Advanced Organic Chemistry

THIRD EDITION

Part B: Reactions and Synthesis

Advanced Organic Chemistry

PART A: Structure and Mechanisms
PART B: Reactions and Synthesis

Advanced Organic Chemistry

THIRD EDITION

Part B: Reactions and Synthesis

FRANCIS A. CAREY
and RICHARD J. SUNDBERG

University of Virginia
Charlottesville, Virginia

PLENUM PRESS • NEW YORK AND LONDON

Library of Congress Cataloging in Publication Data

(Revised for 3rd ed.)
Carey, Francis A., 1937–
Advanced organic chemistry.
Includes bibliographical references.
Contents: pt. A. Structure and mechanisms—pt. B. Reactions and synthesis.
1. Chemistry, Organic. I. Sundberg, Richard J., 1938– . I. Title.
QD251.2.C36 1990 547 90-6851
ISBN 0-306-43440-7 (Part A)
ISBN 0-306-43447-4 (pbk.: Part A)
ISBN 0-306-43456-3 (Part B)
ISBN 0-306-43457-1 (pbk.: Part B)

A Division of Plenum Publishing Corporation
233 Spring Street, New York, N.Y. 10013

Printed in the United States of America

Preface to the Third Edition

The main theme of Part B is the description of synthetically useful reactions and the illustration of their application. We have attempted to update the material to reflect the most important advances in synthetic methodology. Because of the extensive developments in the use of organic derivatives of transition metals, as well as of silicon and tin, we have separated the organometallic material into three chapters. Chapter 7 emphasizes organolithium and organomagnesium chemistry and also considers the group IIB metals. Transition metal chemistry is discussed in Chapter 8, with emphasis on copper and palladium intermediates. In Chapter 9, the carbon–carbon bond-forming reactions of organoboranes, silanes, and stannanes are discussed. The increased importance of free-radical reactions in synthesis has led to the incorporation of a section on radical reactions into Chapter 10, in which carbocations, carbenes, and nitrenes are also discussed.

Certainly a major advance in synthetic chemistry during the 1980s was the development of methods for enantioselective synthesis. We have increased the level of attention to stereochemistry in the discussion of many reactions. In areas in which new stereoselective methods have been well developed, such as in aldol condensations, hydroboration, catalytic reduction, and epoxidation, we discuss these methods.

The final chapter discusses some of the general issues which must be addressed in multistep synthesis and provides some illustrative syntheses which can provide the basis for more detailed study of this aspect of synthetic chemistry.

Each chapter contains problems, some of which are new, drawn from the research literature. The level of difficulty varies. Many will be quite challenging to the student, but they should provide a basis for application of the facts and principles which are discussed in the text.

Francis A. Carey
Richard J. Sundberg

Charlottesville, Virginia

Contents of Part B

Chapter 1. Alkylation of Nucleophilic Carbon. Enolates and Enamines . . . 1

1.1. Generation of Carbanions by Deprotonation 1
1.2. Regioselectivity and Stereoselectivity in Enolate Formation 5
1.3. Other Means of Generating Enolates 10
1.4. Alkylation of Enolates 11
1.5. Generation and Alkylation of Dianions 19
1.6. Medium Effects in the Alkylation of Enolates 20
1.7. Oxygen versus Carbon as the Site of Alkylation 23
1.8. Alkylation of Aldehydes, Esters, Amides, and Nitriles 27
1.9. The Nitrogen Analogs of Enols and Enolates—Enamines and Imine Anions 30
1.10. Alkylation of Carbon Nucleophiles by Conjugate Addition 39
General References 46
Problems 46

Chapter 2. Reactions of Carbon Nucleophiles with Carbonyl Groups . . . 55

2.1. Aldol Condensation 55
2.1.1. The General Mechanism 55
2.1.2. Mixed Aldol Condensations with Aromatic Aldehydes . . 58
2.1.3. Control of Regiochemistry and Stereochemistry of Mixed Aldol Condensations of Aliphatic Aldehydes and Ketones . 60
2.1.4. Intramolecular Aldol Condensations and the Robinson Annulation 75
2.2. Condensation Reactions of Imines and Iminium Ions 80
2.2.1. The Mannich Reaction 80
2.2.2. Amine-Catalyzed Condensation Reactions 83

2.3. Acylation of Carbanions 84
2.4. The Wittig and Related Reactions 95
2.5. Reactions of Carbonyl Compounds with α-Trimethylsilyl Carbanions 102
2.6. Sulfur Ylides and Related Nucleophiles 103
2.7. Nucleophilic Addition–Cyclization 109
General References 110
Problems 111

Chapter 3. Functional Group Interconversion by Nucleophilic Substitution 121

3.1. Conversion of Alcohols to Alkylating Agents 121
3.1.1. Sulfonate Esters 121
3.1.2. Halides 122
3.2. Introduction of Functional Groups by Nucleophilic Substitution at Saturated Carbon 128
3.2.1. General Solvent Effects 128
3.2.2. Nitriles 130
3.2.3. Azides 131
3.2.4. Alkylation of Amines and Amides 132
3.2.5. Oxygen Nucleophiles 133
3.2.6. Sulfur Nucleophiles 136
3.2.7. Phosphorus Nucleophiles 136
3.2.8. Summary of Nucleophilic Substitution at Saturated Carbon 137
3.3. Nucleophilic Cleavage of Carbon–Oxygen Bonds in Ethers and Esters 141
3.4. Interconversion of Carboxylic Acid Derivatives 144
3.4.1. Preparation of Reaction Reagents for Acylation 145
3.4.2. Preparation of Esters 151
3.4.3. Preparation of Amides 152
Problems 156

Chapter 4. Electrophilic Additions to Carbon–Carbon Multiple Bonds . . . 167

4.1. Addition of Hydrogen Halides 167
4.2. Hydration and Other Acid-Catalyzed Additions 170
4.3. Oxymercuration 171
4.4. Addition of Halogens to Alkenes 176
4.5. Electrophilic Sulfur and Selenium Reagents 185
4.6. Addition of Other Electrophilic Reagents 189
4.7. Electrophilic Substitution Alpha to Carbonyl Groups 191
4.8. Additions to Allenes and Alkynes 195
4.9. Addition at Double Bonds via Organoboranes 200
4.9.1. Hydroboration 200

4.9.2. Reactions of Organoboranes 203
4.9.3. Enantioselective Hydroboration 207
4.9.4. Hydroboration of Alkynes 210
General References 212
Problems 212

Chapter 5. Reduction of Carbonyl and Other Functional Groups 219

5.1. Addition of Hydrogen 219
5.1.1. Catalytic Hydrogenation 219
5.1.2. Other Hydrogen-Transfer Reagents 230
5.2. Group III Hydride-Donor Reagents 232
5.2.1. Reduction of Carbonyl Compounds 232
5.2.2. Reduction of Other Functional Groups by Hydride Donors 244
5.3. Group IV Hydride Donors 248
5.4. Hydrogen Atom Donors 250
5.5. Dissolving-Metal Reductions 253
5.5.1. Addition of Hydrogen 253
5.5.2. Reductive Removal of Functional Groups 257
5.5.3. Reductive Carbon–Carbon Bond Formation 261
5.6. Reductive Deoxygenation of Carbonyl Groups 265
General References 269
Problems 270

Chapter 6. Cycloadditions, Unimolecular Rearrangements, and Thermal Eliminations 283

6.1. Cycloaddition Reactions 283
6.1.1. The Diels–Alder Reaction: General Features 284
6.1.2. The Diels–Alder Reaction: Dienophiles 289
6.1.3. The Diels–Alder Reaction: Dienes 296
6.1.4. Intramolecular Diels–Alder Reactions 298
6.2. Dipolar Cycloaddition Reactions 300
6.3. [2 + 2] Cycloadditions and Other Reactions Leading to Cyclobutanes 307
6.4. Photochemical Cycloaddition Reactions 310
6.5. [3,3]-Sigmatropic Rearrangements: Cope and Claisen Rearrangements 316
6.6. [2,3]-Sigmatropic Rearrangements 328
6.7. Ene Reactions 332
6.8. Unimolecular Thermal Elimination Reactions 336
6.8.1. Cheletropic Elimination 336
6.8.2. Decomposition of Cyclic Azo Compounds 339

6.8.3. β-Eliminations Involving Cyclic Transition States 343
General References 349
Problems 350

Chapter 7. Organometallic Compounds of Group I and II Metals 365

7.1. Preparation and Properties of Organolithium and Organomagnesium Compounds 365
7.2. Reactions of Organolithium and Organomagnesium Compounds . 374
7.2.1. Reactions with Alkylating Agents 374
7.2.2. Reactions with Carbonyl Compounds 376
7.3. Organic Derivatives of Group IIB Metals 388
7.3.1. Organozinc Compounds 388
7.3.2. Organocadmium Compounds 392
7.3.3. Organomercury Compounds 392
7.4. Organocerium Compounds 394
General References 394
Problems 395

Chapter 8. Reactions Involving the Transition Metals 401

8.1. Reactions Involving Organocopper Intermediates 401
8.2. Reactions Involving Organopalladium Intermediates 414
8.3. Reactions Involving Organonickel Compounds 422
8.4. Reactions Involving Rhodium, Iron, and Cobalt 425
8.5. Organometallic Compounds with π Bonding 427
General References 431
Problems 431

Chapter 9. Carbon–Carbon Bond-Forming Reactions of Compounds of Boron, Silicon, and Tin 443

9.1. Organoboron Compounds 443
9.1.1. Synthesis of Organoboranes 443
9.1.2. Carbon–Carbon Bond-Forming Reactions 445
9.2. Organosilicon Compounds 464
9.2.1. Synthesis of Organosilanes 464
9.2.2. Carbon–Carbon Bond-Forming Reactions 465
9.3. Organotin Compounds 474
9.3.1. Synthesis of Organostannanes 474
9.3.2. Carbon–Carbon Bond-Forming Reactions 475
General References 484
Problems 485

Chapter 10. Reactions Involving Highly Reactive Electron-Deficient Intermediates . . . 493

10.1. Reactions Involving Carbocation Intermediates . . . 493
10.1.1. Carbon–Carbon Bond-Formation Involving Carbocations . 493
10.1.2. Rearrangements of Carbocations . . . 499
10.1.3. Related Rearrangements . . . 506
10.1.4. Fragmentation Reactions . . . 509
10.2. Reactions Involving Carbenes and Nitrenes . . . 511
10.2.1. Structure and Reactivity of Carbenes . . . 512
10.2.2. Generation of Carbenes . . . 516
10.2.3. Addition Reactions . . . 522
10.2.4. Insertion Reactions . . . 528
10.2.5. Rearrangement Reactions . . . 531
10.2.6. Related Reactions . . . 532
10.2.7. Nitrenes and Related Intermediates . . . 535
10.2.8. Rearrangements to Electron-Deficient Nitrogen . . . 536
10.3. Reactions Involving Free-Radical Intermediates . . . 541
10.3.1. Sources of Radical Intermediates . . . 542
10.3.2. Introduction of Functionality by Radical Reactions . . . 543
10.3.3. Addition Reactions of Radicals with Substituted Alkenes . 545
10.3.4. Cyclization of Free-Radical Intermediates . . . 551
10.3.5. Fragmentation and Rearrangement Reactions . . . 555
General References . . . 558
Problems . . . 559

Chapter 11. Aromatic Substitution Reactions . . . 571

11.1. Electrophilic Aromatic Substitution . . . 571
11.1.1. Nitration . . . 571
11.1.2. Halogenation . . . 572
11.1.3. Friedel–Crafts Alkylations and Acylations . . . 575
11.1.4. Electrophilic Metalation . . . 585
11.2. Nucleophilic Aromatic Substitution . . . 588
11.2.1. Aromatic Diazonium Ions as Synthetic Intermediates . . 588
11.2.2. Substitution by the Addition–Elimination Mechanism . . 596
11.2.3. Substitution by the Elimination–Addition Mechanism . . 599
11.2.4. Copper-Catalyzed Reactions . . . 601
11.3. Aromatic Radical Substitution Reactions . . . 603
11.4. Substitution by the $S_{RN}1$ Mechanism . . . 605
General References . . . 606
Problems . . . 607

Chapter 12. Oxidations . . . 615

12.1. Oxidation of Alcohols to Aldehydes, Ketones, or Carboxylic Acids 615
12.1.1. Transition Metal Oxidants . . . 615
12.1.2. Other Oxidants . . . 620

12.2. Addition of Oxygen at Carbon–Carbon Double Bonds 624
12.2.1. Transition Metal Oxidants 624
12.2.2. Epoxides from Alkenes and Peroxidic Reagents 630
12.2.3. Subsequent Transformations of Epoxides 633
12.2.4. Reactions of Alkenes with Singlet Oxygen 640
12.3. Cleavage of Carbon–Carbon Double Bonds 643
12.3.1. Transition Metal Oxidants 643
12.3.2. Ozonolysis 645
12.4. Selective Oxidative Cleavages at Other Functional Groups 647
12.4.1. Cleavage of Glycols 647
12.4.2. Oxidative Decarboxylation 649
12.5. Oxidation of Ketones and Aldehydes 651
12.5.1. Transition Metal Oxidants 651
12.5.2. Oxidation of Ketones and Aldehydes by Oxygen and Peroxidic Compounds 654
12.5.3. Oxidation with Other Reagents 657
12.6. Allylic Oxidation 658
12.6.1. Transition Metal Oxidants 658
12.6.2. Other Oxidants 659
12.7. Oxidations at Unfunctionalized Carbon 661
General References 664
Problems 664

Chapter 13. Multistep Syntheses 677

13.1. Protective Groups 677
13.1.1. Hydroxyl-Protecting Groups 678
13.1.2. Amino-Protecting Groups 686
13.1.3. Carbonyl-Protecting Groups 689
13.1.4. Carboxylic Acid-Protecting Groups 691
13.2. Synthetic Equivalent Groups 692
13.3. Synthetic Analysis and Planning 699
13.4. Control of Stereochemistry 701
13.5. Illustrative Syntheses 702
13.5.1. Juvabione 703
13.5.2. Longifolene 712
13.5.3. Prelog–Djerassi Lactone 720
13.5.4. Aphidicolin 732
General References 747
Problems 748

References for Problems 763

Index 785

Contents of Part A

Chapter 1. Chemical Bonding and Structure

1.1. Valence Bond Approach to Chemical Bonding
1.2. Bond Energies, Lengths, and Dipoles
1.3. Molecular Orbital Theory and Methods
1.4. Qualitative Application of Molecular Orbital Theory
1.5. Hückel Molecular Orbital Theory
1.6. Perturbation Molecular Orbital Theory
1.7. Interactions between σ and π Systems—Hyperconjugation
General References
Problems

Chapter 2. Stereochemical Principles

2.1. Enantiomeric Relationships
2.2. Diastereomeric Relationships
2.3. Stereochemistry of Dynamic Processes
2.4. Prochiral Relationships
General References
Problems

Chapter 3. Conformational, Steric, and Stereoelectronic Effects

3.1. Steric Strain and Molecular Mechanics
3.2. Conformations of Acyclic Molecules
3.3. Conformations of Cyclohexane Derivatives
3.4. Carbocyclic Rings Other Than Six-Membered

3.5. The Effect of Heteroatoms on Conformational Equilibria
3.6. Molecular Orbital Methods Applied to Conformational Analysis
3.7. Conformational Effects on Reactivity
3.8. Angle Strain and Its Effect on Reactivity
3.9. Relationships between Ring Size and Facility of Ring Closure
3.10. Torsional and Stereoelectronic Effects on Reactivity
General References
Problems

Chapter 4. Study and Description of Organic Reaction Mechanisms

4.1. Thermodynamic Data
4.2. Kinetic Data
4.3. Substituent Effects and Linear Free-Energy Relationships
4.4. Basic Mechanistic Concepts: Kinetic versus Thermodynamic Control, Hammond's Postulate, and the Curtin–Hammett Principle
4.4.1. Kinetic versus Thermodynamic Control
4.4.2. Hammond's Postulate
4.4.3. The Curtin–Hammett Principle
4.5. Isotope Effects
4.6. Isotopes in Labeling Experiments
4.7. Characterization of Reaction Intermediates
4.8. Catalysis by Acids and Bases
4.9. Lewis Acid Catalysis
4.10. Solvent Effects
4.11. Structural Effects in the Gas Phase
4.12. Stereochemistry
4.13. Conclusion
General References
Problems

Chapter 5. Nucleophilic Substitution

5.1. The Limiting Cases—Substitution by the Ionization (S_N1) Mechanism
5.2. The Limiting Cases—Substitution by the Direct Displacement (S_N2) Mechanism
5.3. Detailed Mechanistic Description and Borderline Mechanisms
5.4. Carbocations
5.5. Nucleophilicity and Solvent Effects
5.6. Leaving Group Effects
5.7. Steric and Strain Effects on Substitution and Ionization Rates
5.8. Substituent Effects on Reactivity
5.9. Stereochemistry of Nucleophilic Substitution

5.10. Neighboring-Group Participation
5.11. Rearrangements of Carbocations
5.12. The Norbornyl Cation and Other Nonclassical Carbocations
General References
Problems

Chapter 6. Polar Addition and Elimination Reactions

6.1. Addition of Hydrogen Halides to Alkenes
6.2. Acid-Catalyzed Hydration and Related Addition Reactions
6.3. Addition of Halogens
6.4. Electrophilic Additions Involving Metal Ions
6.5. Additions to Alkynes and Allenes
6.6. The E2, E1, and E1cb Mechanisms
6.7. Orientation Effects in Elimination Reactions
6.8. Stereochemistry of E2 Elimination Reactions
6.9. Dehydration of Alcohols
6.10. Eliminations Not Involving C—H Bonds
General References
Problems

Chapter 7. Carbanions and Other Nucleophilic Carbon Species

7.1. Acidity of Hydrocarbons
7.2. Carbanions Stabilized by Functional Groups
7.3. Enols and Enamines
7.4. Carbanions as Nucleophiles in S_N2 Reactions
General References
Problems

Chapter 8. Reactions of Carbonyl Compounds

8.1. Hydration and Addition of Alcohols to Aldehydes and Ketones
8.2. Addition–Elimination Reactions of Aldehydes and Ketones
8.3. Addition of Carbon Nucleophiles to Carbonyl Groups
8.4. Reactivity of Carbonyl Compounds toward Addition
8.5. Ester Hydrolysis
8.6. Aminolysis of Esters
8.7. Amide Hydrolysis
8.8. Acylation of Nucleophilic Oxygen and Nitrogen Groups
8.9. Intramolecular Catalysis
General References
Problems

Chapter 9. Aromaticity

9.1. The Concept of Aromaticity
9.2. The Annulenes
9.3. Aromaticity in Charged Rings
9.4. Homoaromaticity
9.5. Fused-Ring Systems
9.6. Heterocyclic Rings
General References
Problems

Chapter 10. Aromatic Substitution

10.1 Electrophilic Aromatic Substitution Reactions
10.2. Structure-Reactivity Relationships
10.3. Reactivity of Polycyclic and Heteroaromatic Compounds
10.4. Specific Substitution Mechanisms
10.4.1. Nitration
10.4.2. Halogenation
10.4.3. Protonation and Hydrogen Exchange
10.4.4. Friedel-Crafts Alkylation and Related Reactions
10.4.5. Friedel-Crafts Acylation and Related Reactions
10.4.6. Coupling with Diazonium Compounds
10.4.7. Substitution of Groups Other than Hydrogen
10.5. Nucleophilic Aromatic Substitution by Addition-Elimination
10.6. Nucleophilic Aromatic Substitiution by the Elimination-Addition Mechanism
General References
Problems

Chapter 11. Concerted Reactions

11.1 Electrocyclic Reactions
11.2. Sigmatropic Rearrangements
11.3. Cycloaddition Reactions
General References
Problems

Chapter 12. Free-Radical Reactions

12.1. Generation and Characterization of Free Radicals
12.1.1. Background

12.1.2. Stable and Persistent Free Radicals
12.1.3. Direct Detection of Radical Intermediates
12.1.4. Sources of Free Radicals
12.1.5. Structural and Stereochemical Properties of Radical Intermediates
12.1.6. Charged Radical Species
12.2. Characteristics of Reaction Mechanisms Involving Radical Intermediates
12.2.1. Kinetic Characteristics of Chain Reactions
12.2.2. Structure–Reactivity Relationships
12.3. Free-Radical Substitution Reactions
12.3.1. Halogenation
12.3.2. Oxidation
12.4. Free-Radical Addition Reactions
12.4.1. Addition of Hydrogen Halides
12.4.2. Addition of Halomethanes
12.4.3. Addition of Other Carbon Radicals
12.4.4. Addition of Thiols and Thiocarboxylic Acids
12.5. Intramolecular Free-Radical Reactions
12.6. Rearrangement and Fragmentation Reactions of Free Radicals
12.6.1. Rearrangement Reactions
12.6.2. Fragmentation Reactions
12.7. Electron Transfer Reactions Involving Transition Metal Ions
12.8. $S_{RN}1$ Substitution Processes
General References
Problems

Chapter 13. Photochemistry

13.1. General Principles
13.2. Orbital Symmetry Considerations Related to Photochemical Reactions
13.3. Photochemistry of Carbonyl Compounds
13.4. Photochemistry of Alkenes and Dienes
13.5. Photochemistry of Aromatic Compounds
General References
Problems

References for Problems

Index

List of Figures

1.1. Unsolvated hexameric aggregate of lithium enolate of methyl *t*-butyl ketone
1.2. Potassium enolate of methyl *t*-butyl ketone
1.3. Crystal structure of dimer of lithium derivative of *N*-phenyl imine of methyl *t*-butyl ketone
6.1. Cycloaddition of an alkene and a diene, showing interaction of LUMO of alkene with HOMO of diene
6.2. *Endo* and *exo* addition in a Diels–Alder reaction
6.3. HOMO–LUMO interactions rationalize regioselectivity of Diels–Alder cycloaddition reactions
6.4. Prediction of regioselectivity of 1,3-dipolar cycloaddition
6.5. HOMO–LUMO interactions in the [2 + 2] cycloaddition of an alkene and a ketene
6.6. A concerted ene reaction corresponds to the interaction of a hydrogen atom with the HOMO of an allyl radical and the LUMO of the enophile and is allowed
7.1. Crystal structures of ethylmagnesium bromide
7.2. Crystal structure of tetrameric phenyllithium etherate
8.1. Representation of π bonding in alkene–transition-metal complexes
10.1. Mechanisms for addition of singlet and triplet carbenes to alkenes
10.2. Frontier orbital interpretation of radical substituent effects
10.3. Chain mechanism for radical addition reactions mediated by trialkylstannyl radicals

List of Tables

1.1. Approximate pK values for some carbon acids and some common bases
1.2. Compositions of enolate mixtures
1.3. Relative alkylation rates of sodium diethyl *n*-butylmalonate in various solvents
1.4. Enantioselective alkylation of ketimines
2.1. Stereoselectivity of lithium enolates toward benzaldehyde
2.2. Stereoselectivity of boron enolates toward aldehydes
2.3. Stereoselectivity of ester enolates toward aldehydes
4.1. Stereochemistry of addition of hydrogen halides to alkenes
4.2. Relative reactivity of some alkenes in oxymercuration
4.3. Relative reactivity of alkenes and alkynes
4.4. Regioselectivity of diborane and alkylboranes toward representative alkenes
4.5. Stereoselectivity of hydroboration of cyclic alkenes
5.1. Enantiomeric excess for asymmetric catalytic hydrogenation of substituted acrylic acids
5.2. Relative reactivity of hydride-donor reducing agents
5.3. Stereoselectivity of hydride reducing agents
5.4. Reaction conditions for reductive replacement of halogen and tosylate by hydride donors
6.1. Representative dienophiles
6.2. 1,3-Dipolar compounds
6.3. Relative reactivity of substituted alkenes toward some 1,3-dipoles
6.4. [3,3]-Sigmatropic rearrangements
10.1. Relative rates of addition to alkenes
10.2. Classification of carbenes on the basis of reactivity toward alkenes
11.1. Relative activity of Friedel–Crafts catalysts

List of Schemes

1.1. Generation of carbon nucleophiles by deprotonation
1.2. Resonance in some carbanions
1.3. Generation of specific enolates
1.4. Alkylations of relatively acidic carbon acids
1.5. Synthesis of ketones and carboxylic acid derivatives via alkylation techniques
1.6. Regioselective enolate alkylation
1.7. Generation and alkylation of dianions
1.8. Alkylation of esters, lactones, and nitriles
1.9. Enamine alkylations
1.10. Alkylation of carbon by conjugate addition
1.11. Michael additions under kinetic conditions

2.1. Aldol condensation of simple aldehydes and ketones
2.2. Mixed condensation of aromatic aldehydes with ketones
2.3. Directed aldol condensations
2.4. Addition reactions of carbanions derived from esters, carboxylic acids, amides, and nitriles
2.5. Intramolecular aldol condensations
2.6. The Robinson annulation reaction
2.7. Synthesis and utilization of Mannich bases
2.8. Amine-catalyzed condensations of the Knoevenagel type
2.9. Acylation of nucleophilic carbon by esters
2.10. Acylation of ester enolates with acid halides, anhydrides, and imidazolides
2.11. Acylation of ketones with esters
2.12. The Wittig reaction
2.13. Carbonyl olefination using phosphonate carbanions

2.14. Carbonyl olefination using trimethylsilyl-substituted organolithium reagents
2.15. Reactions of sulfur ylides
2.16. Darzens condensation reactions

3.1. Preparation of alkyl halides
3.2. Transformation of functional groups by nucleophilic substitution
3.3. Cleavage of ethers and esters
3.4. Preparation and reactions of active acylating agents
3.5. Acid-catalyzed esterification
3.6. Synthesis of amides

4.1. Synthesis via mercuration
4.2. Iodolactonization and other cyclizations induced by iodine
4.3. Other sources of positive halogen
4.4. Sulfur and selenium reagents for electrophilic addition
4.5. Cyclizations induced by electrophilic sulfur and selenium reagents
4.6. Addition reactions of other electrophilic reagents
4.7. α-Sulfenylation and α-selenenylation of carbonyl compounds
4.8. Ketones by hydration of alkynes
4.9. Alcohols, ketones, aldehydes, and amines from organoboranes

5.1. Stereochemistry of hydrogenation of some alkenes
5.2. Homogeneous catalytic hydrogenation
5.3. Conditions for catalytic reduction of various functional groups
5.4. Reductions with diimide
5.5. Reduction of other functional groups by hydride donors
5.6. Dehalogenations with stannanes
5.7. Deoxygenation of alcohols via thioesters and related derivatives
5.8. Birch reduction of aromatic rings
5.9. Reductive dehalogenation and deoxygenation
5.10. Reductive removal of functional groups from α-substituted carbonyl compounds
5.11. Reductive carbon–carbon bond formation
5.12. Carbonyl-to-methylene reductions
5.13. Conversion of ketones to alkenes via sulfonylhydrazones

6.1. Diels–Alder reactions of some representative dienophiles
6.2. Enantioselective Diels–Alder reactions
6.3. Intramolecular Diels–Alder reactions
6.4. Typical 1,3-dipolar cycloaddition reactions
6.5. Generation of dipolar intermediates from small rings
6.6. [2 + 2] Cycloadditions of ketenes
6.7. Intramolecular [2 + 2] photochemical cycloaddition reactions of dienes
6.8. Photochemical cycloaddition reactions of enones and alkenes

6.9. Photochemical cycloaddition reactions of carbonyl compounds with alkenes
6.10. Cope rearrangements of 1,5-dienes
6.11. Claisen rearrangements
6.12. Carbon–carbon bond formation via [2,3]-sigmatropic rearrangements of sulfur and nitrogen ylides
6.13. [2,3]-Wittig rearrangements
6.14. Ene Reactions
6.15. Photochemical and Thermal decomposition of cyclic azo compounds
6.16. Eliminations via cyclic transition states
6.17. Thermal eliminations via cyclic transition states

7.1. Organolithium compounds by metalation
7.2. Organolithium reagents by halogen–metal exchange
7.3. Synthetic procedures involving Grignard reagents
7.4. Synthetic procedures involving organolithium reagents
7.5. Condensation of α-halocarbonyl compounds using zinc—the Reformatsky reaction

8.1. Mixed cuprate reagents
8.2. Reactions of organocopper intermediates
8.3. Tandem reactions involving trapping of enolates generated by conjugate addition of organocopper reagents
8.4. Copper-catalyzed reactions of Grignard reagents
8.5. Generation and reactions of alkenylcopper reagents by additions to acetylenes
8.6. Palladium-catalyzed vinylation of aryl and alkenyl halides
8.7. Reactions of cyclobutadiene

9.1. Synthesis via carbonylation of organoboranes
9.2. Some one-carbon donors in alcohol and ketone synthesis using organoboranes
9.3. Alkylation of trialkylboranes by α-halocarbonyl and related compounds
9.4. Alkylation of organoboranes by α,β-unsaturated carbonyl compounds
9.5. Addition reactions of allylic boranes with carbonyl compounds
9.6. Palladium-catalyzed coupling of organoboranes
9.7. Reactions involving electrophilic attack on alkenyl and allylic silanes
9.8. Reactions of silanes with α,β-unsaturated carbonyl compounds
9.9. Reactions of allylic stannanes with carbonyl compounds
9.10. Palladium-catalyzed coupling of stannanes with halides and sulfonates
9.11. Synthesis of ketones by palladium-catalyzed acylation of stannanes

10.1. Polyolefin cyclizations
10.2. Rearrangements promoted by adjacent heteroatoms
10.3. Base-catalyzed rearrangements of α-haloketones

10.4. Fragmentation reactions
10.5. General methods for generation of carbenes
10.6. Cyclopropane formation by carbenoid additions
10.7. Intramolecular carbene-insertion reactions
10.8. Wolff rearrangement of α-diazoketones
10.9. Rearrangement to electron-deficient nitrogen
10.10. Alkylation of alkyl radicals by reaction with alkenes
10.11. Allylation of radical centers using allylstannanes
10.12. Radical cyclizations

11.1. Aromatic nitration
11.2. Aromatic halogenation
11.3. Friedel–Crafts alkylation reactions
11.4. Friedel–Crafts acylation reactions
11.5. Other electrophilic aromatic substitutions related to Friedel–Crafts reactions
11.6. Aromatic substitution via diazonium ions
11.7. Meerwein arylation reactions
11.8. Nucleophilic aromatic substitution
11.9. Some syntheses via benzyne intermediates
11.10. Biaryls by radical substitution
11.11. Aromatic substitution by the $S_{RN}1$ process

12.1. Oxidations with Cr(VI)
12.2. Oxidations of alcohols with manganese dioxide
12.3. Oxidation of alcohols using dimethyl sulfoxide
12.4. Hydroxylation of alkenes
12.5. Enantioselective epoxidation of allylic alcohols
12.6. Synthesis of epoxides from alkenes
12.7. Multistep synthetic transformations via epoxides
12.8. Nucleophilic and solvolytic ring opening of epoxides
12.9. Generation of singlet oxygen
12.10. Oxidation of alkenes with singlet oxygen
12.11. Oxidative cleavage of carbon–carbon double bonds with transition metal oxidants
12.12. Ozonolysis reactions
12.13. Baeyer–Villiger oxidations
12.14. Side-chain oxidation of aromatic compounds

13.1. Protection of hydroxyl groups
13.2. Synthetic sequences used for reaction of acyl anion equivalents
13.3. Reaction sequences involving propanal homoenolate anion synthetic equivalents
13.4. Retrosynthetic analysis of juvabione with disconnection to *p*-methoxyacetophenone

13.5. Juvabione synthesis: K. Mori and M. Matsui
13.6. Juvabione synthesis: K. S. Ayyar and G. S. K. Rao
13.7. Retrosynthetic analysis of juvabione with disconnection to a terpene structure
13.8. Juvabione synthesis: B. A. Pawson, H.-C. Cheung, S. Gurbaxani, and G. Saucy
13.9. Juvabione synthesis: E. Negishi, M. Sabanski, J. J. Katz, and H. C. Brown
13.10. Retrosynthetic analysis of juvabione with alternate disconnections to cyclohexenone
13.11. Juvabione synthesis: J. Ficini, J. D'Angelo, and J. Noiré
13.12. Juvabione synthesis: D. A. Evans and J. V. Nelson
13.13. Juvabione synthesis: A. G. Schultz and J. P. Dittami
13.14. Juvabione synthesis: D. J. Morgans, Jr., and G. B. Feigelson
13.15. Retrosynthesis of longifolene corresponding to the synthesis in Scheme 13.16
13.16. Longifolene synthesis: E. J. Corey, R. B. Mitra, and P. A. Vatakencherry
13.17. Longifolene synthesis: J. E. McMurry and S. J. Isser
13.18. Retrosynthetic analysis corresponding to synthesis in Scheme 13.19
13.19. Longifolene synthesis: R. A. Volkmann, G. C. Andrews, and W. S. Johnson
13.20. Longifolene synthesis: W. Oppolzer and T. Godel
13.21. Longifolene synthesis: A. G. Schultz and S. Puig
13.22. Prelog–Djerassi lactone synthesis: P. A. Grieco, Y. Ohfune, Y. Yokoyama, and W. Owens
13.23. Prelog–Djerassi lactone synthesis: W. C. Still and K. R. Shaw
13.24. Prelog–Djerassi lactone synthesis: S. Masamune, M. Hirama, S. Mori, S. A. Ali, and D. S. Garvey; S. Masamune, S. A. Ali, D. L. Snitman, and D. S. Garvey
13.25. Prelog–Djerassi lactone synthesis: R. W. Hoffman, H.-J. Zeiss, W. Ladner, and S. Tabche
13.26. Prelog-Djerassi lactone synthesis: Y. Nagao, T. Inoue, K. Hashimoto, Y. Hagiwara, M. Ochai, and E. Fujita
13.27. Prelog–Djerassi lactone synthesis: S. Jarosz and B. Fraser-Reid
13.28. Prelog–Djerassi lactone synthesis: N. Kawauchi and H. Hashimoto
13.29. Prelog–Djerassi lactone synthesis: R. E. Ireland and J. P. Daub
13.30. Prelog–Djerassi lactone synthesis: D. A. Evans and J. Bartroli
13.31. Prelog–Djerassi lactone synthesis: S. F. Martin and D. E. Guinn
13.32. Prelog–Djerassi lactone synthesis: M. Honda, T. Katsuki, and M. Yamaguchi
13.33. Aphidicolin synthesis: B. M. Trost, Y. Nishimura, and K. Yamamoto
13.34. Aphicicolin synthesis: J. E. McMurry, A. Andrus, G. M. Ksander, J. H. Musser, and M. A. Johnson
13.35. Aphidicolin synthesis: E. J. Corey, M. A. Tius, and J. Das
13.36. Aphidicolin synthesis: R. E. Ireland, W. C. Low, J. D. Godrey, and S. Thaisrivongs

13.37. Aphidicolin synthesis: R. M. Bettolo, P. Tagliatesta, A. Lupi, and D. Bravetti

13.38. Aphidicolin synthesis: E. E. van Tamelen, S. R. Zawacky, R. K. Russell, and J. G. Carlson

13.39. Aphidicolin synthesis: R. A. Holton, R. M. Kennedy, H.-B. Kim, and M. E. Kraft

1

Alkylation of Nucleophilic Carbon. Enolates and Enamines

Carbon–carbon bond formation is the fundamental basis for the construction of the molecular framework in the synthesis of organic molecules. Many carbon–carbon bond-forming processes involve reaction between a nucleophilic carbon and an electrophilic one. The emphasis in this chapter is on *enolate ions* and *enamines*, two of the most useful kinds of carbon nucleophiles, and on their reactions with *alkylating agents.*

1.1. Generation of Carbanions by Deprotonation

A very general means of generating carbon nucleophiles involves removal of a proton from a carbon by a Brønsted base. The anions produced are *carbanions.* Both the rate of deprotonation and the stability of the resulting carbanion are enhanced by the presence of substituent groups that can stabilize negative charge. A carbonyl group bonded directly to the anionic carbon can delocalize negative charge by resonance, and carbonyl compounds are especially important in carbanion chemistry. The anions formed by deprotonation of the carbon *alpha* to a carbonyl group bear most of their negative charge on oxygen and are commonly referred to as *enolates.* Several typical examples of proton abstraction equilibria are listed in Scheme 1.1. Electron delocalization in the corresponding carbanions is represented by the resonance structures presented in Scheme 1.2.

Scheme 1.1. Generation of Carbon Nucleophiles by Deprotonation

1	$RCH_2\overset{O}{\overset{\parallel}{C}}R' + NH_2^- \rightleftharpoons R\underline{C}H\overset{O}{\overset{\parallel}{C}}R' + NH_3$
2	$RCH_2\overset{O}{\overset{\parallel}{C}}OR' + \bar{N}R''_2 \rightleftharpoons R\underline{C}H\overset{O}{\overset{\parallel}{C}}OR' + HNR''_2$
3	$R'O\overset{O}{\overset{\parallel}{C}}CH_2\overset{O}{\overset{\parallel}{C}}OR' + R'O^- \rightleftharpoons R'O\overset{O}{\overset{\parallel}{C}}\underline{C}H\overset{O}{\overset{\parallel}{C}}OR' + R'OH$
4	$CH_3\overset{O}{\overset{\parallel}{C}}CH_2\overset{O}{\overset{\parallel}{C}}OR' + R'O^- \rightleftharpoons CH_3\overset{O}{\overset{\parallel}{C}}\underline{C}H\overset{O}{\overset{\parallel}{C}}OR' + R'OH$
5	$N{\equiv}CCH_2\overset{O}{\overset{\parallel}{C}}OR' + R'O^- \rightleftharpoons N{\equiv}C\underline{C}H\overset{O}{\overset{\parallel}{C}}OR' + R'OH$
6	$RCH_2NO_2 + HO^- \rightleftharpoons R\underline{C}HNO_2 + H_2O$

Scheme 1.2. Resonance in Some Carbanions

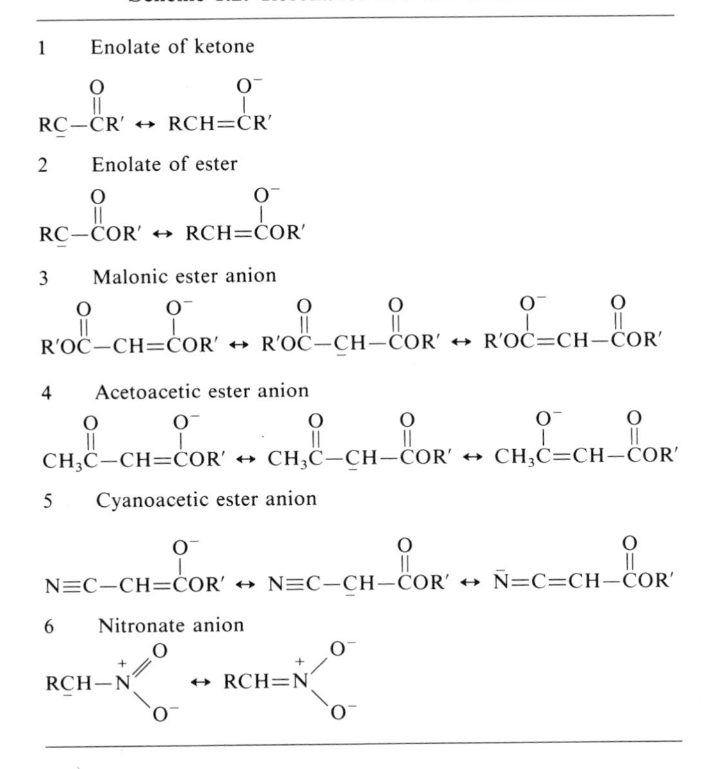

The efficient generation of a significant equilibrium concentration of a carbanion requires choice of a proper Brønsted base. The equilibrium will only favor carbanion formation when the acidity of the carbon acid is greater than that of the conjugate acid of the base used for deprotonation. Acidity is quantitatively expressed as pK_a, which is equal to $-\log K_a$ and applies, by definition, to dilute aqueous solution. Since many important carbon acids are quite weak acids ($pK_a > 15$), accurate measures of their equilibria in aqueous solutions are impossible, and acidities are determined in a variety of organic solvents and referenced to water in an approximate

Table 1.1. Approximate p*K* Values for Some Carbon Acids and Some Common Bases

Carbon acid	p*K*[a]	pK_{DMSO}	Common bases[a,b]	p*K*[c]	pK_{DMSO}
$O_2NCH_2NO_2$	3.6		$CH_3CO_2^-$	4.2	11.6
$CH_3COCH_2NO_2$	5.1				
$PhCH_2NO_2$		12.2[d]			
$CH_3CH_2NO_2$	8.6	16.7[e]			
$CH_3COCH_2COCH_3$	9				
$PhCOCH_2COCH_3$	9.6		PhO^-	9.9	16.4
CH_3NO_2	10.2	17.2[e]			
$CH_3COCH_2CO_2CH_2CH_3$	10.7		$(CH_3CH_2)_3N$	10.7	
$CH_3COCH(CH_3)COCH_3$	11		$(CH_3CH_2)_2NH$	11	
$NCCH_2CN$	11.2	11.1[e]			
$CH_2(SO_2CH_2CH_3)_2$	12.2	14.4[f]			
$CH_2(CO_2CH_2CH_3)_2$	12.7				
Cyclopentadiene	15				
$PhSCH_2COCH_3$		18.7[g]			
$PhCH_2COCH_3$		19.8[e]	CH_3O^-	15.5[i]	29.0[j]
$CH_3CH_2CH(CO_2CH_2CH_3)_2$	15		HO^-	15.7[i]	31.4[j]
$PhSCH_2CN$		20.8[g]			
$PhCH_2CN$		21.9[d]			
$(PhCH_2)_2SO_2$		23.9[e]	$CH_3CH_2O^-$	15.9	29.8[j]
			$(CH_3)_3CO^-$	19	32.2[j]
$PhCOCH_3$	15.8[k]	24.7[e]			
CH_3COCH_3	20	26.5[e]			
$CH_3CH_2COCH_2CH_3$		27.1[e]			
Fluorene	20.5	22.6[e]			
$PhSO_2CH_3$		29.0[e]			
$PhCH_2SOCH_3$		29.0			
CH_3CN	25	31.3[e]			
Ph_3CH	33	30.6[e]	NH_2^-	35	41[h,l]
			$CH_3SOCH_2^-$	35	35.1[e]
			$(CH_3CH_2)_2N^-$	36	
$PhCH_3$		42[h]			
CH_4		55[h]			

a. D. J. Cram, *Fundamentals of Carbanion Chemistry*, Academic Press, New York, 1965, pp. 8–20, 41.
b. H. O. House, *Modern Synthetic Reactions*, Second Edition, W. A. Benjamin, Menlo Park, California, 1972, p. 494.
c. p*K* of the conjugate acid.
d. F. G. Bordwell, J. E. Bares, J. E. Bartmess, G. J. McCollum, M. Van Der Puy, N. R. Vanier, and W. S. Matthews, *J. Org. Chem.* **42**, 321 (1977).
e. W. S. Matthews, J. E. Bares, J. E. Bartmess, F. G. Bordwell, F. J. Cornforth, G. E. Drucker, Z. Margolin, R. J. McCallum, G. J. McCollum, and N. R. Vanier, *J. Am. Chem. Soc.* **97**, 7006 (1975).
f. F. G. Bordwell, J. E. Bartmess, and J. A. Hantala, *J. Org. Chem.* **43**, 3095 (1978).
g. F. G. Bordwell, J. E. Bares, J. E. Bartmess, G. E. Drucker, J. Gerhold, G. J. McCollum, M. Van Der Puy, N. R. Vanier, and W. S. Matthews, *J. Org. Chem.* **42**, 326 (1977).
h. Estimated: D. Algrim, J. E. Bares, J. C. Branca, and F. G. Bordwell, *J. Org. Chem.* **43**, 5024 (1978).
i. True pK_a in water: P. Ballinger and F. A. Long, *J. Am. Chem. Soc.* **82**, 795 (1960).
j. W. N. Olmsted, Z. Margolin, and F. G. Bordwell, *J. Org. Chem.* **45**, 3295 (1980).
k. M. Novak and G. M. Loudon, *J. Org. Chem.* **42**, 2494 (1977).
l. F. G. Bordwell and D. J. Algrim, *J. Am. Chem. Soc.* **110**, 2964 (1988); F. G. Bordwell, G. E. Drucker, and H. E. Fried, *J. Org. Chem.* **46**, 632 (1981).

way. The data produced are not true pK_a's, and their approximate nature is indicated by referring to them as simply pK values, rather than as pK_a's. Table 1.1 presents a listing of pK data for some typical carbon acids. The table also includes a listing of the bases that are frequently used for deprotonation. The strongest acids appear at the top of the table, the strongest bases at the bottom. A favorable equilibrium between a carbon acid and its carbanion will be established if the base which is used appears below the acid in the table. Also included in the table are pK values determined in dimethyl sulfoxide (pK_{DMSO}). The range of acidities which can be directly measured in dimethyl sulfoxide (DMSO) is much greater than can be measured in aqueous media, thereby allowing direct comparisons to be made more confidently. The pK values in DMSO are normally greater than in water because water stabilizes anions more effectively, by hydrogen bonding, than does DMSO. Stated another way, many anions are more basic in DMSO than in water. At the present time, the pK_{DMSO} scale includes the widest variety of structural types of synthetic interest.[1]

From the pK values collected in Table 1.1, an ordering of some substituents with respect to their ability to stabilize carbanions can be established. The order suggested is $NO_2 > COR > CN \approx CO_2R > SO_2R > SOR > Ph \approx SR > H > R$.

By comparing the approximate pK values of the conjugate acids of the bases with those of the carbon acid of interest, it is possible to estimate the position of the acid–base equilibrium for a given reactant–base combination. If we consider the case of a simple alkyl ketone in a protic solvent, for example, it can be seen that hydroxide ion and primary alkoxide ions will convert only a fraction of such a ketone to its anion:

$$\mathrm{R\overset{O}{\overset{\|}{C}}CH_3 + RCH_2O^- \rightleftharpoons R\overset{O^-}{\overset{|}{C}}{=}CH_2 + RCH_2OH} \qquad K < 1$$

The slightly more basic tertiary alkoxides are comparable to the enolates in basicity, and a more favorable equilibrium will be established with such bases:

$$\mathrm{R\overset{O}{\overset{\|}{C}}CH_3 + R_3CO^- \rightleftharpoons R\overset{O^-}{\overset{|}{C}}{=}CH_2 + R_3COH} \qquad K \approx 1$$

Stronger bases, such as amide anion ($^-NH_2$), the conjugate base of DMSO (sometimes referred to as the "dimsyl" anion),[2] and triphenylmethyl anion, are capable of effecting essentially complete conversion of a ketone to its enolate. Lithium diisopropylamide (LDA), generated by addition of *n*-butyllithium to diisopropylamine, is widely used as a strong base in synthetic procedures.[3] It is a

1. W. S. Mathews, J. E. Bares, J. E. Bartmess, F. G. Bordwell, F. J. Cornforth, G. E. Drucker, Z. Margolin, R. J. McCallum, G. J. McCollum, and N. R. Vanier, *J. Am. Chem. Soc.* **97**, 7006 (1975).
2. E. J. Corey and M. Chaykovsky, *J. Am. Chem. Soc.* **87**, 1345 (1965).
3. H. O. House, W. V. Phillips, T. S. B. Sayer, and C.-C. Yau, *J. Org. Chem.* **43**, 700 (1978).

very strong base, yet it is sufficiently bulky so as to be relatively non-nucleophilic, a feature that is important in minimizing side reactions. The lithium and sodium salts of hexamethyldisilazane, $[(CH_3)_3Si]_2NH$, are easily prepared and handled compounds with properties similar to those of lithium diisopropylamide and they also find extensive use in synthesis.[4]

$$\mathrm{R\overset{\overset{\large O}{\|}}{C}CH_3 + [(CH_3)_2CH]_2NLi \rightleftharpoons R\overset{\overset{\large OLi}{|}}{C}{=}CH_2 + [(CH_3)_2CH]_2NH} \qquad K > 1$$

Sodium hydride and potassium hydride can also be used to prepare enolates from ketones. The reactivity of the metal hydrides is somewhat dependent on the means of preparation and purification of the hydride.[5]

For any of the other carbon acids in Table 1.1, similar consideration allows one to estimate the position of the equilibrium with a given base. It is important to keep in mind the position of such equilibria as other aspects of reactions of carbanions are considered.

1.2. Regioselectivity and Stereoselectivity in Enolate Formation

An unsymmetrical dialkyl ketone can form two *regioisomeric* enolates on deprotonation:

$$\mathrm{R_2CH\overset{\overset{\large O}{\|}}{C}CH_2R'} \xrightarrow{B^-} \mathrm{R_2C{=}\overset{\overset{\large O^-}{|}}{C}CH_2R'} \quad \text{or} \quad \mathrm{R_2CH\overset{\overset{\large O^-}{|}}{C}{=}CHR'}$$

In order to exploit fully the synthetic potential of enolate ions, control over the regioselectivity of their formation is required. While it may not be possible to direct deprotonation so as to form one enolate to the exclusion of the other, experimental conditions can often be chosen which will provide a substantial preference for the desired regioisomer. To understand the reason a particular set of experimental conditions leads to the preferential formation of one enolate whereas other conditions leads to the regioisomer, we need to examine the process of enolate generation in more detail.

The composition of an enolate mixture may be governed by kinetic or thermodynamic factors. The enolate ratio is governed by *kinetic control* when the product composition is governed by the relative *rates* of the two competing proton abstraction reactions.

4. E. H. Amonoco-Neizer, R. A. Shaw, D. O. Skovlin, and B. C. Smith, *J. Chem. Soc.* 2997 (1965); C. R. Kruger and E. G. Rochow, *J. Organomet. Chem.* **1**, 476 (1964).
5. C. A. Brown, *J. Org. Chem.* **39**, 1324 (1974); R. Pi, T. Friedl, P. v. R. Schleyer, P. Klusener, and L. Brandsma, *J. Org. Chem.* **52**, 4299 (1987); T. L. Macdonald, K. J. Natalie, Jr., G. Prasad, and J. S. Sawyer, *J. Org. Chem.* **51**, 1124 (1986).

$$R_2CH\overset{O}{\overset{\|}{C}}CH_2R' + B^- \xrightarrow{k_a} R_2C{=}\overset{O^-}{\overset{|}{C}}CH_2R' \ (\mathbf{A}); \quad \xrightarrow{k_b} R_2CH\overset{O^-}{\overset{|}{C}}{=}CHR' \ (\mathbf{B})$$

$$\frac{[\mathbf{A}]}{[\mathbf{B}]} = \frac{k_a}{k_b}$$

Kinetic control of isomeric enolate composition.

One the other hand, if enolates **A** and **B** can be interconverted rapidly, equilibrium will be established and the product composition will reflect the relative thermodynamic stability of the enolates. The enolate ratio is then governed by *thermodynamic control.*

$$R_2CH\overset{O}{\overset{\|}{C}}CH_2R' + B^- \underset{k_{-a}}{\overset{k_a}{\rightleftharpoons}} R_2C{=}\overset{O^-}{\overset{|}{C}}CH_2R' \ (\mathbf{A}); \quad \underset{k_{-b}}{\overset{k_b}{\rightleftharpoons}} R_2CH\overset{O^-}{\overset{|}{C}}{=}CHR' \ (\mathbf{B}); \quad \mathbf{A} \overset{K}{\rightleftharpoons} \mathbf{B}$$

$$\frac{[\mathbf{A}]}{[\mathbf{B}]} = K$$

Thermodynamic control of isomeric enolate composition.

By adjusting the conditions under which an enolate mixture is formed from a ketone, it is possible to establish either kinetic or thermodynamic control. *Ideal conditions for kinetic control of enolate formation are those in which deprotonation is rapid, quantitative, and irreversible.*[6] This ideal is approached experimentally by using a very strong base such as lithium diisopropylamide or hexamethyldisilylamide in an aprotic solvent in the absence of excess ketone. Lithium as the counterion is better than sodium or potassium for regioselective generation of the kinetic enolate. Aprotic solvents are required because protic solvents permit enolate equilibration by allowing reversible protonation–deprotonation, which gives rise to the thermodynamically controlled enolate composition. Excess ketone also catalyzes the equilibration.

The composition of enolate mixtures can be determined by allowing the enolate to react with an electrophile that rapidly traps the enolate. Trimethylsilyl chloride is the most frequently used electrophile.[7]

6. For a review, see J. d'Angelo, *Tetrahedron* **32**, 2979 (1976).
7. H. O. House, M. Gall, and H. D. Olmstead, *J. Org. Chem.* **36**, 2361 (1971).

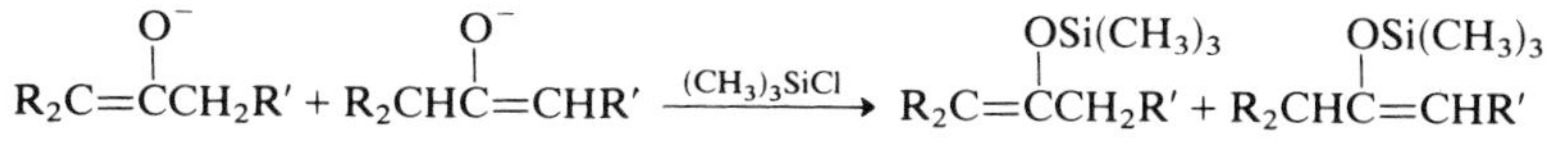

$$R_2C{=}C(O^-)CH_2R' + R_2CHC(O^-){=}CHR' \xrightarrow{(CH_3)_3SiCl} R_2C{=}C(OSi(CH_3)_3)CH_2R' + R_2CHC(OSi(CH_3)_3){=}CHR'$$

The composition of the silyl enol ether mixture is then determined by NMR spectroscopy or by gas chromatography. Table 1.2 shows data for the regioselectivity of enolate formation for several ketones under various reaction conditions.

A quite consistent relationship is found in these and related data. *Conditions of kinetic control usually favor the less substituted enolate.* The principal reason for this result is that removal of the less hindered proton is faster, for steric reasons, than removal of more hindered protons. Removal of the less hindered proton leads to the less substituted enolate. *On the other hand, at equilibrium the more substituted enolate is usually the dominant species.* The stability of carbon–carbon double bonds increases with increasing substitution, and it is this effect that leads to the greater stability of the more substituted enolate.

Kinetically controlled deprotonation of α,β-unsaturated ketones usually occurs preferentially at the α'-carbon adjacent to the carbonyl group. The electron-withdrawing effect of the carbonyl group is probably responsible for the faster deprotonation at this position.

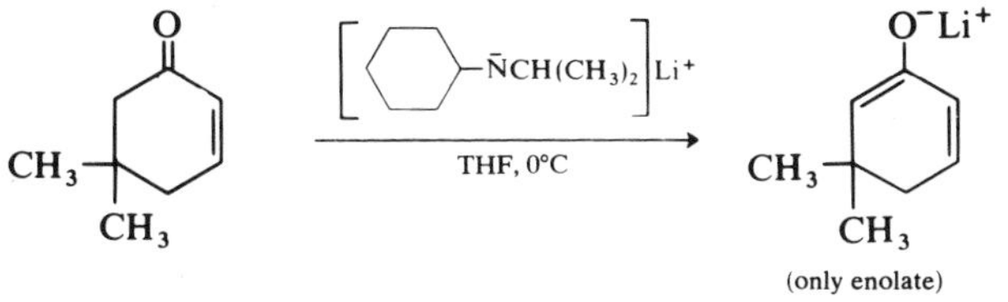

[Scheme: 4,4-dimethylcyclohex-2-enone (CH₃, CH₃) → with [cyclohexyl–$\bar{N}CH(CH_3)_2$] Li^+, THF, 0°C → lithium enolate O^-Li^+ (only enolate)] Ref. 8

Under conditions of thermodynamic control, however, it is the enolate corresponding to deprotonation of the γ-carbon that is present in the greater amount.

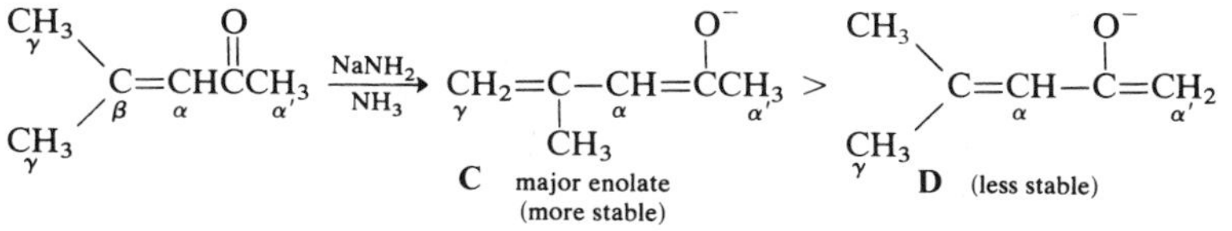

$$(CH_3)_2C{=}CHC(O)CH_3 \xrightarrow[NH_3]{NaNH_2} CH_2{=}C(CH_3){-}CH{=}C(O^-)CH_3 > (CH_3)_2C{=}CH{-}C(O^-){=}CH_2$$

(positions: γ CH₃ groups, β C, α CH, α' CH₃)

C major enolate (more stable)   **D** (less stable)   Ref. 9

These isomeric enolates differ in stability in that **C** is fully conjugated, whereas the π system in **D** is cross-conjugated. In isomer **D** the delocalization of the negative charge is restricted to the oxygen and the α' carbon, whereas in the conjugated system of **C** the negative charge is delocalized on oxygen and both the α- and γ-carbon.

The terms *kinetic control* and *thermodynamic control* are applicable to other reactions besides enolate formation; this concept was covered in general terms in Part A, Section 4.4. In discussions of other reactions in this chapter, it may be stated

8. R. A. Lee, C. McAndrews, K. M. Patel, and W. Reusch, *Tetrahedron Lett.*, 965 (1973).
9. G. Büchi and H. Wüest, *J. Am. Chem. Soc.* **96**, 7573 (1974).

Table 1.2. Compositions of Enolate Mixtures

1[a]	CH_3–cyclopentanone (O) →	CH_3 (O^-) enolate	CH_3 (O^-) enolate
	Kinetic control (Ph_3CLi/dimethoxyethane)	28	72
	Thermodynamic control (Ph_3CLi/equilibration in the presence of excess ketone)	94	6
2[b,c]	CH_3–cyclohexanone (O) →	CH_3 (O^-) enolate	CH_3 (O^-) enolate
	Kinetic control (LDA/dimethoxyethane)	1	99
	Thermodynamic control (Et_3N/DMF)	78	22
3[d]	C_6H_5–cyclohexanone (O) →	C_6H_5 (O^-) enolate	C_6H_5 (O^-) enolate
	Kinetic control (LDA/tetrahydrofuran, −70°C)[d]		Only enolate
	Thermodynamic control (KH, tetrahydrofuran)[e]	Only enolate	
4[a]	H, H decalone (O) →	H, H (O^-) enolate	H, H (O^-) enolate
	Kinetic control (Ph_3CLi/dimethoxyethane)	13	87
	Thermodynamic control (equilibration in the presence of excess ketone)	53	47
5[e]	$CH_3CH_2CH_2C(=O)CH_3$ →	$CH_3CH_2CH_2C(O^-)=CH_2$	
	Kinetic control (LDA/tetrahydrofuran, −78°C)	Only enolate	
6[f]	$CH_3CH_2C(=O)CH_2CH_3$ →	$CH_3(H)C=C(O^-)CH_2CH_3$ *Z*-enolate	$CH_3(H)C=C(CH_2CH_3)O^-$ *E*-enolate
	Kinetic control (lithium 2,2,6,6-tetramethylpiperidide/tetrahydrofuran)	13	87
	Thermodynamic control (equilibration in the presence of excess ketone)	84	16

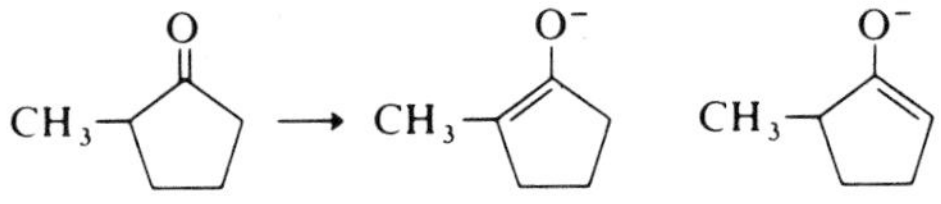

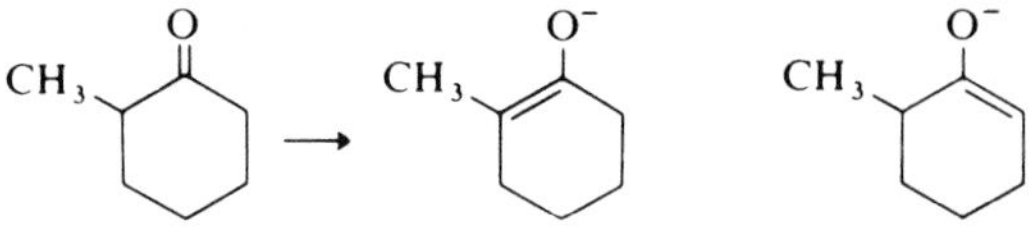

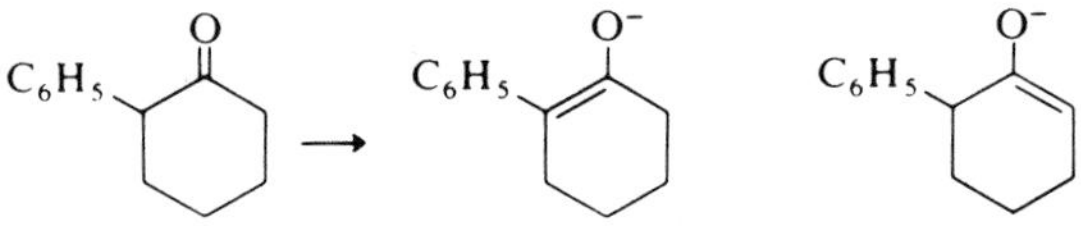

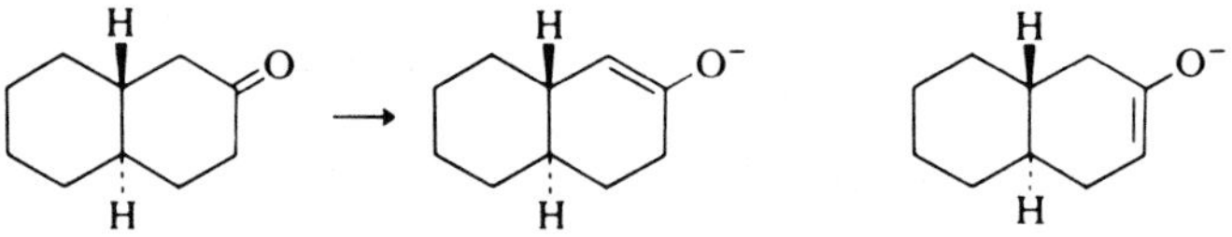

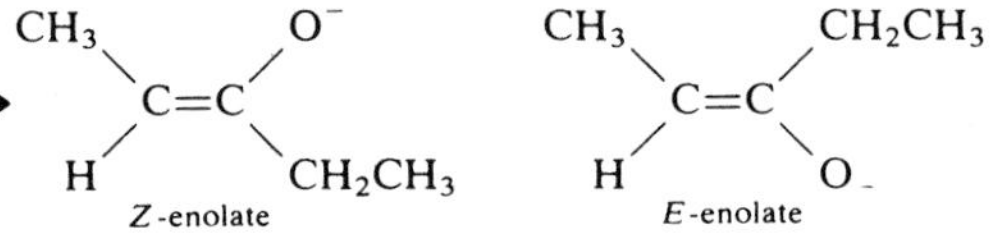

7[g]	$CH_3CH_2C(=O)C(CH_3)_3$ →	$(CH_3)(H)C{=}C(O^-)(C(CH_3)_3)$ *Z*	$(CH_3)(H)C{=}C(C(CH_3)_3)(O^-)$ *E*
Kinetic control (LDA/tetrahydrofuran)		>98	<2
8[g]	$CH_3CH_2C(=O)Ph$ →	$(CH_3)(H)C{=}C(O^-)(Ph)$ *Z*	$(CH_3)(H)C{=}C(Ph)(O^-)$ *E*
Kinetic control (LDA/Tetrahydrofuran)		>98	<2
9[h]	$CH_3(CH_2)_4C(=O)CH_3$ →	$CH_3(CH_2)_4C(O^-){=}CH_2$	$CH_3(CH_2)_3CH{=}C(O^-)CH_3$
Kinetic control (LDA, −78°C)		Only enolate	
Thermodynamic control (KH, tetrahydrofuran, 20°C)		46	54

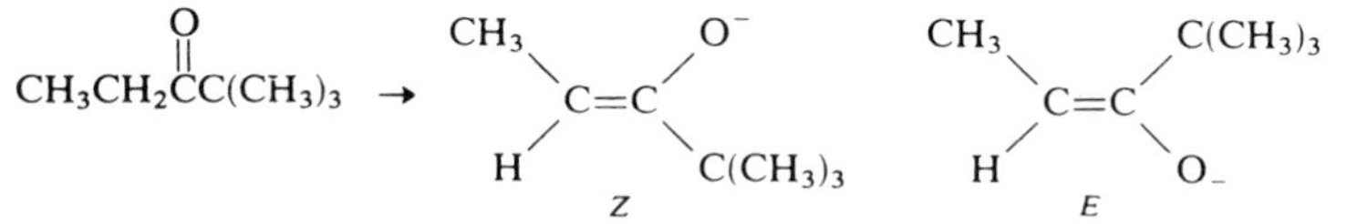

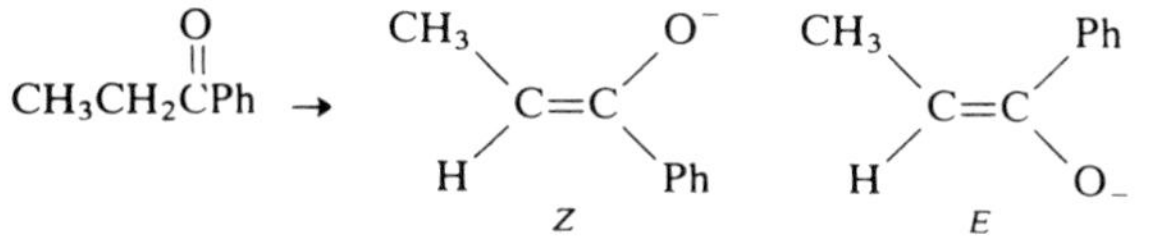

a. H. O. House and B. M. Trost, *J. Org. Chem.* **30**, 1341 (1965).
b. H. O. House, M. Gaol, and H. D. Olmstead, *J. Org. Chem.* **36**, 2361 (1971).
c. H. O. House, L. J. Czuba, M. Gall, and H. D. Olmstead, *J. Org. Chem.* **34**, 2324 (1969).
d. E. Vedejs, *J. Am. Chem. Soc.* **96**, 5944 (1974); H. J. Reich, J. M. Renga, and I. L. Reich, *J. Am. Chem. Soc.* **97**, 5434 (1975).
e. G. Stork, G. A. Kraus, and G. A. Garcia, *J. Org. Chem.* **39**, 3459 (1974).
f. Z. A. Fataftah, I. E. Kopka, and M. W. Rathke, *J. Am. Chem. Soc.* **102**, 3959 (1980).
g. C. H. Heathcock, C. T. Buse, W. A. Kleschick, M. C. Pirrung, J. E. Sohn, and J. Lampe, *J. Org. Chem.* **45**, 1066 (1980).
h. R. D. Clark and C. H. Heathcock, *J. Org. Chem.* **41**, 1396 (1976); C. A. Brown, *J. Org. Chem.* **39**, 3913 (1974).

that a given reagent or set of conditions favors the "thermodynamic product." This statement means that the mechanism operating is such that the various possible products are equilibrated after initial formation. When this is true, the dominant product can be predicted by considering the relative stabilities of the various possible products. On the other hand, if a given reaction is under "kinetic control," prediction or interpretation of the relative amounts of products must be made by analyzing the competing rates of product formation.

For many ketones, *stereoisomeric* as well as regioisomeric enolates can be formed, as is illustrated by entries 6, 7 and 8 of Table 1.2. The *stereoselectivity* of enolate formation, either under conditions of kinetic or thermodynamic control, can also be controlled to some extent. We will return to this topic in more detail in Chapter 2.

1.3. Other Means of Generating Enolates

The recognition of conditions under which lithium enolates are stable and do not equilibrate with regioisomers has permitted the use of other reactions in addition to proton abstraction to generate specific enolates. Several methods are shown in Scheme 1.3.

Scheme 1.3. Generation of Specific Enolates

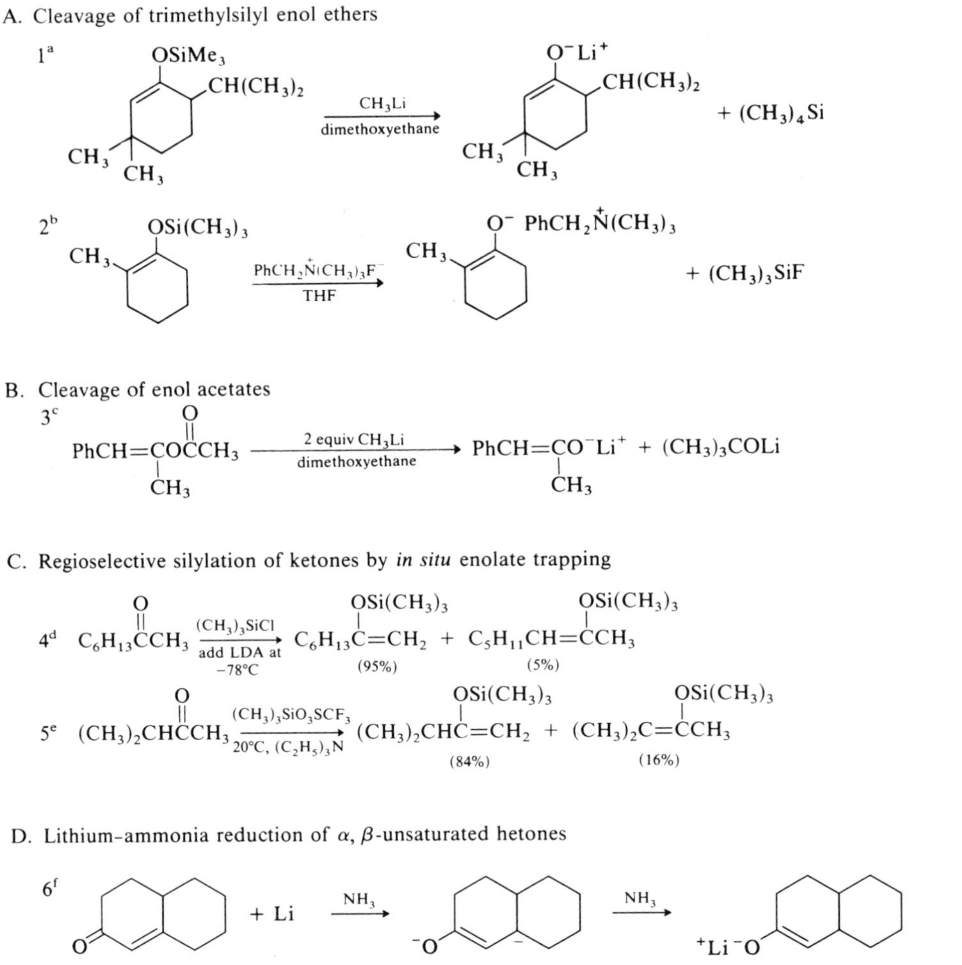

A. Cleavage of trimethylsilyl enol ethers

1[a] $OSiMe_3$, $CH(CH_3)_2$, CH_3, CH_3 $\xrightarrow[\text{dimethoxyethane}]{CH_3Li}$ O^-Li^+, $CH(CH_3)_2$, CH_3, CH_3 + $(CH_3)_4Si$

2[b] $OSi(CH_3)_3$, CH_3 $\xrightarrow[\text{THF}]{PhCH_2\overset{+}{N}(CH_3)_3F^-}$ CH_3, O^- $PhCH_2\overset{+}{N}(CH_3)_3$ + $(CH_3)_3SiF$

B. Cleavage of enol acetates

3[c] $PhCH{=}C(CH_3)OC(O)CH_3$ $\xrightarrow[\text{dimethoxyethane}]{2\text{ equiv }CH_3Li}$ $PhCH{=}C(CH_3)O^-Li^+$ + $(CH_3)_3COLi$

C. Regioselective silylation of ketones by *in situ* enolate trapping

4[d] $C_6H_{13}C(O)CH_3$ $\xrightarrow[\text{add LDA at }-78°C]{(CH_3)_3SiCl}$ $C_6H_{13}C(OSi(CH_3)_3){=}CH_2$ (95%) + $C_5H_{11}CH{=}C(OSi(CH_3)_3)CH_3$ (5%)

5[e] $(CH_3)_2CHC(O)CH_3$ $\xrightarrow[20°C,\ (C_2H_5)_3N]{(CH_3)_3SiO_3SCF_3}$ $(CH_3)_2CHC(OSi(CH_3)_3){=}CH_2$ (84%) + $(CH_3)_2C{=}C(OSi(CH_3)_3)CH_3$ (16%)

D. Lithium-ammonia reduction of α, β-unsaturated hetones

6[f] (octalone) + Li $\xrightarrow{NH_3}$ ^-O (radical anion) $\xrightarrow{NH_3}$ $^+Li\,^-O$ (enolate)

a. G. Stork and P. F. Hudrlik, *J. Am. Chem. Soc.* **90**, 4464 (1968); see also H. O. House, L. J. Czuba, M. Gall, and H. D. Olmstead, *J. Org. Chem.* **34**, 2324 (1969).
b. I. Kuwajima and E. Nakamura, *J. Am. Chem. Soc.* **97**, 3258 (1975).
c. G. Stork and S. R. Dowd, *Org. Synth.* **55**, 46 (1976); see also H. O. House and B. M. Trost, *J. Org. Chem.* **30**, 2502 (1965).
d. E. J. Corey and A. W. Gross, *Tetrahedron Lett.* **25**, 495 (1984).
e. H. Emde, A. Götz, K. Hofmann, and G. Simchen, *Justus Liebigs Ann. Chem.*, 1643 (1981).
f. G. Stork, P. Rosen, N. Goldman, R. V. Coombs and J. Tsuji, *J. Am. Chem. Soc.* **87**, 275 (1965).

Cleavage of enol trimethylsilyl ethers or enol acetates by methyllithium (entries 1 and 3, Scheme 1.3) is a route to specific enolate formation which depends on the availability of these materials in high purity. Trimethylsilyl enol ethers can also be cleaved by tetraalkylammonium fluoride (entry 2, Scheme 1.3). The driving force for this reaction is the formation of the very strong Si—F bond, which has a bond energy of 142 kcal/mol.[10] Preparation of the trimethylsilyl enol ethers from an enolate mixture will reflect the enolate composition. If the enolate formation can be done with high regioselectivity, the corresponding trimethylsilyl enol ether can be obtained in high purity. If not, the silyl enol ether mixture must be separated.

Trimethylsilyl enol ethers can be prepared directly from ketones. One procedure involves reaction with trimethylsilyl chloride and a tertiary amine.[11] This procedure gives the regioisomers in a ratio favoring the thermodynamically stable enol ether. Use of *t*-butyldimethylsilyl chloride with potassium hydride as the base also seems to favor the thermodynamic product.[12] Trimethylsilyl trifluoromethanesulfonate (TMS triflate), which is more reactive, gives primarily the less substituted trimethylsilyl enol ether (entry 5, Scheme 1.3).[13] The best ratio of less substituted to more substituted enol ether is obtained by treating a mixture of ketone and trimethylsilyl chloride with LDA at −78°C.[14] Under these conditions, the kinetically preferred enolate is immediately trapped by reaction with trimethylsilyl chloride (entry 4). Even greater preferences for the less substituted silyl enol ether can be obtained by using the more hindered amide from *t*-octyl-*t*-butylamine.

Lithium–ammonia reduction of α,β-unsaturated ketones (entry 6, Scheme 1.3) provides a very useful method for generating specific enolates.[15] The desired starting materials are often readily available, and the position of the double bond in the enone determines the structure of the resulting enolate. This and other reductive methods for generating enolates from enones will be discussed more fully in Chapter 5. Another very important method for specific enolate generation, the addition of organometallic reagents to enones, will be discussed in Chapter 8.

1.4. Alkylation of Enolates[16]

The alkylation of relatively acidic substances such as β-diketones, β-ketoesters, and esters of malonic acid can be carried out in alcohols as solvents using metal

10. For reviews of the chemistry of *O*-silyl enol ethers, see J. K. Rasmussen, *Synthesis*, 91 (1977); P. Brownbridge, *Synthesis*, 1, 85 (1983): I. Kuwajima and E. Nakamura, *Acc. Chem. Res.* **18**, 181 (1985).
11. H. O. House, L. J. Czuba, M. Gall and H. D. Olmstead, *J. Org. Chem.* **34**, 2324 (1969); R. D. Miller and D. R. McKean, *Synthesis*, 730 (1979).
12. J. Orban, J. V. Turner, and B. Twitchin, *Tetrahedron Lett.* **25**, 5099 (1984).
13. H. Emde, A. Gotz, K. Hofmann and G. Simchen, *Justus Liebigs Ann. Chem.*, 1643 (1981); see also E. J. Corey, H. Cho, C. Rücker, and D. Hua, *Tetrahedron Lett.*, 3455 (1981).
14. E. J. Corey and A. W. Gross, *Tetrahedron Lett.* **25**, 495 (1984).
15. For a review of α,β-enone reduction, see D. Caine, *Org. React.* **23**, 1 (1976).
16. A general review of enolate alkylation is available: D. Caine, in *Carbon–Carbon Bond Formation*, Vol. 1, R. L. Augustine (ed.), Marcel Dekker, New York, 1979, Chapter 2.

alkoxides as bases. The presence of two electron-withdrawing substituents favors formation of the enolate resulting from removal of a proton from the carbon situated between them. Alkylation then occurs by an S_N2 process.

Some examples of alkylation reactions involving relatively acidic carbon acids are shown in Scheme 1.4. These reactions are all mechanistically similar in that a carbanion, formed by deprotonation using a suitable base, reacts with an electrophile in an S_N2 manner. The alkylating agent must be reactive toward nucleophilic displacement. Primary halides and sulfonates, especially allylic and benzylic ones, are the most reactive alkylating agents. Secondary systems react more slowly and often give moderate yields because of competing elimination. Tertiary halides give only elimination products.

Methylene groups can be dialkylated if sufficient base and alkylating agent are used. Dialkylation can be an undesirable side reaction if the monoalkyl derivative

Scheme 1.4. Alkylations of Relatively Acidic Carbon Acids

1[a] $CH_3COCH_2CO_2C_2H_5 + CH_3(CH_2)_3Br \xrightarrow{NaOEt} CH_3COCH(CO_2C_2H_5)(CH_2)_3CH_3$ (69–72%)

2[b] $CH_2(CO_2C_2H_5)_2$ + 3-chlorocyclopentene $\xrightarrow{NaOEt}$ 3-cyclopentenyl-$CH(CO_2C_2H_5)_2$ (61%)

3[c] $CH_3COCH_2COCH_3 + CH_3I \xrightarrow{K_2CO_3} CH_3COCH(CH_3)COCH_3$ (75–77%)

4[d] $CH_3COCH_2CO_2C_2H_5 + ClCH_2CO_2C_2H_5 \xrightarrow{NaOEt} CH_3COCH(CO_2C_2H_5)CH_2CO_2C_2H_5$

5[e] $Ph_2CHCN + KNH_2 \rightarrow Ph_2\overset{-}{C}CN$

$Ph_2\overset{-}{C}CN + PhCH_2Cl \rightarrow Ph_2C(CN)CH_2Ph$ (98–99%)

6[f] $PhCH_2CO_2C_2H_5 + NaNH_2 \rightarrow Ph\overset{-}{C}HCO_2C_2H_5$

$Ph\overset{-}{C}HCO_2C_2H_5 + PhCH_2CH_2Br \rightarrow PhCH(CO_2C_2H_5)CH_2CH_2Ph$ (77–81%)

7[g] $CH_2(CO_2C_2H_5)_2 + BrCH_2CH_2CH_2Cl \xrightarrow{NaOEt}$ 1,1-cyclobutane-$(CO_2C_2H_5)_2$ (53–55%)

8[h] 2-(CO_2CH_3)cyclopentanone $+ BrCH_2(CH_2)_5CO_2C_2H_5 \xrightarrow[DMF]{NaH}$ 2-(CO_2CH_3)-2-[$CH_2(CH_2)_5CO_2C_2H_5$]cyclopentanone (85% on 1-mol scale)

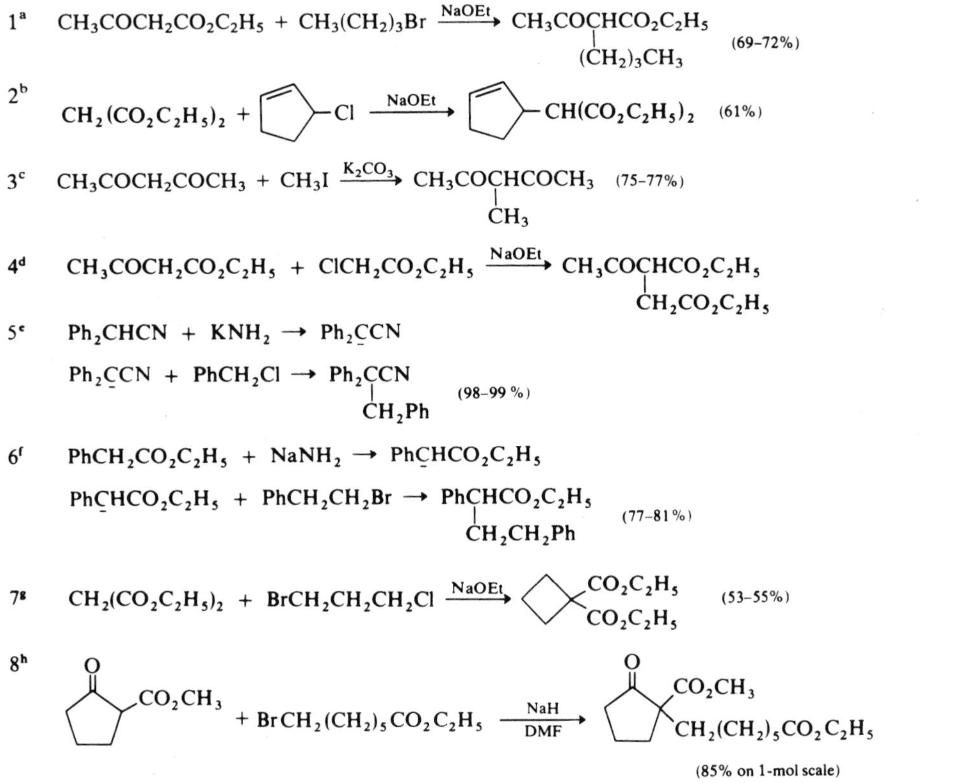

a. C. S. Marvel and F. D. Hager, *Org. Synth.* **I**, 248 (1941).
b. R. B. Moffett, *Org. Synth.* **IV**, 291 (1963).
c. A. W. Johnson, E. Markham, and R. Price, *Org. Synth.* **42**, 75 (1962).
d. H. Adkins, N. Isbell, and B. Wojcik, *Org. Synth.* **II**, 262 (1943).
e. C. R. Hauser and W. R. Dunnavant, *Org. Synth.* **IV**, 962 (1963).
f. E. M. Kaiser, W. G. Kenyon, and C. R. Hauser, *Org. Synth.* **47**, 72 (1967).
g. R. P. Mariella and R. Raube, *Org. Synth.* **IV**, 288 (1963).
h. K. F. Bernardy, J. F. Poletto, J. Nocera, P. Miranda, R. E. Schaub, and M. J. Weiss, *J. Org. Chem.* **45**, 4702 (1980).

is the desired product. Use of dihaloalkanes as the alkylating reagent leads to ring formation, as illustrated by the diethyl cyclobutanedicarboxylate synthesis (entry 7) shown in Scheme 1.4. This example illustrates the synthesis of cyclic compounds by *intramolecular* alkylation reactions. The relative rates of cyclization for haloalkyl malonate esters are 650,000:1:6500:5 for formation of three-, four-, five-, and six-membered rings, respectively.[17]

Scheme 1.5. Synthesis of Ketones and Carboxylic Acid Derivatives via Alkylation Techniques

1[a] $CH_3COCH(CO_2C_2H_5)(CH_2)_3CH_3 \xrightarrow{H_2O,\ ^-OH} CH_3COCH(CO_2^-)(CH_2)_3CH_3 \xrightarrow[\Delta]{H^+} CH_3CO(CH_2)_4CH_3$ (52–61%)

(see Scheme 1.4)

2[b] $CH_2(CO_2C_2H_5)_2 + C_7H_{15}Br \xrightarrow{NaOBu} C_7H_{15}CH(CO_2C_2H_5)_2$

$C_7H_{15}CH(CO_2C_2H_5)_2 \xrightarrow{H_2O,\ ^-OH} \xrightarrow{H^+} C_7H_{15}CH(CO_2H)_2$

$C_7H_{15}CH(CO_2H)_2 \xrightarrow{\Delta} C_8H_{16}CO_2H + CO_2$ (66–75%)

3[c] 1,1-cyclobutane-$(CO_2C_2H_5)_2 \xrightarrow{H_2O,\ ^-OH} \xrightarrow{H^+}$ 1,1-cyclobutane-$(CO_2H)_2 \xrightarrow{\Delta}$ cyclobutane-$CO_2H + CO_2$

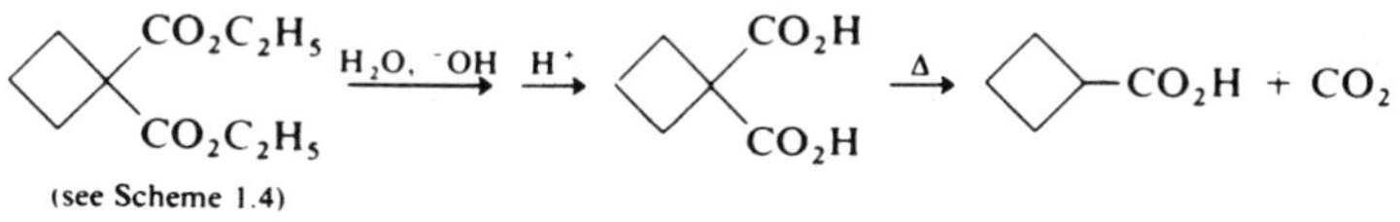

(see Scheme 1.4)

4[d] $NCCH_2CO_2C_2H_5$ + 2-chlorobenzyl-$CH_2Cl \xrightarrow{NaOEt}$ 2-Cl-$C_6H_4CH_2CH(CN)CO_2C_2H_5$

1) $H_2O,\ ^-OH$
2) H^+
3) $\Delta,\ -CO_2$

→ 2-Cl-$C_6H_4CH_2CH_2CN$

5[e] 2-oxocyclopentane-CO_2CH_3 + $PhCH_2Cl \xrightarrow{Na}$ 2-oxo-1-(CH_2Ph)cyclopentane-CO_2CH_3

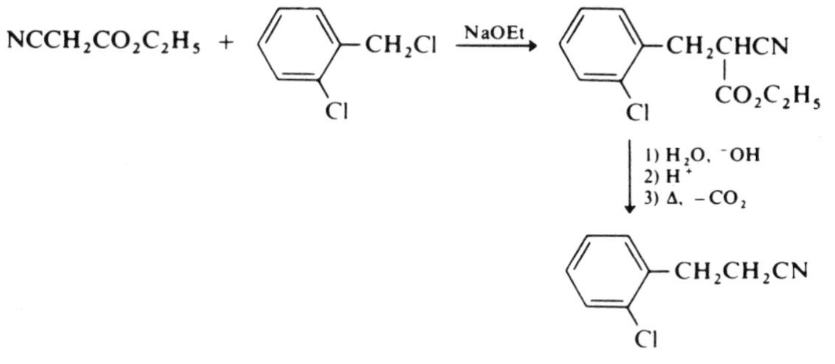

2-oxo-1-(CH_2Ph)cyclopentane-CO_2CH_3 + LiI → 2-(CH_2Ph)cyclopentanone + $CH_3I + CO_2$ (72–76%)

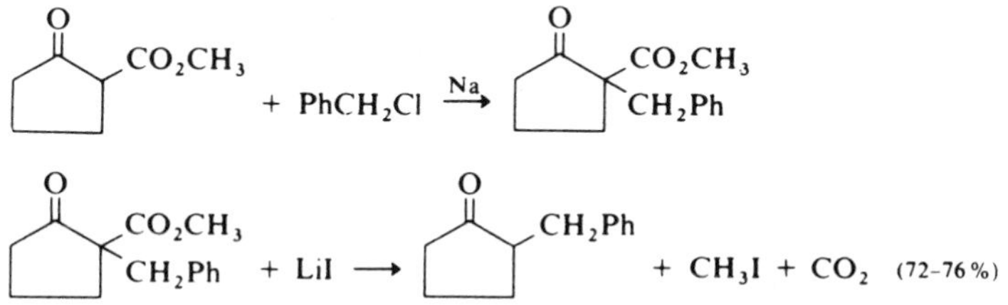

a. J. R. Johnson and F. D. Hager, *Org. Synth.* **I**, 351 (1941).
b. E. E. Reid and J. R. Ruhoff, *Org. Synth.* **II**, 474 (1943).
c. G. B. Heisig and F. H. Stodola, *Org. Synth.* **III**, 213 (1955).
d. J. A. Skorcz and F. E. Kaminski, *Org. Synth.* **48**, 53 (1968).
e. F. Elsinger, *Org. Synth.* **V**, 76 (1973).

17. A. C. Knipe and C. J. Stirling, *J. Chem. Soc., B*, 67 (1968); for a discussion of factors which affect intramolecular alkylation of enolates, see J. Janjatovic and Z. Majerski, *J. Org. Chem.* **45**, 4892 (1980).

Scheme 1.6. Regioselective Enolate Alkylation

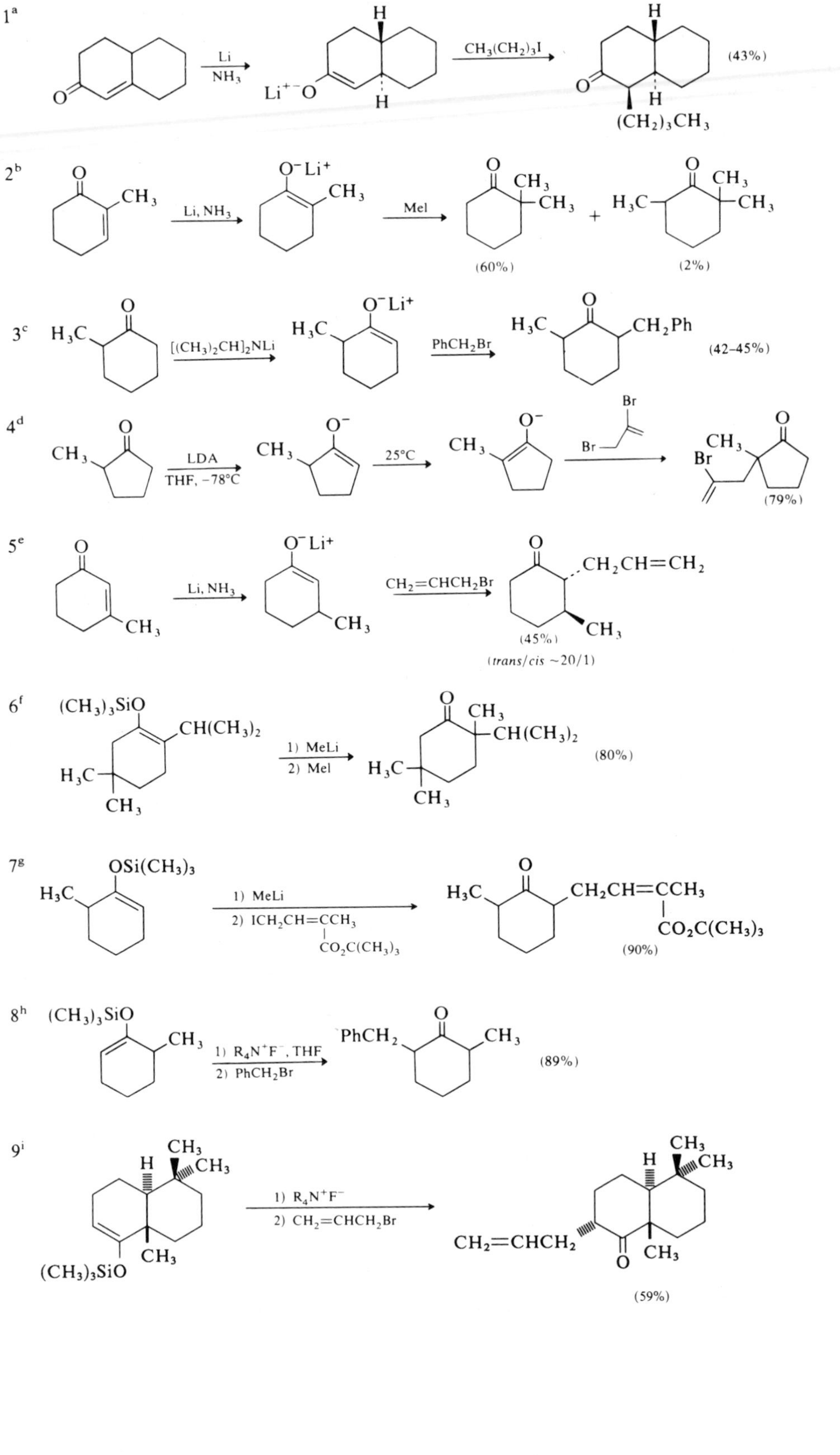

1[a] Li, NH_3; $CH_3(CH_2)_3I$ — (43%)

2[b] Li, NH_3; MeI — (60%) + (2%)

3[c] $[(CH_3)_2CH]_2NLi$; $PhCH_2Br$ — (42–45%)

4[d] LDA, THF, −78°C; 25°C; $BrCH_2C(Br)=CH_2$ — (79%)

5[e] Li, NH_3; $CH_2=CHCH_2Br$ — (45%) (*trans/cis* ~20/1)

6[f] 1) MeLi 2) MeI — (80%)

7[g] 1) MeLi 2) $ICH_2CH=C(CH_3)CO_2C(CH_3)_3$ — (90%)

8[h] 1) $R_4N^+F^-$, THF 2) $PhCH_2Br$ — (89%)

9[i] 1) $R_4N^+F^-$ 2) $CH_2=CHCH_2Br$ — (59%)

Relatively acidic carbon acids such as malonate esters and β-ketoesters were the first class of carbanions for which reliable conditions for alkylation were developed. The reason for this was that these carbanions are formed by easily accessible alkoxide ions. The preparation of 2-substituted β-ketoesters (entries 1, 4, and 8) and 2-substituted derivatives of malonic ester (entries 2 and 7) by the methods illustrated in Scheme 1.4 is useful for the synthesis of ketones and carboxylic acids. Both β-ketoacids and malonic acids undergo facile decarboxylation:

$$\text{X-C(=O)-C(R)(R')-C(=O)OH} \xrightarrow{-CO_2} \text{X-C(OH)=C(R)R'} \longrightarrow \text{X-C(=O)-CH(R)R'}$$

β-keto acid: X = alkyl or aryl = ketone

substituted malonic acid: X = OH = substituted acetic acid

Examples of this approach to the synthesis of ketones and carboxylic acids are presented in Scheme 1.5. In these procedures, an ester group is removed by hydrolysis and decarboxylation after the alkylation step. The malonate and acetoacetate carbanions are the *synthetic equivalents* of the simpler carbanions lacking the ester substituent. In the preparation of 2-heptanone (entry 1), for example, ethyl acetoacetate functions as the synthetic equivalent of acetone. It is also possible to use the dilithium derivative of acetoacetic acid as the synthetic equivalent of acetone enolate.[18] In this case, the hydrolysis step is unnecessary, and decarboxylation can be done directly on the alkylation product.

$$CH_3\overset{O}{\overset{\|}{C}}CH_2CO_2H \xrightarrow{2\ n\text{-BuLi}} CH_3\overset{O^-\ Li^+}{\overset{|}{C}}{=}CH\overset{O}{\overset{\|}{C}}O^-Li^+ \xrightarrow[\substack{2)\ H^+ \\ (-CO_2)}]{1)\ R-X} CH_3\overset{O}{\overset{\|}{C}}CH_2R$$

Similarly, the dilithium dianion of monoethyl malonate is easily alkylated and the product decarboxylates on acidification.[19]

The use of β-ketoesters and malonate ester enolates has, to a significant extent, been supplanted by the development of procedures which are satisfactory for direct

18. R. A. Kjonaas and D. D. Patel, *Tetrahedron Lett.* **25**, 5467 (1984).
19. J. E. McMurry and J. H. Musser, *J. Org. Chem.* **40**, 2557 (1975).

---

a. G. Stork, P. Rosen, N. Goldman, R. V. Coombs, and J. Tsujii, *J. Am. Chem. Soc.* **87**, 275 (1965).
b. H. A. Smith, B. J. L. Huff, W. J. Powers III, and D. Caine, *J. Org. Chem.* **32**, 2851 (1967).
c. M. Gall and H. O. House, *Org. Synth.* **52**, 39 (1972).
d. S. C. Welch and S. Chayabunjonglerd, *J. Am. Chem. Soc.* **101**, 6768 (1979).
e. D. Caine, S. T. Chao, and H. A. Smith, *Org. Synth.* **56**, 52 (1977).
f. G. Stork and P. F. Hudrlik, *J. Am. Chem. Soc.* **90**, 4464 (1968).
g. P. L. Stotter and K. A. Hill, *J. Am. Chem. Soc.* **96**, 6524 (1974).
h. I. Kuwajima, E. Nakamura, and M. Shimizu, *J. Am. Chem. Soc.* **104**, 1025 (1982).
i. A. B. Smith III and R. Mewshaw, *J. Org. Chem.* **49**, 3685 (1984).

alkylation of ketone and ester enolates. When this is not possible, the β-ketoester and malonate procedures are reliable alternatives. Scheme 1.6 illustrates examples of ketone alkylation.

The development of conditions for stoichiometric formation of both kinetically and thermodynamically controlled enolates has permitted the extensive use of enolate alkylation reactions in multistep synthesis of complex molecules. One aspect of the alkylation which is crucial in many cases is the stereoselectivity. First, it must be recognized that the alkylation step will have a stereoelectronic preference for approach of the electrophile perpendicular to the plane of the enolate, since the electrons which are involved in bond formation are the π electrons. A major factor in determining the stereoselectivity of ketone enolate alkylations is the difference in steric hindrance on the two faces of the enolate. The electrophile will approach from the less hindered of the two faces, and the degree of stereoselectivity will depend upon the degree of steric differentiation. Entries 1, 5, and 9 in Scheme 1.6 are cases where stereoselectivity comes into play.

For simple, conformationally biased cyclohexanone enolates such as that from 4-*t*-butycyclohexanone, there is little steric differentiation. The alkylation product is a nearly 1:1 mixture of the *cis* and *trans* isomers.

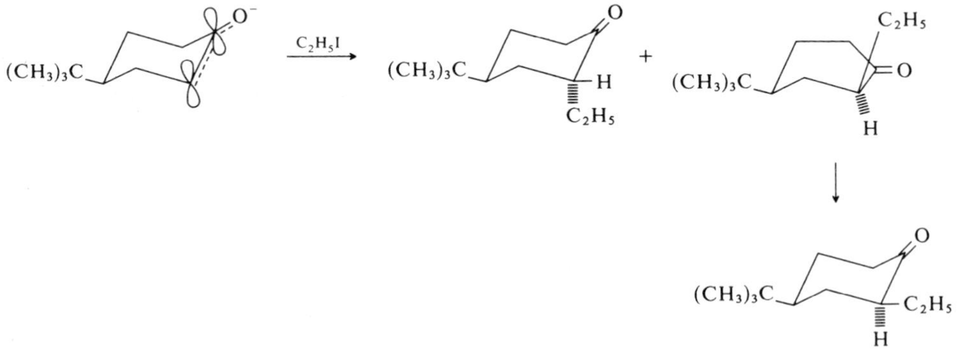

Ref. 20

The *cis* product must be formed through a transition state with a twistlike conformation to adhere to the requirements of stereoelectronic control. The fact that this pathway is not disfavored is consistent with other evidence that in enolate alkylations the transition state is *early* and reflects primarily the structural features of the reactant, not the product. A late transition state should disfavor the formation of the *cis* isomer because of the strain energy associated with the initial nonchair conformation of the product.

The introduction of an alkyl substituent at the α-carbon in the enolate enhances stereoselectivity somewhat. This is attributed to a steric effect in the enolate. To minimize steric interaction with the solvated oxygen, the alkyl group is distorted somewhat from coplanarity. This biases the enolate toward attack from the axial

20. H. O. House, B. A. Terfertiller, and H. D. Olkmstead, *J. Org. Chem.* **33**, 935 (1968).

direction. The alternate approach from the upper face would enhance the steric interaction by forcing the alkyl group to become eclipsed with the enolate oxygen.[21]

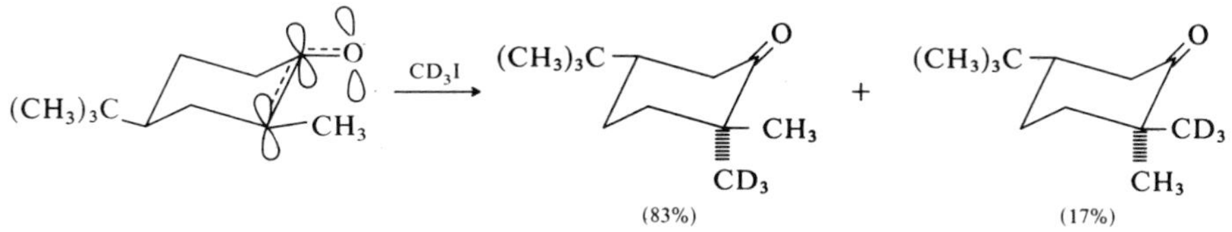

When an additional methyl substituent is placed at C-3, there is a strong preference for alkylation *anti* to the 3-methyl group. This can be attributed to the conformation of the enolate, which will place the methyl group in a pseudoaxial conformation because of allylic strain (see Part A, Section 3.3). The C-3 methyl group then shields the lower face of the enolate.[22]

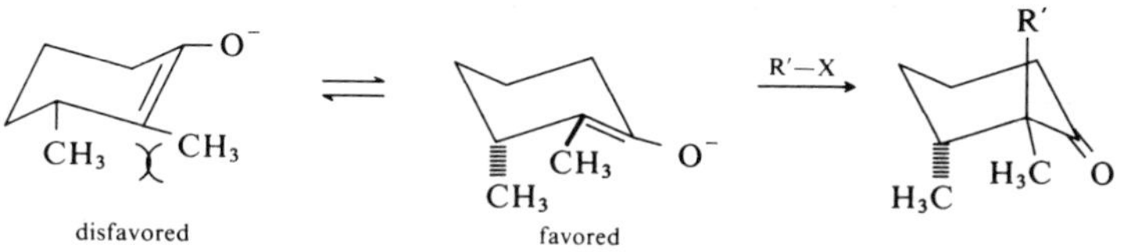

The enolates of 1- and 2-decalone derivatives provide further insights into the factors governing stereoselectivity in enolate alkylations. The 1(9)-enolate of 1-decalone shows a preference for alkylation to give the *cis* ring junction.

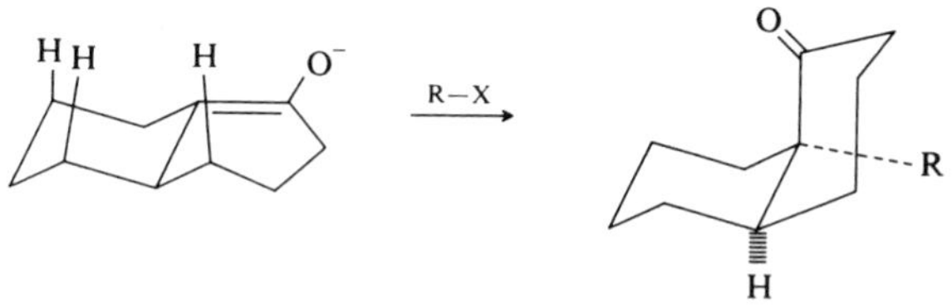

This is believed to be due primarily to a steric effect. The upper face of the enolate presents three hydrogens in a 1,3-diaxial relationship to the approaching electrophile. The corresponding hydrogens on the lower face are equatorial.[23]

The 2(1)-enolate of *trans*-2-decalone is preferentially alkylated by an axial approach of the electrophile.

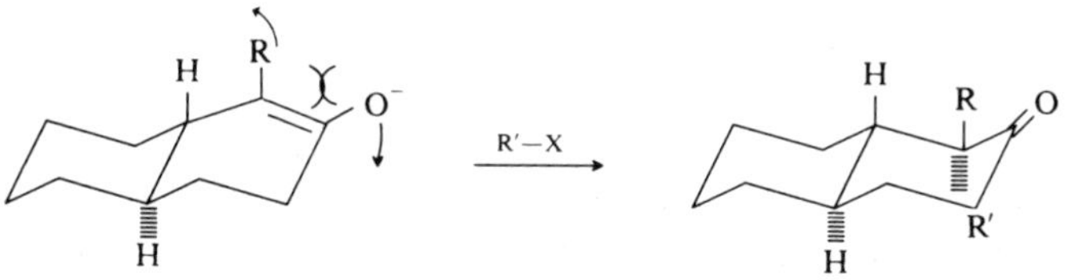

21. H. O. House and M. J. Umen, *J. Org. Chem.* **38**, 1000 (1973).
22. R. K. Boeckman, Jr., *J. Org. Chem.* **38**, 4450 (1973).
23. H. O. House and B. M. Trost, *J. Org. Chem.* **30**, 2502 (1965).

The stereoselectivity is enhanced if there is an alkyl substituent at C-1. The factors operating in this case are similar to those described for 4-*t*-butylcyclohexanone. The *trans*-decalone framework is conformationally rigid. Axial attack from the lower face leads directly to the chair conformation of the product. The 1-alkyl group enhances this stereoselectivity because a steric interaction with the solvated enolate oxygen distorts the enolate in such a way as to favor the axial attack.[24] The placement of an axial methyl group at C-10 in a 2(1)-decalone enolate introduces a 1,3-diaxial interaction with the approaching electrophile. As a result, the preferred alkylation product is that resulting from approach on the other side of the enolate.

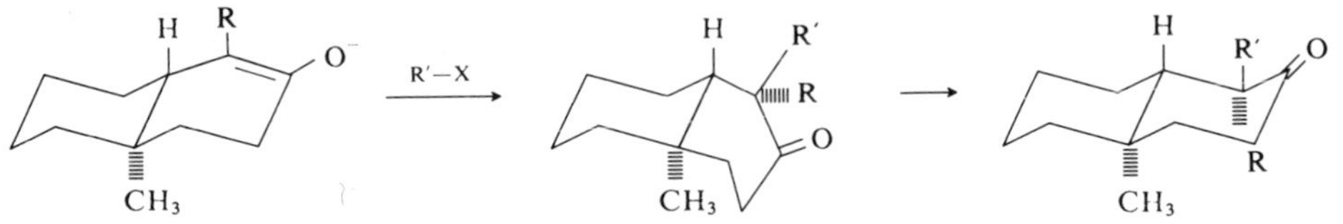

H R O⁻ R′—X H R′ R O CH₃ → H R′ O R CH₃

The prediction and interpretation of alkylation stereochemistry also depends on consideration of conformational effects in the enolate. The decalone enolate **1** is found to have a strong preference for alkylation to give the *cis* ring junction, with alkylation occurring *syn* to the *t*-butyl substituent.[25]

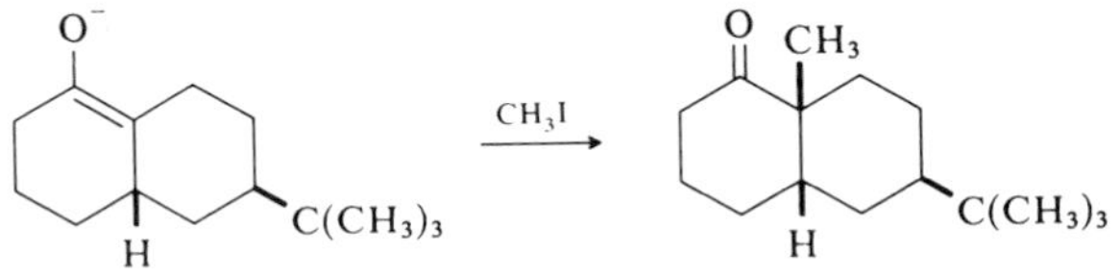

O⁻ H C(CH₃)₃ **1** CH_3I → O CH₃ H C(CH₃)₃

According to molecular mechanics calculations, the minimum-energy conformation of the enolate is a twist boat conformation (because the chair leads to an axial orientation of the *t*-butyl group). The enolate is convex in shape with the second ring shielding the bottom face of the enolate, and alkylation therefore occurs from the top.

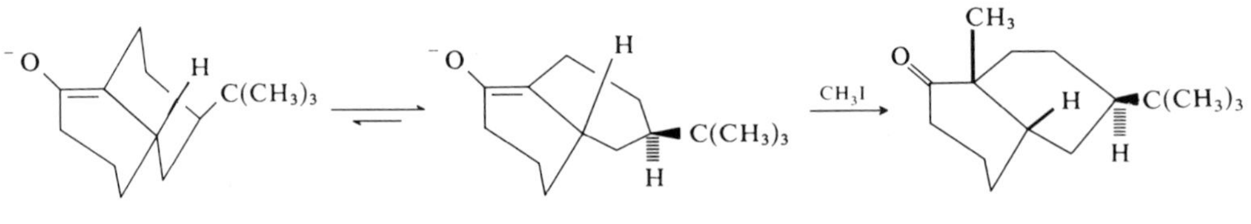

⁻O H C(CH₃)₃ ⇌ ⁻O H C(CH₃)₃ H CH_3I → O CH₃ H C(CH₃)₃ H

If the alkylation is intramolecular, an additional stereoelectronic factor becomes important. This is the conformational restrictions on the direction of approach of the electrophile to the enolate. Baldwin and co-workers have summarized the general

24. R. S. Mathews, S. S. Grigenti, and E. A. Folkers, *J. Chem. Soc., Chem. Commun.*, 708 (1970); P. Lansbury and G. E. DuBois, *Tetrahedron Lett.*, 3305 (1972).
25. H. O. House, W. V. Phillips, and D. Van Derveer, *J. Org. Chem.* **44**, 2400 (1979).

principles that govern the energetics of intramolecular ring-closure reactions.[26] (See Part A, Section 3.9.) The intramolecular alkylation reaction of **2** gives exclusively **3**.[27] The alkylation probably occurs through a transition state like **E**. The transition state **F** for formation of the *trans* ring junction would be more strained because of the necessity to reach across to the opposite face of the enolate π system.

O^- CH_3 $(CH_2)_4OSO_2Ar$ H **2** → CH_3 H O **3**

H H H OSO_2Ar O^- **E** OSO_2Ar H H H O^- **F**

Most enolate alkylations are carried out by deprotonating the ketone under appropriate kinetic or thermodynamic conditions. Alkylation can also be carried out with silyl enol ethers by generating the enolate by fluoride ion.[28] Anhydrous tetraalkylammonium fluoride salts are normally used as the fluoride ion source.

H_3C CH_3 O CH_3 —1) LDA 2) TMSCl→ H_3C CH_3 TMSO CH_3 —$R_4N^+F^-$, $CH_2{=}CHCH_2Br$→ H_3C CH_3 $CH_2{=}CHCH_2$ O CH_3

Ref. 29

1.5. Generation and Alkylation of Dianions

In the presence of a very strong base, such as an alkyllithium, sodium hydride, potassium or sodium amide, or lithium diisopropylamide, 1,3-dicarbonyl compounds may be converted to their *dianions* by two sequential deprotonations.[30] For example, reaction of benzoylacetone with sodium amide leads first to the enolate generated by deprotonation at the methylene group between the two carbonyl groups. A second equivalent of base deprotonates the benzyl methylene group to give a dianion.

$$PhCH_2\overset{O}{\overset{\|}{C}}CH_2\overset{O}{\overset{\|}{C}}CH_3 \xrightarrow{2\ NaNH_2} PhCH{=}\overset{Li^+O^-}{\overset{|}{C}}{-}CH{=}\overset{O^-Li^+}{\overset{|}{C}}{-}CH_3 \xrightarrow{PhCH(Cl)CH_3} PhCH(CHCH_3Ph)\overset{O}{\overset{\|}{C}}CH_2\overset{O}{\overset{\|}{C}}CH_3$$

Ref. 31

26. J. E. Baldwin, R. C. Thomas, L. I. Kruse, and L. Silberman, *J. Org. Chem.* **42**, 3846 (1977).
27. J. M. Conia and F. Rouessac, *Tetrahedron* **16**, 45 (1961).
28. I. Kuwajima, E. Nakamura, and M. Shimizu, *J. Am. Chem. Soc.* **104**, 1025 (1982).
29. A. B. Smith III and R. Mewshaw, *J. Org. Chem.* **49**, 3685 (1984).
30. For reviews, see (a) T. M. Harris and C. M. Harris, *Org. React.* **17**, 155 (1969); (b) E. M. Kaiser, J. D. Petty, and P. L. A. Knutson, *Synthesis*, 509 (1977).
31. D. M. von Schriltz, K. G. Hampton, and C. R. Hauser, *J. Org. Chem.* **34**, 2509 (1969).

Alkylation reactions of dianions occur at the *more basic* carbon. The beauty of this technique is that it allows alkylation of 1,3-dicarbonyl compounds to be carried out cleanly at the less acidic position. Since, as discussed earlier, alkylation of the monoanion occurs at the carbon between the two carbonyl groups, the site of monoalkylation can be controlled by choice of the amount and nature of the base A few examples of the formation and alkylation of dianions are collected in Scheme 1.7.

Scheme 1.7. Generation and Alkylation of Dianions

1[a]

$$CH_3\overset{O}{\overset{\|}{C}}CH_2CHO \xrightarrow[2\ equiv]{KNH_2} CH_2{=}\overset{O^-}{\overset{|}{C}}{-}CH{=}\overset{O^-}{\overset{|}{C}}H \xrightarrow[2)\ H_3O^+]{1)\ PhCH_2Cl} PhCH_2CH_2\overset{O}{\overset{\|}{C}}CH_2CHO\ (80\%)$$

2[b]

$$CH_3\overset{O}{\overset{\|}{C}}CH_2\overset{O}{\overset{\|}{C}}CH_3 \xrightarrow[2\ equiv]{NaNH_2} CH_2{=}\overset{O^-}{\overset{|}{C}}{-}CH{=}\overset{O^-}{\overset{|}{C}}CH_3 \xrightarrow[2)\ H_3O^+]{1)\ BuBr} CH_3(CH_2)_4\overset{O}{\overset{\|}{C}}CH_2\overset{O}{\overset{\|}{C}}CH_3\ (81\text{–}82\%)$$

3[c]

2-methyl-6-(hydroxymethylene)cyclohexanone $\xrightarrow[2\ equiv]{KNH_2}$ dianion (CH_3, CHO^-, O^-) $\xrightarrow{BuBr}$ 2-methyl-2-butyl-6-(CHO^-)cyclohexanone [CH_3, $CH_3(CH_2)_3$] $\xrightarrow{NaOH,\ H_2O}$ 2-methyl-2-butylcyclohexanone [CH_3, $CH_3(CH_2)_3$] (54–74%)

4[d]

$$CH_3\overset{O}{\overset{\|}{C}}CH_2CO_2CH_3 \xrightarrow[2)\ RLi]{1)\ NaH} CH_2{=}\overset{O^-}{\overset{|}{C}}CH{=}\overset{O^-}{\overset{|}{C}}OCH_3 \xrightarrow[2)\ H_3O^+]{1)\ EtBr} CH_3(CH_2)_2\overset{O}{\overset{\|}{C}}CH_2CO_2CH_3\ (84\%)$$

5[e]

$$CH_2{=}\overset{O^-}{\overset{|}{C}}CH{=}\overset{O^-}{\overset{|}{C}}OCH_3 + (CH_3)_2C{=}CHCH_2Br \rightarrow (CH_3)_2C{=}CHCH_2CH_2\overset{O}{\overset{\|}{C}}CH_2CO_2CH_3\ (85\%)$$

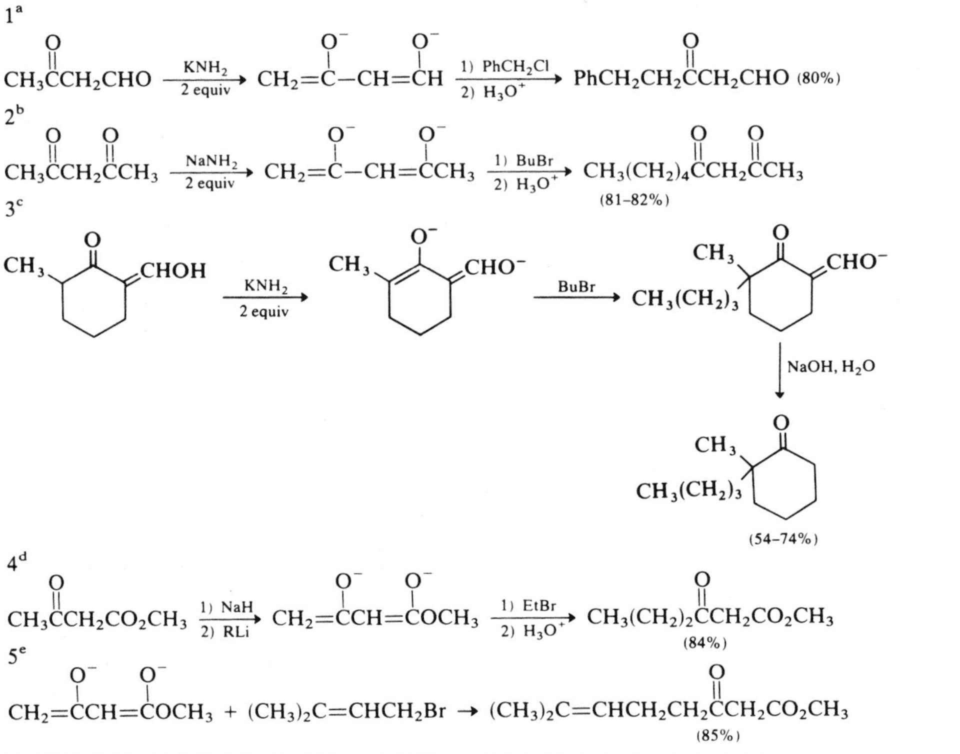

a. T. M. Harris, S. Boatman, and C. R. Hauser, *J. Am. Chem. Soc.* **85**, 3273 (1963); S. Boatman, T. M. Harris, and C. R. Hauser, *J. Am. Chem. Soc.* **87**, **82** (1965); K. G. Hampton, T. M. Harris, and C. R. Hauser, *J. Org. Chem.* **28**, 1946 (1963).
b. K. G. Hampton, T. M. Harris, and C. R. Hauser, *Org. Synth.* **47**, 92 (1967).
c. S. Boatman, T. M. Harris, and C. R. Hauser, *Org. Synth.* **48**, 40 (1968).
d. S. N. Huckin and L. Weiler, *J. Am. Chem. Soc.* **96**, 1082 (1974).
e. F. W. Sum and L. Weiler, *J. Am. Chem. Soc.* **101**, 4401 (1979).

1.6. Medium Effects in the Alkylation of Enolates

The rate of alkylation of enolate ions is strongly dependent on the solvent in which the reaction is carried out.[32] The relative rates of reaction of the sodium enolate of diethyl *n*-butylmalonate with *n*-butyl bromide are shown in Table 1.3.

32. For reviews, see (a) A. J. Parker, *Chem. Rev.* **69**, 1 (1969); (b) L. M. Jackman and B. C. Lange, *Tetrahedron* **33**, 2737 (1977).

Table 1.3. Relative Alkylation Rates of Sodium Diethyl *n*-Butylmalonate in Various Solvents[a]

Solvent	Dielectric constant, ε	Relative rate
Benzene	2·3	1
Tetrahydrofuran	7·3	14
Dimethoxyethane	6·8	80
Dimethylformamide	37	970
Dimethyl sulfoxide	47	1420

a. From H. E. Zaugg, *J. Am. Chem. Soc.* **83**, 837 (1961).

Dimethyl sulfoxide (DMSO) and *N,N*-dimethylformamide (DMF), as Table 1.3 shows, are particularly effective in enhancing the reactivity of enolate ions. Both of these compounds belong to the *polar aprotic* class of solvents. Other members of this class that are used as solvents in reactions between carbanions and alkyl halides include *N*-methylpyrrolidone and hexamethylphosphoric triamide (HMPA). Polar aprotic solvents, as their name implies, are materials which have high dielectric constants but which lack hydroxyl groups or similar hydrogen-bonding functionalities.

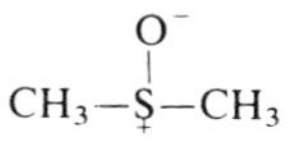

$CH_3-\overset{+}{S}(O^-)-CH_3$
dimethyl sulfoxide (DMSO)
$\varepsilon = 47$

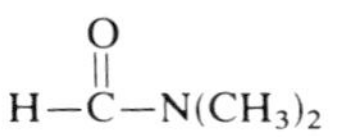

$H-C(=O)-N(CH_3)_2$
N,N-dimethylformamide (DMF)
$\varepsilon = 37$

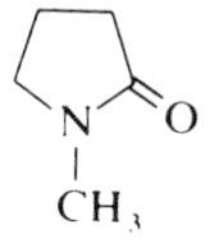

N-methylpyrrolidone ($N-CH_3$)
$\varepsilon = 32$

$O{=}P[N(CH_3)_2]_3$
hexamethylphosphoric triamide (HMPA)
$\varepsilon = 30$

The reactivity of an alkali metal (Li^+, Na^+, K^+) enolate is very sensitive to its state of aggregation, which is, in turn, influenced by the reaction medium. The highest level of reactivity, which can be approached but not achieved in solution, is that of the "bare" unsolvated enolate anion. For an enolate–metal ion pair in solution, the maximum reactivity would be expected in a medium in which the cation was strongly solvated and the enolate was very weakly solvated. Polar aprotic solvents are good cation solvators and poor anion solvators. DMSO, DMF, HMPA, and *N*-methylpyrrolidone each have a negatively polarized oxygen available for coordination to the alkali metal cation. Coordination to the enolate ion is much less effective because the positively polarized atom of these molecules is not nearly as exposed as the oxygen. Thus, these solvents provide a medium in which enolate–metal ion pairs are dissociated to give a less encumbered, more reactive enolate.

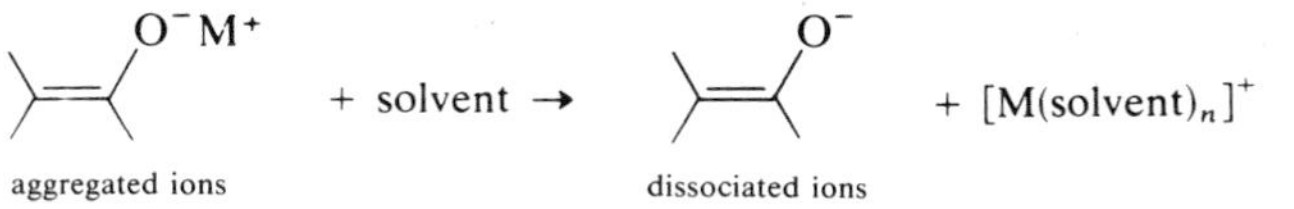

$$\text{enolate } O^-M^+ + \text{solvent} \rightarrow \text{enolate } O^- + [M(\text{solvent})_n]^+$$

aggregated ions — dissociated ions

Polar protic solvents also possess a pronounced ability to separate ion pairs but are less favorable as solvents in enolate alkylation reactions because they can coordinate to both the metal cation and the enolate ion. Coordination to the enolate occurs through hydrogen bonding. The solvated enolate will be relatively less reactive because the solvation lowers its energy. The hydrogen-bonded enolate must shed some of its solvent molecules in order to react with an alkyl halide. Enolates generated in polar protic solvents such as water, alcohols, or ammonia are less reactive than the same enolate in a polar aprotic solvent such as DMSO.

O^-M^+ + solvent–OH → $O^- \cdots$ HO-solvent + $[M(\text{solvent–OH})_n]^+$

solvated ions

Tetrahydrofuran (THF) and dimethoxyethane (DME) are slightly polar solvents which are moderately good cation solvators. Coordination to the metal cation involves the oxygen lone pairs. These solvents, because of their lower dielectric constants, are less effective at separating ion pairs and higher aggregates than are the polar aprotic solvents. The crystal structures of the lithium and potassium enolates of methyl *t*-butyl ketone have been determined by X-ray crystallography. The structures are shown in Figs. 1.1 and 1.2.[33] While these, of course, represent the solid state structural situation, the hexameric clusters which are formed probably give a good indication of the nature of the enolates in relatively weakly coordinating solvents. Despite the somewhat reduced reactivity of the aggregated enolates, THF

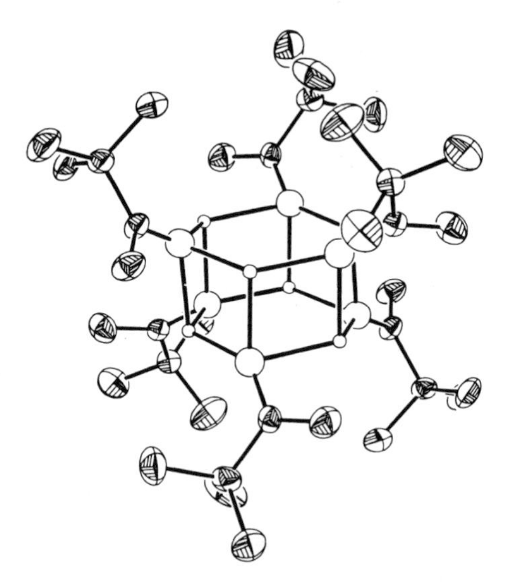

Fig. 1.1. Unsolvated hexameric aggregate of lithium enolate of methyl *t*-butyl ketone; large circles = oxygen, small circles = lithium. (Reproduced with permission from Ref. 33.)

33. P. G. Williard and G. B. Carpenter, *J. Am. Chem. Soc.* **108**, 462 (1986).

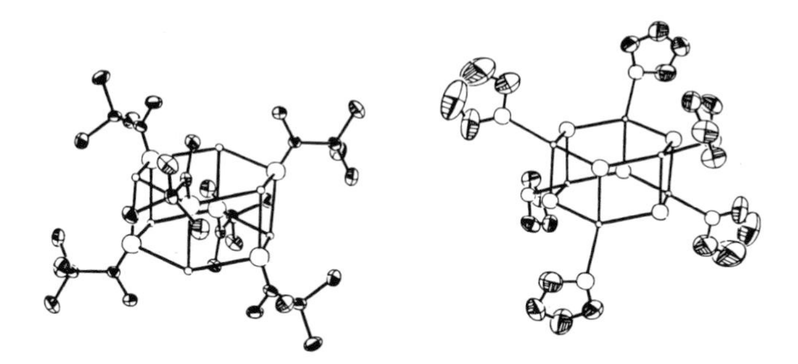

Fig. 1.2. Potassium enolate of methyl *t*-butyl ketone; large circles = oxygen, small circles = potassium. (a) Left-hand plot shows only methyl *t*-butyl ketone residues. (b) Right-hand plot shows only the solvating THF molecules. The crystal is a composite of these two structures. (Reproduced with permission from Ref. 33.)

and DME are the most commonly used solvents for synthetic reactions involving enolate alkylation. They are the most suitable solvents for kinetic enolate generation and also have advantages in terms of product workup and purification over the polar aprotic solvents. Enolate reactivity in these solvents can often be enhanced by adding a reagent which can bind alkali metal cations more strongly. Popular choices are HMPA, tetramethylethylenediamine (TMEDA), and the crown ethers.[34] TMEDA can chelate metal ions through the electron pairs on nitrogen. The crown ethers can coordinate metal ions in structures in which the metal ion is encapsulated by the ether oxygens. The 18-crown-6 structure is of such a size as to allow sodium or potassium ions to fit comfortably in the cavity. The smaller 12-crown-4 binds Li^+ preferentially. The cation-complexing agents lower the degree of aggregation of the enolate–metal cation ion pairs and result in enhanced reactivity.

The reactivity of enolates is also affected by the metal counterion. For the most commonly used ions, the order of reactivity is $Mg^{2+} < Li^+ < Na^+ < K^+$. The factors which are responsible for this order are closely related to those described for solvents. The smaller, harder Mg^{2+} and Li^+ cations are more tightly associated with the enolate than are the Na^+ and K^+ ions. The tighter coordination decreases the reactivity of the enolate and gives rise to more highly associated species.

1.7. Oxygen versus Carbon as the Site of Alkylation

Enolate anions are *ambident nucleophiles*. Alkylation of an enolate may occur at either carbon or oxygen.

C-alkylation $$RC(O^-){=}CH_2 + R'X \rightarrow RC(=O)CH_2R'$$

O-alkylation $$RC(O^-){=}CH_2 + R'X \rightarrow RC(OR'){=}CH_2$$

34. C. L. Liotta and T. C. Caruso, *Tetrahedron Lett.* **26**, 1599 (1985).

Since most of the negative charge of an enolate is on the oxygen atom, it might be supposed that O-alkylation would dominate. A number of factors other than charge density affect the C/O-alkylation ratio, and it is normally possible to establish reaction conditions that favor alkylation on carbon.

O-Alkylation will be most pronounced when the enolate is most free. When the potassium enolate of ethyl acetoacetate is treated with ethyl sulfate in the polar aprotic solvent HMPA, the major product (83%) is the O-alkylated one. In THF, where ion clustering occurs, all of the product is C-alkylated. In *t*-butanol, where the acetoacetate anion is hydrogen-bonded by solvent, again only C-alkylation is observed.[35]

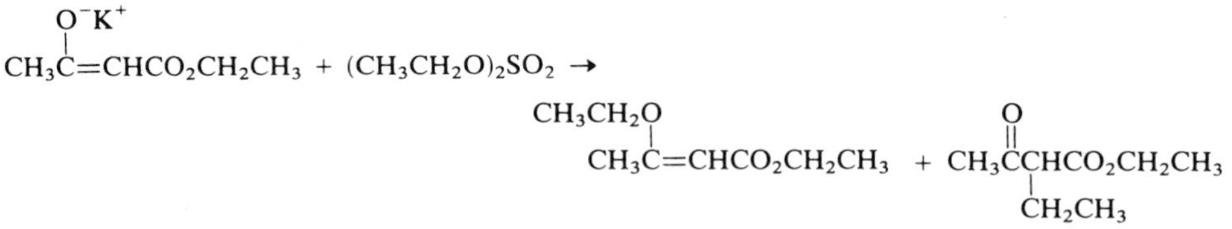

$$CH_3C(O^-K^+){=}CHCO_2CH_2CH_3 + (CH_3CH_2O)_2SO_2 \rightarrow CH_3C(OCH_2CH_3){=}CHCO_2CH_2CH_3 + CH_3C(=O)CH(CH_2CH_3)CO_2CH_2CH_3$$

	O-alkyl	C-alkyl
in hexamethylphosphoric triamide	83%	15% (2% dialkyl)
in *t*-butanol	0%	94% (6% dialkyl)
in tetrahydrofuran	0%	94% (6% dialkyl)

Higher C/O ratios are observed with alkyl halides than with alkyl sulfonates and sulfates. The highest C/O-aklylation ratios are given by alkyl iodides. For ethylation of the potassium enolate of ethyl acetoacetate in HMPA, the product compositions shown below were obtained.[36]

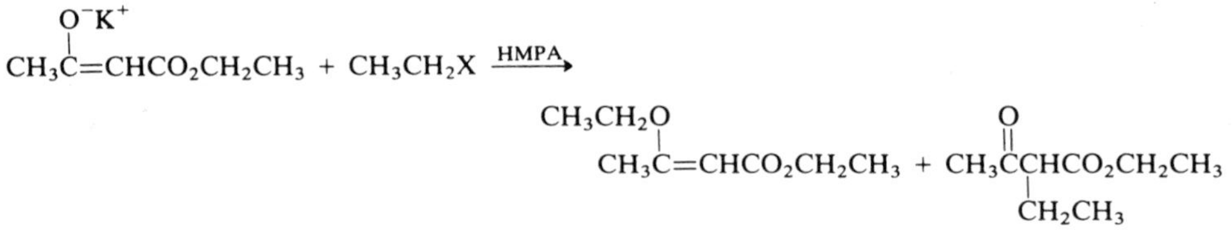

$$CH_3C(O^-K^+){=}CHCO_2CH_2CH_3 + CH_3CH_2X \xrightarrow{HMPA} CH_3C(OCH_2CH_3){=}CHCO_2CH_2CH_3 + CH_3C(=O)CH(CH_2CH_3)CO_2CH_2CH_3$$

	O-alkyl	C-alkyl
X = OTs	88%	11% (1% dialkyl)
X = Cl	60%	32% (8% dialkyl)
X = Br	39%	38% (23% dialkyl)
X = I	13%	71% (16% dialkyl)

Leaving group effects on the ratio of C- to O-alkylation can be correlated by reference to the "hard-soft-acid-base" (HSAB) rationale.[37] Of the two nucleophilic sites in an enolate ion, oxygen is harder than carbon. Nucleophilic substitution reactions of the S_N2 type proceed best when the nucleophile and leaving group are either both hard or both soft.[38] Consequently, ethyl iodide, with the very soft leaving

35. A. L. Kurts, A. Masias, N. K. Genkina, I. P. Beletskaya, and O. A. Reutov, *Dokl. Akad. Nauk. SSSR (Engl. Trans.)* **187**, 595 (1969).
36. A. L. Kurts, N. K. Genkina, A. Masias, I. P. Beletskaya, and O. A. Reutov, *Tetrahedron* **27**, 4777 (1971).
37. T.-L. Ho, *Hard and Soft Acids and Bases Principle in Organic Chemistry*, Academic Press, New York, 1977.
38. R. G. Pearson and J. Songstad, *J. Am. Chem. Soc.* **89**, 1827 (1967).

group iodide, reacts preferentially with the softer carbon site rather than the harder oxygen. Oxygen-containing leaving groups, such as sulfonate and sulfate, are harder, and alkyl sulfonate and sulfate react preferentially at the hard oxygen site of the enolate. The hard–hard combination is favored by an early transition state where the charge distribution is the most important factor. Therefore, conditions which favor a dissociated, more reactive enolate favor O-alkylation. The soft–soft combination is favored by a later transition state where partial bond formation is an important factor. The C-alkylation product is more stable than the O-alkylation product (because the bond energy of C=O + C—C is greater than that of C=C + C—O).

For isopropyl phenyl ketone, the inclusion of one equivalent of 12-crown-4 in a DME solution of the lithium enolate changes the C/O-alkylation ratio from 1.2:1 to 1:3, with the use of methyl sulfate as the alkylating agent.[39] With methyl iodide as the alkylating agent, C-alkylation is strongly favored with or without 12-crown-4.

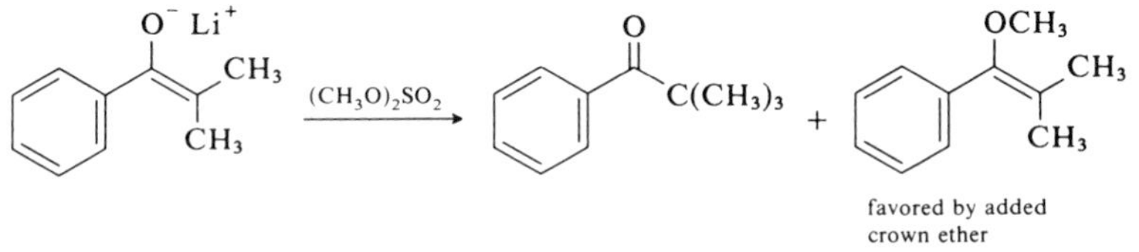
O⁻ Li⁺
CH3
CH3
(CH3O)2SO2
O
C(CH3)3
+
OCH3
CH3
CH3
favored by added
crown ether

To summarize, the amount of *O-alkylation* is maximized by use of an alkyl sulfate or alkyl sulfonate in a polar aprotic solvent. The amount of *C-alkylation* is maximized by use of an alkyl halide in a less polar solvent. The majority of synthetic operations involving ketone enolates are carried out in THF or DME with an alkyl bromide or alkyl iodide as alkylating agent, and C-alkylation is favored.

Intramolecular alkylation of enolates leads to formation of cyclic products. In addition to the other factors which govern C/O-alkylation ratios, the element of stereoelectronic control comes into play in such cases. The following reactions illustrate this point.[40]

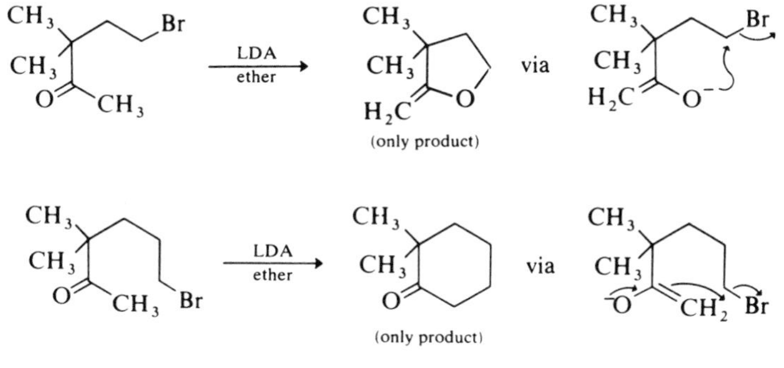
CH3
Br
CH3
O
CH3
LDA
ether
CH3
CH3
H2C
O
(only product)
via
CH3
Br
CH3
H2C
O⁻
CH3
CH3
O
CH3
Br
LDA
ether
CH3
CH3
O
(only product)
via
CH3
CH3
O⁻
CH2
Br

39. L. M. Jackman and B. C. Lange, *J. Am. Chem. Soc.* **103**, 4494 (1981).
40. J. E. Baldwin and L. I. Kruse, *J. Chem. Soc., Chem. Commun.*, 233 (1977).

In order for C-alkylation to occur, the *p*-orbital at the α-carbon must be aligned with the C—Br bond in the linear geometry associated with the S_N2 transition state. When the ring to be closed is six-membered, this geometry is accessible and cyclization to the cyclohexane occurs. When the ring to be closed is five-membered, however, colinearity cannot be achieved easily. Cyclization at oxygen then occurs faster than does cyclopentanone formation. The transition state for O-alkylation can involve an oxygen lone pair orbital and is less strained than the transition state for C-alkylation.

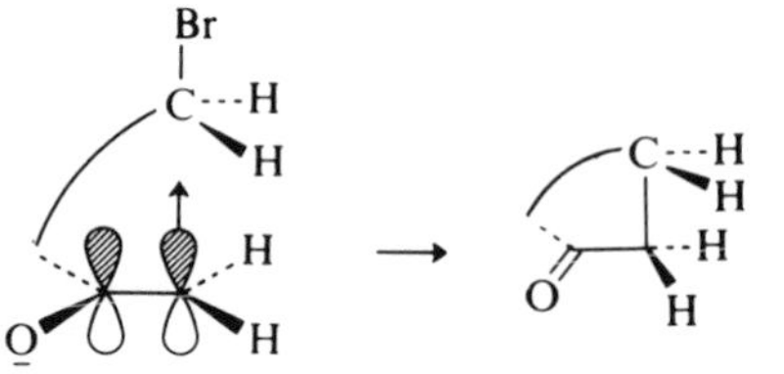

geometry required for intramolecular C-alkylation of enolate

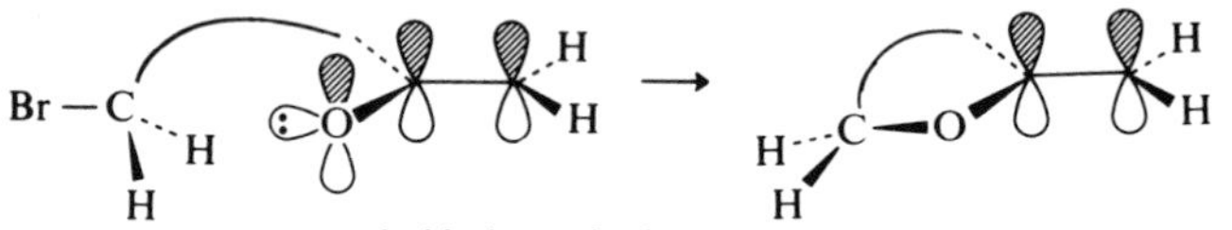

geometry required for intramolecular O-alkylation of enolate

In enolates formed from α,β-unsaturated ketones by proton abstraction from the γ-carbon, there are three potential sites for attack by electrophiles: the oxygen, the α-carbon, and the γ-carbon. The kinetically preferred site for both protonation and alkylation is the α-carbon.

Protonation of the enolate provides a method for converting α,β-unsaturated ketones and esters to the less stable β,γ-unsaturated isomers.

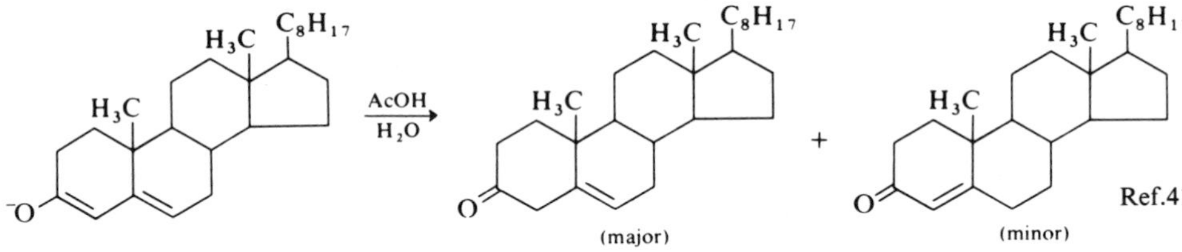

(major) (minor) Ref.41

$$CH_3CH{=}CHCO_2C_2H_5 \xrightarrow{LiNR_2} \xrightarrow{H_2O} CH_2{=}CHCH_2CO_2C_2H_5 \ (87\%) + CH_3CH{=}CHCO_2C_2H_5 \ (13\%)$$ Ref. 42

41. J. H. Ringold and S. K. Malhotra, *Tetrahedron Lett.*, 669 (1962); S. K. Malhotra and H. J. Ringold, *J. Am. Chem. Soc.* **85**, 1538 (1963).
42. M. W. Rathke and D. Sullivan, *Tetrahedron Lett.*, 4249 (1972).

Alkylation also takes place selectively at the α-carbon.[8]

$$(CH_3)_2C{=}CHCOCH_3 + CH_2{=}CHC(CH_3){=}CHCH_2Br \xrightarrow[NH_3]{NaNH_2} CH_2{=}CHC(CH_3){=}CHCH_2{-}CH(COCH_3){-}C(CH_3){=}CH_2 \quad (88\%)$$

Phenoxide ions are a special case related to enolate anions but with a preference for O-alkylation because C-alkylation disrupts aromatic conjugation.

Phenoxides undergo O-alkylation in solvents such as DMSO, DMF, ethers, and alcohols. In water and trifluoroethanol, however, extensive C-alkylation occurs.[43] These latter solvents form particularly strong hydrogen bonds with the oxygen atom of the phenolate anion. This strong solvation decreases the reactivity at oxygen and favors C-alkylation.

+ $PhCH_2Br$ —DMF→ OCH_2Ph (97 %)

—CF_3CH_2OH→ CH_2Ph, OH (85%) + OCH_2Ph (7%)

1.8. Alkylation of Aldehydes, Esters, Amides, and Nitriles

Among the alkylation reactions of compounds capable of forming enolates, those of ketones have been most widely studied and used. Similar reactions of aldehydes, esters, and nitriles have also been developed. Alkylation of aldehyde enolates is not very common. One limitation is the fact that aldehydes are rapidly converted to aldol condensation products by base (see Chapter 2 for more discussion of this reaction). Only when the enolate can be rapidly and quantitatively formed is aldol condensation avoided. Success has been reported using potassium amide in liquid ammonia[44] or potassium hydride in tetrahydrofuran.[45]

43. N. Kornblum, P. J. Berrigan, and W. J. LeNoble, *J. Am. Chem. Soc.* **85**, 1141 (1963); N. Kornblum, R. Seltzer, and P. Haberfield, *J. Am. Chem. Soc.* **85**, 1148 (1963).
44. S. A. G. De Graaf, P. E. R. Oosterhof, and A. van der Gen, *Tetrahedron Lett.*, 1653 (1974).
45. P. Groenewegen, H. Kallenberg, and A. van der Gen, *Tetrahedron Lett.*, 491 (1978).

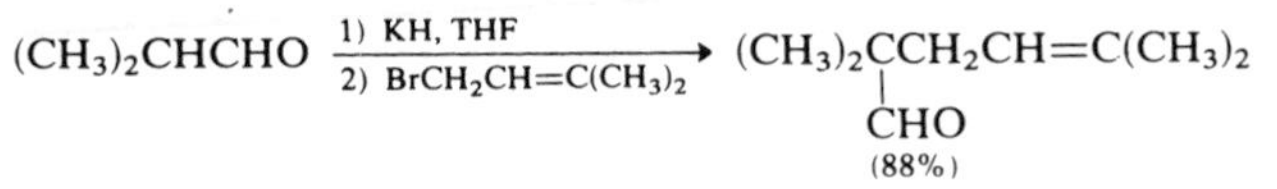

$$(CH_3)_2CHCHO \xrightarrow[\text{2) } BrCH_2CH=C(CH_3)_2]{\text{1) KH, THF}} (CH_3)_2C(CHO)CH_2CH=C(CH_3)_2 \quad (88\%) \qquad \text{Ref. 45}$$

Alkylation via enamines or enamine anions provides a more general method for alkylation of aldehydes. These reactions will be discussed in Section 1.9.

Alkylations of simple esters require strong bases because relatively weak bases such as alkoxides promote condensation reactions (see Chapter 2). The successful formation of ester enolates typically involves amide bases, usually LDA at low temperature.[46] The resulting enolates can be successfully alkylated with alkyl bromides or iodides. Some examples are given in Scheme 1.8.

Carboxylic acids can be directly alkylated by conversion to dianions by two equivalents of LDA. The dianions are alkylated at the α-carbon as would be expected.[47]

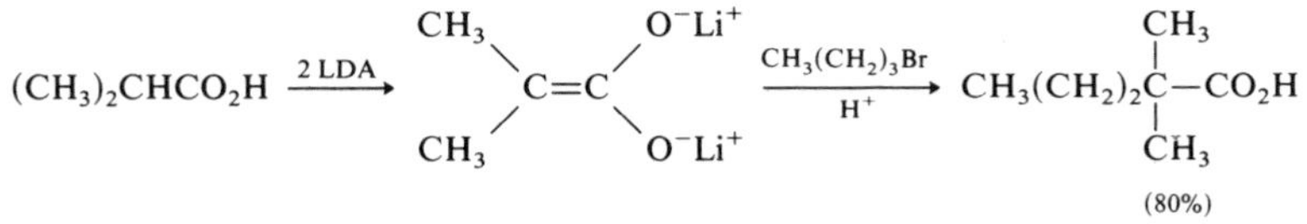

$$(CH_3)_2CHCO_2H \xrightarrow{2\,LDA} (CH_3)_2C=C(O^-Li^+)_2 \xrightarrow[H^+]{CH_3(CH_2)_3Br} CH_3(CH_2)_2C(CH_3)_2-CO_2H \quad (80\%)$$

A method for enantioselective synthesis of carboxylic acid derivatives is based on alkylation of the enolates of N-acyl oxazolidinones.[48] Two enantiomerically pure derivatives which are readily available have received the most study. The lithium enolates have the structures shown because of the tendency for the metal cation to form a chelate.

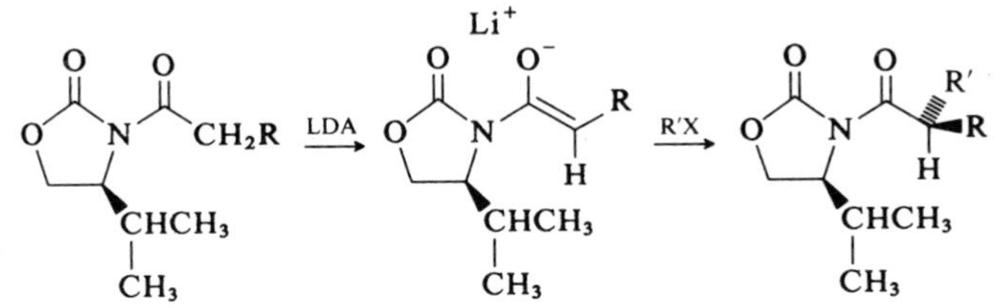

46. (a) M. W. Rathke and A. Lindert, *J. Am. Chem. Soc.* **93**, 2318 (1971); (b) R. J. Cregge, J. L. Herrmann, C. S. Lee, J. E. Richman, and R. H. Schlessinger, *Tetrahedron Lett.*, 2425 (1973); (c) J. L. Herrmann and R. H. Schlessinger, *J. Chem. Soc., Chem. Commun.*, 711 (1973).
47. P. L. Creger, *J. Org. Chem.* **37**, 1907 (1972); P. L. Creger, *J. Am. Chem. Soc.* **89**, 2500 (1967); P. L. Creger, *Org. Synth.* **50**, 58 (1970).
48. D. A. Evans, M. D. Ennis, and D. J. Mathre, *J. Am. Chem. Soc.* **104**, 1737 (1982).

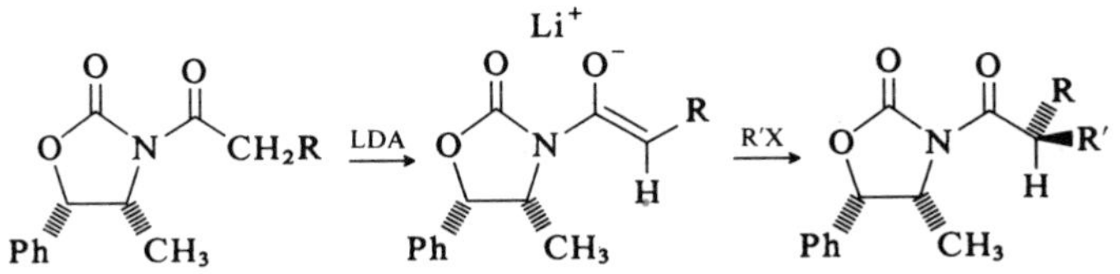

1[a]

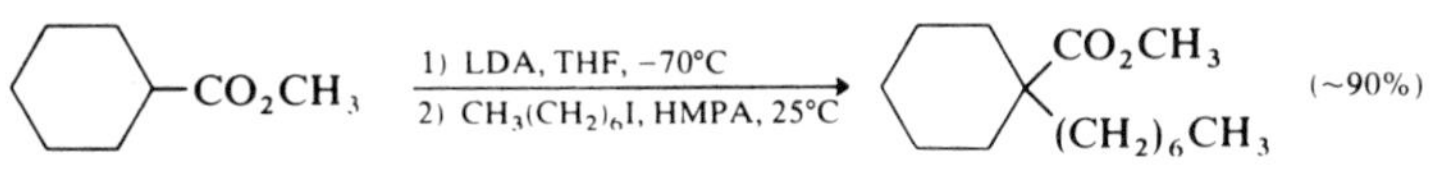

cyclohexyl–CO_2CH_3 → 1) LDA, THF, −70°C; 2) $CH_3(CH_2)_6I$, HMPA, 25°C → 1-CO_2CH_3-1-$(CH_2)_6CH_3$ cyclohexane (~90%)

2[b]

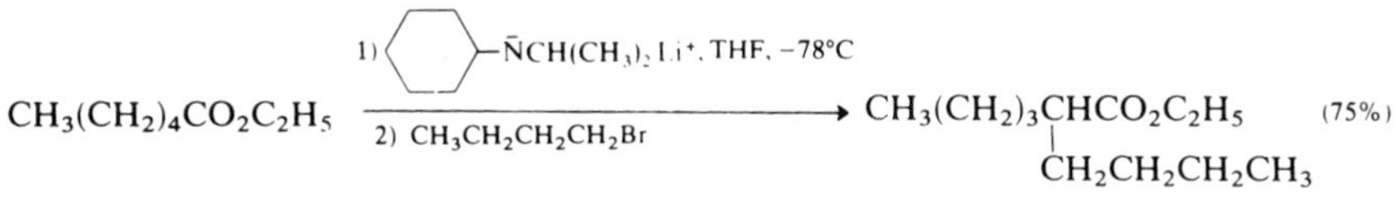

$CH_3(CH_2)_4CO_2C_2H_5$ → 1) cyclohexyl–$\bar{N}CH(CH_3)_2$ Li^+, THF, −78°C; 2) $CH_3CH_2CH_2CH_2Br$ → $CH_3(CH_2)_3CHCO_2C_2H_5$ with $CH_2CH_2CH_2CH_3$ (75%)

3[c]

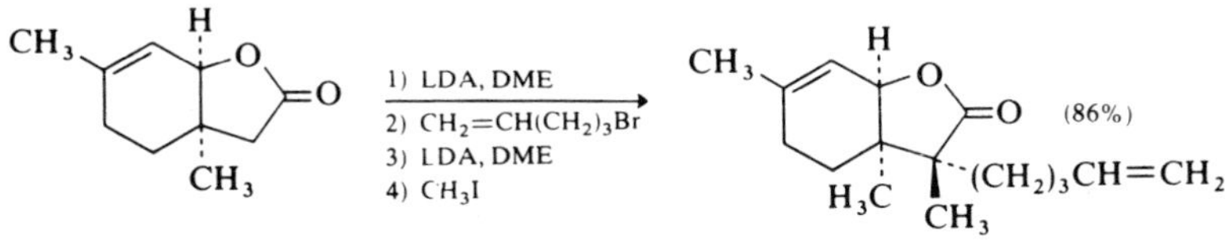

CH_3, H, O, =O, CH_3 → 1) LDA, DME; 2) $CH_2{=}CH(CH_2)_3Br$; 3) LDA, DME; 4) CH_3I → CH_3, H, O, =O, H_3C, $(CH_2)_3CH{=}CH_2$, CH_3 (86%)

4[d]

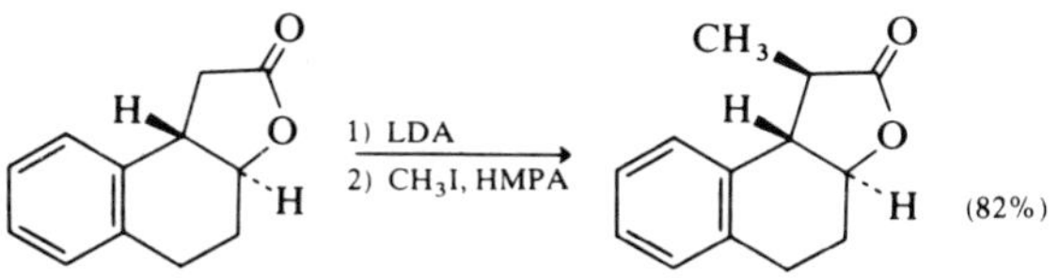

H, O, O, H → 1) LDA; 2) CH_3I, HMPA → CH_3, H, O, O, H (82%)

5[e]

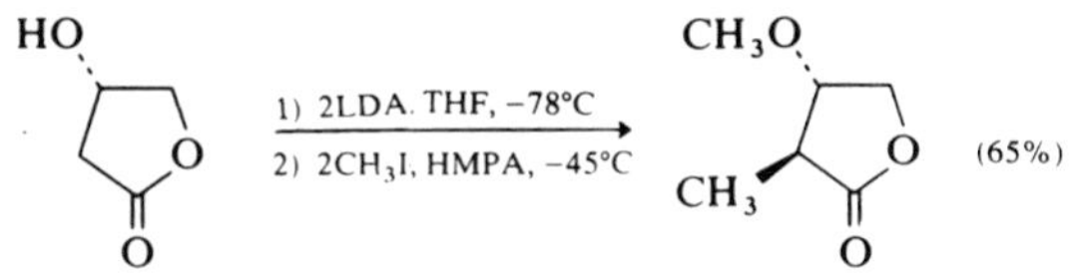

HO, O, O → 1) 2LDA, THF, −78°C; 2) $2CH_3I$, HMPA, −45°C → CH_3O, CH_3, O, O (65%)

6[f]

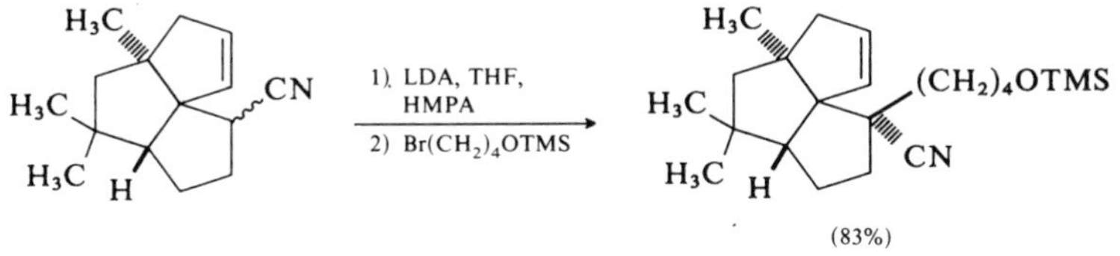

H_3C, H_3C, H_3C, H, CN → 1) LDA, THF, HMPA; 2) $Br(CH_2)_4OTMS$ → H_3C, H_3C, H_3C, H, $(CH_2)_4OTMS$, CN (83%)

7[g]

O, O, CH_2CH_2, CH_2Br, CN, O, O → H, O, O, CN, O, O (83%)

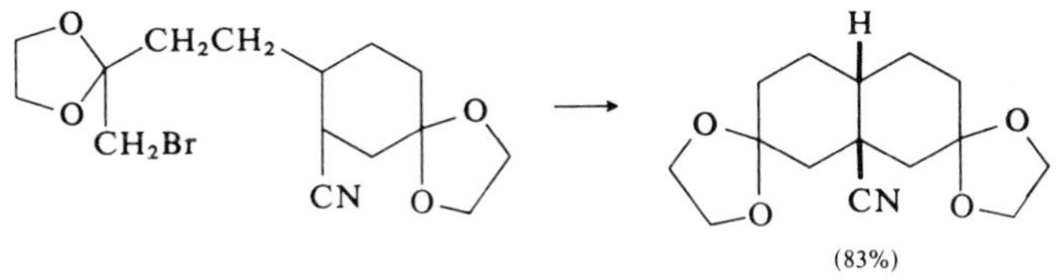

a. T. R. Williams and L. M. Sirvio, *J. Org. Chem.* **45**, 5082 (1980).
b. M. W. Rathke and A. Lindert, *J. Am. Chem. Soc.* **93**, 2320 (1971).
c. S. C. Welch, A. S. C. Prakasa Rao, C. G. Gibbs, and R. Y. Wong, *J. Org. Chem.* **45**, 4077 (1980).
d. W. H. Pirkle and P. E. Adams, *J. Org. Chem.* **45**, 4111 (1980).
e. H.-M. Shieh and G. D. Prestwich, *J. Org. Chem.* **46**, 4319 (1981).
f. L. A. Paquette, M. E. Okazaki, and J.-C. Caille, *J. Org. Chem.* **53**, 477 (1988).
g. G. Stork, J. O. Gardner, R. K. Boeckman, Jr., and K. A. Parker, *J. Am. Chem. Soc.* **95**, 2014 (1973).

In **4** the upper face is shielded by the isopropyl group while in **5** the lower face is shielded by the methyl and phenyl groups. As a result, alkylation of the two derivatives gives products of the opposite configuration. Subsequent hydrolysis or alcoholysis provides acids or esters in enantiomerically enriched form. The initial alkylation product ratios are typically 95:5 in favor of the major isomer. Since the intermediates are diastereomeric mixtures, they can be purified. The final products can then be obtained in >99% enantiomeric purity.

Acetonitrile (pK_{DMSO} 31.3) can be deprotonated, provided a strong non-nucleophilic base such as LDA is used.

$$CH_3C{\equiv}N \xrightarrow[THF]{LDA} LiCH_2C{\equiv}N \xrightarrow[2)\ (CH_3)_3SiCl]{1)\ \text{ethylene oxide}} \underset{(78\%)}{(CH_3)_3SiOCH_2CH_2CH_2C{\equiv}N} \qquad \text{Ref. 49}$$

Phenylacetonitrile (pK_{DMSO} 21.9) is considerably more acidic than acetonitrile. Deprotonation can be done with sodium amide. Dialkylation has been used in the synthesis of meperidine, an analgesic substance.[50]

$$C_6H_5CH_2CN + CH_3N(CH_2CH_2Cl)_2 \xrightarrow{NaNH_2} \text{4-cyano-4-phenyl-1-methylpiperidine (NCH}_3\text{, CN)} \rightarrow \text{4-phenyl-1-methylpiperidine-4-}CO_2CH_2CH_3$$

meperidine

1.9. The Nitrogen Analogs of Enols and Enolates—Enamines and Imine Anions

The nitrogen analogs of ketones and aldehydes are called *imines, azomethines, or Schiff bases.* Imine is the preferred name and will be used here. These compounds can be prepared by condensation of primary amines with ketones and aldehydes.[51]

$$R-\overset{O}{\overset{\|}{C}}-R + RNH_2 \rightarrow R-\overset{N-R}{\overset{\|}{C}}-R + H_2O$$

49. S. Murata and I. Matsuda, *Synthesis*, 221 (1978).
50. O. Eisleb, *Ber.* **74**, 1433 (1941); cited in H. Kagi and K. Miescher, *Helv. Chim. Acta* **32**, 2489 (1949).
51. For general reviews of imines and enamines, see P. Y. Sollenberger and R. B. Martin, in *Chemistry of the Amino Group*, S. Patai (ed.), Wiley-Interscience, New York, 1968, Chapter 7; G. Pitacco and E. Valentin, in *Chemistry of Amino, Nitroso and Nitro Groups and Their Derivatives*, Part 1, S. Patai (ed.), Wiley-Interscience, New York, 1982, Chapter 15; P. W. Hickmott, *Tetrahedron* **38**, 3363 (1982); A. G. Cook (ed.), *Enamines: Synthesis, Structure and Reactions*, Marcel Dekker, New York, 1988.

When secondary amines are heated with ketones or aldehydes in the presence of an acidic catalyst, a related condensation reaction occurs and can be driven to completion by removal of water. This is accomplished by azeotropic distillation or use of molecular sieves. The condensation product is a substituted vinylamine or *enamine.*

$$RC(=O)CHR_2 + R'_2NH \rightleftharpoons R'_2N{-}C(OH)(R){-}CHR_2 \overset{H^+}{\rightleftharpoons} R'_2\overset{+}{N}{=}C(R){-}CHR_2 \overset{-H^+}{\rightleftharpoons} R_2N{-}C(R){=}CR_2$$

There are other methods for preparing enamines from ketones that utilize strong dehydrating reagents to drive the reaction to completion. For example, mixing carbonyl compounds and secondary amines followed by addition of titanium tetrachloride rapidly gives enamines. This method is especially applicable to hindered amines.[52] Another procedure involves converting the secondary amine to its *N*-trimethylsilyl derivative. Because of the higher affinity of silicon for oxygen than nitrogen, enamine formation is favored and takes place under mild conditions.[53]

$$(CH_3)_2CHCH_2CH{=}O + (CH_3)_3SiN(CH_3)_2 \rightarrow (CH_3)_2CHCH{=}CHN(CH_3)_2 \quad (88\%)$$

The β-carbon atom of an enamine is a nucleophilic site because of conjugation with the nitrogen atom. Protonation of enamines takes place at the β-carbon, giving an iminium ion.

$$R'_2N{-}C(R){=}CR_2 \leftrightarrow R'_2\overset{+}{N}{=}C(R){-}\overset{-}{C}R_2 \qquad R'_2N{-}C(R){=}CR_2 \overset{H^+}{\rightarrow} R'_2\overset{+}{N}{=}C(R){-}CHR_2$$

The nucleophilicity of the β-carbon atoms permits enamines to be used in synthetically useful alkylation reactions:

$$R'_2\ddot{N}{-}C(R){=}CR_2 \;\; CH_2(R''){-}X \rightarrow R'_2\overset{+}{N}{=}C(R){-}C(R)(R){-}CH_2R'' \xrightarrow{H_2O} RC(=O){-}C(R)(R){-}CH_2R''$$

The enamines derived from cyclohexanones have been of particular interest. The enamine mixture formed from pyrrolidine and 2-methylcyclohexanone consists predominantly of structure **6.**[54] The pyrrolidine enamine is most frequently used for synthetic applications. The tendency to provide the less substituted enamine is

52. W. A. White and H. Weingarten, *J. Org. Chem.* **32**, 213 (1967); R. Carlson, R. Phan-Tan-Luu, D. Mathieu, F. S. Ahounde, A. Babadjamian, and J. Metzger, *Acta Chem. Scand.* **B32**, 335 (1978); R. Carlson, A. Nilsson, and M. Stromqvist, *Acta Chem. Scand.* **B37**, 7 (1983); R. Carlson and A. Nilsson, *Acta Chem. Scand.* **B38**, 49 (1984).
53. R. Comi, R. W. Franck, M. Reitano, and S. M. Weinreb, *Tetrahedron Lett.*, 3107 (1973).
54. W. D. Gurowitz and M. A. Joseph, *J. Org. Chem.* **32**, 3289 (1968).

quite general. A steric effect is responsible for this preference. Conjugation between the nitrogen atom and the π orbitals of the double bond favors coplanarity of the bonds that are darkened in the structures below. A serious nonbonded repulsion ($A^{1,3}$ strain) in **7** destabilizes this isomer. Furthermore, in isomer **6** the methyl group adopts a quasi-axial conformation to avoid steric interaction with the amine substituents.[55]

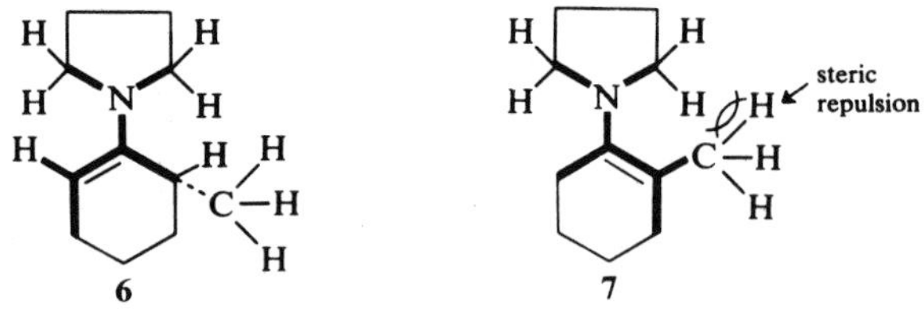

steric repulsion

6 **7**

Because of the predominance of the less substituted enamine, alkylations occur primarily at the less substituted α-carbon. Synthetic advantage can be taken of this selectivity to prepare 2,6-disubstituted cyclohexanones. The iminium ions resulting from C-alkylation are hydrolyzed in the workup procedure.

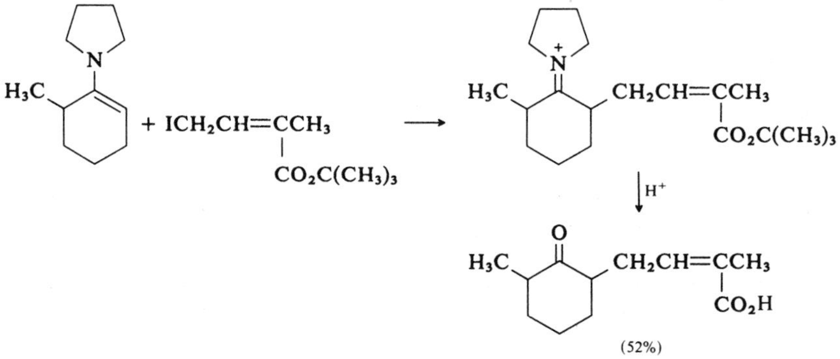

H_3C + $ICH_2CH{=}C(CH_3)CO_2C(CH_3)_3$ → $CH_2CH{=}C(CH_3)CO_2C(CH_3)_3$ →(H^+) $CH_2CH{=}C(CH_3)CO_2H$ (52%) Ref. 56

Alkylation of enamines requires relatively reactive alkylating agents for good results. Methyl iodide, allylic and benzylic halides, α-haloesters, α-haloethers, and α-haloketones are the most successful alkylating agents. Some typical examples of enamine alkylation reactions are shown in Scheme 1.9.

Enamines also react with electrophilic alkenes. This aspect of their chemistry will be described in Section 1.10.

Imines can be deprotonated at the α-carbon by strong bases to give the nitrogen analogs of enolates. Originally, Grignard reagents were used for deprotonation, but LDA is now commonly used. These anions are usually referred to as *imine anions*

55. F. Johnson, L. G. Duquette, A. Whitehead, and L. C. Dorman, *Tetrahedron* **30**, 3241 (1974); K. Muller, F. Previdoli, and H. Desilvestro, *Helv. Chim. Acta* **64**, 2497 (1981); J. E. Anderson, D. Casarini, and L. Lunazzi, *Tetrahedron Lett.* **29**, 3141 (1988).
56. P. L. Stotter and K. A. Hill, *J. Am. Chem. Soc.* **96**, 6524 (1974).

Scheme 1.9. Enamine Alkylations

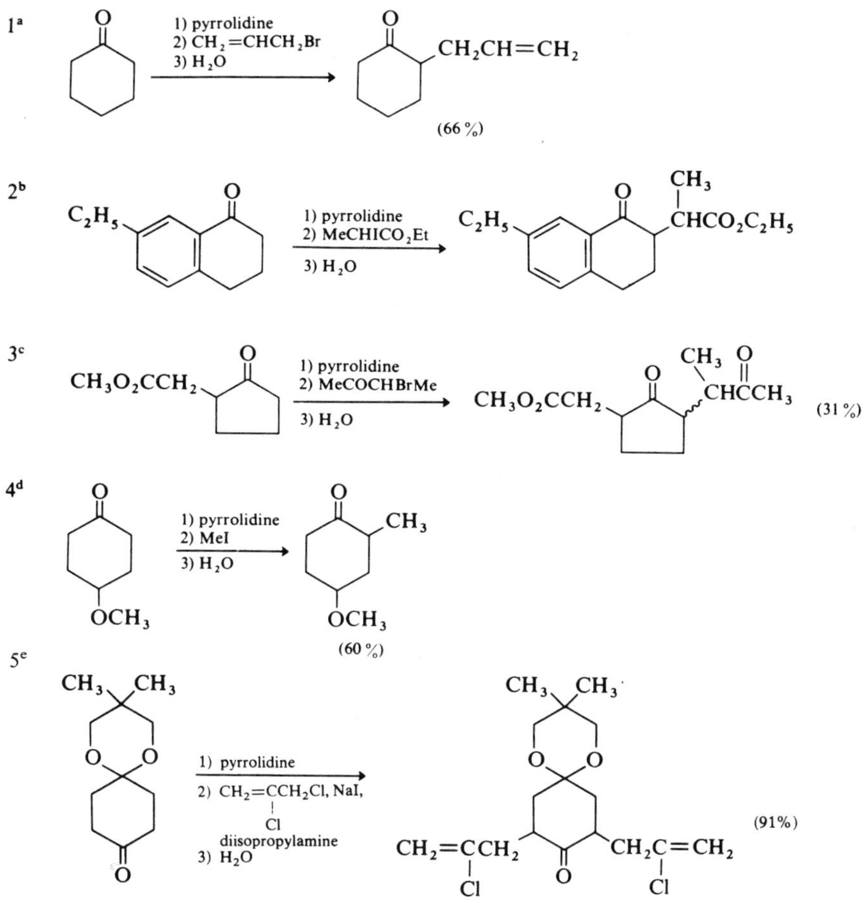

1[a] Cyclohexanone $\xrightarrow{\text{1) pyrrolidine; 2) } CH_2{=}CHCH_2Br\text{; 3) } H_2O}$ 2-($CH_2CH{=}CH_2$)cyclohexanone (66%)

2[b] C_2H_5-substituted tetralone $\xrightarrow{\text{1) pyrrolidine; 2) } MeCHICO_2Et\text{; 3) } H_2O}$ C_2H_5-substituted tetralone with α-$CH(CH_3)CO_2C_2H_5$

3[c] $CH_3O_2CCH_2$-cyclopentanone $\xrightarrow{\text{1) pyrrolidine; 2) MeCOCHBrMe; 3) } H_2O}$ $CH_3O_2CCH_2$-cyclopentanone with α-$CH(CH_3)CCH_3$ (C=O) (31%)

4[d] 4-OCH_3-cyclohexanone $\xrightarrow{\text{1) pyrrolidine; 2) MeI; 3) } H_2O}$ 2-CH_3-4-OCH_3-cyclohexanone (60%)

5[e] Spiro ketal (CH_3, CH_3; O, O) cyclohexanone $\xrightarrow{\text{1) pyrrolidine; 2) } CH_2{=}C(Cl)CH_2Cl\text{, NaI, diisopropylamine; 3) } H_2O}$ 2,6-bis($CH_2C(Cl){=}CH_2$) product (91%)

a. G. Stork, A. Brizzolara, H. Landesman, J. Szmuszkovicz, and R. Terrell, *J. Am. Chem. Soc.* **85**, 207 (1963).
b. D. M. Locke and S. W. Pelletier, *J. Am. Chem. Soc.* **80**, 2588 (1958).
c. K. Sisido, S. Kurozumi, and K. Utimoto, *J. Org. Chem.* **34**, 2661 (1969).
d. G. Stork and S. D. Darling, *J. Am. Chem. Soc.* **86**, 1761 (1964).
e. J. A. Marshall and D. A. Flynn, *J. Org. Chem.* **44**, 1391 (1979).

or *metalloenamines*.[57] Imine anions are isoelectronic and structurally analogous to both enolates and allyl anions and can also be called *azaallyl anions*.

$$\underset{\displaystyle RC{-}CHR''_2}{\overset{NR'}{\|}} \xrightarrow{\text{base}} \underset{\displaystyle RC{-}\bar{C}R''_2}{\overset{NR'}{\|}} \leftrightarrow \underset{\displaystyle RC{=}CR''_2}{\overset{^{-}NR'}{|}}$$

Spectroscopic investigations of the lithium derivatives of cyclohexanone *N*-phenylimine indicate that it exists as a dimer in toluene and that as a better donor solvent, THF, is added, equilibrium with a monomeric structure is established. The

57. For a general review of imine anions, see J. K. Whitesell and M. A. Whitesell, *Synthesis*, 517 (1983).

monomer is favored at high THF concentrations.[58] A crystal structure determination has been done on the lithiated *N*-phenylimine of methyl *t*-butyl ketone. It is a dimeric structure with the lithium cation positioned above the nitrogen and closer to the phenyl ring than to the β-carbon of the imine anion.[59] The structure is shown in Fig. 1.3.

Just as enamines are more nucleophilic than enols, imine anions are more nucleophilic than enolates and react efficiently with alkyl halides. One application of metalloenamines is for the alkylation of aldehydes.

$$(CH_3)_2CHCH{=}NC(CH_3)_3 \xrightarrow{EtMgBr} (CH_3)_2C{=}CH{-}N(MgBr)C(CH_3)_3$$

$$\downarrow PhCH_2Cl \mid H_2O$$

$$(CH_3)_2C(CH_2Ph)CH{=}O \xleftarrow{H_3O^+} (CH_3)_2C(CH_2Ph){-}CH{=}NC(CH_3)_3$$

(80% overall yield)

Ref. 60

$$CH_3CH{=}CH{-}CH{=}N{-}C_6H_{11} \xrightarrow[\text{2) } ICH_2CH_2\text{-}(2,5,5\text{-trimethyl-1,3-dioxan-2-yl}) \quad \text{3) } H_2O]{\text{1) LDA}} CH_3CH{=}C(CH_2CH_2\text{-}(2,5,5\text{-trimethyl-1,3-dioxan-2-yl})){-}CH{=}O$$

Ref. 61

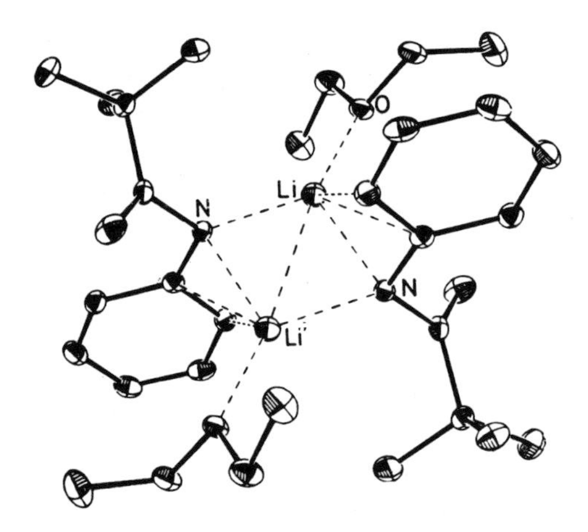

Fig. 1.3. Crystal structure of dimer of lithium derivative of *N*-phenyl imine of methyl *t*-butyl ketone. (Reproduced with permission from Ref. 59.)

58. N. Kallman and D. B. Collum, *J. Am. Chem. Soc.* **109**, 7466 (1987).
59. H. Dietrich, W. Mahdi, and R. Knorr, *J. Am. Chem. Soc.* **108**, 2462 (1986).
60. G. Stork and S. R. Dowd, *J. Am. Chem. Soc.* **85**, 2178 (1963).
61. T. Kametani, Y. Suzuki, H. Furuyama, and T. Honda, *J. Org. Chem.* **48**, 31 (1983).

Ketone imine anions can also be alkylated. The prediction of the regioselectivity of lithioenamine formation is somewhat more complex than for the case of kinetic ketone enolate formation. One of the complicating factors is that there are two imine stereoisomers, each of which can give rise to two regioisomeric imines. The isomers in which the nitrogen substituent R′ is *syn* to the double bond are the more stable.[62]

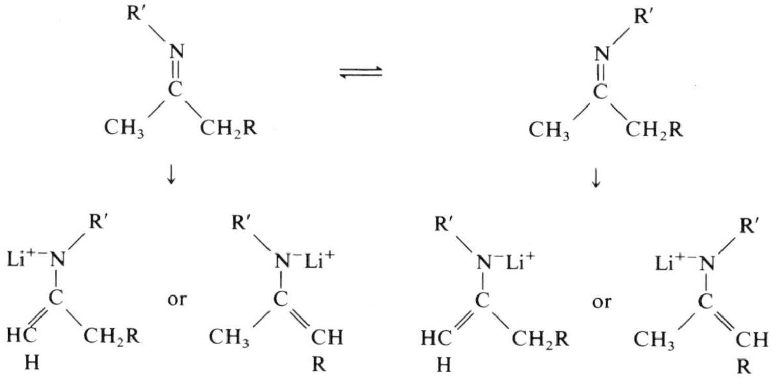

For methyl ketimines, good regiochemical control in favor of methyl deprotonation, regardless of imine stereochemistry, is observed using LDA at −78°C. With larger substituents, deprotonation at 0°C occurs *anti* to the nitrogen substituent.[63]

However, the *syn–anti* isomers of imines are easily thermally equilibrated. They cannot be prepared as single stereoisomers directly from ketones and amines so this method cannot be used to control the regiochemistry of deprotonation. By allowing lithiated ketimines to come to room temperature, the thermodynamic composition is established. The most stable structures are those shown below, which in each case represent the less substituted isomer.

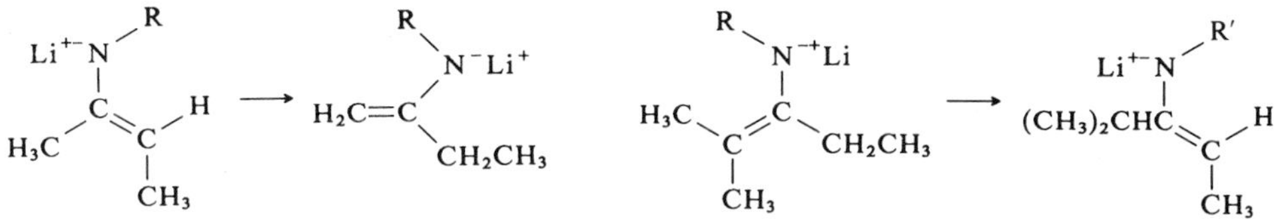

62. K. N. Houk, R. W. Stozier, N. G. Rondan, R. R. Frazier, and N. Chauqui-Ottermans, *J. Am. Chem. Soc.* **102**, 1426 (1980).
63. J. K. Smith, M. Newcomb, D. E. Bergbreiter, D. R. Williams, and A. I. Meyer, *Tetrahedron Lett.* **24**, 3559 (1983); J. K. Smith, D. E. Bergbreiter, and M. Newcomb, *J. Am. Chem. Soc.* **105**, 4396 (1983); A. Hosomi, Y. Arak, and H. Sakurai, *J. Am. Chem. Soc.* **104**, 2081 (1982).

One of the most useful aspects of the imine anions is that they can be readily prepared from enantiomerically pure amines. When imines derived from these amines are alkylated, the new carbon–carbon bond is formed with a bias for one of the two possible stereochemical configurations. Hydrolysis of the imine then leads to enantiomerically enriched ketone. Table 1.4 lists some examples that have been reported.[64]

The interpretation and prediction of the relationship between the configuration of the newly formed chiral center and the configuration of the amine is usually based on steric differentiation of the two faces of the azaallyl anion. Most imine anions which show high stereoselectivity incorporate a substituent which can hold the metal cation in a compact transition state by chelation. In the case of entry 2 in Table 1.4, for example, the transition state **G** rationalizes the observed enantioselectivity.

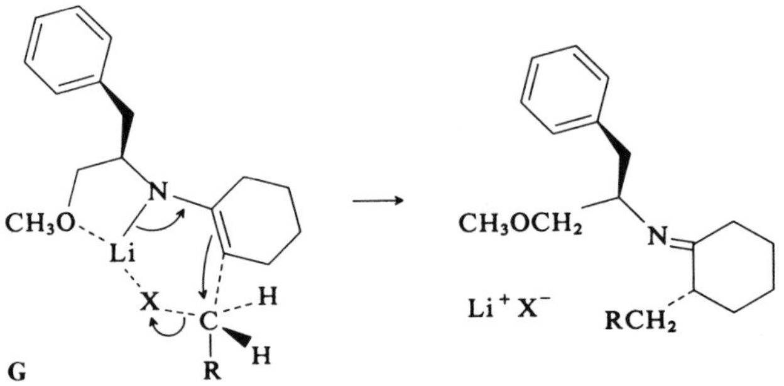

CH_3O ⋯ Li ⋯ X ⋯ C(H)(H)(R) → CH_3OCH_2 … N= … RCH_2 + Li^+X^-

G

The fundamental features of this transition state are (1) the chelation of the methoxy group with the lithium ion, which establishes a rigid transition state; (2) the interaction of the lithium ion with the bromide leaving group; and (3) the steric effect of the benzyl group, which makes the underside the preferred direction of approach for the alkylating agent.

Hydrazones can also be deprotonated to give lithium salts which are reactive toward alkylation at the β-carbon. Hydrazones are more stable than alkylimines and therefore have some advantages in synthesis. The *N,N*-dimethylhydrazones have been investigated most completely.[65] The *N,N*-dimethylhydrazones of methyl ketones are kinetically deprotonated at the methyl group. This regioselectivity is independent of the stereochemistry of the hydrazone.[66] Two successive alkylations of the *N,N*-dimethylhydrazone of acetone provide unsymmetrical ketones.

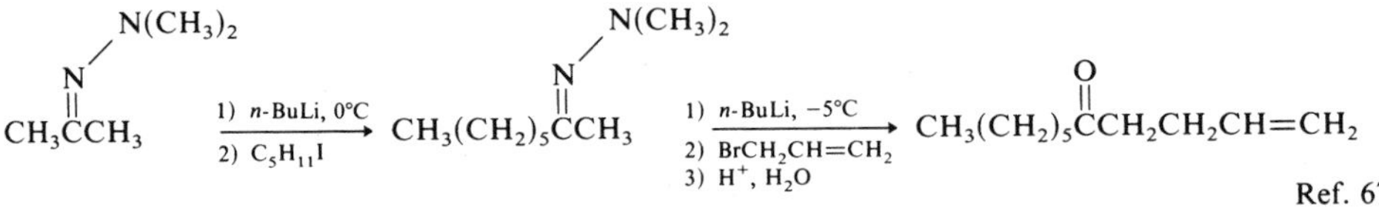

$$CH_3C(=NN(CH_3)_2)CH_3 \xrightarrow[2)\ C_5H_{11}I]{1)\ n\text{-BuLi, }0^\circ C} CH_3(CH_2)_5C(=NN(CH_3)_2)CH_3 \xrightarrow[2)\ BrCH_2CH{=}CH_2;\ 3)\ H^+,\ H_2O]{1)\ n\text{-BuLi, }-5^\circ C} CH_3(CH_2)_5C(=O)CH_2CH_2CH{=}CH_2$$

Ref. 67

64. For a review, see D. E. Bergbreiter and M. Newcomb, in *Asymmetric Synthesis*, Vol. 2, J. D. Morrison (ed.), Academic Press, New York, 1983, Chapter 9.
65. E. J. Corey and D. Enders, *Tetrahedron Lett.*, 3 (1976).
66. D. E. Bergbreiter and M. Newcomb, *Tetrahedron Lett.*, 4145 (1979); M. E. Jung and T. J. Shaw, *Tetrahedron Lett.*, 4149 (1979).
67. M. Yamashita, K. Matsumiya, M. Tanabe, and R. Suetmitsu, *Bull. Chem. Soc. Jpn.* **58**, 407 (1985).

Table 1.4. Enantioselective Alkylation of Ketimines

Amine	Ketone	Alkyl group	Yield	%E.E.	Ref.
$(CH_3)_3C$, H, $(CH_3)_3CO_2C$, NH_2	Cyclohexanone	$CH_2{=}CHCH_2Br$	75	84	a
$PhCH_2$, H, CH_2OCH_3, H_2N	Cyclohexanone	$CH_2{=}CHCH_2Br$	80	>99	b
$(CH_3)_2CH$, H, $(CH_3)_3CO_2C$, NH_2	2-Carbomethoxy-cyclohexanone	CH_3I	57	>99	c
$PhCH_2$, H, CH_2OCH_3, H_2N	5-Nonanone	$CH_2{=}CHCH_2Br$	80	94	d

a. S. Hashimoto and K. Koga, *Tetrahedron Lett.*, 573 (1978).
b. A. I. Meyers, D. R. Williams, G. W. Erickson, S. White, and M. Druelinger, *J. Am. Chem. Soc.* **103**, 3081 (1981).
c. K. Tomioka, K. Ando, Y. Takemasa, and K. Koga, *J. Am. Chem. Soc.* **106**, 2718 (1984).
d. A. I. Meyers, D. R. Williams, S. White, and G. W. Erickson, *J. Am. Chem. Soc.* **103**, 3088 (1981).

The anion of cyclohexanone *N,N*-dimethylhydrazone shows a strong preference for axial alkylation.[68] 2-Methylcyclohexanone *N,N*-dimethylhydrazone is alkylated by methyl iodide to give *cis*-2,6-dimethylcyclohexanone. The methyl group in the hydrazone occupies a pseudoaxial orientation. Alkylation apparently is preferred *anti* to the lithium cation, which is on the face opposite the 2-methyl substituent.

N, $N(CH_3)_2$, CH_3 —LDA→ N, $N(CH_3)_2$, Li, CH_3 → N, $N(CH_3)_2$, H_3C, CH_3 —H_2O→ CH_3, H_3C, O

Chiral hydrazones have also been developed for enantioselective alkylation of ketones. Structure **8**, which is derived from the amino acid proline, is a good example.

68. D. B. Collum, D. Kahne, S. A. Gut, R. T. DePue, F. Mohamadi, R. A. Wanat, J. Clardy, and G. VanDuyne, *J. Am. Chem. Soc.* **106**, 4865 (1984).

The hydrazone can be converted to the lithium salt, alkylated, and then hydrolyzed to give alkylated ketone in good chemical yield and with high enantioselectivity.[69]

$CH_3CH_2C(=O)CH_2CH_3$ + N-amino-2-(methoxymethyl)pyrrolidine (NH_2, OCH_3) ⟶ hydrazone $CH_3CH_2C(=N{-}N)CH_2CH_3$ —LDA→ anion; —$CH_3CH_2CH_2I$→ $CH_3CH_2C(=N)$–C(H)($CH_2CH_2CH_3$)(CH_3) —1) CH_3I 2) H_2O→ $CH_3CH_2C(=O)$–C(H)($CH_2CH_2CH_3$)(CH_3)

Hydrazones are substantially more stable than imines or enamines toward hydrolysis. Several procedures have been developed for conversion of the hydrazones back to ketones.[67–69] In the example above, the hydrazone is N-methylated to facilitate the hydrolysis. Mild conditions are particularly important when stereochemical configuration must be maintained at the enolizable position adjacent to the carbonyl group.

The anions derived from 2,4,4,6-tetramethyl-5,6-dihydro-1,3-oxazine have been used as the basis of several general synthetic procedures. Alkylation of the anion followed by reduction of the heterocyclic imine group and hydrolysis gives aldehydes.[70]

2,4,4,6-tetramethyl-5,6-dihydro-1,3-oxazine —n-BuLi, THF→ N-Li, $=CH_2$ anion —n-BuBr→ 2-$CH_2(CH_2)_3CH_3$ oxazine —1) $NaBH_4$ 2) H_3O^+→ $CH_3(CH_2)_3CH_2CHO$ (65%)

69. D. Enders, H. Eichenauer, U. Bas, H. Schubert, and K. A. M. Kremer, *Tetrahedron* **40**, 1345 (1984); D. Enders, H. Kipphardt, and P. Fey, *Org. Synth.* **65**, 183 (1987).
70. A. I. Meyers, A. Nabeya, H. W. Adickes, I. R. Politzer, G. R. Malone, A. C. Kovelesky, R. L. Nolen, and R. C. Portnoy, *J. Org. Chem.* **38**, 36 (1973).

A procedure for enantioselective synthesis of carboxylic acids is based on sequential alkylation of the anion of the oxazoline **9** via its lithium anion. Chelation by the methoxy group leads preferentially to the transition state in which the lithium is located as shown. The lithium acts as a Lewis acid in directing the approach of the alkyl halide. This is reinforced by a steric effect of the phenyl substituent. As a result, alkylation occurs predominantly from the lower face of the anion. The sequence in which the groups R and R′ are introduced determines the chirality of the product. The enantiomeric purity of disubstituted acetic acids obtained after hydrolysis is in the range of 70–90%.[71]

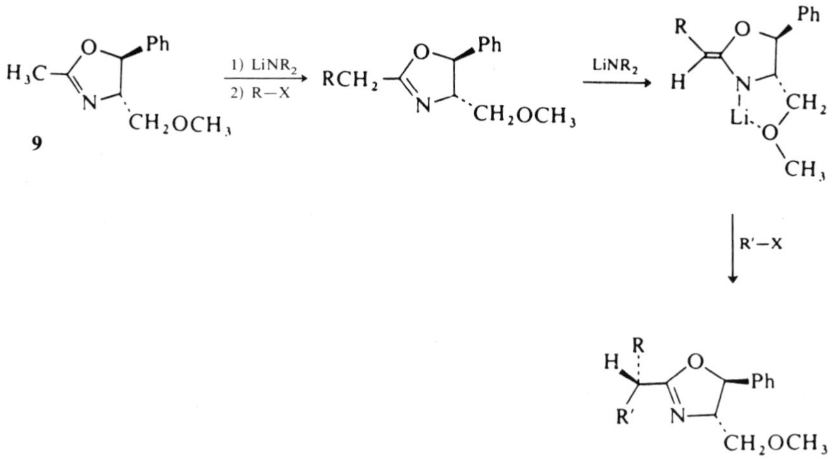

1.10. Alkylation of Carbon Nucleophiles by Conjugate Addition

The previous sections have dealt primarily with reactions in which the new carbon–carbon bond is formed in an S_N2 reaction between the nucleophilic species and the alkylating reagent. There is another important method for alkylation of carbon. This reaction involves the addition of a nucleophilic carbon species to an electrophilic multiple bond. The reaction is applicable to a wide variety of enolates and enamines. The electrophilic reaction partner is typically an α,β-unsaturated ketone, aldehyde, ester, or nitrile, but other electron-withdrawing substituents such as nitro or sulfonyl also activate carbon–carbon double and triple bonds to nucleophilic attack. The reaction is called *conjugate addition* or the *Michael reaction*. Other kinds of nucleophiles such as amines, alkoxides, and sulfide anions also react similarly, but we will focus on the carbon–carbon bond-forming reactions.

71. A. I. Meyers, G. Knaus, K. Kamata, and M. E. Ford, *J. Am. Chem. Soc.* **98**, 567 (1976).

In contrast to the reaction of an enolate anion with an alkyl halide, which requires one equivalent of base, conjugate addition of enolates can be carried out with a catalytic amount of base.

$$\mathrm{R\overset{O}{\overset{\|}{C}}CHR_2 + B^- \rightleftharpoons R\overset{O^-}{\overset{|}{C}}{=}CR_2 + BH}$$

$$\mathrm{R\overset{O^-}{\overset{|}{C}}{=}CR_2 + \;\rangle C{=}C\langle^{X} \rightleftharpoons R\overset{O}{\overset{\|}{C}}{-}\underset{R}{\overset{R}{C}}{-}\overset{|}{\underset{|}{C}}{-}C^-\langle^{X}}$$

$$\mathrm{R\overset{O}{\overset{\|}{C}}{-}\underset{R}{\overset{R}{C}}{-}\overset{|}{\underset{|}{C}}{-}C^-\langle^{X} + BH \rightleftharpoons R\overset{O}{\overset{\|}{C}}{-}\underset{R}{\overset{R}{C}}{-}\overset{|}{\underset{|}{C}}{-}\overset{X}{\underset{|}{C}}{-}H + B^-}$$

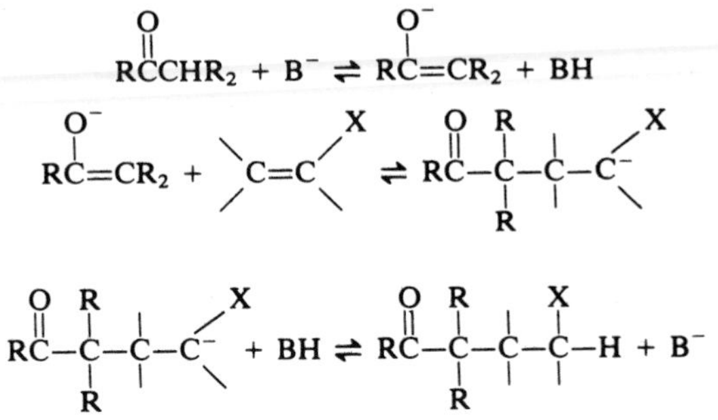

All the steps are reversible. With a catalytic amount of base, the reaction proceeds with thermodynamic control of enolate formation. The most effective nucleophiles under these conditions are carbanions derived from relatively acidic compounds such as β-ketoesters or malonate esters. Scheme 1.10 provides some examples.

The fluoride ion is an effective catalyst for Michael additions involving relatively acidic carbon compounds.[72] The fluoride ion is basic because of the strength of the H—F bond. (Hydrogen fluoride is a weak acid in aqueous solution; pK = 3.45) The reactions are typically done in the presence of excess fluoride, where the formation of the $[\mathrm{F{-}H{-}F}]^-$ ion would occur.

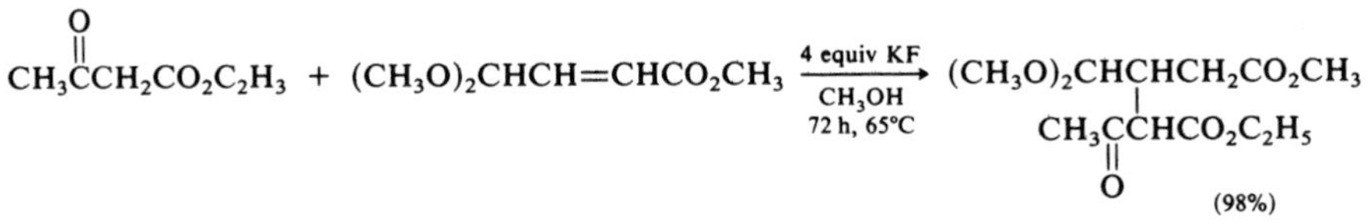

$$\mathrm{CH_3\overset{O}{\overset{\|}{C}}CH_2CO_2C_2H_5 + (CH_3O)_2CHCH{=}CHCO_2CH_3 \xrightarrow[\text{72 h, 65°C}]{\text{4 equiv KF, } CH_3OH} (CH_3O)_2CHCHCH_2CO_2CH_3 \;(\text{with } CH_3\overset{O}{\overset{\|}{C}}CHCO_2C_2H_5 \text{ at C-3})}$$

(98%) Ref. 73

$$\mathrm{(CH_3)_2CHNO_2 + CH_2{=}CHCOCH_3 \xrightarrow[\text{2 h, 25°C}]{\text{0.5 equiv } R_4N^+F^-} O_2N\overset{CH_3}{\underset{CH_3}{C}}CH_2CH_2\overset{O}{\overset{\|}{C}}CH_3}$$

(95%) Ref. 74

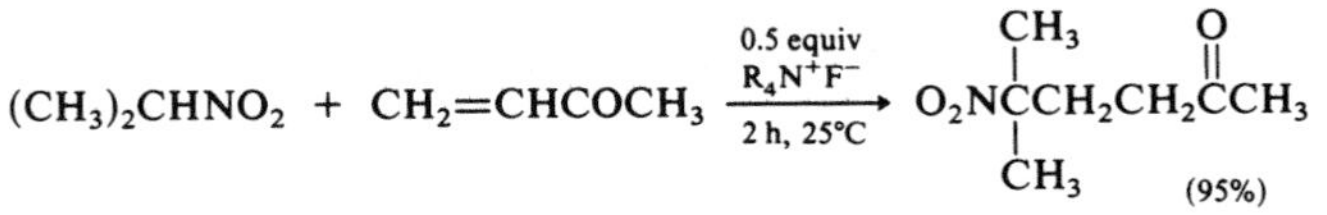

Fluoride ion can also induce reaction of silyl enol ethers with electrophilic alkenes.

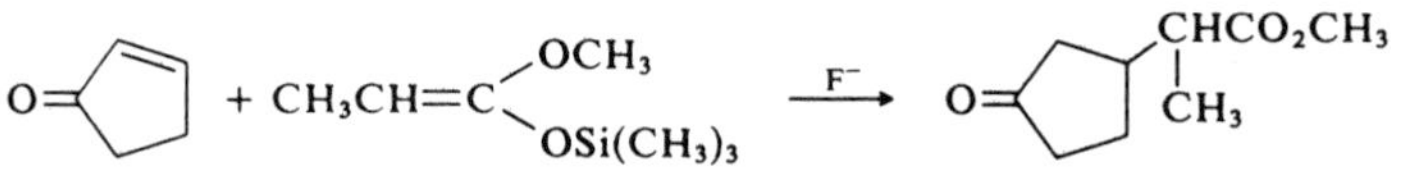

$$\text{cyclopent-2-enone} + \mathrm{CH_3CH{=}C(OCH_3)(OSi(CH_3)_3)} \xrightarrow{F^-} \text{3-}[\mathrm{CH(CH_3)CO_2CH_3}]\text{cyclopentanone}$$

Ref. 75

Conjugate addition can also be carried out by completely forming the nucleophilic enolate under kinetic conditions. Ketone enolates preformed by reac-

72. J. H. Clark, *Chem. Rev.* **80**, 429 (1980).
73. S. Tori, H. Tanaka, and Y. Kobayashi, *J. Org. Chem.* **42**, 3473 (1977).
74. J. H. Clark, J. M. Miller, and K.-H. So, *J. Chem. Soc., Perkin Trans.* 1, 941 (1978).
75. T. V. Rajan Babu, *J. Org. Chem.* **49**, 2083 (1984).

1[a] 2-methylcyclopentanone + $H_2C{=}CHCO_2CH_3$ $\xrightarrow{KOC(CH_3)_3}$ 2-methyl-2-($CH_2CH_2CO_2CH_3$)cyclopentanone (53%)

2[b] $PhCH_2CH(CONH_2)CN + H_2C{=}CHCN \xrightarrow{NH_{3(l)}} PhCH_2C(CN)(CONH_2)CH_2CH_2CN$ (100%)

3[c] $CH_2(CO_2C_2H_5)_2 + H_2C{=}C(Ph)CO_2C_2H_5 \xrightarrow{NaOEt} (H_5C_2O_2C)_2CHCH_2CH(Ph)CO_2C_2H_5$ (55–60%)

4[d] $(CH_3)_2CHNO_2 + CH_2{=}CHCO_2CH_3 \xrightarrow{PhCH_2\overset{+}{N}(CH_3)_3\ ^-OH} O_2NC(CH_3)_2CH_2CH_2CO_2CH_3$ (80–86%)

5[e] $PhCH(CN)CO_2C_2H_5 + CH_2{=}CHCN \xrightarrow[(CH_3)_3COH]{KOH} PhC(CN)(CO_2C_2H_5)CH_2CH_2CN$ (69–83%)

6[f] methyl bicyclo[2.2.1]hepta-2,5-diene-2-carboxylate (CO_2CH_3) + $CH_3\overset{O}{\overset{\|}{C}}CH_2CO_2C_2H_5$ $\xrightarrow{R_4N^+\ ^-OH}$ norbornene bearing $CH(CCH_3)(=O)CO_2C_2H_5$, i.e. $CH(COCH_3)CO_2C_2H_5$, and CO_2CH_3 (86%)

7[g] 1-pyrrolidinyl enamine of 2-methyl-5-isopropylcyclohexanone (CH_3, $CH(CH_3)_2$) + $CH_2{=}CH\overset{O}{\overset{\|}{C}}CH_3$ $\xrightarrow[\text{2) NaOAc, HOAc, } H_2O \text{ reflux}]{\text{1) dioxane, 16 hr}}$ 2-methyl-6-($CH_2CH_2COCH_3$)-5-($CH(CH_3)_2$)cyclohexanone (66%)

a. H. O. House, W. L. Roelofs, and B. M. Trost, *J. Org. Chem.* **31**, 646 (1966).
b. S. Wakamatsu, *J. Org. Chem.* **27**, 1285 (1962).
c. E. M. Kaiser, C. L. Mao, C. F. Hauser, and C. R. Hauser, *J. Org. Chem.* **35**, 410 (1970).
d. R. B. Moffett, *Org. Synth.* **IV**, 652 (1963).
e. E. C. Horning and A. F. Finelli, *Org. Synth.* **IV**, 776 (1963).
f. K. Alder, H. Wirtz, and H. Koppelberg, *Justus Liebigs Ann. Chem.* **601**, 138 (1956).
g. K. D. Croft, E. L. Ghisalberti, P. R. Jeffries, and A. D. Stuart, *Aust. J. Chem.* **32**, 2079 (1979).

tion with LDA in THF react with enones to give 1,4-diketones (entries 1 and 2, Scheme 1.11). Esters of 1,4-dicarboxylic acids are obtained by addition of ester enolates to α,β-unsaturated esters (entry 5, Scheme 1.11). In some cases of kinetically controlled reactions, it is found that the initial product of reaction with an α,β-unsaturated carbonyl compound is the result of 1,2- rather than 1,4-addition. Such

Scheme 1.11. Michael Additions under Kinetic Conditions

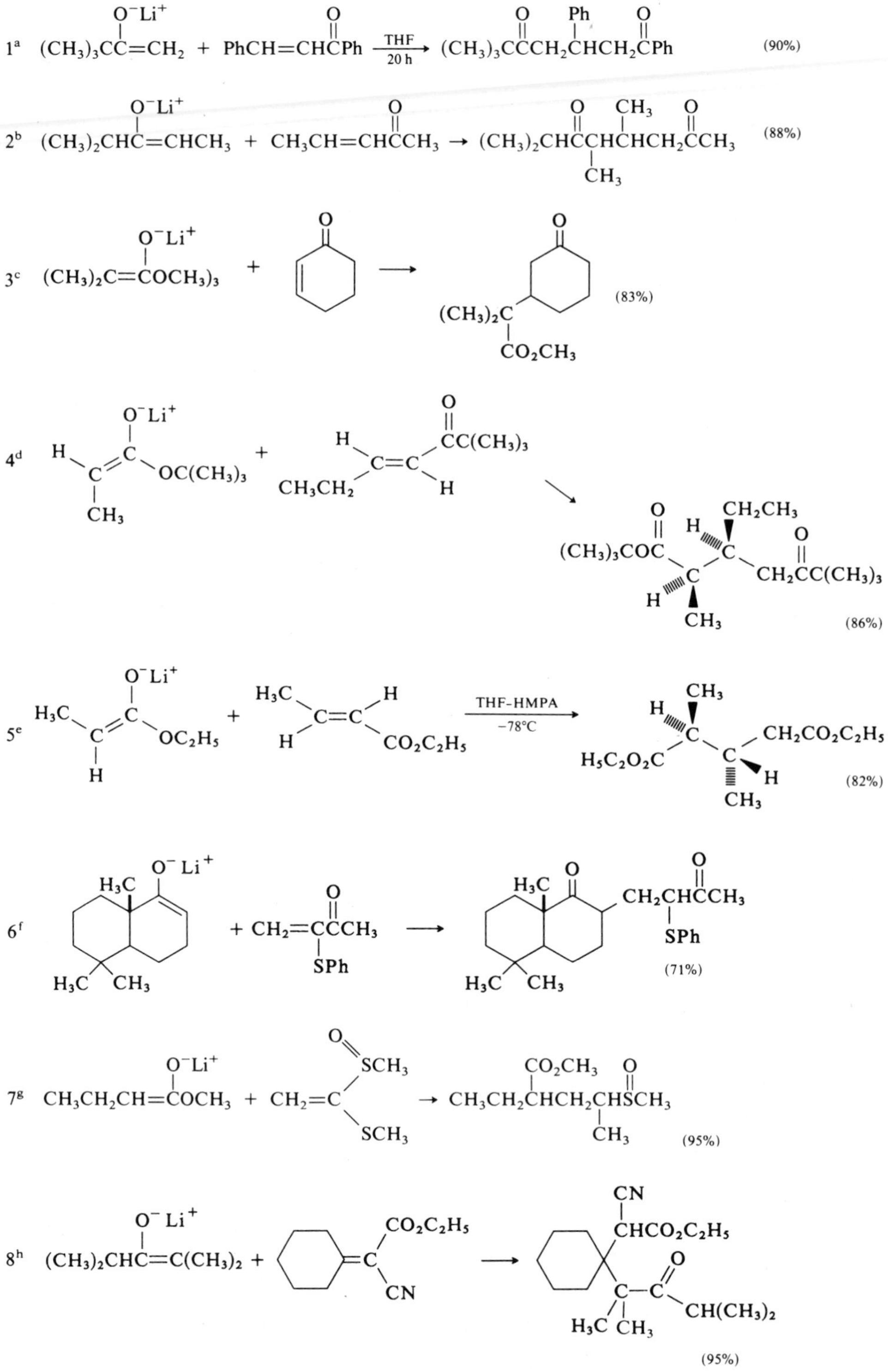

1[a] $(CH_3)_3C(O^-Li^+){=}CH_2$ + $PhCH{=}CHC(O)Ph$ $\xrightarrow[20\,h]{THF}$ $(CH_3)_3CC(O)CH_2CH(Ph)CH_2C(O)Ph$ (90%)

2[b] $(CH_3)_2CHC(O^-Li^+){=}CHCH_3$ + $CH_3CH{=}CHC(O)CH_3$ → $(CH_3)_2CHC(O)CH(CH_3)CH(CH_3)CH_2C(O)CH_3$ (88%)

3[c] $(CH_3)_2C{=}C(O^-Li^+)OCH_3$ + cyclohexenone → 3-[$(CH_3)_2C(CO_2CH_3)$]cyclohexanone (83%)

4[d] $H(CH_3)C{=}C(O^-Li^+)OC(CH_3)_3$ + $CH_3CH_2(H)C{=}C(H)C(O)C(CH_3)_3$ → $(CH_3)_3COC(O)CH(CH_3)CH(CH_2CH_3)CH_2C(O)C(CH_3)_3$ (86%)

5[e] $H_3C(H)C{=}C(O^-Li^+)OC_2H_5$ + $H_3C(H)C{=}C(H)CO_2C_2H_5$ $\xrightarrow[-78^\circ C]{THF-HMPA}$ $H_5C_2O_2CCH(CH_3)CH(CH_3)CH_2CO_2C_2H_5$ (82%)

6[f] enolate (O^-Li^+; H_3C; H_3C, CH_3) + $CH_2{=}C(SPh)C(O)CH_3$ → ketone (H_3C; H_3C, CH_3) bearing $CH_2CH(SPh)C(O)CH_3$ (71%)

7[g] $CH_3CH_2CH{=}C(O^-Li^+)OCH_3$ + $CH_2{=}C(S(O)CH_3)SCH_3$ → $CH_3CH_2CH(CO_2CH_3)CH_2CH(CH_3)S(O)CH_3$ (95%)

8[h] $(CH_3)_2CHC(O^-Li^+){=}C(CH_3)_2$ + cyclohexylidene$=C(CO_2C_2H_5)CN$ → 1-[$CH(CN)CO_2C_2H_5$]-1-[$C(CH_3)_2C(O)CH(CH_3)_2$]cyclohexane (95%)

products frequently are converted to the conjugate addition products under more vigorous reaction conditions.

$$(CH_3)_2C{=}C(O^-Li^+)OCH_3 + \text{cyclohex-2-enone} \longrightarrow \text{1-hydroxy-1-}[C(CH_3)_2CO_2CH_3]\text{cyclohex-2-ene} + \text{3-}[C(CH_3)_2CO_2CH_3]\text{cyclohexanone}$$

	1,2-adduct	1,4-adduct
−78°C	88%	5%
25°C	7%	83%

Ref. 76

$$CH_3CH{=}C(O^-Li^+)C(CH_3)_3 + PhCH{=}CHCOCH_3 \longrightarrow PhCH{=}CHC(OH)(CH_3)CH(CH_3)COC(CH_3)_3 + (CH_3)_3CCOCH(CH_3)CH(Ph)CH_2COCH_3$$

	1,2-adduct	1,4-adduct
30 min	48%	52%
2 h	10%	90%

Ref. 77

The energetic driving force for conjugate addition is the replacement of a π bond by a σ bond. Among Michael acceptors that have been demonstrated to react with ketone and ester enolates under kinetic conditions are methyl α-trimethylsilylvinyl ketone,[78] methyl α-methylthioacrylate,[79] methyl methylthiovinyl sulfoxide,[80] and ethyl α-cyanoacrylate.[81] The latter class of acceptors has been shown to be capable of generating contiguous quaternary carbon centers.

$$(H_3C)_2C{=}C(O^-Li^+)OCH_3 + \text{cyclohexylidene}{=}C(CO_2C_2H_5)CN \longrightarrow \text{1-}[C(NC)HCO_2C_2H_5]\text{-1-}[C(CH_3)_2CO_2CH_3]\text{cyclohexane}$$

Ref. 81

Several examples of conjugate addition of carbanions carried out under kinetically controlled conditions are given in Scheme 1.11.

76. A. G. Schultz and Y. K. Lee, *J. Org. Chem.* **41**, 4045 (1976).
77. J. Berlrand, L. Gorrichon, and P. Maroni, *Tetrahedron* **40**, 4127 (1984).
78. G. Stork and B. Ganem, *J. Am. Chem. Soc.* **95**, 6152 (1973).
79. R. J. Cregge, J. L. Herrmann, and R. H. Schlessinger, *Tetrahedron Lett.*, 2603 (1973).
80. J. L. Herrmann, G. R. Kieczykowski, R. F. Romanet, P. J. Wepple, and R. H. Schlessinger, *Tetrahedron Lett.*, 4711 (1973).
81. R. A. Holton, A. D. Williams, and R. M. Kennedy, *J. Org. Chem.* **51**, 5480 (1986).

---

a. J. Bertrand, L. Gorrichon, and P. Maroni, *Tetrahedron* **40**, 4127 (1984).
b. D. A. Oare and C. H. Heathcock, *Tetrahedron Lett.* **27**, 6169 (1986).
c. A. G. Schultz and Y. K. Yee, *J. Org. Chem.* **41**, 4044 (1976).
d. C. H. Heathcock and D. A. Oare, *J. Org. Chem.* **50**, 3022 (1985).
e. M. Yamaguchi, M. Tsukamoto, S. Tanaka, and I. Hirao, *Tetrahedron Lett.* **25**, 5661 (1984).
f. K. Takaki, M. Ohsugi, M. Okada, M. Yasumura, and K. Negoro, *J. Chem. Soc., Perkin Trans. 1,* 741 (1984).
g. J. L. Herrmann, G. R. Kieczykowski, R. F. Romanet, P. J. Wepplo, and R. H. Schlessinger, *Tetrahedron Lett.*, 4711 (1973).
h. R. A. Holton, A. D. Williams, and R. M. Kennedy, *J. Org. Chem.* **51**, 5480 (1986).

When the conjugate addition is carried out under kinetic conditions with stoichiometric formation of the enolate, the adduct is also an enolate until the reaction mixture is quenched with a proton source. It should therefore be possible to effect a second reaction of the enolate if an electrophile is added prior to protonation of the enolate. This has been done by adding an alkyl halide to the solution of the adduct enolate.

$$H_2C{=}C(O^-Li^+)OC(CH_3)_3 + CH_3CH{=}CHCO_2C_2H_5 \xrightarrow{-78^\circ C} (CH_3)_3CO_2CCH_2CH(CH_3)CH{=}C(O^-Li^+)OC_2H_5 \xrightarrow{CH_3I,\ HMPA} (CH_3)_3CO_2CCH_2CH(CH_3)CH(CH_3)CO_2C_2H_5\ (60\%)$$

Ref. 82

$$H_2C{=}C(CH_3){-}CH{=}C(O^-Li^+)OC_2H_5 + \text{cyclopentenone} \longrightarrow \text{3-[CH(CO}_2\text{C}_2\text{H}_5\text{)C(CH}_3\text{)=CH}_2\text{]cyclopentenolate} \xrightarrow{H_2C=CHCH_2Br} \text{2-(CH}_2\text{CH=CH}_2\text{)-3-[CH(CO}_2\text{C}_2\text{H}_5\text{)C(CH}_3\text{)=CH}_2\text{]cyclopentanone}$$

Ref. 83

Tandem conjugate addition–alkylation has proven to be an efficient means of introducing both α and β substituents at enones.

Trimethylsilyl enol ethers can be caused to react with electrophilic alkenes by use of Lewis acids. These reactions proceed rapidly even at −78°C.[84]

$$PhC(=O)CH{=}C(CH_3)_2 + CH_2{=}C(OSi(CH_3)_3)Ph \xrightarrow{TiCl_4} PhC(=O)CH_2C(CH_3)_2CH_2C(=O)Ph\ (72\text{–}78\%)$$

Ref. 85

Similarly, titanium tetrachloride or stannic tetrachloride induces addition of silyl enol ethers to nitroalkenes. The initial product is a complex of the "aci" tautomer of the nitro group. Hydrolysis of this complex generates the new carbonyl group.[86]

82. M. Yamaguchi, M. Tsukamoto, and I. Hirao, *Tetrahedron Lett.* **26**, 1723 (1985).
83. W. Oppolzer, R. P. Heloud, G. Bernardinelli, and K. Baettig, *Tetrahedron Lett.* **24**, 4975 (1983).
84. K. Narasaka, K. Soai, Y. Aikawa, and T. Mukaiyama, *Bull. Chem. Soc. Jpn.* **49**, 779 (1976).
85. K. Narasaka, *Org. Synth.* **65**, 12 (1987).
86. M. Miyashita, T. Yanimi, T. Kumazawa, and A. Yoshikoshi, *J. Am. Chem. Soc.* **106**, 2149 (1984).

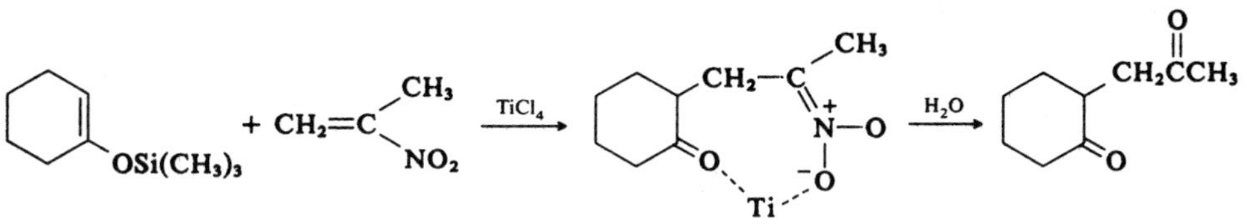

Cyanide ion acts as a carbon nucleophile in the conjugate addition reaction. An alcoholic solution of potassium or sodium cyanide is suitable for simple compounds.

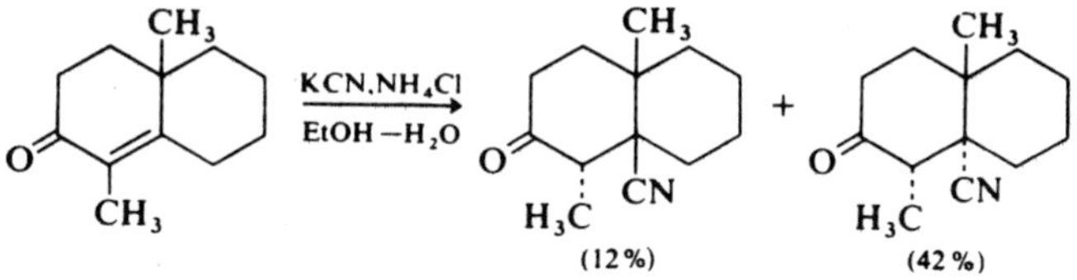

Triethylaluminum–hydrogen cyanide and diethylaluminum cyanide are also useful reagents for conjugate addition of cyanide. The latter reagent is the more reactive of the two. These reactions presumably involve the coordination of the aluminum reagent as a Lewis acid at the carbonyl oxygen.

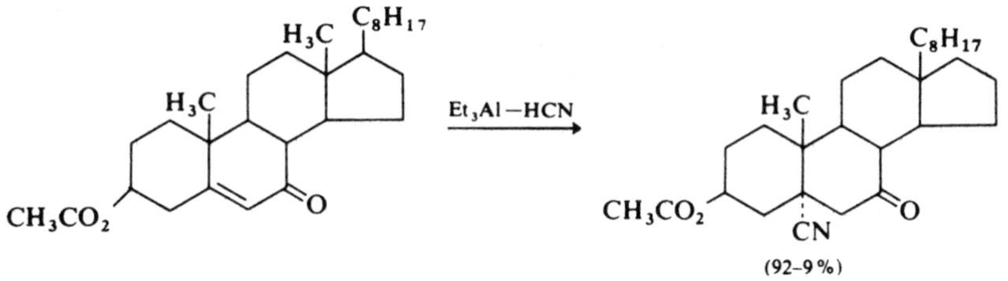

Enamines also react with electrophilic alkenes to give conjugate addition products. The addition reactions of enamines of cyclohexanones show a strong preference for attack from the axial direction.[90] This is anticipated on stereoelectronic grounds since the π orbital of the enamine is the site of nucleophilicity.

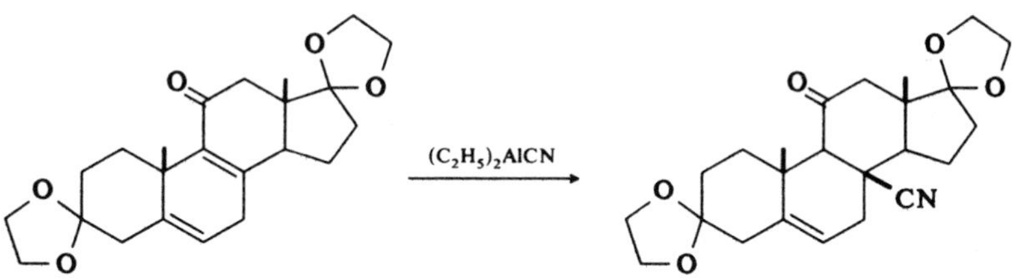

87. O. R. Rodig and N. J. Johnston, *J. Org. Chem.* **34**, 1942 (1969).
88. W. Nagata and M. Yoshioka, *Org. Synth.* **52**, 100 (1972).
89. W. Nagata, M. Yoshioka, and S. Hirai, *J. Am. Chem. Soc.* **94**, 4635 (1972).
90. E. Valentin, G. Pitacco, F. P. Colonna, and A. Risalti, *Tetrahedron* **30**, 2741 (1974); M. Forchiassin, A. Risalti, C. Russo, M. Calligaris, and G. Pitacco, *J. Chem. Soc.*, 660 (1974).

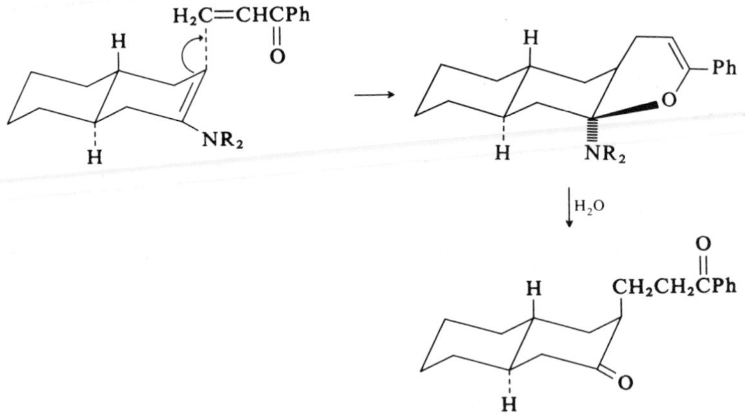

$H_2C{=}CHCPh$ ($C{=}O$)

NR_2 $\longrightarrow$ Ph, O, NR_2

$\downarrow H_2O$

CH_2CH_2CPh ($C{=}O$), O

Another very important method for adding a carbon chain at the β-carbon of an α,β-unsaturated carbonyl system involves organometallic reagents, particularly organocopper intermediates. This reaction will be discussed in Chapter 8.

General References

D. E. Bergbreiter and M. Newcomb, in *Asymmetric Synthesis*, J. D. Morrison (ed.), Academic Press, New York, 1983, Chapter 9.
D. Caine, in *Carbon-Carbon Bond Formation*, Vol. 1, R. L. Augustine (ed.), Marcel Dekker, New York, 1979, Chapter 2.
A. G. Cook (ed.), *Enamines: Synthesis, Structure and Reactions*, Second Edition, Marcel Dekker, New York, 1988.
D. J. Cram, *Fundamentals of Carbanion Chemistry*, Academic Press, New York, 1965.
H. O. House, *Modern Synthetic Reactions*, Second Edition, W. A. Benjamin, Menlo Park, California, 1972, Chapter 9.
J. R. Jones, *The Iionization of Carbon Acids*, Academic Press, New York, 1973.
J. C. Stowell, *Carbanions in Organic Synthesis*, Wiley-Interscience, New York, 1979.

Problems

(*References for these problems will be found on page* 763.)

1. Arrange in order of decreasing acidity:

(a) $CH_3CH_2NO_2$, $(CH_3)_2CHCPh$ (C=O), CH_3CH_2CN, $CH_2(CN)_2$

(b) $[(CH_3)_2CH]_2NH$, $(CH_3)_2CHOH$, $(CH_3)_2CH_2$, $(CH_3)_2CHPh$

(c) $CH_3CCH_2CO_2CH_3$ (C=O), $CH_3CCH_2CCH_3$ (C=O, C=O), CH_3OCCH_2Ph (C=O), CH_3COCH_2Ph (C=O)

(d) $PhCCH_2Ph$ (C=O), $(CH_3)_3CCCH_3$ (C=O), $(CH_3)_3CCCH(CH_3)_2$ (C=O), $PhCCH_2CH_2CH_3$ (C=O)

2. Write the structure of all possible enolates for each ketone. Indicate which you would expect to be favored in a kinetically controlled deprotonation. Which would you expect to be the most stable enolate in each case?

(a) CH_3 ... $C(CH_3)_3$ (b) ... CH_3 (c) $(CH_3)_2CHCCH_2CH_3$ (C=O)

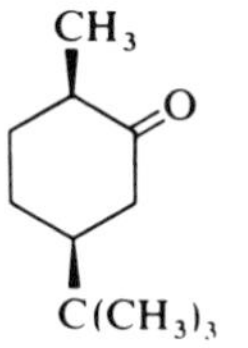

(d) CH_3 ... H_3C CH_3 (e) ... CH_3 CH_3 CH_3 EtO OEt (f) CH_3 ... H_3C CH_3

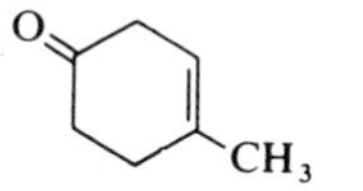

(g) CH_3 ... CH_2 CH_3 (h) ... CH_3

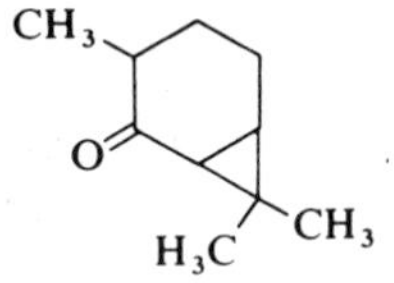

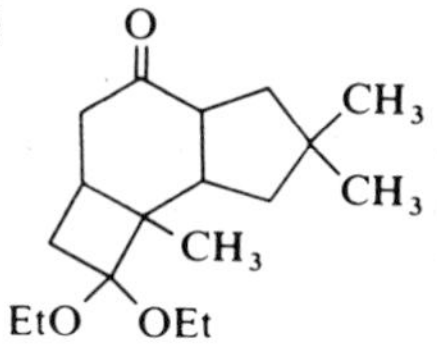

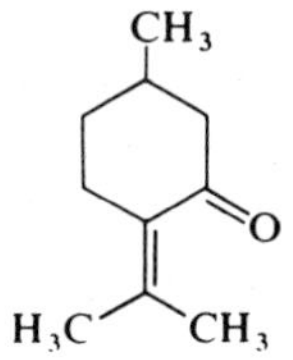

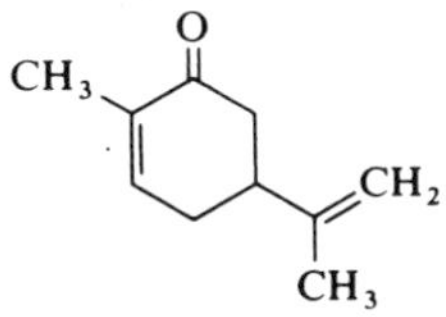

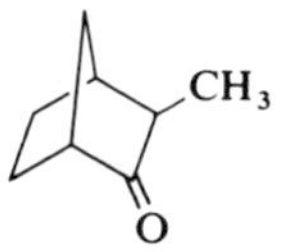

3. Suggest reagents and reaction conditions suitable for effecting each of the following conversions:

(a) 2-methylcyclohexanone to 2-benzyl-6-methylcyclohexanone.

(b) ... CH_3 to CH_3 CH_3 CH_3 ... CH_3

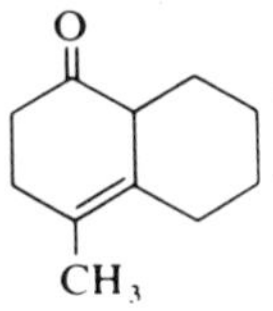

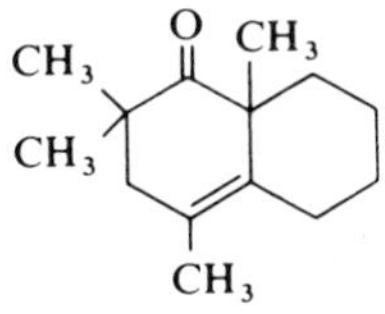

(c) ... Ph to ... Ph CH_2Ph

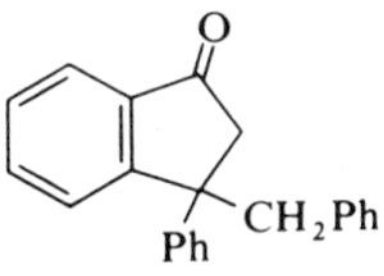

(d) CH_2CN ... N CH_2Ph to CH_3 CHCN ... N CH_2Ph

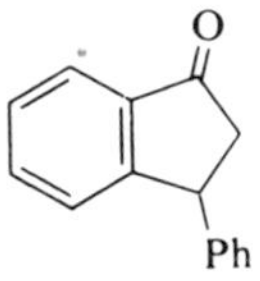

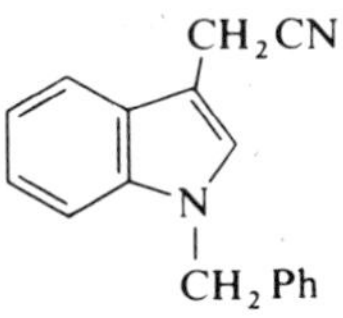

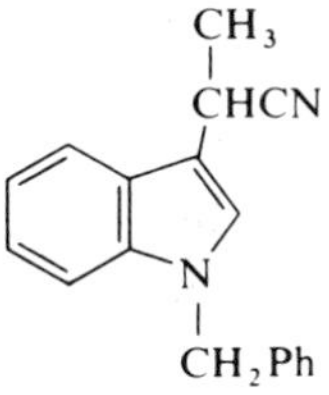

(e) $CH_3\overset{O}{\overset{\|}{C}}CH{=}CH_2$ to $CH_2{=}\overset{OSi(CH_3)_3}{\overset{|}{C}}{-}CH{=}CH_2$

(f) CCH_3 (O) / $CH_2CH_2CH_2Br$ to

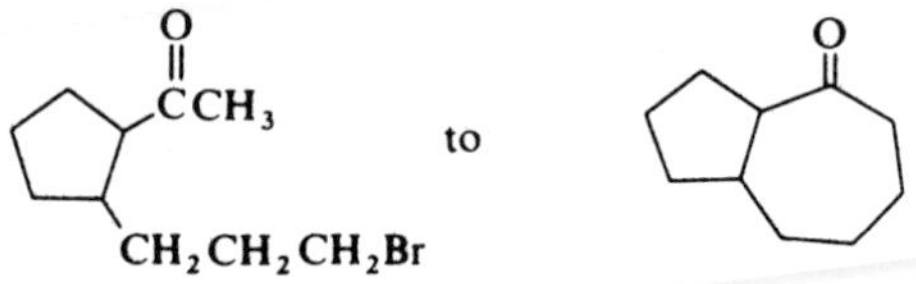

(g) CCH_3 (O) / $CH_2CH_2CH_2Br$ to CH_3

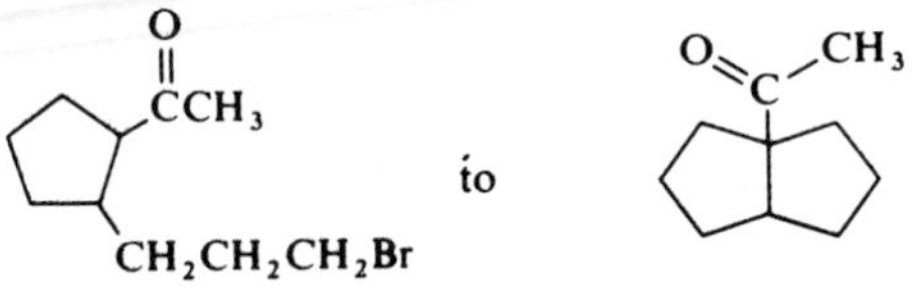

4. Intramolecular alkylation of enolates has been used to advantage in synthesis of bi- and tricyclic compounds. Indicate how such a procedure could be used to synthesize each of the following molecules by drawing the structure of a suitable precursor:

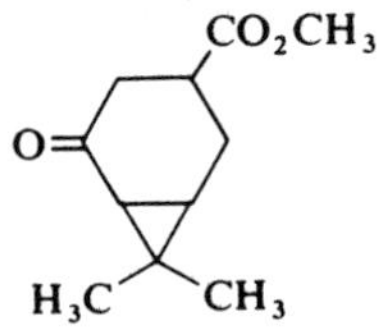

(a) O

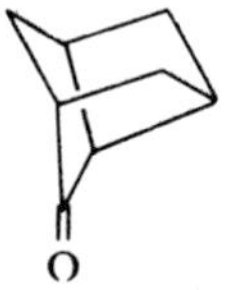

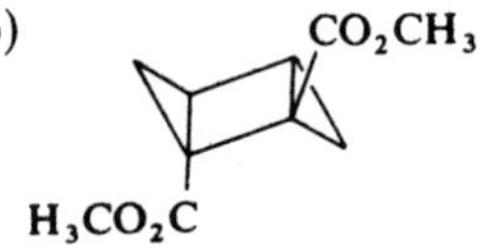

(c) CO_2CH_3, O=, H_3C, CH_3

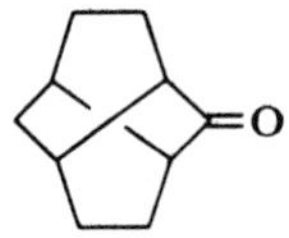

(e) CH_3, OCH_2Ph, O

(b) CO_2CH_3, H_3CO_2C

(d) =O

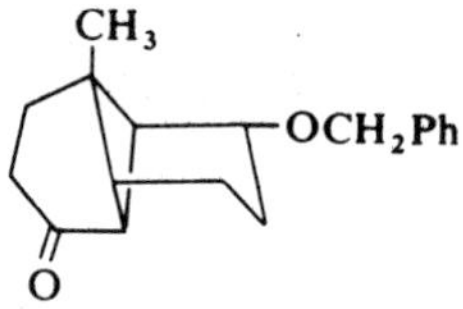

(f) H_3C, CH_3, O

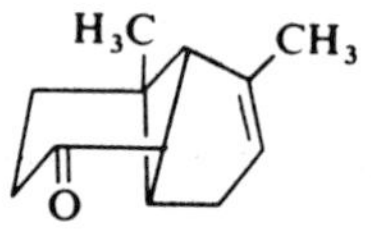

5. Predict the major product of each of the following reactions:

(a) $PhCHCO_2Et$ / CH_2CO_2Et $\xrightarrow[(2)\ CH_3I]{(1)\ 1\ \text{equiv}\ LiNH_2/NH_3}$

(b) $PhCHCO_2Et$ / CH_2CO_2H $\xrightarrow[(2)\ CH_3I]{(1)\ 2\ \text{equiv}\ LiNH_2/NH_3}$

(c) $PhCHCO_2H$ / CH_2CO_2Et $\xrightarrow[(2)\ CH_3I]{(1)\ 2\ \text{equiv}\ LiNH_2/NH_3}$

6. Treatment of 2,3,3-triphenylpropionitrile with one equivalent of potassium amide in liquid ammonia followed by addition of benzyl chloride affords 2-benzyl-2,3,3-triphenylpropionitrile in 97% yield. Use of two equivalents of potassium amide gives an 80% yield of 2,3,3,4-tetraphenylbutyronitrile under the same reaction conditions. Explain.

7. Suggest readily available starting materials and reaction conditions suitable for obtaining each of the following compounds by a procedure involving alkylation of nucleophilic carbon:

(a) $PhCH_2CH_2CH(CN)Ph$

(b) $(CH_3)_2C{=}CHCH_2CH_2C(=O)CH_2CO_2CH_3$

(c) 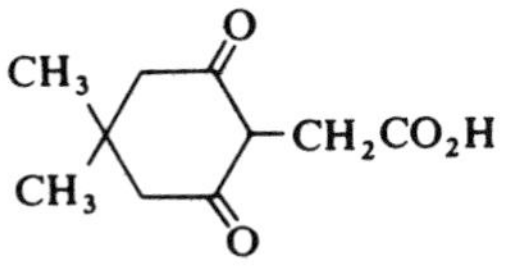 (cyclohexane-1,3-dione with 5,5-dimethyl (CH_3, CH_3) and 2-CH_2CO_2H substituents)

(d) $CH_2{=}CHCH{=}CHCH_2CH_2CO_2H$

(e) 2,3-diphenylpropanoic acid

(f) 2,6-diallylcyclohexanone

(g) 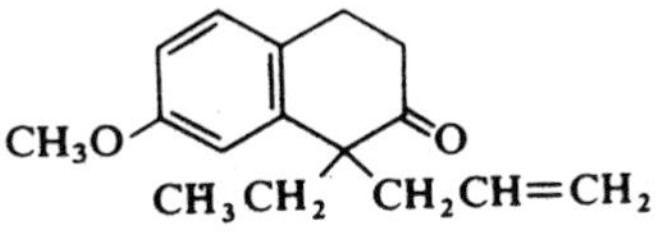 (tetralone with CH_3O, =O, CH_3CH_2 and $CH_2CH{=}CH_2$ substituents)

(h) $H_2C{=}CHCH_2C(CN)(Ph)C(=O)NH_2$

(i) 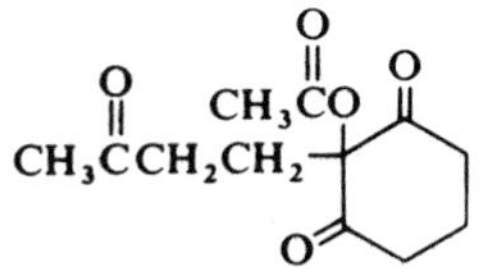 $CH_3C(=O)CH_2CH_2$– (2-acetyl(CH_3CO)-cyclohexanone, O=)

(j) $CH_2{=}CHCH(CO_2CH_2CH_3)CH_2C{\equiv}CH$

8. Suggest starting materials and reaction conditions suitable for obtaining each of the following compounds by a procedure involving a Michael reaction:

(a) 4,4-dimethyl-5-nitropentan-2-one
(b) diethyl 2,3-diphenylglutarate
(c) ethyl 2-benzoyl-4-(2-pyridyl)butyrate
(d) 2-phenyl-3-oxocyclohexaneacetic acid

(e) 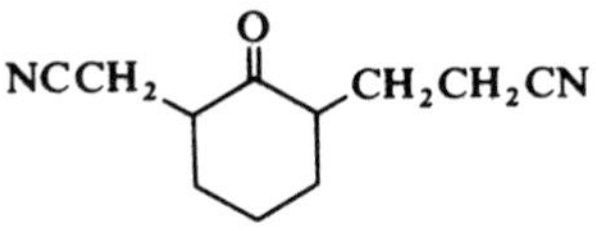 (cyclohexanone with $NCCH_2$ and CH_2CH_2CN substituents)

(f) 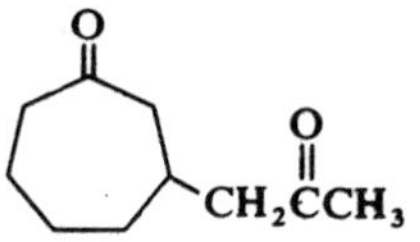 (cycloheptanone with $CH_2C(=O)CH_3$ substituent)

(g) $CH_3CH_2CH(NO_2)CH_2CH_2C(=O)CH_3$

(h) $(CH_3)_2CHCH(CH{=}O)CH_2CH_2CO_2CH_2CH_3$

(i) 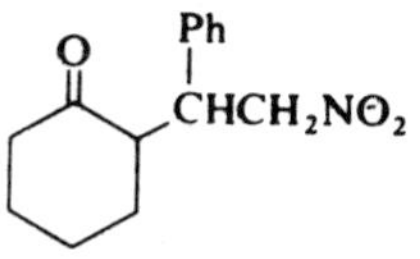 (cyclohexanone with $CH(Ph)CH_2NO_2$ substituent)

(k) (Ph-benzylidene acetal pyranoside with OCH_3, NO_2 and $CH(NO_2)CH_3$ substituents)

(j) $PhCH(CN)CH(Ph)CH_2C(=O)CH_3$

(l) 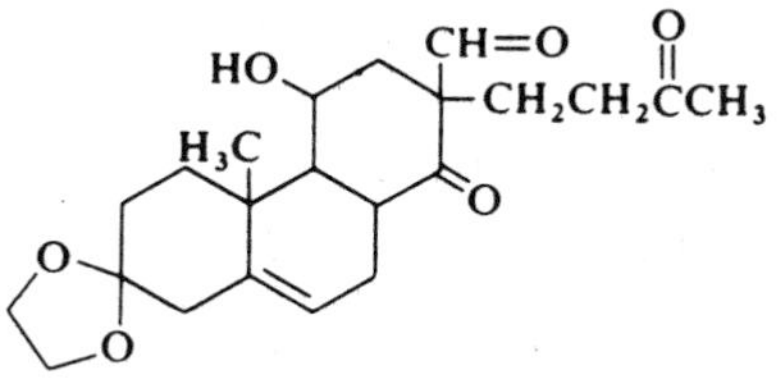 (steroid-type ketal with HO, H_3C, $CH{=}O$, =O and $CH_2CH_2C(=O)CH_3$ substituents)

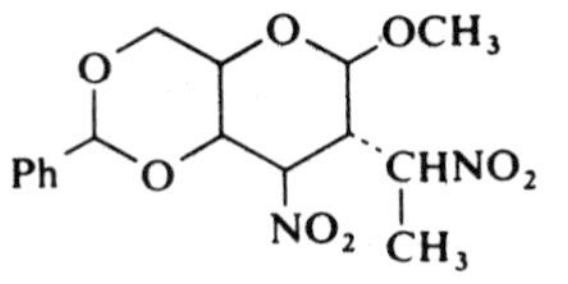

9. In planning a synthesis, the most effective approach is to reason backwards from the target molecule to some readily available starting material. This is called *retrosynthetic analysis* and is indicated by an open arrow of the type shown below. In each of the following problems, the target molecule is shown on the left and the starting material on the right. Determine how you could prepare the target molecule from the indicated starting material using any necessary organic or inorganic reagents. In some cases, more than one step is necessary.

(a) 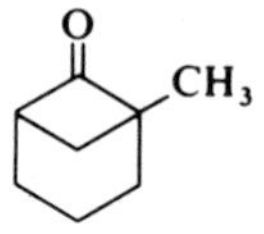 O, CH_3 ⇨ O, $CO_2C_2H_5$

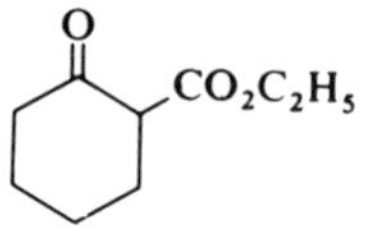

(b) 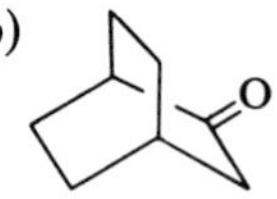 O ⇨ O

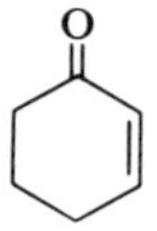

(c) 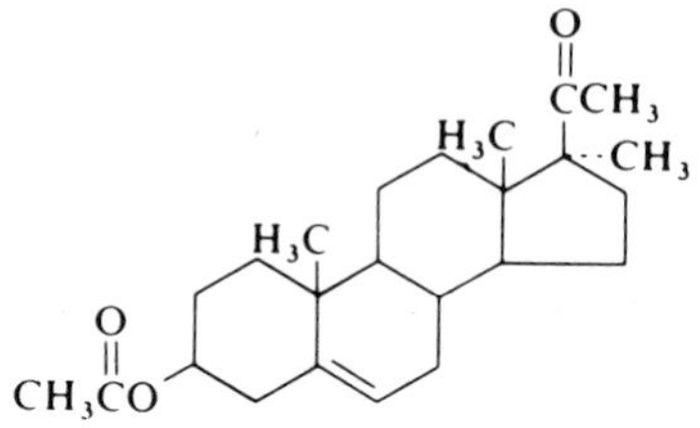 O, $\overset{\|}{C}CH_3$, H_3C, CH_3, H_3C, O, $CH_3\overset{\|}{C}O$ ⇨ O, $\overset{\|}{C}CH_3$, H_3C, H_3C, O, $CH_3\overset{\|}{C}O$

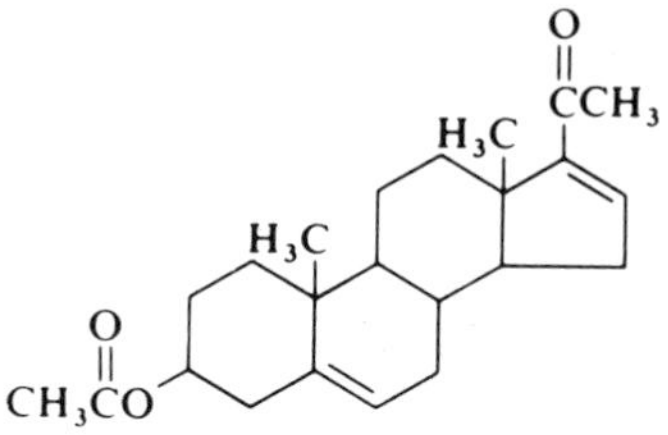

(d) $(CH_3O)_2\overset{O}{\overset{\|}{P}}CH_2\overset{O}{\overset{\|}{C}}(CH_2)_4CH_3$ ⇨ $(CH_3O)_2\overset{O}{\overset{\|}{P}}CH_2\overset{O}{\overset{\|}{C}}CH_3$

(e) $PhCH_2CH_2\underset{Ph}{\underset{|}{C}}HCO_2C_2H_5$ ⇨ $PhCH_2CO_2C_2H_5$

(f) 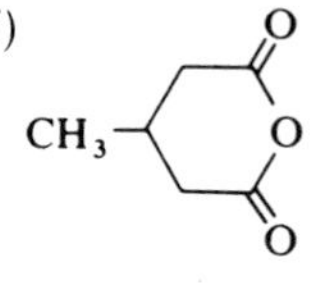 CH_3, O, O, O ⇨ $CH_3CH{=}CHCO_2CH_3$

(g) 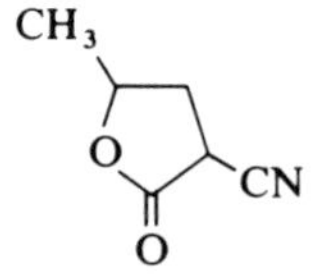 CH_3, O, CN, O ⇨ $NCCH_2CO_2C_2H_5$

(h) 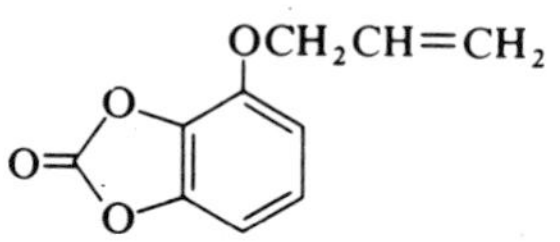 $OCH_2CH{=}CH_2$, O, O=, O ⇨ OH, HO, HO

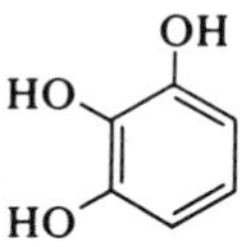

(i) 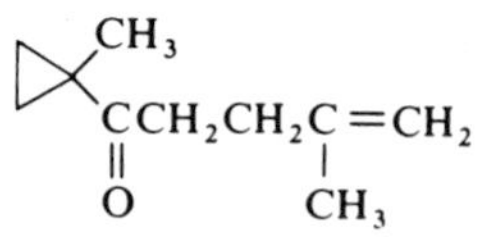 CH_3, $\underset{O}{\underset{\|}{C}}CH_2CH_2\underset{CH_3}{\underset{|}{C}}{=}CH_2$ ⇨ CH_3, $\underset{O}{\underset{\|}{C}}CH_2CO_2CH_2CH_3$

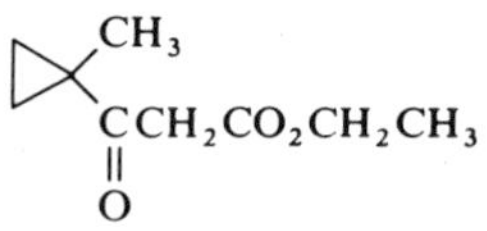

10. In a synthesis of diterpenes via compound **C**, a key intermediate **B** was obtained from carboxylic acid **A**. Suggest a series of reactions for obtaining **B** from **A**.

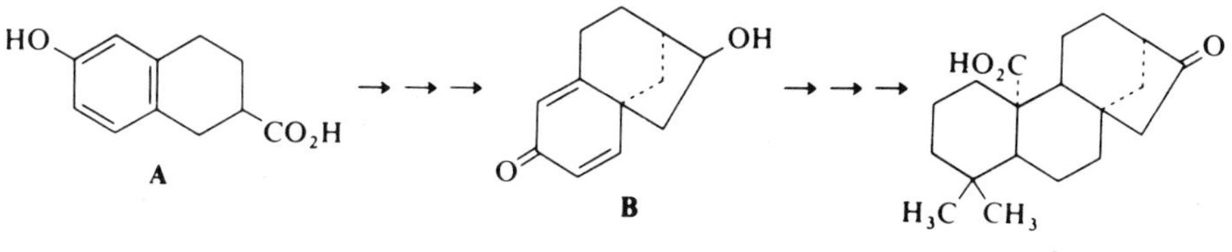

A **B** **C**

11. In a synthesis of the terpene longifolene, the tricyclic intermediate **D** was obtained from a bicyclic intermediate by an intramolecular Michael addition. Deduce the possible structure(s) of the bicyclic precursor.

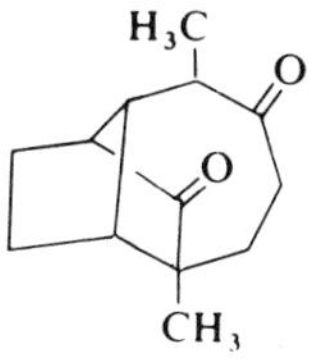

D

12. Substituted acetophenones react with ethyl phenylpropiolate under the conditions of the Michael reaction to give pyrones. Formulate a mechanism.

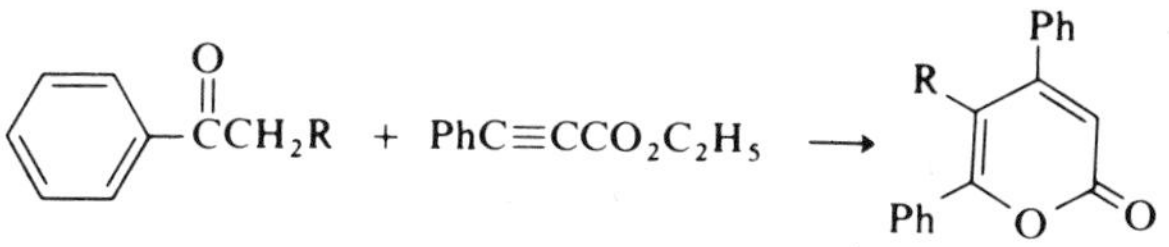

13. The reaction of simple ketones such as 2-butanone or phenylacetone with α,β-unsaturated ketones gives cyclohexenones when the reaction is effected by heating in methanol with potassium methoxide. Explain how the cyclohexenones are formed. What structures are possible for the cyclohexenones? Can you suggest means for distinguishing between possible isomeric cyclohexenones?

14. Analyze the factors which would be expected to control the stereochemistry of the following reactions, and predict the stereochemistry of the product(s).

(a)

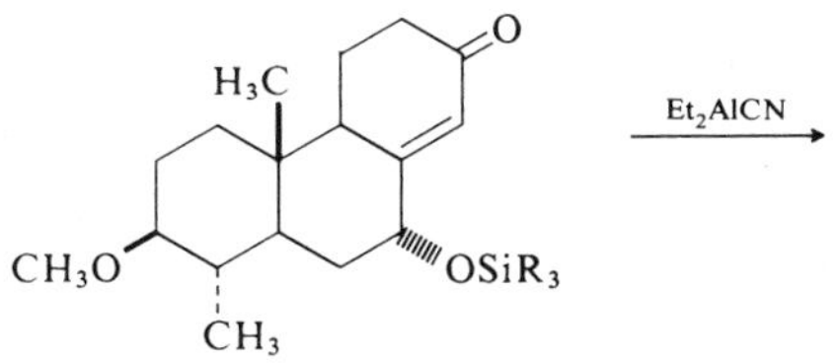

(b) CH_3, $CH_2CH{=}CCH_3$, Cl, O, CH_3 — 1) $K^{+}\,{}^{-}N(SiMe_3)_2$, 25°C; 2) CH_3I

(c) O, O, O, $CH(CH_3)_2$, CN, $PhCH_2OCH_2$, C, RO, $CH_2CH(CH_3)_2$ — 1) NaH; 2) CH_3I; $R = CH_3CH_2OCH(CH_3)-$

(d) CH_3O_2C, N, Ph, O, H_3C — 1) LDA; 2) $BrCH_2CH{=}CH_2$

(e) H_3C, CO_2CH_3, CO_2CH_3, OH, H — 1) NaH; 2) $BrCH_2C(CH_3){=}CH_2$

(f) $N{\equiv}C$, CH_3, O, H — CH_3I, $LiNH_2$

(g) O, O, O, $N\overset{O}{\overset{\|}{C}}CH_2CH_3$, Ph, CH_3 — 1) $NaN[Si(CH_3)_3]_2$; 2) $CH_2{=}CHCH_2I$

(h) Ph_3COCH_2, O, O — 1) LDA/CH_3I; 2) LDA/$CH_2{=}CHCH_2Br$

(i) Ph, CH_3, O, O — 1) LDA/HMPA; 2) C_2H_5I

15. Indicate reaction sequences and approximate conditions which could be used to effect the following transformations. More than one step may be required.

(a)

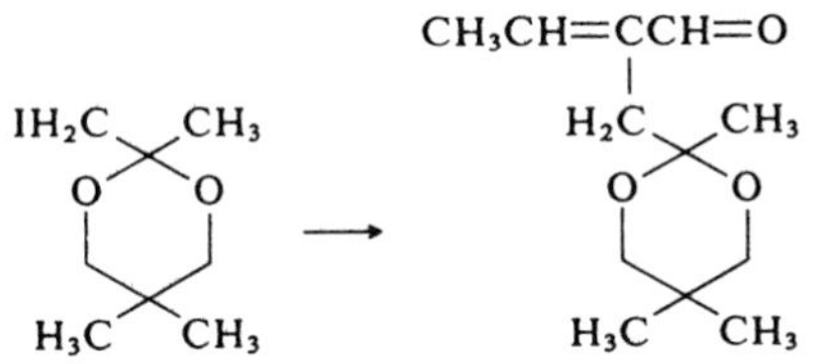

(b) $CH_3\overset{O}{\overset{\|}{C}}CH_2CO_2H \rightarrow CH_3\overset{O}{\overset{\|}{C}}CH_2CH_2CH{=}CH_2$

(c) $(CH_3)_2CHCH_2CH_2\overset{O}{\overset{\|}{C}}CH_2CO_2CH_3 \rightarrow (CH_3)_2CHCH_2\underset{CH_3CH_2CH_2}{\underset{|}{C}H}\overset{O}{\overset{\|}{C}}CH_2CO_2CH_3$

(d)

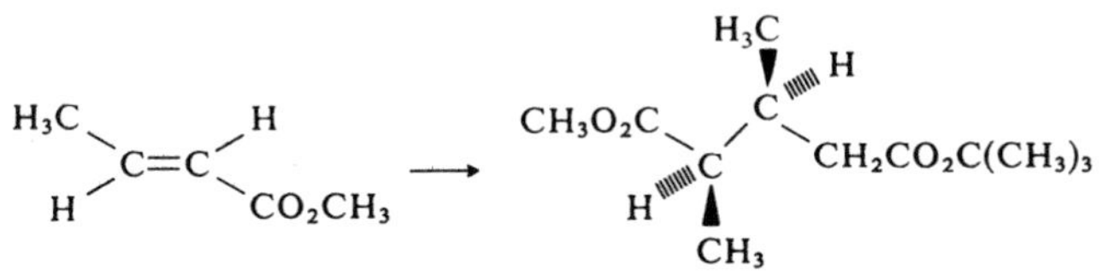

(e)

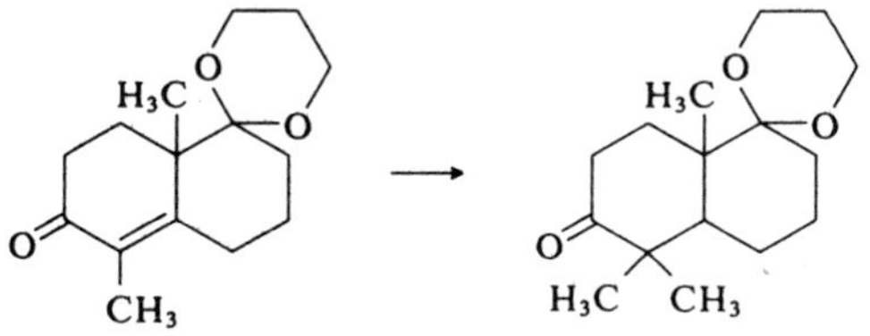

(f)

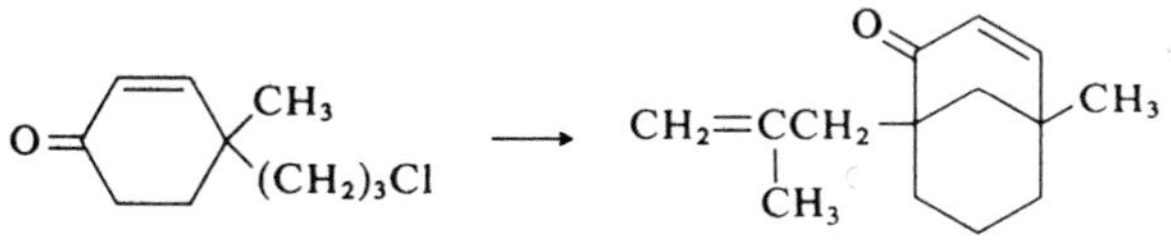

16. One of the compounds shown below undergoes intramolecular cyclization to give a tricyclic ketone on being treated with $[(CH_3)_3Si]_2NNa$. The other does not. Suggest a structure for the product. Explain the difference in reactivity.

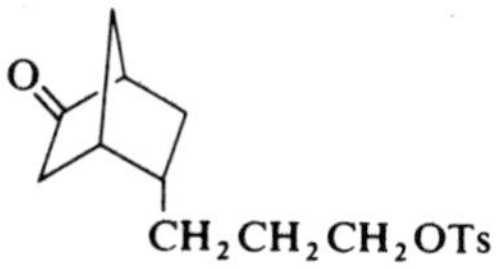

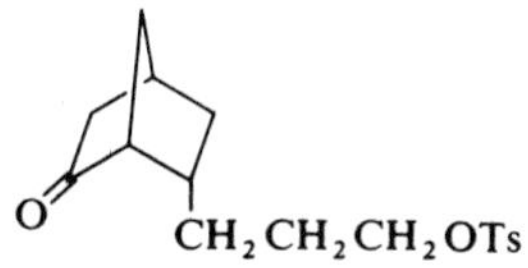

$CH_2CH_2CH_2OTs$ $CH_2CH_2CH_2OTs$

17. The alkylation of 3-methyl-2-cyclohexenone with several dibromides led to the products shown below. Discuss the course of each reaction and suggest an explanation for the dependence of the product structure on the identity of the dihalide.

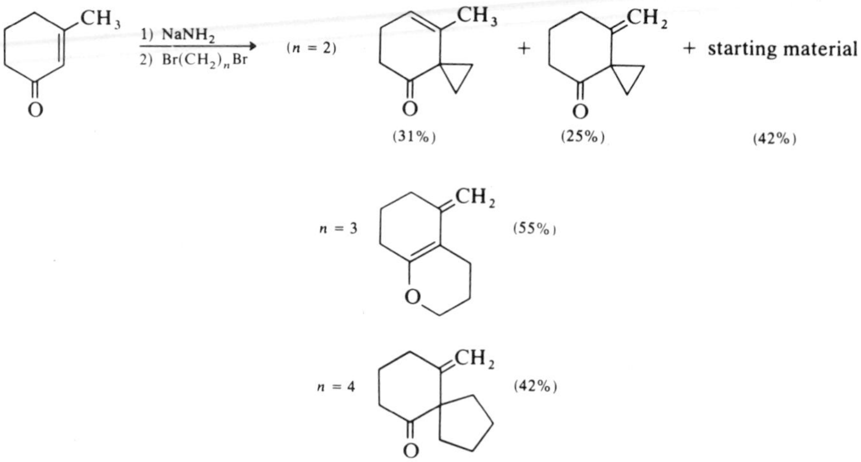

1) $NaNH_2$
2) $Br(CH_2)_nBr$

(n = 2): (31%) + (25%) + starting material (42%)

n = 3: (55%)

n = 4: (42%)

18. Treatment of ethyl 2-azidobutanoate with catalytic quantities of lithium ethoxide in tetrahydrofuran leads to the evolution of nitrogen. On quenching the resulting solution with 3N hydrochloric acid, ethyl 2-oxobutanoate is isolated in 86% yield. Suggest a reasonable mechanism for this process.

$$CH_3CH_2CH(N_3)CO_2CH_2CH_3 \xrightarrow[2)\ H_3O^+]{1)\ LiOEt,\ THF} CH_3CH_2C(=O)CO_2CH_2CH_3 \quad (86\%)$$

19. Suggest a reasonable mechanism for the reaction

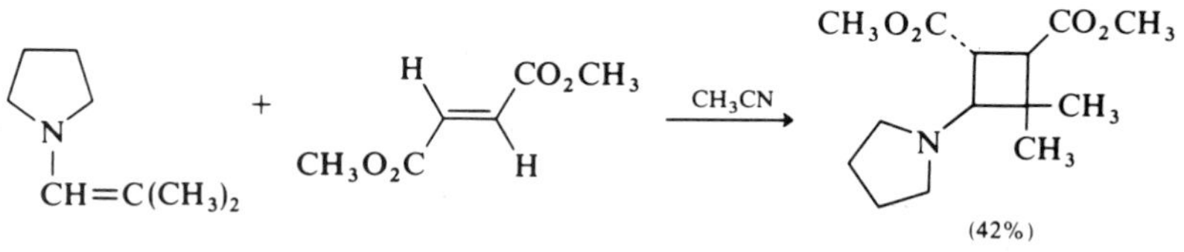

$CH=C(CH_3)_2$ (N-pyrrolidinyl) + CH_3O_2C–CH=CH–CO_2CH_3 $\xrightarrow{CH_3CN}$ (42%)

2

Reactions of Carbon Nucleophiles with Carbonyl Groups

The reactions described in this chapter include some of the most useful synthetic methods for carbon–carbon bond formation: *the aldol and the Claisen condensations, the Robinson annulation, and the Wittig reaction and related olefination reactions.* All of these processes involve the addition of a carbon nucleophile to a carbonyl group. The type of product that is isolated depends on the nature of the substituent (X) on the carbon nucleophile, the substituents (A and B) on the carbonyl group, and the ways in which A, B, and X interact to control the reaction pathways available to the addition intermediate.

$$\text{X}{-}\overset{|}{\underset{|}{\text{C}}}{}^{-} + \text{A}{-}\overset{\text{O}}{\overset{\|}{\text{C}}}{-}\text{B} \rightarrow {-}\overset{\text{X}}{\overset{|}{\underset{|}{\text{C}}}}{-}\overset{\text{O}^-}{\overset{|}{\underset{\underset{\text{A}}{|}}{\text{C}}}}{-}\text{B} \rightarrow \text{product}$$

2.1. Aldol Condensation

2.1.1. The General Mechanism

The aldol condensation reaction is the acid- or base-catalyzed self-condensation of a ketone or aldehyde.[1] Under certain conditions, the reaction product may undergo further transformations, especially dehydration.

1. A. T. Nielsen and W. J. Houlihan, *Org. React.* **16**, 1 (1968); R. L. Reeves, in *Chemistry of the Carbonyl Group*, S. Patai (ed.), Wiley-Interscience, New York, 1966, pp. 580–593; H. O. House, *Modern Synthetic Reactions*, Second Edition, W. A. Benjamin, Menlo Park, California, 1972, pp. 629–682.

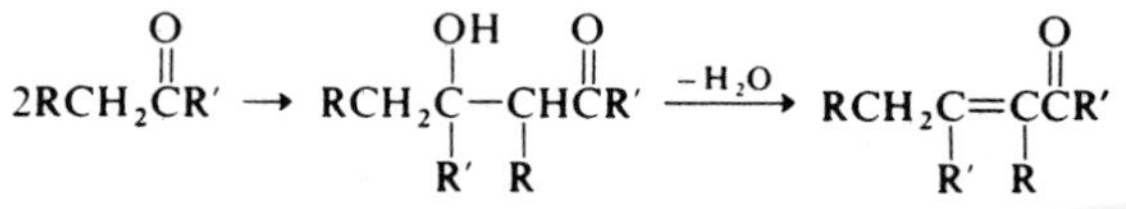

$$2RCH_2C(=O)R' \rightarrow RCH_2C(OH)(R')-CH(R)C(=O)R' \xrightarrow{-H_2O} RCH_2C(R')=C(R)C(=O)R'$$

R' = H or alkyl or aryl

The reaction may also occur between two different carbonyl compounds, in which case the term *mixed aldol condensation* is applied.

The mechanism of the base-catalyzed reaction involves formation of the enolate ion, followed by addition of the enolate to a carbonyl group of the aldehyde or ketone.

Base-Catalyzed Mechanism

1. Addition phase
   a. Enolate formation:

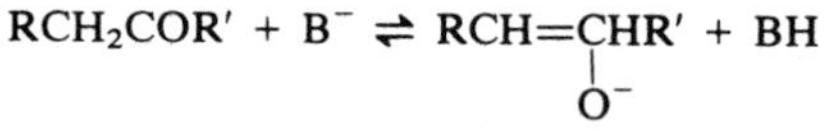

$$RCH_2COR' + B^- \rightleftharpoons RCH=C(O^-)R' + BH$$

   b. Nucleophilic addition:

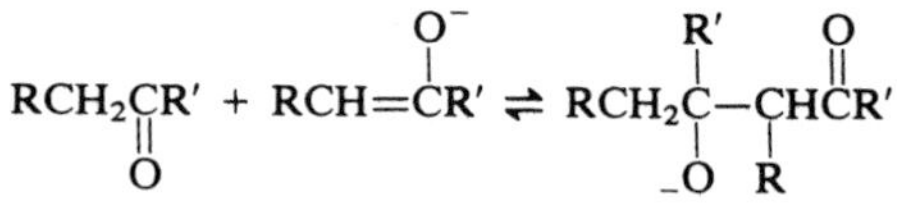

$$RCH_2C(=O)R' + RCH=C(O^-)R' \rightleftharpoons RCH_2C(R')(O^-)-CH(R)C(=O)R'$$

   c. Proton transfer:

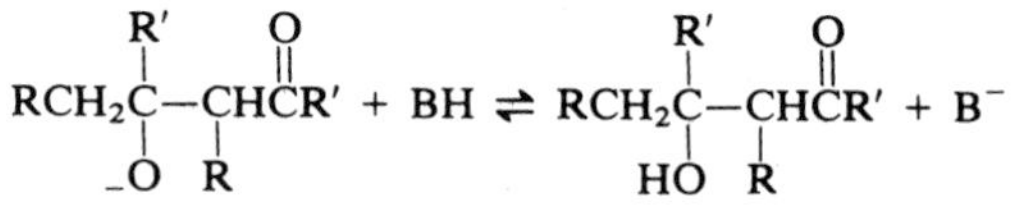

$$RCH_2C(R')(O^-)-CH(R)C(=O)R' + BH \rightleftharpoons RCH_2C(R')(HO)-CH(R)C(=O)R' + B^-$$

2. Dehydration phase

$$RCH_2C(R')(HO)-CH(R)C(=O)R' + B^- \rightarrow (R')(RCH_2)C=C(R)C(=O)R' + BH + HO^-$$

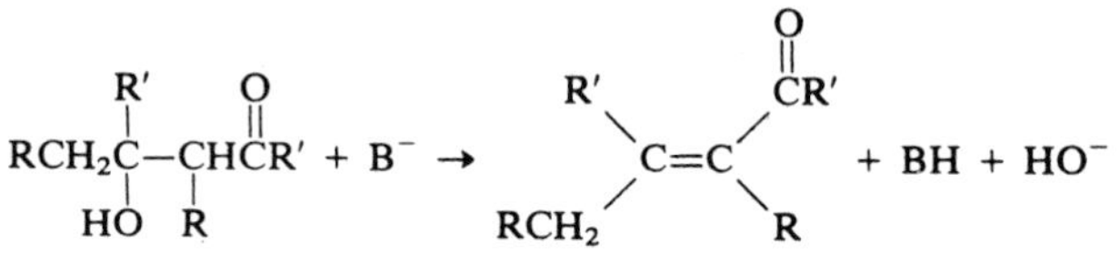

Entries 1 and 2 in Scheme 2.1 illustrate the preparation of aldol condensation products by the base-catalyzed reaction. In entry 1, the product is a β-hydroxy-aldehyde, while in entry 2 dehydration has occurred and the product is an α,β-unsaturated aldehyde.

Under conditions of acid catalysis, it is the enol tautomer of the aldehyde or ketone that functions as the nucleophile. The carbonyl group is activated toward nucleophilic attack by oxygen protonation.

Acid-Catalyzed Mechanism

1. Addition phase
   a. Enolization:

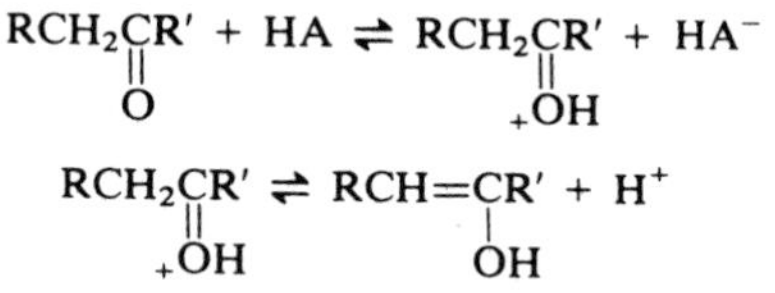

$$RCH_2C(=O)R' + HA \rightleftharpoons RCH_2C(={}^+OH)R' + HA^-$$

$$RCH_2C(={}^+OH)R' \rightleftharpoons RCH=C(OH)R' + H^+$$

b. Nucleophilic addition:

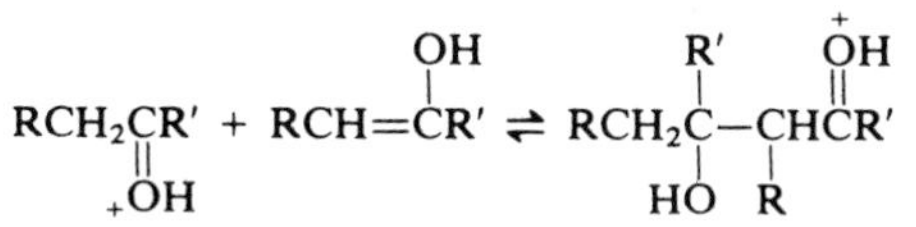

$$\mathrm{RCH_2\overset{\overset{+}{O}H}{\overset{\|}{C}}R' + RCH{=}\overset{OH}{\overset{|}{C}}R' \rightleftharpoons RCH_2\overset{R'}{\overset{|}{\underset{HO}{\underset{|}{C}}}}{-}\underset{R}{\underset{|}{C}}H\overset{\overset{+}{O}H}{\overset{\|}{C}}R'}$$

c. Proton transfer:

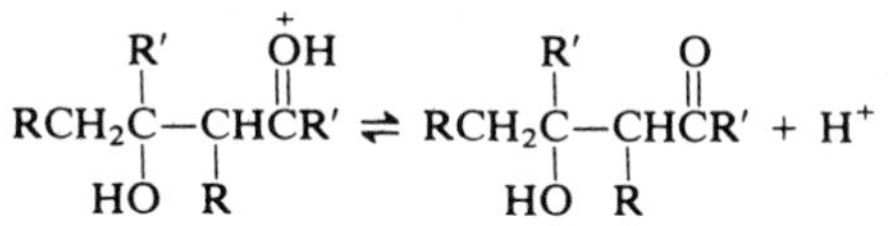

$$\mathrm{RCH_2\overset{R'}{\overset{|}{\underset{HO}{\underset{|}{C}}}}{-}\underset{R}{\underset{|}{C}}H\overset{\overset{+}{O}H}{\overset{\|}{C}}R' \rightleftharpoons RCH_2\overset{R'}{\overset{|}{\underset{HO}{\underset{|}{C}}}}{-}\underset{R}{\underset{|}{C}}H\overset{O}{\overset{\|}{C}}R' + H^+}$$

2. Dehydration phase

$$\mathrm{RCH_2\overset{R'}{\overset{|}{\underset{HO}{\underset{|}{C}}}}{-}\underset{R}{\underset{|}{C}}H\overset{O}{\overset{\|}{C}}R' + H^+ \rightarrow (R')(RCH_2)C{=}C(\overset{O}{\overset{\|}{C}}R')(R) + H_2O}$$

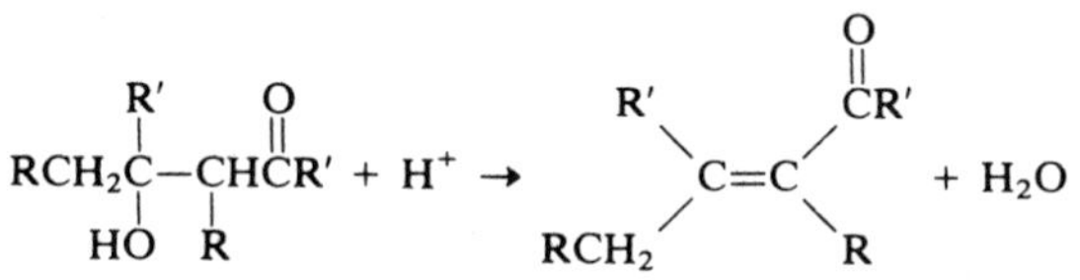

Entries 4 and 5 in Scheme 2.1 depict acid-catalyzed aldol condensation. In entry 4, condensation is accompanied by dehydration. In entry 5, a β-chloroketone is formed by addition of hydrogen chloride to the enone.

In general, the reactions in the addition phase of both the base- and acid-catalyzed mechanisms are reversible. The equilibrium constant for addition is usually unfavorable for acylic ketones. The equilibrium constant for the dehydration phase

Scheme 2.1. Aldol Condensation of Simple Aldehydes and Ketones

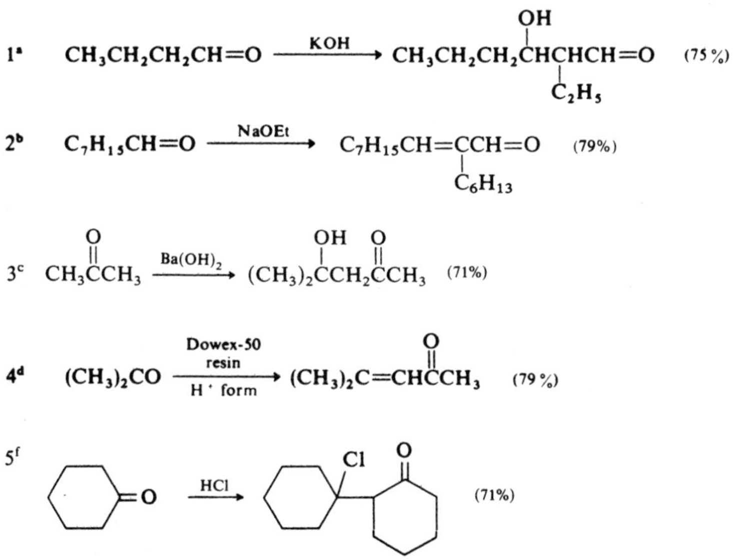

1[a] $\mathrm{CH_3CH_2CH_2CH{=}O \xrightarrow{KOH} CH_3CH_2CH_2\overset{OH}{\overset{|}{C}}H\underset{C_2H_5}{\underset{|}{C}}HCH{=}O}$ (75%)

2[b] $\mathrm{C_7H_{15}CH{=}O \xrightarrow{NaOEt} C_7H_{15}CH{=}\underset{C_6H_{13}}{\underset{|}{C}}CH{=}O}$ (79%)

3[c] $\mathrm{CH_3\overset{O}{\overset{\|}{C}}CH_3 \xrightarrow{Ba(OH)_2} (CH_3)_2\overset{OH}{\overset{|}{C}}CH_2\overset{O}{\overset{\|}{C}}CH_3}$ (71%)

4[d] $\mathrm{(CH_3)_2CO \xrightarrow[H^+\ form]{Dowex\text{-}50\ resin} (CH_3)_2C{=}CH\overset{O}{\overset{\|}{C}}CH_3}$ (79%)

5[f] cyclohexanone $\xrightarrow{HCl}$ 2-(1-chlorocyclohexyl)cyclohexanone (71%)

a. V. Grignard and A. Vesterman, *Bull. Chim. Soc. Fr.* **37**, 425 (1925).
b. F. J. Villani and F. F. Nord, *J. Am. Chem. Soc.* **69**, 2605 (1947).
c. J. B. Conant and N. Tuttle, *Org. Synth.* **I**, 199 (1941).
d. N. B. Lorette, *J. Org. Chem.* **22**, 346 (1957).
e. O. Wallach, *Ber.* **40**, 70 (1907); E. Wenkert, S. K. Bhattacharya, and E. M. Wilson, *J. Chem. Soc.*, 5617 (1964).

is usually favorable, because of the conjugated α,β-unsaturated carbonyl system that is formed. When the reaction conditions are sufficient to cause dehydration, the overall reaction will go to completion, even if the equilibrium constant for the addition phase is unfavorable.

2.1.2. Mixed Aldol Condensations with Aromatic Aldehydes

Aldol condensation reactions involving two different carbonyl compounds are called *mixed aldol condensations.* For such reactions to be useful as a method for synthesis, there must be some basis for controlling which carbonyl component will serve as the electrophile and which will serve as the enolate precursor. One of the most general of the mixed aldol condensations involves the use of aromatic aldehyde with alkyl ketones or aldehydes. Aromatic aldehydes cannot function as the nucleophilic component, because they are incapable of enolization. Furthermore, dehydration is usually especially favorable because the resulting enone is also conjugated with the aromatic ring.

$$\mathrm{ArCH{=}O + RCH_2\overset{O}{\overset{\|}{C}}R' \rightleftharpoons ArCH(OH)CH(R)\overset{O}{\overset{\|}{C}}R' \xrightarrow{-H_2O} ArCH{=}C(R)\overset{O}{\overset{\|}{C}}R'}$$

There are numerous examples of both acid- and base-catalyzed mixed aldol condensations involving aromatic aldehydes. The reaction is sometimes referred to as the *Claisen–Schmidt condensation.* Scheme 2.2 presents some representative examples.

There is a pronounced preference for the formation of a *trans* double bond in the Claisen–Schmidt condensation of methyl ketones. This stereoselectivity arises in the dehydration step. In the transition state for elimination to a *cis* double bond, an unfavorable steric interaction between the ketone substituent (R) and the phenyl group occurs. This interaction is absent in the transition state for elimination to the *trans* double bond.

: base　　: base

less favorable　　more favorable

Additional insight into the factors affecting product structure was obtained by study of the condensation of 2-butanone with benzaldehyde.[2]

$$\mathrm{PhCH{=}C(CH_3)\overset{O}{\overset{\|}{C}}CH_3 \xleftarrow{HCl} PhCHO + CH_3COCH_2CH_3 \xrightarrow{NaOH} PhCH{=}CH\overset{O}{\overset{\|}{C}}CH_2CH_3}$$

2. M. Stiles, D. Wolf, and G. V. Hudson, *J. Am. Chem. Soc.* **81**, 628 (1959); D. S. Noyce and W. L. Reed, *J. Am. Chem. Soc.* **81**, 618, 620, 624 (1959).

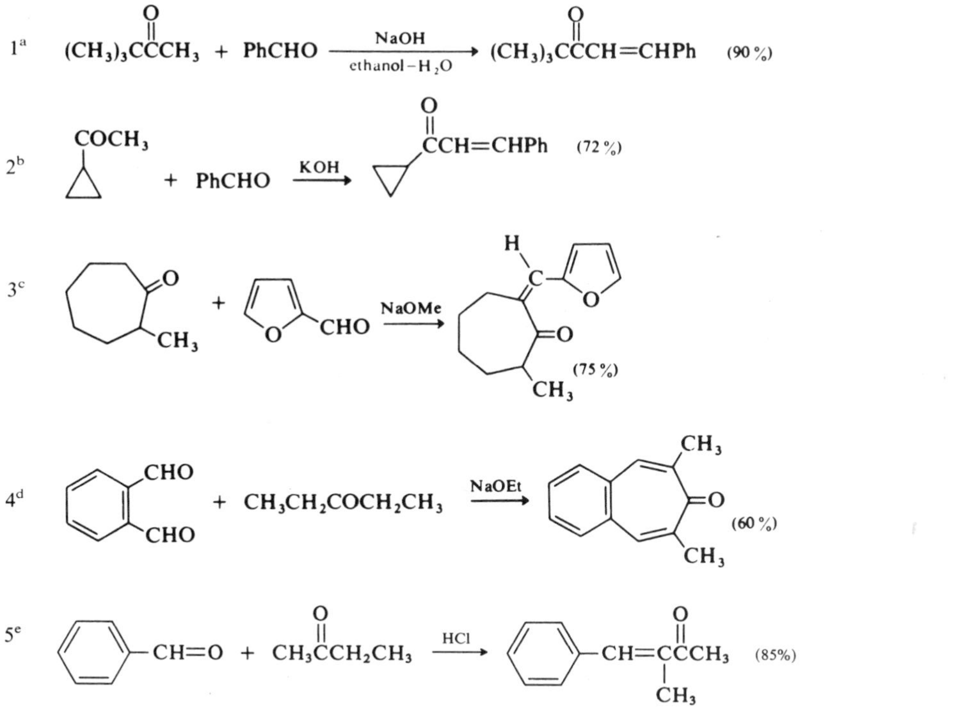

Scheme 2.2. Mixed Condensation of Aromatic Aldehydes with Ketones

1[a] $(CH_3)_3CCCH_3$ (C=O) + PhCHO $\xrightarrow[\text{ethanol}-H_2O]{\text{NaOH}}$ $(CH_3)_3CCCH{=}CHPh$ (C=O) (90%)

2[b] cyclopropyl–$COCH_3$ + PhCHO $\xrightarrow{\text{KOH}}$ cyclopropyl–$CCH{=}CHPh$ (C=O) (72%)

3[c] 2-methylcycloheptanone + furan-2-CHO $\xrightarrow{\text{NaOMe}}$ 2-(furylmethylene)-7-methylcycloheptanone (75%)

4[d] o-phthalaldehyde (CHO, CHO) + $CH_3CH_2COCH_2CH_3$ $\xrightarrow{\text{NaOEt}}$ dimethylbenzocycloheptenone (CH_3, =O, CH_3) (60%)

5[e] C_6H_5–CH=O + $CH_3CCH_2CH_3$ (C=O) $\xrightarrow{\text{HCl}}$ C_6H_5–$CH{=}CCCH_3$ (C=O; CH_3) (85%)

a. G. A. Hill and G. Bramann, *Org. Synth.* **I**, 81 (1941).
b. S. C. Bunce, H. J. Dorsman, and F. D. Popp, *J. Chem. Soc.*, 303 (1963).
c. A. M. Islam and M. T. Zemaity, *J. Am. Chem. Soc.* **79**, 6023 (1957).
d. D. Meuche, H. Strauss, and E. Heilbronner, *Helv. Chim. Acta* **41**, 2220 (1958).
e. M. E. Kronenberg and E. Havinga, *Rec. Trav. Chim.* **84**, 17, 979 (1965).

The results indicate that the product ratio is determined by the competition between the various reaction steps. When catalyzed by base, 2-butanone reacts with benzaldehyde at the methyl group to give 1-phenyl-1-penten-3-one. Under acid-catalyzed conditions, the product is the result of condensation at the methylene group, namely, 3-methyl-4-phenyl-3-buten-2-one. Under the reaction conditions used, it is not possible to isolate the intermediate ketols, because the addition step is rate-limiting. These intermediates can be prepared by alternative methods, and they behave as shown in the following equations:

$$PhCH(OH)CH_2C(=O)CH_2CH_3 \xrightarrow{\text{NaOH}} PhCH{=}CHC(=O)CH_2CH_3 + PhCH{=}O + CH_3CH_2C(=O)CH_3$$

$$PhCH(OH)CH(CH_3)C(=O)CH_3 \xrightarrow{\text{NaOH}} PhCH{=}O + CH_3CH_2C(=O)CH_3$$

$$PhCH(OH)CH_2C(=O)CH_2CH_3 \xrightarrow{HCl} PhCH{=}CHC(=O)CH_2CH_3 + PhCHO + CH_3COCH_2CH_3$$

$$PhCH(OH)CH(CH_3)C(=O)CH_3 \xrightarrow{HCl} PhCH{=}C(CH_3)C(=O)CH_3 + PhCHO + CH_3COCH_2CH_3$$

These results establish that the base-catalyzed dehydration is slow relative to the reverse of the addition phase for the branched-chain isomer. The reason for selective formation of the straight-chain product under conditions of base catalysis is then apparent. In base, the straight-chain ketol is the only intermediate that is dehydrated. The branched-chain ketol reverts to starting material. Under acid conditions, both intermediates are dehydrated. However, under acid-catalyzed conditions the branched-chain ketol is formed most rapidly, because of the preference for acid-catalyzed enolization to give the more substituted enol (see Section 7.3 of Part A).

$$CH_3COCH_2CH_3 \xrightarrow{H^+} \underset{\text{(major)}}{CH_3C(OH){=}CHCH_3} + \underset{\text{(minor)}}{CH_2{=}C(OH)CH_2CH_3}$$

$$CH_3C(OH){=}CHCH_3 + PhCHO \xrightarrow{slow} PhCH(OH)CH(CH_3)C(=O)CH_3 \xrightarrow{fast} PhCH{=}C(CH_3)C(=O)CH_3$$

In general, the product ratio of a mixed aldol condensation will depend upon the individual reaction rates. Most ketones show a pattern similar to butanone in reactions with aromatic aldehydes. Base catalysis favors reaction at a methyl position over a methylene group, whereas acid catalysis gives the opposite preference.

2.1.3. Control of Regiochemistry and Stereochemistry of Mixed Aldol Condensations of Aliphatic Aldehydes and Ketones

The control of mixed aldol condensations between aldehydes and ketones which present several possible sites for enolization is a challenging problem. Such reactions are normally carried out by converting the carbonyl compound which is to serve as the nucleophile to an enolate, silyl enol ether, or metalloenamine. The reactive nucleophile is then allowed to react with the second reaction component. As long as the addition step is faster than proton transfer, or other mechanisms of interconversion of the nucleophilic and electrophilic components, the adduct will have the desired structure. The term *directed aldol condensations* is given to such reactions.[3] Directed aldol condensations must be carried out under conditions designed to ensure that the product structure is that which is desired. Scheme 2.3 illustrates some of the procedures which have been developed to achieve this goal.

3. T. Mukaiyama, *Org. React.* **28**, 203 (1982).

Entries 1–4 represent cases in which the nucleophilic component is converted to the enolate under kinetically controlled conditions, such as those discussed in Section 1.2. Such enolates are usually highly reactive toward aldehydes so that rapid addition occurs when the aldehyde is added, even at low temperature. When the addition step is complete, the reaction is stopped by neutralization and the product is isolated. The guiding mechanistic concept for reactions carried out under these conditions is that they occur through a cyclic transition state in which lithium or another metal cation is coordinated to both the enolate oxygen and the carbonyl oxygen.[4]

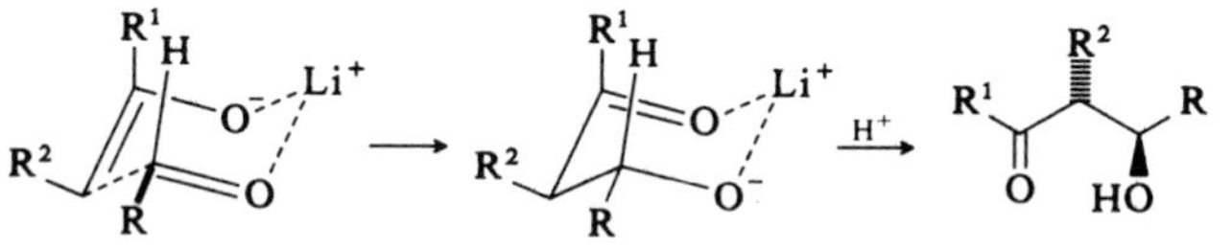

This transition state model has been the basis both for development of other reaction conditions and for the interpretation of the stereochemistry of the reaction.

Most enolates can exist as two stereoisomers.

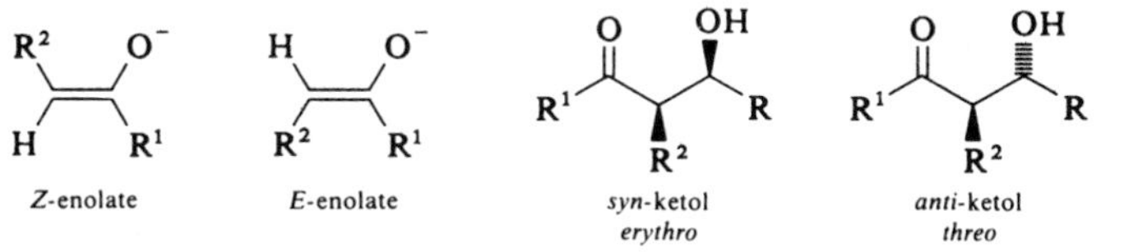

Z-enolate　　*E*-enolate　　*syn*-ketol *erythro*　　*anti*-ketol *threo*

Also, most aldol condensation products formed from a ketone enolate and an aldehyde can have two diastereomeric structures. These can be designated as *erythro* and *threo* or as *syn* and *anti*. We will use the *syn*, *anti* terminology. The cyclic transition state model provides a basis for understanding the relationship between enolate geometry and the stereochemistry of the aldol product.

The enolate formed from 2,2-dimethyl-3-pentanone under kinetically controlled conditions is the *Z*-isomer. When it reacts with benzaldehyde, only the *syn* aldol is formed.[4] This stereochemical relationship is interpreted in terms of a cyclic transition state with a chairlike conformation. The product stereochemistry is correctly predicted if the aldehyde is in a conformation such that the phenyl substituent occupies an equatorial position.

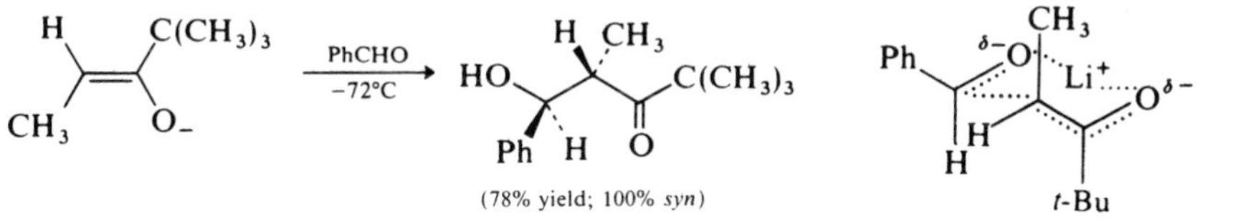

(78% yield; 100% *syn*)

A similar preference for formation of the *syn* aldol is found for other *Z*-enolates derived from ketones in which one of the carbonyl substituents is bulky.[5] Ketone

4. C. H. Heathcock, C. T. Buse, W. A. Kleschick, M. C. Pirrung, J. E. Sohn, and J. Lampe, *J. Org. Chem.* **45**, 1066 (1980).
5. P. Fellman and J. E. Dubois, *Tetrahedron* **34**, 1349 (1978).

Scheme 2.3. Directed Aldol Condensations

A. Condensations of Lithium Enolates under Kinetic Control

1[a] $CH_3CH_2CH_2C(=O)CH_3$ —(LDA, −78°C)→ —(1) $CH_3CH_2CH_2CH{=}O$, 15 min, −78°C; 2) CH_3CO_2H)→ $CH_3CH_2CH_2C(=O)CH_2CH(OH)CH_2CH_2CH_3$ (65%)

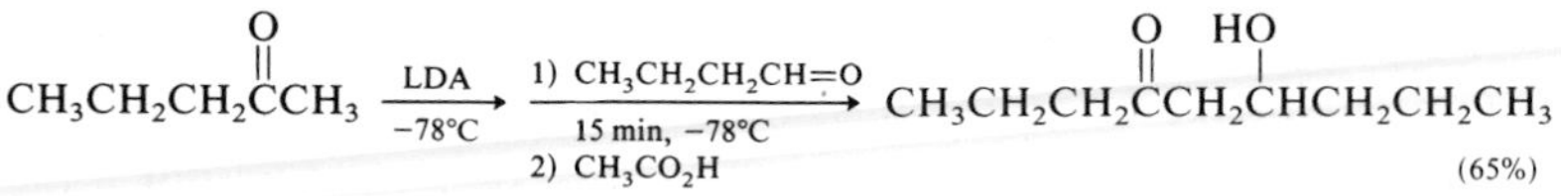

2[b] $CH_3CH{=}C(O^-\,{}^+Li)$(cyclohexyl) + $PhCH_2OCH_2CH(CH_3)CH{=}O$ → $CH_3CH(C(=O)\text{cyclohexyl})CH(OH)CH(CH_3)CH_2OCH_2Ph$ (79%)

3[c] $CH_3CH_2C(=O)C(CH_3)_2(OTMS)$ —(LDA, −78°C)→ —(1) $(CH_3)_2CHCH{=}O$; 2) NH_4Cl)→ $(CH_3)_2CHCH(OH)CH(CH_3)C(=O)C(CH_3)_2(OTMS)$ (61%)

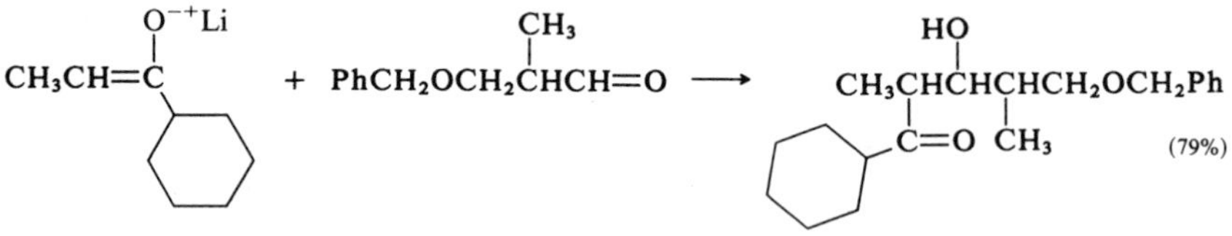

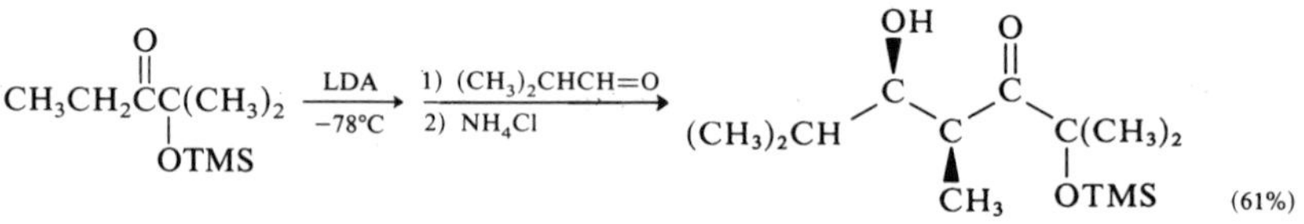

4[d] $CH_3C(=O)C(CH_3)_2OTMS$ —(LDA, −78°C)→ —(1) $CH_3CH_2CH{=}O$, 1.5 h; 2) NH_4Cl)→ $CH_3CH_2CH(OH)CH_2C(=O)C(CH_3)_2OTMS$ (68%)

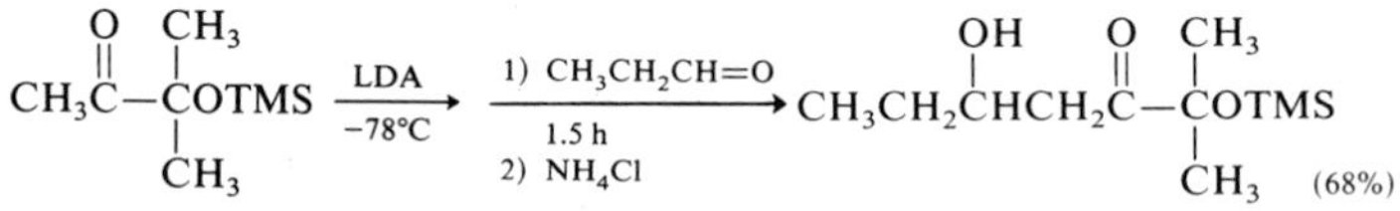

B. Condensations of Magnesium and Zinc Enolates Formed by Dehalogenation

5[e] 2-bromo-2,5,5-trimethylcyclopentanone —(1) Mg; 2) $CH_3CH{=}O$)→ 2-(1-hydroxyethyl)-2,5,5-trimethylcyclopentanone (major and minor diastereomers)

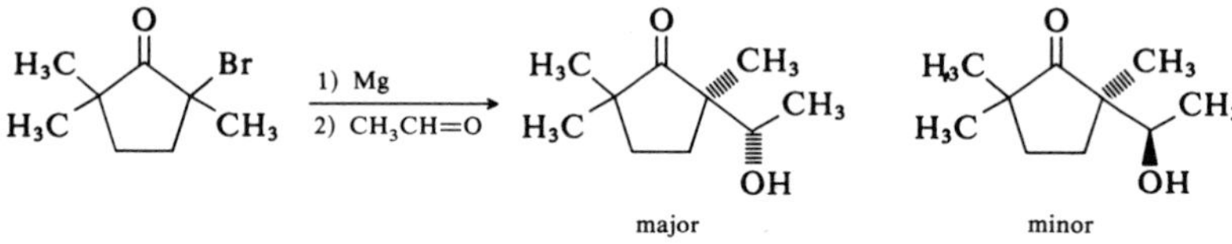

major minor

6[f] 2-bromo-2-methylcyclohexanone + $PhCH{=}O$ —(Zn, Et_2AlCl)→ 2-methyl-2-(CH(OH)Ph)cyclohexanone (100%)

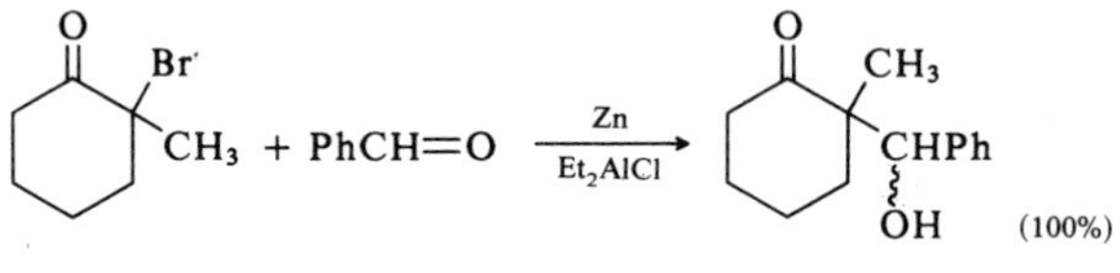

C. Condensations of Boron Enolates

7[g] $PhCH_2CH_2C(=O)CH_3$ —($R_2BO_3SCF_3$, 2,6-lutidine, 78°C, 3 h)→ —(1) $PhCH{=}O$, −78°C, 5 h; 2) H_2O_2, pH 7)→ $PhCH_2CH_2C(=O)CH_2CH(OH)Ph$ (88%)

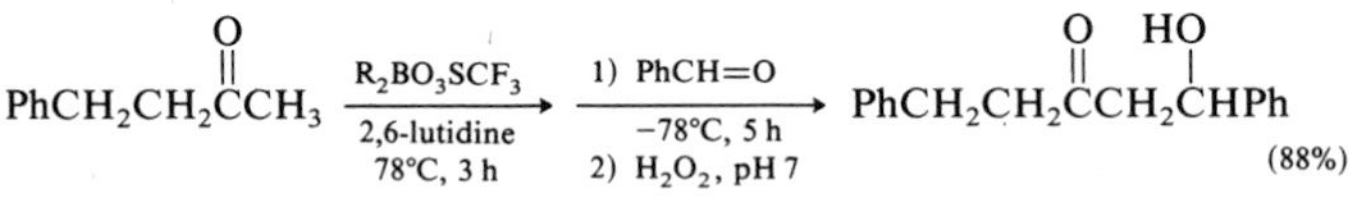

8[h] $C_5H_{11}C(=O)CHN_2$ —(HB(cyclohexyl)$_2$)→ $C_5H_{11}C({=}CH_2)OBR_2$ —(1) O=CH–cyclohexyl; 2) TMS–imidazole)→ $C_5H_{11}C(=O)CH_2CH(OTMS)$–cyclohexyl (60%)

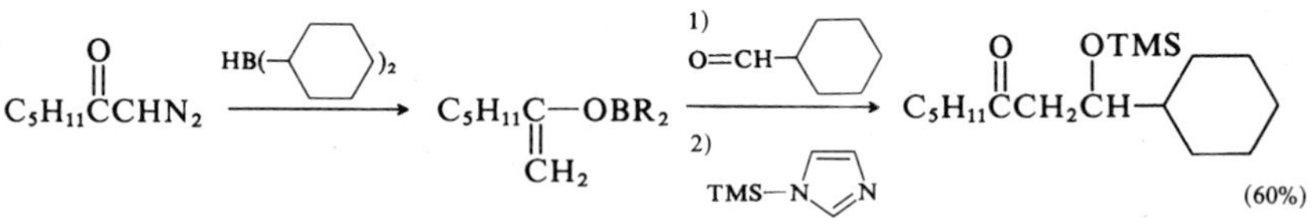

9[i] C_2H_5, OTBDMS, O $\xrightarrow[EtN(i\text{-}Pr)_2]{CF_3SO_3B(C_5H_{11})_2}$ $\xrightarrow{CH_3CH_2CH{=}O}$ C_2H_5, CH_3, OTBDMS, OH, O (70%)

D. Condensations of Silyl Enol Ethers

10[j] CH_3, OTMS + PhCH=O $\xrightarrow[2)\ H_2O]{1)\ R_4N^+{}^-F}$ CH_3, O, OH, Ph (68%)

11[k] $PhCH(CH_3)CH{=}O$ + $CH_2{=}C(OTBDMS)CH_3$ $\xrightarrow{BF_3}$ Ph, CH_3, OH, O, CH_3 (75%)

12[l] TMSO + $CH_3CH{=}O$ $\xrightarrow{TiCl_4}$ O, CH_3, H, OH (96%)

13[m] H_3C, CO_2CH_3, CH_3, CH_3, + CH_3, OTMS $\xrightarrow{TiCl_4}$ H_3C, CO_2CH_3, CH_3, CH_3, CH_3, H, O (90%)

14[n] $PhC(OTMS){=}CH_2$ + $(CH_3)_2C{=}O$ $\xrightarrow{TiCl_4}$ $PhC(=O)CH_2C(OH)(CH_3)_2$ (70–74%)

a. G. Stork, G. A. Kraus, and G. A. Garcia, *J. Org. Chem.* **39**, 3459 (1974).
b. S. Masamune, J. W. Ellingboe, and W. Choy, *J. Am. Chem. Soc.* **104**, 5526 (1982).
c. R. Bal, C. T. Buse, K. Smith, and C. Heathcock, *Org. Synth.* **63**, 89 (1984).
d. P. J. Jerris and A. P. Smith III, *J. Org. Chem.* **46**, 577 (1981).
e. P. Fellmann and J.-E. Dubois, *Tetrahedron* **34**, 1349 (1978).
f. K. Maruoka, S. Hashimoto, Y. Kitagawa, H. Yamamoto, and H. Nozaki, *J. Am. Chem. Soc.* **99**, 7705 (1977).
g. T. Inoue, T. Uchimaru, and T. Mukaiyama, *Chem. Lett.*, 153 (1977).
h. J. Hooz, J. Oudenes, J. L. Roberts, and A. Benderly, *J. Org. Chem.* **52**, 1347 (1987).
i. S. Masamune, W. Choy, F. A. J. Kerdesky, and B. Imperiali, *J. Am. Chem. Soc.* **103**, 1566 (1981).
j. R. Noyori, K. Yokoyama, J. Sakata, I. Kuwajima, E. Nakamura, and M. Shimizu, *J. Am. Chem. Soc.* **99**, 1265 (1977).
k. C. H. Heathcock and L. A. Flippin, *J. Am. Chem. Soc.* **105**, 1667 (1983).
l. T. Yanami, M. Miyashita, and A. Yoshikoshi, *J. Org. Chem.* **45**, 607 (1980).
m. A. S. Kende, S. Johnson, P. Sanfilippo, J. C. Hodges, and L. N. Jungheim, *J. Am. Chem. Soc.* **108**, 3513 (1986).
n. T. Mukaiyama and K. Narasaka, *Org. Synth.* **65**, 6 (1987).

enolates in which the other carbonyl substituent is less bulky show a decreasing stereoselectivity in the order *t*-butyl > *i*-propyl > ethyl.[4]

R	*E*	*Z*	*anti* CH_3	*syn* CH_3
C_2H_5	70	30	36	64
$CH(CH_3)_2$	40	60	18	82
$C(CH_3)_3$	2	98	2	98

The enolates derived from cyclic ketones are necessarily *E*-isomers. The enolate of cyclohexanone reacts with benzaldehyde to give both possible stereoisomeric products under kinetically controlled conditions.

When a more substituted derivative, 2,2-dimethylcyclohexanone, is used, the *anti* stereoisomer is preferred.

13% 87%

This result can also be explained in terms of the cyclic transition state model.[4]

From these and related examples, the following generalizations have been drawn about kinetic stereoselection in aldol condensations.[6] (1) The chair transition state model provides a basis for explaining the stereoselectivity observed in aldol condensation reactions of ketones having one bulky substituent. The preference is *Z*-enolate → *syn* aldol; *E*-enolate → *anti* aldol. (2) When the enolate has no bulky

6. D. A. Evans, J. V. Nelson, and T. R. Taber, *Top. Stereochem.* **13**, 1 (1982); C. H. Heathcock, in *Comprehensive Carbanion Chemistry*, Part B, E. Buncel and T. Durst (eds.), Elsevier, Amsterdam, 1984; C. H. Heathcock, in *Asymmetric Synthesis*, Vol. 3, J. D. Morrison (ed.), Academic Press, New York, 1984.

substituent, stereoselectivity is low. (3) Z-Enolates are more stereoselective than E-enolates. Table 2.1 gives some illustrative data.

Because the aldol reaction is reversible, it is possible to adjust reaction conditions such that the two stereoisomeric aldol products equilibrate. This can be done in the case of lithium enolates by keeping the reaction mixture at room temperature until the product composition reaches equilibrium. This has been done, for example, for the product from the reaction of the enolate of ethyl t-butyl ketone and benzaldehyde.

$$\text{Li}^{+-}\text{O}(\text{(CH}_3)_3\text{C})\text{C}=\text{CH(CH}_3) \xrightarrow[\text{fast}]{\text{PhCH=O}} \text{(CH}_3)_3\text{CCOCH(CH}_3)\text{CH(O}^-\text{Li}^+)\text{Ph}\ (syn) \xrightarrow[\text{slow}]{25^\circ\text{C}} \text{(CH}_3)_3\text{CCOCH(CH}_3)\text{CH(O}^-\text{Li}^+)\text{Ph}\ (anti)$$

The greater stability of the *anti* isomer is attributed to the psuedoequatorial position of the methyl group in the chairlike chelate. With larger substituent groups, the thermodynamic preference for the *anti* isomer is still greater.[7] Thermodynamic equilibrium can be used to control product composition if the desired stereoisomer is significantly more stable than the other.

Table 2.1. Stereoselectivity of Lithium Enolates toward Benzaldehyde[a]

$$Z\text{-enolate (OLi, R}_1) + E\text{-enolate (OLi, R}_1) \xrightarrow{\text{PhCHO}} syn\ \text{PhCH(OH)CH(CH}_3)\text{COR}_1 + anti\ \text{PhCH(OH)CH(CH}_3)\text{COR}_1$$

R^1	Enolate geometry, Z/E	Aldol stereostructure, *syn/anti*
H	100:0	50:50
H	0:100	65:35
Et	30:70	64:36
Et	66:34	77:23
i-Pr	>98:2	90:10
i-Pr	32:68	58:42
i-Pr	0:100	45:55
t-Bu	>98:2	>98:2
1-Adamantyl	>98:2	>98 : 2
Ph	>98:2	88:12
Mesityl	8:92	8:92
Mesityl	87:13	88:12

a. From C. H. Heathcock, in *Asymmetric Synthesis*, Vol. 3, J. D. Morrison (ed.), Academic Press, New York, 1984, Chapter 2.

7. C. H. Heathcock and J. Lampe, *J. Org. Chem.* **48**, 4330 (1983).

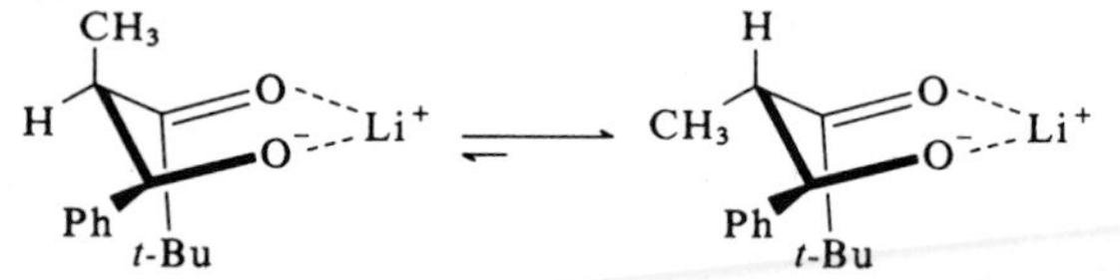

Magnesium enolates of ketones can be prepared from α-bromoketones by reaction with magnesium metal.

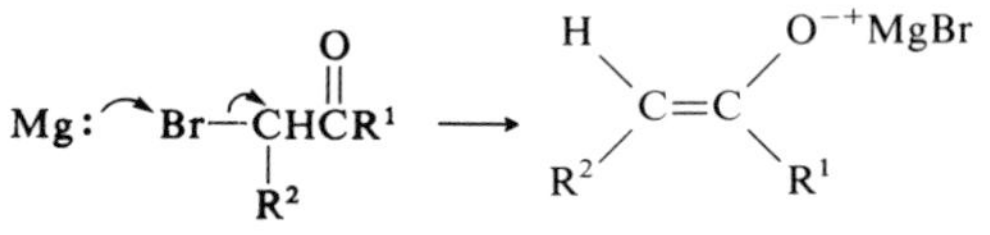

These enolates are similar in their reactivity and stereoselectivity to lithium enolates. In fact, some of the earliest cases of recognition of stereoselectivity in aldol condensations involved studies with magnesium enolates.[8] Entries 5 and 6 in Scheme 2.3 are examples of this type of procedure.

The requirement that an enolate have at least one bulky substituent restricts the types of compounds which can be expected to give highly stereoselective aldol condensations. Furthermore, only the enolate formed by kinetic deprotonation is directly available. Ketones with one tertiary alkyl substituent give mainly the *Z*-enolate.

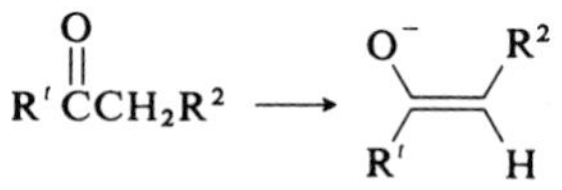

However, less highly substituted ketones usually give mixtures of *E*- and *Z*-enolates.[9] Therefore, efforts aimed at expanding the scope of stereoselective aldol condensations have been directed at two facets of the problem: (1) control of enolate stereochemistry and (2) enhancement of the degree of stereoselectivity in the addition step.

An important modification of the aldol reaction involves the use of boron enolates. The boron enolates react with aldehydes to give aldols. A cyclic transition state is believed to be involved, and, in general, the stereoselectivity is higher than for lithium or magnesium enolates. The O—B bond distances are shorter than those in metal enolates, and this leads to a more compact structure for the transition state. This should magnify the steric interactions which control stereoselectivity.

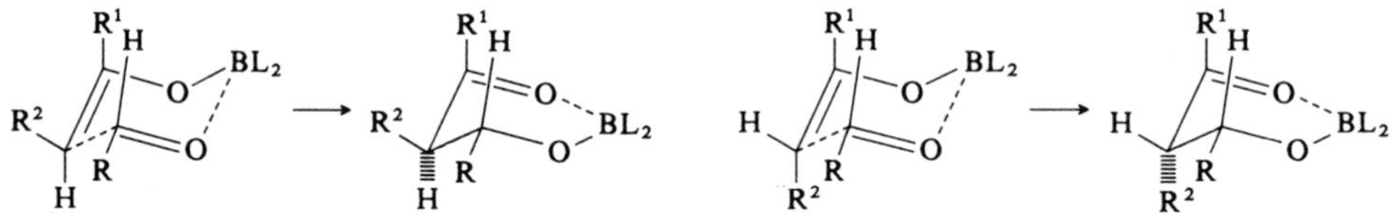

8. J. E. Dubois and J. F. Fort, *Tetrahedron* **28**, 1653, 1665 (1972).
9. R. E. Ireland, R. H. Mueller, and A. K. Willard, *J. Am. Chem. Soc.* **98**, 2868 (1976); W. A. Kleschick, C. T. Buse, and C. H. Heathcock, *J. Am. Chem. Soc.* **99**, 247 (1977); Z. A. Fataftah, I. E. Kopka, and M. W. Rathke, *J. Am. Chem. Soc.* **102**, 3959 (1980).

The boron enolates can be prepared by reaction of the ketone with a dialkylboron trifluoromethanesulfonate (triflate) and a tertiary amine.[10] The Z-stereoisomer is formed preferentially for ethyl ketones with various R^1 substituents. The resulting aldol products are predominantly the *syn* stereoisomers.

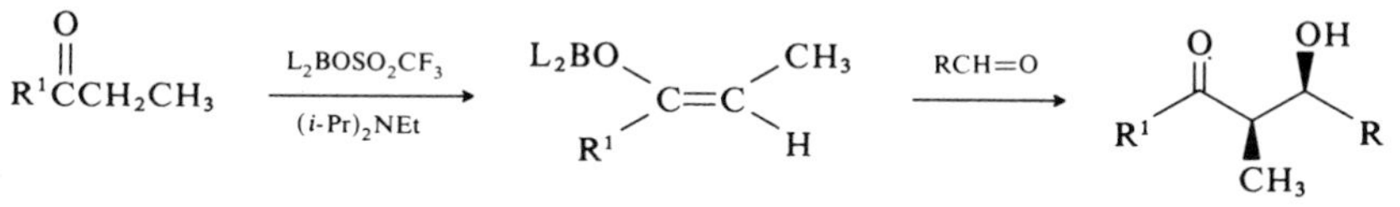

The E boron enolate from cyclohexanone shows a preference for the *anti* aldol product.

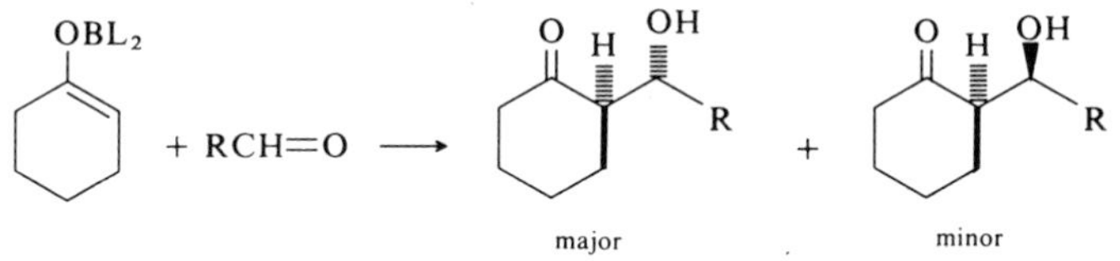

major minor

The exact ratio of stereoisomeric aldols is a function of the substituents on boron and the solvent.

The E boron enolates of acyclic ketones can be obtained by reaction of α-diazo ketones with trialkylboranes.[11] They can be isomerized to the Z-isomer by a catalytic amount of a weak base such as phenoxide ion.

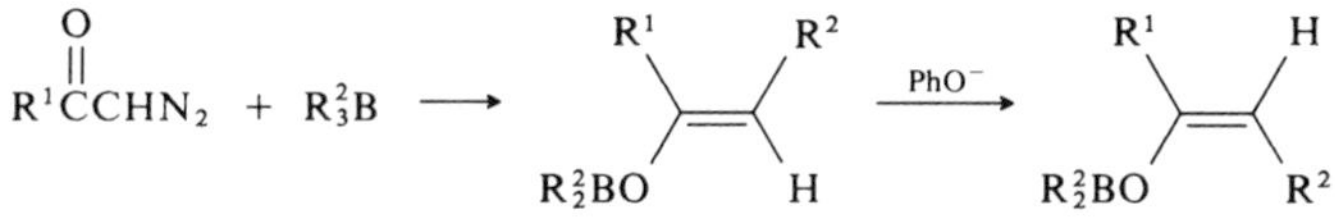

The E boron enolates show a modest preference for formation of the *anti* aldol product.

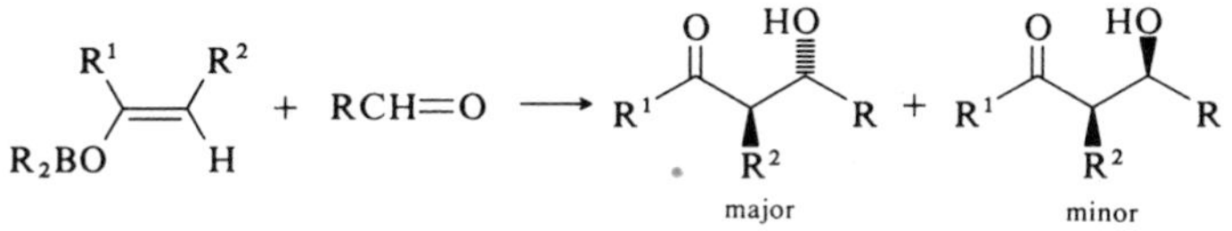

major minor

The general trend then is that boron enolates *parallel* lithium enolates in their stereoselectivity but show *enhanced stereoselectivity*. They also have the advantage of access to both stereoisomeric enol derivatives. Table 2.2 gives a compilation of some of the data on stereoselectivity of aldol reactions with boron enolates.

A number of other atoms can be incorporated into enol type structures, and the effect of such changes on the course of aldol condensations has been extensively investigated. We will discuss here the case of silyl enol ethers. Silyl enol ethers do not react with aldehydes. This is because the silyl enol ether is not a strong enough

10. D. A. Evans, E. Vogel, and J. V. Nelson, *J. Am. Chem. Soc.* **101**, 6120 (1979); D. A. Evans, J. V. Nelson, E. Vogel, and T. R. Taber, *J. Am. Chem. Soc.* **103**, 3099 (1981).
11. S. Masamune, S. Mori, D. Van Horn, and D. W. Brooks, *Tetrahedron Lett.*, 1665 (1979).

Table 2.2. Stereoselectivity of Boron Enolates toward Aldehydes[a]

$\xrightarrow{L_2BOTf}$ OBL_2 (Z) + OBL_2 (E) $\xrightarrow{1.\ R_2CHO}$ R_2CH(OH)CH(CH_3)C(=O)R_1 (*syn*) + R_2CH(OH)CH(CH_3)C(=O)R_1 (*anti*)

R^1	L	R^2	*Z*/*E*	*syn*/*anti*	Ref.
Et	n-C_4H_9	Ph	>97:3	>97:3	b
Et	c-C_5H_9	Ph	82:18	84:16	b
Et	n-C_4H_9	Ph	69:31	72:28	b
Et	n-C_4H_9	n-Pr	>97:3	>97:3	b
Et	n-C_4H_9	t-Pr	>97:3	>97:3	b
Et	n-C_4H_9	$H_2C{=}C(CH_3)$	>97:3	92:8	b
Et	n-C_4H_9	(E)-C_4H_7	>97:3	93:7	b
i-Bu	n-C_4H_9	Ph	>99:1	>97:3	b
i-Bu	c-C_5H_9	Ph	—	84:16	b
i-Pr	n-C_5H_9	Ph	45:55	44:56	b
i-Pr	c-C_5H_9	Ph	19:81	18:82	b
t-Bu	n-C_4H_9	Ph	>99:1	>97:3	b
c-C_6H_{11}	c-C_5H_9	Ph	12:88	14:86	c
c-C_6H_{11}	9-BBN	Ph	>99:1	>97:3	c
Ph	n-C_4H_9	Ph	99:1	>97:3	b
Ph	9-BBN	Ph	>95:5	3:97	d
Ph	9-BBN	i-Pr	>95:5	3:97	d
Ph	9-BBN	$PhCH_2CH_2$	>95:5	3:97	d
Ph	9-BBN	c-C_6H_{11}	>95:5	<3:97	d
Ph	9-BBN	Et	>95:5	<3:97	d
Et	n-C_4H_9	Ph	96:4	95:5	f
n-C_5H_{11}	n-C_4H_9	Ph	95:5	94:6	f
n-C_9H_{19}	n-C_4H_9	Ph	91:9	91:9	f
$PhCH_2$	n-C_4H_9	Ph	95:5	95:5	f
3-C_5H_{11}	n-C_4H_9	Ph	99:1	>99:1	f
c-C_6H_{11}	n-C_4H_9	Ph	98:2	>99:1	f
c-C_6H_{11}	n-C_4H_9	$PhCH_2CH_2$	98:2	>98:2	f

a. From a more complete compilation by C. H. Heathcock, in *Asymmetric Synthesis*, Vol. 3, J. D. Morrison (ed.), Academic Press, New York, 1984, Chap. 3.
b. D. A. Evans, J. V. Nelson, E. Vogel, and T. R. Taber, *J. Am. Chem. Soc.* **103**, 3099 (1981).
c. D. E. Van Horn and S. Masumune, *Tetrahedron Lett.*, 2229 (1979).
d. M. Hirama, D. S. Garvey, L. D.-L. Lu, and S. Masamune, *Tetrahedron Lett.*, 3937 (1979).
e. I. Kuwajima, M. Kato, and A. Mori, *Tetrahedron Lett.* **21**, 4291 (1980).

nucleophile. However, Lewis acids do cause reaction to occur by activating the ketone.[12]

$$PhC(OTMS){=}CH_2 + (CH_3)_2CHCH{=}O \xrightarrow[-78^\circ C]{TiCl_4} PhC(=O)CH_2CH(OH)CH(CH_3)_2 \quad (94\%)$$

12. T. Mukaiyama, K. Banno, and K. Narasaka, *J. Am. Chem. Soc.* **96**, 7503 (1974).

Silyl enol ethers can also be induced to react by fluoride ion sources. These conditions involve liberation of the enolate.[13]

OTMS, CH_3 + PhCH=O $\xrightarrow{R_4N^+F^-}$ CH_3, O, OH, Ph

The stereochemistry of the silyl enol ether-based reactions has not been as extensively explored as for the lithium and boron enolates. Lewis acid-catalyzed condensation of silyl enol ethers of the same structural groups studied in the lithium enolate series showed very little stereoselectivity.[14]

CH_3, OTMS, H, $CH(CH_3)_2$ + PhCH=O $\xrightarrow{BF_3}$ TMSO, O, Ph, $CH(CH_3)_2$, CH_3 *anti* + TMSO, O, Ph, $CH(CH_3)_2$, CH_3 *syn*

An exception is the silyl enol ether of ethyl *t*-butyl ketone, which gave predominantly the *anti* stereoisomer. This is opposite to the stereoselectivity observed with the lithium enolate. The results are considered to be consistent with a noncyclic transition state for the reactions.

$\overset{+}{O}-\overset{-}{B}F_3$, Ph, H, CH_3, $(CH_3)_3SiO$, H, CH_3 ⟶ Ph, H, O, OH, CH_3, H, $(CH_3)_3C$ ≡ O, OH, $(CH_3)_3C$, Ph, CH_3

Entries 10–14 in Scheme 2.3 are examples of aldol condensation reactions proceeding through silyl enol ether intermediates.

Up to this point, we have considered only the effect of enolate geometry on the stereochemistry of the aldol condensation and have considered only achiral aldehydes and enolates. If the aldehyde is chiral, particularly when the chiral center is adjacent to the carbonyl group, the selection between the two diastereotopic faces of the carbonyl group will influence the stereochemical outcome of the reaction. Similarly, when the enolate contains a chiral center, there will be a degree of selectivity between the two faces of the enolate. If both the aldehyde and the enolate are chiral, mutual combinations of stereoselectivity will come into play. One combination should provide complementary, reinforcing stereoselection, while the alterna-

13. E. Nakamura, M. Shimizu, I. Kuwajima, J. Sakata, K. Yokoyama, and R. Noyori, *J. Org. Chem.* **48**, 932 (1983).
14. C. H. Heathcock, K. T. Hug, and L. A. Flippin, *Tetrahedron Lett.* **25**, 5973 (1984).

tive combination would result in opposing preferences and lead to diminished overall stereoselectivity. The combined interactions of chiral centers in both the aldehyde and the enolate component is called *double stereodifferentiation.*[15]

R-enolate ←favored→ *R*-aldehyde   *R*-enolate ←favored→ *S*-aldehyde

S-enolate ←favored→ *S*-aldehyde   or   *S*-enolate ←favored→ *R*-aldehyde

The analysis and prediction of the direction of preferred reaction depend on the same principles as for simple diastereoselectivity and are done by analysis of the attractive and repulsive interactions in the presumed transition state. The enhanced stereoselectivity resulting from interactions in chiral aldehydes and enolates has been useful in the construction of systems with several contiguous chiral centers.

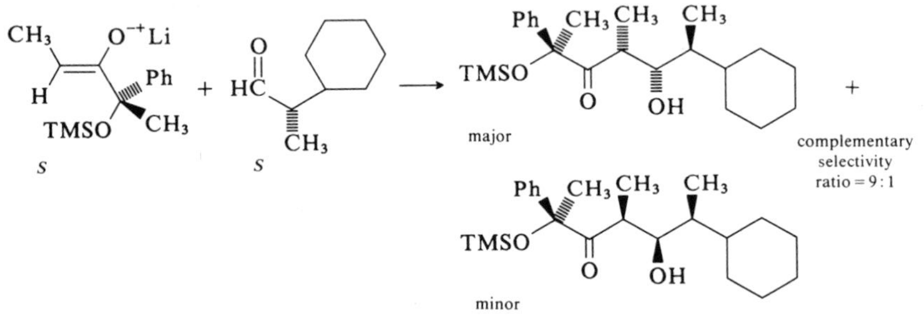

major

minor

complementary selectivity ratio = 9 : 1

Ref. 16

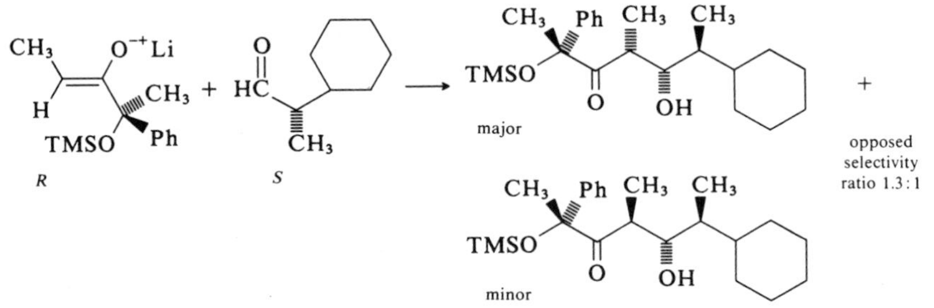

major

minor

opposed selectivity ratio 1.3 : 1

In addition to steric interactions, other structural features may influence the stereoselectivity of aldol condensations. One such factor is chelation by a donor substituent.[17] Several β-alkoxyaldehydes show a preference for *syn* aldol products on reaction with Z-enolates. A chelated transition state can account for the observed stereochemistry.[18] The chelated aldehyde is most easily approached from the face opposite the methyl and R′ substituents.

15. S. Masamune, W. Choy, J. S. Peterson, and L. R. Sita, *Angew. Chem. Int. Ed. Engl.* **24**, 1 (1985).
16. S. Masamune, S. A. Ali, D. L. Snitman, and D. S. Garvey, *Angew. Chem. Int. Ed. Engl.* **19**, 557 (1980).
17. M. T. Reetz, *Angew. Chem. Int. Ed. Engl.* **23**, 556 (1984).
18. S. Masamune, J. W. Ellingboe, and W. Choy, *J. Am. Chem. Soc.* **104**, 5526 (1982).

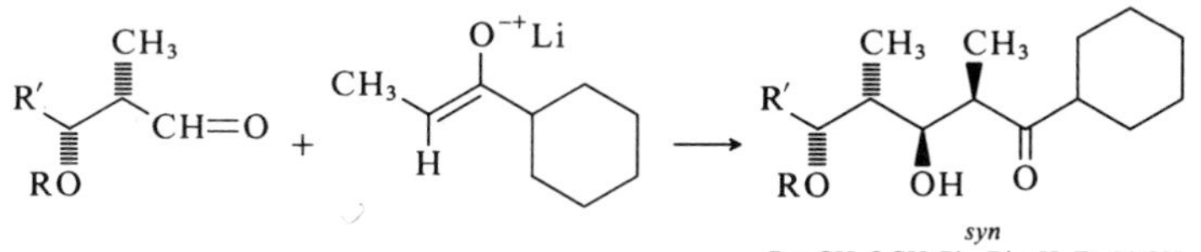

syn

$R = CH_2OCH_2Ph$, $R' = H$, Et, $PhCH_2$

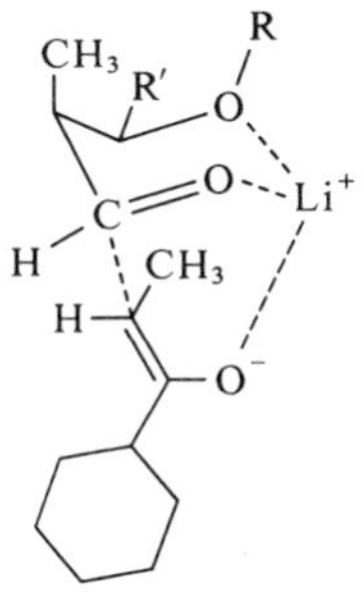

A similar stereoselectivity has been noted for the $TiCl_4$-mediated condensation of β-alkoxyaldehydes with silyl enol ethers.

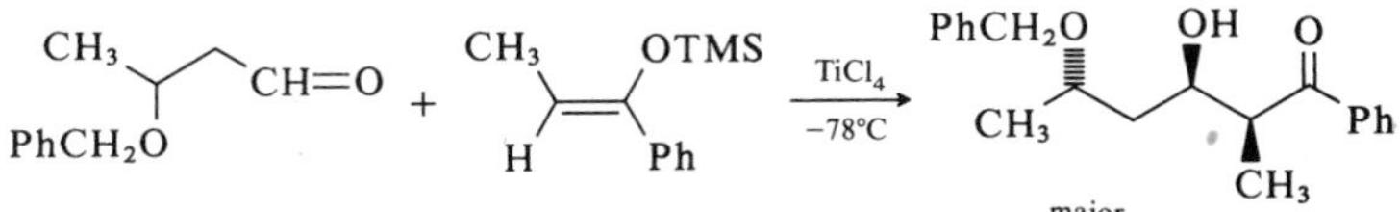

$TiCl_4$, $-78°C$ — major — Ref. 19

The enolates of other carbonyl compounds can be used in mixed aldol condensations. Extensive use has been made of the enolates of esters, thioesters, and amides. Scheme 2.4 gives a selection of such reactions.

Simple alkyl esters show rather low stereoselectivity. However, highly hindered esters derived from 2,6-dimethylphenol or 2,6-di-*t*-butyl-4-methylphenol provide the *anti* stereoisomers.

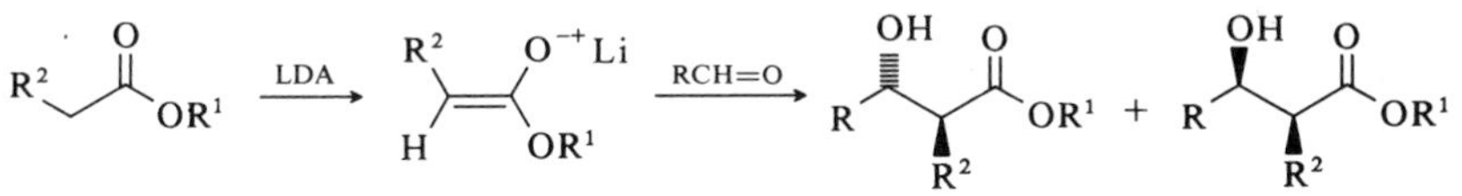

Some illustrative data are given in Table 2.3.

The lithium enolates of α-alkoxy esters have been extensively explored, and several cases where high stereoselectivity is observed have been documented.[20] This stereoselectivity can be explained in terms of a chelated ester enolate which is approached by the aldehyde in such a manner that the aldehyde R group avoids

19. M. T. Reetz and A. Jung, *J. Am. Chem. Soc.* **105**, 4833 (1983).
20. A. I. Meyers and P. Reider, *J. Am. Chem. Soc.* **101**, 2501 (1979); C. H. Heathcock, M. C. Pirrung, S. D. Young, J. P. Hagen, E. T. Jarvi, U. Badertscher, H.-P. Märkl, and S. H. Montgomery, *J. Am. Chem. Soc.* **106**, 8161 (1984).

Scheme 2.4. Addition Reactions of Carbanions Derived from Esters, Carboxylic Acids, Amides, and Nitriles

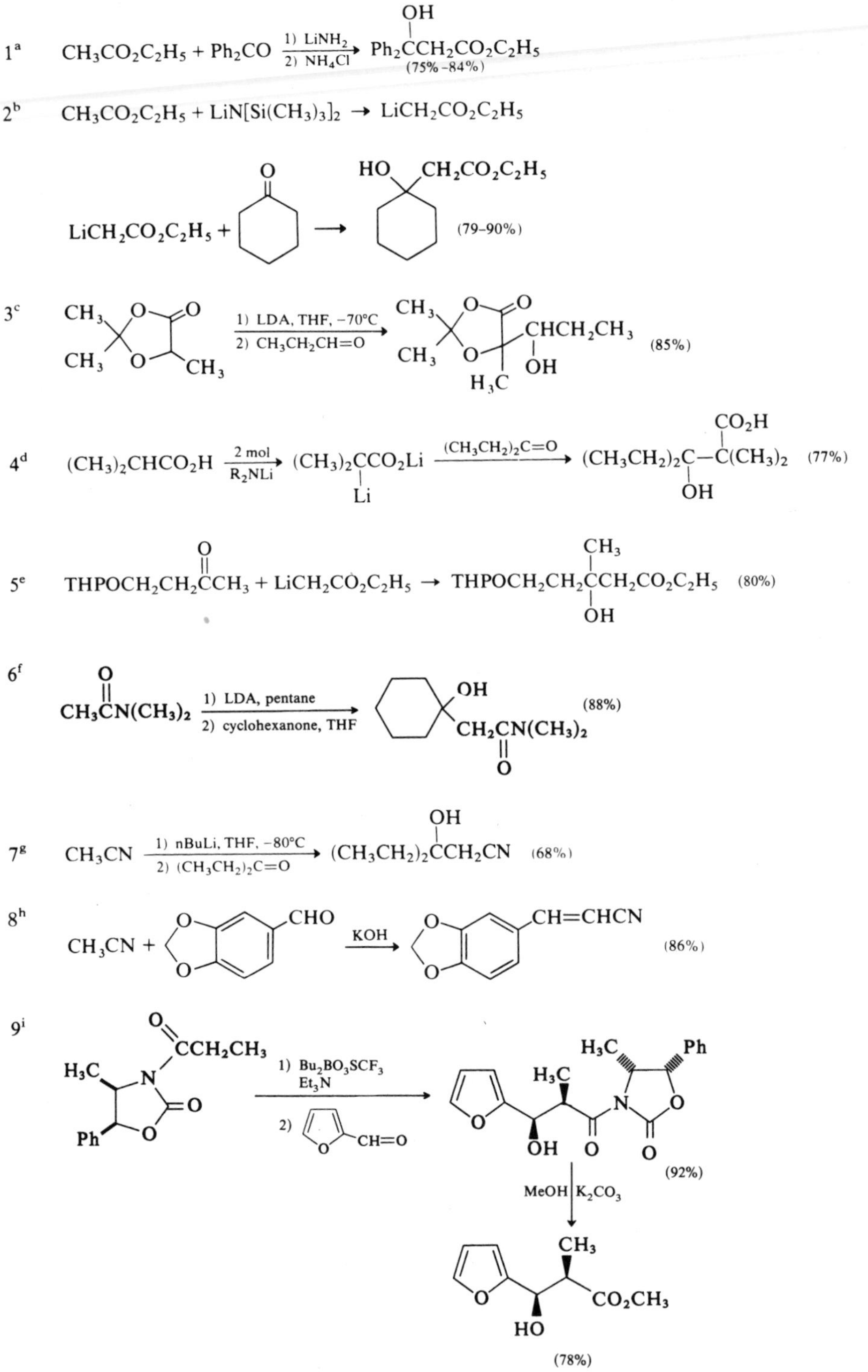

1[a] $CH_3CO_2C_2H_5 + Ph_2CO \xrightarrow[2)\ NH_4Cl]{1)\ LiNH_2} Ph_2C(OH)CH_2CO_2C_2H_5$ (75%–84%)

2[b] $CH_3CO_2C_2H_5 + LiN[Si(CH_3)_3]_2 \rightarrow LiCH_2CO_2C_2H_5$

$LiCH_2CO_2C_2H_5$ + cyclohexanone → 1-hydroxy-1-($CH_2CO_2C_2H_5$)cyclohexane (79–90%)

3[c] 2,2,5-trimethyl-1,3-dioxolan-4-one $\xrightarrow[2)\ CH_3CH_2CH{=}O]{1)\ LDA,\ THF,\ -70^\circ C}$ 5-($CHCH_2CH_3$, OH)-2,2,5-trimethyl-1,3-dioxolan-4-one (85%)

4[d] $(CH_3)_2CHCO_2H \xrightarrow[R_2NLi]{2\ mol} (CH_3)_2C(Li)CO_2Li \xrightarrow{(CH_3CH_2)_2C{=}O} (CH_3CH_2)_2C(OH){-}C(CH_3)_2CO_2H$ (77%)

5[e] $THPOCH_2CH_2C(O)CH_3 + LiCH_2CO_2C_2H_5 \rightarrow THPOCH_2CH_2C(CH_3)(OH)CH_2CO_2C_2H_5$ (80%)

6[f] $CH_3C(O)N(CH_3)_2 \xrightarrow[2)\ cyclohexanone,\ THF]{1)\ LDA,\ pentane}$ 1-hydroxy-1-($CH_2C(O)N(CH_3)_2$)cyclohexane (88%)

7[g] $CH_3CN \xrightarrow[2)\ (CH_3CH_2)_2C{=}O]{1)\ nBuLi,\ THF,\ -80^\circ C} (CH_3CH_2)_2C(OH)CH_2CN$ (68%)

8[h] CH_3CN + 3,4-methylenedioxybenzaldehyde (CHO) $\xrightarrow{KOH}$ 3,4-methylenedioxyphenyl–$CH{=}CHCN$ (86%)

9[i] N-propionyl oxazolidinone (H₃C, Ph; CCH_2CH_3) $\xrightarrow[2)\ \text{furyl}{-}CH{=}O]{1)\ Bu_2BO_3SCF_3,\ Et_3N}$ aldol adduct (furyl, OH, H_3C, N-acyl oxazolidinone with H_3C, Ph) (92%)

$\xrightarrow{MeOH,\ K_2CO_3}$ furyl–CH(HO)–CH(CH_3)–CO_2CH_3 (78%)

10[j]

$$(CH_3)_2CH\text{-oxazolidinone-}N\text{-}C(O)CH_2CH_3 \xrightarrow{R_2BO_3SCF_3,\ EtN(i\text{-}Pr)_2} \xrightarrow{O{=}CH\text{-}CH(CH_3)CH_2OCH_2Ph} \xrightarrow{NaOCH_3,\ CH_3OH} CH_3O_2C\text{-}CH(CH_3)\text{-}CH(OH)\text{-}CH(CH_3)\text{-}CH_2OCH_2Ph$$

a. W. R. Dunnavant and C. R. Hauser, *Org. Synth.* **V**, 564 (1973).
b. M. W. Rathke, *Org. Synth.* **53**, 66 (1973).
c. C. H. Heathcock, S. D. Young, J. P. Hagen, M. C. Pirrung, C. T. White, and D. VanDerveer, *J. Org. Chem.* **45**, 3846 (1980).
d. G. W. Moersch and A. R. Burkett, *J. Org. Chem.* **36**, 1149 (1971).
e. J. D. White, M. A. Avery, and J. P. Carter, *J. Am. Chem. Soc.* **104**, 5486 (1982).
f. R. P. Woodbury and M. W. Rathke, *J. Org. Chem.* **42**, 1688 (1977).
g. E. M. Kaiser and C. R. Hauser, *J. Org. Chem.* **33**, 3402 (1968).
h. S. A. DiBiase, B. A. Lipisko, A. Haag, R. A. Wolak, and G. W. Gokel, *J. Org. Chem.* **44**, 4640 (1979).
i. S. F. Martin and D. E. Guinn, *J. Org. Chem.* **52**, 5588 (1987).
j. D. Seebach, H.-F. Chow, R. F. W. Jackson, K. Lawson, M. A. Sutter, S. Thaisrivongs, and J. Zimmermann, *J. Am. Chem. Soc.* **107**, 5292 (1985).

being between the α-alkoxy group and the methyl group in the ester enolate. When the ester alkyl group, R, becomes very bulky, the stereoselectivity is reversed.

CH₃ R OR / H / R²O O O⁻ Li⁺ — favored

CH₃ H OR / R / R²O O O⁻ Li⁺ — disfavored

A very useful approach for enantioselective aldol condensations has been based on the oxazolidinones **1** and **2**. Both compounds are readily available in enantiomerically pure form.

$(CH_3)_2CH$-oxazolidinone (N–H, C=O) **1** CH_3, Ph-oxazolidinone (N–H, C=O) **2**

These compounds can be acylated and converted to the lithium or boron enolates by the same methods applicable to ketones. The enolates are the Z-stereoisomers.[21]

$$R'\text{-oxazolidinone-}N\text{-}C(O)CH_2R \xrightarrow{L_2BO_3SCF_3} R'\text{-oxazolidinone-}N\text{-}C(OBL_2){=}CHR$$

21. D. A. Evans, J. Bartoli, and T. L. Shih, *J. Am. Chem. Soc.* **103**, 2127 (1981).

Table 2.3. Stereoselectivity of Ester Enolates toward Aldehydes[a]

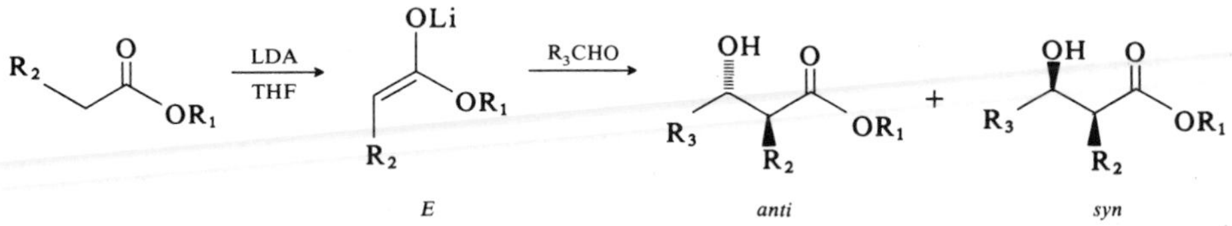

R^{1b}	R^2	R^3	*anti/syn*	Ref.
Me	Me	Ph	55:45	c
Me	Me	*i*-Pr	55:45	c
Me	Me	Me	57:43	c
$MeOCH_2$	Me	*i*-Pr	90:10	c
$MeOCH_2$	Me	Me	67:33	c
DMP	Me	Ph	88:12	d
DMP	$H_2C{=}CHCH_2$	Ph	91:9	d
DMP	Me	*n*-C_5H_{11}	86:14	d
DMP	$H_2C{=}CHCH_2$	Et	84:16	d
DMP	Me	*i*-Pr	>98:2	d
DMP	Et	*i*-Pr	>98:2	d
DMP	$H_2C{=}CHCH_2$	*i*-Pr	>98:2	d
DMP	Me	*t*-Bu	>98:2	d
BHT	Me	Ph	>98:2	d
BHT	$H_2C{=}CHCH_2$	Ph	≥94:6	d
BHT	$H_2C{=}CHCH_2$	Et	>98:2	d
BHT	Me	*i*-Pr	>98:2	d
BHT	$H_2C{=}CHCH_2$	*i*-Pr	>98:2	d

a. From a more extensive compilation by C. H. Heathcock, in *Asymmetric Synthesis*, Vol. 3, J. D. Morrison (ed.), Academic Press, New York, 1984, Chap. 2.
b. DMP = 2,6-dimethylphenyl; BHT = 2,6-di-*t*-butyl-4-methylphenyl.
c. A. I. Meyers and P. Reider, *J. Am. Chem. Soc.* **101**, 2501 (1979).
d. C. H. Heathcock, M. C. Pirrung, S. H. Montgomery, and J. Lampe, *Tetrahedron* **37**, 4087 (1981).

The oxazolidinone substituents R′ then direct the approach of the aldehyde. Because of the opposite steric encumbrance provided by **1** and **2**, the products have the opposite configuration at the reaction site. The acyl oxazolidinones are easily solvolyzed in water or alcohols to give the enantiomeric β-hydroxy acid or ester.

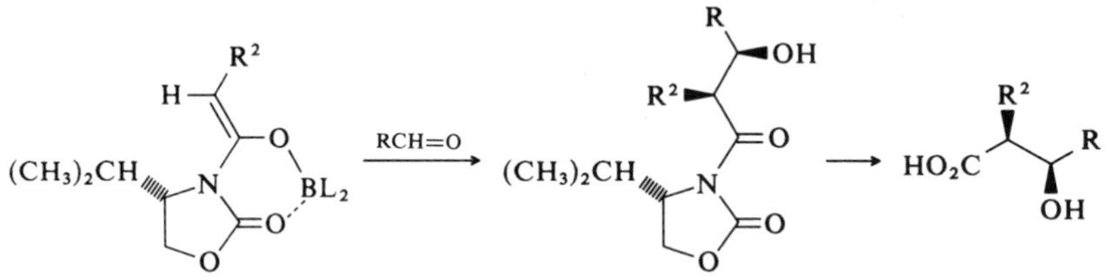

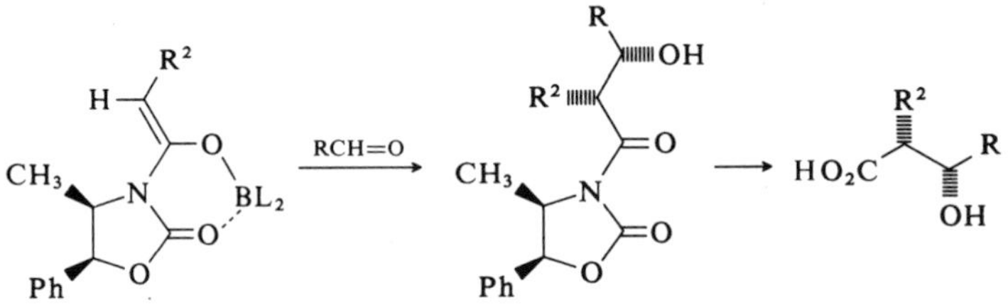

Lithium or magnesium derivatives of imines are also useful nucleophiles in directed aldol condensations. These compounds are especially important when the nucleophilic component must be an aldehyde, since it is difficult to generate aldehyde enolates.

$$CH_3CH_2{=}N{-}C_6H_{11} \xrightarrow[2)\ Ph_2C{=}O]{1)\ LDA} Ph_2C(OH)CH(CH_3)CH{=}N{-}C_6H_{11}\quad 91\%$$ Ref. 22

2.1.4. Intramolecular Aldol Condensations and the Robinson Annulation

The methods for effecting aldol condensation can be applied to dicarbonyl compounds where the two groups are favorably disposed for intramolecular reaction. For formation of five- and six-membered rings, the use of a catalytic amount of a base is frequently satisfactory. With more complex structures, the special techniques required for directed aldol condensations are used. Scheme 2.5 illustrates intramolecular aldol condensations.

A particularly important example is the Robinson annulation, a procedure which constructs a new six-membered ring from a ketone.[23] The reaction sequence starts with conjugate addition of the enolate to methyl vinyl ketone or a similar enone. This is followed by cyclization involving an intramolecular aldol condensation. Dehydration frequently occurs to give a cyclohexenone derivative. Scheme 2.6 shows some examples of Robinson annulation reactions.

$CH_3CCH{=}CH_2$ (O above C) + cyclohexanone enolate $\xrightarrow{\text{conjugate addition}}$ $H_3C{-}C(=O){-}CH_2{-}CH_2$–cyclohexanone $\xrightarrow[\text{dehydration}]{\text{aldol addition and}}$ bicyclic enone

A precursor of methyl vinyl ketone, 4-(trimethylamino)-2-butanone, was used as the reagent in the early examples of the reaction. This compound generates methyl vinyl ketone *in situ*, by β elimination. Other α,β-unsaturated enones can be used, but the reaction is somewhat sensitive to substitution at the β-carbon, and adjustment of the reaction conditions is necessary.[24] Entry 4 in Scheme 2.6 is an example of use of a β-substituted enone.

The original conditions developed for the Robinson annulation reaction are such that the ketone enolate composition is under thermodynamic control. This usually results in the formation of the more substituted enolate, and as illustrated

22. G. Wittig and H. Reiff, *Angew. Chem. Int. Ed. Engl.* **7**, 7 (1968); G. Wittig and P. Suchanek, *Tetrahedron Suppl.* **8**, 347 (1967).
23. E. D. Bergmann, D. Ginsburg, and R. Pappo, *Org. React.* **10**, 179 (1950); J. W. Cornforth and R. Robinson, *J. Chem. Soc.*, 1855 (1949); R. Gawley, *Synthesis*, 777 (1976); M. E. Jung, *Tetrahedron* **32**, 3 (1976); B. P. Mundy, *J. Chem. Educ.* **50**, 110 (1973).
24. C. J. V. Scanio and R. M. Starrett, *J. Am. Chem. Soc.* **93**, 1539 (1971).

Scheme 2.5. Intramolecular Aldol Condensations

1[a] $O{=}CH(CH_2)_3CH(C_3H_7)CH{=}O \xrightarrow[115°]{H_2O}$ CHO, C_3H_7

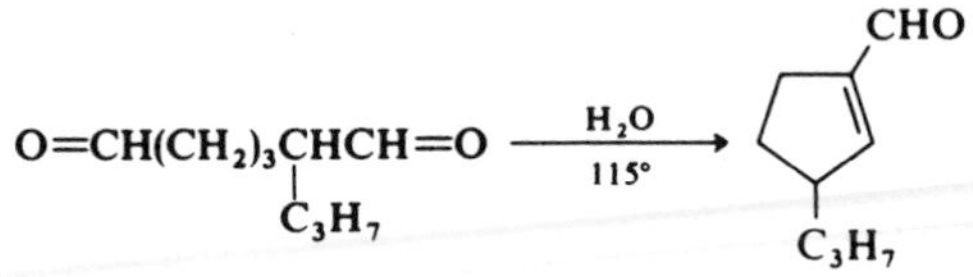

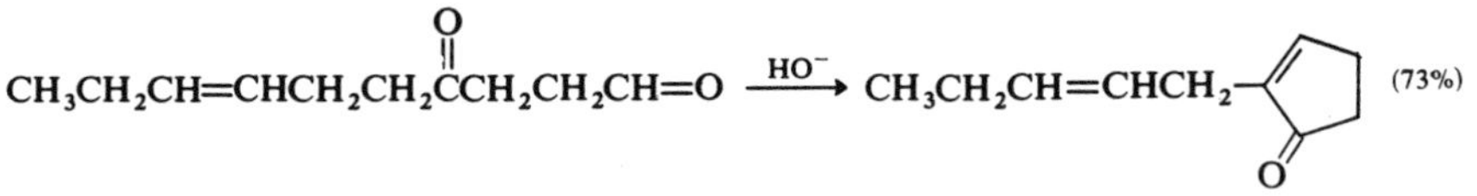

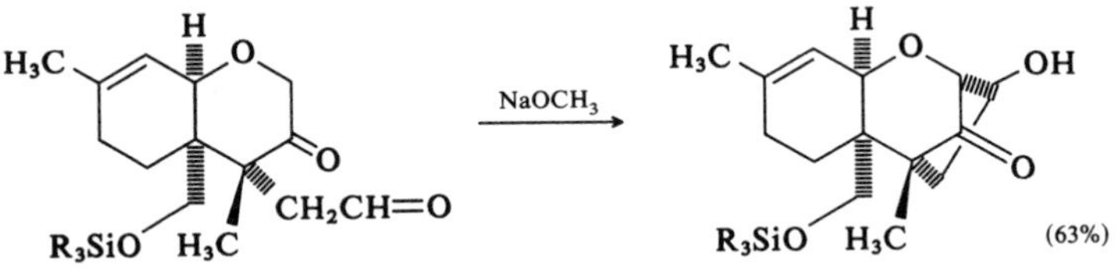

2[b] $CH_3CH_2CH{=}CHCH_2CH_2\overset{O}{\overset{\|}{C}}CH_2CH_2CH{=}O \xrightarrow{HO^-} CH_3CH_2CH{=}CHCH_2-$ (73%)

3[c] H_3C, CH_3C, O, N, O, O $\xrightarrow{NaOH}$ HO, O, H_3C, N, O (80%)

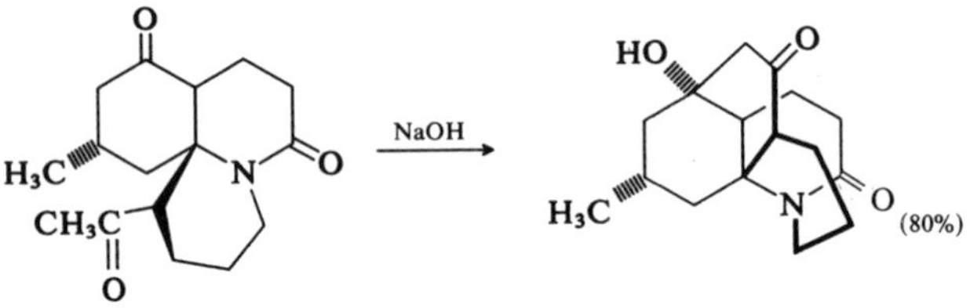

4[d] H_3C, H, O, O, $CH_2CH{=}O$, R_3SiO, H_3C $\xrightarrow{NaOCH_3}$ H_3C, H, O, OH, O, R_3SiO, H_3C (63%)

5[e] TMSO, $=CH_2$, $CH(OCH_3)_2$, H, H, H_3C, CH_3 $\xrightarrow{ZnCl_2}$ O, OCH_3, H, H, H_3C, CH_3 (59%)

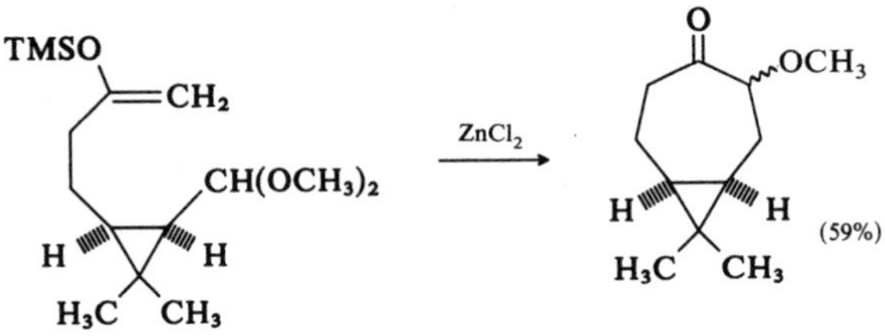

a. J. English and G. W. Barber, *J. Am. Chem. Soc.* **71**, 3310 (1949).
b. A. I. Meyers and N. Nazarenko, *J. Org. Chem.* **38**, 175 (1973).
c. K. Wiesner, V. Musil, and K. J. Wiesner, *Tetrahedron Lett.*, 5643 (1968).
d. G. A. Kraus, B. Roth, K. Frazier, and M. Shimagaki, *J. Am. Chem. Soc.* **104**, 1114 (1982).
e. M. D. Taylor, G. Minaskanian, K. N. Winzenberg, P. Santone, and A. B. Smith III, *J. Org. Chem.* **47**, 3960 (1982).

in entry 4 in Scheme 2.6, gives a product with a substituent at a ring juncture when monosubstituted cyclohexanones are used as reactants. The alternative regiochemistry can be achieved by using an enamine of the ketone. As discussed in Section 1.9, the less substituted enamine is favored, so addition occurs at the less substituted position. Entry 5 illustrates this variation of the reaction.

Scheme 2.6. The Robinson Annulation Reaction

1[a] O, CH_3, O + $CH_2{=}CHCOCH_3$ $\xrightarrow{KOH}$ O, CH_3, $CH_2CH_2COCH_3$, O $\xrightarrow{pyrrolidine}$ O, CH_3, O (63–65%)

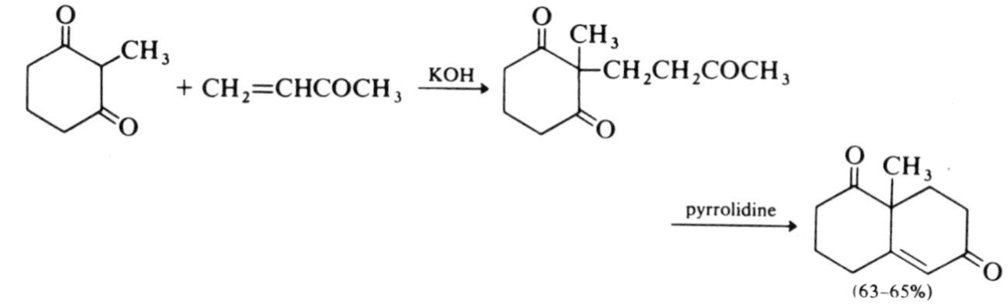

2[b] $CO_2CH_2CH_3$, O + $CH_2{=}CHCOCH_2CH_3$ $\xrightarrow[EtOH]{NaOEt}$ $CO_2CH_2CH_3$, O, CH_3 (59%)

3[c] CH_3, O, CH_3O + $CH_3COCH_2CH_2\overset{+}{N}(CH_3)_3$ $\xrightarrow{^-OEt}$ H_3C, O, CH_3O (71%)

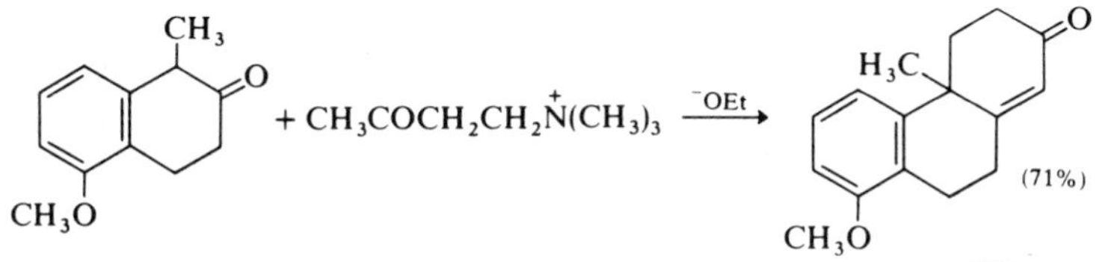

4[d] CH_3, O + $CH_3COCH_2CH_2N(CH_2CH_3)_2$ $\xrightarrow[2)\ H_3O^+]{1)\ Na,\ 135°C}$ CH_3, O (35%)

5[e] CH_3, O ⟶ CH_3, N $\xrightarrow[2)\ HOAc,\ NaOAc,\ H_2O,\ reflux]{1)\ CH_2{=}CHCOCH_3,\ benzene,\ reflux}$ CH_3, O (45%)

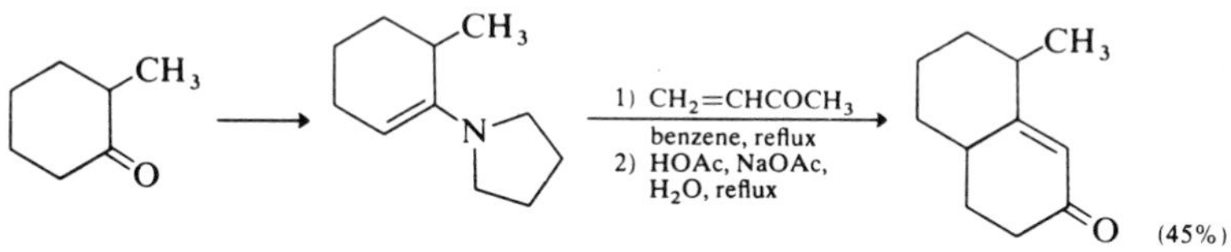

6[f] O, CH_3, O + $CH_2{=}CH\overset{O}{\overset{||}{C}}CH_2CH_3$ $\xrightarrow[PhCO_2H,\ 140°C,\ 24\ h]{1)\ DABCO\ \ 2)\ Et_3N,}$ O, CH_3, O, CH_3 (75%)

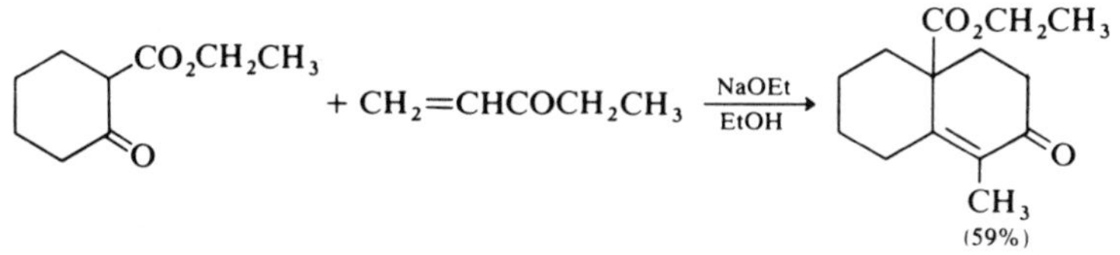

7[g] O, O, OCH_3, OCH_3, O $\xrightarrow[2)\ CH_3CH{=}C(Si(CH_3)_3)\overset{O}{\overset{||}{C}}CH_3\ \ 3)\ MeO^-]{1)\ LDA}$ O, O, OCH_3, OCH_3, O (62%)

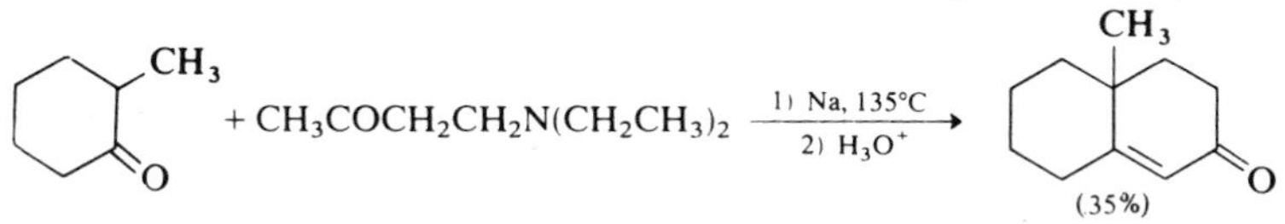

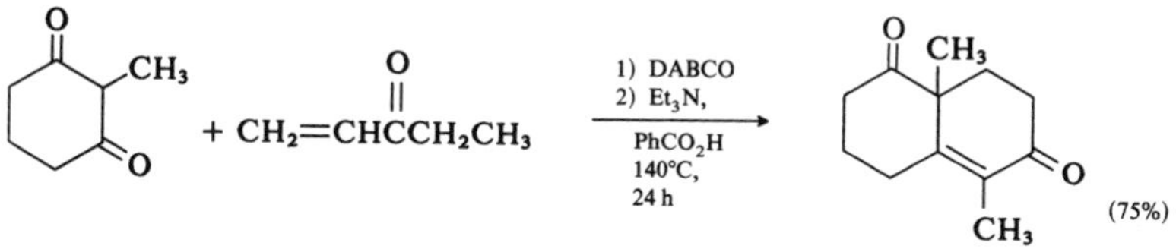

8[h] $O^{-+}Li$ + $CH_2{=}C(SPh)\overset{O}{\overset{||}{C}}CH_3$ $\xrightarrow{-70°C}$ SPh, O, OH (80%)

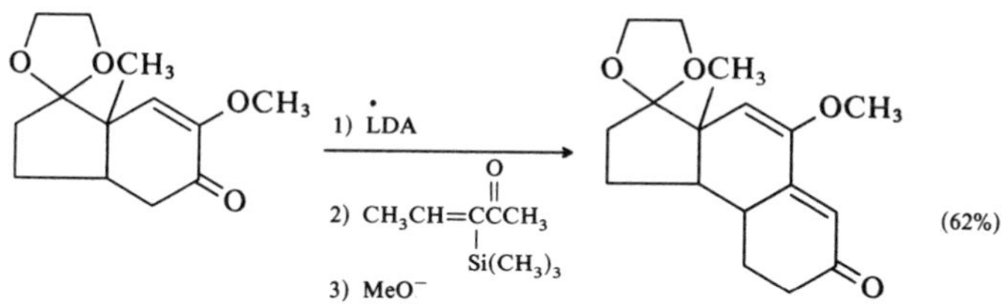

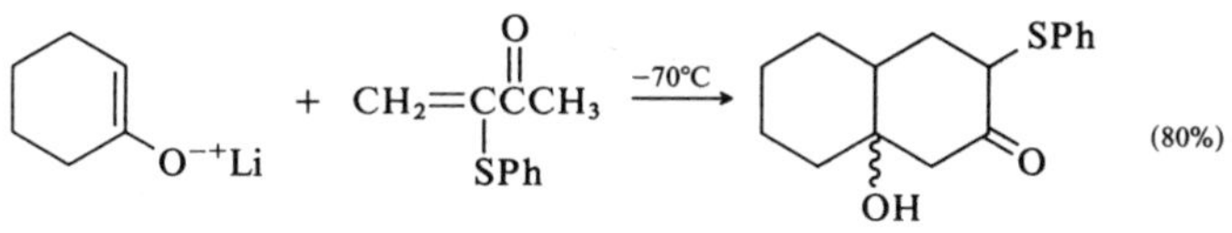

Scheme 2.6—*continued*

9[i]

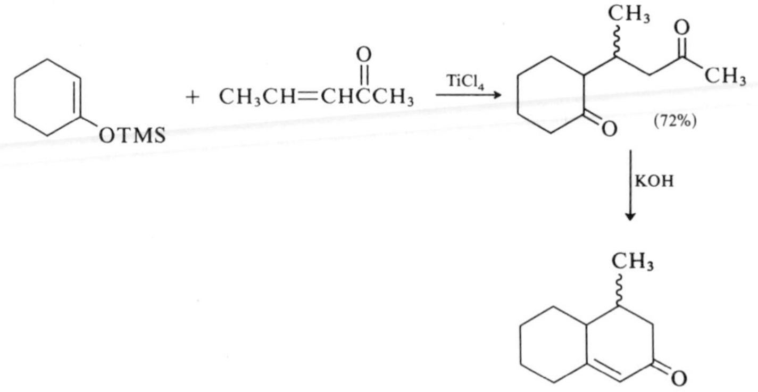

OTMS + $CH_3CH{=}CHCCH_3$ (C=O) $\xrightarrow{TiCl_4}$ (72%) $\xrightarrow{KOH}$

a. S. Ramachandran and M. S. Newman, *Org. Synth.* **41**, 38 (1961).
b. D. L. Snitman, R. J. Himmelsbach, and D. S. Watt, *J. Org. Chem.* **43**, 4578 (1978).
c. J. W. Cornforth and R. Robinson, *J. Chem. Soc.*, 1855 (1949).
d. C. J. V. Scanio and R. M. Starrett, *J. Am. Chem. Soc.* **93**, 1539 (1971).
e. G. Stork, A. Brizzolara, H. Landesman, J. Szmuszkovicz, and R. Terrell, *J. Am. Chem. Soc.* **85**, 207 (1963).
f. F. E. Ziegler, K.-J. Hwang, J. F. Kadow, S. I. Klein, U. K. Pati, and T.-F. Wang, *J. Org. Chem.* **51**, 4573 (1986).
g. G. Stork, J. D. Winkler, and C. S. Shiner, *J. Am. Chem. Soc.* **104**, 3767 (1982).
h. K. Takaki, M. Okada, M. Yamada, and K. Negoro, *J. Org. Chem.* **47**, 1200 (1982).
i. J. W. Huffman, S. M. Potnis, and A. V. Satish, *J. Org. Chem.* **50**, 4266 (1985).

Another version of the Robinson annulation procedure involves the use of methyl 1-trimethylsilylvinyl ketone. The reaction follows the normal sequence of conjugate addition, aldol cyclization, and dehydration.

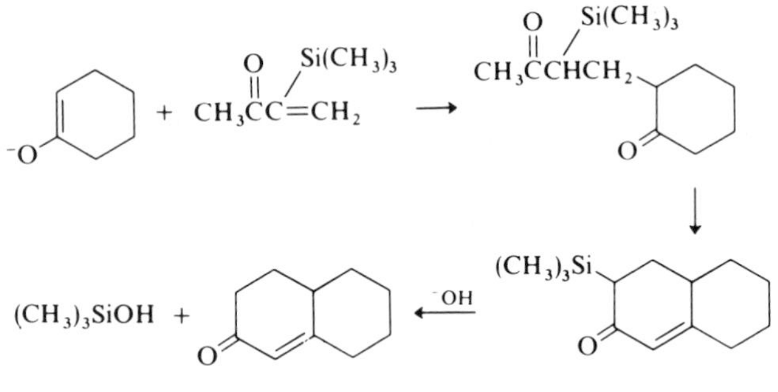

^-O (cyclohexenolate) + $CH_3C(=O)C(Si(CH_3)_3){=}CH_2$ $\longrightarrow$ $CH_3C(=O)CH(Si(CH_3)_3)CH_2$–(cyclohexanone) $\longrightarrow$ $(CH_3)_3Si$–(octalone) $\xrightarrow{^-OH}$ $(CH_3)_3SiOH$ + (octalone)

Ref. 25

The role of the trimethylsilyl group is to stabilize the enolate formed in the conjugate addition. The silyl group is then removed during the dehydration step. The advantage of methyl 1-trimethylsilylvinyl ketone is that it can be used under aprotic conditions which are consistent with regiospecific methods for enolate generation. The direction of annulation of unsymmetrical ketones can therefore be controlled by the method of enolate formation.

25. G. Stork and B. Ganem, *J. Am. Chem. Soc.* **95**, 6152 (1973); G. Stork and J. Singh, *J. Am. Chem. Soc.* **96**, 6181 (1974).

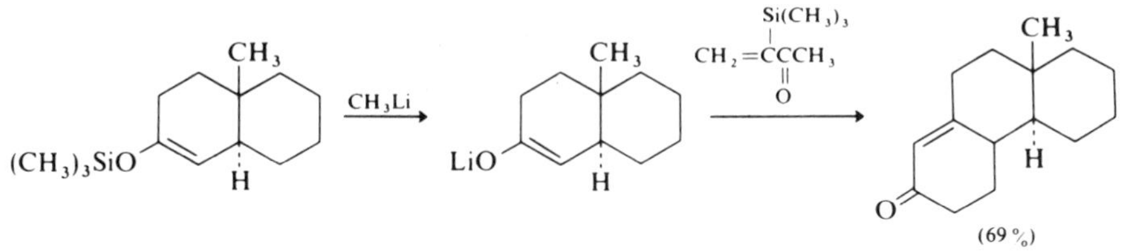

$(CH_3)_3SiO$ ... $\xrightarrow{CH_3Li}$ LiO ... $\xrightarrow{CH_2=C(Si(CH_3)_3)CCH_3(=O)}$ ... (69%) Ref. 26

Methyl 1-phenylthiovinyl ketones can also be used as enones in kinetically controlled Robinson annulation reactions, as illustrated by entry 8 in Scheme 2.6.

The product in entry 1 of Scheme 2.6 is commonly known as the Wieland–Miescher ketone, and it is a useful starting material for the preparation of steroids and terpenes. The Robinson annulation to prepare this ketone can be carried out enantioselectively by using the amino acid L-proline to form an enamine intermediate. The *S*-enantiomer of the product is obtained with high enantiomeric excess.[27] This compound and the corresponding compound obtained from cyclopentane-1,3-dione[28] are key intermediates in the enantioselective synthesis of steroids.[29]

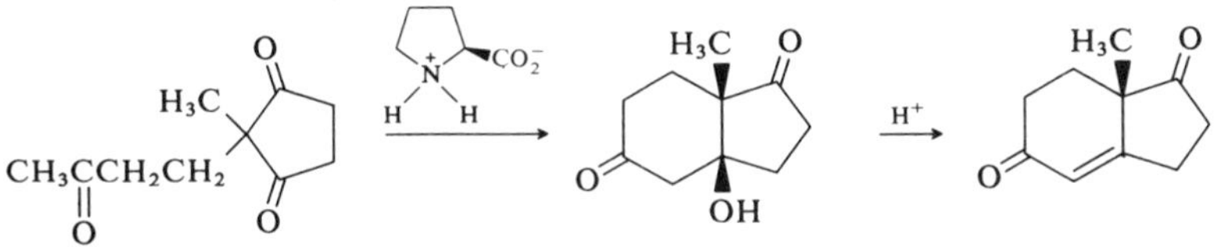

The detailed mechanism of this enantioselective transformation remains under investigation.[30] It is known that the acidic carboxylic group is crucial. The cyclization is believed to occur via the enamine derived from the catalyst and the exocyclic ketone. There is evidence that a second molecule of the catalyst is involved, and it has been suggested that this molecule participates in the proton transfer step that completes the cyclization reaction.[31]

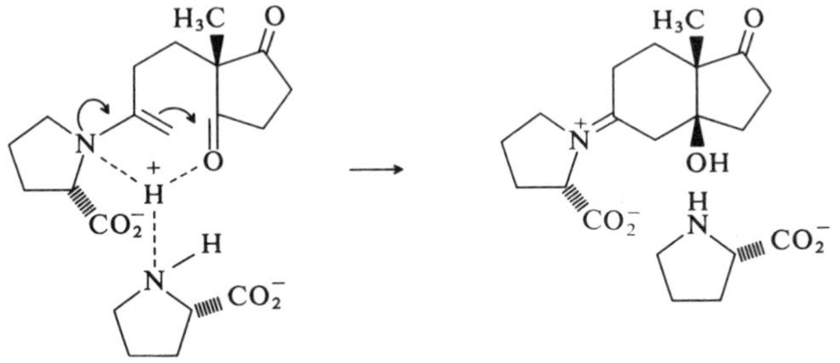

26. R. K. Boeckman, Jr., *J. Am. Chem. Soc.* **96**, 6179 (1974).
27. J. Gutzwiller, P. Buchschacher, and A. Fürst, *Synthesis*, 167 (1977); P. Buchschacher and A. Fürst, *Org. Synth.* **63**, 37 (1984).
28. Z. G. Hajos and D. R. Parrish, *J. Org. Chem.* **39**, 1615 (1974); U. Eder, G. Sauer and R. Wiechert, *Angew. Chem. Int. Ed. Engl.* **10**, 496 (1971). Z. G. Hajos and D. R. Parrish, *Org. Synth.* **63**, 26 (1984).
29. N. Cohen, *Acc. Chem. Res.* **9**, 412 (1976).
30. P. Buchschacher, J.-M. Cassal, A. Fürst, and W. Meier, *Helv. Chim. Acta*, **60**, 2747 (1977); K. L. Brown, L. Damm, J. D. Dunitz, A. Eschenmoser, R. Hobi, and C. Kratky, *Helv. Chim. Acta* **61**, 3108 (1978); C. Agami, F. Meynier, C. Puchot, J. Guilhem, and C. Pascard, *Tetrahedron* **40**, 1031 (1984).
31. C. Agami, J. Levisalles, and C. Puchot, *J. Chem. Soc., Chem. Commun.*, 441 (1985).

2.2. Condensation Reactions of Imines and Iminium Ions

Imines and iminium ions are nitrogen analogs of carbonyl compounds, and they undergo nucleophilic additions like those involved in aldol condensations. The reactivity order is $C{=}NR < C{=}O < [C{=}NR_2]^+ < [C{=}OH]^+$. Because iminium ions are more reactive than imines, condensations involving imines are frequently run under acidic conditions where the imine is protonated.

2.2.1. The Mannich Reaction

The *Mannich reaction* is the condensation of an enolizable carbonyl compound with an iminium ion.[32] The reaction introduces an α dialkylaminomethyl substituent.

$$RCH_2\overset{O}{\overset{\|}{C}}R' + CH_2{=}O + HN(CH_3)_2 \rightarrow (CH_3)_2NCH_2\underset{R}{\underset{|}{C}}H\overset{O}{\overset{\|}{C}}R'$$

The electrophilic species is often generated *in situ* from the amine and formaldehyde.

$$CH_2{=}O + HN(CH_3)_2 \rightleftharpoons HOCH_2N(CH_3)_2 \underset{}{\overset{H^+}{\rightleftharpoons}} H_2O + CH_2{=}\overset{+}{N}(CH_3)_2$$

The reaction is usually limited to secondary amines, since dialkylation can occur with primary amines. The dialkylation reaction can be used advantageously in ring closures.

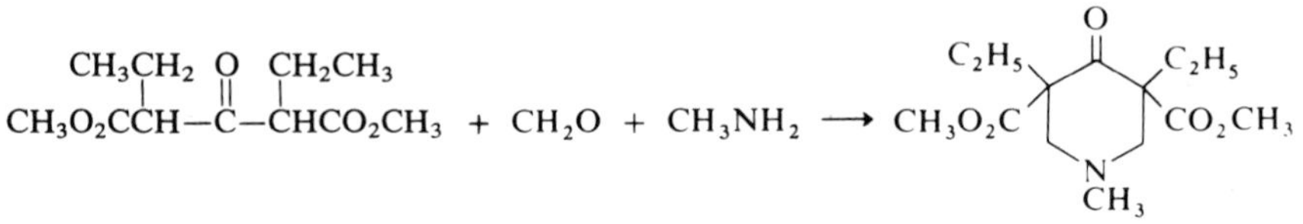

$$CH_3O_2C\underset{}{\overset{CH_3CH_2}{\overset{|}{C}}}H-\overset{O}{\overset{\|}{C}}-\overset{CH_2CH_3}{\overset{|}{C}}HCO_2CH_3 + CH_2O + CH_3NH_2 \longrightarrow \text{3,5-bis}(C_2H_5)\text{-3,5-bis}(CO_2CH_3)\text{-1-}CH_3\text{-piperidin-4-one} \quad \text{Ref. 33}$$

Entries 1 and 2 in Scheme 2.7 show the preparation of "Mannich Bases" from a ketone, formaldehyde, and a dialkylamine following the classical procedure. Alternatively, formaldehyde equivalents may be used, such as bis(dimethylamino)methane in entry 3. On treatment with trifluoroacetic acid, this aminal generates the iminium trifluoroacetate as a reactive electrophile.

$$(CH_3)_2NCH_2N(CH_3)_2 + 2CF_3CO_2H \rightarrow (CH_3)_2\overset{+}{N}{=}CH_2 + H_2\overset{+}{N}(CH_3)_2 + 2CF_3CO_2^-$$

N,N-Dimethylmethyleneammonium iodide is commercially available and is known as "Eschenmoser's salt."[34] This compound is sufficiently electrophilic to react directly

32. F. F. Blicke, *Org. React.* **1**, 303 (1942); J. H. Brewster and E. L. Eliel, *Org. React.* **7**, 99 (1953).
33. C. Mannich and P. Schumann, *Chem. Ber.* **69**, 2299 (1936).
34. J. Schreiber, H. Maag, N. Hashimoto, and A. Eschenmoser, *Angew. Chem. Int. Ed. Engl.* **10**, 330 (1971).

1[a] $PhCOCH_3 + CH_2O + (CH_3)_2\overset{+}{N}H_2Cl^- \rightarrow PhCOCH_2CH_2\overset{H}{\underset{+}{N}}(CH_3)_2Cl^-$ (70%)

2[b] $CH_3COCH_3 + CH_2O + (CH_3CH_2)_2\overset{+}{N}H_2Cl^- \rightarrow CH_3COCH_2CH_2\overset{H}{\underset{+}{N}}(C_2H_5)_2Cl^-$ (66–75%)

3[c] $(CH_3)_2CHCOCH_3 + [(CH_3)_2N]_2CH_2 \xrightarrow{CF_3CO_2H} (CH_3)_2CHCOCH_2CH_2N(CH_3)_2$

4[d] 1-(trimethylsilyloxy)cyclohexene ($OSiMe_3$) $\xrightarrow[THF]{CH_3Li}$ lithium enolate (OLi) $\xrightarrow[2)\ H_2O,\ H^+,\ ^-OH]{1)\ (CH_3)_2\overset{+}{N}{=}CH_2}$ 2-($CH_2N(CH_3)_2$)cyclohexanone (87%)

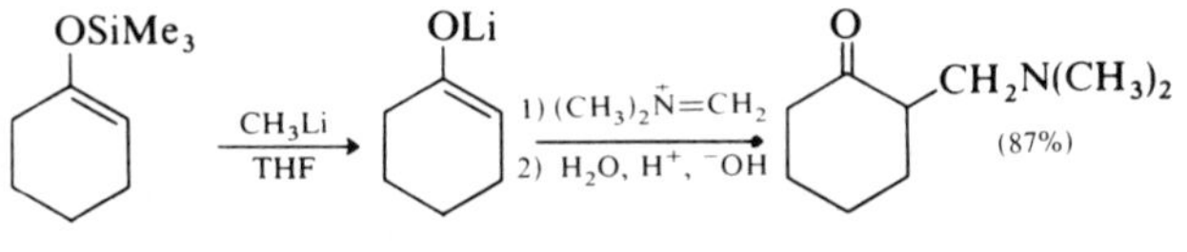

5[e] cyclohexanone $\xrightarrow[THF,\ 0°C]{KH}$ potassium enolate (OK) $\xrightarrow[I^-]{(CH_3)_2\overset{+}{N}{=}CH_2}$ 2-($CH_2N(CH_3)_2$)cyclohexanone (88%)

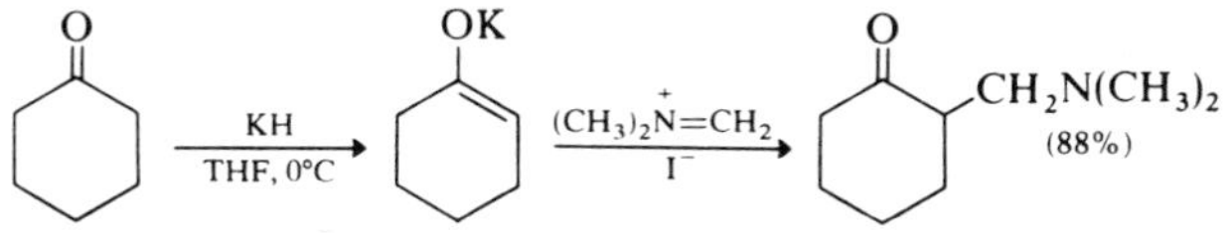

6[f] $CH_3CH_2CH_2CH{=}O + CH_2O + (CH_3)_2\overset{+}{N}H_2Cl^- \xrightarrow[2)\ distill]{1)\ 60°C,\ 6\ h} CH_2{=}C(CH_2CH_3)CH{=}O$ (73%)

7[g] cyclopentanone $+ (CH_2O)_n \xrightarrow[CF_3CO_2^-,\ THF]{PhN^+H_2CH_3}$ 2-methylenecyclopentanone ($=CH_2$) (90%)

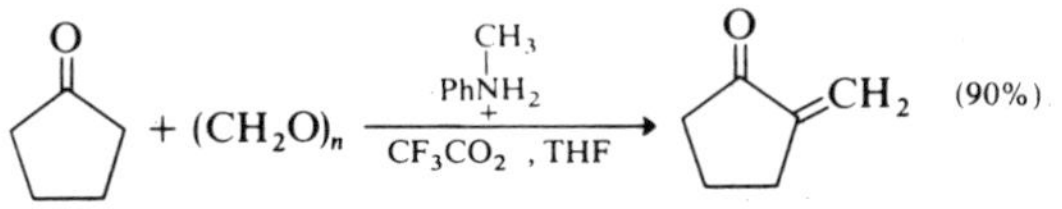

8[h] cyclohexanone $+ PhCOCH_2CH_2N(CH_3)_2 \xrightarrow{NaOH}$ 2-(CH_2CH_2COPh)cyclohexanone (52%)

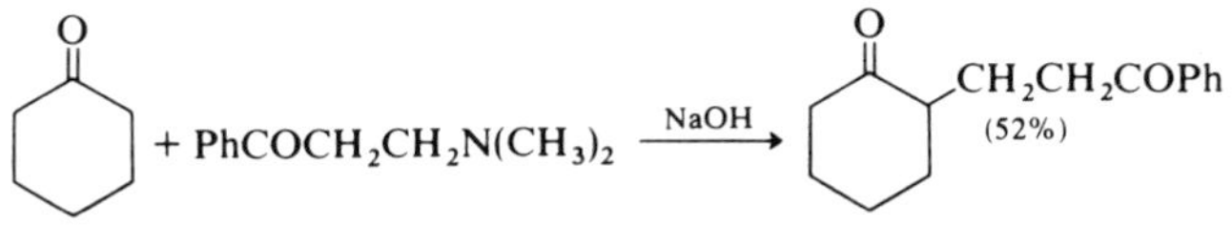

9[i] $PhCOCH_2CH_2N(CH_3)_2 + KCN \rightarrow PhCOCH_2CH_2CN$ (67%)

a. C. E. Maxwell, *Org. Synth.* **III**, 305 (1955).
b. A. L. Wilds, R. M. Nowak, and K. E. McCaleb, *Org. Synth.* **IV**, 281 (1963).
c. M. Gaudry, Y. Jasor, and T. B. Khac, *Org. Synth.* **59**, 153, (1979).
d. S. Danishefsky, T. Kitahara, R. McKee, and P. F. Schuda, *J. Am. Chem. Soc.* **98**, 6715 (1976).
e. J. L. Roberts, P. S. Borromeo, and C. D. Poulter, *Tetrahedron Lett.*, 1621 (1977).
f. C. S. Marvel, R. L. Myers, and J. H. Saunders, *J. Am. Chem. Soc.* **70**, 1694 (1948).
g. J. L. Gras, *Tetrahedron Lett.*, 2111, 2955 (1978).
h. A. C. Cope and E. C. Hermann, *J. Am. Chem. Soc.* **72**, 3405 (1950).
i. E. B. Knott, *J. Chem. Soc.*, 1190 (1947).

with silyl enol ethers in neutral solution.[35] The reagent is added to a solution of an enolate or enolate precursor. This procedure permits the reaction to be carried out under nonacidic conditions. Entries 4 and 5 of Scheme 2.7 illustrate the preparation of Mannich bases with Eschenmoser's salt.

35. S. Danishefsky, T. Kitahara, R. McKee, and P. F. Schuda, *J. Am. Chem. Soc.* **98**, 6715 (1976).

The dialkylaminomethyl ketones formed in the Mannich reaction are useful synthetic intermediates.[36] Thermal decomposition of the amines or the derived quaternary salts provides α-methylene carbonyl compounds.

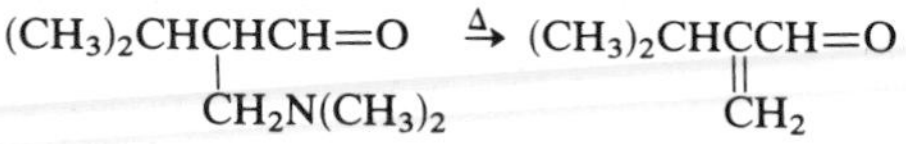

$$(CH_3)_2CHCH(CH_2N(CH_3)_2)CH{=}O \xrightarrow{\Delta} (CH_3)_2CHC({=}CH_2)CH{=}O$$

Ref. 37

These α,β-unsaturated ketones and aldehydes are used as reactants in Michael additions (Section 1.10), Robinson annulations (Section 2.1.4), and in a number of other reactions we will encounter later. Entries 8 and 9 in Scheme 2.7 illustrate Michael reactions carried out by *in situ* generation of α,β-unsaturated carbonyl compounds from Mannich bases.

α-Methylene lactones are present in a number of natural products.[38] The reaction of ester enolates with *N,N*-dimethylmethyleneammonium trifluoroacetate,[39] or Eschenmoser's salt,[40] has been used for introduction of the α-methylene groups in the synthesis of vernolepin, a compound with antileukemic activity.[41,42]

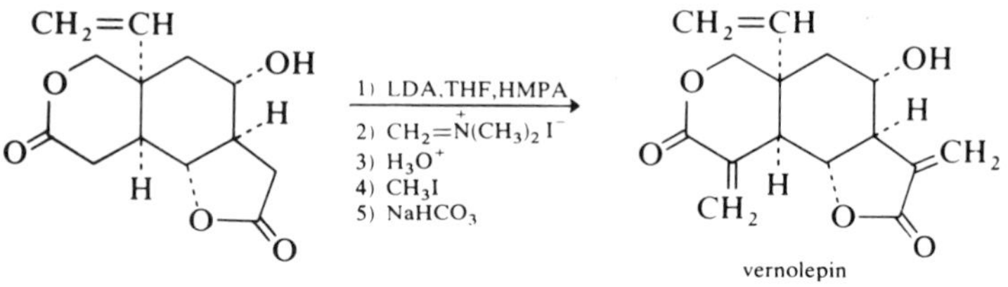

1) LDA,THF,HMPA
2) $CH_2{=}\overset{+}{N}(CH_3)_2\ I^-$
3) H_3O^+
4) CH_3I
5) $NaHCO_3$

vernolepin

Mannich reactions, or a close mechanistic analog, are important in the biosynthesis of many nitrogen-containing natural products. As a result, the Mannich reaction has played an important role in the total synthesis of such compounds, especially in syntheses patterned after the mode of biosynthesis, that is, *biogenetic-type synthesis.* The earliest example of the use of the Mannich reaction in this way was the successful synthesis of tropinone, a derivative of the alkaloid tropine, by Sir Robert Robinson in 1917.

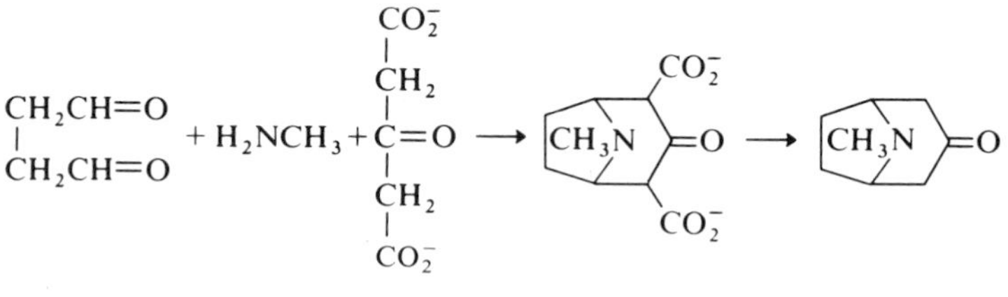

$$CH_2CH{=}O,\ CH_2CH{=}O + H_2NCH_3 + {}^-O_2CCH_2C({=}O)CH_2CO_2^- \longrightarrow \text{(CH}_3\text{N bicyclic ketone with two } CO_2^-) \longrightarrow \text{(CH}_3\text{N bicyclic ketone)}$$

Ref. 43

36. G. A. Gevorgyan, A. G. Agababyan, and O. L. Mndzhoyan, *Russ. Chem. Rev.* (Engl. Transl.), **54**, 495 (1985).
37. C. S. Marvel, R. L. Myers, and J. H. Saunders, *J. Am. Chem. Soc.* **70**, 1694 (1948).
38. S. M. Kupchan, M. A. Eakin, and A. M. Thomas, *J. Med. Chem.* **14**, 1147 (1971).
39. N. L. Holy and Y. F. Wang, *J. Am. Chem. Soc.* **99**, 499 (1977).
40. J. L. Roberts, P. S. Borromes, and C. D. Poulter, *Tetrahedron Lett.*, 1621 (1977).
41. S. Danishefsky, P. F. Schuda, T. Kitahara, and S. J. Etheredge, *J. Am. Chem. Soc.* **99**, 6066 (1977).
42. For reviews of methods for the synthesis of α-methylene lactones, see R. B. Gammill, C. A. Wilson, and T. A. Bryson, *Synth. Commun.* **5**, 245 (1975); J. C. Sarma and R. P. Sharma, *Heterocycles* **24**, 441 (1986); N. Petragnani, H. M. C. Ferraz, and G. V. J. Silva, *Synthesis*, 157 (1986).
43. R. Robinson, *J. Chem. Soc.*, 762 (1917).

Iminium ions are intermediates in a group of reactions which form α,β-unsaturated compounds having structures corresponding to those formed by mixed aldol condensations followed by dehydration. These reactions are catalyzed by amines or buffer systems containing an amine and an acid and are referred to as *Knoevenagel condensations.*[44]

The general mechanism is believed to involve iminium ions as the active electrophiles, rather than the amine simply acting as a base for the aldol condensation. Kinetic evidence in support of this mechanism has been developed in the case of condensation of aromatic aldehydes with nitromethane.[45] Knoevenagel condensation conditions frequently involve both an amine and a weak acid. The reactive electrophile is probably the protonated form of the imine, since this is a more reactive electrophile than the corresponding carbonyl compound.

$$\mathrm{ArCH{=}O + C_4H_9NH_2 \rightleftarrows ArCH{=}NC_4H_9}$$

$$\mathrm{ArCH{=}NC_4H_9 \xrightarrow[\,^-CH_2NO_2]{H^+} ArCH(CH_2NO_2)NHC_4H_9 \longrightarrow ArCH(CHNO_2)\text{–}NHC_4H_9 \xrightarrow{H^+} ArCH{=}CHNO_2}$$

Amine-catalyzed reaction conditions are usually applied to rather acidic compounds containing two electron-attracting substituents. Malonic esters, cyanoacetic esters, and cyanoacetamide are examples of compounds which undergo condensation reactions under Knoevenagel conditions.[46] Nitroalkanes are also effective nucleophilic reactants. The single nitro group sufficiently activates the α-hydrogens to permit deprotonation under the weakly basic conditions. Usually, the product that is isolated is the "dehydrated," that is, α,β-unsaturated, derivative of the original adduct.

$$\mathrm{R_2C(X)\text{–}CH(CO_2R)(CN) \xrightarrow{B} R_2C{=}C(CO_2R)(CN)}$$

X = OH or NR_2

A relatively acidic proton in the potential nucleophile is important for two reasons. First, it permits weak bases, such as amines, to provide a sufficient concentration of the enolate for reaction. A highly acidic proton also facilitates the elimination step which drives the reaction to completion.

44. G. Jones, *Org. React.* **15**, 204 (1967); R. L. Reeves, in *The Chemistry of the Carbonyl Group*, S. Patai (ed.), Wiley-Interscience, New York, 1966, pp. 593–599.
45. T. I. Crowell and D. W. Peck, *J. Am. Chem. Soc.* **75**, 1075 (1953).
46. A. C. Cope, C. M. Hofmann, C. Wykoff, and E. Hardenbergh, *J. Am. Chem. Soc.* **63**, 3452 (1941).

Malonic acid and cyanoacetic acid can also be used as the potential nucleophiles. The mechanism of the addition step is likely to involve iminium ions when secondary amines are used as catalysts. With malonic acid or cyanoacetic acid as reactants, the products usually undergo decarboxylation. This may occur as a concerted decomposition of the adduct.[47]

$$R\overset{O}{\overset{\|}{C}}R + CH_2(CO_2H)_2 \rightarrow R_2\overset{OH}{\overset{|}{C}}-\underset{\underset{\underset{-O}{|}}{C=O}}{\underset{|}{C}}HCO_2H \rightarrow R_2C{=}CHCO_2H$$

Decarboxylative condensations of this type are sometimes carried out in pyridine. Pyridine cannot form an imine intermediate, but it has been shown to catalyze the decarboxylation of arylidene malonic acids.[48] The decarboxylation occurs by concerted decomposition of the adduct between pyridine and the α,β-unsaturated diacid.

$$ArCH{=}C(CO_2H)_2 + C_5H_5NH^+ \longrightarrow Ar\underset{\underset{^+NC_5H_5}{|}}{C}H-\underset{\underset{\underset{O-H}{|}}{C=O}}{|}{C}HCO_2H \longrightarrow ArCH{=}CHCO_2H$$

Scheme 2.8 gives some examples of Knoevenagel condensation reactions.

2.3. Acylation of Carbanions

The reactions to be discussed in this section involve carbanion addition to carbonyl centers with a potential leaving group. The tetrahedral intermediate formed in the addition step then reacts by expulsion of the leaving group.

$$R\overset{O}{\overset{\|}{C}}-X + R'_2\bar{C}Y \rightarrow R\underset{\underset{X}{|}}{\overset{O^-}{\overset{|}{C}}}-\underset{\underset{Y}{|}}{C}R'_2 \rightarrow R\overset{O}{\overset{\|}{C}}\underset{\underset{Y}{|}}{C}R'_2$$

Under these circumstances, the overall transformation results in the acylation of the carbon nucleophile. An important group of these reactions involves acylation by esters, in which case the leaving group is alkloy or aryloxy. The self-condensation of esters is known as the *Claisen condensation.*[49] Ethyl acetoacetate, for example,

47. E. J. Corey, *J. Am. Chem. Soc.* **74**, 5897 (1952).
48. E. J. Corey and G. Fraenkel, *J. Am. Chem. Soc.* **75**, 1168 (1953).
49. C. R. Hauser and B. E. Hudson, Jr., *Org. React.* **1**, 266 (1942).

Scheme 2.8. Amine-Catalyzed Condensations of the Knoevenagel Type

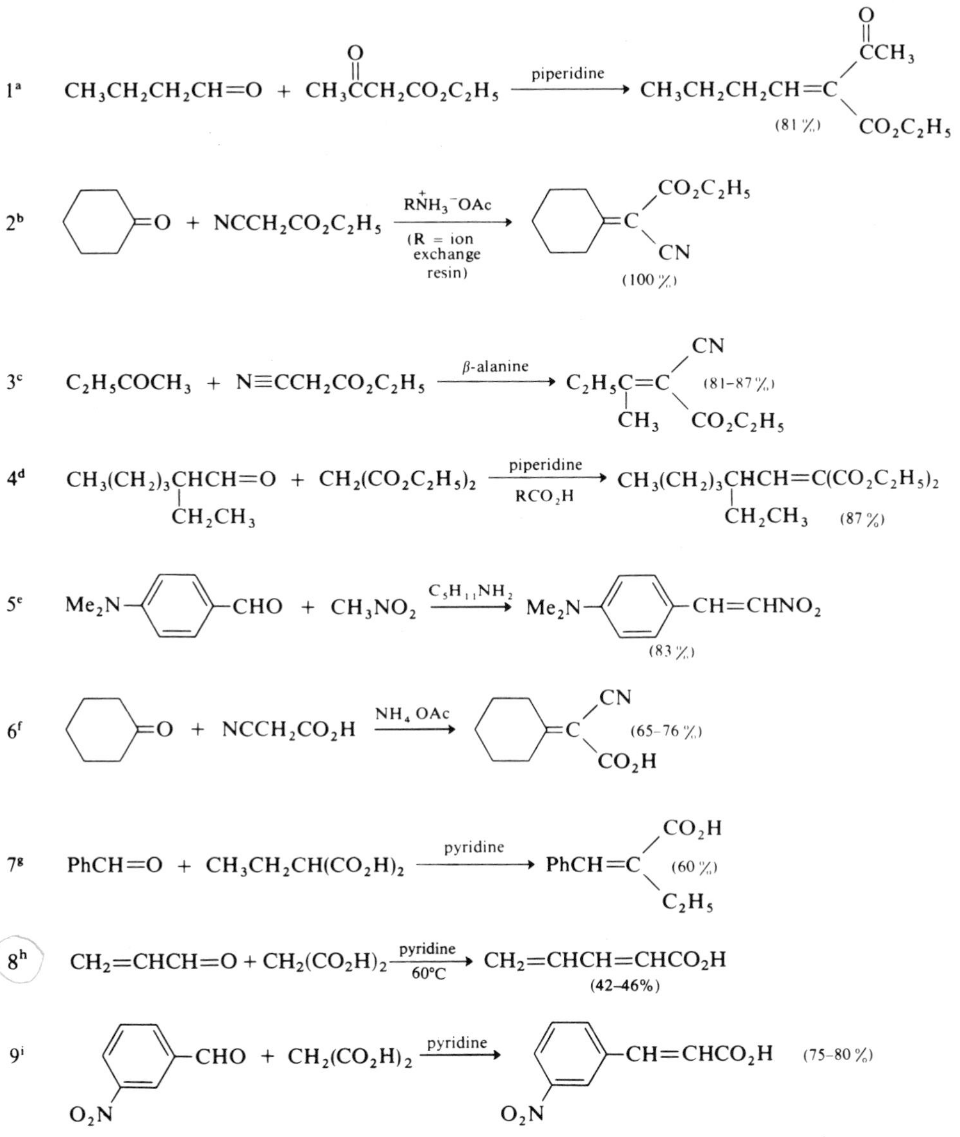

1[a] $CH_3CH_2CH_2CH{=}O + CH_3\overset{O}{\overset{\|}{C}}CH_2CO_2C_2H_5 \xrightarrow{\text{piperidine}} CH_3CH_2CH_2CH{=}C(\overset{O}{\overset{\|}{C}}CH_3)(CO_2C_2H_5)$ (81%)

2[b] cyclohexanone$({=}O) + NCCH_2CO_2C_2H_5 \xrightarrow[\text{(R = ion exchange resin)}]{R\overset{+}{N}H_3\,^{-}OAc}$ cyclohexylidene${=}C(CO_2C_2H_5)(CN)$ (100%)

3[c] $C_2H_5COCH_3 + N{\equiv}CCH_2CO_2C_2H_5 \xrightarrow{\beta\text{-alanine}} C_2H_5C(CH_3){=}C(CN)(CO_2C_2H_5)$ (81–87%)

4[d] $CH_3(CH_2)_3CH(CH_2CH_3)CH{=}O + CH_2(CO_2C_2H_5)_2 \xrightarrow[RCO_2H]{\text{piperidine}} CH_3(CH_2)_3CH(CH_2CH_3)CH{=}C(CO_2C_2H_5)_2$ (87%)

5[e] $Me_2N{-}C_6H_4{-}CHO + CH_3NO_2 \xrightarrow{C_5H_{11}NH_2} Me_2N{-}C_6H_4{-}CH{=}CHNO_2$ (83%)

6[f] cyclohexanone$({=}O) + NCCH_2CO_2H \xrightarrow{NH_4OAc}$ cyclohexylidene${=}C(CN)(CO_2H)$ (65–76%)

7[g] $PhCH{=}O + CH_3CH_2CH(CO_2H)_2 \xrightarrow{\text{pyridine}} PhCH{=}C(CO_2H)(C_2H_5)$ (60%)

8[h] $CH_2{=}CHCH{=}O + CH_2(CO_2H)_2 \xrightarrow[60°C]{\text{pyridine}} CH_2{=}CHCH{=}CHCO_2H$ (42–46%)

9[i] $m\text{-}O_2N{-}C_6H_4{-}CHO + CH_2(CO_2H)_2 \xrightarrow{\text{pyridine}} m\text{-}O_2N{-}C_6H_4{-}CH{=}CHCO_2H$ (75–80%)

a. A. C. Cope and C. M. Hofmann, *J. Am. Chem. Soc.* **63**, 3456 (1941).
b. R. W. Hein, M. J. Astle, and J. R. Shelton, *J. Org. Chem.* **26**, 4874 (1961).
c. F. S. Prout, R. J. Hartman, E. P.-Y. Huang, C. J. Korpics, and G. R. Tichelaar, *Org. Synth.* **IV**, 93 (1963).
d. E. F. Pratt and E. Werble, *J. Am. Chem. Soc.* **72**, 4638 (1950).
e. D. E. Worrall and L. Cohen, *J. Am. Chem. Soc.* **66**, 842 (1944).
f. A. C. Cope, A. A. D'Addieco, D. E. Whyte, and S. A. Glickman, *Org. Synth.* **IV**, 234 (1963).
g. W. J. Gensler and E. Berman, *J. Am. Chem. Soc.* **80**, 4949 (1958).
h. P. J. Jessup, C. B. Petty, J. Roos, and L. E. Overman, *Org. Synth.* **59**, 1 (1979).
i. R. H. Wiley and N. R. Smith, *Org. Synth.* **IV**, 731 (1963).

is prepared by Claisen condensation of ethyl acetate. All of the steps in the mechanism are reversible.

$$CH_3CO_2CH_2CH_3 + CH_3CH_2O^- \rightleftharpoons {}^-CH_2CO_2CH_2CH_3 + CH_3CH_2OH$$

$$CH_3\overset{O}{\overset{\|}{C}}OCH_2CH_3 + {}^-CH_2CO_2CH_2CH_3 \rightleftharpoons CH_3\underset{CH_2CO_2CH_2CH_3}{\overset{O^-}{C}}OCH_2CH_3$$

$$CH_3\underset{CH_2CO_2CH_2CH_3}{\overset{O^-}{C}}{-}OCH_2CH_3 \rightleftharpoons CH_3\overset{O}{\overset{\|}{C}}CH_2CO_2CH_2CH_3 + CH_3CH_2O^-$$

$$CH_3\overset{O}{\overset{\|}{C}}CH_2CO_2CH_2CH_3 + CH_3CH_2O^- \rightarrow CH_3\overset{O}{\overset{\|}{C}}\underset{-}{C}HCO_2CH_2CH_3 + CH_3CH_2OH$$

The final step in the mechanism drives the reaction to completion. Ethyl acetoacetate is more acidic than any of the other species present, and it is converted to its conjugate base in the final step. A full equivalent of base is needed to bring the reaction to completion. The β-ketoester product is obtained after neutralization and workup. As a practical matter, the alkoxide used as the base must be the same as the alcohol portion of the ester to prevent product mixtures resulting from ester interchange. Because the final proton transfer cannot occur when α-substituted esters are used, such compounds do not condense under the normal reaction conditions. This limitation can be overcome by use of a strong base that converts the reactant ester completely to its enolate. Entry 2 of Scheme 2.9 illustrates the use of triphenylmethylsodium for this purpose.

Scheme 2.9. Acylation of Nucleophilic Carbon by Esters

A. Intermolecular Ester Condensations

1[a] $$CH_3(CH_2)_3CO_2C_2H_5 \xrightarrow{NaOEt} CH_3(CH_2)_3CO\underset{CH_2CH_2CH_3}{C}HCO_2C_2H_5 \quad (77\%)$$

2[b] $$CH_3CH_2\underset{CH_3}{C}HCO_2C_2H_5 \xrightarrow{Ph_3C^- Na^+} CH_3CH_2\underset{CH_3}{C}H\overset{O}{\overset{\|}{C}}{-}\underset{CH_3}{\overset{CH_2CH_3}{C}}CO_2C_2H_5 \quad (63\%)$$

B. Cyclization of Diesters

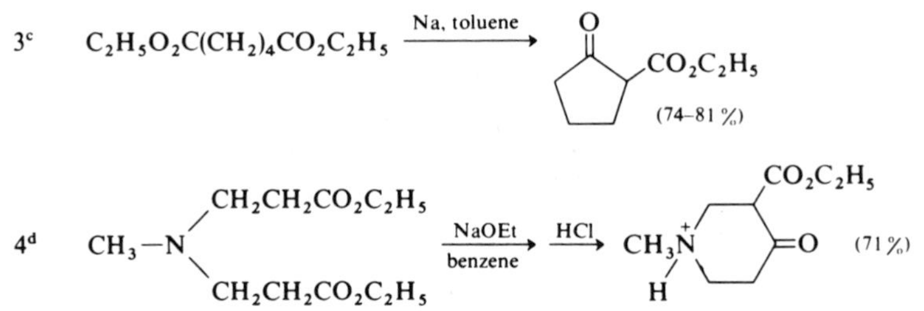

3[c] $$C_2H_5O_2C(CH_2)_4CO_2C_2H_5 \xrightarrow{Na,\ toluene} \text{2-(ethoxycarbonyl)cyclopentanone} \quad (74\text{–}81\%)$$

4[d] $$CH_3{-}N(CH_2CH_2CO_2C_2H_5)_2 \xrightarrow[benzene]{NaOEt} \xrightarrow{HCl} CH_3\overset{+}{N}H \text{ piperidine ring with } {=}O \text{ and } CO_2C_2H_5 \quad (71\%)$$

5[e] $C_2H_5O_2CCH_2CH_2CH(CO_2C_2H_5)CH(CO_2C_2H_5)CH_3$ $\xrightarrow{NaH}$ 2-methyl-3-oxocyclopentane-1,4-dicarboxylate ($CO_2C_2H_5$, CH_3, $C_2H_5O_2C$, O) (92%)

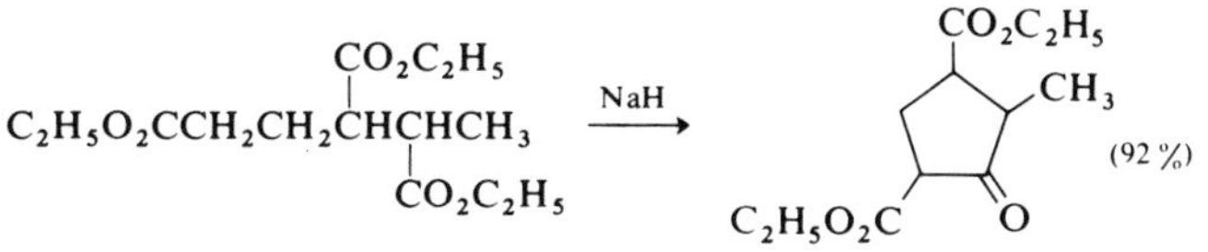

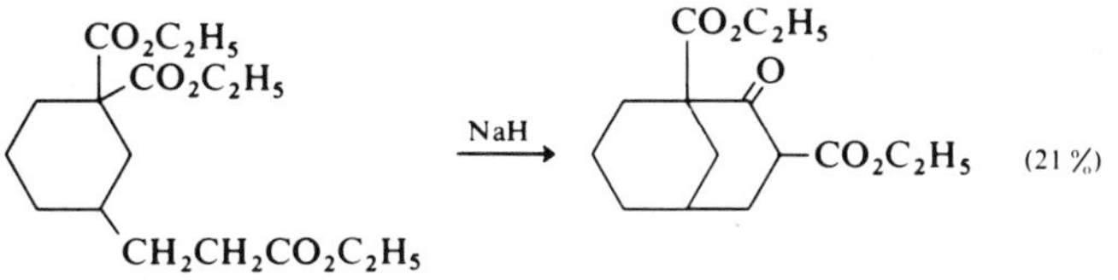

6[f] 1,1-bis($CO_2C_2H_5$)-3-($CH_2CH_2CO_2C_2H_5$)cyclohexane $\xrightarrow{NaH}$ bicyclic keto diester ($CO_2C_2H_5$, O, $CO_2C_2H_5$) (21%)

7[g] $PhCH_2O$, $PhCH_2O$ substituted benzoate ester (O, CH_3, CO_2CH_3, CO_2CH_3) $\xrightarrow[\text{dilute solution}]{[(CH_3)_3Si]_2NNa}$ macrocyclic keto ester ($PhCH_2O$, O, CH_3, CO_2CH_3, $PhCH_2O$, O) (77%)

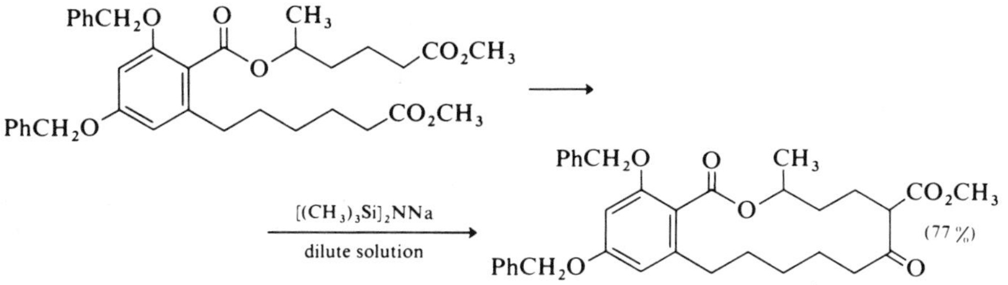

C. Mixed Ester Condensations

8[h] $(CH_2CO_2C_2H_5)_2 + (CO_2C_2H_5)_2 \xrightarrow{NaOEt} COCO_2C_2H_5-CHCO_2C_2H_5-CH_2CO_2C_2H_5$ (86–91%)

9[i] 3-pyridyl-$CO_2C_2H_5$ + $CH_3(CH_2)_2CO_2C_2H_5 \xrightarrow{NaH}$ 3-pyridyl-$COCH(CO_2C_2H_5)CH_2CH_3$ (68%)

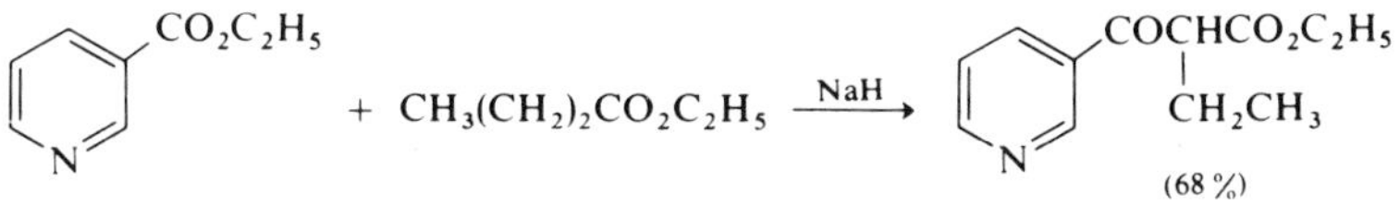

10[j] $C_{17}H_{35}CO_2C_2H_5 + (CO_2C_2H_5)_2 \xrightarrow{NaOEt} C_{16}H_{33}CH(COCO_2C_2H_5)CO_2C_2H_5$ (68–71%)

11[k] $C_6H_5CO_2C_2H_5 + CH_3CH_2CO_2C_2H_5 \xrightarrow{(i\text{-}Pr)_2NMgBr} C_6H_5COCH(CH_3)CO_2C_2H_5$ (51%)

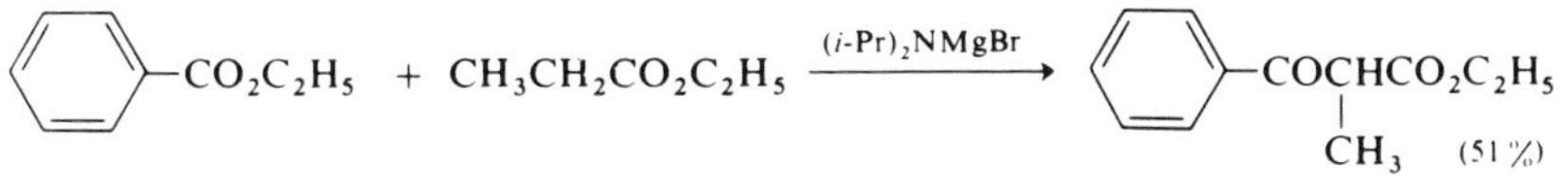

a. R. R. Briese and S. M. McElvain, *J. Am. Chem. Soc.* **55**, 1697 (1933).
b. B. E. Hudson, Jr., and C. R. Hauser, *J. Am. Chem. Soc.* **63**, 3156 (1941).
c. P. S. Pinkney, *Org. Synth.* **II**, 116 (1943).
d. E. A. Prill and S. M. McElvain, *J. Am. Chem. Soc.* **55**, 1233 (1933).
e. M. S. Newman and J. L. McPherson, *J. Org. Chem.* **19**, 1717 (1954).
f. J. P. Ferris and N. C. Miller, *J. Am. Chem. Soc.* **85**, 1325 (1963).
g. R. N. Hurd and D. H. Shah, *J. Org. Chem.* **38**, 390 (1973).
h. E. M. Bottorff and L. L. Moore, *Org. Synth.* **44**, 67 (1964).
i. F. W. Swamer and C. R. Hauser, *J. Am. Chem. Soc.* **72**, 1352 (1950).
j. D. E. Floyd and S. E. Miller, *Org. Synth.* **IV**, 141 (1963).
k. E. E. Royals and D. G. Turpin, *J. Am. Chem. Soc.* **76**, 5452 (1954).

Sodium hydride and a small amount of alcohol is frequently used as the base for ester condensation. It is probable that the reactive base is the sodium alkoxide formed by reaction of sodium hydride with the alcohol released in the condensation.

$$R'OH + NaH \rightarrow R'ONa + H_2$$

The sodium alkoxide is also no doubt the active catalyst in procedures in which sodium metal is used. The alkoxide is formed by reaction of the alcohol that is formed as the reaction proceeds.

The intramolecular version of ester condensation is called the *Dieckmann condensation.*[50] It is an important method for the formation of five- and six-membered rings and has occasionally been used for formation of larger rings. Entries 3–7 in Scheme 2.9 are illustrative.

Since ester condensation is reversible, product structure is governed by thermodynamic control, and in situations where more than one enolate may be formed, the product is derived from the enolate which leads to the more stable product. An example of this effect is the cyclization of the diester **3**.[51] Only **5** is formed, because **4** cannot be converted to a stable enolate. If **4**, synthesized by another method, is subjected to the conditions of the cyclization, it is isomerized to **5** by the reversible condensation mechanism.

$CO_2C_2H_5$ CH_3 **4** $\not\leftarrow$ $C_2H_5O_2CCH_2(CH_2)_3CH(CH_3)CO_2C_2H_5$ **3** $\xrightarrow[\text{xylene}]{\text{NaOEt}}$ CH_3 O^- $CO_2C_2H_5$ **5**

4 $\xrightarrow[\text{xylene}]{\text{NaOEt}}$ **5**

Mixed condensations of esters are subject to the same general restrictions as outlined for mixed aldol condensations (Section 2.1.2). One reactant must act preferentially as the acceptor and another as the nucleophile for good yields to be obtained. Combinations which work most effectively involve one ester than cannot form an enolate and is relatively reactive as an electrophile. Esters of aromatic acids, formic acid, and oxalic acid are especially useful. Some examples are shown in section C of Scheme 2.9.

Acylation of ester enolates can also be carried out with more reactive acylating agents such as acid anhydrides and acyl chlorides. These reactions must be carried out in inert solvents to avoid solvolysis of the acylating agent. The preparation of diethyl benzoylmalonate (entry 1, Scheme 2.10) is an example of the use of an acid anhydride. Entries 2–5 illustrate the use of acid chlorides. The use of these reactive acylating agents can be complicated by competing O-acylation.

Magnesium enolates play an important role in C-acylation reactions. The magnesium enolate of diethyl malonate, for example, can be prepared by reaction

50. J. P. Schaefer and J. J. Bloomfield, *Org. React.* **15**, 1 (1967).
51. N. S. Vul'fson and V. I. Zaretskii, *J. Gen. Chem. USSR* **29**, 2704 (1959).

A. Acylation with Acid Halides and Mixed Anhydrides

1[a] $PhC(=O)OC(=O)OC_2H_5 + C_2H_5OMgCH(CO_2C_2H_5)_2 \rightarrow PhCOCH(CO_2C_2H_5)_2$ (68–75%)

2[b] $CH_3C(O^-)=CHCO_2C_2H_5 + PhCOCl \rightarrow [PhC(=O)-\bar{C}(C(=O)CH_3)CO_2C_2H_5] \rightarrow PhC(=O)CH_2CO_2C_2H_5$ (68–71%)

3[c] o-nitrobenzoyl chloride ($o-NO_2C_6H_4COCl$) $+ C_2H_5OMgCH(CO_2C_2H_5)_2 \rightarrow o-NO_2C_6H_4COCH(CO_2C_2H_5)_2$ (82–88%)

4[d] $CH_3C(O^-)=CHCO_2C_2H_5 + ClC(=O)(CH_2)_3CO_2C_2H_5 \rightarrow C_2H_5O_2C(CH_2)_3C(=O)-CH(C(=O)CH_3)CO_2C_2H_5$ (61–66%)

5[e] $CH_3CO_2C_2H_5 \xrightarrow{R_2NLi} LiCH_2CO_2C_2H_5 \xrightarrow[-78°C]{(CH_3)_3CC(=O)Cl} (CH_3)_3CC(=O)CH_2CO_2C_2H_5$ (70%)

6[f] $CH_3CO_2CH_3 \xrightarrow[\text{2) ClCO(CH}_2)_{12}\text{CH}_3\text{; 3) H}^+]{\text{1) LDA}} CH_3O_2CCH_2C(=O)(CH_2)_{12}CH_3$ (83%)

B. Acylation with Imidazolides

7[g] 4-oxocyclopent-2-enyl-$CH_2C(=O)N$(imidazolyl) $+ Mg(^-O_2CCH_2CO_2C_2H_5)_2 \xrightarrow[\text{2) H}^+]{\text{1) 25°C}}$ 4-oxocyclopent-2-enyl-$CH_2C(=O)CH_2CO_2C_2H_5$ (66%)

8[h] imidazolide (C=O–N-imidazolyl on cyclohexadienyl carbon bearing CH_2CH_2-linked tetrahydrobenzofuran with CH_3) $+ LiCH_2CO_2C(CH_3)_3 \xrightarrow[1\,h]{-78°C} \xrightarrow{H^+}$ product with $C(=O)CH_2CO_2C(CH_3)_3$ (83%)

Scheme 2.10—*continued*

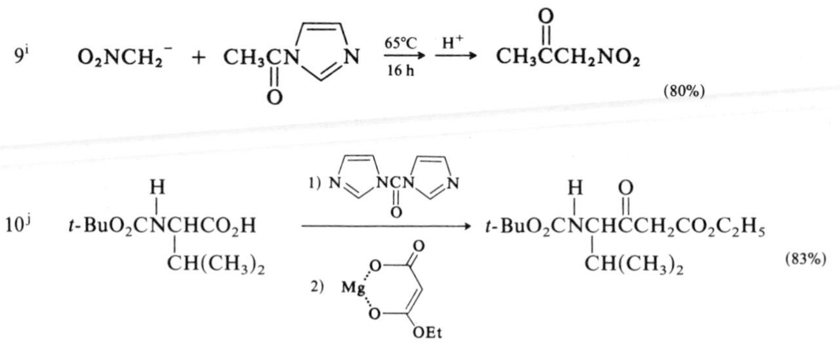

9[i] $O_2NCH_2^-$ + $CH_3C(=O)N$(imidazolyl) $\xrightarrow[16\ h]{65°C}$ $\xrightarrow{H^+}$ $CH_3C(=O)CH_2NO_2$ (80%)

10[j] t-$BuO_2CNHCHCO_2H$ ($CH(CH_3)_2$) $\xrightarrow[2)\ Mg(O_2CCH=C(O)OEt)]{1)\ \text{carbonyldiimidazole}}$ t-$BuO_2CNHCHC(=O)CH_2CO_2C_2H_5$ ($CH(CH_3)_2$) (83%)

a. J. A. Price and D. S. Tarbell, *Org. Synth.* **IV**, 285 (1963).
b. J. M. Straley and A. C. Adams, *Org. Synth.* **IV**, 415 (1963).
c. G. A. Reynolds and C. R. Hauser, *Org. Synth.* **IV**, 708 (1963).
d. M. Guha and D. Nasipuri, *Org. Synth.* **V**, 384 (1973).
e. M. W. Rathke and J. Deitch, *Tetrahedron Lett.*, 2953 (1971).
f. D. F. Taber, P. B. Deker, H. M. Fales, T. H. Jones, and H. A. Lloyd, *J. Org. Chem.* **53**, 2968 (1988).
g. A. Barco, S. Benetti, G. P. Pollini, P. G. Baraldi, and C. Gandolfi, *J. Org. Chem.* **45**, 4776 (1980).
h. E. J. Corey, G. Wess, Y. B. Xiang, and A. K. Singh, *J. Am. Chem. Soc.* **109**, 4717 (1987).
i. M. E. Jung, D. D. Grove, and S. I. Khan, *J. Org. Chem.* **52**, 4570 (1987).
j. J. Maibaum and D. H. Rich, *J. Org. Chem.* **53**, 869 (1988).

with magnesium metal in ethanol. It is soluble in solvents such as ether and undergoes C-acylation by acid anhydrides and acid chlorides (entries 1 and 3 in Scheme 2.10). Monoalkyl esters of malonic acid react with Grignard reagents to give a chelated enolate of the malonate monoanion.

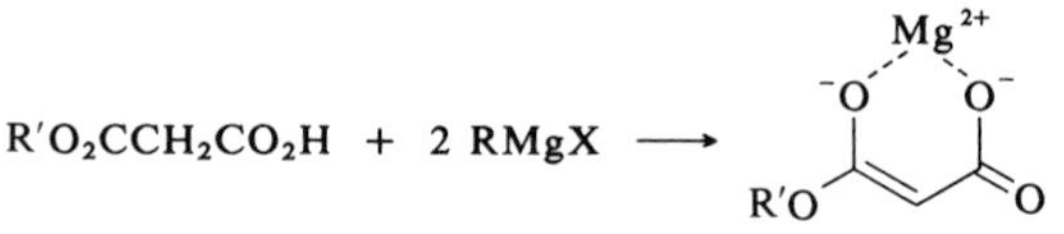

$$R'O_2CCH_2CO_2H + 2\ RMgX \longrightarrow \text{chelate } [R'O(O^-)C=CH-C(=O)O^-]\ Mg^{2+}$$

These carbon nucleophiles react with acid chlorides[52] or acyl imidazolides.[53] The initial products decarboxylate readily so the isolated products are β-ketoesters.

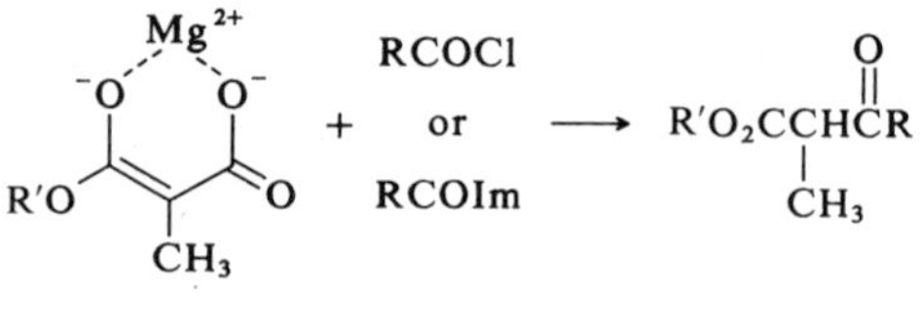

$$[R'O(O^-)C=C(CH_3)-C(=O)O^-]\ Mg^{2+} + \begin{matrix} RCOCl \\ \text{or} \\ RCOIm \end{matrix} \longrightarrow R'O_2CCH(CH_3)C(=O)R$$

Acyl imidazolides have also proven to be useful reagents for acylation of enolates.[54] Acyl imidazolides are more reactive than esters but not as reactive as

52. R. E. Ireland and J. A. Marshall, *J. Am. Chem. Soc.* **81**, 2907 (1959).
53. J. Maibaum and D. H. Rich, *J. Org. Chem.* **53**, 869 (1988); W. H. Moos, R. D. Gless, and H. Rapoport, *J. Org. Chem.* **46**, 5064 (1981).
54. D. W. Brooks, L. D.-L. Lu and S. Masamune, *Angew. Chem. Int. Ed. Engl.* **18**, 72 (1979).

acid halides. β-Ketoesters are formed by reaction of magnesium salts of monoalkyl esters of malonic acid with imidazolides.

$$RC(=O)\text{—N(imidazole)} + Mg(O_2CCH_2CO_2R')_2 \longrightarrow \xrightarrow[-CO_2]{H^+} RC(=O)CH_2CO_2R'$$

Acyl imidazolides have also been used for acylation of ester enolates and nitromethane anion, as illustrated by entries 8 and 9 in Scheme 2.10.

Both diethyl malonate and ethyl acetoacetate can be acylated by acid chlorides with magnesium chloride and an amine as additional reagents.[55]

$$(C_2H_5O_2C)_2CH_2 \xrightarrow[Et_3N]{MgCl_2} \xrightarrow{RC(=O)Cl} (C_2H_5O_2C)_2CHC(=O)R$$

$$C_2H_5O_2CCH_2C(=O)CH_3 \xrightarrow[pyridine]{MgCl_2} \xrightarrow{RC(=O)Cl} C_2H_5O_2CCH(C(=O)R)C(=O)CH_3$$

Rather similar conditions can be used to convert ketones to β-ketoacids by carboxylation.[56]

$$CH_3CH_2C(=O)CH_2CH_3 \xrightarrow[CH_3CN,\ CO_2,\ Et_3N]{MgCl_2,\ NaI} \xrightarrow{H^+} CH_3CH_2C(=O)CH(CO_2H)CH_3$$

Such reactions presumably involve formation of a magnesium chelate of the ketoacid. The β-ketoacid is liberated when the reaction mixture is acidified during workup.

$$Mg^{2+}\ \text{chelate: } {}^-O\text{—}C(R)=C(R)\text{—}C(=O)\text{—}O^-$$

Carboxylation of ketones and esters can be achieved by reaction with the magnesium salt of monomethyl carbonate:

$$C_6H_5C(=O)CH_3 + Mg(O_2COCH_3)_2 \xrightarrow[110°C]{DMF} \xrightarrow{H^+} C_6H_5C(=O)CH_2CO_2H \qquad \text{Ref. 57}$$

55. M. W. Rathke and P. J. Cowan, *J. Org. Chem.* **50**, 2622 (1985).
56. R. E. Tirpak, R. S. Olsen and M. W. Rathke, *J. Org. Chem.* **50**, 4877 (1985).
57. M. Stiles, *J. Am. Chem. Soc.* **81**, 2598 (1959).

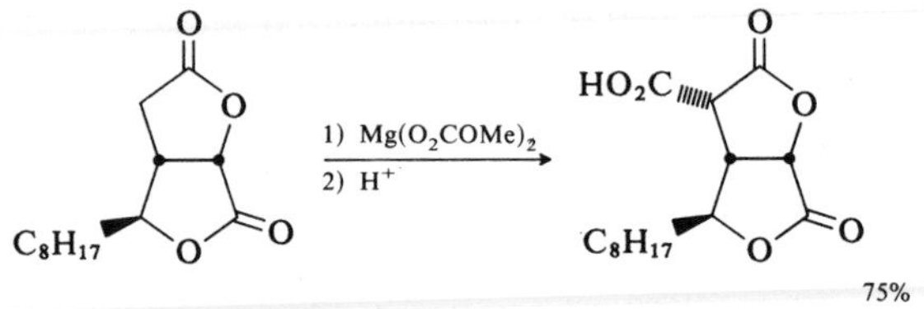

1) $Mg(O_2COMe)_2$
2) H^+

HO_2C C_8H_{17} C_8H_{17} 75% Ref. 58

The enolates of ketones can be acylated by esters and other acylating agents. The acylation of ketones is an important method of enhancing reactivity and selectivity in the synthetic modification of ketones. Some examples of the acylation of ketone enolates are shown in Scheme 2.11. The products of these reactions are

Scheme 2.11. Acylation of Ketones with Esters

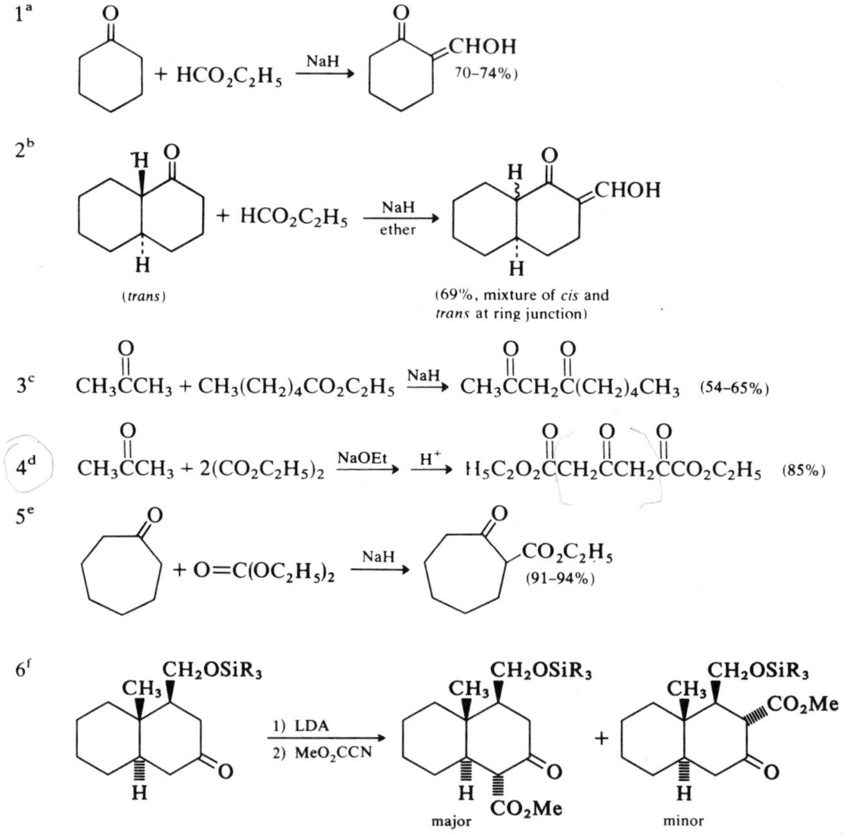

1[a] cyclohexanone + $HCO_2C_2H_5$ $\xrightarrow{NaH}$ 2-(CHOH)cyclohexanone (70–74%)

2[b] (*trans*) decalone + $HCO_2C_2H_5$ $\xrightarrow[ether]{NaH}$ CHOH product (69%, mixture of *cis* and *trans* at ring junction)

3[c] $CH_3CCH_3 + CH_3(CH_2)_4CO_2C_2H_5 \xrightarrow{NaH} CH_3CCH_2C(CH_2)_4CH_3$ (54–65%)

4[d] $CH_3CCH_3 + 2(CO_2C_2H_5)_2 \xrightarrow{NaOEt} \xrightarrow{H^+} H_5C_2O_2CCH_2CCH_2CCO_2C_2H_5$ (85%)

5[e] cycloheptanone + $O{=}C(OC_2H_5)_2$ $\xrightarrow{NaH}$ 2-$CO_2C_2H_5$-cycloheptanone (91–94%)

6[f] CH_2OSiR_3, CH_3, H decalone $\xrightarrow{1)\ LDA;\ 2)\ MeO_2CCN}$ CO_2Me product (major) + CO_2Me product (minor)

a. C. Ainsworth, *Org. Synth.* **IV**, 536 (1963).
b. P. H. Lewis, S. Middleton, M. J. Rosser, and L. E. Stock, *Aust. J. Chem.* **32**, 1123 (1979).
c. N. Green and F. B. La Forge, *J. Am. Chem. Soc.* **70**, 2287 (1948); F. W. Swamer and C. R. Hauser, *J. Am. Chem. Soc.* **72**, 1352 (1950).
d. E. R. Riegel and F. Zwilgmeyer, *Org. Synth.* **II**, 126 (1943).
e. A. P. Krapcho, J. Diamanti, C. Cayen, and R. Bingham, *Org. Synth.* **47**, 20 (1967).
f. F. E. Ziegler, S. I. Klein, U. K. Pati, and T.-F. Wang, *J. Am. Chem. Soc.* **107**, 2730 (1985).

58. W. L. Parker and F. Johnson, *J. Org. Chem.* **38**, 2489 (1973).

all β-dicarbonyl compounds. They are all rather acidic and can be alkylated by the procedures described in Section 1.4.

Reaction of ketone enolates with formate esters gives a β-ketoaldehyde. Since these compounds exist in the enol form, they are referred to as *hydroxymethylene derivatives.* Product formation is under thermodynamic control so the structure of the product can be predicted on the basis of the stability of the various possible product anions.

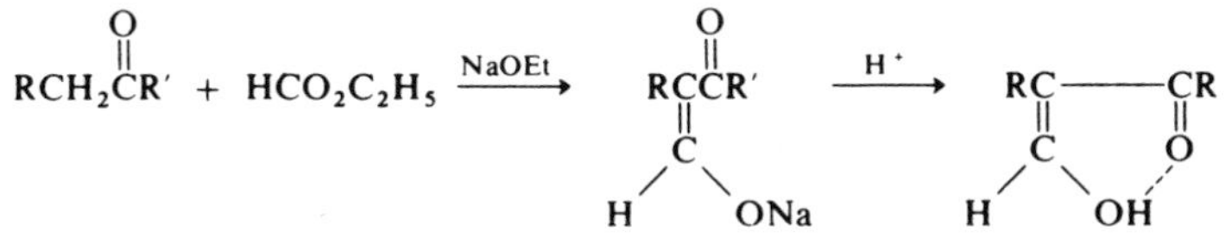

$$RCH_2\overset{O}{\overset{\|}{C}}R' + HCO_2C_2H_5 \xrightarrow{NaOEt} RC(=CH-ONa)\overset{O}{\overset{\|}{C}}R' \xrightarrow{H^+} RC(=CH-OH)-CR(=O)$$

Hydroxymethylene groups can be converted to methyl groups by a two-stage sequence via a β-thioenone:

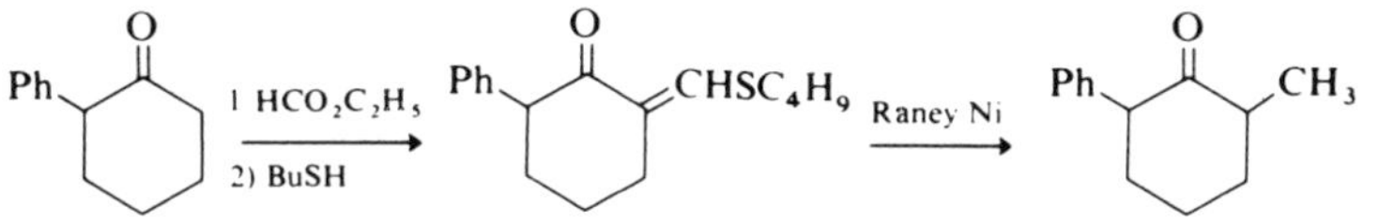

2-Phenylcyclohexanone $\xrightarrow[\text{2) BuSH}]{\text{1) } HCO_2C_2H_5}$ 2-phenyl-6-($CHSC_4H_9$)cyclohexanone $\xrightarrow{\text{Raney Ni}}$ 2-phenyl-6-methylcyclohexanone (Ref. 59)

A sequence involving the dianion of the hydroxymethylene ketone results in alkylation at the α'-carbon.

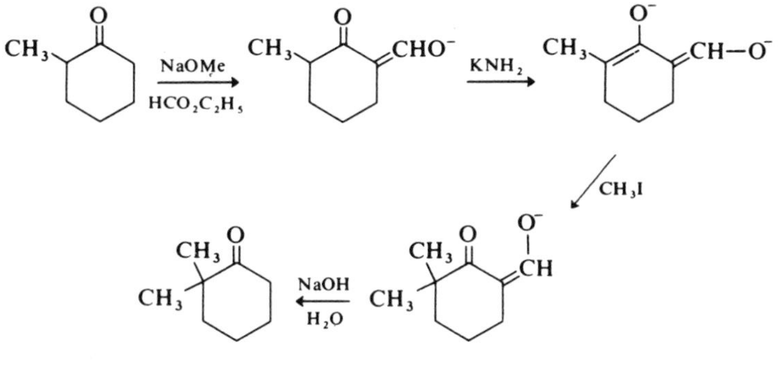

2-Methylcyclohexanone $\xrightarrow[HCO_2C_2H_5]{NaOMe}$ 2-methyl-6-(CHO^-)cyclohexanone $\xrightarrow{KNH_2}$ dianion (O^-, $CH-O^-$) $\xrightarrow{CH_3I}$ 2,2-dimethyl-6-($CH-O^-$)cyclohexanone $\xrightarrow[H_2O]{NaOH}$ 2,2-dimethylcyclohexanone (Ref. 60)

The removal of the hydroxymethylene group in the last step of this sequence occurs by reversal of the acylation.

Ketones are converted to β-ketoesters by acylation with diethyl carbonate or diethyl oxalate. The products are readily alkylated by the procedures discussed in Chapter 1. A convenient procedure for introduction of an ester group alpha to a ketone uses an alkyl cyanoformate as the acylating reagent.[61] These reactions can

59. R. E. Ireland and J. A. Marshall, *J. Org. Chem.* **27**, 1615 (1962); J. D. Metzger, M. W. Baker, and R. J. Morris, *J. Org. Chem.* **37**, 789 (1972).
60. S. Boatman, T. M. Harris, and C. R. Hauser, *Org. Synth.* **48**, 40 (1968).
61. L. N. Mander and S. P. Sethi, *Tetrahedron Lett.* **24**, 5425 (1983).

be used under conditions where a ketone enolate has been formed under kinetic control.

$$\text{2-methylcyclohexanone} \xrightarrow{\text{LDA}} \xrightarrow[\text{THF, HMPA}]{\text{EtO}_2\text{CCN}} \xrightarrow{\text{H}_2\text{O}} \text{2-methyl-6-}(\text{CO}_2\text{C}_2\text{H}_5)\text{cyclohexanone} \quad (86\%)$$

When this type of reaction is quenched with trimethylsilyl chloride, rather than by neutralization, a trimethylsilyl ether of the adduct is isolated. This result shows that the tetrahedral adduct is stable until the reaction mixture is hydrolyzed.

$$\text{cyclohexanone} \xrightarrow[\text{2) EtO}_2\text{CCN}]{\text{1) LDA}} \xrightarrow{(\text{Me})_3\text{SiCl}} \text{2-}[\text{C(OSi(CH}_3)_3)(\text{OC}_2\text{H}_5)(\text{CN})]\text{cyclohexanone}$$ Ref. 62

β-Keto sulfoxides can be prepared by acylation of dimethyl sulfoxide ion with esters.[63]

$$\text{RC(=O)OR}' + {}^{-}\text{CH}_2\text{S(=O)CH}_3 \rightleftharpoons \text{RC(=O)}\bar{\text{C}}\text{HS(=O)CH}_3 + \text{R}'\text{OH}$$

Mechanistically, this reaction is similar to ketone acylation. The β-keto sulfoxides have several synthetic applications. The sulfoxide substituent can be removed reductively, leading to methyl ketones:

$$\text{CH}_3\text{O-C}_6\text{H}_4\text{-C(=O)CH}_2\text{SOCH}_3 \xrightarrow{\text{Zn Hg}} \text{CH}_3\text{O-C}_6\text{H}_4\text{-C(=O)CH}_3$$ Ref. 64

The β-keto sulfoxides can be alkylated via their anions. Inclusion of an alkylation step prior to the reduction provides a route to ketones with longer chains.

$$\text{PhCOCH}_2\text{SOCH}_3 \xrightarrow[\text{2) CH}_3\text{I}]{\text{1) NaH}} \text{PhCOCH(CH}_3)\text{SOCH}_3 \xrightarrow{\text{Zn Hg}} \text{PhCOCH}_2\text{CH}_3$$ Ref. 65

Dimethyl sulfone can be subjected to similar reaction sequences.[66]

62. F. E. Ziegler and T.-F. Wang, *Tetrahedron Lett.* **26**, 2291 (1985).
63. E. J. Corey and M. Chaykovsky, *J. Am. Chem. Soc.* **87**, 1345 (1965); H. D. Becker, G. J. Mikol, and G. A. Russell, *J. Am. Chem. Soc.* **85**, 3410 (1963).
64. G. A. Russell and G. J. Mikol, *J. Am. Chem. Soc.* **88**, 5498 (1966).
65. P. G. Gassman and G. D. Richmond, *J. Org. Chem.* **31**, 2355 (1966).
66. H. O. House and J. K. Larson, *J. Org. Chem.* **33**, 61 (1968).

The Wittig reaction involves phosphorus ylides as the nucleophilic carbon species.[67] An *ylide* is a molecule that has a contributing Lewis structure with opposite charges on adjacent atoms, each of which has an octet of electrons. While this definition includes other classes of compounds, the discussion here will be limited to ylides with the negative charge on carbon. Phosphorus ylides are usually stable, but quite reactive, compounds. They can be represented by two limiting Lewis structures. These are sometimes referred to as the ylide and ylene forms. The ylene form is pentavalent at phosphorus and implies involvement of phosphorus $3d$ orbitals. Using $(CH_3)_3PCH_2$ (trimethylphosphonium methylide) as an example, the two forms are as follows:

$$\underset{\text{ylide}}{(CH_3)_3\overset{+}{P}-CH_2^-} \leftrightarrow \underset{\text{ylene}}{(CH_3)_3P=CH_2}$$

The stability of phosphorus ylides can be ascribed, in part, to resonance of these two structures. NMR spectroscopic studies (^1H, ^{13}C, and ^{31}P), however, are more consistent with the dipolar ylide structure and suggest only a minor contribution from the ylene structure.[68] Theoretical calculations support this view also.[69]

The synthetic potential of phosphorus ylides was initially developed by G. Wittig and his associates at the University of Heidelberg. The reaction of a phosphorus ylide with an aldehyde or ketone introduces a carbon–carbon double bond in place of the carbonyl bond:

$$R_3\overset{+}{P}-\overset{-}{C}R'_2 + R''_2C=O \rightarrow R''_2C=CR'_2 + R_3P=O$$

The mechanism originally proposed is an addition of the nucleophilic ylide carbon to the carbonyl group to yield a dipolar intermediate (a *betaine*), followed by elimination of a phosphine oxide. The elimination is presumed to occur after formation of a four-membered oxaphosphetane intermediate. An alternative mechanism might involve direct formation of the oxaphosphetane.[70] There have been several theoretical studies of these intermediates.[71] Oxaphosphetane intermediates have been observed by NMR studies at low temperature, but direct evidence for the betaine intermediates is less clear.[72]

67. For reviews of the Wittig reaction, see A. Maercker, *Org. React.* **14**, 270 (1965); I. Gosney and A. G. Rowley, in *Organophosphorus Reagents in Organic Synthesis*, J. I. G. Cadogan (ed.), Academic Press, London, 1979, pp. 17–153; B. E. Maryanoff and A. B. Reitz, *Chem. Rev.* **89**, 863 (1989).
68. H. Schmidbaur, W. Bucher, and D. Schentzow, *Chem. Ber.* **106**, 1251 (1973).
69. A. Streitwieser, Jr., A. Rajca, R. S. McDowell, and R. Glaser, *J. Am. Chem. Soc.* **109**, 4184 (1987).
70. E. Vedejs and K. A. J. Snoble, *J. Am. Chem. Soc.* **95**, 5778 (1973).
71. C. Trindle, J.-T. Hwang, and F. A. Carey, *J. Org. Chem.* **38**, 2664 (1973); R. Holler and H. Lischka, *J. Am. Chem. Soc.* **102**, 4632 (1980); F. Volatron and O. Eisenstein, *J. Am. Chem. Soc.* **106**, 6117 (1984).
72. E. Vedejs, G. P. Meier, and K. A. J. Snoble, *J. Am. Chem. Soc.* **103**, 2823 (1981); B. E. Maryanoff, A. B. Reitz, M. S. Mutter, R. R. Inners, H. R. Almond, Jr., R. R. Whittle, and R. A. Olofson, *J. Am. Chem. Soc.* **108**, 7684 (1986).

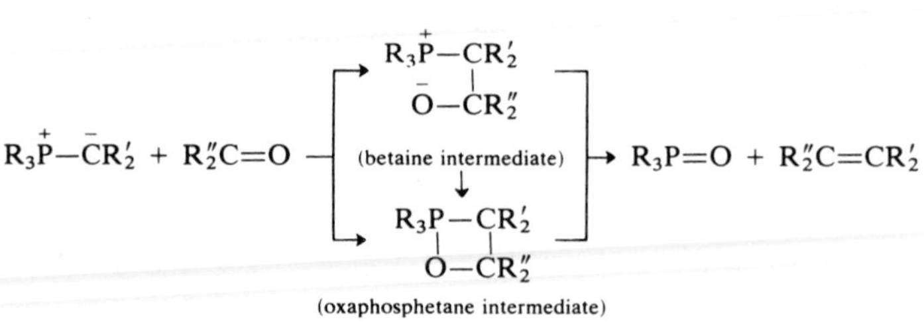

$$R_3\overset{+}{P}-\overset{-}{C}R'_2 + R''_2C{=}O \rightarrow \left[\begin{matrix} R_3\overset{+}{P}-CR'_2 \\ \overset{-}{O}-CR''_2 \\ \text{(betaine intermediate)} \\ \downarrow \\ R_3P-CR'_2 \\ O-CR''_2 \end{matrix}\right] \rightarrow R_3P{=}O + R''_2C{=}CR'_2$$

(oxaphosphetane intermediate)

Phosphorus ylides are usually prepared by deprotonation of phosphonium salts. The phosphonium salts that are used most often are alkyltriphenylphosphonium halides, which can be prepared by the reaction of triphenylphosphine and an alkyl halide:

$$Ph_3P + RCH_2X \rightarrow Ph_3\overset{+}{P}-CH_2R\ X^-$$

X = I, Br, or Cl

$$Ph_3\overset{+}{P}\overset{-}{C}H_2R \xrightarrow{\text{base}} Ph_3P{=}CHR$$

The alkyl halide must be one which is reactive toward S_N2 displacement. Alkyltriphenylphosphonium halides are only weakly acidic, and strong bases are used for deprotonation. These include organolithium reagents, the sodium salt of dimethyl sulfoxide, amide ion, or substituted amide anions such as hexamethyldisilylamide. The ylides are not normally isolated, so the reaction is carried out with the carbonyl compound either present or added after ylide formation. Ylides with nonpolar substituents, for example, H, alkyl, or aryl, are quite reactive toward both ketones and aldehydes. Scheme 2.12 gives some examples of Wittig reactions.

Scheme 2.12. The Wittig Reaction

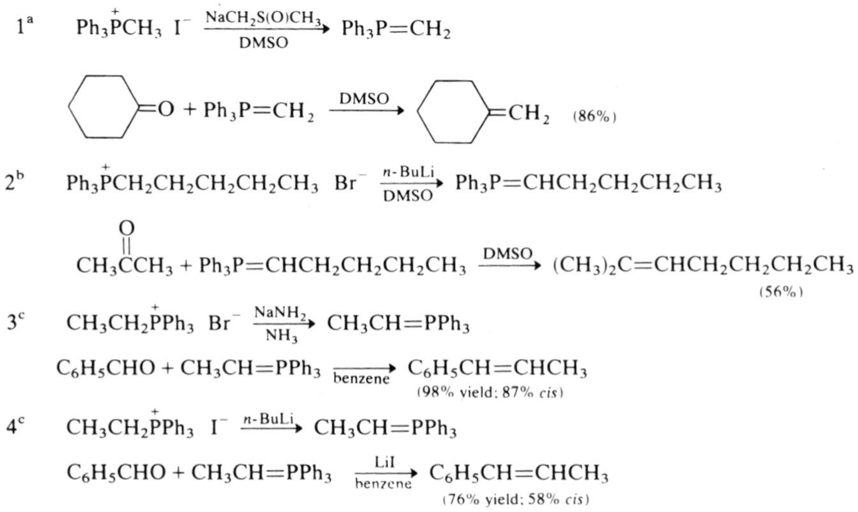

1[a] $Ph_3\overset{+}{P}CH_3\ I^- \xrightarrow[\text{DMSO}]{NaCH_2S(O)CH_3} Ph_3P{=}CH_2$

$$\text{cyclohexanone}{=}O + Ph_3P{=}CH_2 \xrightarrow{\text{DMSO}} \text{cyclohexane}{=}CH_2 \quad (86\%)$$

2[b] $Ph_3\overset{+}{P}CH_2CH_2CH_2CH_2CH_3\ Br^- \xrightarrow[\text{DMSO}]{n\text{-BuLi}} Ph_3P{=}CHCH_2CH_2CH_2CH_3$

$$CH_3\overset{O}{\overset{\|}{C}}CH_3 + Ph_3P{=}CHCH_2CH_2CH_2CH_3 \xrightarrow{\text{DMSO}} (CH_3)_2C{=}CHCH_2CH_2CH_2CH_3 \quad (56\%)$$

3[c] $CH_3CH_2\overset{+}{P}Ph_3\ Br^- \xrightarrow[NH_3]{NaNH_2} CH_3CH{=}PPh_3$

$$C_6H_5CHO + CH_3CH{=}PPh_3 \xrightarrow[\text{benzene}]{} C_6H_5CH{=}CHCH_3$$

(98% yield; 87% *cis*)

4[c] $CH_3CH_2\overset{+}{P}Ph_3\ I^- \xrightarrow{n\text{-BuLi}} CH_3CH{=}PPh_3$

$$C_6H_5CHO + CH_3CH{=}PPh_3 \xrightarrow[\text{benzene}]{\text{LiI}} C_6H_5CH{=}CHCH_3$$

(76% yield; 58% *cis*)

5[d] $CH_3CH_2CH_2CH_2CH_2\overset{+}{P}Ph_3\ Br^- \xrightarrow[THF]{Na^+N(SiMe_3)_2} CH_3CH_2CH_2CH_2CH{=}PPh_3$

$HC(O)(CH_2)_7CH_2OAc + CH_3CH_2CH_2CH_2CH{=}PPh_3 \rightarrow CH_3(CH_2)_3CH{=}CH(CH_2)_7CH_2OAc$

(79% yield; 98% *cis*)

6[e] $Ph_3\overset{+}{P}CH_2CO_2CH_2CH_3\ Br^- \xrightarrow[H_2O]{NaOH} Ph_3P{=}CHCO_2CH_2CH_3$

(stable, isolable ylide)

(benzofuran)-CHO, OH + $Ph_3P{=}CHCO_2CH_2CH_3$ (2 equiv) $\xrightarrow[\text{reflux, 2 hr}]{\text{benzene}}$ (benzofuran)-CH=CH-$CO_2CH_2CH_3$, OH (86%)

7[f] $C_6H_5CHO + Ph_3P{=}CHCO_2CH_2CH_3 \xrightarrow{EtOH} C_6H_5CH{=}CHCO_2CH_2CH_3$

(77% yield; only *trans*)

8[g] $C_6H_5CH_2\overset{+}{P}Ph_3\ Cl^- \xrightarrow[ether]{PhLi} C_6H_5CH{=}PPh_3$

$C_6H_5CHO + C_6H_5CH{=}PPh_3 \rightarrow C_6H_5CH{=}CHC_6H_5$

(82% yield; 70% *trans*)

9[f] (cyclohexanone)=O + $C_6H_5CH{=}PPh_3 \rightarrow$ (cyclohexylidene)=CHC_6H_5 (60%)

10[h] (CH_3, CH_3, CH_3 tricyclic ketone) $\xrightarrow[90°C,\ 30\ min]{Ph_3\overset{+}{P}CH_3Br^-,\ KOCR_3,\ toluene}$ (CH_2 methylene product) (56%)

11[i] (CH_3, CH_3, CH_3 bicyclic ketone, camphor) $\xrightarrow[100°C,\ 2\ h]{Ph_3\overset{+}{P}CH_3Br^-,\ KOCR_3}$ (CH_2 methylene product) (91%)

12[b] $CH_3CH_2CH_2CH_2CHO + CH_3CH{=}PPh_3 \xrightarrow{\text{1) LiBr, THF, }-78°C\text{; 2) BuLi; 3) }CH_2O\text{, 25°C}}$

$(CH_3CH_2CH_2CH_2)(H)C{=}C(CH_2OH)(CH_3)$

a. R. Greenwald, M. Chaykovsky, and E. J. Corey, *J. Org. Chem.* **28**, 1128 (1963).
b. U. T. Bhalerao and H. Rapoport, *J. Am. Chem. Soc.* **93**, 4835 (1971).
c. M. Schlosser and K. F. Christmann, *Justus Liebigs, Ann. Chem.* **708**, 1 (1967).
d. H. J. Bestmann, K. H. Koschatzky, and O. Vostrowsky, *Chem. Ber.* **112**, 1923 (1979).
e. Y. Y. Liu, E. Thom, and A. A. Liebman, *J. Heterocycl. Chem.* **16**, 799 (1979).
f. G. Wittig and W. Haag, *Chem. Ber.* **88**, 1654 (1955).
g. G. Wittig and U. Schöllkopf, *Chem. Ber.* **87**, 1318 (1954).
h. A. B. Smith III and P. J. Jerris, *J. Org. Chem.* **47**, 1845 (1982).
i. L. Fitjer and U. Quabeck, *Synth. Commun.* **15**, 855 (1985).

When a hindered ketone is to be converted to a methylene derivative, the best results are obtained when a potassium *t*-alkoxide is used as a base in a hydrocarbon solvent. Under these conditions, the reaction can be carried out at elevated temperature.[73] Entries 10 and 11 in Scheme 2.12 illustrate this procedure.

β-Ketophosphonium salts are considerably more acidic than alkylphosphonium salts and can be converted to ylides by relatively weak bases. The resulting ylides, which are stabilized by the carbonyl group, are substantially less reactive than unfunctionalized ylides. Vigorous conditions may be required to bring about reactions with ketones. Entries 6 and 7 in Scheme 2.12 involve stabilized ylides.

The stereoselectivity of the Wittig reaction depends strongly on both the structure of the ylide and the reaction conditions. The broadest generalization is that unstabilized ylides give predominantly the *Z*-alkene while stabilized ylides give mainly the *E*-alkene.[74] Use of sodium amide or sodium hexamethyldisilylamide as bases gives higher selectivity for *Z*-alkenes than do ylides prepared with alkyllithium reagents as base (see entries 3 and 5 of Scheme 2.12). The dependence of the stereoselectivity on the nature of the base is attributed to complexes involving the lithium halide salt which is present when alkyllithium reagents are used as bases. Stabilized ylides such as (carboethoxymethylidene)triphenylphosphorane (entries 6 and 7) react with aldehydes to give exclusively *trans* double bonds. Benzylidenetriphenylphosphorane (entry 8) gives a mixture of both *cis*- and *trans*-stilbene on reaction with benzaldehyde.

The stereoselectivity of the Wittig reaction is believed to be the result of steric effects that develop as the two reagents approach one another. The large phenyl substituents on phosphorus impose large steric demands.[75] Reactions of unstabilized phosphoranes are believed to proceed through an early transition state, and steric factors usually make such transition states selective for the *Z*-alkene.[76] The empirical generalization concerning the preference for *Z*-alkenes from nonstabilized ylides under salt-free conditions and *E*-ylides from stabilized ylides serves as a guide to predicting stereoselectivity.

The reaction of nonstabilized ylides with aldehydes can be induced to yield *E*-alkenes with high stereoselectivity by a procedure known as the *Schlosser modification* of the Wittig reaction.[77] In this procedure, the ylide is generated as a lithium halide complex and allowed to react with an aldehyde at low temperature, presumably forming a mixture of diastereomeric betaine–lithium halide complexes. At the temperature at which the addition is carried out, fragmentation to an alkene and triphenylphosphine oxide does not occur. This complex is then treated with an equivalent of strong base such as phenyllithium to form a β-oxido ylide. Addition of *t*-butyl alcohol protonates the β-oxido ylide stereoselectively to give the more

73. J. M. Conia and J. C. Limasset, *Bull. Soc. Chim. Fr.*, 1936 (1967); J. Provin, F. Leyendecker, and J. M. Conia, *Tetrahedron Lett.*, 4053 (1975); S. R. Schow and T. C. Morris, *J. Org. Chem.* **44**, 3760 (1979).
74. M. Schlosser, *Top. Stereochem.* **5**, 1 (1970).
75. M. Schlosser and B. Schaub, *J. Am. Chem. Soc.* **104**, 5821 (1982); H. J. Bestmann and O. Vostrowsky, *Top. Curr. Chem.* **109**, 85 (1983); E. Vedejs, T. Fleck, and S. Hara, *J. Org. Chem.* **52**, 4637 (1987).
76. E. Vedejs and C. F. Marth, *J. Am. Chem. Soc.* **110**, 3948 (1988).
77. M. Schlosser, and K.-F. Christmann, *Justus Liebigs Ann. Chem.* **708**, 1 (1967); M. Schlosser, K.-F. Christmann, and A. Piskala, *Chem. Ber.* **103**, 2814 (1970).

stable *threo*-betaine as a lithium halide complex. Warming the solution causes the *threo*-betaine–lithium halide complex to give *E*-alkene.

$$\underset{\text{LiO}\quad {}^{+}\text{PPh}_3}{\text{RCH—CHR}'}\ \text{Br}^- \xrightarrow{\text{PhLi}} \underset{\text{LiO}\quad {}^{+}\text{PPh}_3}{\overset{\text{Li}}{\text{RCH—CR}'}}\ \text{Br}^- \xrightarrow{t\text{-BuOH}} \text{[LiBr complex of } threo\text{-betaine]}\ \text{Br}^- \longrightarrow (E)\text{-R'CH=CHR} + \text{Ph}_3\text{P=O} + \text{LiBr}$$

LiBr complex of β-oxide ylide

LiBr complex of *threo*-betaine

A useful extension of this method is one in which the β-oxido ylide intermediate, instead of being protonated, is allowed to react with formaldehyde. The β-oxido ylide and formaldehyde react to give, on warming, an allylic alcohol. Entry 12 in Scheme 2.12 is an example of this reaction. The reaction is valuable for the stereoselective synthesis of *Z*-allylic alcohols from aldehydes.[78]

$$\underset{\text{betaine}}{\text{RCH(O}^-\text{)CH(R')}\overset{+}{\text{P}}\text{Ph}_3} \xrightarrow[-25^\circ\text{C}]{\text{RLi}} \underset{\beta\text{-oxido ylide}}{\text{RCH(O}^-\text{)C(R')=PPh}_3} \xrightarrow[2)\ 25^\circ\text{C}]{1)\ \text{CH}_2\text{=O}} \text{RCH=C(R')CH}_2\text{OH}$$

The Wittig reaction can be extended to functionalized ylides.[79] Methoxymethylene and phenoxymethylene ylides lead to vinyl ethers, which can be hydrolyzed to aldehydes.[80]

$$\text{bicyclic ketone (}=\text{O), OCH}_2\text{OCH}_2\text{CH}_2\text{OCH}_3 \xrightarrow{\text{Ph}_3\text{P=CHOMe}} \text{bicyclic =CHOCH}_3\text{, OCH}_2\text{OCH}_2\text{CH}_2\text{OCH}_3$$

Ref. 81

Methyl ketones have been prepared by an analogous reaction.

$$\text{CH}_3(\text{CH}_2)_5\text{CHO} + \text{CH}_3\text{OC(CH}_3\text{)=PPh}_3 \xrightarrow[-40^\circ\text{C}]{\text{DME}} \text{CH}_3(\text{CH}_2)_5\text{CH=C(CH}_3\text{)OCH}_3 \xrightarrow[\text{CH}_3\text{OH},\Delta]{\text{H}_2\text{O,HCl}} \underset{(57\%)}{\text{CH}_3(\text{CH}_2)_5\text{CH}_2\text{C(=O)CH}_3}$$

Ref. 82

78. E. J. Corey and H. Yamamoto, *J. Am. Chem. Soc.* **92**, 226 (1970); E. J. Corey, H. Yamamoto, D. K. Herron, and K. Achiwa, *J. Am. Chem. Soc.* **92**, 6635 (1970); E. J. Corey and H. Yamamoto, *J. Am. Chem. Soc.* **92**, 6636 (1970); E. J. Corey and H. Yamamoto, *J. Am. Chem. Soc.* **92**, 6637 (1970); E. J. Corey, J. I. Shulman, and H. Yamamoto, *Tetrahedron Lett.*, 447 (1970).
79. S. Warren, *Chem. Ind. (London)*, 824 (1980).
80. S. G. Levine, *J. Am. Chem. Soc.* **80**, 6150 (1958); G. Wittig, W. Boll, and K. H. Kruck, *Chem. Ber.* **95**, 2514 (1962).
81. M. Yamazaki, M. Shibasaki, and S. Ikegami, *J. Org. Chem.* **48**, 4402 (1982).
82. D. R. Coulsen, *Tetrahedron Lett.*, 3323 (1964).

Scheme 2.13. Carbonyl Olefination Using Phosphonate Carbanions

1[a] cyclohexanone + $(C_2H_5O)_2P(O)CH_2CO_2C_2H_5$ $\xrightarrow[\text{benzene}]{\text{NaH}}$ cyclohexylidene$=CHCO_2C_2H_5$ (67–77%)

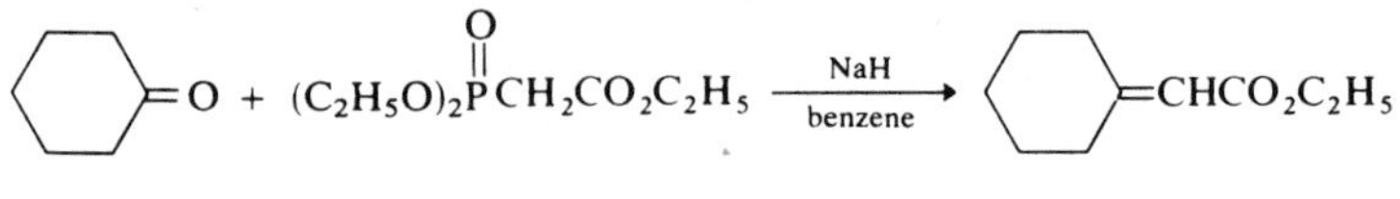

2[b] $CH_2=C(C_2H_5)CH=O$ + $(C_2H_5O)_2P(O)CH_2CO_2C_2H_5$ $\xrightarrow[\text{EtOH}]{\text{NaOEt}}$ $CH_2=C(C_2H_5)CH=CHCO_2C_2H_5$ (66%)

3[c] $C_6H_5CHO + (C_2H_5O)_2P(O)CH_2C_6H_5$ $\xrightarrow[\text{DME}]{\text{NaH}}$ *trans*-$C_6H_5CH=CHC_6H_5$ (63%)

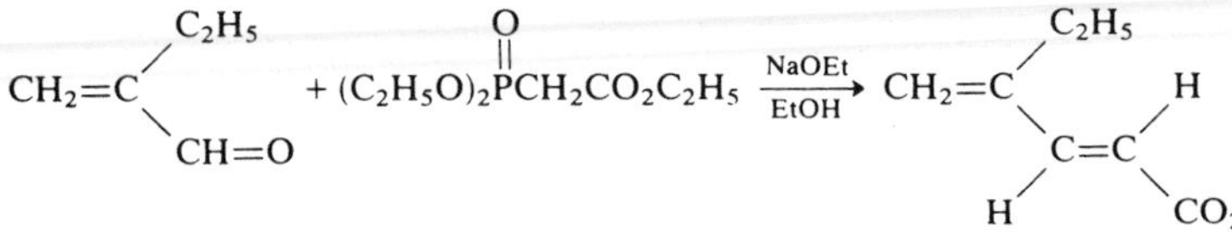

4[d] $(CH_3CH_2CH_2)_2C=O + (C_2H_5O)_2P(O)CH_2CN$ $\xrightarrow[\text{DME}]{\text{NaH}}$ $(CH_3CH_2CH_2)_2C=CHCN$ (74%)

5[e] 2-methoxycyclopentanecarbaldehyde (OCH$_3$, CHO) + $(C_2H_5O)_2P(O)CH_2C(O)(CH_2)_4CH_3$ $\xrightarrow[\text{DMSO}]{\text{NaH}}$ enone product with OCH$_3$ and $(CH_2)_4CH_3$ (55%)

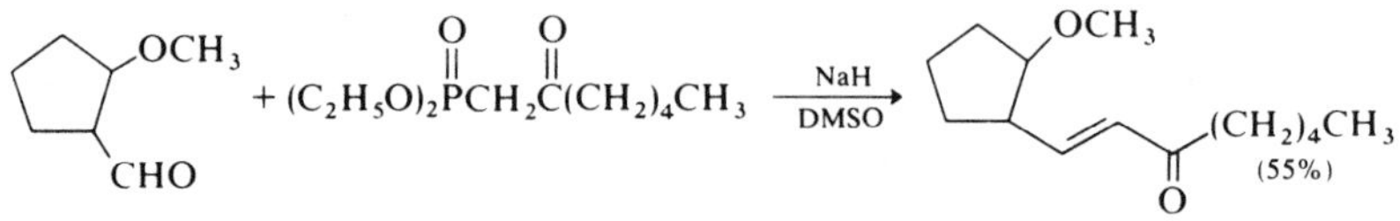

6[f] CHO / $O C(O) CH_2P(O)(OC_2H_5)_2$ $\xrightarrow[\text{benzene-THF-HMPA}]{\text{LiOCH(CH}_3)_2}$ macrocyclic unsaturated lactone (66%)

7[g] $(CH_3O)_2P(O)$ CH$_2$C(O)–, CHO, CH$_3$, lactone $\xrightarrow[\text{DME}]{\text{NaH}}$ macrocyclic dienone lactone with CH$_3$ (70%)

8[h] $PhCH_2O$, R_3SiO, $P(OCH_3)_2$ (=O) ketophosphonate + $O=CH$–diene–O_2CCH_3 $\xrightarrow[\text{CH}_3\text{CN}]{\text{LiCl, DBU, 25°C, 2 h}}$ $PhCH_2O$, R_3SiO trienone–O_2CCH_3 (70%)

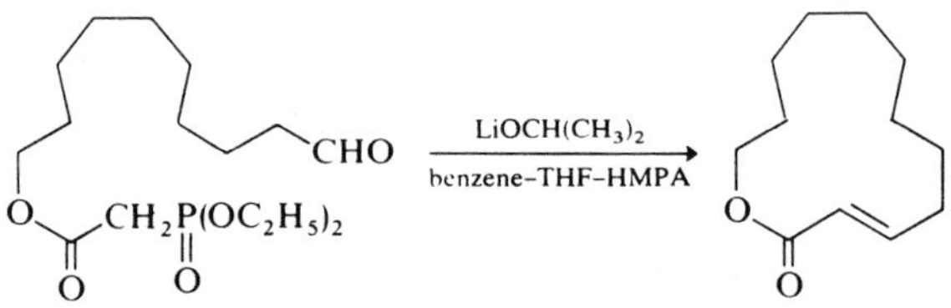

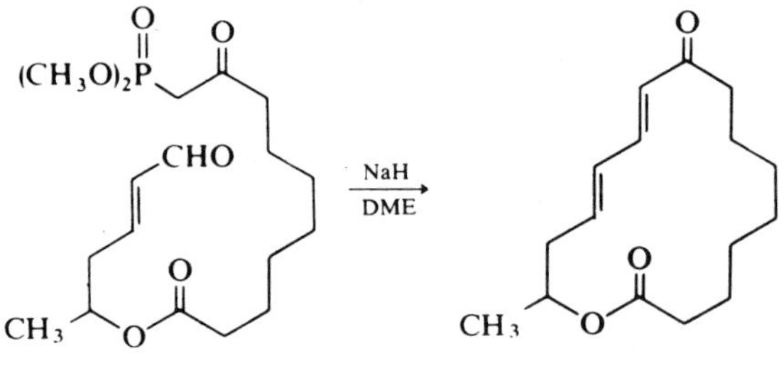

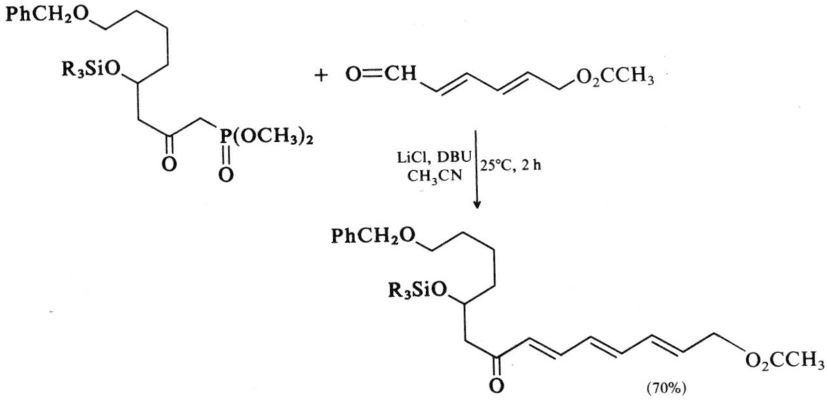

An important complement to the Wittig reaction is the reaction of phosphonate carbanions with carbonyl compounds.[83] Phosphonate carbanions are more nucleophilic than an analogous ylide, and, even when R is a carbanion-stabilizing substituent, they react readily with aldehydes and ketones to give alkenes (Scheme 2.13). Phosphonate carbanions are generated by treating alkylphosphonate esters with bases such as sodium hydride, *n*-butyllithium, or sodium ethoxide. The alkylphosphonate esters are made by the reaction of an alkyl halide, preferably primary, with a phosphite ester.

$$RCH_2X + P(OC_2H_5)_3 \longrightarrow RCH_2\overset{O}{\overset{\|}{P}}(OC_2H_5)_2 + C_2H_5X$$

$$RCH_2\overset{O}{\overset{\|}{P}}(OC_2H_5)_2 \xrightarrow{\text{base}} R\bar{C}H\overset{O}{\overset{\|}{P}}(OC_2H_5)_2$$

$$R\bar{C}H\overset{O}{\overset{\|}{P}}(OC_2H_5)_2 + R'_2C{=}O \rightarrow {}^-O\text{–}CR'_2\text{–}CHR\text{–}\overset{O}{\overset{\|}{P}}(OC_2H_5)_2 \rightarrow R'_2C{=}CHR + (C_2H_5O)_2\overset{O}{\overset{\|}{P}}{-}O^-$$

When R is an electron-attracting group, the reaction exhibits a strong preference for the formation of *trans* double bonds. When the group R is an alkyl group, the addition intermediate is formed but does not undergo elimination.

An alternative procedure for effecting the condensation of phophonates is to carry out the reaction in the presence of lithium chloride and an amine such as *N*,*N*-diisopropyl-*N*-ethylamine. The lithium chelate of the substituted phosphonate is sufficiently acidic to be deprotonated by the amine.[84]

$$(R'O)_2\overset{O}{\overset{\|}{P}}CH_2\overset{O}{\overset{\|}{C}}OR \cdot Li^+ \xrightarrow{R_3N} (R'O)_2\overset{O}{\overset{\|}{P}}CH{=}C(O^-)OR \cdot Li^+ \xrightarrow{R''CH=O} R''CH{=}CHCO_2R$$

Intramolecular Wittig reactions have been used to prepare a variety of cycloalkenes.[85]

83. For reviews of reactions of phosphonate carbanions with carbonyl compounds, see J. Boutagy and R. Thomas, *Chem. Rev.* **74**, 87 (1974); W. S. Wadsworth, Jr., *Org. React.* **25**, 73 (1977); H. Gross and I. Keitels, *Z. Chem.* **22**, 117 (1982).
84. M. A. Blanchette, W. Choy, J. T. Davis, A. P. Essenfeld, S. Masamune, W. R. Roush, and T. Sakai, *Tetrahedron Lett.* **25**, 2183 (1984).
85. K. B. Becker, *Tetrahedron* **36**, 1717 (1980).

---

a. W. S. Wadsworth, Jr. and W. D. Emmons, *Org. Synth.* **45**, 44 (1965).
b. R. J. Sundberg, P. A. Buckowick, and F. O. Holcombe, *J. Org. Chem.* **32**, 2938 (1967).
c. W. S. Wadsworth, Jr. and W. D. Emmons, *J. Am. Chem. Soc.* **83**, 1733 (1961).
d. J. A. Marshall, C. P. Hagan, and G. A. Flynn, *J. Org. Chem.* **40**, 1162 (1975).
e. N. Finch, J. J. Fitt, and I. H. S. Hsu, *J. Org. Chem.* **40**, 206 (1975).
f. G. Stork and E. Nakamura, *J. Org. Chem.* **44**, 4010 (1979).
g. K. C. Nicolaou, S. P. Seitz, M. R. Pavia, and N. A. Petasis, *J. Org. Chem.* **44**, 4010 (1979).
h. M. A. Blanchette, W. Choy, J. T. Davis, A. P. Essenfeld, S. Masamune, W. R. Roush, and T. Sakai, *Tetrahedron Lett.* **25**, 2183 (1984).

$$CH_3\overset{\displaystyle O}{\overset{\|}{C}}(CH_2)_3\overset{\displaystyle O}{\overset{\|}{C}}CH_2\overset{\displaystyle O}{\overset{\|}{P}}(OC_2H_5)_2 \xrightarrow{NaH} \text{3-methylcyclohex-2-enone}$$

Ref. 86

Intramolecular condensation of phosphonate carbanions with carbonyl groups carried out under conditions of high dilution have been utilized in macrocycle synthesis (entries 6 and 7, Scheme 2.13).

Carbanions derived from phosphine oxides also add to carbonyl compounds. The adducts are stable but undergo elimination to form alkene on heating with a base such as sodium hydride. This reaction is known as the *Horner–Wittig* reaction.

$$Ph_2\overset{O}{\overset{\|}{P}}CH_2R \xrightarrow{RLi} Ph_2\overset{O}{\overset{\|}{P}}\underset{Li}{CHR} \xrightarrow{R'CH=O} Ph_2\overset{O}{\overset{\|}{P}}\underset{R}{CH}\overset{O^-}{\underset{H}{C}}R' \rightarrow RCH{=}CHR'$$

The elimination process is *syn* so that the stereochemistry of the alkene formed depends on the stereochemistry of the adduct. It is possible to separate the two diastereomeric adducts in order to prepare the pure alkenes.

$$Ph_2\overset{O}{\overset{\|}{P}}\underset{CH_3}{CH}CH_2CH_2Ph \xrightarrow[2)\ CH_3CH=O]{1)\ BuLi} \text{two diastereomeric adducts } Ph_2P(O)C(CH_3)(CH_2CH_2Ph)CH(OH)CH_3 \text{ (separate)} \xrightarrow{NaH} CH_3(CH_2CH_2Ph)C{=}CH(CH_3) \text{ (each isomer)}$$

Ref. 87

2.5. Reactions of Carbonyl Compounds with α-Trimethylsilyl Carbanions

β-Hydroxyalkyltrimethylsilanes are converted to alkenes in either acidic or basic solution.[88] These eliminations provided the basis for a synthesis of alkenes which begins with the nucleophilic addition of α-trimethylsilyl-substituted carbanion to an aldehyde or ketone.

86. P. A. Grieco and C. S. Pogonowski, *Synthesis* 425 (1973).
87. A. D. Buss and S. Warren, *Tetrahedron Lett.* **24**, 111, 3931 (1983); A. D. Buss and S. Warren, *J. Chem. Soc., Perkin Trans. 1*, 2307 (1985).
88. P. F. Hudrlik and D. Peterson, *J. Am. Chem. Soc.* **97**, 1464 (1975).

$$Me_3Si-\overset{|}{\underset{|}{C}}-\overset{|}{\underset{OH}{C}}- \xrightarrow{HY} Me_3Si-\overset{|}{\underset{|}{C}}-\overset{|}{\underset{^+OH_2}{C}}-\;(^-Y) \rightarrow Me_3SiY + \text{>C=C<} + H_2O$$

$$Me_3Si-\overset{|}{\underset{|}{C}}-\overset{OH}{\underset{|}{C}}- \xrightarrow{base} Me_3Si-\overset{|}{\underset{|}{C}}-\overset{O^-}{\underset{|}{C}}- \rightarrow Me_3SiO^- + \text{>C=C<}$$

For example, the Grignard reagent derived from chloromethyltrimethylsilane adds to an aldehyde or ketone and the intermediate can be converted to a terminal alkene by base.[89]

$$Me_3SiCH_2X \xrightarrow{n\text{-BuLi}} Me_3Si\underset{Li}{\underset{|}{C}}HX \xrightarrow{R_2C=O} R_2C{=}CHX$$

Similarly, organolithium reagents of the type $(CH_3)_3SiCH(Li)X$, where X is a carbanion-stabilizing substituent, can be prepared by deprotonation of $(CH_3)_3SiCH_2X$ with *n*-butyllithium. These reagents usually react with aldehydes and ketones to give substituted alkenes directly. No separate elimination step is necessary since fragmentation of the intermediate occurs spontaneously under the reaction conditions. Several examples of synthesis of substituted alkenes in this way are given in Scheme 2.14.

2.6. Sulfur Ylides and Related Nucleophiles

Sulfur ylides are next to phosphorus ylides in importance as synthetic reagents.[90] Dimethylsulfonium methylide and dimethylsulfoxonium methylide are especially useful.[91] These sulfur ylides are prepared by deprotonation of the corresponding sulfonium salts, both of which are commercially available.

$$(CH_3)_2\overset{+}{S}CH_3\ I^- \xrightarrow[DMSO]{NaCH_2\overset{O}{\overset{||}{S}}CH_3} (CH_3)_2\overset{+}{S}-CH_2^-$$

dimethylsulfonium methylide

$$(CH_3)_2\underset{+}{\overset{O}{\overset{||}{S}}}CH_3\ I^- \xrightarrow[DMSO]{NaH} (CH_3)_2\underset{+}{\overset{O}{\overset{||}{S}}}-CH_2^-$$

dimethylsulfoxonium methylide

There is an important difference between the reactions of these sulfur ylides and those of phosphorus ylides. Whereas phosphorus ylides normally react with carbonyl compounds to give alkenes, dimethylsulfonium methylide and dimethylsulfoxonium methylide yield epoxides. Instead of a four-center elimination, the adducts

89. D. J. Peterson, *J. Org. Chem.* **33**, 780 (1968).
90. B. M. Trost and L. S. Melvin, Jr., *Sulfur Ylides*, Academic Press, New York, 1975; E. Block, *Reactions of Organosulfur Compounds*, Academic Press, New York, 1978.
91. E. J. Corey and M. Chaykovsky, *J. Am. Chem. Soc.* **87**, 1353 (1965).

Scheme 2.14. Carbonyl Olefination Using Trimethylsilyl-Substituted Organolithium Reagents

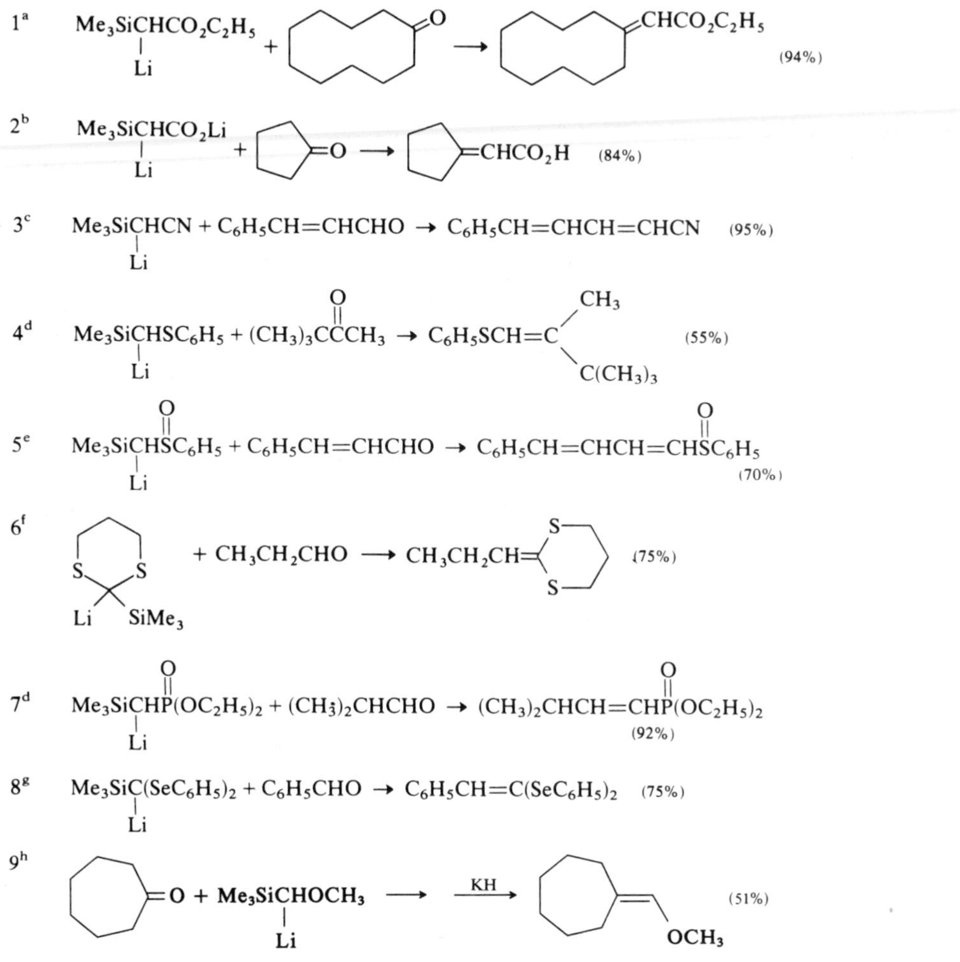

1[a] $Me_3SiCH(Li)CO_2C_2H_5$ + cyclododecanone → cyclododecylidene$=CHCO_2C_2H_5$ (94%)

2[b] $Me_3SiCH(Li)CO_2Li$ + cyclopentanone ($=O$) → cyclopentylidene$=CHCO_2H$ (84%)

3[c] $Me_3SiCH(Li)CN + C_6H_5CH{=}CHCHO \rightarrow C_6H_5CH{=}CHCH{=}CHCN$ (95%)

4[d] $Me_3SiCH(Li)SC_6H_5 + (CH_3)_3CC(=O)CH_3 \rightarrow C_6H_5SCH{=}C(CH_3)C(CH_3)_3$ (55%)

5[e] $Me_3SiCH(Li)S(=O)C_6H_5 + C_6H_5CH{=}CHCHO \rightarrow C_6H_5CH{=}CHCH{=}CHS(=O)C_6H_5$ (70%)

6[f] 2-Li-2-$SiMe_3$-1,3-dithiane $+ CH_3CH_2CHO \longrightarrow CH_3CH_2CH{=}$(1,3-dithian-2-ylidene) (75%)

7[d] $Me_3SiCH(Li)P(=O)(OC_2H_5)_2 + (CH_3)_2CHCHO \rightarrow (CH_3)_2CHCH{=}CHP(=O)(OC_2H_5)_2$ (92%)

8[g] $Me_3SiC(Li)(SeC_6H_5)_2 + C_6H_5CHO \rightarrow C_6H_5CH{=}C(SeC_6H_5)_2$ (75%)

9[h] cycloheptanone ($=O$) + $Me_3SiCH(Li)OCH_3$ ⟶ $\xrightarrow{KH}$ cycloheptylidene$=CHOCH_3$ (51%)

a. K. Shimoji, H. Taguchi, K. Oshima, H. Yamanoto, and H. Hozaki, *J. Am. Chem. Soc.* **96**, 1620 (1974).
b. P. A. Grieco, C. L. J. Wang, and S. D. Burke, *J. Chem. Soc. Chem. Commun.*, 537 (1975).
c. I. Matsuda, S. Murata, and Y. Ishii, *J. Chem. Soc. Perkin Trans. 1*, 26 (1979).
d. F. A. Carey and A. S. Court, *J. Org. Chem.* **37**, 939 (1972).
e. F. A. Carey and O. Hernandez, *J. Org. Chem.* **38**, 2670 (1973).
f. D. Seebach, M. Kolb, and B.-T. Grobel, *Chem. Ber.* **106**, 2277 (1973).
g. B.-T. Grobel and D. Seebach, *Chem. Ber.* **110**, 852 (1977).
h. P. Magnus and G. Roy, *Organometallics* **1**, 553 (1982).

formed from the sulfur ylides undergo intramolecular displacement of the sulfur substituent by oxygen.

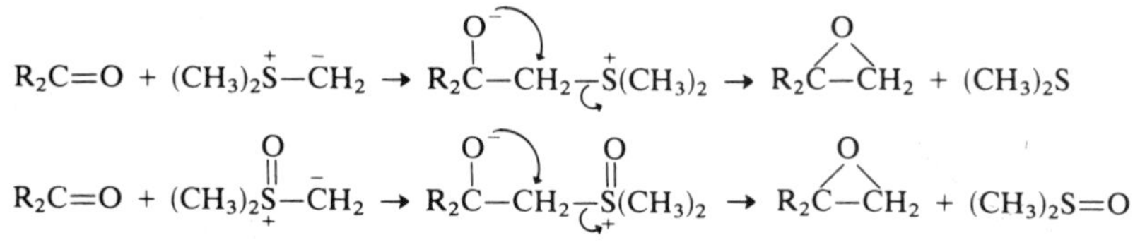

$$R_2C{=}O + (CH_3)_2\overset{+}{S}{-}\overset{-}{C}H_2 \rightarrow R_2C(O^-){-}CH_2{-}\overset{+}{S}(CH_3)_2 \rightarrow R_2\overset{O}{C{-}C}H_2 + (CH_3)_2S$$

$$R_2C{=}O + (CH_3)_2\overset{+}{S}(=O){-}\overset{-}{C}H_2 \rightarrow R_2C(O^-){-}CH_2{-}\overset{+}{S}(=O)(CH_3)_2 \rightarrow R_2\overset{O}{C{-}C}H_2 + (CH_3)_2S{=}O$$

Examples of the use of dimethylsulfonium methylide and dimethylsulfoxonium methylide in the preparation of epoxides are listed in Scheme 2.15. Entries 1-4 illustrate epoxide formation with simple aldehydes and ketones.

Dimethylsulfonium methylide is both more reactive and less stable than dimethylsulfoxonium methylide, so it is generated and used at a lower temperature. A sharp distinction between the two ylides emerges in their reactions with α,β-unsaturated carbonyl compounds. Dimethylsulfonium methylide yields epoxides, while dimethylsulfoxonium methylide reacts by conjugate addition and gives cyclopropanes (entries 5 and 6). It appears that the reason for the difference in their behavior lies in the relative rates of the two reactions available to the betaine intermediate: (a) reversal to starting materials or (b) intramolecular nucleophilic displacement.[92] Presumably both reagents react most rapidly at the carbonyl group. In the case of dimethylsulfonium methylide, the intramolecular displacement step is faster than the reverse of the addition, and epoxide formation takes place.

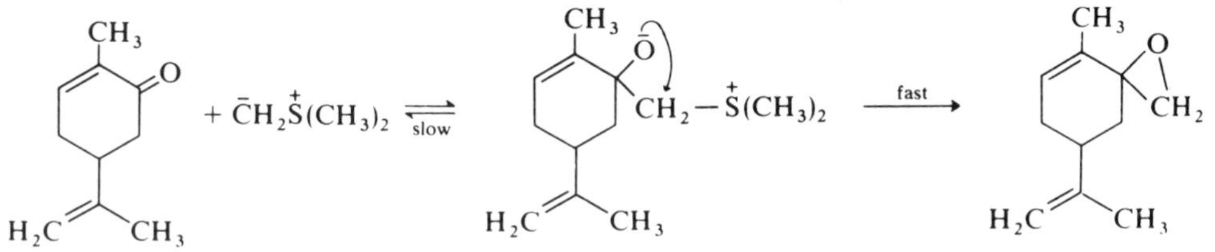

With the more stable dimethylsulfoxonium methylide, the reversal is relatively more rapid, and product formation takes place only after conjugate addition.

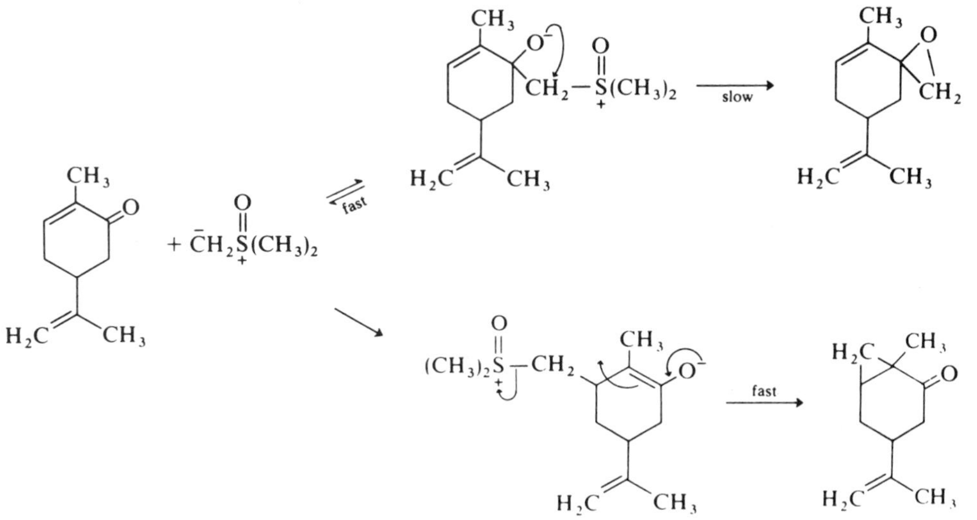

Another difference between dimethylsulfonium methylide and dimethylsulfoxonium methylide concerns the stereoselectivity in formation of epoxides from

92. C. R. Johnson, C. W. Schroeck, and J. R. Shanklin, *J. Am. Chem. Soc.* **95**, 7424 (1973).

Scheme 2.15. Reactions of Sulfur Ylides

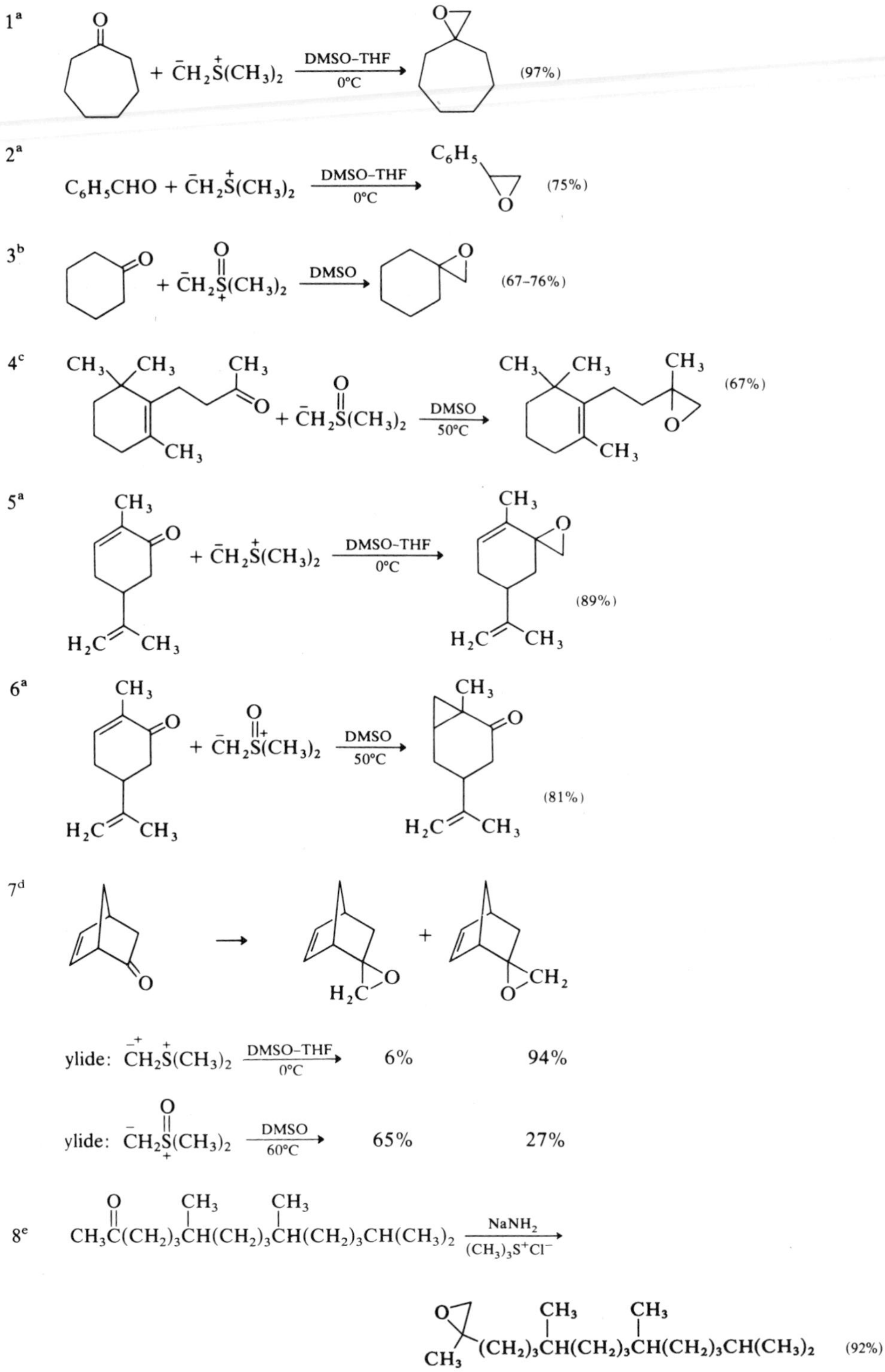

1[a] Cycloheptanone + $\bar{C}H_2\overset{+}{S}(CH_3)_2$ $\xrightarrow[0°C]{\text{DMSO–THF}}$ spiro-epoxide (97%)

2[a] $C_6H_5CHO + \bar{C}H_2\overset{+}{S}(CH_3)_2$ $\xrightarrow[0°C]{\text{DMSO–THF}}$ C_6H_5-oxirane (75%)

3[b] Cyclohexanone + $\bar{C}H_2\overset{+}{S}(O)(CH_3)_2$ $\xrightarrow{\text{DMSO}}$ spiro-epoxide (67–76%)

4[c] CH_3, CH_3, CH_3 ketone, CH_3 + $\bar{C}H_2\overset{+}{S}(O)(CH_3)_2$ $\xrightarrow[50°C]{\text{DMSO}}$ CH_3, CH_3, CH_3 epoxide, CH_3 (67%)

5[a] CH_3 enone, H_2C=C–CH_3 + $\bar{C}H_2\overset{+}{S}(CH_3)_2$ $\xrightarrow[0°C]{\text{DMSO–THF}}$ CH_3 spiro-epoxide, H_2C=C–CH_3 (89%)

6[a] CH_3 enone, H_2C=C–CH_3 + $\bar{C}H_2\overset{+}{S}(O)(CH_3)_2$ $\xrightarrow[50°C]{\text{DMSO}}$ CH_3 cyclopropyl ketone, H_2C=C–CH_3 (81%)

7[d] Bicyclic ketone → epoxide (H_2C–O) + epoxide (O–CH_2)

ylide: $\bar{C}H_2\overset{+}{S}(CH_3)_2$ $\xrightarrow[0°C]{\text{DMSO–THF}}$ 6% 94%

ylide: $\bar{C}H_2\overset{+}{S}(O)(CH_3)_2$ $\xrightarrow[60°C]{\text{DMSO}}$ 65% 27%

8[e] $CH_3C(O)(CH_2)_3CH(CH_3)(CH_2)_3CH(CH_3)(CH_2)_3CH(CH_3)_2$ $\xrightarrow[(CH_3)_3S^+Cl^-]{NaNH_2}$

CH_3-oxirane-$(CH_2)_3CH(CH_3)(CH_2)_3CH(CH_3)(CH_2)_3CH(CH_3)_2$ (92%)

9[i] $(CH_3)_3S^+Cl^-$ / NaOH (87%)

10[g] $+ (CH_3)_2\bar{C}\overset{+}{S}Ph_2$ DME, −50°C (82%)

11[h] CH_3...$CO_2CH_3 + (CH_3)_2\bar{C}\overset{+}{S}Ph_2$ DME, −20°C (72%)

12[i] $+ \triangleright\!\!-\overset{+}{S}Ph_2$ DMSO, 25°C (75%)

13[j] $CH_3\overset{O}{\overset{\|}{C}}(CH_2)_5CH_3 + \triangleright\!\!-\overset{+}{S}Ph_2$ DMSO, 25°C; CH_3, $(CH_2)_5CH_3$ (92%)

a. E. J. Corey and M. Chaykovsky, *J. Am. Chem. Soc.* **87**, 1353 (1965).
b. E. J. Corey and M. Chaykovsky, *Org. Synth.* **49**, 78 (1969).
c. M. G. Fracheboud, O. Shimomura, R. K. Hill, and F. H. Johnson, *Tetrahedron Lett.*, 3951 (1969).
d. R. S. Bly, C. M. DuBose, Jr., and G. B. Konizer, *J. Org. Chem.* **33**, 2188 (1968).
e. G. L. Olson, H.-C. Cheung, K. Morgan, and G. Saucy, *J. Org. Chem.* **45**, 803 (1980).
f. M. Rosenberger, W. Jackson, and G. Saucy, *Helv. Chim. Acta* **63**, 1665 (1980).
g. E. J. Corey, M. Jauetlat, and W. Oppolzer, *Tetrahedron Lett.*, 2325 (1967).
h. E. J. Corey and M. Jautelat, *J. Am. Chem. Soc.* **89**, 3112 (1967).
i. B. M. Trost and M. J. Bogdanowicz, *J. Am. Chem. Soc.* **95**, 5307 (1973).
j. B. M. Trost and M. J. Bogdanowicz, *J. Am. Chem. Soc.* **95**, 5311 (1973).

cyclohexanones. Dimethylsulfonium methylide usually adds from the axial direction while dimethylsulfoxonium methylide favors the equatorial direction. This result may also be due to reversibility of addition in the case of the sulfoxonium methylide.[92] The product from the sulfonium ylide would be the result of the kinetic preference for axial addition by small nucleophiles (see Part A, Section 3.10). In the case of reversible addition of the sulfoxonium ylide, product structure would be determined by the rate of displacement, and this may be faster for the more stable epoxide.

$(CH_3)_3C$–cyclohexanone → $(CH_3)_3C$–(epoxide, H_2C axial) + $(CH_3)_3C$–(epoxide, CH_2 equatorial)

ylide: $\bar{C}H_2\overset{+}{S}(CH_3)_2$	THF, 0°C	83%	17%	Ref. 91
ylide: $\bar{C}H_2\overset{O}{\overset{\|}{S}}(CH_3)_2$	THF, 65°C	not formed	only product	

Sulfur ylides are also available which allow transfer of substituted methylene units, such as isopropylidene (Scheme 2.15, entries 10 and 11) or cyclopropylidene (entries 12 and 13). The oxaspiropentanes formed by reaction of aldehydes and ketones with diphenylsulfonium cyclopropylide are useful intermediates in a number of transformations such as acid-catalyzed rearrangement to cyclobutanones.[93]

$$\text{CH}_3(\text{CH}_2)_5\text{CH}_3\text{-oxaspiropentane} \xrightarrow{\text{H}^+} \text{2-CH}_3\text{-2-(CH}_2)_5\text{CH}_3\text{-cyclobutanone} \quad (92\%)$$

Aside from the methylide and cyclopropylide reagents, the sulfonium ylides are not very stable. A related group of reagents derived from sulfoximines offer greater versatility in alkylidene transfer reactions.[94] The preparation and use of this class of ylides is illustrated by the sequence

$$\text{Ar}\overset{\text{O}}{\overset{\|}{\text{S}}}\text{CH}_2\text{CH}_3 \xrightarrow[\text{H}_2\text{SO}_4,\ \text{CHCl}_3]{\text{NaN}_3} \text{Ar}\overset{\text{O}}{\overset{\|}{\text{S}}}(=\text{NH})\text{CH}_2\text{CH}_3 \xrightarrow{(\text{CH}_3)_3\text{O}^+\text{BF}_4^-} \text{Ar}\overset{\text{O}}{\overset{\|}{\text{S}}}{}^+(\text{N}(\text{CH}_3)_2)\text{CH}_2\text{CH}_3\ \text{BF}_4^-$$

$$\xrightarrow{\text{NaH, DMF}} \text{Ar}\overset{\text{O}}{\overset{\|}{\text{S}}}{}^+(\text{N}(\text{CH}_3)_2)-\bar{\text{C}}\text{HCH}_3 \xrightarrow{\text{C}_6\text{H}_5\text{CHO}} \text{C}_6\text{H}_5\text{CH}\overset{\text{O}}{—}\text{CHCH}_3 \quad (67\%)$$

$$\text{Ar} = p\text{-CH}_3\text{C}_6\text{H}_4—$$

A similar pattern of reactivity has been demonstrated for the anions formed by deprotonation of *S*-alkyl-*N*-*p*-toluenesulfoximines.[95]

$$\text{CH}_3\text{C}_6\text{H}_4-\overset{\text{O}}{\overset{\|}{\text{S}}}{}^+(\text{NMe}_2)-\bar{\text{C}}\text{XY} \qquad \text{CH}_3\text{C}_6\text{H}_4-\overset{\text{O}}{\overset{\|}{\text{S}}}(=\text{NTs})-\bar{\text{C}}\text{XY}$$

dimethylaminooxosulfonium ylide

N-tosylsulfoximine anion

The sulfur atom in both these type of reagents is chiral. These reagents have been utilized in the preparation of enantiomerically enriched epoxides and cyclopropanes.[96]

93. B. M. Trost and M. H. Bogdanowicz, *J. Am. Chem. Soc.* **95**, 5321 (1973).
94. C. R. Johnson, *Acc. Chem. Res.* **6**, 341 (1973).
95. C. R. Johnson, R. A. Kirchoff, R. J. Reischer, and G. F. Katekar, *J. Am. Chem. Soc.* **95**, 4287 (1973).
96. C. R. Johnson and E. R. Janiga, *J. Am. Chem. Soc.* **95**, 7673 (1973).

The pattern of nucleophilic addition at a carbonyl group followed by intramolecular nucleophilic displacement of a leaving group present in the nucleophile can also be recognized in a much older synthetic technique, the *Darzens reaction*.[97] The first step in the reaction is addition of the enolate of an α-haloester to the carbonyl compound. The alkoxide oxygen formed in the addition then effects nucleophilic attack, displacing the halide and forming an α,β-epoxy ester (also called a glycidic ester).

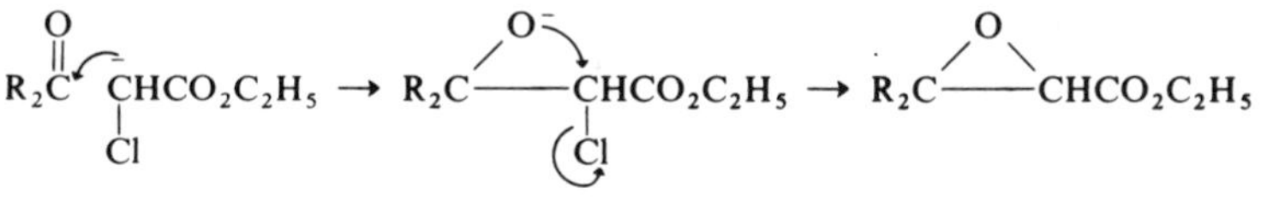

$$R_2C{=}O + {}^-CH(Cl)CO_2C_2H_5 \rightarrow R_2C(O^-)\text{—}CH(Cl)CO_2C_2H_5 \rightarrow R_2C\overset{O}{\text{—}}CHCO_2C_2H_5$$

Scheme 2.16 gives some examples of the Darzens reaction.

Scheme 2.16. Darzens Condensation Reactions

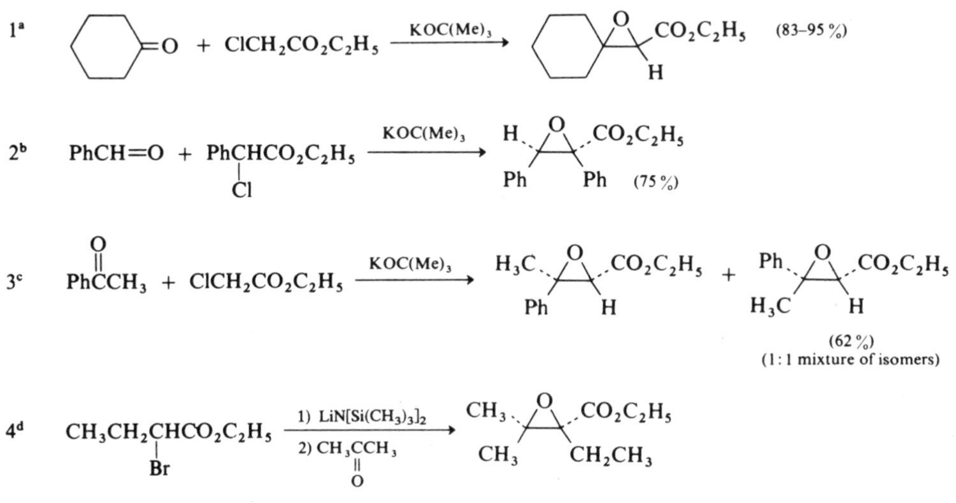

1[a] Cyclohexanone + $ClCH_2CO_2C_2H_5$ $\xrightarrow{KOC(Me)_3}$ spiro epoxide (cyclohexane ring, $CO_2C_2H_5$, H) (83–95%)

2[b] $PhCH{=}O$ + $PhCH(Cl)CO_2C_2H_5$ $\xrightarrow{KOC(Me)_3}$ epoxide (H, $CO_2C_2H_5$; Ph, Ph) (75%)

3[c] $PhC(O)CH_3$ + $ClCH_2CO_2C_2H_5$ $\xrightarrow{KOC(Me)_3}$ epoxide (H_3C, $CO_2C_2H_5$; Ph, H) + epoxide (Ph, $CO_2C_2H_5$; H_3C, H) (62%) (1:1 mixture of isomers)

4[d] $CH_3CH_2CH(Br)CO_2C_2H_5$ $\xrightarrow[2)\ CH_3C(O)CH_3]{1)\ LiN[Si(CH_3)_3]_2}$ epoxide (CH_3, $CO_2C_2H_5$; CH_3, CH_2CH_3)

a. R. H. Hunt, L. J. Chinn, and W. S. Johnson, *Org. Synth.* **IV**, 459 (1963).
b. H. E. Zimmerman and L. Ahramjian, *J. Am. Chem. Soc.* **82**, 5459 (1960).
c. F. W. Bachelor and R. K. Bansal, *J. Org. Chem.* **34**, 3600 (1969).
d. R. F. Borch, *Tetrahedron Lett.*, 3761 (1972).

Trimethylsilylepoxides can be prepared by an addition–cyclization process. Reaction of chloromethyltrimethylsilane with *s*-butyllithium at very low temperature gives an α-chloro lithium reagent that gives an epoxide on reaction with an aldehyde or ketone.[98]

97. M. S. Newman and B. J. Magerlein, *Org. React.* **5**, 413 (1951).
98. C. Burford, F. Cooke, E. Ehlinger, and P. D. Magnus, *J. Am. Chem. Soc.* **99**, 4536 (1977).

$$Me_3SiCH_2Cl \xrightarrow[\text{THF, -78°C}]{s\text{-BuLi}} Me_3Si\underset{\underset{Li}{|}}{C}HCl$$

$$Me_3Si\underset{\underset{Li}{|}}{C}HCl + CH_3CH_2CH_2CHO \rightarrow CH_3CH_2CH_2\underset{\underset{O^-}{|}}{C}H-\overset{\overset{Cl}{|}}{C}HSiMe_3 \rightarrow CH_3CH_2CH_2CH\overset{O}{\text{——}}CHSiMe_3$$

General References

Aldol Condensations

D. A. Evans, J. V. Nelson, and T. R. Taber, *Top. Stereochem.* **13**, 1 (1982).
C. H. Heathcock, in *Comprehensive Carbanion Chemistry*, E. Buncel and T. Durst (ed.), Elsevier, Amsterdam, 1984.
C. H. Heathcock, in *Asymmetric Synthesis*, Vol 3. J. D. Morrison (ed.), Academic Press, New York, 1984.
S. Masamune, W. Choy, J. S. Petersen, and L. R. Sita, *Angew. Chem. Int. Ed. Engl.* **24**, 1 (1985).
A. T. Nielsen and W. T. Houlihan, *Org. React.* **16**, 1 (1968).
T. Mukaiyama, *Org. React.* **28**, 203 (1982).

Annulation Reactions

R. E. Gawley, *Synthesis*, 777 (1976).
M. E. Jung, *Tetrahedron* **32**, 3 (1976).

Mannich Reactions

F. F. Blicke, *Org. React.* **1**, 303 (1942).
H. Böhme and M. Heake, in *Iminium Salts in Organic Chemistry*, H. Böhme and H. G. Viehe (eds.), Wiley-Interscience, New York, 1976, pp. 107–223.

Phosphorus-Stabilized Ylides and Carbanions

J. Boutagy and R. Thomas, *Chem. Rev.* **74**, 87 (1974).
I. Gosney and A. G. Rowley, in *Organophosphorus Reagents in Organic Synthesis*, J. I. G. Cadogan (ed.), Academic Press, London, 1979, pp. 17–153.
A. Maercker, *Org. React.* **14**, 270 (1965).
W. S. Wadsworth, Jr., *Org. React.* **25**, 73 (1977).

Silicon-Stabilized Carbanions

D. J. Ager, *Synthesis*, 384 (1984).

Sulfur Ylides

B. M. Trost and L. S. Melvin, *Sulfur Ylides*, Academic Press, New York, 1975.

Problems

(References for these problems will be found on page 764.)

1. Predict the product formed in each of the following reactions:

(a) γ-butyrolactone + ethyl oxalate $\xrightarrow[\text{2) } H^+]{\text{1) } NaOCH_2CH_3}$

(b) 4-bromobenzaldehyde + ethyl cyanoacetate $\xrightarrow[\text{piperidine}]{\text{ethanol}}$

(c) $CH_3CH_2CH_2C(=O)CH_3$ $\xrightarrow{\text{1) } LiN(iPr)_2,\ -78°C;\ \text{2) } CH_3CH_2CHO,\ 15\ \text{min};\ \text{3) } H_2O}$

(d) 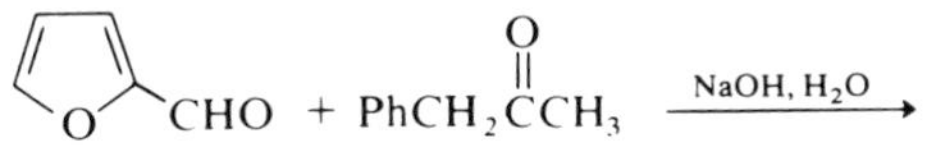 2-furaldehyde (furan-2-CHO) + $PhCH_2C(=O)CH_3$ $\xrightarrow{NaOH,\ H_2O}$

(e) $C_6H_5CH{=}C(OAc)CH_3$ $\xrightarrow{\text{1) } CH_3Li,\ 2\ \text{equiv};\ \text{2) } ZnCl_2;\ \text{3) } n\text{-}C_3H_7CHO}$

(f) 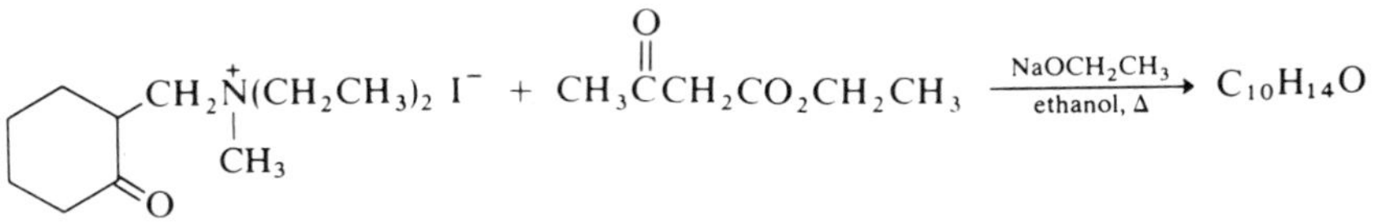 2-[$CH_2\overset{+}{N}(CH_2CH_3)_2(CH_3)$ I^-]cyclohexanone + $CH_3C(=O)CH_2CO_2CH_2CH_3$ $\xrightarrow[\text{ethanol, }\Delta]{NaOCH_2CH_3}$ $C_{10}H_{14}O$

(g) 2-methylcyclopentanone + $HCO_2CH_2CH_3$ $\xrightarrow[\text{ether}]{Na}$

(h) cyclohexyl–$C(=O)CH_3$ + $(CH_3CH_2O)_2C{=}O$ $\xrightarrow[\text{toluene}]{NaNH_2}$

(i) $C_6H_5C(=O)CH_3 + (CH_3CH_2O)_2P(=O)CH_2CN$ $\xrightarrow[\text{THF}]{NaH}$

(j) $CH_3CH_2C(=O)CH_2CH_2CO_2CH_2CH_3$ $\xrightarrow[\text{xylene}]{NaOCH_3}$

(k) 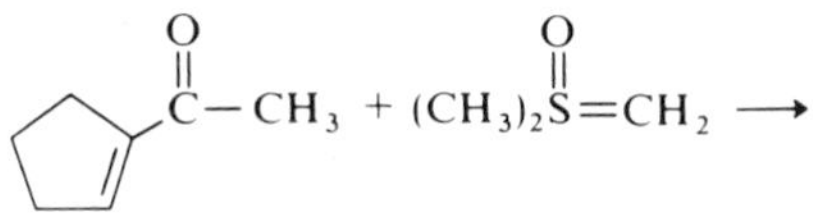 1-cyclopentenyl–$C(=O){-}CH_3$ + $(CH_3)_2S(=O){=}CH_2$ $\longrightarrow$

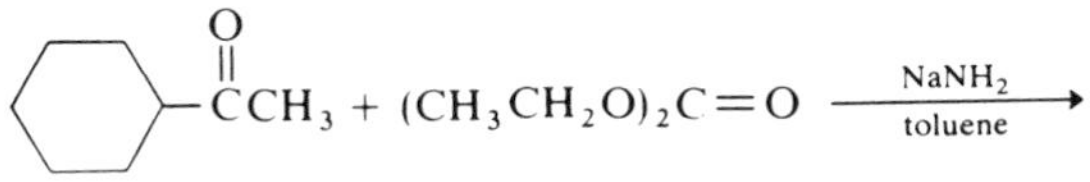

(l) 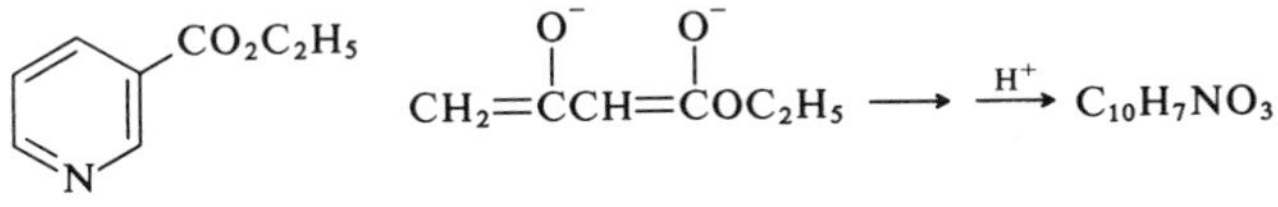 ethyl nicotinate (3-$CO_2C_2H_5$-pyridine) + $CH_2{=}C(O^-)CH{=}C(O^-)OC_2H_5$ $\longrightarrow$ $\xrightarrow{H^+}$ $C_{10}H_7NO_3$

(m) 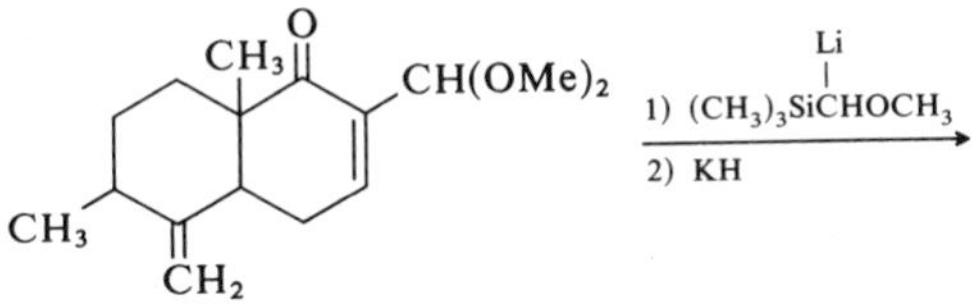 bicyclic enone bearing CH_3, C=O, $CH(OMe)_2$, CH_3, and $=CH_2$ substituents $\xrightarrow{\text{1) } (CH_3)_3SiCH(Li)OCH_3;\ \text{2) } KH}$

2. Indicate reaction conditions or a series of reactions that could effect each of the following synthetic conversations:

(a) $CH_3CO_2C(CH_3)_3 \rightarrow (CH_3)_2C(OH)CH_2CO_2C(CH_3)_3$

(b) tetrahydropyranyl–$O(CH_2)_3CH{=}O$ → tetrahydropyranyl–$O(CH_2)_3CH{=}C(CH_3)CH_2OH$

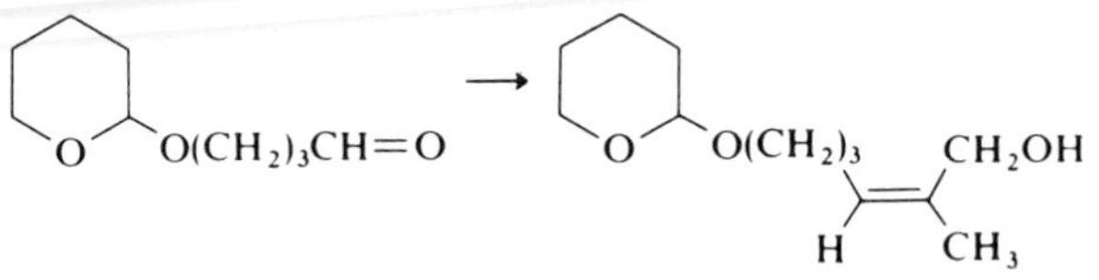

(c) cyclohexanone → 2-(CHOH)cyclohexanone

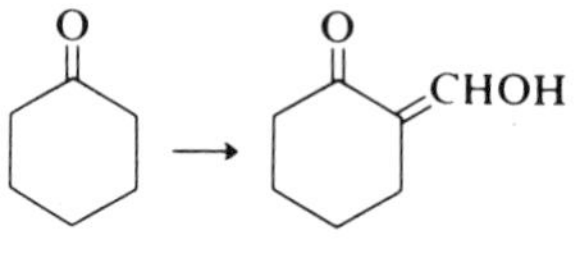

(d) $Ph_2C{=}O \rightarrow$ Ph_2-cyclopropane with CN and $CO_2C_2H_5$

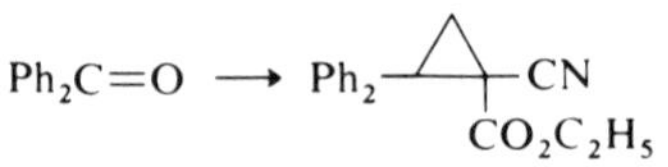

(e) cyclodecanone → 2-($CO_2C_2H_5$)cyclodecanone

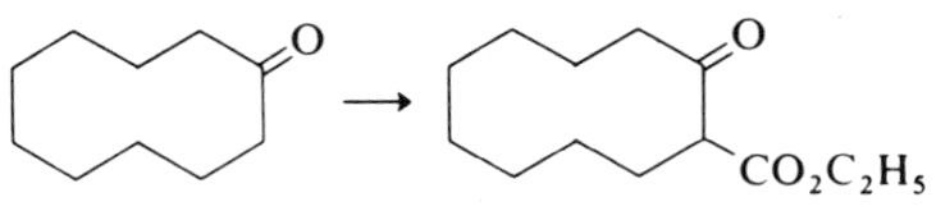

(f) bicyclic lactone (O, =O) → α-(CH_2OH) lactone

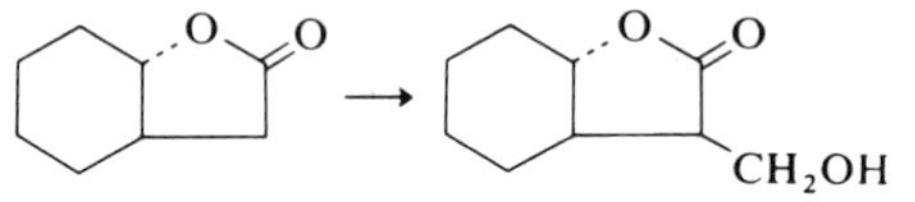

(g) γ-butyrolactone → $HO(CH_2)_3\overset{O}{\overset{\|}{C}}CH_2\overset{O}{\overset{\|}{S}}CH_3$

(h) $H_2C{=}CHCH_2CH_2\overset{O}{\overset{\|}{C}}CH_3 \rightarrow H_2C{=}CHCH_2CH_2\overset{O}{\overset{\|}{C}}CH_2CO_2C_2H_5$

$\downarrow$

$H_2C{=}CHCH_2CH_2\overset{O}{\overset{\|}{C}}CH(CO_2C_2H_5)CH_2CH_2\overset{O}{\overset{\|}{C}}CH_3$

$\leftarrow$ 2-($H_2C{=}CHCH_2$)-3-CH_3-cyclohex-2-enone

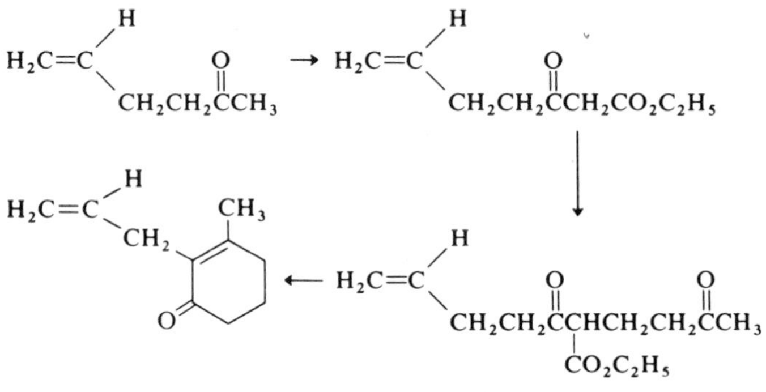

(i) $H_5C_2O_2CCH_2CH_2CO_2C_2H_5 \rightarrow$ cyclohexane-1,4-dione

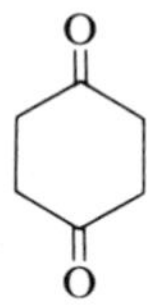

(j) O O C CH₂ C CH₃ → O O O C CH₂ C CH₂ C OCH₃

$C_6H_5C(=O)CH_2C(=O)CH_3 \longrightarrow C_6H_5C(=O)CH_2C(=O)CH_2C(=O)C_6H_4OCH_3$

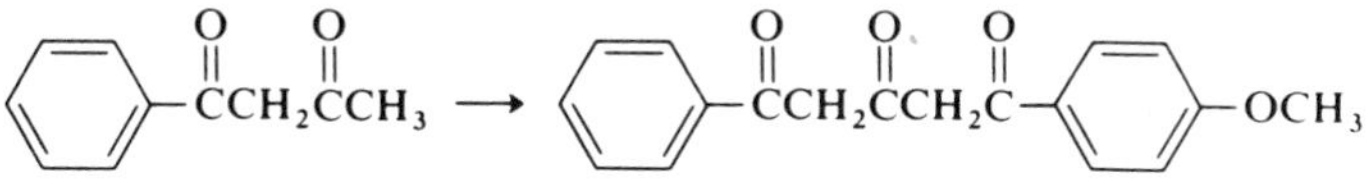

(k) $CH_3O{-}C_6H_4{-}C(=O)CH_3 \longrightarrow CH_3O{-}C_6H_4{-}C(=O)CH{=}CH_2$

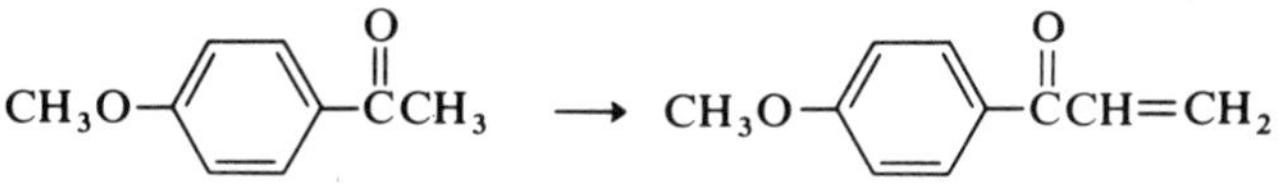

(l) CH_3O, CH_3O, $C(=O)H$ $\longrightarrow$ CH_3O, CH_3O, $CH{=}CHCH{=}CH_2$

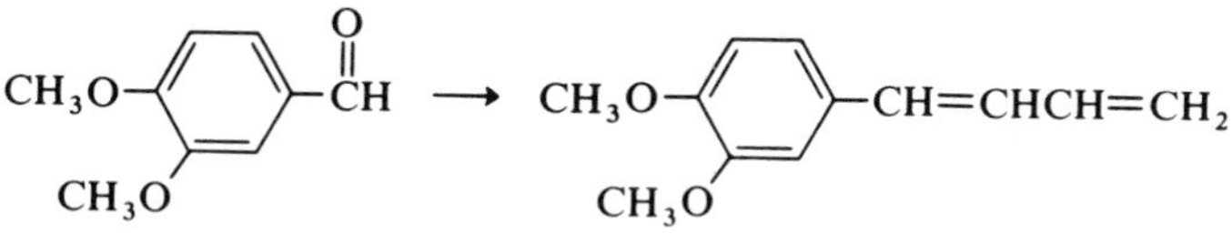

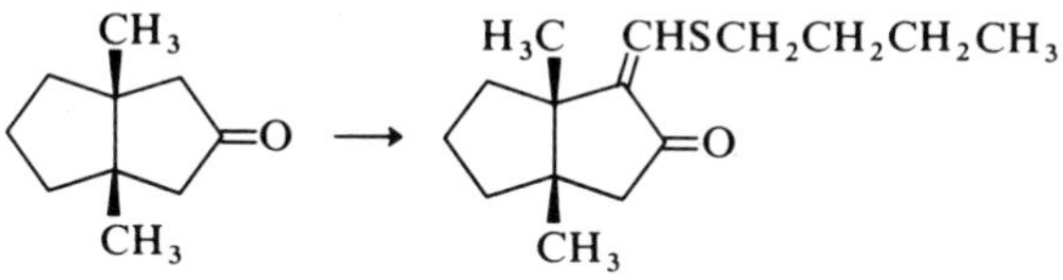

(m) CH_3, $=O$, CH_3 $\longrightarrow$ H_3C, $CHSCH_2CH_2CH_2CH_3$, $=O$, CH_3

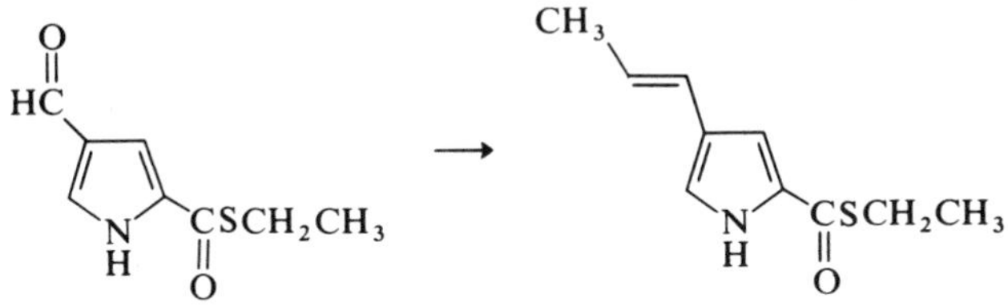

(n) O, HC, N, H, $CSCH_2CH_3$, O $\longrightarrow$ CH_3, N, H, $CSCH_2CH_3$, O

(o) CH_3, CH_3, O $\longrightarrow$ CH_3, CH_3, O

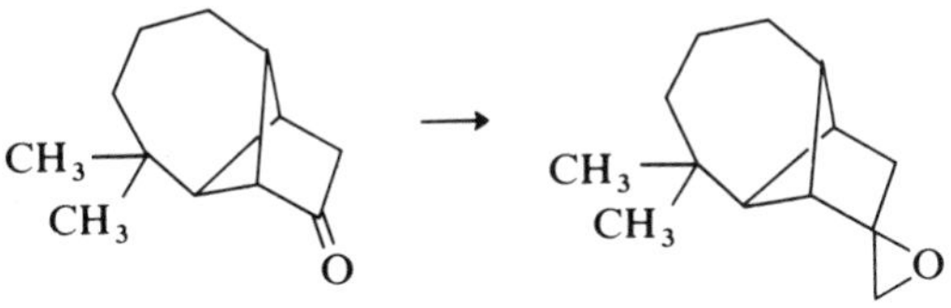

(p) CH_3, $(CH_3)_2CH$, O $\longrightarrow$ CH_3, $(CH_3)_2CH$, $CHOCH_3$

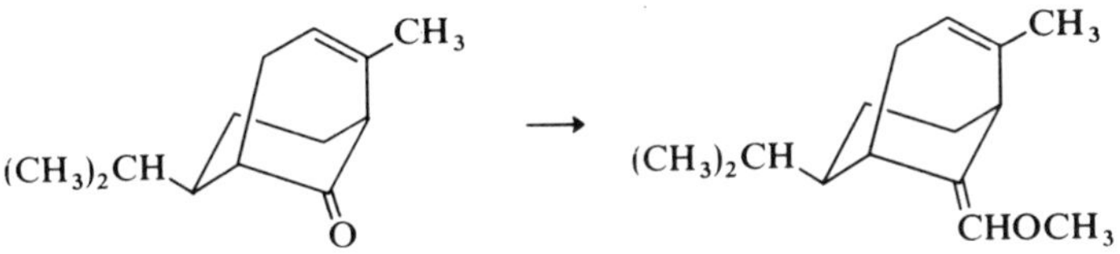

(q) $(CH_3)_3CC(=O)C(CH_3)_3 \rightarrow (CH_3)_3CC(=CH_2)C(CH_3)_3$

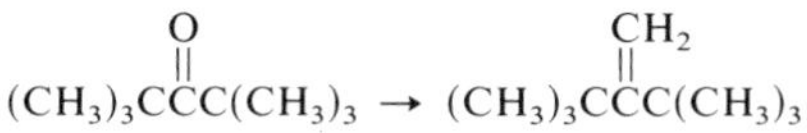

(r) H, CH=O, Ph, Ph $\longrightarrow$ H, H, H, H, Ph, Ph H, Ph

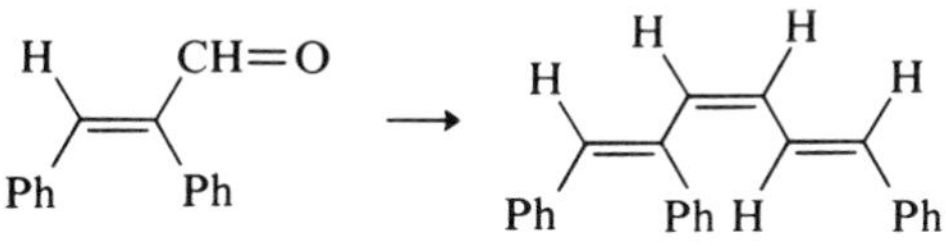

(s) $(CH_3)_2CHCH{=}O \longrightarrow$ H, CO_2CH_3, H, H, $(CH_3)_2CH$, H

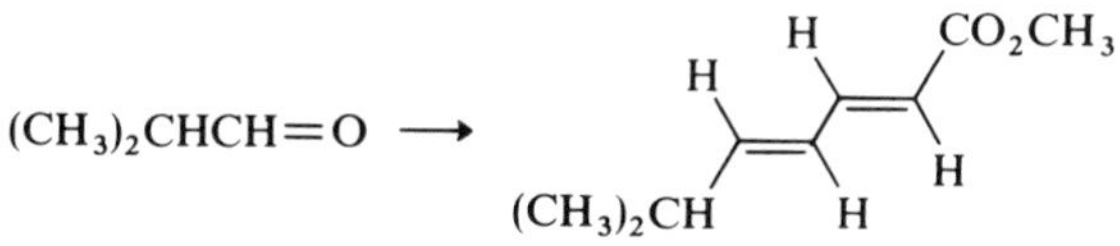

(t)

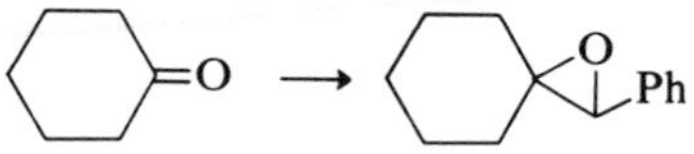

(u) 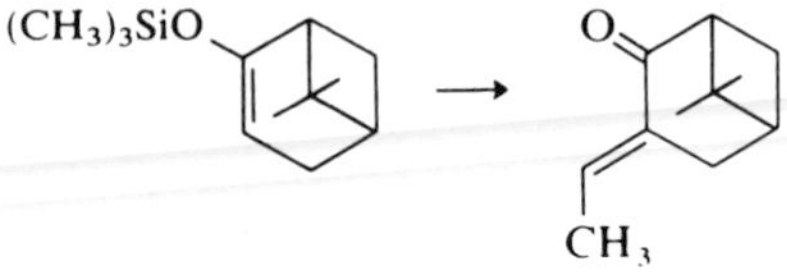 $(CH_3)_3SiO$ → CH_3

3. Step-by-step retrosynthetic analysis of each of the target molecules reveals that they can be efficiently prepared in a few steps from the starting material shown on the right. Show a retrosynthetic analysis and suggest reagents and conditions for carrying out the desired synthesis.

(a)  CH_3, $CH(CH_3)_2$ ⇒ CH_3, $CH(CH_3)_2$

(b) 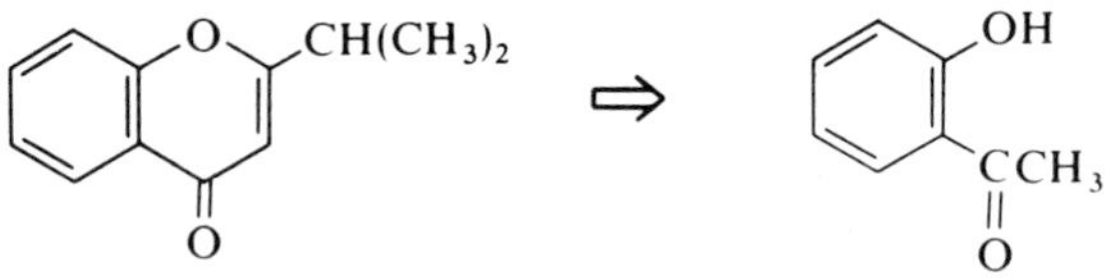 $CH(CH_3)_2$ ⇒ OH, CCH_3 (C=O)

(c) $CH_2{=}C(CH_3)CH{=}CHCH(CH(CH_3)_2)CH_2CH_2CO_2CH_2CH_3$ ⇒ $(CH_3)_2CHCH_2CH{=}O$

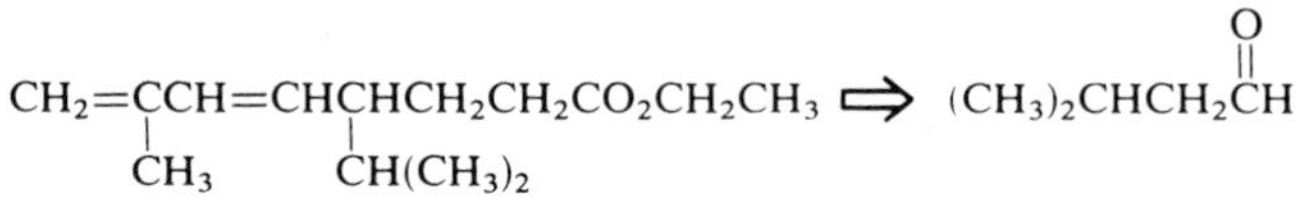

Hint: See Chapter 1, problem 8h.

(d)  C_6H_5 ⇒ $C_6H_5CH{=}CHCH{=}O$

(e) 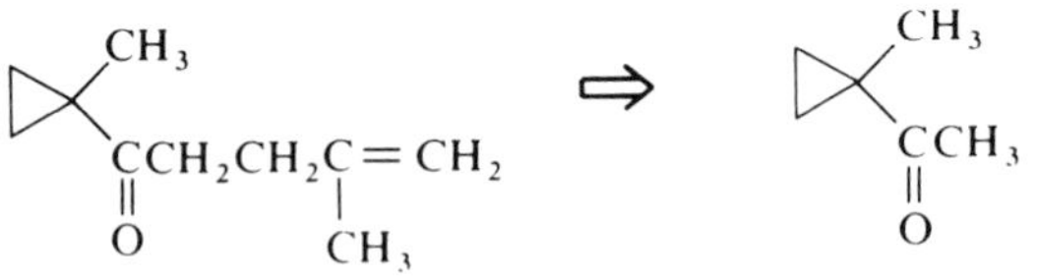 CH_3, $C(=O)CH_2CH_2C(CH_3){=}CH_2$ ⇒ CH_3, $C(=O)CH_3$

(f) 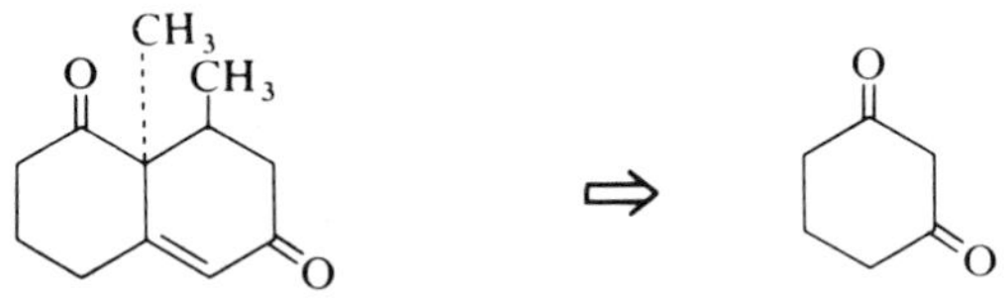 CH_3, CH_3 ⇒

(g) 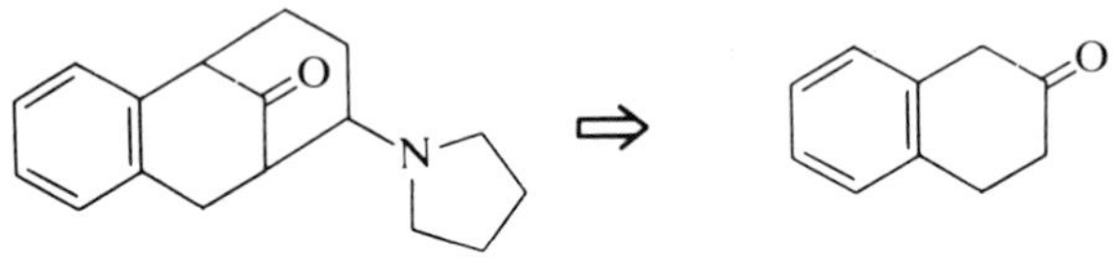 ⇒

(h) $Ph_2C{=}CHCH{=}O$ ⇨ $Ph_2C{=}O$

(i) $CH_3CH_2C(=CH_2)CH{=}CHCO_2C_2H_5$ ⇨ $CH_3CH_2CH_2CH{=}O$, $ClCH_2CO_2C_2H_5$

(j) (1-methyl-4-piperidone) ⇨ CH_3NH_2, $CH_2{=}CHCO_2C_2H_5$

(k) (4,4-dimethylcyclohex-2-enone) ⇨ $(CH_3)_2CHCH{=}O$, $CH_2{=}CHCCH_3$ (C=O)

(l) (2-bromophenyl)–$C(CH_3)(OH)–CH_2NH_2$ ⇨ 2-bromophenyl–CCH_3 (C=O)

(m) (3-oxocyclopentene with $CH_2CO_2CH_3$ and CO_2CH_3) ⇨ $ClCCH_2CH_2CCl$ (both C=O), $CH_3O_2CCH_2CO_2H$

(n) $PhCH_2N$ bicyclic (with CH_3O_2C, CH_2, CH_3O_2C) ⇨ CH_3O_2C–cyclohexanone–CO_2CH_3

4. Offer a mechanism for each of the following transformations:

(a) dimethyl phthalate (CO_2CH_3, CO_2CH_3) $\xrightarrow[\text{NaH, benzene}]{C_2H_5CC_2H_5\ (C=O)}$ 2-methylindane-1,3-dione (CH_3)

(b) CHO, OH, $CH_2P(OCH_3)_2$ (P=O), $(CH_3)_3CO$, CH_3 $\xrightarrow[t\text{-BuOH}]{\text{KO-}t\text{-Bu}}$ H, =O, $(CH_3)_3CO$, CH_3

(c) CH_3CH_2C(=O)–cyclopropyl–C(CH_3)($OCCH_3$, C=O) $\xrightarrow{\text{NaOH, MeOH}}$ 2,3-dimethylcyclopent-2-enone (CH_3, CH_3)

(d) 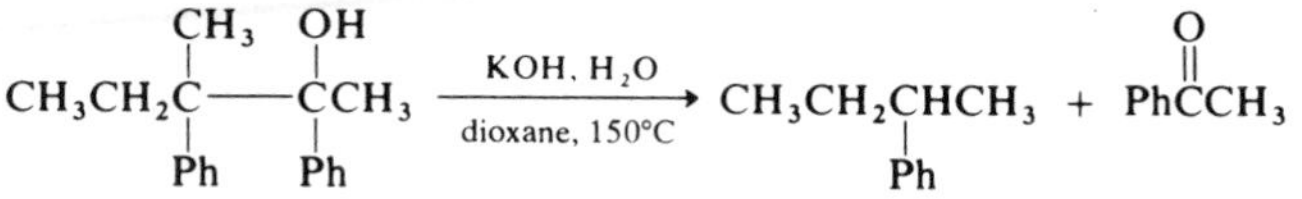 $CH_3CH_2C(CH_3)(Ph)-C(OH)(Ph)CH_3 \xrightarrow[\text{dioxane, 150°C}]{\text{KOH, H}_2\text{O}} CH_3CH_2CH(Ph)CH_3 + PhCCH_3$ (with C=O)

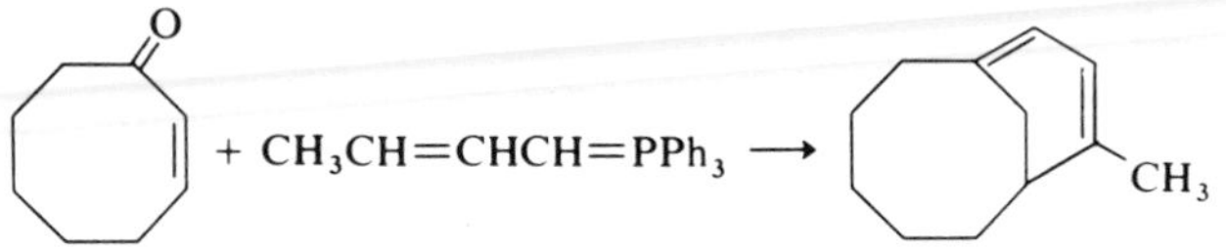

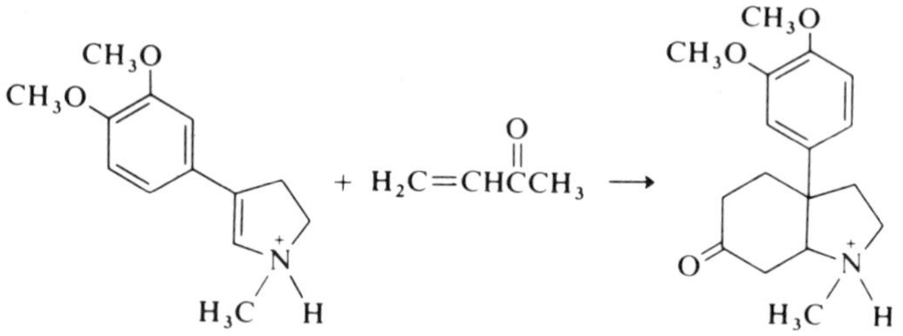

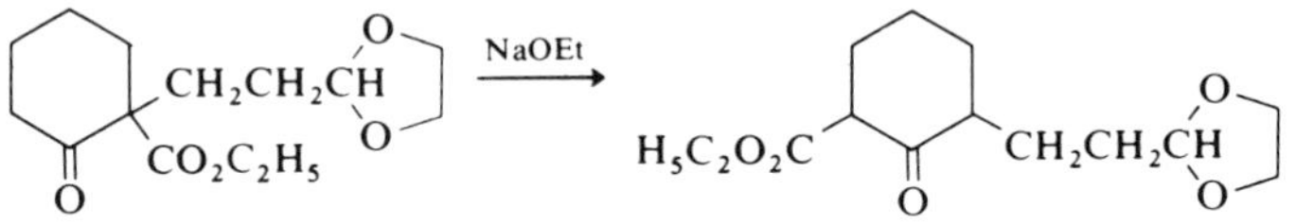

(e) + $CH_3CH{=}CHCH{=}PPh_3$ ⟶ CH_3

(f) CH_3O, CH_3O, + $H_2C{=}CHCCH_3$ (with C=O) ⟶ CH_3O, CH_3O, H_3C, H, N⁺

(g) CH_2CH_2CH, $CO_2C_2H_5$, O $\xrightarrow{\text{NaOEt}}$ $H_5C_2O_2C$, O, CH_2CH_2CH

(h) $CO_2CH_2CO_2C_2H_5$ $\xrightarrow[\text{2) H}^+]{\text{1) LDA, 0°C}}$ $CCHCO_2C_2H_5$ (with C=O), OH

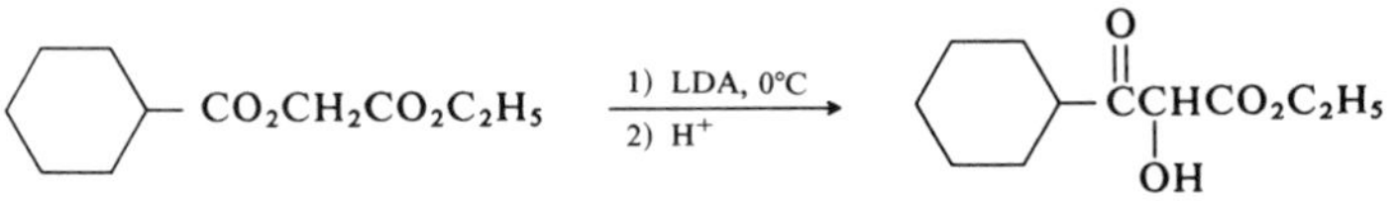

(i) 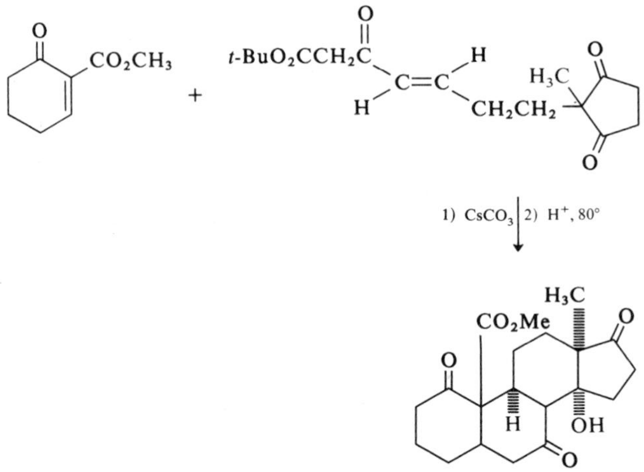 O, CO_2CH_3 + t-BuO_2CCH_2C (with C=O), $C{=}C$, H, H, CH_2CH_2, H_3C, O, O

1) $CsCO_3$ 2) H^+, 80°

H_3C, CO_2Me, O, O, H, OH, O

5. Tetraacetic acid (or a biological equivalent) has been suggested as an intermediate in the biosynthesis of phenolic natural products. Its synthesis has been described, as has its ready conversion to orsellinic acid. Suggest a mechanism for formation of orsellinic acid under the conditions specified:

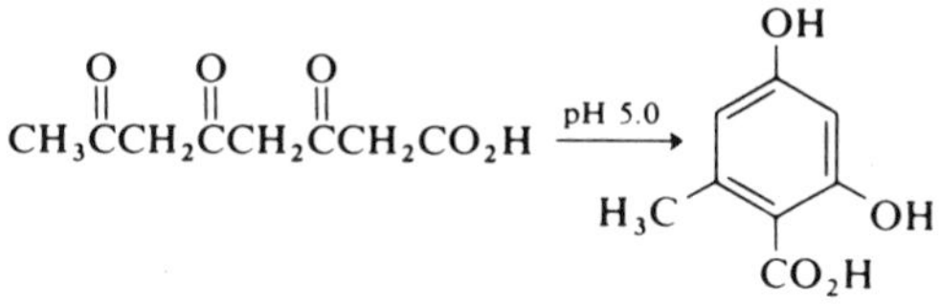

$$CH_3\overset{O}{\overset{\|}{C}}CH_2\overset{O}{\overset{\|}{C}}CH_2\overset{O}{\overset{\|}{C}}CH_2CO_2H \xrightarrow{\text{pH 5.0}} \text{orsellinic acid}$$

(OH, H_3C, OH, CO_2H substituted benzene)

orsellinic acid

6. (a) A stereospecific method for deoxygenating epoxides to alkenes involves reaction of the epoxide with the diphenylphosphide ion, followed by methyl iodide. The method results in overall inversion of the alkene stereochemistry. Thus, *cis*-cyclooctene epoxide gives *trans*-cyclooctene. Propose a mechanism for this process and discuss the relationship of the reaction to the Wittig reaction.

   (b) Reaction of the epoxide of *E*-4-octene (*trans*-2,3-di-*n*-propyloxirane) with trimethylsilylpotassium affords *Z*-4-octene as the only alkene in 93% yield. Suggest a reasonable mechanism for this reaction.

7. (a) A fairly general method for ring closure that involves vinyltriphenylphosphonium halides has been developed. Two examples are shown. Comment on the mechanism of the reaction and suggest two additional types of rings that could be synthesized using vinyltriphenylphosphonium salts.

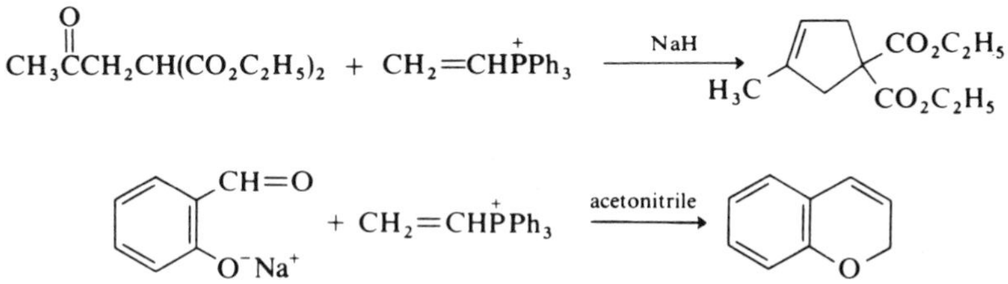

$$CH_3\overset{O}{\overset{\|}{C}}CH_2CH(CO_2C_2H_5)_2 + CH_2{=}CH\overset{+}{P}Ph_3 \xrightarrow{\text{NaH}} \text{(cyclopentene with } H_3C\text{ and } (CO_2C_2H_5)_2)$$

$$\text{(o-}CH{=}O\text{, } O^-Na^+\text{ benzene)} + CH_2{=}CH\overset{+}{P}Ph_3 \xrightarrow{\text{acetonitrile}} \text{(2H-chromene)}$$

   (b) The two phosphonium salts shown have both been used in syntheses of cyclohexadienes. Suggest appropriate coreactants and catalysts that would be expected to lead to cyclohexadienes.

$$H_2C{=}CHCH_2\overset{+}{P}Ph_3 \qquad H_2C{=}CHCH{=}CH\overset{+}{P}Ph_3$$

   (c) The product shown below is formed by the reaction of vinyltriphenylphosphonium bromide, the lithium enolate of cyclohexanone, and 1,3-diphenyl-2-propen-1-one. Formulate a mechanism.

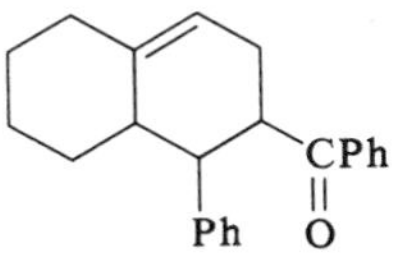

(bicyclic product with Ph and $\overset{O}{\overset{\|}{C}}Ph$ substituents)

Labels: CPh, O, Ph

8. Compounds **A** and **B** are key intermediates in one total synthesis of cholesterol. Rationalize their formation by the routes shown.

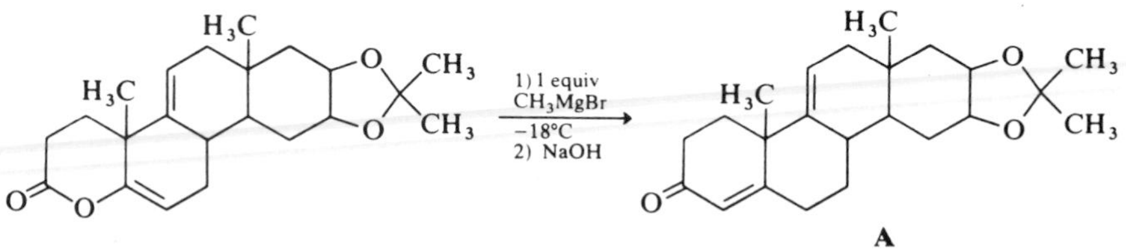

1) 1 equiv CH_3MgBr −18°C
2) NaOH

A

piperidine, acetic acid in benzene

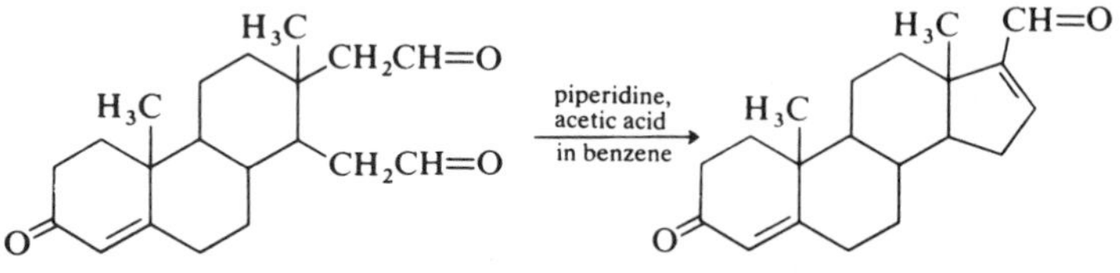

B

9. The first few steps of the synthesis of the alkaloid conessine produce **D** from **C**. Suggest a sequence of reactions for effecting this conversion.

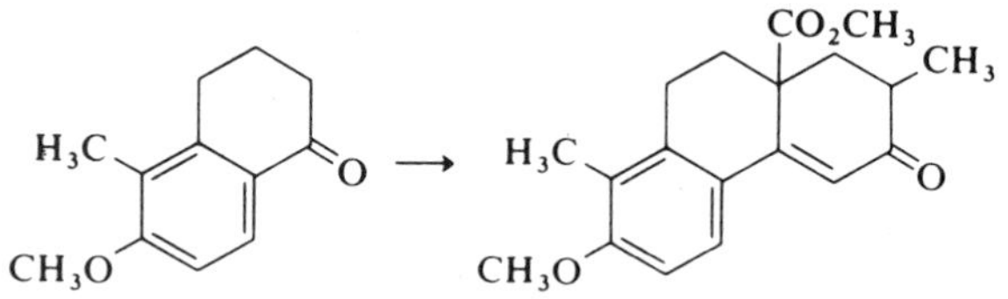

C → **D**

10. A substance known as elastase is a primary cause of arthritis, various inflammations, pulmonary emphysema, and pancreatitis. Elastase activity can be inhibited by a compound known as elasnin, obtained from the culture broth of a particular microorganism. The structure of elasnin is shown. A synthesis of elasnin has been reported which utilizes compound **E** as a key intermediate. Suggest a synthesis of compound **E** from methyl hexanoate and hexanal.

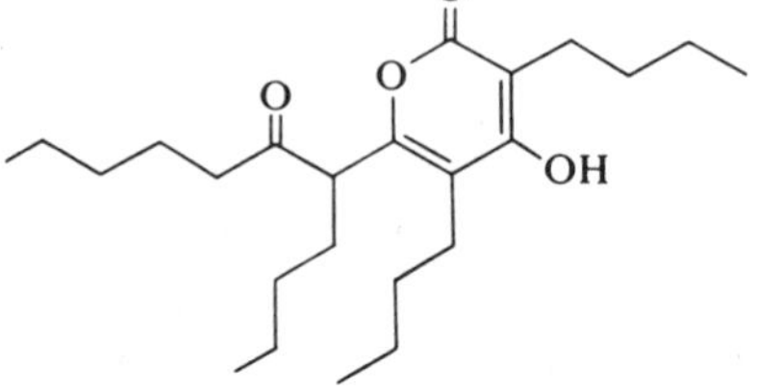

Elasnin

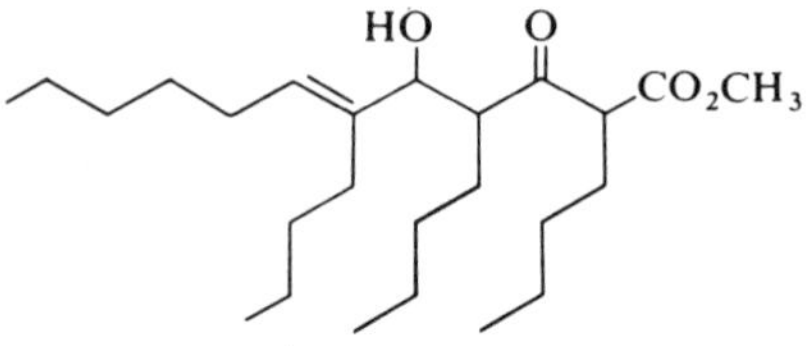

Compound **E**

11. Treatment of compound **F** with lithium diisopropylamide followed by cyclohexanone gives either **G** or **H**. **G** is formed if the aldehyde is added at −78°C,

whereas **H** is formed if the aldehyde is added at 0°C. Furthermore, treatment of **G** with lithium diisopropylamide at 0°C gives **H**. Explain these results.

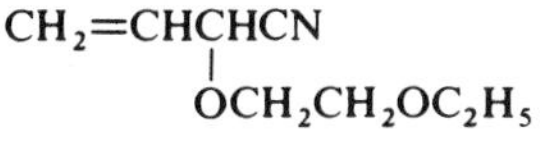

$CH_2{=}CHCH(OCH_2CH_2OC_2H_5)CN$ **F**

HO, CN, $CCH{=}CH_2$, $OCH_2CH_2OC_2H_5$ **G**

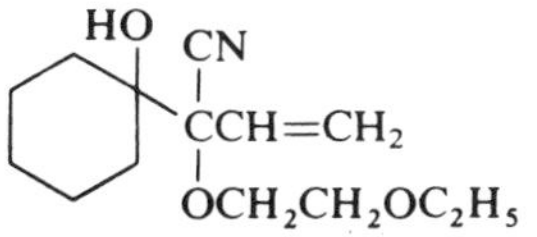

OH, $CH_2CH{=}C(CN)OCH_2CH_2OC_2H_5$ **H**

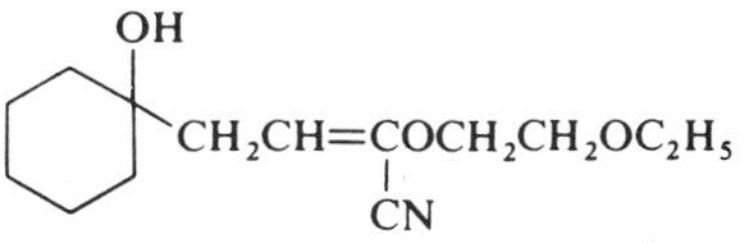

12. Dissect the following molecules into potential precursors by locating all bond connections which could be made by aldol-type condensations. Suggest the structure for potential precursors and conditions for performing the desired condensation.

(a) O

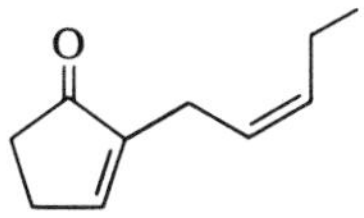

(b) CH_3 O CH_3 CH_3

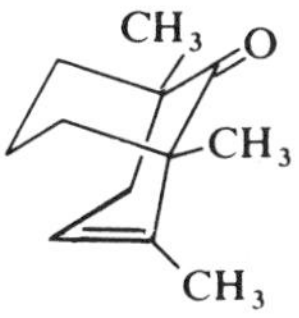

13. Mannich condensations permit one-step reactions to form the following substances from substantially less complex starting materials. By retrosynthetic analysis, identify a potential starting material which could give rise to the product shown in a single step under Mannich reaction conditions.

(a) N CO_2CH_3 O CO_2CH_3

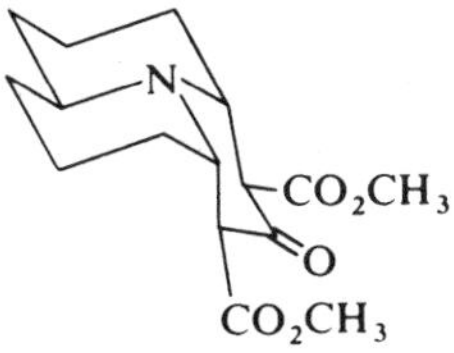

(b) N $PhCH_2OCH_2CH_2CH_2$ CH_3 O

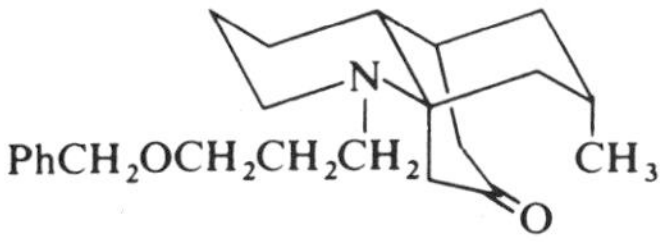

14. (a) The reagent **I** has found use in constructing rather complex molecules from simple precursors; for example, the enolate of 3-pentanone, treated first with **I**, then with benzaldehyde, gives **J** as a 2:1 mixture of stereoisomers. Explain the mechanism by which this synthesis occurs.

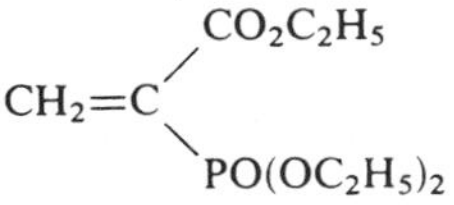

$CH_2{=}C(CO_2C_2H_5)PO(OC_2H_5)_2$

I

$$CH_3CH_2\overset{O}{\overset{\|}{C}}CH_2CH_3 \xrightarrow[2)\ \mathbf{I}]{1)\ LDA,\ -78°C} \xrightarrow[68°C,\ 45\ min]{PhCH{=}O} CH_3CH_2\overset{O}{\overset{\|}{C}}CH(CH_3)CH_2C(CO_2C_2H_5){=}CHPh$$

74%

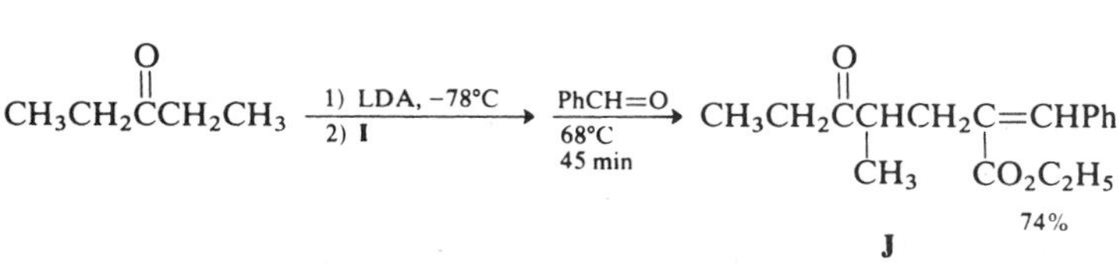

J

(b) The reagent **K** converts enolates of aldehydes into the cyclohexadienyl phosphonates **L**. What is the mechanism of this reaction? What alternative product might have been expected?

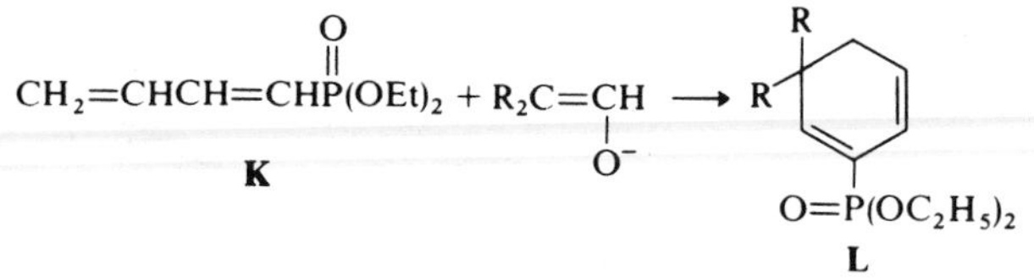

$$CH_2{=}CHCH{=}CHP(O)(OEt)_2 + R_2C{=}CHO^- \longrightarrow$$

K → **L** (4,4-dialkylcyclohexa-1,5-dienyl with $O{=}P(OC_2H_5)_2$)

15. Indicate whether the aldol condensation reactions shown below would be expected to exhibit high stereoselectivity. If high stereoselectivity is to be expected, show the relative configuration which is to be expected for the predominant product.

(a) $Ph_3CC(O)CH_2CH_3$ → 1) BuLi, −50°C (enolate formation); 2) PhCH=O

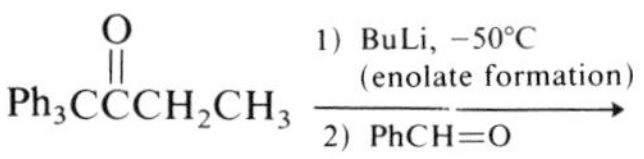

(b) $CH_3CH_2C(O)CH(C_6H_{11})OSi(CH_3)_2C(CH_3)_3$ → 1) $(i\text{-}Pr)_2NC_2H_5$; 2) $Bu_2BOSO_2CF_3$, −78°C → $CH_3CH_2CH{=}O$

(c) $CH_3CH_2C(O)CH_2CH_3$ → 1) LDA, THF, −70°C; 2) $C_6H_5CH{=}O$

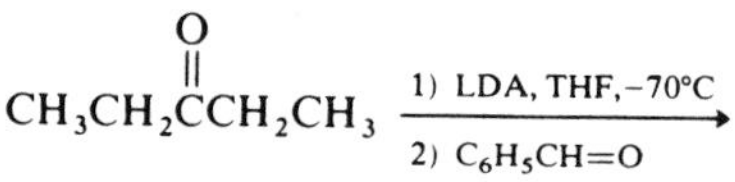

(d) 1-trimethylsilyloxy-2,6-dimethylcyclohexene ($OSi(CH_3)_3$, CH_3, CH_3) → KF, $C_6H_5CH{=}O$

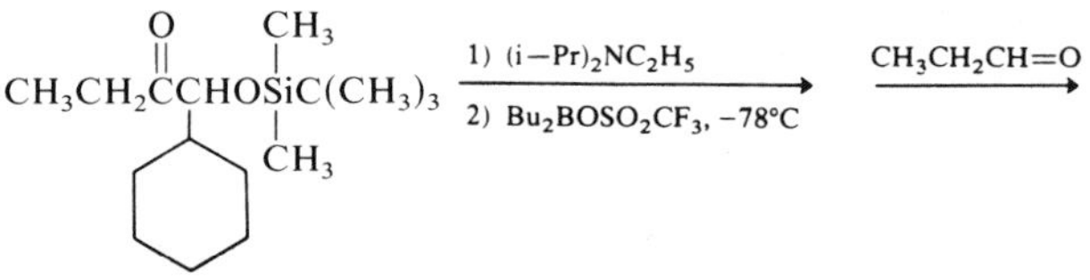

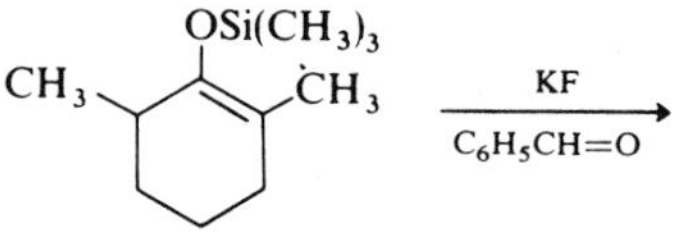

(e) $PhC(O)CH_2CH_3$ → 1) $Bu_2BOSO_2CF_3$, $(iPr)_2NC_2H_5$, −78°C → 2) PhCH=O

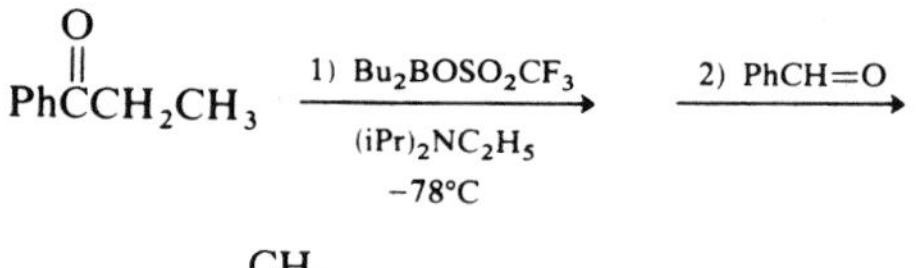

(f) $CH_3CH_2C(O){-}C(CH_3)_2OSi(CH_3)_3$ → 1) LDA, −70°C; 2) $(CH_3)_2CHCH{=}O$

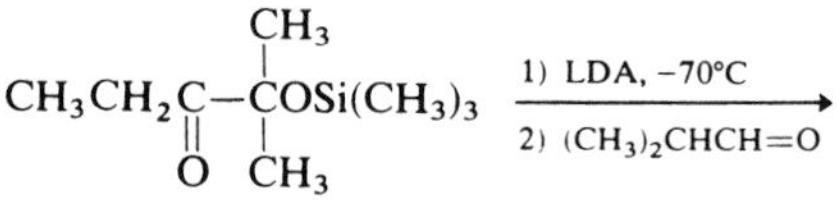

16. The reaction of 3-benzyloxybutanecarboxaldehyde with the trimethylsilyl enol ether of acetophenone is stereoselective for the *anti* diastereomer.

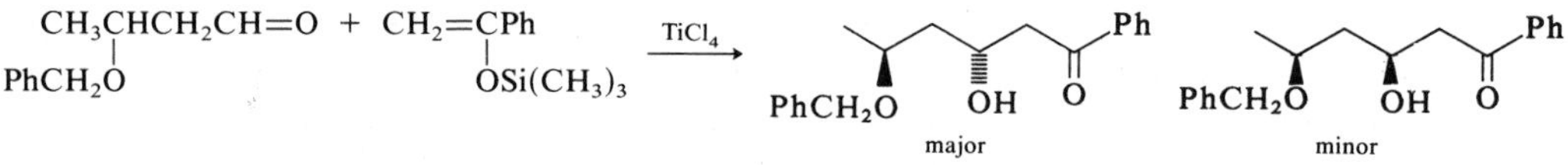

$$CH_3CH(OCH_2Ph)CH_2CH{=}O + CH_2{=}C(OSi(CH_3)_3)Ph \xrightarrow{TiCl_4}$$

PhCH$_2$O, OH, O, Ph — major

PhCH$_2$O, OH, O, Ph — minor

Propose a transition state which would account for the observed stereoselectivity.

3

Functional Group Interconversion by Nucleophilic Substitution

The first two chapters have dealt with formation of new carbon-carbon bonds by processes in which carbon is the nucleophilic atom. The reactions considered in Chapter 1 involve attack by carbon nucleophiles at sp^3 centers, while those discussed in Chapter 2 involve reaction at sp^2 centers, primarily carbonyl groups. In this chapter, we turn our attention to noncarbon nucleophiles. Nucleophilic substitution at both sp^3 and sp^2 centers is used in a variety of synthetic operations, particularly in the interconversion of functional groups. The mechanistic aspects of the reactions were considered in Part A, Chapters 5 and 8.

3.1. Conversion of Alcohols to Alkylating Agents

3.1.1. Sulfonate Esters

The preparation of sulfonate esters from alcohols is an effective way of introducing a reactive leaving group onto an alkyl chain. The reaction is very general, and complications arise only if the resulting sulfonate ester is sufficiently reactive to require special precautions. *p*-Toluenesulfonate (tosylate) and methanesulfonate (mesylate) esters are the most frequently used groups for preparative work, but the very reactive trifluoromethanesulfonates (triflates) are useful when an especially good leaving group is required. The usual method for introducing tosyl or mesyl groups is to allow the alcohol to react with the sulfonyl chloride in pyridine at

0–25°C.[1] An alternative for preparing mesylates and tosylates is to convert the alcohol to a lithium salt which is then allowed to react with the sulfonyl chloride.[2]

$$ROLi + ClSO_2-C_6H_4-CH_3 \longrightarrow ROSO_2-C_6H_4-CH_3$$

Trifluoromethanesulfonates of alkyl and allylic alcohols can be prepared by reaction with trifluoromethanesulfonic anhydride in halogenated solvents in the presence of pyridine.[3]

Since the preparation of sulfonate esters does not disturb the C—O bond, problems of rearrangement or racemization do not arise in the ester formation step. However, sensitive sulfonate esters, such as allylic systems, may be subject to ionization reactions, so that appropriate precautions must be taken. Tertiary alkyl tosylates are not as easily prepared nor as stable as those from primary and secondary alcohols. Under the standard conditions, tertiary alcohols are likely to be converted to the corresponding alkene.

3.1.2. Halides

The prominent role of alkyl halides in formation of carbon–carbon bonds by nucleophilic substitution was evident in Chapter 1. The most common precursors for alkyl halides are the corresponding alcohols, and a variety of procedures for this transformation have been developed. The choice of an appropriate reagent is usually dictated by the sensitivity of the alcohol and any other functional groups present in the molecule. Unsubstituted primary alcohols can be converted to bromides with hot concentrated hydrobromic acid.[4] Alkyl chlorides can be prepared by reaction of primary alcohols with hydrochloric acid–zinc chloride (Lucas reagent).[5] These reactions proceed by an S_N2 mechanism, and elimination and rearrangements are not a problem for primary alcohols. Reactions with tertiary alcohols proceeded by an S_N1 mechanism so these reactions are preparatively useful only when the carbocation is unlikely to give rise to rearranged product.[6] Because of the relative harshness of the conditions in these procedures, they are only applicable to very acid-stable molecules.

Another general method for converting alcohols to halides involves reactions with various halides of nonmettalic elements. Thionyl chloride, phosphorus trichloride, and phosphorus tribromide are the most common examples of this group of reagents. These reagents are suitable for alcohols that are neither acid-sensitive

1. R. S. Tipson, *J. Org. Chem.* **9**, 235 (1944); G. W. Kabalka, M. Varma, R. W. Varma, P. C. Srivastava, and F. F. Knapp, Jr., *J. Org. Chem.* **51**, 2386 (1986).
2. H. C. Brown, R. C. Bernheimer, C. J. Kim, and S. E. Scheppele, *J. Am. Chem. Soc.* **89**, 370 (1967).
3. C. D. Beard, K. Baum, and V. Grakauskas, *J. Org. Chem.* **38**, 3673 (1973).
4. E. E. Reid, J. R. Ruhoff, and R. E. Burnett, *Org. Synth.* **II**, 246 (1943).
5. J. E. Copenhaver and A. M. Wharley, *Org. Synth.* **I**, 142 (1941).
6. J. F. Norris and A. W. Olmsted, *Org. Synth.* **I**, 144 (1941); H. C. Brown and M. H. Rei, *J. Org. Chem.* **31**, 1090 (1966).

nor prone to structural rearrangement. The reaction of alcohols with thionyl chloride initially results in the formation of a chlorosulfite ester. There are two mechanisms by which the chlorosulfite can be converted to a chloride. In nucleophilic solvents, such as dioxane, the solvent participates and can lead to overall retention of configuration[7]:

$$ROH + SOCl_2 \rightarrow ROS(=O)Cl + HCl$$

$$\text{dioxane} + ROS(=O)Cl \longrightarrow \text{dioxane-}\overset{+}{O}{-}R + SO_2 + Cl^- \longrightarrow R{-}Cl + \text{dioxane}$$

In the absence of solvent participation, chloride attack on the chlorosulfite ester leads to product with inversion of configuration:

$$ROH + SOCl_2 \rightarrow ROS(=O)Cl + HCl$$

$$Cl^- \; R{-}OS(=O){-}Cl \rightarrow R{-}Cl + SO_2 + Cl^-$$

The mechanism for the reactions with phosphorus halides can be illustrated using phosphorus tribromide. Initial reaction between the alcohol and phosphorus tribromide leads to a trialkyl phosphate ester by successive displacements of bromide. The reaction stops at this stage if it is run in the presence of an amine which neutralizes the hydrogen bromide that is formed.[8] If the hydrogen bromide is not neutralized, the phosphite ester is protonated and each alkyl group is successively converted to the halide by nucleophilic substitution by bromide ion. The driving force for cleavage of the C—O bond is the formation of a strong phosphoryl bond.

$$ROH + PBr_3 \rightarrow (RO)_3P + 3HBr$$

$$(RO)_3P + HBr \rightarrow RBr + O{=}P(H)(OR)_2$$

$$O{=}P(H)(OR)_2 + HBr \rightarrow R{-}Br + O{=}P(OH)(H)OR$$

$$O{=}P(OH)(H){-}OR + HBr \rightarrow RBr + O{=}P(H)(OH)_2$$

Since C—Br bond formation occurs by back-side attack, inversion of configuration at carbon is anticipated. However, both racemization and rearrangement can be

7. E. S. Lewis and C. E. Boozer, *J. Am. Chem. Soc.* **74**, 308 (1952).
8. A. H. Ford-Moore and B. J. Perry, *Org. Synth.* **IV**, 955 (1963).
9. H. R. Hudson, *Synthesis*, 112 (1969).

observed as competing processes.[9] For example, conversion of homochiral 2-butanol to 2-butyl bromide with PBr_3 is accompanied by 10–13% racemization, and a small amount of *t*-butyl bromide is also formed.[10] The extent of rearrangement increases with increasing chain length and branching:

$$CH_3CH_2\underset{\displaystyle OH}{\underset{|}{C}}HCH_2CH_3 \xrightarrow[\text{ether}]{PBr_3} \underset{(85\text{–}90\%)}{CH_3CH_2\underset{\displaystyle Br}{\underset{|}{C}}HCH_2CH_3} + \underset{(10\text{–}15\%)}{CH_3CH_2CH_2\underset{\displaystyle Br}{\underset{|}{C}}HCH_3} \qquad \text{Ref. 11}$$

$$(CH_3)_3CCH_2OH \xrightarrow[\text{quinoline}]{PBr_3} \underset{(63\%)}{(CH_3)_3CCH_2Br} + \underset{(26\%)}{(CH_3)_2\underset{\displaystyle Br}{\underset{|}{C}}CH_2CH_3} + \underset{(11\%)}{CH_3\underset{\displaystyle Br}{\underset{|}{C}}HCH(CH_3)_2} \qquad \text{Ref. 12}$$

Because of the generation of very acidic solutions, the methods based on thionyl chloride and phosphorus halides are limited to acid-stable molecules. Milder reagents are necessary for most functionally substituted alcohols. A very general and important method for activating alcohols toward nucleophilic substitution is by converting them to alkoxyphosphonium ions.[13] The alkoxyphosphonium ions are very reactive toward nucleophilic attack, with the driving force for substitution being formation of the strong phosphoryl bond.

$$R'_3P + E{-}Y \rightarrow R'_3P\begin{matrix} \diagup E \\ \diagdown Y \end{matrix} \rightleftharpoons R'_3\overset{+}{P}{-}E + Y^-$$

$$R'_3\overset{+}{P}{-}E + ROH \rightarrow R'_3\overset{+}{P}{-}OR + HE \qquad R'_3\overset{+}{P}{-}OR + \mathbf{Nu}^- \rightarrow R_3P{=}O + R{-}\mathbf{Nu}$$

A wide variety of species can provide the electrophile E^+ in the general mechanism. The most useful synthetic procedures for preparation of halides are based on the halogens, positive halogen sources, and diethyl azodicarboxylate.

A 1 : 1 adduct formed from triphenylphosphine and bromine converts alcohols to bromides.[14] The alcohol displaces bromide ion from the pentavalent adduct, giving an alkoxyphosphonium intermediate. The phosphonium ion intermediate then undergoes nucleophilic attack by bromide ion, displacing triphenylphosphine oxide. The strength of the P=O bond formed in this step provides the driving force for the reaction.

$$PPh_3 + Br_2 \rightarrow Br_2PPh_3$$

$$Br_2PPh_3 + ROH \rightarrow RO\overset{+}{P}Ph_3\ Br^- + HBr$$

$$Br^- + RO\overset{+}{P}Ph_3 \rightarrow RBr + O{=}PPh_3$$

10. D. G. Goodwin and H. R. Hudson, *J. Chem. Soc.*, **B**, 1333 (1968); E. J. Coulson, W. Gerrard, and H. R. Hudson, *J. Chem. Soc.*, 2364 (1965).
11. J. Cason and J. S. Correia, *J. Org. Chem.* **26**, 3645 (1961).
12. H. R. Hudson, *J. Chem. Soc.*, 664 (1968).
13. B. P. Castro, *Org. React.* **29**, 1 (1983).
14. G. A. Wiley, R. L. Hershkowitz, B. M. Rein, and B. C. Chung, *J. Am. Chem. Soc.* **86**, 964 (1964).

Since the alkoxyphosphonium intermediate is formed in a reaction that does not break the C—O bond and the second step proceeds by back-side displacement on carbon, the overall stereochemistry of the process is inversion.

$$\text{3-hydroxy steroid (HO, } H_3C\text{, } H_3C\text{, } C_8H_{17}) \xrightarrow{Ph_3PBr_2} \text{3-bromo steroid (Br, inverted)}$$

Ref. 15

Triphenylphosphine dichloride exhibits similar reactivity toward alcohols and has been used to prepare chlorides.[16] The most convenient methods for converting alcohols to chlorides are based on *in situ* generation of chlorophosphonium ions[17] by reaction of triphenylphosphine with various chlorine compounds such as carbon tetrachloride[18] and hexachloroacetone.[19]

$$Ph_3P + CCl_4 \rightarrow Ph_3\overset{+}{P}{-}Cl + {}^-CCl_3$$

$$Ph_3P + Cl_3C\overset{O}{\overset{\|}{C}}CCl_3 \rightarrow Ph_3\overset{+}{P}{-}Cl + {}^-CCl_2\overset{O}{\overset{\|}{C}}CCl_3$$

The chlorophosphonium ion then reacts with the alcohol to give an alkoxyphosphonium ion, which is converted to the chloride:

$$Ph_3\overset{+}{P}{-}Cl + ROH \rightarrow Ph_3\overset{+}{P}{-}OR + HCl$$

$$Ph_3\overset{+}{P}{-}OR + Cl^- \rightarrow Ph_3P{=}O + R{-}Cl$$

There are several procedures for conversion of alcohols to iodides which are based on alkoxyphosphonium intermediates. One involves preparation of a cyclic phosphite ester from the alcohol and *o*-phenylene phosphorochloridite. Treatment of the cyclic phosphite ester with iodine then generates the alkyl iodide[20]:

$$ROH + ClP(O_2C_6H_4) \longrightarrow ROP(O_2C_6H_4) \xrightarrow{I_2} RI + O{=}P(I)(O_2C_6H_4)$$

A related procedure involves heating the alcohol with an adduct formed from triphenyl phosphite and methyl iodide.[21] In this method, a reactive alkoxyphosphonium intermediate is formed by displacement of a phenoxy group at phosphorus:

15. D. Levy and R. Stevenson, *J. Org. Chem.* **30**, 2635 (1965).
16. L. Horner, H. Oediger, and H. Hoffmann, *Justus Liebigs Ann. Chem.* **626**, 26 (1959).
17. R. Appel, *Angew. Chem. Int. Ed. Engl.* **14**, 801 (1975).
18. J. B. Lee and T. J. Nolan, *Can. J. Chem.* **44**, 1331 (1966).
19. R. M. Magid, O. S. Fruchey, W. L. Johnson, and T. G. Allen, *J. Org. Chem.* **44**, 359 (1979).
20. E. J. Corey and J. E. Anderson, *J. Org. Chem.* **32**, 4160 (1967).
21. J. P. H. Verheyden and J. G. Moffatt, *J. Org. Chem.* **35**, 2319 2868 (1970).

$$CH_3I + (PhO)_3P \rightarrow CH_3\overset{+}{P}(OPh)_3 + I^-$$

$$ROH + CH_3\overset{+}{P}(OPh)_3 \rightarrow RO\overset{+}{P}(OPh)_2(CH_3) + PhOH$$

$$I^- + RO\overset{+}{P}(OPh)_2(CH_3) \rightarrow RI + O{=}P(OPh)_2(CH_3)$$

A very mild procedure for converting alcohols to iodides uses triphenylphosphine, diethyl azodicarboxylate, and methyl iodide.[22] This reaction occurs with clean inversion of stereochemistry.[23] The key intermediate is again an alkoxyphosphonium ion.

$$Ph_3P + ROH + C_2H_5O_2CN{=}NCO_2C_2H_5 \rightarrow Ph_3\overset{+}{P}OR + C_2H_5O_2C\overset{-}{N}NHCO_2C_2H_5$$

$$C_2H_5O_2C\overset{-}{N}NHCO_2C_2H_5 + CH_3I \rightarrow C_2H_5O_2CN(CH_3)NHCO_2C_2H_5 + I^-$$

$$Ph_3\overset{+}{P}OR + I^- \rightarrow RI + Ph_3P{=}O$$

The role of the diethyl azodicarboxylate is to activate the triphenylphosphine toward nucleophilic attack by the alcohol. In the course of the reaction, the N=N double bond is reduced. As will be discussed subsequently, this method is applicable for activation of alcohols to nucleophilic attack by other nucleophiles in addition to halide ions. The activation of alcohols to nucleophilic attack by the triphenylphosphine–diethyl azodicarboxylate combination is called the *Mitsunobu reaction.*

There are a number of other useful methods for converting alcohols to halides. A very mild method which is useful for compounds that are prone to allylic rearrangement involves prior conversion of the alcohol to a sulfonate ester followed by nucleophilic displacement with halide ion:

$$(CH_3CH_2CH_2)_2C{=}CHCH_2OH \xrightarrow[\text{2) LiCl, DMF}]{\text{1) }CH_3SO_2Cl} (CH_3CH_2CH_2)_2C{=}CHCH_2Cl \quad (83\%)$$

Ref. 24

Another very mild procedure involves reaction of the alcohol with the heterocyclic 2-chloro-3-ethylbenzoxazolium cation.[25] The alcohol adds to the electrophilic heterocyclic ring, displacing chloride. The alkoxy group is thereby activated toward nucleophilic substitution, which leads to a stable product, 3-ethylbenzoxazolinone:

$$\text{2-chloro-3-ethylbenzoxazolium} + ROH \longrightarrow \text{2-alkoxy-3-ethylbenzoxazolium} + Cl^- \longrightarrow \text{3-ethylbenzoxazolinone} + RCl$$

22. O. Mitsunobu, *Synthesis*, 1 (1981).
23. H. Loibner and E. Zbiral, *Helv. Chim. Acta* **59**, 2100 (1976).
24. E. W. Collington and A. I. Meyers, *J. Org. Chem.* **36**, 3044 (1971).
25. T. Mukaiyama, S. Shoda, and Y. Watanabe, *Chem. Lett.*, 383 (1977); T. Mukaiyama, *Angew. Chem. Int. Ed. Engl.* **18**, 707 (1979).

Scheme 3.1. Preparation of Alkyl Halides

1[a] $(CH_3)_2CHCH_2OH \xrightarrow{PBr_3} (CH_3)_2CHCH_2Br$ (55–60%)

2[b] O CH_2OH $\xrightarrow[\text{pyridine}]{PBr_3}$ O CH_2Br (53–61%)

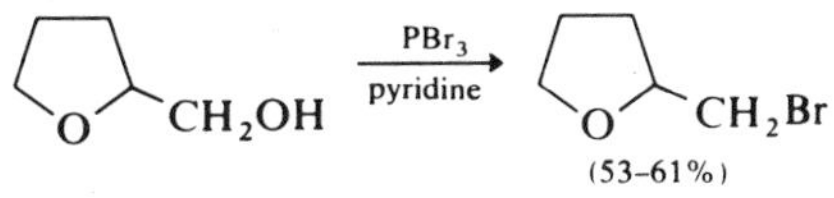

3[c] $(CH_3)_3CCH_2OH \xrightarrow[PPh_3]{Cl_2} (CH_3)_3CCH_2Cl$ (92%)

4[d] H_3C–(cyclohexenyl)–$CHCH_2OH$ with CH_3 $\xrightarrow[PPh_3]{CCl_4}$ H_3C–(cyclohexenyl)–$CHCH_2Cl$ with CH_3 (70%)

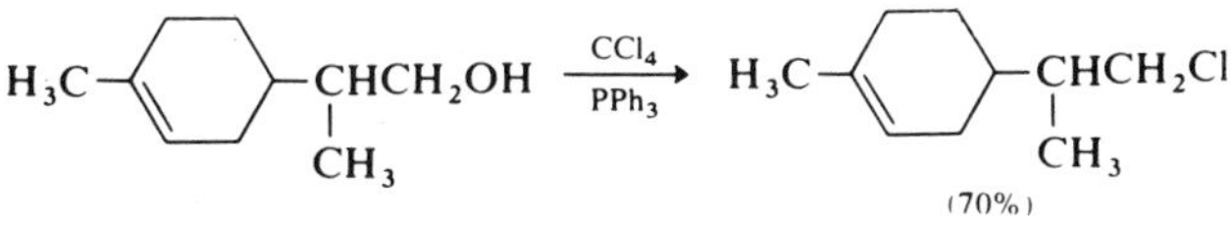

5[e] CH_3, H, C=C, H, CH_2OH $\xrightarrow[PPh_3]{Cl_3CCOCCl_3}$ CH_3, H, C=C, H, CH_2Cl (99%)

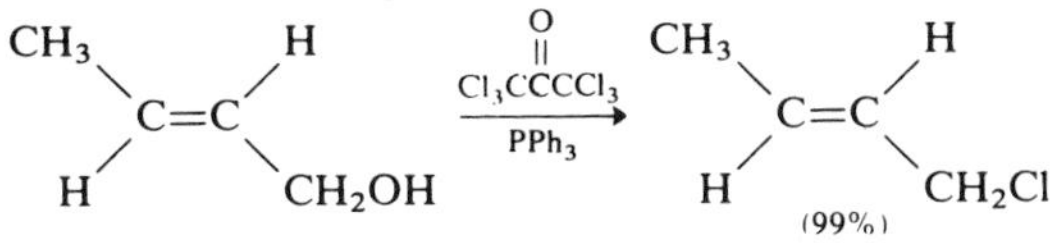

6[f] $Ph_2C{=}CHCH_2CH_2OH \xrightarrow[\text{2) LiBr}]{\text{1) tosyl chloride}} Ph_2C{=}CHCH_2CH_2Br$ (89%)

7[g] CH_2OH $\xrightarrow[\text{2) LiBr, acetone}]{\text{1) tosyl chloride}}$ CH_2Br (94%)

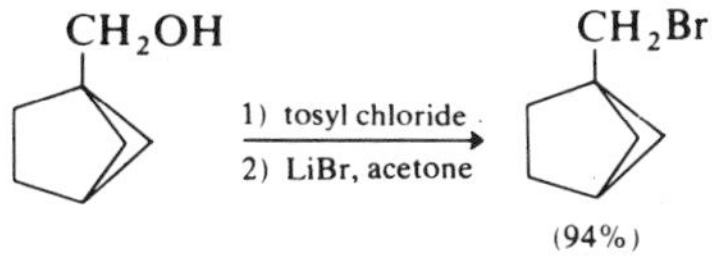

8[h] $CH_3(CH_2)_5CHCH_3$ (OH) + (C_2H_5, ^+N, Cl, O) $\xrightarrow{R_4N^+Cl^-}$ $CH_3(CH_2)_5CHCH_3$ (Cl) (76%)

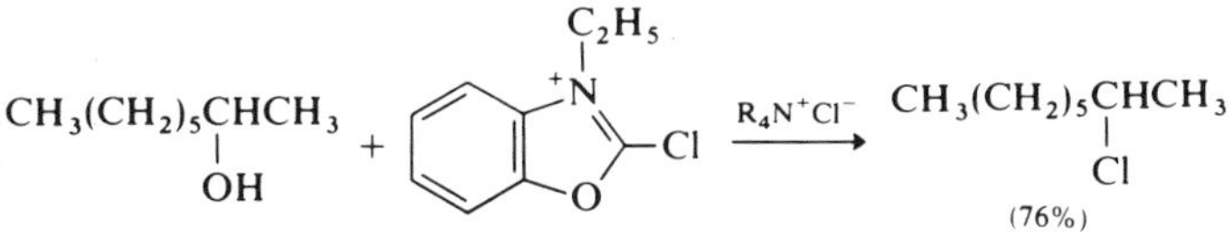

9[i] $(CH_3)_2NCH_2CH_2OH \xrightarrow{SOCl_2} (CH_3)_2\overset{+}{N}(H)CH_2CH_2Cl\ Cl^-$ (90%)

10[j] HO, CH_3 $\xrightarrow[\text{2) } CH_3I]{\text{1) } Ph_3P,\ C_2H_5O_2CN{=}NCO_2C_2H_5}$ CH_3, I (90%)

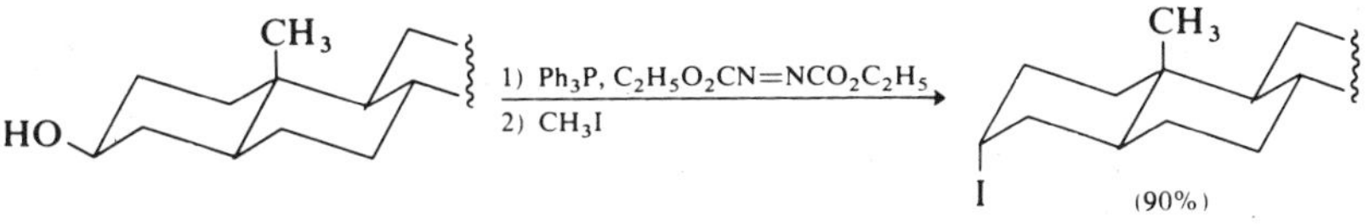

11[k] $PhCH{=}CHCH_2OH \xrightarrow{Ph_3PBr_2} PhCH{=}CHCH_2Br$ (60–70%)

a. C. R. Noller and R. Dinsmore, *Org. Synth.* **II**, 358 (1943).
b. L. H. Smith, *Org. Synth.* **III**, 793 (1955).
c. G. A. Wiley, R. L. Hershkowitz, B. M. Rein, and B. C. Chung, *J. Am. Chem. Soc.* **86**, 964 (1964).
d. B. D. MacKenzie, M. M. Angelo, and J. Wolinsky, *J. Org. Chem.* **44**, 4042 (1979).
e. R. M. Magid, O. S. Fruchy, W. L. Johnson, and T. G. Allen, *J. Org. Chem.* **44**, 359 (1979).
f. M. E. H. Howden, A. Maercker, J. Burdon, and J. D. Roberts, *J. Am. Chem. Soc.* **88**, 1732 (1966).
g. K. B. Wiberg and B. R. Lowry, *J. Am. Chem. Soc.* **85**, 3188 (1963).
h. T. Mukaiyama, S. Shoda, and Y. Watanabe, *Chem. Lett.*, 383 (1977).
i. L. A. R. Hall, V. C. Stephens, and J. H. Burckhalter, *Org. Synth.* **IV**, 333 (1963).
j. H. Loibner and E. Zbiral, *Helv. Chim. Acta* **59**, 2100 (1976).
k. J. P. Schaefer, J. G. Higgins, and P. K. Shenoy, *Org. Synth.* **V**, 249 (1973).

The reaction can be used for making either chlorides or bromides by using the appropriate tetraalkylammonium salt as a halide source.

Scheme 3.1 gives some examples of the various alcohol-to-halide conversions that have been discussed.

3.2. Introduction of Functional Groups by Nucleophilic Substitution at Saturated Carbon

The mechanistic aspects of nucleophilic substitution reactions were treated in detail in Chapter 5 of Part A. That mechanistic understanding has contributed to the development of nucleophilic substitution reactions as important synthetic processes. The S_N2 mechanism, because of its predictable stereochemistry and avoidance of carbocation intermediates, is the most desirable substitution process from a synthetic point of view. In this section, we will discuss the role of S_N2 reactions in the preparation of several classes of compounds. First, however, the important role that solvent plays in S_N2 reactions will be reviewed. The knowledgeable manipulation of solvent and related medium effects has led to significant improvement of many synthetic procedures that proceed by the S_N2 mechanism.

3.2.1. General Solvent Effects

The objective in selecting the reaction conditions for a preparative nucleophilic substitution is to enhance the mutual reactivity of the leaving group and nucleophile so that the desired substitution occurs at a convenient rate and with minimal competition from other possible reactions. The generalized order of leaving group reactivity $RSO_3^- > I^- > Br^- > Cl^-$ pertains for most S_N2 processes. (See Part A, Section 5.6 for more complete data.) Mesylates, tosylates, iodides, and bromides are all widely used in synthesis. Chlorides usually react rather slowly, except in especially reactive systems, such as allyl and benzyl. The overall synthetic objective normally governs the choice of the nucleophile. Optimization of reactivity therefore must be achieved by choice of the reaction conditions, particularly the solvent. Several generalizations about solvents can be made. Hydrocarbons, halogenated hydrocarbons, and ethers are usually unsuitable solvents for reactions involving salts. Acetone and acetonitrile are somewhat more polar, but the solubility of most ionic compounds in these solvents is low. Solubility can be considerably improved by use of salts of cations having substantial nonpolar character, such as those containing tetraalkylammonium ions. Alcohols are reasonably good solvents for salts, but the nucleophilicity of hard anions is relatively low in alcohols because of extensive solvation. The polar aprotic solvents, particularly dimethylformamide (DMF) and dimethyl sulfoxide (DMSO), are good solvents for salts, and, by virtue of selective cation solvation, anions usually show enhanced nucleophilicity in these solvents. The high water solubility of these solvents and their high boiling points can sometimes cause problems in product separation and purification. Hexamethylphosphoric triamide (HMPA), *N,N*-diethylacetamide, and *N*-methylpyrrolidinone are other examples of useful polar aprotic solvents.[26] In addition to enhancing reactivity, polar aprotic solvents also affect the order of reactivity of nucleophilic anions. In DMF, the halides are all of comparable nucleophilicity,[27] whereas in hydroxylic solvents the order is $I^- > Br^- > Cl^-$ and the difference in reactivity are much greater.[28]

In addition to use of solvent effects, there are two other valuable approaches to enhancing reactivity in nucleophilic substitutions. These are use of *crown ethers* as catalysts and the use of *phase transfer catalysts.* The crown ethers are a family of cyclic polyethers, three examples of which are shown below.

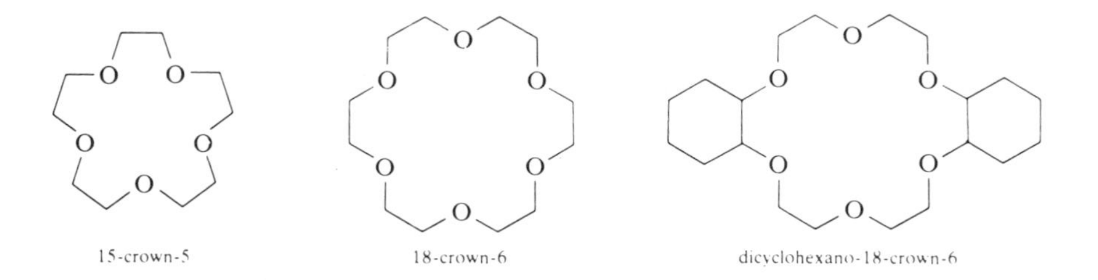

The first number designates the ring size, and the second the number of oxygen atoms in the ring. These materials have specific cation-complexing properties and give rise to catalysis of nucleophilic substitution under many conditions. By complexing the cation in the cavity of the crown ether, these compounds solubilize many salts in nonpolar solvents. Once in solution, the anions are highly reactive as nucleophiles since they are weakly solvated. Tight ion pairing is also precluded by the complexation of the cation by the nonpolar crown ether. As a result, nucleophilicity approaches or exceeds that observed in aprotic polar solvents.[29]

The second method for enhancing nucleophilic substitution processes is to use phase transfer catalysts.[30] The phase transfer catalysts are ionic substances, usually quaternary ammonium or phosphonium salts, in which the size of the hydrocarbon groups in the cation is large enough to confer good solubility of the salt in organic solvents. In other words, the cation must be highly lipophilic. Phase transfer catalysis usually involves carrying out the reaction in a two-phase system. The organic reactant is dissolved in a water-insoluble solvent such as a hydrocarbon or halogenated hydrocarbon. The salt containing the nucleophile is dissolved in water. Even with vigorous mixing, such systems show little tendency to react, since the nucleophile and reactant remain separated in the water and organic phases, respectively. When

26. A. F. Sowinski and G. M. Whitesides, *J. Org. Chem.* **44**, 2369 (1979).
27. W. M. Weaver and J. D. Hutchinson, *J. Am. Chem. Soc.* **86**, 261 (1964).
28. R. G. Pearson and J. Songstad, *J. Org. Chem.* **32**, 2899 (1967).
29. M. Hiraoka, *Crown Compounds. Their Characteristics and Application*, Elsevier, Amsterdam, 1982.
30. E. V. Dehmlow and S. S. Dehmlow, *Phase Transfer Catalysis*, Second Edition, VCH Publishers, New York, 1983; W. P. Weber and G. W. Gokel, *Phase Transfer Catalysis in Organic Synthesis*, Springer-Verlag, New York, 1977; C. M. Stark and C. Liotta, *Phase Transfer Catalysis*, Academic Press, New York, 1978.

a phase transfer catalyst is added, the lipophilic cations are transferred to the nonpolar phase and, to maintain electrical neutrality in this phase, anions are transferred from the water to the organic phase. The anions are only weakly solvated in the organic phase and therefore exhibit enhanced nucleophilicity. As a result, the substitution reactions proceed under relatively mild conditions. The salts of the nucleophile are often used in high concentration in the aqueous solution, and in some procedures the solid salt is used.

3.2.2. Nitriles

The replacement of a halide or tosylate by cyanide ion, extending the carbon chain by one atom and providing an entry to carboxylic acid derivatives, has been a reaction of synthetic importance since the early days of organic chemistry. The classical conditions for preparing nitriles involve heating a halide with a cyanide salt in aqueous alcohol solution:

$$C_6H_5CH_2Cl + NaCN \xrightarrow[\text{reflux 4 h}]{H_2O,\ C_2H_5OH} C_6H_5CH_2CN \ (80\text{–}90\%) \qquad \text{Ref. 31}$$

$$ClCH_2CH_2CH_2Br + KCN \xrightarrow[\text{reflux 1.5 h}]{H_2O,\ C_2H_5OH} ClCH_2CH_2CH_2CN \ (40\text{–}50\%) \qquad \text{Ref. 32}$$

Similar reactions proceed more rapidly in aprotic polar solvents. In DMSO, for example, primary alkyl chlorides are converted to nitriles in 1 h or less at temperatures of 120–140°C.[33] Phase transfer catalysis by hexadecyltributylphosphonium bromide permits conversion of 1-chlorooctane to octyl cyanide in 95% yield in 2 h at 105°C.[34]

$$CH_3CH_2CH_2CH_2Cl \xrightarrow[\text{DMSO, 90–160°C}]{NaCN} CH_3CH_2CH_2CH_2CN \ (93\%)$$

$$CH_3(CH_2)_6CH_2Cl \xrightarrow[\substack{H_2O,\ \text{decane} \\ CH_3(CH_2)_{15}P^+(CH_2CH_2CH_2CH_3)_3 \\ 105°C,\ 2\ h}]{NaCN} CH_3(CH_2)_6CH_2CN \ (95\%)$$

Catalysis by 18-crown-6 of the reaction of solid potassium cyanide with a variety of chlorides and bromides has been demonstrated.[35] With primary bromides, yields are high and reaction times are 15–30 h at reflux in acetonitrile (83°C). Interestingly, the chlorides are more reactive and require reaction times of only about 2 h.

31. R. Adams and A. F. Thal, *Org. Synth.* **I**, 101 (1932).
32. C. F. H. Allen, *Org. Synth.* **I**, 150 (1932).
33. L. Friedman and H. Shechter, *J. Org. Chem.* **25**, 877 (1960); R. A. Smiley and C. Arnold, *J. Org. Chem.* **25**, 257 (1960).
34. C. M. Starks, *J. Am. Chem. Soc.* **93**, 195 (1971); C. M. Starks and R. M. Owens, *J. Am. Chem. Soc.* **95**, 3613 (1973).
35. F. L. Cook, C. W. Bowers, and C. L. Liotta, *J. Org. Chem.* **39**, 3416 (1974).

Secondary halides react more slowly, and yields drop because of competing elimination. Tertiary halides do not react satisfactorily because elimination processes dominate.

3.2.3. Azides

Azides are useful intermediates for synthesis of various nitrogen-containing compounds. They undergo cycloaddition reactions, as will be discussed in Section 6.2, and can also be easily reduced to primary amines. Azido groups are usually introduced into aliphatic compounds by nucleophilic substitution.[36] The most reliable procedures involve heating the appropriate halide with sodium azide in DMSO[37] or DMF.[38] Alkyl azides can also be prepared by reaction in high-boiling alcohols[39]:

$$CH_3(CH_2)_3CH_2I + NaN_3 \xrightarrow[H_2O]{CH_3CH_2OCH_2CH_2OCH_2CH_2OH} CH_3(CH_2)_3CH_2N_3 \ (84\%)$$

Phase transfer conditions have also been used for the preparation of azides[40]:

$$CH_2{=}CH\text{-}CH(CH_3)\text{-}CH_2\text{-}CH(Br)\text{-}CO_2CH_3 \xrightarrow[R_4P^+\,^-Br,\ 4\ h,\ 25°C]{NaN_3} CH_2{=}CH\text{-}CH(CH_3)\text{-}CH_2\text{-}CH(N_3)\text{-}CO_2CH_3$$

There are also useful procedures for preparation of azides directly from alcohols. Reaction of alcohols with 2-fluoro-1-methylpyridinium iodide followed by reaction with lithium azide gives good yields of alkyl azides[41]:

$$ROH + \text{2-fluoro-1-methylpyridinium} \longrightarrow \text{2-OR-1-methylpyridinium} \xrightarrow{N_3^-} \text{1-methyl-2-pyridone} + RN_3$$

These reaction conditions lead to predominant inversion for secondary alcohols.

Diphenylphosphoryl azide reacts with alcohols in the presence of triphenylphosphine and diethyl azodicarboxylate.[42] Hydrazoic acid, HN_3, can also serve as the

36. M. E. C. Biffin, J. Miller, and D. B. Paul, in *The Chemistry of the Azido Group*, S. Patai (ed.), Wiley-Interscience, New York, 1971, Chapter 2.
37. R. Gougarel, A. Cave, L. Tan, and M. Leboeuf, *Bull. Soc. Chim. Fr.*, 646 (1962).
38. E. J. Reist, R. R. Spencer, B. R. Baker, and L. Goodman, *Chem. Ind. (London)*, 1794 (1962).
39. E. Lieber, T. S. Chao, and C. N. R. Rao, *J. Org. Chem.* **22**, 238 (1957); H. Lehmkuhl, F. Rabet, and K. Hauschild, *Synthesis*, 184 (1977).
40. W. P. Reeves and M. L. Bahr, *Synthesis*, 823 (1976); B. B. Snider and J. V. Duncia, *J. Org. Chem.* **46**, 3223 (1981).
41. K. Hojo, S. Kobayashi, K. Soai, S. Ikeda, and T. Mukaiyama, *Chem. Lett.*, 635 (1977).
42. B. Lal, B. N. Pramanik, M. S. Manhas, and A. K. Bose, *Tetrahedron Lett.*, 1977 (1977).

azide ion source under these activating conditions.[43] These reactions are examples of the Mitsunobu reaction discussed earlier.

$$ROH + Ph_3P + C_2H_5O_2CN{=}NCO_2C_2H_5 \rightarrow RO\overset{+}{P}Ph_3 + C_2H_5O_2C\overset{-}{N}NHCO_2C_2H_5$$

$$RO\overset{+}{P}Ph_3 + N_3^- \rightarrow RN_3 + Ph_3P{=}O$$

3.2.4. Alkylation of Amines and Amides

The alkylation of neutral amines by halides is complicated from a synthetic point of view because of the possibility of multiple alkylation, which can proceed to the quaternary ammonium salt in the presence of excess alkyl halide:

$$RNH_2 + R'{-}X \rightarrow R\overset{+}{N}H_2R' + X^-$$

$$R\overset{+}{N}H_2R' + RNH_2 \rightleftharpoons RNHR' + R\overset{+}{N}H_3$$

$$RNHR' + R'{-}X \rightarrow R\overset{+}{N}HR'_2 + X^-$$

$$R\overset{+}{N}HR'_2 + RNH_2 \rightleftharpoons RNR'_2 + R\overset{+}{N}H_3$$

$$RNR'_2 + R'{-}X \rightarrow R\overset{+}{N}R'_3 + X^-$$

Even with a limited amount of the alkylating agent, the equilibria between protonated product and the neutral starting amine are sufficiently fast that a mixture of products may be obtained. For this reason, when monoalkylation of an amine is desired, the reaction is usually best carried out by reductive alkylation, a reaction which will be discussed in Chapter 5. If complete alkylation to the quaternary salt is desired, use of excess alkylating agent and a base to neutralize the liberated acid will normally result in complete reaction.

Amides are only weakly nucleophilic and react very slowly with alkyl halides. The anions of amides are substantially more reactive. The classical Gabriel procedure for synthesis of amines from phthalimide is illustrative.[44]

$$\text{(phthalimide)}N^-K^+ + BrCH_2CH_2Br \longrightarrow \text{(phthalimide)}NCH_2CH_2Br \quad (70\text{–}80\%)$$ Ref. 45

43. J. Schweng and E. Zbiral, *Justus Liebigs Ann. Chem.*, 1089 (1978); M. S. Hadley, F. D. King, B. McRitchie, D. H. Turner, and E. A. Watts, *J. Med. Chem.* **28**, 1843 (1985).
44. M. S. Gibson and R. N. Bradshaw, *Angew. Chem. Int. Ed. Engl.* **7**, 919 (1968).
45. P. L. Salzberg and J. V. Supniewski, *Org. Synth.* **I**, 114 (1932).

The enhanced acidity of the NH group in phthalimide permits formation of an anion which is readily alkylated by alkyl halides or tosylates. The amine can then be liberated by reaction of the substituted phthalimide with hydrazine:

$$CH_3O_2CCH(NPhth)CH_2CH(NPhth)CO_2CH_3 \xrightarrow[CH_3OH]{NH_2NH_2} \xrightarrow{HCl} HO_2CCH(NH_2)CH_2CH(NH_2)CO_2H$$ Ref. 46

Secondary amides can be alkylated by the use of sodium hydride for proton abstraction, followed by reaction with an alkyl halide[47]:

$$\text{caprolactam (NH)} \xrightarrow[CH_3I]{\text{NaH, benzene}} \text{N-methylcaprolactam (NCH}_3\text{)}$$

Neutral tertiary and secondary amides react with very reactive alkylating agents, such as triethyloxonium tetrafluoroborate, to give O-alkylation products.[48] The same reaction occurs, but more slowly, with tosylates and dimethyl sulfate. Neutralization of the resulting salt provides iminoethers:

$$RC(=O)NHR' \xrightarrow[2)\ ^-OH]{1)\ (CH_3O)_2SO_2} RC(OCH_3)=NR'$$

3.2.5. Oxygen Nucleophiles

The oxygen nucleophiles that are of primary interest in synthesis are the hydroxide ion (or water), alkoxide ions, and carboxylate anions, which lead, respectively, to alcohols, ethers, and esters. Since each of these nucleophiles can also act as a base, reaction conditions must be selected to favor substitution over elimination.

Frequently, a given alcohol is more easily obtained than the corresponding halide so the halide-to-alcohol transformations are not extensively used for synthesis.

46. J. C. Sheehan and W. A. Bolhofer, *J. Am. Chem. Soc.* **72**, 2786 (1950).
47. W. S. Fones, *J. Org. Chem.* **14**, 1099 (1949); R. M. Moriarty, *J. Org. Chem.* **29**, 2748 (1964).
48. L. Weintraub, S. R. Oles, and N. Kalish, *J. Org. Chem.* **33**, 1679 (1968); H. Meerwein, E. Battenberg, H. Gold, E. Pfeil, and G. Willfang, *J. Prakt. Chem.* **154**, 83 (1939).

The hydrolysis of benzyl halides to the corresponding alcohols proceeds in good yield. This can be a useful synthetic transformation, since benzyl halides are available either by side-chain halogenation or by the chloromethylation reaction (Section 11.1.3).

$$NC-C_6H_4-CH_2Cl \xrightarrow[H_2O,\ 100°C,\ 2.5\ h]{K_2CO_3} NC-C_6H_4-CH_2OH\ (85\%)$$ Ref. 49

Ether formation from alkoxides and alkylating reagents is a reaction of wide synthetic importance. The conversion of phenols to methoxyaromatics, for example, is a very common reaction. Methyl iodide, methyl tosylate, and dimethyl sulfate are the usual alkylating agents. The reaction proceeds in the presence of a weak base, such as Na_2CO_3 or K_2CO_3, which serves to deprotonate the phenol. The conjugate bases of alcohols are considerably more basic than phenoxides, and therefore β elimination can become a problem. Fortunately, the most useful and commonly encountered ethers are methyl and benzyl ethers, where elimination is not a problem and the corresponding halides are especially reactive. Entries 12–15 in Scheme 3.2 provide some typical examples of ether preparations.

Two methods for converting carboxylic acids to esters fall into the mechanistic group under discussion. One of these methods is the reaction of carboxylic acids with diazo compounds, especially diazomethane. The second is alkylation of carboxylate anions by halides or sulfonates. The esterification of carboxylic acids with diazomethane is a very quick and clean reaction.[50] The alkylating agent is the extremely reactive methyl diazonium ion, which is generated by proton transfer from the carboxylic acid to diazomethane. The collapse of the resulting ion pair with loss of nitrogen is extremely rapid:

$$RCO_2H + CH_2N_2 \rightarrow [RCO_2^- \ CH_3\overset{+}{N}_2] \rightarrow RCO_2CH_3 + N_2$$

The main drawback to this reaction is the toxicity of diazomethane and some of its precursors. One possible alternative is the use of alkyltriazenes as reactive alkylating agents.[51] Alkyltriazenes are readily prepared from primary amines and aryl diazonium salts.[52] The triazenes, on being protonated by the carboxylic acid, generate a reactive alkylating agent that is equivalent, if not identical, to the alkyl diazonium ions generated from diazoalkanes.

$$RCO_2H + Ar{-}N{=}NNHR' \rightleftharpoons Ar{-}\overset{+}{N}H_2{-}N{=}N{-}R' \ \ {}^-O_2CR \rightarrow RCO_2R' + ArNH_2 + N_2$$

49. J. N. Ashley, H. J. Barber, A. J. Ewins, G. Newbery, and A. D. Self, *J. Chem. Soc.*, 103 (1942).
50. T. H. Black, *Aldrichimica Acta* **16**, 3 (1983).
51. E. H. White, H. Maskill, D. J. Woodcock, and M. A. Schroeder, *Tetrahedron Lett.*, 1713 (1969).
52. E. H. White and H. Scherrer, *Tetrahedron Lett.*, 758 (1961).

Especially for large-scale work, esters may be more safely and efficiently prepared by reaction of carboxylate salts with alkyl halides or tosylates. Carboxylate anions are not very reactive nucleophiles so the best results are obtained in polar aprotic solvents[53] or with crown ether catalysts.[54] Cesium carboxylates are especially useful in polar aprotic solvents. The reactivity order is $Na^+ < K^+ < Rb^+ < Cs^+$. The enhanced reactivity of the cesium salts is due both to high solubility and to the absence of ion pairing with the anion.[55] Acetone has been found to be a good solvent for reaction of carboxylate anions with alkyl iodides.[56] Carboxylate ion alkylation procedures are particularly advantageous for preparation of hindered esters, which can be relatively difficult to prepare by the acid-catalyzed esterification (Fischer esterification) method, which will be discussed in Section 3.4. Sections F and G of Scheme 3.2 give some specific examples of preparation of ester by alkylations using diazomethane and by carboxylate alkylation.

In the course of synthesis, it is sometimes necessary to invert the configuration at an oxygen-substituted center. One of the best ways of doing this is to activate the hydroxyl group to substitution by a carboxylate anion. The activation is frequently done with the Mitsunobu combination of reagents. Hydrolysis of the resulting ester gives the alcohol of inverted configuration:

Ph_3P, $EtO_2CN{=}NCO_2Et$, $PhCO_2H$ (89%) Ref. 57

Ph_3P, $EtO_2CN{=}NCO_2Et$, $PhCO_2H$ (74%) Ref. 58

Use of zinc tosylate in place of the carboxylic acid gives a tosylate of inverted configuration:

Ph_3P, $EtO_2CN{=}NCO_2Et$, $Zn(O_3SAr)_2$ (96%) Ref. 59

53. P. E. Pfeffer, T. A. Foglia, P. A. Barr, I. Schmeltz, and L. S. Silbert, *Tetrahedron Lett.*, 4063 (1972); J. E. Shaw, D. C. Kunerth, and J. J. Sherry, *Tetrahedron Lett.*, 689 (1973); J. Grundy, B. G. James, and G. Pattenden, *Tetrahedron Lett.*, 757 (1972).
54. C. L. Liotta, H. P. Harris, M. McDermott, T. Gonzalez, and K. Smith, *Tetrahedron Lett.* 2417 (1974).
55. G. Dijkstra, W. H. Kruizinga, and R. M. Kellog, *J. Org. Chem.* **52**, 4230 (1987).
56. G. G. Moore, T. A. Foglia, and T. J. McGahan, *J. Org. Chem.* **44**, 2425 (1979).
57. M. J. Arco, M. H. Trammel, and J. D. White, *J. Org. Chem.* **41**, 2075 (1976).
58. C.-T. Hsu, N.-Y. Wang, L. H. Latimer, and C. J. Sih, *J. Am. Chem. Soc.* **105**, 593 (1983).
59. I. Galynker and W. C. Still, *Tetrahedron Lett.*, 4461 (1982).

3.2.6. Sulfur Nucleophiles

Anions derived from thiols are very nucleophilic and can easily be alkylated by halides.

$$CH_3S^-Na^+ + ClCH_2CH_2OH \xrightarrow{C_2H_5OH} \underset{(75-80\%)}{CH_3SCH_2CH_2OH} \qquad \text{Ref. 60}$$

Neutral sulfur compounds are also good nucleophiles. Sulfides and thioamides readily form salts with methyl iodide, for example:

$$(CH_3)_2S + CH_3I \xrightarrow[12-16\text{ h}]{25°C} (CH_3)_3\overset{+}{S}\ I^- \qquad \text{Ref. 61}$$

$$\text{1-methyl-2-pyrrolidinethione} + CH_3I \xrightarrow[12\text{ h}]{25°C} \text{1-methyl-2-(methylthio)-1-pyrrolinium } (N^+\text{–}CH_3,\ SCH_3) \qquad \text{Ref. 62}$$

Even sulfoxides, where nucleophilicity is decreased by the additional oxygen, can be alkylated by methyl iodide. The sulfoxonium salts formed have useful synthetic applications, as discussed in Section 2.6.

$$(CH_3)_2S{=}O + CH_3I \xrightarrow[72\text{ h}]{25°C} (CH_3)_3\overset{+}{S}{=}O\ I^- \qquad \text{Ref. 63}$$

3.2.7. Phosphorus Nucleophiles

Both neutral and anionic phosphorus compounds are good nucleophiles toward alkyl halides. Examples of these reactions were already encountered in Chapter 2 in connection with the preparation of the valuable phosphorane and phosphonate intermediates used for Wittig reactions:

$$Ph_3P + CH_3Br \xrightarrow[2\text{ days}]{\text{room temp}} Ph_3\overset{+}{P}CH_3\ Br^- \qquad \text{Ref. 64}$$

$$[(CH_3)_2CHO]_3P + CH_3I \rightarrow [(CH_3)_2CHO]_2\overset{O}{\overset{\|}{P}}CH_3 + (CH_3)_2CHI \qquad \text{Ref. 65}$$

The reaction with phosphite esters is known as the Michaelis–Arbuzov reaction and proceeds through an unstable trialkoxyphosphonium intermediate. The second stage

60. W. Windus and P. R. Shildneck, *Org. Synth.* **II**, 345 (1943).
61. E. J. Corey and M. Chaykovsky, *J. Am. Chem. Soc.* **87**, 1353 (1965).
62. R. Gompper and W. Elser, *Org. Synth.* **V**, 780 (1973).
63. R. Kuhn and H. Trischmann, *Justus Liebigs Ann. Chem.* **611**, 117 (1958).
64. G. Wittig and U. Schoellkopf, *Org. Synth.* **V**, 751 (1973).
65. A. H. Ford-Moore and B. J. Perry, *Org. Synth.* **IV**, 325 (1963).

in the reaction is another example of the great tendency of alkoxyphosphonium ions to react with nucleophiles to break the O—C bond, resulting in formation of a phosphoryl bond.

$$R'{-}X + P(OR)_3 \rightarrow (RO)_2\overset{+}{\underset{R'}{P}}{-}O{-}R\ \ ^{-}X \rightarrow (RO)_2\overset{O}{\overset{\|}{P}}R' + RX$$

The reaction of α-bromoketones with phosphines and phosphites can take an alternative course in which phosphorus attacks at bromine. In protic solvents, the enolate intermediate is protonated so that the overall course of the reaction is dehalogenation of the ketone[66]:

$$R_3P + BrCH_2\overset{O}{\overset{\|}{C}}R' \rightarrow R_3\overset{+}{P}Br + H_2C{=}\overset{O^-}{\overset{|}{C}}R' \xrightarrow{ROH} R_3P{=}O + CH_3\overset{O}{\overset{\|}{C}}R' + RBr$$

When phosphite esters are used as nucleophiles, the stable product is an enol phosphate[67]:

$$(RO)_3P + BrCH_2\overset{O}{\overset{\|}{C}}R' \rightarrow H_2C{=}\overset{\overset{+}{O}P(OR)_3}{\overset{|}{C}}R' \ + Br^- \rightarrow H_2C{=}\overset{\overset{O}{\overset{\|}{O P}}(OR)_2}{\overset{|}{C}}R' + RBr$$

3.2.8. Summary of Nucleophilic Substitution at Saturated Carbon

In the preceding sections, some of the nucleophilic substitution reactions at sp^3 carbon which are most valuable for synthesis have been outlined. These reactions all fit into the general mechanistic patterns that were discussed in Chapter 5 of Part A. The order of reactivity of alkylating groups is benzyl $\approx$ allyl $>$ methyl $>$ primary $>$ secondary. Tertiary halides and sulfonates are generally not satisfactory reactants because of the preference for ionization processes over S_N2 substitution. Because of their high reactivity toward nucleophilic substitution, α-haloesters, α-haloketones, and α-halonitriles are usually favorable reactants for substitution reactions. The reactivity of leaving groups is sulfonate $>$ iodide $>$ bromide $>$ chloride. Steric hindrance can greatly decrease the rate of nucleophilic substitution. Thus, projected synthetic steps involving nucleophilic substitution must be evaluated for potential steric problems. Scheme 3.2 gives some representative examples of nucleophilic substitution processes drawn from *Organic Synthesis* and from recent synthetic efforts.

66. N. Kreutzkamp and H. Kayser, *Chem. Ber.* **9**, 1614 (1956).
67. I. J. Borowitz and H. Parnes, *J. Org. Chem.* **32**, 3560 (1967).

Scheme 3.2. Transformation of Functional Groups by Nucleophilic Substitution

A. Nitriles

1[a] CH_3CHCH_2OH (octahydroanthracenyl) → 1) CH_3SO_2Cl, pyridine; 2) NaCN, DMF, 40–60°C, 3 h → CH_3CHCH_2CN (85%)

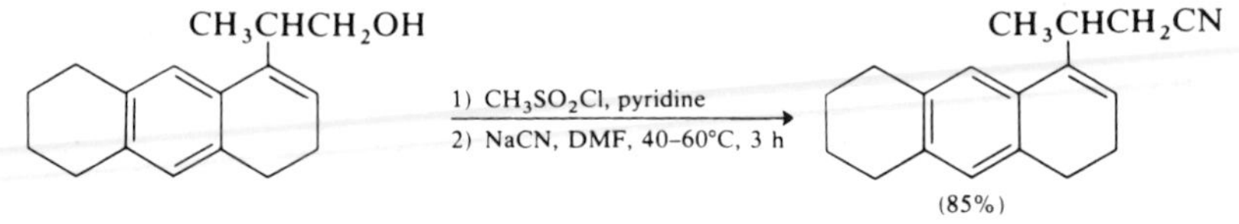

2[b] CH_3–$CHCH_2OH$ (4-methylcyclohexenyl, H_3C) → 1) $ArSO_2Cl$; 2) NaCN, DMSO, 90°C, 5 h → CH_3–$CHCH_2CN$ (80%)

3[c] CH_2OH / CH_2OH (cyclohexene) → 1) $ArSO_2Cl$; 2) NaCN, DMSO → CH_2CN / CH_2CN

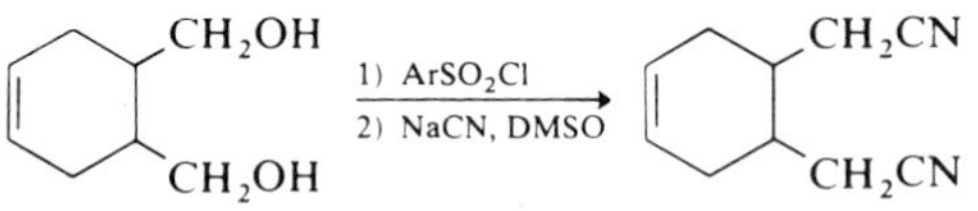

B. Azides

4[d] $CH_3CH_2CH_2CH_2Br + NaN_3 \xrightarrow[H_2O,\ 100°C,\ 6\ h]{R_4\overset{+}{N}Cl^-} CH_3CH_2CH_2CH_2N_3$ (97%)

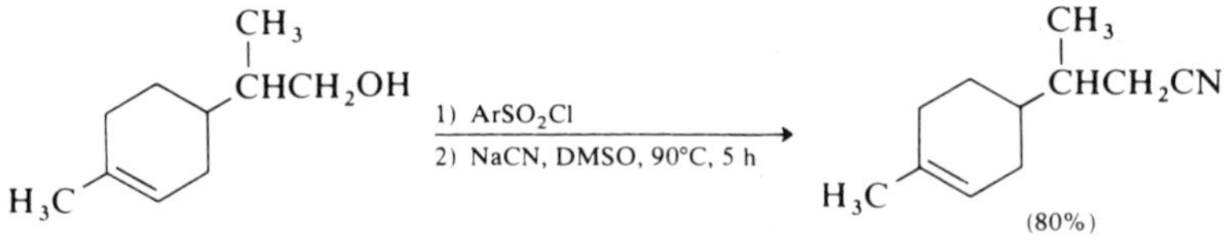

5[e] OH, $CH(CH_3)_2$, CH_3 (cyclohexanol) → $Ph_2\overset{O}{\overset{\|}{P}}N_3$, Ph_3P, $C_2H_5O_2CN{=}NCO_2C_2H_5$ → N_3, $CH(CH_3)_2$, CH_3 (90%)

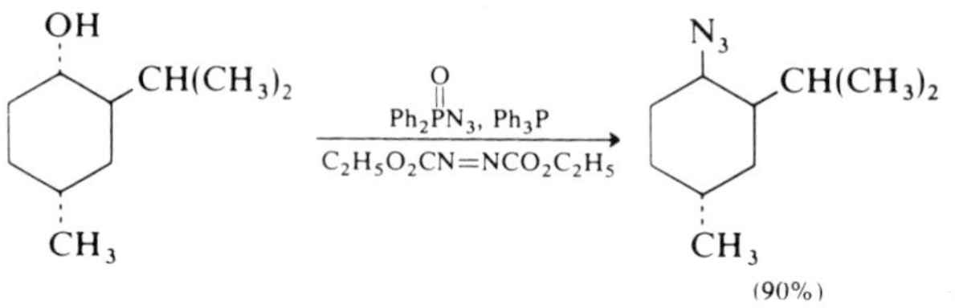

6[f] CH_3, HO, N, H (quinolizidine) → Ph_3P, $C_2H_5O_2CN{=}NCO_2C_2H_5$, $Ph_2\underset{O}{\underset{\|}{P}}N_3$ → CH_3, N_3, N, H (60%)

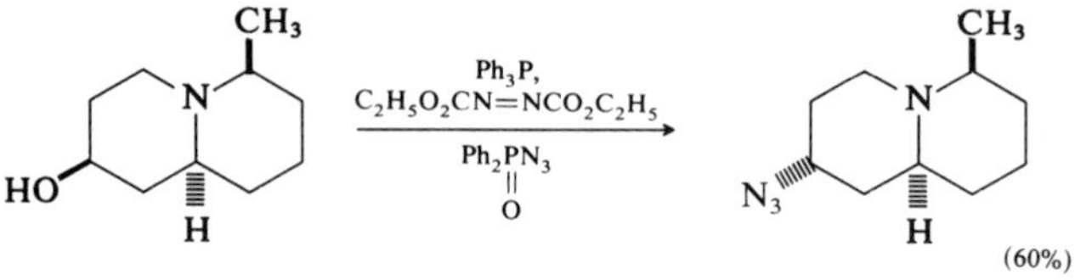

C. Amines and Amides

7[g] pyrrolidine NH + $CH_3\underset{Br}{CH}CO_2C_2H_5$ → $NCHCO_2C_2H_5$ (CH_3) (80–90%)

8[h] $HN\ \ \overset{+}{N}H_2$ (piperazinium) $\xrightarrow{PhCH_2Cl} \xrightarrow{^-OH}$ $PhCH_2N\ \ NH$ (65–75%)

9[i] caprolactam (=O, NH) $\xrightarrow[benzene]{(CH_3O)_2SO_2} \xrightarrow{K_2CO_3}$ $-OCH_3$, N (60–70%)

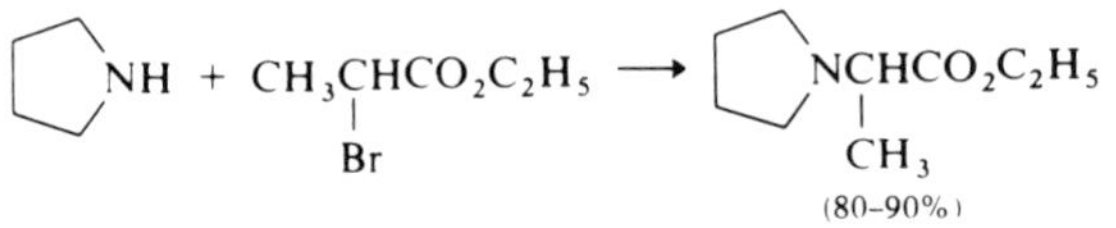

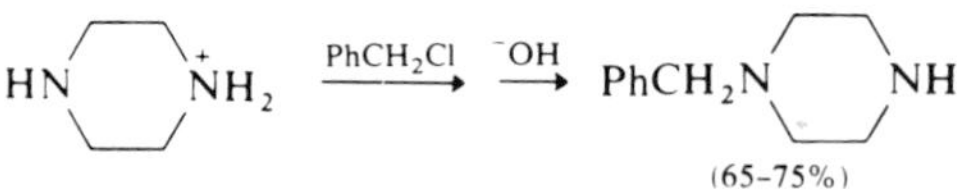

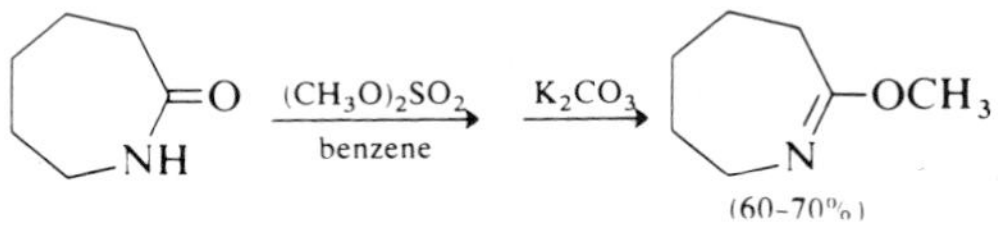

D. Hydrolysis of Alkyl Halides

10[j] $H_3C-C_6H_4-C(=O)CH(Cl)-C_6H_5 \xrightarrow[4\ h,\ 25^\circ C]{NaOH,\ H_2O} H_3C-C_6H_4-C(=O)CH(OH)-C_6H_5$ (92%)

11[k] $(CH_3O)_2C_6H_3-CH(Br)-CH(Br)CO_2CH_3 \xrightarrow[10\ min]{H_2O,\ 100^\circ C} (CH_3O)_2C_6H_3-CH(OH)-CH(Br)CO_2CH_3$ (92%)

E. Ethers by Base-Catalyzed Alkylation

12[l] (diacetone glucose, HO) $\xrightarrow[PhCH_2Cl,\ heat,\ 3\ h]{NaH,\ THF,\ DMSO,}$ ($PhCH_2O$ derivative) (95%)

13[m] (glucose: HO, HO, HO, OH, OH) $\xrightarrow[NaOH]{(CH_3O)_2SO_2} \xrightarrow[H_2O]{HCl}$ (CH_3O, CH_3O, CH_3O, OCH_3, OH)

14[n] o-$O_2NC_6H_4OH + CH_3CH_2CH_2CH_2Br \xrightarrow{K_2CO_3} o$-$O_2NC_6H_4OCH_2CH_2CH_2CH_3$ (75–80%)

15[o] 2,4-$(HO)_2C_6H_3COCH_3 \xrightarrow[K_2CO_3]{CH_3I}$ 2-HO-4-$CH_3OC_6H_3COCH_3$ (55–65%)

F. Esterification by Diazoalkanes and Triazenes

16[p] $\triangleright-CH_2CO_2H + CH_2N_2 \longrightarrow \triangleright-CH_2CO_2CH_3$ (79%)

17[q] 3,5-$(O_2N)_2C_6H_3CO_2H + H_3C-C_6H_4-N{=}NNHCH_3 \longrightarrow$ 3,5-$(O_2N)_2C_6H_3CO_2CH_3$ (70–90%)

G. Esterification by Nucleophilic Substitution with Carboxylate Salts

18[r] $(CH_3)_3CCO_2^- + BrCH_2C(=O)-C_6H_4-Br \xrightarrow{18\text{-crown-6}} (CH_3)_3CCO_2CH_2C(=O)-C_6H_4-Br$ (95%)

Scheme 3.2—*continued*

19[s] 2,4,6-trimethylbenzoate (H_3C–$C_6H_2(CH_3)_2$–CO_2^-) + $CH_3CH(I)(CH_2)_5CH_3$ $\xrightarrow[56°C]{acetone}$ H_3C–$C_6H_2(CH_3)_2$–$CO_2CH(CH_3)(CH_2)_5CH_3$ (100%)

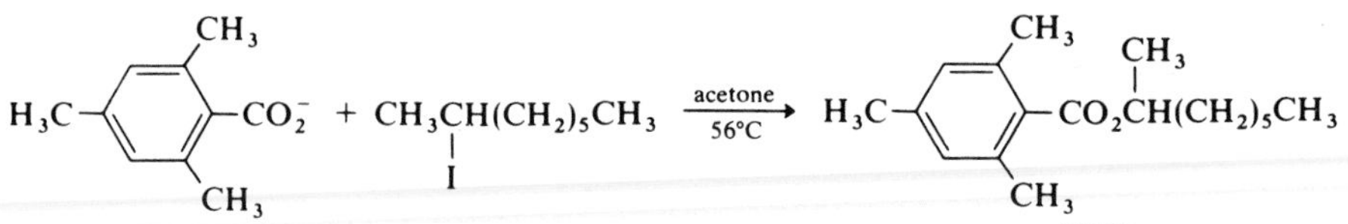

20[t] diacetone sugar acid (CO_2H) $\xrightarrow[18\ h]{CH_3I,\ KF,\ DMF,\ 25°C}$ methyl ester (CO_2CH_3) (84%)

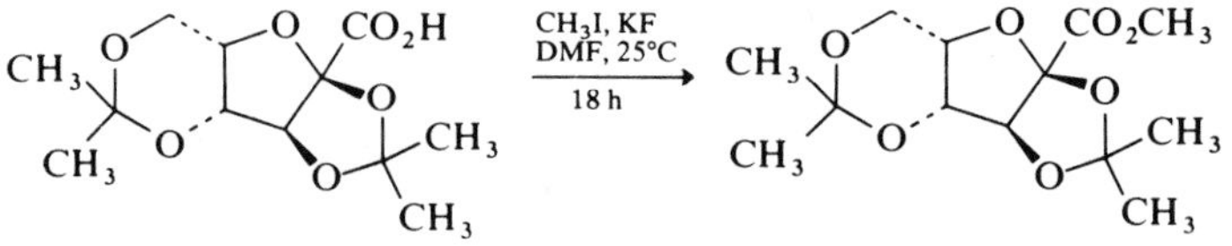

H. Phosphorus Nucleophiles

21[u] $Ph_3P + BrCH_2CH_2OPh \rightarrow Ph_3\overset{+}{P}CH_2CH_2OPh\ Br^-$

22[v] $[(CH_3)_2CHO]_3P + CH_3I \rightarrow [(CH_3)_2CHO]_2\overset{O}{\overset{\|}{P}}CH_3 + (CH_3)_2CHI$ (85–90%)

I. Sulfur Nucleophiles

23[w] $CH_3(CH_2)_{10}CH_2Br + S{=}C(NH_2)_2 \rightarrow \xrightarrow[H_2O]{NaOH} CH_3(CH_2)_{10}CH_2SH$ (80%)

24[x] $Na^+\ {}^-SCH_2CH_2S^-\ Na^+ + BrCH_2CH_2Br \longrightarrow$ 1,4-dithiane (S, S ring) (55–60%)

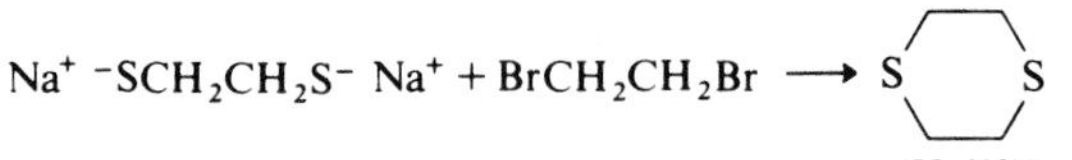

25[y] N-methylpyrrolidine-2-thione (N–CH_3, =S) $\xrightarrow[2)\ (CH_3)_3CO^-K^+]{1)\ CH_3I}$ 1-methyl-2-(SCH_3)-2-pyrroline (N–CH_3) (62%)

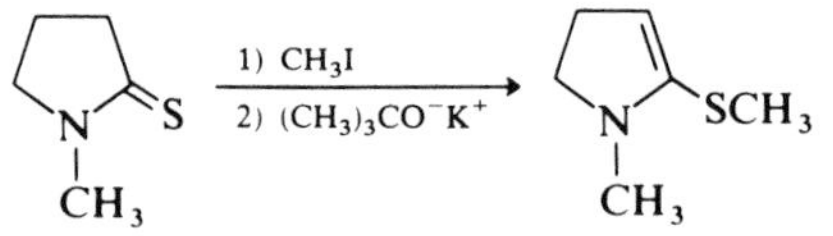

a. M. S. Newman and S. Otsuka, *J. Org. Chem.* **23**, 797 (1958).
b. B. A. Pawson, H.-C. Cheung, S. Gurbaxani, and G. Saucy, *J. Am. Chem. Soc.* **92**, 336 (1970).
c. J. J. Bloomfield and P. V. Fennessey, *Tetrahedron Lett.*, 2273 (1964).
d. W. P. Reeves and M. L. Bahr, *Synthesis*, 823 (1976).
e. B. Lal, B. N. Pramanik, M. S. Manhas, and A. K. Bose, *Tetrahedron Lett.*, 1977 (1977).
f. M. S. Hadley, F. D. King, B. McRitchie, D. H. Turner, and E. A. Watts, *J. Med. Chem.* **28**, 1843 (1985).
g. R. B. Moffett, *Org. Synth.* **IV**, 466 (1963).
h. J. C. Craig and R. J. Young, *Org. Synth.* **V**, 88 (1973).
i. R. E. Benson and T. L. Cairns, *Org. Synth.* **IV**, 588 (1963).
j. R. N. McDonald and P. A. Schwab, *J. Am. Chem. Soc.* **85**, 4004 (1963).
k. E. Adler and K. J. Bjorkquist, *Acta Chem. Scand.* **5**, 241 (1951).
l. C. H. Heathcock, C. T. White, J. J. Morrison, and D. VanDerveer, *J. Org. Chem.* **46**, 1296 (1981).
m. E. S. West and R. F. Holden, *Org. Synth.* **III**, 800 (1955).
n. C. F. H. Allen and J. W. Gates, Jr., *Org. Synth.* **III**, 140 (1955).
o. G. N. Vyas and M. N. Shah, *Org. Synth.* **IV**, 836 (1963).
p. L. I. Smity and S. McKenzie, Jr., *J. Org. Chem.* **15**, 74 (1950); A. I. Vogel, *Practical Organic Chemistry*, third edition, Wiley (1956), p. 973.
q. E. H. White, A. A. Baum, and D. E. Eitel, *Org. Synth.* **V**, 797 (1973).
r. H. D. Durst, *Tetrahedron Lett.*, 2421 (1974).
s. G. G. Moore, T. A. Foglia, and T. J. McGahan, *J. Org. Chem.* **44**, 2425 (1979).
t. C. H. Heathcock, C. T. White, J. Morrison, and D. VanDerveer, *J. Org. Chem.* **46**, 1296 (1981).
u. E. E. Schweizer and R. D. Bach, *Org. Synth.* **V**, 1145 (1973).
v. A. H. Ford-Moore and B. J. Perry, *Org. Synth.* **IV**, 325 (1963).
w. G. G. Urquhart, J. W. Gates, Jr., and R. Conor, *Org. Synth.* **III**, 363 (1965).
x. R. G. Gillis and A. B. Lacey, *Org. Synth.* **IV**, 396 (1963).
y. R. Gompper and W. Elser, *Org. Synth.* **V**, 780 (1973).

The cleavage of carbon-oxygen bonds in ethers or esters by nucleophilic substitution is frequently a useful synthetic transformation.

$$R{-}O{-}CH_3 + \mathbf{Nu}^- \rightarrow RO^- + CH_3{-}\mathbf{Nu}$$

$$RC(=O){-}O{-}CH_3 + \mathbf{Nu}^- \rightarrow RCO_2^- + CH_3{-}\mathbf{Nu}$$

The classical ether cleavage conditions involving concentrated hydrogen halides are much too strenuous for most polyfunctional molecules, so several milder reagents have been developed for effecting these transformations.[68] These reagents include boron tribromide,[69] dimethylboron bromide,[70] trimethylsilyl iodide,[71] and boron trifluoride in the presence of thiols.[72] The mechanism for ether cleavage with boron tribromide involves attack of bromide ion on an adduct formed from the ether and the electrophilic boron reagent. The cleavage step can occur by either an S_N2 or S_N1 process, depending on the nature of the alkyl group.

$$R{-}O{-}R + BBr_3 \rightarrow R{-}\overset{+}{O}(BBr_3^-){-}R \rightleftharpoons R{-}\overset{+}{O}(BBr_2){-}R + Br^-$$

$$R{-}\overset{+}{O}(BBr_2){-}R \;\; Br^- \rightarrow R{-}O{-}BBr_2 + RBr$$

$$R{-}O{-}BBr_2 + 3H_2O \rightarrow ROH + B(OH)_3 + 2HBr$$

Generally, good yields are observed, especially for methyl ethers. The combination of boron tribromide with dimethyl sulfide has been found to be particularly effective for cleaving aryl methyl ethers.[73]

Trimethylsilyl iodide cleaves methyl ethers in a period of a few hours at room temperature.[71] Benzyl and *t*-butyl systems are cleaved very rapidly, whereas secondary systems require longer times. The reaction presumably proceeds via an initially formed silyl oxonium ion.

$$R{-}O{-}R + (CH_3)_3SiI \rightarrow R{-}\overset{+}{O}(Si(CH_3)_3){-}R + I^- \rightarrow R{-}O{-}Si(CH_3)_3 + RI$$

The direction of cleavage in unsymmetrical ethers is determined by the relative ease of O—R bond breaking by either S_N2 (methyl, benzyl) or S_N1 (*t*-butyl) processes.

68. M. V. Bhatt and S. U. Kulkarni, *Synthesis*, 249 (1983).
69. J. F. W. McOmie, M. L. Watts, and D. E. West, *Tetrahedron* **24**, 2289 (1968).
70. Y. Guindon, M. Therien, Y. Girard, and C. Yoakim, *J. Org. Chem.* **52**, 1680 (1987).
71. M. E. Jung and M. A. Lyster, *J. Org. Chem.* **42**, 3761 (1977).
72. M. Node, H. Hori, and E. Fujita, *J. Chem. Soc., Perkin Trans. 1*, 2237 (1976); K. Fuji, K. Ichikawa, M. Node, and E. Fujita, *J. Org. Chem.* **44**, 1661 (1979).
73. P. G. Williard and C. R. Fryhle, *Tetrahedron Lett.* **21**, 3731 (1980).

Because trimethylsilyl iodide is rather expensive, alternative procedures which generate the reagent *in situ* have been devised:

$$(CH_3)_3SiCl + NaI \xrightarrow{CH_3CN} (CH_3)_3SiI + NaCl \qquad \text{Ref. 74}$$

$$PhSi(CH_3)_3 + I_2 \rightarrow (CH_3)_3SiI + PhI \qquad \text{Ref. 75}$$

Diiodosilane, SiH_2I_2, is an especially effective reagent for cleaving secondary alkyl ethers.[76]

Trimethylsilyl iodide also effects rapid cleavage of esters. The first products formed are trimethylsilyl esters, but these are hydrolyzed rapidly on exposure to water.[77]

$$RC(=O)O{-}R' + (CH_3)_3SiI \rightarrow RC(={}^{+}OSi(CH_3)_3)O{-}R' + I^- \rightarrow RC(=O)OSi(CH_3)_3 + R'I$$

$$RC(=O)OSi(CH_3)_3 + H_2O \rightarrow RCO_2H + (CH_3)_3SiOH$$

Benzyl, methyl, and *t*-butyl esters are rapidly cleaved, but secondary esters react more slowly. In the case of *t*-butyl esters, the initial silylation is followed by a rapid ionization to form the *t*-butyl cation.

The boron trifluoride–alkyl thiol combination reagent also operates on the basis of nucleophilic attack on an oxonium ion generated by reaction of the ether with boron trifluoride[78]:

$$R{-}O{-}R + BF_3 \rightarrow R{-}\overset{+}{O}({}^{-}BF_3){-}R$$

$$R{-}\overset{+}{O}({}^{-}BF_3){-}R + R'SH \rightarrow RO\overset{-}{B}F_3 + RSR' + H^+$$

Ether cleavage can also be effected by reaction with acetic anhydride and Lewis acids such as BF_3, $FeCl_3$, and $MgBr_2$.[79] Mechanistic investigations have pointed to acylium ions generated from the anhydride and Lewis acid as the reactive electrophiles:

74. T. Morita, Y. Okamoto, and H. Sakurai, *J. Chem. Soc., Chem. Commun.*, 874 (1978); G. A. Olah, S. C. Narang, B. G. B. Gupta, and R. Malhotra, *Synthesis*, 61 (1979).
75. T. L. Ho and G. A. Olah, *Synthesis*, 417 (1977); R. A. Benkeser, E. C. Mozdzen, and C. L. Muth, *J. Org. Chem.* **44**, 2185 (1979).
76. E. Keinan and D. Perez, *J. Org. Chem.* **52**, 4846 (1987).
77. T. L. Ho and G. A. Olah, *Angew. Chem. Int. Ed. Engl.* **15**, 774 (1976); M. E. Jung and M. A. Lyster, *J. Am. Chem. Soc.* **99**, 968 (1977).
78. K. Fuji, K. Ichikawa, M. Node, and E. Fujita, *J. Org. Chem.* **44**, 1661 (1979).
79. C. R. Narayanan and K. N. Iyer, *J. Org. Chem.* **30**, 1734 (1965); B. Ganem and V. R. Small, Jr., *J. Org. Chem.* **39**, 3728 (1974); D. J. Goldsmith, E. Kennedy, and R. G. Campbell, *J. Org. Chem.* **40**, 3571 (1975).

$$(RCO)_2O + MX_n \rightarrow RC\equiv\overset{+}{O} + [MX_nO_2CR]^-$$

$$RC\equiv\overset{+}{O} + R'{-}O{-}R' \rightarrow R'{-}\overset{+}{\underset{\underset{R-C=O}{|}}{O}}{-}R'$$

$$R'{-}\overset{+}{\underset{\underset{R-C=O}{|}}{O}}{-}R' + X^- \rightarrow R'{-}X + RCO_2R'$$

Scheme 3.3 gives some specific examples of ether and ester cleavage reactions.

Scheme 3.3. Cleavage of Ethers and Esters

1[a] 3,3'-dimethoxybiphenyl (CH_3O, OCH_3) $\xrightarrow{BBr_3}$ $\xrightarrow{H_2O}$ 3,3'-dihydroxybiphenyl (HO, OH) (75–85%)

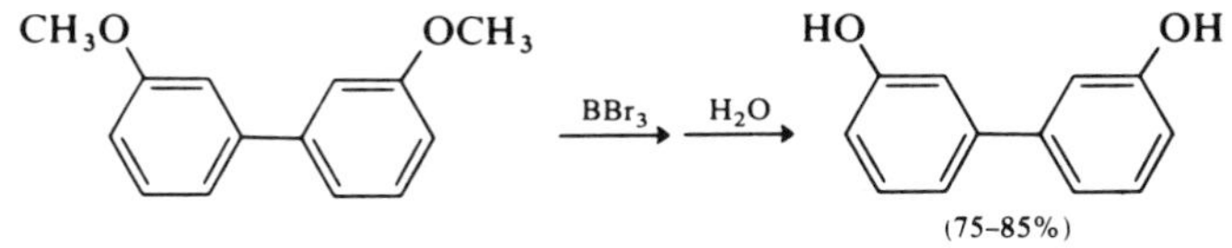

2[b] $CH_2{=}CH$, CH_2OCH_3, $CH_3O_2CCH_2$, H $\xrightarrow[-78°C]{BBr_3,\ CH_2Cl_2}$ lactone with $CH{=}CH_2$, H (88%)

3[c] cyclohexyl–OCH_3 $\xrightarrow{(CH_3)_3SiI}$ cyclohexyl–OH (83–89%)

4[d] cyclohexyl–CO_2CH_3 $\xrightarrow[NaI,\ CH_3CN]{(CH_3)_3SiCl}$ cyclohexyl–$CO_2Si(CH_3)_3$ + CH_3I (86%)

5[e] aryl with OCH_3, OCH_3, CH_3 $\xrightarrow{(CH_3)_3SiI}$ $\xrightarrow{H_2O}$ aryl with OCH_3, OH, CH_3

6[f] CH_3–aryl(Br)–O–CH_2–phenyl $\xrightarrow[C_2H_5SH]{BF_3}$ CH_3–aryl(Br)–OH (90%)

7[g] polycyclic ether (H_3C, H, OCH_3, H_3C, H, CH_3) $\xrightarrow[C_2H_5SH]{BF_3}$ polycyclic alcohol (H_3C, H, OH, H_3C, H, CH_3) (75%)

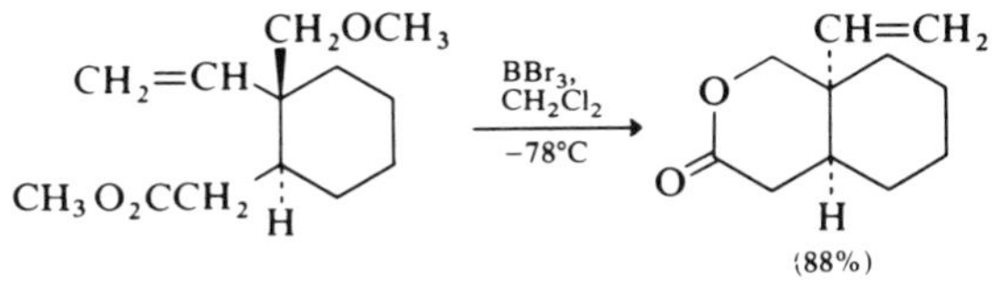

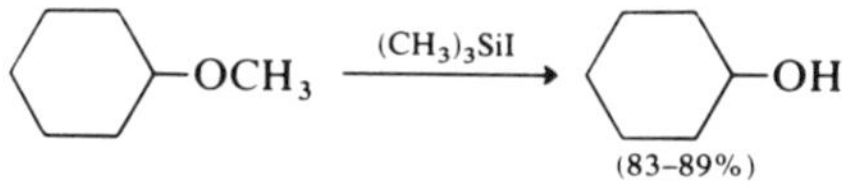

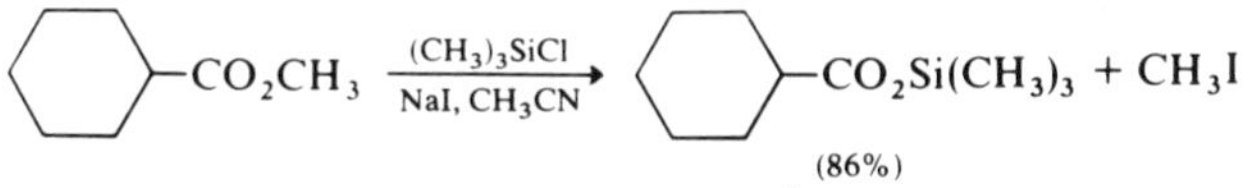

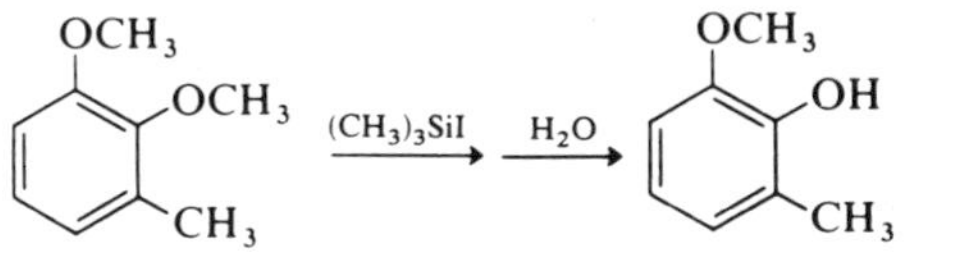

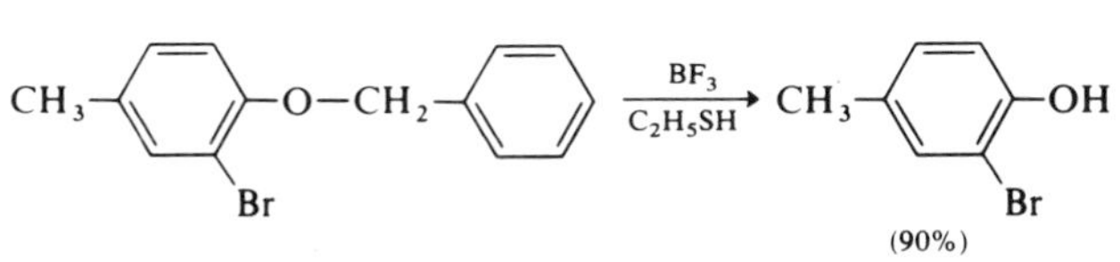

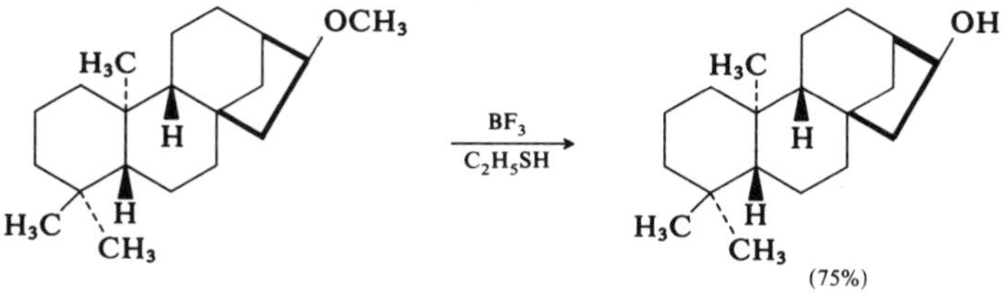

8[h] (structure) $\xrightarrow[\text{NaOAc}]{BF_3 \cdot OEt_2,\ EtSH}$ (structure) (61%)

9[i] (structure) $\xrightarrow{(CH_3)_2BBr}$ Br---(cyclohexane)—OH (85%)

10[j] $(CH_3)_2CHOCH(CH_3)_2 \xrightarrow[(CH_3CO)_2O]{FeCl_3} (CH_3)_2CHO_2CCH_3$ (83%)

a. J. F. W. McOmie and D. E. West, *Org. Synth.* **V**, 412 (1973).
b. P. A. Grieco, K. Hiroi, J. J. Reap, and J. A. Noguez, *J. Org. Chem.* **40**, 1450 (1975).
c. M. E. Jung and M. A. Lyster, *Org. Synth.* **59**, 35 (1980).
d. T. Morita, Y. Okamoto, and H. Sakurai, *J. Chem. Commun.*, 874 (1978).
e. E. H. Vickery, L. F. Pahler, and E. J. Eisenbraun, *J. Org. Chem.* **44**, 4444 (1979).
f. K. Fuji, K. Ichikawa, M. Node, and E. Fujita, *J. Org. Chem.* **44**, 1661 (1979).
g. M. Nobe, H. Hori, and E. Fujita, *J. Chem. Soc. Perkin Trans. 1*, 2237 (1976).
h. A. B. Smith III, N. J. Liverton, N. J. Hrib, H. Sivaramakrishnan, and K. Winzenberg, *J. Am. Chem. Soc.* **108**, 3040 (1986).
i. Y. Guindon, M. Therien, Y. Girard, and C. Yoakim, *J. Org. Chem.* **52**, 1680 (1987).
j. B. Ganem and V. R. Small, Jr., *J. Org. Chem.* **39**, 3728 (1974).

3.4. Interconversion of Carboxylic Acid Derivatives

The classes of compounds that are conveniently considered together as derivatives of carboxylic acids include the carboxylic acid anhydrides, acid chlorides, esters, and amides. In the case of simple aliphatic and aromatic acids, synthetic transformations among these derivatives are usually a straightforward matter involving such fundamental reactions as ester saponification, formation of acid chlorides, and the reactions of amines with acid anhydrides or acid chlorides:

$$RCO_2CH_3 \xrightarrow[H_2O]{^-OH} RCO_2^- + CH_3OH$$

$$RCO_2H + SOCl_2 \rightarrow RCOCl + HCl + SO_2$$

$$RCOCl + R'_2NH \rightarrow RCONR'_2$$

When a multistep synthesis is being undertaken with other sensitive functional groups present in the molecule, milder reagents and reaction conditions may be necessary. As a result, many alternative methods for effecting interconversion of the carboxylic acid derivatives have been developed, and some of the most common will be considered in the succeeding sections.

3.4.1. Preparation of Reactive Reagents for Acylation

The traditional method for transforming carboxylic acids into reactive acylating agents capable of converting alcohols to esters or amines to amides is by formation of the acid chloride. Molecules devoid of acid-sensitive functional groups can be converted to acid chlorides with thionyl chloride or phosphorus pentachloride. When milder conditions are necessary, the reaction of the acid or its sodium salt with oxalyl chloride provides the acid chloride. When a salt is used, the reaction solution remains essentially neutral.

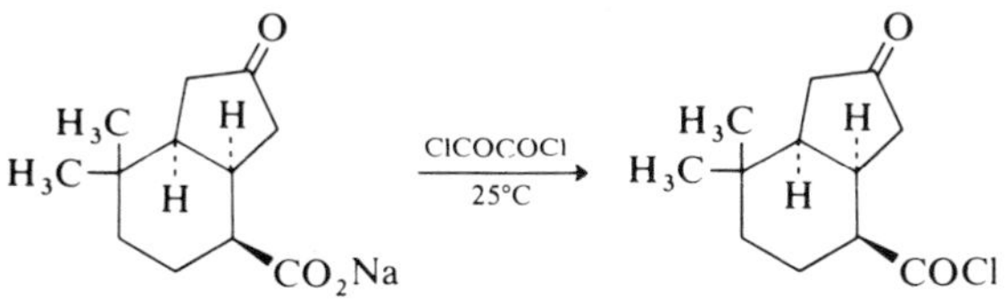

H_3C, H_3C, H, H, O, CO_2Na —(ClCOCOCl, 25°C)→ H_3C, H_3C, H, H, O, COCl   Ref. 80

Acid chlorides are highly reactive acylating agents and react very rapidly with amines. For alcohols, preparative procedures often call for use of pyridine as a catalyst. Pyridine catalysis involves initial formation of an acyl pyridinium ion, which then reacts with the alcohol. Pyridine is a better nucleophile than the neutral alcohol, but the acylpyridinium ion reacts more rapidly with the alcohol than the acid chloride.[81]

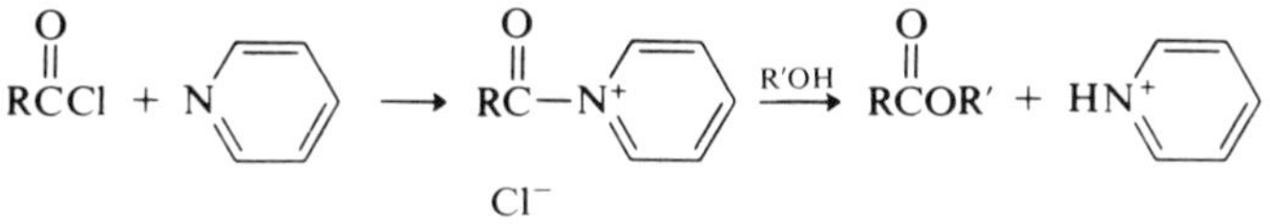

$$\text{RCCl (O)} + \text{N(pyridine)} \longrightarrow \text{RC(O)}{-}\text{N}^+\text{(pyridinium)}\ \text{Cl}^- \xrightarrow{\text{R'OH}} \text{RCOR' (O)} + \text{HN}^+\text{(pyridinium)}$$

An even stronger catalytic effect is obtained when 4-dimethylaminopyridine (DMAP) is used as a nucleophilic catalyst.[82] The dimethylamino function acts as an electron donor substituent, increasing both the nuclephilicity and basicity of the pyridine nitrogen.

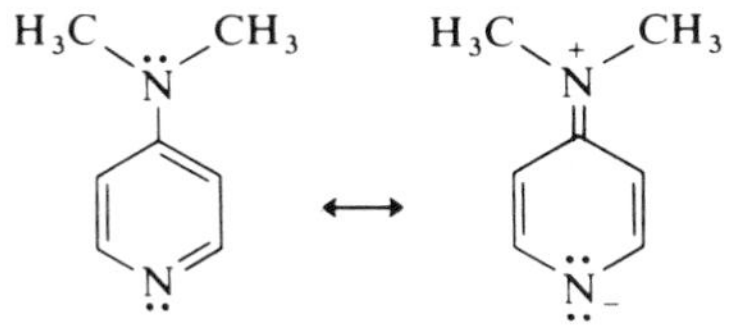

H_3C, CH_3, N: (DMAP) ⟷ H_3C, CH_3, N^+, N^-

The inclusion of DMAP to the extent of 5–10 mol % in acylations by acid anhydrides and acid chlorides increases acylation rates by up to four orders of magnitude and permits successful acylation of tertiary and other hindered alcohols.

80. M. Miyano and C. R. Dorn, *J. Org. Chem.* **37**, 268 (1972).
81. A. R. Fersht and W. P. Jencks, *J. Am. Chem. Soc.* **92**, 5432, 5442 (1970).
82. G. Hofle, W. Steglich, and H. Vorbruggen, *Angew. Chem. Int. Ed. Engl.* **17**, 569 (1978); E. F. V. Scriven, *Chem. Soc. Rev.* **12**, 129 (1983).

The mechanism of catalysis by DMAP is considered to involve an *N*-acylpyridinium ion. However, the identity of the anion also affects the reactivity so that a complete formulation requires that the ion pair characteristics of the acylpyridinium ion be taken into account. Interestingly, in the presence of DMAP, acetic anhydride is a more reactive acylating agent than acetyl chlorode. This reversal of the normal reactivity order can be explained if the acetate ion, a stronger base than chloride, is involved in deprotonating the alcohol[83]:

$(CH_3)_2N$–$C_5H_4N^+$–C(=O)CH_3 + $^-$O–C(=O)R + H–O–R′ → $(CH_3)_2N$–$C_5H_4N^+$–C(CH_3)(OR′)–O^- → CH_3COR' (C=O) + $(CH_3)_2N$–C_5H_4N

There are other activation procedures which generate acyl halides *in situ* in the presence of the nucleophile. Refluxing a carboxylic acid, triphenylphosphine, bromotrichloromethane, and an amine gives rise to the corresponding amide[84]:

$$RCO_2H + R'NH_2 \xrightarrow{PPh_3,\ CBrCl_3} RC(=O)NHR'$$

This reaction presumably proceeds via the acid bromide, since it is known that triphenylphosphine and carbon tetrachloride convert acids to the corresponding acid chloride.[85] Similarly, carboxylic acids react with the triphenylphosphine-bromine adduct to give acyl bromides.[86] These reactions are mechanistically analogous to the alcohol-to-halide conversions that were discussed in Section 3.1.2.

$$RCO_2H + Ph_3\overset{+}{P}Br \rightarrow RC(=O){-}O{-}\overset{+}{P}Ph_3 + HBr$$

$$Br^- + RC(=O){-}O{-}\overset{+}{P}Ph_3 \rightarrow RC(=O)Br + O{=}PPh_3$$

In addition to acid chlorides and acid bromides, there are a number of milder and more selective acylating agents which can be readily prepared from carboxylic acids. Imidazolides, the *N*-acyl derivatives of imidazole, are examples.[87] Two factors are responsible for the high reactivity of the imidazolides as acylating reagents. One is the relative weakness of the "amide" bond. Because of the aromatic character of imidazole, there is little of the N→C=O delocalization that characterizes normal

83. E. Guibe-Jampel, G. Le Corre, and M. Wakselman, *Tetrahedron Lett.*, 1157 (1979).
84. L. E. Barstow and V. J. Hruby, *J. Org. Chem.* **36**, 1305 (1971).
85. J. B. Lee, *J. Am. Chem. Soc.* **88**, 3440 (1966).
86. H. J. Bestmann and L. Mott, *Justus Liebigs Ann. Chem.* **693**, 132 (1966).
87. H. A. Staab and W. Rohr, *Newer Methods Prep. Org. Chem.* **5**, 61 (1968).

amides. The second factor is the enhancement in reactivity that results from protonation of the other imidazole nitrogen, which makes the imidazole ring a better leaving group.

$$\mathbf{Nu} + RC(=O){-}N\text{(imidazole)} \xrightarrow{H^+} \mathbf{Nu}{-}C(=O)R + \text{imidazole (NH)}$$

Imidazolides are isolable substances and can be prepared directly from the carboxylic acid by reaction with carbonyldiimidazole.

$$RCO_2H + \text{Im}{-}C(=O){-}\text{Im} \longrightarrow RC(=O){-}\text{Im} + \text{HIm} + CO_2$$

Imidazolides react with alcohols on heating to give esters and react at room temperature with amines to give amides. Imidazolides are particularly appropriate for acylation of acid-sensitive materials.

Dicyclohexylcarbodiimide (DCC) is another example of a reagent that converts carboxylic acids to reactive acylating agents. This compound has been particularly widely applied in the acylation step in the synthesis of polypeptides from amino acids.[88] The reactive species is an acyl isourea. The acyl group is highly reactive in this environment because the cleavage of the acyl–oxygen bond converts the carbon–nitrogen double bond of the isourea to a more stable carbonyl group.[89]

$$RCO_2H + RN{=}C{=}NR \rightarrow RC(=O){-}O{-}C(=NR)NHR$$

$$RC(=O){-}O{-}C(=NR)NHR + \mathbf{Nu} \xrightarrow{H^+} RC(=O)\mathbf{Nu} + RNHC(=O)NHR$$

The combination of carboxyl activation by cyclohexylcarbodiimide and catalysis by DMAP provides a useful method for *in situ* activation of carboxylic acids for reaction with alcohols. The reaction proceeds at room temperature[90]:

$$Ph_2CHCO_2H + C_2H_5OH \xrightarrow[DMAP]{DCC} Ph_2CHCO_2C_2H_5$$

2-Chloropyridinium[91] and 3-chloroisoxazolium[92] cations also activate carboxyl groups toward nucleophilic attack. In each instance, the halide is displaced from

88. F. Kurzer and K. Douraghi-Zadeh, *Chem. Rev.* **67**, 107 (1967).
89. D. F. DeTar and R. Silverstein, *J. Am. Chem. Soc.* **88**, 1013, 1020 (1966); D. F. DeTar, R. Silverstein, and F. F. Rogers, Jr., *J. Am. Chem. Soc.* **88**, 1024 (1966).
90. A. Hassner and V. Alexanian, *Tetrahedron Lett.*, 4475 (1978); B. Neises and W. Steglich, *Angew. Chem. Int. Ed. Engl.* **17**, 522 (1978).
91. T. Mukaiyama, M. Usui, E. Shimada, and K. Saigo, *Chem. Lett.*, 1045 (1975).
92. K. Tomita, S. Sugai, T. Kobayashi, and T. Murakami, *Chem. Pharm. Bull.* **27**, 2398 (1979).

the heterocycle by the carboxylate via an addition–elimination mechanism. Nucleophilic attack on the activated carbonyl group results in elimination of the heterocyclic ring, with the departing oxygen becoming part of an amide-like structure. The positive charge on the heterocyclic ring accelerates both the initial addition step and the subsequent elimination of the heterocycle.

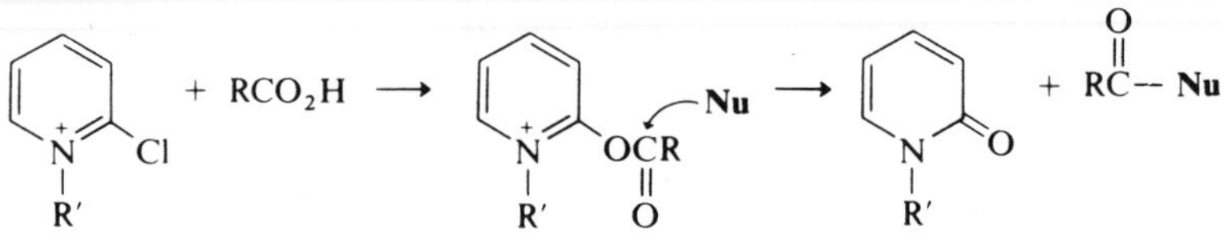

Carboxylic acid esters of thiols are considerably more reactive as acylating reagents than are the esters of alcohols. Particularly reactive are esters of pyridine-2-thiol since there is an additional driving force—the formation of the more stable pyridine-2-thione tautomer:

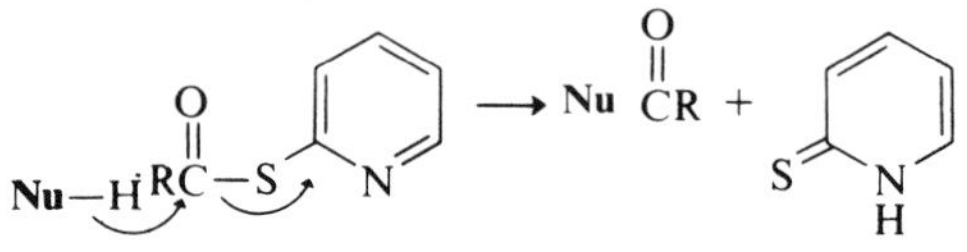

Additional acceleration of the rate of acylation can be obtained by inclusion of cupric salts which coordinate at the pyridine nitrogen. This modification is especially useful for the preparation of highly hindered esters.[93] Pyridine-2-thiol esters can be prepared by reaction of the carboxylic acid with 2,2′-dipyridyl disulfide and triphenylphosphine[94] or directly from the acid and 2-pyridyl thiochloroformate.[95]

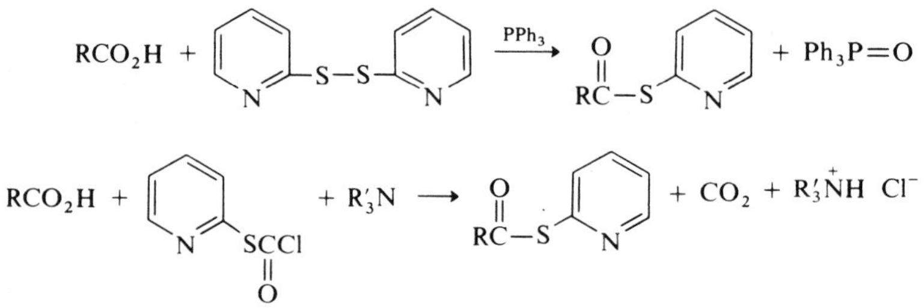

The 2-pyridyl and related 2-imidazoyl disulfides have found special use in the closure of large lactone rings.[96] This type of structure is encountered in a number of antibiotics which, because of the presence of numerous other sensitive functional groups, require mild conditions for cyclization. It has been suggested that the pyridyl

93. S. Kim and J. I. Lee, *J. Org. Chem.* **49**, 1712 (1984).
94. T. Mukaiyama, R. Matsueda, and M. Suzuki, *Tetrahedron Lett.*, 1901 (1970).
95. E. J. Corey and D. A. Clark, *Tetrahedron Lett.*, 2875 (1979).
96. E. J. Corey and K. C. Nicolaou, *J. Am. Chem. Soc.* **96**, 5614 (1974); K. C. Nicolaou, *Tetrahedron* **33**, 683 (1977).

and imidazyl thioesters function by a mechanism in which the heterocyclic nitrogen acts as a base, deprotonating the alcohol group:

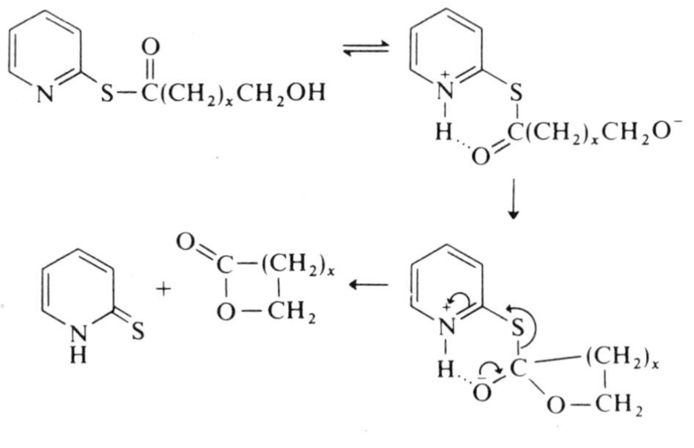

This provides an intermediate in which hydrogen bonding can enhance the reactivity of the carbonyl group.[97] Excellent yields of large ring lactones are achieved by this method.

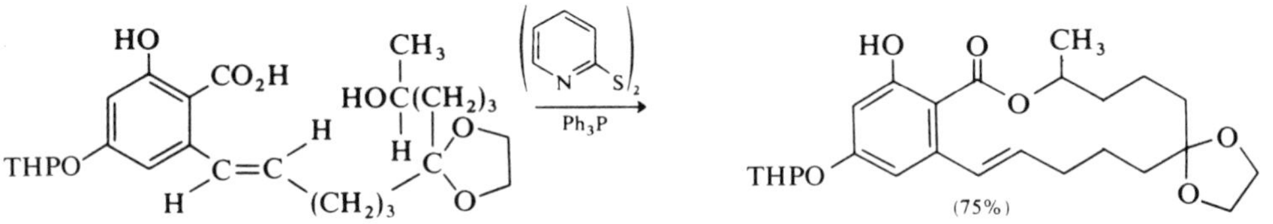

Ref. 96

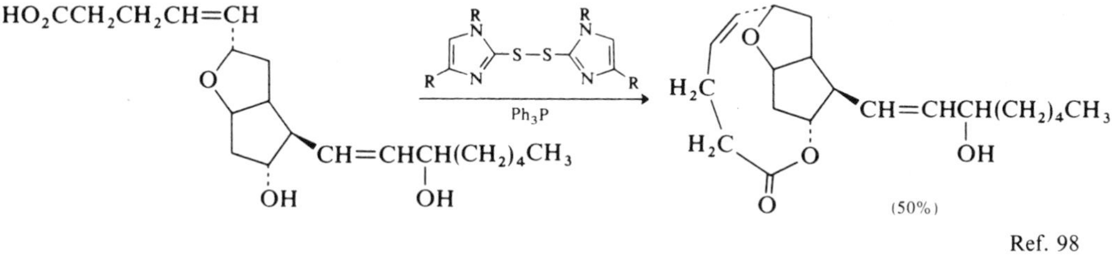

Ref. 98

Intramolecular lactonization can also be carried out with dicyclohexylcarbodiimide and dimethylaminopyridine. As with other macrolactonizations, the reactions must be carried out in dilute solution to promote the intramolecular transformation in competition with intermolecular reaction, which leads to dimers or higher oligomers. A study with 15-hydroxypentadecanoic acid demonstrated that a proton source is beneficial under these conditions and found the hydrochloride of DMAP to be convenient.[99]

97. E. J. Corey, K. C. Nicolaou, and L. S. Melvin, Jr., *J. Am. Chem. Soc.* **97**, 654 (1975); E. J. Corey, D. J. Brunelle, and P. J. Stork, *Tetrahedron Lett.*, 3405 (1976).
98. E. J. Corey, H. L. Pearce, I. Szekely, and M. Ishiguro, *Tetrahedron Lett.*, 1023 (1978).
99. E. P. Boden and G. E. Keck, *J. Org. Chem.* **50**, 2394 (1985).

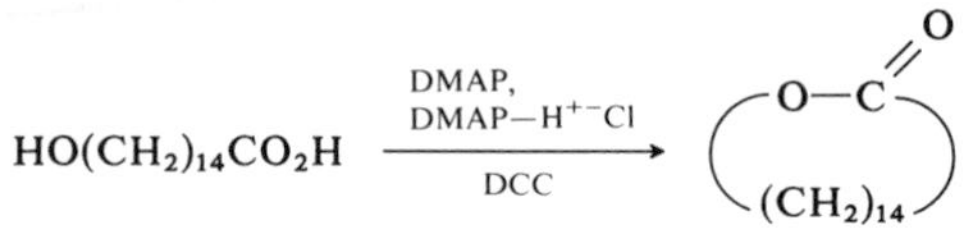

$$HO(CH_2)_{14}CO_2H \xrightarrow[\text{DCC}]{\text{DMAP, DMAP-H}^{+-}\text{Cl}} \text{(cyclic lactone: O–C(=O)–(CH}_2)_{14}\text{)}$$

Scheme 3.4 gives some typical examples of preparation and use of active acylating agents from carboxylic acids.

Scheme 3.4. Preparation and Reactions of Active Acylating Agents

1[a] (bicyclic acid bearing CH_3, CH_3, CH_3, CH_3CO_2 and CO_2H) $\xrightarrow[25°C]{ClCOCOCl}$ (corresponding acid chloride, COCl)

2[b] $(CH_3)_2C{=}CHCH_2CH_2C(CH_3){=}CHCH_2CH_2CH_2CO_2^-\ Na^+ \xrightarrow{ClCOCOCl} (CH_3)_2C{=}CHCH_2CH_2C(CH_3){=}CHCH_2CH_2CH_2COCl$

3[c] $PhCO_2H \xrightarrow{\text{N,N'-carbonyldiimidazole}} \xrightarrow{CH_2{=}CHCH(OH)CH_2{-}N\triangleleft} PhCO_2CH(CH{=}CH_2)CH_2{-}N\triangleleft$ (60%)

4[d] $C_6H_5CO_2H + HO{-}C_6H_4{-}NO_2 \xrightarrow{\text{dicyclohexyl-carbodiimide}} C_6H_5C(=O){-}O{-}C_6H_4{-}NO_2$

5[e] $PhCH_2CO_2H \xrightarrow{\text{2-chloro-1-methylpyridinium}} \xrightarrow{PhCH(CH_3)OH} PhCH_2C(=O)OCH(CH_3)Ph$ (88%)

6[f] $CH_3CH{=}CHCH{=}CHCO_2H \xrightarrow{\text{2-pyridyl-S-C(=O)Cl}} CH_3CH{=}CHCH{=}CHC(=O)S{-}\text{(2-pyridyl)}$

7[g] (lactone bearing HCO_2CH_2, $HCO_2(CH_2)_3CH(CH_3)$, OH and $=CH_2$) $\xrightarrow[\text{4-dimethylamino-pyridine}]{(CH_2{=}C(CH_3)CO)_2O}$ (corresponding ester $OC(=O)C(CH_3){=}CH_2$)

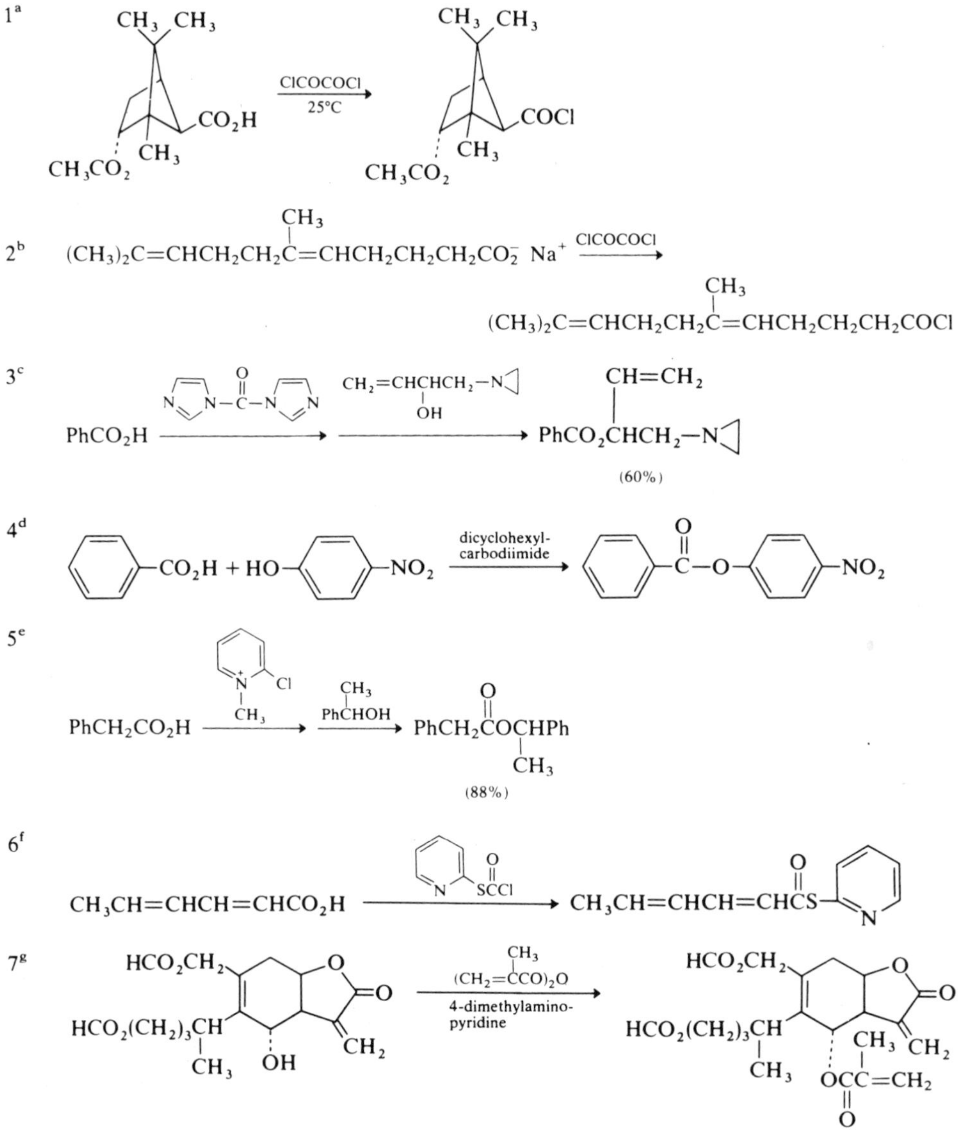

Scheme 3.4—*continued*

8[h]

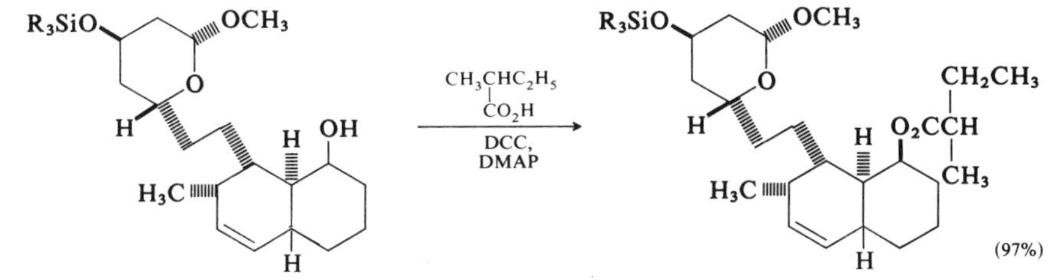
R3SiO
OCH3
O
H
H
OH
H3C
H
CH3CHC2H5
CO2H
DCC,
DMAP
R3SiO
OCH3
O
CH2CH3
H
H
O2CCH
CH3
H3C
H
(97%)

9[i]

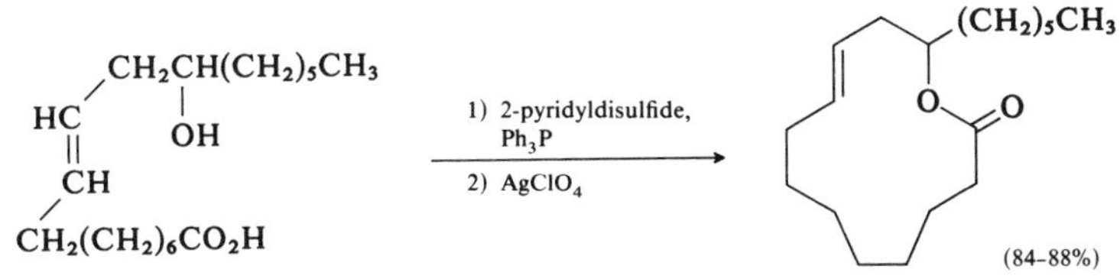
CH2CH(CH2)5CH3
HC
OH
CH
CH2(CH2)6CO2H
1) 2-pyridyldisulfide,
Ph3P
2) AgClO4
(CH2)5CH3
O
O
(84–88%)

a. J. Meinwald, J. C. Shelton, G. L. Buchanan, and A. Courtain, *J. Org. Chem.* **33**, 99 (1968).
b. U. T. Bhalerao, J. J. Plattner, and H. Rapaport, *J. Am. Chem. Soc.* **92**, 3429 (1970).
c. H. A. Staab and W. Rohr, *Chem. Ber.* **95**, 1298 (1962).
d. S. Neelakantan, R. Padmasani, and T. R. Seshadri, *Tetrahedron* **21**, 3531 (1965).
e. T. Mukaiyama, M. Usui, E. Shimada, and K. Saigo, *Chem. Lett.*, 1045 (1975).
f. E. J. Corey and D. A. Clark, *Tetrahedron Lett.*, 2875 (1979).
g. P. A. Grieco, T. Oguri, S. Gilman, and G. DeTitta, *J. Am. Chem. Soc.* **100**, 1616 (1978).
h. Y.-L. Yang, S. Manna, and J. R. Falck, *J. Am. Chem. Soc.* **106**, 3811 (1984).
i. A. Thalman, K. Oertle, and H. Gerlach, *Org. Synth.* **63**, 192 (1984).

3.4.2. Preparation of Esters

As mentioned in the preceding section, one of the most general methods of synthesis of esters is by reaction of alcohols with an acid chloride or other activated carboxylic acid derivative. Section 3.2.5 included a discussion of two other important methods, namely, reactions of carboxylic acids with diazoalkanes and reactions of carboxylate salts with alkyl halides or sulfonate esters. There remains to be mentioned the acid-catalyzed reaction of carboxylic acids with alcohols, which is frequently referred to as Fischer esterification:

$$RCO_2H + R'OH \xrightarrow{H^+} RCO_2R' + H_2O$$

This is an equilibrium process, and there are two techniques which are used to drive the reaction to completion. One is to use a large excess of the alcohol. This is feasible for simple and relatively inexpensive alcohols. The second method is to drive the reaction forward by irreversible removal of water. Azeotropic distillation is one method for doing this. Entries 1–4 in Scheme 3.5 are examples of acid-catalyzed esterifications. Entry 5 is the preparation of a diester starting with an anhydride. This is a closely related reaction in which the initial opening of the anhydride ring is followed by an acid-catalyzed esterification.

Scheme 3.5. Acid-Catalyzed Esterification

1[a] $CH_3CO_2H + HOCH_2CH_2CH_2Cl \xrightarrow[\text{benzene, azeotropic removal of water}]{ArSO_3H} CH_3CO_2CH_2CH_2CH_2Cl$ (93%–95%)

2[b] $HO_2CC{\equiv}CCO_2H + CH_3OH \xrightarrow[\text{25°C, 4 days}]{H_2SO_4} CH_3O_2CC{\equiv}CCO_2CH_3$ (72–88%)

3[c] $CH_3CH{=}CHCO_2H + CH_3CH(OH)CH_2CH_3 \xrightarrow[\text{benzene, azeotropic removal of water}]{H_2SO_4} CH_3CH{=}CHCO_2CH(CH_3)CH_2CH_3$ (85–90%)

4[d] $PhCH(OH)CO_2H + C_2H_5OH \xrightarrow[\text{78°C, 5 h}]{HCl} PhCH(OH)CO_2C_2H_5$ (82–86%)

5[e] 3-methylenecyclobutane-1,2-dicarboxylic anhydride $+ CH_3OH \xrightarrow[\text{67–68°C, 40 h}]{ArSO_3H}$ $H_2C{=}$cyclobutane-1,2-(CO_2CH_3)$_2$ (80–90%)

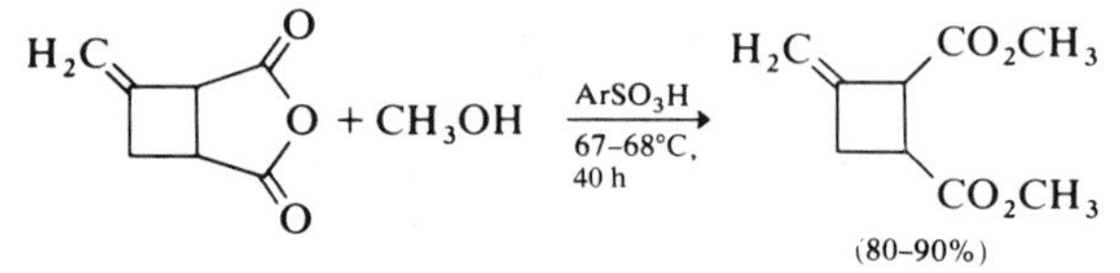

a. C. F. H. Allen and F. W. Spangler, *Org. Synth.* **III**, 203 (1955).
b. E. H. Huntress, T. E. Lesslie, and J. Bornstein, *Org. Synth.* **IV**, 329 (1963).
c. J. Munch-Petersen, *Org. Synth.* **V**, 762 (1973).
d. E. L. Eliel, M. T. Fisk, and T. Prosser, *Org. Synth.* **IV**, 169 (1963).
e. H. B. Stevenson, H. N. Cripps, and J. K. Williams, *Org. Synth.* **V**, 459 (1973).

3.4.3. Preparation of Amides

By far the most common method for preparation of amides is the reaction of ammonia or a primary or secondary amine with one of the reactive acylating reagents described in Section 3.4.1. When acid halides are used, some provision for neutralizing the hydrogen halide is necessary, since it will otherwise tie up some of the reagent amine as the corresponding salt. Acid anhydrides give rapid acylation of most amines and are convenient if available. The Schotten–Baumann conditions, involving shaking an amine with excess anhydride or acid chloride and an alkaline aqueous solution, provide a very satisfactory method for preparation of simple amides.

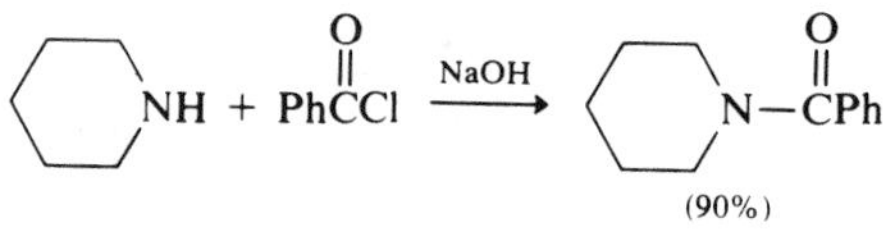

$$\text{(piperidine)NH} + PhC(=O)Cl \xrightarrow{NaOH} \text{(piperidine)}N{-}C(=O)Ph \quad (90\%)$$ Ref. 100

A great deal of work has been done on the *in situ* activation of carboxylic acids toward nucleophilic substitution by amines. This type of reaction forms the backbone

100. C. S. Marvel and W. A. Lazier, *Org. Synth.* **I**, 99 (1941).

of the methods for synthesis of peptides and proteins. Dicyclohexylcarbodiimide is very widely used for coupling carboxylic acids and amines to give amides. Since amines are better nucleophiles than alcohols, the leaving group in the acylation reagent need not be as reactive as is necessary for alcohols. The *p*-nitrophenyl[101] and 2,4,5-trichlorophenyl[102] esters of amino acids are sufficiently reactive toward amines to be useful in peptide synthesis. Acyl derivatives of *N*-hydroxysuccinimide are also useful for synthesis of peptides and other types of amides.[103] Like the *p*-nitrophenyl esters, the acylated *N*-hydroxysuccinimides can be isolated and purified but react rapidly with free amino groups.

$$XC(=O)NHCH(R^2)C(=O)O{-}N(\text{succinimide}) + H_2NCH(R^1)C(=O)Y \longrightarrow XC(=O)NHCH(R^2)C(=O)NHCH(R^1)C(=O)Y + HO{-}N(\text{succinimide})$$

The *N*-hydroxysuccinimide that is liberated is easily removed because of its solubility in dilute base. The relative stability of the anion of *N*-hydroxysuccinimide is also responsible for the acyl derivative being reactive toward nucleophilic attack by an amino group.

Carboxylic acids can also be activated by formation of mixed anhydrides with various phosphoric acid derivatives. Diphenylphosphoryl azide, for example, is an effective reagent for conversion of amines to amides.[104] The postulated mechanism involves formation of the acyl azide as a reactive intermediate.

$$RCO_2^- + (PhO)_2P(=O)N_3 \rightarrow RC(=O){-}O{-}P(=O)(OPh)_2 + N_3^-$$

$$RC(=O){-}O{-}P(=O)(OPh)_2 + N_3^- \rightarrow RC(=O)N_3 + {}^-O_2P(OPh)_2$$

$$RC(=O)N_3 + R'NH_2 \rightarrow RC(=O)NHR' + HN_3$$

Another useful reagent for amide formation is compound **1**. The reaction with this reagent also proceeds via a mixed carboxylic phosphoric anhydride.

$$RCO_2^- + \underset{\mathbf{1}}{(\text{oxazolidinon-3-yl})_2P(=O)Cl} \longrightarrow RC(=O){-}O{-}P(=O)(\text{oxazolidinon-3-yl})_2$$

101. M. Bodanszky and V. DuVigneaud, *J. Am. Chem. Soc.* **81**, 5688 (1959).
102. J. Pless and R. A. Boissonnas, *Helv. Chim. Acta* **46**, 1609 (1963).
103. G. W. Anderson, J. E. Zimmerman, and F. M. Callahan, *J. Am. Chem. Soc.* **86**, 1839 (1964).
104. T. Shiori and S. Yamada, *Chem. Pharm. Bull.* **22**, 849, 855, 859 (1974).

Scheme 3.6. Synthesis of Amides

A. From Acid Chlorides and Anhydrides

1[a] $(CH_3)_2CHCO_2H \xrightarrow[2)\ NH_3]{1)\ SOCl_2} (CH_3)_2CHC(=O)NH_2$ (70%)

2[b] $C_6H_{11}CO_2H \xrightarrow[2)\ (CH_3)_2NH]{1)\ SOCl_2} C_6H_{11}C(=O)N(CH_3)_2$ (85–90%)

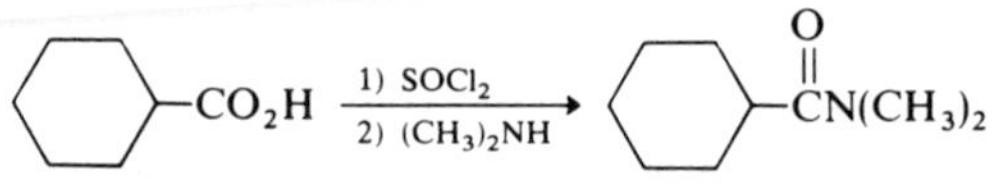

3[c] $(CH_3CO)_2O + H_2NCH_2CO_2H \rightarrow CH_3C(=O)NHCH_2CO_2H$ (90%)

B. From Esters

4[d] $NCCH_2CO_2C_2H_5 \xrightarrow{NH_4OH} NCCH_2C(=O)NH_2$

5[e] phenyl salicylate (o-HOC$_6$H$_4$CO$_2$Ph) + o-toluidine (o-CH$_3$C$_6$H$_4$NH$_2$) $\xrightarrow[185\text{–}200°C]{\text{trichlorobenzene}}$ o-HOC$_6$H$_4$C(=O)–NH–C$_6$H$_4$-o-CH$_3$ (75%)

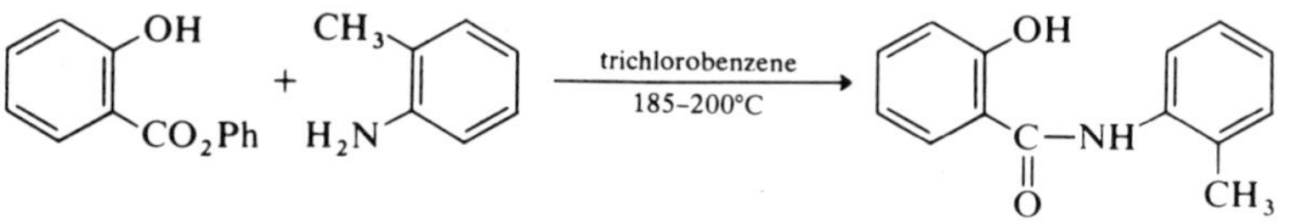

C. From Carboxylic Acids

6[f] 4-azidopyrrolidine-2-CO_2CH_3 (N–H) $\xrightarrow[Et_3N]{PhCO_2H,\ DCC,}$ N-benzoyl (N–C(=O)Ph) 4-azidopyrrolidine-2-CO_2CH_3 (63%)

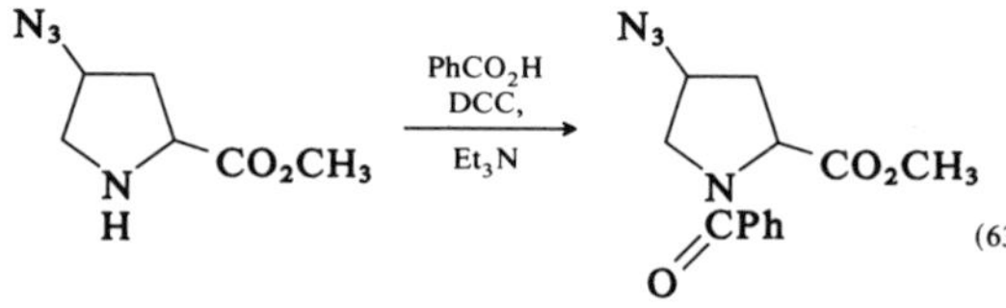

7[g] 2-(CH_2CN)C$_6$H$_4$CO$_2$H + C$_6$H$_5$NH$_2$ $\xrightarrow{Et_3N}$ [bis(2-oxo-3-oxazolidinyl)phosphinic chloride, $(C_3H_4NO_2)_2P(O)Cl$] → 2-(CH_2CN)C$_6$H$_4$C(=O)NH–C$_6$H$_5$

8[h] 2,3-dimethoxybenzoic acid (OCH_3, OCH_3, CO_2H) + $H_2N(CH_2)_2CO_2H$ $\xrightarrow{DCC}$ (N-hydroxysuccinimide, NOH) → 2,3-dimethoxy-C$_6$H$_3$–$CONH(CH_2)_2CO_2H$ (82%)

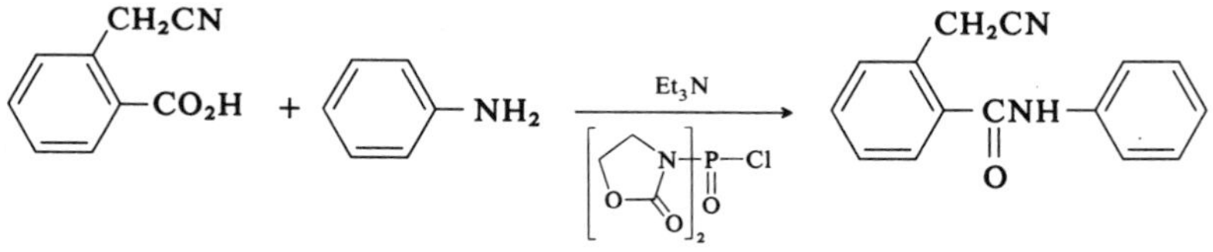

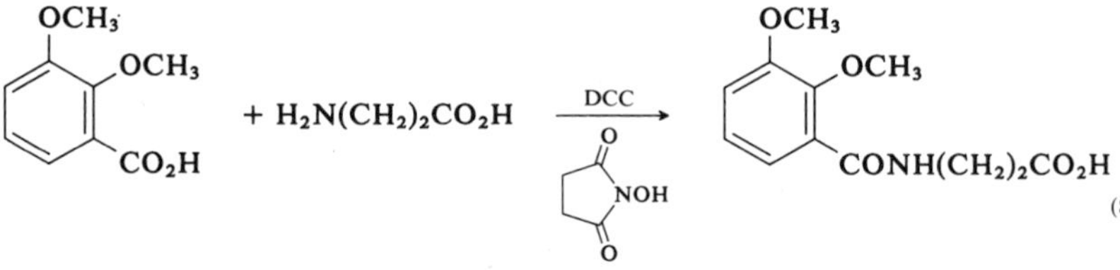

D. From Nitriles

9[i] $C_6H_5CH_2CN \xrightarrow[40\text{–}50°C,\ 1\ hr]{HCl,\ H_2O} C_6H_5CH_2C(=O)NH_2$ (80%)

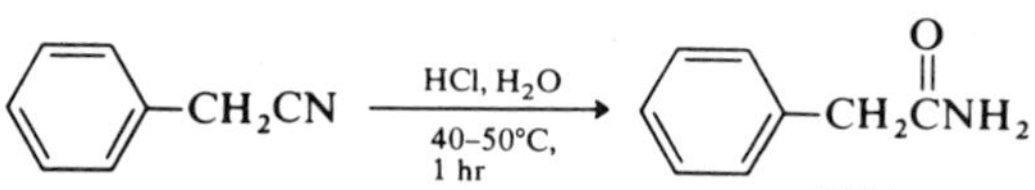

10[j] o-methylbenzonitrile (CH_3, CN) $\xrightarrow[40\text{–}50°C,\ 4\ hr]{30\%\ H_2O_2,\ NaOH}$ o-methylbenzamide (CH_3, CNH_2 with C=O) (90%)

a. R. E. Kent and S. M. McElvain, *Org. Synth.* **III**, 490 (1955).
b. A. C. Cope and E. Ciganek, *Org. Synth.* **IV**, 339 (1963).
c. R. M. Herbst and D. Shemin, *Org. Synth.* **II**, 11 (1943).
d. B. B. Corson, R. W. Scott, and C. E. Vose, *Org. Synth.* **I**, 179 (1941).
e. C. F. H. Allen and J. Van Allan, *Org. Synth.* **III**, 765 (1955).
f. D. J. Abraham, M. Mokotoff, L. Sheh, and J. E. Simmons, *J. Med. Chem.* **26**, 549 (1983).
g. J. Diago-Mesenguer, A. L. Palamo-Coll, J. R. Fernandez-Lizarbe, and A. Zugaza-Bilbao, *Synthesis*, 547 (1980).
h. R. J. Bergeron, S. J. Kline, N. J. Stolowich, K. A. McGovern, and P. S. Burton, *J. Org. Chem.* **46**, 4524 (1981).
i. W. Wenner, *Org. Synth.* **IV**, 760 (1963).
j. C. R. Noller, *Org. Synth.* **II**, 586 (1943).

The preparation of amides directly from alkyl esters is also feasible but is usually too slow for preparative convenience. Entries 4 and 5 in Scheme 3.6 are successful examples. The reactivity of ethyl cyanoacetate (entry 4) is higher than that of unsubstituted aliphatic esters because of the inductive effect of the cyano group.

A recently developed method for converting esters to amides involves aluminum amides, which can be prepared from trimethylaluminum and the amine. These reagents convert esters directly to amides at room temperature.

$$C_6H_{11}CO_2CH_3 \xrightarrow{(CH_3)_2AlN(H)CH_2Ph} C_6H_{11}CNHCH_2Ph\ (C{=}O)\quad (78\%)$$ Ref. 105

The driving force for this reaction is the greater strength of the aluminum-oxygen bond relative to the aluminum-nitrogen bond. Trialkyltin amides and tetrakis(dimethylamino)titanium show similar reactivity.[106]

The cyano group is at the carboxylic acid oxidation level so nitriles are potential precursors of primary amides. Partial hydrolysis is sometimes possible.[107]

$$PhCH_2C{\equiv}N \xrightarrow[40\text{–}50°C,\ 1\ h]{HCl,\ H_2O} PhCH_2\overset{O}{\overset{\|}{C}}NH_2$$

A milder procedure involves the reaction of a nitrile with an alkaline solution of hydrogen peroxide.[108] The strongly nucleophilic hydrogen peroxide adds to the

105. A. Basha, M. Lipton, and S. M. Weinreb, *Tetrahedron Lett.*, 4171 (1977).
106. G. Chandra, T. A. George, and M. F. Lappert, *J. Chem. Soc., C*, 2565 (1969).
107. W. Wenner, *Org. Synth.* **IV**, 760 (1963).
108. C. R. Noller, *Org. Synth.* **II**, 586 (1943); J. S. Buck and W. S. Ide, *Org. Synth.* **II**, 44 (1943).

nitrile, and the resulting adduct gives the amide. There are several possible mechanisms for the subsequent decomposition of the peroxycarboximidic adduct.[109]

$$RC{\equiv}N + {}^{-}O_2H \rightarrow R\overset{NH}{\overset{\|}{C}}OO^- \underset{}{\overset{H_2O}{\rightleftharpoons}} R\overset{NH}{\overset{\|}{C}}OOH + H_2O_2 \rightarrow R\overset{O}{\overset{\|}{C}}NH_2 + O_2 + H_2O$$

In all the mechanisms, the hydrogen peroxide is converted to oxygen and water, leaving the organic substrate hydrolyzed but at the same oxidation level. Scheme 3.6 includes two specific examples of conversion of nitriles to amides.

Problems

(*References for these problems will be found on page* 766.)

1. Give the products which would be expected to be formed under the specified reaction conditions. Be sure to specify all aspects of stereochemistry.

(a) 5-ethyl-γ-butyrolactone (CH_3CH_2-substituted tetrahydrofuran-2-one) $\xrightarrow[CH_3OH]{HCl}$

(b) $CH_3(CH_2)_4CH_2OH + ClCH_2OCH_3 \xrightarrow[CH_2Cl_2]{CH_3CH_2N(i\text{-}Pr)_2}$

(c) (S)-$CH_3(CH_2)_3\underset{OH}{CHCH_3}$ + 3-ethyl-2-chlorobenzoxazolium (^+N-C_2H_5) $\xrightarrow[Et_4N^+Cl^-]{Et_3N}$

(d) $C_2H_5O_2CCH_2\underset{OH}{CH}CO_2C_2H_5 \xrightarrow[2)\ C_2H_5O_2CN{=}NCO_2C_2H_5]{1)\ Ph_3P,\ HN_3}$

(e) purine nucleoside with $N(\overset{O}{\overset{\|}{C}}R)_2$ at C-6, $HOCH_2$ at C-5′, RCO_2 and O_2CR at C-2′/C-3′ $\xrightarrow[DMF,\ 20°C,\ 10\ min]{(PhO)_3\overset{+}{P}CH_3\ I^-}$

(f) 4-methylcyclohexene bearing CH_3 on the alkene carbon and CH_3CHCH_2OH at C-4 $\xrightarrow[CCl_4]{PPh_3}$

(g) 2-methoxyphenyl with stereocenter C_2H_5, H, CH_2CO_2H (OCH$_3$ ortho) $\xrightarrow[0°C,\ 1\ h,\ CH_2Cl_2]{BBr_3,\ -78°C}$

(h) $Ph\overset{CO_2CH_3}{C}(H)(OH)$ $\xrightarrow[2)\ PhS^-Na^+]{1)\ p\text{-toluenesulfonyl chloride}}$

109. K. B. Wiberg, *J. Am. Chem. Soc.* **75**, 3961 (1953); **77**, 2519 (1955); J. E. McIsaac, Jr., R. E. Ball, and E. J. Behrman, *J. Org. Chem.* **36**, 3048 (1971).

(i) $(C_6H_5)_2CHBr + P(OCH_3)_3 \rightarrow$

(j) [Structure: hydrophenanthrene bearing two $CH_3SO_2OCH_2$ groups] $\xrightarrow[\text{HMPA}]{Na_2S}$

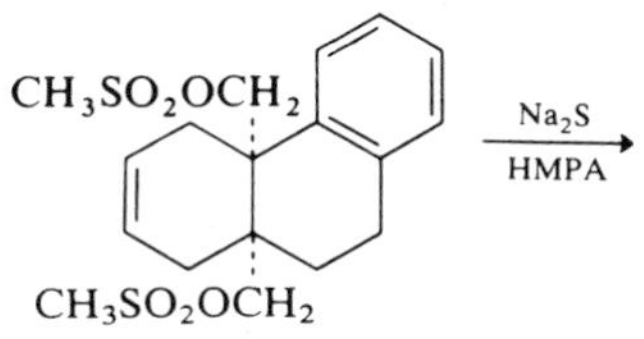

(k) [Structure: 6,7-dimethoxy-2-benzyl tetrahydroisoquinolin-4-one, CH_3O, CH_3O, O, $NCH_2C_6H_5$] $\xrightarrow[\text{heat}]{\text{48\% HBr}}$

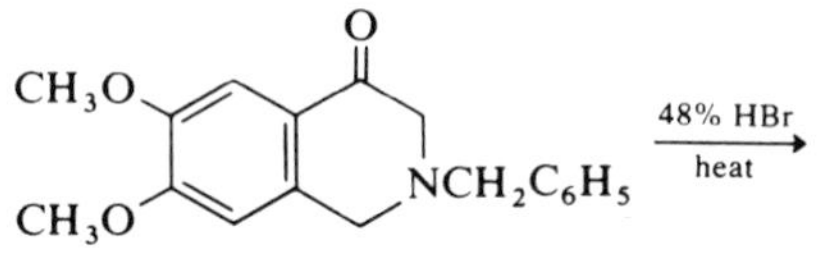

(l) [Structure: H(C$_2$H$_5$O$_2$C)C=C(CO$_2$H)H] $\xrightarrow[\text{DCC, DMAP}]{t\text{-BuOH}}$

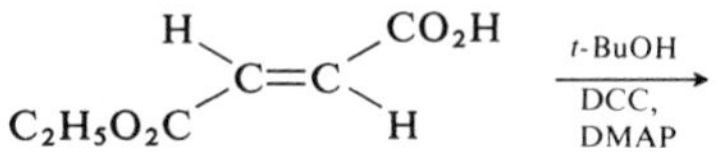

2. When (R)-(−)-5-hexen-2-ol was treated with triphenylphosphine in refluxing carbon tetrachloride, (+)-5-chloro-1-hexene was obtained. Conversion of (R)-(−)-5-hexen-2-ol to its *p*-bromobenzenesulfonate ester and subsequent reaction with lithium chloride gave (+)-5-chloro-1-hexene. Reaction of (S)-(+)-5-hexen-2-ol with phosphorus pentachloride in ether gave (−)-5-chloro-1-hexene.
   (a) Write chemical equations for each of the reactions described above and specify whether each one proceeds with net retention of configuration or inversion of configuration.
   (b) What is the sign of rotation of (R)-5-chloro-1-hexene?

3. A careful investigation of the extent of isomeric products formed by reaction of several alcohols with thionyl chloride has been reported. The product compositions for several of the alcohols are given below. Show how each of the rearranged products arises and discuss the structural features which promote isomerization.

$$\text{ROH} \xrightarrow[100°C]{SOCl_2} \text{RCl}$$

R	% un-rearranged RCl	Structure and amount of rearranged RCl
$CH_3CH_2CH_2CH_2-$	100	
$(CH_3)_2CHCH_2-$	99.7	$CH_3CH_2CH(Cl)CH_3$ (0.3%)
$(CH_3)_2CHCH_2CH_2-$	100	
$CH_3CH_2CH(CH_3)CH_2-$	78	$CH_3CH(Cl)CH_2CH_2CH_3$ (1%), $CH_3CH_2CH(Cl)CH_2CH_3$ (11%), $CH_3CH_2C(Cl)(CH_3)_2$ (10%)
$(CH_3)_3CCH_2-$	2	$CH_3CH_2C(Cl)(CH_3)_2$ (98%)
$CH_3CH_2CH_2CH(-)CH_3$	98	$CH_3CH_2CH(Cl)CH_2CH_3$ (2%)
$CH_3CH_2CH(-)CH_2CH_3$	90	$CH_3CH_2CH_2CH(Cl)CH_3$ (10%)
$(CH_3)_2CHCH(-)CH_3$	5	$CH_3CH_2C(Cl)(CH_3)_2$ (95%)

4. Give a reaction mechanism which will explain the following observations and transformations.

(a) Kinetic measurements reveal that solvolytic displacement is about 5×10^5 faster for **B** than for **A**.

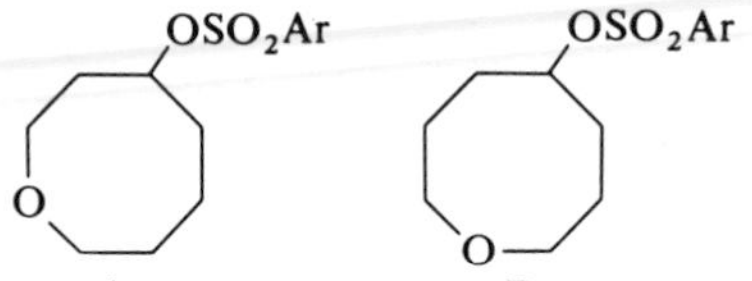

OSO_2Ar — **A** OSO_2Ar — **B**

(b) H_2N, Br, H, $CH_2CH_2CO_2CH_3$, S, H $\xrightarrow{HOAc}$ Br, H, H, N, O, S, H

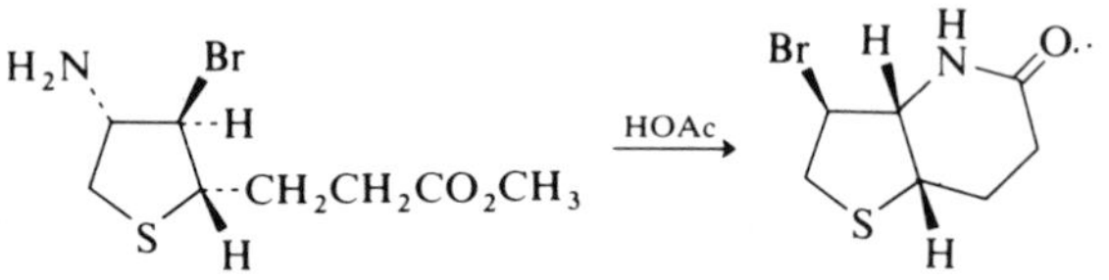

(c) CH_3, H_3C, H_3C, S, O, C_6H_5 $\xrightarrow[2)\ NaCN]{1)\ (CH_3)_3O^+PF_6^-}$ $CH_3SCH(CH_3)CH_2C(CH_3)_2{-}O{-}CH(CN)C_6H_5$

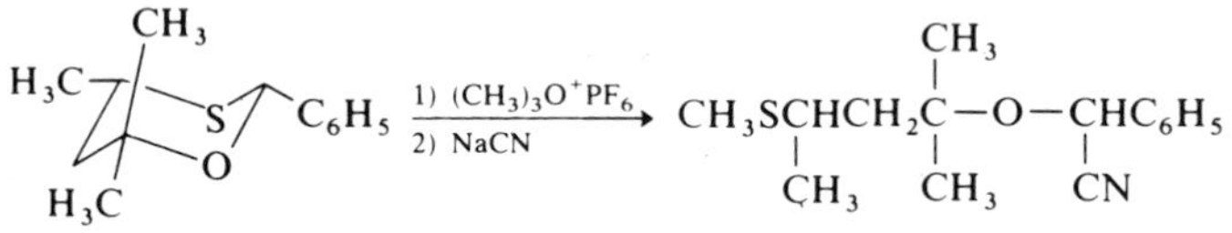

(d) $C_6H_5CH_2SCH_2CH(OH)CH_2SCH_2C_6H_5 \xrightarrow{SOCl_2} C_6H_5CH_2SCH_2CH(SCH_2C_6H_5)CH_2Cl$

(e) HO, $CO_2CH_2CH_3$, CN, N $\xrightarrow[t\text{-BuOH}]{KOH}$ HO, O, NH, O, N

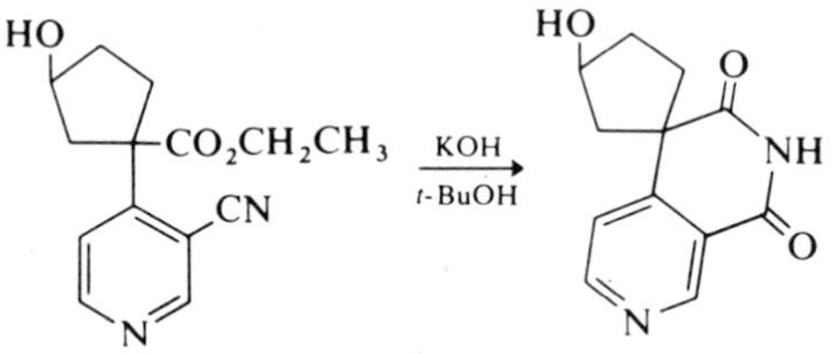

(f) $CH_3(CH_2)_6CO_2H + PhCH_2NH_2 \xrightarrow[Bu_3P,\ 25°C]{o\text{-nitrophenyl-isothiocyanate}} CH_3(CH_2)_6C(=O)NHCH_2Ph$
(99%)

5. Substances such as carbohydrates, amino acids, and other small molecules available from natural sources are valuable starting materials for the synthesis of stereochemically defined substances. Suggest a sequence of reactions which could effect the following transformations, taking particular care to ensure that the product would be obtained stereochemically pure.

(a) $(CH_3)_2NC(=O)$–CH(OCH_3)–CH(OCH_3)–$C(=O)N(CH_3)_2$ from CH_3O_2C–CH(OH)–CH(OH)–CO_2CH_3

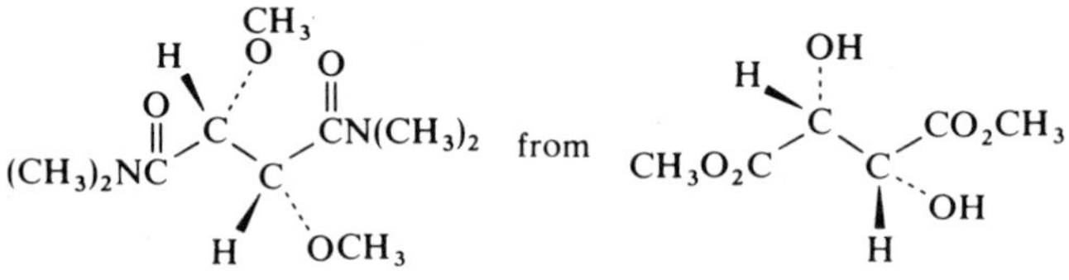

(b) 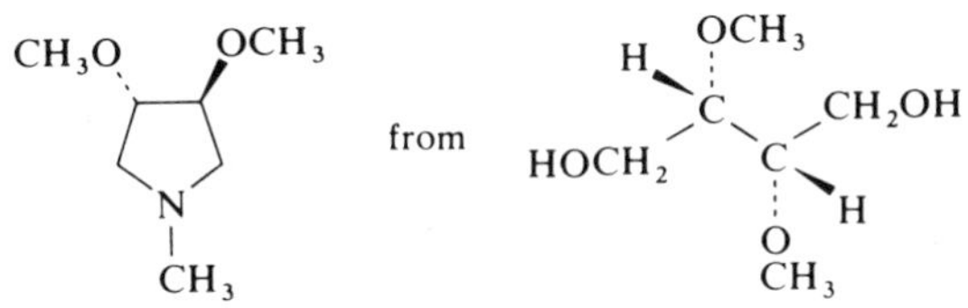 CH_3O OCH_3 N CH_3 from H OCH_3 C CH_2OH $HOCH_2$ C H O CH_3

(c) Ph_2P N CH_2PPh_2 $(CH_3)_3COC{=}O$ from HO N CH_2OH $(CH_3)_3COC{=}O$

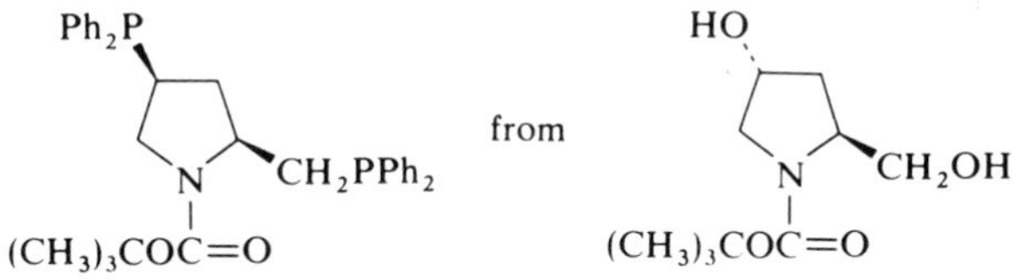

(d) 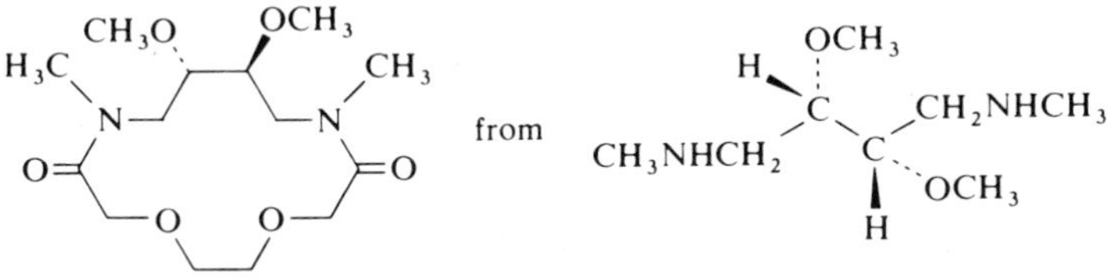 CH_3O OCH_3 H_3C N CH_3 N O= =O O O from H OCH_3 C CH_2NHCH_3 CH_3NHCH_2 C OCH_3 H

(e) 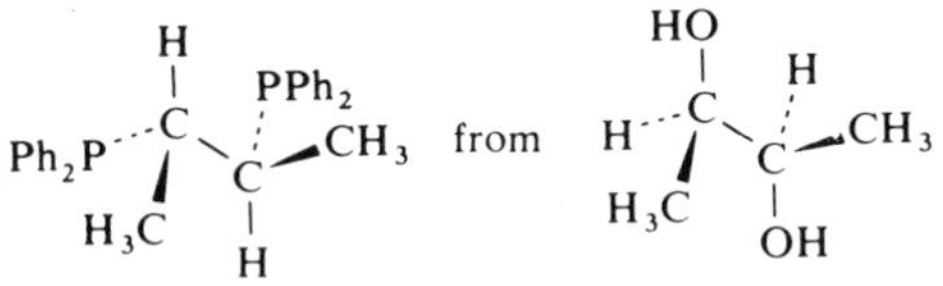 H PPh_2 Ph_2P C CH_3 C H_3C H from HO H H C C CH_3 H_3C OH

(f) 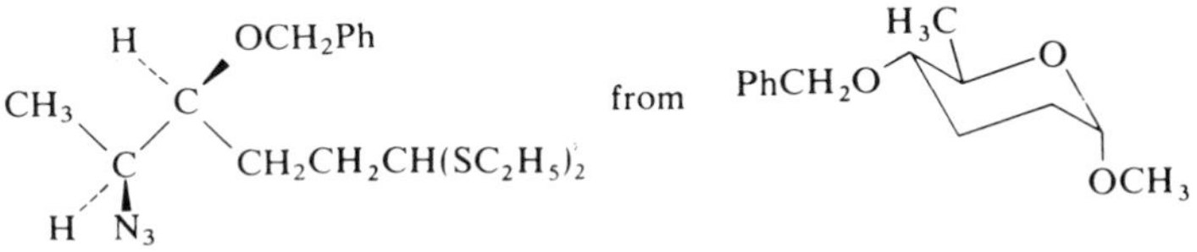 H OCH_2Ph CH_3 C C $CH_2CH_2CH(SC_2H_5)_2$ H N_3 from H_3C O $PhCH_2O$ OCH_3

(g) 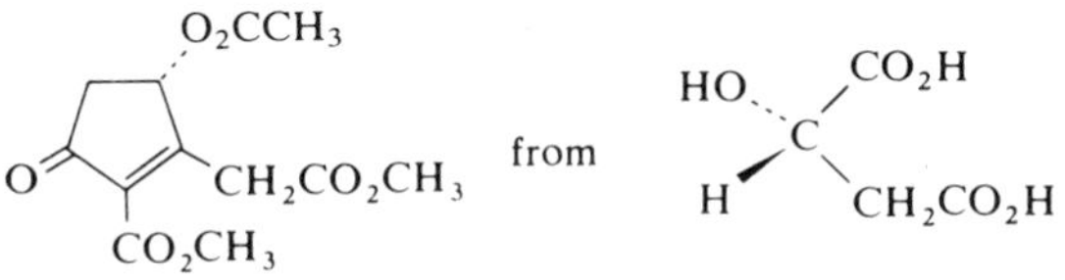 O_2CCH_3 O $CH_2CO_2CH_3$ CO_2CH_3 from HO CO_2H C H CH_2CO_2H

6. Suggest reagents and reaction conditions which could be expected to effect the following conversions.

(a) 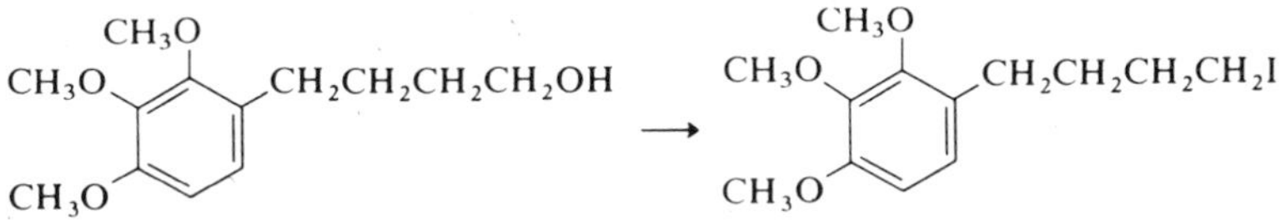 CH_3O CH_3O $CH_2CH_2CH_2CH_2OH$ CH_3O → CH_3O CH_3O $CH_2CH_2CH_2CH_2I$ CH_3O

(b) 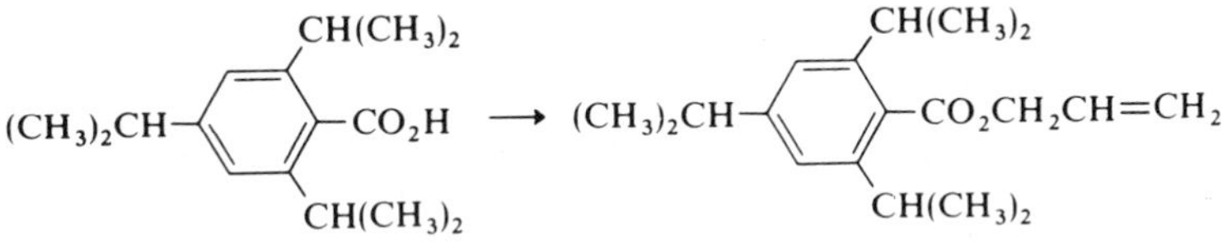 $CH(CH_3)_2$ $(CH_3)_2CH$ CO_2H $CH(CH_3)_2$ → $CH(CH_3)_2$ $(CH_3)_2CH$ $CO_2CH_2CH{=}CH_2$ $CH(CH_3)_2$

(c) 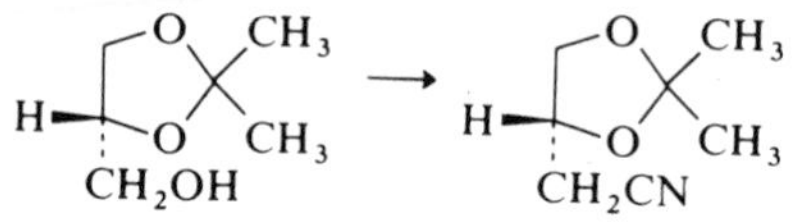 [dioxolane with H, CH_2OH, CH_3, CH_3] → [dioxolane with H, CH_2CN, CH_3, CH_3]

(d) 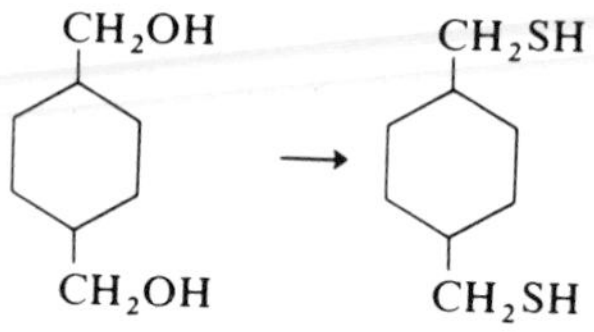 1,4-bis(CH_2OH)cyclohexane → 1,4-bis(CH_2SH)cyclohexane (more than one step is required)

(e) 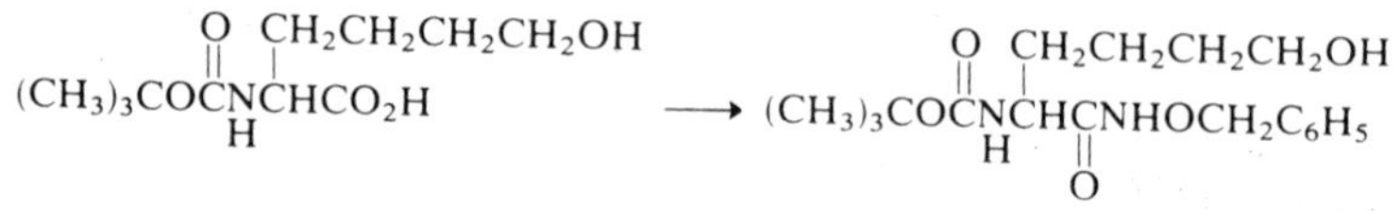 $(CH_3)_3COC(=O)NHCH(CH_2CH_2CH_2CH_2OH)CO_2H$ → $(CH_3)_3COC(=O)NHCH(CH_2CH_2CH_2CH_2OH)C(=O)NHOCH_2C_6H_5$

(f) 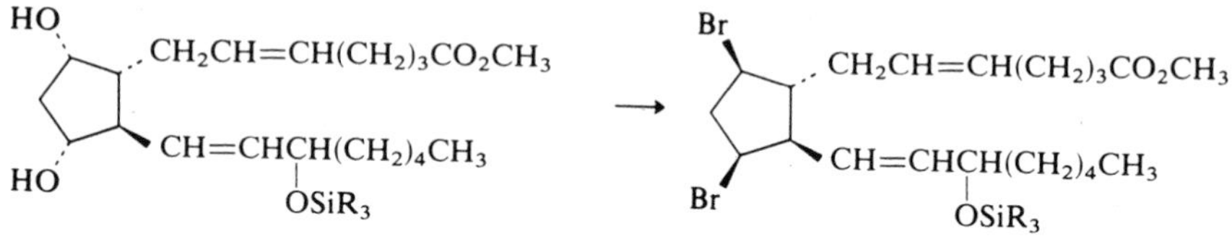 [cyclopentane with HO, HO, $CH_2CH{=}CH(CH_2)_3CO_2CH_3$, $CH{=}CHCH(OSiR_3)(CH_2)_4CH_3$] → [cyclopentane with Br, Br, $CH_2CH{=}CH(CH_2)_3CO_2CH_3$, $CH{=}CHCH(OSiR_3)(CH_2)_4CH_3$]

(g) 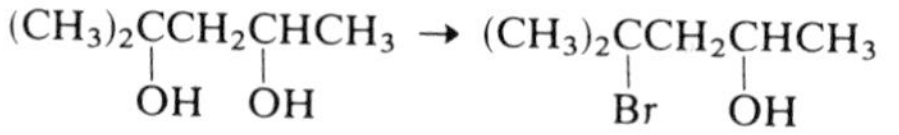 $(CH_3)_2C(OH)CH_2CH(OH)CH_3$ → $(CH_3)_2C(Br)CH_2CH(OH)CH_3$

7. Provide a mechanistic interpretation for each of the following observations.

(a) A procedure for inverting the configuration of alcohols has been developed and demonstrated using cholesterol as a substrate:

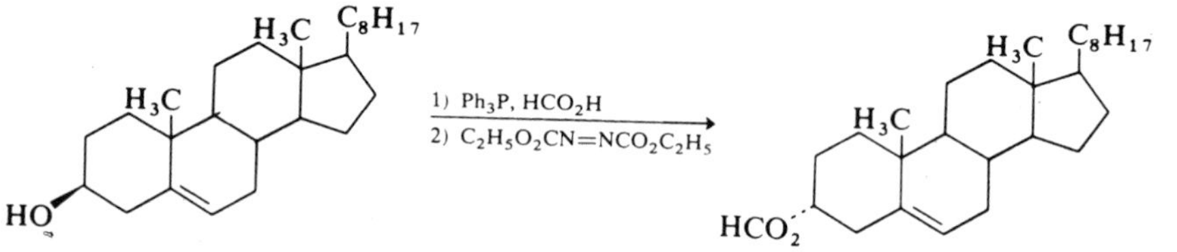

[cholesterol (HO, H_3C, H_3C, C_8H_{17})] $\xrightarrow[\text{2) } C_2H_5O_2CN{=}NCO_2C_2H_5]{\text{1) } Ph_3P,\ HCO_2H}$ [inverted formate ester (HCO_2, H_3C, H_3C, C_8H_{17})]

Show the details of the mechanism of the key step which converts cholesterol to the inverted formate ester.

(b) It has been found that triphenylphosphine oxide reacts with trifluoromethylsulfonic anhydride to give an ionic substance with the composition of a simple 1:1 adduct. When this substance is added to a solution containing a carboxylic acid, followed by addition of an amine, amides are formed in good yield. Similarly, esters are formed by treating carboxylic acids first with the reagent and then with an alcohol. What is a likely structure for this ionic substance and how can it effect the activation of the carboxylic acids?

(c) Sulfonate esters having quaternary nitrogen substituents, such as **A** and **B**, show exceptionally high reactivity toward nucleophilic displacement reactions. Discuss factors which might contribute to the reactivity of these substances.

$ROSO_2CH_2CH_2\overset{+}{N}(CH_3)_3$
A

$ROS(=O)_2-C_6H_4-\overset{+}{N}(CH_3)_3$
B

(d) Alcohols react with hexachloroacetone in the presence of dimethylformamide to give alkyl trichloroacetates in high yield. Primary alcohols react fastest. Tertiary alcohols do not react. Suggest a reasonable mechanism for this reaction.

8. Short synthetic sequences have been used to accomplish synthesis of the material at the left from that on the right. Suggest appropriate methods. No more than three separate steps should be required.

(a) $PhCH(OCH_3)C(=O)NHCH(CH_3)CH_2C_6H_5 \Leftarrow PhCH(OH)CO_2H$ (with retention of configuration)

(b) $(CH_3)_2CHCH_2CH{=}CHCH(CH_3)CH_2CO_2C_2H_5 \Leftarrow$ H_3C-substituted lactone (O, =O)

(c) TsO, N-($CH_3C{=}O$), CO_2CH_3 $\Leftarrow$ HO, N–H, CO_2H

(d) $(CH_3)_2CH(H)C{=}C(H)–(H)C{=}C(CH_3)CH_2CN \Leftarrow (CH_3)_2CH(H)C{=}C(H)–(H)C{=}C(CH_3)CH_2OH$

(e) CH_3O, CH_3O, $CH(CH_3)CH(CO_2H)(CH_2)_4CH_3 \Leftarrow$ CH_3O, CH_3O, $CH(OH)CH_3$

9. Amino acids can be converted to epoxides in high enantiomeric purity by the following reaction sequence. Analyze the sterochemistry at each step of the reaction.

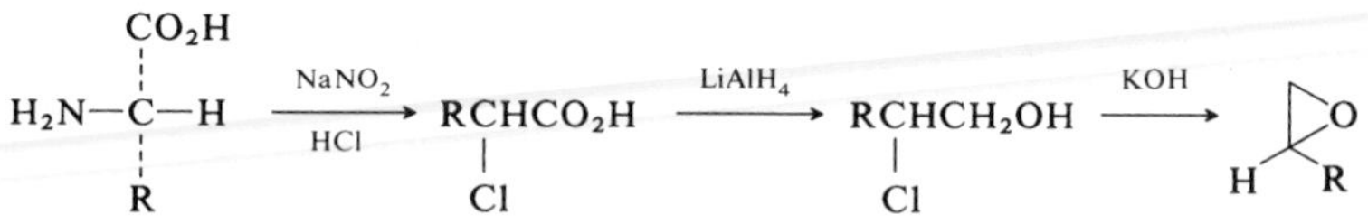

$$H_2N-C(CO_2H)(R)-H \xrightarrow[HCl]{NaNO_2} RCH(Cl)CO_2H \xrightarrow{LiAlH_4} RCH(Cl)CH_2OH \xrightarrow{KOH} \text{epoxide (H, R)}$$

10. A reagent which has been found to be useful for introduction of the benzyloxycarbonyl group onto amino groups of nucleosides is prepared by allowing benzyl chloroformate to react first with imidazole and then with trimethyloxonium tetrafluoroborate. What is the structure of the resulting reagent (a salt), and why is it an especially reactive acylating reagent?

11. (a) Write the equilibrium expression for phase transfer involving a tetraalkylammonium salt, $R_4N^+X^-$, NaOH, a water phase, and a nonaqueous phase.
    (b) The concentration of ^-OH in the nonaqueous phase under phase transfer conditions is a function of the anion X^-. What structural characteristics of X^- would be expected to influence the position of the equilibrium?
    (c) It has been noted in a comparison of 15% aqueous NaOH versus 50% NaOH that the extent of transfer of ^-OH to the nonaqueous phase is less for 50% NaOH than for lower concentrations. What could be the cause of this?

12. The scope of the reaction of triphenylphosphine-hexachloroacetone with allylic alcohols has been studied. Primary and some secondary alcohols such as **1** and **2** give good yields of unrearranged halides. Certain other alcohols, such as **3** and **4**, give more complex mixtures. Discuss structural features which are probably important in determining how cleanly a given alcohol is converted to halide.

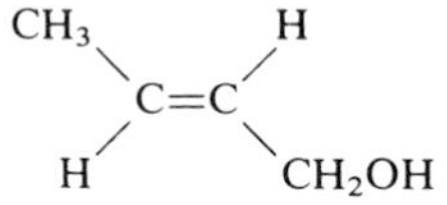

1: $(CH_3)(H)C{=}C(H)(CH_2OH)$

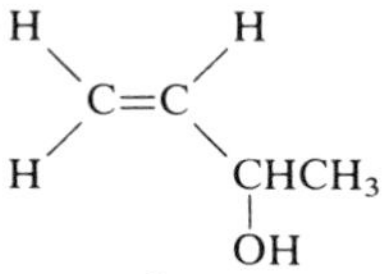

2: $(H)(H)C{=}C(H)(CH(OH)CH_3)$

3: $(H)(H)C{=}C(H)(C(CH_3)_2OH)$ $\xrightarrow[Cl_3CCOCCl_3]{Ph_3P}$ $CH_2{=}CHC(Cl)(CH_3)_2$ (21%) + $ClCH_2CH{=}C(CH_3)_2$ (43%) + $CH_2{=}CHC(CH_3){=}CH_2$ (18%)

4: $(H)(H)C{=}C(H)(CH(OH)CH(CH_3)_2)$ $\xrightarrow[Cl_3CCOCCl_3]{Ph_3P}$ $CH_2{=}CHCH(Cl)CH(CH_3)_2$ (27%) + $ClCH_2CH{=}CHCH(CH_3)_2$ (15%) + $CH_2{=}CHCH{=}C(CH_3)_2$ (58%)

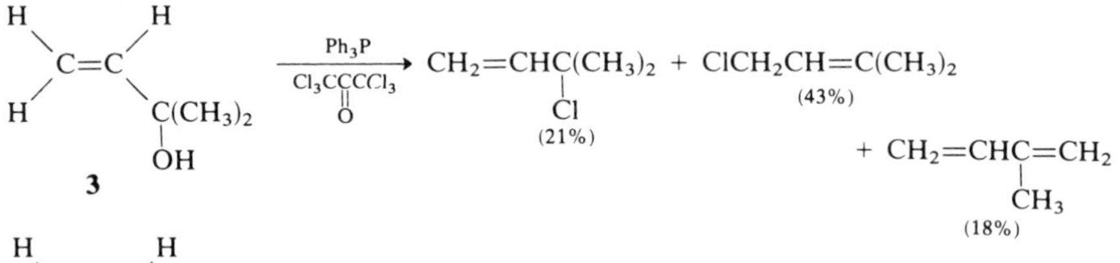

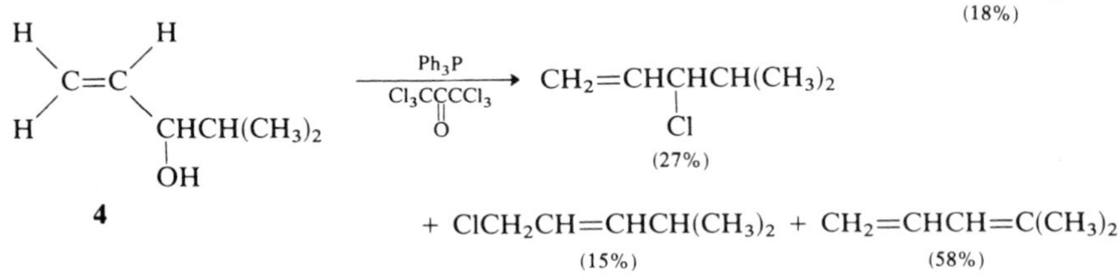

13. Two heterocyclic ring systems which have found some use in the formation of amides under mild conditions are *N*-alkyl-5-arylisoxazolium salts (structure **A**) and *N*-acyloxy-2-alkoxydihydroquinolines (structure **B**).

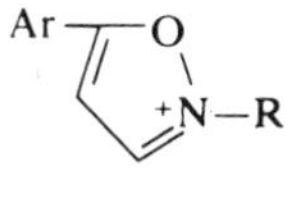

Ar, O, +N—R, **A**; N, OR, O=COR, **B**

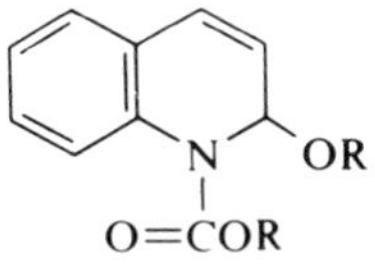

A typical set of reactions conditions is indicated below for each reagent. Consider mechanisms by which these heterocyclic molecules might function to activate the carboxylic acid group under these conditions, and outline the mechanisms you consider to be most likely.

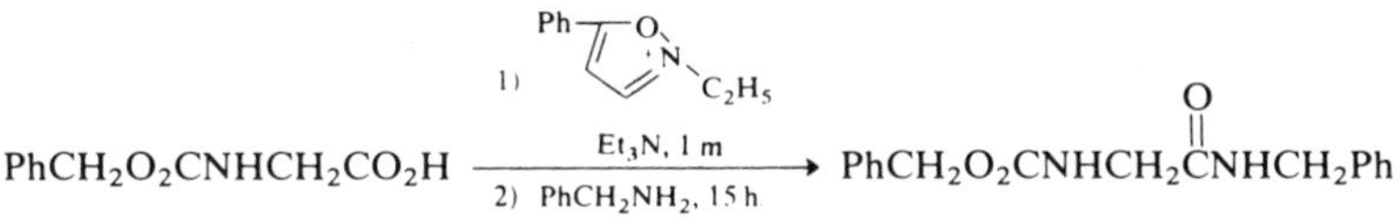

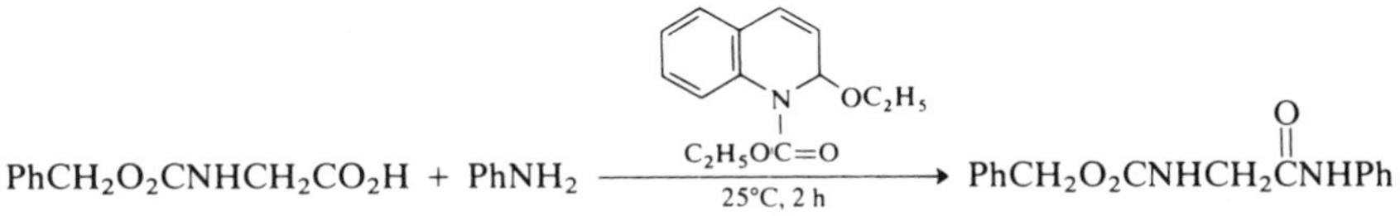

$$PhCH_2O_2CNHCH_2CO_2H \xrightarrow[\text{2) } PhCH_2NH_2\text{, 15 h}]{\text{1) Ph-isoxazolium-}N\text{-}C_2H_5\text{, } Et_3N\text{, 1 m}} PhCH_2O_2CNHCH_2\overset{O}{\overset{\|}{C}}NHCH_2Ph$$

$$PhCH_2O_2CNHCH_2CO_2H + PhNH_2 \xrightarrow[25^\circ C\text{, 2 h}]{\text{quinoline-}N(C_2H_5OC{=}O)\text{-}OC_2H_5} PhCH_2O_2CNHCH_2\overset{O}{\overset{\|}{C}}NHPh$$

14. Either because of potential interferences with other functional groups present in the molecule or because of special structural features, the following reactions would require expecially careful selection of reagents and reaction conditions. Identify the special requirements of each substrate and suggest appropriate conditions for effecting the desired transformation.

(a) RO, OR, H, CHC, H, C—CO_2H, H, H, C=C, H, $CH_2CH_2CH_2CHCH_3$, OH → RO, OR, H, O, CH_3, O, H; R= (tetrahydropyranyl)

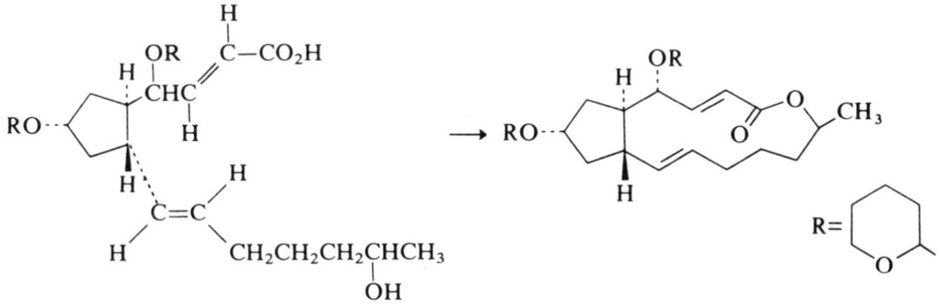

(b) $HOCH_2$, O, N, NH_2, N, N, N, HO → $ClCH_2$, O, N, NH_2, N, N, N, HO

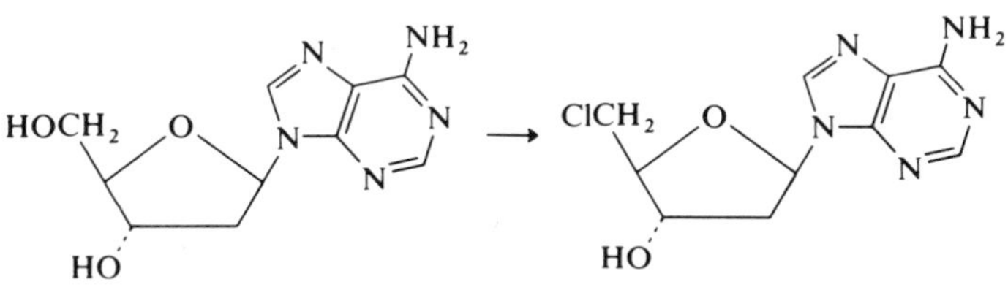

(c) $(CH_3)_3CSC(=O)CH_2CH(CH_3)-C(OH)(CH_3)-CO_2H \longrightarrow (CH_3)_3CSC(=O)CH_2CH(CH_3)-C(OC(=O)CH_3)(CH_3)-CO_2H$

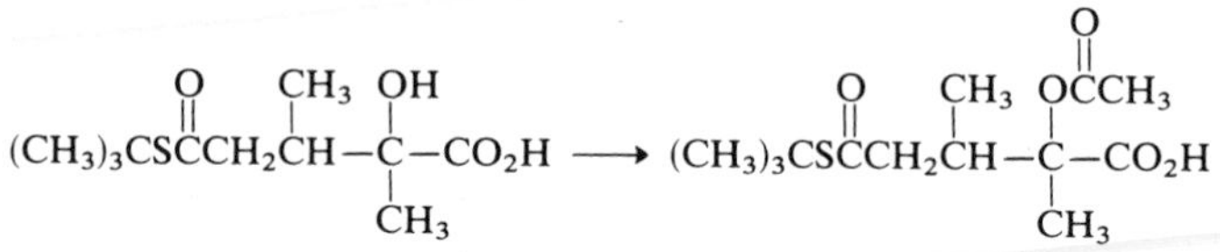

(d) 2,6-bis$((CH_3)_2CH)$-3,5-bis$(N(CH_3)_2)$-substituted benzene $-CO_2H \longrightarrow$ same $-CO_2C_2H_5$

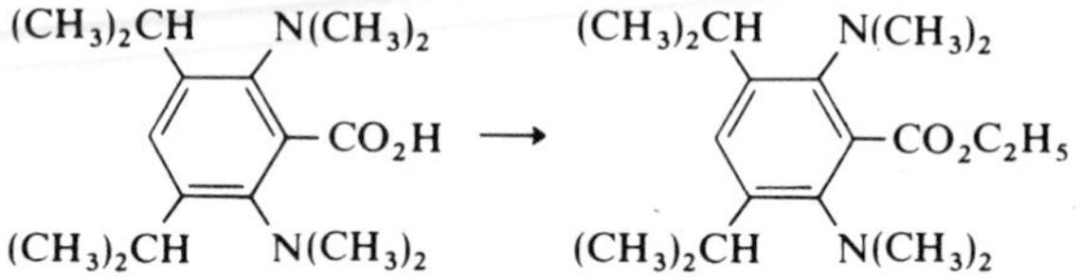

15. The preparation of nucleosides by reaction of carbohydrates and heterocyclic bases is fundamental to the study of the important biological activity of such substances. Several methods have been developed for accomplishing this reaction.

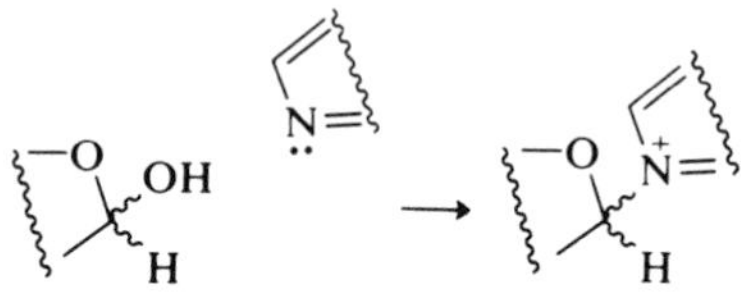

Application of 2-chloro-3-ethylbenzoxazolium chloride to this problem has been investigated using 2,3,4,6-tetra-*O*-acetyl-*β*-D-glucopyranose as the carbohydrate derivative. Good yields were observed and, furthermore, the process was stereoselective, giving the *β*-nucleoside. Suggest a mechanism and explain the stereochemistry.

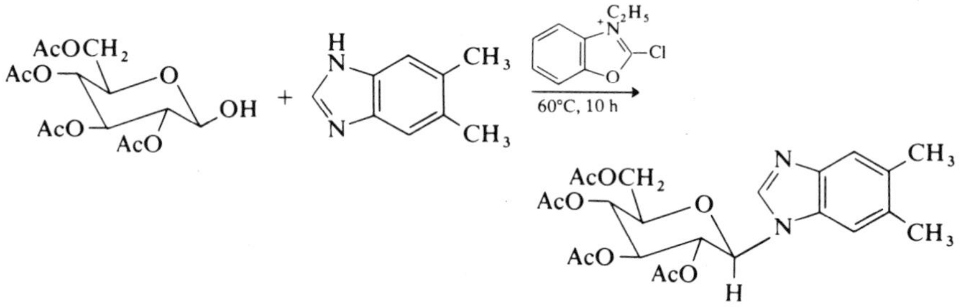

60°C, 10 h

16. A route to *α*-glycosides has been described in which 2,3,4,6-tetra-*O*-benzyl-*α*-D-glucopyranosyl bromide is treated with an alcohol and tetraethylammonium bromide and diisopropylethylamine in dichloromethane.

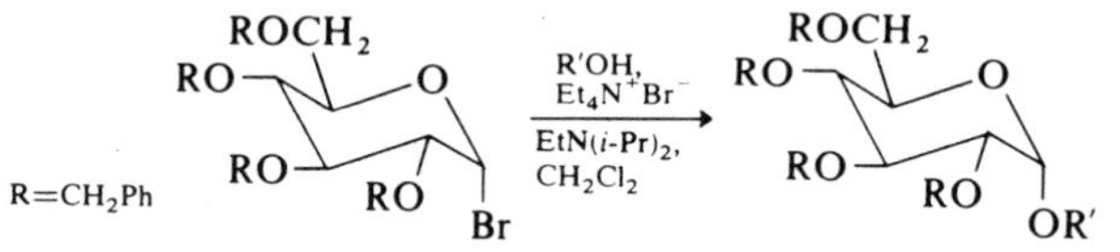

$R=CH_2Ph$; $R'OH$, $Et_4N^+Br^-$, $EtN(i\text{-}Pr)_2$, CH_2Cl_2

Suggest an explanation for the stereochemical course of this reaction.

17. Write the mechanism for the formation of 2-pyridylthio esters by the following reactions.

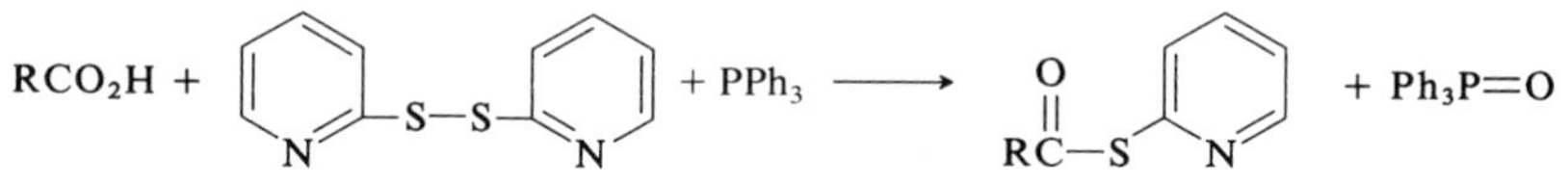

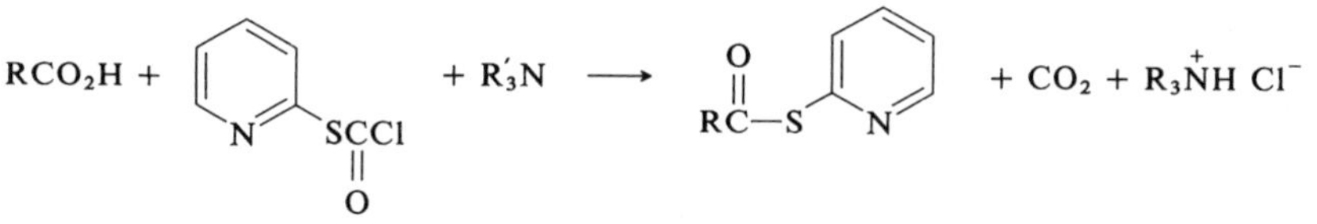

$$RCO_2H + \text{(2-pyridyl)–S–S–(2-pyridyl)} + PPh_3 \longrightarrow RC(=O)\text{–S–(2-pyridyl)} + Ph_3P{=}O$$

$$RCO_2H + \text{(2-pyridyl)–S–C(=O)Cl} + R'_3N \longrightarrow RC(=O)\text{–S–(2-pyridyl)} + CO_2 + R_3\overset{+}{N}H\ Cl^-$$

4

Electrophilic Additions to Carbon–Carbon Multiple Bonds

One of the most general and useful reactions of alkenes and acetylenes for synthetic purposes is the addition of electrophilic reagents. This chapter is restricted to reactions that proceed through polar intermediates or transition states. Several other classes of addition reactions are also of importance, and these are discussed elsewhere. Nucleophilic additions to electrophilic alkenes were covered in Chapter 1, and cycloadditions involving concerted mechanisms will be encountered in Chapter 6. Free-radical addition reactions are considered in Chapter 10.

4.1. Addition of Hydrogen Halides

Hydrogen chloride and hydrogen bromide react with alkenes to give addition products. Many years ago, it was observed that addition usually takes place to give the product in which the halogen atom is attached to the more substituted carbon of the double bond. This behavior was sufficiently general that the name *Markownikoff's rule* was given to the statement describing this mode of addition. A very rudimentary picture of the mechanism of addition reveals the basis of Markownikoff's rule. The addition involves either initial protonation of the alkene or addition involving a partial transfer of a proton to the double bond. The relative stability of the two possible carbocations from an unsymmetrical alkene favors formation of the more highly substituted intermediate. Addition is completed when the carbocation reacts with a halide ion.

$$R_2C{=}CH_2 + HX \rightarrow R_2\overset{+}{C}{-}CH_3 + X^- \rightarrow R_2C(X)CH_3$$

A more complete discussion of the mechanism of ionic addition of hydrogen halides to alkenes is given in Chapter 6 of Part A. In particular, the extent to which discrete carbocations are involved is considered there.

The terms *regioselective* and *regiospecific* are used to describe addition reactions that proceed selectively or exclusively in one direction with unsymmetrical alkenes.[1] Markownikoff's rule describes a specific case of regioselectivity that is based on the stabilizing effect that alkyl and aryl substituents have on carbocations.

$$^{+}CH_2CH_2R \qquad CH_3\overset{+}{C}HR \qquad CH_3\overset{+}{C}HAr \qquad CH_3\overset{+}{C}R_2 \qquad CH_3\overset{+}{C}(Ar)_2$$

——increasing stability——→

The term regiospecific can be used when a single product is formed. The addition of hydrogen bromide to styrene, for example, is regiospecific because the phenyl group strongly stabilizes cationic character.

$$C_6H_5{-}CH{=}CH_2 \xrightarrow{HBr} C_6H_5{-}CH(Br)CH_3$$

In nucleophilic solvents, products that arise from reaction of the solvent with the cationic intermediate may be encountered. For example, reaction of cyclohexene with hydrogen bromide in acetic acid gives cyclohexyl acetate as well as cyclohexyl bromide. This occurs because acetic acid acts as a nucleophile in competition with the bromide ion.

$$\text{cyclohexene} + HBr \xrightarrow[40^\circ C]{AcOH} \text{cyclohexyl-Br } (85\%) + \text{cyclohexyl-OAc } (15\%)$$

Ref. 2

When carbocations are involved as intermediates, carbon skeleton rearrangement can occur during electrophilic addition reactions. Reaction of *t*-butylethylene with hydrogen chloride in acetic acid gives both rearranged and unrearranged product.[3] The rearranged acetate may also be formed, but it is unstable under the reaction conditions and is converted to the rearranged chloride.

$$(CH_3)_3CCH{=}CH_2 \xrightarrow[HCl]{AcOH} (CH_3)_3CCH(Cl)CH_3 \; (35\text{–}40\%) + (CH_3)_2C(Cl)CH(CH_3)_2 \; (40\text{–}50\%) + (CH_3)_3CCH(OAc)CH_3 \; (15\text{–}20\%)$$

1. A. Hassner, *J. Org. Chem.* **33**, 2684 (1968).
2. R. C. Fahey and R. A. Smith, *J. Am. Chem. Soc.* **86**, 5035 (1964).
3. R. C. Fahey and C. A. McPherson, *J. Am. Chem. Soc.* **91**, 3865 (1969).

The stereochemistry of addition of hydrogen halides to alkenes is dependent on the structure of the alkene and also on the reaction conditions. Addition of hydrogen bromide to cyclohexene and to *E*- and *Z*-2-butene is *anti*.[4] The addition of hydrogen chloride to 1-methylcyclopentene is entirely *anti* when carried out at 25°C in nitromethane.[5]

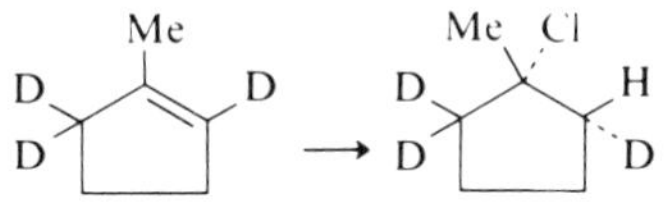

1,2-Dimethylcyclohexene is an example of an alkene for which the stereochemistry of hydrogen chloride addition is dependent on the solvent and temperature. At −78°C in dichloromethane, 88% of the product is the result of *syn* addition, whereas at 0°C in ether, 95% of the product results from *anti* addition.[6] *Syn* addition is particularly common with alkenes having an aryl substituent. Table 4.1 lists examples

Table 4.1. Stereochemistry of Addition of Hydrogen Halides to Alkenes

Alkene	Hydrogen halide	Stereochemistry	Ref.
1,2-Dimethylcyclohexene	HBr	*anti*	a
1,2-Dimethylcyclohexene	HCl	solvent- and temperature-dependent	a
Cyclohexene	HBr	*anti*	b
cis-2-Butene	DBr	*anti*	c
trans-2-Butene	DBr	*anti*	c
1,2-Dimethylcyclopentene	HBr	*anti*	d
1-Methylcyclopentene	HCl	*anti*	e
Norbornene	HBr	*syn* and rearrangement	f
Norbornene	HCl	*syn* and rearrangement	g
trans-1-Phenylpropene	HBr	*syn* (9 : 1)	h
cis-1-Phenylpropene	HBr	*syn* (8 : 1)	h
Bicyclo[3.1.0]hex-2-ene	DCl	*syn*	i
1-Phenyl-4-*t*-butylcyclohexene	DCl	*syn*	j

a. G. S. Hammond and T. D. Nevitt, *J. Am. Chem. Soc.* **76**, 4121 (1954); R. C. Fahey and C. A. McPherson, *J. Am. Chem. Soc.* **93**, 2445 (1971); K. B. Becker and C. A. Grob, *Synthesis*, 789 (1973).
b. R. C. Fahey and R. A. Smith, *J. Am. Chem. Soc.* **86**, 5035 (1964).
c. D. J. Pasto, G. R. Meyer, and B. Lepeska, *J. Am. Chem. Soc.* **96**, 1858 (1974).
d. G. S. Hammond and C. H. Collins, *J. Am. Chem. Soc.* **82**, 4323 (1960).
e. Y. Pocker and K. D. Stevens, *J. Am. Chem. Soc.* **91**, 4205 (1969).
f. H. Kwart and J. L. Nyce, *J. Am. Chem. Soc.* **86**, 2601 (1964).
g. J. K. Stille, F. M. Sonnenberg, and T. H. Kinstle, *J. Am. Chem. Soc.* **88**, 4922 (1966).
h. M. J. S. Dewar and R. C. Fahey, *J. Am. Chem. Soc.* **85**, 3645 (1963).
i. P. K. Freeman, F. A. Raymond, and M. F. Grostic, *J. Org. Chem.* **32**, 24 (1967).
j. K. D. Berlin, R. O. Lyerla, D. E. Gibbs, and J. P. Devlin, *J. Chem. Soc., Chem. Commun.*, 1246 (1970).

4. D. J. Pasto, G. R. Meyer, and S. Kang, *J. Am. Chem. Soc.* **91**, 4205 (1969).
5. Y. Pocker and K. D. Stevens, *J. Am. Chem. Soc.* **91**, 4205 (1969).
6. K. B. Becker and C. A. Grob, *Synthesis*, 789 (1973).

of alkenes for which the stereochemistry of addition of hydrogen chloride or hydrogen bromide has been studied.

The stereochemistry of addition depends on the details of the mechanism. Two general mechanisms have been encountered for alkenes. The addition can proceed through an ion pair intermediate formed by an initial protonation step:

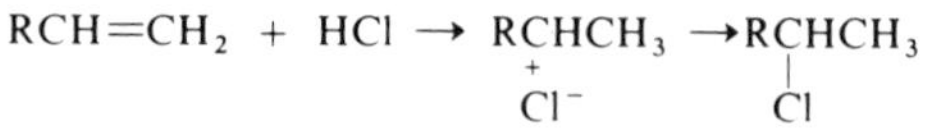

$$RCH{=}CH_2 + HCl \rightarrow \underset{Cl^-}{RC^{+}HCH_3} \rightarrow \underset{Cl}{\underset{|}{RCHCH_3}}$$

Most alkenes, however, react via a transition state that involves the alkene, hydrogen halide, and a third species which delivers the nucleophile. This termolecular mechanism is generally pictured as a nucleophilic attack on the alkene-hydrogen halide complex. This mechanism bypasses a discrete carbocation.

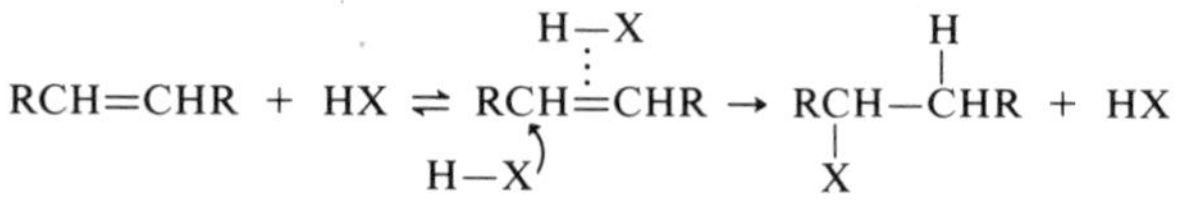

$$RCH{=}CHR + HX \rightleftharpoons \underset{H-X}{\overset{H-X}{RCH{=}CHR}} \rightarrow \underset{X}{RCH}-\overset{H}{CHR} + HX$$

The major factor in determining which mechanism is followed is the stability of the carbocation intermediate. Alkenes that can give rise to a particularly stable carbocation are likely to react via the ion pair mechanism. The ion pair mechanism would not be expected to be stereospecific, since the carbocation intermediate permits loss of stereochemistry relative to the reactant alkene. It might be expected that the ion pair mechanism would lead to a preference for *syn* addition, since at the instant of formation of the ion pair, the halide is necessarily on the same side of the alkene as the proton being added. If the lifetime of the ion pair is long, or if the ion pair dissociates, a mixture of *syn* and *anti* addition products would be expected. The termolecular mechanism would be expected to give *anti* addition. Attack by the nucleophile would occur at the opposite side of the double bond from proton addition:

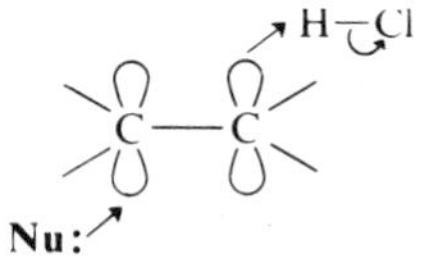

H—Cl

C—C

Nu:

Section 6.1 of Part A gives further discussion of the structural features that affect the competition between the two possible mechanisms.

4.2. Hydration and Other Acid-Catalyzed Additions

In addition to halide ions, various other nucleophilic species can be added to double bonds under acidic conditions. A fundamental example is the hydration of alkenes in acidic aqueous solution:

$$R_2C{=}CH_2 + H^+ \longrightarrow R_2\overset{+}{C}CH_3 \xrightarrow{H_2O} R_2C(\overset{+}{O}H_2)CH_3 \xrightarrow{-H^+} R_2C(OH)CH_3$$

Addition of a proton occurs to give the more substituted carbocation so that addition is regioselective and in accord with Markownikoff's rule. A more detailed discussion of the reaction mechanism is given in Section 6.2 of Part A. The reaction is occasionally applied to the synthesis of tertiary alcohols:

$$(CH_3)_2C{=}CHCH_2CH_2\overset{O}{\overset{\|}{C}}CH_3 \xrightarrow[H_2O]{H_2SO_4} (CH_3)_2C(OH)CH_2CH_2CH_2\overset{O}{\overset{\|}{C}}CH_3 \qquad \text{Ref. 7}$$

Because of the strongly acidic and rather vigorous conditions required to effect hydration of most alkenes, the reaction is only applicable to molecules that have no other acid-sensitive functional groups. Also, because of the involvement of cationic intermediates, rearrangements can occur in systems where a more stable cation would result by aryl, alkyl, or hydrogen migration. A much milder and more general procedure for alkene hydration is discussed in the next section.

Addition of nucleophilic solvents such as alcohols and carboxylic acids can be effected by use of strong acids as catalysts[8]:

$$(CH_3)_2C{=}CH_2 + MeOH \xrightarrow{HBF_4} (CH_3)_3COCH_3$$

$$CH_3CH{=}CH_2 + CH_3CO_2H \xrightarrow{HBF_4} (CH_3)_2CHOCOCH_3$$

Trifluoroacetic acid is a sufficiently strong acid to react with alkenes under relatively mild conditions.[9] The addition is regiospecific in the direction predicted by Markownikoff's rule.

$$ClCH_2CH_2CH_2CH{=}CH_2 \xrightarrow[\Delta]{CF_3CO_2H} ClCH_2CH_2CH_2CH(O_2CCF_3)CH_3$$

Ring strain enhances alkene reactivity. Norbornene, for example, undergoes rapid addition at 0°C.[10]

4.3. Oxymercuration

The addition reactions that were discussed in Sections 4.1 and 4.2 are initiated by interaction of a proton with the alkene, which causes nucleophilic attack on the

7. J. Meinwald, *J. Am. Chem. Soc.* **77**, 1617 (1955).
8. R. D. Morin and A. E. Bearse, *Ind. Eng. Chem.* **43**, 1596 (1951); D. T. Dalgleish, D. C. Nonhebel, and P. L. Pauson, *J. Chem. Soc., C*, 1174 (1971).
9. P. E. Peterson, R. J. Bopp, D. M. Chevli, E. L. Curran, D. E. Dillard, and R. J. Kamat, *J. Am. Chem. Soc.* **89**, 5902 (1967).
10. H. C. Brown, J. H. Kawakami, and K.-T. Liu, *J. Am. Chem. Soc.* **92**, 5536 (1970).

double bond. The role of the initial electrophile can be played by metal cations as well. Mercuric ion is the reactive electrophile in several synthetically valuable procedures.[11] The most commonly used reagent is mercuric acetate, but the trifluoroacetate or nitrate salts are preferable in some applications. A general mechanism depicts a mercurinium ion as the first intermediate.[12] Such species can be detected by physical measurements when alkenes react with mercuric ions in nonnucleophilic solvents.[13] Depending on the structure of the particular alkene, the mercurinium ion may be predominantly bridged or open. The addition is completed by attack of a nucleophile.

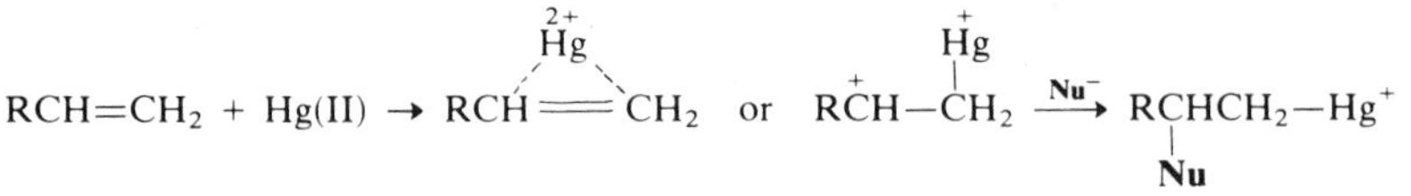

$$RCH{=}CH_2 + Hg(II) \rightarrow RCH\overset{Hg^{2+}}{\overline{\;\;\;}}CH_2 \text{ or } R\overset{+}{C}H{-}CH_2(Hg^+) \xrightarrow{Nu^-} RCH(Nu)CH_2{-}Hg^+$$

The nucleophiles that are used for synthetic purposes include water, alcohols, carboxylate ions, hydroperoxides, amines, and nitriles. After the addition step is complete, the mercury is usually reductively removed by sodium borohydride. The net result is the addition of hydrogen and the nucleophile to the alkene. The regioselectivity is excellent and is in the same sense as is observed for proton-initiated additions.[14] Scheme 4.1 includes examples of these reactions. Electrophilic attack by mercuric ion can effect cyclization by intramolecular capture of a nucleophilic functional group, as illustrated by entries 9–11.

Scheme 4.1. Synthesis via Mercuration

Alcohols

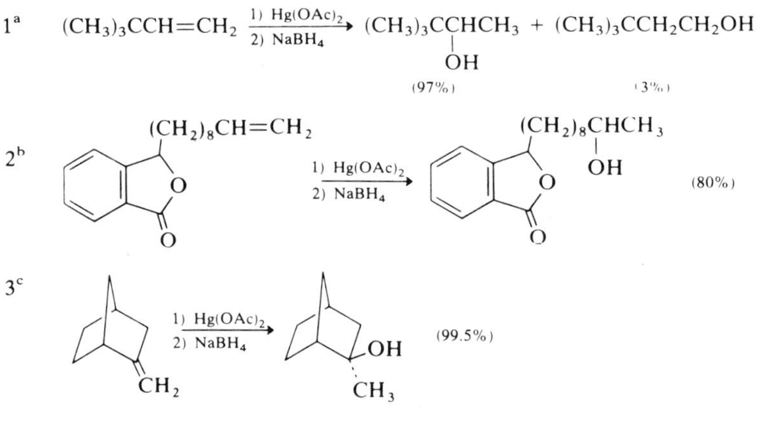

1[a] $(CH_3)_3CCH{=}CH_2 \xrightarrow[2)\ NaBH_4]{1)\ Hg(OAc)_2} (CH_3)_3CCH(OH)CH_3 + (CH_3)_3CCH_2CH_2OH$
(97%) (3%)

2[b] 3-[$(CH_2)_8CH{=}CH_2$]phthalide $\xrightarrow[2)\ NaBH_4]{1)\ Hg(OAc)_2}$ 3-[$(CH_2)_8CH(OH)CH_3$]phthalide (80%)

3[c] 2-methylenenorbornane (CH_2) $\xrightarrow[2)\ NaBH_4]{1)\ Hg(OAc)_2}$ 2-methyl-2-norbornanol (OH, CH_3) (99.5%)

11. R. C. Larock, *Angew. Chem. Int. Ed. Engl.* **17**, 27 (1978); W. Kitching, *Organomet. Chem. Rev.* **3**, 61 (1968).
12. S. J. Cristol, J. S. Perry, Jr., and R. S. Beckley, *J. Org. Chem.* **41**, 1912 (1976); D. J. Pasto and J. A. Gontarz, *J. Am. Chem. Soc.* **93**, 6902 (1971).
13. G. A. Olah and P. R. Clifford, *J. Am. Chem. Soc.* **95**, 6067 (1973); G. A. Olah and S. H. Yu, *J. Org. Chem.* **40**, 3638 (1975).
14. H. C. Brown and P. J. Geoghegan, Jr., *J. Org. Chem.* **35**, 1844 (1970).

Ethers

4[d] cyclohexene $\xrightarrow[\text{2) NaBH}_4]{\text{1) Hg(O}_2\text{CCF}_3)_2\text{, (CH}_3)_2\text{CHOH}}$ cyclohexyl–$OCH(CH_3)_2$ (98%)

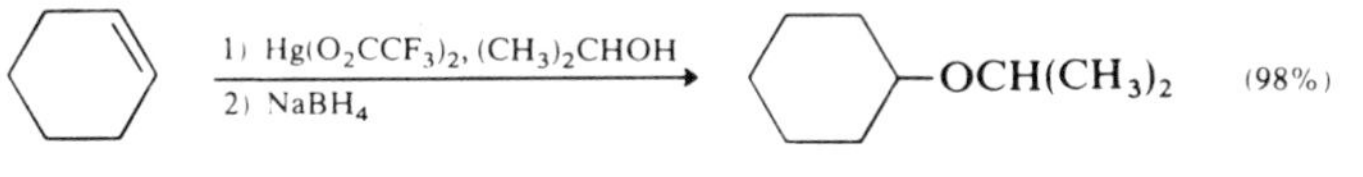

5[e] $CH_3(CH_2)_3CH{=}CH_2 \xrightarrow[\text{EtOH}]{\text{Hg(O}_2\text{CCF}_3)_2} CH_3(CH_2)_3CH(OC_2H_5)CH_3$ (97%)

Amides

6[f] $CH_3(CH_2)_3CH{=}CH_2 \xrightarrow[\text{2) NaBH}_4\text{, H}_2\text{O}]{\text{1) Hg(NO}_3)_2\text{, CH}_3\text{CN}} CH_3CH_2CH_2CH_2CH(HNCOCH_3)CH_3$ (92%)

Peroxides

7[g] $CH_3(CH_2)_4CH{=}CHCH_3 \xrightarrow[\text{2) NaBH}_4]{\text{1) Hg(OAc)}_2\text{, }t\text{-BuOOH}} CH_3(CH_2)_4CH(OOC(CH_3)_3)CH_2CH_3$ (40%)

Amines

8[f] CH_3O–C_6H_4–$CH_2CH{=}CH_2$ + $PhCH_2NH_2$ $\xrightarrow[\text{2) NaBH}_4]{\text{1) Hg(ClO}_4)_2}$ CH_3O–C_6H_4–$CH_2CH(HNCH_2Ph)CH_3$ (70%)

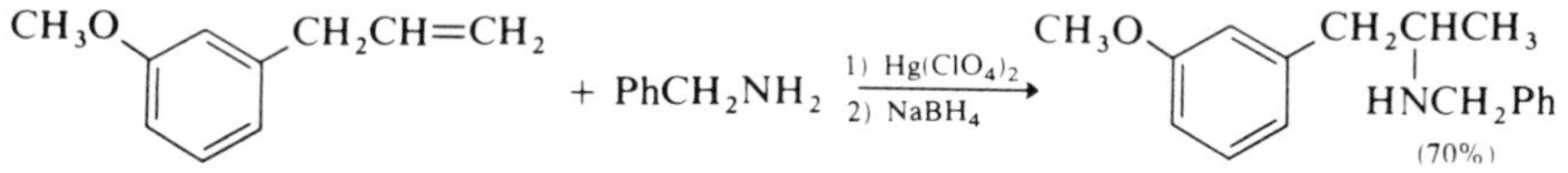

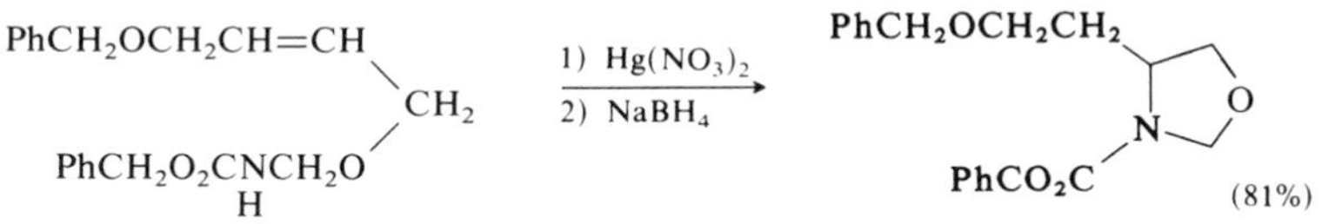

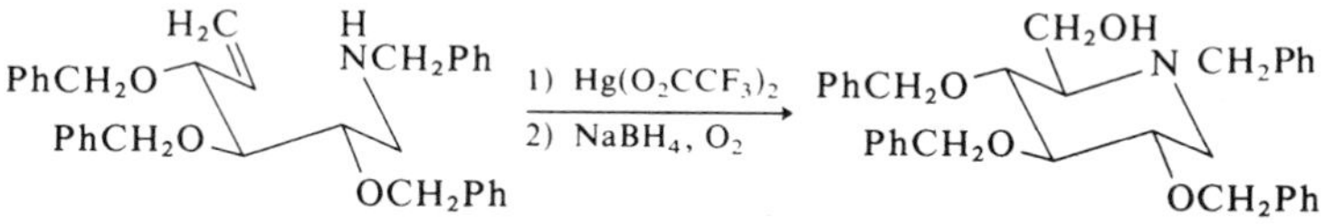

9[i] $PhCH_2OCH_2CH{=}CH$–CH_2–$OCH_2NHCO_2CH_2Ph$ ($PhCH_2O_2CNCH_2O$, H) $\xrightarrow[\text{2) NaBH}_4]{\text{1) Hg(NO}_3)_2}$ $PhCH_2OCH_2CH_2$-substituted oxazolidine, N-$PhCO_2C$ (81%)

10[j] H_2C, $PhCH_2O$, $PhCH_2O$, H, NCH_2Ph, OCH_2Ph $\xrightarrow[\text{2) NaBH}_4\text{, O}_2]{\text{1) Hg(O}_2\text{CCF}_3)_2}$ CH_2OH, $PhCH_2O$, $PhCH_2O$, N CH_2Ph, OCH_2Ph

11[k] H_2C, HO, OH, H, O, N, O, N, O, OCH_2OCH_2Ph, CH_2OCH_2Ph $\xrightarrow[\text{2) NaBH}_4\text{, O}_2]{\text{1) Hg(OAc)}_2}$ $HOCH_2$, O, OH, O, N, O, N, O, H, OCH_2OCH_2Ph, CH_2OCH_2Ph

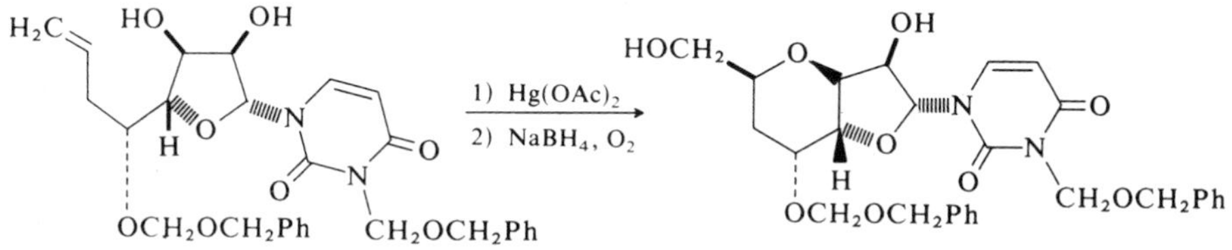

a. H. C. Brown and P. J. Geoghegan, Jr., *J. Org. Chem.* **35**, 1844 (1970).
b. H. L. Wehrmeister and D. E. Robertson, *J. Org. Chem.* **33**, 4173 (1968).
c. H. C. Brown and W. J. Hammar, *J. Am. Chem. Soc.* **89**, 1524 (1967).
d. H. C. Brown and M.-H. Rei, *J. Am. Chem. Soc.* **91**, 5646 (1969).
e. H. C. Brown, J. T. Kurek, M.-H. Rei, and K. L. Thompson, *J. Org. Chem.* **50**, 1171 (1985).
f. H. C. Brown and J. T. Kurek, *J. Am. Chem. Soc.* **91**, 5647 (1969).
g. D. H. Ballard and A. J. Bloodworth, *J. Chem. Soc. C*, 945 (1971).
h. R. C. Griffith, R. J. Gentile, T. A. Davidson, and F. L. Scott, *J. Org. Chem.* **44**, 3580 (1979).
i. K. E. Harding and D. R. Hollingsworth, *Tetrahedron Lett.* **29**, 3789 (1988).
j. R. C. Bernotas and B. Ganem, *Tetrahedron Lett.* **26**, 1123 (1985).
k. S. Hanessian, J. Kloss, and T. Sugawara, *J. Am. Chem. Soc.* **108**, 2758 (1986).

The reductive replacement of mercury by hydrogen with the use of sodium borohydride is a free-radical process.[15]

$$RHgX + NaBH_4 \rightarrow RHgH$$

$$RHgH \rightarrow R\cdot + Hg^{(I)}H$$

$$R\cdot + RHgH \rightarrow RH + Hg^{(0)} + R\cdot$$

The evidence for this mechanism includes the fact that the course of the reaction can be diverted by oxygen, an efficient radical scavenger. In the presence of oxygen, the mercury is replaced by a hydroxy group. Also, the stereochemistry of the reduction, as studied by using $NaBD_4$ as the reducing agent, is consistent with a radical intermediate.[16] For example, a 50 : 50 mixture of *erythro*- and *threo*-3-*deuterio*-2-butanol is obtained by oxymercuration of either *cis*- or *trans*-2-butene, followed by reduction with $NaBD_4$:

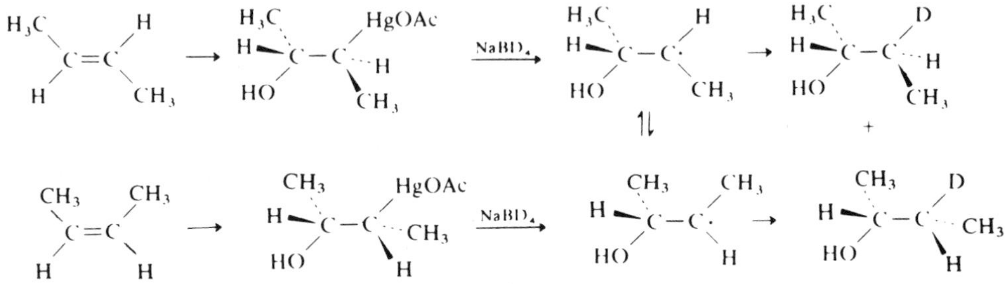

H3C H C=C H CH3 → H3C H C–C HgOAc H HO CH3 —NaBD4→ H3C H C–C· H HO CH3 → H3C H C–C D H HO CH3 +

CH3 CH3 C=C H H → CH3 H C–C HgOAc CH3 HO H —NaBD4→ CH3 H C–C· CH3 HO H → CH3 H C–C D CH3 HO H

Also consistent with occurrence of a free-radical intermediate is the formation of cyclic products when 5-hexenylmercury compounds are reduced with sodium borohydride.[17] In the presence of oxygen, no cyclic product is formed, indicating that O_2 trapping of the radical is much faster than cyclization.

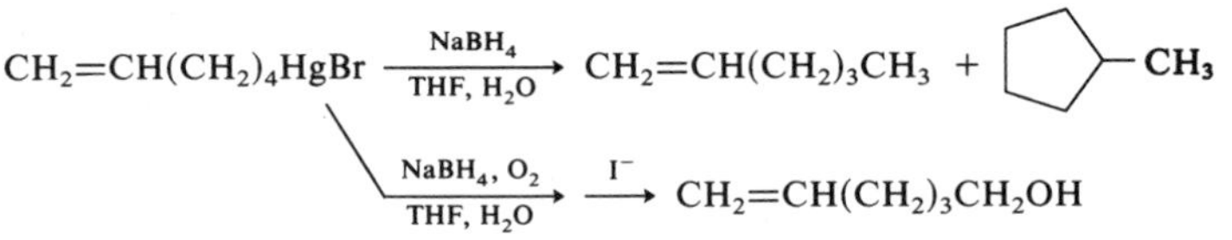

$$CH_2{=}CH(CH_2)_4HgBr \xrightarrow[\text{THF, } H_2O]{NaBH_4} CH_2{=}CH(CH_2)_3CH_3 + \text{(methylcyclopentane)}$$

$$CH_2{=}CH(CH_2)_4HgBr \xrightarrow[\text{THF, } H_2O]{NaBH_4,\ O_2} \xrightarrow{I^-} CH_2{=}CH(CH_2)_3CH_2OH$$

The trapping of the radical intermediate by oxygen has been exploited as a method for introduction of an additional hydroxyl substituent. The example below and entries 10 and 11 in Scheme 4.1 illustrate this reaction.

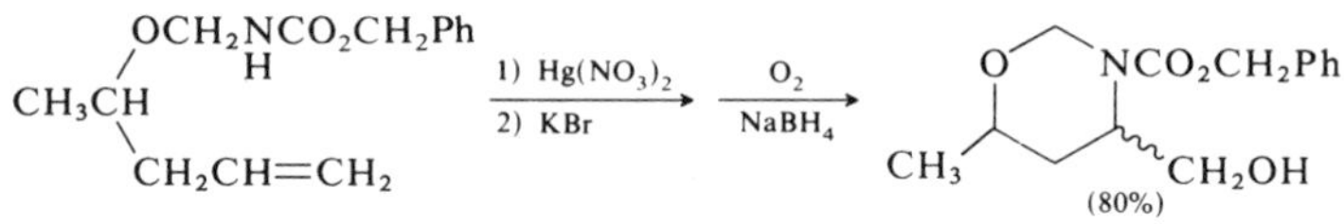

$CH_3CH(OCH_2NHCO_2CH_2Ph)CH_2CH{=}CH_2$ $\xrightarrow[2)\ KBr]{1)\ Hg(NO_3)_2}$ $\xrightarrow[NaBH_4]{O_2}$ cyclic product: O, NCO_2CH_2Ph, CH_3, CH_2OH (80%) Ref. 18

15. C. L. Hill and G. M. Whitesides, *J. Am. Chem. Soc.* **96**, 870 (1974).
16. D. J. Pasto and J. A. Gontarz, *J. Am. Chem. Soc.* **91**, 719 (1969); G. A. Gray and W. R. Jackson, *J. Am. Chem. Soc.* **91**, 6205 (1969).
17. R. P. Quirk and R. E. Lea, *J. Am. Chem. Soc.* **98**, 5973 (1976).
18. K. E. Harding, T. H. Marman, and D. Nam, *Tetrahedron Lett.* **29**, 1627 (1988).

An alternative reagent for demercuration is sodium amalgam in a protic solvent. Here the evidence is that free radicals are not involved, and the mercury is replaced with complete retention of configuration[19]:

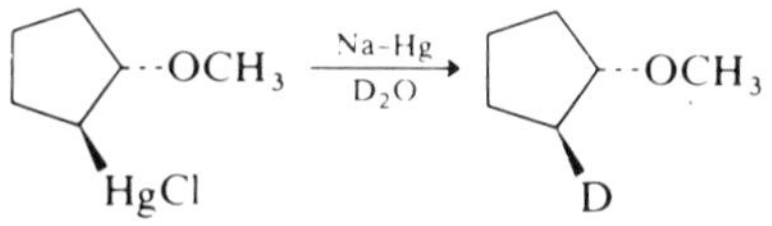

The stereochemistry of oxymercuration has been examined in a number of systems. Conformationally biased cyclic alkenes such as 4-*t*-butycyclohexene and 4-*t*-butyl-1-methycyclohexene give exclusively the product of *anti* addition, which is consistent with a mercurinium ion intermediate.[16,20]

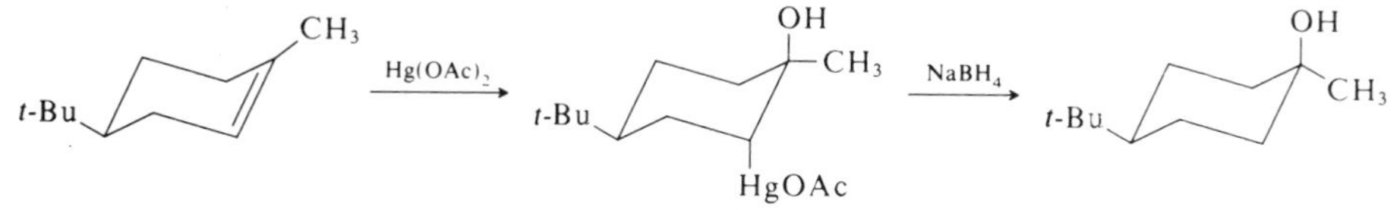

With hindered or strained alkenes, where the transition state for *anti* addition is disfavored, a *syn* mode of addition is also accessible.[21] Norbornene, for example, in which an *anti* addition is opposed by both steric (difficulty of *endo* approach) and torsional (prohibition of an *anti*-periplanar transition state) factors, reacts by *syn–exo* addition.

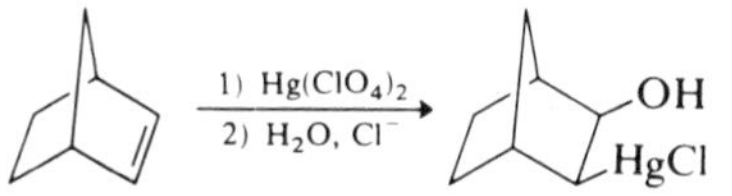

Ref. 22

The reactivity of different alkenes toward mercuration spans a considerable range and is governed by a combination of steric and electronic factors.[23] Terminal double bonds are more reactive than internal ones. Disubstituted terminal alkenes, however, are more reactive than monosubstituted cases, as would be expected for electrophilic attack. The differences in relative reactivities are large enough that selectivity can be achieved in certain dienes:

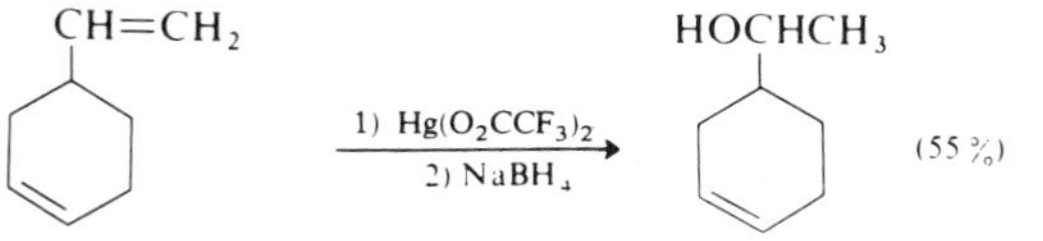

(55%) Ref. 23b

19. F. R. Jensen, J. J. Miller, S. J. Cristol, and R. S. Beckley, *J. Org. Chem.* **37**, 434 (1972); R. P. Quirk, *J. Org. Chem.* **37**, 3554 (1972); W. Kitching, A. R. Atkins, G. Wickham, and V. Alberts, *J. Org. Chem.* **46**, 563 (1981).
20. H. C. Brown, G. J. Lynch, W. J. Hammar, and L. C. Liu, *J. Org. Chem.* **44**, 1910 (1979).
21. W. L. Waters, T. G. Traylor, and A. Factor, *J. Org. Chem.* **38**, 2306 (1973).
22. T. G. Traylor and A. W. Baker, *Tetrahedron. Lett.*, No. 19, 14 (1959).
23. (a) H. C. Brown and P. J. Geoghegan, *J. Org. Chem.* **37**, 1937 (1972); (b) H. C. Brown, P. J. Geoghegan, Jr., G. J. Lynch, and J. T. Kurek, *J. Org. Chem.* **37**, 1941 (1972); (c) H. C. Brown, P. J. Geoghegan, Jr., and J. T. Kurek, *J. Org. Chem.* **46**, 3810 (1981).

Table 4.2. Relative Reactivity of Some Alkenes in Oxymercuration[a]

1-Pentene	6.6
2-Methyl-1-pentene	48
cis-2-Pentene	0.56
trans-2-Pentene	0.17
2-Methyl-2-pentene	1.24

a. Relative to cyclohexene; Data from H. C. Brown and P. J. Geoghegan, Jr., *J. Org. Chem.* **37**, 1937 (1972).

The relative reactivity data for some pentene derivatives are given in Table 4.2.

Diastereoselectivity has been observed in oxymercuration of alkenes with nearby oxygen substituents. Terminal allylic alcohols show a preference for formation of the *anti* 2,3-diols.

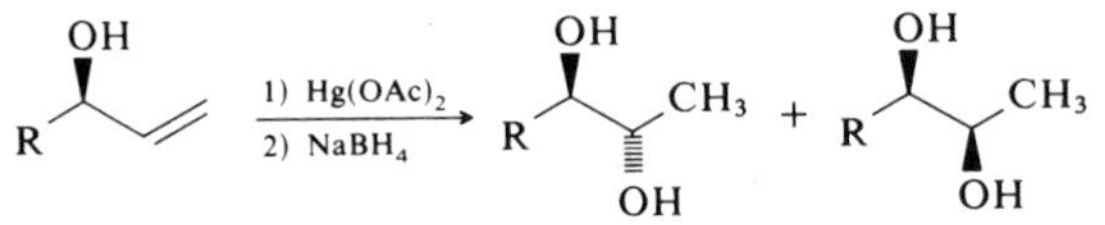

1) $Hg(OAc)_2$
2) $NaBH_4$

R	*anti*	*syn*
Et	76	24
i-Pr	80	20
t-Bu	98	2
Ph	88	12

The approach of the mercuric ion is directed by the hydroxyl group. There is a preference for transition state **A** over **B** for steric reasons, and the selectivity increases with the size of the substituent R.[24]

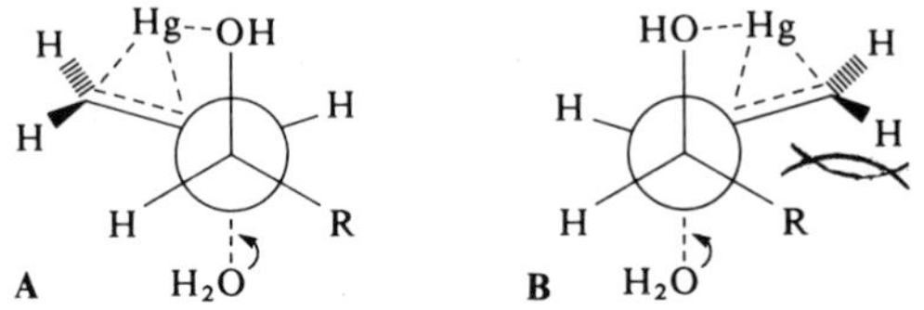

When the hydroxyl group is acetylated, the *syn* isomer is preferred. This result is attributed to direct nucleophilic participation by the carbonyl oxygen of the ester.

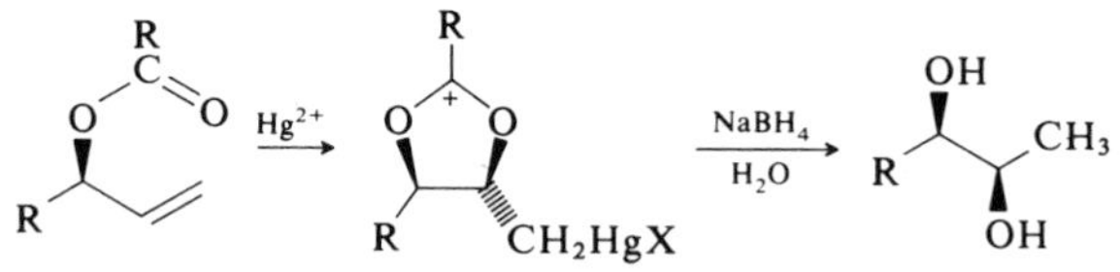

4.4. Addition of Halogens to Alkenes

The addition of chlorine and bromine to alkenes is a very general reaction. Considerable insight has been gained into the mechanism of halogen addition by

24. B. Giese and D. Bartmann, *Tetrahedron Lett.* **26**, 1197 (1985).

studies on the stereochemistry of the reaction. Most types of alkenes are known to add bromine in a stereospecific manner, giving the product of *anti* addition. Among the alkenes that are known to give *anti* addition products are maleic and fumaric acid, *cis*-2-butene, *trans*-2-butene, and a number of cycloalkenes.[25] Cyclic, positively charged bromonium ion intermediates offer an attractive explanation for the observed stereospecificity:

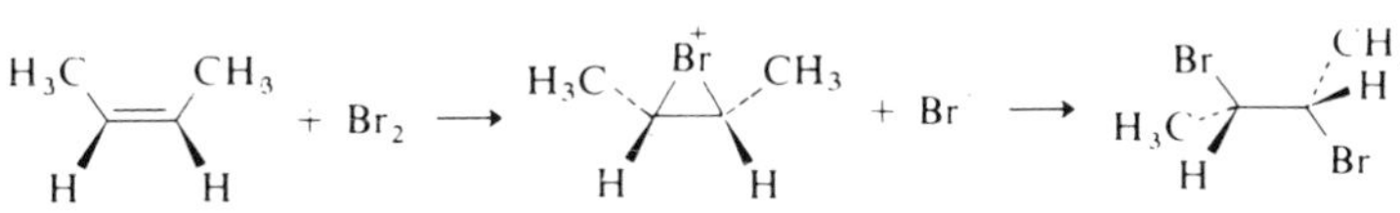

The bridging bromine prevents rotation about the remaining bond, and back-side nucleophilic opening of the ring by bromide ion would lead to the observed *anti* addition. Direct evidence for the existence of bromonium ions has been obtained from NMR measurements.[26] A bromonium ion salt (with Br_3^- as the counterion) has been isolated from the reaction of bromine with the very hindered alkene adamantylideneadamantane.[27]

Substantial amounts of *syn* addition have been observed for *cis*-1-phenylpropene (27–80% *syn* addition), *trans*-1-phenylpropene (17–29% *syn* addition), and *cis*-stilbene (up to 90% *syn* addition in polar solvents).

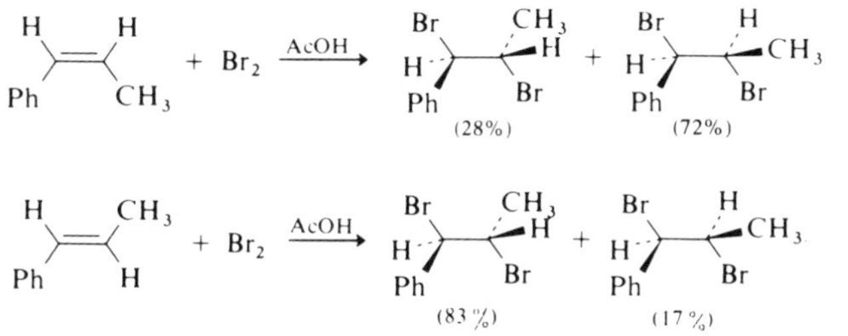

(28%) (72%)

Ref. 28

(83%) (17%)

A common feature of the compounds that give extensive *syn* addition is the presence of at least one phenyl substituent on the double bond. The presence of a phenyl substituent diminishes the necessity for strong bromonium ion bridging by stabilizing the cationic center. A weakly bridged structure in equilibrium with an open benzylic cation can account for the loss in stereospecificity:

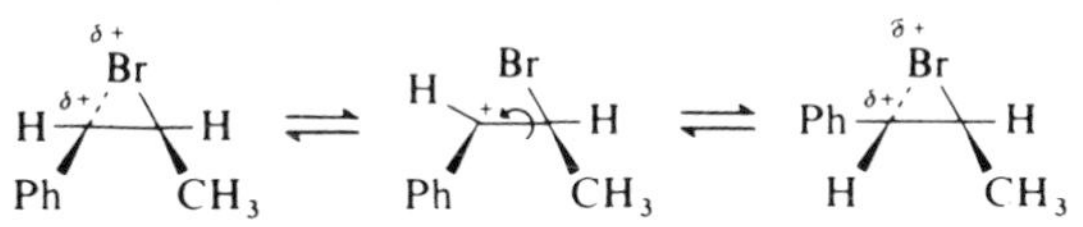

The diminished stereospecificity is similar to that noted for hydrogen halide addition to phenyl-substituted alkenes.

25. J. H. Rolston and K. Yates, *J. Am. Chem. Soc.* **91**, 1469, 1477 (1969).
26. G. A. Olah, J. M. Bollinger, and J. Brinich, *J. Am. Chem. Soc.* **90**, 2587 (1968); G. A. Olah, P. Schilling, P. W. Westerman, and H. C. Lin, *J. Am. Chem. Soc.* **96**, 3581 (1974).
27. J. Strating, J. H. Wierenga, and H. Wynberg, *J. Chem. Soc., Chem. Commun.*, 907 (1969).
28. J. H. Rolston and K. Yates, *J. Am. Chem. Soc.* **91**, 1469, 1477 (1969).

Although chlorination of aliphatic alkenes usually gives *anti* addition, *syn* addition is often dominant for phenyl-substituted alkenes[29]:

$$\text{cis-}CH_3CH{=}CHCH_3 + Cl_2 \xrightarrow{AcOH} \text{(dichloride)} \quad \text{only dichloride formed}$$

$$\text{cis-}PhCH{=}CHCH_3 + Cl_2 \xrightarrow{AcOH} \text{(major)} + \text{(minor)}$$

These results, too, reflect a difference in the extent of bridging in the intermediates. With unconjugated alkenes, there is strong bridging and high *anti* stereospecificity. Phenyl substitution leads to greater cationic character at the benzylic site, and there is a preference for *syn* addition. Because of its smaller size and lesser polarizability, chlorine is not as effective as bromine in maintaining bridging for any particular alkene. Bromination therefore generally gives a higher degree of *anti* addition than chlorination, all other factors being the same.[30]

Chlorination can be accompanied by other reactions that are indicative of carbocation intermediates. Branched alkenes can give products that are the result of elimination of a proton from a cationic intermediate:

$$(H_3C)_2C{=}CH_2 \xrightarrow{Cl_2} (CH_3)_2\overset{+}{C}{-}CH_2Cl \rightarrow H_2C{=}C(CH_3){-}CH_2Cl \quad (80\%)$$

$$(H_3C)_2C{=}C(CH_3)_2 \xrightarrow{Cl_2} (CH_3)_2\overset{+}{C}{-}CCl(CH_3)_2 \rightarrow H_2C{=}C(CH_3)CCl(CH_3)_2 \quad (99\%)$$

Skeletal rearrangements have also been observed in systems that are prone toward migration:

$$(CH_3)_3CCH{=}CHC(CH_3)_3 \xrightarrow{Cl_2} CH_3C(=CH_2)CH(CH_3)CHClC(CH_3)_3 \qquad \text{Ref. 31}$$

$$Ph_3CCH{=}CH_2 \xrightarrow{Br_2} Ph_3CCHBrCH_2Br + Ph_2C{=}C(Ph)CH_2Br \qquad \text{Ref. 32}$$

Since halogenation involves electrophilic attack, substituents on the double bond that increase electron density increase the rate of reaction, whereas electron-withdrawing substituents have the opposite effect. Bromination of simple alkenes

29. M. L. Poutsma, *J. Am. Chem. Soc.* **87**, 2161, 2172 (1965); R. C. Fahey, *J. Am. Chem. Soc.* **88**, 4681 (1966); R. C. Fahey and C. Shubert, *J. Am. Chem. Soc.* **87**, 5172 (1965).
30. R. J. Abraham and J. R. Monasterios, *J. Chem. Soc., Perkin Trans. 1*, 1446 (1973).
31. M. L. Poutsma, *J. Am. Chem. Soc.* **87**, 4285 (1965).
32. R. O. C. Norman and C. B. Thomas, *J. Chem. Soc., B*, 598 (1967).

is an extremely fast reaction. Some specific rate data are tabulated and discussed in Section 6.3 of Part A.

In nucleophilic solvents, the solvent can compete with halide ion for the positively charged intermediate. For example, the bromination of styrene in acetic acid leads to substantial amounts of the acetoxybromo derivative:

$$PhCH{=}CH_2 + Br_2 \xrightarrow{AcOH} \underset{(80\%)}{PhCH(Br)CH_2Br} + \underset{(20\%)}{PhCH(OAc)CH_2Br} \qquad \text{Ref. 33}$$

The acetoxy group is introduced exclusively at the benzylic carbon. This is in accord with the intermediate being a weakly bridged species or a benzylic cation. In an intermediate with C—Br$^+$ bonds of equal strength to both carbons, preferential attack at the less hindered carbon would be anticipated.

[Structure: weakly bridged bromonium intermediate (H, Ph, δ+ on carbon, δ+ Br) attacked by AcO–H at the benzylic carbon → Ph(H)(AcO)C–C(Br)(H)(H)]

The addition of bromide salts to the reaction mixture diminishes the amount of acetoxy compound formed by tipping the competition between acetic acid and bromide ion for the electrophile in favor of the bromide ion.

Chlorination in nucleophilic solvents can also lead to solvent incorporation, as for example in the chlorination of phenylpropene in methanol.[34]

$$PhCH{=}CHCH_3 + Cl_2 \xrightarrow{CH_3OH} \underset{(82\%)}{PhCH(OCH_3)CH(Cl)CH_3} + \underset{(18\%)}{PhCH(Cl){-}CH(Cl)CH_3}$$

From a synthetic point of view, the participation of water in brominations, leading to bromohydrins, is probably the most important example of nucleophilic participation by solvent. In the case of unsymmetrical alkenes, water reacts at the more substituted carbon, which is the carbon with the greatest cationic character. If it is desired to favor introduction of water, it is necessary to keep the concentration of the bromide ion as low as possible. One method for accomplishing this is to use *N*-bromosuccinimide (NBS) as the brominating reagent.[35,36] High yields of bromohydrins are obtained by use of NBS in aqueous DMSO.[37] The reaction is a stereospecific *anti* addition. As in bromination, a bromonium ion intermediate can explain the *anti* stereospecificity. It has been shown that the reactions in dimethyl

33. J. H. Rolston and K. Yates, *J. Am. Chem. Soc.* **91**, 1469 (1969).
34. M. L. Poutsma and J. L. Kartch, *J. Am. Chem. Soc.* **89**, 6595 (1967).
35. A. J. Sisti and M. Meyers, *J. Org. Chem.* **38**, 4431 (1973).
36. C. O. Guss and R. Rosenthal, *J. Am. Chem. Soc.* **77**, 2549 (1955).
37. D. R. Dalton, V. P. Dutta, and D. C. Jones, *J. Am. Chem. Soc.* **90**, 5498 (1968).

sulfoxide involve initial nucleophilic attack by the sulfoxide oxygen. The resulting intermediate reacts with water to give the bromohydrin.

$$RCH{=}CH_2 \xrightarrow{Br^+} R{-}\overset{Br^+}{CH{-}CH_2} \xleftarrow{} (CH_3)_2S{=}O \longrightarrow (CH_3)_2\overset{+}{S}{-}O{-}\overset{R}{\overset{|}{C}}HCH_2Br \; (H_2O,\ H^+) \xrightarrow{H_2O} HO\overset{R}{\overset{|}{C}}HCH_2Br$$

In accord with the Markownikoff rule, the hydroxyl group is introduced at the carbon best able to support positive charge:

$$(CH_3)_3C\overset{CH_3}{\overset{|}{C}}{=}CH_2 \xrightarrow[DMSO,\ H_2O]{NBS} (CH_3)_3C\overset{CH_3}{\underset{OH}{C}}{-}CH_2Br \quad (60\%) \qquad \text{Ref. 37}$$

$$PhCH_2CH{=}CH_2 \xrightarrow[H_2O,\ DMSO]{NBS} PhCH_2\underset{OH}{C}HCH_2Br \quad (89\%) \qquad \text{Ref. 38}$$

Because of its great reactivity, special precautions must be used in reactions of fluorine, and its use is somewhat specialized.[39] Nevertheless, there is some basis for comparison with the less reactive halogens. Addition of fluorine to *Z*- and *E*-1-propenylbenzene is not stereospecific, but *syn* addition is somewhat favored.[40] This result is consistent with formation of a cationic intermediate.

$$PhCH{=}CHCH_3 \xrightarrow{F_2} PhCHF{-}CHFCH_3 \ (\text{diastereomer}) + PhCHF{-}CHFCH_3 \ (\text{diastereomer})$$

In methanol, the solvent incorporation product is formed, as would be expected for a cationic intermediate:

$$PhCH{=}CHCH_3 \xrightarrow[MeOH]{F_2} Ph\underset{MeO}{C}H\underset{F}{C}HCH_3$$

These results are consistent with the expectation that fluoride would not be an effective bridging atom.

There are other reagents, such as CF_3OF and CH_3CO_2F, which appear to transfer an electrophilic fluorine to double bonds and form an ion pair which collapses to an addition product.

$$PhCH{=}CHPh + CF_3OF \rightarrow Ph\underset{CF_3O}{C}H\underset{F}{C}HPh \qquad \text{Ref. 41}$$

38. A. W. Langman and D. R. Dalton, *Org. Synth.* **59**, 16 (1979).
39. H. Vypel, *Chimia* **39**, 305 (1985).
40. R. F. Merritt, *J. Am. Chem. Soc.* **89**, 609 (1967).
41. D. H. R. Barton, R. H. Hesse, G. P. Jackson, L. Ogankoya, and M. M. Pechet, *J. Chem. Soc., Perkin Trans. 1*, 739 (1974).

$$CH_3(CH_2)_9CH{=}CH_2 + CH_3CO_2F \rightarrow CH_3(CH_2)_9\underset{\displaystyle CH_3CO_2}{\underset{|}{C}}HCH_2F \quad (30\%)$$ Ref. 42

Addition of iodine to alkenes can be accomplished by a photochemically initiated reaction. Elimination of iodine is catalyzed by excess iodine radicals, but the diiodo compounds can be obtained if unreacted iodine is removed[43]:

$$RCH{=}CHR + I_2 \rightleftharpoons R\underset{I}{\underset{|}{C}}H{-}\underset{I}{\underset{|}{C}}HR$$

The diiodo compounds are very sensitive to light and have not been used very often in synthesis.

Iodine is a very good electrophile for effecting intramolecular nucleophilic addition to alkenes, as exemplified by the iodolactonization reaction. Reaction of iodine with carboxylic acids having carbon–carbon double bonds placed to permit intramolecular reaction results in formation of iodolactones.[44] The reaction shows a preference for formation of five-membered rings over six-membered[45] and is a strictly *anti* stereospecific addition when carried out under basic conditions.

CH_2CO_2H ; $CH{=}CH_2$ $\xrightarrow[NaHCO_3]{I_2,\ I^-}$ I ; O ; O ; $CH{=}CH_2$ Ref. 46

The *anti* addition is kinetically controlled and results from irreversible back-side opening of an iodonium ion intermediate by the carboxylate nucleophile.

When iodolactonization is carried out under nonbasic conditions, the addition step becomes reversible and the product is then the thermodynamically favored one.[47] This usually results in the formation of the stereoisomeric lactone that has adjacent substituents *trans* with respect to one another.

H ; CH_2 ; R ; CH_2CO_2H $\xrightarrow[CH_3CN]{I_2}$ ICH_2 ; H ; O ; $\overset{+}{O}H$; R ; H $\rightleftharpoons$ ICH_2 ; H ; O ; $\overset{+}{O}H$; R ; H $\xrightarrow{-H^+}$ ICH_2 ; H ; O ; =O ; R ; H

Several other nucleophilic functional groups can be induced to participate in iodocyclization reactions. *t*-Butyl carbonate esters cyclize to diol carbonates[48]:

42. S. Rozen, O. Lerman, M. Kol, and D. Hebel, *J. Org. Chem.* **50**, 4753 (1985).
43. P. S. Skell and R. R. Pavlis, *J. Am. Chem. Soc.* **86**, 2956 (1964); R. L. Ayres, C. J. Michejda, and E. P. Rack, *J. Am. Chem. Soc.* **93**, 1389 (1971).
44. M. D. Dowle and D. I. Davies, *Chem. Soc. Rev.* **8**, 171 (1979).
45. S. Ranganathan, D. Ranganathan, and A. K. Mehrotra, *Tetrahedron* **33**, 807 (1977).
46. L. A. Paquette, G. D. Crouse, and A. K. Sharma, *J. Am. Chem. Soc.* **102**, 3972 (1980).
47. P. A. Bartlett and J. Myerson, *J. Am. Chem. Soc.* **100**, 3950 (1978).
48. P. A. Bartlett, J. D. Meadows, E. G. Brown, A. Morimoto, and K. K. Jernstedt, *J. Org. Chem.* **47**, 4013 (1982).

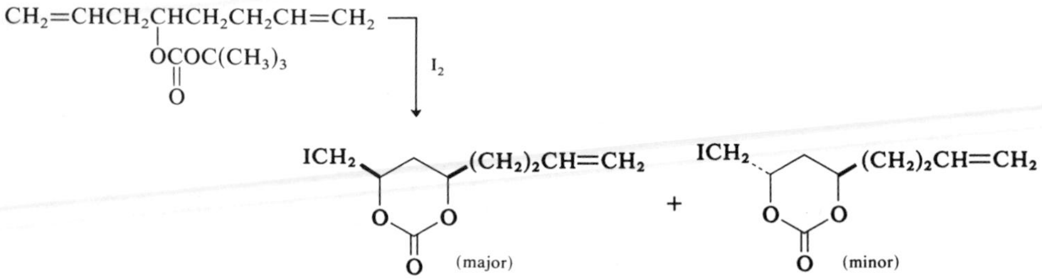

$CH_2{=}CHCH_2CH(OCOC(CH_3)_3)CH_2CH_2CH{=}CH_2$ $\xrightarrow{I_2}$ (major) + (minor)

Lithium salts of carbonate monoesters can also be cyclized[49]:

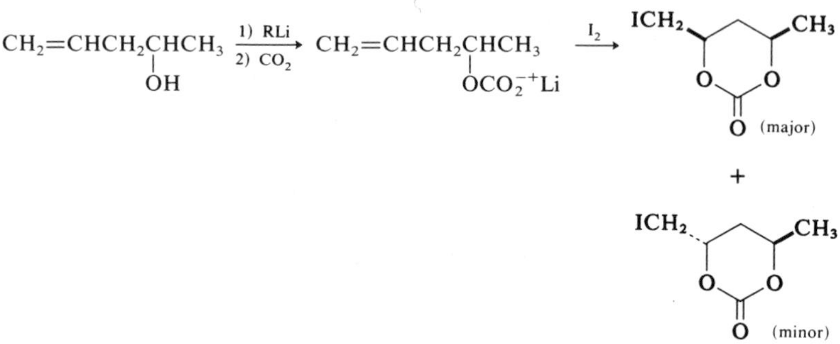

$CH_2{=}CHCH_2CH(OH)CH_3 \xrightarrow[2)\ CO_2]{1)\ RLi} CH_2{=}CHCH_2CH(OCO_2^-\,{}^+Li)CH_3 \xrightarrow{I_2}$ (major) + (minor)

Because the iodocylization products have a potentially nucleophilic oxygen substituent β to the iodide, they are useful in stereospecific syntheses of epoxides and diols:

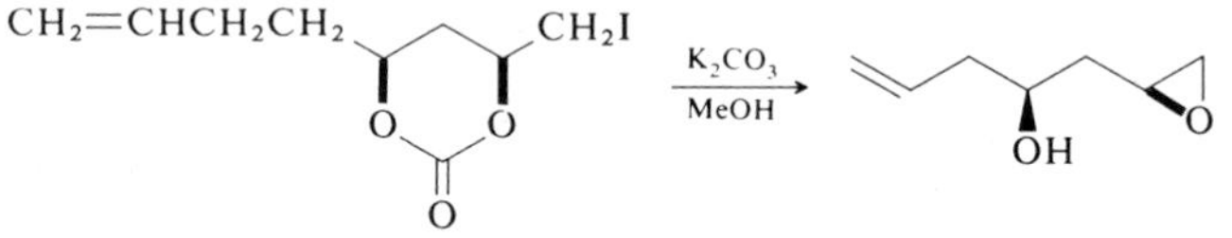

$\xrightarrow[MeOH]{K_2CO_3}$ Ref. 48

$\xrightarrow{I_2}$ $\xrightarrow[MeOH]{Na_2CO_3}$ Ref. 50

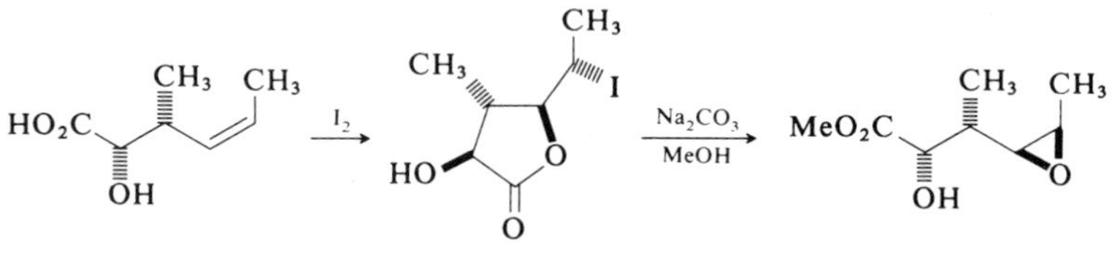

Lactams can be obtained by iodocyclization of *O,N*-trimethylsilyl imidates[51]:

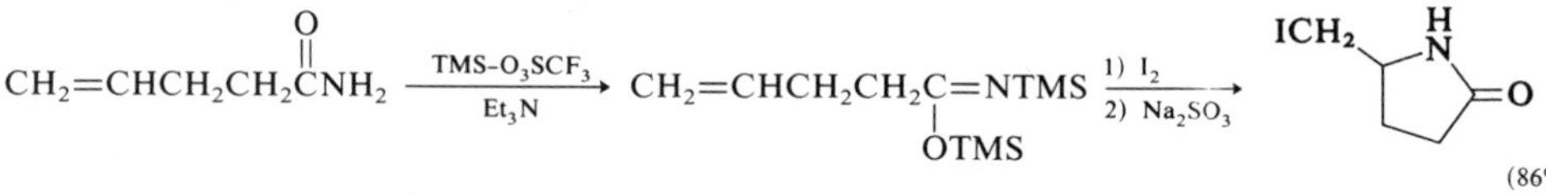

$CH_2{=}CHCH_2CH_2CONH_2 \xrightarrow[Et_3N]{TMS\text{-}O_3SCF_3} CH_2{=}CHCH_2CH_2C(OTMS){=}NTMS \xrightarrow[2)\ Na_2SO_3]{1)\ I_2}$ (86%)

49. A. Bogini, G. Cardillo, M. Orena, G. Porzi, and S. Sandri, *J. Org. Chem.* **47**, 4626 (1982).
50. C. Neukome, D. P. Richardson, J. H. Myerson, and P. A. Bartlett, *J. Am. Chem. Soc.* **108**, 5559 (1986).
51. S. Knapp, K. E. Rodriquez, A. T. Levorse, and R. M. Ornat, *Tetrahedron Lett.* **26**, 1803 (1985); S. Knapp and A. T. Levorse, *J. Org. Chem.* **53**, 4006 (1988).

Scheme 4.2. Iodolactonization and Other Cyclizations Induced by Iodine

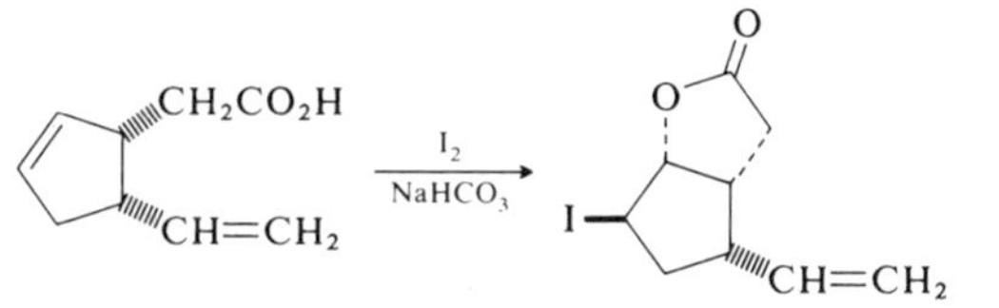

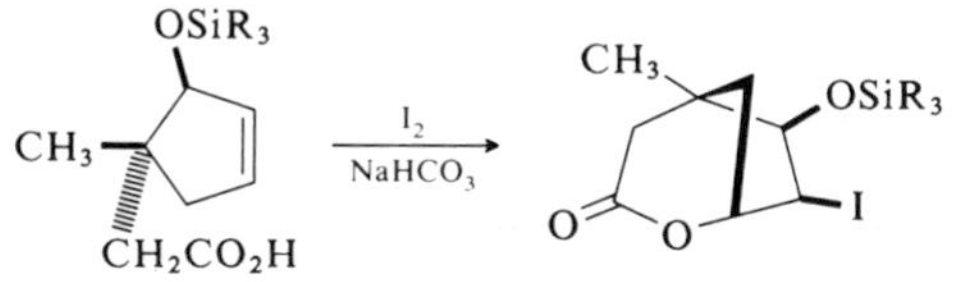

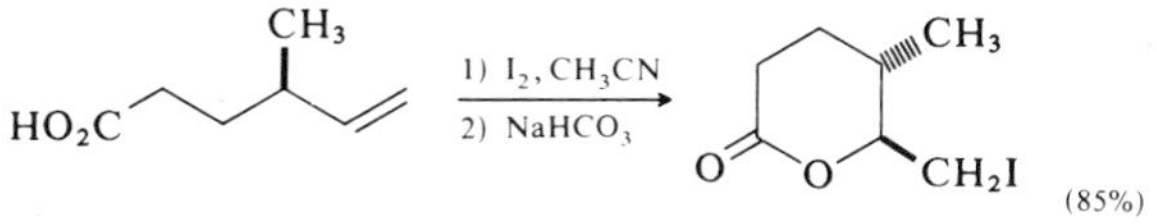

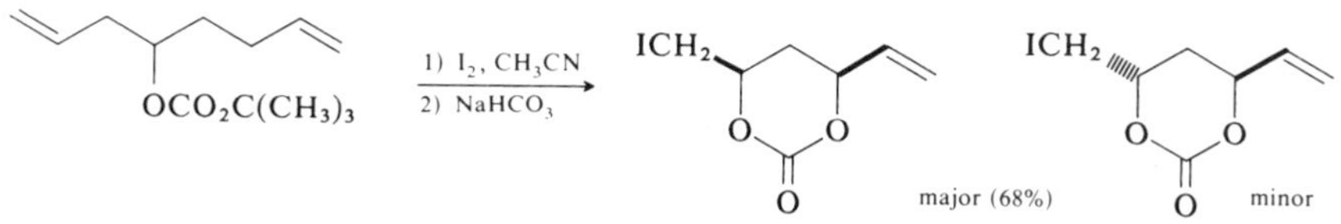

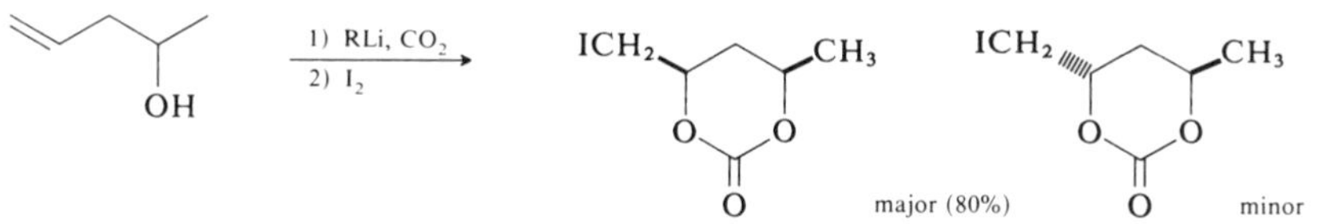

1[a] I_2, $NaHCO_3$

2[b] I_2, $NaHCO_3$

3[c] 1) I_2, CH_3CN 2) $NaHCO_3$ (85%)

4[d] 1) I_2, CH_3CN 2) $NaHCO_3$ major (68%) minor

5[e] 1) RLi, CO_2 2) I_2 major (80%) minor

6[f] *N*-iodosuccinimide (79%)

7[g] $Me_3SiO_3SCF_3$, Et_3N; 1) I_2, THF 2) Na_2SO_3 (88%)

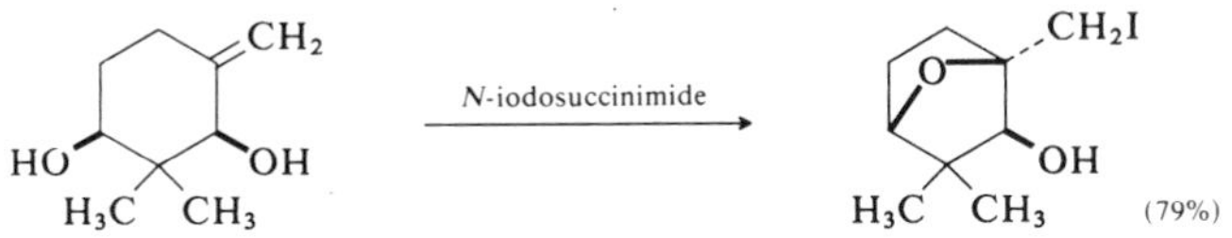

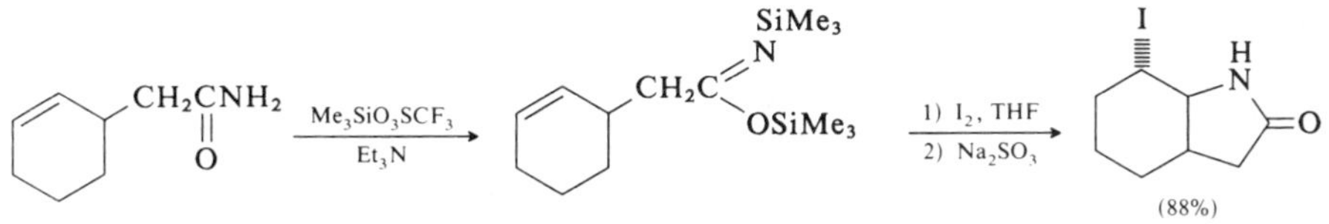

a. L. A. Paquette, G. D. Crouse, and A. K. Sharma, *J. Am. Chem. Soc.* **102**, 3972 (1980).
b. A. J. Pearson and S.-Y. Hsu, *J. Org. Chem.* **51**, 2505 (1986).
c. A. G. M. Barrett, R. A. E. Carr, S. V. Attwood, G. Richardson, and N. D. A. Walshe, *J. Org. Chem.* **51**, 4840 (1986).
d. P. A. Bartlett, J. D. Meadows, E. G. Brown, A. Morimoto, and K. K. Jernstedt, *J. Org. Chem.* **47**, 4013 (1982).
e. A. Bongini, G. Cardillo, M. Orena, G. Porzi, and S. Sandri, *J. Org. Chem.* **47**, 4626 (1982).
f. A. Murai, N. Tanimoto, N. Sakamoto, and T. Masamune, *J. Am. Chem. Soc.* **110**, 1985 (1988).
g. S. Knapp and A. T. Levorse, *J. Org. Chem.* **53**, 4006 (1988).

Examples of iodolactonization and related iodocyclizations can be found in Scheme 4.2.

Conjugated dienes often give mixtures of products from addition of halogens. The case of isoprene is an example[52]:

$$CH_2{=}C(CH_3){-}CH{=}CH_2 \xrightarrow[\text{chloroform}]{Br_2,\ 0^\circ C} CH_2{=}C(CH_3){-}CHBr{-}CH_2Br\ (3\%) + BrCH_2{-}CBr(CH_3){-}CH{=}CH_2\ (21\%) + (BrCH_2)(CH_3)C{=}CH(CH_2Br)\ (5\%) + (BrCH_2)(CH_3)C{=}CH(CH_2Br)\ (71\%)$$

The product mixture includes both 1,2- and 1,4-addition products. The formation of this type of product mixture can be explained in terms of an allylic cation intermediate. Such a cation would be sufficiently stable to minimize the need for bromine bridging, and nucleophilic bromide would attack at both positive centers in the allylic cation.

$$H_2C{=}C(H_3C){-}CH{=}CH_2 \xrightarrow{Br_2} BrH_2C{-}\overset{+}{C}(H_3C){-}CH{=}CH_2 \longleftrightarrow BrH_2C{-}C(H_3C){=}CH{-}\overset{+}{C}H_2$$

Chlorination of conjugated dienes also leads to mixtures of 1,2- and 1,4-addition products.[53]

The elemental halogens are not the only source of electrophilic halogen atoms, and, for some synthetic purposes, other "positive halogen" compounds may be preferable sources of the desired electrophile. The utility of *N*-bromosuccinimide in formation of bromohydrins was mentioned earlier. Other compounds that are useful for specific purposes are indicated in Scheme 4.3. Pyridinium hydrotribromide (pyridinium hydrobromide perbromide) and dioxane–bromine complex are examples of complexes of bromine in which its reactivity is somewhat attenuated, resulting in increased selectivity. *N*-Chlorosuccinimide and *N*-bromosuccinimide transfer electrophilic halogen with the succinimide anion acting as the leaving group. This anion is subsequently protonated to give the weak nucleophile succinimide. These reagents therefore favor nucleophilic additions by solvent and cyclization reactions, since there is no competition from a nucleophilic anion. In tetrabromo-

52. V. L. Heasley, C. L. Frye, R. T. Gore, Jr., and P. Š. Wilday, *J. Org. Chem.* **33**, 2342 (1968).
53. M. L. Poutsma, *J. Org. Chem.* **31**, 4167 (1966); V. L. Heasley and S. K. Taylor, *J. Org. Chem.* **34**, 2779 (1969).

Scheme 4.3. Other Sources of Positive Halogen

Source	Synthetic applications[a]
A. Chlorinating Agents	
Sodium hypochlorite solution	Formation of chlorohydrins from alkenes.
N-Chlorosuccinimide	Chlorination with solvent participation and cyclization.
Antimony pentachloride	Controlled chlorination of acetylenes.
B. Brominating Agents	
Pyridinium hydrotribromide (pyridinium hydro-bromide perbromide)	Substitute for bromine when increased selectivity or mild reaction conditions are required
Dioxane-bromine complex	Same as for pyridinium hydrotribromide.
N-Bromosuccinimide	Substitute for bromine when low Br^- concentration is required.
2,4,4,6-Tetrabromocyclohexadienone	Selective bromination of polyolefins and cyclization induced by Br^+.

a. For specific examples, consult M. Fieser and L. F. Fieser, *Reagents for Organic Synthesis*, Vols. 1–8, John Wiley and Sons, New York, 1979.

cyclohexadienone, the leaving group is 2,4,6-tribromophenoxide ion. This reagent is a very selective source of electrophilic bromine.

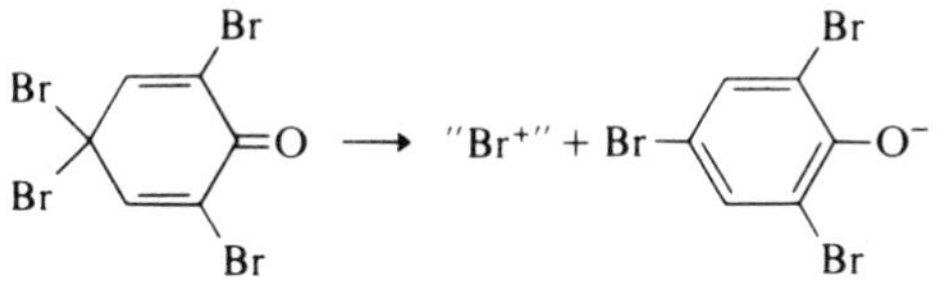

$$\text{2,4,4,6-tetrabromocyclohexadienone} \longrightarrow \text{"}Br^{+}\text{"} + \text{2,4,6-}Br_3C_6H_2O^-$$

4.5. Electrophilic Sulfur and Selenium Reagents

Compounds in which sulfur and selenium atoms are bound to more electrophilic elements can react with alkenes to give addition products. The mechanism is similar to that in halogenation with a bridged cationic intermediate being involved.

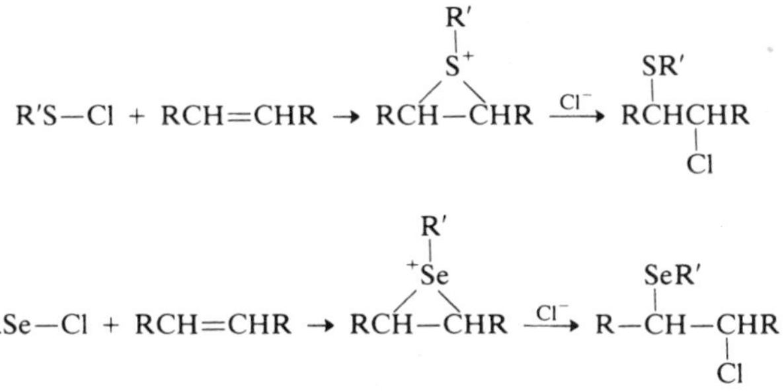

$$R'S{-}Cl + RCH{=}CHR \rightarrow RCH{-}CHR\ (\text{bridged by } S^+{-}R') \xrightarrow{Cl^-} RCH(SR')CH(Cl)R$$

$$RSe{-}Cl + RCH{=}CHR \rightarrow RCH{-}CHR\ (\text{bridged by } {}^+Se{-}R') \xrightarrow{Cl^-} R{-}CH(SeR'){-}CH(Cl)R$$

In many synthetic applications, the sulfur or selenium substituent is subsequently removed by elimination as will be discussed in Chapter 6. Arenesulfenyl halides,

ArSCl, are the most commonly used of the sulfur reagents. A veriety of electrophilic selenium reagents have been employed, and several examples are given in Scheme 4.4.

Scheme 4.4. Sulfur and Selenium Reagents for Electrophilic Addition

	Reagent	Product
1[a,b]	CH_3SCl	$RCH(SCH_3)CH(Cl)R$
2[a]	PhSCl	RCH(SPh)CH(Cl)R
3[c]	PhSeCl	RCH(SePh)CH(Cl)R
4[d]	$PhSeO_2CCF_3$	$RCH(SePh)CH(O_2CCF_3)R$
5[e]	PhSe—N-phthalimide, H_2O	RCH(SePh)CH(OH)R
6[f]	$PhSeO_2H$, H_3PO_2	RCH(SePh)CH(OH)R
7[g]	PhSeCN, Cu(II), R′OH	RCH(SePh)CH(OR′)R
8[h]	PhSeCl, $AgBF_4$, $H_2NCO_2C_2H_5$	$RCH(SePh)CH(HNCO_2C_2H_5)R$

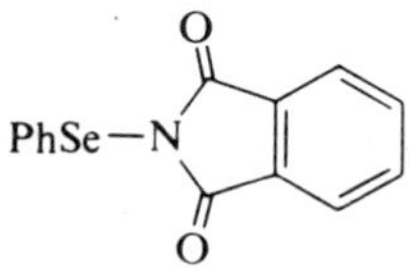

a. W. M. Mueller and P. E. Butler, *J. Am. Chem. Soc.* **90**, 2075 (1968).
b. W. A. Thaler, *J. Org. Chem.* **34**, 871 (1969).
c. K. B. Sharpless and R. F. Lauer, *J. Org. Chem.* **39**, 429 (1974); D. Liotta and G. Zima, *Tetrahedron Lett.* 4977 (1978).
d. H. J. Reich, *J. Org. Chem.* **39**, 428 (1974).
e. K. C. Nicolaou, D. A. Claremon, W. E. Barnette, and S. P. Seitz, *J. Am. Chem. Soc.* **101**, 3704 (1979).
f. D. Labar, A. Krief, and L. Hevesi, *Tetrahedron Lett.* 3967 (1978).
g. A. Toshimitsu, T. Aoai, S. Uemura, and M. Okano, *J. Org. Chem.* **45**, 1953 (1980).
h. C. G. Francisco, E. I. León, J. A. Salazar, and E. Suárez, *Tetrahedron Lett.* **27**, 2513 (1986).

Mechanistic studies have been most thorough with the sulfenyl halides.[54] The reactions show moderate sensitivity to alkene structure, with electron-releasing groups on the alkene accelerating the reaction. The addition can occur in either the Markownikoff or anti-Markownikoff sense.[55] The regioselectivity can be understood by focusing attention on the sulfur-bridged intermediate, which may range from being a sulfonium ion to a less electrophilic chlorosulfurane:

R′–S⁺ bridging R–C(H)–C(H)–H; R′–S(Cl) bridging R–C(H)–C(H)–H

Compared to the C-Br bonds in the bromonium ion, the C—S bonds will be stronger so that steric interactions which dictate access by the nucleophile become a more important factor in determining the direction of addition. For reactions involving phenylsulfenyl chloride or methylsulfenyl chloride, the intermediate is a fairly stable species, and ease of approach by the nucleophile is the major factor in determining the direction of ring opening. In these cases, the product has the anti-Markownikoff orientation.[56]

$$CH_2{=}CHCH(CH_3)_2 \xrightarrow{CH_3SCl} ClCH_2CH(SCH_3)CH(CH_3)_2\ (94\%) + CH_3SCH_2CH(Cl)CH(CH_3)_2\ (6\%)$$ Ref. 57

$$CH_3CH_2CH{=}CH_2 \xrightarrow{p\text{-}ClPhSCl} ClCH_2CH(SAr)CH_2CH_3\ (77\%) + ArSCH_2CH(Cl)CH_2CH_3\ (23\%)$$ Ref. 58

The stereospecific *anti* addition of phenylsulfenyl chloride to norbornene is a particularly interesting example of the stability of the intermediate. Neither rearrangement nor *syn* addition products, which are observed with many electrophilic reagents, are formed.[54] This result indicates that the intermediate must be quite stable and reacts only by nucleophilic attack.[55]

norbornene + PhSCl → (SPh, Cl adduct)

54. W. A. Smit, N. S. Zefirov, I. V. Bodrikov, and M. Z. Krimer, *Acc. Chem. Res.* **12**, 282 (1979); G. H. Schmid and D. G. Garratt, *The Chemistry of Double-Bonded Functional Groups*, S. Patai (ed., Wiley-Interscience, New York, 1977, Chapter 9.
55. W. H. Mueller and P. E. Butler, *J. Am. Chem. Soc.* **90**, 2075 (1968); G. H. Schmid and D. I. Macdonald, *Tetrahedron Lett.* **25**, 157 (1984).
56. G. H. Schmid, M. Strukelj, S. Dalipi, and M. D. Ryan, *J. Org. Chem.* **52**, 2403 (1987).
57. W. H. Mueller and P. E. Butler, *J. Am. Chem. Soc.* **90**, 2075 (1968).
58. G. H. Schmid, C. L. Dean, and D. G. Garratt, *Can. J. Chem.* **54**, 1253 (1976).

When non-nucleophilic salts, for example, $LiClO_4$, are included in the reaction medium, products indicative of a more reactive intermediate with carbocationic character are observed:

ArSCl, $LiClO_4$, CH_3CO_2H; SAr; SAr; O_2CCH_3; ArS; O_2CCH_3; Ref. 59

Ar = O_2N–C_6H_3(NO_2)–

These contrasting results have been interpreted in terms of a relatively unreactive species, perhaps a chlorosulfurane, being the main intermediate in the absence of the salt. The presence of the lithium salt gives rise to a more reactive species such as the episulfonium ion structure, as the result of ion pairing with the chloride ion.

Ar, S, Cl + Li^+ ⇌ Ar, S^+ + $Li^{+-}Cl$

Terminal alkenes react with selenenyl halides with anti-Markownikoff regioselectivity.[60] However, the β-selenenyl halide addition product can readily rearrange to isomeric products[61]:

$$R_2C{=}CHR + ArSeX \rightarrow R_2\underset{X}{\overset{SeAr}{C}}CHR \rightleftharpoons R_2\underset{X}{C}\overset{ArSe}{C}HR$$

Electrophilic selenium reagents are very effective in promoting cyclization of unsaturated molecules containing potentially nucleophilic substituents. Unsaturated carboxylic acids, for example, give selenolactones, and this reaction has been termed *selenolactonization.*[62]

CH_2CO_2H —PhSeCl→ PhSe, O, O (93%)

N-Phenylselenophthalimide is an excellent reagent for this process and permits the formation of large-ring lactones.[63] The advantage of the reagent in this particular application is that the phthalimide anion, by virtue of its low nucleophilicity, does not compete with the remote internal nucleophile.

59. N. S. Zefirov, N. K. Sadovaja, A. M. Maggerramov, I. V. Bodrikov, and V. E. Karstashov, *Tetrahedron* **31**, 2949 (1975).
60. D. Liotta and G. Zima, *Tetrahedron Lett.*, 4977 (1978); P. T. Ho and R. J. Holt, *Can. J. Chem.* **60**, 663 (1982).
61. S. Raucher, *J. Org. Chem.* **42**, 2950 (1977).
62. K. C. Nicolaou, S. P. Seitz, W. J. Sipio, and J. F. Blount, *J. Am. Chem. Soc.* **101**, 3884 (1979).
63. K. C. Nicolaou, D. A. Claremon, W. E. Barnette, and S. P. Seitz, *J. Am. Chem. Soc.* **101**, 3704 (1979).

The reaction of phenylselenenyl chloride or *N*-phenylselenophthalimide with unsaturated alcohols leads to formation of β-phenylselenenyl ethers:

$$\text{(cyclopentenyl)}-CH_2CH_2OH + PhSeN(\text{phthalimide}) \longrightarrow \text{PhSe-(bicyclic ether)}$$ Ref. 64

Scheme 4.5 gives some examples of cyclizations induced by sulfur and selenium electrophiles.

Reaction of alcohol solutions of alkenes with potassium selenocyanate and cupric chloride gives β-selenocyanato ethers:

$$RCH{=}CHR + {}^{-}SeCN + Cu(II) \xrightarrow{ROH} RCH(SeCN)CH(R'O)R$$ Ref. 65

This reaction proceeds by the generation of an electrophilic species, probably N≡C-Se-Cl, which is formed by oxidation by Cu(II).

$$2\,CuCl_2 + {}^{-}SeC{\equiv}N \rightarrow 2\,CuCl + Cl{-}SeC{\equiv}N + Cl^{-}$$

This reagent then gives a cyclic selenonium ion intermediate which captures the solvent. Thiocyanate salts behave similarly under these reaction conditions.[66]

4.6. Addition of Other Electrophilic Reagents

Many small halogen-containing compounds react with alkenes to give addition products by mechanisms similar to that of halogenation. A complex is formed, and the electrophilic portion of the reagent is transferred to the alkene to generate a cationic intermediate. This may be a symmetrically bridged ion or an unsymmetrically bridged species, depending on the ability of the reacting carbon atoms of the alkene to accommodate positive charge. The direction of opening of the bridged intermediate is usually governed by electronic factors. That is, the addition is completed by attack of the nucleophile at the more positive carbon atom of the bridged intermediate. The orientation of addition therefore follows Markownikoff's rule. The stereochemistry of addition is usually *anti*, again because of the involvement of a bridged intermediate.[67] Several reagents of this type are listed in Scheme 4.6.

64. K. C. Nicolaou, R. L. Magolda, W. J. Sipio, W. E. Barnette, Z. Lysenko, and M. M. Joullie, *J. Am. Chem. Soc.* **102**, 3784 (1980).
65. A. Toshimitsu, Y. Kozawa, S. Uemura, and M. Okano, *J. Chem. Soc., Perkin Trans. 1,* 1273 (1978).
66. A. Onoe, S. Uemura, and M. Okano, *Bull. Chem. Soc. Jpn.* **47**, 2818 (1974).
67. A. Hassner and C. Heathcock, *J. Org. Chem.* **30**, 1748 (1965).

Scheme 4.5. Cyclizations Induced by Electrophilic Sulfur and Selenium Reagents

1[a] $CH_2{=}CH(CH_2)_4OH$ —PhSCl, EtN(i-Pr)$_2$→ $PhSCH_2$-tetrahydropyran (85%)

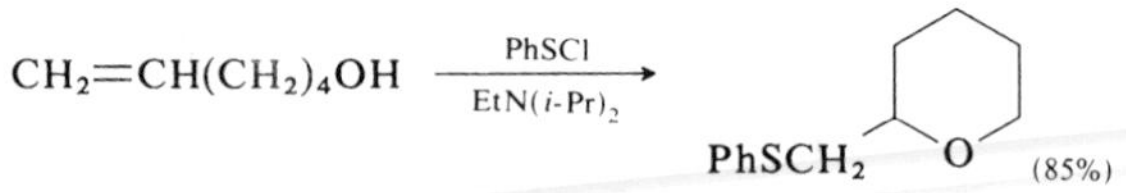

2[b] o-allylphenol ($CH_2CH{=}CH_2$, OH) —PhSCl→ 2-(CH_2SPh)-2,3-dihydrobenzofuran (35%)

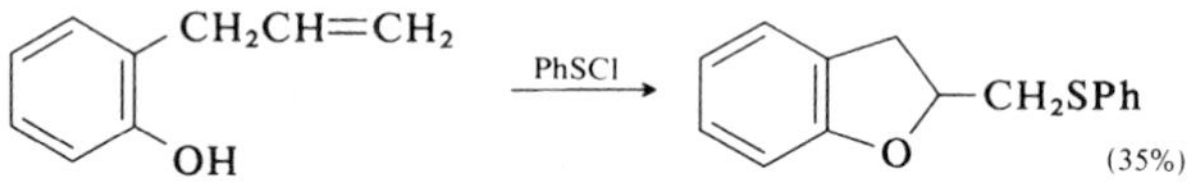

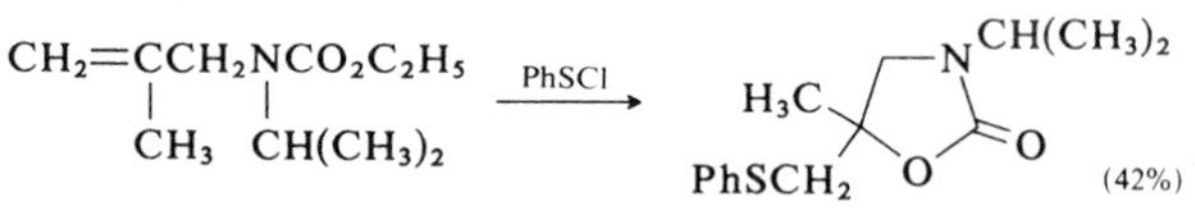

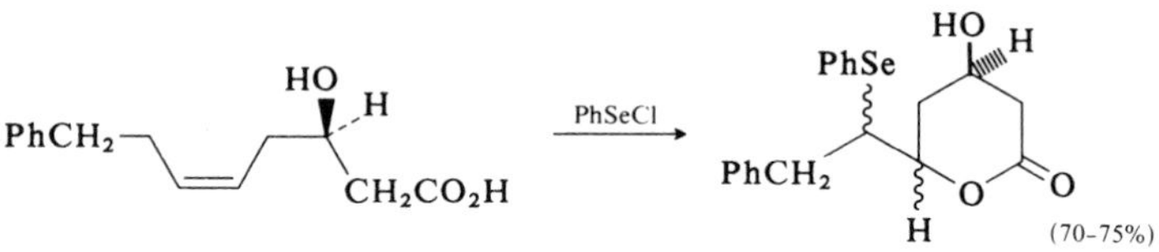

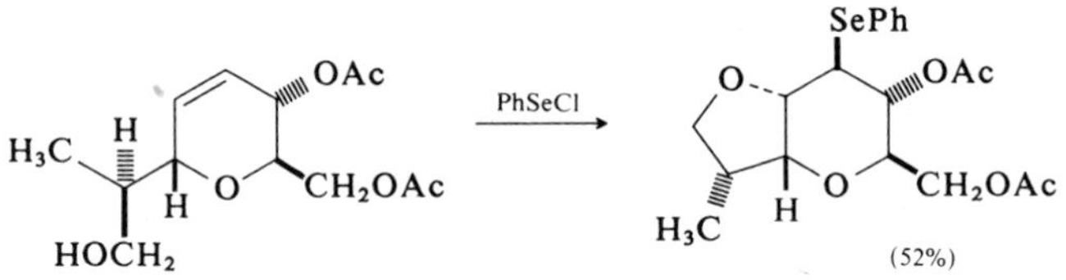

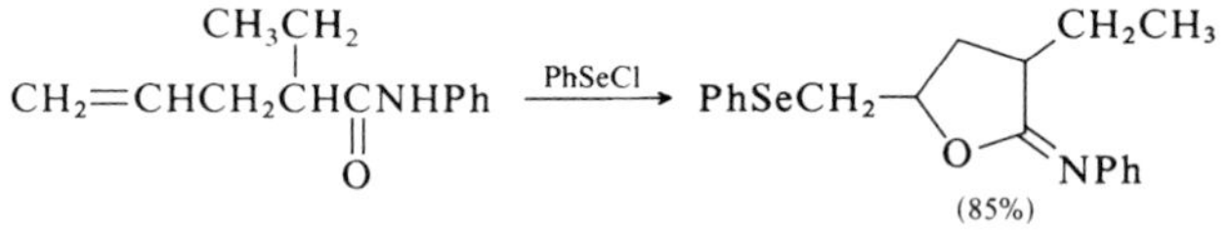

3[c] $CH_2{=}C(CH_3)CH_2N(CH(CH_3)_2)CO_2C_2H_5$ —PhSCl→ oxazolidinone with H_3C, $PhSCH_2$, N-$CH(CH_3)_2$, =O (42%)

4[d] $PhCH_2$–CH=CH–CH$_2$–CH(OH)(H)–CH_2CO_2H —PhSeCl→ lactone with PhSe, $PhCH_2$, HO, H, O, =O (70–75%)

5[e] H_3C, H, HOCH$_2$, H, O, OAc, CH_2OAc —PhSeCl→ SePh, O, OAc, CH_2OAc, H_3C, H, O (52%)

6[f] $CH_2{=}CHCH_2CH(CH_3CH_2)C(=O)NHPh$ —PhSeCl→ $PhSeCH_2$-tetrahydrofuran with CH_2CH_3, =NPh (85%)

7[g] 5-($CH_2CH_2CH{=}CH_2$)-pyrrolidin-2-one (O, N, H) —PhSeBr→ pyrrolizidinone with CH_2SePh (68%)

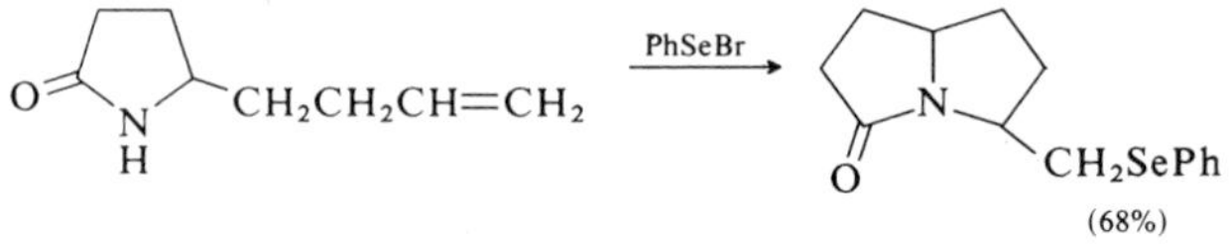

a. S. M. Tuladhar and A. G. Fallis, *Tetrahedron Lett.* **28**, 523 (1987).
b. M. Muehlstaedt, C. Schubert, and E. Kleinpeter, *J. Prakt. Chem.* **327**, 270 (1985).
c. M. Muehlstaedt, R. Widera, and B. Olk, *J. Prakt. Chem.* **324**, 362 (1982).
d. F. Bennett and D. W. Knight, *Tetrahedron Lett.* **29**, 4625 (1988).
e. S. J. Danishefsky, S. DeNinno, and P. Lartey, *J. Am. Chem. Soc.* **109**, 2082 (1987).
f. A. Toshimitsu, K. Terao, and S. Uemura, *J. Org. Chem.* **52**, 2018 (1987).
g. A. Toshimitsu, K. Terao, and S. Uemura, *J. Org. Chem.* **51**, 1724 (1986).

Certain of the reagents shown in Scheme 4.6 exhibit special features which should be noted. The addition of nitrosyl chloride or nitrosyl formate to alkenes is accompanied by subsequent reaction, unless the nitroso group is attached to a tertiary carbon. The nitroso compound may dimerize or rearrange to a more stable oxime tautomer:

+ NOCl → (2-chlorocyclohexyl N=O) —SO_2, −30°C→ (dimer, $^{-}O-\overset{+}{N}=\overset{+}{N}-O^{-}$, with 2-chlorocyclohexyl groups); —HCl→ (2-chlorocyclohexanone oxime, NOH, Cl)

Ref. 68

In the case of thiocyanogen chloride and thiocyanogen, the formal electrophile is $N{\equiv}C{-}S^{+}$. The presumed intermediate is a cyanosulfonium ion. The thiocyanate anion is an ambident nucleophile, and both carbon-sulfur and carbon-nitrogen bond formation can be observed, depending upon the reaction conditions (see entry 9 in Scheme 4.6).

4.7. Electrophilic Substitution Alpha to Carbonyl Groups

Although the reaction of ketones and other carbonyl compounds with electrophiles such as bromine leads to substitution rather than addition, it is mechanistically closely related to electrophilic additions to alkenes. An enol or enolate derived from the carbonyl compound is the reactive species, and the initial attack is similar to the electrophilic attack on alkenes. The reaction is completed by restoration of the carbonyl bond, rather than by addition of a nucleophile. The acid- and base-catalyzed halogenation of ketones, which were discussed briefly in Part A, Chapter 7, are the most studied examples of the reaction.

$$R_2CHC(=O)R' \xrightleftharpoons{H^+} R_2C{=}C(OH)R' \xrightarrow{Br_2} R_2C{=}C(OH)R' \;(Br{-}Br) \rightarrow R_2C(Br)C(=O)R'$$

$$R_2CHC(=O)R' \xrightleftharpoons{^-OH} R_2C{=}C(O^-)R' \xrightarrow{Br_2} R_2C{=}C(O^-)R' \;(Br{-}Br) \rightarrow R_2C(Br)C(=O)R'$$

68. B. W. Ponder, T. E. Walton, and W. J. Pollock, *J. Org. Chem.* **33**, 3957 (1968); M. Ohno, N. Naruse, S. Torimitsu, and M. Oiamoto, *Bull. Chem. Soc. Jpn.* **39**, 1119 (1966).

Scheme 4.6. Addition Reactions of Other Electrophilic Reagents

	Reagent	Preparation	Product
1[a]	I—N=C=O	AgCNO, I_2	RCH(I)—CHR(NCO)
2[b]	$Br—N=\overset{+}{N}=N^-$	HN_3, Br_2	RCH(Br)—CHR(N_3)
3[c]	$I—N=\overset{+}{N}=N^-$	NaN_3, ICl	RCH(I)—CHR(N_3)
4[d]	I—S—C≡N	$(NCS)_2$, I_2	RCH(I)—CHR(S—C≡N)
5[e]	$I—ONO_2$	$AgNO_3$, ICl	RCH(I)—CHR(ONO_2)
6[f]	O=N—Cl		RC(=NOH)—CHR(Cl)
7[g]	$O=N—O_2CH$	$(CH_3)_2CHCH_2CH_2ON=O$, HCO_2H	RC(=NOH)—CHR(O_2CH)
8[h]	Cl—SCN	$Pb(SCN)_2$, Cl_2	RCH(Cl)—CHR(SCN)
9[i]	N≡CS—SC≡N	$Pb(SCN)_2$, Br_2	RCH(SC≡N)—CHR(SC≡N) and RCH(SC≡N)—CHR(N=C=S)

a. A. Hassner, R. P. Hoblitt, C. Heathcock, J. E. Kropp, and M. Lorber, *J. Am. Chem. Soc.* **92**, 1326 (1970); A. Hassner, M. E. Lorber, and C. Heathcock, *J. Org. Chem.* **32**, 540 (1967).
b. A. Hassner, F. P. Boerwinkle, and A. B. Levy, *J. Am. Chem. Soc.* **92**, 4879 (1970).
c. F. W. Fowler, A. Hassner, and L. A. Levy, *J. Am. Chem. Soc.* **89**, 2077 (1967).
d. R. J. Maxwell and L. S. Silbert, *Tetrahedron Lett.* 4991 (1978).
e. J. W. Lown and A. V. Joshua, *J. Chem. Soc. Perkin Trans. 1*, 2680 (1973).
f. J. Meinwald, Y. C. Meinwald, and T. N. Baker III, *J. Am. Chem. Soc.* **86**, 4074 (1967).
g. H. C. Hamann and D. Swern, *J. Am. Chem. Soc.* **90**, 6481 (1968).
h. R. G. Guy and I. Pearson, *J. Chem. Soc. Perkin Trans. 1*, 281 (1973); *J. Chem. Soc. Perkin Trans. 2*, 1359 (1973).
i. R. J. Maxwell, L. S. Silbert, and J. R. Russell, *J. Org. Chem.* **42**, 1510 (1977).

The most common preparative procedures involve use of the halogen, usually bromine, in acetic acid. Other suitable halogenating agents include *N*-bromosuccinimide, tetrabromocyclohexadienone, and sulfuryl chloride.

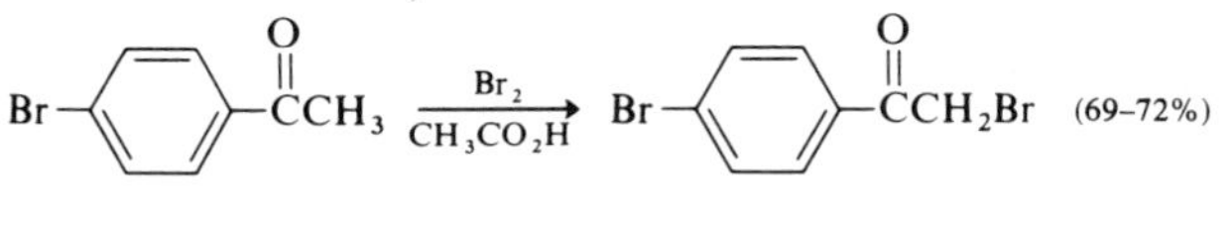

$$Br-C_6H_4-\overset{O}{\overset{\|}{C}}CH_3 \xrightarrow[CH_3CO_2H]{Br_2} Br-C_6H_4-\overset{O}{\overset{\|}{C}}CH_2Br \quad (69\text{–}72\%)$$ Ref. 69

69. W. D. Langley, *Org. Synth.* **I**, 122 (1932).

Cycloheptanone $\xrightarrow[\text{CCl}_4]{N\text{-bromosuccinimide}}$ 2-bromocycloheptanone — Ref. 70

2-Methylcyclohexanone $\xrightarrow{SO_2Cl_2}$ 2-chloro-2-methylcyclohexanone (83–85%) — Ref. 71

C_6H_5–CH=CHCCH$_3$ (O above the carbonyl C) $\xrightarrow{\text{2,4,4,6-tetrabromocyclohexa-2,5-dienone}}$ C_6H_5–CH=CHCCH$_2$Br (91%) — Ref. 72

The reactions involving bromine or chlorine generate hydrogen halide and are autocatalytic. Reactions with *N*-bromosuccinimide or tetrabromocyclohexadienone form no hydrogen bromide and may therefore be preferable reagents in the case of acid-sensitive compounds.

As was pointed out in Part A, Section 7.3, under many conditions halogenation is faster than enolization. When this is true, the position of substitution in unsymmetrical ketones is governed by the relative rates of formation of the isomeric enols. In general, mixtures are formed with unsymmetrical ketones. The presence of a halogen substituent decreases the rate of acid-catalyzed enolization and therefore retards the introduction of a second halogen at the same site. Monohalogenation can therefore usually be carried out satisfactorily. In contrast, in basic solution halogenation tends to proceed to polyhalogenated products because the inductive effect of a halogen accelerates base-catalyzed enolization. With methyl ketones, base-catalyzed reaction with iodine or bromine leads eventually to cleavage to a carboxylic acid.[73] The reaction can also be effected with hypochlorite ion.

$$(CH_3)_2C{=}CH\overset{O}{\overset{\|}{C}}CH_3 + {}^{-}OCl \rightarrow (CH_3)_2C{=}CHCO_2H \quad (49\text{–}53\%)$$ Ref. 74

A preparatively useful procedure for monohalogenation of ketones involves reaction with cupric chloride or cupric bromide.[75]

Cyclodecanone $\xrightarrow[\text{HCCl}_3,\ \text{CH}_3\text{CO}_2\text{C}_2\text{H}_5]{\text{CuBr}_2}$ 2-bromocyclodecanone — Ref. 76

70. E. J. Corey, *J. Am. Chem. Soc.* **75**, 2301 (1954).
71. E. W. Warnhoff, D. G. Martin, and W. S. Johnson, *Org. Synth.* **IV**, 162 (1963).
72. V. Calo, L. Lopez, G. Pesce, and P. E. Todesco, *Tetrahedron* **29**, 1625 (1973).
73. S. J. Chakrabartty, in *Oxidations in Organic Chemistry*, Part C, W. Trahanovsky (ed.), Academic Press, New York, 1978, Chapter V.
74. L. J. Smith, W. W. Prichard, and L. J. Spillane, *Org. Synth.* **III**, 302 (1955).
75. E. M. Kosower, W. J. Cole, G.-S. Wu, D. E. Cardy, and G. Meisters, *J. Org. Chem.* **28**, 630 (1963); E. M. Kosower and G.-S. Wu, *J. Org. Chem.* **28**, 633 (1963).
76. D. P. Bauer and R. S. Macomber, *J. Org. Chem.* **40**, 1990 (1975).

Instead of direct halogenation of ketones, reactions with more reactive ketone derivatives such as silyl enol ethers and enamines have advantages in certain cases.

OSi(CH$_3$)$_3$ → 1) I_2, AgOAc; $R_4N^+F^-$ → (84%) Ref. 77

Cl$_2$, −78°C; H_2O → (65%) Ref. 78

There are also procedures in which the enolate is generated and allowed to react with a halogenating agent. Among the sources of halogen that have been used under these conditions are bromine,[79] *N*-chlorosuccinimide,[80] trifluoromethanesulfonyl chloride,[81] and hexachloroethane.[82]

1) LDA; 2) CF_3SO_2Cl

α-Fluoroketones have been made primarily by reactions of enol acetates or silyl enol ethers with fluorinating agents such as CF_3OF[83], XeF_2,[84] and dilute F_2.[85]

Another example of α-halogenation which has synthetic utility is the α-halogenation of acid chlorides. The mechanism is presumed to be similar to that of ketone halogenation and to proceed through an enol. The reaction can be effected in thionyl chloride as solvent to give α-chloro, α-bromo, or α-iodo acid chlorides, with, respectively, *N*-chlorosuccinimide, *N*-bromosuccinimide, or molecular iodine as the halogenating agent.[86] Because thionyl chloride rapidly converts carboxylic acids to acid chlorides, the acid can be used as the starting material.

77. G. M. Rubottom and R. C. Mott, *J. Org. Chem.* **44**, 1731 (1979); G. A. Olah, L. Ohannesian, M. Arvanaghi, and G. K. S. Prakash, *J. Org. Chem.* **49**, 2032 (1984).
78. W. Seufert and F. Effenberger, *Chem. Ber.* **112**, 1670 (1979).
79. T. Woolf, A. Trevor, T. Baille, and N. Castagnoli, Jr., *J. Org. Chem.* **49**, 3305 (1984).
80. A. D. N. Vaz and G. Schoellmann, *J. Org. Chem.* **49**, 1286 (1984).
81. P. A. Wender and D. A. Holt, *J. Am. Chem. Soc.* **107**, 7771 (1985).
82. M. B. Glinski, J. C. Freed, and T. Durst, *J. Org. Chem.* **52**, 2749 (1987).
83. W. J. Middleton and E. M. Bingham, *J. Am. Chem. Soc.* **102**, 4845 (1980).
84. B. Zajac and M. Zupan, *J. Chem. Soc., Chem. Commun.*, 759 (1980).
85. S. Rozen and Y. Menahem, *Tetrahedron Lett.*, 725 (1979).
86. D. N. Harpp, L. Q. Bao, C. J. Black, J. G. Gleason, and R. A. Smith, *J. Org. Chem.* **40**, 3420 (1975); Y. Ogata, K. Adachi, and F.-C. Chen, *J. Org. Chem.* **48**, 4147 (1983).

$$CH_3(CH_2)_3CH_2CO_2H \xrightarrow[SOCl_2]{N\text{-chlorosuccinimide}} CH_3(CH_2)_3\underset{\displaystyle Cl}{\underset{|}{C}}HCOCl \quad (87\%)$$

$$PhCH_2CH_2CO_2H \xrightarrow[SOCl_2]{I_2} PhCH_2\underset{\displaystyle I}{\underset{|}{C}}HCOCl \quad (95\%)$$

The α-sulfenylation[87] and α-selenenylation[88] of carbonyl compounds are important reactions, since the products can subsequently be oxidized to sulfoxides and selenoxides. The sulfoxides and selenoxides readily undergo elimination (see Section 6.8.3), generating the corresponding α,β-unsaturated carbonyl compound. Sulfenylations and selenenylations are usually carried out under conditions where the enolate of the carbonyl compound is the reactive species. Scheme 4.7 gives some specific examples of these types of reactions. The most general procedure involves generating the enolate by deprotonation, or one of the alternative methods, followed by reaction with the sulfenylation or selenenylation reagent. Disulfides are the most common sulfenylation reagents, whereas diselenides or selenenyl halides are used for selenenylation. As entries 6 and 7 indicate, the selenenylation of ketones can also be effected by reactions of enol acetates or enol silyl ethers. If a specific enolate is generated by one of the methods described in Chapter 1, the position of sulfenylation or selenenylation can be controlled.[89]

4.8. Additions to Allenes and Alkynes

Both allenes and alkynes require special consideration with regard to mechanisms of electrophilic addition. The attack by a proton on allene can conceivably lead to the allyl cation or the 2-propenyl cation:

$$^{+}CH_2{-}CH{=}CH_2 \xleftarrow{H^+} CH_2{=}C{=}CH_2 \xrightarrow{H^+} CH_3{-}\overset{+}{C}{=}CH_2$$

An immediate presumption that the more stable allyl ion will be formed ignores the stereoelectronic facets of the reaction. Protonation at the center carbon without rotation of one of the terminal methylene groups leads to a primary carbocation that is not stabilized by resonance, since the adjacent π bond is orthogonal to the empty p orbital.

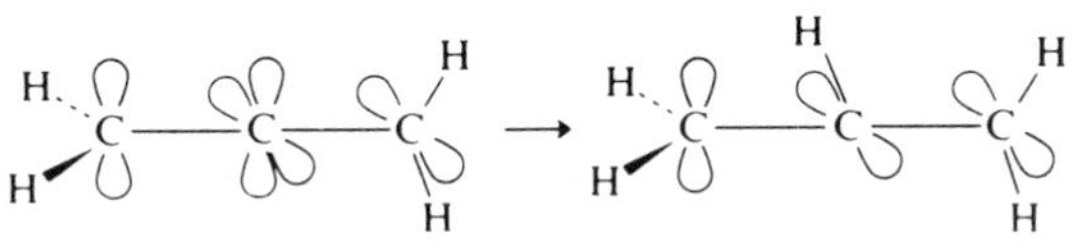

87. B. M. Trost, *Chem. Rev.* **78**, 363 (1978).
88. H. J. Reich, *Acc. Chem. Res.* **12**, 22 (1979); H. J. Reich, J. M. Renga, and I. L. Reich, *J. Am. Chem. Soc.* **97**, 5434 (1975).
89. P. G. Gassman, D. P. Gilbert, and S. M. Cole, *J. Org. Chem.* **42**, 3233 (1977).

Scheme 4.7. α-Sulfenylation and α-Selenenylation of Carbonyl Compounds

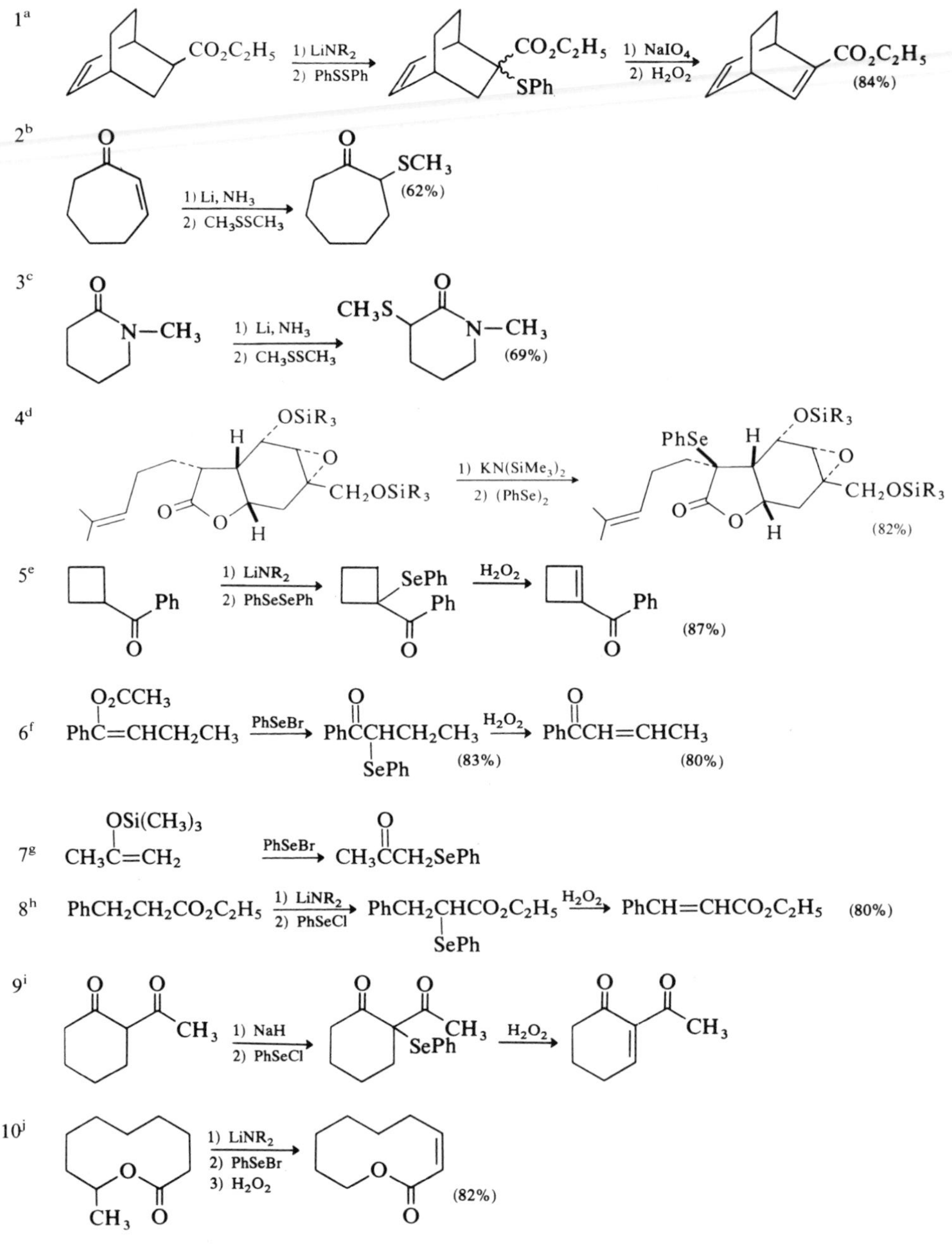

1[a] $CO_2C_2H_5$ — 1) $LiNR_2$ 2) PhSSPh → $CO_2C_2H_5$, SPh — 1) $NaIO_4$ 2) H_2O_2 → $CO_2C_2H_5$ (84%)

2[b] 1) Li, NH_3 2) CH_3SSCH_3 → SCH_3 (62%)

3[c] $N-CH_3$ — 1) Li, NH_3 2) CH_3SSCH_3 → CH_3S, $N-CH_3$ (69%)

4[d] $OSiR_3$, H, O, CH_2OSiR_3, O, H — 1) $KN(SiMe_3)_2$ 2) $(PhSe)_2$ → PhSe, H, $OSiR_3$, O, CH_2OSiR_3, O, H (82%)

5[e] Ph, O — 1) $LiNR_2$ 2) PhSeSePh → SePh, Ph, O — H_2O_2 → Ph, O (87%)

6[f] $PhC(O_2CCH_3){=}CHCH_2CH_3$ —PhSeBr→ $PhC(O)CH(SePh)CH_2CH_3$ (83%) —H_2O_2→ $PhC(O)CH{=}CHCH_3$ (80%)

7[g] $CH_3C(OSi(CH_3)_3){=}CH_2$ —PhSeBr→ $CH_3C(O)CH_2SePh$

8[h] $PhCH_2CH_2CO_2C_2H_5$ —1) $LiNR_2$ 2) PhSeCl→ $PhCH_2CH(SePh)CO_2C_2H_5$ —H_2O_2→ $PhCH{=}CHCO_2C_2H_5$ (80%)

9[i] O, O, CH_3 — 1) NaH 2) PhSeCl → O, O, CH_3, SePh — H_2O_2 → O, O, CH_3

10[j] O, CH_3, O — 1) $LiNR_2$ 2) PhSeBr 3) H_2O_2 → O, O (82%)

a. B. M. Trost, T. N. Salzmann, and K. Hiroi, *J. Am. Chem. Soc.* **98**, 4887 (1976).
b. P. G. Gassman, D. P. Gilbert, and S. M. Cole, *J. Org. Chem.* **42**, 3233 (1977).
c. P. G. Gassman and R. J. Balchunis, *J. Org. Chem.* **42**, 3236 (1977).
d. A. B. Smith III and R. E. Richmond, *J. Am. Chem. Soc.* **105**, 575 (1983).
e. H. J. Reich, J. M. Renga, and I. L. Reich, *J. Am. Chem. Soc.* **97**, 5434 (1975).
f. H. J. Reich, I. L. Reich, and J. M. Renga, *J. Am. Chem. Soc.* **95**, 5813 (1973).
g. I. Ryu, S. Murai, I. Niwa, and N. Sonoda, *Synthesis*, 874 (1977).
h. J. M. Renga and H. J. Reich, *Org. Synth.* **59**, 58 (1979).
i. T. Wakamatsu, K. Akasaka, and Y. Ban, *J. Org. Chem.* **44**, 2008 (1979).

The addition of HCl, HBr, and HI to allene has been studied in some detail.[90] In each case, a 2-halopropene is formed, corresponding to protonation at a terminal carbon. The initial product can undergo a second addition, giving rise to 2,2-dihalopropanes. Dimers are also formed, but we will not consider them.

$$CH_2{=}C{=}CH_2 + HX \rightarrow CH_3C(X){=}CH_2 + H_3CC(X)_2CH_3$$

The presence of a phenyl group results in the formation of products from protonation at the center carbon[91]:

$$PhCH{=}C{=}CH_2 \xrightarrow[HOAc]{HCl} PhCH{=}CHCH_2Cl$$

Two alkyl substituents, as in 1,1-dimethylallene, also lead to protonation at the center carbon[92]:

$$(CH_3)_2C{=}C{=}CH_2 \rightarrow (CH_3)_2C{=}CHCH_2Cl$$

These substituent effects are due to the stabilization of the carbocation resulting from protonation at the center carbon. Even if allylic conjugation is not available in the transition state, the aryl and alkyl substituents make the terminal carbocation more stable than the alternative, a secondary vinyl cation.

Alkynes, although not as accessible as alkenes, have a number of important uses in synthesis. In general, alkynes are somewhat less reactive than alkenes toward many electrophiles. A major reason for this difference in reactivity is the substantially higher energy of the vinyl cation intermediate that is formed by an electrophilic attack on an alkyne. It is estimated that vinyl cations are about 10 kcal/mol less stable than alkyl cations with similar substitution. The observed differences in rate of addition in direct comparisons between alkenes and alkynes depend upon the

Table 4.3. Relative Reactivity of Alkenes and Alkynes[a]

	Ratio of second-order rate constants (alkene/alkyne)		
	Bromination, acetic acid	Chlorination, acetic acid	Acid-catalyzed hydration, water
$CH_3CH_2CH_2CH_2CH{=}CH_2$ $CH_3CH_2CH_2CH_2C{\equiv}CH$	1.8×10^5	5.3×10^5	3.6
trans-$CH_3CH_2CH{=}CHCH_2CH_3$ $CH_3CH_2C{\equiv}CCH_2CH_3$	3.4×10^5	$\sim 1 \times 10^5$	16.6
$PhCH{=}CH_2$ $PhC{\equiv}CH$	2.6×10^3	7.2×10^2	0.65

a. From data tabulated in Ref. 93.

90. K. Griesbaum, W. Naegele, and G. G. Wanless, *J. Am. Chem. Soc.* **87**, 3151 (1965).
91. T. Okuyama, K. Izawa, and T. Fueno, *J. Am. Chem. Soc.* **95**, 6749 (1973).
92. T. L. Jacobs and R. N. Johnson, *J. Am. Chem. Soc.* **82**, 6397 (1960).

specific electrophile and the reaction conditions.[93] Table 4.3 summarizes some specific rate comparisons. A more complete discussion of the mechanistic aspects of addition to alkynes can be found in Section 6.5 of Part A.

Acid-catalyzed additions to alkynes follow the Markownikoff rule.

$$CH_3(CH_2)_6C{\equiv}CH \xrightarrow{Et_4N^+HBr_2^-} CH_3(CH_2)_6\underset{\displaystyle Br}{\underset{|}{C}}{=}CH_2 \quad (77\%) \qquad \text{Ref. 94}$$

The initial addition products are not always stable, however. Addition of acetic acid, for example, results in the formation of enol acetates, which are easily converted to the corresponding ketone under the reaction conditions[95]:

$$H_5C_2C{\equiv}CC_2H_5 \xrightarrow[CH_3CO_2H]{H^+} H_5C_2\underset{\displaystyle O_2CCH_3}{\underset{|}{C}}{=}CHCH_2CH_3 \rightarrow H_5C_2\underset{\displaystyle O}{\underset{\|}{C}}CH_2CH_2CH_3$$

The most synthetically valuable method for converting alkynes to ketones is by mercuric-ion-catalyzed hydration. Terminal alkynes give methyl ketones, in accordance with the Markownikoff rule. Internal alkynes will give mixtures of ketones unless some structural feature directs regioselectivity. Scheme 4.8 gives some examples of alkyne hydrations.

Addition of chlorine to 1-butyne is slow in the absence of light. When addition is initiated by light, the major product is *E*-1,2-dichlorobutene when butyne is present in large excess[96]:

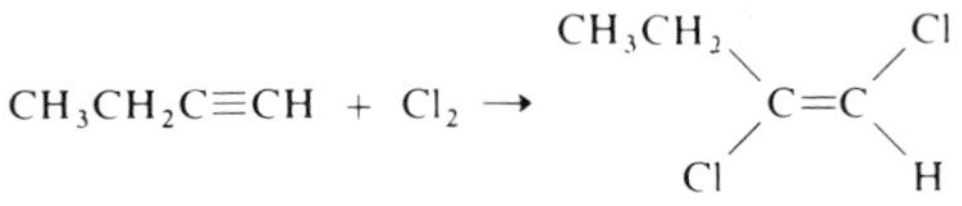

$$CH_3CH_2C{\equiv}CH + Cl_2 \rightarrow (CH_3CH_2)(Cl)C{=}C(Cl)(H)$$

In acetic acid, both 1-pentyne and 1-hexyne give the *syn* addition product. With 2-butyne and 3-butyne, the major products are β-chlorovinyl acetates of *E*-configuration.[97] Some of the dichloro compounds are also formed, with more of the *E*- than the *Z*-isomer being observed.

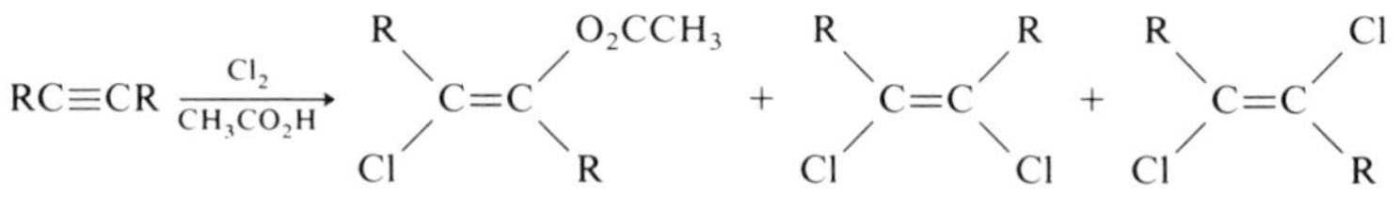

$$RC{\equiv}CR \xrightarrow[CH_3CO_2H]{Cl_2} (R)(Cl)C{=}C(O_2CCH_3)(R) + (R)(Cl)C{=}C(R)(Cl) + (R)(Cl)C{=}C(Cl)(R)$$

The reactions of the internal alkynes are considered to involve a cyclic halonium ion intermediate, whereas the terminal alkynes seem to react by a rapid collapse of a vinyl cation.

93. K. Yates, G. H. Schmid, T. W. Regulski, D. G. Garratt, H. W. Leung, and R. McDonald, *J. Am. Chem. Soc.* **95**, 160 (1973).
94. J. Cousseau, *Synthesis*, 805 (1980).
95. R. C. Fahey and D.-J. Lee, *J. Am. Chem. Soc.* **90**, 2124 (1968).
96. M. L. Poutsma and J. L. Kartch, *Tetrahedron* **22**, 2167 (1966).
97. K. Yates and T. A. Go, *J. Org. Chem.* **45**, 2385 (1980).

1[a]

$$CH_3(CH_2)_3C{\equiv}CH \xrightarrow[HgSO_4]{H_2SO_4} CH_3(CH_2)_3\overset{O}{\overset{\|}{C}}CH_3 \quad (79\%)$$

2[b]

$$PhC{\equiv}CH \xrightarrow[HOAc\text{-}H_2O]{H_2SO_4} Ph\overset{O}{\overset{\|}{C}}CH_3$$

3[c]

1-ethynylcyclohexanol (HO, C≡CH) $\xrightarrow[H_2O]{HgSO_4,\ H_2SO_4}$ 1-acetylcyclohexanol (HO, $\overset{O}{\overset{\|}{C}}CH_3$) (65–67%)

4[d]

ketal-protected cyclohexanone bearing CH_3 and $CH_2C{\equiv}CH$ $\xrightarrow[H_2SO_4]{Hg^{2+},\ Dowex\ 50}$ product bearing CH_3 and $CH_2\overset{O}{\overset{\|}{C}}CH_3$ (100%)

5[e]

hydrindane with CH_3, OH, H, CH_3CO_2, $C{\equiv}C\text{–}H$, $CH(CH_3)_2$ $\xrightarrow[2)\ H_2S]{1)\ Hg^{++},\ H_2SO_4,\ H_2O}$ product with $CH_3C{=}O$ in place of $C{\equiv}CH$ (~60%)

a. R. J. Thomas, K. N. Campbell, and G. F. Hennion, *J. Am. Chem. Soc.* **60**, 718 (1938).
b. R. W. Bott, C. Eaborn, and D. R. M. Walton, *J. Chem. Soc.* 384 (1965).
c. G. N. Stacy and R. A. Mikulec, *Org. Synth.* **IV**, 13 (1963).
d. W. G. Dauben and D. J. Hart, *J. Org. Chem.* **42**, 3787 (1977).
e. D. Caine and F. N. Tuller, *J. Org. Chem.* **38**, 3663 (1973).

Alkynes react with bromine by an electrophilic addition mechanism. A bridged bromonium ion type intermediate has been postulated for alkyl-substituted acetylenes, while vinyl cations are suggested for aryl-substituted examples.[98] 1-Phenylpropyne gives mainly the *trans* addition product in acetic acid, but some of the *cis* isomer is formed.[99] The proportion of dibromide formed is enhanced when lithium bromide is added to the reaction mixture.

$$PhC{\equiv}CCH_3 \xrightarrow[CH_3CO_2H]{Br_2} \underset{Ph}{\overset{Br}{}}C{=}C\underset{Br}{\overset{CH_3}{}} + \underset{Ph}{\overset{Br}{}}C{=}C\underset{CH_3}{\overset{Br}{}} + Ph\overset{AcO}{\overset{|}{C}}{=}\overset{Br}{\overset{|}{C}}CH_3 \text{ (both isomers)}$$

no LiBr	59%	14%	21%
LiBr added	98%	0.2%	1.5%

98. G. H. Schmid, A. Modro, and K. Yates, *J. Org. Chem.* **45**, 665 (1980).
99. J. A. Pincock and K. Yates, *J. Am. Chem. Soc.* **90**, 5643 (1968).

Some of the most useful reactions of acetylenes are with organometallic reagents. These reactions, which can lead to carbon–carbon bond formation, will be discussed in Chapter 8.

4.9. Addition at Double Bonds via Organoboranes

4.9.1. Hydroboration

Borane, BH_3, is an avid electron pair acceptor because only six valence electrons are present at boron. Pure borane exists as a dimer in which two hydrogens bridge the borons. In aprotic solvents that can act as electron donors, such as ethers, tertiary amines, and sulfides, diborane forms Lewis acid–base adducts.

$$R_2\overset{+}{O}-\overset{-}{B}H_3 \qquad R_3\overset{+}{N}-\overset{-}{B}H_3 \qquad R_2\overset{+}{S}-\overset{-}{B}H_3$$

Borane dissolved in THF or dimethyl sulfide undergoes addition reactions rapidly with most alkenes. This reaction, which is known as hydroboration, has been extensively studied, and a variety of useful synthetic processes have been developed, largely through the work of H. C. Brown and his associates.

Hydroboration is highly regioselective and is stereospecific. The boron becomes bonded primarily to the *less substituted* carbon atom of the alkene. A combination of steric and electronic effects work together to favor this orientation. Borane is an electrophilic reagent. The reaction with substituted styrenes exhibits a weakly negative ρ value (-0.5).[100] Compared with the value for bromination $(\rho^+ = -4.3)$,[101] this is a small substituent effect but does favor addition of the electrophilic boron at the less substituted end of the double bond. In contrast to the case of addition of acids to alkenes, it is the boron atom, not hydrogen, that is the electrophilic center. This electronic effect is reinforced by steric factors. Hydroboration is usually done under conditions where the borane eventually reacts with three alkene molecules to give a trialkylborane. The second and third alkyl groups would result in severe steric repulsion if the boron were added at the internal carbon.

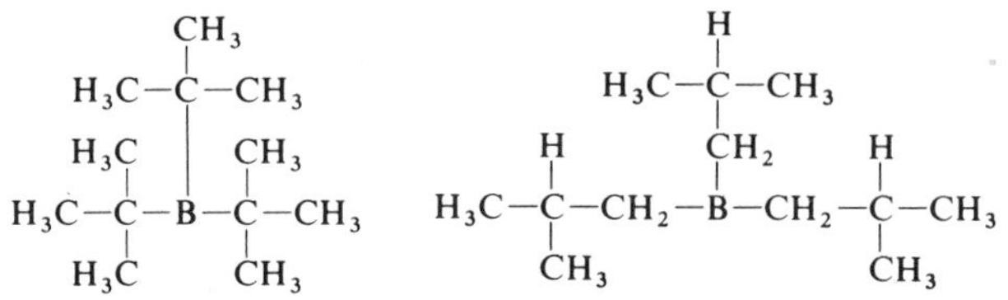

```
        CH3                                   H
        |                                     |
   H3C—C—CH3                            H3C—C—CH3
        |                                     |
   H3C  |   CH3                   H          CH2         H
     |  |   |                     |           |          |
H3C—C—B—C—CH3              H3C—C—CH2—B—CH2—C—CH3
     |      |                     |                      |
   H3C      CH3                  CH3                    CH3
```

severe nonbonded repulsions — nonbonded repulsions reduced

Table 4.4 provides some data on the regioselectivity of addition of diborane and several of its derivatives to representative alkenes.

100. L. C. Vishwakarma and A. Fry, *J. Org. Chem.* **45**, 5306 (1980).
101. J. A. Pincock and K. Yates, *Can. J. Chem.* **48**, 2944 (1970).

The table includes data for some monoalkyl and dialkylboranes which show even higher regioselectivity than diborane itself. These derivatives have been widely used in synthesis and are frequently referred to by the shortened names shown below with the structures.

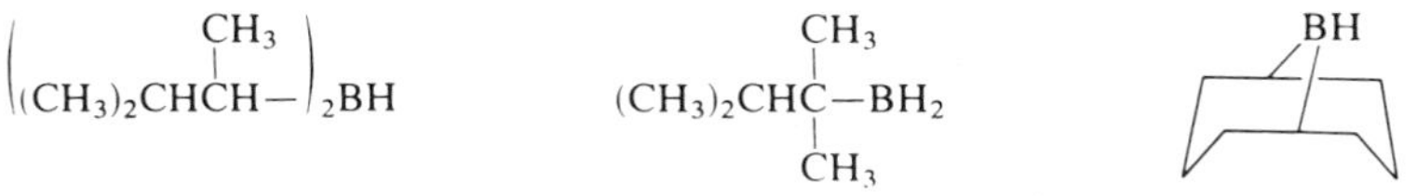

$$\left((CH_3)_2CH\overset{CH_3}{\overset{|}{C}}H-\right)_2BH \qquad (CH_3)_2CH\underset{CH_3}{\underset{|}{\overset{CH_3}{\overset{|}{C}}}}-BH_2$$

disiamylborane
bis(3-methyl-2-butyl)borane

thexylborane
1,1,2-trimethylpropylborane

BH

9-BBN
9-borabicyclo[3.3.1]nonane

These reagents are prepared by hydroboration of the appropriate alkene, using control of stoichiometry to terminate the hydroboration at the desired degree of alkylation.

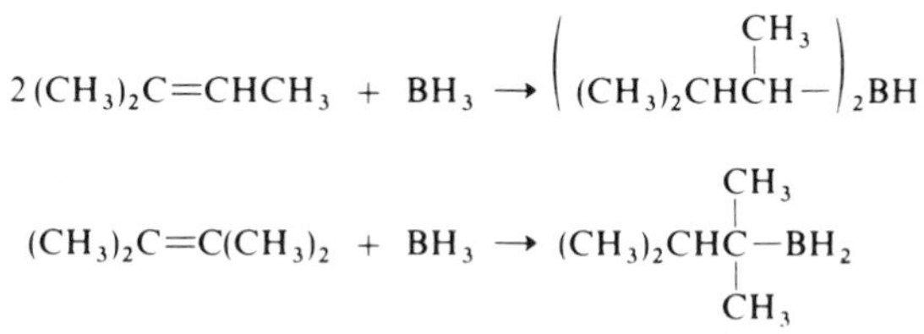

$$2(CH_3)_2C{=}CHCH_3 + BH_3 \rightarrow \left((CH_3)_2CH\overset{CH_3}{\overset{|}{C}}H-\right)_2BH$$

$$(CH_3)_2C{=}C(CH_3)_2 + BH_3 \rightarrow (CH_3)_2CH\underset{CH_3}{\underset{|}{\overset{CH_3}{\overset{|}{C}}}}-BH_2$$

Hydroboration is a stereospecific *syn* addition. The addition occurs through a four-center transition state with essentially simultaneous bonding to boron and hydrogen. Both the new C—B and C—H bonds are therefore formed from the same

Table 4.4. Regioselectivity of Diborane and Alkylboranes toward Representative Alkenes

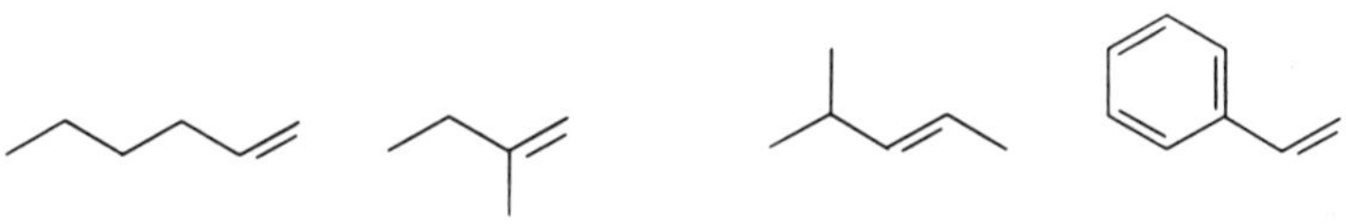

Hydroborating reagent	Percent of Boron Added at Less Substituted Carbon			
	1-Hexene	2-Methyl-1-butene	4-Methyl-2-pentene	Styrene
Diborane[a]	94	99	57	80
Chloroborane-dimethyl sulfide[b]	99	99.5	—	98
Disiamylborane[a]	99	—	97	98
Thexylborane[c]	94	—	66	95
Thexylchloroborane-[d] dimethyl sulfide	99	99	97	99
9-BBN[e]	99.9	99.8*	99.8	98.5

* data for 2-methyl-1-pentene

a. G. Zweifel and H. C. Brown, *Org. React.* **13**, 1 (1963).
b. H. C. Brown, N. Ravindran, and S. U. Kulkarni, *J. Org. Chem.* **44**, 2417 (1969); H. C. Brown and U. S. Racherla, *J. Org. Chem.* **51**, 895 (1986).
c. H. C. Brown and G. Zweifel, *J. Am. Chem. Soc.* **82**, 4708 (1960).
d. H. C. Brown, J. A. Sikorski, S. U. Kulkarni, and H. D. Lee, *J. Org. Chem.* **45**, 4540 (1980).
e. H. C. Brown, E. F. Knights, and C. G. Scouten, *J. Am. Chem. Soc.* **96**, 7765 (1974).

side of the double bond. In molecular orbital terms, the addition is viewed as taking place by interaction of the filled alkene π orbital with the empty p orbital on boron, accompanied by concerted C—H bond formation[102]:

As is true for most reagents, there is a preference for approach of the borane from the less hindered side of the molecule. Since diborane itself is a relatively small molecule, the stereoselectivity is not great for unhindered molecules. Table 4.5 gives some data comparing the direction of approach for three cyclic alkenes. The products in all cases result from *syn* addition, but the mixtures result both from the low regioselectivity and from addition to both faces of the double bond. Even the quite hindered 7,7-dimethylnorbornene shows only modest preference for *endo* addition with diborane. The selectivity is enhanced with the bulkier reagent 9-BBN.

The haloboranes BH_2Cl, BH_2Br, $BHCl_2$, and $BHBr_2$ are also useful hydroborating reagents.[103] These compounds are somewhat more regioselective than borane itself but otherwise show similar reactivity. The most useful aspects of the chemistry of the haloboranes is their application in sequential introduction of substituents at boron. The halogens can be replaced by alkoxide or by hydride. When halogen is replaced by hydride, a second hydroboration step can be carried out.

$$R_2BX + NaOR' \rightarrow R_2BOR'$$
$$R_2BX + LiAlH_4 \rightarrow R_2BH$$
$$RBX_2 + LiAlH_4 \rightarrow RBH_2$$
$$X = Cl, Br$$

Table 4.5. Stereoselectivity of Hydroboration of Cyclic Alkenes[a]

Hydroborating reagent	Product composition[b]								
	3-Methylcyclopentene			3-Methylcyclohexene				7,7-Dimethylnorbornene	
	trans-2	*cis*-3	*trans*-3	*cis*-2	*trans*-2	*cis*-3	*trans*-3	*exo*	*endo*
Borane	45	55 (*cis*-3 + *trans*-3)		16	34	18	32	22	78[c]
Disiamylborane	40	60 (*cis*-3 + *trans*-3)		18	30	27	25	—	—
9-BBN	25	50	25	0	20	40	40	3	97

a. Data from H. C. Brown, R. Liotta, and L. Brener, *J. Am. Chem. Soc.* **99**, 3427 (1977), except where noted otherwise.
b. Product composition refers to methylcycloalkanol formed by subsequent oxidation.
c. H. C. Brown, J. H. Kawakami, and K.-T. Liu, *J. Am. Chem. Soc.* **95**, 2209 (1973).

102. D. J. Pasto, B. Lepeska, and T.-C. Cheng, *J. Am. Chem. Soc.* **94**, 6083 (1972); P. R. Jones, *J. Org. Chem.* **37**, 1886 (1972); S. Nagase, K. N. Ray, and K. Morokuma, *J. Am. Chem. Soc.* **102**, 4536 (1980).
103. H. C. Brown and S. U. Kulkarni, *J. Organomet. Chem.* **239**, 23 (1982).

Application of these transformations will be discussed in Chapter 9, where carbon–carbon bond-forming reactions of boranes are covered.

Hydroboration is thermally reversible. At 160°C and above, B—H molecules are eliminated from alkylboranes, but the equilibrium is still in favor of the addition products. This provides a mechanism for migration of the boron group along the carbon chain by a series of eliminations and additions:

$$R_2CH-CH(B<)-CH_3 \rightleftharpoons R_2C=CH-CH_3 + H-B< \;;\; R_2CH-CH=CH_2 + H-B< \rightleftharpoons R_2CH-CH_2-CH_2-B<$$

Migration cannot occur past a quaternary carbon, however, since the required elimination is blocked. At equilibrium, the major trialkylborane is the least substituted terminal isomer that is accessible, since this is the isomer which minimizes unfavorable steric interactions.

$$[\text{(H}_3\text{C, H)-pinanyl}]_3B \xrightarrow{160°C} [\text{(H, CH}_2\text{–)-pinanyl}]_3B \qquad \text{Ref. 104}$$

$$CH_3(CH_2)_{13}CH=CH(CH_2)_{13}CH_3 \xrightarrow[2)\ 80°C,\ 14\ h]{1)\ B_2H_6} [CH_3(CH_2)_{29}]_3B \qquad \text{Ref. 105}$$

Bulky substituents on boron facilitate the migration. Bis(bicyclo[2.2.2]oct-2-yl)borane, in which there are no complications from migrations in the bicyclic substituent, has been found to be a particularly useful reagent for equilibration.

$$(\text{bicyclo[2.2.2]octyl})_2B-H + (CH_3)_2C=CHCH_3 \xrightarrow{\Delta} (\text{bicyclo[2.2.2]octyl})_2BCH_2CH_2CH(CH_3)_2 \qquad \text{Ref. 106}$$

4.9.2. Reactions of Organoboranes

The organoboranes have proven to be very useful intermediates in organic synthesis. In this section, we will discuss methods by which the boron atom can be

104. G. Zweifel and H. C. Brown, *J. Am. Chem. Soc.* **86**, 393 (1964).
105. K. Maruyama, K. Terada, and Y. Yamamoto, *J. Org. Chem.* **45**, 737 (1980).
106. H. C. Brown and U. S. Racherla, *J. Am. Chem. Soc.* **105**, 6506 (1983).

efficiently replaced by hydroxyl, halogen, or amino groups. There are also processes that use alkylboranes in the formation of new carbon-carbon bonds. These reactions will be discussed in Section 9.1.

The most widely used reaction of organoboranes is the oxidation to alcohols. Alkaline hydrogen peroxide is the reagent usually employed to effect the oxidation. The mechanism is outlined below.

$$R_3B + HOO^- \rightarrow R-\overset{R}{\underset{R}{\overset{|}{\underset{|}{B^-}}}}-O-OH \rightarrow R-\overset{R}{\overset{|}{B}}-OR + {}^-OH$$

$$R_2BOR + HOO^- \rightarrow R-\overset{R-O}{\underset{R}{\overset{|}{\underset{|}{B^-}}}}-O-O-H \rightarrow R-\overset{RO}{\underset{RO}{\overset{|}{\underset{|}{B}}}} + {}^-OH$$

$$(RO)_2BR + HOO^- \rightarrow (RO)_2\underset{R}{\underset{|}{B^-}}-O-O-H \rightarrow (RO)_3B + {}^-OH$$

$$(RO)_3B + H_2O \rightarrow ROH + B(OH)_3$$

The R—O—B bonds are hydrolyzed in the alkaline aqueous solution, generating the alcohol. The oxidation mechanism involves a series of boron-to-oxygen migrations of the alkyl groups. The stereochemical outcome is replacement of the C—B bond by a C—O bond with *retention of configuration.* In combination with the stereospecific *syn* hydroboration, this allows the structure and stereochemistry of the alcohols to be predicted with confidence. The preference for hydroboration at the less substituted carbon of a double bond results in the alcohol being formed with regiochemistry that is complementary to that observed for direct hydration or oxymercuration.

Conditions that permit oxidation of organoboranes to alcohols using molecular oxygen[107] or amine oxides[108] as oxidants have also been developed.

$$[CH_3(CH_2)_6]_3B + 3(CH_3)_3\overset{+}{N}-O^- \xrightarrow[2)\ H_2O]{1)\ heat} CH_3(CH_2)_6OH$$

The oxidation by amine oxides provides a basis for selection among nonequivalent groups on boron. In acyclic organoboranes, the order of reaction is *tertiary > secondary > primary.* In cyclic boranes, stereoelectronic factors dominate. With 9-BBN derivatives, for example, preferential migration of a C—B bond that is part of the bicyclic ring structure occurs.

107. H. C. Brown, M. M. Midland, and G. W. Kabalka, *J. Am. Chem. Soc.* **93**, 1024 (1971).
108. G. W. Kabalka and H. C. Hedgecock, Jr., *J. Org. Chem.* **40**, 1776 (1975); R. Koster and Y. Monta, *Justus Liebigs Ann. Chem.* **704**, 70 (1967).

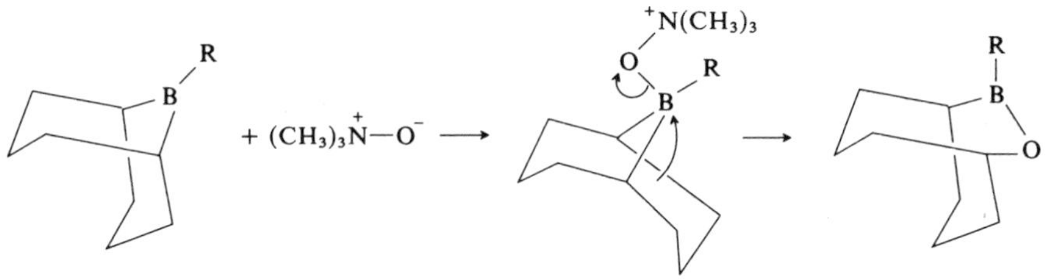

This is attributed to the unfavorable steric interactions that arise in the transition state for antiperiplanar migration of the exocyclic substituent.[109]

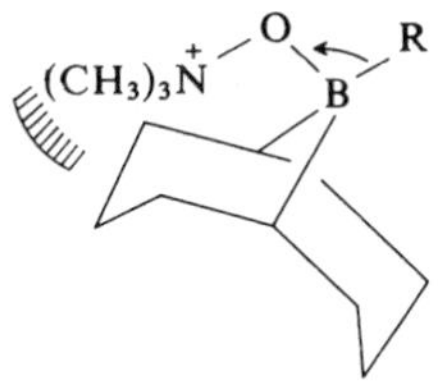

Some examples of synthesis of alcohols by hydroboration–oxidation are included in Scheme 4.9.

More vigorous oxidizing agents effect replacement of boron and oxidation to the carbonyl level.[110]

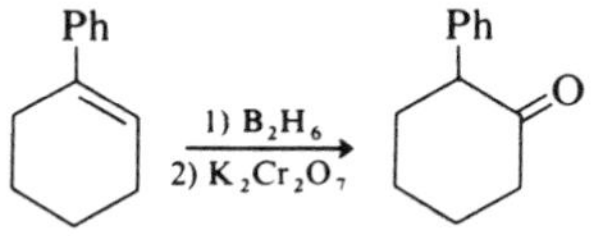

The boron atom can also be replaced by an amino group.[111] The reagents that effect this conversion are chloramine or hydroxylamine-*O*-sulfonic acid. The mechanism of these reactions is very similar to that of the hydrogen peroxide oxidation of organoboranes. The nitrogen-containing reagent initially reacts as a nucleophile by adding at boron, and then rearrangement with expulsion of chloride or sulfate ion follows. As in the oxidation, the migration step occurs with retention of configuration. The amine is freed by hydrolysis.

$$R_3B + NH_2X \rightarrow R_2\bar{B}(R)-NH-X \rightarrow R_2B-NH(R) \xrightarrow{H_2O} RNH_2$$

$X = Cl$ or OSO_3

Secondary amines are formed by reaction of trisubstituted boranes with alkyl or aryl azides. The most efficient borane intermediates to use are monoalkyldi-

109. J. A. Soderquist and M. R. Najafi, *J. Org. Chem.* **51**, 1330 (1986).
110. H. C. Brown and C. P. Garg, *J. Am. Chem. Soc.* **83**, 2951 (1961).
111. M. W. Rathke, N. Inoue, K. R. Varma, and H. C. Brown, *J. Am. Chem. Soc.* **88**, 2870 (1966); G. W. Kabalka, K. A. R. Sastry, G. W. McCollum, and H. Yoshioka, *J. Org. Chem.* **46**, 4296 (1981).

Scheme 4.9. Alcohols, Ketones, Aldehydes, and Amines from Organoboranes

A. Alcohols

1[a]

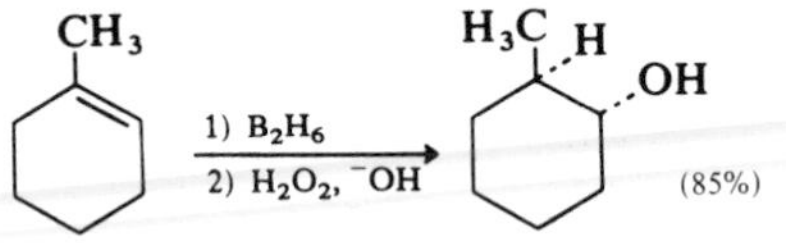

CH_3 — 1) B_2H_6 2) H_2O_2, ^-OH → H_3C H OH (85%)

2[b]

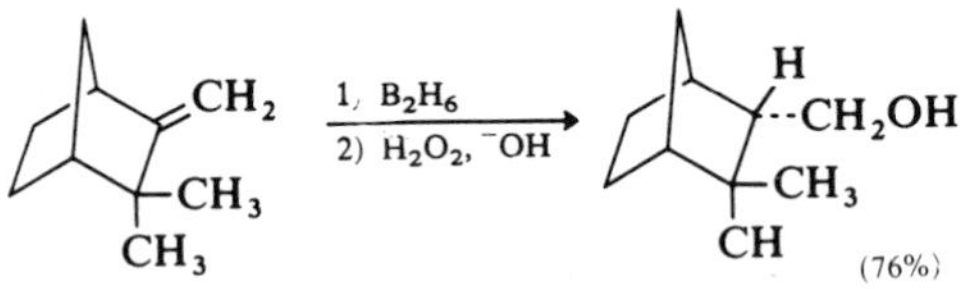

CH_2 CH_3 CH_3 — 1) B_2H_6 2) H_2O_2, ^-OH → H CH_2OH CH_3 CH (76%)

3[c]

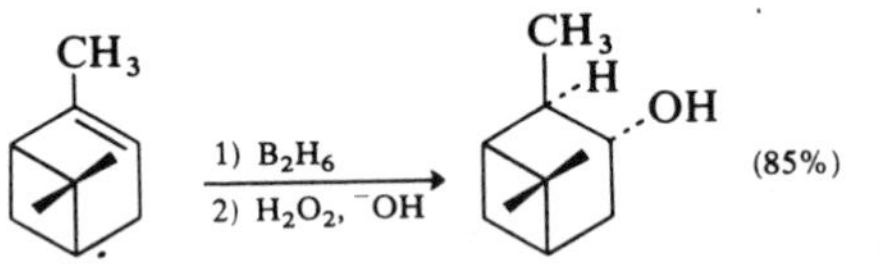

CH_3 — 1) B_2H_6 2) H_2O_2, ^-OH → CH_3 H OH (85%)

4[d]

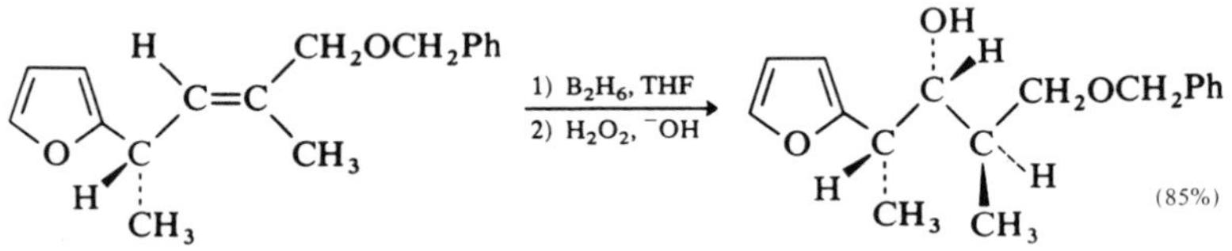

H CH_2OCH_2Ph C=C O C H CH_3 CH_3 — 1) B_2H_6, THF 2) H_2O_2, ^-OH → OH H C CH_2OCH_2Ph O C C H H CH_3 CH_3 (85%)

B. Ketones

5[e]

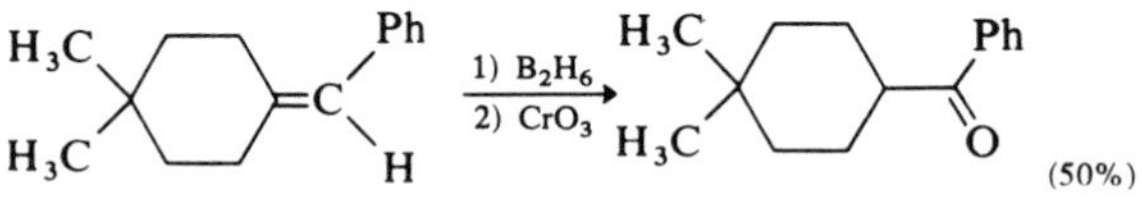

H_3C H_3C Ph C H — 1) B_2H_6 2) CrO_3 → H_3C H_3C Ph O (50%)

6[f]

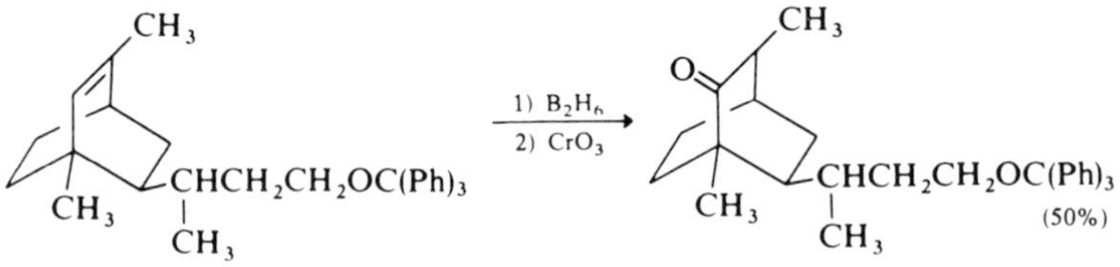

CH_3 CH_3 $CHCH_2CH_2OC(Ph)_3$ CH_3 — 1) B_2H_6 2) CrO_3 → O CH_3 CH_3 $CHCH_2CH_2OC(Ph)_3$ CH_3 (50%)

7[g]

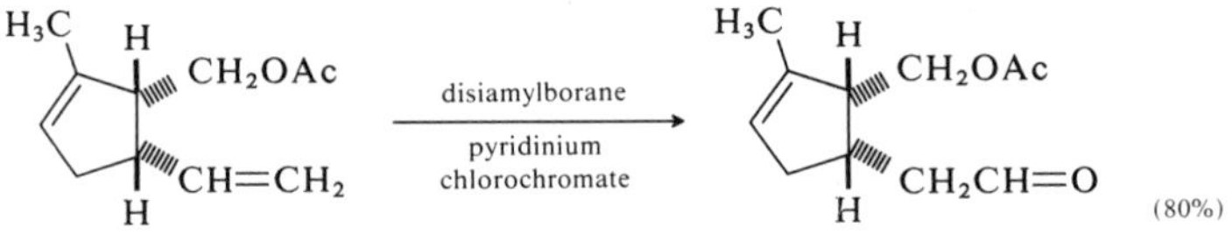

H_3C H CH_2OAc $CH{=}CH_2$ H — disiamylborane; pyridinium chlorochromate → H_3C H CH_2OAc $CH_2CH{=}O$ H (80%)

C. Amines

8[h]

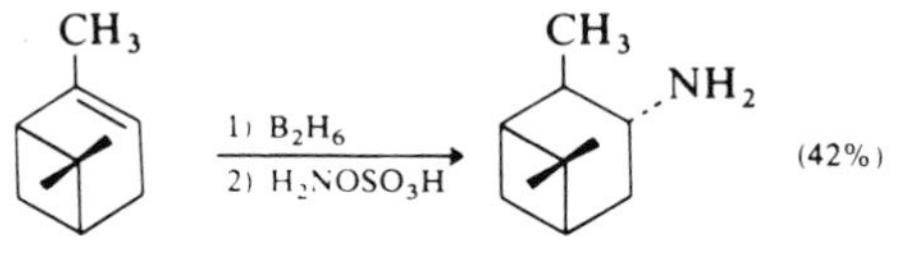

CH_3 — 1) B_2H_6 2) H_2NOSO_3H → CH_3 NH_2 (42%)

9[i]

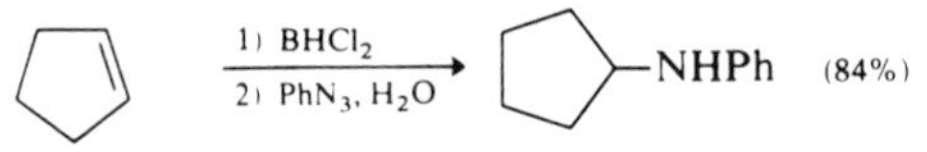

1) $BHCl_2$ 2) PhN_3, H_2O → NHPh (84%)

chloroboranes, which are generated by reaction of an alkene with $BHCl_2 \cdot Et_2O$.[112] The entire sequence of steps and the mechanism of the final stages are summarized by the equations below.

$$BHCl_2{\cdot}Et_2O + RCH{=}CH_2 \rightarrow RCH_2CH_2BCl_2$$

$$RCH_2CH_2BCl_2 + R'{-}N_3 \rightarrow Cl_2\overset{-}{B}(CH_2CH_2R){-}N(R'){-}\overset{+}{N}{\equiv}N \rightarrow Cl_2BN(R')CH_2CH_2R$$

$$Cl_2BN(R')CH_2CH_2R \xrightarrow{H_2O} R'NHCH_2CH_2R$$

Secondary amines can also be made using the *N*-chloro derivatives of primary amines[113]:

$$(CH_3CH_2)_3B + HN(Cl)(CH_2)_7CH_3 \rightarrow CH_3CH_2N(H)(CH_2)_7CH_3 \quad (90\%)$$

Organoborane intermediates can also be used to synthesize alkyl halides. Replacement of boron by iodine is rapid in the presence of base.[114] The best yields are obtained with sodium methoxide in methanol.[115] If less basic conditions are desirable, the use of iodine monochloride and sodium acetate gives good yields.[116] As is the case in hydroboration-oxidation, the regioselectivity of hydroboration-halogenation is opposite to that observed for direct addition of hydrogen halides to alkenes. Terminal alkenes give primary halides.

$$RCH{=}CH_2 \xrightarrow[2)\ Br_2,\ NaOH]{1)\ B_2H_6} RCH_2CH_2Br$$

4.9.3. Enantioselective Hydroboration

Several alkylboranes are available in enantiomerically enriched or pure form, and they can be used to prepare homochiral alcohols and other compounds available

112. H. C. Brown, M. M. Midland, and A. B. Levy, *J. Am. Chem. Soc.* **95**, 2394 (1973).
113. G. W. Kabalka, G. W. McCollum, and S. A. Kunda, *J. Org. Chem.* **49**, 1656 (1984).
114. H. C. Brown, M. W. Rathke, and M. M. Rogic, *J. Am. Chem. Soc.* **90**, 5038 (1968).
115. N. R. De Lue and H. C. Brown, *Synthesis*, 114 (1976).
116. G. W. Kabalka and E. E. Gooch III, *J. Org. Chem.* **45**, 3578 (1980).

---

a. H. C. Brown and G. Zweifel, *J. Am. Chem. Soc.* **83**, 2544 (1961).
b. R. Dulou, Y. Chretien-Bessiere, *Bull. Soc. Chim. France*, 1362 (1959).
c. G. Zweifel and H. C. Brown, *Org. Synth.* **52**, 59 (1972).
d. G. Schmid, T. Fukuyama, K. Akasaka, and Y. Kishi, *J. Am. Chem. Soc.* **101**, 259 (1979).
e. W. B. Farnham, *J. Am. Chem. Soc.* **94**, 6857 (1972).
f. R. N. Mirrington, and K. J. Schmalzl, *J. Org. Chem.* **37**, 2871 (1972).
g. H. C. Brown, S. U. Kulkarni, and C. G. Rao, *Synthesis*, 151 (1980); T. H. Jones and M. S. Blum, *Tetrahedron Lett.* **22**, 4373 (1981).
h. M. W. Rathke and A. A. Millard, *Org. Synth.* **58**, 32 (1978).
i. H. C. Brown, M. M. Midland, and A. B. Levy, *J. Am. Chem. Soc.* **95**, 2394 (1973).

via organoborane intermediates.[117] One route to homochiral boranes is by hydroboration of readily available terpenes that occur naturally in enantiomerically enriched or pure form. The most thoroughly investigated of these is bis(isopinocampheyl)borane, which can be prepared in 100% enantiomeric purity from the readily available terpene α-pinene.[118] Both enantiomers are available.

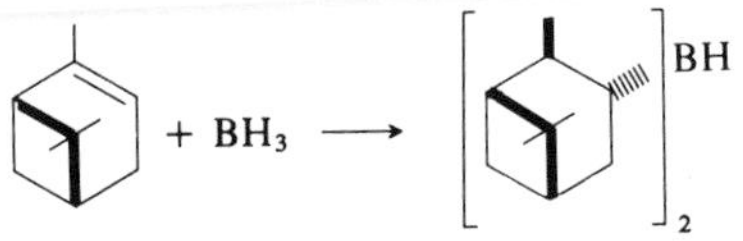

+ BH_3 ⟶ [ ]BH$_2$

Other examples of chiral organoboranes derived from terpenes are **C**, **D**, and **E**, which are derived from longifolene,[119] 2-carene,[120] and limonene,[121] respectively.

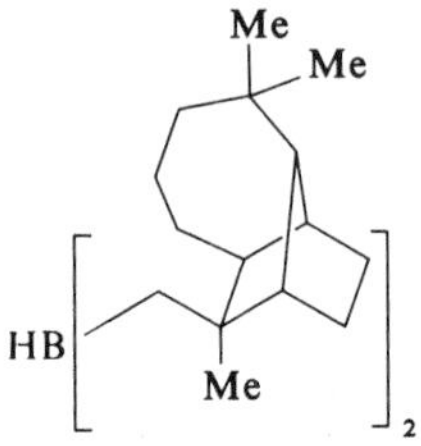

Me Me HB Me 2 **C**   HB Me Me Me 2 **D**   Me HB Me **E**

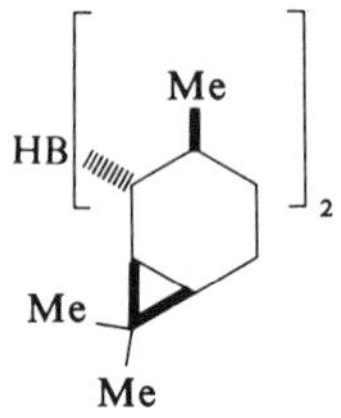

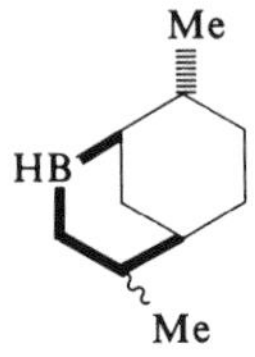

Bis(isopinocampheyl)borane adopts a conformation which minimizes steric interactions. This conformation can be represented schematically as in **F** and **G**, where the S, M, and L substituents are, respectively, the 3—H, 4—CH_2, and 2—$CHCH_3$ groups of the carbocyclic structure. The steric environment at boron in this conformation is such that *Z*-alkenes encounter less steric encumberance in transition state **G** than in **F**.

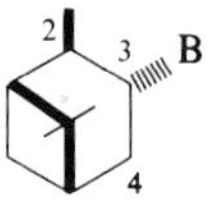

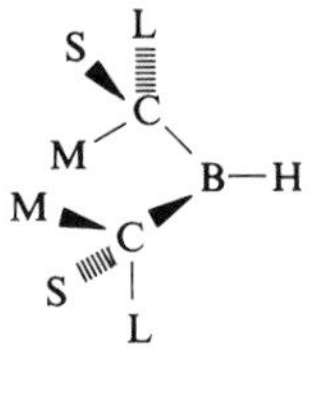

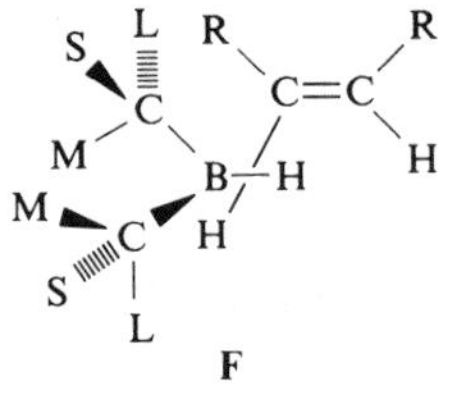

F 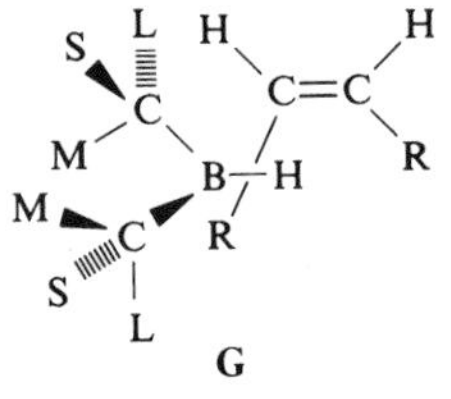   **G**

The degree of enantioselectivity of bis(isopinocampheyl)borane is not high for all simple alkenes. *Z*-Disubstituted alkenes give good enantioselectivity (75–90%), but *E*-alkenes and simple cycloalkenes give low enantioselectivity (5–30%).

117. H. C. Brown and B. Singaram, *Acc. Chem. Res.* **21**, 287 (1988); D. S. Matteson, *Acc. Chem. Res.* **21**, 294 (1988).
118. H. C. Brown, P. K. Jadhav, and A. K. Mandal, *Tetrahedron* **37**, 3547 (1981); H. C. Brown and P. K. Jadhav, in *Asymmetric Synthesis*, Vol. 2, J. D. Morrison (ed.), Academic Press, New York, 1983, Chapter 1.
119. P. K. Jadhav and H. C. Brown, *J. Org. Chem.* **46**, 2988 (1981).
120. H. C. Brown, J. V. N. Vara Prasad, and M. Zaidlewics, *J. Org. Chem.* **53**, 2911 (1988).
121. P. K. Jadhav and S. U. Kulkarni, *Heterocycles* **18**, 169 (1982).

A promising hydroborating reagent has been reported recently.[122] The *trans*-dimethylborolane **1** is prepared and can be separated from the contaminating *cis* isomer. The *trans* isomer can be resolved by formation of an adduct with a chiral aminoalcohol. Finally, reaction with $LiAlH_4$ generates the dialkylborane in high enantiomeric purity.

Et_2NBCl_2 + $CH_3CH(MgBr)CH(MgBr)CH_3$ → (2,5-dimethylborolane)$BNEt_2$ —(chiral aminoalcohol)→ adduct —($LiAlH_4$)→ **1**

This dialkylborane effects hydroboration of a variety of internal alkenes with good enantioselectivity, but terminal alkenes give low enantioselectivity. The main drawback with this reagent is that its preparation requires several steps.

Mono(isocampheyl)borane can be prepared in enantiomerically pure form by purification of a TMEDA adduct.[123] When this monoalkylborane reacts with a prochiral alkene, two diastereomeric products can be formed. One is normally formed in excess and can be obtained in high enantiomeric purity by an appropriate separation.[124] Oxidation of the borane then provides the corresponding alcohol in the same enantiomeric purity achieved for the borane.

$IpcBH_2$ + $R^3(H)C{=}C(R^1)R^2$ → $IpcB(H)$–$C(R^3)(H)$–$C(R^1)(R^2)H$ or $IpcB(H)$–$C(R^3)(H)$–$C(R^1)(R^2)H$

Since oxidation also converts the original chiral terpene-derived group to an alcohol, it is not directly reusable as a chiral auxiliary. While this is not a problem with inexpensive materials, the overall efficiency of generation of enantiomerically pure product is improved by procedures that can regenerate the original terpene. This can be done by heating the dialkylborane intermediate with acetaldehyde. The α-pinene is released and a diethoxyborane is produced.[125]

122. S. Masamune, B. M. Kim, J. S. Petersen, T. Sato, S. J. Veenstra, and T. Imai, *J. Am. Chem. Soc.* **107**, 4549 (1985).
123. H. C. Brown, J. R. Schwier, and B. Singaram, *J. Org. Chem.* **43**, 4395 (1978); H. C. Brown, A. K. Mandal, N. M. Yoon, B. Singaram, J. R. Schwier, and P. K. Jadhav, *J. Org. Chem.* **47**, 5069 (1982).
124. H. C. Brown and B. Singaram, *J. Am. Chem. Soc.* **106**, 1797 (1984); H. C. Brown, P. K. Jadhav, and A. K. Mandal, *J. Org. Chem.* **47**, 5074 (1982).
125. H. C. Brown, B. Singaram, and T. E. Cole, *J. Am. Chem. Soc.* **107**, 460 (1985); H. C. Brown, T. Imai, M. C. Desai, and B. Singaram, *J. Am. Chem. Soc.* **107**, 4980 (1985).

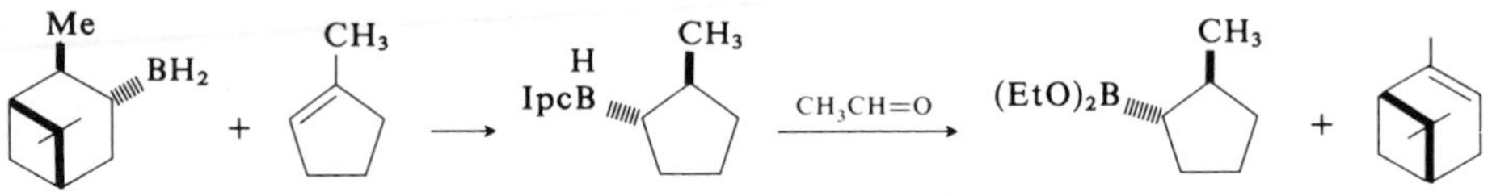

The usual oxidation conditions then convert this boronic ester to an alcohol.[126]

Procedures for enantioselective synthesis of chiral amines[127] and halides[128] based on alkylboranes have been developed by applying the methods discussed earlier to homochiral organoborane intermediates.

4.9.4. Hydroboration of Alkynes

Alkynes are reactive toward hydroboration reagents. The most useful procedures involve addition of a disubstituted borane to the alkyne. This avoids complications that can occur with borane itself which lead to cross-linked structures. Catecholborane (1,3,2-benzodioxaborole), which is prepared from equimolar amounts of catechol (1,2-dihydroxybenzene) and borane, is a particularly useful reagent for hydrocarboration of alkynes.[129] Protonolysis with acetic acid of the adduct of an alkyne with this reagent results in reduction of the alkyne to the corresponding *Z*-alkene. Oxidative workup with hydrogen peroxide gives ketones via enol intermediates.

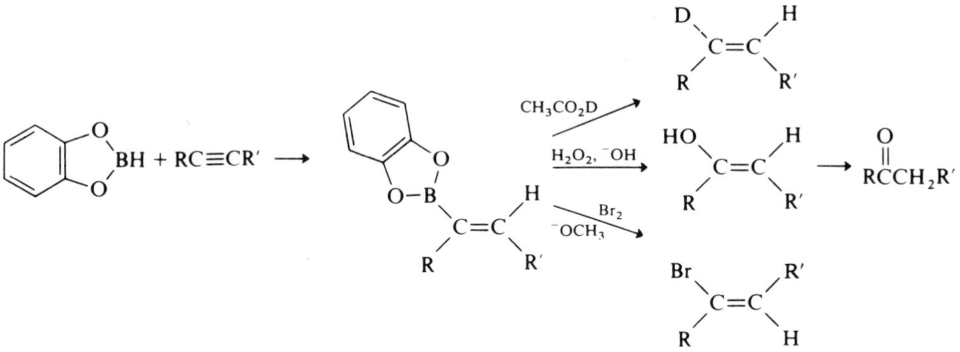

Treatment of the vinylborane with bromine and base leads to vinyl bromides. The reaction occurs with net *anti* addition. The stereoselectivity has been rationalized

126. D. S. Matteson and K. M. Sadhu, *J. Am. Chem. Soc.* **105**, 2077 (1983).
127. L. Verbit and P. J. Heffron, *J. Org. Chem.* **32**, 3199 (1967); H. C. Brown, K.-W. Kim, T. E. Cole, and B. Singaram, *J. Am. Chem. Soc.* **108**, 6761 (1986).
128. H. C. Brown, N. R. De Lue, G. W. Kabalka, and H. C. Hedgecock, Jr., *J. Am. Chem. Soc.* **98**, 1290 (1976).
129. H. C. Brown, T. Hamaoka, and N. Ravindran, *J. Am. Chem. Soc.* **95**, 6456 (1973); C. F. Lane and G. W. Kabalka, *Tetrahedron* **32**, 981 (1976).

on the basis of *anti* addition of bromine followed by *anti* elimination of bromide and boron:

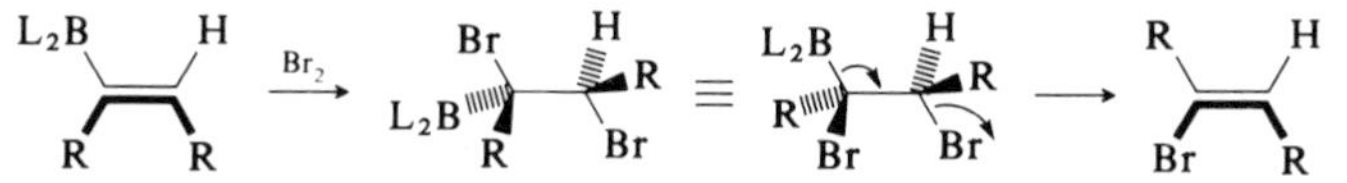

Exceptions to this stereoselectivity have been noted.[130]

The adducts derived from catecholborene are hydrolyzed by water to vinylboronic acids. These materials are useful intermediates for preparation of terminal vinyl iodides. Since the hydroboration is a *syn* addition and the iodinolysis occurs with retention of the alkene geometry, the iodides have the E-configuration.[131]

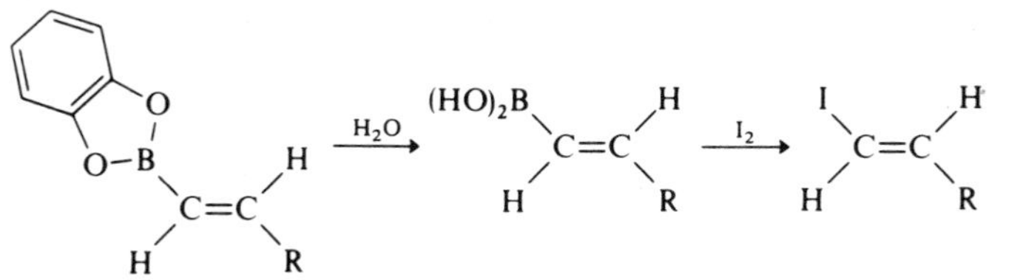

Other disubstituted boranes have also been used for selective hydroboration of alkynes. 9-BBN can be used to hydroborate internal alkynes. Protonolysis can be carried out with methanol, and this forms a convenient method for formation of a disubstituted Z-alkene.[132]

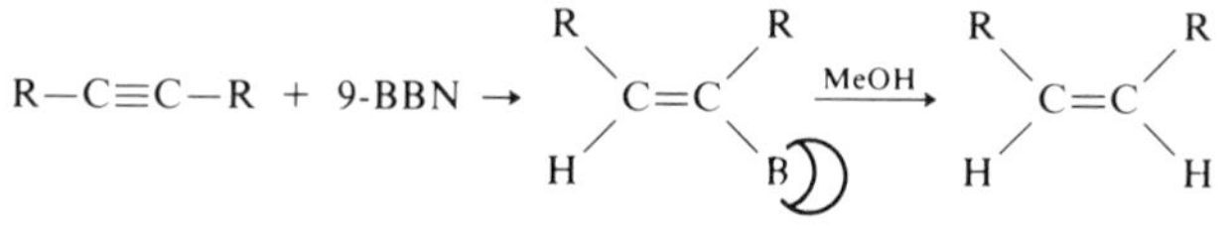

$$R{-}C{\equiv}C{-}R + \text{9-BBN} \rightarrow$$

The dimethyl sulfide complex of dibromoborane is also useful for synthesis of E-vinyl iodides from terminal alkynes[133]:

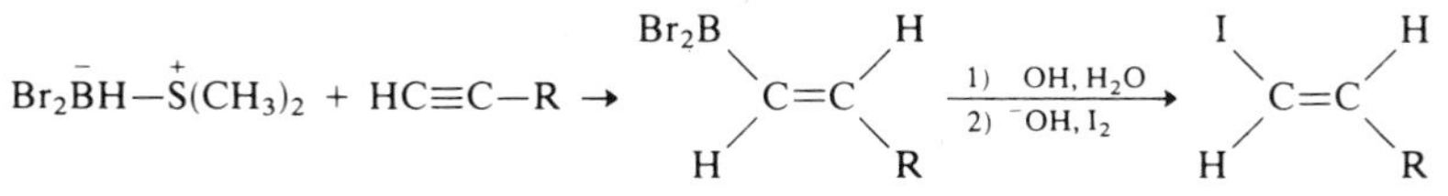

$$Br_2\overset{-}{B}H{-}\overset{+}{S}(CH_3)_2 + HC{\equiv}C{-}R \rightarrow$$

A large number of procedures that involve carbon–carbon bond formation have developed around organoboranes. These reactions are considered in Chapter 9.

130. J. R. Wiersig, N. Waespe-Sarcevic, and C. Djerassi, *J. Org. Chem.* **44**, 3374 (1979).
131. H. C. Brown, T. Hamaoka, and N. Ravindran, *J. Am. Chem. Soc.* **95**, 5786 (1973).
132. H. C. Brown and G. A. Molander, *J. Org. Chem.* **51**, 4512 (1986); H. C. Brown and K. K. Wang, *J. Org. Chem.* **51**, 4514 (1986).
133. H. C. Brown and J. B. Campbell, Jr., *J. Org. Chem.* **45**, 389 (1980).

General References

Addition of Hydrogen Halide, Halogens, and Related Electrophiles

P. B. De la Mare and R. Bolton, *Electrophilic Addition to Unsaturated Systems*, Elsevier, Amsterdam, 1982.
R. C. Fahey, *Top. Stereochem.* **2**, 237 (1968).

Solvomercuration

W. Kitching, *Organomet. Chem. Rev.* **3**, 61 (1968).
R. C. Larock, *Angew. Chem. Int. Ed. Engl.* **12**, 27 (1978).

Addition of Sulfur and Selenium Reagents

D. J. Clive, *Tetrahedron* **34**, 1049 (1978).
D. Liotta (ed.), *Organoselenium Chemistry*, Wiley, New York, 1987.
S. Patai and Z. Rappoport (ed.), *The Chemistry of Organic Selenium and Tellurium Compounds*, Wiley, New York, 1986.
C. Paulmier, *Selenium Reagents and Intermediates in Organic Synthesis*, Pergamon, Oxford, 1986.

Additions to Acetylenes and Allenes

T. F. Rutledge, *Acetylenes and Allenes*, Reinhold, New York, 1969.
G. H. Schmid, in *The Chemistry of the Carbon–Carbon Triple Bond*, S. Patai (ed.), Wiley, New York, 1978, Chapter 8.

Organoboranes as Synthetic Intermediates

H. C. Brown, *Organic Synthese via Boranes*, Wiley, New York, 1975.
G. Cragg, *Organoboranes in Organic Synthesis*, Marcel Dekker, New York, 1973.
A. Pelter, K. Smith, and H. C. Brown, *Borane Reagents*, Academic Press, New York, 1988.

Problems

(*References for these problems will be found on page* 767.)

1. Predict the direction of addition and structure of the product for each of the following reactions.

(a) $CH_3CH{=}CH_2 + O_2N\text{-}C_6H_3(NO_2)\text{-}SCl \longrightarrow$ (2,4-dinitrophenyl)SCl, with NO_2 ortho to SCl

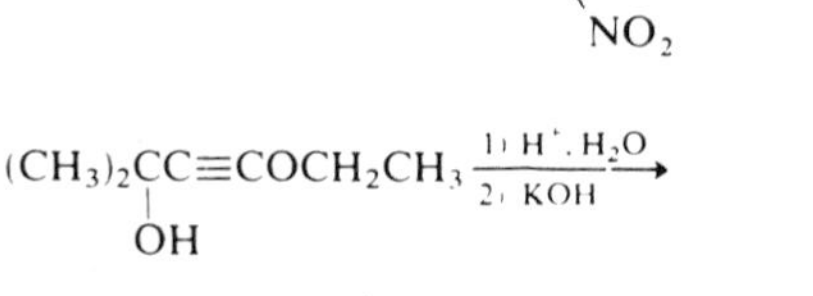

(b) $(CH_3)_2C(OH)C{\equiv}COCH_2CH_3 \xrightarrow[\text{2) KOH}]{\text{1) H}^+,\ H_2O}$

(c) cyclohexylidene$=CH_2 \xrightarrow{\text{HOBr}}$

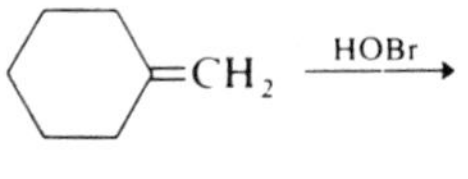

(d) $(CH_3)_2C{=}CHCH_3 \xrightarrow[\text{2) } H_2O_2,\ HO^-]{\text{1) disiamylborane}}$

(e) $CH_3CH_2CH_2CH_2CH{=}CH_2 \xrightarrow{IN_3}$

(f) $(CH_3)_3CCH{=}CHCH_3 \xrightarrow{IN_3}$

(g) $C_6H_5CH{=}CHCH(OCH_3)_2 \xrightarrow{IN_3}$

(h) 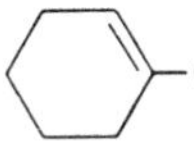 1-(trimethylsilyloxy)cyclohexene, $-OSi(CH_3)_3 \xrightarrow[\text{ether, } -78°C]{PhSeBr} \xrightarrow[H_2O]{NaHCO_3}$

(i) $HC{\equiv}CCH_2CH_2CO_2H \xrightarrow[CH_2Cl_2]{Hg(OAc)_2} \xrightarrow[\text{10 min}]{H_2O,\ NaHCO_3}$

(j) $H_2C{=}CHCH_2CH_2CH_2CH_2OH \xrightarrow[\text{2) } NaBH_4]{\text{1) } Hg(OAc)_2}$

(k)  cyclohexene $\xrightarrow[\text{CHCl}_3,\text{ crown ether}]{I_2,\ NaN_3}$

(l) 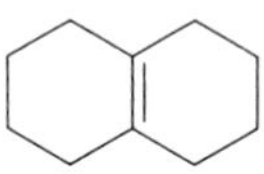 $\Delta^{9,10}$-octalin $\xrightarrow{NOCl}$

(m) 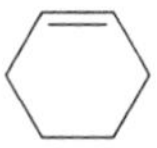 cyclohexene $\xrightarrow[LiClO_4,\ CH_3CN]{PhSCl,\ Hg(OAc)_2}$

(n) $PhCH(CH{=}CH_2)CH_2CO_2H \xrightarrow[CH_3CN]{I_2}$

2. Bromination of 4-*t*-butylcyclohexene in methanol gives a 45:55 mixture of two compounds, each of composition $C_{11}H_{21}BrO$. Predict the structure and stereochemistry of these two products. How would you confirm your prediction?

3. Hydroboration-oxidation of $PhCH{=}CHOC_2H_5$ gives **A** as the major product if the hydroboration step is of short duration (7 s), but **B** is the major product if the hydroboration is allowed to proceed for a longer time (2 h). Explain.

$PhCH(OH)CH_2OC_2H_5$ — **A**

$PhCH_2CH_2OH$ — **B**

4. Oxymercuration of 4-*t*-butycyclohexene, followed by $NaBH_4$ reduction, gives *cis*-4-*t*-butylcyclohexanol and *trans*-3-*t*-butylcyclohexanol in approximately equal amounts. 1-Methyl-4-*t*-butylcyclohexene under similar conditions gives only *cis*-4-*t*-butyl-1-methylcyclohexanol. Formulate a mechanism for the oxymercuration-reduction process that is consistent with this stereochemical result.

5. Treatment of compound **C** with *N*-bromosuccinimide in acetic acid containing sodium acetate gives a product $C_{13}H_{19}BrO_3$. Propose the structure, including stereochemistry, of the product and explain the basis for your proposal.

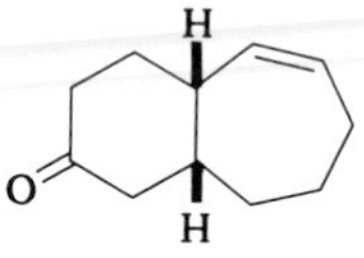

C

6. The hydration of 5-undecyn-2-one with mercuric sulfate and sulfuric acid in methanol is regioselective, giving 2,5-undecadione in 85% yield. Suggest an explanation for the high selectivity.

7. A procedure for the preparation of allylic alcohols has been devised in which the elements of phenylselenenic acid are added to an alkene, then the reaction mixture is treated with *t*-butyl hydroperoxide. Suggest a mechanistic rationale for this process.

$$CH_3CH_2CH_2CH{=}CHCH_2CH_2CH_3 \xrightarrow[2)\ t\text{-BuOOH}]{1)\ \text{"}C_6H_5SeOH\text{"}} CH_3CH_2CH_2\underset{\displaystyle OH}{\underset{|}{C}}HCH{=}CHCH_2CH_3 \quad (88\%)$$

8. Suggest synthetic sequences that could accomplish each of the following transformations.

(a)

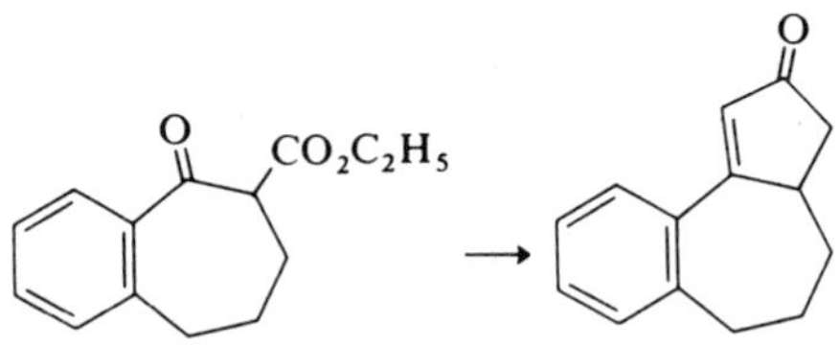

$CO_2C_2H_5$ →

(b)

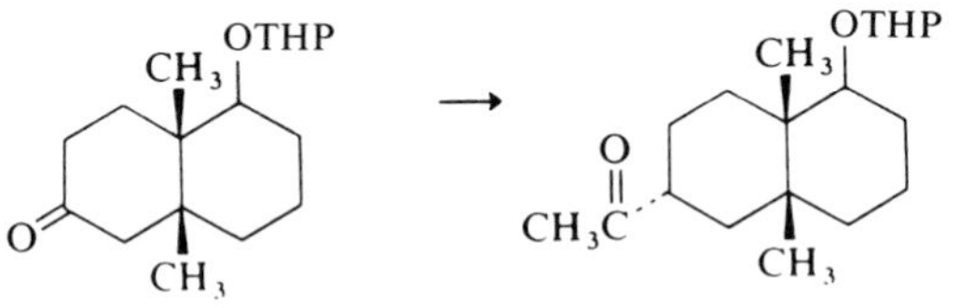

CH_3 OTHP → CH_3 OTHP, CH_3C (=O), CH_3

(c)

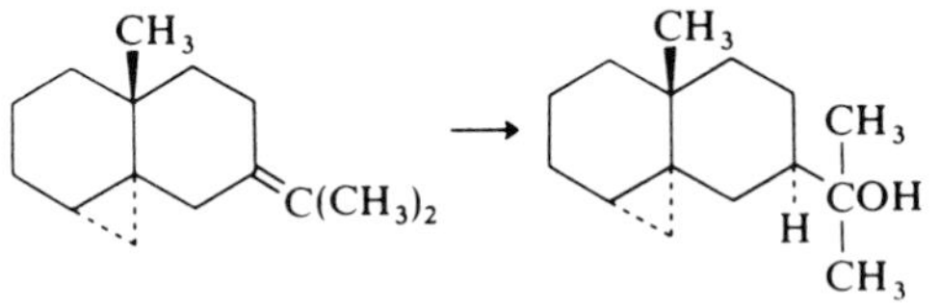

CH_3, $C(CH_3)_2$ → CH_3, H, CH_3–COH–CH_3

(d)

CH_3 → OH, CH_3

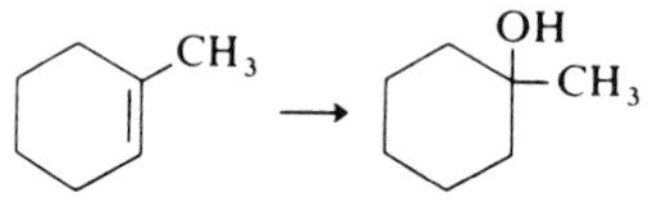

(e) $CH_3CH_2CH_2CH_2C{\equiv}CH \longrightarrow$ $CH_3CH_2CH_2CH_2$(H)C=C(H)(I)

(f) $\longrightarrow$ $HC(CH_3)_2$

(g) $CH_3\overset{O}{\overset{\|}{C}}CH_2CH_2CH{=}C(CH_3)_2 \longrightarrow$ $(CH_3)_2CH$

(h) H_3C, C=CH_2, CH=CH_2, CH_3 $\longrightarrow$ H_3C, C=CH_2, CH_2CH_2OH, CH_3

(i) $CH_2CH{=}CH_2 \longrightarrow (CH_2)_3I$

(j) CH_3, C≡CH $\longrightarrow$ CH_3, $CH_2CH{=}O$

(k) OH, C≡CH, MeO $\longrightarrow$ OH, CH=CHI, MeO

(l) $CH_3CH_2\underset{}{C}HCH_2CH{=}CH_2$ with $HC(OCH_3)_2$ $\rightarrow$ $CH_3CH_2CHCH_2CH_2CH_2Br$ with $HC(OCH_3)_2$

(m) $CH_3(CH_2)_5C{\equiv}CH \rightarrow CH_3(CH_2)_5$(H)C=C(H)(Br)

9. Three methods for the preparation of nitroalkenes are outlined as shown. Describe in mechanistic terms how each of these transformations might occur.

(a) cyclohexene $\xrightarrow[\text{2) NaOH}]{\text{1) } HgCl_2\text{, } NaNO_2}$ 1-nitrocyclohexene ($-NO_2$)

(b) 1-(trimethylstannyl)cycloheptene [$Sn(CH_3)_3$] + $C(NO_2)_4$ $\longrightarrow$ 1-nitrocycloheptene (NO_2)

(c) cyclohexene $\xrightarrow{NO_2^+BF_4^-}$ 1-nitrocyclohexene ($-NO_2$)

10. Hydroboration-oxidation of 1,4-di-*t*-butylcyclohexene gave three alcohols: C (77%), D (20%), and E (3%). Oxidation of C gave ketone F, which was readily converted in either acid or base to an isomeric ketone G. Ketone G was the only oxidation product of alcohols D and E. What are the structures of compounds C—G?

11. Show, using enolate chemistry and organoselenium reagents, how you could convert 2-phenylcyclohexanone regiospecifically to either 2-phenyl-2-cyclohexen-1-one or 6-phenyl-2-cyclohexen-1-one.

12. On the basis of the mechanistic picture of oxymercuration involving a mercurinium ion, predict the structure and stereochemistry of the major alcohols to be expected by application of the oxymercuration-demercuration sequence to each of the following substituted cyclohexenes.

(a) $C(CH_3)_3$ (b) CH_3 (c) CH_3

13. Reaction of the unsaturated acid **C** and I_2 in acetonitrile (no base) gives rise in 89% yield to a 20:1 mixture of two stereoisomeric iodolactones. Formulate the complete stereochemistry of both the major and the minor product to be expected under these conditions.

$H_2C{=}CH$, CH_2, CO_2H, C, C, H_3C, H, H, CH_3

C

14. Some synthetic transformations are shown in the retrosynthetic format. Propose a short series of reactions (no more than three steps should be necessary) which could effect the synthetic conversion.

(a) [oxazolidinone: O, O, N, O, CH$_2$Ph; C(NH$_2$)CH(CH$_3$)$_2$] ⇒ [oxazolidinone: O, O, N, O, CH$_2$Ph; CH$_2$CH(CH$_3$)$_2$]

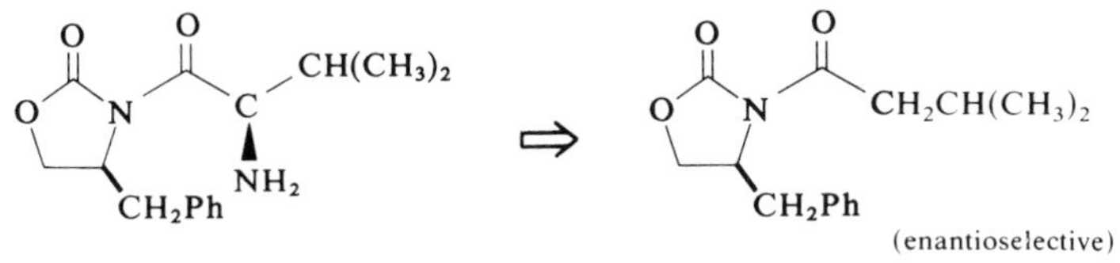

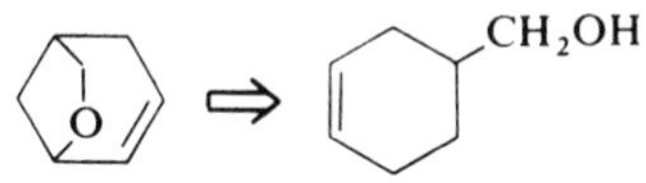

(enantioselective)

(b) [bicyclic ether: O] ⇒ [cyclohexene–CH$_2$OH]

(c) I—CHCH$_2$CH$_2$CH$_2$CO$_2$CH$_3$; S; H; C=C; H; CH(CH$_2$)$_4$CH$_3$; RO; OR ⇒ HO; CH$_2$; H C=C; CH$_2$CH$_2$CH$_2$CO$_2$CH$_3$; C H; H; C=C; RO; H; CH(CH$_2$)$_4$CH$_3$; OR

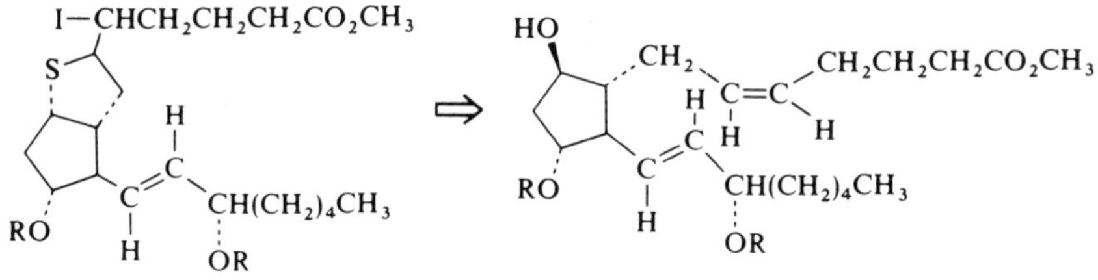

(d) CH$_3$; PhCHN; O; O; Br ⇒ O; O CH$_3$; CNCHCHPh; H

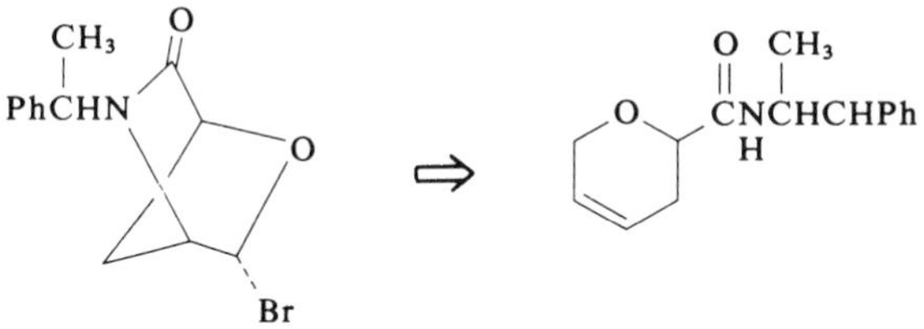

(e) CH$_3$; CH$_2$O$_2$CCH$_3$; C=CH$_2$; CH$_3$ ⇒ CH$_3$; CH$_2$O$_2$CCH$_3$; CH$_3$CHCH=O

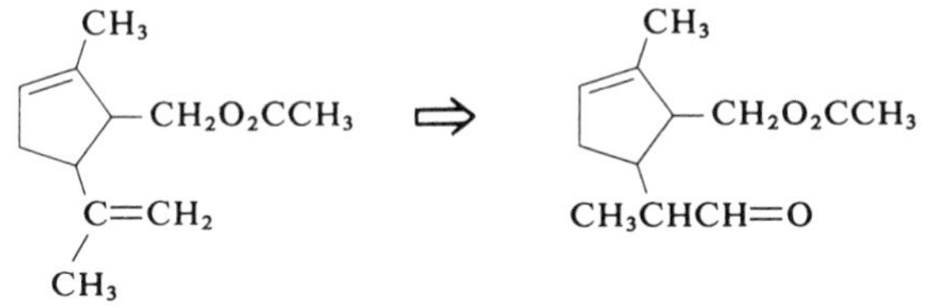

15. Write detailed mechanisms for the following reactions.

(a) H$_3$C; H$_3$C; O; O $\xrightarrow{\text{NaOCl}}$ HO$_2$CCH$_2$C(CH$_3$)$_2$CH$_2$CO$_2$H

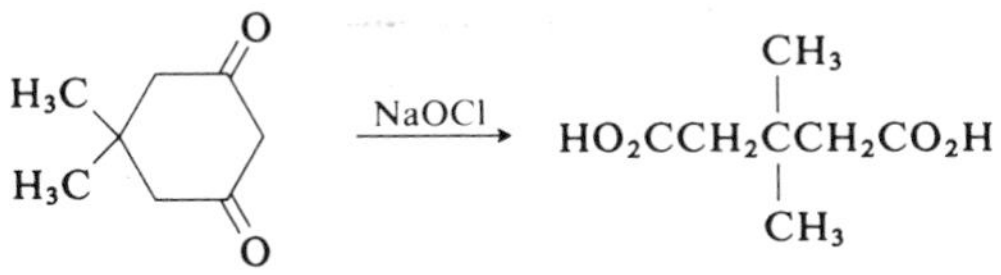

(b) CH$_2$=CHCH$_2$CH(CH$_3$)OCH$_2$NHCO$_2$CH$_2$Ph $\xrightarrow[\text{2) KBr}]{\text{1) Hg(NO}_3)_2}$ $\xrightarrow[\text{O}_2]{\text{NaBH}_4}$ O; NCO$_2$CH$_2$Ph; H$_3$C; CH$_2$OH

3:1 *cis* : *trans*

(c) N; CO$_2$H; Ph; CH$_3$ $\xrightarrow[\text{2) NaOCH}_3]{\text{1) NBS}}$ N; CO$_2$CH$_3$; O; Ph; CH$_3$

>90% enantiomerically pure

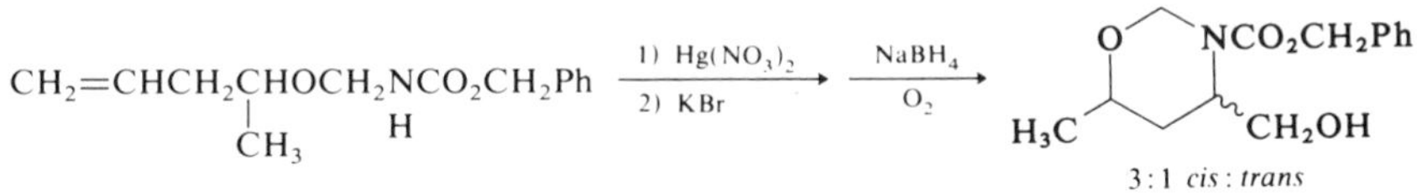

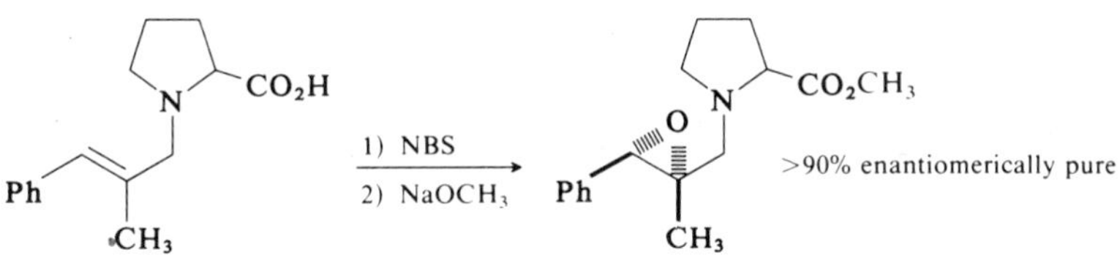

16. It has been observed that 4-pentenyl amides such as **1** cyclize to lactams **2** on reaction with phenylselenenyl bromide. The 3-butenyl compound **3**, on the other hand, cyclizes to an imino ether, **4**. What is the basis for the different course of these reactions?

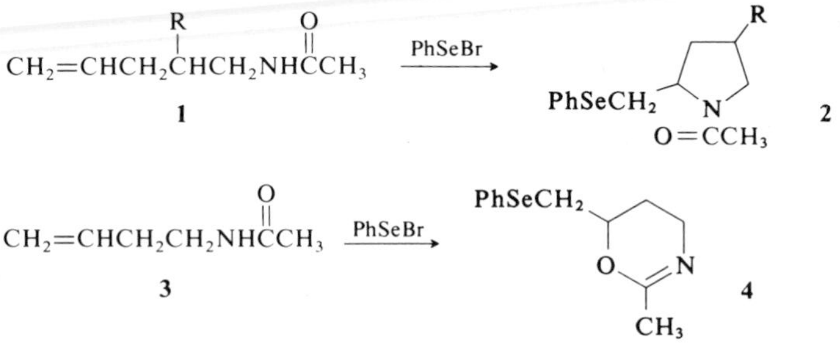

$CH_2{=}CHCH_2CH(R)CH_2NHC(O)CH_3$ (**1**) $\xrightarrow{PhSeBr}$ **2** ($PhSeCH_2$-substituted pyrrolidine bearing R, N-acetyl, $O{=}CCH_3$)

$CH_2{=}CHCH_2CH_2NHC(O)CH_3$ (**3**) $\xrightarrow{PhSeBr}$ **4** ($PhSeCH_2$-substituted six-membered ring O–C(=N)–CH_3)

17. Procedures for enantioselective synthesis of derivatives of α-bromoacids based on reaction of compounds **A** and **B** with N-bromosuccinimide have been developed. Predict the absolute configuration at the halogenated carbon in each product. Explain the basis of your prediction.

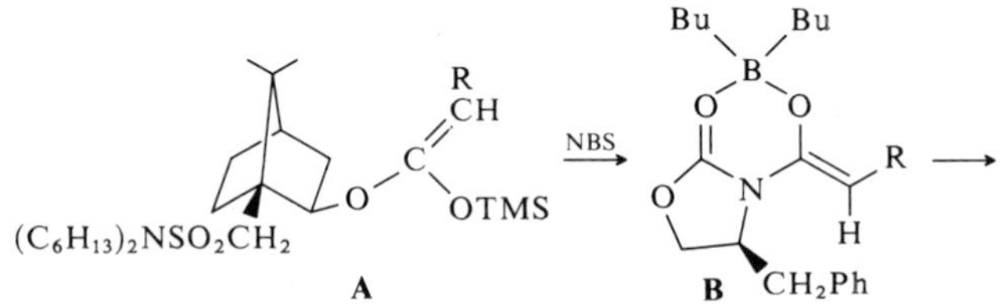

A: $(C_6H_{13})_2NSO_2CH_2$-substituted bornyl–O–C(OTMS)=CHR $\xrightarrow{NBS}$

B: Bu, Bu, B, O, O, N, O, R, H, CH_2Ph $\longrightarrow$

18. The stereochemical outcome of the hydroboration-oxidation of 1,1′-bicyclohexenyl depends on the amount of diborane used in the hydroboration. When 1.1 equiv. is used the product is a 3 : 1 mixture of **C** and **D**. When the ratio is increased to 2.1 : 1 the ratio is changed so that **C** is formed nearly exclusively. Offer an explanation of these results.

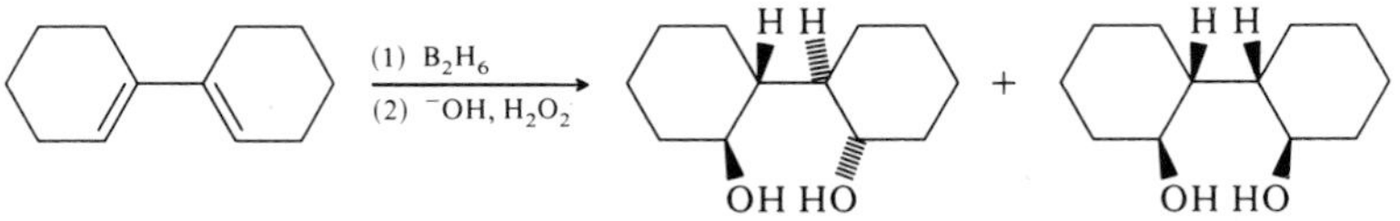

1,1′-bicyclohexenyl $\xrightarrow[(2)\ ^-OH,\ H_2O_2]{(1)\ B_2H_6}$ (H H, OH HO) + (H H, OH HO)

19. Predict the absolute configuration of the product obtained from the following reactions based on asymmetric hydroboration.

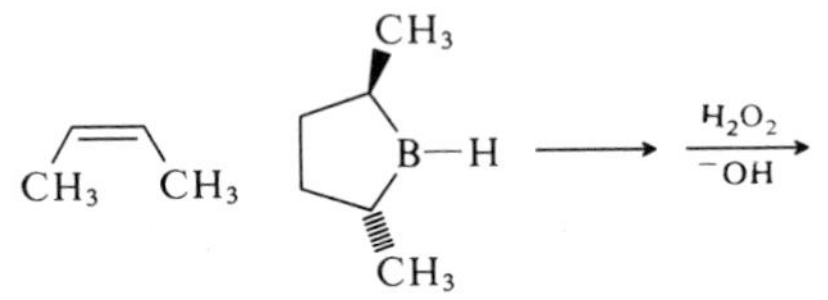

1-methylcyclopentene (CH_3) + H_2B–(pinanyl) $\xrightarrow{CH_3CH{=}O}$ $\xrightarrow[^-OH]{H_2O_2}$

cis-2-butene (CH_3, CH_3) + 2,5-dimethylborolane (CH_3, B—H, CH_3) $\longrightarrow$ $\xrightarrow[^-OH]{H_2O_2}$

5

Reduction of Carbonyl and Other Functional Groups

5.1. Addition of Hydrogen

5.1.1. Catalytic Hydrogenation

The most widely used method of adding the elements of hydrogen to carbon–carbon double bonds is catalytic hydrogenation. Except for very sterically hindered alkenes, this reaction usually proceeds rapidly and cleanly. The most common catalysts are various forms of transition metals, particularly platinum, palladium, rhodium, ruthenium, and nickel. Both the metals, as finely dispersed solids or adsorbed on inert supports such as carbon or alumina, and certain soluble complexes of these metals exhibit catalytic activity. Depending upon conditions and catalyst, other functional groups are also subject to catalytic reduction.

$$RCH{=}CHR + H_2 \xrightarrow{\text{catalyst}} RCH_2CH_2R$$

The mechanistic description of alkene hydrogenation has remained somewhat vague, partly because the reactive sites on the metal surface are not as readily described as small-molecule reagents in solution. As understanding of the chemistry of soluble hydrogenation catalysts has developed, it has become possible to extrapolate some mechanistic concepts to heterogeneous systems. It is known that hydrogen is adsorbed onto the metal surface, presumably forming metal–hydrogen bonds like those in transition metal hydride complexes. The alkenes are also adsorbed on the metal surface, and at least three types of intermediates have been implicated in the process of hydrogenation. The initially formed intermediate is pictured as attached at both carbon atoms of the double bond by π-type bonding, as shown in **A**. The bonding is regarded as an interaction between the alkene π and π^* orbitals and acceptor and donor orbitals of the metal. A hydrogen can be added to the adsorbed group, leading to **B**, which involves a σ-type carbon–metal bond. This species can react with another hydride to give alkane, which is desorbed from the surface. A third intermediate species, shown as **C**, accounts for double-bond isomerization and the exchange of hydrogen that sometimes accompanies hydrogenation. This intermediate is equivalent to an allyl radical bound to the metal surface by π bonds. It can be formed from adsorbed alkene by abstraction of an allylic hydrogen atom by the metal. In Chapter 8, the reactions of transition metals with organic compounds will be discussed. Well-characterized examples of structures corresponding to each of the intermediates **A**, **B**, and **C** are involved in those reactions.

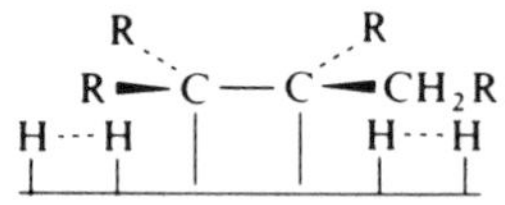

A

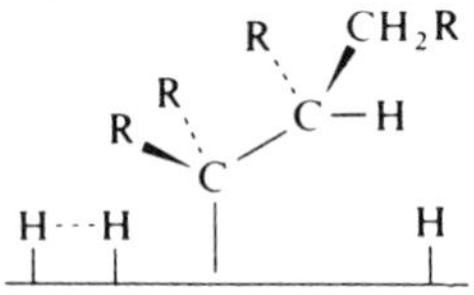

B 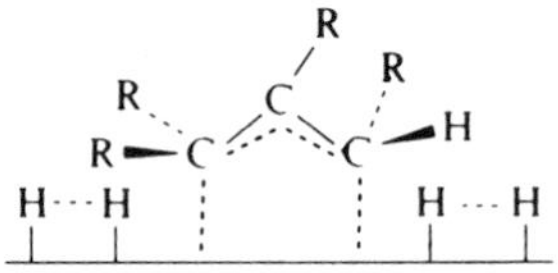 **C**

In most cases, both hydrogen atoms are added to the same side of the reactant (*syn* addition). If hydrogenation occurs by addition of hydrogen in two steps, as implied by the mechanism above, the intermediate must remain bonded to the metal surface in such a way that the stereochemical relationship is maintained. Adsorption to the catalyst surface normally involves the less sterically congested side of the double bond. Scheme 5.1 illustrates some hydrogenations in which the *syn* addition from the less hindered side is observed. Some exceptions are also included. There are many hydrogenations where hydrogen addition is not entirely *syn*, and independent corroboration of the stereochemistry is normally necessary.

The stereochemistry of hydrogenation is affected by the presence of polar functional groups that can govern the mode of adsorption to the catalyst surface. For instance, there are a number of examples where the presence of a hydroxyl group results in the hydrogen being introduced from the side of the molecule occupied by the hydroxyl group. This implies that the hydroxyl group is involved in the interaction with the catalyst surface. This behavior can be illustrated with the alcohol **1a** and the ester **1b**.[1] Although the overall shapes of the two molecules are similar, the alcohol gives mainly the product with a *cis* ring juncture (**2a**), while the ester gives a product with *trans* stereochemistry (**3b**). The stereoselectivity of

1. H. W. Thompson, *J. Org. Chem.* **36**, 2577 (1971); H. W. Thompson, E. McPherson, and B. L. Lences, *J. Org. Chem.* **41**, 2903 (1976).

Scheme 5.1. Stereochemistry of Hydrogenation of Some Alkenes

A. Examples of Preferential *syn* Addition from Less Hindered Side

1[a] H_3C, H, $=CH_2$ → (Pt, H_2) → H_3C, H, CH_3, H (70%) + H_3C, H, H, CH_3 (30%)

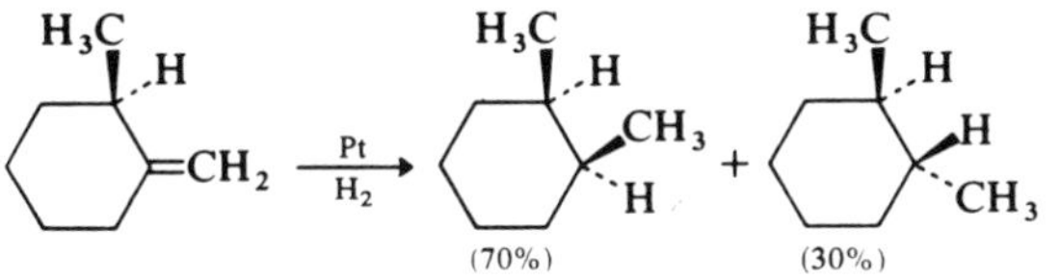

2[b] CH_3, CH_3 → (Pt, H_2) → CH_3, CH_3 (70–85%) + CH_3, CH_3 (30–15%)

3[c] CH_3, CH_3 → H, H, H_3C, CH_3

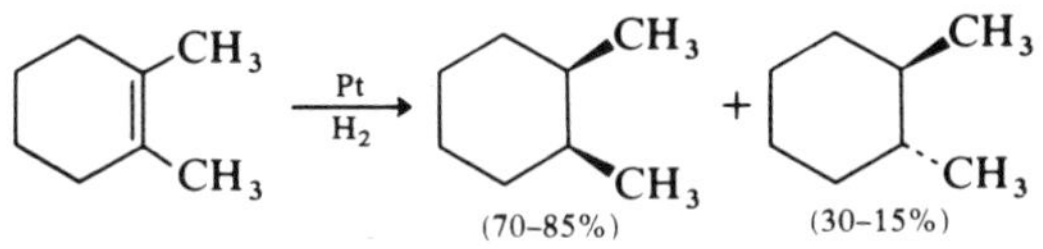

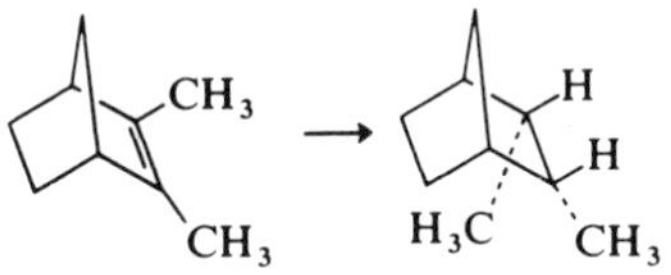

4[b] CH_2 → (Pt(BH_4^-)—C, H_2) → CH_3, H

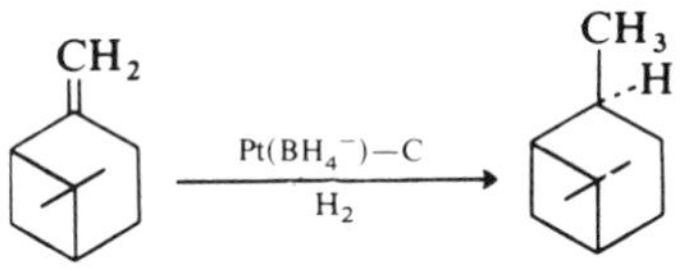

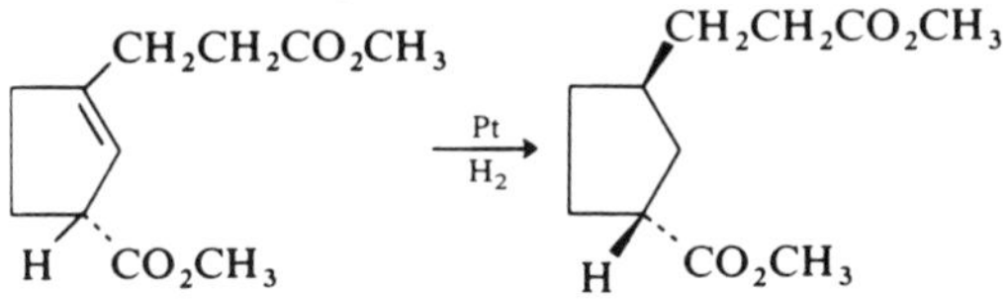

B. Exceptions

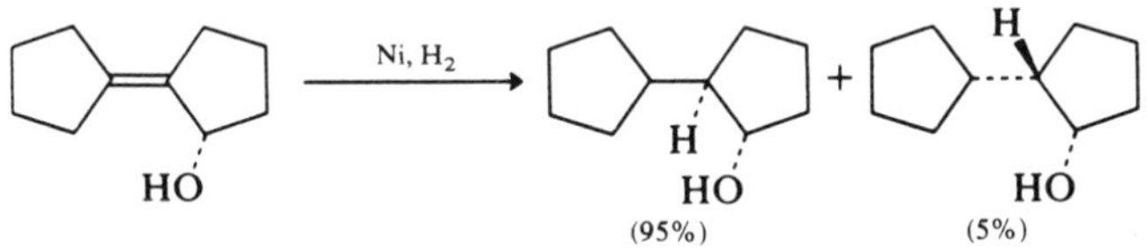

5[a] CH_3, CH_3 → (Pd, H_2) → CH_3, CH_3 (75%) + CH_3, CH_3 (25%)

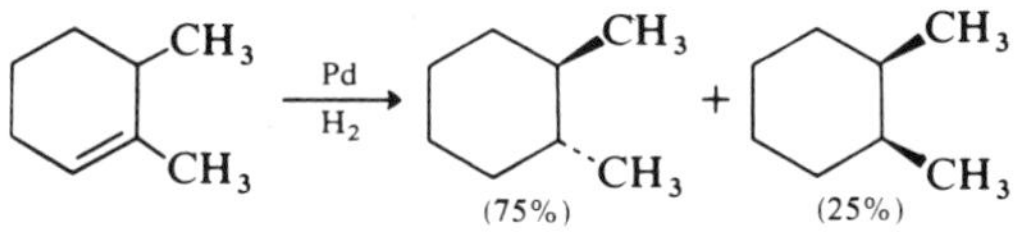

6[d] $CH_2CH_2CO_2CH_3$, H, CO_2CH_3 → (Pt, H_2) → $CH_2CH_2CO_2CH_3$, H, CO_2CH_3

7[e] HO → (Ni, H_2) → H, HO (95%) + H, HO (5%)

8[f] CH_3 → (Pt, H_2, acetic acid) → H, CH_3 (80%) + H, CH_3 (20%)

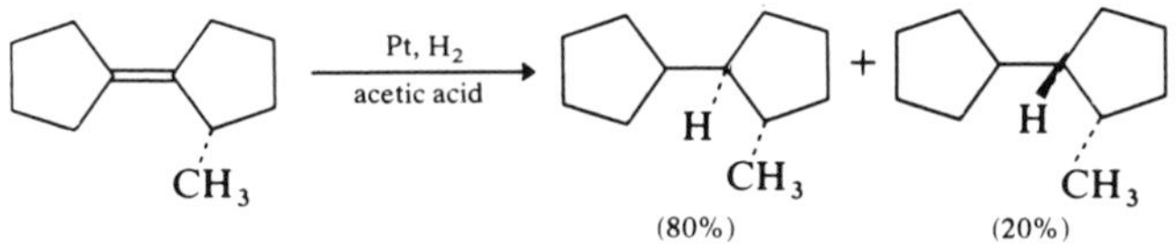

a. S. Siegel and G. V. Smith, *J. Am. Chem. Soc.* **82**, 6082, 6087 (1960).
b. C. A. Brown, *J. Am. Chem. Soc.* **91**, 5901 (1969).
c. K. Alder and W. Roth, *Chem. Ber.* **87**, 161 (1954).
d. J. P. Ferris and N. C. Miller, *J. Am. Chem. Soc.* **88**, 3522 (1966).
e. S. Mitsui, Y. Senda and H. Saito, *Bull. Chem. Soc. Jpn.* **39**, 694 (1966).
f. S. Siegel and J. R. Cozort, *J. Org. Chem.* **40**, 3594 (1975).

hydroxyl-directed hydrogenation is a function of solvent and catalyst. The ratio of *cis to trans* product from the alcohol **4** changes with solvent. The *cis* isomer is the main product in hexane.

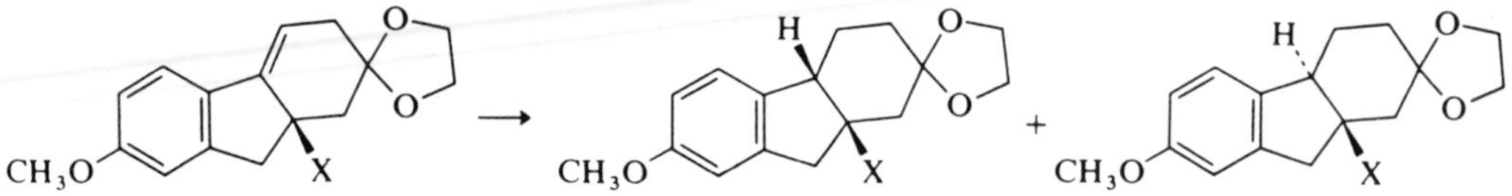

1a x = CH_2OH
1b x = CO_2CH_3

2b 15%

3a 6%
3b 85%

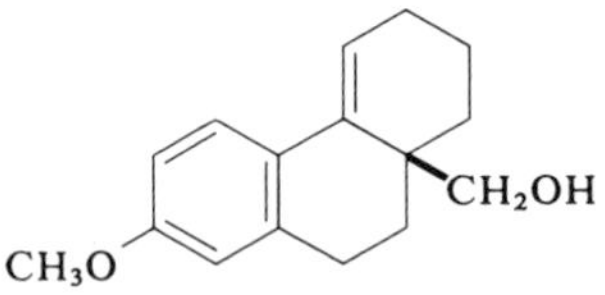

4

Solvent	% *cis*	% *trans*
Hexane	61	39
DME	20	80
EtOH	6	94

Catalytic hydrogenations are usually extremely clean reactions with little by-product formation, unless reduction of other groups is competitive. Careful study, however, sometimes reveals that double-bond migration can take place in competition with reduction. For example, hydrogenation of 1-pentene over Raney nickel is accompanied by some isomerization to both *E*- and *Z*-2-pentene.[2] The isomerized products are converted to pentane, but at a slower rate than is 1-pentene. Exchange of hydrogen atoms between the reactant and adsorbed hydrogen can be detected by observing exchange of deuterium for hydrogen. Allylic positions undergo such exchange particularly rapidly.[3] Both the isomerization and allylic hydrogen exchange can be explained by the intervention of the π-allyl intermediate **C**. If this intermediate, adds a hydrogen at the alternative end of the allyl system on being desorbed, an isomeric alkene is formed. Hydrogen exchange occurs if a hydrogen from the metal surface, rather than the original hydrogen, is involved in desorption.

Besides solid transition metals, certain soluble transition metal complexes are active hydrogenation catalysts.[4] The most commonly used example is tris(triphenylphosphine)chlororhodium, which is known as *Wilkinson's catalyst.*[5] This and related homogeneous catalysts usually minimize exchange and isomerization processes. Hydrogenation by homogeneous catalysts is believed to take place by initial formation of a π complex, followed by transfer of hydrogen from rhodium to carbon.

$$\text{Rh—H} + \text{RCH=CHR} \rightarrow \underset{\text{RCH=CHR}}{\text{RhH}} \longrightarrow \underset{\text{RCH}_2\text{CHR}}{\text{Rh}} \xrightarrow{H_2} \underset{\text{RCH}_2\text{CHR}}{\text{RhH}} \rightarrow \text{RCH}_2\text{CH}_2\text{R}$$

2. H. C. Brown and C. A. Brown, *J. Am. Chem. Soc.* **85**, 1005 (1963).
3. G. V. Smith and J. R. Swoap, *J. Org. Chem.* **31**, 3904 (1966).
4. A. J. Birch and D. H. Williamson, *Org. React.* **24**, 1 (1976); B. R. Jones, *Homogeneous Hydrogenation*, Wiley, New York, 1973.
5. J. A. Osborn, F. H. Jardine, J. F. Young, and G. Wilkinson, *J. Chem. Soc., A*, 1711 (1966).

The phosphine ligands serve both to provide a soluble complex and to adjust the reactivity of the metal center. Scheme 5.2 gives some examples of hydrogenations carried out with homogeneous catalysts. One potential advantage of homogeneous catalysts is the ability to achieve a high degree of selectivity among different functional groups. Entries 3 and 5 are examples of such selectivity.

The stereochemistry of reduction by homogeneous catalysts is often controlled by functional groups in the reactant. Homogeneous iridium catalysts have been found to be influenced not only by hydroxyl groups, but also by amide, ester, and ether substituents.

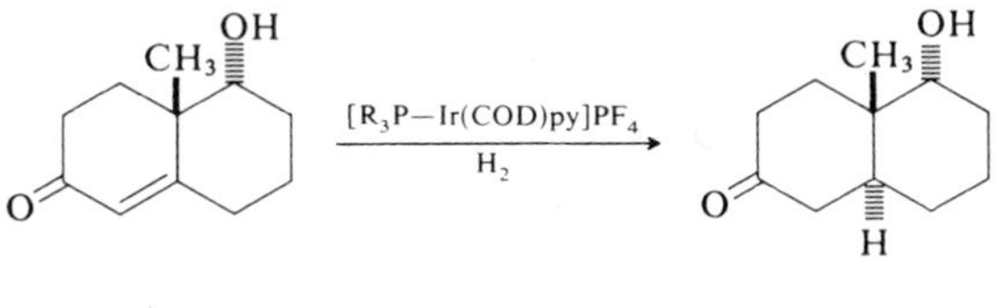

$[R_3P{-}Ir(COD)py]PF_4$, H_2 — Ref. 6

$[R_3P{-}Ir(COD)py]PF_4$, H_2 — Ref. 7

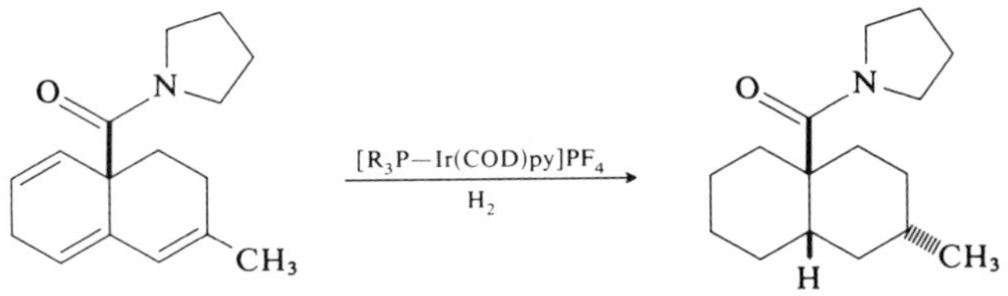

Delivery of hydrogen occurs *syn* to the functional group. Presumably, the stereoselectivity is the result of prior coordination of iridium by the functional group. The crucial property required for a catalyst to be stereodirective is that it be able to coordinate with both the directive group and the double bond and still accommodate the metal hydride bond necessary for hydrogenation. In the iridium catalyst illustrated above, the cyclooctadiene ligand (COD) in the catalysts is released upon coordination of the reactant.

A number of chiral ligands, especially phosphines, have been explored in order to develop highly enantioselective hydrogenation catalysts.[8] Some of the most successful catalysts are derived from chiral 1,1′-binaphthyldiphosphines. These ligands are chiral by virtue of the sterically restricted rotation of the two naphthyl rings.

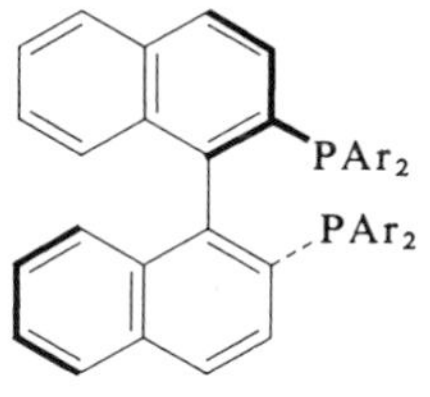

PAr_2

PAr_2

6. G. Stork and D. E. Kahne, *J. Am. Chem. Soc.* **105**, 1072 (1983).
7. A. G. Schultz and P. J. McCloskey, *J. Org. Chem.* **50**, 5905 (1985).
8. B. Bosnich and M. D. Fryzuk, *Top. Stereochem.* **12**, 119 (1981); W. S. Knowles, W. S. Chrisopfel, K. E. Koenig, and C. F. Hobbs, *Adv. Chem. Ser.* **196**, 325 (1982); W. S. Knowles, *Acc. Chem. Res.* **16**, 106 (1983).

Scheme 5.2. Homogeneous Catalytic Hydrogenation

1[a] $(Ph_3P)_3RhBr$, D_2; H^+, H_2O (56%)

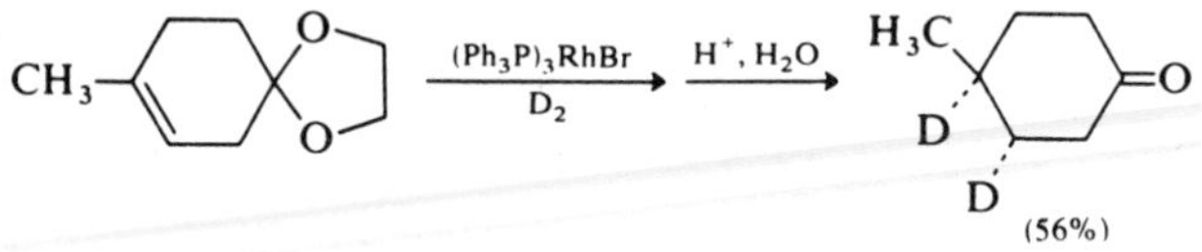

2[b] $(Ph_3P)_3RhCl$, H_2 (90%)

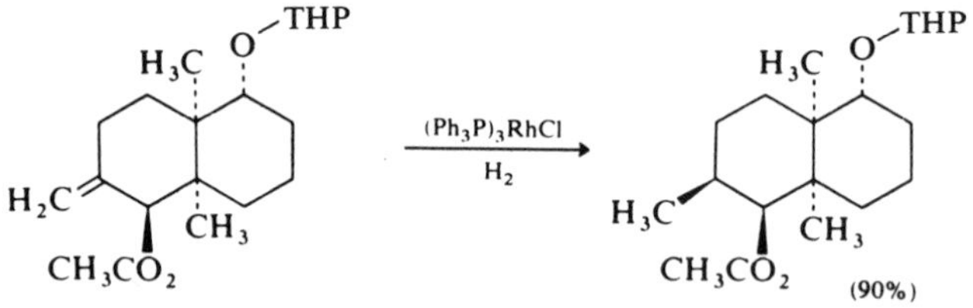

3[c] $(Ph_3P)_3RhCl$, H_2 (94%)

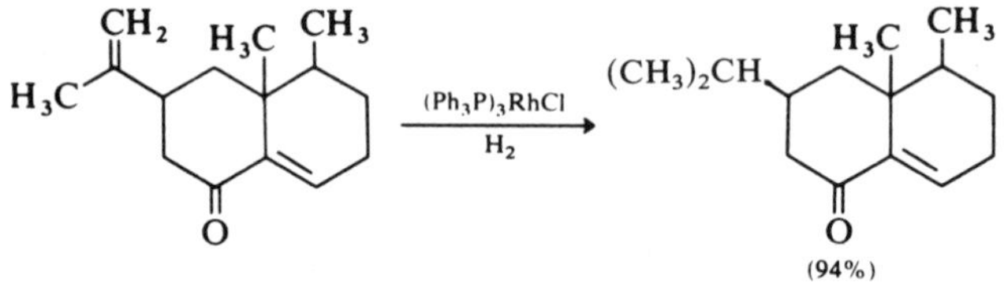

4[d] $CH_3O-C_6H_4-CH{=}CHNO_2 \xrightarrow[H_2]{(Ph_3P)_3RhCl} CH_3O-C_6H_4-CH_2CH_2NO_2$ (90%)

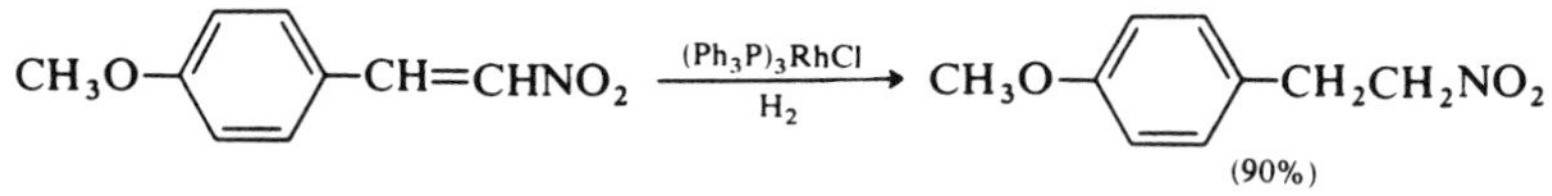

5[e] $(Ph_3P)_3RhCl$, H_2 (90–94%)

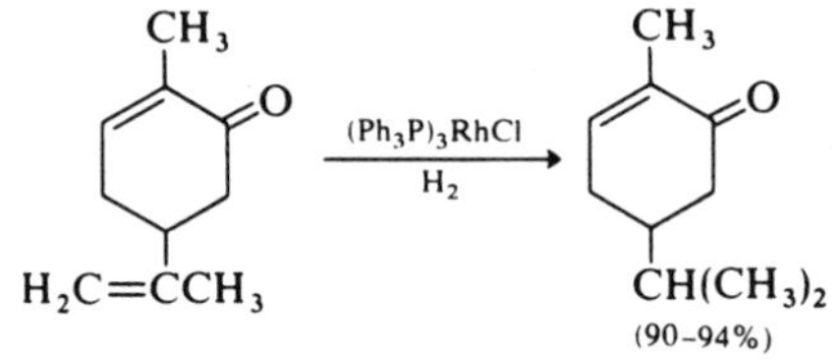

6[f] $[R_3R-Ir(COD)pyr]BF_4$ (67%)

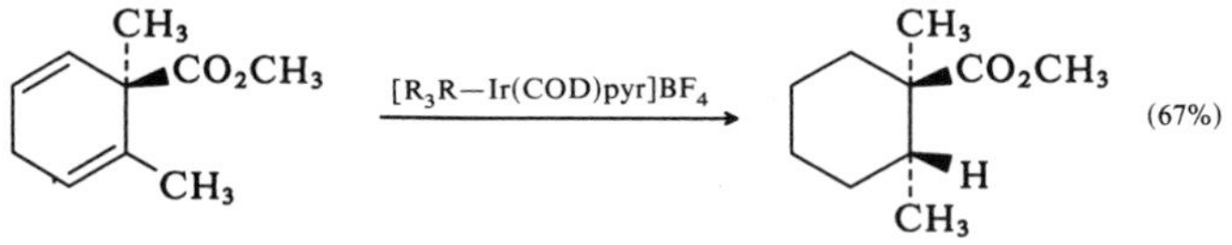

7[g] $[Rh(NBD)(diphosp)]BF_4$ (95%)

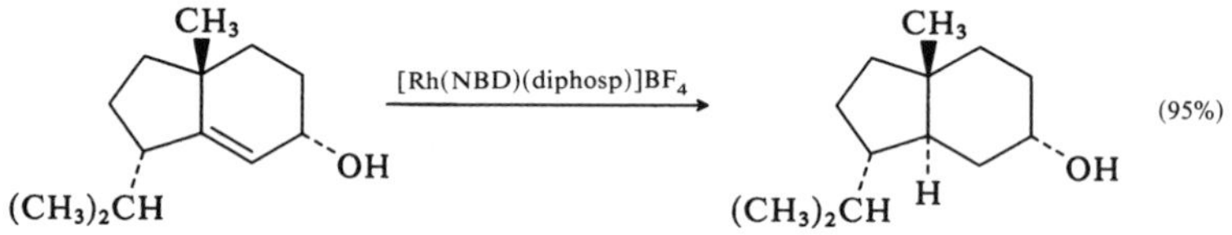

8[h] $[P_3R-Ir(COD)pyr]$ (100%)

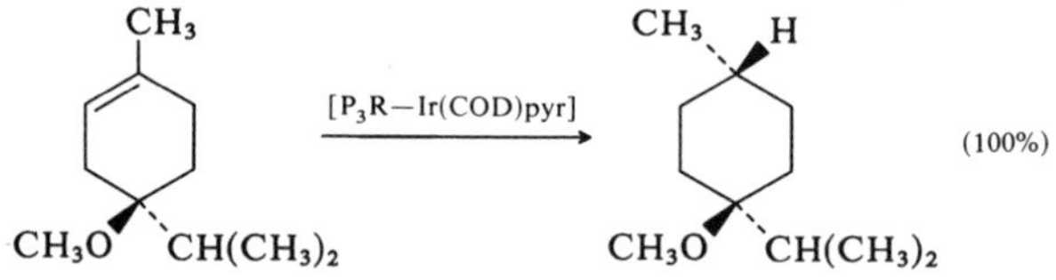

a. W. C. Agosta and W. L. Schreiber, *J. Am. Chem. Soc.* **93**, 3947 (1971).
b. E. Piers, W. de Waal, and R. W. Britton, *J. Am. Chem. Soc.* **93**, 5113 (1971).
c. M. Brown and L. W. Piszkiewicz, *J. Org. Chem.* **32**, 2013 (1967).
d. R. E. Harmon, J. L. Parsons, D. W. Cooke, S. K. Gupta, and J. Schoolenberg, *J. Org. Chem.* **34**, 3684 (1969).
e. R. E. Ireland and P. Bey, *Org. Synth.* **53**, 63 (1973).
f. A. G. Schultz and P. J. McCloskey, *J. Org. Chem.* **50**, 5905 (1985).
g. D. A. Evans and M. M. Morrissey, *J. Am. Chem. Soc.* **106**, 3866 (1984).
h. R. H. Crabtree and M. W. Davis, *J. Org. Chem.* **51**, 2655 (1986).

Ruthenium complexes containing this phosphine ligand are able to reduce a variety of double bonds with enantiomeric excesses above 95%. In order to achieve high enantioselectivity, the compound to be reduced must show a strong preference for a specific orientation when complexed with the catalyst. This ordinarily requires the presence of a functional group that can coordinate with the metal. The ruthenium binaphthyldiphosphine catalyst has been used successfully with unsaturated amides,[9] allylic and homoallylic alcohols,[10] and unsaturated carboxylic acids.[11]

An especially important case is the enantioselective hydrogenation of α-amidoacrylic acids, which leads to α-aminoacids. A particularly detailed study has been carried out on the mechanism of reduction of methyl Z-α-acetamidocinnamate by a rhodium catalyst with a chiral diphosphine.[12] It has been concluded that the reactant can bind reversibly to the catalyst to give either of two complexes. Addition of hydrogen at rhodium then leads to a reactive rhodium hydride and eventually to product. Interestingly, the addition of hydrogen occurs most rapidly in the minor isomeric complex, and the enantioselectivity is due to this kinetic preference.

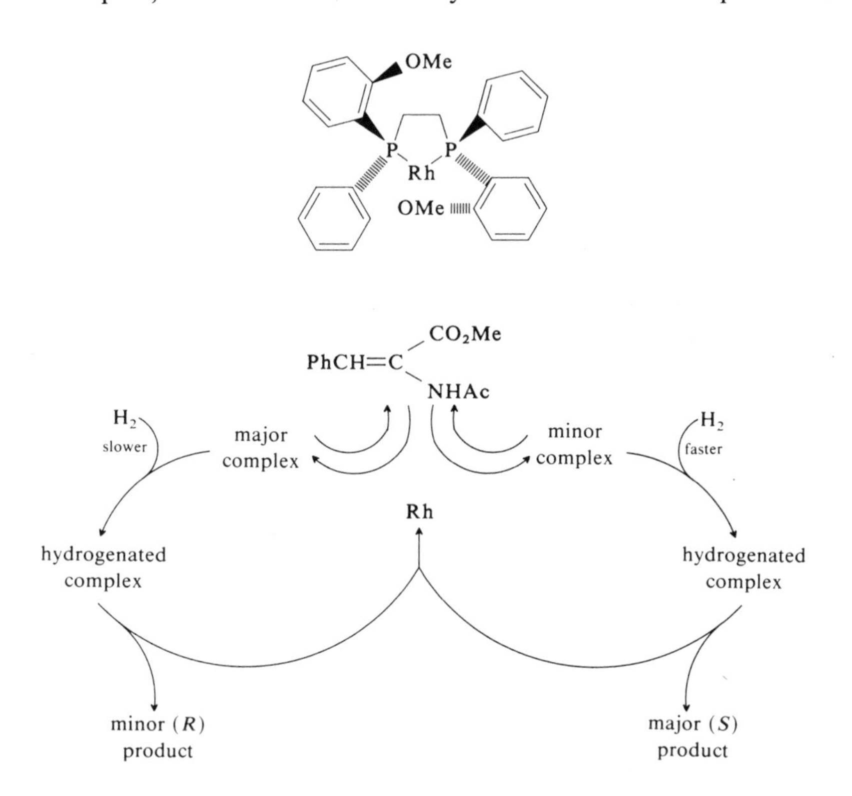

Table 5.1 gives the enantioselectivity of some other hydrogenations of unsaturated acids and amides.

9. R. Noyori, M. Ohta, Y. Hsiao, M. Kitamura, T. Ohta, and H. Takaya, *J. Am. Chem. Soc.* **108**, 7117 (1986).
10. H. Takaya, T. Ohta, N. Sayo, H. Kumobayashi, S. Akutagawa, S. Inoue, I. Kasahara, and R. Noyori, *J. Am. Chem. Soc.* **109**, 1596 (1987).
11. T. Ohta, H. Takaya, M. Kitamura, K. Nagai, and R. Noyori, *J. Org. Chem.* **52**, 3176 (1987).
12. C. R. Landis and J. Halpern, *J. Am. Chem. Soc.* **109**, 1746 (1987).

Table 5.1. Enantiomeric Excess (E.E.) for Asymmetric Catalytic Hydrogenation of Substituted Acrylic Acids

Substrate	Catalyst	Product	Configuration	% E.E.	Ref.
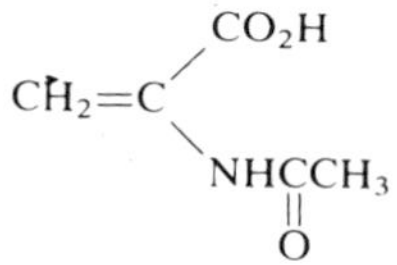 $CH_2{=}C(CO_2H)NHC(O)CH_3$	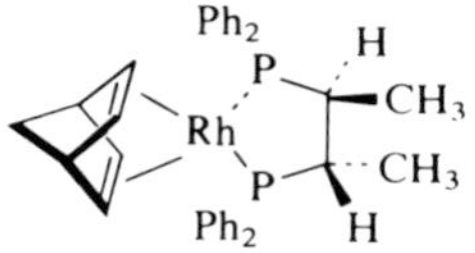 Norbornadiene–Rh complex with chelating Ph_2P–PPh_2 ligand (ring substituents H, CH_3, CH_3, H)	$CH_3CH(CO_2H)NHC(O)CH_3$	*R*	90	a
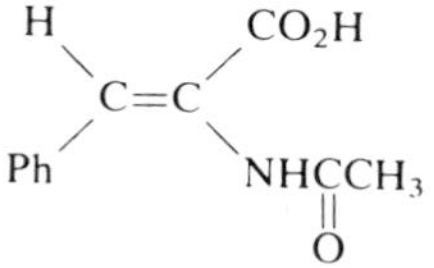 $(H)(Ph)C{=}C(CO_2H)NHC(O)CH_3$	Same as above	$PhCH_2CH(CO_2H)NHC(O)CH_3$	*R*	95	a
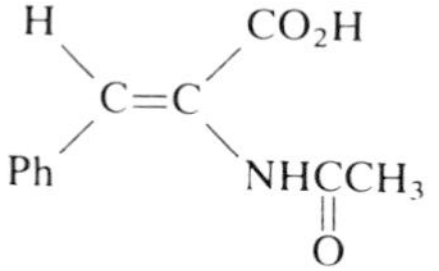 $(H)(Ph)C{=}C(CO_2H)NHC(O)CH_3$	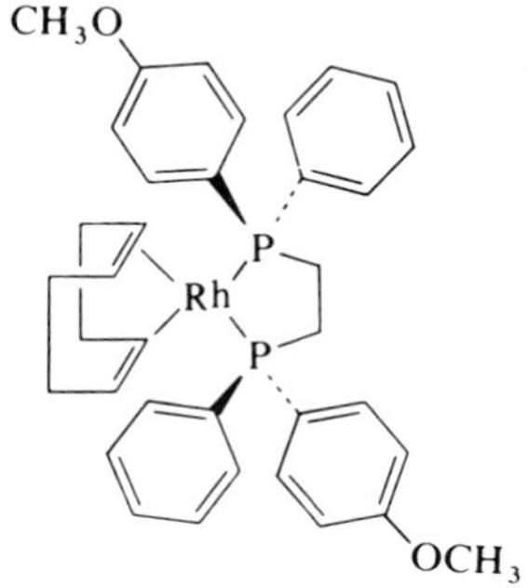 Cyclooctadiene–Rh complex with P–CH_2CH_2–P ligand bearing phenyl and CH_3O-substituted phenyl groups (CH_3O, OCH_3)	$PhCH_2CH(CO_2H)NHC(O)CH_3$	*S*	94	b
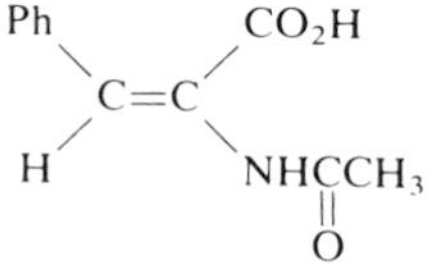 $(Ph)(H)C{=}C(CO_2H)NHC(O)CH_3$	Same as above	$PhCH_2CH(CO_2H)NHC(O)CH_3$	*S*	47	b

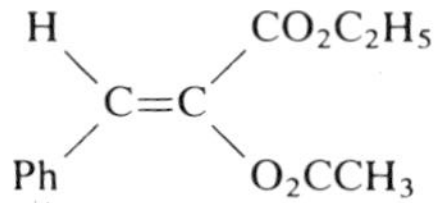	Same as above	$PhCH_2CHCO_2C_2H_5$ O_2CCH_3	*S*	90	b
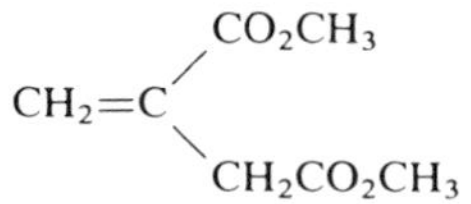	Same as above	$CH_3CHCO_2CH_3$ $CH_2CO_2CH_3$	*R*	88	c
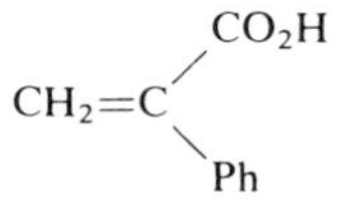	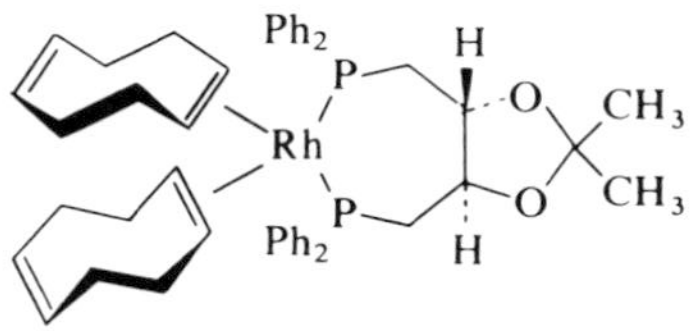	CH_3CHCO_2H Ph	*S*	64	d
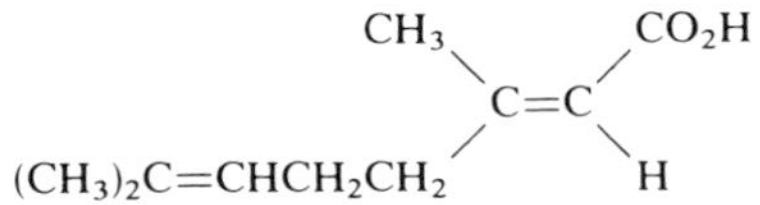	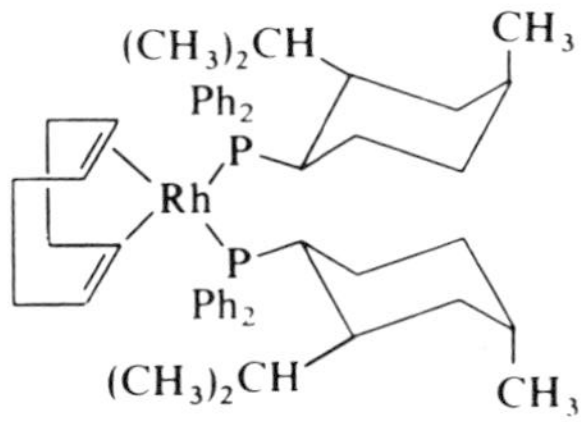	CH_3 $(CH_3)_2C{=}CHCH_2CH_2CHCH_2CO_2H$	*R*	79	e

a. M. D. Fryzuk and B. Bosnich, *J. Am. Chem. Soc.* **99**, 6262 (1977).
b. B. D. Vineyard, W. S. Knowles, M. J. Sabacky, G. L. Bachman, and D. J. Weinkauff, *J. Am. Chem. Soc.* **99**, 5946 (1977).
c. W. C. Christopfel and B. D. Vineyard, *J. Am. Chem. Soc.* **101**, 4406 (1979).
d. H. B. Kagan and T.-P. Dang, *J. Am. Chem. Soc.* **94**, 6429 (1972).
e. D. Valentine, Jr., K. K. Johnson, W. Priester, R. C. Sun, K. Toth, and G. Saucy, *J. Org. Chem.* **45**, 3698 (1980).

Partial reduction of alkynes to Z-alkenes is another important synthetic application of selective hydrogenation catalysts. The transformation can be carried out under heterogeneous or homogeneous conditions. Among heterogeneous catalysts, the one that is most successful is *Lindlar's catalyst*, which is a lead-modified palladium-$CaCO_3$ catalyst.[13] A nickel-boride catalyst prepared by reduction of nickel salts with sodium borohydride is also useful.[14] A homogeneous rhodium catalyst has also been reported to show good selectivity.[15]

Many other functional groups are also reactive under conditions of catalytic hydrogenation. The reduction of nitro compounds to amines, for example, usually proceeds very rapidly. Ketones, aldehydes, and esters can all be reduced to alcohols, but in most cases these reactions are slower than alkene reductions. For most synthetic applications, the hydride transfer reagents to be discussed in Section 5.2 are used for reduction of carbonyl groups. Amides and nitriles can be reduced to amines. Hydrogenation of amides requires extreme conditions and is seldom used in synthesis, but reduction of nitriles is quite useful. Scheme 5.3 gives a summary of the approximate conditions for catalytic reduction of some common functional groups.

Scheme 5.3. Conditions for Catalytic Reduction of Various Functional Groups[a]

Functional group	Reduction product	Common catalysts	Typical reaction conditions
>C=C<	−CH−CH− (H, H added)	Pd, Pt, Ni, Ru, Rh	Rapid at R.T. and 1 atm except for highly substituted or hindered cases
−C≡C−	(H)C=C(H) (cis)	Pd	R.T. and low pressure, quinoline or lead added to deactivate catalyst
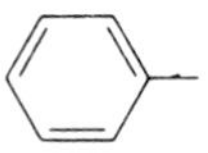 phenyl (benzene ring)	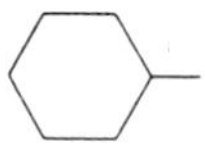 cyclohexyl	Rh, Pt	Moderate pressure (5–10 atm), 50–100°C
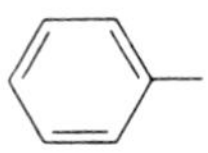 phenyl (benzene ring)	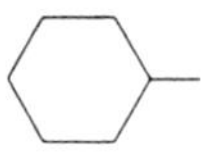 cyclohexyl	Ni, Pd	High pressure (100–200 atm), 100–200°C
RC(=O)R	RCH(OH)R	Pt, Ru	Moderate rate at R.T. and 1–4 atm, acid-catalyzed
RC(=O)R	RCH(OH)R	Cu–Cr, Ni	High pressure, 50–100°C

13. H. Lindlar and R. Dubuis, *Org. Synth.* **V**, 880 (1973).
14. H. C. Brown and C. A. Brown, *J. Am. Chem. Soc.* **85**, 1005 (1963); E. J. Corey, K. Achiwa, and J. A. Katzenellenbogen, *J. Am. Chem. Soc.* **91**, 4318 (1969).
15. R. R. Schrock and J. A. Osborn, *J. Am. Chem. Soc.* **98**, 2143 (1976).

Scheme 5.3—*continued*

Functional group	Reduction product	Common catalysts	Typical reaction conditions
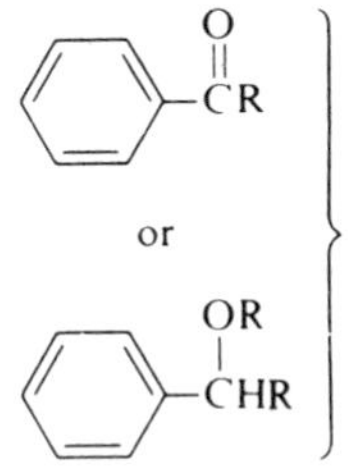 $C_6H_5C(=O)R$ or $C_6H_5CH(OR)R$	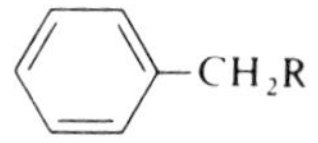 $C_6H_5CH_2R$	Pd	R.T., 1–4 atm, acid-catalyzed
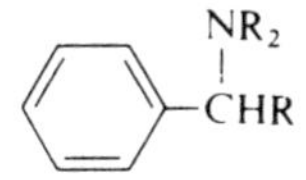 $C_6H_5CH(NR_2)R$	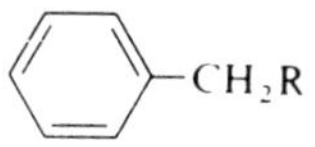 $C_6H_5CH_2R$	Pd, Ni	50–100°C, 1–4 atm
$RC(=O)Cl$	$RC(=O)H$	Pd	R.T., 1 atm, quinoline or other catalyst moderator used
$RC(=O)OH$	RCH_2OH	Pd, Ni, Ru	Very strenuous conditions required
$RC(=O)OR$	RCH_2OH	Cu–Cr, Ni	200°C, high pressure
$RC{\equiv}N$	RCH_2NH_2	Ni, Rh	50–100°C, usually high pressure, NH_3 added to increase yield of primary amine
$RC(=O)NH_2$	RCH_2NH_2	Cu–Cr	Very strenuous conditions required
RNO_2	RNH_2	Pd, Ni, Pt	R.T., 1–4 atm.
$RC(=NR)R$	R_2CHNHR	Pd, Pt	R.T., 4–100 atm
R–Cl, R–Br, R–I	R–H	Pd	Order of reactivity: I > Br > Cl > F, bases promote reaction for R = alkyl
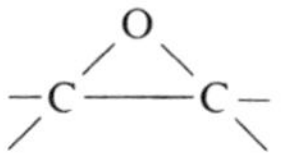 epoxide ($-C(-O-)C-$)	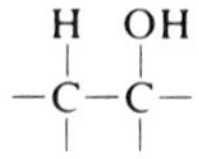 $-CH-C(OH)-$	Pt, Pd	Proceeds slowly at R.T., 1–4 atm, acid-catalyzed

a. General references: M. Freifelder, *Catalytic Hydrogenation in Organic Synthesis: Procedures and Commentary*, Wiley-Interscience, New York, 1978; P. N. Rylander, *Catalytic Hydrogenation in Organic Syntheses*, Academic Press, New York, 1979; R. L. Augustine, *Catalytic Hydrogenation*, Marcel Dekker, New York, 1965.

In certain cases, functional groups can be entirely removed and replaced by hydrogen. This is called *hydrogenolysis*. For example, aromatic halogen substituents are usually removed by hydrogenation over transition metal catalysts. Aliphatic halogens are somewhat less reactive, but hydrogenolysis is promoted by base.[16] The most useful class of hydrogenolysis reactions involves removal of functional groups at benzylic and allylic positions.[17]

$$C_6H_5{-}CH_2OR \xrightarrow{H_2,\ Pd} C_6H_5{-}CH_3 + HOR$$

This facile cleavage of the benzyl-oxygen bond has made the benzyl group a useful "protecting group" in multistep synthesis. A particularly important example is the use of the carbobenzyloxy group in peptide synthesis. The protecting group is removed by hydrogenolysis. The substituted carbamic acid generated by the hydrogenolysis decarboxylates spontaneously to provide the amine.

$$PhCH_2OC(=O)NHR \rightarrow PhCH_3 + HOC(=O)NHR \rightarrow CO_2 + H_2NR$$

5.1.2. Other Hydrogen-Transfer Reagents

Catalytic hydrogenation transfers the elements of molecular hydrogen through a series of complexes and intermediates. Diimide, HN=NH, an unstable hydrogen donor which can only be generated *in situ*, finds some specialized application in the reduction of carbon–carbon double bonds. Simple alkenes are reduced efficiently by diimide, but other easily reduced functional groups, such as nitro and cyano, are unaffected. The mechanism of the reaction is pictured as a transfer of hydrogen via a nonpolar cyclic transition state:

$$HN{=}NH + {>}C{=}C{<} \longrightarrow [\,{-}C{-}C{-}\ \cdots H\ H \cdots N{=}N\,] \longrightarrow H{-}C{-}C{-}H + N{\equiv}N$$

In agreement with this mechanism is the fact that the stereochemistry of addition is *syn*.[18] The rate of reaction with diimide is influenced by torsional and angle strain in the alkene. More strained double bonds react at accelerated rates.[19] For example,

16. A. R. Pinder, *Synthesis*, 425 (1980).
17. W. H. Hartung and R. Simonoff, *Org. React.* **7**, 263 (1953); P. N. Rylander, *Catalytic Hydrogenation over Platinum Metals*, Academic Press, New York, 1967, Chapter 25; P. N. Rylander, *Catalytic Hudrogenation in Organic Synthesis*, Academic Press, New York, 1979, Chapter 15; P. N. Rylander, *Hydrogenation Methods*, Academic Press, Orlando, Florida, 1985, Chapter 13.
18. E. J. Corey, D. J. Pasto, and W. L. Mock, *J. Am. Chem. Soc.* **83**, 2957 (1961).
19. E. W. Garbisch, Jr., S. M. Schildcrout, D. B. Patterson, and C. M. Sprecher, *J. Am. Chem. Soc.* **87**, 2932 (1965).

the more strained *trans* double bond is selectively reduced in *Z,E*-1,5-cyclodecadiene.

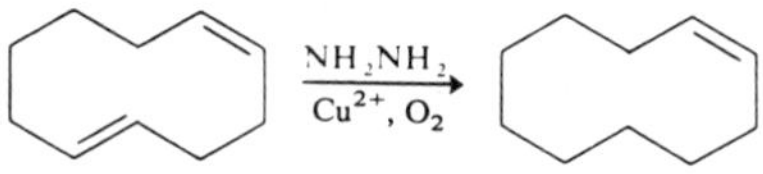

$\xrightarrow[Cu^{2+}, O_2]{NH_2NH_2}$ Ref. 20

Scheme 5.4. Reductions with Diimide

1[a] $CH_2{=}CHCH_2OH \xrightarrow[RCO_2H,\ 25°C]{\overset{+}{Na}\,^{-}O_2C{-}N{=}N{-}CO_2^{-}\,\overset{+}{Na}} CH_3CH_2CH_2OH$ (78%)

2[b] $(CH_2{=}CHCH_2)_2S \xrightarrow{C_7H_7SO_2NHNH_2,\ heat} (CH_3CH_2CH_2)_2S$ (93–100%)

3[c] $\xrightarrow{NH_2NH_2,\ O_2,\ Cu(II)}$

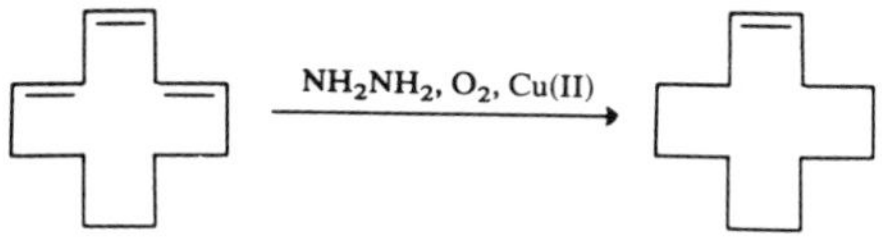

4[d] $O_2N{-}C_6H_4{-}CH{=}CHCO_2H \xrightarrow[NH_2OH]{NH_2OSO_3^-} O_2N{-}C_6H_4{-}CH_2CH_2CO_2H$ (87%)

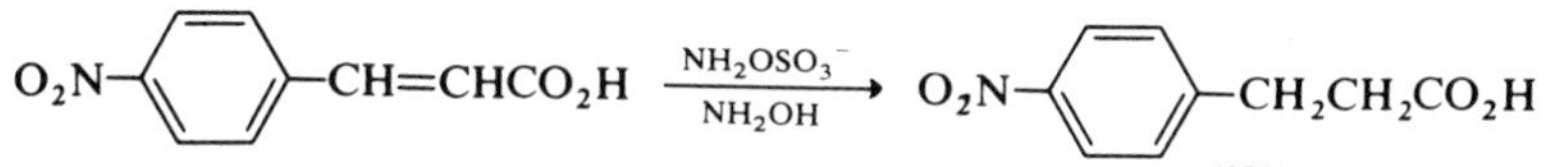

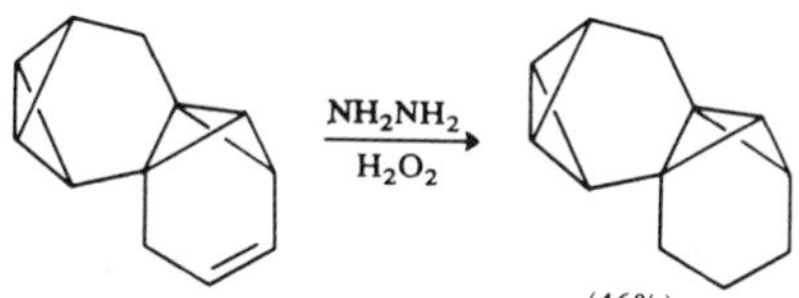

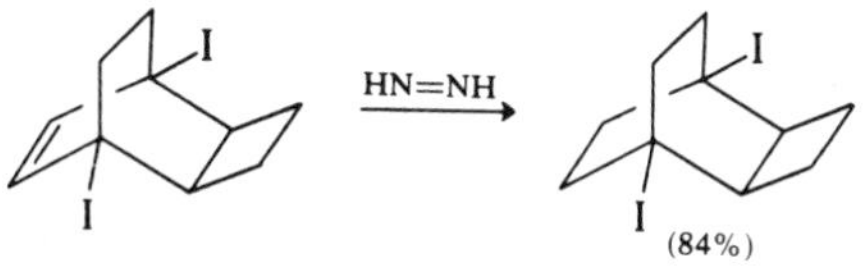

5[e] $\xrightarrow[H_2O_2]{NH_2NH_2}$ (46%)

6[f] I, I $\xrightarrow{HN{=}NH}$ I, I (84%)

7[g] O, $CO_2C_2H_5$, NO_2 $\xrightarrow[MeOH,\ HOAc]{KO_2CN{=}NCO_2K}$ O, $CO_2C_2H_5$, NO_2 (95%)

8[h] O, PhS, N, N, $ArSO_2$ $\xrightarrow[THF,\ H_2O,\ NaOAc]{C_7H_7SO_2NHNH_2}$ O, PhS, N, N, $ArSO_2$ (99%)

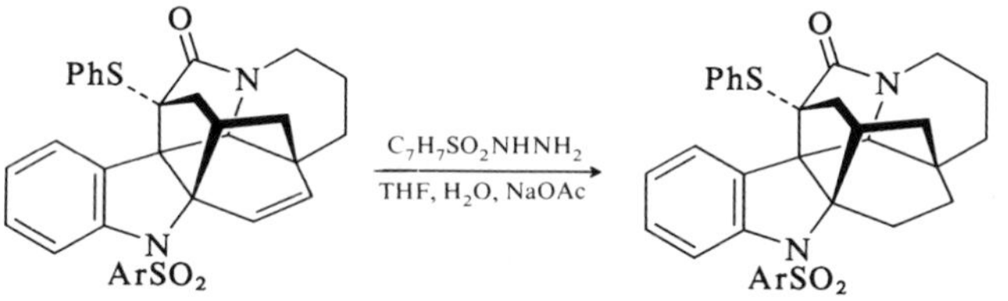

a. E. E. van Tamelen, R. S. Dewey, and R. J. Timmons, *J. Am. Chem. Soc.* **83**, 3725 (1961).
b. E. E. van Tamelen, R. S. Dewey, M. F. Lease, and W. H. Pirkle, *J. Am. Chem. Soc.* **83**, 4302 (1961).
c. M. Ohno and M. Okamoto, *Org. Synth.* **49**, 30 (1969).
d. W. Durckheimer, *Justus Liebigs Ann. Chem.* **721**, 240 (1969).
e. L. A. Paquette, A. R. Browne, E. Chamot, and J. F. Blount, *J. Am. Chem. Soc.* **102**, 643 (1980).
f. J. R. Wiseman and J. J. Vanderbilt, *J. Am. Chem. Soc.* **100**, 7730 (1978).
g. P. A. Grieco, R. Lis, R. E. Zelle, and J. Finn, *J. Am. Chem. Soc.* **108**, 5908 (1986).
h. P. Magnus, T. Gallagher, P. Brown, and J. C. Huffman, *J. Am. Chem. Soc.* **106**, 2105 (1984).

20. J. G. Traynham, G. R. Franzen, G. A. Kresel, and D. J. Northington, Jr., *J. Org. Chem.* **32**, 3285 (1967).

Diimide selectively reduces terminal over internal double bonds in polyunsaturated systems.[21] There are several methods for generation of diimide, and they are summarized in Scheme 5.4.

5.2. Group III Hydride-Donor Reagents

5.2.1. Reduction of Carbonyl Compounds

Most reductions of carbonyl compounds are done with reagents that transfer a hydride from boron or aluminum. The numerous reagents of this type that are available provide a considerable degree of selectivity and stereochemical control. Sodium borohydride and lithium aluminum hydride are the most widely used of these reagents. Sodium borohydride is a mild reducing agent which reacts rapidly with aldehydes and ketones but quite slowly with esters. Lithium aluminum hydride

Table 5.2. Relative Reactivity of Hydride-Donor Reducing Agents

	Reduction products[a]						
Hydride donors	Iminium ion	Acyl halide	Aldehyde	Ketone	Ester	Amide	Carboxylate salt
$LiAlH_4$[b]	Amine	Alcohol	Alcohol	Alcohol	Alcohol	Amine	Alcohol
$LiAlH_2(OCH_2CH_2OCH_3)_2$[c]		Alcohol	Alcohol	Alcohol	Alcohol	Amine	Alcohol
$LiAlH[OC(CH_3)_3]_3$[d]		Aldehyde[e]	Alcohol	Alcohol	Alcohol[f]	Aldehyde[f]	NR
$NaBH_4$[b]	Amine		Alcohol	Alcohol	Alcohol[f]	NR	NR
$NaBH_3CN$[g]	Amine		Alcohol[f]	NR	NR	NR	NR
B_2H_6[h]			Alcohol	Alcohol	NR	Amine	Alcohol[i]
AlH_3[j]		Alcohol	Alcohol	Alcohol	Alcohol	Amine	Alcohol
$[(CH_3)_2CHCH(CH_3)-]_2BH$[k]			Alcohol	Alcohol	NR	Aldehyde[e]	NR
$[(CH_3)_2CHCH_2-]_2AlH$[l]			Alcohol	Alcohol	Aldehyde[e]	Aldehyde[e]	Alcohol

a. Products shown are the usual products of synthetic operations. Where no entry is given, the combination has not been studied or is not of major synthetic utility.
b. See the general references at the end of the chapter.
c. J. Malék, *Org. React.* **34**, 1 (1985); **36**, 249 (1989).
d. H. C. Brown and R. F. McFarlin, *J. Am. Chem. Soc.* **78**, 752 (1956); **80**, 5372 (1958); H. C. Brown and B. C. Subba Rao, *J. Am. Chem. Soc.* **80**, 5377 (1958); H. C. Brown and A. Tsukamoto, *J. Am. Chem. Soc.* **86**, 1089 (1964).
e. Reaction must be controlled by use of a stoichiometric amount of reagent and low temperature.
f. Reaction occurs slowly.
g. C. F. Lane, *Synthesis*, 135 (1975).
h. H. C. Brown, P. Heim, and N. M. Yoon, *J. Am. Chem. Soc.* **92**, 1637 (1970); N. M. Yoon, C. S. Park, H. C. Brown, S. Krishnamurthy, and T. P. Stocky, *J. Org. Chem.* **38**, 2786 (1973); H. C. Brown and P. Heim, *J. Org. Chem.* **38**, 912 (1973).
i. Reaction occurs via the triacyl borate.
j. H. C. Brown and N. M. Yoon, *J. Am. Chem. Soc.* **88**, 1464 (1966).
k. H. C. Brown, D. B. Bigley, S. K. Arora, and N. M. Yoon, *J. Am. Chem. Soc.* **92**, 7161 (1970); H. C. Brown and V. Varma, *J. Org. Chem.* **39**, 1631 (1974).
l. E. Winterfeldt, *Synthesis*, 617 (1975); H. Reinheckel, K. Haage, and D. Jahnke, *Organomet. Chem. Res.* **4**, 47 (1969); N. M. Yoon and Y. S. Gyoung, *J. Org. Chem.* **50**, 2443 (1985).

21. E. J. Corey, H. Yamamoto, D. K. Herron, and K. Achiwa, *J. Am. Chem. Soc.* **92**, 6635 (1970); E. J. Corey and H. Yamamoto, *J. Am. Chem. Soc.* **92**, 6636, 6637 (1970).

is a much more powerful hydride-donor reagent. It will rapidly reduce esters, acids, nitriles, and amides, as well as aldehydes and ketones. Neither sodium borohydride nor lithium aluminum hydride reacts with isolated carbon-carbon double bonds. The reactivity of these reagents and some related reducing reagents is summarized in Table 5.2.

The mechanism by which the group III hydrides effect reduction involves nucleophilic transfer of hydride to the carbonyl group. Activation of the carbonyl group by coordination with a metal cation is probably involved under most conditions.

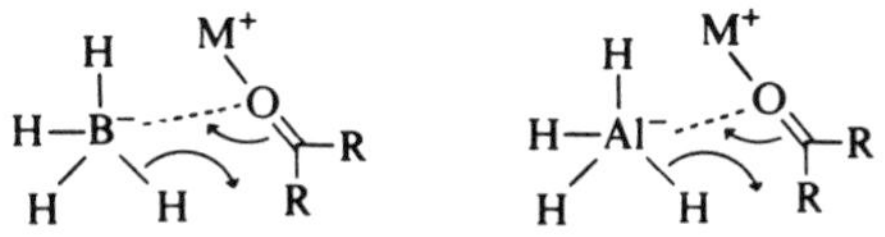

Since all four of the hydrides can eventually be transferred, there are actually several distinct reducing agents functioning during the course of the reaction.[22] While this somewhat complicates interpretation of rates and stereoselectivity, it has not detracted from the synthetic utility of these reagents. Reduction with $NaBH_4$ is usually done in aqueous or alcoholic solution, and the alkoxyboranes formed as intermediates are rapidly solvolyzed.

$$BH_4^- + R_2CO \rightarrow R_2CHO\bar{B}H_3$$

$$R_2CHO\bar{B}H_3 + R_2CO \rightarrow [R_2CHO]_2\bar{B}H_2$$

$$[R_2CHO]_2\bar{B}H_2 + R_2CO \rightarrow [R_2CHO]_3\bar{B}H$$

$$[R_2CHO]_3\bar{B}H + R_2CO \rightarrow [R_2CHO]_4\bar{B}$$

The mechanism for reduction by $LiAlH_4$ is very similar. However, since $LiAlH_4$ reacts very rapidly with protic solvents to release molecular hydrogen, reductions with this reagent must be carried out in aprotic solvents, usually ether or tetrahydrofuran. The products are liberated by hydrolysis of the aluminum alkoxide at the end of the reaction.

Hydride reduction of esters to alcohols involves elimination steps in addition to hydride transfer:

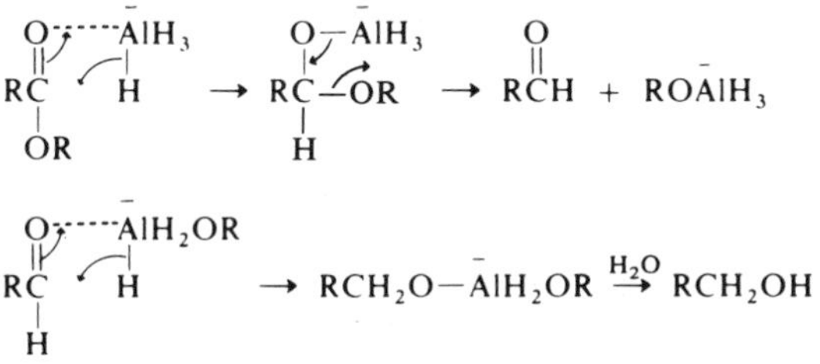

$$\rightarrow RCH + RO\bar{A}lH_3$$

$$\rightarrow RCH_2O-\bar{A}lH_2OR \xrightarrow{H_2O} RCH_2OH$$

22. B. Rickborn and M. T. Wuesthoff, *J. Am. Chem. Soc.* **92**, 6894 (1970).

Amides are reduced to amines because the nitrogen is a poorer leaving group than oxygen at the intermediate stage of the reduction. Primary and secondary amides are rapidly deprotonated by the strongly basic $LiAlH_4$, so the addition step involves the conjugate base.

$$RC(=O)\bar{N}H + H-\bar{A}lH_3 \rightarrow RCH(O-\bar{A}lH_3)\bar{N}H \rightarrow RC(H)=NH + H-AlH_2O^- \rightarrow RCH_2\bar{N}AlH_2O^-(H) \xrightarrow{H_2O} RCH_2NH_2$$

Reduction of amides is an important method of synthesis of amines:

$$C_6H_{11}CON(CH_3)_2 \xrightarrow[\text{35°C, 15 h}]{LiAlH_4,\ \text{ether,}} C_6H_{11}CH_2N(CH_3)_2 \quad (88\%) \qquad \text{Ref. 23}$$

$$\text{5,5-dimethyl-2-pyrrolidinone} \xrightarrow[\text{65°C, 8 h}]{LiAlH_4,\ \text{THF,}} \text{2,2-dimethylpyrrolidine} \quad (67\text{–}79\%) \qquad \text{Ref. 24}$$

Several factors affect the reactivity of the boron and aluminum hydrides. These include the nature of the metal cation present and the ligands, in addition to hydride, in the complex hydride. Some of these effects can be illustrated by considering the reactivity of ketones and aldehydes toward various hydride-transfer reagents. Comparison of $LiAlH_4$ and $NaAlH_4$ has shown the former to be more reactive.[25] This can be attributed to the greater Lewis acid strength and hardness of the lithium cation. Substances that complex metal cations decrease the rate of reduction because they prevent the cation from activating the carbonyl oxygen.[26]

Both $LiBH_4$ and $Ca(BH_4)_2$ are more reactive than sodium borohydride. This enhanced reactivity is due to the greater Lewis acid complexing power of Li^+ and Ca^{2+}, compared with Na^+. Both of these reagents can reduce esters and lactones efficiently.

$$3\text{-}NC\text{-}C_6H_4CO_2C_2H_5 \xrightarrow{Ca(BH_4)_2} 3\text{-}NC\text{-}C_6H_4CH_2OH \quad (70\%) \qquad \text{Ref. 27}$$

$$\gamma\text{-}C_7H_{15}\text{-butyrolactone} \xrightarrow{LiBH_4} C_7H_{15}CH(OH)CH_2CH_2CH_2OH \quad (45\%) \qquad \text{Ref. 28}$$

23. A. C. Cope and E. Ciganek, *Org. Synth.* **IV**, 339 (1963).
24. R. B. Moffett, *Org. Synth.* **IV**, 354 (1963).
25. E. C. Ashby and J. R. Boone, *J. Am. Chem. Soc.* **98**, 5524 (1976).
26. H. Handel and J.-L. Pierre, *Tetrahedron Lett.*, 741 (1976); K. E. Wiegers and S. G. Smith, *J. Org. Chem.* **43**, 1126 (1978).
27. H. C. Brown, S. Narashimhan, and Y. M. Choi, *J. Org. Chem.* **47**, 4702 (1982).
28. K. Soai and S. Ookawa, *J. Org. Chem.* **51**, 4000 (1986).

An extensive series of aluminum hydrides in which one or more of the hydrides is replaced by an alkoxide ion have been prepared by addition of the calculated amount of the appropriate alcohol:

$$LiAlH_4 + 2\,ROH \rightarrow LiAlH_2(OR)_2 + 2\,H_2$$

$$LiAlH_4 + 3\,ROH \rightarrow LiAlH(OR)_3 + 3\,H_2$$

These reagents generally show increased solubility, particularly at low temperatures, in organic solvents and are useful in certain selective reductions.[29] Lithium tri-*t*-butoxyaluminum hydride and sodium bis(2-methoxyethoxy)aluminum hydride (Red-Al)[30] are examples of these types of reagents that have come into wide synthetic use. Their reactivity toward typical functional groups is included in Table 5.2. Sodium cyanoborohydride[31] is a useful derivative of sodium borohydride. The electron-attracting cyano substituent reduces reactivity, and only iminium groups are rapidly reduced by this reagent.

A group of alkylated borohydrides are also used as reducing agents. These reagents are much more bulky than the borohydride ion and therefore are more stereoselective in situations where steric factors are controlling.[32] These compounds are prepared by reaction of trialkylboranes with lithium, sodium, or potassium hydride.[33] Several of the compounds are available commercially under the trade name Selectrides.[34]

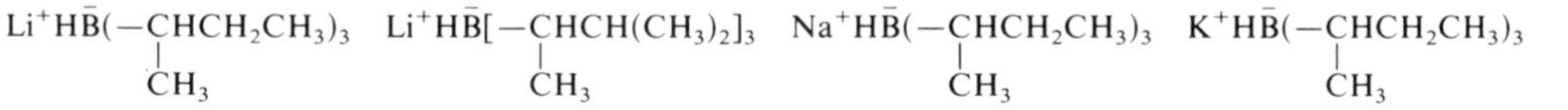

$$Li^+H\bar{B}(-CH(CH_3)CH_2CH_3)_3 \qquad Li^+H\bar{B}[-CH(CH_3)CH(CH_3)_2]_3 \qquad Na^+H\bar{B}(-CH(CH_3)CH_2CH_3)_3 \qquad K^+H\bar{B}(-CH(CH_3)CH_2CH_3)_3$$

L-Selectride LS-Selectride N-Selectride K-Selectride

Closely related to, but distinct from, the anionic boron and aluminum hydrides are the neutral boron (borane, BH_3) and aluminum (alane, AlH_3) hydrides. These molecules also contain hydrogen that can be transferred as hydride. Borane and alane differ from the anionic hydrides in being electrophilic species by virtue of having a vacant *p* orbital. Reduction by these molecules occurs by an intramolecular hydride transfer in a Lewis acid–base complex of the reactant and reductant.

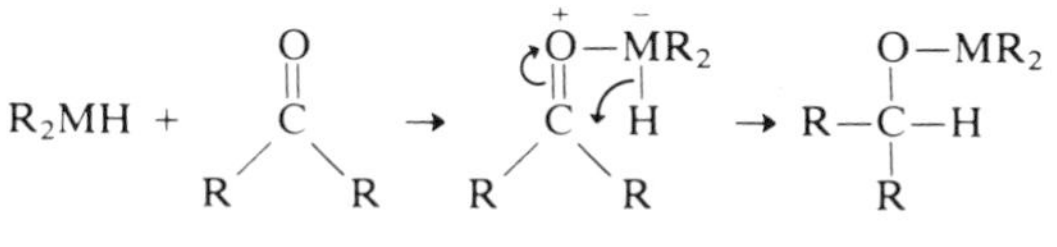

$$R_2MH + R_2C{=}O \rightarrow R_2C{=}\overset{+}{O}{-}\overset{-}{M}R_2(H) \rightarrow R{-}CH(R){-}O{-}MR_2$$

29. J. Malék and M. Cerny, *Synthesis*, 217 (1972); J. Malék, *Org. React.* **34**, 1 (1985); J. Malék, *Org. React.* **36**, 249 (1988).
30. Red-Al is a trademark of Aldrich Chemical Company.
31. C. F. Lane, *Synthesis*, 135 (1975).
32. H. C. Brown and S. Krishnamurthy, *J. Am. Chem. Soc.* **94**, 7159 (1972); S. Krishnamurthy and H. C. Brown, *J. Am. Chem. Soc.* **98**, 3383 (1976).
33. H. C. Brown, S. Krishnamurthy, and J. L. Hubbard, *J. Am. Chem. Soc.* **100**, 3343 (1978); J. A. Soderquist and J. Rivera, *Tetrahedron Lett.* **29**, 3195 (1988); J. L. Hubbard, *Tetrahedron Lett.* **29**, 3197 (1988).
34. Selectride is a trade name of the Aldrich Chemical Company.

Alkyl derivatives of boron and alane can function as reducing reagents in a similar fashion. Two reagents of this group are included in Table 5.2. Diisobutylaluminum hydride (DIBAL) is an especially useful reagent.

In synthesis, the principal factors that affect the choice of a reducing agent are selectivity among functional groups and stereochemistry. Let us consider first the matter of selectivity. Selectivity can involve two issues. It may be desired to effect a *partial reduction* of a particular functional group or it may be necessary to *reduce one group in preference to another.* The reagents in Table 5.2 are arranged in approximate order of decreasing reactivity.[35] The relative ordering of reducing agents with respect to particular functiinal groups can permit selection of the appropriate reagent.

One of the more difficult partial reductions to accomplish is the conversion of a carboxylic acid derivative to an aldehyde without overreduction to the alcohol. Several approaches have been used to achieve this selectivity. One is to replace some of the hydrogens in the group III hydride with more bulky groups, thus modifying reactivity by steric factors. Lithium tri-*t*-butoxyaluminum hydride is an example of this approach. Lithium tri-*t*-butoxyaluminum hydride can be used to reduce acid chlorides to aldehydes without overreduction to the alcohol.[36] The excellent solubility properties of sodium bis(2-methoxyethoxy)aluminum hydride make it a useful reagent for selective reductions. The reagent is soluble in toluene even at −70°C. It is possible to reduce esters to aldehydes and lactones to lactols with this reagent.

$CH_3O-C_6H_4-CH_2CH_2CO_2CH_3 \xrightarrow[\text{HN(CH}_2\text{CH}_2)_2\text{NCH}_3]{NaAlH_2(OCH_2CH_2OCH_3)_2} CH_3O-C_6H_4-CH_2CH_2CH{=}O$

Ref. 37

Lactone (THPO, $(CH_2)_4CO_2C(CH_3)_3$, OTHP) $\xrightarrow{NaAlH_2(OCH_2CH_2OCH_3)_2}$ lactol (OH, THPO, $(CH_2)_4CO_2C(CH_3)_3$, OTHP)

Ref. 38

Probably the most widely used reagent for partial reduction of esters and lactones at the present time is diisobutylaluminum hydride.[39] By use of a controlled amount of the reagent at low temperature, partial reduction can be reliably achieved. The selectivity results from the relative stability of the intermediate that is formed.

35. For more complete discussion of functional group selectivity of hydride reducing agents, see E. R. H. Walter, *Chem. Soc. Rev.* **5**, 23 (1976).
36. H. C. Brown and B. C. Subba Rao, *J. Am. Chem. Soc.* **80**, 5377 (1958).
37. R. Kanazawa and T. Tokoroyama, *Synthesis*, 526 (1976).
38. H. Disselnkötter, F. Liob, H. Oedinger, and D. Wendisch, *Liebigs Ann. Chem.*, 150 (1982).
39. F. Winterfeldt, *Synthesis*, 617 (1975); N. M. Yoon and Y. G. Gyoung, *J. Org. Chem.* **50**, 2443 (1985).

The aldehyde is not liberated until the hydrolytic workup and is therefore not subject to overreduction. At higher temperatures, where the intermediate undergoes elimination, diisobutylaluminum hydride reduces esters to primary alcohols.

CH_3O, CH_3O, N, C_2H_5, $CH_2CO_2C_2H_5$ $\xrightarrow[-60°C,\ 2\ h]{(i\text{-}Bu)_2AlH,\ toluene}$ CH_3O, CH_3O, N, C_2H_5, $CH_2CH{=}O$ (83%) Ref. 40

O, O, $CH{=}CHCHC_5H_{11}$, OTHP, OTHP $\xrightarrow[-60°C]{(i\text{-}Bu)_2AlH,\ toluene}$ OH, O, $CH{=}CHCHC_5H_{11}$, OTHP, OTHP Ref. 41

Another useful approach to aldehydes is by partial reduction of nitriles to imines. The imines are then hydrolysed to the aldehyde. Diisobutylaluminum hydride seems to be the best reagent for this purpose.[42,43]

$$CH_3CH{=}CHCH_2CH_2CH_2C{\equiv}N \xrightarrow[2)\ H^+,\ H_2O]{1)\ (i\text{-}Bu)_2AlH} CH_3CH{=}CHCH_2CH_2CH_2CH{=}O \quad (64\%)$$

A second type of selectivity arises in the context of the need to reduce one functional group in the presence of another. If the group to be reduced is more reactive than the one to be left unchanged, it is simply a matter of choosing an appropriate reducing reagent. Sodium borohydride, for example, is very useful in this respect since it will reduce ketones and aldehydes much more rapidly than esters. Sodium cyanoborohydride is used to reduce imines to amines. This reagent is only reactive toward protonated imines. At pH 6–7, $NaBH_3CN$ is essentially unreactive toward carbonyl groups. When an amine and ketone are mixed together, equilibrium is established with the imine. At mildly acidic pH, only the protonated imine is reactive toward $NaBH_3CN$.[44]

$$R_2C{=}O + R'NH_2 + H^+ \rightleftarrows R_2C{=}\overset{+}{N}(H)R'$$

$$R_2C{=}\overset{+}{N}(H)R' + BH_3CN^- \rightarrow R_2CHNHR'$$

40. C. Szantay, L. Toke, and P. Kolonits, *J. Org. Chem.* **31**, 1447 (1966).
41. E. J. Corey, N. M. Weinshenker, T. K. Schaaf, and W. Huber, *J. Am. Chem. Soc.* **91**, 5675 (1969).
42. N. A. LeBel, M. E. Post, and J. J. Wang, *J. Am. Chem. Soc.* **86**, 3759 (1964).
43. R. V. Stevens and J. T. Lai, *J. Org. Chem.* **37**, 2138 (1972); S. Trofimenko, *J. Org. Chem.* **29**, 3046 (1964).
44. R. F. Borch, M. D. Berstein, and H. D. Durst, *J. Am. Chem. Soc.* **93**, 2897 (1971).

Diborane is another reagent that has a useful pattern of selectivity. It will reduce carboxylic acids to primary alcohols under mild conditions where esters are unchanged.[45] Nitro and cyano groups are also relatively unreactive toward diborane. The rapid reaction between carboxylic acids and diborane is the result of formation of a triacyloxyborane intermediate as a result of protonolysis of the B—H bonds. This compound is essentially a mixed anhydride of the carboxylic acid and boric acid, and the carbonyl groups have enhanced reactivity.

$$3\,RCO_2H + BH_3 \rightarrow (RCO_2)_3B + 3\,H_2$$

$$RC(=O)-O-B(O_2CR)_2 \leftrightarrow RC(=O)-\overset{+}{O}=\overset{-}{B}(O_2CR)_2$$

Diborane is also a useful reagent for reducing amides. Tertiary and secondary amides are easily reduced, but primary amides react only slowly.[46] The electrophilicity of diborane is involved in the reduction of amides. The boron complexes at the carbonyl oxygen, enhancing the reactivity of the carbonyl center.

Amides require vigorous reaction conditions for reduction by $LiAlH_4$ so that little selectivity can be achieved with this reagent. Diborane, however, reduces amides in the presence of ester and nitro groups.

Alane is also a useful group for reducing amides, and it too can be used to reduce amides to amines in the presence of ester groups.

Ref. 47

Again, the electrophilicity of alane is the basis for the selective reaction with the amide group. Alane is also useful for reducing azetidinones to azetidines. Most conventional reducing agents give ring-opened products.

Ref. 48

45. M. N. Yoon, C. S. Pak, H. C. Brown, S. Krishnamurthy, and T. P. Stocky, *J. Org. Chem.* **38**, 2786 (1973).
46. H. C. Brown and P. Heim, *J. Org. Chem.* **38**, 912 (1973).
47. S. F. Martin, H. Rüeger, S. A. Williamson, and S. Grejszczak, *J. Am. Chem. Soc.* **109**, 6124 (1987).
48. M. B. Jackson, L. N. Mander, and T. M. Spotswood, *Aust. J. Chem.* **36**, 779 (1983).

Another approach to reduction of an amide group in the presence of more easily reduced groups is to convert the amide to a more reactive species. One such method is conversion of the amide to an *O*-alkyl derivative with a positive charge on nitrogen.[49] This method has proven successful for tertiary and secondary but not primary amides.

$$\mathrm{RC(=O)NR_2 + Et_3O^+ \rightarrow RC(OEt)=\overset{+}{N}R_2}$$

$$\mathrm{RC(OEt)=\overset{+}{N}R_2 + NaBH_4 \rightarrow RCH_2NR_2}$$

Other compounds that can be derived from amides and are more reactive toward hydride reducing agents are α-alkylthioimmonium ions[50] and α-chloroimmonium ions.[51]

An important case of selectivity arises in the reduction of α,β-unsaturated carbonyl compounds. Reduction can occur at the carbonyl group, giving an allylic alcohol, or at the double bond, giving a saturated ketone. In the latter case, a second reduction may give the saturated alcohol. These alternative reaction modes are called 1,2- and 1,4-reduction, respectively. If a hydride is added at the β position, the initial product is an enolate. In protic solvents, this leads to the ketone, which can be reduced to the saturated alcohol. If hydride is added at the carbonyl group, the allylic alcohol is usually not susceptible to further reduction. Both $NaBH_4$ and $LiAlH_4$ have been observed to give both types of product, although the extent of reduction to saturated alcohol is usually greater with $NaBH_4$.[52]

1,2-reduction

$$\mathrm{R_2C{=}CHC(=O)R' + [H^-] \rightarrow R_2C{=}CHC(O^-)(H)R' \xrightarrow{H^+} R_2C{=}CHCH(OH)R'}$$

1,4-reduction leading to saturated alcohol

$$\mathrm{R_2C{=}CHC(=O)R' + [H^-] \rightarrow R_2CH{-}CH{=}C(O^-)R' \xrightarrow{H^+} R_2CHCH_2C(=O)R'}$$

$$\mathrm{R_2CHCH_2C(=O)R' + [H^-] \rightarrow R_2CHCH_2CH(O^-)R' \xrightarrow{H^+} R_2CHCH_2CH(OH)R'}$$

Several reagents have been developed which lead to exclusive 1,2- or 1,4-reduction. Use of $NaBH_4$ in combination with cerium chloride results in clean 1,2-reduction.[53] Diisobutylaluminum hydride[54] and the dialkylborane 9-BBN[55] also

49. R. F. Borch, *Tetrahedron Lett.*, 61 (1968).
50. S. Raucher and P. Klein, *Tetrahedron Lett.*, 4061 (1980); R. J. Sundberg, C. P. Walters, and J. D. Bloom, *J. Org. Chem.* **46**, 3730 (1981).
51. M. E. Kuehne and P. J. Shannon, *J. Org. Chem.* **42**, 2082 (1977).
52. M. R. Johnson and B. Rickborn, *J. Org. Chem.* **35**, 1041 (1970); W. R. Jackson and A. Zurqiyah, *J. Chem. Soc.*, 5280 (1965).
53. J.-L. Luche, *J. Am. Chem. Soc.* **100**, 2226 (1978); J.-L. Luche, L. Rodrequez-Hahn, and P. Crabbe, *J. Chem. Soc., Chem. Commun.*, 601 (1978).
54. K. E. Wilson, R. T. Seidner, and S. Masamune, *J. Chem. Soc., Chem. Commun.*, 213 (1970).
55. K. Krishnamurthy and H. C. Brown, *J. Org. Chem.* **42**, 1197 (1977).

give exclusive carbonyl reduction. Selective reduction of the carbon–carbon double bond can usually be achieved by catalytic hydrogenation. A series of reagents prepared from a hydride reducing agent and copper salts also give primarily the saturated ketone.[56] Similar reagents have been shown to reduce α,β-unsaturated esters[57] and nitriles[58] to the corresponding saturated compounds. The mechanistic details are not known with certainty, but it is likely that "copper hydrides" are the active reducing agents and that they form an organocopper intermediate by conjugate addition.

$$\text{"H–Cu–H"} + \text{RCH=CHC(=O)R} \rightarrow \text{–Cu–CH(R)–CH}_2\text{C(=O)R} \rightarrow \text{RCH}_2\text{CH}_2\text{C(=O)R}$$

Another reagent combination that selectively reduces the carbon–carbon double bond is Wilkinson's catalyst and triethylsilane. The initial product is the silyl enol ether.[59]

$$(CH_3)_2C{=}CH(CH_2)_2C(CH_3){=}CHCH{=}O \xrightarrow[(Ph_3P)_3RhCl]{Et_3SiH} (CH_3)_2C{=}CH(CH_2)_2CH(CH_3)CH{=}CHOSiEt_3 \xrightarrow{H_2O} (CH_3)_2C{=}CH(CH_2)_2CH(CH_3)CH_2CH{=}O$$

Unconjugated double bonds are unaffected by this reducing system.[60]

The enol ethers of β-dicarbonyl compounds are reduced to α,β-unsaturated ketones by $LiAlH_4$, followed by hydrolysis.[61] Reduction initially stops at the allylic alcohol, but subsequent acid hydrolysis of the enol ether and dehydration leads to the isolated product. This reaction is a useful method for synthesis of substituted cyclohexenones.

OC₂H₅, Ph, Ph, O —LiAlH₄→ OC₂H₅, Ph, Ph, ⁻O, H —H⁺→ O, Ph, Ph

56. S. Masamune, G. S. Bates, and P. E. Georghiou, *J. Am. Chem. Soc.* **96**, 3686 (1974); E. C. Ashby, J.-J. Lin, and R. Kovar, *J. Org. Chem.* **41**, 1939 (1976); E. C. Ashby, J.-J. Lin, and A. B. Goel, *J. Org. Chem.* **43**, 183 (1978); W. S. Mahoney, D. M. Brestensky, and J. M. Stryker, *J. Am. Chem. Soc.* **110**, 291 (1988); D. S. Brestensky, D. E. Huseland, C. McGettigan, and J. M. Stryker, *Tetrahedron Lett.* **29**, 3749 (1988).
57. M. F. Semmelhack, R. D. Stauffer, and A. Yamashita, *J. Org. Chem.* **42**, 3180 (1977).
58. M. E. Osborn, J. F. Pegues, and L. A. Paquette, *J. Org. Chem.* **45**, 167 (1980).
59. I. Ojima, T. Kogure, and Y. Nagai, *Tetrahedron Lett.*, 5035 (1972); I. Ojima, M. Nihonyanagi, T. Kogure, M. Kumagai, S. Horiuchi, K. Nakatsugawa, and Y. Nagai, *J. Organomet. Chem.* **94**, 449 (1973).
60. H.-J. Liu and E. N. C. Browne, *Can. J. Chem.* **59**, 601 (1981); T. Rosen and C. H. Heathcock, *J. Am. Chem. Soc.* **107**, 3731 (1985).
61. H. E. Zimmerman and D. I. Schuster, *J. Am. Chem. Soc.* **84**, 4527 (1962); W. F. Gannon and H. O. House, *Org. Synth.* **40**, 14 (1960).

A very important aspect of reductions by hydride transfer reagents is their stereoselectivity. The stereochemistry of hydride reduction has been studied most thoroughly with conformationally biased cyclohexanone derivatives. Some reagents give predominantly axial cyclohexanols while others give the equatorial isomer. Axial alcohols are likely to be formed when the reducing agent is a sterically hindered hydride donor. This is because the equatorial direction of approach is more open and is preferred by bulky reagents. This is an example of *steric approach control*.[62]

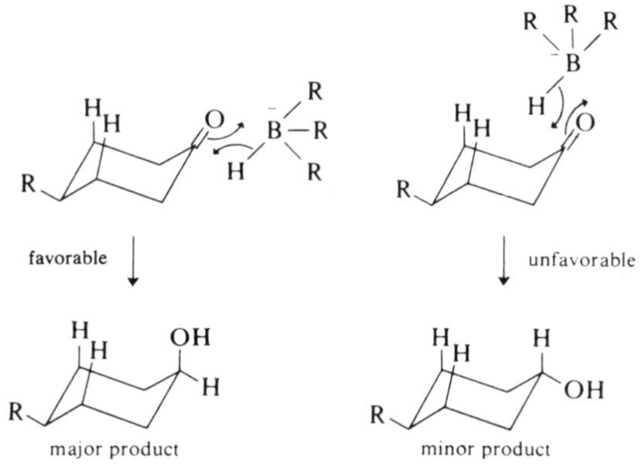

favorable | unfavorable

major product | minor product

Steric Approach Control

With less hindered hydride donors, particularly $NaBH_4$ and $LiAlH_4$, cyclohexanones give predominantly the equatorial alcohol. The equatorial alcohol is normally the more stable of the two isomers. However, hydride reductions are exothermic reactions with low activation energies. The transition state should resemble starting ketone, so product stability should not control the reaction stereochemistry. One explanation of the preference for formation of the equatorial isomer involves the torsional strain that develops in formation of the axial alcohol.[63]

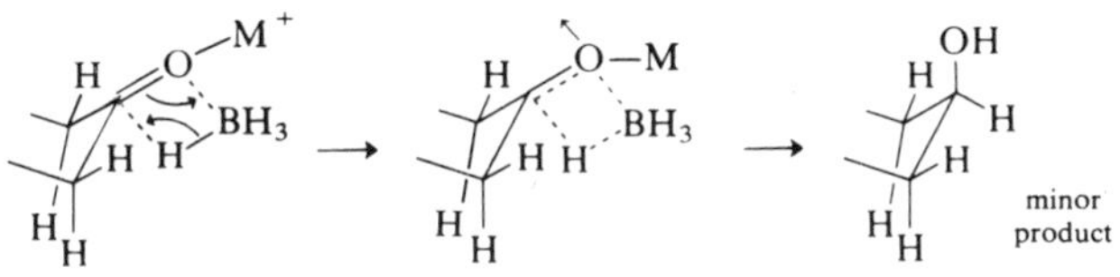

minor product

Torsional strain as oxygen passes through an eclipsed conformation

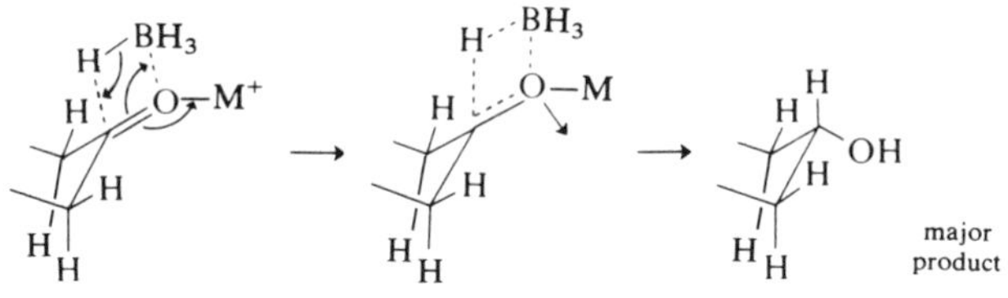

major product

Oxygen moves away from equatorial hydrogens; no torsional strain

62. W. G. Dauben, G. J. Fonken, and D. S. Noyce, *J. Am. Chem. Soc.* **78**, 2579 (1956).
63. M. Cherest, H. Felkin, and N. Prudent, *Tetrahedron Lett.*, 2205 (1968); M. Cherest and H. Felkin, *Tetrahedron Lett.*, 383 (1971).

An alternative suggestion is that the carbonyl group π-antibonding orbital which acts as the LUMO in the reaction has a greater density on the axial face.[64] It is not entirely clear at the present time how important such orbital effects are. Most of the stereoselectivities that have been reported can be reconciled with torsional and steric effects being dominant.[65] See Section 3.10 of Part A for further discussion of this issue.

When a ketone is relatively hindered, as for example in the bicyclo[2.2.1]heptan-2-one system, steric factors govern stereoselectivity even for unhindered hydride donors.

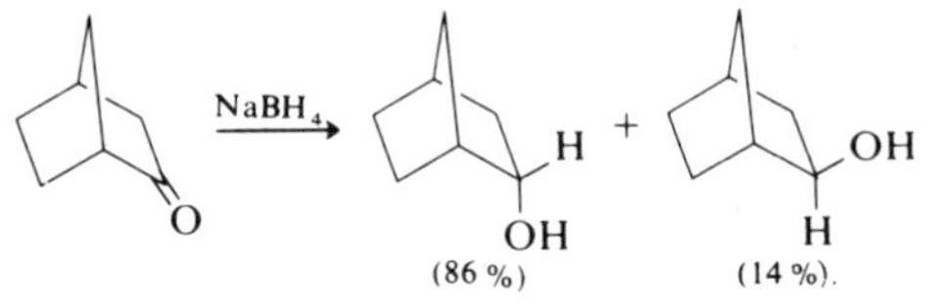

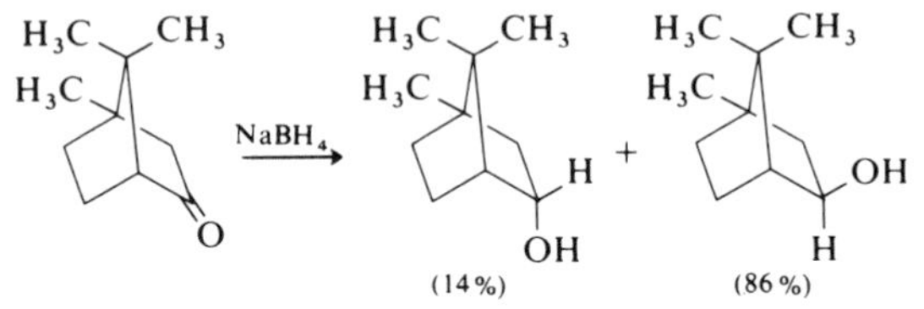

NaBH$_4$ → (86 %) + (14 %).

NaBH$_4$ → (14 %) + (86 %)

A large amount of data has been accumulated on the stereoselectivity of reduction of cyclic ketones.[66] Table 5.3 compares the stereochemistry of reduction of several ketones by hydride donors of increasing steric bulk. The trends in the table illustrate the increasing importance of steric approach control as both the hydride reagent and the ketone become more highly substituted. The alkyl-substituted borohydrides have especially high selectivity for the least hindered direction of approach.

The stereochemistry of reduction of acyclic aldehydes and ketones is a function of the substitution on the adjacent carbon atom and can be predicted on the basis of a conformational model of the transition state.[63]

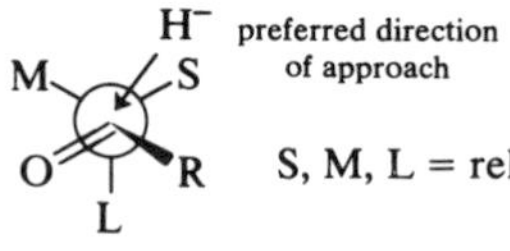

H^- preferred direction of approach

S, M, L = relative size of substituents

This model is rationalized in terms of a combination of steric and stereoelectronic effects. From a purely steric standpoint, an approach involving minimal steric

64. J. Klein, *Tetrahedron Lett.*, 4307 (1973); N. T. Ahn, O. Eisenstein, J.-M. Lefour, and M. E. Tran Huu Dau, *J. Am. Chem. Soc.* **95**, 6146 (1973).
65. W. T. Wipke and P. Gund, *J. Am. Chem. Soc.* **98**, 8107 (1976); J.-C. Perlburger and P. Müller, *J. Am. Chem. Soc.* **99**, 6316 (1977); D. Mukherjee, Y.-D. Wu, F. R. Fronczek, and K. N. Houk, *J. Am. Chem. Soc.* **110**, 3328 (1988).
66. D. C. Wigfield, *Tetrahedron* **35**, 449 (1979); D. C. Wigfield and D. J. Phelps, *J. Org. Chem.* **41**, 2396 (1976).

Table 5.3. Stereoselectivity of Hydride Reducing Agents[a]

Reducing agent	Percentage alcohol favored by steric approach control				
	$(CH_3)_3C$–cyclohexanone	CH_3–cyclohexanone	H_3C, CH_3, CH_3–cyclohexanone	norbornanone	H_3C, CH_3, CH_3–camphor
	% axial	% axial	% axial	% *endo*	% *exo*
$NaBH_4$	20[b]	25[c]	58[c]	86[d]	86[d]
$LiAlH_4$	8	24	83	89	92
$LiAl(OMe)_3H$	9	69		98	99
$LiAl(t\text{-}BuO)_3H$	9[e]	36[f]	95	94[f]	94[f]
$(CH_3CH_2CH(CH_3)-)_3\bar{B}H\ Li^+$	93[g]	98[g]	99.8[g]	99.6[g]	99.6[g]
$[(CH_3)_2CHCH(CH_3)]_3\bar{B}H\ Li^+$	>99[h]	>99[h]		>99[h]	NR[h]

a. Except where otherwise noted, data are those given by H. C. Brown and W. C. Dickason, *J. Am. Chem. Soc.* **92**, 709 (1970). Data for many other cyclic ketones and reducing agents are given by A. V. Kamernitzky and A. A. Akhrem, *Tetrahedron* **18**, 705 (1962) and W. T. Wipke and P. Gund, *J. Am. Chem. Soc.* **98**, 8107 (1976).
b. P. T. Lansbury and R. E. MacLeay, *J. Org. Chem.* **28**, 1940 (1963).
c. B. Rickborn and W. T. Wuesthoff, *J. Am. Chem. Soc.* **92**, 6894 (1970).
d. H. C. Brown and J. Muzzio, *J. Am. Chem. Soc.* **88**, 2811 (1966).
e. J. Klein, E. Dunkelblum, E. L. Eliel, and Y. Senda, *Tetrahedron Lett.*, 6127 (1968).
f. E. C. Ashby, J. P. Sevenair, and F. R. Dobbs, *J. Org. Chem.* **36**, 197 (1971).
g. H. C. Brown and S. Krishnamurthy, *J. Am. Chem. Soc.* **94**, 7159 (1972).
h. S. Krishnamurthy and H. C. Brown, *J. Am. Chem. Soc.* **98**, 3383 (1976).

interaction with the groups L and M, that is, an approach from the direction of the smallest substituent, is favorable. The stereoelectronic effect involves the interaction between the approaching hydride ion and the LUMO of the carbonyl group. This orbital, which accepts the electrons of the incoming nucleophile, is stabilized when the group L is perpendicular to the plane of the carbonyl group.[67] This conformation permits a favorable interaction between the LUMO and the antibonding σ^* orbital associated with the C—L bond.

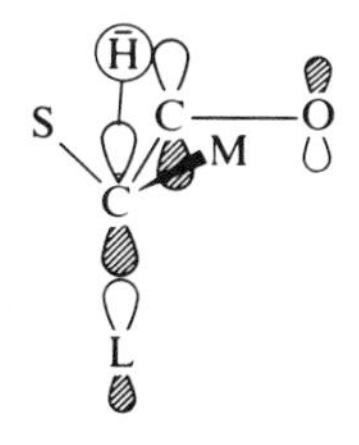

Steric factors arising from groups that are more remote from the center undergoing reduction can also influence the stereoselectivity of reduction. Such steric factors are magnified by use of bulky reducing agents. For example, a 4.5:1

67. N. T. Ahn, *Top. Curr. Chem.* **88**, 145 (1980).

preference for stereoisomer **E** over **F** is achieved by using the trialkylborohydride **D** as the reducing agent in the reduction of a prostaglandin intermediate.[68]

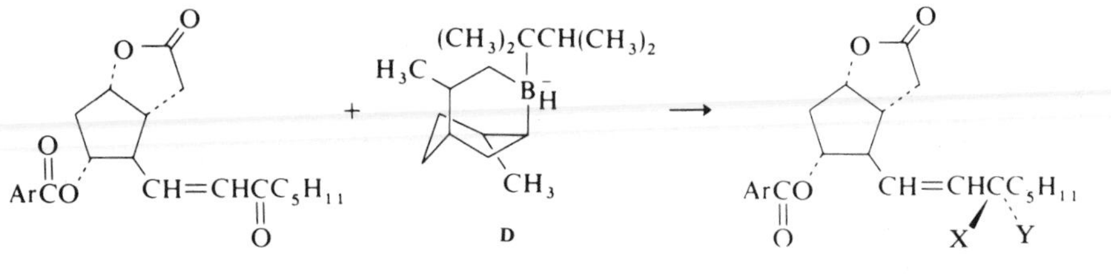

$(CH_3)_2CCH(CH_3)_2$ H$_3$C B$^-$H CH$_3$ ArCO CH=CHCC$_5$H$_{11}$ X Y

D

E: X = H, Y = OH 82

F: X = OH, Y = H 18

The stereoselectivity of reduction of carbonyl groups is affected by the same combination of steric and stereoelectronic factors that control the addition of other nucleophiles, such as enolates and organometallic reagents, to carbonyl groups. A general discussion of these factors with regard to addition of hydride is given in Section 3.10 of Part A.

5.2.2. Reduction of Other Functional Groups by Hydride Donors

Although reductions of the common carbonyl and carboxylic acid derivatives are the most prevalent cases for application of hydride donors, these reagents carry out reduction of a number of other groups in ways that are of synthetic utility. Scheme 5.5 illustrates some of these other applications of the hydride donors. Halogen and sulfonate leaving groups can undergo replacement by hydride. Both the aluminum and boron hydrides exhibit this reactivity. Lithium trialkylborohydrides can also be used.[69] The reduction is particularly rapid and efficient in polar aprotic solvents such as dimethyl sulfoxide, dimethylformamide, and hexamethylphosphoric triamide. Table 5.4 gives some indication of the reaction conditions. The normal factors in susceptibility to nucleophilic attack govern reactivity, with I > Br > Cl being the order in terms of the leaving group and benzyl ≈ allyl > primary > secondary > tertiary in terms of the substitution site.[70] For alkyl groups, it is likely that the reaction proceeds by an S_N2 mechanism. However, the range of halides that can be reduced includes aryl halides and bridgehead halides, which cannot react by the S_N2 mechanism.[71] There is loss of stereochemical integrity in the reduction of vinyl halides, suggesting the involvement of radical intermediates.[72] Formation and subsequent dissociation of a radical anion by one-electron transfer

68. E. J. Corey, S. M. Albonico, U. Koelliker, T. K. Schaaf, and R. K. Varma, *J. Am. Chem. Soc.* **93**, 1491 (1971).
69. S. Krishnamurthy and H. C. Brown, *J. Org. Chem.* **45**, 849 (1980).
70. S. Krishnamurthy and H. C. Brown, *J. Org. Chem.* **47**, 276 (1982).
71. C. W. Jefford, D. Kirkpatrick, and F. Delay, *J. Am. Chem. Soc.* **94**, 8905 (1972).
72. S. K. Chung, *J. Org. Chem.* **45**, 3513 (1980).

Halides

1[a] $CH_3(CH_2)_5CH(Cl)CH_3 \xrightarrow[\text{DMSO}]{\text{NaBH}_4} CH_3(CH_2)_6CH_3$ (67%)

2[b] $CH_3(CH_2)_8CH_2I \xrightarrow[\text{HMPA}]{\text{NaBH}_3\text{CN}} CH_3(CH_2)_8CH_3$ (88–90%)

3[c] 1-Bromonaphthalene $\xrightarrow[\text{THF, reflux}]{\text{LiAlH}_4}$ naphthalene (79%)

Sulfonates

4[d] Cyclooctenyl–$CH_2OSO_2C_7H_7 \xrightarrow{\text{LiAlH}_4}$ cyclooctenyl–CH_3 (33%)

5[e] 2-($CH_2OSO_2CH_3$)-2-(H_3C)cyclopentanone $\xrightarrow{\text{LiAlH}_4}$ 2,2-dimethylcyclopentanol (H_3C, CH_3, OH)

6[f] Cyclohexenyl–$OSO_2C_7H_7 \xrightarrow{\text{LiCuHC}_4\text{H}_9}$ cyclohexene (75%)

Epoxides

7[g] Methylcyclohexene oxide (CH_3) $\xrightarrow{\text{LiAlH}_4}$ 2-methylcyclohexanol (OH, CH_3) (89%)

Acetylenes

8[h] $CH_3CH_2C{\equiv}CCH_2CH_3 \xrightarrow[\text{120–150°C, 4.5 h}]{\text{LiAlH}_4}$ $(CH_3CH_2)(H)C{=}C(H)(CH_2CH_3)$ (90%)

9[i] 1-[$CH(OCH_3)C{\equiv}CCH_3$]-cyclohexanol (HO) $\xrightarrow[\text{NaOCH}_3,\ 65°\text{C, 45 min}]{\text{LiAlH}_4}$ 1-[$CH(OCH_3)CH{=}CHCH_3$]-cyclohexanol (HO; H, H, CH_3) (85%)

a. R. O. Hutchins, D. Hoke, J. Keogh, and D. Koharski, *Tetrahedron Lett.*, 3495 (1969); H. M. Bell, C. W. Vanderslice, and A. Spehar, *J. Org. Chem.* **34**, 3923 (1969).
b. R. O. Hutchins, C. A. Milewski, and B. E. Maryanoff, *Org. Synth.* **53**, 107 (1973).
c. H. C. Brown and S. Krishnamurthy, *J. Org. Chem.* **34**, 3918 (1969).
d. A. C. Cope and G. L. Woo, *J. Am. Chem. Soc.* **85**, 3601 (1963).
e. A. Eschenmoser and A. Frey, *Helv. Chim. Actá* **35**, 1660 (1952).
f. S. Masamune, G. S. Bates, and P. E. Geoghiou, *J. Am. Chem. Soc.* **96**, 3686 (1974).
g. B. Rickborn and W. E. Lamke II, *J. Org. Chem.* **32**, 537 (1967).
h. E. F. Magoon and L. H. Slaugh, *Tetrahedron* **23**, 4509 (1967).
i. D. A. Evans and J. V. Nelson, *J. Am. Chem. Soc.* **102**, 774 (1980).

is a likely mechanism for reductive dehalogenation of compounds that cannot react by an S_N2 mechanism.

$$R{-}X + e^- \rightarrow R{-}X^{\cdot -}$$

$$R{-}X^{\cdot -} \rightarrow R\cdot + X^-$$

$$R\cdot + H^- \rightarrow R{-}H + e^-$$

One experimental test for the involvement of radical intermediates is to study 5-hexenyl systems and look for the characteristic cyclization to cyclopentane derivatives (see Section 12.2 of Part A). When 5-hexenyl bromide or iodide reacts with $LiAlH_4$, no cyclization products are observed. However, the more hindered 2,2-dimethyl-5-hexenyl iodide gives mainly cyclic product.[73]

$$CH_2{=}CH(CH_2)_3CH_2I + LiAlH_4 \xrightarrow[1\,h]{24°C} CH_2{=}CH(CH_2)_3CH_3 \quad (94\%)$$

$$CH_2{=}CH(CH_2)_2C(CH_3)_2CH_2I + LiAlH_4 \xrightarrow[24\,h]{24°C} CH_2{=}CH(CH_2)_2C(CH_3)_2CH_3 \; (3\%) + \text{1,1,3-trimethylcyclopentane } (81\%)$$

Some cyclization also occurs with the bromide but not with the chloride or the tosylate. The occurrence of a radical intermediate is also indicated in the reduction

Table 5.4. Reaction Conditions for Reductive Replacement of Halogen and Tosylate by Hydride Donors

	Approximate conditions for complete reduction	
	Halides	Tosylates
$NaBH_3CN$[a]	1-Iodododecane, HMPA, 25°C, 4 h	1-Dodecyl tosylate, HMPA, 70°C, 8 h
$NaBH_4$[b]	1-Bromododecane, DMSO, 85°C, 1.5 h	1-Dodecyl tosylate, DMSO, 85°C, 2 h
$LiAlH_4$[c,d]	1-Bromooctane, THF, 25°C, 1 h	1-Octyl tosylate, DME, 25°C, 6 h
$LiB(C_2H_5)_3H$[c]	1-Bromooctane, THF, 25°C, 3 h	

a. R. O. Hutchins, D. Kandasamy, C. A. Maryanoff, D. Masilamani, and B. E. Maryanoff, *J. Org. Chem.* **42**, 82 (1977).
b. R. O. Hutchins, D. Kandasamy, F. Dux III, C. A. Maryanoff, D. Rotstein, B. Goldsmith, W. Burgoyne, F. Cistone, J. Dalessandro, and J. Puglis, *J. Org. Chem.* **43**, 2259 (1978).
c. S. Krishnamurthy and H. C. Brown, *J. Org. Chem.* **45**, 849 (1980).
d. S. Krishnamurthy, *J. Org. Chem.* **45**, 2550 (1980).

73. E. C. Ashby, R. N. DePriest, A. B. Goel, B. Wenderoth, and T. N. Pham, *J. Org. Chem.* **49**, 3545 (1984).

of 2-octyl iodide by $LiAlD_4$ since for this compound, in contrast to the other halides, extensive racemization accompanies reduction.

The presence of transition metal ions has a catalytic effect on reduction of halides and tosylates by $LiAlH_4$.[74] Various "copper hydride" reducing agents are effective for removal of halide and tosylate groups.[75] The primary synthetic value of these reductions is for the removal of a hydroxyl function after conversion to a halide or tosylate. Scheme 5.5 includes an example of the use of this type of reaction in synthesis.

Epoxides are converted to alcohols by $LiAlH_4$. The reaction occurs by nucleophilic attack, and hydride addition at the less hindered carbon of the epoxide is usually observed.

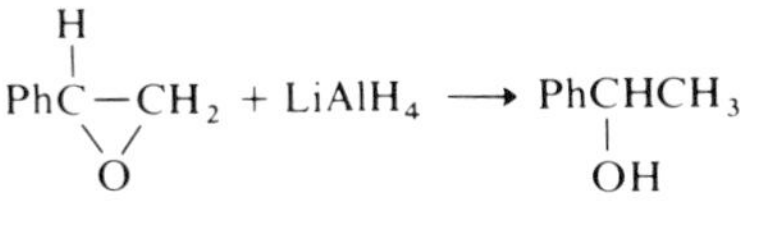

$$PhCH(-O-)CH_2 + LiAlH_4 \longrightarrow PhCH(OH)CH_3$$

Cyclohexene epoxides are preferentially reduced by an axial approach by the nucleophile.[76]

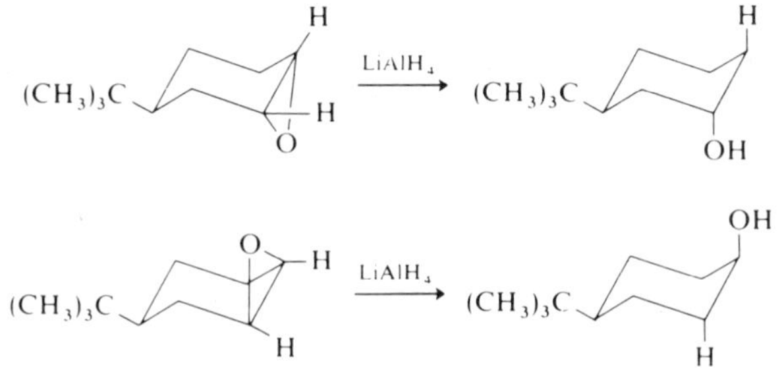

$$(CH_3)_3C\text{-cyclohexene epoxide} \xrightarrow{LiAlH_4} (CH_3)_3C\text{-cyclohexanol (axial OH)}$$

$$(CH_3)_3C\text{-cyclohexene epoxide} \xrightarrow{LiAlH_4} (CH_3)_3C\text{-cyclohexanol (axial OH)}$$

Lithium triethylborohydride is a superior reagent for reduction of epoxides that are relatively unreactive or prone to rearrangement.[77]

Alkynes are reduced to *E*-alkenes by $LiAlH_4$.[78] This stereochemistry is complementary to that of partial hydrogenation, which gives *Z*-isomers. Alkyne reduction by $LiAlH_4$ is greatly accelerated by a nearby hydroxyl group. Typically, propargylic alcohols react in ether or tetrahydrofuran over a period of several hours,[79] whereas forcing conditions are required for isolated triple bonds.[80] (Compare entries

74. E. C. Ashby and J. J. Lin, *J. Org. Chem.* **43**, 1263 (1978).
75. S. Masamune, G. S. Bates, and P. E. Georghiou, *J. Am. Chem. Soc.* **96**, 3686 (1974); E. C. Ashby, J. J. Lin, and A. B. Goel, *J. Org. Chem.* **43**, 183 (1978).
76. B. Rickborn and J. Quartucci, *J. Org. Chem.* **29**, 3185 (1964); B. Rickborn and W. E. Lamke II, *J. Org. Chem.* **32**, 537 (1967); D. K. Murphy, R. L. Alumbaugh, and B. Rickborn, *J. Am. Chem. Soc.* **91**, 2649 (1969).
77. H. C. Brown, S. C. Kim, and S. Krishnamurthy, *J. Org. Chem.* **45**, 1 (1980); H. C. Brown, S. Narasimhan, and V. Somayaji, *J. Org. Chem.* **48**, 3091 (1983).
78. E. F. Magoon and L. H. Slaugh, *Tetrahedron* **23**, 4509 (1967).
79. N. A. Porter, C. B. Ziegler, Jr., F. F. Khouri, and D. H. Roberts, *J. Org. Chem.* **50**, 2252 (1985).
80. H. C. Huang, J. K. Rehmann, and G. R. Gray, *J. Org. Chem.* **47**, 4018 (1982).

8 and 9 in Scheme 5.5). This is presumably the result of coordination of the hydroxyl group at aluminum and formation of a cyclic intermediate. The involvement of intramolecular Al—H addition has been demonstrated by use of $LiAlD_4$ as the reductant. When reduction by $LiAlD_4$ is followed by quenching with normal water, propargylic alcohol gives *Z*-2-^{2}H-2-propenol. Quenching with D_2O gives 2-^{2}H-3-^{2}H-2-propenol.[81]

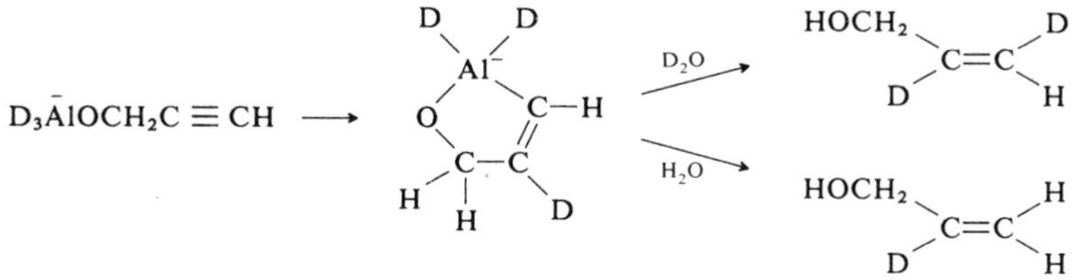

$$D_3\bar{A}lOCH_2C\equiv CH \longrightarrow \text{cyclic Al intermediate} \xrightarrow{D_2O} HOCH_2(D)C{=}C(D)H \qquad \xrightarrow{H_2O} HOCH_2(D)C{=}CH_2$$

The efficiency and stereospecificity of reduction is improved by using a 1:2 $LiAlH_4$–$NaOCH_3$ mixture as the reducing agent.[82] The mechanistic basis of this effect has not been explored in detail.

5.3. Group IV Hydride Donors

Both Si—H and C—H compounds can function as hydride donors under certain circumstances. The silicon-hydrogen bond is capable of transferring a hydride to carbocations. Alcohols that can be ionized by trifluoroacetic acid are reduced to hydrocarbons in the presence of a silane.

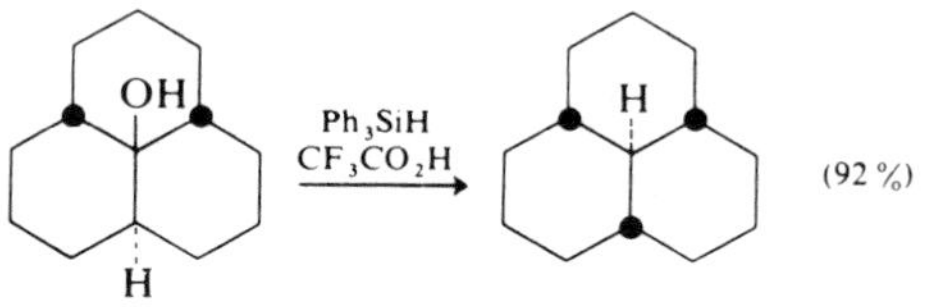

$$\xrightarrow[CF_3CO_2H]{Ph_3SiH} \quad (92\%) \qquad \text{Ref. 83}$$

Aromatic aldehydes and ketones are reduced to alkyl derivatives.[84]

$$ArC(=O)R + H^+ \rightleftarrows Ar\overset{+}{C}(OH)R \xrightarrow{R_3SiH} ArCH(OH)R \rightleftharpoons Ar\overset{+}{C}HR + H_2O$$

$$Ar\overset{+}{C}HR + R_3SiH \rightarrow ArCH_2R$$

81. J. E. Baldwin and K. A. Black, *J. Org. Chem.* **48**, 2778 (1983).
82. E. J. Corey, J. A. Katzenellenbogen, and G. H. Posner, *J. Am. Chem. Soc.* **89**, 4245 (1967); B. B. Molloy and K. L. Hauser, *J. Chem. Soc., Chem. Commun.*, 1017 (1968).
83. F. A. Carey and H. S. Tremper, *J. Org. Chem.* **36**, 758 (1971).
84. C. T. West, S. J. Donelly, D. A. Kooistra, and M. P. Doyle, *J. Org. Chem.* **38**, 2675 (1973); M. P. Doyle, D. J. DeBruyn, and D. A. Kooistra, *J. Am. Chem. Soc.* **94**, 3659 (1972); M. P. Doyle and C. T. West, *J. Org. Chem.* **40**, 3821 (1975).

Aliphatic ketones can be reduced to hydrocarbons by triethylsilane and gaseous BF_3.[85] The BF_3 is a sufficiently strong Lewis acid to promote formation of a carbocation from the intermediate alcohol.

$$\overset{+O-BF_3}{\underset{\|}{RCR}} \xrightarrow{Et_3SiH} R-\underset{H}{\overset{O\bar{B}F_3}{C}}-R \longrightarrow \underset{H}{\overset{R\quad R}{\overset{+}{C}}} \xrightarrow{Et_3SiH} RCH_2R$$

Aromatic carboxylic acids and esters are reduced to methyl groups in a process in which trichlorosilane is the reductant.[86]

$$ArCO_2H \text{ or } ArCO_2CH_3 \xrightarrow[\text{2) } Cl_3SiH,\ R_3N]{\text{1) } (CH_3)_3SiI} ArCH_2SiCl_3 \xrightarrow{^-OH} ArCH_3$$

The mechanism of this reaction has not been delineated in detail.

There is also a group of reactions in which hydride is transferred from carbon. The carbon-hydrogen bond has little intrinsic tendency to break in the way required for hydride transfer. Frequently, these reactions proceed through a cyclic transition state in which a new C—H bond is formed simultaneously with the C—H cleavage. Hydride transfer is facilitated by high charge density at the carbon atom.

Aluminum alkoxides catalyze transfer of hydride from an alcohol to a ketone. This is generally an equilibrium process, and the reaction can be driven to completion if the ketone is removed from the system, by distillation, for example. This process is called the Meerwein-Pondorff-Verley reduction.[87]

$$3\,R_2C{=}O + Al[OCH(CH_3)_2]_3 \rightarrow [R_2CHO]_3Al + 3\,CH_3\underset{\underset{O}{\|}}{C}CH_3$$

The reaction proceeds via a cyclic transition state involving coordination of both the alcohol and ketone oxygens to the aluminum. Hydride donation usually takes place at the less hindered face of the carbonyl group.[88]

Al, O, O, H_3C-C, H_3C, H, C, R, R

Certain lanthanide alkoxides, such as t-BuOSmI_2, have also been found to catalyze hydride exchange between alcohols and ketones.[89] Isopropanol can serve

85. J. L. Frey, M. Orfanopoulos, M. G. Adlington, W. R. Dittman, Jr., and S. B. Silverman, *J. Org. Chem.* **43**, 374 (1978).
86. R. A. Benkeser, E. C. Mozdzen, and C. L. Muth, *J. Org. Chem.* **44**, 2185 (1979).
87. A. L. Wilds, *Org. React.* **2**, 178 (1944).
88. F. Nerdel, D. Frank, and G. Barth, *Chem. Ber.* **102**, 395 (1969).
89. J. L. Namy, J. Souppe, J. Collin, and H. B. Kagan, *J. Org. Chem.* **49**, 2045 (1984).

as the reducing agent for aldehydes and ketones that are thermodynamically better hydride acceptors than acetone.

$$O_2N-C_6H_4-CH{=}O \xrightarrow[t\text{-BuOSmI}_2]{CH_3CH(OH)CH_3} O_2N-C_6H_4-CH_2OH \quad (94\%)$$

Another reduction process, catalyzed by iridium chloride and characterized by very high axial: equatorial product ratios for cyclohexanones, apparently involves hydride transfer from isopropanol.[90]

$$(CH_3)_3C\text{-cyclohexanone} \xrightarrow[(CH_3)_2CHOH]{IrCl_4,\ HCl,\ (CH_3O)_3P,\ H_2O} (CH_3)_3C\text{-cyclohexanol (axial OH)}$$

Formic acid can also act as a donor of hydrogen. The driving force in this case is the formation of carbon dioxide. A useful application is the Clark-Eschweiler reductive alkylation of amines. Heating a primary or secondary amine with formaldehyde and formic acid results in complete methylation to the tertiary amine.[91]

$$RNH_2 + CH_2{=}O + HCO_2H \rightarrow RN(CH_3)_2$$

The hydride acceptor is the iminium ion resulting from condensation of the amine with formaldehyde.

$$R_2\overset{+}{N}{=}CH_2 \quad H-O-C(H){=}O$$

5.4. Hydrogen Atom Donors

Reduction by hydrogen atom donors involves intermediates with unpaired electrons. Tri-*n*-butyltin hydride is the most important example of this type of reducing agent. It is able to reductively replace halogen in many types of organic compounds. Mechanistic studies have established a free-radical chain mechanism.[92] The order of reactivity for the halides is RI > RBr > RCl > RF, which reflects the relative ease of the halogen atom abstraction.[93]

$$In\cdot + Bu_3SnH \rightarrow In{-}H + Bu_3Sn\cdot$$

$$Bu_3Sn\cdot + R{-}X \rightarrow R\cdot + Bu_3SnX$$

$$R\cdot + Bu_3SnH \rightarrow RH + Bu_3Sn\cdot$$

90. E. L. Eliel, T. W. Doyle, R. O. Hutchins, and E. C. Gilbert, *Org. Synth.* **50**, 13 (1970).
91. M. L. Moore, *Org. React.* **5**, 301 (1949); S. H. Pine and B. L. Sanchez, *J. Org. Chem.* **36**, 829 (1971).
92. L. W. Menapace and H. G. Kuivila, *J. Am. Chem. Soc.* **86**, 3047 (1964).
93. H. G. Kuivila and L. W. Menapace, *J. Org. Chem.* **28**, 2165 (1963).

Tri-*n*-butyltin hydride shows substantial selectivity toward polyhalogenated compounds, permitting partial dehalogenation. The reason for the greater reactivity of more highly halogenated carbons toward reduction lies in the stabilizing effect that the remaining halogen has on the radical intermediate. This selectivity has been used, for example, to reduce dihalocyclopropanes to monohalocyclopropanes, as in entry 4 in Scheme 5.6. A procedure that is catalytic in Bu_3SnH and uses $NaBH_4$ as the stoichiometric reagent has been developed.[94] Tributylstannane is regenerated by $NaBH_4$ as the reaction proceeds. This procedure has advantages in the isolation and purification of product. Entry 5 in Scheme 5.6 is an example of this procedure.

Scheme 5.6. Dehalogenations with Stannanes

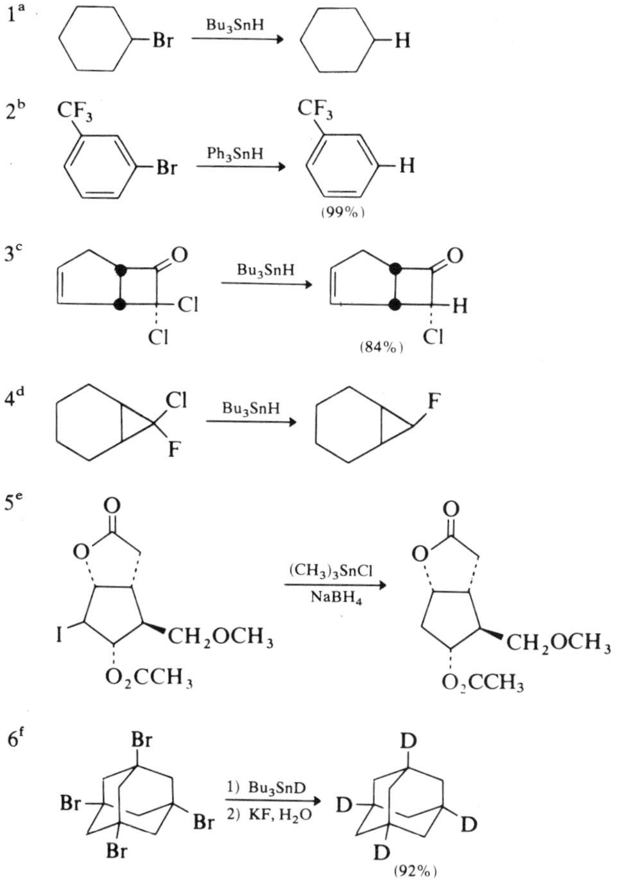

1[a] Br $\xrightarrow{Bu_3SnH}$ H

2[b] CF_3 Br $\xrightarrow{Ph_3SnH}$ CF_3 H (99%)

3[c] O Cl Cl $\xrightarrow{Bu_3SnH}$ O H Cl (84%)

4[d] Cl F $\xrightarrow{Bu_3SnH}$ F

5[e] O O I CH_2OCH_3 O_2CCH_3 $\xrightarrow[NaBH_4]{(CH_3)_3SnCl}$ O O CH_2OCH_3 O_2CCH_3

6[f] Br Br Br Br $\xrightarrow[2)\ KF,\ H_2O]{1)\ Bu_3SnD}$ D D D D (92%)

a. H. G. Kuivila, L. W. Menapace, and C. R. Warner, *J. Am. Chem. Soc.* **84**, 3584 (1962).
b. D. H. Lorenz, P. Shapiro, A. Stern, and E. I. Becker, *J. Org. Chem.* **28**, 2332 (1963).
c. W. T. Brady and E. F. Hoff, Jr. *J. Org. Chem.* **35**, 3733 (1970).
d. T. Ando, F. Namigata, H. Yamanaka, and W. Funasaka, *J. Am. Chem. Soc.* **89**, 5719 (1967).
e. E. J. Corey and J. W. Suggs, *J. Org. Chem.* **40**, 2554 (1975).
f. J. E. Leibner and J. Jacobus, *J. Org. Chem.* **44**, 449 (1979).

94. E. J. Corey and J. W. Suggs, *J. Org. Chem.* **40**, 2554 (1975).

Scheme 5.7. Deoxygenation of Alcohols via Thioesters and Related Derivatives

1[a]

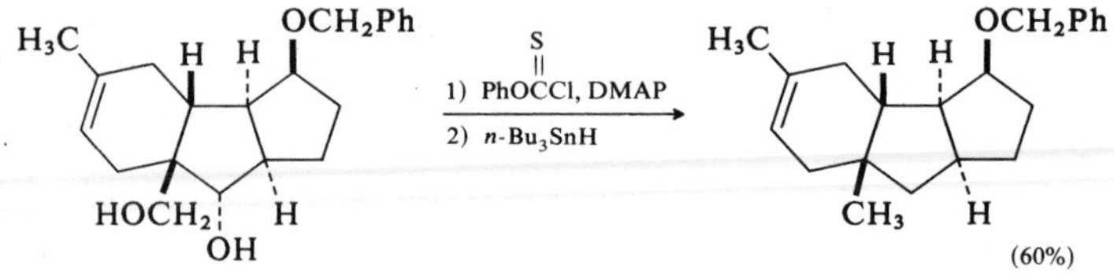

H_3C … OCH_2Ph, H, H, $HOCH_2$, OH, H → 1) $PhOC(=S)Cl$, DMAP; 2) n-Bu_3SnH → H_3C … OCH_2Ph, H, H, CH_3, H (60%)

2[b]

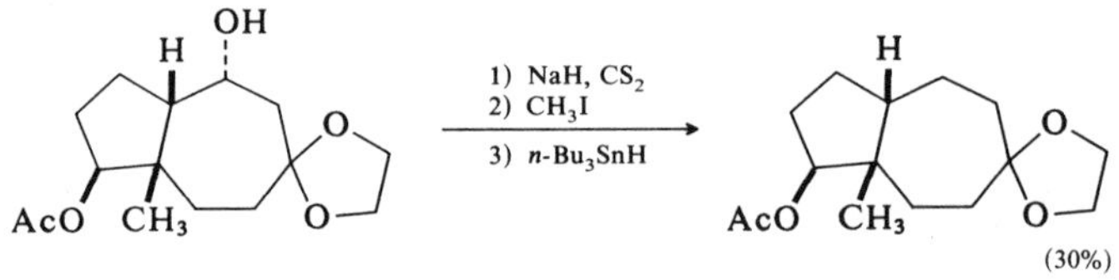

OH, H, AcO, CH_3 → 1) NaH, CS_2; 2) CH_3I; 3) n-Bu_3SnH → H, AcO, CH_3 (30%)

3[c]

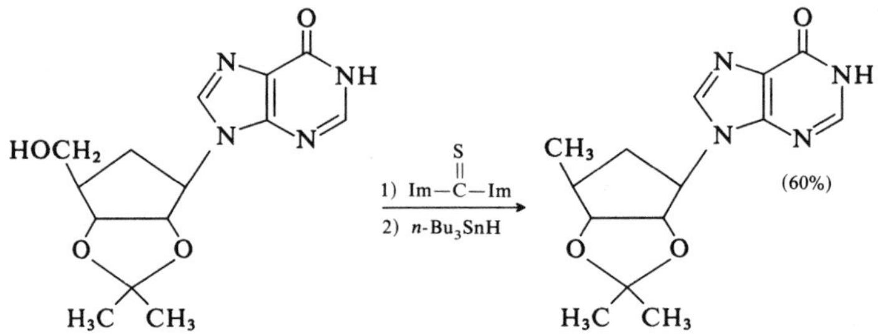

$HOCH_2$, H_3C, CH_3 → 1) $Im-C(=S)-Im$; 2) n-Bu_3SnH → CH_3, H_3C, CH_3 (60%)

4[d]

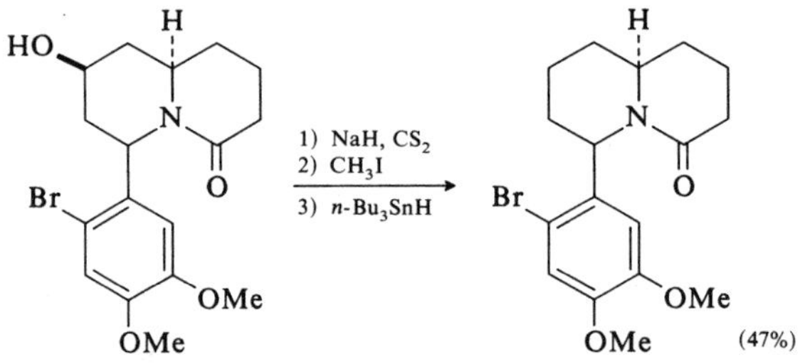

HO, H, Br, OMe, OMe → 1) NaH, CS_2; 2) CH_3I; 3) n-Bu_3SnH → H, Br, OMe, OMe (47%)

5[e]

$PhCO_2CH_2$, HO, OCH_3, $PhCO_2$, O_2CPh → 1) $Im-C(=S)-Im$; 2) n-Bu_3SnH → $PhCO_2CH_2$, OCH_3, $PhCO_2$, O_2CPh (92%)

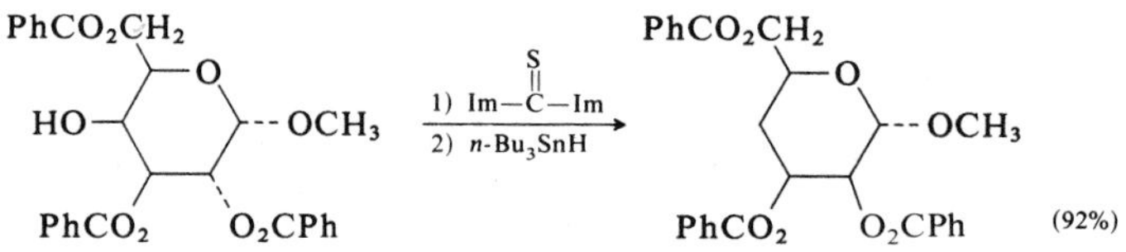

a. H. J. Liu and M. G. Kulkarni, *Tetrahedron Lett.* **26**, 4847 (1985).
b. C. M. Tice and C. H. Heathcock, *J. Org. Chem.* **46**, 9 (1981).
c. O. Miyashita, F. Kasahara, T. Kusaka, and R. Marumoto, *J. Antibiot.* **38**, 981 (1985).
d. D. J. Hart and K. Kanai, *J. Org. Chem.* **47**, 1555 (1982).
e. J. R. Rasmussen, C. J. Slinger, R. J. Kordish, and D. D. Newman-Evans, *J. Org. Chem.* **46**, 4843 (1981).

Tri-*n*-butyltin hydride also serves as a hydrogen atom donor in radical-mediated methods for reductive deoxygenation of alcohols.[95] The alcohol is converted to a thiocarbonyl derivative. These thiono esters undergo a radical reaction with tri-*n*-butyltin hydride. This procedure gives good yields with secondary alcohols and, with appropriate adjustment of conditions, can also be adapted to primary alcohols.[96] Scheme 5.7 illustrates some of the conditions that have been developed for the reductive deoxygenation of alcohols.

$$\mathrm{R{-}O\overset{\overset{S}{\|}}{C}X + BuSn\cdot \rightarrow RO\overset{\overset{S-SnBu_3}{|}}{\dot{C}}X \rightarrow R\cdot + X\overset{\overset{O}{\|}}{C}S{-}SnBu_3}$$

$$\mathrm{R\cdot + Bu_3SnH \rightarrow R{-}H + BuSn\cdot}$$

5.5. Dissolving-Metal Reductions

Another group of synthetically useful reductions employs a metal as the reducing agent. The organic substrate under these conditions accepts one or more electrons from the metal. The subsequent course of the reaction depends on the structure of the reactant and reaction conditions. Three broad classes of reactions can be recognized, and these will be discussed separately. These include reactions in which the overall change involves (a) net addition of hydrogen, (b) reductive removal of a functional group, and (c) formation of carbon–carbon bonds.

5.5.1. Addition of Hydrogen

Although metals have been supplanted for synthetic purposes by hydride donors in reactions involving addition of hydrogen, the reduction of ketones to alcohols in ammonia or alcohols provides some mechanistic insight into this group of reactions. The overall course of the reaction of ketones with metal reductants is determined by the fate of the initial ketyl formed by a single-electron transfer. The intermediate, depending on its structure and the medium, maybe protonated, may disproportionate, or may dimerize.[97] In hydroxylic solvents such as liquid ammonia or in the presence of alcohols, the protonation process dominates over dimerization.

95. D. H. R. Barton and S. W. McCombie, *J. Chem. Soc., Perkin Trans. 1,* 1574 (1975); for a review of this method, see W. Hartwig, *Tetrahedron* **39**, 2609 (1983).
96. D. H. R. Barton, W. B. Motherwell, and A. Stange, *Synthesis*, 743 (1981).
97. V. Rautenstrauch and M. Geoffroy, *J. Am. Chem. Soc.* **99**, 6280 (1977); J. W. Huffman and W. W. McWhorter, *J. Org. Chem.* **44**, 594 (1979); J. W. Huffman, P. C. Desai, and J. E. LaPrade, *J. Org. Chem.* **48**, 1474 (1983); J. W. Huffman, W.-P. Liao, and R. H. Wallace, *Tetrahedron Lett.* **28**, 3315 (1987); J. W. Huffman, *Acc. Chem. Res.* **16**, 399 (1983).

As will be discussed in Section 5.5.3, dimerization may become the dominant process under other conditions.

$$RCH_2-C(=O)-R' \xrightarrow{e^-} RCH_2-\dot{C}(O^-)-R' \text{ (ketyl)}$$

$$\xrightarrow[S-H]{\text{protonation}} RCH_2-\dot{C}(OH)-R' \xrightarrow[S-H]{e} RCH_2C(O^-)(H)-R'$$

$$\xrightarrow{\text{dimerization}} RCH_2-C(O^-)(R')-C(O^-)(R')-CH_2R$$

$$\xrightarrow{\text{disproportionation}} RCH_2-CH(O^-)R' + RCH=C(O^-)R'$$

$$RCH_2-CH(O^-)R' \xrightarrow{\text{protonation, SH}} RCH_2-CH(OH)R'$$

α,β-Unsaturated carbonyl compounds are cleanly reduced to the enolate of the corresponding saturated ketone on reduction with lithium in ammonia.[98] Usually, an alcohol is added to the reduction solution to serve as the proton source.

$$R_2C=CH-C(=O)-R \xrightarrow{e^-} R_2\dot{C}-CH=C(O^-)-R \xrightarrow[S-H]{e^-} R_2CH-CH=C(O^-)-R$$

As mentioned in Chapter 1, this is one of the best methods for generating a specific enolate of a ketone. The enolate generated by conjugate reduction can undergo the characteristic reactions that were discussed in Chapters 1 and 2. When this is the objective of the reduction, it is important to use only one equivalent of the proton donor. Ammonia, being a weaker acid than an aliphatic ketone, does not act as a proton donor toward an enolate. If the saturated ketone is the desired product, the enolate is protonated either by use of excess proton donor during the reduction or on workup.

3-Methylcyclohex-2-enone $\xrightarrow[\text{1 equiv } H_2O]{Li, NH_3}$ $\xrightarrow{CH_2=CHCH_2Br}$ 2-($CH_2CH=CH_2$)-3-CH_3-cyclohexanone (43–47%) + 2-($CH_2CH=CH_2$)-3-CH_3-cyclohexanone (2–2.5%) Ref. 99

Octalone $\xrightarrow[\text{2) } n\text{-}C_4H_9I]{\text{1) Li, } NH_3}$ trans-decalone with H_9C_4 (47%) Ref. 100

98. D. Caine, *Org. React.* **23**, 1 (1976).
99. D. Caine, S. T. Chao, and H. A. Smith, *Org. Synth.* **56**, 52 (1977).
100. G. Stork, P. Rosen, and N. L. Goldman, *J. Am. Chem. Soc.* **83**, 2965 (1961).

The stereochemistry of conjugate reduction is established by the proton transfer to the β-carbon. In the well-studied case of $\Delta^{1,9}$-2-octalones, the ring junction is usually *trans.*[101]

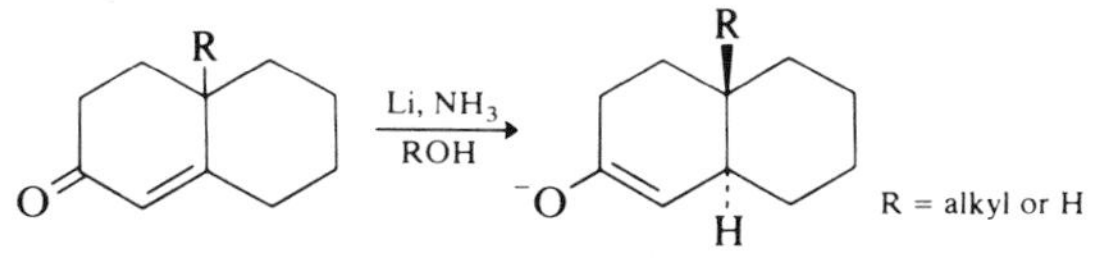

Exceptions to the preference for formation of the *trans* ring fusion by axial protonation can usually be traced to unfavorable steric interactions in the chair–chair conformation of the reduction intermediate.[98] For example, 6-β-*t*-butyl-$\Delta^{1,9}$-2-octalone gives predominantly the *cis* ring junction because a chair–chair conformation is precluded by the bulky *t*-butyl substituent.

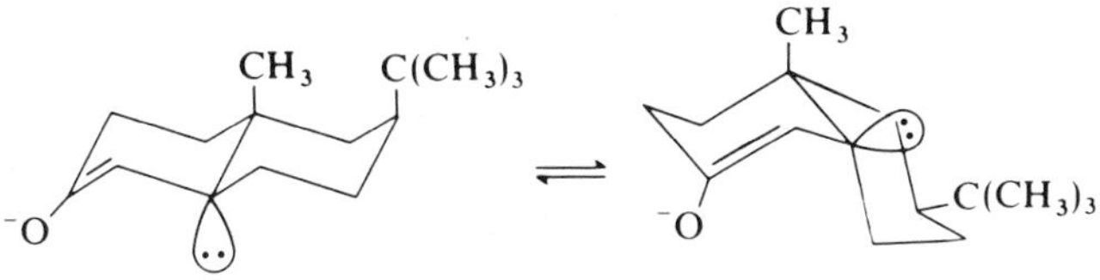

Stereochemical results in other series usually reflect two basic requirements. There is a stereoelectronic preference for protonation perpendicular to the enolate system and, given that this requirement is met, the stereochemistry will normally correspond to protonation of the most stable conformation of the dianion intermediate from its least hindered side.

Dissolving-metal systems constitute the most general method for partial reduction of aromatic rings. The reaction is called the *Birch reduction.*[102] The usual reducing medium is lithium or sodium in liquid ammonia. The first step is electron transfer, and the radical anion is then protonated by the solvent.

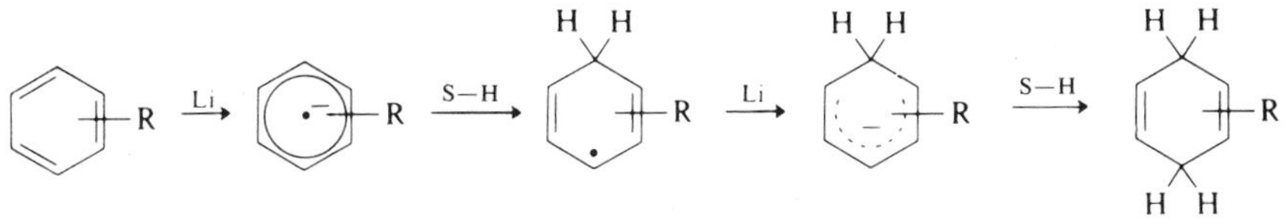

The isolated double bonds in the dihydro structure are much less easily reduced than the aromatic ring, so the reduction stops at the dihydro stage. The rate of reduction is affected in a predictable way by substituent groups. Electron-releasing groups retard electron transfer whereas electron-withdrawing groups facilitate reduction. Alkyl- and alkoxyaromatics, phenols, and benzoate anions are the most useful

101. G. Stork, P. Rosen, N. Goldman, R. V. Coombs, and J. Tsuji, *J. Am. Chem. Soc.* **87**, 275 (1965); M. J. T. Robinson, *Tetrahedron* **21**, 2475 (1965).
102. A. J. Birch and G. Subba Rao, *Adv. Org. Chem.* **8**, 1 (1972); R. G. Harvey, *Synthesis*, 161 (1980); J. M. Hook and L. N. Mander, *Nat. Prod. Rep.* **3**, 35 (1986); P. W. Rabideau, *Tetrahedron* **45**, 1599 (1989).

reactants for Birch reduction. Many other functional groups, for example, ketone and nitro, are reduced in preference to the aromatic ring. The substituents also govern the position of protonation. Alkyl- and alkoxyaromatics normally give the 2,5-dihydro derivative. Benzoate anions give 1,4-dihydro derivatives.

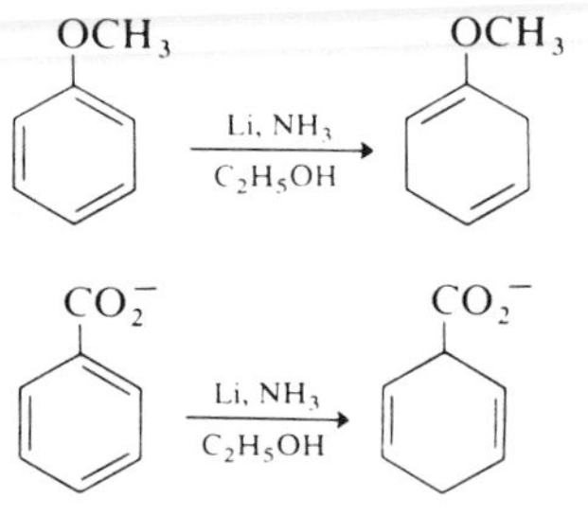

OCH_3 (anisole) $\xrightarrow[C_2H_5OH]{Li,\ NH_3}$ OCH_3 (1-methoxy-1,4-cyclohexadiene)

CO_2^- (benzoate) $\xrightarrow[C_2H_5OH]{Li,\ NH_3}$ CO_2^- (2,5-cyclohexadiene-1-carboxylate)

The structure of the products is determined by the site of protonation of the radical anion intermediate formed after the first electron-transfer step. In general, electron-releasing substituents favor protonation at the *ortho* position, whereas electron-attracting groups favor protonation at the *para* position.[103] Addition of a second electron gives a pentadienyl anion, which is protonated at the center carbon. As a result, 2,5-dihydro products are formed with alkyl or alkoxy substituents and 1,4-reduction products are formed with carboxylate substituents. It has been suggested that the reason the pentadienyl anion is protonated at the center carbon is that this causes a smaller change in bond orders than protonation at one of the terminal positions.[104] The reduction of methoxybenzenes is of importance in the synthesis of cyclohexenones via hydrolysis of the intermediate enol ethers:

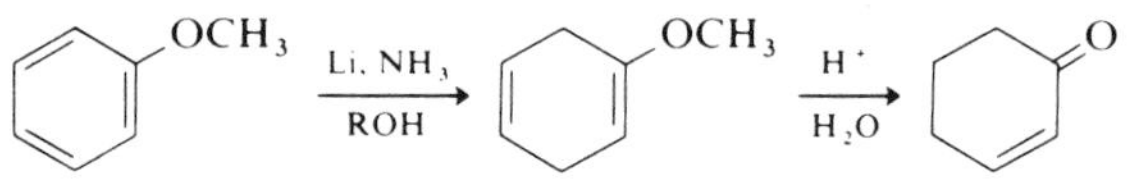

OCH_3 $\xrightarrow[ROH]{Li,\ NH_3}$ OCH_3 $\xrightarrow[H_2O]{H^+}$ O

Scheme 5.8 lists some examples of the use of the Birch reduction.

Reduction of acetylenes with sodium in ammonia,[105] with lithium in low-molecular-weight amines,[106] or with sodium in hexamethylphosphoric triamide containing *t*-butanol as a proton source[107] all lead to the *E*-alkene. The reaction is assumed to involve successive electron-transfer and proton-transfer steps.

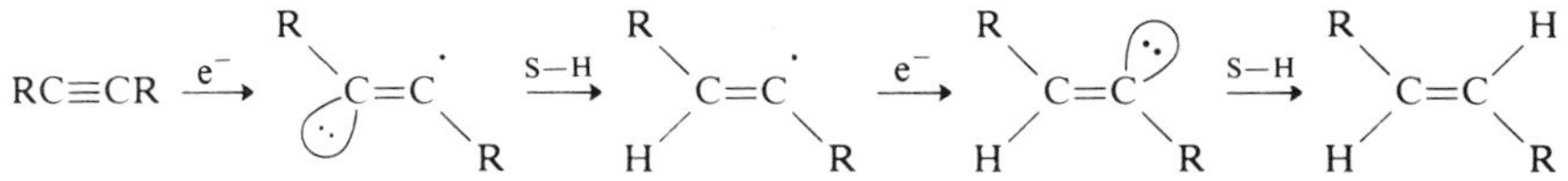

$$RC{\equiv}CR \xrightarrow{e^-} \overset{R}{}\dot{C}{=}\dot{C}R \xrightarrow{S-H} R(H)C{=}\dot{C}R \xrightarrow{e^-} R(H)C{=}\ddot{C}R \xrightarrow{S-H} R(H)C{=}C(H)R$$

103. A. J. Birch, A. L. Hinde, and L. Radom, *J. Am. Chem. Soc.* **102**, 2370 (1980).
104. P. W. Rabideau and D. L. Huser, *J. Org. Chem.* **48**, 4266 (1983).
105. K. N. Campbell and T. L. Eby, *J. Am. Chem. Soc.* **63**, 216, 2683 (1941); A. L. Henne and K. W. Greenlee, *J. Am. Chem. Soc.* **65**, 2020 (1943).
106. R. A. Benkeser, G. Schroll, and D. M. Sauve, *J. Am. Chem. Soc.* **77**, 3378 (1955).
107. H. O. House and E. F. Kinloch, *J. Org. Chem.* **39**, 747 (1974).

1[a] OCH_3 / $C(CH_3)_3$ —Li, NH_3→ OCH_3 / $C(CH_3)_3$ (63%)

2[b] $C(CH_3)_3$ / $C(CH_3)_3$ —Li, $C_2H_5NH_2$→ $C(CH_3)_3$ / $C(CH_3)_3$ (56%)

3[c] H_3C, OCH_3 —1) Li, NH_3; 2) H^+, H_2O→ H_3C, O (80%)

4[d] CO_2H —Na, NH_3, C_2H_5OH→ CO_2H (90%)

5[e] OH —Li, NH_3, C_2H_5OH→ OH (97–99%)

6[f] OC_2H_5 —Na, C_2H_5OH→ OC_2H_5

a. D. A. Bolon, *J. Org. Chem.* **35**, 715 (1970).
b. H. Kwart and R. A. Conley, *J. Org. Chem.* **38**, 2011 (1973).
c. E. A. Braude, A. A. Webb and M. U. S. Sultanbawa, *J. Chem. Soc.*, 3328 (1958); W. C. Agosta and W. L. Schreiber, *J. Am. Chem. Soc.* **93**, 3947 (1971).
d. M. E. Kuehne and B. F. Lambert, *Org. Synth.* **V**, 400 (1973).
e. C. D. Gutsche and H. H. Peter, *Org. Synth.* **IV**, 887 (1963).
f. M. D. Soffer, M. P. Bellis, H. E. Gellerson, and R. A. Stewart, *Org. Synth.* **IV**, 903 (1963).

5.5.2. Reductive Removal of Functional Groups

The reductive removal of halogen can be accomplished with lithium or sodium. Tetrahydrofuran containing *t*-butanol is a useful solvent medium. Good results have also been achieved with polyhalogenated compounds by using sodium in ethanol.

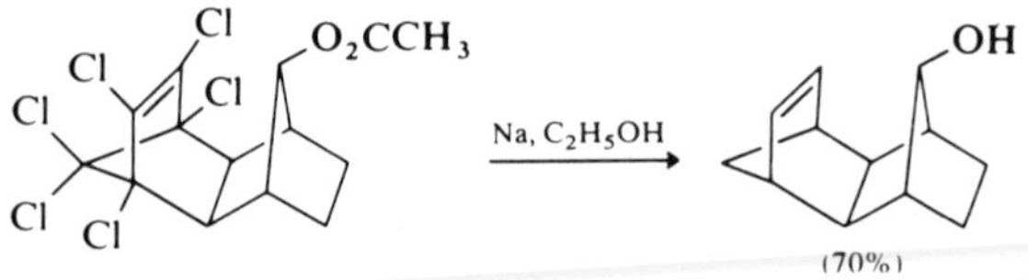

Cl, Cl, Cl, Cl, Cl, Cl, O_2CCH_3 $\xrightarrow{Na,\ C_2H_5OH}$ OH (70%) Ref. 108

An important synthetic application of this reaction is in dehalogenation of dichloro- and dibromocyclopropanes. The dihalocyclopropanes are accessible via carbene addition reactions (see Section 10.2.3). Reductive dehalogenation can also be used to introduce deuterium at a specific site. Some examples of these types of reactions are given in Scheme 5.9. The mechanism of the reaction presumably involves electron transfer to form a radical anion which then fragments with loss of a halide ion. The resulting radical is reduced to a carbanion by a second electron transfer and subsequently protonated.

$$R{-}X \xrightarrow{e^-} R{-}\dot{X}^- \xrightarrow{-X^-} R\cdot \xrightarrow{e^-} \bar{R}: \xrightarrow{S-H} R{-}H$$

Phosphate ester groups can also be removed by dissolving-metal reduction. Reductive removal of vinyl phosphate groups is one of the better methods for conversion of a carbonyl compound to an alkene.[109] The required vinyl phosphate esters are obtained by phosphorylation of the enolate with diethyl phosphorochloridate or *N,N,N′,N′*-tetramethyldiamidophosphorochloridate.[110]

$$RCH_2\overset{O}{\overset{\|}{C}}R' \xrightarrow[(X)_2POCl]{LiNR_2} RCH{=}\overset{OPO(X)_2}{\overset{|}{C}}R' \xrightarrow[t\text{-}BuOH]{Li,\ RNH_2} RCH{=}CHR'$$

$$X = OEt \text{ or } NMe_2$$

Reductive removal of oxygen from aromatic rings can also be achieved by reductive cleavage of diethyl aryl phosphate esters.

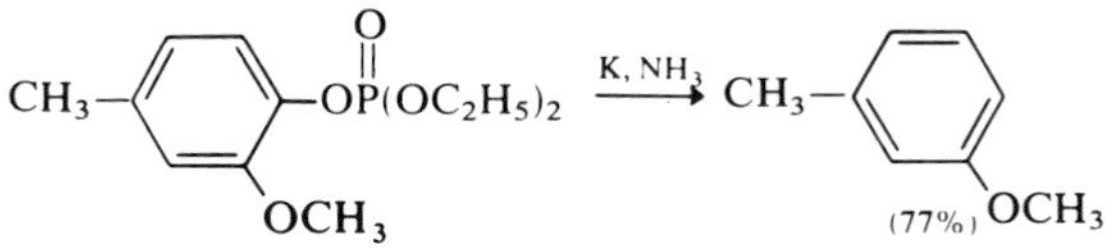

$$CH_3{-}C_6H_3(OCH_3){-}O\overset{O}{\overset{\|}{P}}(OC_2H_5)_2 \xrightarrow{K,\ NH_3} CH_3{-}C_6H_4{-}OCH_3 \ (77\%)$$ Ref. 111

There are also examples where phosphate esters of saturated alcohols are reductively deoxygenated.[112] Mechanistic studies of the cleavage of dialkyl aryl phosphates have indicated that the crucial C—O bond cleavage occurs after transfer of two electrons.[113]

$$ArO\overset{O}{\overset{\|}{P}}(OC_2H_5)_2 \xrightarrow{2e^-} [ArOPO(OEt)_2]^{2-} \rightarrow Ar^- + (EtO)_2PO_2^-$$

108. B. V. Lap and M. N. Paddon-Row, *J. Org. Chem.* **44**, 4979 (1979).
109. R. E. Ireland and G. Pfister, *Tetrahedron Lett.*, 2145 (1969).
110. R. E. Ireland, D. C. Muchmore, and U. Hengartner, *J. Am. Chem. Soc.* **94**, 5098 (1972).
111. R. A. Rossi and J. F. Bunnett, *J. Org. Chem.* **38**, 2314 (1973).
112. R. R. Muccino and C. Djerassi, *J. Am. Chem. Soc.* **96**, 556 (1974).
113. S. J. Shafer, W. D. Closson, J. M. F. vanDijk, O. Piepers, and H. M. Buck, *J. Am. Chem. Soc.* **99**, 5118 (1977).

A. Dehalogenation

1[a]

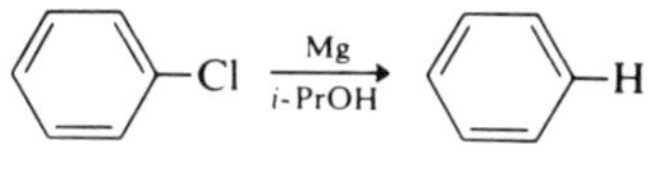
Cl
Mg
i-PrOH
H

2[b]

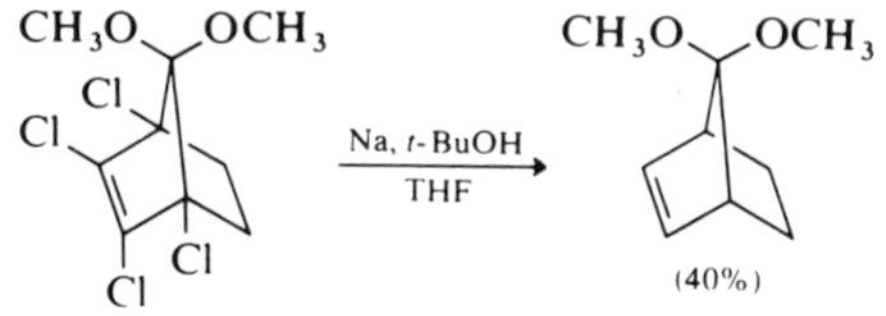
CH3O OCH3
Cl
Cl
Cl
Cl
Na, t-BuOH
THF
CH3O OCH3
(40%)

3[c]

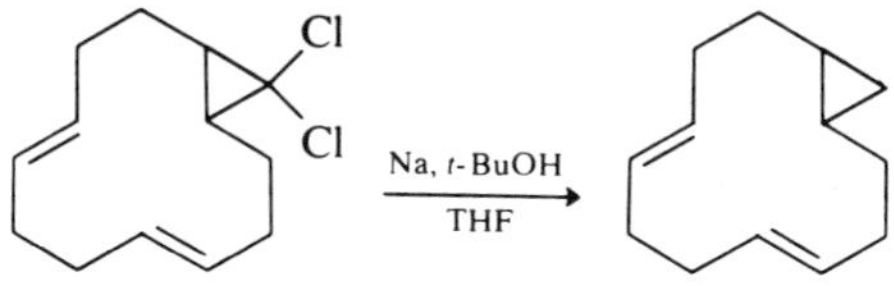
Cl
Cl
Na, t-BuOH
THF

4[d]

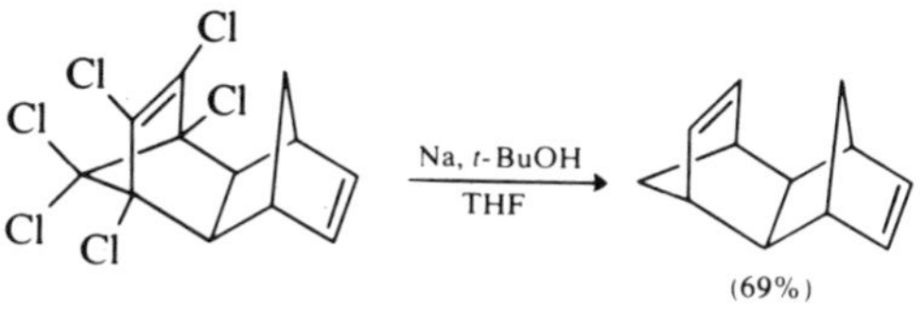
Cl
Cl
Cl
Cl
Cl
Cl
Na, t-BuOH
THF
(69%)

B. Deoxygenation

5[e]

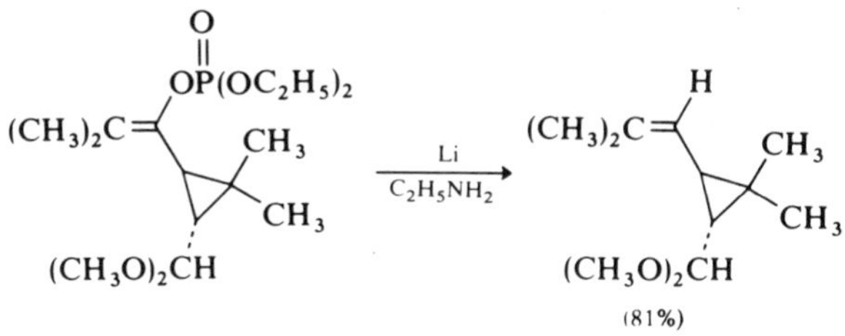
O
OP(OC2H5)2
(CH3)2C
CH3
CH3
(CH3O)2CH
Li
C2H5NH2
H
(CH3)2C
CH3
CH3
(CH3O)2CH
(81%)

6[f]

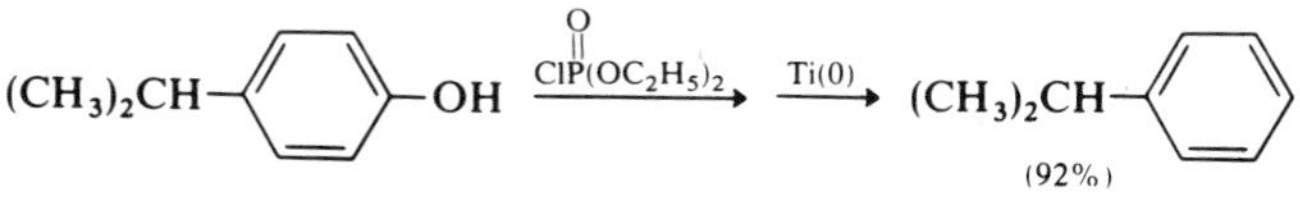
(CH3)2CH
OH
O
ClP(OC2H5)2
Ti(0)
(CH3)2CH
(92%)

7[g]

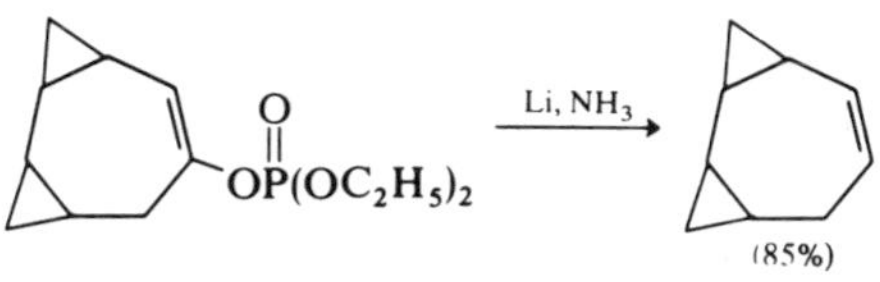
O
OP(OC2H5)2
Li, NH3
(85%)

a. D. Bryce-Smith and B. J. Wakefield, *Org. Synth.* **47**, 103 (1967).
b. P. G. Gassman and J. L. Marshall, *Org. Synth.* **48**, 68 (1968).
c. P. G. Gassman, J. Seter, and F. J. Williams, *J. Am. Chem. Soc.* **93**, 1673 (1971).
d. B. V. Lap and M. N. Paddon-Row, *J. Org. Chem.* **44**, 4979 (1979).
e. S. C. Welch and T. A. Valdes, *J. Org. Chem.* **42**, 2108 (1977).
f. S. C. Welch and M. E. Walter, *J. Org. Chem.* **43**, 4797 (1978).
g. M. R. Detty and L. A. Paquette, *J. Am. Chem. Soc.* **99**, 821 (1977).

For preparative purposes, titanium metal can be used in place of sodium or lithium in liquid ammonia for both the vinyl phosphate[114] and aryl phosphate[115] cleavages. The titanium metal is generated *in situ* from $TiCl_3$ by reduction with potassium metal in tetrahydrofuran.

Both metallic zinc and aluminum amalgam are milder reducing agents than the alkali metals. These reductants selectively remove oxygen and sulfur functional groups alpha to carbonyl groups. The mechanistic picture that seems most generally applicable is a net two-electron reduction with expulsion of the oxygen or sulfur group as an anion. The reaction must be a concerted process since the isolated functional groups are not reduced under these conditions.

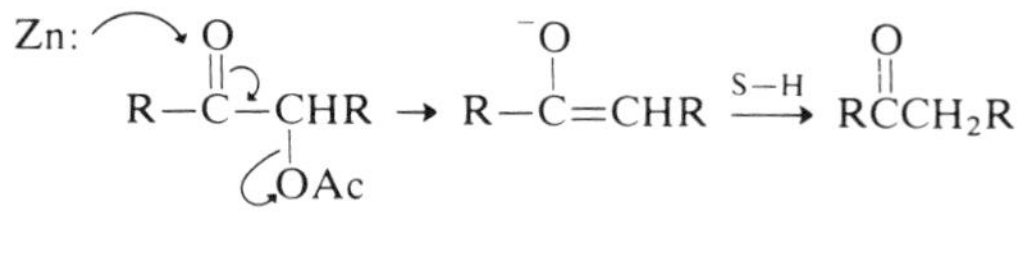

$$\text{Zn:} \; R-\overset{O}{\overset{\|}{C}}-\underset{OAc}{\underset{|}{C}}HR \rightarrow R-\overset{^-O}{\overset{|}{C}}=CHR \xrightarrow{S-H} R\overset{O}{\overset{\|}{C}}CH_2R$$

Some examples of this type of reaction are given in Scheme 5.10. Vinylogous oxygen substituents are also subject to reductive elimination by zinc or aluminum amalgam (see entry 8, Scheme 5.10).

Scheme 5.10. Reductive Removal of Functional Groups from α-Substituted Carbonyl Compounds

1[a] CH_3, O_2CCH_3 —Zn, $(CH_3CO)_2O$→ CH_3 (63%)

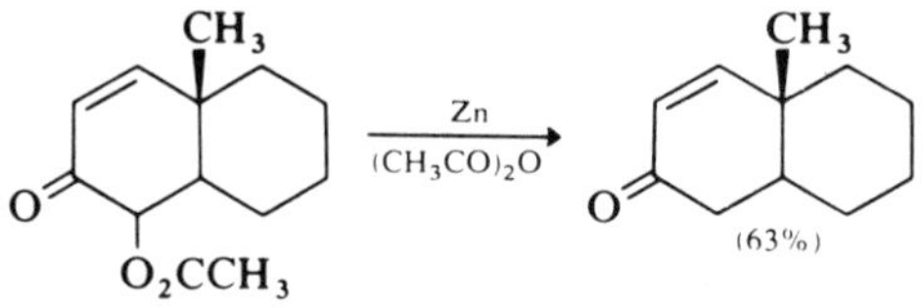

2[b] CH_3, CH_3CO_2 —Ca, NH_3→ CH_3 (80%)

3[c] O, OH —Zn, HCl, CH_3CO_2H→ O (75%)

4[d] H_3C OSO_2CH_3, O, CH_3 —Zn, NH_4Cl→ CH_3, O, CH_3

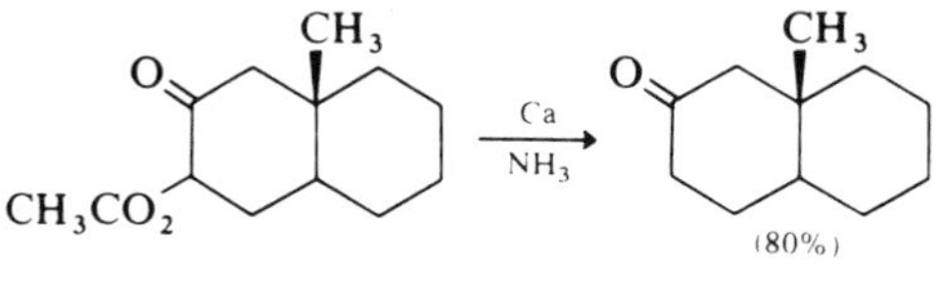

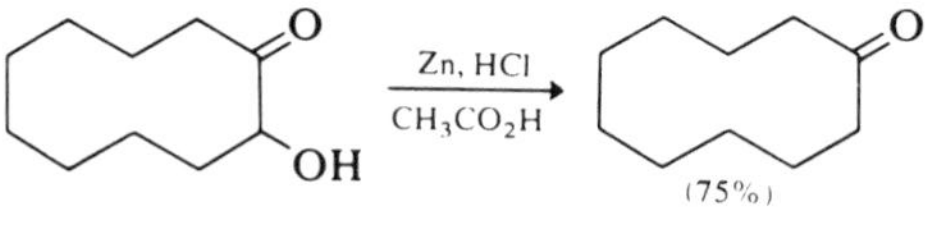

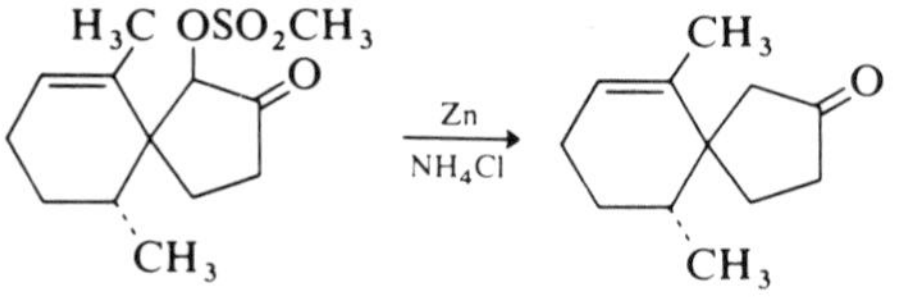

114. S. C. Welch and M. E. Walters, *J. Org. Chem.* **43**, 2715 (1978).
115. S. C. Welch and M. E. Walters, *J. Org. Chem.* **43**, 4797 (1978).

5[e] $\xrightarrow{\text{Al-Hg}}$ (75%)

6[f] $(CH_3)_3CSi(Ph)_2-O$ ... $\xrightarrow{\text{Al-Hg}}$ ... CO_2H (75%)

7[g] $CH_3O-C_6H_4-\overset{O}{\overset{\|}{C}}CH_2SO_2CH_3 \xrightarrow{\text{Al-Hg}} CH_3O-C_6H_4-\overset{O}{\overset{\|}{C}}CH_3$ (98%)

8[h] $H_{11}C_5$ ... $CO_2CH_3 \xrightarrow{\text{Zn}}$ $H_{11}C_5$ ... CO_2CH_3

a. R. B. Woodward, F. Sondheimer, D. Taub, K. Heusler, and W. M. McLamore, *J. Am. Chem. Soc.* **74**, 4223 (1952).
b. J. A. Marshall and H. Roebke, *J. Org. Chem.* **34**, 4188 (1969).
c. A. C. Cope, J. W. Barthel, and R. D. Smith, *Org. Synth.* **IV**, 218 (1963).
d. T. Ibuka, K. Hayashi, H. Minakata, and Y. Inubushi, *Tetrahedron Lett.*, 159 (1979).
e. E. J. Corey, E. J. Trybulski, L. S. Melvin, Jr., K. C. Nicolaou, J. A. Secrist, R. Lett, P. W. Sheldrake, J. R. Falck, D. J. Brunelle, M. F. Haslanger, S. Kim, and S. Yoo, *J. Am. Chem. Soc.* **100**, 4618 (1978).
f. P. A. Grieco, E. Williams, H. Tanaka, and S. Gilman, *J. Org. Chem.* **45**, 3537 (1980).
g. E. J. Corey and M. Chaykovsky, *J. Am. Chem. Soc.* **86**, 1639 (1964).
h. L. E. Overman and C. Fukaya, *J. Am. Chem. Soc.* **102**, 1454 (1980).

5.5.3. Reductive Carbon–Carbon Bond Formation

Since reductions by metal atoms often occur as one-electron processes, radicals are involved as intermediates. When the reaction conditions are adjusted so that coupling competes favorably with other processes, the formation of a carbon–carbon bond will occur. The reductive coupling of acetone to 2,3-dimethylbutane-2,3-diol (pinacol) is an example of such a process.

$$(CH_3)_2CO \xrightarrow{\text{Mg-Hg}} (CH_3)_2\underset{HO}{\underset{|}{C}}-\underset{OH}{\underset{|}{C}}(CH_3)_2$$ Ref. 116

116. R. Adams and E. W. Adams, *Org. Synth.* **I**, 448 (1932).

The most general reagent for effecting this type of reaction is a combination of $TiCl_4$ and a magnesium amalgam.[117] The active reductant is presumably titanium metal formed by reduction of $TiCl_4$.

Mg–Hg, $TiCl_4$ (95%)

$CH_3CCH_2CH_2CCH_3$ (O, O) → Mg–Hg, $TiCl_4$ → H_3C, OH, OH, H_3C (81%)

Titanium metal generated by stronger reducing agents, such as $LiAlH_4$ or lithium or potassium metal, effects complete removal of oxygen with formation of alkene.[118]

$TiCl_4$, Li (85%)

Both unsymmetrical diols and alkenes can be prepared by applying these methods to mixtures of two different carbonyl compounds. An excess of one component can be used to achieve a high conversion of the more valuable reactant.

A version of titanium-mediated reductive coupling in which $TiCl_3$–Zn–Cu serves as the reductant is efficient in closing large rings.

$O{=}CH(CH_2)_{12}CH{=}O$ → $TiCl_3$, Zn–Cu → (71%) Ref. 119

The mechanism of the titanium-mediated reductive couplings is presumably generally similar to that of reduction by other metals, but titanium is uniquely efficient in reductive coupling of carbonyl compounds. The strength of Ti—O bonds is probably the basis for this efficiency. Titanium-mediated reductive couplings are normally heterogeneous, and it is likely the reaction takes place at the metal surface.[120] The partially reduced intermediates are probably bound to the metal surface, and this may account for the effectiveness of the reaction in forming medium and large rings.

117. E. J. Corey, R. L. Danheiser, and S. Chandrasekaran, *J. Org. Chem.* **41**, 260 (1976).
118. J. E. McMurry and M. P. Fleming, *J. Org. Chem.* **41**, 896 (1976); J. E. McMurry and L. R. Krepski, *J. Org. Chem.* **41**, 3929 (1976); J. E. McMurry, M. P. Fleming, K. L. Kees, and L. R. Krepski, *J. Org. Chem.* **43**, 3255 (1978); J. E. McMurry, *Acc. Chem. Res.* **16**, 405 (1983).
119. J. E. McMurry, J. R. Matz, K. L. Kees, and P. A. Bock, *Tetrahedron Lett.* **23**, 1777 (1982).
120. R. Dams, M. Malinowski, I. Westdrop, and H. Y. Geise, *J. Org. Chem.* **47**, 248 (1982).

$$R_2C{=}O \;\; R_2C{=}O \;\; -Ti^0-Ti^0-Ti^0- \longrightarrow R_2\dot{C}-O^- \;\; R_2\dot{C}-O^- \;\; -Ti^{+1}-Ti^0-Ti^{+1}- \longrightarrow R_2C(O^-)-C(O^-)R_2 \;\; -Ti^{+1}-Ti^{+1}-Ti^0 \longrightarrow R_2C{=}CR_2 \;\; -Ti^{+2}(=O)-Ti^{+2}(=O)-Ti^0$$

Another important reductive coupling is the conversion of esters to α-hydroxyketones (acyloins).[121] This reaction is usually carried out with sodium metal in an inert solvent. Diesters undergo intramolecular reactions, and this is also an important method for preparation of medium and large carbocyclic rings.

$$CH_3O_2C(CH_2)_8CO_2CH_3 \xrightarrow[2)\ CH_3CO_2H]{1)\ Na} \text{2-hydroxycyclodecanone (C=O, OH)} \qquad \text{Ref. 122}$$

There has been considerable discussion concerning the mechanism of the acyloin condensation. A simple formulation of the mechanism envisages coupling of radicals generated by one-electron transfer.

$$RC(=O)OR' + Na\cdot \longrightarrow R\dot{C}(O^-)OR' \longrightarrow RC(O^-)(OR')-C(O^-)(OR')R \longrightarrow RC(=O)-C(=O)R$$

$$RC(=O)-C(=O)R + 2\,Na\cdot \longrightarrow RC(O^-)=C(O^-)R \xrightarrow{H^+} RC(=O)CH(OH)R$$

An alternative mechanism bypasses the postulated α-diketone intermediate since its involvement is doubtful.[123]

$$RCO_2R' + Na\cdot \longrightarrow R\dot{C}(O^-)OR' \xrightarrow{RCO_2R'} R\dot{C}(OR')-O-C(OR')(O^-)R \xrightarrow{Na\cdot} R\bar{C}(OR')-O-C(OR')(O^-)R \longrightarrow$$

$$RC(OR')(-O-)CR(O^-) \xrightarrow{Na} RC(OR')(O^-)-\dot{C}R(O^-) \xrightarrow{Na\cdot} RC(OR')(O^-)-\bar{C}R(O^-) \longrightarrow RC(O^-)=C(O^-)R$$

Regardless of the details of the mechanism, the product prior to neutralization is the dianion of the final α-hydroxyketone, namely, a enediolate. It has been found

121. J. J. Bloomfield, D. C. Owsley, and J. M. Nelke, *Org. React.* **23**, 259 (1976).
122. N. Allinger, *Org. Synth.* **IV**, 840 (1963).
123. J. J. Bloomfield, D. C. Owsley, C. Ainsworth, and R. E. Robertson, *J. Org. Chem.* **40**, 393 (1975).

Scheme 5.11. Reductive Carbon-Carbon Bond Formation

A. Pinacol Formation

1[a]

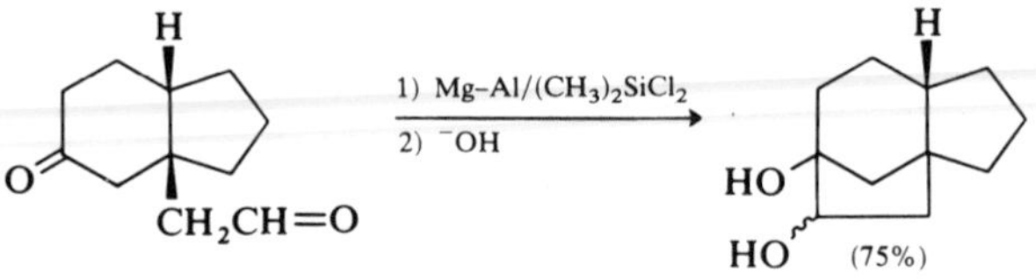

$CH_2CH{=}O$

1) $Mg\text{-}Al/(CH_3)_2SiCl_2$
2) ^{-}OH

H, HO, HO (75%)

2[b]

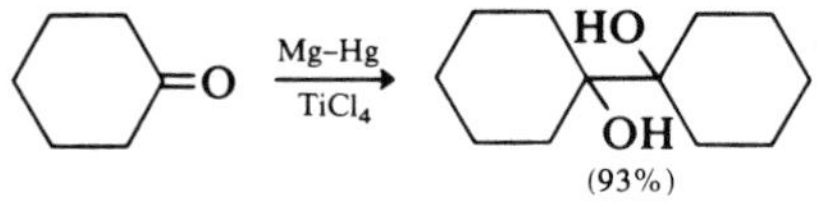

=O

Mg–Hg, $TiCl_4$

HO, OH (93%)

B. Alkene Formation

3[c]

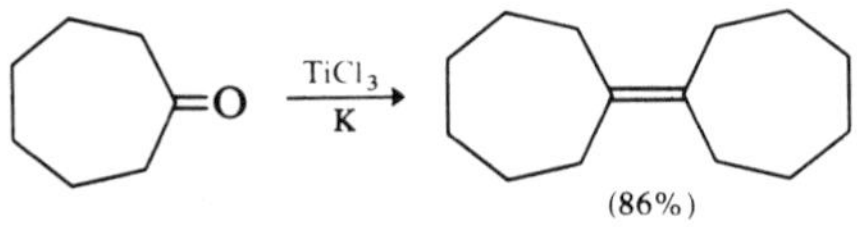

=O

$TiCl_3$, K

(86%)

4[c]

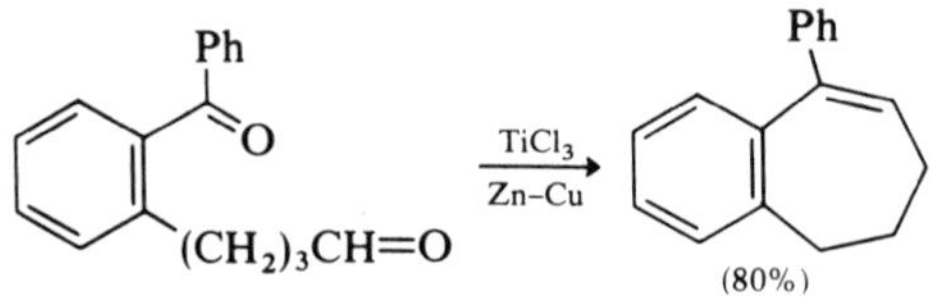

Ph, O, $(CH_2)_3CH{=}O$

$TiCl_3$, Zn–Cu

Ph (80%)

5[d]

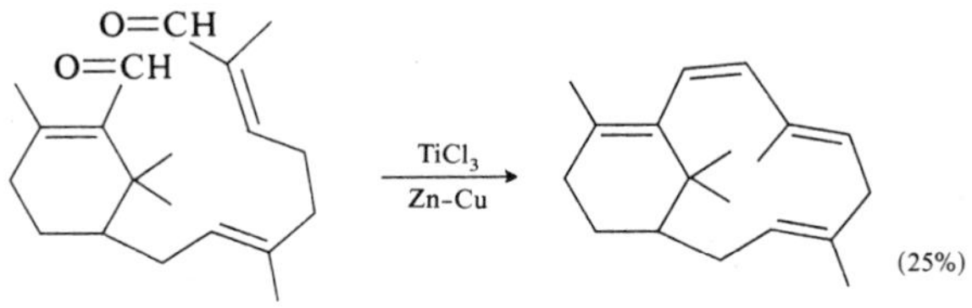

O=CH, O=CH

$TiCl_3$, Zn–Cu

(25%)

C. Acyloin Formation

5[d]

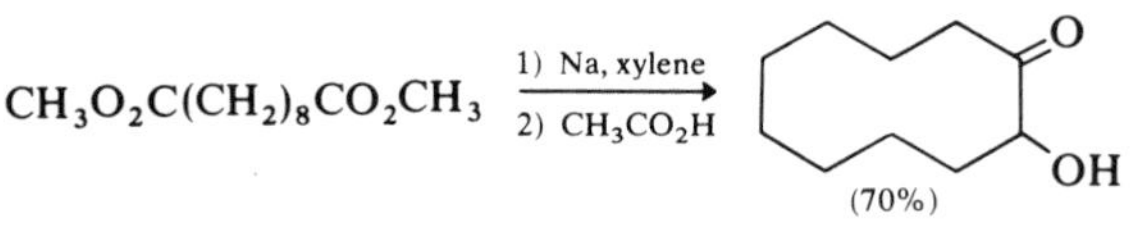

$CH_3O_2C(CH_2)_8CO_2CH_3$

1) Na, xylene
2) CH_3CO_2H

O, OH (70%)

6[e]

$C_2H_5O_2CH_2CH_2CO_2C_2H_5$

1) Na, $(CH_3)_3SiCl$ toluene
2) CH_3OH

O, OH (85%)

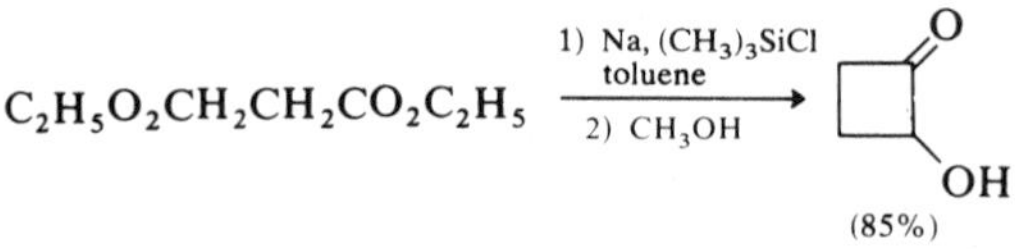

a. E. J. Corey and R. L. Carney, *J. Am. Chem. Soc.* **93**, 7318 (1971).
b. E. J. Corey, R. L. Danheiser, and S. Chandrasekaran, *J. Org. Chem.* **41**, 260 (1976).
c. J. E. McMurry, M. P. Fleming, K. L. Kees, and L. R. Krepski, *J. Org. Chem.* **43**, 3255 (1978).
d. C. B. Jackson and G. Pattenden, *Tetrahedron Lett.* **26**, 3393 (1985).
e. N. L. Allinger, *Org. Synth.* **IV**, 840 (1963).
f. J. J. Bloomfield and J. M. Nelke, *Org. Synth.* **57**, 1 (1977).

that the overall yields are greatly improved if trimethylsilyl chloride is present during the reduction to trap these dianions as trimethylsilyl ethers.[124] The latter are much more stable under the reaction conditions than the enediolates. Hydrolysis during workup gives the acyloin product. This modified version of the reaction has been applied to cyclizations leading to small, medium, and large rings, as well as to intermolecular couplings.

Two examples of acyloin formation from esters are given in Scheme 5.11.

5.6. Reductive Deoxygenation of Carbonyl Groups

Several methods are available for reductive removal of carbonyl groups from organic molecules. Complete reduction to methylene groups or conversion to alkenes can be achieved. Some examples of both types of reactions are given in Scheme 5.12.

Zinc and hydrochloric acid is a classical reagent combination for conversion of carbonyl groups to methylene groups. The reaction is known as the *Clemmensen reduction.*[125] The corresponding alcohols are not reduced under the conditions of the reaction, so they are evidently not intermediates. The mechanism is not known in detail, but it may involve carbon–zinc bonds at the metal surface.

The reaction is commonly carried out in hot concentrated hydrochloric acid with ethanol as a co-solvent. These conditions preclude the presence of acid-sensitive or hydrolyzable functional groups. The Clemmensen reaction works best for aryl ketones and is less reliable with unconjugated ketones. A modification in which the reaction is run in ether saturated with dry hydrogen chloride gave good results in the reaction of steroidal ketones.[126]

The *Wolff–Kishner reaction*[127] is the reduction of carbonyl groups to methylene groups by base-catalyzed decomposition of the hydrazone of the carbonyl compound. Alkyldiimides are believed to be formed and then collapse with loss of nitrogen:

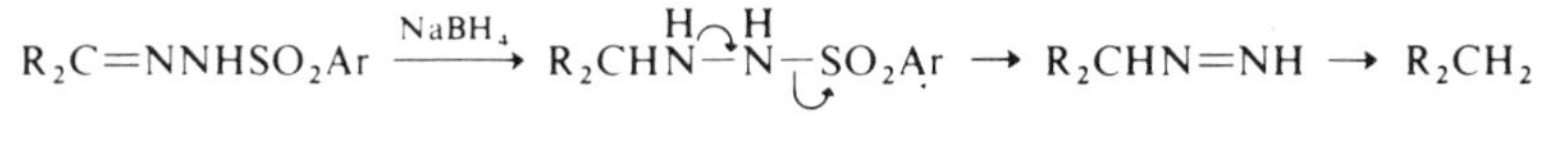

$$R_2C{=}N{-}NH_2 + \bar{O}H \rightleftharpoons R_2C{=}N{-}\bar{N}H \rightarrow R_2\underset{H}{C}{-}N{=}N{-}H \xrightarrow{-N_2} R_2CH_2$$

The reduction of tosylhydrazones by $LiAlH_4$ or $NaBH_4$ also convertes carbonyl groups to methylene groups.[128] It is believed that a diimide is involved, as in the Wolff–Kishner reaction.

$$R_2C{=}NNHSO_2Ar \xrightarrow{NaBH_4} R_2CHN(H){-}N(H){-}SO_2Ar \rightarrow R_2CHN{=}NH \rightarrow R_2CH_2$$

124. K. Ruhlmann, *Synthesis*, 236 (1971).
125. E. Vedejs, *Org. React.* **22**, 401 (1975).
126. M. Toda, M. Hayashi, Y. Hirata, and S. Yamamura, *Bull. Chem. Soc. Jpn.* **45**, 264 (1972).
127. D. Todd, *Org. React.* **4**, 378 (1948); Huang-Minlon, *J. Am. Chem. Soc.* **68**, 2847 (1946).
128. L. Caglioti, *Tetrahedron* **22**, 487 (1966).

Excellent yields can also be obtained using $NaBH_3CN$ as the reducing agent.[129] The $NaBH_3CN$ can be added to a mixture of the carbonyl compound and *p*-toluensulfonylhydrazide. Hydrazone formation is faster than reduction of the carbonyl group by $NaBH_3CN$, and the tosylhydrazone is reduced as it is formed. Another reagent which can reduce tosylhydrazones to give methylene groups is $CuBH_4(PPh_3)_2$.[130]

Reduction of tosylhydrazones of α,β-unsaturated ketones by $NaBH_3CN$ gives alkenes with the double bond located between the former carbonyl carbon and the α-carbon.[131] This reaction is believed to proceed by an initial conjugate reduction, followed by decomposition of the resulting vinylhydrazine to a vinyldiimide.

$$RCH{=}CH\overset{NNHSO_2Ar}{\overset{\|}{C}}R' \xrightarrow{NaBH_3CN} RCH_2CH{=}\overset{NHNHSO_2Ar}{\overset{|}{C}}R' \rightarrow RCH_2CH{=}\overset{N=NH}{\overset{|}{C}}R' \xrightarrow{-N_2} RCH_2CH{=}CHR'$$

Catecholborane or sodium borohydride in acetic acid can also be used as reducing reagents in this reaction.[132]

Carbonyl groups can be converted to methylene groups by desulfurization of thioketals. The cyclic thioketal from ethanedithiol is commonly used. Reaction with excess Raney nickel causes hydrogenolysis of both C—S bonds.

H$_3$C, H$_3$C, H$_3$C, O $\xrightarrow[HSCH_2CH_2SH]{BF_3}$ H$_3$C, H$_3$C, H$_3$C, S, S $\xrightarrow{Ni}$ H$_3$C, H$_3$C, H$_3$C (81%) Ref. 133

Tri-*n*-butyltin hydride is an alternative reagent for desulfurization.[134]

The conversion of ketone *p*-toluenesulfonylhydrazones to alkenes takes place on treatment with strong bases such as an alkyllithium or lithium dialkylamide.[135] This is known as the *Shapiro reaction.*[136] The reaction proceeds through the anion of a vinyldiimide, which decomposes to a vinyllithium reagent. Exposure of this intermediate to a proton source gives the alkene.

$$R\overset{NNHSO_2Ar}{\overset{\|}{C}}CH_2R' \xrightarrow{2RLi} R\overset{Li^+\bar{N}NSO_2Ar}{\overset{\|}{C}}\underset{Li}{\underset{|}{C}}HR' \xrightarrow{-LiSO_2AR} R\overset{N=N^-Li^+}{\overset{|}{C}}{=}CHR' \xrightarrow{-N_2} R\overset{Li}{\overset{|}{C}}{=}CHR'$$

Scheme 5.13 provides some examples of this very useful reaction.

129. R. O. Hutchins, C. A. Milewski, and B. E. Maryanoff, *J. Am. Chem. Soc.* **95**, 3662 (1973).
130. B. Milenkov and M. Hesse, *Helv. Chim. Acta* **69**, 1323 (1986).
131. R. O. Hutchins, M. Kacher, and L. Rua, *J. Org. Chem.* **40**, 923 (1975).
132. G. W. Kabalka, D. T. C. Yang, and J. D. Baker, Jr., *J. Org. Chem.* **41**, 574 (1976); R. O. Hutchins and N. R. Natale, *J. Org. Chem.* **43**, 2299 (1978).
133. F. Sondheimer and S. Wolfe, *Can. J. Chem.* **37**, 1870 (1959).
134. C. G. Gutierrez, R. A. Stringham, T. Nitasaka, and K. G. Glasscock, *J. Org. Chem.* **45**, 3393 (1980).
135. R. H. Shapiro and M. J. Heath, *J. Am. Chem. Soc.* **89**, 5734 (1967).
136. R. H. Shapiro, *Org. React.* **23**, 405 (1976); R. M. Adington and A. G. M. Barrett, *Acc. Chem. Res.* **16**, 53 (1983).

Clemmensen

1[a] OH, OCH_3, CH=O → Zn (Hg), HCl → OH, OCH_3, CH_3 (60–67%)

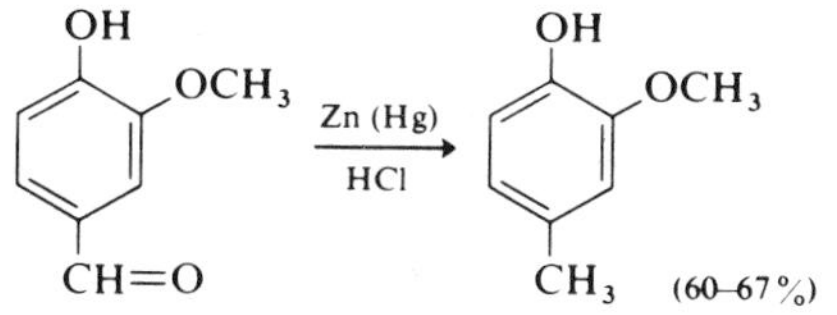

2[b] OH, $CO(CH_2)_5CH_3$ → Zn (Hg), HCl → OH, $CH_2(CH_2)_5CH_3$ (81–86%)

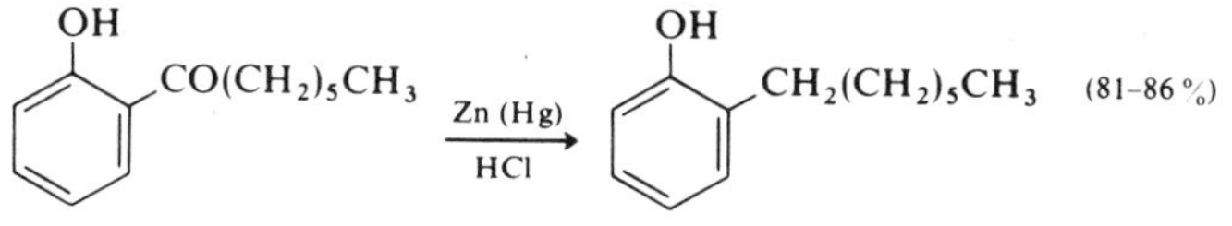

Wolff–Kishner

3[c] $HO_2C(CH_2)_4CO(CH_2)_4CO_2H \xrightarrow[KOH]{NH_2NH_2} HO_2C(CH_2)_9CO_2H$ (87–93%)

4[d] $Ph{-}C(=NNH_2){-}Ph \xrightarrow[DMSO]{KOC(CH_3)_3} PhCH_2Ph$ (90%)

Tosylhydrazone Reduction

5[e] $CH{=}NNHSO_2C_7H_7$ → $LiAlH_4$ → CH_3 (70%)

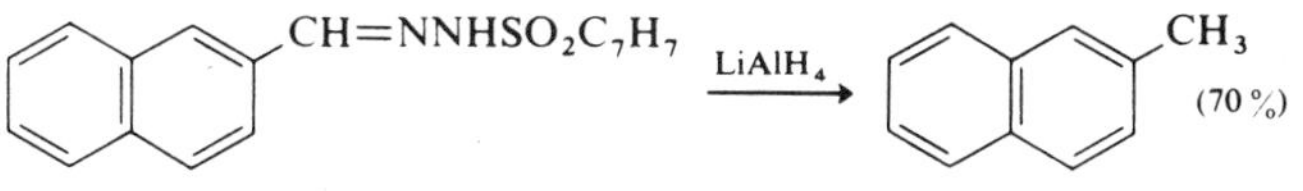

6[f] $(CH_3)_3C$—C=O $\xrightarrow[NaBH_3CN]{C_7H_7SO_2NHNH_2}$ $(CH_3)_3C$— (77%)

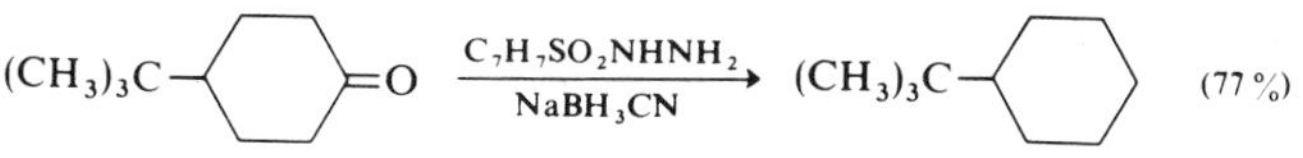

Thioketal Desulfurization

7[g] $H_5C_2O_2C$—(S, S, S, S)—$CO_2C_2H_5$ → Raney Ni → $H_5C_2O_2C$—$CO_2C_2H_5$

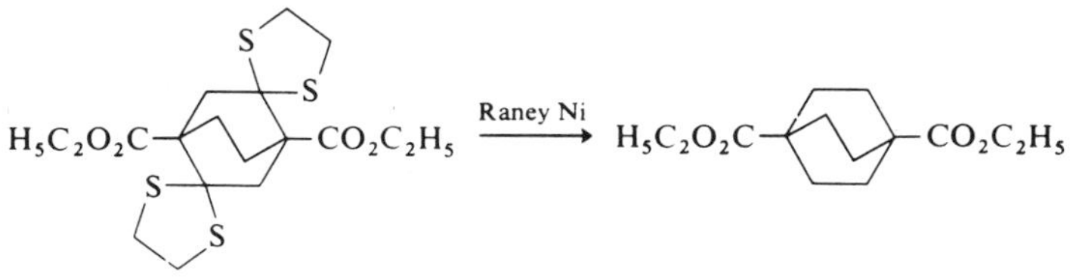

8[h] CH_3, O, $(CH_3)_2CH$, CH_3 → 1) $HSCH_2CH_2SH$, BF_3; 2) Raney Ni → CH_3, $(CH_3)_2CH$, CH_3 (58%)

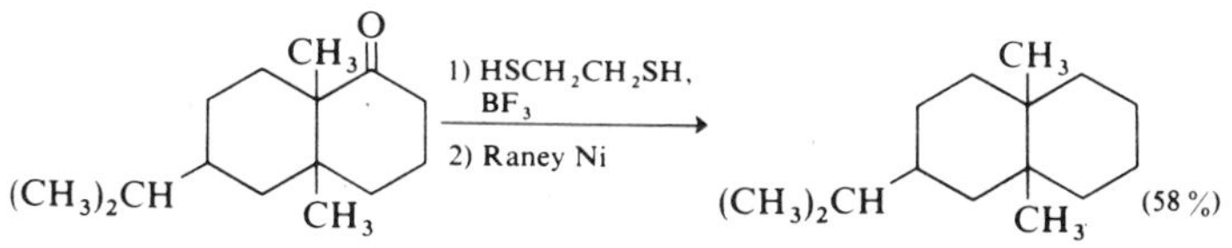

a. R. Schwarz and H. Hering, *Org. Synth.* **IV**, 203 (1963).
b. R. R. Read and J. Wood, Jr., *Org. Synth.* **III**, 444 (1955).
c. L. J. Durham, D. J. McLeod, and J. Cason, *Org. Synth.* **IV**, 510 (1963).
d. D. J. Cram, M. R. V. Sahyun, and G. R. Knox, *J. Am. Chem. Soc.* **84**, 1734 (1962).
e. L. Caglioti and M. Magi, *Tetrahedron* **19**, 1127 (1963).
f. R. O. Hutchins, B. E. Maryanoff, and C. A. Milewski, *J. Am. Chem. Soc.* **93**, 1793 (1971).
g. J. D. Roberts and W. T. Moreland, Jr., *J. Am. Chem. Soc.* **75**, 2167 (1953).
h. P. N. Rao, *J. Org. Chem.* **36**, 2426 (1971).

Scheme 5.13. Conversion of Ketones to Alkenes via Sulfonylhydrazones

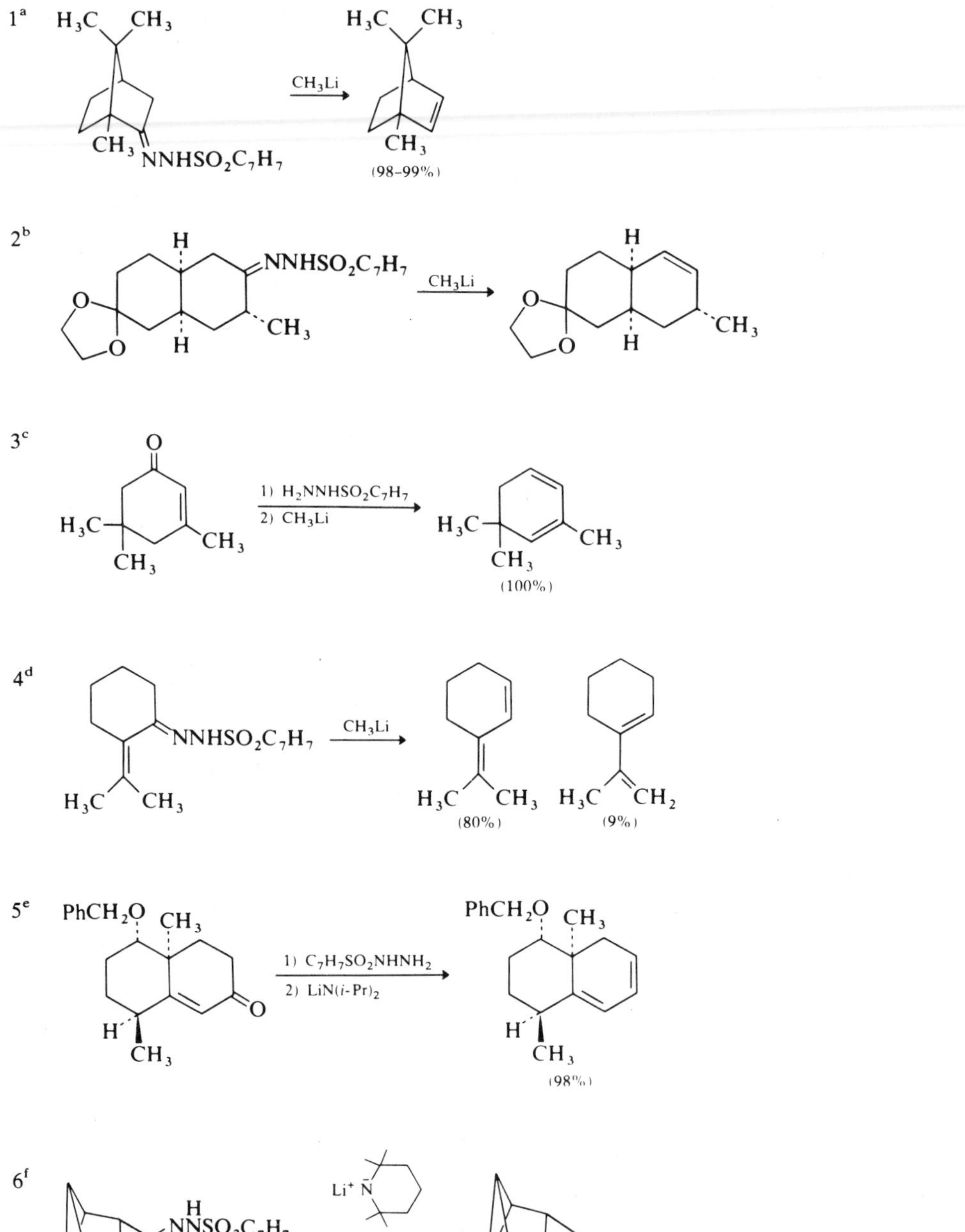

1[a] H_3C CH_3 / CH_3 $NNHSO_2C_7H_7$ —CH_3Li→ H_3C CH_3 / CH_3 (98–99%)

2[b] H / $NNHSO_2C_7H_7$ / O / O / H / CH_3 —CH_3Li→ H / O / O / H / CH_3

3[c] O / H_3C / CH_3 / CH_3 —1) $H_2NNHSO_2C_7H_7$; 2) CH_3Li→ H_3C / CH_3 / CH_3 (100%)

4[d] $NNHSO_2C_7H_7$ / H_3C CH_3 —CH_3Li→ H_3C CH_3 (80%) + H_3C CH_2 (9%)

5[e] $PhCH_2O$ CH_3 / O / H / CH_3 —1) $C_7H_7SO_2NHNH_2$; 2) $LiN(i\text{-}Pr)_2$→ $PhCH_2O$ CH_3 / H / CH_3 (98%)

6[f] H / $NNSO_2C_7H_7$ —Li^+ $\bar{N}$ (2,2,6,6-tetramethylpiperidide)→ (35–55%)

a. R. H. Shapiro and J. H. Duncan, *Org. Synth.* **51**, 66 (1971).
b. W. L. Scott and D. A. Evans, *J. Am. Chem. Soc.* **94**, 4779 (1972).
c. W. G. Dauben, M. E. Lorber, N. D. Vietmeyer, R. H. Shapiro, J. H. Duncan, and K. Tomer, *J. Am. Chem. Soc.* **90**, 4762 (1968).
d. W. G. Dauben, G. T. Rivers, and W. T. Zimmerman, *J. Am. Chem. Soc.* **99**, 3414 (1977).
e. P. A. Grieco, T. Oguri, C.-L. J. Wang, and E. Williams, *J. Org. Chem.* **42**, 4113 (1977).
f. L. R. Smith, G. E. Gream, and J. Meinwald, *J. Org. Chem.* **42**, 927 (1977).

The Shapiro reaction has been particularly useful for cyclic ketones, but the scope of the reaction also includes acylclic systems. In the case of unsymmetrical acyclic ketones, questions of both regiochemistry and stereochemistry arise. 1-Octene is the exclusive product from 2-octanone.[137]

$$\underset{\displaystyle CH_3\overset{\|}{C}(CH_2)_5CH_3}{C_7H_7SO_2N\overset{H}{N}} \xrightarrow{2LiNR_2} CH_2{=}CH(CH_2)_5CH_3$$

This regiospecificity has been shown to depend on the stereochemistry of the C—N bond in the starting hydrazone. There is evidently a strong preference for abstracting the proton *syn* to the arenesulfonyl group, probably because this permits chelation with the lithium ion.

$$ArSO_2N^- \text{—}N{=}CH_3CCH_2R \longrightarrow ArSO_2N^-\cdots Li \text{ chelate } CH_2CCH_2R \longrightarrow \xrightarrow{H^+} CH_2{=}CHCH_2R$$

The Shapiro reaction converts the *p*-toluenesulfonylhydrazones of α,β-unsaturated ketones to dienes (see entries 3-5 in Scheme 5.13).[138]

General References

R. L. Augustine (ed.), *Reduction Techniques and Applications in Organic Synthesis*, Marcel Dekker, New York, 1968.
M. Hudlicky, *Reductions in Organic Chemistry*, Halstead Press, New York, 1984.

Catalytic Reduction

M. Freifelder, *Catalytic Hydrogenation in Organic Synthesis, Procedures and Commentary*, Wiley, New York, 1978.
B. R. James, *Homogeneous Hydrogenation*, Wiley, New York, 1973.
P. N. Rylander, *Hydrogenation Methods*, Academic Press, Orlando, Florida, 1985.
P. N. Rylander, *Hydrogenation in Organic Synthesis*, Academic Press, New York, 1979.

Metal Hydrides

A. Hajos, *Complex Hydrides and Related Reducing Agents in Organic Synthesis*, Elsevier, New York, 1979.
J. Malek, *Org. React.* **34**, 1 (1985); **36**, 249 (1988).

Dissolving-Metal Reductions

A. A. Akhrem, I. G. Rshetova, and Y. A. Titov, *Birch Reduction of Aromatic Compounds*, IFGI/Plenum, New York, 1972.

137. K. J. Kolonko and R. H. Shapiro, *J. Org. Chem.* **43**, 1404 (1978).
138. W. G. Dauben, G. T. Rivers, and W. T. Zimmerman, *J. Am. Chem. Soc.* **99**, 3414 (1977).

Problems

(References for these problems will be found on page 768.)

1. Give the product(s) to be expected from the following reactions. Be sure to specify all facets of stereochemistry.

(a) $(CH_3)_2CHCH{=}CHCH{=}CHCO_2CH_3 \xrightarrow{(i\text{-Bu})_2AlH}$

(b) 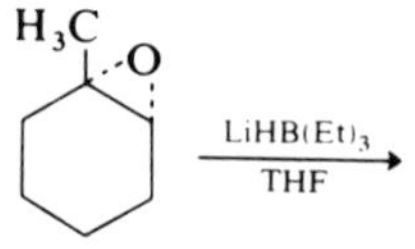

(c) 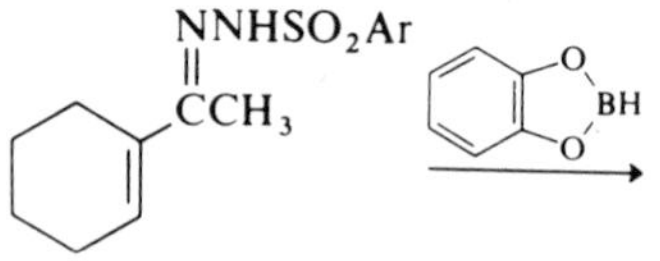

(d) 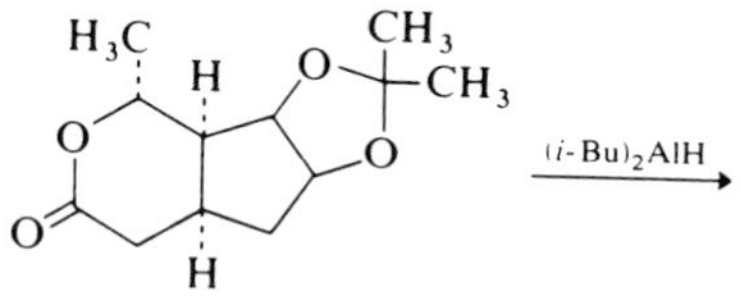

(e) 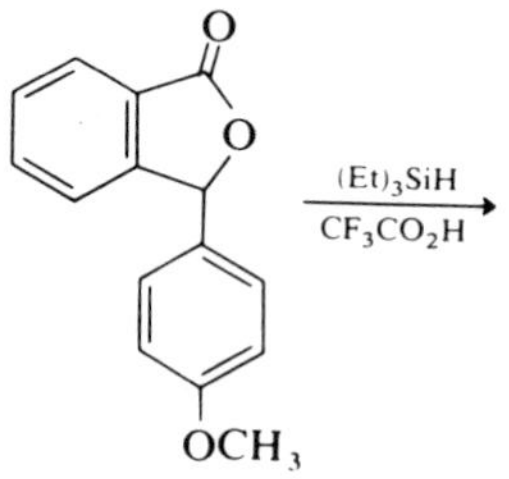

(f) 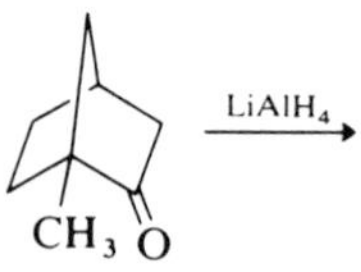

(g) 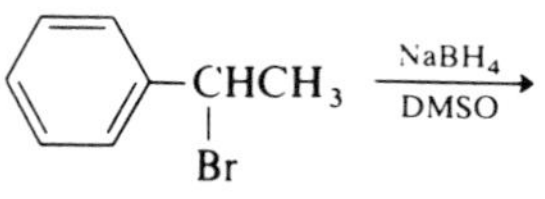

(h) 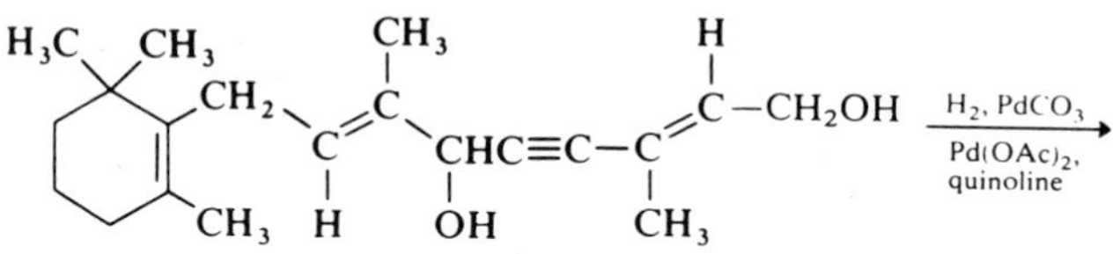

(i) 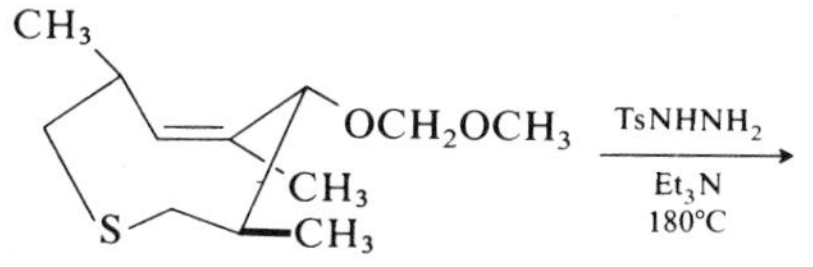 CH_3, OCH_2OCH_3, CH_3, CH_3, S $\xrightarrow[\text{Et}_3\text{N, 180°C}]{\text{TsNHNH}_2}$

(j) 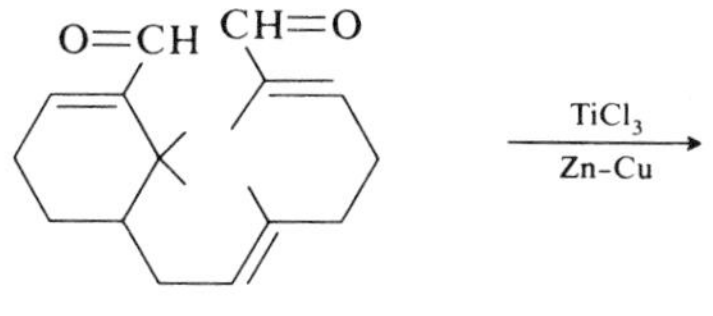 O=CH CH=O $\xrightarrow[\text{Zn-Cu}]{\text{TiCl}_3}$

2. Indicate the stereochemistry of the major alcohol that would be formed by sodium borohydride reduction of each of the cyclohexanone derivatives shown:

(a) 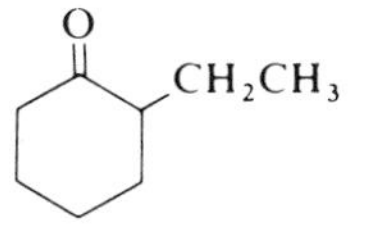 O, CH_2CH_3

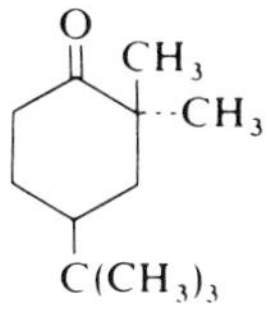

(b) 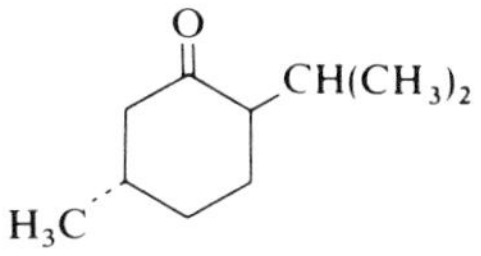 O, $CH(CH_3)_2$, H_3C

(c) O, CH_3, CH_3, $C(CH_3)_3$

(d) 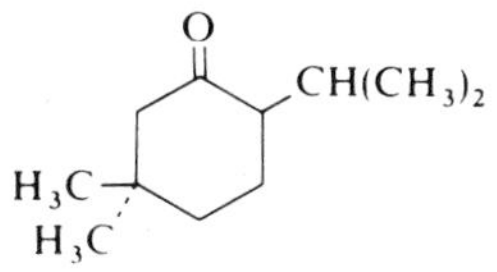 O, $CH(CH_3)_2$, H_3C, H_3C

3. Indicate reaction conditions that would accomplish each of the following transformations in one step.

(a) 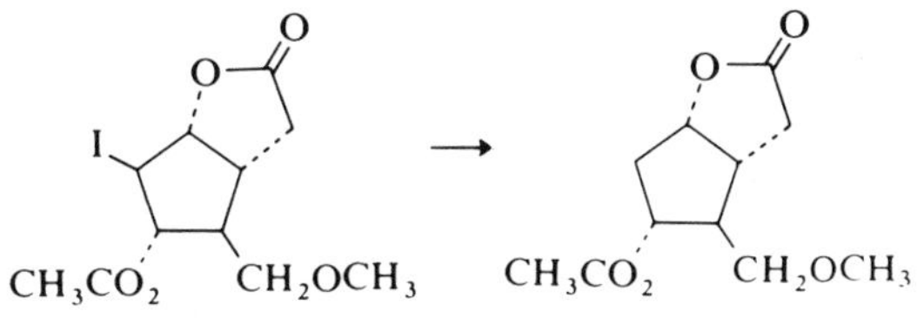 I, O, O, CH_3CO_2, CH_2OCH_3 → O, O, CH_3CO_2, CH_2OCH_3

(b) 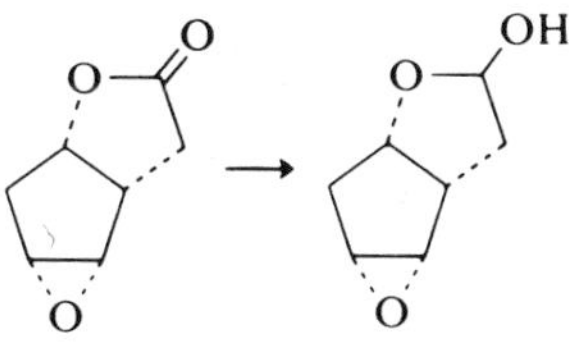 O, O, O → OH, O, O

(c) 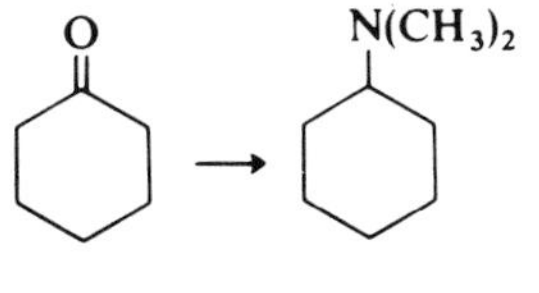 O → $N(CH_3)_2$

(d) O, H, H → OH, H, H

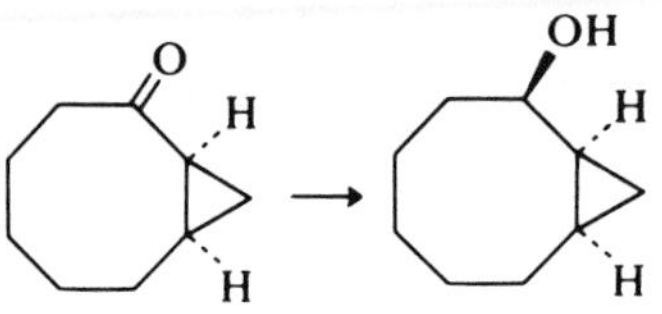

(e) CH_3O–C≡N → CH_3O–CH=O

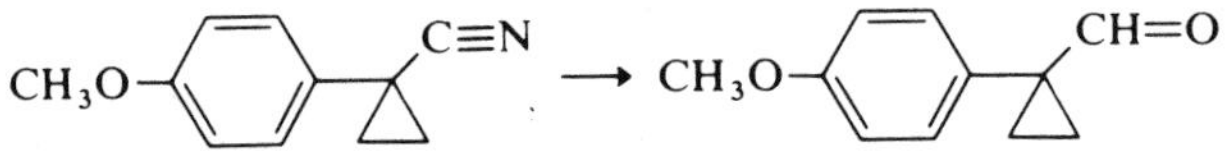

(f) CH_3, O, CH_3, O, O, OH, O, CH_3, O, CH_3 → CH_3, O, CH_3, O, O, O, CH_3, O, CH_3

(g) H_3C, CH_2CN, H_3C, O, H → H_3C, CH_2CN, H_3C, HO, H, H

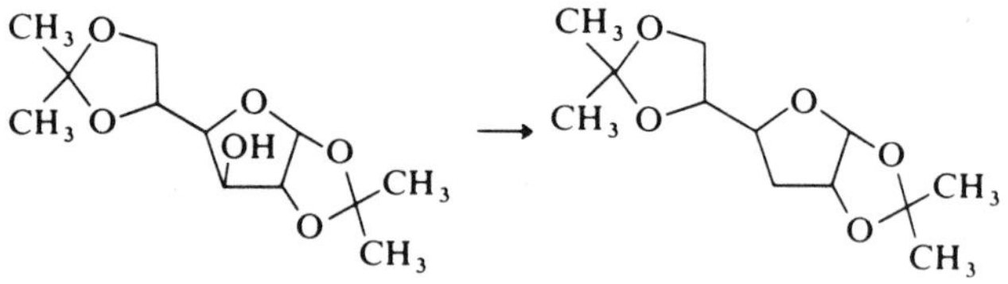

(h) HO, CH_3, O, H, CH_3 → HO, CH_3, H, CH_3

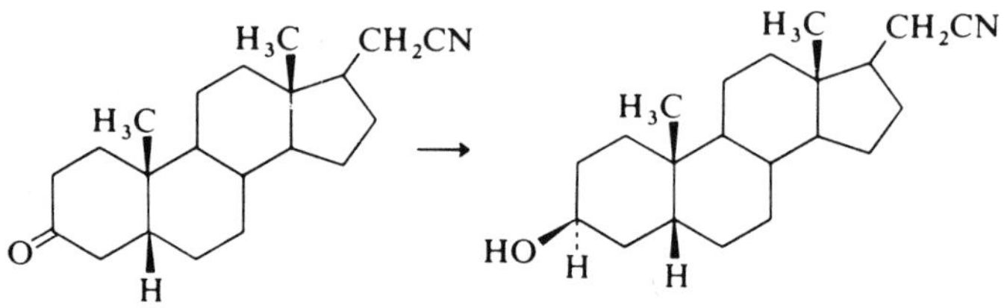

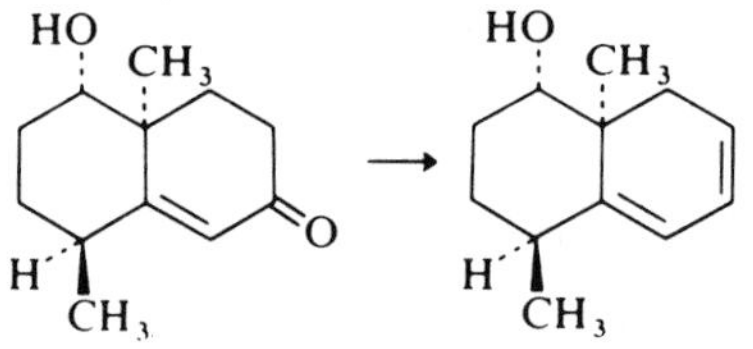

(i) OCH_2Ph, CH_3, CH_3, $CH_2OPO[N(CH_3)_2]_2$, C_2H_5O, OC_2H_5 → OH, CH_3, CH_3, CH_3, C_2H_5O, OC_2H_5

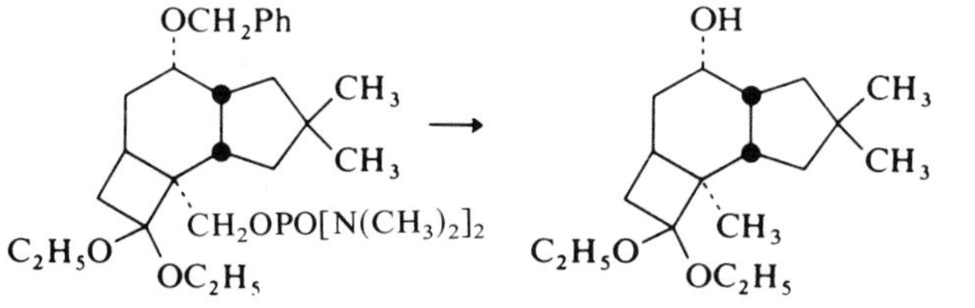

(j) O_2N–C(=O)N$(CH_3)_2$ → O_2N–$CH_2N(CH_3)_2$

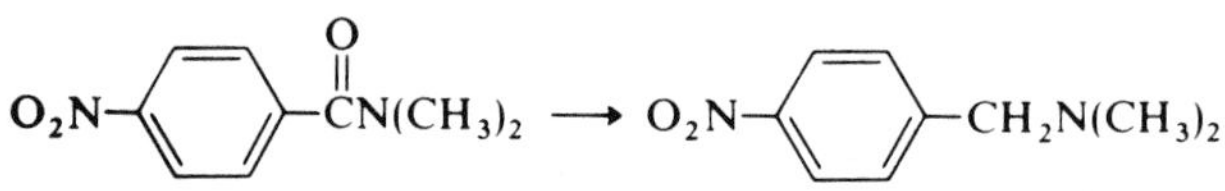

(k) O, H_3C, CH_3, CH_3 → H_3C, CH_3, CH_3

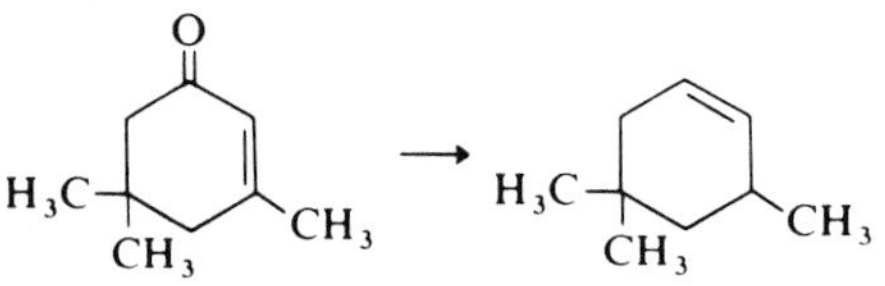

(l)

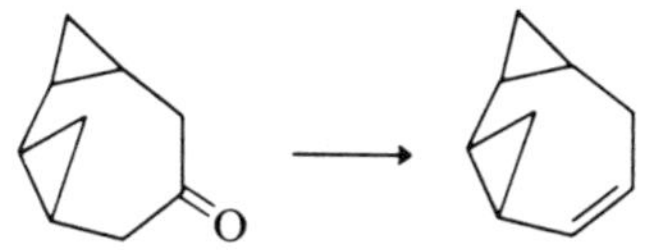

(m) $-CO_2CH_3 \rightarrow$ $-CO_2CH_3$

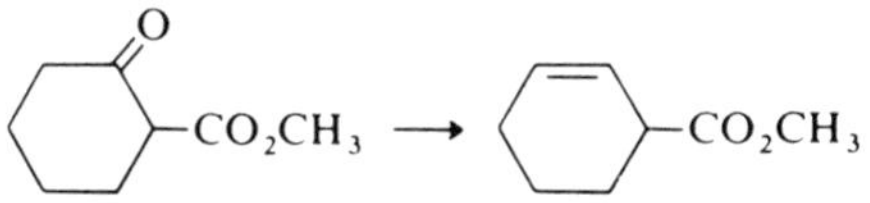

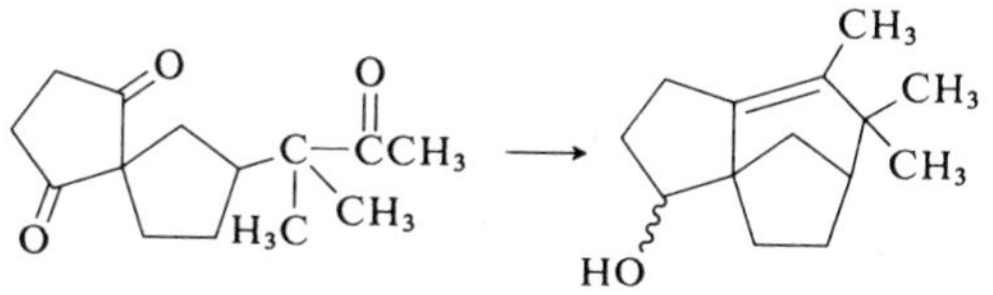

(n) O, O, O, C–CCH$_3$, H_3C, CH_3 $\rightarrow$ CH_3, CH_3, CH_3, HO

(o) $H_3C\ (CH_2)_3C{\equiv}C(CH_2)_3OTHP \rightarrow H_3C\ (CH_2)_3$ H, $(CH_2)_3OTHP$, C=C, H

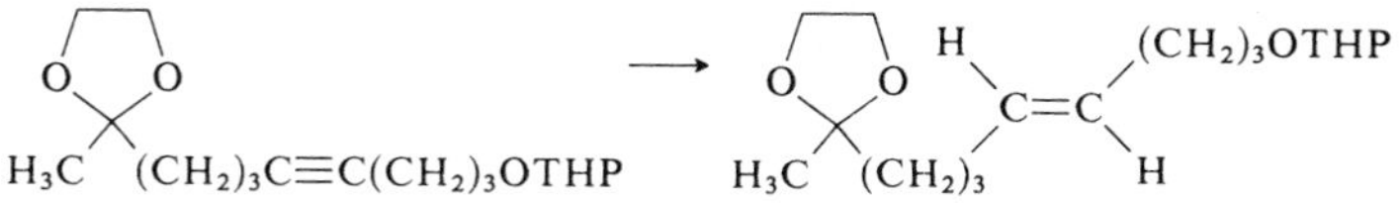

4. Predict the stereochemistry of the products from the following reactions and justify your prediction.

(a) CH_3, CH_3 $\xrightarrow[H_2O]{KBH_4}$

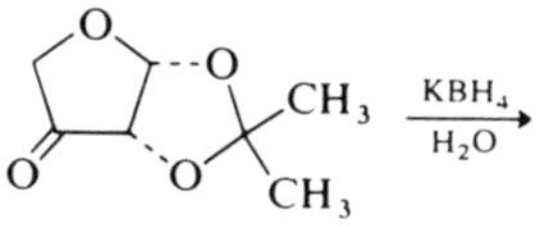

(b) CH_3, H, C–C, O, Ph, Ph $\xrightarrow[Et_2O]{LiAlH_4}$

(c) H, O, H $\xrightarrow{LiHB(Et)_3}$

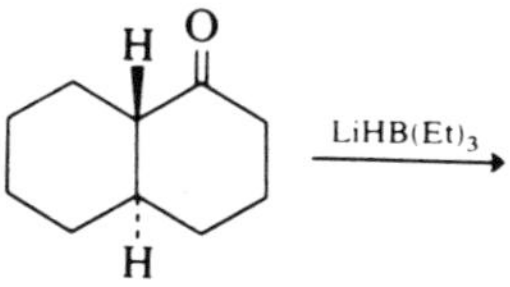

(d) $(CH_3)_2CH$, HO, CH_3 $\xrightarrow[\text{ethanol}]{Pt,\ H_2}$

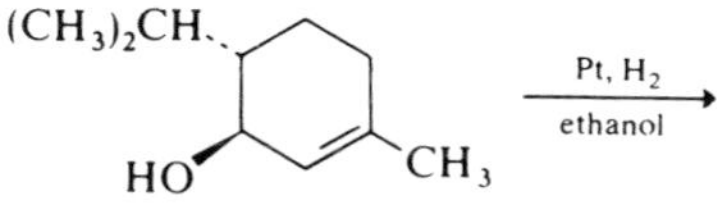

(e) H_3C, O, O, H_2C, CH_3, O_2CCH_3 $\xrightarrow[(PPh_3)_3RhCl]{H_2}$

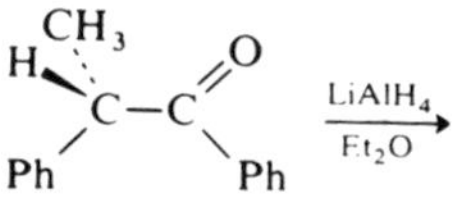

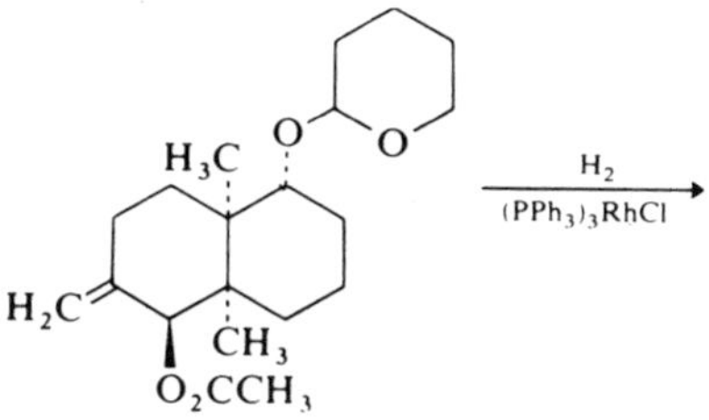

(f) $C(CH_3)_3$ / H_3C / CH_3 / OH $\xrightarrow[Rh/Al_2O_3]{H_2}$

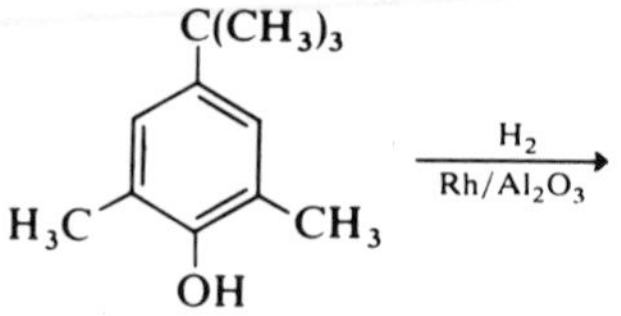

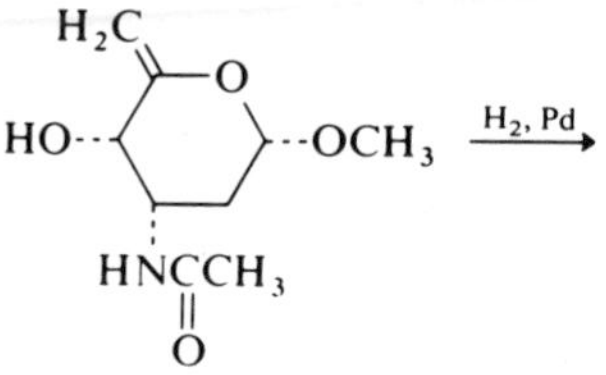

(g) H_2C / O / HO / OCH_3 / $HNCCH_3$ / O $\xrightarrow{H_2, Pd}$

(h) CH_3 / O / O / O / CH_3 $\xrightarrow{H_2/Pd\text{-}C}$

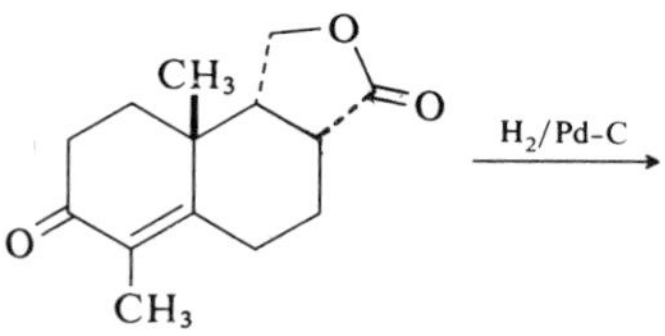

(i) OCH_2OCH_3 / $PhCH_2O$ / CH_3OCH_2O / O / OMe $\xrightarrow{Zn(BH_4)_2}$

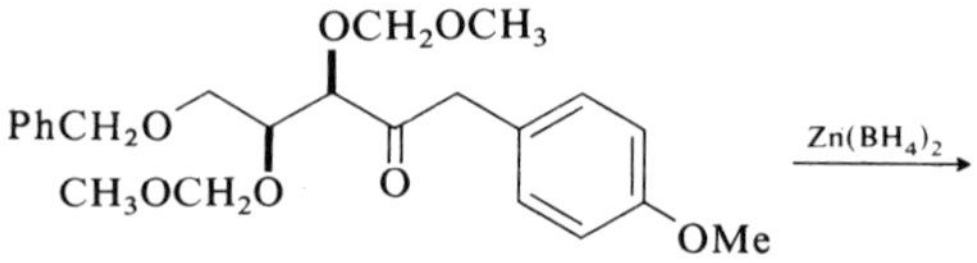

(j) CH_3 / OH $\xrightarrow{[NBD-Rh(P\ P)]^{+1}}$

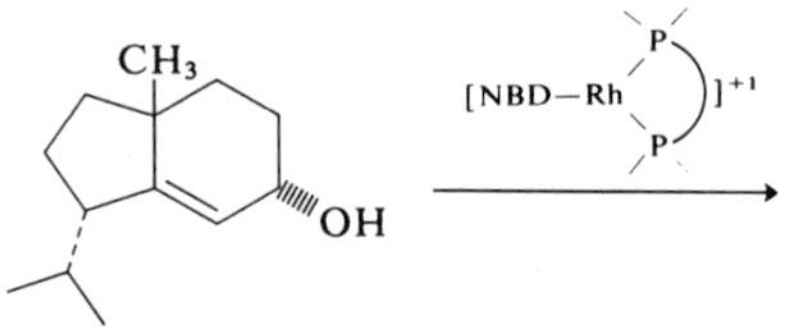

(k) $CH_3(CH_2)_4C{\equiv}CCH_2OH \xrightarrow[ether]{LiAlH_4}$

(l) CH_3 / OH / CH_3 $\xrightarrow{[COD\text{-}Ir\text{-}Pyr, PR_3]^{+1}}$

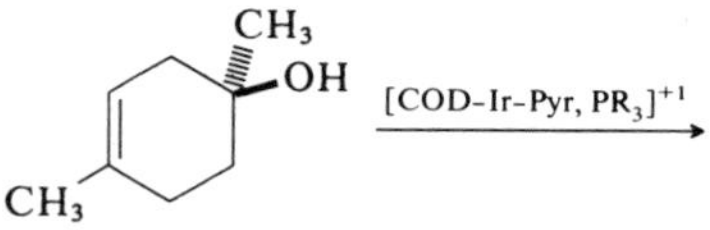

(m) $PhCH(CH_3)C(=O)CH_2CH_3 \xrightarrow{L\text{-}Selectride}$

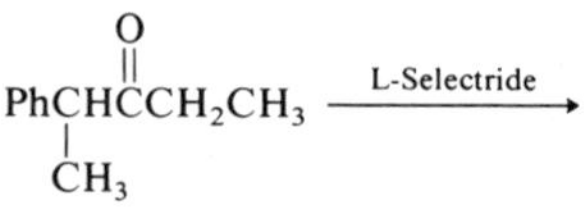

5. Suggest a convenient method for carrying out the following syntheses. The compound on the left is to be synthesized from the one on the right (retrosynthetic notation). No more than three steps should be necessary.

(a) 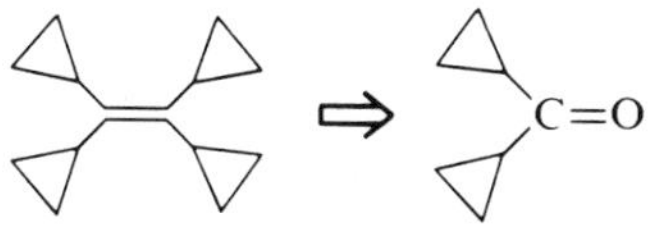 C=O

(b) CO_2CH_3 ⇒ CO_2CH_3, O

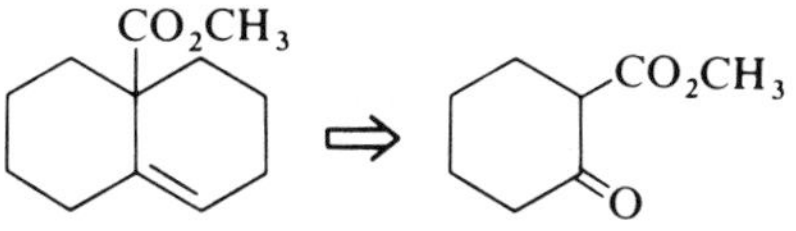

(c) H_3C, O, O, O, H_3C, H, CH_2OH ⇒ H_3C, O, O, O, CH_3

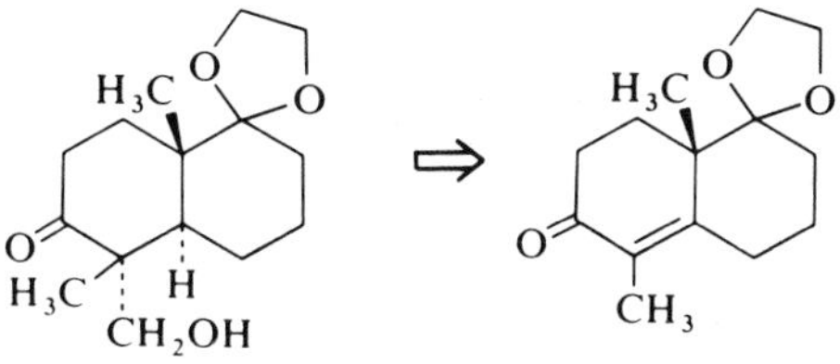

(d) 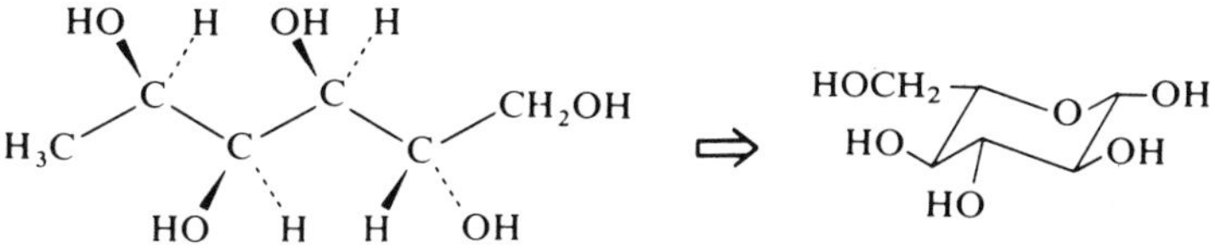 HO H OH H, C, C, H_3C, C, C, CH_2OH, HO H H OH ⇒ $HOCH_2$, O, OH, HO, OH, HO

(e) 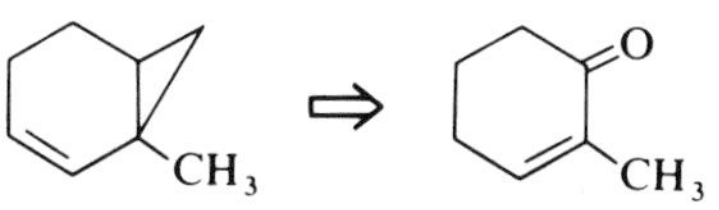 CH_3 ⇒ O, CH_3

(f) 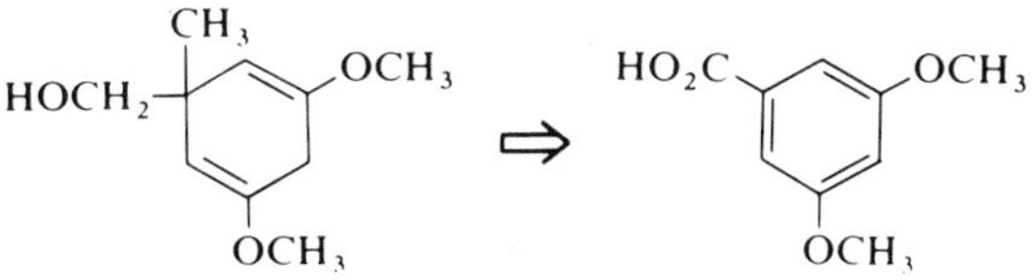 CH_3, $HOCH_2$, OCH_3, OCH_3 ⇒ HO_2C, OCH_3, OCH_3

(g) 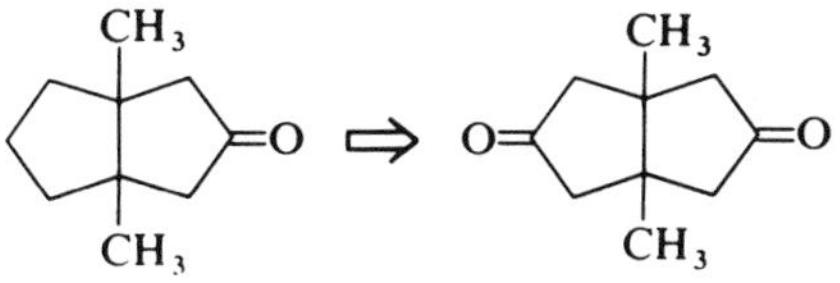 CH_3, =O, CH_3 ⇒ O=, CH_3, =O, CH_3

(h) *meso*-$(CH_3)_2CHCH(OH)CH(OH)CH(CH_3)_2$ ⇒ $(CH_3)_2CHCO_2CH_3$

(i) 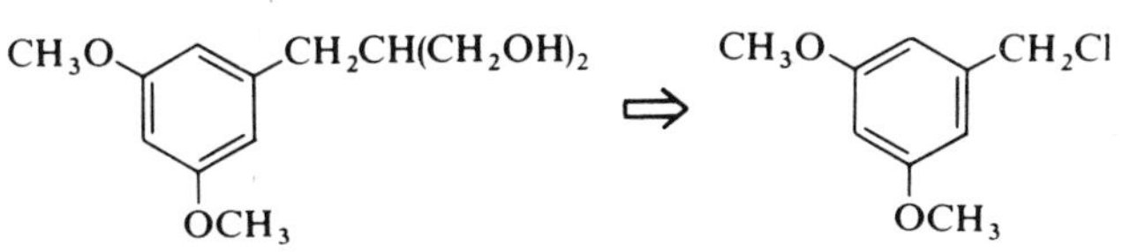 CH_3O, $CH_2CH(CH_2OH)_2$, OCH_3 ⇒ CH_3O, CH_2Cl, OCH_3

(j) $C_6H_5CH(SC_6H_5)CH_2CH(OH)CH_3$ ⇒ $C_6H_5CH{=}CHC(=O)CH_3$

6. Offer an explanation to account for the observed differences in rate which are described.
   (a) $LiAlH_4$ reduces the ketone camphor about 30 times faster than does $NaAlH_4$.
   (b) The rate of reduction of camphor by $LiAlH_4$ is decreased by a factor of about 4 when a crown ether is added to the reaction mixture.
   (c) For reduction of cyclohexanones by lithium tri-*t*-butoxyaluminum hydride, the addition of one methyl group at C-3 has little effect on the rate but a second group has a large effect. The addition of a third methyl group at C-5 has no effect. The effect of a fourth group is also rather small.

	Rate:
cyclohexanone	439
3-methylcyclohexanone	280
3,3-dimethylcyclohexanone	17.5
3,3,5-trimethylcyclohexanone	17.4
3,3,5,5-tetramethylcyclohexanone	8.9

7. Suggest reaction conditions appropriate for stereoselectively converting the octalone shown to each of the diastereomeric decalones.

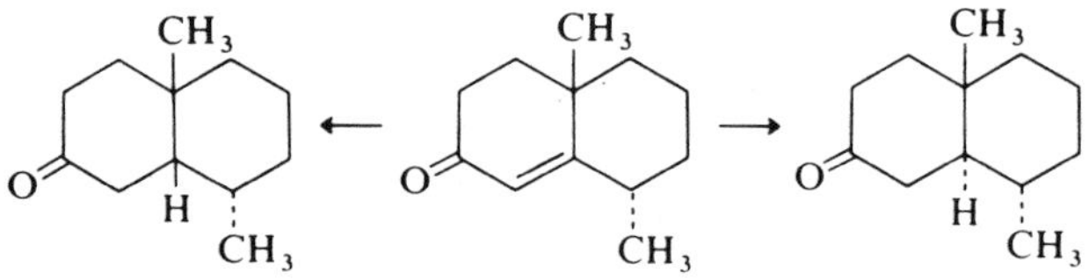

CH3 CH3 CH3
O H CH3 O CH3 O H CH3

8. The fruit of a shrub which grows in Sierra Leone is very toxic and has been used as a rat poison. The toxic principle has been identified as *Z*-18-fluoro-9-octadecenoic acid. Suggest a synthesis for this material from 8-fluorooctanol, 1-chloro-7-iodoheptane, acetylene, and any other necessary organic or inorganic reagents.

9. Each of the following molecules contains more than one potentially reducible group. Indicate a reducing agent which would be suitable for effecting the desired selective reduction. Explain the basis for the expected selectivity.

(a)
H CO2CH3 H CO2CH3
O O
O H O H
CH3 CH2 CH3 CH3

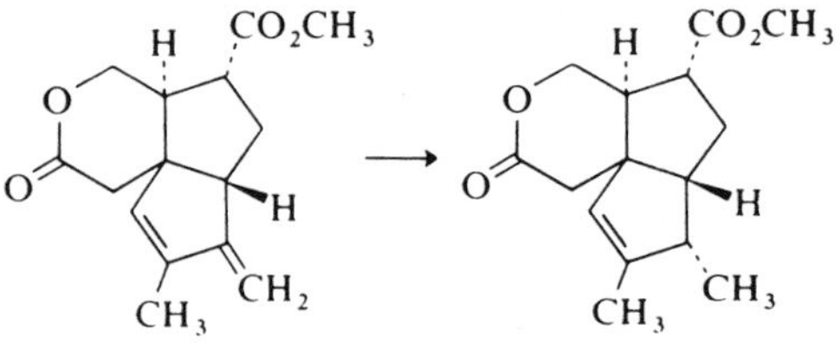

(b) [structures: ketone → alcohol; lactone ring with trimethoxyphenyl group]

CH_3O OCH_3 OCH_3 → CH_3O OCH_3 OCH_3

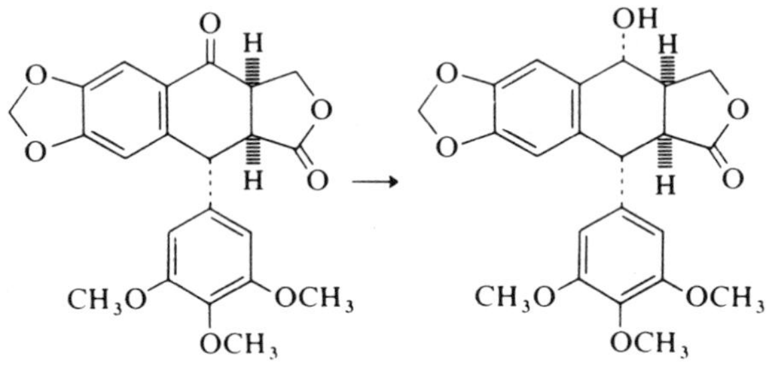

(c) HO_2C–(acetonide-protected furanose)–$CH_2CO_2C_2H_5$ → $HOCH_2$–(acetonide-protected furanose)–$CH_2CO_2C_2H_5$

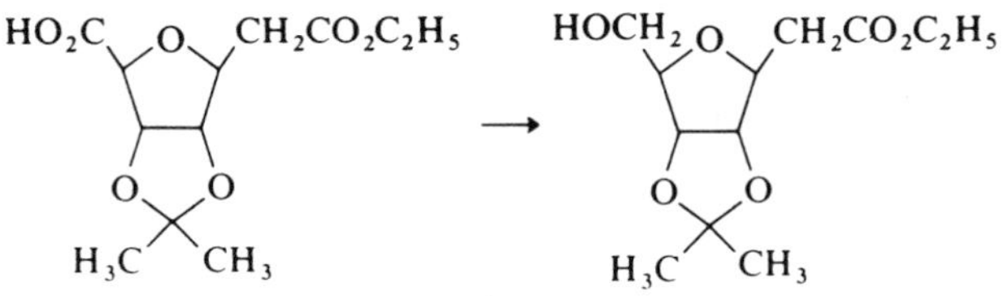

(d) $CH_3CH_2C\equiv CCH_2C\equiv CCH_2OH \longrightarrow CH_3CH_2C\equiv CCH_2C(H)=C(H)CH_2OH$

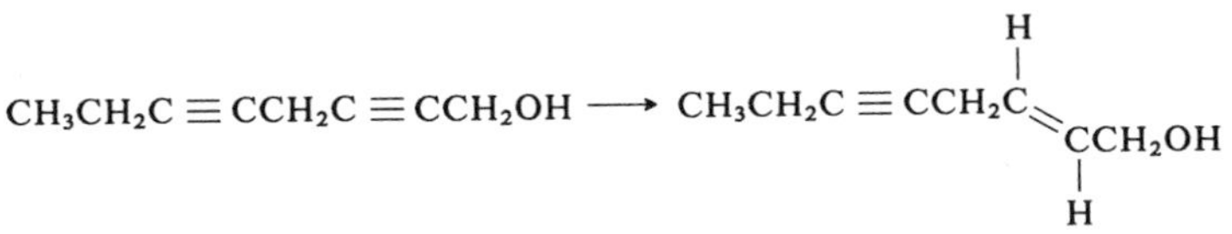

(e) [structure with $CH_3CH_2CH(CH_3)CO_2$, H, CH_3, O, $CH_2CCH_2CO_2CH_3$, OTMS] → [structure with $CH_3CH_2CH(CH_3)CO_2$, H, CH_3, $CH_2CH_2\overset{O}{\overset{\|}{C}}CH_2CHCH_2CO_2CH_3$, OTMS]

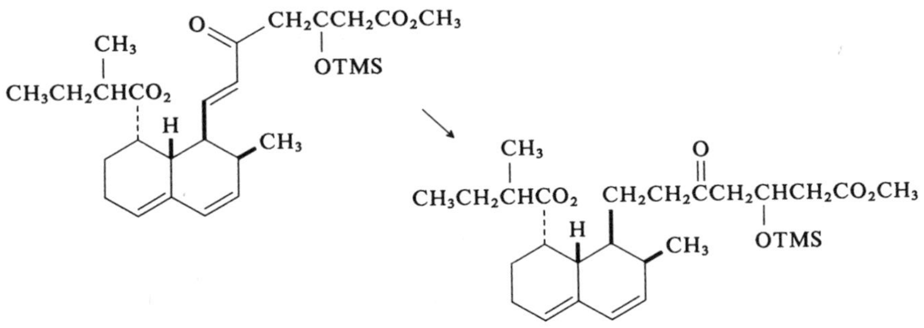

(f) $CH_3(CH_2)_3\overset{O}{\overset{\|}{C}}(CH_2)_4\overset{O}{\overset{\|}{C}}Cl \longrightarrow CH_3(CH_2)_3\overset{O}{\overset{\|}{C}}(CH_2)_4CH{=}O$

(g) [structure with H_3C, H, H, OCH_2Ph, HOH_2C, OH, H] → [structure with H_3C, H, OCH_2Ph, H_3C, OH, H]

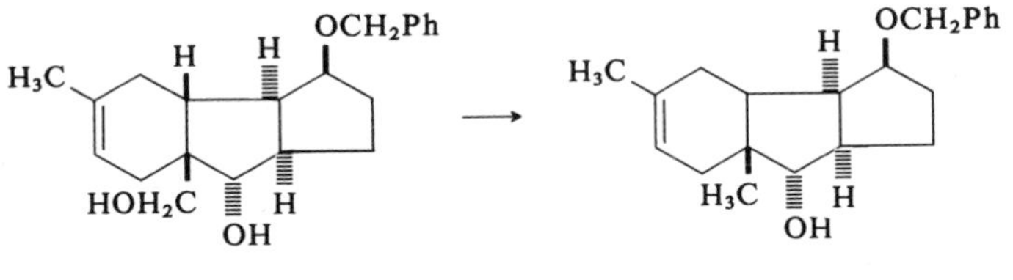

10. Explain the basis of the observed stereoselectivity for the following reductions:

(a) [2-methylcyclohexanone, CH_3] —(borohydride reagent, H–B⁻)→ [cis-2-methylcyclohexanol, OH, CH_3]

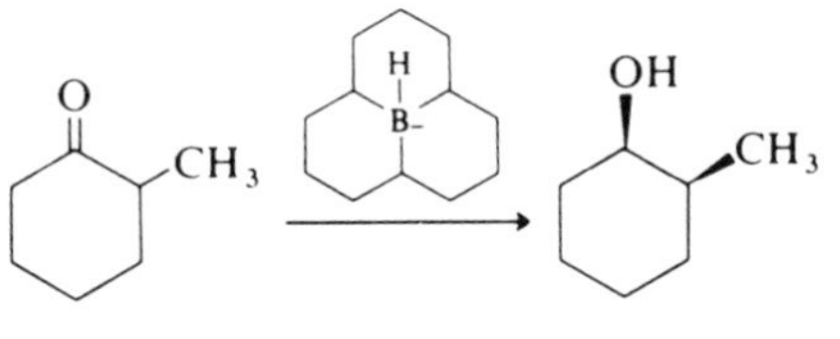

(b) H Br, Br, H $\xrightarrow{Bu_3SnH}$ H Br, H, H

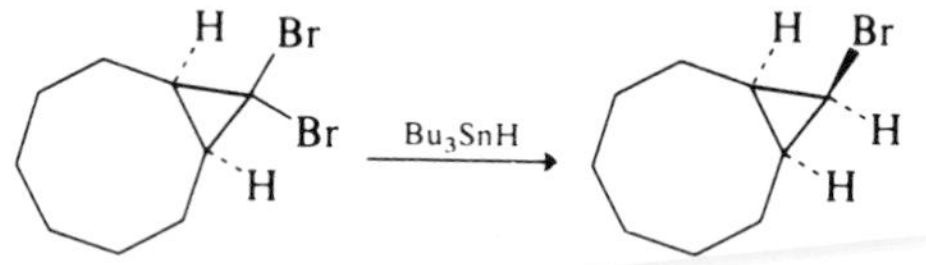

(c) H_3C, OCH_3, O $\xrightarrow[EtOH]{Li, NH_3}$ H_3C, OCH_3, O, H

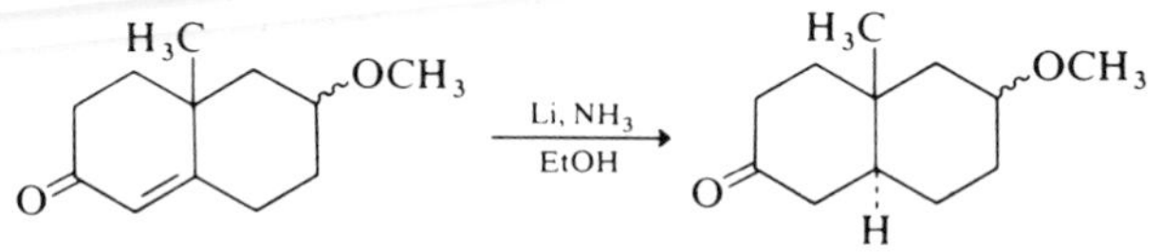

11. A valuable application of sodium cyanoborohydride is in the synthesis of amines by reductive amination. What combination of carbonyl and amine components would you choose to prepare the following amines by this route? Explain your choices.

(a) $N(CH_3)_2$ (b) NH_2 (c) N, N H

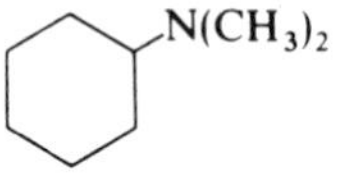

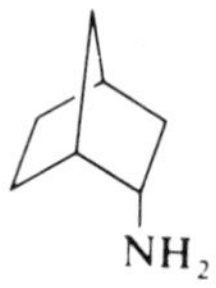

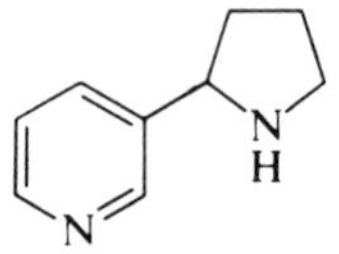

12. The reduction of *o*-bromophenyl allyl ether by $LiAlH_4$ has been studied in several solvents. In ether two products are formed. The ratio **A**:**B** increases with increasing $LiAlH_4$ concentration. When $LiAlD_4$ is used as the reductant, about half of the product **B** is a monodeuterated derivative. Provide a mechanistic rationale for these results. What is the most likely location of the deuterium atom in the deuterated product? Why is the product not completely deuterated?

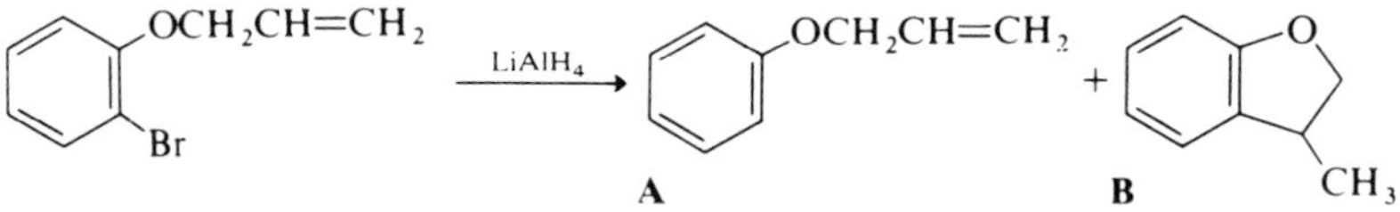

$OCH_2CH{=}CH_2$, Br $\xrightarrow{LiAlH_4}$ $OCH_2CH{=}CH_2$ (**A**) + O, CH_3 (**B**)

13. A simple synthesis of 2-substituted cyclohexanones has been developed. Although the yields are only 25-30%, it is carried out as a "one-pot" process using the sequence of reactions shown below. Explain the mechanistic basis of this synthesis and identify the intermediate present after each stage of the reaction.

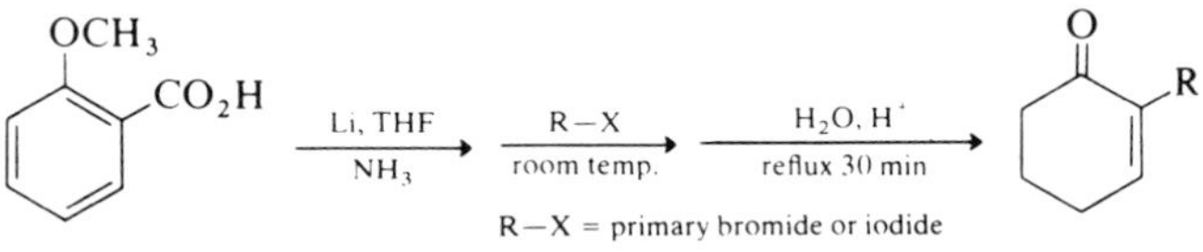

OCH_3, CO_2H $\xrightarrow[NH_3]{Li, THF}$ $\xrightarrow[\text{room temp.}]{R-X}$ $\xrightarrow[\text{reflux 30 min}]{H_2O, H^+}$ O, R

R—X = primary bromide or iodide

14. Birch reduction of 3,4,5-trimethoxybenzoic acid gives in 94% yield a dihydrobenzoic acid which bears only *two* methoxysubstituents. Suggest a plausible structure for this product based on the mechanism of the Birch reduction.

15. In a multistep synthetic sequence it was necessary to remove selectively one of two secondary hydroxyl groups.

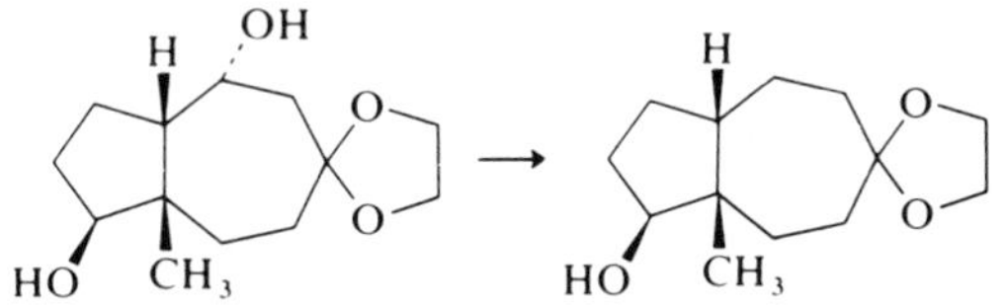

H OH O O HO CH_3 $\rightarrow$ H O O HO CH_3

Consider several (at least three) methods by which this transformation might be accomplished. Discuss the relative merits of the various possibilities and recommend one as the most likely to succeed or most convenient.

16. Wolff-Kishner reduction of ketones that bear other functional groups sometimes give products other than the corresponding methylene compound. Some examples are given. Indicate a mechanism for each of the reactions.

(a) $(CH_3)_3CC(=O)CH_2OPh \longrightarrow (CH_3)_3CCH{=}CH_2$

(b) 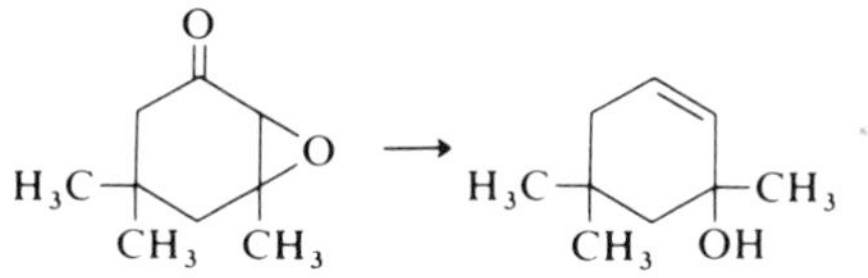 O O H_3C CH_3 CH_3 $\rightarrow$ H_3C CH_3 CH_3 OH

(c) 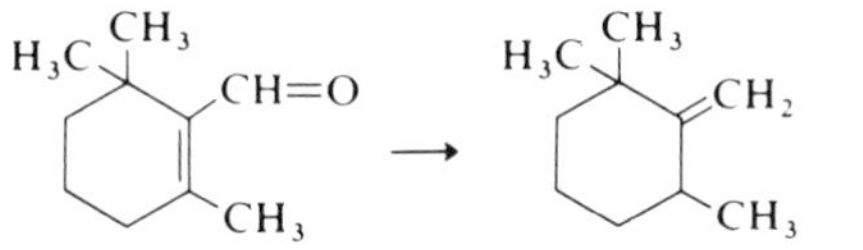 H_3C CH_3 $CH{=}O$ CH_3 $\rightarrow$ H_3C CH_3 CH_2 CH_3

(d) $PhCH{=}CHCH{=}O \rightarrow$ Ph N N H H

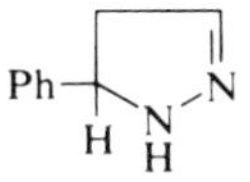

17. Suggest reagents and reaction conditions that would be suitable for each of the following selective or partial reductions:

(a) $HO_2C(CH_2)_4CO_2C_2H_5 \rightarrow HOCH_2(CH_2)_4CO_2C_2H_5$

(b) 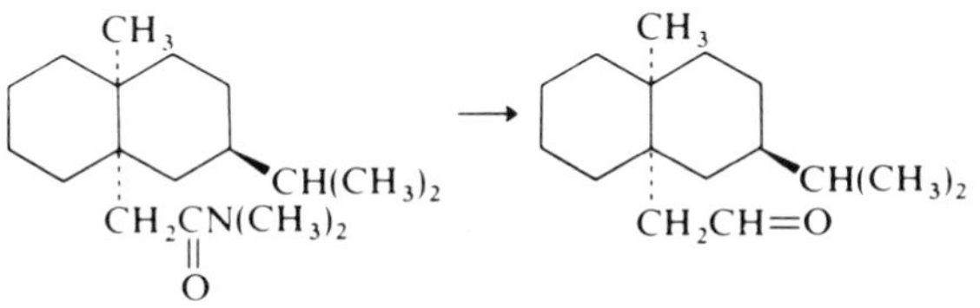 CH_3 $CH(CH_3)_2$ $CH_2C(=O)N(CH_3)_2$ $\rightarrow$ CH_3 $CH(CH_3)_2$ $CH_2CH{=}O$

(c) $CH_3C(=O)(CH_2)_2CO_2C_8H_{17} \rightarrow CH_3(CH_2)_3CO_2C_8H_{17}$

(d) $CH_3CONH-C_6H_4-CO_2CH_3 \rightarrow CH_3CH_2NH-C_6H_4-CO_2CH_3$

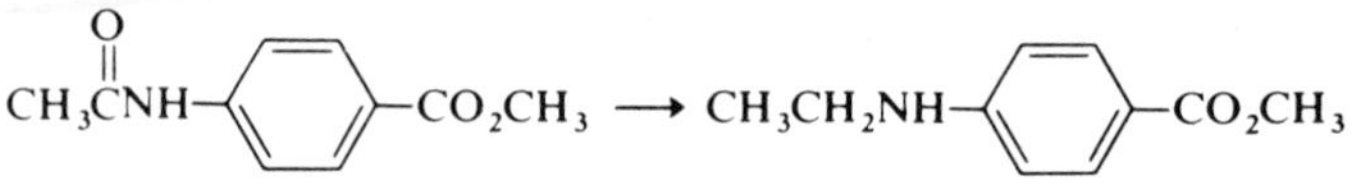

(e) $O_2N-C_6H_4-CO-C_6H_5 \rightarrow O_2N-C_6H_4-CH_2-C_6H_5$

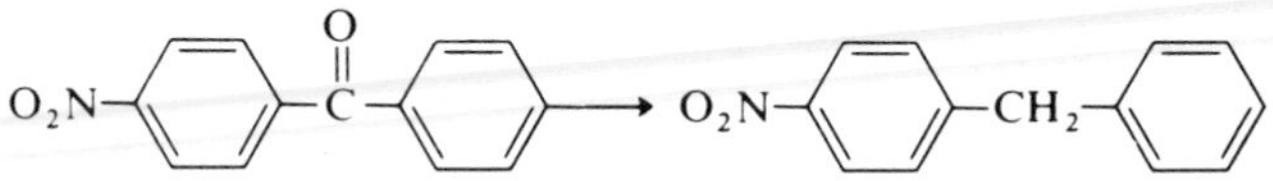

(f) 3-methylcyclohex-2-enone (CH_3, =O) → 3-methylcyclohex-2-enol (CH_3, –OH)

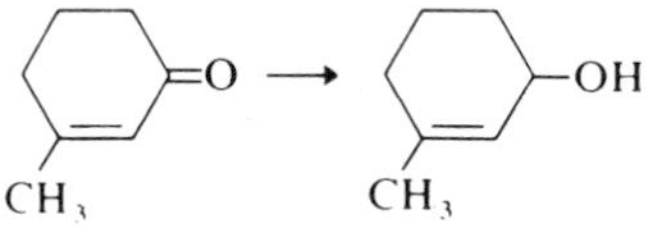

(g) O–O, O → O–O, O

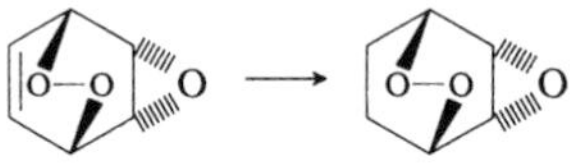

18. In the reduction of the ketone **A**, product **B** is favored with *increasing* stereoselectivity in the order $NaBH_4 < LiAlH_2(OCH_2CH_2OCH_3)_2 < Zn(BH_4)_2$. With L-Selectride, stereoisomer **C** is favored. Account for the dependence of the stereoselectivity on the various reducing agents.

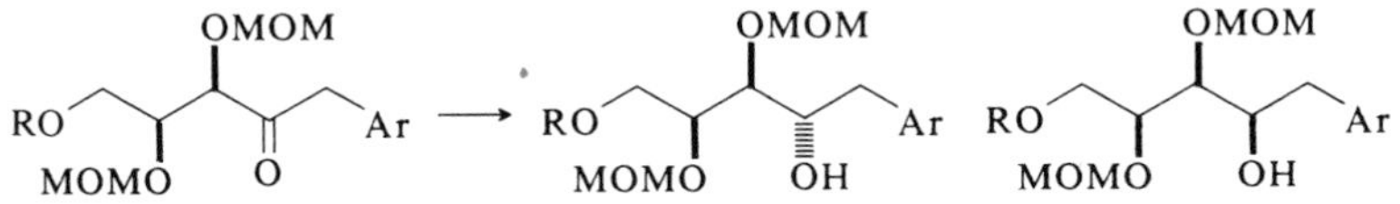

OMOM, RO, Ar, MOMO, O → OMOM, RO, Ar, MOMO, OH; OMOM, RO, Ar, MOMO, OH

Ar = 4-methoxyphenyl
R = benzyl
MOM = methoxymethyl

A **B** **C**

19. The following reducing agents effect enantioselective reduction of ketones. Propose a mechanism and transition state structure which would be in accord with the observed enantioselectivity.

(a) $C_6H_5COCO_2CH_3$ + (CH_3, B) → *R*-α-hydroxyester in 90% e.e.

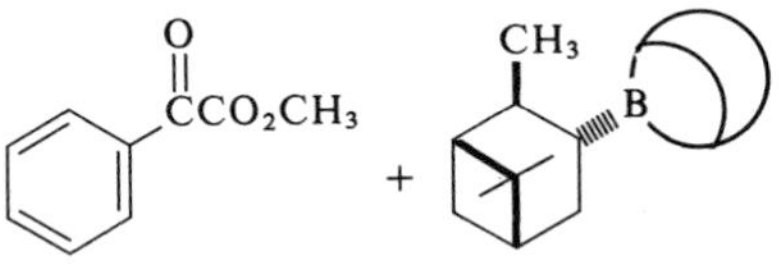

(b) $C_6H_5COCH_2CH_3$ + BH_3 (0.6 equiv) + (H, Ph, Ph, O, N–B, Me) (0.1 equiv) → *R*-alcohol in 97% e.e.

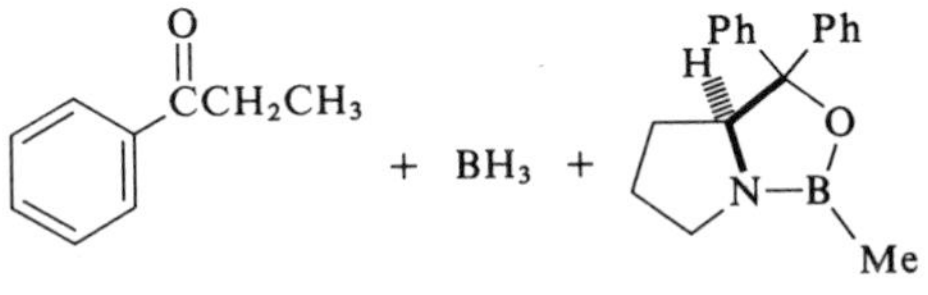

(c) $p\text{-}BrC_6H_4COCH_3$ + ( )$_2$BCl → *S*-alcohol in 97% e.e.

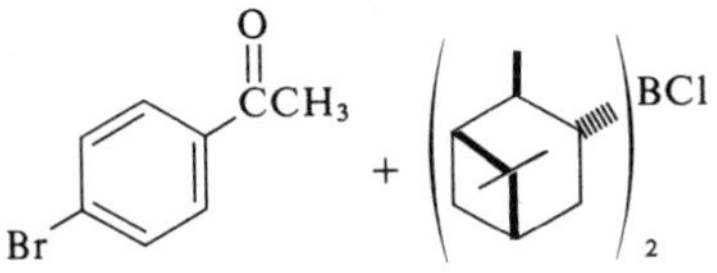

20. Devise a sequence of reactions which would accomplish the following synthesis:

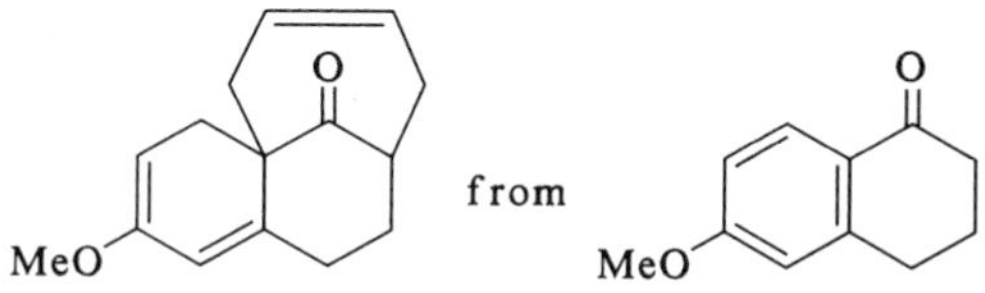

21. A group of topologically unique molecules known as "betweenenes" have been synthesized. Successful synthesis of such molecules depends on effective means of closing large rings. Suggest an overall strategic approach (details are not required) to synthesize such molecules. Suggest reaction types which might be considered for formation of the large rings.

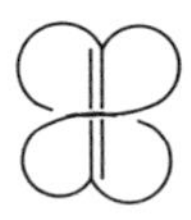

6

Cycloadditions, Unimolecular Rearrangements, and Thermal Eliminations

Most of the reactions described to this point involve polar or polarizable reactants and proceed through polar intermediates or transition states. Carbanion alkylations, nucleophilic additions to carbonyl groups, and electrophilic additions to alkenes are all examples of such reactions. The reactions to be examined in the present chapter, on the other hand, occur by a reorganization of valence electrons through activated complexes that are not much more polar than the reactants. These reactions usually proceed through cyclic transition states, and little separation of charge occurs during these processes. The energy necessary to attain the transition state is usually provided by thermal or photochemical excitation of the reactant(s), and frequently no other reagents are involved. Many of the reactions fall into the category of *concerted processes* and can be analyzed mechanistically in terms of the orbital symmetry concepts discussed in Chapter 11 of Part A. We will also discuss some reactions which effect closely related transformations but which, on mechanistic scrutiny, are found to proceed through discrete intermediates.

6.1. Cycloaddition Reactions

Cycloaddition reactions result in the formation of a new ring from two reacting molecules. A concerted mechanism requires that a single transition state, and therefore no intermediate, lie on the reaction path between reactants and adduct.

Two important examples of cycloadditions that usually occur by concerted mechanisms are the *Diels–Alder reaction,*

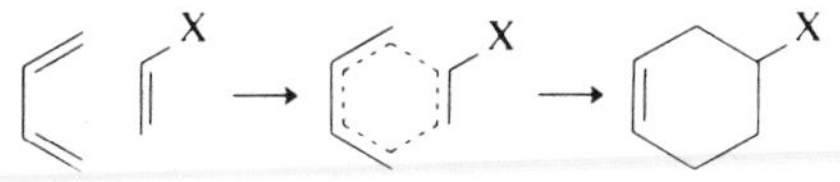

and *1,3-dipolar cycloaddition*:

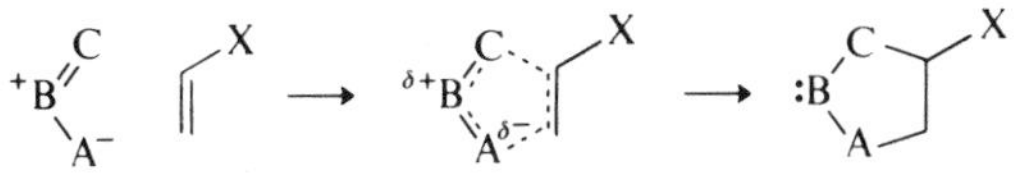

A firm understanding of concerted cycloaddition reactions developed as a result of the formulation of the mechanism within the framework of molecular orbital theory. Consideration of the molecular orbitals of reactants and products revealed that in some cases a smooth transformation of the orbitals of the reactants to those of products is possible. In other cases, reactions that appear feasible if no consideration is given to the symmetry and spatial orientation of the orbitals are found to require high-energy transition states when the orbitals are considered in detail. (Review Section 11.3 of Part A for a discussion of the orbital symmetry analysis of cycloaddition reactions.) These considerations permit description of various types of cycloaddition reactions as "allowed" or "forbidden" and permit conclusions as to whether specific reactions are likely to be energetically feasible. In this chapter, the synthetic applications of cycloaddition reactions will be emphasized. The same orbital symmetry relationships that are informative as to the feasibility of a reaction are often predictive of the regiochemistry and stereochemistry of the process. This predictability is an important feature for synthetic purposes. Another attractive feature of cycloaddition reactions is the fact that *two* new bonds are formed in a single reaction. This can enhance the efficiency of a synthetic process.

6.1.1. The Diels–Alder Reaction: General Features

The cycloaddition of alkenes and dienes is a very useful method for forming substituted cyclohexenes. This reaction is known as the *Diels–Alder reaction.*[1] The concerted nature of the mechanism was generally accepted and the stereospecificity of the reaction was firmly established before the importance of orbital symmetry was recognized. In the terminology of orbital symmetry classification, the Diels–Alder reaction is a $[_{\pi}4_s + {_{\pi}2_s}]$ cycloaddition, an allowed process. The transition state for a concerted reaction requires that the diene adopt the *s-cis* conformation.

1. L. W. Butz and A. W. Rytina, *Org. React.* **5**, 136 (1949); M. C. Kloetzel, *Org. React.* **4**, 1 (1948); A. Wasserman, *Diels–Alder Reactions*, Elsevier, New York, 1965; R. Huisgen, R. Grashey, and J. Sauer, in *Chemistry of Alkenes*, S. Patai (ed.), Wiley-Interscience, New York, 1964, pp. 878–928.

The diene and alkene (which is called the *dienophile*) approach each other in approximately parallel planes. The symmetry properties of the orbitals permit stabilizing interactions between C-1 and C-4 of the diene and the two carbons of the dienophile. The interaction between the frontier orbitals is depicted in Fig. 6.1.

For an unsymmetrical dienophile, there are two possible stereochemical orientations with respect to the diene. The two possible orientations, called *endo* and *exo*, are illustrated in Fig. 6.2. In the *endo* transition state, the reference substituent on the dienophile is oriented toward the π orbitals of the diene. In the *exo* transition state, the substituent is oriented away from the π system.

Whether the products of *endo* and *exo* addition will be different depends on the substitution pattern in the diene. Except for symmetrically substituted butadiene derivatives, the two transition states will lead to two different stereoisomeric products. The *endo* mode of addition is usually preferred when an unsaturated substituent, such as a carbonyl group, is present on the dienophile. The empirical statement

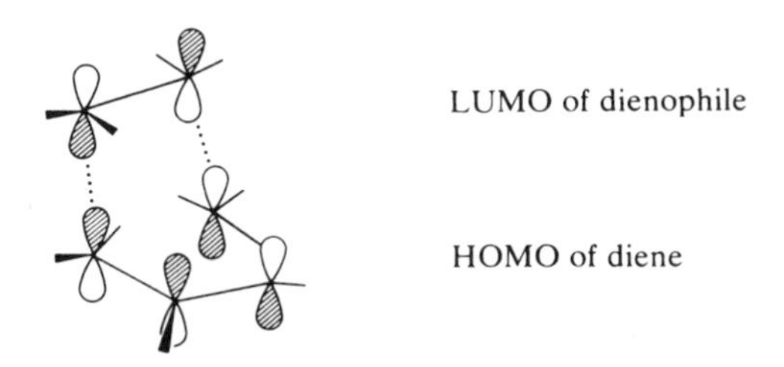

Fig. 6.1. Cycloaddition of an alkene and a diene, showing interaction of LUMO of alkene with HOMO of diene.

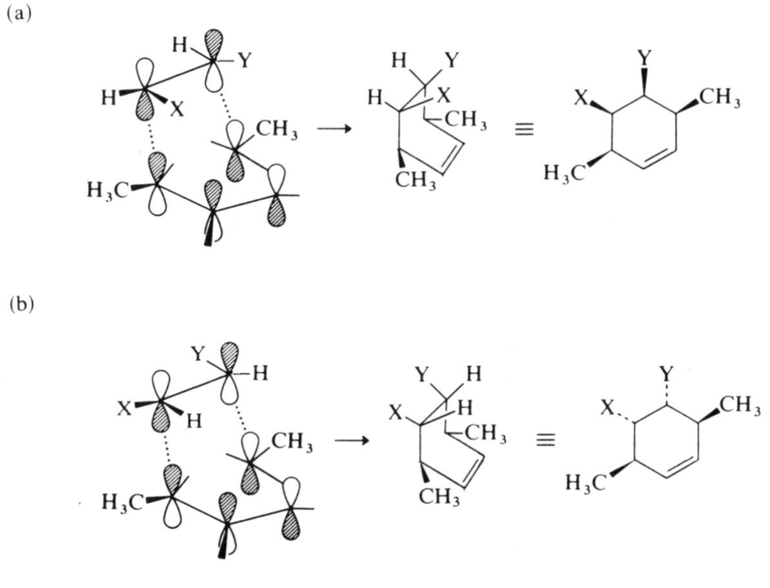

Fig. 6.2. *Endo* (a) and *exo* (b) addition in a Diels–Alder reaction.

that describes this preference is called the *Alder rule.* Frequently, a mixture of both stereoisomers is formed and sometimes the *exo* product predominates, but the Alder rule is a useful initial guide to prediction of the stereochemistry of the Diels–Alder reaction. The *endo* product is often the more sterically congested. The preference for the *endo* transition state is the result of interaction between the dienophile substituent and the π electrons of the diene. Dipolar attractions and van der Waals attractions may also be involved.[2]

There is a strong electronic substituent effect in the Diels–Alder addition. The most reactive alkenes toward simple dienes are those bearing electron-attracting groups. Thus, among the most reactive dienophiles are quinones, maleic anhydride, and nitroalkenes. α,β-Unsaturated esters, ketones, and nitriles are also effective dienophiles. It is significant that if a relatively electron-poor diene is utilized, the preference is reversed and electron-rich alkenes are the best dienophiles. Such reactions are called *inverse electron demand* Diels–Alder reactions. These relationships are readily understood in terms of frontier orbital theory. Electron-rich dienes have high-energy HOMOs and interact strongly with the LUMOs of electron-poor dienophiles. When the substituent pattern is reversed and the diene is electron-poor, the strongest interaction is between the dienophile HOMO and the diene LUMO.

A question of regioselectivity arises when both the diene and alkene are unsymmetrically substituted. Generally, there is a preference for the "*ortho*" and "*para*" orientations, respectively, as in the examples shown below.[3]

$N(CH_2CH_3)_2$ diene + $CO_2CH_2CH_3$ alkene $\xrightarrow{20°C}$ $N(CH_2CH_3)_2$, $CO_2CH_2CH_3$ cyclohexene

"*ortho*"-like
only product (94%)

CH_3CH_2O diene + alkene CO_2CH_3 $\xrightarrow{160°C}$ CH_3CH_2O cyclohexene CO_2CH_3

"*para*"-like
only product (50%)

This preference can also be understood in terms of frontier orbital theory.[4] In the most common cases, the dienophile bears an electron-withdrawing substituent and the diene an electron-releasing one. The strongest interaction is then between the HOMO of the diene and the LUMO of the dienophile. Because of this interaction, the reactants will be oriented so that the carbons having the highest coefficients in the two frontier orbitals will begin the bonding process. This leads to the observed regiochemical preference, as illustrated in Fig. 6.3.

2. Y. Kobuke, T. Sugimoto, J. Furukawa, and T. Funco, *J. Am. Chem. Soc.* **94**, 3633 (1972); K. L. Williamson and Y.-F. L. Hsu, *J. Am. Chem. Soc.* **92**, 7385 (1970).
3. J. Sauer, *Angew. Chem. Int. Ed. Engl.* **6**, 16 (1967).
4. K. N. Houk, *Acc. Chem. Res.* **8**, 361 (1975); I. Fleming, *Frontier Orbitals and Organic Chemical Reactions*, Wiley-Interscience, New York, 1976; O. Eisenstein, J. M. LeFour, N. T. Anh, and R. F. Hudson, *Tetrahedron* **33**, 523 (1977).

(a) Coefficient of C-2 is higher than coefficient of C-1 in LUMO of dienophile bearing an electron-withdrawing substituent.

EWG is a π acceptor such as —C(O)R, —NO_2, —CN

(b) Coefficient of C-4 is higher than coefficient of C-1 in HOMO of diene bearing an electron-releasing substituent at C-1.

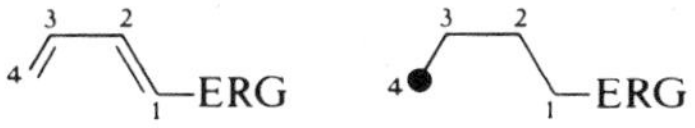

ERG is a π donor such as —OR, —SR, —$OSiMe_3$

(c) Coefficient of C-1 is higher than coefficient of C-4 in HOMO of diene bearing an electron-releasing substituent at C-2.

(d) Regioselectivity of Diels–Alder addition corresponds to that given by matching carbon atoms having the largest coefficients in the frontier orbitals.

"*ortho*"-like orientation:

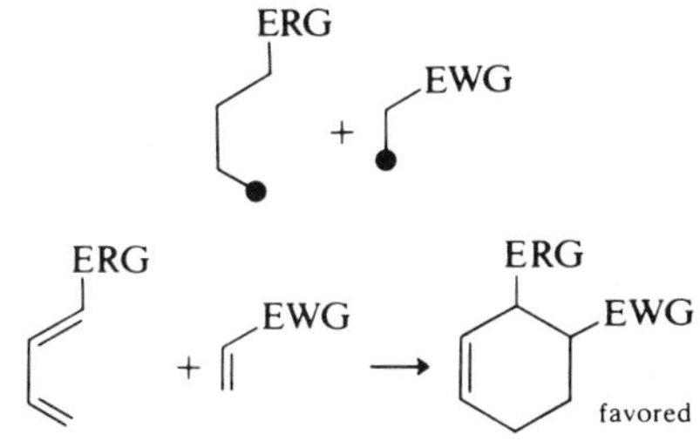

favored

"*para*"-like orientation:

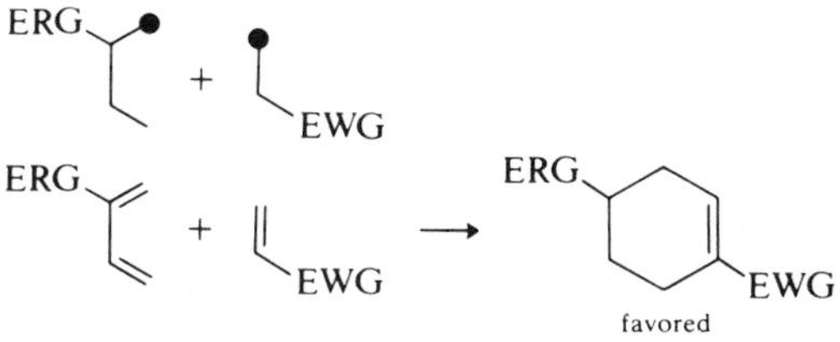

favored

Fig. 6.3. HOMO-LUMO interactions rationalize regioselectivity of Diels–Alder cycloaddition reactions.

Diels–Alder cycloadditions are sensitive to steric effects of two major types. Bulky substituents on the dienophile or on the termini of the diene can hinder approach of the two components to each other and decrease the rate of reaction. This effect can be seen in the relative reactivity of 1-substituted butadienes toward maleic anhydride.[5]

5. D. Craig, J. J. Shipman, and R. B. Fowler, *J. Am. Chem. Soc.* **83**, 2885 (1961).

R	k_{rel} (25°C)
$-H$	1
$-CH_3$	4.2
$-C(CH_3)_3$	<0.05

Substitution of hydrogen by methyl results in a slight rate *increase*, as a result of the electron-releasing effect of the methyl group. A *t*-butyl substituent produces a large rate *decrease*, because the steric effect is dominant.

The other type of steric effect has to do with interactions between diene substituents. Adoption of the *s-cis* conformation of the diene in the transition state brings the *cis*-oriented 1- and 4-substituents on the diene close together. *trans*-1,3-Pentadiene is 10^3 times more reactive than 4-methyl-1,3-pentadiene toward the very reactive dienophile tetracyanoethylene. This is because of the unfavorable interaction between the additional methyl substituent and the C-1 hydrogen in the *s-cis* conformation.[6]

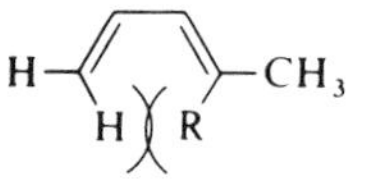

R	k_{rel}
$-H$	1
$-CH_3$	10^{-3}

Relatively small substituents at C-2 and C-3 of the diene exert little steric influence on the rate of Diels–Alder addition. 2,3-Dimethylbutadiene reacts with maleic anhydride about 10 times faster than butadiene does, and this is because of the electronic effect of the methyl groups. 2-*t*-Butyl-1,3-butadiene is 27 times more reactive than butadiene. This is because the *t*-butyl substituent makes the *s-cis* conformation more stable, relative to the *s*-trans conformation.

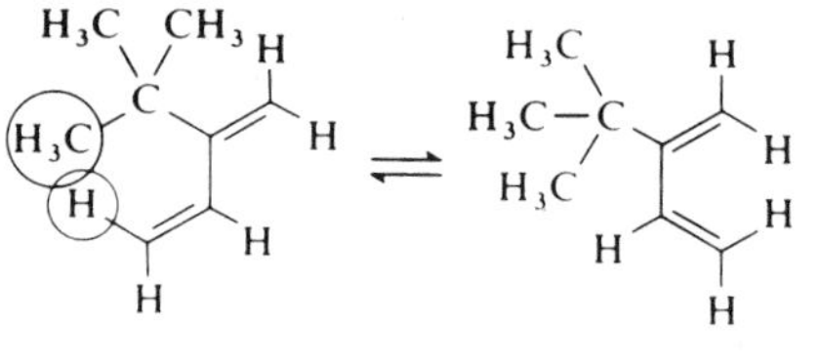

The presence of a *t*-butyl substituent on *both* C-2 and C-3, however, prevents attainment of the *s-cis* conformation, and Diels–Alder reactions of 2,3-di-*t*-butyl-1,3-butadiene have not been observed.[7]

Lewis acids such as zinc chloride, aluminum chloride, and diethylaluminum chloride catalyze Diels–Alder reactions.[8] The catalytic effect is the result of coordination of the Lewis acid with the dienophile.

6. C. A. Stewart, Jr., *J. Org. Chem.* **28**, 3320 (1963).
7. H. J. Backer, *Recl. Trav. Chim. Pays-Bas* **58**, 643 (1939).
8. P. Yates and P. Eaton, *J. Am. Chem. Soc.* **82**, 4436 (1960); T. Inukai and M. Kasai, *J. Org. Chem.* **30**, 3567 (1965); T. Inukai and T. Kojima, *J. Org. Chem.* **32**, 869, 872 (1967); F. Fringuelli, F. Pizzo, A. Taticchi, and E. Wenkert, *J. Org. Chem.* **48**, 2802 (1983); F. K. Brown, K. N. Houk, D. J. Burnell, and Z. Valenta, *J. Org. Chen.* **52**, 3050 (1987).

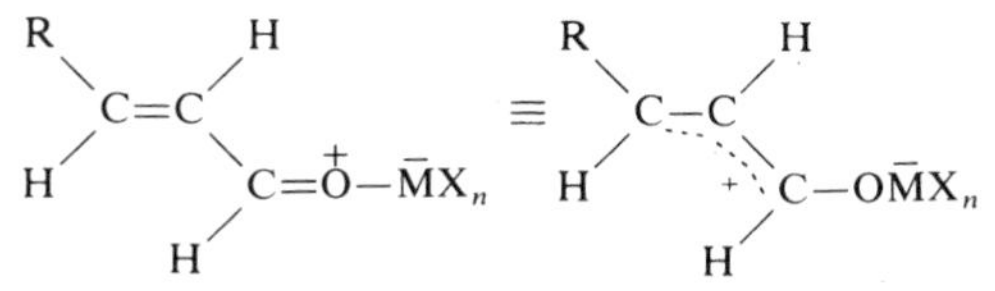

The complexed dienophile is then more electrophilic and more reactive toward electron-rich dienes. The mechanism of the addition is believed to still be concerted, and high stereospecificity is observed.[9] Lewis acid catalysts also usually increase the regioselectivity of the reaction.

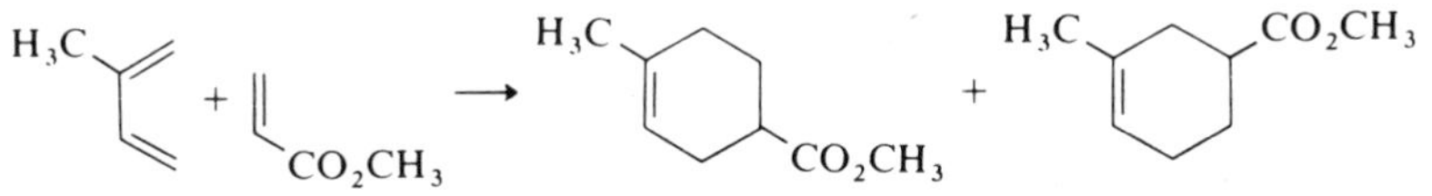

Ref. 10

	"para"-like	"meta"-like
	Product ratio	
Uncatalyzed reaction: 120°C, 6 h	70%	30%
Aluminum-chloride-catalyzed: 20°C, 3 h	95%	5%

6.1.2. The Diels–Alder Reaction: Dienophiles

Examples of some compounds that exhibit a high level of reactivity as dienophiles are collected in Table 6.1. Scheme 6.1 presents some typical Diels–Alder reactions. Each of the reactive dienophiles has at least one strongly electron-attracting substituent on the double or triple carbon–carbon bond. Ethylene, acetylene, and their alkyl derivatives are notoriously poor dienophiles.

Diels–Alder reactions have long played an important role in synthetic organic chemistry. The reaction of a substituted benzoquinone and 1,3-butadiene, for example, was the first step in one of the early syntheses of steroids. The angular methyl group was introduced by the methyl group on the quinone, and the other functional groups were used for further elaboration.

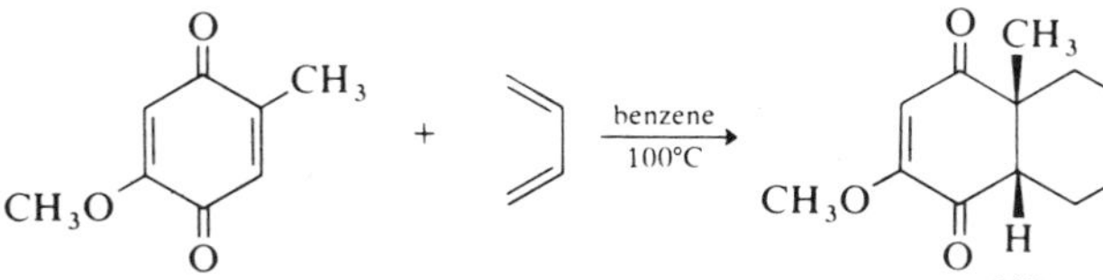

benzene, 100°C (86%) Ref. 11

The synthetic utility of the Diels–Alder reaction can be significantly expanded by the use of dienophiles that contain *masked functionality* and are the *synthetic equivalent* of unreactive or inaccessible species. For example, α-chloroacrylonitrile

9. K. N. Houk, *J. Am. Chem. Soc.* **95**, 4094 (1973).
10. T. Inukai and T. Kojima, *J. Org. Chem.* **31**, 1121 (1966).
11. R. B. Woodward, F. Sondheimer, D. Taub, K. Heusler, and W. M. McLamore, *J. Am. Chem. Soc.* **74**, 4223 (1952).

Table 6.1. Representative Dienophiles

A. Substituted Alkenes

1[a] Maleic anhydride

2[b] Benzoquinone

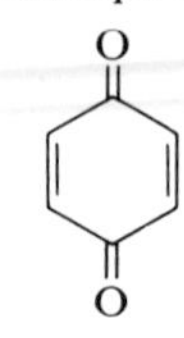

3[c] Diethyl vinylphosphonate

$H_2C{=}CHP(O)(OC_2H_5)_2$

4[d] Methyl vinyl sulfone

$H_2C{=}CHS(O)_2CH_3$

5[e] Tetracyanoethylene

$(NC)_2C{=}C(CN)_2$

6[f] Vinyl ketones, acrolein, acrylate esters, acrylonitrile, nitroalkenes, etc.

$RCH{=}CH{-}X$

$X = C(O)R,\ C(O)OR,\ C{\equiv}N,\ NO_2$

B. Substituted Alkynes

7[f] Esters of acetylenedicarboxylic acid

$H_3CO_2CC{\equiv}CCO_2CH_3$

8[g] Hexafluoro-2-butyne

$F_3CC{\equiv}CCF_3$

9[h] Dibenzoylacetylene

$PhC(O)C{\equiv}CC(O)Ph$

10[i] Dicyanoacetylene

$N{\equiv}CC{\equiv}CC{\equiv}N$

C. Heteroatomic Dienophiles

11 Esters of azodicarboxylic acid

$H_3CO_2CN{=}NCO_2CH_3$

12[k] 4-phenyl-1,2,4-triazoline-3,5-dione

13[l] Iminourethanes

a. M. C. Kloetzel, *Org. React.* **4**, 1 (1948).
b. L. W. Butz and A. W. Rytina, *Org. React.* **5**, 136 (1949).
c. W. M. Daniewski and C. E. Griffin, *J. Org. Chem.* **31**, 3236 (1966).
d. J. C. Philips and M. Oku, *J. Org. Chem.* **37**, 4479 (1972).
e. W. J. Middleton, R. E. Heckert, E. L. Little, and C. G. Krespan, *J. Am. Chem. Soc.* **80**, 2783 (1958); E. Ciganek, W. J. Linn, and O. W. Webster, *The Chemistry of the Cyano Group*, Z. Rappoport (ed.), Interscience, New York, 1970, pp. 423-638.
f. H. L. Holmes, *Org. React.* **4**, 60 (1948).
g. R. E. Putnam, R. J. Harder, and J. E. Castle, *J. Am. Chem. Soc.* **83**, 391 (1961); C. G. Krespan, B. C. McKusick, and T. L. Cairns, *J. Am. Chem. Soc.* **83**, 3428 (1961).
h. J. D. White, M. E. Mann, H. D. Kirshenbaum, and A. Mitra, *J. Org. Chem.* **36**, 1048 (1971).
i. C. D. Weis, *J. Org. Chem.* **28**, 74 (1963).
j. B. T. Gillis and P. E. Beck, *J. Org. Chem.* **28**, 3177 (1963).
k. B. T. Gillis and J. D. Hagarty, *J. Org. Chem.* **32**, 330 (1967).
l. M. P. Cava, C. K. Wilkins, Jr., D. R. Dalton, and K. Bessho, *J. Org. Chem.* **30**, 3772 (1965); G. Krow, R. Rodebaugh, R. Carmosin, W. Figures, H. Pannella, G. De Vicaris, and M. Grippi, *J. Am. Chem. Soc.* **95**, 5273 (1973).

Scheme 6.1. Diels–Alder Reactions of Some Representative Dienophiles

1[a] Maleic Anhydride

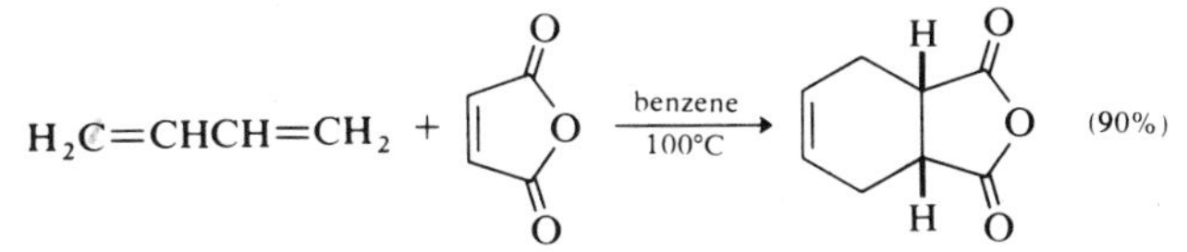

$H_2C{=}CHCH{=}CH_2$ + maleic anhydride $\xrightarrow[100°C]{benzene}$ (90%)

2[b] Benzoquinone

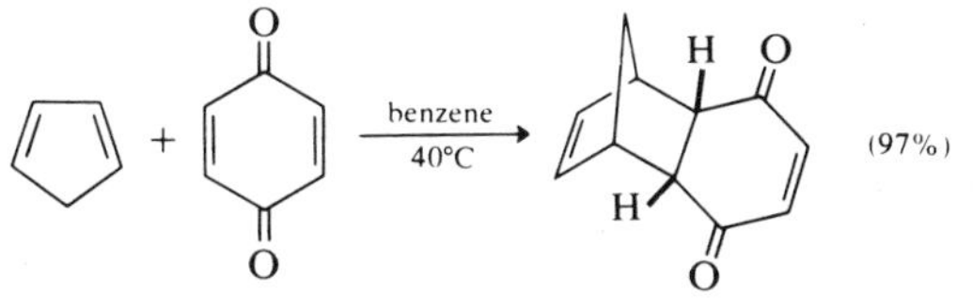

cyclopentadiene + benzoquinone $\xrightarrow[40°C]{benzene}$ (97%)

3[c] Methyl Vinyl Ketone

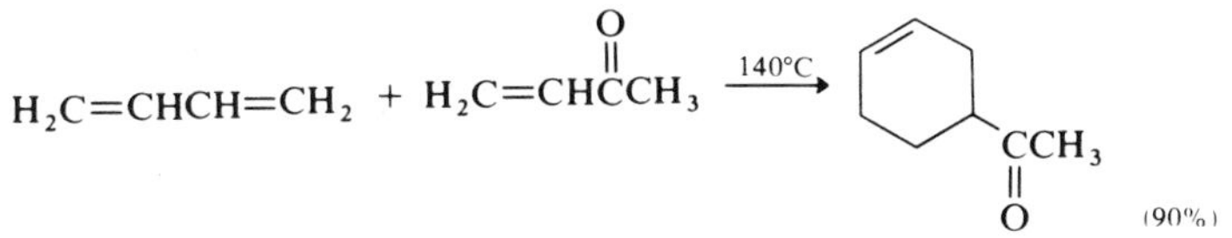

$H_2C{=}CHCH{=}CH_2 + H_2C{=}CHCCH_3$ (C=O) $\xrightarrow{140°C}$ 4-acetylcyclohexene (CCH_3, C=O) (90%)

4[d] Methyl Acrylate

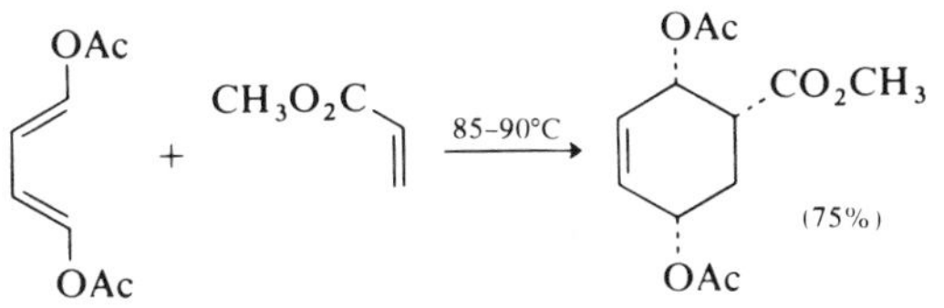

OAc diene (OAc) + CH_3O_2C (vinyl) $\xrightarrow{85–90°C}$ OAc, CO_2CH_3, OAc (75%)

5[e] Acrolein

$H_2C{=}C(OCH_3)CH{=}CH_2 + H_2C{=}CHCH{=}O \longrightarrow$ CH_3O, $CH{=}O$ (75%)

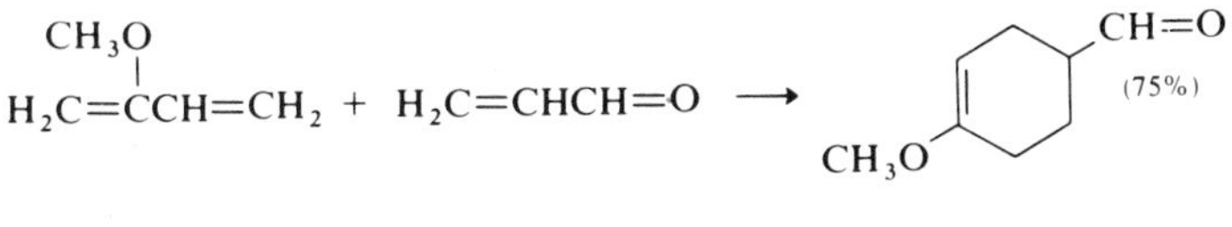

6[f] Tetracyanoethylene

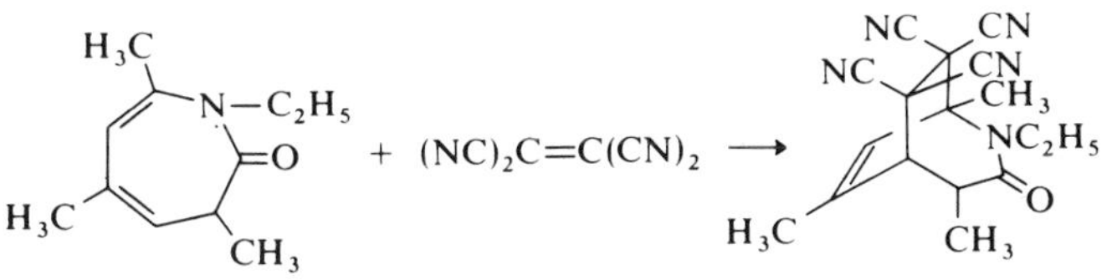

H_3C, $N{-}C_2H_5$, $=O$, H_3C, CH_3 + $(NC)_2C{=}C(CN)_2 \longrightarrow$ NC, CN, NC, CN, CH_3, NC_2H_5, O, H_3C, CH_3

a. L. F. Fieser and F. C. Novello, *J. Am. Chem. Soc.* **64**, 802 (1942).
b. A. Wassermann, *J. Chem. Soc.*, 1511 (1935).
c. W. K. Johnson, *J. Org. Chem.* **24**, 864 (1959).
d. R. McCrindle, K. H. Overton, and R. A. Raphael, *J. Chem. Soc.* 1560 (1960); R. K. Hill and G. R. Newkome, *Tetrahedron Lett.*, 1851 (1968).
e. J. I. DeGraw, L. Goodman, and B. R. Baker, *J. Org. Chem.* **26**, 1156 (1961).
f. L. A. Paquette, *J. Org. Chem.* **29**, 3447 (1964).

shows satisfactory reactivity as a dienophile. The α-chloronitrile functionality in the adduct can be hydrolyzed to a carbonyl group. Thus, α-chloroacrylonitrile can function as the equivalent of ketene, $CH_2{=}C{=}O$. Ketene is not a suitable dienophile because it has a tendency to react with dienes by [2 + 2] cycloaddition, rather than the desired [4 + 2] fashion.

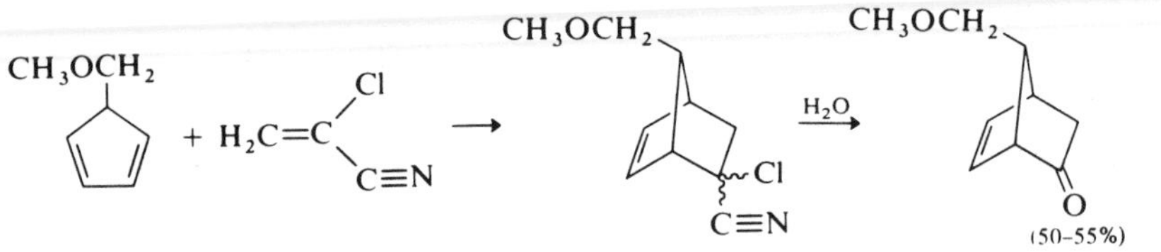

Ref. 12

Nitroalkenes are good dienophiles, and the variety of transformations that are available for nitro groups make them versatile intermediates.[13] Nitro groups can be coverted to carbonyl groups by reductive hydrolysis, so nitroethylene can be used as a ketene equivalent.[14]

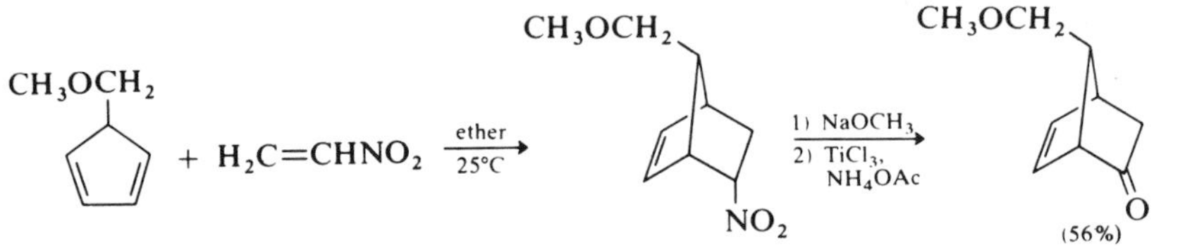

Ref. 15

Vinyl sulfones are reactive as dienophiles. The sulfonyl group can be removed reductively with sodium amalgam. In this two-step reaction sequence, the vinyl sulfone functions as an ethylene equivalent. The sulfonyl group also permits alkylation of the adduct, via the carbanion. This three-step sequence permits the vinyl sulfone to serve as the synthetic equivalent of a terminal alkene.[16]

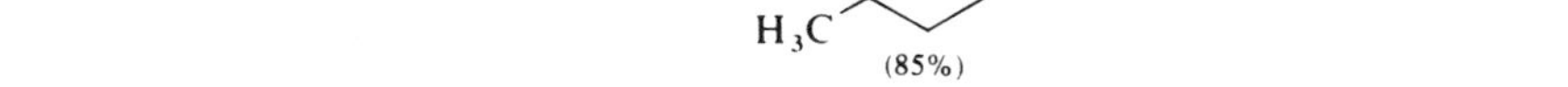

12. E. J. Corey, N. M. Weinshenker, T. K. Schaaf, and W. Huber, *J. Am. Chem. Soc.* **91**, 5675 (1969).
13. D. Ranganathan, C. B. Rao, S. Ranganathan, A. K. Mehrotra, and R. Iyengar, *J. Org. Chem.* **45**, 1185 (1980).
14. For a review of ketene equivalents, see S. Ranganathan, D. Ranganathan, and A. K. Mehrotra, *Synthesis*, 289 (1977).
15. S. Ranganathan, D. Ranganathan, and A. K. Mehrotra, *J. Am. Chem. Soc.* **96**, 5261 (1974).
16. R. V. C. Carr and L. A. Paquette, *J. Am. Chem. Soc.* **102**, 853 (1980); R. V. C. Carr, R. V. Williams, and L. A. Paquette, *J. Org. Chem.* **48**, 4976 (1983); W. A. Kinney, G. O. Crouse, and L. A. Paquette, *J. Org. Chem.* **48**, 4986 (1983).

Phenyl vinyl sulfoxide is a useful acetylene equivalent. Its Diels–Alder adducts can undergo elimination of benzenesulfenic acid.

$$\text{C}_5\text{Cl}_6 + \text{PhS(O)CH=CH}_2 \xrightarrow{100°\text{C}} \text{adduct (S=O, Ph)} \xrightarrow{100°\text{C}} \text{product (83\%)}$$

Ref. 17

The *cis* and *trans* isomers of bis(benzenesulfonyl)ethene are also acetylene equivalents. The two sulfonyl groups undergo reductive elimination on reaction with sodium amalgam.

$$\text{cyclohexadiene} + \text{PhSO}_2\text{CH=CHSO}_2\text{Ph} \longrightarrow \text{adduct (SO}_2\text{Ph, SO}_2\text{Ph)} \xrightarrow[\text{MeOH}]{\text{Na-Hg}} \text{bicyclo[2.2.2]octadiene (69\%)}$$

Ref. 18

Vinylphosphonium salts are reactive as dienophiles as a result of the electron-withdrawing capacity of the phosphonium substituent. The Diels–Alder adducts can be deprotonated to give ylides which undergo the Wittig reaction to introduce an exocyclic double bond. This sequence of reactions corresponds to a Diels–Alder reaction employing allene as the dienophile.[19]

$$\text{cyclohexadiene} + \text{H}_2\text{C=CH}\overset{+}{\text{P}}\text{Ph}_3 \quad \text{adduct}\sim\overset{+}{\text{P}}\text{Ph}_3 \ (96\%) \xrightarrow{1)\ \text{LiNR}_2;\ 2)\ \text{CH}_2\text{=O}} \text{=CH}_2 \ (50\%)$$

The use of 2-vinyldioxolane, the ethylene glycol acetal of acrolein, as a dienophile represents application of the masked functionality concept in a different way. The acetal itself would not be expected to be a reactive dienophile, but in the presence of a catalytic amount of acid, the acetal is in equilibrium with the highly reactive oxycarbocation.

$$\text{CH}_2\text{=CH}-\text{(1,3-dioxolan-2-yl)} + \text{H}^+ \rightleftharpoons \text{CH}_2\text{=CH}-\text{CH=}\overset{+}{\text{O}}-\text{CH}_2\text{CH}_2\text{OH}$$

17. L. A. Paquette, R. E. Moerck, B. Harirchian, and P. D. Magnus, *J. Am. Chem. Soc.* **100**, 1597 (1978).
18. O. DeLucchi, V. Lucchini, L. Pasquato, and G. Modena, *J. Org. Chem.* **49**, 596 (1984).
19. R. Bonjouklian and R. A. Ruden, *J. Org. Chem. Chem.* **42**, 4095 (1977).

Scheme 6.2. Enantioselective Diels–Alder Reactions

1[a]

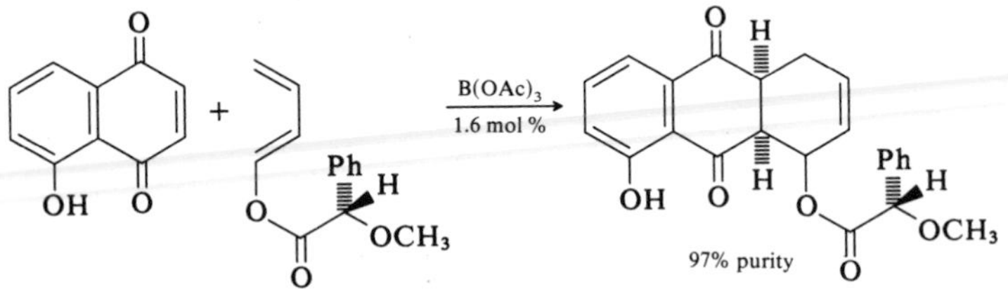

OH O + Ph H OCH$_3$ — $B(OAc)_3$ 1.6 mol % → O H, OH O H, Ph H OCH$_3$

97% purity

2[b]

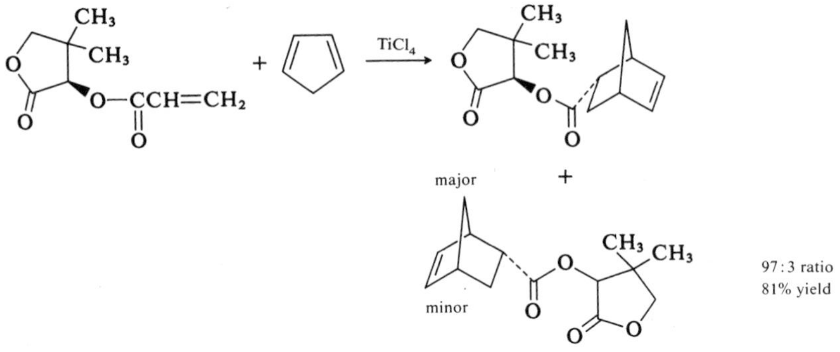

CH_3 CH_3 O–$\overset{O}{\overset{\|}{C}}$CH=$CH_2$ + cyclopentadiene — $TiCl_4$ → CH_3 CH_3

major + minor CH_3 CH_3

97 : 3 ratio
81% yield

3[c]

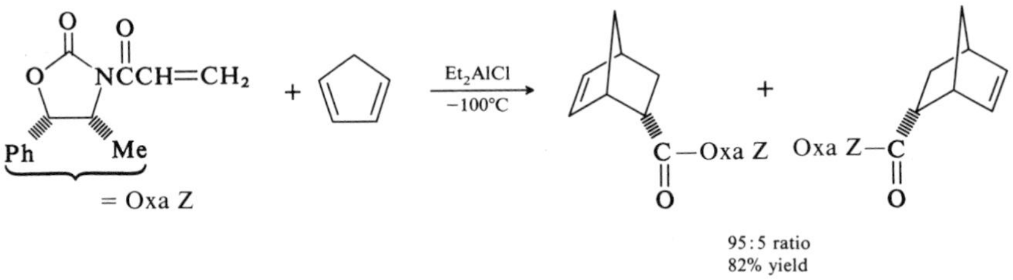

O O NCCH=CH_2, Ph Me = Oxa Z + cyclopentadiene — Et_2AlCl, −100°C → C–Oxa Z + Oxa Z–C

95 : 5 ratio
82% yield

4[d]

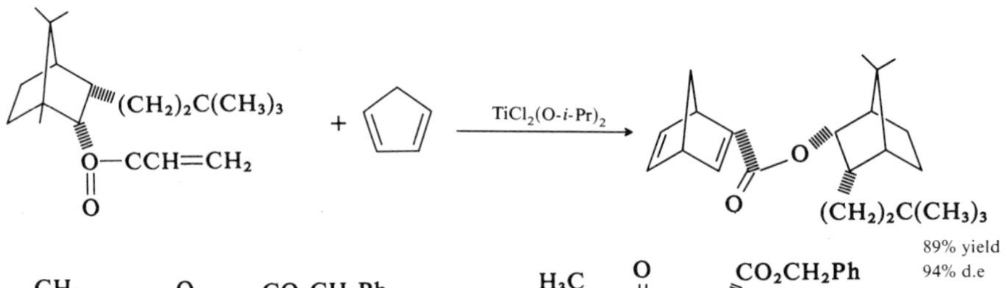

$(CH_2)_2C(CH_3)_3$, O–CCH=CH_2 + cyclopentadiene — $TiCl_2(O\text{-}i\text{-}Pr)_2$ → $(CH_2)_2C(CH_3)_3$

89% yield
94% d.e

5[e]

CH_3, CH_3 + O, N, CO_2CH_2Ph — TiCl → H_3C, C–N, CO_2CH_2Ph, CH_3

42% yield
80% d.e

+

H_3C, C–N, CO_2CH_2Ph, CH_3

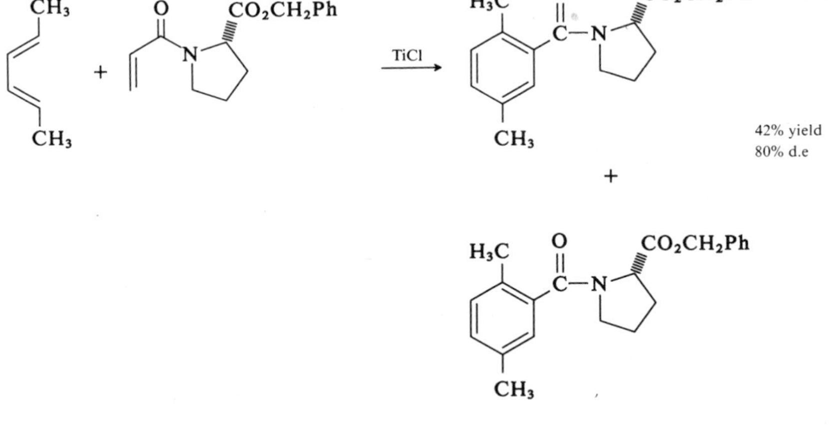

Diels–Alder addition occurs through this cationic intermediate at room temperature.[20] Similar reactions occur with substituted alkenyldioxolanes.

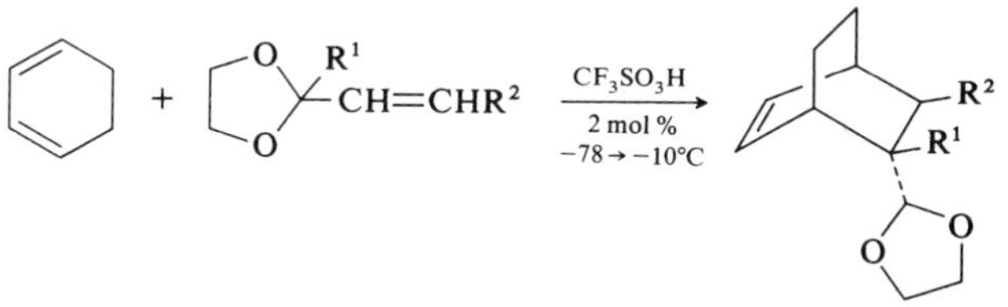

CF_3SO_3H 2 mol % −78 → −10°C

Chiral substituents in dienophiles can be used to make Diels–Alder reactions enantioselective.[21] Some examples of chiral dienophiles that have been used successfully are collected in Scheme 6.2. Because of the lower temperature required and the greater stereoselectivity observed in Lewis acid catalyzed reactions, the best enantioselectivity is often observed in the catalyzed reactions. Chiral esters and amides of acrylic acid are particularly useful because the chiral auxiliary can be easily recovered by hydrolysis of the adduct. The cycloaddition proceeds to give two diastereomeric products, which can be separated and purified. Hydrolysis then gives enantiomerically pure carboxylic acid.

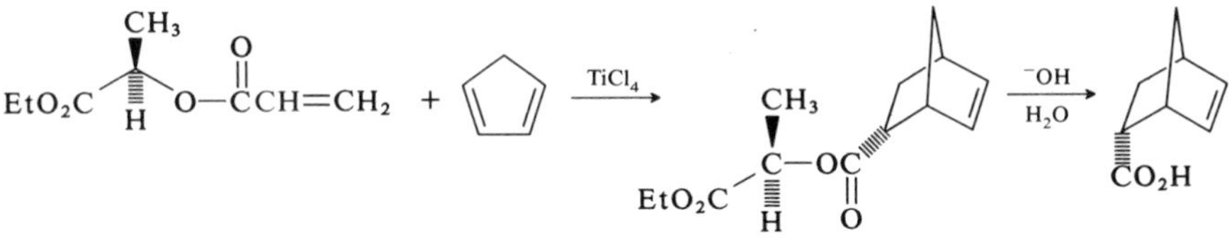

$TiCl_4$ → ; ^-OH, H_2O → Ref. 22

Prediction and analysis of diastereoselectivity is based on postulation of a transition state structure on the basis of steric, stereoelectronic, and complexing interactions.[23]

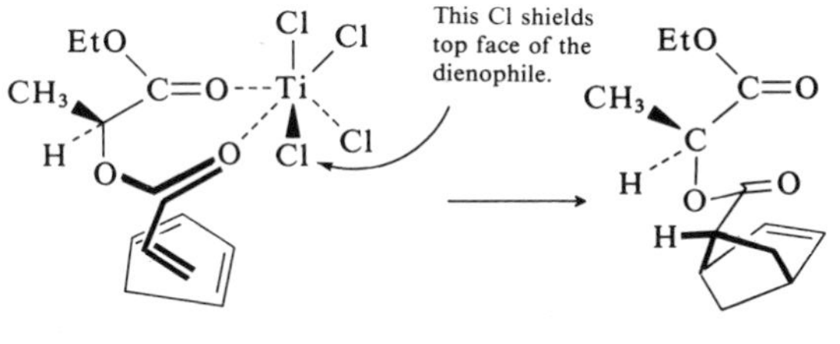

This Cl shields top face of the dienophile.

20. P. G. Gassman, D. A. Singleton, J. J. Wilwerding, and S. P. Chavan, *J. Am. Chem. Soc.* **109**, 2182 (1987).
21. W. Oppolzer, *Angew. Chem. Int. Ed. Engl.* **23**, 876 (1984); H. Wurziger, *Kontakte (Darmstadt)*, 3 (1984).
22. T. Poll, G. Helmchen, and B. Bauer, *Tetrahedron Lett.* **25**, 2191 (1984).
23. For example, see T. Poll, A. Sobczak, H. Hartmann, and G. Helmchen, *Tetrahedron Lett.* **26**, 3095 (1985).

---

a. B. M. Trost, D. O'Krongly, and J. L. Belletire, *J. Am. Chem. Soc.* **102**, 7595 (1980).
b. T. Poll, A. Sobczak, H. Hartmann, and G. Helmchen, *Tetrahedron Lett.* **26**, 3095 (1985).
c. D. A. Evans, K. T. Chapman, and J. Bisaha, *J. Am. Chem. Soc.* **110**, 1238 (1988).
d. W. Oppolzer and C. Chapuis, *Tetrahedron Lett.* **25**, 5383 (1984).
e. H. Waldmann, *J. Org. Chem.* **53**, 6133 (1988).

6.1.3. The Diels–Alder Reaction: Dienes

Simple dienes react readily with good dienophiles in Diels–Alder reactions. As discussed earlier, steric effects can play a role in their reactivity. Functionalized dienes have become important in organic synthesis. One example which illustrates the versatility of such reagents is 1-methoxy-3-trimethylsiloxy-1,3-butadiene (*Danishefsky's diene*).[24] Its Diels–Alder adducts are trimethylsilyl enol ethers and can be readily hydrolyzed to ketones. The β-methoxy group is often eliminated after hydrolysis.

OCH₃ / Me₃SiO diene + $H_2C{=}C(CH_3)CHO$ —benzene, Δ→ adduct (OCH₃, CHO, CH₃, Me₃SiO) —H_3O^+→ enone (CHO, CH₃, O) (72%)

Related transformations of the adduct with dimethyl acetylenedicarboxylate lead to dimethyl 4-hydroxyphthalate.

OCH₃ / Me₃SiO diene + $CH_3OCC{\equiv}CCOCH_3$ (with C=O groups) —benzene, Δ→ adduct (OCH₃, CO_2CH_3, CO_2CH_3, Me₃SiO) —H_3O^+→ HO-phenyl with CO_2CH_3, CO_2CH_3

Unstable dienes can also be generated *in situ* in the presence of a dienophile. Among the most useful examples of this type of diene are the quinodimethanes. These compounds are exceedingly reactive as dienes because the cycloaddition reestablishes a benzenoid ring and results in aromatic stabilization.[25]

quinodimethane (CH_2, CH_2) + $CH_2{=}CHX$ → tetrahydronaphthalene (X)

quinodimethane

24. S. Danishefsky and T. Kitahara, *J. Am. Chem. Soc.* **96**, 7807 (1974).
25. W. Oppolzer, *Angew. Chem. Int. Ed. Engl.* **16**, 10 (1977); T. Kametani and K. Fukumoto, *Heterocycles* **3**, 29 (1975); J. J. McCullough, *Acc. Chem. Res.* **13**, 270 (1980); W. Oppolzer, *Synthesis*, 793 (1978); J. L. Charlton and M. M. Alauddin, *Tetrahedron* **43**, 2873 (1987); H. N. C. Wong, K.-L. Lau, and K. F. Tam, *Top. Curr. Chem.* **133**, 85 (1986).

There are several general routes to quinodimethanes. One is pyrolysis of benzocyclobutenes.[26]

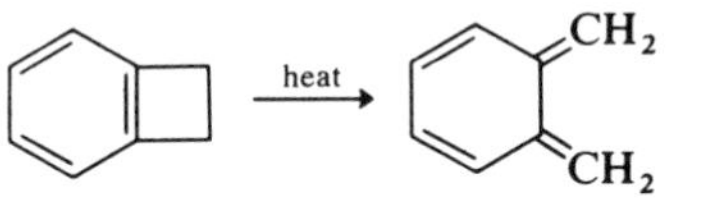

Eliminations from α,α'-*ortho*-disubstituted benzenes can be carried out with various potential leaving groups.

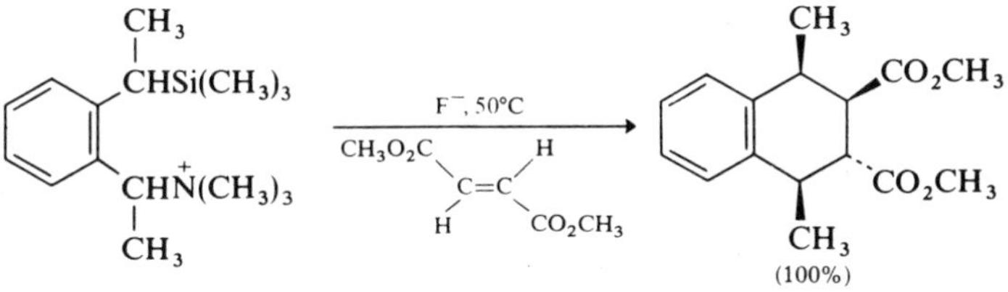

Ref. 27

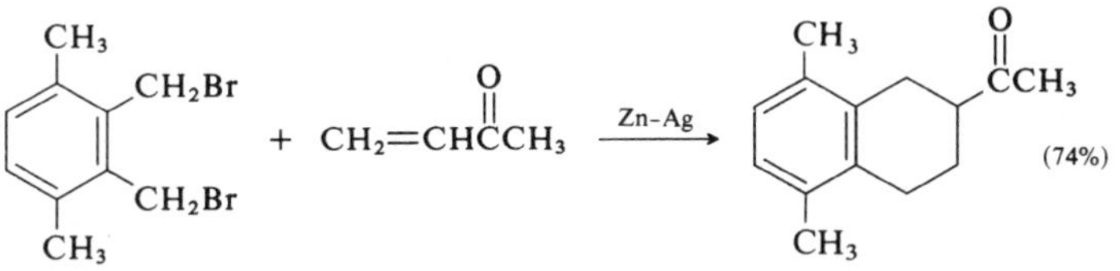

Ref. 28

Quinodimethanes have been especially useful in intramolecular Diels–Alder reactions, as will be illustrated in Section 6.1.4.

Another group of dienes with extraordinarily high reactivity are derivatives of benzo[*c*]furan (isobenzofuran).[29]

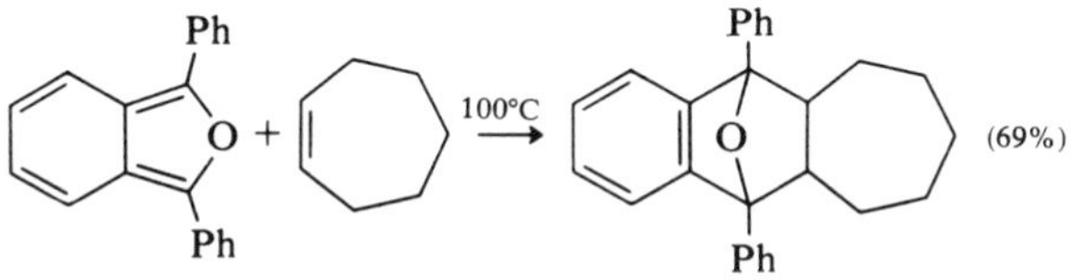

Ref. 30

Here again, the high reactivity can be traced to the aromatic stabilization of the adduct.

Polycyclic aromatic hydrocarbons are moderately reactive as the diene component of Diels–Alder reactions. Anthracene forms adducts with a number of reactive dienophiles. The addition occurs at the center ring. There is no net loss of resonance

26. M. P. Cava and M. J. Mitchell, *Cyclobutadiene and Related Compounds*, Academic Press, New York, 1967, Chapter 6; I. L. Klundt, *Chem. Rev.* **70**, 471 (1970); R. P. Thummel, *Acc. Chem. Res.* **13**, 70 (1980).
27. Y. Ito, M. Nakatsuka, and T. Saegusa, *J. Am. Chem. Soc.* **104**, 7609 (1982).
28. G. M. Rubottom and J. E. Wey, *Synth. Commun.* **14**, 507 (1984).
29. M. J. Haddadin, *Heterocycles* **9**, 865 (1978); W. Friedrichsen, *Adv. Heterocycl. Chem.* **26**, 135 (1980).
30. G. Wittig and T. F. Burger, *Justus Liebigs Ann. Chem.* **632**, 85 (1960).

stabilization, since the anthracene ring (resonance energy = 1.60 eV) is replaced by two benzenoid rings (total resonance energy = 2 × 0.87 = 1.74 eV).[31]

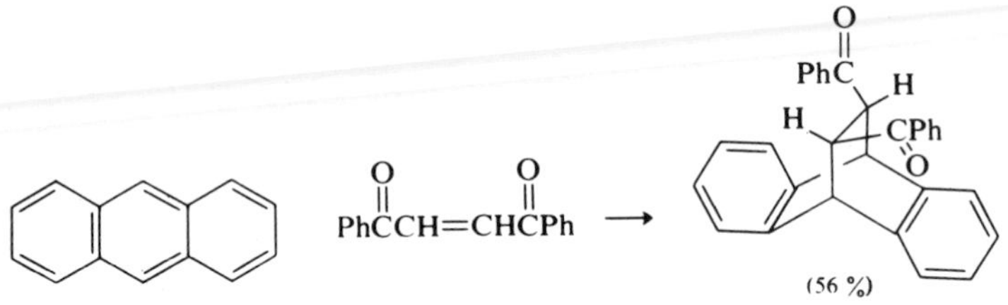

PhCCH=CHCPh → (56 %) Ref. 32

The naphthalene ring is much less reactive. Polymethylnaphthalenes are more reactive than the parent molecule, and 1,2,3,4-tetramethylnaphthalene gives an adduct with maleic anhydride in 82% yield. Reaction occurs exclusively in the substituted ring.[33] This is because the steric repulsions between the methyl groups, which are relieved in the nonplanar adduct, exert an accelerating effect.

With benzenoid compounds, Diels–Alder addition is rare and occurs only with very reactive dienophiles. Formation of an adduct between benzene and dicyanoacetylene in the presence of $AlCl_3$ has been reported, for example.[34]

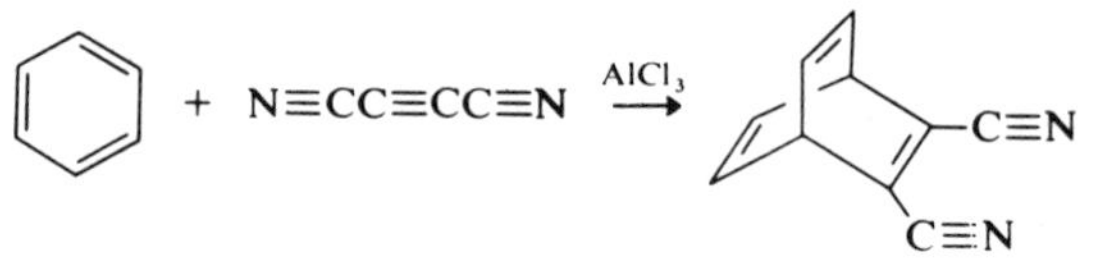

+ N≡CC≡CC≡N $\xrightarrow{AlCl_3}$ C≡N, C≡N

6.1.4. Intramolecular Diels–Alder Reactions

Intramolecular Diels–Alder reactions have proven very useful in the synthesis of polycyclic compounds.[35] Some examples are given in Scheme 6.3.

In entry 1, the dienophilic portion bears a carbonyl substituent, and cycloaddition occurs easily. Two stereoisomeric products are formed but both have the *cis* ring fusion. This is the stereochemistry expected for an *endo* transition state.

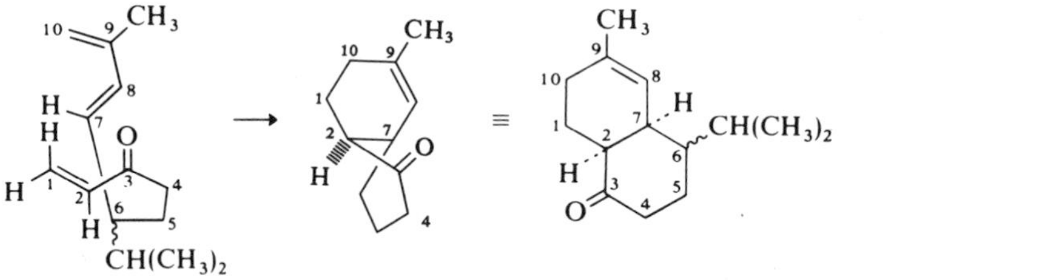

CH_3 … $CH(CH_3)_2$ → CH_3 ≡ CH_3 … $CH(CH_3)_2$

31. M. J. S. Dewar and D. de Llano, *J. Am. Chem. Soc.* **91**, 789 (1969).
32. D. M. McKinnon and J. Y. Wong, *Can. J. Chem.* **49**, 3178 (1971).
33. A. Oku, Y. Ohnishi, and F. Mashio, *J. Org. Chem.* **37**, 4264 (1972).
34. E. Ciganek, *Tetrahedron Lett.*, 3321 (1967).
35. W. Oppolzer, *Angew. Chem. Int. Ed. Engl.* **16**, 10 (1977); G. Brieger and J. N. Bennett, *Chem. Rev.* **80**, 63 (1980); E. Ciganek, *Org. React.* **32**, 1 (1984); D. F. Taber, *Intramolecular Diels–Alder and Alder Ene Reactions*, Springer-Verlag, Berlin, 1984.

1[a]

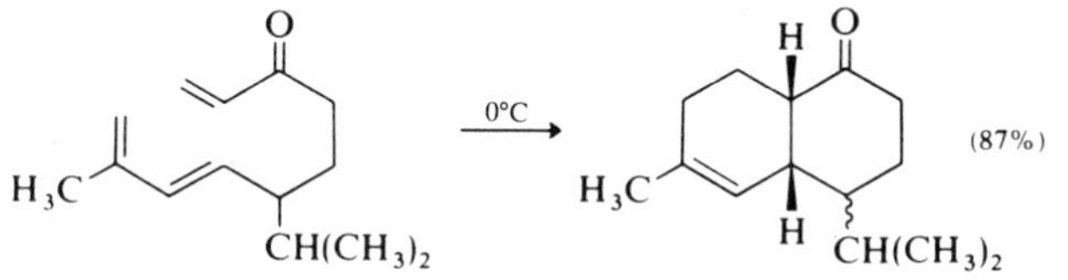

H_3C ... $CH(CH_3)_2$ —0°C→ H, O, H_3C, H, $CH(CH_3)_2$ (87%)

2[b]

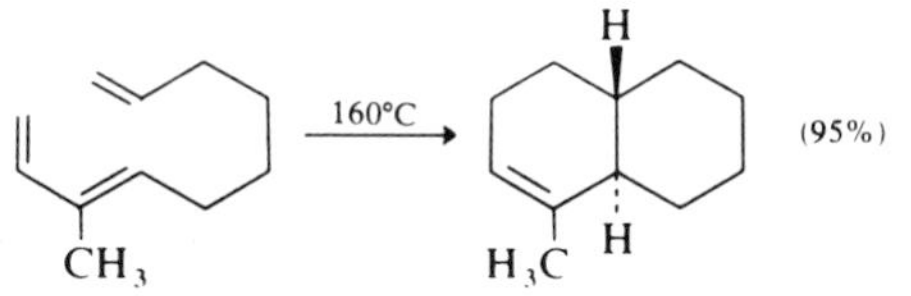

CH_3 —160°C→ H, H_3C, H (95%)

3[c]

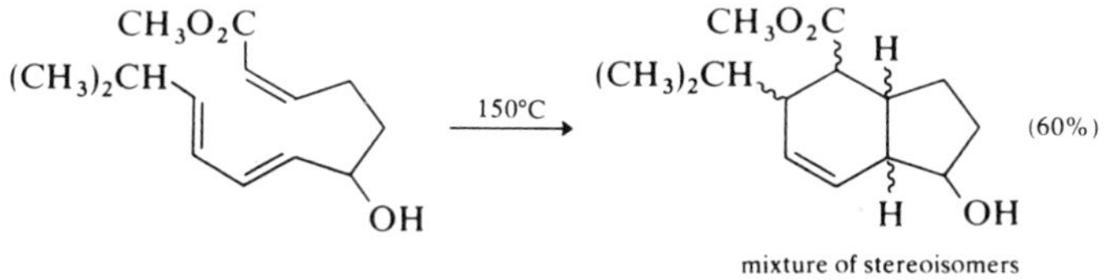

CH_3O_2C, $(CH_3)_2CH$, OH —150°C→ CH_3O_2C, H, $(CH_3)_2CH$, H, OH (60%)

mixture of stereoisomers

4[d]

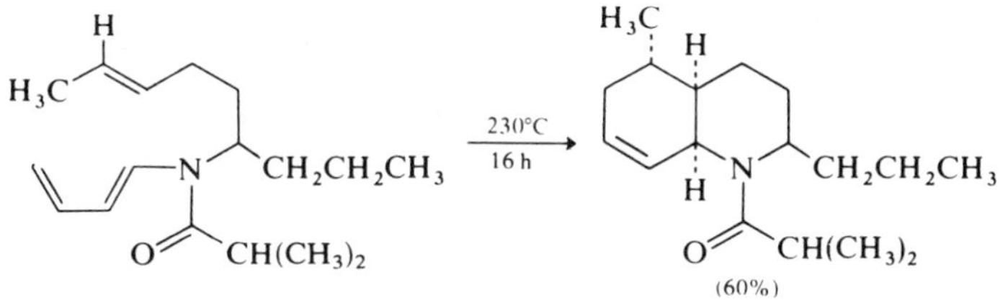

H, H_3C, N, $CH_2CH_2CH_3$, O, $CH(CH_3)_2$ —230°C, 16 h→ H_3C, H, H, N, $CH_2CH_2CH_3$, O, $CH(CH_3)_2$ (60%)

5[e]

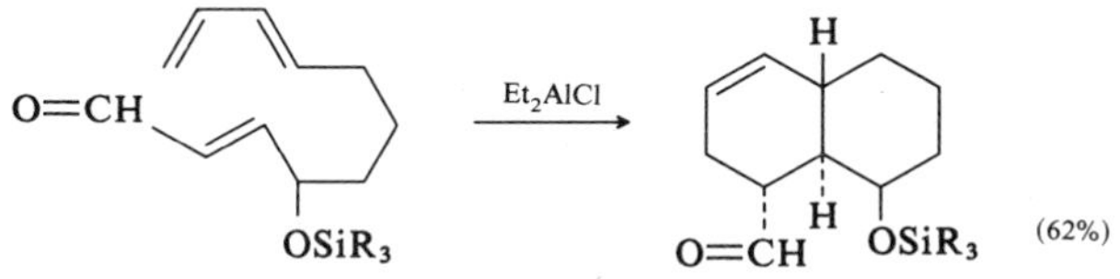

O=CH, $OSiR_3$ —Et_2AlCl→ H, H, O=CH, $OSiR_3$ (62%)

6[f]

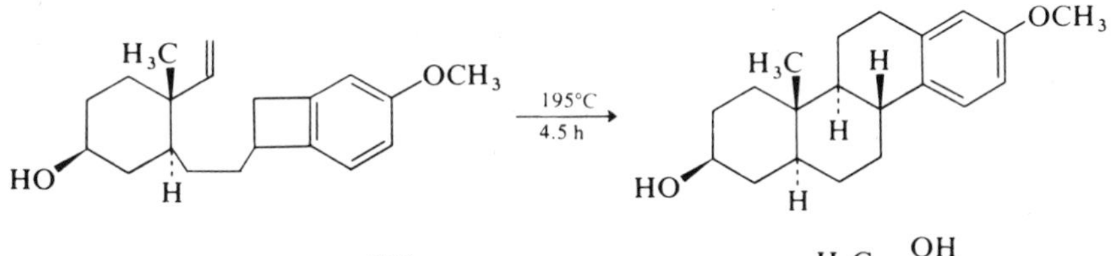

H_3C, OCH_3, HO, H —195°C, 4.5 h→ OCH_3, H_3C, H, H, HO, H

7[g]

H_3C, OH, CH_3O, H —200°C, 7.5 h→ H_3C, OH, H, H, H, CH_3O (91%)

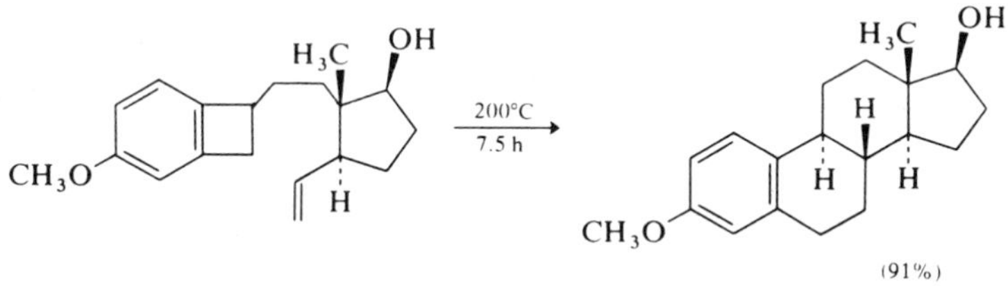

a. D. F. Taber and B. P. Gunn, *J. Am. Chem. Soc.* **101**, 3992 (1979).
b. S. R. Wilson and D. T. Mao, *J. Am. Chem. Soc.* **100**, 6289 (1978).
c. W. R. Roush, *J. Am. Chem. Soc.* **102**, 1390 (1980).
d. W. Oppolzer and E. Flaskamp, *Helv. Chim. Acta.* **60**, 204 (1977).
e. J. A. Marshall, J. E. Audia, and J. Grote, *J. Org. Chem.* **49**, 5277 (1984).
f. T. Kametani, K. Suzuki, and H. Nemoto, *J. Org. Chem.* **45**, 2204 (1980); *J. Am. Chem. Soc.* **103**, 2890 (1981).
g. P. A. Grieco, T. Takigawa, and W. J. Schillinger, *J. Org. Chem.* **45**, 2247 (1980).

In entry 2, a similar triene that lacks the activating carbonyl group undergoes reaction but a higher temperature is required. In this case, the ring junction is *trans*. In entry 3, the dienophilic double bond bears an electron-withdrawing group, but a higher temperature than for entry 1 is required because the connecting chain contains one less methylene group and this leads to a more strained transition state. A mixture of stereoisomers is formed reflecting a conflict between the Alder rule, which favors *endo* addition, and conformational factors which favor the *exo* transition state. The stereoselectivity of a number of intramolecular Diels–Alder reactions has been analyzed, and conformational factors in the transition state seem to play the dominant role in determining product structure.[36]

Lewis acid catalysis frequently improves the stereoselectivity of intramolecular Diels–Alder reactions, just as it does in intermolecular cases. For example, the thermal cyclization of **1** at 160°C gives a 50:50 mixture of two stereoisomers, but the use of $(C_2H_5)_2AlCl$ as a catalyst permits the reaction to proceed at room temperature and *endo* addition is favored by 8:1.[37]

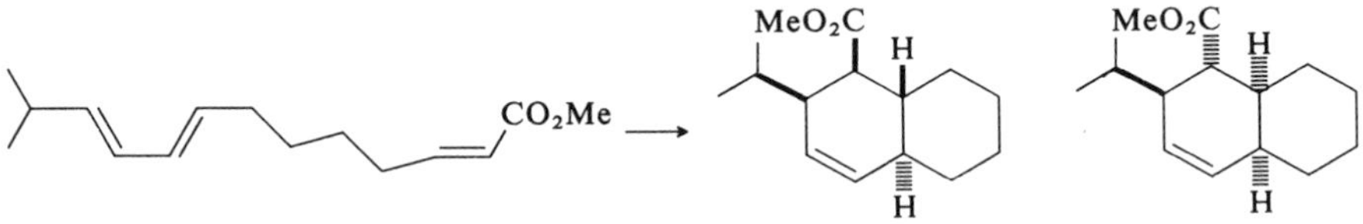

CO_2Me → MeO_2C H H ; MeO_2C H H

	endo T.S.	*exo* T.S.
thermal (160°)	50%	50%
Et_2AlCl (23°)	88%	12%

It has also been noted in certain systems that the stereoselectivity is a function of the activating substituent on the double bond, both for thermal and Lewis acid catalyzed reactions.[38] The general trend in these systems is in agreement with frontier orbital interactions and conformational effects being the main factors in determining stereoselectivity. Because the conformational interactions depend on the substituent pattern in the specific case, no general rules for stereoselectivity can be put forward.

6.2. Dipolar Cycloaddition Reactions

In Section 11.3 of Part A, the relationship of 1,3-dipolar cycloadditions to the general topic of concerted cycloadditions was discussed. Dipolar cycloaddition reactions are useful both for the synthesis of heterocyclic compounds and for carbon–carbon bond formation. Table 6.2 lists some of the types of molecules that are capable of dipolar cycloaddition. These molecules, which are called *1,3-dipoles,* are isoelectronic with allyl anion. They have four π electrons, and each has at least

36. W. R. Roush, A. I. Ko, and H. R. Gillis, *J. Org. Chem.* **45**, 4264 (1980); R. K. Boeckman, Jr., and S. K. Ko, *J. Am. Chem. Soc.* **102**, 7146 (1980); W. R. Roush and S. E. Hall, *J. Am. Chem. Soc.* **103**, 5200 (1981); K. A. Parker and T. Iqbal, *J. Org. Chem.* **52**, 4369 (1987).
37. W. R. Roush and H. R. Gillis, *J. Org. Chem.* **47**, 4825 (1982).
38. J. A. Marshall, J. E. Audia, and J. Grote, *J. Org. Chem.* **49**, 5277 (1984); W. R. Roush, A. P. Eisenfeld, and J. S. Warmus, *Tetrahedron Lett.* **28**, 2447 (1987); T.-C. Wu and K. N. Houk, *Tetrahedron Lett.* **26**, 2293 (1985).

one charge-separated resonance structure with opposite charges in a 1,3-relationship. It is this structural feature that leads to the name *1,3-dipolar cycloaddition reactions* for this class of reactions.[39] The other reactant in a dipolar cycloaddition, usually an alkene or alkyne, is referred to as the *dipolarophile.* Other multiply bonded functional groups such as imines, azo groups, and nitroso groups can also act as dipolarophiles.

Mechanistic studies have shown that the transition state for 1,3-dipolar cycloaddition is not very polar. The rate of reaction is not strongly sensitive to solvent polarity. There is general agreement that the reaction is a concerted $[_{\pi}4_s + _{\pi}2_s]$ cycloaddition.[40] The destruction of charge separation that is implied is more apparent than real, because most 1,3-dipolar compounds are not highly polar. The polarity implied by any single structure is balanced by other contributing structures.

Table 6.2. 1,3-Dipolar Compounds

Structures	Name
$:\overset{+}{N}{=}\ddot{N}{-}\overset{-}{\underset{..}{C}}R_2 \leftrightarrow :N{\equiv}\overset{+}{N}{-}\overset{-}{\underset{..}{C}}R_2$	Diazoalkane
$:\overset{+}{N}{=}\ddot{N}{-}\overset{-}{\ddot{N}}R \leftrightarrow :N{\equiv}\overset{+}{N}{-}\overset{-}{\ddot{N}}R$	Azide
$R\overset{+}{C}{=}N{=}\overset{-}{C}R_2 \leftrightarrow RC{\equiv}\overset{+}{N}{-}\overset{-}{C}R_2$	Nitrile ylide
$R\overset{+}{C}{=}N{-}\overset{-}{\ddot{N}}R \leftrightarrow RC{\equiv}\overset{+}{N}{-}\overset{-}{\ddot{N}}R$	Nitrile imine
$R\overset{+}{C}{=}N{-}\overset{-}{\ddot{O}}: \leftrightarrow RC{\equiv}\overset{+}{N}{-}\overset{-}{\ddot{O}}:$	Nitrile oxide
$R_2\overset{+}{C}{-}\ddot{N}(R){-}\overset{-}{C}R_2 \leftrightarrow R_2C{=}\overset{+}{N}(R){-}\overset{-}{C}R_2$	Azomethine ylide
$R_2\overset{+}{C}{-}\ddot{N}(R){-}\overset{-}{\ddot{O}}: \leftrightarrow R_2C{=}\overset{+}{N}(R){-}\overset{-}{\ddot{O}}:$	Nitrone
$R_2\overset{+}{C}{-}\ddot{O}{-}\overset{-}{\ddot{O}}: \leftrightarrow R_2C{=}\overset{+}{O}{-}\overset{-}{\ddot{O}}:$	Carbonyl oxide

39. For comprehensive reviews of 1,3-dipolar cycloaddition reactions, see G. Bianchi, C. DeMicheli, and R. Gandolfi, in *The Chemistry of Double Bonded Functional Groups, Part I, Supplement A*, S. Patzi (ed.), Wiley-Interscience, New York, 1977, pp. 369–532; A. Padwa (ed.), *1,3-Dipolar Cycloaddition Chemistry,* Wiley, New York, 1984. For a review of intramolecular 1,3-dipolar cycloaddition reactions, see A. Padwa, *Angew. Chem. Int. Ed. Engl.* **15**, 123 (1976).
40. P. K. Kadaba, *Tetrahedron* **25**, 3053 (1969); R. Huisgen, G. Szeimies, and L. Mobius, *Chem. Ber.* **100**, 2494 (1967); P. Scheiner, J. H. Schomaker, S. Deming, W. J. Libbey, and G. P. Nowack, *J. Am. Chem. Soc.* **87**, 306 (1965).

Two questions are of principal interest for predicting the structure of 1,3-dipolar cycloaddition products: (1) What is the regioselectivity? (2) What is the stereoselectivity? Many specific examples demonstrate that 1,3-dipolar cycloaddition is a stereospecific *syn* addition with respect to the dipolarophile. This is what would be expected for a concerted process.

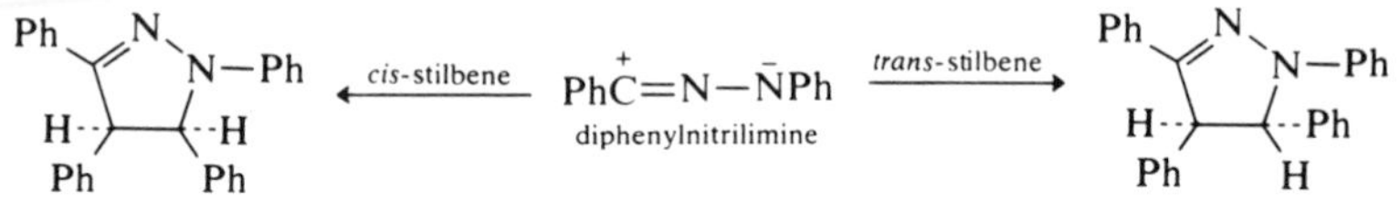
cis-stilbene
trans-stilbene
diphenylnitrilimine

Ref. 41

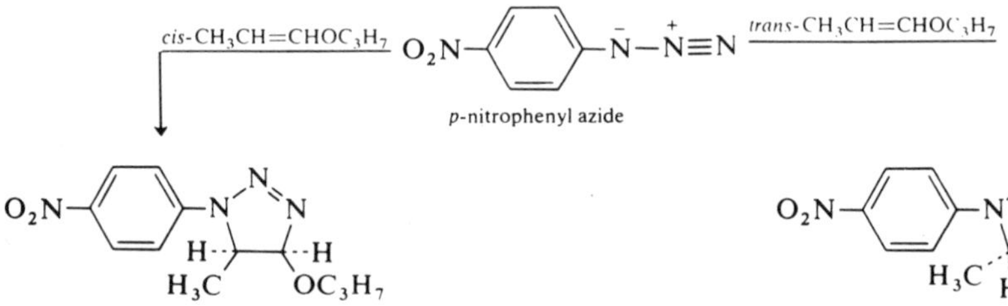
cis-CH3CH=CHOC3H7
trans-CH3CH=CHOC3H7
p-nitrophenyl azide

Ref. 42

With some 1,3-dipoles, two possible stereoisomers can be formed by *syn* addition. These result from two differing orientations of the reacting molecules, which are analogous to the *endo* and *exo* transition states in Diels–Alder reactions. Diazoalkanes, for example, can add to unsymmetrical dipolarophiles to give two diastereomers.

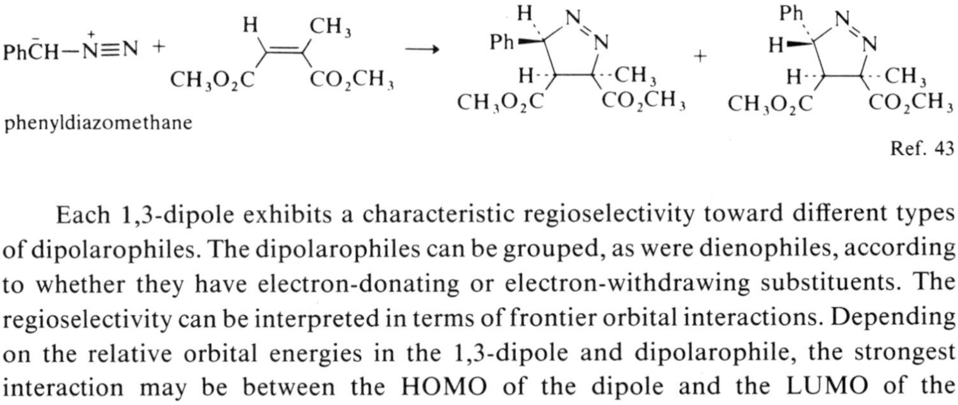
phenyldiazomethane

Ref. 43

Each 1,3-dipole exhibits a characteristic regioselectivity toward different types of dipolarophiles. The dipolarophiles can be grouped, as were dienophiles, according to whether they have electron-donating or electron-withdrawing substituents. The regioselectivity can be interpreted in terms of frontier orbital interactions. Depending on the relative orbital energies in the 1,3-dipole and dipolarophile, the strongest interaction may be between the HOMO of the dipole and the LUMO of the dipolarophile or vice versa. Usually, for dipolarophiles with electron-attracting groups the dipole-HOMO/dipolarophile-LUMO interaction is dominant. The reverse is true for dipolarophiles with donor substituents. In some circumstances, the magnitudes of the two interactions may be comparable.[44]

41. R. Huisgen, M. Seidel, G. Wallbillich, and H. Knupfer, *Tetrahedron* **17**, 3 (1965).
42. R. Huisgen and G. Szeimies, *Chem. Ber.* **98**, 1153 (1965).
43. R. Huisgen and P. Eberhard, *Tetrahedron Lett.*, 4343 (1971).
44. K. N. Houk, J. Sims, B. E. Duke, Jr., R. W. Strozier, and J. K. George, *J. Am. Chem. Soc.* **95**, 7287 (1973); I. Fleming, *Frontier Orbitals and Organic Chemical Reactions*, Wiley, New York, 1977; K. N. Houk, in *Pericyclic Reactions*, Vol. II, A. P. Marchand and R. E. Lehr (eds.), Academic Press, New York, 1977, pp. 181–271.

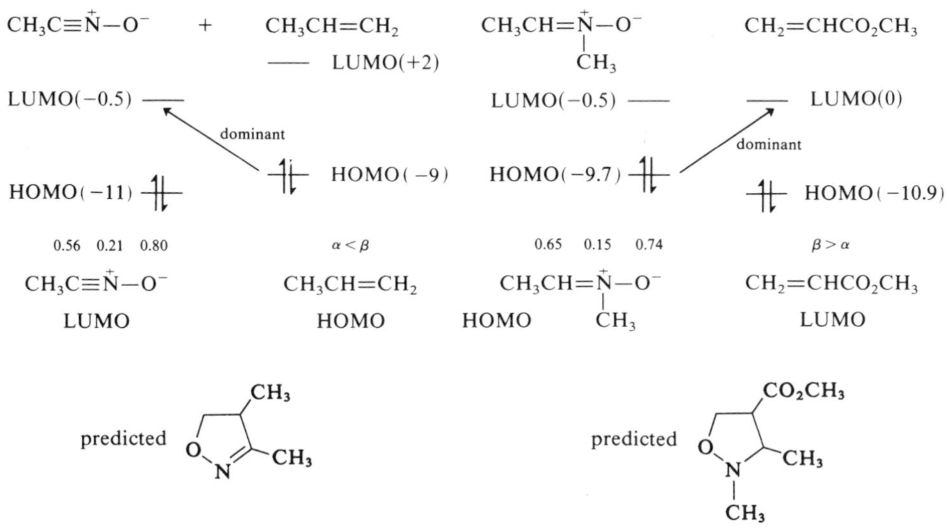

$CH_3C{\equiv}\overset{+}{N}{-}O^-$ + $CH_3CH{=}CH_2$ — LUMO(+2); LUMO(−0.5); HOMO(−11); HOMO(−9); dominant

$CH_3CH{=}\overset{+}{N}(CH_3){-}O^-$ + $CH_2{=}CHCO_2CH_3$ — LUMO(−0.5); LUMO(0); HOMO(−9.7); HOMO(−10.9); dominant

0.56 0.21 0.80 $CH_3C{\equiv}\overset{+}{N}{-}O^-$ LUMO; $\alpha < \beta$ $CH_3CH{=}CH_2$ HOMO

0.65 0.15 0.74 $CH_3CH{=}\overset{+}{N}(CH_3){-}O^-$ HOMO; $\beta > \alpha$ $CH_2{=}CHCO_2CH_3$ LUMO

predicted (isoxazoline with CH_3, CH_3) predicted (isoxazolidine with CO_2CH_3, CH_3, N–CH_3)

Fig. 6.4. Prediction of regioselectivity of 1,3-dipolar cycloaddition.

The prediction of regiochemistry requires estimation or calculation of the energies of the orbitals involved. This permits identification of the frontier orbitals. The energies and orbital coefficients for the most common dipoles and dipolarophiles have been summarized.[44] Figure 11.13 in Part A gives the orbital coefficients of some representative 1,3-dipoles. Regioselectivity is determined by the preference for the orientation that results in bond formation between the atoms having the largest coefficients in the two frontier orbitals. This analysis illustrated in Fig. 6.4.

In addition to the role of substituents in determining regioselectivity, several other structural features affect the reactivity of dipolarophiles. Strain increases reactivity. Norbornene, for example, is consistently more reactive than cyclohexene in 1,3-dipolar cycloadditions. Conjugated functional groups also usually increase reactivity. This increased reactivity has most often been demonstrated with electron-attracting substituents, but, in the case of some 1,3-dipoles, enamines, enol ethers, and other alkenes with donor substituents are also quite reactive. Some reactivity data for a series of alkenes with a few 1,3-dipoles are given in Table 6.3. Scheme 6.4 gives some examples of 1,3-dipolar cycloaddition reactions.

Dipolar cycloadditions are an important means of synthesis of a wide variety of heterocyclic molecules. Some of these are useful intermediates in multistage synthesis. Pyrazolines, which are formed from alkenes and diazo compounds, for example, can be pyrolyzed or photolyzed to give cyclopropanes.

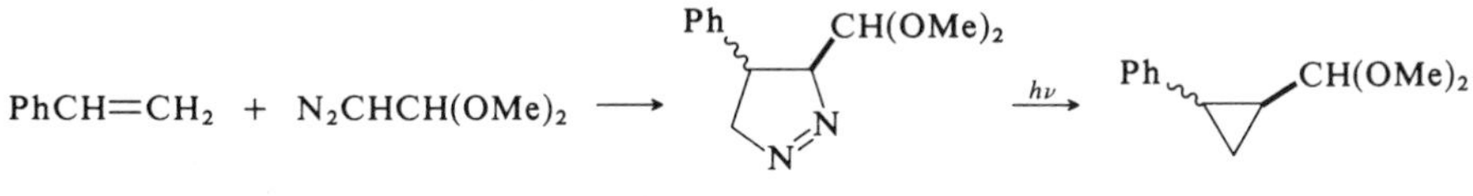

$PhCH{=}CH_2$ + $N_2CHCH(OMe)_2$ → (pyrazoline with Ph, $CH(OMe)_2$) $\xrightarrow{h\nu}$ (cyclopropane with Ph, $CH(OMe)_2$) Ref. 45

45. P. Carrie, *Heterocycles* **14**, 1529 (1980).

Table 6.3. Relative Reactivity of Substituted Alkenes toward Some 1,3-Dipoles[a,b]

Substituted alkene	Ph_2CN_2	PhN_3	$Ph\overset{+}{N}{=}N{-}\overset{-}{N}Ph$	$PhC{\equiv}\overset{+}{N}{-}\overset{-}{O}$	$PhC(H){=}\overset{+}{N}(O^-){-}CH_3$
Dimethyl fumarate	100	31	283	94	18.3
Dimethyl maleate	27.8	1.25	7.9	1.61	6.25
Norbornene	1.15	700	3.1	97	0.13
Ethyl acrylate	28.8	36.5	48	66	11.1 (methyl ester)
Butyl vinyl ether	—	1.5	—	15	—
Styrene	0.57	1.5	1.6	9.3	0.32
Ethyl crotonate	1.0	1.0	1.0	1.0	1.0
Cyclopentene	—	6.9	0.13	1.04	0.022
Terminal alkene	—	0.89 (heptene)	0.15 (heptene)	2.6 (hexene)	0.072 (heptene)
Cyclohexene	—	—	0.011	0.055	—

a. Data are selected from those compiled by R. Huisgen, R. Grashey, and J. Sauer, in *Chemistry of Alkenes*, S. Patai (ed.), Interscience, New York, 1964, pp. 806–877.
b. Conditions such as solvent and temperature vary for each 1,3-dipole, so comparison from dipole to dipole is not possible. Following Huisgen, Grashey, and Sauer,[a] ethyl crotonate is assigned reactivity = 1.0 for each 1,3-dipole.

Scheme 6.4. Typical 1,3-Dipolar Cycloaddition Reactions

A. Intermolecular Cycloaddition

1[a] $O_2N{-}C_6H_4{-}N{=}\overset{+}{N}{=}\overset{-}{N}$ + norbornene → triazoline (92%)

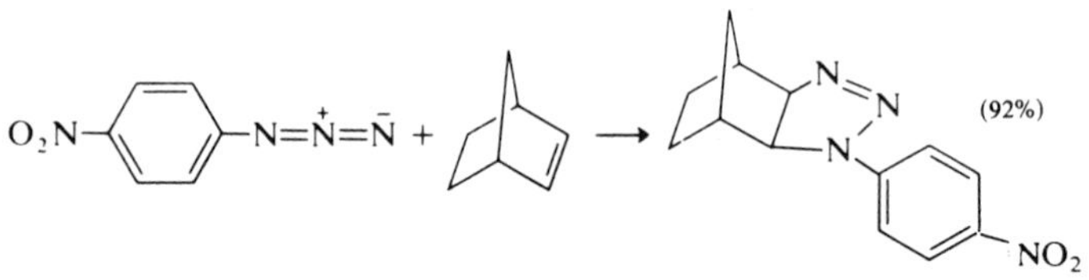

2[b] $C_6H_5{-}N{=}\overset{+}{N}{=}\overset{-}{N} + H_3CO_2CC{\equiv}CCO_2CH_3$ → triazole with H_3CO_2C, CO_2CH_3 (87%)

3[c] $CH_2N_2 + H_2C{=}CH$–(1,3-dioxolan-2-yl) → pyrazoline (80%)

4[d] $PhCH{=}\overset{+}{N}(O^-)CH_3$ + $H_2C{=}CHC{\equiv}N$ → isoxazolidine with CH_3, H, Ph, $C{\equiv}N$ (91%)

5[e] cyclopentanone with R and CH_2NO_2 $\xrightarrow[Et_3N]{PhNCO}$ cyclopentanone with R and $C{\equiv}\overset{+}{N}{-}O^-$

$R = -(CH_2)_6\overset{O}{\overset{\|}{C}}OCH_2CH_2CH_2CH_3$

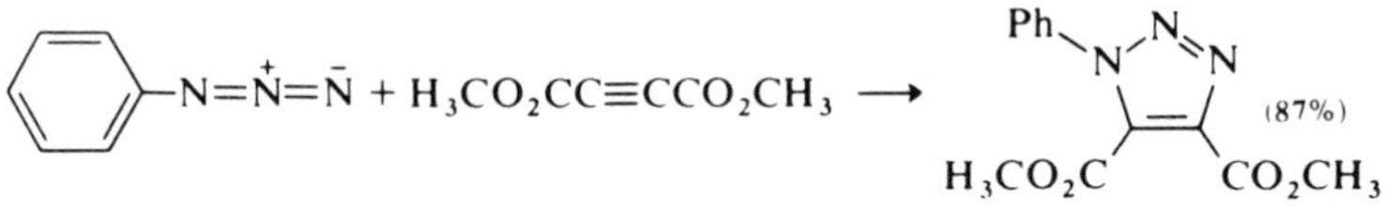

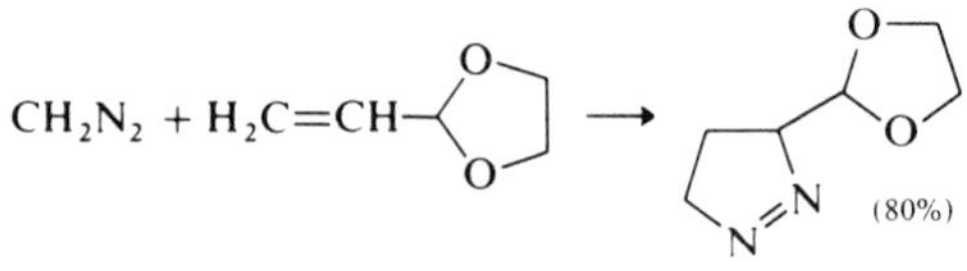

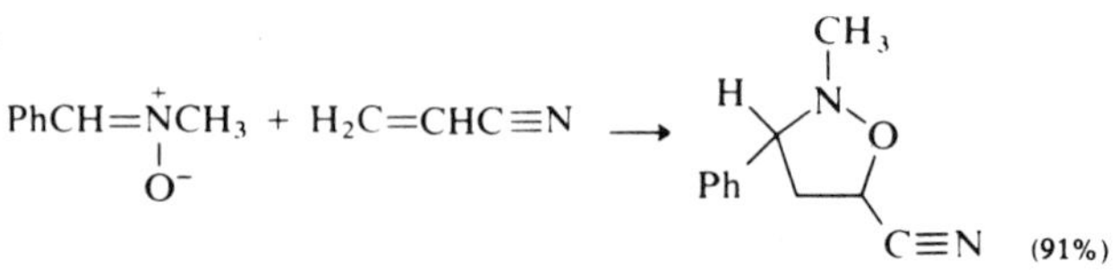

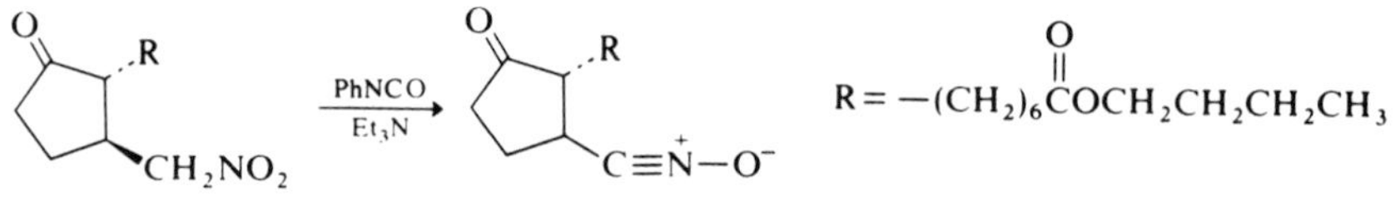

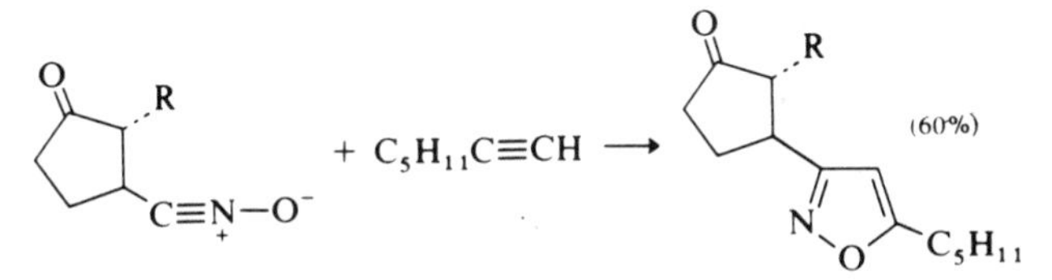

(60%)

B. Intramolecular Cycloaddition

6[f]

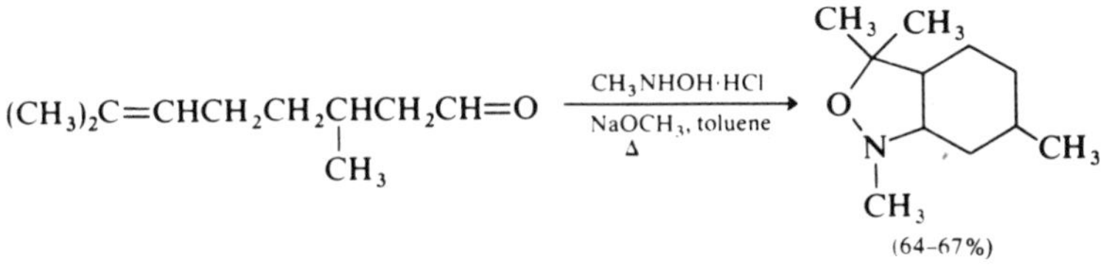

$(CH_3)_2C{=}CHCH_2CH_2CH(CH_3)CH_2CH{=}O \xrightarrow[\Delta]{CH_3NHOH{\cdot}HCl,\ NaOCH_3,\ toluene}$ (64–67%)

7[g]

toluene, Δ; 1) H_2, Pd/C 2) CH_2O, HCO_2H

8[h]

$O{=}CHCH_2S$ —($PhCH_2NHOH$, acetonitrile)→ (66%) —(Zn, HOAc, H_2O)→ (96%)

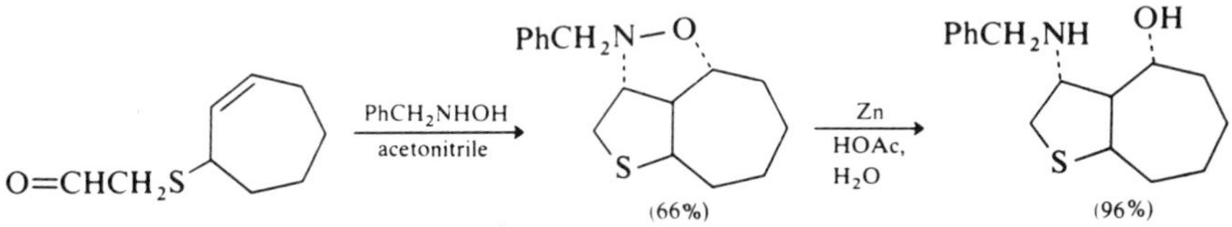

a. P. Scheiner, J. H. Schomaker, S. Deming, W. J. Libbey, and G. P. Nowack, *J. Am. Chem. Soc.* **87**, 306 (1965).
b. R. Huisgen, R. Knorr, L. Mobius, and G. Szeimies, *Chem. Ber.* **98**, 4014 (1965).
c. J. M. Stewart, C. Carlisle, K. Kem, and G. Lee, *J. Org. Chem.* **35**, 2040 (1970).
d. R. Huisgen, H. Hauck, R. Grashey, and H. Seidl, *Chem. Ber.* **101**, 2568 (1968).
e. A. Barco, S. Benetti, G. P. Pollini, P. G. Baraldi, M. Guarneri, D. Simoni, and C. Gandolfi, *J. Org. Chem.* **46**, 4518 (1981).
f. N. A. LeBel and D. Hwang, *Org. Synth.* **58**, 106 (1978).
g. J. J. Tufariello, G. B. Mullen, J. J. Tegeler, E. J. Trybulski, S. C. Wong, and S. A. Ali, *J. Am. Chem. Soc.* **101**, 2435 (1979).
h. P. N. Confalone, G. Pizzolato, D. L. Confalone, and M. R. Uskokovic, *J. Am. Chem. Soc.* **102**, 1954 (1980).

Intramolecular 1,3-dipolar cycloadditions have proven to be particularly useful in synthesis. The addition of nitrones to alkenes serves both to form a carbon–carbon bond and to introduce functionality.[46] Entry 6 in Scheme 6.4 is an example. The nitrone **A** is generated by condensation of the aldehyde group with *N*-methylhydroxylamine.

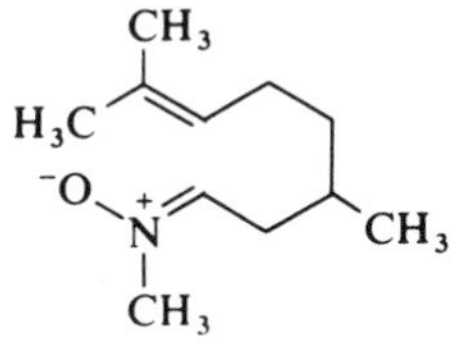

A

46. For reviews of nitrone cycloadditions, see D. St. C. Black, R. F. Crozier, and V. C. Davis, *Synthesis*, 205 (1975); J. J. Tufariello, *Acc. Chem. Res.* **12**, 396 (1979); P. N. Confalone and E. M. Huie, *Org. React.* **36**, 1 (1988).

The products of nitrone–alkene cycloadditions are isoxazolines, and the oxygen–nitrogen bond can be cleaved by reduction, leaving both an amino and a hydroxy function in place. A number of imaginative syntheses have employed this strategy. Entry 7 in Scheme 6.4 is the synthesis of the alkaloid pseudotropine. The proper stereochemical orientation of the hydroxyl group is assured by the structure of the isoxazoline from which it is formed. Entry 8 in Scheme 6.4 portrays the early stages of the synthesis of the biologically important molecule biotin.

Nitrile oxides, which are usually formed by dehydration of nitroalkanes, are also useful 1,3-dipoles. They are highly reactive and must be generated *in situ.*[47] They react with alkenes and alkynes. Entry 5 in Scheme 6.4 is an example in which the cycloaddition product (an isoxazole) was eventually converted to a prostaglandin derivative.

An interesting variation of the 1,3-dipolar cycloaddition involves generation of 1,3-dipoles from three-membered rings. As an example, aziridines **2** and **4** give adducts derived from apparent formation of 1,3-dipoles **3** and **5**, respectively.[48]

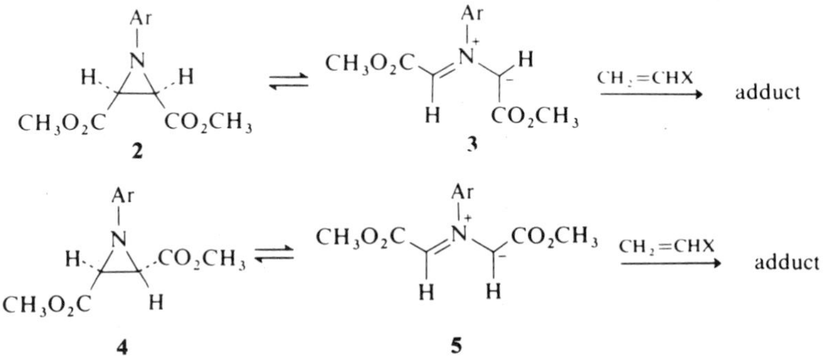

Scheme 6.5. Generation of Dipolar Intermediates from Small Rings

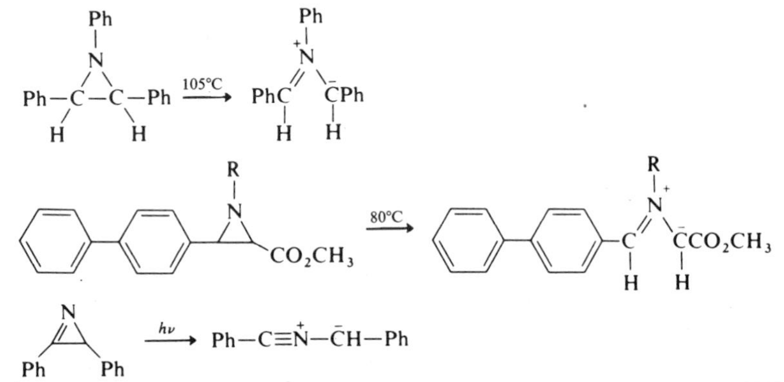

a. H. W. Heine, R. Peavy, and A. J. Durbetaki, *J. Org. Chem.* **31**, 3924 (1966).
b. P. B. Woller and N. H. Cromwell, *J. Org. Chem.* **35**, 888 (1970).
c. A. Padwa, M. Dharan, J. Smolanoff, and S. I. Wetmore, Jr., *J. Am. Chem. Soc.* **95**, 1945, 1954 (1973).

47. K. Torssell, *Nitrile Oxides, Nitrones and Nitronates in Organic Synthesis,* VCH Publishers, New York, 1988.
48. R. Huisgen and H. Mader, *J. Am. Chem. Soc.* **93**, 1777 (1971).

The evidence for the involvement of 1,3-dipoles as discrete intermediates includes the observation that the reaction rates are independent of dipolarophile concentration. This fact indicates that the ring opening is the rate-determining step in the reaction. Ring opening is most facile for aziridines that have an electron-attracting substituent to stabilize the carbanion center in the dipole. Scheme 6.5 gives some examples of formation of 1,3-dipoles in ring opening reactions.

Cyclopropanones are also reactive toward certain types of cycloadditions. It is suspected that a dipolar species resulting from reversible cleavage of the cyclopropanone ring is the reactive species. These intermediates, which are known as oxyallyl cations, can also be generated by a number of other reaction processes.[49]

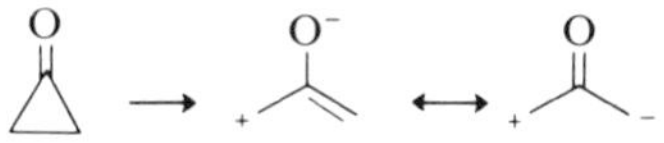

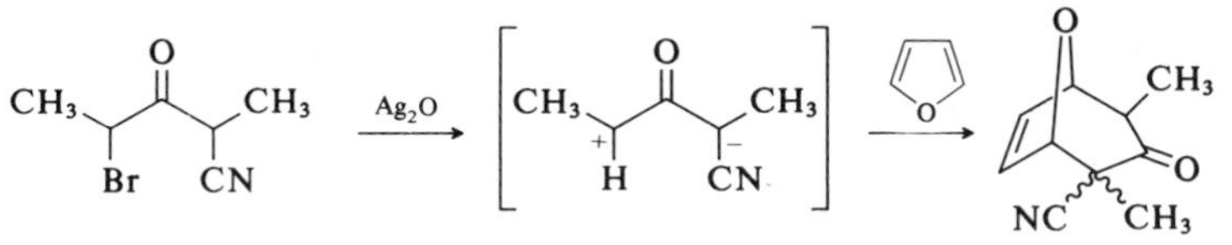

Ag_2O Ref. 50

6.3. [2 + 2] Cycloadditions and Other Reactions Leading to Cyclobutanes

Among the cycloaddition reactions that have been shown to have general synthetic utility are the [2 + 2] cycloadditions of ketenes and alkenes.[51] The stereoselectivity of ketene–alkene cycloaddition can be analyzed in terms of the Woodward–Hoffmann rules.[52] To be an allowed process, the $[_{\pi}2 + _{\pi}2]$ cycloaddition must be suprafacial in one component and antarafacial in the other. Figure 6.5 illustrates the transition state. The ketene, utilizing the low-lying LUMO, is the antarafacial component and interacts with the HOMO of the alkene. The stereoselectivity of ketene cycloadditions can be rationalized in terms of steric effects in this transition state. Minimization of interaction between the substituents R and R′ leads to a cyclobutanone in which these substituents are *cis*. This is the stereochemistry observed in these reactions.

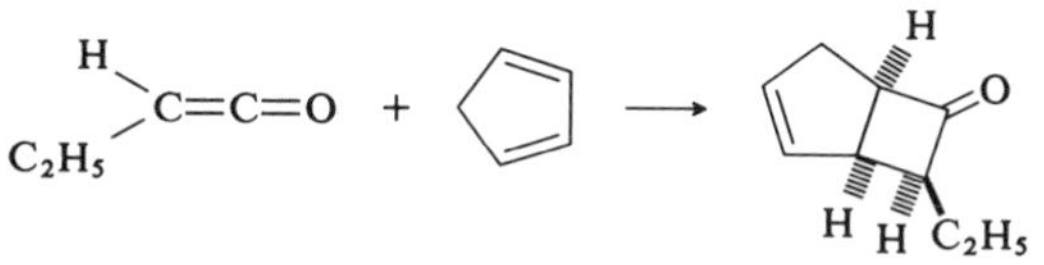

Ref. 53

49. N. J. Turro, S. S. Edelson, J. R. Williams, T. R. Darling, and W. B. Hammond, *J. Am. Chem. Soc.* **91**, 2283 (1969); S. S. Edelson and N. J. Turro, *J. Am. Chem. Soc.* **92**, 2770 (1970); N. J. Turro, *Acc. Chem. Res.* **2**, 25 (1969); J. Mann, *Tetrahedron* **42**, 4611 (1986).
50. B. Foehlisch, D. Lutz, W. Gottstein, and U. Dukek, *Justus Liebigs Ann. Chem.*, 1847 (1977).
51. For reviews, see W. T. Brady, in *The Chemistry of Ketenes, Allenes, and Related Compounds*, S. Patai (ed.), Wiley-Interscience, New York, 1980, Chapter 8; W. T. Brady, *Tetrahedron* **37**, 2949 (1981).
52. R. B. Woodward and R. Hoffmann, *Angew. Chem. Int. Ed. Engl.* **8**, 781 (1969).
53. M. Rey, S. M. Roberts, A. S. Dreiding, A. Roussel, H. Vanlierde, S. Toppert, and L. Ghosez, *Helv. Chim. Acta* **65**, 703 (1982).

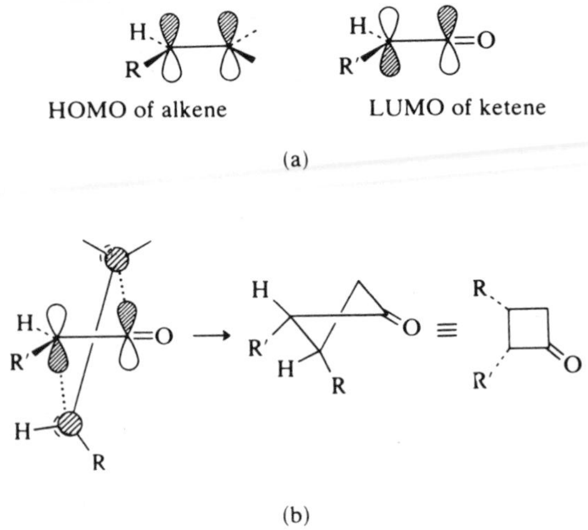

HOMO of alkene LUMO of ketene

(a)

(b)

Fig. 6.5. HOMO-LUMO interactions in the [2 + 2] cycloaddition of an alkene and a ketene. (a) Frontier orbitals of alkene and ketene. (b) Transition state required for suprafacial addition to alkene and antarafacial addition to ketene, leading to R and R′ in *cis* orientation in cyclobutanone products.

Ketenes are especially reactive in [2 + 2] cycloadditions, and an important reason is that they offer a low degree of steric interactions in the $[{}_{\pi}2_s + {}_{\pi}2_a]$ transition state. Another reason is the electrophilic character of the ketone LUMO. The best yields are obtained in reactions in which the ketene has an electronegative substituent, such as halogen. Simple ketenes are not very stable and must usually be generated *in situ*. The most convenient method for generating ketenes for synthesis is by dehydrohalogenation of acid chorides. This is usually done with an amine such as triethylamine. Ketene itself and certain alkyl derivatives can be generated by pyrolysis of carboxylic anhydrides.[54] Scheme 6.6 gives some examples of ketone–alkene cycloadditions.

Intramolecular ketene cycloadditions are possible if the ketene and alkene functionalities can achieve an appropriate orientation.[55]

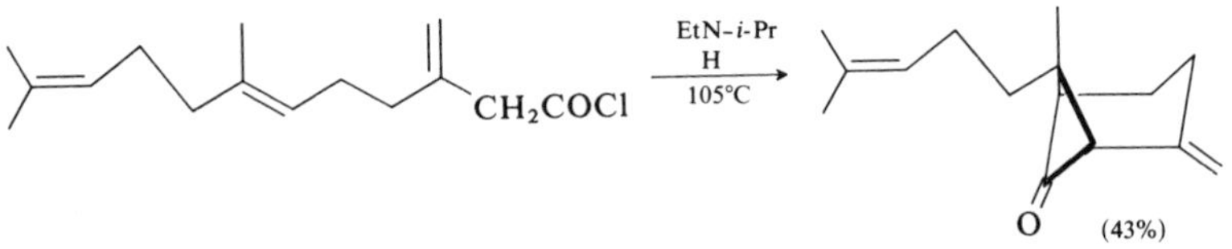

CH_2COCl $\xrightarrow[105°C]{EtN\text{-}i\text{-}Pr \atop H}$ (43%) Ref. 56

Cyclobutanes can also be formed by nonconcerted processes involving zwitterionic intermediates. The combination of an electron-rich alkene (enamine, enol

54. G. J. Fisher, A. F. MacLean, and A. W. Schnizer, *J. Org. Chem.* **18**, 1055 (1953).
55. B. B. Snider, R. A. H. F. Hui, and Y. S. Kulkarni, *J. Am. Chem. Soc.* **107**, 2194 (1985); B. B. Snider and R. A. H. F. Hui, *J. Org. Chem.* **50**, 5167 (1985); W. T. Brady and Y. F. Giang, *J. Org. Chem.* **50**, 5177 (1985).
56. E. J. Corey and M. C. Desai, *Tetrahedron Lett.* **26**, 3535 (1985).

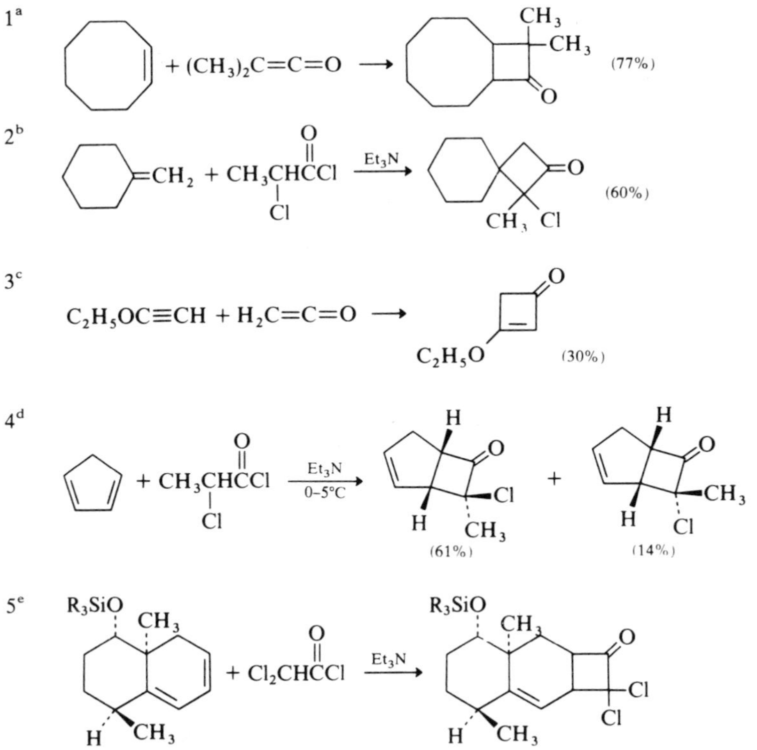

1[a]

$$\text{cyclooctene} + (CH_3)_2C{=}C{=}O \longrightarrow \text{bicyclic cyclobutanone (}CH_3, CH_3, O\text{)} \quad (77\%)$$

2[b]

$$\text{methylenecyclohexane} + CH_3CH(Cl)C(O)Cl \xrightarrow{Et_3N} \text{spiro cyclobutanone (}CH_3, Cl, =O\text{)} \quad (60\%)$$

3[c]

$$C_2H_5OC{\equiv}CH + H_2C{=}C{=}O \longrightarrow C_2H_5O\text{-cyclobutenone} \quad (30\%)$$

4[d]

$$\text{cyclopentadiene} + CH_3CH(Cl)C(O)Cl \xrightarrow[0\text{–}5^\circ C]{Et_3N} \text{(Cl exo, }CH_3\text{ endo)} \ (61\%) + \text{(}CH_3\text{ exo, Cl endo)} \ (14\%)$$

5[e]

$$\text{diene (}R_3SiO, CH_3, H, CH_3\text{)} + Cl_2CHC(O)Cl \xrightarrow{Et_3N} \text{dichlorocyclobutanone adduct}$$

a. A. P. Krapcho and J. H. Lesser, *J. Org. Chem.* **31**, 2030 (1966).
b. W. T. Brady and A. D. Patel, *J. Org. Chem.* **38**, 4106 (1973).
c. H. H. Wasserman, J. U. Piper, and E. V. Dehmlow, *J. Org. Chem.* **38**, 1451 (1973).
d. W. T. Brady and R. Roe, *J. Am. Chem. Soc.* **93**, 1662 (1971).
e. P. A. Grieco, T. Oguri, and S. Gilman, *J. Am. Chem. Soc.* **102**, 5886 (1980).

ether) and an electrophilic one (nitro- or polycyanoalkene) is required for such processes.

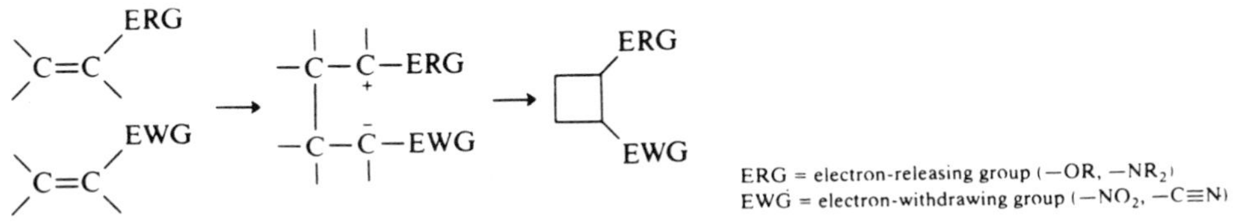

$$\begin{matrix} >C{=}C<^{ERG} \\ >C{=}C<^{EWG} \end{matrix} \longrightarrow \begin{matrix} -\overset{|}{C}-\overset{+}{C}-ERG \\ -\underset{|}{C}-\overset{-}{C}-EWG \end{matrix} \longrightarrow \text{cyclobutane (ERG, EWG)}$$

ERG = electron-releasing group ($-OR$, $-NR_2$)
EWG = electron-withdrawing group ($-NO_2$, $-C{\equiv}N$)

Two examples of this reaction type are shown below.

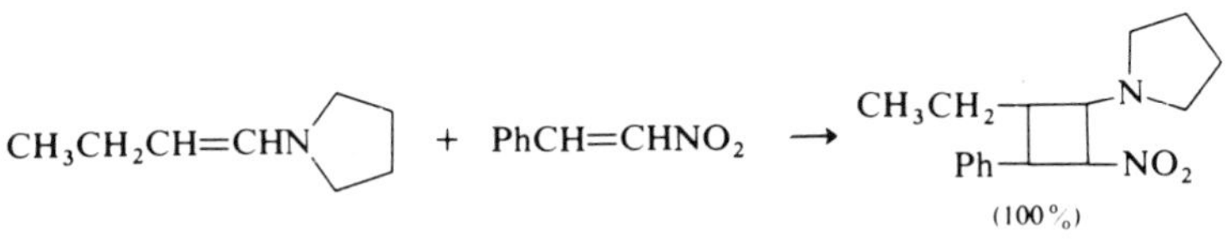

$$CH_3CH_2CH{=}CHN(\text{pyrrolidine}) + PhCH{=}CHNO_2 \longrightarrow \text{cyclobutane (}CH_3CH_2, N\text{-pyrrolidinyl, Ph, }NO_2\text{)} \quad (100\%)$$

Ref. 57

57. M. E. Kuehne and L. Foley, *J. Org. Chem.* **30**, 4280 (1965).

$$H_2C{=}CHOCH_3 + (NC)_2C{=}C(CN)_2 \longrightarrow$$ 3-methoxy-1,1,2,2-tetracyanocyclobutane (90%) Ref. 58

The stereochemistry of these reactions depends on the lifetime of the dipolar intermediate, which, in turn, is influenced by the polarity of the solvent. In the reactions of enol ethers with tetracyanoethylene, the stereochemistry of the enol ether portion is retained in nonpolar solvents. In polar solvents, cycloaddition is nonstereospecific, as a result of a longer lifetime for the zwitterionic intermediate.[59]

6.4. Photochemical Cycloaddition Reactions

Photochemical cycloaddition provides a method that is often complementary to thermal cycloadditions with regard to the types of compounds that can be prepared. The theoretical basis for this complementary relationship between thermal and photochemical modes of reaction lies in orbital symmetry relationships, as discussed in Chapter 11 of Part A. The reaction types permitted by photochemical excitation that are particularly useful for synthesis are $[2+2]$ additions between two carbon–carbon double bonds and $[2+2]$ additions of alkenes and carbonyl groups to form oxetanes. Photochemical cycloadditions are not always concerted processes because in some cases the reactive excited state is a triplet. In this case, the initial adduct is a triplet 1,4-diradical which must undergo spin inversion before product formation is complete. Stereospecificity is lost if the intermediate 1,4-diradical undergoes bond rotation faster than ring closure. As a result, photochemical $[2+2]$ cycloadditions are not always stereospecific, in contrast to the concerted thermal cycloadditions.

$$>C{=}C< \xrightarrow{h\nu} {}^1[>C{=}C<]^* \xrightarrow{\text{intersystem crossing}} {}^3[>C{=}C<]^*$$

$${}^3[>C{=}C<]^* + >C{=}C< \rightarrow >\overset{\uparrow}{C}-\underset{|}{\overset{|}{C}}-\underset{|}{\overset{|}{C}}-\overset{\uparrow}{C}<$$

$$>\overset{\uparrow}{C}-\underset{|}{\overset{|}{C}}-\underset{|}{\overset{|}{C}}-\overset{\uparrow}{C}< \rightarrow >\overset{\uparrow}{C}-\underset{|}{\overset{|}{C}}-\underset{|}{\overset{|}{C}}-\overset{\downarrow}{C}< \rightarrow \text{cyclobutane}$$

Intermolecular photocycloadditions of alkenes can be carried out by photosensitization with mercury or directly with short wavelength light.[60] Relatively little

58. J. K. Williams, D. W. Wiley, and B. C. McKusick, *J. Am. Chem. Soc.* **84**, 2210 (1962).
59. R. Huisgen, *Acc. Chem. Res.* **10**, 117, 199 (1977).
60. H. Yamazaki and R. J. Cvetanovic, *J. Am. Chem. Soc.* **91**, 520 (1969).

preparative use has been made of this reaction for simple alkenes. Dienes can be photosensitized using benzophenone, 2,3-butanedione, and acetophenone.[61] The photodimerization of derivatives of cinnamic acid was among the earliest photochemical reactions to be studied.[62] Good yields of dimers are obtained when irradiation is carried out in the crystalline state. In solution, *cis–trans* isomerization is the dominant reaction.

$$PhCH{=}CHCO_2H \xrightarrow[H_2O]{h\nu}$$ (56%)

The presence of Cu(I) salts promotes intermolecular photocycloaddition of simple alkenes. Copper (I) triflate is especially effective.[63] It is believed that the photoreactive species is a 2:1 alkene-Cu(I) complex in which the two alkene molecules are brought together prior to photoexcitation.[64]

$$2\ RCH{=}CH_2 + Cu^I \rightleftharpoons \ldots\ \xrightarrow{h\nu}$$

$$\xrightarrow[h\nu]{CuO_3SCF_3}$$

Intramolecular [2 + 2] photocycloaddition of alkenes is an important method of formation of compounds containing four-membered rings.[65] Direct irradiation of simple nonconjugated dienes leads to cyclobutanes.[66] Strain makes the reaction unfavorable for 1,4-dienes, but when the alkene units are separated by at least two carbon atoms, cycloaddition becomes possible.

Ref. 67

The most widely exploited photochemical cycloadditions involve irradiation of dienes in which the two double bonds are fairly close and result in formation of

61. G. S. Hammond, N. J. Turro, and R. S. H. Liu, *J. Org. Chem.* **28**, 3297 (1963).
62. A. Mustafa, *Chem. Rev.* **51**, 1 (1951); D. G. Farnum and A. J. Mostashari, *Org. Photochem. Synth.* **1**, 103 (1971).
63. R. G. Salomon, *Tetrahedron* **39**, 485 (1983).
64. R. G. Salomon, K. Folking, W. E. Streib, and J. K. Kochi, *J. Am. Chem. Soc.* **96**, 1145 (1974).
65. P. de Mayo, *Acc. Chem. Res.* **4**, 41 (1971).
66. R. Srinivasan, *J. Am. Chem. Soc.* **84**, 4141 (1962); *J. Am. Chem. Soc.* **90**, 4498 (1968).
67. J. Meinwald and G. W. Smith, *J. Am. Chem. Soc.* **89**, 4923 (1967); R. Srinivasan and K. H. Carlough, *J. Am. Chem. Soc.* **89**, 4932 (1967).

Scheme 6.7. Intramolecular [2 + 2] Photochemical Cycloaddition Reactions of Dienes

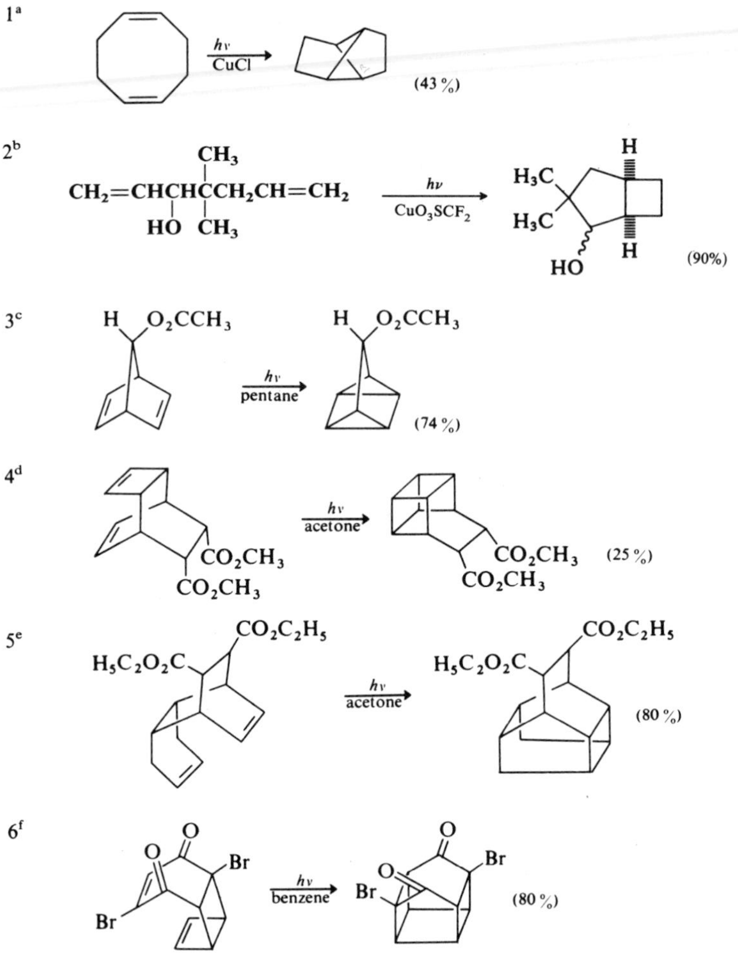

1[a] $h\nu$, CuCl (43%)

2[b] $CH_2{=}CHCH(OH)C(CH_3)_2CH_2CH{=}CH_2$ $\xrightarrow{h\nu,\ CuO_3SCF_3}$ (90%)

3[c] $h\nu$, pentane (74%)

4[d] $h\nu$, acetone (25%)

5[e] $h\nu$, acetone (80%)

6[f] $h\nu$, benzene (80%)

a. P. Srinivasan, *Org. Photochem. Synth.* **1**, 101 (1971); *J. Am. Chem. Soc.* **86**, 3318 (1964).
b. R. G. Salomon and S. Ghosh, *Org. Synth.* **62**, 125 (1984).
c. P. G. Gassman and D. S. Patton, *J. Am. Chem. Soc.* **90**, 7276 (1968).
d. W. G. Dauben, C. H. Schallhorn, and D. L. Whalen, *J. Am. Chem. Soc.* **93**, 1446 (1971).
e. B. M. Jacobson, *J. Am. Chem. Soc.* **95**, 2579 (1973).
f. J. C. Barborak, L. Watts, and R. Pettit, *J. Am. Chem. Soc.* **88**, 1328 (1966).

polycyclic cage compounds. Some examples are given in Scheme 6.7. Copper(I) triflate facilitates these intramolecular additions, as was the case for intermolecular reactions.

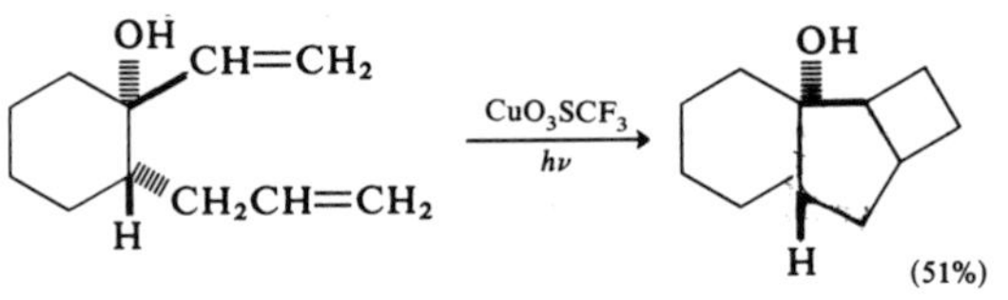

$\xrightarrow{CuO_3SCF_3,\ h\nu}$ (51%) Ref. 68

68. K. Avasthi and R. G. Salomon, *J. Org. Chem.* **51**, 2556 (1986).

Another class of molecules that undergo photochemical cycloadditions is α,β-unsaturated ketones.[69] The reactive excited state is believed to be a π–π^* triplet.[70] Since the initial intermediate is a triplet diradical, these reactions need not be stereospecific with respect to the alkene component.[71] The reaction is most successful with cyclopentenones and cyclohexenones. The excited states of acylic enones and larger ring compounds are deactivated by *cis–trans* isomerization and do not readily add to alkenes. Photoexcited enones can also add to alkynes.[72] Unsymmetrical alkenes can undergo two regioisomeric modes of addition. It is generally observed that alkenes with donor groups are oriented such that the substituted carbon becomes bound to the β-carbon, whereas with acceptor substituents the other orientation is preferred.[73]

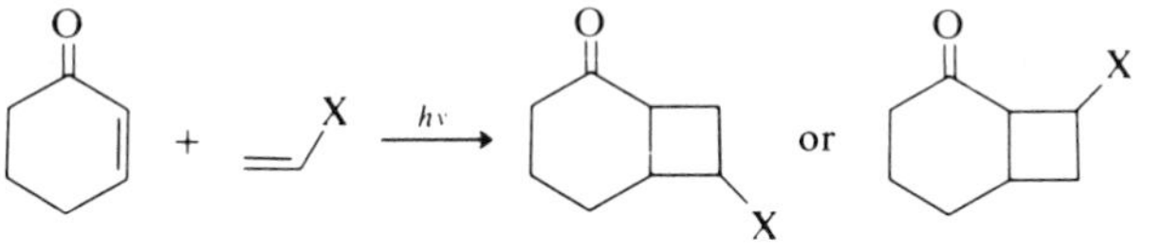

favored for
X = electron donor

favored for
X = electron acceptor

Selectivity is low for alkenes without strong donor or acceptor substituents.[74]

Intramolecular enone–alkene cycloadditions are also possible.

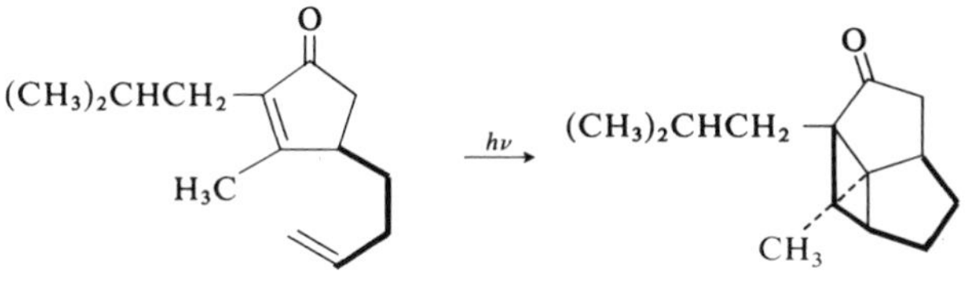

Ref. 75

In the case of β-(5-pentenyl) substituents, there is a general preference for *exo*-type cyclization to form a five-membered ring.[76]

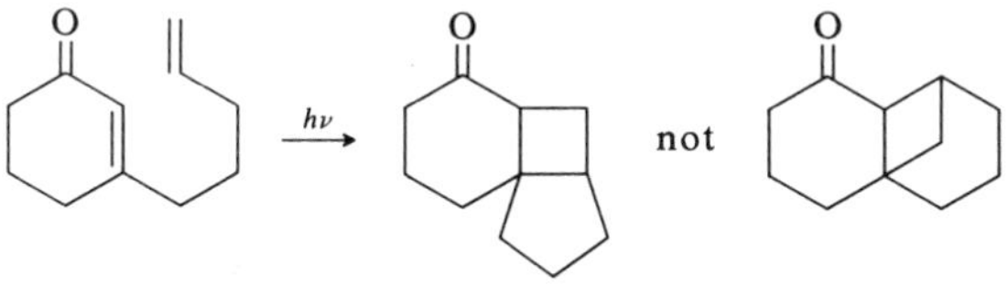

69. A. C. Weedon, in *Synthetic Organic Photochemistry*, W. M. Horspool (ed.), Plenum, New York, 1984, Chapter 2.
70. D. I. Schuster, M. M. Greenberg, I. M. Nunez, and P. C. Tucker, *J. Org. Chem.* **48**, 2615 (1983).
71. E. J. Corey, J. D. Bass, R. Le Mahieu, and R. B. Mitra, *J. Am. Chem. Soc.* **86**, 5570 (1964).
72. R. L. Cargill, T. Y. King, A. B. Sears, and M. R. Willcott, *J. Org. Chem.* **36**, 1423 (1971); W. C. Agosta and W. W. Lowrance, *J. Org. Chem.* **35**, 3851 (1970).
73. E. J. Corey, J. D. Bass, R. Le Mahieu, and R. B. Mitra, *J. Am. Chem. Soc.* **86**, 5570 (1984).
74. J. D. White and D. N. Gupta, *J. Am. Chem. Soc.* **88**, 5364 (1966); P. E. Eaton, *Acc. Chem. Res.* **1**, 50 (1968).
75. P. J. Connolly and C. H. Heathcock, *J. Org. Chem.* **50**, 4135 (1985).
76. W. C. Agosta and S. Wolff, *J. Org. Chem.* **45**, 3139 (1980); M. C. Pirrung, *J. Am. Chem. Soc.* **103**, 82 (1981); P. J. Connolly and C. H. Heathcock, *J. Org. Chem.* **50**, 4135 (1985).

Scheme 6.8. Photochemical Cycloaddition Reactions of Enones and Alxenes

1[a]

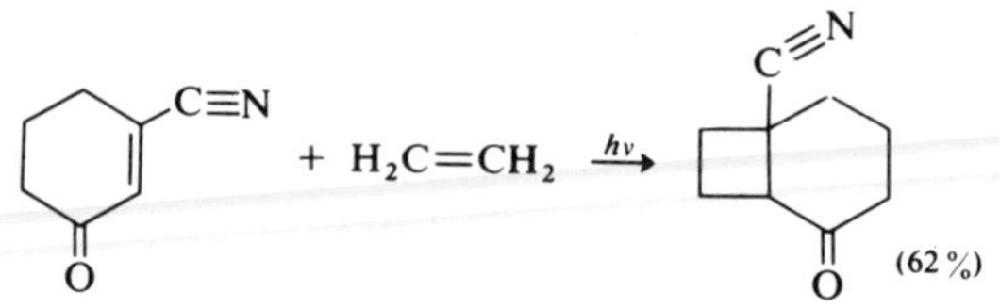
C≡N
+ H₂C=CH₂
hν
C≡N
O
O
(62%)

2[b]

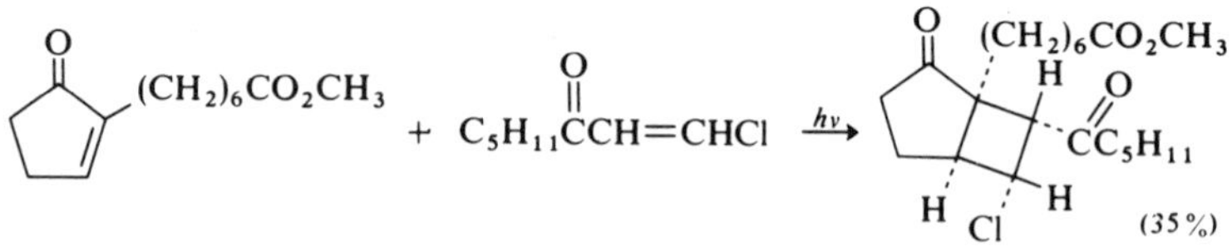
O
(CH₂)₆CO₂CH₃
+ C₅H₁₁CCH=CHCl
hν
O (CH₂)₆CO₂CH₃
H
CC₅H₁₁
H
H Cl
(35%)

3[c]

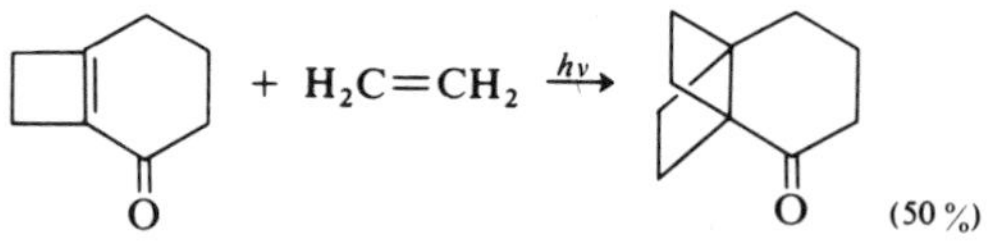
+ H₂C=CH₂
hν
O
O
(50%)

4[d]

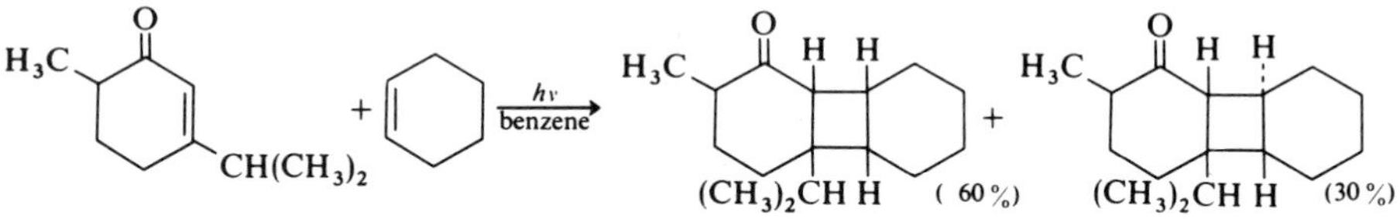
H₃C
O
CH(CH₃)₂
+
hν
benzene
H₃C
O H H
(CH₃)₂CH H
(60%)
+
H₃C
O H H
(CH₃)₂CH H
(30%)

5[e]

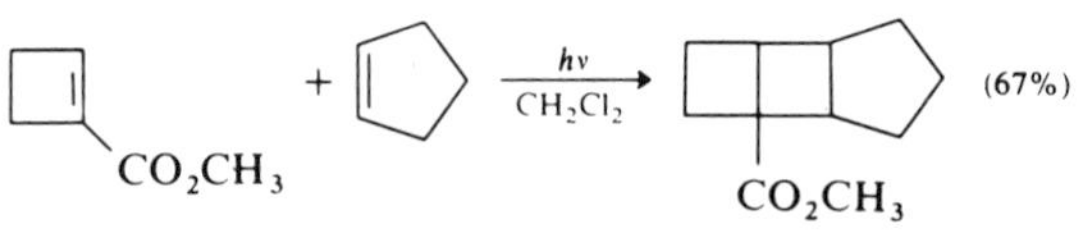
CO₂CH₃
+
hν
CH₂Cl₂
(67%)
CO₂CH₃

6[f]

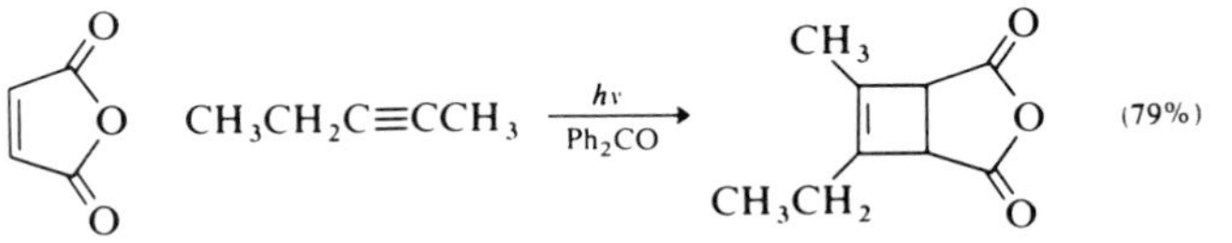
O
O
O
CH₃CH₂C≡CCH₃
hν
Ph₂CO
CH₃
O
O
O
CH₃CH₂
(79%)

7[g]

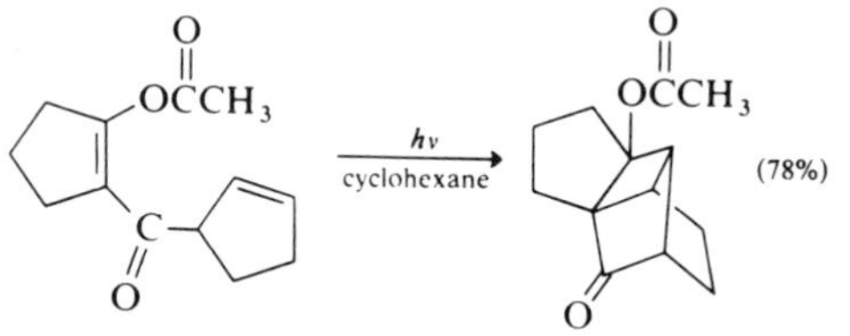
O
OCCH₃
C
O
hν
cyclohexane
O
OCCH₃
O
(78%)

8[h]

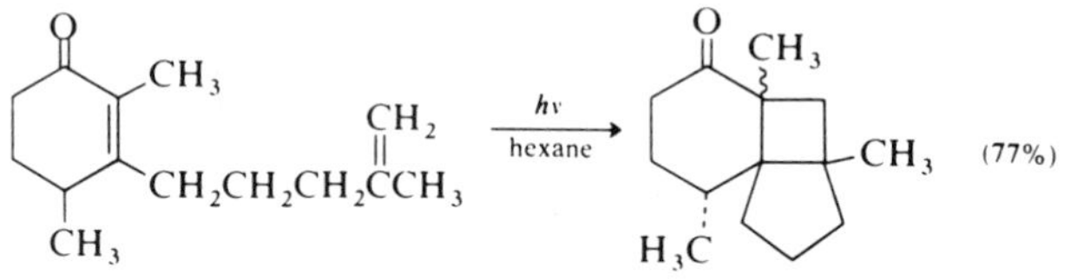
O
CH₃
CH₂
CH₂CH₂CH₂CCH₃
CH₃
hν
hexane
O CH₃
CH₃
H₃C
(77%)

a. W. C. Agosta and W. W. Lowrance, Jr., *J. Org. Chem.* **35**, 3851 (1970).
b. J. F. Bagli and T. Bogri, *J. Org. Chem.* **37**, 2132 (1972).
c. P. E. Eaton and K. Nyi, *J. Am. Chem. Soc.* **93**, 2786 (1971).
d. P. Singh, *J. Org. Chem.* **36**, 3334 (1971).
e. P. A. Wender and J. C. Lechleiter, *J. Am. Chem. Soc.* **99**, 267 (1977).
f. R. M. Scarborough, Jr., B. H. Toder, and A. B. Smith III, *J. Am. Chem. Soc.* **102**, 3904 (1980).
g. W. Oppolzer and T. Godel, *J. Am. Chem. Soc.* **100**, 2583 (1978).
h. M. C. Pirrung, *J. Am. Chem. Soc.* **101**, 7130 (1979).

Some examples of photochemical enone–alkene cycloadditions are given in Scheme 6.8.

With other ketones and aldehydes, reaction between the photoexcited carbonyl chromophore and alkene can result in formation of four-membered cyclic ethers (oxetanes). This reaction is often referred to as the *Paterno–Büchi reaction.*[77]

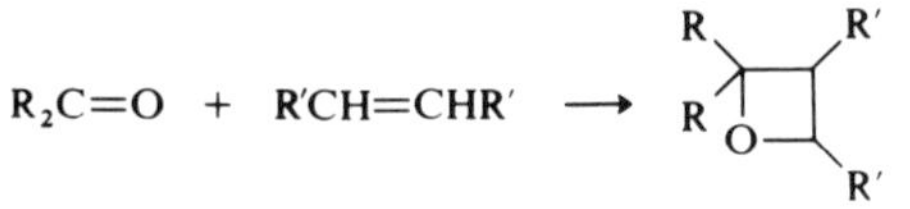

$$R_2C{=}O \;+\; R'CH{=}CHR' \;\longrightarrow\; \text{oxetane (2,2-R}_2\text{-3,4-R}'_2\text{)}$$

The reaction is stereospecific for at least some aliphatic ketones but not for aromatic carbonyls.[78] This result suggests that the reactive excited state is a singlet for aliphatics and a triplet for aromatics. With aromatic ketones, the regioselectivity of addition can usually be predicted on the basis of formation of the more stable of the two possible radicals by bond formation between oxygen and the alkene. Some examples of Paterno–Büchi reactions are given in Scheme 6.9.

Scheme 6.9. Photochemical Cycloaddition Reactions of Carbonyl Compounds with Alkenes

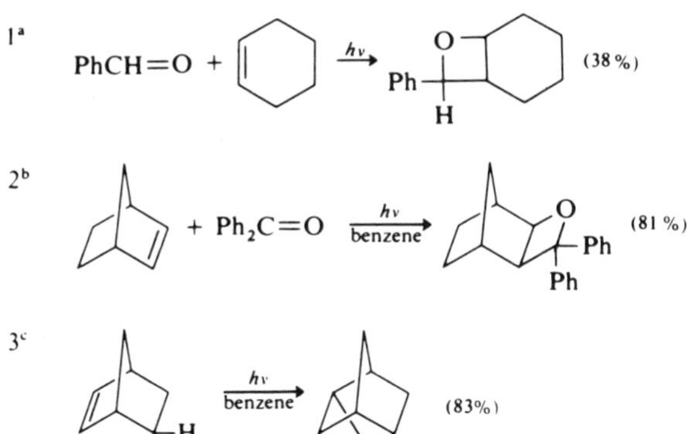

1[a] PhCH=O + cyclohexene $\xrightarrow{h\nu}$ oxetane (Ph, H) (38%)

2[b] norbornene + $Ph_2C{=}O$ $\xrightarrow[\text{benzene}]{h\nu}$ oxetane (Ph, Ph) (81%)

3[c] norbornenyl CCH_3 (=O), H $\xrightarrow[\text{benzene}]{h\nu}$ intramolecular oxetane, CH_3 (83%)

4[d] PhCH=O + $PhCH{=}CH_2$ $\xrightarrow{h\nu}$ oxetane (Ph, Ph) (31%)

a. J. S. Bradshaw, *J. Org. Chem.* **31**, 237 (1966).
b. D. R. Arnold, A. H. Glick, and V. Y. Abraitys, *Org. Photochem. Synth.* **1**, 51 (1971).
c. R. R. Sauers, W. Schinski, and B. Sickles, *Org. Photochem. Synth.* **1**, 76 (1971).
d. H. A. J. Carless, A. K. Maitra, and H. S. Trivedi, *J. Chem. Soc. Chem. Commun.*, 984 (1979).

77. D. R. Arnold, *Adv. Photochem.* **6**, 301 (1968); H. A. J. Carless, in *Synthetic Organic Photochemistry*, W. M. Horspool (ed.), Plenum, New York, 1984, Chapter 8.
78. N. C. Yang and W. Eisenhardt, *J. Am. Chem. Soc.* **93**, 1277 (1971); D. R. Arnold, R. L. Hinman, and A. H. Glick, *Tetrahedron Lett.*, 1425 (1964); N. J. Turro and P. A. Wriede, *J. Am. Chem. Soc.* **90**, 6863 (1968); J. A. Barltrop and H. A. J. Carless, *J. Am. Chem. Soc.* **94**, 8761 (1972).

6.5. [3,3]-Sigmatropic Rearrangements: Cope and Claisen Rearrangements

The mechanistic basis of sigmatropic rearrangements was introduced in Part A, Chapter 11. The sigmatropic process that is most widely applied in synthetic methodology is the [3,3]-sigmatropic rearrangement. The principles of orbital symmetry establish that concerted [3,3]-sigmatropic rearrangements are allowed processes. Stereochemical predictions and analyses are based on the transition state implied by a concerted reaction mechanism. Some of the various [3,3]-sigmatropic rearrangements that are used in synthesis are presented in outline form in Table 6.4.[79]

The Cope rearrangement is the conversion of a 1,5-hexadiene derivative to an isomeric 1,5-hexadiene by the [3,3]-sigmatropic mechanism. The reaction is both stereospecific and stereoselective. It is stereospecific in that a *Z*- or *E*-configurational relationship at either double bond is maintained in the transition state and governs the stereochemical relationship at the newly formed single bond in the product.[80]

Table 6.4. [3,3]-Sigmatropic Rearrangements

1[a] Cope Rearrangement

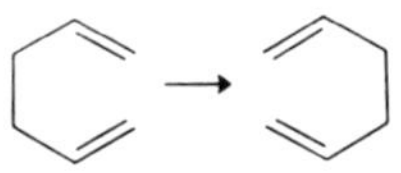

2[b] Oxy-Cope Rearrangement

HO → HO → O

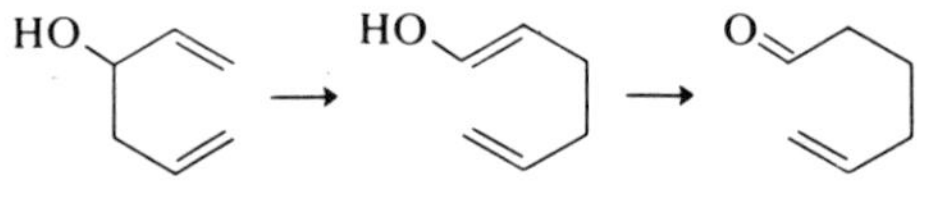

3[c] Anionic Oxy-Cope Rearrangement

$\bar{O}$ → $\bar{O}$ $\xrightarrow{H^+}$ O

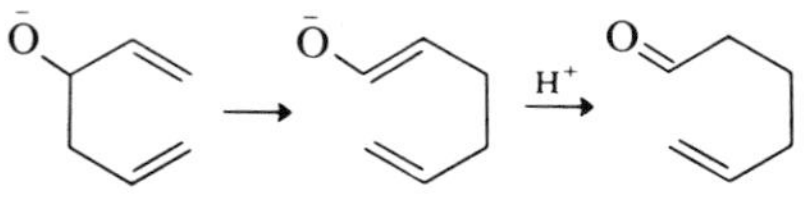

4[d] Claisen Rearrangement of Allyl Vinyl Ethers

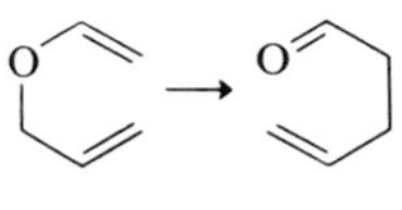

O → O

5[d] Claisen Rearrangement of Allyl Phenyl Ethers

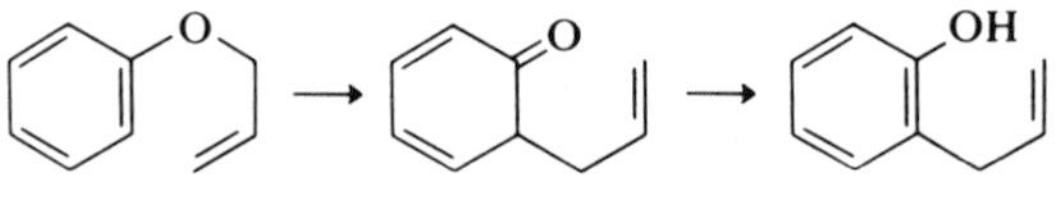

O → O → OH

79. For reviews of synthetic application of [3,3]-sigmatropic rearrangements, see G. B. Bennett, *Synthesis*, 589 (1977); F. E. Zielger, *Acc. Chem. Res.* **10**, 227 (1977); S. J. Rhoads and N. E. Raulins, *Org. React.* **22**, 1 (1974).
80. W. von E. Doering and W. R. Roth, *Tetrahedron* **18**, 67 (1962).

6[e] Orthoester Claisen Rearrangement

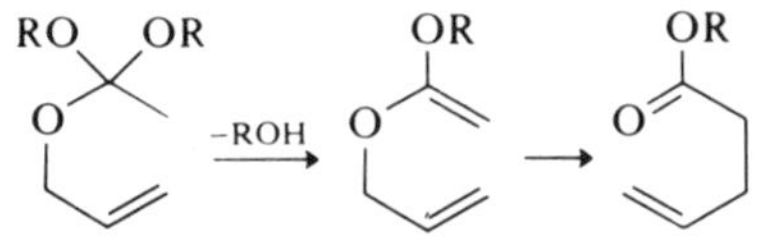

7[f] Claisen Rearrangement of *O*-Allyl-*O*′-trimethylsilyl Ketene Acetals

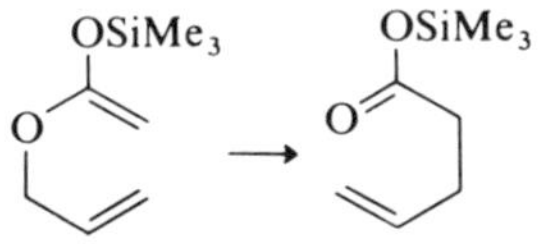

8[g] Ester Enolate Claisen Rearrangement

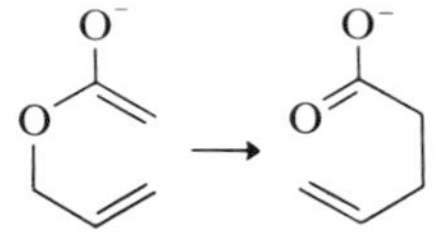

9[h] Claisen Rearrangement of *O*-Allyl-*N*,*N*-dialkyl Ketene Aminals

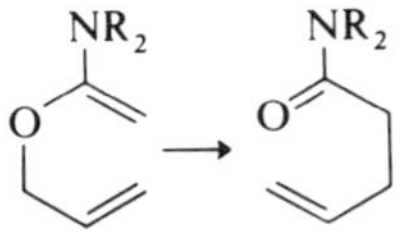

a. S. J. Rhoads and N. R. Raulins, *Org. React.* **22**, 1 (1975).
b. J. A. Berson and M. Jones, Jr., *J. Am. Chem. Soc.* **86**, 5019 (1964).
c. D. A. Evans and A. M. Golob, *J. Am. Chem. Soc.* **97**, 4765 (1975).
d. D. S. Tarbell, *Org. React.* **2**, 1 (1944).
e. W. S. Johnson, L. Werthemann, W. R. Bartlett, T. J. Brocksom, T. Li, D. J. Faulkner, and M. R. Petersen, *J. Am. Chem. Soc.* **92**, 741 (1970).
f. R. E. Ireland and R. H. Mueller, *J. Am. Chem. Soc.* **94**, 5898 (1972).
g. R. E. Ireland, R. H. Mueller, and A. K. Willard, *J. Am. Chem. Soc.* **98**, 2868 (1976).
h. D. Felix, K. Gschwend-Steen, A. E. Wick, and A. Eschenmoser, *Helv. Chim. Acta.* **52**, 1030 (1969).

In general, when both *E*- and *Z*-stereoisomers are possible for the product, there will be stereoselectivity in favor of one. The stereochemical aspects of the Cope rearrangement are consistent with a chairlike transition state in which the larger substituent at C-3 (or C-4) adopts an equatorial-like conformation.

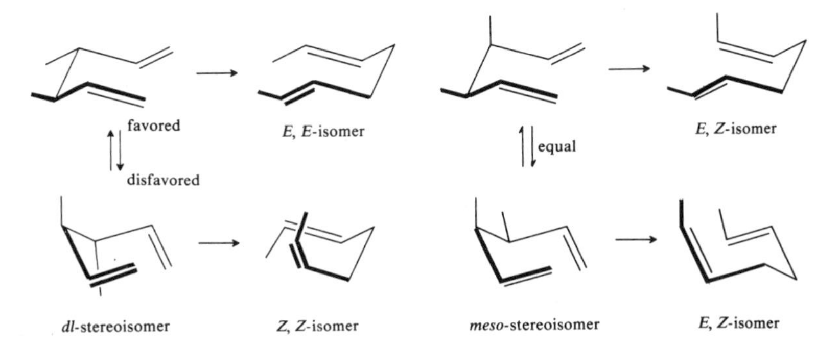

Because of the concerted nature of the mechanism, chirality at C-3 (or C-4) leads to enantiospecific formation of the new chiral center at C-1 (or C-6).[81] These relationships are illustrated in the example below. Both the configuration of the new chiral center and the new double bond are those expected on the basis of a chairlike transition state. Since there are two stereogenic centers, the double bond and the chiral carbon, there are four possible stereoisomers of the product. Only two are formed. The *E*-double bond isomer has the *S*-configuration at C-4 whereas the *Z*-isomer has the *R*-configuration at C-4. The stereochemistry of the new double bond is determined by the competition between the chair transition states. The amount of the two products is determined by the relative stability of the two chair transition states. Transition state **B** is less favorable because of the axial placement of the larger phenyl substituent.

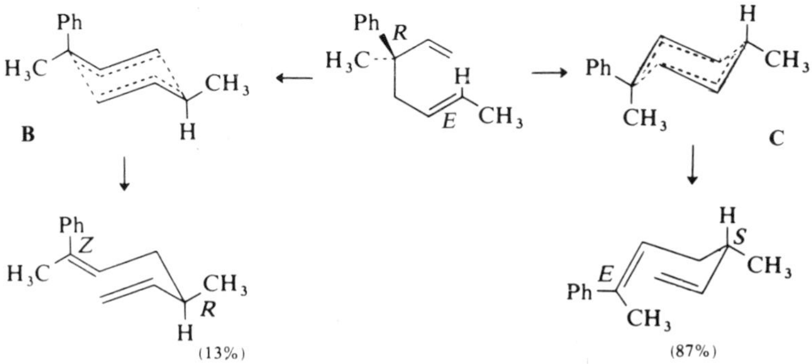

(13%) (87%)

The products corresponding to boatlike transition states are usually not observed for acylic dienes:

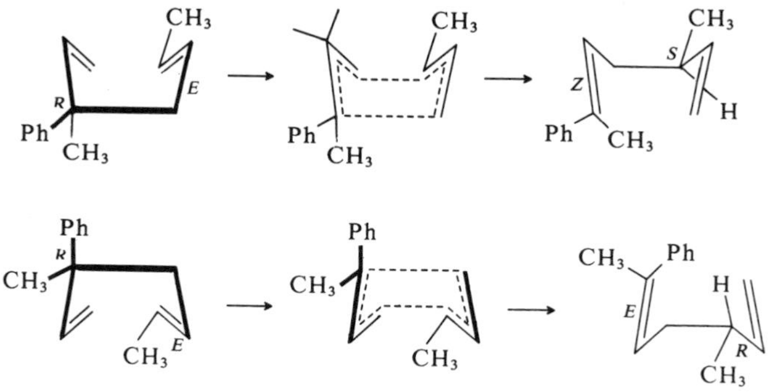

Cope rearrangements are generally reversible processes, and, since there are no changes in the number or types of bonds as a result of the reaction, to a first approximation the total bond energy is unchanged. The position of the final equilibrium is governed by the relative stability of the starting material and product. In the example just cited, the equilibrium is favorable because the product is stabilized by conjugation with the phenyl ring.

81. R. K. Hill and N. W. Gilman, *J. Chem. Soc., Chem. Commun.*, 619 (1967); R. K. Hill, in *Asymmetric Synthesis*, Vol. 3, J. D. Morrison (ed.), Academic Press, New York, 1984, pp. 503–572.

Scheme 6.10. Cope Rearrangements of 1,5-Dienes

A. Thermal

1[a] 350°C, 1 h (100%)

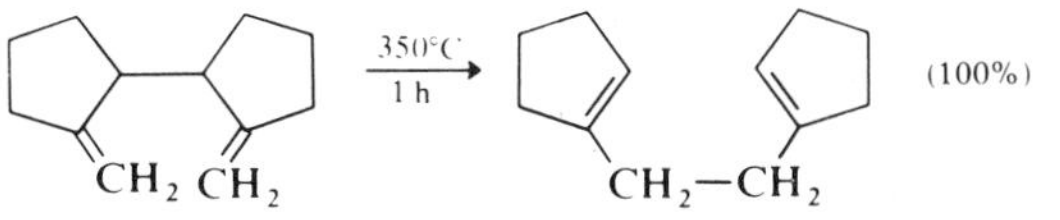

CH_2 CH_2 → CH_2—CH_2

2[b] $K = 0.25$, 275°C

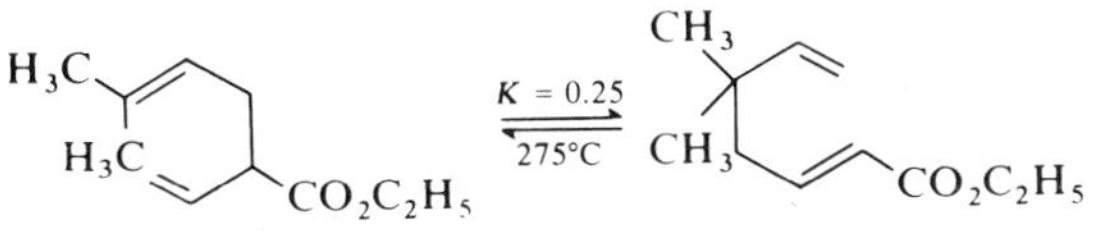

H_3C, H_3C, $CO_2C_2H_5$ ⇌ CH_3, CH_3, $CO_2C_2H_5$

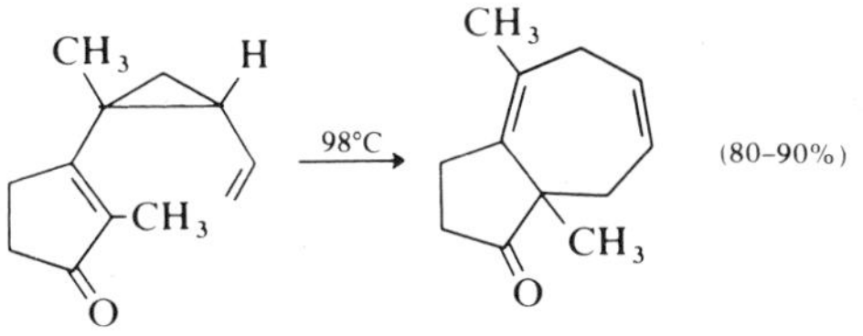

3[c] 98°C (80–90%)

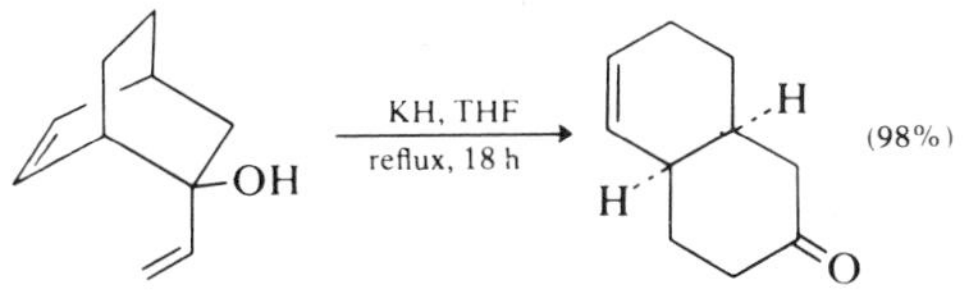

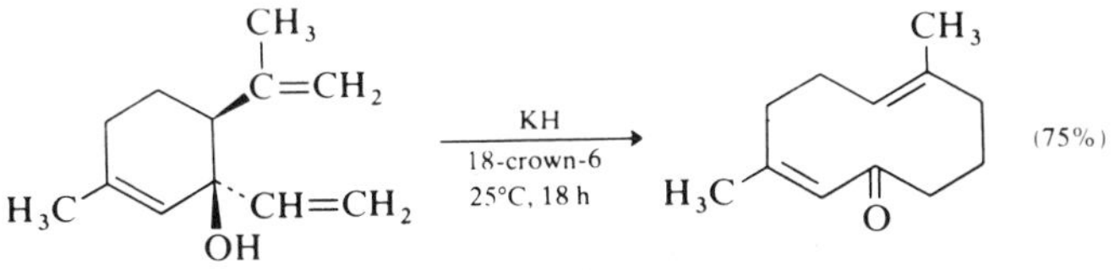

CH_3, H, CH_3, O → CH_3, CH_3, O

4[d] 320°C (90%)

OH, $CH=CH_2$, $CH=CH_2$ → O

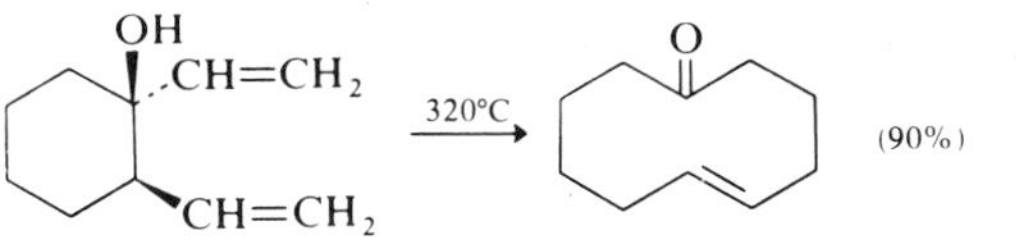

B. Anionic Oxy-Cope

5[e] KH, THF, reflux, 18 h (98%)

OH → H, H, O

6[f] KH, 18-crown-6, 25°C, 18 h (75%)

CH_3, $C=CH_2$, H_3C, $CH=CH_2$, OH → CH_3, H_3C, O

7[g] KH, $Bu_4N^+I^-$

H, O, HO, O, $CH_2=CH$, H, CH_3, CH_3C, CH, O, CH_3, H → H, O, O, O, H, CH_3, O, CH_3, CH_3

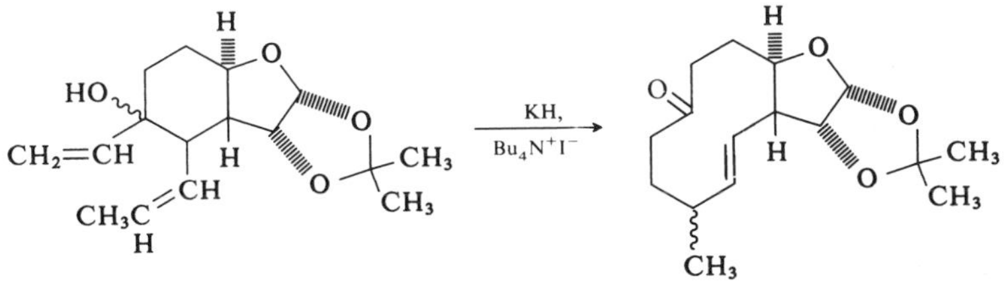

a. K. J. Shea and R. B. Phillips, *J. Am. Chem. Soc.* **102**, 3156 (1980).
b. F. E. Zeigler and J. J. Piwinski, *J. Am. Chem. Soc.* **101**, 1612 (1979).
c. P. A. Wender, M. A. Eissenstat, and M. P. Filosa, *J. Am. Chem. Soc.* **101**, 2196 (1979).
d. E. N. Marvell and W. Whalley, *Tetrahedron Lett.*, 509 (1970).
e. D. A. Evans, A. M. Golob, N. S. Mandel, and G. S. Mandel, *J. Am. Chem. Soc.* **100**, 8170 (1978).
f. W. C. Still, *J. Am. Chem. Soc.* **99**, 4186 (1977).
g. M. Georges, T.-F. Tam, and B. Fraser-Reid, *J. Org. Chem.* **50**, 5747 (1985).

Some other examples of Cope rearrangements are given in Scheme 6.10. In entry 1, the equilibrium is biased toward product by the fact that the double bonds in the product are more highly substituted, and therefore more stable, than those in the reactant. In entry 2, a gain in conjugation is offset by the formation of a less highly substituted double bond, and the equilibrium mixture contains both dienes. When ring strain is relieved, Cope rearrangements can occur at much lower temperatures and with complete conversion to ring-opened products. A classic example of such a process is the conversion of *cis*-divinylcyclopropane to 1,4-cycloheptadiene, as reaction which occurs readily at temperatures below −40°C.[82]

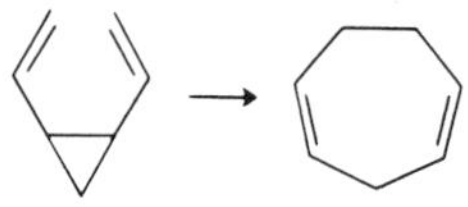

Entry 3 in Scheme 6.10 illustrates the application of a *cis*-divinylcyclopropane rearrangement in the preparation of an intermediate for the synthesis of pseudoguiane-type natural products.

Several transition metal species, especially Pd(II) salts, have been found to catalyze Cope rearrangements.[83] The catalyst that has been adopted for synthetic purposes is $PdCl_2(CH_3CN)_2$. With this catalyst, the rearrangement of **6** to **7** and **8** occurs at room temperature, as contrasted to 240°C in its absence.[84] The catalyzed reaction shows enhanced stereoselectivity which is consistent with a chairlike transition state structure.

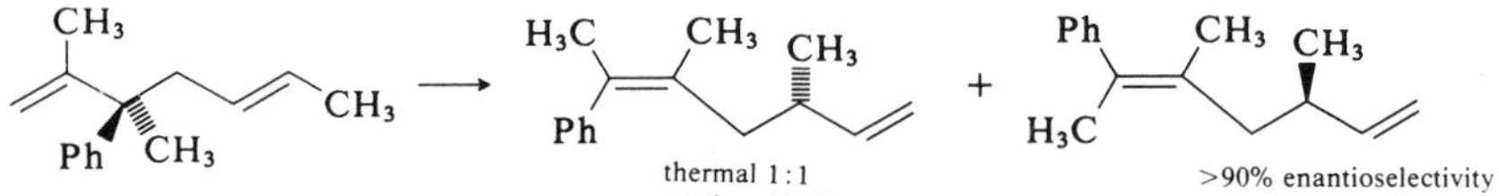

CH₃ Ph CH₃ CH₃ **6** → H₃C CH₃ CH₃ Ph **7** thermal 1:1 catalyzed 7:3 + Ph CH₃ CH₃ H₃C **8** >90% enantioselectivity under both conditions

The mechanism for catalysis can be formulated as a stepwise process in which the electrophilic character of Pd(II) is important.

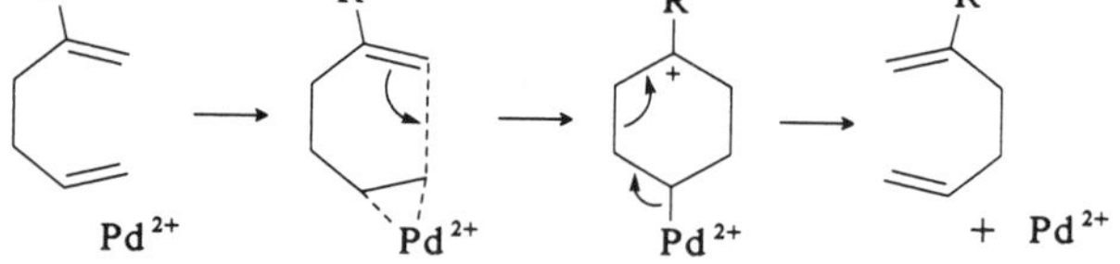

R Pd²⁺ → R Pd²⁺ → R + Pd²⁺ → R + Pd²⁺

When there is a hydroxyl substituent at C-3 of the diene system, the Cope rearrangement product is an enol which is subsequently converted to the

82. W. von. E. Doering and W. R. Roth, *Tetrahedron* **19**, 715 (1963).
83. R. P. Lutz, *Chem. Rev.* **84**, 205 (1984).
84. L. E. Overman and F. M. Knoll, *J. Am. Chem. Soc.* **102**, 865 (1980); L. E. Overman and E. J. Jacobsen, *J. Am. Chem. Soc.* **104**, 7225 (1982).

corresponding carbonyl compound. This is called the *oxy-Cope* rearrangement. The formation of the carbonyl compound provides a net driving force for the reaction.[85] Entry 4 in Scheme 6.10 illustrates the use of the oxy-Cope rearrangement in formation of a medium-sized ring.

An important improvement in the oxy-Cope reaction was made when it was found that the reactions are markedly catalyzed by base.[86] When the C-3 hydroxyl group is converted to its alkoxide, the reaction is accelerated by factors of 10^{10}–10^{17}.

$^{-}$O ⟶ $^{-}$O $\xrightarrow{H^+}$ O H

These base-catalyzed reactions are called *anionic oxy-Cope rearrangements.* The rates of anionic oxy-Cope rearrangements depend on the degree of cation coordination at the oxy anion. The reactivity trend is $K^+ > Na^+ > Li^+$. Catalytic amounts of tetra-*n*-butylammonium salts lead to accelerated rates in some cases. This presumably results from the dissociation of less reactive ion pair species which is promoted by the tetra-*n*-butylammonium ion.[87] Entries 5, 6, and 7 in Scheme 6.10 illustrate the mild conditions under which anionic oxy-Cope rearrangement occurs.

The [3,3]-sigmatropic rearrangement of allyl vinyl ethers is known as the *Claisen rearrangement.* It is mechanistically analogous to the Cope rearrangement. Because the product is a carbonyl compound, the equilibrium is usually favorable. The reactants are usually made from allylic alcohols by mercuric-ion-catalyzed exchange with ethyl vinyl ether.[88] The allyl vinyl ether need not be isolated but is usually prepared under conditions which lead to its rearrangement. The simplest of all Claisen rearrangements, the conversion of allyl vinyl ether to 4-pentenal, typifies the process.

$$CH_2{=}CHCH_2OH + CH_2{=}CHOCH_2CH_3 \xrightarrow[\Delta]{Hg(OAc)_2} [CH_2{=}CHCH_2OCH{=}CH_2] \rightarrow CH_2{=}CHCH_2CH_2\overset{O}{\overset{\|}{C}}H \ (96\%) \qquad \text{Ref. 89}$$

Acid-catalyzed exchange can also be used to prepare the allyl vinyl ethers.

$$RCH{=}CHCH_2OH + CH_3CH_2OCH{=}CH_2 \xrightarrow{H^+} RCH{=}CHCH_2OCH{=}CH_2 \qquad \text{Ref. 90}$$

85. A. Viola, E. J. Iorio, K. K. Chen, G. M. Glover, U. Nayak, and P. J. Kocienski, *J. Am. Chem. Soc.* **89**, 3462 (1967).
86. D. A. Evans and A. M. Golob, *J. Am. Chem. Soc.* **97**, 4765 (1975); D. A. Evans, D. J. Baillargeon, and J. V. Nelson, *J. Am. Chem. Soc.* **100**, 2242 (1978).
87. M. George, T.-F. Tam, and B. Fraser-Reid, *J. Org. Chem.* **50**, 5747 (1985).
88. W. H. Watanabe and L. E. Conlon, *J. Am. Chem. Soc.* **79**, 2828 (1957).
89. S. E. Wilson, *Tetrahedron Lett.*, 4651 (1975).
90. G. Saucy and R. Marbet, *Helv. Chim. Acta* **50**, 2091 (1967); R. Marbet and G. Saucy, *Helv. Chim. Acta* **50**, 2095 (1967).

Some representative Claisen rearrangements are shown in Scheme 6.11. Entry 1 illustrates the application of the Claisen rearrangement in introduction of a substituent at the junction of two six-membered rings. Introduction of a substituent at this type of position is frequently necessary in the synthesis of steroids and terpenes. In entry 2, rearrangement of a 2-propenyl ether leads to formation of a methyl ketone. Entry 3 illustrates the use of 3-methoxyisoprene to form the allylic ether. The rearrangement of this type of ether leads to introduction of isoprene structural units.

There are several variants of the Claisen rearrangement that increase its versatility and have made it a powerful synthetic tool. The orthoester modification of the Claisen rearrangement allows carboalkoxymethyl groups to be introduced at the γ position of allylic alcohols.[91] A mixed orthoester is formed as an intermediate and undergoes sequential elimination and rearrangement.

$$RCH{=}CHCH_2OH + CH_3C(OCH_3)_3 \rightleftharpoons RCH{=}CHCH_2O\underset{\displaystyle OCH_3}{\overset{\displaystyle OCH_3}{\overset{|}{\underset{|}{C}}}}CH_3 \rightleftharpoons RCH{=}CHCH_2O\underset{\displaystyle OCH_3}{\underset{|}{C}}{=}CH_2 \rightarrow R\underset{}{\overset{\displaystyle CH_2CO_2CH_3}{\overset{|}{C}}}HCH{=}CH_2$$

Both the exchange and elimination are catalyzed by addition of a small amount of a weak acid, such as propionic acid. Entries 5–8 in Scheme 6.11 are representative examples.

The mechanism and stereochemistry of the orthoester Claisen rearrangement is analogous to that of the Cope rearrangement. The reaction is stereospecific with respect to the double bond present in the initial allylic alcohol. In acyclic molecules, the stereochemistry of the product can usually be predicted on the basis of a chairlike transition state. When steric effects or ring geometry preclude a chairlike structure, the reaction can proceed through a boatlike transition state.[92]

High levels of enantioselectivity have been observed in the rearrangement of chiral reactants. This reaction can be used to establish the configuration of the newly formed carbon–carbon bond on the basis of the chirality of the C—O bond in the starting allylic alcohol. Treatment of (2*R*,3*E*)-3-penten-2-ol with ethyl orthoacetate gives the ethyl ester of (3*R*,4*E*)-3-methyl-4-hexenoic acid in 90% enantiomeric excess.[93] The configuration of the new chiral center is that predicted by a chairlike transition state with the methyl group occupying a pseudoequatorial position.

91. W. S. Johnson, L. Werthemann, W. R. Bartlett, T. J. Brocksom, T. Li, D. J. Faulkner, and M. R. Petersen, *J. Am. Chem. Soc.* **92,** 741 (1970).
92. R. J. Cave, B. Lythgoe, D. A. Metcalf, and I. Waterhouse, *J. Chem. Soc., Perkin Trans. 1,* 1218 (1977); G. Büchi and J. E. Powell, Jr., *J. Am. Chem. Soc.* **92**, 3126 (1970); J. J. Gajewski and J. L. Jiminez, *J. Am. Chem. Soc.* **108**, 468 (1986).
93. R. K. Hill, R. Soman, and S. Sawada, *J. Org. Chem.* **37**, 3737 (1972); **38**, 4218 (1973).

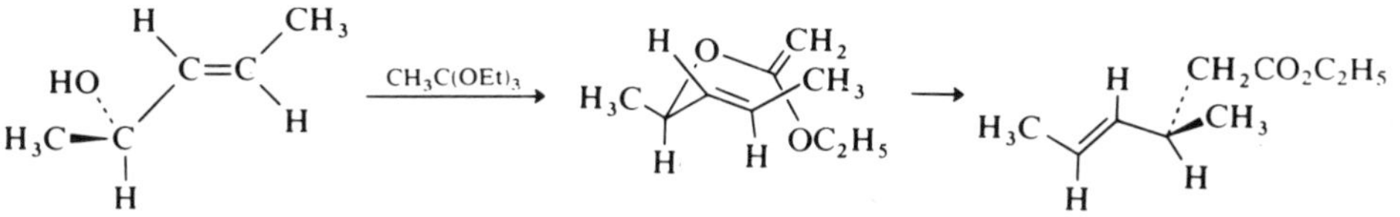

Esters of allylic alcohols can be rearranged to γ,δ-unsaturated carboxylic acids via the *O*-trimethylsilyl ethers of the ester enolate.[94] This rearrangement takes place under much milder conditions than for the orthoester method. The reaction occurs at room temperature or slightly above. Entries 10, 11, and 13 of Scheme 6.11 are examples. The example in entry 12 is a rearrangement on the enolate without intervention of the silyl enol ether.

The stereochemistry of the ester silyl enol ether Claisen rearrangement can be controlled not only by the stereochemistry of the double bond in the allylic alcohol, but also by the stereochemistry of the silyl enol ether. The chair transition state predicts that the configuration at the newly formed C—C bond will be determined by the *E*- or *Z*-configuration of the silyl enol ether.

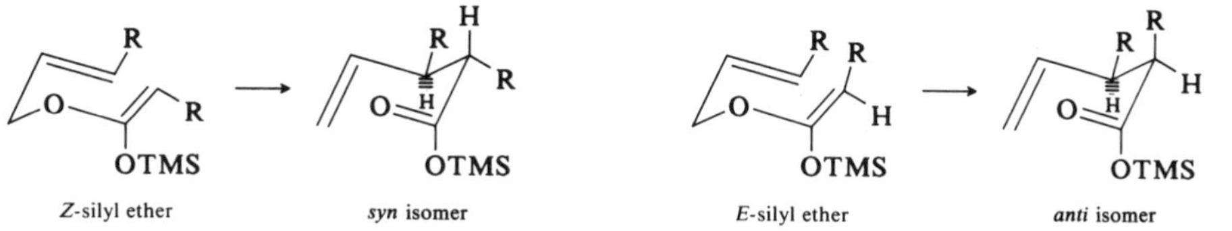

Z-silyl ether → *syn* isomer   *E*-silyl ether → *anti* isomer

The stereochemistry of the silyl enol ether can be controlled by the conditions of preparation. The base that is usually used for enolate formation is lithium diisopropylamide. If the enolate is prepared in pure THF, the *E*-enolate is generated and this stereochemistry is maintained in the silylated derivative. If HMPA is included in the solvent, the *Z*-enolate predominates.[95] The preferential formation of the *E*-enolate can be explained in terms of a cyclic transition state in which the proton is abstracted from the stereoelectronically preferred orientation.

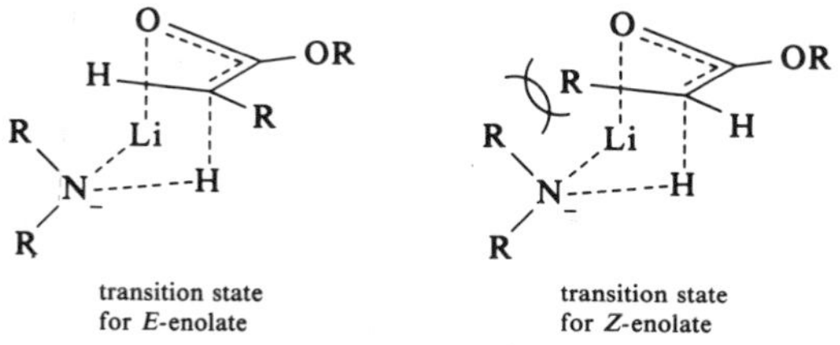

transition state for *E*-enolate   transition state for *Z*-enolate

The switch to the *Z*-enolate in HMPA can be attributed to a noncyclic transition state being favored as the result of strong solvation of the lithium ion by HMPA.

94. R. E. Ireland, R. H. Mueller, and A. K. Willard, *J. Am. Chem. Soc.* **98**, 2868 (1976).
95. R. E. Ireland, R. H. Mueller, and A. K. Willard, *J. Am. Chem. Soc.* **98**, 2868 (1972); R. E. Ireland and A. K. Willard, *Tetrahedron Lett.*, 3975 (1975).

Scheme 6.11. Claisen Rearrangements

A. Rearrangements of Allyl Vinyl Ethers

1[a] (bicyclic allyl vinyl ether with $OCH{=}CH_2$) $\xrightarrow{195°C}$ (product with angular $CH_2CH{=}O$) (87%)

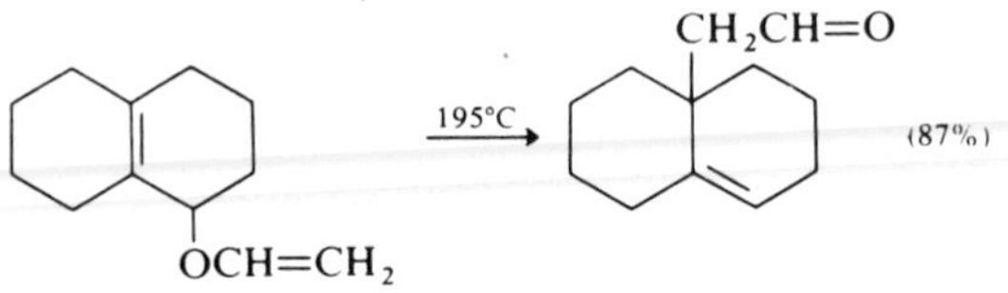

2[b] $(CH_3)_2C(OH)CH{=}CH_2 + H_2C{=}C(CH_3)OCH_3 \xrightarrow[125°C]{H^+} (CH_3)_2C{=}CHCH_2CH_2C(=O)CH_3$ (94%)

3[c] $CH_2{=}C(CH_3)CH(OH)CH_2CH_3 + CH_2{=}C(CH_3)C(OCH_3){=}CH_2 \xrightarrow[H^+]{110°C} CH_2{=}C(CH_3)C(=O)CH_2CH_2C(CH_3){=}CHCH_2CH_3$ (~70%)

4[d] $CH_3CH{=}CHCH_2OC(CH_3){=}C(H)CO_2CH_2CH{=}CHCH_3 \xrightarrow{140–145°C} CH_3C(=O)CH(CO_2CH_2CH{=}CHCH_3)CH(CH_3)CH{=}CH_2$ (61%)

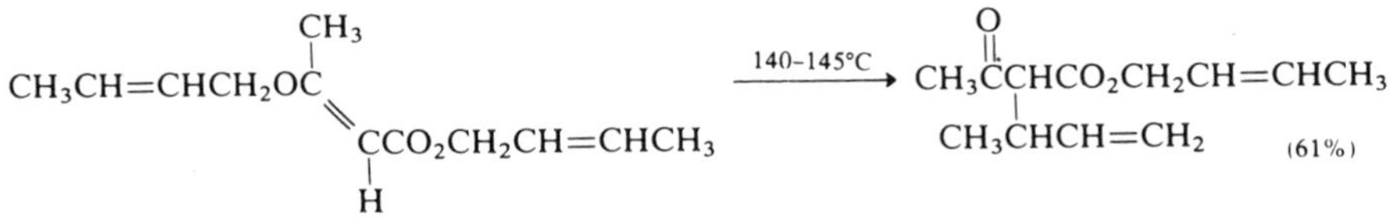

B. Rearrangements via Orthoesters and Orthoamides

5[e] $H_2C{=}CHCH_2CH_2CH(OH)C(CH_3){=}CH_2 \xrightarrow[H^+,\ 140°C]{CH_3C(OC_2H_5)_3} H_2C{=}CHCH_2CH_2C(H){=}C(CH_3)CH_2CH_2CO_2C_2H_5$ (83–88%)

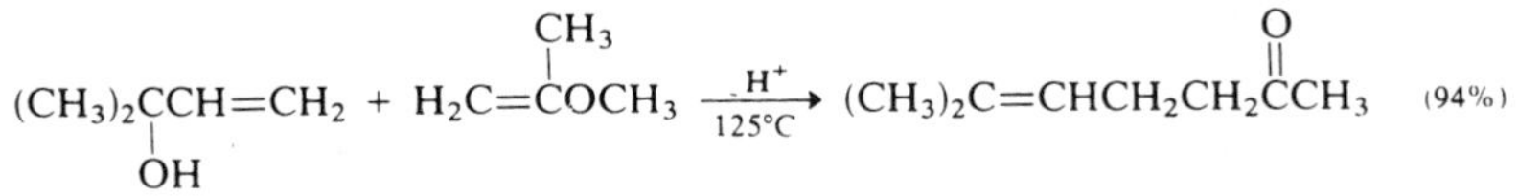

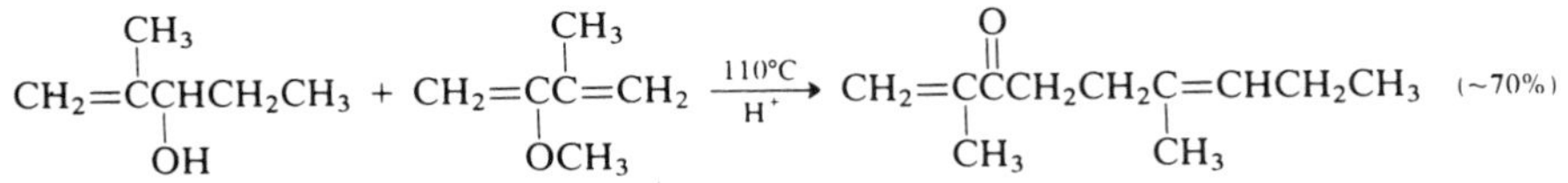

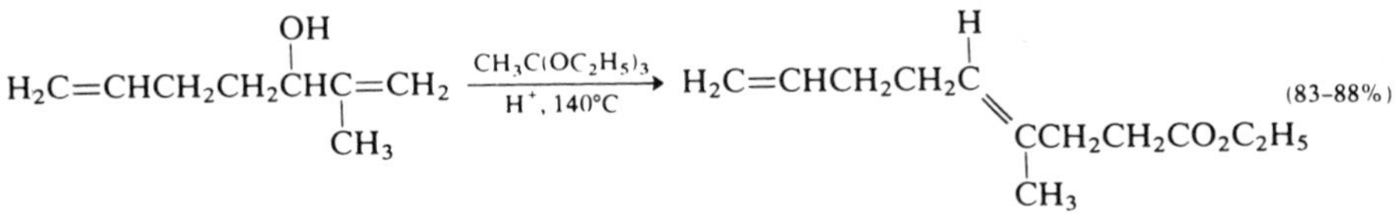

6[f] $CH_2{=}C(C_2H_5)CH(OH)CH_2CH_2C(CH_3){=}C(H)CO_2CH_3 \xrightarrow[110°C]{CH_3C(OCH_3)_3} CH_3O_2CCH_2CH_2C(C_2H_5){=}C(H)CH_2CH_2C(CH_3){=}C(H)CO_2CH_3$ (85%)

7[g] (N-CH_2Ph tetrahydropyridine with HO and CH_2CH_3) $\xrightarrow[140°C]{CH_3C(OC_2H_5)_3}$ (N-CH_2Ph tetrahydropyridine with CH_2CH_3 and $CH_2CO_2C_2H_5$) (74%)

8[h] (N-CH_3 lactam with CH_2OH) $\xrightarrow[145°C]{CH_3CH_2CH_2C(OCH_3)_3}$ (N-CH_3 lactam with $=CH_2$ and $CH_3CH_2CHCO_2CH_3$) (96%)

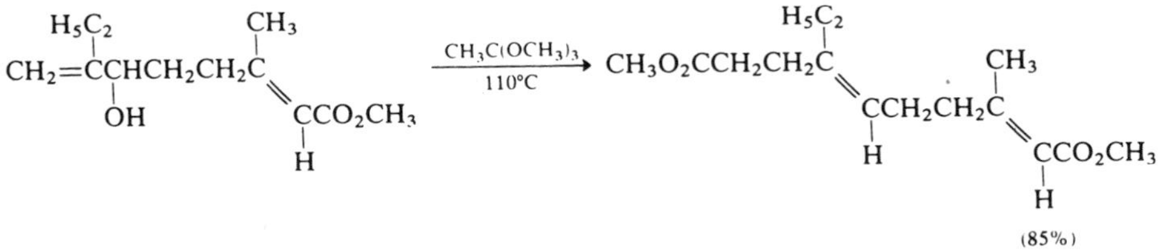

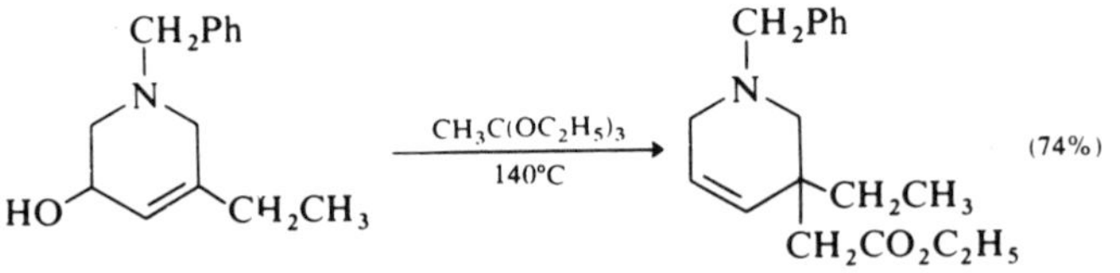

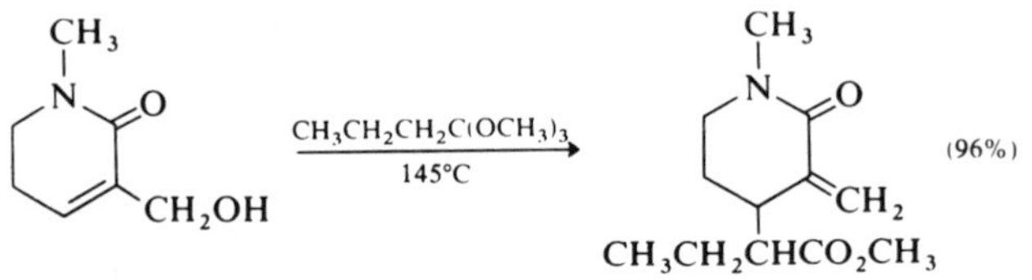

9[g]

$(CH_3)_2NC(OCH_3)_2$ / CH_3 ; diglyme, 160°C — (45%)

C. Rearrangements of Ester Enolates and Silyl Enol Ethers

10[i] 1) 67°C 2) CH_3OH 3) HO → $CH_2{=}CHCH(CH_3)CH_2CO_2H$ (70%)

11[j] $CH_3(CH_2)_5CH(OC(OSiMe_3){=}CH_2)C(CH_3){=}CH_2$ 1) 70°C 2) H_3O → $CH_3(CH_2)_5CH{=}C(CH_3)CH_2CH_2CO_2H$ (53%)

12[i] 1) $Li^+[(CH_3)_2CH\bar{N}C_6H_{11}]$ 2) 25°C, 3 h → $CH_2{=}C(CH_3)CH(CH_2)_5CH_3$ / CH_3CHCO_2H (71%)

13[k] 1) LDA, TMS—Cl 2) CH_2N_2 — 80%

a. A. W. Burgstahler and I. C. Nordin, *J. Am. Chem. Soc.* **83**, 198 (1961).
b. G. Saucy and R. Marbet, *Helv. Chim. Acta.* **50**, 2091 (1967).
c. D. J. Faulkner and M. R. Petersen, *J. Am. Chem. Soc.* **95**, 553 (1973).
d. J. W. Ralls, R. E. Lundin, and G. F. Bailey, *J. Org. Chem.* **28**, 3521 (1963).
e. R. I. Trust and R. E. Ireland, *Org. Synth.* **53**, 116 (1973).
f. C. A. Henrick, R. Schaub, and J. B. Siddall, *J. Am. Chem. Soc.* **94**, 5374 (1972).
g. F. E. Ziegler and G. B. Bennett, *J. Am. Chem. Soc.* **95**, 7458 (1973).
h. J. J. Plattner, R. D. Glass, and H. Rapoport, *J. Am. Chem. Soc.* **94**, 8614 (1972).
i. R. E. Ireland, R. H. Mueller, and A. K. Willard, *J. Am. Chem. Soc.* **98**, 2868 (1976).
j. J. A. Katzenellenbogen and K. J. Christy, *J. Org. Chem.* **39**, 3315 (1974).
k. R. E. Ireland and D. W. Norbeck, *J. Am. Chem. Soc.* **107**, 3279 (1985).

A number of steric effects on the rate of rearrangement have been observed and can be accommodated by the chairlike transition state model.[96] The *E*-silyl enol ethers rearrange somewhat more slowly than the corresponding *Z*-isomers. This is interpreted as resulting from the pseudoaxial placement of the methyl group in the transition state for rearrangement of the *E*-isomer.

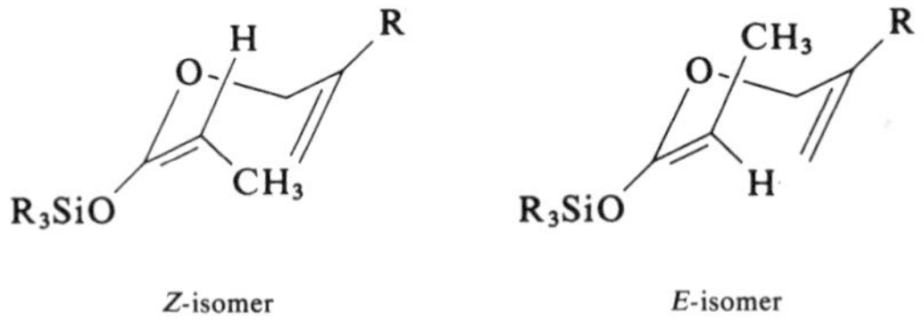

Z-isomer *E*-isomer

The size of the substituent R also influences the rate, with the rate increasing somewhat for both stereoisomers as R becomes larger. It is proposed that steric interactions with R are relieved as the C—O bond stretches. The rate acceleration would reflect the higher ground state energy resulting from these steric interactions.

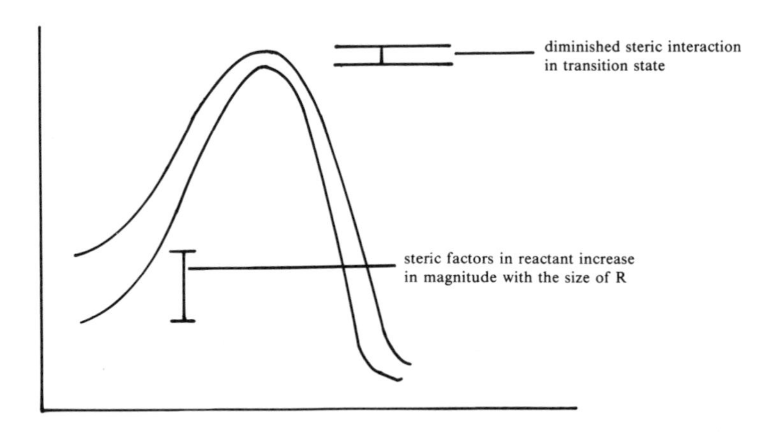

The enolates of α-alkoxy esters adopt the *Z*-configuration because of chelation by the alkoxy substituent.

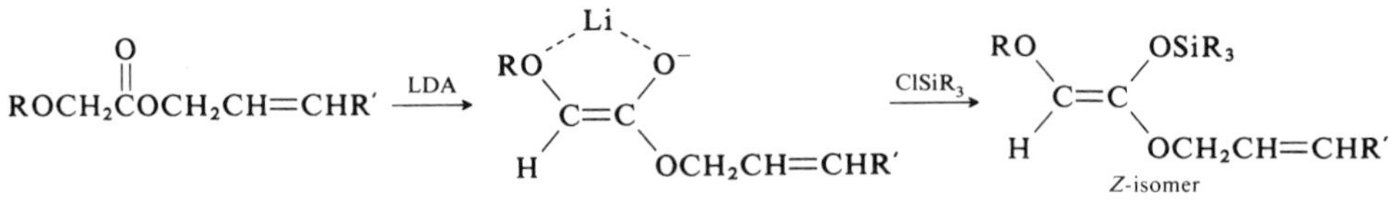

$$ROCH_2COCH_2CH{=}CHR' \xrightarrow{LDA} \text{Li-chelated enolate} \xrightarrow{ClSiR_3} (RO)(H)C{=}C(OSiR_3)(OCH_2CH{=}CHR')$$

Z-isomer

The configuration at the newly formed C—C bond is then controlled by the stereochemistry of the double bond in the allylic alcohol. The *E*-isomer gives a *syn* orientation while the *Z*-isomer gives rise to *anti* stereochemistry.[97]

96. C. S. Wilcox and R. E. Babston, *J. Am. Chem. Soc.* **108**, 6636 (1986).
97. T. J. Gould, M. Balestra, M. D. Wittman, J. A. Gary, L. T. Rossano, and J. Kallmerten, J. Org. Chem. **52**, 3889 (1987); S. D. Burke, W. F. Fobare, and G. J. Pacofsky, *J. Org. Chem.* **48**, 5221 (1983); P. A. Bartlett, D. J. Tanzella, and J. F. Barstow, *J. Org. Chem.* **47**, 3941 (1982).

A reaction which is related to the orthoester Claisen rearrangement utilizes an amide acetal, such as dimethylacetamide dimethyl acetal, rather than an orthoester. The reaction is carried out by heating the allylic alcohol and the amide acetal. The reaction occurs by exchange of the allylic alcohol for one of the alkoxy groups of the amide acetal, followed by elimination.[98] The stereochemistry of the reaction is analogous to that of the other variants of the Claisen rearrangement.[99]

$$RCH{=}CHCH_2OH + (CH_3)_2N\underset{OCH_3}{\overset{OCH_3}{C}}CH_3 \rightleftharpoons (CH_3)_2N\underset{CH_3}{\overset{OCH_3}{C}}OCH_2CH{=}CHR \rightleftharpoons (CH_3)_2N\underset{CH_2}{\underset{\|}{C}}OCH_2CH{=}CHR \rightarrow (CH_3)_2N\overset{O}{\overset{\|}{C}}CH_2\underset{R}{CH}CH{=}CH_2$$

The rearrangement of enolates of α-allyloxy ketones gives α-hydroxy-α-allyl ketones.[100]

$$Ph\overset{O}{\overset{\|}{C}}\underset{CH_3}{C}HOCH_2CH{=}CH_2 \rightarrow Ph\overset{O^-}{C}{=}C\begin{matrix} CH_3 \\ OCH_2CH{=}CH_2\end{matrix} \rightarrow Ph\underset{CH_2CH=CH_2}{\overset{O^-}{C}}{-}\overset{O}{\overset{\|}{C}}CH_3$$

In analogy with the oxy-Cope rearrangement, it has been called an anionic oxy-Claisen rearrangement. Like the anionic oxy-Cope reaction, this reaction shows a high sensitivity to the metal cation and solvent. The reaction rate is in the order $K^+ > Na^+ > Li^+$ and THF > toluene.

98. A. E. Wick, D. Felix, K. Steen, and A. Eschenmoser, *Helv. Chim. Acta* **47**, 2425 (1964); D. Felix, K. Gschwend-Steen, A. E. Wick, and A. Eschenmoser, *Helv. Chim. Acta* **52**, 1030 (1969).
99. W. Sucrow, M. Slopianka, and P. P. Calderia, *Chem. Ber.* **108**, 1101 (1975).
100. M. Koreeda and J. I. Luengo, *J. Am. Chem. Soc.* **107**, 5572 (1985).

Claisen rearrangements of allyl phenyl ethers to *ortho*-allyl phenols were the first [3,3]-sigmatropic rearrangements to be thoroughly studied.[101] The reaction proceeds through a cyclohexadienone which enolizes to the stable phenol.

If both *ortho* positions are substituted, the allyl group undergoes a second migration, giving the *para*-substituted phenol:

$OCH_2CH{=}CH_2$, CH_3O, OCH_3 → 180°C → OH, CH_3O, OCH_3, $CH_2CH{=}CH_2$ (88%)

[CH_3O, O, CH_2, HC, H_2C, OCH_3 → CH_3O, O, OCH_3, H, $CH_2CH{=}CH_2$] Ref. 102

6.6. [2,3]-Sigmatropic Rearrangements

The [2,3]-sigmatropic class of rearrangements is represented as

Y, X, + → Y, X or $RCH{=}CHCH_2{-}X{-}CHZ$ → $RCHCH{=}CH_2$, $ZCHX^-$

Neutral Anionic

The rearrangements of allylic sulfoxides, selenoxides, and amine oxides are the most important examples of the first type while rearrangements of carbanions of allyl ethers are the major examples of the anionic type.

101. S. J. Rhoads, in *Molecular Rearrangements*, Vol. 1, P. de Mayo (ed.), Wiley-Interscience, New York, 1963, pp. 655–684.
102. I. A. Pearl, *J. Am. Chem. Soc.* **70**, 1746 (1948).

The rearrangement of allylic sulfoxides to allylic sulfenates first received study in connection with the mechanism of racemization of allyl aryl sulfoxides.[103] While the allyl sulfoxide structure is strongly favored at equilibrium, rearrangement through the achiral allyl sulfenate provides a low-energy pathway for racemization.

$$R-\overset{+}{S}(O^-)-CH_2-CH=CH_2 \rightleftharpoons RS-O-CH_2-CH=CH_2$$

The synthetic utility of the allyl sulfoxide–allyl sulfenate rearrangement is as a method of preparation of allylic alcohols.[104] The reaction is carried out in the presence of some reagent, such as phenylthiolate or trimethyl phosphite, which reacts with the sulfenate to cleave the S—O bond:

$$(CH_3)_3C\text{-}C_6H_{10}{=}CHCH_2\overset{+}{S}(O^-)Ph \longrightarrow (CH_3)_3C\text{-}C_6H_{10}(OSPh)(CH{=}CH_2) \xrightarrow{PhS^-} (CH_3)_3\text{-}C_6H_{10}(OH)(CH{=}CH_2)\ (95\%)$$

Ref. 105

A comparable transformation occurs with allylic selenoxides when they are generated *in situ* by oxidation of allylic seleno ethers.[106]

$$PhCH_2CH_2CH(SePh)CH{=}CHCH_3 \xrightarrow{H_2O_2} PhCH_2CH_2CH{=}CHCH(OH)CH_3$$

Allylic sulfonium ylides readily undergo [2,3]-sigmatropic rearrangement.[107]

$$(CH_3)_2C{=}CHCH_2-\overset{+}{S}(CH_3)-\overset{-}{C}HCH{=}C(CH_3)_2 \rightarrow (CH_3)_2C{=}CHCH(SCH_3)C(CH_3)_2CH{=}CH_2\ (95\%)$$

This reaction results in carbon–carbon bond formation. It has found synthetic application in ring-expansion sequences for generation of medium-sized rings. The

103. R. Tang and K. Mislow, *J. Am. Chem. Soc.* **92**, 2100 (1970).
104. D. A. Evans and G. C. Andrews, *Acc. Chem. Res.* **7**, 147 (1974).
105. D. A. Evans, G. C. Andrews, and C. L. Sims, *J. Am. Chem. Soc.* **93**, 4956 (1971).
106. H. J. Reich, *J. Org. Chem.* **40**, 2570 (1975); D. L. J. Clive, G. Chittatu, N. J. Curtis, and S. M. Menchen, *J. Chem. Soc., Chem. Commun.*, 770 (1978).
107. J. E. Baldwin, R. E. Hackler, and D. P. Kelly, *J. Chem. Soc., Chem. Commun.*, 537 (1968).

reaction proceeds best when the ylide has a carbanion-stabilizing substituent. Part A of Scheme 6.12 shows some examples of the reaction.

The corresponding nitrogen ylides can also be generated when one of the nitrogen substituents has an electron-withdrawing group on the α-carbon. Entries 4 and 5 of Scheme 6.11 are examples. Entry 4 illustrates the use of the reaction for ring expansion.

Scheme 6.12. Carbon-Carbon Bond Formation via [2,3]-Sigmatropic Rearrangements of Sulfur and Nitrogen Ylides

A. Sulfonium Ylides

1[a] $(CH_3)_2C{=}CHCH_2{-}\overset{+}{S}(CH_3){-}CH_2CO_2C_2H_5 \xrightarrow{Na_2CO_3} (CH_3)_2C(SCH_3)CH(CO_2C_2H_5)$ … $CH{=}CH_2$ (91%)

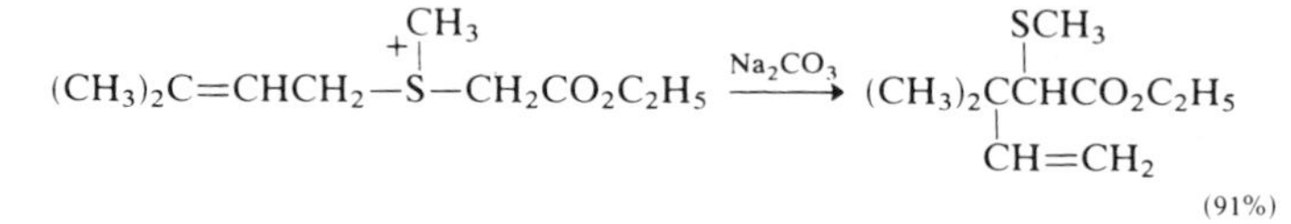

2[b] S-methyl-2-vinylthiacycloheptanium $\xrightarrow[-40°C]{K^+\ OC(CH_3)_3}$ thiacyclononene (85%)

3[c] $\xrightarrow[20°C]{DBU}$ (40%)

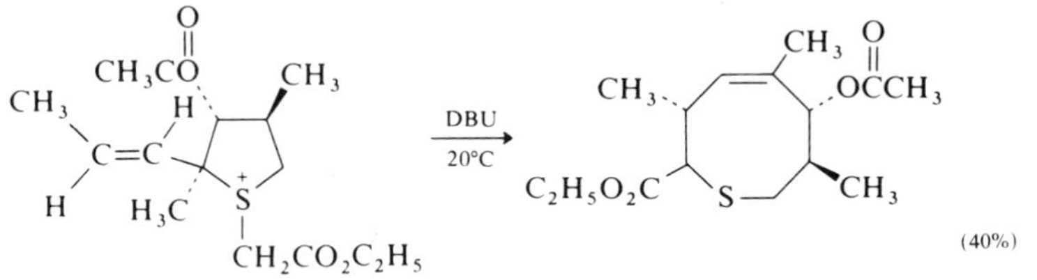

B. Ammonium Ylides

4[d] N-benzyl-N-($CH_2CO_2C_2H_5$)-2-vinylpiperidinium $\xrightarrow[20°C]{DBU}$ $PhCH_2$-N azacyclononene, $CO_2C_2H_5$ (90%)

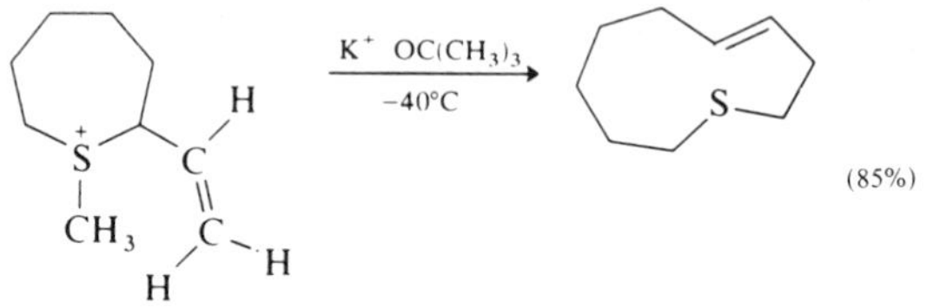

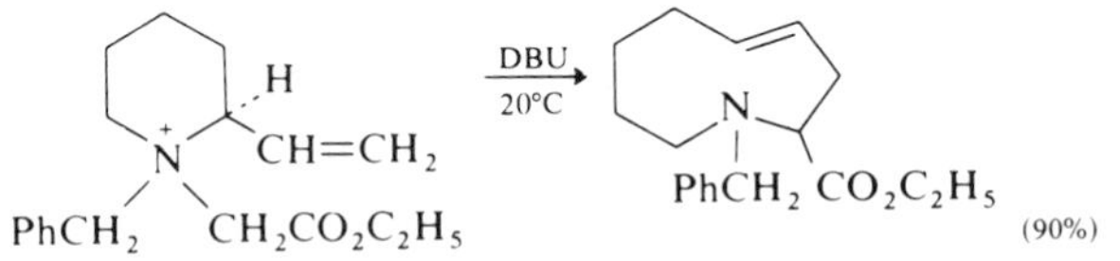

5[e] $(CH_3)_3C$-cyclohexylidene$=CHCH_2\overset{+}{N}$(pyrrolidinium)$CH_2CN \xrightarrow{K^{+-}O\text{-}t\text{-}Bu} (CH_3)_3C$-cyclohexyl-C(CHCN-N-pyrrolidinyl)($CH{=}CH_2$) (94%)

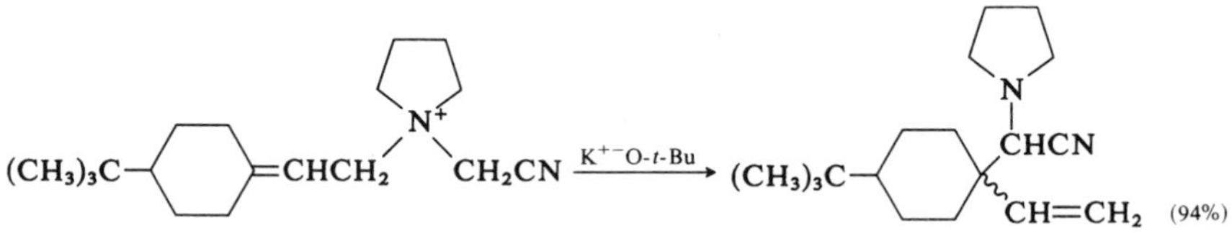

a. K. Ogura, S. Furukawa, and G. Tsuchihashi, *J. Am. Chem. Soc.* **102**, 2125 (1980).
b. V. Cere, C. Paolucci, S. Pollicino, E. Sandri, and A. Fava, *J. Org. Chem.* **43**, 4826 (1978).
c. E. Vedejs and M. J. Mullins, *J. Org. Chem.* **44**, 2947 (1979).
d. E. Vedejs, M. J. Arco, D. W. Powell, J. M. Renga, and S. P. Singer, *J. Org. Chem.* **43**, 4831 (1978).
e. L. N. Mander and J. V. Turner, *Aust. J. Chem.* **33**, 1559 (1980).

N-Allylamine oxides represent the general structure pattern for [2,3]-sigmatropic rearrangement, with $X = N^+$ and $Y = O^-$. The rearrangement proceeds readily to provide *O*-allyl hydroxylamine derivatives.

$$R-\overset{+R}{\underset{O^-}{N}}-CH_2CH=CH_2 \rightarrow R_2N-OCH_2CH=CH_2$$

A useful method for *ortho*-alkylation of aromatic amines is based on [2,3]-sigmatropic rearrangement of *S*-anilinosulfonium ylides. These ylides are generated from anilinosulfonium ions, which can be prepared from *N*-chloroanilines and sulfides.[108]

This method is the basis for synthesis of nitrogen-containing heterocyclic compounds when Z is a carbonyl-containing group.[109]

The [2,3]-sigmatropic rearrangement pattern is also observed with anionic species. The most important for synthetic purposes is the *Wittig rearrangement*, in which a strong base converts allylic ethers to α-allylalkoxides.

Since the deprotonation at the α'-carbon must compete with deprotonation of the α-carbon in the allyl group, most examples involve a conjugated or electron-withdrawing substituent Z.[110]

The stereochemistry of the Wittig rearrangement can, in general, be predicted in terms of a cyclic five-membered transition state.[111]

108. P. G. Gassman and G. D. Gruetzmacher, *J. Am. Chem. Soc.* **96**, 5487 (1974); P. G. Gassman and H. R. Drewes, *J. Am. Chem. Soc.* **100**, 7600 (1978).
109. P. G. Gassman, T. J. van Bergen, D. P. Gilbert, and B. W. Cue, Jr., *J. Am. Chem. Soc.* **96**, 5495 (1974); P. G. Gassman and T. J. van Bergen, *J. Am. Chem. Soc.* **96**, 5508 (1974); P. G. Gassman, G. Gruetzmacher, and T. J. van Bergen, *J. Am. Chem. Soc.* **96**, 5512 (1974).
110. For reviews of [2,3]-sigmatropic rearrangement of allyl ethers, see T. Nakai and K. Mikami, *Chem. Rev.* **86**, 885 (1986).
111. K. Mikami, Y. Kimura, N. Kishi, and T. Nakai, *J. Org. Chem.* **48**, 279 (1983); K. Mikami, K. Azuma, and T. Nakai, *Tetrahedron* **40**, 2303 (1984); R. W. Hoffmann, *Angew. Chem. Int. Ed. Engl.* **18**, 563 (1979).

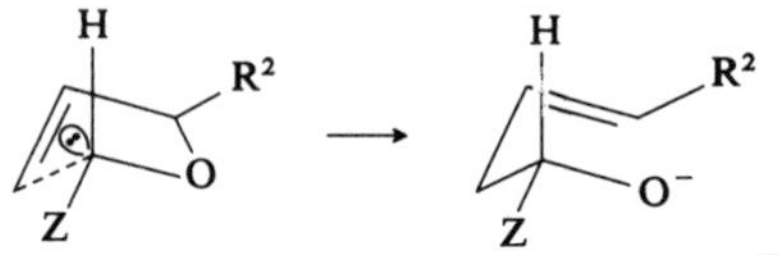

A consistent feature of the observed stereochemistry is a preference for *E*-stereochemistry at the newly formed double bond. The reaction can also show stereoselectivity at the newly formed single bond. This stereoselectivity has been carefully studied for the case where the substituent Z is an acetylenic group.[112]

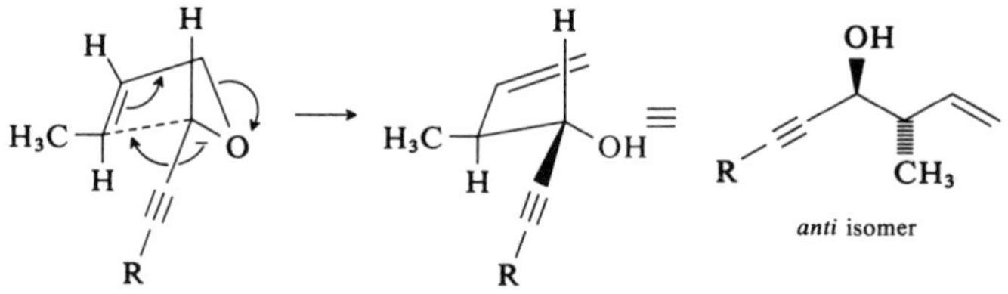

E-isomer

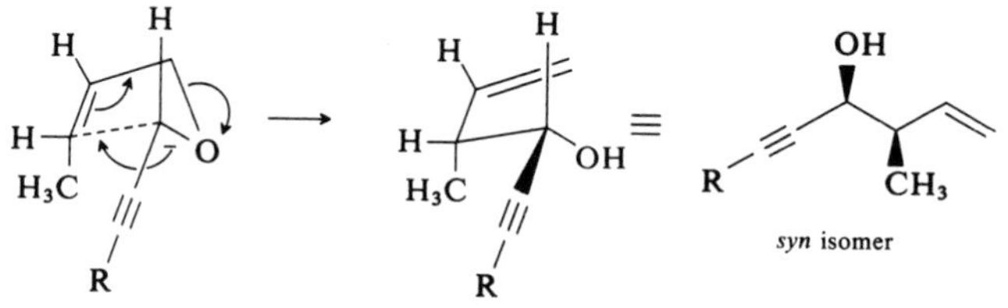

anti isomer

Z-isomer

syn isomer

The preferred stereochemistry arises from the transition state that minimizes interaction between the Z and R^2 groups. This stereoselectivity is revealed in the rearrangement of **9** to **10**.

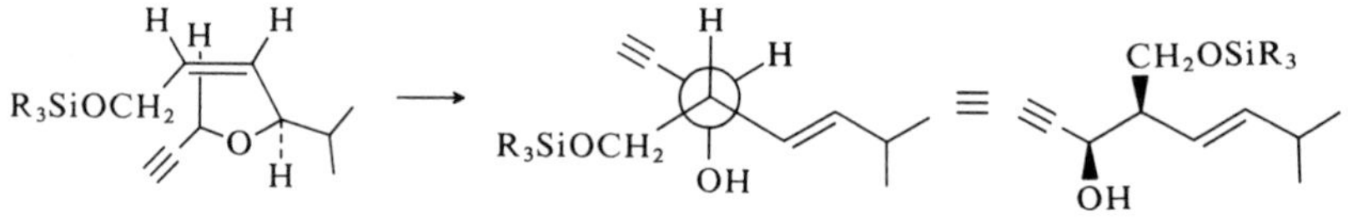

9

10

The [2,3]-Wittig rearrangement has proven useful for ring contraction in the synthesis of a number of medium-ring unsaturated structures, as illustrated by entry 3 in Scheme 6.13.

6.7. Ene Reactions

Certain electrophilic carbon–carbon and carbon–oxygen double bonds can undergo an addition reaction with alkenes in which an allylic hydrogen is transferred

112. M. M. Midland and J. Gabriel, *J. Org. Chem.* **50**, 1143 (1985).

1[a]

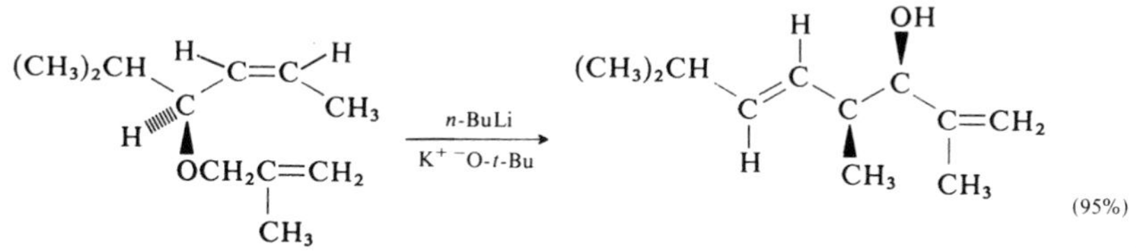

$(CH_3)_2CH$, H, C=C, H, CH_3, H, C, $OCH_2C=CH_2$, CH_3 → (n-BuLi, K^+ ^-O-t-Bu) → $(CH_3)_2CH$, H, C=C, H, C, OH, C, CH_3, $C=CH_2$, CH_3 (95%)

2[b]

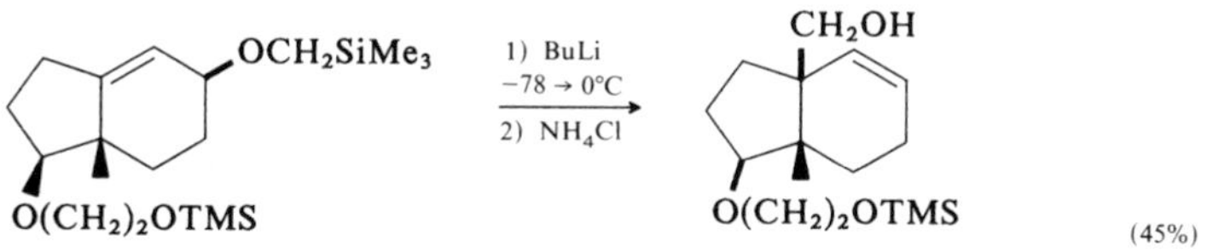

OCH_2SiMe_3, $O(CH_2)_2OTMS$ → (1) BuLi, −78 → 0°C; 2) NH_4Cl) → CH_2OH, $O(CH_2)_2OTMS$ (45%)

3[c]

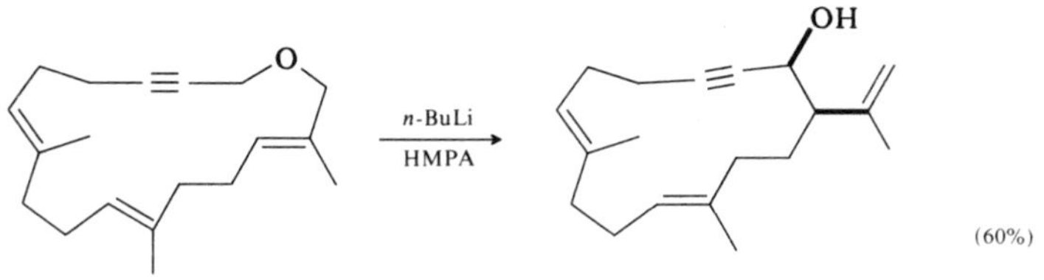

O → (n-BuLi, HMPA) → OH (60%)

4[d]

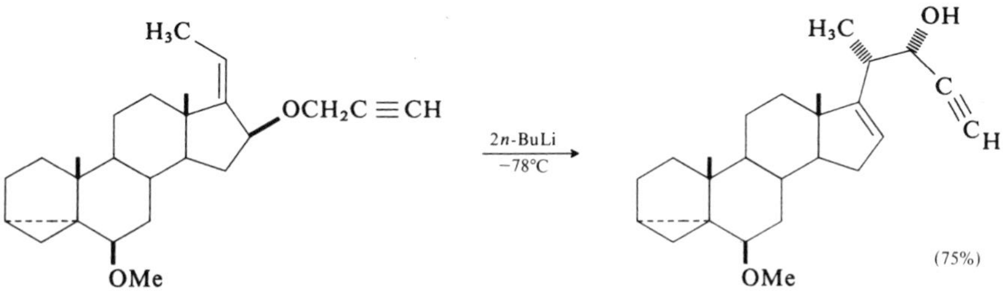

H_3C, $OCH_2C\equiv CH$, OMe → (2n-BuLi, −78°C) → H_3C, OH, C≡CH, OMe (75%)

a. D. J.-S. Tsai and M. M. Midland, *J. Am. Chem. Soc.* **107**, 3915 (1985).
b. T. Sugimura and L. A. Paquette, *J. Am. Chem. Soc.* **109**, 3017 (1987).
c. J. A. Marshall, T. M. Jenson, and B. S. De Hoff, *J. Org. Chem.* **51**, 4316 (1986).
d. K. Mikami, K. Kawamoto, and T. Nakai, *Tetrahedron Lett.* **26**, 5799 (1985).

to the electrophile. This process is called the *ene reaction* and the electrophile is called an *enophile.*[113]

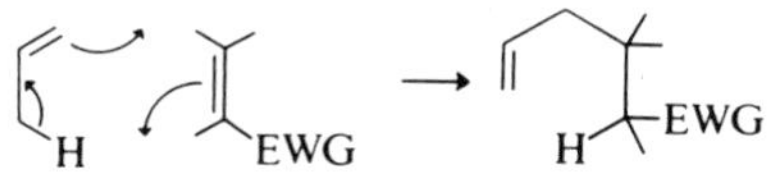

H, EWG → H, EWG

ene + enophile

113. For review of the ene reaction, see H. M. R. Hoffmann, *Angew. Chem. Int. Ed. Engl.* **8**, 556 (1969); W. Oppolzer, *Pure Appl. Chem.* **53**, 1181 (1981).

The concerted mechanism shown above is allowed by the Woodward–Hoffmann rules. The transition state involves the π electrons of the alkene and enophile and the σ electrons of the C—H bond, as illustrated in Fig. 6.6.

Ene reactions have relatively high activation energies, and intermolecular reaction is observed only for strongly electrophilic enophiles. Some examples are given in Scheme 6.14. The ene reaction is strongly catalyzed by Lewis acids such as aluminum chloride and diethylaluminum chloride.[114] Coordination by the aluminum at the carbonyl group increases the electrophilicity of the conjugated system and allows reaction to occur below room temperature, as illustrated in entry 6. Intramolecular ene reactions can be carried out under either thermal (entry 5) or catalyzed (entry 6) conditions.[115]

Certain carbonyl compounds also react with alkenes according to the ene reaction pattern. Formaldehyde in acidic solution reacts to form homoallylic alcohols or the corresponding esters as in entry 3.

Diethyl oxomalonate is another example of a carbonyl compound that gives ene reaction products with a number of alkenes.[116]

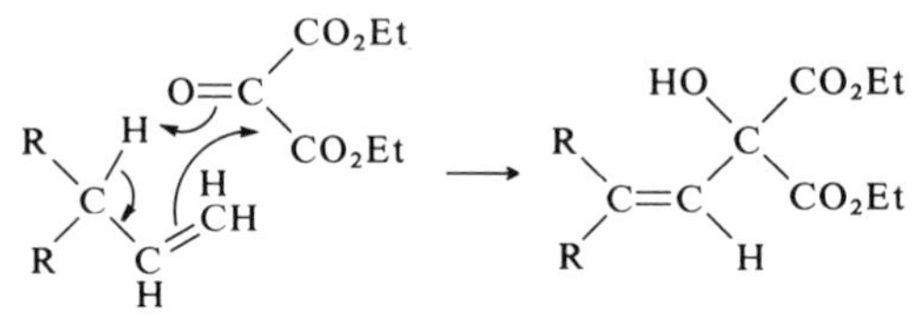

Mechanistic studies have been designed to determine if the concerted cyclic transition state is a good representation of the mechanism. The reaction is only moderately sensitive to electronic effects. The ρ value for a series of 1-arycyclopentenes is -1.2, which would indicate that there is little charge development in the transition state. The reaction shows a primary kinetic isotope effect indicative of C—H bond breaking in the rate-determining step.[117] These observations are consistent with a concerted

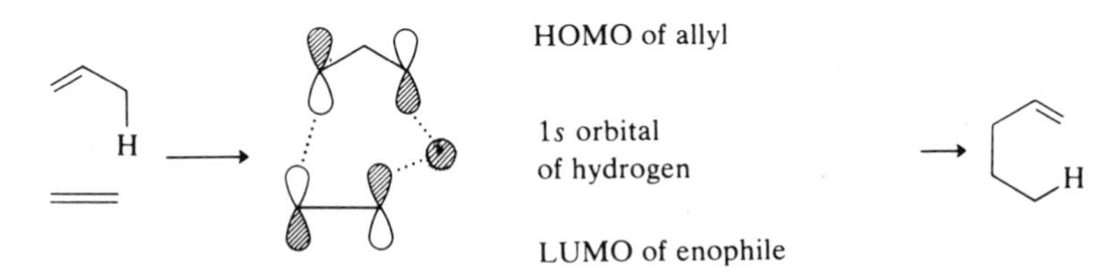

Fig. 6.6. A concerted ene reaction corresponds to the interaction of a hydrogen atom with the HOMO of an allyl radical and the LUMO of the enophile and is allowed.

114. B. B. Snider, *Acc. Chem. Res.* **13**, 426 (1980).
115. W. Oppolzer and V. Snieckus, *Angew. Chem. Int. Ed. Engl.* **17**, 476 (1978).
116. M. F. Salomon, S. N. Pardo, and R. G. Salomon, *J. Org. Chem.* **49**, 2446 (1984); *J. Am. Chem. Soc.* **106**, 3797 (1984).
117. O. Achmatowicz and J. Szymoniak, *J. Org. Chem.* **45**, 4774 (1980); H. Kwart and M. Brechbiel, *J. Org. Chem.* **47**, 3353 (1982).

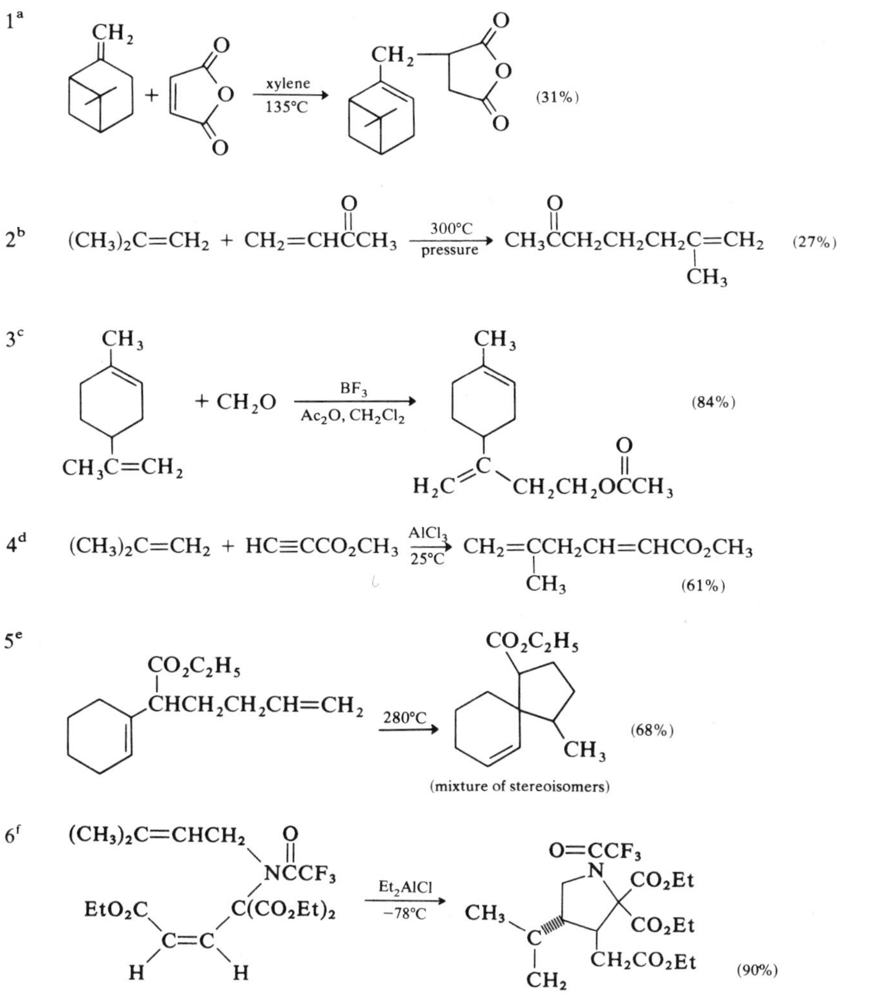

1[a] (CH_2-substituted bicyclic alkene) + (maleic anhydride) $\xrightarrow[135°C]{\text{xylene}}$ (CH_2-succinic anhydride adduct) (31%)

2[b] $(CH_3)_2C{=}CH_2 + CH_2{=}CH\overset{O}{\overset{\|}{C}}CH_3 \xrightarrow[\text{pressure}]{300°C} CH_3\overset{O}{\overset{\|}{C}}CH_2CH_2CH_2\underset{CH_3}{\underset{|}{C}}{=}CH_2$ (27%)

3[c] (CH_3-cyclohexene with $CH_3C{=}CH_2$ substituent) $+ CH_2O \xrightarrow[Ac_2O,\ CH_2Cl_2]{BF_3}$ (CH_3-cyclohexene with $H_2C{=}C\text{—}CH_2CH_2O\overset{O}{\overset{\|}{C}}CH_3$ substituent) (84%)

4[d] $(CH_3)_2C{=}CH_2 + HC{\equiv}CCO_2CH_3 \xrightarrow[25°C]{AlCl_3} CH_2{=}\underset{CH_3}{\underset{|}{C}}CH_2CH{=}CHCO_2CH_3$ (61%)

5[e] (cyclohexenyl)$\underset{}{\overset{CO_2C_2H_5}{\overset{|}{C}}}HCH_2CH_2CH{=}CH_2 \xrightarrow{280°C}$ (spiro compound bearing $CO_2C_2H_5$ and CH_3) (68%)

(mixture of stereoisomers)

6[f] $(CH_3)_2C{=}CHCH_2\text{—}N(\overset{O}{\overset{\|}{C}}CF_3)\text{—}C(CO_2Et)_2\text{—}C(H){=}C(H)\text{—}CO_2Et$ (EtO$_2$C, H / H) $\xrightarrow[-78°C]{Et_2AlCl}$ (pyrrolidine with N–$O{=}CCF_3$, CO_2Et, CO_2Et, CH_2CO_2Et, and $CH_3\text{—}C{=}CH_2$ substituents) (90%)

a. R. T. Arnold and J. S. Showell, *J. Am. Chem. Soc.* **79**, 419 (1957).
b. C. J. Albisetti, N. G. Fisher, M. J. Hogsed, and R. M. Joyce, *J. Am. Chem. Soc.* **78**, 2637 (1956).
c. A. T. Blomquist and R. J. Himics, *J. Org. Chem.* **33**, 1156 (1968).
d. B. B. Snider, D. J. Rodini, R. S. E. Conn, and S. Sealfon, *J. Am. Chem. Soc.* **101**, 5283 (1979).
e. W. Oppolzer, K. K. Mahalanabis, and K. Battig, *Helv. Chim. Acta.* **60**, 2388 (1977).
f. W. Oppolzer and C. Robbiani, *Helv. Chim. Acta* **63**, 2010 (1980).

process. Mechanistic analysis of Lewis acid catalyzed reactions indicates that they may more closely resemble an electrophile substitution process related to Friedel-Crafts acylation reactions that will be discussed in Section 10.11.[118]

118. B. B. Snider, D. M. Roush, D. J. Rodini, D. M. Gonzalez, and D. Spindell, *J. Org. Chem.* **45**, 2773 (1980); J. V. Duncia, P. T. Lansbury, Jr., T. Miller, and B. B. Snider, *J. Org. Chem.* **47**, 4538 (1982); B. B. Snider and G. B. Phillips, *J. Org. Chem.* **48**, 464 (1983); B. B. Snider and E. Ron, *J. Am. Chem. Soc.* **107**, 8160 (1985).

Electrophilic azo compounds such as diethyl azodicarboxylate and triazoline-3,5-dione also give ene-type products on reaction with alkenes.

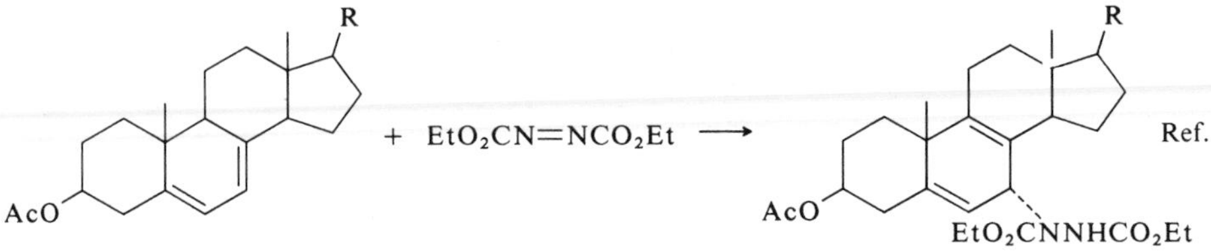

R
AcO
+ $EtO_2CN{=}NCO_2Et$ ⟶
R
AcO
$EtO_2CNNHCO_2Et$
Ref. 119

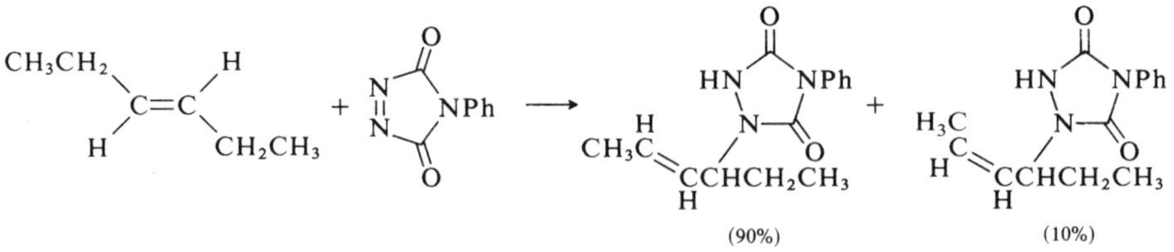

CH_3CH_2 H
C=C
H CH_2CH_3
+ N=N, O, NPh, O ⟶
O, HN, NPh, N, O
H
CH_3C
$CCHCH_2CH_3$
H
(90%)
+
O, HN, NPh, N, O
H_3C
C
H
$CCHCH_2CH_3$
H
(10%)
Ref. 120

6.8. Unimolecular Thermal Elimination Reactions

There are several thermal elimination reactions that find use in synthesis. Some of these are concerted processes. The transition state energy requirements and stereochemistry of concerted elimination processes can be analyzed in terms of orbital symmetry considerations. We will also consider an important group of unimolecular β-elimination reactions in Section 6.8.3.

6.8.1. Cheletropic Elimination

Cheletropic processes are defined as reactions in which two bonds are broken at a single atom. Concerted cheletropic reactions are subject to orbital symmetry restrictions in the same way that cycloadditions and sigmatropic processes are.

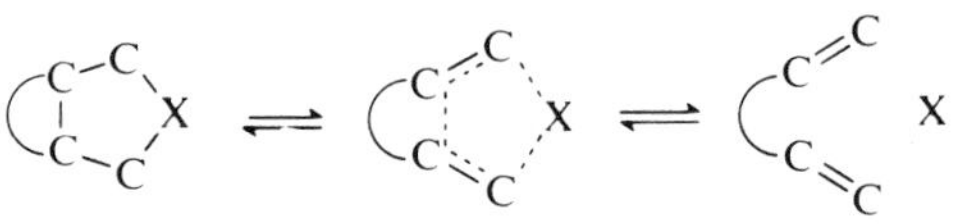

C C C C X ⇌ C C C C X ⇌ C C C C X

In the elimination processes of interest here, the atom X is normally bound to other atoms in such a way that elimination will give rise to a stable molecule. The most common examples involve five-membered rings.

119. M. Anastasia, A. Fiocchi, and G. Galli, *J. Org. Chem.* **46**, 3421 (1981).
120. C.-C. Cheng, C. A. Seymour, M. A. Petti, F. D. Greene, and J. F. Blount, *J. Org. Chem.* **49**, 2910 (1984).

A good example of a concerted cheletropic elimination is the reaction of 3-pyrroline with *N*-nitrohydroxylamine, which gives rise to the diazene **E**, which then undergoes elimination of nitrogen.

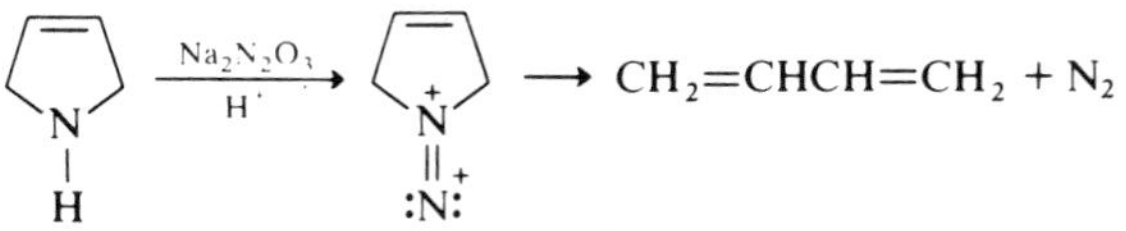

$$\mathbf{D} \xrightarrow[H^+]{Na_2N_2O_3} \mathbf{E} \longrightarrow CH_2{=}CHCH{=}CH_2 + N_2$$

Use of substituted systems has shown that the reaction is completely stereospecific.[121] The groups on C-2 and C-5 of the pyrroline ring rotate in the disrotatory mode on going to product. This stereochemistry is consistent with conservation of orbital symmetry.

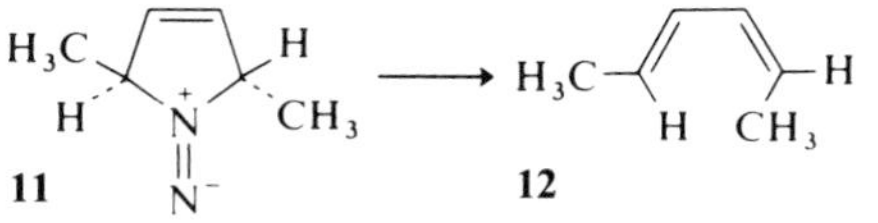

$$\mathbf{11} \longrightarrow \mathbf{12}$$

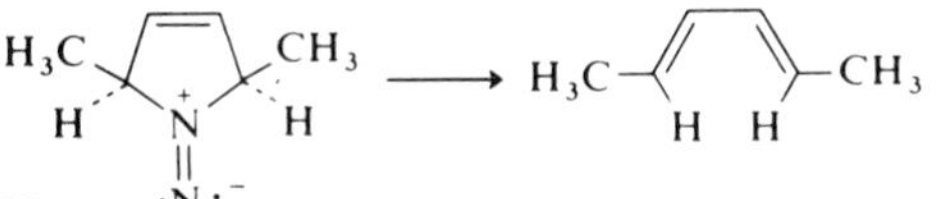

$$\mathbf{13} \longrightarrow \mathbf{14}$$

The most synthetically useful cheletropic elimination involves 2,5-dihydrothiophene-1,1-dioxides (sulfolene dioxides). At elevated temperatures, these compounds fragment to give dienes and sulfur dioxide.[122] The reaction is stereospecific. For example, the dimethyl derivatives **15** and **16** give the *E,E*- and *Z,E*-isomers of 2,4-hexadiene, respectively, at temperatures of 100–150°C.[123] This stereospecificity corresponds to disrotatory elimination.

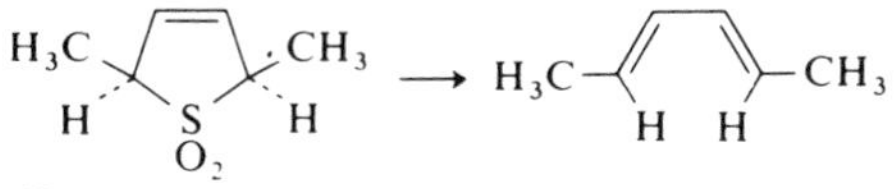

$$\mathbf{15} \longrightarrow H_3C{-}CH{=}CH{-}CH{=}CH{-}CH_3$$

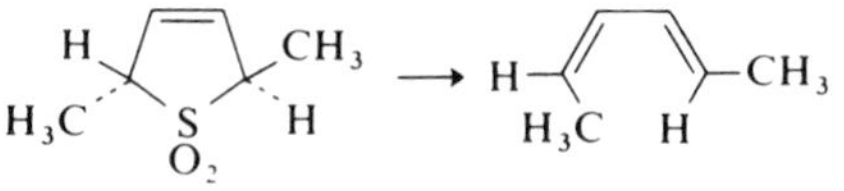

$$\mathbf{16} \longrightarrow H_3C{-}CH{=}CH{-}CH{=}CH{-}CH_3$$

121. D. M. Lemal and S. D. McGregor, *J. Am. Chem. Soc.* **88**, 1335 (1966).
122. W. L. Mock, in *Pericyclic Reactions*, Vol. II, A. P. Marchand and R. E. Lehr (eds.), Academic Press, New York, 1977, Chapter 3.
123. W. L. Mock, *J. Am. Chem. Soc.* **88**, 2857 (1966); S. D. McGregor and D. M. Lemal, *J. Am. Chem. Soc.* **88**, 2858 (1966).

Elimination of sulfur dioxide has proven to be a useful method for generating dienes which can undergo subsequent Diels–Alder addition.

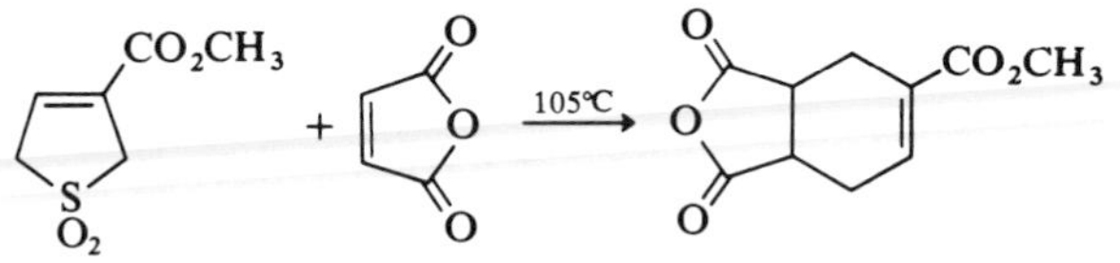

Ref. 124

The method is particularly useful in formation of *o*-quinodimethanes.

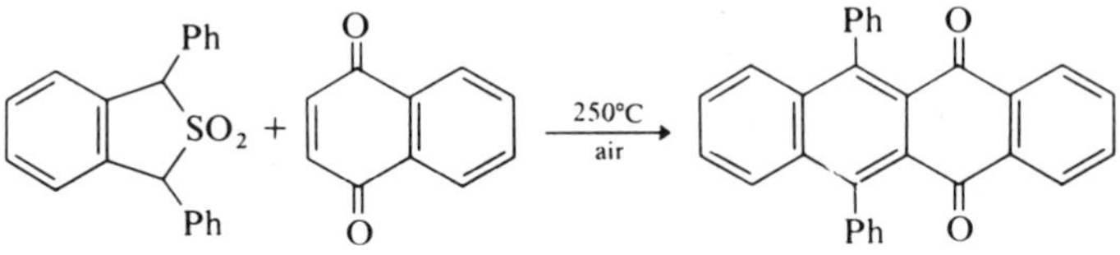

(oxidation product of initial adduct) (42%) Ref. 125

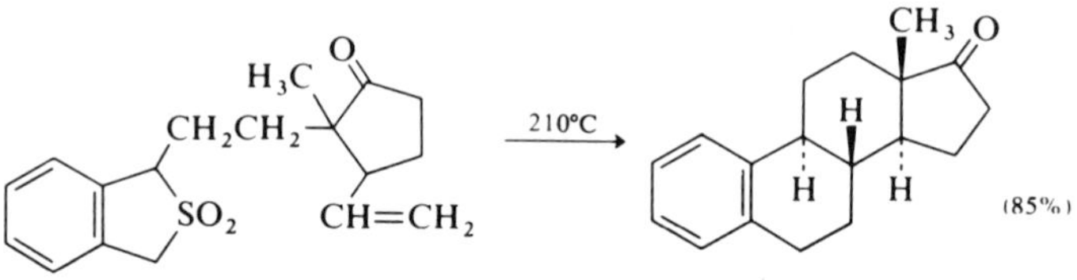

(85%) Ref. 126

The elimination of carbon monoxide can occur by a concerted process in some cyclic ketones. The elimination of carbon monoxide from bicyclo[2.2.1]heptadien-7-ones is very facile. In fact, generation of bicyclo[2.2.1]heptadien-7-ones is usually accompanied by spontaneous elimination of carbon monoxide.

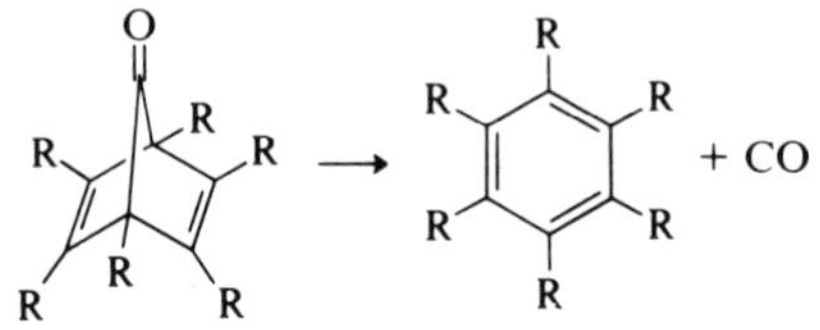

The ring system can be generated by Diels–Alder addition of a substituted cyclopentadienone and an alkyne. A reaction sequence involving addition followed by CO elimination can be used for the synthesis of highly substituted benzene rings.[127]

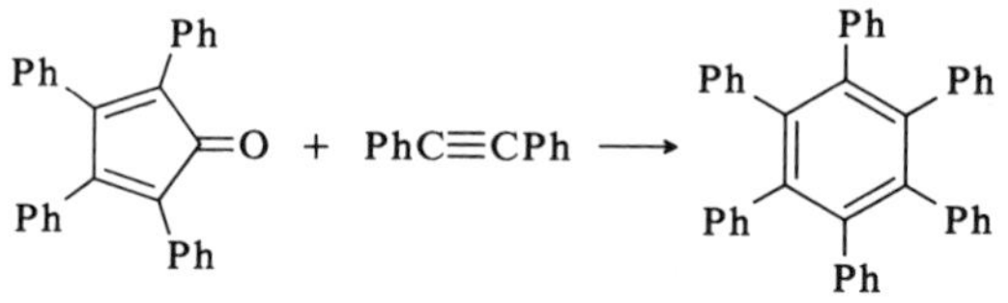

Ref. 128

124. J. M. McIntosh and R. A. Sieler, *J. Org. Chem.* **43**, 4431 (1978).
125. M. P. Cava, M. J. Mitchell, and A. A. Deana, *J. Org. Chem.* **25**, 1481 (1960).
126. K. C. Nicolaou, W. E. Barnette, and P. Ma, *J. Org. Chem.* **45**, 1463 (1980).
127. M. A. Ogliaruso, M. G. Romanelli, and E. I. Becker, *Chem. Rev.* **65**, 261 (1965).
128. L. F. Fieser, *Org. Synth.* **V**, 604 (1973).

Exceptionally facile elimination of CO also takes place from **17**, in which homoaromaticity can stabilize the transition state:

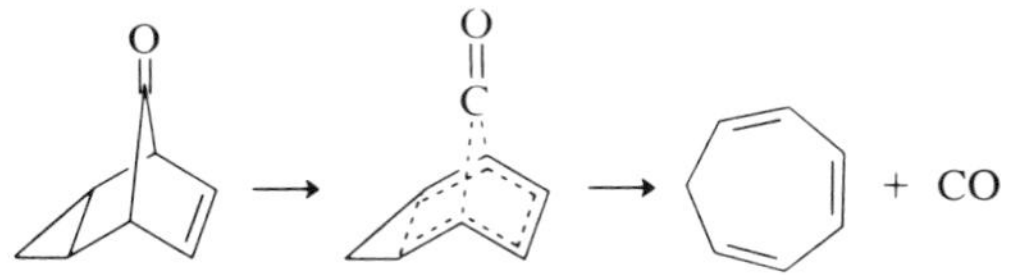

+ CO   Ref. 129

7

6.8.2. Decomposition of Cyclic Azo Compounds

Another significant group of elimination reactions involves processes in which a small molecule is eliminated from a ring system and the two reactive sites that remain react to re-form a ring.

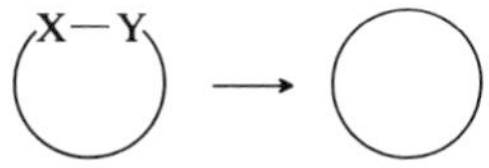

X—Y

The most widely studied example is azo compounds, where —X—Y— is —N=N—.[130] The elimination of nitrogen from cyclic azo compounds can be carried out either photochemically or thermally. Although the reaction generally does not proceed by a concerted mechanism, there are some special cases where concerted elimination is possible. We will consider some of these cases first and then consider the more general case.

An interesting illustration of the importance of orbital symmetry effects is the contrasting stability of azo compounds **18** and **19**. Compound **18** decomposes to norbornene and nitrogen only above 100°C. In contrast, **19** eliminates nitrogen immediately on preparation, even at −78°C.[131]

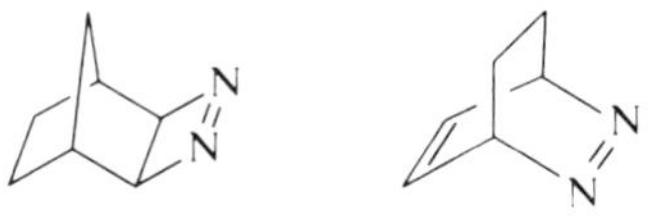

18 **19**

The reason for this difference is that if **18** were to undergo a concerted elimination, it would have to follow the forbidden (high-energy) $[_{\pi}2_s + {}_{\pi}2_s]$ pathway. For **19**, the elimination can take place by the allowed $[_{\pi}2_s + {}_{\pi}4_s]$ pathway. Thus, these reactions are the reverse of the [2 + 2] and [4 + 2] cycloadditions, respectively, and only the latter is an allowed concerted process. The temperature at which **18**

129. B. A. Halton, M. A. Battiste, R. Rehberg, C. L. Deyrup, and M. E. Brennan, *J. Am. Chem. Soc.* **89**, 5964 (1967).
130. P. S. Engel, *Chem. Rev.* **80**, 99 (1980).
131. N. Rieber, J. Alberts, J. A. Lipsky, and D. M. Lemal, *J. Am. Chem. Soc.* **91**, 5668 (1969).

decomposes is fairly typical for strained azo compounds, and the reaction presumably proceeds by a nonconcerted biradical mechanism. Because a C—N bond must be broken without concomitant compensation by carbon–carbon bond formation, the activation energy is much higher than for a concerted process.

Although the concerted mechanism described in the preceding paragraph is available only to those azo compounds with appropriate orbital arrangements, the nonconcerted mechanism occurs at low enough temperatures to be synthetically useful. The elimination can also be carried out photochemically. These reactions presumably occur by stepwise elimination of nitrogen.

$$R{-}N{=}N{-}R' \xrightarrow{\text{slow}} R{-}N{=}N\cdot + \cdot R' \xrightarrow{\text{fast}} R\cdot \quad N{\equiv}N \quad \cdot R' \rightarrow R{-}R'$$

The stereochemistry of the nonconcerted reaction has been a topic of considerable study and discussion. Frequently, there is not complete randomization as would be expected for a long-lived diradical intermediate. The details vary from case to case, and both preferential inversion and retention of relative stereochemistry have been observed.

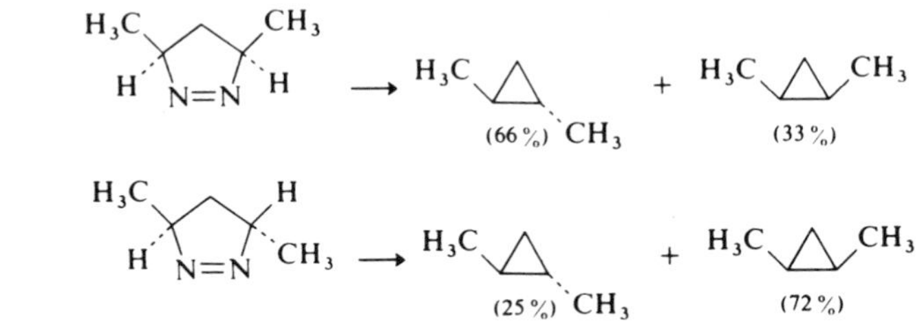

(66%) (33%)

(25%) (72%)

Ref. 132

145°

(43%) (2.5%)

(3.5%) (42%)

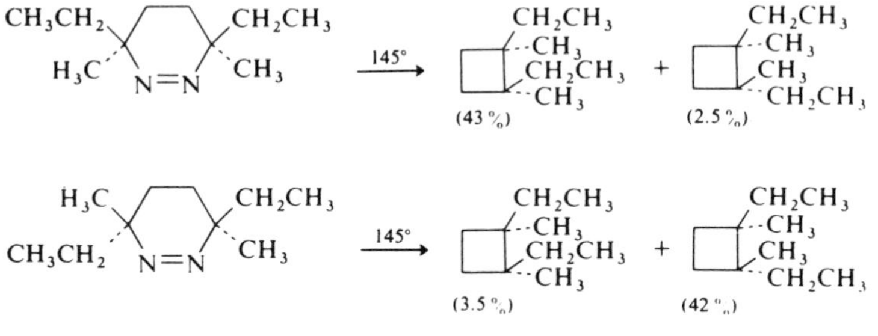

Ref. 133

These results can be interpreted in terms of competition between recombination of the diradical intermediate and conformational equilibration, which would destroy the stereochemical relationships present in the azo compound.

The main synthetic application of azo compound decomposition is in the synthesis of cyclopropanes and other strained ring systems. Some of the required azo compounds can be made by dipolar cycloadditions of diazo compounds. Some examples of decomposition of cyclic azo compounds are given in Scheme 6.15.

132. R. J. Crawford and A. Mishra, *J. Am. Chem. Soc.* **88**, 3963 (1966).
133. P. D. Bartlett and N. A. Porter, *J. Am. Chem. Soc.* **90**, 5317 (1968).

1[a] H_3C, CH_3 / H, CO_2CH_3 C=C + CH_2N_2 ⟶ pyrazoline (H_3C, CH_3, $CO_2C_2H_5$, N=N)

↓ hν

H_3C, CH_3, $CO_2C_2H_5$ cyclopropane (major) + H_3C, $CO_2C_2H_5$, CH_3 cyclopropane (minor)

2[b] $CF_3CH{=}CH_2$ + CH_2N_2 ⟶ F_3C-pyrazoline (N, N–H) $\xrightarrow{260°C}$ F_3C-cyclopropane (51%)

3[c] Ph-pyrazoline (CH_3, OCCH_3 (=O), N=N) $\xrightarrow{200°C}$ Ph-cyclopropane (CH_3, OCCH_3 (=O))

4[d] H_3C, H_3C, CH_3, CH_3 pyrazolenine (N=N) $\xrightarrow{h\nu}$ H_3C, CH_3, H_3C, CH_3 cyclopropene (74%)

5[e] bicyclic azo compound (N=N) $\xrightarrow{h\nu}$ bicyclo[2.1.0]pentane

6[f] CH_3CO_2C-indazole derivative (CH_3, CH_3, N=N) $\xrightarrow{h\nu}$ CH_3CO_2C-benzocyclopropene derivative (CH_3, CH_3)

7[g] polycyclic azo compound (N=N) $\xrightarrow{h\nu}$ polycyclic hydrocarbon (35%)

a. T. V. Van Auken and K. L. Rinehart, Jr., *J. Am. Chem. Soc.* **84**, 3736 (1962).
b. F. Misani, L. Speers, and A. M. Lyon, *J. Am. Chem. Soc.* **78**, 2801 (1956).
c. J. P. Freeman, *J. Org. Chem.* **29**, 1379 (1964).
d. G. L. Closs, W. A. Böll, H. Heyn, and V. Dev, *J. Am. Chem. Soc.* **90**, 173 (1968).
e. C. G. Overberger, N. R. Byrd, and R. B. Mesrobian, *J. Am. Chem. Soc.* **78**, 1961 (1956).
f. R. Anet and F. A. L. Anet, *J. Am. Chem. Soc.* **86**, 525 (1964).
g. F. M. Moriarty, *J. Org. Chem.* **28**, 2385 (1963).

The facile elimination of nitrogen can be used in conjunction with Diels–Alder additions to construct aromatic rings.[134] Pyridazine-3,6-dicarboxylate esters, for example, react with electron-rich alkenes to give adducts that undergo subsequent elimination of nitrogen.[135]

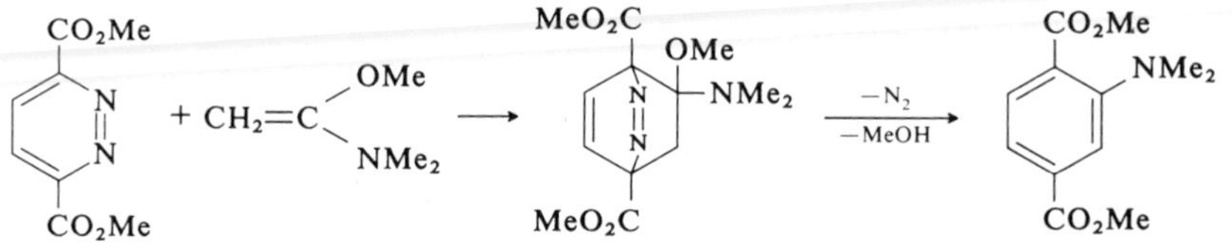

CO_2Me N N CO_2Me + $CH_2{=}C(OMe)NMe_2$ ⟶ MeO_2C OMe N N NMe_2 MeO_2C $\xrightarrow[-MeOH]{-N_2}$ CO_2Me NMe_2 CO_2Me

Similar reactions have been developed for 1,2,4-triazines and 1,2,4,5-tetrazines.

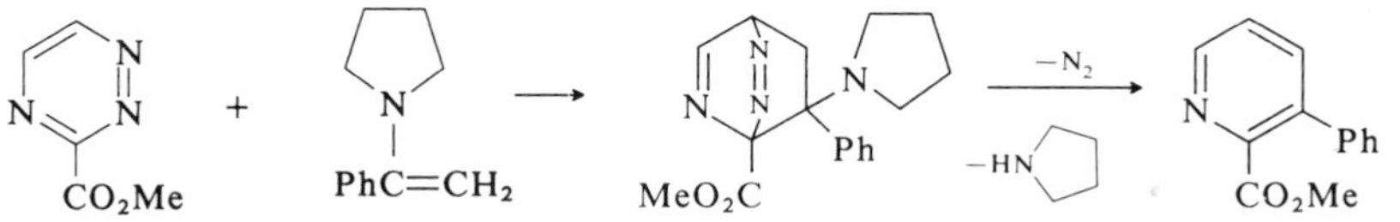

N N N CO_2Me + N $PhC{=}CH_2$ ⟶ N N N N Ph MeO_2C $\xrightarrow[-HN]{-N_2}$ N Ph CO_2Me Ref. 136

CO_2Me N N N N CO_2Me + $(MeO)_2C{=}CHCCH_3$ (O) ⟶ MeO_2C O CCH_3 N N N N H OMe OMe CO_2Me $\xrightarrow[-MeOH]{-N_2}$ MeO_2C O CCH_3 N N OMe CO_2Me Ref. 137

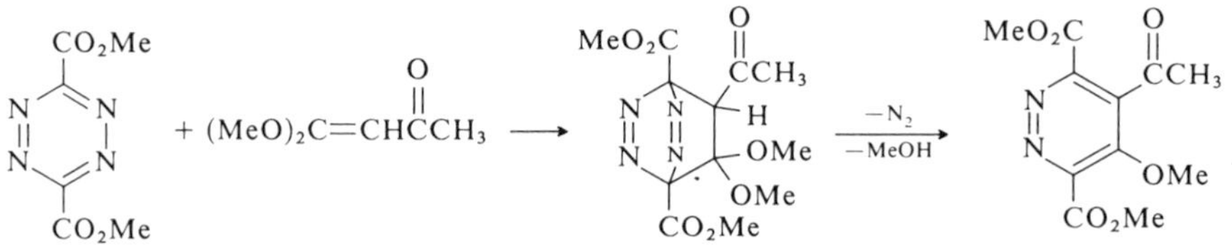

Diels–Alder reactions of pyrones can be used in a related way. The adducts are unstable toward elimination of carbon dioxide. In the examples shown below, there is a further elimination leading to the aromatic product.

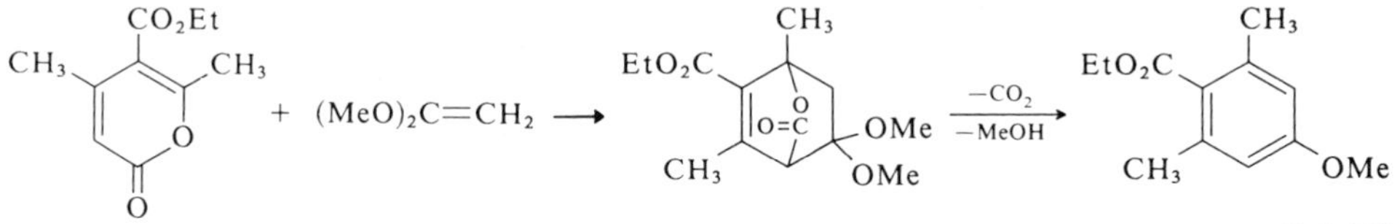

CO_2Et CH_3 CH_3 O O + $(MeO)_2C{=}CH_2$ ⟶ CH_3 EtO_2C O O=C OMe CH_3 OMe $\xrightarrow[-MeOH]{-CO_2}$ CH_3 EtO_2C CH_3 OMe

Ref. 138

CO_2Et CH_3 CH_3 O O + PhCH=CH—N O ⟶ CH_3 EtO_2C Ph O O=C CH_3 N O $\xrightarrow[HN\ O]{-CO_2}$ CH_3 EtO_2C Ph CH_3

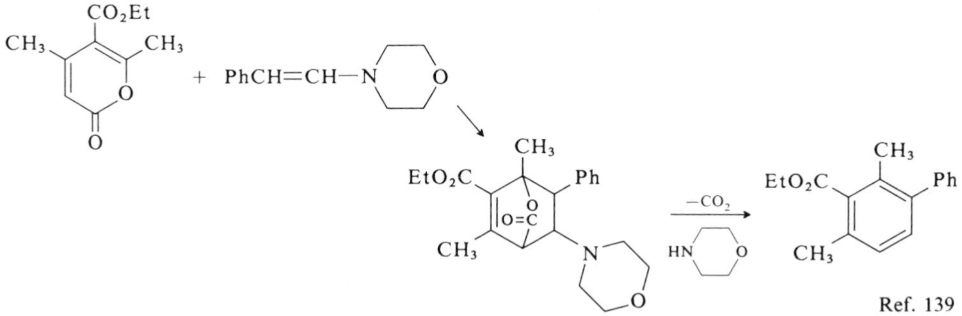

Ref. 139

134. D. L. Boger, *Chem. Rev.* **86**, 781 (1986).
135. H. Neunhoeffer and G. Werner, *Justus Liebigs Ann. Chem.* **437**, 1955 (1973).
136. D. L. Boger and J. S. Panek, *J. Am. Chem. Soc.* **107**, 5745 (1985).
137. D. L. Boger and R. S. Coleman, *J. Am. Chem. Soc.* **109**, 2717 (1987).
138. M. E. Jung and J. A. Hagenah, *J. Org. Chem.* **52**, 1889 (1987).
139. H. L. Gingrich, D. M. Roush, and W. A. Van Saum, *J. Org. Chem.* **48**, 4869 (1983).

It should be noted that the Diels–Alder reactions of both the nitrogen heterocycles and the pyrones are of the inverse electron demand type.

6.8.3. β Eliminations Involving Cyclic Transition States

Another important family of elimination reactions has as the common mechanistic feature cyclic transition states in which an intramolecular proton transfer accompanies elimination to form a new carbon–carbon double bond. Scheme 6.16 depicts examples of the most important of these reaction types, which normally do not involve acidic or basic catalysts. There is, however, a wide variation in the temperature at which elimination proceeds at a convenient rate. The cyclic transition states dictate that elimination will occur with *syn* stereochemistry. The reactions, as a group, are often referred to as *thermal syn eliminations.*

Scheme 6.16. Eliminations via Cyclic Transition States

	Reactant	Transition state	Product	Temp. range	Ref.
1[a]	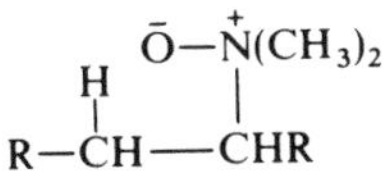 $R-CH(H)-CHR-\overset{+}{N}(CH_3)_2-\bar{O}$	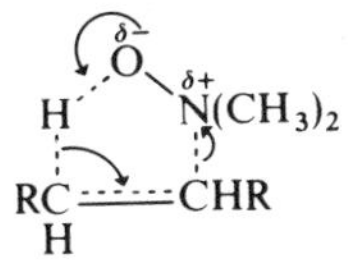 $\delta-$ O ··· H, $\delta+$ $N(CH_3)_2$, RC(H) ⋯ CHR	$RCH{=}CHR$ + $HON(CH_3)_2$	100–150°C	a
2[b]	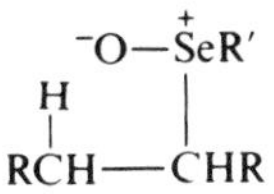 $RCH(H)-CHR-\overset{+}{Se}R'-O^-$	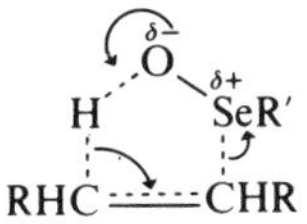 $\delta-$ O ··· H, $\delta+$ SeR', RHC ⋯ CHR	$RCH{=}CHR$ + $HOSeR'$	0–100°C	b
3[c]	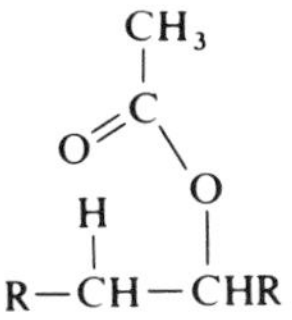 $R-CH(H)-CHR-O-C(=O)CH_3$	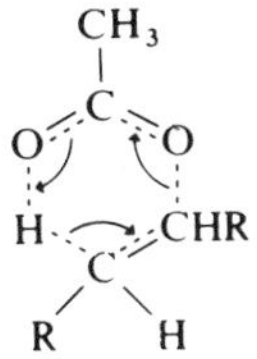 CH_3–C(O···H)(O), H ⋯ C(R)(H) ⋯ CHR	$RCH{=}CHR$ + CH_3CO_2H	400–600°C	c
4[d]	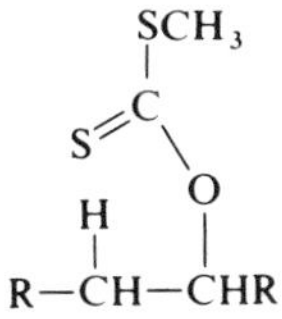 $R-CH(H)-CHR-O-C(=S)SCH_3$	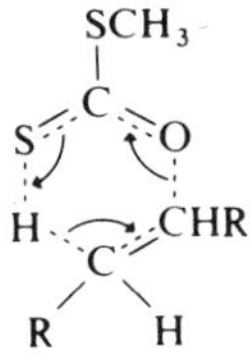 SCH_3–C(S···H)(O), H ⋯ C(R)(H) ⋯ CHR	$RCH{=}CHR$ + CH_3SH + SCO	150–250°C	d

a. A. C. Cope and E. R. Trumbull, *Org. React.* **11**, 317 (1960).
b. D. L. J. Clive, *Tetrahedron* **34**, 1049 (1978).
c. C. H. De Puy and R. W. King, *Chem. Rev.* **60**, 431 (1960).
d. H. R. Nace, *Org. React.* **12**, 57 (1962).

Amine oxide pyrolysis occurs at temperatures of 100–150°C. The reaction can proceed at room temperature in DMSO.[140] If more than one type of β-hydrogen can attain the eclipsed conformation of the cyclic transition state, a mixture of alkenes will be formed. The product ratio should parallel the relative stability of the competing transition states. Usually, more of the E-alkene is formed because of the additional eclipsed interactions present in the transition state leading to the Z-alkene. The selectivity is usually not high, however.

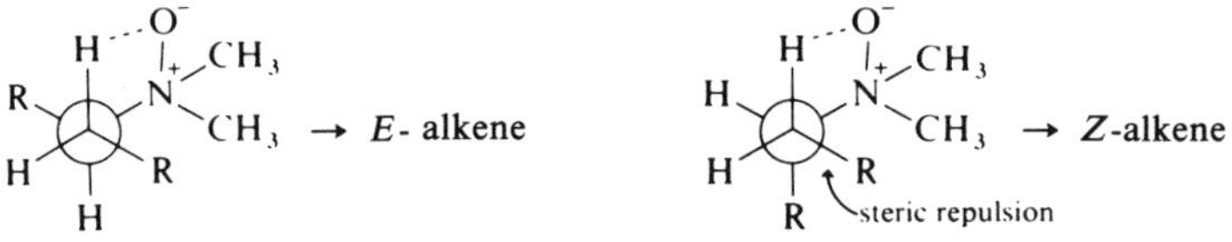

H···O⁻, N⁺(CH₃)₂ → E-alkene
more favorable

H···O⁻, N⁺(CH₃)₂ → Z-alkene; steric repulsion
less favorable

In cyclic systems, conformational effects and the requirement for a cyclic transition state will determine the product composition. Elimination to give a double bond conjugated with an aromatic ring is especially favorable. This presumably reflects both the increased acidity of the proton alpha to the phenyl ring and the stabilizing effect of the developing conjugation in the transition state. Amine oxides can be readily prepared from amines by oxidation with hydrogen peroxide or a peroxycarboxylic acid. Some typical examples are given in section A of Scheme 6.17.

Scheme 6.17. Thermal Eliminations via Cyclic Transition States

A. Amine Oxide Pyrolyses

1[a] $PhCH(CH_3)CH(CH_3)\overset{+}{N}(CH_3)_2O^-$ → $PhC(CH_3)=CHCH_3$ (92%) + $PhCH(CH_3)CH=CH_2$ (8%)

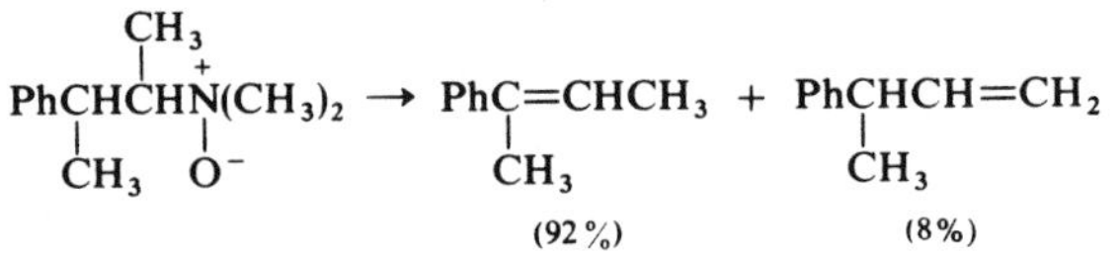

2[b] trans-2-phenylcyclohexyl-$\overset{+}{N}(CH_3)_2O^-$ → 1-phenylcyclohexene (85%) + 3-phenylcyclohexene (15%)

3[c] cis-2-phenylcyclohexyl-$\overset{+}{N}(CH_3)_2O^-$ → 1-phenylcyclohexene (2%) + 3-phenylcyclohexene (98%)

4[d] cyclohexyl-$CH_2\overset{+}{N}(CH_3)_2O^-$ → methylenecyclohexane ($=CH_2$) (85%)

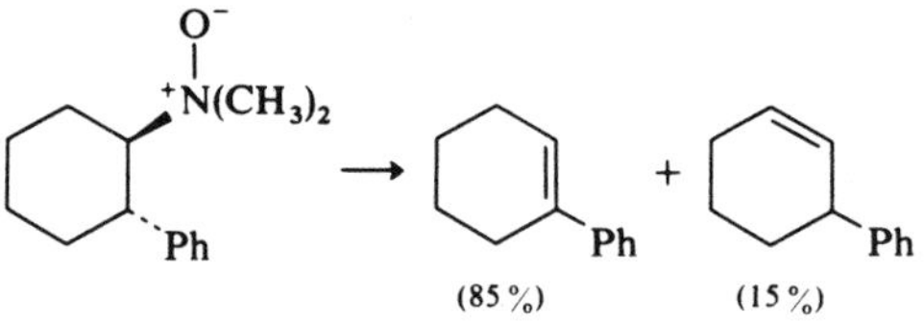

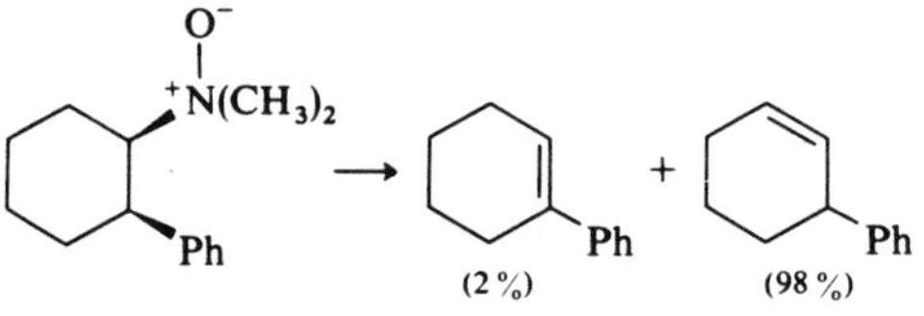

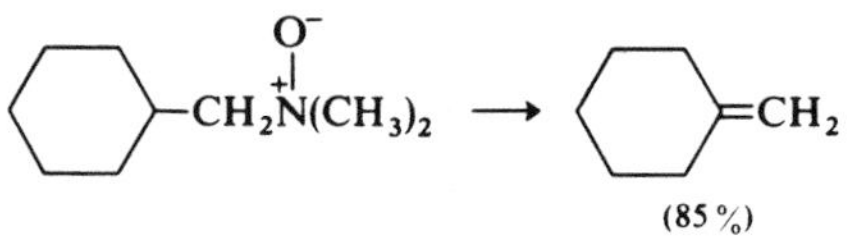

140. D. J. Cram, M. R. V. Sahyun, and G. R. Knox, *J. Am. Chem. Soc.* **84**, 1734 (1962).

5[e] $\rightarrow$ (67%)

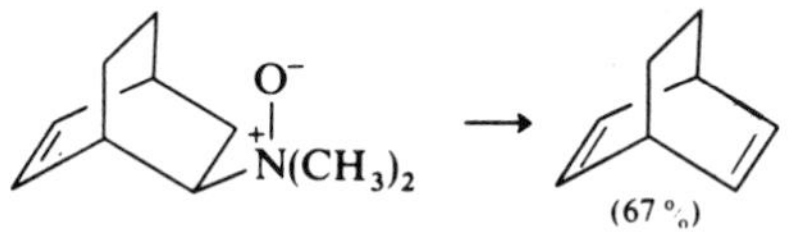

6[f] $CH_3(CH_2)_6CH_2\overset{+}{N}(CH_3)_2(O^-) \longrightarrow CH_3(CH_2)_5CH{=}CH_2$ (87%)

B. Selenoxide Elimination

7[g] $CH_2CH_2OSO_2Ar$ $\xrightarrow{\text{1) PhSe}^-\text{; 2) O}_3}$ $CH_2CH_2\overset{O}{\overset{\|}{Se}}Ph$ $\xrightarrow{\text{77°C, CCl}_4\text{, 10 min}}$ $CH{=}CH_2$ (60%)

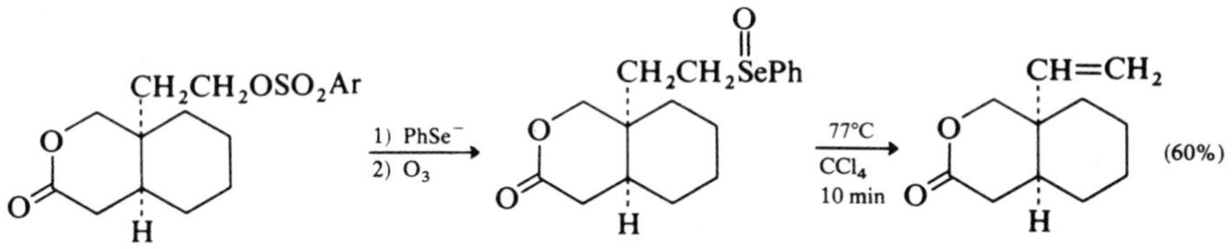

8[h] $\xrightarrow{\text{1) PhSe}^-\text{; 2) O}_3}$ CO_2H, $CH_2\overset{O}{\overset{\|}{Se}}Ph$ $\xrightarrow{\text{3) pyridine; 4) H}^+}$ CO_2H, CH_2

9[i] CH_2CO_2H $\xrightarrow{\text{PhSeCl, Et}_3\text{N; CH}_2\text{Cl}_2}$ PhSe (93%) $\xrightarrow{\text{H}_2\text{O}_2\text{; THF}}$ (92%)

10[j] OH, H $\xrightarrow{\text{1) (o-NO}_2\text{C}_6\text{H}_4\text{)SeCN, Ph}_3\text{P; 2) O}_3}$ $O{=}Se$, O_2N $\xrightarrow{\text{DBU}}$ (5%)

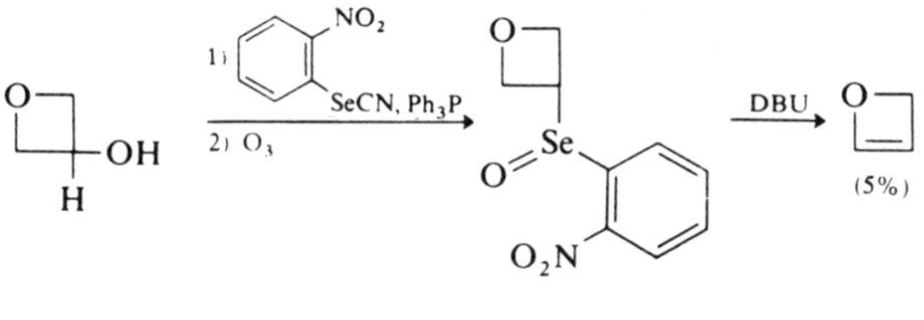

C. Acetate Pyrolyses

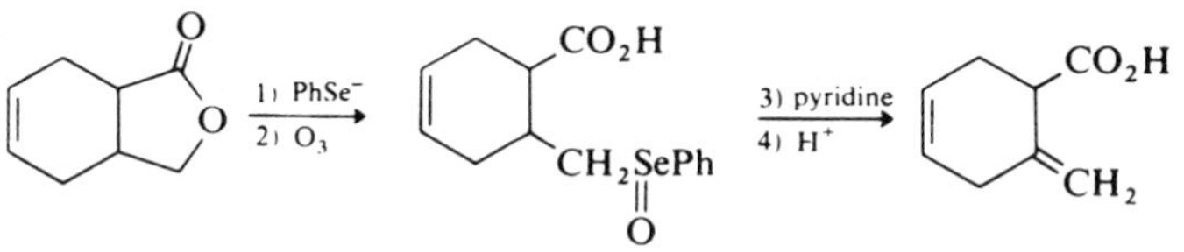

11[k] $N{\equiv}C{-}C_6H_4{-}CH(O_2CCH_3)CH_3 \xrightarrow{575\text{–}600°C} N{\equiv}C{-}C_6H_4{-}CH{=}CH_2$ (76%)

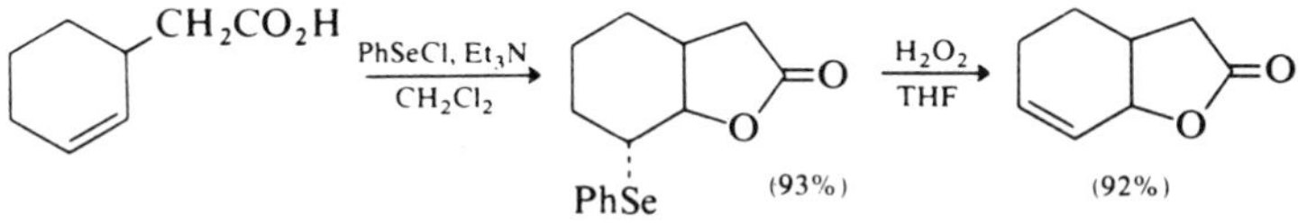

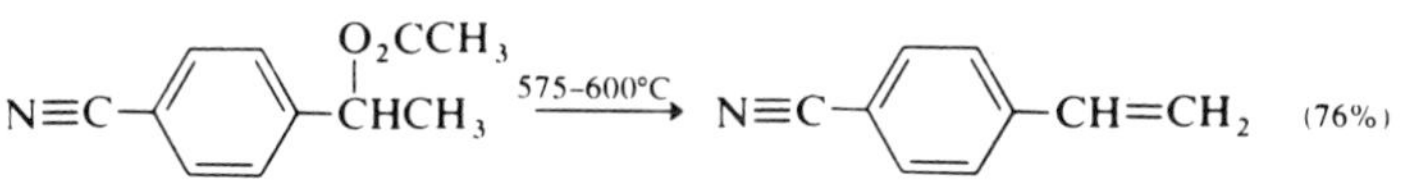

12[l] $CH_2O_2CCH_3$, $CH_2O_2CCH_3$ $\xrightarrow{555°C}$ CH_2, CH_2 (61%) + CH_2, $CH_2O_2CCH_3$

13[m] $CH_2CH_2O_2CCH_3$, N, NH $\xrightarrow{500°C}$ $CH{=}CH_2$, N, NH (50%)

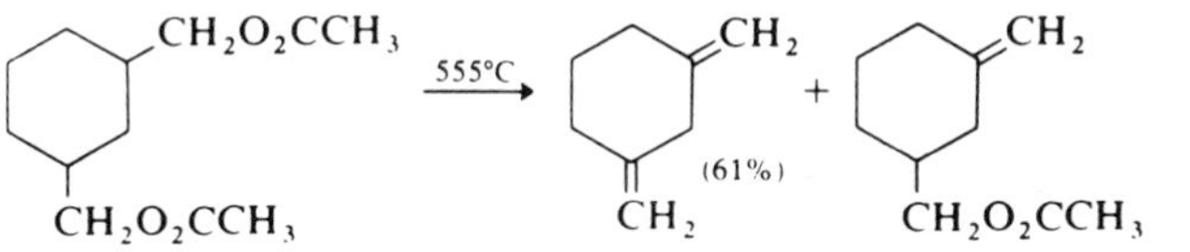

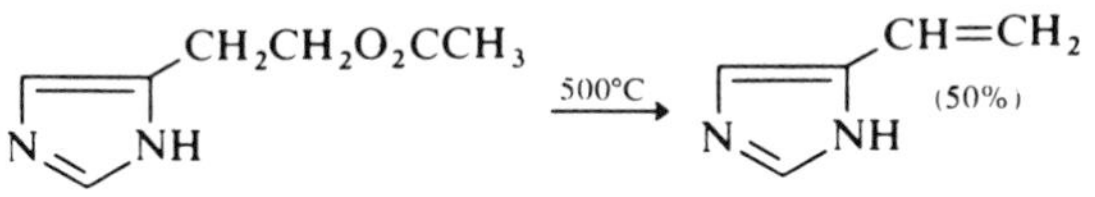

Scheme 6.17—*continued*

14[n] CH_3, O, CH_3, CH_3, $CH_2O_2CCH_3$ $\xrightarrow{400°C}$ CH_3, O, CH_3, CH_3, CH_2

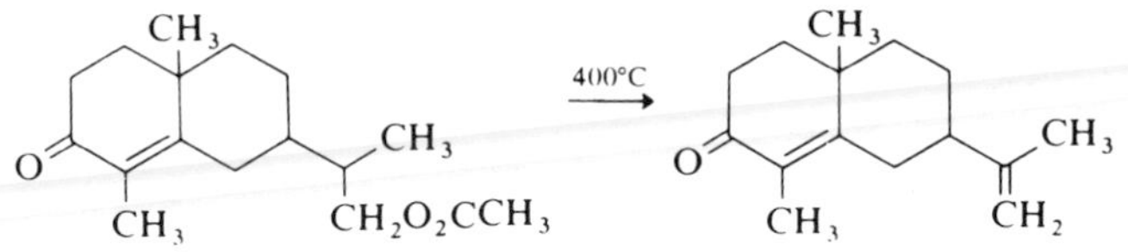

D. Xanthate Ester Pyrolyses

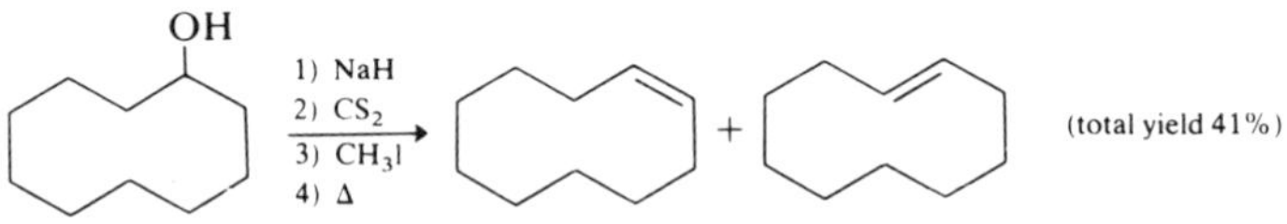

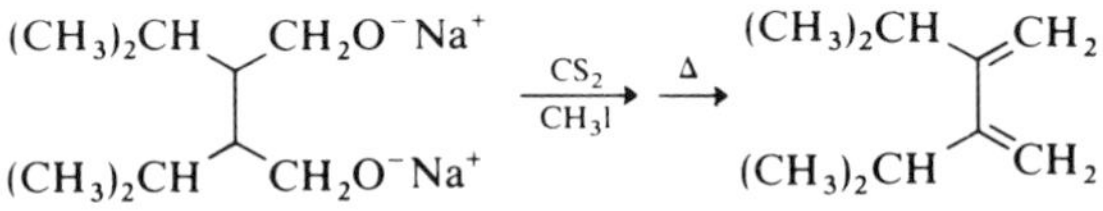

15[o] $PhCH(CH_3)CH(OH)CH_3$ $\xrightarrow{1)\ K;\ 2)\ CS_2;\ 3)\ CH_3I;\ 4)\ \Delta}$ $PhC(CH_3){=}CHCH_3$ (91%)

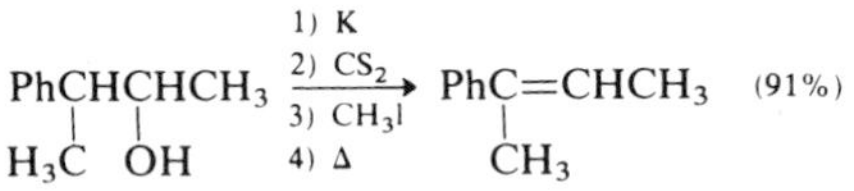

16[p] OH (cyclodecanol) $\xrightarrow{1)\ NaH;\ 2)\ CS_2;\ 3)\ CH_3I;\ 4)\ \Delta}$ (cyclodecene) + (cyclodecene) (total yield 41%)

17[q] $(CH_3)_2CH$, $CH_2O^-Na^+$, $(CH_3)_2CH$, $CH_2O^-Na^+$ $\xrightarrow[CH_3I]{CS_2}$ $\xrightarrow{\Delta}$ $(CH_3)_2CH$, CH_2, $(CH_3)_2CH$, CH_2

18[r] OH $\xrightarrow{1)\ NaH;\ 2)\ CS_2;\ 3)\ CH_3I;\ 4)\ \Delta}$ (71%)

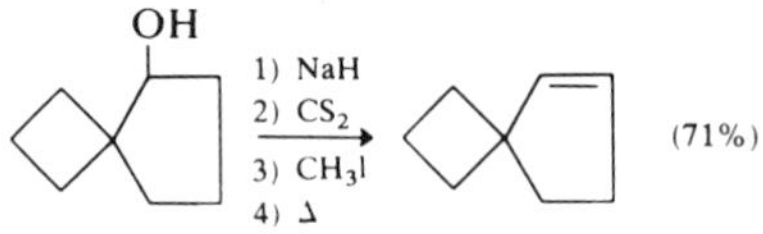

a. D. J. Cram and J. E. McCarty, *J. Am. Chem. Soc.* **76**, 5740 (1954).
b. A. C. Cope and C. L. Bumgardner, *J. Am. Chem. Soc.* **79**, 960 (1957).
c. A. C. Cope, C. L. Bumgardner, and E. C. Schweizer, *J. Am. Chem. Soc.* **79**, 4729 (1957).
d. A. C. Cope, E. Ciganek, and N. A. LeBel, *J. Am. Chem. Soc.* **81**, 2799 (1959).
e. A. C. Cope and C. L. Bumgardner, *J. Am. Chem. Soc.* **78**, 2812 (1956).
f. J. I. Roberts, P. S. Borromeo, and C. D. Poulter, *Tetrahedron Lett.* 1299 (1977).
g. R. D. Clark and C. H. Heathcock, *J. Org. Chem.* **41**, 1396 (1976).
h. D. Liotta and H. Santiesteban, *Tetrahedron Lett.* 4369 (1977); R. M. Scarborough, Jr. and A. B. Smith, III, *Tetrahedron Lett.* 4361 (1977).
i. K. C. Nicolaou and Z. Lysenko, *J. Am. Chem. Soc.* **99**, 3185 (1977).
j. L. E. Friedrich and P. Y. S. Lam, *J. Org. Chem.* **46**, 306 (1981).
k. C. G. Overberger and R. E. Allen, *J. Am. Chem. Soc.* **68**, 722 (1946).
l. W. J. Bailey and J. Economy, *J. Org. Chem.* **23**, 1002 (1958).
m. C. G. Overberger and N. Vorchheimer, *J. Am. Chem. Soc.* **85**, 951 (1963).
n. E. Piers and K. F. Cheng, *Can. J. Chem.* **46**, 377 (1968).
o. D. J. Cram, *J. Am. Chem. Soc.* **71**, 3883 (1949).
p. A. T. Blomquist and A. Goldstein, *J. Am. Chem. Soc.* **77**, 1001 (1955).
q. A. de Groot, B. Evenhuis, and H. Wynberg, *J. Org. Chem.* **33**, 2214 (1968).
r. C. F. Wilcox, Jr. and G. C. Whitney, *J. Org. Chem.* **32**, 2933 (1967).

Selenoxides are even more reactive than amine oxides toward β elimination. In fact, many selenoxides react spontaneously when generated at room temperature. Synthetic procedures based on selenoxide eliminations usually involve synthesis of the corresponding selenide followed by oxidation and *in situ* elimination. We have already discussed examples of these procedures in Section 4.7, where the conversion of ketones and esters to their α,β-unsaturated derivatives was considered. Selenides can also be prepared by electrophilic addition of selenenyl halides and related compounds to alkenes (see section 4.5). Selenide anions are powerful nucleophiles and can displace halides or tosylates and open epoxides.[141] Selenide substituents stabilize an adjacent carbanion so that α-selenenyl carbanions can be prepared. One versatile procedure involves conversion of a ketone to a bis-selenoketal, which can then be cleaved by n-butyllithium.[142] The carbanions in turn add to aldehydes or ketones to give β-hydroxyselenides.[143] Elimination gives an allylic alcohol.

$$RCH_2C(=O)R' + 2PhSeH \rightarrow RCH_2C(SePh)_2R' \xrightarrow{BuLi} RCH_2-C(Li)(SePh)R' \xrightarrow{R''CH=O} RCH_2C(R')(SePh)-CH(OH)R''$$

$$\xrightarrow{[o]} RCH=C(R')-CH(OH)R''$$

Alcohols can be converted to o-nitrophenyl selenides by reaction with o-nitrophenyl selenocyanate and tri-n-butylphosphine.[144]

$$RCH_2OH + o\text{-}O_2NC_6H_4SeCN \xrightarrow{Bu_3P} RCH_2Se\text{-}C_6H_4\text{-}o\text{-}NO_2$$

The selenides prepared by any of these methods can be converted to selenoxides by such oxidants as hydrogen peroxide, sodium metaperiodate, peroxycarboxylic acids, t-butyl hydroperoxide, or ozone.

Like amine oxide elimination, selenoxide eliminations normally favor formation of the E-isomer in acyclic structures. In cyclic systems, the stereochemical requirements of the cyclic transition state govern the product structure. Section B of Scheme 6.17 gives some examples of selenoxide eliminations.

A third category of *syn* eliminations involves pyrolytic decomposition of esters with elimination of a carboxylic acid. The pyrolysis is usually done with acetate esters, and temperatures above 400°C are normally required. The pyrolysis is therefore usually a vapor phase reaction. In the laboratory, this can be carried out by using a glass tube in the heating zone of a small furnace. The vapors of the reactant are swept through the hot chamber by an inert gas and into a cold trap. Similar reactions occur with esters derived from long-chain acids. If the boiling point of the ester is above the decomposition temperature, the reaction can be carried out in the liquid phase.

141. D. L. J. Clive, *Tetrahedron* **34**, 1049 (1978).
142. W. Dumont, P. Bayet, and A. Krief, *Angew. Chem. Int. Ed. Engl.* **13**, 804 (1974).
143. D. Van Ende, W. Dumont, and A. Krief, *Angew. Chem. Int. Ed. Engl.* **14**, 700 (1975); W. Dumont and A. Krief, *Angew. Chem. Int. Ed. Engl.* **14**, 350 (1975).
144. P. A. Grieco, S. Gilman, and M. Nishizawa, *J. Org. Chem.* **41**, 1485 (1976).

Ester pyrolysis has been shown to be a *syn* elimination by use of deuterium labels. In the case of stilbene formation, deuterium was introduced by stereospecific reduction of *cis*- and *trans*-stilbene oxide by $LiAlD_4$. The *syn* elimination is demonstrated by retention of deuterium in the product from the *trans* epoxide and its elimination in the product from the *cis* epoxide.[145]

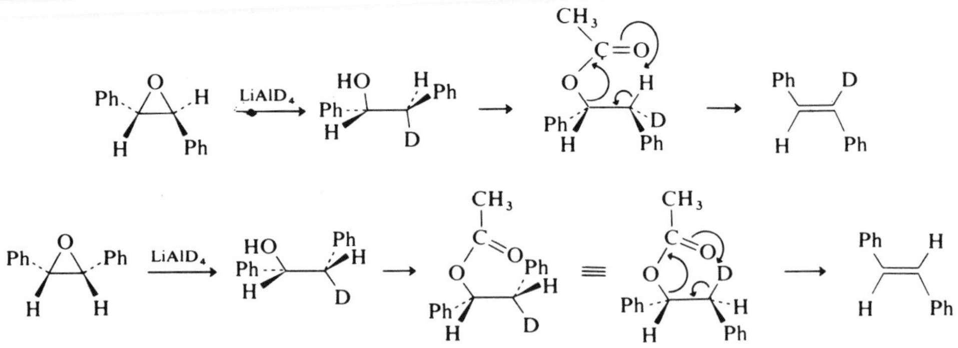

An alternative view of the mechanism has been presented. Although the existence of the concerted cyclic mechanism is not disputed, it has been proposed that most preparative pyrolyses proceed as surface-catalyzed reactions.[146]

Mixtures of alkenes are formed when more than one type of β-hydrogen is present. In acyclic compounds, the product composition often approaches that expected on a statistical basis from the number of each type of hydrogen. The E-alkene usually predominates over the Z-alkene for a given isomeric pair. In cyclic structures, elimination is in the direction in which the cyclic mechanism can operate most favorably.

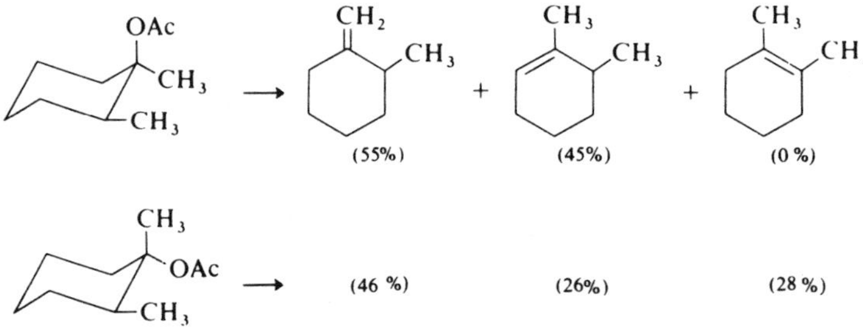

(55%) (45%) (0 %) Ref. 147

(46 %) (26%) (28 %)

Alcohols can be dehydrated via xanthate esters at temperatures that are much lower than those required for acetate pyrolysis. The preparation of xanthate esters involves reaction of the alkoxide with carbon disulfide. The resulting salt is alkylated with methyl iodide.

$$RO^-Na^+ + CS_2 \rightarrow RO\overset{S}{\overset{\|}{C}}S^-Na^+ \xrightarrow{CH_3I} RO\overset{S}{\overset{\|}{C}}SCH_3$$

145. D. Y. Curtin and D. B. Kellom, *J. Am. Chem. Soc.* **75**, 6011 (1953).
146. D. H. Wertz and N. L. Allinger, *J. Org. Chem.* **42**, 698 (1977).
147. D. H. Froemsdorf, C. H. Collins, G. S. Hammond, and C. H. DePuy, *J. Am. Chem. Soc.* **81**, 643 (1959).

The elimination is often effected simply by distillation:

$$R-CH(H)-C(H)(R)-O-C(=S)-SCH_3 \xrightarrow{\Delta} RCH{=}CHR + \left[HSC(=O)SCH_3\right] \rightarrow CH_3SH + COS$$

Product mixtures are observed when more than one type of β-hydrogen can participate in the reaction. As with the other *syn* thermal eliminations, there are no intermediates that are prone to skeletal rearrangement.

General References

A. P. Marchand and R. E. Lehr (eds.), *Pericyclic Reactions*, Vols. I and II, Academic Press, New York, 1977.

R. B. Woodward and R. Hoffmann, *The Conservation of Orbital Symmetry*, Academic Press, New York, 1970.

Diels–Alder Reactions

G. Brieger and J. N. Bennett, *Chem. Rev.* **80**, 63 (1980).

E. Ciganek, *Org. React.* **32**, 1 (1984).

W. Oppolzer, *Angew. Chem. Int. Ed. Engl.* **23**, 876 (1984).

W. Oppolzer, *Synthesis*, 793 (1978).

Cycloaddition Reactions

A. Padwa (ed.), *1,3-Dipolar Cycloaddition Chemistry*, Wiley, New York, 1984.

Sigmatropic Rearrangements

E. Block, *Reactions of Organosulfur Compounds*, Academic Press, New York, 1978, Chapter 7.

H.-J. Hansen, in *Mechanisms of Molecular Migrations*, Vol. 3, B. S. Thyagarajan (ed.), Wiley-Interscience, New York, 1971, pp. 177–236.

R. K. Hill, in *Asymmetric Synthesis*, Vol. 3, J. D. Morrison (ed.), Academic Press, New York, 1984, Chapter 8.

S. J. Rhoads and N. R. Raulins, *Org. React.* **22**, 1 (1975).

T. S. Stevens and W. E. Watts, *Selected Molecular Rearrangements*, Van Nostrand Reinhold, London, 1973, Chapter 8.

B. M. Trost and L. S. Melvin, Jr., *Sulfur Ylides*, Academic Press, New York, 1975, Chapter 7.

Elimination Reactions

W. H. Saunders, Jr., and A. F. Cockerill, *Mechanisms of Elimination Reactions*, Wiley, New York, 1973, Chapter VIII.

Problems

(*References for these problems will be found on page* 770.)

1. Predict the product of each of the following reactions, clearly showing stereochemistry where appropriate.

(a) OAc, CH_2CH_3 + CH_2=CHCHO $\xrightarrow[\text{toluene, }-10°C]{BF_3\cdot Et_2O}$

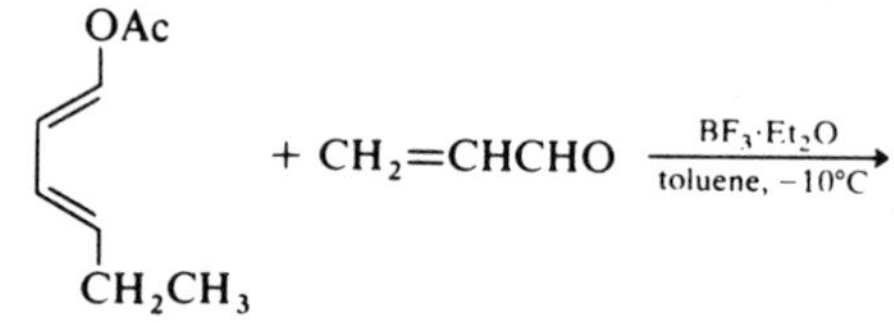

(b) $OSiMe_3$, CH_3, CH_3 + CH_2=CHCHO $\longrightarrow$

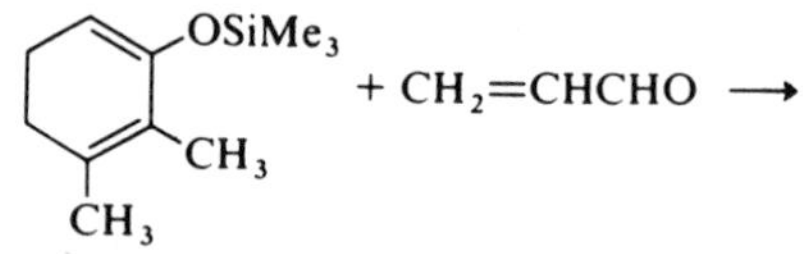

(c) $NHCO_2C_2H_5$ + (*E*)-CH_3CH=CHCHO $\xrightarrow{110°C}$

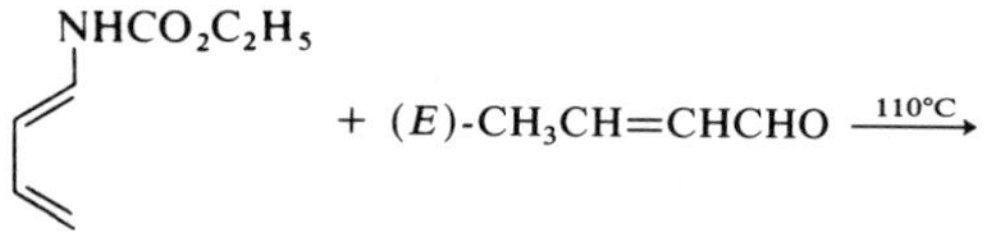

(d) H_2C=CH, H, H $\xrightarrow{60°C}$

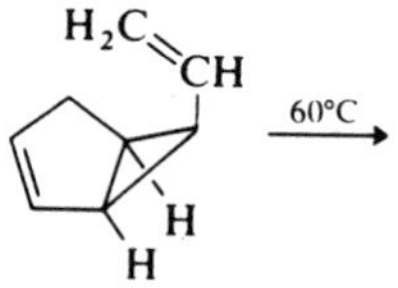

(e) CH_3, CH_3, CH_3, H, O, O $\xrightarrow{200°C}$

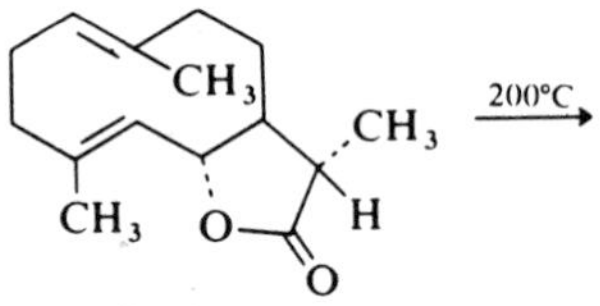

(f) H_3C, O, C, CH_3; C, H, CO_2CH_3 $\xrightarrow{230°C}$

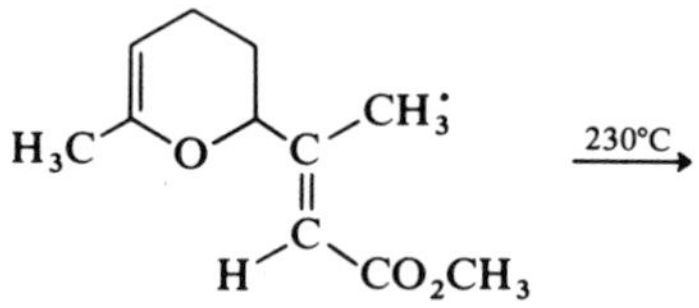

(g) $CH_3\overset{O}{\overset{\|}{C}}CH_2CH_2CH_2CH$=$CH_2$ + $CH_3NHOH\cdot HCl$ $\rightarrow$

(h) OH, H_3C, CH_3 $\xrightarrow[\text{2) 210°C}]{\text{1) } C_2H_5OCH=CH_2\text{, } Hg^{2+}}$

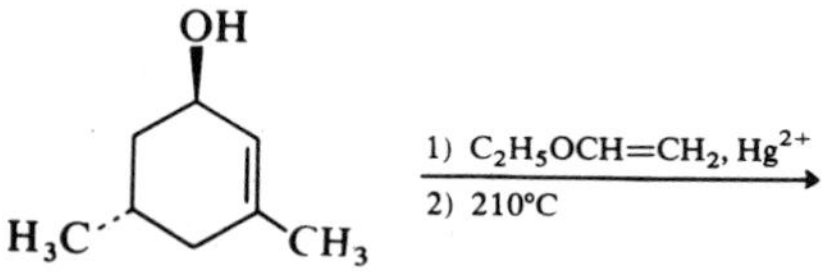

(i) OH, $-\overset{|}{C}CH_2CH$=CH_2, CH_3 $\xrightarrow[\text{dimethoxyethane, 80°C}]{KH}$

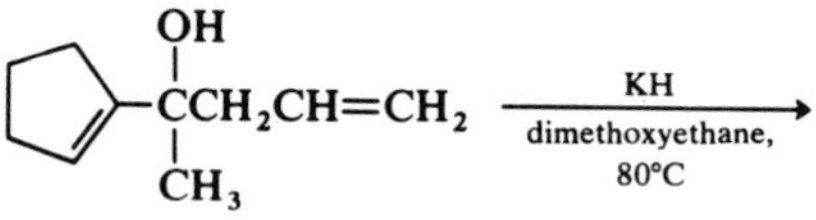

(j) [structure: 2,4-dimethyl-3-($CH_2CH_2CH_2C(CH_3)=CH_2$)cyclohex-2-enone] $\xrightarrow{h\nu}$

CH_3 $CH_2CH_2CH_2C{=}CH_2$ CH_3 CH_3

(k) $C_6H_5CH(SeCH_3)_2 \xrightarrow[\text{2) 1,2-epoxybutane; 3) } H_2O_2]{\text{1) } n\text{-BuLi}}$

(l) [cyclonona-diene with H, OH] $\xrightarrow[25°C]{KH, THF}$

(m) HO H; CH_3 C; C=C $CH_2CH(CH_3)_2$; H H $\xrightarrow[\Delta]{(CH_3)_2NC(CH_3)(OCH_3)_2}$

(n) CH_3; CH_3; O $\xrightarrow{100°C}$

(o) CH_3; CH_3; O $\xrightarrow{Li^+\ ^-N(\text{2,2,6,6-tetramethylpiperidide})}$

(p) CH_3OCH_2O; H_3C; O; O; H $\xrightarrow[\text{3) } 105°C]{\text{1) LDA, HMPA; 2) } t\text{-BuMe}_2\text{SiCl}}$

(q) H_3C CH_3; O O; O; $CH_2{=}CH(CH_2)_3$ $\xrightarrow{h\nu}$

(r) O OMe CH_2Br; CH_2Br; O OMe $+ \ CH_2{=}CHCCH_3$ (C=O) $\xrightarrow{Zn}$

(s) $CH_2{=}C(CH_3)CH_2CH_2CH_2OCH_2CO_2H \xrightarrow[\text{2) } Et_3N]{\text{1) ClCOCOCl}}$

2. Intramolecular cycloaddition reactions occur under the reaction conditions specified for each of the following reactants. Show the structure of the product, including all aspects of its stereochemistry, and indicate the structures of any intermediates which are involved in the reactions.

(a) NHOH, H, H, CH_3 $\xrightarrow{CH_2=O}$

(b) C_6H_5, N, CH_3, $CH_2CH_2CH_2CH=CH_2$ $\xrightarrow{h\nu}$

(c) O=, H N, N H, S, S, $CH_2\overset{O}{\overset{\|}{C}}NHNHCH_3$ $\xrightarrow[80°C,\ 1\ hr]{C_6H_5CH=O}$

(d) Ph, Ph, H, H, H, $CCH_2N\overset{O}{\overset{\|}{C}}$, CH_3, H, H, Ph $\xrightarrow{90°C}$

(e) CH_3O, CN, $(CH_2)_4CH=CH_2$ $\xrightarrow{\Delta}$

(f) $CH_2=CH$, H, H, H, $CH(CH_3)_2$, C=C, C=C, H, $CH_2CH_2CH_2C$, O, H $\xrightarrow{\Delta}$

(g) CH_3, O, $CH_2=CCH_2CH_2C$, AcO $\xrightarrow{h\nu}$

3. Indicate the mechanistic type to which each of the following reactions belongs.

(a) $(CH_3)_2C=CHN(CH_3)_2$ + H, NO_2, C=C, C_6H_5, H $\longrightarrow$ CH_3, C_6H_5, H_3C, $(CH_3)_2N$, NO_2

(b) $CH_3CH{=}CHCH_2Br$ + $\xrightarrow[\text{2) } K_2CO_3]{\text{1) } CH_3SCH_2\overset{O}{\overset{\|}{C}}Ph}$ $CH_3CHCH{=}CH_2$ | $CH_3SCHCPh$ ‖ O

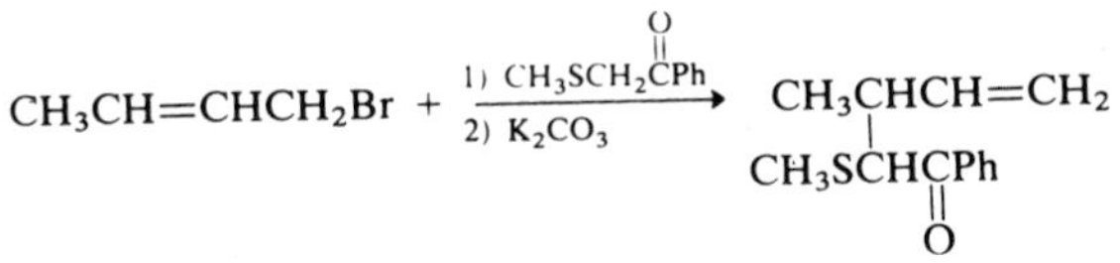

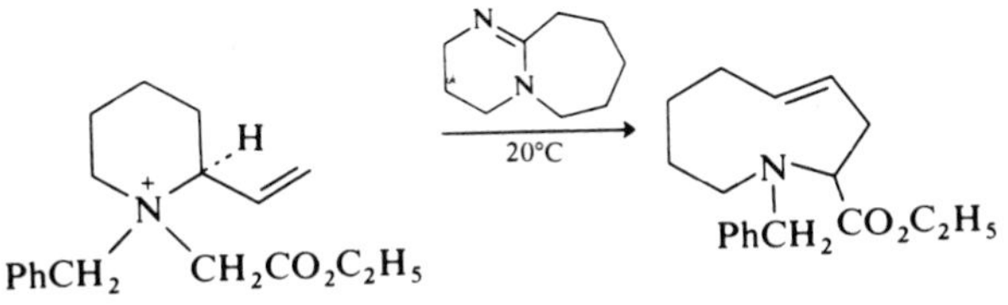

(c) H, $\overset{+}{N}$, $PhCH_2$, $CH_2CO_2C_2H_5$ —N, N (amidine base), 20°C→ N, $PhCH_2$, $CO_2C_2H_5$

(d) $\overset{+}{N}$, O^- + CO_2CH_3, CH_3SO_2O → CO_2CH_3, $CH_2CH_2OSO_2CH_3$, N–O

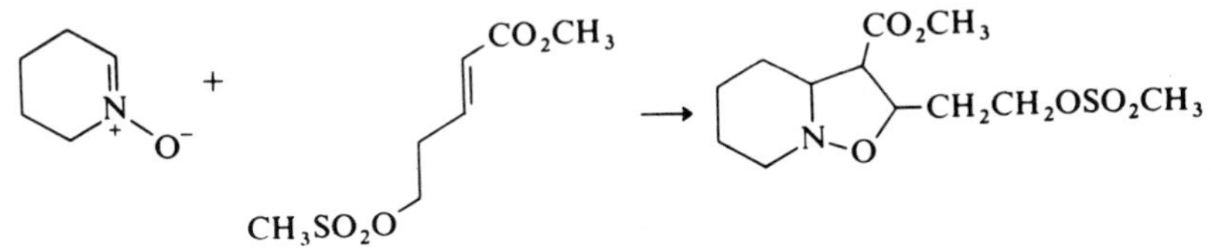

(e) $CH_2{=}CHCH_2CH_3$ + $C_2H_5O_2CN{=}NCO_2C_2H_5$ → $CH_3CH{=}CHCH_2NNHCO_2C_2H_5$ | $CO_2C_2H_5$

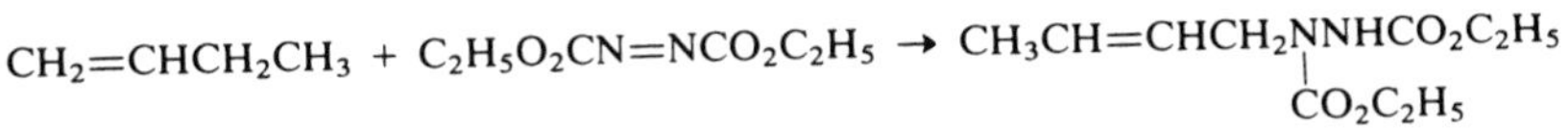

(f) $(CH_3)_2C{=}CHCH_2CH_2\underset{CH_3}{\underset{|}{C}}HCH_2CH_2CO_2CH_3$ + $HC{\equiv}CCO_2CH_3$

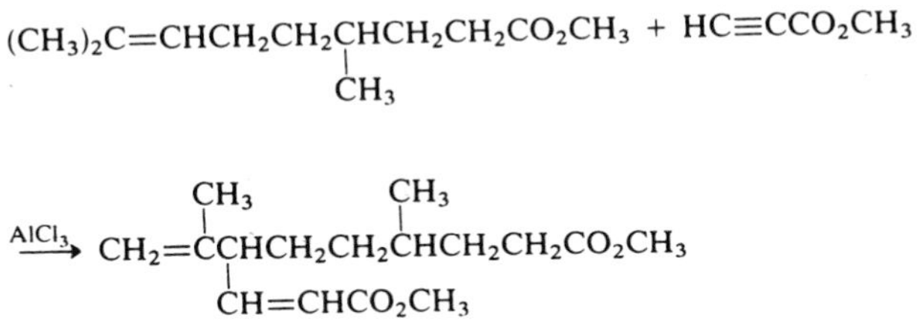

$\xrightarrow{AlCl_3}$ $CH_2{=}\overset{CH_3}{\overset{|}{C}}\underset{CH{=}CHCO_2CH_3}{\underset{|}{C}}HCH_2CH_2\overset{CH_3}{\overset{|}{C}}HCH_2CH_2CO_2CH_3$

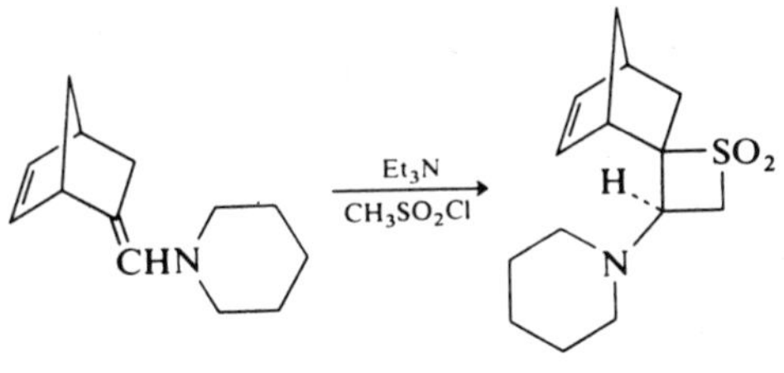

(g) CHN $\xrightarrow[CH_3SO_2Cl]{Et_3N}$ SO_2, H, N

(h) PhN_3 + $NCH{=}CHPh$ → Ph, N, N, N, N, Ph

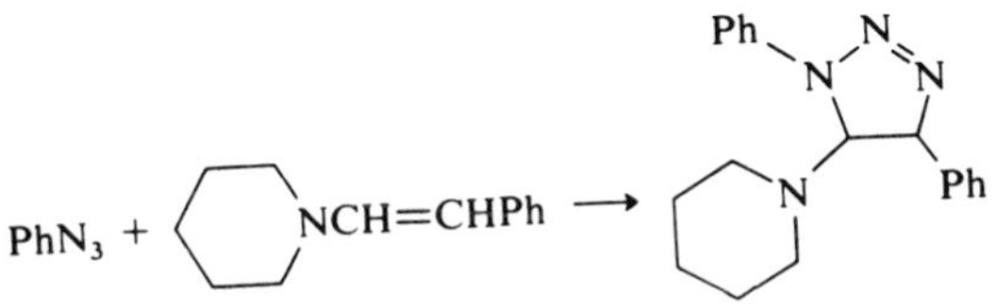

(i) $(CH_3)_2CH\underset{CN}{\underset{|}{C}}HOCH_2CH{=}C(CH_3)_2$ $\xrightarrow[\text{2) } H_3O^+]{\text{1) } LiNR_2}$ $(CH_3)_2CH\overset{O}{\overset{\|}{C}}{-}\overset{CH_3}{\underset{CH_3}{C}}CH{=}CH_2$

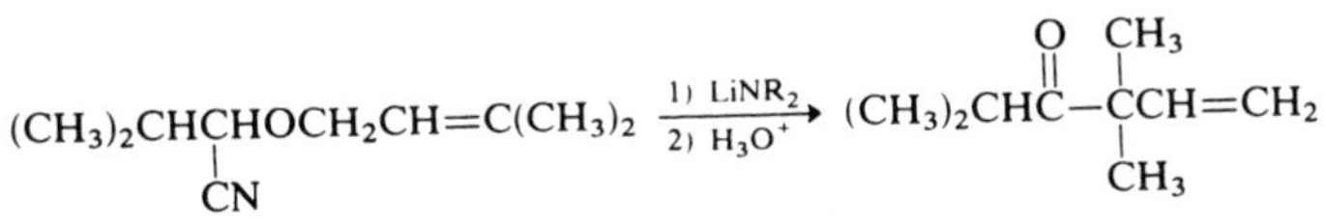

(j) O, O, CO_2CH_3 + $CH_2{=}C(OCH_3)_2$ $\xrightarrow{110°C}$ OCH_3, CO_2CH_3

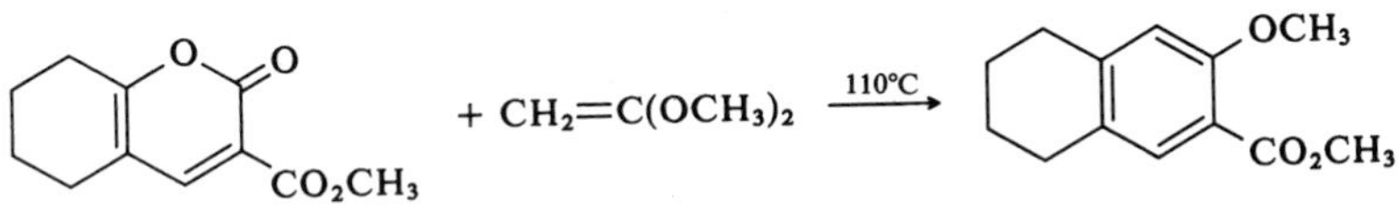

4. By applying the principles of retrosynthetic analysis, show how each of the indicated target molecules could be prepared from the starting material(s) given. No more than three separate transformations are necessary in any of the syntheses.

(a) CH_3O OCH_3, Cl, Cl, Cl, Cl ⇨ Cl Cl Cl Cl CH_3O OCH_3 + cyclooctatetraene

(b) O CH_3 CO_2CH_3 CH_3 CO_2CH_3 ⇨ H_3C O CH_3 + dimethyl acetylenedicarboxylate

(c) $N(C_2H_5)_2$ C(O)Ph C(O)Ph ⇨ crotonaldehyde, diethylamine, and *trans*-1,2-dibenzoylethylene

(d) $CH_3CH_2C{\equiv}CCH{=}CHCH_2CH_2CO_2CH_3$ ⇨

$CH_3CH_2C{\equiv}CH$ + $CH_2{=}CHCHO$ + $CH_3C(OCH_3)_3$

(e) OH HO HO $CH_2CH{=}CH_2$ ⇨ OH O O=C O + $H_2C{=}CHCH_2Br$

(f) H_3C H_3C H Ph $CO_2C_2H_5$ $CO_2C_2H_5$ Ph H ⇨ *trans*-stilbene, diethyl malonate, and acetone

(g) CO_2CH_3 H_3C CO_2CH_3 H_3C CH_3 ⇨ CH_3 CHO H_3C CH_3 and any other necessary reagents

(h) NO_2 O CO_2CH_3 ⇨ (E)-$O_2NCH{=}CHCO_2CH_3$ and any other necessary reagents

(i)  [structure: 3,3-dimethylcyclopentenyl diene with $CH_2CH_2CO_2C_2H_5$] ⇒ [structure: 4,4-dimethylcyclopentene-1-carbaldehyde, H_3C, H_3C, CHO] and any other necessary reagents

5. Reaction of α-pyrone (**A**) with methyl acrylate at reflux for extended periods gives a mixture of stereoisomers of **B**. Account for the formation of this product.

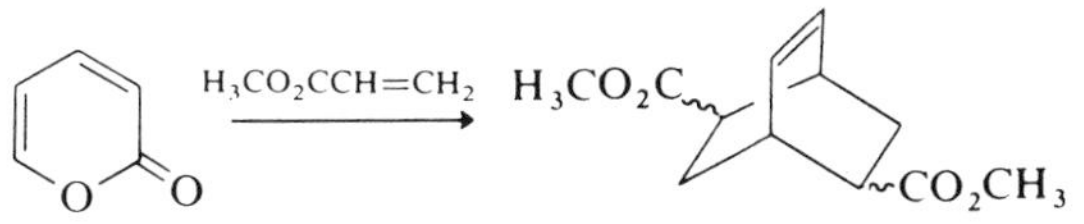

A $\xrightarrow{H_3CO_2CCH=CH_2}$ **B** (H_3CO_2C, CO_2CH_3)

6. When 2-methylpropene and acrolein are heated at 300°C under pressure, 3-methylenecyclohexanol and 6,6-dimethyldihydropyran are formed. Explain the formation of these products.

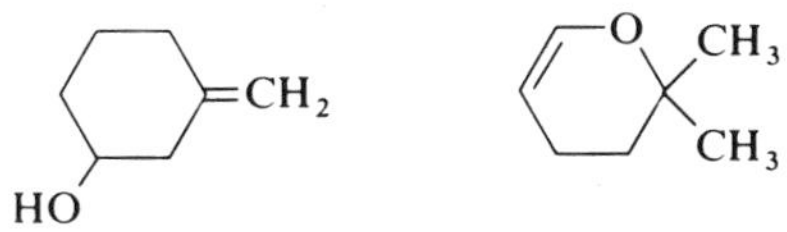

3-methylenecyclohexanol (HO, $=CH_2$)   6,6-dimethyldihydropyran (O, CH_3, CH_3)

7. Vinylcyclopropane, when irradiated with benzophenone or benzaldehyde, gives a mixture of two types of products. Suggest the mechanism by which product of type **C** is formed.

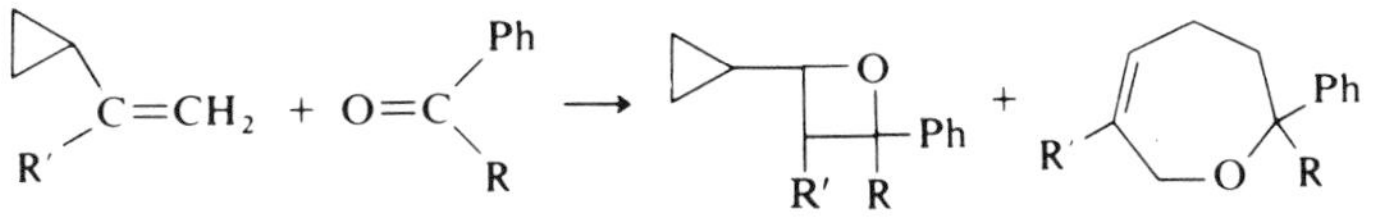

$$\text{(cyclopropyl)(R')C=CH}_2 + \text{O=C(Ph)(R)} \longrightarrow \text{oxetane (Ph, R', R)} + \textbf{C} \text{ (R', Ph, R)}$$

8. The addition reaction of tetracyanoethylene and ethyl vinyl ether in acetone gives 94% of the 2 + 2 adduct and 6% of an adduct having the composition: tetracyanoethylene + ethyl vinyl ether + acetone. If the 2 + 2 adduct is kept in contact with acetone for several days, it is completely converted to the minor product. Suggest a structure for this product, and indicate its mode of formation (a) in the initial reaction and (b) on standing in acetone.

9. A convenient preparation of 2-allycyclohexanone involves simply heating the diallyl ketal of cyclohexanone in toluene containing a trace of p-toluenesulfonic acid and collecting a distillate consisting of toluene and allyl alcohol. Distillation of the residue gives a 90% yield of 2-allylcyclohexanone. Outline the mechanism of this reaction.

10. The preparation of a key intermediate in an imaginative synthesis of prephenic acid is depicted below. Write a series of equations showing the important steps and intermediates in this process. Indicate the reagents required to bring about the desired transformations where other than thermal reactions are involved.

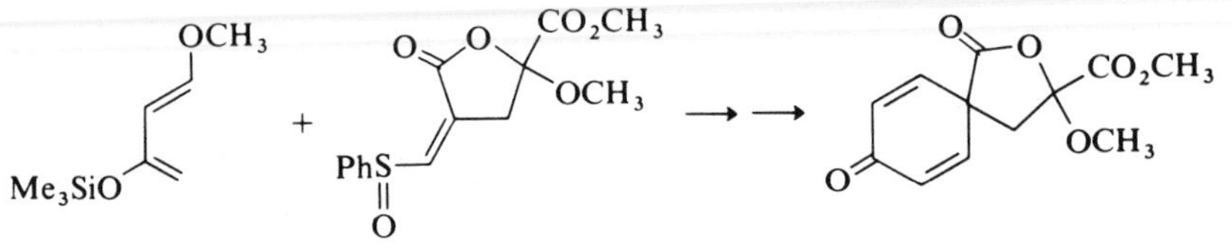

OCH_3 ... Me_3SiO + PhS(=O) ... CO_2CH_3, OCH_3 → → ... CO_2CH_3, OCH_3

11. A route to hasubanan alkaloids has been described involving reaction of 1-butadienyl phenyl sulfoxide with the tetrahydrobenzindole **D**. Treatment of the resulting adduct with sodium sulfide in refluxing methanol gave **F**. Suggest a structure for **E**, and rationalize the formation of **F** from **E**.

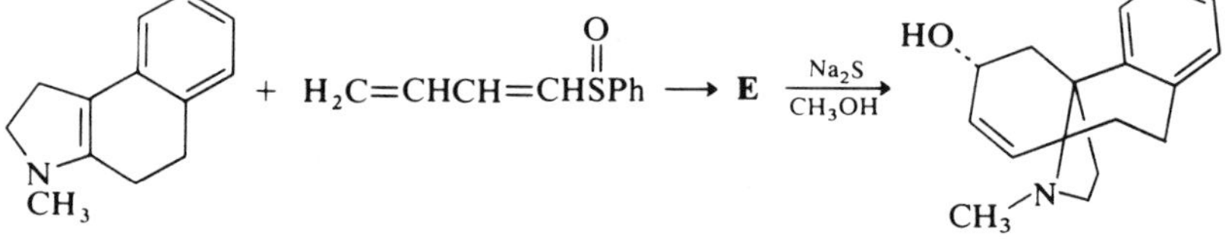

D (N–CH_3) + $H_2C{=}CHCH{=}CH\overset{O}{\overset{\|}{S}}Ph$ → **E** $\xrightarrow[CH_3OH]{Na_2S}$ **F** (HO, CH_3–N)

12. A solution of 2-butenal, 2-acetoxypropene, and dimethyl acetylenedicarboxylate refluxed in the presence of a small amount of an acidic catalyst gives an 80% yield of dimethyl phthalate. Explain the course of this reaction.

13. Irradiation of the dienone shown generates three isomeric saturated ketones, all of which contain cyclobutane rings. Postulate reasonable structures.

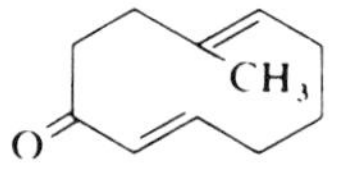

(CH_3, O)

14. Irradiation of *o*-methylbenzaldehyde in the presence of maleic anhydride gives **G**. The same compound is obtained when **H** is heated with maleic anhydride. Both reactions give only the stereoisomer shown. Formulate a mechanism.

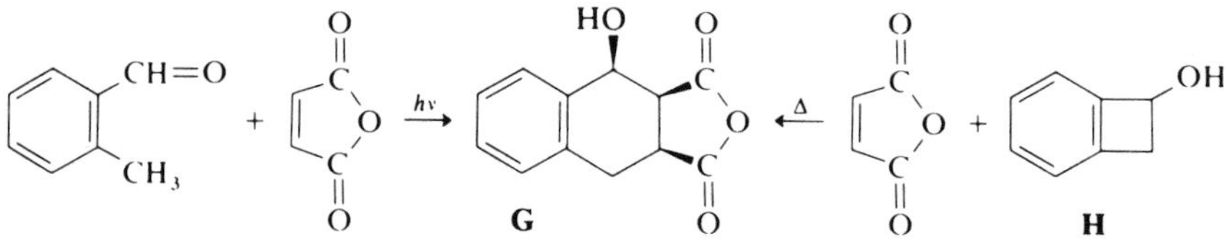

CH=O, CH_3 + maleic anhydride $\xrightarrow{h\nu}$ **G** (HO) $\xleftarrow{\Delta}$ maleic anhydride + **H** (OH)

15. Photolysis of **I** gives an isomeric compound **J** in 83% yield. Alkaline hydrolysis of **J** affords a hydroxy carboxylic acid **K**, $C_{25}H_{32}O_4$. Treatment of **J** with silica gel in hexane yields **L**, $C_{24}H_{28}O_2$. **L** is converted by sodium periodate-potassium permanganate to a mixture of **M** and **N**. What are the structures of **J**, **K**, and **L**?

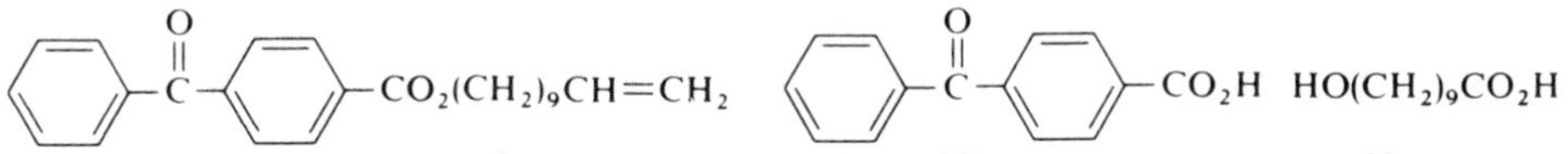

$C_6H_5C(=O)C_6H_4CO_2(CH_2)_9CH{=}CH_2$ **I** $C_6H_5C(=O)C_6H_4CO_2H$ **M** $HO(CH_2)_9CO_2H$ **N**

16. (a) 1,2,4,5-Tetrazines react with alkenes to give dihydropyridazines, as illustrated in the equation below. Suggest a mechanism.

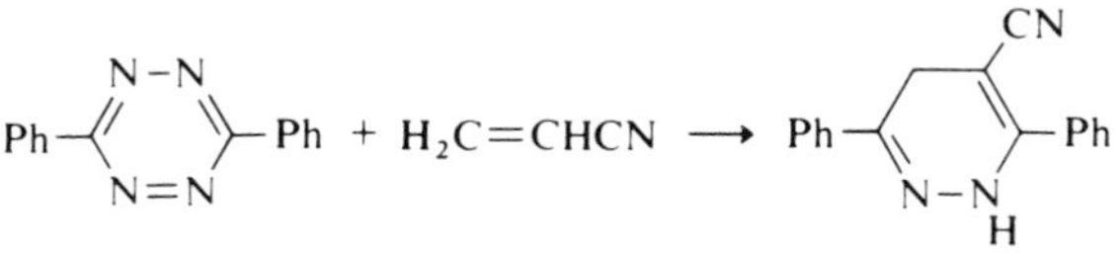

3,6-diphenyl-1,2,4,5-tetrazine + $H_2C{=}CHCN$ → 3,6-diphenyl-4-cyano-1,4-dihydropyridazine (N–N, H)

(b) Compounds **O** and **P** are both unstable toward loss of nitrogen at room temperature. Both compounds give **Q** as the product of decomposition. Account for the formation of **Q**.

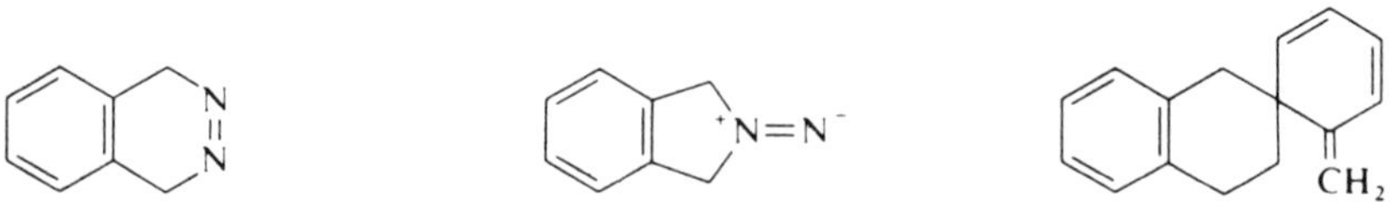

O **P** **Q**

17. In each part below, the molecule shown has been employed as a synthetic equivalent in cycloaddition reactions. Show the sequence of reactions by which the adduct could be converted to the adduct that cannot be obtained directly.

(a) $RC(NO_2){=}CHSO_2Ph$ as an acetylene equivalent
(b) $PhSO_2CH{=}CHSiMe_3$ as an acetylene equivalent
(c) $CH_2{=}C(O_2CCH_3)CN$ as a ketene equivalent
(d) $CH_2{=}CHNO_2$ as a ketene equivalent

18. Provide a detailed mechanistic explanation for each of the following synthetically useful transformations.

(a) 1-cyclohexenylmethanol (CH_2OH) $\xrightarrow{\text{1) } p\text{-}O_2NC_6H_4SeCN;\ \text{2) } Bu_3P;\ \text{3) } H_2O_2}$ 2-methylenecyclohexanol (CH_2, OH)

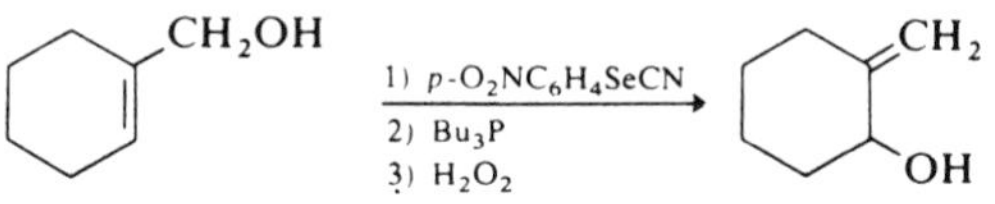

(b) $HC{\equiv}CCH(NHCOCH_3)CH(OH)CH{=}CH_2 \xrightarrow[2)\ \Delta]{1)\ Cl_3CCN} HC{\equiv}CCH(NHCOCH_3)CH{=}CHCH_2NHCOCCl_3$

(c) 2-(phenylthiomethyl)cyclopent-2-enone (CH_2SPh) $\xrightarrow[CsF,\ PhCH{=}O]{Me_3SiCH_2O_3SCF_3}$ 2-methylene-3-(CH_2SPh)cyclopentanone + $PhCH{-}CH_2$ (epoxide, O)

(d) 2-(2-furyl)cyclobutanol (OH) $\xrightarrow[25°C,\ 16\ h]{KH}$ 4-hydroxy-4,5,6,7-tetrahydrobenzofuran (OH, O)

(e) $CH_2{=}C(CH_3)CH_2SePh + H_2NCH_2CO_2C_2H_5 \xrightarrow{N\text{-chlorosuccinimide (N–Cl)}} CH_2{=}C(CH_3)CH_2NHCO_2C_2H_5$

(f) $CH_2{=}CHCH_2CH_2$–(1-H_3C)cyclohexane–CO_2H $\xrightarrow[2)\ Et_3N]{(1)\ \text{2-fluoro-1-methylpyridinium (}\overset{+}{N}\text{, F, Me)}}$ tricyclic cyclobutanone (O)

(g) cyclohexyl(H)$CH{=}\overset{+}{N}(O^-)C(CH_3)_3 \xrightarrow[2)\ Et_3N]{1)\ CH_3O_2CCl} \xrightarrow{H_2O}$ 1-(O_2COCH_3)cyclohexane-1-$CH{=}O$

(h) $CH_3(H)C{=}C(H)CH_2Br + CH_3SCH_2\overset{O}{\overset{\|}{C}}Ph \xrightarrow{K_2CO_3} Ph\overset{O}{\overset{\|}{C}}CH(SCH_3)CH(CH_3)CH{=}CH_2$

(i) $CH_3CH_2CH_2CH(OTMS)$–(2,5-dihydrothiophene SO_2)–$CH_2CH_2\overset{O}{\overset{\|}{C}}\underset{H}{N}CH_2O_2CCH_3 \xrightarrow{380°C}$ $CH_3CH_2CH_2CH(OTMS)$–indolizidinone (N, O)

(j) [Reaction scheme: 1-(3-hydroxy-4,4-dimethyl-7-octen-1-ynyl)-2-methylcyclobutene, $C{\equiv}CCHOHC(CH_3)_2CH_2CH{=}CH_2$, treated with PhSCl, Et_3N, 25°C, 38 h, gives a tricyclic diene bearing O=SPh, CH_3, and two CH_3 groups]

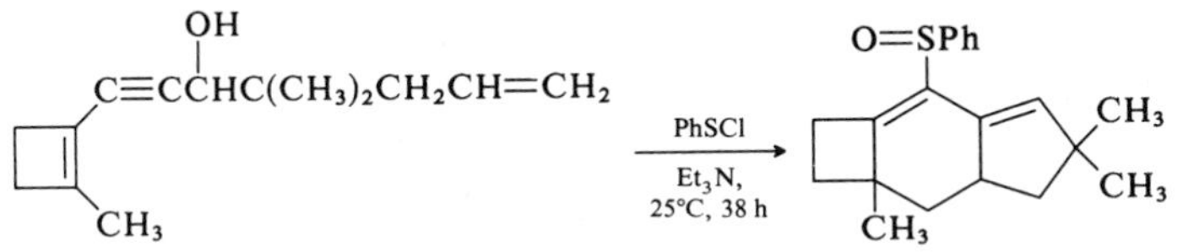

(k) [Reaction scheme: bicyclic lactam bearing $CH_2Si(CH_3)_3$, CH_3, H, $PhCH_2$, $CH_3C(=O)$–N, OAc, and a thiane ketone with a $H\!C{=}C(H)CH_2Cl$ side chain, treated with NaI, K_2CO_3, CH_3CN, gives the macrocyclic product]

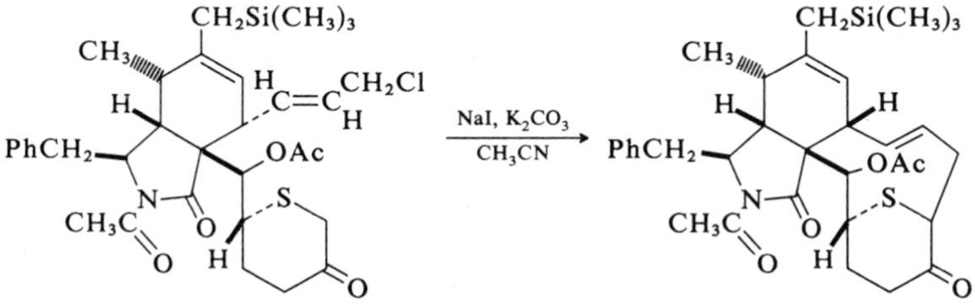

19. Suggest sequences of reactions for accomplishing each of the following synthetic transformations:

(a) squalene from succinaldehyde, isopropyl bromide, and 3-methoxy-2-methyl-1,3-butadiene

(b) [bicyclic amino alcohol with $(CH_3)_2N$ and HO substituents] from [3-cyclohexenecarboxaldehyde, CH=O]

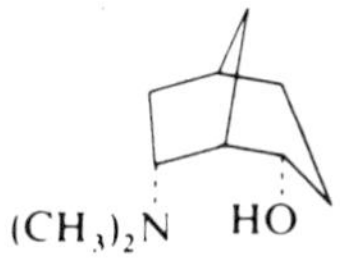

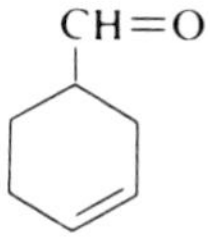

(c) $H_2C{=}CH(H_3C)C{=}CH\,CH_2(H)C{=}C(CH_3)CH_2CH_2(H)C{=}C(CH{=}O)CH_3$

from $H_2C{=}CH(H_3C)C{=}CH\,CH_2CH(C(CH_3){=}CH_2)C(CH_3)(CH{=}CH_2)CH{=}O$

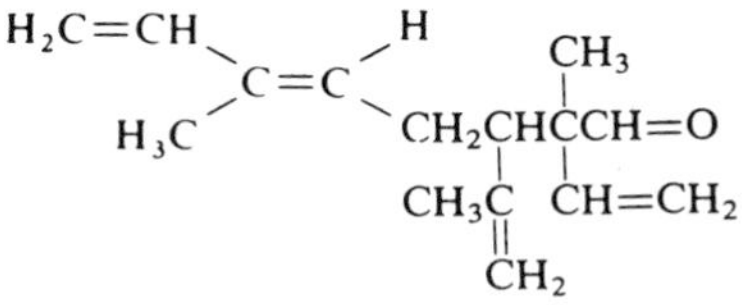

(d) $H_5C_2O_2CCH_2CH_2C(C_2H_5){=}CHCH_2CH_2Cl$ from $H_2C{=}C(C_2H_5)CH(OH)CH_2CH_2Cl$

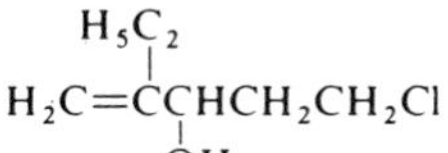

(e) [2-cyclopentenone with H, C_5H_{11}, CH_3, H substituents] from $H_5C_2CO_2CH_2C(OC_2H_5){=}CHC_5H_{11}$

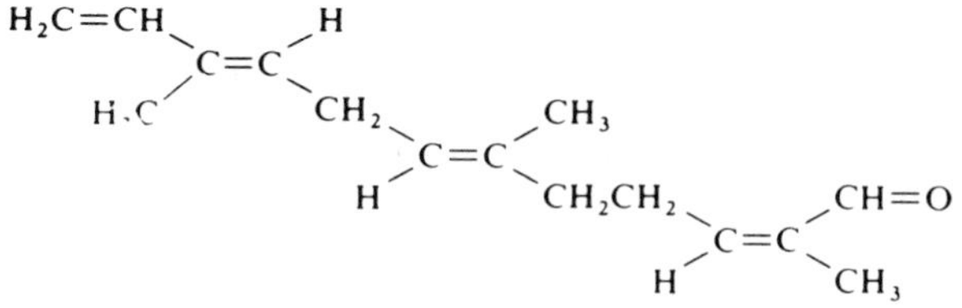

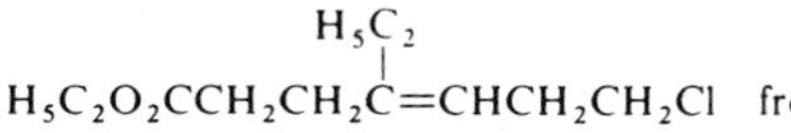

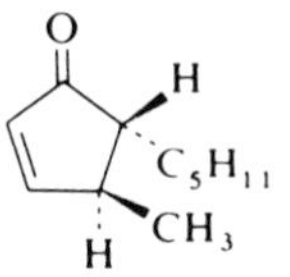

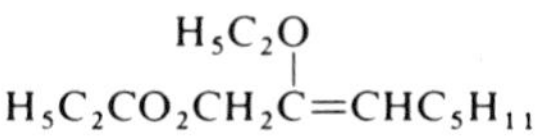

(f) 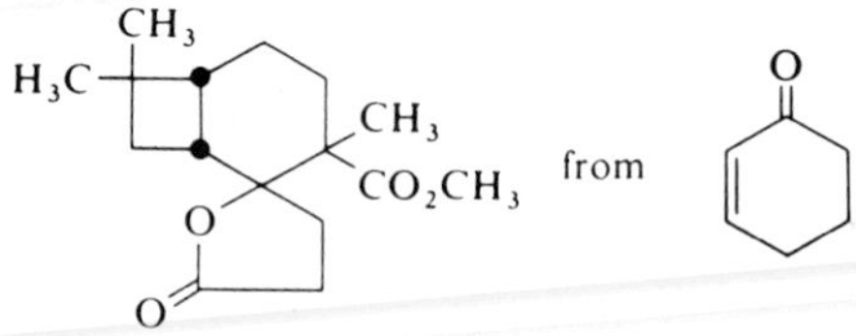 from

(g) 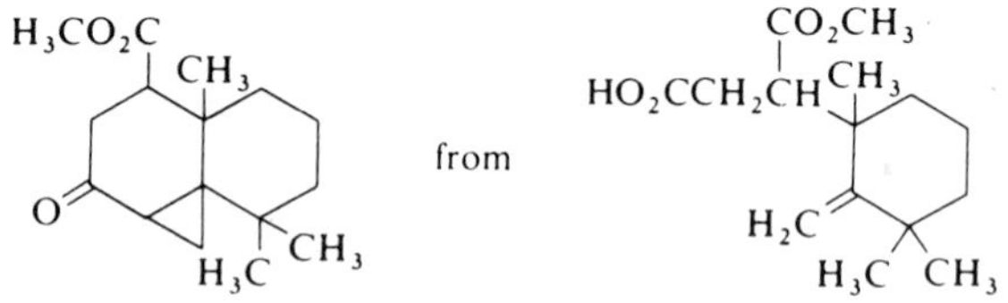 from

(h) $(CH_3)_2C{=}CHCH(CH{=}CH_2)C(CH_3){=}CH_2$ from $(CH_3)_2C{=}CHCH{=}CHCH_2OH$

(i) $(CH_3)_2CHCH_2COCH_2C(CH_3){=}CH_2$ from $H_2C{=}CHOCH_2CH{=}CHCH_3$ and $(CH_3)_2CHCH_2Br$

(j) 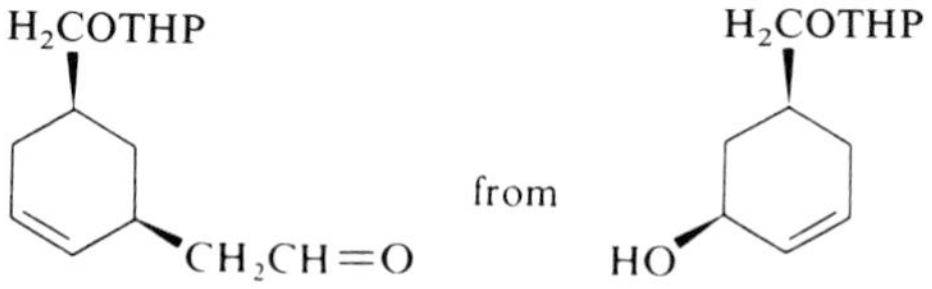 from

(k) 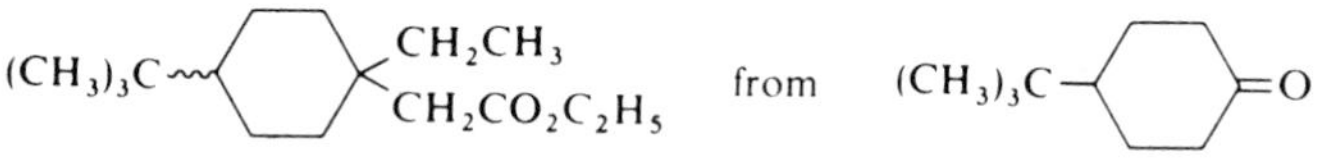 from

(l) 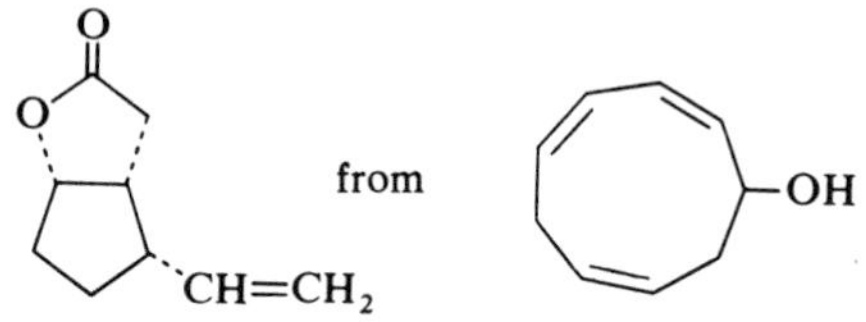 from

(m) 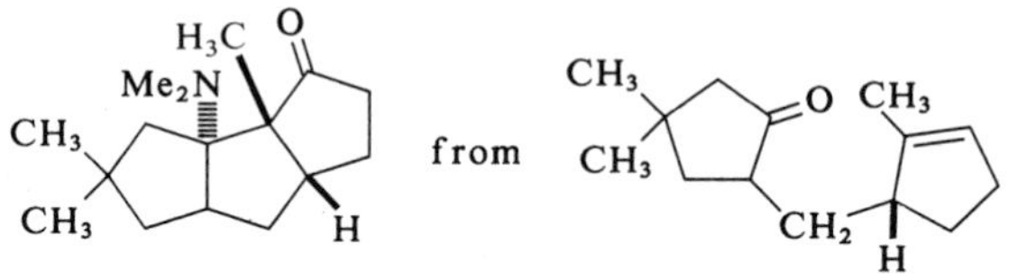 from

(n) 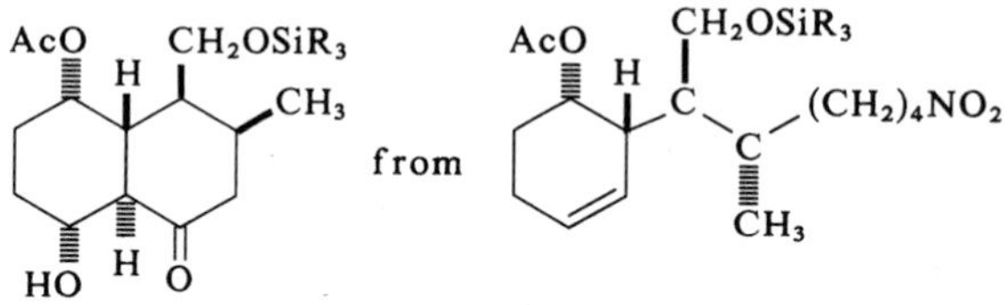 from

(o) 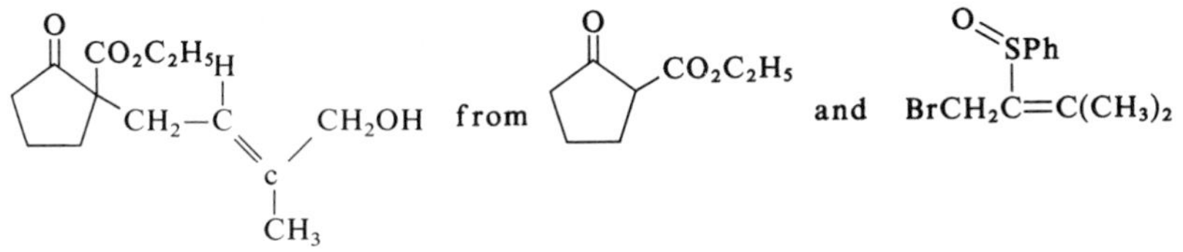 [structure] $CO_2C_2H_5$ H CH_2—C CH_2OH c CH_3 from [structure] $CO_2C_2H_5$ and O SPh $BrCH_2C{=}C(CH_3)_2$

(p) 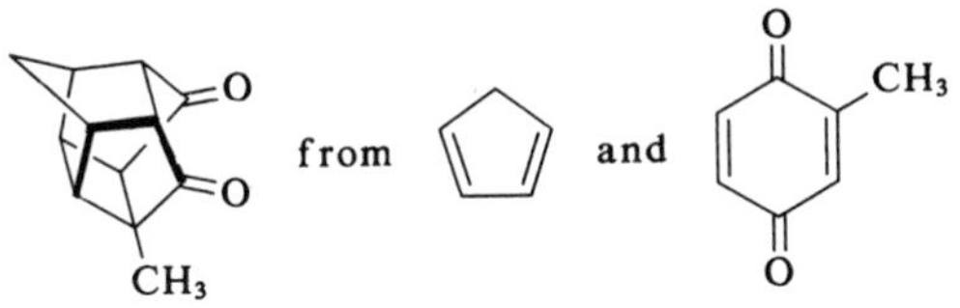 [structure] O O CH_3 from [structure] and [structure] O CH_3 O

(q) 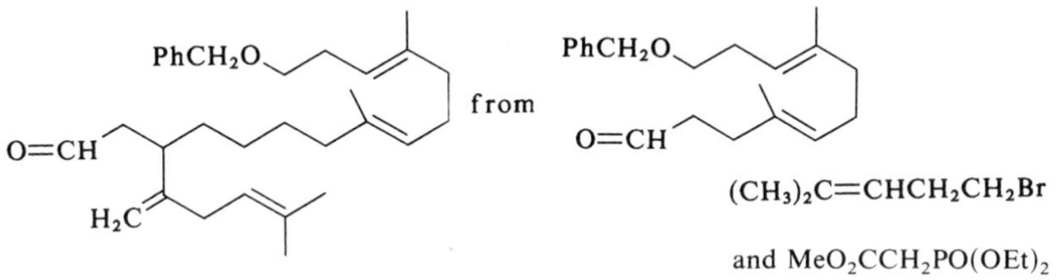 $PhCH_2O$ O=CH H_2C from $PhCH_2O$ O=CH

$(CH_3)_2C{=}CHCH_2CH_2Br$

and $MeO_2CCH_2PO(OEt)_2$

20. Predict the major product and its stereochemistry for each of the following reactions. Provide a structure for the transition state, and indicate the features that control the stereochemistry.

(a) 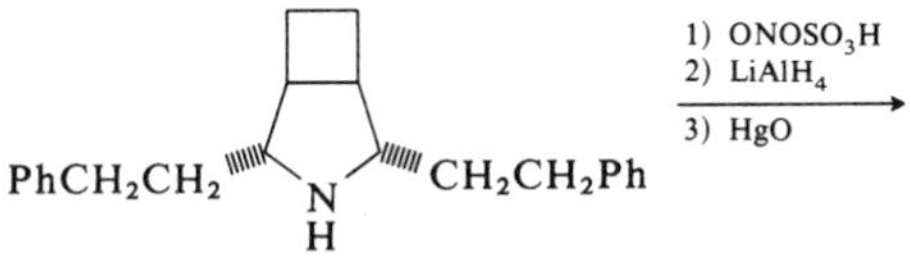 $PhCH_2CH_2$ N H CH_2CH_2Ph — 1) $ONOSO_3H$ 2) $LiAlH_4$ 3) HgO →

(b) 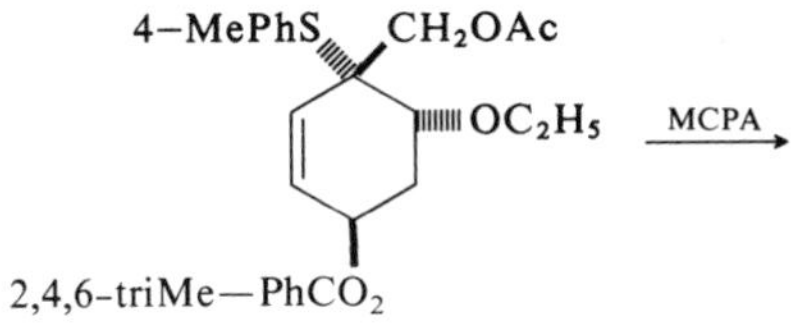 4–MePhS CH_2OAc OC_2H_5 2,4,6-triMe—$PhCO_2$ — MCPA →

(c) $OSiR_3$ O—CH_2 NC CH_3 — 165°C 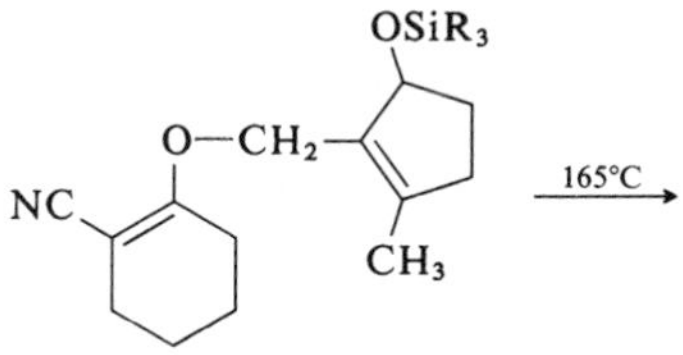 →

(d) 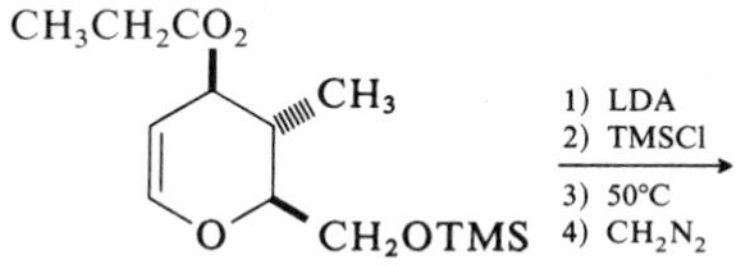 $CH_3CH_2CO_2$ CH_3 O CH_2OTMS — 1) LDA 2) TMSCl 3) 50°C 4) CH_2N_2 →

(e) $CH_2O_2CCH_2CH_3$; $C(CH_3)_2$ — 1) LDA, −78°C, THF 2) t-BuMe$_2$SiCl, HMPA 3) 50°C

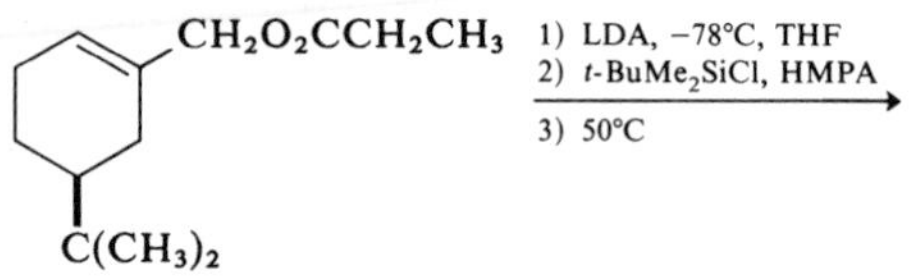

(f) $(CH_3)_2CH$, CH_3, $CH_2{=}CCH_2O$ — n-BuLi, $KOC(CH_3)_3$

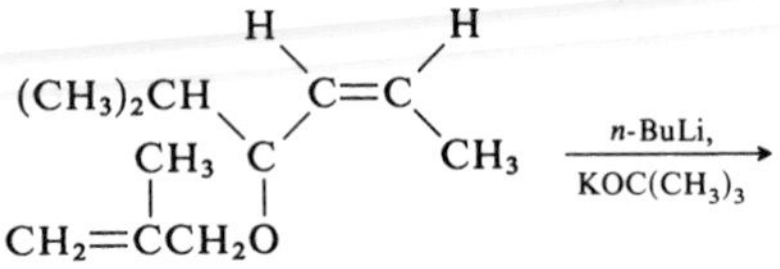

(g) $CH_2{=}CCH_3$, S, CH_3, OCH_2Ph, CH_3 — 1) $EtO_2CCH(CH_3)O_3SCF_3$ 2) DBU

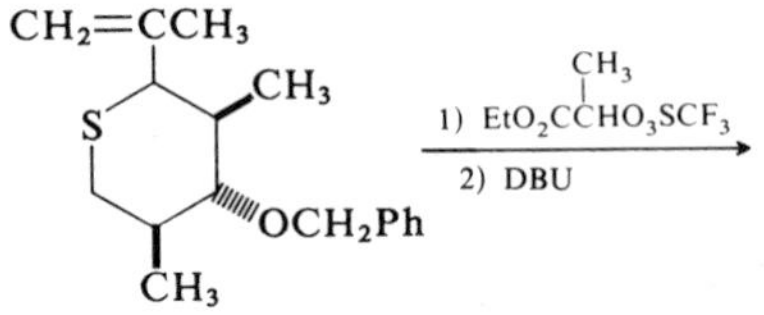

21. When the lactone enol silyl ether **R** is heated to 135°C, a mixture of four stereoisomers is obtained. Although the major one is that expected for a [3,3]-sigmatropic rearrangement, lesser amounts of the other possible C-4a and C-5 epimers are also present. When the reactant is heated to 100°C, partial conversion to the same mixture of stereoisomers is observed, but most of the product at this temperature is an acyclic triene ester.

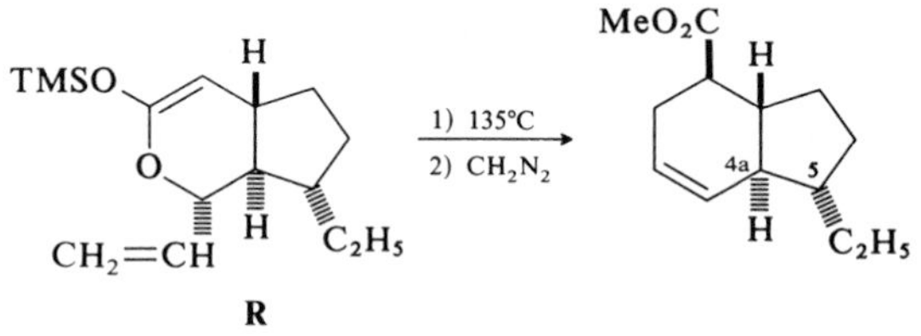

TMSO, O, $CH_2{=}CH$, H, H, C_2H_5 — 1) 135°C 2) CH_2N_2 → MeO_2C, H, 4a, 5, H, C_2H_5

R

Suggest a structure for the triene ester and show how it could be formed. Discuss the significance of the observation of the triene ester for the incomplete stereospecificity in the rearrangement.

22. Provide a detailed mechanistic description of the following transformations:

(a) O, O, CN — $CH_3S^{+}(O^{-})CH_2Li$, 10 min 0°C, then add MeO_2C–CH=CH–(furan)–OMe — 0°C, 1.25 h, then H^+, H_2O → OH, CO_2Me, OH, O, OMe

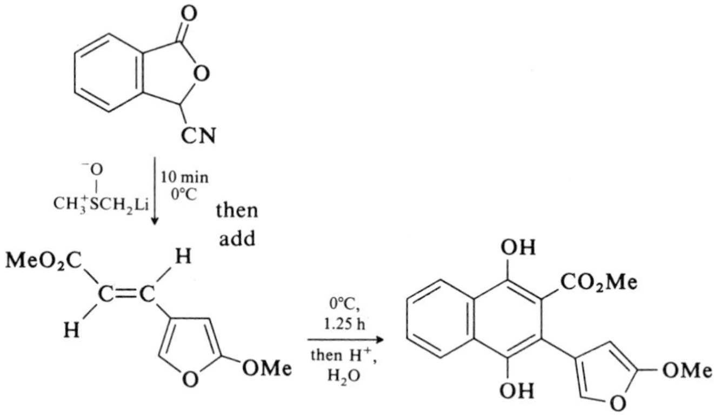

(b)

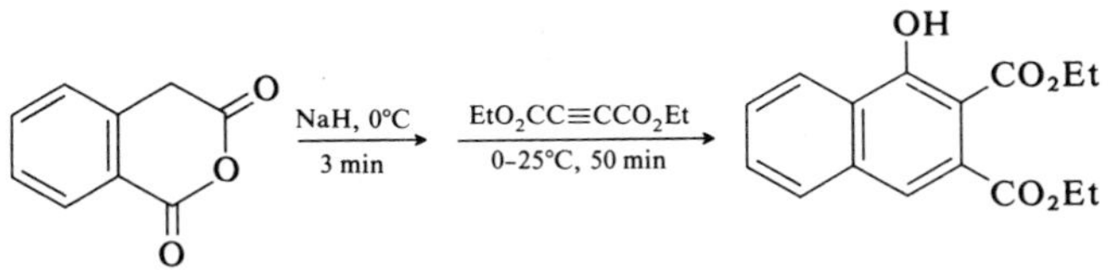

O, O, O; NaH, 0°C, 3 min; $EtO_2CC{\equiv}CCO_2Et$, 0–25°C, 50 min

OH, CO_2Et, CO_2Et

23. Each of the following cycloaddition reactions exhibits a good degree of diastereoselectivity. Provide a rationalization of the observed diastereoselectivity in terms of a preferred transition state structure.

(a)

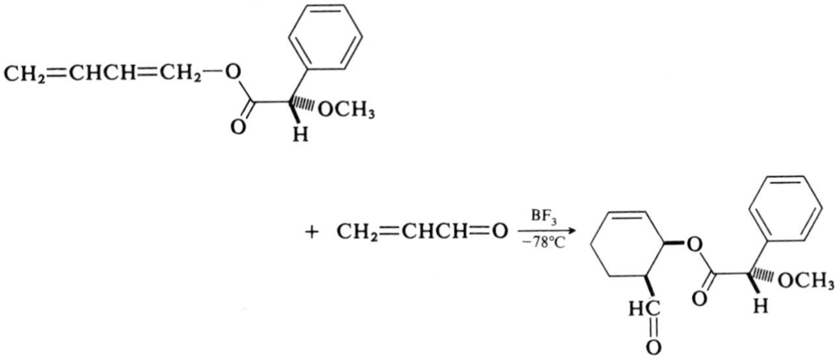

$CH_2{=}CHCH{=}CH_2{-}O$, O, OCH_3, H

+ $CH_2{=}CHCH{=}O$ $\xrightarrow[-78°C]{BF_3}$ O, O, OCH_3, HC, O, H

d.e. = 88%

(b)

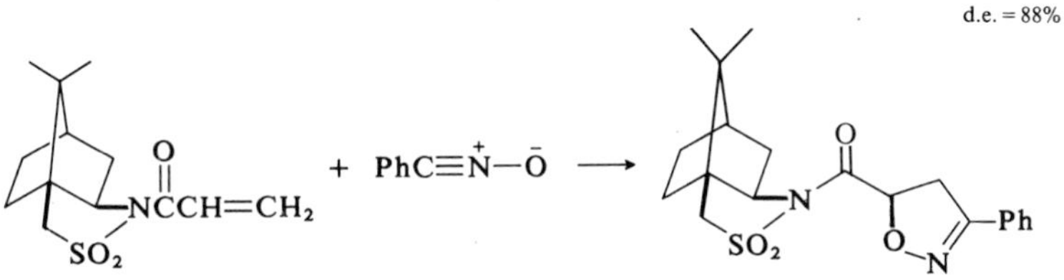

O, $NCCH{=}CH_2$, SO_2 + $PhC{\equiv}\overset{+}{N}{-}\overset{-}{O}$ ⟶ O, N, SO_2, O, N, —Ph

d.e. = 90%

(c)

O, $CH{=}CH_2$, N, $PhCH_2O_2C$ + $\xrightarrow{TiCl_4}$ O, N, $PhCH_2O_2C$

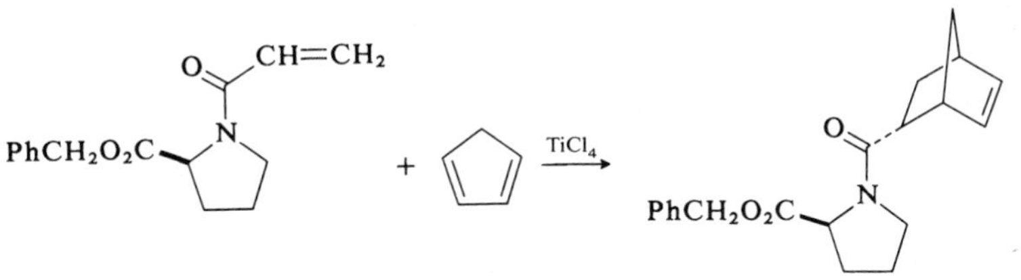

d.e. = 88%

7

Organometallic Compounds of Group I and II Metals

The use of organometallic reagents in organic synthesis had its beginning around 1900 with the work of Victor Grignard, who discovered that alkyl and aryl halides reacted with magnesium metal to give homogeneous solutions. The "Grignard reagents" proved to be reactive carbon nucleophiles and have remained very useful synthetic reagents since that time. Organolithium reagents came into synthetic use somewhat later. In the last 25 years, the synthetic utility of reactions involving metal ions and organometallic compounds has expanded enormously. Certain of the transition metals, such as copper, palladium, and nickel, have gained important places in synthetic methodology. In addition to providing reagents for organic synthesis, the systematic study of the reactions of organic compounds with metal ions and complexes has created a large number of organometallic compounds, many having unique structures and reactivity. In this chapter, we will discuss the Grignard reagents and organolithium compounds. In Chapter 8, the role of transition metals in organic synthesis will be given attention.

7.1. Preparation and Properties of Organolithium and Organomagnesium Compounds

The compounds of lithium and magnesium are the most important of the group IA and IIA organometallics. The metals in these two groups are the most electropositive of the elements. The polarity of the metal–carbon bond is such as to place high electron density on carbon. This electronic distribution is responsible for the strong nucleophilicity and basicity that characterize these compounds.

The reaction of magnesium metal with an alkyl or aryl halide in diethyl ether is the classical method of synthesis of Grignard reagents.

$$RX + Mg \rightarrow RMgX$$

The order of reactivity of the halides is RI > RBr > RCl. Solutions of some Grignard reagents such as methylmagnesium bromide, ethylmagnesium bromide, and phenylmagnesium bromide are available commercially. Some Grignard reagents are formed in tetrahydrofuran more rapidly than in ether. This is true of vinylmagnesium bromide, for example.[1] The solubility of Grignard reagents in ethers is the result of strong Lewis acid–base complex formation between the ether molecules and the magnesium atom. A number of Grignard reagents have been subjected to X-ray structure determination. Ethylmagnesium bromide has been observed in both monomeric and dimeric forms in crystal structures.[2] Figure 7.1a shows the crystal structure of the monomer with two diethyl ether molecules coordinated at magnesium. Figure 7.1b shows a dimeric structure with one diisopropyl ether molecule at each magnesium.

Organic halides that are unreactive toward magnesium shavings can often be induced to react by using an extremely reactive form of magnesium which is obtained by reducing magnesium salts with sodium or potassium metal.[3] Even alkyl fluorides, which are normally unreactive, form Grignard reagents under these conditions.

The mechanism of the Grignard reaction has not been specified precisely. The reaction takes place at the metal surface. Reaction probably commences with an

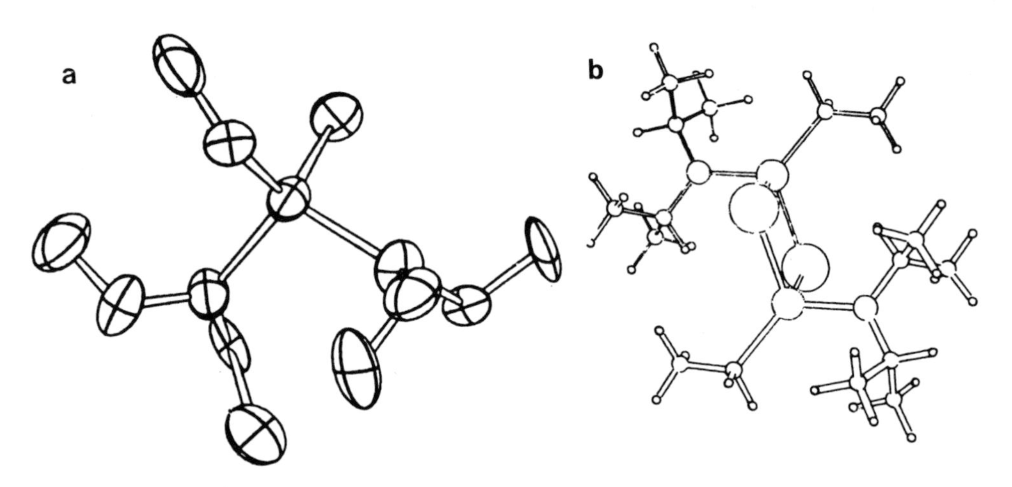

Fig. 7.1. Crystal structures of ethylmagnesium bromide. (a) Monomeric $C_2H_5MgBr[O(C_2H_5)_2]_2$; reproduced from Ref. 2a. (b) Dimeric $C_2H_5MgBr[O(i\text{-}C_3H_7)_2]$; reproduced from Ref. 2b.

1. D. Seyferth and F. G. A. Stone, *J. Am. Chem. Soc.* **79**, 515 (1957); H. Normant, *Adv. Org. Chem.* **2**, 1 (1960).
2. (a) L. J. Guggenberger and R. E. Rundle, *J. Am. Chem. Soc.* **90**, 5375 (1968); (b) A. L. Spek, P. Voorbergen, G. Schat, C. Blomberg, and F. Bickelhaupt, *J. Organomet. Chem.* **77**, 147 (1974).
3. R. D. Rieke and S. E. Bales, *J. Am. Chem. Soc.* **96**, 1775 (1974); R. D. Rieke, *Acc. Chem. Res.* **10**, 301 (1977).

electron transfer, followed by rapid combination of the organic group with a magnesium ion. The carbon–bromine bond must break either prior to or during the formation of the carbon–magnesium bond.[4]

$$\mathrm{R{-}Br + Mg \rightarrow R{-}Br^{\cdot -} + Mg(I)}$$

$$\mathrm{R{-}Br^{\cdot -} + Mg(I) \rightarrow RMgBr}$$

One test for the involvement of radical intermediates is to determine if cyclization occurs in the 6-hexenyl system, where radical cyclization is rapid (see Part A, Section 12.2.2). Small amounts of cyclized products are obtained after the preparation of Grignard reagents from 5-hexenyl bromide.[5] This indicates that cyclization of the intermediate radical competes to a small extent with combination of the radical with the metal.

The preparation of Grignard reagents from alkyl halides normally occurs with stereochemical randomization at the site of the reaction. Stereoisomeric halides give rise to organomagnesium compounds of identical composition.[6] The main exceptions to this generalization are cyclopropyl and alkenyl systems, which can be prepared with partial retention of configuration.[7] Once formed, secondary alkylmagnesium compounds undergo stereochemical inversion only slowly. *Endo*- and *exo*- norbornylmagnesium bromide, for example, require one day at room temperature to reach equilibrium.[8] NMR studies have demonstrated that inversion of configuration is quite slow, on the NMR time scale, even up to 170°C.[9] In contrast, the inversion of configuration of primary alkylmagnesium halides can be shown to be very fast.[10] This difference between the primary and secondary systems may be the result of a mechanism for inversion that involves exchange of alkyl groups between magnesium atoms:

$$\mathrm{X{-}Mg{-}C(R)_2H \;\; + \;\; R_2HC{-}Mg{-}X \;\rightleftharpoons\; R_2HC{-}Mg(\mu\text{-}X)_2Mg{-}CHR_2}$$

4. H. R. Rogers, C. L. Hill, Y. Fujuwara, R. J. Rogers, H. L. Mitchell, and G. M. Whitesides, *J. Am. Chem. Soc.* **102**, 217 (1980); J. F. Garst, J. E. Deutch, and G. M. Whitesides, *J. Am. Chem. Soc.* **108**, 2490 (1986); E. C. Ashby and J. Oswald, *J. Org. Chem.* **53**, 6068 (1988).
5. R. C. Lamb, P. W. Ayers, and M. K. Toney, *J. Am. Chem. Soc.* **85**, 3483 (1963); R. C. Lamb and P. W. Ayers, *J. Org. Chem.* **27**, 1441 (1962); C. Walling and A. Cioffari, *J. Am. Chem. Soc.* **92**, 6609 (1970); H. W. H. J. Bodewitz, C. Blomberg, and F. Bickelhaupt, *Tetrahedron* **31**, 1053 (1975).
6. N. G. Krieghoff and D. O. Cowan, *J. Am. Chem. Soc.* **88**, 1322 (1966).
7. T. Yoshino and Y. Manabe, *J. Am. Chem. Soc.* **85**, 2860 (1963); H. M. Walborsky and A. E. Young, *J. Am. Chem. Soc.* **86**, 3288 (1964); H. M. Walborsky and B. R. Banks, *Bull. Soc. Chim. Belg.* **89**, 849 (1980).
8. F. R. Jensen and K. L. Nakamaye, *J. Am. Chem. Soc.* **88**, 3437 (1966); N. G. Krieghoff and D. O. Cowan, *J. Am. Chem. Soc.* **88**, 1322 (1966).
9. E. Pechold, D. G. Adams, and G. Fraenkel, *J. Org. Chem.* **36**, 1368 (1971).
10. G. M. Whitesides, M. Witanowski, and J. D. Roberts, *J. Am. Chem. Soc.* **87**, 2854 (1965); G. M. Whitesides and J. D. Roberts, *J. Am. Chem. Soc.* **87**, 4878 (1965); G. Fraenkel and D. T. Dix, *J. Am. Chem. Soc.* **88**, 979 (1966).

If bridged intermediates are involved, the larger steric bulk of secondary systems would retard the reaction. Steric restrictions may be further enhanced by the fact that organomagnesium reagents are often present as clusters (see below).

The usual designation of Grignard reagents as RMgX is a basically correct representation of the composition of the compounds in ether solution, but an equilibrium exists with magnesium bromide and the dialkylmagnesium.

$$2\,RMgX \rightleftharpoons R_2Mg + MgX_2$$

The position of the equilibrium depends upon the solvent and the identity of the specific organic group but lies far to the left in ether for simple aryl-, alkyl-, and alkenylmagnesium halides.[11]

Solutions of organomagnesium compounds in diethyl ether contain aggregated species.[12] Dimers predominate in ether solutions of alkylmagnesium chlorides.

$$2\,RMgCl \rightleftharpoons R{-}Mg\langle\begin{smallmatrix}Cl\\ \\Cl\end{smallmatrix}\rangle Mg{-}R$$

The corresponding bromides and iodides show concentration-dependent behavior, and in very dilute solutions they exist as monomers. In tetrahydrofuran, there is less tendency to aggregate, and several alkyl and aryl Grignard reagents have been found to be monomeric in this solvent.

Most simple organolithium reagents can be prepared by reaction of the appropriate halide with lithium metal.

$$R{-}X + 2\,Li \rightarrow RLi + LiX$$

As with organomagnesium reagents, there is usually loss of stereochemical integrity at the site of reaction in the preparation of alkyllithium compounds.[13] Alkenyllithium reagents can usually be prepared with retention of configuration of the double bond.[14]

The simple alkyllithium reagents exist mainly as hexamers in hydrocarbon solvents.[15] In ethers, tetrameric structures are usually dominant.[16] The tetramers, in

11. G. E. Parris and E. C. Ashby, *J. Am. Chem. Soc.* **93**, 1206 (1971); P. E. M. Allen, S. Hagias, S. F. Lincoln, C. Mair, and E. H. Williams, *Ber. Bunsenges. Phys. Chem.* **86**, 515 (1982).
12. E. C. Ashby and M. B. Smith, *J. Am. Chem. Soc.* **86**, 4363 (1964); F. W. Walker and E. C. Ashby, *J. Am. Chem. Soc.* **91**, 3845 (1969).
13. W. H. Glaze and C. M. Selman, *J. Org. Chem.* **33**, 1987 (1968).
14. J. Millon, R. Lorne, and G. Linstrumelle, *Synthesis*, 434 (1975).
15. G. Fraenkel, W. E. Beckenbaugh, and P. P. Yang, *J. Am. Chem. Soc.* **98**, 6878 (1976); G. Fraenkel, M. Henrichs, J. M. Hewitt, B. M. Su, and M. J. Geckle, *J. Am. Chem. Soc.* **102**, 3345 (1980).
16. H. L. Lewis and T. L. Brown, *J. Am. Chem. Soc.* **92**, 4664 (1970); P. West and R. Waack, *J. Am. Chem. Soc.* **89**, 4395 (1967); J. F. McGarrity and C. A. Ogle, *J. Am. Chem. Soc.* **107**, 1085 (1985); D. Seebach, R. Hässig, and J. Gabriel, *Helv. Chim. Acta* **66**, 308 (1983); T. L. Brown, *Adv. Organometal. Chem.* **3**, 365 (1965); W. N. Setzer and P. von R. Schleyer, *Adv. Organomet. Chem.* **24**, 354 (1985).

turn, are solvated with ether molecules.[17] Phenyllithium is tetrameric in cyclohexane and dimeric in tetrahydrofuran.[18]

The crystal structures of many organolithium compounds have been determined. Phenyllithium has been crystallized as an ether solvate. The structure is tetrameric with lithium and carbon atoms at alternating corners of a strongly distorted cube. Each carbon is 2.33 Å from the three neighboring lithium atoms. An ether molecule is coordinated to each lithium atom. Figure 7.2a shows the Li—C cluster, while Fig. 7.2b shows the complete array of atoms, except for hydrogen.[19] Section 7.1 of Part A provides additional information on the structure of organolithium compounds.

There are several other methods that are very useful for preparing organometallic reagents. The first of these is hydrogen-metal exchange or metalation. This reaction is the usual method for preparing alkynylmagnesium and alkynyllithium reagents. The reaction proceeds readily because of the relative acidity of the hydrogen bound to *sp* carbon.

$$\mathrm{H{-}C{\equiv}C{-}R + R'MgBr \rightarrow BrMgC{\equiv}C{-}R + R'{-}H}$$
$$\mathrm{H{-}C{\equiv}C{-}R + R'Li \rightarrow LiC{\equiv}C{-}R + R'{-}H}$$

Although of limited utility in the synthesis of other types of Grignard reagents, metalation is an important means of preparing a variety of organolithium compounds. The position of lithiation is determined by the relative acidities of the

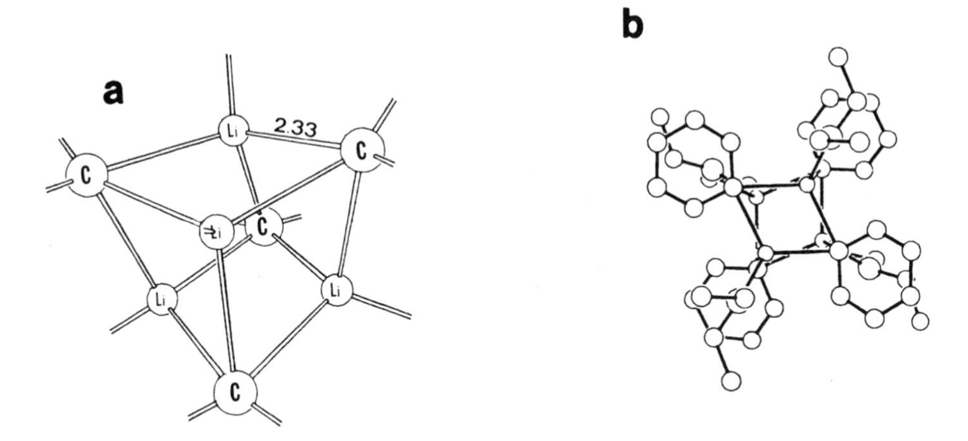

Fig. 7.2. Crystal structure of tetrameric phenyllithium etherate. (a) Tetrameric cluster. (b) Complete structure except for hydrogens. (Reproduced with permission from Ref. 19.)

17. P. D. Bartlett, C. V. Goebel, and W. P. Weber, *J. Am. Chem. Soc.* **91**, 7425 (1969).
18. L. M. Jackman and L. M. Scarmoutzos, *J. Am. Chem. Soc.* **106**, 4627 (1984).
19. H. Hope and P. P. Power, *J. Am. Chem. Soc.* **105**, 5320 (1983).

available hydrogens and the directing effect of substituent groups. Benzylic and allylic hydrogens are relatively reactive toward lithiation because of the resonance stabilization of the resulting anions. Substituents that can coordinate to the lithium ion, such as alkoxy, amido, and sulfonyl, have a powerful influence on the position and rate of lithiation of aromatic compounds.[20] In heterocyclic compounds, the preferred site for lithiation is usually adjacent to the heteroatom. Scheme 7.1 gives some examples of the preparation of organolithium compounds by this method.

Scheme 7.1. Organolithium Compounds by Metalation

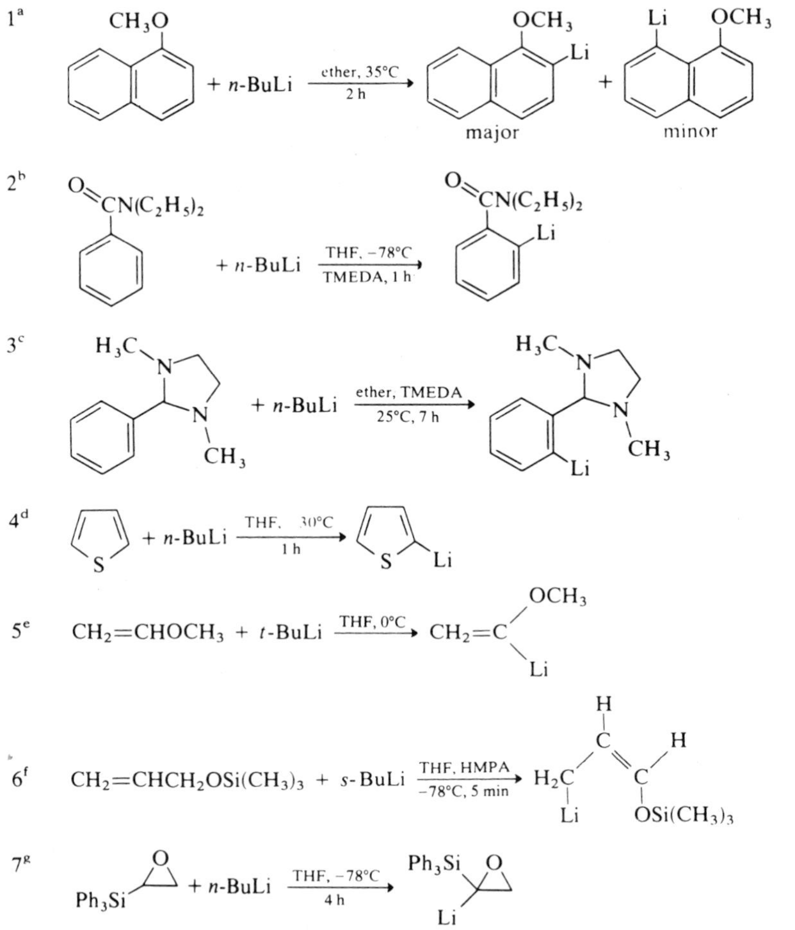

1[a] 1-Methoxynaphthalene (CH_3O) + n-BuLi $\xrightarrow[\text{2 h}]{\text{ether, 35°C}}$ 1-methoxy-2-lithionaphthalene (OCH_3, Li) (major) + 1-methoxy-8-lithionaphthalene (Li, OCH_3) (minor)

2[b] $C_6H_5C(=O)N(C_2H_5)_2$ + n-BuLi $\xrightarrow[\text{TMEDA, 1 h}]{\text{THF, −78°C}}$ 2-lithio-$C_6H_4C(=O)N(C_2H_5)_2$

3[c] 1,3-dimethyl-2-phenylimidazolidine (H_3C–N, N–CH_3) + n-BuLi $\xrightarrow[\text{25°C, 7 h}]{\text{ether, TMEDA}}$ ortho-Li derivative

4[d] thiophene + n-BuLi $\xrightarrow[\text{1 h}]{\text{THF, 30°C}}$ 2-lithiothiophene (S, Li)

5[e] $CH_2{=}CHOCH_3$ + t-BuLi $\xrightarrow{\text{THF, 0°C}}$ $CH_2{=}C(OCH_3)Li$

6[f] $CH_2{=}CHCH_2OSi(CH_3)_3$ + s-BuLi $\xrightarrow[\text{−78°C, 5 min}]{\text{THF, HMPA}}$ $H_2C(Li)$–$C(H){=}C(H)OSi(CH_3)_3$

7[g] Ph_3Si-oxirane + n-BuLi $\xrightarrow[\text{4 h}]{\text{THF, −78°C}}$ $Ph_3Si(Li)$-oxirane

a. B. M. Graybill and D. A. Shirley, *J. Org. Chem.* **31**, 1221 (1966).
b. P. A. Beak and R. A. Brown, *J. Org. Chem.* **42**, 1823 (1977); **44**, 4463 (1979).
c. T. D. Harris and G. P. Roth, *J. Org. Chem.* **44**, 2004 (1979).
d. E. Jones and I. M. Moodie, *Org. Synth.* **50**, 104 (1970).
e. J. E. Baldwin, G. A. Höfle, and O. W. Lever, Jr., *J. Am. Chem. Soc.* **96**, 7125 (1974).
f. W. C. Still and T. L. Macdonald, *J. Org. Chem.* **41**, 3620 (1976).
g. J. J. Eisch and J. E. Galle, *J. Am. Chem. Soc.* **98**, 4646 (1976).

20. D. W. Slocum and C. A. Jennings, *J. Org. Chem.* **41**, 3653 (1976); J. M. Mallan and R. C. Rebb, *Chem. Rev.* **69**, 693 (1969). H. W. Gschwend and H. R. Rodriguez, *Org. React.* **26**, 1 (1979).

Reaction conditions can be modified to accelerate the rate of lithiation when necessary. Addition of tertiary amines, especially tetramethylethylenediamine (TMEDA), accelerates lithiation by coordination at the lithium, which promotes dissociation of aggregated structures.[21] Hydrocarbons lacking directing substituents are not very reactive toward metalation, but it has been found that a mixture of *n*-butyllithium and potassium *t*-butoxide[22] is sufficiently reactive to give allyl anions from alkenes such as isobutene and 2,3-dimethyl-1,3-butadiene.[23]

$$CH_2{=}C(CH_3)_2 \xrightarrow[KOC(CH_3)_3]{n\text{-BuLi}} CH_2{=}C(CH_3)CH_2Li$$

$$H_2C{=}C(CH_3){-}C(CH_3){=}CH_2 \xrightarrow[KOC(CH_3)_3]{n\text{-BuLi}} H_2C{=}C(CH_2Li){-}C(CH_2Li){=}CH_2$$

(2) Metal–halogen exchange is a second important method for preparation of organolithium reagents. This reaction proceeds in the direction of forming the more stable organolithium reagent, that is, the one derived from the more acidic organic structure.

$$RX + R'Li \rightarrow RLi + R'X$$

Thus, by use of the very basic organolithium compound *n*-butyl- or *t*-butyllithium, halogen substituents at more acidic carbons are readily exchanged to give the corresponding lithium compounds. Halogen–metal exchange is particularly useful for converting aryl and alkenyl halides to the corresponding lithium compounds. The driving force of the reaction is the greater stability of sp^2 carbanions in comparison with sp^3 carbanions. Scheme 7.2 gives some examples of these reactions.

$$\text{Ph(H)C}{=}\text{C(H)Br} \xrightarrow[\text{pentane-THF-Et}_2\text{O}]{t\text{-BuLi, }-120^\circ\text{C}} \text{Ph(H)C}{=}\text{C(H)Li} \qquad \text{Ref. 24}$$

$$m\text{-MeOC}_6\text{H}_4\text{Br} \xrightarrow[-78^\circ\text{C}]{n\text{-BuLi}} m\text{-MeOC}_6\text{H}_4\text{Li} \qquad \text{Ref. 25}$$

21. G. G. Eberhardt and W. A. Butte, *J. Org. Chem.* **29**, 2928 (1964); R. West and P. C. Jones, *J. Am. Chem. Soc.* **90**, 2656 (1968); S. Akiyama and J. Hooz, *Tetrahedron Lett.*, 4115 (1973).
22. L. Lochmann, J. Pospisil, and D. Lim, *Tetrahedron Lett.*, 257 (1966).
23. M. Schlosser and J. Hartmann, *Angew. Chem. Int. Ed. Engl.* **12**, 508 (1973); J. J. Bahl, R. B. Bates, and B. Gordon III, *J. Org. Chem.* **44**, 2290 (1979); M. Schlosser and G. Rauchshwalbe, *J. Am. Chem. Soc.* **100**, 3258 (1978).
24. N. Neumann and D. Seebach, *Tetrahedron Lett.*, 4839 (1976).
25. T. R. Hoye, S. J. Martin, and D. R. Peck, *J. Org. Chem.* **47**, 331 (1982).

Metal–halogen exchange is a rapid reaction and is usually carried out at −60 to −120°C. This makes it possible to prepare aryllithium compounds containing functional groups, such as cyano and nitro, which would react under the conditions required for preparation from lithium metal. Entries 5 and 6 in Scheme 7.2 are examples.

Retention of configuration is usually observed when organolithium compounds are prepared by metal–halogen exchange. The degree of retention is low for exchange of most alkyl systems,[26] but it is normally high for cyclopropyl and alkenyl halides.[27] Once formed, both cyclopropyl- and alkenyllithium reagents retain their configuration at room temperature.

A third useful method of preparing organolithium reagents involves metal–metal exchange. The reaction between two organometallic compounds will proceed in the

Scheme 7.2. Organolithium Reagents by Halogen-Metal Exchange

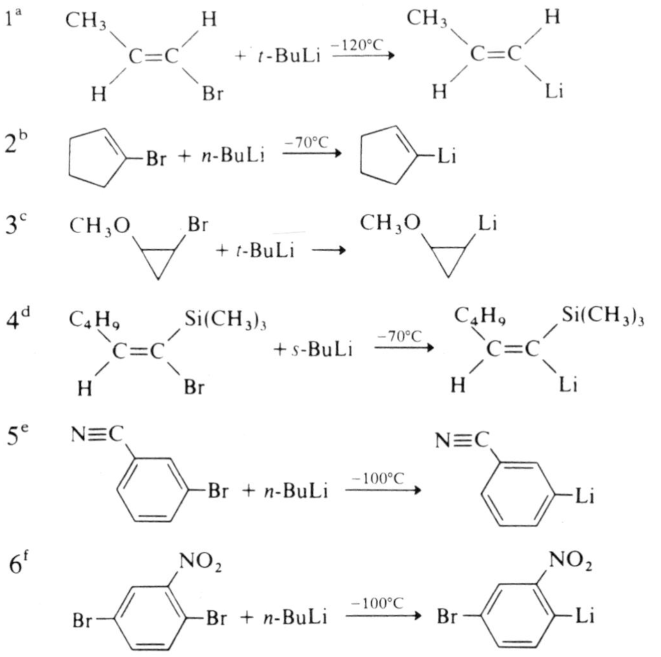

1[a] $CH_3(H)C{=}C(H)Br$ + t-BuLi $\xrightarrow{-120°C}$ $CH_3(H)C{=}C(H)Li$

2[b] 1-Bromocyclopentene (–Br) + n-BuLi $\xrightarrow{-70°C}$ 1-cyclopentenyl–Li

3[c] CH_3O-, Br-substituted cyclopropane + t-BuLi $\longrightarrow$ CH_3O-, Li-substituted cyclopropane

4[d] $C_4H_9(H)C{=}C(Si(CH_3)_3)Br$ + s-BuLi $\xrightarrow{-70°C}$ $C_4H_9(H)C{=}C(Si(CH_3)_3)Li$

5[e] N≡C–C_6H_4–Br + n-BuLi $\xrightarrow{-100°C}$ N≡C–C_6H_4–Li

6[f] Br–$C_6H_3(NO_2)$–Br + n-BuLi $\xrightarrow{-100°C}$ Br–$C_6H_3(NO_2)$–Li

a. H. Neuman and D. Seebach, *Tetrahedron Lett.*, 4839 (1976).
b. J. Millon, R. Lorne and G. Linstrumelle, *Synthesis*, 434 (1975).
c. E. J. Corey and P. Ulrich, *Tetrahedron Lett.*, 3685 (1975).
d. R. B. Miller and G. McGarvey, *J. Org. Chem.* **44**, 4623 (1979).
e. W. E. Parham and L. D. Jones, *J. Org. Chem.* **41**, 1187 (1976).
f. W. E. Parham and R. M. Piccirilli, *J. Org. Chem.* **42**, 257 (1977).

26. R. L. Letsinger, *J. Am. Chem. Soc.* **72**, 4842 (1950); D. Y. Curtin and W. J. Koehl, Jr., *J. Am. Chem. Soc.* **84**, 1967 (1962).
27. H. M. Walborsky, F. J. Impastato, and A. E. Young, *J. Am. Chem. Soc.* **86**, 3283 (1964); M. J. S. Dewar and J. M. Harris, *J. Am. Chem. Soc.* **91**, 3652 (1969); E. J. Corey and P. Ulrich, *Tetrahedron Lett.*, 3685 (1975); N. Neumann and D. Seebach, *Tetrahedron Lett.*, 4839 (1976); R. B. Miller and G. McGarvey, *J. Org. Chem.* **44**, 4623 (1979).

direction of placing the more electropositive metal on the more acidic carbon position. Exchanges between organotin reagents and alkyllithium reagents are particularly significant from a synthetic point of view.

$$HC{\equiv}CCH_2OTHP \xrightarrow{Bu_3SnH} (Bu_3Sn)(H)C{=}C(H)(CH_2OTHP) \xrightarrow{n\text{-}BuLi} (Li)(H)C{=}C(H)(CH_2OTHP) \qquad \text{Ref. 28}$$

$$RCH(SnBu_3)OR' + n\text{-}BuLi \xrightarrow{-78°C} RCH(Li)OR' \qquad \text{Ref. 29}$$

$$R_2NCH_2SnBu_3 + n\text{-}BuLi \xrightarrow{0°C} R_2NCH_2Li \qquad \text{Ref. 30}$$

The tributyltin compounds are used most frequently in these procedures. The α-stannyl derivatives needed for the latter two examples are readily available.

$$RCH{=}O + Bu_3SnLi \rightarrow RCH(SnBu_3){-}O^- \xrightarrow{R'X} RCH(SnBu_3)OR'$$

$$R_2NCH_2SPh + Bu_3SnLi \rightarrow R_2NCH_2SnBu_3$$

The exchange reactions of α-alkoxystannanes occur with retention of configuration at the carbon–metal bond.[31]

$$RCH_2C(H)(OCH_2OR')SnBu_3 \xrightarrow{RLi} RCH_2C(H)(OCH_2OR')Li$$

Alkyllithium reagents can also be generated by reduction of sulfides. This technique is especially useful for the preparation of α-lithio ethers, sulfides, and silanes.[32]

$$\text{2-(phenylthio)tetrahydropyran (SPh)} \xrightarrow{LDMAN} \text{2-lithiotetrahydropyran (Li)}$$

$$PhSC(CH_3)_2Si(CH_3)_3 \xrightarrow{LDMAN} LiC(CH_3)_2Si(CH_3)_3$$

The lithium radical anion of naphthalene or dimethylaminonaphthalene (LDMAN) is used as the reducing agent. This method can also be applied to prepare unsub-

28. E. J. Corey and R. H. Wollenberg, *J. Org. Chem.* **40**, 2265 (1975).
29. W. C. Still, *J. Am. Chem. Soc.* **100**, 1481 (1978).
30. D. J. Peterson, *J. Am. Chem. Soc.* **93**, 4027 (1971).
31. W. C. Still and C. Sreekumar, *J. Am. Chem. Soc.* **102**, 1201 (1980); J. S. Sawyer, A. Kucerovy, T. L. Macdonald, and G. J. McGarvey, *J. Am. Chem. Soc.* **110**, 842 (1988).
32. T. Cohen and J. R. Matz, *J. Am. Chem. Soc.* **102**, 6900 (1980); T. Cohen, J. P. Sherbine, J. R. Matz, R. R. Hutchins, B. M. McHenry, and P. R. Wiley, *J. Am. Chem. Soc.* **106**, 3245 (1984); T. Cohen and M. Bhupathy, *Acc. Chem. Res.* **22**, 152 (1989).

stituted alkyllithium reagents, although in this case there is no special advantage over the conventional procedure.

$$\mathrm{PhCH_2CH_2SPh} \xrightarrow[\mathrm{Li^+Naph^-}]{\mathrm{Li\ or}} \mathrm{PhCH_2CH_2Li}$$ Ref. 33

Alkenyllithium compounds are intermediates in the Shapiro reaction, which was discussed in Section 5.6. The reaction can be run in such a way that the organolithium compound is generated in high yield and subsequently reacted with a variety of electrophiles.[34] This method provides a route to vinyllithium compounds starting from a ketone.

NNHTs; C_2H_5; H_3C $\xrightarrow[\text{TMEDA}]{2\ n\text{-BuLi}}$ Li; C_2H_5; H_3C Ref. 35

7.2. Reactions of Organolithium and Organomagnesium Compounds

7.2.1. Reactions with Alkylating Agents

The organometallic compounds of the group IA and IIA metals are strongly basic and nucleophilic. Although the alkylation of these compounds might seem to offer a general synthetic method for formation of carbon–carbon bonds, in fact this reaction is somewhat limited in scope. Several mechanistic studies of the reaction between simple alkyllithium compounds and alkyl halides has led to the conclusion that radicals are involved. Radicals can be detected during the reaction by EPR and CIDNP techniques.[36] The alkylation product is also accompanied by products formed by disproportionation and coupling of radical intermediates and hydrogen abstraction form the solvent.[37]

$$n\text{-}C_4H_9Br + n\text{-}C_4H_9Li \xrightarrow{\text{HC(CH}_3)_2\text{-C}_6\text{H}_5} \underset{(43\%)}{CH_3(CH_2)_6CH_3} + \underset{(3\%)}{CH_3CH_2CH{=}CH_2} + \underset{(19\%)}{CH_3CH_2CH_2CH_3}$$

$$+ \ \mathrm{C_6H_5C(CH_3)_2C(CH_3)_2C_6H_5}\ (18\%)$$

33. C. G. Screttas and M. Micha-Screttas, *J. Org. Chem.* **43**, 1064 (1978); C. G. Screttas and M. Micha-Screttas, *J. Org. Chem.* **44**, 113 (1979).
34. F. T. Bond and R. A. DiPietro, *J. Org. Chem.* **46**, 1315 (1981); T. H. Chan, A. Baldassarre, and D. Massuda, *Synthesis*, 801 (1976); B. M. Trost and T. N. Nanninga, *J. Am. Chem. Soc.* **107**, 1293 (1985).
35. W. Barth and L. A. Paquette, *J. Org. Chem.* **50**, 2438 (1985).
36. G. A. Russell and D. W. Lamson, *J. Am. Chem. Soc.* **91**, 3967 (1969); H. R. Ward and R. G. Lawler, *J. Am. Chem. Soc.* **89**, 5518 (1967); A. R. Lepley and R. L. Landau, *J. Am. Chem. Soc.* **91**, 748 (1969); H. R. Ward, R. G. Lawler, and R. A. Cooper, *J. Am. Chem. Soc.* **91**, 746 (1969).
37. D. Bryce-Smith, *J. Chem. Soc.*, 1603 (1956).

Alkylation by allylic halides is a more satisfactory reaction. Alkylation in this case may often proceed through a cyclic mechanism.[38] For example, when allyl 1-^{14}C chloride reacts with phenyllithium, about three-fourths of the product has the labeled carbon at the terminal methylene group.

$$H_2C{=}CH{-}\overset{*}{C}H_2Cl + PhLi \longrightarrow PhCH_2CH{=}\overset{*}{C}H_2$$

Organolithium reagents in which the carbanion is delocalized also appear to constitute an exception to the radical-coupling mechanism. Allyllithium and benzyllithium reagents can be alkylated, and with secondary alkyl bromides a high degree of inversion of configuration is observed.[39]

$$PhCH_2Li + CH_3CH_2CH(CH_3)Br \longrightarrow PhCH_2{-}C(H)(CH_3)CH_2CH_3$$

(58% yield, 100% inversion)

Alkenyllithium reagents can be alkylated in good yields by alkyl iodides and bromides.[40]

$$(CH_3)(Br)C{=}C(CH_3)(H) \xrightarrow[2)\ CH_3(CH_2)_3I]{1)\ Li} (CH_3)(CH_3(CH_2)_3)C{=}C(CH_3)(H)$$

(77%)

The reaction of 1,3-, 1,4-, and 1,5-diiodides with *t*-butyllithium is an effective means of ring closure, but 1,6-diiodides give very little cyclization.[41]

$$1,2\text{-bis}(CH_2I)\text{cyclohexane} \xrightarrow{t\text{-BuLi}} \text{bicyclo[4.2.0]octane}$$

(97%)

The alkylation of Grignard reagents is of some synthetic value, especially when methyl, allyl, or benzyl halides are involved.

$$\text{6-bromo-1,1,4,4-tetramethyltetralin} \xrightarrow[2)\ CH_2{=}CHCH_2Br]{1)\ Mg} \text{6-}(CH_2CH{=}CH_2)\text{-1,1,4,4-tetramethyltetralin}$$

(79%) Ref. 42

38. R. M. Magid and J. G. Welch, *J. Am. Chem. Soc.* **90**, 5211 (1968); R. M. Magid, E. C. Nieh, and R. D. Gandour, *J. Org. Chem.* **36**, 2099 (1971); R. M. Magid and E. C. Nieh, *J. Org. Chem.* **36**, 2105 (1971).
39. L. H. Sommer and W. D. Korte, *J. Org. Chem.* **35**, 22 (1970).
40. J. Milon, R. Lorne, and G. Linstrumelle, *Synthesis*, 434 (1975).
41. W. F. Bailey, R. P. Gagnier, and J. J. Patricia, *J. Org. Chem.* **49**, 2098 (1984).
42. J. Eustache, J.-M. Bernardon, and B. Shroot, *Tetrahedron Lett.* **28**, 4681 (1987).

Synthetically useful alkylation of Grignard reagents can also be carried out with alkyl sulfonates and sulfates:

$$PhCH_2MgCl + CH_3CH_2CH_2CH_2OSO_2C_7H_7 \longrightarrow \underset{(50\text{–}59\,\%)}{PhCH_2CH_2CH_2CH_2CH_3} \quad \text{Ref. 43}$$

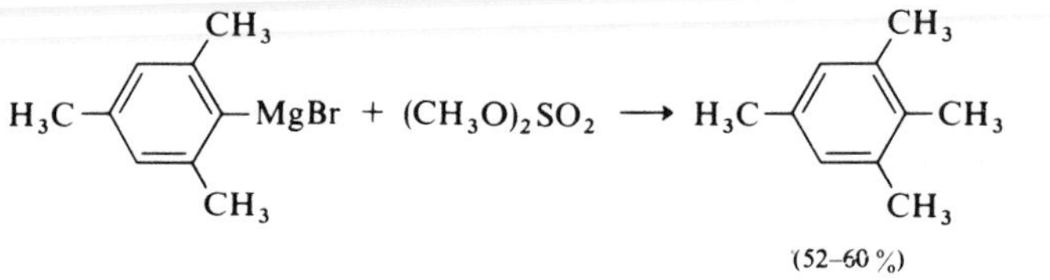

$$\text{2,4,6-}(CH_3)_3C_6H_2MgBr + (CH_3O)_2SO_2 \longrightarrow \underset{(52\text{–}60\,\%)}{\text{1,2,3,5-}(CH_3)_4C_6H_2} \quad \text{Ref. 44}$$

7.2.2. Reactions with Carbonyl Compounds

The most important type of reaction of Grignard reagents involves addition to carbonyl groups. The transition state for addition of Grignard reagents is often represented as a cyclic array containing the carbonyl group and two molecules of Grignard reagent. There is considerable evidence favoring this mechanism involving a termolecular complex.[45]

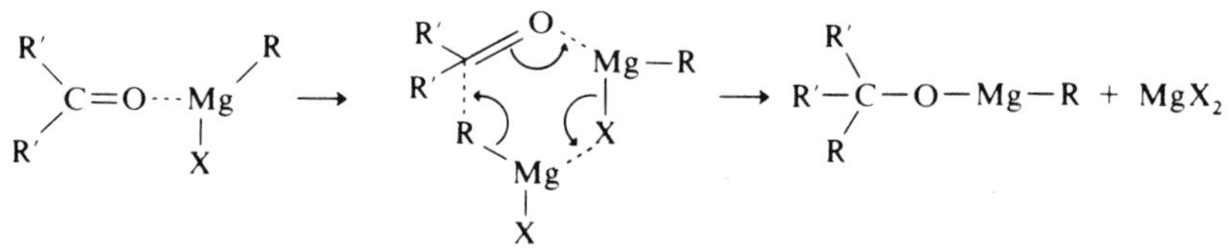

$$R'_2C{=}O\cdots Mg(R)X \longrightarrow [\text{cyclic transition state: } R'_2C{=}O\cdots Mg{-}R,\ R{-}Mg{-}X] \longrightarrow R'{-}\underset{R}{\overset{R'}{C}}{-}O{-}Mg{-}R + MgX_2$$

When the carbonyl carbon carries a substituent that can act as a leaving group, the initial adduct can break down to regenerate a C=O bond and a second addition step can occur. Esters, for example, usually are converted to tertiary alcohols, rather than ketones, in reactions with Grignard reagents.

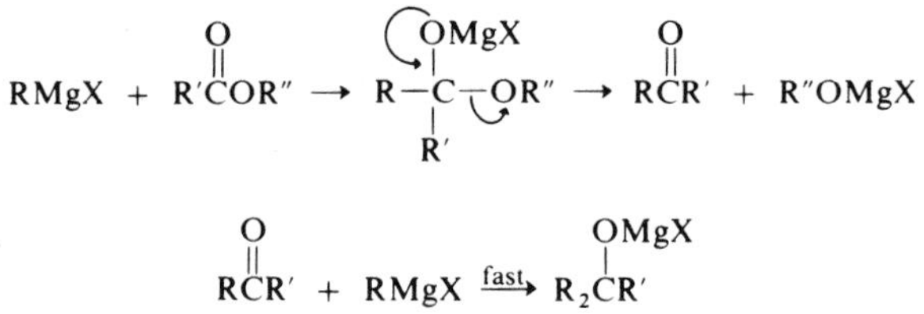

$$RMgX + R'\overset{O}{\overset{\|}{C}}OR'' \longrightarrow R{-}\underset{R'}{\overset{OMgX}{C}}{-}OR'' \longrightarrow R\overset{O}{\overset{\|}{C}}R' + R''OMgX$$

$$R\overset{O}{\overset{\|}{C}}R' + RMgX \xrightarrow{\text{fast}} R_2\overset{OMgX}{C}R'$$

The addition of Grignard reagents to aldehydes, ketones, and esters is the basis for synthesis of a wide variety of alcohols, and a number of examples are given in Scheme 7.3.

43. H. Gilman and J. Robinson, *Org. Synth.* **II**, 47 (1943).
44. L. I. Smith, *Org. Synth.* **II**, 360 (1943).
45. E. C. Ashby, R. B. Duke, and H. M. Neuman, *J. Am. Chem. Soc.* **89**, 1964 (1967); E. C. Ashby, *Pure Appl. Chem.* **52**, 545 (1980).

Grignard reagents add to nitriles, and, after hydrolysis of the reaction mixture, a ketone is obtained. Hydrocarbons are the preferred solvents for this reaction.[46]

$$RMgX + R'C{\equiv}N \rightarrow RC(=NMgX)R' \xrightarrow{H_2O} RC(=O)R'$$

Ketones can also be prepared from acid chlorides by reaction at low temperature with an excess of acid chloride. Tetrahydrofuran is the preferred solvent.[47] The reaction conditions must be controlled to prevent formation of tertiary alcohol by addition of Grignard reagent to the ketone as it is formed.

$$CH_3(CH_2)_5MgBr + CH_3CH_2CH_2C(=O)Cl \xrightarrow{-30°C} CH_3(CH_2)_5C(=O)(CH_2)_2CH_3 \quad (92\%)$$

THIS INTO THF

2-Pyridinethiolate esters, which are easily prepared from acid chlorides, also react with Grignard reagents to give ketones.[48]

$$\text{2-methyl-1,3-dithiolan-2-yl}-CH_2CH_2C(=O)S\text{-}C_6H_5 + BrMgCH_2CH{=}CHCH_2CH_3 \; (cis) \longrightarrow \text{2-methyl-1,3-dithiolan-2-yl}-CH_2CH_2C(=O)CH_2CH{=}CHCH_2CH_3 \; (cis) \quad (93\%)$$

Aldehydes can be obtained by reaction of Grignard reagents with triethyl orthoformate. The addition step is preceded by elimination of one of the alkoxy groups to generate an electrophilic carbon. The elimination is promoted by the magnesium ion acting as a Lewis acid.[49] The acetals formed by the addition are stable under the reaction conditions but are hydrolyzed to aldehydes on contact with aqueous acid.

$$(C_2H_5O)_2HC{-}OC_2H_5 \cdots RMg{-}X \rightarrow HC^{+}(OC_2H_5)_2 \xrightarrow{RMgX} RCH(OC_2H_5)_2 + C_2H_5OMgR + X^-$$

46. P. Canonne, G. B. Foscolos, and G. Lemay, *Tetrahedron Lett.*, 155 (1980).
47. F. Sato, M. Inoue, K. Oguro, and M. Sato, *Tetrahedron Lett.*, 4303 (1979).
48. T. Mukaiyama, M. Araki, and H. Takei, *J. Am. Chem. Soc.* **95**, 4763 (1973); M. Araki, S. Sakata, H. Takei, and T. Mukaiyama, *Bull. Chem. Soc. Jpn.* **47**, 1777 (1974).
49. E. L. Eliel and F. W. Nader, *J. Am. Chem. Soc.* **92**, 584 (1970).

A. Primary Alcohols from Formaldehyde

1[a] $C_6H_{11}MgCl + CH_2O \longrightarrow \xrightarrow[H^+]{H_2O} C_6H_{11}CH_2OH$ (64–69 %)

B. Primary Alcohols from Ethylene Oxide

2[b] $CH_3(CH_2)_3MgBr + H_2C\underset{O}{—}CH_2 \longrightarrow CH_3(CH_2)_5OH$ (60–62 %)

C. Secondary Alcohols from Aldehydes

3[c] $PhCH{=}CHCH{=}O + HC{\equiv}CMgBr \longrightarrow \xrightarrow[H^+]{H_2O} HC{\equiv}CCH(OH)CH{=}CHPh$ (58–69 %)

4[d] $m\text{-}ClC_6H_4MgBr + CH_3CH{=}O \longrightarrow \xrightarrow{H_2O} m\text{-}ClC_6H_4CHOHCH_3$ (82–85%)

5[e] $CH_3CH{=}CHCH{=}O + CH_3MgCl \longrightarrow \xrightarrow{H_2O} CH_3CH{=}CHCH(OH)CH_3$ (81–86 %)

6[f] $(CH_3)_2CHMgBr + CH_3CH{=}O \rightarrow (CH_3)_2CHCH(OH)CH_3$ (53–54%)

D. Secondary Alcohols from Formate Esters

7[g] $2\,CH_3(CH_2)_3MgBr + HCO_2C_2H_5 \longrightarrow \xrightarrow[H^+]{H_2O} (CH_3CH_2CH_2CH_2)_2CHOH$ (83–85 %)

E. Tertiary Alcohols from Esters and Lactones

8[h] $3\,C_2H_5MgBr + (C_2H_5O)_2CO \longrightarrow \xrightarrow[NH_4Cl]{H_2O} (C_2H_5)_3COH$ (82–88 %)

9[i] $2\,PhMgBr + PhCO_2C_2H_5 \longrightarrow \xrightarrow{H_2O} Ph_3COH$ (89–93%)

10[j] $CH_3(CH_2)_4$-γ-butyrolactone $+ 2\,CH_3MgBr \longrightarrow \xrightarrow[H^+]{H_2O} CH_3(CH_2)_4CH(OH)(CH_2)_2C(OH)(CH_3)_2$ (57 %)

F. Aldehydes from Ethyl Orthoformate

11[k] 9-phenanthryl-MgBr $+ HC(OC_2H_5)_3 \longrightarrow \xrightarrow[H^+]{H_2O}$ 9-phenanthryl-CH=O (40–42 %)

12[l] $CH_3(CH_2)_4MgBr + HC(OC_2H_5)_3 \longrightarrow \xrightarrow[H^+]{H_2O} CH_3(CH_2)_4CH{=}O$ (45–50%)

G. Ketones from Nitriles

13[m] 9-cyanophenanthrene ($C{\equiv}N$) + $CH_3MgI \longrightarrow \xrightarrow[HCl]{H_2O}$ 9-acetylphenanthrene ($\overset{O}{\overset{\|}{C}}CH_3$) (52–59%)

14[n] $CH_3OCH_2C{\equiv}N + PhMgBr \longrightarrow \xrightarrow[HCl]{H_2O} Ph\overset{O}{\overset{\|}{C}}CH_2OCH_3$ (71–78%)

H. Carboxylic Acids by Carbonation

15[o] 2,4,6-$(CH_3)_3C_6H_2$–MgBr + $CO_2 \longrightarrow \xrightarrow[H^+]{H_2O}$ 2,4,6-$(CH_3)_3C_6H_2$–CO_2H (86–87%)

16[p] $CH_3CH_2CH(MgBr)CH_3 + CO_2 \longrightarrow \xrightarrow[H^+]{H_2O} CH_3CH_2CH(CO_2H)CH_3$ (76–86%)

17[q] 1-chloronorbornane (Cl) $\xrightarrow[3)\ H^+,\ H_2O]{1)\ \text{active Mg}\quad 2)\ CO_2}$ norbornane-1-carboxylic acid (CO_2H) 60–70%

I. Amines from Imines

18[r] $PhCH{=}NCH_3 + PhCH_2MgCl \longrightarrow \xrightarrow{H_2O} PhCH(NHCH_3)CH_2Ph$ (96%)

J. Alkenes after Dehydration of Intermediate Alcohols

19[s] $PhCH{=}CHCH{=}O + CH_3MgBr \longrightarrow \xrightarrow{H_2SO_4} PhCH{=}CHCH{=}CH_2$ (75%)

20[t] $2\ PhMgBr + CH_3CO_2C_2H_5 \longrightarrow \xrightarrow[H_2O]{H^+} Ph_2C{=}CH_2$ (67–70%)

a. H. Gilman and W. E. Catlin, *Org. Synth.* **I**, 182 (1932).
b. E. E. Dreger, *Org. Synth.* **I**, 299 (1932).
c. L. Skattebøl, E. R. H. Jones, and M. C. Whiting, *Org. Synth.* **IV**, 792 (1963).
d. C. G. Overberger, J. H. Saunders, R. E. Allen, and R. Gander, *Org. Synth.* **III**, 200 (1955).
e. E. R. Coburn, *Org. Synth.* **III**, 696 (1955).
f. N. L. Drake and G. B. Cooke, *Org. Synth.* **II**, 406 (1943).
g. G. H. Coleman and D. Craig, *Org. Synth.* **II**, 179 (1943).
h. W. W. Moyer and C. S. Marvel, *Org. Synth.* **II**, 602 (1943).
i. W. E. Bachman and H. P. Hetzner, *Org. Synth.* **III**, 839 (1955).
j. J. Colonge and R. Marey, *Org. Synth.* **IV**, 601 (1963).
k. C. A. Dornfeld and G. H. Coleman, *Org. Synth.* **III**, 701 (1955).
l. G. B. Bachman, *Org. Synth.* **II**, 323 (1943).
m. J. E. Callen, C. A. Dornfeld, and G. H. Coleman, *Org. Synth.* **III**, 26 (1955).
n. R. B. Moffett and R. L. Shriner, *Org. Synth.* **III**, 562 (1955).
o. D. M. Bowen, *Org. Synth.* **III**, 553 (1955).
p. H. Gilman and R. H. Kirby, *Org. Synth.* **I**, 353 (1932).
q. R. D. Rieke, S. E. Bales, P. M. Hudnall, and G. S. Poindexter, *Org. Synth.* **59**, 85 (1977).
r. R. B. Moffett, *Org. Synth.* **IV**, 605 (1963).
s. O. Grummitt and E. I. Becker, *Org. Synth.* **IV**, 771 (1963).
t. C. F. H. Allen and S. Converse, *Org. Synth.* **I**, 221 (1932).

Carboxylic acids are obtained from Grignard reagents by reaction with carbon dioxide. Scheme 7.3 includes some specific examples of procedures described in *Organic Syntheses.*

$$\mathrm{RMgX} + \mathrm{CO_2} \rightarrow \mathrm{R\overset{O}{\overset{\|}{C}}OMgX} \xrightarrow[\mathrm{H_2O}]{\mathrm{H^+}} \mathrm{RCO_2H}$$

It is important to recognize that the use of Grignard reagents is quite restricted in terms of the types of functional groups that can be present in either the organometallic or the carbonyl compound. Alkene, ether, and acetal functionality usually causes no difficulty, but unprotected OH, NH, SH, or carbonyl groups cannot be present, and CN and NO_2 groups cause problems in many cases.

Grignard additions are sensitive to steric effects, and with hindered ketones a competing process involving reduction of the carbonyl group is observed. A cyclic transition state similar to that proposed for the Meerwein–Pondorff–Verley reduction (p. 249) is involved.

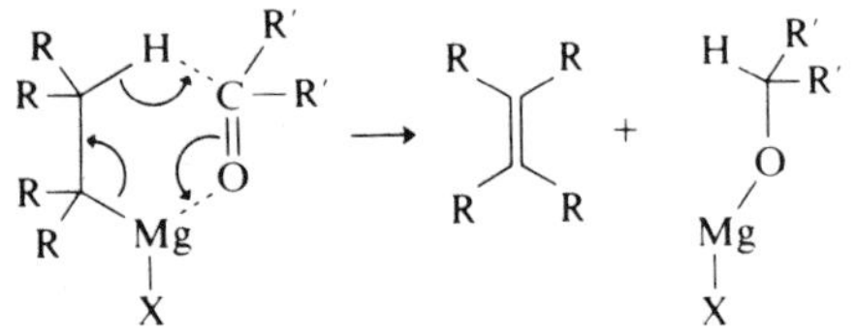

The extent of this reaction increases with the steric bulk of the ketone and the Grignard reagent. For example, no addition occurs between diisopropyl ketone and isopropylmagnesium bromide, and the reduction product diisopropylcarbinol is formed in 70% yield.[50] Competing reduction can be minimized in troublesome cases by using benzene or toluene as the solvent.[51]

Enolization of the ketone is also sometimes a competing reaction. Since the enolate is unreactive toward nucleophilic addition, the ketone is recovered unchanged after hydrolysis. Enolization has been shown to be especially important when a considerable portion of the Grignard reagent is present as an alkoxide.[52] Alkoxides are formed as the addition reaction proceeds. They also can be present as the result of oxidation of some of the Grignard reagent by oxygen during preparation or storage. As with reduction, enolization is most seriously competitive in cases where addition is retarded by steric factors.

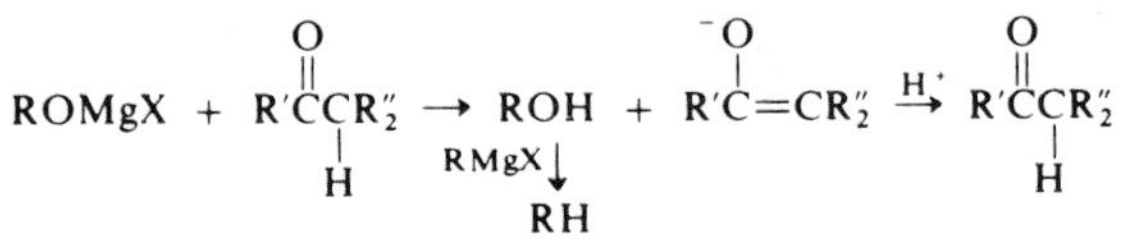

$$\mathrm{ROMgX} + \mathrm{R'\overset{O}{\overset{\|}{C}}\underset{H}{\underset{|}{C}}R''_2} \rightarrow \underset{\downarrow \mathrm{RMgX}}{\mathrm{ROH}} + \mathrm{R'\overset{^-O}{\overset{|}{C}}{=}CR''_2} \xrightarrow{\mathrm{H^+}} \mathrm{R'\overset{O}{\overset{\|}{C}}\underset{H}{\underset{|}{C}}R''_2}$$

$$\mathrm{RH}$$

Structural rearrangements are not encountered with saturated Grignard reagents. Allylic and homoallylic systems can give products resulting from isomeriz-

50. D. O. Cowan and H. S. Mosher, *J. Org. Chem.* **27**, 1 (1962).
51. P. Canonne, G. B. Foscolos, and G. Lemay, *Tetrahedron Lett.*, 4383 (1979).
52. H. O. House and D. D. Traficante, *J. Org. Chem.* **28**, 355 (1963).

ation. NMR studies indicate that allylmagnesium bromide exists as a σ-bonded structure in which there is rapid equilibration of the two terminal carbons.[53] 2-Butenylmagnesium bromide and 1-methylpropenylmagnesium bromide are in equilibrium in solution, with the latter constituent being the minor constituent.[54]

$$CH_3CH{=}CHCH_2MgBr \rightleftharpoons CH_2{=}CHCH(CH_3)MgBr$$

Addition products incorporate the 1-methylpropenyl group. Addition is believed to occur through a cyclic process that leads to an allylic shift.

$$R_2C{=}O \cdots Mg(Br)CH_2CH{=}CHCH_3 \rightarrow R_2C(OMgBr)CH(CH_3)CH{=}CH_2$$

This mode of addition is supplanted by reaction at the primary carbon when hindered ketones are involved. This is undoubtedly the result of a steric effect.

$$(CH_3)_3CC(=O)C(CH_3)_3 + BrMgCH_2CH{=}CHCH_3 \rightarrow [(CH_3)_3C]_2C(OMgBr)CH_2CH{=}CHCH_3$$

The adducts of allylic Grignard reagents with ketones can undergo isomerization to less hindered isomers by a reversal of the addition step.[55]

$$(CH_3)_3CC(OMgBr)(CH(CH_3)CH{=}CH_2)C(CH_3)_3 \xrightarrow[50^\circ C]{6\ h} (CH_3)_3CC(OMgBr)(CH_2CH{=}CHCH_3)C(CH_3)_3$$

3-Butenylmagnesum bromide is in equilibrium with a small amount of cyclopropylmethylmagnesium bromide. The existence of the mobile equilibrium has been established by deuterium-labeling techniques.[56] Cyclopropylmethylmagnesium bromide[57] (and cyclopropylmethyllithium[58]) can be prepared by working at low temperature. At room temperature, the ring-opened 3-butenyl reagents are formed.

$$CH_2{=}CHCH_2CD_2MgBr \rightleftharpoons \underset{CD_2}{\overset{CH_2}{|}}\!\!>CHCH_2MgBr \rightleftharpoons BrMgCH_2CD_2CH{=}CH_2$$

53. M. Schlosser and N. Stähle, *Angew. Chem. Int. Ed. Engl.* **19**, 487 (1980); M. Stähle and M. Schlosser, *J. Organomet. Chem.* **220**, 277 (1981).
54. R. A. Benkeser, W. G. Young, W. E. Broxterman, D. A. Jones, Jr., and S. J. Piaseczynski, *J. Am. Chem. Soc.* **91**, 132 (1969).
55. R. A. Benkeser, M. P. Siklosi, and E. C. Mozdzen, *J. Am. Chem. Soc.* **100**, 2134 (1978).
56. M. E. H. Howden, A. Maercker, J. Burdon, and J. D. Roberts, *J. Am. Chem. Soc.* **88**, 1732 (1966).
57. D. J. Patel, C. L. Hamilton, and J. D. Roberts, *J. Am. Chem. Soc.* **87**, 5144 (1965).
58. P. T. Lansbury, V. A. Pattison, W. A. Clement, and J. D. Sidler, *J. Am. Chem. Soc.* **86**, 2247 (1964).

When the double bond is further removed, as in 5-hexenylmagnesium bromide, there is no evidence of a similar equilibrium.[59]

$$CH_2{=}CHCH_2CH_2CH_2CH_2MgBr \not\rightleftharpoons BrMgCH_2\text{-cyclopentyl}$$

The corresponding lithium reagent does undergo cyclization.

The benzyl Grignard reagent exhibits a distinctive reactivity which has been carefully studied in the case where formaldehyde is the electrophile. Three products are formed, and the "normal" product is a minor one.[60]

$C_6H_5CH_2MgCl$ + $CH_2{=}O$ → A (o-$CH_3C_6H_4CH_2OH$) + B ($C_6H_5CH_2CH_2OH$) + C (o-$HOCH_2CH_2C_6H_4CH_2OH$)

A 30–50% B 5–8% C 0–15%

A cyclic mechanism can account for the formation of the major product **A**. The diol **C** can be formed by reaction of formaldehyde with an intermediate in this process.

CH₂—Mg(Cl)···O=CH₂ (cyclic) → H, CH₂OMgCl-substituted methylenecyclohexadiene → o-$CH_3C_6H_4CH_2OMgCl$ → A

methylenecyclohexadiene + $CH_2{=}O$ → o-$(ClMgOCH_2CH_2)C_6H_4CH_2OMgCl$ → C

The reactivity of organolithium reagents toward carbonyl compounds is generally similar to that of Grignard reagents. The lithium reagents are less likely to undergo the competing reduction with ketones, however. This is illustrated by the comparison of the reaction of ethyllithium and ethylmagnesium bromide with adamantone. An 83% yield of the tertiary alcohol is obtained with ethyllithium, whereas the Grignard reagent gives mainly the reduction product 2-adamantol.[61]

59. R. C. Lamb, P. W. Ayers, M. K. Toney, and J. F. Garst, *J. Am. Chem. Soc.* **88**, 4261 (1966).
60. R. A. Benkeser and D. C. Snyder, *J. Org. Chem.* **47**, 1243 (1982).
61. J. L. Fry, E. M. Engler, and P. v. R. Schleyer, *J. Am. Chem. Soc.* **94**, 4628 (1972).

One reaction that is quite efficient with lithium reagents but poor with Grignard reagents is the synthesis of ketones from carboxylic acids.[62] The success of the reaction depends upon the stability of the dilithio adduct that is formed. This intermediate does not break down until hydrolysis, at which point the ketone is liberated. Some examples of this reaction are shown in section B of Scheme 7.4.

$$RLi + R'\overset{O}{\overset{\|}{C}}O^-Li^+ \rightarrow R'\underset{R}{\overset{O^-Li^+}{C}}O^-Li^+ \xrightarrow[H_2O]{H^+} R'\underset{R}{\overset{OH}{C}}OH \rightarrow R\overset{O}{\overset{\|}{C}}R'$$

A study aimed at optimizing yields in this reaction found that carbinol formation was a major competing process if the reaction was not carried out in such a way that all of the lithium compound had been consumed prior to hydrolysis.[63] Any excess lithium reagent that is present reacts extremely rapidly with the ketone as it is formed by hydrolysis. Another way to avoid the problem of carbinol formation is to quench the reaction mixture with trimethylsilyl chloride.[64]

$$\text{cyclohexyl-}CO_2H \xrightarrow[\text{3) } H_2O,\ H^+]{\text{1) 4 equiv MeLi; 2) TMS—Cl}} \text{cyclohexyl-}\overset{O}{\overset{\|}{C}}CH_3 \quad (92\%)$$

Scheme 7.4 illustrates some of the important synthetic reactions in which organolithium reagents act as nucleophiles. In addition to this type of reactivity, the lithium reagents have enormous importance in synthesis as bases and as lithiating reagents. The commercially available methyl, *n*-butyl, and *t*-butyl reagents are used most frequently in this context.

The stereochemistry of the addition of both organomagnesium and organolithium compounds to cyclohexanones is similar.[65] With unhindered ketones the stereoselectivity is not high, but there is generally a preference for attack from the equatorial direction to give the axial alcohol. This preference for the equatorial approach increases with the size of the alkyl group. With alkyllithium reagents, added salts improve the stereoselectivity. For example, one equivalent of $LiClO_4$ enhances the proportion of the axial alcohol in the addition of methyllithium to 4-*t*-butylcyclohexanone.[66]

$$t\text{-Bu-cyclohexanone} \xrightarrow{CH_3Li} t\text{-Bu-cyclohexane(OH ax, }CH_3\text{ eq)} + t\text{-Bu-cyclohexane(}CH_3\text{ ax, OH eq)}$$

	axial OH, equatorial CH_3	axial CH_3, equatorial OH
no $LiClO_4$	(65%)	(35%)
1 equiv $LiClO_4$	(92%)	(8%)

62. M. J. Jorgenson, *Org. React.* **18**, 1 (1971).
63. R. Levine, M. J. Karten, and W. M. Kadunce, *J. Org. Chem.* **40**, 1770 (1975).
64. G. M. Rubottom and C. Kim, *J. Org. Chem.* **48**, 1550 (1983).
65. E. C. Ashby and J. T. Laemmle, *Chem. Rev.* **75**, 521 (1975).
66. E. C. Ashby and S. A. Noding, *J. Org. Chem.* **44**, 4371 (1979).

Scheme 7.4. Synthetic Procedures Involving Organolithium Reagents

A. Alkylation

1[a] 3,5-(MeO)$_2$C$_6$H$_3$–OSiR$_3$ $\xrightarrow[\text{2) BrCH}_2\text{CH=C(CH}_3)_2]{\text{1) } n\text{-BuLi}}$ $(CH_3)_2C{=}CHCH_2$–C$_6$H$_2$(MeO)$_2$–OSiR$_3$ (60%)

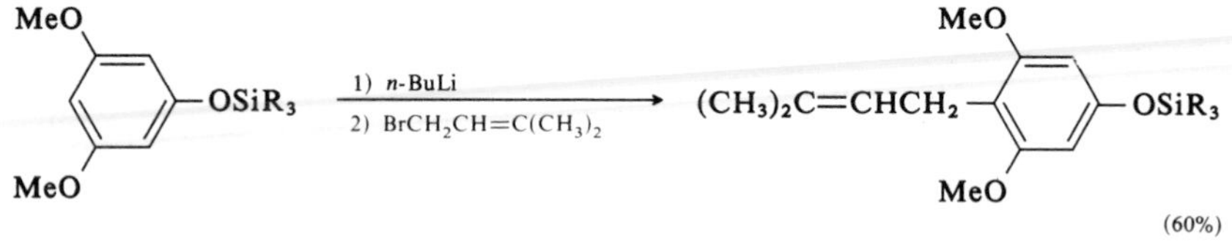

2[b] $(CH_3)_3CO\underset{\text{Li}}{C}HCH{=}CH_2 + CH_3(CH_2)_5I \rightarrow (CH_3)_3CO(H)C{=}C(H)(CH_2)_6CH_3$ (83%)

3[c] Bicyclic allylic bromide (CH$_2$Br, H$_3$C, CH$_3$, H$_3$C CH$_3$) + CH$_3$O, Li, CH$_2$OSiR$_3$, THPO, CH$_2$OSiR$_3$ substituted aryllithium → coupled product (CH$_3$O, CH$_2$OSiR$_3$, CH$_2$OSiR$_3$, OTHP, CH$_2$, H$_3$C, CH$_3$, H$_3$C CH$_3$) (65%)

4[d] Cyclohexene oxide + C_4H_9Li/BF_3 (3 equiv) → trans-2-(CH$_2$)$_3$CH$_3$-cyclohexanol (OH) (97%)

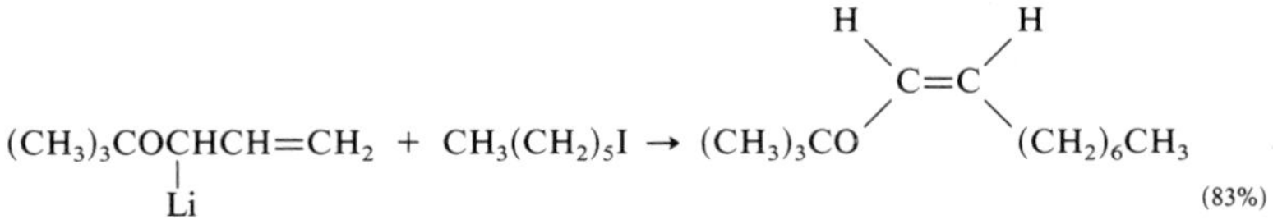

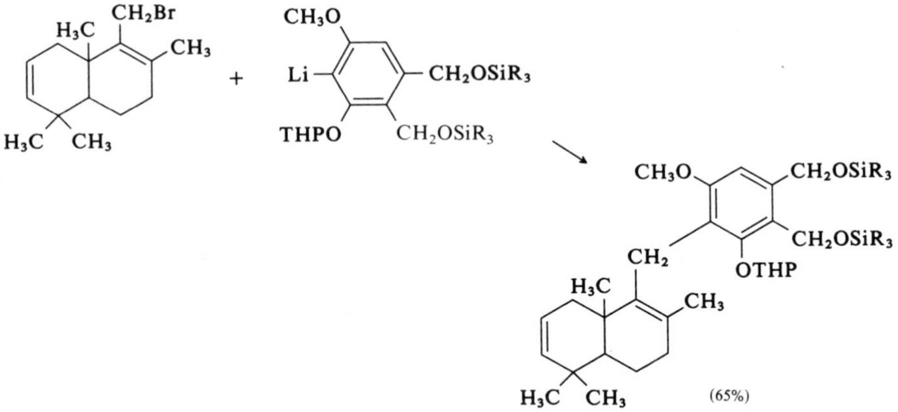

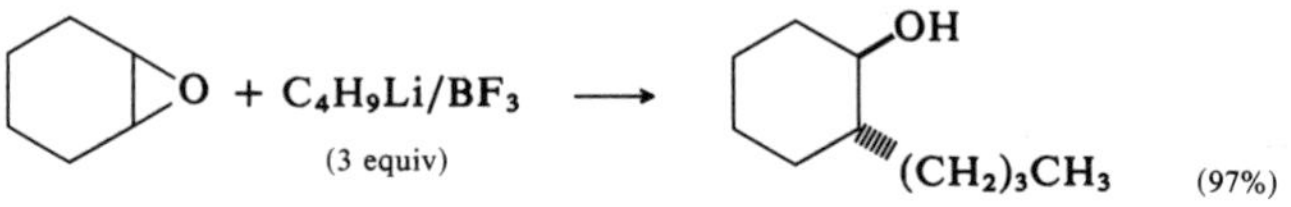

B. Reactions with Aldehydes and Ketones to Give Alcohols

5[e] $CH_2{=}CHCH_2Li + CH_3\overset{O}{\overset{\|}{C}}CH_2CH(CH_3)_2 \rightarrow CH_2{=}CHCH_2\overset{OH}{\underset{CH_3}{C}}CH_2CH(CH_3)_2$ (70–72%)

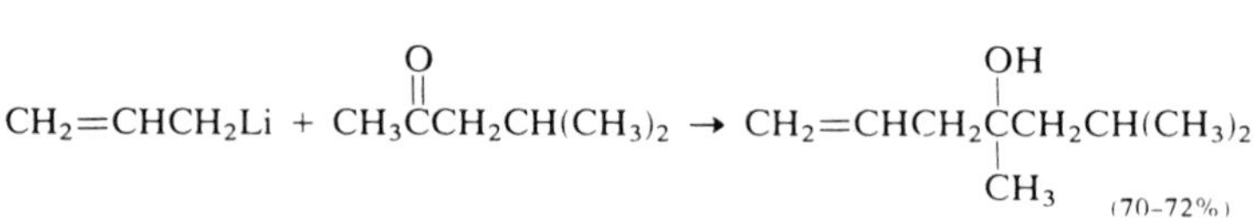

6[f] C_4H_9Li + cyclohexanone → 1-(CH$_2$CH$_2$CH$_2$CH$_3$)cyclohexanol (OH) (89%)

7[g] 2-methylpyridine (CH$_3$) $\xrightarrow{\text{PhLi}}$ 2-(CH$_2$Li)pyridine $\xrightarrow{CH_3CH{=}O}$ 2-(CH$_2\underset{OH}{C}$HCH$_3$)pyridine (44–50%)

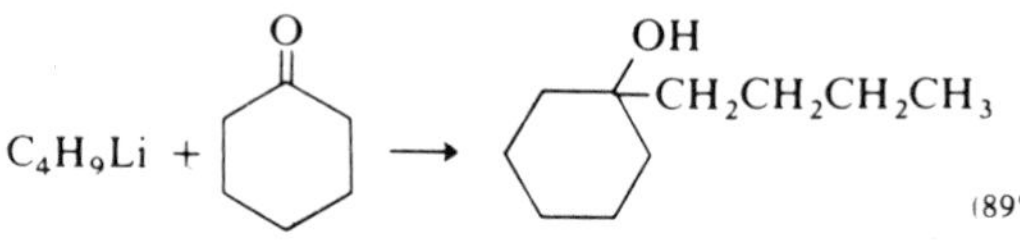

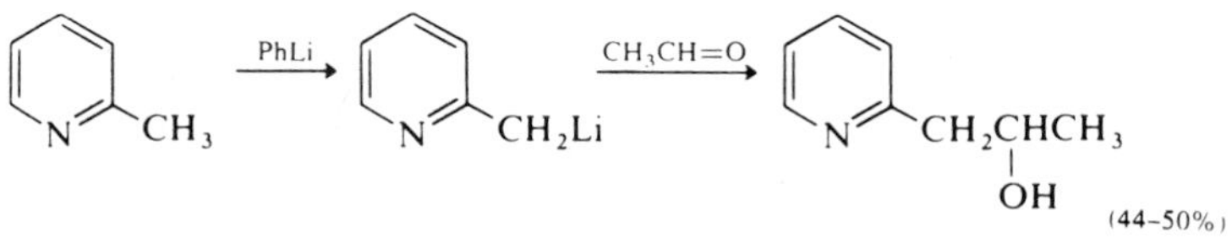

8[h] $CH_3CH{=}CHBr \xrightarrow[-120°C]{2\ t\text{-BuLi}} CH_3CH{=}CHLi \xrightarrow{PhCH=O} CH_3CH{=}CHCH(OH)Ph$ (72%)

9[i] 3,5-dimethoxy-$C_6H_3C(O)N(C_2H_5)_2$ $\xrightarrow{s\text{-BuLi}}$ 2-lithio-3,5-dimethoxy-$C_6H_2C(O)N(C_2H_5)_2$ $\xrightarrow{\text{2-}CH_3O\text{-4-}CH_3\text{-5-}OCH_3C_6H_2CH=O}$ product bearing CH_3O, $(C_2H_5)_2NC{=}O$, CH_3O, OH, OCH_3, CH_3, OCH_3 (63%)

C. Reactions with Carboxylic Salts, Acid Chlorides, and Acid Anhydrides to Give Ketones

10[j] $C_6H_{11}CO_2Li + CH_3Li \longrightarrow \xrightarrow{H_2O} C_6H_{11}CCH_3(=O)$ (91%)

11[k] $(CH_3)_3CCO_2H + 2\,PhLi \longrightarrow \xrightarrow{H_2O} PhCC(CH_3)_3(=O)$ (65%)

12[l] H, Ph, CO_2H, CH_3-cyclopropane $\xrightarrow{2CH_3Li}$ H, Ph, $CCH_3(=O)$, CH_3-cyclopropane (90%)

13[m] CH_3O–C_6H_3(Br)(CO_2H) $\xrightarrow[-100°C]{2\ n\text{-BuLi}}$ CH_3O–C_6H_3(Li)(CO_2Li) $\xrightarrow{m\text{-}CH_3OC_6H_4COCl}$ CH_3O–C_6H_3(CO_2H)–C(=O)–C_6H_4–OCH_3 (78%)

14[n] NC–C_6H_4–Br $\xrightarrow[2)\ \text{phthalic anhydride}]{1)\ n\text{-BuLi},\ -100°C}$ NC–C_6H_4–C(=O)–C_6H_4–CO_2H (71%)

Scheme 7.4—*continued*

D. Reactions with Carbon Dioxide to Give Carboxylic Acids

15[o] 3-(OCH_2OCH_3)-H_3C-benzene $\xrightarrow{\text{1) } t\text{-BuLi, 0°C; 2) } CO_2}$ H_3C-(OCH_2OCH_3)-benzene-CO_2H (90%)

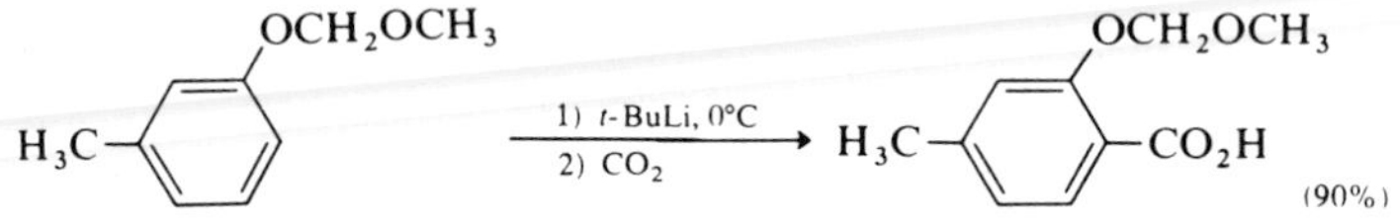

16[p] 2-methylpyridine (CH_3) $\xrightarrow{\text{1) PhLi; 2) } CO_2}$ 2-pyridyl-CH_2CO_2Li

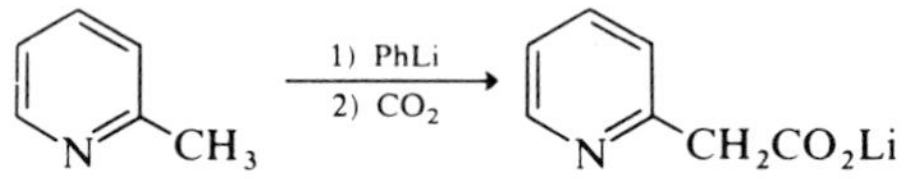

E. Other Reactions

17[q] $CH_2OSi(CH_3)_2C(CH_3)_3$, C_2H_5, OCH_3 (dihydronaphthalene) $\xrightarrow{\text{1) } n\text{-BuLi, TMEDA; 2) } HCN(CH_3)_2\text{ (C=O), 0°C}}$ $CH_2OSi(CH_3)_2C(CH_3)_3$, C_2H_5, O=CH, OCH_3 (80%)

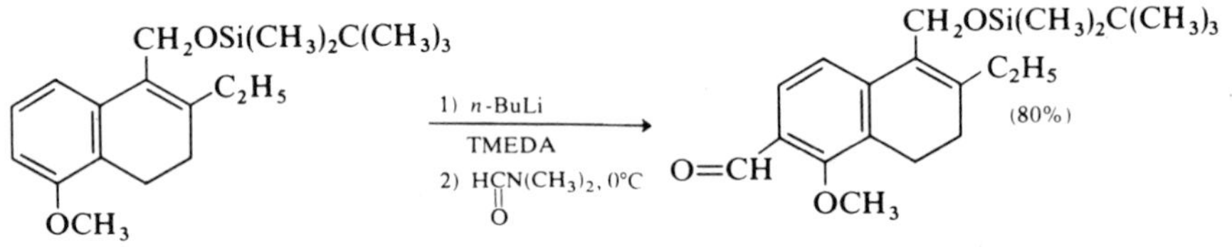

18[r] $(CH_3)_2C{=}CHC(O)N(i\text{-Pr})_2$ $\xrightarrow{\text{1) } LiN(i\text{-Pr})_2\text{; 2) } CH_3I}$ $H_2C{=}C(CH_3)CH(CH_3)C(O)N(i\text{-Pr})_2$ (98%)

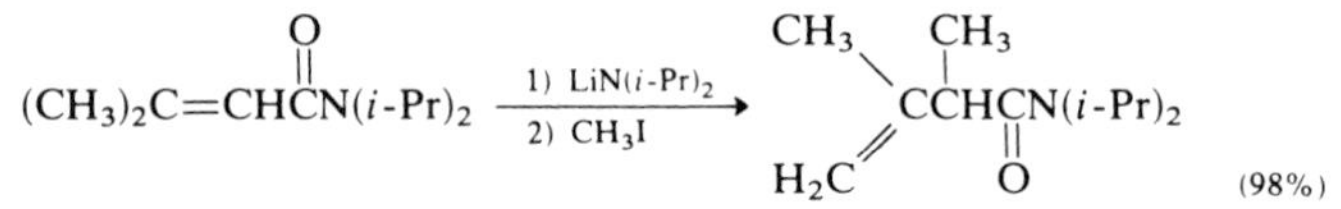

a. T. L. Shih, M. J. Wyvratt, and H. Mrozik, *J. Org. Chem.* **52**, 2029 (1987).
b. D. A. Evans, G. C. Andrews, and B. Buckwalter, *J. Am. Chem. Soc.* **96**, 5560 (1974).
c. J. E. McMurry and M. D. Erion, *J. Am. Chem. Soc.* **107**, 2712 (1985).
d. M. J. Eis, J. E. Wrobel, and B. Ganem, *J. Am. Chem. Soc.* **106**, 3693 (1984).
e. D. Seyferth and M. A. Weiner, *Org. Synth.* **V**, 452 (1973).
f. J. D. Buhler, *J. Org. Chem.* **38**, 904 (1973).
g. L. A. Walker, *Org. Synth.* **III**, 757 (1955).
h. H. Neumann and D. Seebach, *Tetrahedron Lett.*, 4839 (1976).
i. S. O. deSilva, M. Watanabe, and V. Snieckus, *J. Org. Chem.* **44**, 4802 (1979).
j. T. M. Bare and H. O. House, *Org. Synth.* **49**, 81 (1969).
k. R. Levine and M. J. Karten, *J. Org. Chem.* **41**, 1176 (1976).
l. C. H. DePuy, F. W. Breitbeil, and K. R. DeBruin, *J. Am. Chem. Soc.* **88**, 3347 (1966).
m. W. E. Parham, C. K. Bradsher, and K. J. Edgar, *J. Org. Chem.* **46**, 1057 (1981).
n. W. E. Parham and R. M. Piccirilli, *J. Org. Chem.* **41**, 1268 (1976).
o. R. C. Ronald, *Tetrahedron Lett.*, 3973 (1975).
p. R. B. Woodward and E. C. Kornfeld, *Org. Synth.* **III**, 413 (1955).
q. A. S. Kende and J. R. Rizzi, *J. Am. Chem. Soc.* **103**, 4247 (1981).
r. M. Majewski, G. B. Mpango, M. T. Thomas, A. Wu, and V. Snieckus, *J. Org. Chem.* **46**, 2029 (1981).

Bicyclic ketones react with organometallic reagents to give the products of addition from the less hindered face of the carbonyl group.

The stereochemistry of addition of organometallic reagents to acyclic carbonyl compounds parallels the behavior of the hydride reducing agents, as discussed in Section 5.2.1. Organometallic compounds were included in the early studies that established the preference for addition according to Cram's rule.[67]

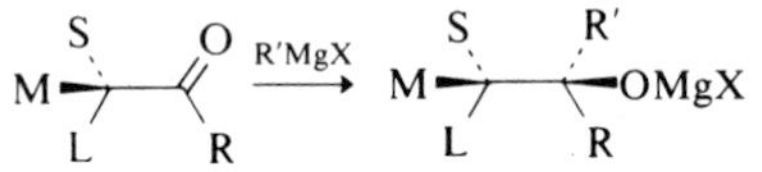

M(S)(L)C–C(=O)R $\xrightarrow{R'MgX}$ M(S)(L)C–C(R')(R)OMgX   S, M, L = relative size of substituents

67. D. J. Cram and F. A. A. Elhafez, *J. Am. Chem. Soc.* **74**, 5828 (1952).

The interpretation of the basis for this stereoselectivity can be made in terms of the steric, torsional, and stereoelectronic effects discussed in connection with reduction by hydrides. It has been found that crown ethers enhance stereoselectivity in the reaction of both Grignard reagents and alkyllithium compounds.[68]

For ketones and aldehydes in which adjacent substituents permit the possibility of chelation with the metal ion in the transition state, the stereochemistry can often be interpreted in terms of the steric requirements of the chelated transition state. In the case of α-alkoxyketones, for example, an assumption that the alkoxy substituent will be coordinated with the metal ion and that addition will occur from the less hindered side of this structure correctly predicts the stereochemistry of addition. The predicted product dominates by as much as 100 to 1 for several Grignard reagents.[69]

R″ XMg R′O O R H R → R′O OMgX R R″ H R

R CH$_3$ H R′O O + C_4H_9MgBr $\xrightarrow[-78°C]{THF}$ R CH$_3$ H C_4H_9 R′O OH

R = C_7H_{15} R′ = CH_2OCH_3
$CH_2OCH_2CH_2OCH_3$
CH_2Ph
CH_2OCH_2Ph

A similar study with β-alkoxyaldehydes revealed that neither Grignard reagents nor organolithium compounds exhibit high stereoselection in addition reactions.[70]

An alternative to preparing organometallic reagents and then carrying out reaction with a carbonyl compound is to generate the organometallic intermediate *in situ* in the presence of the carbonyl compound. The organometallic compound then reacts immediately with the carbonyl compound. This procedure is referred to as the *Barbier reaction.*[71] This technique has no advantage over the conventional one for most cases. However, when the organometallic reagent is very unstable, it can be a useful method. Allylic halides, which are difficult to convert to Grignard reagents in good yield, frequently give excellent results in the Barbier procedure. Since solid metals are used, one of the factors affecting the rate of the reaction is the physical state of the metal. Ultrasonic irradiation has been found to have a

68. Y. Yamamoto and K. Maruyama, *J. Am. Chem. Soc.* **107**, 6411 (1985).
69. W. C. Still and J. H. McDonald III, *Tetrahedron Lett.*, 1031 (1980).
70. W. C. Still and J. A. Schneider, *Tetrahedron Lett.*, 1035 (1980).
71. C. Blomberg and F. A. Hartog, *Synthesis*, 18 (1977).

favorable effect on the Barbier reaction, presumably by accelerating the generation of reactive sites on the metal surface.[72]

$$(CH_3)_2CHCH_2CH{=}O + CH_2{=}\underset{}{\overset{CH_3}{\overset{|}{C}}}CH_2Cl \xrightarrow[\text{ether}]{\text{Mg}} (CH_3)_2CHCH_2\overset{OH}{\overset{|}{C}}HCH_2\overset{CH_3}{\overset{|}{C}}{=}CH_2 \quad (92\%)$$

7.3. Organic Derivatives of Group IIB Metals

In this section, we will discuss organometallic derivatives of zinc, cadmium, and mercury. These group IIB metals have the d^{10} electronic configuration in the +2 oxidation state. Because of the filled *d* level, the +2 oxidation state is quite stable, and most reactions do not involve changes in oxidation level. This property makes the reactivity patterns of the group IIB organometallics more similar to those of the organometallics of groups IA and IIA than to those of organometallics of transition metals with vacancies in the *d* levels. The IIB metals, however, are much less electropositive than the IA and IIA metals so the nucleophilicity of the organometallics is much less than for organolithium or organomagnesium compounds. Many of the synthetic applications of these organometallics are based on this attentuated reactivity.

7.3.1. Organozinc Compounds

Organozinc compounds can be prepared by reaction of Grignard or organolithium reagents with zinc salts. A one-pot process in which the organic halide, magnesium metal, and zinc chloride are sonicated has proven to be a convenient method for the preparation.[73] Organozinc compounds can also be prepared from organic halides by reaction with highly reactive zinc metal.[74] Simple alkylzinc compounds can also be prepared from alkyl halides and Zn–Cu couple.[75] They are distillable liquids.

When prepared *in situ* from $ZnCl_2$ and Grignard reagents, organozinc reagents add to carbonyl compounds to give carbinols.[76] This must reflect some activation of the carbonyl group by Lewis acid catalysis by magnesium ion, since ketones are less reactive toward pure dialkylzinc reagents and tend to react by reduction rather than addition.[77]

72. J.-L. Luche and J.-C. Damiano, *J. Am. Chem. Soc.* **102**, 7926 (1980).
73. J. Boersma, *Comprehensive Organometallic Chemistry*, G. Wilkinson (ed.), Pergamon Press, Oxford, 1982, Chapter 16; G. E. Coates and K. Wade, *Organometallic Compounds*, Vol. 1, Third Edition, Methuen, London, 1967, pp. 121–128.
74. R. D. Rieke, P. T.-J. Li, T. P. Burns, and S. T. Uhm, *J. Org. Chem.* **46**, 4323 (1981).
75. C. R. Noller, *Org. Synth.* **II**, 184 (1943).
76. P. R. Jones, W. J. Kauffman, and E. J. Goller, *J. Org. Chem.* **36**, 186 (1971); P. R. Jones, E. J. Goller, and W. J. Kaufmann, *J. Org. Chem.* **36**, 3311 (1971).
77. G. Giacomelli, L. Lardicci, and R. Santi, *J. Org. Chem.* **39**, 2736 (1974).

High degrees of enantioselectivity have been observed when alkylzinc reagents react with aldehydes in the presence of chiral aminoalcohols.[78] The enantioselectivity is the result of chelation by the aminoalcohol. The aminoalcohols also serve to activate the organozinc reagent.

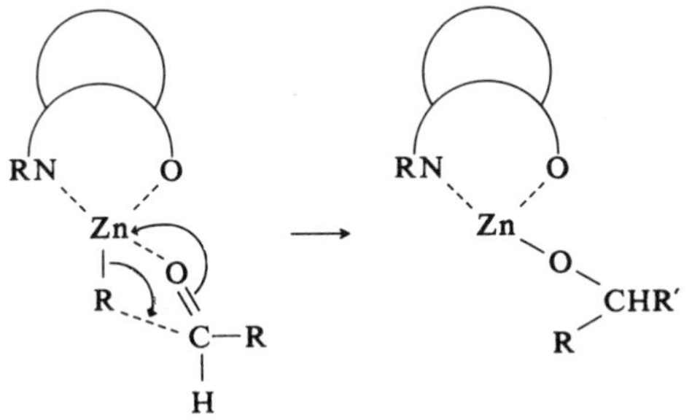

A frequently used reaction involving zinc is the Reformatsky reaction, in which zinc, an α-haloester, and a carbonyl compound react to give a β-hydroxyester.[79] The zinc and α-haloester react to form an organozinc reagent. Since the carboxylate group can stabilize the carbanionic center, the product is essentially the zinc enolate of the dehalogenated ester.[80] The enolate can then carry out a nucleophilic attack on the carbonyl group.

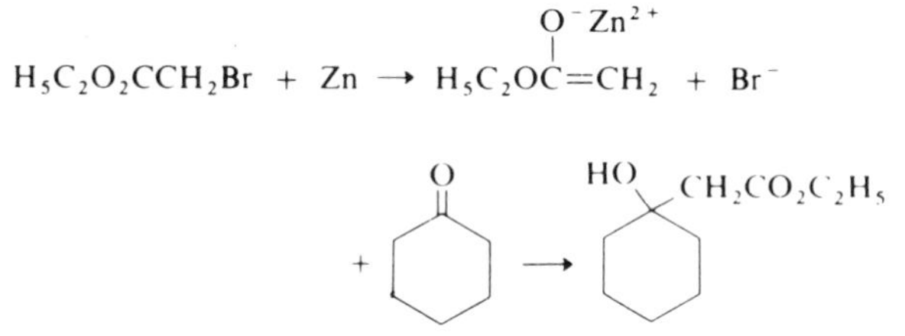

$$H_5C_2O_2CCH_2Br + Zn \rightarrow H_5C_2OC(O^-Zn^{2+})=CH_2 + Br^-$$

Several techniques have been used to "activate" the zinc metal and improve yields. For example, pretreatment of zinc dust with a solution of copper acetate gives a more reactive zinc–copper couple.[81] Exposure to trimethylsilyl chloride also activates the zinc.[82] Scheme 7.5 gives some examples of the Reformatsky reaction.

Zinc enolates prepared from α-haloketones can be used as nucleophiles in mixed aldol condensations (see Section 2.1.3). Entry 7 in Scheme 7.5 is an example. This reaction can be conducted in the presence of the Lewis acid diethylaluminum chloride, in which case addition occurs at $-20°C$.[83]

78. K. Soai, A. Ookawa, T. Kaba, and K. Ogawa, *J. Am. Chem. Soc.* **109**, 7111 (1987); M. Kitamura, S. Suga, K. Kawai, and R. Noyori, *J. Am. Chem. Soc.* **108**, 6071 (1986); W. Oppolzer and R. N. Rodinov, *Tetrahedron Lett.* **29**, 5645 (1988); M. Kitamura, S. Okada, and R. Noyori, *J. Am. Chem. Soc.* **111**, 4028 (1989).
79. R. L. Shriner, *Org. React.* **1**, 1 (1942); M. W. Rathke, *Org. React.* **22**, 423 (1975).
80. W. R. Vaughan and H. P. Knoess, *J. Org. Chem.* **35**, 2394 (1970).
81. E. Le Goff, *J. Org. Chem.* **29**, 2048 (1964); L. R. Krepski, L. E. Lynch, S. M. Heilmann, and J. K. Rasmussen, *Tetrahedron Lett.* **26**, 981 (1985).
82. G. Picotin and P. Miginiac, *J. Org. Chem.* **52**, 4796 (1987).
83. K. Maruoka, S. Hashimoto, Y. Kitagawa, H. Yamamoto, and H. Nozaki, *J. Am. Chem. Soc.* **99**, 7705 (1977).

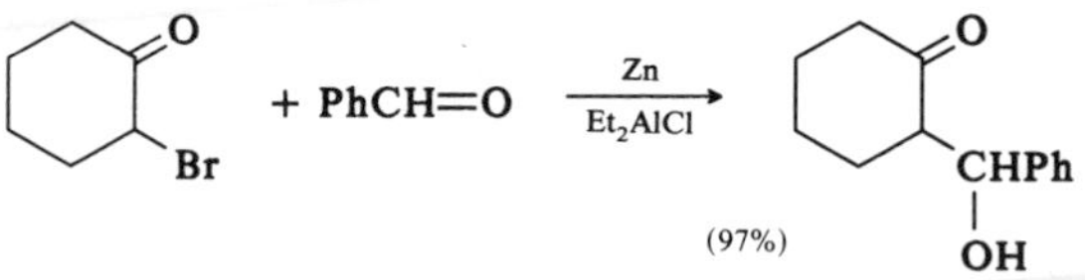

$$\text{2-bromocyclohexanone} + \text{PhCH=O} \xrightarrow[\text{Et}_2\text{AlCl}]{\text{Zn}} \text{2-(CH(OH)Ph)cyclohexanone} \quad (97\%)$$

Zinc enolates prepared under Reformatsky conditions undergo addition reactions with nitriles. The initial products are β-amino-α,β-unsaturated esters, which can be readily hydrolyzed to β-ketoesters.[84]

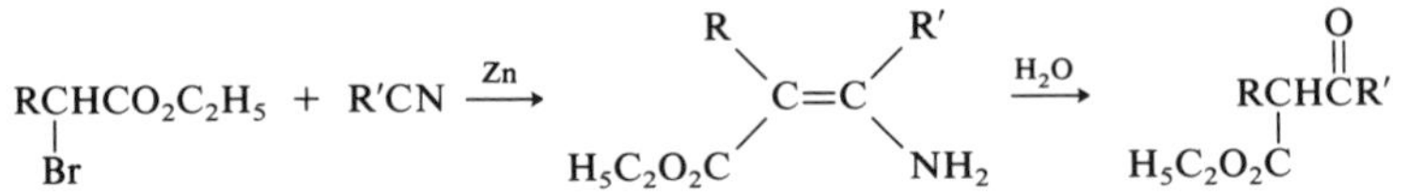

$$\text{RCH(Br)CO}_2\text{C}_2\text{H}_5 + \text{R'CN} \xrightarrow{\text{Zn}} (\text{R})(\text{H}_5\text{C}_2\text{O}_2\text{C})\text{C=C}(\text{R'})(\text{NH}_2) \xrightarrow{\text{H}_2\text{O}} \text{RCH(CO}_2\text{C}_2\text{H}_5)\text{C(=O)R'}$$

Scheme 7.5. Condensation of α-Halocarbonyl Compounds Using Zinc—The Reformatsky Reaction

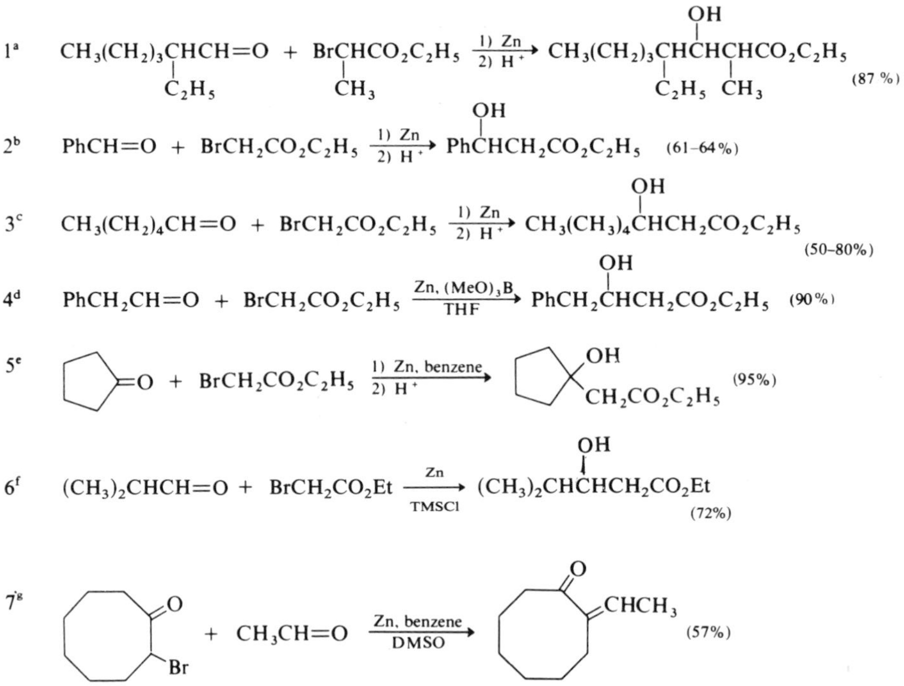

1[a] $$CH_3(CH_2)_3CH(C_2H_5)CH{=}O + BrCH(CH_3)CO_2C_2H_5 \xrightarrow[2)\ H^+]{1)\ Zn} CH_3(CH_2)_3CH(C_2H_5)CH(OH)CH(CH_3)CO_2C_2H_5 \quad (87\%)$$

2[b] $$PhCH{=}O + BrCH_2CO_2C_2H_5 \xrightarrow[2)\ H^+]{1)\ Zn} PhCH(OH)CH_2CO_2C_2H_5 \quad (61\text{–}64\%)$$

3[c] $$CH_3(CH_2)_4CH{=}O + BrCH_2CO_2C_2H_5 \xrightarrow[2)\ H^+]{1)\ Zn} CH_3(CH_3)_4CH(OH)CH_2CO_2C_2H_5 \quad (50\text{–}80\%)$$

4[d] $$PhCH_2CH{=}O + BrCH_2CO_2C_2H_5 \xrightarrow[THF]{Zn,\ (MeO)_3B} PhCH_2CH(OH)CH_2CO_2C_2H_5 \quad (90\%)$$

5[e] $$\text{cyclopentanone} + BrCH_2CO_2C_2H_5 \xrightarrow[2)\ H^+]{1)\ Zn,\ benzene} \text{1-hydroxy-1-(}CH_2CO_2C_2H_5\text{)cyclopentane} \quad (95\%)$$

6[f] $$(CH_3)_2CHCH{=}O + BrCH_2CO_2Et \xrightarrow[TMSCl]{Zn} (CH_3)_2CHCH(OH)CH_2CO_2Et \quad (72\%)$$

7[g] $$\text{2-bromocyclooctanone} + CH_3CH{=}O \xrightarrow[DMSO]{Zn,\ benzene} \text{2-ethylidenecyclooctanone }(=CHCH_3) \quad (57\%)$$

a. K. L. Rinehart, Jr., and E. G. Perkins, *Org. Synth.* **IV**, 444 (1963).
b. C. R. Hauser and D. S. Breslow, *Org. Synth.* **III**, 408 (1955).
c. J. W. Frankenfeld and J. J. Werner, *J. Org. Chem.* **34**, 3689 (1969).
d. M. W. Rathke and A. Lindert, *J. Org. Chem.* **35**, 3966 (1970).
e. J. F. Ruppert and J. D. White, *J. Org. Chem.* **39**, 269 (1974).
f. G. Picotin and P. Miginiac, *J. Org. Chem.* **52**, 4796 (1987).
g. T. A. Spencer, R. W. Britton, and D. S. Watt, *J. Am. Chem. Soc.* **89**, 5727 (1967).

84. S. M. Hannick and Y. Kishi, *J. Org. Chem.* **48**, 3833 (1983).

Organozinc compounds in which the ester group is farther removed from the carbanion center can also be prepared. This would not be possible for the more reactive lithium or magnesium reagents. The β-, γ-, and δ-esters can be made by reaction of the corresponding iodo esters with zinc–copper couple[85]:

$$C_2H_5O_2C(CH_2)_nI + Zn\text{-}Cu \rightarrow C_2H_5O_2C(CH_2)_nZnI \quad n = 2\text{-}4$$

Bis(β-carboalkoxyalkyl)zinc reagents can also be prepared by ring opening of the cyclopropane ring of mixed cyclopropanone acetals[86]:

$$\text{(1-OSiMe}_3\text{-1-OR-cyclopropane)} + ZnCl_2 \rightarrow (RO_2CCH_2CH_2)_2Zn$$

These organozinc compounds react with various electrophiles, including acid chlorides, allylic tosylates, and aldehydes in combination with TMS-Cl.

$$C_2H_5O_2C(CH_2)_nZnI \xrightarrow{RCOCl} RC(=O)(CH_2)_nCO_2C_2H_5$$

$$C_2H_5O_2C(CH_2)_nZnI \xrightarrow{RCH=O,\ TMS-Cl} RCH(OTMS)(CH_2)_nCO_2C_2H_5$$

$$C_2H_5O_2C(CH_2)_nZnI \xrightarrow{RCH=CHCH_2X} C_2H_5O_2C(CH_2)_nCH_2CH=CHR$$

The reagent combination $Zn/CH_2Br_2/TiCl_4$ gives rise to an organometallic reagent which is called Lombardo's reagent. It converts ketones to methylene groups.[87] The active reagent is presumed to be a dimetalated species which adds to the ketone under the influence of the Lewis acidity of titanium. β-Elimination then generates the methylene group.

$$R_2C=O\cdots Ti + -Zn-CH_2-Zn- \longrightarrow R_2C(O-Ti-)(CH_2-Zn) \longrightarrow R_2C=CH_2$$

Use of esters and 1,1-dibromoalkanes as reactants gives enol ethers[88]:

$$C_4H_9CO_2CH_3 + (CH_3)_2CHCHBr_2 \xrightarrow[TiCl_4,\ TMEDA]{Zn} (C_4H_9)(CH_3O)C=CH(CH(CH_3)_2) \quad (95\%)$$

85. Y. Tamaru, H. Ochiai, T. Nakaumura, and Z. Yoshida, *Tetrahedron Lett.* **26**, 5559 (1985); H. Ochiai, Y. Tamaru, K. Tsubaki, and Z. Yoshida, *J. Org. Chem.* **52**, 4418 (1987).
86. E. Nakamura, S. Aoki, K. Sekiya, H. Oshino, and I. Kuwajima, *J. Am. Chem. Soc.* **109**, 8056 (1987).
87. K. Oshima, K. Takai, Y. Hotta, and H. Nozaki, *Tetrahedron Lett.*, 2417 (1978); L. Lombardo, *Tetrahedron Lett.* **23**, 4293 (1982).
88. T. Okazoe, K. Takai, K. Oshima, and K. Utimoto, *J. Org. Chem.* **52**, 4410 (1987).

A similar procedure starting with trimethylsilyl esters generates trimethylsilyl enol ethers[89]:

$$PhCO_2Si(CH_3)_3 + CH_3CHBr_2 \xrightarrow[\text{TMEDA}]{\text{Zn, TiCl}_4} PhC(OSi(CH_3)_3)=CHCH_3$$

Organozinc reagents are also used in conjunction with palladium in a number of carbon–carbon bond-forming processes which will be discussed in Section 8.2.

7.3.2. Organocadmium Compounds

Organocadmium compounds can be prepared from Grignard reagents or organolithium compounds by reaction with Cd(II) salts.[90] Organocadmium compounds can also be prepared directly from alkyl, benzyl, and aryl halides by reaction with highly reactive cadmium metal generated by reduction of Cd(II) salts[91]:

$$NC-C_6H_4-CH_2Br \xrightarrow{\text{Cd}} NC-C_6H_4-CH_2CdBr$$

The reactivity of these reagents is similar to that of the corresponding organozinc compounds.

The most common application of organocadmium compounds has been in the preparation of ketones by reaction with acid chlorides.

$$[(CH_3)_2CHCH_2CH_2]_2Cd + ClCCH_2CH_2CO_2CH_3 \rightarrow (CH_3)_2CHCH_2CH_2COCH_2CH_2CO_2CH_3 \quad (73\text{–}75\%)$$

(with C=O on the ClC carbon) Ref. 92

$$\text{bicyclic ketone (H}_3\text{C, H}_3\text{C, COCl)} + (CH_3)_2Cd \rightarrow \text{bicyclic ketone (H}_3\text{C, H}_3\text{C, CCH}_3\text{=O)} \quad (60\%)$$

Ref. 93

7.3.3. Organomercury Compounds

There are several useful means for preparation of organomercury compounds. The general metal–metal exchange reaction between mercury(II) salts and organolithium or organomagnesium compounds is applicable. The oxymercuration reaction discussed in Section 4.3 provides a means of acquiring certain functionalized

89. K. Takai, Y. Kataoka, T. Okazoe, and K. Utimoto, *Tetrahedron Lett.* **29**, 1065 (1988).
90. P. R. Jones and P. J. Desio, *Chem. Rev.* **78**, 491 (1978).
91. E. R. Burkhardt and R. D. Rieke, *J. Org. Chem.* **50**, 416 (1985).
92. J. Cason and F. S. Prout, *Org. Synth.* **III**, 601 (1955).
93. M. Miyano and B. R. Dorn, *J. Org. Chem.* **37**, 268 (1972).

organomercury reagents. Organomercury compounds can also be obtained by reaction of mercuric salts with trialkylboranes, although only primary alkyl groups react readily.[94] Other organoboron compounds, such as boronic acids and boronate esters, also react with mercuric salts.

$$R_3B + 3\,Hg(O_2CCH_3)_2 \rightarrow 3\,RHgO_2CCH_3$$
$$RB(OH)_2 + Hg(O_2CCH_3)_2 \rightarrow RHgO_2CCH_3$$
$$RB(OR')_2 + Hg(O_2CCH_3)_2 \rightarrow RHgO_2CCH_3$$

Alkenylmercury compounds, for example, can be prepared by hydroboration of an acetylene with catecholborane, followed by reaction with mercuric acetate.[95]

$$RC{\equiv}CR + HB(O_2C_6H_4) \longrightarrow (R)(H)C{=}C(R)(B(O_2C_6H_4)) \xrightarrow{Hg(O_2CCH_3)_2} (R)(H)C{=}C(R)(HgO_2CCH_3)$$

The organomercury compounds can be used *in situ* or they can be isolated as organomercuric halides.

Organomercury compounds are very weakly nucleophilic. They readily undergo electrophilic substitution by halogens.

$$CH_3(CH_2)_6CH{=}CH_2 \xrightarrow[\substack{2)\ Hg(O_2CCH_3)_2 \\ 3)\ Br_2}]{1)\ B_2H_6} CH_3(CH_2)_8Br \quad (69\%)$$ Ref. 94

$$o\text{-}HOC_6H_4HgCl + I_2 \longrightarrow o\text{-}HOC_6H_4I$$ Ref. 96

Organomercury reagents do not react with ketones or aldehydes, but Lewis acids cause reaction with acid chlorides.[97] With alkenyl mercury compounds, the reaction probably proceeds by electrophilic attack on the double bond, with the regiochemistry being directed by the stabilization of the β-carbocation by the mercury.[98]

$$RCH{=}CH{-}HgCl + R'\overset{O}{\overset{\|}{C}}Cl \xrightarrow{AlCl_3} R\overset{+}{C}H{-}\underset{}{CH}(\overset{O}{\overset{\|}{C}}R'){-}HgCl \rightarrow RCH{=}CH\overset{O}{\overset{\|}{C}}R'$$

94. R. C. Larock and H. C. Brown, *J. Am. Chem. Soc.* **92**, 2467 (1970); J. J. Turariello and M. M. Hovey, *J. Am. Chem. Soc.* **92**, 3221 (1970).
95. R. C. Larock, S. K. Gupta, and H. C. Brown, *J. Am. Chem. Soc.* **94**, 4371 (1972).
96. F. C. Whitmore and E. R. Hanson, *Org. Synth.* **I**, 326 (1941).
97. A. L. Kurts, I. P. Beletskaya, I. A. Savchenko, and O. A. Reutov, *J. Organomet. Chem.* **17**, P21 (1969).
98. R. C. Larock and J. C. Bernhardt, *J. Org. Chem.* **43**, 710 (1978).

The majority of the synthetic applications of organomercury compounds are in transition metal-catalyzed processes in which the organic substituent is transferred from mercury to the transition metal in the course of the reaction. Examples of this type of reaction will be considered in Chapter 8.

7.4. Organocerium Compounds

Recent years have seen the development of synthetic procedures involving lanthanide metals, especially cerium. In the synthetic context, one of the most important examples is the conversion of organolithium compounds to organocerium derivatives by $CeCl_3$.[99] The organocerium compounds are useful for addition to carbonyl compounds that are prone to enolization. The organocerium reagents retain strong nucleophilicity but show a much reduced tendency to effect deprotonation. For example, in addition of trimethylsilylmethyllithium to relatively acidic ketones such as 2-indanone, the yield was substantially increased by use of the organocerium intermediate.[100]

$(CH_3)_3SiCH_2Li$: 6% yield

OH, CH_2SiMe_3

$(CH_3)_3SiCH_2CeCl_2$: 83% yield

Organocerium compounds were also found to be the preferred organometallic reagent for addition to hydrazones in an enantioselective synthesis of amines[101]:

CH_2OCH_3

$R'CH_2CH{=}NN$

RLi $\xrightarrow{CeCl_3}$ $RCeCl_2$ $\longrightarrow$ $\xrightarrow{ClCO_2CH_3}$ $R'CH_2CH{-}N{-}N$

R CO_2CH_3

CH_2OCH_3

H_2, Raney Ni

$R'CH_2CHNH_2$

R

General References

R. C. Larock, *Organomercury Compounds in Organic Synthesis*, Springer-Verlag, Berlin, 1985.
B. J. Wakefield, *The Chemistry of Organolithium Compounds*, Pergamon, Oxford, 1974.
B. J. Wakefield, *Organolithium Methods*, Academic Press, Orlando, Florida, 1988.

99. T. Imamoto, T. Kusumoto, Y. Tawarayama, Y. Sugiura, T. Mita, Y. Hatanaka, and M. Yokoyama, *J. Org. Chem.* **49**, 3904 (1984).
100. C. R. Johnson and B. D. Tait, *J. Org. Chem.* **52**, 281 (1987).
101. S. E. Denmark, T. Weber, and D. W. Piotrowski, *J. Am. Chem. Soc.* **109**, 2224 (1987).

(*References for these problems will be found on page 772.*)

1. Predict the product of each of the following reactions. Be sure to specify all elements of stereochemistry.

(a) 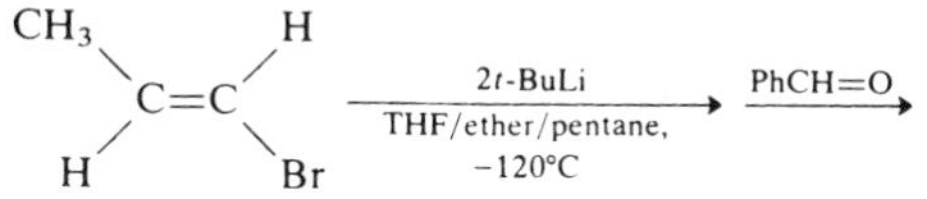 $CH_3(H)C{=}C(H)Br$ (cis-1-bromopropene: CH_3 and Br trans; H and H) $\xrightarrow[\text{THF/ether/pentane, } -120°C]{2t\text{-BuLi}}$ $\xrightarrow{PhCH{=}O}$

(b) 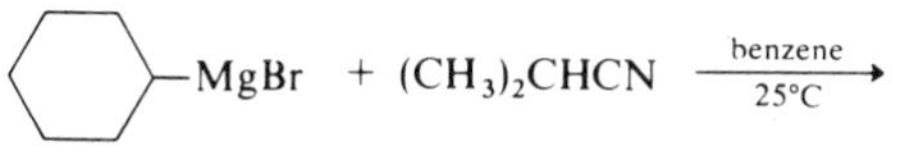 cyclohexyl–MgBr + $(CH_3)_2CHCN$ $\xrightarrow[25°C]{\text{benzene}}$

(c) 3,5-(MeO)$_2$C$_6$H$_3$–$OSi(CH_3)_2C(CH_3)_3$ $\xrightarrow[\text{2) } BrCH_2C{=}C(CH_3)_2]{\text{1) } n\text{-BuLi}}$

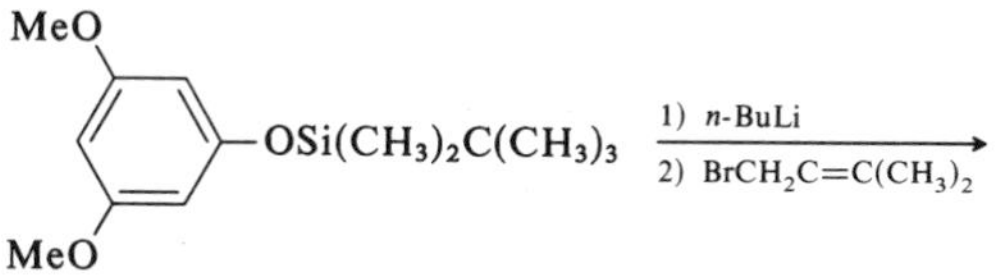

(d) $THPO(CH_2)_3C{\equiv}CH$ $\xrightarrow[\text{2) } CH_3C{\equiv}CCH_2Br]{\text{1) EtMgBr}}$

(e) 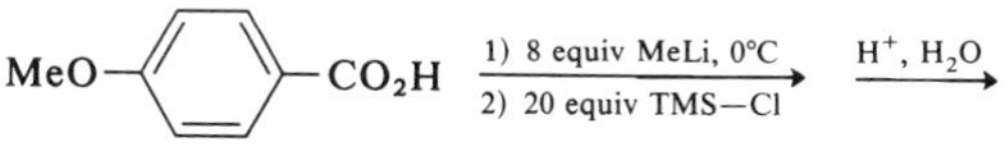 MeO–C$_6$H$_4$–CO_2H $\xrightarrow[\text{2) 20 equiv TMS—Cl}]{\text{1) 8 equiv MeLi, 0°C}}$ $\xrightarrow{H^+, H_2O}$

(f) 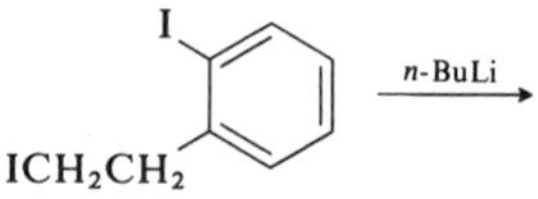 1-iodo-2-(ICH_2CH_2)benzene $\xrightarrow{n\text{-BuLi}}$

(g) $PhCO_2CH_3$ + CH_3CHBr_2 $\xrightarrow[\text{TMEDA, 25°C}]{Zn, TiCl_4}$

(h) $CH_3(CH_2)_4CH{=}O$ + $BrCH_2CO_2Et$ $\xrightarrow[\text{benzene}]{\text{Zn dust}}$

(i) 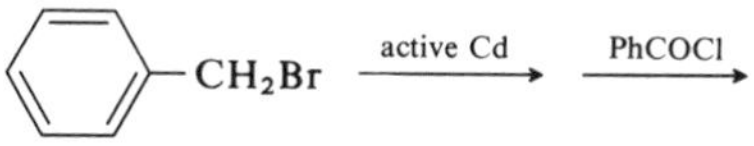 C$_6$H$_5$–CH_2Br $\xrightarrow{\text{active Cd}}$ $\xrightarrow{PhCOCl}$

2. Reaction of the epoxide of 1-butene with methyllithium gives 3-pentanol in 90% yield. In contrast, methylmagnesium bromide under similar conditions gives the array of products shown below. Explain the difference in the reactivity of the two organometallic compounds toward this epoxide.

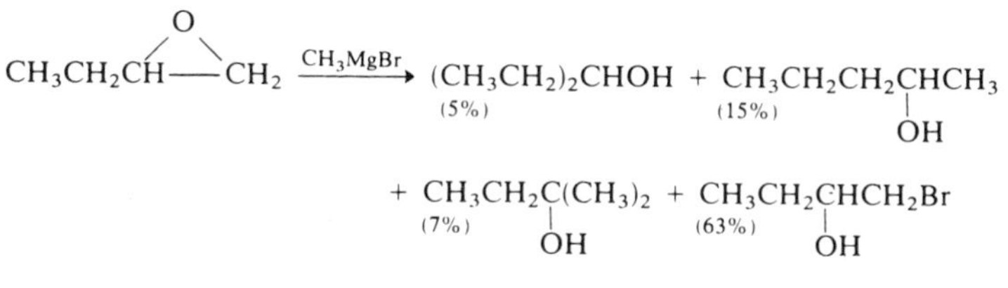

$$CH_3CH_2\overset{O}{CH{-}CH_2} \xrightarrow{CH_3MgBr} \underset{(5\%)}{(CH_3CH_2)_2CHOH} + \underset{(15\%)}{CH_3CH_2CH_2CH(OH)CH_3}$$

$$+ \underset{(7\%)}{CH_3CH_2C(OH)(CH_3)_2} + \underset{(63\%)}{CH_3CH_2CH(OH)CH_2Br}$$

3. Devise an efficient synthesis for the following organometallic compounds from the starting material specified.

(a) 1-lithio-1-(methoxymethoxy)cyclohexane (Li, OCH_2OCH_3) from cyclohexanone

(b) $(CH_3)_2C(Li)OCH_3$ from $(CH_3)_2C(OCH_3)_2$

(c) $CH_3OCH_2OCH_2Li$ from Bu_3SnCH_2OH

(d) bicyclic vinyllithium (CH_3, $H_2C=$, Li) from bicyclic ketone (CH_3, $H_2C=$, O)

(e) $LiCH_2C(OSi(CH_3)_3)=NSi(CH_3)_3$ from $CH_3C(=O)NH_2$

(f) $PhCH_2OCH_2OCH_2CH(CH_3)CH(Li)OCH_2OCH_3$ from $PhCH_2OCH_2OCH_2CH(CH_3)CH=O$

(g) $(CH_3)_3Si(H)C=C(H)Li$ from $(CH_3)_3SiC\equiv CH$

4. Each of the following compounds gives products in which one or more lithium atoms have been introduced under the conditions specified. Predict the structures of the lithiated product on the basis of structural features known to promote lithiation and/or stabilization of lithiated species. The number of lithium atoms introduced is equal to the number of moles of lithium reagent used in each case.

(a) $H_2C=C(CH_3)C(=O)NHC(CH_3)_3 \xrightarrow[\text{TMEDA, THF, }-20°C]{2\ n\text{-BuLi}}$

(b) $(CH_3)_2C=CH_2 \xrightarrow[\text{TMEDA, hexane}]{n\text{-BuLi}}$

(c) $C_6H_5CH_2N(CH_3)_2 \xrightarrow[\text{ether, 25°C, 24 h}]{n\text{-BuLi}}$

(d) 1-methoxynaphthalene (OCH_3) $\xrightarrow[\text{ether, 38°C, 20 h}]{n\text{-BuLi}}$

(e) $HC\equiv CCO_2CH_3 \xrightarrow[\text{THF/pentane/ether}]{n\text{-BuLi, }-120°C}$

(f) $C_6H_5NHC(=O)C(CH_3)_3 \xrightarrow[\text{THF, 0°C, 2 h}]{2\ n\text{-BuLi}}$

(g) $(CH_3)_2CH$–C₆H₄–OCH_3 $\xrightarrow[\text{TMEDA, ether}]{n\text{-BuLi}}$

(h) Ph(H)C=C(H)CN $\xrightarrow[-113°C]{\text{LDA}}$

(i) $CH_2=C(CH_3)CH_2OH$ $\xrightarrow[2\ n\text{-BuLi, 0°C}]{2\ K^{+-}O\text{-}t\text{-Bu}}$

(j) N-$PhSO_2$-indole $\xrightarrow[-5°C]{2\ t\text{-BuLi}}$

5. Each of the following compounds can be prepared from reactions of organometallic reagents and readily available starting materials. Identify the appropriate organometallic reagent in each case and show how it would be prepared. Show how the desired material would be made.

(a) $H_2C=CHCH_2CH_2CH_2OH$

(b) $H_2C=C(CH_3)C(OH)(CH_2CH_2CH_2CH_3)_2$

(c) $PhC(OH)(CH_2OCH_3)_2$

(d) 2-$N(CH_3)_2$-5-CH_3-C₆H₃–$C(OH)Ph_2$

(e) $(CH_3)_3CCH(CO_2C_2H_5)_2$

(f) $H_2C=CHCH=CHCH=CH_2$

6. Identify an organometallic reagent system which will permit formation of the product on the left of each equation from the starting material specified in a "one-pot" process.

(a) [bicyclic ketal with CH_2, H, H, CH_3, OSiTBDMS] $\Rightarrow$ [bicyclic ketal with O, H, H, CH_3, OSiTBDMS]

(b) cyclohexyl–$C(OSiMe_3)=CH(CH_3)$ (H) $\Rightarrow$ cyclohexyl–CO_2H

(c) $PhCCH_2CH_2CO_2C_2H_5$ (C=O) $\Rightarrow$ $PhCOCl$

(d) $(CH_3)_2CH(CH_2)_2C(=O)(CH_2)_6CO_2C_2H_5 \Rightarrow ClC(=O)(CH_2)_6CO_2C_2H_5$

7. The solvomercuration reaction (Section 4.3) provides a convenient source of such organomercury compounds as **1** and **2**. How could these be converted to functionalized lithium reagents such as **3** and **4**?

$$\underset{\mathbf{1}}{\mathrm{HOCH(R)CH_2HgBr}} \qquad \underset{\mathbf{2}}{\mathrm{PhN(H)CH_2CH_2HgBr}} \qquad \underset{\mathbf{3}}{\mathrm{LiOCH(R)CH_2Li}} \qquad \underset{\mathbf{4}}{\mathrm{PhN(Li)CH_2CH_2Li}}$$

Would the procedure you have suggested also work for the following transformation? Explain your reasoning.

$$\mathrm{CH_3OCH(R)CH_2HgBr} \rightarrow \mathrm{CH_3OCH(R)CH_2Li}$$

8. Predict the stereochemical outcome of the following reactions and indicate the basis of your predictions.

(a) 2-(OCH_2OCH_2Ph)cyclohexanone $\xrightarrow{CH_3MgCl}$

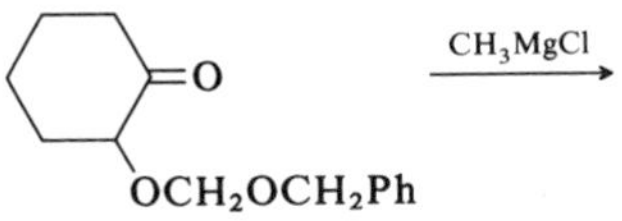

(b) $CH_3(CH_2)_6C(H)(OCH_2OCH_2CH_2OCH_3)CCH_3$ (ketone, C=O) $\xrightarrow[\text{THF}]{n\text{-BuMgBr}}$

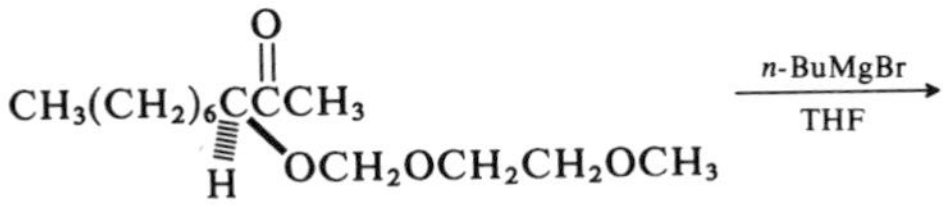

(c) trans-decalone (H, H ring-fusion, C=O) $\xrightarrow{CH_3MgI}$

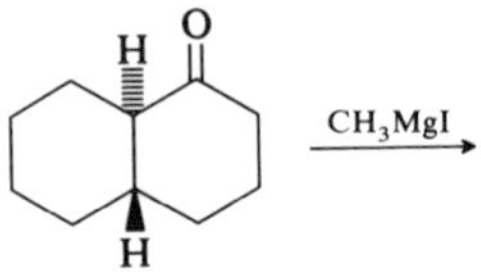

9. Tertiary amides **1, 2**, and **3** are lithiated at the β-carbon, rather than the α-carbon, when treated with s-BuLi/TMEDA. It is estimated, however, that the intrinsic acidity of the α position exceeds that of the β position by ~9 pK units. What would cause the β deprotonation to be kinetically preferred?

$$\mathrm{CH_3CH(CH_2R)C(=O)N(\mathit{i}\text{-}Pr)_2} \rightarrow \mathrm{CH_3CH(CHRLi)C(=O)N(\mathit{i}\text{-}Pr)_2}$$

1 R = Ph
2 R = CH=CH$_2$
3 R = SPh

10. The following reaction sequence converts esters to bromomethyl ketones. Show the intermediates that are involved in each of the steps in the sequence.

$$\mathrm{CH_2Br_2} \xrightarrow[-90°\mathrm{C}]{\mathrm{LDA}} \xrightarrow[-90°\mathrm{C}]{\mathrm{RCO_2Et}} \xrightarrow[-90°\mathrm{C}]{n\text{-BuLi}} \xrightarrow[-78°\mathrm{C}]{\mathrm{H^+}} \mathrm{RCCH_2Br}\ (\mathrm{C{=}O})$$

11. Normally, the reaction of an ester with one equivalent of a Grignard reagent leads to formation of a mixture of tertiary alcohol, ketone, and unreacted ester. However, if one equivalent of LDA is present along with the Grignard reagent, good yields of ketones are obtained. What is the role of LDA in this process?

12. Several examples of intramolecular additions to carbonyl groups by organolithium reagents generated by halogen–metal exchange have been reported, such as the examples shown below. What relative reactivity relationships must hold in order for such procedures to succeed?

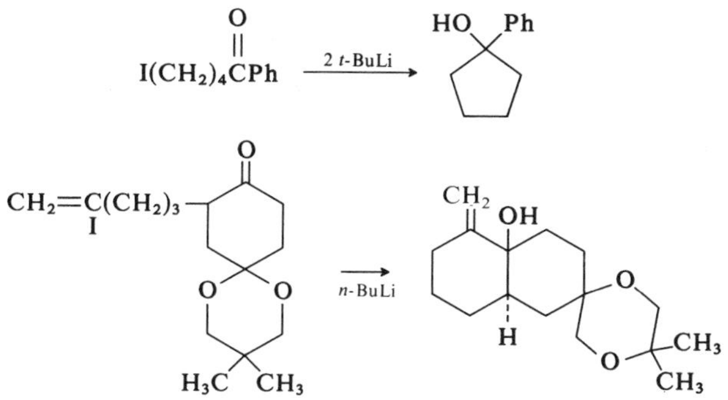

$I(CH_2)_4\overset{O}{\overset{\|}{C}}Ph \xrightarrow{2\ t\text{-BuLi}}$ HO Ph

$CH_2{=}\underset{I}{C}(CH_2)_3$— O O H₃C CH₃ $\xrightarrow{n\text{-BuLi}}$ CH₂ OH O H CH₃ O CH₃

13. Short synthetic sequences (three steps or less) involving functionally substituted organometallic compounds as key reagents can effect the following synthetic transformations. Suggest reaction sequences which would be effective for each transformation. Indicate how the required organometallic reagent could be prepared.

(a) $CH_3(CH_2)_4$ O O $\longrightarrow$ $CH_3(CH_2)_4$ O O CH₃ CH₃

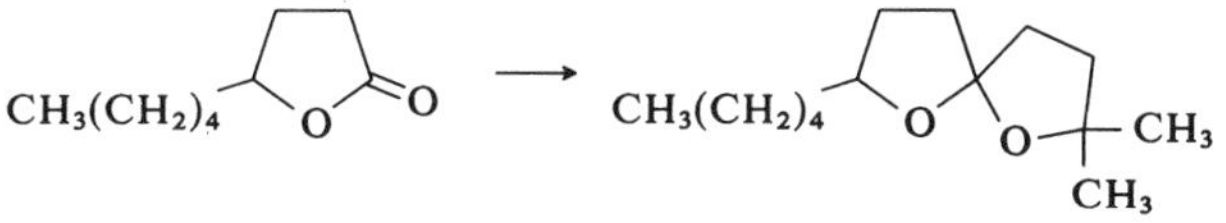

(b) $CH_3CH{=}O \longrightarrow$ CH₃ O O CH₃

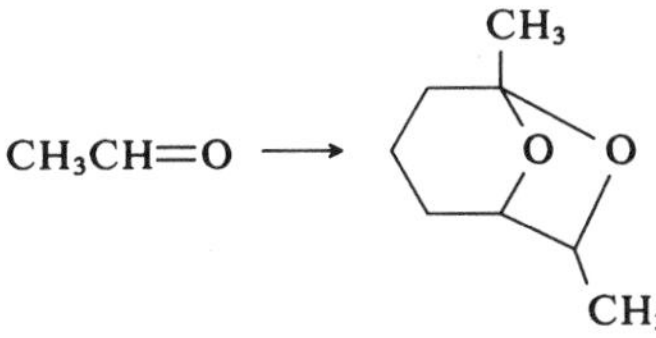

(c) MeO CH=O $\longrightarrow$ O O OMe

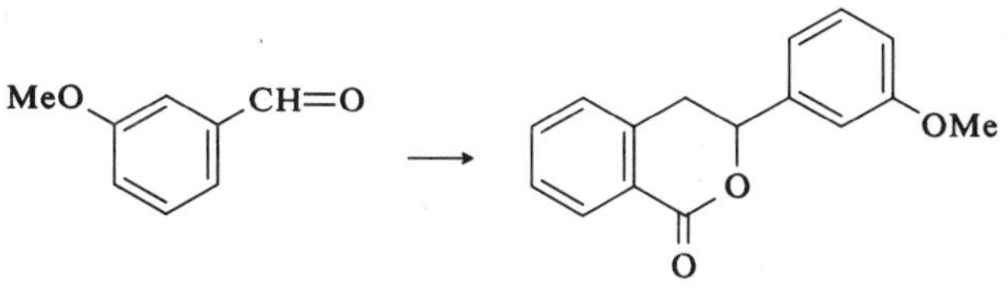

(d) CH=O O $\longrightarrow$ $\overset{CH_3}{\overset{|}{CHOCHOCH_2CH_3}}$ $\underset{CH_3}{|}$ O

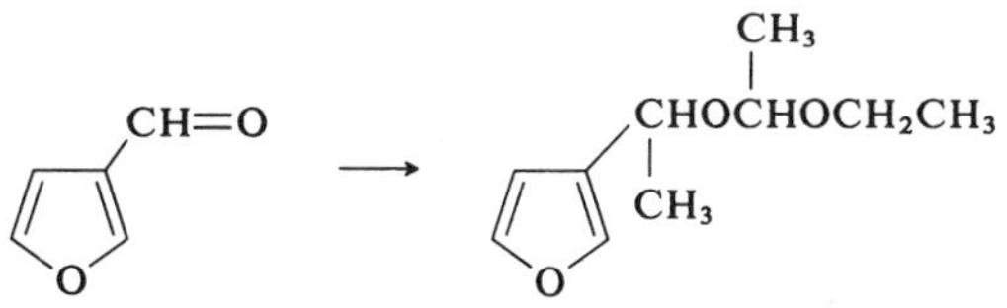

(e) $THPOCH_2CH_2C{\equiv}CH \rightarrow$ (H)(THPOCH₂CH₂)C=C(H)(CH₂CH=CH₂)

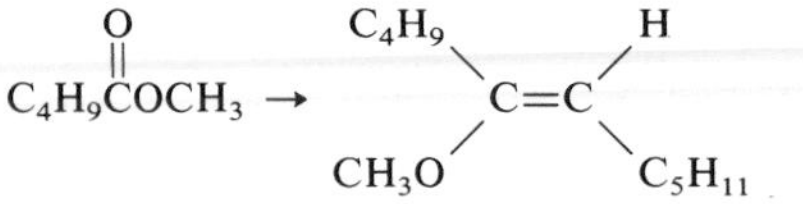

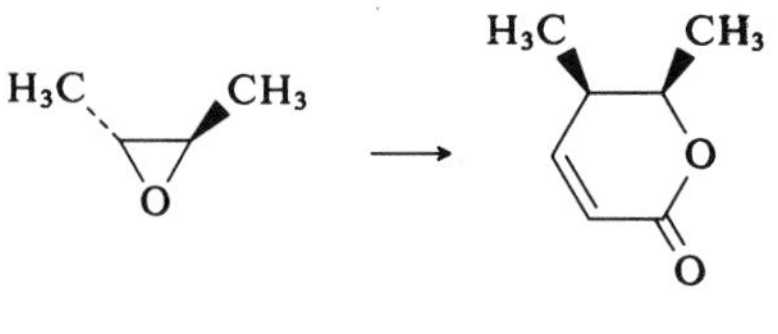

(f) $C_4H_9\overset{O}{\overset{\|}{C}}OCH_3 \rightarrow$ $(C_4H_9)(CH_3O)C{=}C(H)(C_5H_{11})$

(g) [epoxide with H_3C and CH_3] → [dihydropyranone with H_3C and CH_3]

(h) [aromatic aldehyde: CH=O, OCH_3, CH_3O, CH_3] → [phthalide: CH_3O, O, O, H, CH_3O; aryl: OCH_3, CH_3O, CH_3]

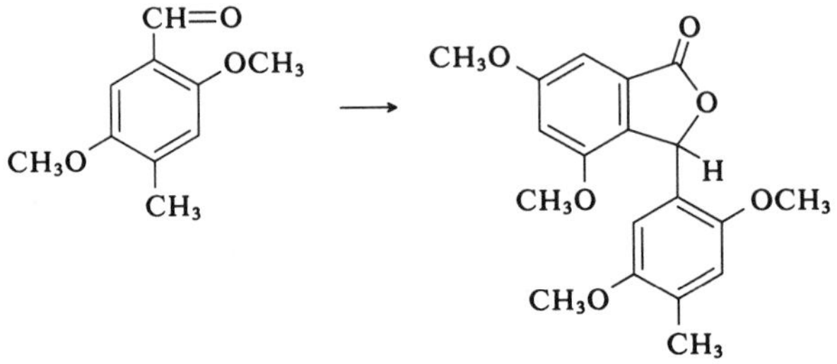

14. Chiral aminoalcohols both catalyze reactions of simple dialkylzinc reagents with aldehydes and also induce a high degree of enantioselectivity, even when used in only catalytic amounts. Two examples are given below. Indicate how the aminoalcohols can have a catalytic effect. Suggest transition states for the examples show which would be in accord with the observed enantioselectivity.

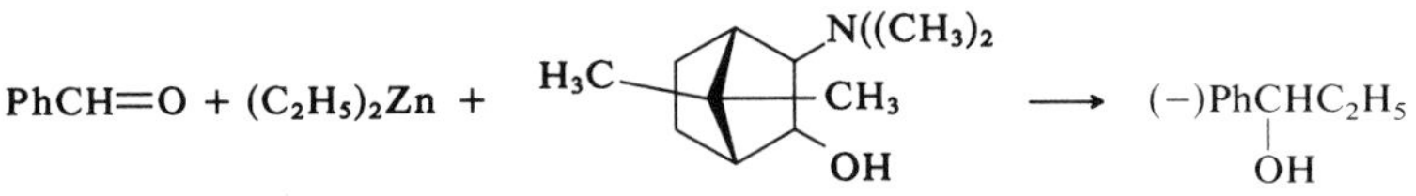

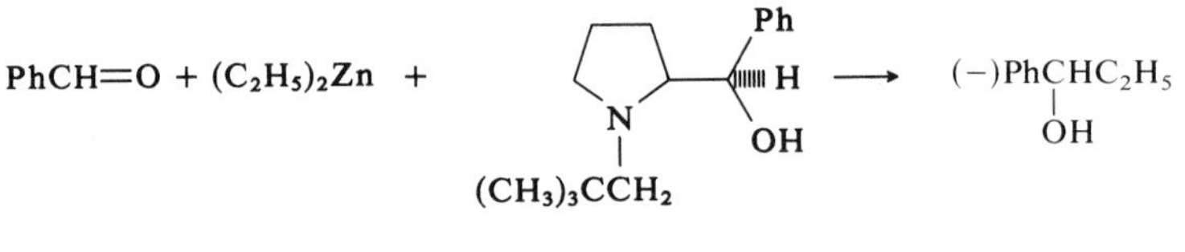

$$PhCH{=}O + (C_2H_5)_2Zn + \text{[bicyclic aminoalcohol: } H_3C\text{, } N(CH_3)_2\text{, } CH_3\text{, } OH] \longrightarrow (-)PhCH(OH)C_2H_5$$

$$PhCH{=}O + (C_2H_5)_2Zn + \text{[pyrrolidine: } N\text{-}CH_2C(CH_3)_3\text{, } C(Ph)(H)(OH)] \longrightarrow (-)PhCH(OH)C_2H_5$$

8

Reactions Involving the Transition Metals

While the main-group metals, especially magnesium and lithium, were the first metals to have a prominent role in organic synthesis, several of the transition metals have also become very important. In this chapter, we will discuss the reactions of transition metal compounds and intermediates that are important in the repertoire of synthetic organic chemistry. In contrast to the reactions involving lithium and magnesium, where the reagents are used in stoichiometric quantities, many of the transition metal reactions will be found to be catalytic processes.

8.1. Reactions Involving Organocopper Intermediates

Development of the chemistry of organocopper compounds has put many important new reactions at the disposal of synthetic chemists. These advances received initial impetus from the study of the catalytic effect of copper salts on reactions of Grignard reagents with α,β-unsaturated ketones.[1] While Grignard reagents normally add to such compounds to give the 1,2-addition product, the presence of catalytic amounts of Cu(I) results in conjugate addition. Mechanistic study of this effect pointed to a very fast reaction by an organocopper intermediate.

$$CH_3CH{=}CHCOCH_3 \xrightarrow{CH_3MgBr} \xrightarrow[H^+]{H_2O} CH_3CH{=}CHC(OH)(CH_3)_2$$

$$CH_3CH{=}CHCOCH_3 \xrightarrow{CuI,\ CH_3MgBr} \xrightarrow[H^+]{H_2O} (CH_3)_2CHCH_2C(=O)CH_3$$

1. H. O. House, W. L. Respess, and G. M. Whitesides, *J. Org. Chem.* **31**, 3128 (1966).

Many subsequent studies have led to the characterization of several organocopper compounds that result from reaction of organolithium reagents with copper salts.[2]

$$RLi + Cu(I) \rightarrow RCu + Li^+$$

$$2\,RLi + Cu(I) \rightarrow [R_2CuLi] + Li^+$$

$$3\,RLi + Cu(I) \rightarrow [R_3CuLi_2] + Li^+$$

The species from the 2:1 molar ratio are known as cuprates, and they have been used most frequently as synthetic reagents. In solution, lithium dimethylcuprate exists as a dimer, $[LiCu(CH_3)_2]_2$,[3] but the precise structure is not known. The compound is often represented as four methyl groups attached to a tetrahedral cluster of lithium and copper atoms.

Cu
H₃C CH₃
Li Li
H₃C CH₃
Cu

Discrete R_2Cu^- anions have been observed in crystals in which the lithium cation is complexed by crown ethers.[4]

Cuprates in which there are two different copper substituents have been developed. These compounds can have important advantages in cases in which one of the substituents is derived from a valuable synthetic intermediate. Scheme 8.1 presents some of these mixed cuprate reagents.

An important group of mixed cuprates are prepared from a 2:1 ratio of alkyllithium and CuCN.[5] These compounds are called *higher-order cyanocuprates*. The composition of the major species in THF solution is $R_2CuCNLi_2$, but it is thought that most of the molecules probably are present as dimers. NMR studies have established that individual alkyl groups can exchange between copper centers.[6]

$$2\,RLi + CuCN \rightarrow [R_2CuCN]^{2-}2Li^+ \rightleftharpoons [(R_2CuCN)_2]^{4-}4Li^+$$

These reagents are similar to other cuprates in reactivity, but they are more stable than the dialkylcuprates. Because higher-order cuprates usually transfer only one of the two organic groups, it is useful to have a group that normally does not transfer. The 2-thienyl group has been shown to be useful for this purpose.[7] In a

2. E. C. Ashby and J. J. Lin, *J. Org. Chem.* **42**, 2805 (1977); E. C. Ashby and J. J. Watkins, *J. Am. Chem. Soc.* **99**, 5312 (1977).
3. R. G. Pearson and C. D. Gregory, *J. Am. Chem. Soc.* **98**, 4098 (1976); B. H. Lipshutz, J. A. Kozlowski, and C. M. Breneman, *J. Am. Chem. Soc.* **107**, 3197 (1985).
4. H. Hope, M. M. Olmstead, P. P. Power, J. Sandell, and X. Xu, *J. Am. Chem. Soc.* **107**, 4337 (1985).
5. B. H. Lipshutz, R. S. Wilhelm, and J. Kozlowski, *Tetrahedron* **40**, 5005 (1984); B. H. Lipshutz, *Synthesis*, 325 (1987).
6. B. H. Lipshutz, J. A. Kozlowski, and R. S. Wilhelm, *J. Org. Chem.* **49**, 3943 (1984).
7. B. H. Lipshutz, J. A. Kozlowski, D. A. Parker, S. L. Nguyen, and K. E. McCarthy, *J. Organomet. Chem.* **285**, 437 (1985); B. H. Lipshutz, M. Koerner, and D. A. Parker, *Tetrahedron Lett.* **28**, 945 (1987).

Scheme 8.1. Mixed Cuprate Reagents

	Mixed cuprate	Reactivity and properties
1[a]	[RC≡C—Cu—R]Li	Conjugate addition to α,β-unsaturated ketones and certain esters.
2[b,c]	[ArS—Cu—R]Li	Nucleophilic substitution and conjugate addition to unsaturated ketones. Ketones from acid chlorides.
3[b]	[$(CH_3)_3CO$—Cu—R]Li	Nucleophilic substitution and conjugate addition to unsaturated ketones.
4[d]	[$(c-C_6H_{11})_2N$—Cu—R]Li	Normal range of nucleophilic reactivity, improved thermal stability.
5[d]	[Ph_2P—Cu—R]Li	Normal range of nucleophilic reactivity, improved thermal stability.
6[e]	[MeS(=O)CH_2—Cu—R]Li	Normal range of nucleophilic reactivity, ease of preparation, thermal stability.
7[f]	[N≡C—Cu—R]Li	Efficient opening of epoxides.
8[g]	[R—Cu—BF_3]	Conjugate addition including addition to acrylate esters and acrylonitriles. S_N2' displacement of allylic halides.

a. H. O. House and M. J. Umen, *J. Org. Chem.* **38**, 3893 (1973); E. J. Corey, D. Floyd and B. H. Lipshutz, *J. Org. Chem.* **43**, 3418 (1978).
b. G. H. Posner, C. E. Whitten, and J. J. Sterling, *J. Am. Chem. Soc.* **95**, 7788 (1973).
c. G. H. Posner and C. E. Whitten, *Org. Synth.* **55**, 122 (1975).
d. S. H. Bertz, G. Dabbagh, and G. M. Villacorta, *J. Am. Chem. Soc.* **104**, 5824 (1982).
e. C. R. Johnson and D. S. Dhanoa, *J. Org. Chem.* **52**, 1885 (1987).
f. R. D. Acker, *Tetrahedron Lett.*, 3407 (1977); J. P. Marino and N. Hatanaka, *J. Org. Chem.* **44**, 4467 (1979).
g. K. Maruyama and Y. Yamamoto, *J. Am. Chem. Soc.* **99**, 8068 (1977); Y. Yamamoto and K. Maruyama, *J. Am. Chem. Soc.* **100**, 3240 (1978).

mixed alkyl-thienyl cyanocuprate, only the alkyl substituent is normally transferred as a nucleophile.

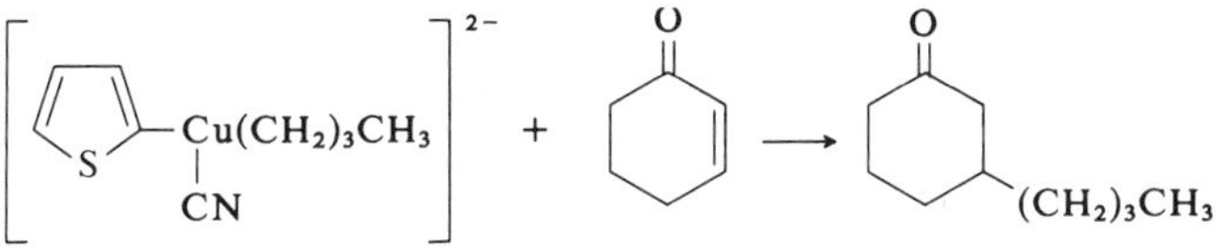

$$\left[\text{(2-thienyl)}\underset{\displaystyle CN}{\underset{|}{Cu}}(CH_2)_3CH_3\right]^{2-} + \text{cyclohexenone} \longrightarrow \text{3-}(CH_2)_3CH_3\text{-cyclohexanone}$$

The 1 : 1 organocopper reagents can be prepared directly from a halide and highly reactive copper metal obtained by reducing Cu(I) salts with lithium naphthalenide.[8] This method is advantageous for preparation of compounds containing substituents that would be incompatible with organolithium intermediates. For example, nitrophenyl and cyanophenyl copper reagents can be prepared in this way. Alkylcopper reagents having ester and cyano substituents have also been prepared.[9]

There has been much study on the effect of solvents and other reaction conditions on the stability and reactivity of organocuprates.[10] These studies have found, for

8. G. W. Ebert and R. D. Rieke, *J. Org. Chem.* **49**, 5280 (1984); *J. Org. Chem.* **53**, 4482 (1988).
9. R. M. Wehmeyer and R. D. Rieke, *J. Org. Chem.* **52**, 5056 (1987); T.-C. Wu, R. M. Wehmeyer, and R. D. Rieke, *J. Org. Chem.* **52**, 5059 (1987); R. M. Wehmeyer and R. D. Rieke, *Tetrahedron Lett.* **29**, 4513 (1988).
10. R. H. Schwartz and J. San Filippo, Jr., *J. Org. Chem.* **44**, 2705 (1979).

example, that $(CH_3)_2S\text{-}CuBr$, a readily prepared and purified complex of CuBr, is an especially reliable source of Cu(I) in cuprate preparation.[11] Copper(I) cyanide and iodide are also generally effective and are preferable in some cases.[12]

The most important reactions of organocuprate reagents are nucleophile displacements on halides and sulfonates, epoxide ring opening, conjugate additions to α,β-unsaturated carbonyl compounds, and additions to acetylenes.[13] Scheme 8.2 gives some examples of each of these reaction types, and they are discussed in more detail in the following paragraphs.

Corey and Posner discovered that lithium dimethylcuprate could replace iodine or bromine by methyl in a wide variety of compounds, including alkenyl and aryl derivatives. This halogen displacement reaction is more general and gives higher yields than displacements with Grignard or lithium reagents.[14]

$$C_6H_5I + (CH_3)_2CuLi \longrightarrow C_6H_5CH_3 \quad (90\%)$$

$$PhCH{=}CHBr + (CH_3)_2CuLi \rightarrow PhCH{=}CHCH_3 \quad (81\%)$$

Allylic halides usually give both S_N2 and S_N2' products although the mixed organocopper reagent $RCu\text{-}BF_3$ is reported to give mainly the S_N2' product.[15] Allylic acetates undergo displacement with an allylic shift (S_N2' mechanism).[16] The allylic substitution process may involve initial coordination with the double bond.[17]

$$R_2Cu^{I} + CH_2{=}CHCH_2X \longrightarrow \underset{CH_2-X}{\overset{R_2Cu}{H_2C{=}CH}}$$

$$\downarrow$$

$$R_2Cu^{III}{-}CH_2CH{=}CH_2 \rightarrow R{-}CH_2CH{=}CH_2 + RCu^{I}$$

11. H. O. House, C.-Y. Chu, J. M. Wilkins, and M. J. Umen, *J. Org. Chem.* **40**, 1460 (1975).
12. B. H. Lipshutz, R. S. Wilhelm, and D. M. Floyd, *J. Am. Chem. Soc.* **103**, 7672 (1981); S. H. Bertz, C. P. Gibson, and G. Dabbagh, *Tetrahedron Lett.* **28**, 4251 (1987); B. H. Lipshutz, S. Whitney, J. A. Kozlowski, and C. M. Breneman, *Tetrahedron Lett.* **27**, 4273 (1986).
13. For reviews of the reactions of organocopper reagents, see G. H. Posner, *Org. React.* **19**, 1 (1972); **22**, 253 (1975); G. H. Posner, *An Introduction to Synthesis Using Organocopper Reagents*, Wiley, New York, 1980.
14. E. J. Corey and G. H. Posner, *J. Am. Chem. Soc.* **89**, 3911 (1967).
15. K. Maruyama and Y. Yamamoto, *J. Am. Chem. Soc.* **99**, 8068 (1977).
16. R. J. Anderson, C. A. Henrick, and J. B. Siddall, *J. Am. Chem. Soc.* **92**, 735 (1970); E. E. van Tamelen and J. P. McCormick, *J. Am. Chem. Soc.* **92**, 737 (1970).
17. H. L. Goering and S. S. Kantner, *J. Org. Chem.* **49**, 422 (1984).

The reaction shows a preference for *anti* stereochemistry in cyclic systems.[18]

CH_3 CH_3

$\xrightarrow{[Me_2Cu]Li}$

O_2CCH_3 H_3C

It has been suggested that the preference for the *anti* stereochemistry is the result of simultaneous overlap of a *d* orbital on copper with both the π^* and σ^* orbitals of the allyl system.

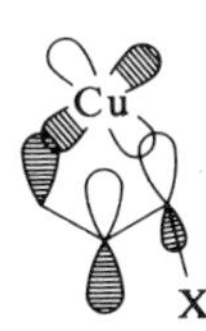

Propargylic acetates, halides, and sulfonates also react with a double bond shift to give allenes.[19] Some direct substitution product can be formed, as well. The highest proportion of allenic product is found with CH_3Cu–LiBr–MgBrI, which is prepared by addition of methylmagnesium bromide to a 1:1 LiBr–CuI mixture.[20]

O_2CCH_3

$-C{\equiv}CCHC_5H_{11}$ + CH_3Cu-LiBr-MgBrI $\longrightarrow$ $C{=}C{=}C$ (H)(C_5H_{11})(H_3C) (100%)

Saturated epoxides are opened in good yield by lithium dialkylcuprate.[21] The alkyl group is introduced at the less hindered carbon of the epoxide ring:

CH_3CH_2(epoxide O) + $(CH_3)_2CuLi \rightarrow CH_3CH_2CH(OH)CH_2CH_3$ (88%)

Epoxides with alkenyl substituents undergo alkylation at the double bond with a double bond shift accompanying ring opening.[22]

$(CH_3)_2CuLi$ + $H_2C{=}C(CH_3)$(epoxide O)CH_3 $\rightarrow$ $CH_3CH_2C(CH_3){=}CHCH(OH)CH_3$

18. H. L. Goering and V. D. Singleton, Jr., *J. Am. Chem. Soc.* **98**, 7854 (1976); H. L. Goering and C. C. Tseng, *J. Org. Chem.* **48**, 3986 (1983).
19. P. Rona and P. Crabbe, *J. Am. Chem. Soc.* **90**, 4733 (1968); R. A. Amos and J. A. Katzenellenbogen, *J. Org. Chem.* **43**, 555 (1978); D. J. Pasto, S.-K. Chou, E. Fritzen, R. H. Shults, A. Waterhouse, and G. F. Hennion, *J. Org. Chem.* **43**, 1389 (1978).
20. T. L. Macdonald, D. R. Reagan, and R. S. Brinkmeyer, *J. Org. Chem.* **45**, 4740 (1980).
21. C. R. Johnsin, R. W. Herr, and D. M. Wieland, *J. Org. Chem.* **38**, 4263 (1973).
22. R. J. Anderson, *J. Am. Chem. Soc.* **92**, 4978 (1970); R. W. Herr and C. R. Johnson, *J. Am. Chem. Soc.* **92**, 4979 (1970).

Scheme 8.2. Reactions of Organocopper Intermediates

A. Conjugate Addition Reactions

1[a]

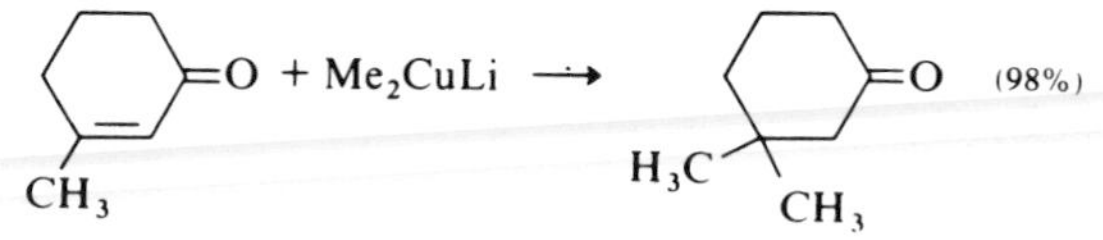

$=O$ + Me_2CuLi → $=O$ (98%)

CH_3 ; H_3C CH_3

2[b]

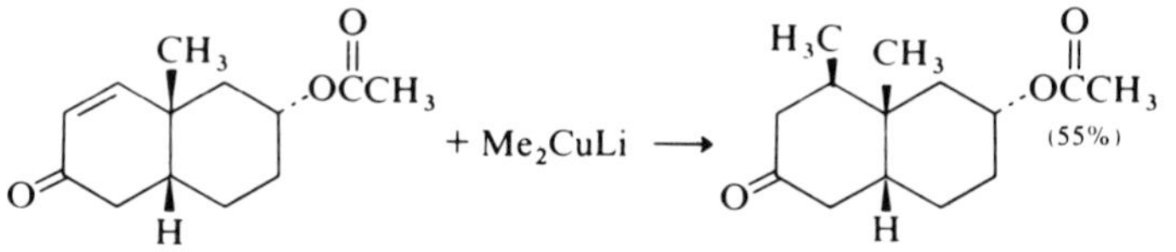

CH_3 ; O ; $OCCH_3$; O ; H

+ Me_2CuLi →

H_3C CH_3 ; O ; $OCCH_3$ (55%) ; O ; H

3[c]

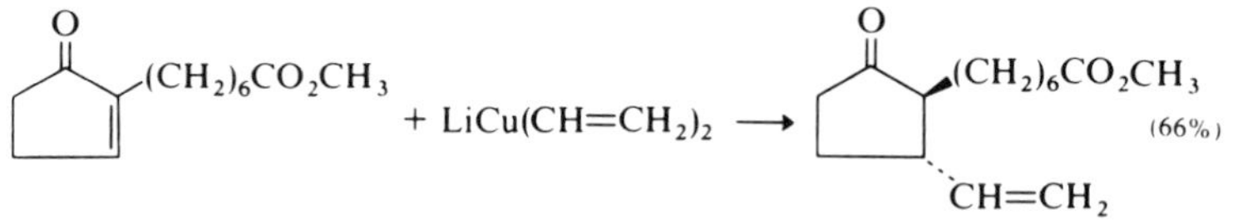

O ; $(CH_2)_6CO_2CH_3$

+ $LiCu(CH{=}CH_2)_2$ →

O ; $(CH_2)_6CO_2CH_3$ (66%) ; $CH{=}CH_2$

4[d]

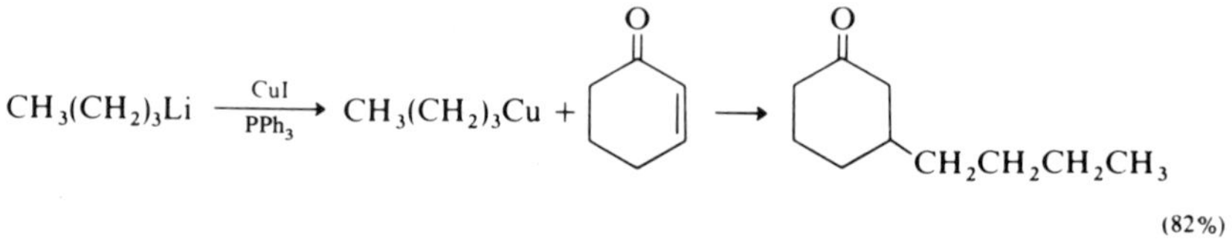

$CH_3(CH_2)_3Li \xrightarrow[PPh_3]{CuI} CH_3(CH_2)_3Cu$ + O → O ; $CH_2CH_2CH_2CH_3$

(82%)

5[e]

$CH_3{-}N$; O $\xrightarrow{Ph_2CuLi}$ $CH_3{-}N$; Ph ; O

(75%)

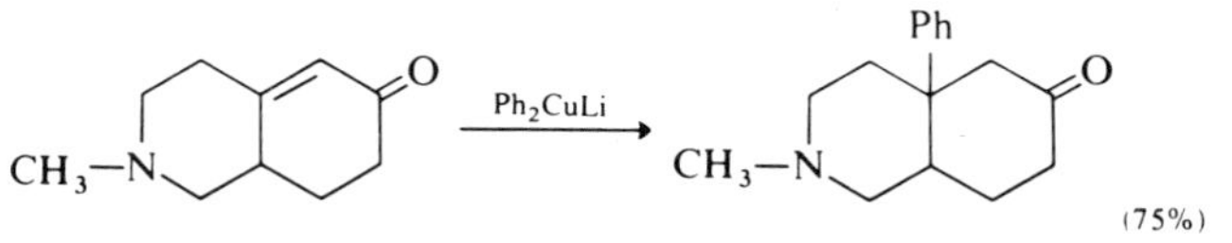

B. Halide Substitution

6[f]

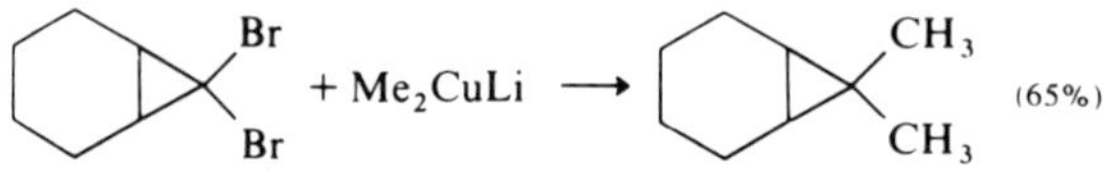

Br ; Br + Me_2CuLi → CH_3 ; CH_3 (65%)

7[g]

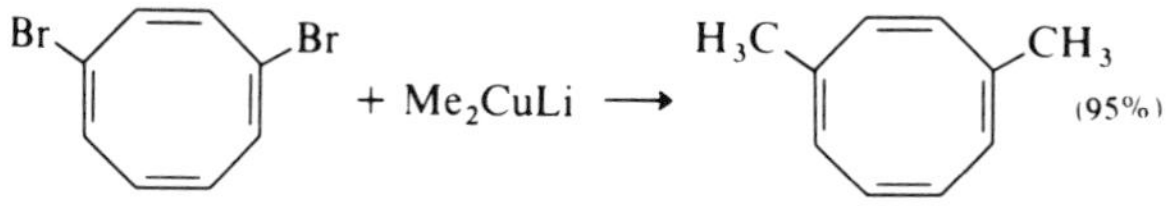

Br ; Br + Me_2CuLi → H_3C ; CH_3 (95%)

8[h]

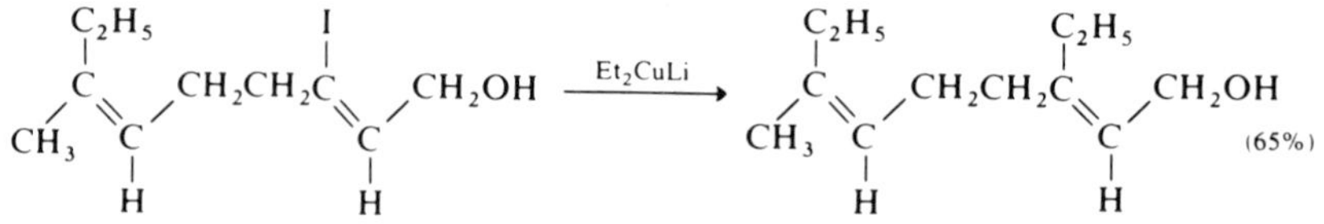

C_2H_5 ; I ; CH_3 ; C ; C ; CH_2CH_2C ; C ; CH_2OH ; H ; H $\xrightarrow{Et_2CuLi}$ C_2H_5 ; C_2H_5 ; CH_3 ; C ; C ; CH_2CH_2C ; C ; CH_2OH ; H ; H

(65%)

9[i]

CH_2 ; CH_2 ; CH_3 ; H_3C ; Cl ; Cl ; CH_2OCPh_3 ; H

$\xrightarrow{Me_2CuLi}$ H_3C ; C_2H_5 ; C_2H_5 ; CH_3 ; CH_2OCPh_3 ; H ; H ; H

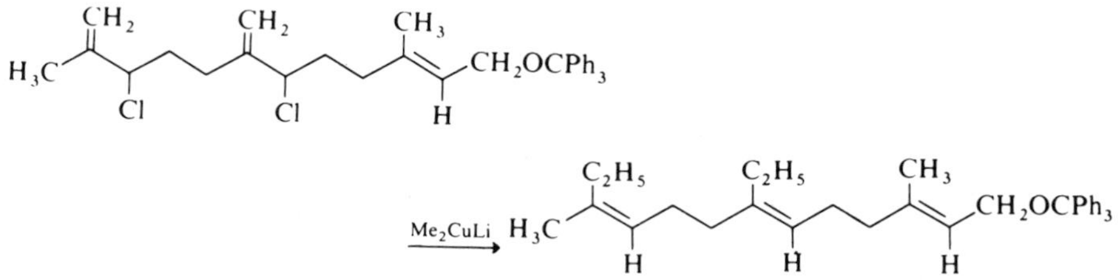

10[j]

$$\underset{H}{\overset{CH_3}{}}\!\!C{=}C\!\!\underset{Cl}{\overset{H}{}} \xrightarrow{Li(1\%Na)} \underset{H}{\overset{CH_3}{}}\!\!C{=}C\!\!\underset{Li}{\overset{H}{}} \xrightarrow[CH_3(CH_2)_6CH_2I]{CuI,\ -78°C} \underset{H}{\overset{CH_3}{}}\!\!C{=}C\!\!\underset{CH_2(CH_2)_6CH_3}{\overset{H}{}}$$

(90–93%)

C. Displacement of Allylic Acetates

11[k]

$$CH_3C(=CH_2)(O_2CCH_3)CHCH_2CH_2C(CH_3){=}CHCH_2CH_2C(CH_3){=}CHCO_2CH_3 \xrightarrow[ether,\ -10°C]{LiCu(CH_3)_2}$$

$$CH_3C(CH_3CH_2){=}CHCH_2CH_2C(CH_3){=}CHCH_2CH_2C(CH_3){=}CHCO_2CH_3$$

(87 %)

12[l]

$$\text{3-methylcyclohex-2-enyl... } CH_3\text{-cyclohexenyl}{-}O_2CCH_3 \xrightarrow{LiCu(CH_3)_2} \text{3,5-dimethylcyclohexene}$$

(90–95%)

D. Ketones from Acid Chlorides

13[m]

$$PhCCl(=O) + [(CH_3)_3CCuSPh]Li \rightarrow PhC(=O)C(CH_3)_3$$

(84–87%)

14[n]

$$ClC(=O)CH_2CH_2\text{-steroid (}H_3C,\ OC(CH_3)_3\text{)} \xrightarrow{(C_2H_5)_2CuLi} C_2H_5C(=O)CH_2CH_2\text{-steroid (}H_3C,\ OC(CH_3)_3\text{)}$$

(65%)

a. H. O. House, W. L. Respess, and G. M. Whitesides, *J. Org. Chem.* **31**, 3128 (1966).
b. J. A. Marshall and G. M. Cohen, *J. Org. Chem.* **36**, 877 (1971).
c. F. S. Alvarez, D. Wren, and A. Prince, *J. Am. Chem. Soc.* **94**, 7823 (1972).
d. M. Suzuki, T. Suzuki, T. Kawagishi, and R. Noyori, *Tetrahedron Lett.*, 1247 (1980).
e. N. Finch, L. Blanchard, R. T. Puckett, and L. H. Werner, *J. Org. Chem.* **39**, 1118 (1974).
f. E. J. Corey and G. H. Posner, *J. Am. Chem. Soc.* **89**, 3911 (1967).
g. W. E. Konz, W. Hechtl and R. Huisgen, *J. Am. Chem. Soc.* **92**, 4104 (1970).
h. E. J. Corey, J. A. Katzenellenbogen, N. W. Gilman, S. A. Roman, and B. W. Erickson, *J. Am. Chem. Soc.* **90**, 5618 (1968).
i. E. E. van Tamelen and J. P. McCormick, *J. Am. Chem. Soc.* **92**, 737 (1970).
j. G. Linstrumelle, J. K. Krieger, and G. M. Whitesides, *Org. Synth* **55**, 103 (1976).
k. R. J. Anderson, C. A. Henrick, J. B. Siddall, and R. Zurfluh, *J. Am. Chem. Soc.* **94**, 5379 (1972).
l. H. L. Goering and V. D. Singleton, Jr., *J. Am. Chem. Soc.* **98**, 7854 (1976).
m. G. Posner and C. E. Whitten, *Org. Synth.* **55**, 122 (1976).
n. W. G. Dauben, G. Ahlgren, T. J. Leitereg, W. C. Schwarzel, and M. Yoshioko, *J. Am. Chem. Soc.* **94**, 8593 (1972).

All of the above reactions illustrate the powerful nucleophilicity that organocuprates exhibit toward carbon. Secondary tosylates react with inversion of stereochemistry, as in the classical S_N2 substitution reaction.[23] The overall mechanism probably consists of two steps. First, an oxidative addition to the metal occurs. The formal oxidation state of copper in the intermediate from this addition step is +3. This step is followed by migration of one of the alkyl groups from copper. The addition of halides and tosylates to transition metal species with low oxidation states is a common reaction type in transition metal chemistry.

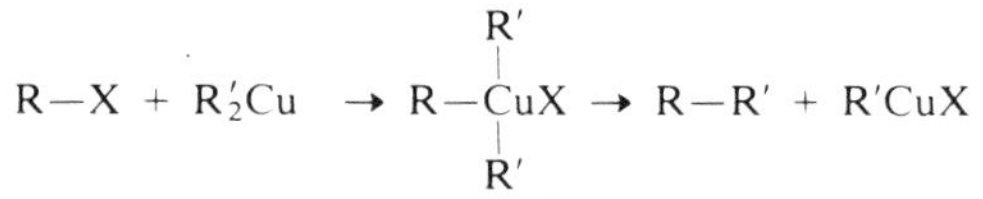

$$\mathrm{R{-}X + R'_2Cu \rightarrow R{-}\overset{R'}{\underset{R'}{Cu}}X \rightarrow R{-}R' + R'CuX}$$

All of the types of mixed cuprate reagents described in Scheme 8.1 react with conjugated enones. A number of improvements in the methodology for carrying out the conjugate addition reactions have been introduced. The efficiency of the reaction is improved by the addition of trialkylphosphines to the reaction mixture.[24]

The conjugate addition reactions probably occur by a mechanism similar to that for substitution on halides. There may be a fast electron transfer step, but, if so, the radicals must combine faster than they diffuse apart since there is no evidence that *free* radicals are generated.[25]

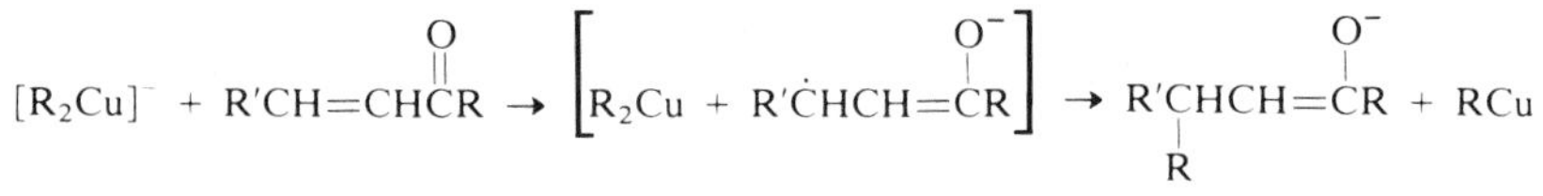

$$\mathrm{[R_2Cu]^- + R'CH{=}CH\overset{O}{\overset{\|}{C}}R \rightarrow \left[R_2Cu + R'\dot{C}HCH{=}\overset{O^-}{\overset{|}{C}}R\right] \rightarrow R'\underset{R}{\underset{|}{C}}HCH{=}\overset{O^-}{\overset{|}{C}}R + RCu}$$

There is a correlation between the reduction potential of the carbonyl compounds and the ease of reaction with cuprate reagents.[26] The more easily the compound is reduced, the more reactive it is toward cuprate reagents. Compounds such as α,β-unsaturated esters and nitriles, which are not as easily reduced as the corresponding ketones, do not react as readily with dialkylcuprates, even though they are good acceptors in classical Michael reactions with carbanions. α,β-Unsaturated esters are borderline in terms of reactivity toward standard dialkylcuprate reagents. β-Substitution retards reactivity. The RCu–BF_3 reagent combination is more reactive toward conjugated esters and nitriles.[27] Additions to hindered α,β-unsaturated ketones are also accelerated by BF_3.[28] The presence of trimethylsilyl chloride also

23. C. R. Johnson and G. A. Dutra, *J. Am. Chem. Soc.* **95**, 7783 (1973).
24. M. Suzuki, T. Suzuki, T. Kawagishi, and R. Noyori, *Tetrahedron Lett.*, 1247 (1980).
25. H. O. House, *Acc. Chem. Res.* **9**, 59 (1976); H. O. House and P. D. Weeks, *J. Am. Chem. Soc.* **97**, 2770, 2778 (1975); H. O. House and K. A. J. Snoble, *J. Org. Chem.* **41**, 3076 (1976); S. H. Bertz, G. Dabbagh, J. M. Cook, and V. Honkan, *J. Org. Chem.* **49**, 1739 (1984).
26. H. O. House and M. J. Umen, *J. Org. Chem.* **38**, 3893 (1973); B. H. Lipshutz, R. S. Wilhelm, S. T. Nugent, R. D. Little, and M. M. Baizer, *J. Org. Chem.* **48**, 3306 (1983).
27. Y. Yamamoto and K. Maruyama, *J. Am. Chem. Soc.* **100**, 3240 (1978).
28. A. B. Smith III and P. J. Jerris, *J. Am. Chem. Soc.* **103**, 194 (1981).

accelerates the addition of cuprates to enones. Under these conditions, the initial product is a silyl enol ether. The rate enhancement is attributed to silylation of a reversibly formed complex between the enone and the cuprate.[29]

This technique also greatly improves yields of conjugate addition of cuprates to α,β-unsaturated esters and amides.[30]

Several mixed cuprate reagents containing chiral anionic ligands have been explored to determine the degree of enantioselectivity that can be achieved in conjugate addition. With the aminoalkoxide **A**, 85–92% enantioselectivity was observed in additions to cyclohexenone.[31]

R	e.e.
Et	92
n-Bu	89
$(CH_3)_3COCH_2$	85

Several other nitrogen anions of chiral amines have also been examined.[32]

Prior to protonolysis, the products of conjugate addition to unsaturated carbonyl compounds are enolates and, therefore, potential nucleophiles. A useful extension of the conjugate addition method is to combine it with an alkylation step that can add a substituent at the α position.[33] Several examples of this tandem conjugate addition/alkylation procedure are given in Scheme 8.3.

29. E. J. Corey and N. W. Boaz, *Tetrahedron Lett.* **26**, 6019 (1985); E. Nakamura, S. Matsuzawa, Y. Horiguchi, and I. Kuwajima, *Tetrahedron Lett.* **27**, 4029 (1986); C. R. Johnson and T. J. Marren, *Tetrahedron Lett.* **28**, 27 (1987).
30. A. Alexakis, J. Berlan, and Y. Besace, *Tetrahedron Lett.* **27**, 1047 (1986).
31. E. J. Corey, R. Neaf, and F. J. Hannon, *J. Am. Chem. Soc.* **108**, 7144 (1986).
32. R. K. Dieter and M. Tokles, *J. Am. Chem. Soc.* **109**, 2040 (1987); S. H. Bertz, G. Dabbagh, and G. Sundararajan, *J. Org. Chem.* **51**, 4953 (1986).
33. For a review of such reactions, see R. J. K. Taylor, *Synthesis*, 364 (1985).

Scheme 8.3. Tandem Reactions Involving Trapping of Enolates Generated by Conjugate Addition of Organocopper Reagents

1[a] TMSO, H, TMSO, H, O → 1) $[(CH_2{=}CH)_2Cu]Li$ 2) CH_3I → TMSO, $CH{=}CH_2$, H, CH_3, O, H, TMSO (85%)

2[b] $PhCH_2O$, H, H, $PhCH_2O$, O → 1) $[EtOCHO(CH_2)_3CuSPh]Li$ (CH_3) 2) CH_3I → $PhCH_2O$, $(CH_2)_3OCHOEt$ (CH_3), H, CH_3, O, H, $PhCH_2O$ (80%)

3[c] O, CH_3, CH_3 → 1) $R_3P{\cdot}Cu$(H)C=C(H)$CH(CH_2)_4CH_3$ (OTBDMS) 2) ICH_2(H)C=C(H)$(CH_2)_3CO_2CH_3$ → O, CH_3, CH_3, CH_2, $(CH_2)_3CO_2CH_3$, C=C, H, H, H, C=C, H, $CH(CH_2)_4CH_3$, OTBDMS (64%)

4[d] O, CH_3, $(CH_2)_2$ → 1) $[PhSCH_2CH_2CHCH{=}CHCu]$ ($OC(CH_3)_2OCH_3$) 2) CH_3I 3) H^+ → CH_3, O, CH_3, $(CH_2)_2$, H, $CHCH_2CH_2SPh$, OH

5[f] O, CH_3 → 1) Cu(H)C=C(H)$CHCH_2CH(CH_3)_2$ (OP) 2) $CH_2{=}C(SiMe_3)CCH_3$ (=O) 3) NaOMe 4) HCl → O, OH (58%)

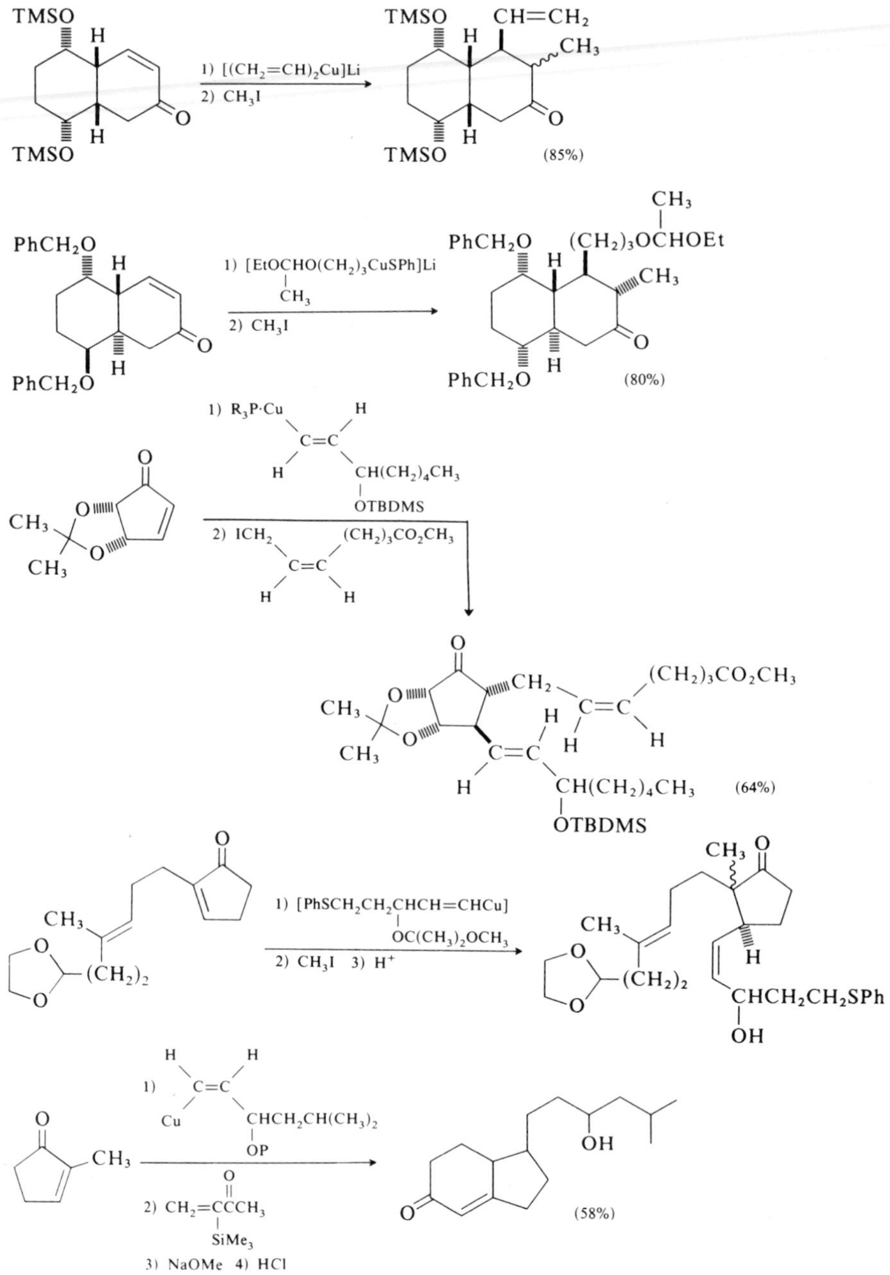

$[P{=}C(CH_3)_2OCH_3]$

a. N. N. Girotra, R. A. Reamer, and N. L. Wendler, *Tetrahedron Lett.* **25**, 5371 (1984).
b. N.-Y. Wang, C.-T. Hsu, and C. J. Sih, *J. Am. Chem. Soc.* **103**, 6538 (1981).
c. C. R. Johnson and T. D. Penning, *J. Am. Chem. Soc.* **110**, 4726 (1988).
d. T. Takahashi, H. Okumoto, J. Tsuji, and N. Harada, *J. Org. Chem.* **49**, 948 (1984).
e. T. Takahashi, K. Shimizu, T. Doi, and J. Tsuji, *J. Am. Chem. Soc.* **110**, 2674 (1988).

Conjugated acetylenic esters react readily with cuprate reagents, with *syn* addition being kinetically preferred.[34]

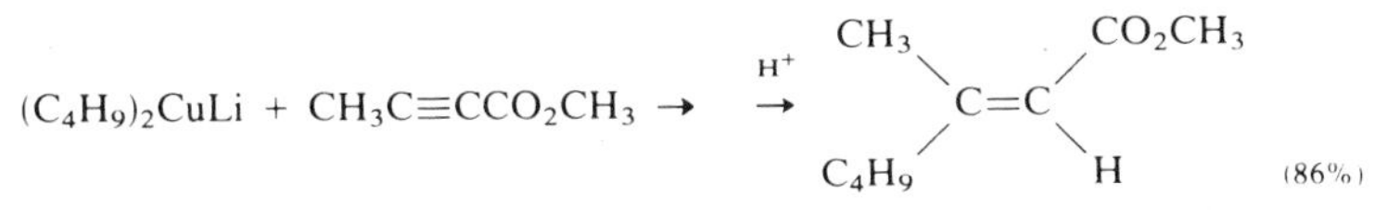

$$(C_4H_9)_2CuLi + CH_3C{\equiv}CCO_2CH_3 \rightarrow \xrightarrow{H^+} (CH_3)(C_4H_9)C{=}C(CO_2CH_3)(H) \quad (86\%)$$

The intermediate adduct can be substituted at the α position by a variety of electrophiles, including acid chlorides, epoxides, aldehydes, and ketones.[35]

The cuprate reagents which have been discussed in the preceding paragraphs are normally prepared by reaction of an organolithium reagent with a copper(I) salt in a 2:1 ratio. There are also valuable synthetic procedures which involve organocopper intermediates that are generated in the reaction system by use of only a catalytic amount of a copper(I) species.[36] Conjugate addition to α,β-unsaturated esters can often be effected by copper-catalyzed reaction with a Grignard reagent. Other reactions, such as epoxide ring opening, can also be carried out under catalytic conditions. Some examples of catalyzed additions and alkylations are given in Scheme 8.4.

Mixed copper–magnesium reagents related to the lithium cuprates can be prepared.[37] These compounds are often called *Normant reagents.* The precise structural nature of these compounds has not been determined. Individual species with differing Mg:Cu ratios may be in equilibrium.[38] These reagents undergo addition to terminal acetylenes to generate alkenylcopper reagents. The addition is stereospecifically *syn.*

$$C_2H_5MgBr + CuBr \rightarrow C_2H_5CuMgBr_2$$

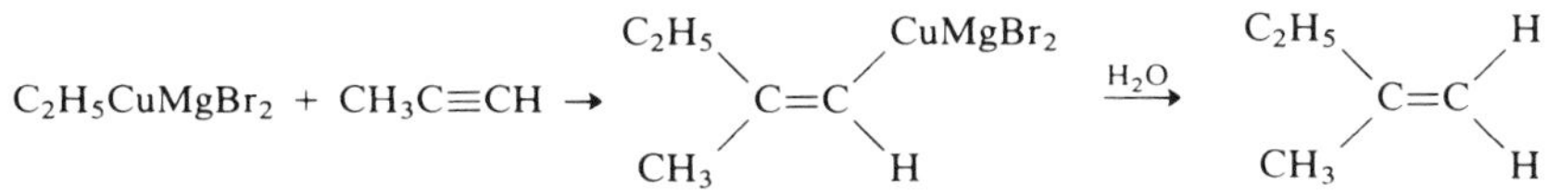

$$C_2H_5CuMgBr_2 + CH_3C{\equiv}CH \rightarrow (C_2H_5)(CH_3)C{=}C(CuMgBr_2)(H) \xrightarrow{H_2O} (C_2H_5)(CH_3)C{=}C(H)(H)$$

The alkenylcopper adducts can be worked up by protonolysis or they can be subjected to further elaboration by alkylation or electrophilic substitution. Some examples are given in Scheme 8.5.

Organocopper intermediates are also involved in several procedures for coupling of two organic reactants to form a new carbon–carbon bond. A classical example of this type of reaction is the *Ullman coupling* of aryl halides, which is done by

35. J. P. Marino and R. G. Lindeman, *J. Org. Chem.* **48**, 4621 (1983).
36. For a review, see E. Erdiky, *Tetrahedron* **40**, 641 (1984).
37. J. F. Normant and M. Bourgain, *Tetrahedron Lett.*, 2583 (1971); J. F. Normant, G. Cahiez, M. Bourgain, C. Chuit, and J. Villieras, *Bull. Soc. Chim. Fr.*, 1656 (1974); H. Westmijze, J. Meier, H. J. T. Bos, and P. Vermeer, *Recl. Trav. Chim. Pays-Bas* **95**, 299, 304 (1976).
38. E. C. Ashby, R. S. Smith, and A. B. Goel, *J. Org. Chem.* **46**, 5133 (1981); E. C. Ashby and A. B. Goel, *J. Org. Chem.* **48**, 2125 (1983).

Scheme 8.4. Copper-Catalyzed Reactions of Grignard Reagents

A. Alkylations

1[a] $n\text{-}C_8H_{17}Br + CH_2{=}CHC(MgCl){=}CH_2 \xrightarrow[2\text{ mol }\%]{Li_2CuCl_4} CH_2{=}CHC((CH_2)_7CH_3){=}CH_2$ (80%)

2[b] $PhCH{=}CHCH(O_2C(CH_3)_3)CH_3 + n\text{-}C_4H_9MgBr \xrightarrow[1\text{ mol }\%]{CuCN} PhCH((CH_2)_3CH_3)CH{=}CHCH_3$ (95%)

3[c] 5-$(CH_3)_3CCO_2$-1-($CH_2CH(OMe)_2$)cyclopentene + t-BuMgCl $\xrightarrow[3\text{ mol }\%]{CuCN}$ 5-$C(CH_3)_3$-1-($CH_2CH(OMe)_2$)cyclopentene (87%)

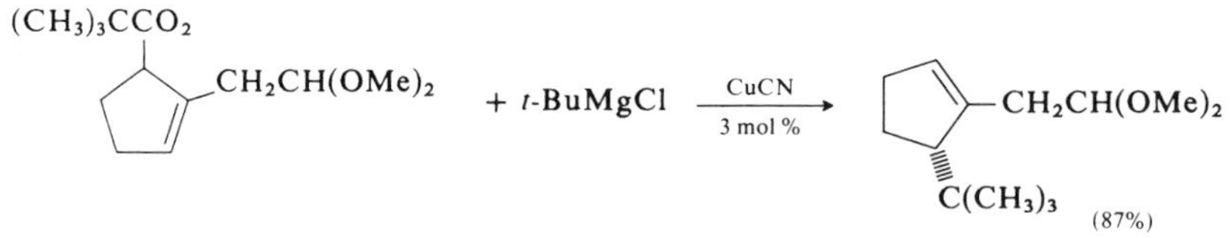

4[d] $n\text{-}C_4H_9MgCl$ + ethylene oxide $\xrightarrow[10\text{ mol }\%]{CuBr}$ $CH_3(CH_2)_5OH$ (88%)

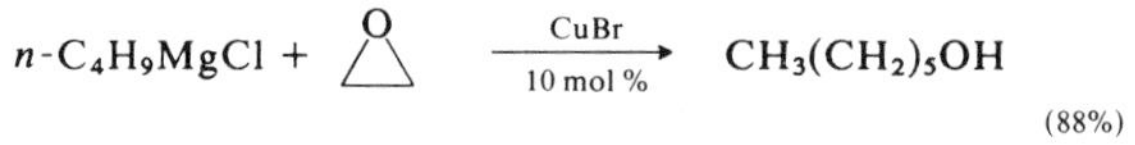

B. Conjugate Additions

5[e] $(CH_3)_2C{=}C(CO_2CH_3)_2 + CH_3MgBr \xrightarrow[2\text{ mol }\%]{CuCl} \xrightarrow{H_2O} (CH_3)_3CCH(CO_2CH_3)_2$ (84–94%)

6[f] cyclohexyl-MgBr + $CH_2{=}CHCO_2C_2H_5 \xrightarrow[1\text{ mol }\%]{CuCl}$ cyclohexyl-$CH_2CH_2CO_2C_2H_5$ (68%)

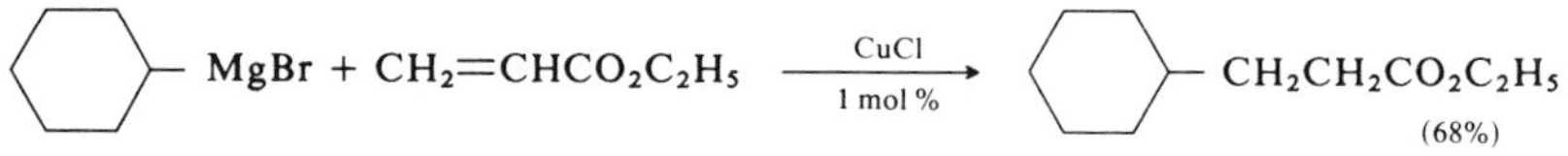

7[g] $CH_3CH{=}CHCO_2CH(CH_3)CH_2CH_3 + CH_3(CH_2)_3MgBr \xrightarrow{CuCl\ 1.4\text{ mol }\%} CH_3(CH_2)_3CH(CH_3)CH_2CO_2CH(CH_3)CH_2CH_3$ (51%)

8[h] $H_5C_2O_2CCH{=}CHCO_2C_2H_5 + (CH_3)_2CHMgBr \xrightarrow{CuCl} H_5C_2O_2CCH(CH(CH_3)_2)CH_2CO_2C_2H_5$ (81%)

C. Other Reactions

9[i] $C_6H_5C{\equiv}N + (CH_3)_3CMgCl \xrightarrow[2\text{ mol }\%]{CuBr} C_6H_5C({=}NH)C(CH_3)_3$ (95%)

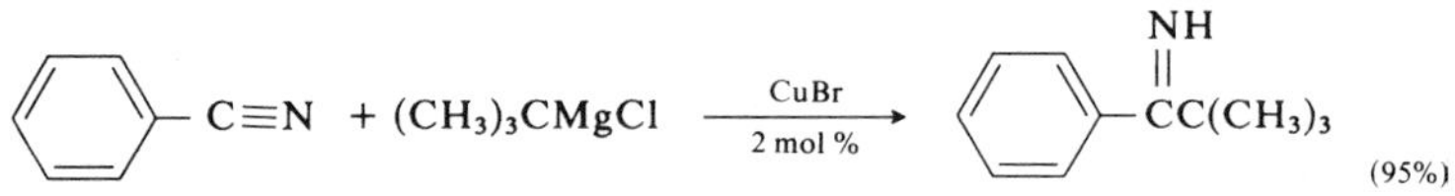

a. S. Numoto, Y. Kawatami, and Y. Yarnashita, *J. Org. Chem.* **48**, 1912 (1983).
b. C. C. Tseng, S. D. Paisley, and H. L. Goering, *J. Org. Chem.* **51**, 2884 (1986).
c. E. J. Corey and A. V. Gavai, *Tetrahedron Lett.* **29**, 3201 (1988).
d. G. Huynh, F. Derguini-Boumechal, and G. Linstrumelle, *Tetrahedron Lett.*, 1503 (1979).
e. E. L. Eliel, R. O. Hutchins, and M. Knoeber, *Org. Synth.* **50**, 38 (1970).
f. S.-H. Liu, *J. Org. Chem.* **42**, 3209 (1977).
g. T. Kindt-Larsen, V. Bitsch, I. G. K. Andersen, A. Jart, and J. Munch-Petersen, *Acta Chem. Scand.* **17**, 1426 (1963).
h. V. K. Andersen and J. Munch-Petersen, *Acta Chem. Scand.* **16**, 947 (1962).
i. F. J. Weiberth and S. S. Hall, *J. Org. Chem.* **52**, 3901 (1987).

1[a] $C_2H_5MgBr + CuBr + C_4H_9C{\equiv}CH \rightarrow (C_2H_5)(C_4H_9)C{=}C(Cu)(H) \xrightarrow{H^+} (C_2H_5)(C_4H_9)C{=}CH_2$ (82%)

2[b] $C_2H_5Cu(SMe_2)MgBr_2 + C_6H_{13}C{\equiv}CH \rightarrow (C_2H_5)(C_6H_{13})C{=}C(Cu)(H) \xrightarrow{I_2} (C_2H_5)(C_6H_{13})C{=}C(I)(H)$ (63%)

3[c] $[(n\text{-}C_4H_9)_2Cu]Li + HC{\equiv}CH \rightarrow (CH_3(CH_2)_3)(H)C{=}C(Cu)(H) \xrightarrow{I_2} (CH_3(CH_2)_3)(H)C{=}C(I)(H)$ (65-75%)

4[d] $(C_5H_{11})_2CuLi + HC{\equiv}CH \rightarrow (C_5H_{11})(H)C{=}C(Cu)(H) \xrightarrow{HC{\equiv}CCO_2C_2H_5} (C_5H_{11})(H)C{=}C(H)\text{–}C(H){=}C(CO_2C_2H_5)(H)$ (78%)

5[e] $(CH_3)_2CHCuMgBr_2 + C_4H_9C{\equiv}CH \rightarrow ((CH_3)_2CH)(C_4H_9)C{=}C(Cu)(H) \xrightarrow{Cl-C{\equiv}N} ((CH_3)_2CH)(C_4H_9)C{=}C(C{\equiv}N)(H)$ (92%)

6[f] $C_2H_5Cu(SMe_2)MgBr_2 + C_6H_{13}C{\equiv}CH \rightarrow (C_2H_5)(C_6H_{13})C{=}C(Cu)(H) \xrightarrow{H_2C{=}CHCH_2Br} (C_2H_5)(C_6H_{13})C{=}C(CH_2CH{=}CH_2)(H)$ (85%)

6[f] $C_3H_7Cu(SMe_2)MgBr_2 + CH_3C{\equiv}CH \rightarrow (CH_3)(C_3H_7)C{=}C(Cu)(H) \xrightarrow{LiC{\equiv}CC_3H_7,\ \text{ethylene oxide}} (CH_3)(C_3H_7)C{=}C(CH_2CH_2OH)(H)$ (95%)

a. J.-F. Normant, G. Cahiez, M. Bourgain, C. Chuit, and J. Villieras, *Bull. Chim. Soc. France*, 1656 (1974).
b. N. J. LaLima, Jr. and A. B. Levy, *J. Org. Chem.* **43**, 1279 (1978).
c. A. Alexakis, G. Cahiez, and J. F. Normant, *Org. Synth.* **62**, 1 (1984).
d. A. Alexakis, J. Normant, and J. Villieras, *Tetrahedron Lett.*, 3461 (1976).
e. H. Westmijze and P. Vermeer, *Synthesis*, 784 (1977).
f. R. S. Iyer and P. Helquist, *Org. Synth.* **64**, 1 (1985).
g. P. R. McGuirk, A. Marfat, and P. Helquist, *Tetrahedron Lett.*, 2465 (1978).

heating an aryl halide with a copper-bronze alloy.[39] Good yields by this method are limited to halides with electron-attracting substituents.

Cu-bronze, 220°C — Ref. 40

Mechanistic studies have established the involvement of arylcopper intermediates. Soluble Cu(I) salts, particularly the triflate, effect coupling of aryl halides at much lower temperatures and under homogeneous conditions.[41]

CuO_3SCF_3, NH_3, 24 h, 25°C

Arylcopper intermediates can be generated from organolithium compounds as in the preparation of cuprates.[42] These compounds react with a second aryl halide to provide unsymmetrical biaryls. This reaction is essentially a variant of the cuprate alkylation process discussed earlier in the chapter (p. 404).

8.2. Reactions Involving Organopalladium Intermediates

For the most part, organic reactions involving palladium do not involve the preparation of stoichiometric organopalladium reagents. Instead, organopalladium species are usually generated *in situ* in the course of the reaction. Indeed, in the most useful processes, only a *catalytic amount* of palladium is used. Three types of organopalladium intermediates are of primary importance in the reactions that have found synthetic application. Alkenes react with Pd(II) to give π complexes, which are subject to nucleophilic attack. These reactions are closely related to the solvomercuration reactions discussed in Section 4.3. The products that are derived from the resulting intermediates depend upon specific reaction conditions. The palladium can be replaced by hydrogen under reductive conditions (path a). In the absence of a reducing agent, an elimination of Pd(0) and a proton occurs, leading to net

39. P. E. Fanta, *Chem. Rev.* **64**, 613 (1964); *Synthesis*, 9 (1974).
40. R. C. Fuson and E. A. Cleveland, *Org. Synth.* **III**, 339 (1955).
41. T. Cohen and I. Christea, *J. Am. Chem. Soc.* **98**, 748 (1976).
42. F. E. Ziegler, I. Chliwner, K. W. Fowler, S. J. Kanfer, S. J. Kuo, and N. D. Sinha, *J. Am. Chem. Soc.* **102**, 790 (1980).

substitution of a vinyl hydrogen by the nucleophile (path b). We will return to specific examples of these reactions shortly.

$$RCH{=}CH_2 + Pd(II) \rightleftharpoons RCH{=}CH_2 \cdot Pd^{2+}$$

$$\mathbf{Nu} + RCH{=}CH_2 \cdot Pd^{2+} \rightarrow \mathbf{Nu}{-}CH(R)CH_2Pd^{2+}$$

$$\mathbf{Nu}{-}CH(R)CH_2Pd^{2+} \xrightarrow{[H]} \mathbf{Nu}{-}CH(R)CH_3 \quad \text{(path a)}$$

$$\mathbf{Nu}{-}CH(R)CH_2Pd^{2+} \xrightarrow{-Pd(0),\ -H^+} \mathbf{Nu}{-}C(R){=}CH_2 \quad \text{(path b)}$$

The second type of organopalladium intermediates are π-allyl complexes. These complexes can be obtained from Pd(II) salts and allylic acetates and other compounds with potential leaving groups in an allylic position.[43] The same type of π-allyl complexes can be prepared from alkenes by reaction with $PdCl_2$ or $Pd(O_2CCF_3)_2$.[44] The reaction is formulated as an electrophilic attack on the π electrons followed by loss of a proton. The proton loss probably proceeds via an unstable species in which the hydrogen is bound to palladium.[45]

$$\text{alkene–H} + Pd^{2+} \rightleftharpoons \text{alkene}{-}Pd^{2+}\text{ (H)} \xrightarrow{-H} \pi\text{-allyl}{-}Pd^{2+}$$

The complexes can be isolated as halide-bridged dimers.

These π-allyl complexes are electrophilic in character and undergo reaction with a variety of nucleophiles.

$$RCH_2CH{=}CH_2 \text{ or } RCH(O_2CCH_3)CH{=}CH_2 \rightarrow \pi\text{-allyl}{-}Pd^{2+} \xrightarrow{Nu} RCH(Pd^{2+})CH_2CH_2{-}\mathbf{Nu} \xrightarrow[-H^+]{-Pd^0} RCH{=}CHCH_2\mathbf{Nu} + Pd^0H$$

After nucleophilic addition occurs, the resulting organopalladium intermediate usually breaks down by elimination of Pd(0) and H^+.

The third general type of reactivity involves the reaction of Pd(0) species with halides by oxidative addition, generating reactive intermediates having the organic

43. R. Huttel, *Synthesis*, 225 (1970); B. M. Trost, *Tetrahedron* **33**, 2615 (1977).
44. B. M. Trost and P. J. Metzner, *J. Am. Chem. Soc.* **102**, 3572 (1980); B. M. Trost, P. E. Strege, L. Weber, T. J. Fullerton, and T. J. Dietsche, *J. Am. Chem. Soc.* **100**, 3407 (1978).
45. D. R. Chrisope, P. Beak, and W. H. Saunders, Jr., *J. Am. Chem. Soc.* **110**, 230 (1988).

group attached to Pd by a σ bond. The oxidative addition reaction is very useful for aryl and alkenyl halides, but the products from saturated alkyl halides usually decompose by elimination.

$$RCH{=}CHX + Pd(0) \rightarrow RCH{=}CH{-}Pd^{II}{-}X \qquad ArX + Pd(0) \rightarrow Ar{-}Pd^{II}{-}X$$

The σ-bonded species formed by oxidative addition can react with alkenes and other unsaturated compounds to form new carbon–carbon bonds. Specific examples of this type of reaction will be discussed below.

In considering the mechanisms involved in organopalladium chemistry, several general points should be kept in mind. Frequently, reactions involving organopalladium intermediates are done in the presence of phosphine ligands. These ligands coordinate at palladium and can play a key role in the course of the reaction by influencing the degree of reactivity. Another general point concerns the relative weakness of the C—Pd bond and, especially, the instability of alkylpalladium species in which there is a β hydrogen. The final stage in many palladium-mediated reactions is the elimination of Pd(0) and H^+ to generate a carbon–carbon double bond. This tendency toward elimination distinguishes organopalladium species from the organometallic species we have discussed to this point. Finally, organopalladium species with two organic substituents show the same tendency to decompose with recombination of the organic groups that was exhibited by copper(III) intermediates.

An important industrial process based on Pd–alkene complexes is the *Wacker reaction*, a catalytic method for conversion of ethylene to acetaldehyde. The first step is addition of water to the Pd-activated alkene. The addition intermediate undergoes the characteristic elimination of Pd(0) and H^+ to generate the enol of acetaldehyde.

$$CH_2{=}CH_2 + Pd(II) \rightarrow CH_2{=}CH_2{\cdot}Pd^{2+} \xrightarrow{H_2O} HO{-}CH_2CH_2{-}Pd^{2+} \rightarrow (HO)(H)C{=}CH_2 + Pd(^0) + H^+$$

$$(HO)(H)C{=}CH_2 \rightarrow CH_3CH{=}O$$

The reaction is run with only a catalytic amount of Pd. The co-reagents $CuCl_2$ and O_2 serve to reoxidize the Pd(0) to Pd(II). The net reaction consumes only alkene and oxygen.

½ O_2
$2Cu^{1+}$
^-OH
$2Cu^{2+}$
Pd^{2+}
$CH_2{=}CH_2$
Pd^{2+} | $CH_2{=}CH_2$
H_2O
$HOCH_2CH_2Pd^{2+}$
$HOCH{=}CH_2$
$Pd^0 + H^+$

When the Wacker conditions are applied to terminal alkenes, methyl ketones are formed.[46]

$$CH_2{=}CHCH_2C(CH_3)_2CH{=}O \xrightarrow[H_2O,\ DMF,\ O_2]{CuCl_2,\ PdCl_2} CH_3C({=}O)CH_2C(CH_3)_2CH{=}O \quad (78\%)$$ Ref. 46c

π-Allyl palladium species are involved in a number of useful reactions which result in allylation of nucleophiles. These reactions can be applied to carbon–carbon bond formation with relatively stable carbanions, such as those derived from malonate esters and β-ketoesters. The π-allyl complexes can be synthesized and used in stoichiometric amount[47] or they can be generated *in situ* by reaction of an allylic acetate with a catalytic amount of tetrakis(triphenylphosphine)palladium.[48] The reactive Pd(0) species is regenerated in an elimination step.

$$CH_2{=}CHCH_2OAc + Pd^0 \longrightarrow \pi\text{-allyl}\ Pd^{2+} \xrightarrow[H^+]{Nu^-} NuH_2C{-}CH(H){-}CH_2{-}Pd^{2+} \longrightarrow NuCH_2CH{=}CH_2 + Pd^0 + H^+$$

$$CH_3O_2C\text{-cyclohexenyl-}O_2CCH_3 \xrightarrow[NaCH(CO_2Et)_2]{Pd(PPh_3)_4} CH_3O_2C\text{-cyclohexenyl-}CH(CO_2Et)_2 \quad (57\%)$$ Ref. 49

The reaction has also been used to form rings. β-Sulfonylesters have proven particularly useful for formation of both medium and large rings. In some cases, medium-sized rings are formed in preference to six-and seven-membered rings.[50]

$$PhSO_2CH(CO_2CH_3)CH_2C({=}O)OCH_2CH_2CH{=}CHCH_2O_2CCH_3 \xrightarrow[NaH]{Pd(PPh_3)_4} \text{lactone ring with } PhSO_2,\ CO_2CH_3 \quad (54\%)$$

+5% of *trans*-isomer

46. (a) J. Tsuji, I. Shimizu, and K. Yamamoto, *Tetrahedron Lett.*, 2975 (1976); (b) J. Tsuji, H. Nagashima, and H. Nemoto, *Org. Synth.* **62**, 9 (1984); (c) D. Pauley, F. Anderson, and T. Hudlicky, *Org. Synth.* **67**, 12 (1988); (d) K. Januszkiewicz and H. Alper, *Tetrahedron Lett.* **24**, 5159 (1983); (e) K. Januszkiewicz and D. J. H. Smith, *Tetrahedron Lett.* **26**, 2263 (1985).
47. B. M. Trost, W. P. Conway, P. E. Strege, and T. J. Dietsche, *J. Am. Chem. Soc.* **96**, 7165 (1974); B. M. Trost, L. Weber, P. E. Strege, T. J. Fullerton, and T. J. Dietsche, *J. Am. Chem. Soc.* **100**, 3416 (1978); B. M. Trost, *Acc. Chem. Res.* **13**, 385 (1980).
48. B. M. Trost and T. R. Verhoeven, *J. Am. Chem. Soc.* **102**, 4730 (1980).
49. B. M. Trost and P. E. Strege, *J. Am. Chem. Soc.* **99**, 1649 (1977).
50. B. M. Trost and T. R. Verhoeven, *J. Am. Chem. Soc.* **102**, 4743 (1980); B. M. Trost and S. J. Brickner, *J. Am. Chem. Soc.* **105**, 568 (1983).

O_2CCH_3 C_2H_5O CH_2 CH_3 O O CH_2SO_2Ph $\xrightarrow{Pd(PPh_3)_4}$ C_2H_5O CH_3 SO_2Ph O O (60%)

The sulfonyl substituent can be removed by reduction after the ring closure (see Section 5.5.2).

The third important type of reactivity of palladium, namely, oxidative addition to Pd(0), is the foundation for several methods of forming carbon–carbon bonds. Aryl[51] and alkenyl[52] halides react with alkenes in the presence of catalytic amounts of palladium to give net substitution of the halide by the alkenyl group. The reaction is quite general and has been observed for simple alkenes, aryl-substituted alkenes, electrophilic alkenes such as acrylate esters, and *N*-vinylamides.[53] The reaction is carried out in the presence of a phosphine ligand, with tri-(*o*-tolyl)phosphine being preferred in many cases.

$$C_6H_5\text{-}Br + CH_2{=}CHCO_2CH_3 \xrightarrow[R_3N]{(Ph_3P)_2Pd(OAc)_2} C_6H_5\text{-}CH{=}CHCO_2CH_3 \quad (85\%)$$

The reaction is initiated by oxidative addition of the halide to a Pd(0) species generated *in situ* from the Pd(II) catalyst. The arylpalladium intermediate forms a complex with the alkene. The complex decomposes with carbon–carbon bond formation and regeneration of Pd(0), presumably through a σ-bonded intermediate.

$$C_6H_5\text{-}Br + Pd(PPh_3)_2 \longrightarrow C_6H_5\text{-}Pd(PPh_3)_2Br \xrightarrow{CH_2{=}CH{-}Y} C_6H_5\text{-}Pd(PPh_3)Br(CH_2{=}CHY)$$

$$C_6H_5\text{-}Pd(PPh_3)Br(CH_2{=}CHY) \longrightarrow C_6H_5CH_2{-}CHY{-}Pd(PPh_3)Br \longrightarrow C_6H_5\text{-}CH{=}CHY + Pd(PPh_3)_2 + HBr$$

A number of modified reaction conditions have been developed. One involves addition of silver salts, which serve to activate the halide toward displacement.[54]

51. H. A. Dieck and R. F. Heck, *J. Am. Chem. Soc.* **96**, 1133 (1974); R. F. Heck, *Acc. Chem. Res.* **12**, 146 (1979).
52. B. A. Patel and R. F. Heck, *J. Org. Chem.* **43**, 3898 (1978); B. A. Patel, J. I. Kim, D. D. Bender, L. C. Kao, and R. F. Heck, *J. Org. Chem.* **46**, 1061 (1981); J. I. Kim, B. A. Patel, and R. F. Heck, *J. Org. Chem.* **46**, 1067 (1981).
53. C. B. Ziegler, Jr., and R. F. Heck, *J. Org. Chem.* **43**, 2941 (1978); W. C. Frank, Y. C. Kim, and R. F. Heck, *J. Org. Chem.* **43**, 2947 (1978); C. B. Ziegler, Jr., and R. F. Heck, *J. Org. Chem. Soc.* **96**, 1133 (1974).
54. M. M. Abelman, T. Oh, and L. E. Overman, *J. Org. Chem.* **52**, 4130 (1987); M. M. Abelman and L. E. Overman, *J. Am. Chem. Soc.* **110**, 2328 (1988).

Use of sodium bicarbonate or sodium carbonate in the presence of a phase transfer catalyst permits especially mild conditions to be used for many systems.[55] Pretreatment with nickel bromide causes normally unreactive aryl chlorides to undergo Pd-catalyzed substitution.[56] Solid phase catalysts in which the palladium is complexed by polymer-bound phosphine groups have also been developed.[57] Aryl triflates have also been found to be excellent substrates for Pd-catalyzed vinylations.[58] Scheme 8.6 illustrates the vinylation reaction.

Scheme 8.6. Palladium-Catalyzed Vinylation of Aryl and Alkenyl Halides

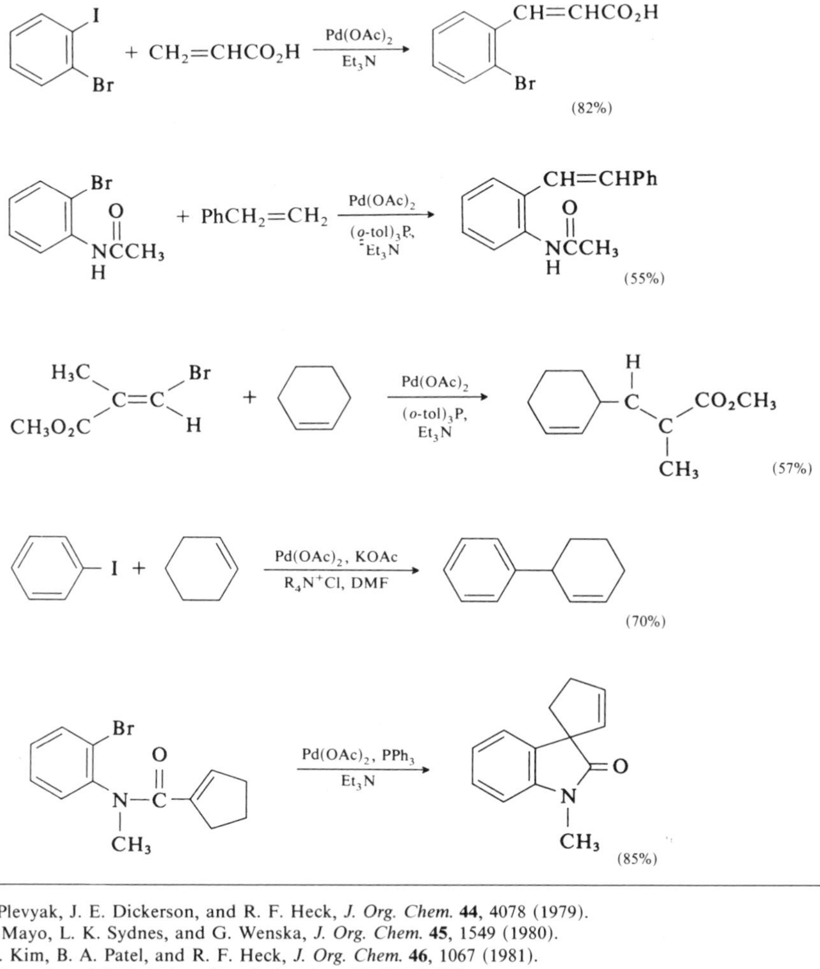

1[a] (2-bromoiodobenzene) + $CH_2{=}CHCO_2H$ $\xrightarrow[Et_3N]{Pd(OAc)_2}$ 2-bromo-C_6H_4–$CH{=}CHCO_2H$ (82%)

2[b] 2-bromo-C_6H_4–$NHCOCH_3$ + $PhCH_2{=}CH_2$ $\xrightarrow[(o\text{-tol})_3P,\ Et_3N]{Pd(OAc)_2}$ 2-($CH{=}CHPh$)-C_6H_4–$NHCOCH_3$ (55%)

3[c] $(H_3C)(CH_3O_2C)C{=}C(Br)(H)$ + cyclohexene $\xrightarrow[(o\text{-tol})_3P,\ Et_3N]{Pd(OAc)_2}$ cyclohexenyl–$CH{=}C(CH_3)CO_2CH_3$ (57%)

4[d] C_6H_5–I + cyclohexene $\xrightarrow[R_4N^+Cl^-,\ DMF]{Pd(OAc)_2,\ KOAc}$ 3-phenylcyclohexene (70%)

5[e] 2-bromo-C_6H_4–$N(CH_3)$–$C(=O)$–cyclopentenyl $\xrightarrow[Et_3N]{Pd(OAc)_2,\ PPh_3}$ spiro[cyclopentene-oxindole], N–CH_3 (85%)

a. J. E. Plevyak, J. E. Dickerson, and R. F. Heck, *J. Org. Chem.* **44**, 4078 (1979).
b. P. de Mayo, L. K. Sydnes, and G. Wenska, *J. Org. Chem.* **45**, 1549 (1980).
c. J.-I. I. Kim, B. A. Patel, and R. F. Heck, *J. Org. Chem.* **46**, 1067 (1981).
d. R. C. Larock and B. E. Baker, *Tetrahedron Lett.* **29**, 905 (1988).
e. M. M. Abelman, T. Oh, and L. E. Overman, *J. Org. Chem.* **52**, 4130 (1987).

55. T. Jeffery, *J. Chem. Soc., Chem. Commun.*, 1287 (1984); *Tetrahedron Lett.* **26**, 2667 (1985); *Synthesis*, 70 (1987); R. C. Larock and S. Babu, *Tetrahedron Lett.* **28**, 5291 (1987).
56. J. J. Bozell and C. E. Vogt, *J. Am. Chem. Soc.* **110**, 2655 (1988).
57. C.-M. Andersson, K. Karabelas, A. Hallberg, and C. Andersson, *J. Org. Chem.* **50**, 3891 (1985).
58. A. M. Echavarren and J. K. Stille, *J. Am. Chem. Soc.* **109**, 5478 (1987).

Tetrakis(triphenylphosphine)palladium catalyzes coupling of alkenyl halides with Grignard reagents and organolithium reagents.

$$\underset{C_6H_{13}}{\overset{H}{}}\!\!>\!C{=}C\!<\!\!\underset{I}{\overset{H}{}} + \underset{H}{\overset{BrMg}{}}\!\!>\!C{=}C\!<\!\!\underset{H}{\overset{H}{}} \xrightarrow{Pd(PPh_3)_4} \underset{C_6H_{13}}{\overset{H}{}}\!\!>\!C{=}C\!<\!\!\underset{CH=CH_2}{\overset{H}{}} \; (75\%) \qquad \text{Ref. 59}$$

$$\underset{C_4H_9}{\overset{H}{}}\!\!>\!C{=}C\!<\!\!\underset{Br}{\overset{H}{}} + C_4H_9Li \xrightarrow{Pd(PPh_3)_4} \underset{C_4H_9}{\overset{H}{}}\!\!>\!C{=}C\!<\!\!\underset{C_4H_9}{\overset{H}{}} \; (63\%) \qquad \text{Ref. 60}$$

These processes are similar to the halide-alkene couplings except that the reactive disubstituted Pd intermediate is generated by transfer of the nucleophilic organic group to the Pd center from the organometallic reagent.

$$\underset{R}{\overset{H}{}}\!\!>\!C{=}C\!<\!\!\underset{\overset{2+}{Pd}-X}{\overset{H}{}} + R'M \rightarrow \underset{R}{\overset{H}{}}\!\!>\!C{=}C\!<\!\!\underset{\overset{2+}{Pd}-R'}{\overset{H}{}} + MX \rightarrow \underset{R}{\overset{H}{}}\!\!>\!C{=}C\!<\!\!\underset{R'}{\overset{H}{}} + Pd^{(0)}$$

Organozinc compounds are also useful in palladium-catalyzed coupling with aryl and alkenyl halides. Procedures employing arylzinc,[61] alkenylzinc,[62] and alkylzinc[63] reagents have been developed. A ferrocenyldiphosphine has been found to be an especially good Pd ligand for these reactions.[64]

$$2\text{-}CH_3C_6H_4ZnCl + Br\text{-}C_6H_4\text{-}NO_2 \xrightarrow{Pd(PPh_3)_4} 2\text{-}CH_3C_6H_4\text{-}C_6H_4\text{-}NO_2 \; (78\%) \qquad \text{Ref. 61}$$

$$CH_2{=}CHCH_2CH_2ZnCl + \underset{H}{\overset{I}{}}\!\!>\!C{=}C\!<\!\!\underset{(CH_2)_3CH_3}{\overset{CH_3}{}} \xrightarrow{Pd(PPh_3)_4} \underset{H}{\overset{CH_2=CHCH_2CH_2}{}}\!\!>\!C{=}C\!<\!\!\underset{(CH_2)_3CH_3}{\overset{CH_3}{}} \; (81\%) \qquad \text{Ref. 63}$$

$$\underset{H_5C_2}{\overset{Ph}{}}\!\!>\!C{=}C\!<\!\!\underset{Ph}{\overset{Br}{}} + MeO\text{-}C_6H_4\text{-}ZnCl \xrightarrow{Pd(PPh_3)_4} \underset{H_5C_2}{\overset{Ph}{}}\!\!>\!C{=}C\!<\!\!\underset{Ph}{\overset{C_6H_4\text{-}OMe}{}} \; (84\%) \qquad \text{Ref. 65}$$

59. M. P. Dang and G. Linstrumelle, *Tetrahedron Lett.*, 191 (1978).
60. M. Yamamura, I. Moritani, and S. Murahashi, *J. Organometal. Chem.* **91**, C39 (1975).
61. E. Negishi, A. O. King, and N. Okukado, *J. Org. Chem.* **42**, 1821 (1977); E. Negishi, T. Takahashi, and A. O. King, *Org. Synth.* **66**, 67 (1987).
62. U. H. Lauk, P. Skrabal, and H. Zollinger, *Helv. Chim. Acta* **68**, 1406 (1985).
63. E. Negishi, L. F. Valente, and M. Kobayashi, *J. Am. Chem. Soc.* **102**, 3298 (1980).
64. T. Hayashi, M. Konishi, Y. Kobori, M. Kumada, T. Higuchi, and K. Hirotsu, *J. Am. Chem. Soc.* **106**, 158 (1984).
65. R. B. Miller and M. I. Al-Hassan, *J. Org. Chem.* **50**, 2121 (1985).

A combination of $Pd(PPh_3)_4$ and Cu(I) effects coupling of terminal alkynes with vinyl or aryl halides.[66] The alkyne is presumably converted to the copper acetylide. The halide reacts with Pd(0) by oxidative addition. Transfer of the acetylide group to Pd results in reductive elimination and formation of the observed product.

$$HC{\equiv}CR \xrightarrow[R_3N]{Cu(I)} CuC{\equiv}CR$$

$$R'X + Pd^0 \longrightarrow R'Pd^{II}X$$

$$CuC{\equiv}CR + R'Pd^{II}X \longrightarrow R'Pd^{II}(C{\equiv}CR) \longrightarrow R'C{\equiv}CR + Pd^0$$

Use of alkenyl halides in this reaction has proven to be an effective method for the synthesis of enynes.[67]

Organopalladium intermediates generated from halides by oxidative addition react with carbon monoxide in the presence of alcohols to give esters.[68]

$$(C_2H_5)(H)C{=}C(C_2H_5)(I) + CO \xrightarrow[n\text{-BuOH}]{PdI_2(PPh_3)_2} (C_2H_5)(H)C{=}C(C_2H_5)(CO_2C_4H_9) \quad (74\%)$$

Complexes of alkenes are also reactive toward carbon monoxide. A catalytic process which includes copper(II) results in concomitant addition of nucleophilic solvent. The copper(II) reoxidizes Pd(0) to the Pd(II) state.[69]

$$Pd^{2+} + CH_2{=}CHR \rightarrow Pd^{2+}(CH_2{=}CHR) \xrightarrow{MeOH} Pd^{2+}{-}CH_2CH(OMe)R \xrightarrow{CO} Pd^{2+}{-}C(=O)CH_2CH(OMe)R \xrightarrow{MeOH} MeOC(=O)CH_2CH(OMe)R + Pd^0$$

$$Pd^0 + 2\,Cu^{2+} \rightarrow Pd^{2+} + 2\,Cu^{1+}$$

Both of these reactions depend on a carbonyl insertion step. This is a common reaction for palladium and certain other metals and takes place by migration of the organic group from the metal to the coordinated carbon monoxide.

$$R{-}Pd{-}C{\equiv}O^+ \rightarrow Pd{-}C(=O){-}R \xrightarrow{R'OH} Pd^0 + R'O{-}C(=O){-}R + H^+$$

66. K. Sonogashira, Y. Tohda, and N. Hagihara, *Tetrahedron Lett.*, 4467 (1975).
67. V. Ratoveloma and G. Linstrumelle, *Synth. Commun.* **11**, 917 (1981); L. Crombie and M. A. Horsham, *Tetrahedron Lett.* **28**, 4879 (1987); G. Just and B. O'Connor, *Tetrahedron Lett.* **29**, 753 (1988); D. Guillerm and G. Linstrumelle, *Tetrahedron Lett.* **27**, 5857 (1986).
68. A. Schoenberg, I. Bartoletti, and R. F. Heck, *J. Org. Chem.* **39**, 3318 (1974).
69. D. E. James and J. K. Stille, *J. Am. Chem. Soc.* **98**, 1810 (1976).

The detailed mechanisms of such reactions have been shown to involve addition and elimination of phosphine ligands. The efficiency of individual reactions can often be improved by careful study of the effect of added ligands.

Application of the carbonylation reaction to halides with appropriately placed hydroxyl groups leads to lactone formation. In this case, the acylpalladium intermediate is trapped intramolecularly.

$$(CH_3)(I)C{=}C(H)CH(OH)CH_3 \xrightarrow[C\equiv O]{Pd(PPh_3)_2Cl_2} \text{lactone } (CH_3, CH_3) \quad (99\%)$$ Ref. 70

8.3. Reactions Involving Organonickel Compounds

The most useful synthetic processes that have been developed utilizing nickel involve the coupling of halides. Allylic halides react with nickel carbonyl, $Ni(CO)_4$, to give π-allyl complexes. These complexes react with a variety of halides to give coupling products.[71]

$$2\ CH_2{=}CHCH_2Br + 2\ Ni(CO)_4 \rightarrow [(\pi\text{-allyl})NiBr]_2$$ Ref. 72

$$CH_2{=}CHBr + [(CH_2{\cdots}CH{\cdots}CH_2)NiBr]_2 \rightarrow CH_2{=}CHCH_2CH{=}CH_2 \quad (70\%)$$

$$C_6H_{11}I + [(CH_2{\cdots}CH{\cdots}CH_2)NiBr]_2 \rightarrow C_6H_{11}CH_2CH{=}CH_2 \quad (91\%)$$

These coupling reactions are believed to involve Ni(I) and Ni(III) intermediates in a chain process which is initiated by formation of a small amount of a Ni(I) species.[73]

Ni^{II} ⇌ —Ni^{I}— ; e$^-$ → Ni^{I} ; Ni^{I} + R—X → R—Ni^{III} → RCH$_2$CH=CH$_2$

70. A. Cowell and J. K. Stille, *J. Am. Chem. Soc.* **102**, 4193 (1980).
71. M. F. Semmelhack, *Org. React.* **19**, 115 (1972).
72. E. J. Corey and M. F. Semmelhack, *J. Am. Chem. Soc.* **89**, 2755 (1967).
73. L. S. Hegedus and D. H. P. Thompson, *J. Am. Chem. Soc.* **107**, 5663 (1985).

Nickel carbonyl effects coupling of allylic halides when the reaction is carried out in very polar solvents such as DMF or DMSO. This coupling reaction has been used intramolecularly to bring about cyclization of bis-allylic halides and has been found useful in the preparation of large rings.

$BrCH_2CH{=}CH(CH_2)_{12}CH{=}CHCH_2Br \xrightarrow{Ni(CO)_4}$ (76–84%) Ref. 74

$BrCH_2CH{=}CHCH_2CH_2C(=O)O\text{–}CH_2CH_2CH_2CH{=}CHCH_2Br \xrightarrow{Ni(CO)_4}$ (70–75%) Ref. 75

Nickel carbonyl is an extremely toxic compound, and a number of other nickel reagents with generally similar reactivity can be used in its place. The Ni(0) complex of 1,5-cyclooctadiene, $Ni(COD)_2$, has been found to bring about coupling of allylic, alkenyl, and aryl halides.

$PhCH{=}CHBr \xrightarrow{Ni(COD)_2} PhCH{=}CH\text{–}CH{=}CHPh$ (46%) Ref. 76

$N{\equiv}C\text{–}C_6H_4\text{–}Br \xrightarrow{Ni(COD)_2} N{\equiv}C\text{–}C_6H_4\text{–}C_6H_4\text{–}C{\equiv}N$ (81%) Ref. 77

Tetrakis(triphenylphoshphine)nickel(0) is an effective reagent for coupling aryl halides.[78] Large rings can be formed in intramolecular reactions.

$CH_3O\text{–}C_6H_3(I)\text{–}CH_2CH_2N(CH_3)CH_2CH_2\text{–}C_6H_3(I)\text{–}OCH_3 \xrightarrow{Ni(PPh_3)_4}$ Ref. 79

The coupling of aryl halides can be made catalytic in nickel by using zinc as a reductant for *in situ* regeneration of the active Ni(0) species.[80]

$O{=}CH\text{–}C_6H_4\text{–}Cl \xrightarrow[NiCl_2\ (5\ mol\ \%),\ PPh_3\ (5\ mol\ \%)]{Zn,\ NaBr} O{=}CH\text{–}C_6H_4\text{–}C_6H_4\text{–}CH{=}O$ (62%)

74. E. J. Corey and E. K. W. Wat, *J. Am. Chem. Soc.* **89**, 2757 (1967).
75. E. J. Corey and H. A. Kirst, *J. Am. Chem. Soc.* **94**, 667 (1972).
76. M. F. Semmelhack, P. M. Helquist, and J. D. Gorzynski, *J. Am. Chem. Soc.* **94**, 9234 (1972).
77. M. F. Semmelhack, P. M. Helquist, and L. D. Jones, *J. Am. Chem. Soc.* **93**, 5908 (1971).
78. A. S. Kende, L. S. Liebeskind, and D. M. Braitsch, *Tetrahedron Lett.*, 3375 (1975).
79. S. Brandt, A. Marfat, and P. Helquist, *Tetrahedron Lett.*, 2193 (1979).
80. M. Zembayashi, K. Tamao, J. Yoshida, and M. Kumada, *Tetrahedron Lett.*, 4089 (1977); I. Colon and D. R. Kelly, *J. Org. Chem.* **51**, 2627 (1986).

Mechanistic study of the aryl couplings has revealed the importance of the changes in redox state that are involved in the reaction.[81] Ni(I), Ni(II), and Ni(III) states are believed to be involved. Changes in the degree of coordination by phosphine ligands are also believed to be involved, but these have been omitted in the mechanism shown here. The detailed kinetics of the reaction are inconsistent with a mechanism involving only formation and decomposition of a biarylnickel(II) intermediate.

initiation by electron transfer $$ArNi(II)X + ArX \rightarrow ArNi(III)X^+ + Ar\cdot + X^-$$

propagation $$ArNi(III)X^+ + ArNi(II)X \rightarrow Ar_2Ni(III)X + Ni^{2+} + X^-$$

$$Ar_2Ni(III)X \rightarrow Ar{-}Ar + Ni(I)X$$

$$Ni(I)X + ArX \rightarrow ArNi(III)X^+ + X^-$$

The key aspects of the mechanism are (1) the reductive elimination occurs via a diaryl Ni(III) intermediate and (2) the oxidative addition involves a Ni(I) species.

Nickel(II) salts are able to catalyze the coupling of Grignard reagents with alkenyl and aryl halides. A soluble bis-phosphine complex, Ni $(Ph_2PCH_2CH_2PPh_2)_2Cl_2$, is a particularly effective catalyst.[82]

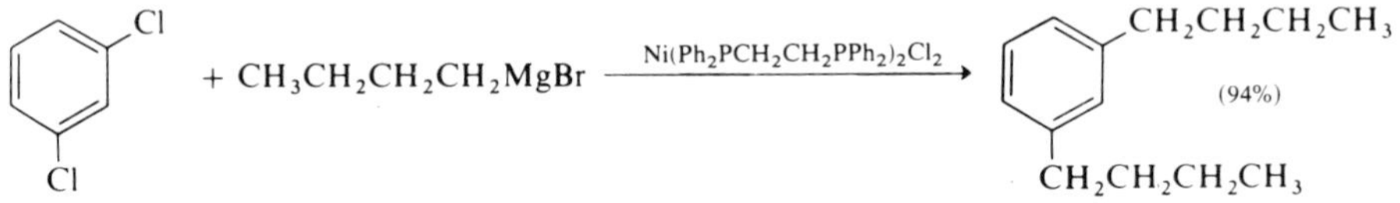

$$m\text{-}C_6H_4Cl_2 + CH_3CH_2CH_2CH_2MgBr \xrightarrow{Ni(Ph_2PCH_2CH_2PPh_2)_2Cl_2} m\text{-}C_6H_4(CH_2CH_2CH_2CH_3)_2 \quad (94\%)$$

The reaction has been applied to the synthesis of cyclophane-type structures by use of dihaloarenes and Grignard reagents from α,ω-dihalides.

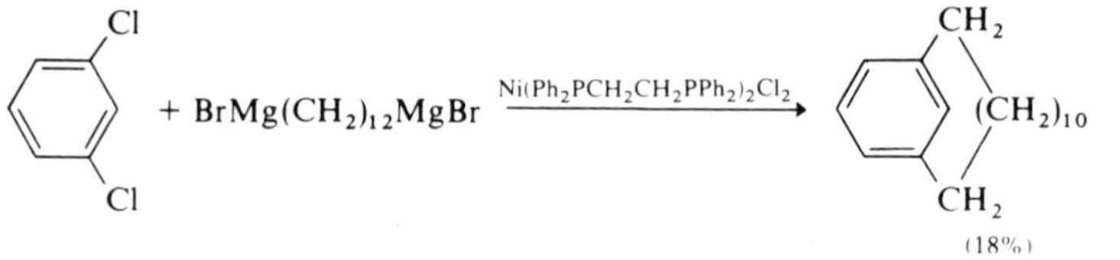

$$m\text{-}C_6H_4Cl_2 + BrMg(CH_2)_{12}MgBr \xrightarrow{Ni(Ph_2PCH_2CH_2PPh_2)_2Cl_2} m\text{-}C_6H_4[CH_2(CH_2)_{10}CH_2] \quad (18\%)$$ Ref. 83

When secondary Grignard reagents are used, the coupling product sometimes is derived from the corresponding primary alkyl group.[84] This transformation can occur by reversible formation of a nickel–alkene complex from the σ-bonded alkyl group. Re-formation of the σ-bonded structure will be preferred at the less hindered primary position.

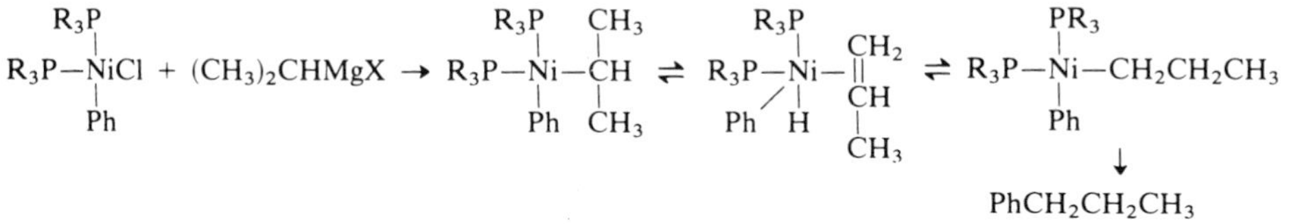

$$(R_3P)_2Ni(Ph)Cl + (CH_3)_2CHMgX \rightarrow (R_3P)_2Ni(Ph){-}CH(CH_3)_2 \rightleftharpoons (R_3P)_2Ni(Ph)(H)(CH_2{=}CHCH_3) \rightleftharpoons (R_3P)_2Ni(Ph){-}CH_2CH_2CH_3 \rightarrow PhCH_2CH_2CH_3$$

81. T. T. Tsou and J. K. Kochi, *J. Am. Chem. Soc.* **101**, 7547 (1979).
82. K. Tamao, K. Sumitani, and M. Kumada, *J. Am. Chem. Soc.* **94**, 4374 (1972).
83. K. Tamao, S. Kodama, T. Nakatsuka, Y. Kiso, and A. Kumadak *J. Am. Chem. Soc.* **97**, 4405 (1975).
84. K. Tamao, Y. Kiso, K. Sumitani, and M. Kumada, *J. Am. Chem. Soc.* **94**, 9268 (1972).

The metals rhodium, iron, and cobalt each participate in several reactions that are of value in organic synthesis. Rhodium and cobalt are active catalysts for the reaction of alkenes with hydrogen and carbon monoxide to give aldehydes. This reaction is called *hydroformylation.*[85]

$$\text{cyclohexene} + CO + H_2 \xrightarrow[\text{100°C, 60–150 atm}]{Rh_2O_3} \text{C}_6\text{H}_{11}\text{CH=O} \quad (82\text{–}84\%) \qquad \text{Ref. 86}$$

The key steps in the reaction are addition of hydridometal to the double bond of the alkene and migration of the alkyl group to the complexed carbon monoxide.

$$\text{cyclohexene} + HRh(CO)$$
$$\downarrow$$
$$C_6H_{11}{-}Rh{-}C{\equiv}O^+ \longrightarrow Rh{-}\overset{O}{\overset{\|}{C}}{-}C_6H_{11} \xrightarrow{H_2} Rh{-}H + H\overset{O}{\overset{\|}{C}}{-}C_6H_{11}$$

The steps in the hydroformulation reaction are closely related to those that occur in the *Fischer–Tropsch* process. The Fischer–Tropsch process is the reductive conversion of carbon monoxide to alkanes. It occurs by a series of carbonylation, migration, and reduction steps.

$$M + CO \rightarrow M{-}CO \xrightarrow{+H_2} M{-}CH_3 \xrightarrow{+CO} OC{-}M{-}CH_3$$

$$OC{-}M{-}CH_3 \rightarrow M{-}\overset{O}{\overset{\|}{C}}{-}CH_3 \xrightarrow{+H_2} M{-}CH_2CH_3 \qquad \text{etc.}$$

The Fischer–Tropsch process is of interest because of its potential for conversion of carbon monoxide to synthetic hydrocarbons fuels.

The key carbonylation step which is involved in both hydroformylation and the Fischer–Tropsch reaction can be reversible. Under appropriate conditions, rhodium catalysts can be used for the decarbonylation of aldehydes[87] and acid chlorides.[88]

$$R\overset{O}{\overset{\|}{C}}H + Rh(PPh_3)_3Cl \rightarrow RH \qquad R\overset{O}{\overset{\|}{C}}Cl + Rh(PPh_3)_3Cl \rightarrow RCl$$

85. R. L. Pruett, *Adv. Organometal. Chem.* **17**, 1 (1979); H. Siegel and W. Himmele, *Angew. Chem. Int. Ed. Engl.* **19**, 178 (1980); J. Falbe, *New Syntheses with Carbon Monoxide*, Springer-Verlag, Berlin, 1980.
86. P. Pino and C. Botteghi, *Org. Synth.* **57**, 11 (1977).
87. J. A. Kampmeier, S. H. Harris, and D. K. Wedgaertner, *J. Org. Chem.* **45**, 315 (1980).
88. J. K. Stille and M. T. Regan, *J. Am. Chem. Soc.* **96**, 1508 (1974); J. K. Stille and R. W. Fries, *J. Am. Chem. Soc.* **96**, 1514 (1974).

An acylrhodium intermediate is involved in both cases. The elimination of the hydrocarbon or halide occurs by reductive elimination.[89]

$$\mathrm{RC(=O)X + Rh(PPh_3)_3Cl \rightarrow RC(=O){-}Rh(Cl)(X)(PPh_3)_2 \rightarrow R{-}Rh(Cl)(X)(PPh_3)_2 + CO}$$

$$\mathrm{R{-}Rh(Cl)(X)(PPh_3)_2 \rightarrow R{-}X + Rh(PPh_3)_2COCl}$$

X = H, Cl

There are related reversible carbonylations involved in organoiron chemistry. These reactions can be illustrated by discussion of the chemistry of sodium tetracarbonylferrate, $Na_2Fe(CO)_4$.[90] The formal oxidation state of iron in this compound is −2. The carbon monoxide ligands serve to stabilize this very low oxidation state, but the iron is very reactive toward oxidative addition by halides and tosylates. The adducts can undergo carbonylation, leading to the formation of aldehydes, ketones, and carboxylic acid derivatives.

$$\mathrm{R{-}X + Fe(CO)_4^{2-} \rightarrow [R{-}Fe(CO)_4(X)]^{2-} \rightarrow [R{-}C(=O){-}Fe(CO)_3]^{-} + X^{-}}$$

$$\mathrm{[R{-}C(=O){-}Fe(CO)_3] \xrightarrow{H^+} RCH{=}O}$$

$$\mathrm{[R{-}C(=O){-}Fe(CO)_3] \xrightarrow{O_2,\ R'OH} RCO_2R'}$$

$$\mathrm{[R{-}C(=O){-}Fe(CO)_3] \xrightarrow{R'{-}X,\ PPh_3} R{-}C(=O){-}Fe(R')(PPh_3)(CO)_3 \rightarrow R{-}C(=O){-}R'}$$

This particular reagent illustrates several of the fundamental reaction types which make transition metal complexes useful in organic synthesis. The first step in each path is oxidative addition. This is followed by carbonylation and migration of the alkyl group from iron to a coordinated carbon monoxide molecule. This reaction is promoted by coordination of a phosphine ligand at iron. The acyliron intermediates are reactive toward protons (giving an aldehyde), toward O_2 and other oxidants (giving carboxylic acids and esters), and toward alkyl halides (giving ketones).

The acyliron intermediates also react with electrophilic alkenes such as acrylonitrile, acrylate esters, and α,β-unsaturated ketones.[91]

$$\mathrm{[RFe(CO)_4]^- + CH_2{=}CH{-}Z \rightarrow R{-}C(=O){-}CH_2CH_2Z}$$

89. J. E. Baldwin, T. C. Barden, R. L. Pugh, and W. C. Widdison, *J. Org. Chem.* **52**, 3303 (1987).
90. J. P. Collman, *Acc. Chem. Res.* **8**, 342 (1975); R. G. Finke and T. N. Sorrell, *Org. Synth.* **59**, 102 (1979).
91. M. P. Cooke, Jr., and R. M. Parlman, *J. Am. Chem. Soc.* **99**, 5222 (1977).

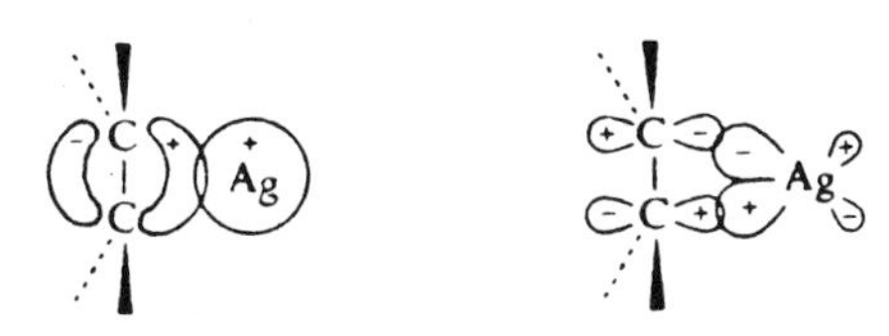

Fig. 8.1. Representation of π bonding in alkene-transition-metal complexes.

8.5. Organometallic Compounds with π Bonding

The organometallic intermediates in the previous sections involved, in most cases, carbon–metal σ bonds, although examples of π bonding with alkenes and allyl groups were also encountered. The compounds that are emphasized in this section involve organic groups that are bound to the metal through delocalized π systems. Among the classes of organic compounds that serve as π ligands are alkenes, allyl groups, dienes, the cyclopentadienide anion, and aromatic compounds. There are a very large number of such compounds, and we will illustrate only a few representative examples.

The bonding in π complexes of alkenes is the result of two major contributions. The filled π orbital acts as an electron donor to the metal ion. There is also a contribution to bonding, called "back-bonding," from a filled metal orbital interacting with the alkene π^* orbital. These two types of bonding are represented in Fig. 8.1.

These same general bonding concepts apply to all the other π organometallics. The details of structure and reactivity of an individual compound depend on such factors as (a) how many electrons can be accommodated by the metal orbitals; (b) the oxidation level of the metal; and (c) the electronic character of other ligands on the metal.

Alkene–metal complexes are usually prepared by a process in which some other ligand is first dissociated from the metal. Both thermal and photochemical reactions are used.

$(C_6H_5CN)_2PdCl_2$ + 2 RCH=CH$_2$ → [RCH=CH$_2$, Cl, Pd, Cl, Pd, Cl, Cl, CH$_2$=CHR] Ref. 92

[(OC)$_2$Rh(μ-Cl)$_2$Rh(CO)$_2$] + 2 [cyclooctadiene] → [(cyclooctadiene)Rh(μ-Cl)$_2$Rh(cyclooctadiene)] Ref. 93

92. M. S. Kharasch, R. C. Seyler, and F. R. Mayo, *J. Am. Chem. Soc.* **60**, 882 (1938).
93. J. Chatt and L. M. Venanzi, *J. Chem. Soc.*, 4735 (1957).

π-Allyl complexes of nickel can be prepared either by oxidative addition to Ni(0) or by ligation of a Ni(II) salt.

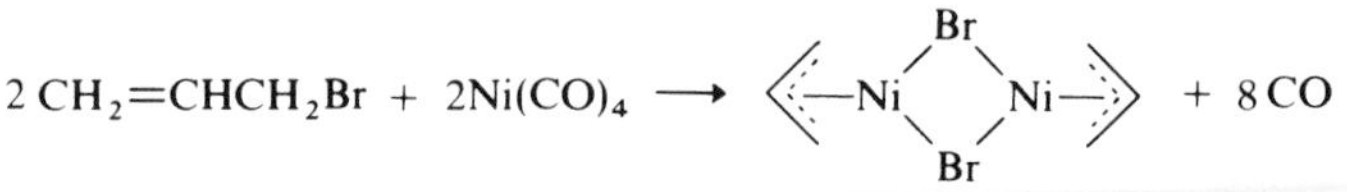

$$2\,CH_2{=}CHCH_2Br + 2Ni(CO)_4 \longrightarrow (\pi\text{-allyl})Ni(\mu\text{-Br})_2Ni(\pi\text{-allyl}) + 8\,CO$$ Ref. 94

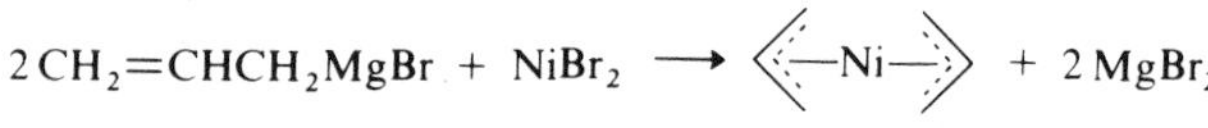

$$2\,CH_2{=}CHCH_2MgBr + NiBr_2 \longrightarrow (\pi\text{-allyl})_2Ni + 2\,MgBr_2$$ Ref. 95

Organic ligands having a cyclic array of four carbon atoms have been of particular interest in connection with the chemistry of cyclobutadiene. Organometallic compounds containing cyclobutadiene as a ligand were first prepared in 1965.[96] The carbocyclic ring in the cyclobutadiene–iron tricarbonyl complex reacts as an aromatic ring and can undergo electrophilic substitutions.[97] Subsequent studies provided evidence that oxidative decomposition of the complex could liberate cyclobutadiene, which could be trapped by appropriate reactants.[98] Some examples of these reactions are given in Scheme 8.7.

Scheme 8.7. Reactions of Cyclobutadiene

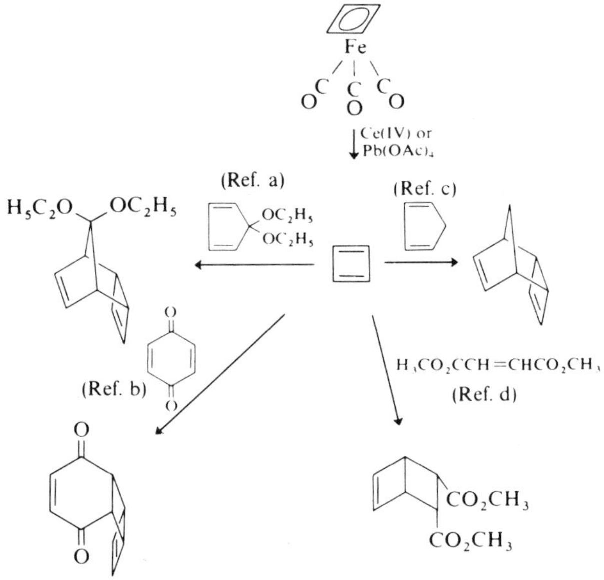

Fe, C, O, C, O, C, O
Ce(IV) or Pb(OAc)$_4$
(Ref. a)
OC_2H_5
OC_2H_5
H_5C_2O OC_2H_5
(Ref. c)
O
O
(Ref. b)
O
O
$H_3CO_2CCH{=}CHCO_2CH_3$
(Ref. d)
CO_2CH_3
CO_2CH_3

a. J. C. Barborak and R. Pettit, *J. Am. Chem. Soc.* **89**, 3080 (1967).
b. J. C. Barborak, L. Watts, and R. Pettit, *J. Am. Chem. Soc.* **88**, 1328 (1966).
c. L. Watts, J. D. Fitzpatrick, and R. Pettit, *J. Am. Chem. Soc.* **88**, 623 (1966).
d. P. Reeves, J. Henery, and R. Pettit, *J. Am. Chem. Soc.* **91**, 5889 (1969).

94. E. J. Corey and M. F. Semmelhack, *J. Am. Chem. Soc.* **89**, 2755 (1967).
95. D. Walter and G. Wilke, *Angew. Chem. Int. Ed. Engl.* **5**, 151 (1966).
96. G. F. Emerson, L. Watts, and R. Pettit, *J. Am. Chem. Soc.* **87**, 131 (1965); R. Pettit and J. Henery, *Org. Synth.* **50**, 21 (1970).
97. J. D. Fitzpatrick, L. Watts, G. F. Emerson, and R. Pettit, *J. Am. Chem. Soc.* **87**, 3254 (1965).
98. R. H. Grubbs and R. A. Grey, *J. Am. Chem. Soc.* **95**, 5765 (1973).

One of the best known of the π-organometallic compounds is ferrocene. It is a neutral compound which can be readily prepared from cyclopentadienide anion and iron(II).[99]

$$2\,[C_5H_5]^- + FeCl_2 \longrightarrow Fe(C_5H_5)_2$$

Many related compounds have been prepared. The total number of electrons contributed by the ligands (six for each cyclopentadiene ion) plus the number of valence electrons on the metal atom or ion usually equals 18, to satisfy the effective atomic number rule.[100]

	$C_5H_5Mn(CO)_3$	C_5H_5NiNO	$(C_5H_5)_2Ti(CO)_2$
Metal	6	9	2
Ligands	12	9	16
Total	18	18	18

Numerous chemical reactions have been carried out on ferrocene and its derivatives. The molecule behaves as an electron-rich aromatic system, and substitution reactions occur readily. Reagents that are relatively strong oxidizing agents, such as the halogens, effect oxidation at iron and destroy the compound.

The most useful π complexes of aromatic compounds from the synthetic point of view are the chromium complexes obtained by heating benzene or other aromatics with $Cr(CO)_6$.

$$C_6H_6 + Cr(CO)_6 \longrightarrow C_6H_6Cr(CO)_3 \qquad \text{Ref. 101}$$

$$C_6H_5Cl + Cr(CO)_6 \longrightarrow (C_6H_5Cl)Cr(CO)_3 \qquad \text{Ref. 102}$$

99. G. Wilkinson, *Org. Synth.* **IV**, 473, 476 (1963).
100. M. Tsutsui, M. N. Levy, A. Nakamura, M. Ichikawa, and K. Mori, *Introduction to Metal π-Complex Chemistry*, Plenum Press, New York, 1970, pp. 44–45; J. P. Collman, L. S. Hegedus, J. R. Norton, and R. G. Finke, *Principles and Applications of Organotransition Metal Chemistry*, University Science Books, Mill Valley, California, 1987, pp. 166–173.
101. W. Strohmeier, *Chem. Ber.* **94**, 2490 (1961).
102. J. F. Bunnett and H. Hermann, *J. Org.Chem.* **36**, 4081 (1971).

The $Cr(CO)_3$ unit in these compounds is strongly electron-withdrawing and activates the ring to nucleophilic attack. Reactions with certain carbanions result in arylation.[103]

$$(OC)_3Cr\text{-}C_6H_5Cl + (CH_3)_2CCN \longrightarrow (OC)_3Cr\text{-}C_6H_5C(CH_3)_2C{\equiv}N \; (78\%) \longrightarrow C_6H_5C(CH_3)_2C{\equiv}N$$

In compounds where the aromatic ring does not have a leaving group, addition occurs. The intermediate can be oxidized by I_2.

$$(OC)_3Cr\text{-}C_6H_6 + LiC(CH_3)_2CO_2C(CH_3)_3 \longrightarrow \text{adduct}\; Cr(CO)_3 \xrightarrow{I_2} C_6H_5C(CH_3)_2CO_2C(CH_3)_3 \; (91\%)$$

Ref. 104

Existing substituent groups such as CH_3, OCH_3, and $^{+}N(CH_3)_3$ exert a directive effect, often resulting in a major amount of the *meta* substitution product.[105] The intermediate adducts can be converted to cyclohexadiene derivatives if the adduct is protonolyzed.[106]

$$CH_3O\text{-}C_6H_5\text{-}Cr(CO)_3 + LiC(CH_3)_2C{\equiv}N \longrightarrow \text{adduct}\; Cr(CO)_3 \xrightarrow{CF_3CO_2H} CH_3O\text{-cyclohexadienyl-}C(CH_3)_2C{\equiv}N$$

Not all carbon nucleophiles will add to arenechromiumtricarbonyl complexes. For example, alkyllithium reagents[107] and simple ketone enolates do not give adducts.

Organometallic chemistry is a very active field of research, and new types of compounds, new reactions, and useful catalysts are being discovered at a rapid rate. These developments have had a major impact on organic synthesis, and future developments can be expected to continue to do so.

103. M. F. Semmelhack and H. T. Hall, *J. Am. Chem. Soc.* **96**, 7091 (1974).
104. M. F. Semmelhack, H. T. Hall, M. Yoshifuji, and G. Clark, *J. Am. Chem. Soc.* **97**, 1247 (1975); M. F. Semmelhack, H. T. Hall, Jr., R. Farina, M. Yoshifuji, G. Clark, T. Bargar, K. Hirotsu, and J. Clardy, *J. Am. Chem. Soc.* **101**, 3535 (1979).
105. M. F. Semmelhack, G. R. Clark, R. Farina, and M. Saeman, *J. Am. Chem. Soc.* **101**, 217 (1979).
106. M. F. Semmelhack, J. J. Harrison, and Y. Thebtaranonth, *J. Org. Chem.* **44**, 3275 (1979).
107. R. J. Card and W. S. Trahanovsky, *J. Org. Chem.* **45**, 2555, 2560 (1980).

General References

H. Alper (ed.), *Transition Metal Organometallics in Organic Synthesis*, Vols. I and II, Academic Press, New York, 1978.
J. P. Collman, L. S. Hegedus, J. R. Norton, and R. G. Finke, *Principles and Applications of Organotransition Metal Chemistry*, University Science Books, Mill Valley, California, 1987.
H. M. Colquhoun, J. Holton, D. J. Thomson, and M. V. Twigg, *New Pathways for Organic Synthesis*, Plenum, New York, 1984.
S. G. Davies, *Organo-Transition Metal Chemistry: Applications to Organic Synthesis*, Pergamon, Oxford, 1982.
J. K. Kochi, *Organometallic Mechanisms and Catalysis*, Academic Press, New York, 1979.
E. Negishi, *Organometallics in Organic Synthesis*, Wiley, New York, 1980.
M. Tsutsui, M. N. Levy, A. Nakamura, M. Ichikawa, and K. Mori, *Introduction to Metal π-Complex Chemistry*, Plenum, New York, 1970.

Organocopper Reactions

G. Posner, *Org. React.* **19**, 1 (1972).
G. Posner, *Org. React.* **22**, 253 (1975).
G. Posner, *An Introduction to Synthesis Using Organocopper Reagents*, Wiley-Interscience, New York, 1975.

Organopalladium Reactions

R. F. Heck, *Palladium Reagents in Organic Synthesis*, Academic Press, Orlando, Florida, 1985.
R. F. Heck, *Org. React.* **27**, 345 (1982).
J.Tsuji, *Organic Synthesis with Palladium Compounds*, Springer-Verlag, Berlin, 1980.

Problems

(*References for these problems will be found on page* 773.)

1. Predict the product of the following reactions. Be sure to specify all elements of stereochemistry.

(a) $H_2C{=}C(CH_3)MgBr + CH_3{-}CH{-}CH_2$ (epoxide, O bridging CH and CH_2) $\xrightarrow{Cu(I)}$

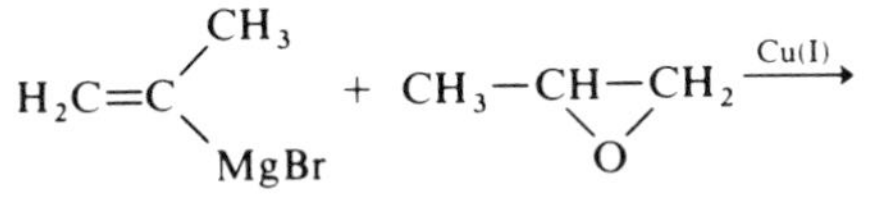

(b) $C_2H_5MgBr \xrightarrow[\text{2) } C_6H_{13}C\equiv CH \text{ 3) } I_2]{\text{1) } CuBr{-}S(CH_3)_2,\ -45°C}$

(c) (benzene ring with ortho CH_2Br and CH_2OH) $\xrightarrow[CO]{(Ph_3P)_2PdCl_2}$

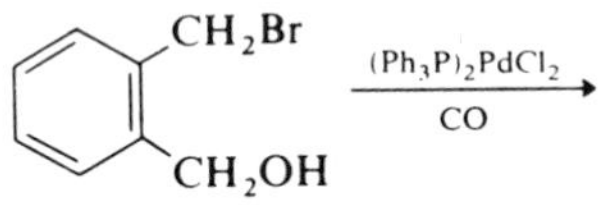

(d) $H_2C{=}CHCH_2MgBr$ + (β-lactone with $HC{\equiv}C$ substituent) $\xrightarrow{Cu(I)}$

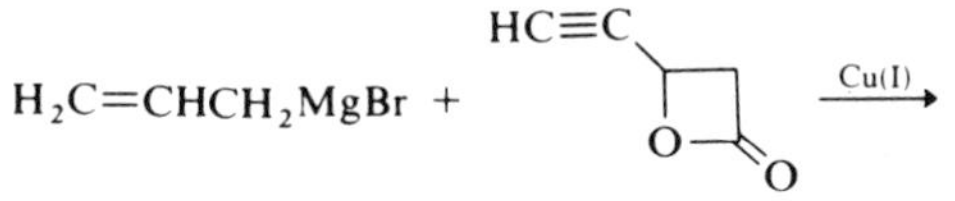

(e) CH_3CH_2MgBr + (H)(C_6H_{13})C=C(H)(I) $\xrightarrow{Pd(PPh_3)_4}$

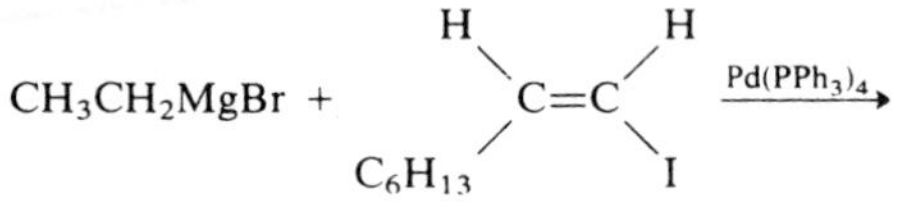

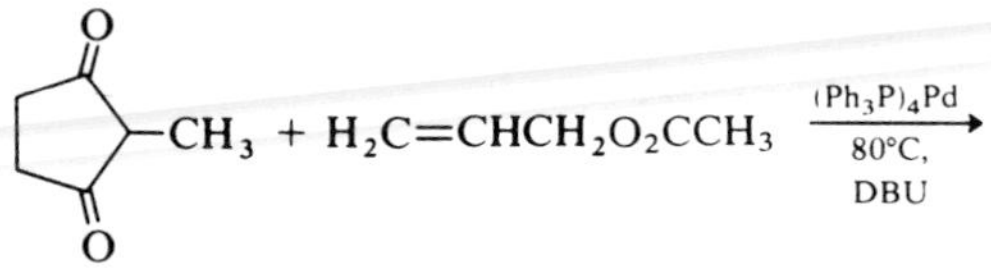

(f) (O, O, cyclopentane-dione)–CH_3 + $H_2C{=}CHCH_2O_2CCH_3$ $\xrightarrow[80°C,\ DBU]{(Ph_3P)_4Pd}$

(g) (cyclohexane, Cl, C≡CH) + $[CH_3(CH_2)_3]_2CuLi$ ⟶

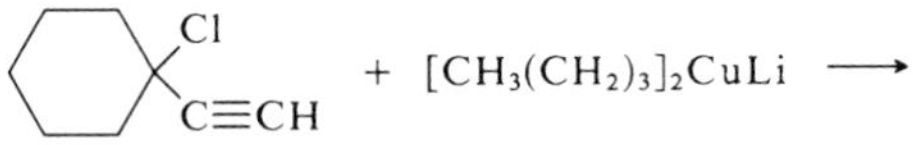

(h) CH_3, O, $C(CH_3)_3$ + $[(C_2H_5)_2CuCN]Li_2$ ⟶

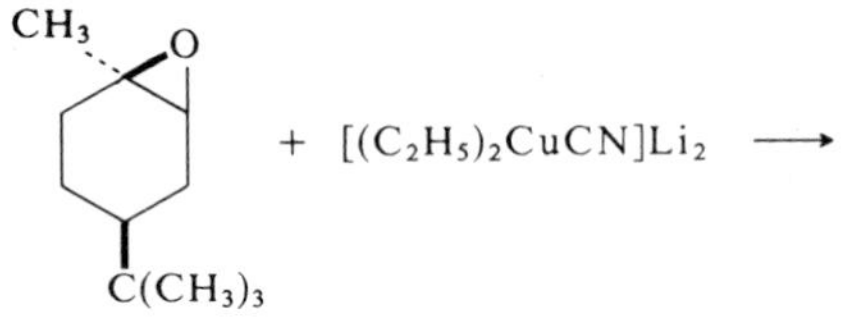

(i) $PhCH_2OCH_2$, CH_3, =O, O + $(CH_3)CuLi$ ⟶

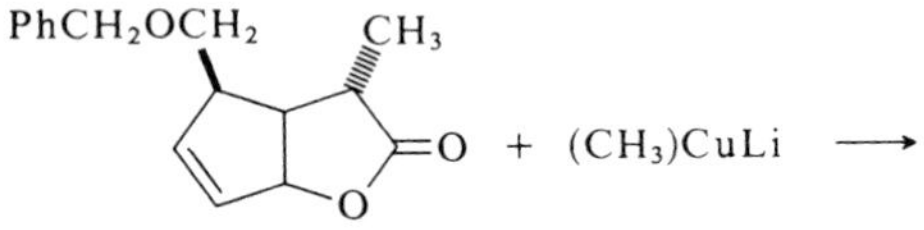

(j) (O, O)CH_2CH_2, $CH_2CH{=}CH_2$, =O, CH_3, CH_3 $\xrightarrow[O_2,\ DMF,\ H_2O]{PdCl_2,\ CuCl_2}$

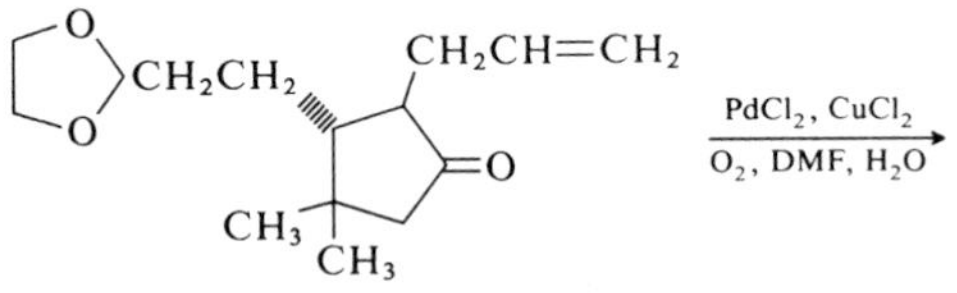

(k) C_4H_9Li $\xrightarrow[\text{2) } HC{\equiv}CH \quad \text{3) } I_2]{\text{1) } CuBr{\cdot}SMe_2}$

(l) $PhOCH_2CH{=}CHCH_2CH_2\overset{O}{\overset{\|}{C}}CH_2CO_2CH_3$ $\xrightarrow[PPh_3]{Pd(OAc)_2}$

(m) (naphthalene, Br, CH_3O) + $CH_2{=}CHMgBr$ $\xrightarrow{Ni(dmpe)Cl_2}$

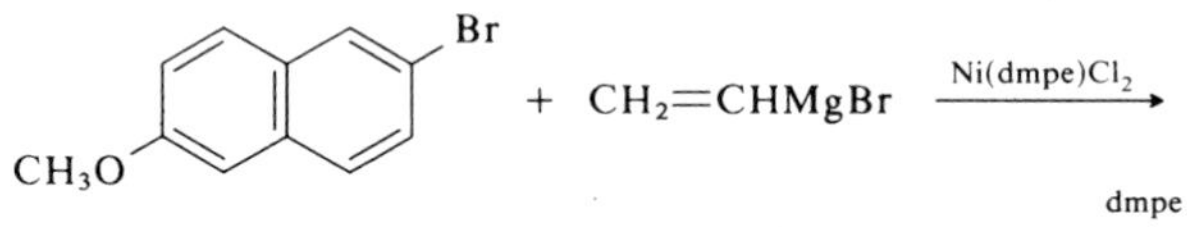

dmpe = 1,2-bis(dimethylphosphino)ethane

(n) (PhN, O, O, N, N) $\xrightarrow[Rh_2(CO)_4Cl_2,\ Ph_3P]{CO,\ H_2}$

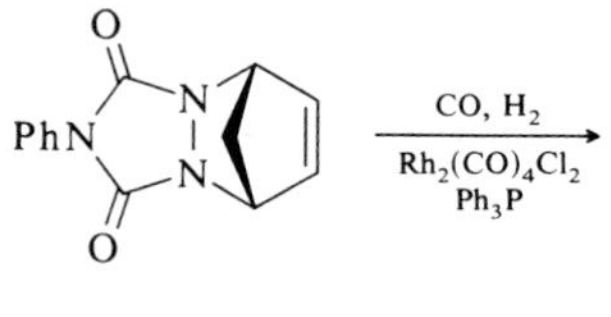

(o) 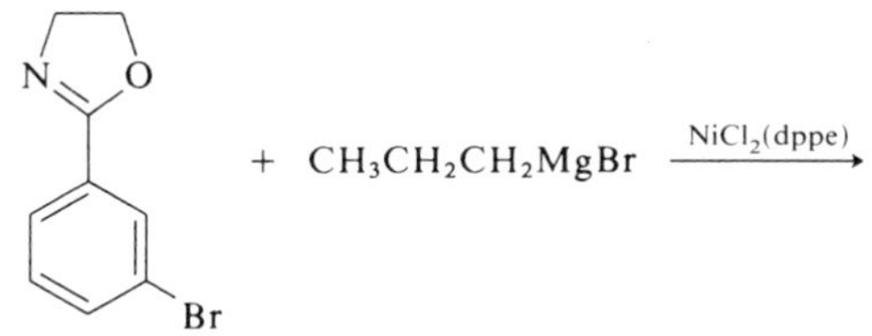 [2-(3-bromophenyl)-2-oxazoline] + $CH_3CH_2CH_2MgBr$ $\xrightarrow{NiCl_2(dppe)}$

dppe = 1,2-bis(diphenylphosphino)ethane

2. Give the products to be expected from each of the following reactions involving mixed or higher-order cuprate reagents.

(a) 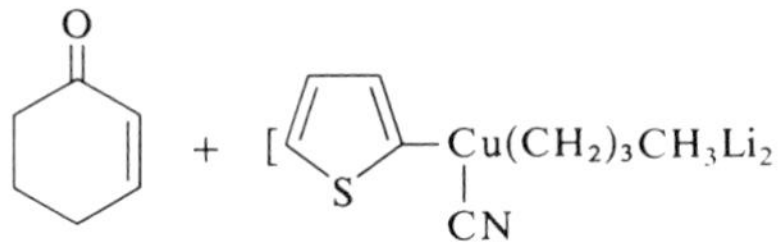 [2-cyclohexen-1-one] + [(2-thienyl)Cu(CN)$(CH_2)_3CH_3Li_2$]

(b) 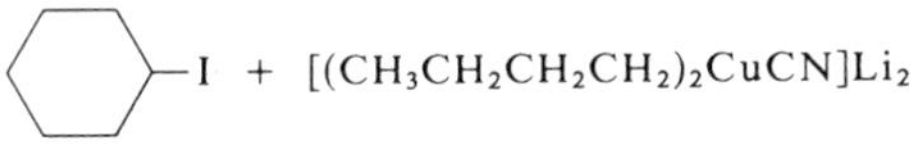 [cyclohexyl]–I + $[(CH_3CH_2CH_2CH_2)_2CuCN]Li_2$

(c) 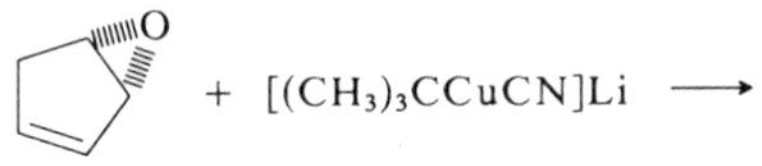 [cyclopentadiene monoepoxide] + $[(CH_3)_3CCuCN]Li$ $\longrightarrow$

(d) 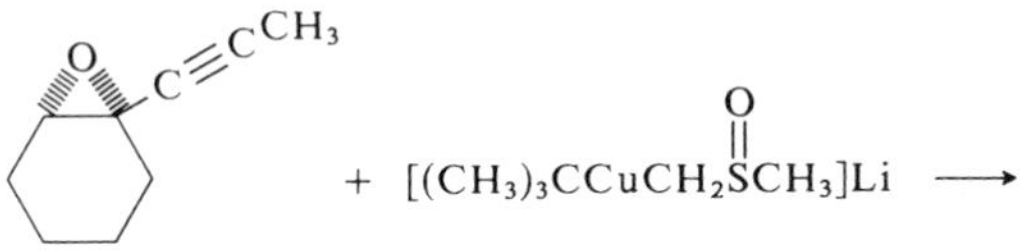 [1-(propynyl)cyclohexene oxide, $C\equiv CCH_3$] + $[(CH_3)_3CCuCH_2\overset{O}{\overset{\|}{S}}CH_3]Li$ $\longrightarrow$

3. Write a mechanism for each of the following transformations which accounts for the observed product and is in accord with other information which is available concerning the reaction.

(a) 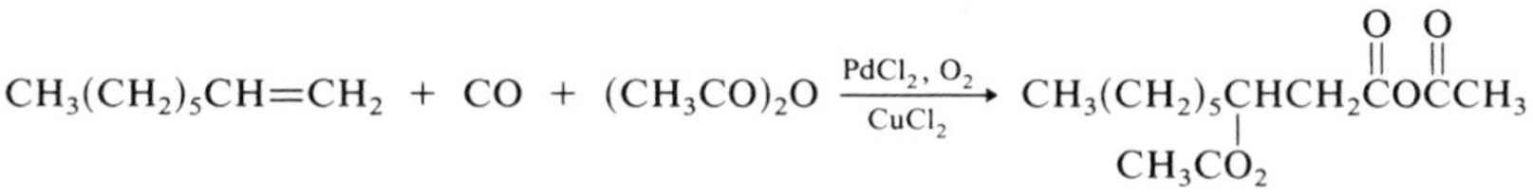 $CH_3(CH_2)_5CH{=}CH_2 + CO + (CH_3CO)_2O \xrightarrow[CuCl_2]{PdCl_2,\ O_2} CH_3(CH_2)_5\underset{CH_3CO_2}{\underset{|}{C}}HCH_2\overset{O}{\overset{\|}{C}}O\overset{O}{\overset{\|}{C}}CH_3$

(b) $CH_2{=}CHCH_2CH_2\overset{OSi(CH_3)_3}{\overset{|}{C}}{=}CH_2 \xrightarrow[10\ h,\ 25°C]{Pd(OAc)_2}$ [3-methylenecyclopentanone... H_2C, O] + Pd^0

(c) $PhCH_2O$–[CH_3, CH_3, OH, vinyl chain] $\xrightarrow[Rh_2(OAc)_2,\ Ph_3P,\ 100°C]{CO,\ H_2}$ $PhCH_2O$–[CH_3, H, CH_3, δ-lactone, O, O]

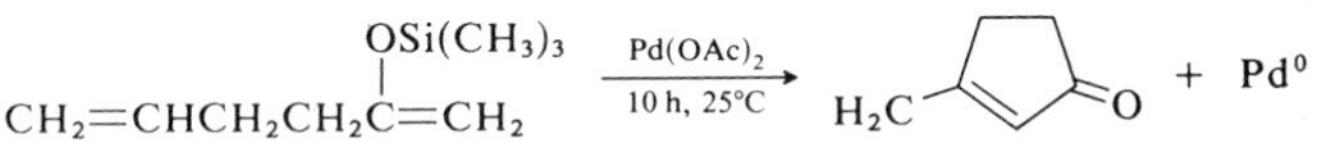

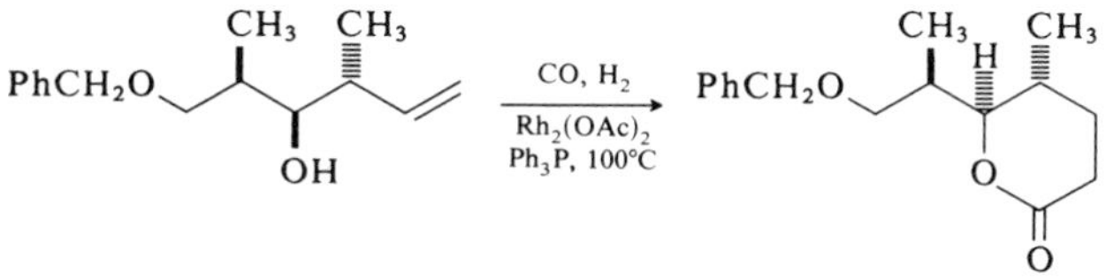

4. Indicate appropriate conditions and reagents for effecting the following transformations. "One-pot" processes are possible in all cases.

(a)

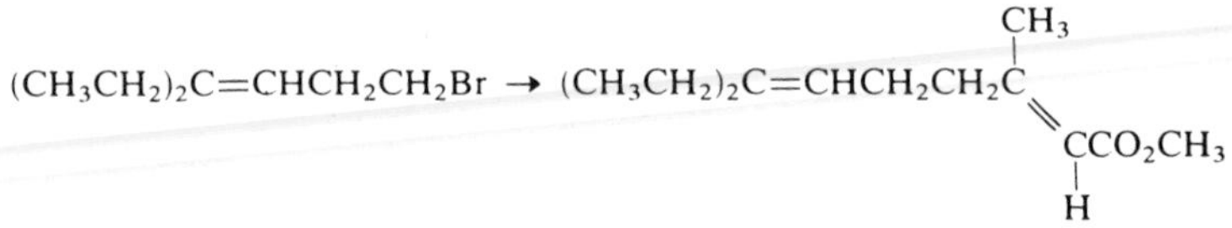

$(CH_3CH_2)_2C{=}CHCH_2CH_2Br \rightarrow (CH_3CH_2)_2C{=}CHCH_2CH_2C(CH_3){=}CHCO_2CH_3$

(b)

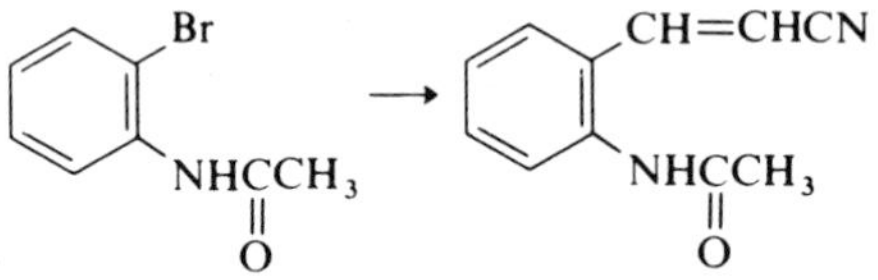

Br, $NHCCH_3$ (=O) → $CH{=}CHCN$, $NHCCH_3$ (=O)

(c)

$CH_3CH_2CH_2CH_2Br + CH_3O_2CC{\equiv}CCO_2CH_3 \rightarrow$ CH_3O_2C, CO_2CH_3, C=C, $CH_3(CH_2)_3$, H

(d)

O, H_3C, CH_3 → O, H_3C, H_3C, $CH_2CH_2CH{=}CH_2$

(e)

O_2CCH_3, H, CH_3, H, $CHCH_2CH_2C{=}C$, C=C, CH_2Br, H, $(CH_3)_2CHCH$, H, C, CH_2CH_2C, CH_2Br, CH_3 → $(CH_3)_2CH$, O_2CCH_3, CH_3, H_3C

(f)

H_3CO_2C, CO_2CH_3, C=C, H, H → CO_2CH_3, CO_2CH_3

(g)

$(CH_3)_2C{=}C(Br)CH_3 \rightarrow (CH_3)_2C{=}C(CH_3)CH{=}CHCO_2H$

(h)

O → O, CH_3

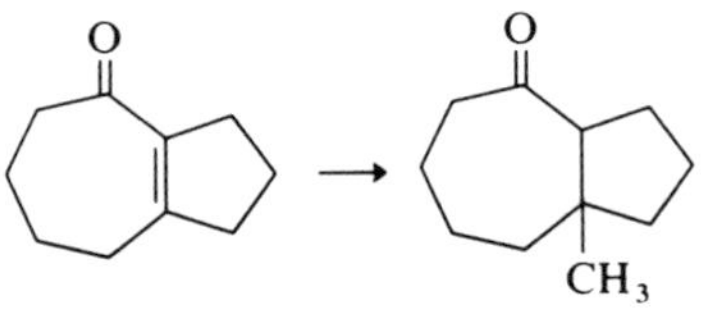

(i)

$OCH_2OCH_2CH_2Si(CH_3)_3$, N, Br → $OCH_2OCH_2CH_2Si(CH_3)_3$, N, N, $(CH_3)_3SiCH_2CH_2OCH_2O$

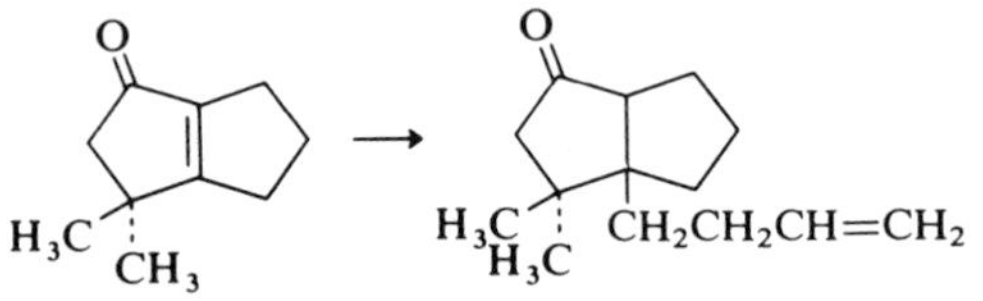

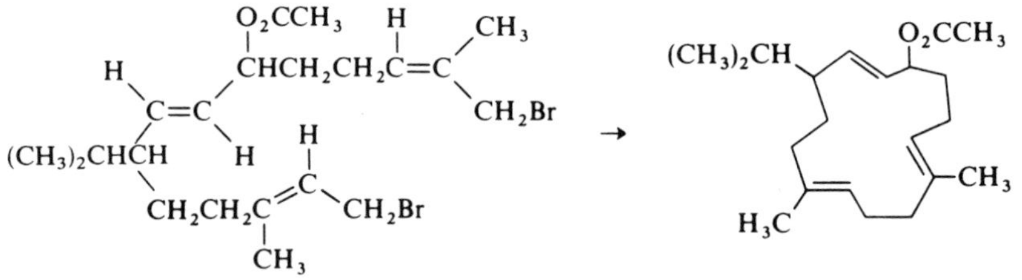

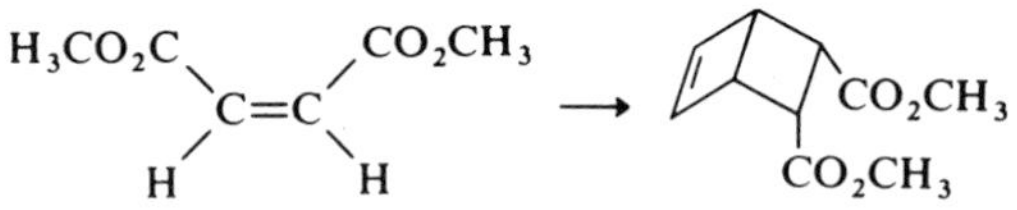

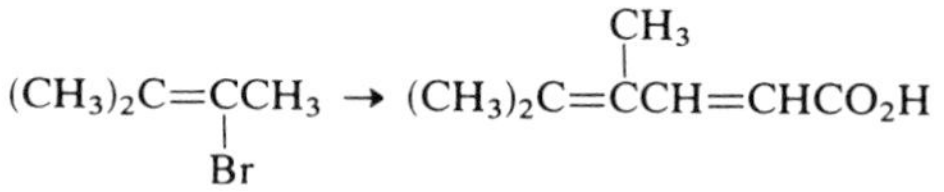

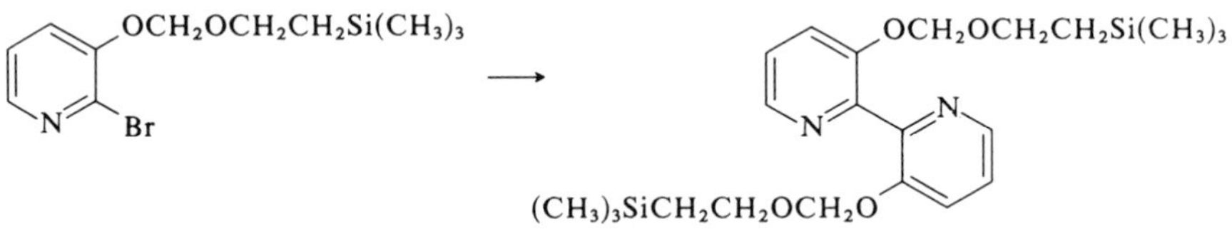

5. Vinyl triphenylphosphonium ion has been found to react with cuprate reagents by nucleophilic addition, generating an ylide structure. This intermediate can then be treated with an aldehyde to give an alkene by the Wittig reaction. Show how organocuprate intermediates could be used in conjunction with vinyl triphenylphosphonium ion to generate the following products from the specified starting material.

(a) 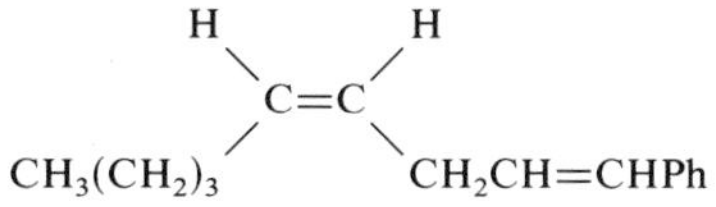 $CH_3(CH_2)_3(H)C{=}C(H)CH_2CH{=}CHPh$ from $CH_3(CH_2)_3(H)C{=}C(H)I$

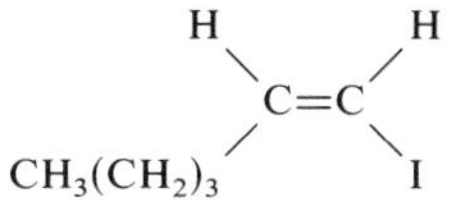

(b) $PhCH_2CH{=}CH(CH_2)_3CH_3$ from C_6H_5I

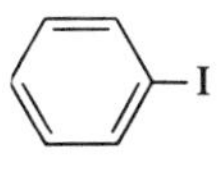

6. It has been observed that the reaction of $[(C_2H_5)_2Cu]Li$ or $[(C_2H_5)_2CuCN]Li_2$ with 2-iodooctane proceeds with racemization in both cases. On the other hand, the corresponding bromide reacts with nearly complete inversion of configuration with both reagents. When 6-halo-2-heptenes are used in similar reactions with dimethylcuprate, the iodide gives mainly the cyclic product 1-ethyl-2-methylcyclopentane whereas the bromide gives mainly 6-methyl-1-heptene. Provide a mechanism which accounts for the different behavior of the iodides as compared with the bromides.

7. Short synthetic sequences involving no more than three steps can be used to prepare the compound shown on the left from the potential starting material on the right. Suggest an appropriate series of reactions for each transformation.

(a) 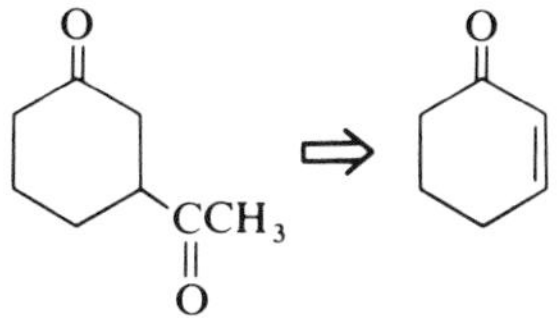 3-acetylcyclohexanone (CCH_3, C=O) ⇒ cyclohex-2-enone and $H_2C{=}CHOCH_3$

(b) 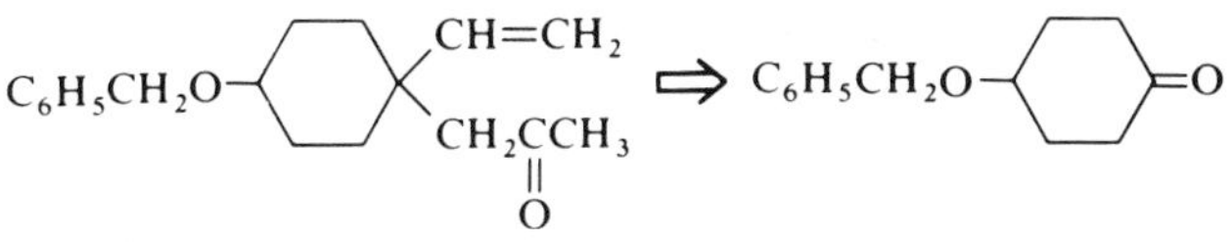 $C_6H_5CH_2O$–cyclohexane bearing $CH{=}CH_2$ and CH_2CCH_3 (C=O) ⇒ $C_6H_5CH_2O$–cyclohexanone (=O)

(c) bicyclic cyclopentenone fused to cyclooctene (O=) ⇒ cyclooctenone with $CH_2{=}CHCH_2$ substituent

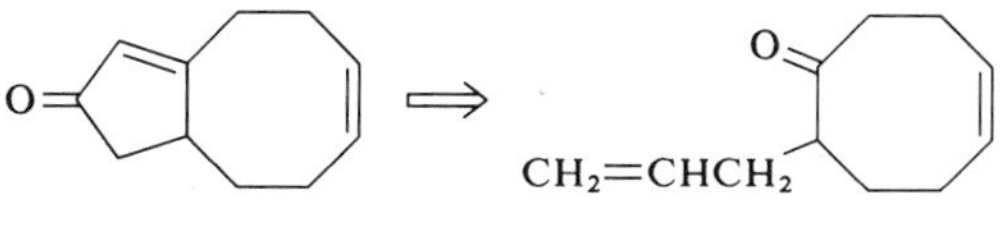

8. The conversions shown were carried out in a multistep, but "one-pot," synthetic process in which none of the intermediates needs to be isolated. Show how you could perform the transformation by suggesting a sequence of organic and inorganic reagents to be employed and the approximate reaction conditions.

(a)

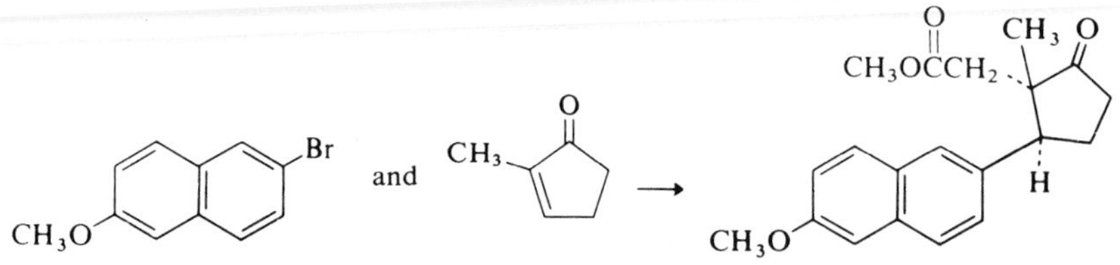

(b) and $HC{\equiv}C(CH_2)_5CH_3$ →

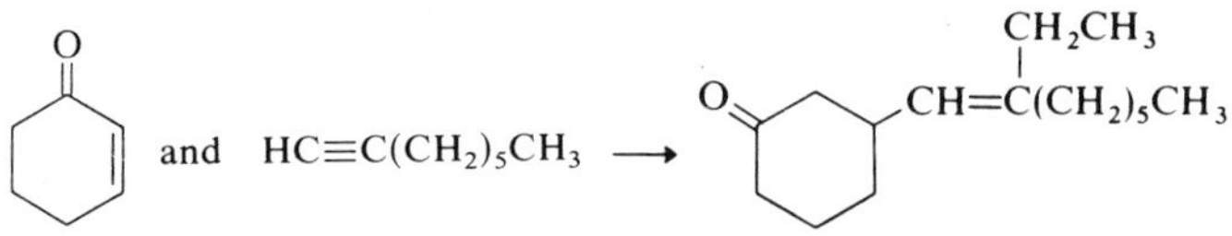

(c) and $BrCH_2C{\equiv}CSi(CH_3)_3$ →

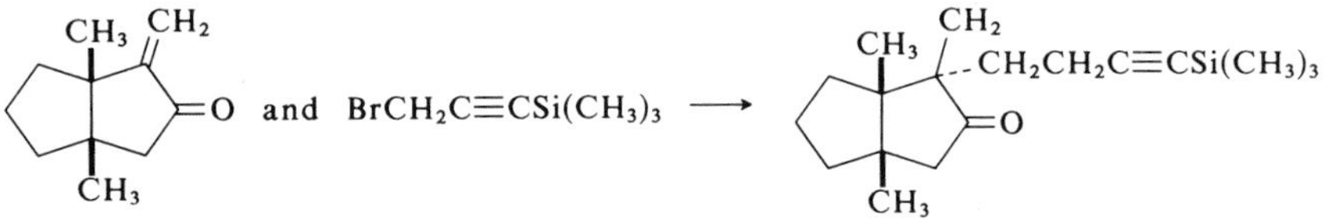

(d) and $HC{\equiv}CH$, C_2H_5I →

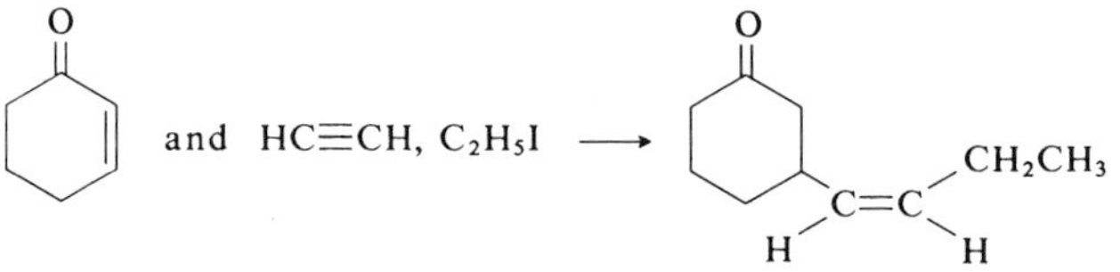

(e) and →

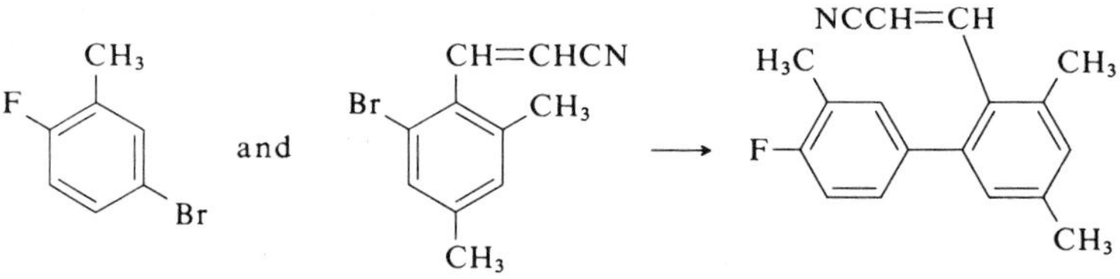

9. A number of syntheses of medium and large ring compounds which involve transition metal reagents have been described. Suggest an organometallic reagent or metal complex which could bring about the following conversions:

(a)

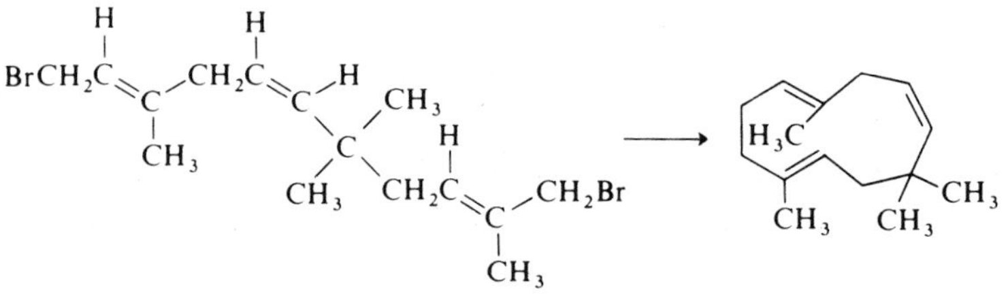

(b)

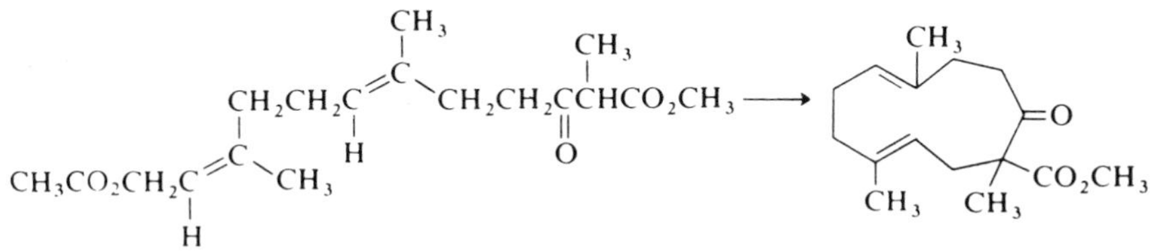

$CH_3CO_2CH_2C(H)=C(CH_3)CH_2CH_2C(H)=C(CH_3)CH_2CH_2C(CH_3)(C=O)CHCO_2CH_3 \longrightarrow$ (ring product with CH_3, CH_3, CH_3, $=O$, CO_2CH_3)

(c)

$(PhSO_2)_2CH(CH_2)_{10}CO_2(CH_2)_{10}$–(epoxide, O)–$CH{=}CH_2 \longrightarrow$ (macrocycle: O=OC, $(CH_2)_{10}$, $(CH_2)_{10}$, $(PhSO_2)_2C$, H–C=C–H, CH–OH)

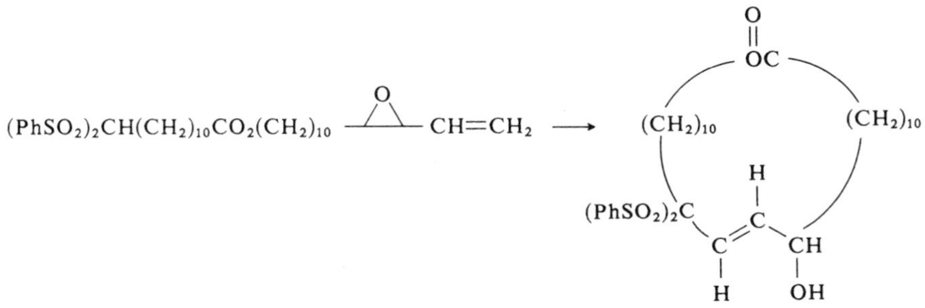

(d)

MeO, I–C$_6$H$_3$–$CH_2CH_2N(CH_3)CH_2CH_2$–C$_6$H$_3$(OMe)(I) $\longrightarrow$ MeO–(dibenzazecine with N–CH_3)–OMe

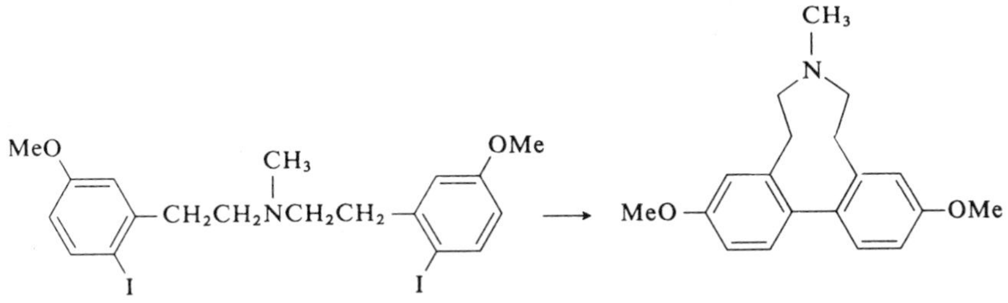

10. The cyclobutadiene complex **1** can be prepared in optically active form. When the complex is reacted with an oxidizing agent and a compound capable of trapping cyclobutadienes, the products are racemic. When the reaction is carried only to partial completion, the recovered complex remains optically active. Discuss the relevance of these results to the following question. "In oxidative decomposition of cyclobutadiene complexes, is the cyclobutadiene liberated from the complex before or after it has reacted with the trapping reagent?"

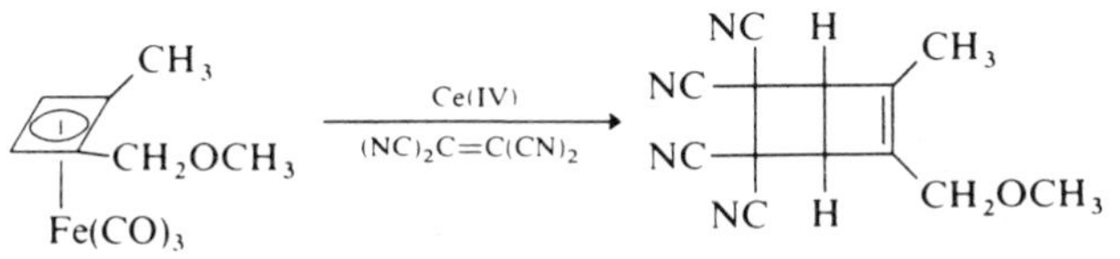

CH_3, CH_2OCH_3, $Fe(CO)_3$ **1** $\xrightarrow[(NC)_2C=C(CN)_2]{Ce(IV)}$ NC, H, NC, NC, NC, H, CH_3, CH_2OCH_3

11. When the isomeric acetates **A** and **B** react with dialkylcuprates, both give a very similar product mixture containing mainly **C** with small amounts of **D**. Only trace amounts of the corresponding *Z*-isomers are found. Suggest a mechanism to account for the formation of essentially the same product mixture from both reactants.

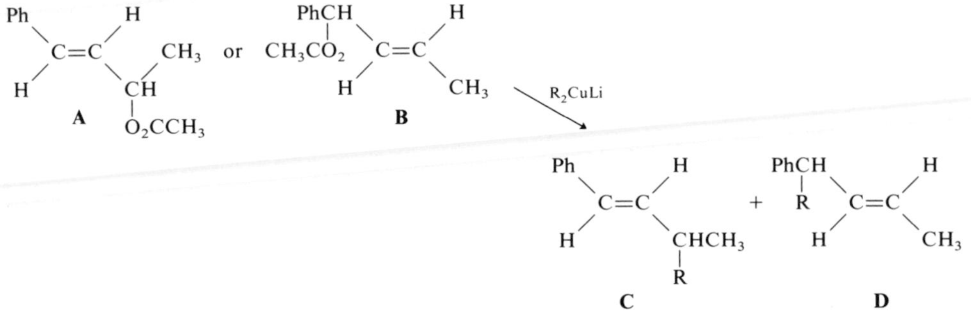

A: Ph(H)C=C(H)–CH(O$_2$CCH$_3$)CH$_3$ or B: PhCH(CH$_3$CO$_2$)(H)C=C(H)CH$_3$ $\xrightarrow{R_2CuLi}$ C: Ph(H)C=C(H)–CHCH$_3$(R) + D: PhCH(R)(H)C=C(H)CH$_3$

12. The compound shown below is a constituent of the pheromone of the codling moth. It has been synthesized using *n*-propy bromide, propyne, 1-pentyne, ethylene oxide, and CO_2 as the source of the carbon atoms. Devise a route for such a synthesis. Hint: Extensive use of the chemistry of organocopper reagents is the basis for the existing synthesis.

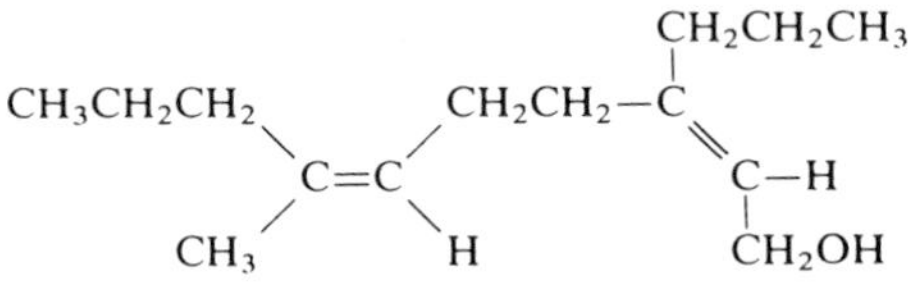

CH$_3$CH$_2$CH$_2$(CH$_3$)C=C(H)–CH$_2$CH$_2$–C(CH$_2$CH$_2$CH$_3$)=C(H)–CH$_2$OH

13. *S*-3-Hydroxy-2-methylpropanoic acid, **1**, can be obtained enantiomerically pure from isobutyric acid by a microbiological oxidation. The aldehyde **2** is available from a natural product, pulegone, also in enantiomerically pure form.

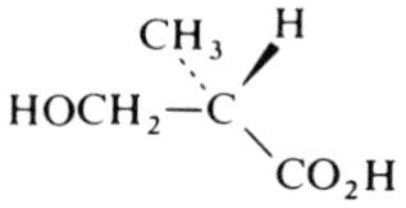

1: HOCH$_2$–C(CH$_3$)(H)–CO$_2$H

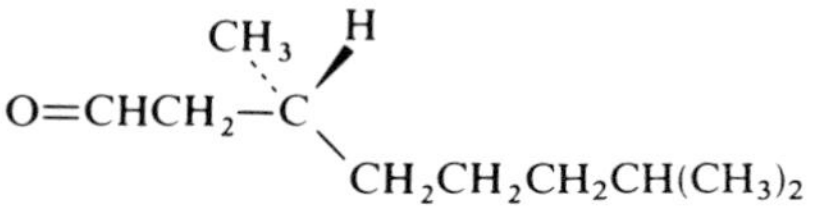

2: O=CHCH$_2$–C(CH$_3$)(H)–CH$_2$CH$_2$CH$_2$CH(CH$_3$)$_2$

Devise a synthesis of enantiomerically pure **3**, a compound of interest as a starting material for the synthesis of α-tocopherol (vitamin E).

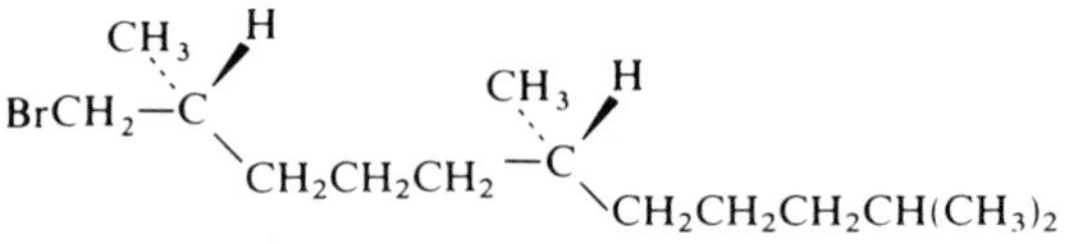

3: BrCH$_2$–C(CH$_3$)(H)–CH$_2$CH$_2$CH$_2$–C(CH$_3$)(H)–CH$_2$CH$_2$CH$_2$CH(CH$_3$)$_2$

14. Each of the following transformations can be carried out in good yield under optimized conditions. Consider the special factors in each case, and discuss the most appropriate reagent and reaction conditions to obtain good yields.

(a) cyclopent-2-enone → 3-substituted cyclopentanone (substituent: CH$_3$OCH$_2$O–CHCH(CH$_3$)$_2$)

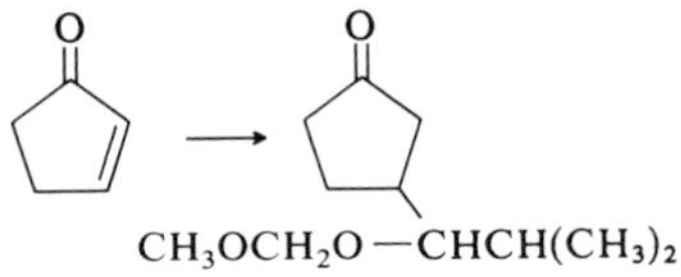

(b) O, H_3C CH_3 $\longrightarrow$ O, H_3C CH_3 $CH_2CH{=}CH_2$

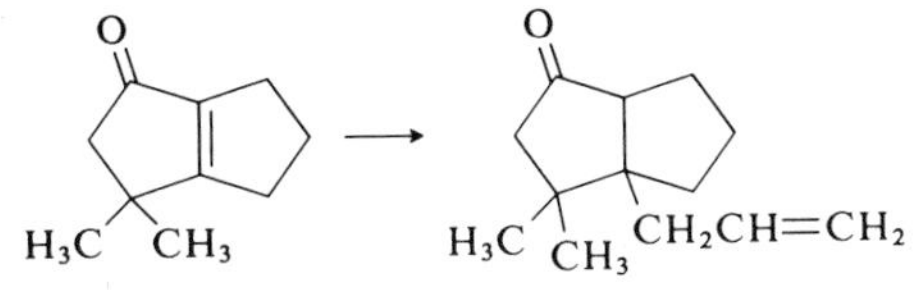

(c) $(CH_3)_2C{=}CHCO_2C_2H_5 \rightarrow CH_3(CH_2)_3C(CH_3)_2CH_2CO_2C_2H_5$

(d) O, CH_3, H, CH_2OCH_3, O, O $\longrightarrow$ O, CH_3, H, $CH{=}CH_2$, CH_2OCH_3, O, O

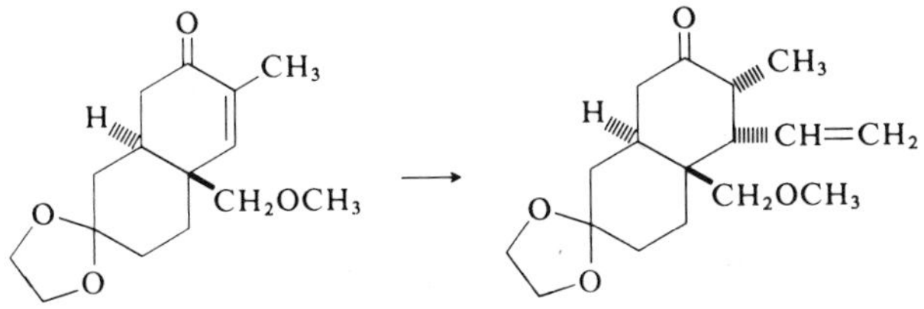

15. Each of the following synthetic transformations can be accomplished by use of organometallic reagents and/or catalysts. Indicate a sequence of reactions which would permit each of the syntheses to be completed.

(a) N, Br $+ CH_2{=}C(SiMe_3)Br \longrightarrow$ N, $C{=}CH_2$, $SiMe_3$

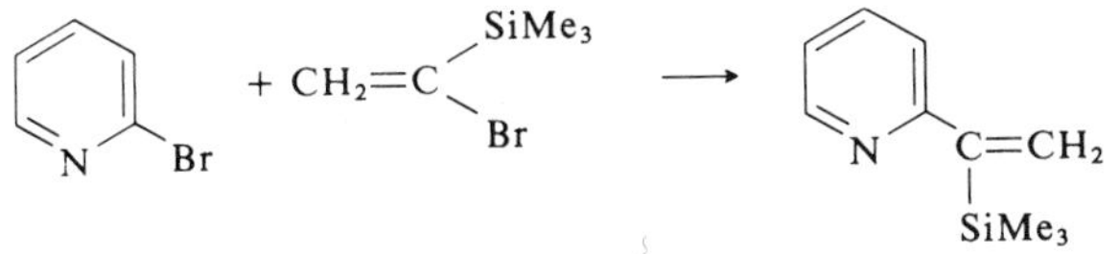

(b) O, O, Br + MeO, MeO, MeO, I, $CH{=}N$ $\searrow$ N, HC, O, O, MeO, OMe, OMe

(c) CH_3, CH_3, O, H, =O, O, MeO $\longrightarrow$ CH_3, CH_3, O, H, CO_2Me, O, MeO

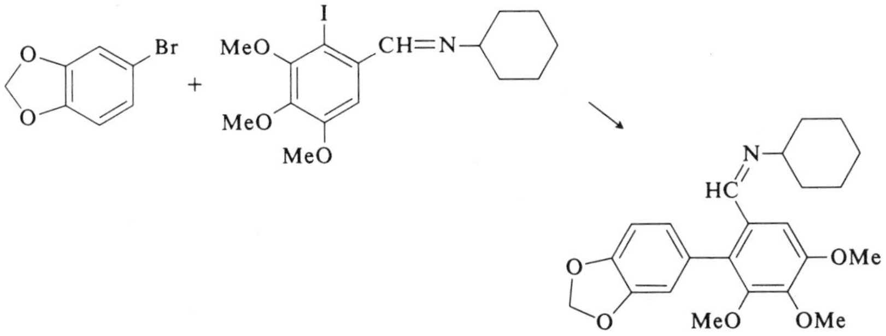

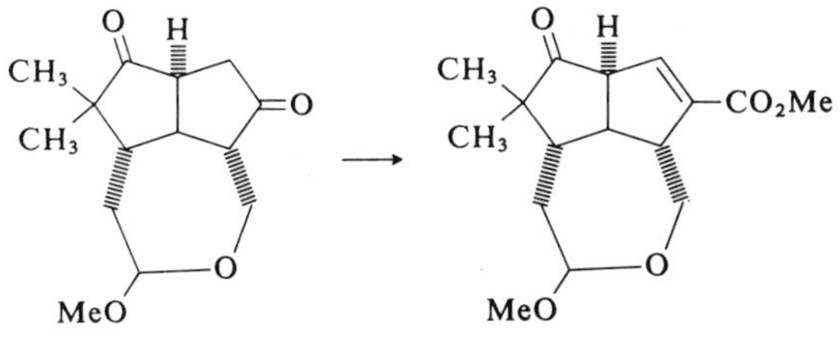

(d) [structures: MeO_2C … Br (cyclopentylidene acetal) + $HC{\equiv}C{-}CH{=}CH{-}CH_2OTBS$ → MeO_2C … $CH_2OTBDMS$]

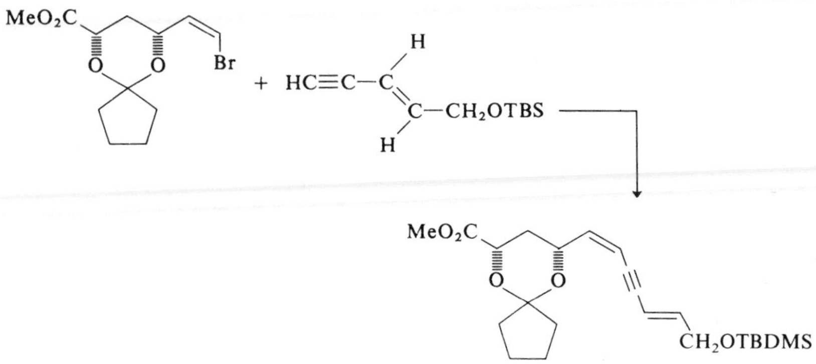

(e) [structures: MeO, CH_3-aryl-$CHCH_2CH_2CH_2CN$ with $CH(CH_3)_2$ → CH_3, O, NC, $(CH_3)_2CH$ spiro compound]

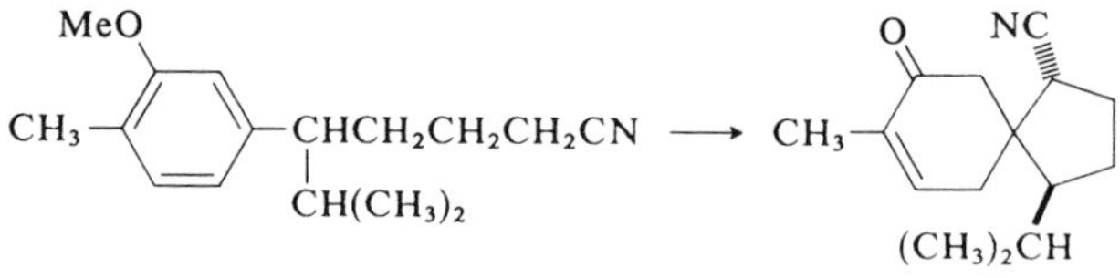

(f) [structures: MeO, MeO quinolinone (N–H) + $C_6H_5NHCO_2C(CH_3)_3$ → MeO, MeO quinoline with $NCO_2C(CH_3)_3$ phenyl]

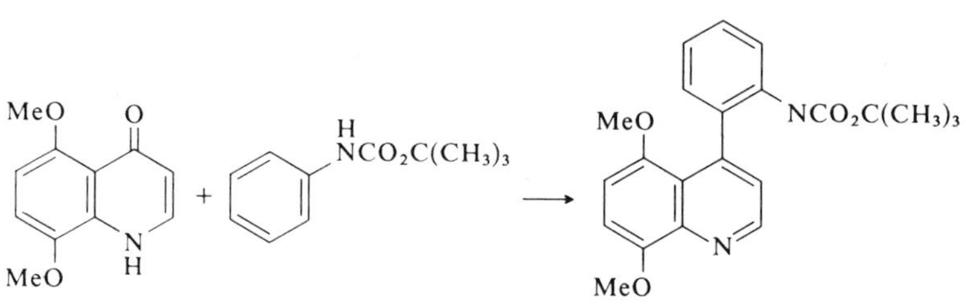

16. Each of the following reactions is accomplished with a palladium reagent or catalyst. Write a detailed mechanism for each reaction. The number of equivalents of each reagent which is used is given in parentheses. Be sure your mechanism accounts for the regeneration of a catalytically active species in those reactions which are catalytic in palladium.

(a) 1,3-cyclohexadiene $\xrightarrow[\text{benzoquinone (0.25), } MnO_2 \text{ (1.2)}]{Pd(OAc)_2 \text{ (0.05); LiOAc (1.0)}}$ CH_3CO_2–cyclohexene–O_2CCH_3

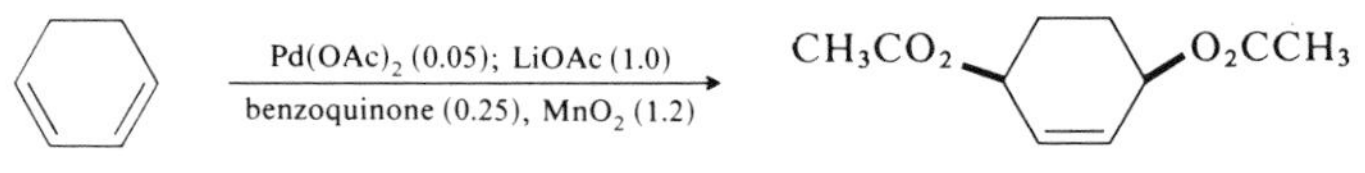

(b) $(CH_3)_2CHCH_2CH(OH)(CH_2)_3CH{=}CHCH_3$ $\xrightarrow[\text{CO (excess); } CH_3OH \text{ (excess)}]{PdCl_2 \text{ (0.1); } CuCl_2 \text{ (3.0)}}$ $(CH_3)_2CHCH_2$–tetrahydropyran–$CH(CH_3)CO_2CH_3$ (O, H)

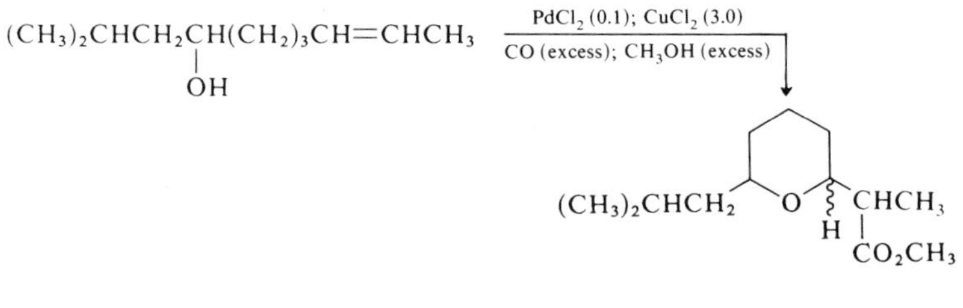

(c) 2-bromo-5-methoxyacetanilide (CH_3O, Br, NH, $CH_3C{=}O$) $\xrightarrow[\text{CO (excess); } Bu_3N \text{ (1.2), } H_2O \text{ (excess)}]{Pd(PPh_3)_2Cl_2 \text{ (0.005); } PPh_3 \text{ (0.02)}}$ CH_3O–aryl–CO_2H, $\overset{O}{\overset{\|}{N}}CCH_3$ H

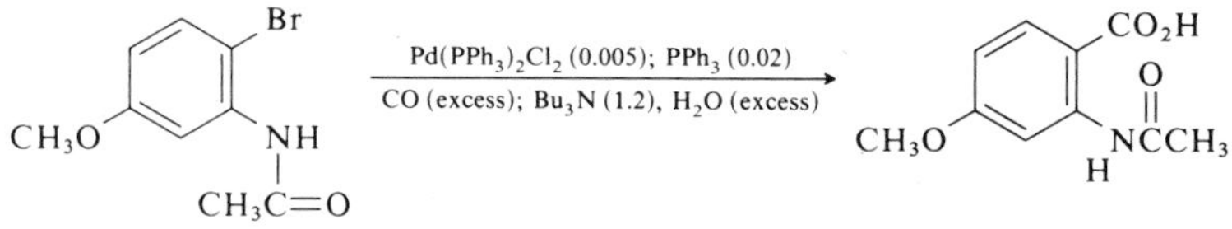

(d) [chemical structure: $CH_2{=}CHCH_2CH_2$-substituted cyclopentene with $(CH_3)_2$, $CH_2CO_2C_2H_5$, $OSi(CH_3)_3$] $\xrightarrow{Pd(OAc)\ (1.0)}$ [bicyclic products with CH_3, CH_3, $=CH_2$, O, $CH_2CO_2CH_3$] + [bicyclic product with CH_3, CH_3, CH_3, O, $CH_2CO_2CH_3$]

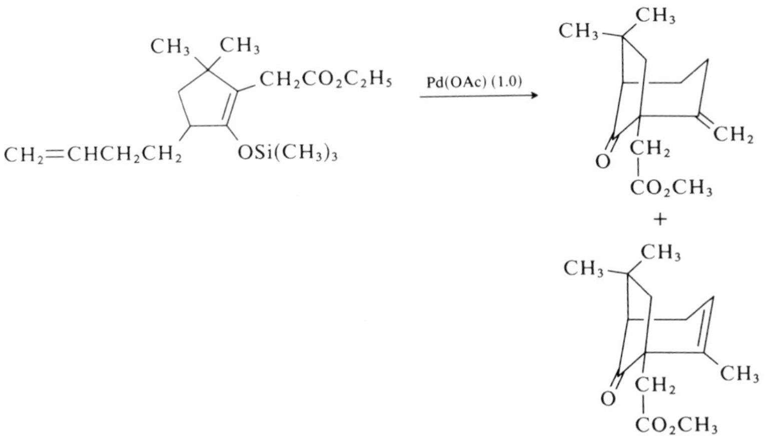

17. The following transformations have been carried out *enantiospecifically* by synthetic sequences involving organometallic reagents. Devise a scheme by which each desired material could be prepared in high enantiomeric purity from the specified starting material.

(a) $CH_2{=}CHCH_2\text{–}C(OH)\text{–}CH_2CH_2CH{=}CH_2 \Rightarrow CH_3O_2C\text{–}C(OH)\text{–}CH_2CO_2CH_3$

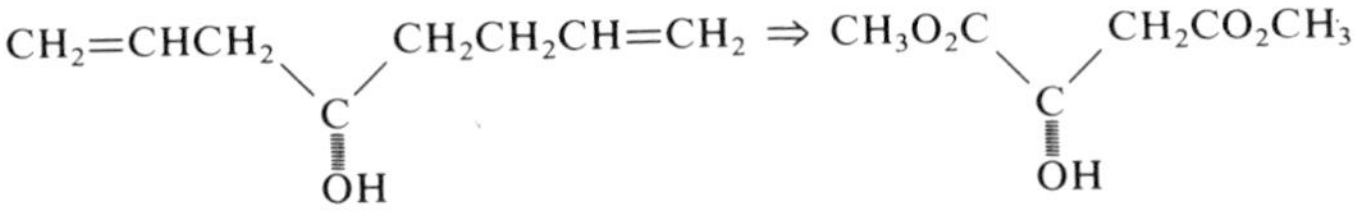

(b) $(CH_3)_2C{=}CHCH_2\text{–}C(OH)\text{–}CO_2C_2H_5 \Rightarrow HOCH_2\text{–}C(NH_2)\text{–}CO_2H$

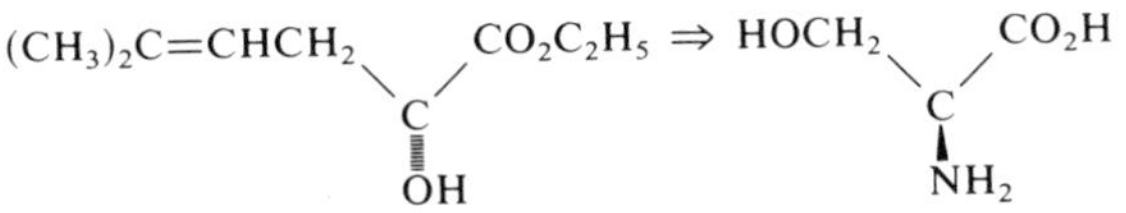

9

Carbon–Carbon Bond-Forming Reactions of Compounds of Boron, Silicon, and Tin

In this chapter, we will discuss examples of the use of the compounds of boron, silicon, and tin in organic synthesis to form carbon–carbon bonds. These elements are at the boundary of the metals and nonmetals, with boron being the most electronegative and tin the least electronegative of the three. The neutral alkyl derivatives of boron have the formula R_3B, whereas those of silicon and tin have the formulas R_4Si and R_4Sn. These compounds are relatively volatile, nonpolar substances which exist as discrete molecules and in which the carbon–metal bonds are largely covalent. The synthetically important reactions of these compounds involve transfer of a carbon substituent with one (radical equivalent) or two (carbanion equivalent) electrons to a reactive carbon center. This chapter will emphasize the nonradical reactions. In contrast to the reactions of the transition metals, during which there is often a change in oxidation level at the metal, there is usually no oxidation level change at the heteroatom during the reactions of boron, silicon, and tin compounds.

9.1. Organoboron Compounds

9.1.1. Synthesis of Organoboranes

The most widely used route to organoboranes is hydroboration, which was discussed in Section 4.9.1. Hydroboration provides access to both alkyl- and alkenylboranes. Aryl-, methyl-, and benzylboranes cannot be prepared by hydroboration. One route to methyl and aryl derivatives is by reaction of a dialkylborane, such as 9-BBN, with a cuprate reagent.[1]

$$\text{(9-BBN)BH} + R_2CuLi \longrightarrow \text{(9-BBN)B—R} + [RCuH]^- Li^+$$

These reactions are formulated as occurring by oxidative addition at copper, followed by decomposition of the Cu(III) intermediate.

$$R'_2B{-}H + {}^-Cu^IR_2 \rightarrow R'_2B{-}Cu^{III}(H)(R)R \rightarrow R'_2B{-}R + RCu^IH$$

Two successive reactions with organocuprates can convert thexylborane to an unsymmetrical trialkylborane.[2]

$$\text{ThexylBH}_2 \xrightarrow{R^1_2CuLi} \xrightarrow{R^2_2CuLi} \text{ThexylB}(R^1)(R^2)$$

Alkyl, aryl, and allyl derivatives of boron can be prepared directly from the corresponding halides, BF_3, and magnesium metal. This process presumably involves *in situ* generation of a Grignard reagent, which then displaces fluoride from boron.[3]

$$3\,R{-}X + BF_3 + 3\,Mg \rightarrow R_3B + 3\,MgXF$$

Alkoxy groups can be displaced from boron by both alkyl and aryllithium reagents. The reaction of diisopropoxyboranes with organolithium reagents, for example, provides good yields of unsymmetrically disubstituted isopropoxyboranes.[4]

$$RB(O{-}i\text{-}Pr)_2 + R'Li \rightarrow RR'B{-}O{-}i\text{-}Pr$$

Alkoxyboron compounds can also be named as esters. Compounds with one alkoxy group are esters of borinic acids and are called borinates. Compounds with two alkoxy groups are called boronate esters.

R_2BOH	R_2BOR'	$RB(OH)_2$	$RB(OR')_2$
borinic acid	borinate ester	boronic acid	boronate ester

1. C. G. Whiteley, *J. Chem. Soc., Chem. Commun.*, 5 (1981).
2. C. G. Whiteley, *Tetrahedron Lett.* **25**, 5563 (1984).
3. H. C. Brown and U. S. Racherla, *J. Org. Chem.* **51**, 427 (1986).
4. H. C. Brown, T. E. Cole, and M. Srebnik, *Organometallics* **4**, 1788 (1985).

Organometallic displacement reactions on haloboranes can also provide access to boranes that cannot be obtained by hydroboration.[5]

$$[\text{Ipc}]_2\text{BCl} \xrightarrow{CH_2{=}CHCH_2MgBr} [\text{Ipc}]_2\text{BCH}_2\text{CH}{=}\text{CH}_2$$

9.1.2. Carbon–Carbon Bond-Forming Reactions

The reactions of organoboranes that were discussed in Chapter 4 are valuable methods for introducing functional groups into alkenes. In this section, we will discuss carbon–carbon bond-forming reactions of organoboranes.[6] Trivalent organoboranes are not very nucleophilic, but they are reactive Lewis acids. Most reactions in which carbon–carbon bonds are formed involve a tetracoordinate intermediate with a negative charge on boron. Adduct formation weakens the boron–carbon bonds and permits a transfer of a carbon substituent with its electrons. A general mechanistic pattern is

$$R_3B + :Nu \rightarrow R_3\bar{B}{-}\overset{+}{Nu}$$

$$R_3\bar{B}{-}\overset{+}{Nu} + E^+ \rightarrow R_2B{-}\overset{+}{Nu} + R{-}E$$

The electrophilic center is sometimes generated from the Lewis base by formation of the adduct.

$$R_3B + :Nu{=}X \rightarrow R{-}\bar{B}R_2{-}\overset{+}{Nu}{=}X \quad \longrightarrow \quad R_2B{-}{}^{+}Nu(R){-}X^-$$

An important group of reactions of this type are the reactions of organoboranes with carbon monoxide. Carbon monoxide forms Lewis acid–base complexes with organoboranes. In these adducts, the boron bears a formal negative charge and carbon is electrophilic because of the triple bond to oxygen bearing a formal positive charge. The adducts undergo boron-to-carbon migration of the boron substituents. The reaction can be controlled so that it results in the migration of one, two, or all three of the substituents on boron.[7]

$$R_3\bar{B}{-}C{\equiv}O^+$$

$$\xrightarrow{NaBH_4} R_2B{-}CH(OH)R \xrightarrow{H_2O,\ ^-OH} RCH_2OH$$

$$\xrightarrow{100°C,\ H_2O} RB(OH){-}C(OH)R_2 \xrightarrow{H_2O_2} RC(=O)R$$

$$\xrightarrow{100\text{–}125°C} \text{“}O{=}B{-}CR_3\text{”} \xrightarrow{H_2O_2,\ ^-OH} R_3COH$$

5. H. C. Brown and P. K. Jadhav, *J. Am. Chem. Soc.* **105**, 2092 (1983).
6. For a review of this topic, see E. Negishi and M. Idacavage, *Org. React.* **33**, 1 (1985).
7. H. C. Brown and M. W. Rathke, *J. Am. Chem. Soc.* **89**, 2737 (1967).

If the organoborane is heated with carbon monoxide to 100–125°C, all of the groups migrate and a tertiary alcohol is obtained after workup by oxidation. The presence of water in the reaction mixture causes the reaction to cease after migration of two groups from boron to carbon. Oxidation of the reaction mixture at this stage gives a ketone.[8] Primary alcohols are obtained when the carbonylation is carried out in the presence of sodium borohydride or lithium borohydride.[9] The product of the first migration step is reduced, and subsequent hydrolysis gives a primary alcohol.

In this synthesis of primary alcohols, only one of the three groups in the organoborane is converted to product. This disadvantage can be overcome by using a dialkylborane, particularly 9-BBN, in the initial hydroboration. After carbonylation and boron-to-carbon migration, the reaction mixture can be processed to give an aldehydes, alcohol, or the homologated 9-alkyl-9-BBN.[10] The utility of 9-BBN in these procedures is the result of the minimal tendency of the bicyclic ring to undergo migration.

H
B
RCH=CH$_2$
B—CH$_2$CH$_2$R
CO | KBH(O-*i*-Pr)$_3$
$^-$OH
HOCH$_2$CH$_2$CH$_2$R
OH
B—CHCH$_2$CH$_2$R
LiAlH$_4$
– 20°C
B—CH$_2$CH$_2$CH$_2$R
H$_2$O$_2$
O=CHCH$_2$CH$_2$R

Unsymmetrical ketones can be made by using either thexylborane or thexychloroborane.[11] Thexylborane works well when one of the desired carbonyl substituents is derived from a moderately hindered alkene. Under these circumstances, a clean monoalkylation of thexylborane can be accomplished. This is followed by reaction with a second alkene and carbonylation.

$$(CH_3)_2CH\overset{CH_3}{\underset{CH_3}{C}}-BH_2 \xrightarrow{RCH=CHR} \xrightarrow{R'CH=CH_2} \xrightarrow[2)\ H_2O,\ 100°C \quad 3)\ H_2O_2]{1)\ CO} RCH_2\underset{R}{C}H\overset{O}{\overset{\|}{C}}CH_2CH_2R'$$

8. H. C. Brown and M. W. Rathke, *J. Am. Chem. Soc.* **89**, 2738 (1967).
9. M. W. Rathke and H. C. Brown, *J. Am. Chem. Soc.* **89**, 2740 (1967).
10. H. C. Brown, E. F. Knights, and R. A. Coleman, *J. Am. Chem. Soc.* **91**, 2144 (1969); H. C. Brown, J. L. Hubbard, and K. Smith, *Synthesis* 701 (1979); H. C. Brown, T. M. Ford, and J. L. Hubbard, *J. Org. Chem.* **45**, 4067 (1980).
11. H. C. Brown and E. Negishi, *J. Am. Chem. Soc.* **89**, 5285 (1967); S. U. Kulkarni, H. D. Lee, and H. C. Brown, *J. Org. Chem.* **45**, 4542 (1980).

Thexychloroborane can be alkylated and then converted to a dialkylborane by a reducing agent such as $KBH[OCH(CH_3)_2]_3$.

$$(CH_3)_2CHC(CH_3)_2-BHCl \xrightarrow[2)\ KBH[OCH(CH_3)_2]_3]{1)\ CH_3CH_2CH=CH_2} (CH_3)_2CHC(CH_3)_2-BH-CH_2CH_2CH_2CH_3$$

$$\downarrow \text{4-vinylcyclohexene}$$

$$(CH_3)_2CHC(CH_3)_2-B(CH_2CH_2CH_2CH_3)(CH_2CH_2\text{-cyclohexenyl})$$

$$\xrightarrow[2)\ (CF_3CO)_2O,\ -78^\circ C;\ 3)\ NaOH,\ H_2O_2]{1)\ NaCN} \text{cyclohexenyl}-CH_2CH_2\overset{O}{\overset{\|}{C}}CH_2CH_2CH_2CH_3 \quad (67\%)$$

The success of both of these approaches depends upon the thexyl group being noncompetitive with the other groups in the migration steps. A related procedure begins with dibromoborane (see entry 7 in Scheme 9.1). Several other examples of these carbonylation reactions are given in Scheme 9.1.

A number of other procedures have been developed in which other reagents besides carbon monoxide serve as the electrophilic migration terminus. Perhaps the most generally applicable of these methods involves the use of cyanide ion and trifluoroacetic anhydride as illustrated by entries 3 and 5 in Scheme 9.2. In this reaction, the borane initially forms an adduct with cyanide ion. The migration is induced by acylation of the cyano group by trifluoroacetic anhydride.

$$R_3B + {}^-CN \rightarrow R_3\bar{B}-C\equiv N \xrightarrow{(CF_3CO)_2O} R_3\bar{B}-C\equiv\overset{+}{N}-\overset{O}{\overset{\|}{C}}CF_3$$

$$R_3\bar{B}-C\equiv\overset{+}{N}-\overset{O}{\overset{\|}{C}}CF_3 \rightarrow R_2B-\underset{R}{C}=N-\overset{O}{\overset{\|}{C}}CF_3 \rightarrow R-B\langle O-C(CF_3)=N-CR_2 \rangle \xrightarrow{H_2O_2} R\overset{O}{\overset{\|}{C}}R$$

The remaining entries in Scheme 9.2 illustrate other methods which proceed by a generally similar mechanism involving adduct formation and a boron-to-carbon rearrangement. Problem 3 deals with the mechanisms of these reactions.

An efficient process for one-carbon homologation to aldehydes is based on cyclic boronate esters.[12] These can be prepared by hydroboration of an alkene with dibromoborane, followed by displacement of the labile bromines. The homologation

12. H. C. Brown and T. Imai, *J. Am. Chem. Soc.* **105**, 6285 (1983).

Scheme 9.1. Synthesis via Carbonylation of Organoboranes

1[a] $\left(CH_3CH_2\underset{}{\overset{CH_3}{|}}{CH}-\right)_3B \xrightarrow[2)\ H_2O_2,\ ^-OH]{1)\ CO,\ 125°C} \left(CH_3CH_2\overset{CH_3}{|}{CH}-\right)_3COH$ (87%)

2[b] $(C_6H_{11})_3B \xrightarrow[2)\ H_2O_2,\ ^-OH]{1)\ CO,\ 125°C,\ H_2O} C_6H_{11}-\overset{O}{\overset{\|}{C}}-C_6H_{11}$ (90%)

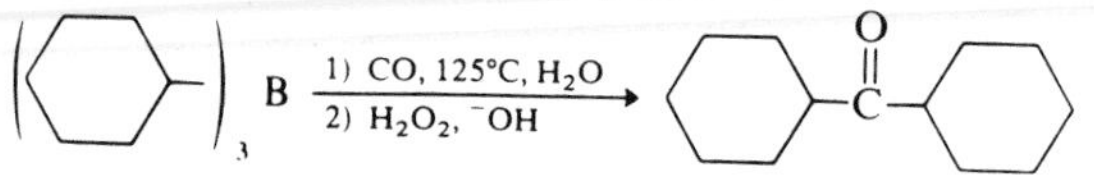

3[c] 1-methylcyclopentene $\xrightarrow{9\text{-BBN}}$ $\xrightarrow[2)\ CO;\ 3)\ H_2O_2,\ ^-OH]{1)\ KBH(O\text{-}i\text{-}Pr)_3}$ trans-2-methylcyclopentanecarboxaldehyde (CH_3, $CH{=}O$) (96%)

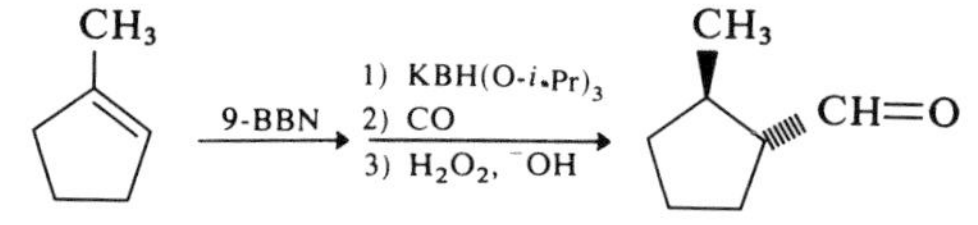

4[d] $(CH_3)_2C{=}CH_2 \xrightarrow{\text{thexylborane}} \xrightarrow{CH_2{=}CHCO_2C_2H_5} \xrightarrow[2)\ H_2O_2,\ ^-OAc]{1)\ CO,\ 50°C} (CH_3)_2CHCH_2\overset{O}{\overset{\|}{C}}CH_2CH_2CO_2C_2H_5$ (81%)

5[e] $CH_3CH{=}CHCH_3 \xrightarrow{B_2H_6} \xrightarrow[2)\ CO]{1)\ LiAlH(OCH_3)_3} \xrightarrow[2)\ H_2O_2,\ ^-OH]{1)\ H_2O,\ H^+} CH_3CH_2\underset{CH_3}{CH}\overset{OH}{CH}\underset{CH_3}{CH}CH_2CH_3$ (82%)

6[f] $CH_3(CH_2)_5CH{=}CH_2 \xrightarrow[2)\ KBH(OR)_3]{1)\ \text{thexylchloroborane}} \xrightarrow{CH_2{=}CH(CH_2)_7CH_3} \xrightarrow[2)\ (CF_3CO)_2O]{1)\ NaCN} CH_3(CH_2)_7\overset{O}{\overset{\|}{C}}(CH_2)_9CH_3$ (74%)

7[g] 1-vinylcyclohexene $\xrightarrow[2)\ CO,\ 50°C;\ 3)\ H_2O_2]{1)\ \text{thexylborane}}$ cis-hydrindanone (H, H, O) (54%)

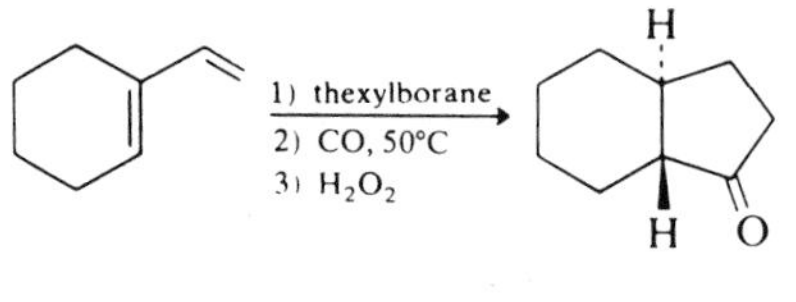

8[h] $CH_2{=}CH$, CH_3, CH_3O substrate $\xrightarrow[2)\ CO,\ 50°C;\ 3)\ H_2O_2]{1)\ \text{siamylborane}}$ product (CH_3, O, H, CH_3O) (53%)

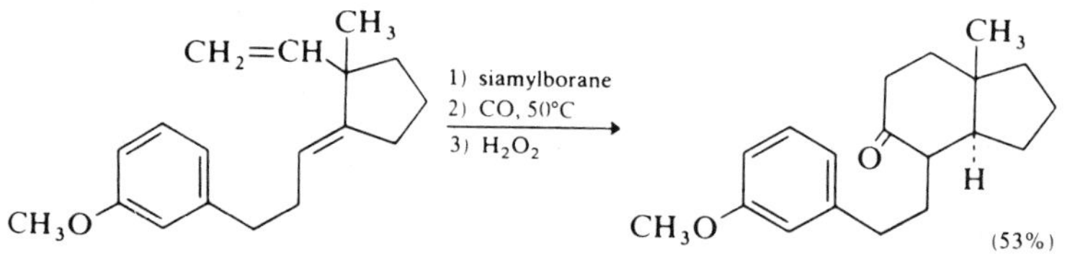

a. H. C. Brown and M. W. Rathke, *J. Am. Chem. Soc.* **89**, 2737 (1967).
b. H. C. Brown and M. W. Rathke, *J. Am. Chem. Soc.* **89**, 2738 (1967).
c. H. C. Brown, J. L. Hubbard, and K. Smith, *Synthesis*, 701 (1979).
d. H. C. Brown and E. Negishi, *J. Am. Chem. Soc.* **89**, 5285 (1967).
e. J. L. Hubbard and H. C. Brown, *Synthesis*, 676 (1978).
f. S. U. Kulkarni, H. D. Lee, and H. C. Brown, *J. Org. Chem.* **45**, 4542 (1980).
g. H. C. Brown and E. Negishi, *J. Am. Chem. Soc.* **89**, 5477 (1967).
h. T. A. Bryson and W. E. Pye, *J. Org. Chem.* **42**, 3214 (1977).

Scheme 9.2. Some One-Carbon Donors in Alcohol and Ketone Synthesis Using Organoboranes

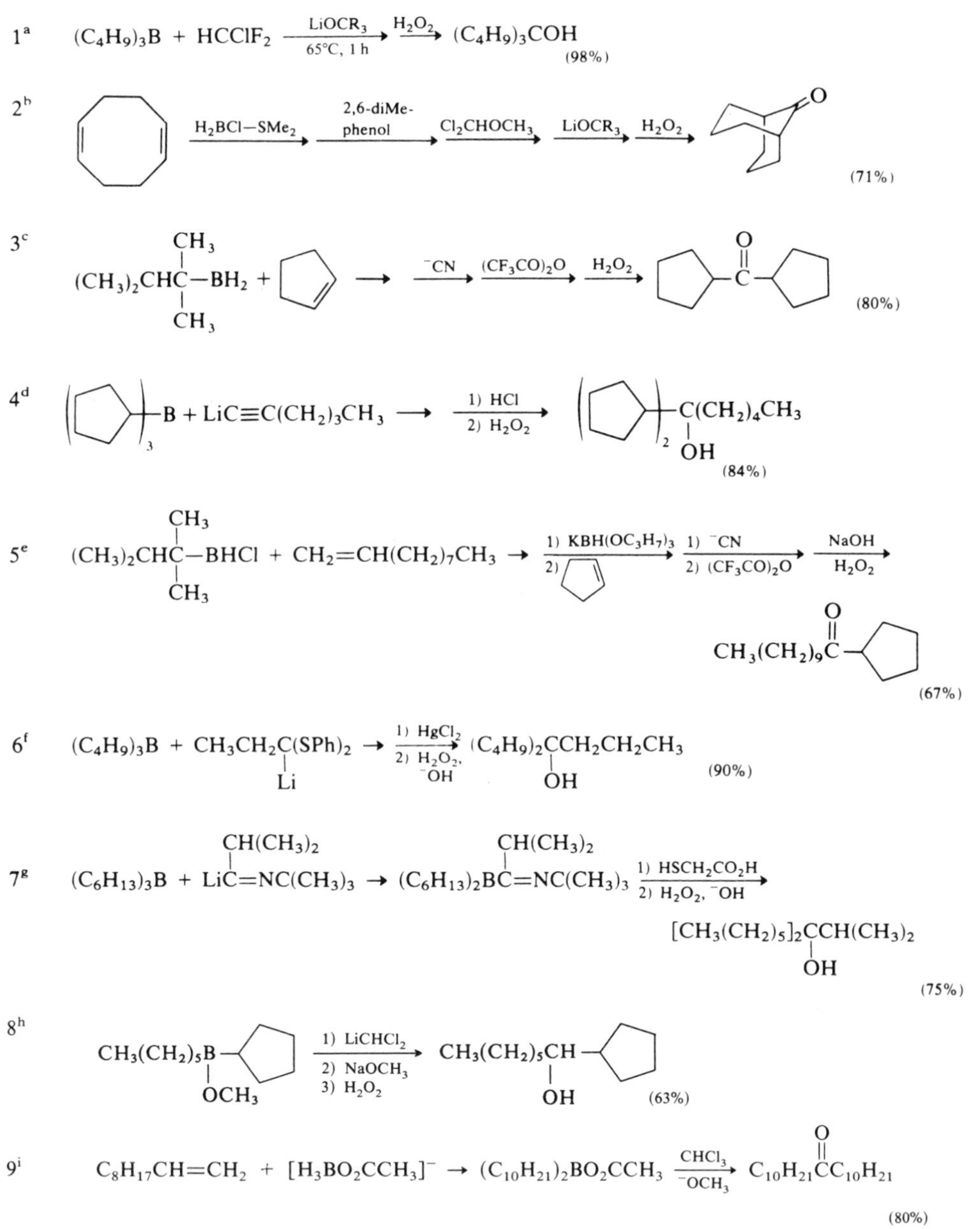

1[a] $(C_4H_9)_3B + HCClF_2 \xrightarrow[65°C, 1 h]{LiOCR_3} \xrightarrow{H_2O_2} (C_4H_9)_3COH$ (98%)

2[b] cyclooctadiene $\xrightarrow{H_2BCl-SMe_2} \xrightarrow{2,6\text{-diMe-phenol}} \xrightarrow{Cl_2CHOCH_3} \xrightarrow{LiOCR_3} \xrightarrow{H_2O_2}$ bicyclic ketone (71%)

3[c] $(CH_3)_2CHC(CH_3)_2-BH_2$ + cyclopentene $\rightarrow \xrightarrow{^-CN} \xrightarrow{(CF_3CO)_2O} \xrightarrow{H_2O_2}$ $(C_5H_9)_2C=O$ (80%)

4[d] $(C_5H_9)_3B + LiC\equiv C(CH_2)_3CH_3 \rightarrow \xrightarrow[2)\ H_2O_2]{1)\ HCl} (C_5H_9)_2C(OH)(CH_2)_4CH_3$ (84%)

5[e] $(CH_3)_2CHC(CH_3)_2-BHCl + CH_2{=}CH(CH_2)_7CH_3 \rightarrow \xrightarrow[2)\ \text{cyclopentene}]{1)\ KBH(OC_3H_7)_3} \xrightarrow[2)\ (CF_3CO)_2O]{1)\ ^-CN} \xrightarrow[H_2O_2]{NaOH}$ $CH_3(CH_2)_9C(=O)-C_5H_9$ (67%)

6[f] $(C_4H_9)_3B + CH_3CH_2C(SPh)_2Li \rightarrow \xrightarrow[2)\ H_2O_2,\ ^-OH]{1)\ HgCl_2} (C_4H_9)_2C(OH)CH_2CH_2CH_3$ (90%)

7[g] $(C_6H_{13})_3B + LiC(CH(CH_3)_2){=}NC(CH_3)_3 \rightarrow (C_6H_{13})_2BC(CH(CH_3)_2){=}NC(CH_3)_3 \xrightarrow[2)\ H_2O_2,\ ^-OH]{1)\ HSCH_2CO_2H}$ $[CH_3(CH_2)_5]_2C(OH)CH(CH_3)_2$ (75%)

8[h] $CH_3(CH_2)_5B(OCH_3)-C_5H_9 \xrightarrow[2)\ NaOCH_3;\ 3)\ H_2O_2]{1)\ LiCHCl_2} CH_3(CH_2)_5CH(OH)-C_5H_9$ (63%)

9[i] $C_8H_{17}CH{=}CH_2 + [H_3BO_2CCH_3]^- \rightarrow (C_{10}H_{21})_2BO_2CCH_3 \xrightarrow[^-OCH_3]{CHCl_3} C_{10}H_{21}C(=O)C_{10}H_{21}$ (80%)

a. H. C. Brown, B. A. Carlson, and R. H. Prager, *J. Am. Chem. Soc.* **93**, 2070 (1971).
b. H.C. Brown and S. U. Kulkarni, *J. Org. Chem.* **44**, 2422 (1979).
c. A. Pelter, K. Smith, M. G. Hutchings, and K. Rowe, *J. Chem. Soc. Perkin Trans. 1*, 129 (1975).
d. M. M. Midland and H. C. Brown, *J. Org. Chem.* **40**, 2845 (1975).
e. S. U. Kulkarni, H. D. Lee, and H. C. Brown, *J. Org. Chem.* **45**, 4542 (1980).
f. R. J. Hughes, S. Ncube, A. Pelter, K. Smith, E. Negishi, and T. Yoshida, *J. Chem. Soc., Perkin Trans. 1,* 1172 (1977); S. Ncube, A. Pelter, and K. Smith, *Tetrahedron Lett.* 1895 (1979).
g. Y. Yamamoto, K. Kondo and I. Moritani, *J. Org. Chem.* **40**, 3644 (1975).
h. H. C. Brown, T. Imai, P. T. Perumal, and B. Singaram, *J. Org. Chem.* **50**, 4032 (1985).
i. C. Narayana and M. Periasamy, *Tetrahedron Lett.* **26**, 6361 (1985).

step is carried out by addition of methoxy(phenylthio)methyllithium to the boronate ester. The migration step is induced by mercuric ion.

$$RCH{=}CH_2 + HBBr_2 \longrightarrow RCH_2CH_2BBr_2 \xrightarrow{Me_3SiO(CH_2)_3OSiMe_3} RCH_2CH_2B(O(CH_2)_3O)$$

$$RCH_2CH_2B(O(CH_2)_3O) + LiCH(OCH_3)SPh \longrightarrow RCH_2CH_2\overset{-}{B}(O(CH_2)_3O)CH(SPh)OCH_3 \xrightarrow{Hg^{2+}} RCH_2CH_2C(H)(OCH_3){-}B(O(CH_2)_3O) \xrightarrow{H_2O_2,\ pH\ 8} RCH_2CH_2CH{=}O$$

Organoboranes can also be used to construct carbon-carbon bonds by several other types of reactions involving migration of a boron substituent to carbon. One such reaction involves α-halocarbonyl compounds.[13] For example, ethyl bromoacetate reacts with trialkylboranes in the presence of base to give alkylated acetic acid derivatives in excellent yield. The reaction is most efficiently carried out with a 9-BBN derivative.

$$\text{(9-BBN)}B{-}R + BrCH_2CO_2R' \xrightarrow{^-OC(CH_3)_3} RCH_2CO_2R'$$

These reactions can also be effected with *B*-alkenyl derivatives of 9-BBN to give β,γ-unsaturated esters.[14]

The mechanism of these alkylations involves a tetracoordinate boron intermediate formed by addition of the enolate of the α-bromoester to the organoborane. The migration then occurs with displacement of bromide ion. In agreement with this mechanism, retention of configuration of the migrating group is observed.[15]

$$R_3B + \overset{-}{C}H(Br)CO_2C_2H_5 \rightarrow R{-}\overset{-}{B}(R)_2{-}CH(Br)CO_2C_2H_5 \rightarrow R{-}B(R){-}CH(R)CO_2C_2H_5 \xrightarrow{RO^-} RCH_2CO_2C_2H_5$$

α-Haloketones and α-halonitriles undergo similar reactions.[16]

A closely related reaction employs α-diazoesters or α-diazoketones.[17] In these compounds, molecular nitrogen acts as the leaving group in the migration step. The best results are achieved using dialkylchloroboranes or monoalkyldichloroboranes.

$$RBCl_2 + N_2CHCO_2CH_3 \rightarrow RCH_2CO_2CH_3$$

A number of these alkylation reactions are illustrated in Scheme 9.3.

13. H. C. Brown, M. M. Rogić, M. W. Rathke, and G. W. Kabalka, *J. Am. Chem. Soc.* **90**, 818 (1968); H. C. Brown and M. M. Rogić, *J. Am. Chem. Soc.* **91**, 2146 (1969).
14. H. C. Brown, N. G. Bhat, and J. B. Cambell, Jr., *J. Org. Chem.* **51**, 3398 (1986).
15. H. C. Brown, M. M. Rogić, M. W. Rathke, and G. W. Kabalka, *J. Am. Chem. Soc.* **91**, 2151 (1969).
16. H. C. Brown, M. M. Rogić, H. Nambu, and M. W. Rathke, *J. Am. Chem. Soc.* **91**, 2147 (1969); H. C. Brown, H. Nambu, and M. M. Rogić, *J. Am. Chem. Soc.* **91**, 6853, 6855 (1969).
17. H. C. Brown, M. M. Midland, and A. B. Levy, *J. Am. Chem. Soc.* **94**, 3662 (1972); J. Hooz, J. N. Bridson, J. G. Calzada, H. C. Brown, M. M. Midland, and A. B. Levy, *J. Org. Chem.* **38**, 2574 (1973).

Scheme 9.3. Alkylation of Trialkylboranes by α-Halocarbonyl and Related Compounds

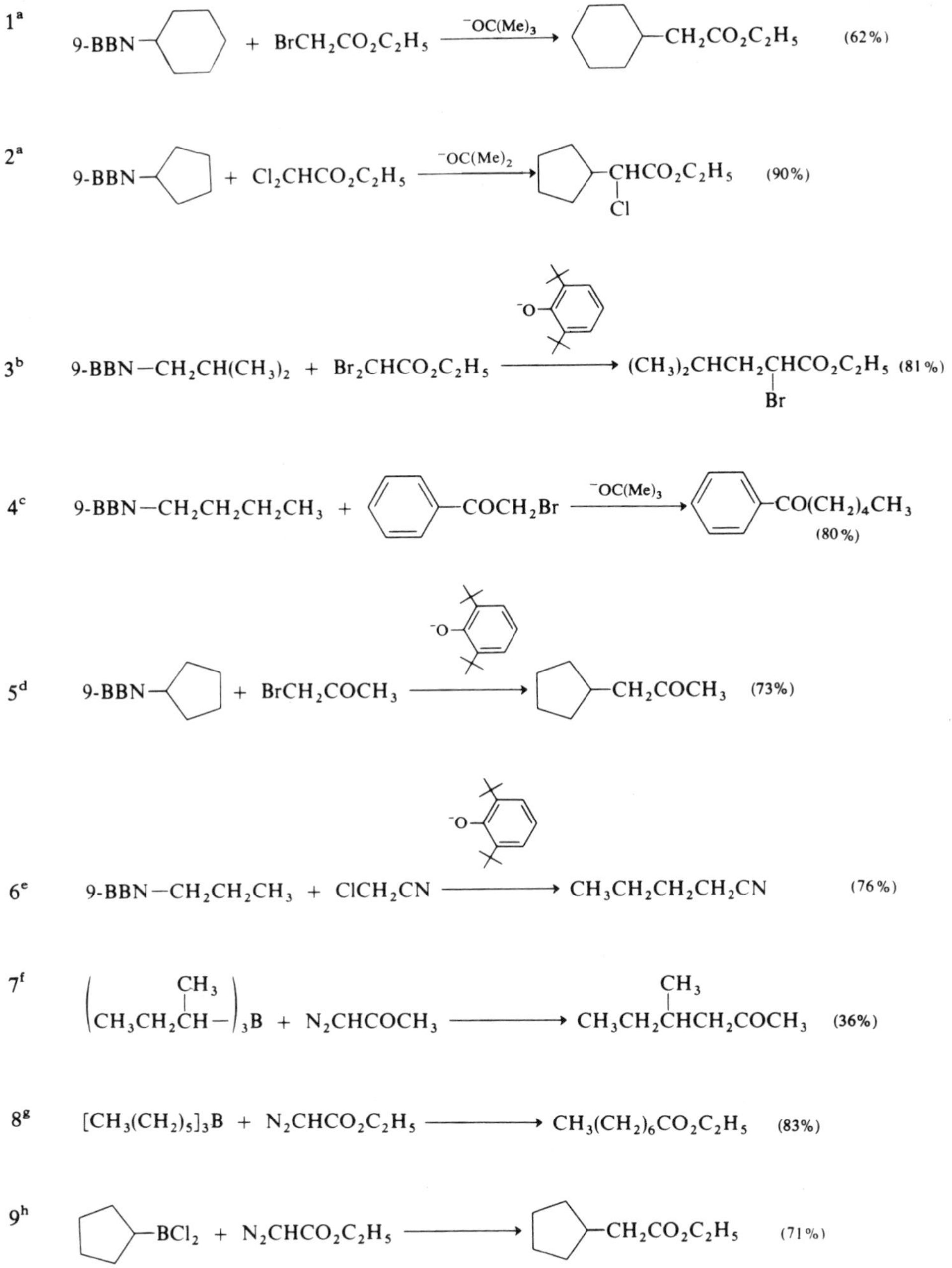

1[a] 9-BBN–(cyclohexyl) + $BrCH_2CO_2C_2H_5$ $\xrightarrow{^-OC(Me)_3}$ (cyclohexyl)–$CH_2CO_2C_2H_5$ (62%)

2[a] 9-BBN–(cyclopentyl) + $Cl_2CHCO_2C_2H_5$ $\xrightarrow{^-OC(Me)_2}$ (cyclopentyl)–$CHClCO_2C_2H_5$ (90%)

3[b] 9-BBN–$CH_2CH(CH_3)_2$ + $Br_2CHCO_2C_2H_5$ $\xrightarrow{\text{2,6-di-}t\text{-butylphenoxide}}$ $(CH_3)_2CHCH_2CHBrCO_2C_2H_5$ (81%)

4[c] 9-BBN–$CH_2CH_2CH_2CH_3$ + $C_6H_5COCH_2Br$ $\xrightarrow{^-OC(Me)_3}$ $C_6H_5CO(CH_2)_4CH_3$ (80%)

5[d] 9-BBN–(cyclopentyl) + $BrCH_2COCH_3$ $\xrightarrow{\text{2,6-di-}t\text{-butylphenoxide}}$ (cyclopentyl)–CH_2COCH_3 (73%)

6[e] 9-BBN–$CH_2CH_2CH_3$ + $ClCH_2CN$ $\xrightarrow{\text{2,6-di-}t\text{-butylphenoxide}}$ $CH_3CH_2CH_2CH_2CN$ (76%)

7[f] $(CH_3CH_2CH(CH_3)-)_3B$ + $N_2CHCOCH_3$ ⟶ $CH_3CH_2CH(CH_3)CH_2COCH_3$ (36%)

8[g] $[CH_3(CH_2)_5]_3B$ + $N_2CHCO_2C_2H_5$ ⟶ $CH_3(CH_2)_6CO_2C_2H_5$ (83%)

9[h] (cyclopentyl)–BCl_2 + $N_2CHCO_2C_2H_5$ ⟶ (cyclopentyl)–$CH_2CO_2C_2H_5$ (71%)

a. H. C. Brown and M. M. Rogić, *J. Am. Chem. Soc.* **91**, 2146 (1969).
b. H. C. Brown, H. Nambu, and M. M. Rogić, *J. Am. Chem. Soc.* **91**, 6855 (1969).
c. H. C. Brown, M. M. Rogić, H. Nambu, and M. W. Rathke, *J. Am. Chem. Soc.* **91**, 2147 (1968).
d. H. C. Brown, H. Nambu, and M. M. Rogić, *J. Am. Chem. Soc.* **91**, 6853 (1969).
e. H. C. Brown, H. Nambu, and M. M. Rogić, *J. Am. Chem. Soc.* **91**, 6855 (1969).
f. J. Hooz and S. Linke, *J. Am. Chem. Soc.* **90**, 5936 (1968).
g. J. Hooz and S. Linke, *J. Am. Chem. Soc.* **90**, 6891 (1968).
h. J. Hooz, J. N. Bridson, J. G. Caldaza, H. C. Brown, M. M. Midland, and A. B. Levy, *J. Org. Chem.* **38**, 2574 (1973).

Organoboranes also undergo alkylation reactions with certain α,β-unsaturated carbonyl compounds. A radical mechanism has been proposed for these reactions.[18]

$$\text{In}\cdot + \text{RCH{=}CHCH{=}O} \rightarrow \text{InCH(R)}{-}\dot{\text{C}}\text{HCH{=}O} \xrightarrow{R'_3B} \text{InCH(R)CH{=}CHOBR}'_2 + \text{R}'\cdot$$

$$\left.\begin{array}{l} \text{R}'\cdot + \text{RCH{=}CHCH{=}O} \rightarrow \text{R}'\text{CH(R)}{-}\dot{\text{C}}\text{HCH{=}O} \\ \text{R}'\text{CH(R)}{-}\dot{\text{C}}\text{HCH{=}O} + \text{R}'_3\text{B} \rightarrow \text{R}'\text{CH(R)CH{=}CHOBR}'_2 + \text{R}'\cdot \end{array}\right\} \text{chain process}$$

$$\text{R}'\text{CH(R)CH{=}CHOBR}'_2 + \text{H}_2\text{O} \rightarrow \text{R}'\text{CH(R)CH}_2\text{CH{=}O}$$

Since only one of the boron substituents is transferred, an improved method that uses the starting alkene more efficiently has been developed. Hetereocyclic dialkylboranes (borinanes) prepared from 1,4-dienes have proved satisfactory for this purpose. (See entry 6 of Scheme 9.4.)

When the reaction is applied to β-alkoxyenones, the β-alkoxy group is eliminated, providing 2,4-dienones.[19]

$$>\text{B}{-}\text{CH{=}CHR} + \text{CH}_3\text{OCH{=}CHC(=O)CH}_3 \longrightarrow \text{RCH{=}CHCH{=}CHC(=O)CH}_3$$

A similar reaction with *B*-alkynyl-9-BBN compounds gives 2-en-4-ynones.[20] The elimination of the β-alkoxy group may be assisted by the Lewis acid character of boron and may take place through a cyclic transition state.

$$>\text{B}{-}\text{C}{\equiv}\text{CR} \longrightarrow [\text{RC}{\equiv}\text{C}{-}\text{B} \cdots \text{O{=}CH{-}CH{=}CH{-}OR}] \longrightarrow [\text{cyclic intermediate}] \longrightarrow \text{RC}{\equiv}\text{CCH{=}CHCH{=}O}$$

Terminal acetylenes can also be alkylated by organoboranes. Adducts are formed between a lithium acetylide and a trialkylborane. Reaction with iodine induces migration and results in the formation of the alkylated acetylene.[21]

$$\text{Li}^+\ (\text{C}_5\text{H}_9)_3\bar{\text{B}}{-}\text{C}{\equiv}\text{C(CH}_2)_3\text{CH}_3 \xrightarrow[-78^\circ\text{C}]{\text{I}_2} \text{C}_5\text{H}_9{-}\text{C}{\equiv}\text{C(CH}_2)_3\text{CH}_3 \ (100\%)$$

18. G. W. Kabalka, H. C. Brown, A. Suzuki, S. Honma, A. Arase, and M. Itoh, *J. Am. Chem. Soc.* **92**, 710 (1970); G. W. Kabalka, *Intra-Sci. Chem. Rept.* **7**, 57 (1973).
19. G. A. Molander, B. Singaram, and H. C. Brown, *J. Org. Chem.* **49**, 5024 (1984).
20. G. A. Molander and H. C. Brown, *J. Org. Chem.* **42**, 3106 (1977).
21. A. Suzuki, N. Miyaura, S. Abiko, H. C. Brown, J. A. Sinclair, and M. M. Midland, *J. Am. Chem. Soc.* **95**, 3080 (1973); A. Suzuki, N. Miyaura, S. Abiko, M. Itoh, M. M. Midland, J. A. Sinclair, and H. C. Brown, *J. Org. Chem.* **51**, 4507 (1986).

Scheme 9.4. Alkylation of Organoboranes by α,β-Unsaturated Carbonyl Compounds

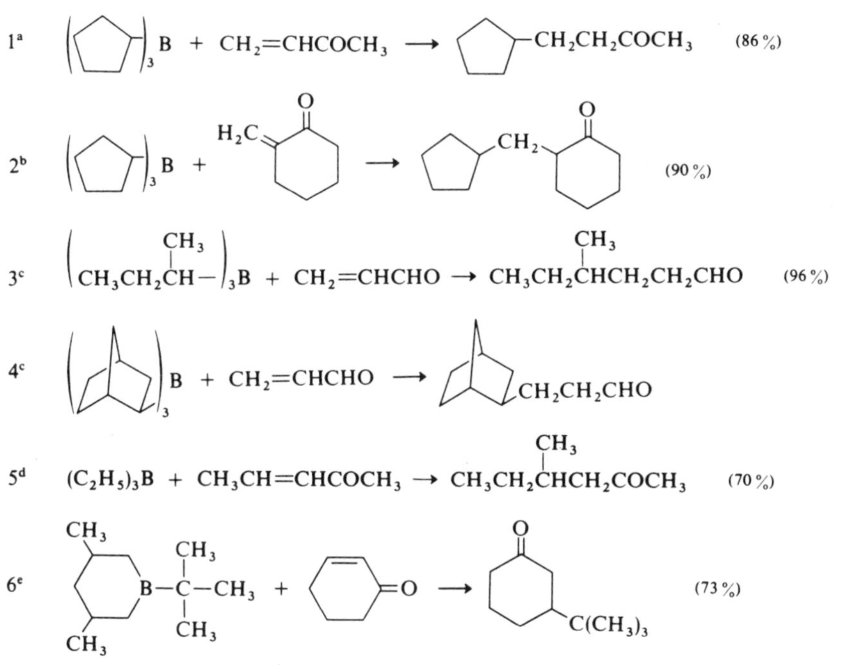

1[a] (cyclopentyl)$_3$B + $CH_2{=}CHCOCH_3$ → cyclopentyl–$CH_2CH_2COCH_3$ (86%)

2[b] (cyclopentyl)$_3$B + 2-methylenecyclohexanone → 2-(cyclopentyl-CH_2)cyclohexanone (90%)

3[c] $(CH_3CH_2CH(CH_3)-)_3B$ + $CH_2{=}CHCHO$ → $CH_3CH_2CH(CH_3)CH_2CH_2CHO$ (96%)

4[c] (norbornyl)$_3$B + $CH_2{=}CHCHO$ → norbornyl–CH_2CH_2CHO

5[d] $(C_2H_5)_3B$ + $CH_3CH{=}CHCOCH_3$ → $CH_3CH_2CH(CH_3)CH_2COCH_3$ (70%)

6[e] (3,5-dimethylboracyclohexyl)B–$C(CH_3)_3$ + cyclohexenone → 3-$C(CH_3)_3$-cyclohexanone (73%)

a. A. Suzuki, A. Arase, H. Matsumoto, M. Itoh, H. C. Brown, M. M. Rogić, and M. W. Rathke, *J. Am. Chem. Soc.* **89**, 5708 (1967).
b. H. C. Brown, M. W. Rathke, G. W. Kabalka, and M. M. Rogić, *J. Am. Chem. Soc.* **90**, 4166 (1968).
c. H. C. Brown, M. M. Rogić, M. W. Rathke, and G. W. Kabalka, *J. Am. Chem. Soc.* **89**, 5709 (1967).
d. H. C. Brown and G. W. Kabalka, *J. Am. Chem. Soc.* **92**, 714 (1970).
e. E. Negishi and H. C. Brown, *J. Am. Chem. Soc.* **95**, 6757 (1973).

The mechanism involves electrophilic attack by iodine on the triple bond, which induces migration of an alkyl group from boron. This is followed by elimination of dialkyliodoboron.

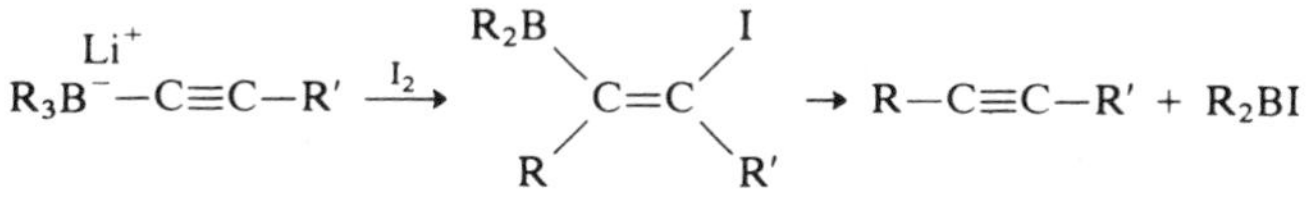

$$Li^+\ R_3B^- {-} C{\equiv}C{-}R' \xrightarrow{I_2} (R_2B)(R)C{=}C(I)(R') \rightarrow R{-}C{\equiv}C{-}R' + R_2BI$$

Related procedures have been developed for the synthesis of both Z- and E-alkenes. Treatment of alkenyldialkylboranes with iodine results in the formation of the Z-alkene.[22]

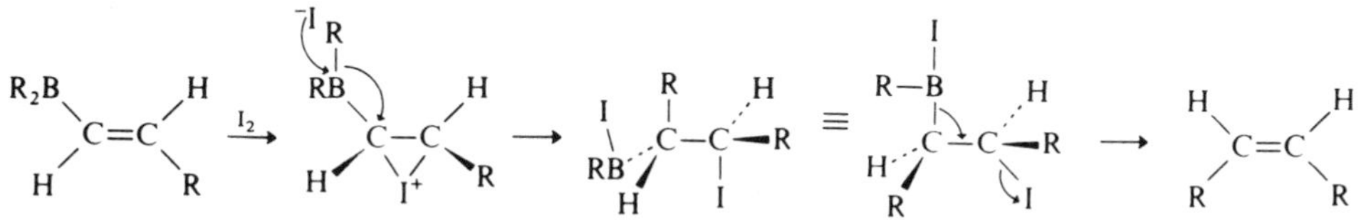

$$(R_2B)(H)C{=}C(H)(R) \xrightarrow{I_2} \text{iodonium ion} \rightarrow RB(I)(H)C(R){-}C(H)(I)(R) \equiv \rightarrow (H)(R)C{=}C(H)(R)$$

22. G. Zweifel, H. Arzoumanian, and C. C. Whitney, *J. Am. Chem. Soc.* **89**, 3652 (1967); G. Zweifel, R. P. Fisher, J. T. Snow, and C. C. Whitney, *J. Am. Chem. Soc.* **93**, 6309 (1971).

Similarly, alkenyllithium reagents add to dimethyl boronates to give adducts which decompose to Z-alkenes on treatment with iodine.[23]

$$RB(OCH_3)_2 + \underset{H}{\overset{Li}{}}C{=}C\underset{R'}{\overset{H}{}} \rightarrow R\overset{OCH_3}{\underset{OCH_3}{\overset{|^-}{\underset{|}{B}}}}{-}CH{=}CHR' \xrightarrow{I_2} \underset{H}{\overset{R}{}}C{=}C\underset{H}{\overset{R'}{}}$$

The stereoselectivity of these reactions arises from a base-induced *anti* elimination after the migration.

$$\text{RB(OMe)}_2\text{–CH=CHR'} \longrightarrow \text{(MeO)}_2\text{B(R)–C(H)–C(H)(R')}\ (\text{I}^+ \text{bridged}) \longrightarrow \text{(MeO)}_2\text{B–C(R)(H)–C(H)(I)–R'} \equiv \text{(MeO)}_2\text{B–C(H)(R)–C(H)(R')–I} \xrightarrow{MeO^-} \underset{R}{\overset{H}{}}C{=}C\underset{R'}{\overset{H}{}}$$

The synthesis of Z-alkenes can also be carried out starting with an alkylbromoborane. In this case, migration presumably follows replacement of the bromide by methoxide.[24]

$$R'BHBr + HC{\equiv}CR \rightarrow \underset{H}{\overset{R'BBr}{}}C{=}C\underset{R}{\overset{H}{}} \xrightarrow[I_2]{MeO^-} \underset{R'}{\overset{H}{}}C{=}C\underset{R}{\overset{H}{}}$$

E-Alkenes can be prepared by several related reactions.[25] Hydroboration of a bromoalkyne generates an α-bromoalkenylborane. On treatment with methoxide ion, these intermediates undergo boron-to-carbon migration to give an alkylated alkenylborane. Protonolysis generates an E-alkene.

$$RC{\equiv}CBr + R'_2BH \rightarrow \underset{H}{\overset{R}{}}C{=}C\underset{BR'_2}{\overset{Br}{}} \xrightarrow{^-OMe} \underset{H}{\overset{R}{}}C{=}C\underset{R'}{\overset{MeOBR'}{}} \xrightarrow{CH_3CO_2H} \underset{H}{\overset{R}{}}C{=}C\underset{R'}{\overset{H}{}}$$

The dialkylboranes can be prepared from thexylchloroborane. The thexyl group does not normally migrate.

$$\text{Thx–BHCl} + R'CH{=}CH_2 \longrightarrow \text{Thx–}\underset{Cl}{\underset{|}{B}}CH_2CH_2R' \xrightarrow{KBH(OR)_3} \text{Thx–}\underset{H}{\underset{|}{B}}CH_2CH_2R'$$

23. D. A. Evans, T. C. Crawford, R. C. Thomas, and J. A. Walker, *J. Org. Chem.* **41**, 3947 (1976).
24. H. C. Brown, D. Basavaiah, S. U. Kulkarni, N. G. Bhat, and J. V. N. Vara Prasad, *J. Org. Chem.* **53**, 239 (1988).
25. H. C. Brown, D. Basavaiah, S. U. Kulkarni, H. P. Lee, E. Negishi, and J.-J. Katz, *J. Org. Chem.* **51**, 5270 (1986).

A similar strategy involves initial hydroboration by $BrBH_2$.[26]

$$R'CH{=}CH_2 + BrBH_2 \rightarrow R'CH_2CH_2BBr(H) \xrightarrow{BrC\equiv CR} (R'CH_2CH_2B(Br))(Br)C{=}C(H)(R) \xrightarrow{^-OMe} (R'CH_2CH_2)((MeO)_2B)C{=}C(H)(R) \xrightarrow{H^+} (R'CH_2CH_2)(H)C{=}C(H)(R)$$

The boron-to-carbon migration can also be induced by other types of electrophiles. Trimethylsilyl chloride and trimethylsilyl triflate both induce a stereospecific migration to form β-trimethylsilylalkenylboranes with the silicon and boron substituents *cis*.[27] It has been suggested that the stereospecificity arises from a silicon-bridged intermediate.

$$R_3B{-}C{\equiv}C{-}R' + (CH_3)_3Si{-}X \longrightarrow [R_2B(R)C{=}C(R')\text{, bridged by } Si(CH_3)_3\cdots X] \longrightarrow (R)(R_2B)C{=}C(R')(Si(CH_3)_3)$$

Tributyltin chloride induces migration and also gives the product in which the C—Sn bond is *syn* to the C—B bond. Protonolysis of both the C—Sn and C—B bonds by acetic acid gives the Z-alkene.[28]

$$R_3\bar{B}{-}C{\equiv}CR' + ClSnR''_3 \rightarrow (R)(R_2B)C{=}C(R')(SnR''_3) \xrightarrow{CH_3CO_2H} (R)(H)C{=}C(R')(H)$$

As can be judged from the preceding discussion, organoboranes are versatile intermediates for formation of carbon–carbon bonds. An important aspect of all of these synthetic procedures involving boron-to-carbon migration is that they involve *retention* of the configuration of the group that migrates. Very effective procedures for enentioselective hydroboration have been developed (see Section 4.9.3). These homochiral organoboranes offer the opportunity for enantioselective synthesis.

A sequence for enantioselective formation of ketones starts with hydroboration of monoisopinocampheylborane, which can be obtained in high enantiomeric

26. H. C. Brown, T. Imai, and N. G. Bhat, *J. Org. Chem.* **51**, 5277 (1986); H. C. Brown, D. Basavaiah, and S. U. Kulkarni, *J. Org. Chem.* **47**, 3808 (1982).
27. P. Binger and R. Köster, *Synthesis*, 309 (1973); E. J. Corey and W. L. Seibel, *Tetrahedron Lett.* **27**, 905 (1986).
28. K. K. Wang and K.-H. Chu, *J. Org. Chem.* **49**, 5175 (1984).

purity.[29] The hydroboration of a prochiral alkene establishes a new stereocenter. A third alkyl group can be introduced by a second hydroboration step.

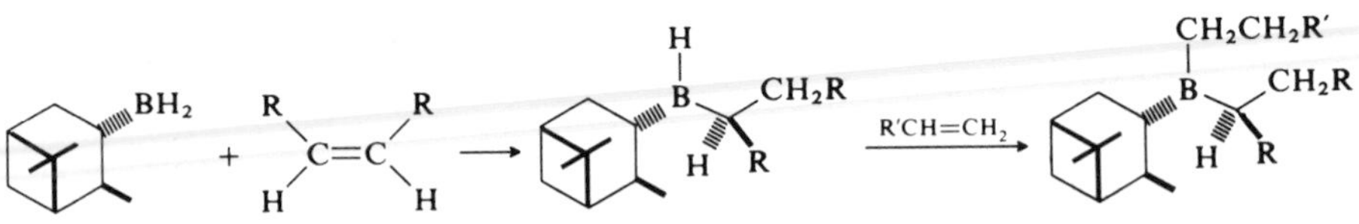

The trialkylborane can be transformed to an ethyl borinate by heating with acetaldehyde, which releases the original chiral α-pinene. Finally, application of one of the carbonylation procedures outlined in Scheme 9.1 or 9.2 gives a chiral ketone.[30] The enantiomeric excess observed for ketones prepared in this way ranges from 60% to 90%.

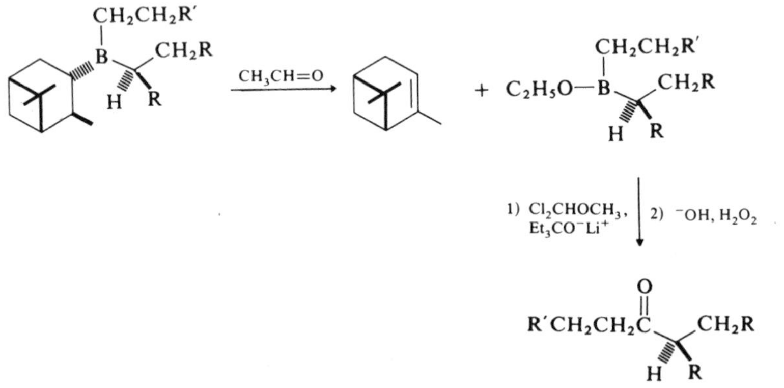

Higher enantiomeric purity can be obtained by a modified procedure in which the monoalkylborane intermediate is prepared.[31]

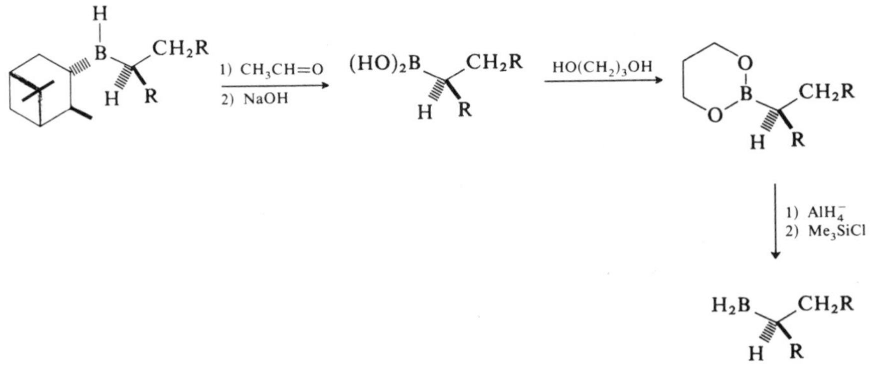

29. H. C. Brown, P. K. Jadhav, and A. K. Mandal, *J. Org. Chem.* **47**, 5074 (1982).
30. H. C. Brown, P. K. Jadhav, and M. C. Desai, *Tetrahedron* **40**, 1325 (1984).
31. H. C. Brown, R. K. Bakshi, and B. Singaram, *J. Am. Chem. Soc.* **110**, 1529 (1988); H. C. Brown, M. Srebnik, R. K. Bakshi, and T. E. Cole, *J. Am. Chem. Soc.* **109**, 5420 (1987).

Subsequent steps involve introduction of a thexyl group and then the second ketone substituent. Finally, the ketone is formed by the cyanide–trifluoroacetic anhydride method (see p. 41).

H_2B, CH_2R, H, R — $(CH_3)_2C{=}C(CH_3)_2$ → thexyl–BH, CH_2R, H, R — $R'CH{=}CH_2$ → thexyl–B(CH_2CH_2R'), CH_2R, H, R — 1) NaCN, 2) $(CF_3CO)_2O$, 3) H_2O_2 → $R'CH_2CH_2C(=O)$–C(H)(R)CH_2R

Bis(isopinocampheyl)borane has been used in the enantioselective preparation of alkynes by the reaction sequence for alkylation of alkynes (see page 452).

$[\text{Ipc}]_2B$–C(H)(CH_2CH_3)(CH_3) — 1) $LiC{\equiv}CR$, 2) I_2, 3) NaOH, H_2O_2 → $RC{\equiv}C$–C(H)(CH_2CH_3)(CH_3)

In general, any of the synthetic processes involving boron-to-carbon migration should be adaptable to enantioselective syntheses.

Allylboranes such as 9-allyl-9-BBN react with aldehydes and ketones to give allylic carbinols. The reaction proceeds through a cyclic transition state. Bond formation takes place at the γ carbon of the allyl group, and the double bond shifts.[33]

$R_2C{=}O \cdots B$, CH_2, $H_2C{=}C$–H (cyclic) → $R_2C(CH_2CH{=}CH_2)$–O–B< — $HO(CH_2)_2NH_2$ → $R_2C(OH)CH_2CH{=}CH_2$

This reaction probably begins with Lewis acid–base complexation at the carbonyl oxygen. Such complexation both increases the electrophilicity of the carbonyl group and weakens the C—B bond of the allyl group. After the reaction is complete, the carbinol product can be liberated from the borinate ester by displacement with ethanolamine. Yields for a series of aldehydes and ketones were usually above 90% for 9-allyl-9-BBN.

The cyclic mechanism would predict that the allyl addition reaction would be stereospecific with respect to the geometry of the double bond in the allylic group. This has been demonstrated to be the case. The *E*- and *Z*-2-butenyl cyclic boronate

32. C. A. Brown, M. C. Desai, and P. K. Jadhav, *J. Org. Chem.* **51**, 162 (1986).
33. G. W. Kramer and H. C. Brown, *J. Org. Chem.* **42**, 2292 (1977).

esters **1** and **2** were synthesized and allowed to react with aldehydes. The *E*-boronate gave the carbinol having *anti* stereochemistry whereas the *Z*-boronate gave the *syn* product.[34]

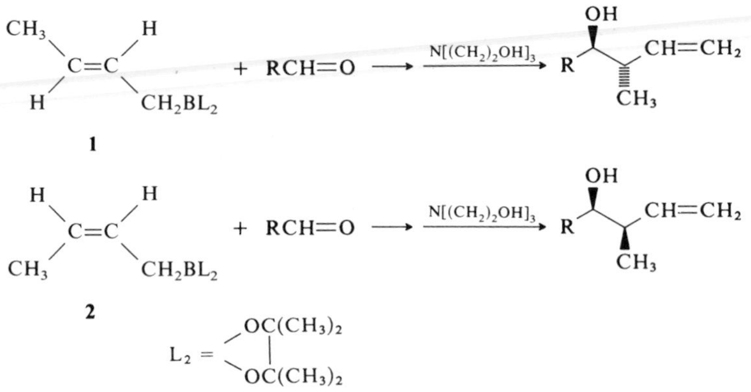

This stereochemistry is that predicted for a cyclic transition state in which the aldehyde substituent occupies an equatorial position.

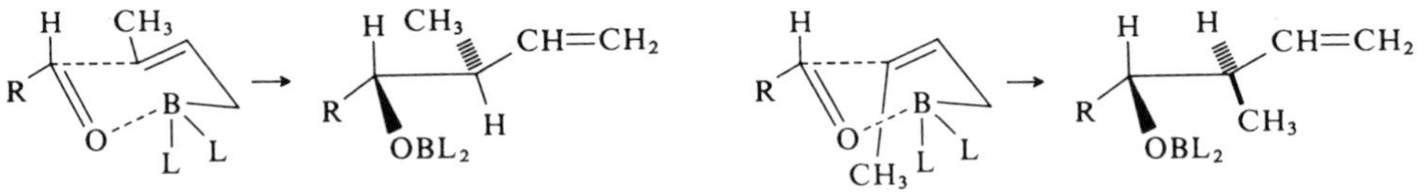

The diastereoselectivity observed in simple systems led to investigation of enantiomerically pure aldehydes. It was found that the *E*- and *Z*-2-butenylboronates both exhibit high diastereoselectivity with chiral α-substituted aldehydes. However, only the *Z*-isomer also exhibited high enantioselectivity.[35]

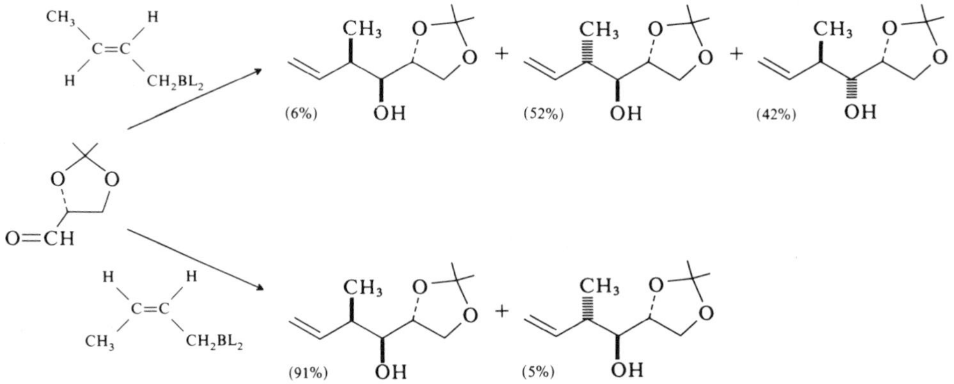

Addition reactions of allylic boranes have proven to be quite general and useful. Several methods for synthesis of allylic boranes esters have been developed.[36] The

34. R. W. Hoffmann and H.-J. Zeiss, *J. Org. Chem.* **46**, 1309 (1981); K. Fujita and M. Schlosser, *Helv. Chim. Acta* **65**, 1258 (1982).
35. W. R. Roush, M. A. Adam, A. E. Walts, and D. J. Harris, *J. Am. Chem. Soc.* **108**, 3422 (1986).
36. P. G. M. Wuts, P. A. Thompson, and G. R. Callen, *J. Org. Chem.* **48**, 5398 (1983); E. Moret and M. Schlosser, *Tetrahedron Lett.* **25**, 4491 (1984).

reaction has found application in the stereospecific synthesis of complex structures.

MeO, B(OMe)₂ + HC(=O), O, O, CH₃ ⟶ MeO, O, O, OH, CH₃ Ref. 37

The allylation reaction has also been extended to enantiomerically pure allylic boranes. For example, the 3-methyl-2-butenyl derivative of bis-(isopinocampheyl)borane reacts with aldehydes to give carbinols of >90% enantiomeric excess in most cases.[38]

$BCH_2CH{=}C(CH_3)_2$ 1) $(CH_3)_2C{=}CHCH{=}O$ 2) NaOH, H_2O_2 ⟶ CH_3, H, OH, $CH{=}CH_2$, CH_3, CH_3 CH_3 85% yield 96% e.e.

B-Allylbis(isopinocampheyl)borane exhibits high stereoselectivity in reactions with chiral α-substituted aldehydes.[39]

$PhCH_2O$, CH_3, O, H; $BCH_2CH{=}CH_2$ ⟶ $PhCH_2O$, CH_3, OH (94%) + $PhCH_2O$, CH_3, OH (6%); $BCH_2CH{=}CH_2$ ⟶ (4%) (96%)

Other enantiomerically pure *B*-allylboranes also show excellent stereoselectivity in these reactions.[40] Allyl and 2-butenyl derivatives of the cyclic boronate ester **3**, derived from tartaric acid, also give enantioselective additions to aldehydes.[41]

CH_3–CH=CH–CH_2–B(O)(O), CO_2–i-Pr, CO_2–i-Pr (**3**) + O=CH, O, O, CH_3 ⟶ CH_3, O, O, OH, CH_3 major stereoisomer

37. W. R. Roush, M. R. Michaelides, D. F. Tai, and W. K. M. Chong, *J. Am. Chem. Soc.* **109**, 7575 (1987).
38. H. C. Brown and P. K. Jadhav, *Tetrahedron Lett.* **25**, 1215 (1984); H. C. Brown, P. K. Jadhav, and K. S. Bhat, *J. Am. Chem. Soc.* **110**, 1535 (1988).
39. H. C. Brown, K. S. Bhat, and R. S. Randad, *J. Org. Chem.* **52**, 319 (1987).
40. H. C. Brown and P. K. Jadhav, *J. Org. Chem.* **49**, 4089 (1984); H. C. Brown and K. S. Bhat, *J. Am. Chem. Soc.* **108**, 5919 (1986); J. Garcia, B. M. Kim, and S. Masamune, *J. Org. Chem.* **52**, 4831 (1987).
41. W. R. Roush, A. E. Walts, and L. K. Hoong, *J. Am. Chem. Soc.* **107**, 8186 (1985); W. R. Roush and R. L. Halterman, *J. Am. Chem. Soc.* **108**, 294 (1986).

Scheme 9.5. Addition Reactions of Allylic Boranes with Carbonyl Compounds

1[a] H, H, H_3C, C=C, CH_2B, O, O + O=CH, O, O → CH_3, O, O, OH

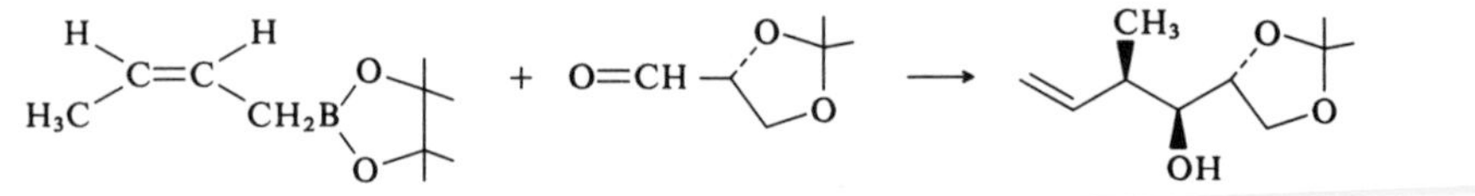

2[b] H, BCH_2C, CCH_3, H + $CH_3\overset{O}{\overset{\|}{C}}CO_2CH_3$ → CH_2=CHCHCCO$_2$CH$_3$ (CH_3; H_3C OH)

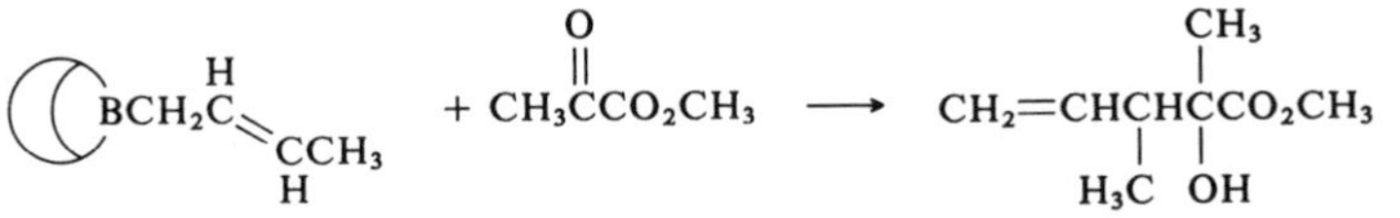

96% yield
73:27 *anti–syn* mixture

3[c] $Si(CH_3)_3$, B—CH, H, C=C, H, CH_3 + C_3H_7CH=O → C_3H_7, CH_3, $Si(CH_3)_3$, OH

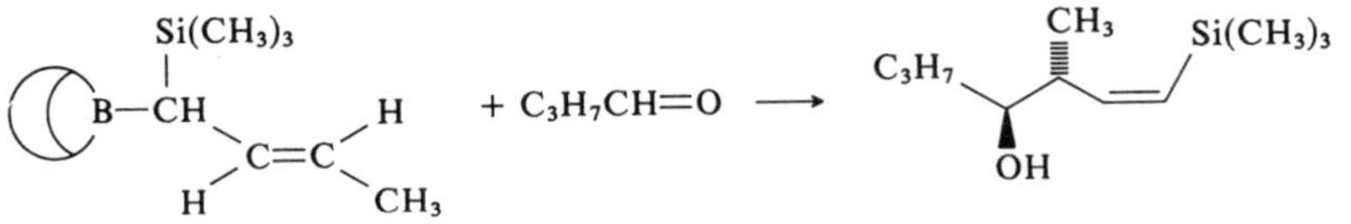

4[d] O=CH, O, O, CH_3 + CH_3OCH_2CH=CHB(OCH$_3$)$_2$ → MeO, O, O, OH, CH_3 (83%)

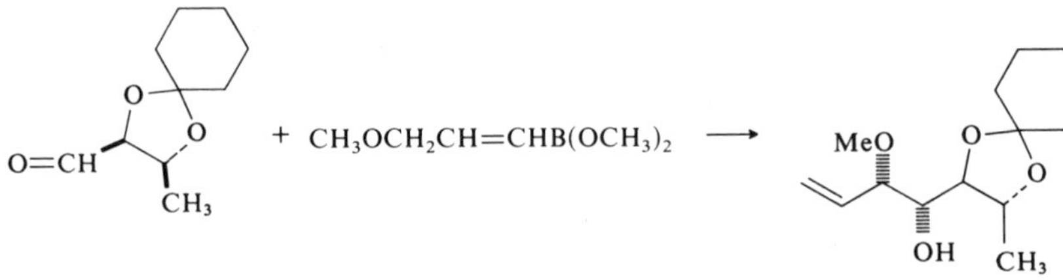

5[e] $(Ipc)_2BCH_2$, H, C=C, H, Ph + O=CH, O, O, CH_3, CH_3 → OH, Ph, O, O, CH_3, CH_3

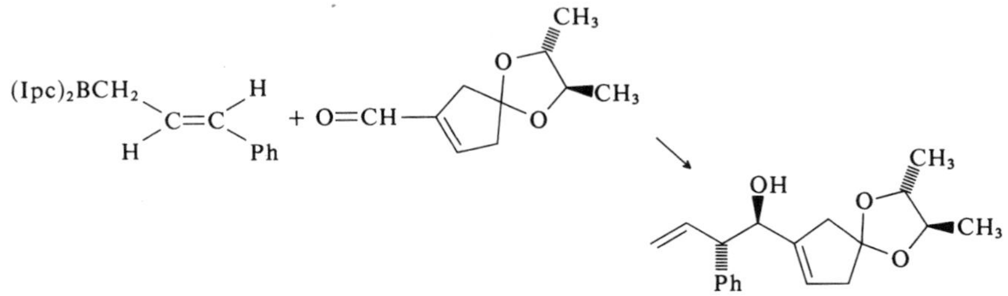

96% d.e.
89% e.e.

6[f] H_3C, C=C, H, CH_2—B, O, O, H + PhCH=O $\xrightarrow{-78°C}$ OH, CH_3

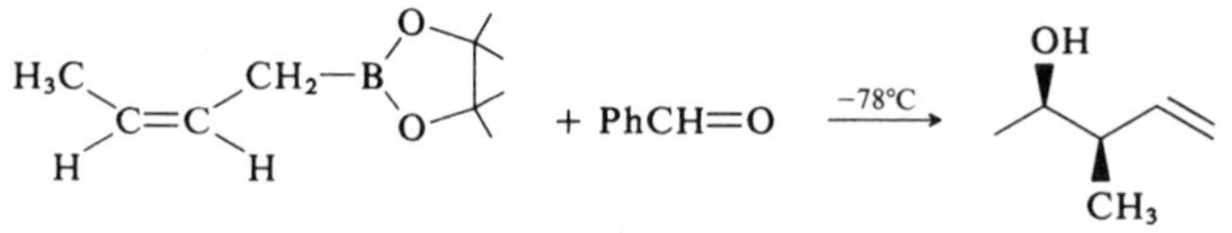

diastereosolectivity > 95%

7[f] H, C=C, H_3C, CH_2—B, O, O, H + PhCH=O $\xrightarrow{-78°C}$ OH, CH_3

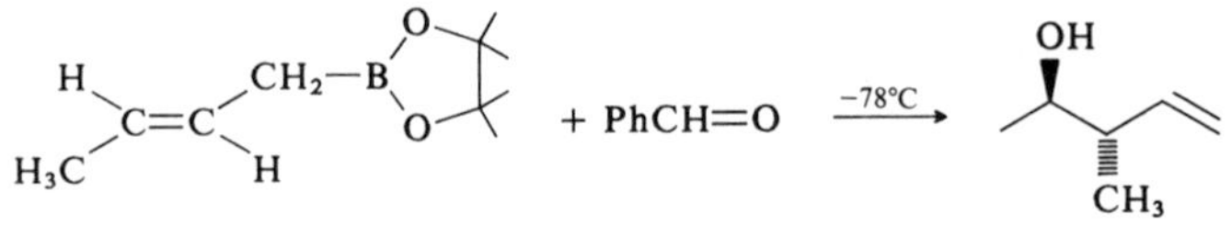

diastereoselectivity > 95%

8[g]

$$i\text{-}PrO_2C,\ i\text{-}PrO_2C\text{-dioxaborolane}\text{-}BCH_2CH{=}CH_2 + O{=}CH\text{-dioxolane} \longrightarrow \text{homoallylic alcohol (OH)}$$

(91%)
96:4 diastereoselectivity

9[h]

$$[\text{caranyl}]_2BCH_2CH{=}CH_2 + CH_3CH_2CH{=}O \longrightarrow CH_3CH_2C(OH)(H)CH_2CH{=}CH_2$$

76% yield
91% e.e. (*R*)

a. W. R. Roush and A. E. Walts, *Tetrahedron Lett.* **26**, 3427 (1985); W. R. Roush, M. A. Adam, and D. J. Harris, *J. Org. Chem.* **50**, 2000 (1985).
b. Y. Yamamoto, K. Maruyama, T. Komatsu, and W. Ito, *J. Org. Chem.* **51**, 886 (1986).
c. Y. Yamamoto, H. Yatagai, and K. Maruyama, *J. Org. Chem.* **103**, 3229 (1981).
d. W. R. Roush, M. R. Michaelides, D. F. Tai, and W. K. M. Chong, *J. Am. Chem. Soc.* **109**, 7575 (1987).
e. L. K. Truesdale, D. Swanson, and R. C. Sun, *Tetrahedron Lett.* **26**, 5009 (1985).
f. R. W. Hoffmann and H.-J. Zeiss, *J. Org. Chem.* **46**, 1309 (1981).
g. W. R. Roush, A. E. Walts, and L. K. Hoong, *J. Am. Chem. Soc.* **107**, 8186 (1985).
h. H. C. Brown and P. K. Jadhav, *J. Org. Chem.* **49**, 4089 (1984).

Scheme 9.5 illustrates some examples of synthesis of allylic carbinols via allylic boranes and boronate esters.

B-Alkynyl derivatives of 9-BBN act as mild sources of nucleophilic acetylenic groups. Reaction occurs with both aldehydes and ketones, but the rate is at least 100 times faster for aldehydes.[42]

$$(CH_3)_3CC{\equiv}C{-}BL_2 + CH_3CH_2CH{=}O \rightarrow \xrightarrow{HOCH_2CH_2NH_2} (CH_3)_3CC{\equiv}CC(OH)(H)CH_2CH_3 \quad (83\%)$$

$BL_2 = $ 9-BBN

Ethanolamine is used to displace the adduct from the borinate ester. The facility with which the transfer of acetylenic groups occurs is associated with the relative stability of the *sp*-hybridized carbon. This reaction is an alternative to the addition of magnesium or lithium salts of acetylides to aldehydes.

Arylboronic acids and esters serve as aryl group donors for palladium-catalyzed arylation procedures.[43] The boronic acids are readily prepared from an aryllithium compound and trimethyl borate, followed by hydrolysis. The boronic acids react

42. H. C. Brown, G. A. Molander, S. M. Singh, and U. S. Racherla, *J. Org. Chem.* **50**, 1577 (1985).
43. N. Miyaura, T. Yanagi, and A. Suzuki, *Synth. Commun.* **11**, 513 (1981); W. J. Thompson and J. J. Gaudino, *J. Org. Chem.* **49**, 5237 (1984); M. J. Sharp and V. Snieckus, *Tetrahedron Lett.* **26**, 5997 (1985).

with aryl bromides to give unsymmetrical biaryls in the presence of the palladium catalyst.

$$\text{ArLi} + (\text{MeO})_3\text{B} \xrightarrow{\text{H}_2\text{O}} \text{ArB(OH)}_2$$

$$\text{ArB(OH)}_2 + \text{Ar'Br} \xrightarrow{\text{Pd(PPh}_3)_4} \text{Ar—Ar'}$$

Based on the general mechanism for Pd-catalyzed carbon–carbon bond formation (see p. 418), the boronic acid must deliver an aryl anion equivalent to Pd at some stage in the reaction.

Alkenylboronic acids, alkenyl boronate esters, and alkenylboranes can be coupled with alkenyl halides by palladium catalysts to give dienes.[44]

$$(R)(H)C{=}C(H)(BX_2) + (R')(H)C{=}C(H)(Y) \xrightarrow{\text{Pd(PPh}_3)_4} (R)(H)C{=}C(H)\text{–}C(H){=}C(H)(R')$$

X = OH, OR, R
Y = Br, I

These reactions proceed with retention of double-bond configuration in both the boron derivative and the alkenyl halide. The basic steps must involve oxidative addition at the alkenyl halide, transfer of an alkenyl group from boron to palladium, and reductive elimination. However, the details of the mechanism are not established.

$$(R')(H)C{=}C(H)(Y) \xrightarrow{\text{Pd}^0} (R')(H)C{=}C(H)(\text{Pd}^{II}) + (R)(H)C{=}C(H)(BX_2) \rightarrow (R')(H)C{=}C(H)\text{–Pd}^{II}\text{–}C(H){=}C(R)(H) \rightarrow (R')(H)C{=}C(H)\text{–}C(H){=}C(H)(R)$$

Alkyl substituents on boron in 9-BBN derivatives can be coupled with both vinyl and aryl halides through Pd catalysis.[45] This is an especially interesting reaction because of its ability to effect coupling of saturated alkyl groups. Palladium-catalyzed coupling of alkyl groups by other methods frequently fails because of the tendency for β elimination.

$$\begin{array}{c}\text{Ar—X}\\ \text{or}\\ \text{R'CH=CHX}\end{array} + \text{RBL}_2 \xrightarrow[\text{NaOMe}]{\text{Pd}} \begin{array}{c}\text{Ar—R}\\ \text{or}\\ \text{R'CH=CHR}\end{array}$$

Scheme 9.6 gives some examples of Pd-catalyzed couplings of organoboranes.

44. N. Miyaura, K. Yamada, H. Suginome, and A. Suzuki, *J. Am. Chem. Soc.* **107**, 972 (1985); N. Miyaura, M. Satoh, and A. Suzuki, *Tetrahedron Lett.* **27**, 3745 (1986); F. Björkling, T. Norin, C. R. Unelius, and R. B. Miller, *J. Org. Chem.* **52**, 292 (1987); J. Uenishi, J.-M. Beau, R. W. Armstrong, and Y. Kishi, *J. Am. Chem. Soc.* **109**, 4756 (1987).
45. N. Miyaura, T. Ishiyama, M. Ishikawa, and A. Suzuki, *Tetrahedron Lett.* **27**, 6369 (1986).

Scheme 9.6. Palladium-Catalyzed Coupling of Organoboranes

1[a] $CH_3(CH_2)_3CH{=}CH{-}B(O_2C_6H_4)$ + $HC(Br){=}CH(Ph)$ $\xrightarrow{Pd(PPh_3)_4,\ NaOEt}$ $CH_3(CH_2)_3CH{=}CH{-}CH{=}CHPh$ (86%)

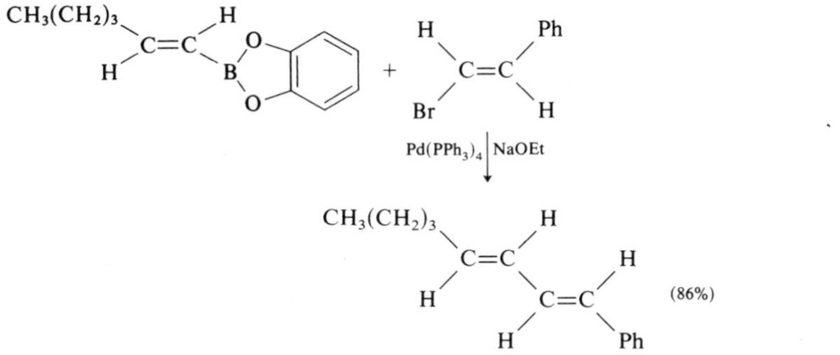

2[b] $CH_3(CH_2)_5CH{=}CH{-}B(O\text{-}i\text{-}Pr)_2$ + $I{-}C_6H_5$ $\xrightarrow{Pd(PPh_3)_4,\ NaOEt}$ $CH_3(CH_2)_5CH{=}CHPh$ (98%)

3[c] $CH_3CH_2CH{=}CH{-}BR_2$ + $BrCH{=}CH(CH_2)_8OTHP$ $\xrightarrow{Pd(PPh_3)_4,\ NaOEt}$ $CH_3CH_2CH{=}CH{-}CH{=}CH(CH_2)_8OTHP$ (73%)

4[d] CH_3OCH_2, R_3SiO, R_3SiO, R_3SiO, OH, I + $(HO)_2B$, $OSiR_3$, $OSiR_3$, R_3SiO, $OSiR_3$ $\xrightarrow{Pd(PPh_3)_4,\ TlOH}$ (73%) CH_3OCH_2, R_3SiO, R_3SiO, R_3SiO, OH, R_3SiO, $OSiR_3$, $OSiR_3$, $OSiR_3$

5[e] $C_6H_5CONHC_2H_5$ $\xrightarrow{1)\ s\text{-BuLi};\ 2)\ B(OMe)_3;\ 3)\ H^+}$ $2\text{-}(CONHC_2H_5)C_6H_4B(OH_2)$ + 2-bromothiophene (Br, S) $\xrightarrow{Pd(PPh_3)_4}$ $2\text{-}(CONHC_2H_5)C_6H_4$-(2-thienyl) (92%)

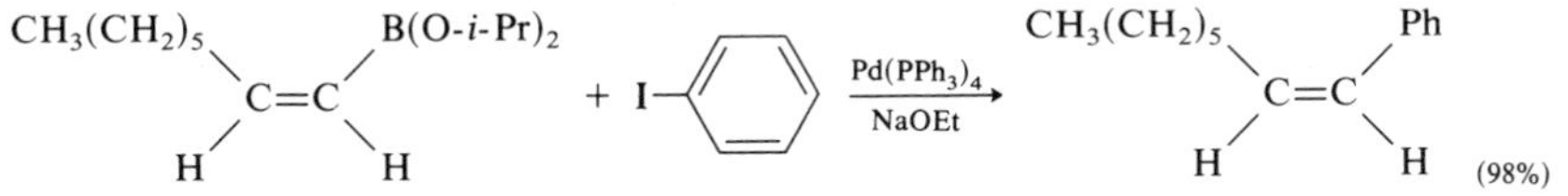

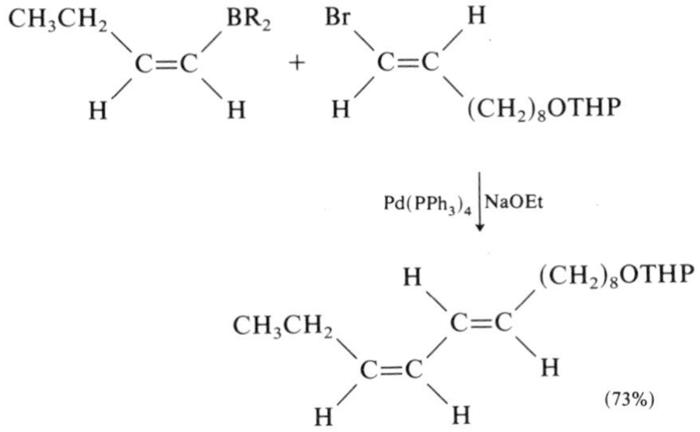

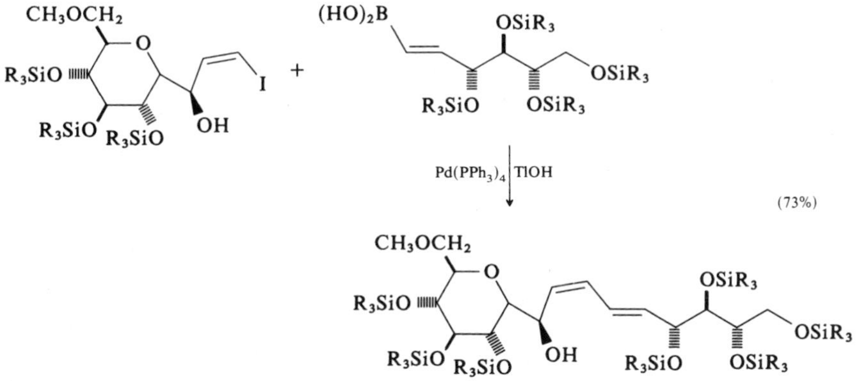

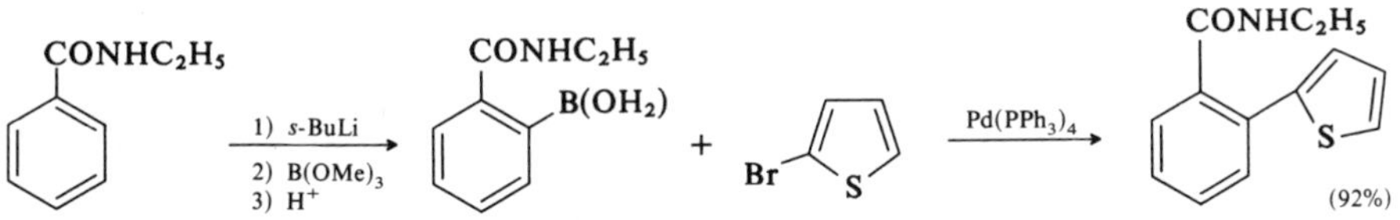

Scheme 9.6—*continued*

6[f] 2-iodoanisole + $CH_3(CH_2)_7B$(9-BBN) $\xrightarrow[\text{NaOH}]{PdCl_2(dppf)}$ 2-$(CH_2)_7CH_3$-anisole (90%)

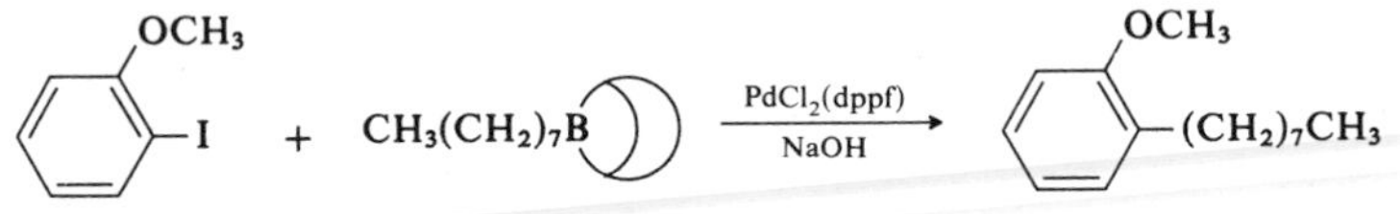

dppf = bis-(diphenylphorphine) ferrocene

7[g] 3-bromopyridine $\xrightarrow[\text{2) } Et_2BOMe]{\text{1) } n\text{-BuLi}}$ 3-$B(C_2H_5)_2$-pyridine + 4-bromoanisole $\xrightarrow{Pd(PPh_3)_4}$ 3-(4-OCH_3-phenyl)pyridine (75%)

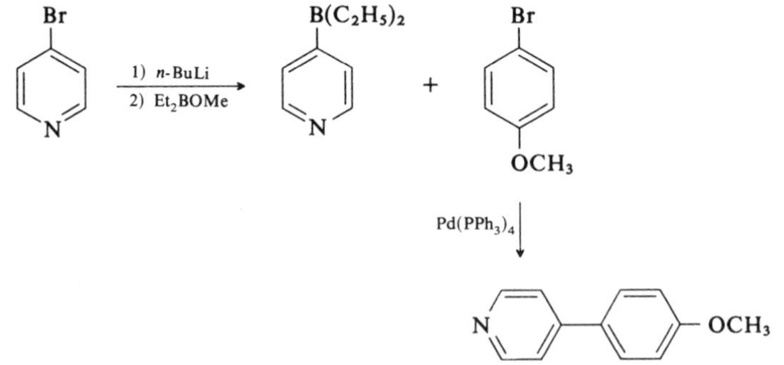

a. N. Miyaura, K. Yamada, H. Suginome, and A. Suzuki, *J. Am. Chem. Soc.* **107**, 972 (1985).
b. N. Miyaura, M. Satoh, and A. Suzuki, *Tetrahedron Lett.* **27**, 3745 (1986).
c. F. Björkling, T. Norin, C. R. Unelius, and R. B. Miller, *J. Org. Chem.* **52**, 292 (1987).
d. J. Uenishi, J.-M. Beau, R. W. Armstrong, and Y. Kishi, *J. Am. Chem. Soc.* **109**, 4756 (1987).
e. M. J. Sharp and V. Snieckus, *Tetrahedron Lett.* **26**, 5997 (1985).
f. N. Miyaura, T. Ishiyama, M. Ishikawa, and A. Suzuki, *Tetrahedron Lett.* **27**, 6369 (1986).
g. M. Ishikura, T. Ohta, and M. Terashima, *Chem. Pharm. Bull.* **33**, 4755 (1985).

9.2. Organosilicon Compounds

9.2.1. Synthesis of Organosilanes

The two most general means of synthesis of organosilanes for use in synthesis are nucleophilic displacement of halogen from a halosilane by an organometallic reagent and addition of silanes at multiple bonds (*hydrosilation*). Organomagnesium and organolithium compounds react with trimethylsilyl chloride to give the corresponding tetrasubstituted silanes.

$$CH_2{=}CHMgBr + (CH_3)_3SiCl \rightarrow CH_2{=}CHSi(CH_3)_3 \qquad \text{Ref. 46}$$

$$CH_2{=}C(Li)OC_2H_5 + (CH_3)_3SiCl \rightarrow CH_2{=}C(Si(CH_3)_3)OC_2H_5 \qquad \text{Ref. 47}$$

The carbon–silicon bond is quite strong, and trimethylsilyl groups are stable to many of the reactions that are typically used in organic synthesis. Thus, much of

46. R. K. Boeckman, Jr., D. M. Blum, B. Ganem, and N. Halvey, *Org. Synth.* **58**, 152 (1978).
47. R. F. Cunico and C.-P. Kuan, *J. Org. Chem.* **50**, 5410 (1985).

the repertoire of synthetic organic chemistry can be used for elaboration of organosilanes.

Silicon substituents can also be introduced into alkenes and alkynes by hydrosilation.[48] This reaction, in contrast to hydroboration, does not occur spontaneously, but it can be carried out in the presence of catalysts, the most common of which is H_2PtCl_6, hexachloroplatinic acid. Other catalysts are also available.[49] Silanes that carry one or more halogen substituents are more reactive than trialkylsilanes.[50]

$$\text{C}_6\text{H}_{10}{=}\text{CH}_2 \xrightarrow[\text{H}_2\text{PtCl}_6]{\text{CH}_3\text{SiCl}_2\text{H}} \text{C}_6\text{H}_{11}\text{–CH}_2\text{SiCl}_2\text{CH}_3$$

With more substituted alkenes, similar reaction conditions are often accompanied by double-bond migrations which eventually lead to the formation of an alkyltrichlorosilane with a primary alkyl group.[51]

9.2.2. Carbon–Carbon Bond-Forming Reactions

The carbon-silicon bond to saturated alkyl groups is not very reactive. Most of the valuable synthetic procedures based on organosilanes involve either alkenyl or allylic silicon substituents. The dominant reactivity pattern involves attack by an electrophilic carbon intermediate at the double bond.

$$\text{(R}_3\text{Si)(H)C{=}CHR} + {-}\overset{|}{\underset{}{C}}{}^{\delta+} \longrightarrow \text{H–C(R}_3\text{Si)(–C–)–}\overset{+}{\text{C}}\text{HR} \longrightarrow \text{(–C–)(H)C{=}CHR}$$

$$-\overset{|}{\underset{|}{C}}{}^{\delta+}\text{----CH}_2{=}\text{CHCH}_2\text{SiR}_3 \longrightarrow -\overset{|}{\underset{|}{C}}\text{–CH}_2\overset{+}{\text{C}}\text{HCH}_2\text{SiR}_3 \longrightarrow -\overset{|}{\underset{|}{C}}\text{–CH}_2\text{CH{=}CH}_2$$

Attack on alkenylsilanes takes place at the α carbon and results in replacement of the silicon substituent by the electrophile. Attack on allylic groups is at the γ carbon and results in replacement of the silicon substituent and an allylic shift of the double bond. The crucial influence on the reactivity pattern in both cases is the *very high stabilization that silicon provides for carbocationic character at the β-carbon atom.*

48. J. L. Speier, *Adv. Organomet. Chem.* **17**, 407 (1979); E. Lukenvics, *Russ. Chem. Rev.* (*Engl. transl.*), **46**, 264 (1977).
49. A. Onopchenko and E. T. Sabourin, *J. Org. Chem.* **52**, 4118 (1987); H. M. Dickens, R. N. Hazeldine, A. P. Mather, and R. V. Parish, *J. Organomet. Chem.* **161**, 9 (1978); A. J. Cornish and M. F. Lappert, *J. Organomet. Chem.* **271**, 153 (1984).
50. T. G. Selin and R. West, *J. Am. Chem. Soc.* **84**, 1863 (1962).
51. R. A. Benkeser, S. Dunny, G. S. Li, P. G. Nerlekar, and S. D. Work, *J. Am. Chem. Soc.* **90**, 1871 (1968).

This stabilization is attributed primarily to a hyperconjugation with the C—Si bond.[52]

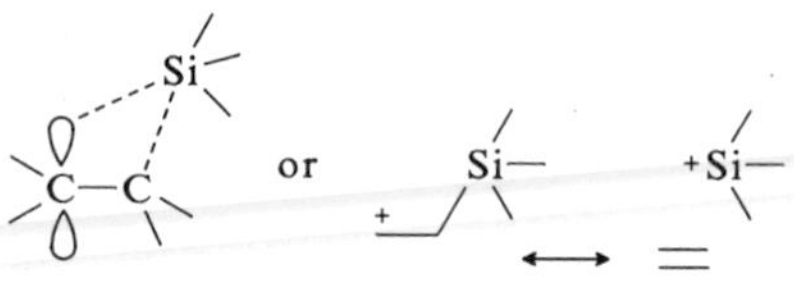

The most useful electrophiles from a synthetic standpoint are carbonyl compounds, iminium ions, and electrophilic alkenes.

The nucleophilic reactivity of allylic silanes can also be induced by fluoride ion. Fluoride adds at silicon to form a hypervalent structure with enhanced nucleophilicity.

$$CH_2{=}CHCH_2SiR_3 + F^- \rightarrow CH_2{=}CHCH_2\overset{|}{\underset{/\backslash}{Si^-}}{-}F$$

Most reactions of alkenylsilanes require strong carbon electrophiles, and Lewis acid catalysts are often involved. Reaction with acid chlorides is catalyzed by aluminum chloride or stannic chloride.[53]

$$RCH{=}CHSi(CH_3)_3 + RCOCl \xrightarrow[\text{or } SnCl_4]{AlCl_3} RCH{=}CH\overset{O}{\overset{\|}{C}}R$$

Titanium tetrachloride induces reaction with dichloromethyl methyl ether.[54]

$$RCH{=}CHSi(CH_3)_3 + Cl_2CHOCH_3 \xrightarrow{TiCl_4} RCH{=}CHCH{=}O$$

Similar conditions are used to effect reactions of allylsilanes with acid halides.[55]

$$Ph\overset{O}{\overset{\|}{C}}Cl + CH_2{=}CHCH_2Si(CH_3)_3 \xrightarrow{AlCl_3} Ph\overset{O}{\overset{\|}{C}}CH_2CH{=}CH_2$$

The reactions of allylic silanes with carbonyl compounds also require the use of Lewis acid catalysts such as $TiCl_4$ and BF_3.[56]

$$CH_2{=}CHCH_2SiR_3 + R_2C{=}O \xrightarrow[\text{or } BF_3]{TiCl_4} R_2\overset{OH}{\overset{|}{C}}CH_2CH{=}CH_2$$

52. S. G. Wierschke, J. Chandrasekhar, and W. L. Jorgensen, *J. Am. Chem. Soc.* **107**, 1496 (1985); J. B. Lambert, G. Wang, R. B. Finzel, and D. H. Teramura, *J. Am. Chem. Soc.* **109**, 7838 (1987).
53. I. Fleming and A. Pearce, *J. Chem. Soc., Chem. Commun.*, 633 (1975); W. E. Fristad, D. S. Dime, T. R. Bailey, and L. A. Paquette, *Tetrahedron Lett.*, 1999 (1979).
54. K. Yamamoto, O. Nunokawa, and J. Tsumi, *Synthesis*, 721 (1977).
55. J.-P. Pillot, G. Déléris, J. Dunoguès, and R. Calas, *J. Org. Chem.* **44**, 3397 (1979); R. Calas, J. Dunoguès, J.-P. Pillot, C. Biran, F. Pisciotti, and B. Arreguy, *J. Organomet. Chem.* **85**, 149 (1975).
56. A. Hosomi and H. Sakurai, *Tetrahedron Lett.*, 1295 (1976).

These reactions involve activation of the carbonyl group by the Lewis acid. A nucleophile, either a ligand from the Lewis acid or from the solution, assists in the desilylation step.

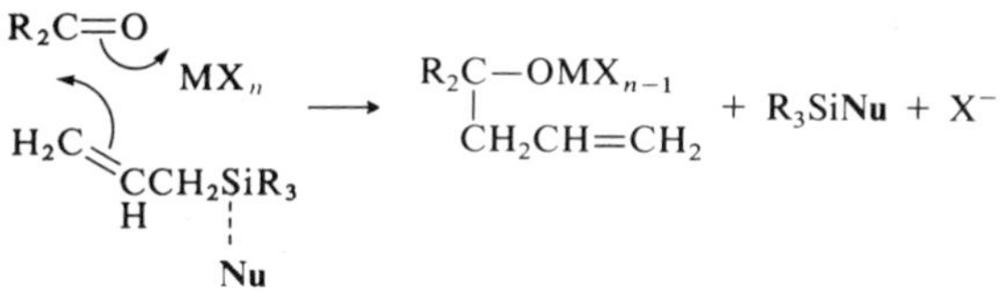

$$R_2C{=}O \cdots MX_n + H_2C{=}CHCH_2SiR_3 \;(\cdots Nu) \longrightarrow R_2C(OMX_{n-1})CH_2CH{=}CH_2 + R_3SiNu + X^-$$

While this reaction is formally analogous to the addition of allylboranes to carbonyl derivatives, it does not appear to occur through a cyclic transition state. This is because, in contrast to the boranes, the silicon in allylic silanes has no Lewis acid character and would not be expected to coordinate at the carbonyl oxygen. The stereochemistry of addition of allylic silanes to carbonyl compounds is consistent with an acyclic transition state. Both the *E*- and *Z*-stereoisomers of 2-butenyl(trimethyl)silane react with aldehydes to give the product in which the newly formed hydroxyl group is *syn* to the methyl substituent.[57] The preferred orientation of approach by the silane minimizes interaction between the aldehyde substituent R and the terminal methyl group.

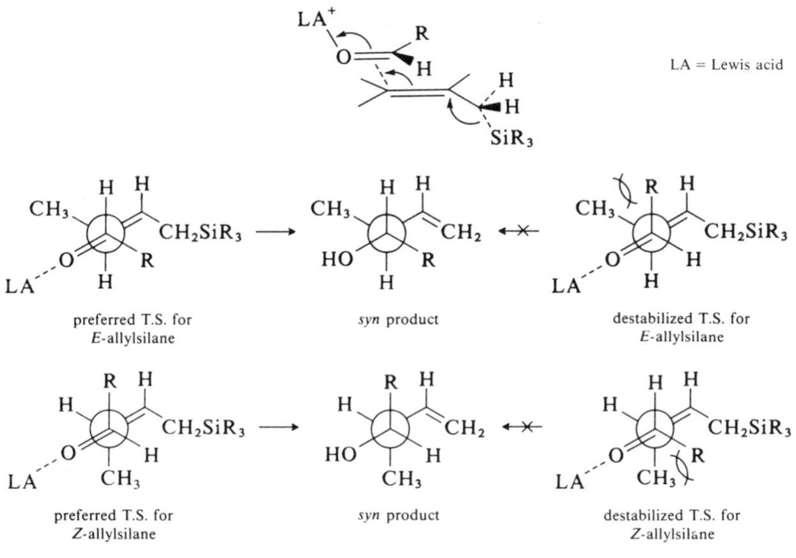

LA = Lewis acid

preferred T.S. for *E*-allylsilane — *syn* product — destabilized T.S. for *E*-allylsilane

preferred T.S. for *Z*-allylsilane — *syn* product — destabilized T.S. for *Z*-allylsilane

When chiral aldehydes such as **4** are used, there is a modest degree of diastereoselectivity in the direction predicted by Cram's rule.

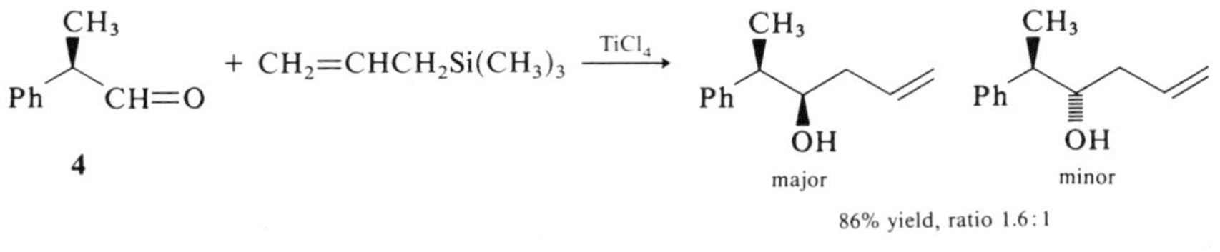

$$\textbf{4} + CH_2{=}CHCH_2Si(CH_3)_3 \xrightarrow{TiCl_4} \text{major} + \text{minor}$$

86% yield, ratio 1.6 : 1

57. T. Hayashi, K. Kabeta, I. Hamachi, and M. Kumada, *Tetrahedron Lett.* **24**, 2865 (1983).

Aldehydes with donor substituents that can form a chelate with the Lewis acid catalyst react by approach from the less hindered side of the chelate structure. The best electrophile in this case is $SnCl_4$.[58]

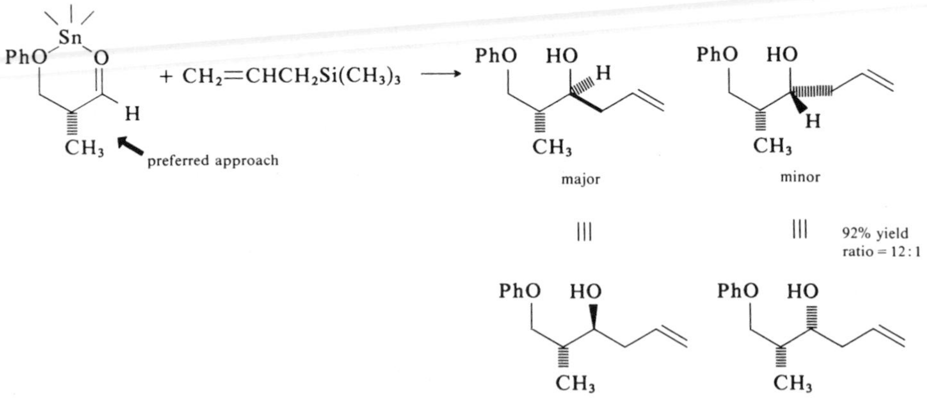

Sn PhO O H CH_3 preferred approach + $CH_2{=}CHCH_2Si(CH_3)_3$ ⟶

PhO HO H CH_3 major PhO HO H CH_3 minor

92% yield
ratio = 12 : 1

PhO HO CH_3 PhO HO CH_3

Carbon electrophiles can also be generated from acetals. Trimethylsilyl iodide causes addition to occur.[59] The trimethylsilyl iodide can be used in catalytic quantity because it is regenerated by recombination of iodide with silicon in the desilylation step.

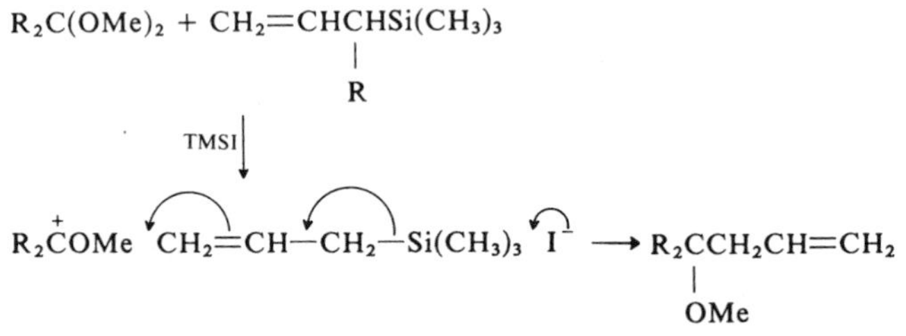

$$R_2C(OMe)_2 + CH_2{=}CHCH(R)Si(CH_3)_3$$

TMSI

$$R_2\overset{+}{C}OMe \;\; CH_2{=}CH{-}CH_2{-}Si(CH_3)_3 \;\; I^- \longrightarrow R_2C(OMe)CH_2CH{=}CH_2$$

Application of this reaction to carbohydrate derivatives has been used for the extension of the carbon chain.[60]

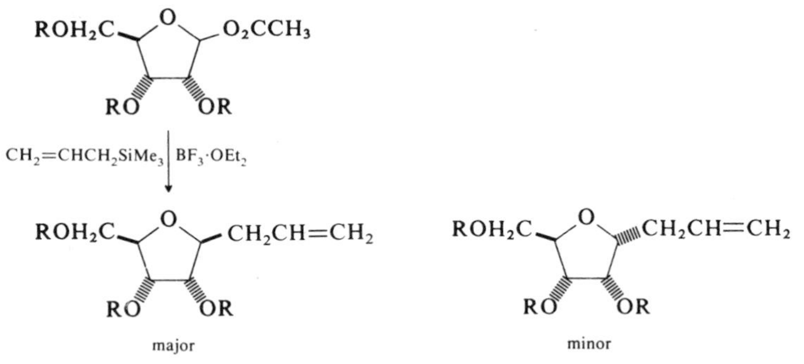

ROH_2C O O_2CCH_3 RO OR

$CH_2{=}CHCH_2SiMe_3$ $BF_3 \cdot OEt_2$

ROH_2C O $CH_2CH{=}CH_2$ RO OR major

ROH_2C O $CH_2CH{=}CH_2$ RO OR minor

58. C. H. Heathcock, S. Kiyooka, and T. Blumenkopf, *J. Org. Chem.* **49**, 4214 (1984).
59. H. Sakurai, K. Sasaki, and A. Hosmoni, *Tetrahedron Lett.* **22**, 745 (1981).
60. A. P. Kozikowski, K. Sorgi, B. C. Wang, and Z. Xu, *Tetrahedron Lett.* **24**, 1563 (1983).

Reaction of allylic silanes with enantiomerically pure 1,3-dioxanes has been found to proceed with high enantioselectivity. The observed stereochemistry can be explained in terms of concerted attack on the most stable acetal conformer.[61]

$$\text{(1,3-dioxane acetal, R)} + \text{CH}_2\text{=CHCH}_2\text{SiMe}_3 \longrightarrow \text{(ether product, OH)} \xrightarrow[2)\ ^-\text{OH}]{1)\ \text{oxid.}} \text{H–C(R)(OH)CH}_2\text{CH=CH}_2$$

The homoallylic alcohol can be liberated by oxidation followed by base-catalyzed β elimination. The alcohols obtained in this way are formed in $70 \pm 5\%$ enantiomeric excess. A similar reaction occurs with alkynylsilanes to give propargylic alcohols in 70–90% enantiomeric excess.[62]

$$\text{(2-R-4,6-dimethyl-1,3-dioxane)} + \text{R}'\text{C}\equiv\text{CSi(CH}_3)_3 \xrightarrow{\text{TiCl}_4} \xrightarrow[2)\ ^-\text{OH}]{1)\ \text{oxid}} \text{RCH(OH)C}\equiv\text{CR}'$$

Scheme 9.7 gives some additional examples of Lewis acid mediated reactions of allylic silanes with aldehydes and acetals.

Reaction of allylic silanes with aldehydes and ketones can also be induced by fluoride ion, which is usually supplied by the THF-soluble salt tetrabutylammonium fluoride (TBAF). These reactions proceed via a pentavalent fluoride adduct which transfers a nucleophilic allylic anion.[63] Unsymmetrical allylic anions generated in this way react with ketones at their less substituted terminus.

$$\text{CH}_2\text{=CHCH}_2\text{Si(CH}_3)_3 + \text{F}^- \rightarrow \text{CH}_2\text{=CHCH}_2\bar{\text{Si}}\text{F(CH}_3)_3 \xrightarrow{\text{RCH=O}} \text{RCH(OH)CH}_2\text{CH=CH}_2$$

$$\text{(CH}_3)_2\text{C=CHCH}_2\text{Si(CH}_3)_3 + \text{Ph}_2\text{C=O} \xrightarrow{\text{TBAF}} \xrightarrow{\text{H}_2\text{O}} \text{(CH}_3)_2\text{C=CHCH}_2\text{C(OH)Ph}_2 \quad (87\%)$$

An allylic silane which serves as a reagent for introduction of isoprenoid structures has been developed.[64]

$$\text{(CH}_3)_2\text{C=CHCH=O} + \text{(CH}_3)_3\text{SiCH}_2\text{C(=CH}_2\text{)CH=CH}_2 \xrightarrow{\text{R}_4\text{N}^+\text{F}^-} \text{(CH}_3)_2\text{C=CHCH(OH)CH}_2\text{C(=CH}_2\text{)CH=CH}_2 \quad (70\%)$$

61. P. A. Bartlett, W. S. Johnson, and J. D. Elliott, *J. Am. Chem. Soc.* **105**, 2088 (1983).
62. W. S. Johnson, R. Elliott, and J. D. Elliott, *J. Am. Chem. Soc.* **105**, 2904 (1983).
63. A. Hosomi, A. Shirahata, and H. Sakurai, *Tetrahedron Lett.*, 3043 (1978).
64. A. Hosomi, Y. Araki, and H. Sakurai, *J. Org. Chem.* **48**, 3122 (1983).

Fluoride-induced desilylation has also been used to effect ring closures.[65]

$$\text{bicyclic ketone with } CH_2CCH_2Si(CH_3)_3 (=CH_2) \text{ and } SO_2Ph \xrightarrow{RyN^+F^-} \text{tricyclic alcohol with } =CH_2 \text{ and } SO_2Ph \; (94\%)$$

Iminium ions are reactive electrophiles toward both alkenyl and allylic silanes. Useful techniques for closing nitrogen-containing rings have been based on *in situ* generation of iminium ions from amines and formaldehyde.[66]

$$PhCH_2NHCH_2CH_2C(=CH_2)CH_2Si(CH_3)_3 \xrightarrow[H_2C=O]{CF_3CO_2H} PhCH_2\overset{+}{N}(=CH_2)CH_2CH_2C(=CH_2)CH_2Si(CH_3)_3 \longrightarrow PhCH_2N\langle\text{piperidine}\rangle{=}CH_2 \; (73\%)$$

When primary amines are employed, the initially formed 3-butenylamine undergoes a further reaction, forming a 4-piperidinol.[67]

$$PhCH_2\overset{+}{N}H_3 + CH_2{=}CHCH_2Si(CH_3)_3 + CH_2{=}O \longrightarrow PhCH_2N\langle\text{piperidine}\rangle{-}OH$$

Reactions of this type can also be observed with 4-(trimethylsilyl)alkenylamines.[68]

$$R'NHCH_2CH_2CH{=}C(Si(CH_3)_3)R^3 \xrightarrow[H^+]{CH_2=O} \text{1-}R'\text{-3-}R^3\text{-1,2,5,6-tetrahydropyridine}$$

Mechanistic investigation in this case has shown that there is an equilibration between an alkenyl silane and an allylic silane by a rapid 3,3-sigmatropic process. The cyclization occurs through the more reactive allylic silane.

$$R'NH(CH_2)_2CH{=}C(Si(CH_3)_3)R^3 + CH_2{=}O \longrightarrow R'\overset{+}{N}(=CH_2)CH_2CH_2CH{=}C(Si(CH_3)_3)R^3 \rightleftharpoons R'\overset{+}{N}{=}CH_2 \cdots (CH_3)_3Si \; R^3 \longrightarrow R'\overset{+}{N}\langle\text{ring}\rangle R^3$$

Section C of Scheme 9.7 gives some examples of cyclization involving iminium ions as electrophiles in reactions with unsaturated silanes.

65. B. M. Trost and J. E. Vincent, *J. Am. Chem. Soc.* **102**, 5680 (1980); B. M. Trost and D. P. Curran, *J. Am. Chem. Soc.* **103**, 7380 (1981).
66. P. A. Grieco and W. F. Fobare, *Tetrahedron Lett.* **27**, 5067 (1986).
67. S. D. Larsen, P. A. Grieco, and W. F. Fobare, *J. Am. Chem. Soc.* **108**, 3512 (1986).
68. C. Flann, T. C. Malone, and L. E. Overman, *J. Am. Chem. Soc.* **109**, 6097 (1987).

Scheme 9.7. Reactions Involving Electrophilic Attack on Alkenyl and Allylic Silanes

A. Reactions with Carbonyl Compounds

1[a]

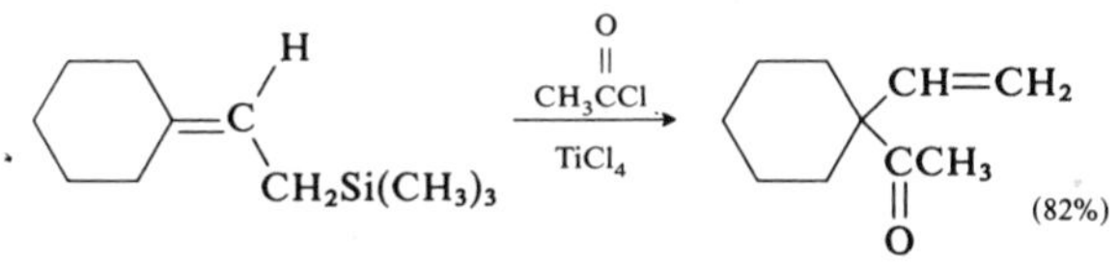

(cyclohexylidene)C(H)$CH_2Si(CH_3)_3$ $\xrightarrow[TiCl_4]{CH_3CCl(=O)}$ 1-(CH=CH_2)-1-(CCH_3(=O))cyclohexane (82%)

2[b]

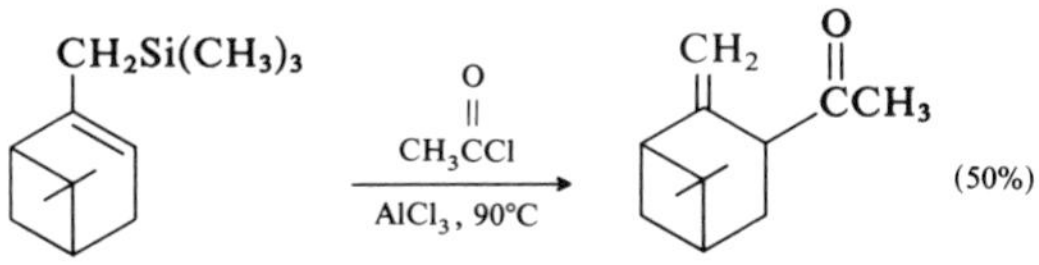

$CH_2Si(CH_3)_3$ (pinene) $\xrightarrow[AlCl_3,\ 90°C]{CH_3CCl(=O)}$ CH_2 / CCH_3(=O) (50%)

3[c]

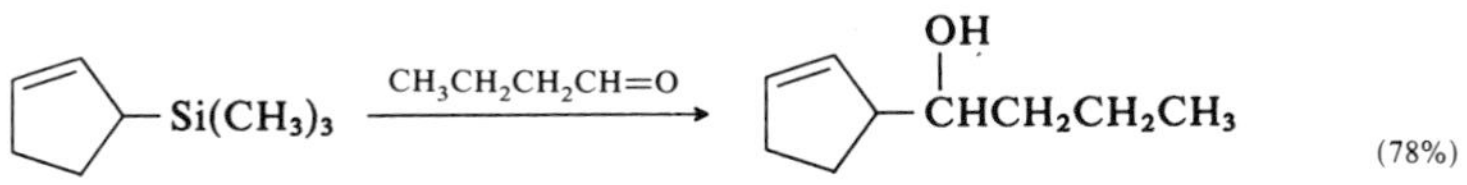

(cyclopentenyl)–$Si(CH_3)_3$ $\xrightarrow{CH_3CH_2CH_2CH=O}$ (cyclopentenyl)–$CHCH_2CH_2CH_3$(OH) (78%)

4[d]

CH_2=$CHCH_2Si(CH_3)_3$ + O=CH–(furanose)–OCH_3 (O, O acetonide) $\xrightarrow{BF_3}$ CH_2=$CHCH_2$–CH(OH)–(furanose)–OCH_3 (80%)

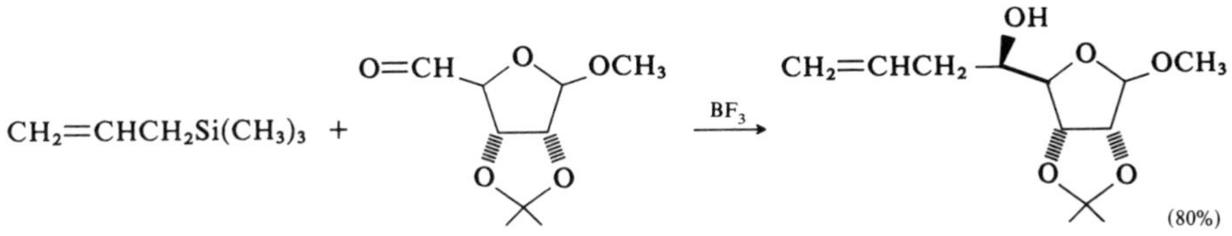

B. Reactions with Acetals and Related Compounds

5[e]

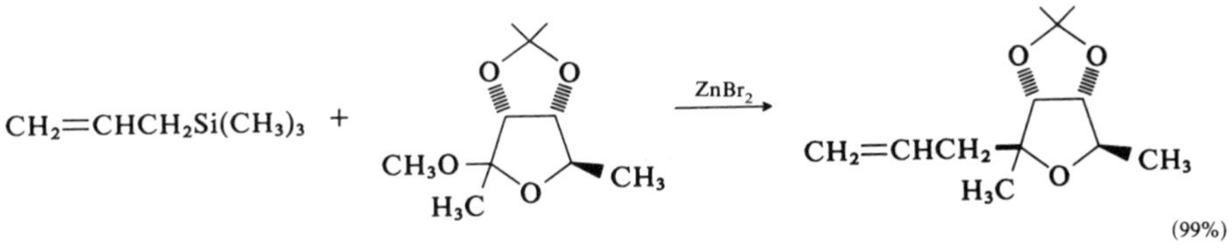

CH_2=$CHCH_2Si(CH_3)_3$ + CH_3O, H_3C, O, CH_3 (acetonide) $\xrightarrow{ZnBr_2}$ CH_2=$CHCH_2$, H_3C, O, CH_3 (99%)

6[f]

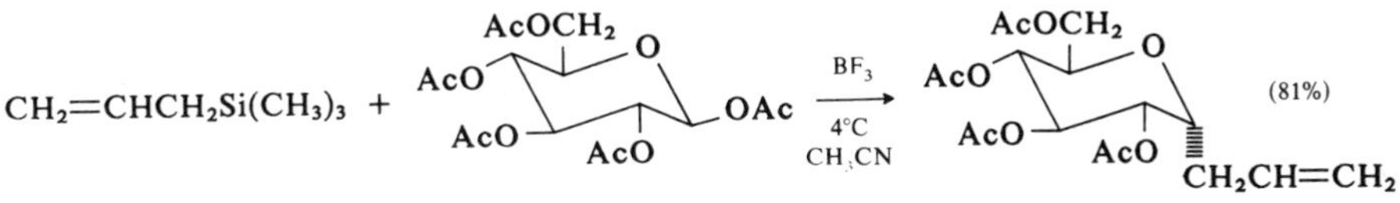

CH_2=$CHCH_2Si(CH_3)_3$ + $AcOCH_2$, AcO, AcO, AcO, O, OAc $\xrightarrow[4°C,\ CH_3CN]{BF_3}$ $AcOCH_2$, AcO, AcO, AcO, O, CH_2CH=CH_2 (81%)

7[g]

CH_2=$CHCH_2Si(CH_3)_3$ + $ROCH_2$, OR, RO, RO, O, OAc, OCH_3 (R = CH_2Ph) $\xrightarrow{Me_3SiO_3SCF_3}$ $ROCH_2$, OR, RO, RO, O, CH_2CH=CH_2 (87%)

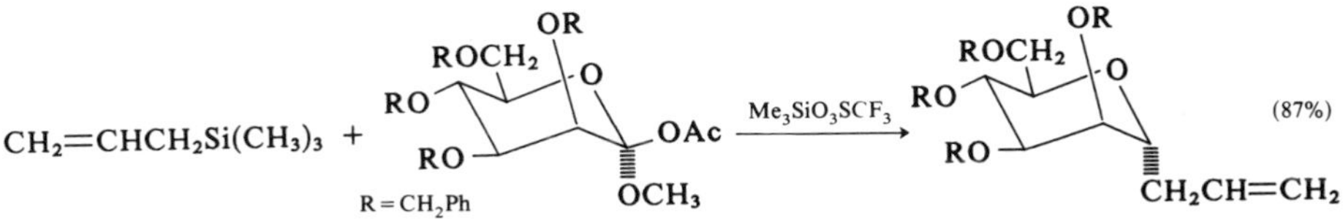

C. Iminium Ions

8[h]

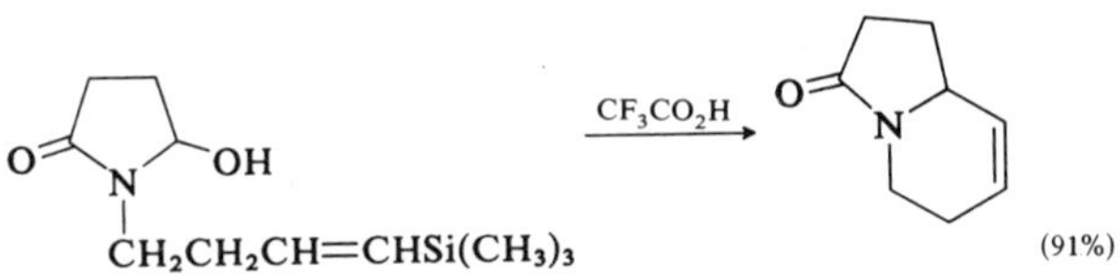

O=(pyrrolidinone)–OH, N–CH_2CH_2CH=$CHSi(CH_3)_3$ $\xrightarrow{CF_3CO_2H}$ O=(bicyclic lactam, N) (91%)

Scheme 9.7—*continued*

9[i]

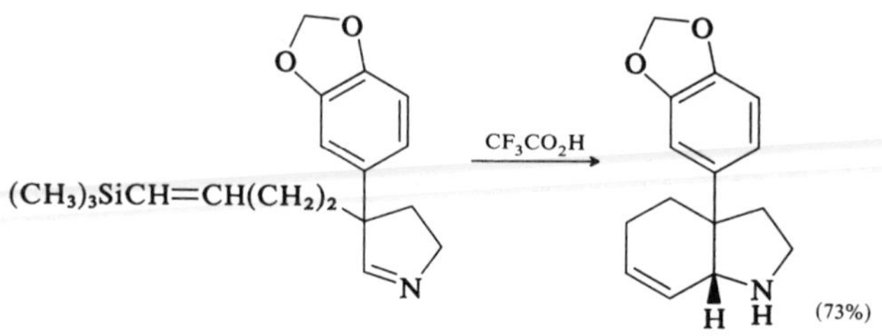

$(CH_3)_3SiCH{=}CH(CH_2)_2$ —— CF_3CO_2H ——→ (73%)

a. I. Fleming and I. Paterson, *Synthesis*, 446 (1979).
b. J. P. Pillot, G. Déléris, J. Dunoguès, and R. Calas, *J. Org. Chem.* **44**, 3397 (1979).
c. I. Ojima, M. Kumagai, and Y. Miyazawa, *Tetrahedron Lett.*, 1385 (1977).
d. S. Danishefsky and M. De Ninno, *Tetrahedron Lett.* **26**, 823 (1985).
e. H. Suh and C. S. Wilcox, *J. Am. Chem. Soc.* **110**, 470 (1988).
f. A. Giannis and K. Sanshoff, *Tetrahedron Lett.* **26**, 1479 (1985).
g. A. Hosomi, Y. Sakata, and H. Sakurai, *Tetrahedron Lett.* **25**, 2383 (1984).
h. C. Flann, T. C. Malone, and L. E. Overman, *J. Am. Chem. Soc.* **109**, 6097 (1987).
i. L. E. Overman and R. M. Burk, *Tetrahedron Lett.* **25**, 5739 (1984).

Allylic silanes act as nucleophilic species toward α,β-unsaturated ketones in the presence of Lewis acids such as $TiCl_4$.[69]

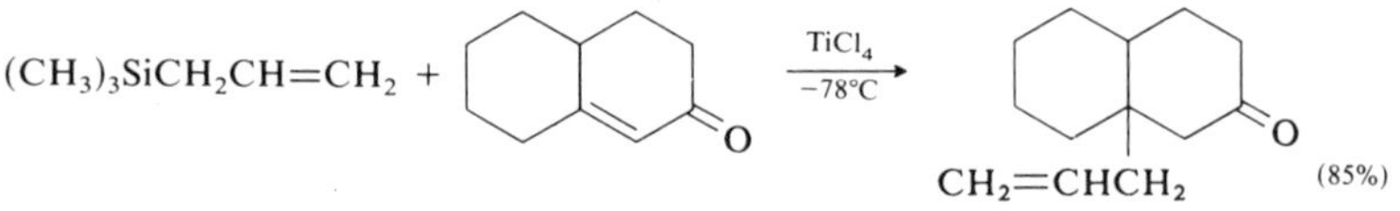

$(CH_3)_3SiCH_2CH{=}CH_2$ + [enone] $\xrightarrow[-78^\circ C]{TiCl_4}$ [product] $CH_2{=}CHCH_2$ (85%)

The stereochemistry of this reaction in cyclic systems is in accord with expectations for stereoelectronic control. The allylic group approaches along a trajectory that is appropriate for interaction with the LUMO of the conjugated system.[70]

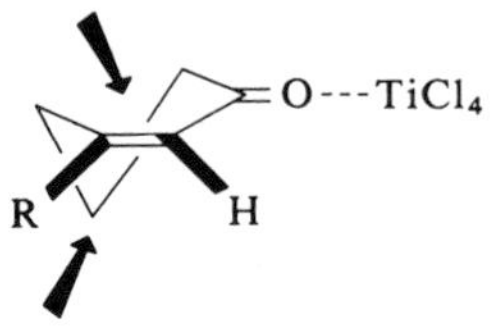

O---$TiCl_4$ R H

The stereoselectivity then depends on the conformation of the enone and the location of substituents which establish a steric bias for one of the two potential directions of approach. In the ketone **5**, the preferred approach is from the β side, since this permits maintaining a chair conformation as the reaction proceeds.[71]

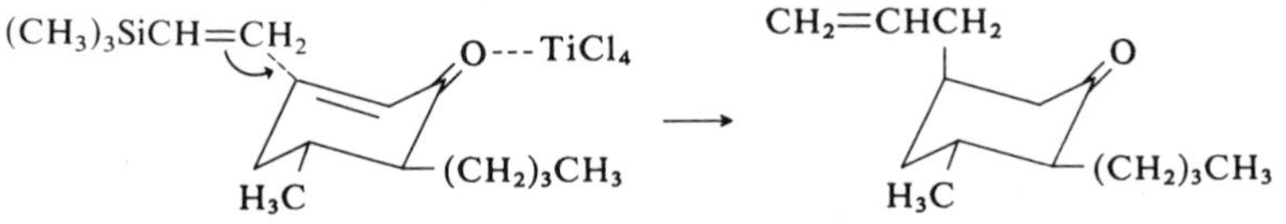

$(CH_3)_3SiCH{=}CH_2$ O---$TiCl_4$ H_3C $(CH_2)_3CH_3$ **5** ——→ $CH_2{=}CHCH_2$ O H_3C $(CH_2)_3CH_3$

69. A. Hosomi and H. Sakurai, *J. Am. Chem. Soc.* **99**, 1673 (1977).
70. T. A. Blumenkopf and C. H. Heathcock, *J. Am. Chem. Soc.* **105**, 2354 (1983).
71. W. R. Roush and A. E. Walts, *J. Am. Chem. Soc.* **106**, 721 (1984).

Intramolecular conjugate addition of allylic silanes can also be used to construct new rings, as illustrated by entry 6 in Scheme 9.8.

Scheme 9.8. Reactions of Silanes with α,β-Unsaturated Carbonyl Compounds

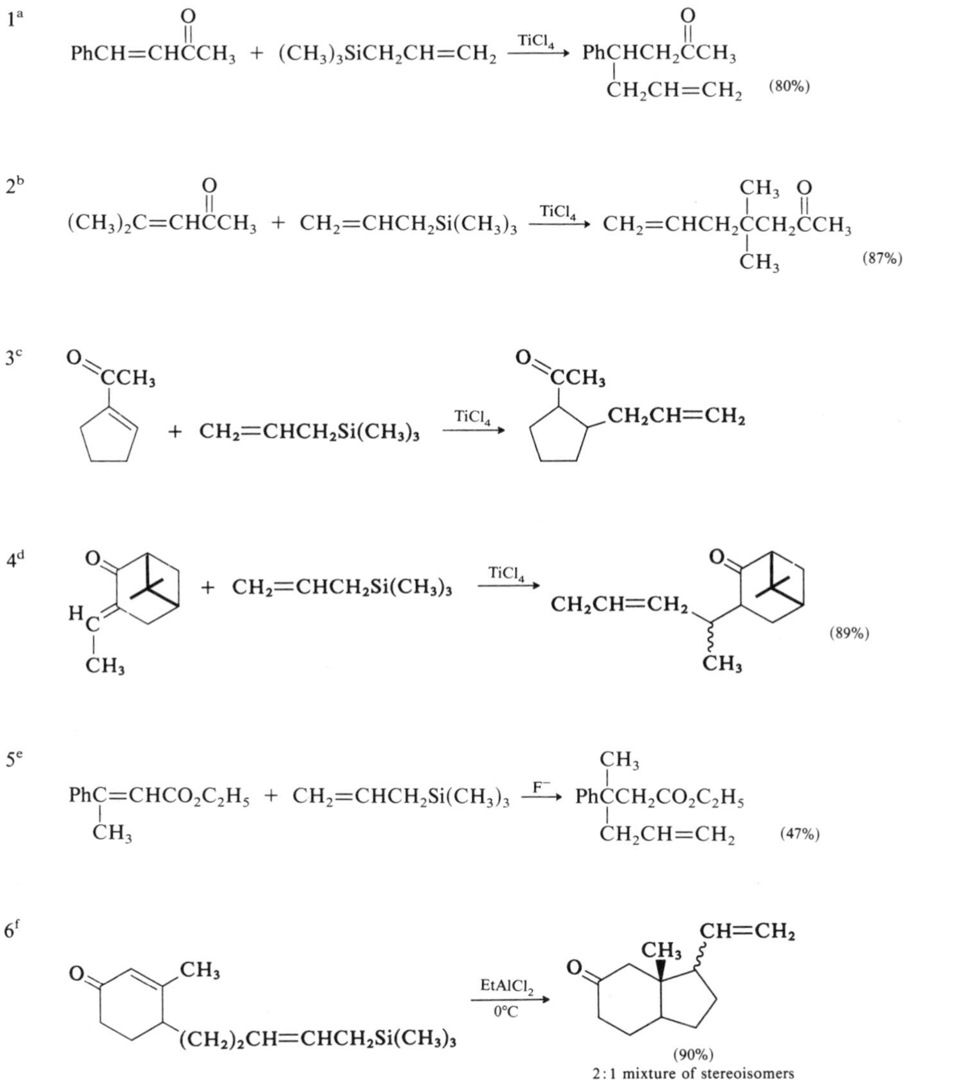

1[a]

$$PhCH{=}CHC(O)CH_3 + (CH_3)_3SiCH_2CH{=}CH_2 \xrightarrow{TiCl_4} PhCH(CH_2CH{=}CH_2)CH_2C(O)CH_3 \quad (80\%)$$

2[b]

$$(CH_3)_2C{=}CHC(O)CH_3 + CH_2{=}CHCH_2Si(CH_3)_3 \xrightarrow{TiCl_4} CH_2{=}CHCH_2C(CH_3)_2CH_2C(O)CH_3 \quad (87\%)$$

3[c]

$$\text{1-acetylcyclopentene} + CH_2{=}CHCH_2Si(CH_3)_3 \xrightarrow{TiCl_4} \text{1-acetyl-2-}(CH_2CH{=}CH_2)\text{cyclopentane}$$

4[d]

$$\text{ethylidene bicyclic ketone} + CH_2{=}CHCH_2Si(CH_3)_3 \xrightarrow{TiCl_4} CH_2CH{=}CH_2\text{-CH(}CH_3\text{)-substituted bicyclic ketone} \quad (89\%)$$

5[e]

$$PhC(CH_3){=}CHCO_2C_2H_5 + CH_2{=}CHCH_2Si(CH_3)_3 \xrightarrow{F^-} PhC(CH_3)(CH_2CH{=}CH_2)CH_2CO_2C_2H_5 \quad (47\%)$$

6[f]

$$\text{3-methyl-4-}[(CH_2)_2CH{=}CHCH_2Si(CH_3)_3]\text{cyclohex-2-enone} \xrightarrow[0^\circ C]{EtAlCl_2} \text{bicyclic ketone with } CH_3 \text{ and } CH{=}CH_2$$

(90%)
2:1 mixture of stereoisomers

a. H. Sakurai, A. Hosomi, and J. Hayashi, *Org. Synth.* **62**, 86 (1984).
b. D. H. Hua, *J. Am. Chem. Soc.* **108**, 3835 (1986).
c. H. O. House, P. C. Gaa, and D. VanDerveer, *J. Org. Chem.* **48**, 1661 (1983).
d. T. Yanami, M. Miyashita, and A. Yoshikoshi, *J. Org. Chem.* **45**, 607 (1980).
e. G. Majetich, A. Casares, D. Chapman, and M. Behnke, *J. Org. Chem.* **51**, 1745 (1986).
f. D. Schinzer, S. Sólyom, and M. Becker, *Tetrahedron Lett.* **26**, 1831 (1985).

Conjugate addition can also be carried out by fluoride-mediated desilylation. A variety of α,β-unsaturated esters and amides have been found to undergo this reaction.[72]

$$PhCH=CHCO_2C_2H_5 + CH_2=CHCH_2Si(CH_3)_3 \xrightarrow{F^-} PhCH(CH_2CH=CH_2)CH_2CO_2C_2H_5$$

With unsaturated aldehydes, 1,2-addition occurs; with ketones, both the 1,2- and 1,4-products are formed.

$$PhCH=CHCOCH_3 + CH_2=CHCH_2Si(CH_3)_3 \xrightarrow[HMPA]{TBF} \underset{(25\%)}{PhCH(CH_2CH=CH_2)CH_2CCH_3(=O)} + \underset{(50\%)}{PhCH=CHC(OH)(CH_3)CH_2CH=CH_2}$$

9.3. Organotin Compounds

9.3.1. Synthesis of Organostannanes

The readily available organotin compounds include trisubstituted tin hydrides (stannanes) and chlorides. The tin hydrides can be added to carbon–carbon double and triple bonds. The reaction is normally carried out by a radical chain process.[73] Addition is facilitated by the presence of radical-stabilizing substituents.

$$(C_2H_5)_3SnH + CH_2=CHCN \rightarrow (C_2H_5)_3SnCH_2CH_2CN \qquad \text{Ref. 74}$$

$$(C_4H_9)_3SnH + CH_2=C(CO_2CH_3)(Ph) \rightarrow (C_4H_9)_3SnCH_2CH(Ph)CO_2CH_3 \qquad \text{Ref. 75}$$

With terminal alkynes, the stannyl group is added at the terminal carbon and the Z-stereoisomer is initially formed but is thermally isomerized to the E-isomer.[76]

$$HC\equiv CCH_2OTHP \xrightarrow[AIBN]{(C_4H_9)_3SnH} \underset{Z}{(C_4H_9)_3Sn(H)C=C(H)CH_2OTHP} \rightarrow \underset{E}{(C_4H_9)_3Sn(H)C=C(H)CH_2OTHP}$$

72. G. Majetich, A. Casares, D. Chapman, and M. Behnke, *J. Org. Chem.* **51**, 1745 (1986).
73. H. G. Kuivila, *Adv. Organomet. Chem.* **1**, 47 (1964).
74. A. J. Leusinsk and J. G. Noltes, *Tetrahedron Lett.*, 335 (1966).
75. I. Fleming and C. J. Urch, *Tetrahedron Lett.* **24**, 4591 (1983).
76. E. J. Corey and R. H. Wollenberg, *J. Org. Chem.* **40**, 2265 (1975).

The reaction with internal acetylenes can lead to both a mixture of regioisomers and a mixture of stereoisomers.[77] Palladium-catalyzed procedures have also been developed for addition of stannanes to alkynes.[78]

$$PhC{\equiv}CCH_3 + (C_4H_9)_3SnH \xrightarrow{PdCl_2\text{-}PPh_3} (C_4H_9)_3Sn(Ph)C{=}C(CH_3)H$$

The other major route for synthesis of stannanes is reaction of an organometallic reagent with a trisubstituted halostannane. This is the normal route for preparation of arylstannanes.

$$CH_3O\text{-}C_6H_4\text{-}MgBr + BrSn(CH_3)_3 \longrightarrow CH_3O\text{-}C_6H_4\text{-}Sn(CH_3)_3 \qquad \text{Ref. 79}$$

$$H(Ph)C{=}C(Li)OCH_3 + (CH_3)_3SnCl \rightarrow H(Ph)C{=}C(Sn(CH_3)_3)OCH_3 \qquad \text{Ref. 80}$$

Deprotonated trialkylstannanes are potent nucleophiles. Addition to carbonyl groups or iminium intermediates provides routes to α-alkoxy- and α-aminoalkylstannanes.

$$RCH{=}O + (C_4H_9)_3SnLi \rightarrow RCH(O^-)Sn(C_4H_9)_3 \xrightarrow{R'X} RCH(OR')Sn(C_4H_9)_3 \qquad \text{Ref. 81}$$

$$R_2NCH_2SPh + (C_4H_9)_3SnLi \rightarrow R_2NCH_2Sn(C_4H_9)_3 \qquad \text{Ref. 82}$$

9.3.2. Carbon–Carbon Bond-Forming Reactions

As with the silanes, some of the most useful procedures for synthetic applications of stannanes involve electrophilic attack on alkenyl and allylic derivatives. The stannanes are more reactive than the corresponding silanes because there is more anionic character on carbon in the C—Sn bond and it is a weaker bond. There are also a number of synthetic procedures in which organotin compounds act as carbanion donors in transition metal-catalyzed reactions. Organotin compounds are also used in free-radical reactions, as will be discussed in Chapter 10.

77. H. E. Ensley, R. R. Buescher, and K. Lee, *J. Org. Chem.* **47**, 404 (1982).
78. H. X. Zhang, F. Guibé, and G. Balavoine, *Tetrahedron Lett.* **29**, 619 (1988).
79. C. Eaborn, A. R. Thompson, and D. R. M. Walton, *J. Chem. Soc. C*, 1364 (1967); C. Eaborn, H. L. Hornfeld, and D. R. M. Walton, *J. Chem. Soc., B*, 1036 (1967).
80. J. A. Soderquist and G. J.-H. Hsu, *Organometallics* **1**, 830 (1982).
81. W. C. Still, *J. Am. Chem. Soc.* **100**, 1481 (1978).
82. D. J. Peterson, *J. Am. Chem. Soc.* **93**, 4027 (1971).

Tetrasubstituted organotin compounds are not sufficiently reactive to add directly to aldehydes and ketones, although reactions with aldehydes do occur with heating.

$$Cl-C_6H_4-CH{=}O + CH_2{=}CHCH_2Sn(C_2H_5)_3 \xrightarrow[4\ h]{100^\circ C} Cl-C_6H_4-CH(OSn(C_2H_5)_3)CH_2CH{=}CH_2 \quad (90\%)$$ Ref. 83

Use of Lewis acid catalysts allows allylic stannanes to react under mild conditions. As was the case with allylic silanes, a double-bond migration occurs.

$$PhCH{=}O + (CH_3)(H)C{=}C(H)(CH_2Sn(C_4H_9)_3) \xrightarrow{BF_3} PhCH(OH)CH(CH_3)CH{=}CH_2 \quad (92\%)$$ Ref. 84

Various allylstannyl halides can transfer allyl groups to carbonyl compounds. In this case, the reagent acts as both a Lewis acid and the source of the nucleophilic allyl group.

$$PhCH_2CH_2C(=O)CH_3 + (CH_2{=}CHCH_2)_2SnBr_2 \rightarrow PhCH_2CH_2C(OH)(CH_3)CH_2CH{=}CH_2$$ Ref. 85

The stannyl halides can be generated *in situ* by reactions of allylic halides with tin metal or with stannous halides.

$$PhCH{=}O + CH_2{=}CHCH_2I \xrightarrow{Sn} \xrightarrow{H_2O} PhCH(OH)CH_2CH{=}CH_2$$ Ref. 85

$$PhCH{=}CHCH{=}O + CH_2{=}CHCH_2I \xrightarrow{SnF_2} PhCH{=}CHCH(OH)CH_2CH{=}CH_2$$ Ref. 86

Reactions of tetrallylstannanes with aldehydes catalyzed by $SnCl_4$ also appear to involve a stannyl halide intermediate. It can be demonstrated by NMR spectroscopy that there is a rapid-redistribution of the allyl group.[87]

$$(CH_2{=}CHCH_2)_4Sn + SnCl_4 \rightarrow CH_2{=}CHCH_2SnCl_3 + (CH_2{=}CHCH_2)_3SnCl$$

The allylation reaction can be adapted to the synthesis of terminal dienes by using 1-bromo-3-iodopropene and stannous chloride.

$$PhCH{=}O + ICH_2CH{=}CHBr \xrightarrow{SnCl_2} PhCH(OH)CH(Br)CH{=}CH_2 \rightarrow PhCH{=}CHCH{=}CH_2$$ Ref. 88

83. K. König and W. P. Neumann, *Tetrahedron Lett.*, 495 (1967).
84. H. Yatagai, Y. Yamamoto, and K. Maruyama, *J. Am. Chem. Soc.* **102**, 4548 (1980); Y. Yamamoto, *Acc. Chem. Res.* **20**, 243 (1987).
85. T. Mukaiyama and T. Harada, *Chem. Lett.*, 1527 (1981).
86. T. Mukaiyama, T. Harada, and S. Shoda, *Chem. Lett.* 1507 (1980).
87. S. E. Denmark, T. Wilson, and T. M. Wilson, *J. Am. Chem. Soc.* **110**, 984 (1988).
88. J. Auge, *Tetrahedron Lett.* **26**, 753 (1985).

The elinination step is a reductive elimination of the type discussed in Section 6.10 of Part A. Excess stannous chloride acts as the reducing agent.

The stereoselectivity of addition to aldehydes and ketones has been of considerable interest. With benzaldehyde, the addition of 2-butenylstannanes catalyzed by BF_3 gives the *syn* isomer, irrespective of the stereochemistry of the butenyl group.[89]

$H_3C(H)C{=}C(H)CH_2Sn(C_4H_9)_3$ —PhCH=O→ Ph–CH(OH)–CH(CH_3)–CH=CH_2 ←PhCH=O— $H(H_3C)C{=}C(H)CH_2Sn(C_4H_9)_3$

This stereochemistry is the same as that observed for allylic silanes and can be interpreted in terms of an acyclic transition state in which the *anti* orientation of the aldehyde substituent and the methyl group is dictated by steric factors.

CH_2SnBu_3 ; H, R, H_3C, H ; F_3B–O Bu_3SnCH_2 ; H, R, H_3C, H ; F_3B–O

When $TiCl_4$ is used as the catalyst, the stereoselectivity depends on the order of addition of the reagents. When *E*-2-butenylstannane is added to a $TiCl_4$–aldehyde mixture, *syn* stereoselectivity is observed. When the aldehyde is added to a premixed solution of the 2-butenylstannane and $TiCl_4$, the *anti* isomer predominates.[90]

$TiCl_4$ + cyclohexyl–CH=O —add $CH_2{=}CHCH_2SnBu_3$→ cyclohexyl–CH(OH)–CH(CH_3)–CH=CH_2 (*syn*)

$TiCl_4$ + $CH_2{=}CHCH_2SnBu_3$ —add cyclohexyl–CH=O→ cyclohexyl–CH(OH)–CH(CH_3)–CH=CH_2 (*anti*)

The formation of the *anti* stereoisomer is attributed to involvement of a butenyltitanium intermediate formed by rapid exchange with the butenylstannane. Because of its Lewis acidity, this intermediate reacts through a cyclic transition state.

When an aldehyde subject to "chelation control" is used, the *syn* stereoisomer dominates with $TiCl_4$ or $MgBr_2$ as the Lewis acid.[91] Allylic stannanes with γ-oxygen

89. Y. Yamamoto, H. Yatagi, H. Ishihara, and K. Maruyama, *Tetrahedron* **40**, 2239 (1984).
90. G. E. Keck, D. E. Abbott, E. P. Boden, and E. J. Enholm, *Tetrahedron Lett.* **25**, 3927 (1984).
91. G. E. Keck and E. P. Boden, *Tetrahedron Lett.* **25**, 265 (1984); G. E. Keck, D. E. Abbott and M. R. Wiley, *Tetrahedron Lett.*, **28**, 139 (1987).

substituents have been used to build up polyoxygenated carbon chains. For example, **6** reacts with the stannane **7** to give a dioxygenated chain, with a high preference for the stereoisomer in which the two oxygen substituents are *anti*. This stereoselectivity is consistent with chelation control.

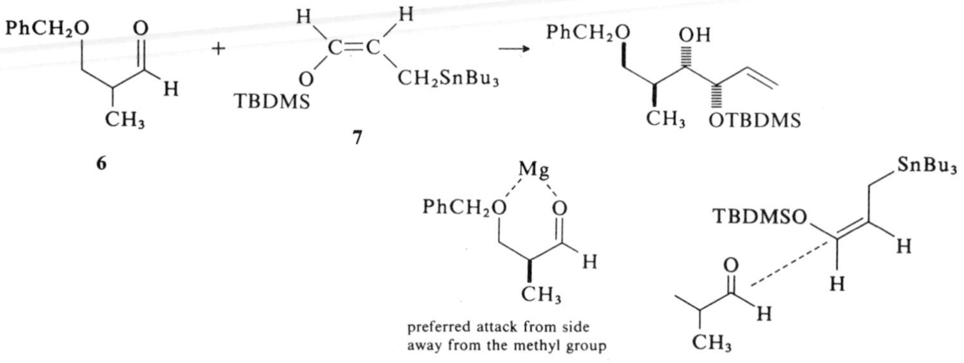

PhCH$_2$O, O, H, CH$_3$ (**6**) + H, H, C=C, O, TBDMS, CH$_2$SnBu$_3$ (**7**) → PhCH$_2$O, OH, CH$_3$, OTBDMS

Mg, PhCH$_2$O, O, H, CH$_3$ — preferred attack from side away from the methyl group

SnBu$_3$, TBDMSO, H, O, H, H, CH$_3$

Scheme 9.9 gives some other examples of Lewis acid catalyzed reactions of allylic stannanes with carbonyl compounds.

Lewis acid mediated ionization of acetals also generates electrophilic carbon intermediates which react readily with allylic stannanes.[92] Dithioacetals are activated by the sulfonium salt $[(CH_3)_2SSCH_3]^+BF_4^-$.[93]

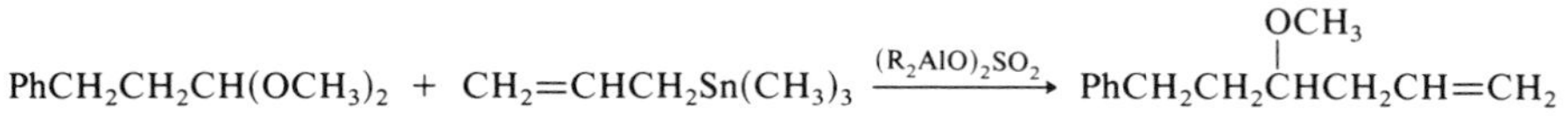

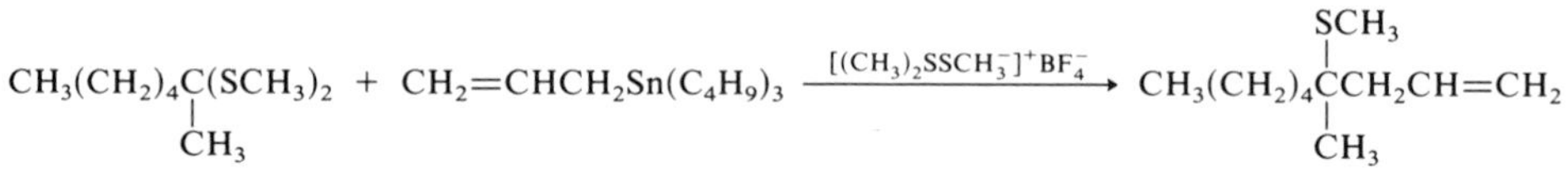

$$PhCH_2CH_2CH(OCH_3)_2 + CH_2{=}CHCH_2Sn(CH_3)_3 \xrightarrow{(R_2AlO)_2SO_2} PhCH_2CH_2\underset{}{\overset{OCH_3}{C}}HCH_2CH{=}CH_2$$

$$CH_3(CH_2)_4\underset{CH_3}{C}(SCH_3)_2 + CH_2{=}CHCH_2Sn(C_4H_9)_3 \xrightarrow{[(CH_3)_2SSCH_3]^+BF_4^-} CH_3(CH_2)_4\overset{SCH_3}{\underset{CH_3}{C}}CH_2CH{=}CH_2$$

Another broad class of carbon-carbon bond-forming reactions in which organostannanes participate are those involving transition metals, especially palladium. One general reaction is the cross coupling of organic halides with stannanes, which couples the organic group from the halide with one of the stannane ligands. The overall nature of such reactions involves a transfer of a carbon ligand from tin to palladium. The carbon-carbon bond formation proceeds by a reductive elimination step. The palladium is recycled through an oxidative addition reaction with the halide.[94]

$$R'{-}Pd{-}X + R_4Sn \rightarrow R'{-}Pd{-}R \rightarrow R{-}R' + Pd^0$$

$$Pd^0 + R'X \rightarrow R'{-}Pd{-}X$$

Details of the mechanism include specification of the oxidation state of palladium during the catalytic cycle and the timing of the ligand transfer from tin. Phosphines are normally present and presumably function as reversibly bound Pd ligands. The

92. A. Hosomi, H. Iguchi, M. Endo, and H. Sakurai, *Chem. Lett.*, 977 (1979).
93. B. M. Trost and T. Sato, *J. Am. Chem. Soc.* **107**, 719 (1985).
94. J. K. Stille, *Angew. Chem. Int. Ed. Engl.* **25**, 508 (1986).

Scheme 9.9. Reactions of Allylic Stannanes with Carbonyl Compounds

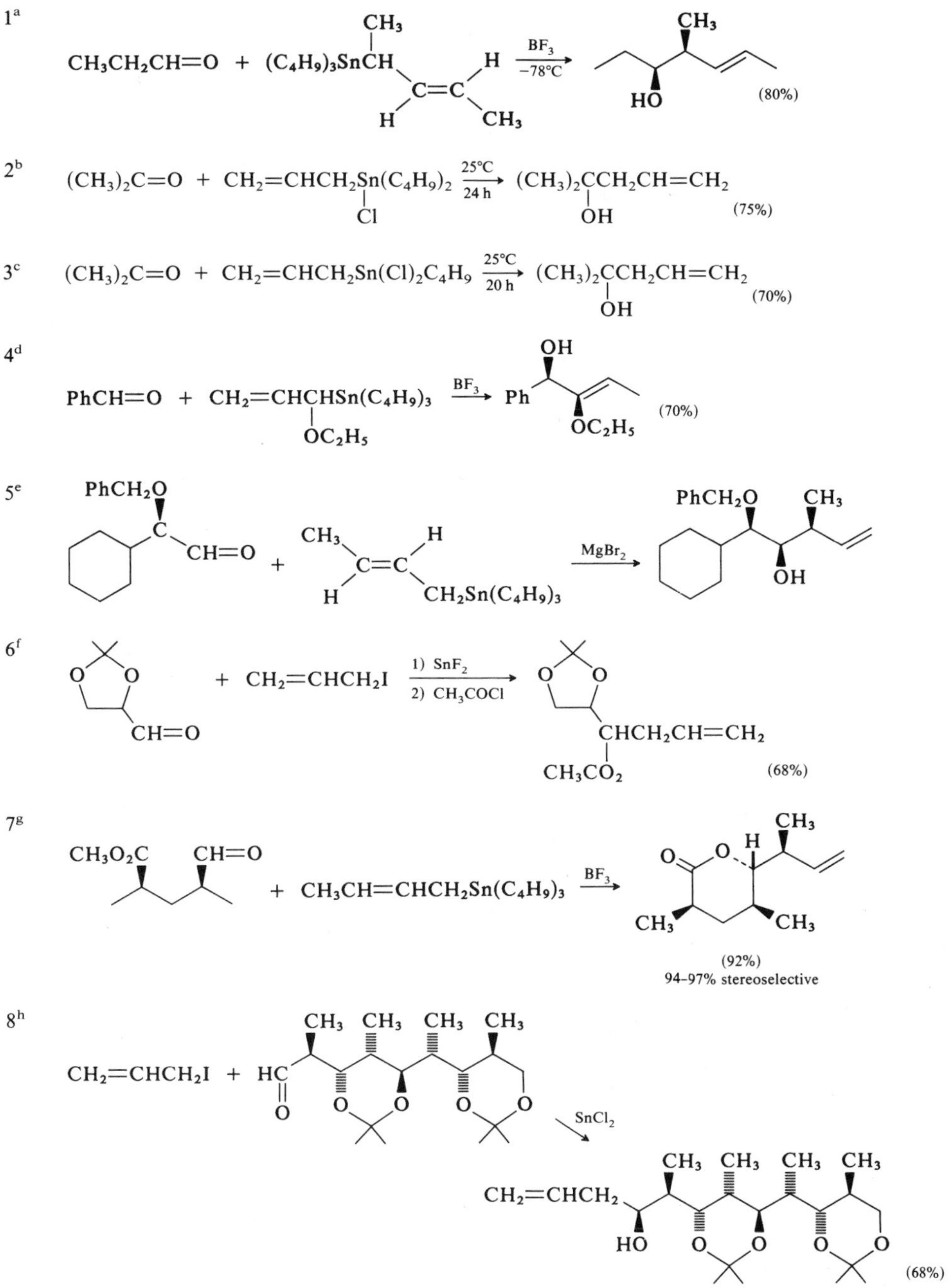

1[a]

$CH_3CH_2CH{=}O$ + $(C_4H_9)_3SnCH(CH_3)CH{=}CHCH_3$ $\xrightarrow[-78°C]{BF_3}$ CH_3, HO (80%)

2[b]

$(CH_3)_2C{=}O$ + $CH_2{=}CHCH_2Sn(Cl)(C_4H_9)_2$ $\xrightarrow[24\ h]{25°C}$ $(CH_3)_2C(OH)CH_2CH{=}CH_2$ (75%)

3[c]

$(CH_3)_2C{=}O$ + $CH_2{=}CHCH_2Sn(Cl)_2C_4H_9$ $\xrightarrow[20\ h]{25°C}$ $(CH_3)_2C(OH)CH_2CH{=}CH_2$ (70%)

4[d]

$PhCH{=}O$ + $CH_2{=}CHCH(OC_2H_5)Sn(C_4H_9)_3$ $\xrightarrow{BF_3}$ Ph, OH, OC_2H_5 (70%)

5[e]

$PhCH_2O$, C, $CH{=}O$ + CH_3, H, C=C, H, $CH_2Sn(C_4H_9)_3$ $\xrightarrow{MgBr_2}$ $PhCH_2O$, CH_3, OH

6[f]

O, O, $CH{=}O$ + $CH_2{=}CHCH_2I$ $\xrightarrow[2)\ CH_3COCl]{1)\ SnF_2}$ O, O, $CHCH_2CH{=}CH_2$, CH_3CO_2 (68%)

7[g]

CH_3O_2C, $CH{=}O$ + $CH_3CH{=}CHCH_2Sn(C_4H_9)_3$ $\xrightarrow{BF_3}$ O, O, H, CH_3, CH_3, CH_3

(92%)
94–97% stereoselective

8[h]

$CH_2{=}CHCH_2I$ + HC(=O), CH_3 CH_3 CH_3 CH_3, O, O, O, O $\xrightarrow{SnCl_2}$

$CH_2{=}CHCH_2$, HO, CH_3 CH_3 CH_3 CH_3, O, O, O, O (68%)

a. M. Koreeda and Y. Tanaka, *Chem. Lett.*, 1297 (1982).
b. V. Peruzzo and G. Tagliavini, *J. Organomet. Chem.* **162**, 37 (1978).
c. A. Gambaro, V. Peruzzo, G. Plazzogna, and G. Tagliavini, *J. Organomet. Chem.* **197**, 45 (1980).
d. J.-P. Quintard, B. Elissondo, and M. Pereyre, *J. Org. Chem.* **48**, 1559 (1983).
e. G. E. Keck and E. P. Boden, *Tetrahedron Lett.* **25**, 1879 (1984).
f. T. Harada and T. Mukaiyama, *Chem. Lett.*, 1109 (1981).
g. K. Maruyama, Y. Ishihara, and Y. Yamamoto, *Tetrahedron Lett.* **22**, 4235 (1981).
h. H. Nagaoka and Y. Kishi, *Tetrahedron* **37**, 3873 (1981).

reaction has proven to be very general with respect to both the halides and the types of stannanes that can be used. Benzylic, aryl, alkenyl, and allylic halides can all be used.[95] The groups that can be transferred from tin include alkyl, alkenyl, aryl, and alkynyl. The approximate order of effectiveness of transfer of groups from tin is alkynyl > alkenyl > aryl > benzyl > methyl > alkyl, so unsaturated groups are normally transferred selectively.[96]

Vinyltri-*n*-butylstannane reacts with aryl halides to give good yields of styrene derivatives.

$$Ar{-}Br + CH_2{=}CHSn(C_4H_9)_3 \xrightarrow{Pd(PPh_3)_4} ArCH{=}CH_2 \qquad \text{Ref. 97}$$

Various substituted alkenyl trialkylstannanes also react with alkenyl iodides to give dienes. These reactions proceed with retention of configuration at both double bonds.

$$CH_3(CH_2)_3CH{=}CHSn(CH_3)_3 + ICH{=}CH(CH_2)_4OH \xrightarrow{PdCl_2} CH_3(CH_2)_3CH{=}CHCH{=}CH(CH_2)_4OH$$

Ref. 98

The coupling reaction is very general with respect to the functionality that can be carried both in the halide and in the tin reagent. Groups such as ester, nitrile, nitro, cyano, and formyl can be present. This permits applications involving "masked functionality." For example, when the coupling reaction is applied to 1-alkoxy-2-butenylstannanes, the double-bond shift leads to a vinyl ether which can be hydrolyzed to an aldehyde.

$$CH_3C_6H_4Br + (C_4H_9)_3SnCH(OC_2H_5)CH{=}CHCH_3 \longrightarrow CH_3C_6H_4CH(CH_3)CH{=}CHOC_2H_5 \xrightarrow{H^+} CH_3C_6H_4CH(CH_3)CH_2CH{=}O$$

Ref. 99

The versatility of Pd-catalyzed couplings of stannanes has been extended by the demonstration that alkenyl trifluoromethylsulfonates (vinyl triflates) are also reactive.[100]

$$\text{6-methyl-1-cyclohexenyl-}OSO_2CF_3 + (CH_3)_3SnCH{=}CHSi(CH_3)_3 \xrightarrow{Pd(PPh_3)_4} \text{6-methyl-1-cyclohexenyl-}CH{=}CHSi(CH_3)_3$$

95. F. K. Sheffy, J. P. Godschalx, and J. K. Stille, *J. Am. Chem. Soc.* **106**, 4833 (1984); I. P. Beltskaya, *J. Organomet. Chem.* **250**, 551 (1983); J. K. Stille and B. L. Groh, *J. Am. Chem. Soc.* **109**, 813 (1987).
96. J. W. Labadie and J. K. Stille, *J. Am. Chem. Soc.* **105**, 6129 (1983).
97. D. R. McKean, G. Parrinello, A. F. Renaldo, and J. K. Stille, *J. Org. Chem.* **52**, 422 (1987).
98. J. K. Stille and B. L. Groh, *J. Am. Chem. Soc.* **109**, 813 (1987).
99. A. Duchene and J.-P. Quintard, *Synth. Commun.* **15**, 873 (1985).
100. W. J. Scott, G. T. Crisp, and J. K. Stille, *J. Am. Chem. Soc.* **106**, 4630 (1984); W. J. Scott and J. E. McMurry, *Acc. Chem. Res.* **21**, 47 (1988).

The alkenyl triflates can be prepared from ketones.[101] Methods for regioselective preparation of alkenyl triflates from unsymmetrical ketones are available.[102]

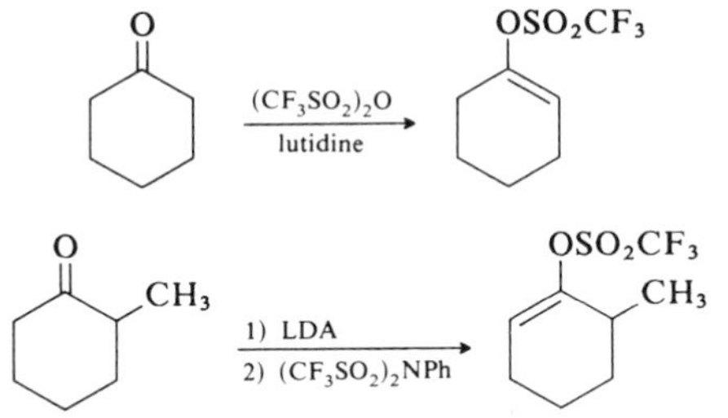

$$\text{cyclohexanone} \xrightarrow[\text{lutidine}]{(CF_3SO_2)_2O} \text{cyclohexenyl-}OSO_2CF_3$$

$$\text{2-methylcyclohexanone} \xrightarrow[\text{2) } (CF_3SO_2)_2NPh]{\text{1) LDA}} \text{6-methylcyclohex-1-enyl-}OSO_2CF_3$$

Some examples of Pd-catalyzed coupling of organostannanes with halides and triflates are given in Scheme 9.10.

Scheme 9.10. Palladium-Catalyzed Coupling of Stannanes with Halides and Sulfonates

A. Aryl Halides

1[a]
$$CH_3O\text{-}C_6H_4\text{-}Br + CH_2{=}CHCH_2Sn(C_4H_9)_3 \xrightarrow[120°C,\ 20\ h]{Pd(PPh_3)_4} CH_3O\text{-}C_6H_4\text{-}CH_2CH{=}CH_2 \quad (96\%)$$

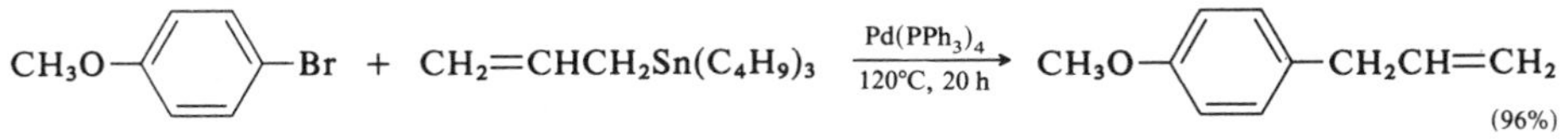

2[b]
$$O_2N\text{-}C_6H_4\text{-}Br + CH_2{=}CHSn(C_4H_9)_3 \xrightarrow[105°C,\ 4\ h]{Pd(PPh_3)_4} O_2N\text{-}C_6H_4\text{-}CH{=}CH_2 \quad (80\%)$$

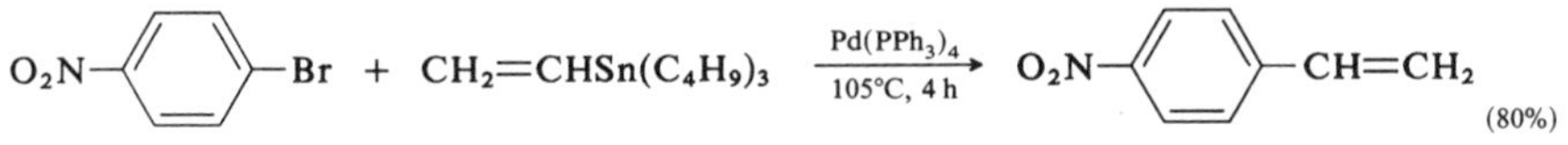

3[c]
$$CH_3O_2C\text{-}C_6H_4\text{-}I + \text{2-pyridyl-}Sn(CH_3)_3 \xrightarrow{PdCl_2(PPh_3)_4} \text{2-pyridyl-}C_6H_4\text{-}CO_2CH_3 \quad (95\%)$$

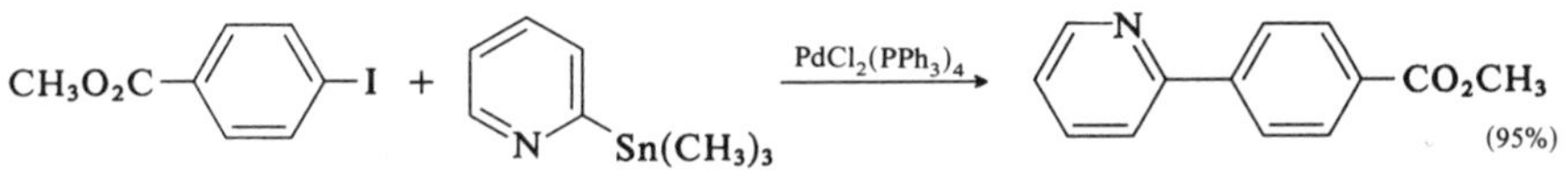

B. Alkenyl Halides and Sulfonates

4[d]
$$PhCH{=}CHI + CH_2{=}CHSn(C_4H_9)_3 \xrightarrow[25°C,\ 0.1\ h]{PdCl_2(CH_3CN)_2} PhCH{=}CHCH{=}CH_2 \quad (85\%)$$

5[e]
$$\text{(5,5-dimethyl-2-methylcyclopentenyl)-}OSO_2CF_3 + (CH_3)_3Sn(H)C{=}C(H)Si(CH_3)_3 \xrightarrow{Pd(PPh_3)_4} \text{(5,5-dimethyl-2-methylcyclopentenyl)-}C(H){=}C(H)Si(CH_3)_3 \quad (100\%)$$

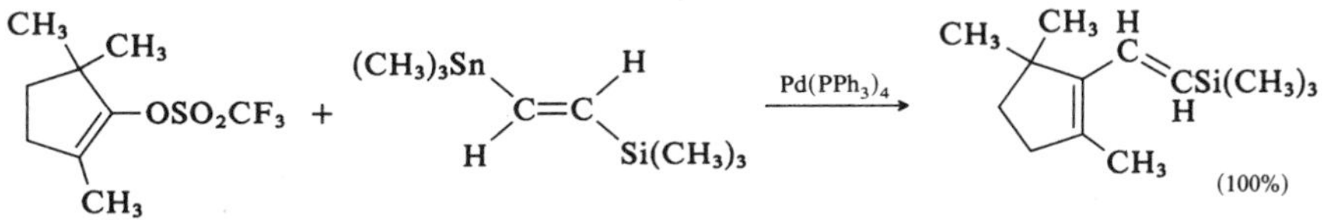

101. P. J. Stang, M. Hanack, and L. R. Subramanian, *Synthesis*, 85 (1982).
102. J. E. McMurry and W. J. Scott, *Tetrahedron Lett.* **24**, 979 (1983).

C. Allylic and Benzylic Halides

6[f] $BrCH_2(CH_3O)C{=}CH(CO_2CH_3)$ + $CH_3CH{=}CHSn(C_4H_9)_3$ —Pd(dba)$_2$, PPh_3→ $CH_3CH{=}CHCH_2(CH_3O)C{=}CH(CO_2CH_3)$ (86%)

dba = dibenzylideneacetone

7[g] $C_6H_5CH_2Br$ + $(C_4H_9)_3Sn(H)C{=}C(CH_3)(CH_2)_2OCH_2Ph$ —Pd$(PPh_3)_4$→ $C_6H_5CH_2(H)C{=}C(CH_3)(CH_2)_2OCH_2Ph$ (81%)

a. M. Kosugi, K. Sasazawa, Y. Shimizu, and T. Migata, *Chem. Lett.*, 301 (1977).
b. D. R. McKean, G. Parrinello, A. F. Renaldo, and J. K. Stille, *J. Org. Chem.* **52**, 422 (1987).
c. T. R. Bailey, *Tetrahedron Lett.* **27**, 4407 (1986).
d. J. K. Stille and B. L. Groh, *J. Am. Chem. Soc.* **109**, 813 (1987).
e. W. J. Scott, G. T. Crisp, and J. K. Stille, *J. Am. Chem. Soc.* **106**, 4630 (1984).
f. F. K. Sheffy, J. P. Godschalx, and J. K. Stille, *J. Am. Chem. Soc.* **106**, 4833 (1984).
g. J. Hibino, S. Matsubara, Y. Morizawa, K. Oshima, and H. Nozaki, *Tetrahedron Lett.* **25**, 2151 (1984).

Procedures for synthesis of ketones based on coupling of organostannanes with acid chlorides have also been developed.[103] The catalytic cycle is similar to that involved in the coupling with organic halides. The scope of the tin compounds to which the procedure can be applied is wide and includes successful results with tetra-*n*-butyltin. This implies that the reductive elimination step (step c) in the mechanism occurs faster than β elimination of the butylpalladium intermediate.

Pd^0 (a) + RCOCl → (b) $RC(=O){-}Pd^{2+}{-}Cl$; + R'_4Sn → R'_3SnCl' + $RC(=O){-}Pd^{2+}{-}R'$ (c) → RC(=O)R' + Pd^0

103. D. Milstein and J. K. Stille, *J. Org. Chem.* **44**, 1613 (1979); J. W. Labadie and J. K. Stille, *J. Am. Chem. Soc.* **105**, 6129 (1983).

Scheme 9.11. Synthesis of Ketones by Palladium-Catalyzed Acylation of Stannanes

A. From Acid Halides

1[a]

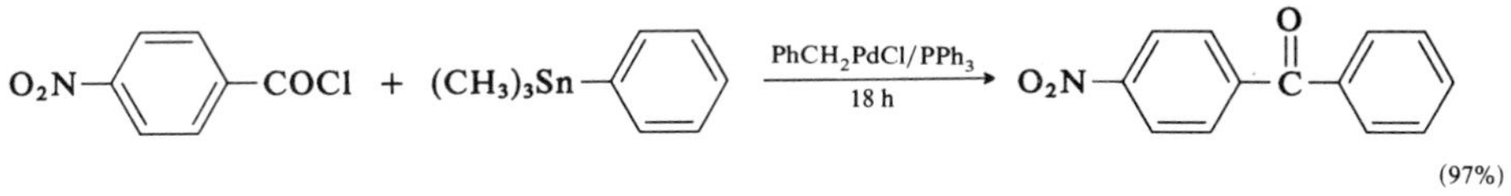

$$O_2N\text{-}C_6H_4\text{-}COCl + (CH_3)_3Sn\text{-}C_6H_5 \xrightarrow[18\ h]{PhCH_2PdCl/PPh_3} O_2N\text{-}C_6H_4\text{-}\overset{O}{\overset{\|}{C}}\text{-}C_6H_5$$

(97%)

2[a]

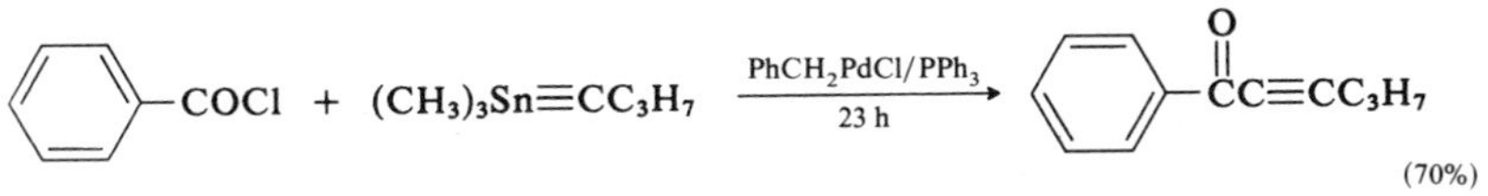

$$C_6H_5\text{-}COCl + (CH_3)_3Sn{\equiv}CC_3H_7 \xrightarrow[23\ h]{PhCH_2PdCl/PPh_3} C_6H_5\text{-}\overset{O}{\overset{\|}{C}}C{\equiv}CC_3H_7$$

(70%)

3[b]

$$(CH_3)_2C{=}CHCOCl + (C_4H_9)_3Sn\text{-}(1\text{-cyclopentenyl}) \xrightarrow{PdCl_2,\ PPh_3} (CH_3)_2C{=}CH\overset{O}{\overset{\|}{C}}\text{-}(1\text{-cyclopentenyl})$$

(85%)

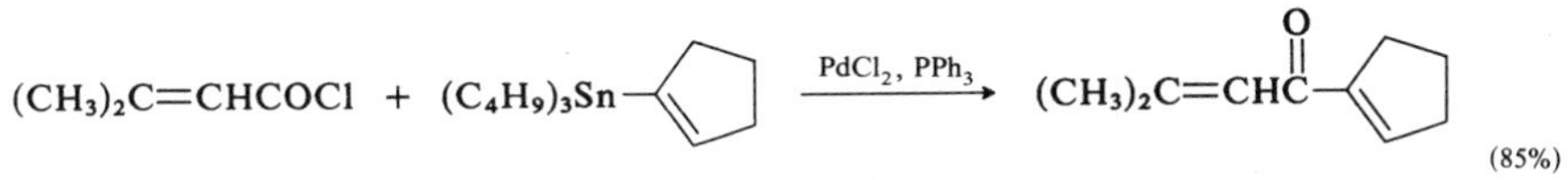

4[c]

$$CH_3\overset{O}{\overset{\|}{C}}NH\underset{CO_2C_2H_5}{\underset{|}{C}}H(CH_2)_5COCl + (CH_2{=}CH)_4Sn \xrightarrow{PhCH_2PdCl/PPh_3} CH_3\overset{O}{\overset{\|}{C}}NH\underset{CO_2C_2H_5}{\underset{|}{C}}H(CH_2)_5\overset{O}{\overset{\|}{C}}CH{=}CH_2$$

(70%)

B. By Carbonylation

5[d]

$$CH_2{=}CHCH_2Cl + (CH_3)_3SnCH_2CH{=}C(CH_3)_2 \xrightarrow[CO]{PdCl_2(CH_3CN)} CH_2{=}CHCH_2\overset{O}{\overset{\|}{C}}CH_2CH{=}C(CH_3)_2$$

(52%)

6[b]

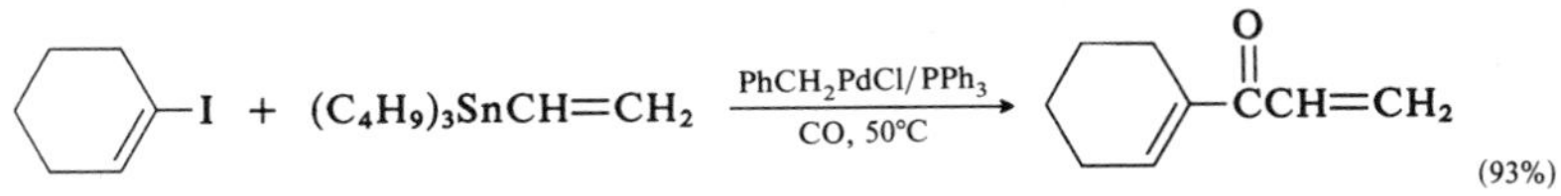

$$(1\text{-cyclohexenyl})\text{-}I + (C_4H_9)_3SnCH{=}CH_2 \xrightarrow[CO,\ 50°C]{PhCH_2PdCl/PPh_3} (1\text{-cyclohexenyl})\text{-}\overset{O}{\overset{\|}{C}}CH{=}CH_2$$

(93%)

7[e]

$$(CH_3)_2C{=}CHCH_2Cl + 3\text{-}(Sn(CH_3)_3)\text{furan} \xrightarrow[CO]{PhCH_2PdCl/PPh_3} 3\text{-}(\overset{O}{\overset{\|}{C}}CH_2CH{=}C(CH_3)_2)\text{furan}$$

(75%)

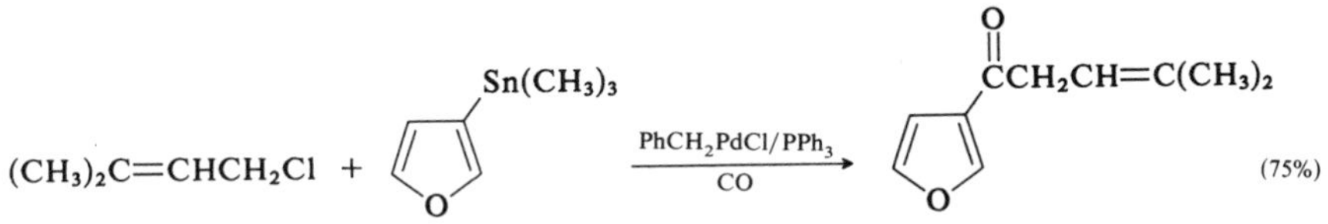

a. J. W. Labadie, D. Tueting, and J. K. Stille, *J. Org. Chem.* **48**, 4634 (1983).
b. W. F. Goure, M. E. Wright, P. D. Davis, S. S. Labadie, and J. K. Stille, *J. Am. Chem. Soc.* **106**, 6417 (1984).
c. D. H. Rich, J. Singh, and J. H. Gardner, *J. Org. Chem.* **48**, 432 (1983).
d. J. H. Merrifield, J. P. Godschalx, and J. K. Stille, *Organometallics* **3**, 1108 (1984).
e. F. K. Sheffy, J. P. Godschalx, and J. K. Stille, *J. Am. Chem. Soc.* **106**, 4833 (1984).

Coupling of organotin compounds with halides in a carbon monoxide atmosphere leads to formation of ketones by incorporation of a carbonylation step.[104]

$$\mathrm{Pd^0} + \mathrm{R'X} \rightarrow \mathrm{R'{-}Pd^{2+}{-}X} \xrightarrow{\mathrm{CO}} \mathrm{R'{-}Pd^{2+}(CO){-}X} \xrightarrow{\mathrm{R_4Sn}} \mathrm{R'{-}Pd^{2+}(R){-}C{\equiv}O} + \mathrm{R_3SnCl} \rightarrow \mathrm{RC(=O)R'} + \mathrm{Pd^0}$$

Procedures for carbonylation of alkenyl triflates are also successful.

$$\text{(trimethylbicyclooctenyl triflate, } \mathrm{OSO_2CF_3}) + \mathrm{(CH_3)_3Sn{-}CH{=}CH{-}Si(CH_3)_3} \xrightarrow[\mathrm{CO,\ LiCl}]{\mathrm{Pd(PPh_3)_4}} \text{alkenyl}\mathrm{{-}C(=O){-}CH{=}CH{-}Si(CH_3)_3}\ (86\%)$$

Ref. 105

Scheme 9.11 gives some examples of these methods of ketone synthesis.

General References

Organoboron Compounds

H. C. Brown, *Organic Synthesis via Boranes*, Wiley, New York, 1975.
E. Negishi and M. Idacavage, *Org. React.* **33**, 1 (1985).
A. Pelter, K. Smith, and H. C. Brown, *Borane Reagents*, Academic Press, New York, 1988.
A. Pelter, in *Rearrangements in Ground and Excited States*, P. de Mayo (ed.), Vol. 2, Academic Press, New York, 1980, Chapter 8.

Organosilicon Compounds

T. H. Chan and I. Fleming, *Synthesis*, 761 (1979).
E. W. Colvin, *Silicon in Organic Synthesis*, Butterworths, London, 1981.
I. Fleming, J. Dunogvès, and R. Smithers, *Org. React.* **37**, 57, 1989.
W. P. Weber, Silicon Reagents for Organic Synthesis, Springer-Verlag, Berlin, 1983.

Organotin Compounds

M. Peyeyre, J.-P. Quintard, and A. Rahm, *Tin in Organic Synthesis*, Butterworths, London, 1987.

104. M. Tanaka, *Tetrahedron Lett.*, 2601 (1979).
105. G. T. Crisp, W. J. Scott, and J. K. Stille, *J. Am. Chem. Soc.* **106**, 7500 (1984).

(*References for these problems will be found on page 775.*)

1. Give the expected product(s) for the following reactions.

(a) $CH_3CH_2CH_2CH_2(H)C{=}C(H)B(O_2C_6H_4)$ (catecholborane) + $Br(H)C{=}C(H)Ph$ $\xrightarrow[NaOC_2H_5]{Pd(PPh_3)_4}$

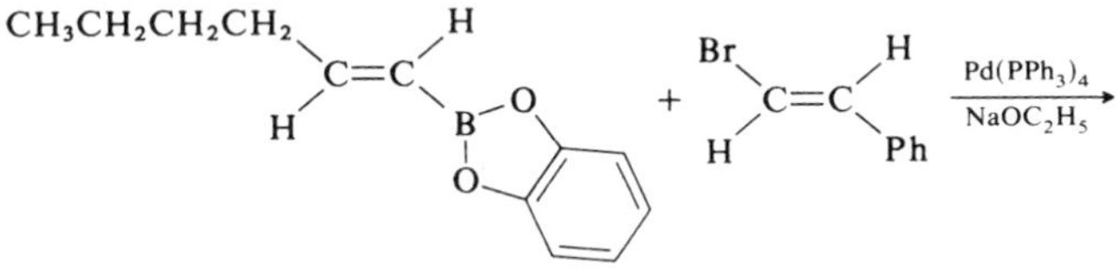

(b) $(C_4H_9)_3Sn(H)C{=}C(H)CH_2OTHP$ + $CH_3O(BrCH_2)C{=}C(CO_2C_2H_5)H$ $\xrightarrow[bis(dba)Pd]{CO,\ 55\ psi}$

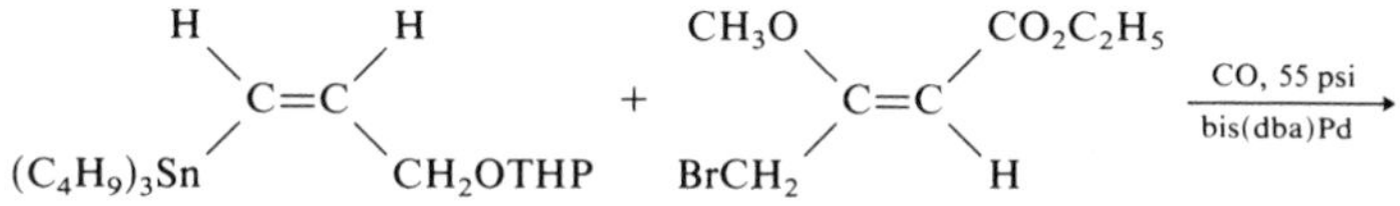

dba = dibenzylideneacetone

(c) $(CH_3)_3CC{\equiv}CCH_3 \xrightarrow[2)\ H_2C{=}CHCCH_3,\ 65^\circ C]{1)\ 9\text{-BBN},\ 0^\circ C,\ 16\ h}$ (C=O) $\xrightarrow[H_2O_2]{NaOH,}$

(d) $PhCH{=}CHCOCl + (CH_3)_4Sn \xrightarrow[HMPA]{PhCH_2PdCl/Ph_3P}$

(e) $[CH_3CO_2(CH_2)_5]_3B + LiC{\equiv}C(CH_2)_3CH_3 \longrightarrow \xrightarrow{I_2}$

(f) $PhCH{=}O$ + $CH_3(H)C{=}C(CH_2Sn(C_4H_9)_3)H$ $\xrightarrow{BF_3}$

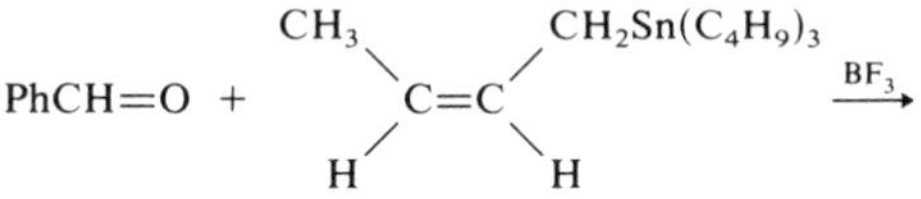

(g) $CH_3O{-}C_6H_4{-}CH{=}CHCH_2Si(CH_3)_3$ + $CH_3CH_2CH(OCH_3){-}C_6H_4{-}OCH_3$ $\xrightarrow{TiCl_4}$

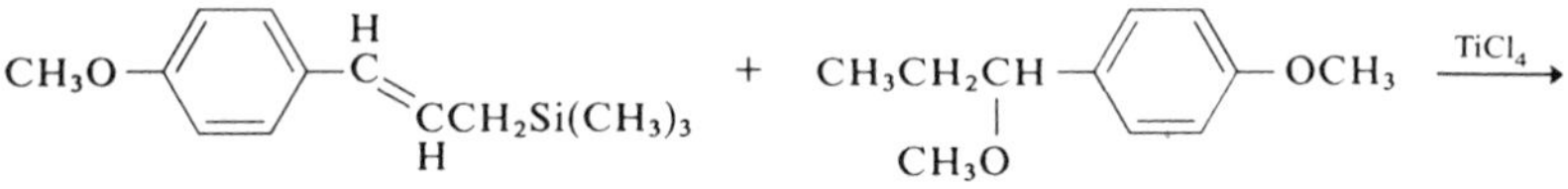

(h) $(CH_3O_2C)_2C(Si(CH_3)_3)CH(CH_3)CH{=}CHSi(CH_3)_3 + PhCH{=}O \xrightarrow{F^-}$

(i) 2-methylcyclopentyl-B(1,3,2-dioxaborinane) (CH₃ and B trans) $\xrightarrow[\substack{2)\ HgCl_2 \\ 3)\ H_2O_2,\ pH\ 8}]{1)\ LiCHOCH_3(SPh)}$

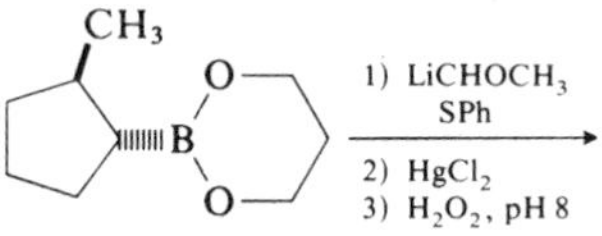

(j) 9-BBN-$B{-}C(H){=}C(H)(CH_2)_3CH_3$ + $ClCH_2CN$ $\xrightarrow{K^{+}\ {}^{-}O{-}C_6H_3(t\text{-}Bu)_2\ (2,6)}$

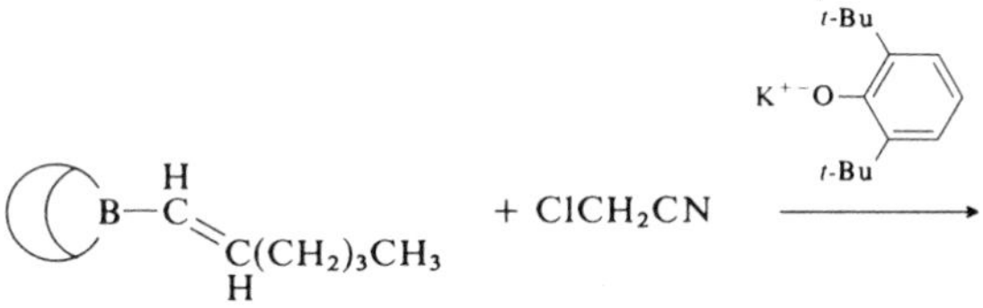

2. Starting with an alkene $RCH{=}CH_2$, indicate how organoborane intermediate could be used for each of the following synthetic transformations:

(a) $RCH{=}CH_2 \rightarrow RCH_2CH_2CH_2C(=O)C_6H_5$

(b) $RCH{=}CH_2 \rightarrow RCH_2CH_2CH{=}O$

(c) $RCH{=}CH_2 \rightarrow RCH_2CH_2CH(CH_3)CH_2C(=O)CH_3$

(d) $RCH{=}CH_2 \rightarrow$ $(RCH_2CH_2)HC{=}CH(CH_3)$ (RCH₂CH₂ and CH₃ cis; H and H cis)

(e) $RCH{=}CH_2 \rightarrow RCH_2CH_2C(=O)CH_2CH_2R$

(f) $RCH{=}CH_2 \rightarrow RCH_2CH_2CH_2CO_2C_2H_5$

3. In Scheme 9.2 there are described reactions of organoboranes with cyanide ion, chlorodifluoromethane, dichloromethyl methyl ether, acetylides, α-lithiated diphenylthioacetals, and α-lithioimines. Compare the structures of these reagents and the final reaction products from the various reagents. Develop a general mechanistic outline which encompasses all these reactions and discuss the structural features which these reagents have in common with one another and with carbon monoxide.

4. Give appropriate reagents, other organic reactants, and approximate reaction conditions for effecting the following syntheses in a "one-pot" process.

(a) $CH_3(H)C{=}C(H)(CH_2)_5O_2CCH_3$ (CH₃ and H trans; H and (CH₂)₅O₂CCH₃) from $IC{\equiv}CCH_2CH_3$ and $CH_2{=}CH(CH_2)_3O_2CCH_3$

(b) $CH_3(CH_2)_3C{\equiv}C(CH_2)_7CH_3$ from $CH_2{=}CH(CH_2)_5CH_3$ and $HC{\equiv}C(CH_2)_3CH_3$

(c) $CH_3(CH_2)_{11}C(=O)$–cyclopentyl from $CH_2{=}CH(CH_2)_9CH_3$ and cyclopentene

(d) bicyclic hydroazulenone (H, CH₃ with wavy bond, =O, TBSO, CH₃) from substituted cyclopentane (H, $C(CH_3){=}CH_2$, $CH{=}CH_2$, TBSO, CH_3)

5. Give a mechanism for each of the following reactions.

(a)

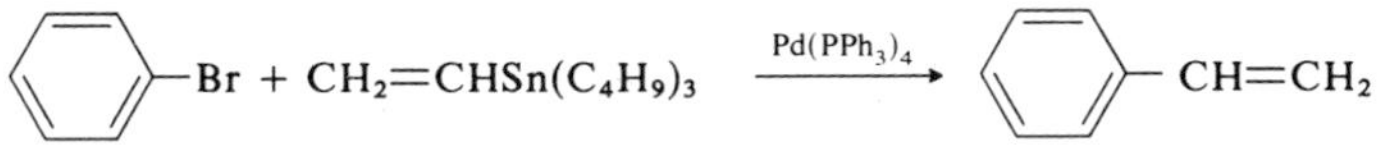

$$C_6H_5Br + CH_2{=}CHSn(C_4H_9)_3 \xrightarrow{Pd(PPh_3)_4} C_6H_5CH{=}CH_2$$

(b)

$$CH_3(CH_2)_6C(OSi(CH_3)_3){=}CH_2 + PhBr + Bu_3SnF \xrightarrow[3\ mol\ \%]{PdCl_2[P(tol)_3]} CH_3(CH_2)_6\overset{O}{\overset{\|}{C}}CH_2Ph$$

(c)

$$(C_5H_9)_3\bar{B}{-}C{\equiv}C(CH_2)_5CH_3 \xrightarrow[2)\ H_2O_2,\ ^-OAc]{1)\ ICH_2CN} C_5H_9{-}\overset{O}{\overset{\|}{C}}{-}CH(CH_2CN)(CH_2)_5CH_3$$

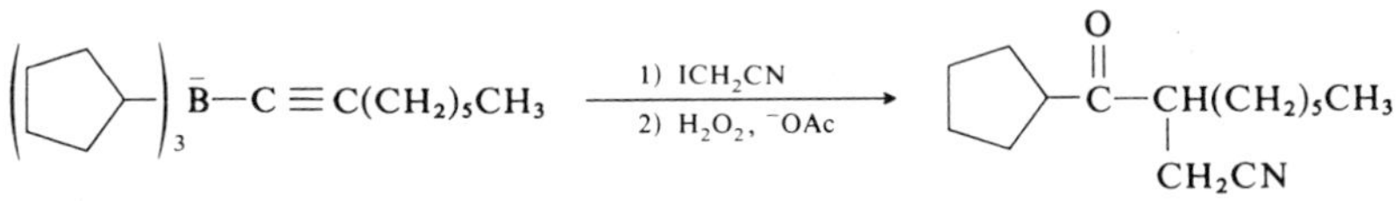

(d)

$$(CH_3)_3\bar{B}C{\equiv}C(CH_2)_3CH_3 \xrightarrow[ether/THF]{2.5\ equiv\ CH_3SO_3H} HO{-}C(CH_3)_2(CH_2)_4CH_3$$

(e)

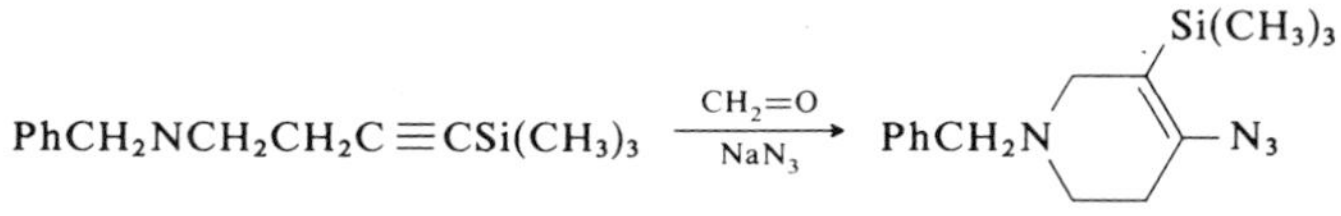

$$PhCH_2NCH_2CH_2C{\equiv}CSi(CH_3)_3 \xrightarrow[NaN_3]{CH_2{=}O} PhCH_2N\text{(ring)}{-}Si(CH_3)_3,\ N_3$$

(f)

$$\text{(9-BBN)}B{-}Cl \xrightarrow[3)\ (C_2H_5)_3CO^-Li^+;\ 4)\ H_2O_2,\ ^-OH]{1)\ 2,6\text{-dimethylphenol};\ 2)\ HCCl_2OCH_3} \text{bicyclo[3.3.1]nonan-9-one}$$

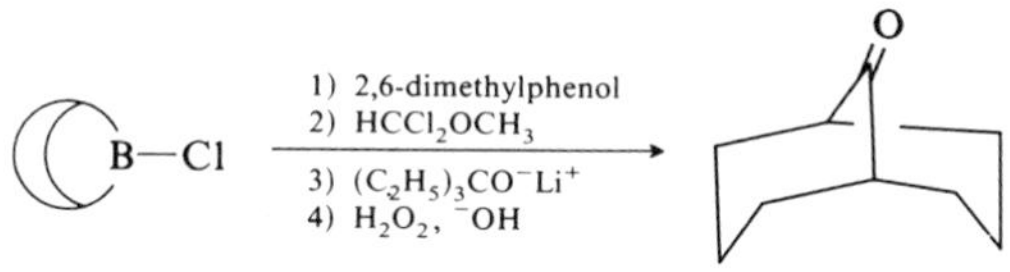

6. Offer a detailed mechanistic explanation for the following observations.

(a) When the E- and Z-isomers of the 2-butenyl boradioxolane **A** are caused to react with the aldehyde **B**, the Z-isomer gives the *syn* product with >90% stereoselectivity. The E-isomer, however, gives a nearly 1:1 mixture of **D** and **E**.

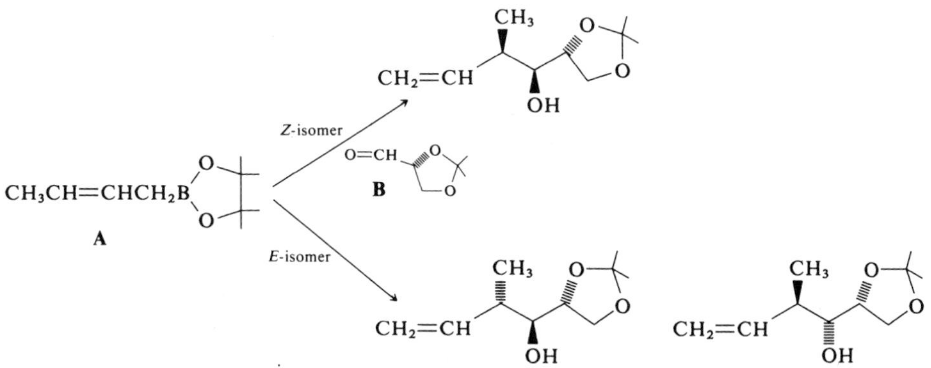

$CH_3CH{=}CHCH_2B(OC(CH_3)_2)_2$ **A** + O=CH–(dioxolane) **B**

Z-isomer → $CH_2{=}CH$–CH(CH_3)–CH(OH)–(dioxolane)

E-isomer → $CH_2{=}CH$–CH(CH_3)–CH(OH)–(dioxolane) + $CH_2{=}CH$–CH(CH_3)–CH(OH)–(dioxolane)

(b) The reaction of a variety of $\Delta^{2,3}$-pyranyl acetates with allylsilanes under the influence of Lewis acid catalysts gives 2-allyl-$\Delta^{3,4}$-pyrans. The stereochemistry of the methyl group is dependent on whether the E- or Z-allylsilane is used. Predict which stereoisomer is formed in each case and explain the mechanistic basis of your prediction.

O_2CCH_3, O_2CCH_3, $CH_2O_2CCH_3$ (pyranyl acetate) $+ CH_2{=}CHCH_2Si(CH_3)_3 \longrightarrow$ $CH_2{=}CH\overset{CH_3}{C}H$–(pyran, H, O) with O_2CCH_3, $CH_2O_2CCH_3$

(c) When trialkylboranes react with α-diazoketones or α-diazoesters in D_2O, the resulting products are monodeuterated. Formulate the reaction mechanism in sufficient detail to account for this fact.

$$R_3B + N_2CH\overset{O}{\overset{\|}{C}}R' \xrightarrow{D_2O} R\underset{D}{\underset{|}{C}}H\overset{O}{\overset{\|}{C}}R'$$

7. A number of procedures for stereoselective syntheses of alkenes involving alkenylboranes have been developed. For each of the procedures given below, give the structures of the intermediates and describe the mechanism in sufficient detail to account for the observed stereoselectivity.

(a)

$$R^1C{\equiv}CBr \xrightarrow[\text{2) } HO(CH_2)_3OH]{\text{1) } BHBr_2{\cdot}SMe_2} A \xrightarrow{R^2Li} B \xrightarrow[\text{2) } I_2,\ MeOH;\ \text{3) } NaOH]{\text{1) } R^3Li} (R^1)(H)C{=}C(R^2)(R^3)$$

(b)

$$R^1C{\equiv}CBr \xrightarrow[\text{2) } HO(CH_2)_3OH]{\text{1) } BHBr_2{\cdot}SMe_2} A \xrightarrow{R^2Li} B \xrightarrow[\text{2) } NaOMe,\ MeOH]{\text{1) } Br_2} C \xrightarrow[\text{2) } (i\text{-}PrO)_3B;\ \text{3) } HCl]{\text{1) } s\text{-}BuLi} D$$

$$D \xrightarrow[HO(CH_2)_3OH]{H_2O} E \xrightarrow[\text{2) } I_2,\ MeOH;\ \text{3) } NaOH]{\text{1) } R^3Li} (R^1)(H)C{=}C(R^3)(R^2)$$

(c)

$$R^1C{\equiv}CH \xrightarrow{BBr_3} A \xrightarrow[PdCl_2(PPh_3)_3]{R^2ZnCl} B \xrightarrow[\text{2) } R^3I]{\text{1) } LiOCH_3} (R^2)(R^1)C{=}C(R^3)(H)$$

(d)

$$R^2_2BCl \xrightarrow[R^1C{\equiv}CH]{LiAlH_4} A \xrightarrow[I_2]{NaOCH_3} (R^1)(H)C{=}C(R^2)(H)$$

(e)

$$R^1C{\equiv}CH \xrightarrow[\text{2) } HO(CH_2)_2OH]{\text{1) } BHBr_2{\cdot}SMe_2} A \xrightarrow[\text{2) } I_2,\ MeOH;\ \text{3) } NaOH]{\text{1) } R^2Li} (R_1)(H)C{=}C(R^2)(H)$$

8. Suggest reagents which could be effective for the following cyclization reactions.

(a)

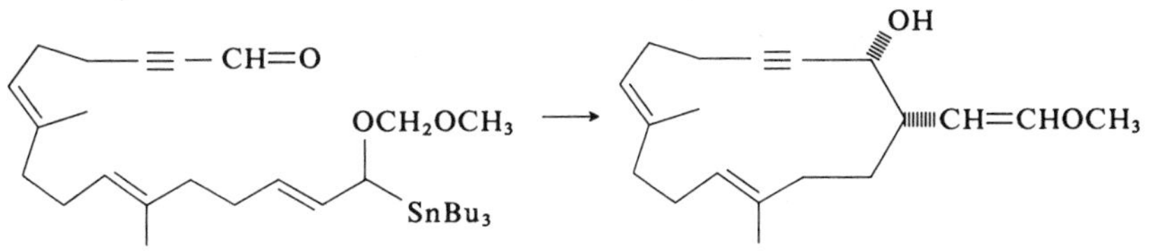

[Structure: acyclic polyene with ≡—CH=O terminus, OCH_2OCH_3 and $SnBu_3$ substituents → macrocycle bearing OH, ≡, and $CH{=}CHOCH_3$]

(b)

[Structure: lactone-tethered allylstannane ($SnBu_3$) with CH_3S SCH_3 dithioketal → macrocyclic lactone bearing CH_2, C_2H_5, SCH_3]

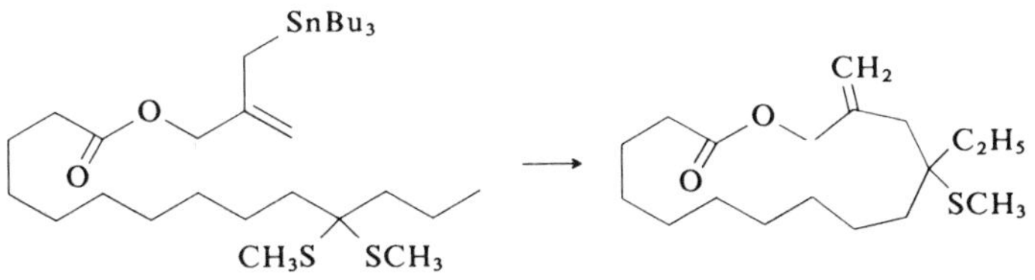

9. Show how the following silanes and stannanes could be synthesized from the suggested starting material.

(a) 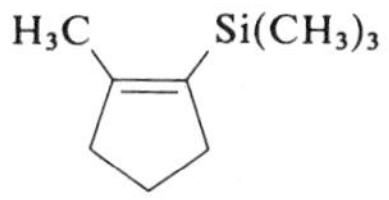 1-H_3C-2-$Si(CH_3)_3$-cyclopentene from 2-methylcyclopentanone (H_3C, O)

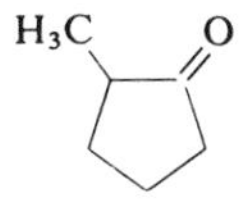

(b) 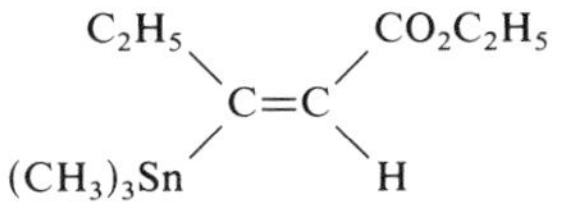 $(CH_3)_3Sn(C_2H_5)C{=}C(H)CO_2C_2H_5$ from $CH_3CH_3C{\equiv}CCO_2C_2H_5$

(c) $Bu_3SnCH{=}CHSnBu_3$ from $HC{\equiv}CH$, Bu_3SnCl, and Bu_3SnH

(d) 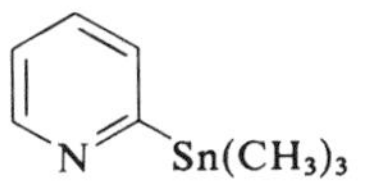 2-$Sn(CH_3)_3$-pyridine from 2-Cl-pyridine

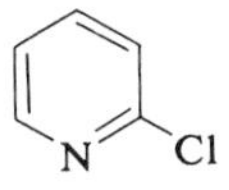

(e) $RC(=CH_2)CH_2Si(CH_3)_3$ from $RCOCl$ or RCO_2R'

(f) $CH_2{=}C(CH_2Si(CH_3)_3)CH{=}CH_2$ from $CH_2{=}C(Cl)CH{=}CH_2$

10. Each of the cyclic amines shown below has been synthesized by reaction of an amino-substituted allylic silane with formaldehyde in the presence of trifluoroacetic acid. Identify the appropriate precursor of each amine and suggest a method for its synthesis.

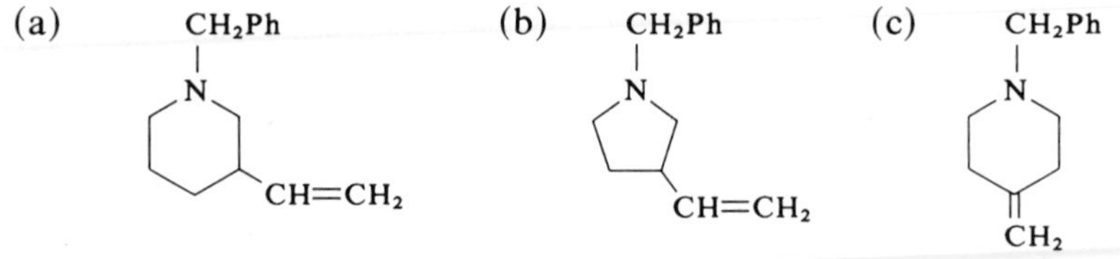

(a) CH_2Ph N CH=CH$_2$ (b) CH_2Ph N CH=CH$_2$ (c) CH_2Ph N CH_2

11. Both the *E*- and *Z*-stereoisomers of the terpene γ-bisabolene can be isolated from natural sources. Recently, stereoselective syntheses of these compounds were developed which rely heavily on borane intermediates.

E-γ-bisabolene **A** *Z*-γ-bisabolene **B**

The synthesis of the *E*-isomer, **A**, proceeds from **D** through **C** as a key intermediate. Show how **C** could be obtained from **D** and how it could be converted to **A**.

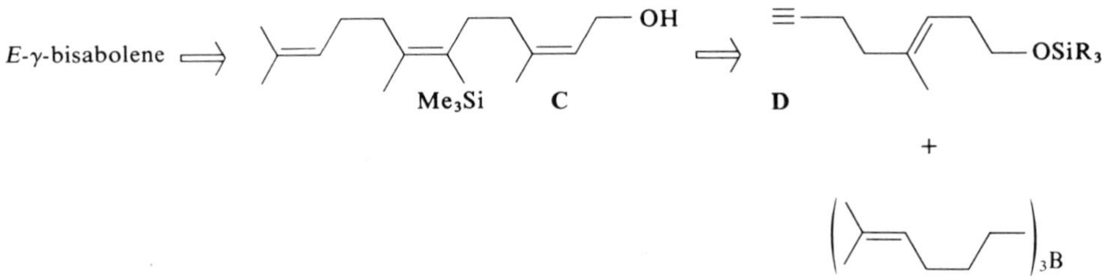

E-γ-bisabolene ⇒ Me_3Si **C** OH ⇒ **D** $OSiR_3$ + $(\ldots)_3B$

The synthesis of the *Z*-isomer employs the bromide **F** and the borane **G** as starting materials. Devise a method for synthesis of **B** from **F** and **G**.

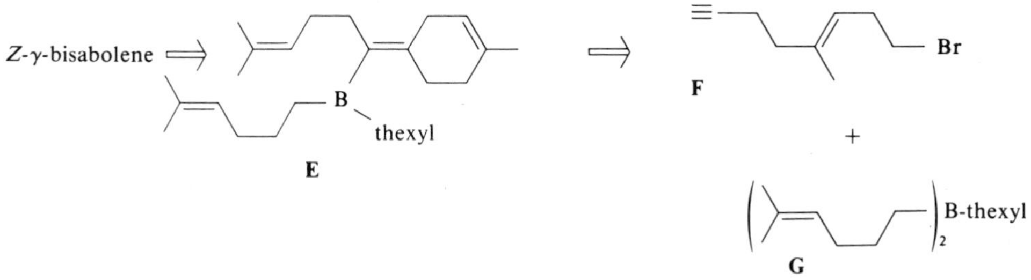

Z-γ-bisabolene ⇒ B thexyl **E** ⇒ **F** Br + $(\ldots)_2$B-thexyl **G**

The crucial element of these syntheses is control of the stereochemistry of the exocyclic double bond on the basis of the stereochemistry of migration of a substituent from boron to carbon. Discuss the requirements for a stereoselective synthesis and suggest how these requirements might be met.

12. Devise a sequence of reactions which would provide the desired compound from the suggested starting material(s).

(a) [macrocyclic structure with OH, ≡, CHOMOM] from $HC{\equiv}CCH_2CH_2C(H){=}C(CH_3)CH_2CH_2C(H){=}C(CH_3)CH_2CH_2C(H){=}C(H)CH{=}O$

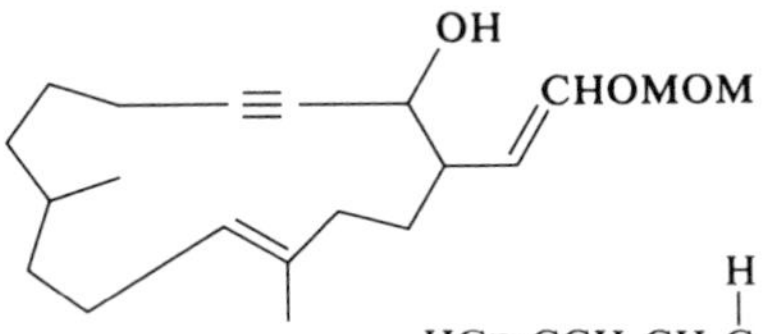

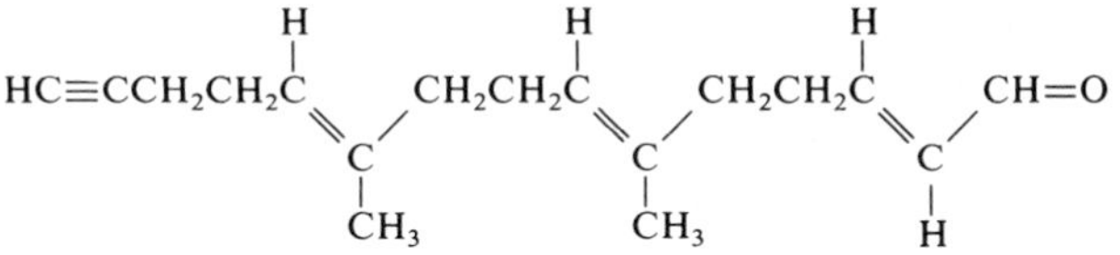

(b) $CH_3(CH_2)_3$, CH_3, OH, OTHP from $CH_3(CH_2)_3CH{=}O$ and $CH_3(H)C{=}C(H)(CH_2)_2OTHP$

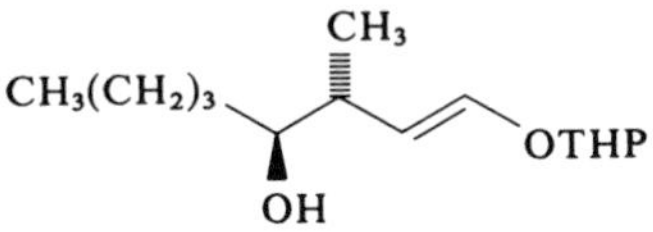

(c) H_3C, CH_3, O, O, $CH_2C(H){=}C(H)Ph$, $CH{=}O$ from H_3C, CH_3, O, O, $CH{=}O$ and $[\ ]_2BCH_2CH{=}CHPh$

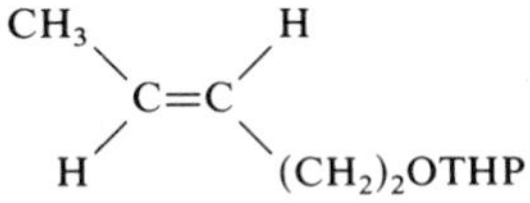

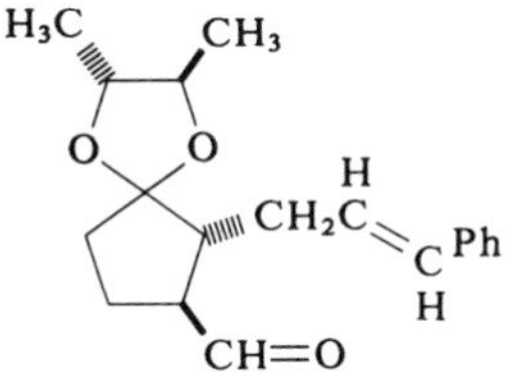

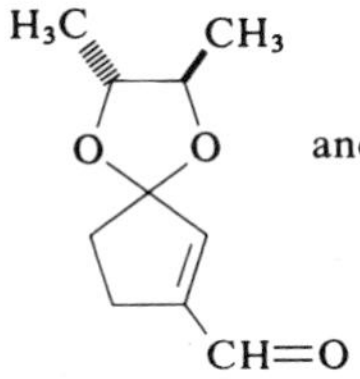

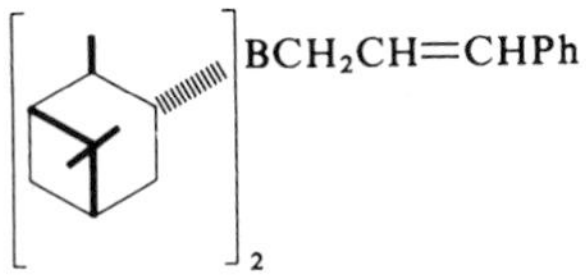

(d) CH_2, O, O, C_2H_5, SCH_3 from $(C_4H_9)_3SnCH_2C(=CH_2)CH_2OH$ and $CH_3CH_2C(SCH_3)_2(CH_2)_9CO_2H$

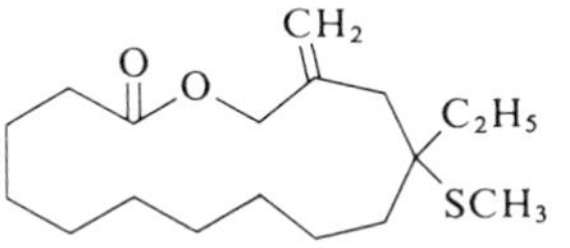

(e) $THPO(CH_2)_8(H)C{=}C(H)C(H){=}C(H)CH_2CH_3$ from $BrC{\equiv}CCH_2CH_3$ and $THPO(CH_2)_8C{\equiv}CH$

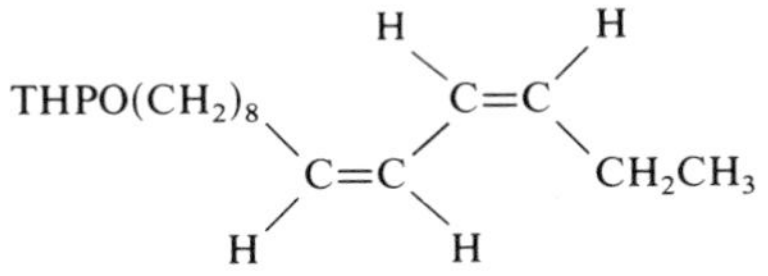

(f) $CH_3(CH_2)_3(H)C{=}C(H)C(H){=}C(H)(CH_2)_4OH$ from $(H)(I)C{=}C(H)(CH_2)_4OH$

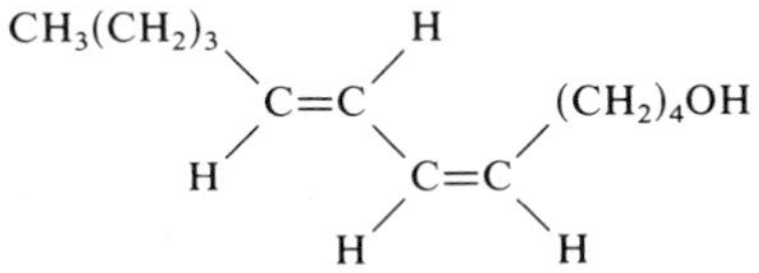

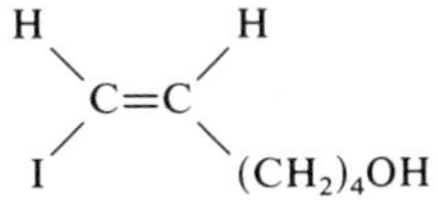

(g) 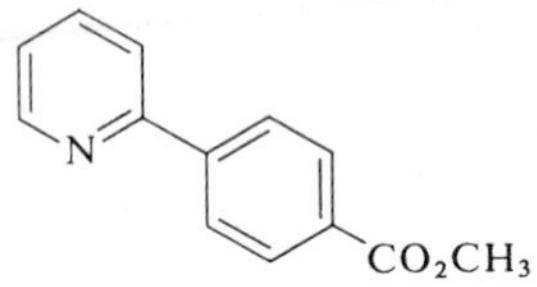 [structure: 2-(4-methoxycarbonylphenyl)pyridine, CO_2CH_3] from [2-chloropyridine, N, Cl] and I–[C_6H_4]–CO_2CH_3

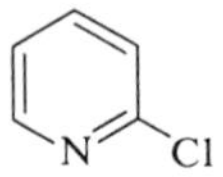

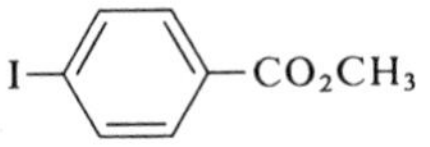

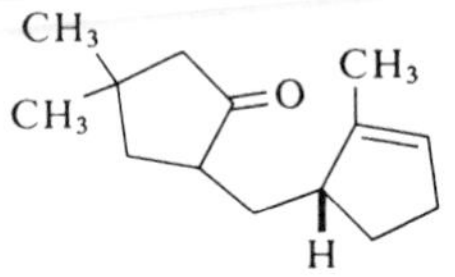

(h) [structure: CH_3, CH_3, =O, CH_3, H] from [structure: CH_3, CH_3, O, CH_2]

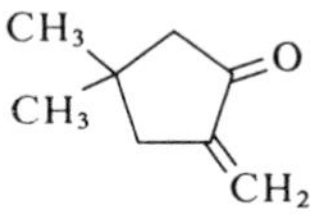

(i) 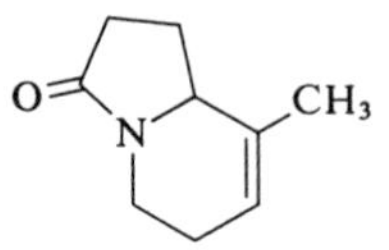 [structure: O, N, CH_3] from [structure: O, N, O, $CH_2CH_2CH{=}C(CH_3)Si(CH_3)_3$]

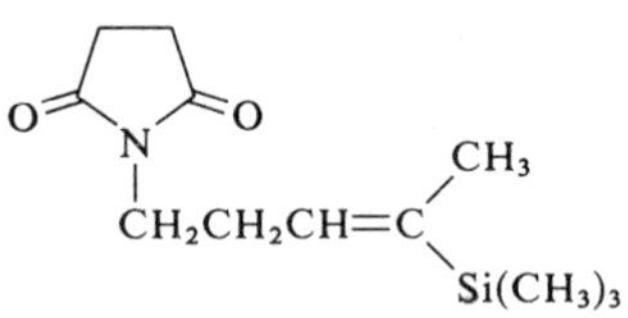

(j) 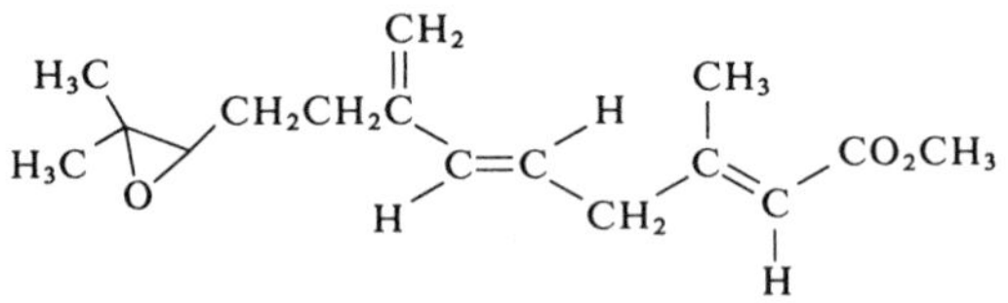 [structure: H_3C, H_3C, O, $CH_2CH_2C(=CH_2)$, H, C=C, H, CH_2, CH_3, C=C, CO_2CH_3, H] from

[structure: H_3C, H_3C, O, $CH_2CH_2\overset{O}{\overset{\|}{C}}CH_3$], $BrCH_2\overset{CH_3}{\overset{|}{C}}{=}CHCO_2CH_3$

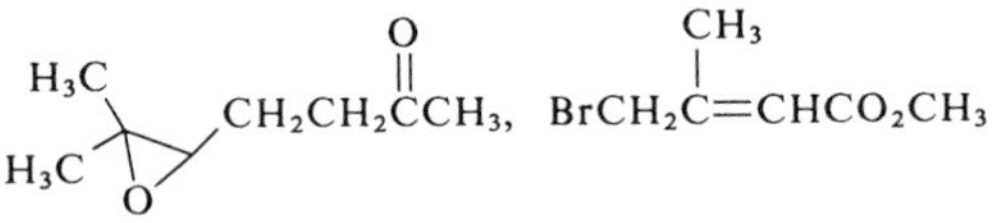

and $(C_4H_9)_3SnCH{=}CHSn(C_4H_9)_3$

(k) 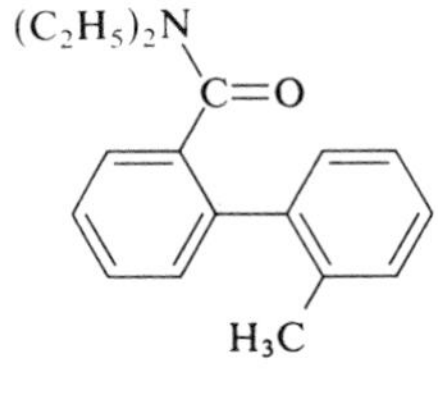 [structure: $(C_2H_5)_2N$, C=O, H_3C] from [C_6H_5–$\overset{O}{\overset{\|}{C}}N(C_2H_5)_2$]

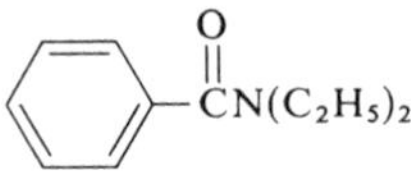

13. Show how the following compounds can be prepared in high enantiomeric purity using homochiral boranes as reagents.

(a) 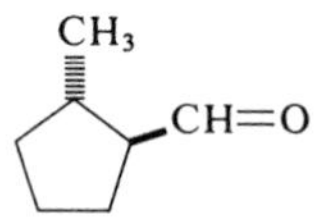 [structure: CH_3, CH=O]

(b) 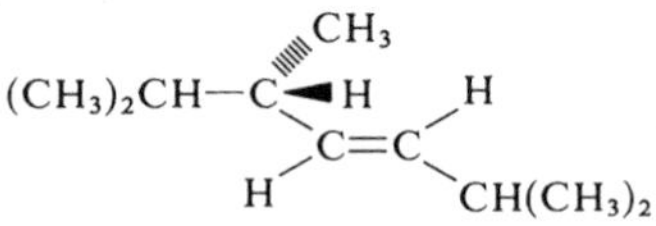 [structure: $(CH_3)_2CH$–C(CH_3)(H)–C(H)=C(H)–$CH(CH_3)_2$]

(c) [structure: Ph, $\overset{O}{\overset{\|}{C}}(CH_2)_4CH_3$]

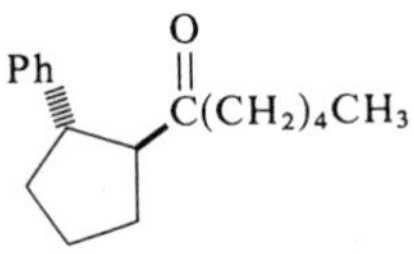

(d) 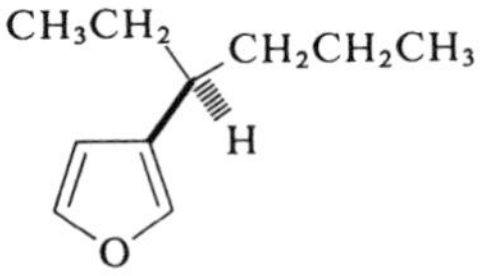 [structure: CH_3CH_2, $CH_2CH_2CH_3$, H, O]

10

Reactions Involving Highly Reactive Electron-Deficient Intermediates

10.1. Reactions Involving Carbocation Intermediates

In this section, we will discuss reactions which involve carbocation intermediates. These include carbon–carbon bond-forming reactions and also rearrangements and fragmentation reactions. The discussion will also include processes which are closely related in reactivity pattern but which avoid "free" carbocation intermediates.

10.1.1. Carbon–Carbon Bond Formation Involving Carbocations

The formation of carbon–carbon bonds by electrophilic attack on the π system is an important reaction in aromatic chemistry, with both Friedel-Crafts alkylation and acylation following this pattern. There also are valuable synthetic procedures in which carbon–carbon bond formation results from electrophilic attack by a carbocation on an alkene. The reaction of a carbocation with an alkene to form a new carbon–carbon bond is both kinetically and thermodynamically favorable.

$$>\overset{+}{C}- \; + \; >C=C< \;\rightarrow\; -\overset{|}{\underset{|}{C}}-\overset{|}{\underset{|}{C}}-\overset{+}{C}<$$

There are, however, important problems that must be overcome in the application of this reaction to synthesis. One is that the product is a new carbocation, which can react further. Repetitive addition to alkene molecules leads to polymerization. Indeed, this is the mechanism of acid-catalyzed polymerization of alkenes. There is also the possibility of rearrangement. A key requirement for adapting the reaction of carbocations with alkenes to the synthesis of small molecules is control of the reactivity of the newly formed carbocation intermediate. We have already encountered one successful strategy in the reaction of alkenyl and allylic silanes and stannanes with electrophilic carbon. In those reactions, the silyl or stannyl substituent is eliminated and a stable alkene is formed.

$$>\overset{+}{C}< \; + \; >C{=}C(M)< \;\rightarrow\; -C-C(M)-C^{+} \;\rightarrow\; -C-C{=}C<$$

The enhanced reactivity of the silyl- and stannyl-substituted alkenes is also favorable to the synthetic utility of carbocation–alkene reactions. Silyl enol ethers also show enhanced reactivity. Electrophilic attack is followed by desilylation to give an α-substituted carbonyl compound. The carbocations can be generated from tertiary chlorides and a Lewis acid, such as $TiCl_4$. This reaction provides a method for introducing tertiary alkyl groups alpha to a carbonyl. This transformation cannot be achieved by base-catalyzed alkylation because of the strong tendency for tertiary halides to undergo elimination.

$$\text{1-cyclohexenyl-}OSi(CH_3)_3 + (CH_3)_3CCl \xrightarrow[25^\circ C]{TiCl_4} \text{2-}C(CH_3)_3\text{-cyclohexanone} \;(40\text{–}50\%) \qquad \text{Ref. 1}$$

$$\text{1-cyclopentenyl-}OSi(CH_3)_3 + (CH_3)_2C(Cl)CH_2CH_3 \xrightarrow[-50^\circ C]{TiCl_4} \text{2-}[C(CH_3)_2CH_2CH_3]\text{-cyclopentanone} \;(62\%) \qquad \text{Ref. 2}$$

Slightly less reactive halides such as secondary benzyl bromides, allylic bromides, and α-chloroethers can undergo analogous reactions with the use of $ZnBr_2$ as the catalyst.[3]

Alkenes react with acid halides or acid anhydrides in the presence of a Lewis acid catalyst. The reaction works better with cyclic alkenes than acyclic ones.

$$M{-}\overset{+}{O}{=}C(R)X + H{-}C{-}C{=}C< \;\rightarrow\; M{-}\overset{+}{O}{-}C(R)(X){-}C{-}C{=}C<\;(H) \;\rightarrow\; O{=}C(R){-}C{-}C{=}C< \; + \; MX + H^{+}$$

1. T. H. Chan, I. Paterson, and J. Pinsonnault, *Tetrahedron Lett.*, 4183 (1977).
2. M. T. Reetz, I. Chatziiosifidis, U. Löwe, and W. F. Maier, *Tetrahedron Lett.*, 1427 (1979); M. T. Reetz, I. Chatziiosifidis, F. Hübner, and H. Heimbach, *Org. Synth.* **62**, 95 (1984).
3. I. Paterson, *Tetrahedron Lett.*, 1519 (1979).

A mechanistically significant feature of this reaction is the kinetic preference for formation of β,γ-unsaturated ketones. It has been suggested that this regiochemistry results from an intramolecular deprotonation, as shown in the mechanism above.[4] A related reaction occurs between alkenes and acylium ions, as exemplified by the reaction between 2-methylpropene and the acetylium ion.[5] The reaction leads regiospecifically to β,γ-enones. A concerted "ene reaction" mechanism has been suggested.

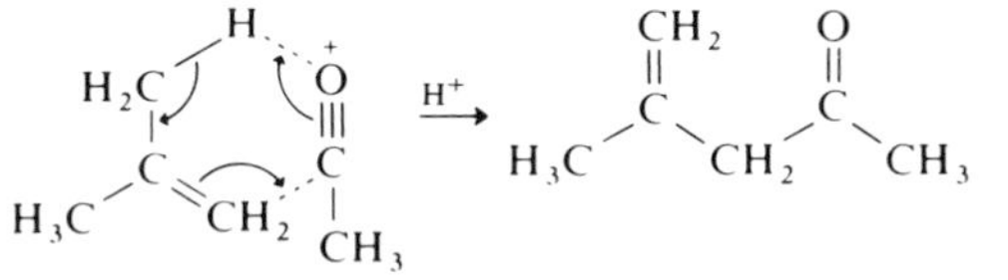

A variety of other reaction conditions have been examined for acylation of alkenes by acid chlorides. With the use of Lewis acid catalysts, reaction typically occurs to give both enones and β-haloketones.[6] This reaction has been most synthetically useful in intramolecular reactions. The following examples are illustrative.

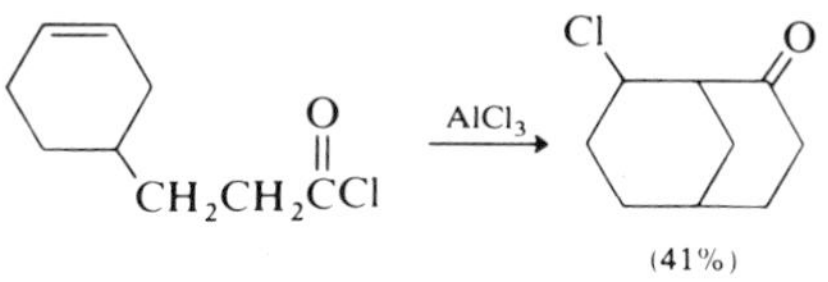

Ref. 7

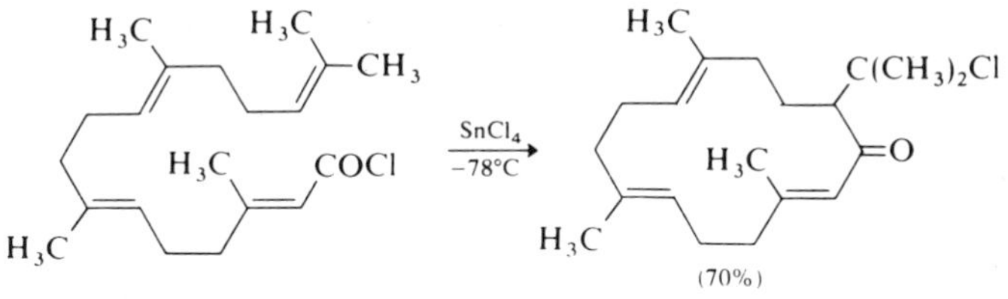

(41%)

Ref. 8

(70%)

Lewis acid catalyzed cyclization of unsaturated aldehydes is also an effective reaction. Stannic chloride is the usual catalyst for this cyclization.

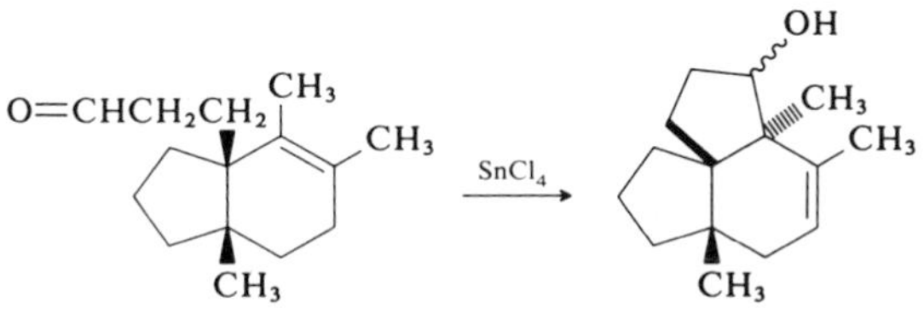

Ref. 9

4. P. Beak and K. R. Berger, *J. Am. Chem. Soc.* **102**, 3848 (1980).
5. H. M. R. Hoffmann and T. Tsushima, *J. Am. Chem. Soc.* **99**, 6008 (1977).
6. For example: T. S. Cantrell, J. M. Harless, and B. L. Strasser, *J. Org. Chem.* **36**, 1191 (1971); L. Rand and R. J. Dolinski, *J. Org. Chem.* **31**, 3063 (1966).
7. E. N. Marvell, R. S. Knutson, T. McEwen, D. Sturmer, W. Federici, and K. Salisbury, *J. Org. Chem.* **35**, 391 (1970).
8. T. Kato, M. Suzuki, T. Kobayashi, and B. P. Moore, *J. Org. Chem.* **45**, 1126 (1980).
9. L. A. Paquette and Y.-K. Han, *J. Am. Chem. Soc.* **103**, 1835 (1981).

In those cases where the molecular geometry makes it possible, the proton transfer step is intramolecular.[10]

Treatment of diazomethyl ketones with boron trifluoride generates an electrophilic reagent which undergoes cyclization when there is an appropriately located site of unsaturation.

CH_2CCHN_2 (O) / CH_3 $\xrightarrow{BF_3}$ =O / CH_3 (57%) Ref. 11

CF_3CO_2 / $CCHN_2$ (O) / CH_3O $\xrightarrow{BF_3}$ $\xrightarrow{^-OH}$ OH / =O / CH_3O (60%) Ref. 12

The reactive electrophilic species in this process can be formulated as a complexed β-oxyvinyl cation:

$$R-\overset{O}{\overset{\|}{C}}-CH=\overset{+}{N}=N^- \rightarrow R-\overset{\bar{B}F_3-O}{C}=CH-\overset{+}{N}\equiv N \xrightarrow{-N_2} R-\overset{\bar{B}F_3-O}{C}=\overset{+}{C}H$$

Perhaps most useful from a synthetic point of view are reactions of polyenes having two or more double bonds positioned in such a way that successive bond-forming steps can occur. This process, called *polyolefin cyclization*, has proven to be an effective way of making polycyclic compounds containing six-membered and, in some cases, five-membered rings. The reaction proceeds through an electrophilic attack and requires that the double bonds that participate in the cyclization be properly positioned. For example, compound **1** is converted quantitatively to **2** on treatment with formic acid. The reaction is initiated by protonation and ionization of the allylic alcohol.

HO / $(CH_2)_2$ / CH_3 / C−H / H−C / H **1** $\xrightarrow[-H_2O]{H}$ CH_2 / CH_3 CH_2 / CH_2=C−H $\rightarrow$ H / + / CH_3 $\xrightarrow{HCO_2H}$ H / OCH (O) / CH_3 **2** Ref. 13

10. N. H. Andersen and D. W. Ladner, *Synth. Commun.*, 449 (1978).
11. A. B. Smith III, B. H. Toder, S. J. Branca, and R. K. Dieter, *J. Am. Chem. Soc.* **103**, 1996 (1981).
12. T. R. Klose and L. M. Mander, *Aust. J. Chem.* **27**, 1287 (1974).
13. W. S. Johnson, P. J. Neustaedter, and K. K. Schmiegel, *J. Am. Chem. Soc.* **87**, 5148 (1965).

More extended polyenes can cyclize to tricyclic systems:

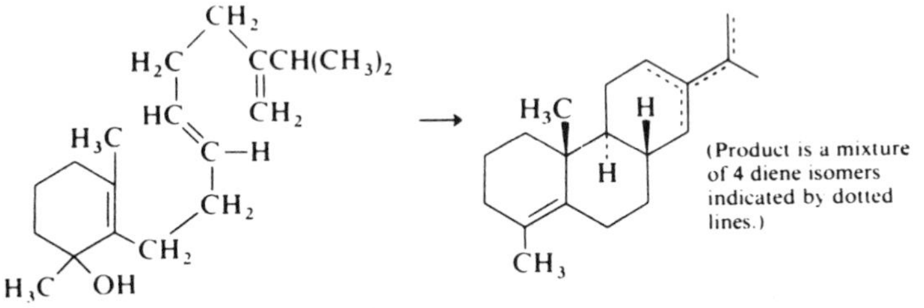

(Product is a mixture of 4 diene isomers indicated by dotted lines.)

Ref. 14

These cyclizations are usually highly stereoselective, with the stereochemical outcome being predictable on the basis of reactant conformation.[15] The stereochemistry of cyclization products in the decalin family can be predicted by assuming that cyclizations will occur through conformations that resemble chair cyclohexane rings. The stereochemistry at ring junctures is that expected from *anti* attack at the participating double bonds:

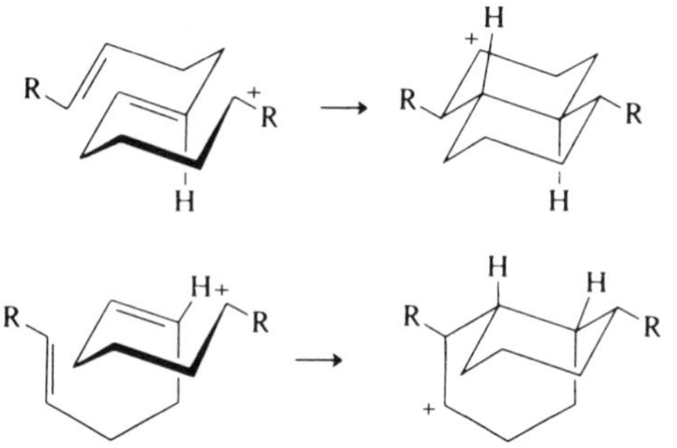

In order for the reaction to be of maximum synthetic value, the generation of the cationic site that initiates cyclization must involve mild reaction conditions. Formic acid and stannic chloride have proved to be effective reagents for cyclization of polyunsaturated allylic alcohols. Acetals generate α-alkoxy carbocations in acidic solution, and they can also be used to initiate the cyclization of polyenes[16]:

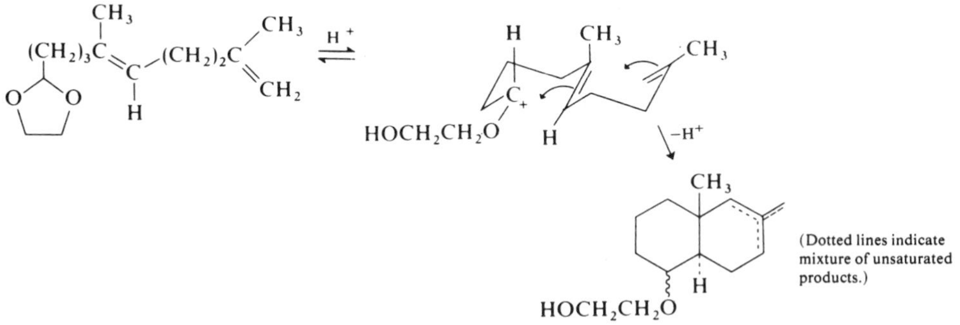

(Dotted lines indicate mixture of unsaturated products.)

14. W. S. Johnson, N. P. Jensen, J. Hooz, and E. J. Leopold, *J. Am. Chem. Soc.* **90**, 5872 (1968).
15. W. S. Johnson, *Acc. Chem. Res.* **1**, 1 (1968); P. A. Bartlett, in *Asymmetric Synthesis*, Vol. 3, J. D. Morrison (ed.), Academic Press, New York, 1984, Chapter 5.
16. A van der Gen, K. Wiedhaup, J. J. Swoboda, H. C. Dunathan, and W. S. Johnson, *J. Am. Chem. Soc.* **95**, 2656 (1973).

Another significant method for generating the electrophilic site is acid-catalyzed epoxide ring opening.[17]

Mercuric ion has been found to be capable of inducing cyclization of polyenes.

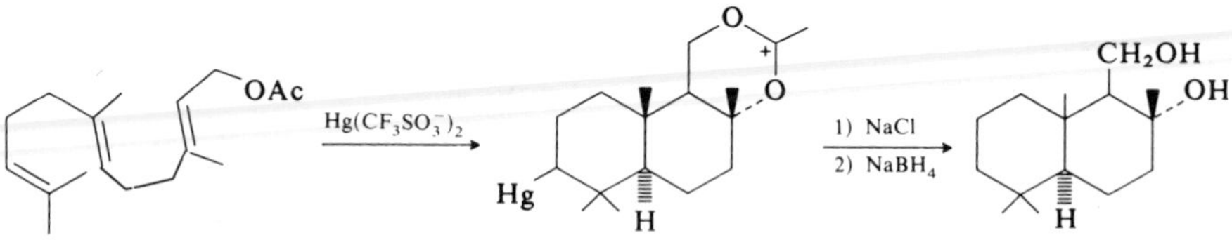

The particular example shown also has a special mechanism for stabilization of the cyclized carbocation. The adjacent acetoxy group is captured to form a stabilized dioxanylium cation. After reductive demercuration (see Section 4.3) and hydrolysis, a diol is isolated.

Since the immediate product of the polyolefin cyclization is a carbocation, it is often found that the final product is a mixture of closely related compounds resulting from the competing modes of reaction of the carbocation, that is, from capture by solvent or other nucleophile or deprotonation to form an alkene. Polyolefin cyclizations can be carried out on reactants which have special structural features in place that facilitate transformation of the carbocation to a stable product. Allylic silanes, for example, are stabilized by desilylation.[19]

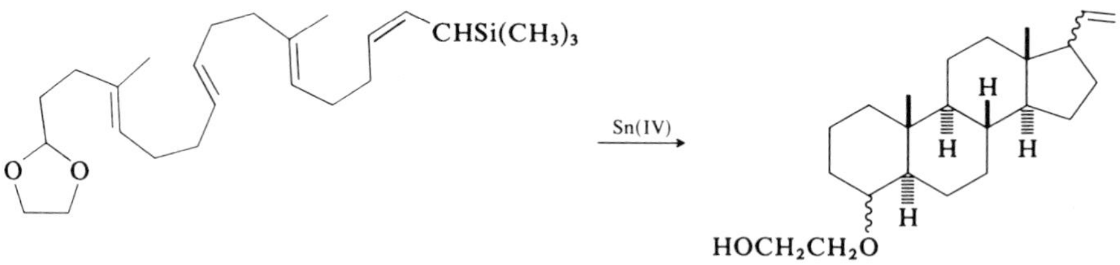

With terminal acetylenes, vinyl cations are formed. Capture of water leads to formation of a ketone.[20]

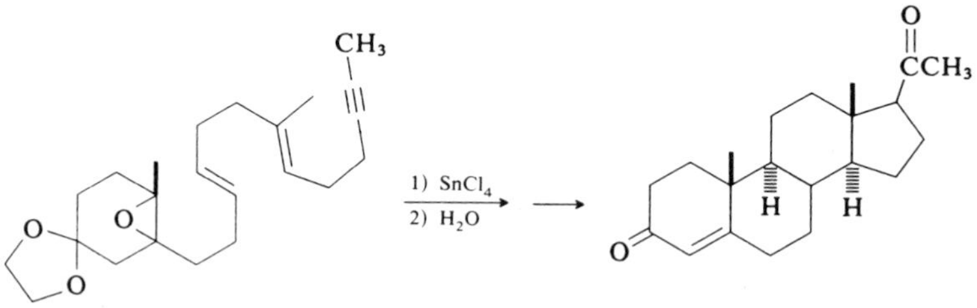

Polyolefin cyclizations have been of substantial value in the synthesis of polycyclic natural products of the terpene type. These syntheses resemble the processes by which the polycyclic compounds are assembled in nature. The most dramatic

17. E. E. van Tamelen and R. G. Nadeau, *J. Am. Chem. Soc.* **89**, 176 (1967).
18. M. Nishizawa, H. Takenaka, and Y. Hayashi, *J. Org. Chem.* **51**, 806 (1986); E. J. Corey, J. G. Reid, A. G. Myers, and R. W. Hahl, *J. Am. Chem. Soc.* **109**, 918 (1987).
19. W. S. Johnson, Y.-Q. Chen, and M. S. Kellogg, *J. Am. Chem. Soc.* **105**, 6653 (1983).
20. E. E. van Tamelen and J. R. Hwu, *J. Am. Chem. Soc.* **105**, 2490 (1983).

example of biological synthesis of a polycyclic skeleton from an acyclic intermediate is the conversion of squalene oxide to the steroid lanosterol. In the biological reaction, the enzyme presumably functions not only to induce the cationic cyclization but also to hold the substrate in a conformation corresponding to the stereochemistry of the polycyclic product.

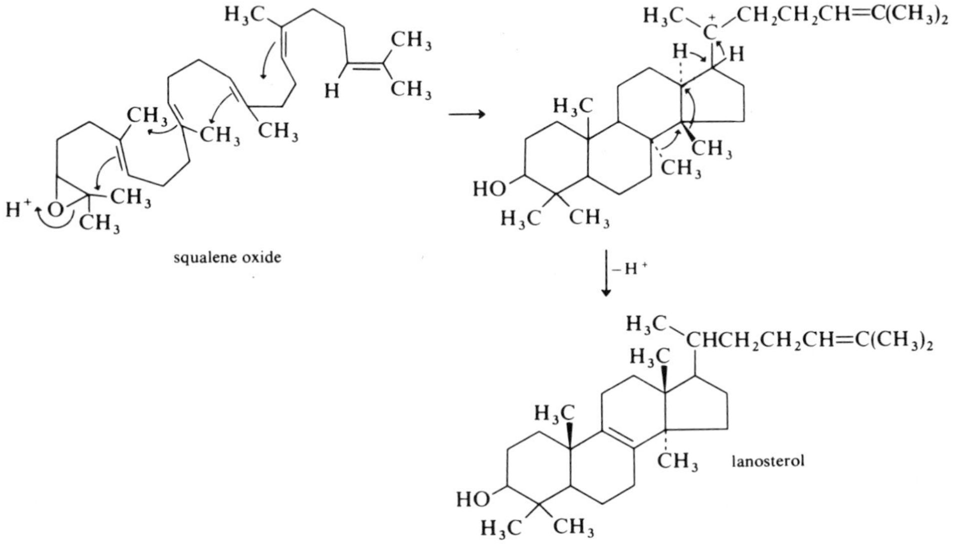

squalene oxide $\longrightarrow$ $\xrightarrow{-H^+}$ lanosterol

Scheme 10.1 gives some representative examples of laboratory syntheses involving polyene cyclization.

10.1.2. Rearrangements of Carbocations

Carbocations can be stabilized by the migration of hydrogen, alkyl or aryl groups, and, occasionally, other substituents. A mechanistic discussion of these reactions is given in Section 5.11 of Part A. Reactions involving rearrangements of carbocations are usually avoided in planning syntheses, because of the potential for complications resulting from competing rearrangement pathways. Rearrangements can become highly specific and, therefore, reliable synthetic reactions when the structural situation is such as to strongly favor a particular reaction. One case where this arises is in the reaction of carbocations having a hydroxyl group on an adjacent carbon. The formation of a carbonyl group is then possible:

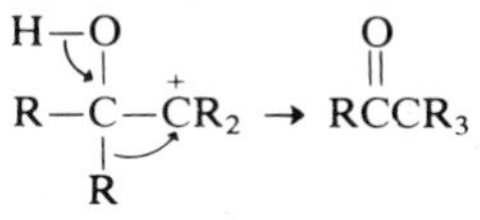

$$R-\overset{\overset{\displaystyle H-O}{|}}{\underset{\underset{\displaystyle R}{|}}{C}}-\overset{+}{C}R_2 \rightarrow R\overset{\overset{\displaystyle O}{\|}}{C}CR_3$$

A reaction that follows this pattern is the acid-catalyzed conversion of diols to ketones, which is known as the *pinacol rearrangement*.[21] The classic example of this

21. C. J. Collins, *Q. Rev.* **14**, 357 (1960).

Scheme 10.1. Polyolefin Cyclizations

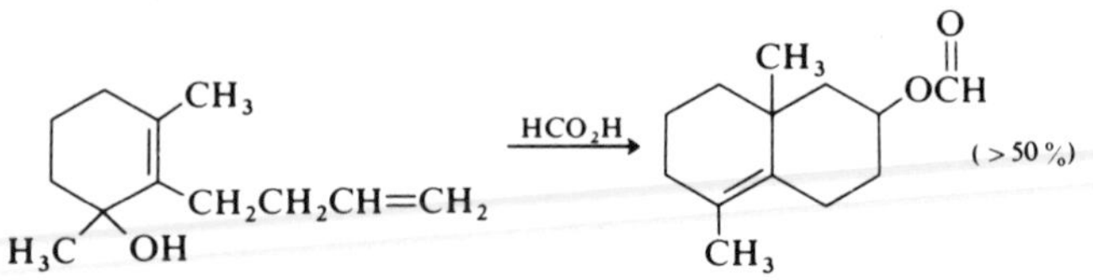

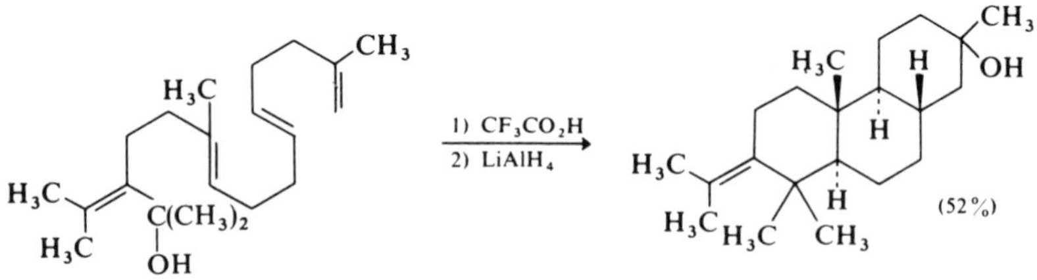

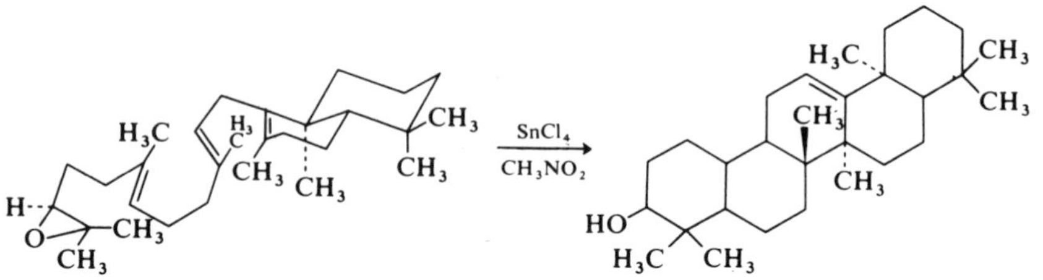

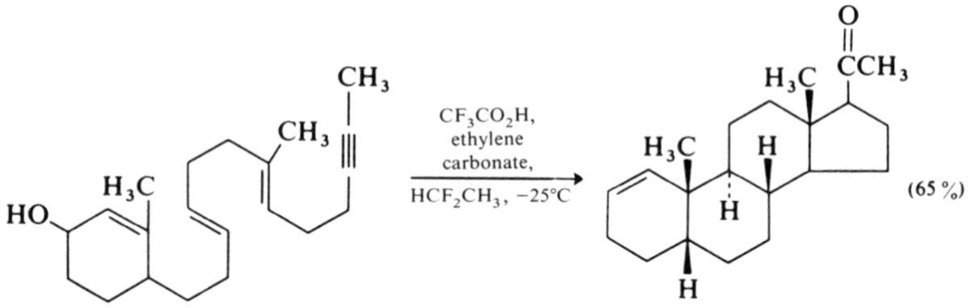

1[a] $\xrightarrow{HCO_2H}$ (> 50%)

2[b] 1) CF_3CO_2H 2) $LiAlH_4$ (52%)

3[c] $SnCl_4$, CH_3NO_2

4[d] CF_3CO_2H, ethylene carbonate, HCF_2CH_3, −25°C (65%)

5[e] $BF_3\cdot OEt_2$, Et_3N, −78°C (25-35%)

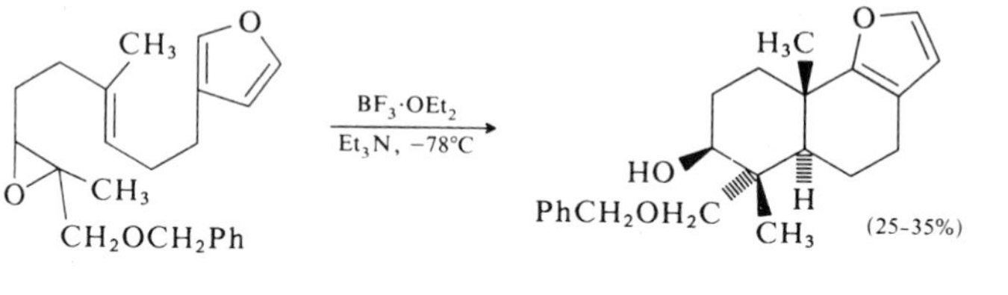

a. J. A. Marshall, N. Cohen, and A. R. Hochstetler, *J. Am. Chem. Soc.* **88**, 3408 (1966).
b. W. S. Johnson and T. K. Schaaf, *Chem. Commun.* 611 (1969).
c. E. E. van Tamelen, R. A. Holton, R. E. Hopla, and W. E. Konz, *J. Am. Chem. Soc.* **94**, 8228 (1972).
d. B. E. McCarry, R. L. Markezich, and W. S. Johnson, *J. Am. Chem. Soc.* **95**, 4416 (1973).
e. S. P. Tanis, Y.-H. Chuang, and D. B. Head, *J. Org. Chem.* **53**, 4929 (1988).

reaction is the conversion of 2,3-dimethylbutane-2,3-diol (pinacol) to methyl *t*-butyl ketones (pinacolone)[22]:

$$(CH_3)_2C(OH){-}C(OH)(CH_3)_2 \xrightarrow{H^+} CH_3C(=O)C(CH_3)_3 \quad (67\text{–}72\%)$$

22. G. A. Hill and E. W. Flosdorf, *Org. Synth.* **I**, 451 (1932).

The mechanism involves carbocation formation and substituent migration assisted by the hydroxyl group:

$$R_2C(OH)-CR_2(OH) \xrightarrow{H^+} RC(\overset{\delta+}{O}H)\cdots(R^{\delta+})\cdots CR_2 \longrightarrow RC(=\overset{+}{O}H)-CR_3 \longrightarrow RC(=O)CR_3 + H^+$$

The traditional conditions for carrying out the pinacol rearrangement involve treating the glycol with a strong acid. Under these conditions, the more easily ionized C—O bond generates the carbocation, and migration of one of the groups from the adjacent carbon ensues. Both stereochemistry and "migratory aptitude" can be factors in determining the extent of migration of the different groups.

Another method for carrying out the same net rearrangement involves synthesis of a glycol monosulfonate ester. These compounds rearrange under the influence of base.

$$B^- \; R_2C(HO)-CR_2(OSO_2R') \longrightarrow RC(O^-)-CR_2(OSO_2R)\;(R \text{ migrates}) \longrightarrow RC(=O)CR_3$$

These conditions permit greater control over the course of the rearrangement, since ionization will take place only at the sulfonylated alcohol. These reactions have been of value in rearranging ring systems, especially in the synthesis of terpenes, as illustrated by entries 3 and 4 in Scheme 10.2.

In cyclic systems that enforce sufficient structural rigidity or conformational bias, the course of the rearrangement is controlled by stereoelectronic factors. The carbon substituent that is *anti* to the leaving group is the one that undergoes migration. In cyclic systems such as **3**, for example, selective migration of the ring fusion bond occurs because of this stereoelectronic effect.

CH_3SO_2O, CH_3, $PhCH_2O$, CH_3, H, O, CH_3, O — **3** $\longrightarrow$ CH_3, +, $PhCH_2O$, CH_3, H, O, CH_3, O $\xrightarrow{-H^+}$ $PhCH_2O$, CH_3, H, O, CH_3, O — **4**

Ref. 23

23. M. Ando, A. Akahane, H. Yamaoka, and K. Takase, *J. Org. Chem.* **47**, 3909 (1982).

Scheme 10.2. Rearrangements Promoted by Adjacent Heteroatoms

A. Pinacol-Type Rearrangements

1[a]

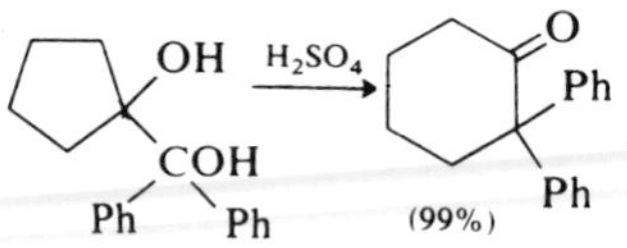

OH H_2SO_4 → O Ph Ph
COH
Ph Ph (99%)

2[b]

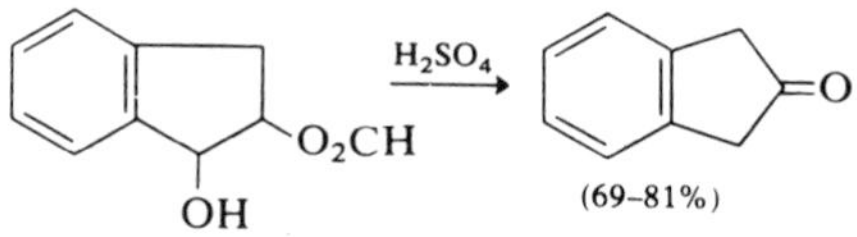

H_2SO_4 → =O
O_2CH
OH (69–81%)

3[c]

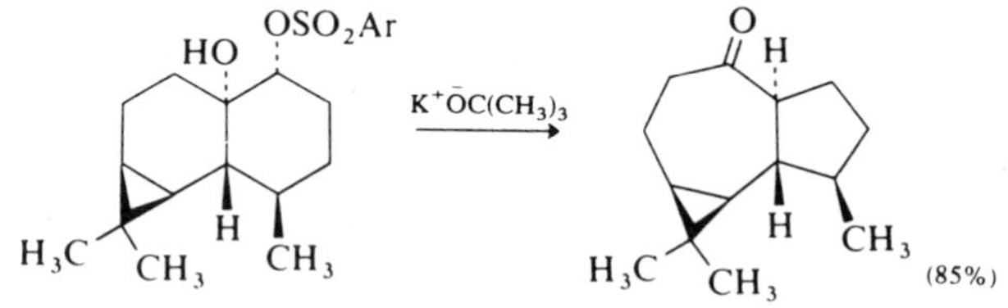

OSO_2Ar HO $K^+\bar{O}C(CH_3)_3$ → O H
H H_3C CH_3 CH_3 H CH_3 H_3C CH_3 (85%)

4[d]

CH_3SO_2O OH CH_3 CH_3 pyridine Et_3N → H O CH_3 CH_3
H_3CO_2C H_3CO_2C (91%)

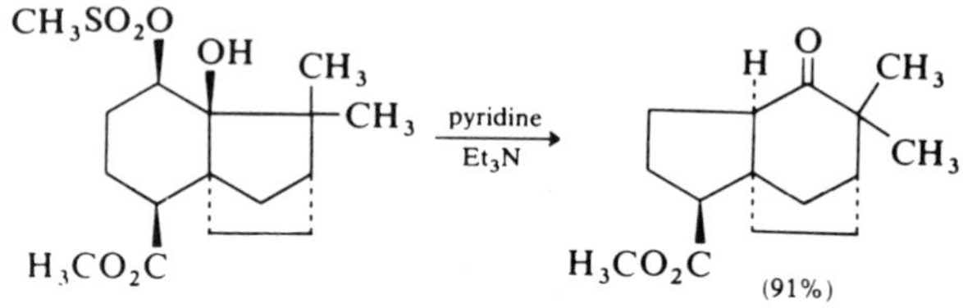

B. Rearrangement of β-Aminoalcohols by Diazotization

5[e]

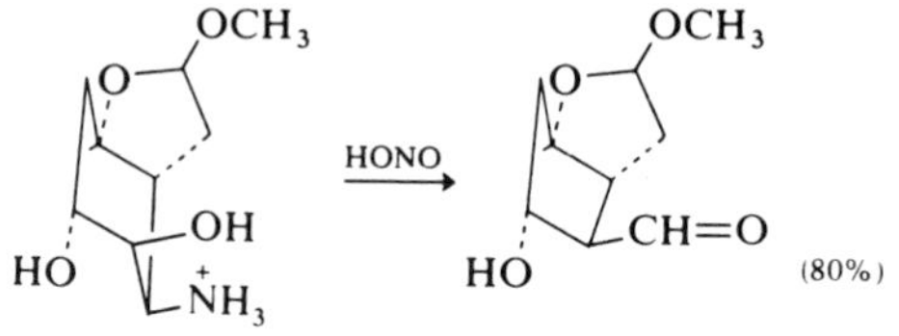

OCH_3 O OH HO $\overset{+}{N}H_3$ HONO → OCH_3 O HO CH=O (80%)

6[f]

H_3C H_3C =O CH_3 1) $(CH_3)_3SiCN$ 2) $LiAlH_4$ 3) HNO_2 → H_3C H_3C O CH_3 (75–85%) + H_3C H_3C O CH_3 (15–25%) (total yield 70%)

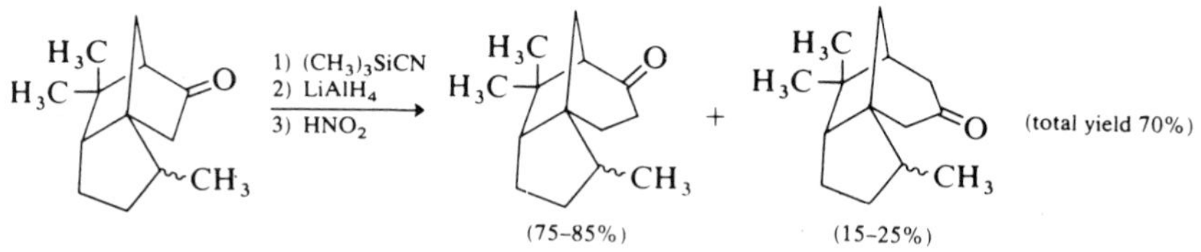

C. Ring Expansion of Cyclic Ketones with Diazo Compounds

7[g]

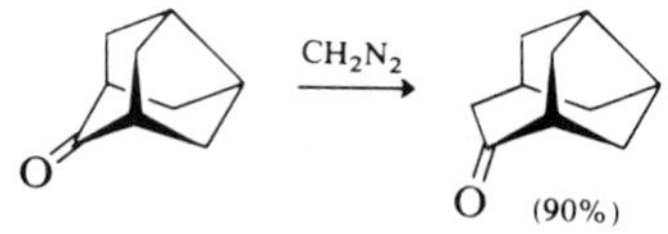

O CH_2N_2 → O (90%)

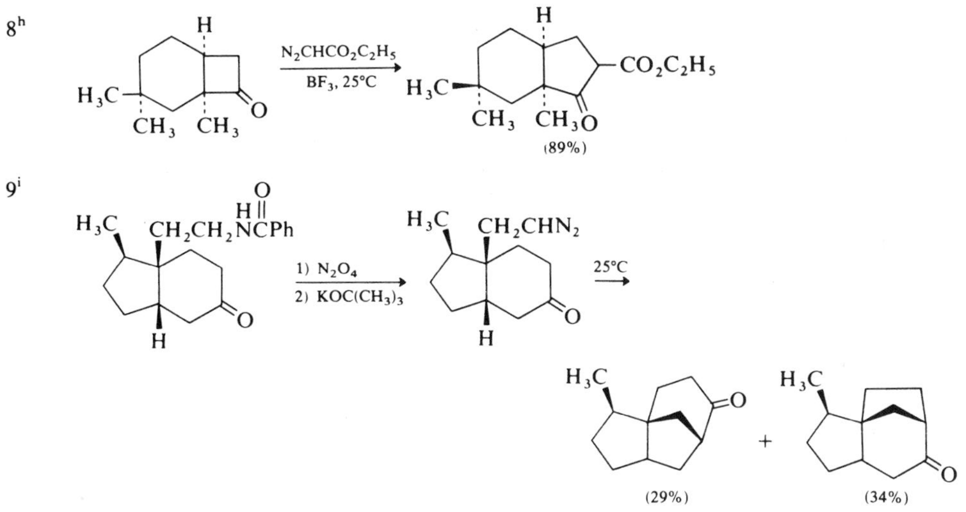

8[h]

H, H_3C, CH_3 CH_3, O — $N_2CHCO_2C_2H_5$, BF_3, 25°C → H, H_3C, $-CO_2C_2H_5$, CH_3 CH_3O (89%)

9[i]

H_3C $CH_2CH_2\overset{H\,\|}{N}CPh$ (O), H, O — 1) N_2O_4 2) $KOC(CH_3)_3$ → H_3C CH_2CHN_2, H, O — 25°C →

H_3C, =O (29%) + H_3C, O (34%)

a. H. E. Zaugg, M. Freifelder, and B. W. Horrom, *J. Org. Chem.* **15**, 1191 (1950).
b. J. E. Horan and R. W. Schiessler, *Org. Synth.* **41**, 53 (1961).
c. G. Büchi, W. Hofheinz, and J. V. Paukstelis, *J. Am. Chem. Soc.* **88**, 4113 (1966).
d. D. F. MacSweeney and R. Ramage, *Tetrahedron* **27**, 1481 (1971).
e. R. B. Woodward, J. Gosteli, I. Ernest, R. J. Friary, G. Nestler, H. Raman, R. Sitrin, C. Suter, and J. K. Whitesell, *J. Am. Chem. Soc.* **95**, 6853 (1973).
f. E. G. Breitholle and A. G. Fallis, *J. Org. Chem.* **43**, 1964 (1978).
g. Z. Majerski, S. Djigas, and V. Vinkovic, *J. Org. Chem.* **44**, 4064 (1979).
h. H. J. Liu and T. Ogino, *Tetrahedron Lett.*, 4937 (1973).
i. P. R. Vettel and R. M. Coates, *J. Org. Chem.* **45**, 5430 (1980).

Similarly, **5** gives **6** by antiperiplanar migration.

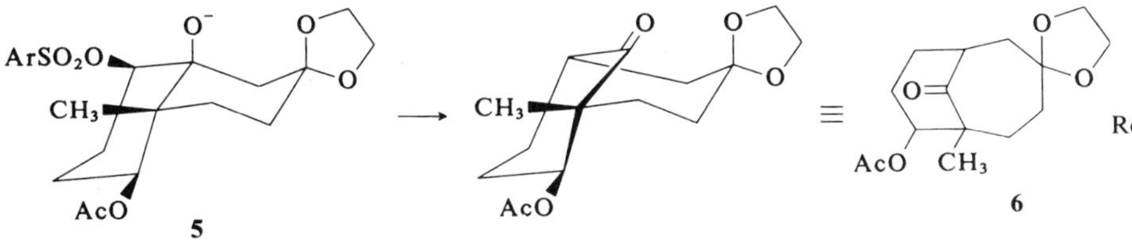

O^-, O, O, $ArSO_2O$, CH_3, AcO, **5** → O, O, O, CH_3, AcO ≡ O, O, O=, AcO CH_3, **6** Ref. 24

There is kinetic evidence that the migration step in these base-catalyzed rearrangements is concerted with ionization. Thus, in cyclopentane derivatives, the rate of reaction depends on the nature of the *trans* substituent R[25]:

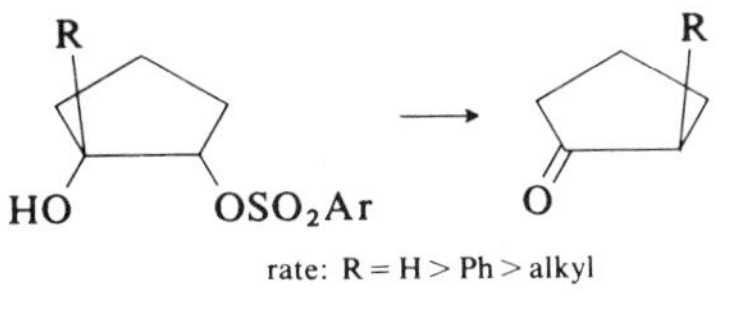

R, HO, OSO_2Ar → R, O

rate: R = H > Ph > alkyl

This implies that the migration is part of the rate-determining step.

24. C. H. Heathcock, E. G. Del Mar, and S. L. Graham, *J. Am. Chem. Soc.* **104**, 1907 (1982).
25. E. Wistuba and C. Rüchardt, *Tetrahedron Lett.* **22**, 4069 (1981).

Aminomethylcarbinols yield ketones when treated with nitrous acid. This reaction has been used to form ring-expanded cyclic ketones, a procedure which is called the *Tiffeneau–Demjanov reaction.*[26]

$$R_2C(OH)CH_2NH_2 \xrightarrow{HONO} R_2C(OH)CH_2\overset{+}{N}{\equiv}N \rightarrow RC(OH)(R)-\overset{+}{C}H_2 \rightarrow RC(=O)CH_2R$$

Ref. 27

$$\text{1-(aminomethyl)cycloheptanol (HO, } CH_2NH_2\text{)} \xrightarrow{HONO} \text{cyclooctanone (61\%)}$$

The diazotization reaction generates the same type of β-hydroxycarbocation that is involved in the pinacol rearrangement.

One method for obtaining the required aminomethylcarbinols involves cyanohydrin formation, followed by reduction:

$$\text{cyclohexanone} \xrightarrow{HC{\equiv}N} \text{1-hydroxycyclohexane-1-carbonitrile (HO, } C{\equiv}N\text{)} \rightarrow \text{1-(aminomethyl)cyclohexanol (HO, } CH_2NH_2\text{)}$$

The reaction can also be carried out on trimethylsilyl ethers of cyanohydrins, which can be obtained from the ketone and trimethylsilyl cyanide. The cyano ether is reduced to the aminomethylcarbinol by $LiAlH_4$:

$$\text{cyclohexanone} + (CH_3)_3SiCN \xrightarrow{ZnI_2} \text{(N}{\equiv}C\text{, } OSi(CH_3)_3\text{)cyclohexane} \xrightarrow{LiAlH_4} \text{(HO, } CH_2NH_2\text{)cyclohexane}$$

Ref. 28

Another method for synthesis of aminomethyl cycloalkanols involves reduction of the adducts of nitromethane and cyclic ketones:

$$\text{cyclohexanone} \xrightarrow[-OH]{CH_3NO_2} \text{(HO, } CH_2NO_2\text{)cyclohexane} \xrightarrow{H_2,\ Ni} \text{(HO, } CH_2NH_2\text{)cyclohexane}$$

Ref. 29

The reaction of ketones with diazoalkanes sometimes leads to a ring-expanded ketone in synthetically useful yields.[30] The reaction occurs by addition of the diazoalkane, followed by elimination of nitrogen and migration:

$$\text{C=O ring } (CH_2)_x + CH_2N_2 \rightarrow {}^{-}O\text{–C(}CH_2\overset{+}{N}{\equiv}N\text{) ring } (CH_2)_x \rightarrow \text{C=O ring } (CH_2)_{x+1}$$

26. P. A. S. Smith and D. R. Baer, *Org. React.* **11**, 157 (1960).
27. F. F. Blicke, J. Azuara, N. J. Dorrenbos, and E. B. Hotelling, *J. Am. Chem. Soc.* **75**, 5418 (1953).
28. D. A. Evans, G. L. Carroll, and L. K. Truesdale, *J. Org. Chem.* **39**, 914 (1974).
29. W. E. Noland, J. F. Kneller, and D. E. Rice, *J. Org. Chem.* **22**, 695 (1957).
30. C. D. Gutsche, *Org. React.* **8**, 364 (1954).

The rearrangement proceeds via essentially the same intermediate as is involved in the Tiffeneau–Demjanov reaction. Since the product is also a ketone, a subsequent addition of diazomethane can lead to higher homologs. The best yields are obtained when the starting ketone is more reactive than the product. For this reason, strained ketones work especially well. Higher diazoalkanes can also be used in place of diazomethane. The reaction is found to be accelerated by alcoholic solvents. This effect probably involves the hydroxyl group being hydrogen-bonded to the carbonyl oxygen and acting as a proton donor in the addition step.[31]

$$R_2C{=}O\cdots H{-}O{-}R \;+\; {:}CH_2N_2 \longrightarrow R{-}\underset{R}{\overset{OH}{C}}{-}CH_2\overset{+}{N}{\equiv}N$$

Ketones react with esters of diazoacetic acid in the presence of Lewis acids such as BF_3 and SbF_5.[32]

$$\text{cyclohexanone} + N_2CHCO_2C_2H_5 \xrightarrow{SbCl_5} \text{2-}(CO_2C_2H_5)\text{cycloheptanone}$$

In unsymmetrical ketones, the direction of migration can be controlled by the presence of a halogen substituent. The electronegative halogen retards migration, and the migration of the nonhalogenated substituent is preferred.

Intramolecular reactions between diazo groups and carbonyl centers can be used to construct bicyclic ring systems:

$$\text{2-}[(CH_2)_3N(N{=}O)CCH_3(=O)]\text{cyclopentanone} \xrightarrow{CH_3O^-} \text{bicyclic ketone} \quad (65\%)$$

Ref. 33

The pinacol rearrangement pattern can also be seen in a procedure for ketone ring expansion which begins with addition of a lithio sulfone[34]:

$$\underset{(CH_2)_n}{C{=}O} + LiCHSO_2Ph\,(X) \longrightarrow \underset{(CH_2)_n}{C}(OLi)(CHXSO_2Ph) \xrightarrow{Et_2AlCl} \underset{(CH_2)_n}{C(=O)CHX}$$

X = SPh or OMe

The rearrangement step is induced by diethylaluminum chloride, which promotes rearrangement by complexation at the sulfonyl group.

31. J. N. Bradley, G. W. Cowell, and A. Ledwith, *J. Chem. Soc.*, 4334 (1964).
32. H. J. Liu and T. Ogino, *Tetrahedron Lett.*, 4937 (1973); W. T. Tai and E. W. Warnhoff, *Can. J. Chem.* **42**, 1333 (1964); W. L. Mock and M. E. Hartman, *J. Org. Chem.* **42**, 459 (1977); V. Dave and E. W. Warnhoff, *J. Org. Chem.* **48**, 2590 (1983).
33. C. D. Gutsche and D. M. Bailey, *J. Org. Chem.* **28**, 607 (1963).
34. B. M. Trost and G. K. Mikhail, *J. Am. Chem. Soc.* **109**, 4124 (1987).

10.1.3. Related Rearrangements

α-Haloketones undergo a skeletal change when treated with base. The overall structural change is similar to that caused by rearrangement of α-diazoketones. The most commonly used bases are alkoxide ions, which lead to esters as the reaction products:

$$\mathrm{RCH_2C(=O)CH(X)R'} \xrightarrow{\mathrm{CH_3O^-}} \mathrm{CH_3OC(=O)CH(CH_2R)R'} + \mathrm{X^-}$$

This reaction is known as the *Favorskii rearrangement.*[35] If the ketone is cyclic, a ring contraction occurs.

$$\text{2-chlorocyclohexanone} \xrightarrow{\mathrm{Na^+\ ^-OMe}} \text{cyclopentyl-}\mathrm{CO_2Me} \qquad \text{Ref. 36}$$

The reaction has been subjected to extensive mechanistic study. There is strong evidence that the rearrangement involves the open 1,3-dipolar isomers of cyclopropanone and/or cyclopropanones as reaction intermediates.[37]

$$\mathrm{RCH_2C(=O)CH(X)R'} \overset{^-\mathrm{OR}^-}{\rightleftarrows} \mathrm{R\bar{C}HC(=O)CH(X)R'} \rightarrow \mathrm{R\bar{C}HC(=O)\overset{+}{C}HR'} \leftrightarrow \mathrm{R\overset{+}{C}HC(=O)\bar{C}HR'}$$

$$\downarrow$$

$$\mathrm{RCH(CO_2R'')CH_2R'} + \mathrm{RCH_2CH(CO_2R'')R'} \longleftarrow \underset{\text{RHC—CHR'}}{\mathrm{C(O^-)(OR'')}} \xleftarrow{\ ^-\mathrm{OR''}\ } \underset{\text{RHC—CHR'}}{\mathrm{C{=}O}}$$

There is also a related mechanism that can operate in the absence of an acidic α-hydrogen. This process is called the "semibenzilic" rearrangement:

$$\mathrm{RC(=O)CH(X)R'} \xrightarrow{\mathrm{R''O^-}} \mathrm{RC(O^-)(OR'')\text{—}CH(X)R} \rightarrow \mathrm{R''O\text{—}C(=O)\text{—}CH(R)R'}$$

The net structural change is the same for both mechanisms. The energy requirements of the cyclopropanone and semibenzilic mechanisms may be fairly closely balanced, because cases of operation of the semibenzilic mechanism have been reported even for compounds with a hydrogen available for enolization.[38] Included in the evidence that the cyclopropanone mechanism usually operates is the demonstration that a

35. A. S. Kende, *Org. React.* **11**, 261 (1960); A. A. Akhrem, T. K. Ustynyuk, and Y. A. Titov, *Russ. Chem. Rev.* (Engl. transl.), **39**, 732 (1970).
36. D. W. Goheen and W. R. Vaughan, *Org. Synth.* **IV**, 594 (1963).
37. F. G. Bordwell, T. G. Scamehorn, and W. R. Springer, *J. Am. Chem. Soc.* **91**, 2087 (1969); F. G. Bordwell and J. G. Strong, *J. Org. Chem.* **38**, 579 (1973).
38. E. W. Warnhoff, C. M. Wong, and W. T. Tai, *J. Am. Chem. Soc.* **90**, 514 (1968).

symmetrical intermediate is involved. The isomeric chloroketones **7** and **8**, for example, lead to the same ester:

$$\underset{\mathbf{7}}{PhCH(Cl)C(=O)CH_3} \xrightarrow{CH_3O^-} \underset{\mathbf{9}}{PhCH_2CH_2CO_2CH_3} \xleftarrow{CH_3O^-} \underset{\mathbf{8}}{PhCH_2C(=O)CH_2Cl} \quad \text{Ref. 37}$$

It has been shown that the two ketones are not interconverted under the conditions of the rearrangement. A common intermediate in the rearrangements of the two ketones can explain these observations, and the cyclopropanone intermediate would meet this requirement. The involvement of a symmetrical intermediate has also been demonstrated by ^{14}C labeling in he case of α-chlorocyclohexanone[39]:

$* = {}^{14}C$ label

Numbers refer to percentage of label at each carbon.

Because of the operation of the cyclopropanone mechanism, the structure of the ester product cannot be predicted directly from the structure of the reacting haloketone. Instead, the identity of the product is governed by the direction of ring opening of the cyclopropanone intermediate. The dominant mode of ring opening would be expected to be that which forms the more stable of the two possible ester enolates. For this reason, a phenyl substituent favors breaking of the bond to the substituted carbon, but an alkyl group directs the cleavage to the less substituted carbon.[40] That both **7** and **8** above give the same ester, **9**, is illustrative of the directing effect that the phenyl group can have on the ring-opening step.

α-Alkoxyketones are common by-products of Favorskii rearrangements catalyzed by alkoxide ions. Generally, these by-products are not formed by a direct S_N2 displacement, since a symmetrical intermediate appears to be involved. The most satisfactory general mechanism suggests that the enol form of the chloroketone is the precursor of the alkoxyketones[41]:

$$RCH_2C(=O)CHXR' \overset{RO^-}{\rightleftharpoons} RCH{=}C(OH)CHXR' \qquad RCHX{-}C(OH){=}CHR' \overset{RO^-}{\rightleftharpoons} RCHXC(=O)CH_2R'$$

$$\downarrow \qquad\qquad \downarrow$$

$$RCH{=}C(OH){-}\overset{+}{C}HR' \leftrightarrow R\overset{+}{C}H{-}C(OH){=}CHR' \xrightarrow{ROH} RCH_2C(=O)CH(OR)R' + RCH(OR)C(=O)CH_2R'$$

39. R. B. Loftfield, *J. Am. Chem. Soc.* **73**, 4707 (1951).
40. C. Rappe, L. Knutsson, N. J. Turro, and R. B. Gagosian, *J. Am. Chem. Soc.* **92**, 2032 (1970).
41. F. G. Bordwell and M. W. Carlson, *J. Am. Chem. Soc.* **92**, 3377 (1970).

The Favorskii reaction has been used to effect ring contraction in the course of synthesis of strained ring compounds. Entry 4 in Scheme 10.3 illustrates this application of the reaction. With α,α'-dihaloketones, the rearrangement is accompanied by dehydrohalogenation to yield an α,β-unsaturated ester, as illustrated by entry 3 in Scheme 10.3.

α-Halosulfones undergo a related rearrangement known as the *Ramberg–Bäcklund reaction.* The carbanion formed by deprotonation forms an unstable thiirane dioxide.[42] The thiirane dioxide decomposes with elimination of sulfur dioxide.

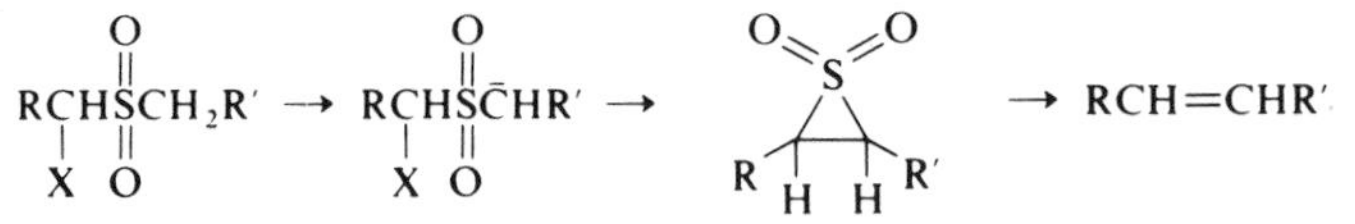

$$RCH(X)SO_2CH_2R' \rightarrow RCH(X)SO_2\bar{C}HR' \rightarrow \text{thiirane dioxide (R, H, H, R')} \rightarrow RCH{=}CHR'$$

The reaction is useful for the synthesis of certain types of alkenes:

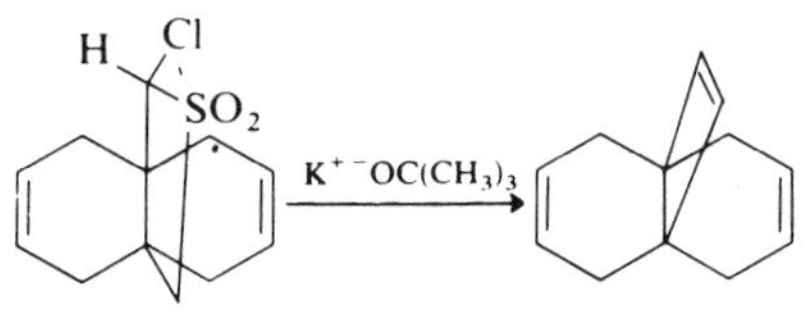

H Cl SO$_2$ $\xrightarrow{K^{+}\,{}^{-}OC(CH_3)_3}$ Ref. 43

Scheme 10.3. Base-Catalyzed Rearrangements of α-Haloketones

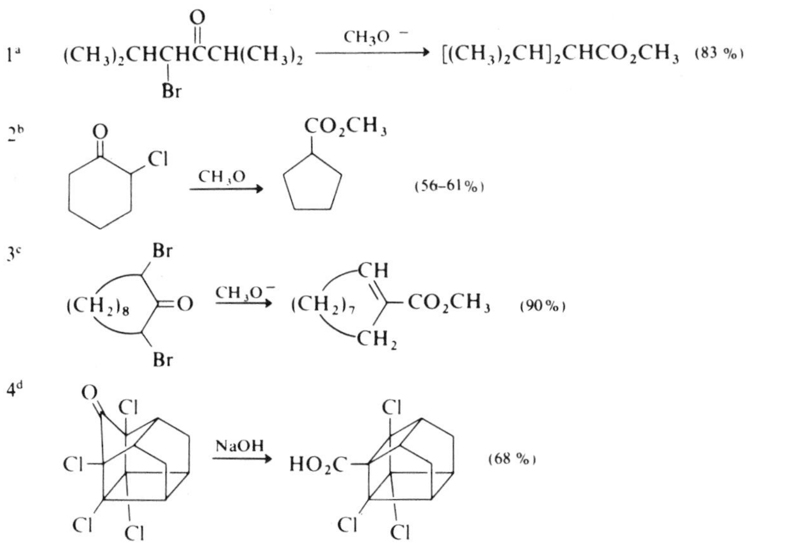

1[a] $(CH_3)_2CHCH(Br)C(O)CH(CH_3)_2 \xrightarrow{CH_3O^-} [(CH_3)_2CH]_2CHCO_2CH_3$ (83%)

2[b] 2-chlorocyclohexanone $\xrightarrow{CH_3O^-}$ CO_2CH_3 (methyl cyclopentanecarboxylate) (56–61%)

3[c] Br, $(CH_2)_8$, =O, Br $\xrightarrow{CH_3O^-}$ $(CH_2)_7$, CH, CO_2CH_3, CH_2 (90%)

4[d] O, Cl, Cl, Cl, Cl $\xrightarrow{NaOH}$ HO_2C, Cl, Cl, Cl (68%)

a. S. Sarel and M. S. Newman, *J. Am. Chem. Soc.* **78**, 416 (1956).
b. D. W. Goheen and W. R. Vaughan, *Org. Synth.* **IV**, 594 (1963).
c. E. W. Garbisch, Jr., and J. Wohllebe, *J. Org. Chem.* **33**, 2157 (1968).
d. R. J. Stedman, L. S. Miller, L. D. Davis, and J. R. E. Hoover, *J. Org. Chem.* **35**, 4169 (1970).

42. L. A. Paquette, *Acc. Chem. Res.* **1**, 209 (1968); L. A. Paquette, in *Mechanism of Molecular Migrations*, Vol. 1, B. S. Thyagaragan (ed.), Wiley-Interscience, New York, 1968, Chapter 3.
43. L. A. Paquette, J. C. Philips, and R. E. Wingard, Jr., *J. Am. Chem. Soc.* **93**, 4516 (1971).

10.1.4. Fragmentation Reactions

The classification "fragmentation" applies to reactions in which a carbon–carbon bond is broken when an electron deficiency develops. A structural feature that permits fragmentation to occur readily is the presence of a carbon that can accommodate carbocationic character *beta* to the developing electron deficiency. Fragmentation occurs particularly readily when the γ atom is a heteroatom, such as nitrogen or oxygen, having an unshared electron pair available for stabilization of the new cationic center.[44]

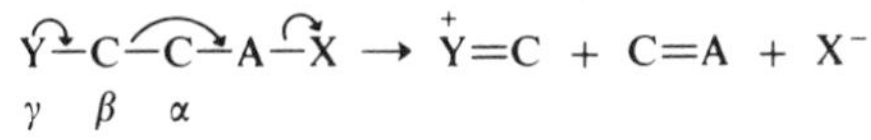

$$\ddot{Y}-C-C-A-X \rightarrow \overset{+}{Y}=C + C=A + X^-$$
$$\gamma \quad \beta \quad \alpha$$

The fragmentation can be concerted or stepwise. The concerted mechanism is restricted to molecular geometry that is appropriate for continuous overlap of the participating orbitals. An example is the solvolysis of 4-chloropiperidine, which is more rapid than the solvolysis of chlorocyclohexane and occurs with ring opening by fragmentation of the C-2—C-3 bond.[45]

A particularly useful type of substrate for fragmentation is diols or hydroxyethers in which the two oxygen substituents are in a 1,3 relationship. If the diol or hydroxyether is converted to a monotosylate, the remaining oxygen function can serve to promote fragmentation.

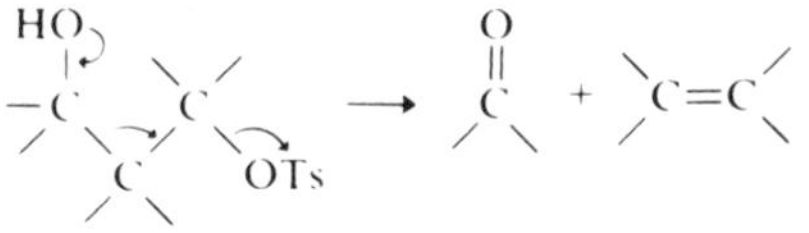

This reaction pattern can be seen in a fragmentation used to construct the ring structure found in the taxane group of diterpenes:

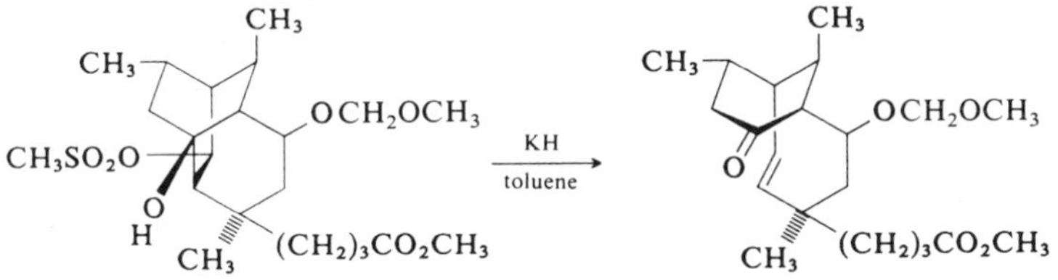

Ref. 46

44. C. A. Grob, *Angew. Chem. Int. Ed. Engl.* **8**, 535 (1969).
45. R. D'Arcy, C. A. Grob, T. Kaffenberger, and V. Krasnobajew, *Helv. Chim. Acta* **49**, 185 (1966).
46. H. Nagaoka, K. Ohsawa, T. Takata, and Y. Yamad, *Tetrahedron Lett.* **25**, 5389 (1984).

Scheme 10.4. Fragmentation Reactions

A. Heteroatom-Promoted Fragmentation

1[a] CH_3SO_2O ... O^- ... CH_3 — $K^+{}^-OC(CH_3)_3$ → $H_2C{=}CH$ ... O ... CH_3 (64%)

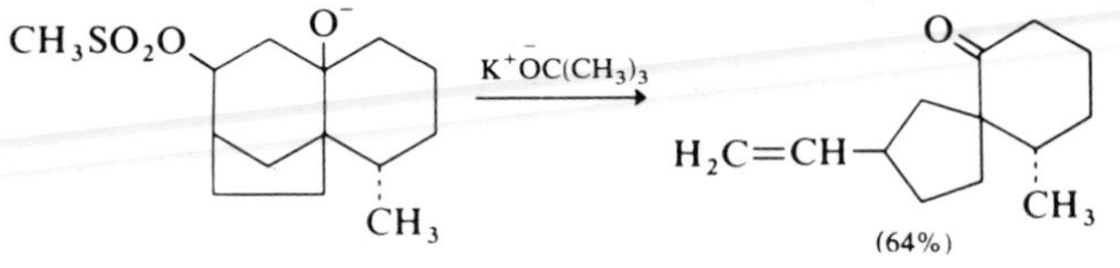

2[b] H_3C, OSO_2CH_3, OH — $LiAlH_4$ → H_3C, OH (71%)

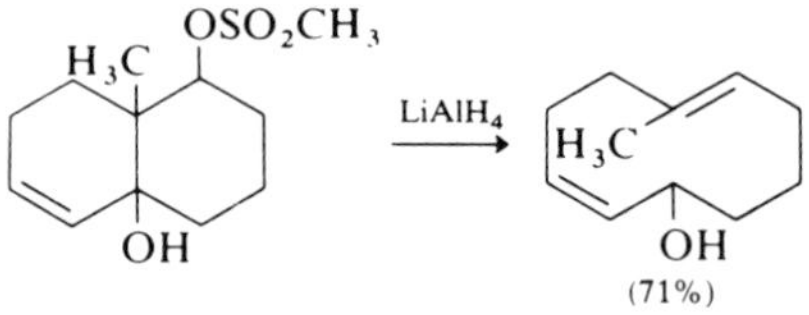

3[c] OCH_3, CCH_3, O — H^+, CH_3CO_2H, 25°C, 0.5 h → O, $CH_2CH_2\overset{O}{\overset{\|}{C}}CH_3$ (70%)

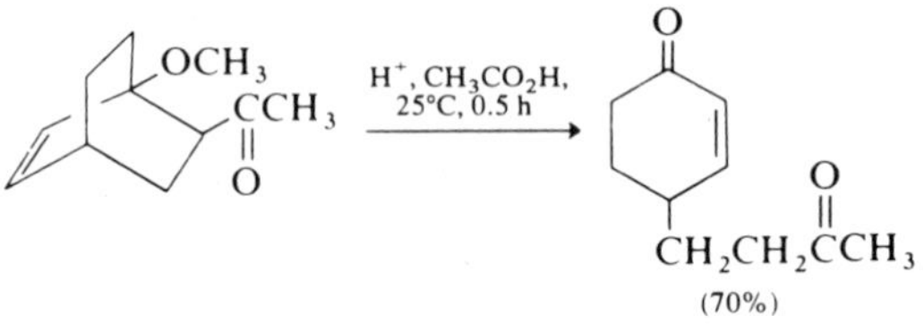

4[d] OSO_2Ar, N, CH_2Ph — solvolysis in the presence of $NaBH_4$ → N, CH_2Ph (44–58%)

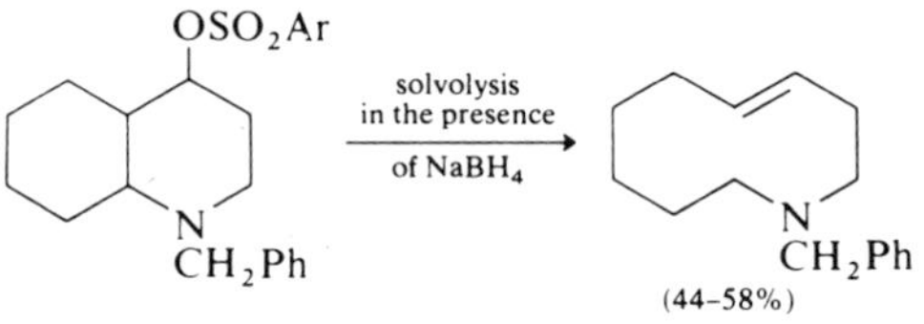

B. Boronate Fragmentations

5[e] OSO_2CH_3, CH_3, H_3C, CH_3 — 1) B_2H_6 2) H_2O_2, ^-OH → H_3C, CH_3, CH_3 (70%)

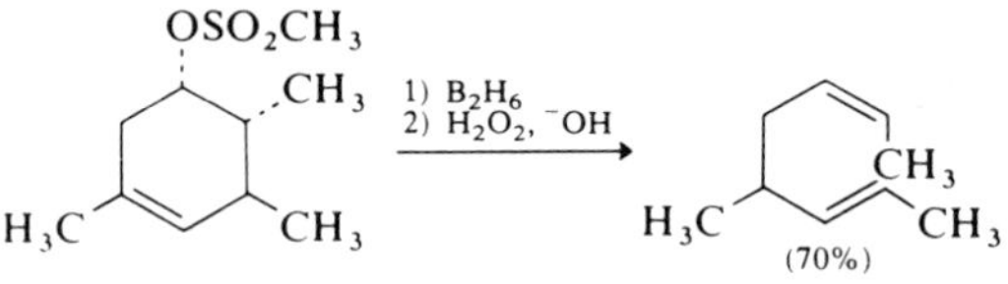

C. δ-Tosyloxy Enolate Fragmentation

6[f] O, CH_3, CH_3, $OSO_2C_6H_5$ — 1) $LiNR_2$ 2) R_2AlH, 25°C, 2 h → OH, CH_3, CH_3, H (93%)

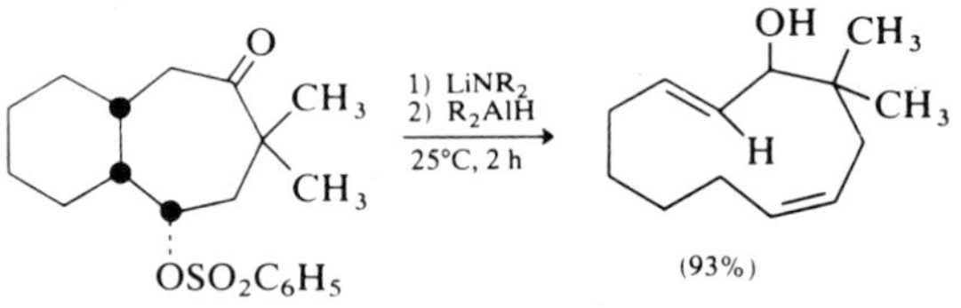

a. J. A. Marshall and S. F. Brady, *J. Org. Chem.* **35**, 4068 (1970).
b. J. A. Marshall, W. F. Huffman, and J. A. Ruth, *J. Chem. Soc.* **94**, 4691 (1972).
c. A. J. Birch and J. S. Hill, *J. Chem. Soc. C*, 419 (1966).
d. J. A. Marshall and J. H. Babler, *J. Org. Chem.* **34**, 4186 (1969).
e. J. A. Marshall and J. H. Babler, *Tetrahedron Lett.*, 3861 (1970).
f. D. A. Clark and P. L. Fuchs, *J. Am. Chem. Soc.* **101**, 3567 (1979).

Similarly, a carbonyl group at the fifth carbon from a leaving group, reacting as the enolate, promotes fragmentation with formation of an enone[47]:

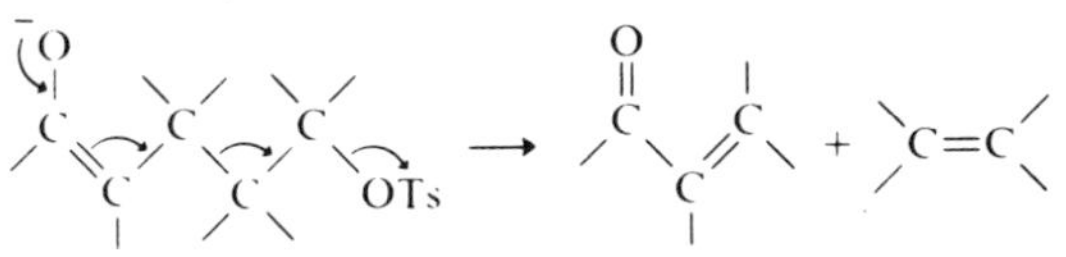

Organoboranes have been shown to undergo fragmentation if a good leaving group is present on the δ-carbon.[48] The reactive intermediate is the tetrahedral species formed by addition of hydroxide ion at boron.

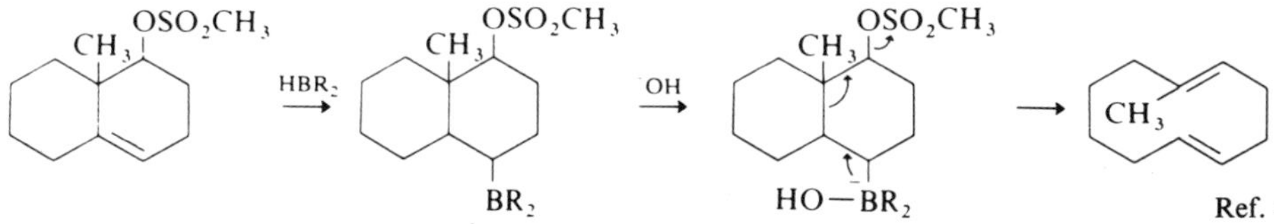

Ref. 49

The usual synthetic objective of a fragmentation reaction is the construction of a medium-sized ring from a fused ring system. Furthermore, because the fragmentation reactions of the type being discussed are usually concerted processes, the stereochemistry is predictable. In 3-hydroxy tosylates, the fragmentation is favorable only for geometry in which the carbon–carbon bond being broken can be in an antiperiplanar relationship to the leaving group.[50] Other stereochemical relationships in the molecule are retained during the concerted fragmentation. In the case below, for example, the newly formed double bond has the E-configuration:

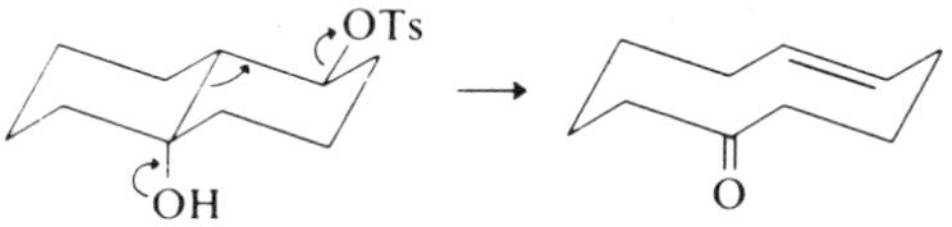

Scheme 10.4 provides some additional examples of fragmentation reactions that have been employed in a synthetic context.

10.2. Reactions Involving Carbenes and Nitrenes

Carbenes can be classed with carbanions, carbocations, and carbon-centered radicals as among the fundamental intermediates in the reactions of carbon compounds. Carbenes are neutral, divalent derivatives of carbon. Depending on whether

47. J. M. Brown, T. M. Cresp, and L. N. Mander, *J. Org. Chem.* **42**, 3984 (1977); D. A. Clark and P. L. Fuchs, *J. Am. Chem. Soc.* **101**, 3567 (1979).
48. J. A. Marshall, *Synthesis*, 229 (1971); J. A. Marshall and G. L. Bundy, *J. Chem. Soc., Chem. Commun.*, 854 (1967); P. S. Wharton, C. E. Sundin, D. W. Johnson, and H. C. Kluender, *J. Org. Chem.* **37**, 34 (1972).
49. J. A. Marshall and G. L. Bundy, *J. Am. Chem. Soc.* **88**, 4291 (1966).
50. P. S. Wharton and G. A. Hiegel, *J. Org. Chem.* **30**, 3254 (1965); C. H. Heathcock and R. A. Badger, *J. Org. Chem.* **37**, 234 (1972).

the nonbonding electrons are of the same or opposite spin, they are triplet or singlet species.

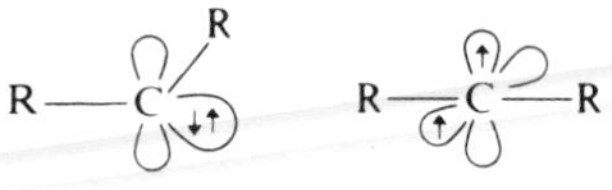

singlet triplet

As would be expected from the electron-deficient nature of these species, they are highly reactive.

Carbenes are often generated by α-elimination reactions.

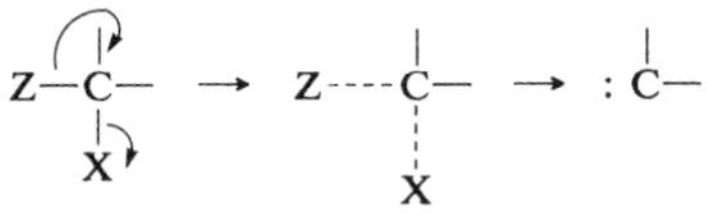

Under some circumstances, the question arises as to whether the carbene has a finite lifetime, and in some cases there is evidence that a completely free carbene structure is never attained. When reaction appears to involve a species that reacts as expected for a carbene but must still be at least partially bound to other atoms, the term *carbenoid* is applied. Some reactions which involve carbene-like processes involve transition metal ions. In many of these reactions, the divalent carbene is bound to the transition metal. Some compounds of this type are stable whereas others exist only as transient intermediates.

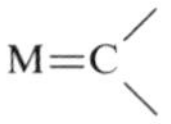

metal-bound carbene

Carbenes and carbenoids can add to double bonds to form cyclopropanes or insert into C—H bonds. These reactions have low activation energies when the intermediate is a "free" carbene. When this is the case, the reactions are inherently nonselective. The course of intramolecular reactions is frequently controlled by the proximity of the reacting groups.[51]

Nitrenes are neutral, monovalent nitrogen analogs of carbenes. The term *nitrenoid* is applied to nitrene-like intermediates which transfer a monosubstituted nitrogen fragment.

In this section, we will also consider a number of rearrangement reactions that probably do not involve carbene or nitrene intermediates but give overall transformations that correspond to those characteristic of a carbene or nitrene.

10.2.1. Structure and Reactivity of Carbenes

Depending upon the mode of generation, a carbene may initially be formed in either the singlet or triplet state, no matter which is lower in energy. These two

51. S. D. Burke and P. A. Grieco, *Org. React.* **26**, 361 (1979).

electronic configurations possess different geometries and chemical reactivities. Because of their paramagnetic character, triplet carbenes can be observed by electron spin resonance spectroscopy, provided they have sufficient lifetime. A rough picture of the bonding in the singlet assumes sp^2 hybridization at carbon, with the two unshared electrons in an sp^2 orbital. The p orbital is unoccupied. The R—C—R angle would be expected to be contracted slightly from the normal 120° because of the electronic repulsions between the unshared electron pair and the electrons in the two bonding orbitals. The bonds in the corresponding triplet carbene structure are formed from sp orbitals, with the unpaired electrons being in two equivalent p orbitals. For this bonding arrangement, a linear structure would be predicted.

Both theroetical and experimental studies have provided more detailed information about carbene structure. Molecular orbital calculations lead to the prediction of H—C—H angles for methylene of ~135° for the triplet and ~105° for the singlet. The triplet is calculated to be about 8 kcal/mol lower in energy than the singlet.[52] Experimental determinations of the geometry of CH_2 tend to confirm the theoretical results. The H—C—H angle of the triplet state, as determined from the EPR spectrum, is 125–140°. The H—C—H angle of the singlet state is found to be 102° by electronic spectroscopy. All the evidence is consistent with the triplet being the ground state.

Substituents have the effect of perturbing the relative energies of the singlet and triplet states. In general, alkyl groups resemble hydrogen as a substituent, and dialkylcarbenes are ground state triplets. Substituents that act as electron-pair donors stabilize the singlet state more than the triplet state by delocalization of an electron pair into the empty p-orbital.[53,54]

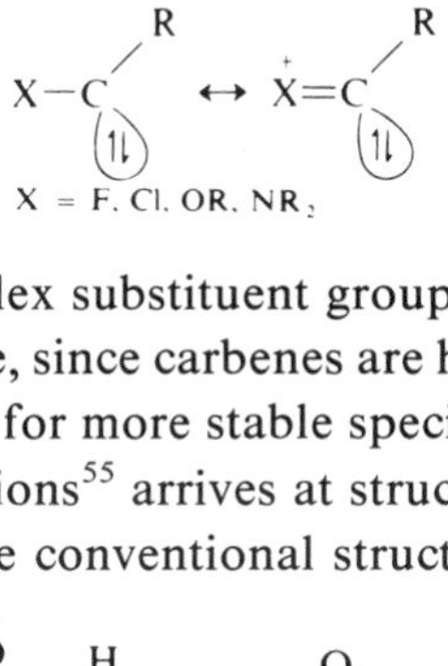

X = F, Cl, OR, NR_2

The presence of more complex substituent groups complicates the description of carbene structure. Furthermore, since carbenes are high-energy species, structural entities that would be unrealistic for more stable species must be considered. As an example, one set of MO calculations[55] arrives at structure **A** as a better description of carbomethoxycarbene than the conventional structure **B**.

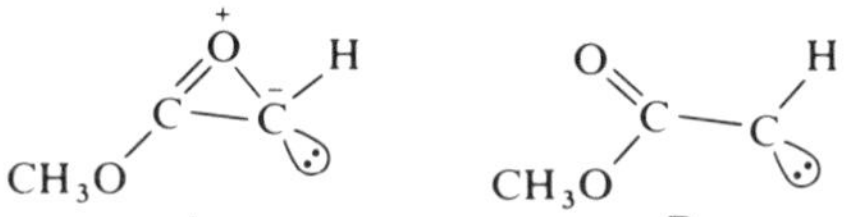

A **B**

52. J. F. Harrison, *Acc. Chem. Res.* **7**, 378 (1974); P. Saxe, H. F. Schaefer, III, and N. C. Hardy, *J. Phys. Chem.* **85**, 745 (1981); C. C. Hayden, M. Newmark, K. Shobatake, R. K. Sparks, and Y. T. Lee, *J. Chem. Phys.* **76**, 3607 (1982); R. K. Lengel and R. N. Zare, *J. Am. Chem. Soc.* **100**, 739 (1978); C. W. Bauschlicher, Jr., and I. Shavitt, *J. Am. Chem. Soc.* **100**, 739 (1978); A. R. W. M. Kellar, P. R. Bunker, T. J. Sears, K. M. Evenson, R. Saykally and S. R. Langhoff, *J. Chem. Phys.* **79**, 5251 (1983).
53. N. C. Baird and K. F. Taylor, *J. Am. Chem. Soc.* **100**, 1333 (1978).
54. J. F. Harrison, R. C. Liedtke, and J. F. Liebman, *J. Am. Chem. Soc.* **101**, 7162 (1979).
55. R. Noyori and M. Yamanaka, *Tetrahedron Lett.*, 2851 (1980).

From the point of view of both synthetic and mechanistic interest, much attention has been focused on the addition reaction between carbenes and alkenes to give cyclopropanes. Characterization of the reactivity of individual substituted carbenes in addition reactions has emphasized stereochemistry and selectivity. The reactivity of singlet and triplet states should be different. The triplet state is a diradical and would be expected to exhibit a selectivity similar to that of free radicals and other species with unpaired electrons. The singlet state, with its unfilled *p* orbital, should be electrophilic and exhibit reactivity similar to that of other electrophiles. Also, a triplet addition process must go through an intermediate that has two unpaired electrons of the same spin. In contrast, a singlet carbene can go to a cyclopropane in a single concerted step (see Fig. 10.1). As a result, it was predicted[56] that additions of singlet carbenes would be stereospecific while those of triplet carbenes would not be. This prediction has been confirmed, and the stereoselectivity of addition reactions with alkenes has come to be used as a test for the involvement of the singlet versus the triplet carbene in specific reactions.[57]

The radical versus electrophilic character of triplet and singlet carbenes also shows up in relative reactivity patterns, shown in Table 10.1. The relative reactivity of singlet dibromocarbene toward the three alkenes in Table 10.1 is more similar to that of electrophiles (bromination, epoxidation) than to that of radicals ($\cdot CCl_3$).

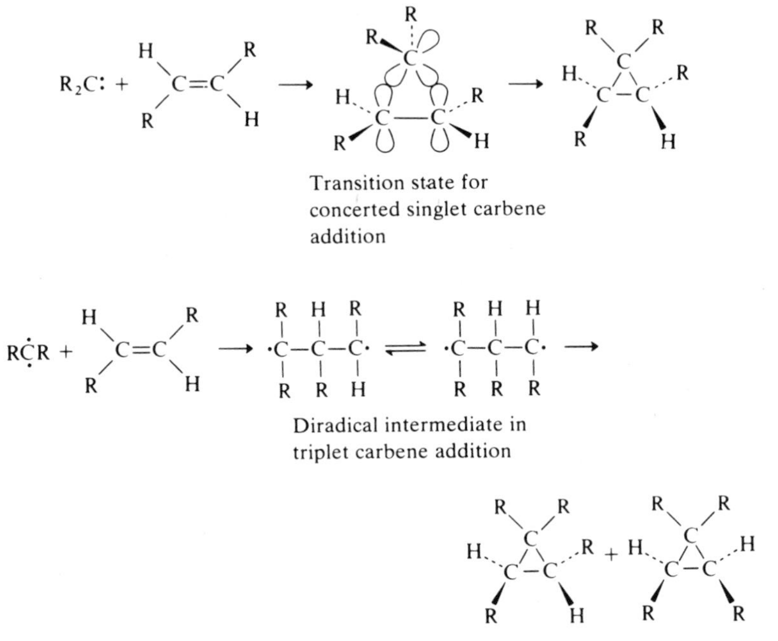

Transition state for concerted singlet carbene addition

Diradical intermediate in triplet carbene addition

Fig. 10.1. Mechanisms for addition of singlet and triplet carbenes to alkenes.

56. P. S. Skell and A. Y. Garner, *J. Am. Chem. Soc.* **78**, 5430 (1956).
57. R. C. Woodworth and P. S. Skell, *J. Am. Chem. Soc.* **81**, 3383 (1959); P. S. Skell, *Tetrahedron* **41**, 1427 (1985).

Carbene reactivity is strongly affected by substituents.[58] Various singlet carbenes have been characterized as nucleophilic, ambiphilic, or electrophilic as shown in Table 10.2. This classification is based on relative reactivity toward a series of different alkenes containing both nucleophilic alkenes, such as tetramethylethylene, and electrophilic ones, such as acrylonitrile. The principal structural feature which determines the reactivity of the carbene is the ability of the substituent to act as an electron donor. For example, dimethoxycarbene is devoid of electrophilicity toward alkenes[59] because of electron donation by the methoxy groups:

$$CH_3{-}\ddot{O}{-}\ddot{C}{-}\ddot{O}{-}CH_3 \longleftrightarrow CH_3{-}\overset{+}{\dot{O}}{=}\overset{-}{\ddot{C}}{-}\ddot{O}{-}CH_3 \longleftrightarrow CH_3{-}\ddot{O}{-}\overset{-}{\ddot{C}}{=}\overset{+}{\dot{O}}{-}CH_3$$

π Delocalization involving divalent carbon in conjugated cyclic systems has been studied in the interesting species cyclopropenylidene (**C**)[60] and cycloheptatrienylidene (**D**).[61] In these molecules, the empty p orbital on the carbene carbon can be part of the aromatic π system and is therefore delocalized over the entire

Table 10.1. Relative Rates of Addition to Alkenes[a]

Alkene	$\cdot CCl_3$	$:CBr_2$	Br_2	Epoxidation
Isobutylene	1.00	1.00	1.00	1.00
Styrene	>19	0.4	0.6	0.1
2-Methylbutene	0.17	3.2	1.9	13.5

a. P. S. Skell and A. Y. Garner, *J. Am. Chem. Soc.* **78**, 5430 (1956).

Table 10.2. Classification of Carbenes on the Basis of Reactivity toward Alkenes[a]

Nucleophilic	Ambiphilic	Electrophilic
$CH_3O\ddot{C}OCH_3$	$CH_3O\ddot{C}Cl$	$Cl\ddot{C}Cl$
$CH_3O\ddot{C}N(CH_3)_2$	$CH_3O\ddot{C}F$	$Ph\ddot{C}Cl$
		$CH_3\ddot{C}Cl$
		$Br\ddot{C}CO_2C_2H_5$

a. R. A. Moss and R. C. Munjal, *Tetrahedron Lett.*, 4721 (1979); R. A. Moss, *Acc. Chem. Res.* **13**, 58 (1980).

58. A comprehensive review of this topic is given by R. A. Moss, in *Carbenes*, M. Jones, Jr., and R. A. Moss (eds.), Wiley, New York, 1973, pp. 153–304; more recent work is reviewed in the series *Reactive Intermediates*, R. A. Moss and M. Jones, Jr. (eds.), Wiley, New York and by R. A. Mosss, *Acc. Chem. Res.* **22**, 15 (1989).
59. D. M. Lemal, E. P. Gosselink, and S. D. McGregor, *J. Am. Chem. Soc.* **88**, 582 (1966).
60. H. P. Reisenauer, G. Maier, A. Reimann, and R. W. Hoffmann, *Angew. Chem. Int. Ed. Engl.* **23**, 641 (1984); J. M. Bofill, J. Farras, S. Olivella, A. Sole, and J. Vilarrasa, *J. Am. Chem. Soc.* **110**, 1694 (1988); T. J. Lee, A. Bunge, and H. F. Schaefer III, *J. Am. Chem. Soc.* **107**, 137 (1985).
61. R. J. McMahon and O. L. Chapman, *J. Am. Chem. Soc.* **108**, 1713 (1986); M. Kusaz, H. Lüerssen, and C. Wentrup, *Angew. Chem. Int. Ed. Engl.* **25**, 480 (1986); C. L. Janssen and H. F. Schaefer III, *J. Am. Chem. Soc.* **109**, 5030 (1987).

ring. Currently available data indicate that the ground state configuration for **C** is the singlet, but the issue is not entirely settled for **D**, although theoretical calculations support a singlet ground state.

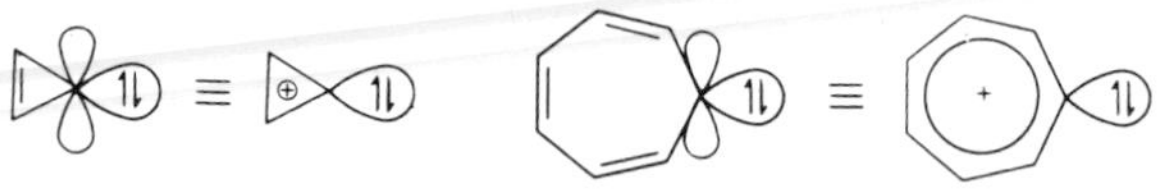

C **D**

10.2.2. Generation of Carbenes

There are numerous ways of generating carbene intermediates. Some of the most general routes are summarized in Scheme 10.5 and will be discussed in the succeeding paragraphs.

Decomposition of diazo compounds to carbenes is a quite general reaction. Examples of compounds that can be used include diazomethane and other diazoalkanes, diazoalkenes, and diazo compounds with aryl and acyl substituents. The

Scheme 10.5. General Methods for Generation of Carbenes

	Precursor	Condition	Products
1[a]	$R_2C{=}\overset{+}{N}{=}N^-$ Diazoalkanes	Photolysis, thermolysis, or metal-ion catalysis	$R_2C{:} + N_2$
2[b]	$R_2C{=}N{-}\bar{N}SO_2Ar$ Salts of sulfonylhydrazones	Photolysis or thermolysis: diazoalkanes are intermediates	$R_2C{:} + N_2 + ArSO_2^-$
3[c]	 R_2C(N=N) Diazirines	Photolysis	$R_2C{:} + N_2$
4[d]	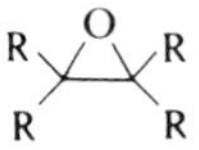 R_2C–O–CR_2 Epoxides	Photolysis	$R_2C{:} + R_2C{=}O$
5[e]	$R_2CH{-}X$ Halides	Strong base or organometallic compounds	$R_2C{:} + BH + X^-$
6[f]	$R_2C(X)HgR'$ α-Halomercury compounds	Thermolysis	$R_2C{:} + R'HgX$

a. W. J. Baron, M. R. DeCamp, M. E. Hendrick, M. Jones, Jr., R. H. Levin, and M. B. Sohn, in *Carbenes*, M. Jones, Jr., and R. A. Moss (eds.), Wiley, New York, 1973, pp. 1–151.
b. W. R. Bamford and T. S. Stevens, *J. Chem. Soc.*, 4735 (1952).
c. H. M. Frey, *Adv. Photochem.* **4**, 225 (1966); R. A. G. Smith and J. R. Knowles, *J. Chem. Soc., Perkin Trans. 2*, 686 (1975).
d. G. W. Griffin and N. R. Bertoniere, in *Carbenes*, M. Jones, Jr., and R. A. Moss (eds.), Wiley, New York, 1973, pp. 318–332.
e. W. Kirmse, *Carbene Chemistry*, Academic Press, New York, 1971, pp. 96–109, 129–149.
f. D. Seyferth, *Acc. Chem. Res.* **5**, 65 (1972).

main restrictions on this method are the methods of synthesis for and stability of the diazo compounds. The lower diazoalkanes are toxic and unstable. They are usually prepared immediately before use. The most general synthetic route involves base-catalyzed decomposition of *N*-nitroso derivatives of amides or sulfonamides. These reactions are illustrated by several reactions used for the preparation of diazomethane.

$$CH_3N(N{=}O){-}C({=}NH)NHNO_2 \xrightarrow{KOH} CH_2N_2 \qquad \text{Ref. 62}$$

$$CH_3N(N{=}O){-}C({=}O)NH_2 \xrightarrow[^-OR]{^-OH} CH_2N_2 \qquad \text{Ref. 63}$$

$$CH_3N(N{=}O){-}C({=}O){-}C_6H_4{-}C({=}O){-}N(N{=}O)CH_3 \xrightarrow{NaOH} CH_2N_2 \qquad \text{Ref. 64}$$

$$CH_3N(N{=}O){-}SO_2Ph \xrightarrow{KOH} CH_2N_2 \qquad \text{Ref. 65}$$

The details of the base-catalyzed decompositions vary somewhat, but the mechanisms involve two essential steps.[66] The initial reactants undergo a base-catalyzed elimination to form an alkyl diazoate. This is followed by a deprotonation of the α-carbon and elimination of the oxygen.

$$RCH_2N(N{=}O){-}Y \xrightarrow{-Y^+} RCH_2{-}N{=}N{-}O{-}H^+ \xrightarrow[-H_2O]{^-OH} RCH{=}\overset{+}{N}{=}\ddot{N}^-$$

Diazo compounds can also be obtained by oxidation of the corresponding hydrazone. This route is employed most frequently when one of the substitutents is an aromatic ring.

$$Ph_2C{=}NNH_2 \xrightarrow{HgO} Ph_2C{=}\overset{+}{N}{=}\ddot{N}^- \qquad \text{Ref. 67}$$

α-Diazoketones are an especially useful group of diazo compounds in synthesis. There are several methods of preparation. Reaction of diazomethane with an acid chloride results in formation of a diazomethyl ketones:

$$RC({=}O)Cl + H_2C{=}\overset{+}{N}{=}N^- \rightarrow RC({=}O)CH{=}\overset{+}{N}{=}N^-$$

62. M. Neeman and W. S. Johnson, *Org. Synth.* **V**, 245 (1973).
63. F. Arndt, *Org. Synth.* **II**, 165 (1943).
64. Th. J. de Boer and H. J. Backer, *Org. Synth.* **IV**, 250 (1963).
65. J. A. Moore and D. E. Reed, *Org. Synth.* **V**, 351 (1973).
66. W. M. Jones, D. L. Muck, and T. K. Tandy, Jr., *J. Am. Chem. Soc.* **88**, 3798 (1966); R. A. Moss, *J. Org. Chem.* **31**, 1082 (1966); D. E. Applequist and D. E. McGreer, *J. Am. Chem. Soc.* **82**, 1965 (1960); S. M. Hecht and J. W. Kozarich, *J. Org. Chem.* **38**, 1821 (1973); E. H. White, J. T. DePinto, A. J. Polito, I. Bauer, and D. F. Roswell, *J. Am. Chem. Soc.* **110**, 3708 (1988).
67. L. I. Smith and K. L. Howard, *Org. Synth.* **III**, 351 (1955).

The HCl generated in this reaction destroys one equivalent of diazomethane. This can be avoided by including a base, such as triethylamine, to neutralize the acid.[68] Cyclic α-diazoketones, which are not available from acid chlorides, can be prepared by reaction of an enolate equivalent with a sulfonyl azide. Several types of compounds can act as the carbon nucleophile. These include the anion of the hydroxymethylene derivative of the ketone[69] or the dialkylaminomethylene derivative of the ketone.[70]

$$RC(=O)C(R')=CHY + ArSO_2N_3 \rightarrow RC(=O)C(R')=\overset{+}{N}=N^-$$

$Y = O^-$ or NR_2

This reaction is called *diazo transfer*. Various arenesulfonyl azides[71] and methanesulfonyl azide[72] are used most frequently. α-Diazoketones can also be made by first converting the ketone to an α-oximino derivative by nitrosation and then allowing the oximinoketone to react with chloramine.[73]

1) RONO, $KOC(CH_3)_3$; 2) NH_3, CaOCl → (70%) Ref. 74

The driving force for decomposition of diazo compounds to carbenes is the formation of the very stable nitrogen molecule. Activation energies for decomposition of diazoalkanes in the gas phase are in the neighborhood of 30 kcal/mol. The requisite energy can also be supplied by photochemical excitation. It is often possible to control the photochemical process to give predominantly singlet or triplet carbene. Direct photolysis leads to the singlet intermediate when the dissociation of the excited diazoalkane is faster than intersystem crossing to the triplet state. The triplet carbene is the principal intermediate in photosensitized decomposition of diazoalkanes.

Reaction of diazo compounds with a variety of transition metal compounds leads to evolution of nitrogen and formation of products of the same general type as those formed by thermal and photochemical decomposition of diazoalkanes.

68. M. S. Newman and P. Beall III, *J. Am. Chem. Soc.* **71**, 1506 (1949); M. Berebom and W. S. Fones, *J. Am. Chem. Soc.* **71**, 1629 (1949); L. T. Scott and M. A. Minton, *J. Org. Chem.* **42**, 3757 (1977).
69. M. Regitz and G. Heck, *Chem. Ber.* **97**, 1482 (1964); M. Regitz, *Angew. Chem. Int. Ed. Engl.* **6**, 733 (1967).
70. M. Rosenberger, P. Yates, J. B. Hendrickson, and W. Wolf, *Tetrahedron Lett.*, 2285 (1964); K. B. Wiberg, B. L. Furtek, and L. K. Olli, *J. Am. Chem. Soc.* **101**, 7675 (1979).
71. J. B. Hendrickson and W. A. Wolf, *J. Org. Chem.* **33**, 3610 (1968); J. S. Baum, D. A. Shook, H. M. L. Davies, and H. D. Smith, *Synth. Commun.* **17**, 1709 (1987).
72. D. F. Taber, R. E. Ruckle Jr., and M. J. Hennessy, *J. Org. Chem.* **51**, 4077 (1986).
73. T. N. Wheeler and J. Meinwald, *Org. Synth.* **52**, 53 (1972).
74. T. Sasaki, S. Eguchi, and Y. Hirako, *J. Org. Chem.* **42**, 2981 (1977).

These transition metal-catalyzed reactions in general appear to involve carbenoid intermediates in which the carbene becomes bound to the metal.[75]

The second method listed in Scheme 10.5, thermal or photochemical decomposition of salts of arenesulfonylhydrazones, is actually a variation of the diazoalkane method, since diazo compounds are intermediates. The conditions of the decomposition are usually such that the diazo compound reacts immediately on formation.[76] The nature of the solvent plays an important role in the outcome of sulfonylhydrazone decompositions. In protic solvents, the diazoalkane can be diverted to a carbocation by protonation.[77]

$$RC(=O)R' + NH_2NHSO_2Ar \rightarrow R_2C{=}N\text{-}N(H)SO_2Ar \xrightarrow{\text{base}} R_2C{=}N{-}\bar{N}SO_2Ar$$

$$R_2C{=}N{-}\bar{N}SO_2Ar \xrightarrow[\text{or } \Delta]{h\nu} R_2C{=}\overset{+}{N}{=}\overset{-}{N}$$

$$R_2C{=}\overset{+}{N}{=}N^- \xrightarrow{XOH} R_2C(H){-}\overset{+}{N}{\equiv}N \rightarrow R_2\overset{+}{C}H + N_2$$

Aprotic solvents favor decomposition via the carbene pathway.

The diazirine precursors of carbenes (entry 3, Scheme 10.5) are cyclic isomers of diazo compounds. The strain of the small ring makes these compounds highly reactive toward loss of nitrogen on photoexcitation. They are, in general, somewhat less easily available than diazo compounds or arenesulfonylhydrazones. However, there are several useful synthetic routes.[78]

CH₂OR / O / O= / OR / OMe / OR, R = TMS $\xrightarrow[\text{2) } I_2]{\text{1) } NH_3,\ NH_2OSO_3H}$ OR / O / N=N (diazirine) / OR / OMe / OR — Ref. 79

$$CH_3O_2C{-}C_6H_4{-}C(=NH)NH_2 \xrightarrow{\text{NaOCl, LiCl}} CH_3O_2C{-}C_6H_4{-}C(N{=}N)Cl$$ Ref. 80

Carbenes are also generated when aryl epoxides are photolyzed (entry 4, Scheme 10.5). The other product formed is a carbonyl compound. The photodecomposition

75. W. R. Moser, *J. Am. Chem. Soc.* **91**, 1135, 1141 (1969); M. P. Doyle, *Chem. Rev.* **86**, 919 (1986); M. Brookhart, *Chem. Rev.* **87**, 411 (1987).
76. G. M. Kaufman, J. A. Smith, G. G. Vander Stouw, and H. Shechter, *J. Am. Chem. Soc.* **87**, 935 (1965).
77. J. H. Bayless, L. Friedman, F. B. Cook, and H. Shechter, *J. Am. Chem. Soc.* **90**, 531 (1968).
78. For reviews of synthesis of diazirines, see E. Schmitz, *Dreiringe mit Zwei Heteroatomen*, Springer-Verlag, Berlin, 1967, pp. 114–121; E. Schmitz, *Adv. Heterocycl. Chem.* **24**, 63 (1979); H. W. Heine, in *Chemistry of Heterocyclic Compounds*, Vol. 42, Pt. 2, A. Hassner (ed.), Wiley-Interscience, New York, 1983, pp. 547–628.
79. G. Kurz, J. Lehmann, and R. Thieme, *Carbohydr. Res.* **136**, 125 (1983).
80. D. F. Johnson and R. K. Brown, *Photochem. Photobiol.* **43**, 601 (1986).

of epoxides is not a single-step process.[81] The intermediates are carbonyl ylides, which are valence isomers of epoxides.[82] It is believed that the second step in the process is also a photoreaction:

$$R'_2C\overset{O}{—}CR_2 \xrightarrow{h\nu} R'_2C{=}\overset{+}{O}{-}\overset{-}{C}R_2 \longleftrightarrow R'_2\overset{-}{C}{-}\overset{+}{O}{=}CR_2 \xrightarrow{h\nu} R'_2C{=}O + R\ddot{C}R$$

Unsymmetrical epoxides can conceivably give rise to two carbenes and two carbonyl compounds. The nature of the substituent groups ordinarily favors one mode of reaction over the other. When R = aryl and R′ = alkyl, the aliphatic ketone is generated and the aryl-substituted carbon is released as the carbene fragment. Electron-withdrawing substituents such as cyano or carbomethoxy favor carbene formation, but a methoxy group directs carbene formation to the other oxirane carbon. These substituent effects can be understood by considering which of the two possible resonance structures will be the principal contributor to the carbonyl ylide. The relative contribution from each resonance structure is reflected in the bond order of the C—O bonds. The C—O bond with the greatest double bond character becomes the carbonyl group, while the weaker C—O bond is cleaved.

$$\underset{\text{minor}}{Ar_2C{=}\overset{+}{O}{-}\overset{-}{C}(Ar){-}OCH_3} \longleftrightarrow \underset{\text{major}}{Ar_2\overset{-}{C}{-}\overset{+}{O}{=}C(Ar){-}OCH_3} \rightarrow Ar_2C\text{:} + ArCO_2CH_3$$

$$\underset{\text{major}}{Ar_2C{=}\overset{+}{O}{-}\overset{-}{C}(Ar){-}CN} \longleftrightarrow \underset{\text{minor}}{Ar_2\overset{-}{C}{-}\overset{+}{O}{=}C(Ar){-}CN} \rightarrow Ar_2C{=}O + Ar\ddot{C}CN$$

The α elimination of hydrogen halide induced by strong base (entry 5, Scheme 10.5) is restricted to reactants which do not have β-hydrogens, since dehydrohalogenation by β elimination dominates when it can occur. The classic example of this method of carbene generation is the generation of dichlorocarbene by base-catalyzed decomposition of chloroform.[83]

$$HCCl_3 + {}^{-}OR \rightleftharpoons \text{:}\overset{-}{C}Cl_3 \rightarrow \text{:}CCl_2 + Cl^-$$

81. R. S. Becker, R. O. Bost, J. Kolc, N. R. Bertoniere, R. L. Smith, and G. W. Griffin, *J. Am. Chem. Soc.* **92**, 1302 (1970).
82. T. Do-Minh, A. M. Trozzolo, and G. W. Griffin, *J. Am. Chem. Soc.* **92**, 1402 (1970).
83. J. Hine, *J. Am. Chem. Soc.* **72**, 2438 (1950); J. Hine and A. M. Dowell, Jr., *J. Am. Chem. Soc.* **76**, 2688 (1954).

Both phase transfer and crown ether catalysis have been used to promote α-elimination reactions of chloroform and other haloalkanes.[84]

$$Ph_2C{=}CH_2 + CHCl_3 \xrightarrow[50\%\ NaOH]{PhCH_2\overset{+}{N}(C_2H_5)_3} \text{1,1-dichloro-2,2-diphenylcyclopropane} \qquad \text{Ref. 85}$$

Dichlorocarbene can also be generated by sonication of a solution of chloroform with powdered KOH.[86]

α Elimination also occurs in the reaction of dichloromethane and benzyl chlorides with alkyllithium reagents.

$$H_2CCl_2 + RLi \rightarrow RH + LiCHCl_2 \rightarrow :CHCl + LiCl \qquad \text{Ref. 87}$$

$$ArCH_2X + RLi \rightarrow RH + Ar\overset{Li}{C}HX \rightarrow Ar\ddot{C}H + LiX \qquad \text{Ref. 88}$$

The reactive intermediates under some conditions may be the carbenoid α-haloalkyllithium compounds or carbene–lithium halide complexes.[89] In the case of the trichloromethyllithium $\rightarrow$ dichlorocarbene conversion, the equilibrium lies heavily to the side of trichloromethyllithium at $-100°C$.[90] The addition reaction with alkenes seems to involve dichlorocarbene, however, since the pattern of reactivity toward different alkenes is identical to that observed for the free carbene in the gas phase.[91]

Lithium dialkylamides of hindered amines can generate aryl-substituted carbenes from benzyl halides.[92] Reaction of α,α-dichlorotoluene or α,α-dibromotoluene with potassium *t*-butoxide in the presence of 18-crown-6 generates the corresponding α-halophenylcarbene.[93] The relative reactivity data for carbenes generated under these latter conditions suggest that they are "free." The potassium cation would be expected to be strongly solvated by the crown ether and is evidently not involved in the carbene-generating step.

A method that provides an alternative route to dichlorocarbene is the decarboxylation of trichloroacetic acid.[94] The decarboxylation generates the trichloromethyl

84. W. P. Weber and G. W. Gokel, *Phase Transfer Catalysis in Organic Synthesis*, Springer-Verlag, New York, 1977, Chapters 2–4.
85. E. V. Dehmlow and J. Schönefeld, *Justus Liebigs Ann. Chem.* **744**, 42 (1971).
86. S. L. Regen and A. Singh, *J. Org. Chem.* **47**, 1587 (1982).
87. G. Köbrich, H. Trapp, K. Flory, and W. Drischel, *Chem. Ber.* **99**, 689 (1966); G. Köbrich and H. R. Merkle, *Chem. Ber.* **99**, 1782 (1966).
88. G. L. Closs and L. E. Closs, *J. Am. Chem. Soc.* **82**, 5723 (1960).
89. G. Köbrich, *Angew. Chem. Int. Ed. Engl.* **6**, 41 (1967).
90. W. T. Miller, Jr., and D. M. Whalen, *J. Am. Chem. Soc.* **86**, 2089 (1964); D. F. Hoeg, D. I. Lusk, and A. L. Crumbliss, *J. Am. Chem. Soc.* **87**, 4147 (1965).
91. P. S. Skell and M. S. Cholod, *J. Am. Chem. Soc.* **91**, 6035, 7131 (1969); *J. Am. Chem. Soc.* **92**, 3522 (1970).
92. R. A. Olofson and C. M. Dougherty, *J. Am. Chem. Soc.* **95**, 581 (197).
93. R. A. Moss and F. G. Pilkiewicz, *J. Am. Chem. Soc.* **96**, 5632 (1974).
94. W. E. Parham and E. E. Schweizer, *Org. React.* **13**, 55 (1963).

anion, which decomposes to the carbene. Treatment of alkyl trichloroacetates with an alkoxide also generates dichlorocarbene.

$$^{-}O-\overset{O}{\overset{\|}{C}}-CCl_3 \xrightarrow{-CO_2} :\bar{C}Cl_3 \leftarrow Cl_3C-\overset{O^{-}}{\underset{OR'}{\overset{|}{\underset{|}{C}}}}OR \leftarrow Cl_3C\overset{O}{\overset{\|}{C}}OR \quad {}^{-}OR'$$

$$:\bar{C}Cl_3 \rightarrow :CCl_2 + Cl^{-}$$

The applicability of these methods is restricted to polyhalogenated compounds, since the inductive effect of the halogen atoms is necessary for facilitating formation of the trichloromethyl anion.

The principle underlying the use or organomercury compounds for carbene generation (entry 6, Scheme 10.5) is also the α-elimination mechanism. The carbon-mercury bond is much more covalent then the C—Li bond, however, so the mercury reagents are generally stable at room temperature and can be isolated. They then decompose to the carbene on heating.[95] Addition reactions occur in the presence of alkenes. The decomposition rate is not greatly influenced by the alkene. This observation implies that a free carbene is generated from the organomercury precursor.[96]

$$PhHg-\underset{Cl}{\overset{Cl}{\underset{|}{\overset{|}{C}}}}-Br \rightarrow :CCl_2 + PhHgBr$$

A variety of organomercury compounds that can serve as precursors of substituted carbenes have been synthesized. For example, carbenes with carbomethoxy or trifluoromethyl substituents can be generated in this way[97]:

$$PhHg-\underset{CCF_3}{\overset{Cl}{\underset{|}{\overset{|}{C}}}}-Br \rightarrow Cl\ddot{C}CF_3$$

$$PhHgCCl_2CO_2CH_3 \rightarrow Cl\ddot{C}CO_2CH_3$$

The addition reactions between alkenes and phenylmercuric bromide typically occur at about 80°C. Phenylmercuric iodides are somewhat more reactive and may be advantageous in reactions with relatively unstable alkenes.[98]

10.2.3. Addition Reactions

The addition reaction with alkenes to form cyclopropanes is the best-studied reaction of carbene intermediates, both from the point of view of understanding

95. D. Seyferth, J. M. Burlitch, R. J. Minasz, J. Y.-P. Mui, H. D. Simmons, Jr., A. J. Treiber, and S. R. Dowd, *J. Am. Chem. Soc.* **87**, 4259 (1965).
96. D. Seyferth, J. Y.-P. Mui, and J. M. Burlitch, *J. Am. Chem. Soc.* **89**, 4953 (1967).
97. D. Seyferth, D. C. Mueller, and R. L. Lambert, Jr., *J. Am. Chem. Soc.* **91**, 1562 (1969).
98. D. Seyferth and C. K. Haas, *J. Org. Chem.* **40**, 1620 (1975).

carbene mechanisms and for synthetic applications. A concerted mechanism is possible for singlet carbenes. As a result, the stereochemistry present in the alkene is retained in the cyclopropane. With triplet carbenes, an intermediate diradical is involved. Closure to cyclopropane requires a spin inversion. The rate of spin inversion is slow relative to rotation about single bonds, so that mixtures of the two possible stereoisomers are obtained from either alkene stereoisomer.

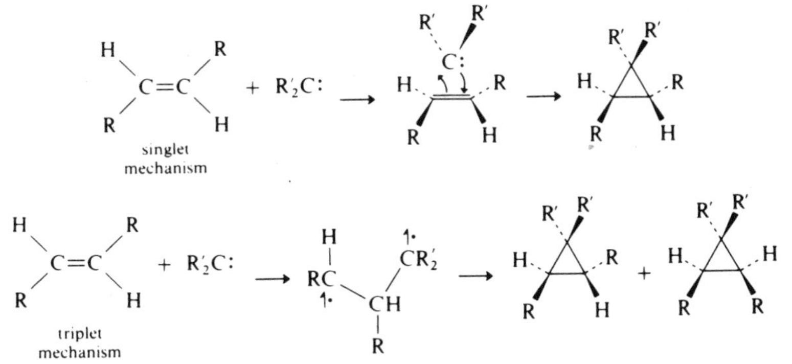

singlet mechanism

triplet mechanism

Reactions involving free carbenes are very exothermic since two new σ bonds are formed and only the alkene π bond is broken. The reaction is very fast, and, in fact, theoretical treatment of the addition of singlet methylene to ethylene suggests that there is no activation barrier.[99] Thus, the slow step in carbene addition reactions under most circumstances is generation of the carbene.

The addition of carbenes to alkenes is an important method for synthesis of many types of cyclopropanes, and several of the methods for carbene generation listed in Scheme 10.5 have been adapted for use in synthesis. A number of specific synthetic examples are given in Scheme 10.6.

Scheme 10.6. Cyclopropane Formation by Carbenoid Additions

A. Cyclopropanes by Methylene Transfer

1[a] + CH_2I_2 $\xrightarrow[\text{CuCl}]{\text{Zn dust}}$ (92%)

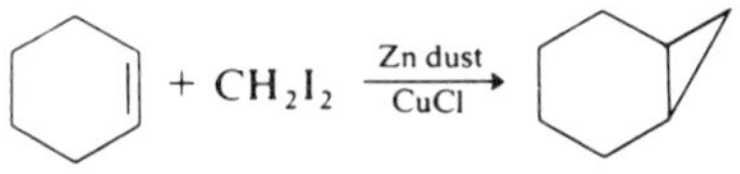

2[b] + CH_2I_2 $\xrightarrow{\text{Cu-Zn}}$ (66%)

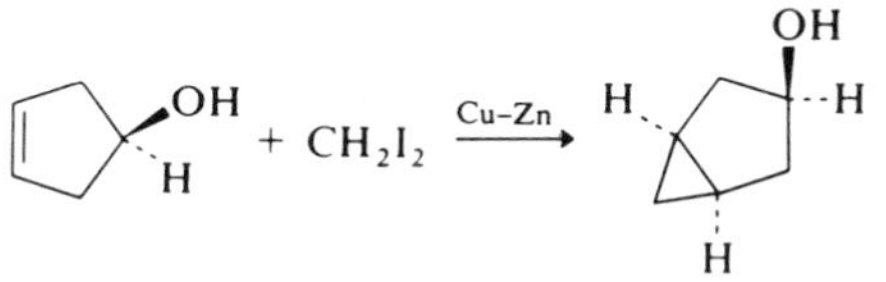

99. B. Zurawaski and W. Kutzelnigg, *J. Am. Chem. Soc.* **100**, 2654 (1978).

3[c] HO CH$_3$ → (CH_2I_2, Cu-Zn) HO CH$_3$ (76%)

4[d] OH, CH_3O, CH_3 → ($(C_2H_5)_2Zn$, CH_2I_2) OH, CH_3O, CH_3 (99%)

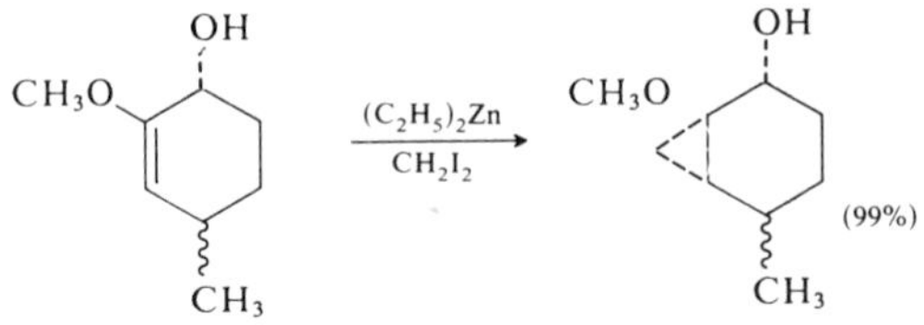

B. Catalytic Cyclopropanation by Diazo Compounds and Metal Salts

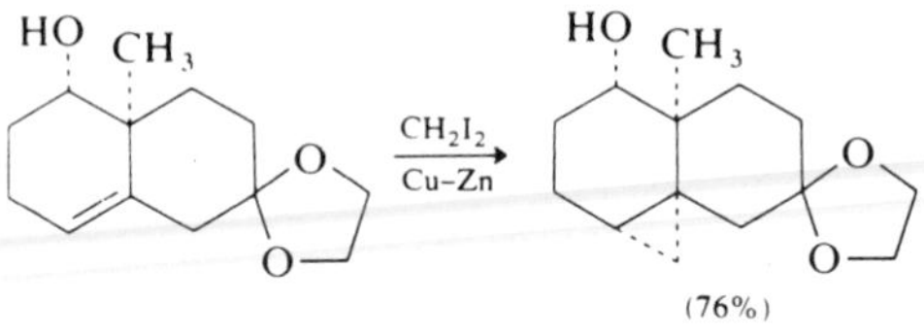

5[e] + $N_2CHCO_2C_2H_5$ →(CuCN) —$CO_2C_2H_5$ (58%)

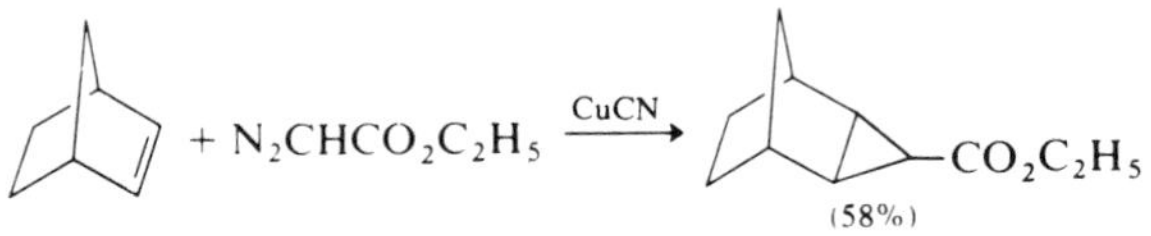

6[f] H_3C(H)C=C(H)CH_3 + $N_2CHCO_2C_2H_5$ →(CuO_3SCF_3) $CO_2C_2H_5$, CH_3, H_3C (51%)

7[g] $H_2C{=}CHO_2CCH_3$ + $N_2CHCO_2C_2H_5$ →($Rh(O_2CCH_3)_4$) $CO_2C_2H_5$, O_2CCH_3 (77%)

8[h] $(CH_3)_3C$—⟨⟩=CH_2 + $N_2CH\overset{O}{\overset{\|}{C}}CO_2C_2H_5$ →($Cu(acac)_2$) $(CH_3)_3C$—, $\overset{O}{\overset{\|}{C}}CO_2C_2H_5$ (45%)

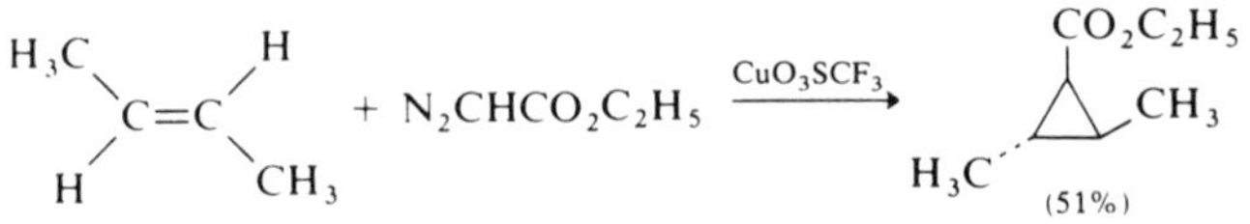

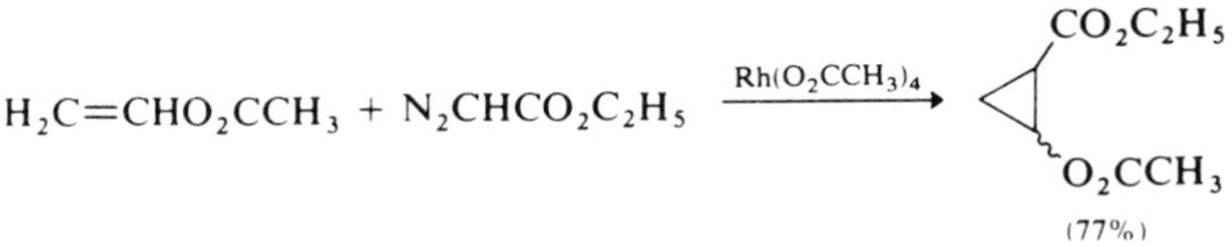

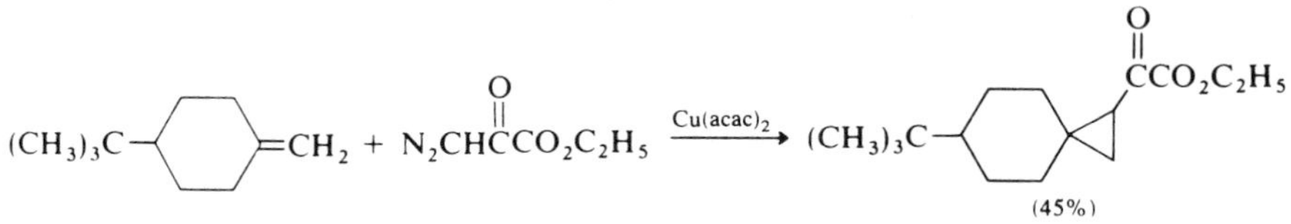

C. Cyclopropane Formation Using Haloalkylmercurials

9[i] + $PhHgCBr_2CO_2CH_3$ → CO_2CH_3, Br + Br, CO_2CH_3 (50% total yield)

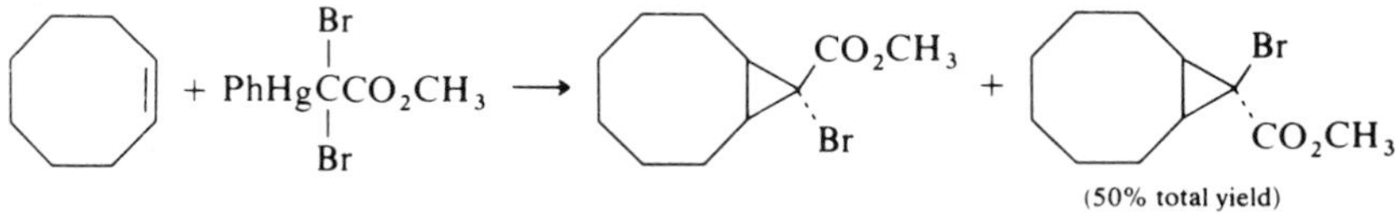

10[j] $(CH_3)_2C{=}C(CH_3)_2$ + $PhHgC(CF_3)BrCl$ → Cl, CF_3, H_3C, CH_3, H_3C, CH_3 (58%)

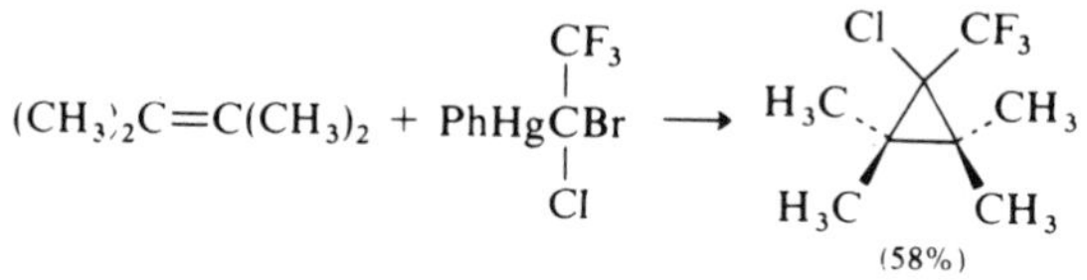

D. Reactions of Carbenes Generated by α Elimination

11[k]

$+ HCBr_3 + K^{+-}OC(CH_3)_3 \longrightarrow$ (79%)

12[l]

$CH_3O-C_6H_4-CHBr_2 + (H_3C)HC{=}CH(CH_3) \xrightarrow{n\text{-BuLi}}$ (55%)

13[m]

$(CH_3)_2C{=}CHCH_3 + CFBr_3 \xrightarrow{n\text{-BuLi}}$ (55%)

14[n]

$+ CHCl_3 \xrightarrow[\text{50\% NaOH, benzene}]{PhCH_2\overset{+}{N}(C_2H_5)_3}$

E. Intramolecular Addition Reactions

15[o]

$\xrightarrow{Rh_2(OAc)_4}$ (37%)

16[p]

$\xrightarrow{h\nu}$ (50%)

a. R. J. Rawson and I. T. Harrison, *J. Org. Chem.* **35**, 2057 (1970).
b. S. Winstein and J. Sonnenberg, *J. Am. Chem. Soc.* **83**, 3235 (1961).
c. P. A. Grieco, T. Oguir, C.-L. J. Wang, and E. Williams, *J. Org. Chem.* **42**, 4113 (1977).
d. R. C. Gadwood, R. M. Lett, and J. E. Wissinger, *J. Am. Chem. Soc.* **108**, 6343 (1986).
e. R. R. Sauers and P. E. Sonnett, *Tetrahedron* **20**, 1029 (1964).
f. R. G. Salomon and J. K. Kochi, *J. Am. Chem. Soc.* **95**, 3300 (1973).
g. A. J. Anciaux, A. J. Hubert, A. F. Noels, N. Petiniot, and P. Teyssie, *J. Org. Chem.* **45**, 695 (1980).
h. M. E. Alonso, P. Jano, and M. I. Hernandez, *J. Org. Chem.* **45**, 5299 (1980).
i. D. Seyferth, D. C. Mueller, and R. L. Lambert, Jr., *J. Am. Chem. Soc.* **91**, 1562 (1969).
j. D. Seyferth and D. C. Mueller, *J. Am. Chem. Soc.* **93**, 3714 (1971).
k. L. A. Paquette, S. E. Wilson, R. P. Henzel, and G. R. Allen, Jr., *J. Am. Chem. Soc.* **94**, 7761 (1972).
l. G. L. Closs and R. A. Moss, *J. Am. Chem. Soc.* **86**, 4042 (1964).
m. D. J. Burton and J. L. Hahnfeld, *J. Org. Chem.* **42**, 828 (1977).
n. T. T. Sasaki, K. Kanematsu, and N. Okamura, *J. Org. Chem.* **40**, 3322 (1975).
o. P. Dowd, P. Garner, R. Schappert, H. Irngartinger, and A. Goldman, *J. Org. Chem.* **47**, 4240 (1982).
p. B. M. Trost, R. M. Cory, P. H. Scudder, and H. B. Neubold, *J. Am. Chem. Soc.* **95**, 7813 (1973).

A very effective means for conversion of alkenes to cyclopropanes by transfer of a CH_2 unit involves the system methylene iodide and zinc-copper couple, commonly referred to as the *Simmons-Smith reagent.*[100] The active species is believed to be iodomethylzinc iodide in equilibrium with bis(iodomethyl)zinc.[101]

$$2ICH_2ZnI \rightleftharpoons (ICH_2)_2Zn + ZnI_2$$

The transfer of methylene occurs stereospecifically. Free $:CH_2$ is not an intermediate. In molecules with hydroxyl groups, the CH_2 unit is introduced on the side of the double bond *syn* to the hydroxyl group. This indicates that the reagent is complexed to the hydroxyl group and that the complexation directs the addition. Entries 2 and 3 in Scheme 10.6 illustrate the stereodirective effect of the hydroxyl group.

A modified version of the Simmons-Smith reaction uses dibromomethane and *in situ* generation of the Cu-Zn couple.[102] Sonication is used in this procedure to promote reaction at the metal surface.

CH_2Br_2
Zn-Cu
sonication

(50%)

Ref. 103

Cyclopropanation can also be effected with a combination of methylene iodide and an alkylzinc reagent.

CO_2CH_3
$C_2H_5ZnI,$
CH_2I_2
CO_2CH_3

(51%)

Ref. 104

The transition metal-catalyzed decomposition of diazo compounds is a very useful reaction for formation of substituted cyclopropanes. The reaction has been carried out with several copper salts.[105] A wide variety of other transition metal

100. H. E. Simmons and R. D. Smith, *J. Am. Chem. Soc.* **80**, 5323 (1958); **81**, 4256 (1959); H. E. Simmons, T. L. Cairns, S. A. Vladuchick, and C. M. Hoiness, *Org. React.* **20**, 1 (1973).
101. E. P. Blanchard and H. E. Simmons, *J. Am. Chem. Soc.* **86**, 1337 (1964); H. E. Simmons, E. P. Blanchard, and R. D. Smith, *J. Am. Chem. Soc.* **86**, 1347 (1964).
102. E. C. Friedrich, J. M. Demek, and R. Y. Pong, *J. Org. Chem.* **50**, 4640 (1985).
103. S. Sawada and Y. Inouye, *Bull. Chem. Soc. Jpn.* **42**, 2669 (1969); N. Kawabata, T. Nakagawa, T. Nakao, and S. Yamashita, *J. Org. Chem.* **42**, 3031 (1977); J. Furukawa, N. Kawabata, and J. Nishimura, *Tetrahddron* **24**, 53 (1968).
104. J. S. Swenton, K. A. Burdett, D. M. Madigan, T. Johnson, and P. D. Rosso, *J. Am. Chem. Soc.* **97**, 3428 (1975).
105. W. von E. Doering and W. R. Roth, *Tetrahedron* **19**, 715 (1963); J. P. Chesick, *J. Am. Chem. Soc.* **84**, 3250 (1962); H. Nozaki, H. Takaya, S. Moriuti, and R. Noyori, *Tetrahedron* **24**, 3655 (1968); R. G. Salomon and J. H. Kochi, *J. Am. Chem. Soc.* **95**, 3300 (1973); M. E. Alonso, P. Jano, and M. I. Hernandez, *J. Org. Chem.* **45**, 5299 (1980); T. Hudlicky, F. J. Koszyk, T. M. Kutchean, and J. P. Sheth, *J. Org. Chem.* **45**, 5020 (1980); M. P. Doyle and M. L. Truell, *J. Org. Chem.* **49**, 1196 (1984).

complexes are also useful including rhodium,[106] palladium,[107] and molybdenum[108] compounds. The catalytic cycle can be generally represented as below[109]:

$$L_nM \xrightarrow{R_2CN_2,\ -N_2} L_nM{=}CR_2 \xrightarrow{\text{alkene}} \text{cyclopropane} + L_nM$$

The metal–carbene complexes are electrophilic in character. They can, in fact, be represented as metal-stabilized carbocations.

$$\ddot{M} + R_2\bar{C}{-}\overset{+}{N}{\equiv}N \xrightarrow{-N_2} \ddot{M}{-}\overset{+}{C}R_2 \leftrightarrow M{=}CR_2$$

In most transition metal-catalyzed reactions, one of the carbene substituents is a carbonyl group, which further enhances the electrophilicity of the intermediate. There are two general mechanisms that can be considered for cyclopropane formation. One involves formation of a four-membered ring intermediate that incorporates the metal. The alternative represents an electrophilic attack giving a polar species which undergoes 1,3-bond formation.

$$M{=}CR_2 + {>}C{=}C{<} \longrightarrow \text{M–CR}_2\text{–C–C (four-membered ring)} \longrightarrow \text{cyclopropane}$$

or

$$M{=}CR_2 + {>}C{=}C{<} \longrightarrow \ddot{M}{-}CR_2{-}C{-}\overset{+}{C} \longrightarrow M: + \text{cyclopropane}$$

The observation that the additions are normally stereospecific with respect to the alkene indicates that if an open-chain intermediate is involved, it must collapse to product more rapidly than single-bond rotations which would destroy the stereoselectivity.

106. S. Bien and Y. Segal, *J. Org. Chem.* **42**, 1685 (1977); A. J. Anciaux, A. J. Hubert, A. F. Noels, N. Petiniot, and P. Teyssie, *J. Org. Chm.* **45**, 695 (1980); M. P. Doyle, W. H. Tamblyn, and V. Baghari, *J. Org. Chem.* **46**, 5094 (1981); D. F. Taber and R. E. Ruckle, Jr., *J. Am. Chem. Soc.* **108**, 7686 (1986).
107. R. Paulissen, A. J. Hubert, and P. Teyssie, *Tetrahedron Lett.*, 1465 (1972); U. Mende, B. Radüchel, W. Skuballa, and H. Vorbrüggen, *Tetrahedron Lett.*, 629 (1975); M. Suda, *Synthesis*, 714 (1981); M. P. Doyle, L. C. Wang, and K.-L. Loh, *Tetrahedron Lett.* **25**, 4087 (1984).
108. M. P. Doyle and J. G. Davidson, *J. Org. Chem.* **45**, 1538 (1980); M. P. Doyle, R. L. Dorow, W. E. Buhro, J. H. Tamblyn, and M. L. Trudell, *Organometallics* **3**, 44 (1984).
109. M. P. Doyle, *Chem. Rev.* **86**, 919 (1986).

Entries 5–8 in Scheme 10.6 are examples of transition metal-catalyzed carbene addition reactions.

Haloalkylmercury compounds are also useful in synthesis. The addition reactions are usually carried out by heating the organomercury compound with the alkene. Two typical examples are given in section C of Scheme 10.6.

The addition of dichlorocarbene, generated from chloroform, to alkenes is a useful synthesis of cyclopropanes. The procedures based on lithiated halogen compounds have been less generally used in synthesis. Section D of Scheme 10.6 gives a few examples of addition reactions of carbenes generated by α elimination.

Intramolecular carbene addition reactions have a special importance in the synthesis of strained ring compounds. Because of the high reactivity of carbene or carbenoid species, the formation of highly strained bonds is possible. The strategy for synthesis is to construct a potential carbene precursor, such as a diazo compound or di- or trihalo compound, which can undergo intramolecular addition to the desired structure. Section E of Scheme 10.6 gives some representative examples.

The high reactivity of carbenes is also essential to the addition reaction that occurs with aromatic compounds.[110] The resulting adducts are in thermal equilibrium with the corresponding cycloheptatriene. The position of the equilibrium depends on the nature of the substituent (see Section 11.1 of Part A).

$$\text{benzene} + CH_2N_2 \xrightarrow{h\nu} \text{norcaradiene} \rightleftharpoons \text{cycloheptatriene} \qquad \text{Ref. 111}$$

$$\text{benzene} + (N{\equiv}C)_2CN_2 \xrightarrow{80^\circ C} \text{7,7-di}(C{\equiv}N)\text{norcaradiene} \qquad \text{Ref. 112}$$

$$\text{benzene} + H_5C_2O_2CHN_2 \xrightarrow{\text{heat}} \text{7-}(H)(CO_2C_2H_5)\text{cycloheptatriene} \qquad \text{Ref. 113}$$

10.2.4. Insertion Reactions

Insertion reactions are processes in which a reactive intermediate, in this case a carbene, interposes itself into an existing bond. In terms of synthesis, this usually involves C—H bonds. Many singlet carbenes are sufficiently reactive that this insertion can occur as a one-step process.

$$CH_3{-}CH_2{-}CH_3 + {:}CH_2 \rightarrow CH_3{-}\underset{\substack{|\\ CH_3}}{CH}{-}CH_3$$

110. E. Ciganek, *J. Am. Chem. Soc.* **93**, 2207 (1971).
111. G. A. Russell and D. G. Hendry, *J. Org. Chem.* **28**, 1933 (1963).
112. E. Ciganek, *J. Am. Chem. Soc.* **89**, 1454 (1967).
113. J. E. Baldwin and R. A. Smith, *J. Am. Chem. Soc.* **89**, 1886 (1967).

The same products can be formed by a two-step hydrogen abstraction and recombination involving a triplet carbene.

$$CH_3-CH_2-CH_3 + \cdot\dot{C}H_2 \rightarrow CH_3-\dot{C}H-CH_3 + CH_3\cdot \rightarrow CH_3-\underset{\displaystyle CH_3}{\underset{|}{C}H}-CH_3$$

It is sometimes a difficult task to clearly distinguish between these mechanisms, but determination of reaction stereochemistry provides one approach. The true one-step insertion must occur with complete *retention* of configuration. The results for the two-step process will depend on the rate of recombination in competition with stereorandomization of the radical intermediate.

Because of the very high reactivity of the intermediates which are involved, intermolecular carbene insertion reactions are not very selective. The distribution of insertion products from the photolysis of diazomethane in heptane, for example, is almost exactly that which would be expected on a statistical basis.[114]

$$CH_3CH_2CH_2CH_2CH_2CH_2CH_3 \xrightarrow[h\nu]{CH_2N_2} \underset{(38\%)}{CH_3(CH_2)_6CH_3} + \underset{(25\%)}{CH_3\underset{\displaystyle CH_3}{\underset{|}{C}H}(CH_2)_4CH_3}$$

$$+ \underset{(24\%)}{CH_3CH_2\underset{\displaystyle CH_3}{\underset{|}{C}H}(CH_2)_3CH_3}$$

$$+ \underset{(13\%)}{(CH_3CH_2CH_2)_2CHCH_3}$$

There is some increase in selectivity with functionally substituted carbenes, but the selectivity is still not high enough to prevent formation of mixtures. Carbethoxycarbene, for example, inserts at tertiary C—H bonds about three times as fast as at primary C—H bonds in simple alkanes.[115] For this reason, intermolecular insertion reactions are seldom useful in synthesis.

Intramolecular insertion reactions are of considerably more use. Intramolecular insertion reactions usually occur at the C—H bond that is closest to the carbene, and good yields can frequently be obtained. Intramolecular insertion reactions can provide routes to highly strained structures that would be difficult to approach in other ways.

Rhodium carboxylates have been found to be effective catalysts for intramolecular C—H insertion reactions of α-diazoketones and esters.[116] In flexible systems, five-membered rings are formed in preference to six-membered ones. Insertion into a methine carbon-hydrogen bond is preferred to insertion at a

114. D. B. Richardson, M. C. Simmons, and I. Dvoretzky, *J. Am. Chem. Soc.* **83**, 1934 (1961).
115. W. von E. Doering and L. H. Knox, *J. Am. Chem. Soc.* **83**, 1989 (1961).
116. D. F. Taber and E. H. Petty, *J. Org. Chem.* **47**, 4808 (1982); D. F. Taber and R. E. Ruckle, Jr., *J. Am. Chem. Soc.* **108**, 7686 (1986).

Scheme 10.7. Intramolecular Carbene-Insertion Reactions

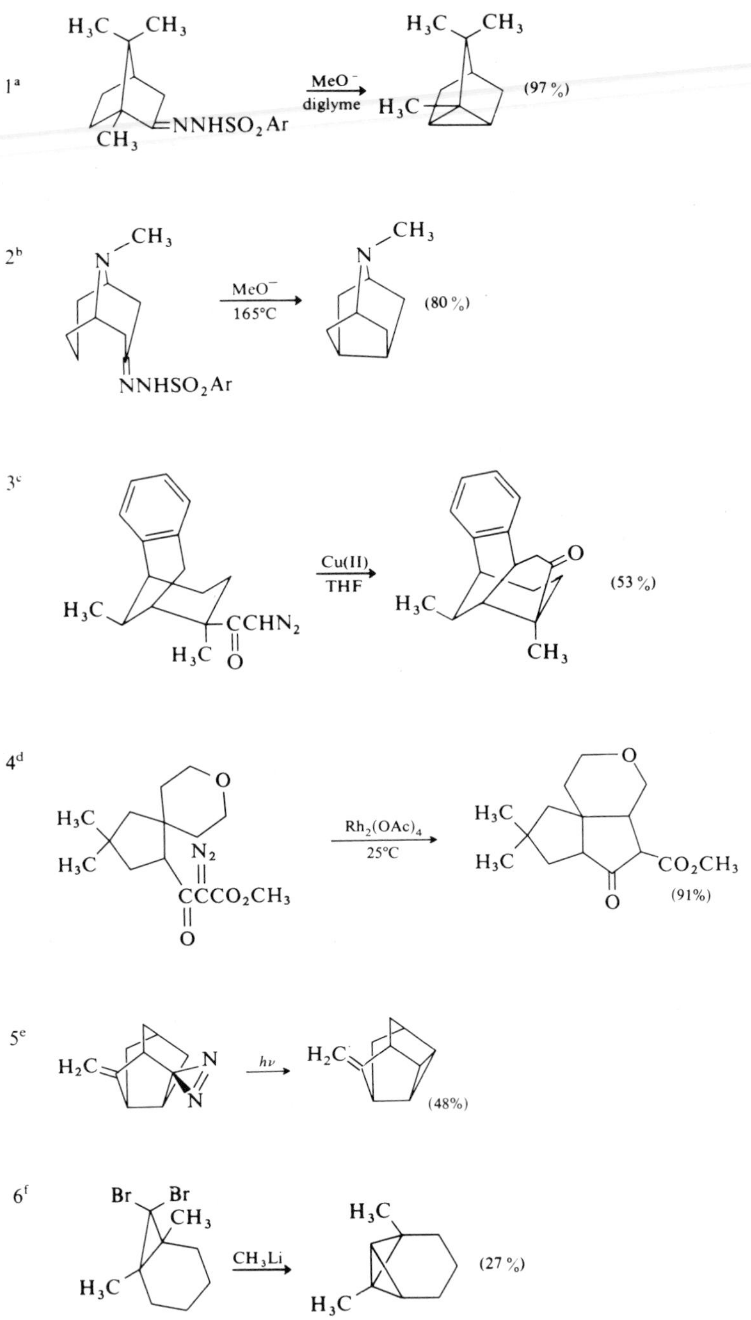

1[a] H_3C CH_3; CH_3; $=NNHSO_2Ar$ —(MeO⁻, diglyme)→ H_3C CH_3; H_3C (97%)

2[b] N–CH_3; $NNHSO_2Ar$ —(MeO⁻, 165°C)→ N–CH_3 (80%)

3[c] H_3C; $CCHN_2$ (C=O); H_3C —(Cu(II), THF)→ H_3C; O; CH_3 (53%)

4[d] O; H_3C; H_3C; N_2; $CCCO_2CH_3$ (C=O) —($Rh_2(OAc)_4$, 25°C)→ O; H_3C; H_3C; CO_2CH_3; O (91%)

5[e] H_2C; N=N —($h\nu$)→ H_2C (48%)

6[f] Br Br; CH_3; H_3C —(CH_3Li)→ H_3C; H_3C (27%)

a. R. H. Shapiro, J. H. Duncan, and J. C. Clopton, *J. Am. Chem. Soc.* **89**, 1442 (1967).
b. T. Sasaki, S. Eguchi, and T. Kiriyama, *J. Am. Chem. Soc.* **91**, 212 (1969).
c. U. R. Ghatak and S. Chakrabarty, *J. Am. Chem. Soc.* **94**, 4756 (1972).
d. D. F. Taber and J. L. Schuchardt, *J. Am. Chem. Soc.* **107**, 5289 (1985).
e. Z. Majerski, Z. Hamersak, and R. Sarac-Arneri, *J. Org. Chem.* **53**, 5053 (1988).
f. L. A. Paquette, S. E. Wilson, R. P. Henzel, and G. R. Allen, Jr., *J. Am. Chem. Soc.* **94**, 7761 (1972).

methylene group. Intramolecular insertion can be competitive with intramolecular addition:

$$(CH_3CH_2CH_2)(CH_2{=}CHCH_2)CH{-}\overset{O}{\overset{\|}{C}}{-}\underset{N_2}{\underset{\|}{C}}CO_2CH_3 \xrightarrow{Rh_2(OAc)_4}$$

addition product ($CH_3CH_2CH_2$, O, CO_2CH_3)

+

insertion product ($CH_2{=}CHCH_2$, O, CO_2CH_3, CH_3)

Scheme 10.7 gives some additional examples of intramolecular insertion reactions.

10.2.5. Rearrangement Reactions

The most common rearrangement reaction of carbenes is the shift of hydrogen, generating an alkene. This mode of stabilization predominates to the exclusion of most intermolecular reactions of aliphatic carbenes and often competes with intramolecular insertion reactions. For example, the carbene generated by decomposition of the tosylhydrazone of 2-methylcyclohexanone gives mainly 1- and 3-methylcyclohexene rather than the intramolecular insertion product.

2-methylcyclohexanone $NNHSO_2Ar$ $\xrightarrow[180°C]{NaOCH_3}$ 1-methylcyclohexene (38%) + 3-methylcyclohexene (16%) + bicyclo[4.1.0]heptane (trace) Ref. 117

Carbenes can also be stabilized by migration of alkyl or aryl groups. 2-Methyl-2-phenyl-1-diazopropane provides a case in which both phenyl and methyl migration, as well as intramolecular insertion, are observed.

$$PhC(CH_3)_2CHN_2 \xrightarrow{60°C} \underset{(50\%)}{(CH_3)_2C{=}CHPh} + \underset{(9\%)}{PhC(CH_3){=}CHCH_3} + \underset{(41\%)}{Ph{-}C(CH_3)\langle CH_2CH_2 \rangle}$$ Ref. 118

117. J. W. Wilt and W. J. Wagner, *J. Org. Chem.* **29**, 2788 (1964).
118. H. Philip and J. Keating, *Tetrahedron Lett.*, 523 (1961).

Carbene centers adjacent to double bonds (vinylcarbenes) usually cyclize to cyclopropenes.[119]

$$(CH_3CH_2)(CH_3)C{=}C(CH_3)CH{=}NNHSO_2Ar \longrightarrow$$ (1-H, 1-CH_3, 3-CH_2CH_3, 3-CH_3 cyclopropene) Ref. 120

Cyclopropylidenes undergo ring opening to give allenes. Reactions that would be expected to generate a cyclopropylidenes therefore lead to allenes, often in preparatively useful yields.

(2,3-diphenylcyclopropyl)$N(N{=}O)CNH_2$ (C=O) $\xrightarrow{LiOC_2H_5}$ [2,3-diphenylcyclopropylidene] $\rightarrow$ $PhHC{=}C{=}CHPh$ (79%) Ref. 121

1-methyl-1-propyl-2,2-dichlorocyclopropane $\xrightarrow{BuLi}$ $CH_3CH{=}C{=}CHCH_2CH_2CH_3$ Ref. 122

10.2.6. Related Reactions

There are several reactions which are conceptually related to carbene reactions but which do not involve carbene, or even carbenoid, intermediates. Usually, these are reactions in which the generation of a carbene is circumvented by a concerted rearrangement process. Important examples of this type of reaction are the thermal and photochemical reactions of α-diazoketones. When α-diazoketones are decomposed thermally or photochemically, they usually rearrange to ketenes. This reaction is known as the *Wolff rearrangement.* If this reaction can proceed in a concerted fashion, a carbene intermediate is avoided. Mechanistic studies have been aimed at determining if migration is concerted with loss of nitrogen. The conclusion that has emerged is that a carbene is generated in photochemical reactions but that the reaction can be concerted under thermal condition. A related issue is whether the carbene, when it is involved, is in equilibrium with a ring-closed isomer, an oxirine.[123] This aspect of the reaction has been probed by isotopic labeling of the carbonyl carbon. If a symmetrical oxirene is formed, the label should be distributed between both the carbonyl and the α-carbon. A concerted reaction or a carbene intermediate

119. G. L. Closs, L. E. Closs, and W. A. Böll, *J. Am. Chem. Soc.* **85**, 3796 (1963).
120. E. J. York, W. Dittmar, J. R. Stevenson, and R. G. Bergman, *J. Am. Chem. Soc.* **95**, 5680 (1973).
121. W. M. Jones, J. W. Wilson, Jr., and F. B. Tutwiler, *J. Am. Chem. Soc.* **85**, 3309 (1963).
122. W. R. Moore and H. R. Ward, *J. Org. Chem.* **25**, 2073 (1960).
123. M. Torres, E. M. Lown, H. E. Gunning, and O. P. Strausz, *Pure Appl. Chem.* **52**, 1623 (1980); E. G. Lewars, *Chem. Rev.* **83**, 519 (1983).

that did not equilibrate with the oxirene should have label only in the carbonyl carbon.

$$R-\overset{O}{\overset{\|}{C}}-\overset{+}{C}H-N\equiv N \rightarrow O=C=CHR \quad \text{concerted mechanism}$$

$$R-\overset{O}{\overset{\|}{C}}-CH=\overset{+}{N}=\ddot{N}^- \rightarrow R-\overset{O}{\overset{\|}{C}}-\ddot{C}H \rightarrow O=C=CHR \quad \text{carbene mechanism}$$

$$\rightleftharpoons \quad \text{oxirene}$$

The extent to which the oxirene is formed depends on the structure of the diazo compound. For diazoacetaldehyde, photolysis leads to only 8% migration of label, which would correspond to formation of 16% of the product through the oxirene.[124]

$$H-\overset{O}{\overset{\|}{C^*}}-CHN_2 \rightarrow H-\overset{O}{\overset{\|}{C^*}}-\ddot{C}H \rightleftharpoons H\overset{*}{C}\overset{O}{=}CH \rightleftharpoons H\overset{*}{\ddot{C}}-\overset{O}{\overset{\|}{C}}H$$

$$CH_2=\overset{*}{C}=O \qquad {}^*CH_2=C=O$$

$$CH_3CO_2R \quad (8\%)\ (92\%) \text{ distribution of label}$$

The diphenyl analog shows about 20–30% rearrangement.[125] α-Diazocyclohexanone gives no evidence of an oxirene intermediate, since all the label remains at the carbonyl carbon.[126]

$$\text{(2-diazocyclohexanone, }{}^*C=O) \xrightarrow{h\nu} \text{(cyclopentylidene)}=\overset{*}{C}=O \xrightarrow{H_2O} \text{(cyclopentyl)}-\overset{*}{C}O_2H$$

One synthetic application of the Wolff rearrangement is for the one-carbon homologation of carboxylic acids.[127] In this procedure, a diazomethyl ketone is synthesized from an acid chloride. The rearrangement is then carried out in a nucleophilic solvent which traps the ketene to form a carboxylic acid (in water) or an ester (in alcohols). Silver oxide is often used as a catalyst, since it seems to promote the rearrangement over carbene formation.[128]

124. K.-P. Zeller, *Tetrahedron Lett.*, 707 (1977).
125. K.-P. Zeller, H. Meier, H. Kolshorn, and E. Müller, *Chem. Ber.* **105**, 1875 (1972).
126. U. Timm, K.-P. Zeller, and H. Meier, *Tetrahedron* **33**, 453 (1977).
127. W. E. Bachmann and W. S. Stuve, *Org. React.* **1**, 38 (1942); L. L. Rodina and I. K. Korobitsyna, *Russ. Chem. Rev.* (Engl. Transl.), **36**, 260 (1967); W. Ando, in *Chemistry of Diazonium and Diazo Groups*, S. Patai (ed.), Wiley, New York, 1978, pp. 458–475; H. Meier and K.-P. Zeller, *Angew. Chem. Int. Ed. Engl.* **14**, 32 (1975).
128. T. Hudlicky and J. P. Sheth, *Tetrahedron Lett.*, 2667 (1979).

The photolysis of cyclic α-diazoketones results in ring contraction to a ketene, which is usually isolated as the corresponding ester.

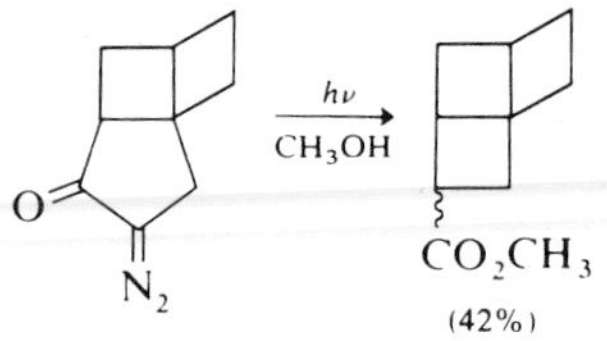

$h\nu$, CH_3OH → CO_2CH_3 (42%) Ref. 129

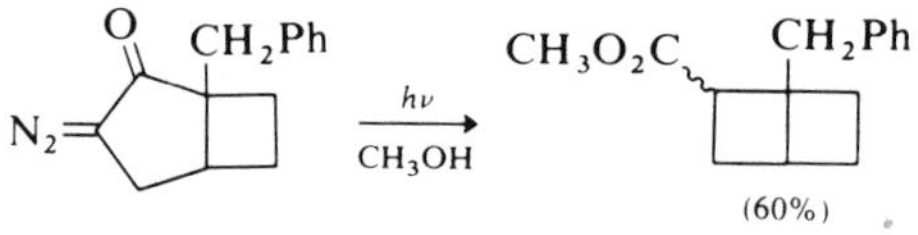

N_2, O, CH_2Ph → $h\nu$, CH_3OH → CH_3O_2C, CH_2Ph (60%) Ref. 130

Scheme 10.8 gives some other examples of Wolff rearrangement reactions.

Scheme 10.8. Wolff Rearrangement of α-Diazoketones

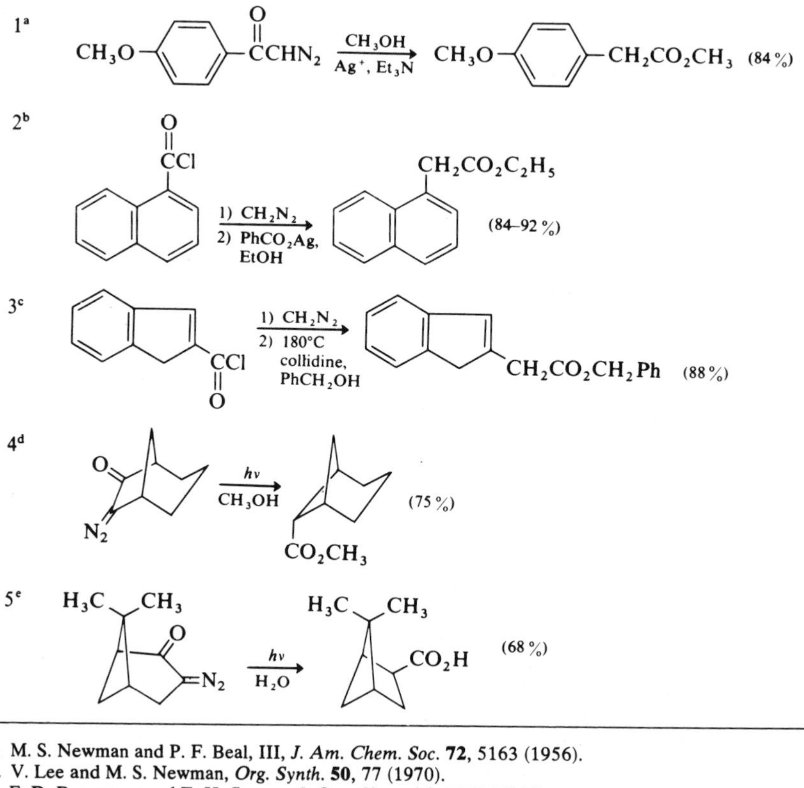

1[a] $CH_3O-C_6H_4-\overset{O}{\overset{\|}{C}}CHN_2 \xrightarrow[Ag^+, Et_3N]{CH_3OH} CH_3O-C_6H_4-CH_2CO_2CH_3$ (84%)

2[b] 1-naphthyl-$\overset{O}{\overset{\|}{C}}Cl$ → 1) CH_2N_2, 2) $PhCO_2Ag$, EtOH → 1-naphthyl-$CH_2CO_2C_2H_5$ (84–92%)

3[c] indenyl-$\underset{O}{\underset{\|}{C}}Cl$ → 1) CH_2N_2, 2) 180°C collidine, $PhCH_2OH$ → indenyl-$CH_2CO_2CH_2Ph$ (88%)

4[d] O, N_2 → $h\nu$, CH_3OH → CO_2CH_3 (75%)

5[e] H_3C, CH_3, O, N_2 → $h\nu$, H_2O → H_3C, CH_3, CO_2H (68%)

a. M. S. Newman and P. F. Beal, III, *J. Am. Chem. Soc.* **72**, 5163 (1956).
b. V. Lee and M. S. Newman, *Org. Synth.* **50**, 77 (1970).
c. E. D. Bergmann and E. Hoffmann, *J. Org. Chem.* **26**, 3555 (1961).
d. K. B. Wiberg and B. A. Hess, Jr., *J. Org. Chem.* **31**, 2250 (1966).
e. J. Meinwald and P. G. Gassman, *J. Am. Chem. Soc.* **82**, 2857 (1960).

129. K. B. Wiberg, L. K. Olli, N. Golembeski, and R. D. Adams, *J. Am. Chem. Soc.* **102**, 7467 (1980).
130. K. B. Wiberg, B. L. Furtek, and L. K. Olli, *J. Am. Chem. Soc.* **101**, 7675 (1979).

The nitrogen analogs of carbenes are called nitrenes. As with carbenes, both singlet and triplet electronic states are possible. The triplet state is usually the ground state, but either species can be involved in reactions. The most common method for generating nitrene intermediates is by thermolysis or photolysis of azides.[131] This method is analogous to formation of carbenes from diazo compounds.

$$R-\bar{\ddot{N}}-\overset{+}{N}\equiv N \xrightarrow[\text{or } h\nu]{\Delta} R-\ddot{N} + N_2$$

The types of azides which have been used for generation of nitrenes include alkyl,[132] aryl,[133] acyl,[134] and sulfonyl[135] derivatives.

The characteristic reaction of alkylnitrenes is migration of one of the substituents to nitrogen, giving an imine:

$$R_3C-\bar{\ddot{N}}-\overset{+}{N}\equiv N \xrightarrow[\text{or } h\nu]{\Delta} R_2C=N-R$$

R = H or alkyl

Intramolecular insertion and addition reactions are almost unknown for alkyl-nitrenes. In fact, it is not clear that the nitrenes are formed as discrete species. The migration may be concerted with elimination, as in the case of the Wolff rearrangement.[136]

Arylnitrenes also generally rearrange rather than undergo addition or insertion reactions.[137]

N_3 → :N: → N → (Nu) → Nu, NH

$Nu = HNR_2$, etc.

131. E. F. V. Scriven (ed.), *Azides and Nitrenes; Reactivity and Utility*, Academic Press, Orlando, Florida, 1984.
132. F. D. Lewis and W. H. Saunders, Jr., in *Nitrenes*, W. Lwoski (ed.), Wiley-Interscience, New York, 1970, pp. 47–98; E. P. Kyba, in *Azides and Nitrenes*, E. F. V. Scriven (ed.), Academic Press, Orlando, Florida, 1984, pp. 2–34.
133. P. A. Smith, in *Nitrenes*, W. Lwoski (ed.), Wiley-Interscience, New York, 1970, pp. 99–162; P. A. S. Smith, in *Azides and Nitrenes*, E. F. V. Scriven (ed.), Academic Press, Orlando, Florida, 1984, pp. 95–204.
134. W. Lwowski, in *Nitrenes*, W. Lwowski (ed.), Wiley-Interscience, New York, 1970, pp. 185–224; W. Lwowski, in *Azides and Nitrenes*, E. F. V. Scriven (ed.), Academic Press, Orlando, Florida, 1984, pp. 205–245.
135. D. S. Breslow, in *Nitrenes*, W. Lwowski (ed.), Wiley-Interscience, New York, 1970, pp. 245-303; R. A. Abramovitch and R. G. Sutherland, *Fortschr. Chem. Forsch.* **16**, 1 (1970).
136. R. M. Moriarty and R. C. Reardon, *Tetrahedron* **26**, 1379 (1970); R. A. Abramovitch and E. P. Kyba, *J. Am. Chem. Soc.* **93**, 1537 (1971); R. M. Moriarty and P. Serridge, *J. Am. Chem. Soc.* **93**, 1534 (1971).
137. O. L. Chapman and J.-P. LeRoux, *J. Am. Chem. Soc.* **100**, 282 (1978); O. L. Chapman, R. S. Sheridan, and J.-P. LeRoux, *Recl. Trav. Chim. Pay-Bas* **98**, 334 (1979); R. J. Sundberg, S. R. Suter, and M. Brenner, *J. Am. Chem. Soc.* **94**, 573 (1972).

A few intramolecular insertion reactions, especially in aromatic systems, go in good yield.[138]

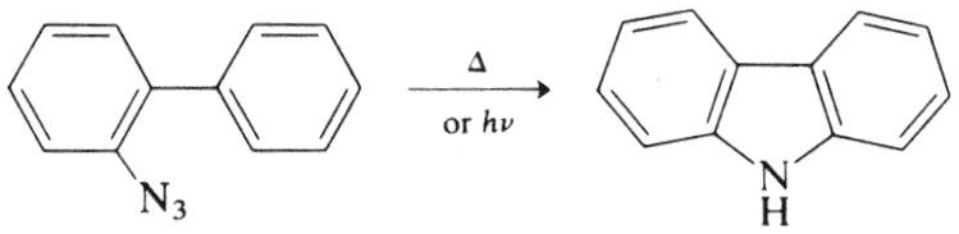

N_3 $\xrightarrow[\text{or } h\nu]{\Delta}$ N H

The nitrenes that most consistently give addition and insertion reactions are carboalkoxynitrenes generated from alkyl azidoformates:

$$RO-\overset{O}{\overset{\|}{C}}-N_3 \xrightarrow[\text{or } h\nu]{\Delta} RO-\overset{O}{\overset{\|}{C}}-N:$$

These intermediates undergo addition reactions with alkenes and aromatic compounds and insertion reactions with saturated hydrocarbons.[139]

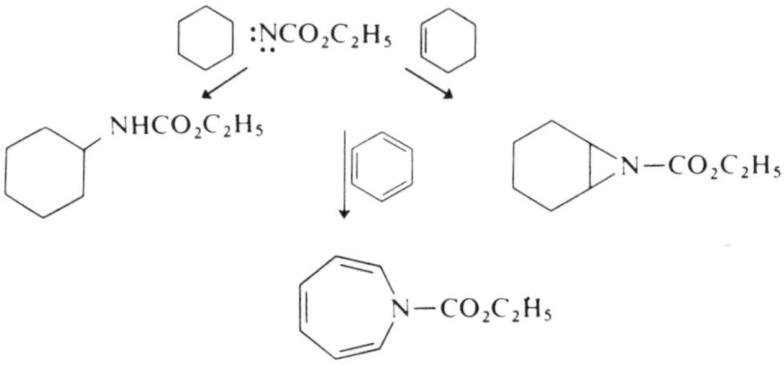

$:\ddot{N}CO_2C_2H_5$

$NHCO_2C_2H_5$ $N-CO_2C_2H_5$

$N-CO_2C_2H_5$

Carboalkoxynitrenes are somwehat more selective than the corresponding carbenes, showing selectivities of roughly 1:10:40 for the primary, secondary, and tertiary positions in 2-methylbutane in insertion reactions.

Sulfonylnitrenes are formed by thermal decomposition of sulfonyl azides. Insertion reactions occur with saturated hydrocarbons.[140] With aromatic rings, the main products are formally insertion products, but they are believed to be formed through addition intermediates.

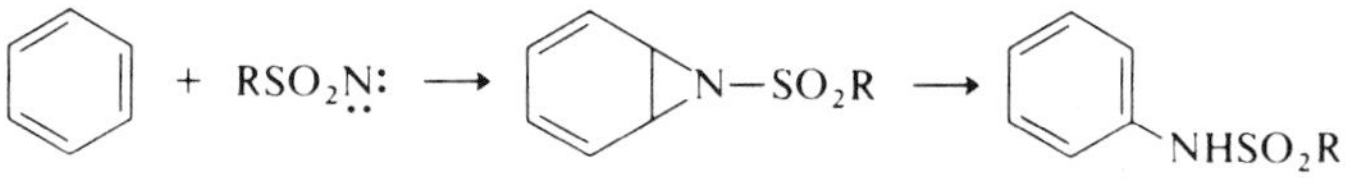

$+ \; RSO_2\ddot{N}: \longrightarrow$ $N-SO_2R \longrightarrow$ $NHSO_2R$ Ref. 141

10.2.8. Rearrangements to Electron-Deficient Nitrogen

In contrast to the somewhat limited synthetic utility of nitrenes, there is an important group of reactions in which migration occurs to electron-deficient nitrogen.

138. P. A. S. Smith and B. B. Brown, *J. Am. Chem. Soc.* **73**, 2435, 2438 (1951); J. S. Swenton, T. J. Ikeler, and B. H. Williams, *J. Am. Chem. Soc.* **92**, 3103 (1970).
139. W. Lwowski, *Angew. Chem. Int. Ed. Engl.* **6**, 897 (1967).
140. D. S. Breslow, M. F. Sloan, N. R. Newburg, and W. B. Renfrow, *J. Am. Chem. Soc.* **91**, 2273 (1969).
141. R. A. Abramovitch, G. N. Knaus, and V. Uma, *J. Org. Chem.* **39**, 1101 (1974).

One of the most useful of these reactions is the *Curtius rearrangement.*[142] This reaction has the same relationship to acylnitrene intermediates as the Wolff rearrangement has to acylcarbenes. The initial product is an isocyanate that can be isolated or trapped by a nucleophilic solvent.

$$\text{RC(=O)}\ddot{N}^{-}\text{–}N^{+}\equiv N \rightarrow \text{R–C(=O)}\cdots N\text{:}\cdots N\equiv N \rightarrow O{=}C{=}N\text{–}R \xrightarrow{H_2O} [R\text{–}NH\text{–}C(=O)\text{–}OH] \rightarrow RNH_2 + CO_2$$

$$O{=}C{=}N\text{–}R \xrightarrow{R'OH} R\text{–}NH\text{–}C(=O)\text{–}OR'$$

This reaction is usually considered to be a concerted process in which migration accompanies loss of nitrogen.[143] The migrating group retains it stereochemical configuration. The temperature required for reaction is in the vicinity of 100°C.

The acyl azide intermediates are prepared either by reaction of sodium azide with a reactive acylating agent or by diazotization of an acid hydrazide. An especially convenient version of the former process is to treat the carboxylic acid with ethyl chloroformate to form a mixed anhydride, which then reacts with azide ion.[144]

$$RCO_2H \xrightarrow{ClC(=O)OEt} RC(=O)OC(=O)OEt \xrightarrow{N_3^-} RC(=O)N_3$$

$$RC(=O)NHNH_2 \xrightarrow[H^+]{NaNO_2} RCON_3$$

The reaction can also be carried out on the acid using diphenylphosphoryl azide[145]:

$$RCO_2H + (PhO)_2P(=O)N_3 \rightarrow RC(=O)N_3 \xrightarrow{R'OH} RNHC(=O)OR'$$

Some examples of the Curtius reaction are given in Scheme 10.9.

Another reaction that can be used for conversion of carboxylic acids to the corresponding amines with loss of carbon dioxide is the *Hofmann rearrangement.* The reagent is hypobromite ion, which reacts to form an *N*-bromoamide intermediate. Like the Curtius reaction, the rearrangement is believed to be a concerted process:

$$RC(=O)NH_2 + {}^{-}OBr \rightarrow RC(=O)NHBr + {}^{-}OH \rightleftharpoons RC(=O)\bar{N}Br + H_2O$$

$$R\text{–}C(=O)\text{–}\bar{N}\text{–}Br \rightarrow O{=}C{=}N\text{–}R + Br^{-} \xrightarrow{H_2O} NH_2R + CO_2$$

142. P. A. S. Smith, *Org. React.* **3**, 337 (1946).
143. S. Linke, G. T. Tisue, and W. Lwowski, *J. Am. Chem. Soc.* **89**, 6308 (1967).
144. J. Weinstock, *J. Org. Chem.* **26**, 3511 (1961).
145. D. Kim and S. M. Weinreb, *J. Org. Chem.* **43**, 125 (1978).

Scheme 10.9. Rearrangement to Electron-Deficient Nitrogen

A. Beckmann Rearrangement Reactions

1[a] NOH, $C_6H_5C(=NOH)C(CH_3)_3$ → (HCl, CH_3CO_2H) → $(CH_3)_3CCONHC_6H_5$ (94%)

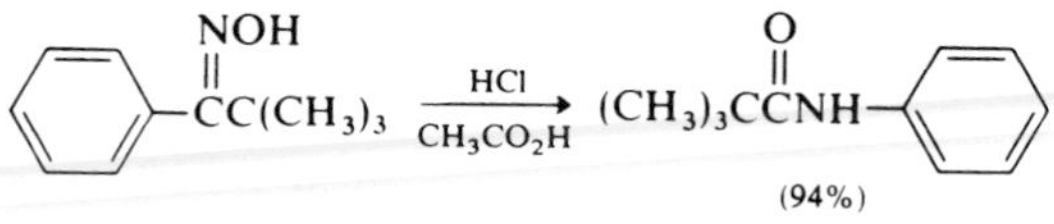

2[b] H_3C–C=NOH (decalin) → (*p*-toluenesulfonyl chloride, pyridine) → NHCCH$_3$ (O) (decalin), H (92%)

3[c] H_3C, H, NOH → (polyphosphoric acid) → H_3C, NH, H, O (92%)

4[d] O, =NOH → (*p*-toluenesulfonyl chloride, pyridine) → O, O, NH (91%)

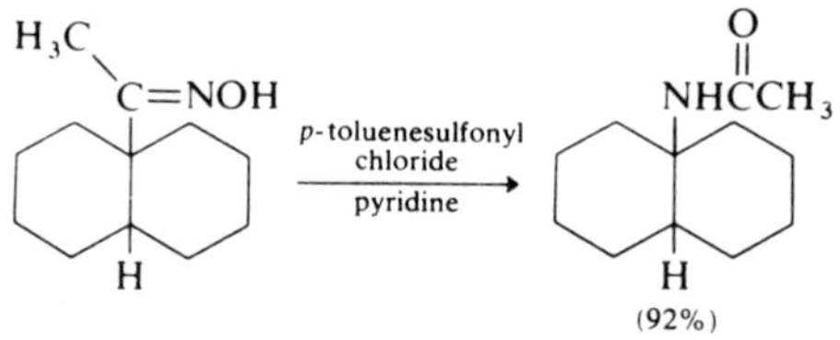

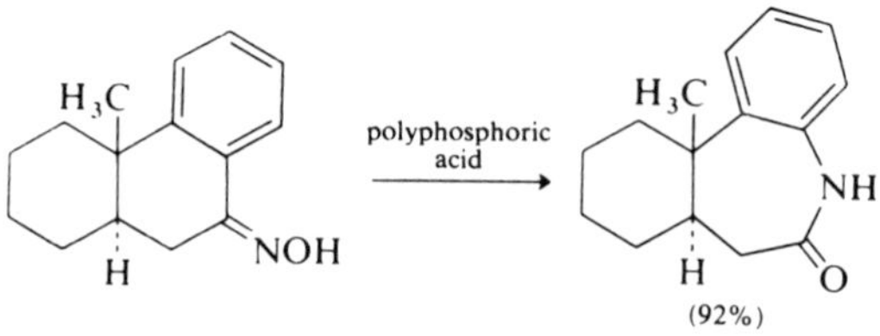

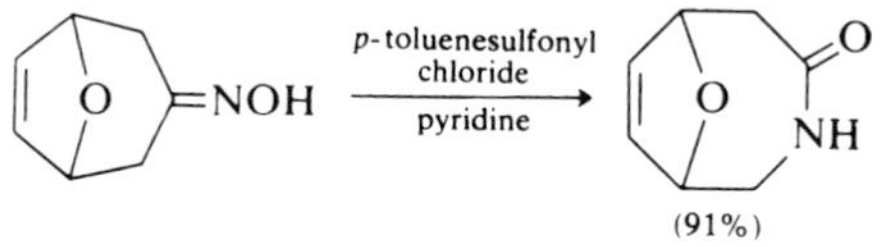

B. Curtius Rearrangement Reactions

5[e] $CH_3(CH_2)_{10}CCl$ (C=O) → (1) NaN_3 2) benzene, 70°C) → $CH_3(CH_2)_{10}N{=}C{=}O$

6[f] $H_5C_2O_2C(CH_2)_4CO_2C_2H_5$ → (1) N_2H_4 2) HNO_2 3) Δ 4) H^+, H_2O) → $Cl^-H_3\overset{+}{N}(CH_2)_4\overset{+}{N}H_3Cl^-$

7[g] Ph, CO_2H (cyclopentane) → (1) EtOCCl (C=O) 2) NaN_3 3) heat 4) H^+, H_2O) → Ph, $\overset{+}{N}H_3$ Cl^- (cyclopentane) (76–81%)

8[h] H_3C, CH_2CH_3, Ph, C–CO_2H → (1) $SOCl_2$, pyridine 2) NaN_3, xylene) → H_3C, CH_2CH_3, Ph, C–NH_2 (66%)

9[i] CH_3O_2C, CH_3, N, Ph, CH_3O_2C, CO_2H (pyridine) → (1) $(PhO)_2PN_3$ (P=O), 80°C 2) MeOH) → CH_3O_2C, CH_3, N, Ph, CH_3O_2C, $NHCO_2CH_3$ (100%)

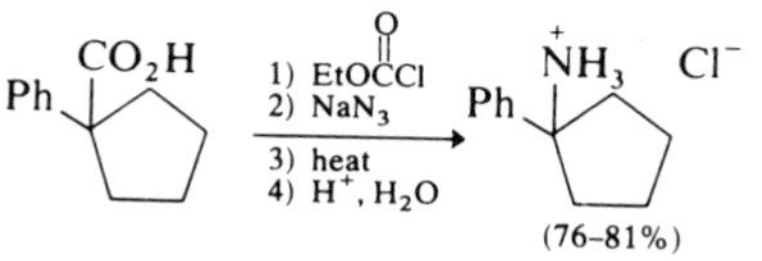

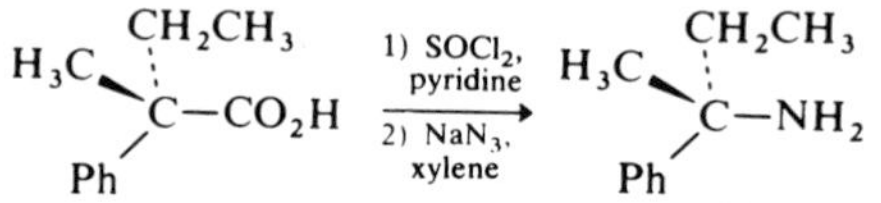

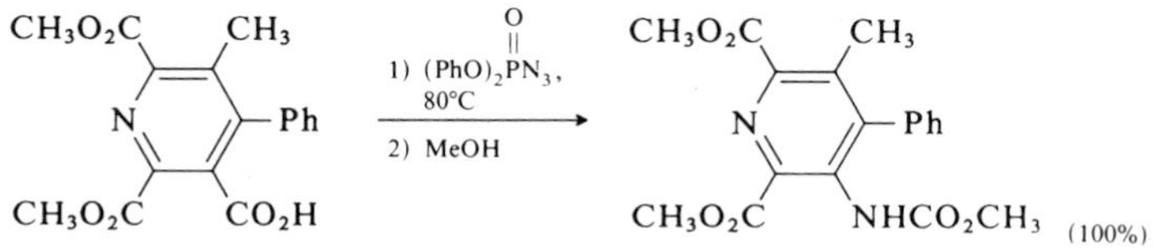

C. Schmidt Reactions

10[j] $PhCH_2CO_2H \xrightarrow[\text{polyphosphoric acid}]{NaN_3} PhCH_2NH_2$

11[k] 3,5-$((CH_3)_3C)_2C_6H_3CO_2H \xrightarrow[NaN_3]{H_2SO_4}$ 3,5-$((CH_3)_3C)_2C_6H_3NH_3^+$ (93%)

12[l] adamantanone $\xrightarrow[CF_3CO_2H]{NaN_3}$ lactam (NH) (59%)

a. R. F. Brown, N. M. van Gulick, and G. H. Schmid, *J. Am. Chem. Soc.* **77**, 1094 (1955).
b. R. K. Hill and O. T. Chortyk, *J. Am. Chem. Soc.* **84**, 1064 (1962).
c. R. A. Barnes and M. T. Beachem, *J. Am. Chem. Soc.* **77**, 5388 (1955).
d. S. R. Wilson, R. A. Sawicki, and J. C. Huffman, *J. Org. Chem.* **46**, 3887 (1981).
e. C. F. H. Allen and A. Bell, *Org. Synth.* **III**, 846 (1955).
f. P. A. S. Smith, *Org. Synth.* **IV**, 819 (1963).
g. C. Kaiser and J. Weinstock, *Org. Synth.* **51**, 48 (1971).
h. D. J. Cram and J. S. Bradshaw, *J. Am. Chem. Soc.* **85**, 1108 (1963).
i. D. Kim and S. M. Weinreb, *J. Org. Chem.* **43**, 125 (1978).
j. R. M. Palmere and R. T. Conley, *J. Org. Chem.* **35**, 2703 (1970).
k. J. W. Elder and R. P. Mariella, *Can. J. Chem.* **41**, 1653 (1963).
l. T. Sasaki, S. Eguchi, and T. Toru, *J. Org. Chem.* **35**, 4109 (1970).

The reaction has been useful in the conversion of aromatic carboxylic acids to aromatic amines.

2-fluoro-3-pyridinecarboxamide (CNH_2 with C=O) $\xrightarrow[KOH]{Br_2}$ 3-amino-2-fluoropyridine (NH_2, F, N) — Ref. 146

Carboxylic acids and esters can also be converted to amines wity loss of the carbonyl group by reaction with hydrazoic acid, HN_3. This is known as the *Schmidt reaction.*[147] The mechanism is related to that of the Curtius reaction. An azido intermediate is generated by addition of hydrazoic acid to the carbonyl group. The migrating group retains its stereochemical configurations.

$$RCO_2H + HN_3 \rightarrow HO{-}\underset{OH}{\overset{R}{C}}{-}\underset{H}{N}{-}\overset{+}{N}{\equiv}N \rightarrow HO\overset{O}{\overset{\|}{C}}\underset{H}{N}R + N_2 \xrightarrow{H^+} R\overset{+}{N}H_3 + CO_2$$

146. G. C. Finger, L. D. Starr, A. Roe, and W. J. Link, *J. Org. Chem.* **27**, 3965 (1962).
147. H. Wolff, *Org. React.* **3**, 307 (1946); P. A. S. Smith, in *Molecular Rearrangements*, P. de Mayo (ed.), Vol. 1, Wiley-Interscience, New York, 1963, pp. 507–522.

The reaction of hydrazoic acids converts ketones to amides:

$$RC(=O)R + HN_3 \rightleftharpoons RC(OH)(R)-N^{-}-\overset{+}{N}\equiv N \xrightarrow{H^+} R-C(OH)(R)-N(H)-\overset{+}{N}\equiv N \rightarrow RC(=O)NHR$$

$$\downarrow -H_2O$$

$$R-C(R)=N-\overset{+}{N}\equiv N \rightarrow R-\overset{+}{N}\equiv C-R \xrightarrow{H_2O} RC(=O)NHR$$

Unsymmetrical ketones can give mixtures of products because it is possible for either group to migrate.

$$RC(=O)R' \xrightarrow{HN_3} RC(=O)NHR' + RNHC(=O)R'$$

Section C of Scheme 10.9 includes some examples of the Schmidt reaction.

Another important reaction involving migration to electron-deficient nitrogen is the *Beckmann rearrangement*, in which oximes are converted to amides[148]:

$$R-C(=N-OH)-R' \rightarrow R-N(H)-C(=O)-R'$$

A variety of protic acids, Lewis acids, acid anhydrides, and acid halides can cause the reaction to occur. The mechanism involves conversion of the oxime hydroxyl group to a leaving group. Ionization and migration then occur as a concerted process, with the group which is *anti* to the oxime leaving group migrating. This results in formation of a nitrilium ion, which captures a nucleophile. Eventually, hydrolysis leads to the amide.

$$RC(R')=N-OH \xrightarrow{X^+} RC(R')=N-\overset{+}{O}(H)X \rightarrow [R\cdots N(=C-R')\cdots\overset{\delta+}{O}(H)-X]^{\delta+} \rightarrow R-\overset{+}{N}\equiv C-R' \xrightarrow{YOH} R-N=C(O-Y)R' \rightarrow RNHC(=O)R'$$

The migrating group retains its configuration. Some reaction conditions can lead to *syn-anti* isomerization occurring at a rate exceeding that of rearrangement. When this occurs, a mixture of products will be formed. The reagents which have been found least likely to cause competing isomerization are phosphorus pentachloride and *p*-toluenesulfonyl chloride.[149]

148. L. G. Donaruma and W. Z. Heldt, *Org. React.* **11**, 1 (1960); P. A. S. Smith, *Open Chain Nitrogen Compounds*, Vol. II, W. A. Benjamin, New York, 1966, pp. 47–54; P. A. S. Smith, in *Molecular Rearrangements*, Vol. 1, P. de Mayo (ed.), Wiley-Interscience, New York, 1973, pp. 483–507; G. R. Krow, *Tetrahedron* **37**, 1283 (1981); R. E. Gawley, *Org. React.* **35**, 1 (1988).

149. R. F. Brown, N. M. van Gulick, and G. H. Schmid, *J. Am. Chem. Soc.* **77**, 1094 (1955); J. C. Craig and A. R. Naik, *J. Am. Chem. Soc.* **84**, 3410 (1962).

A fragmentation reaction occurs if one of the oxime substituents can give rise to a relatively stable carbocation. Fragmentation is very likely to occur if X is a nitrogen, oxygen, or sulfur atom.

$$X-C-C(R)=N-OY \rightarrow \overset{+}{X}=C + RC\equiv N + {}^{-}OY$$

$$\text{1,1-dimethyl-2-tetralone oxime (NOH; } H_3C, CH_3) \xrightarrow{PCl_5} \text{o-}(CH_2CH_2C\equiv N)C_6H_4C(CH_3)=CH_2 \quad (93\%)$$

Ref. 150

Section A of Scheme 10.9 provides some examples of the Beckmann rearrangement.

10.3. Reactions Involving Free-Radical Intermediates

The mechanistic basis of free-radical reactions was considered in Chapter 12 of Part A. Several mechanistic points are crucial in development of free-radical reactions for synthetic applications.[151] Successful free-radical reactions are usually chain processes. The lifetimes of the intermediate radicals are very short. To meet the requirement of synthesis for high selectivity and efficiency, all steps in a desired process must be fast in comparison with competing reactions. Because of the requirement that all steps be quite fast, only steps that are exothermic or very slightly endothermic can participate in chain processes. Comparison of two sets of radical processes can illustrate this point. Let us compare the enthalpy of addition of a radical to a carbon–carbon double bond with addition to a carbonyl group:

$$-\dot{C}\cdot + C=C \rightarrow -C-C-C\cdot$$

$$\Delta H = C-C \; - \; C^{\pi}-C^{\pi} = -81-(-64) = -17$$

$$-\dot{C}\cdot + C=O \rightarrow -C-C-O\cdot$$

$$\Delta H = C-C \; - \; C^{\pi}-O^{\pi} = -81-(-94) = +13$$

This comparison suggests that of these two similar reactions, only the former is likely to be a part of an efficient radical chain reaction. Radical additions to carbon–carbon double bonds should be further enhanced by radical-stabilizing groups on the carbon–carbon double bond. A similar bond energy comparison can be made for abstraction of hydrogen from carbon as opposed to oxygen:

$$-C\cdot + H-C- \rightarrow -C-H + \cdot C-$$

$$\Delta H = 0$$

$$-C\cdot + H-O-C- \rightarrow -C-H + \cdot O-C-$$

$$\Delta H = C-H - O-H = -98-(-109) = +11$$

150. R. T. Conley and R. J. Lange, *J. Org. Chem.* **28**, 210 (1963).
151. C. Walling, *Tetrahedron* **41**, 3887 (1985).

We conclude that while abstraction of a hydrogen atom from carbon may be a feasible step in a chain process, abstraction of a hydrogen atom from a hydroxyl group is unlikely. Homolytic cleavage of an O—H bond is likely only if the resulting oxygen radical is stabilized in some way.

10.3.1. Sources of Radical Intermediates

A discussion of some of the radical sources used for mechanistic studies was given in Section 12.1.4 of Part A. Some of the reactions discussed there, particularly those involving the use of azo compounds and peroxides as reaction initiators, are also important in synthetic chemistry. One of the most useful sources of free radicals in preparative chemistry is the reaction of halides with stannyl radicals:

$$\text{initiation} \quad In\cdot + R'_3SnH \rightarrow R'_3Sn\cdot + In{-}H$$

$$R{-}X + R'Sn\cdot \rightarrow R\cdot + R'_3Sn{-}X$$

$$\text{propagation} \quad R\cdot + X{=}Y \rightarrow R{-}X{-}Y\cdot$$

$$R{-}X{-}Y\cdot + R'_3Sn{-}H \rightarrow R{-}X{-}Y{-}H + R'_3Sn\cdot$$

This generalized reaction sequence consumes the halide, the stannane, and the other reactant X=Y and effects addition of the organic radical and a hydrogen atom to the X=Y bond. The order of reactivity of organic halides toward stannyl radicals is iodides > bromides > chlorides.

The esters of *N*-hydroxypyridine-2-thione have proven to be a versatile source of radicals. The radical is formed by fragmentation of an adduct resulting from attack at sulfur by the chain-carrying radical.[152] The generalized chain sequence is as follows:

$$X\cdot + \text{(N-acyloxypyridine-2-thione, } S{=}C\text{–N–OC(=O)R)} \rightarrow \text{(X–S adduct radical, N–O–C(=O)–R)} \rightarrow \text{2-(X–S)pyridine} + CO_2 + R\cdot$$

$$R\cdot + X{-}Y \rightarrow R{-}Y + X\cdot$$

When X—Y is $R_3Sn{-}H$, the net reaction is decarboxylation and reduction of the original acyloxy group. When X—Y is $Cl_3C{-}Cl$, the final product is a chloride.[153] Use of $Cl_3C{-}Br$ gives the corresponding bromide.[154]

$$\text{(N-acyloxypyridine-2-thione, } S{=}C\text{–N–OC(=O)R)} \xrightarrow{CCl_4} R{-}Cl + \text{2-}(Cl_3CS)\text{pyridine} + CO_2$$

152. D. H. R. Barton, D. Crich, and W. B. Motherwell, *Tetrahedron* **41**, 3901 (1985); D. H. R. Barton, D. Bridson, I. Fernandez-Picot, and S. Z. Zard, *Tetrahedron* **43**, 2733 (1987); D. H. R. Barton, D. Crich, and G. Kretzchmar, *J. Chem. Soc., Perkin Trans. 1*, 39 (1986).
153. D. H. R. Barton, D. Crich, and W. B. Motherwell, *Tetrahedron Lett.* **24**, 4979 (1983).
154. D. H. R. Barton, R. Lacher, and S. Z. Zard, *Tetrahedron Lett.* **26**, 5939 (1985).

Selenyl groups can be abstracted from acyl selenides to generate radicals on reaction with stannyl radicals.[155] The resulting acyl radicals can be reduced to aldehydes or undergo decarbonylation prior to reduction.

$$R'_3Sn\cdot + PhSe\overset{O}{\overset{\|}{C}}R \rightarrow R\overset{O}{\overset{\|}{C}}\cdot + R'_3SnSePh$$

$$R\overset{O}{\overset{\|}{C}}\cdot \rightarrow R\cdot + CO$$

The reductive decomposition of alkylmercury compounds is also a useful source of radicals. The organomercury compounds are available by oxymercuration (Section 4.3) or from an organometallic compound as a result of metal–metal exchange (Section 7.3.3). The mercuric hydride formed by reduction undergoes chain decomposition to generate alkyl radicals.

$$RHgX + NaBH_4 \rightarrow RHgH$$

$$RHgH \rightarrow R\cdot + HgH$$

$$\text{propagation} \qquad R\cdot + RHgH \rightarrow R{-}H + RHg$$

$$RHg \rightarrow R\cdot + Hg^0$$

10.3.2. Introduction of Functionality by Radical Reactions

The introduction of halogen substituents by free-radical substitution was discussed in Section 12.3 of Part A. Halogenation is a fairly general method for functionalization but is limited to the site in the molecule which is most reactive toward halogen atoms. Halogenations at benzylic and allylic positions are therefore the most useful synthetic reactions.

In this section, we will focus on intramolecular functionalization. Such reactions normally achieve selectivity on the basis of proximity of the reacting centers. In acyclic molecules, intramolecular functionalization normally involves hydrogen atom abstraction via six-membered cyclic transition states. The net result is introduction of functionality at the δ atom in relation to the radical site.

[Scheme: C–C–C–C(H)–C–C–X· → (1,5-H transfer) C–C–C–C·–C–C–X–H $\xrightarrow{Y-Z}$ C–C–C–C(Y)–C–C–X–H + Z·]

One example of this type of reaction is the photolytically initiated decomposition of N-chloroamines in acidic solution, which is known as the *Hofmann–Loeffler*

155. J. Pfenninger, C. Heuberger, and W. Graf, *Helv. Chim. Acta* **63**, 2328 (1980).

reaction.[156] The initial products are δ-chloroamines, but these are usually converted to pyrrolidines by intramolecular nucleophilic substituents.

$$RCH_2CH_2CH_2CH_2\overset{+}{N}(Cl)HCH_3 \xrightarrow{h\nu} RCH_2CH_2CH_2CH_2\overset{+}{\dot{N}}HCH_3 + Cl\cdot$$

$$RCH_2CH_2CH_2CH_2\overset{+}{\dot{N}}HCH_3 \longrightarrow R\dot{C}HCH_2CH_2CH_2\overset{+}{N}H_2CH_3$$

$$R\dot{C}HCH_2CH_2CH_2\overset{+}{N}H_2CH_3 + RCH_2CH_2CH_2CH_2\overset{+}{N}(Cl)HCH_3 \longrightarrow RCH(Cl)CH_2CH_2CH_2\overset{+}{N}H_2CH_3 + RCH_2CH_2CH_2CH_2\overset{+}{\dot{N}}HCH_3$$

$$RCH(Cl)CH_2CH_2CH_2\overset{+}{N}H_2CH_3 \xrightarrow{NaOH} \text{2-R-N-methylpyrrolidine}$$

A closely related procedure results in formation of γ-lactones. Amides are converted to *N*-iodoamides by reaction with iodine and *t*-butyl hypochlorite. Photolysis of the *N*-iodoamides gives lactones via iminolactone intermediates.[157]

$$RCH_2(CH_2)_2\overset{O}{\overset{\|}{C}}NHI \xrightarrow{h\nu} RCH(I)(CH_2)_2\overset{O}{\overset{\|}{C}}NH_2 \longrightarrow \text{iminolactone } (=\overset{+}{N}H_2\ I^-) \xrightarrow{H_2O} \gamma\text{-lactone}$$

Steps similar to those involved in the Hofmann–Loeffler reaction are also involved in cyclization of *N*-alkylmethanesulfonamides by oxidation with $Na_2S_2O_4$ in the presence of cupric ion[158]:

$$RCH_2(CH_2)_3NHSO_2CH_3 \xrightarrow[-H^+]{-e^-} RCH_2(CH_2)_3\dot{N}SO_2CH_3 \rightarrow R\dot{C}H(CH_2)_3N(H)SO_2CH_3$$

$$R\dot{C}H(CH_2)_3NHSO_2CH_3 \xrightarrow{Cu^{2+}} R\overset{+}{C}H(CH_2)_3NHSO_2CH_3 \rightarrow \text{2-R-N-}(SO_2CH_3)\text{pyrrolidine}$$

There are also useful intramolecular functionalization methods which involve hydrogen atom abstraction by oxygen radicals. The conditions that were initially developed involved thermal or photochemical dissociation of alkoxy derivatives of Pb(IV) generated by exchange with $Pb(OAc)_4$[159]:

$$RCH_2(CH_2)_3OH \xrightarrow{Pb(OAc)_4} RCH_2(CH_2)_3O{-}Pb(OAc)_3 \rightarrow RCH_2(CH_2)_3O\cdot + Pb(OAc)_3$$

$$RCH_2(CH_2)_3O\cdot \rightarrow R\dot{C}H(CH_2)_3OH \xrightarrow{-e^-} R\overset{+}{C}H(CH_2)_3OH \rightarrow \text{2-R-tetrahydrofuran}$$

156. M. E. Wolff, *Chem. Rev.* **63**, 55 (1963).
157. D. H. R. Barton, A. L. J. Beckwith, and A. Goosen, *J. Chem. Soc.*, 181 (1965).
158. G. I. Nikishin, E. I. Troyansky, and M. Lazareva, *Tetrahedron Lett.* **26**, 1877 (1985).
159. K. Heusler, *Tetrahedron Lett.*, 3975 (1964).

The oxidation of the radical to a carbocation is effected by Pb(IV) or Pb(III). Current procedures include iodine and are believed to involve a hypoiodite intermediate.[160]

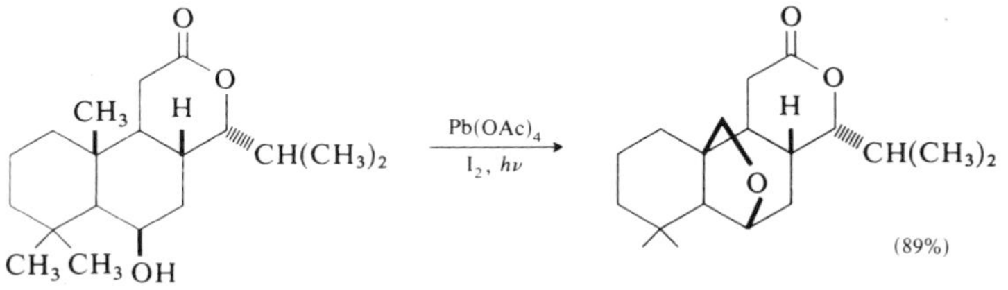

$\xrightarrow[I_2,\ h\nu]{Pb(OAc)_4}$ (89%) Ref. 161

Alkoxy radicals are also the active hydrogen-abstracting species in a procedure which involves photolysis of nitrite esters.

$$RCH_2(CH_2)_3ON{=}O \rightarrow RCH_2(CH_2)_3O\cdot \quad \cdot N{=}O \rightarrow R\dot{C}H(CH_2)_3OH \quad \cdot N{=}O$$

$$\downarrow$$

$$RCH(N{=}O)(CH_2)_3OH \rightarrow RC(=NOH)(CH_2)_3OH$$

This reaction was originally developed as a method for functionalization of methyl groups in steroids.[162]

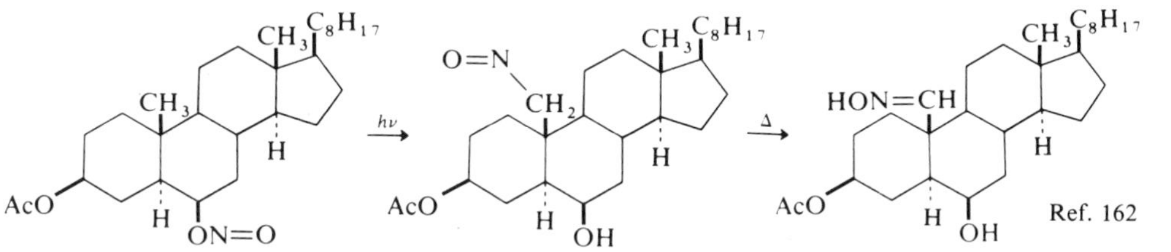

$\xrightarrow{h\nu}$ $\xrightarrow{\Delta}$ Ref. 162

The reaction has found other synthetic applications.

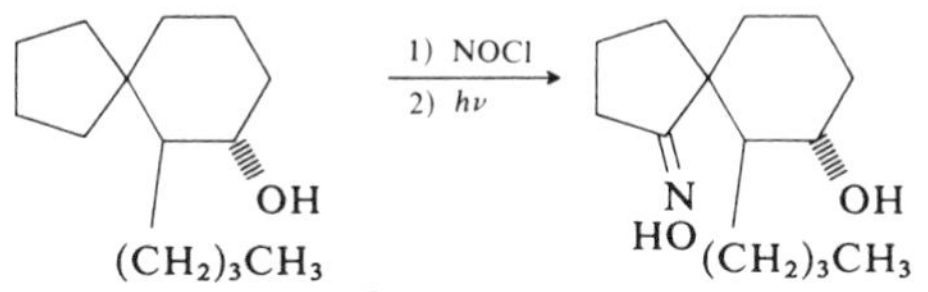

$\xrightarrow[2)\ h\nu]{1)\ NOCl}$ Ref. 163

10.3.3. Addition Reactions of Radicals with Substituted Alkenes

The most general method for formation of new carbon–carbon bonds via radical intermediates involves addition of the radical to an alkene. The addition reaction

160. K. Heusler, P. Wieland, and C. Meystre, *Org. Synth.* **V**, 692 (1973); K. Heusler and J. Kalvoda, *Angew. Chem. Int. Ed. Engl.* **3**, 525 (1964).
161. S. D. Burke, L. A. Silks III, and S. M. S. Strickland, *Tetrahedron Lett.* **29**, 2761 (1988).
162. D. H. R. Barton, J. M. Beaton, L. E. Geller, and M. M. Pechet, *J. Am. Chem. Soc.* **83**, 4076 (1961).
163. E. J. Corey, J. F. Arnett, and G. N. Widiger, *J. Am. Chem. Soc.* **97**, 430 (1975).

generates a new radical which can propagate a chain reaction. The preferred alkenes for trapping alkyl radicals are often ethylene derivatives with electron-attracting groups, such as cyano, ester, or other carbonyl substituents.[164] There are two factors which make such compounds particularly useful: (1) alkyl radicals are relatively nucleophilic, and they react at enhanced rates with alkenes having electron-withdrawing substituents; (2) alkenes with such substituents exhibit a good degree of regioselectivity, resulting from a combination of steric and radical-stabilizing effects of the substituent. The "nucleophilic" versus "electrophilic" character of radicals can be understood in terms of the MO description of substituent effects on radicals. The three most important cases are outlined in Fig. 10.2.

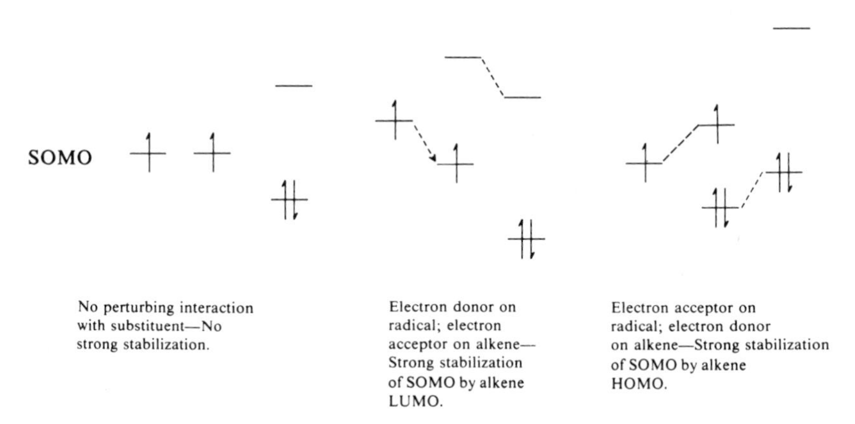

SOMO
No perturbing interaction with substituent—No strong stabilization.
Electron donor on radical; electron acceptor on alkene—Strong stabilization of SOMO by alkene LUMO.
Electron acceptor on radical; electron donor on alkene—Strong stabilization of SOMO by alkene HOMO.

Fig. 10.2. Frontier orbital interpretation of radical substituent effects.

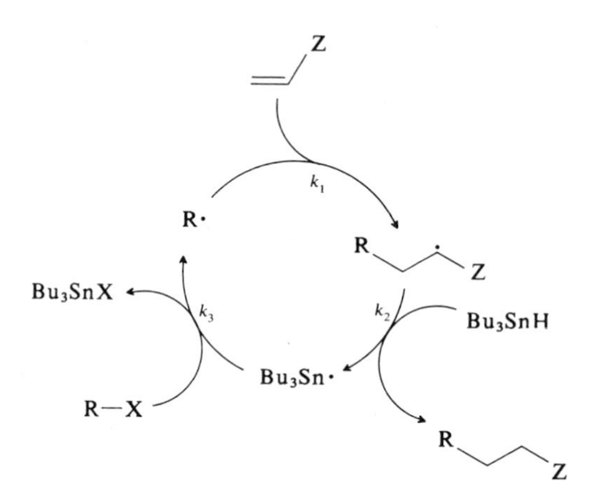

Fig. 10.3. Chain mechanism for radical addition reactions mediated by trialkylstannyl radicals.

164. B. Giese, *Angew. Chem. Int. Ed. Engl.* **22**, 753 (1983); B. Giese, *Angew. Chem. Int. Ed. Engl.* **24**, 553 (1985).

Radicals for addition reactions can be generated by halogen atom abstraction by stannyl radicals. The chain mechanism for alkylation of alkyl halides by reaction with a substituted alkene is outlined in Fig. 10.3. There are three reactions in the propagation cycle of this chain mechanism. The rate of each of these steps must exceed those of competing chain termination reactions in order for good yields to be obtained. The most important competitions are between the addition step k_1 and reaction of the intermediate R· with Bu_3SnH and between the H-abstraction step k_2 and addition to another molecule of the alkene. In the case of step k_1, if this step is not fast enough, the radical R· will abstract H from the stannane and the overall reaction will simply be dehalogenation. If step k_2 is not fast relative to a successive addition step, formation of oligomers containing several alkene units will occur. For good yields to be obtained, R· must be more reactive toward the substituted alkene than is $RCH_2\dot{C}HZ$, and $RCH_2\dot{C}HZ$ must be more reactive toward Bu_3SnH than is R·. These requirements are met when Z is an electron-attracting group. Yields are also improved if the concentration of Bu_3SnH is kept low to minimize the reductive dehalogenation. This can be done by adding the stannane slowly as the reaction proceeds. Another method is to use only a small amount of the trialkyltin hydride along with a reducing agent, such as $NaBH_4$ or $NaBH_3CN$, which can regenerate the reactive stannane.[165]

$$\text{C}_6\text{H}_{11}\text{–I} + CH_2{=}CHCN \xrightarrow[1.3\ NaBH_4]{0.2\ \text{equiv}\ Bu_3SnH,\ h\nu} \text{C}_6\text{H}_{11}\text{–}CH_2CH_2CN$$

Radicals formed by fragmentation of xanthate and related thiono esters can also be trapped by reactive alkenes.[165] The mechanism of radical generation from thiono esters was discussed in connection with the Barton deoxygenation method on p. 253.[166]

$$R_2CHOC(=S)SCH_3 + CH_2{=}CHCN \xrightarrow{Bu_3SnH} R_2CHCH_2CH_2CN$$

Organomercury compounds are also sources of alkyl radicals. Organomercurials can be prepared either by solvomercuration (Section 4.3) or from organometallic reagents (Section 7.3.3).

$$RCH{=}CH_2 + HgX_2 \xrightarrow{SH} RCH(S)CH_2HgX$$

$$RLi + HgX_2 \rightarrow RHgX + LiX$$

Alkylmercury reagents can also be prepared from alkylboranes.

$$R_3B + 3\,Hg(OAc)_2 \rightarrow 3\,RHgOAc \qquad \text{Ref. 167}$$

165. B. Giese, J. A. Gonzalez-Gomez, and T. Witzel, *Angew. Chem. Int. Ed. Engl.* **23**, 69 (1984).
166. D. H. R. Barton and S. W. McCombie, *J. Chem. Soc., Perkin Trans 1*, 1574 (1975).
167. R. C. Larock and H. C. Brown, *J. Am. Chem. Soc.* **92**, 2467 (1970).

α-Acetoxyalkylmercury compounds can be prepared from hydrazones by treatment with mercuric oxide and mercuric acetate.

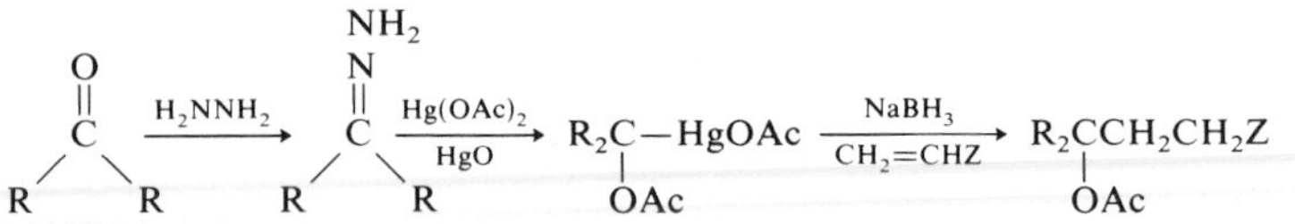

$$R_2C{=}O \xrightarrow{H_2NNH_2} R_2C{=}NNH_2 \xrightarrow[HgO]{Hg(OAc)_2} R_2C(OAc){-}HgOAc \xrightarrow[CH_2{=}CHZ]{NaBH_3} R_2C(OAc)CH_2CH_2Z$$

Ref. 168

Radicals are generated by reduction of the organomercurial by $NaBH_4$ or a similar reductant. These techniques have been applied to β-hydroxy-,[169] β-alkoxy-,[170] and β-amido-[171] alkylmercury derivatives.

Scheme 10.10 illustrates addition reaction of radicals with alkenes. A variety of methods for radical generation are included in these examples.

Scheme 10.10. Alkylation of Alkyl Radicals by Reaction with Alkenes

A. With Radical Generation Using Trisubstituted Stannanes

1[a] I (bicyclic lactone) $\xrightarrow[CH_2{=}CHCO_2CH_3]{Bu_3SnH,\ AIBN}$ $CH_3O_2CCH_2CH_2$ (bicyclic lactone) (55%)

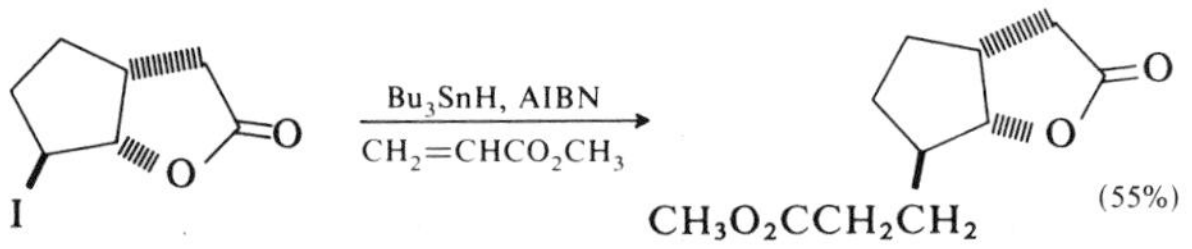

2[b] Br (pyranose, OCH₂Ph, PhCH₂O, OCH₂Ph) $\xrightarrow[CH_2{=}CHSO_2Ph]{Bu_3SnCl,\ NaBH_3CN}$ $PhSO_2CH_2CH_2$ (pyranose, OCH₂Ph, PhCH₂O, OCH₂Ph) (80%)

3[c] $CH_3O_2CCH{=}CH$ (furanose acetonide, CH_3, CH_3, R_3Si) $\xrightarrow[ICH_2CONH_2]{Bu_3SnH,\ h\nu}$ $H_2NC(O)CH_2$, CH_3O_2C–CH_2CH (furanose acetonide, CH_3, CH_3, R_3Si) (64%)

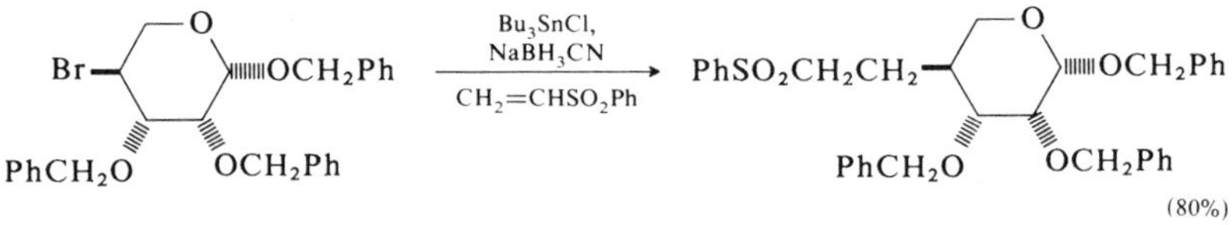

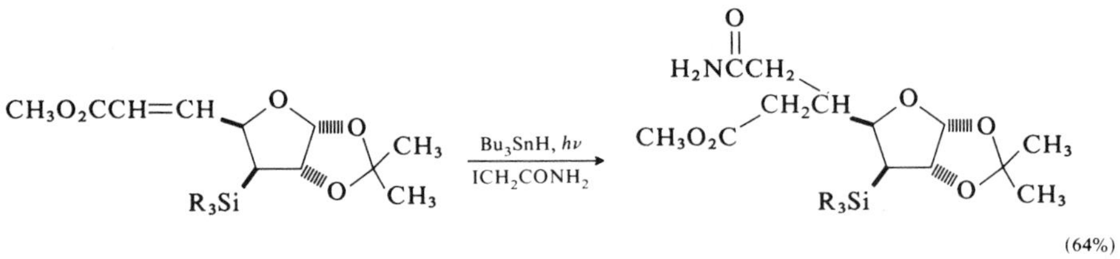

4[d] H_3C, H_3C (dioxolane)–(furanose acetonide, CH_3, CH_3), HO $\xrightarrow{1)\ CS_2,\ NaH;\ 2)\ CH_3I;\ 3)\ Bu_3SnH,\ CH_2{=}CHCN}$ H_3C, H_3C (dioxolane)–(furanose acetonide, CH_3, CH_3), $NCCH_2CH_2$ (40%)

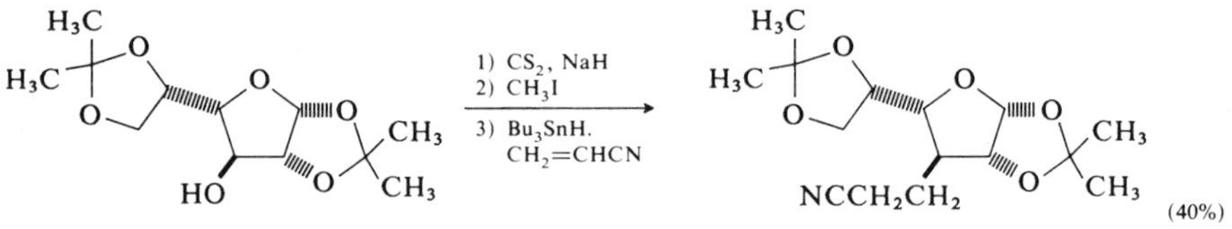

168. B. Giese and U. Erfort, *Chem. Ber.* **116**, 1240 (1983).
169. A. P. Kozikowski, T. R. Nieduzak, and J. Scripko, *Organometallics* **1**, 675 (1982).
170. B. Giese and K. Heuck, *Chem. Ber.* **112**, 3759 (1979); B. Giese and U. Lüning, *Synthesis*, 735 (1982).
171. A. P. Kozikowski and J. Scripko, *Tetrahedron Lett.* **24**, 2051 (1983).

B. Using Other Methods of Radical Generation

5[e] (2-methoxycyclohexyl)mercuric chloride $\xrightarrow[CH_2{=}CHCN]{NaBH(OCH_3)_3}$ 2-methoxycyclohexyl-CH_2CH_2CN (77%)

6[f] (2-acetamidocyclopentyl)mercuric chloride ($HgCl$, $NHCCH_3$ with C=O) $\xrightarrow[CH_2{=}C(Cl)CN]{NaBH_4}$ 2-acetamidocyclopentyl-$CH_2CH(Cl)CN$ (49%)

7[g] $CH_3(CH_2)_8C(OH)(CH_2OTHP)CH_2HgBr \xrightarrow[CH_2{=}C(CH_3)CN]{NaBH(OMe)_3} CH_3(CH_2)_8C(OH)(CH_2OTHP)CH_2CH_2CH(CH_3)CN$ (49%)

8[h] norbornyl ($HgOAc$, O_2CCH_2) $\xrightarrow[CH_2{=}CHCO_2CH_3]{NaBH_4}$ norbornyl ($CH_2CH_2CO_2CH_3$, O_2CCH_2) (75%)

a. S. D. Burke, W. B. Fobare, and D. M. Arminsteadt, *J. Org. Chem.* **47**, 3348 (1982).
b. M. V. Rao and M. Nagarajan, *J. Org. Chem.* **53**, 1432 (1988).
c. G. Sacripante, C. Tan, and G. Just, *Tetrahedron Lett.* **26**, 5643 (1985).
d. B. Giese, J. A. Gonzalez-Gomez, and T. Witzel, *Angew. Chem. Int. Ed. Engl.* **23**, 69 (1984).
e. B. Giese and K. Heuck, *Chem. Ber.* **112**, 3759 (1979).
f. R. Henning and H. Urbach, *Tetrahedron Lett.* **24**, 5343 (1983).
g. A. P. Kozikowski, T. R. Nieduzak, and J. Scripko, *Organometallics* **1**, 675 (1982).
h. B. Giese and U. Erfort, *Chem. Ber.* **116**, 1240 (1983).

One of the procedures for formation of carbon-carbon bonds from organoboranes which was discussed in Section 9.1.2 involves a radical chain sequence[172]:

$$R\cdot + CH_2{=}CH\overset{O}{\overset{\|}{C}}CH_3 \rightarrow RCH_2\dot{C}H\overset{O}{\overset{\|}{C}}CH_3 \leftrightarrow RCH_2CH{=}\overset{O\cdot}{\overset{|}{C}}CH_3$$

$$RCH_2CH{=}\overset{O\cdot}{\overset{|}{C}}CH_3 + R_3B \rightarrow \left[RCH_2CH{=}\overset{O\cdot BR_3}{\overset{|}{C}}CH_3\right] \rightarrow RCH_2CH{=}\overset{OBR_2}{\overset{|}{C}}CH_3 + R\cdot$$

The intermediate radical shown in brackets is an unusual type of radical in that the unpaired electron is involved in a one-electron bond in an electron-deficient borane. The radical that propagates the chain is split off from this radical when a two-electron bond is formed between oxygen and boron.

172. H. C. Brown and M. M. Midland, *Angew. Chem. Int. Ed. Engl.* **11**, 692 (1972).

Another class of compounds which undergo addition reactions with alkyl radicals are allylstannanes. The chain is propagated by elimination of the trialkylstannyl radical.[173]

$$R{-}X + Bu_3Sn\cdot \rightarrow R\cdot + Bu_3SnX$$

$$R\cdot + CH_2{=}CHCH_2SnBu_3 \rightarrow [RCH_2\dot{C}HCH_2SnBu_3] \rightarrow RCH_2CH{=}CH_2 + \cdot SnBu_3$$

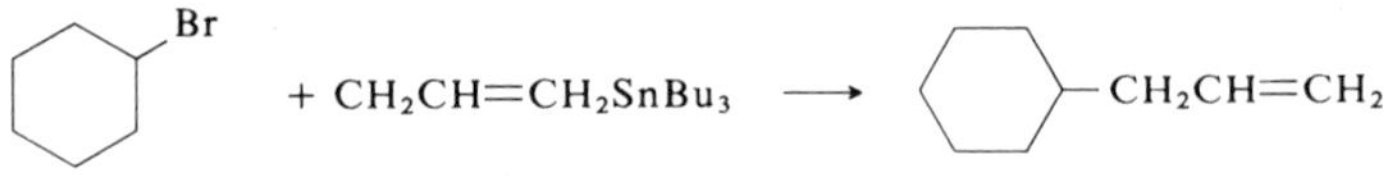

Br

$+ CH_2CH{=}CH_2SnBu_3 \longrightarrow$ $-CH_2CH{=}CH_2$

The radical source must have some functional group X which can be abstracted by trialkylstannyl radicals. In addition to halides, both thiono esters[174] and selenides[175] are reactive. Scheme 10.11 illustrates allylation by reaction of radical intermediates with allylstannanes.

Scheme 10.11. Allylation of Radical Centers Using Allylstannanes

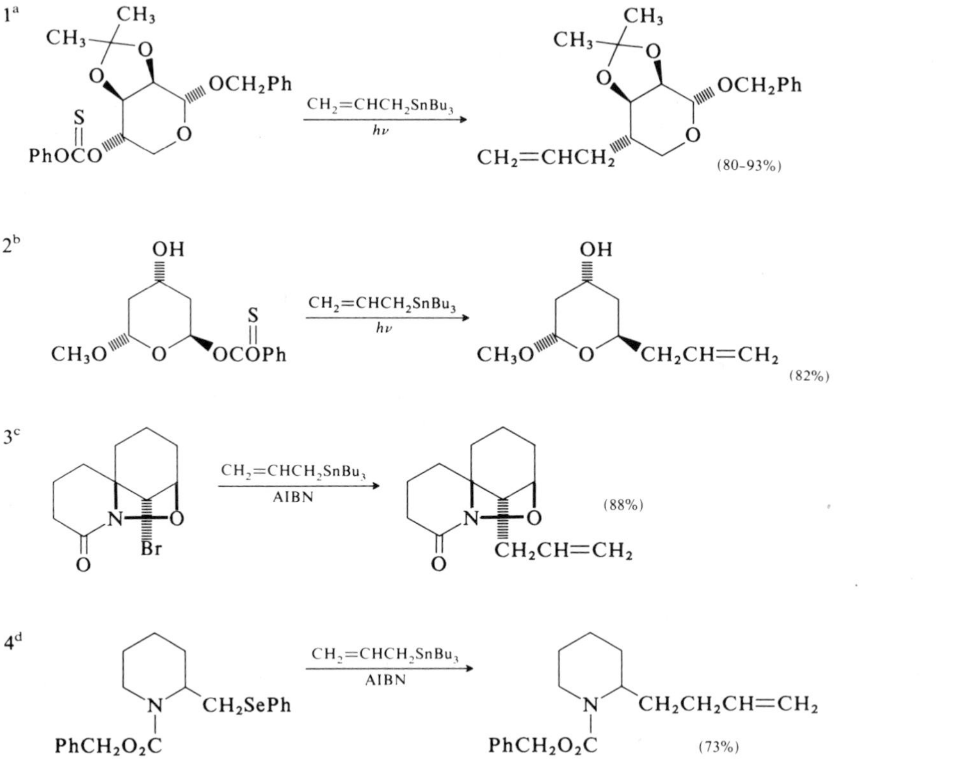

1[a] CH_3, CH_3, O, O, OCH_2Ph, S, PhOCO, O $\xrightarrow[h\nu]{CH_2{=}CHCH_2SnBu_3}$ CH_3, CH_3, O, O, OCH_2Ph, $CH_2{=}CHCH_2$, O (80–93%)

2[b] OH, CH_3O, O, S, OCOPh $\xrightarrow[h\nu]{CH_2{=}CHCH_2SnBu_3}$ OH, CH_3O, O, $CH_2CH{=}CH_2$ (82%)

3[c] N, O, Br, O $\xrightarrow[AIBN]{CH_2{=}CHCH_2SnBu_3}$ N, O, $CH_2CH{=}CH_2$, O (88%)

4[d] N, CH_2SePh, $PhCH_2O_2C$ $\xrightarrow[AIBN]{CH_2{=}CHCH_2SnBu_3}$ N, $CH_2CH_2CH{=}CH_2$, $PhCH_2O_2C$ (73%)

a. G. E. Keck, D. F. Kachensky, and E. J. Enholm, *J. Org. Chem.* **50**, 4317 (1985).
b. G. E. Keck and D. F. Kachensky, *J. Org. Chem.* **51**, 2487 (1986).
c. G. E. Keck and J. B. Yates, *J. Org. Chem.* **47**, 3590 (1982).
d. R. R. Webb II and S. Danishefsky, *Tetrahedron Lett.* **24**, 1357 (1983).

173. G. E. Keck and J. B. Yates, *J. Am. Chem. Soc.* **104**, 5829 (1982).

There are also reactions in which electrophilic radicals react with relatively nucleophilic alkenes. These reactions are perhaps best represented by a group of procedures in which a radical intermediate is formed by oxidation of the enol of a readily enolized compound. This reaction was initially developed for β-ketoacids.[176] The method has been extended to β-diketones, malonic acids, and cyanoacetic acid.[177]

$$HO_2CCH_2CN + Mn^{3+} \rightarrow HO_2C\dot{C}HCN$$

$$HO_2C\dot{C}HCN + CH_2{=}CHR \rightarrow HO_2CCH(CN)CH_2\dot{C}HR \xrightarrow{Mn^{3+}} HO_2CCH(CN)CH_2\overset{+}{C}HR \rightarrow$$

The radicals formed by the addition step are rapidly oxidized to cations, which give rise to the final product by intramolecular capture of a nucleophilic carboxylate group. With diones such as 1,3-cyclohexadienone, one of the carbonyl oxygens serves as the internal nucleophile:

$$\text{1,3-cyclohexanedione} + CH_2{=}C(CH_3)CH_2CH_2CH_3 \xrightarrow{Mn(OAc)_3} \text{(dihydrofuranone product, } CH_3, CH_2CH_2CH_3)$$

Ref. 177

10.3.4. Cyclization of Free-Radical Intermediates

There have been a number of useful procedures developed for cyclization of radical intermediates. The key step in these procedures involves addition of a radical center to an unsaturated functional group. The radical formed by the cyclization must then give rise to a new radical which can propagate the chain. An important group of such reactions involves halides as the source of the radical intermediate. The radicals are normally generated by halogen atom abstraction with a trialkylstannane as the reagent and azoisobutyronitrile (AIBN) as the initiator. The cyclization step must be fast relative to hydrogen abstraction from the stannane. The chain is propagated when the cyclized radical abstracts hydrogen from the stannane.

$$In\cdot + Bu_3Sn{-}H \rightarrow In{-}H + Bu_3Sn\cdot$$

$$Bu_3Sn\cdot + X{-}CH_2 \frown CH{=}CH_2 \longrightarrow Bu_3Sn{-}X + \cdot CH_2 \frown CH{=}CH_2 \longrightarrow CH_2{-}CH{-}CH_2\cdot$$

$$CH_2{-}CH{-}CH_2\cdot + Bu_3Sn{-}H \longrightarrow CH_2CH{-}CH_3 + Bu_3Sn\cdot$$

174. G. E. Keck, D. F. Kachensky, and E. J. Enholm, *J. Org. Chem.* **49**, 1462 (1984).
175. R. R. Webb and S. Danishefsky, *Tetrahedron Lett.* **24**, 1357 (1983).
176. E. Heiba and R. M. Dessau, *J. Org. Chem.* **39**, 3456 (1974).
177. E. J. Corey and M. C. Kang, *J. Am. Chem. Soc.* **106**, 5384 (1984); E. J. Corey and A. W. Gross, *Tetrahedron Lett.* **26**, 4291 (1985); W. E. Fristad and S. S. Hershberger, *J. Org. Chem.* **50**, 1026 (1985).

From a synthetic point of view, the regiochemistry and stereochemistry of the cyclization are of paramount importance. As discussed in Section 12.5 of Part A, there is usually a preference for ring formation in the order $5 > 6 > 7$. The other major influence on the direction of cyclization is the presence of substituents. Attack at a less hindered position is favored, both by steric effects and by the stabilizing effect that most substituents have on a radical center. For relatively rigid cyclic structures, proximity factors determined by the specific geometry of the ring system will be a major factor. Theoretical analysis of radical addition indicates that the major interaction of the attacking radical is with the alkene LUMO.[178] The preferred direction of attack is not directly perpendicular to the π system but, instead, at an angle of about 110°.

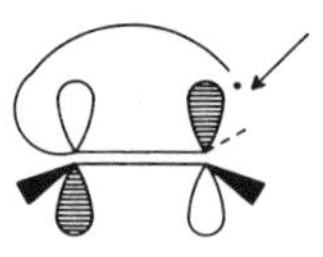

Five-membered rings can be expected to be fused onto other rings in a *cis* manner in order to minimize strain. When cyclization is followed by hydrogen abstraction, the hydrogen atoms is normally delivered from the less hindered side of the molecule. The example below illustrates these principles. The initial tetrahydrofuran ring closure gives the *cis*-fused ring. The subsequent hydrogen abstraction is from the less hindered axial direction.[179]

Reaction conditions have been developed under which the cyclized radical can react in some manner other than hydrogen atom abstraction. One such reaction is abstraction of an iodine atom. The cyclization of 2-iodo-2-methyl-6-heptyne is a

178. A. L. J. Beckwith and C. H. Schiesser, *Tetrahedron* **41**, 3925 (1985); D. C. Spellmeyer and K. N. Houk, *J. Org. Chem.* **52**, 959 (1987).
179. M. J. Begley, H. Bhandal, J. H. Hutchinson, and G. Pattenden, *Tetrahedron Lett.* **28**, 1317 (1987).

structurally simple example:

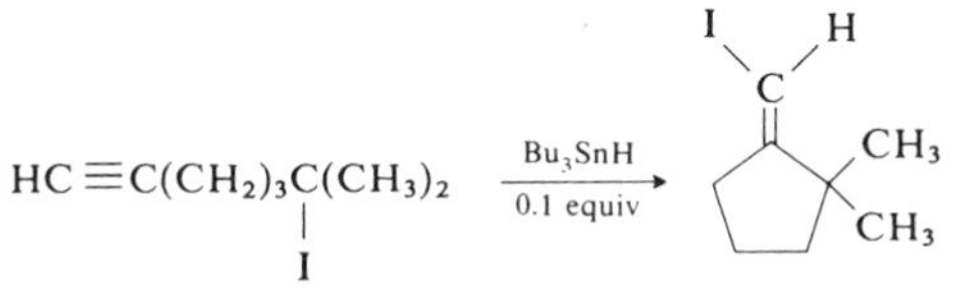

$$HC{\equiv}C(CH_2)_3C(CH_3)_2(I) \xrightarrow[0.1\ equiv]{Bu_3SnH} \text{(iodomethylene)-2,2-dimethylcyclopentane} \quad \text{Ref. 180}$$

In this reaction, the trialkylstannane serves to initiate chains, but it is present in low concentration to minimize the rate or H-atom abstraction from the stannane. Under these conditions, the chain is propagated by iodine atom abstraction.

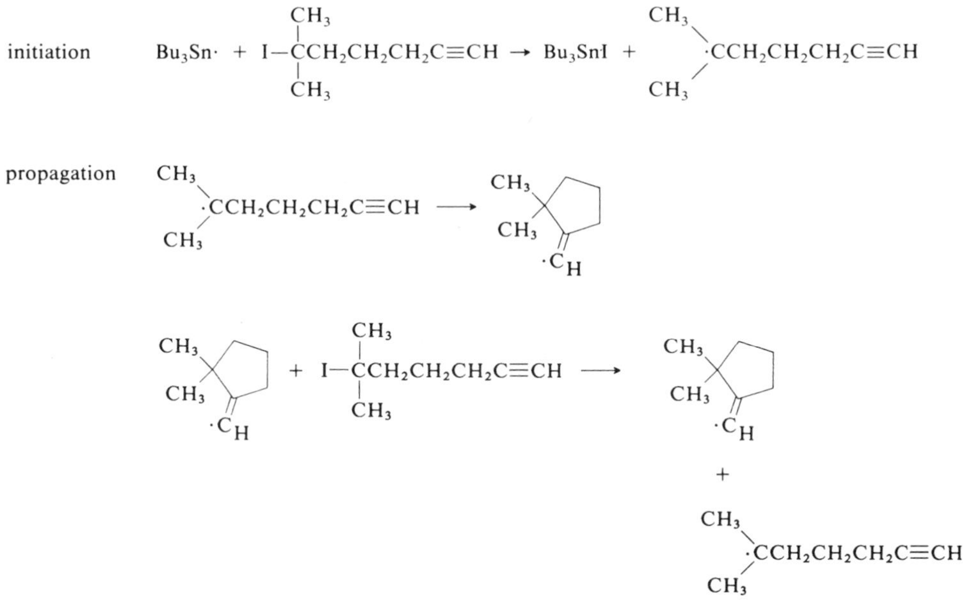

initiation

$$Bu_3Sn\cdot + I{-}C(CH_3)_2CH_2CH_2CH_2C{\equiv}CH \rightarrow Bu_3SnI + (CH_3)_2\dot{C}CH_2CH_2CH_2C{\equiv}CH$$

propagation

$$(CH_3)_2\dot{C}CH_2CH_2CH_2C{\equiv}CH \longrightarrow \text{2,2-dimethylcyclopentylidene-}\dot{C}H$$

$$\text{2,2-dimethylcyclopentylidene-}\dot{C}H + I{-}C(CH_3)_2CH_2CH_2CH_2C{\equiv}CH \longrightarrow \text{2,2-dimethylcyclopentylidene-}\dot{C}H + (CH_3)_2\dot{C}CH_2CH_2CH_2C{\equiv}CH$$

The fact that the cyclization is directed to an acetylenic group and leads to formation of an alkenyl radical is significant. Formation of a saturated iodide would be expected to lead to a more complex product mixture because the cyclized product could undergo iodine abstraction and proceed to add to a second unsaturated center.

Radicals react with tertiary isonitriles by abstraction of the cyano group. In the initial example of this process, irradiation of hexaphenyldistannane was used to generate the radical because tributylstannane was more reactive toward the radical intermediate than was the isonitrile.

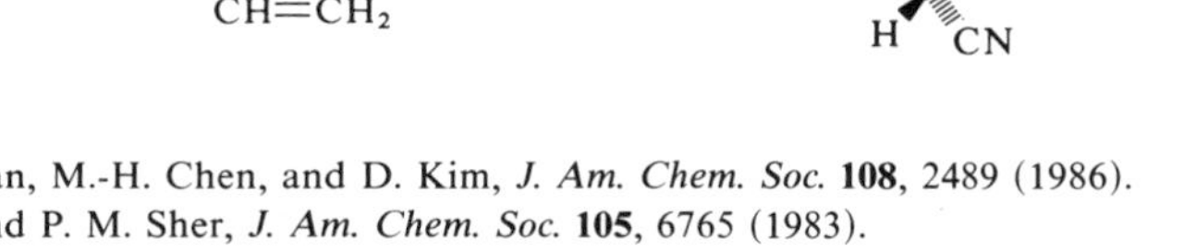

$$CH_3(CH_2)_4CH(CH{=}CH_2)OCH(OC_2H_5)CHBr \xrightarrow[(CH_3)_3CN{\equiv}C]{Ph_3SnSnPh_3} \text{2-ethoxy-4-cyano-3-pentyltetrahydrofuran} \quad \text{Ref. 181}$$

180. D. P. Curran, M.-H. Chen, and D. Kim, *J. Am. Chem. Soc.* **108**, 2489 (1986).
181. G. Stork and P. M. Sher, *J. Am. Chem. Soc.* **105**, 6765 (1983).

An improved procedure involves maintenance of a low concentration of tri-*n*-butylstannane by using $NaBH_3CN$ to reduce the stannyl iodide that is formed as the reaction proceeds.

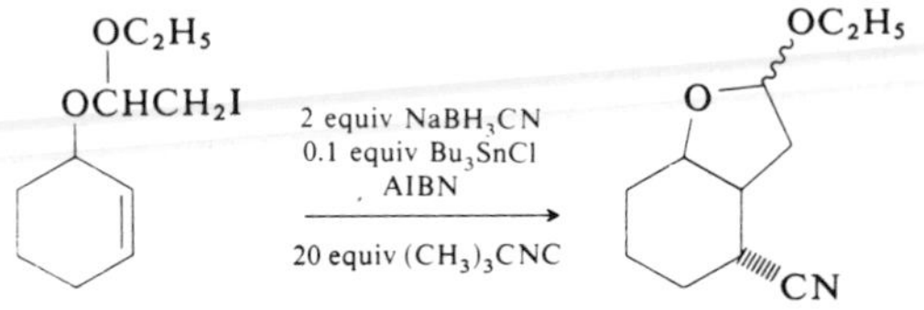

Ref. 182

Radicals formed by intramolecular addition can be further extended by trapping the intermediate cyclic radical with an electrophilic alkene.

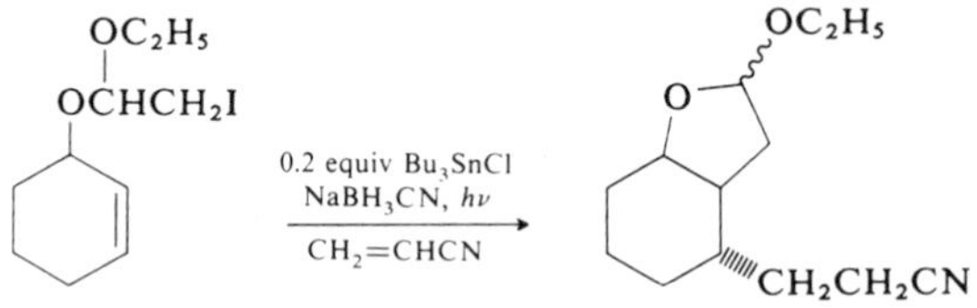

Ref. 182

The radical intermediate can also add to another double bond in the same molecule. The use of vinyl radicals in cyclizations of this type is particularly promising. Addition of a vinyl radical to a double blnd is usually favorable thermodynamically because a more stable alkyl radical results. The vinyl radical can be generated by dehalogenation of vinyl bromides or iodides. An early study provided examples of both five and six-membered rings being formed[183]:

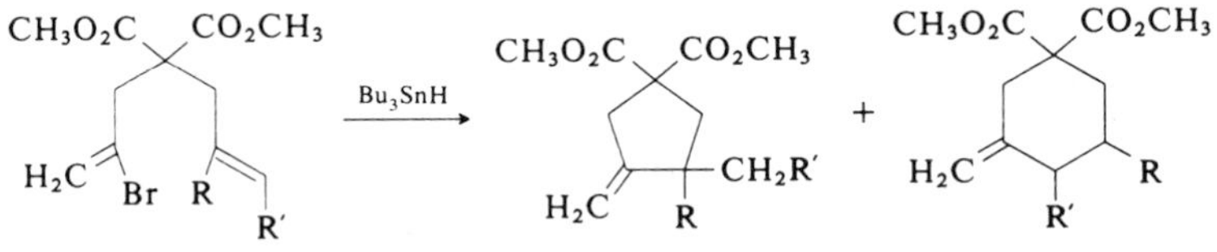

R	R′	Product ratio **E**:**F**
H	H	3:1
CH_3	H	**F** exclusively
H	CH_3	2:1

Vinyl radicals generated by addition of trialkylstannyl radicals to terminal alkynes can also undergo cyclization with a nearby double bond.

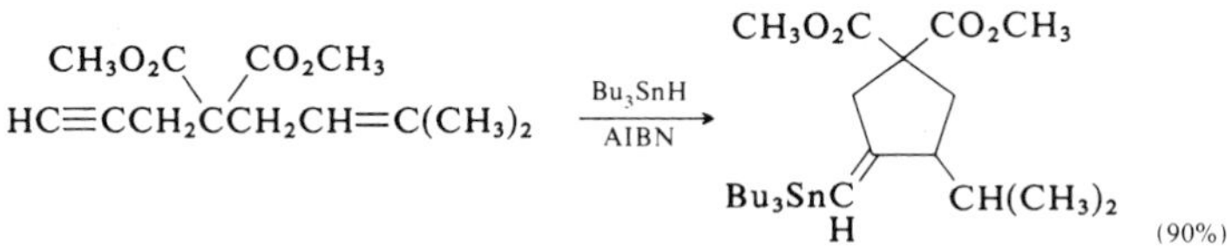

Ref. 184

182. G. Stork and P. M. Sher, *J. Am. Chem. Soc.* **108**, 303 (1986).
183. G. Stork and N. H. Baine, *J. Am. Chem. Soc.* **104**, 2321 (1982).
184. G. Stork and R. Mook, Jr., *J. Am. Chem. Soc.* **109**, 2829 (1987).

Vinyl radicals generated by intramolecular addition can add to a nearby double bond in a tandem cyclization process.

CH_3 OCH_2CH_2Cl $CH_2CH_2C{\equiv}CCH_2OCHCH_2Br$ → (1.1 equiv Bu_3SnH, AIBN, 80°C) CH_3 OCH_2CH_2Cl H (75%) Ref. 185

Cyclization of both alkyl and acyl radicals generated by selenide abstraction has also been observed.

Ph C C HO CN $CH_2CHSePh$ → (Bu_3SnH) H OH CPh CN (91%) Ref. 186

O CH_2CH_2CSePh → (Bu_3SnH) H H O Ref. 187

Scheme 10.12 gives some additional examples of cyclization reactions involving radical intermediates.

10.3.5. Fragmentation and Rearrangement Reactions

Fragmentation is the reverse of radical addition. Fragmentation of radicals is often observed to be fast when the overall transformation is exothermic.
radical rearrangements proceeding through addition–elimination.

$$\cdot Y-\overset{|}{\underset{|}{C}}-\overset{|}{\underset{|}{C}}-X \rightarrow Y{=}C\langle \; + \; \rangle\dot{C}-X$$

Rearrangements of radicals frequently occur by a series of addition–elimination steps which incorporate fragmentation. The fragmentation of alkoxyl radicals is especially common because the formation of a carbonyl bond frequently makes such reactions exothermic. The following two reaction sequences are examples of

185. G. Stork and R. Mook, Jr., *J. Am. Chem. Soc.* **105**, 3720 (1983).
186. D. L. J. Clive, T. L. B. Boivin, and A. G. Angoh, *J. Org. Chem.* **52**, 4943 (1987).
187. D. L. Boger and R. J. Mathvink, *J. Org. Chem.* **53**, 3377 (1988).

Scheme 10.12. Radical Cyclizations

A. Cyclizations Terminated by Hydrogen Atom Abstraction

1[a]

Br
$CH_2CH_2CH{=}C(CH_3)_2$
CH_3
O
Bu_3SnH
$CH(CH_3)_2$
CH_3
O
(62%)

2[b]

CH_3
H
$OCHCH_2Br$
OC_2H_5
Bu_3SnH
CH_3
H
O
H
C_2H_5O

3[c]

H_3C
$CH_2CH{=}C$
CH_3
Br
O
CH_3
Bu_3SnH
AIBN, 80°C
H_3C
O
CH_3
(70%)

4[d]

O
PhSeCO
CH_3
H
CH_2OSiR_3
Bu_3SnH
AIBN
CH_3
H
CH_2OSiR_3
O
H
O
(73%)

B. Cyclization with Tandem Alkylation

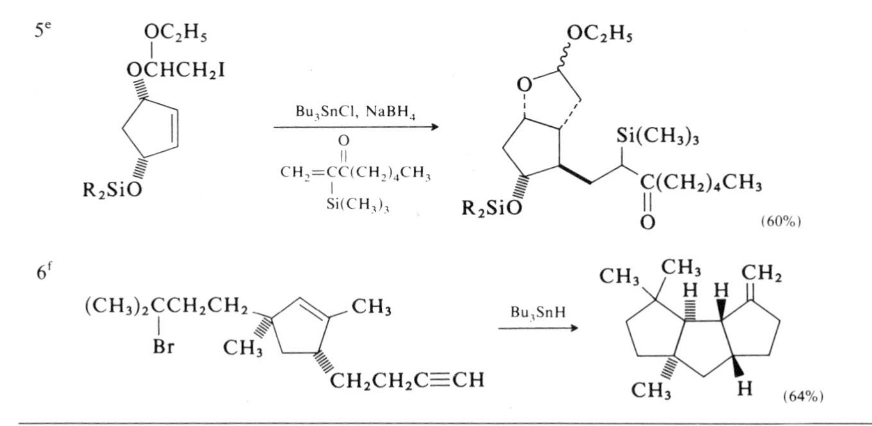

a. P. Bakuzis, O. O. S. Campos, and M. L. F. Bakuzis, *J. Org. Chem.* **41**, 3261 (1976).
b. G. Stork and M. Kahn, *J. Am. Chem. Soc.* **107**, 500 (1985).
c. G. Stork and N. H. Baine, *Tetrahedron Lett.* **26**, 5927 (1985).
d. A. K. Singh, R. K. Bakshi, and E. J. Corey, *J. Am. Chem. Soc.* **109**, 6187 (1987).
e. G. Stork, P. M. Sher, and H.-L. Chen, *J. Am. Chem. Soc.* **108**, 6384 (1986).
f. D. P. Curran and D. M. Rakewicz, *Tetrahedron* **41**, 3943 (1985).

radical rearrangements proceeding through addition–elimination.

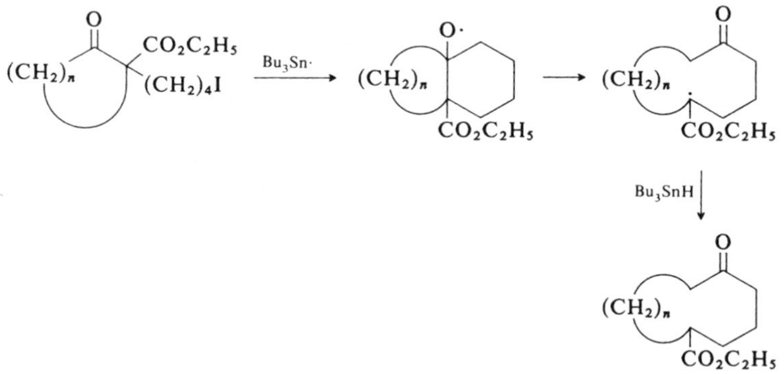

Ref. 188

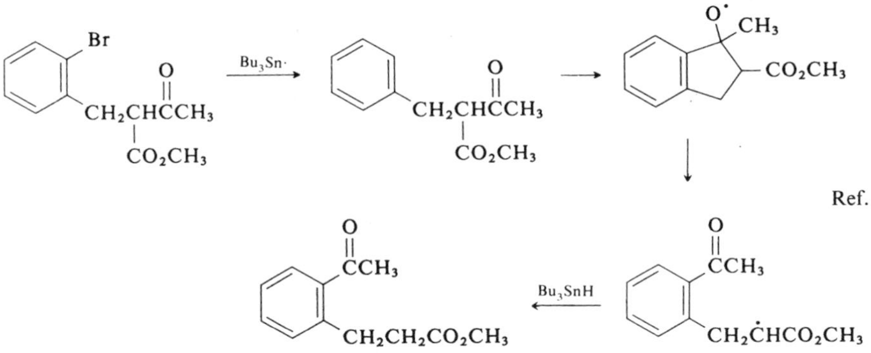

Ref. 189

Both of these transformations feature addition of a carbon-centered radical to a carbonyl group, followed by fragmentation to a more stable radical. The rearranged radicals then abstract hydrogen from the co-reactant n-Bu_3SnH. The addition step must be fast relative to hydrogen abstraction, since if this were not the case, simple reductive dehalogenation would occur. The fragmentation step is considered to be irreversible. There are two reasons: (1) the reverse addition is endothermic; (2) the stabilized radical substituted by electron-withdrawing alkoxycarbonyl groups is unreactive toward addition to carbonyl bonds.

The two reactions above are examples of a more general reactivity pattern[190]:

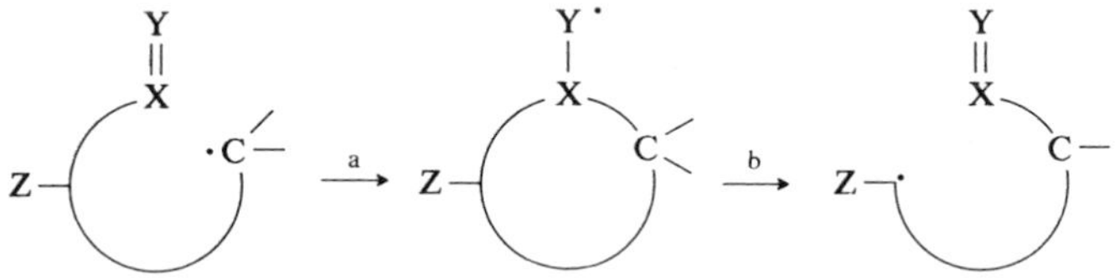

188. P. Dowd and S.-C. Choi, *J. Am. Chem. Soc.* **109**, 6548 (1987).
189. A. L. J. Beckwith, D. M. O'Shea, and S. W. Westwood, *J. Am. Chem. Soc.* **110**, 2565 (1988).
190. A. L. J. Beckwith, D. M. O'Shea, and S. W. Westwood, *J. Am. Chem. Soc.* **110**, 2565 (1988); R. Tsang, J. K. Pickson, Jr., H. Pak, R. Walton, and B. Fraser-Reid, *J. Am. Chem. Soc.* **109**, 3484 (1987).

The unsaturated group X=Y that is "transferred" by the rearrangement process can be C=C, C=O, C≡N, or other groups which fulfill the following general criteria: (1) the addition step (step a) must be fast relative to other potentially competing reactions; (2) the group Z must stabilize the product radical so that the overall process is energetically favorable.

A direct comparison of the ease with which unsaturated groups migrate by cyclization–fragmentation has been made for the case of net 1,2-migration:

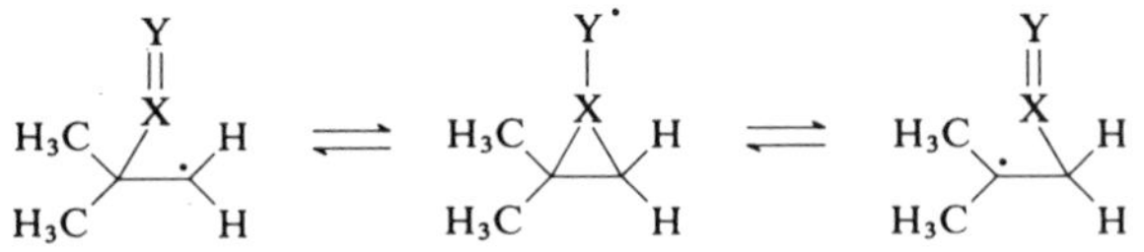

In this system, the overall driving force is the conversion of a primary radical to a tertiary one ($\Delta H \approx 5$ kcal), and the activation barrier would incorporate the strain associated with formation of the three-membered ring. Rates and activation energies determined for several migrating groups are given below.[191]

	X=Y				
	$HC{=}CH_2$	$(CH_3)_3C{-}C{=}O$	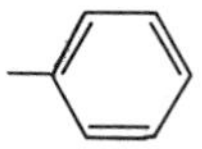 $-C_6H_5$	$-C\equiv CC(CHH_3)_3$	$C\equiv N$
k_r (sec^{-1})	10^7	1.7×10^5	7.6×10^3	93	0.9
E_a (kcal)	5.7	7.8	11.8	12.8	16.4

General References

S. P. McManus (ed.), *Organic Reactive Intermediates*, Academic Press, New York, 1973.

Carbocation Cyclizations and Rearrangements

P. A. Bartlett, in *Asymmetric Synthesis*, Vol. 3, J. D. Morrison (ed.), Academic Press, New York, 1984, Chapter 5.
P. de Mayo (ed.), *Molecular Rearrangements*, Vols. 1 and 2, New York, 1963.
W. S. Johnson, *Angew. Chem. Int. Ed. Engl.* **15**, 9 (1976).
D. Redmore and C. D. Gutsche, *Adv. Alicyclic Chem.* **3**, 1 (1971).
B. S. Thyagarajan (ed.), *Mechanisms of Molecular Migrations*, Vols. 1–4, Wiley-Interscience, New York, 1968–1971.

191. Data from D. A. Lindsay, J. Lusztyk, and K. U. Ingold, *J. Am. Chem. Soc.* **106**, 7087 (1984).

G. L'Abbe, *Chem. Rev.* **69**, 345 (1969).
R. A. Abramovitch and E. P. Kyba, in *The Chemistry of the Azido Group*, S. Patai (ed.), Wiley-Interscience, New York, 1971, pp. 331-395.
D. Bethell, *Adv. Phys. Org. Chem.* **7**, 153 (1968).
R. E. Gawley, *Org. React.* **35**, 1 (1988).
M. Jones, Jr., and R. A. Moss (eds.), Vols. I and II, Wiley, New York, 1973, 1975.
W. Kirmse, *Carbene Chemistry*, Academic Press, New York, 1971.
W. Lwowski (ed.), *Nitrenes*, Wiley-Interscience, New York, 1970.
E. F. V. Scriven (ed.), *Azides and Nitrenes*, Academic Press, New York, 1984.

Free Radicals

A. L. J. Beckwith and K. U. Ingold, in *Rearrangements in Ground and Excited States*, Vol. 1, P. de Mayo (ed.), Academic Press, 1980, Chapter 4.
B. Giese, *Radicals in Organic Synthesis: Fromation of Carbon–Carbon Bonds*, Pergamon, Oxford, 1986.
B. Giese, *Angew. Chem. Int. Ed. Engl.* **24**, 553 (1985).
J. M. Tedder, *Angew. Chem. Int. Ed. Engl.* **21**, 401 (1982).

Problems

(*References for these problems will be found on page* 776.)

1. Indicate the major product to be expected in the following reactions:

(a) cyclohexene + $CHCl_3$ $\xrightarrow[\text{NaOH, } H_2O]{PhCH_2\overset{+}{N}(C_2H_5)_3Cl}$

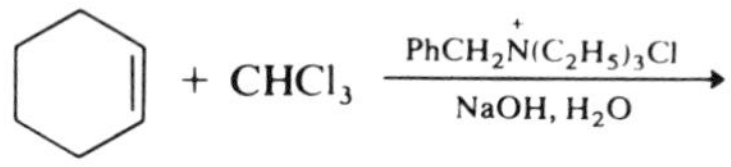

(b) adamantane + $CH_3OC(=O)N_3$ $\xrightarrow{\Delta}$

(c) $(CH_3)_2C{=}C(CH_3)_2$ + $CFCl_3$ $\xrightarrow[-120°C]{n\text{-BuLi}}$

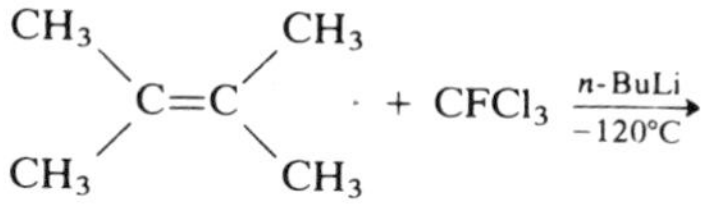

(d) cyclooctene + $PhHgCF_3$ $\xrightarrow[12\text{ hr}]{80°C}$

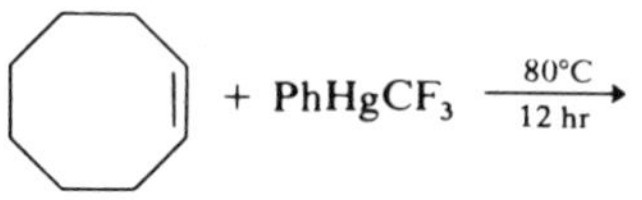

(e) $(CH_3)_2C{=}C(CH_3)CH{=}NNHTs$ $\xrightarrow{NaOCH_3}$

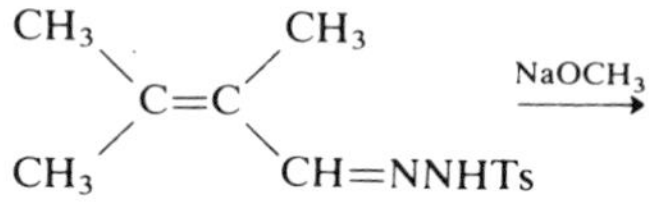

(f) 1,3-cyclohexadiene + $N_2CHC(=O)C(=O)OC(CH_3)_3$ $\xrightarrow{Rh(OAc)_2}$

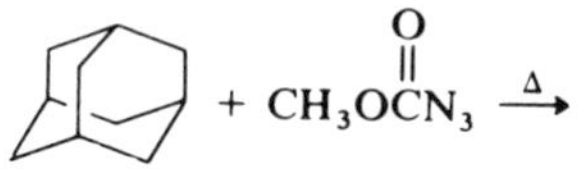

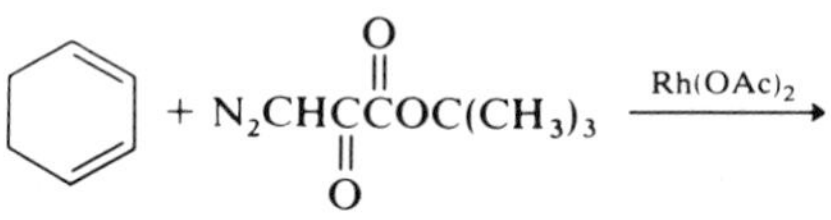

(g) $PhCH_2C(=O)CH(Br)CH_3 + CH_3CH_2O^- \rightarrow$

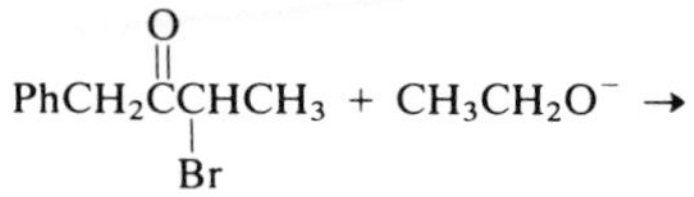

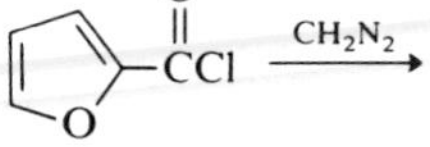

(h) 2-furoyl chloride (furan–CCl, C=O) $\xrightarrow{CH_2N_2}$

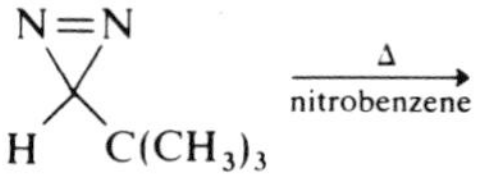

(i) 3-H-3-$C(CH_3)_3$-diazirine (N=N) $\xrightarrow[\text{nitrobenzene}]{\Delta}$

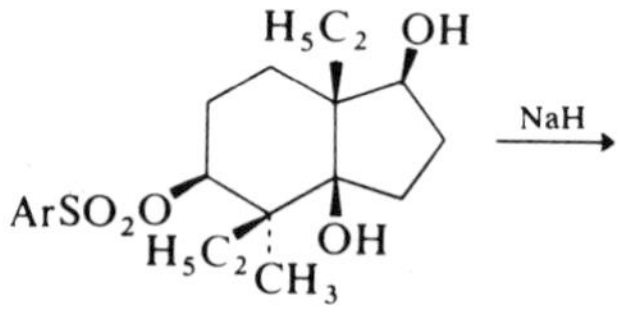

(j) H_5C_2, OH, $ArSO_2O$, H_5C_2, CH_3, OH $\xrightarrow{NaH}$

(k) $(CH_3)_2C{=}CHCH_2CH_2(CH_3)C{=}C(H)CH_2CH_2(CH_3)C{=}C(H)CH_2O_2CCH_3$ $\xrightarrow[\text{3) } NaBH_4]{\text{1) } Hg(O_3SCF_3)_2/PhN(CH_3)_2 \text{ 2) } NaCl}$

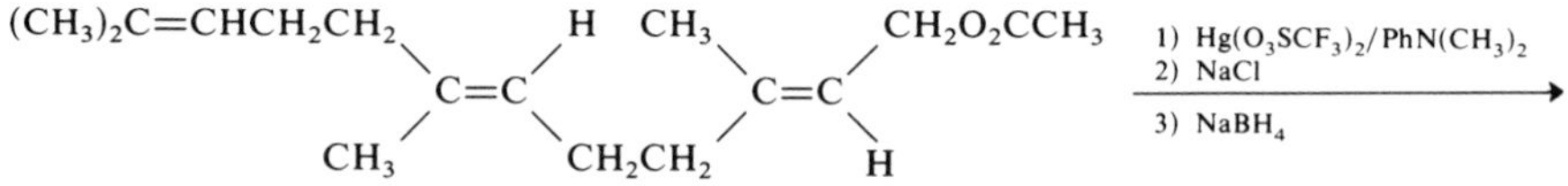

(l) $CH_3C(=O)O{-}N{=}$ (cyclohexylidene with C_5H_{11} and $Si(CH_3)_3$) $\xrightarrow{(CH_3)_3SiO_3SCF_3}$ $C_{11}H_{19}N$

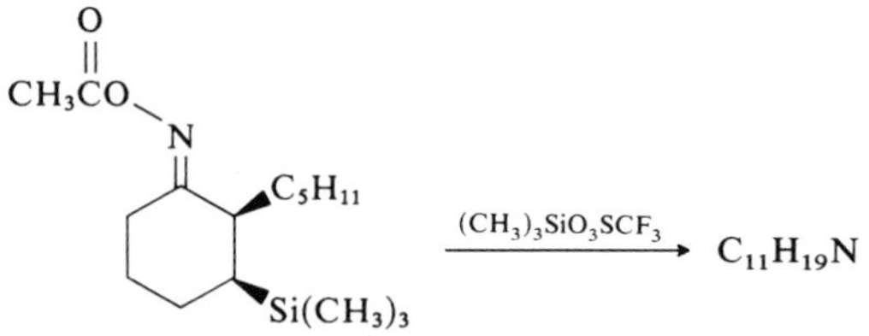

(m) $ArSO_2O$, CH_3, OH $\xrightarrow{K^{+-}OC(CH_3)_3}$

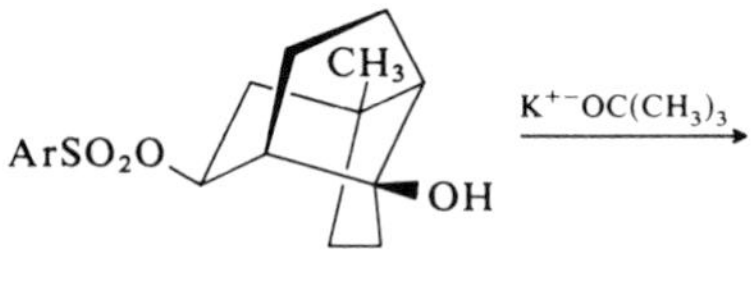

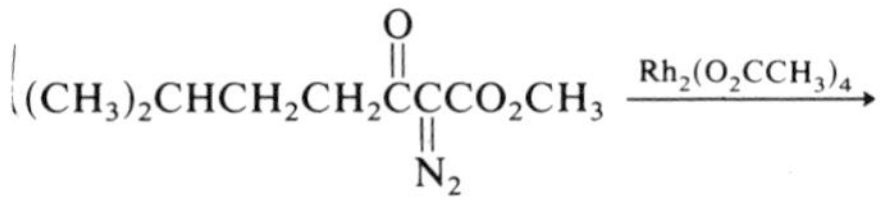

(n) $(CH_3)_2CHCH_2CH_2C(=O)C(=N_2)CO_2CH_3$ $\xrightarrow{Rh_2(O_2CCH_3)_4}$

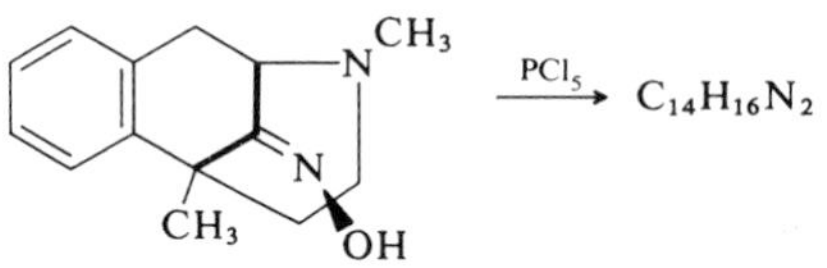

(o) CH_3, N, N, CH_3, OH $\xrightarrow{PCl_5}$ $C_{14}H_{16}N_2$

(p) H, O, O, H, CH_3, CO_2CH_3, O_2CCH_3, CH_2OH $\xrightarrow[I_2,\ h\nu]{Pb(OAc)_4}$

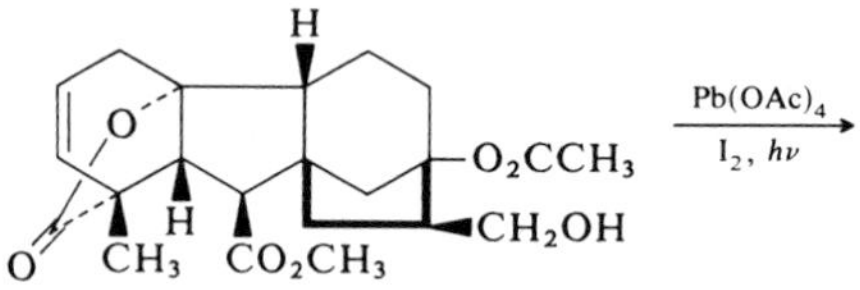

(q)

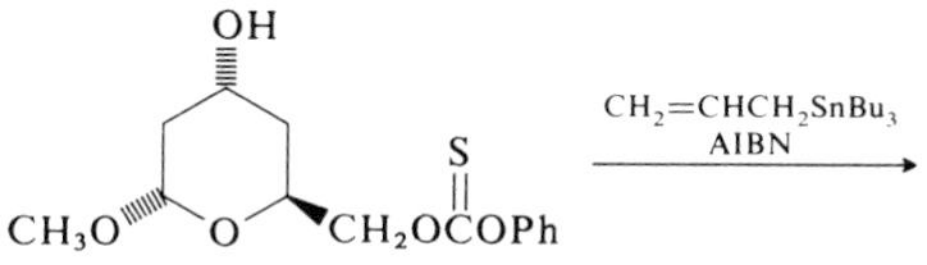

$CH_2{=}CHCH_2SnBu_3$, AIBN

OH; CH_3O; O; CH_2OCOPh (C=S)

(r)

CH_3; O=C; $CO_2C(CH_3)_3$; NCH; CO—N; O; S; $h\nu$

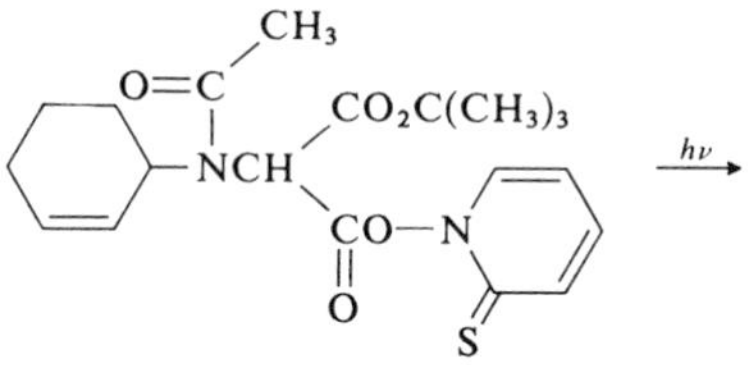

2. Indicate appropriate reagents and conditions or a short reaction sequence which could be expected to effect the following transformations:

(a) 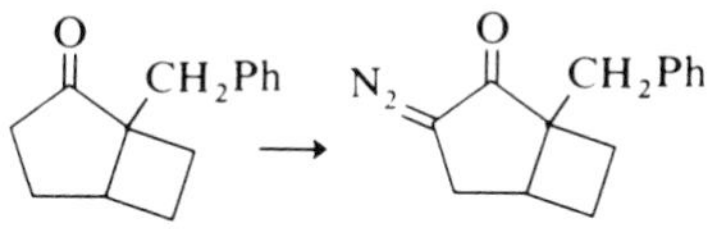 O, CH_2Ph → N_2, O, CH_2Ph

(b) 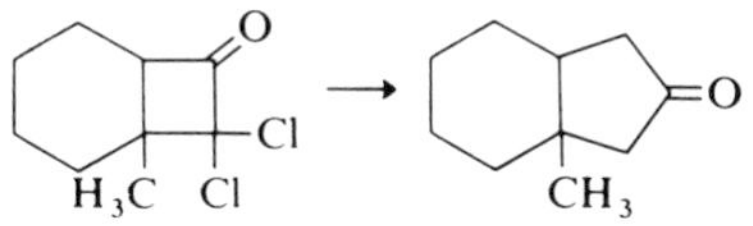 O, Cl, H_3C, Cl → O, CH_3

(c) 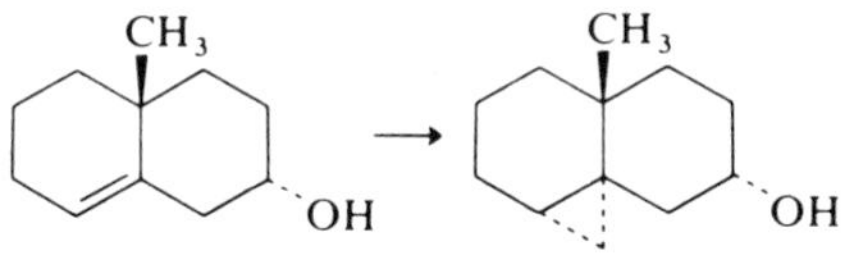 CH_3, OH → CH_3, OH

(d) 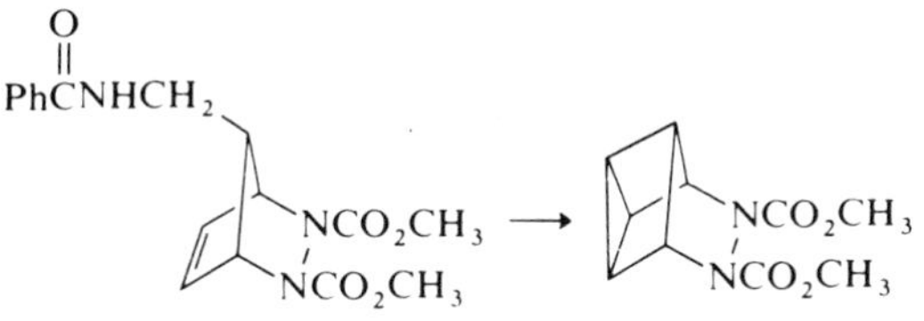 O, $PhCNHCH_2$, NCO_2CH_3, NCO_2CH_3 → NCO_2CH_3, NCO_2CH_3

(e) 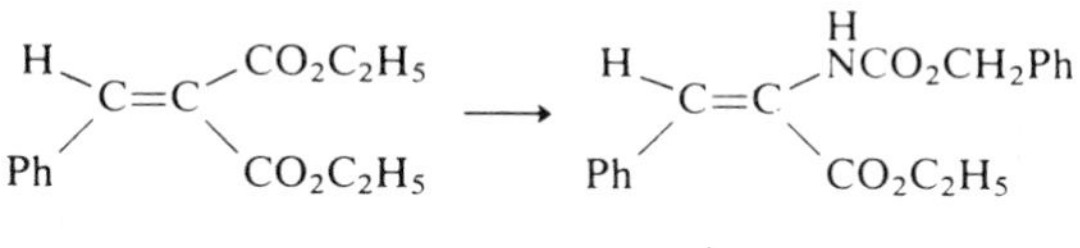 H, Ph, C=C, $CO_2C_2H_5$, $CO_2C_2H_5$ → H, Ph, C=C, $\overset{H}{N}CO_2CH_2Ph$, $CO_2C_2H_5$

(f) 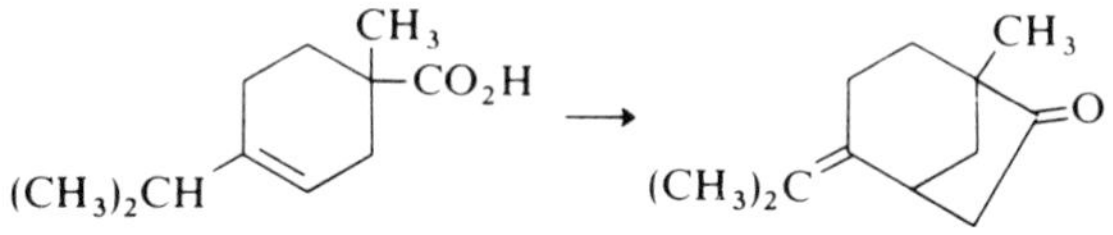 CH_3, CO_2H, $(CH_3)_2CH$ → CH_3, O, $(CH_3)_2C$

(g) $CH_2{=}CHCH{=}CHCO_2H \rightarrow CH_2{=}CHCH{=}CH\overset{H}{N}CO_2CH_2Ph$

(h)

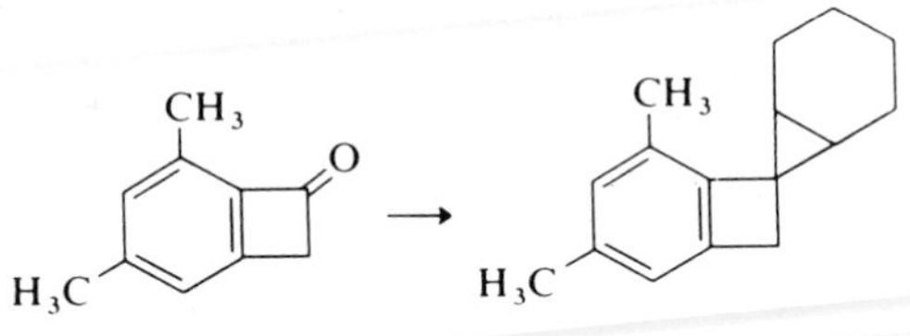

CH_3 H_3C → CH_3 H_3C

(i)

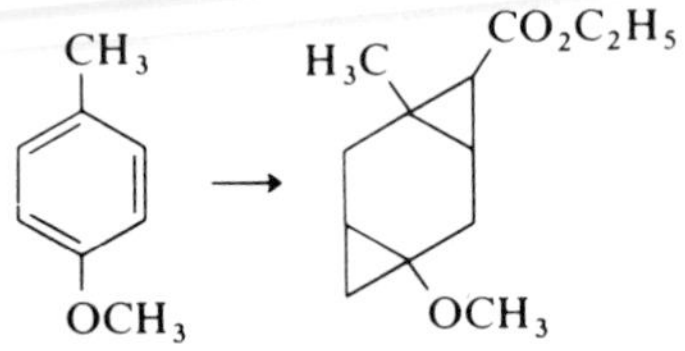

CH_3 OCH_3 → H_3C $CO_2C_2H_5$ OCH_3

(j)

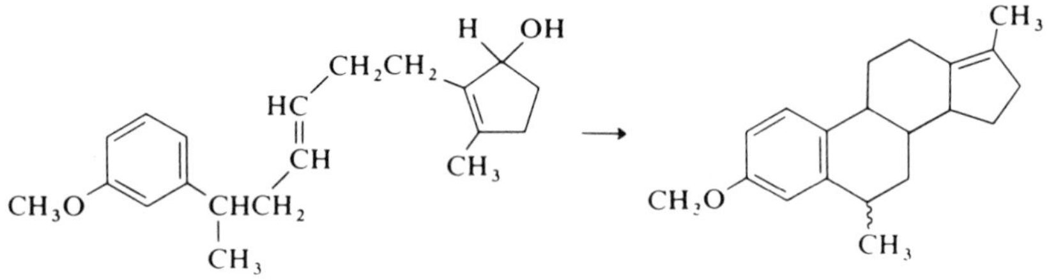

CH_3O $CHCH_2$ CH_3 HC=CH CH_2CH_2 H OH CH_3 → CH_3 CH_3O CH_3

(k)

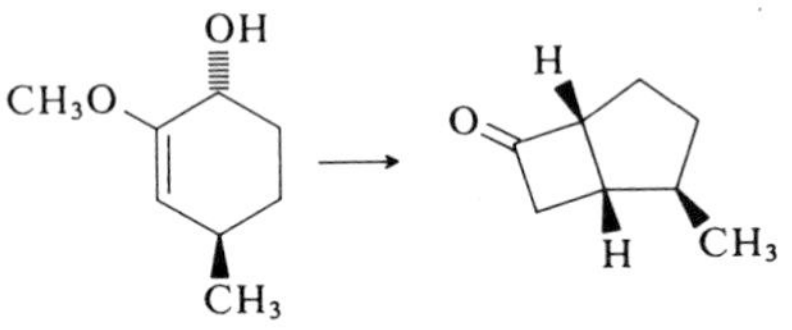

OH CH_3O CH_3 → O H H CH_3

(l)

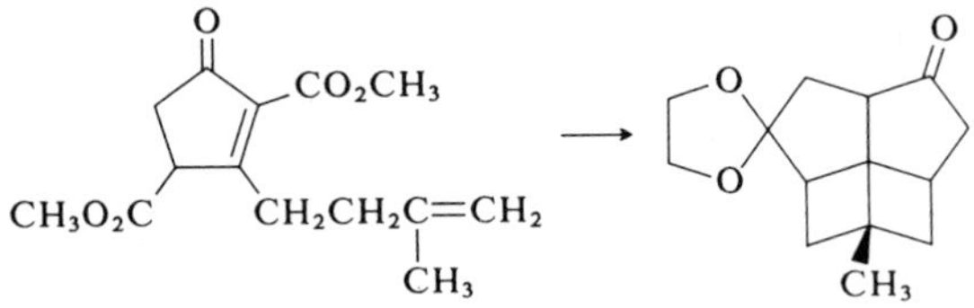

O CO_2CH_3 CH_3O_2C $CH_2CH_2C{=}CH_2$ CH_3 → O O O CH_3

(m) $AcOCH_2$ O AcO AcO OAc I → $AcOCH_2$ O AcO AcO OAc O NC_2H_5 O

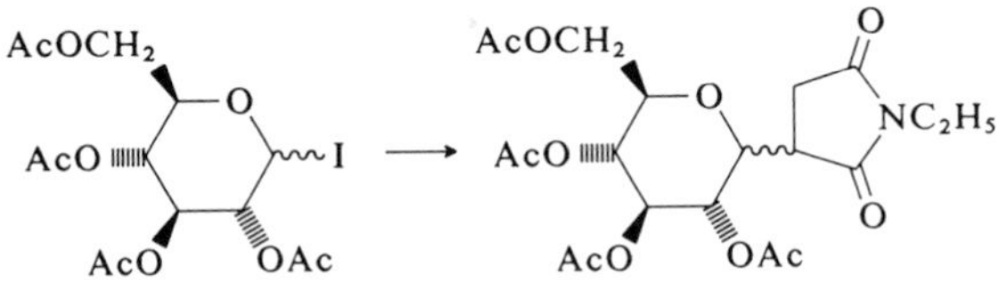

(n) $CH_3(CH_2)_7CO_2H \rightarrow CH_3(CH_2)_9CN$

(o)

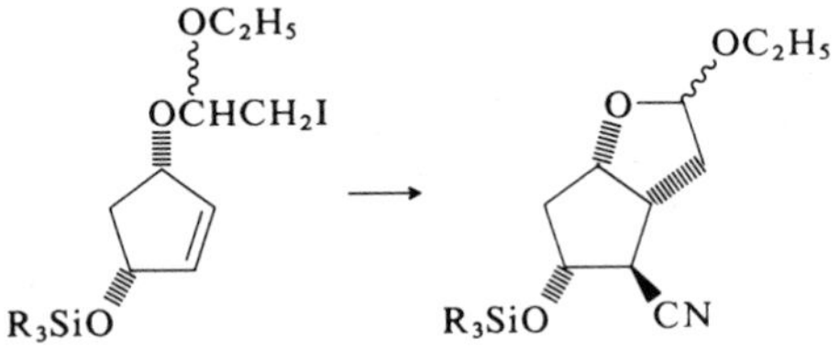

OC_2H_5 $OCHCH_2I$ R_3SiO → OC_2H_5 O R_3SiO CN

(p) [structure: 3-cyclopentenyl-$CH_2COCH_2CO_2H$ → tricyclic keto lactone]

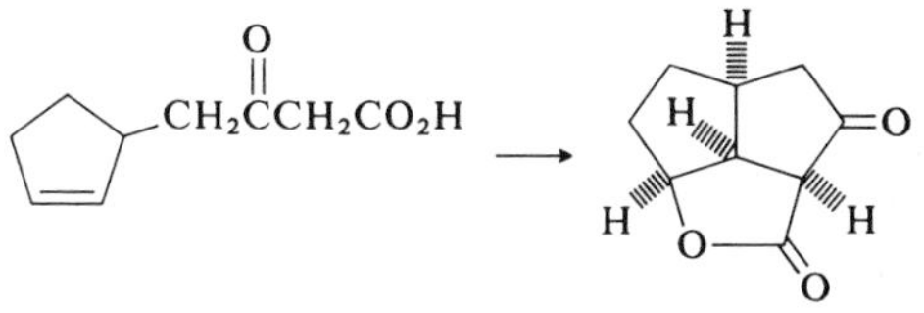

(q) [structure: acetonide-protected pyranoside with OCH_2Ph and HO → same with $CH_2{=}CHCH_2$]

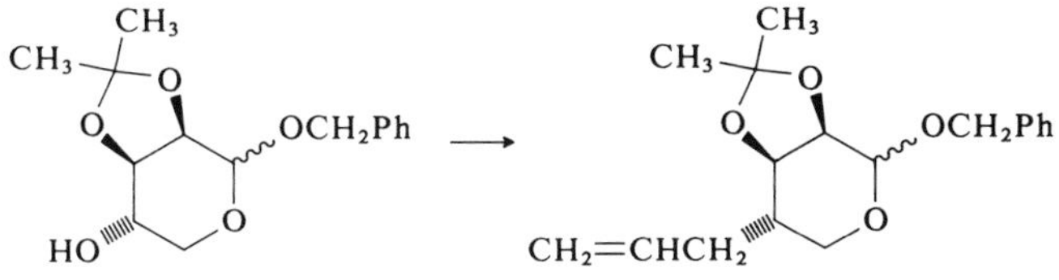

(r) [structure: tri-O-acetyl glycal ($AcOCH_2$, AcO, OAc) → fused bicyclic tetrahydrofuran with exocyclic CH_2]

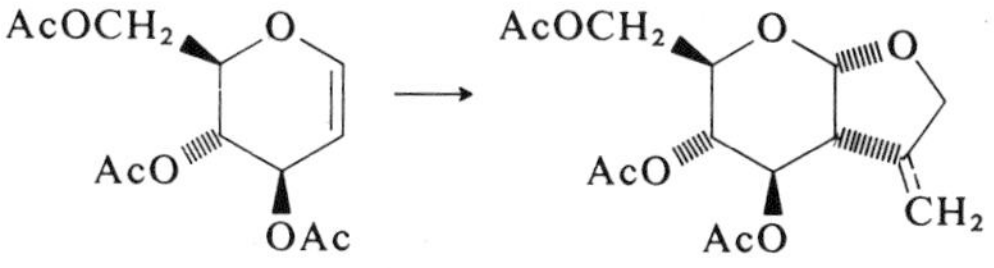

3. Each of the following carbenes has been predicted to have a singlet ground state, either as the result of qualitative structural considerations or on the basis of theoretical calculations. Indicate what structural feature in each case might lead to stabilization of the singlet state.

(a) 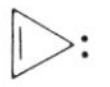 cyclopropenylidene

(b) 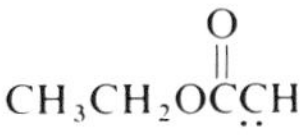 $CH_3CH_2O\overset{O}{\overset{\|}{C}}\ddot{C}H$

(c) cycloheptatrienylidene

(d) 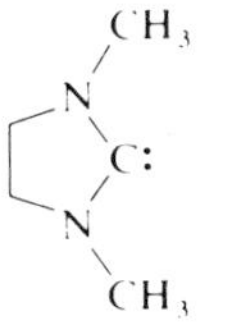 1,3-dimethylimidazolidin-2-ylidene (CH_3–N, C:, N–CH_3)

4. The hydroxyl group in *trans*-cycloocten-3-ol determines the stereochemistry of reaction of this compound with the Simmons-Smith reagent. By examining a model, predict the stereochemistry of the resulting product.

5. Discuss the significance of the relationships between substrate stereochemistry and product composition exhibited by the reactions shown below:

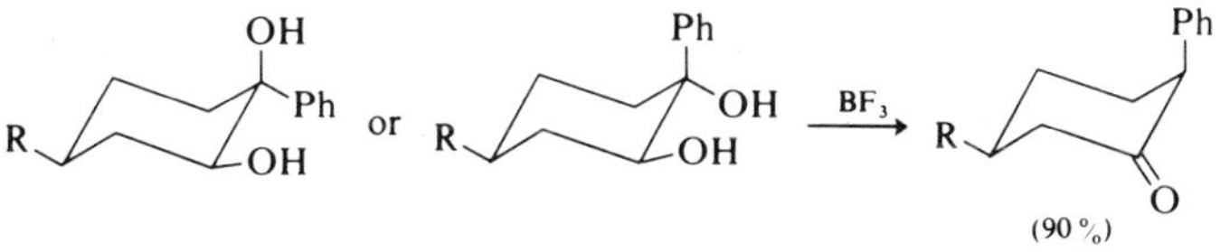

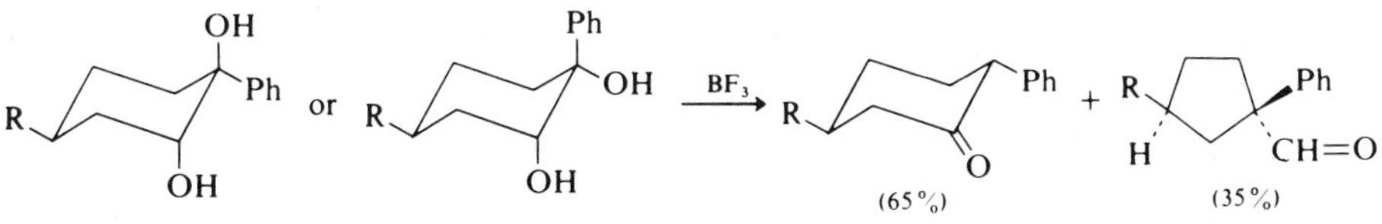

[structures: diols (R, OH, Ph, OH) or (R, Ph, OH, OH) $\xrightarrow{BF_3}$ ketone (R, Ph, O) (90%)]

[structures: diols (R, OH, Ph, OH) or (R, Ph, OH, OH) $\xrightarrow{BF_3}$ ketone (R, Ph, O) (65%) + cyclopentane (R, H, Ph, CH=O) (35%)]

R = *t*-butyl

6. Suggest a mechanistic rationalization for the following reactions. Point out the structural features which contribute to the unusual or abnormal course of the reaction. What product might have been expected if the reaction followed a "normal" course?

(a) NOH ... S $\xrightarrow{SOCl_2}$ C≡N, CH_2SCH_2Cl

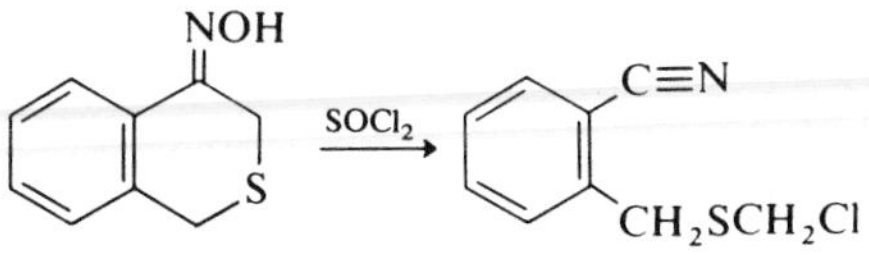

(b) H_3C, CH_3, C, $\overset{O}{\overset{\|}{C}}CHN_2$ $\xrightarrow[CH_3OH]{Cu(acac)_2}$ $C(CH_3)_2$, $CH_2CO_2CH_3$

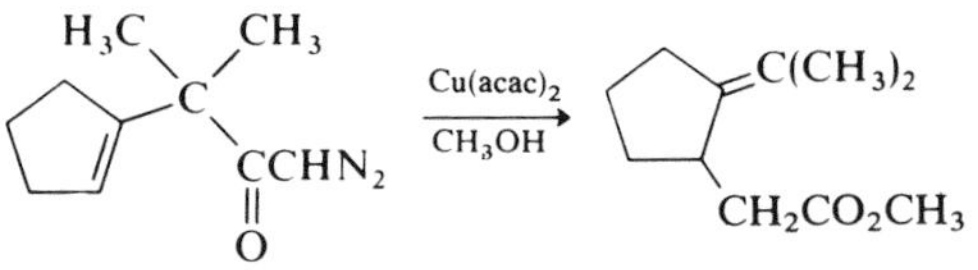

(c) $\overset{O}{\overset{\|}{C}}\underset{\underset{N_2}{\|}}{C}CO_2C_2H_5$, $OCH_2CH{=}CH_2$ $\xrightarrow{Rh_2(O_2CCH_3)_4}$ O, $CO_2C_2H_5$, $CH_2CH{=}CH_2$, O

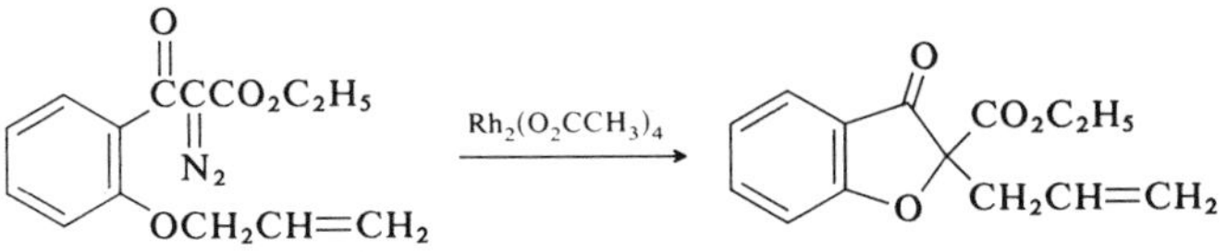

7. Give the structure of the expected Favorskii rearrangement product of compound **A**:

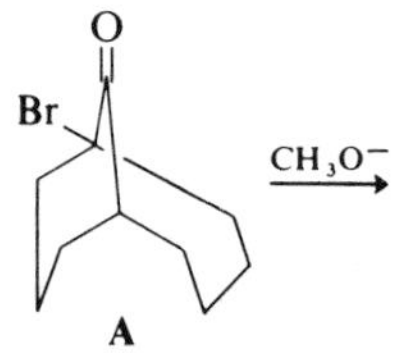

O, Br $\xrightarrow{CH_3O^-}$

A

Experimentally, it has been found that the above ketone can rearrange by either the cyclopropanone or the semibenzilic mechanism, depending on the reaction conditions. Devise two experiments that would permit you to determine which mechanism was operating under a given set of circumstances.

8. Predict the major product of the following reactions:

(a) $\overset{O}{\overset{\|}{C}}CHN_2$, H $\xrightarrow[\text{toluene, 105°C, 2 h}]{CuSO_4}$

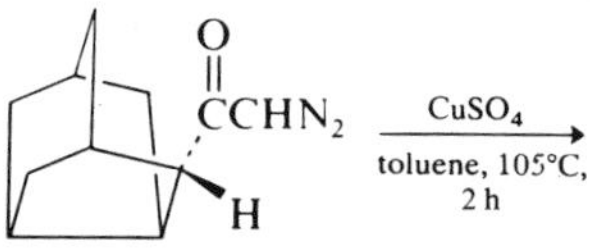

(b) O, OSO_2Ar $\xrightarrow[2)\ H^+]{1)\ OH,\ 110°C}$

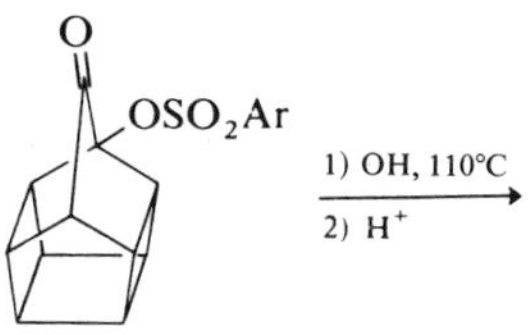

(c) 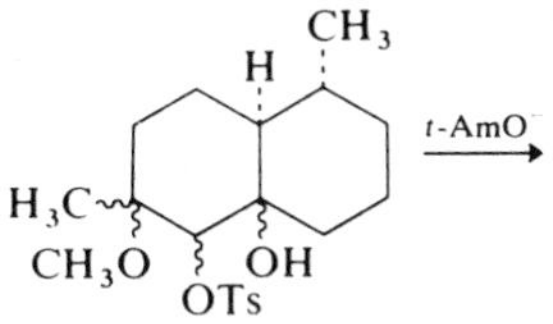 H_3C, CH_3O, OTs, OH, H, CH_3 $\xrightarrow{t\text{-AmO}^-}$

(d) 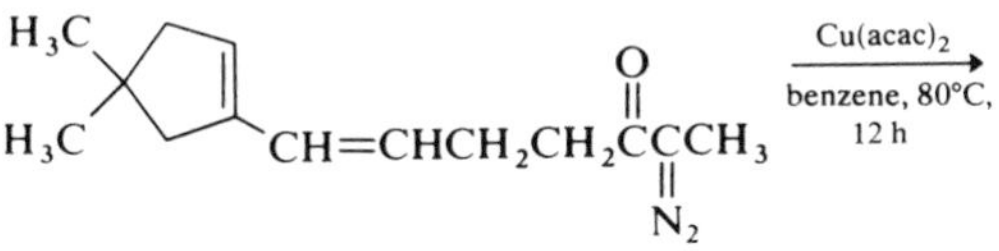 H_3C, H_3C, $CH{=}CHCH_2CH_2\overset{O}{\overset{\|}{C}}\underset{N_2}{\underset{\|}{C}}CH_3$ $\xrightarrow[\text{benzene, 80°C, 12 h}]{Cu(acac)_2}$

(e) 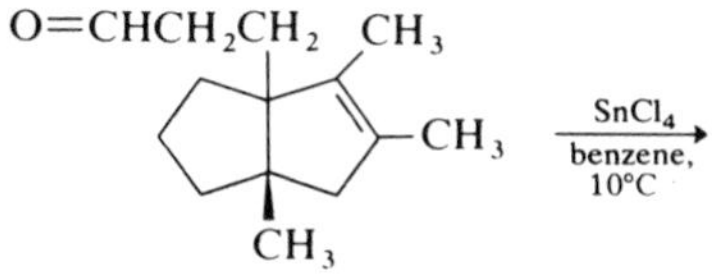 $O{=}CHCH_2CH_2$, CH_3, $-CH_3$, CH_3 $\xrightarrow[\text{benzene, 10°C}]{SnCl_4}$

(f) 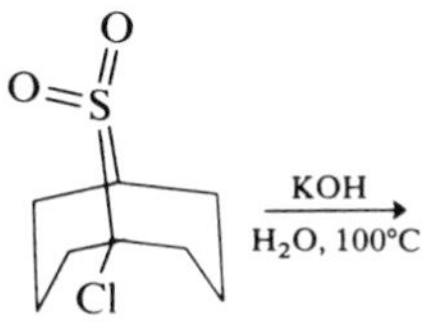 O=S=O, Cl $\xrightarrow[H_2O,\ 100°C]{KOH}$

9. Short reaction series can effect formation of the desired material on the left from the starting material on the right. Devise an appropriate reaction sequence.

(a) 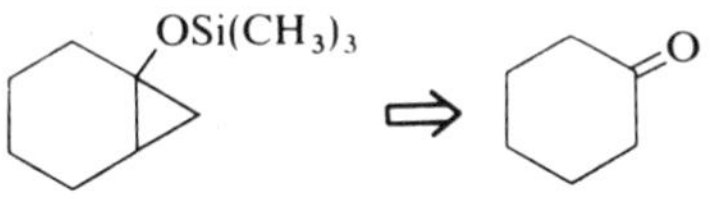 $OSi(CH_3)_3$ $\Rightarrow$ O

(b) 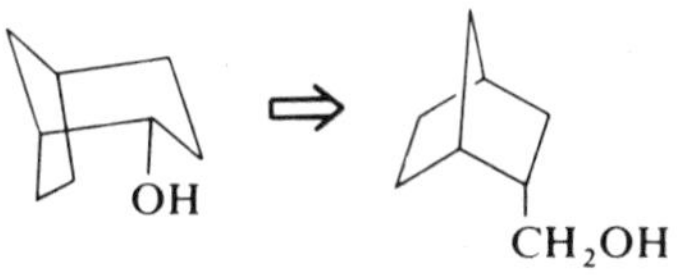 OH $\Rightarrow$ CH_2OH

(c) 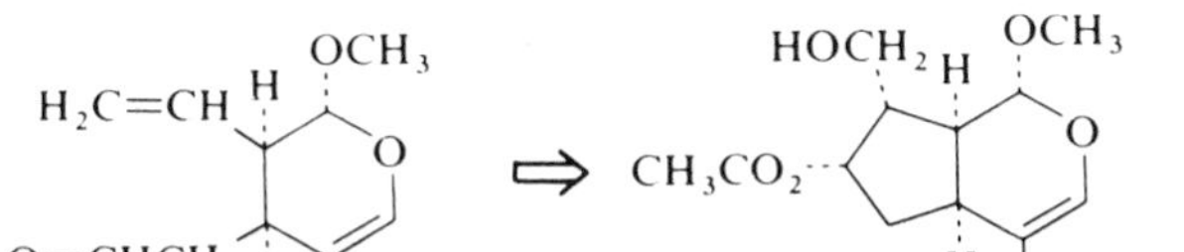 $H_2C{=}CH$, H, OCH_3, O, $O{=}CHCH_2$, H, CO_2CH_3 $\Rightarrow$ $HOCH_2$, H, OCH_3, O, CH_3CO_2, H, CO_2CH_3

(d) 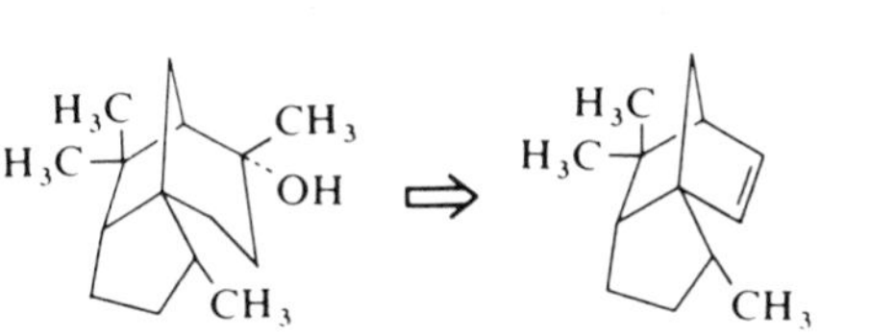 H_3C, H_3C, CH_3, OH, CH_3 $\Rightarrow$ H_3C, H_3C, CH_3

(e) CH_3, CO_2CH_3, CH_3 ⟹ CH_3, CH_2OH, CH_3

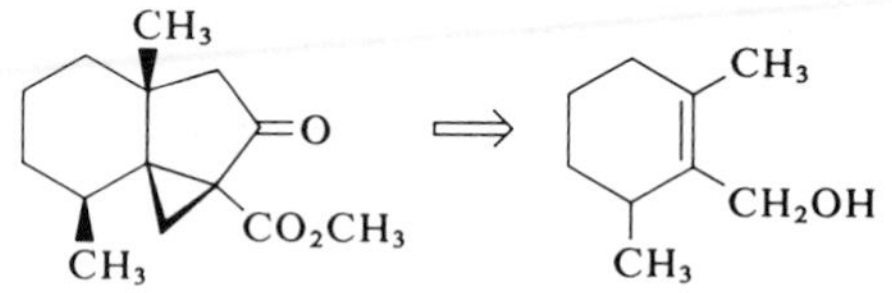

(f) H H H O H H H O ⟹ H H O H H

(g) CH_2OCH_3, O, H, Ph, O, OCH_2Ph ⟹ O, O, OH, H, Ph, O, OCH_2Ph

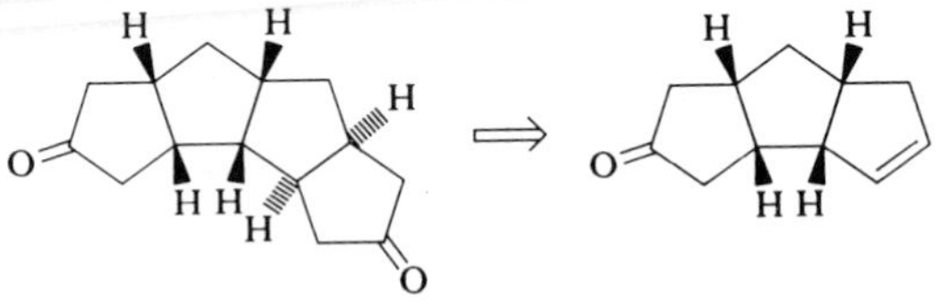

(h) ⟹ O O O

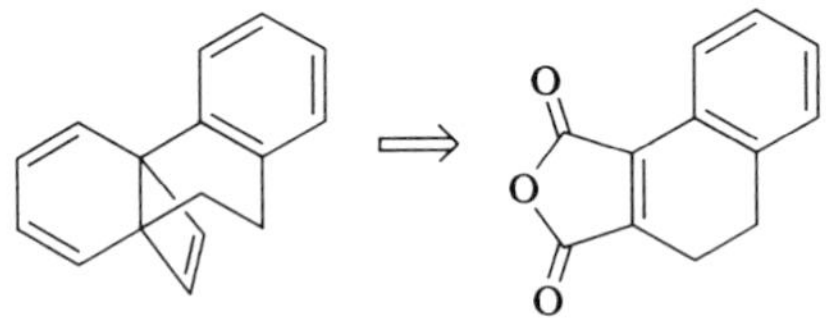

(i) HO, CH_2, CH_3, CH_2CH_2O, CH_2OH, OH, H ⟹ CH=O, H_3C, $CHCH_2CH_2OCH_2Ph$, O, O, CH_3, CH_3, H

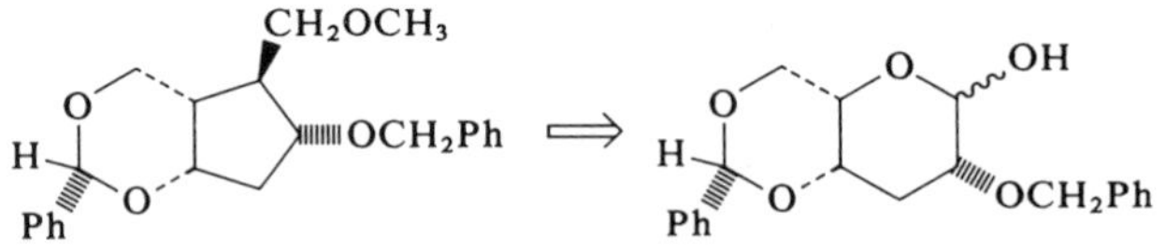

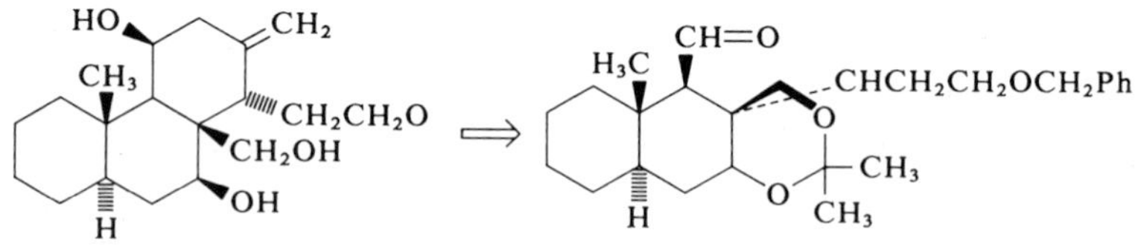

(j) O, H, C=C, CH_3, H ⟹ H, $CH_2CH_2CH_2CO_2H$, C=C, H, H, C=C, CH_3, H

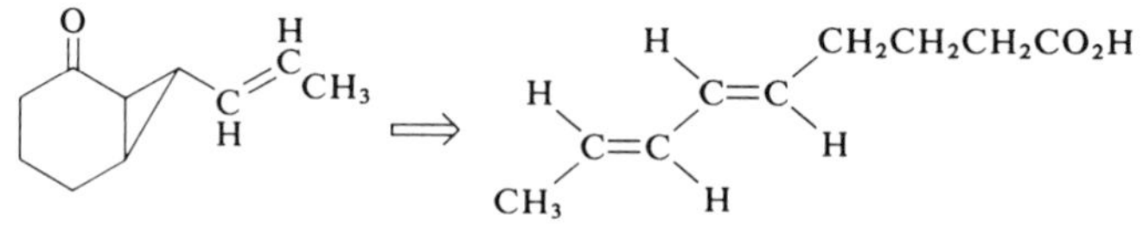

(k) CH_2, H, H_3C, O, O, O, H, H ⟹ $CH_2Si(CH_3)_3$, H_3C, O, O, O, H

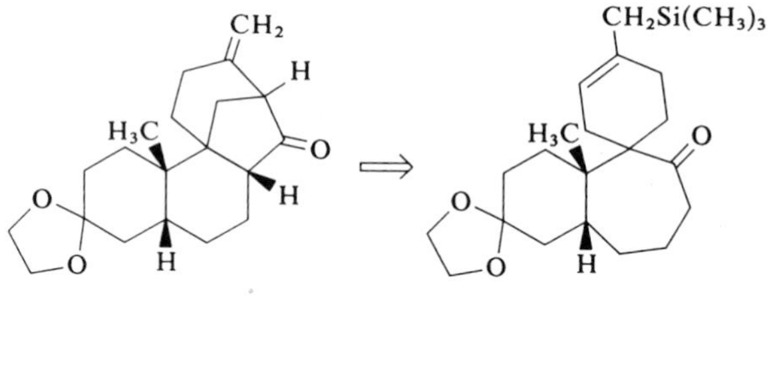

(l)

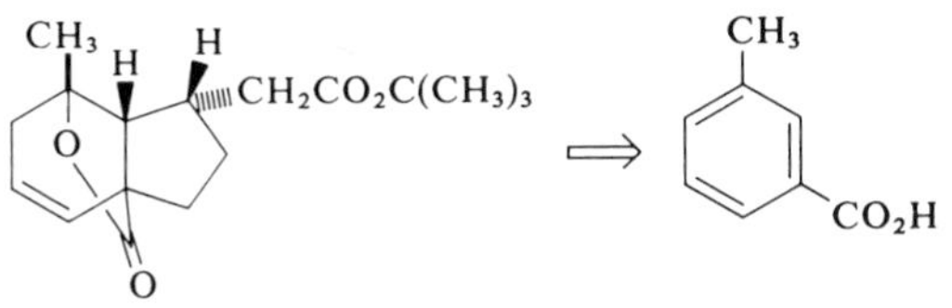

$\Longrightarrow$

10. Formulate mechanisms for the following reactions:

(a) $(CH_3)_2C{=}CHCH_2CH_2$ … $\xrightarrow{NaNH_2}$ $(CH_3)_2C{=}CHCH_2CH_2$, CNH_2 … CH_3

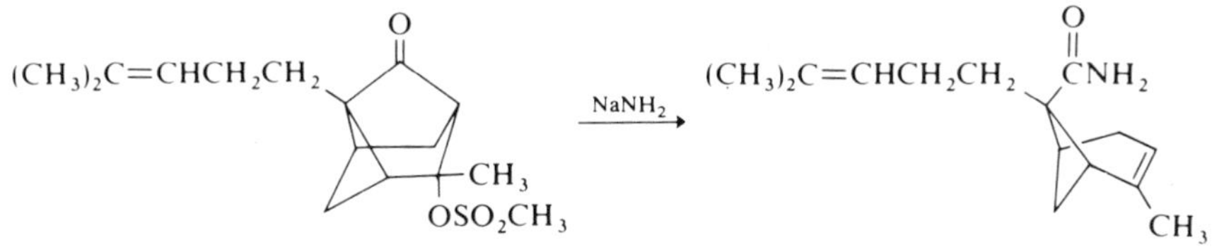

(b) $\xrightarrow{\text{1) KOH; 2) } H^+}$ CO_2H

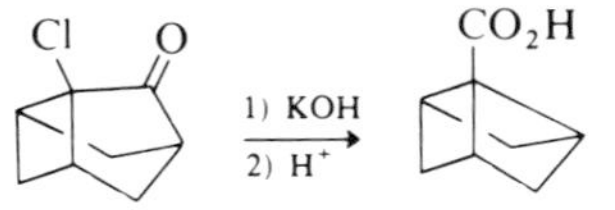

(c) $\xrightarrow{\text{1) KOH; 2) } H^+}$ $H_2C{=}CHCH_2C(=CH_2)CO_2H$

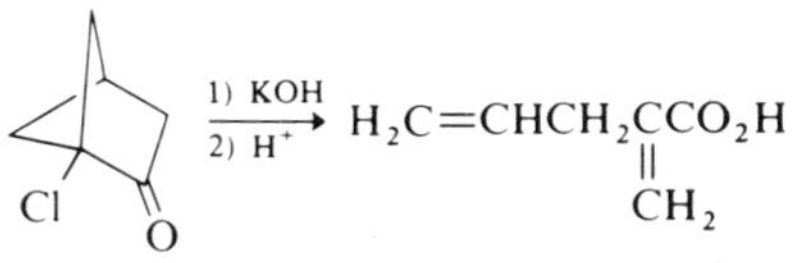

(d) $CH_2{=}CHCH{=}CH_2 + N_2CHCCO_2C_2H_5$ $\xrightarrow{Rh_2(OAc)_4}$ $CH_2{=}CH$ … $CO_2C_2H_5$ + … $CO_2C_2H_5$

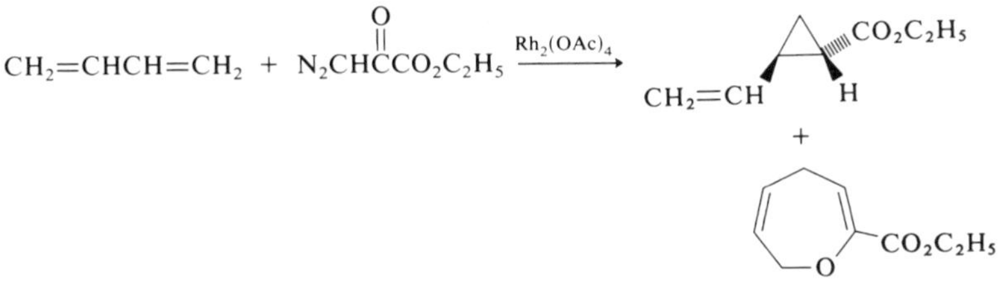

(e) $CH_3CH{=}CHCH_2CH_2CH_2C(=O)CH(CO_2CH_3)CH_2CH{=}CH_2$ $\xrightarrow[Cu(OAc)_2]{Mn(OAc)_3}$ … CO_2CH_3, $=CH_2$, CH_3

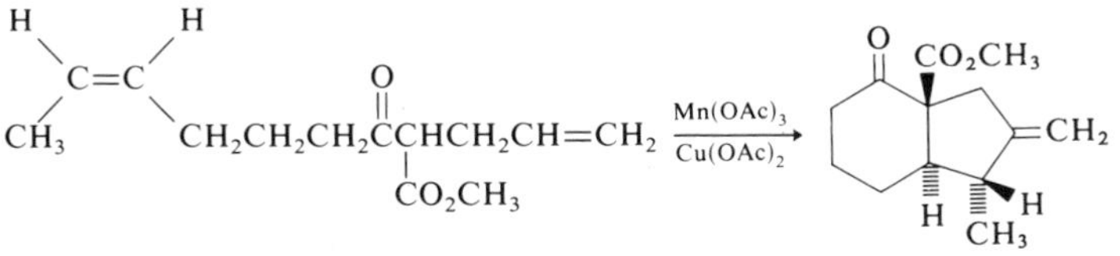

(f) $CH_3CH(I)CO_2SnBu_3 + C_4H_9CH{=}CH_2$ $\xrightarrow{AIBN}$ … CH_3, C_4H_9

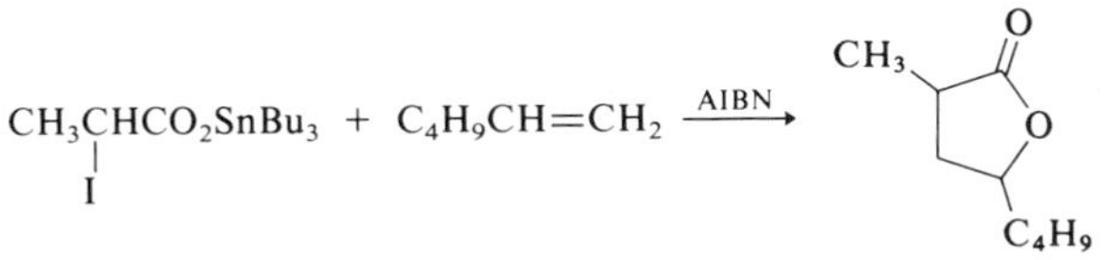

11. A sequence of reactions for converting acyclic and cyclic ketones to α,β-unsaturated ketones with an additional $=CCH_3$ unit has been developed.

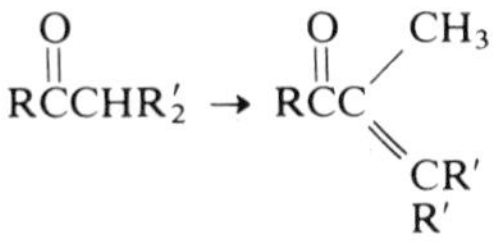

$$RCCHR'_2 \rightarrow RCC(CH_3){=}CR'R'$$

The method utilizes a carbenoid reagent, 1-lithio-1,1-dichloroethane, as a key reagent. The overall sequence involves three steps, one of them before and one of them after the carbenoid reaction. Attempt, by analysis of the bond changes and with your knowledge of carbene chemistry, to devise such a reaction sequence.

12. The synthesis of globulol from the octalin derivative shown proceeds in four stages. These include, not necessarily in sequence, addition of a carbene, fragmentation, and an acid-catalyzed cyclization of a cyclodeca-2,7-dienol. The final stage of the process converts a dibromocyclopropane to a demethylcyclopropane using dimethylcuprate. Working back from globulol, attempt to discover the appropriate sequence of reactions and suggest appropriate reagents for each step.

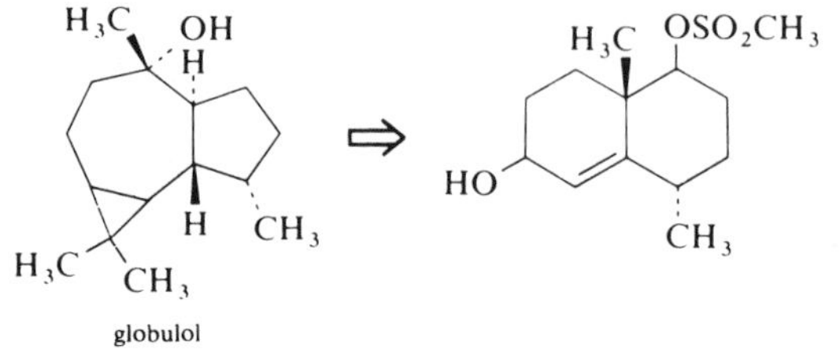

H_3C OH H H CH_3 H_3C CH_3 ⇨ H_3C OSO_2CH_3 HO CH_3

globulol

13. The three decahydroquinolines shown below each gives a different product composition on solvolysis. One gives 9-methylamino-*trans*-5-nonenal, one gives 9-methylamino-*cis*-nonenal, and the third gives a mixture of the two quinoline derivatives **D** and **E**. Deduce which compound gives rise to which product. Explain your reasoning.

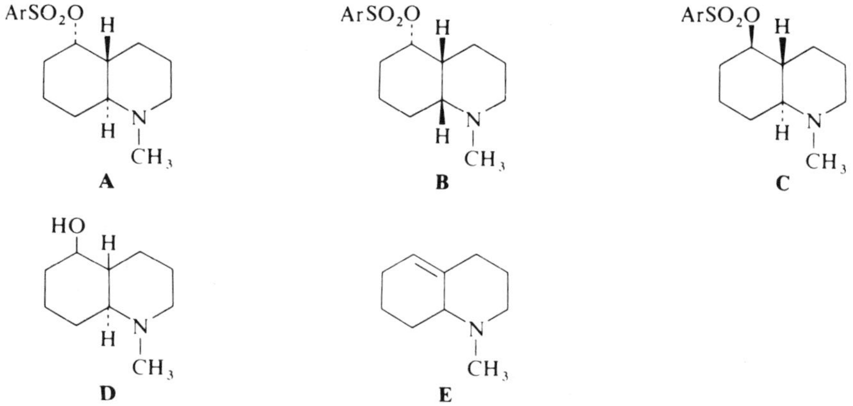

$ArSO_2O$ H H N CH_3 **A**

$ArSO_2O$ H H N CH_3 **B**

$ArSO_2O$ H H N CH_3 **C**

HO H H N CH_3 **D**

N CH_3 **E**

14. Normally, the dominant reaction between acyldiazo compounds and simple α,β-unsaturated carbonyl compounds is a cycloaddition.

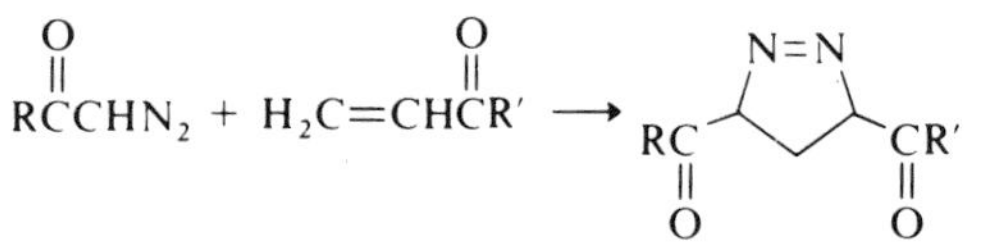

$$\text{RCCHN}_2\ (\text{C=O}) + \text{H}_2\text{C=CHCR}'\ (\text{C=O}) \longrightarrow \text{pyrazoline: N=N ring bearing RC(=O) and CR}'\text{(=O)}$$

If, however, the reaction is run in the presence of a Lewis acid, particularly antimony pentafluoride, the reaction takes a different course, giving a diacyl cyclopropane.

$$RC(=O)CHN_2 + H_2C{=}CHC(=O)R' \xrightarrow{SbF_5} \text{1,2-}(RC(=O))(R'C(=O))\text{cyclopropane}$$

Formulate a mechanism to account for the altered course of the reaction in the presence of SbF_5.

15. Compound **A** on reaction with Bu_3SnH in the presence of AIBN gives **B** rather than **C**. How is **B** formed? Why is **C** not formed? What relationship do these results have to the rate data given on p. 558?

$$\underset{\mathbf{A}}{I(CH_2)_3CH(CH_2CH{=}O)(CH_2CH{=}CH_2)} \xrightarrow[AIBN]{Bu_3SnH} \underset{\mathbf{B}}{\text{1-OH-3-}(CH_2CH{=}CH_2)\text{cyclohexane}} \text{ but not } \underset{\mathbf{C}}{\text{1-}CH_3\text{-3-}(CH_2CH{=}O)\text{cyclohexane}}$$

11

Aromatic Substitution Reactions

This chapter is concerned with reactions that introduce substituent groups on aromatic rings. The most important group of such reactions is the electrophilic aromatic substitutions, but there are also important reactions which occur by nucleophilic substitution. The mechanism of electrophilic aromatic substitution has been studied in great detail, and much information is available about structure-reactivity relationships. These areas were discussed in Chapter 10 of Part A. In this chapter, the synthetic aspects of aromatic substitution will be emphasized.

11.1. Electrophilic Aromatic Substitution

11.1.1. Nitration

Nitration is the most important method for introduction of nitrogen functionality onto aromatic rings. The nitro compounds can easily be reduced to the corresponding amino derivatives. The amino group provides access to diazonium ions, which are useful intermediates for the introduction of a variety of functional groups. There are several reagent systems that are capable of effecting nitration. A major factor in the choice of reagent is the reactivity of the ring to be nitrated. Concentrated nitric acid can effect nitration, but it is not as reactive as a mixture of nitric acid with sulfuric acid. The active nitrating species in both media is the nitronium ion, NO_2^+. The NO_2^+ *ion* is formed by protonation and dissociation of nitric acid. The concentration of NO_2^+ is higher in the more acidic sulfuric acid than in nitric acid.

$$HNO_3 + 2\,H^+ \rightleftarrows H_3O^+ + NO_2^+$$

Nitration can also be carried out in organic solvents, with acetic acid and nitromethane being common examples. In these solvents, the formation of the NO_2^+ ion is often the rate-controlling step[1]:

$$2\,HNO_3 \rightleftarrows H_2NO_3^+ + NO_3^-$$

$$H_2NO_3^+ \xrightarrow{\text{slow}} NO_2^+ + H_2O$$

$$ArH + NO_2^+ \xrightarrow{\text{fast}} ArNO_2 + H^+$$

Another useful medium for nitration is a solution prepared by dissolving nitric acid in acetic anhydride. This generates acetyl nitrate:

$$HNO_3 + (CH_3CO)_2O \rightleftarrows CH_3\overset{O}{\overset{\|}{C}}ONO_2 + CH_3CO_2H$$

This reagent tends to give high *ortho*:*para* ratios for some nitrations.[2] A related procedure involves reaction of the aromatic in chloroform or dichloromethane with a nitrate salt and trifluoroacetic anhydride.[3] Presumably, trifluoroacetyl nitrate is generated under these conditions.

$$NO_3^- + (CF_3CO)_2O \rightarrow CF_3\overset{O}{\overset{\|}{C}}ONO_2 + CF_3CO_2^-$$

Salts containing the nitronium ion can be prepared, and they are reactive nitrating agents. The tetrafluoroborate has been used most frequently,[4] but the trifluoromethanesulfonate can also be prepared readily.[5] Pyridine and quinoline form *N*-nitro salts on reaction with NO_2BF_4.[6] These *N*-nitro heterocycles in turn can act as nitrating reagents. This is called "transfer nitration."

Nitration is a very general reaction, and satisfactory conditions can normally be developed for both activated and deactivated aromatic compounds. Since each successive nitro group that is introduced reduces the reactivity of the ring, it is usually possible to control conditions to obtain a mononitration product. If polynitration is desired, more vigorous conditions are used. Scheme 11.1 gives some examples of nitration reactions.

11.1.2. Halogenation

The introduction of the halogens onto aromatic rings by electrophilic substitution is an important synthetic procedure. Chlorine and bromine are reactive toward

1. E. D. Hughes, C. K. Ingold, and R. I. Reed, *J. Chem. Soc.*, 2400 (1950); J. G. Hoggett, R. B. Moodie, and K. Schofield, *J. Chem. Soc., B*, 1 (1969); K. Schofield, *Aromatic Nitration*, Cambridge University Press, Cambridge, 1980, Chapter 2.
2. A. K. Sparks, *J. Org. Chem.* **31**, 2299 (1966).
3. J. V. Crivello, *J. Org. Chem.* **46**, 3056 (1981).
4. S. J. Kuhn and G. A. Olah, *J. Am. Chem. Soc.* **83**, 4564 (1961); G. A. Olah and S. J. Kuhn, *J. Am. Chem. Soc.* **84**, 3684 (1962); G. A. Olah, S. C. Narang, J. A. Olah, and K. Lammertsma, *Proc. Natl. Acad. Sci. U.S.A.* **79**, 4487 (1982).
5. C. L. Coon, W. G. Blucher, and M. E. Hill, *J. Org. Chem.* **38**, 4243 (1973).
6. G. A. Olah, S. C. Narang, J. A. Olah, R. L. Pearson, and C. A. Cupas, *J. Am. Chem. Soc.* **102**, 3507 (1980).

aromatic hydrocarbons, but Lewis acid catalysts are normally used to achieve desirable rates. Elemental fluorine reacts very exothermically, and very careful control of conditions is required. Iodine can effect substitution only on very reactive aromatics, but a number of useful iodination reagents have been developed.

Scheme 11.1. Aromatic Nitration

1[a] CH_2CN (benzene) $\xrightarrow{HNO_3, H_2SO_4}$ CH_2CN, NO_2 (para) (50–54%)

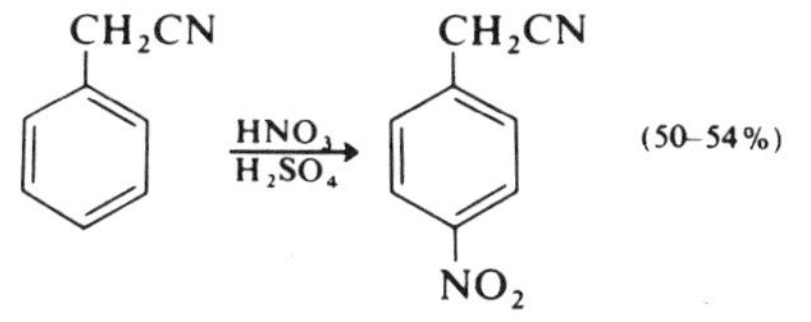

2[b] $N(CH_3)_2$ (benzene) $\xrightarrow{H_2SO_4, HNO_3}$ $N(CH_3)_2$, NO_2 (meta) (56–63%)

3[c] CO_2H (benzene) $\xrightarrow{H_2SO_4, HNO_3}$ CO_2H, O_2N, NO_2 (3,5-dinitro) (54–58%)

4[d] CH_3O, CH_3O, $CH{=}O$ $\xrightarrow{HNO_3}$ CH_3O, CH_3O, $CH{=}O$, NO_2 (73–79%)

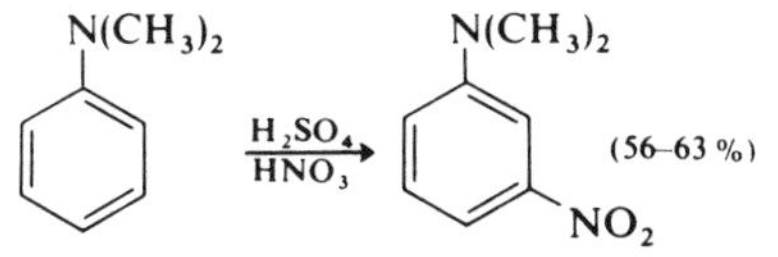

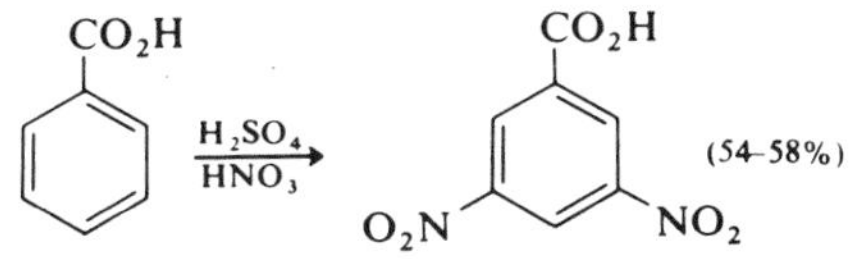

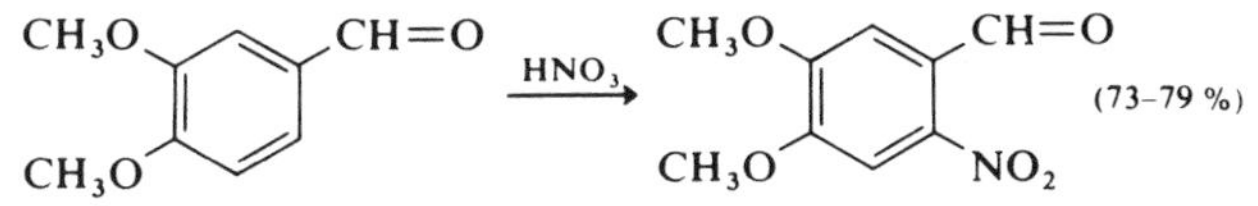

5[e] $CH{=}CHCH{=}O$ (benzene) $\xrightarrow{HNO_3, Ac_2O}$ $CH{=}CHCH{=}O$, NO_2 (ortho) (36–46%)

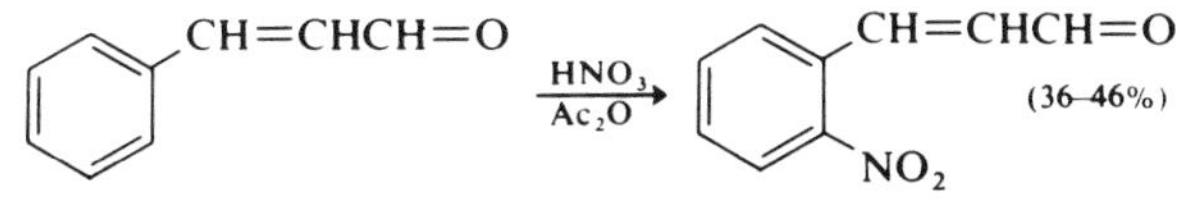

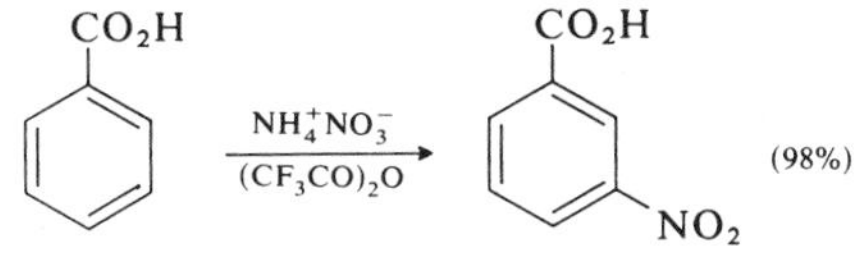

6[f] CO_2H (benzene) $\xrightarrow{NH_4^+NO_3^-, (CF_3CO)_2O}$ CO_2H, NO_2 (meta) (98%)

7[g] CH_3 (benzene) + 2-CH_3-pyridinium ^+N–NO_2 $\longrightarrow$ CH_3, NO_2 $o:m:p$ (100%) 64:3:33

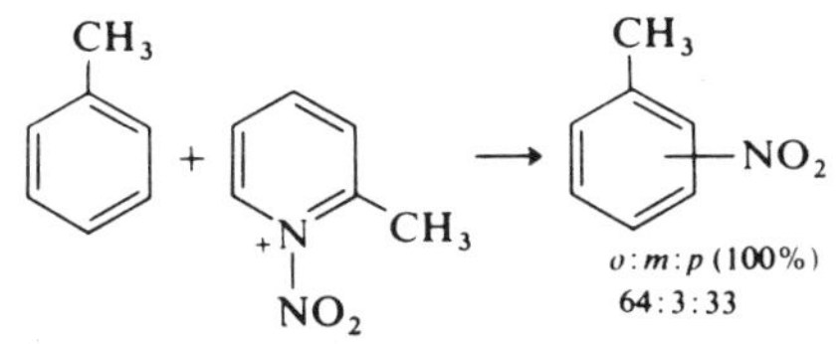

a. G. R. Robertson, *Org. Synth.* **I**, 389 (1932).
b. H. M. Fitch, *Org. Synth.* **III**, 658 (1955).
c. R. Q. Brewster, B. Williams, and R. Phillips, *Org. Synth.* **III**, 337 (1955).
d. C. A. Fetscher, *Org. Synth.* **IV**, 735 (1963).
e. R. E. Buckles and M. P. Bellis, *Org. Synth.* **IV**, 722 (1963).
f. J. V. Crivello, *J. Org. Chem.* **46**, 3056 (1981).
g. C. A. Cupas and R. L. Pearson, *J. Am. Chem. Soc.* **90**, 4742 (1968).

Rate studies show that chlorination is subject to acid catalysis, although the kinetics are frequently complex.[7] The proton is believed to assist Cl—Cl bond breaking in a reactant-Cl_2 complex. Chlorination is much more rapid in polar than in nonpolar solvents.[8] Bromination exhibits similar mechanistic features.

Cl—Cl H—A → (Cl, H, +) → product

+ HCl + A^-

For preparative reactions, Lewis acid catalysts are used. Zinc chloride or ferric chloride can be used in chlorination, and metallic iron, which generates ferric bromide, is often used in bromination. The Lewis acid facilitates cleavage of the halogen–halogen bond.

$$MX_n + X_2 \rightleftharpoons X{-}X\cdots MX_n$$

X—X···MX_n → (X, H, +) → (X) + H^+

A wide variety of aromatic compounds can be brominated. Highly reactive ones, such as anilines and phenols, may undergo bromination at all activated positions. Moderately activated compounds such as anilides, haloaromatics, and hydrocarbons can be readily brominated, with the usual directing effects controlling the regiochemistry. Use of Lewis acid catalysts permits bromination of rings with strongly deactivating substituents, such as nitro and cyano. Scheme 11.2 gives some specific examples of aromatic halogenation reactions.

Halogenations are strongly catalyzed by mercuric acetate or trifluoroacetate. In these reactions, acetyl or trifluoroacetyl hypohalites are formed *in situ*, and these are the reactive halogenating agents. The trifluoroacetyl hypohalites are very reactive reagents. Even nitrobenzene, for example, is readily brominated by trifluoroacetyl hypobromite.[9]

$$Hg(O_2CR)_2 + X_2 \rightleftharpoons HgX(O_2CR) + RCO_2X$$

A solution of bromine in CCl_4 containing sulfuric acid and mercuric oxide is also a reactive brominating agent.[10]

Fluorination can be carried out with the use of fluorine gas diluted with an inert gas. However, great care is necessary to avoid uncontrolled reaction.[11] Several

7. L. M. Stock and F. W. Baker, *J. Am. Chem. Soc.* **84**, 1661 (1962); L. J. Andrews and R. M. Keefer, *J. Am. Chem. Soc.* **81**, 1063 (1959); R. M. Keefer and L. J. Andrews, *J. Am. Chem. Soc.* **82**, 4547 (1960); L. J. Andrews and R. M. Keefer, *J. Am. Chem. Soc.* **79**, 5169 (1957).
8. L. M. Stock and A. Himoe, *J. Am. Chem. Soc.* **83**, 4605 (1961).
9. J. R. Barnett, L. J. Andrews, and R. M. Keefer, *J. Am. Chem. Soc.* **94**, 6129 (1972).
10. S. A. Khan, M. A. Munawar, and M. Siddiq, *J. Org. Chem.* **53**, 1799 (1988).
11. F. Cacace, P. Giacomello, and A. P. Wolf, *J. Am. Chem. Soc.* **102**, 3511 (1980).

other reagents have been devised which are capable of aromatic fluorination.[12] Acetyl hypofluorite can be prepared *in situ* from fluorine and sodium acetate.[13] This reagent effects fluorination of activated aromatics. While this procedure does not avoid the special precautions necessary for manipulation of elemental fluorine, it does provide a system with much greater selectivity. Acetyl hypofluorite shows a strong preference for *o*-fluorination of alkoxy- and acetamido-substituted rings. *N*-Fluoro-bis(trifluoromethanesulfonyl)amine displays similar reactivity. It can fluorinate benzene and activated aromatics and shows a high *ortho* : *para* selectivity.[14]

CH_3O–(benzene) + $(CF_3SO_2)_2NF$ ⟶ CH_3O–(*o*-F-benzene) (69%) + CH_3O–(benzene)–F (24%)

Iodinations can be carried out by mixtures of iodide and cupric salts.[15] A mixture of an iodide salt and ceric ammonium nitrate, $Ce(NH_3)_2(NO_3)_6$, is also a reactive iodinating system.[16]

p-xylene (CH_3, CH_3) + CuI + $CuCl_2$ ⟶ 2-iodo-*p*-xylene (CH_3, I, CH_3) (~70%)

Oxidation of iodide to iodine is undoubtely involved in these reactions, but the details of the mechanism have not been explored. Iodination of moderately reactive aromatics can be effected by mixtures of iodine and silver or mercuric salts.[17] Hypoiodites are presumably the active iodinating species. Some examples of iodination procedures are included in Scheme 11.2.

11.1.3. Friedel–Crafts Alkylations and Acylations

Friedel-Crafts reactions are one of the most important methods for introducing carbon substituents onto aromatic rings. The reactive electrophiles can be either discrete carbocations or acylium ions or polarized complexes that still contain the

12. S. T. Purrington, B. S. Kagan, and T. B. Patrick, *Chem. Rev.* **86**, 997 (1986).
13. O. Lerman, Y. Tor, and S. Rozen, *J. Org. Chem.* **46**, 4629 (1981); O. Lerman, Y. Tor, D. Hebel, and S. Rozen, *J. Org. Chem.* **49**, 806 (1984).
14. S. Singh, D. D. DesMarteau, S. S. Zuberi, M. Whitz, and H.-N. Huang, *J. Am. Chem. Soc.* **109**, 7194 (1987).
15. W. C. Baird, Jr., and J. H. Surridge, *J. Org. Chem.* **35**, 3436 (1970).
16. T. Sugiyama, *Bull. Chem. Soc. Jpn.* **54**, 2847 (1981).
17. Y. Kobayashi, I. Kumadaki, and T. Yoshida, *J. Chem. Res.* (Synopses), 215 (1977); R. N. Hazeldine and A. G. Sharpe, *J. Chem. Soc.*, 993 (1952); W. Minnis, *Org. Synth.* **II**, 357 (1943); D. E. Janssen and C. V. Wilson, *Org. Synth.* **IV**, 547 (1963).

Scheme 11.2. Aromatic

A. Chlorination

1[a]

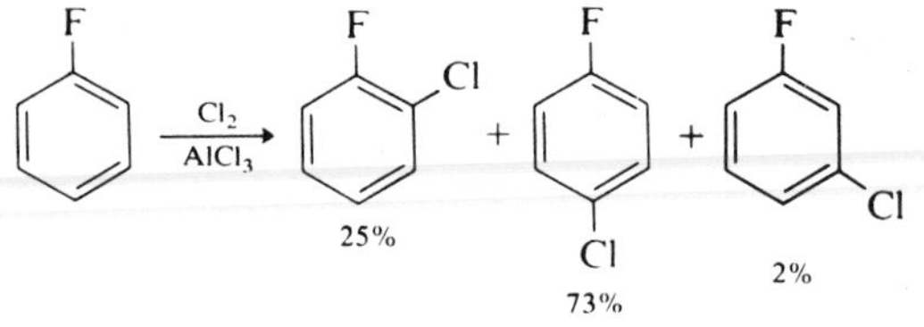

F, Cl_2, $AlCl_3$, F, Cl, 25%, +, F, Cl, 73%, +, F, Cl, 2%

2[b]

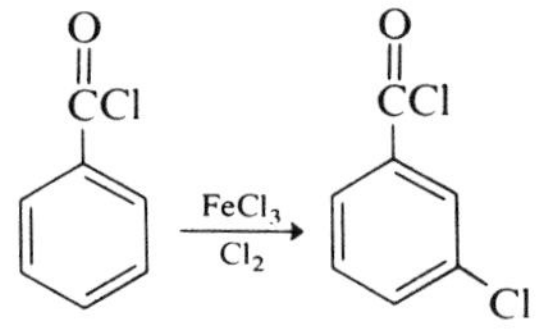

O, CCl, $FeCl_3$, Cl_2, O, CCl, Cl

B. Bromination

3[c]

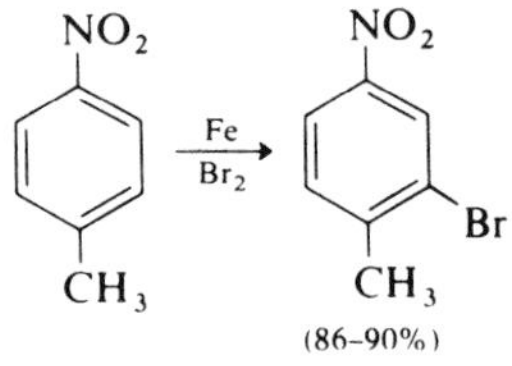

NO_2, CH_3, Fe, Br_2, NO_2, Br, CH_3, (86–90%)

4[d]

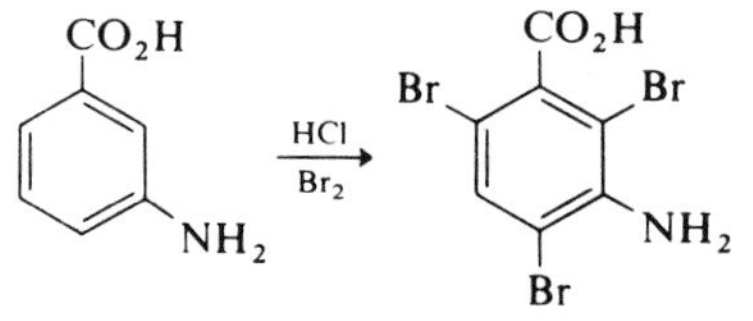

CO_2H, NH_2, HCl, Br_2, CO_2H, Br, Br, NH_2, Br

5[e]

CH_3, NO_2, CO_2H, Br, O, Br, N, N, O, N, O, H, H_2SO_4, CH_3, Br, NO_2, CO_2H, (98%)

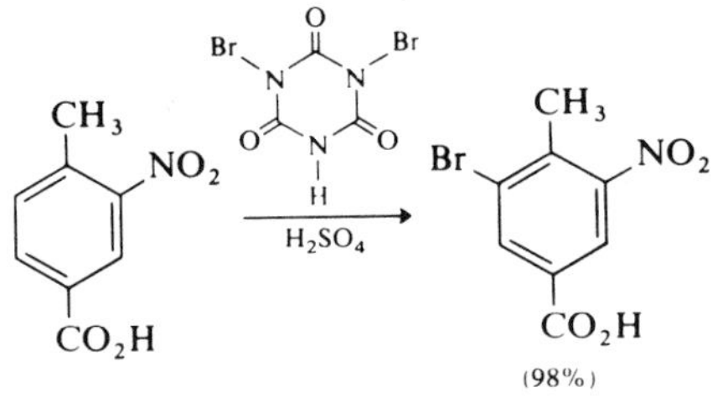

a. G. A. Olah, S. J. Kuhn, and B. A. Hardie, *J. Am. Chem. Soc.* **86**, 1055 (1964).
b. E. Hope and G. F. Riley, *J. Chem. Soc.* **121**, 2510 (1922).
c. M. M. Robison and B. L. Robison, *Org. Synth.* **IV**, 947 (1963).
d. W. A. Wisansky and S. Ansbacher, *Org. Synth.* **III**, 138 (1955).
e. A. R. Leed, S. D. Boettger, and B. Ganem, *J. Org. Chem.* **45**, 1098 (1980).

6[f] $C_6H_5CO_2CH_3$ $\xrightarrow[H^+]{Br_2,\ HgO}$ m-$BrC_6H_4CO_2CH_3$ (80%)

C. Iodination

6[f] 2-$H_2NC_6H_4CO_2H$ $\xrightarrow{ICl}$ 2-amino-5-iodobenzoic acid (76–84%)

7[g] 3,4,5-$(CH_3O)_3C_6H_2CH_2OH$ $\xrightarrow[Hg(O_2CCH_3)_2]{I_2}$ 2-iodo-3,4,5-trimethoxybenzyl alcohol (76%)

8[h] 1,2-$(CH_3O)_2C_6H_4$ $\xrightarrow[AgO_2CCF_3]{I_2}$ 4-iodo-1,2-dimethoxybenzene (85–91%)

9[i] 1,2,4,5-$(CH_3)_4C_6H_2$ $\xrightarrow[HIO_4]{I_2}$ 3-iodo-1,2,4,5-tetramethylbenzene (80–81%)

11[k] $C_6H_5OCH_3$ $\xrightarrow[Ce(NH_3)_2(NO_3)_6]{n\text{-}Bu_4I}$ p-$IC_6H_4OCH_3$ (84%)

f. S. A. Khan, M. A. Munawar, and M. Siddiq, *J. Org. Chem.* **53**, 1799 (1988).
g. V. H. Wallingord and P. A. Krueger, *Org. Synth.* **II**, 349 (1943).
h. F. E. Ziegler and J. A. Schwartz, *J. Org. Chem.* **43**, 985 (1978).
i. D. E. Janssen and C. V. Wilson, *Org. Synth.* **IV**, 547 (1963).
j. H. Suzuki, *Org. Synth.* **51**, 94 (1971).
k. T. Sugiyama, *Bull. Chem. Soc. Jpn.* **54**, 2847 (1981).

leaving group. Various combinations of reagents can be used to generare alkylating species. These include alkyl halides with Lewis acids and alcohols or alkenes with strong acids.

$$R{-}X + AlCl_3 \rightleftharpoons R{-}\overset{+}{X}{-}\overset{-}{Al}Cl_3 \rightleftharpoons R^+ + X\overset{-}{Al}Cl_3$$

$$R{-}OH + H^+ \rightleftharpoons R{-}\overset{+}{O}H(H) \rightleftharpoons R^+ + H_2O$$

$$RCH{=}CH_2 + H^+ \rightleftharpoons R\overset{+}{C}HCH_3$$

Because of the involvement of carbocations, Friedel–Crafts alkylations are often accompanied by rearrangement of the alkylating group. For example, isopropyl groups are often introduced when *n*-propyl reactants are used.[18]

$$C_6H_6 + CH_3CH_2CH_2Cl \xrightarrow{AlCl_3} C_6H_5CH(CH_3)_2$$

Under a variety of reaction conditions, alkylation of benzene with either 2-chloro- or 3-chloropentane gives rise to a mixture of both 2-pentyl- and 3-pentylbenzene.[19] Rearrangement can also occur after the initial alkylation. The reaction of 2-chloro-2-methylbutane with benzene under Friedel–Crafts conditions is an example of this behavior.[20] With relatively mild Friedel–Crafts catalysts such as BF_3 or $FeCl_3$, the main product is **A**. With $AlCl_3$, equilibration of **A** and **B** occurs, and the equilibrium favors **B**. Thus, the apparent rearrangement of a tertiary carbocation to a less stable secondary one is actually the result of product equilibration.

$$C_6H_6 + (CH_3)_2C(Cl)CH_2CH_3 \longrightarrow \underset{\mathbf{A}}{C_6H_5C(CH_3)_2CH_2CH_3} + \underset{\mathbf{B}}{C_6H_5CH(CH_3)CH(CH_3)_2}$$

Groups can also migrate from one position to another on the ring.[21] Such migrations are thermodynamically controlled and proceed in the direction of minimizing

18. S. H. Sharman, *J. Am. Chem. Soc.* **84**, 2945 (1962).
19. R. M. Roberts, S. E. McGuire, and J. R. Baker, *J. Org. Chem.* **41**, 659 (1976).
20. A. A. Khalaf and R. M. Roberts, *J. Org. Chem.* **35**, 3717 (1970); R. M. Roberts and S. E. McGuire, *J. Org. Chem.* **35**, 102 (1970).
21. R. M. Roberts and D. Shiengthong, *J. Am. Chem. Soc.* **86**, 2851 (1964).

Table 11.1. Relative Activity of Friedel-Crafts Catalysts[a]

Very active	Moderately active	Weak
$AlCl_3$, $AlBr_3$, $GaCl_3$, $GaCl_2$, SbF_5, $MoCl_5$	$InCl_3$, $LnBr_3$, $SbCl_5$, $FeCl_3$, $AlCl_3$–CH_3NO_2, SbF_5–CH_3NO_2	BCl_3, $SnCl_4$, $TiCl_4$, $TiBr_4$, $FeCl_2$

a. G. A. Olah, S. Kobayashi, and M. Tashiro, *J. Am. Chem. Soc.* **94**, 7448 (1972).

interactions between substituents.

$$p\text{-}CH_3C_6H_4CH_3 + CH_3CH(Cl)CH_3 \xrightarrow[50^\circ C]{AlCl_3} 2\text{-}(CH(CH_3)_2)\text{-}1,4\text{-}(CH_3)_2C_6H_3 + 3\text{-}(CH(CH_3)_2)\text{-}1,5\text{-}(CH_3)_2C_6H_3$$

The relative reactivity of various Friedel–Crafts catalysts has not been described in a quantitative way, but comparative studies with a series of benzyl halides have resulted in the qualitative groupings shown in Table 11.1. Proper choice of catalyst can minimize subsequent product equilibrations.

The Friedel–Crafts alkylation reaction does not proceed successfully with aromatic substrates having electron-attracting groups. Another limitation is that the alkyl group introduced in the reaction increases the reactivity of the ring toward further substitution, so polyalkylation can be a problem. Polyalkylation can be minimized by using the aromatic reactant in excess.

Besides the alkyl halide–Lewis acid combination, two other sources of carbocations are used in Friedel–Crafts reactions. Alcohols can serve as carbocation precursors in strong acids such as sulfuric or phosphoric acid. Alkylation can also be effected by alcohols in combination with BF_3 and $AlCl_3$.[22] Alkenes can also serve as alkylating agents when protic acids, especially H_2SO_4, H_3PO_4 or HF, or Lewis acids, such as BF_3 and $AlCl_3$, are used as catalysts.[23]

Friedel–Crafts alkylation can occur intramolecularly to form a new fused ring. It is somewhat easier to form six-membered rings than five-membered ones in such reactions. Thus, while 4-phenyl-1-butanol gives a 50% yield of a cyclized product

22. A. Schriesheim, in *Friedel–Crafts and Related Reactions*, Vol. II, G. Olah (ed.), Wiley-Interscience, New York, 1964, Chapter XVIII.
23. S. H. Patinkin and B. S. Friedman, in *Friedel–Crafts and Related Reactions*, Vol. II, G. Olah (ed.), Wiley-Interscience, New York, 1964, Chapter XIV.

in phosphoric acid, 3-phenyl-1-propanol is mainly dehydrated to alkene.[24]

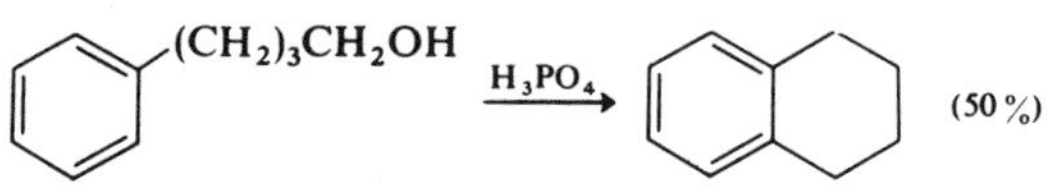

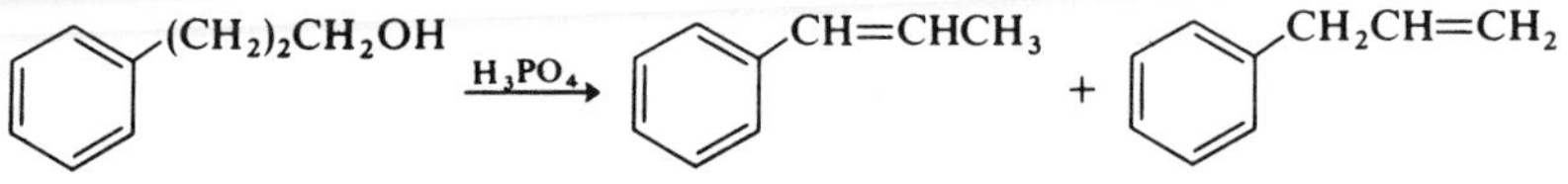

$$C_6H_5(CH_2)_3CH_2OH \xrightarrow{H_3PO_4} \text{tetralin} \quad (50\%)$$

$$C_6H_5(CH_2)_2CH_2OH \xrightarrow{H_3PO_4} C_6H_5CH{=}CHCH_3 + C_6H_5CH_2CH{=}CH_2$$

If a potential carbocation intermediate can undergo a hydride or alkyl shift, this shift will occur in preference to closure of the five-membered ring:

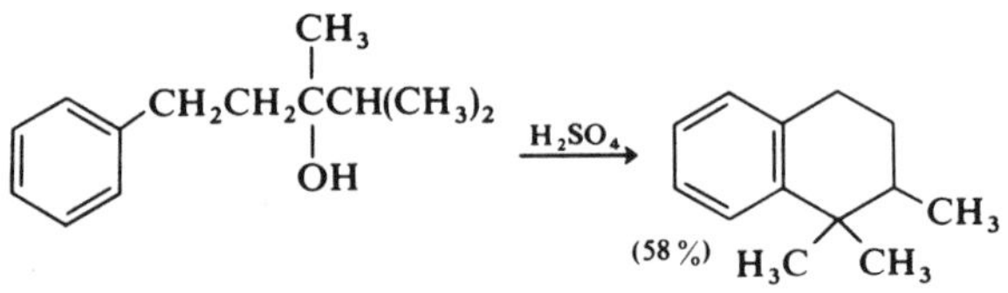

$$C_6H_5CH_2CH_2C(CH_3)(OH)CH(CH_3)_2 \xrightarrow{H_2SO_4} \text{1,1,2-trimethyltetralin} \quad (58\%)$$

Ref. 25

Intramolecular Friedel–Cradts reactions are an important method for constructing polycyclic hydrocarbon frameworks. Entry 5 in Scheme 11.3 is an example of this type of reaction.

Friedel-Crafts acylation generally involves reaction of an acid halide and a Lewis acid such as $AlCl_3$, SbF_5, or BF_3. Acid anhydrides can also be used in some cases. As in the alkylation reaction, the reactive intermediate can be a dissociated acylium ion or a complex of the acid chloride and Lewis acid.[26]

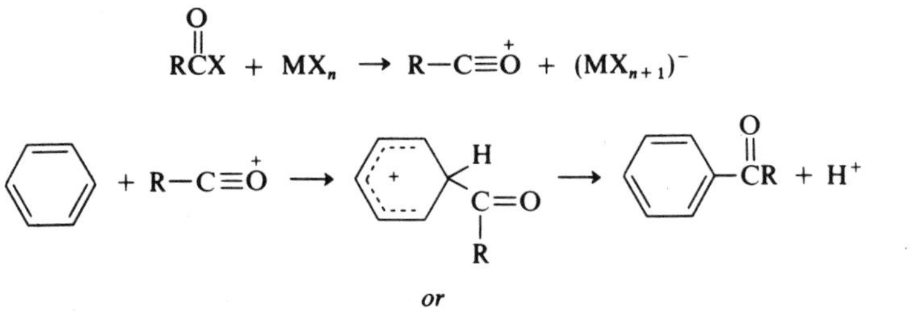

$$RCX(=O) + MX_n \rightarrow R{-}C{\equiv}\overset{+}{O} + (MX_{n+1})^-$$

$$C_6H_6 + R{-}C{\equiv}\overset{+}{O} \rightarrow [C_6H_6(H)(C(=O)R)]^+ \rightarrow C_6H_5C(=O)R + H^+$$

or

$$RCX(=O) + MX_n \rightarrow R{-}C(={}^{+}O{-}\overset{-}{M}X_n){-}X$$

$$C_6H_6 + RC({=}{}^{+}O{-}\overset{-}{M}X_n){-}X \rightarrow [C_6H_6(H)(C(=O)R)]^+ + (MX_{n+1})^- \rightarrow C_6H_5C(=O)R + H^+$$

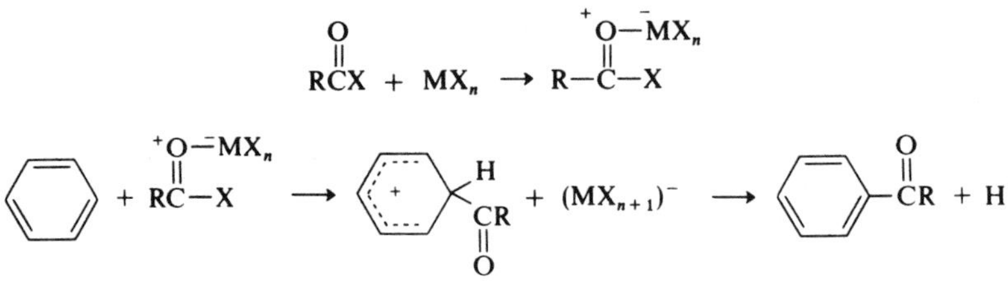

24. A. A. Khalaf and R. M. Roberts, *J. Org. Chem.* **34**, 3571 (1969).
25. A. A. Khalaf and R. M. Roberts, *J. Org. Chem.* **37**, 4227 (1972).
26. F. R. Jensen and G. Goldman, in *Friedel-Crafts and Related Reactions*, Vol. III, G. Olah (ed.), Wiley-Interscience, New York, 1964, Chapter XXXVI.

A. Intermolecular Reactions

1[a] $PhCH(Br)COCH_3$ + benzene $\xrightarrow{AlCl_3}$ $Ph_2CHCOCH_3$ (53–57%)

2[b] $CH_3C(CH_2Cl)=CH_2$ + benzene $\xrightarrow{H_2SO_4}$ $PhC(CH_3)_2CH_2Cl$ (70–73%)

3[c] $PhCH(Br)CH(Br)CO_2H$ + benzene $\xrightarrow{AlCl_3}$ $Ph_2CHCH(Ph)CO_2H$ (66–78%)

B. Intramolecular Friedel–Crafts Cyclizations

4[d] $PhCH_2CH_2CH(CH_3)C(OH)(CH_3)_2$ $\xrightarrow{H_2SO_4}$ tetrahydronaphthalene with CH_3, H_3C, CH_3 (60%)

5[e] HO, CH_3, CH_3, H, O, CH_3O $\xrightarrow{\text{polyphosphoric acid}}$ H_3C, H, O, CH_3, CH_3O (87%)

a. E. M. Schultz and S. Mickey, *Org. Synth.* **III**, 343 (1955).
b. W. T. Smith, Jr., and J. T. Sellas, *Org. Synth.* **IV**, 702 (1963).
c. C. P. Krimmol and L. E. Thielen, E. A. Brown, and W. J. Heidtke, *Org. Synth.* **IV**, 960 (1963).
d. A. A. Khalaf and R. M. Roberts, *J. Org. Chem.* **37**, 4227 (1972).
e. R. E. Ireland, S. W. Baldwin, and S. C. Welch, *J. Am. Chem. Soc.* **94**, 2056 (1972).

Orientation of the incoming acyl group in Friedel–Crafts acylations can be quite sensitive to the reaction solvent and other procedural variables.[27] In general, *para* attack predominates for alkylbenzenes.[28] The percentage of *ortho* attack increases with the electrophilicity of the acylium ion, and as much as 50% *ortho* product is observed with the formyl and 2,4-dinitrobenzoyl ions.[29] Rearrangement of the acyl group is not a problem in Friedel–Crafts acylation. Neither is polyacylation, because the first acyl group serves to deactivate the ring to further attack.

27. For example, see L. Friedman and R. J. Honour, *J. Am. Chem. Soc.* **91**, 6344 (1969).
28. H. C. Brown, G. Marino, and L. M. Stock, *J. Am. Chem. Soc.* **81**, 3310 (1959); H. C. Brown and G. Marino, *J. Am. Chem. Soc.* **81**, 5611 (1959); G. A. Olah, M. E. Moffatt, S. J. Kuhn, and B. A. Hardie, *J. Am. Chem. Soc.* **86**, 2198 (1964).
29. G. A. Olah and S. Kobayashi, *J. Am. Chem. Soc.* **93**, 6964 (1971).

Intramolecular acylations are quite common. The normal procedure involving an acid halide and Lewis acid can be used. One useful alternative is to dissolve the carboxylic acid in polyphosphoric acid (PPA) and heat to effect cyclization. This procedure probably involves formation of a mixed phosphoric–carboxylic anhydride.[30] Cyclizations can also be carried out with an esterified oligomer of phosphoric acid called "polyphosphate ester." This material is chloroform-soluble.[31]

$(CH_2)_3CO_2H$ PPA

Neat methanesulfonic acid is also an effective reagent for intramolecular Friedel–Crafts acylation.[32]

A procedure for fusing a six-membered ring to an aromatic ring uses succinic anhydride or a derivative. An intermolecular acylation is followed by reduction and an intramolecular acylation. The reduction is necessary to provide a more reactive ring for the second acylation.

CH_3 CH_3 + H_3C O O O $\xrightarrow{AlCl_3}$ CH_3 CH_3 O CCH_2CHCO_2H CH_3

Pd, H_2 — Ref. 33

CH_3 CH_3 $(CH_2)_2CHCO_2H$ CH_3 $\xleftarrow{PPA}$ CH_3 CH_3 O CH_3

Scheme 11.4 shows some other representative Friedel–Crafts acylation reactions.

There are a number of other variations of the Friedel–Crafts reaction that are useful in synthesis. The introduction of chloromethyl substituents is brought about by reaction with formaldehyde in concentrated hydrochloric acid in the presence of halide salts, especially zinc chloride.[34] The reaction proceeds with benzene and

30. W. E. Bachmann and W. J. Horton, *J. Am. Chem. Soc.* **69**, 58 (1947).
31. Y. Kanaoka, O. Yonemitsu, K. Tanizawa, and Y. Ban, *Chem. Pharm. Bull.* **12**, 773 (1964); T. Kametani, S. Takano, S. Hibino, and T. Terui, *J. Heterocycl. Chem.* **6**, 49 (1969).
32. V. Premasagar, V. A. Palaniswamy, and E. J. Eisenbraun, *J. Org. Chem.* **46**, 2974 (1981).
33. E. J. Eisenbraun, C. W. Hinman, J. M. Springer, J. W. Burnham, T. S. Chou, P. W. Flanagan, and M. C. Hamming, *J. Org. Chem.* **36**, 2480 (1971).
34. R. C. Fuson and C. H. McKeever, *Org. React.* **1**, 63 (1942); G. A. Olah and S. H. Yu, *J. Am. Chem. Soc.* **97**, 2293 (1975).

derivatives with electron-releasing groups. The reactive electrophile is probably protonated choromethyl alcohol.

$$CH_2{=}O + HCl + H^+ \rightleftharpoons H_2\overset{+}{O}CH_2Cl$$

$$C_6H_6 + H_2\overset{+}{O}CH_2Cl \longrightarrow C_6H_5CH_2Cl$$

Chloromethylation can also be carried out by using various chloromethyl ethers and $SnCl_4$.[35]

$$p\text{-}CH_3C_6H_4CH_3 \xrightarrow[SnCl_4]{ClCH_2O(CH_2)_4OCH_2Cl} 2,5\text{-}(CH_3)_2C_6H_3CH_2Cl$$

Carbon monoxide, hydrogen cyanide, and nitriles also react with aromatic compounds in the presence of strong acids or other Friedel–Crafts catalysts. These reactions introduce formyl or acyl substituents. The general outlines of the mechanisms of these reactions are given below:

a. Formylation with carbon monoxide:

$$\overset{-}{C}{\equiv}\overset{+}{O} + H^+ \rightleftharpoons H{-}C{\equiv}\overset{+}{O}$$

$$ArH + H{-}C{\equiv}\overset{+}{O} \rightarrow ArCH{=}O + H^+$$

b. Formylation with hydrogen cyanide:

$$H{-}C{\equiv}N + H^+ \rightleftharpoons H{-}C{\equiv}\overset{+}{N}{-}H$$

$$ArH + H{-}C{\equiv}\overset{+}{N}{-}H \rightarrow Ar\underset{H}{C}{=}\overset{+}{N}H_2 \xrightarrow{H_2O} ArCH{=}O$$

c. Acylation with nitriles:

$$R{-}C{\equiv}N + H^+ \rightleftharpoons R{-}C{\equiv}\overset{+}{N}{-}H$$

$$ArH + R{-}C{\equiv}\overset{+}{N}{-}H \rightarrow R{-}\underset{Ar}{C}{=}\overset{+}{N}H_2 \xrightarrow{H_2O} Ar\overset{O}{\overset{\|}{C}}R$$

Many specific examples of these reactions can be found in reviews in the *Organic Reactions* series.[36] Dichloromethyl ethers are also precursors of the formyl group.[37]

35. G. A. Olah, D. A. Beal, and J. A. Olah, *J. Org. Chem.* **41**, 1627 (1976); G. A. Olah, D. A. Bell, S. H. Yu, and J. A. Olah, *Synthesis*, 560 (1974).
36. N. N. Crounse, *Org. React.* **5**, 290 (1949); W. E. Truce, *Org. React.* **9**, 37 (1957); P. E. Spoerri and A. S. DuBois, *Org. React.* **5**, 387 (1949); see also G. A. Olah, L. Ohannesian, and M. Arvanaghi, *Chem. Rev.* **87**, 671 (1987).
37. P. E. Sonnet, *J. Med. Chem.* **15**, 97 (1972).

Scheme 11.4. Friedel-Crafts

A. Intermolecular Reactions

1[a] Bromobenzene + $(CH_3CO)_2O$ $\xrightarrow{AlCl_3}$ p-$BrC_6H_4C(=O)CH_3$ (69–79%)

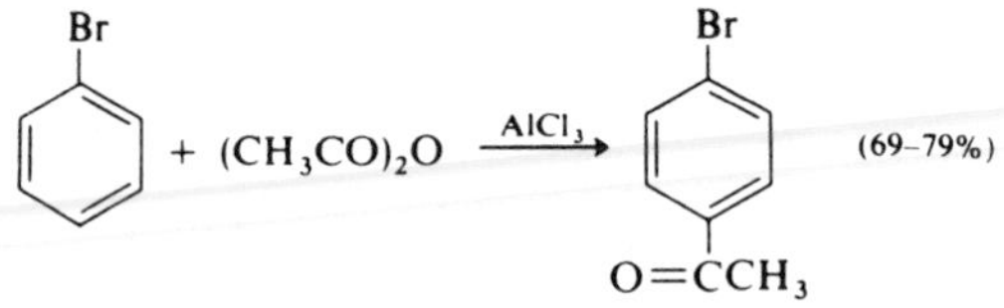

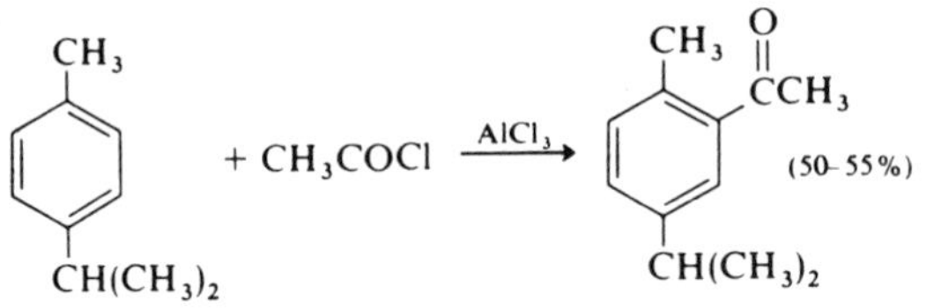

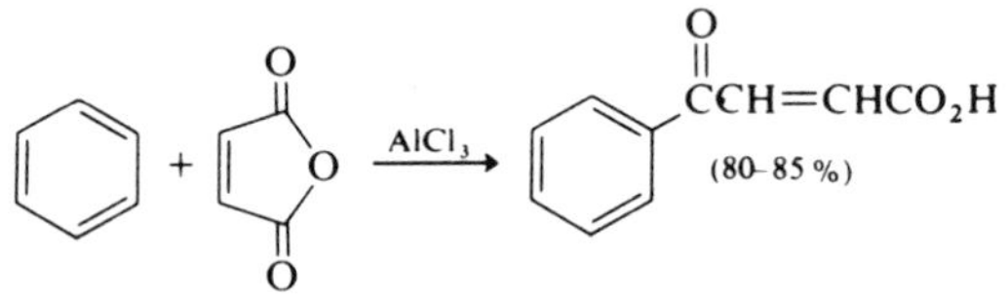

2[b] p-Cymene (CH_3, $CH(CH_3)_2$) + CH_3COCl $\xrightarrow{AlCl_3}$ 2-acetyl-p-cymene (CH_3, $\overset{O}{\parallel}CCH_3$, $CH(CH_3)_2$) (50–55%)

3[c] Benzene + maleic anhydride $\xrightarrow{AlCl_3}$ $C_6H_5\overset{O}{\overset{\parallel}{C}}CH{=}CHCO_2H$ (80–85%)

4[d] Acetanilide ($NHCOCH_3$) + $ClCCH_2Cl$ (O) $\xrightarrow{AlCl_3}$ p-$CH_3CONHC_6H_4C(=O)CH_2Cl$ (79–83%)

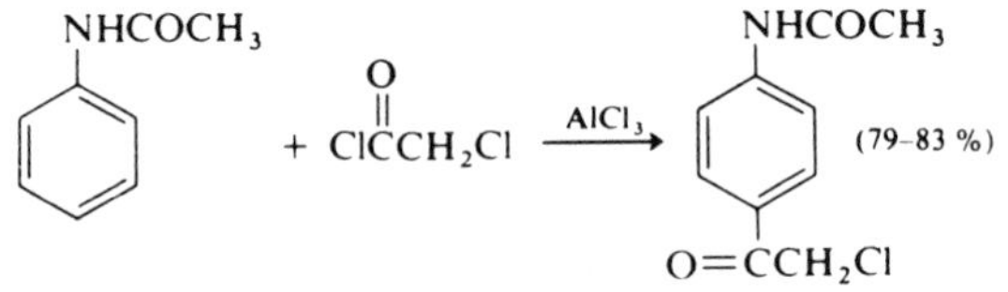

a. R. Adams and C. R. Noller, *Org. Synth.* **I**, 109 (1941).
b. C. F. H. Allen, *Org. Synth.* **II**, 3 (1943).
c. O. Grummitt, E. I. Becker, and C. Miesse, *Org. Synth.* **III**, 109 (1955).
d. J. L. Leiserson and A. Weissberger, *Org. Synth.* **III**, 183 (1955).

Alkylation is catalyzed by $SnCl_4$. The dichloromethyl group is hydrolyzed to a formyl group.

$$Ar{-}H \xrightarrow[SnCl_4]{Cl_2CHOR} ArCHCl_2 \xrightarrow{H_2O} ArCH{=}O$$

Another useful method for introducing formyl and acyl groups is the Vilsmeier-Haack reaction.[38] An *N,N*-dialkylamide and phosphorus oxychloride react to give a chloroiminium ion, which is the reactive electrophile.

$$R\overset{O}{\overset{\parallel}{C}}N(CH_3)_2 + POCl_3 \rightarrow R\overset{Cl}{\overset{|}{C}}{=}\overset{+}{N}(CH_3)_2$$

38. G. Martin and M. Martin, *Bull. Soc. Chim. Fr.*, 1637 (1963); S. Seshadri, *J. Sci. Ind. Res.* **32**, 128 (1973); C. Just, in *Iminium Salts in Organic Chemistry*, H. Böhme and H. G. Viehe (eds.), Vol. 9 in *Advances in Organic Chemistry: Methods and Results*, Wiley-Interscience, New York, 1976, pp. 225–342.

B. Intramolecular Friedel–Crafts Acylations

5[e] $C_6H_5(CH_2)_3CO_2H$ —PPA→ 1-tetralone (75–86%)

6[f] $C_6H_5(CH_2)_3COCl$ —$AlCl_3$→ 1-tetralone (74–91%)

7[g] benzene + γ-butyrolactone —$AlCl_3$→ 1-tetralone (91–96%)

8[h] m-$CH_3OC_6H_4CHClCH_2COCl$ —1) $AlCl_3$, 2) pyridine→ methoxyindenone isomers (85% total yield)

9[i] $(CH_3O)_2C_6H_2(CH(CH_3)_2)CH_2CH(CH_3)CH_2CO_2H$ —polyphosphate ester, CH_2Cl_2→ 6,7-dimethoxy-5-isopropyl-3-methyl-1-tetralone

e. L. Arsenijevic, V. Arsenijevic, A. Horeau, and J. Jacques, *Org. Synth.* **53**, 5 (1973).
f. E. L. Martin and L. F. Fieser, *Org. Synth.* **II**, 569 (1943).
g. C. E. Olson and A. F. Bader, *Org. Synth.* **IV**, 898 (1963).
h. M. B. Floyd and G. R. Allen, Jr., *J. Org. Chem.* **35**, 2647 (1970).
i. M. C. Venuti, *J. Org. Chem.* **46**, 3124 (1981).

This species acts as an electrophile in the absence of any added Lewis acid, but only rings with electron-releasing substituents are reactive.

Scheme 11.5 gives some examples of these acylation reactions.

11.1.4. Electrophilic Metalation

Aromatic compounds react with mercuric salts to give arylmercury compounds.[39] The reaction shows substituent effects that are characteristic of electrophilic aromatic substitution.[40] Mercuration is one of the few electrophilic aromatic

39. W. Kitching, *Organomet. Chem. Rev.* **3**, 35 (1968).
40. H. C. Brown and C. W. McGary, Jr., *J. Am. Chem. Soc.* **77**, 2300, 2310 (1955); A. J. Kresge and H. C. Brown, *J. Org. Chem.* **32**, 756 (1967); G. A. Olah, I. Hashimoto, and H. C. Lin, *Proc. Natl. Acad. Sci. U.S.A.* **74**, 4121 (1977).

Scheme 11.5. Other Electrophilic Aromatic

Chloromethylation

1[a]

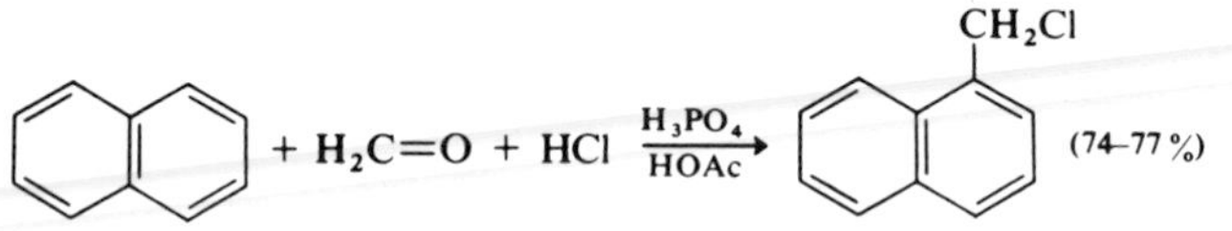

naphthalene + $H_2C{=}O$ + HCl $\xrightarrow[\text{HOAc}]{H_3PO_4}$ 1-(chloromethyl)naphthalene (CH_2Cl) (74–77 %)

Formylation with Carbon Monoxide

2[b]

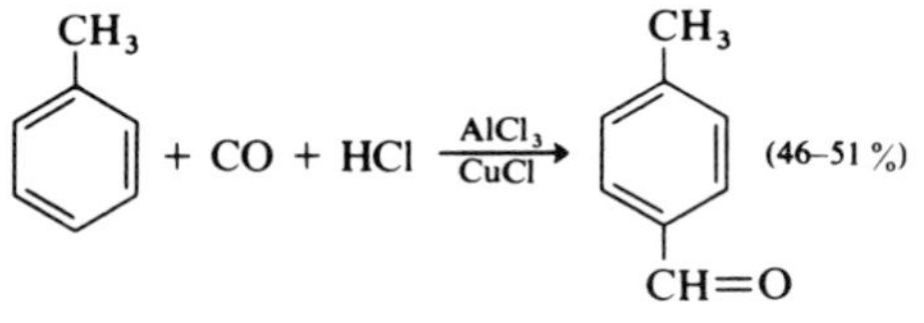

toluene (CH_3) + CO + HCl $\xrightarrow[\text{CuCl}]{AlCl_3}$ 4-methylbenzaldehyde (CH_3, $CH{=}O$) (46–51 %)

Acylation with Cyanide and Nitriles

3[c]

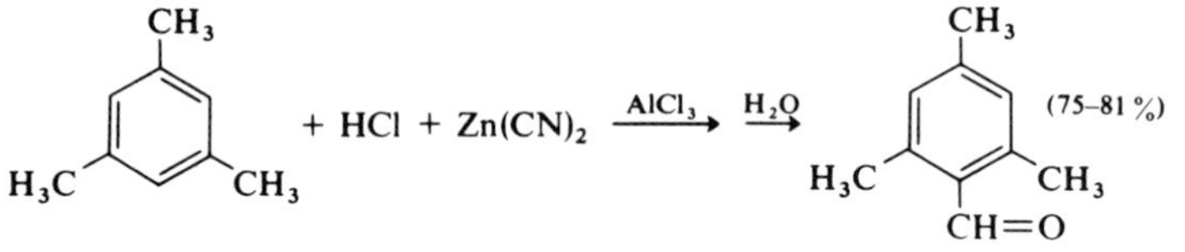

mesitylene (CH_3, H_3C, CH_3) + HCl + $Zn(CN)_2$ $\xrightarrow{AlCl_3}$ $\xrightarrow{H_2O}$ 2,4,6-trimethylbenzaldehyde (CH_3, H_3C, CH_3, $CH{=}O$) (75–81 %)

4[d]

phloroglucinol (OH, HO, OH) + CH_3CN $\xrightarrow[Zn(CN)_2]{HCl}$ $\xrightarrow{H_2O}$ 2,4,6-trihydroxyacetophenone (OH, $\overset{O}{\|}{C}CH_3$, HO, OH) (74–87 %)

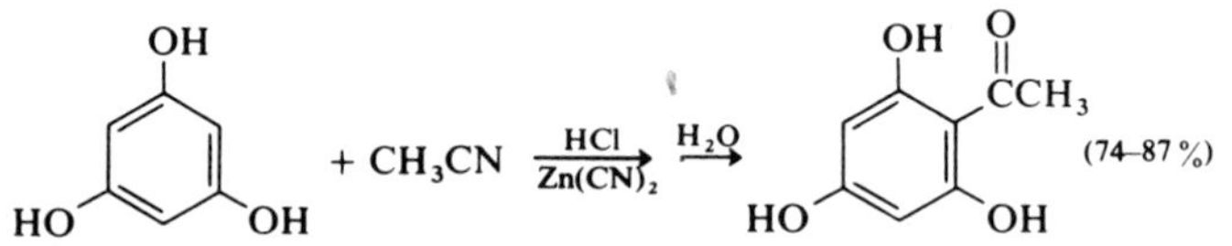

substitutions in which proton loss from the σ complex is rate-determining. Mercuration of benzene shows an isotope effect $k_H/k_D = 6$.[41] This result indicates that the σ complex must be reversibly formed.

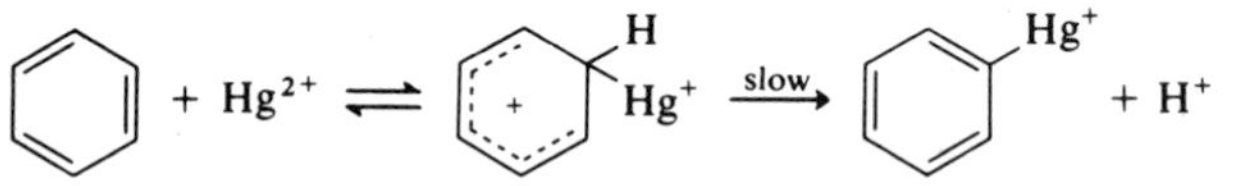

$$C_6H_6 + Hg^{2+} \rightleftharpoons [\sigma\text{ complex: } C_6H_6^{+}(H)(Hg^+)] \xrightarrow{\text{slow}} C_6H_5Hg^+ + H^+$$

Mercuric acetate or mercuric trifluoroacetate is the usual reagent.[42] The synthetic utility of the mercuration reaction derives from subsequent transformations of the arylmercury compounds. As indicated in Section 7.3.3, arylmercury compounds are only weakly nucleophilic, but the carbon–mercury bond is reactive toward various

41. C. Perrin and F. H. Westheimer, *J. Am. Chem. Soc.* **85**, 2773 (1963); A. J. Kresge and J. F. Brennan, *J. Org. Chem.* **32**, 752 (1967); C. W. Fung, M. Khorramdel-Vahad, R. J. Ranson, and R. M. G. Roberts, *J. Chem. Soc., Perkin Trans.* 2, 267 (1980).
42. A. J. Kresge, M. Dubeck and H. C. Brown, *J. Org. Chem.* **32**, 745 (1967); H. C. Brown and R. A. Wirkkala, *J. Am. Chem. Soc.* **88**, 1447, 1453, 1456 (1966).

Vilsmeier–Haack Acylation

5[e] $C_6H_5N(CH_3)_2 + HCON(CH_3)_2 \xrightarrow{POCl_3} \xrightarrow{H_2O} (CH_3)_2N\text{-}C_6H_4\text{-}CH{=}O$ (80–84%)

6[f] $C_6H_5N(CH_3)_2 + PhCONHPh \xrightarrow{POCl_3} \xrightarrow{H_2O} (CH_3)_2N\text{-}C_6H_4\text{-}C(=O)\text{-}C_6H_5$ (72–77%)

7[g] 2-Ethoxynaphthalene (OC_2H_5) $+ HCONHPh \xrightarrow{POCl_3} \xrightarrow{H_2O}$ 2-ethoxy-1-naphthaldehyde (CHO, OC_2H_5) (74–84%)

a. O. Grummitt and A. Buck, *Org. Synth.* **III**, 195 (1955).
b. G. H. Coleman and D. Craig, *Org. Synth.* **II**, 583 (1955).
c. R. C. Fuson, E. C. Horning, S. P. Rowland, and M. L. Ward, *Org. Synth.* **III**, 549 (1955).
d. K. C. Gulati, S. R. Seth, and K. Venkataraman, *Org. Synth.* **II**, 522 (1943).
e. E. Campaigne and W. L. Archer, *Org. Synth.* **IV**, 331 (1963).
f. C. D. Hurd and C. N. Webb, *Org. Synth.* **I**, 217 (1941).
g. J. H. Wood and R. W. Bost, *Org. Synth.* **IV**, 98 (1955).

electrophiles. The nitroso group can be introduced by reaction with nitrosyl chloride[43] or nitrosonium tetrafluoroborate[44] as the electrophile. Arylmercury compounds are also useful in certain palladium-catalyzed reactions discussed in Section 8.2.

Thallium(III), particularly as the trifluoroacetate salt, is also a reactive electrophilic metalating species, and a variety of synthetic schemes based on arylthallium intermediates have been devised.[45] Arylthallium compounds are converted to chlorides or bromides by reaction with the appropriate cupric halide.[46] Reaction with potassium iodide gives aryl iodides.[47] Fluorides are prepared by successive

43. L. I. Smith and F. L. Taylor, *J. Am. Chem. Soc.* **57**, 2460 (1935); S. Terabe, S. Kuruma, and R. Konaka, *J. Chem. Soc., Perkin Trans.* 2, 1252 (1973).
44. L. M. Stock and T. L. Wright, *J. Org. Chem.* **44**, 3467 (1979).
45. E. C. Taylor and A. McKillop, *Acc. Chem. Res.* **3**, 338 (1970).
46. S. Uemura, Y. Ikeda, and K. Ichikawa, *Tetrahedron* **28**, 5499 (1972).
47. A. McKillop, J. D. Hunt, M. J. Zelesko, J. S. Fowler, E. C. Taylor, G. McGillivray, and F. Kienzle, *J. Am. Chem. Soc.* **93**, 4841 (1971).

treatment with potassium fluoride and boron trifluoride.[48] Procedures for converting arylthallium compounds to nitriles and phenols have also been described.[49]

The thallium intermediates can be useful in directing substitution to specific positions when the site of thallation can be controlled in an advantageous way. The two principal means of control are chelation and the ability to effect thermal equilibration of arylthallium intermediates. Oxygen-containing groups normally direct thallation to the *ortho* position by a chelation effect. The thermodynamically favored position is normally the *meta* position, and heating the thallium derivatives of alkylbenzenes gives a predominance of the *meta* isomer.[50]

Both mercury and thallium compounds are very toxic, so great care is needed in their manipulation.

11.2. Nucleophilic Aromatic Substitution

Many synthetically important substitutions of aromatic compounds are effected by nucleophilic reagents. Unlike nucleophilic substitution at saturated carbon, aromatic nucleophilic substitution does not occur by a single-step mechanism. Three broad mechanistic classes can be recognized: addition–elimination, elimination–addition, and radical or electron-transfer processes. The most broadly useful intermediates for nucleophilic aromatic substitution are the aryl diazonium salts, and these compounds will be the first topic.

11.2.1. Aromatic Diazonium Ions as Synthetic Intermediates

Aryl diazonium ions are usually prepared by reaction of an aniline with nitrous acid, which is generated *in situ* from a nitrite salt.[51] Unlike aliphatic diazonium ions, which decompose very rapidly to molecular nitrogen and a carbocation (see Section 10.1), aryl diazonium ions are stable enough to exist in solution at room temperature and below. They can also be isolated as salts with non-nucleophilic anions, such as tetrafluoroborate. The steps in forming a diazonium ion involve

48. E. C. Taylor, E. C. Bigham, and D. K. Jonson, *J. Org. Chem.* **42**, 362 (1977).
49. S. Uemura, Y. Ikeda, and K. Ichikawa, *Tetrahedron* **28**, 3025 (1972); E. C. Taylor, H. W. Atland, R. H. Danforth, G. McGillivray, and A. McKillop, *J. Am. Chem. Soc.* **92**, 3520 (1970).
50. A. McKillop, J. D. Hunt, M. J. Zelesko, J. S. Fowler, E. C. Taylor, G. McGillivray, and F. Kienzle, *J. Am. Chem. Soc.* **93**, 4841 (1971).
51. H. Zollinger, *Azo and Diazo Chemistry*, Wiley-Interscience, New York, 1961; S. Patai (ed.), *The Chemistry of Diazonium and Diazo Groups*, Wiley, New York, 1978, Chapters 8, 11, and 14; H. Saunders and R. L. M. Allen, *Aromatic Diazo Compounds*, Third Edition, Edward Arnold, London, 1985.

addition of ^+NO to the amino group, followed by elimination of water.

$$ArNH_2 + HONO \xrightarrow{H^+} ArN(H){-}N{=}O + H_2O$$

$$ArN(H){-}N{=}O \rightarrow ArN{=}N{-}OH \xrightarrow{H^+} Ar\overset{+}{N}{\equiv}N + H_2O$$

In alkaline solution, diazonium ions are converted to diazoate anions, which are in equilibrium with diazooxides.[52]

$$Ar\overset{+}{N}{\equiv}N + 2\ ^-OH \rightarrow ArN{=}N{-}O^- + H_2O$$

$$ArN{=}N{-}O^- + Ar\overset{+}{N}{\equiv}N \rightleftharpoons ArN{=}N{-}O{-}N{=}NAr$$

In addition to the classical techniques for diazotization in aqueous solution, diazonium ions can be generated in organic solvents by reaction with alkyl nitrites.

$$RO{-}N{=}O + ArNH_2 \rightarrow ArN(H){-}N{=}O + ROH$$

$$ArN(H){-}N{=}O \rightleftharpoons ArN{=}N{-}OH \xrightarrow{H^+} Ar\overset{+}{N}{\equiv}N + H_2O$$

The great usefulness of aryl diazonium ions as synthetic intermediates results from the excellence of N_2 as a leaving group. There are at least three general mechanisms by which substitution can occur. One involves unimolecular decomposition of the diazonium ion, followed by capture of the resulting aryl cation by a nucleophile. The phenyl cation is very unstable (see Part A, Section 5.4) and therefore highly unselective.[53] Either the solvent or an anion can act as the nucleophile.

$$C_6H_5{-}\overset{+}{N}{\equiv}N \longrightarrow C_6H_5^+ + N_2$$

$$C_6H_5^+ + X^- \longrightarrow C_6H_5{-}X$$

Another possible mechanism for substitution is adduct formation followed by collapse of the adduct with loss of nitrogen.

$$C_6H_5{-}\overset{+}{N}{\equiv}N + X^- \rightleftharpoons C_6H_5{-}N{=}N{-}X \longrightarrow C_6H_5{-}X + N_2$$

52. E. S. Lewis and M. P. Hanson, *J. Am. Chem. Soc.* **89**, 6268 (1967).
53. C. G. Swain, J. E. Sheats, and K. G. Harbison, *J. Am. Chem. Soc.* **97**, 783 (1975).

The third mechanism involves electron-transfer processes. This mechanism is particularly likely to operate in reactions where copper salts are used as catalysts.[54]

$$C_6H_5-\overset{+}{N}\equiv N + Cu(I)X_2^- \longrightarrow C_6H_5-\ddot{N}=\dot{N}: + Cu(II)X_2$$

$$C_6H_5-\ddot{N}=\dot{N}: + Cu(II)X_2 \longrightarrow C_6H_5-X + Cu(I)X + N_2$$

Examples of the three mechanistic types are, respectively: (a) hydrolysis of diazonium salts to phenols[55]; (b) reaction with azide ion to form aryl azides[56]; and (c) reaction with cuprous halides to form aryl chlorides or bromides.[57] In the paragraphs which follow, the synthetically useful reactions of diazonium intermediates are considered. The reactions are organized on the basis of the group which is introduced, rather than on the mechanism involved. It will be seen that the reactions which are discussed fall into one of the three general mechanistic types.

Replacement of a nitro or amino group by hydrogen is sometimes required as a sequel to a synthetic operation in which the substituent has been used to control the position selectivity of a transformation. The best reagents for reductive dediazonation are hypophoshorous acid, H_3PO_2,[58] and $NaBH_4$.[59] The reduction by hypophosphorous acid is substantially improved by catalysis by cuprous oxide.[60] Reductive dediazonation by H_3PO_2 proceeds by one-electron reduction followed by loss of nitrogen and formation of the phenyl radical.[61]

initiation $$Ar\overset{+}{N}\equiv N + e^- \rightarrow Ar\cdot + N_2$$

propagation $$Ar\cdot + H_3PO_2 \rightarrow Ar{-}H + [H_2PO_2\cdot]$$

$$Ar\overset{+}{N}\equiv N + [H_2PO_2\cdot] \rightarrow Ar\cdot + N_2 + [H_2PO_2^+]$$

$$[H_2PO_2^+] + H_2O \rightarrow H_3PO_3 + H^+$$

An alternative method for reductive dediazonation involves *in situ* diazotization by an alkyl nitrite in dimethylformamide.[62] This is probably a chain reaction with the

54. T. Cohen, R. J. Lewarchik, and J. Z. Tarino, *J. Am. Chem. Soc.* **97**, 783 (1975).
55. E. S. Lewis, L. D. Hartung, and B. M. McKay, *J. Am. Chem. Soc.* **91**, 419 (1969).
56. C. D. Ritchie and D. J. Wright, *J. Am. Chem. Soc.* **93**, 2429 (1971); C. D. Ritchie and P. O. I. Virtanen, *J. Am. Chem. Soc.* **94**, 4966 (1972).
57. J. K. Kochi, *J. Am. Chem. Soc.* **79**, 2942 (1957); S. C. Dickerman, K. Weiss, and A. K. Ingberman, *J. Am. Chem. Soc.* **80**, 1904 (1958).
58. N. Kornblum, *Org. React.* **2**, 262 (1944).
59. J. B. Hendrickson, *J. Am. Chem. Soc.* **83**, 1251 (1961).
60. S. Korzeniowski, L. Blum, and G. W. Gokel, *J. Org. Chem.* **42**, 1469 (1977).
61. N. Kornblum, G. D. Cooper, and J. E. Taylor, *J. Am. Chem. Soc.* **72**, 3013 (1950).
62. M. P. Doyle, J. F. Dellaria, Jr., B. Siegfried, and S. W. Bishop, *J. Org. Chem.* **42**, 3494 (1977).

solvent acting as a hydrogen atom donor.

$$\text{initiation} \qquad Ar\overset{+}{N}{\equiv}N + e^- \rightarrow Ar\cdot + N_2$$

$$\text{propagation} \qquad Ar\cdot + HC(=O)N(CH_3)_2 \rightarrow Ar{-}H + \cdot C(=O)N(CH_3)_2$$

$$Ar\overset{+}{N}{\equiv}N \; \cdot C(=O)N(CH_3)_2 \rightarrow Ar\cdot + N_2 + C{\equiv}O + CH_3\overset{+}{N}H{=}CH_2$$

Aryl diazonium ions can be converted to phenols by heating in water. Under these conditions, formation of a phenyl cation probably occurs.

$$Ar\overset{+}{N}{\equiv}N \rightarrow Ar^+ + N_2 \xrightarrow{H_2O} ArOH$$

By-products from capture of nucleophilic anions may also be observed.[53] Phenols can be formed under milder conditions by an alternative redox mechanism.[63] The reaction is initiated by cuprous oxide, which effects reduction and decomposition to an aryl radical. The reaction is run in the presence of Cu(II) salts. The radical is captured by Cu(II) and oxidized to the phenol. This procedure is very rapid and gives good yields of phenols over a range of structural types.

$$Ar\overset{+}{N}{\equiv}N + Cu(I) \rightarrow Ar\cdot + N_2 + Cu(II)$$

$$Ar\cdot + Cu(II) \rightarrow [Ar{-}Cu]^{2+} \xrightarrow{H_2O} ArOH + Cu(I) + H^+$$

Replacement of diazonium groups by halide is a valuable alternative to direct halogenation for preparation of aryl halides. Aryl bromides and chlorides are usually prepared by reaction with the appropriate Cu(I) salt, a process which is known as the Sandmeyer reaction. Under the classic conditions, the diazonium salt is added to a hot acidic solution of the cuprous halide.[64] It is also possible to convert anilines to aryl halides by generating the diazonium ion *in situ*. Reaction of anilines with alkyl nitrites and Cu(II) halides in acetonitrile gives good yields of aryl chlorides and bromides.[65] Examples of these reactions are given in section C of Scheme 11.6.

The Sandmeyer reaction is formulated as proceeding by reduction of the diazonium ion by Cu(I) and halide transfer from copper.

$$Ar\overset{+}{N}{\equiv}N + [CuX_2]^- \rightarrow Ar{-}CuX_2 + N_2$$

$$Ar{-}CuX_2 \rightarrow ArX + CuX$$

Diazonium salts can also be converted to halides by processes involving aryl free radicals. In basic solution, aryl diazonium ions are converted to radicals via

63. T. Cohen, A. G. Dietz, Jr., and J. R. Miser, *J. Org. Chem.* **42**, 2053 (1977).
64. W. A. Cowdrey and D. S. Davies, *Q. Rev. Chem. Soc.* **6**, 358 (1952); H. H. Hodgson, *Chem. Rev.* **40**, 251 (1947).
65. M. P. Doyle, B. Siegfried, and J. F. Dellaria, Jr., *J. Org. Chem.* **42**, 2426 (1977).

A. Replacement by Hydrogen

1[a] 2,4,6-tribromoaniline → 1) HONO; 2) C_2H_5OH, Δ → 1,3,5-tribromobenzene (74–77%)

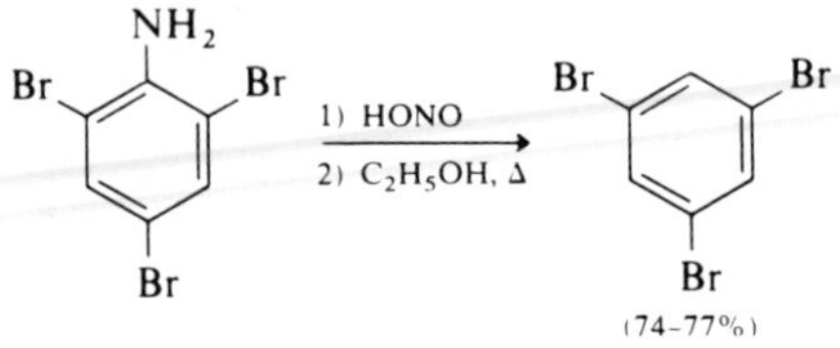

2[b] 3,3'-dimethylbenzidine → 1) HONO; 2) H_3PO_2 → 3,3'-dimethylbiphenyl (76–82%)

3[c] 2,5-dichlorobenzenediazonium $\overset{+}{N}_2BF_4$ → H_3PO_2, Cu_2O → 1,3-dichlorobenzene (97%)

4[d] 4,5-dichloro-2-nitroaniline → $(CH_3)_3CONO$, DMF → 3,4-dichloronitrobenzene (68%)

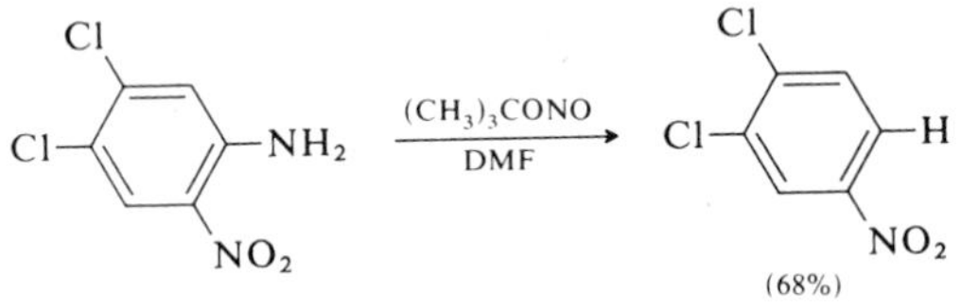

B. Replacement by Hydroxyl

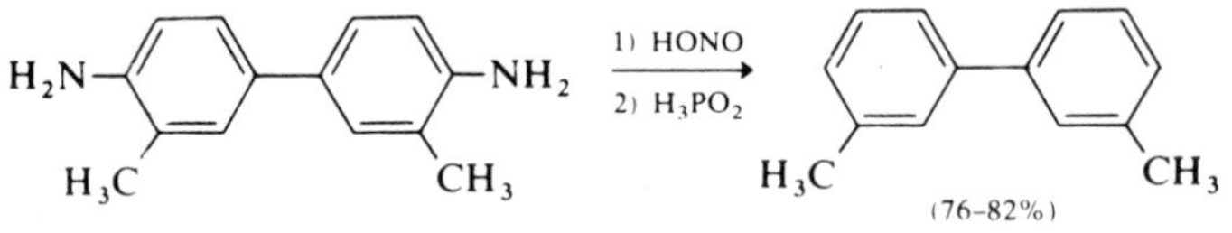

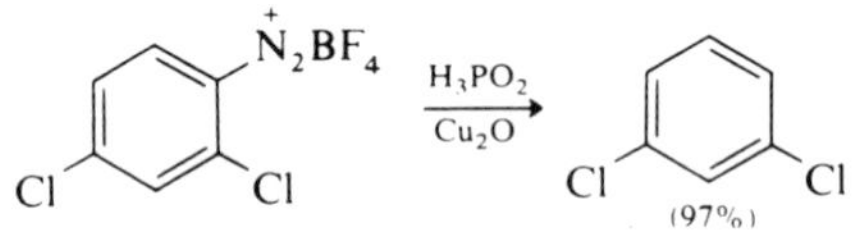

4[d] 2-bromo-4-methylaniline → 1) HONO; 2) H_2O, Δ → 2-bromo-4-methylphenol (80–92%)

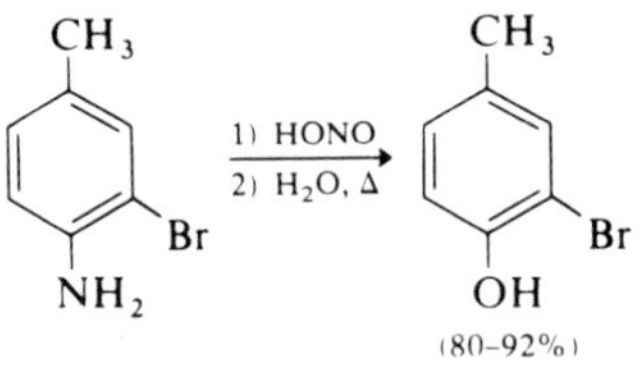

5[e] 2-methyl-5-nitroaniline → 1) HONO; 2) $Cu(NO_3)_2$, CuO → 2-methyl-5-nitrophenol (95%)

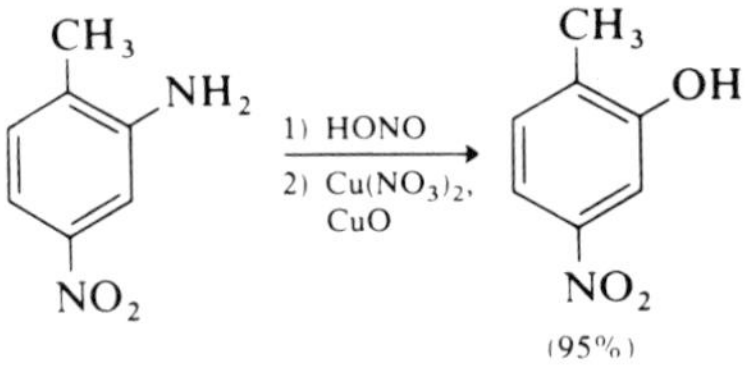

C. Replacement by Halogen

6[f] 3-aminobenzaldehyde (CH=O) → 1) HONO; 2) Cu_2Cl_2 → 3-chlorobenzaldehyde (75–79%)

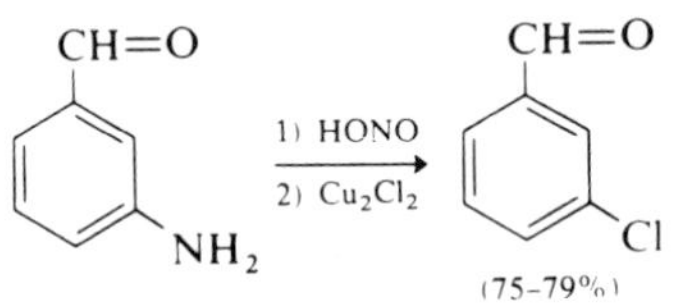

7[g] 2,6-dinitroaniline → 1) HONO; 2) Cu_2Cl_2 → 2-chloro-1,3-dinitrobenzene (71–74%)

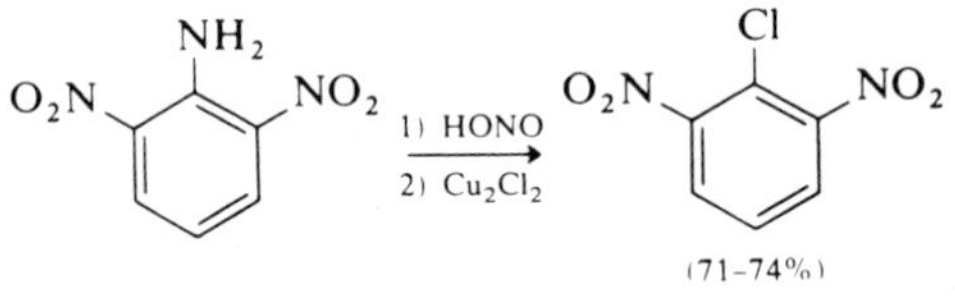

8[h] $H_3C-C_6H_4-NH_2 \xrightarrow[CuBr_2]{RONO} H_3C-C_6H_4-Br$ (76%)

9[i] $3-ClC_6H_4\overset{+}{N}_2BF_4^- \xrightarrow[BrCCl_3]{NaOAc,\ 18\text{-crown-6}} 3-ClC_6H_4Br$ (88%)

10[j] 2-Bromoaniline $\xrightarrow[2)\ KI]{1)\ HONO}$ 1-bromo-2-iodobenzene (72–83%)

11[k] 2-Bromoaniline $\xrightarrow[2)\ HPF_6;\ 3)\ \Delta]{1)\ HONO}$ 1-bromo-2-fluorobenzene (73–75%)

12[l] $H_2N-C_6H_4-C_6H_4-NH_2 \xrightarrow[2)\ HBF_4;\ 3)\ \Delta]{1)\ HONO} F-C_6H_4-C_6H_4-F$ (54–56%)

D. Replacement by Other Anions

13[m] 2-Methylaniline (NH_2, CH_3) $\xrightarrow[2)\ CuCN]{1)\ HONO}$ 2-methylbenzonitrile ($C{\equiv}N$, CH_3) (64–70%)

14[n] 2-Aminobiphenyl (NH_2) $\xrightarrow[2)\ NaN_3]{1)\ HONO}$ 2-azidobiphenyl (N_3) (88%)

a. G. H. Coleman and W. F. Talbot, *Org. Synth.* **II**, 592 (1943).
b. N. Kornblum, *Org. Synth.* **III**, 295 (1955).
c. S. H. Korzeniowski, L. Blum, and G. W. Gokel, *J. Org. Chem.* **42**, 1469 (1977).
d. M. P. Doyle, J. F. Dellaria, Jr., B. Siegfried, and S. W. Bishop, *J. Org. Chem.* **42**, 3494 (1977).
e. H. E. Ungnade and E. F. Orwoll, *Org. Synth.* **III**, 130 (1955).
f. T. Cohen, A. G. Dietz, Jr., and J. R. Miser, *J. Org. Chem.* **42**, 2053 (1977).
g. J. S. Buck and W. S. Ide, *Org. Synth.* **II**, 130 (1943).
h. F. D. Gunstone and S. H. Tucker, *Org. Synth.* **IV**, 160 (1963).
i. M. P. Doyle, B. Siegfried, and J. F. Dellaria, Jr., *J. Org. Chem.* **42**, 2426 (1977).
j. S. H. Korzeniowski and G. W. Gokel, *Tetrahedron Lett.*, 3519 (1977).
k. H. Heaney and I. T. Millar, *Org. Synth.* **40**, 105 (1960).
l. K. G. Rutherford and W. Redmond, *Org. Synth.* **43**, 12 (1963).
m. G. Schiemann and W. Winkelmuller, *Org. Synth.* **II**, 188 (1943).
n. H. T. Clarke and R. R. Read, *Org. Synth.* **I**, 514 (1941).
o. P. A. S. Smith and B. B. Brown, *J. Am. Chem. Soc.* **73**, 2438 (1957).

diazoxides.[66]

$$2\,Ar\overset{+}{N}{\equiv}N + 2\,^{-}OH \rightarrow ArN{=}N{-}O{-}N{=}NAr + H_2O$$

$$ArN{=}N{-}O{-}N{=}NAr \rightarrow ArN{=}N{-}O\cdot + Ar\cdot + N_2$$

$$Ar\cdot + S{-}X \rightarrow ArX + S\cdot$$

The reaction can be carried out efficiently by using aryl diazonium tetrafluoroborates with crown ethers, polyethers, or phase transfer catalysts.[67] In solvents that can act as halogen atom donors, the radicals react to give aryl halides. Bromotrichloromethane gives aryl bromides and methyl iodide gives iodides.[68] Diazonium ions can also be generated by *in situ* methods. Under these conditions, bromoform and bromotrichloromethane have been used as bromine donors and carbon tetrachloride is the best chlorine donor.[69]

Fluorine substituents can also be introduced via diazonium ions. One procedure is to isolate aryl diazonium tetrafluoroborates. These decompose thermally to give aryl fluorides.[70] This reaction probably involves formation of an aryl cation, which abstracts fluoride ion from the tetrafluoroborate anion.[71]

$$Ar\overset{+}{N}{\equiv}N + BF_4^- \rightarrow ArF + N_2 + BF_3$$

Hexafluorophosphate salts behave similarly.[72] The diazonium tetrafluoroborates can be prepared either by precipitation from an aqueous solution by fluoroboric acid[73] or by anhydrous diazotization in ether, THF, or acetonitrile with *t*-butyl nitrate and boron trifluoride.[74]

Aryl diazonium ions are converted to iodides in high yield by reaction with iodide salts. This reaction is probably initiated by reduction of the diazonium ion by iodide. The aryl radical then abstracts iodine from either I_2 or I_3^-. A chain mechanism then proceeds which consumes I^- and ArN_2^+.[75] Evidence for the involvement of radicals includes the isolation of cyclized products from *o*-allyl derivatives.

$$Ar\overset{+}{N}{\equiv}N + I^- \rightarrow Ar\cdot + N_2 + I\cdot \qquad 2I\cdot \rightarrow I_2$$

$$Ar\cdot + I_3^- \rightarrow ArI + I_2^{\overline{\cdot}}$$

$$Ar\overset{+}{N}{\equiv}N + I_2^{\overline{\cdot}} \rightarrow Ar\cdot + N_2 + I_2 \qquad I_2 + I^- \rightarrow I_3^-$$

66. C. Rüchardt and B. Freudenberg, *Tetrahedron Lett.*, 3623 (1964); C. Rüchardt and E. Merz, *Tetrahedron Lett.*, 2431 (1964).
67. S. H. Korzeniowski and G. W. Gokel, *Tetrahedron Lett.*, 1637 (1977).
68. S. H. Korzeniowski and G. W. Gokel, *Tetrahedron Lett.*, 3519 (1977); R. A. Bartsch and I. W. Wang, *Tetrahedron Lett.*, 2503 (1979).
69. J. I. G. Cadogan, D. A. Roy, and D. M. Smith, *J. Chem. Soc., C*, 1249 (1966).
70. A. Roe, *Org. React.* **5**, 193 (1949).
71. C. G. Swain and R. J. Rogers, *J. Am. Chem. Soc.* **97**, 799 (1975).
72. M. S. Newman and R. H. B. Galt, *J. Org. Chem.* **25**, 214 (1960).
73. E. B. Starkey, *Org. Synth.* **II**, 225 (1943); G. Schiemann and W. Winkelmuller, *Org. Synth.* **II**, 299 (1943).
74. M. P. Doyle and W. J. Bryker, *J. Org. Chem.* **44**, 1572 (1979).
75. P. R. Singh and R. Kumar, *Aust. J. Chem.* **25**, 2133 (1972); A. Abeywickrema and A. L. J. Beckwith, *J. Org. Chem.* **52**, 2568 (1987).

Cyano and azido groups are also readily introduced via diazonium intermediates. The former process involves a copper-catalyzed reaction analogous to the Sandmeyer reaction. Reaction of diazonium salts with azide ion gives adducts which smoothly decompose to nitrogen and the aryl azide.[56]

$$\mathrm{Ar\overset{+}{N}{\equiv}N} + \mathrm{{}^{-}N{=}\overset{+}{N}{=}N^{-}} \rightarrow \mathrm{ArN{=}N{-}N{=}\overset{+}{N}{=}N^{-}} \rightarrow \mathrm{ArN{=}\overset{+}{N}{=}N^{-}} + \mathrm{N_2}$$

Aryl thiolates react with aryl diazonium ions to give diaryl sulfides. This reaction is believed to be a radical chain process, with a mechanism similar to that for the reaction of diazonium ions with iodide ion.[76]

initiation

$$\mathrm{Ar\overset{+}{N}{\equiv}N} + \mathrm{PhS^{-}} \rightarrow \mathrm{ArN{=}NSPh}$$

$$\mathrm{ArN{=}NSPh} \rightarrow \mathrm{Ar\cdot} + \mathrm{N_2} + \mathrm{PhS\cdot}$$

propagation

$$\mathrm{Ar\cdot} + \mathrm{PhS^{-}} \rightarrow \mathrm{Ar\overset{\cdot-}{S}Ph}$$

$$\mathrm{Ar\overset{\cdot-}{S}Ph} + \mathrm{Ar\overset{+}{N}{\equiv}N} \rightarrow \mathrm{ArSPh} + \mathrm{Ar\cdot} + \mathrm{N_2}$$

Scheme 11.6 gives some examples of the various substitution reactions of aryl diazonium ions.

Aryl diazonium ions can also be used to form certain types of carbon–carbon bonds. The copper-catalyzed reaction of aryl diazonium ions with conjugated alkenes results in arylation of the alkene. This is known as the *Meerwein arylation reaction.*[77] The reaction sequence is initiated by reduction of the diazonium ion by Cu(I). The aryl radical adds to the alkene to give a new arylated radical. The final step is an oxidation/ligand transfer that takes place in the copper coordination sphere. An alternative course is oxidation/deprotonation, which gives a styrene derivative.

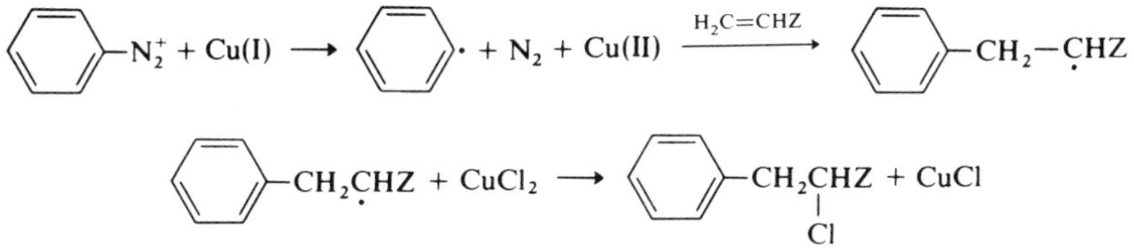

$$\mathrm{C_6H_5{-}N_2^{+}} + \mathrm{Cu(I)} \longrightarrow \mathrm{C_6H_5\cdot} + \mathrm{N_2} + \mathrm{Cu(II)} \xrightarrow{\mathrm{H_2C{=}CHZ}} \mathrm{C_6H_5{-}CH_2{-}\dot{C}HZ}$$

$$\mathrm{C_6H_5{-}CH_2\dot{C}HZ} + \mathrm{CuCl_2} \longrightarrow \mathrm{C_6H_5{-}CH_2CH(Cl)Z} + \mathrm{CuCl}$$

The reaction gives better yields with dienes, styrenes, or alkenes substituted with electron-withdrawing groups than with simple alkenes. These groups increase the rate of capture of the aryl radical. The standard conditions for the Meerwein arylation reaction employ aqueous solutions of diazonium ions prepared in the usual way. Conditions for *in situ* diazotization by *t*-butyl nitrite in the presence of $CuCl_2$ and acrylonitrile or styrene are also effective.[78]

Reduction of aryl diazonium ions by Ti(III) in the presence of α,β-unsaturated ketones and aldehydes leads to β arylation and formation of the saturated ketone

76. A. N. Abeywickerma and A. L. J. Beckwith, *J. Am. Chem. Soc.* **108**, 8227 (1986).
77. C. S. Rondestvedt, Jr., *Org. React.* **11**, 189 (1960); **24**, 225 (1976); A. V. Dombrovskii, *Russ. Chem. Rev.* (Engl. Transl.), **53**, 943 (1984).
78. M. P. Doyle, B. Siegfried, R. C. Elliot, and J. F. Dellaria, Jr., *J. Org. Chem.* **42**, 2431 (1977).

Scheme 11.7. Meerwein Arylation Reactions

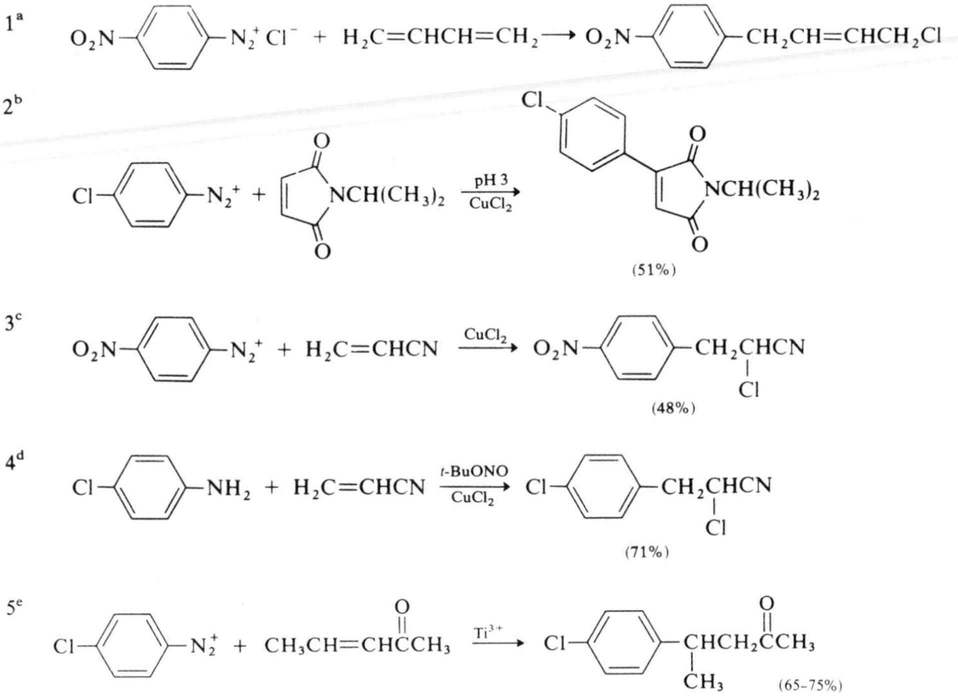

1[a] $O_2N-C_6H_4-N_2^+Cl^- + H_2C{=}CHCH{=}CH_2 \longrightarrow O_2N-C_6H_4-CH_2CH{=}CHCH_2Cl$

2[b] $Cl-C_6H_4-N_2^+$ + N-isopropylmaleimide ($NCH(CH_3)_2$) $\xrightarrow[CuCl_2]{pH\ 3}$ 3-(4-chlorophenyl)-N-isopropylmaleimide ($NCH(CH_3)_2$) (51%)

3[c] $O_2N-C_6H_4-N_2^+ + H_2C{=}CHCN \xrightarrow{CuCl_2} O_2N-C_6H_4-CH_2CH(Cl)CN$ (48%)

4[d] $Cl-C_6H_4-NH_2 + H_2C{=}CHCN \xrightarrow[CuCl_2]{t\text{-}BuONO} Cl-C_6H_4-CH_2CH(Cl)CN$ (71%)

5[e] $Cl-C_6H_4-N_2^+ + CH_3CH{=}CHCCH_3(=O) \xrightarrow{Ti^{3+}} Cl-C_6H_4-CH(CH_3)CH_2C(=O)CH_3$ (65–75%)

a. G. A. Ropp and E. C. Coyner, *Org. Synth.* **IV**, 727 (1963).
b. C. S. Rondestvedt, Jr., and O. Vogl, *J. Am. Chem. Soc.* **77**, 2313 (1955).
c. C. F. Koelsch, *J. Am. Chem. Soc.* **65**, 57 (1943).
d. M. P. Doyle, B. Siegfried, R. C. Elliott, and J. F. Dellaria, Ur., *J. Org. Chem.* **42**, 2431 (1977).
e. A. Citterio and E. Vismara, *Synthesis*, 291 (1980); A. Citterio, *J. Org. Synth.* **62**, 67 (1984).

or aldehyde. The early steps in this reaction parallel those in the copper-catalyzed reaction. However, rather than being oxidized, the radical formed by the addition step is reduced by Ti(III).[79]

$$Ar\overset{+}{N}{\equiv}N + Ti(III) \rightarrow Ar\cdot + N_2 + Ti(IV)$$

$$Ar\cdot + RCH{=}CHCR(=O) \rightarrow ArCH(R)\dot{C}HCR(=O) \xrightarrow[H^+]{Ti(III)} ArCH(R)CH_2CR(=O)$$

Scheme 11.7 illustrates some typical arylations of alkenes by diazonium ions.

11.2.2. Substitution by the Addition–Elimination Mechanism

The addition of a nucleophile to an aromatic ring, followed by elimination of a substituent, results in nucleophilic substitution. The highest energetic requirement

79. A. Citterio and E. Vismara, *Synthesis*, 191 (1980); A. Citterio, A. Cominelli, and F. Bonavoglia, *Synthesis*, 308 (1986).

for this mechanism is formation of the addition intermediate. The addition step is greatly facilitated by strongly electron-attracting substituents, so that nitroaromatics are the best substrates for nucleophilic aromatic substitution. Other electron-attracting groups such as cyano, acetyl, and trifluoromethyl also enhance reactivity.

$$O_2N-C_6H_4-X + Y^- \longrightarrow {}^-O_2\overset{+}{N}{=}C_6H_4(X)(Y) \longrightarrow O_2N-C_6H_4-Y + X^-$$

Nucleophilic substitution occurs when there is a potential leaving group present at the carbon at which addition occurs. Although halides are the most common leaving groups, alkoxy, cyano, nitro, and sulfonyl groups can also be displaced. The ordering of leaving group ability does not necessarily parallel that found for nucleophilic substitution at saturated carbon. As a particularly striking example, fluoride is often a better leaving group than the other halogens in nucleophilic aromatic substitution. The relative reactivity of the *p*-halonitrobenzenes toward sodium methoxide at 50°C is F (312) $\gg$ Cl (1) $>$ Br (0.74) $>$ I (0.36).[80] A principal reason for the order I $>$ Br $>$ Cl $>$ F in S_N2 reactions is that the carbon–halogen bond strength increases from I to F. The carbon–halogen bond strength is not so important a factor in nucleophilic aromatic substitution because bond breaking is not ordinarily part of the rate-determining step. Furthermore, the highly electronegative fluorine stabilizes the addition transition state more effectively than the other halogens.

There are not many successful examples of arylation of carbanions by nucleophilic aromatic substitution. A major limitation is the fact that aromatic nitro compounds often react with carbanions by electron-transfer processes.[81]

2-Halopyridines are excellent reactants for nucleophilic aromatic substitution.[82] Substitution reactions also occur readily for other heterocyclic systems, such as 2-haloquinolines and 1-haloisoquinolines, in which a potential leaving group is adjacent to a pyridine-type nitrogen. 4-Halopyridines and related heterocyclic compounds can also undergo substitution by nucleophilic addition–elimination.

$$\text{2-chloropyridine} + NaOC_2H_5 \longrightarrow \text{2-}OC_2H_5\text{-pyridine}$$ Ref. 83

Scheme 11.8 gives some examples of nucleophilic aromatic substitution reactions.

80. G. P. Briner, J. Mille, M. Liveris, and P. G. Lutz, *J. Chem. Soc.*, 1265 (1954).
81. R. D. Guthrie, in *Comprehensive Carbanion Chemistry, Part A*, E. Buncel and T. Durst (ed.), Elsevier, Amsterdam, 1980, Chapter 5.
82. H. E. Mertel, in *Heterocyclic Compounds*, Vol. 14, Part 2, E. Klingsberg (ed.), Wiley-Interscience, New York, 1961; M. M. Boudakian, in *Heterocyclic Compounds*, Vol. 14, Part 2, Supplement, R. A. Abramovitch (ed.), Wiley-Interscience, New York, 1974, Chapter 6; B. C. Uff, in *Comprehensive Heterocyclic Chemistry*, Vol. 2A, A. J. Boulton and A. McKillop (ed.), Pergamon Press, Oxford, 1984, Chapter 2.06.
83. N. Al-Awadi, J. Ballam, R. R. Hemblade, and R. Taylor, *J. Chem. Soc., Perkin Trans. 2*, 1175 (1982).

Scheme 11.8. Nucleophilic Aromatic Substitution

1[a]
NO_2 Cl + H N → NO_2 N (94%)

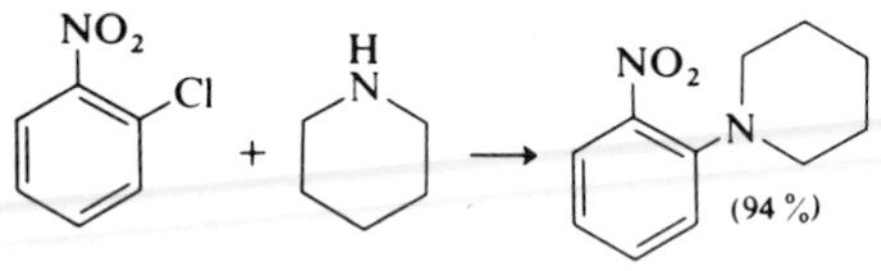

2[b]
NO_2 F + H N CH_3 → O_2N– –N CH_3 (85%)

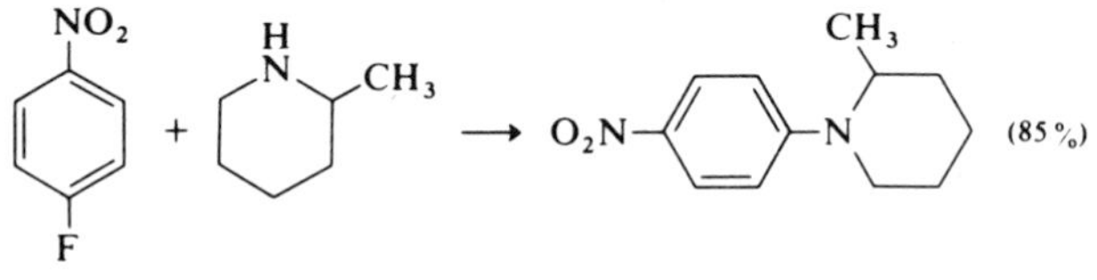

3[c]
$F-\overset{O}{\overset{\|}{C}}CH_3 + (CH_3)_2NH \longrightarrow (CH_3)_2N-\overset{O}{\overset{\|}{C}}CH_3$ (96%)

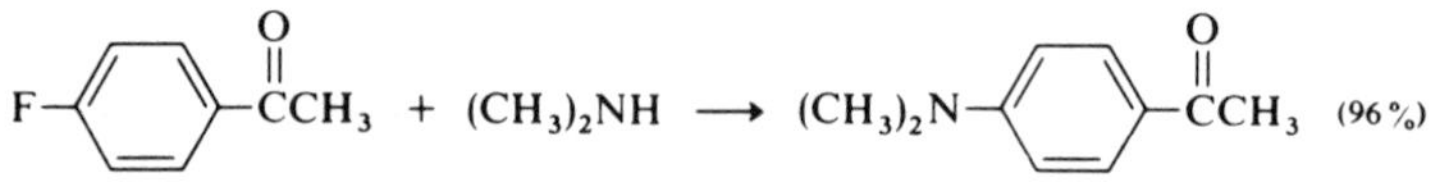

4[d]
NO_2 $N\equiv C$ NO_2 + CH_3O^- → OCH_3 $N\equiv C$ NO_2

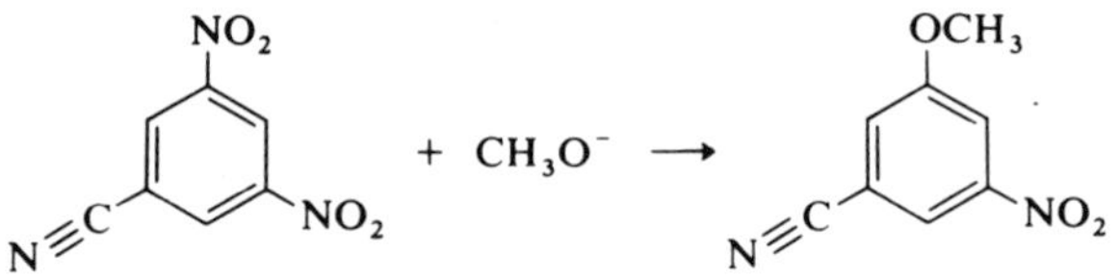

5[e]
O_2N– –Cl + –OH $\xrightarrow{KOH}$ O_2N– –O– (80–82%)

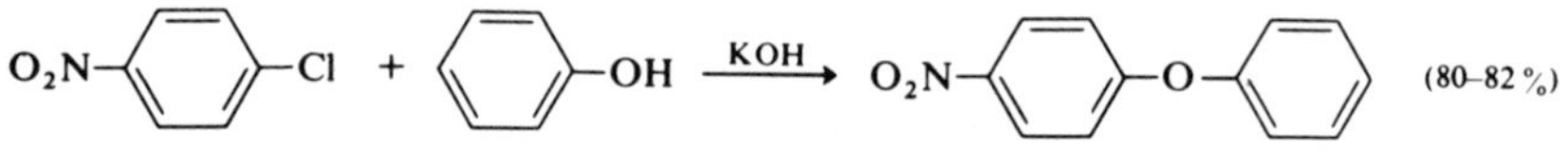

6[f]
N + Cl– –NO_2 O_2N → O NO_2 NO_2 (92%)

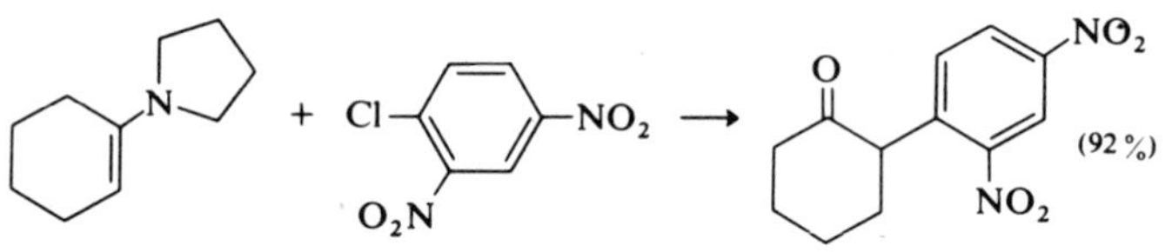

7[g]
$N\equiv C\underset{-}{C}HCO_2C_2H_5$ + Cl– –NO_2 O_2N → $H_5C_2O_2C\overset{N\equiv C}{\overset{|}{C}}H$– –$NO_2$ O_2N

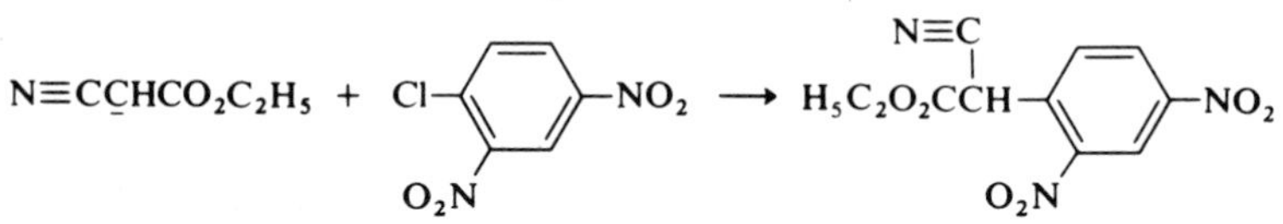

a. S. D. Ross and M. Finkelstein, *J. Am. Chem. Soc.* **85**, 2603 (1963).
b. F. Pietra and F. Del Cima, *J. Org. Chem.* **33**, 1411 (1968).
c. H. Bader, A. R. Hansen, and F. J. McCarty, *J. Org. Chem.* **31**, 2319 (1966).
d. E. J. Fendler, J. H. Fendler, N. L. Arthur, and C. E. Griffin, *J. Org. Chem.* **37**, 812 (1972).
e. R. O. Brewster and T. Groening, *Org. Synth.* **II**, 445 (1943).
f. M. E. Kuehne, *J. Am. Chem. Soc.* **84**, 837 (1962).
g. H. R. Snyder, E. P. Merica, C. G. Force, and E. G. White, *J. Am. Chem. Soc.* **80**, 4622 (1958).

The elimination-addition mechanism involves a highly unstable intermediate, which is called *dehydrobenzene* or *benzyne*.[84]

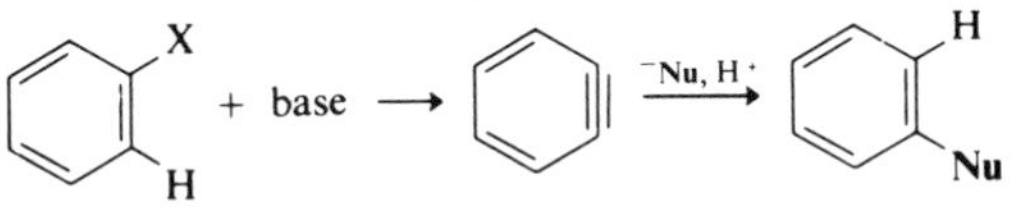

A characteristic feature of this mechanism is that the entering nucleophile does not necessarily become bound to the carbon to which the leaving group was bound.

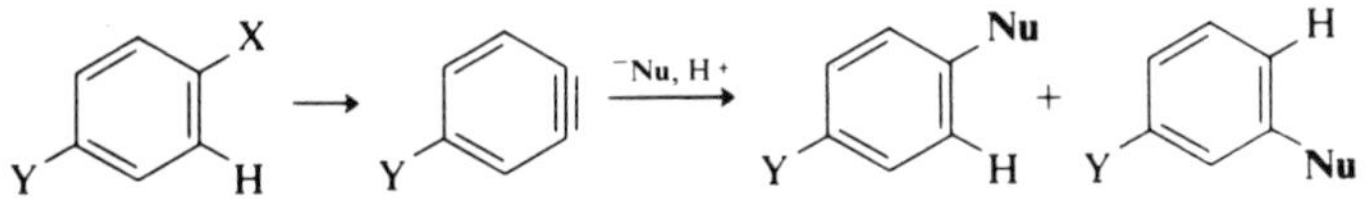

The elimination-addition mechanism is facilitated by electronic effects that favor removal of a hydrogen from the ring as a proton. Relative reactivity also depends on the leaving group. The order Br > I > Cl ≫ F has been established in the reaction of aryl halides with KNH_2 in liquid ammonia.[85] This order has been interpreted as representing a balance of two effects. The inductive order favoring proton removal would be F > Cl > Br > I, but this effect is largely overwhelmed by the ease of bond breaking, which is in the order I > Br > Cl > F. Thus, under these conditions, carbon–halogen bond breaking must be part of the rate-determining step. With organolithium reagents in aprotic solvents, the order of reactivity is F > Cl > Br > I, which indicates that the acidity of the ring hydrogen is the dominant factor governing reactivity.[86]

Addition of nucleophiles such as ammonia or alcohols, or their conjugate bases, to benzynes takes place very rapidly. The addition is believed to involve capture of the nucleophile by benzyne, followed by protonation to give the substitution product.[87] Electron-attracting groups tend to favor addition of the nucleophile at the more distant end of the "triple bond," since this permits maximum stabilization of the developing negative charge. Selectivity is usually not high, however, and formation of both possible products from monosubstituted benzynes is common.[88]

There are several methods for generation of benzyne in addition to base-catalyzed elimination of hydrogen halide from a halobenzene, and some of these are more widely applicable for preparative work. Probably the most useful method is diazotization of *o*-aminobenzoic acids.[89] Loss of nitrogen and carbon dioxide

84. R. W. Hoffmann, *Dehydrobenzene and Cycloalkynes*, Academic Press, New York, 1967.
85. F. W. Bergstrom, R. E. Wright, C. Chandler, and W. A. Gilkey, *J. Org. Chem.* **1**, 170 (1936).
86. R. Huisgen and J. Sauer, *Angew. Chem.* **72**, 91 (1960).
87. J. F. Bunnett, D. A. R. Happer, M. Patsch, C. Pyun, and H. Takayama, *J. Am. Chem. Soc.* **88**, 5250 (1966); J. F. Bunnett and J. K. Kim, *J. Am. Chem. Soc.* **95**, 2254 (1973).
88. E. R. Biehl, E. Nieh, and K. C. Hsu, *J. Org. Chem.* **34**, 3595 (1969).
89. M. Stiles, R. G. Miller, and U. Burckhardt, *J. Am. Chem. Soc.* **85**, 1792 (1963); L. Friedman and F. M. Logullo, *J. Org. Chem.* **34**, 3089 (1969).

follows diazotization and generates benzyne. This method permits generation of benzyne in the presence of a number of molecules with which it can react.

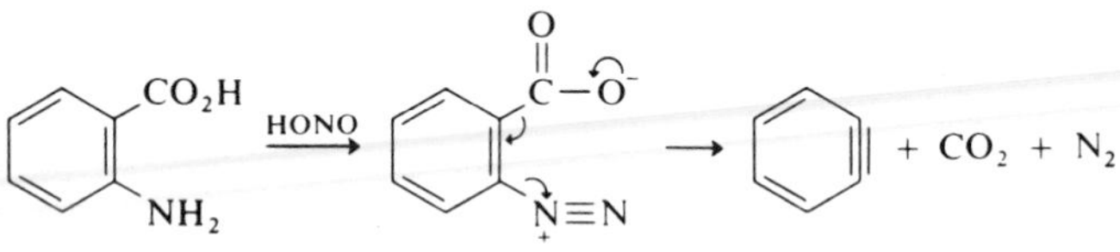

Oxidation of 1-aminobenzotriazole also serves as a source of benzyne under mild conditions. An oxidized intermediate decomposes with loss of two molecules of nitrogen.[90]

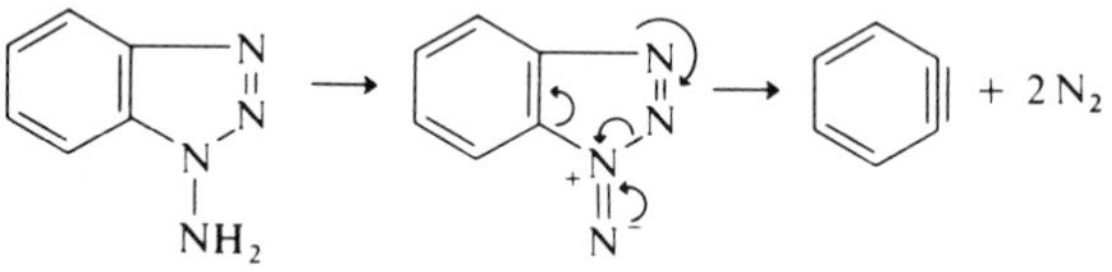

Another heterocyclic molecule that can serve as a benzyne precursor is benzothiadiazole-1,1-dioxide, which decomposes with elimination of nitrogen and sulfur dioxide.[91]

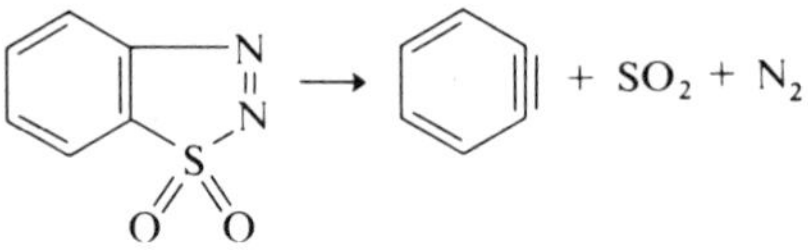

Benzyne can also be generated from *o*-dihaloaromatics. Reaction with lithium amalgam or magnesium results in the formation of transient organometallic compounds that decompose with elimination of lithium halide. *o*-Fluorobromobenzene is the usual starting material in this procedure.[92]

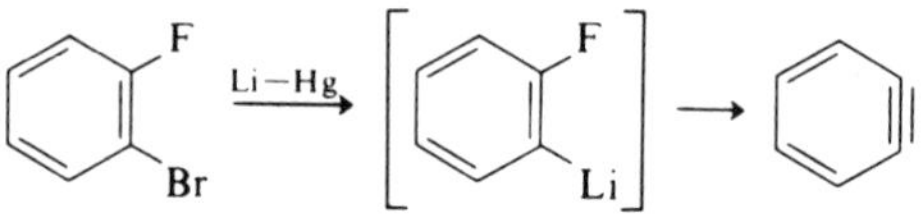

When benzyne is generated in the absence of another reactive molecule, it dimerizes to biphenylene.[93] In the presence of dienes, benzyne is a very reactive dienophile, and [4 + 2] cycloaddition products are formed.

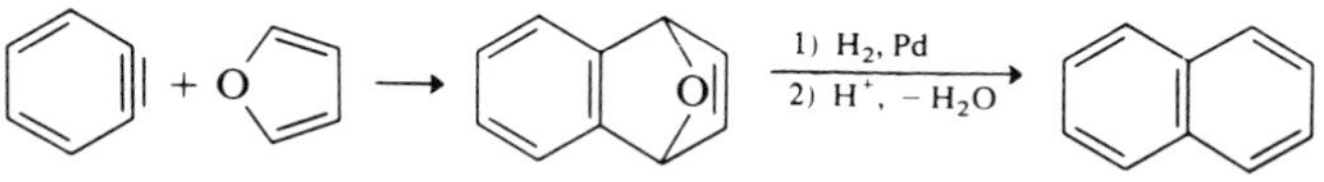

Ref. 94

90. C. D. Campbell and C. W. Rees, *J. Chem. Soc. C*, 742, 752 (1969); S. E. Whitney and B. Rickborn, *J. Org. Chem.* **53**, 5595 (1988).
91. G. Wittig and R. W. Hoffmann, *Org. Synth.* **47**, 4 (1967); G. Wittig and R. W. Hoffmann, *Chem. Ber.* **95**, 2718, 2729 (1962).
92. G. Wittig and L. Pohmer, *Chem. Ber.* **89**, 1334 (1956); G. Wittig, *Org. Synth.* **IV**, 964 (1963).
93. F. M. Logullo, A. H. Seitz, and L. Friedman, *Org. Synth.* **V**, 54 (1973).
94. G. Wittig and L. Pohmer, *Angew. Chem.* **67**, 348 (1955).

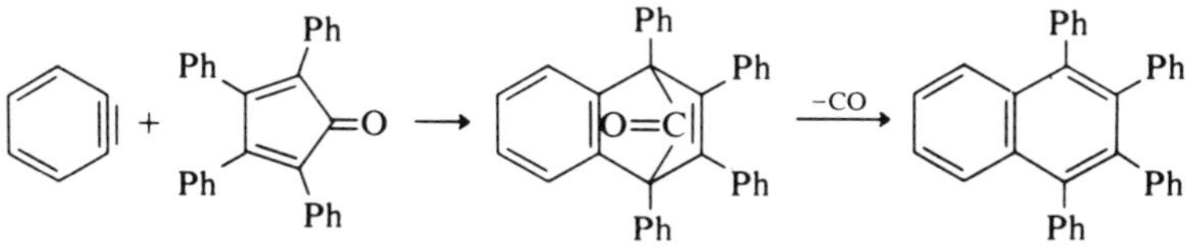

Ref. 95

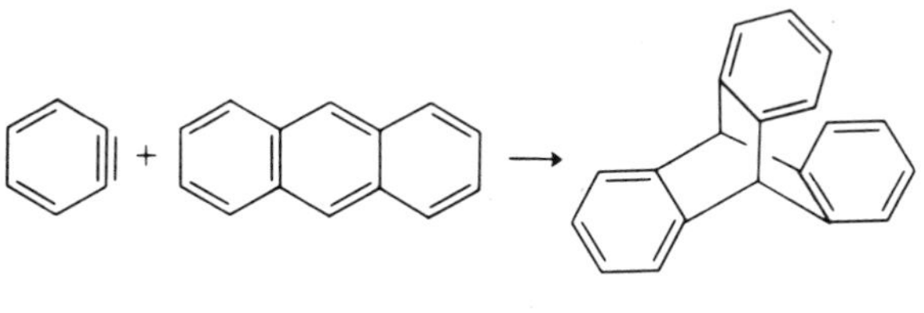

Ref. 96

Benzyne gives both [2 + 2] cycloaddition and ene reaction products with simple alkenes.[97]

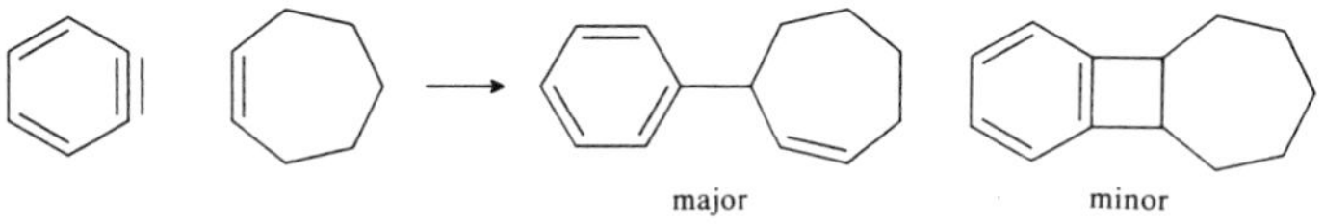

major minor

Scheme 11.9 illustrates some of the types of compounds that can be prepared via benzyne intermediates.

11.2.4. Copper-Catalyzed Reactions

It has been known for a long time that the nucleophilic substitution of aromatic halides is often strongly catalyzed by the presence of copper salts. Perhaps the most useful of the synthetic procedures based on this observation is the synthesis of aryl nitriles by reaction of aryl bromides with Cu(I)CN. The reaction is usually carried out at elevated temperatures in DMF or a similar solvent.

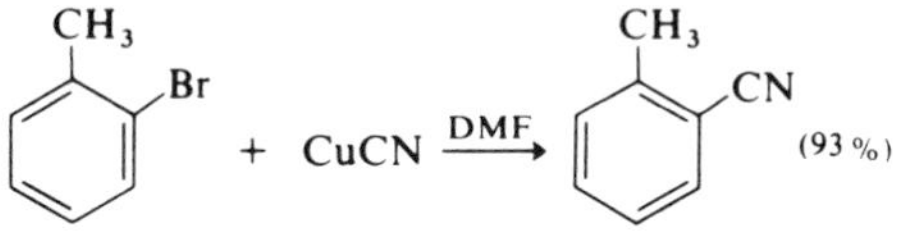

CH_3 Br + CuCN $\xrightarrow{DMF}$ CH_3 CN (93%) Ref. 98

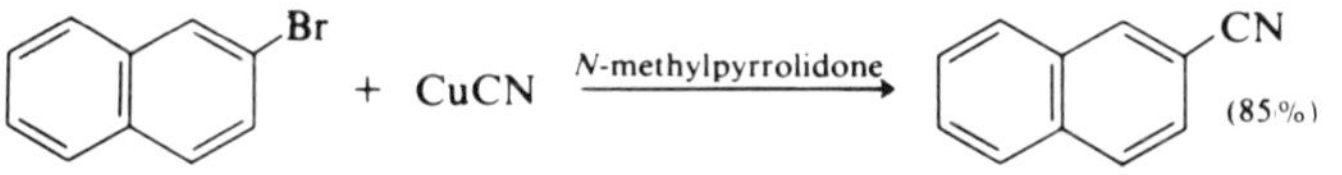

Br + CuCN $\xrightarrow{N\text{-methylpyrrolidone}}$ CN (85%) Ref. 99

95. L. F. Fieser and M. J. Haddadin, *Org. Synth.* **V**, 1037 (1973).
96. L. Friedman and F. M. Logullo, *J. Org. Chem.* **34**, 3089 (1969).
97. P. Crews and J. Beard, *J. Org. Chem.* **38**, 522 (1973).
98. L. Friedman and H. Shechter, *J. Org. Chem.* **26**, 2522 (1961).
99. M. S. Newman and H. Boden, *J. Org. Chem.* **26**, 2525 (1961).

Scheme 11.9. Some Syntheses via Benzyne Intermediates

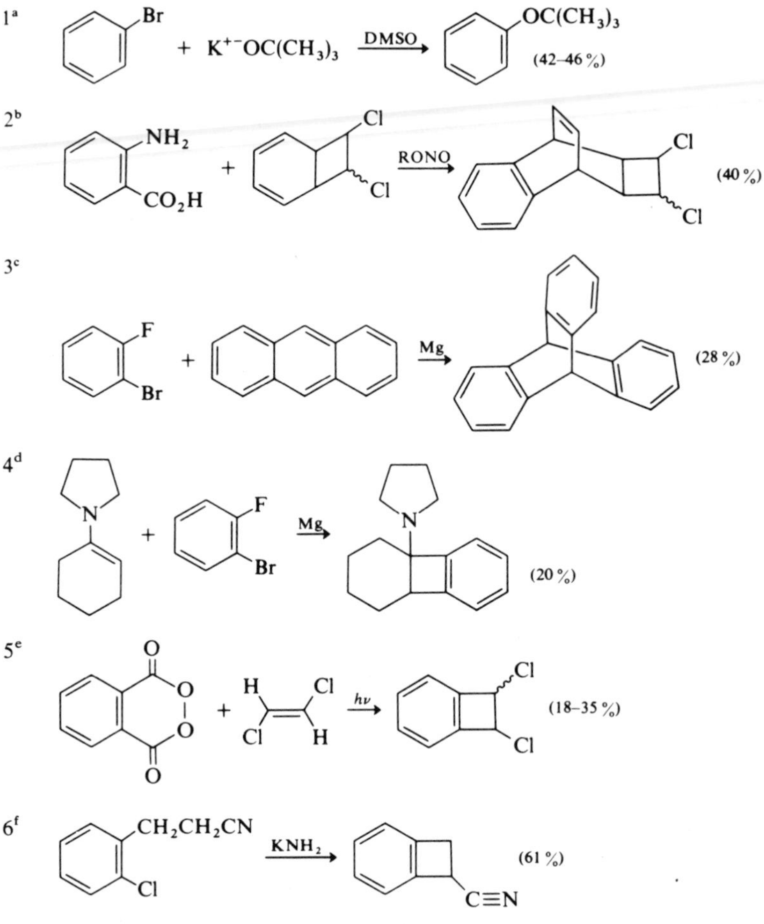

1[a] Bromobenzene + $K^{+-}OC(CH_3)_3$ $\xrightarrow{DMSO}$ $C_6H_5OC(CH_3)_3$ (42–46%)

2[b] Anthranilic acid (NH_2, CO_2H) + dichlorobicyclooctadiene $\xrightarrow{RONO}$ adduct (Cl, Cl) (40%)

3[c] 1-Bromo-2-fluorobenzene + anthracene $\xrightarrow{Mg}$ triptycene (28%)

4[d] 1-Pyrrolidinocyclohexene + 1-bromo-2-fluorobenzene $\xrightarrow{Mg}$ benzocyclobutene amine (N) (20%)

5[e] Phthaloyl peroxide + trans-ClCH=CHCl $\xrightarrow{h\nu}$ 1,2-dichlorobenzocyclobutene (18–35%)

6[f] o-ClC$_6$H$_4$CH$_2$CH$_2$CN $\xrightarrow{KNH_2}$ 1-cyanobenzocyclobutene (C≡N) (61%)

a. M. R. Sahyun and D. J. Cram, *Org. Synth.* **45**, 89 (1965).
b. L. A. Paquette, M. J. Kukla, and J. C. Stowell, *J. Am. Chem. Soc.* **94**, 4920 (1972).
c. G. Wittig, *Org. Synth.* **IV**, 964 (1963).
d. M. E. Kuehne, *J. Am. Chem. Soc.* **84**, 837 (1962).
e. M. Jones, Jr. and M. R. DeCamp, *J. Org. Chem.* **36**, 1536 (1971).
f. J. F. Bunnett and J. A. Skorcz, *J. Org. Chem.* **27**, 3836 (1962).

A general mechanistic description of the copper-promoted nucleophilic substitution pictures an oxidative addition of the aryl halide at Cu(I) followed by collapse of the arylcopper intermediate with a ligand transfer (reductive elimination).[100]

$$\mathrm{Ar{-}X + Cu(I)Z \rightarrow Ar{-}\underset{\underset{X}{|}}{Cu(III)}{-}Z \rightarrow Ar{-}Z + CuX}$$

X = halide
Z = nucleophile

100. T. Cohen, J. Wood, and A. G. Dietz, *Tetrahedron Lett.*, 3555 (1974).

Many kinds of nucleophiles can be arylated by copper-catalyzed substitution.[101] Among the reactive nuclophiles are carboxylate ions,[102] alkoxide ions,[103] amines,[104] phthalimide anions,[105], thiolate anions,[106] and acetylides.[107] In some of these reactions, there is a competitive reduction of the aryl halide to the dehalogenated arene. This is attributed to protonolysis of the arylcopper intermediate. Traditionally, most of these reactions have been carried out under heterogeneous conditions with copper powder or copper bronze as the catalyst. The general mechanism would suggest that these catalysts act as sources of Cu(I) ions. Homogeneous reactions have been carried out with soluble Cu(I) salts, particularly Cu(I)O_3SCF_3.[108]

11.3. Aromatic Radical Substitution Reactions

Aromatic rings are moderately reactive toward addition of free radicals (see Part A, Section 12.2) and several synthetically useful reactions involve free-radical substitution. One example is the synthesis of biaryls.[109]

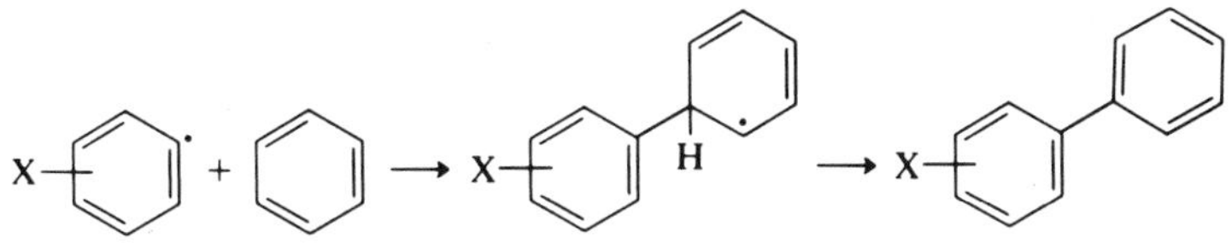

There are some inherent limits to the usefulness of such reactions. Radical substitutions are only moderately sensitive to substituent directing effects, so that substituted reactants usually give a mixture of products. This means that the practical utility is limited to symmetrical reactants, such as benzene, where the position of attack is immaterial. The best sources of aryl radicals for the reaction are aryl diazonium ions and *N*-nitrosoacetanilides. In the presence of base, aryl diazonium ions form diazooxides, which decompose to aryl radicals.[110]

$$\mathrm{Ar\overset{+}{N}{\equiv}N + 2\,{}^{-}OH \rightarrow ArN{=}N{-}O{-}N{=}NAr + H_2O}$$

$$\mathrm{ArN{=}N{-}O{-}N{=}NAr \rightarrow Ar\cdot + N_2 + \cdot O{-}N{=}NAr}$$

101. For a review of this reaction, see J. Lindley, *Tetrahedron* **48**, 1433 (1984).
102. T. Cohen and A. H. Lewin, *J. Am. Chem. Soc.* **88**, 4521 (1966).
103. R. G. R. Bacon and S. C. Rennison, *J. Chem. Soc., C*, 312 (1969).
104. A. J. Paine, *J. Am. Chem. Soc.* **109**, 1496 (1987).
105. R. G. R. Bacon and A. Karim, *J. Chem. Soc., Perkin Trans. 1*, 272 (1973).
106. H. Suzuki, H. Abe and A. Osuka, *Chem. Lett.*, 1303 (1980); R. G. R. Bacon and H. A. O. Hill, *J. Chem. Soc.*, 1108 (1964).
107. C. E. Castro, R. Havlin, V. K. Honwad, A. Malte, and S. Moje, *J. Am. Chem. Soc.* **91**, 6464 (1969).
108. T. Cohen and J. G. Tirpak, *Tetrahedron Lett.*, 143 (1975).
109. W. E. Bachmann and R. A. Hoffman, *Org. React.* **2**, 224 (1944); D. H. Hey, *Adv. Free Radical Chem.* **2**, 47 (1966).
110. C. Rüchardt and B. Freudenberg, *Tetrahedron Lett.*, 3623 (1964); C. Rüchardt and E. Merz, *Tetrahedron Lett.*, 2431 (1964).

Scheme 11.10. Biaryls by Radical Substitution

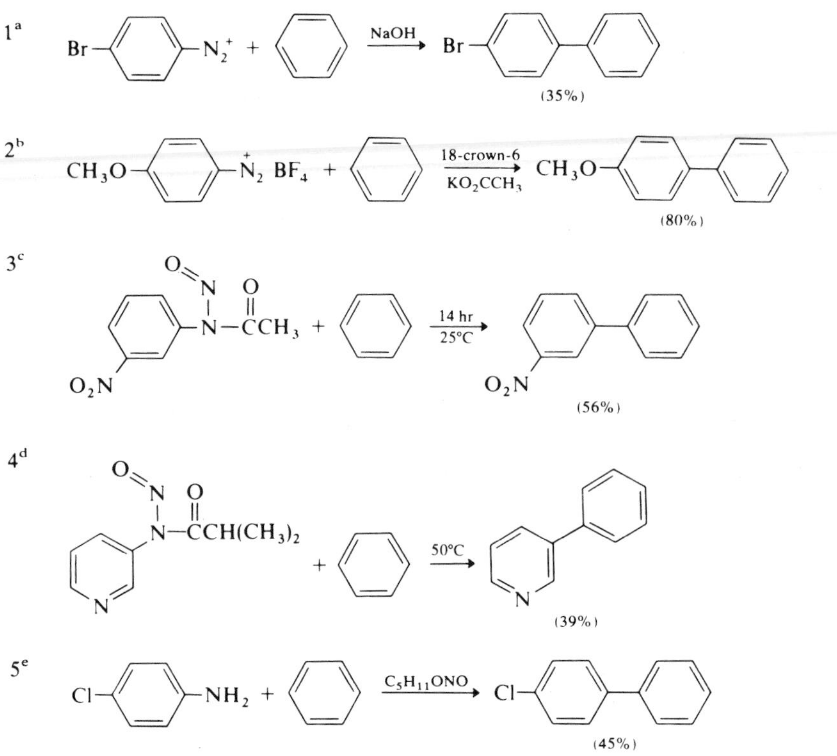

1[a] $Br-C_6H_4-N_2^+ + C_6H_6 \xrightarrow{NaOH} Br-C_6H_4-C_6H_5$ (35%)

2[b] $CH_3O-C_6H_4-\overset{+}{N_2}\ BF_4^- + C_6H_6 \xrightarrow[KO_2CCH_3]{18\text{-crown-6}} CH_3O-C_6H_4-C_6H_5$ (80%)

3[c] $3\text{-}O_2N-C_6H_4-N(N{=}O)-C(=O)CH_3 + C_6H_6 \xrightarrow[25°C]{14\ hr} 3\text{-}O_2N-C_6H_4-C_6H_5$ (56%)

4[d] $3\text{-}C_5H_4N-N(N{=}O)-C(=O)CH(CH_3)_2 + C_6H_6 \xrightarrow{50°C} 3\text{-phenylpyridine}$ (39%)

5[e] $Cl-C_6H_4-NH_2 + C_6H_6 \xrightarrow{C_5H_{11}ONO} Cl-C_6H_4-C_6H_5$ (45%)

a. M. Gomberg and W. E. Bachmann, *Org. Synth.* **I**, 113 (1941).
b. S. H. Korzeniowski, L. Blum, and G. W. Gokel, *Tetrahedron Lett.*, 1871 (1977); J. R. Beadle, S. H. Korzeniowski, D. E. Rosenberg, B. J. Garcia-Slanga, and G. W. Gokel, *J. Org. Chem.* **49**, 1594 (1984).
c. W. E. Bachmann and R. A. Hoffman, *Org. React.* **2**, 249 (1944).
d. H. Rapoport, M. Lock, and G. J. Kelly, *J. Am. Chem. Soc.* **74**, 6293 (1952).
e. J. I. G. Cadogan, *J. Chem. Soc.*, 4257 (1962).

In the classical procedure, base is added to a two-phase mixture of the aqueous diazonium salt and an excess of the aromatic that is to be substituted. Improved yields have been obtained by using polyethers or phase transfer catalysts with solid aryl diazonium tetrafluoroborate salts in an excess of the reactant.[111] *N*-Nitrosoacetanilides rearrange to diazonium acetates, which also give rise to aryl radicals via diazooxides.[112]

$$ArN(N{=}O)C(=O)CH_3 \rightarrow ArN{=}N-OC(=O)CH_3$$

$$2\,ArN{=}N-OC(=O)CH_3 \rightarrow ArN{=}N-O-N{=}NAr + (CH_3CO)_2O$$

111. J. R. Beadle, S. H. Korzeniowski, D. E. Rosenberg, G. J. Garcia-Slanga, and G. W. Gokel, *J. Org. Chem.* **49**, 1594 (1984).
112. J. I. G. Cadogan, *Acc. Chem. Res.* **4**, 186 (1971); *Adv. Free Radical Chem.* **6**, 185 (1980).

1[a]

$$C_6H_5Br + H_2C{=}C(O^-)CH_3 \xrightarrow[h\nu]{NH_3} C_6H_5CH_2C(O)CH_3$$

(86%)

2[b]

$$C_6H_5Br + H_2C{=}C(O^-)CH(CH_3)_2 \xrightarrow[h\nu]{NH_3} C_6H_5CH_2C(O)CH(CH_3)_2$$

(79%)

3[c]

$$2,6\text{-}(CH_3)_2\text{-}4\text{-}H_3C\text{-}C_6H_2Br + H_2C{=}C(O^-)CH{=}C(O^-)CH_3 \xrightarrow[h\nu]{NH_3} 2,6\text{-}(CH_3)_2\text{-}4\text{-}H_3C\text{-}C_6H_2\text{-}CH_2C(O)CH_2C(O)CH_3$$

4[d]

$$CH_3O\text{-}C_6H_4\text{-}I + {}^-OP(OC_2H_5)_2 \xrightarrow[h\nu]{NH_3} CH_3O\text{-}C_6H_4\text{-}P(O)(OC_2H_5)_2$$

(65%)

5[e]

$$\text{2-chloropyridine} + H_2C{=}C(O^-)CH_3 \xrightarrow[h\nu]{NH_3} \text{2-pyridyl-}CH_2C(O)CH_3$$

(84%)

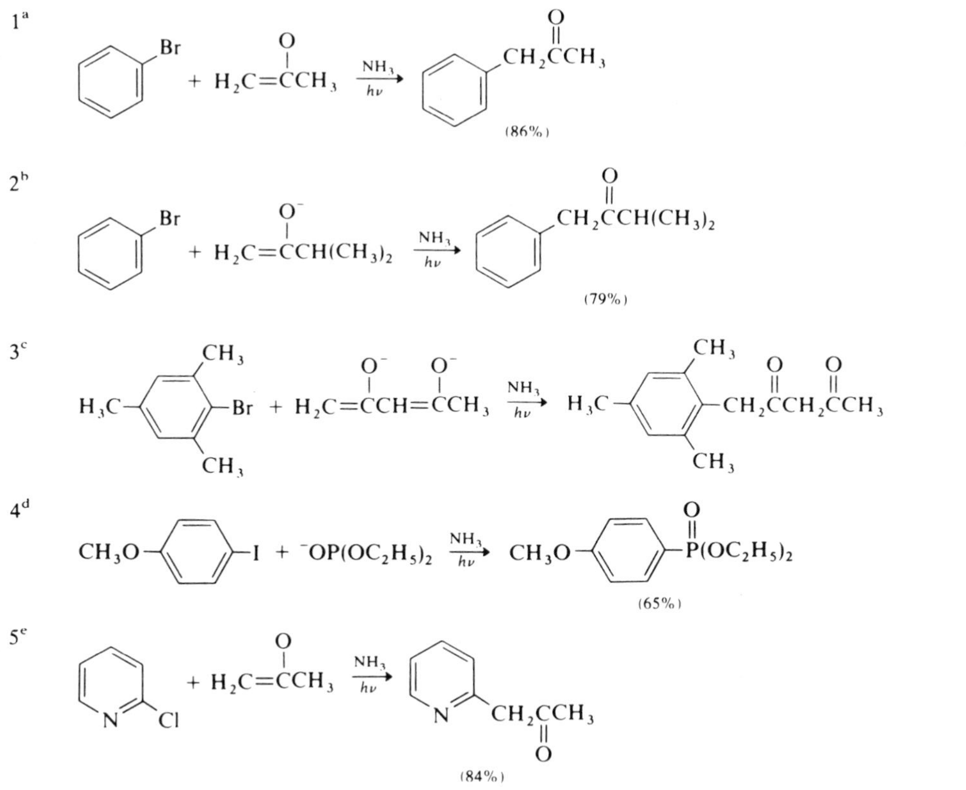

a. R. A. Rossi and J. F. Bunnett, *J. Org. Chem.* **38**, 1407 (1973).
b. M. F. Semmelhack and T. Bargar, *J. Am. Chem. Soc.* **102**, 7765 (1980).
c. J. F. Bunnett and J. E. Sundberg, *J. Org. Chem.* **41**, 1702 (1976).
d. J. F. Bunnett and X. Creary, *J. Org. Chem.* **39**, 3612 (1974).
e. A. P. Komin and J. F. Wolfe, *J. Org. Chem.* **42**, 2481 (1977).

A procedure for arylation involving *in situ* diazotization has also been developed.[113] Scheme 11.10 gives some representative preparative methods.

11.4. Substitution by the $S_{RN}1$ Mechanism

The distinctive feature of the $S_{RN}1$ mechanism is an electron transfer between the nucleophile and the aryl halide.[114] The overall reaction is normally a chain process.

113. J. I. G. Cadogan, *J. Chem. Soc.*, 4257 (1962).
114. J. F. Bunnett, *Acc. Chem. Res.* **11**, 413 (1978); R. A. Rossi and R. H. de Rossi, *Aromatic Substitution by the $S_{RN}1$ Mechanism*, ACS Monograph Series, No. 178, American Chemical Society, Washington, D.C., 1983.

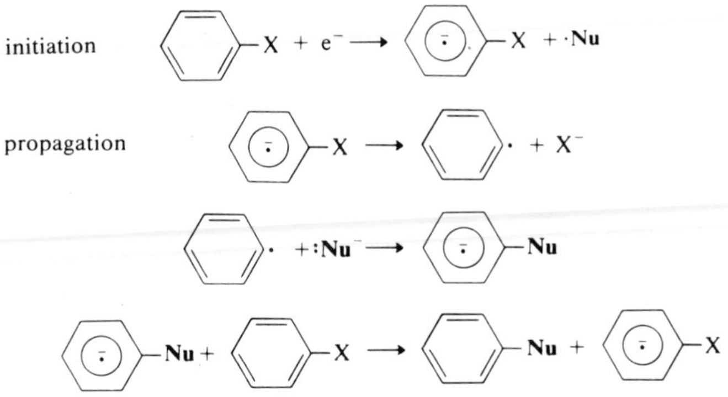

initiation

propagation

The potential advantage of the $S_{RN}1$ mechanism is that it is not particularly sensitive to the nature of other aromatic ring substituents. Chloropyridines and chloroqinolines are also excellent reactants.[115] A variety of nucleophiles undergo the reaction, although not always in high yield. The nucleophiles which have been found to participate in $S_{RN}1$ substitution include ketone enolates,[116] 2,4-pentanedione dianion,[117] amide enolates,[118] pentadienyl and indenyl carbanions,[119] phenolates,[120] diethyl phosphite anion,[121] and thiolates.[122] The reactions are frequently initiated by light, which accelerates the initiation step. As for other radical chain processes, the reaction is sensitive to substances that can intercept the propagation intermediates. Scheme 11.11 provides some examples of the preparative use of the $S_{RN}1$ reaction.

General References

Electrophilic Aromatic Substitution

J. G. Hoggett, R. B. Moodie, J. R. Penton, and K. S. Schofield, *Nitration and Aromatic Reactivity*, Cambridge University Press, Cambridge, UK, 1971.

G. A. Olah, *Friedel–Crafts Chemistry*, Wiley-Interscience, New York, 1973.

G. A. Olah (ed.), *Friedel–Crafts and Related Reactions*, Vols. I–IV, Wiley-Interscience, New York, 1962–1964.

R. O. C. Norman and R. Taylor, *Electrophilic Substitution in Benzenoid Compounds*, Elsevier, Amsterdam, 1965.

R. M. Roberts and A. A. Khalaf, *Friedel–Crafts Alkylation Chemistry*, Marcel Dekker, New York, 1984.

K. Schofield, *Aromatic Nitration*, Cambridge University Press, Cambridge, UK, 1980.

L. M. Stock, *Atomatic Substitution Reactions*, Prentice-Hall, Englewood Cliffs, New Jersey, 1968.

115. J. V. Hay, T. Hudlicky, and J. F. Wolfe, *J. Am. Chem. Soc.* **97**, 374 (1975); J. V. Hay and J. F. Wolfe, *J. Am. Chem. Soc.* **97**, 3702 (1975); A. P. Komin and J. F. Wolfe, *J. Org. Chem.* **42**, 2481 (1977); R. Beugelmans, M. Bois-Choussy, and B. Boudet, *Tetrahedron* **24**, 4153 (1983).
116. M. F. Semmelhack and T. Bargar, *J. Am. Chem. Soc.* **102**, 7765 (1980).
117. J. F. Bunnett and J. E. Sundberg, *J. Org. Chem.* **41**, 1702 (1976).
118. R. A. Rossi and R. A. Alonso, *J. Org. Chem.* **45**, 1239 (1980).
119. R. A. Rossi and J. F. Bunnett, *J. Org. Chem.* **38**, 3020 (1973).
120. A. B. Pierini, M. T. Baumgartner, and R. A. Rossi, *Tetrahedron Lett.* **29**, 3429 (1988).
121. J. F. Bunnett and X. Creary, *J. Org. Chem.* **39**, 3612 (1974).
122. J. F. Bunnett and X. Creary, *J. Org. Chem.* **39**, 3173, 3611 (1974); *J. Org. Chem.* **40**, 3740 (1975).

Nucleophilic Aromatic Substitution

R. W. Hoffman, *Dehydrobenzene and Cycloalkynes*, Academic Press, New York, 1967.
J. Miller, *Aromatic Nucleophilic Substitution*, Elsevier, Amsterdam, 1968.
S. Patai (ed.), *The Chemistry of Diazonium and Diazo Groups*, Wiley, New York, 1978.
K. H. Saunders and R. L. M. Allen, *Aromatic Diazo Compounds*, Edward Arnold, London, 1985.
H. Zollinger, *Azo and Diazo Chemistry*, Interscience, New York, 1961.

Problems

(*References for these problems will be found on page* 777.)

1. Give reaction conditions that would accomplish each of the following transformations. Multistep schemes are not necessary. Be sure to choose conditions that would afford the desired isomer as the principal product.

(a) $H_3C-C_6H_4-Br \longrightarrow H_3C-C_6H_4-C{\equiv}N$

(b) $C_6H_5CO_2CH_3 \longrightarrow$ o-$IC_6H_4CO_2CH_3$

(c) $(CH_3)_3C-C_6H_4-C(CH_3)_3 \longrightarrow (CH_3)_3C-C_6H_4-CCH_3$ (with C=O)

(d) 3,4,5-$(CH_3O)_3C_6H_2CH_2CH(CO_2C_2H_5)C(=O)CO_2C_2H_5 \longrightarrow$ indene with CH_3O, CH_3O, CH_3O, $CO_2C_2H_5$, $CO_2C_2H_5$

(e) $C_6H_5CH(CH_3)_2 \longrightarrow$ m-$IC_6H_4CH(CH_3)_2$

(f) p-$NO_2C_6H_4NH_2 \longrightarrow$ p-$NO_2C_6H_4HC{=}CHCH{=}CH_2$

(g) chroman with O_2CCH_3 at C-8 and C-3 $\longrightarrow$ same with $CH_3C(=O)$ at C-6 (O_2CCH_3, O, O_2CCH_3, CH_3C, O)

(h) 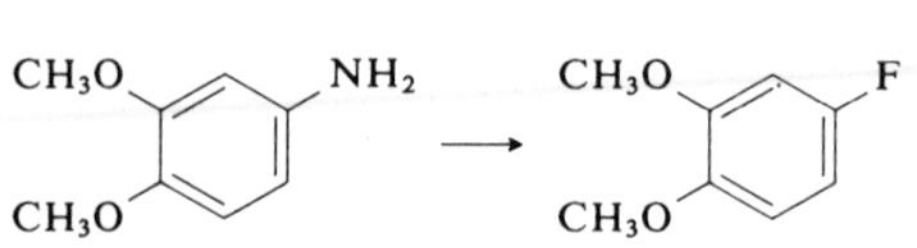 [structures: 3,4-dimethoxyaniline (CH_3O, CH_3O, NH_2) → 3,4-dimethoxyfluorobenzene (CH_3O, CH_3O, F)]

(i) 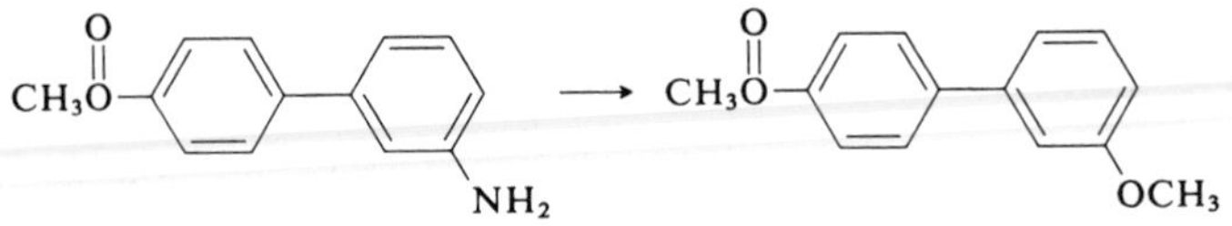 [structures: $CH_3\overset{O}{\overset{\|}{C}}O$-biphenyl with NH_2 → $CH_3\overset{O}{\overset{\|}{C}}O$-biphenyl with OCH_3]

2. Suggest a short series of reactions which could be expected to transform the material on the right into the desired product shown on the left.

(a) 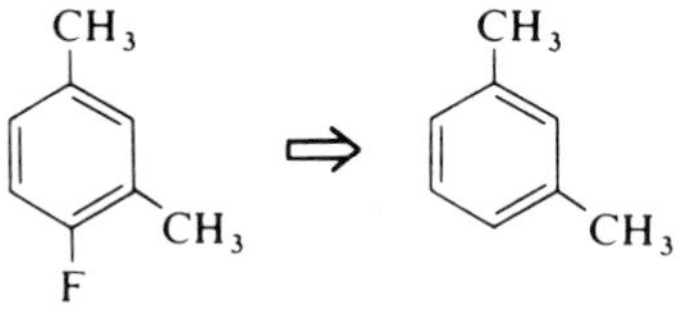 [structures: CH_3, CH_3, F-substituted benzene ⇨ CH_3, CH_3 (m-xylene)]

(b) 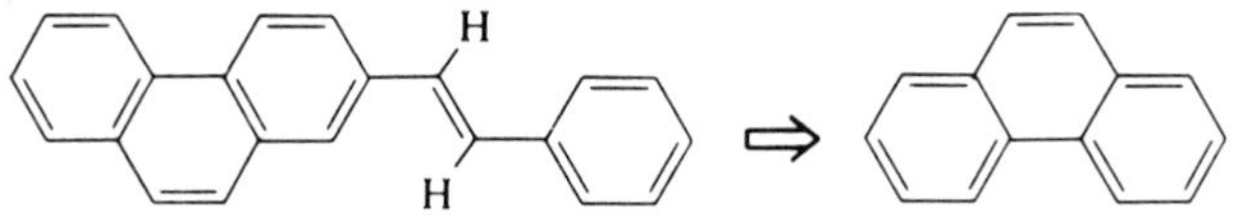 [structures: 2-styrylphenanthrene (H, H) ⇨ phenanthrene]

(c) 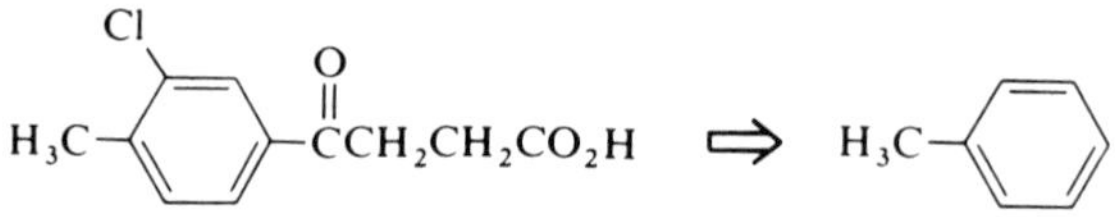 [structures: Cl, H_3C-C₆H₃-$\overset{O}{\overset{\|}{C}}CH_2CH_2CO_2H$ ⇨ H_3C-C₆H₅]

(d) 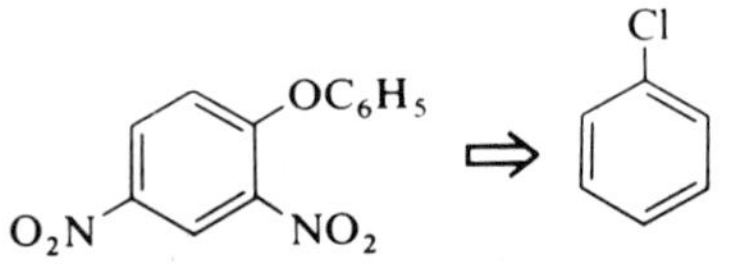 [structures: OC_6H_5, O_2N, NO_2-substituted benzene ⇨ Cl-benzene]

(e) 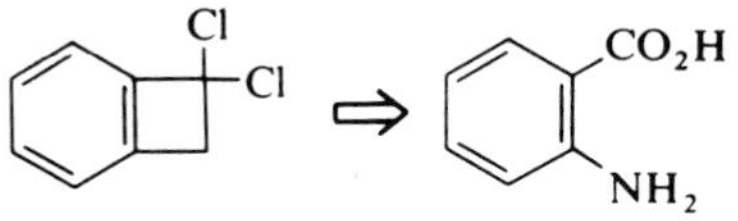 [structures: benzocyclobutene with Cl, Cl ⇨ benzene with CO_2H, NH_2]

3. Write mechanisms that would account for the following reactions:

(a) 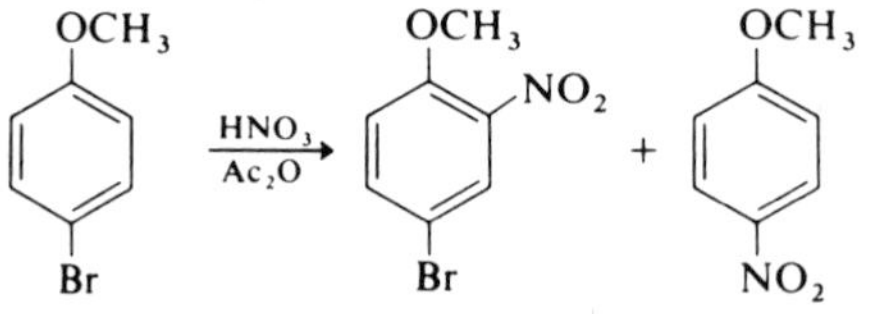 [structures: p-bromoanisole (OCH_3, Br) $\xrightarrow[Ac_2O]{HNO_3}$ (OCH_3, NO_2, Br) + (OCH_3, NO_2)]

(b) 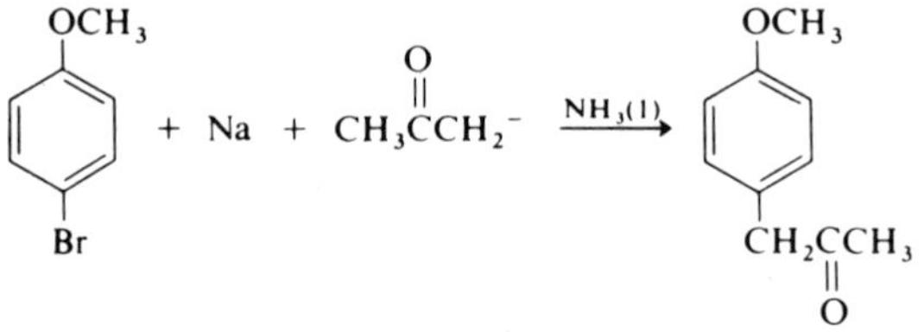 [structures: p-bromoanisole (OCH_3, Br) + Na + $CH_3\overset{O}{\overset{\|}{C}}CH_2^-$ $\xrightarrow{NH_3(l)}$ (OCH_3, $CH_2\underset{O}{\underset{\|}{C}}CH_3$)]

(c) [o-carboxybenzenediazonium, CO_2^-, N_2^+] + $H_3C-CH=CH-CH=CH-CH_3$ →

→ [1,4-dimethyl-1,4-dihydronaphthalene, cis] (~74%) + $H_2C=C(H)-CH(Ph)-C(H)=C(H)-CH_3$ (~6%)

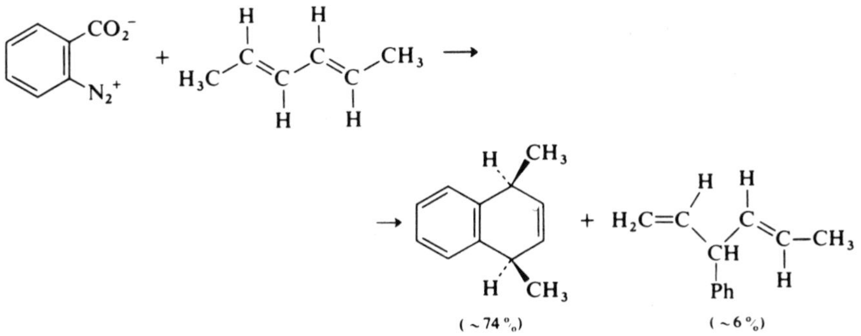

(d) [1-OCH_3, 5-CH_3O, 4-$O_2CC_2H_5$ naphthalene] $\xrightarrow{BF_3}$ [OCH_3, CH_3O, OH, $\overset{O}{\parallel}CC_2H_5$ naphthalene]

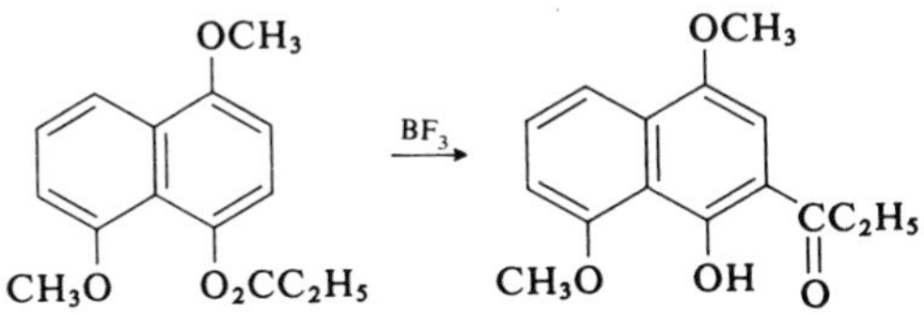

(e) [3-cyanophthalide, CN, O, O] $\xrightarrow[-78°C]{LDA}$ $\xrightarrow[-40 \rightarrow 25°c]{\text{2-bromo-1,4-dimethoxybenzene } (CH_3O, Br, CH_3O)}$ [1,4-dimethoxyanthraquinone; O, OCH_3, O, OCH_3]

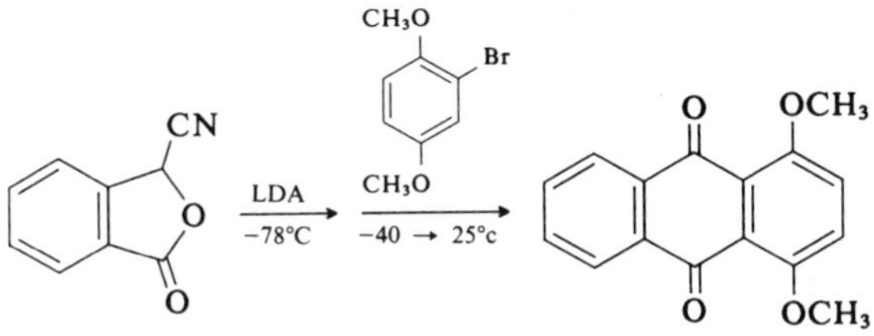

(f) [CH_3O, CO_2CH_3, OCH_3, CH_3, CH_3, CH_3, CH_3, O] $\xrightarrow[SnCl_4]{MeOCH_2COCl}$ [O, O, CH_3O, OCH_3, CH_3, CH_3, O, CH_3, CH_3]

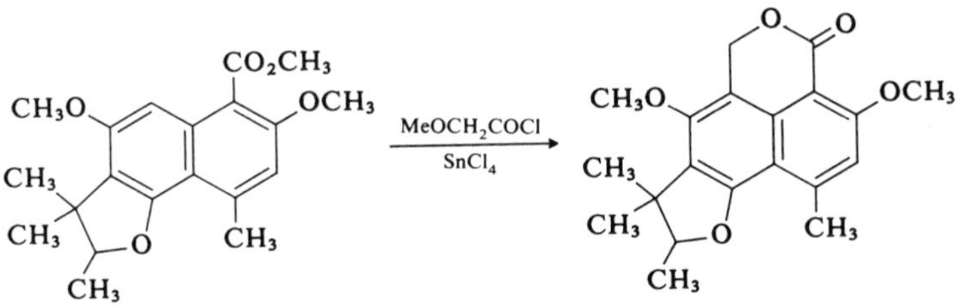

4. Predict the product(s) of the following reactions. If more than one product is expected, indicate which will be major and which will be minor.

(a) $PhC(=O)-C_6H_4-NH_2$ $\xrightarrow[DMF,\ 65°C]{(CH_3)_3CONO}$

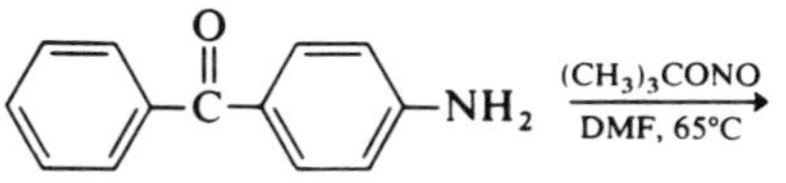

(b) [2-nitro-6-methylaniline: NO_2, NH_2, CH_3] $\xrightarrow[\text{2) } Cu(NO_3)_2,\ CuO,\ H_2O]{\text{1) } H_2SO_4,\ NaNO_2,\ 0°C}$

(c) $Cl-C_6H_4-NH_2 + PhCH=CH_2$ $\xrightarrow[CuCl]{(CH_3)_3CONO}$

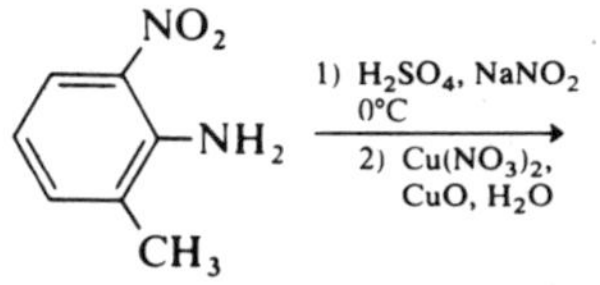

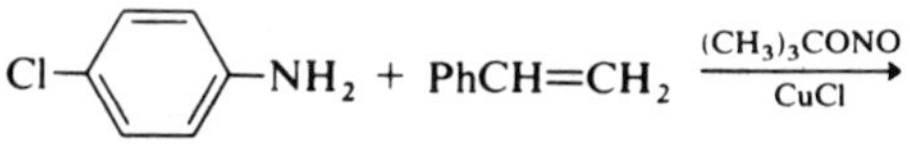

(d) H_3C–(C$_6H_3$, I)–SO_3H $\xrightarrow[HgSO_4]{H_2SO_4,\ H_2O}$

(e) CH_3O, CH_3O (C$_6H_3$) CH_2CH_2OH $\xrightarrow[0°C]{HNO_3,\ AcOH}$

(f) Cl, Cl, Cl (C$_6H_2$) –NH_2 $\xrightarrow[CH_3CN]{(CH_3)_3CONO,\ CuCl}$

(g) NO_2, H_3C, O_2N (C$_6H_2$) + $HSCH_2CO_2CH_3$ $\xrightarrow[HMPA]{LiOH}$

(h) Cl (naphthalene) $\xrightarrow[CHCl_3]{CH_3\overset{O}{\overset{\|}{C}}Cl,\ AlCl_3}$

(i) $CH_3\overset{O}{\overset{\|}{C}}CH_2$ (benz[a]anthracene) $\xrightarrow{CH_3SO_3H}$

(j) $(CH_3)_2Si$–(C$_6H_4$)–$CH_2\underset{\displaystyle NH\overset{O}{\overset{\|}{C}}CH_3}{CH}CO_2CH_3$ $\xrightarrow{I_2,\ AgBF_4}$

(k) Br (C$_6H_5$) $\xrightarrow[H^+]{Br_2,\ HgO}$

(l) (C$_6H_5$)–CN + F–(C$_6H_4$)–N_2^+ $\xrightarrow[KOAc]{18\text{-crown-6}}$

5. Suggest efficient syntheses of *ortho*-, *meta*-, and *para*-fluoropropiophenone from benzene and any other necessary organic or inorganic reagents.

6. Treatment of compound **A** in dibromomethane with one equivalent of aluminum bromide yielded **B** as the only product in 78% yield. When three equivalents of aluminum bromide were used, however, compounds **C** and **D** were obtained in a combined yield of 97%. Suggest an explanation for these observations.

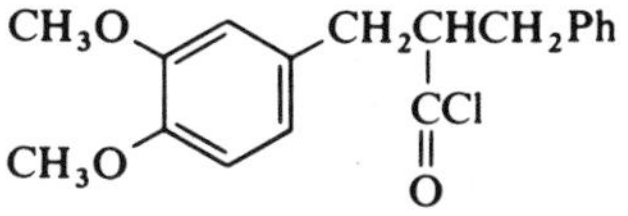

CH_3O, CH_3O, CH_2CHCH_2Ph, CCl, O

A

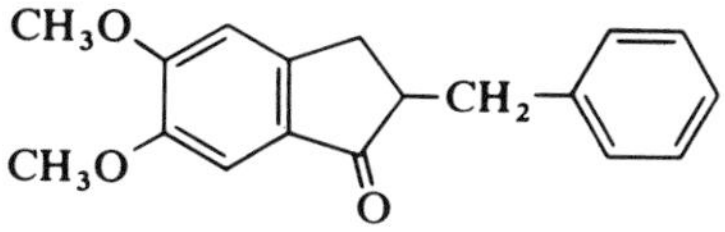

CH_3O, CH_3O, CH_2, O

B

RO, R'O, CH_2, O

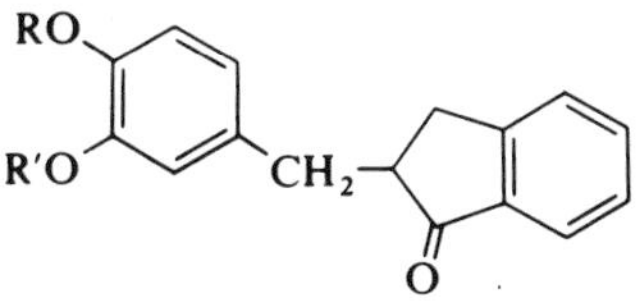

C R = CH_3, R' = H
D R = H, R' = CH_3

7. Some data for the alkylation of naphthalene by isopropyl bromide under various conditions are given.

Reaction medium A: $AlCl_3$–CS_2
Reaction medium B: $AlCl_3$–CH_3NO_2

Reaction time (min)	$\alpha:\beta$ Ratio A	$\alpha:\beta$ Ratio B
5	4:96	83:17
15	2.5:97.5	74:26
45	2:98	70:30

What factors are responsible for the difference in the product ratio for the two reaction media and why might the ratio change with reaction time?

8. Addition of a solution of bromine and potassium bromide to a solution of the carboxylate salt **A** results in the precipitation of a neutral compound having the formula $C_{11}H_{13}BrO_3$. Various spectroscopic data show that the compound is nonaromatic. Suggest a structure and discuss the significance of the formation of this product.

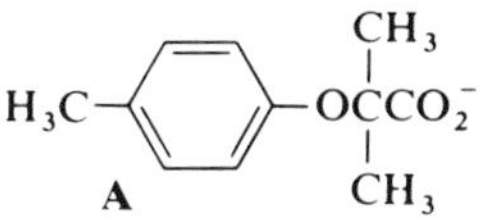

H_3C, CH_3, $OCCO_2^-$, CH_3

A

9. Benzaldehyde, benzyl methyl ether, benzoic acid, methyl benzoate, and phenylacetic acid all undergo thallation initially in the *ortho* position. Explain this observation.

10. Reaction of 3,5,5-trimethyl-2-cyclohexenone with $NaNH_2$ (3 equiv) in THF generates its enolate. When bromobenzene is then added and this solution stirred for 4 h, the product **A** is isolated in 30% yield. Formulate a mechanism for this transformation.

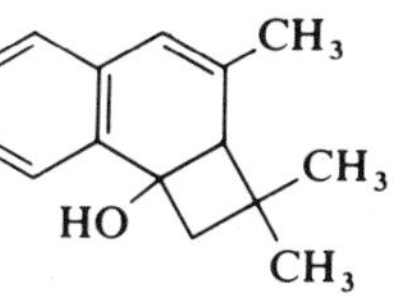

CH_3, CH_3, HO, CH_3

A

11. When phenylacetonitrile is converted to its anion in the presence of an excess of LDA and then allowed to react with a brominated aromatic ether such as 2-bromo-4-methylmethoxybenzene, the product is the result of both cyanation and benzylation. Propose a mechanism for this reaction.

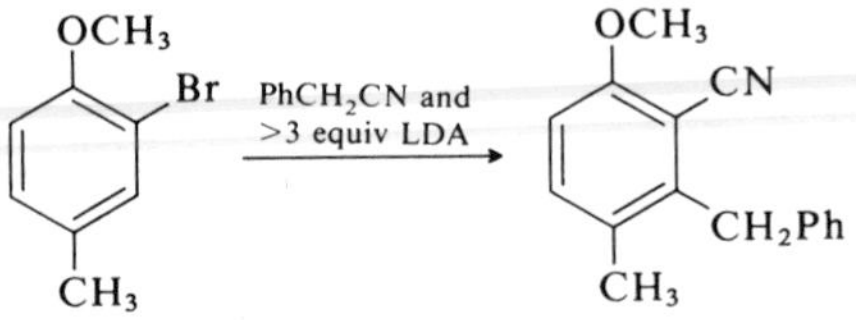

OCH_3 Br CH_3 — $PhCH_2CN$ and >3 equiv LDA → OCH_3 CN CH_2Ph CH_3

12. Suggest reaction sequences that would permit synthesis of the following aromatic compounds from the starting material indicated on the right.

(a) H_2N NH_2 O_2N NO_2 ⇒ Cl Cl

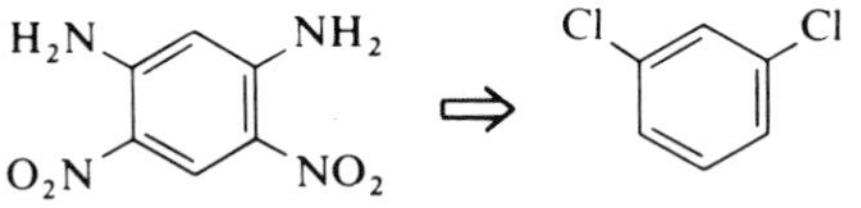

(b) NH_2 O_2N NO_2 ⇒ Cl

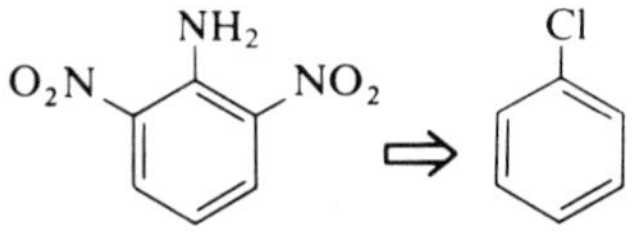

(c) Cl Cl ⇒ Cl CO_2H

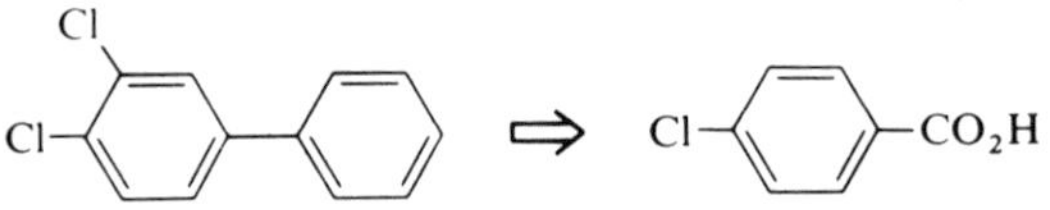

(d) CH_3O $CH_2C{\equiv}N$ ⇒ CH_3O

(e) Br F ⇒ Br NH_2

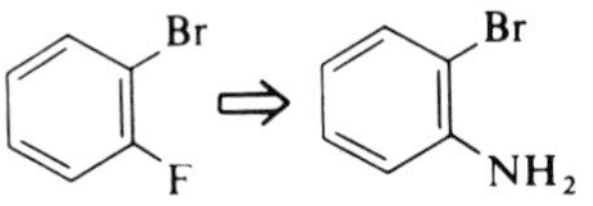

(f) CO_2H Br Br Br ⇒ CO_2H

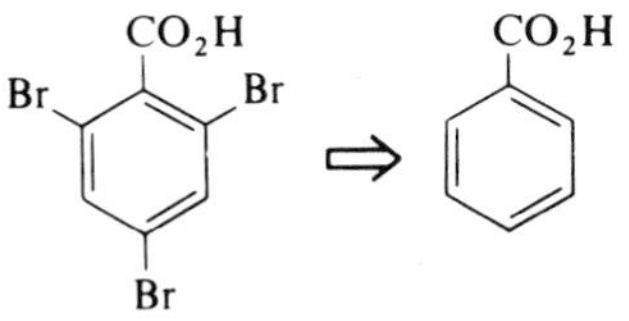

(g) $S(CH_2)_3CH_3$ $S(CH_2)_3CH_3$ ⇒ Br Br

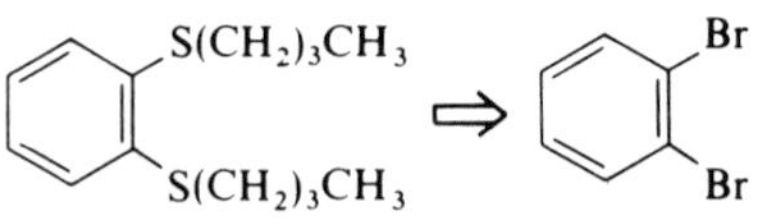

(h) $CH_3C(=O)$ CH_2CHCO_2H Br ⇒ $CH_3C(=O)$ NH_2

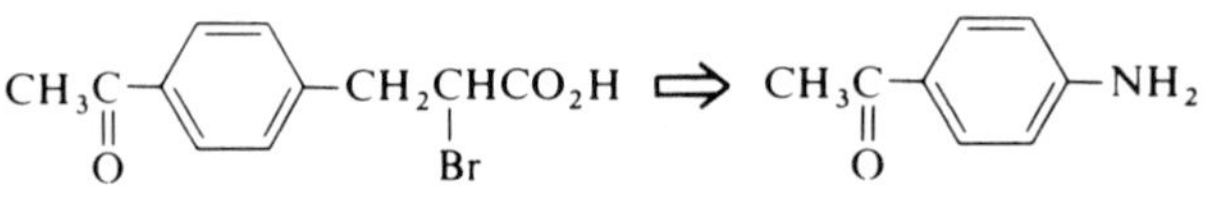

13. Aromatic substitution reactions are key steps in multistep synthetic sequences that effect the following transformations. Suggest reaction sequences that might accomplish the desired syntheses.

(a) ... from H_3C-cyclohexane-1,3-dione, CH_3O-C_6H_4-$(CH_2)_3\overset{O}{\overset{\|}{C}}CH{=}CH_2$

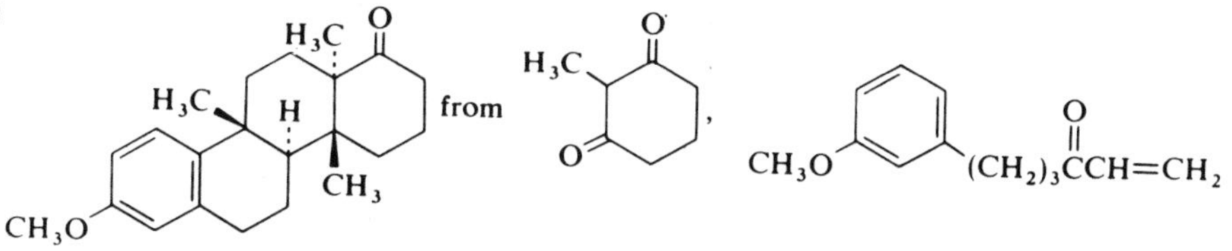

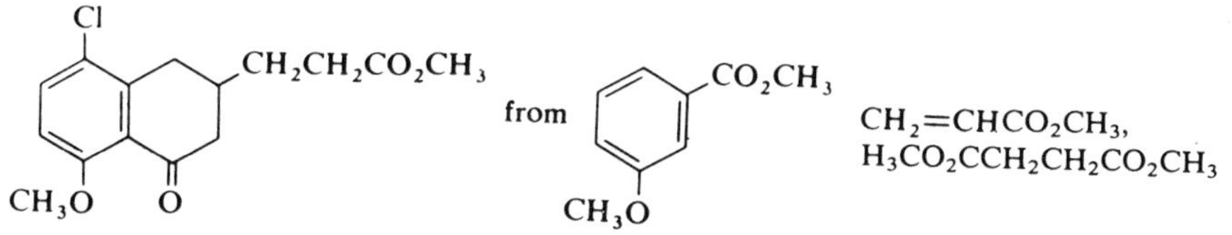

(b) ... $CH_2CH_2CO_2CH_3$ ... from CO_2CH_3, CH_3O ... $CH_2{=}CHCO_2CH_3$, $H_3CO_2CCH_2CH_2CO_2CH_3$

(c) CH_3O CN ... CH_3O O from OCH_3 $CH{=}O$ OCH_3, $CO_2C_2H_5$

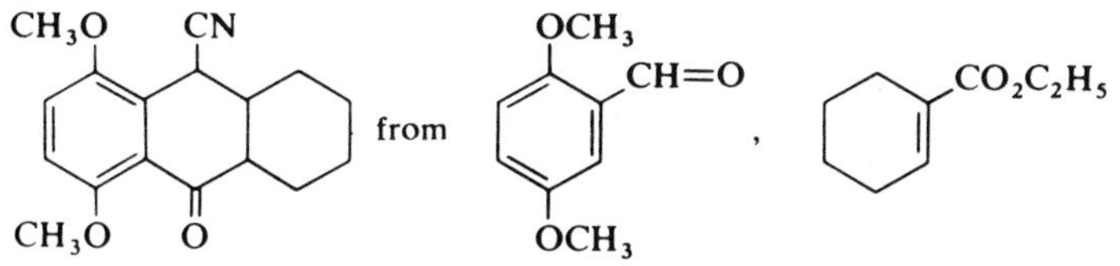

(d) CH_3O H N O CH_3 CH_3O N O H C=C H CH_3 from CO_2H CH_3O OCH_3 CH_3, H CH_3 C=C H $H_5C_2O_2C$ N H

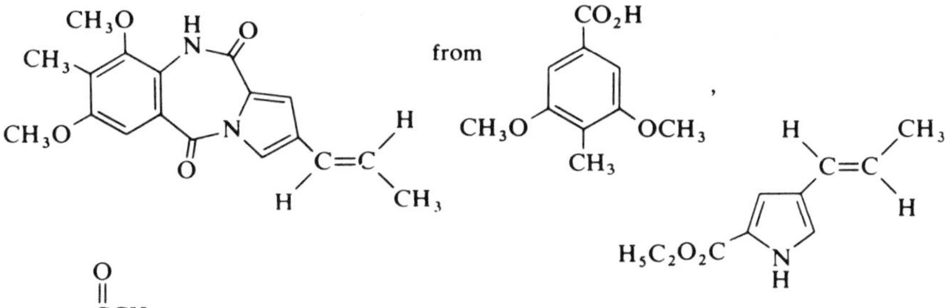

(e) O $\overset{\|}{C}CH_3$ CH_3 CH_3 from

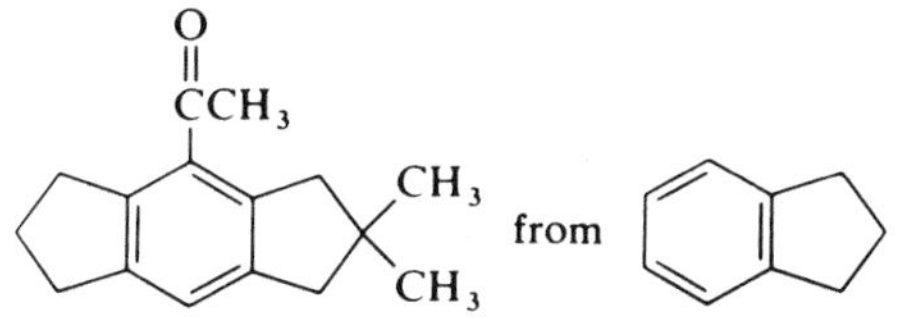

(f) CH_3O CH_3O $CO_2C_2H_5$ CH_3 from OCH_3 CH_3O $CH{=}O$

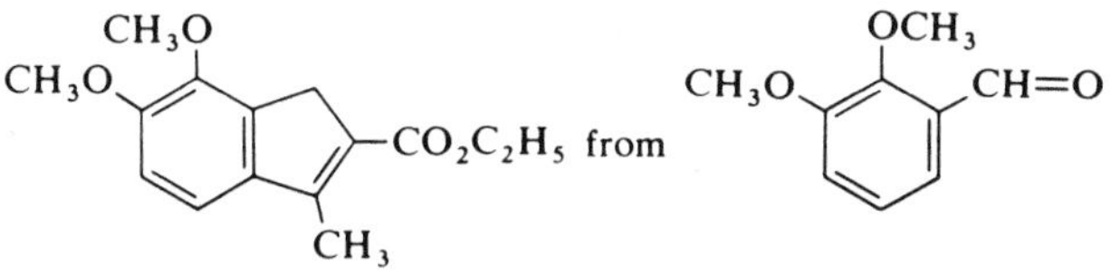

14. In the cyclization of **A** by an intramolecular Friedel–Crafts acylation, it is observed that the product formed depends on the amount of Lewis acid catalyst used in the reaction. Offer a mechanistic explanation of this effect.

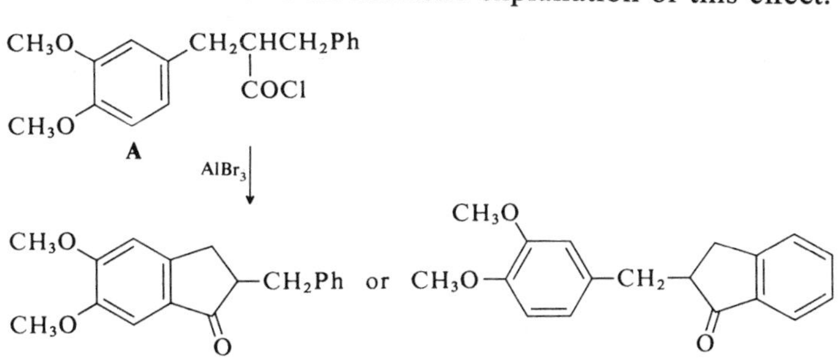

CH_3O $CH_2\underset{COCl}{\underset{|}{C}}HCH_2Ph$ CH_3O

A

$AlBr_3$

CH_3O CH_3O O $-CH_2Ph$ or CH_3O CH_3O- $-CH_2-$ O

formed with 1 equiv of $AlBr_3$

formed with 3 equiv of $AlBr_3$

15. The use of aryltrimethylsilanes as intermediates has been found to be a useful complement to direct thallation in the preparation of arylthallium intermediates. The advantages are (a) position specificity, because the thallium always replaces the silane substituent, and (b) improved ability to effect thallation of certain deactivated rings, such as those having trifluoromethyl substituents. What role does the silyl substituent play in these reactions?

16. The Pschorr reaction is a method of synthesis of phenanthrenes from diazotized *cis*-2-aminostilbene derivatives. A traditional procedure involves heating with a copper catalyst. Improved yields are frequently observed, however, if the diazonium salt is allowed to react with iodide ion. What might be the mechanism of the iodide-catalyzed reaction?

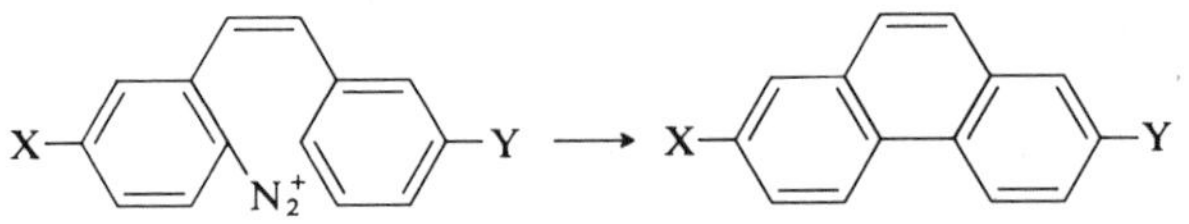

X— —Y → X— —Y (N_2^+)

17. When compound **B** is dissolved in FSO_3H at −78°C, NMR spectroscopy shows that a carbocation is formed. If the solution is then allowed to warm to −10°C, a different ion forms. The first ion gives compound **C** when quenched with base, while the second gives **D**. What are the structures of the two carbocations and why do they give different products on quenching?

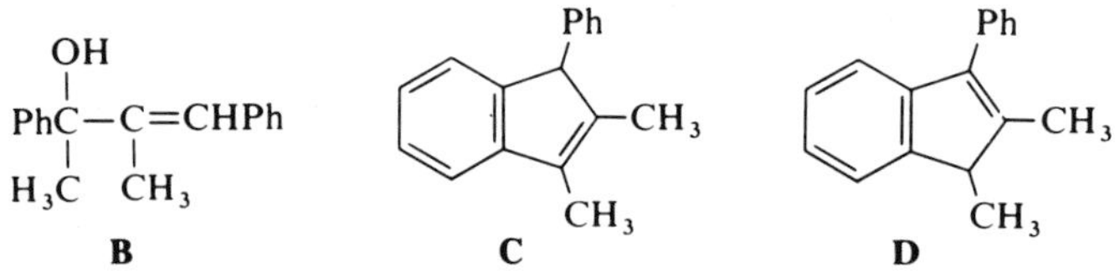

B: PhC(OH)(CH$_3$)—C(CH$_3$)=CHPh   C (Ph, CH$_3$, CH$_3$)   D (Ph, CH$_3$, CH$_3$)

18. Various phenols can be selectively hydroxymethylated at the *ortho* position by heating with paraformaldehyde and phenylboronic acid.

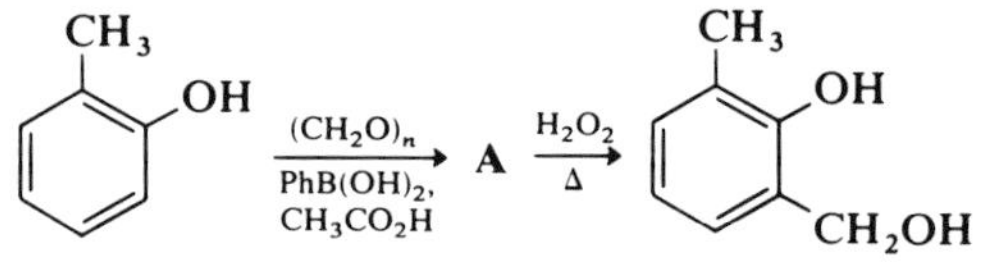

$$\text{2-CH}_3\text{C}_6\text{H}_4\text{OH} \xrightarrow[\text{PhB(OH)}_2,\ \text{CH}_3\text{CO}_2\text{H}]{(\text{CH}_2\text{O})_n} \textbf{A} \xrightarrow[\Delta]{\text{H}_2\text{O}_2} \text{(CH}_3\text{, OH, CH}_2\text{OH)}$$

An intermediate **A**, having the formula $C_{14}H_{13}O_2B$ for the case above, can be isolated after the first step. Postulate a structure for the intermediate and comment on its role in the reaction.

12

Oxidations

This chapter is concerned with reactions which transform a functional group to a more highly oxidized derivative. There is a very large number of such processes, and the reactions have been chosen for discussion on the basis of general utility in organic synthesis. As the reactions are considered, it will become evident that the material in this chapter spans a wider variety of mechanistic patterns than is true for most of the earlier chapters. Because of this range in mechanisms, the chapter has been organized by the functional-group transformation that is accomplished. This organization facilitates comparison of the methods available for effecting a given synthetic transformation. In general, oxidants have been grouped into three classes: transition metal drivatives; oxygen, ozone, and peroxides; and other reagents.

12.1. Oxidation of Alcohols to Aldehydes, Ketones, or Carboxylic Acids

12.1.1. Transition Metal Oxidants

The most widely employed of the transition metal oxidants are Cr(VI)-based reagents. The form of Cr(VI) in aqueous solution depends upon concentration and pH. In dilute solution, the monomeric acid chromate ion is present; as concentration increases, the dichromate ion dominates.

$$2\,HO{-}\overset{\displaystyle O}{\overset{\|}{\underset{\underset{\displaystyle O}{\|}}{Cr}}}{-}O^- \rightleftharpoons {}^-O{-}\overset{\displaystyle O}{\overset{\|}{\underset{\underset{\displaystyle O}{\|}}{Cr}}}{-}O{-}\overset{\displaystyle O}{\overset{\|}{\underset{\underset{\displaystyle O}{\|}}{Cr}}}{-}O^- + H_2O$$

The extent of protonation of these ions depends on the pH. In acetic acid, Cr(VI) is present as mixed anhydrides of acetic acid and chromic acid.[1]

$$CH_3CO_2H + CrO_3 \rightarrow CH_3CO_2\overset{O}{\overset{\|}{\underset{\underset{O}{\|}}{Cr}}}-OH \rightleftharpoons CH_3CO_2\overset{O}{\overset{\|}{\underset{\underset{O}{\|}}{Cr}}}O_2CCH_3$$

In pyridine, an adduct involving Cr—N bonding is formed:

$$C_5H_5N + CrO_3 \longrightarrow C_5H_5\overset{+}{N}-\overset{O}{\overset{\|}{\underset{\underset{O}{\|}}{Cr}}}-O^-$$

The oxidation state of Cr in each of these species is (VI), and they are powerful oxidants. The precise reactivity, however, depends on the solvent and the chromium ligands, so substantial selectivity can be achieved by choice of the particular reagent and conditions.

The transformation that is most often effected with CrO_3-based oxidants is the conversion of an alcohol to the corresponding ketone or aldehyde. The general mechanism of alcohol oxidation is outlined below:

$$R_2CHOH + HCrO_4^- + H^+ \rightarrow R_2CHOCrO_3H + H_2O$$
$$R_2C(H)-O-CrO_3H \xrightarrow{\text{slow}} R_2C{=}O + HCrO_3^- + H^+$$

The kinetics of the reaction also indicate a contribution from a variant that includes an additional proton in the rate-determining step. An important piece of evidence pertinent to identification of the rate-determining step is the fact that a large isotope effect is observed when the α-hydrogen is replaced by deuterium.[2] The Cr(IV) that is produced in the initial step is not stable, and this species is capable of a further one-electron oxidation step. It is believed that a part of the substrate is oxidized via a free-radical intermediate resulting from oxidation by Cr(IV).[3] This scheme also includes Cr(V) as a participant in the mechanism:

$$R_2CHOH + Cr(IV) \rightarrow R_2\dot{C}OH + Cr(III) + H^+$$
$$R_2\dot{C}OH + Cr(VI) \rightarrow R_2C{=}O + Cr(V) + H^+$$
$$R_2CHOH + Cr(V) \rightarrow R_2C{=}O + Cr(III) + 2H^+$$

Because of the participation of radical intermediates, a competing fragmentation reaction is sometimes observed with molecules that can generate a

1. K. B. Wiberg, *Oxidation in Organic Chemistry*, Part A, Academic Press, New York, 1965, pp. 69–72.
2. F. H. Westheimer and N. Nicolaides, *J. Am. Chem. Soc.* **71**, 25 (1949).
3. M. Rahman and J. Rocek, *J. Am. Chem. Soc.* **93**, 5462 (1971); P. M. Nave and W. S. Trahanovsky, *J. Am. Chem. Soc.* **92**, 1120 (1970); K. B. Wiberg and S. K. Mukherjee, *J. Am. Chem. Soc.* **96**, 1884 (1974); M. Doyle, R. J. Swedo, and J. Rocek, *J. Am. Chem. Soc.* **95**, 8352 (1973).

relatively stable radical by homolytic fragmentation. This process can be formulated as involving an alkoxy radical formed by one-electron oxidation. (See Section 10.3.5 for other examples of this type of bond cleavage.)

$$\mathrm{PhC(H)(-CH_2Ph)-O-Cr} \rightarrow \mathrm{PhCH{=}O} + \cdot\mathrm{CH_2Ph}$$

A variety of experimental conditions have been used for oxidations of alcohols by Cr(VI) on a synthetic scale. For simple unfunctionalized alcohols, oxidation can be done by addition of an acidic aqueous solution containing chromic acid (known as *Jones' reagent*) to an acetone solution of the alcohol. Oxidation normally occurs rapidly, and overoxidation is minimal. Often, the reduced chromium salts precipitate, and the acetone solution can be decanted. Entries 2 and 4 in Scheme 12.1 are examples of this technique.

The chromium trioxide–pyridine complex is useful in situations where other functional groups might be susceptible to oxidation or when the molecule is sensitive to strong acid.[4] A procedure for utilizing the CrO_3-pyridine complex which was originated by Collins and co-workers[5] has become quite widely accepted. The CrO_3-pyridine complex is isolated and dissolved in dichloromethane. With an excess of the reagent, oxidation of simple alcohols is complete in a few minutes, giving the aldehyde or ketone in good yield. A procedure that avoids isolation of the complex can further simplify the experimental operations.[6] Chromium trioxide is added to pyridine in dichloromethane. Subsequent addition of the alcohol to this solution results in oxidation to high yield. Other modifications for use of the CrO_3-pyridine complex have been developed.[7]Entries 5–8 in Scheme 12.1 demonstrate the excellent results that have been reported using the CrO_3–pyridine complex in dichloromethane.

Another very useful Cr(VI) reagent is pyridinium chlorochromate (PCC), which is prepared by dissolving CrO_3 in hydrochloric acid and adding pyridine to obtain a solid reagent having the composition CrO_3ClH-pyridine.[8] This reagent can be used in close to the stoichiometric ratio. Entries 10 and 11 in Scheme 12.1 are examples of the use of this reagent. Reaction of pyridine with CrO_3 in a small amount of water gives pyridinium dichromate, which is also a useful oxidant.[9] Dissolved in DMF or as a suspension in dichloromethane, this reagent oxidizes secondary alcohols to ketones. Allylic primary alcohols give the corresponding

4. G. I. Poos, G. E. Arth, R. E. Beyler, and L. H. Sarett, *J. Am. Chem. Soc.* **75**, 422 (1953); W. S. Johnson, W. A. Vredenburgh, and J. E. Pike, *J. Am. Chem. Soc.* **82**, 3409 (1960); W. S. Allen, S. Bernstein, and R. Little, *J. Am. Chem. Soc.* **76**, 6116 (1954).
5. J. C. Collins, W. W. Hess, and F. J. Frank, *Tetrahedron Lett.*, 3363 (1968).
6. R. Ratcliffe and R. Rodehorst, *J. Org. Chem.* **35**, 4000 (1970).
7. J. Herscovici, M.-J. Egron, and K. Antonakis, *J. Chem. Soc., Perkin Trans. 1,* 1967 (1982); E. J. Corey and G. Schmidt, *Tetrahedron Lett.*, 399 (1979); S. Czernecki, C. Georgoulis, C. L. Stevens, and K. Vijayakumaran, *Tetrahedron Lett.* **26**, 1699 (1985).
8. E. J. Corey and J. W. Suggs, *Tetrahedron Lett.*, 2647 (1975); G. Piancatelli, A. Scettri, and M. D'Auria, *Synthesis*, 245 (1982).
9. E. J. Corey and G. Schmidt, *Tetrahedron Lett.*, 399 (1979).

Scheme 12.1. Oxidations with Cr(VI)

A. Chromic Acid Solutions

1[a] $CH_3CH_2CH_2OH \xrightarrow[H_2O]{H_2CrO_4} CH_3CH_2CH{=}O$ (45–49%)

2[b] cyclooctanol $\xrightarrow[\text{acetone}]{H_2CrO_4}$ cyclooctanone (92–96%)

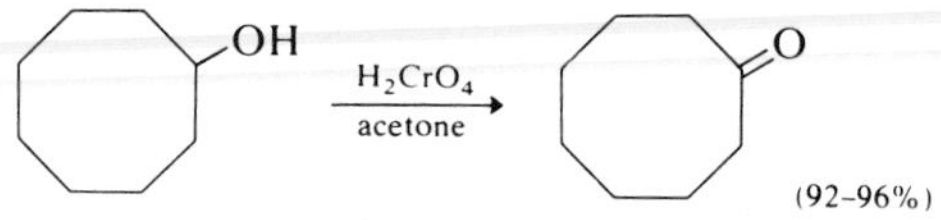

3[c] $\xrightarrow[\text{ether}]{H_2CrO_4}$ (84%)

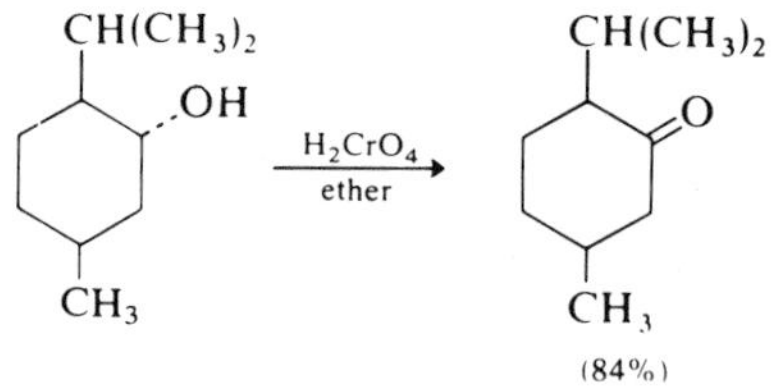

4[d] $\xrightarrow[\text{acetone}]{H_2CrO_4}$ (79–88%)

B. Chromium Trioxide–Pyridine

5[e] $CH_3(CH_2)_5CH_2OH \xrightarrow[CH_2Cl_2]{CrO_3\text{-pyridine}} CH_3(CH_2)_5CH{=}O$ (70–84%)

6[f] $CH_3CH_2CH(CH_3)(CH_2)_4CH_2OH \xrightarrow[CH_2Cl_2]{CrO_3\text{-pyridine}} CH_3CH_2CH(CH_3)(CH_2)_4CH{=}O$ (69%)

7[g] $\xrightarrow[CH_2Cl_2]{CrO_3\text{-pyridine}}$ (95%)

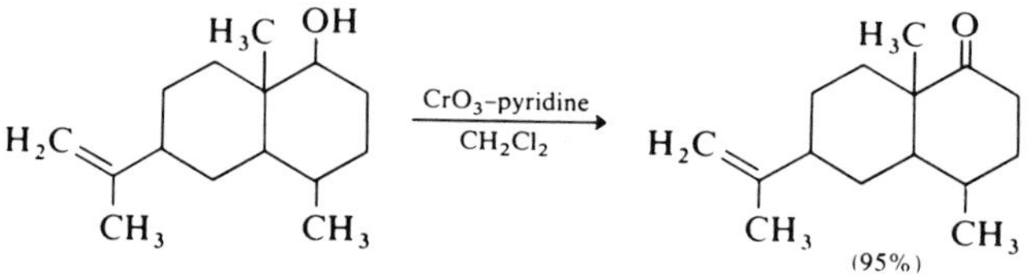

8[h] $\xrightarrow[CH_2Cl_2]{CrO_3\text{-pyridine}}$

9[i] $\xrightarrow[\text{mol. sieves, } CH_3CO_2H]{CrO_3\text{-pyridine}}$ (96%)

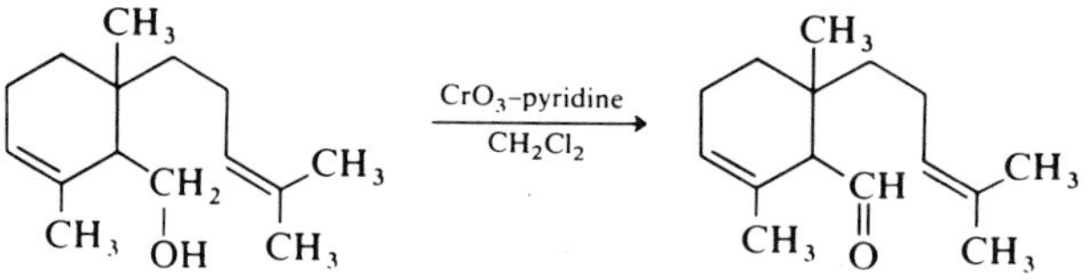

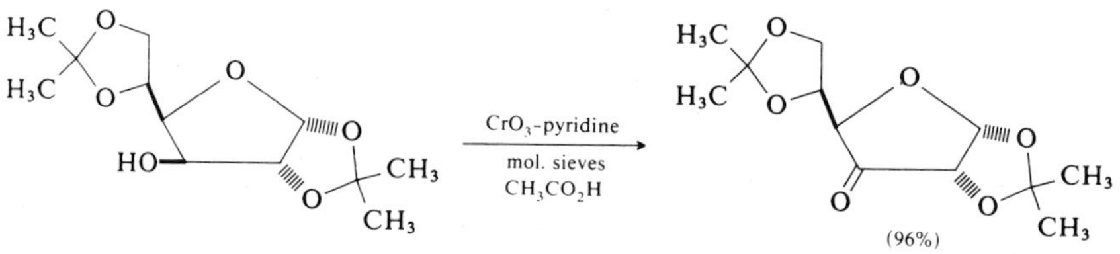

C. Pyridinium Chlorochromate

10[j] $(CH_3)_2C{=}CHCH_2CH_2CH(CH_3)CH_2CH_2OH \xrightarrow{PCC} (CH_3)_2C{=}CHCH_2CH_2CH(CH_3)CH_2CH{=}O$ (82%)

11[k] $HOCH_2CH_2C(CH_3)_2CH_2CH{=}CHCO_2CH_3 \xrightarrow{PCC} O{=}CHCH_2C(CH_3)_2CH_2CH{=}CHCO_2CH_3$ (83%)

a. C. D. Hurd and R. N. Meinert, *Org. Synth.* **II**, 541 (1943).
b. E. J. Eisenbraun, *Org. Synth.* **IV**, 310 (1973).
c. H. C. Brown, C. P. Garg, and K.-T. Liu, *J. Org. Chem.* **36**, 387 (1971).
d. J. Meinwald, J. Crandall, and W. E. Hymans, *Org. Synth.* **45**, 77 (1965).
e. J. C. Collins and W. W. Hess, *Org. Synth.* **52**, 5 (1972).
f. J. I. DeGraw and J. O. Rodin, *J. Org. Chem.* **36**, 2902 (1971).
g. R. Ratcliffe and R. Rodehorst, *J. Org. Chem.* **35**, 4000 (1970).
h. M. A. Schwartz, J. D. Crowell, and J. H. Musser, *J. Am. Chem. Soc.*, **94**, 4361 (1972).
i. C. Czernecki, C. Georgoulis, C. L. Stevens, and K. Vijayakumaran, *Tetrahedron Lett.* **26**, 1699 (1985).
j. E. J. Corey and J. W. Suggs, *Tetrahedron Lett.*, 2647 (1975).
k. R. D. Little and G. W. Muller, *J. Am. Chem. Soc.* **103**, 2744 (1981).

aldehydes. Depending upon the conditions, saturated primary alcohols give either an aldehyde or the corresponding carboxylic acid.

Potassium permanganate has found relatively little application in the oxidation of alcohols to ketones and aldehydes. The reagent is less selective than Cr(VI), and overoxidation is a problem. On the other hand, manganese(IV) dioxide is quite useful.[10] This reagent preferentially attacks allylic and benzylic hydroxyl groups and therefore possesses a useful degree of selectivity. Manganese dioxide is prepared by reaction of manganese(II) sulfate with potassium permanganate and sodium hydroxide. The precise reactivity of MnO_2 depends on its mode of preparation and the extent of drying.[11] Scheme 12.2 illustrates the various classes of alcohols that are most susceptible to MnO_2 oxidation.

Another reagent which finds application in oxidations of alcohols to ketones is ruthenium tetroxide. For example, the oxidation of **1** to **2** was successfully achieved with this reagent after a number of other methods failed.

HO … 1 → O … 2 (O, O) Ref. 12

10. D. G. Lee, in *Oxidation*, Vol. 1, R. L. Augustine (ed.), Marcel Dekker, New York, 1969, pp. 66–70; A. J. Fatiadi, *Synthesis*, 65, 133 (1976).
11. J. Attenburrow, A. F. B. Cameron, J. H. Chapman, R. M. Evans, A. B. A. Jansen, and T. Walker, *J. Chem. Soc.*, 1094 (1952); I. M. Goldman, *J. Org. Chem.* **34**, 1979 (1969).
12. R. M. Moriarty, H. Gopal, and T. Adams, *Tetrahedron Lett.*, 4003 (1970).

Scheme 12.2. Oxidations of Alcohols with Manganese Dioxide

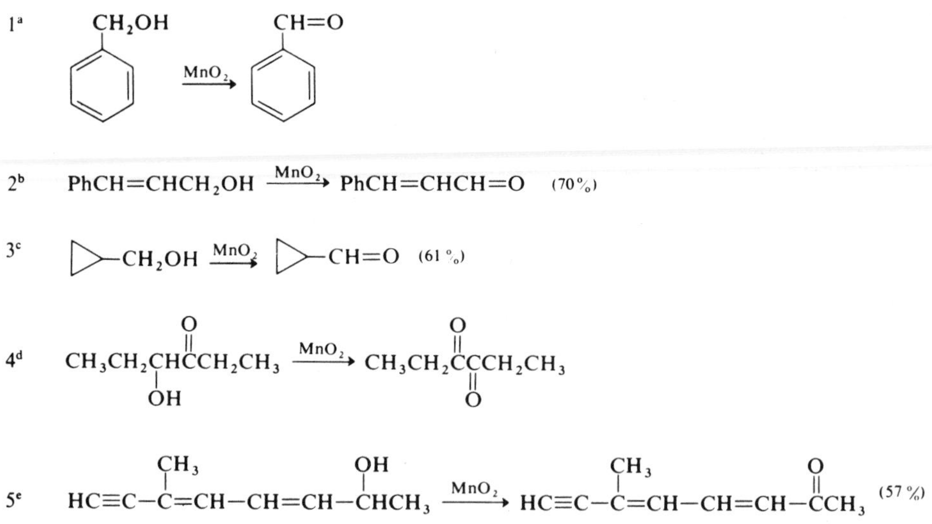

1[a] $C_6H_5CH_2OH \xrightarrow{MnO_2} C_6H_5CH{=}O$

2[b] $PhCH{=}CHCH_2OH \xrightarrow{MnO_2} PhCH{=}CHCH{=}O$ (70%)

3[c] $c\text{-}C_3H_5\text{-}CH_2OH \xrightarrow{MnO_2} c\text{-}C_3H_5\text{-}CH{=}O$ (61%)

4[d] $CH_3CH_2CH(OH)C(=O)CH_2CH_3 \xrightarrow{MnO_2} CH_3CH_2C(=O)C(=O)CH_2CH_3$

5[e] $HC{\equiv}C{-}C(CH_3){=}CH{-}CH{=}CH{-}CH(OH)CH_3 \xrightarrow{MnO_2} HC{\equiv}C{-}C(CH_3){=}CH{-}CH{=}CH{-}C(=O)CH_3$ (57%)

a. E. F. Pratt and J. F. Van De Castle, *J. Org. Chem.* **26**, 2973 (1961).
b. I. M. Goldman, *J. Org. Chem.* **34**, 1979 (1969).
c. L. Crombie and J. Crossley, *J. Chem. Soc.* 4983 (1963).
d. E. P. Papadopoulos, A. Jarrar, and C. H. Issidorides, *J. Org. Chem.* **31**, 615 (1966).
e. J. Attenburrow, A. F. B. Cameron, J. H. Chapman, R. M. Evans, B. A. Hems, A. B. A. Jansen, and T. Walker, *J. Chem. Soc.* 1094 (1952).

This compound is a potent oxidant, however, and it readily attacks carbon–carbon double bonds.[13] Procedures for *in situ* generation of RuO_4 from RuO_2 using periodate or hypochlorite as oxidants are available.[14]

12.1.2. Other Oxidants

A very useful group of procedures for oxidation of alcohols to ketones have been developed which involve dimethyl sulfoxide (DMSO) and any one of a number of electrophilic molecules, such as dicyclohexylcarbodiimide, acetic anhydride, trifluoroacetic anhydride, oxalyl chloride, and sulfur trioxide.[15] The initial work involved DMSO and dicyclohexylcarbodiimide.[16] The primary utility of this method is for oxidation of molecules that are sensitive to the more powerful transition metal oxidants. The mechanism of the oxidation involves formation of intermediate **A** by

13. J. L. Courtney and K. F. Swansborough, *Rev. Pure Appl. Chem.* **22**, 47 (1972); D. G. Lee and M. van den Engh, in *Oxidation*, Part B, W. S. Trahanovsky (ed.), Academic Press, New York, 1973, Chapter IV.
14. P. E. Morris, Jr., and D. E. Kiely, *J. Org. Chem.* **52**, 1149 (1987).
15. A. J. Mancuso and D. Swern, *Synthesis*, 165 (1981).
16. K. E. Pfitzner and J. G. Moffatt, *J. Am. Chem. Soc.* **87**, 5661, 5670 (1965).

nucleophilic attack of DMSO on the carbodiimide, followed by reaction of this intermediate with the alcohol.[17]

$$RN{=}C{=}NR \;+\; {}^{-}O{-}\overset{+}{S}(CH_3)_2 \;\underset{}{\overset{H^+}{\rightleftharpoons}}\; \underset{\mathbf{A}}{RNH{-}\underset{O-\overset{+}{S}(CH_3)_2}{C}{=}NR} \xrightarrow{R_2CHOH} \underset{\mathbf{B}}{R_2CH{-}O{-}S(CH_3)(CH_2{-}H)\cdots O{-}C(NHR){=}NR} \longrightarrow \underset{\mathbf{C}}{R_2\underset{H}{C}{-}O{-}\overset{+}{S}(CH_3)\overset{-}{C}H_2}$$

$$\downarrow$$

$$R_2C{=}O + (CH_3)_2S + RNH\overset{O}{\overset{\|}{C}}NHR$$

The activation of DMSO toward the addition step can be accomplished by other electrophiles. All of these reagents are believed to form a sulfoxonium species by electrophilic attack at the sulfoxide oxygen. The addition of the alcohol and the departure of the sulfoxide oxygen as part of a leaving group generates an intermediate comparable to **C** in the above mechanism.

$$(CH_3)_2\overset{+}{S}{-}O^- + X^+ \rightarrow (CH_3)_2\overset{+}{S}{-}O{-}X$$

$$R_2CHOH + (CH_3)_2\overset{+}{S}{-}O{-}X \rightarrow R_2CHO{-}\underset{CH_3}{\overset{CH_3}{S}}{-}O{-}X \rightarrow R_2CHO{-}\overset{+}{S}(CH_3)_2 + {}^{-}OX$$

$$R_2CHO{-}\overset{+}{S}(CH_3)_2 \rightarrow R_2C{=}O + (CH_3)_2S + H^+$$

Preparatively useful procedures based on acetic anhydride,[18] trifluoroacetic anhydride,[19] and oxalyl chloride[20] have been developed. The latter method, known as *Swern oxidation,* is currently the most popular and is frequently used as an alternative to the Cr(VI) oxidation. Scheme 12.3 gives some representative examples of these methods. Entry 4 is an example of the use of a water-soluble carbodiimide as the activating reagent. The modified carbodiimide facilitates product purification by providing for easy removal of the urea by-product.

Oxidation of alcohols under extremely mild conditions can be achieved by a procedure that is mechanistically related to the DMSO methods. Dimethyl sulfide is converted to a chlorosulfonium ion by reaction with *N*-chlorosuccinimide. This sulfonium ion reacts with alcohols, generating the same kind of alkoxysulfonium

17. J. G. Moffatt, *J. Org. Chem.* **36**, 1909 (1971).
18. J. D. Albright and L. Goldman, *J. Am. Chem. Soc.* **89**, 2416 (1967).
19. J. Yoshimura, K. Sato, and H. Hashimoto, *Chem. Lett.*, 1327 (1977); K. Omura, A. K. Sharma, and D. Swern, *J. Org. Chem.* **41**, 957 (1976); S. L. Huang, K. Omura, and D. Swern, *J. Org. Chem.* **41**, 3329 (1976).
20. A. J. Mancuso, S.-L. Huang, and D. Swern, *J. Org. Chem.* **43**, 2480 (1978).

Scheme 12.3. Oxidation of Alcohols Using Dimethyl Sulfoxide

1[a] (steroid CH_2OH → $CH{=}O$) DMSO, dicyclohexylcarbodiimide (84%)

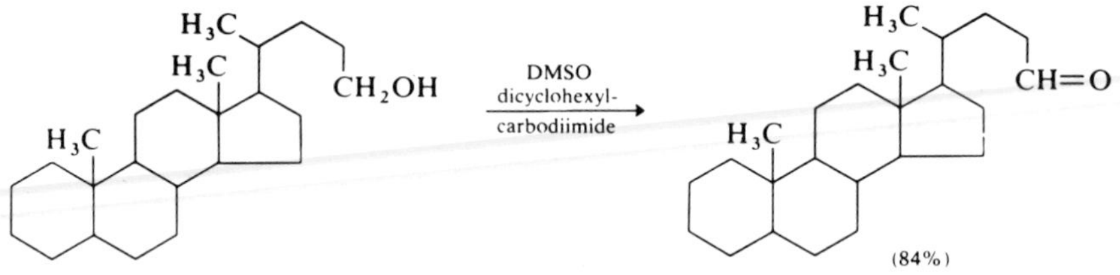

2[b] DMSO, SO_3 (44%)

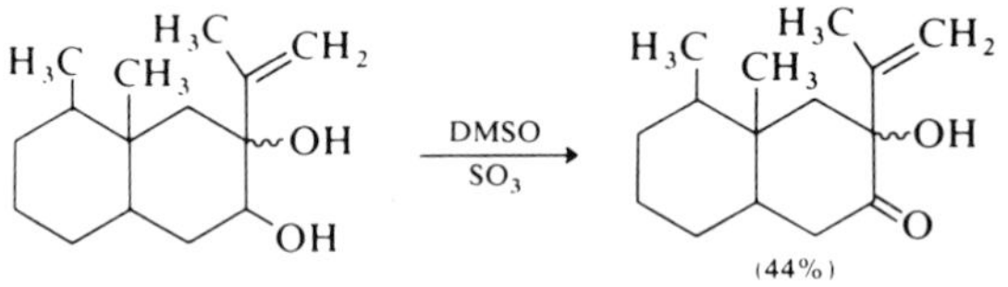

3[c] indole–$C(CH_3)_2$–CH_2OH → indole–$C(CH_3)_2$–$CH{=}O$ DMSO, $(CH_3CO)_2O$ (60%)

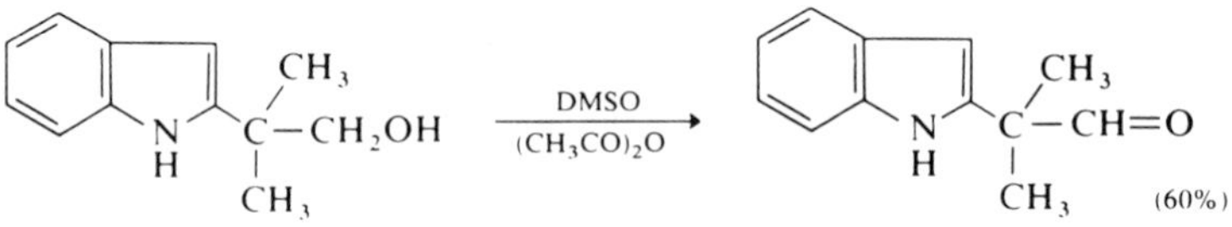

4[d] $PhSCH_2ON{=}$cyclopentane with $(CH_2)_6CO_2CH_3$, CH_2OH, O-THP → $CH{=}O$; DMSO, cyclohexyl–$N{=}C{=}NCH_2CH_2N^+$(morpholinium, CH_3) (98%)

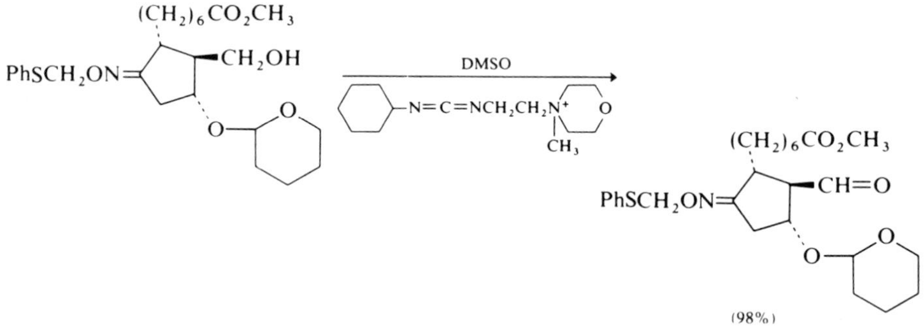

5[e] $(CH_3)_2CHCH{=}CHCH{=}CHCH_2OH \xrightarrow[\text{ClCOCOCl}]{\text{DMSO}} (CH_3)_2CHCH{=}CHCH{=}CHCH{=}O$ (93%)

6[f] CN, OH, $CHCH$, OCH_3, CH_3, CH_3 → CHC=O; DMSO, $(CF_3CO)_2O$ (90%)

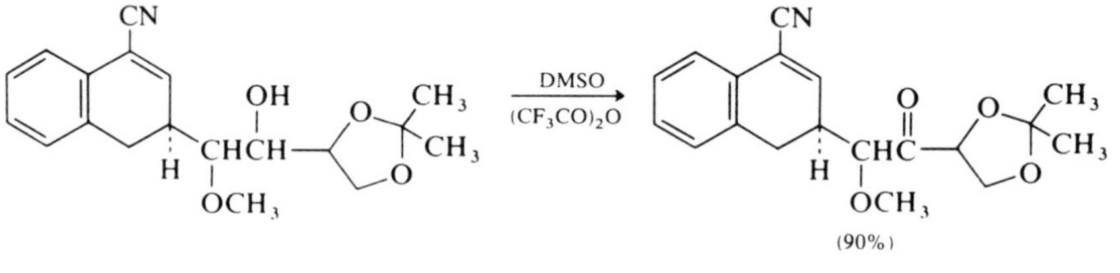

7[g] $CH_3CH{=}CHCH_2CH_2CH(OH)CH_2CO_2CH_3 \xrightarrow[\text{Et}_3\text{N}]{\text{DMSO, P}_2\text{O}_5} CH_3CH{=}CHCH_2CH_2C(O)CH_2CO_2CH_3$ (85%)

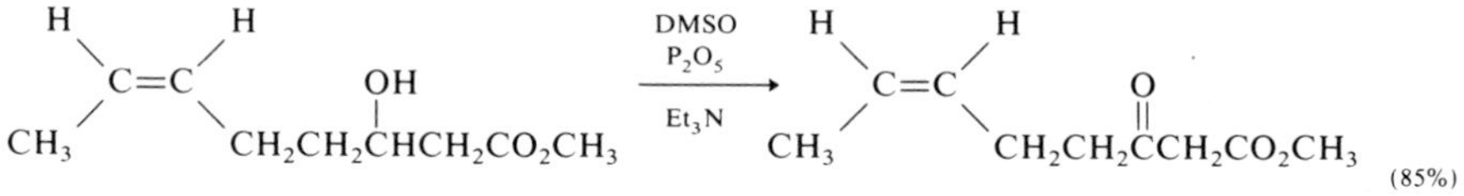

a. J. G. Moffatt, *Org. Synth.* **47**, 25 (1967).
b. J. A. Marshall and G. M. Cohen, *J. Org. Chem.* **36**, 877 (1971).
c. E. Houghton and J. E. Saxton, *J. Chem. Soc. C*, 595 (1969).
d. N. Finch, L. D. Vecchia, J. J. Fitt, R. Stephani, and I. Vlattas, *J. Org. Chem.* **38**, 4412 (1973).
e. W. R. Roush, *J. Am. Chem. Soc.* **102**, 1390 (1980).
f. R. W. Franck and T. V. John, *J. Org. Chem.* **45**, 1170 (1980).
g. D. F. Taber, J. C. Amedio, Jr., and K.-Y. Jung, *J. Org. Chem.* **52**, 5621 (1987).

ion that is involved in the DMSO procedures. In the presence of a mild base, elimination of dimethyl sulfide completes the oxidation.[21]

$$\text{N-chlorosuccinimide} + (CH_3)_2S \rightarrow (CH_3)_2\overset{+}{S}{-}Cl \xrightarrow{R_2CHOH} R_2C(H){-}O{-}\overset{+}{S}(CH_3)_2 \;(B:) \rightarrow R_2C{=}O + (CH_3)_2S$$

Similarly, reaction of chlorine and DMSO at low temperature gives an adduct that reacts with alcohols to give the ketone and DMSO.[22]

$$(CH_3)_2S{=}O \xrightarrow{Cl_2} (CH_3)_2\overset{+}{S}(=O){-}Cl \xrightarrow{R_2CHOH} (CH_3)_2\overset{+}{S}(=O){-}O{-}CR_2(H) \xrightarrow{Et_3N} (CH_3)_2S{=}O + R_2C{=}O$$

The development of the CrO_3-pyridine and DMSO-based methods has decreased the number of instances in which older oxidation techniques are used. One such method, the *Oppenauer oxidation*,[23] is the reverse of the Meerwein–Pondorff–Verley reduction (Chapter 5). It involves heating the alcohol to be oxidized with an aluminum alkoxide in the presence of a carbonyl compound, which acts as the hydrogen acceptor. The reaction is an equilibrium process and proceeds through a cyclic transition state.

$$R_2CHOAl[OCH(CH_3)_2]_2 + (CH_3)_2C{=}O \rightleftarrows [R_2C(H){-}O{-}Al[OCH(CH_3)_2]_2 \cdots O{=}C(CH_3)_2]^{\ddagger} \rightleftarrows R_2C{=}O + Al[OCH(CH_3)_2]_3$$

The reaction can be driven in the desired direction by using a carbonyl compound that is a good acceptor of hydrogen, that is, one that is easily reduced. Quinone and fluorenone have been utilized for this purpose. Alternatively, an excess of a hydrogen acceptor can be used. Since the reaction conditions are nonacidic, this method is valuable for very acid-sensitive molecules.

$$\text{benzobicyclic alcohol (OH)} \xrightarrow[\text{Al(O-}t\text{-Bu)}_3]{\text{quinone}} \text{ketone (90\%)}$$ Ref. 24

21. E. J. Corey and C. U. Kim, *J. Am. Chem. Soc.* **94**, 7586 (1972).
22. E. J. Corey and C. U. Kim, *Tetrahedron Lett.*, 919 (1973).
23. C. Djerassi, *Org. React.* **6**, 207 (1951).
24. P. D. Bartlett and W. P. Giddings, *J. Am. Chem. Soc.* **82**, 1240 (1960).

Probably related mechanistically to the Oppenauer oxidations are several methods for oxidation that involve transfer of hydrogen to trichloroacetaldehyde. The reaction is mediated by alumina and is carried out by simply mixing the alcohol to be oxidized, the hydrogen acceptor, and alumina in an inert solvent.[25] This reaction is suitable for selective oxidation of secondary alcohols in the presence of primary alcohols (which do not react) and also for the oxidation of compounds containing other easily oxidized functional groups.

SePh, OH → ($Cl_3CCH{=}O$, Al_2O_3, 55°C, 24 h) SePh, O   Ref. 26

12.2. Addition of Oxygen at Carbon–Carbon Double Bonds

12.2.1. Transition Metal Oxidants

Compounds of certain transition metals in which the metals are in high oxidation states such as the permanganate ion and osmium tetroxide, are effective reagents for addition of atoms at a carbon–carbon double bond. Under mild reaction conditions, potassium permanganate can effect conversion of alkenes to glycols. This oxidant is, however, capable of further oxidizing the glycol with cleavage of the carbon–carbon bond, so careful control of the reaction conditions is necessary. A cyclic manganese ester is an intermediate in these oxidations. Because of the cyclic nature of this intermediate, the glycols are formed by *syn* addition.

$$RCH{=}CHR + MnO_4^- \rightarrow \text{cyclic Mn ester} \xrightarrow[OH^-]{H_2O} \underset{HO\ \ OH}{RCHCHR}$$

Ketols are also observed as products of permanganate oxidation of alkenes. The ratio between the two products is pH dependent, with alkaline pH favoring the glycol. The ketols are believed to be formed as a result of oxidation of the cyclic intermediate.[27]

$$\text{R–CH–O, R–CH–O (cyclic Mn ester, Mn(=O)O}^-) \rightarrow \text{R–CH–O, R–C(H)–O (Mn ester)} \rightarrow \text{R–CH–OH, R–C{=}O} + MnO_2$$

25. G. H. Posner, *Angew. Chem. Int. Ed. Engl.* **17**, 487 (1978); G. H. Posner, R. B. Perfetti, and A. W. Runquist, *Tetrahedron Lett.*, 3499 (1976).
26. G. H. Posner and M. J. Chapdelaine, *Tetrahedron Lett.*, 3227 (1977).
27. S. Wolfe, C. F. Ingold, and R. U. Lemieux, *J. Am. Chem. Soc.* **103**, 938 (1981).

It is also possible to carry out permanganate oxidation in organic solvents by using crown ethers or phase transfer agents to effect solubilization of permanganate salts. Studies with 5-decene and cyclodecene revealed that mixtures of carboxylic acids, diones, ketols, and diols were obtained but that the ratio of the various products could be influenced by the amount of oxidant used.

Cyclodecene —(3.3 mol $KMnO_4$, benzene, polyether)→ $HO_2C(CH_2)_8CO_2H$ (59%) + cyclodecane-1,2-dione (16%) Ref. 28

Cyclodecene —(1.0 mol $KMnO_4$, benzene, $R_4N^+Cl^-$)→ cyclodecane-1,2-diol (50%) Ref. 29

Permanganate ion can be used to oxidize acetylenes to diones.

$$PhC{\equiv}CCH_2CH_2CH_3 \xrightarrow[R_4N^+,\ CH_2Cl_2]{KMnO_4} PhC(=O){-}C(=O)CH_2CH_2CH_3 \quad (81\%)$$ Ref. 30

A mixture of $NaIO_4$ and RuO_2 in a heterogeneous solvent system is also effective.

$$PhC{\equiv}CCH_3 \xrightarrow[NaIO_4]{RuO_2} PhC(=O)C(=O)CH_3 \quad (80\%)$$ Ref. 31

Osmium tetroxide is a highly selective oxidant which gives glycols by a stereospecific *syn* addition.[32] The reaction occurs through a cyclic osmate ester.

$$RCH{=}CHR + OsO_4 \rightarrow \text{cyclic osmate ester} \rightarrow RCH(OH)CH(OH)R$$

The reagent is quite expensive, but this disadvantage can be minimized by procedures that use only a catalytic amount of osmium tetroxide. A very useful procedure

28. D. G. Lee and V. S. Chang, *J. Org. Chem.* **43**, 1532 (1978).
29. W. P. Weber and J. P. Shepherd, *Tetrahedron Lett.*, 4907 (1972).
30. D. G. Lee and V. S. Chang, *J. Org. Chem.* **44**, 2726 (1979).
31. R. Zibuck and D. Seebach, *Helv. Chim. Acta* **71**, 237 (1988).
32. M. Schröder, *Chem. Rev.* **80**, 187 (1980).

involves an amine oxide, such as morpholine-*N*-oxide, as the stoichiometric oxidant.[33]

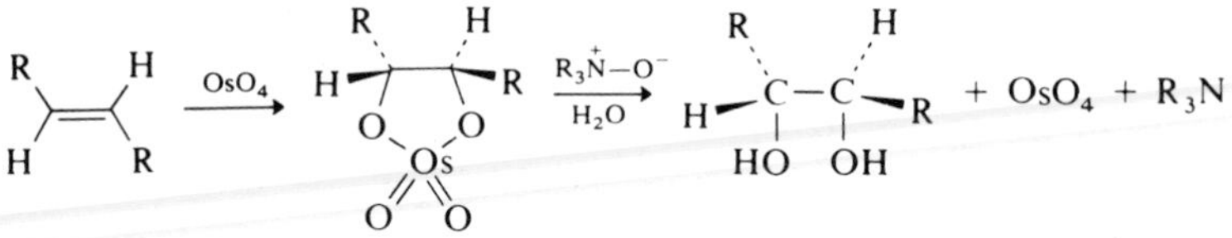

t-Butyl hydroperoxide[34] or barium chlorate[35] can also be used as oxidants in catalytic procedures. Scheme 12.4 provides some examples of oxidations of alkenes to glycols by both permanganate and osmium tetroxide.

Scheme 12.4. Hydroxylation of Alkenes

A. Potassium Permanganate

1[a] $CH_2{=}CHCH(OC_2H_5)_2 + KMnO_4 \rightarrow HOCH_2CH(OH)CH(OC_2H_5)_2$ (67%)

2[b] (58%)

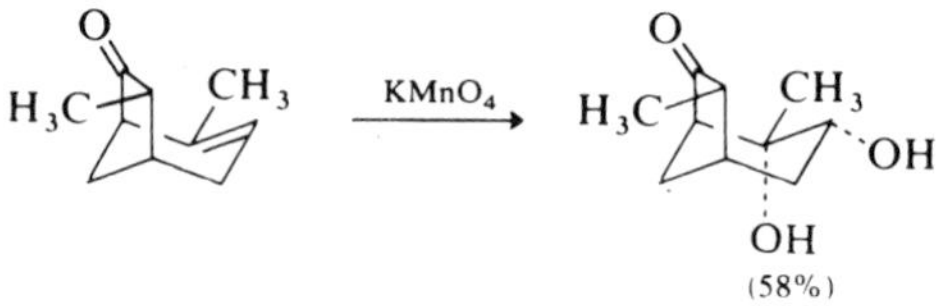

B. Osmium Tetroxide

3[c] $CH_3CH_2CH{=}CHC{\equiv}CCH_3 \xrightarrow[OsO_4]{\text{N-methylmorpholine N-oxide}} CH_3CH_2CH(OH)CH(OH)C{\equiv}CCH_3$ (65%)

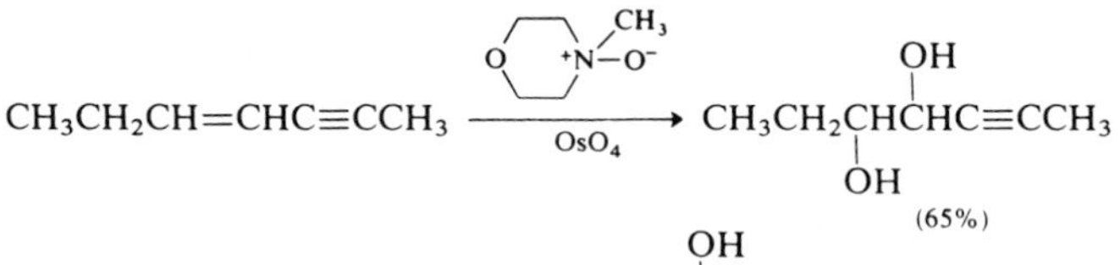

4[d] $CH_3CH{=}CHCO_2C_2H_5 \xrightarrow[t\text{-BuOOH}]{OsO_4} CH_3CH(OH)CH(OH)CO_2C_2H_5$ (72%)

5[e] $\xrightarrow[BaClO_3]{OsO_4}$ (84%)

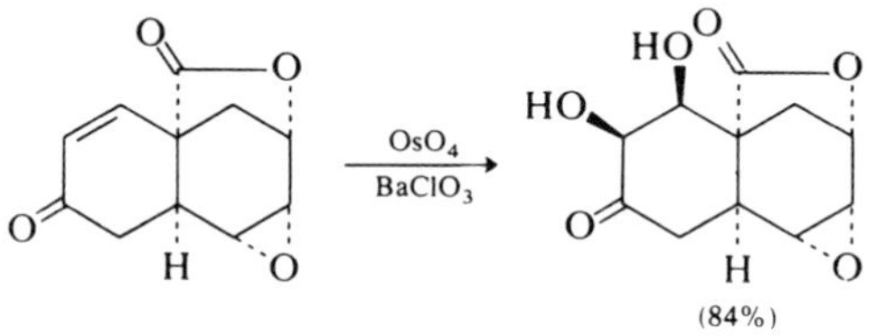

a. E. J. Witzeman, W. L. Evans, H. Haas, and E. F. Schroeder, *Org. Synth.* **II**, 307 (1943).
b. S. D. Larsen and S. A. Monti, *J. Am. Chem. Soc.* **99**, 8015 (1977).
c. E. J. Corey, P. B. Hopkins, S. Kim, S. Yoo, K. P. Nambiar, and J. R. Falck, *J. Am. Chem. Soc.* **101**, 7131 (1979).
d. K. Akashi, R. E. Palermo, and K. B. Sharpless, *J. Org. Chem.* **43**, 2063 (1978).
e. S. Danishefsky, P. F. Schuda, T. Kitahara, and S. J. Etheredge, *J. Am. Chem. Soc.* **99**, 6066 (1977).

33. V. Van Rheenen, R. C. Kelly, and D. Y. Cha, *Tetrahedron Lett.*, 1973 (1976).
34. K. B. Sharpless and K. Akashi, *J. Am. Chem. Soc.* **98**, 1986 (1976); K. Akashi, R. E. Palermo, and K. B. Sharpless, *J. Org. Chem.* **43**, 2063 (1978).
35. L. Plaha, J. Weichert, J. Zvacek, S. Smolik, and B. Kakac, *Collect. Czech. Chem. Commun.* **25**, 237 (1960); A. S. Kende, T. V. Bentley, R. A. Mader, and D. Ridge, *J. Am. Chem. Soc.* **96**, 4332 (1974).

Other transition metal oxidants convert alkenes to epoxides. The most useful procedures involve *t*-butyl hydroperoxide as the stoichiometric oxidant in combination with vanadium, molybdenum, or titanium compounds. The most useful substrates for oxidation are allylic alcohols. The hydroxyl group of the alcohol plays both an activating and a stereodirecting role in these reactions. *t*-Butyl hydroperoxide and a catalytic amount of VO(acac) convert allylic alcohols to the corresponding epoxides in good yields.[36] The reaction proceeds through a complex in which the allylic alcohol is coordinated to vanadium through the hydroxyl group. In cyclic alcohols, this results in epoxidation *cis* to the hydroxyl group. In acyclic alcohols, the observed stereochemistry is consistent with a transition state in which the double bond is oriented at an angle of about 50° to the coordinated hydroxy group. This transition state leads preferentially to formation of an alcohol with *erythro* or *syn* stereochemistry from allylic alcohols with a terminal double bond. When the double bond is disubstituted the stereoselectivity is lower. The *Z*-isomer favors the *threo* isomer while for the *E*-isomer the *erythro* isomer is the major product.[37]

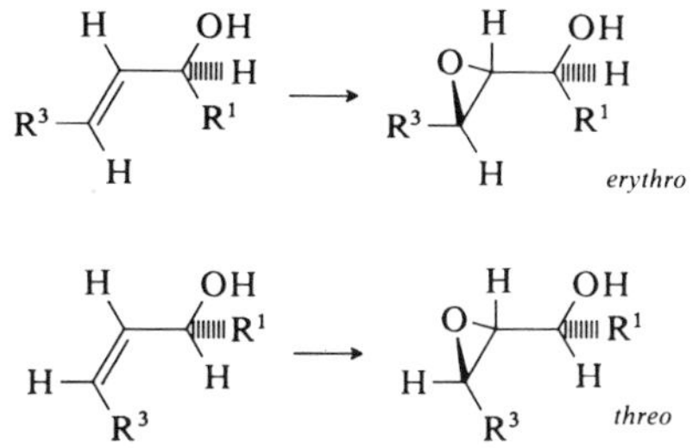

The epoxidation of allylic alcohols can also be effected by *t*-butyl hydroperoxide and titanium tetraisopropoxide. When enantiomerically pure tartrate esters are included in the system, the reaction is highly enantioselective. This reaction is called the *Sharpless asymmetric epoxidation*.[38] Either the (+)- or (−)-tartrate ester can be used so either enantiomer of the desired product can be obtained.

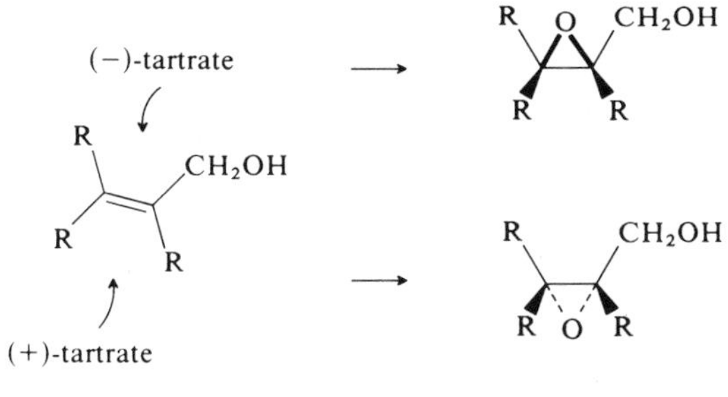

The mechanism by which the enantioselective oxidation occurs is generally similar to that for the vanadium-catalyzed oxidations. The allylic hydroxyl group serves to coordinate the reactant to titanium. The tartrate esters are also coordinated

36. K. B. Sharpless and R. C. Michaelson, *J. Am. Chem. Soc.* **95**, 6136 (1973).
37. E. D. Mihelich, *Tetrahedron Lett.*, 4729 (1979); B. E. Rossiter, T. R. Verhoeven, and K. B. Sharpless, *Tetrahedron Lett.*, 4733 (1979).
38. For a review, see A. Pfenninger, *Synthesis*, 89 (1986).

at titanium, which creates a chiral environment. Oxidation occurs through an intermediate in which both the allylic alcohol and *t*-butyl hydroperoxide are complexed to the titanium ion.

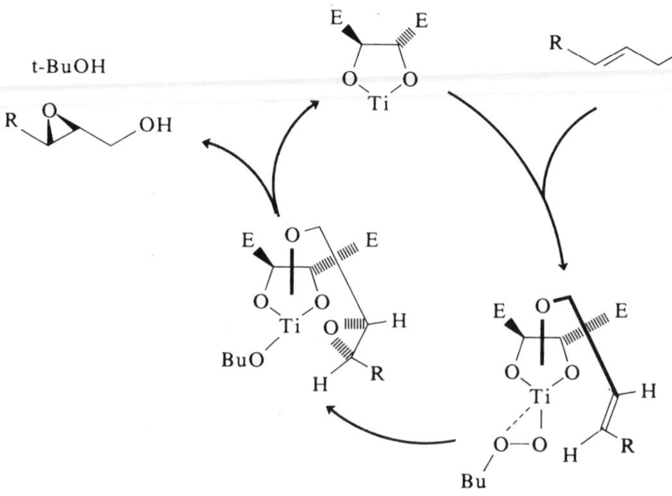

The orientation of the reactive ligands is governed by the chirality of the tartrate ester. In the transition state, an oxygen atom from the peroxide is transferred to the double bond. The enantioselectivity is consistent with a transition state such as that shown below.[39]

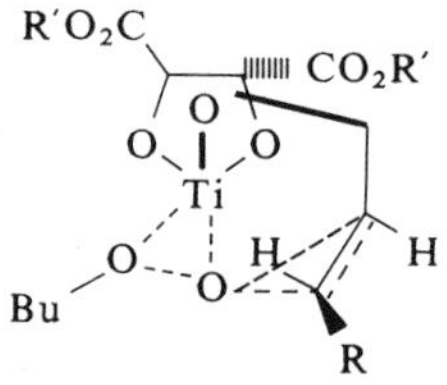

This method has proven to be an extremely useful means of synthesizing enantiomerically enriched compounds. Various improvements in the methods for carrying out the Sharpless oxidation have been developed.[40] The reaction can be done with catalytic amounts of titanium isopropoxide and the tartrate ester.[41] This procedure uses molecular sieves to sequester water, which has a deleterious effect on both the rate and enantioselectivity of the reaction. Scheme 12.5 gives some examples of enantioselective oxidation of allylic alcohols.

39. V. S. Martin, S. S. Woodard, T. Katsuki, Y. Yamada, M. Ikeda, and K. B. Sharpless, *J. Am. Chem. Soc.* **103**, 6237 (1981); K. B. Sharpless, S. S. Woodard, and M. F. Finn, *Pure Appl. Chem.* **55**, 1823 (1983); M. G. Finn and K. B. Sharpless, in *Asymmetric Synthesis*, Vol. 5, J. D. Morrison (ed.), Academic Press, New York, 1985, Chapter 8.
40. J. G. Hill, B. E. Rossiter, and K. B. Sharpless, *J. Org. Chem.* **48**, 3607 (1983); L. A. Reed III, S. Masamune, and K. B. Sharpless, *J. Am. Chem. Soc.* **104**, 6468 (1982).
41. R. M. Hanson and K. B. Sharpless, *J. Org. Chem.* **51**, 1922 (1986); Y. Gao, R. M. Hanson, J. M. Klunder, S. Y. Ko, H. Masamune, and K. B. Sharpless, *J. Am. Chem. Soc.* **109**, 5765 (1987).

Scheme 12.5. Enantioselective Epoxidation of Allylic Alcohols

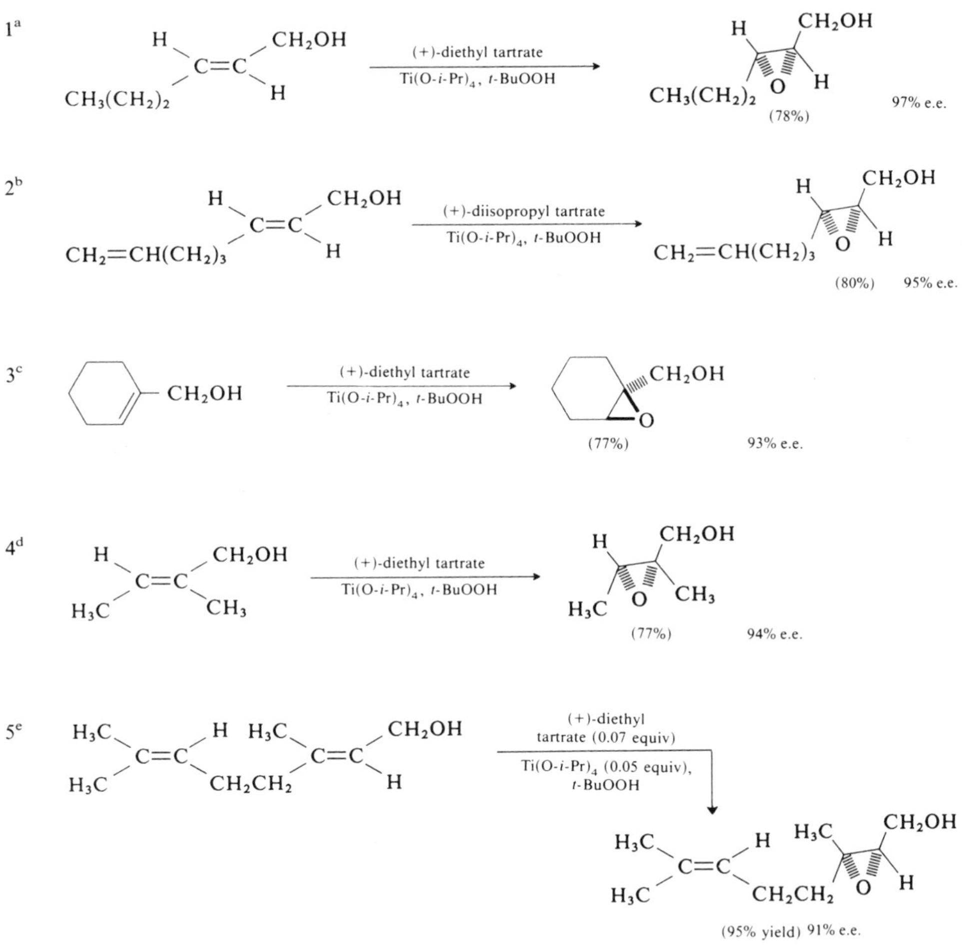

1[a]

$CH_3(CH_2)_2$(H)C=C(H)CH_2OH —(+)-diethyl tartrate, Ti(O-*i*-Pr)$_4$, *t*-BuOOH→ epoxide: H, CH_2OH, $CH_3(CH_2)_2$, O, H (78%) 97% e.e.

2[b]

$CH_2{=}CH(CH_2)_3$(H)C=C(H)CH_2OH —(+)-diisopropyl tartrate, Ti(O-*i*-Pr)$_4$, *t*-BuOOH→ epoxide: H, CH_2OH, $CH_2{=}CH(CH_2)_3$, O, H (80%) 95% e.e.

3[c]

1-cyclohexenyl-CH_2OH —(+)-diethyl tartrate, Ti(O-*i*-Pr)$_4$, *t*-BuOOH→ epoxycyclohexyl-CH_2OH (O) (77%) 93% e.e.

4[d]

H_3C(H)C=C(CH_3)CH_2OH —(+)-diethyl tartrate, Ti(O-*i*-Pr)$_4$, *t*-BuOOH→ epoxide: H, CH_2OH, H_3C, O, CH_3 (77%) 94% e.e.

5[e]

$(H_3C)_2$C=C(H)CH_2CH_2(H_3C)C=C(H)CH_2OH —(+)-diethyl tartrate (0.07 equiv), Ti(O-*i*-Pr)$_4$ (0.05 equiv), *t*-BuOOH→ $(H_3C)_2$C=C(H)CH_2CH_2-epoxide (H_3C, CH_2OH, O, H) (95% yield) 91% e.e.

a. J. G. Hill and K. B. Sharpless, *Org. Synth.* **63**, 66 (1985).
b. B. E. Rossiter, T. Katsuki, and K. B. Sharpless, *J. Am. Chem. Soc.* **103**, 464 (1981).
c. Y. Gao, R. M. Hanson, J. M. Klunder, S. Y. Ko, H. Masamune, and K. B. Sharpless, *J. Am. Chem. Soc.* **109**, 5765 (1987).
d. D. A. Evans, S. L. Bender, and J. Morris, *J. Am. Chem. Soc.* **110**, 2506 (1988).
e. R. M. Hanson and K. B. Sharpless, *J. Org. Chem.* **51**, 1922 (1986).

Because of the importance of the allylic hydroxyl group in coordinating the reactant to the titanium, the structural relationship between the double bond and the hydroxyl group is crucial. Homoallylic alcohols can be oxidized but the degree of enantioselectivity is reduced. Interestingly, the facial selectivity is reversed from that observed with allylic alcohols.[42] Compounds lacking a coordinating functional group are not reactive under these conditions.

42. B. E. Rossiter and K. B. Sharpless, *J. Org. Chem.* **49**, 3707 (1984).

12.2.2. Epoxides from Alkenes and Peroxidic Reagents

The most general reagents for conversion of alkenes to epoxides are peroxycarboxylic acids.[43] *m*-Chloroperoxybenzoic acid[44] is a particularly convenient reagent, but it is not commercially available at the present time. The magnesium salt of monoperoxyphthalic acid has been recommended as a replacement.[45] Potassium hydrogen peroxysulfate, which is sold commercially as "oxone," is a convenient reagent for epoxidations that can be done in aqueous methanol.[46] Peroxyacetic acid, peroxybenzoic acid, and peroxytrifluoroacetic acid have also been used frequently for epoxidation.

It has been demonstrated that ionic intermediates are not involved in the epoxidation reaction. The reaction rate is not very sensitive to solvent polarity.[47] Stereospecific *syn* addition is consistently observed. The oxidation is therefore believed to be a concerted process. A representation of the transition state is shown below.

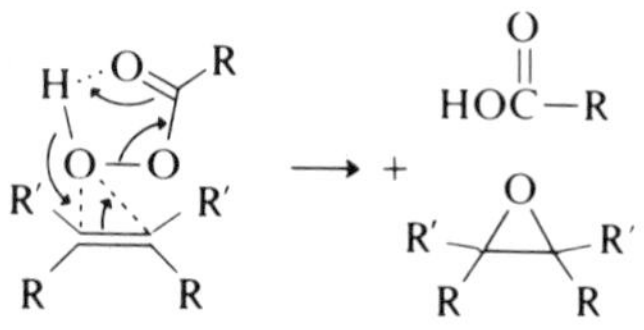

$$\text{[transition state: R}'\text{RC=CRR}' \cdots \text{O–O(H)–C(O)R]} \longrightarrow \mathrm{HO\overset{O}{\overset{\|}{C}}-R} + \text{R}'\text{R C–C R R}' \text{ (epoxide)}$$

The rate of epoxidation is increased by alkyl groups and other electron-donating substituents on the alkene, and the reactivity of the peroxyacids is increased by electron-accepting substituents.[48] These structure-reactivity relationships demonstrate that the peroxyacid acts as an electrophile in the reaction. Very low reactivity is exhibited by double bonds that are conjugated with strongly electron-attracting substituents, and strongly oxidizing peroxyacids, such as trifluoroacetic acid, are required for oxidation of such compounds.[49] Electron-poor alkenes can also be epoxidized by alkaline solutions of hydrogen peroxide or *t*-butyl hydroperoxide. A quite different mechanism, involving conjugate nucleophilic addition, operates in this case[50]:

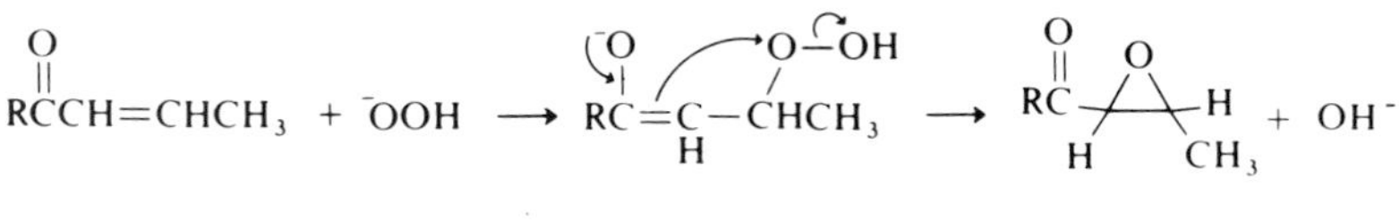

$$\mathrm{R\overset{O}{\overset{\|}{C}}CH{=}CHCH_3} + {}^{-}\mathrm{OOH} \longrightarrow \mathrm{R\overset{O^-}{C}{=}\underset{H}{C}-\overset{O-OH}{CHCH_3}} \longrightarrow \mathrm{R\overset{O}{\overset{\|}{C}}}\text{–}\overset{O}{\text{C(H)–C(H)CH}_3} + \mathrm{OH^-}$$

The stereoselectivity of epoxidation with peroxycarboxylic acids has been well studied. Addition of oxygen occurs preferentially from the less hindered side of the

43. D. Swern, *Organic Peroxides*, Vol. II, Wiley-Interscience, New York, 1971, pp. 355–533; B. Plesnicar, in *Oxidation in Organic Chemistry*, Part C, W. Trahanovsky (ed.), Academic Press, New York, 1978, pp. 211–253.
44. R. N. McDonald, R. N. Steppel, and J. E. Dorsey, *Org. Synth.* **50**, 15 (1970).
45. P. Brougham, M. S. Cooper, D. A. Cummerson, H. Heaney, and N. Thompson, *Synthesis*, 1015 (1987).
46. R. Bloch, J. Abecassis, and D. Hassan, *J. Org. Chem.* **50**, 1544 (1985).
47. N. N. Schwartz and J. N. Blumbergs, *J. Org. Chem.*, 29, 1976 (1964).
48. B. M. Lynch and K. H. Pausacker, *J. Chem. Soc.*, 1525 (1955).
49. W. D. Emmons and A. S. Pagano, *J. Am. Chem. Soc.* **77**, 89 (1955).
50. C. A. Bunton and G. J. Minkoff, *J. Chem. Soc.*, 665 (1949).

molecule. Norbornene, for example, gives a 96:4 *exo*:*endo* ratio.[51] In molecules where two potential modes of approach are not greatly different, a mixture of products is to be expected. For example, the unhindered exocyclic double bond in 4-*t*-butylmethylenecyclohexane gives both stereoisomeric products.[52]

$(CH_3)_3C$–C$_6$H$_{10}$=CH_2 $\xrightarrow[CH_2Cl_2]{m\text{-chloroperoxybenzoic acid}}$ (69%) + (31%)

Hydroxyl groups exert a directive effect on epoxidation and favor approach from the side of the double bond closest to the hydroxyl group.[53] Hydrogen bonding between the hydroxyl group and the reagent evidently stabilizes this transition state.

OH $\xrightarrow{\text{peroxybenzoic acid}}$ HO, H, H, O, H

This is a strong directive effect which can exert stereochemical control even when steric effects are opposed. Several examples of epoxidation reactions are given in Scheme 12.6. Entries 4 and 5 illustrate the hydroxyl directive effect.

A process that is effective for epoxidation and avoids acidic conditions involves reaction of an alkene, a nitrile, and hydrogen peroxide.[54] The nitrile and hydrogen peroxide react, forming a peroxyimidic acid, which epoxidizes the alkene, presumably by a mechanism similar to that for peroxyacids. An important contribution to the reactivity of the peroxyimidic acids comes from the formation of the stable amide carbonyl group.

$$R'C{\equiv}N + H_2O_2 \longrightarrow R'C({=}NH){-}O{-}OH$$

$$R'C({=}NH){-}O{-}OH + R_2C{=}CR_2 \longrightarrow R'C({=}O)NH_2 + \text{R}_2\text{C}\overset{\text{O}}{-}\text{CR}_2$$

A variety of other reagents have been examined with the objective of activating H_2O_2 to generate a good epoxidizing agent. In principle, any species which can convert one of the hydroxyl groups in hydrogen peroxide to a good leaving group would generate a reactive epoxidizing reagent:

$$H{-}O{-}O{-}X + {>}C{=}C{<} \longrightarrow {-}\overset{\text{O}}{C{-}C}{-} + HO{-}X$$

51. H. Kwart and T. Takeshita, *J. Org. Chem.* **28**, 670 (1963).
52. R. G. Carlson and N. S. Behn, *J. Org. Chem.* **32**, 1363 (1967).
53. H. B. Henbest and R. A. L. Wilson, *J. Chem. Soc.*, 1958 (1957).
54. G. B. Payne, *Tetrahedron* **18**, 763 (1962); R. D. Bach and J. W. Knight, *Org. Synth.* **60**, 63 (1981); L. A. Arias, S. Adkins, C. J. Nagel, and R. D. Bach, *J. Org. Chem.* **48**, 888 (1983).

Scheme 12.6. Synthesis of Epoxides from Alkenes

A. Oxidation of Alkenes with Peroxyacids

1[a] $C_6H_5CH=CH_2$ —peroxybenzoic acid→ (69–75%)

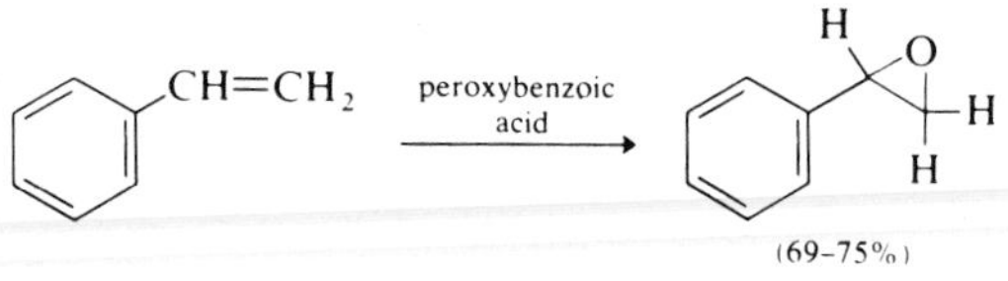

2[b] —peroxybenzoic acid→ (72%)

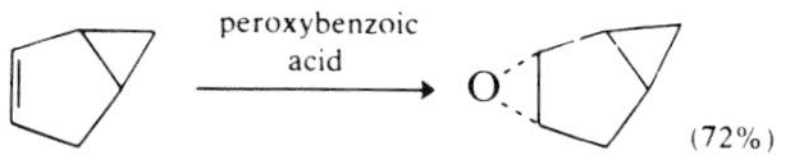

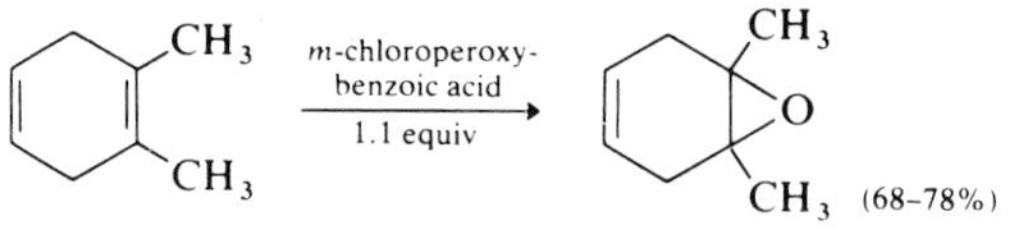

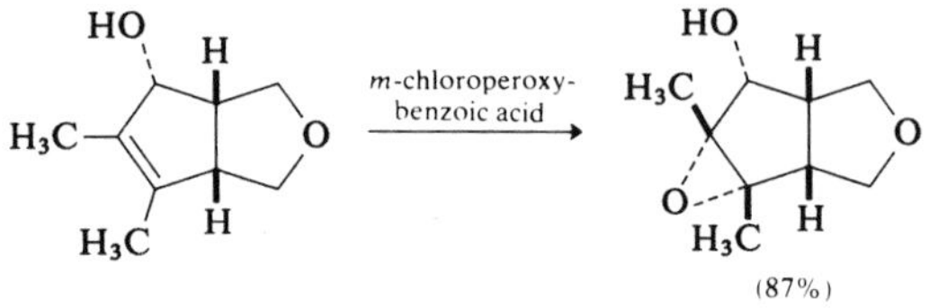

3[c] —*m*-chloroperoxybenzoic acid, 1.1 equiv→ (68–78%)

4[d] —*m*-chloroperoxybenzoic acid→ (87%)

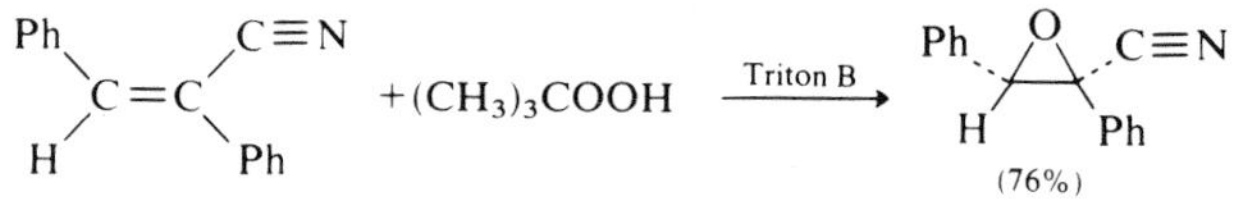

5[e] —*m*-chloroperoxybenzoic acid→ (78%)

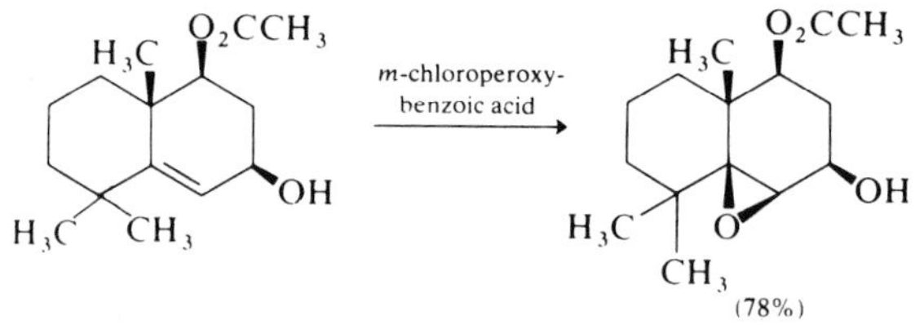

B. Epoxidation of Electrophilic Alkenes

6[f] —H_2O_2, ^-OH→ (70–72%)

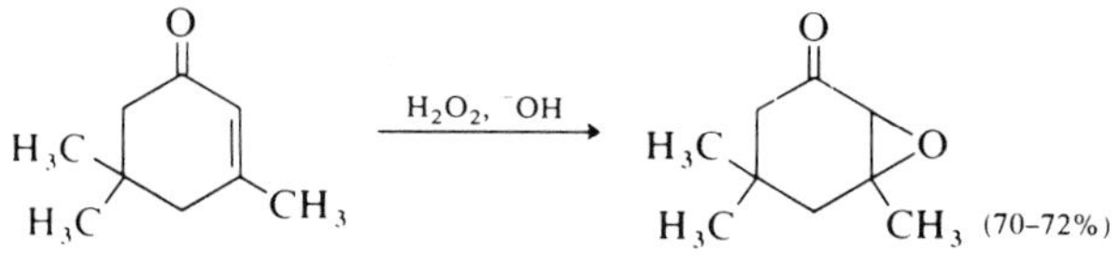

7[g] Ph(H)C=C(C≡N)Ph + $(CH_3)_3COOH$ —Triton B→ (76%)

8[h] $CH_3CH=CHCO_2C_2H_5$ —CF_3CO_3H→ (73%)

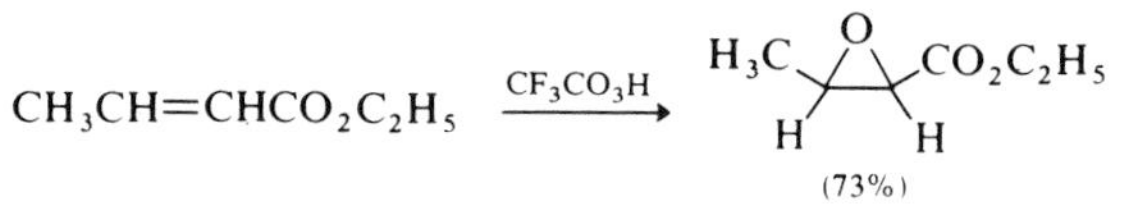

a. H. Hibbert and P. Burt, *Org. Synth.* **I**, 481 (1932).
b. E. J. Corey and R. L. Dawson, *J. Am. Chem. Soc.* **85**, 1782 (1963).
c. L. A. Paquette and J. H. Barrett, *Org. Synth.* **49**, 62 (1969).
d. R. M. Scarborough, Jr., B. H. Toder, and A. B. Smith, III, *J. Am. Chem. Soc.* **102**, 3904 (1980).
e. M. Miyashita and A. Yoshikoshi, *J. Am. Chem. Soc.* **96**, 1917 (1974).
f. R. L. Wasson and H. O. House, *Org. Synth.* **IV**, 552 (1963).
g. G. B. Payne and P. H. Williams, *J. Org. Chem.* **26**, 651 (1961).
h. W. D. Emmons and A. S. Pagano, *J. Am. Chem. Soc.* **77**, 89 (1955).

In practice, promising results have been obtained for several systems. Fair to good yields of epoxides are obtained when a two-phase system consisting of alkene and ethyl chloroformate is stirred with a buffered basic solution of hydrogen peroxide. The active oxidant is presumed to be *O*-ethyl peroxycarbonic acid.[55]

$$H_2O_2 + C_2H_5OC(=O)Cl \rightarrow C_2H_5OC(=O){-}OH + HCl$$

$$C_2H_5OC(=O){-}OH + RCH{=}CHR \rightarrow C_2H_5OH + CO_2 + RCH\overset{O}{\frown}CHR$$

Similarly, the adduct of hydrogen peroxide with hexafluoroacetone is an active epoxidizing reagent.[56]

$$CF_3C(OH)(O{-}OH)CF_3 + RCH{=}CHR \rightarrow CF_3C(OH)_2CF_3 + RCH\overset{O}{\frown}CHR$$

Although neither of these reagents, nor similar combinations, have been as generally useful as the peroxycarboxylic acids, they serve to illustrate that epoxidizing activity is not unique to the peroxyacids.

12.2.3. Subsequent Transformations of Epoxides

Epoxides are useful synthetic intermediates, and the conversion of an alkene to an epoxide is often part of a more extensive molecular transformation. In many instances, advantage is taken of the high reactivity of the epoxide ring to introduce additional functionality. Such two- or three-step operations can accomplish specific oxidative transformations of an alkene that may not easily be accomplished in a single step. Scheme 12.7 provides a preview of the type of reactivity to be discussed.

Epoxidation may be preliminary to solvolytic or nucleophilic ring opening in synthetic sequences. In acidic aqueous solution, epoxides are opened to give diols by an *anti* addition process. In cyclic systems, ring opening occurs to give the diaxial diol.

CH$_3$, O, CH$_3$ $\xrightarrow[H_2O]{H^+}$ OH, CH$_3$, CH$_3$, OH — Ref. 57

Base-catalyzed reactions, in which the nucleophile provides the driving force for ring opening, usually involve breaking of the epoxide bond at the less substituted

55. R. D. Bach, M. W. Klein, R. A. Ryntz, and J. W. Holubka, *J. Org. Chem.* **44**, 2569 (1979).
56. R. P. Heggs and B. Ganem, *J. Am. Chem. Soc.* **101**, 2484 (1979).
57. B. Rickborn and D. K. Murphy, *J. Org. Chem.* **34**, 3209 (1969).

Scheme 12.7. Multistep Synthetic Transformations via Epoxides

A. Epoxidation Followed by Nucleophilic Ring Opening

$$\rangle C{=}C\langle \longrightarrow -\overset{O}{C{-}C}- \xrightarrow{NuH} -\underset{Nu}{\overset{|}{C}}-\overset{OH}{\underset{|}{C}}-$$

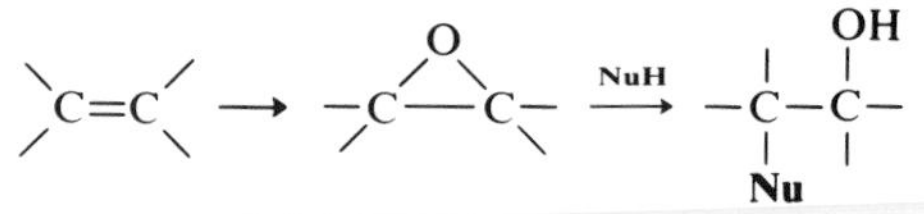

B. Epoxidation Followed by Rearrangement to a Carbonyl Compound

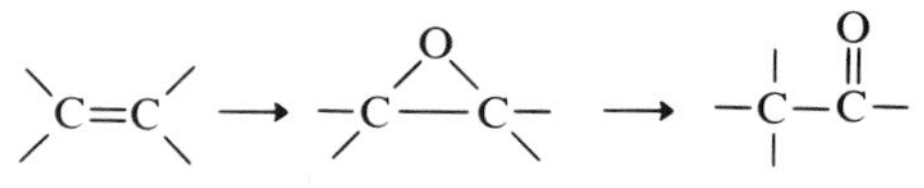

$$\rangle C{=}C\langle \longrightarrow -\overset{O}{C{-}C}- \longrightarrow -\underset{|}{\overset{|}{C}}-\overset{O}{\overset{\|}{C}}-$$

C. Epoxidation Followed by Ring Opening to an Allyl Alcohol

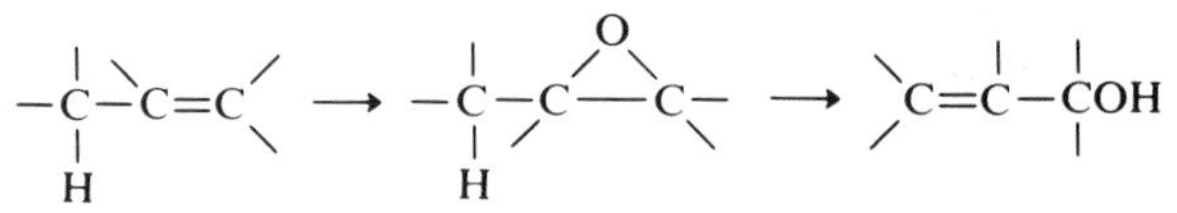

$$-\underset{H}{\overset{|}{C}}-C{=}C\langle \longrightarrow -\underset{H}{\overset{|}{C}}-\overset{O}{C{-}C}- \longrightarrow \rangle C{=}\overset{|}{C}-\underset{|}{\overset{|}{C}}OH$$

D. Epoxidation Followed by Ring Opening and Elimination

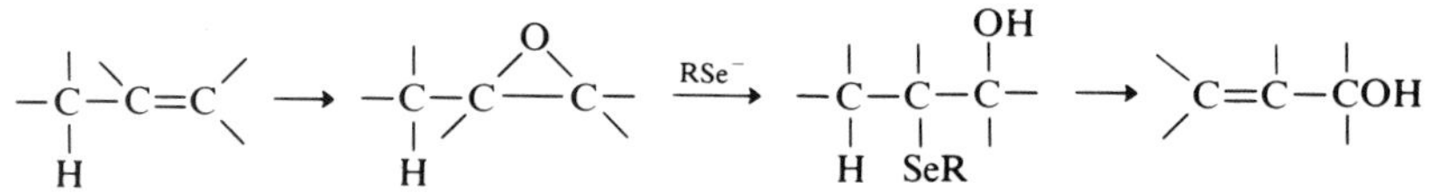

$$-\underset{H}{\overset{|}{C}}-C{=}C\langle \longrightarrow -\underset{H}{\overset{|}{C}}-\overset{O}{C{-}C}- \xrightarrow{RSe^-} -\underset{H}{\overset{|}{C}}-\underset{SeR}{\overset{|}{C}}-\overset{OH}{\underset{|}{C}}- \longrightarrow \rangle C{=}\overset{|}{C}-\underset{|}{\overset{|}{C}}OH$$

E. Epoxidation Followed by Reductive Ring Opening

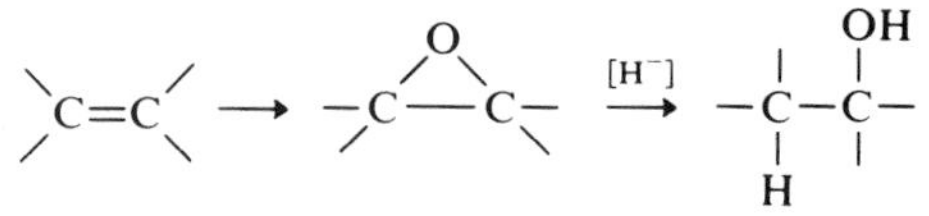

$$\rangle C{=}C\langle \longrightarrow -\overset{O}{C{-}C}- \xrightarrow{[H^-]} -\underset{H}{\overset{|}{C}}-\overset{OH}{\underset{|}{C}}-$$

carbon, since this is the position most open to nucleophilic attack.[58] The situation in acid-catalyzed reactions is more complex. The bonding of a proton to the oxygen weakens the C—O bonds and facilitates rupture by weak nucleophiles. If the C—O bonds are largely intact at the transition state, the nucleophile will become attached to the less substituted position for the same steric reasons that were cited for nucleophilic ring opening. If, on the other hand, C—O rupture is nearly complete when the transition state is reached, the opposite orientation is observed. This change in regiochemistry results from the ability of the more substituted carbon to stabilize the developing positive charge.

58. R. E. Parker and N. S. Isaacs, *Chem. Rev.* **59**, 737 (1959).

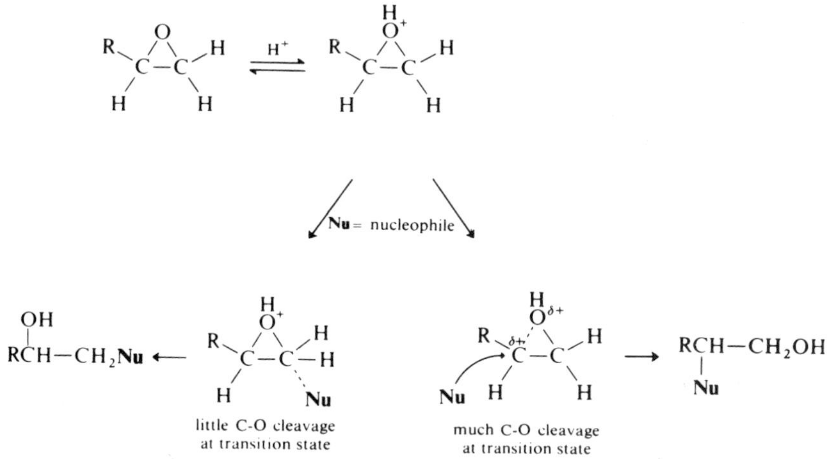

Nu = nucleophile

little C-O cleavage at transition state

much C-O cleavage at transition state

When simple aliphatic epoxides such as propylene oxide react with hydrogen halides, the dominant mode of reaction introduces halide at the less substituted primary carbon.[59]

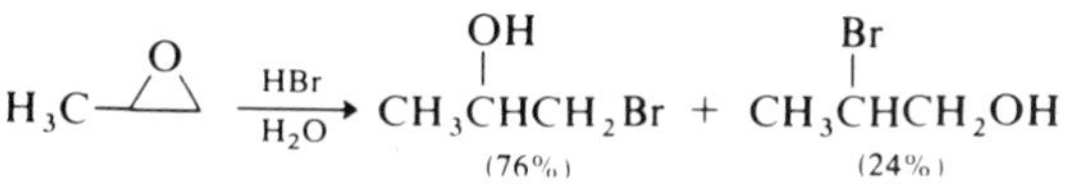

$$H_3C\text{-epoxide} \xrightarrow[H_2O]{HBr} CH_3CH(OH)CH_2Br\ (76\%) + CH_3CH(Br)CH_2OH\ (24\%)$$

Substituents that further stabilize a carbocation intermediate lead to reversal of the mode of addition.[60] Scheme 12.8 gives some examples of both acid-catalyzed and nucleophilic ring openings of epoxides.

Epoxides can be isomerized to carbonyl compounds by Lewis acids.[61] Boron trifluoride is frequently used as the reagent. Carbocation intermediates appear to be involved, and the structure and stereochemistry of the product are determined by the factors that govern substituent migration in the carbocation. Clean, high-yield reactions can be expected only where structural or conformational factors promote a very selective rearrangement.

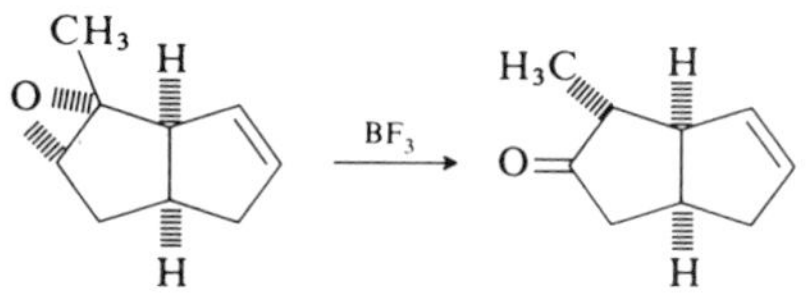

$\xrightarrow{BF_3}$ Ref. 62

Double bonds having oxygen and halogen substituents are susceptible to epoxidation, and the reactive epoxides that are thereby generated serve as intermedi-

59. C. A. Stewart and C. A. VanderWerf, *J. Am. Chem. Soc.* **76**, 1259 (1954).
60. S. Winstein and L. L. Ingraham, *J. Am. Chem. Soc.* **74**, 1160 (1952).
61. J. N. Coxon, M. P. Hartshorn, and W. J. Rae, *Tetrahedron* **26**, 1091 (1970).
62. J. K. Whitesell, R. S. Matthews, M. A. Minton, and A. M. Helbling, *J. Am. Chem. Soc.* **103**, 3468 (1981).

Scheme 12.8. Nucleophilic and Solvolytic Ring Opening of Epoxides

A. Oxidation with Solvolysis of the Intermediate Epoxide

1[a] cyclohexene $\xrightarrow[HCO_2H]{H_2O_2}$ trans-1,2-cyclohexanediol (65-73%)

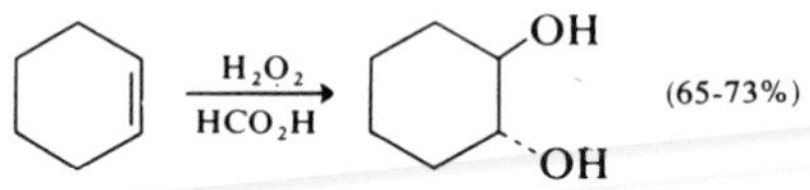

2[b] $\xrightarrow[\text{2) NaOH}]{\text{1) } HCO_2H, H_2O_2}$ (65-73%)

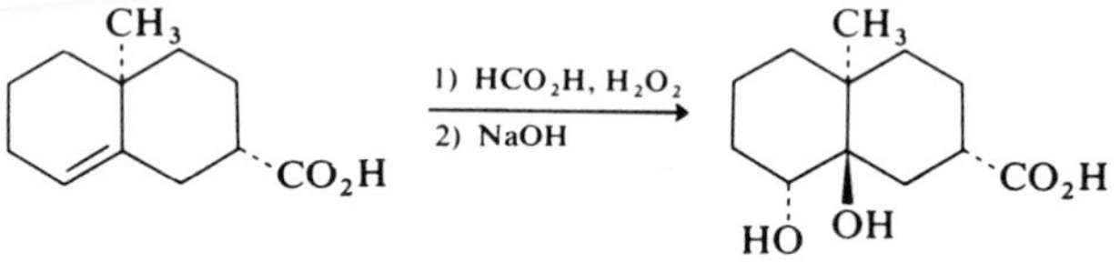

B. Acid-Catalyzed Solvolytic Ring Opening

3[c] $(H_3C)_2C(O)CHCH_3$ epoxide $\xrightarrow[MeOH]{H_2SO_4}$ $(CH_3)_2C(OCH_3)-CH(OH)CH_3$ (76%)

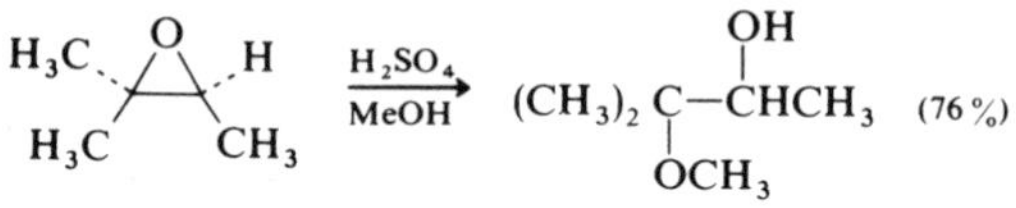

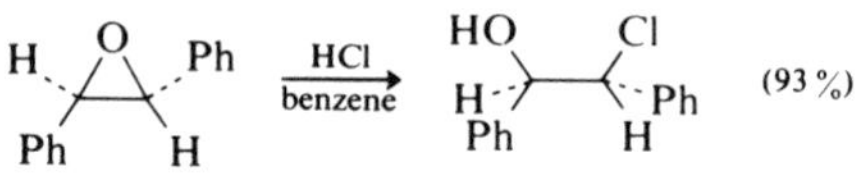

4[d] $\xrightarrow[\text{benzene}]{HCl}$ (93%)

5[e] $\xrightarrow[H_2O]{HClO_4}$ (100%)

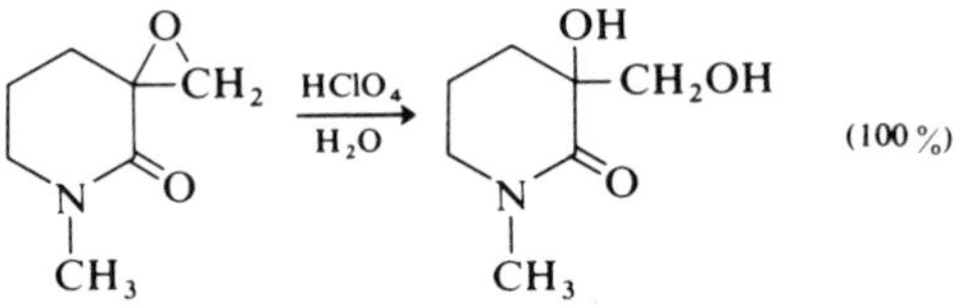

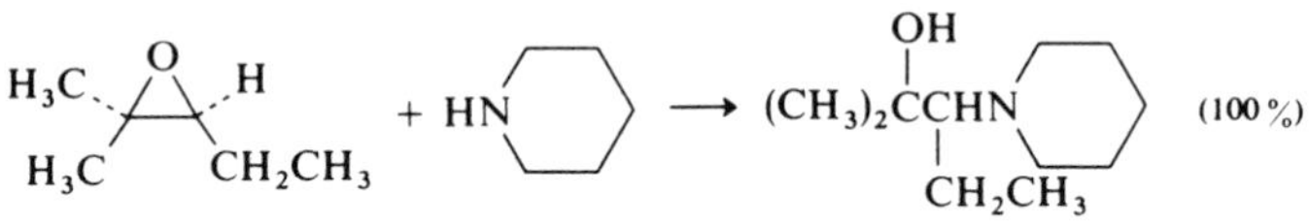

C. Nucleophilic Ring-Opening Reactions

6[f] + CH_3O^- ⟶ $(CH_3)_2C(OH)CH(OCH_3)CH_3$ (53%)

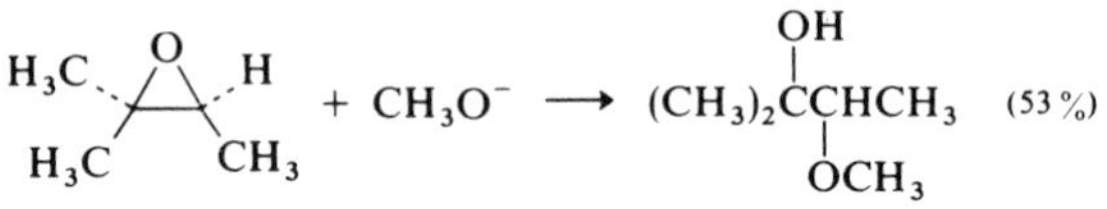

7[g] + HN(piperidine) ⟶ $(CH_3)_2C(OH)CH(CH_2CH_3)N$(piperidine) (100%)

8[h] $CH_2N(C_2H_5)_2$ epoxide + ^-SH ⟶ $HSCH_2CH(OH)CH_2N(C_2H_5)_2$ (63%)

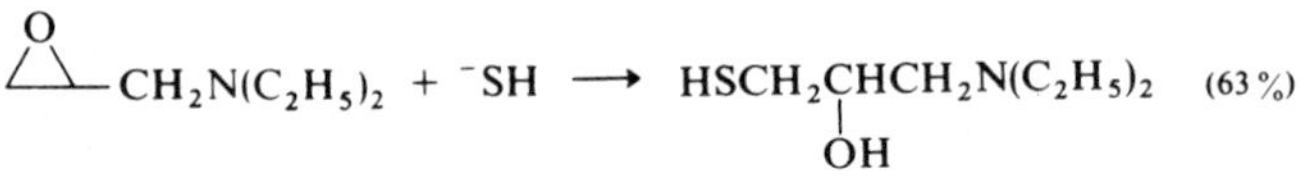

a. A. Roebuck and H. Adkins, *Org. Synth.* **III**, 217 (1955).
b. T. R. Kelly, *J. Org. Chem.* **37**, 3393 (1972).
c. S. Winstein and L. L. Ingrahm, *J. Am. Chem. Soc.* **74**, 1160 (1952).
d. G. Berti, F. Bottari, P. L. Ferrarini, and B. Macchia, *J. Org. Chem.* **30**, 4091 (1965).
e. M. L. Rueppel and H. Rapoport, *J. Am. Chem. Soc.* **94**, 3877 (1972).
f. T. Colclough, J. I. Cunneen, and C. G. Moore, *Tetrahedron* **15**, 187 (1961).
g. D. M. Burness and H. O. Bayer, *J. Org. Chem.* **28**, 2283 (1963).

ates in some useful synthetic transformations. Vinyl chlorides furnish haloepoxides, which can rearrange to α-haloketones:

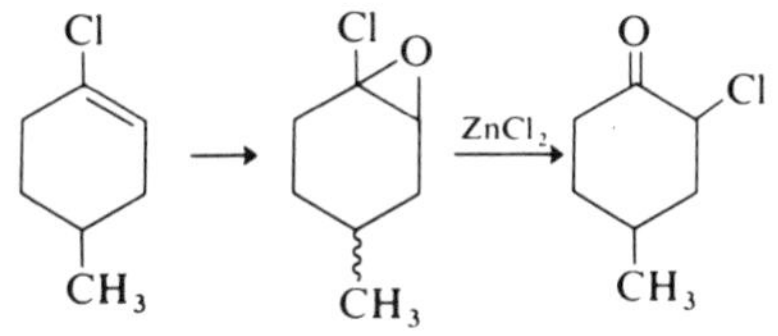

Cl CH$_3$ → Cl O CH$_3$ $\xrightarrow{ZnCl_2}$ O Cl CH$_3$ Ref. 63

Enol acetates form epoxides that can rearrange to α-acetoxyketones:

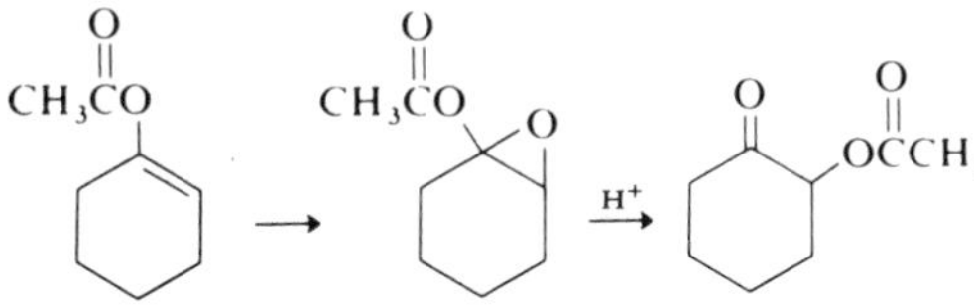

$CH_3\overset{O}{\overset{\|}{C}}O$ → $CH_3\overset{O}{\overset{\|}{C}}O$ O $\xrightarrow{H^+}$ O $O\overset{O}{\overset{\|}{C}}CH_3$ Ref. 64

The stereochemistry of the rearrangement of the acetoxy epoxides involves inversion at the carbon to which the acetoxy group migrates.[65] The reaction probably proceeds through a cyclic transition state:

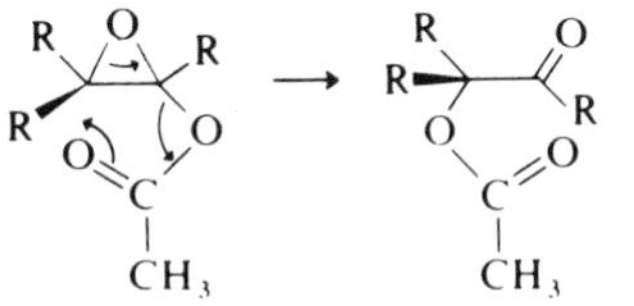

R O R R O O C CH_3 → R R O R O C O CH_3

A more synthetically useful version of this reaction involves epoxidation of trimethylsilyl enol ethers of ketones. Epoxidation of the silyl enol ethers, followed by aqueous workup, gives α-hydroxyketones and α-hydroxyaldehydes.[66]

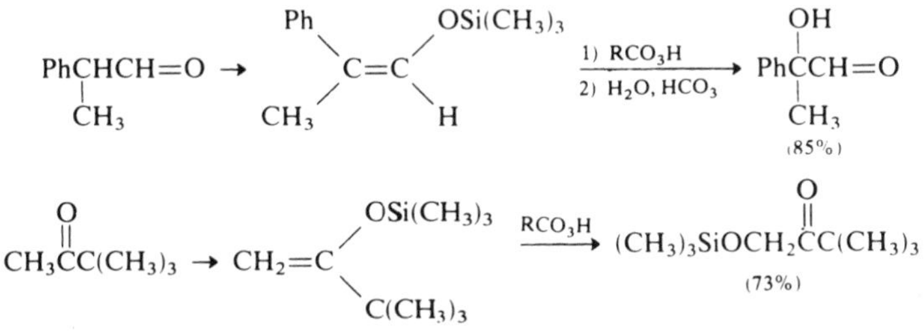

$PhCH(CH_3)CH{=}O \rightarrow (Ph)(CH_3)C{=}C(OSi(CH_3)_3)(H) \xrightarrow[2)\ H_2O,\ HCO_3^-]{1)\ RCO_3H} PhC(OH)(CH_3)CH{=}O$ (85%)

$CH_3\overset{O}{\overset{\|}{C}}C(CH_3)_3 \rightarrow CH_2{=}C(OSi(CH_3)_3)(C(CH_3)_3) \xrightarrow{RCO_3H} (CH_3)_3SiOCH_2\overset{O}{\overset{\|}{C}}C(CH_3)_3$ (73%)

The oxidation of silyl enol ethers with the osmium tetroxide–amine oxide combination also leads to α-hydroxyketones in generally good yields.[67]

63. R. N. McDonald and T. E. Tabor, *J. Am. Chem. Soc.* **89**, 6573 (1967).
64. K. L. Williamson, J. I. Coburn, and M. F. Herr, *J. Org. Chem.* **32**, 3934 (1967).
65. K. L. Williamson and W. S. Johnson, *J. Org. Chem.* **26**, 4563 (1961).
66. A. Hassner, R. H. Reuss, and H. W. Pinnick, *J. Org. Chem.* **40**, 3427 (1975).
67. J. P. McCormick, W. Tomasik, and M. W. Johnson, *Tetrahedron Lett.*, 607 (1981).

Epoxides derived from vinylsilanes are converted by mildly acidic conditions into ketones or aldehydes.[68]

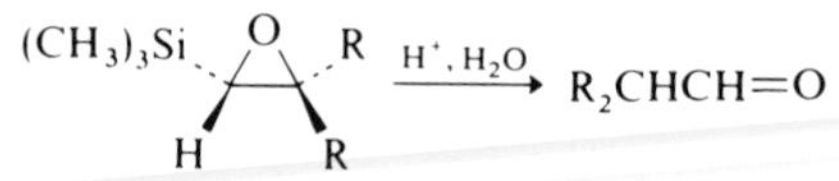

The ring opening of the silyl epoxides is facilitated by the stabilizing effect that silicon has on a positive charge at the β position. This facile transformation permits vinylsilanes to serve as the equivalent of carbonyl groups in multistep synthesis.[69]

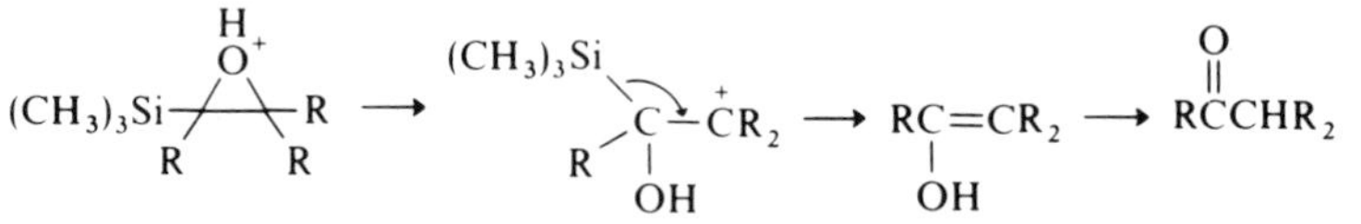

Base-catalyzed ring opening of epoxides constitutes a route to allylic alcohols:

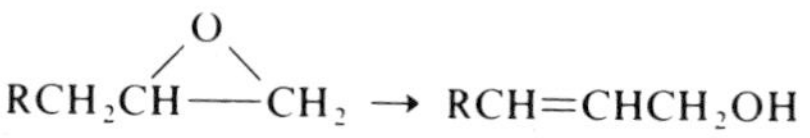

Strongly basic reagents, such as lithium salts of dialkylamines, are required to promote the reaction. The stereochemistry of the ring opening has been investigated by deuterium labeling. A proton *cis* to the epoxide ring is selectively removed.[70]

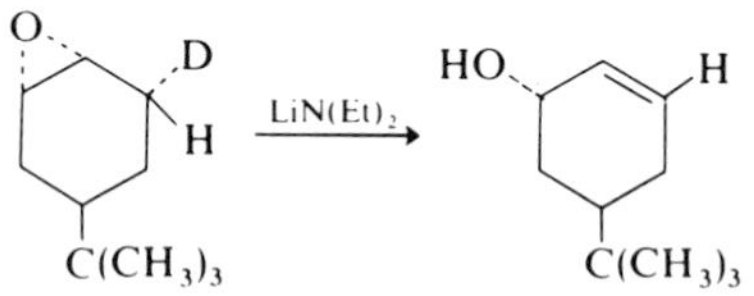

A transition state represented by structure **D** could account for this stereochemistry. Such an arrangement would be favored by ion pairing, which would bring the amide anion and lithium cation into close proximity. Simultaneous coordination of the lithium ion at the epoxide would result in a *syn* elimination.

D

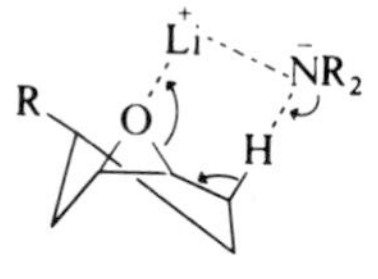

Among other reagents that effect epoxide ring opening are diethylaluminum 2,2,6,6-tetramethylpiperidide and magnesium *N*-cyclohexyl-*N*-2-propylamide.

68. G. Stork and E. Colvin, *J. Am. Chem. Soc.* **93**, 2080 (1971).
69. G. Stork and M. E. Jung, *J. Am. Chem. Soc.* **96**, 3682 (1974).
70. R. P. Thummel and B. Rickborn, *J. Am. Chem. Soc.* **92**, 2064 (1970).

$\xrightarrow[\text{0°C, 3 h}]{\text{NAl}(C_2H_5)_2}$ (90%) Ref. 71

$\xrightarrow[\text{0–23°C, 2 h}]{\text{BrMgN(cyclohexyl)CH}(CH_3)_2}$ (70%) Ref. 72

These latter reagents are appropriate even for very sensitive molecules. Their efficacy is presumably due to the strong Lewis acid effect of the aluminum and magnesium ions. The hindered nature of the amide bases minimizes competition from nucleophilic ring opening.

Allylic alcohols can also be obtained from epoxides by ring opening with a selenide anion followed by elimination via the selenoxide (see Section 6.8.3 for discussion of selenoxide elimination). The elimination occurs regiospecifically away from the hydroxy group.[73]

$$RCH_2\overset{O}{CH{-}CH}R' + PhSe^- \rightarrow RCH_2\underset{PhSe}{CH}{-}\overset{OH}{CH}R' \xrightarrow{H_2O_2} RCH{=}CH\overset{OH}{CH}R'$$

Epoxides can also be converted to allylic alcohols with electrophilic reagents. The treatment of epoxides with trisubstituted silyl iodides and an organic base gives the silyl ether of the corresponding allylic alcohols.[74]

$\xrightarrow{1)\ (CH_3)_2SiC(CH_3)_3I}$ $OSi(CH_3)_2C(CH_3)_3$ (70–80%)

71. A. Yasuda, S. Tanaka, K. Oshima, H. Yamamoto, and H. Nozaki, *J. Am. Chem. Soc.* **96**, 6513 (1974).
72. E. J. Corey, A. Marfat, J. R. Falck, and J. O. Albright, *J. Am. Chem. Soc.* **102**, 1433 (1980).
73. K. B. Sharpless and R. F. Lauer, *J. Am. Chem. Soc.* **95**, 2697 (1973).
74. M. R. Detty, *J. Org. Chem.* **45**, 924 (1980); M. R. Detty and M. D. Seidler, *J. Org. Chem.* **46**, 1283 (1981).

Each of these procedures for epoxidation and ring opening is the equivalent of an allylic oxidation of a double bond with migration of the double bond:

$$R_2CHCH{=}CHR' \rightarrow R_2C{=}CH{-}\underset{}{\overset{OH}{\overset{|}{C}}}HR'$$

In Section 12.2.4, alternative means of effecting this transformation will be described.

Epoxides can also be reduced to saturated alcohols. Lithium aluminum hydride acts as a nucleophilic reducing agent, and the hydride is added at the less substituted carbon atom of the epoxide ring. Lithium triethylborohydride is more reactive than $LiAlH_4$ and is superior for epoxides that are resistant to reduction.[75] Reduction by dissolving metals, such as lithium in ethylenediamine,[76] also gives good yields.

Diborane in THF reduces epoxides, but the yields are low, and other products are formed by pathways that result from the electrophilic nature of diborane.[77] Better yields are obtained when BH_4^- is included in the reaction system, but the electrophilic nature of diborane is still evident because the dominant product is that resulting from addition of the hydride at the more substituted carbon.[78]

$$CH_3\underset{CH_3}{\underset{|}{C}}\overset{O}{\frown}CHCH_3 \xrightarrow[BH_4^-]{BH_3} (CH_3)_2CH\overset{OH}{\overset{|}{C}}HCH_3 + (CH_3)_2\overset{OH}{\overset{|}{C}}CH_2CH_3$$

(78%) (22%)

The overall transformation of alkenes to alcohols that is accomplished by epoxidation and reduction corresponds to alkene hydration. This reaction sequence is therefore an alternative to the hydration methods discussed in Chapter 4 for converting alkenes to alcohols.

12.2.4. Reaction of Alkenes with Singlet Oxygen

Also among the oxidative reactions that add oxygen at carbon–carbon double bonds is the reaction with singlet oxygen.[79] For most alkenes, this reaction proceeds with the specific removal of an allylic hydrogen and shift of the double bond to provide an allylic hydroperoxide as the initial product.

O=O H → O O H → O–OH

75. S. Krishnamurthy, R. M. Schubert, and H. C. Brown, *J. Am. Chem. Soc.*, **95**, 8486 (1973).
76. H. C. Brown, S. Ikegami, and J. H. Kawakami, *J. Org. Chem.* **35**, 3243 (1970).
77. D. J. Pasto, C. C. Cumbo, and J. Hickman, *J. Am. Chem. Soc.* **88**, 2201 (1966).
78. H. C. Brown and N. M. Yoon, *J. Am. Chem. Soc.* **90**, 2686 (1968).
79. H. H. Wasserman and R. W. Murray (eds.), *Singlet Oxygen*, Academic Press, New York, 1979; A. A. Frimer, *Chem. Rev.* **79**, 359 (1979).

Singlet oxygen is usually generated from oxygen by dye-sensitized photoexcitation. A number of alternative methods of generating singlet oxygen are summarized in Scheme 12.9.

Singlet oxygen decays to the ground state triplet oxygen at a rate which is strongly dependent on the solvent.[80] Measured lifetimes range from about 700 μs in carbon tetrachloride to 2 μs in water. The choice of solvent can therefore have a pronounced effect on the efficiency of oxidation; the longer the excited state lifetime, the more likely it is that reaction with the alkene can occur.

The reactivity order of alkenes is that expected for attack by an electrophilic reagent. Reactivity increases with the number of alkyl substituents on the alkene.[81] Terminal alkenes are relatively inert. Steric effects govern the direction of approach of the oxygen, so that the hydroperoxy group is introduced on the less hindered face of the double bond.

Many alkenes present several different allylic hydrogens, and in this type of situation it is important to be able to predict the degree of selectivity. A useful

Scheme 12.9. Generation of Singlet Oxygen

1[a] Photosensitizer + $h\nu \rightarrow$ ^{1}[Photosensitizer]*

^{1}[Photosensitizer]* $\rightarrow$ ^{3}[Photosensitizer]*

^{3}[Photosensitizer]* + $^3O_2 \rightarrow {}^1O_2$ + Photosensitizer

2[b] $H_2O_2 + {}^-OCl \rightarrow {}^1O_2 + H_2O + Cl^-$

3[c] $(RO)_3P + O_3 \rightarrow (RO)_3P\langle O_3 \rangle \rightarrow (RO)_3P{=}O + {}^1O_2$

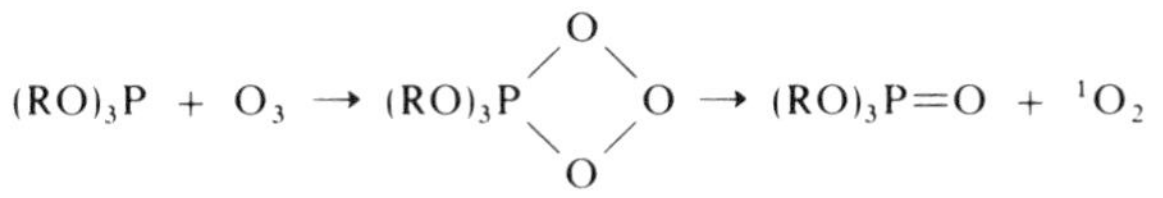

4[d] 9,10-diphenylanthracene endoperoxide (Ph, O–O bridge, Ph) $\rightarrow$ 9,10-diphenylanthracene + $^1O_2$

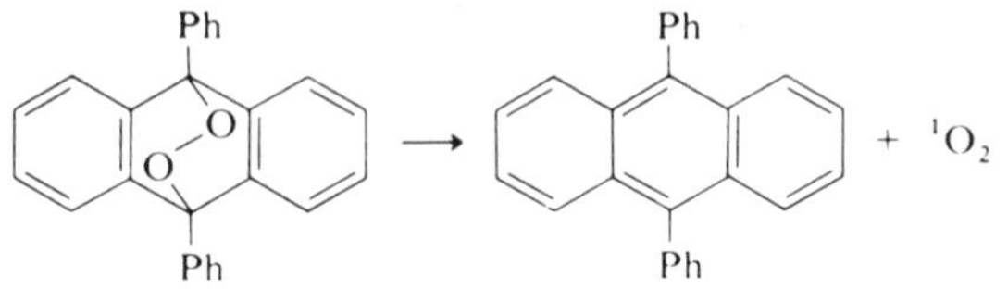

5[e] $(C_2H_5)_3SiH + O_3 \rightarrow (C_2H_5)_3SiOOOH \rightarrow (C_2H_5)_3SiOH + {}^1O_2$

a. C. S. Foote and S. Wexler, *J. Am. Chem. Soc.* **86**, 3880 (1964).
b. C. S. Foote and S. Wexler, *J. Am. Chem. Soc.* **86**, 3879 (1964).
c. R. W. Murray and M. L. Kaplan, *J. Am. Chem. Soc.* **90**, 537 (1964).
d. H. H. Wasserman, J. R. Scheffer, and J. L. Cooper, *J. Am. Chem. Soc.* **94**, 4991 (1972).
e. E. J. Corey, M. M. Mehotra, and A. U. Khan, *J. Am. Chem. Soc.* **108**, 2472 (1986).

80. P. B. Merkel and D. R. Kearns, *J. Am. Chem. Soc.* **94**, 1029, 7244 (1972); P. R. Ogilby and C. S. Foote, *J. Am. Chem. Soc.* **105**, 3423 (1983); J. R. Hurst, J. D. McDonald, and G. B. Schuster, *J. Am. Chem. Soc.* **104**, 2065 (1982).
81. K. R. Kopecky and H. J. Reich, *Can. J. Chem.* **43**, 2265 (1965); C. S. Foote and R. W. Denny, *J. Am. Chem. Soc.* **93**, 5162 (1971); A. Nickon and J. F. Bagli, *J. Am. Chem. Soc.* **83**, 1498 (1961).

generalization is that *there is a preference for abstraction of hydrogen from the more congested side of the double bond.*[82]

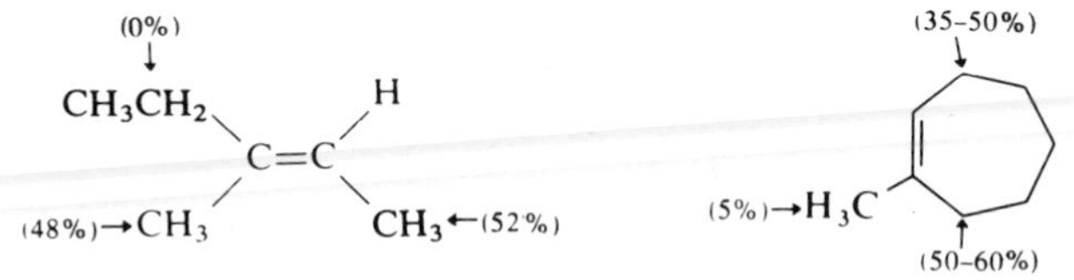

$$\underset{(48\%)\rightarrow CH_3}{\overset{\overset{(0\%)}{\downarrow}}{CH_3CH_2}}\!\!>C=C<\!\!\underset{CH_3\leftarrow(52\%)}{\overset{H}{}}$$

(35–50%), (5%)→H_3C, (50–60%) [1-methylcycloheptene]

The allylic hydroperoxides generated by singlet oxygen oxidation are normally reduced to the corresponding allylic alcohols. The net synthetic transformation is then formation of an allylic alcohols with transposition of the double bond. Scheme 12.10 gives some examples of oxidations by singlet oxygen.

Certain compounds react with singlet oxygen in a different manner, giving dioxetanes as products[83]:

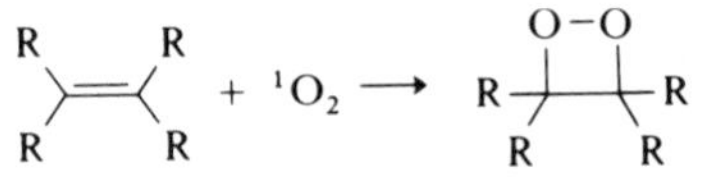

$$R_2C{=}CR_2 + {}^1O_2 \longrightarrow \text{R}_2\overset{\text{O}-\text{O}}{\text{C}-\text{C}}\text{R}_2$$

This reaction is not usually a major factor with alkenes bearing only alkyl groups but is important for vinyl ethers and other alkenes with donor substituents. Enaminoketones undergo a clean oxidation to α-diketones, presumably through a dioxetane intermediate[84]:

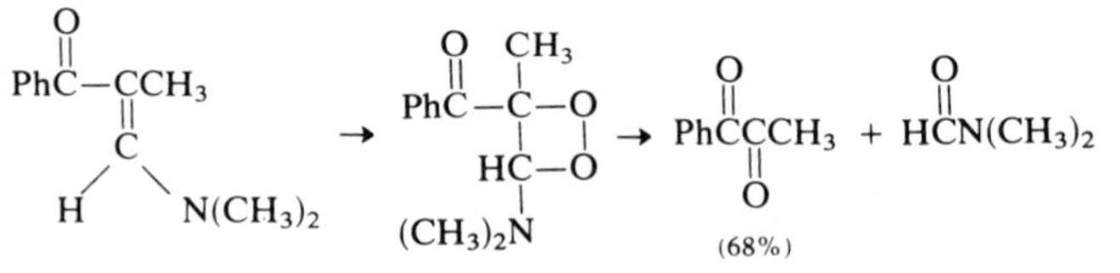

$$PhC(=O)C(CH_3){=}CHN(CH_3)_2 \rightarrow PhC(=O)\!-\!\underset{|}{\overset{CH_3}{C}}\!-\!O\text{—}O\!-\!\underset{(CH_3)_2N}{HC} \rightarrow \underset{(68\%)}{PhC(=O)C(=O)CH_3} + HC(=O)N(CH_3)_2$$

Singlet oxygen undergoes [4 + 2] cycloaddition with dienes:

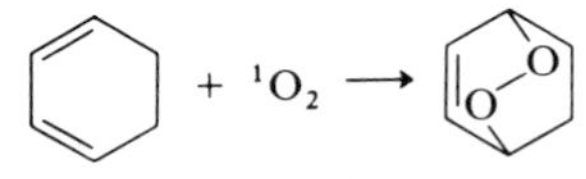

$$\text{1,3-cyclohexadiene} + {}^1O_2 \longrightarrow \text{endoperoxide} \qquad \text{Ref. 85}$$

$$\text{cyclohexadiene epoxide} + {}^1O_2 \longrightarrow \text{endoperoxide epoxide} \qquad \text{Ref. 86}$$

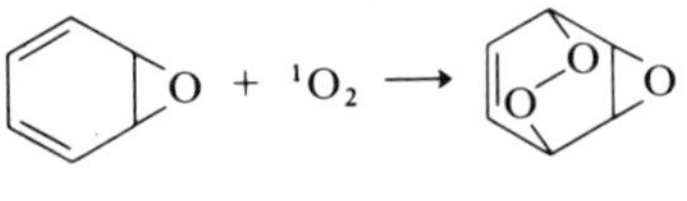

82. M. Orfanopoulos, M. B. Grdina, and L. M. Stephenson, *J. Am. Chem. Soc.* **101**, 275 (1079); K. H. Schulte-Elte, B. L. Muller, and V. Rautenstrauch, *Helv. Chim. Acta* **61**, 2777 (1978); K. H. Schulte-Elte and V. Rautenstrauch, *J. Am. Chem. Soc.* **102**, 1738 (1980).
83. W. Fenical, D. R. Kearns, and P. Radlick, *J. Am. Chem. Soc.* **91**, 3396 (1969); S. Mazur and C. S. Foote, *J. Am. Chem. Soc.* **92**, 3225 (1970); P. D. Bartlett and A. P. Schaap, *J. Am. Chem. Soc.* **92**, 3223 (1970).
84. H. H. Wasserman and J. L. Ives, *J. Am. Chem. Soc.* **98**, 7868 (1976).
85. C. S. Foote, S. Wexler, W. Ando, and R. Higgins, *J. Am. Chem. Soc.* **90**, 975 (1968).
86. C. H. Foster and G. A. Berchtold, *J. Am. Chem. Soc.* **94**, 7939 (1972).

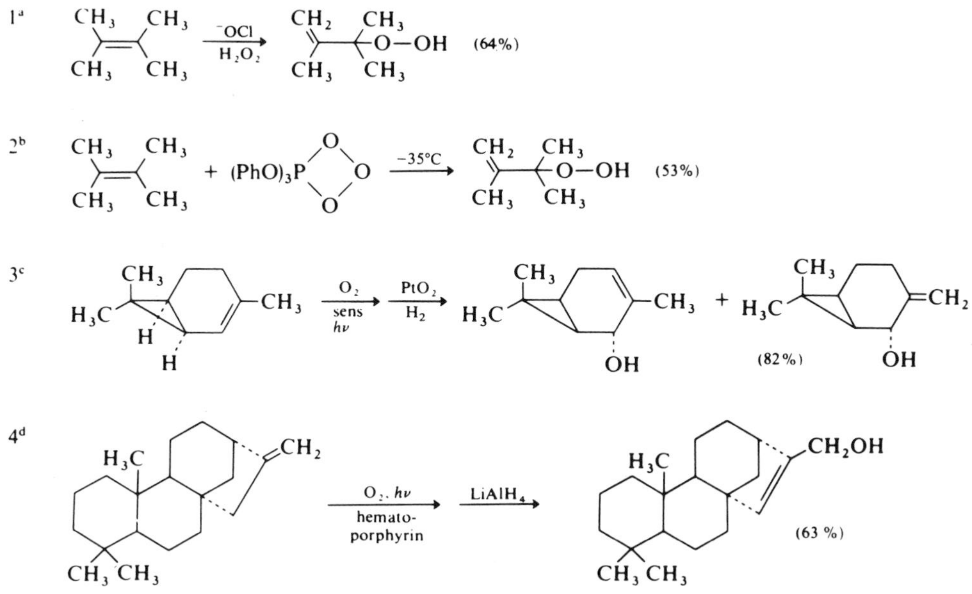

Scheme 12.10. Oxidation of Alkenes with Singlet Oxygen

1[a] $(CH_3)_2C{=}C(CH_3)_2 \xrightarrow[H_2O_2]{^-OCl} CH_2{=}C(CH_3){-}C(CH_3)_2{-}O{-}OH$ (64%)

2[b] $(CH_3)_2C{=}C(CH_3)_2 + (PhO)_3P\text{(ozonide)} \xrightarrow{-35°C} CH_2{=}C(CH_3){-}C(CH_3)_2{-}O{-}OH$ (53%)

3[c] $\xrightarrow[\text{sens}, h\nu]{O_2} \xrightarrow[H_2]{PtO_2}$ (82%)

4[d] $\xrightarrow[\text{hemato-porphyrin}]{O_2, h\nu} \xrightarrow{LiAlH_4}$ (63 %)

a. C. S. Foote, S. Wexler, W. Ando, and R. Higgins, *J. Am. Chem. Soc.* **90**, 975 (1968).
b. R. W. Murray and M. L. Kaplan, *J. Am. Chem. Soc.* **91**, 5358 (1968).
c. K. Gollnick and G. Schade, *Tetrahedron Lett.* 2335 (1966).
d. R. A. Bell, R. E. Ireland, and L. N. Mander, *J. Org. Chem.* **31**, 2536 (1966).

12.3. Cleavage of Carbon–Carbon Double Bonds

12.3.1. Transition Metal Oxidants

The most selective methods for cleaving organic molecules at carbon–carbon double bonds are based on procedures in which glycols are intermediates. Oxidation of alkenes to glycols was discussed in Section 12.2.1. Cleavage of alkenes can be carried out in one operation under mild conditions by using a solution containing periodate ion and a catalytic amount of permanganate ion.[87] The permanganate ion effects the hydroxylation, and the glycol is then cleaved by reaction with periodate. A cyclic intermediate is believed to be involved in the periodate oxidation. Permanganate is regenerated by the oxidizing action of periodate.

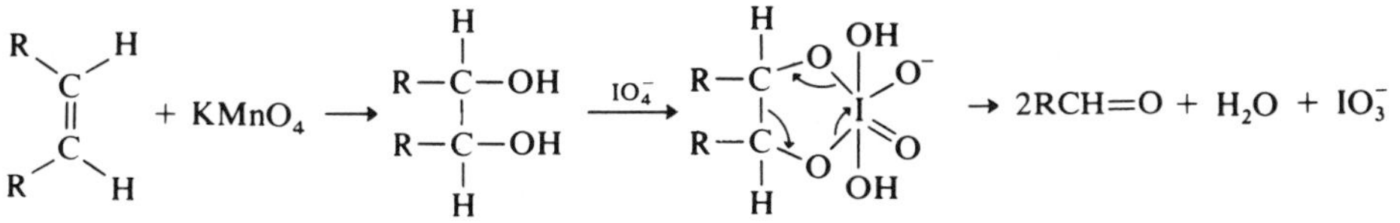

$$RCH{=}CHR + KMnO_4 \longrightarrow R{-}CH(OH){-}CH(OH){-}R \xrightarrow{IO_4^-} \text{[cyclic periodate ester]} \rightarrow 2RCH{=}O + H_2O + IO_3^-$$

87. R. U. Lemieux and E. von Rudloff, *Can. J. Chem.* **33**, 1701, 1710 (1955); E. von Rudloff, *Can. J. Chem.* **33**, 1714 (1955).

Scheme 12.11. Oxidative Cleavage of Carbon–Carbon Double Bonds with Transition Metal Oxidants

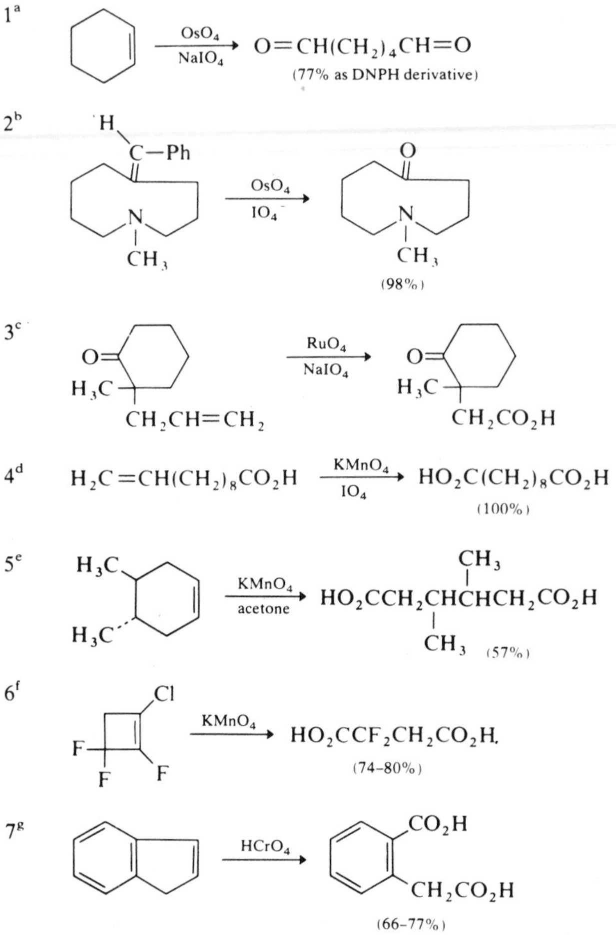

1[a] cyclohexene $\xrightarrow[NaIO_4]{OsO_4}$ $O{=}CH(CH_2)_4CH{=}O$ (77% as DNPH derivative)

2[b] H, C–Ph, N, CH_3 $\xrightarrow[IO_4^-]{OsO_4}$ O, N, CH_3 (98%)

3[c] O, H_3C, $CH_2CH{=}CH_2$ $\xrightarrow[NaIO_4]{RuO_4}$ O, H_3C, CH_2CO_2H

4[d] $H_2C{=}CH(CH_2)_8CO_2H$ $\xrightarrow[IO_4]{KMnO_4}$ $HO_2C(CH_2)_8CO_2H$ (100%)

5[e] H_3C, H_3C $\xrightarrow[acetone]{KMnO_4}$ $HO_2CCH_2CH(CH_3)CH(CH_3)CH_2CO_2H$ (57%)

6[f] Cl, F, F, F $\xrightarrow{KMnO_4}$ $HO_2CCF_2CH_2CO_2H$. (74–80%)

7[g] indene $\xrightarrow{HCrO_4}$ CO_2H, CH_2CO_2H (66–77%)

a. R. U. Lemieux and E. von Rudloff, *Can. J. Chem.* **33**, 1701 (1955).
b. M. G. Reinecke, L. R. Kray, and R. F. Francis, *J. Org. Chem.* **37**, 3489 (1972).
c. A. A. Asselin, L. G. Humber, T. A. Dobson, J. Komlossy, and R. R. Martel, *J. Med. Chem.* **19**, 787 (1976).
d. R. Pappo, D. S. Allen, Jr., R. U. Lemieux, and W. S. Johnson, *J. Org. Chem.* **21**, 478 (1956).
e. W. C. M. C. Kokke and F. A. Varkvisser, *J. Org. Chem.*, **39**, 1535 (1974).
f. N. S. Raasch and J. E. Castle, *Org. Synth.* **42**, 44 (1962).
g. O. Grummitt, R. Egan, and A. Buck, *Org. Synth.* **III**, 449 (1955).

Osmium tetroxide used in combination with sodium periodate can also effect alkene cleavage cleanly.[88] Successful oxidative cleavage of double bonds using ruthenium tetroxide and sodium periodate has also been reported.[89] In these procedures, the osmium or ruthenium can be used in substoichiometric amounts because the perio-

88. R. Pappo, D. S. Allen, Jr., R. U. Lemieux, and W. S. Johnson, *J. Org. Chem.* **21**, 478 (1956); H. Vorbrueggen and C. Djerassi, *J. Am. Chem. Soc.* **84**, 2990 (1962).
89. W. G. Dauben and L. E. Friedrich, *J. Org. Chem.* **37**, 241 (1972); B. E. Rossiter, T. Katsuki, and K. B. Sharpless, *J. Am. Chem. Soc.* **103**, 464 (1981); J. W. Patterson, Jr., and D. V. Krishna Murthy, *J. Org. Chem.* **48**, 4413 (1983).

date reoxidizes the metal to the tetroxide state. Entries 1-4 in Scheme 12.11 are examples of these procedures.

The strong oxidants Cr(VI) and MnO_4^- can also be used for oxidative cleavage of double bonds, provided there are no other sensitive groups in the molecule. The permanganate oxidation proceeds first to the diols and ketols, as described earlier (p. 624), and these are further oxidized to carboxylic acids or ketones. Good yields can be obtained, provided care is taken to prevent subsequent oxidative degradation of the products. Entries 5 and 6 in Scheme 12.11 are illustrative.

The oxidation of cyclic alkenes by Cr(VI) reagents can be a useful method for formation of dicarboxylic acids. The initial oxidation step appears to yield an epoxide, which then undergoes solvolytic ring opening to a glycol or glycol ester which is then oxidatively cleaved.[90] Two possible complications that can be encountered are competing allylic attack and skeletal rearrangement. Allylic attack can lead to eventual formation of a dicarboxylic acid that has lost one carbon atom. Pinacol-type rearrangements of the epoxide or glycol intermediates can give rise to rearranged products.

$$RCH{=}CHR \xrightarrow{Cr(VI)} RCH\overset{O}{\text{—}}CHR \xrightarrow{H^+} R_2CHCH{=}O \xrightarrow{Cr(VI)} R_2CHCO_2H$$

12.3.2. Ozonolysis

The reaction of alkenes with ozone constitutes an important method of cleaving carbon-carbon double bonds.[91] Application of low-temperature spectroscopic techniques has provided information about the rather unstable species that are intermediates in the ozonization process. These studies, along with isotope labeling results, have provided an understanding of the reaction mechanism.[92]

The two key intermediates in ozonolysis appear to be the 1,2,3-trioxolane, or initial ozonide, and the 1,2,4-trioxolane, or ozonide. The first step of the reaction is a cycloaddition to give the 1,2,3-trioxolane. This is followed by a fragmentation and recombination to give the isomeric 1,2,4-trioxolane. The mechanistic pattern of the first step is that of a 1,3-dipolar cycloaddition reaction. Ozone is expected to be a very electrophilic 1,3-dipole because of the accumulation of electronegative oxygen atoms in the ozone molecule. The initial cycloaddition, fragmentation, and

90. J. Rocek and J. C. Drozd, *J. Am. Chem. Soc.* **92**, 6668 (1970); A. K. Awasthy and J. Rocek, *J. Am. Chem. Soc.* **91**, 991 (1969).
91. P. S. Bailey, *Ozonization in Organic Chemistry*, Vol. 1, Academic Press, New York, 1978.
92. R. P. Lattimer, R. L. Kuczkowski, and C. W. Gillies, *J. Am. Chem. Soc.* **96**, 348 (1974); C. W. Gillies, R. P. Lattimer, and R. L. Kuczkowski, *J. Am. Chem. Soc.* **96**, 1536 (1974); G. Klopman and C. M. Joiner, *J. Am. Chem. Soc.* **97**, 5287 (1975); P. S. Bailey and T. M. Ferrell, *J. Am. Chem. Soc.* **100**, 899 (1978); I. C. Hisatsune, K. Shinoda, and J. Heicklen, *J. Am. Chem. Soc.* **101**, 2524 (1979); J.-I. Choe, M. Srinivasan, and R. L. Kuczkowski, *J. Am. Chem. Soc.* **105**, 4703 (1983); R. L. Kuczkowski, in *1,3-Dipolar Cycloaddition Chemistry*, A. Padwa (ed.), Wiley-Interscience, New York, Vol. 2, 1984, Chapter 11.

recombination are all predicted to be exothermic on the basis of thermochemical considerations.[93]

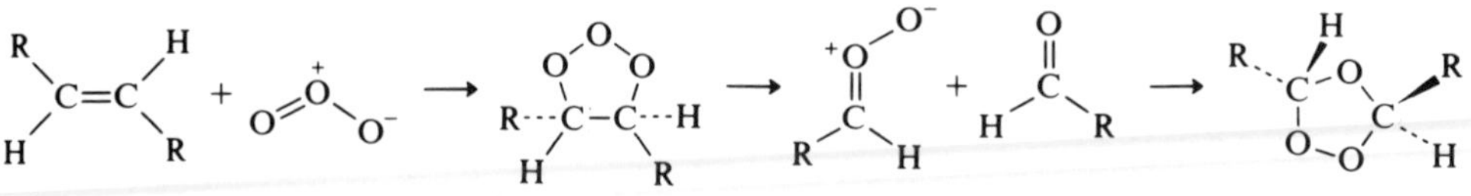

$$RHC{=}CHR + O{=}\overset{+}{O}{-}O^{-} \longrightarrow \text{(R}\cdots\text{C(H)}{-}\text{C(R)}\cdots\text{H, 1,2,3-trioxolane)} \longrightarrow RCH{=}\overset{+}{O}{-}O^{-} + HC(=O)R \longrightarrow \text{(R, H / R, H 1,2,4-trioxolane)}$$

The fragmentation–recombination may take place more rapidly than diffusion, or it may be represented as a concerted process.

The actual products isolated after ozonolysis depend upon the conditions of workup. Simple hydrolysis leads to the carbonyl compounds and hydrogen peroxide, and these can react to give secondary oxidation products. It is usually preferable to include a mild reducing agent that is capable of reducing peroxidic products. The current practice is to use dimethyl sulfide, though numerous other reducing agents have been used, including zinc,[94] trivalent phosphorus compounds,[95] and sodium sulfite.[96]

When ozonolysis is done in alcohol solvents, the carbonyl oxide fragmentation product can be trapped as an α-hydroperoxy ether.[97] Recombination is then prevented, and the carbonyl compound formed in the fragmentation step can also be isolated. If the reaction mixture is then treated with dimethyl sulfide, the hydroperoxide is reduced and the second carbonyl compound is also formed in good yield.[98] This procedure prevents oxidation of the aldehyde by the peroxidic compounds present at the conclusion of ozonolysis.

$$R_2C{=}\overset{+}{O}{-}O^{-} + CH_3OH \longrightarrow R_2\underset{\displaystyle OCH_3}{C}OOH$$

$$PhCH{=}CH_2 \xrightarrow[CH_3OH]{O_3} \underset{(31\%)}{PhCH(OCH_3)OOH} + \underset{(23\%)}{CH_2(OCH_3)OOH} + \underset{(26\%)}{PhCH{=}O} + \underset{(27\%)}{CH_2{=}O}$$

If the alcohols resulting from the reduction of the carbonyl cleavage products are desired, the reaction mixture can be reduced with a hydride reducing agent.[99] Carboxylic acids are formed in good yields from aldehydes when the ozonolysis reaction mixture is worked up in the presence of excess hydrogen peroxide.[100] Scheme 12.12 illustrates some cases where ozonolysis reactions have been used in the course of synthesis.

93. P. S. Nangia and S. W. Benson, *J. Am. Chem. Soc.* **102**, 3105 (1980).
94. S. M. Church, F. C. Whitmore, and R. V. McGrew, *J. Am. Chem. Soc.* **56**, 176 (1934).
95. W. S. Knowles and Q. E. Thompson, *J. Org. Chem.* **25**, 1031 (1960).
96. R. H. Callighan and M. H. Wilt, *J. Org. Chem.* **26**, 4912 (1961).
97. W. P. Keaveney, M. G. Berger, and J. J. Pappas, *J. Org. Chem.* **32**, 1537 (1967).
98. J. J. Pappas, W. P. Keaveney, E. Gancher, and M. Berger, *Tetrahedron Lett.*, 4273 (1966).
99. F. L. Greenwood, *J. Org. Chem.* **20**, 803 (1955).
100. A. L. Henne and P. Hill, *J. Am. Chem. Soc.* **65**, 752 (1943).

A. Reductive Workup

1[a] 6-methyl-2-vinylpyridine (H_3C, $CH{=}CH_2$) $\xrightarrow[\text{2) } Na_2SO_3]{\text{1) } O_3}$ 6-methylpyridine-2-carbaldehyde (H_3C, $CH{=}O$) (80%)

2[b] 2-nitrophenyl-$CH_2CH{=}CHCH_2Cl$ (NO_2) $\xrightarrow[\text{2) NaI}]{\text{1) } O_3}$ 2-nitrophenyl-$CH_2CH{=}O$ (NO_2) (89%)

3[c] 2,6-dimethylnaphthalene (H_3C, CH_3) $\xrightarrow[\text{2) } Ph_3P]{\text{1) } O_3}$ 4-methylphthalaldehyde (H_3C, $CH{=}O$, $CH{=}O$) (51%)

4[d] 3-methylene-1-methyl-2-piperidone (CH_2, N–CH_3, O) $\xrightarrow[\text{2) } Me_2S]{\text{1) } O_3}$ 1-methylpiperidine-2,3-dione (O, N–CH_3, O) (66%)

5[e] (H_3C, H_2C, Cl bicyclic alkene) $\xrightarrow[\text{2) } (CH_3)_2S]{\text{1) } O_3}$ (H_3C, O, Cl bicyclic ketone) (84%)

B. Oxidative Workup

6[f] tetrahydrophthalic anhydride $\xrightarrow[\text{2) } H_2O_2]{\text{1) } O_3\text{, } HCO_3H}$ HO_2C, CO_2H, HO_2C, CO_2H (butanetetracarboxylic acid) (95%)

7[g] $PhP(O)(CH_2CH{=}CH_2)_2$ $\xrightarrow[\text{2) } HCO_2H\text{, } H_2O_2]{\text{1) } O_3}$ $PhP(O)(CH_2CO_2H)_2$

a. R. H. Callighan and M. H. Wilt, *J. Org. Chem.* **26**, 4912 (1961).
b. W. E. Noland and J. H. Sellstedt, *J. Org. Chem.* **31**, 345 (1966).
c. J. J. Pappas, W. P. Keaveney, M. Berger, and R. V. Rush, *J. Org. Chem.* **33**, 787 (1968).
d. M. L. Rueppel and H. Rapoport, *J. Am. Chem. Soc.* **94**, 3877 (1972).
e. J. V. Paukstelis and B. W. Macharia, *J. Org. Chem.* **38**, 646 (1973).
f. J. E. Franz, W. S. Knowles, and C. Osuch, *J. Org. Chem.* **30**, 4328 (1965).
g. J. L. Eichelberger and J. K. Stille, *J. Org. Chem.* **36**, 1840 (1971).

12.4. Selective Oxidative Cleavages at Other Functional Groups

12.4.1. Cleavage of Glycols

As discussed in connection with cleavage of double bonds by permanganate-periodate or osmium tetroxide-periodate (see p. 643), the glycol unit is susceptible to mild oxidative cleavage. The most commonly used reagent for this oxidative cleavage is the periodate ion.[101] The fragmentation is believed to occur via a cyclic adduct of the glycol and the oxidant.

$$\text{R-CH(OH)-CH(OH)-R} \xrightarrow{IO_4^-} \text{cyclic periodate ester (R-CH-O)(R-CH-O)I(=O)(O^-)(OH)(OH)} \longrightarrow 2\text{RCH=O} + H_2O + IO_3^-$$

Structural features which retard formation of the cyclic intermediate decrease the reaction rate. For example, *cis*-1,2-dihydroxycyclohexane is substantially more reactive than the *trans* isomer.[102] Glycols for which the geometry of the molecule precludes the possibility of a cyclic intermediate are essentially inert to periodate.

Certain other combinations of adjacent functional groups are also cleaved be periodate. Diketones are cleaved to carboxylic acids, and it has been proposed that a reactive cyclic intermediate is formed by nucleophilic attack on the diketone[103]:

$$H_3C\text{-CO-CO-}CH_3 + IO_4^- \xrightarrow{^-OH} H_3C\text{-C(OH)(OIO}_4^{2-}\text{)-C(=O)-}CH_3 \xrightarrow{H_2O} \text{cyclic } H_3C\text{-C(OH)(O-)-C(OH)(O-)-}CH_3\ IO_4H^- \rightarrow 2CH_3CO_2H + IO_3^-$$

α-Hydroxyketones and α-aminoalcohols are also subject to oxidative cleavage, presumably by a similar mechanism.

Lead tetraacetate is an alternative reagent to periodate for glycol cleavage. It is particularly useful for glycols that have low solubility in the aqueous media used for periodate reactions. A cyclic intermediate is suggested by the same kind of stereochemistry-reactivity relationships discussed for periodate.[104] Unlike the situation with periodate, however, glycols that cannot form cyclic intermediates are eventually oxidized. For example, *trans*-9,10-dihydroxydecalin is oxidized, although the rate is 100 times less than for the *cis* isomer.[105] Thus, while a mechanism involving cyclic transition state appears to provide the lowest-energy pathway for this oxidative cleavage, it is not the only possible mechanism.

101. C. A. Bunton, in *Oxidation in Organic Chemistry*, Part A, K. B. Wiberg (ed.), Academic Press, New York, 1965, pp. 367–388; A. S. Perlin, in *Oxidation*, Vol. 1, R. L. Augustine (ed.), Marcel Dekker, New York, 1969, pp. 189–204.
102. C. C. Price and M. Knell, *J. Am. Chem. Soc.* **64**, 552 (1942).
103. C. A. Bunton and V. J. Shiner, *J. Chem. Soc.*, 1593 (1960).
104. C. A. Bunton, in *Oxidation in Organic Chemistry*, K. Wiberg (ed.), Academic Press, New York, 1965, pp. 398–405; W. S. Trahanovsky, J. R. Gilmore, and P. C. Heaton, *J. Org. Chem.* **38**, 760 (1973).
105. R. Criegee, E. Höger, G. Huber, P. Kruck, F. Marktscheffel, and H. Schellenberger, *Justus Liebigs Ann. Chem.* **599**, 81 (1956).

Both the periodate cleavage and lead tetraacetate oxidation can be applied synthetically to the generation of medium-sized rings when the glycol is at the junction of two rings.

Ref. 106

12.4.2. Oxidative Decarboxylation

Carboxylic acids are oxidized by lead tetraacetate. Decarboxylation occurs, and the product may be an alkene, alkane, or acetate ester, or, under modified conditions, a halide. A radical mechanism operates, and the product composition depends on the fate of the radical intermediate.[107] The reaction is catalyzed by cupric salts, which function by oxidizing the intermediate radical to a carbocation (step 3 in the mechanism). Cu(II) is more reactive than $Pb(OAc)_4$ in this step.

$$Pb(OAc)_4 + RCO_2H \rightleftharpoons RCO_2Pb(OAc)_3 + CH_3CO_2H$$

$$RCO_2Pb(OAc)_3 \rightarrow R\cdot + CO_2 + Pb(OAc)_3$$

$$R\cdot + Pb(OAc)_4 \rightarrow R^+ + Pb(OAc)_3 + CH_3CO_2^-$$

and

$$R\cdot + Pb(OAc)_3 \rightarrow R^+ + Pb(OAc)_2 + CH_3CO_2^-$$

Alkanes are formed when the intermediate radical abstracts hydrogen from solvent faster than it is oxidized to the carbocation. This reductive process is promoted by good hydrogen-donor solvents. It is also most favorable for primary alkyl radicals because of the higher activation energy associated with formation of primary carbocations. The most favorable conditions for alkane formation involve photochemical decomposition of the carboxylic acid in chloroform, which is a relatively good hydrogen donor.

CO_2H $\xrightarrow[h\nu]{CHCl_3,\ Pb(OAc)_4}$ (65%) Ref. 108

Normally, the dominant products are the alkene and the acetate ester. These arise from the carbocation intermediate by, respectively, elimination of a proton and

106. T. Wakamatsu, K. Akasaka, and Y. Ban, *Tetrahedron Lett.*, 2751, 2755 (1977).
107. R. A. Sheldon and J. K. Kochi, *Org. React.* **19**, 279 (1972).
108. J. K. Kochi and J. D. Bacha, *J. Org. Chem.* **33**, 2746 (1968).

capture of an acetate ion. The presence of copper acetate increases the alkene : ester ratio.[109]

CO_2H CO_2H $\xrightarrow[KOAc]{Pb(OAc)_2,\ Cu(OAc)_2}$ O_2CCH_3 CH_3CO_2 (93%)

Ref. 110

CH_3 H_3C H_3C CO_2H O $\xrightarrow[Cu(OAc)_2]{Pb(OAc)_4}$ CH_3 H_3C H_3C O + CH_3 H_3C H_3C O

Ref. 111

In the presence of lithium chloride, the product is the corresponding chloride.[112]

$$\underset{CH_3CHCH_2CO_2CH_3}{Cl_2CCO_2H} \xrightarrow[LiCl]{Pb(OAc)_4} \underset{CH_3CHCH_2CO_2CH_3}{Cl_3C} \quad (77\%)$$

Ref. 113

A related method for conversion of carboxylic acids to bromides with decarboxylation is the *Hunsdiecker reaction.*[114] The most convenient method for carrying out this transformation involves heating the carboxylic acid with mercuric oxide and bromine:

$$\triangleright\!-CO_2H \xrightarrow[Br_2]{HgO} \triangleright\!-Br \quad (41\text{–}46\%)$$

Ref. 115

The overall transformation can also be accomplished by reaction of thallium(I) carboxylate with bromine.[116]

1,2-Dicarboxylic acids undergo bis-decarboxylation on reaction with lead tetraacetate to give alkenes. This reaction has been of occasional use, when the required diacid is available, for the synthesis of strained alkenes.

O O O O O O $\xrightarrow[Pyridine,\ 80°C]{Pb(OAc)_4}$ O O O (39%)

Ref. 117

109. J. D. Bacha and J. K. Kochi, *Tetrahedron* **24**, 2215 (1968).
110. P. Caluwe and T. Pepper, *J. Org. Chem.* **53**, 1786 (1988).
111. D. D. Sternbach, J. W. Hughes, D. E. Bardi and B. A. Banks, *J. Am. Chem. Soc.* **107**, 2149 (1985).
112. J. K. Kochi, *J. Org. Chem.* **30**, 3265 (1965).
113. S. E. de Laszlo and P. G. Williard, *J. Am. Chem. Soc.* **107**, 199 (1985).
114. C. V. Wilson, *Org. React.* **9**, 332 (1957).
115. J. S. Meek and D. T. Osuga, *Org. Synth.* **V**, 126 (1973).
116. A. McKillop, D. Bromley, and E. C. Taylor, *J. Org. Chem.* **34**, 1172 (1969).
117. E. Grovenstein, Jr., D. V. Rao, and J. W. Taylor, *J. Am. Chem. Soc.* **83**, 1705 (1961).

The reaction can be formulated as occurring by a concerted process initiated by a two-electron oxidation:

$$H-O-\overset{O}{\overset{\|}{C}}-\underset{R}{\overset{R}{C}}-\underset{R}{\overset{R}{C}}-\overset{O}{\overset{\|}{C}}-O-Pb(OAc)_3 \rightarrow 2\,CO_2 + R_2C{=}CR_2 + Pb(OAc)_2 + CH_3CO_2H$$

A concerted mechanism is also possible for α-hydroxycarboxylic acids, and these compounds readily undergo oxidative decarboxylation to ketones[118]:

$$R_2C(OH)-\overset{O}{\overset{\|}{C}}-O-Pb(OAc)_3 \rightarrow R_2C{=}O + CO_2 + Pb(OAc)_2 + CH_3CO_2H$$

γ-Ketocarboxylic acids are oxidatively decarboxylated to enones.[119] This reaction is presumed to proceed through the usual oxidative decarboxylation, with the carbocation intermediate being efficiently deprotonated because of the developing conjugation.

HO_2C, CH_3 $\xrightarrow[Cu(OAc)_2]{Pb(OAc)_4}$ CH_3 (78%) Ref. 119

12.5. Oxidation of Ketones and Aldehydes

12.5.1. Transition Metal Oxidants

Ketones are cleaved oxidatively by Cr(VI) or Mn(VII) reagents. The reaction is sometimes of utility in the synthesis of difunctional molecules by ring cleavage. The mechanism for both reagents is believed to involve cleavage of an enolic intermediate.[120] A study involving both kinetic data and quantitative product studies has permitted a fairly complete description of the Cr(VI) oxidation of benzyl phenyl ketone.[121] The products include both oxidative-cleavage products and benzil, **3**, which results from oxidation alpha to the carbonyl. In addition, the dimeric product **4**, which is suggestive of radical intermediates, is formed under some conditions.

$$PhCH_2\overset{O}{\overset{\|}{C}}Ph \xrightarrow{Cr(VI)} \underset{\mathbf{3}}{PhCOCOPh} + PhCHO + PhCO_2H + \underset{\mathbf{4}}{PhCOCH(Ph)-CH(Ph)COPh}$$

118. R. Criegee and E. Büchner, *Chem. Ber.* **73**, 563 (1940).
119. J. E. McMurry and L. C. Blaszczak, *J. Org. Chem.* **39**, 2217 (1974).
120. K. B. Wiberg and R. D. Geer, *J. Am. Chem. Soc.* **87**, 5202 (1965); J. Rocek and A. Riehl, *J. Am. Chem. Soc.* **89**, 6691 (1967).
121. K. B. Wiberg, O. Aniline, and A. Gatzke, *J. Org. Chem.* **37**, 3229 (1972).

Both the diketone and the cleavage products were shown to arise from an α-hydroxyketone intermediate (benzoin), **5**:

$$PhCH_2\overset{O}{\overset{\|}{C}}Ph \rightleftharpoons PhCH{=}\underset{OH}{\underset{|}{C}}Ph \xrightarrow{H_2CrO_4} Ph{-}CH{=}\underset{OCrO_3H}{\underset{|}{C}}H{-}Ph \rightarrow \underset{OH}{\underset{|}{PhCH}}{-}\underset{O}{\underset{\|}{C}}Ph + Cr(IV) \rightarrow \text{products}$$

5

The coupling product is considered to involve a radical intermediate formed by one-electron oxidation, probably effected by Cr(IV).

The oxidation of cyclohexanone involves 2-hydroxycyclohexanone and 1,2-cyclohexanedione as intermediates.[122]

$$\text{cyclohexanone} \xrightarrow{Cr(VI)} \text{2-hydroxycyclohexanone} \rightarrow \text{1,2-cyclohexanedione} \rightarrow HO_2C(CH_2)_4CO_2H$$

Because of the efficient oxidation of alcohols to ketones, the alcohols can be used as the starting materials in oxidative cleavages. The conditions required for oxidative cleavage are more vigorous than for the alcohol → ketone transformation.

Aldehydes can be oxidized to carboxylic acids by both Mn(VII) and Cr(VI). Fairly detailed mechanistic studies have been carried out for Cr(VI). A chromate ester of the aldehyde hydrate is believed to be formed, and this species decomposes in the rate-determining step by a mechanism similar to that which operates in alcohol oxidations.[123]

$$RCH{=}O + H_2CrO_4 \rightleftharpoons R\overset{OH}{\underset{H}{C}}{-}OCrO_3H \rightarrow RCO_2H + HCrO_3^- + H^+$$

Effective conditions for oxidation of aldehydes to carboxylic acids with $KMnO_4$ involve use of *t*-butanol and an aqueous NaH_2PO_4 buffer as the reaction medium.[124] Another reagent for carrying out the aldehyde → carboxylic acid oxidation synthetically is silver oxide:

$$\text{4-HO-3-CH}_3\text{O-C}_6\text{H}_3\text{CH=O} \xrightarrow[\text{2) HCl}]{\text{1) Ag}_2\text{O, NaOH}} \text{4-HO-3-CH}_3\text{O-C}_6\text{H}_3\text{CO}_2\text{H} \quad (83\text{–}95\%)$$

Ref. 125

The reaction of aldehydes with MnO_2 in the presence of cyanide ion in an alcohol solvent is a convenient method of converting aldehydes directly to esters.[126]

122. J. Rocek and A. Riehl, *J. Org. Chem.* **32**, 3569 (1967).
123. K. B. Wiberg, *Oxidation in Organic Chemistry*, Part A, Academic Press, New York, 1965, pp. 172–178.
124. A. Abiko, J. C. Roberts, T. Takemasa, and S. Masamune, *Tetrahedron Lett.* **27**, 4537 (1986).
125. I. A. Pearl, *Org. Synth.* **IV**, 972 (1963).
126. E. J. Corey, N. W. Gilman, and B. E. Ganem, *J. Am. Chem. Soc.* **90**, 5616 (1968).

This reaction involves the cyanohydrin as an intermediate. The initial product is an acyl cyanide, which is solvolyzed under the reaction conditions.

$$RCH{=}O + {}^{-}CN + H^{+} \rightleftharpoons RCH(OH)CN$$

$$RCH(OH)CN + MnO_2 \rightarrow RC(=O)CN \xrightarrow{R'OH} RC(=O)OR'$$

Lead tetraacetate can effect oxidation of carbonyl groups, leading to formation of α-acetoxyketones.[127] The yields are seldom high, however. Boron trifluoride can be used to catalyze these oxidations. It is presumed to function by catalyzing the formation of the enol, which is assumed to be the reactive species.[128] With unsymmetrical ketones, products from oxidation at both α-methylene groups are found.[129]

$$R_2CHC(=O)R' \rightleftharpoons R_2C{=}C(OH)R' \xrightarrow{Pb(OAc)_4} [CH_3C(=O)O{-}Pb(OAc)_2{-}O{-}C(R'){=}CR_2] \rightarrow R_2C(O_2CCH_3)C(=O)R'$$

Introduction of oxygen alpha to a ketone function can also be carried out via the silyl enol ether. Lead tetraacetate gives the α-acetoxyketone[130]:

$$\text{2,2,6-trimethyl-1-(trimethylsilyloxy)cyclohexene} \xrightarrow{Pb(OAc)_4} \text{2-acetoxy-2,6,6-trimethylcyclohexanone}\ (56\%)$$

α-Hydroxyketones can be obtained from silyl enol ethers by oxidation using a catalytic amount of OsO_4 with an amine oxide serving as the stochiometric oxidant.[131]

$$(CH_3)_3SiO\text{-cyclohexene}(CH_3)_2(OSiR_3) \xrightarrow[\text{N-methylmorpholine N-oxide}]{OsO_4} O{=}\text{cyclohexane}(OH)(CH_3)_2(OSiR_3) \quad \text{Ref. 132}$$

The silyl enol ethers of ketones are also oxidized to α-hydroxyketones by m-chloroperoxybenzoic acid. If the reaction woikup includes acylation, α-

127. R. Criegee, in *Oxidation in Organic Chemistry*, Part A, K. B. Wiberg (ed.), Academic Press, New York, 1965, pp. 305–312.
128. J. D. Cocker, H. B. Henbest, G. H. Philipps, G. P. Slater, and D. A. Thomas, *J. Chem. Soc.*, 6 (1965).
129. S. Moon and H. Bohm, *J. Org. Chem.* **37**, 4338 (1972).
130. G. M. Rubottom, J. M. Gruber, and K. Kincaid, *Synth. Commun.* **6**, 59 (1976); G. M. Rubottom and J. M. Gruber, *J. Org. Chem.* **42**, 1051 (1977); G. M. Rubottom and H. D. Juve, Jr., *J. Org. Chem.* **48**, 422 (1983).
131. J. P. McCormick, W. Tomasik, and M. W. Johnson, *Tetrahedron Lett.* **22**, 607 (1981).
132. R. K. Boeckman, Jr., J. E. Starrett, Jr., D. G. Nickell, and P.-E. Sun, *J. Am. Chem. Soc.* **108**, 5549 (1986).

acyloxyketones are obtained.[133] These reactions proceed by initial epoxidation of the silyl enol ether, which then undergoes ring opening. Subsequent transfer of either the O-acyl or O-TMS substituent occurs, depending on the reaction conditions.

$OSi(CH_3)_3$ $\xrightarrow{RCO_3H}$ $(CH_3)_3SiO$ ... O $\xrightarrow{RCO_2H}$ RCO_2, $OSi(CH_3)_3$, OH $\downarrow$ O, $OSi(CH_3)_3$ or O, O_2CR

Other procedures for α oxidation of ketones are based on prior generation of the enolate. The most useful oxidant in these procedures is a molybdenum compound, MoO_5·pyridine·HMPA, which is prepared by dissolving MoO_3 in hydrogen peroxide, followed by addition of HMPA. This reagent oxidizes the enolates of aldehydes, ketones, esters, and lactones to the corresponding α-hydroxy compounds.[134]

$\xrightarrow[\text{2) } MoO_5\text{-pyridine HMPA}]{\text{1) LDA}}$ (85%) Ref. 135

12.5.2. Oxidation of Ketones and Aldehydes by Oxygen and Peroxidic Compounds

In the presence of acid catalysts, peroxy compounds are capable of oxidizing carbonyl compounds in a manner involving formal insertion of an oxygen atom into one of the carbon–carbon bonds at the carbonyl group. This is known as the *Baeyer–Villiger oxidation.*[136] The insertion of oxygen is accomplished by a sequence

133. G. M. Rubottom, J. M. Gruber, R. K. Boeckman, Jr., M. Ramaiah, and J. B. Medwick, *Tetrahedron Lett.*, 4603 (1978); G. M. Rubottom and J. M. Gruber, *J. Org. Chem.* **43**, 1599 (1978); G. M. Rubottom, M. A. Vazquez, and D. R. Pelegrina, *Tetrahedron Lett.*, 4319 (1974).
134. E. Vedejs, *J. Am. Chem. Soc.* **96**, 5945 (1974); E. Vedejs, D. A. Engler, and J. E. Telschow, *J. Org. Chem.* **43**, 188 (1978); E. Vedejs and S. Larsen, *Org. Synth.* **64**, 127 (1985).
135. S. P. Tanis and K. Nakanishi, *J. Am. Chem. Soc.* **101**, 4398 (1979).
136. C. H. Hassall, *Org. React.* **9**, 73 (1957).

of steps involving addition to the carbonyl group and migration to oxygen, as outlined in the mechanism below.

$$RC(=O)R + R'C(=O)OOH \rightleftarrows R-C(OH)(R)-O-O-C(=O)R' \rightarrow RC(=O)OR + R'CO_2H$$

The concerted O—O heterolysis-migration is usually the rate-determining step.[137]

When the reaction involves an unsymmetrical ketone, the structure of the product depends on which group migrates. A number of studies have been directed at ascertaining the basis of migratory aptitude in the Baeyer–Villiger oxidation. From these studies, a general order of likelihood of migration has been established: *tert*-alkyl, *sec*-alkyl > benzyl, phenyl > *pri*-alkyl > cyclopropyl > methyl.[138] Thus, methyl ketones are uniformly found to give acetate esters resulting from migration of the larger group.[139] A major factor in determining which group migrates is the ability to accommodate partial positive charge. In *para*-substituted phenyl groups, for example, electron-donor substituents favor migration.[140] Steric and conformational factors probably also come into play.[141] As is generally true in cases of migration to an electron-deficient center, the configuration of the migrating group is retained in Baeyer–Villiger oxidations.

Some typical examples of Baeyer–Villiger oxidations are shown in Scheme 12.13.

Although ketones are essentially inert to molecular oxygen, enolate anions are susceptible to oxidation. The combination of oxygen and a strong base has found some utility in the introduction of an oxygen function at carbanionic sites.[142] Hydroperoxides are the initial products of such oxidations, but when DMSO or some other substance capable of reducing the hydroperoxide is present, the corresponding alcohol is isolated. A procedure that has met with considerable success involves oxidation in the presence of a trialkyl phosphite.[143] The intermediate hydroperoxide is efficiently reduced by the phosphite ester.

O OCH₃ O CCH₃ … O OCH₃ → (NaO—t-Bu, O_2, DMF, $P(OEt)_3$) O OCH₃ O CCH₃ OH … O OCH₃ (55%) Ref. 144

137. Y. Ogata and Y. Sawaki, *J. Org. Chem.* **37**, 2953 (1972).
138. H. O. House, *Modern Synthetic Reactions*, Second Edition, W. A. Benjamin, Menlo Park, California, 1972, p. 325.
139. P. A. S. Smith, in *Molecular Rearrangements*, P. de Mayo (ed.), Wiley-Interscience, New York, 1963, pp. 84–587.
140. W. E. Doering and L. Speers, *J. Am. Chem. Soc.* **72**, 5515 (1950).
141. M. F. Hawthorne, W. D. Emmons, and K. S. McCallum, *J. Am. Chem. Soc.* **80**, 6393 (1958); J. Meinwald and E. Frauenglass, *J. Am. Chem. Soc.* **82**, 5235 (1960).
142. J. N. Gardner, T. L. Popper, F. E. Carlon, O. Gnoj, and J. L. Herzog, *J. Org. Chem.* **33**, 3695 (1968).
143. J. N. Gardner, F. E. Carlon, and O. Gnoj, *J. Org. Chem.* **33**, 3294 (1968).
144. F. A. J. Kerdesky, R. J. Ardecky, M. V. Lashmikanthan, and M. P. Cava, *J. Am. Chem. Soc.* **103**, 1992 (1981).

Scheme 12.13. Baeyer-Villiger Oxidations

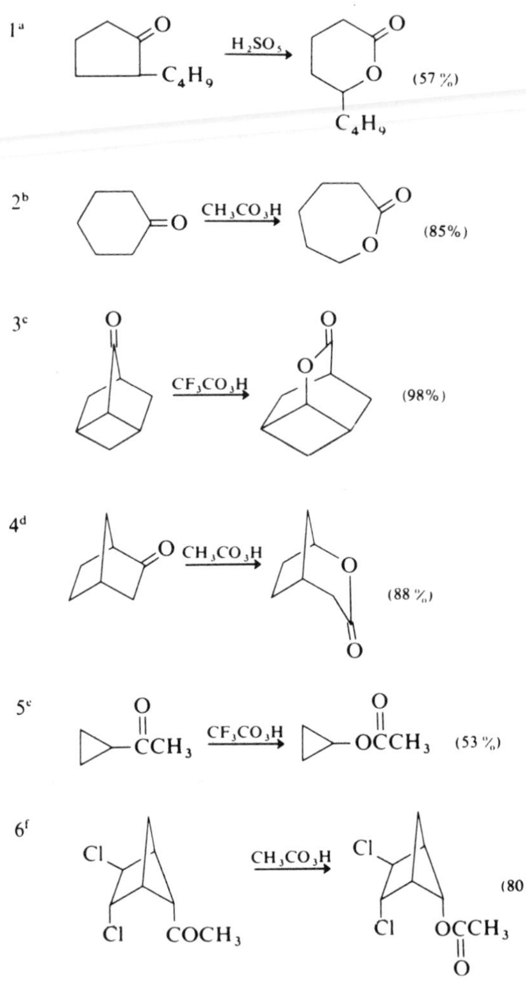

1[a] H_2SO_5 (57%)

2[b] CH_3CO_3H (85%)

3[c] CF_3CO_3H (98%)

4[d] CH_3CO_3H (88%)

5[e] CF_3CO_3H (53%)

6[f] CH_3CO_3H (80%)

a. T. H. Parliment, M. W. Parliment, and I. S. Fagerson, *Chem. Ind.* 1845 (1966).
b. P. S. Starcher and B. Phillips, *J. Am. Chem. Soc.* **80**, 4079 (1958).
c. S. A. Monti and S.-S. Yuan, *J. Org. Chem.* **36**, 3350 (1971).
d. J. Meinwald and E. Frauenglass, *J. Am. Chem. Soc.* **82**, 5235 (1960).
e. W. D. Emmohs and G. B. Lucas, *J. Am. Chem. Soc.* **77**, 2287 (1955).
f. K. B. Wiberg and R. W. Ubersax, *J. Org. Chem.* **37**, 3827 (1972).

This oxidative process has been successful with ketones,[144] esters,[145] and lactones.[146] Hydrogen peroxide can also be used as the oxidant, in which case the alcohol is formed directly.[147]

145. E. J. Corey and H. E. Ensley, *J. Am. Chem. Soc.* **97**, 6908 (1975).
146. J. J. Plattner, R. D. Gless, and H. Rapoport, *J. Am. Chem. Soc.* **94**, 8613 (1972); R. Volkmann, S. Danishefsky, J. Eggler, and D. M. Solomon, *J. Am. Chem. Soc.* **93**, 5576 (1971).
147. G. Büchi, K. E. Matsumoto, and H. Nishimura, *J. Am. Chem. Soc.* **93**, 3299 (1971).

The mechanism for the oxidation of enolates by oxygen is usually a radical chain autoxidation in which the propagation step involves electron transfer from the carbanion to a hydroperoxy radical[148]:

$$RC(O^-){=}CR_2 + O_2 \rightarrow RC(=O)-\dot{C}R_2 + O_2^{\cdot -}$$

$$RC(=O)\dot{C}R_2 + O_2 \rightarrow RC(=O)CR_2(O{-}O\cdot)$$

$$RC(O^-){=}CR_2 + RC(=O)CR_2(O{-}O\cdot) \rightarrow RC(=O)\dot{C}R_2 + RC(=O)CR_2(O{-}O^-)$$

Arguments for a non-chain reaction between the enolate and oxygen to give the hydroperoxide anion directly have also been advanced.[149]

12.5.3. Oxidation with Other Reagents

Selenium dioxide can be used to oxidize ketones and aldehydes to α-dicarbonyl compounds. The reaction often gives high yields of products when there is a single type of CH_2 group adjacent to the carbonyl group. In unsymmetrical ketones, oxidation usually occurs at the CH_2 that is most readily enolized.[150]

Cyclohexanone $\xrightarrow{SeO_2}$ cyclohexane-1,2-dione (60%) — Ref. 151

$C_6H_5CCH_3$ (C=O) $\xrightarrow{SeO_2}$ $C_6H_5C(=O)CH{=}O$ (69–72%) — Ref. 152

The oxidation is regarded as taking place by an electrophilic attack of selenium dioxide (or selenous acid, H_2SeO_3, the hydrate) on the enol of the ketone or aldehyde. This is followed by hydrolytic elimination of the selenium.[153]

$$RC(OH){=}CHR' \xrightarrow{SeO_2} RC(=O)-CHR'(Se(=O)OH) \xrightarrow{-H_2O} RC(=O)-C(=Se{=}O)R' \xrightarrow[-H_2SeO]{H_2O} RC(=O)-C(=O)R'$$

148. G. A. Russell and A. G. Bemix, *J. Am. Chem. Soc.* **88**, 5491 (1966).
149. H. R. Gersmann and A. F. Bickel, *J. Chem. Soc., B*, 2230 (1971).
150. E. N. Trachtenberg, in *Oxidation*, Vol. 1, R. L. Augustine (ed.), Marcel Dekker, New York, 1969, Chapter 3.
151. C. C. Hach, C. V. Banks, and H. Diehl, *Org. Synth.* **IV**, 229 (1963).
152. H. A. Riley and A. R. Gray, *Org. Synth.* **II**, 509 (1943).
153. K. B. Sharpless and K. M. Gordon, *J. Am. Chem. Soc.* **98**, 300 (1976).

Methyl ketones are degraded to the next lower carboxylic acid by reaction with hypochlorite or hypobromite ions. The initial step in these reactions involves base-catalyzed halogenation. The haloketones are more reactive than their precursors, and rapid halogenation to the trihalo compound results. Trihalomethyl ketones are susceptible to alkaline cleavage because of the inductive stabilization provided by the halogen atoms.

$$RC(=O)CH_3 \underset{}{\overset{slow}{\rightleftharpoons}} RC(O^-)=CH_2 \xrightarrow{^-OBr} RC(=O)CH_2Br \xrightarrow{^-OH} RC(O^-)=CHBr \xrightarrow{fast} RC(=O)CBr_3$$

$$RC(=O)CBr_3 + {}^-OH \rightleftharpoons RC(O^-)(OH)CBr_3 \rightarrow RCO_2H + {}^-CBr_3 \rightleftharpoons RCO_2^- + HCBr_3$$

$$(CH_3)_3CC(=O)CH_3 \xrightarrow[Br_2]{NaOH} (CH_3)_3CCO_2H \quad (71\text{–}74\%)$$ Ref. 154

$$(CH_3)_2C{=}CHCOCH_3 \xrightarrow{KOCl} \xrightarrow{H^+} (CH_3)_2C{=}CHCO_2H \quad (49\text{–}53\%)$$ Ref. 155

12.6. Allylic Oxidation

12.6.1. Transition Metal Oxidants

Carbon-carbon double bonds, besides being susceptible to addition of oxygen or cleavage, can also react at allylic positions. Synthetic utility requires that there be a good degree of selectivity between the possible modes of reaction. Among the transition metal oxidants, the CrO-pyridine reagent in methylene chloride is the most satisfactory for allylic oxidation.[156] A related complex in which 3,5-dimethylpyrazole is used in place of pyridine has also been found to be effective for allylic oxidations.[157]

H_3C CH_3 (tricyclic alkene) $\xrightarrow{CrO_3\text{-3,5-dimethylpyrazole}}$ H_3C CH_3 (tricyclic enone, O=) Ref. 158

Several pieces of mechanistic evidence indicative that allylic radicals or cations are intermediates in these oxidations. Thus, ^{14}C in cyclohexene is distributed in the

154. L. T. Sandborn and E. W. Bousquet, *Org. Synth.* **I**, 512 (1932).
155. L. I. Smith, W. W. Prichard, and L. J. Spillane, *Org. Synth.* **III**, 302 (1955).
156. W. G. Dauben, M. Lorber, and D. S. Fullerton, *J. Org. Chem.* **34**, 3587 (1969).
157. W. G. Salmond, M. A. Barta, and J. L. Havens, *J. Org. Chem.* **43**, 2057 (1978); R. H. Schlessinger, J. L. Wood, A. J. Poos, R. A. Nugent, and W. H. Parson, *J. Org. Chem.* **48**, 1146 (1983).
158. A. B. Smith III and J. P. Konopelski, *J. Org. Chem.* **49**, 4094 (1984).

product cyclohexenone in a manner indicating that a symmetrical allylic intermediate is involved at some stage[159]:

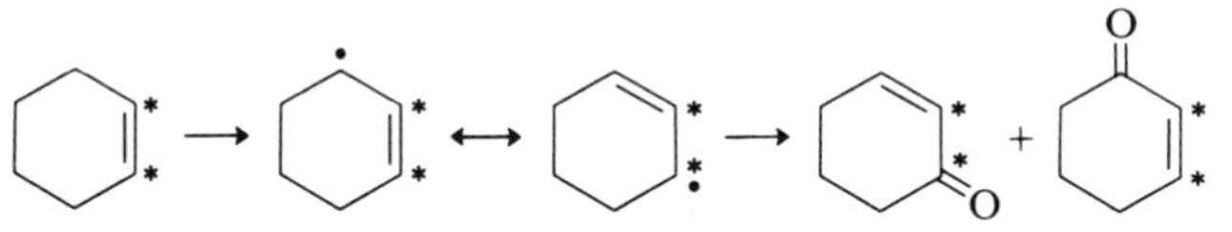

In many allylic oxidations, the double bond is found in a position indicating that an "allylic shift" occurs during the oxidation.

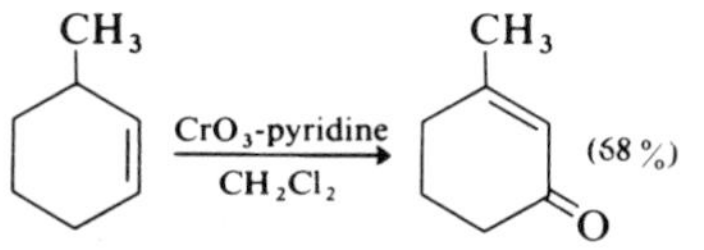

CrO_3-pyridine, CH_2Cl_2 (68%) Ref. 156

Detailed mechanistic understanding of the allylic oxidation has not been developed. One possibility is that an intermediate oxidation state of Cr, such as Cr(IV), acts as the key reagent by abstracting hydrogen.[160]

12.6.2. Other Oxidants

Selenium dioxide is a very useful reagent for allylic oxidation of alkenes. The products are either carbonyl compounds, allylic alcohols, or allylic esters, depending on the reaction conditions. The basic mechanism consists of three essential steps: (a) an electrophilic "ene" reaction with SeO_2, (b) a sigmatropic rearrangement which restores the original location of the double bond, and (c) breakdown of the resulting selenium ester[161]:

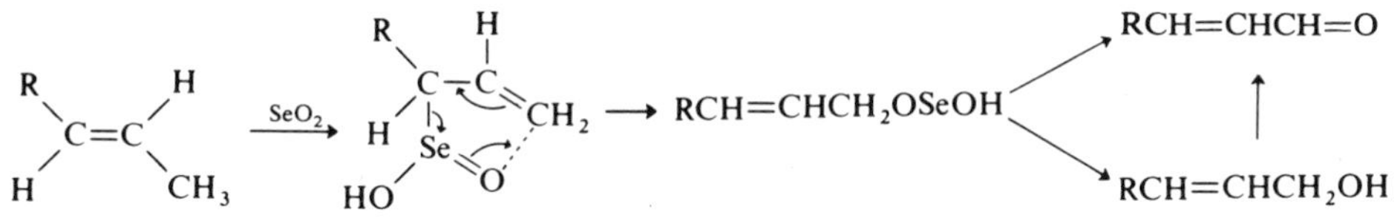

$$RCH{=}CHCH_3 \xrightarrow{SeO_2} [\text{ene adduct}] \rightarrow RCH{=}CHCH_2OSeOH \rightarrow RCH{=}CHCH{=}O \text{ or } RCH{=}CHCH_2OH$$

The alcohols that are the initial oxidation products can be further oxidized to carbonyl groups by SeO_2, and it is usually the carbonyl compound that is isolated. If the alcohol is the desired product, the oxidation can be run in acetic acid as solvent, in which case acetate esters are formed.

Although the traditional conditions for effecting SeO_2 oxidations involve use of a stoichiometric or excess amount of SeO_2, it is also possible to carry out the reaction with 1.5–2 mol % SeO_2, using *t*-butyl hydroperoxide as a stoichiometric

159. K. B. Wiberg and S. D. Nielsen, *J. Org. Chem.* **29**, 3353 (1964).
160. P. Müller and J. Rocek, *J. Am. Chem. Soc.* **96**, 2836 (1974).
161. K. B. Sharpless and R. F. Lauer, *J. Am. Chem. Soc.* **94**, 7154 (1972).

oxidant. Under these conditions, the allylic alcohol is the principal reaction product. The use of a catalytic amount of SeO_2 and excess *t*-butyl hydroperoxide permits good yields of allylic alcohols, even from alkenes that are poorly reactive under the traditional conditions.[162]

Selenium dioxide reveals a very high and useful selectivity when applied to trisubstituted *gem*-dimethyl alkenes. The products are always predominantly the *E*-allylic alcohol or unsaturated aldehydes.[163]

$$(CH_3)_2C{=}CH(CH_2CH_3)\ \xrightarrow{SeO_2}\ O{=}CH(CH_3)C{=}CH(CH_2CH_3)\ (45\%)$$

This stereoselectivity can be explained by a cyclic transition state for the sigmatropic rearrangement step. The observed stereochemistry results if the alkyl substituent adopts a pseudoequatorial conformation:

$$HOSe(=O){-}CH(C_2H_5)C({=}CH_2){-}CH_3 \longrightarrow [\text{cyclic TS}] \longrightarrow HOSeOCH_2(CH_3)C{=}CH(C_2H_5)$$

Trisubstituted alkenes are oxidized selectively at the more substituted end of the carbon–carbon double bond, indicating that the ene reaction step is electrophilic in character.

$$RCH_2(R)C{=}CH(CH_2R) \longrightarrow RCH_2(R)C^{\delta+}{-}CH(Se(O^{\delta-})O)(CH_2R) \longrightarrow RCH(OSeOH)(R)C{=}CH(CH_2R) \longrightarrow RCH(OH)(R)C{=}CH(CH_2R)$$

Thus, trisubstituted alkenes are oxidized at one of the allylic groups at the disubstituted carbon.

$$\text{4-isopropyl-1-methylcyclohexene} \xrightarrow[CH_3CO_2H,\ (CH_3CO)_2O]{SeO_2} \text{allylic } O_2CCH_3\ (35\%) + \text{allylic } OH\ (18\%) + \text{enone}\ (8\%)$$

Ref. 164

162. M. A. Umbreit and K. B. Sharpless, *J. Am. Chem. Soc.* **99**, 5526 (1977).
163. U. T. Bhalerao and H. Rapoport, *J. Am. Chem. Soc.* **93**, 4835 (1971); G. Büchi and H. Wüest, *Helv. Chim. Acta* **50**, 2440 (1967).
164. T. Suga, M. Sugimoti, and T. Matsuura, *Bull. Chem. Soc. Jpn.* **36**, 1363 (1963).

The equivalent to allylic oxidation of alkenes, but with allylic transposition of the carbon–carbon double bond, can be carried out by an indirect oxidative process involving addition of an electrophilic arylselenenyl reagent, followed by oxidative elimination of selenium. In one procedure, addition of an arylselenenyl halide is followed by solvolysis and oxidation:

$$\text{cyclohexene} \xrightarrow{\text{PhSeBr}} \text{1-bromo-2-(phenylseleno)cyclohexane} \xrightarrow[\text{2) } H_2O_2]{\text{1) } CH_3CO_2H} \text{3-acetoxycyclohexene } (O_2CCH_3)$$ Ref. 165

This reaction depends upon the facile solvolysis of β-haloselenides and the oxidative elimination of selenium, which was discussed in Section 6.8.3. An alternative method, which is experimentally simpler, involves reaction of alkenes with a mixture of diphenyl diselenide and phenylseleninic acid.[166] The two selenium reagents generate an electrophilic selenium species, phenylselenenic acid, PhSeOH.

$$RCH_2CH{=}CHR' \xrightarrow{\text{PhSeOH}} RCH_2\underset{\text{PhSe}}{CH}\overset{\text{OH}}{CH}R' \xrightarrow{t\text{-BuOOH}} RCH{=}CH\underset{\text{OH}}{CH}R'$$

The elimination is promoted by oxidation to the selenoxide by t-butyl hydroperoxide. The regioselectivity in this reaction is such that the hydroxyl group becomes bound at the more substituted end of the carbon–carbon double bond. The origin of this orientation is that the addition follows Markownikoff's rule with "PhSe$^+$" acting as the electrophile. The elimination step specifically proceeds away from the oxygen functionality.

12.7. Oxidations at Unfunctionalized Carbon

Attempts to achieve selective oxidations of hydrocarbons or other compounds when the desired site of attack is remote from an activating functional group are faced with difficulties. With powerful transition metal oxidants, the initial oxidation products are almost always more susceptible to oxidation than the starting material. Once a hydrocarbon is attacked, it is likely to be oxidized to a carboxylic acid, with chain cleavage by successive oxidation of alcohol and carbonyl intermediates. There are a few circumstances under which oxidations of hydrocarbons can be synthetically useful processes. One group involves catalytic industrial processes. Much work has been expended on the development of selective catalytic oxidation processes, and several have economic importance. Since the mechanisms are often obscured by

165. K. B. Sharpless and R. F. Lauer, *J. Org. Chem.* **39**, 429 (1974); D. L. J. Clive, *J. Chem. Soc., Chem. Commun.*, 100 (1974).
166. T. Hori and K. B. Sharpless, *J. Org. Chem.* **43**, 1689 (1978).

limited understanding of heterogeneous catalysis, however, we will not devote additional attention to these reactions.

Perhaps the most familiar and useful hydrocarbon oxidation is the oxidation of side chains on aromatic rings. Two factors contribute to making this a high-yield procedure, despite the use of strong oxidants. First, the benzylic site is activated to oxidation. Either radical or carbocation intermediates can be stabilized by resonance. Second, the aromatic ring is resistant to attack by Mn(VII) and Cr(VI) reagents which oxidize the side chain. Scheme 12.14 provides some examples of the familiar oxidation of aromatic alkyl substituents to carboxylic acid groups.

Scheme 12.14. Side-Chain Oxidation of Aromatic Compounds

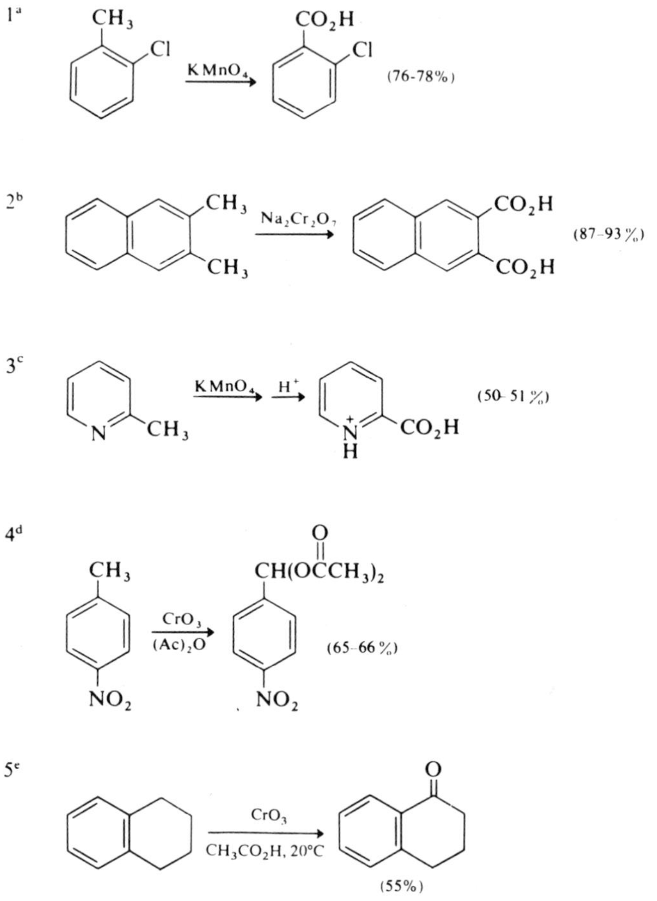

1[a] CH_3, Cl → $KMnO_4$ → CO_2H, Cl (76-78%)

2[b] CH_3, CH_3 → $Na_2Cr_2O_7$ → CO_2H, CO_2H (87-93%)

3[c] N, CH_3 → $KMnO_4$, H^+ → $\overset{+}{N}$, H, CO_2H (50-51%)

4[d] CH_3, NO_2 → CrO_3, $(Ac)_2O$ → $CH(OCCH_3)_2$ (with O above C), NO_2 (65-66%)

5[e] → CrO_3, CH_3CO_2H, 20°C → O (55%)

a. H. T. Clarke and E. R. Taylor, *Org. Synth.* **II**, 135 (1943).
b. L. Friedman, *Org. Synth.* **43**, 80 (1963); L. Friedman, D. L. Fishel, and H. Shechter, *J. Org. Chem.* **30**, 1453 (1965).
c. A. W. Singer and S. M. McElvain, *Org. Synth.* **III**, 740 (1955).
d. T. Nishimura, *Org. Synth.* **IV**, 713 (1963).
e. J. W. Burnham, W. P. Duncan, E. J. Eisenbraun, G. W. Keen, and M. C. Hamming, *J. Org. Chem.* **39**, 1416 (1974).

Partial oxidations of side chains on aromatic compounds have been achieved using tetralkylammonium permanganate in organic solvents.[167]

$$C_6H_5CH_2CH(CH_3)_2 \xrightarrow[R_4N^+MnO_4^-]{CH_3CO_2H} C_6H_5\overset{O}{\overset{\|}{C}}CH(CH_3)_2 + C_6H_5\underset{O_2CCH_3}{\underset{|}{C}}HCH(CH_3)_2$$

These reagents, however, must be used with caution because of a potential danger of explosion.

Benzeneselenic anhydride is an effective reagent for oxidizing methyl groups on aromatic rings to aldehydes[168]:

$$H_3C{-}C_6H_4{-}CH_3 \xrightarrow[\Delta,\ 10\ h]{PhSe(O)OSe(O)Ph} H_3C{-}C_6H_4{-}CH{=}O \quad (66\%)$$

A second type of hydrocarbon in which selective oxidations are possible is certain bicyclic hydrocarbons.[169] Here, the bridgehead position is the preferred site of initial attack because of the order of reactivity of C—H bonds, which is tertiary > secondary > primary. The tertiary alcohols that are the initial oxidation products, however, are not easily further oxidized. The geometry of the bicyclic rings (Bredt's rule) prevents dehydration of the alcohol. Tertiary bridgehead hydroxyls cannot be converted to ketones. Therefore, oxidation that begins at a bridgehead position stops at the alcohol stage. Chromic acid oxidation has been the most useful reaction for functionalizing unstrained bicyclic hydrocarbons. The reaction fails for strained bicyclic compounds such as norbornane because the reactivity of the bridgehead position is lowered by the unfavorable energy of radical or carbocation intermediates.

$$\xrightarrow[HOAc,\ Ac_2O]{CrO_3} \text{OH} \quad (40\text{–}50\%)$$

Other successful selective oxidations of hydrocarbons by Cr(VI) have been reported—for example, the oxidation of *cis*-decalin to the corresponding alcohol—but careful attention to reaction conditions is required to obtain satisfactory yields.

$$\xrightarrow[8^\circ C,\ 30\ min]{HCrO_4^-} \text{OH} \quad (15\%) \qquad \text{Ref. 170}$$

167. H. Jäger, J. Lütolf, and M. W. Meyer, *Angew. Chem. Int. Ed. Engl.* **18**, 756 (1979); H. J. Schmidt and H. J. Schmidt and H. J. Schäfer, *Angew. Chem. Int. Ed. Engl.* **18**, 787 (1979); H. J. Schmidt and H. J. Schäfer, *Angew. Chem. Int. Ed. Engl.* **18**, 68 (1979).
168. D. H. R. Barton, R. A. H. F. Hus, D. J. Lester, and S. V. Ley, *Tetrahedron Lett.*, 3331 (1979).
169. R. C. Bingham and P. v. R. Schleyer, *J. Org. Chem.* **36**, 1198 (1971).
170. K. B. Wiberg and G. Foster, *J. Am. Chem. Soc.* **83**, 423 (1961).

General References

R. L. Augustine (ed.), *Oxidations*, Vol. 1, Marcel Dekker, New York, 1969.
R. L. Augustine and D. J. Trecker (eds.), *Oxidations*, Vol. 2, Marcel Dekker, New York, 1971.
P. S. Bailey, *Ozonization in Organic Synthesis*, Vols. I and II, Academic Press, New York, 1978, 1982.
G. Cainelli and G. Cardillo, *Chromium Oxidations in Organic Chemistry*, Springer-Verlag, New York, 1984.
L. J. Chinn, *Selection of Oxidants in Synthesis*, Marcel Dekker, New York, 1971.
A. A. Frimer, *Chem. Rev.* **79**, 359 (1979).
A. H. Haines, *Methods for the Oxidation of Organic Compounds; Alkanes, Alkenes, Alkynes and Arynes*, Academic Press, Orlando, Florida, 1985.
W. J. Mijs and C. R. H. de Jonge, *Organic Synthesis by Oxidation with Metal Compounds*, Plenum, New York, 1986.
W. Trahanovsky (ed.), *Oxidations in Organic Chemistry*, Parts B–D, Academic Press, New York, 1973–1982.
H. H. Wasserman and R. W. Murray (eds.), *Singlet Oxygen*, Academic Press, New York, 1979.
K. B. Wiberg (ed.), *Oxidations in Organic Chemistry*, Part A, Academic Press, New York, 1965.

Problems

(*References for these problems will be found on page* 779.)

1. Indicate an appropriate oxidant for carrying out the following transformations:

(a) $(CH_3)_2C{=}CHCH_2CH_2CH(CH_3)CH_2CN \rightarrow CH_2{=}C(CH_3)CH(OH)CH_2CH_2CH(CH_3)CH_2CN$

(b) $PhCH_2OCH_2$ … CO_2R $\rightarrow$ $PhCH_2OCH_2$ … OH, CO_2R

(c) H, H_3C, H_3C, $C{=}C(CH_3)_2$, H, H, $CH_2O_2CCH_3$ $\rightarrow$ H, $CH{=}O$, H_3C, H_3C, $C{=}C$, CH_3, H, H, $CH_2O_2CCH_3$

(d) O, Ph $\rightarrow$ HO, O, Ph

(e) $CH_3CH_2CH_2(H)C{=}C(H)CH_2CH_2CH_3 \rightarrow CH_3CH_2CH_2CH(OH)CH(OH)CH_2CH_2CH_3$

(f) $PhCOCH_2CH_3 \rightarrow PhCOCH(OH)CH_3$

(g) CH_2 ; CH_3 ; $CHCH_2CH_2OCPh_3$; CH_3 → O ; CH_3 ; $CHCH_2CH_2OCPh_3$; CH_3

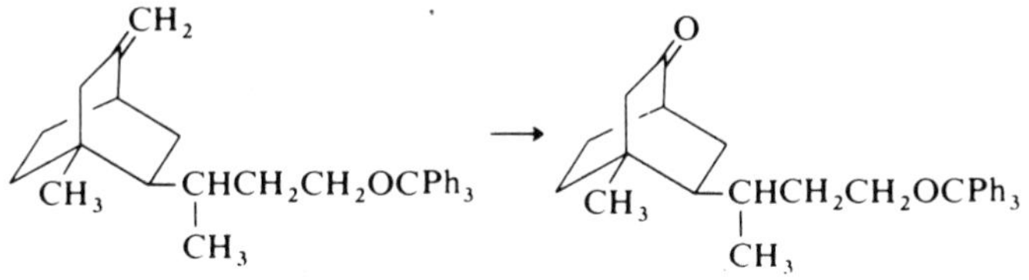

(h) → $OCCH_3$; O

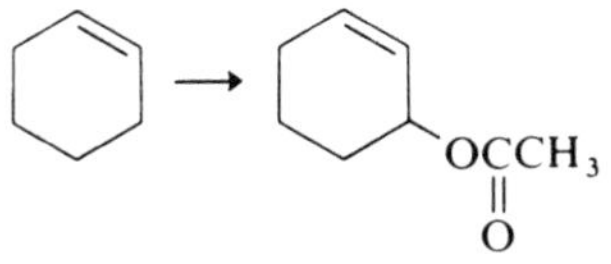

(i) H, CH_3, H ; C=C ; $C{=}C(CH_3)_2$; $PhSO_2$; CH_2CH_2 → H, CH_3, H, CH=O ; C=C ; C=C ; $PhSO_2$; CH_2CH_2 ; CH_3

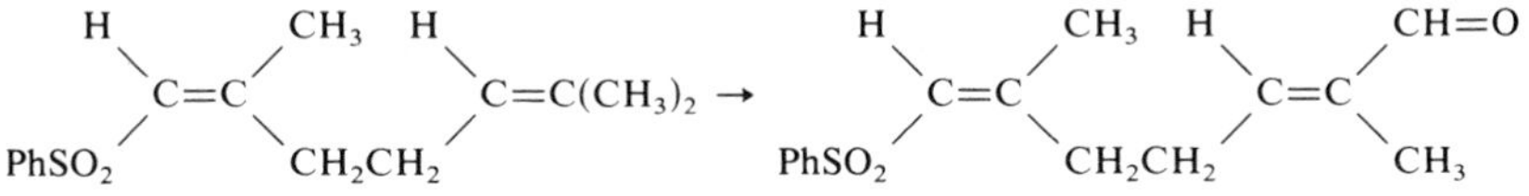

(j) H_3C CH_3 ; O → H_3C CH_3 ; O ; O

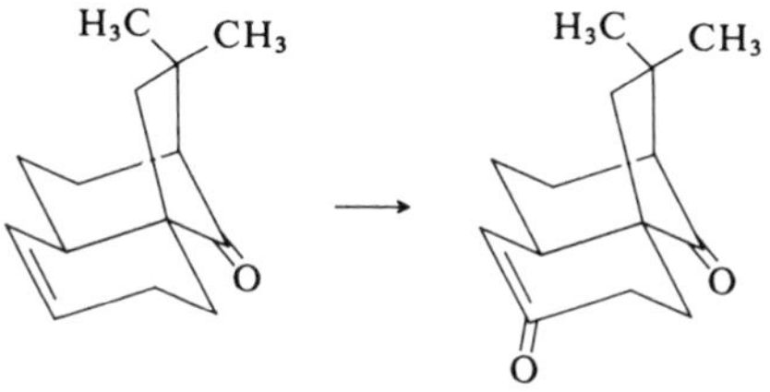

(k) CH_3 ; O= ; $OSiR_3$; CH_3 → CH_3 ; O= ; $OSiR_3$; HO ; CH_3

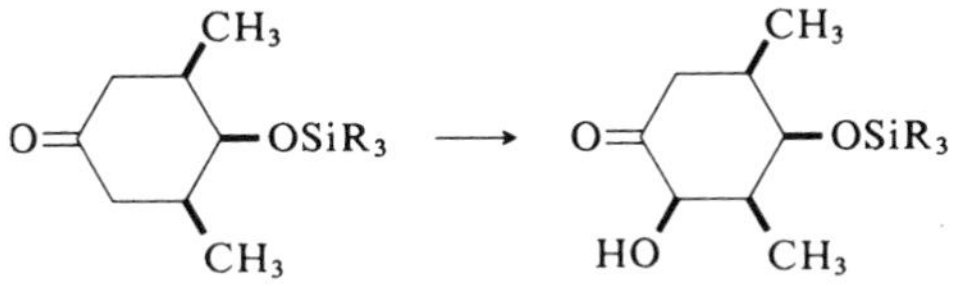

(l) H ; O ; OH ; H → H ; OCH_2OCH_3 ; CH=O ; H

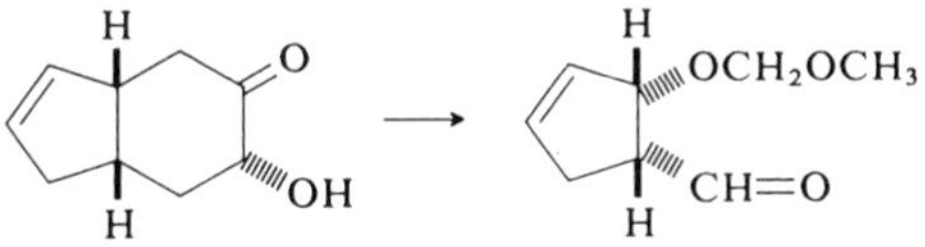

(m) O ; OCH_2OCH_3 ; CH_3 ; OCH_2OCH_3 ; CH_3 ; CH_3OCH_2O ; CH_3 ; CH_2CH_3 → O ; O ; OCH_2OCH_3 ; CH_3 ; OCH_2OCH_3 ; CH_3 ; CH_3OCH_2O ; CH_3 ; CH_2CH_3

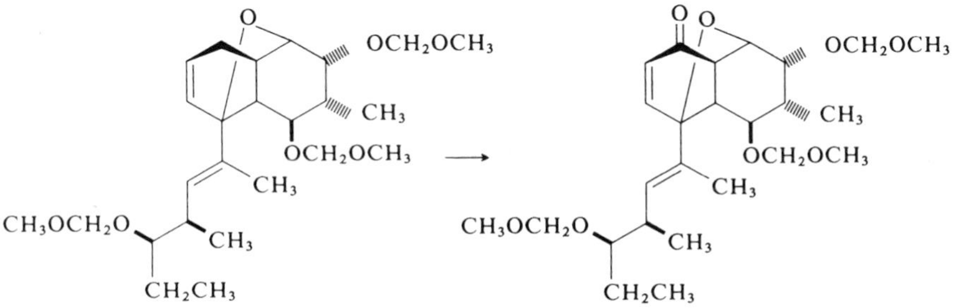

(n) H, CH_2OH ; C=C ; H_3C, CH_3 → H, CH_2OH ; CH_3CH_2 ; O ; CH_3

(o)

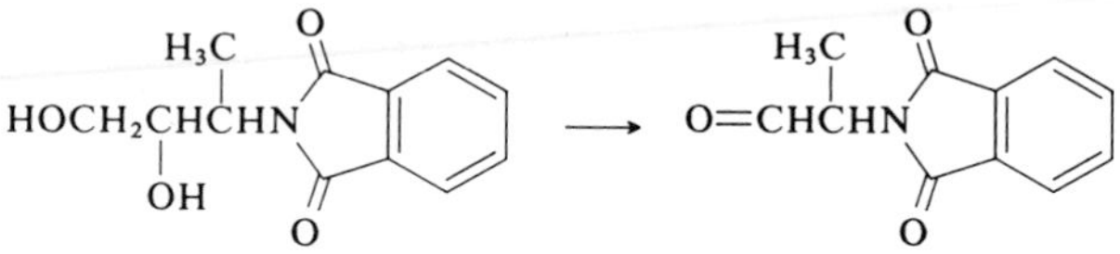

H_3C O $HOCH_2CHCHN$ OH O $\longrightarrow$ H_3C O $O{=}CHCHN$ O

(p)

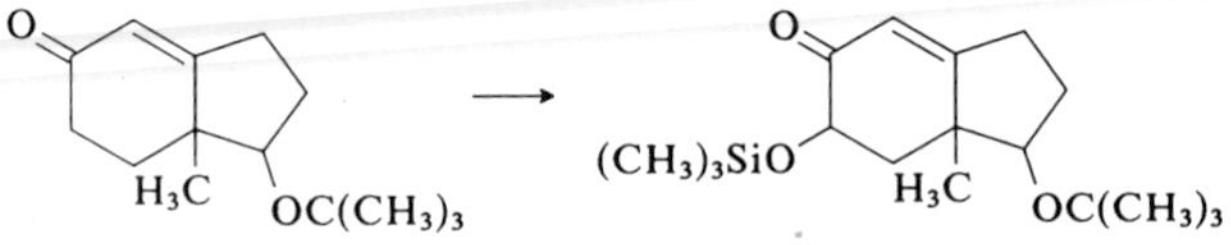

O H_3C $OC(CH_3)_3$ $\longrightarrow$ O $(CH_3)_3SiO$ H_3C $OC(CH_3)_3$

(q)

O CH_3 HO CH_3 CH_3 O O CH_3 CH_3 $\longrightarrow$ O O CH_3 HO CH_3 CH_3 O O CH_3 CH_3

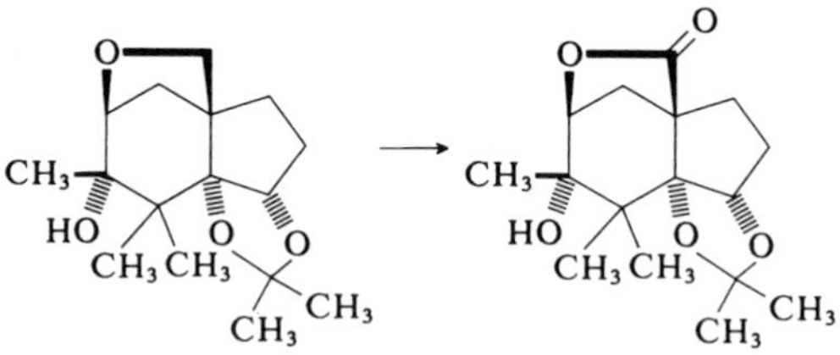

2. Predict the products of the following reactions. Be careful to consider all stereochemical aspects.

(a) $OSi(CH_3)_3$ $\xrightarrow{OsO_4}$ (N-methylmorpholine N-oxide: O, ^+N, CH_3, O^-)

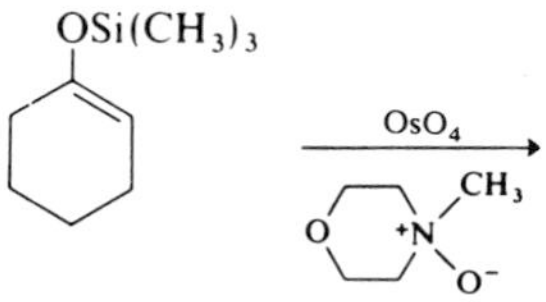

(b) H_3C CH_3 H_2C $\xrightarrow{m\text{-chloroperoxybenzoic acid}}$

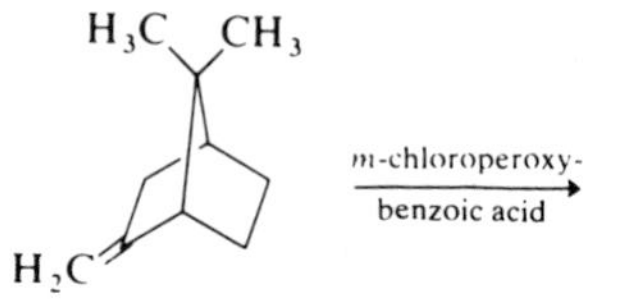

(c) HO H H_3C H_3C H O $\xrightarrow{m\text{-chloroperoxybenzoic acid}}$

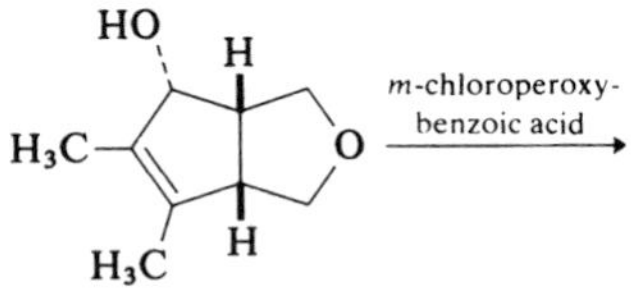

(d) CH_3 O CH_3 $\xrightarrow{LiClO_4}$

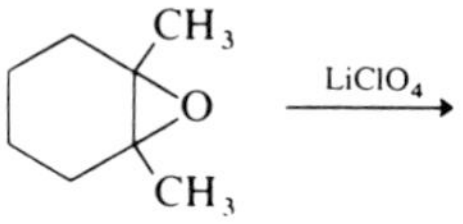

(e) H_3C CO_2CH_3 O CH_3 H O $\xrightarrow{BF_3}$

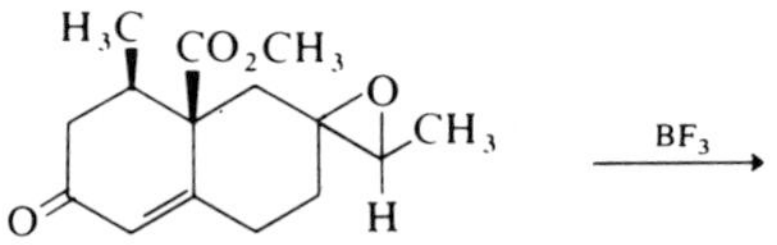

(f) $\xrightarrow[(CH_3)_3COOH]{Mo(CO)_6}$

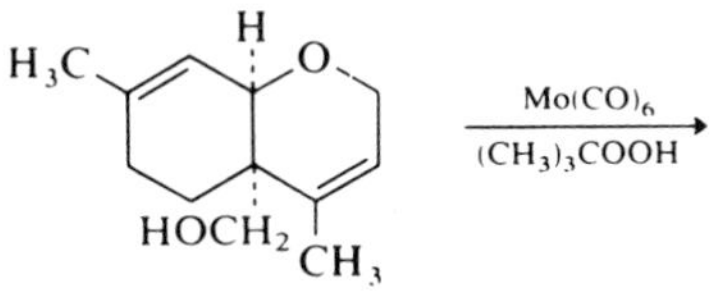

H, O, H_3C, $HOCH_2$, CH_3

(g) $\xrightarrow{OsO_4}$

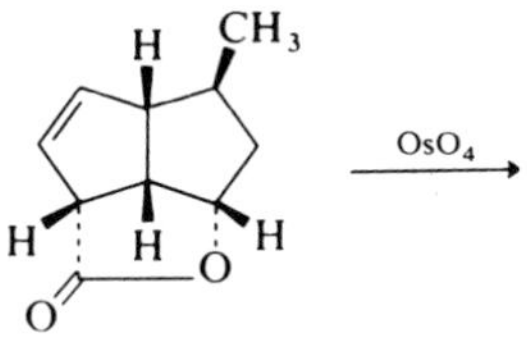

H, CH_3, H, H, H, O, O

(h) H_3C, H, C=C, H_3C, $CH_2CH_2CHCH_2$, CH_3, O, O $\xrightarrow{SeO_2}$

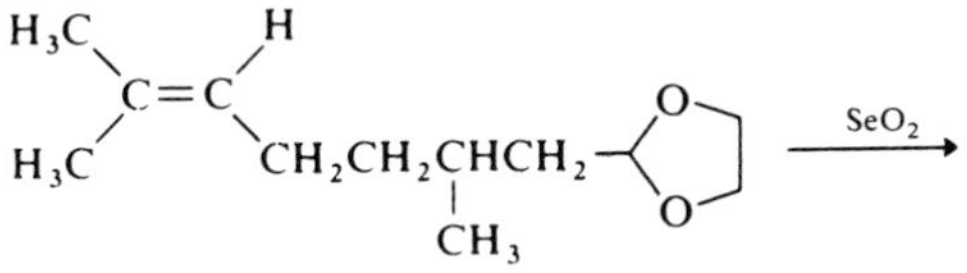

(i) CH_2OCH_3, O, O — Collins reagent (excess) $\longrightarrow$

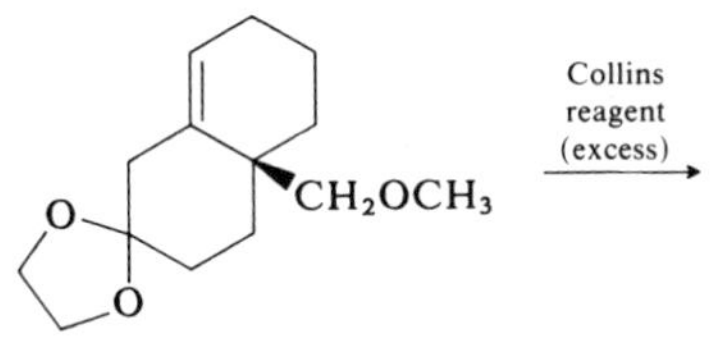

(j) CH_2CN, CH_3 $\xrightarrow[NaIO_4]{RuO_2}$

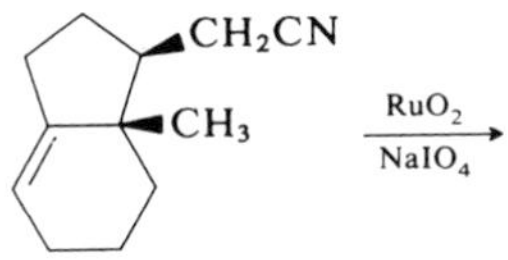

(k) OH, CH_3 $\xrightarrow[VO(acac)_2]{t\text{-}BuOOH}$

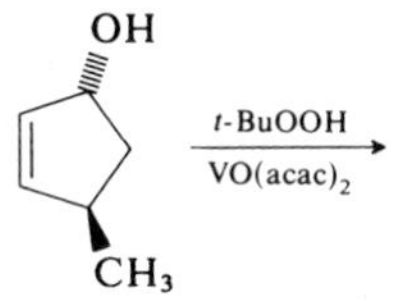

(l) CH_3, CO_2CH_3, O, O, CH_3 $\xrightarrow{BF_3}$

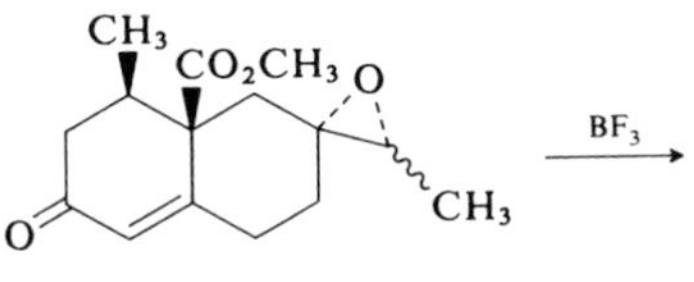

(m) CH_3, $ArSO_2NCH_2CH_2$, $OSiR_3$, $PhSO_2$ — *m*-chloroperoxy-benzoic acid $\longrightarrow$

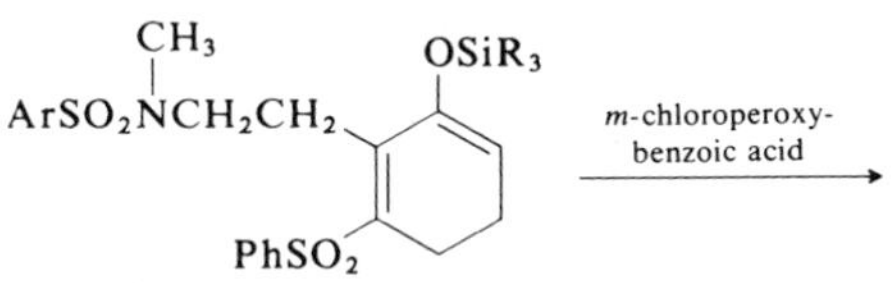

3. In chromic acid oxidation of isomeric cyclohexanols, it is usually found that axial hydroxyl groups react more rapidly than equatorial groups. For example, *trans*-4-*t*-butylcyclohexanol is less reactive (by a factor of 3.2) than the *cis* isomer. An even larger difference is noted with *cis*- and *trans*-3,3,5-trimethylcyclohexanol. The *trans* alcohol is more than 35 times more reactive than the *cis*. Are these data compatible with the mechanism given on p. 616? What additional detail do these data provide about the reaction mechanism? Explain.

4. Predict the products from opening of the two stereoisomeric epoxides derived from limonene by reaction with (a) acetic acid, (b) dimethylamine, and (c) lithium aluminum hydride.

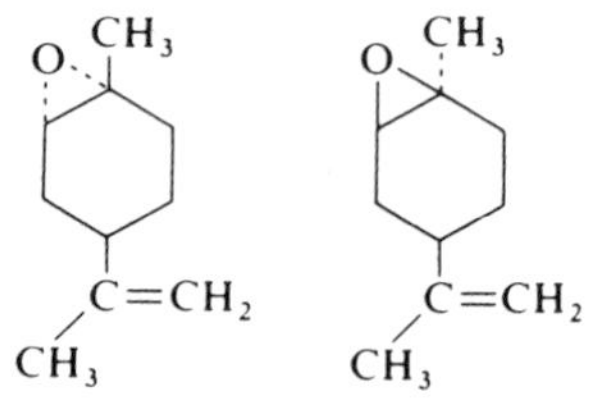

5. The direct oxidative conversion of primary halides or tosylates to aldehydes can be carried out by reaction with dimethyl sulfoxide under alkaline conditions. Formulate a mechanism for this general reaction.

6. A method for synthesis of ozonides that involves no ozone has been reported. It consists of photosensitized oxidation of solutions of diazo compounds and aldehydes. Suggest a mechanism.

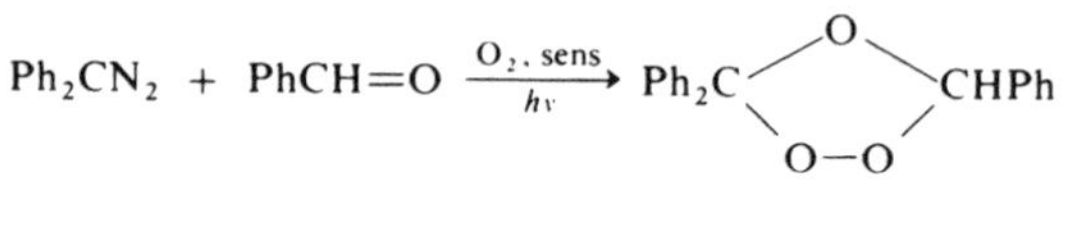

$$Ph_2CN_2 + PhCH{=}O \xrightarrow[h\nu]{O_2,\ sens} Ph_2C\langle O,\ O{-}O \rangle CHPh$$

7. Overoxidation of carbonyl products during ozonolysis can be prevented by addition of tetracyanoethylene to the reaction mixture. The stoichiometry of the reaction is then

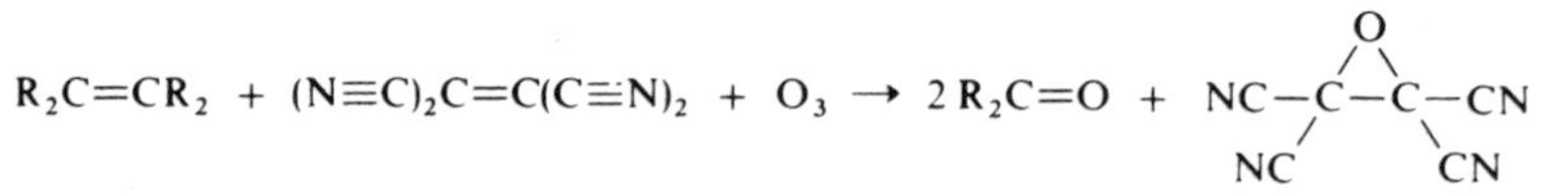

$$R_2C{=}CR_2 + (N{\equiv}C)_2C{=}C(C{\equiv}N)_2 + O_3 \rightarrow 2\,R_2C{=}O + (NC)_2C\overset{O}{-}C(CN)_2$$

Propose a reasonable mechanism that would account for the effect of tetracyanoethylene. Does your mechanism suggest that tetracyanoethylene would be a particularly effective alkene for this purpose? Explain.

8. Suggest a mechanism by which the "abnormal" oxidations shown below might occur.

(a) $^1O_2$

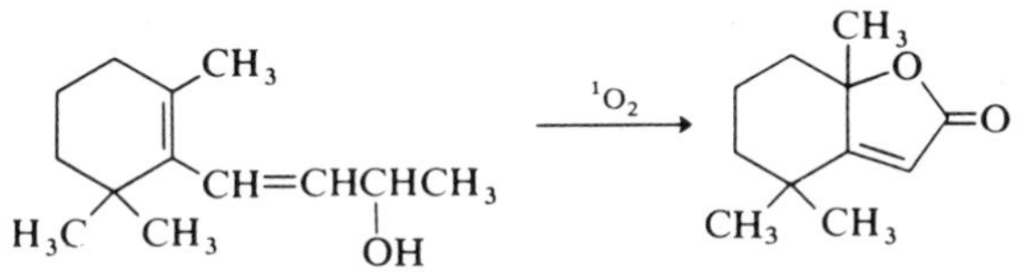

(b) $PhC(=O)C(CH_3)_2OH \xrightarrow[H_2O_2]{^-OH} PhCO_2H + (CH_3)_2C{=}O$

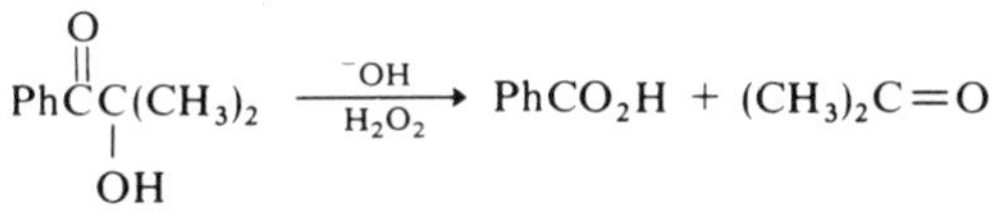

(c) O_2, $NaOCH_3$

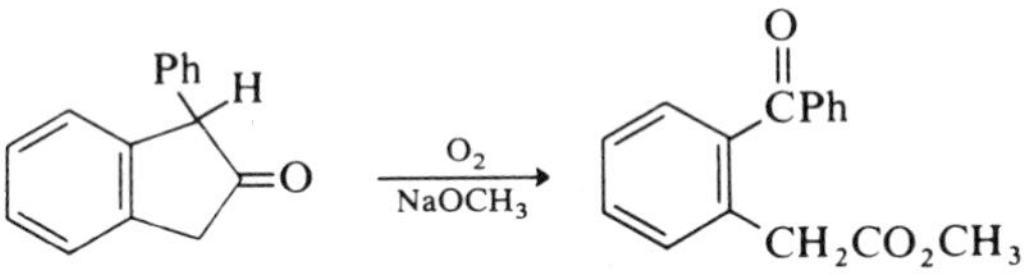

(d) 1) H_2O_2, ^-OH 2) H^+

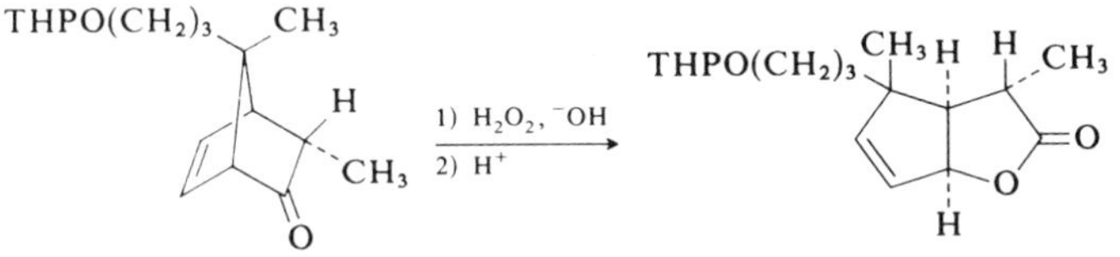

(e) O_2, $h\nu$, sens

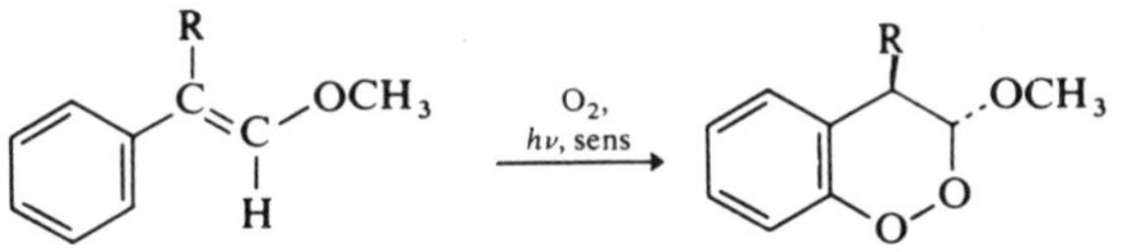

(f) H_3C–C6H4–OH + $CH_3S(O)CH_3$ → (dicyclohexylcarbodiimide, H_3PO_4) (None of the *para* isomer is formed.)

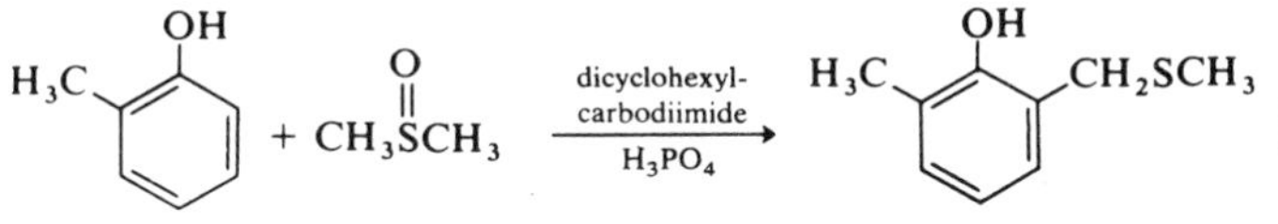

(g) 1) O_3, MeOH, −78° 2) CF_3CO_2H

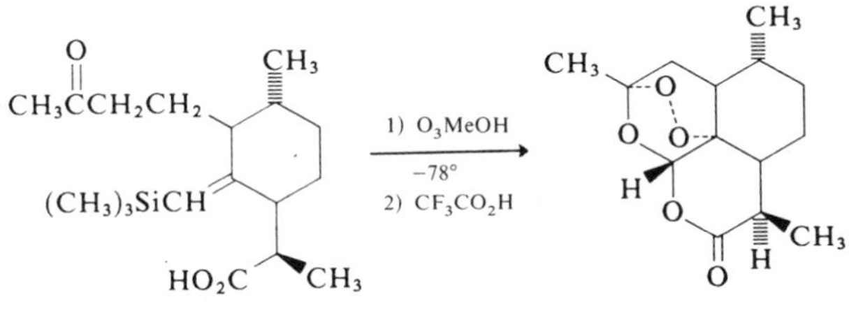

9. Indicate one or more satisfactory oxidants for effecting the following transformations. Each molecule poses problems of selectivity or the need to preserve a potentially sensitive functional group. In most cases, a "single-pot" process is possible and in no case are more than three steps required. Explain the basis for your choice of reagent.

(a) CH_3, CH_3O_2CN, OH, S → CH_3, CH_3O_2CN, O, S

(b) O, CH_3, CH_3, H_3C, CH_3, O, CH_3CHCCO_2H, OH → O, CH_3, CH_3, H_3C, O, CH_3CHCCH_3, O

(c) H_3C, H, H → H_3C, H, O, H

(d) O, OCH_3, O, CCH_3, O, OCH_3 → O, OCH_3, O, CCH_3, OH, O, OCH_3

(e) CH_3CO_2, O, O, OH, H → CH_3CO_2, O, O, O, H

(f) $CH_2CH_2O_2CCH_3$, H, H, N, H, H, HO → $CH_2CH_2O_2CCH_3$, H, H, N, H, H, O

(g) H_3C, CH_3, CO_2CH_3, H_3C, CH_3 → H_3C, CH_3, CO_2CH_3, O, H_3C, CH_3

(h) OCH_3, HO → OCH_3, O

(i)

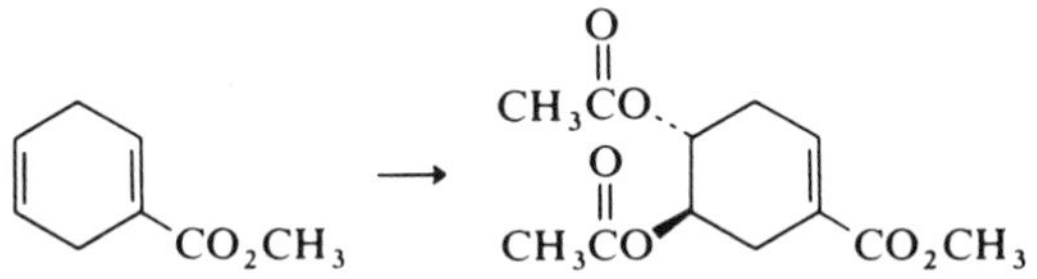

(j)

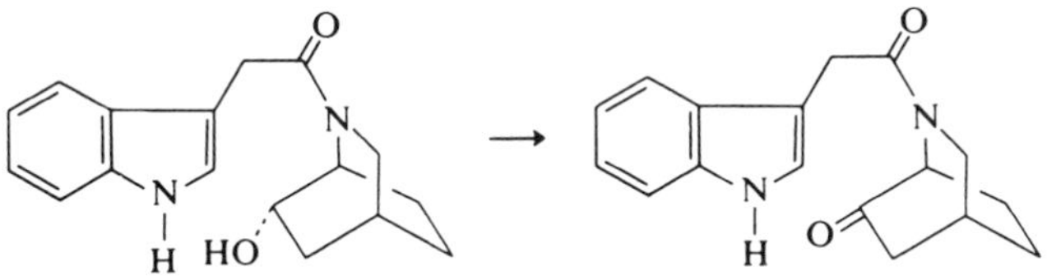

(k)

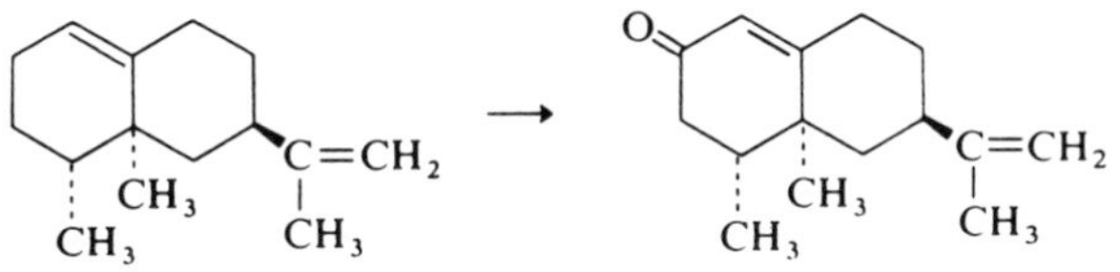

(l)

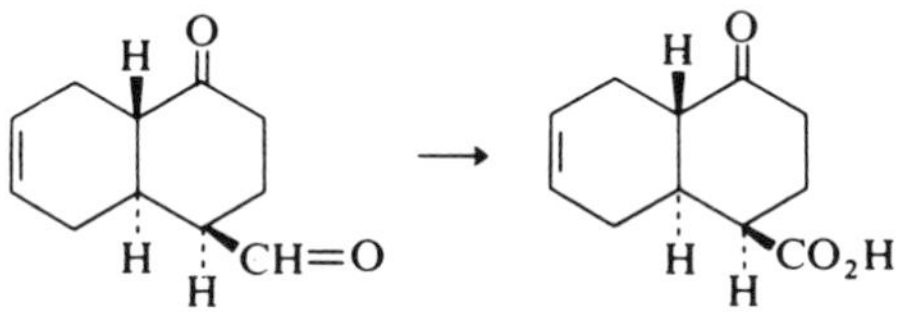

(m)

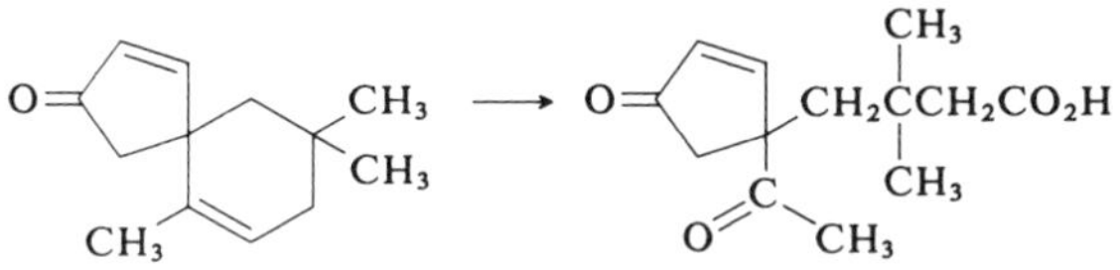

(n)

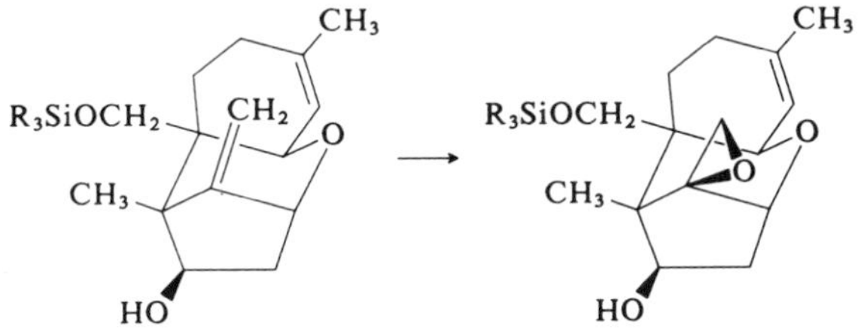

10. It has been noted that when unsymmetrical olefins are ozonized in methanol, there is often a large preference for one cleavage mode over the other. For example,

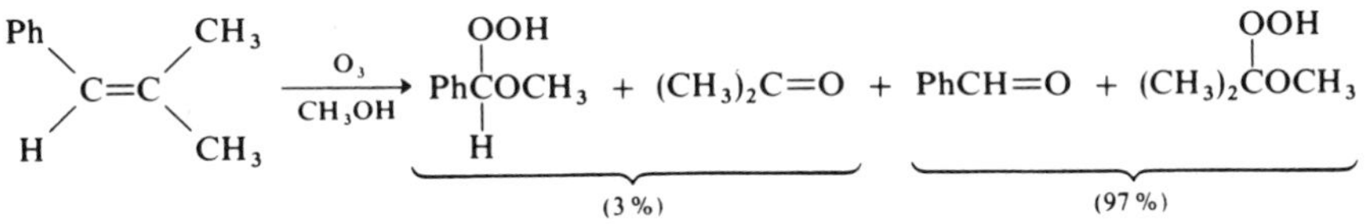

$$\mathrm{PhCH{=}C(CH_3)_2} \xrightarrow[\mathrm{CH_3OH}]{\mathrm{O_3}} \underbrace{\mathrm{PhCH(OOH)OCH_3} + \mathrm{(CH_3)_2C{=}O}}_{(3\%)} + \underbrace{\mathrm{PhCH{=}O} + \mathrm{(CH_3)_2C(OOH)OCH_3}}_{(97\%)}$$

How would you explain this example of regioselective cleavage?

11. A method for oxidative cleavage of cyclic ketones involves a four-stage process. First, the ketone is converted to an α-phenylthio derivative (see Section 4.7). The ketone is then converted to an alcohol, either by reduction or addition of an organolithium reagent. This compound is then treated with lead tetraacetate to give an oxidation product in which the hydroxyl group has been acetylated and an additional oxygen added to the β-thioalcohol. Aqueous hydrolysis of this intermediate in the presence of Hg^{2+} gives a dicarbonyl compound. Formulate a likely structure for the product of each reaction step and an overall mechanism for this process.

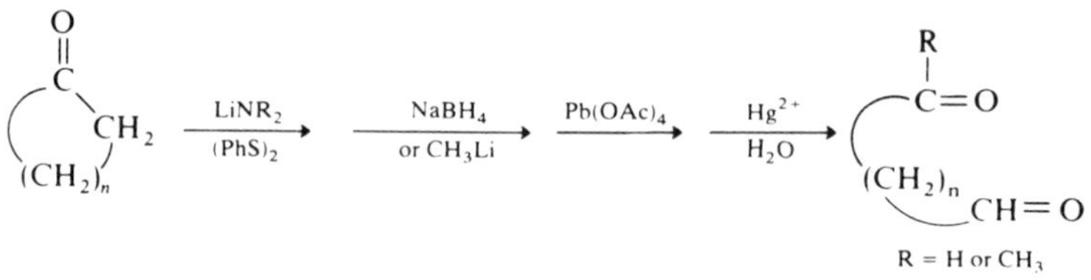

$$\text{cyclic ketone } C(=O)CH_2(CH_2)_n \xrightarrow[(PhS)_2]{LiNR_2} \xrightarrow[\text{or } CH_3Li]{NaBH_4} \xrightarrow{Pb(OAc)_4} \xrightarrow[H_2O]{Hg^{2+}} RC(=O)(CH_2)_nCH{=}O$$

R = H or CH_3

12. Certain thallium salts, particularly $Tl(NO_3)_3$, effect oxidation with an accompanying rearrangement. Especially good yields are found when the thallium salt is supported on inert material. Two examples are given. Formulate a mechanistic rationalization.

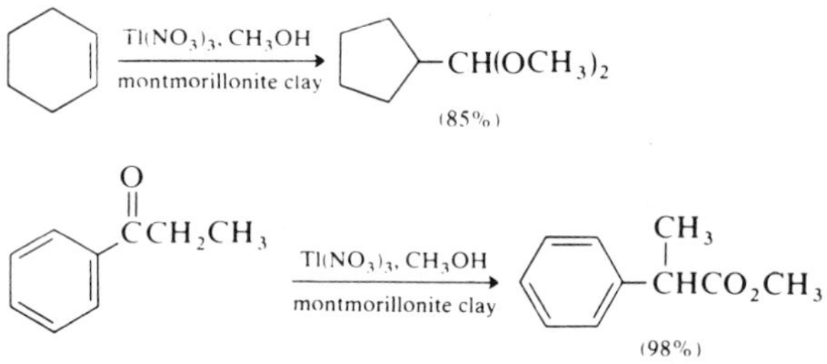

$$\text{cyclohexene} \xrightarrow[\text{montmorillonite clay}]{Tl(NO_3)_3,\ CH_3OH} \text{cyclopentyl}{-}CH(OCH_3)_2 \quad (85\%)$$

$$PhCCH_2CH_3 (C{=}O) \xrightarrow[\text{montmorillonite clay}]{Tl(NO_3)_3,\ CH_3OH} PhCH(CH_3)CO_2CH_3 \quad (98\%)$$

13. The two transformations shown below have been carried out by short reaction sequences involving several oxidative steps. Deduce a series of steps which could effect these transformations, and suggest reagents which might be suitable for each step.

(a) [1-methyl-4-isopropenylcyclohexene (CH_3, $CH_3C{=}CH_2$)] → [epoxy ketone with CH_3, O]

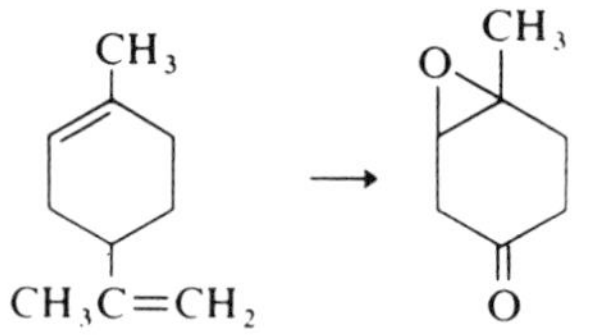

(b) [bicyclic oxa-ketone] → [$HOCH_2$, O, CH_2CO_2H tetrahydrofuran acetonide with O, O, H_3C, CH_3]

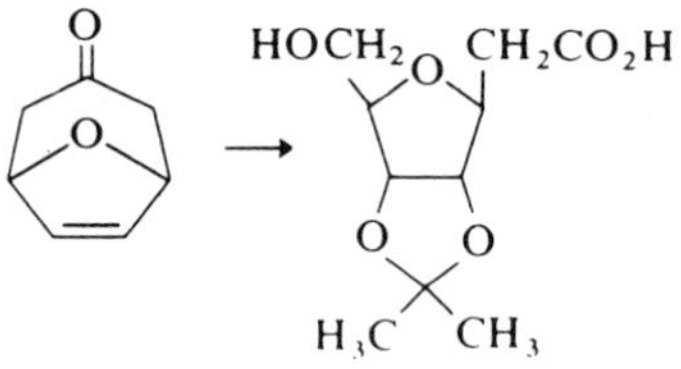

14. Provide mechanistic interpretations of the following reactions.

(a) Account for the products formed under the following conditions. How does the inclusion of cupric acetate affect the course of the reaction?

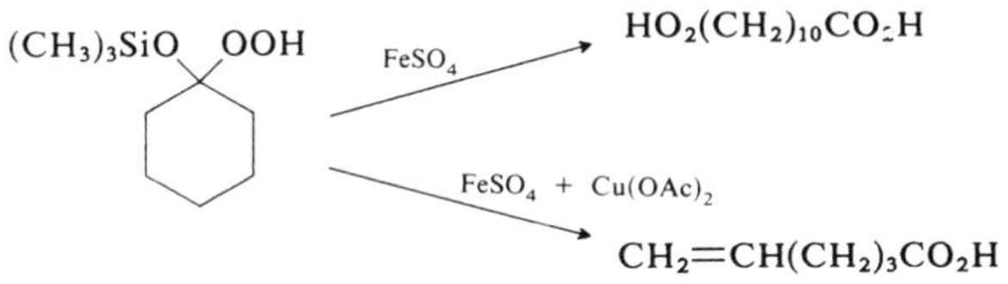

$(CH_3)_3SiO$ OOH (cyclohexane) $\xrightarrow{FeSO_4}$ $HO_2(CH_2)_{10}CO_2H$

$\xrightarrow{FeSO_4 + Cu(OAc)_2}$ $CH_2{=}CH(CH_2)_3CO_2H$

(b) CH_3N CN CH_3 $\xrightarrow{\text{1) LDA; 2) } O_2\text{; 3) } NaSO_3}$ CH_3N O CH_3

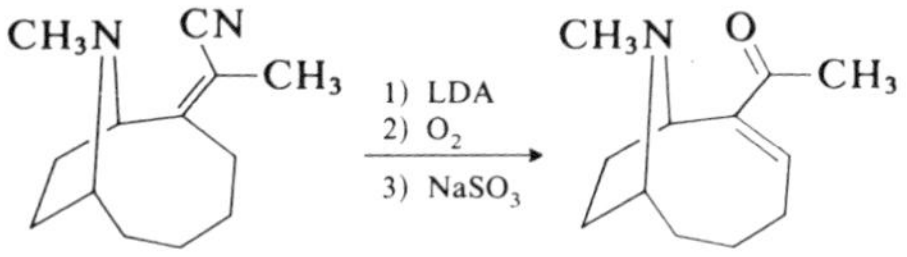

(c) $R_3MCHCHCHCO_2H$ (R′, R″) $\xrightarrow{Pb(OAc)_4}$ $R'CH{=}CHR''$

M = Si, Sn

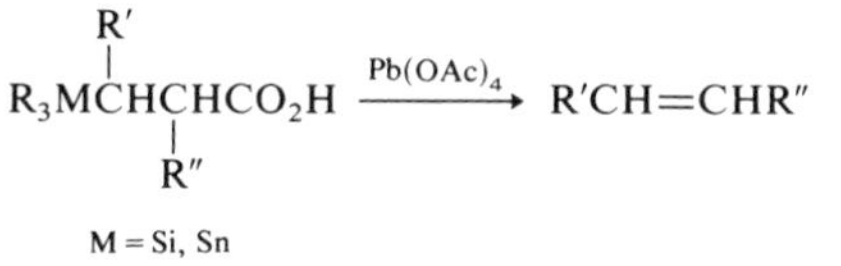

15. Devise a sequence of reactions which could accomplish the formation of the structure on the left from the potential precursor on the right. Pay close attention to stereochemical requirements.

(a) CH_3O, OCH_2Ph, CH_3, O, O, CH_3, H, $HOCH_2$, CH_3, H $\Rightarrow$ CH_3O, OCH_2Ph, CH_3, O, OCH_3, C, H_3C, H C=C H, H, CH_2OH

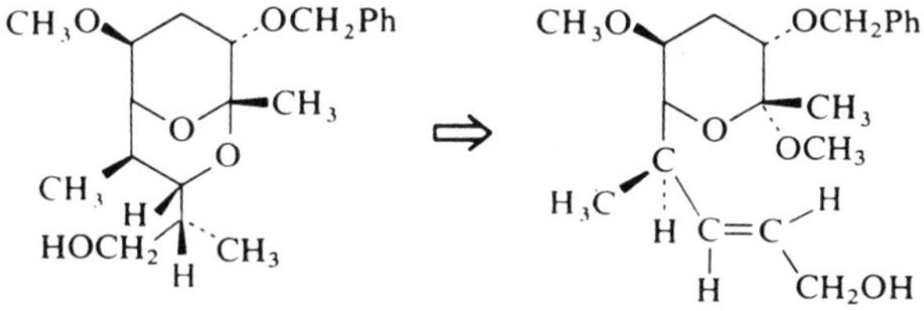

(b) HO, H, H_3C, O, H_3C, H $\Rightarrow$ CH_3CH_2, H, O, O, H_3C, H, O

(c) $CH_3O_2C(CH_2)_4CH{=}CH_2$ $\Rightarrow$ O, $CH_2Si(CH_3)_3$

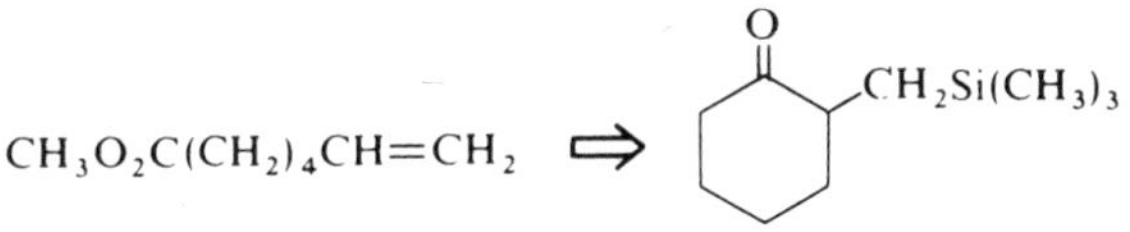

(d) O, O, CH_3, O $\Rightarrow$ $CH_3O_2C(CH_2)_4CO_2CH_3$

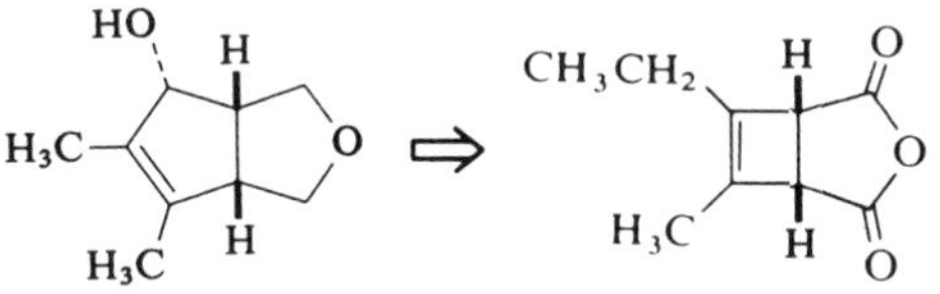

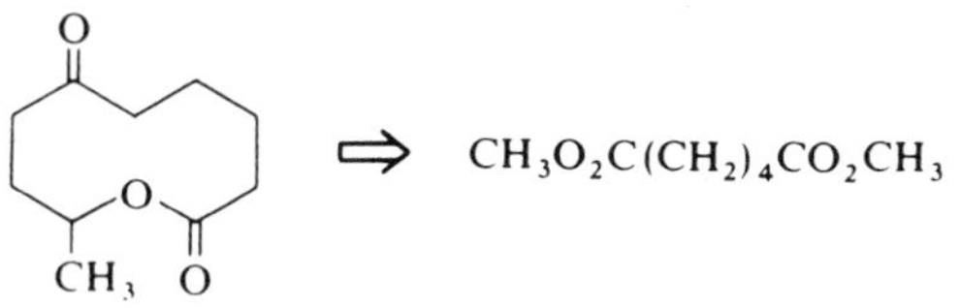

(e) $HOCH_2CH_2(H)C=C(CH_3)CH_2CH_2OH \Rightarrow$ p-$CH_3C_6H_4OCH_3$

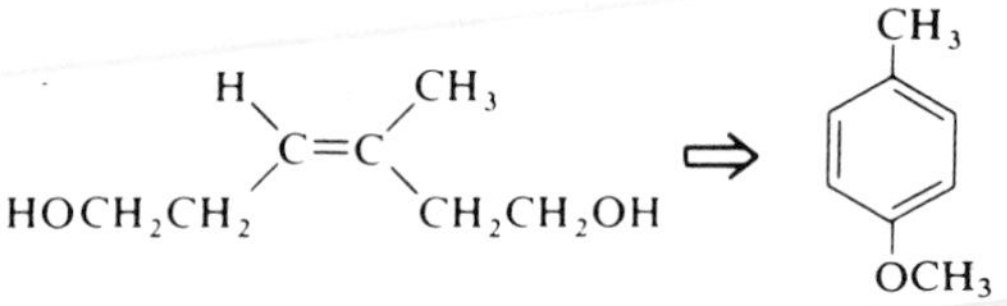

(f) $CH_3CH_2CH_2CH_2C(=O)C\equiv CC_6H_5 \Rightarrow C_6H_5C\equiv CH,\quad CH_3CH_2CH_2CH_2CH_2OH$

(g) OH, CH_2OH $\Rightarrow$ OH

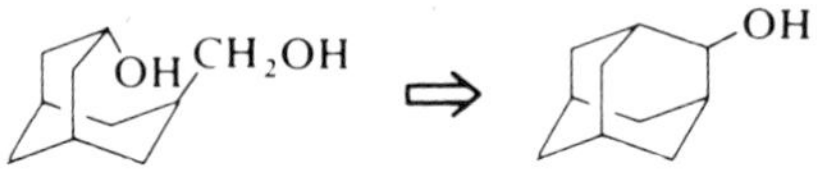

(h) O $\Rightarrow$

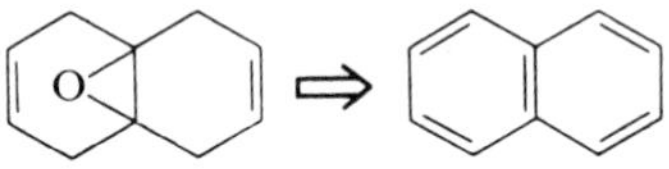

(i) $C_6H_5CH(OH)CH_2-N\lhd \Rightarrow C_6H_5CH=CH_2$

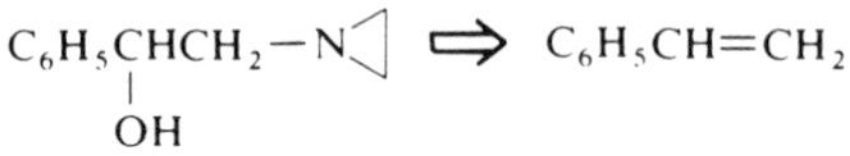

(j) $CH_3CH_2CH-C(CH_3)CH_2OH$ (epoxide, O) $\Rightarrow CH_3CH_2CH_2OH$

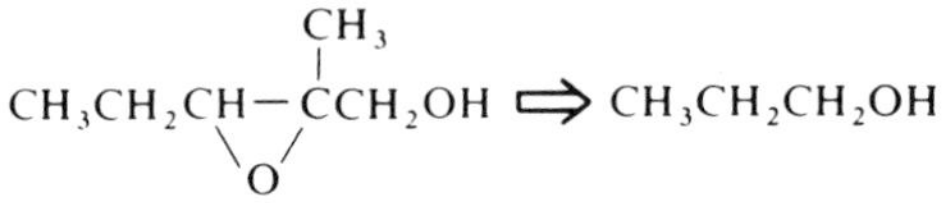

(k) CH_3, CH_3, CH_3, O, O, $Si(CH_3)_3$ $\Rightarrow$ CH_3, CH_2, =O, CH_3

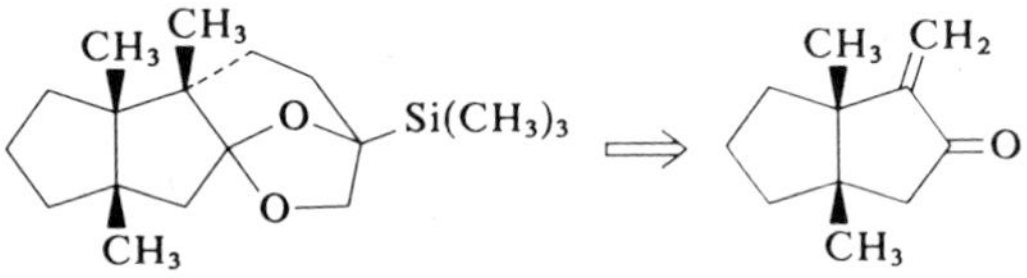

(l) O, O, CH_3 $\Rightarrow$ $(CH_3)_3SiO$, $(CH_3)_3SiO$

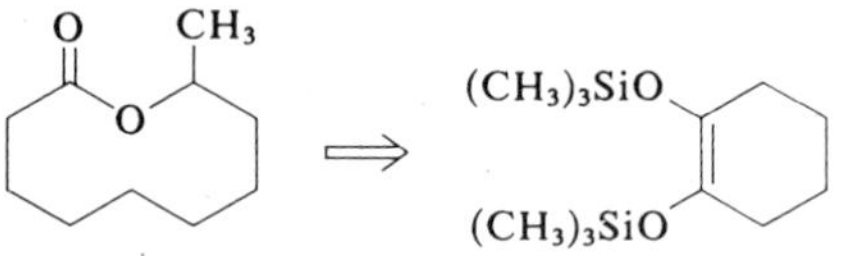

(m) $FCH_2CH(OH)CH_2OCH_2Ph \Rightarrow CH_2=CHCH_2OCH_2Ph$

(n) $CH_3O_2CC=CHOCH_3$, H, H, $CH=CH_2$, O, O, O_2CCH_3, H $\Rightarrow$ $CH_3O_2CC=CHOCH_3$, O

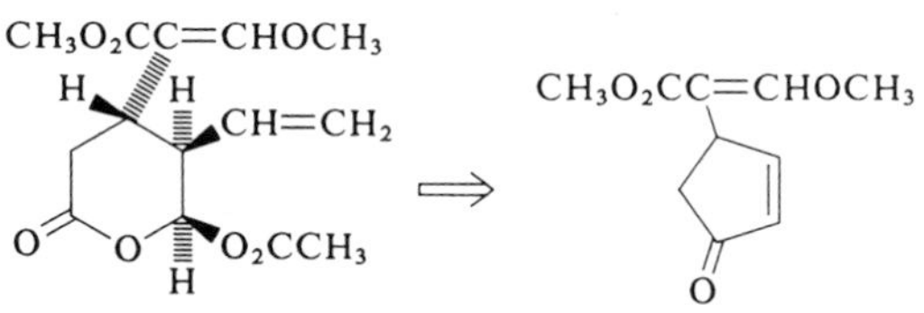

(o) H, O, C_2H_5O, C_2H_5O, H, CH_3 $\Rightarrow$ O, CO_2CH_3, CH_3

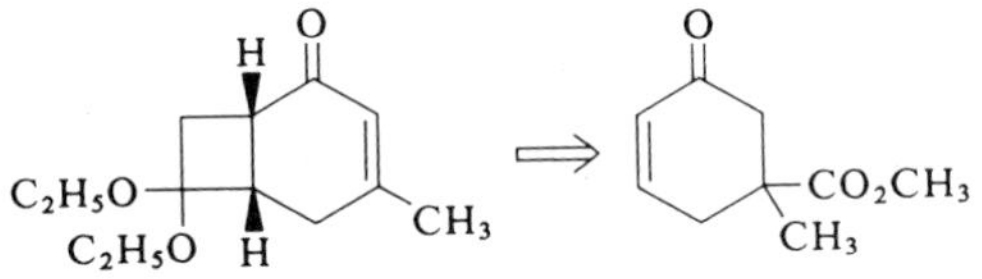

(p)

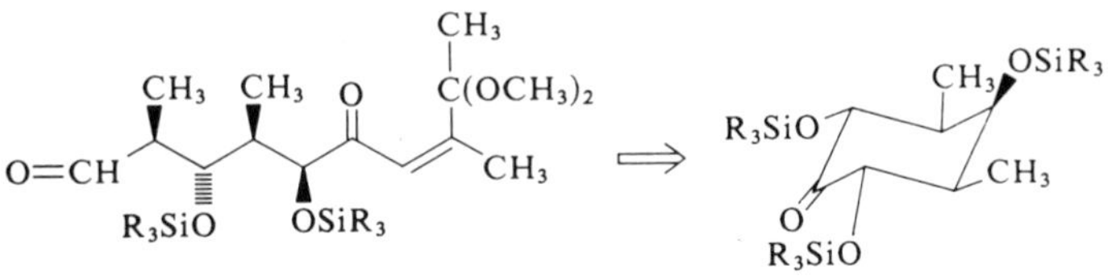

$O{=}CH$–CH(CH_3)–CH($OSiR_3$)–CH(CH_3)–CH($OSiR_3$)–C(=O)–CH=C(CH_3)–C(OCH_3)$_2$–CH_3 $\Longrightarrow$ cyclohexanone bearing R_3SiO, R_3SiO, CH_3, $OSiR_3$, CH_3

16. Tomexetine and fluoroxetine are antidepressants. Both enantiomers of each compound can be prepared enantiospecifically, starting from cinnamyl alcohol. Give a reaction sequence which would accomplish this objective.

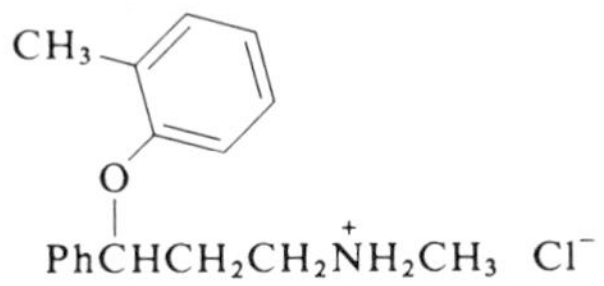

$PhCHCH_2CH_2\overset{+}{N}H_2CH_3\ Cl^-$ (O-2-methylphenyl)

tomoxetine

$PhCHCH_2CH_2\overset{+}{N}H_2CH_3\ Cl^-$ (O-4-fluorophenyl)

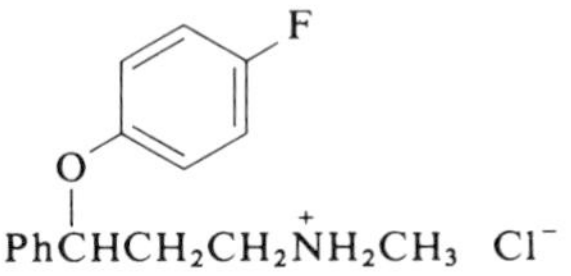

fluoxetine

17. The irradiation of **A** in the presence of rose bengal, oxygen, and acetaldehyde yields the mixture of products shown. Account for the formation of each product.

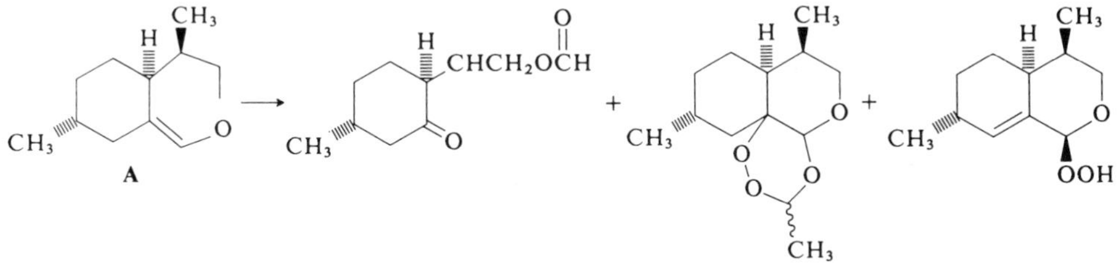

A $\longrightarrow$ (product with $CHCH_2OCH$=O side chain and ketone) + (trioxolane with CH_3) + (hydroperoxide, OOH)

13

Multistep Syntheses

13.1. Protective Groups

The reactions which have been discussed to this point provide the tools for synthesis of organic compounds. When the synthetic target is a relatively complex molecule, a sequence of such reactions that would lead to the desired product must be devised. At the present time, syntheses requiring 15-20 steps are common, and many that are even longer have been developed. In the planning and execution of such multistep syntheses, an important consideration is the compatibility of the functional groups that are already present in the molecule with the reaction conditions required for subsequent steps. It is frequently necessary to modify a functional group in order to prevent interference with some reaction in the synthetic sequence. One way to do this is by use of a *protective group*. A protective group is some derivative that can be put in place, and then subsequently removed, in order to prevent such problems. For example, alcohols are often protected as trisubstituted silyl ethers and aldehydes as acetals. The silyl group replaces the labile proton of the hydroxyl group, and the acetal group prevents unwanted nucleophilic additions at an aldehyde.

$$R{-}OH + R'_3SiX \rightarrow R{-}O{-}SiR'_3$$

$$RCH{=}O + R'OH \rightarrow RCH(OR')_2$$

Protective groups play a passive role in synthesis. Each operation of introduction and removal of a protective group adds steps to the synthetic sequence. It is thus desirable to minimize the number of such operations.

Three considerations are important in choosing an appropriate protective group: (1) the nature of the group requiring protection; (2) the reaction conditions under which the protective group must be stable; and (3) the conditions that can be tolerated for removal of the protective group. No universal protective groups exist.

The state of the art has been brought to a high level, however, and the many mutually complementary protective groups that have been developed provide a great degree of flexibility in the design of syntheses of complex molecules.[1]

An alternative approach to the problem of interfering functionality is to introduce the group in a modified form called a *masked functionality*. For example, vinyl ethers can be readily hydrolyzed to carbonyl compounds. A synthesis which introduces a vinyl ether by a Wittig reaction provides a carbonyl group in a masked form which can be hydrolyzed at a later stage of the synthesis.

$$RCH{=}O + Ph_3P{=}CHOR' \rightarrow RCH{=}CHOR' \xrightarrow[H_2O]{H^+} RCH_2CH{=}O$$

13.1.1. Hydroxyl-Protecting Groups

A common requirement in synthesis is that a hydroxyl group be masked as a derivative lacking an active hydrogen. An example of this requirement is in reactions involving Grignard or other organometallic reagents. The acidic hydrogen of a hydroxyl group will destroy one equivalent of a strongly basic organometallic reagent and possibly adversely affect the reaction in other ways. Conversion to alkyl or silyl ethers is the most common means of protecting hydroxyl groups. The choice of the most appropriate ether group is largely dictated by the conditions that can be tolerated in subsequent removal of the protecting group. An important method that is applicable when mildly acidic hydrolysis is an appropriate method for deprotection is to form a tetrahydropyranyl ether (THP group).[2]

ROH + (dihydropyran) $\xrightarrow{H^+}$ RO–(tetrahydropyranyl)

This protective group is introduced by an acid-catalyzed addition of the alcohol to the vinyl ether moiety in dihydropyran. *p*-Toluenesulfonic acid or its pyridinium salt is recommended as a catalyst,[3] although other catalysts are advantageous in special cases. The THP group can be removed by dilute aqueous acid. The chemistry involved in both the introduction and deprotection stages is the reversible acid-catalyzed formation and hydrolysis of an acetal (see Part A, Section 8.1).

introduction: ROH + (protonated dihydropyran oxonium ion) $\rightleftharpoons$ RO^+H–(tetrahydropyranyl) $\rightleftharpoons$ RO–(tetrahydropyranyl) + H^+

1. The book by T. W. Greene, *Protective Groups in Organic Synthesis*, which is listed in the general references, provides a thorough survey of protective groups and the conditions for their introduction and removal.
2. W. E. Parham and E. L. Anderson, *J. Am. Chem. Soc.* **70**, 4187 (1948).
3. J. H. van Boon, J. D. M. Herschied, and C. B. Reese, *Synthesis*, 169 (1973); M. Miyashita, A. Yoshikoshi, and P. A. Grieco, *J. Org. Chem.* **42**, 3772 (1977).

removal: $\text{RO-THP} \xrightarrow[H_2O]{H^+} ROH + \text{HO-THP} \rightleftharpoons ROH + O{=}CH(CH_2)_3CH_2OH$

The THP group, like other acetals and ketals, is inert to nucleophilic reagents and is unchanged under the conditions of hydride reduction, organometallic reactions, or base-catalyzed reactions in aqueous solution. It also protects the hydroxy group against oxidation under most conditions.

A disadvantage of the THP group is the fact that a chiral center is produced at C-2 of the tetrahydropyran ring. This presents no difficulties if the alcohol is achiral, since a racemic mixture results. However, if the alcohol is chiral, the reaction may give a mixture of diastereomeric ethers, which can complicate purification and characterization. One way of avoiding this problem is to use methyl 2-propenyl ether in place of dihydropyran. No new chiral center is introduced, and this acetal offers the further advantage of being hydrolyzed under somewhat milder conditions than those required for THP ethers.[4]

$$ROH + CH_2{=}C(CH_3){-}OCH_3 \xrightarrow{H^+} ROC(CH_3)_2OCH_3$$

Ethyl vinyl ether is also useful for hydroxyl group protection. The resulting derivative (1-ethoxyethyl ether) is abbreviated as the EE group.[5] As with the THP group, the EE group contains a chiral center.

The methoxymethyl (MOM) and β-methoxyethoxymethyl (MEM) groups are used to protect alcohols and phenols as formaldehyde acetals. The groups are normally introduced by reaction of an alkali metal salt of the alcohol with methoxymethyl chloride or β-methoxyethoxymethyl chloride.

$$RO^-M^+ \xrightarrow{CH_3OCH_2Cl} ROCH_2OCH_3$$
$$RO^-M^+ \xrightarrow{CH_3OCH_2CH_2OCH_2Cl} ROCH_2OCH_2CH_2OCH_3$$

An attractive feature of the MEM group is the ease with which it can be removed under nonaqueous conditions. Lewis acids such as zinc bromide, titanium tetrachloride, dimethylboron bromide, or trimethylsilyl iodide permit its removal.[6] The MEM group is cleaved in preference to the MOM or THP groups under these conditions. Conversely, the MEM group is more stable to acidic aqueous hydrolysis than the THP group. These relative reactivity relationships allow the THP and MEM groups to be used in a complementary fashion when two hydroxyl groups must be deprotected at different points in a synthetic sequence.

4. A. F. Kluge, K. G. Untch, and J. H. Fried, *J. Am. Chem. Soc.* **94**, 7827 (1972).
5. H. J. Sims, H. B. Parseghian, and P. L. DeBenneville, *J. Org. Chem.* **23**, 724 (1958).
6. E. J. Corey, J.-L. Gras, and P. Ulrich, *Tetrahedron Lett.*, 809 (1976); Y. Quindon, H. E. Morton, and C. Yoakim, *Tetrahedron Lett.* **24**, 3969 (1983); J. H. Rigby and J. Z. Wilson, *Tetrahedron Lett.* **25**, 1429 (1984).

$$\text{(THPO, OMEM-substituted vinyl hydrindene)} \xrightarrow[35^\circ C,\ 40\ h]{CH_3CO_2H,\ H_2O,\ THF} \text{(HO, OMEM-substituted vinyl hydrindene)}$$

CH$_2$=CH ... THPO ... OMEM → CH$_2$=CH ... HO ... OMEM Ref. 7

The methylthiomethyl (MTM) group is a related alcohol-protecting group. There are several methods for introducing the MTM group. Alkylation of an alcoholate by methylthiomethyl chloride is efficient if catalyzed by iodide ion.[8] Alcohols are also converted to MTM ethers by reaction with dimethyl sulfoxide in the presence of acetic acid and acetic anhydride[9] or with benzoyl peroxide and dimethyl sulfide.[10] The latter two methods involve the generation of the methylthiomethylium ion by ionization of an acyloxysulfonium ion.

$$RO^-M^+ + CH_3SCH_2Cl \xrightarrow{I^+} ROCH_2SCH_3$$

$$ROH + CH_3SOCH_3 \xrightarrow[(CH_3CO)_2O]{CH_3CO_2H} ROCH_2SCH_3$$

$$ROH + (CH_3)_2S + (PhCO_2)_2 \rightarrow ROCH_2SCH_3$$

The MTM group is selectively removed under nonacidic conditions in aqueous solutions containing Ag^+ or Hg^{2+} salts. The THP and MOM groups are stable under these conditions.[8] The MTM group can also be removed by reaction with methyl iodide, followed by hydrolysis of the resulting sulfonium salt in moist acetone.[9]

The simple alkyl groups are generally not very useful for protection of alcohols as ethers. Although they can be introduced readily by alkylation, subsequent cleavage requires strongly electrophilic reagents such as boron tribromide (see Section 3.3). The *t*-butyl group is an exception and has found some use as a protecting group. Because of the stability of the *t*-butyl cation, *t*-butyl ethers can be cleaved under moderately acidic conditions. Trifluoroacetic acid in an inert solvent is frequently used.[11] *t*-Butyl ethers can also be cleaved by acetic anhydride–$FeCl_3$ in ether.[12] The *t*-butyl group is normally introduced by reaction of the alcohol with isobutylene in the presence of an acid catlyst.[11,13] Acidic ion exchange resins are effective catalysts.[14]

$$ROH + CH_2{=}C(CH_3)_2 \xrightarrow{H^+} ROC(CH_3)_3$$

The triphenylmethyl (trityl, abbreviated Tr) group is removed under even milder conditions than the *t*-butyl group and is an important hydroxyl-protecting group,

7. E. J. Corey, R. L. Danheiser, S. Chandrasekaran, P. Siret, G. E. Keck, and J.-L. Gras, *J. Am. Chem. Soc.* **100**, 8031 (1978).
8. E. J. Corey and M. G. Bock, *Tetrahedron Lett.*, 3269 (1975).
9. P. M. Pojer and S. J. Angyal, *Tetrahedron Lett.*, 3067 (1976).
10. J. C. Modina, M. Salomon, and K. S. Kyler, *Tetrahedron Lett.* **29**, 3773 (1988).
11. H. C. Beyerman and G. J. Heiszwolf, *J. Chem. Soc.*, 755 (1963).
12. B. Ganem and V. R. Small, Jr., *J. Org. Chem.* **39**, 3728 (1974).
13. J. L. Holcombe and T. Livinghouse, *J. Org. Chem.* **51**, 111 (1986).
14. A. Alexakis, M. Gardette, and S. Colin, *Tetrahedron Lett.* **29**, 2951 (1988).

especially in carbohydrate chemistry. This group is introduced by reaction of the alcohol with triphenylmethyl chloride by an S_N1 mechanism. Hot aqueous acetic acid suffices to remove the trityl group. The ease of removal can be increased by addition of electron-releasing substituents. The *p*-methoxy derivatives are used in this way.[15] Because of their steric bulk, triarylmethyl groups can usually be introduced only at primary hydroxyl groups.

The benzyl group can serve as an alcohol-protecting group when acidic conditions for ether cleavage cannot be tolerated. The benzyl C—O bond is cleaved by catalytic hydrogenolysis[16] or with sodium in liquid ammonia.[17] Benzyl ethers can also be cleaved with the use of formic acid, cyclohexene, or cyclohexadiene as hydrogen sources in transfer hydrogenolysis catalyzed by platinum or palladium.[18]

Several nonreductive methods for cleavage of benzyl groups have also been developed. Treatment with *s*-butyllithium followed by reaction with trimethyl borate and then hydrogen peroxide liberates the alcohol.[19] The lithiated ether forms an alkyl boronate, which is oxidized as discussed in Section 4.9.2.

$$\mathrm{ROCH_2Ph} \xrightarrow{s\text{-BuLi}} \mathrm{ROCH(Li)Ph} \xrightarrow{\mathrm{B(OCH_3)_3}} \mathrm{ROCH(Ph)B(OCH_3)_2} \xrightarrow{\mathrm{H_2O_2}} \mathrm{ROCH(Ph)OB(OCH_3)_2} \rightarrow \mathrm{ROH} + \mathrm{PhCH{=}O}$$

Lewis acids such as $FeCl_3$ and $SnCl_4$ also cleave benzyl ethers.[20]

Benzyl groups having 4-methoxy or 3,5-dimethoxy substituents can be removed oxidatively by dichlorodicyanoquinone.[21] These reactions presumably proceed through a benzyl cation, and the methoxy substituent is necessary to facilitate the oxidation.

$$\mathrm{CH_3O{-}C_6H_4{-}CH_2OR} \xrightarrow[-\mathrm{H^+}]{-2e^-} \mathrm{CH_3O{-}C_6H_4{-}\overset{+}{C}H{-}OR} \xrightarrow{\mathrm{H_2O}} \mathrm{CH_3O{-}C_6H_4{-}CH(OH){-}OR} \longrightarrow \mathrm{ROH}$$

These reaction conditions do not affect most of the other common hydroxyl-protecting groups, and the methoxybenzyl groups are therefore useful in synthetic sequences that require selective deprotection of different hydroxyl groups.

15. M. Smith, D. H. Rammler, I. H. Goldberg, and H. G. Khorana, *J. Am. Chem. Soc.* **84**, 430 (1962).
16. W. H. Hartung and R. Simonoff, *Org. React.* **7**, 263 (1953).
17. E. J. Reist, V. J. Bartuska, and L. Goodman, *J. Org. Chem.* **29**, 3725 (1964).
18. B. El Amin, G. M. Anantharamiah, G. P. Royer, and G. E. Means, *J. Org. Chem.* **44**, 3442 (1979); A. M. Felix, E. P. Heimer, T. J. Lambros, C. Tzougraki, and J. Meienhofer, *J. Org. Chem.* **43**, 4194 (1978); A. E. Jackson and R. A. W. Johnstone, *Synthesis*, 685 (1976); G. M. Anantharamaiah and K. M. Sivandaiah, *J. Chem. Soc., Perkin Trans. 1*, 490 (1977).
19. D. A. Evans, C. E. Sacks, W. A. Kleschick, and T. R. Taber, *J. Am. Chem. Soc.* **101**, 6789 (1979).
20. M. H. Park, R. Takeda, and K. Nakanishi, *Tetrahedron Lett.* **28**, 3823 (1987).
21. Y. Oikawa, T. Yoshioka, and O. Yonemitsu, *Tetrahedron Lett.* **23**, 885 (1982); Y. Oikawa, T. Tanaka, K. Horita, T. Yoshioka, and O. Yonemitsu, *Tetrahedron Lett.* **25**, 5393 (1984); N. Nakajima, T. Hamada, T. Tanaka, Y. Oikawa, and O. Yonemitsu, *J. Am. Chem. Soc.* **108**, 4645 (1986).

Benzyl groups are usually introduced by the Williamson reaction (Section 3.2.5). They can also be introduced under nonbasic conditions if necessary. Benzyl ethers are converted to trichloroacetimidates by reaction with trichloroacetonitrile. These then react with an alcohol to transfer the benzyl group.[22]

$$ArCH_2OH + Cl_3CCN \rightarrow ArCH_2O\overset{\overset{NH}{\|}}{C}CCl_3 \xrightarrow{ROH} ROCH_2Ar + Cl_3C\overset{\overset{O}{\|}}{C}NH_2$$

Allyl ethers can be cleaved by conversion to propenyl ethers, followed by acidic hydrolysis of the resulting enol ether.

$$ROCH_2CH{=}CH_2 \rightarrow ROCH{=}CHCH_3 \xrightarrow{H_3O^+} ROH + CH_3CH_2CH{=}O$$

The isomerization of an allyl ether to a propenyl ether can be achieved either by treatment with potassium *t*-butoxide in dimethyl sulfoxide[23] or by Wilkinson's catalyst, tris(triphenylphosphine)chlororhodium.[24]

Silyl ethers play a very important role as hydroxyl-protecting groups.[25] Alcohols can be easily converted to trimethylsilyl (TMS) ethers by reaction with trimethylsilyl chloride in the presence of an amine or by heating with hexamethyldisilazane. *t*-Butyldimethylsilyl (TBDMS) ethers are also very useful. The increased steric bulk of the TBDMS group improves the stability of the group toward such reactions as hydride reduction and Cr(VI) oxidation. The TBDMS group is normally introduced by the reaction of the alcohol with *t*-butyldimethylsilyl chloride with the use of imidazole as a catalyst. Cleavage of the TBDMS group is slow under hydrolytic conditions, but fluoride ion (from anhydrous tetra-*n*-butylammonium fluoride),[26] aqueous hydrogen fluoride,[27] or boron trifluoride[28] can be used for its removal. Other highly substituted silyl groups, such as dimethyl(1,2,2-trimethylpropyl)silyl[29] and tris(isopropyl)silyl,[30] are even more sterically hindered than the TBDMS group and can be used when added stability is required.

Diols represent a special case in terms of applicable protecting groups. 1,2-Diols and 1,3-diols easily form cyclic acetals with aldehydes and ketones, unless cyclization is precluded by molecular geometry. The isopropylidene derivatives (acetonides) formed by reaction with acetone are a common example.

22. H.-P. Wessel, T. Iverson, and D. R. Bundle, *J. Chem. Soc., Perkin Trans. 1*, 2247 (1985); N. Nakajima, K. Horita, R. Abe, and O. Yonemitsu, *Tetrahedron Lett.* **29**, 4139 (1988); S. J. Danishefsky, S. DeNinno, and P. Lartey, *J. Am. Chem. Soc.* **109**, 2082 (1987).
23. R. Griggs and C. D. Warren, *J. Chem. Soc., C*, 1903 (1968).
24. E. J. Corey and J. W. Suggs, *J. Org. Chem.* **38**, 3224 (1973).
25. J. F. Klebe, in *Advances in Organic Chemistry, Methods and Results*, Vol. 8, E. C. Taylor (ed.), Wiley-Interscience, New York, 1972, pp. 97–178; A. E. Pierce, *Silylation of Organic Compounds*, Pierce Chemical Company, Rockford, Illinois, 1968.
26. E. J. Corey and A. Venkateswarlu, *J. Am. Chem. Soc.* **94**, 6190 (1972).
27. R. F. Newton, D. P. Reynolds, M. A. W. Finch, D. R. Kelly, and S. M. Roberts, *Tetrahedron Lett.*, 3981 (1979).
28. D. R. Kelly, S. M. Roberts, and R. F. Newton, *Synth. Commun.* **9**, 295 (1979).
29. H. Wetter and K. Oertle, *Tetrahedron Lett.* **26**, 5515 (1985).
30. R. F. Cunico and L. Bedell, *J. Org. Chem.* **45**, 4797 (1980).

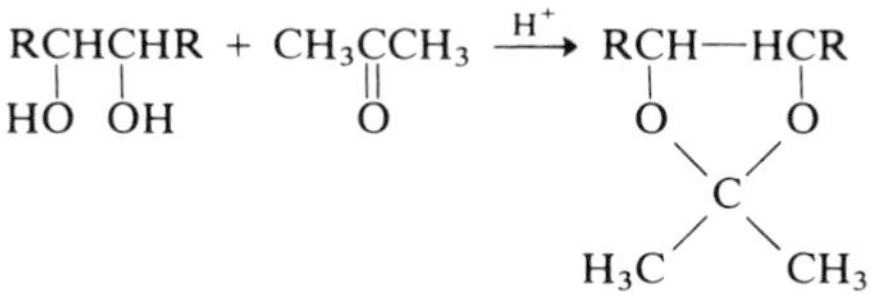

$$\underset{\text{HO}\ \ \text{OH}}{\text{RCHCHR}} + \text{CH}_3\overset{}{\underset{\text{O}}{\overset{\|}{\text{C}}}}\text{CH}_3 \xrightarrow{\text{H}^+} \text{RCH—HCR (O–C(CH}_3)_2\text{–O ring)}$$

The isopropylidene group can also be introduced by acid-catalyzed exchange with 2,2-dimethoxypropane.[31]

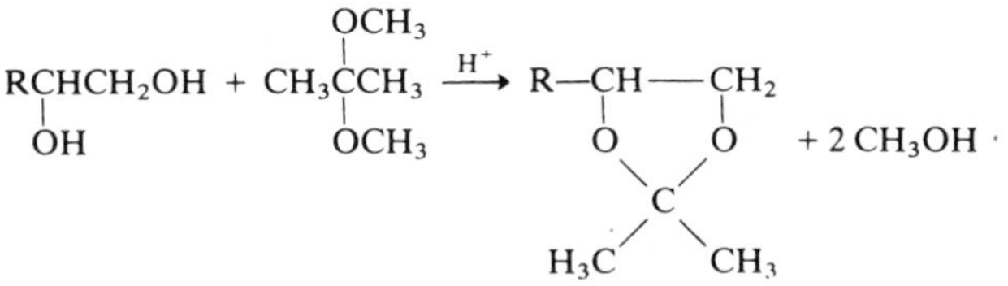

$$\underset{\text{OH}}{\text{RCHCH}_2\text{OH}} + \text{CH}_3\text{C(OCH}_3)_2\text{CH}_3 \xrightarrow{\text{H}^+} \text{R—CH—CH}_2 \text{ (O–C(CH}_3)_2\text{–O ring)} + 2\,\text{CH}_3\text{OH}$$

This ketal protective group is resistant to basic and nucleophilic reagents but is readily removed by aqueous acid. Formaldehyde, acetaldehyde, and benzaldehyde are also used as the carbonyl component in the formation of cyclic acetals. They function in the same manner as acetone. A disadvantage in the case of acetaldehyde and benzaldehyde is the possibility of forming a mixture of diastereomers, because of the new chiral center at the acetal carbon.

Protection of an alcohol function by esterification sometimes offers advantages over use of acetal or alkyl groups. Generally, ester groups are more stable under acidic conditions. Esters are especially useful for protection during oxidations. Acetates and benzoates are the most commonly used ester derivatives. They can be conveniently prepared by reaction of unhindered alcohols with acetic anhydride or benzoyl chloride, respectively, in the presence of pyridine or other tertiary amines. The use of *N*-acylimidazoles (see Section 3.4.1) allows the acylation reaction to be carried out in the absence of added base.[32] Imidazolides are less reactive than the corresponding acid chlorides and can exhibit a higher degree of selectivity in reactions with a molecule possessing several hydroxyl groups:

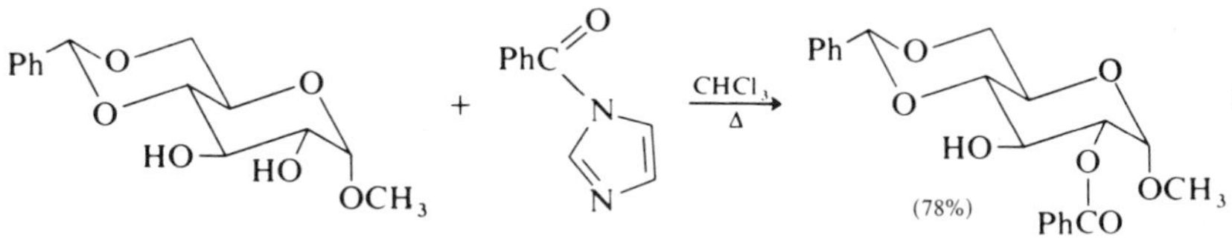

Ph, O, O, O, HO, HO, OCH_3 + PhC(=O)–N-imidazole $\xrightarrow[\Delta]{CHCl_3}$ Ph, O, O, O, HO, O–COPh, OCH_3 (78%) Ref. 33

Hindered hydroxyl groups may require special acylation procedures. One method is to increase the reactivity of the hydroxyl group by converting it to an alkoxide ion with strong base (e.g., *n*-BuLi or KH). When this conversion is not feasible, more reactive acylating reagents are used. Highly reactive acylating agents are generated *in situ* when carboxylic acids are mixed with trifluoroacetic anhydride.

31. M. Tanabe and B. Bigley, *J. Am. Chem. Soc.* **83**, 756 (1961).
32. H. A. Staab, *Angew. Chem.* **74**, 407 (1962).
33. F. A. Carey and K. O. Hodgson, *Carbohydr. Res.* **12**, 463 (1970).

Scheme 13.1. Protection of Hydroxyl Groups

1. Tetrahydropyranyl ether[a]

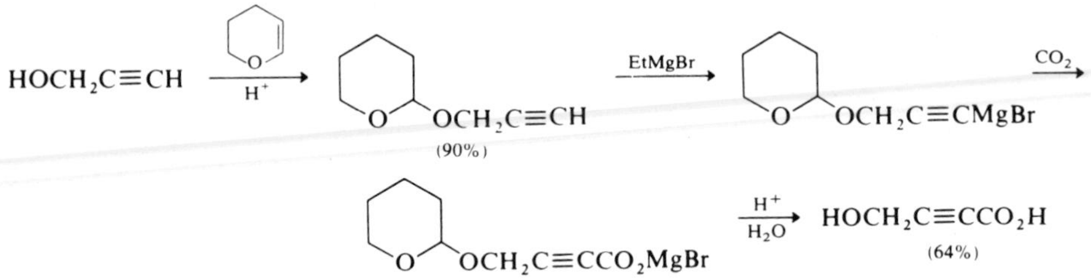

$HOCH_2C{\equiv}CH$ → (dihydropyran, H^+) → THP–$OCH_2C{\equiv}CH$ (90%) → ($EtMgBr$) → THP–$OCH_2C{\equiv}CMgBr$ → (CO_2) → THP–$OCH_2C{\equiv}CCO_2MgBr$ → (H^+, H_2O) → $HOCH_2C{\equiv}CCO_2H$ (64%)

2. Methoxymethyl ether[b]

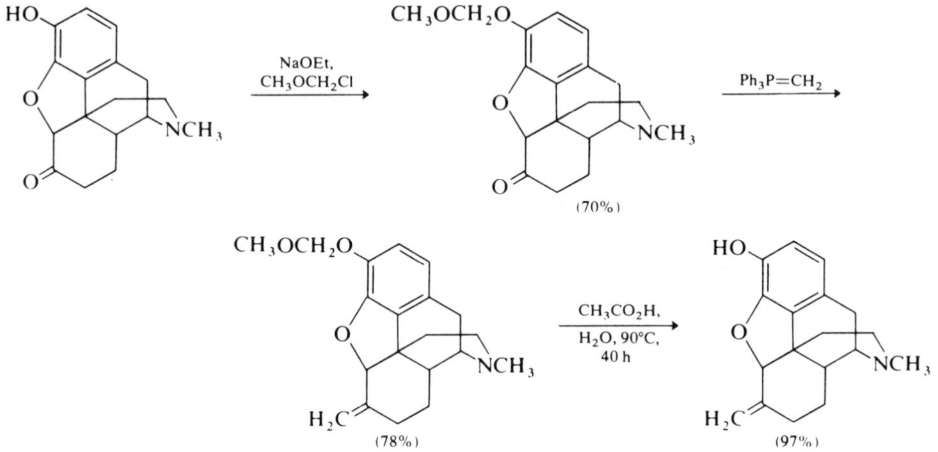

HO-morphinan ketone (NCH_3) → ($NaOEt$, CH_3OCH_2Cl) → CH_3OCH_2O ether (70%) → ($Ph_3P{=}CH_2$) → $H_2C{=}$ derivative (78%) → (CH_3CO_2H, H_2O, 90°C, 40 h) → HO, $H_2C{=}$ product (97%)

3. Triarylmethyl ether[c]

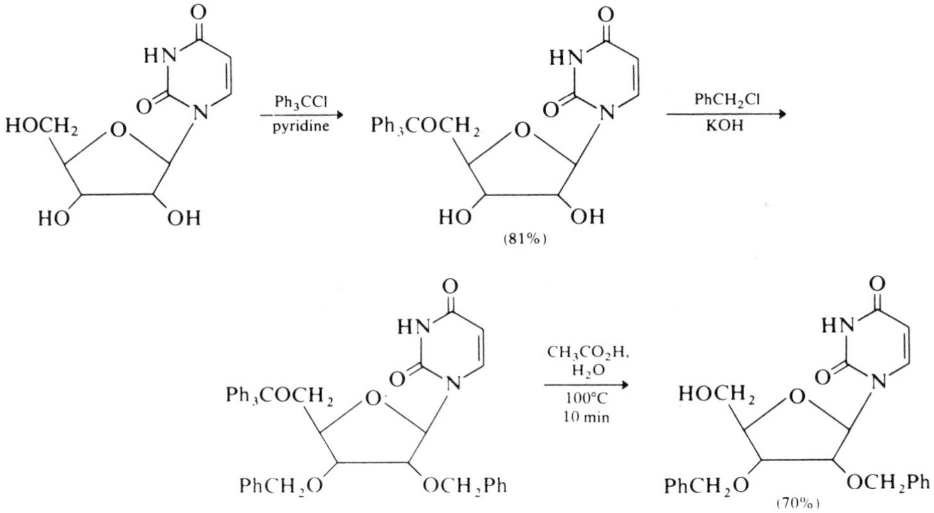

Uridine ($HOCH_2$, HO, OH) → (Ph_3CCl, pyridine) → Ph_3COCH_2 derivative (81%) → ($PhCH_2Cl$, KOH) → Ph_3COCH_2, $PhCH_2O$, OCH_2Ph derivative → (CH_3CO_2H, H_2O, 100°C, 10 min) → $HOCH_2$, $PhCH_2O$, OCH_2Ph product (70%)

4. Benzyl ether[d]

PhCH$_2$O–(cholest-5-ene, C_8H_{17}) $\xrightarrow[\text{2) } H^+, Cr(VI)]{\text{1) } PhCO_3H}$ PhCH$_2$O–(5-hydroxy-6-keto steroid, HO, O, C_8H_{17}) (96%)

$\xrightarrow[\text{2) } H_2, Pt]{\text{1) } PhCO_3H}$ HO–(seco keto acid, O, CH_2CO_2H, C_8H_{17}) (76%)

5. Glycol protection by isopropylidene derivative[e]

$$CH_3O_2C(CH_2)_7\underset{HO}{CH}\underset{OH}{CH}(CH_2)_5CH_2OH \xrightarrow[H^+]{\text{acetone}} CH_3O_2C(CH_2)_7HC\text{—}CH(CH_2)_5CH_2OH \text{ (O,O-isopropylidene: } O{-}C(CH_3)_2{-}O)$$

$$\xrightarrow[CH_3CO_2H]{KMnO_4} CH_3O_2C(CH_2)_7HC\text{—}CH(CH_2)_5CO_2H \text{ (O,O-isopropylidene: } O{-}C(CH_3)_2{-}O) \quad (90\%)$$

a. H. B. Henbest, E. R. H. Jones, and I. M. S. Walls, *J. Chem. Soc.*, 3646 (1950).
b. M. A. Abdel-Rahman, H. W. Elliott, R. Binks, W. Küng, and H. Rapoport, *J. Med. Chem.* **9**, 1 (1965).
c. A. M. Michelson and A. Todd, *J. Chem. Soc.*, 3459 (1956).
d. L. Knof, *Justus Liebigs Ann. Chem.* **656**, 183 (1962).
e. S. D. Sabnis, H. H. Mathur, and S. C. Bhattacharyya, *J. Chem. Soc.*, 2477 (1963).

The mixed anhydride exhibits increased reactivity because of the high reactivity of the trifluoroacetate ion as a leaving group.[34] Dicyclohexylcarbodiimide is another reagent that serves to activate carboxy groups by forming the iminoanhydride **A** (see Section 3.4.1).

$$R\overset{O}{\overset{\|}{C}}{-}O{-}\underset{NC_6H_{11}}{\underset{\|}{C}}{-}NHC_6H_{11} \qquad \mathbf{A}$$

Ester groups can be removed readily by base-catalyzed hydrolysis. When basic hydrolysis is inappropriate, special acyl groups are required. Trichloroethyl carbonate esters, for example, can be reductively removed with zinc.[35]

$$RO\underset{O}{\underset{\|}{C}}OCH_2CCl_3 \xrightarrow{Zn} ROH + CH_2{=}CCl_2$$

34. R. C. Parish and L. M. Stock, *J. Org. Chem.* **30**, 927 (1965); J. M. Tedder, *Chem. Rev.* **55**, 787 (1955).
35. T. B. Windholz and D. B. R. Johnston, *Tetrahedron Lett.*, 2555 (1967).

Cyclic carbonate esters are easily prepared from 1,2- and 1,3-diols, by reaction with *N,N'*-carbonyldiimidazole[36] or by transesterification with diethyl carbonate.

Scheme 13.1 depicts some short synthetic sequences that illustrate the use of several of the important hydroxyl-protecting groups.

13.1.2. Amino-Protecting Groups

Primary and secondary amino groups are sites of both nucleophilicity and a weakly acidic hydrogen. If either of these types of reactivity will cause a problem, the amino group must be protected. The most general way of masking nucleophilicity is by acylation. A most useful group for this purpose is the carbobenzyloxy (CBz) group.[37] The utility of this group lies in the ease with which it can be removed. Because of the lability of the benzyl C—O bond toward hydrogenolysis, the amine can be regenerated from a carbobenzyloxy derivative by hydrogenation, which is accompanied by spontaneous decarboxylation of the resulting carbamic acid.

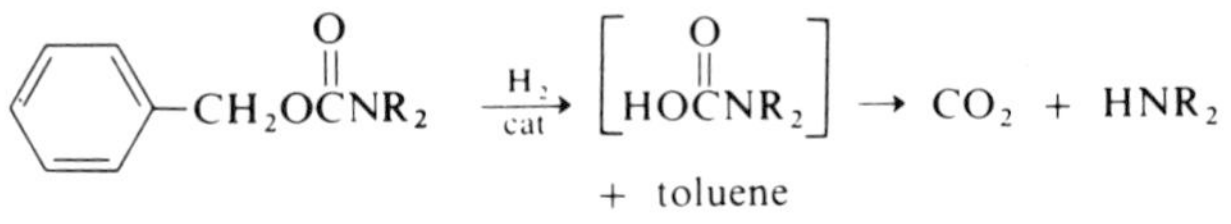

$$C_6H_5\text{-}CH_2O\overset{O}{\overset{\|}{C}}NR_2 \xrightarrow[\text{cat}]{H_2} \left[HO\overset{O}{\overset{\|}{C}}NR_2\right] \rightarrow CO_2 + HNR_2$$

+ toluene

In addition to standard catalytic hydrogenolysis, methods for transfer hydrogenolysis using such hydrogen donors as ammonium formate or formic acid with Pd/C catalyst are available.[38] The *t*-butoxycarbonyl (TBOC) group is another useful amino-protecting group. The removal in this case is done with an acid such as trifluoroacetic acid or *p*-toluenesulfonic acid.[39] *t*-Butoxycarbonyl groups are introduced by reaction of amines with *t*-butyl pyrocarbonate or the mixed carbonate ester known as "BOC-ON."[40]

$$(CH_3)_3CO\overset{O}{\overset{\|}{C}}O\overset{O}{\overset{\|}{C}}OC(CH_3)_3$$

t-butyl pyrocarbonate

$$(CH_3)_3CO\overset{O}{\overset{\|}{C}}ON{=}\overset{CN}{\overset{|}{C}}Ph \quad \text{"BOC-ON"}$$

2-(*t*-butoxycarbonyloxyimino)-2-phenylacetonitrile

Other carbamates in which the *O*-alkyl bond can be easily cleaved can serve as amino-protecting groups. Allyloxycarbonyl groups are cleaved by a combination of tributyltin hydride and a Pd(II) catalyst.[41] This reaction involves liberation of the carbamic acid by oxidative addition to the palladium. The allylpalladium species is then reductively cleaved by the stannane.

36. J. P. Kutney and A. H. Ratcliffe, *Synth. Commun.* **5**, 47 (1975).
37. W. H. Hartung and R. Simonoff, *Org. React.* **7**, 263 (1953).
38. S. Ram and L. D. Spicer, *Tetrahedron Lett.* **28**, 515 (1987); B. ElAmin, G. Anatharamaiah, G. Royer, and G. Means, *J. Org. Chem.* **44**, 3442 (1979).
39. E. Wünsch, *Methoden der Organischen Chemie*, Vol. 15, Fourth Edition, Thieme, Stuttgart, 1975.
40. O. Keller, W. Keller, G. van Look, and G. Wersin, *Org. Synth.* **63**, 160 (1984); W. J. Paleveda, F. W. Holly, and D. F. Weber, *Org. Synth.* **63**, 171 (1984).
41. O. Dangles, F. Guibe, G. Balavoine, S. Lavielle, and A. Marquet, *J. Org. Chem.* **52**, 4984 (1987).

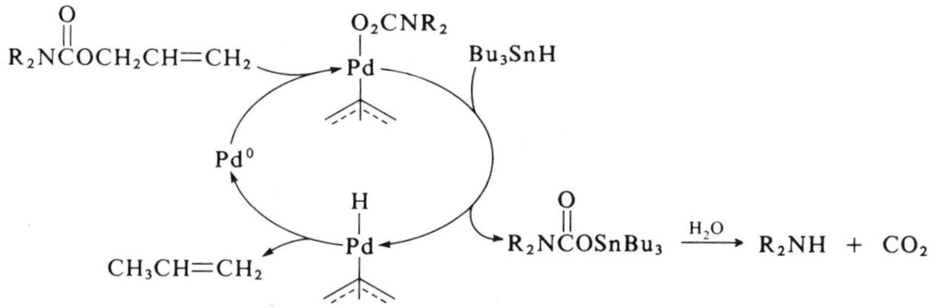

There are other Pd-catalyzed procedures for deallylation.[42] Allyl groups attached directly to amine or amide nitrogen can be removed by isomerization and hydrolysis. Catalysts which have been found to be effective include Wilkinson's catalyst,[43] other rhodium catalysts, and iron pentacarbonyl.[44] 2,2,2-Trichloroethylcarbamates can be reductively cleaved by zinc.[45]

Simple amides are satisfactory protecting groups only if the rest of the molecule can resist the vigorous acidic or alkaline hydrolysis necessary for their removal. For this reason, only amides that can be removed under mild conditions have been found useful as amino-protecting groups. Phthalimides are used to protect primary amino groups. The phthalimides can be cleaved by treatment with hydrazine. This reaction proceeds by initial nucleophilic addition at an imide carbonyl, followed by an intramolecular acyl transfer.

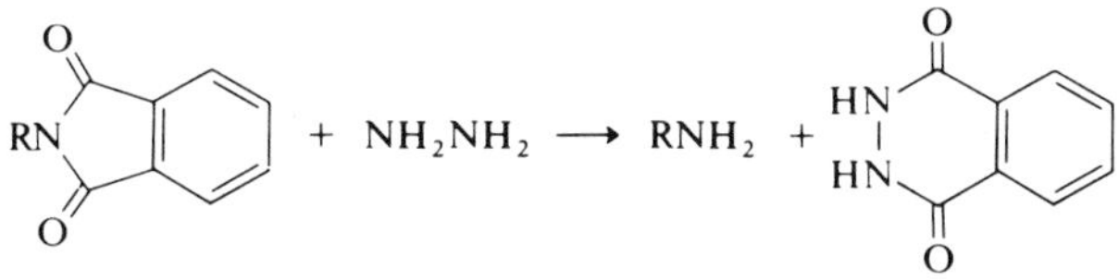

Reduction by $NaBH_4$ in aqueous ethanol is an alternative method for deprotection of phthalimides. This reaction involves formation of an *o*-hydroxybenzamide in the reduction step. Intramolecular displacement of the amino group follows.[46]

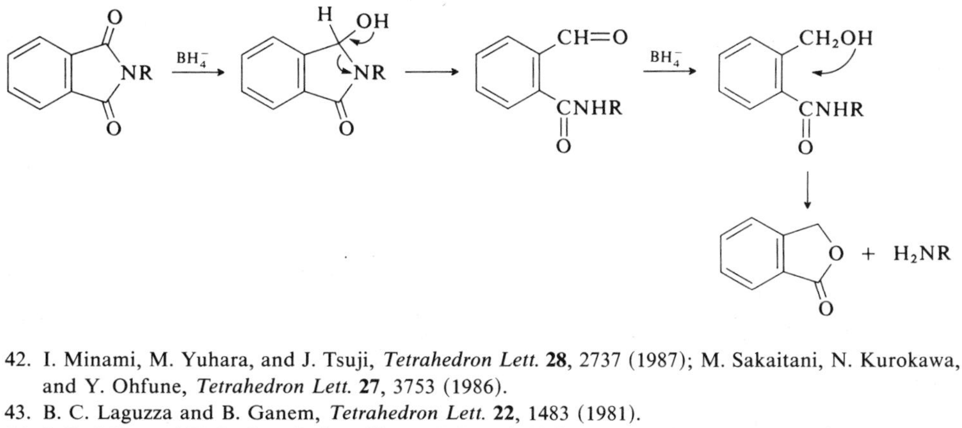

42. I. Minami, M. Yuhara, and J. Tsuji, *Tetrahedron Lett.* **28**, 2737 (1987); M. Sakaitani, N. Kurokawa, and Y. Ohfune, *Tetrahedron Lett.* **27**, 3753 (1986).
43. B. C. Laguzza and B. Ganem, *Tetrahedron Lett.* **22**, 1483 (1981).
44. J. K. Stille and Y. Becker, *J. Org. Chem.* **45**, 2139 (1980); R. J. Sundberg, G. S. Hamilton, and J. P. Laurino, *J. Org. Chem.* **53**, 976 (1988).
45. G. Just and K. Grozinger, *Synthesis*, 457 (1976).
46. J. O. Osborn, M. G. Martin, and B. Ganem, *Tetrahedron Lett.* **25**, 2093 (1984).

Because of the strong electron-withdrawing effect of the trifluoromethyl group, trifluoroacetamides are subject to hydrolysis under mild conditions. This has permitted trifluoroacetyl groups to be used as amino-protecting groups in some situations.

$$\text{(AcO)}_4\text{-glucopyranosyl-O-}C_6H_4\text{-}CH_2CH(NHCOCF_3)CO_2C_2H_5 \xrightarrow[25^\circ C]{Ba(OH)_2,\ 18\ h} \text{(HO)}_4\text{-glucopyranosyl-O-}C_6H_4\text{-}CH_2CH(NH_2)CO_2H\ (88\%)$$

Ref. 47

Some amides can be cleaved by partial reduction. If the reduction proceeds only to the carbinolamine stage, hydrolysis can liberate the deprotected amine. Trichloroacetamides are readily cleaved by sodium borohydride in alcohols by this mechanism.[48] Benzamides, and probably other simple amides, can be cleaved by careful partial reduction with diisobutylaluminum hydride.[49]

$$R_2NC(=O)Ph \xrightarrow{R_2AlH} R_2NCH(OAlR_2)Ph \xrightarrow[H_2O]{H^+} R_2NH + PhCHO$$

Simple sulfonamides are very difficult to hydrolyze. However, a photoactivated reductive method for desulfonylation has been developed.[50] Sodium borohydride is used in conjunction with 1,2- or 1,4-dimethoxybenzene or 1,5-dimethoxynaphthalene. The photoexcited aromatic serves as an electron donor toward the sulfonyl group, which then fragments to give the deprotected amine. The $NaBH_4$ reduces the sulfonyl radical.

$$R_2NSO_2Ar + CH_3O\text{-}C_6H_4\text{-}OCH_3 \xrightarrow{h\nu} R_2N^- + [CH_3O\text{-}C_6H_4\text{-}OCH_3]^{\cdot +} + RSO_2\cdot$$

Amide nitrogens can be protected by 4- or 2,4-dimethoxyphenyl groups. The protecting group can be removed by oxidation with ceric ammonium nitrate.[51]

47. A. Taurog, S. Abraham, and I. Chaikoff, *J. Am. Chem. Soc.* **75**, 3473 (1953).
48. F. Weygand and E. Frauendorder, *Chem. Ber.* **103**, 2437 (1970).
49. J. Gutzwiller and M. Uskokovic, *J. Am. Chem. Soc.* **92**, 204 (1970); K. Psotta and A. Wiechers, *Tetrahedron* **35**, 255 (1979).
50. T. Hamada, A. Nishida, and O. Yonemitsu, *Heterocycles* **12**, 647 (1979); T. Hamada, A. Nishida, and O. Yonemitsu, *J. Am. Chem. Soc.* **108**, 140 (1986).
51. M. Yamaura, T. Suzuki, H. Hashimoto, J. Yoshimura, T. Okamoto, and C. Shin, *Bull. Chem. Soc. Jpn.* **58**, 1413 (1985); R. M. Williams, R. W. Armstrong, and J.-S. Dung, *J. Med. Chem.* **28**, 733 (1985).

2,4-Dimethoxybenzyl groups can be removed with the use of anhydrous trifluoroacetic acid.[52]

13.1.3. Carbonyl-Protecting Groups

Conversion to acetals or ketals is a very general method for protecting aldehydes and ketones against addition by nucleophiles or reduction. Ethylene glycol, which gives a cyclic dioxolane derivative, is the most frequently employed reagent for this purpose. The dioxolane derivative is usually prepared by heating the carbonyl compound with ethylene glycol in the presence of an acid catalyst, with provision for azeotropic removal of water.

$$RC(=O)R' + HOCH_2CH_2OH \xrightarrow{H^+} \underset{R'}{\overset{R}{>}}C\underset{O-CH_2}{\overset{O-CH_2}{<}}\big| + H_2O$$

Dimethyl or diethyl acetals and ketals can be conveniently prepared by acid-catalyzed exchange with a ketal such as 2,2-dimethoxypropane or an orthoester.[53]

$$RC(=O)R' + HC(OCH_3)_3 \xrightarrow{H^+} R-C(OCH_3)_2-R' + HCO_2CH_3$$

$$RC(=O)R' + (CH_3O)_2C(CH_3)_2 \xrightarrow{H^+} R-C(OCH_3)_2-R' + (CH_3)_2C{=}O$$

Acetals and ketals can be prepared under very mild conditions by reaction of the carbonyl compound with an alkoxytrimethylsilane, by the use of trimethylsilyl trifluoromethylsulfonate as the catalyst.[54]

$$R_2C{=}O + 2R'OSi(CH_3)_3 \xrightarrow{Me_3SiO_3SCF_3} R_2C(OR')_2 + (CH_3)_3SiOSi(CH_3)_3$$

Dioxolanes and other acetals and ketals are generally inert to powerful nucleophiles, including organometallic reagents and hydride-transfer reagents. The carbonyl group can be deprotected by acid-catalyzed hydrolysis by the general mechanism for acetal hydrolysis (Part A, Section 8.1). Hydrolysis is also promoted by $LiBF_4$ in acetonitrile.[55]

52. R. H. Schlessinger, G. R. Bebernitz, P. Lin, and A. J. Poss, *J. Am. Chem. Soc.* **107**, 1777 (1985); P. DeShong, S. Ramesh, V. Elango, and J. J. Perez, *J. Am. Chem. Soc.* **107**, 5219 (1985).
53. C. A. MacKenzie and J. H. Stocker, *J. Org. Chem.* **20**, 1695 (1955); E. C. Taylor and C. S. Chiang, *Synthesis*, 467 (1977).
54. T. Tsunoda, M. Suzuki, and R. Noyori, *Tetrahedron Lett.* **21**, 1357 (1980).
55. B. H. Lipshutz and D. F. Harvey, *Synth. Commun.* **12**, 267 (1982).

If the carbonyl group must be regenerated under nonhydrolytic conditions, β-haloalcohols such as 3-bromodihydroxypropane or 2,2,2-trichloroethanol can be used for acetal formation. These groups can be removed by reduction with zinc, which promotes β elimination.

$$\text{BrCH}_2\text{-(2,2-R}_2\text{-1,3-dioxolan-4-yl)} \xrightarrow{\text{Zn}} R_2C{=}O + HOCH_2CH{=}CH_2$$

Ref. 56

$$RC(R')(OCH_2CCl_3)_2 \xrightarrow[\text{THF}]{\text{Zn}} RC(=O)R' + CH_2{=}CCl_2$$

Ref. 57

Another type of carbonyl protecting group is the 1,3-oxathiolane derivative, which can be prepared by reaction with mercaptoethanol in the presence of BF_3[58] or by heating with an acid catalyst with azeotropic removal of water.[59] The 1,3-oxathiolanes are useful when nonacidic conditions are required for deprotection. The 1,3-oxathiolane group can be removed by treatment with Raney nickel in alcohol, even under slightly alkaline conditions.[60] Deprotection can also be accomplished by treating with a mild halogenating agent, such as chloramine-T.[61] This reagent oxidizes the sulfur in the 1,3-oxathiolane group, yielding a chlorosulfonium salt and activating the ring to hydrolytic cleavage.

$$\text{R}_2\text{C(1,3-oxathiolane)} + H_3C\text{-}C_6H_4\text{-}SO_2\bar{N}(Cl)\,Na^+ \;(\mathbf{1}) \longrightarrow \text{R}_2\text{C(1,3-oxathiolane, }\overset{+}{S}\text{-X)} \xrightarrow{H_2O} R_2C{=}O$$

X = Cl or NSO_2Ar

Dithioketals, especially the cyclic dithiolanes and dithianes, are also useful carbonyl-protecting groups. These can be formed from the corresponding dithiols by Lewis acid catalyzed reactions. The catalysts that are used include BF_3, $Mg(O_3SCF_3)_2$, and $Zn(O_3SCF_3)_2$.[62] *S*-Trimethylsilyl ethers of thiols and dithiols also react with ketones to form dithioketals.[63]

$$R_2C{=}O + 2R'SSi(CH_3)_3 \xrightarrow{ZnI_2} R_2C(SR')_2 + (CH_3)_3SiOSi(CH_3)_3$$

56. E. J. Corey and R. A. Ruden, *J. Org. Chem.* **38**, 834 (1973).
57. J. L. Isidor and R. M. Carlson, *J. Org. Chem.* **38**, 544 (1973).
58. G. E. Wilson, Jr., M. G. Huang, and W. W. Scholman, Jr., *J. Org. Chem.* **33**, 2133 (1968).
59. C. Djerassi and M. Gorman, *J. Am. Chem. Soc.* **75**, 3704 (1953).
60. C. Djerassi, E. Batres, J. Romo, and G. Rosenkranz, *J. Am. Chem. Soc.* **74**, 3634 (1952).
61. D. W. Emerson and H. Wynberg, *Tetrahedron Lett.*, 3445 (1971).
62. L. F. Fieser, *J. Am. Chem. Soc.* **76**, 1945 (1954); E. J. Corey and K. Shimoji, *Tetrahedron Lett.* **24**, 169 (1983).
63. D. A. Evans, L. K. Truesdale, K. G. Grimm, and S. L. Nesbitt, *J. Am. Chem. Soc.* **99**, 5009 (1977).

The regeneration of carbonyl compounds from dithioacetals and dithio ketals is done best with reagents that activate the sulfur as a leaving group and facilitate hydrolysis. Among the reagents which have been found effective are nitrous acid ($X = NO^+$), *t*-butyl hypochlorite ($X = Cl^+$), and cupric salts ($X = Cu^{2+}$).[64]

$$R_2C(SR')_2 + X^+ \rightarrow R_2\underset{\substack{|\\ +SR'\\ X}}{C}{-}SR' \rightarrow R_2C{=}\overset{+}{S}R' \xrightarrow{H_2O} R_2\underset{\substack{|\\ OH}}{C}{-}SR' \rightarrow R_2C{=}O$$

13.1.4. Carboxyl Acid-Protecting Groups

If only the O—H, as opposed to the carbonyl, group of a carboxyl group needs to be masked, this can be readily accomplished by esterification. Alkaline hydrolysis is the usual way for regenerating the acid. *t*-Butyl esters, which are readily cleaved by acid, can be used if alkaline conditions must be avoided. 2,2,2-Trichloroethyl esters, which can be reductively cleaved with zinc, are another possibility.[65]

The more difficult problem of protecting the carbonyl group can be accomplished by conversion to an oxazoline derivative. The most commonly used example is the 4,4-dimethyl derivative, which can be prepared from the acid by reaction with 2-amino-2-methylpropanol or with 2,2-dimethylaziridine.[66]

$$RCO_2H + HOCH_2\underset{\substack{|\\ NH_2}}{C}(CH_3)_2 \longrightarrow \text{R–C(=N–C(CH}_3)_2\text{–CH}_2\text{–O) (4,4-dimethyloxazoline)}$$

$$RCO_2H + HN\text{(2,2-dimethylaziridine)} \longrightarrow R\overset{O}{\overset{\|}{C}}{-}N\text{(2,2-dimethylaziridine)} \xrightarrow{H^+} \text{RC(–OH)=N–CH}_2\text{–}\overset{+}{C}(CH_3)_2 \longrightarrow \text{R–(4,4-dimethyloxazoline)}$$

The heterocyclic derivative successfully protects the acid from attack by Grignard reagents or hydride-transfer reagents. The carboxylic acid group can be regenerated by acidic hydrolysis or converted to an ester by acid-catalyzed reaction with the appropriate alcohol.

Carboxylic acids can also be protected as orthoesters. Orthoesters derived from simple alcohols are easily hydrolyzed. A more useful orthoester protecting group is the 4-methyl-2,6,7-trioxabicyclo[2.2.2]octane structure. This bicyclic ortho ester

64. M. T. M. El-Wassimy, K. A. Jorgensen, and S. O. Lawesson, *J. Chem. Soc., Perkin Trans. 1,* 2201 (1983); J. Lucchetti and A. Krief, *Synth. Commun.* **13**, 1153 (1983).
65. R. B. Woodward, K. Heusler, J. Gostelli, P. Naegeli, W. Oppolzer, R. Ramage, S. Ranganathan, and H. Vorbrüggen, *J. Am. Chem. Soc.* **88**, 852 (1966).
66. A. I. Meyers, D. L. Temple, D. Haidukewych, and E. Mihelich, *J. Org. Chem.* **39**, 2787 (1974).

can be prepared by exchange with other orthoesters, by reaction with iminoethers, or by rearrangement of the ester derived from 3-hydroxymethyl-3-methyloxetane.

$RC(OCH_3)_3 \xrightarrow{(HOCH_2)_3CCH_3}$ Ref. 67

$RC(=NH)OR' \xrightarrow{(HOCH_2)_3CCH_3}$ R–(orthoester)–CH_3 Ref. 68

$RC(=O)OCH_2$–(3-methyloxetane) $\xrightarrow{BF_3}$ Ref. 69

Lactones can be protected as their dithioketals by a method which is analogous to ketone protection. The required reagent is readily prepared from trimethylaluminum and ethanedithiol.

$$\text{lactone} + (CH_3)_2AlSCH_2CH_2SAl(CH_3)_2 \longrightarrow \text{spiro dithioketal}$$ Ref. 70

Acyclic esters react with this reagent to give ketenethioacetals.

$$R_2CHCO_2R' + (CH_3)_2AlSCH_2CH_2SAl(CH_2)_2 \longrightarrow R_2C{=}\text{(1,3-dithiolane)}$$

In general, the methods for protection and deprotection of carboxylic acids and esters are not as convenient as those for alcohols, aldehydes, and ketones. It is therefore common to carry potential carboxylic acids through synthetic schemes in the form of protected primary alcohols or aldehydes. The carboxylic acid can then be formed at a late stage in the synthesis by an appropriate oxidation. This strategy allows the wide variety of alcohol and aldehyde protective groups available to be utilized indirectly for carboxylic acid protection.

13.2. Synthetic Equivalent Groups

The protective groups discussed in the previous section play only a passive role during a synthetic sequence. The groups are introduced and removed at appropriate stages but do not directly contribute to bond formation. It is often advantageous to combine the need for masking of a functional group with a change in the reactivity

67. M. P. Atkins, B. T. Golding, D. A. Howe and P. J. Sellars, *J. Chem. Soc., Chem. Commun.*, 207 (1980).
68. E. J. Corey and K. Shimoki, *J. Am. Chem. Soc.* **105**, 1662 (1983).
69. E. J. Corey and N. Raju, *Tetrahedron Lett.* **24**, 5571 (1983).
70. E. J. Corey and D. J. Beames, *J. Am. Chem. Soc.* **95**, 5829 (1973).

of the functionality in question. As an example, suppose the transformation shown below was to be accomplished:

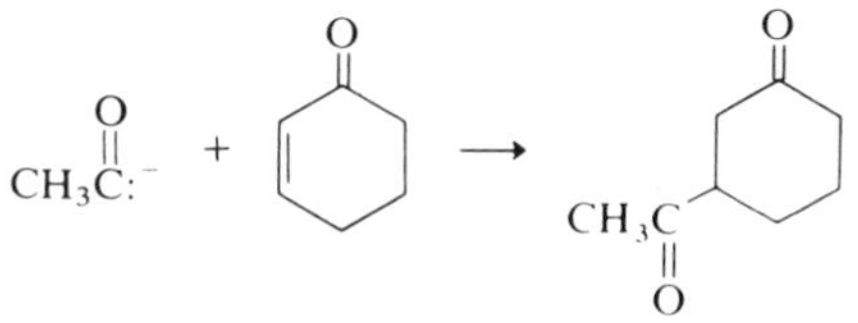

$$CH_3\overset{O}{\overset{\|}{C}}:^- + \text{cyclohexenone} \longrightarrow \text{3-acetylcyclohexanone } (CH_3C(=O)-)$$

The electrophilic α,β-unsaturated ketone is reactive toward nucleophiles, but the nucleophile that is required, an acyl anion, is not normally an accessible entity. As will be discussed, however, there are several potential reagents which could introduce the desired acyl anion in a masked form. The masked functionality used in place of an inaccessible species is termed a *synthetically equivalent group.* Often the concept of "umpolung" is involved in devising synthetic equivalent groups. The term *umpolung* refers to the formal reversal of the normal polarity of a functional group.[71] For instance, acyl groups are normally electrophilic, but often, as in the example above, a synthetic operation may require the transfer of an acyl group as a nucleophile.

Because of the great importance of carbonyl groups in synthesis, a substantial effort has been devoted to developing nucleophilic equivalents for introduction of acyl groups.[72] One successful method involves a three-step sequence in which an aldehyde is converted to an O-protected cyanohydrin. This α-alkoxynitrile is then deprotonated, generating a nucleophilic carbanion.[73] After carbon–carbon bond formation, the carbonyl group can be regenerated by hydrolysis of the cyanohydrin. This sequence has been used to solve the problem of introducing an acetyl group at the β position of cyclohexenone.[74]

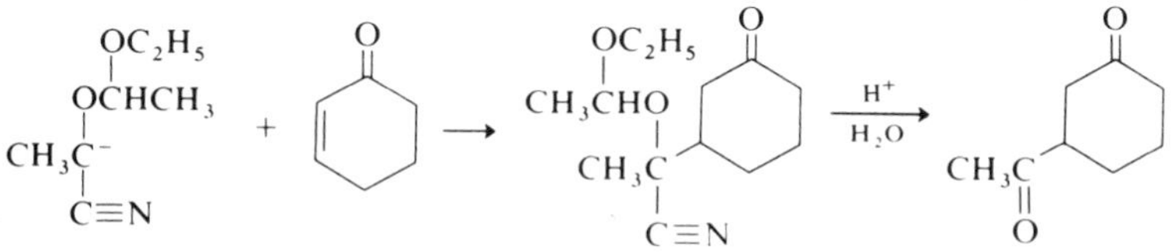

$$CH_3\overset{OCHCH_3(OC_2H_5)}{\underset{C\equiv N}{C^-}} + \text{cyclohexenone} \longrightarrow \text{3-[}CH_3C(C\equiv N)(OCH(CH_3)OC_2H_5)\text{]cyclohexanone} \xrightarrow[H_2O]{H^+} \text{3-acetylcyclohexanone} \quad \text{Ref. 73}$$

α-Lithiovinyl ethers provide another group of acyl anion equivalents.

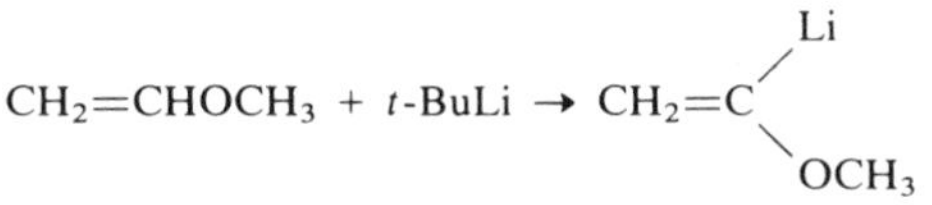

$$CH_2{=}CHOCH_3 + t\text{-BuLi} \rightarrow CH_2{=}C(Li)(OCH_3) \quad \text{Ref. 75}$$

71. For a general discussion and many examples of the use of the umpolung concept, see D. Seebach, *Angew. Chem. Int. Ed. Engl.* **18**, 239 (1979).
72. For reviews of acyl anion synthons, see T. A. Hase and J. K. Koskimies, *Aldrichimica Acta* **15**, 35 (1982).
73. G. Stork and L. Maldonado, *J. Am. Chem. Soc.* **93**, 5286 (1971); *J. Am. Chem. Soc.* **96**, 5272 (1974).
74. For futher discussion of synthetic applications of the carbanions of O-protected cyanohydrins, see J. D. Albright, *Tetrahedron* **39**, 3207 (1983).
75. J. E. Baldwin, G. A. Höfle, and O. W. Lever, Jr., *J. Am. Chem. Soc.* **96**, 7125 (1974).

$$CH_2{=}CHOC_2H_5 \xrightarrow[2)\ CuI]{1)\ t\text{-}BuLi,\ -65°C} CH_2{=}C(Cu)(OC_2H_5)$$ Ref. 76

These reagents are capable of adding the α-alkoxyvinyl group to electrophilic centers. Subsequent hydrolysis can generate the carbonyl group and complete the desired transformation.

$$CH_2{=}C(Li)(OCH_3) + \text{cyclopentanone} \rightarrow CH_2{=}C(OCH_3)\text{-}C(OH)\text{(cyclopentyl)}\ (88\%) \xrightarrow[H_2O]{H^+} CH_3C(=O)\text{-}C(OH)\text{(cyclopentyl)}\ (86\%)$$ Ref. 75

$$(CH_2{=}C(OC_2H_5))_2CuLi + \text{5,5-dimethylcyclohex-2-enone} \rightarrow \text{3-}(CH_2{=}C(OC_2H_5))\text{-5,5-dimethylcyclohexanone}\ (67\%) \xrightarrow[H_2O]{H^+} \text{3-}(CH_3C(=O))\text{-5,5-dimethylcyclohexanone}\ (74\%)$$ Ref. 76

Sulfur compounds have also proven to be useful as nucleophilic acyl equivalents. The first reagent of this type to find use was the 1,3-dithiane ring, which on lithiation provides a nucleophilic acyl equivalent. The lithio derivative is a reactive nucleophile toward alkyl halides and carbonyl compounds.[77]

$$\text{2-methyl-1,3-dithiane} \xrightarrow{n\text{-BuLi}} \text{2-lithio-2-methyl-1,3-dithiane} \xrightarrow{O{=}\text{cyclohexenone}} \text{2-methyl-2-(1-hydroxycyclohex-2-enyl)-1,3-dithiane} \xrightarrow[H_2O,\ CaCO_3]{Hg^{2+}} CH_3C(=O)\text{-}C(OH)\text{(cyclohex-2-enyl)}$$

Closely related procedures are based on α-alkylthiosulfoxides, with ethyllthiomethyl ethyl sulfoxide being a particularly convenient example.[78]

$$CH_3CH_2S(=O)\text{-}\bar{C}(R)\text{-}SCH_2CH_3 \xrightarrow{R'X} CH_3CH_2S(=O)\text{-}C(R)(R')\text{-}SC_2H_5 \rightarrow RC(=O)R'$$

$$CH_3CH_2S(=O)\text{-}\bar{C}(R)\text{-}SCH_2CH_3 \xrightarrow{R_2'C{=}O} CH_3CH_2S(=O)\text{-}C(R)(SC_2H_5)\text{-}C(OH)R_2' \rightarrow RC(=O)\text{-}C(OH)R_2'$$

$$CH_3CH_2S(=O)\text{-}\bar{C}(R)\text{-}SCH_2CH_3 \xrightarrow{R'CH{=}CHC(=O)R''} CH_3CH_2S(=O)\text{-}C(R)(SC_2H_5)\text{-}CH(R')CH_2C(=O)R'' \rightarrow RC(=O)CH(R')CH_2C(=O)R''$$

76. R. K. Boeckman, Jr., and K. J. Bruza, *J. Org. Chem.* **44**, 4781 (1979).
77. D. Seebach and E. J. Corey, *J. Org. Chem.* **40**, 231 (1975).
78. J. E. Richman, J. L. Herrmann, and R. H. Schlessinger, *Tetrahedron Lett.*, 3267 (1973); J. L. Herrmann, J. E. Richman, and R. H. Schlessinger, *Tetrahedron Lett*, 3271 (1973); J. L. Herrmann, J. E. Richman, P. J. Wepplo, and R. H. Schlessinger, *Tetrahedron Lett.*, 4707 (1973).

Scheme 13.2. Synthetic Sequences Used for Reaction of Acyl Anion Equivalents

1[a]

$$\underset{\mathrm{OEE}}{\mathrm{RCHCN}} \xrightarrow{\mathrm{LDA}} \overset{\mathrm{Li}}{\underset{\mathrm{OEE}}{\mathrm{RCCN}}} \xrightarrow{\mathrm{R'X}} \overset{\mathrm{OEE}}{\underset{\mathrm{CN}}{\mathrm{RCR'}}} \xrightarrow[\mathrm{H_2O}]{\mathrm{H^+}} \overset{\mathrm{O}}{\mathrm{RCR'}}$$

2[b]

$$\text{2-R-1,3-dithiane} \xrightarrow{n\text{-BuLi}} \text{2-R-2-Li-1,3-dithiane} \xrightarrow{\mathrm{R'X}} \text{2-R-2-R'-1,3-dithiane} \longrightarrow \overset{\mathrm{O}}{\mathrm{RCR'}}$$

3[c]

$$R_2C{=}CHSC_2H_5 \xrightarrow{s\text{-BuLi}} R_2C{=}\underset{\mathrm{Li}}{C}SC_2H_5 \xrightarrow{\mathrm{R'I}} R_2C{=}\underset{\mathrm{R'}}{C}SC_2H_5 \xrightarrow[\mathrm{H_2O}]{\mathrm{HgCl_2}} R_2CH\overset{\mathrm{O}}{C}R'$$

4[d]

$$R_2C{=}C(SPh)_2 \xrightarrow{\mathrm{LiNaphth^-}} R_2C{=}\underset{\mathrm{Li}}{C}SPh \xrightarrow{\mathrm{R'CH{=}O}} R_2C{=}\overset{\mathrm{SPh}}{C}\underset{\mathrm{OH}}{C}HR' \rightarrow R_2CH\overset{\mathrm{O}}{C}\underset{\mathrm{OH}}{C}HR'$$

5[e]

$$C_2H_5S\overset{\mathrm{O}}{\underset{\mathrm{CH_3}}{C}}SC_2H_5 + CH_2{=}C\overset{\mathrm{O}}{C}CH_2CH_3 \text{ (with } CH_2SCH_3 \text{ on C-2)} \rightarrow CH_3\overset{C_2H_5S}{\underset{C_2H_5S(O)}{C}}CH_2\underset{CH_2SCH_3}{C}H\overset{\mathrm{O}}{C}CH_2CH_3$$

$$\xrightarrow{1)\ H^+,\ H_2O\quad 2)\ NaOH} \text{2,3-dimethyl-5-(}CH_2SCH_3\text{)cyclopent-2-enone}$$

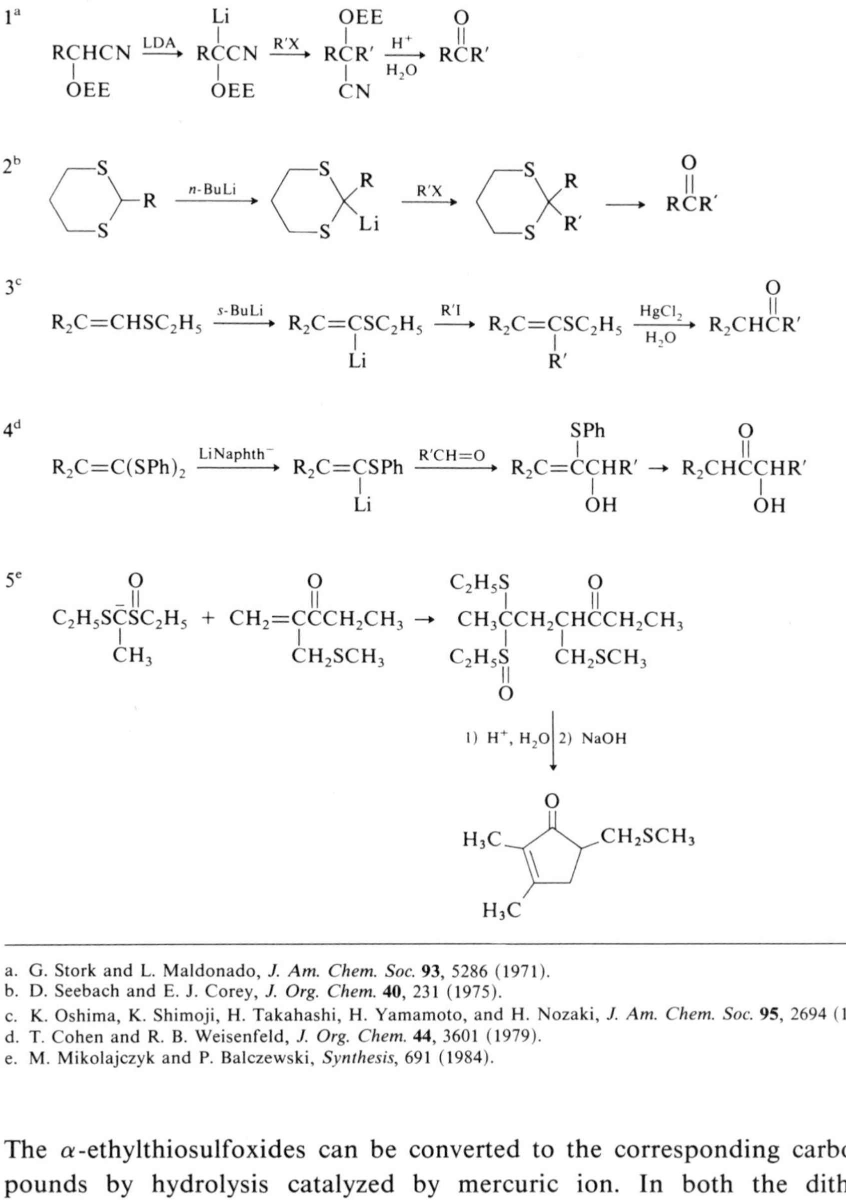

a. G. Stork and L. Maldonado, *J. Am. Chem. Soc.* **93**, 5286 (1971).
b. D. Seebach and E. J. Corey, *J. Org. Chem.* **40**, 231 (1975).
c. K. Oshima, K. Shimoji, H. Takahashi, H. Yamamoto, and H. Nozaki, *J. Am. Chem. Soc.* **95**, 2694 (1973).
d. T. Cohen and R. B. Weisenfeld, *J. Org. Chem.* **44**, 3601 (1979).
e. M. Mikolajczyk and P. Balczewski, *Synthesis*, 691 (1984).

The α-ethylthiosulfoxides can be converted to the corresponding carbonyl compounds by hydrolysis catalyzed by mercuric ion. In both the dithiane and alkylthiomethylsulfoxide systems, an umpolung is achieved on the basis of the carbanion-stabilizing ability of the sulfur substituents. Scheme 13.2 summarizes some examples of synthetic sequences which employ acyl anion equivalents.

Another group of synthetic equivalents has been developed which correspond to the propanal "homoenolate," $^-CH_2CH_2CH{=}O$.[79] This structure is the umpolung equivalent of an important electrophilic reagent, the α,β-unsaturated aldehyde

79. For reviews of homoenolate anions, see J. C. Stowell, *Chem. Rev.* **84**, 409 (1984); N. H. Werstiuk, *Tetrahedron* **39**, 205 (1983).

Scheme 13.3. Reaction Sequences Involving Propanal Homoenolate Anion Synthetic Equivalents

	Reagent	Reaction sequence
1[a]		

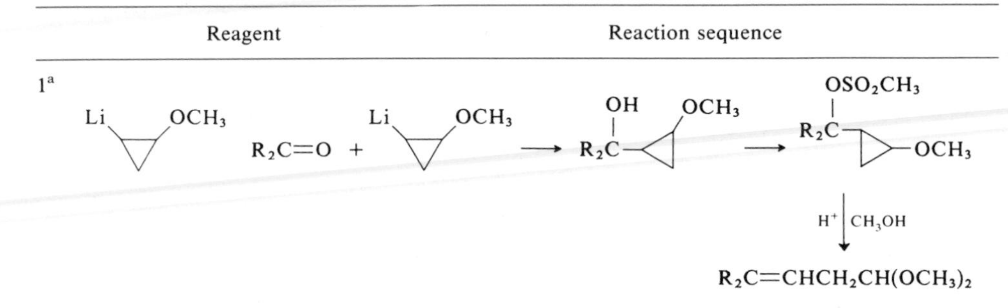

2[b] $CH_2{=}CHCH(Li)OCH_3$

$$R_2C{=}O + CH_2{=}CHCH(Li)OCH_3 \rightarrow R_2C(OH)CH_2CH{=}CHOCH_3 \xrightarrow[H_2O]{H^+} R_2C(OH)CH_2CH_2CH{=}O$$

3[c] $LiCH_2CH{=}CHNR'_2$

$$RX + LiCH_2CH{=}CHNR'_2 \rightarrow RCH_2CH{=}CHNR'_2 \rightarrow RCH_2CH_2CH{=}O$$

4[d] $CH_2{=}CHCH(Li)OSi(CH_3)_3$

$$RX + CH_2{=}CHCH(Li)OSi(CH_3)_3 \rightarrow RCH_2CH{=}CHOSi(CH_3)_3 \xrightarrow[H_2O]{H^+} RCH_2CH_2CH{=}O$$

5[e] $PhSO_2CH(Li)CH_2CH(OR')_2$

$$RX + PhSO_2CH(Li)CH_2CH(OR')_2 \rightarrow PhSO_2CH(R)CH_2CH(OR')_2 \xrightarrow[2)\ H^+,\ H_2O]{1)\ Na/Hg} RCH_2CH_2CH{=}O$$

6[f] $LiCH_2CH{=}CHS^-$

$$R_2C{=}O + LiCH_2CH{=}CHS^- \rightarrow R_2C(OH)CH_2CH{=}CHS^- \xrightarrow{CH_3I} R_2C(OH)CH_2CH{=}CHSCH_3$$

7[g] $LiCH_2CH{=}CHSi(CH_3)_3$

$$R_2C{=}O + LiCH_2CH{=}CHSi(CH_3)_3 \rightarrow R_2C(OH)CH_2CH{=}CHSi(CH_3)_3$$

$$\xrightarrow{1)\ RCO_3H\quad 2)\ BF_3,\ MeOH}$$

2,2-R,R-5-OCH_3-tetrahydrofuran (R, R, O, OCH_3)

Scheme 13.3—*continued*

	Reagents	Reaction sequence
8[h]	CuBr/BrMgCH$_2$CH$_2$CH(OR')$_2$	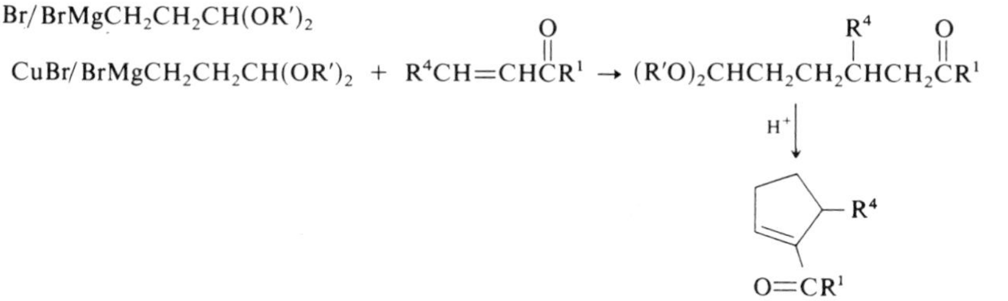 CuBr/BrMgCH$_2$CH$_2$CH(OR')$_2$ + R^{4}CH=CHC(=O)R^1 → (R'O)$_2$CHCH$_2$CH$_2$CH(R^4)CH$_2$C(=O)R^1 —H$^+$→ cyclopentene bearing R^4 and O=CR^1

a. E. J. Corey and P. Ulrich, *Tetrahedron Lett.*, 3685 (1975).
b. D. A. Evans, G. C. Andrews, and B. Buckwalter, *J. Am. Chem. Soc.* **96**, 5560 (1974).
c. H. Ahlbrect and J. Eichler, *Synthesis*, 672 (1974); S. F. Martin and M. T. DuPriest, *Tetrahedron Lett.*, 3925 (1977); H. Ahlbrect, G. Bonnet, D. Enders, and G. Zimmerman, *Tetrahedron Lett.* **21**, 3175 (1980).
d. W. C. Still and T. L. Macdonald, *J. Am. Chem. Soc.* **96**, 5561 (1974).
e. M. Julia and B. Badet, *Bull. Soc. Chim. Fr.*, 1363 (1975); K. Kondo and D. Tunemoto, *Tetrahedron Lett.*, 1007 (1975).
f. K.-H. Geiss, B. Seuring, R. Pieter, and D. Seebach, *Angew. Chem. Int. Ed. Engl.* **13**, 479 (1974); K.-H. Geiss, D. Seebach, and B. Seuring, *Chem. Ber.* **110**, 1833 (1977).
g. E. Ehlinger and P. Magnus, *J. Am. Chem. Soc.* **102**, 5004 (1980).
h. A. Marfat and P. Helquist, *Tetrahedron Lett.*, 4217 (1978); A. Leone-Bay and L. A. Paquette, *J. Org. Chem.* **47**, 4172 (1982).

acrolein. Scheme 13.3 illustrates some of the propanal homoenolate equivalents which have been developed. In general, the reagents are reactive toward such electrophiles as alkyl halides and carbonyl compounds. Several general points can be made about the reagents in Scheme 13.3. First, it should be noted that all deliver the aldehyde functionlity in a masked form, such as an acetal or enol ether. The aldehyde must be liberated in a final step from the protected precursor. Several of the reagents involve delocalized allylic anions. This gives rise to the possibility of electrophilic attack at either the α or γ position of the allylic group. In most cases, the γ attack, which is necessary for the anion to function as a propanal homoenolate, is dominant.

The concept of developing reagents that are the synthetic equivalents of inaccessible species can be taken a step further by considering dipolar species. For example, structures **C** and **D** incorporate both electrophilic and nucleophilic centers. Such reagents might be incorporated into ring-forming schemes, since they have the ability, at least in a formal sense, to undergo cycloaddition reactions.

$$H_5C_2OC(=O)\overset{-}{C}HCH_2\overset{+}{C}H_2 \quad (\mathbf{C}) \qquad {}^{-}C(=O)CH_2CH_2{}^{+} \quad (\mathbf{D})$$

Among the real chemical species which have been developed along these lines are the cyclopropanes **2** and **3**.

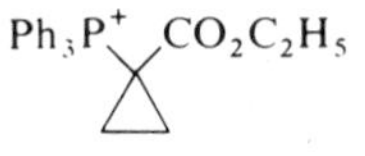

$$\text{1-}(Ph_3P^+)\text{-1-}(CO_2C_2H_5)\text{cyclopropane} \quad (\mathbf{2}) \qquad \text{1-}(Ph_3P^+)\text{-1-}(SPh)\text{cyclopropane} \quad (\mathbf{3})$$

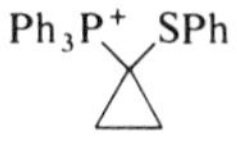

The phosphonium salt **2** reacts with β-ketoesters and β-ketoaldehydes to give excellent yields of cyclopentenecarboxylate esters.

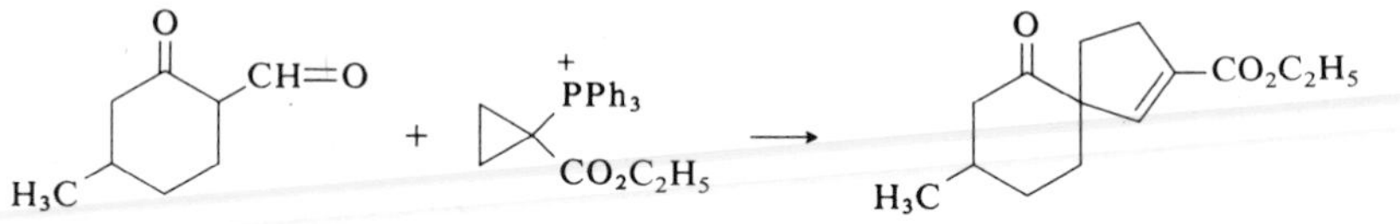

Ref. 80

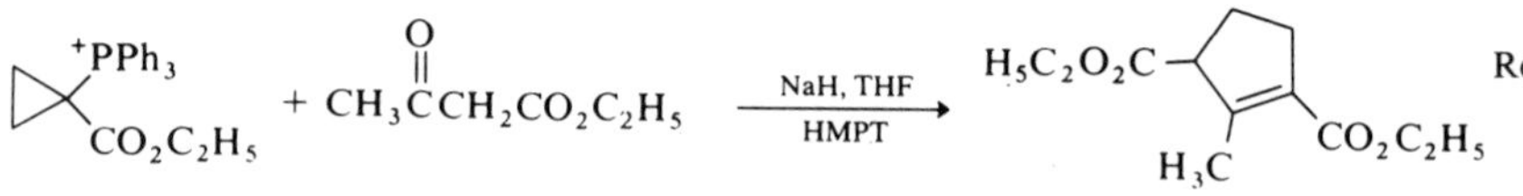

Ref. 81

Several steps are involved. First, the enolate opens the cyclopropane ring. The polarity of this process corresponds to that in the formal synthon **C**. The product of this step is a stabilized Wittig ylide, which goes on to react with the carbonyl group.

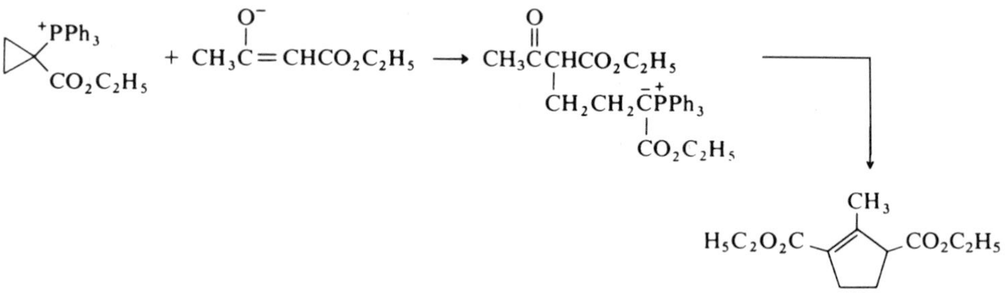

The phosphonium salt **3** reacts similarly with enolates to give vinyl sulfides. The vinyl sulfide group can then be hydrolyzed to a ketone. The overall transformation corresponds to the reactivity of the dipolar synthon **D**.

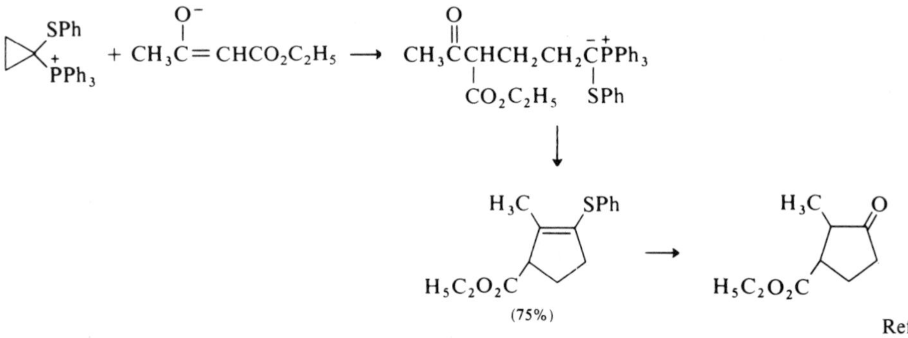

Ref. 82

Many other examples of synthetic equivalent groups have been developed. For example, in Chapter 6 the use of dienes and dienophiles with masked functionality

80. W. G. Dauben and D. J. Hart, *J. Am. Chem. Soc.* **99**, 7307 (1977).
81. P. L. Fuchs, *J. Am. Chem. Soc.* **96**, 1607 (1974).
82. J. P. Marino and R. C. Landick, *Tetrahedron Lett.*, 4531 (1975).

in the Diels–Alder reaction was discussed. It should be recognized that there is no absolute difference between what is termed a "reagent" and a "synthetic equivalent group." For example, we tend to think of potassium cyanide as a reagent, but the cyanide ion is a nucleophilic equivalent of a carboxyl group. This reactivity is evident in the classical preparation of carboxylic acids from alkyl halides via nitrile intermediates.

$$RX + KCN \rightarrow RCN \xrightarrow[H^+]{H_2O} RCO_2H$$

The important point is that synthetic analysis and planning should not be restricted to the specific functionalities that is to appear in the target molecules. These groups can be incorporated as masked equivalents by methods that would not be possible for the functional group itself.

13.3. Synthetic Analysis and Planning

The material covered to this point has been a description of the tools at the disposal of the synthetic chemist, consisting of the extensive catalog of reactions and the associated information on such issues as stereochemistry and mutual reactivity. This knowledge permits a judgment of the applicability of a particular reaction in a synthetic sequence. Broad mechanistic insight is also crucial to synthetic analysis. The relative position of functional groups in a potential reactant may lead to special reactions. Mechanistic concepts are also the basis for developing new reactions which may be necessary in a particular situation. In this chapter, tactical tools of synthesis such as protective groups and synthetic equivalent groups have been considered. The objective of synthetic analysis and planning is to develop a reaction sequence that will efficiently complete the desired synthesis within the constraints which apply.

The planning of a synthesis involves a critical comparative evaluation of alternative reaction sequences that could reasonably be expected to lead to the desired structure from appropriate starting materials. In general, the complexity of any synthetic plan will increase with the size of the molecule and with increasing numbers of functional groups and chiral centers. The goal of synthetic analysis is to recognize possible pathways to the target compound and to develop a suitable sequence of synthetic steps.

The restrictions which apply to a synthesis will depend on the reason the synthesis is being conducted. A synthesis of a material to be prepared in substantial quantity may impose a limitation as to cost of starting materials. Syntheses for commercial production must meet such criteria as economic feasibility, acceptability of by-products, and safety. Synthesis of complex structures with several chiral centers must deal with the problem of stereoselectivity. If an enantiomerically pure material is to be synthesized, the means of controlling enantioselectivity must be considered. The development of a satisfactory plan is the intellectual challenge to the chemist. The task puts a premium on creativity and ingenuity. There is no single correct solution. While there is no established routine by which a synthetic plan can be formulated, general principles that should guide synthetic analysis and planning have been described.[83]

The initial step in creating a synthetic plan should involve a *retrosynthetic analysis*. The structure of the molecule is dissected step by step along reasonable pathways to successively simpler compounds until molecules that are acceptable as starting materials are reached. Several factors enter into this process, and all are closely interrelated. The recognition of *bond disconnections* allows the molecule to be broken down into *key intermediates*. Such disconnections must be made in such a way that it is feasible to form the bonds by some synthetic process. Thus, the relative placement of potential functionality will strongly influence which bond disconnections are preferred. To emphasize that these bond disconnections must correspond to transformations that can be conducted in the synthetic sense, they are sometimes called *antisynthetic transforms*, that is, the reverse of synthetic steps. The overall synthetic plan will consist of a sequence of reactions designed to construct the total molecular framework from the key intermediates. The plan should take into account the advantages of a *convergent synthesis*. The objective of making a synthesis convergent is to decrease its overall length. In general, it is desirable to construct the molecule from a few key fragments that can be combined late in the synthesis rather than to build the molecule step by step from a single starting material. The reason for this is that the overall yield is the multiplication product of the yields for all the individual steps. Overall yields will decrease as the number of steps to which the original starting material is subjected increases.[84]

Convergent synthesis:

$$A + B \rightarrow C \xrightarrow{D} G \searrow$$
$$\rightarrow G{-}H \xrightarrow{I} G{-}H{-}I$$
$$E + F \rightarrow H \nearrow$$

Linear synthesis: $A + B \rightarrow C \xrightarrow{D} G \xrightarrow{E} G{-}E \xrightarrow{F} G{-}H \xrightarrow{I} G{-}H{-}I$

After a plan for assembly of the key intermediates into the molecular framework has been developed, the details of incorporation and transformation of functional groups must be considered. It is frequently necessary to interconvert functional groups. This may be to develop a particular kind of reactivity at a center or to avoid interference with a particular reaction step. Protective groups and synthetic equivalent groups are important for planning of functional group transformations. Achieving the final array of functionality is often less difficult than establishing the overall molecular skeleton and stereochemistry, because of the large number of procedures which exist for interconverting the common functional groups.

The care with which a synthesis is analyzed and planned will have a great impact on its likelihood of success. A single flaw can cause failure. The investment

83. E. J. Corey and X.-M. Cheng, *The Logic of Chemical Synthesis*, Wiley, New York, 1989.
84. A formal analysis of the concept of convergency has been presented by J. B. Hendrickson, *J. Am. Chem. Soc.* **99**, 5439 (1977).

of material and effort that is made when the synthesis is begun may be lost if the plan is faulty. Even with the best of planning, however, unexpected problems are frequently encountered. This circumstance again tests the ingenuity of the chemist to devise a modified plan that can expeditiously overcome the unanticipated obstacle.

13.4. Control of Stereochemistry

The degree of control of stereochemistry that is necessary depends on the nature of the molecule and the objective of the synthesis. The issue becomes of critical importance when the target molecule has several stereogenic centers, such as double bonds, ring junctions, and chiral centers. The number of possible stereoisomers is 2^n, where n is the number of stereogenic centers. Failure to control stereochemistry of intermediates in the synthesis of a compound with several centers of stereochemistry will result in a mixture of stereoisomers, which will, at best, lead to a reduced yield of the desired product and may generate inseparable mixtures.

We have considered in the earlier chapters the means of stereochemical control of many of the synthetic methods discussed. In ring compounds, for example, stereoselectivity can frequently be predicted on the basis of conformational analysis of the reactant and consideration of the steric and stereoelectronic factors that will control reagent approach. In the stereoselective synthesis of a chiral material in racemic form, it is necessary to control the relative configuration of all stereogenic centers. Thus, in planning a synthesis, the stereochemical outcome of all reactions which form new double bonds, ring junctions, and chiral centers must be incorporated into the synthetic plan. In a completely stereoselective synthesis, each successive center is introduced in the proper relationship to existing stereocenters. This ideal is often difficult to achieve. When a reaction is not completely stereoselective, the product will contain one or more diastereomers of the desired product. This requires either a purification or some manipulation to correct the stereochemistry. Fortunately, diastereomers are usually separable, but the overall efficiency of the synthesis is decreased with each such separation. Thus, high stereoselectivity is an important goal of synthetic planning.

If the compound is to be obtained in enantiomerically pure form, an enantioselective synthesis must be developed. (Review Section 2.1 of Part A for the basis of enantiomeric relationships.) There are four general approaches that are commonly used to obtain enantiomerically pure material by synthesis. One is based on incorporating a *resolution* into the synthetic plan. This approach involves use of racemic or achiral starting materials and resolving some intermediate in the synthesis. In a synthesis based on a resolution, the steps subsequent to the resolution step must meet two criteria: (1) they must not disturb the configuration at existing centers of stereochemistry and (2) all new centers of stereochemistry must be introduced with the correct configuration relative to the existing centers.

A second general approach is to use a starting material that is enantiomerically pure. There are a number of naturally occurring materials, or substances derived from them, which are available in enantiomerically pure form.[85] Again, a completely stereoselective synthesis must be capable of controlling the stereochemistry of all newly introduced stereogenic centers so that they have the proper relationship to the chiral centers existing in the starting material. When this is not achieved, the desired stereoisomer must be separated and purified.

A third method for enantioselective synthesis involves the use of a stoichiometric amount of a *chiral auxiliary*. This is an enantiomerically pure material that can control the stereochemistry of one or more reaction steps in such a way as to give product having the desired configuration. Once the chiral auxiliary has achieved its purpose, it can be eliminated from the molecule. As in syntheses involving resolution or enantiomerically pure starting materials, subsequent steps must be controlled to give the correct configuration of newly created chiral centers.

A fourth approach to enantioselective synthesis is to use a chiral catalyst in a reaction which creates one or more chiral centers. If the catalyst operates with complete efficiency, an enantiomerically pure material will be obtained. Subsequent steps must control the configuration of newly introduced chiral centers.

In practice, any of these four approaches might be the most effective for a given synthesis. If they are judged on the basis of absolute efficiency in the use of chiral material, the ranking is resolution < natural source < chiral auxiliary < enantioselective catalyst. A resolution process inherently employs only half of the original racemic material. A starting material from a natural source can, in principle, be used with 100% efficiency, but it is consumed and cannot be reused. A chiral auxiliary can, in principle, be reused, but it must be used in stoichiometric amount. A chiral catalyst, in principle, can produce an unlimited amount of an enantiomerically pure material.

The key issue for synthesis of pure stereoisomers, in either racemic or homochiral form, is that the configuration at newly created chiral centers be controlled in some way. This may be accomplished by several means. An existing functional group may control the approach of a reagent by coordination. An existing chiral center may control reactant conformation, and thereby the direction of approach of a reagent. Whatever the detailed mechanism, the synthetic plan must include the means by which the required stereochemical control is to be achieved. When this cannot be done, the price to be paid is a separation of stereoisomers and the resulting reduction in overall yield.

13.5. Illustrative Syntheses

In the remainder of this chapter, we will consider several syntheses of four illustrative compounds. We will examine the retrosynthetic plans and discuss crucial bond-forming steps and, where appropriate, the means of stereochemical control.

85. For a discussion of this approach to enantioselective synthesis, see S. Hanessian, *Total Synthesis of Natural Products, the Chiron Approach*, Pergamon Press, New York, 1983.

In this discussion, we have the benefit of hindsight by being able to look at successfully completed syntheses. This retrospective analysis can serve to illustrate the issues that arise in planning a synthesis and provide examples of solutions that have been developed.

13.5.1. Juvabione

Juvabione is a terpene-derived ketoester that has been isolated from various plant sources. It exhibits "juvenile hormone" activity in insects; that is, it can modify the process of metamorphosis.

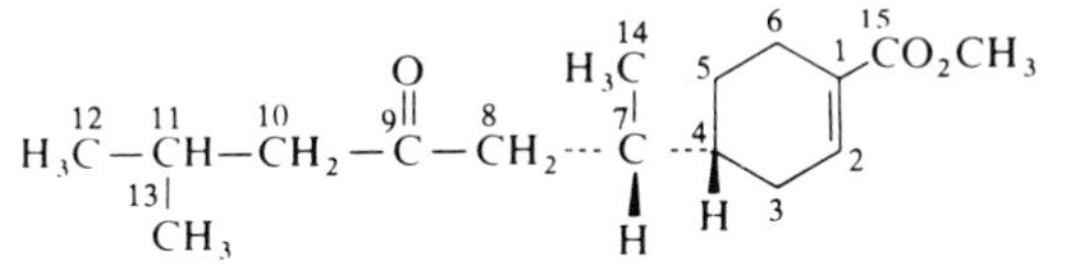

erythro-juvabione

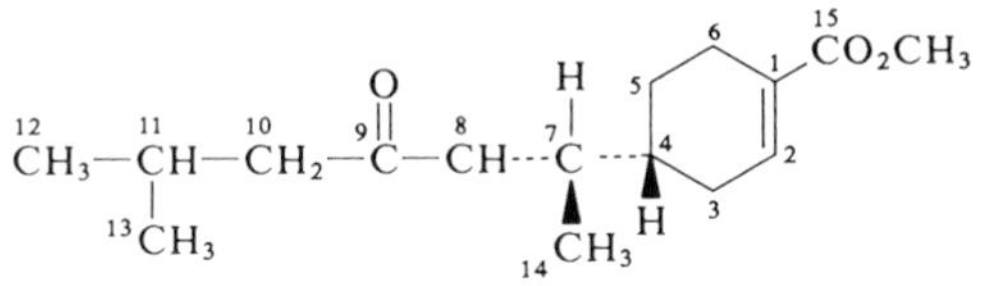

threo-juvabione

In considering the retrosynthetic analysis of juvabione, two factors draw special attention to the bond between C-4 and C-7. First, this bond establishes the only stereochemical feature of the molecule. The two carbons it connects are both chiral, and their relative configuration determines which diastereomeric structure will be obtained. Both stereoisomers have been characterized and are referred to as *erythro*- and *threo*-juvabione. In a stereocontrolled synthesis, it will be necessary to establish the desired stereochemistry at C-4 and C-7. The C-4—C-7 bond also connects the side chain to the cyclohexene ring. Because a compound incorporating the ring would make a logical candidate for one key intermediate, the C-4—C-7 bond is a potential bond disconnection. Other bonds that merit attention are those connecting C-7 through C-11. These could all be formed by one of the many methods for synthesis of ketones. The only other point of functionality is the conjugated unsaturated ester. This functionality is remote from the stereochemical centers and the ketone functionality, and in most of the reported synthesis it does not play a key role.

Some of the existing syntheses use similar starting materials. Those in Schemes 13.5 and 13.6 lead back to a *para*-substituted aromatic ether. The syntheses in Schemes 13.8 and 13.9 begin with an accessible terpene intermediate. The syntheses in Schemes 13.11, 13.12, 13.13, and 13.14 start with cyclohexenone or 3-ethoxycyclohexenone.

Scheme 13.4 presents a retrosynthetic analysis leading to the key intermediates used by the syntheses in Schemes 13.5 and 13.6. The first disconnection is that of the ester functionality. This corresponds to a decision that the ester group can be

Scheme 13.4. Retrosynthetic Analysis of Juvabione with Disconnection to *p*-Methoxy-acetophenone

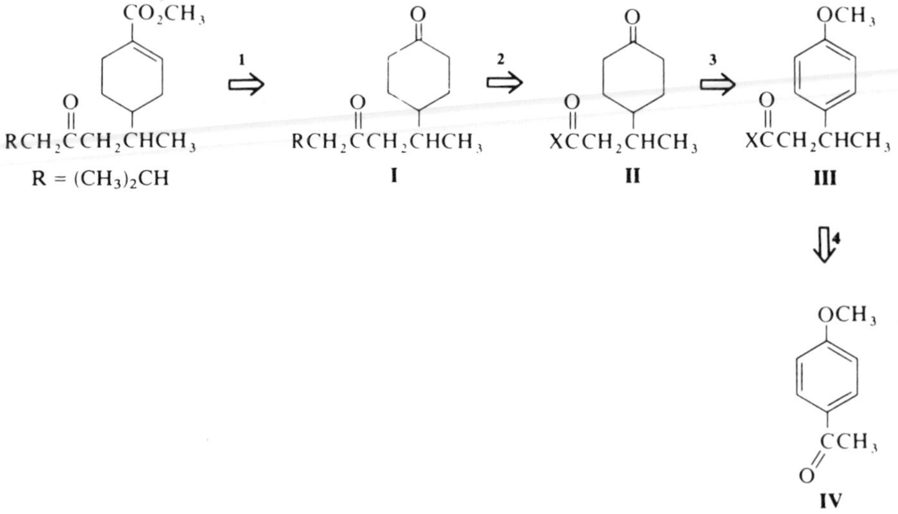

CO_2CH_3 ; $RCH_2\overset{O}{\overset{\|}{C}}CH_2CHCH_3$; $R = (CH_3)_2CH$ ⇒(1) $RCH_2\overset{O}{\overset{\|}{C}}CH_2CHCH_3$ **I** ⇒(2) $X\overset{O}{\overset{\|}{C}}CH_2CHCH_3$ **II** ⇒(3) OCH_3 ; $X\overset{O}{\overset{\|}{C}}CH_2CHCH_3$ **III** ⇓(4) OCH_3 ; CCH_3 ; O **IV**

added late in the synthesis. Disconnection **2** identifies the C-9—C-10 bond as one that can be readily formed by addition of some nucleophilic group corresponding to C-10—C-13 to the carbonyl center at C-9. The third retrosynthetic transform recognizes that the cyclohexanone ring might be obtained by a Birch reduction of an appropriately substituted aromatic ether. The methoxy substituent would provide for eventual introduction of the cyclic carbonyl group. The final disconnection identifies a simple starting material, 4-methoxyacetophenone.

A synthesis corresponding to this pattern is shown in Scheme 13.5. It relies on well-known reaction types. The C-4—C-7 bond is formed by a Reformatsky reaction and is followed by hydrogenolysis. Steps **B–D** introduce the C-10–C-13 isobutyl group. The C-9—C-10 bond connection is made in step **D** by a Grignard addition reaction. In this synthesis, the relative configuration at C-4 and C-7 is established by the hydrogenation in step **F**. In principle, this reaction could be diastereoselective if the adjacent chiral center at C-7 strongly influenced the direction of addition of hydrogen. In practice, the reduction is not very selective, and a mixture of isomers was obtained. Steps **G** and **H** introduce the C-1 ester group. A key feature of this synthesis is the use of the methoxybenzene ring as a synthetic equivalent for the cyclohexanone structure.

The synthesis in Scheme 13.6 also makes use of an aromatic starting material and follows a retrosynthetic plan corresponding to that in Scheme 13.4. This synthesis is somewhat more convergent in that the entire side chain, except for C-14, is introduced as a single unit. The C-14 methyl is introduced by a copper-catalyzed conjugate addition in step **B**.

Scheme 13.7 is a retrosynthetic outline of the syntheses in Schemes 13.8 and 13.9. The common feature of these syntheses is the use of terpene-derived starting materials. The use of such a starting material is suggested by the terpenoid structure of juvabione, which can be divided into "isoprene units."

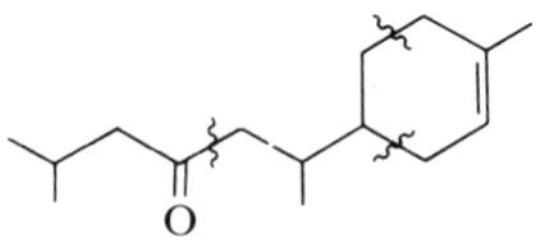

isoprene units in juvabione

The synthesis shown in Scheme 13.8 used limonene as the starting material ($R = CH_3$ in Scheme 13.7) whereas that in Scheme 13.9 uses an aldehyde (R =

Scheme 13.5. Juvabione Synthesis: K. Mori and M. Matsui[a]

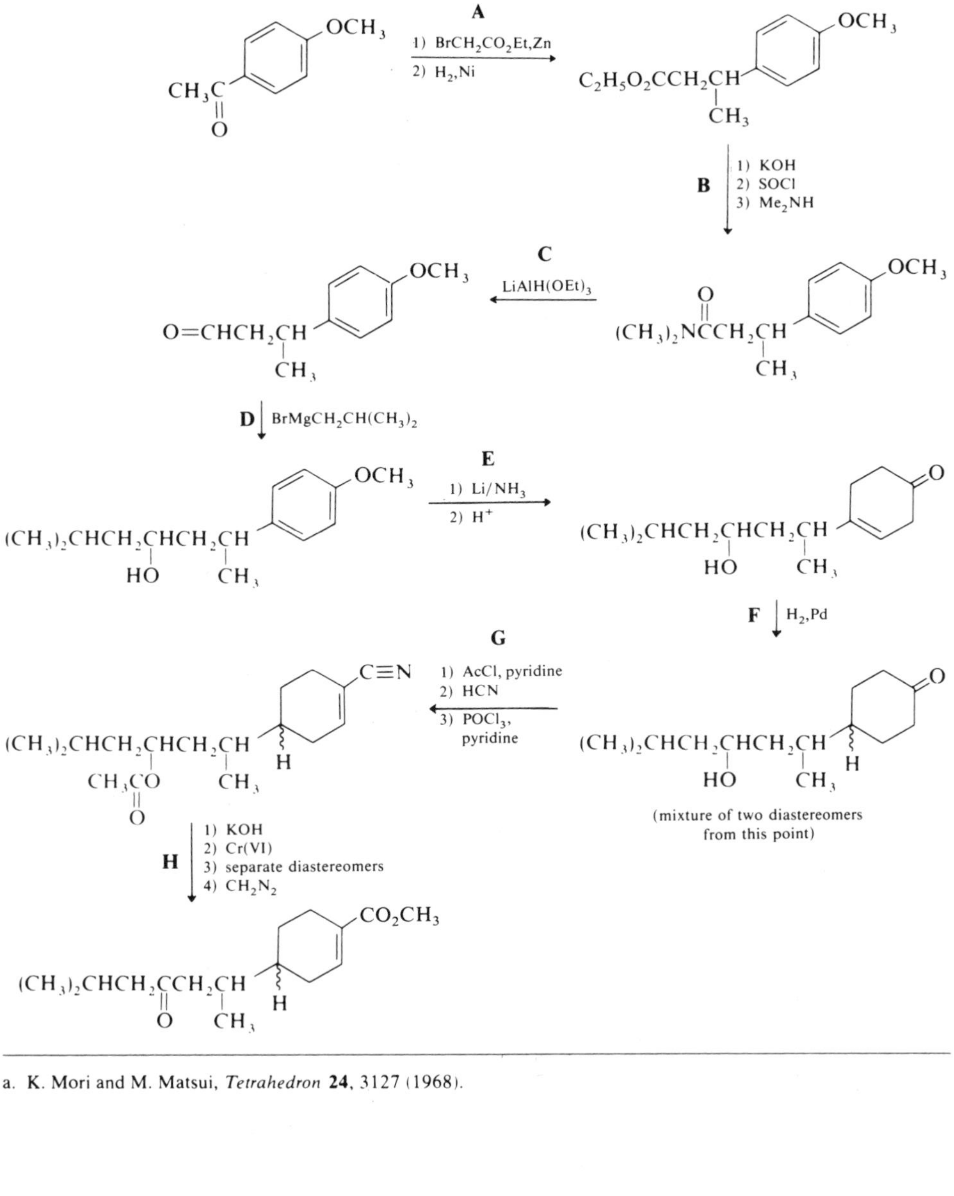

A: 1) $BrCH_2CO_2Et$,Zn; 2) H_2,Ni

B: 1) KOH; 2) SOCl; 3) Me_2NH

C: $LiAlH(OEt)_3$

D: $BrMgCH_2CH(CH_3)_2$

E: 1) Li/NH_3; 2) H^+

F: H_2,Pd

(mixture of two diastereomers from this point)

G: 1) AcCl, pyridine; 2) HCN; 3) $POCl_3$, pyridine

H: 1) KOH; 2) Cr(VI); 3) separate diastereomers; 4) CH_2N_2

a. K. Mori and M. Matsui, *Tetrahedron* **24**, 3127 (1968).

Scheme 13.6. Juvabione Synthesis: K. S. Ayyar and G. S. K. Rao[a]

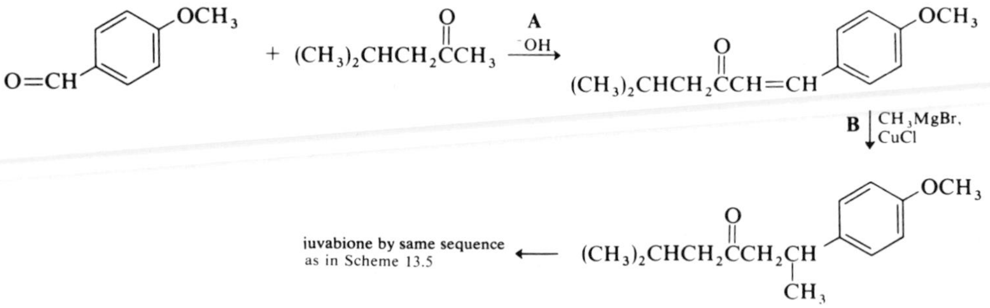

$O{=}CH-C_6H_4-OCH_3$ + $(CH_3)_2CHCH_2C(=O)CH_3$ $\xrightarrow[\ ^-OH]{A}$ $(CH_3)_2CHCH_2C(=O)CH{=}CH-C_6H_4-OCH_3$

B: CH_3MgBr, CuCl

$(CH_3)_2CHCH_2C(=O)CH_2CH(CH_3)-C_6H_4-OCH_3$ ⟶ juvabione by same sequence as in Scheme 13.5

a. K. S. Ayyar and G. S. K. Rao, *Can. J. Chem.* **46**, 1467 (1968).

Scheme 13.7. Retrosynthetic Analysis of Juvabione with Disconnection to a Terpene Structure

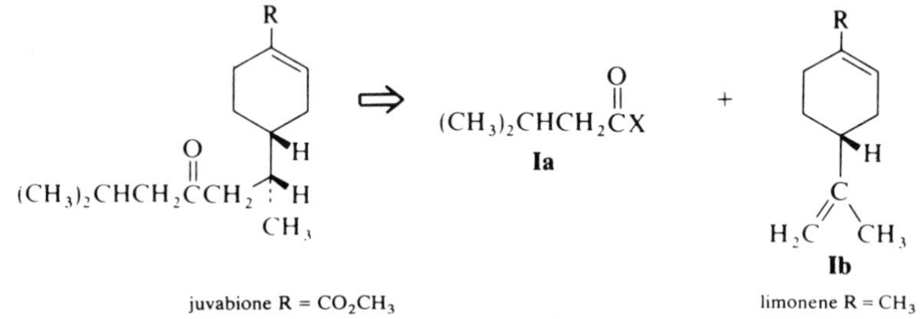

juvabione R = CO_2CH_3 ⇒ $(CH_3)_2CHCH_2C(=O)X$ **Ia** + **Ib** limonene R = CH_3

CH=O). The use of these starting materials focuses attention on the means of attaching the C-9–C-13 side chain. Furthermore, since the starting material is an enantiomerically pure terpene, enantiospecificity controlled by the chiral center at C-4 of the starting material might be feasible. In the synthesis in Scheme 13.8, the C-4—C-7 stereochemistry is established in the hydroboration which is the first step of the synthesis. Unfortunately, this reaction shows only very modest stereoselectivity, and a 3:2 mixture of diastereomers was obtained and separated. The subsequent steps do not affect these chiral centers. The synthesis in Scheme 13.8 uses a three-step oxidation sequence to oxidize the C-15 methyl group at step **D**. The first reaction is oxidation by singlet oxygen to give a mixture of hydroperoxides, with oxygen bound mainly at C-2. The mixture is reduced to the corresponding alcohols, which are then oxidized to the acid via an aldehyde intermediate.

In Scheme 13.9, the side chain is added in one step by a borane carbonylation reaction. This synthesis is very short, and the first four steps are used to transform the aldehyde group in the starting material to a methyl ester. The stereochemistry at C-4—C-7 is established by the hydroboration in which the C-7—H bond is formed. A 1:1 mixture of diastereomers was formed, indicating the configuration at C-4 did not influence the direction of approach of the borane reagent.

Scheme 13.8. Juvanione Synthesis: B. A. Pawson, H.-C. Cheung, S. Gurbaxani, and G. Saucy[a]

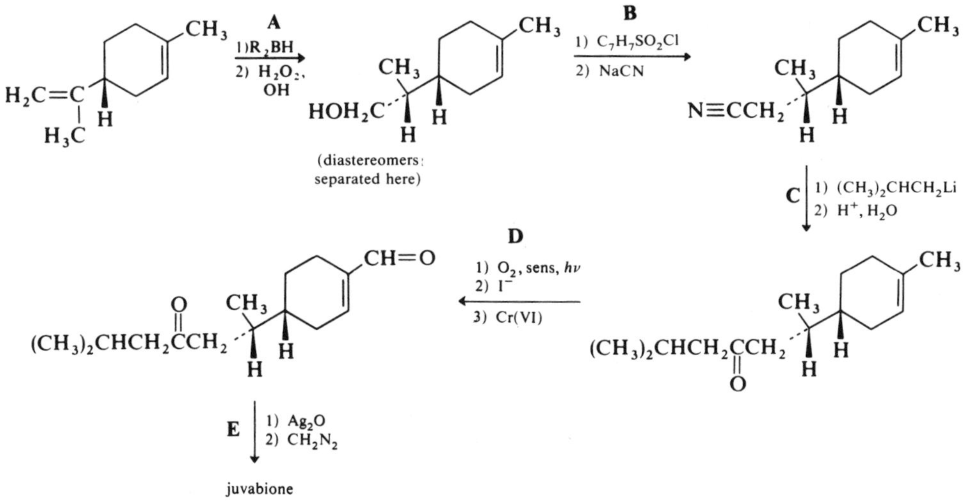

A: 1) R_2BH; 2) H_2O_2, ^-OH (diastereomers separated here)

B: 1) $C_7H_7SO_2Cl$; 2) NaCN

C: 1) $(CH_3)_2CHCH_2Li$; 2) H^+, H_2O

D: 1) O_2, sens, $h\nu$; 2) I^-; 3) Cr(VI)

E: 1) Ag_2O; 2) CH_2N_2

juvabione

a. B. A. Pawson, H.-C. Cheung, S. Gurbaxani, and G. Saucy, *J. Am. Chem. Soc.* **92**, 336 (1970).

Scheme 13.9. Juvabione Synthesis: E. Negishi, M. Sabanski, J. J. Katz, and H. C. Brown[a]

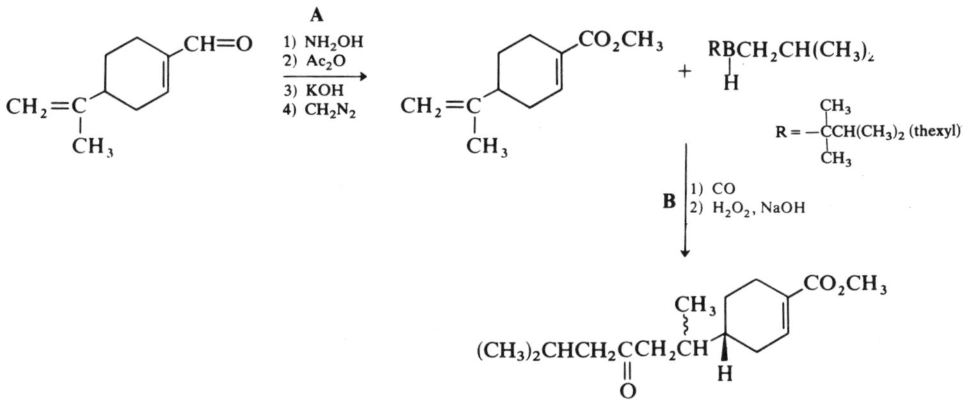

A: 1) NH_2OH; 2) Ac_2O; 3) KOH; 4) CH_2N_2

+ $RBHCH_2CH(CH_3)_2$, R = $-C(CH_3)_2CH(CH_3)_2$ (thexyl)

B: 1) CO; 2) H_2O_2, NaOH

a. E. Negishi, M. Sabanski, J. J. Katz, and H. C. Brown, *Tetrahedron* **32**, 925 (1976).

The first stereocontrolled syntheses of juvabione are described in Schemes 13.11 and 13.12. Scheme 13.10 is a retrosynthetic analysis corresponding to these syntheses. These syntheses have certain similarities. Both start with cyclohexenone. There is a general similarity in the fragments which were utilized, but the order of construction differs.

In the synthesis in Scheme 13.11, the crucial stereochemical control occurs in step **B**. The first intermediate is constructed by a [2 + 2] cycloaddition between reagents of complementary polarity, the electron-rich ynamine and the electron-poor

Scheme 13.10. Retrosynthetic Analysis of Juvabione with Alternate Disconnections to Cyclohexenone

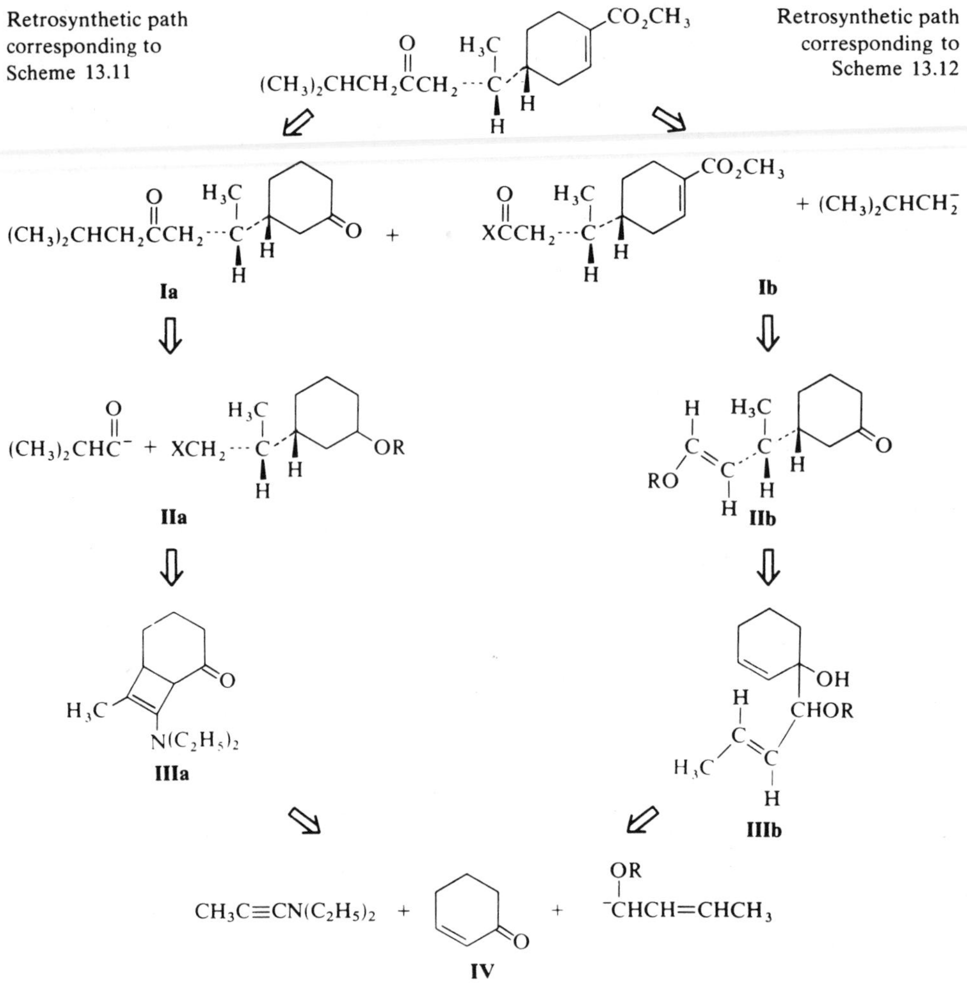

Retrosynthetic path corresponding to Scheme 13.11

Retrosynthetic path corresponding to Scheme 13.12

$(CH_3)_2CHCH_2CCH_2$ + $(CH_3)_2CHCH_2^-$

Ia **Ib**

$(CH_3)_2CHC^-$ + XCH_2 OR

IIa **IIb**

H_3C $N(C_2H_5)_2$ OH CHOR

IIIa **IIIb**

$CH_3C{\equiv}CN(C_2H_5)_2$ + **IV** + $^-$CHCH=CHCH$_3$ (OR)

enone. The cyclobutane ring is then opened in a process which corresponds to retrosynthetic step **IIa** ⇒ **IIIa** in Scheme 13.10. The stereoselectivity of this step results from preferential protonation of the enamine from the less hindered side of the bicyclic intermediate. The cyclobutane ring is then cleaved by hydrolysis of the enamine and resulting β-diketone.

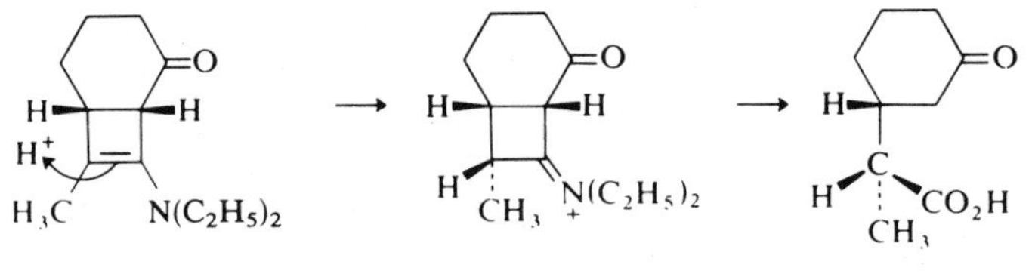

$CH_3C{\equiv}CN(C_2H_5)_2$ + (cyclohexenone) —A→ (H_3C, $N(C_2H_5)_2$ cyclobutene-fused cyclohexanone) —B (H^+, H_2O)→ (H_3C, HO_2C, H, H, O)

C: 1) H_2, Pt; 2) $H_2C{=}CHOEt$ → (CH_3, HO_2C, H, H, OEE)

D: 1) $LiAlH_4$; 2) CBr_4, PPh_3 → (CH_3, $BrCH_2$, H, H, OEE)

E: $CH_3CHOC_2H_5$–O–$(CH_3)_2CHCH_2\overset{-}{C}CN$ → EEO, $(CH_3)_2CHCH_2C(CN)CH_2$, CH_3, H, H, OEE

F: H^+, H_2O → $(CH_3)_2CHCH_2C(=O)CH_2$, CH_3, H, H, OH

G: 1) $HOCH_2CH_2OH$, H^+; 2) Cr(VI) → $(CH_3)_2CHCH_2$, CH_2, CH_3, H, H, O (ketal, O, O)

H: $(CH_3O)_2C{=}O$, NaH → CO_2CH_3, $(CH_3)_2CHCH_2$, CH_2, CH_3, H, H, O (ketal)

I: 1) $NaBH_4$; 2) $C_7H_7SO_2Cl$; 3) NaOMe → CO_2CH_3, $(CH_3)_2CHCH_2$, CH_2, CH_3, H, H (ketal)

J: H^+, H_2O → juvabione

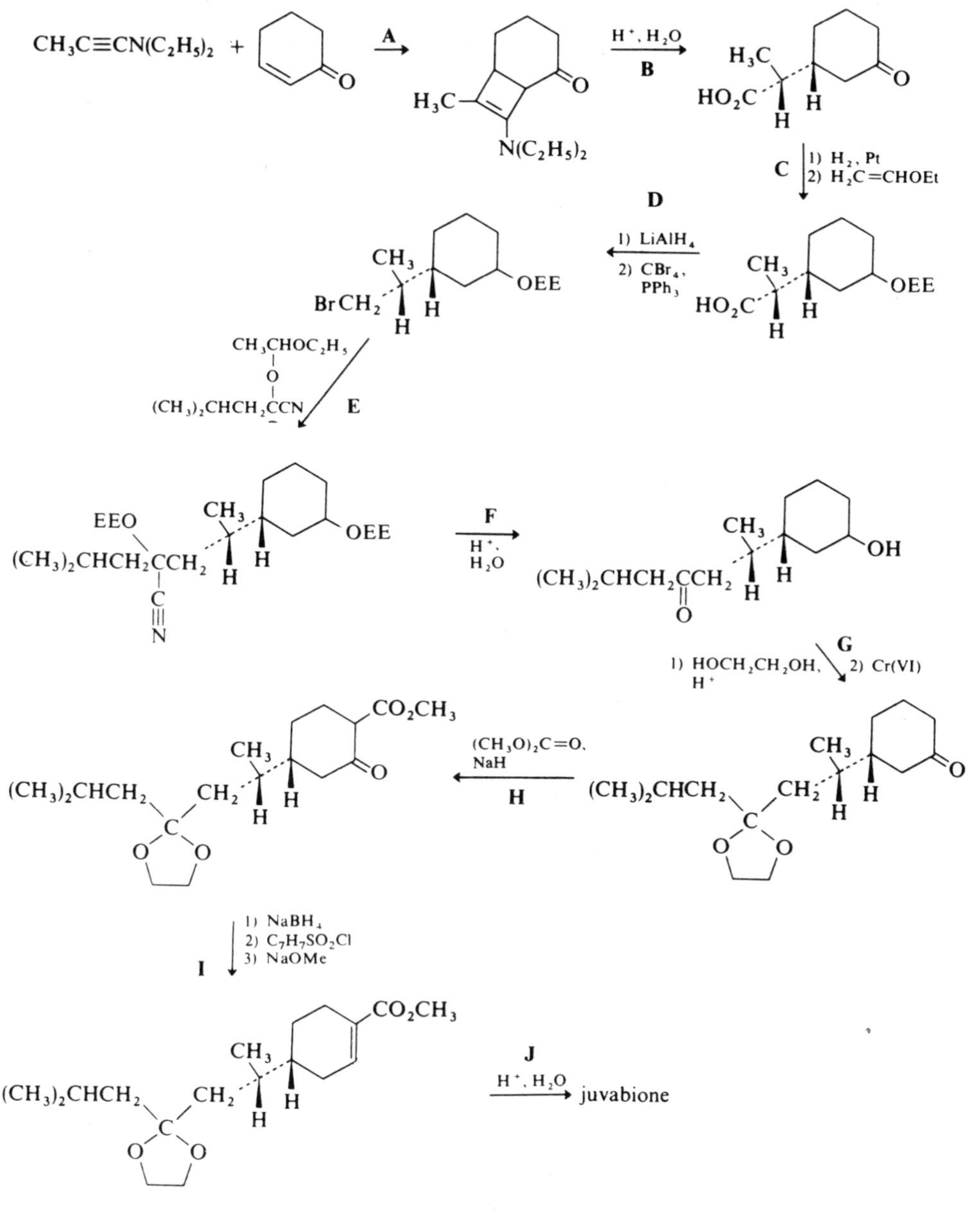

a. J. Ficini, J. D'Angelo, and J. Noiré, *J. Am. Chem. Soc.* **96**, 1213 (1974).

The relative configuration of the chiral centers is unaffected by subsequent reaction steps, so the overall sequence is stereoselective. Another key step in this synthesis is step **E**. This and the following steps, **F** and **G**, correspond to the transformation **Ia** ⇒ **IIa** in the retrosynthesis. A protected cyanohydrin is used as a nucleophilic acyl anion equivalent in this step. Steps **H**, **I**, and **J** of the synthesis in Scheme 13.11

Scheme 13.12. Juvabione Synthesis: D. A. Evans and J. V. Nelson[a]

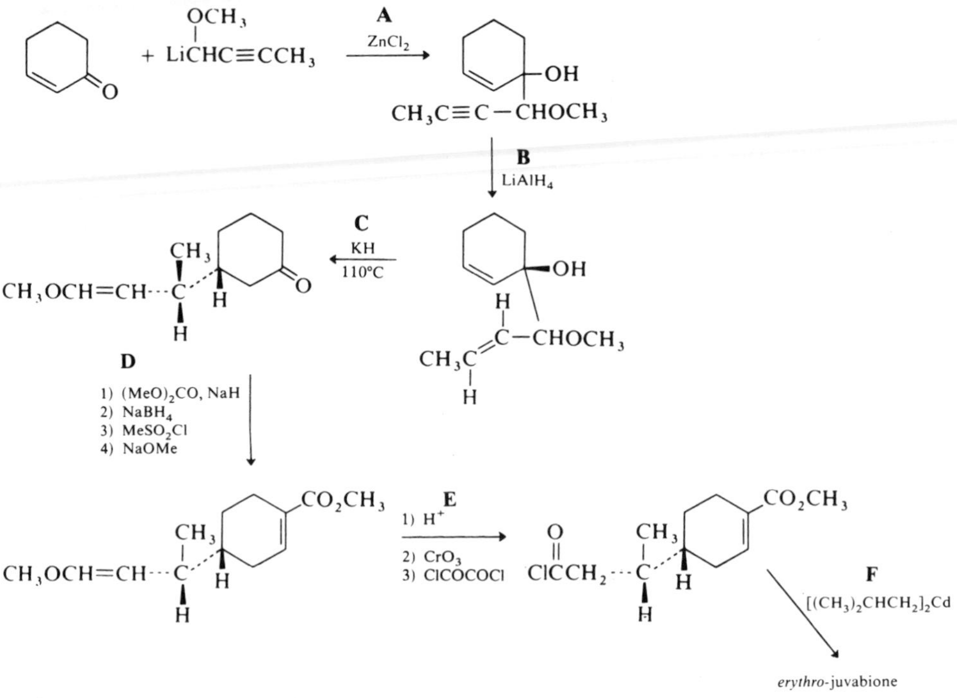

A $ZnCl_2$

B $LiAlH_4$

C KH 110°C

D
1) $(MeO)_2CO$, NaH
2) $NaBH_4$
3) $MeSO_2Cl$
4) NaOMe

E
1) H^+
2) CrO_3
3) ClCOCOCl

F
$[(CH_3)_2CHCH_2]_2Cd$

erythro-juvabione

a. D. A. Evans and J. V. Nelson, *J. Am. Chem. Soc.* **102**, 774 (1980).

employ the C-2 carbonyl group to introduce the carboxy group and the C-1—C-2 double bond.

The stereoselectivity achieved in the synthesis in Scheme 13.12 is the result of a preferred conformation for the base-catalyzed oxy-Cope rearrangement in step **C**. Although the intermediate used in step **C** is a mixture of stereoisomers, both give predominantly the desired relative stereochemistry at C-4 and C-7. The stereoselectivity is based on the preferred chair conformation for the transition state of the concerted rearrangement.

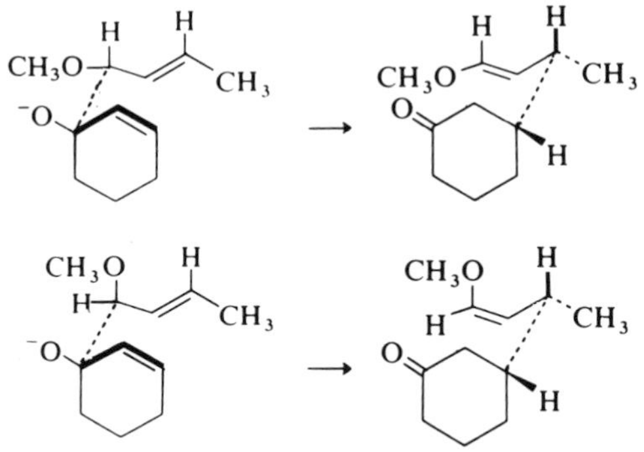

Scheme 13.13. Juvabione Synthesis: A. G. Schultz and J. P. Dittami[a]

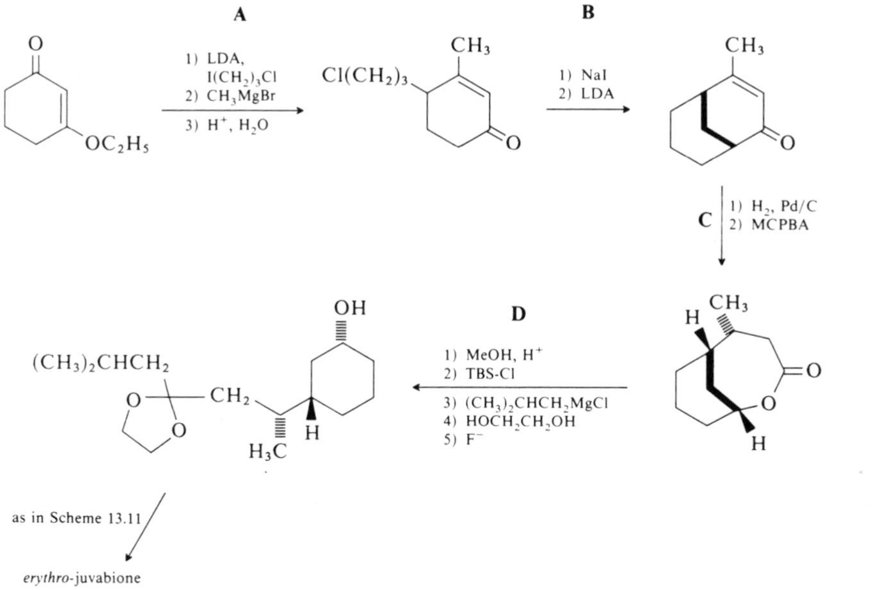

A: 1) LDA, $I(CH_2)_3Cl$ 2) CH_3MgBr 3) H^+, H_2O
B: 1) NaI 2) LDA
C: 1) H_2, Pd/C 2) MCPBA
D: 1) MeOH, H^+ 2) TBS-Cl 3) $(CH_3)_2CHCH_2MgCl$ 4) $HOCH_2CH_2OH$ 5) F^-

as in Scheme 13.11 → *erythro*-juvabione

a. A. G. Schultz and J. P. Dittami, *J. Org. Chem.* **49**, 2615 (1984).

The various [3,3]-sigmatropic reactions related to the Cope and Claisen rearrangements have proved to be very useful for stereoselective formation of carbon–carbon bonds.

The synthesis in Scheme 13.13 leads to the *erythro* stereoisomer. The relative configuration of C-4 and C-7 is established by the hydrogenation in step **C**. The hydrogen is added from the less hindered *exo* face of the bicyclic enone. This synthesis is an example of the use of geometric constraints built in to a ring system to control relative stereochemistry.

The *threo* stereoisomer is the major one obtained by the synthesis in Scheme 13.14. This stereochemistry is established by the conjugate addition in step **A**, where a significant (4:1–6:1) diastereoselectivity is observed. The C-4—C-7 stereochemical relationship is retained through the remainder of the synthesis. The other special features of this synthesis are in steps **B** and **C**. The mercuric acetate-mediated cyclopropane ring opening is facilitated by the alkoxy substituent.[86] The reduction by $NaBH_4$ accomplishes both demercuration and reduction of the aldehyde group.

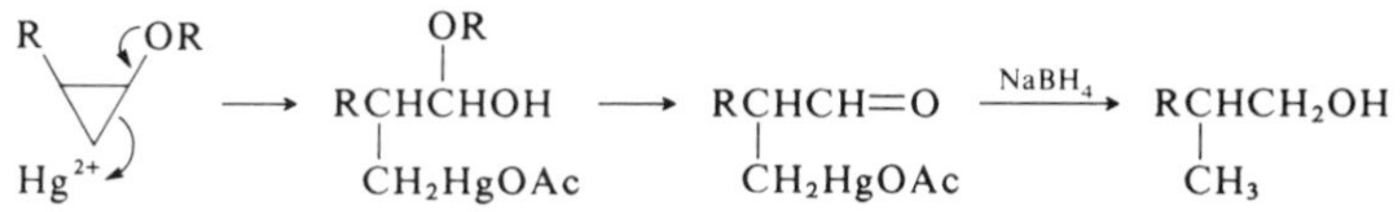

$$\text{R-cyclopropyl-OR} + Hg^{2+} \longrightarrow \underset{CH_2HgOAc}{RCHCH(OR)OH} \longrightarrow \underset{CH_2HgOAc}{RCHCH{=}O} \xrightarrow{NaBH_4} \underset{CH_3}{RCHCH_2OH}$$

86. A. DeBoer and C. H. DePuy, *J. Am. Chem. Soc.* **92**, 4008 (1970).

Scheme 13.14. Juvabione Synthesis: D. J. Morgans, Jr., and G. B. Feigelson[a]

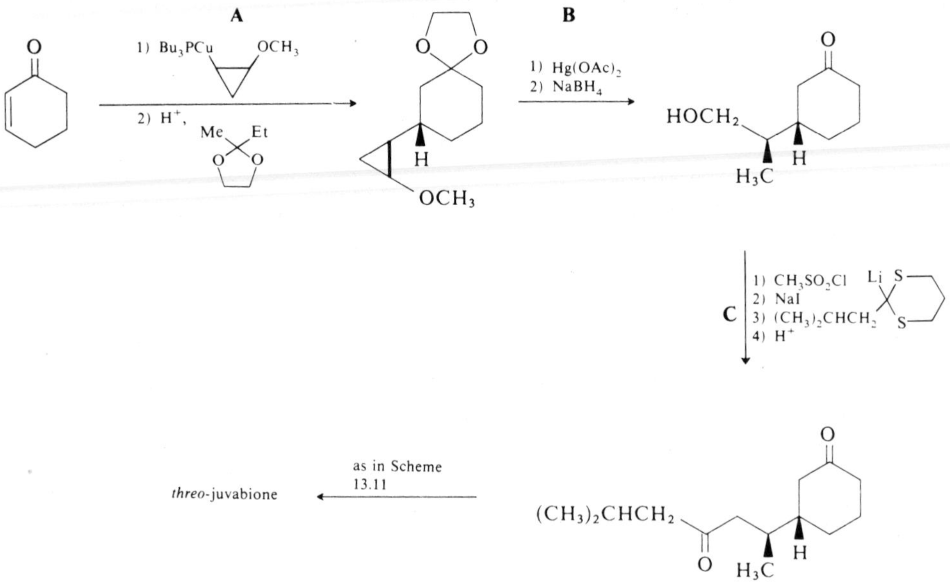

A: 1) Bu_3PCu (methoxycyclopropyl); 2) H^+, Me, Et dioxolane

B: 1) $Hg(OAc)_2$ 2) $NaBH_4$

C: 1) CH_3SO_2Cl 2) NaI 3) $(CH_3)_2CHCH_2$ (2-lithio-1,3-dithiane) 4) H^+

threo-juvabione ← as in Scheme 13.11

a. D. J. Morgans, Jr., and G. B. Feigelson, *J. Am. Chem. Soc.* **105**, 5477 (1983).

In step **C** a dithiane anion is used as a nucleophilic acyl anion equivalent to introduce the C-10—C-13 isobutyl group.

Several other syntheses of juvabione have also been completed.[87]

13.5.2. Longifolene

Lnogifolene is a tricyclic terpene. It is a typical terpene in terms of the structural complexity. Schemes 13.15 through 13.21 describe five separate syntheses of longifolene. We wish to particularly emphasize the methods for carbon-carbon bond formation used in these syntheses.

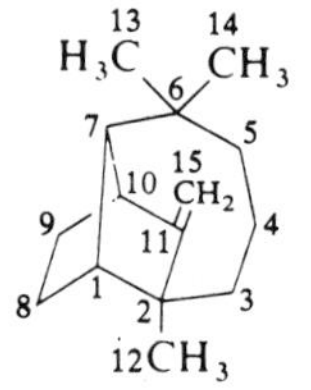

87. A. A. Drabkina and Y. S. Tsizin, *J. Gen. Chem. USSR* (Engl. Transl.), **43**, 422, 691 (1973); R. J. Crawford, U.S. Patent, 3,676,506; *Chem. Abstr.* **77**, 113889e (1972); A. J. Birch, P. L. Macdonald, and V. H. Powell, *J. Chem. Soc., C*, 1469 (1970).

There are four chiral carbons in longifolene, but they are not independent of one another since the geometry of the ring system requires that they have a specific relative relationship. That does not mean stereochemistry can be ignored, however, since the formation of the various rings will fail if the reactants do not have the proper stereochemistry.

The first successful synthesis of longifolene was described in detail by Corey and co-workers in 1964. Scheme 13.15 presents a retrosynthetic analysis corresponding to this route. A key disconnection is made on going from **I** ⇒ **II**. This transformation simplifies the tricyclic skeleton to a bicyclic one. For this disconnection to correspond to a reasonable synthetic step, the functionality in the intermediate to be cyclized must engender mutual reactivity between C-7 and C-10. This is achieved in diketone **II**, since an enolate generated by deprotonation at C-10 can undergo an intramolecular Michael addition to C-7. Retrosynthetic step **II** ⇒ **III** is attractive, since it suggests a decalin derivative as a key intermediate. Methods for preparing this type of structure have been well developed, since such structures are useful intermediates in the synthesis of other terpenes and also steroids. Can a chemical reaction be recognized which would permit **III** ⇒ **II** to proceed in the synthetic sense? The hydroxyl → carbonyl transformation with migration would correspond to the pinacol rearrangement (Section 10.1.2). The retrosynthetic transformation **II** ⇒ **III** corresponds to a workable synthetic step if the group X in **III** is a leaving group which could promote the rearrangement. The other transformations in the retrosynthetic plan, **III** ⇒ **IV** ⇒ **V**, are straightforward in concept and lead to identification of **V** as a potential starting material.

Scheme 13.15. Retrosynthesis of Longifolene Corresponding to the Synthesis in Scheme 13.16

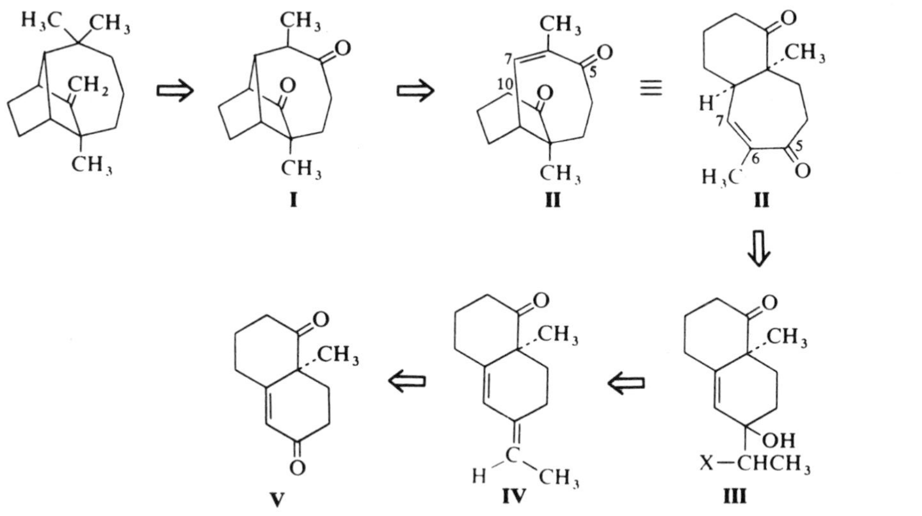

The synthesis was carried out as is shown in Scheme 13.16. The key intramolecular Michael addition was accomplished using triethylamine under high-temperature conditions.

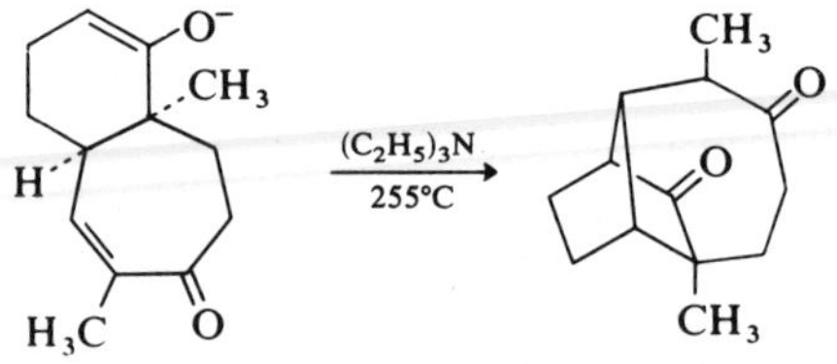
O⁻
CH₃
H
H₃C
O
(C₂H₅)₃N
255°C
CH₃
O
O
CH₃

The cyclization requires that the intermediate have a *cis* ring fusion. The stereochemistry of the ring junction is established when the double bond is moved into conjugation in step **D**. This product was not stereochemically characterized, and need not be, because the stereochemically important site at C-1 can be epimerized under the basic cyclization conditions. Thus, the equilibration of the ring junction through a dienol can allow the cyclization to proceed to completion from either stereoisomer. Step **C** is the pinacol rearrangement corresponding to **II** ⇒ **III** in the retrosynthesis. A diol is formed and selectively tosylated at the secondary hydroxyl group (step **B**). Base then promotes the skeletal rearrangement in step **C**. The remaining transformations effect the addition of the remaining methyl and methylene

Scheme 13.16. Longifolene Synthesis: E. J. Corey, R. B. Mitra, and P. A. Vatakencherry[a]

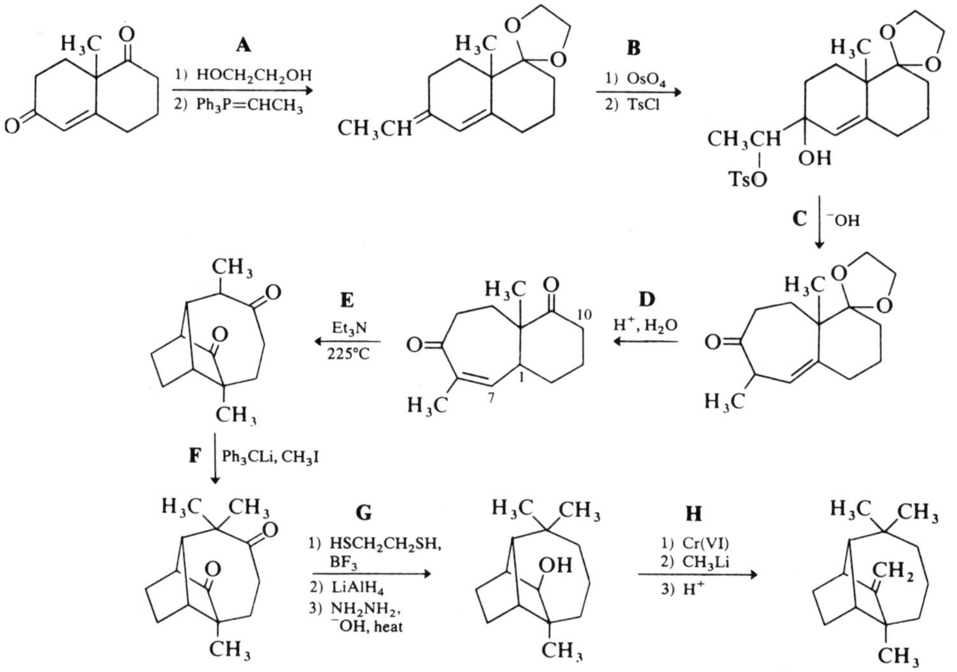
A
1) HOCH₂CH₂OH
2) Ph₃P=CHCH₃
B
1) OsO₄
2) TsCl
CH₃CH
TsO
OH
C
⁻OH
D
H⁺, H₂O
E
Et₃N
225°C
10
1
7
F
Ph₃CLi, CH₃I
G
1) HSCH₂CH₂SH, BF₃
2) LiAlH₄
3) NH₂NH₂, ⁻OH, heat
H
1) Cr(VI)
2) CH₃Li
3) H⁺
CH₂

a. E. J. Corey, R. B. Mitra, and P. A. Vatakencherry, *J. Am. Chem. Soc.* **86**, 478 (1964).

groups by well-known methods. Step **G** accomplishes a selective reduction of one of the two carbonyl groups to a methylene by taking advantage of the difference in the steric environment of the two carbonyls. Selective protection of the less hindered C-5 carbonyl was done using a thioketal. The C-11 carbonyl was then reduced to give the alcohol, and finally C-5 was reduced to a methylene group under Wolff–Kishner conditions. The hydroxyl group at C-11 provides the reactive center necessary to introduce the C-15 methylene group in step **H**.

The key bond closure in Scheme 13.17 is somewhat similar to that used in Scheme 13.16 but is performed on a bicyclo[4.4.0]decane ring system. The ring

Scheme 13.17. Longifolene Synthesis: J. E. McMurry and S. J. Isser[a]

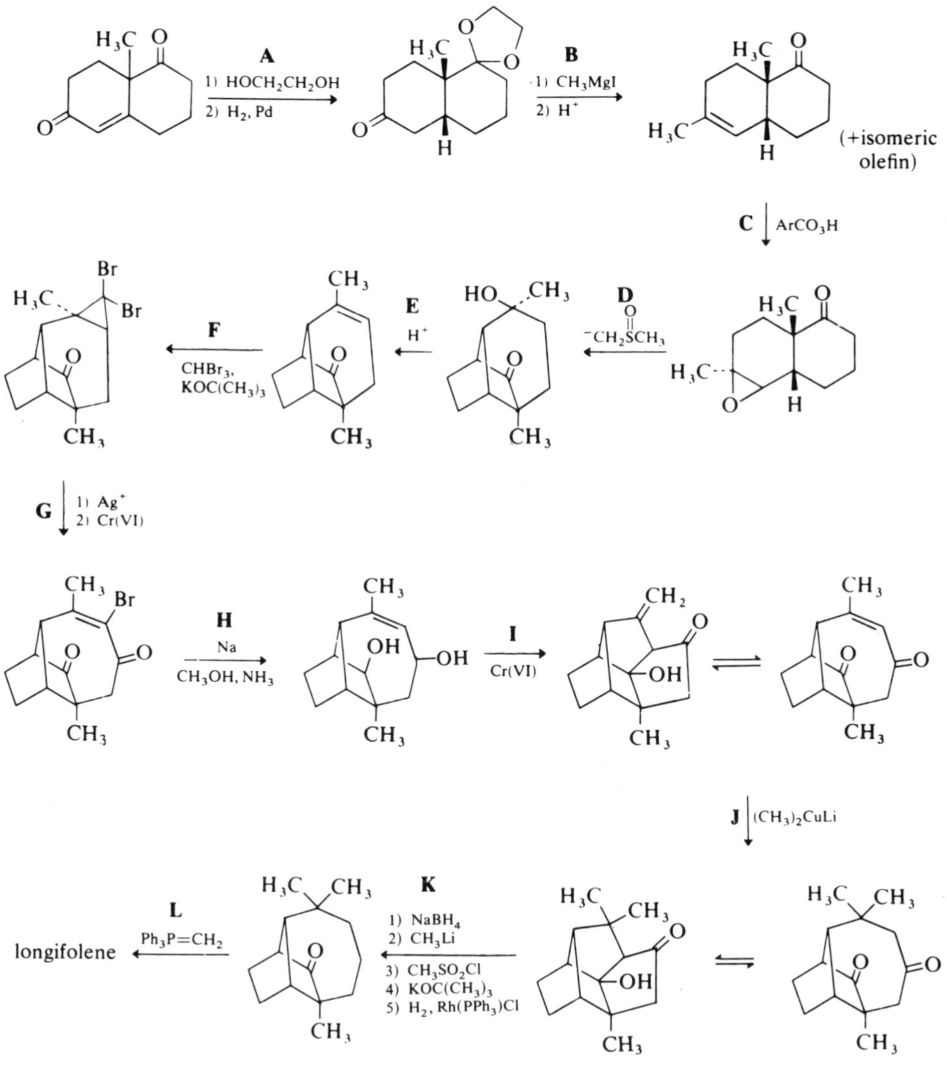

A: 1) $HOCH_2CH_2OH$; 2) H_2, Pd
B: 1) CH_3MgI; 2) H^+ (+isomeric olefin)
C: $ArCO_3H$
D: $^-CH_2S(O)CH_3$
E: H^+
F: $CHBr_3$, $KOC(CH_3)_3$
G: 1) Ag^+; 2) Cr(VI)
H: Na, CH_3OH, NH_3
I: Cr(VI)
J: $(CH_3)_2CuLi$
K: 1) $NaBH_4$; 2) CH_3Li; 3) CH_3SO_2Cl; 4) $KOC(CH_3)_3$; 5) H_2, $Rh(PPh_3)Cl$
L: $Ph_3P{=}CH_2$ → longifolene

a. J. E. McMurry and S. J. Isser, *J. Am. Chem. Soc.* **94**, 7132 (1972).

juncture must be *cis* to permit the intramolecular epoxide ring opening. The required *cis* ring fusion is established during the catalytic hydrogenation in step **A**.

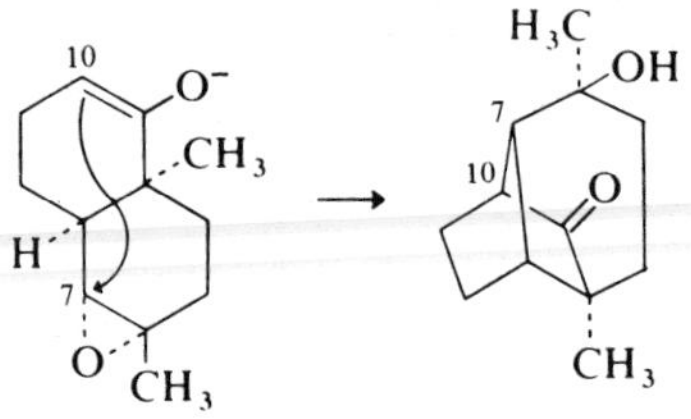

The cyclization is followed by a sequence of steps **F–H**, which effect a ring expansion via a carbene addition and cyclopropyl halide solvolysis. The products of steps **I** and **J** are interesting in that the tricyclic structures are largely converted to tetracyclic derivatives by intramolecular aldol reactions. The extraneous bond is broken in step **K**. First, a diol is formed by $NaBH_4$ reduction, and this is converted to a monomesylate. The resulting β-hydroxy mesylate is capable of a concerted fragmentation, which occurs on treatment with potassium *t*-butoxide.

A retrosynthetic analysis corresponding to the synthesis in Scheme 13.19 is given in Scheme 13.18. The striking feature of this synthesis is the structural simplicity of the key intermediate **IV**. A synthesis according to this scheme would generate the tricyclic skeleton in a single step from a monocyclic intermediate. The disconnection **III** ⇒ **IV** corresponds to a cationic cyclization of a highly symmetric cation, **IVa**.

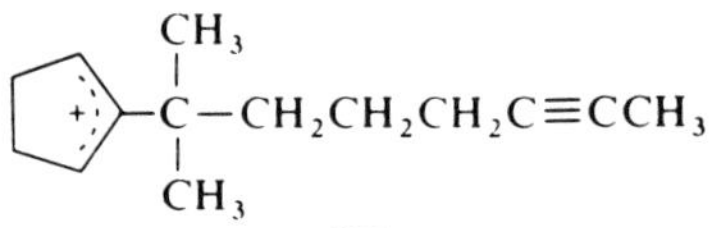

IVa

Scheme 13.18. Retrosynthetic Analysis Corresponding to Synthesis in Scheme 13.19.

I **II** **III** **IV**

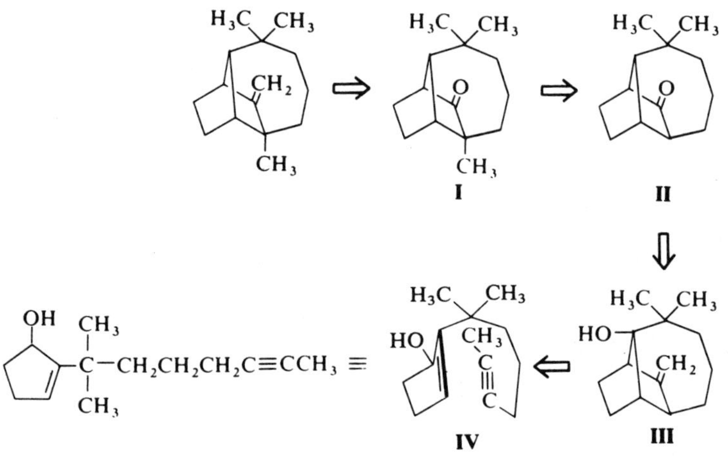

No issues of stereochemistry would arise until the carbon skeleton had been formed, at which point all of the chiral centers would be in the proper relationship. The structures of the successive intermediates, assuming a stepwise mechanism for the cationic cyclization, are shown below.

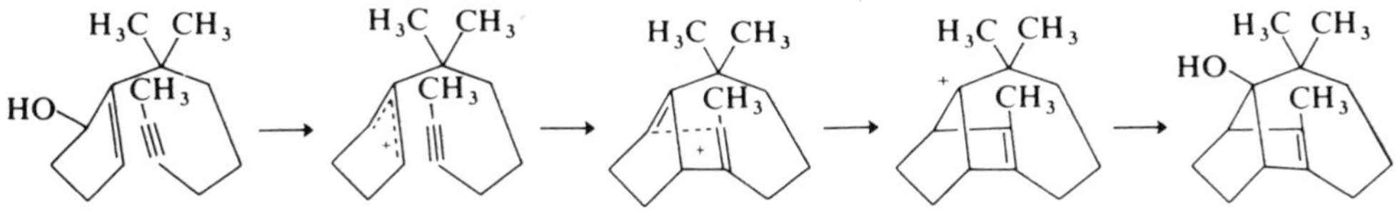

Evidently, these or closely related intermediates are accessible and reactive, since the synthesis was successfully achieved as outlined in Scheme 13.19. In addition to the key cationic cyclization in step **D**, interesting transformations are carried out in step **E**, where a bridgehead tertiary alcohol is reductively removed, and in step **F**, where a methylene group, which is eventually reintroduced, must be removed. The endocyclic double bond, which is strained because of its bridgehead location, is isomerized to the exocyclic position and then cleaved with RuO_4/IO_4^-. The enolate of the ketone is then used to introduce the C-12 methyl group.

Scheme 13.19. Longifolene Synthesis: R. A. Volkmann, G. C. Andrews, and W. S. Johnson[a]

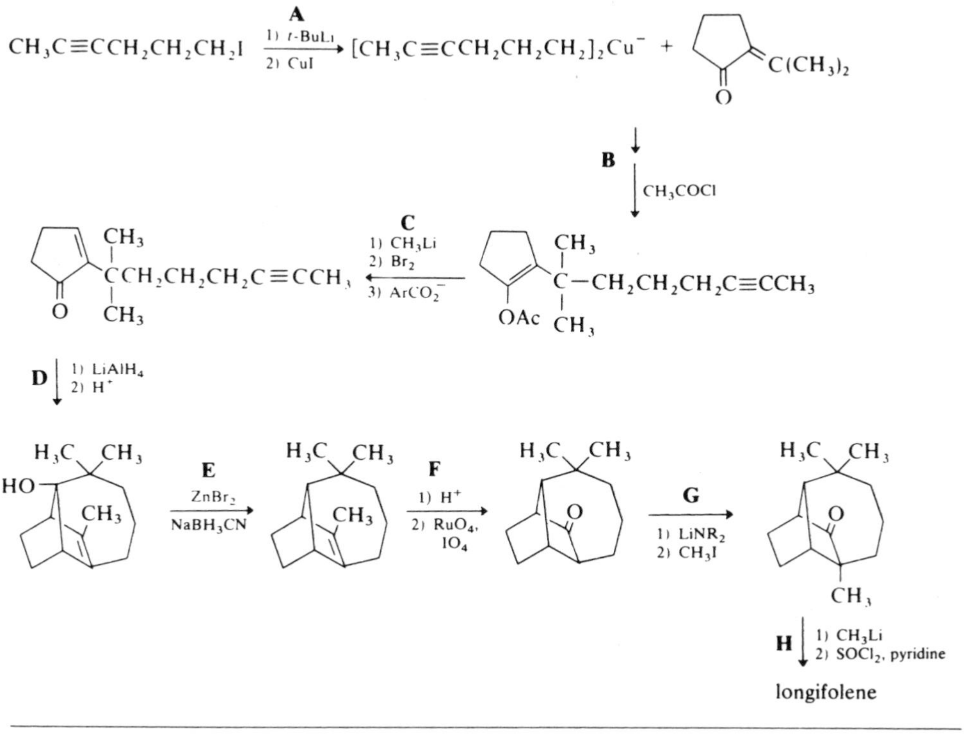

$CH_3C{\equiv}CCH_2CH_2CH_2I$ —**A**: 1) *t*-BuLi, 2) CuI→ $[CH_3C{\equiv}CCH_2CH_2CH_2]_2Cu^-$ + (2-isopropylidenecyclopentanone, $=C(CH_3)_2$)

B: CH_3COCl

C: 1) CH_3Li, 2) Br_2, 3) $ArCO_3^-$

D: 1) $LiAlH_4$, 2) H^+

E: $ZnBr_2$, $NaBH_3CN$

F: 1) H^+, 2) RuO_4, IO_4^-

G: 1) $LiNR_2$, 2) CH_3I

H: 1) CH_3Li, 2) $SOCl_2$, pyridine

longifolene

a. R. A. Volkmann, G. C. Andrews, and W. S. Johnson, *J. Am. Chem. Soc.* **97**, 4777 (1975).

The synthesis in Scheme 13.20 also uses a remarkably simple starting material to achieve the construction of the tricyclic skeleton. A partial retrosynthesis is outlined below:

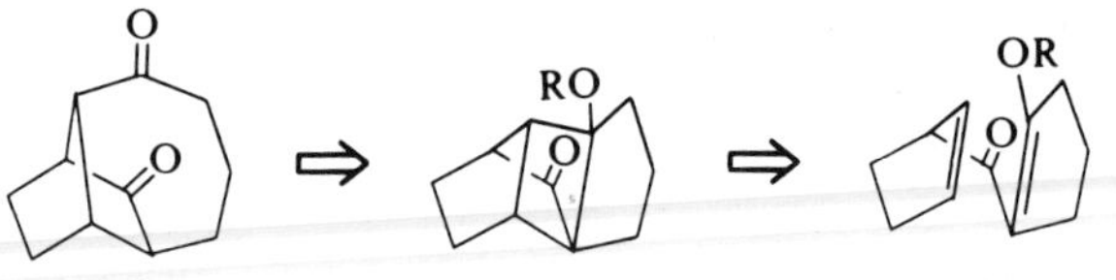

I ⇒ II ⇒ III

Intermediate **I** contains the tricyclic skeleton of longifolene, shorn of its substituent groups, but containing carbonyl groups suitably placed so that the methyl groups at C-2 and C-6 and the C-15 methylene can eventually be introduced. The retrosynthetic step **I** ⇒ **II** corresponds to an intramolecular aldol condensation. However, **II** clearly is strained relative to **I**, so **II** (with OR = OH) should open to **I**.

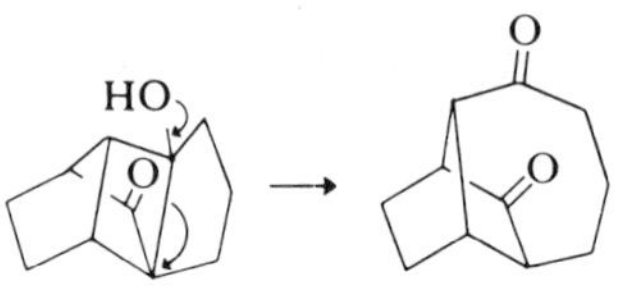

How might **II** be obtained? The four-membered ring suggests that a [2 + 2] (photochemical) cycloaddition might be useful, and this, in fact, was successful (step **B** in Scheme 13.20). After liberation of the hydroxyl group by hydrogenolysis in step

Scheme 13.20. Longifolene Synthesis: W. Oppolzer and T. Godel

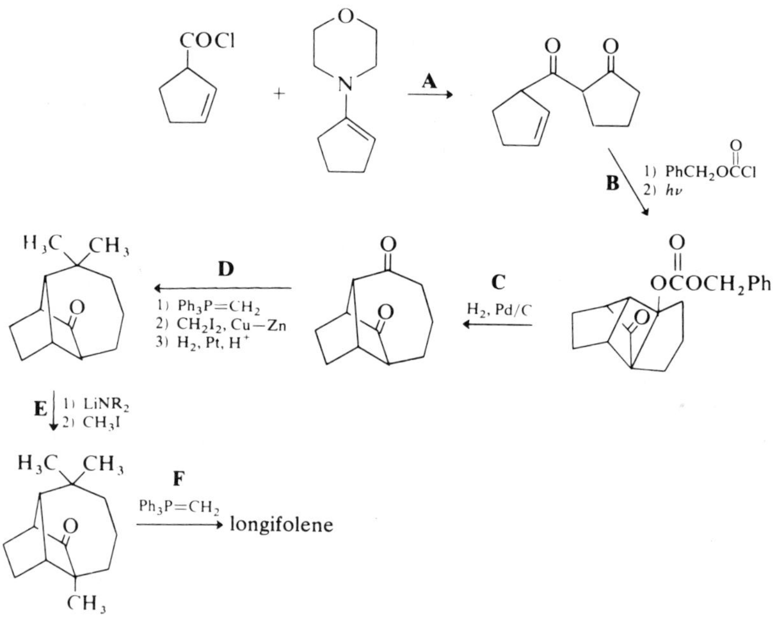

A: COCl-cyclopentene + morpholine enamine of cyclopentanone →

B: 1) $PhCH_2OCCl$ (chloroformate) 2) $h\nu$ → $OCOCH_2Ph$ adduct

C: H_2, Pd/C

D: 1) $Ph_3P{=}CH_2$ 2) CH_2I_2, Cu–Zn 3) H_2, Pt, H^+ → (H_3C, CH_3)

E: 1) $LiNR_2$ 2) CH_3I → (H_3C, CH_3, CH_3)

F: $Ph_3P{=}CH_2$ → longifolene

a. W. Oppolzer and T. Godel, *J. Am. Chem. Soc.* **100**, 2583 (1978).

C, the extra carbon–carbon bond between C-2 and C-6 was broken by a spontaneous retro-aldol reaction. Step **D** in this synthesis is an interesting way of introducing the geminal dimethyl groups. It proceeds through a cyclopropane intermediate, which is cleaved by hydrogenolysis.

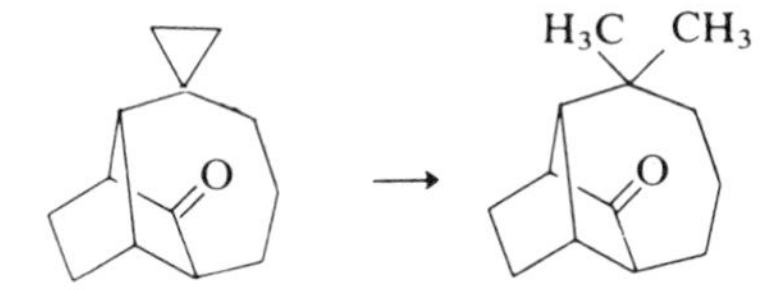

H$_3$C CH$_3$

O → O

The synthesis of longifolene in Scheme 13.21 commences with a Birch reduction and tandem alkylation of methyl 2-methoxybenzoate (see Section 5.5.1). Step **C** is an intramolecular cycloaddition of a diazoalkane, which is generated from an aziridinohydrazine intermediate:

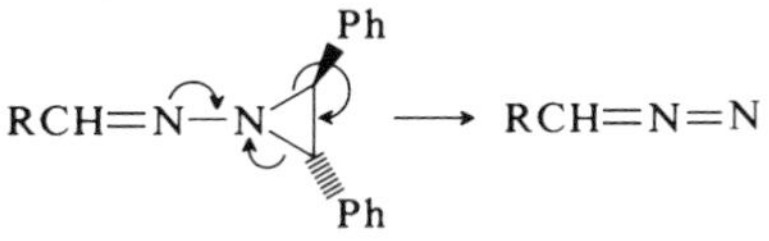

Ph

$$RCH{=}N{-}N \longrightarrow RCH{=}N{=}N$$

Ph

Scheme 13.21. Longifolene Synthesis: A. G. Schultz and S. Puig[a]

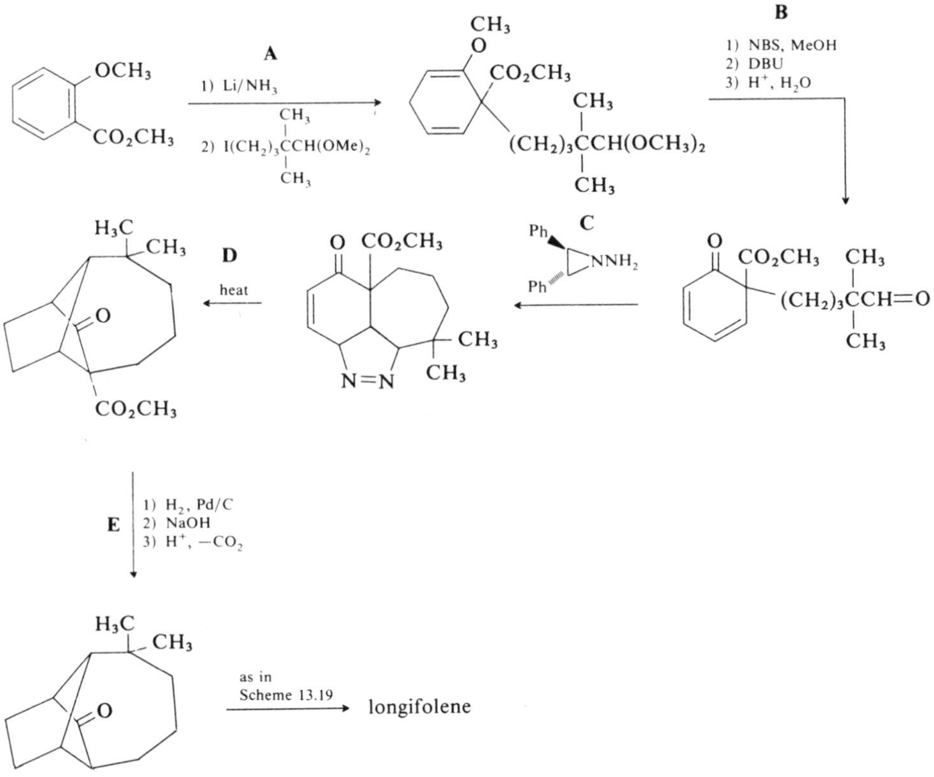

OCH$_3$ CO$_2$CH$_3$

A 1) Li/NH$_3$ 2) I(CH$_2$)$_3$C(CH$_3$)$_2$CH(OMe)$_2$ →

CH$_3$ O CO$_2$CH$_3$ (CH$_2$)$_3$C(CH$_3$)$_2$CH(OCH$_3$)$_2$

B 1) NBS, MeOH 2) DBU 3) H$^+$, H$_2$O →

O CO$_2$CH$_3$ (CH$_2$)$_3$C(CH$_3$)$_2$CH=O

C Ph, Ph, NNH$_2$ →

O CO$_2$CH$_3$ CH$_3$ CH$_3$ N=N

D heat →

H$_3$C CH$_3$ O CO$_2$CH$_3$

E 1) H$_2$, Pd/C 2) NaOH 3) H$^+$, $-$CO$_2$ →

H$_3$C CH$_3$ O

as in Scheme 13.19 → longifolene

a. A. G. Schultz and S. Puig, *J. Org. Chem.* **50**, 915 (1985).

The thermolysis in step **D** generates a diradical (or the corresponding dipolar intermediate), which then closes to generate the desired carbon skeleton.

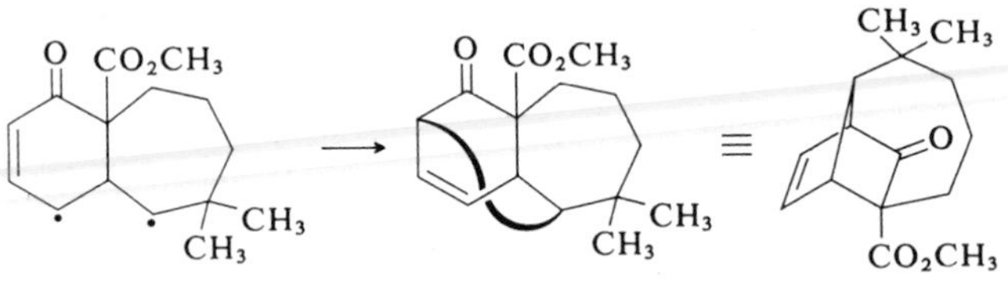

The cyclization product is then converted to an intermediate that was used in the longifolene synthesis described in Scheme 13.19.

The synthesis in Scheme 13.21 has also been done in such a way as to give enantiomerically pure longifolene. A starting material, whose chirality is derived from the amino acid L-proline, was enantioselectively converted to the product of step **A**.

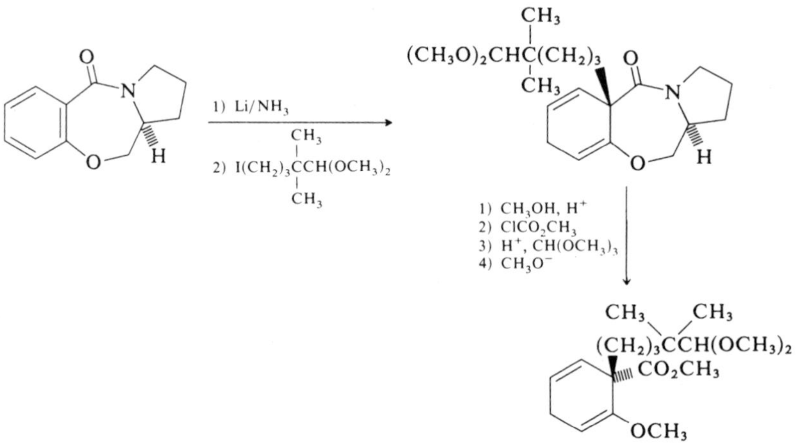

1) Li/NH$_3$
2) $I(CH_2)_3C(CH_3)_2CH(OCH_3)_2$

1) CH_3OH, H^+
2) $ClCO_2CH_3$
3) H^+, $CH(OCH_3)_3$
4) CH_3O^-

This chiral intermediate, when carried through the reaction sequence in Scheme 13.21, generates the enantiomer of natural longifolene. Therefore, D-proline would have to be used to generate the natural enantiomer.

These syntheses of longifolene provide good examples of the approaches that are available for construction of ring compounds of this type. On each case, a set of functionalities that have the potential for *intramolecular* reaction was assembled. After assembly of the carbon framework, the final functionality changes were effected. *It is the necessity for the formation of the carbon skeleton that determines the functionalities that are present at the ring-closure stage.*

13.5.3. Prelog–Djerassi Lactone

The Prelog–Djerassi lactone (abbreviated as P–D-lactone) was originally isolated as a degradation product during structural investigation of antibiotics. Its open-chain precursor, structure **1**, is typical of methyl-branched carbon chains which occur frequently in macrolide and polyether antibiotics.

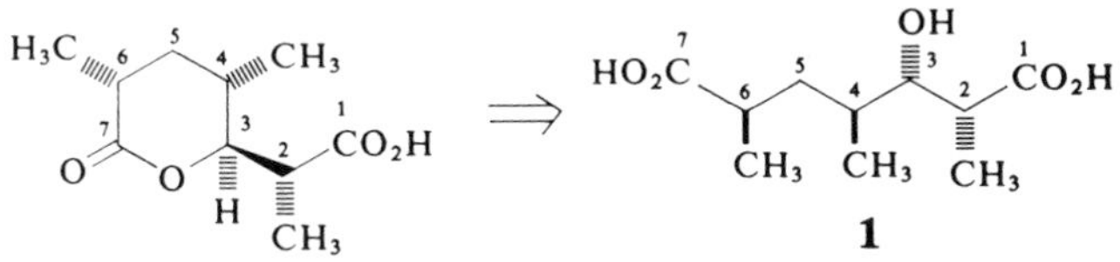

There have been about 20 different syntheses of P–D-lactone.[88] We will focus here on some of those that provide enantiomerically pure product, since they illustrate several of the methods for enantioselective synthesis.[89]

The synthesis in Scheme 13.22 is based on a starting material which can be prepared in enantiomerically pure form. A resolution is carried out during the preparation of the norbornenone derivative. In the synthesis in Scheme 13.22, C-7 of the nornornenone starting material becomes C-4 of P–D-lactone. The configuration of the C-3 hydroxyl and the methyl groups at C-2 and C-6 must then be established relative to the C-4 stereochemistry. The alkylation in step **A** establishes the configuration at C-2. The basis for the stereoselectivity is the preference for *exo* versus *endo* approach in the alkylation. The Baeyer–Villiger oxidation in step **B** is followed by a Lewis acid mediated allylic rearrangement. This rearrangement is suprafacial, This stereochemistry is evidently dictated by the preference for maintaining a *cis* ring juncture at the five-membered ring.

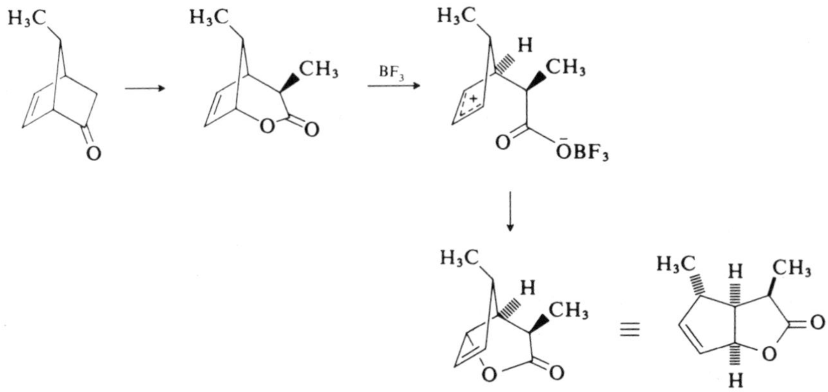

The stereochemistry of the C-3 hydroxyl is established in step **E**. The Baeyer–Villiger oxidation proceeds with retention of configuration of the migrating group (see

88. For references to these syntheses, see S. F. Martin and D. G. Guinn, *J. Org. Chem.* **52**, 5588 (1987); H. F. Chow and I. Fleming, *Tetrahedron Lett.* **26**, 397 (1985).
89. For other syntheses of enantiomerically pure Prelog–Djerassi lactone, see F. E. Ziegler, A. Kneisley, J. K. Thottathil, and R. T. Wester, *J. Am. Chem. Soc.* **110**, 5434 (1988); A. Nakano, S. Takimoto, J. Inanaga, T. Katsuki, S. Ouchida, K. Inoue, M. Aiga, N. Okukado, and M. Yamaguchi, *Chem. Lett.*, 1019 (1979); K. Suzuki, K. Tomooko, T. Matsumoto, E. Katayama, and G. Tsuchihashi, *Tetrahedron Lett.* **26**, 3711 (1985); M. Isobo, Y. Ichikawa, and T. Goto, *Tetrahedron Lett.* **22**, 4287 (1981); M. Mori, T. Chuman, and K. Kato, *Carbohydr. Res.* **129**, 73 (1984).

Scheme 13.22. Prelog–Djerassi Lactone Synthesis: P. A. Grieco, Y. Ohfune, Y. Yokoyama, and W. Owens[a]

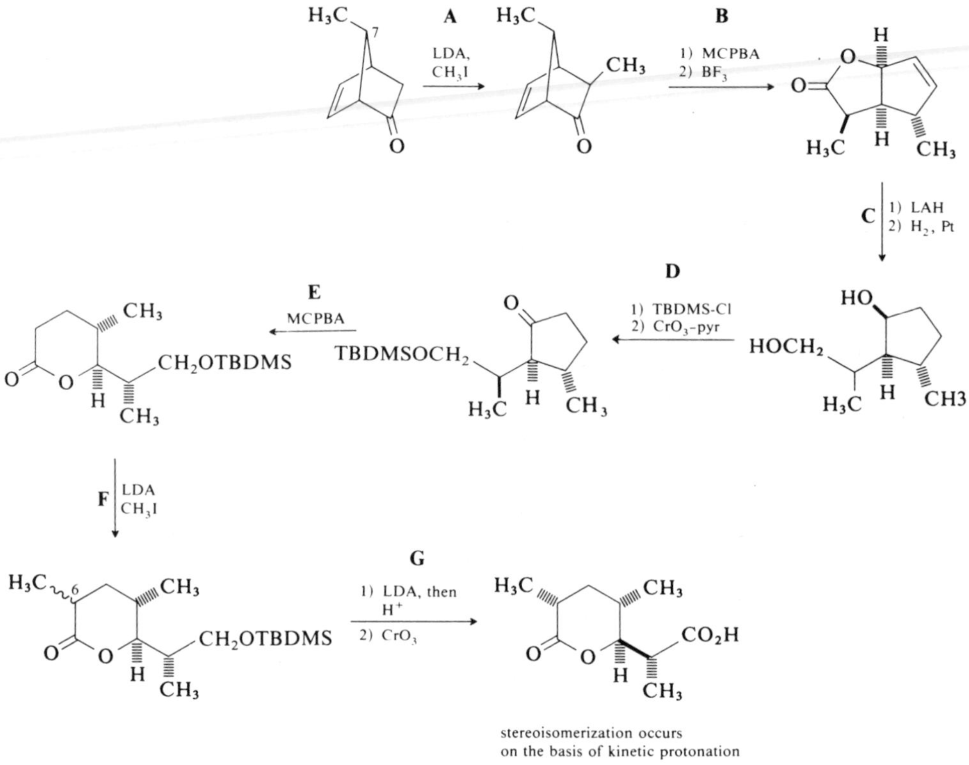

A: LDA, CH_3I

B: 1) MCPBA 2) BF_3

C: 1) LAH 2) H_2, Pt

D: 1) TBDMS-Cl 2) CrO_3-pyr

E: MCPBA

F: LDA, CH_3I

G: 1) LDA, then H^+ 2) CrO_3

stereoisomerization occurs on the basis of kinetic protonation

a. P. A. Grieco, Y. Ohfune, Y. Yokoyama, and W. Owens, *J. Am. Chem. Soc.* **101**, 4749 (1979).

Section 12.5.2), so the correct stereochemistry is established for the C—O bond. The final center for which configuration must be established is C-6. The methyl group is introduced at C-6 by an enolate alkylation which is not highly stereoselective. However, since this center is adjacent to the lactone carbonyl, it can be epimerized through the enolate. The enolate is formed and quenched with acid. The kinetically preferred protonation from the axial direction provides the correct stereochemistry at C-6.

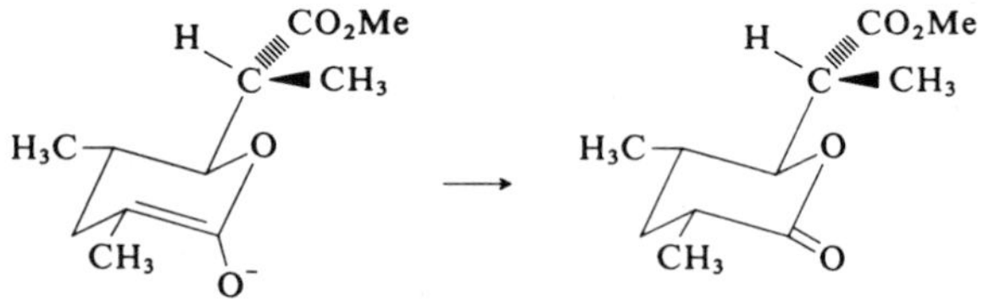

Another synthesis of P-D-lactone based on a resolved starting material is shown in Scheme 13.23. The chiral carbon in the starting material is destined to become C-4 in the final product. Steps **A** and **B** serve to extend the chain to provide a

Scheme 13.23. Prelog–Djerassi Lactone Synthesis: W. C. Still and K. R. Shaw[a]

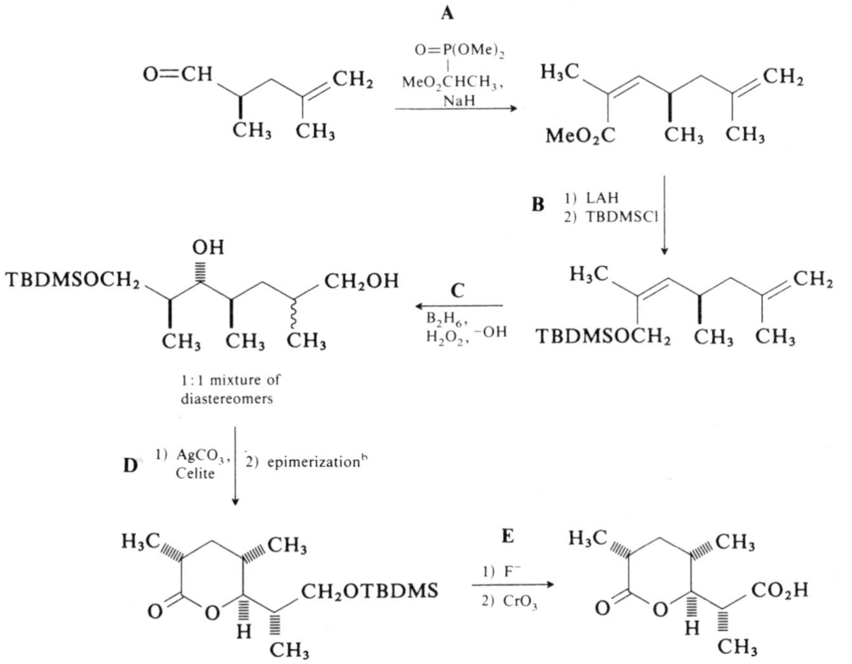
A
O=CH
CH₂
CH₃ CH₃
O=P(OMe)₂
MeO₂CHCH₃,
NaH
H₃C
CH₂
MeO₂C CH₃ CH₃
B 1) LAH
2) TBDMSCl
OH
TBDMSOCH₂
CH₂OH
CH₃ CH₃ CH₃
C
B₂H₆,
H₂O₂, ⁻OH
H₃C
CH₂
TBDMSOCH₂ CH₃ CH₃
1:1 mixture of
diastereomers
D 1) AgCO₃, Celite 2) epimerization[b]
H₃C
CH₃
CH₂OTBDMS
O O H CH₃
E
1) F⁻
2) CrO₃
H₃C
CH₃
CO₂H
O O H CH₃

a. W. C. Still and K. R. Shaw, *Tetrahedron Lett.* **22**, 3725 (1981).
b. Epimerization carried out as in Scheme 13.22.

seven-carbon diene. The configuration of two of the three remaining chiral centers is controlled by the hydroboration step. Hydroboration is a stereospecific *syn* addition (Section 4.9.1). In 1,5-dienes of this type, an intramolecular hydroboration occurs and establishes the configurations of the two newly formed C—B and C—H bonds.

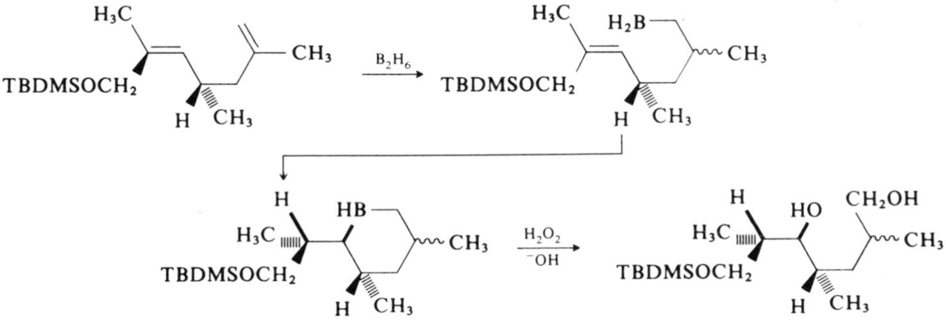
H₃C
CH₃
TBDMSOCH₂
H CH₃
B₂H₆
H₃C
H₂B
CH₃
TBDMSOCH₂
H CH₃
H
HB
H₃C
CH₃
TBDMSOCH₂
H CH₃
H₂O₂
⁻OH
H
HO
CH₂OH
H₃C
CH₃
TBDMSOCH₂
H CH₃

There is, however, no significant selectivity in the initial hydroboration of the terminal double bond. As a result, both configurations are formed at C-6. This problem was overcome using the epimerization process from Scheme 13.22.

The syntheses in Schemes 13.24 and 13.25 are conceptually related. The starting material is prepared by reduction of the half-ester of *meso*-2,4-dimethylglutaric acid. The use of the *meso*-diacid ensures the correct relative configuration of the C-4 and C-6 methyl substituents. The half-acid is resolved, and the correct enantiomer is reduced to the aldehyde.

The synthesis in Scheme 13.24 uses stereoselective aldol condensation methodology. Both the lithium enolate and the boron enolate methods were employed. The enol derivatives were used in enantiomerically pure form so the condensations are examples of *double stereodifferentiation* (Section 2.1.3). The stereoselectivity observed in the reaction is that predicted by a cyclic transition state for the aldol condensations.

The synthesis in Scheme 13.25 also relies on *meso*-2,4-dimethylglutaric acid as the starting material. Both the resolved aldehyde employed in Scheme 13.24 and a resolved half-amide were successfully used as intermediates. The configuration at C-2 and C-3 was controlled by addition of a butenylborane to an aldehyde (see Section 9.1.2). The boronate ester was used in enantiomerically pure form so that stereoselectivity was enhanced by double stereodifferentiation. The allylic additions carried out by the butenyl boronates do not appear to have been quite as highly stereoselective as the aldol condensations used in Scheme 13.24, since a minor diastereomer was formed in the boronate addition reactions.

Scheme 13.24. Prelog–Djerassi Lactone Synthesis: S. Masamune, M. Hirama, S. Mori, S. A. Ali, and D. S. Garvey; S. Masamune, S. A. Ali, D. L. Snitman, and D. S. Garvey[a]

MeO₂C CH=O CH₃ CH₃ — A: H₃C OLi OTMS CH₃ C₆H₁₁ → MeO₂C OH O OTMS CH₃ CH₃ CH₃ CH₃

B: 1) H⁺ 2) Zn(BH₄)₂ → H₃C CH₃ OH OTMS O O H CH₃ CH₃

C: oxid. → H₃C CH₃ O O H CH₃ CO₂H

A′: H₃C OB OTBDMS H C₆H₁₁ → MeO₂C OH O OTBDMS H CH₃ CH₃ CH₃

B′: 1) HF 2) IO_4^-

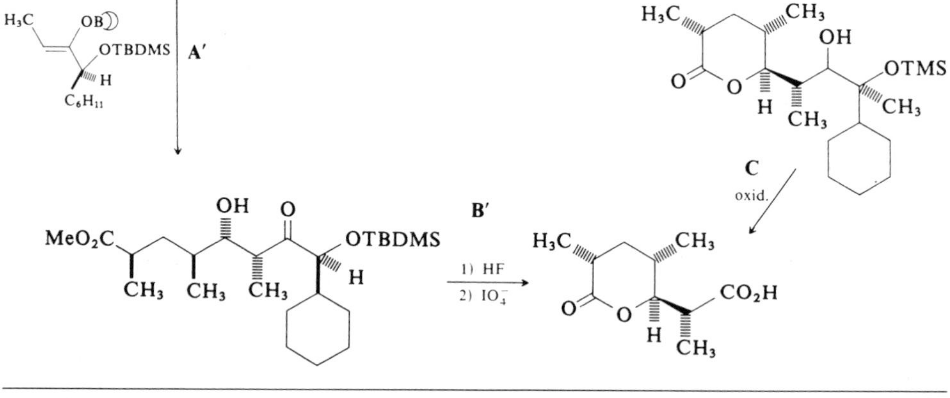

a. S. Masamune, S. A. Ali, D. L. Snitman, and D. S. Garvey, *Angew. Chem. Int. Ed. Engl.* **19**, 557 (1980); S. Masamune, M. Hirama, S. Mori, S. A. Ali, and D. S. Garvey, *J. Am. Chem. Soc.* **103**, 1568 (1981).

Scheme 13.25. Prelog–Djerassi Lactone Synthesis: R. W. Hoffmann, H.-J. Zeiss, W. Ladner, and S. Taber[a]

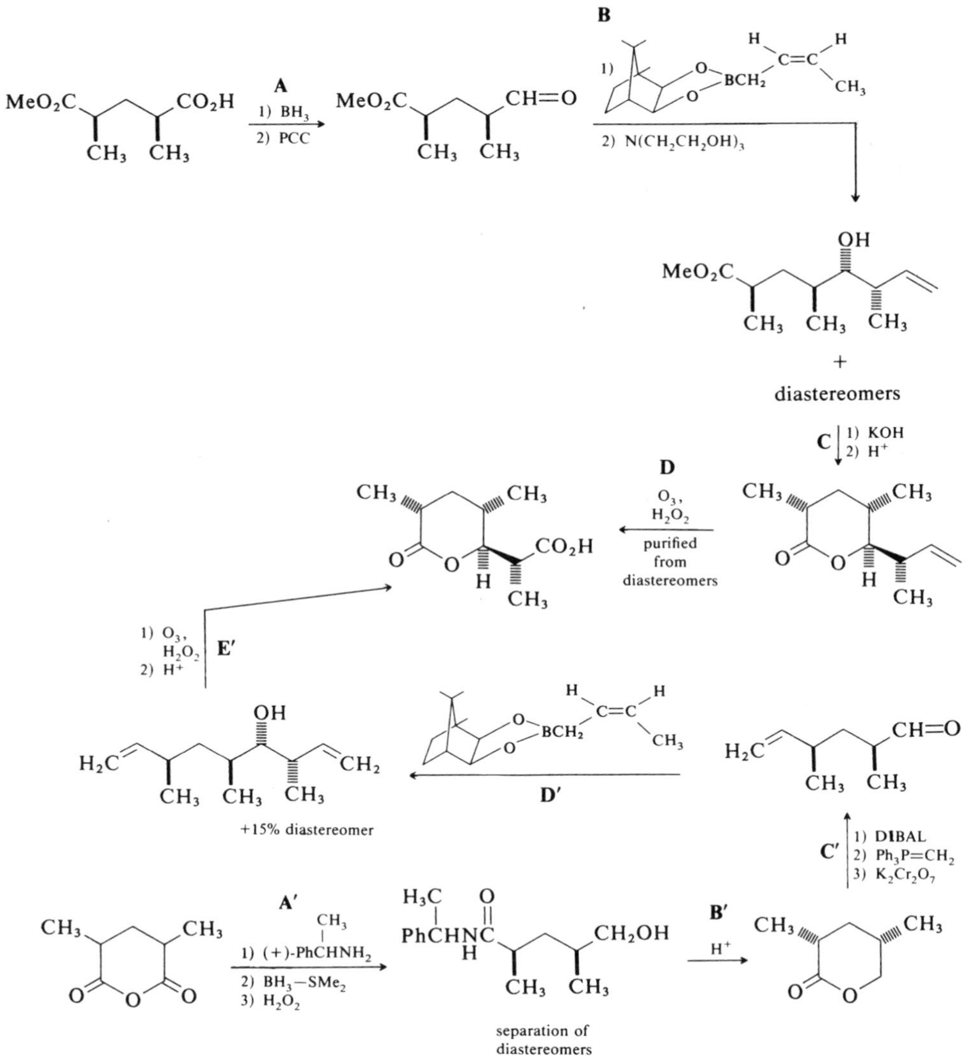

a. R. W. Hoffmann, H.-J. Zeiss, W. Ladner, and S. Tabche, *Chem. Ber.* **115**, 2357 (1982).
b. Resolved via α-phenylethylamine salt; S. Masamune, S. A. Ali, D. L. Snitman, and D. S. Garvey, *Angew. Chem. Int. Ed. Engl.* **19**, 557 (1980).

The synthesis in Scheme 13.26 is based on an interesting kinetic differentiation in the reactivity of two centers which are structurally identical but are diastereomeric. A bis-amide of *meso*-2,4-dimethylglutaric acid and a homochiral thiazoline is formed in step **A**. The thiazoline is derived from the amino acid cysteine. The two amide carbonyls in this bis-amide are nonequivalent by virtue of the diastereomeric relationship established by the chiral centers at C-2 and C-4 in the glutaric acid portion of

Scheme 13.26. Prelog–Djerassi Lactone Synthesis: Y. Nagao, T. Inoue, K. Hashimoto, Y. Hagiwara, M. Ochai, and E. Fujita[a]

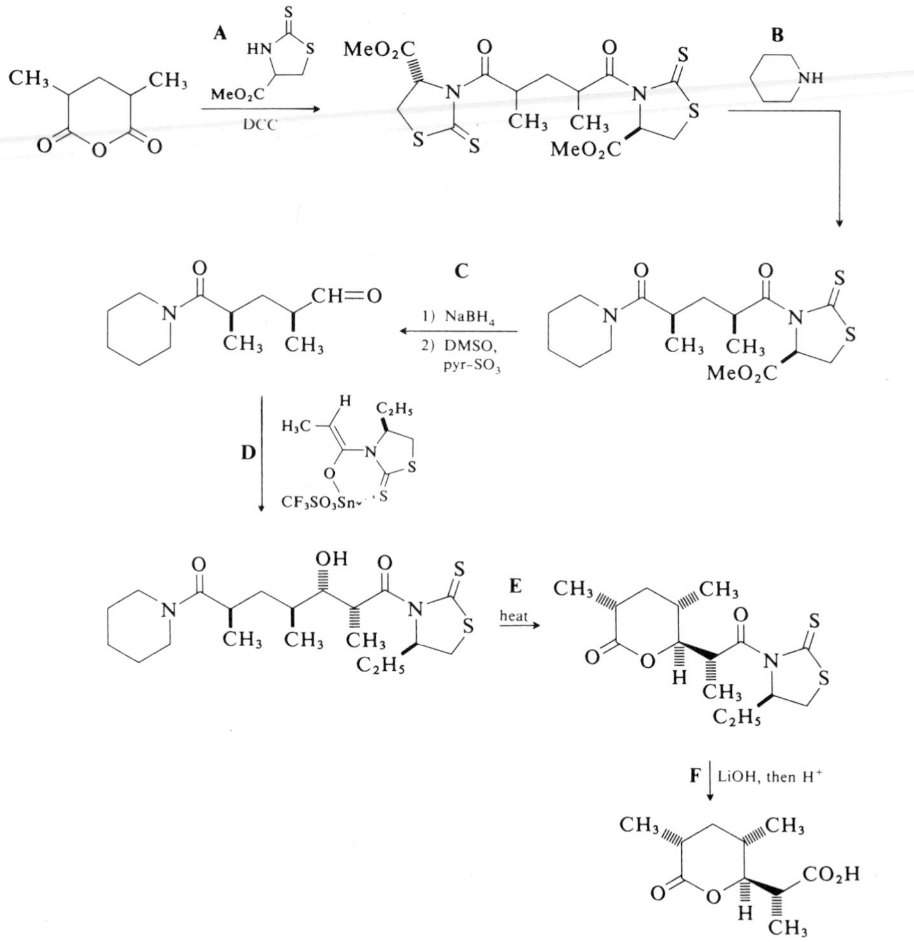

A: MeO_2C-thiazolidinethione, DCC
B: piperidine (NH)
C: 1) $NaBH_4$ 2) DMSO, pyr-SO_3
D: CF_3SO_3Sn enolate, C_2H_5
E: heat
F: LiOH, then H^+

a. Y. Nagao, T. Inoue, K. Hashimoto, Y. Hagiwara, M. Ochai, and E. Fujita, *J. Chem. Soc., Chem. Commun.*, 1419 (1985).

the structure. One of the centers reacts with a 97 : 3 preference with the achiral amine piperidine.

Two amide bonds are in nonequivalent stereochemical environments

S—R—S—S

more reactive ↗ ↖ less reactive

In step **D** a *chiral auxiliary*, also derived from cysteine, is used to achieve double stereodifferentiation in an aldol condensation. A tin enolate was used. The stereoselectivity of this reaction parallels that of aldol condensations carried out with lithium or zinc enolates. Once the configuration of all the centers has been established, the synthesis proceeds to P–D-lactone by functional group modifications.

There have been several syntheses of P–D-lactone which are based on carbohydrate-derived starting materials. The starting material used in Scheme 13.27 had been prepared from carbohydrate in earlier work.[90] The relative stereochemistry at C-4 and C-6 was established by the hydrogenation in step **B**. This *syn* hydrogenation is not completely stereoselective but provides a 4:1 mixture favoring the desired isomer. The stereoselectivity is presumably the result of preferential adsorption from the less hindered β face of the molecule. The configuration at C-2 is established by protonation during the hydrolysis of the enol ether in step **D**. This step is not stereoselective, and so a separation of diastereomers after the oxidation in step **E** was required.

The synthesis in Schemes 13.28 also begins with carbohydrate-derived starting material and also uses catalytic hydrogenation to establish the stereochemical relationship between the methyl groups at C-4 and C-6. As was the case in Scheme 13.27, the configuration at C-2 is not controlled in this synthesis, and separation of the diastereomeric products was necessary.

The synthesis in Scheme 13.29 is also based on a carbohydrate-derived starting material. It controls the stereochemistry at C-2 by means of the stereoselectivity of the Ireland–Claisen rearrangement in step **A** (see Section 6.5). The ester enolate is formed under conditions where the *E*-enolate is expected to predominate. Heating

Scheme 13.27. Prelog–Djerassi Lactone Synthesis; S. Jarosz and B. Fraser-Reid[a]

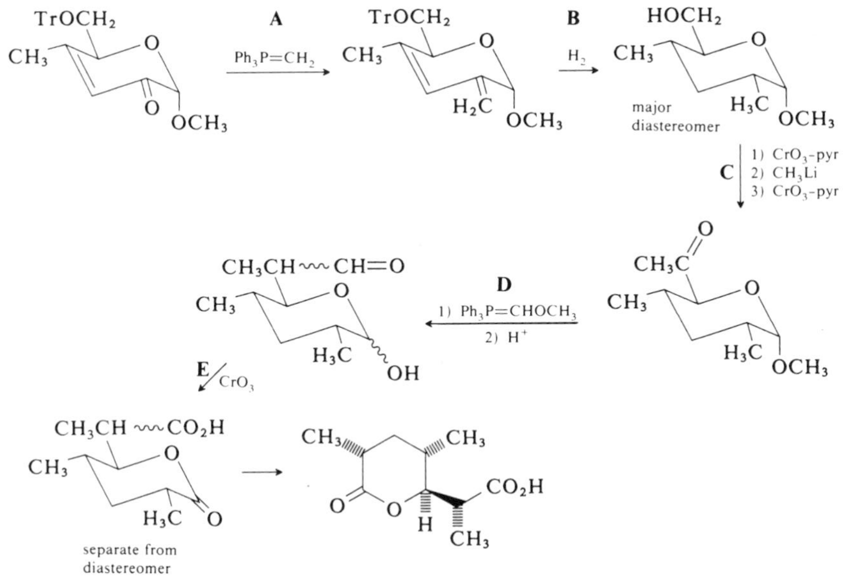

a. S. Jarosz and B. Fraser-Reid, *Tetrahedron Lett.* **22**, 2533 (1981).

90. M. B. Yunker, D. E. Plaumann, and B. Fraser-Reid, *Can. J. Chem.* **55**, 4002 (1977).

Scheme 13.28. Prelog–Djerassi Lactone Synthesis: N. Kawachi and H. Hashimoto[a]

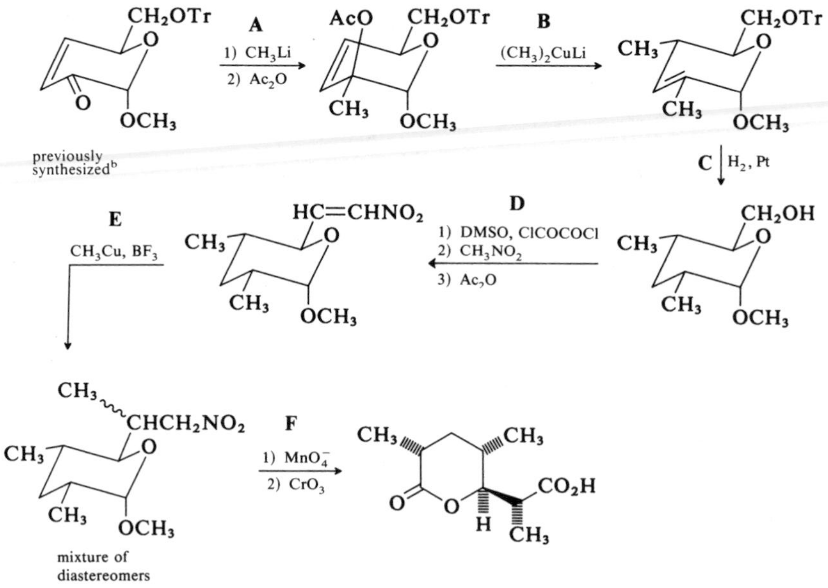

a. N. Kawauchi and H. Hashimoto, *Bull. Chem. Soc. Jpn.* **60**, 1441 (1987).
b. N. L. Holder and B. Fraser-Reid, *Can. J. Chem.* **51**, 3357 (1973).

the resulting silyl enol ether gave a 9:1 preference for the expected stereoisomer. The preferred transition state, which is boatlike, minimizes the steric interaction between the bulky silyl substituent and the ring structure.

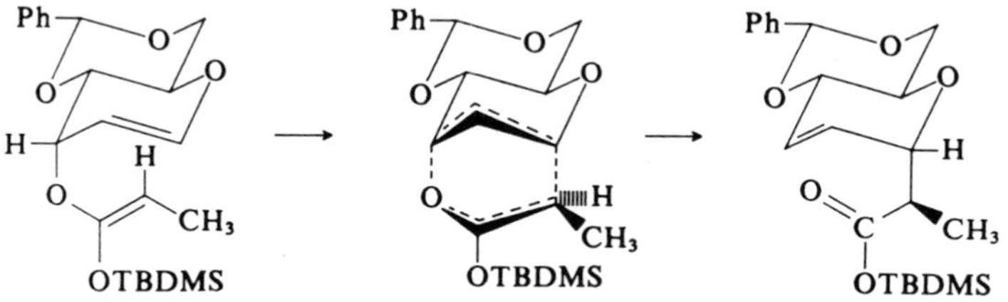

The stereochemistry at C-4 and C-6 is then established. The cuprate addition in step **C** occurs *anti* to the substituent at C-2 of the pyran ring. After a Wittig methylenation, the catalytic hydrogenation in step **D** establishes the stereochemistry at C-6.

The syntheses in Schemes 13.30 and 13.31 illustrate the use of chiral auxiliaries in enantioselective synthesis. Step **A** in Scheme 13.30 establishes the configuration at the carbon that becomes C-4 in the product. This is an enolate alkylation in which the steric effect of the oxazolidinone substituents directs the approach of the alkylating group. Step **C** also uses the oxazolinone structure. In this case, the enol borinate is formed and condensed with the aldehyde intermediate. This stereoselec-

Scheme 13.29. Prelog–Djerassi Lactone Synthesis: R. E. Ireland and J. P. Daub[a]

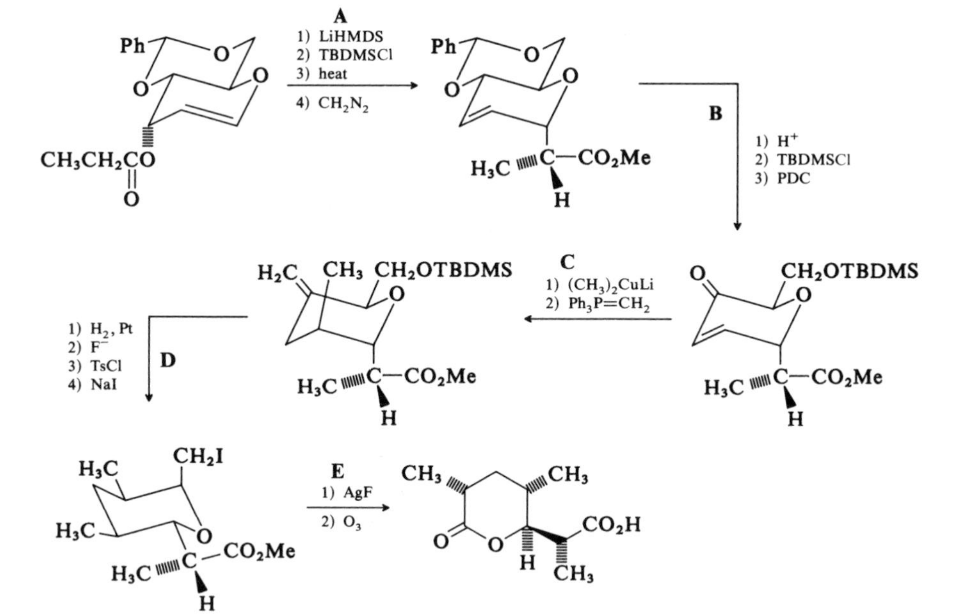

A: 1) LiHMDS 2) TBDMSCl 3) heat 4) CH_2N_2

B: 1) H^+ 2) TBDMSCl 3) PDC

C: 1) $(CH_3)_2CuLi$ 2) $Ph_3P{=}CH_2$

D: 1) H_2, Pt 2) F^- 3) TsCl 4) NaI

E: 1) AgF 2) O_3

a. R. E. Ireland and J. P. Daub, *J. Org. Chem.* **46**, 479 (1981).

Scheme 13.30. Prelog–Djerassi Lactone Synthesis: D. A. Evans and J. Bartroli[a]

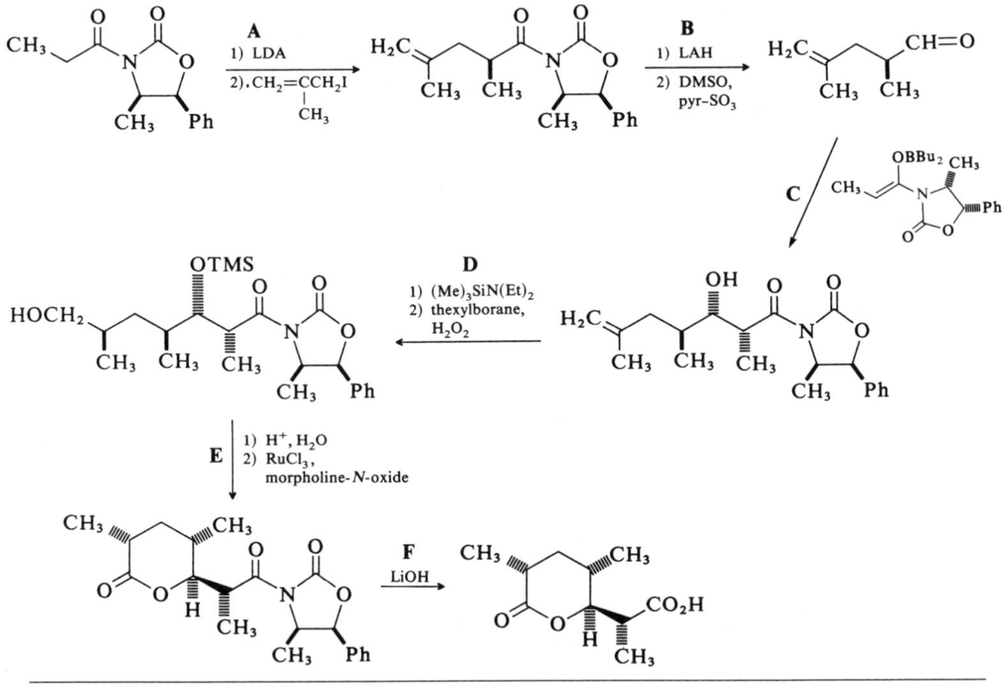

A: 1) LDA 2) $CH_2{=}C(CH_3)CH_2I$

B: 1) LAH 2) DMSO, pyr-SO_3

C

D: 1) $(Me)_3SiN(Et)_2$ 2) thexylborane, H_2O_2

E: 1) H^+, H_2O 2) $RuCl_3$, morpholine-*N*-oxide

F: LiOH

a. D. A. Evans and J. Bartroli, *Tetrahedron Lett.* **23**, 807 (1982).

Scheme 13.31. Prelog–Djerassi Lactone Synthesis: S. F. Martin and D. E. Guinn[a]

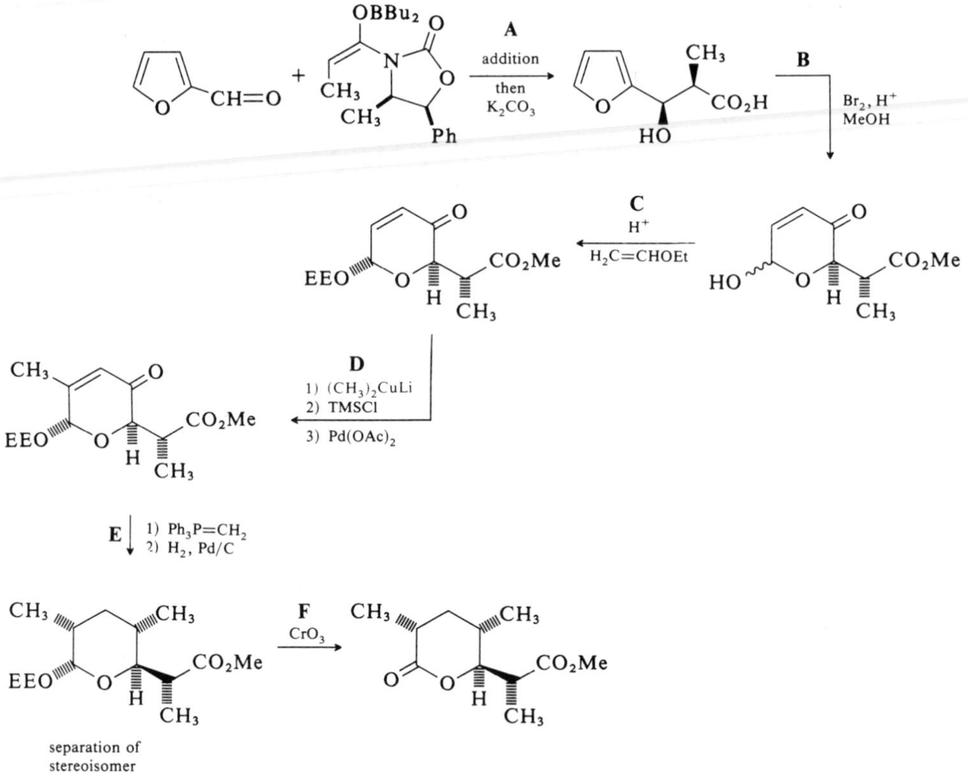

A: addition, then K_2CO_3

B: Br_2, H^+, MeOH

C: H^+, $H_2C{=}CHOEt$

D: 1) $(CH_3)_2CuLi$ 2) TMSCl 3) $Pd(OAc)_2$

E: 1) $Ph_3P{=}CH_2$ 2) H_2, Pd/C

separation of stereoisomer

F: CrO_3

a. S. F. Martin and D. E. Guinn, *J. Org. Chem.* **52**, 5588 (1987).

tive aldol condensation establishes the configuration at C-2 and C-3. The configuration at the final chiral center is established by the hydroboration in step **D**. The selectivity for the desired stereoisomer is 85 : 15. Stereoselectivity in the same sense has been observed for a number of other 2-methylalkenes in which the remainder of the alkene constitutes a relatively bulky group.[91] A transition state such as **A** can rationalize this result.

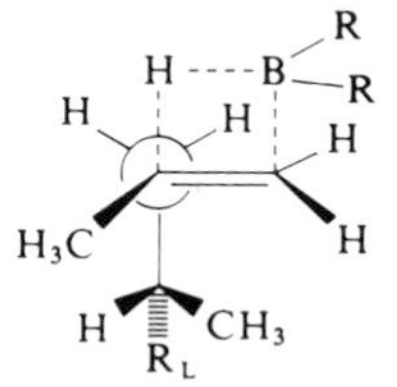

A

91. D. A. Evans, J. Bartroli, and T. Godel, *Tetrahedron Lett.* **23**, 4577 (1982).

In the synthesis in Scheme 13.31, a stereoselective aldol condensation is used to establish the configuration at C-2 and C-3 in step **A**. The furan ring is then subjected to an electrophilic addition and solvolytic rearrangement in step **B**.

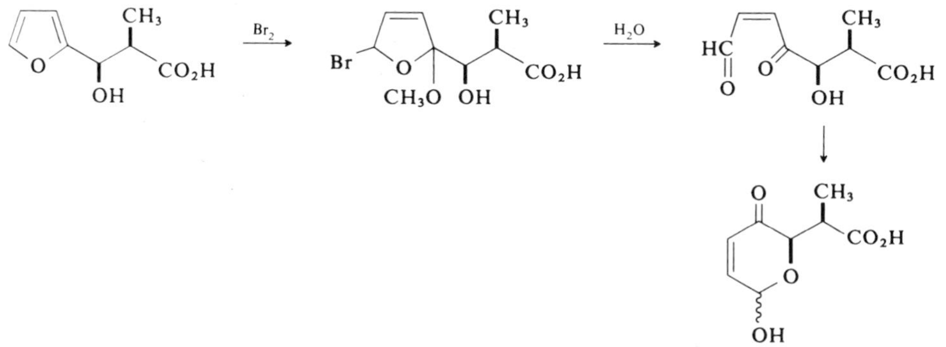

The protection of the hemiacetal hydroxyl in step **C** is followed by a purification of the dominant stereoisomer. The C-6 methyl group is introduced in step **C** by conjugate addition of dimethylcuprate. The enolate is trapped as the silyl enol ether and oxidized to the enone by palladium acetate. The enone from step **D** is then subjected to a Wittig reaction. As in several of the other syntheses, the hydrogenation in step **E** is used to establish the configuration at C-4 and C-6.

The synthesis in Scheme 13.32 features a catalytic asymmetric epoxidation (see Section 12.2.1). By use of *meso*-2,4-dimethylglutaric anhydride as the starting material, the proper relative configuration at C-4 and C-6 is ensured. The epoxidation directed by the homochiral tartrate catalyst controls the configuration established

Scheme 13.32. Prelog–Djerassi Lactone Synthesis: M. Honda, T. Katsuki, and M. Yamaguchi[a]

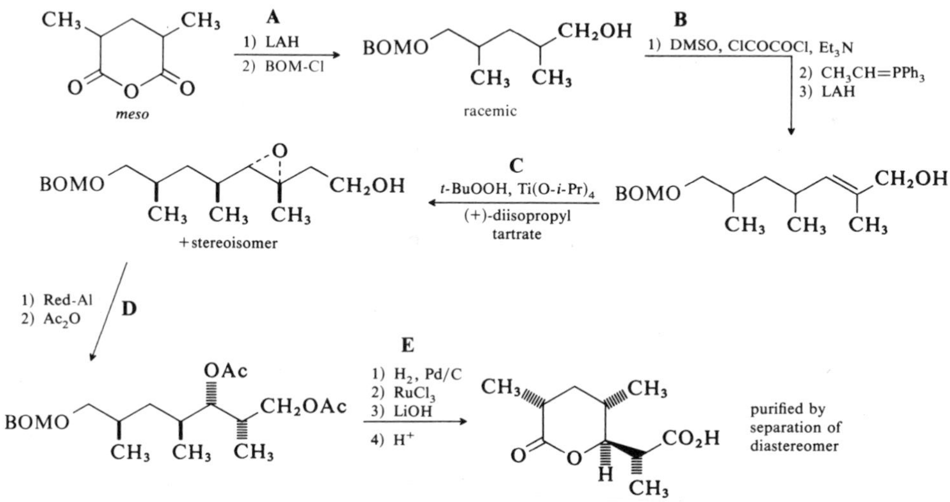

a. M. Honda, T. Katsuki, and M. Yamaguchi, *Tetrahedron Lett.* **25**, 3857 (1984).

at C-2 and C-3 by the epoxidation. While the epoxidation is highly selective in establishing the configuration at C-2 and C-3, the configuration at C-4 and C-6 does not strongly influence the reaction, and a mixture of diastereomeric products is formed and must be separated at a later stage of the synthesis. The reductive ring opening in step **D** occurs with dominant inversion to establish the necessary configuration at C-2. The preference for 1,3-diol formation is characteristic of reductive ring opening by Red-Al of epoxides derived from allylic alcohols.[92] Presumably, initial coordination at the hydroxyl group and intramolecular delivery of hydride is responsible for this stereoselectivity.

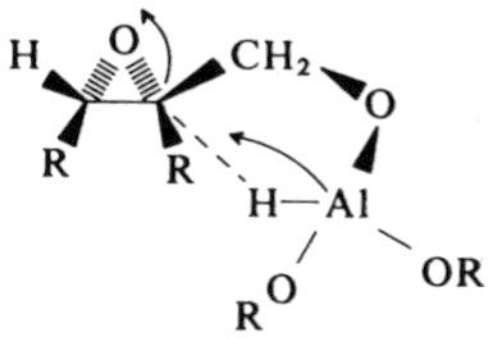

13.5.4. Aphidicolin

Aphidicolin is an antibiotic isolated from a fungus. It has antimitotic activity and inhibits the growth of *Herpes simplex* virus. It also has antitumor activity. The molecule has a somewhat more complex structure than those we have discussed up to this point. There are a total of eight chiral centers. Five of these are associated with ring junctions so a major part of the planning of a synthesis of aphidicolin revolves around the stereochemistry of the intermediates for ring formation.

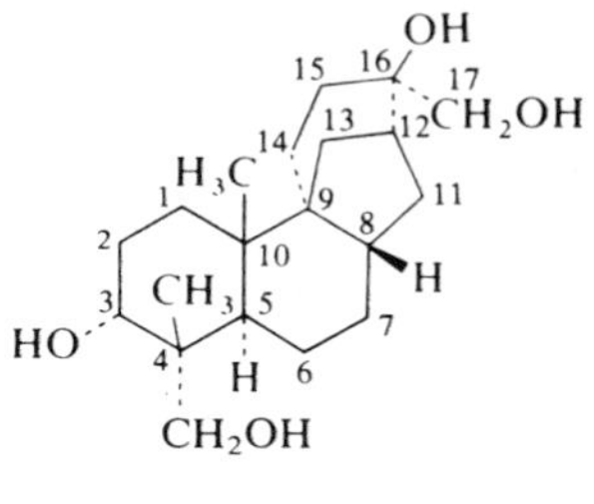

The syntheses of Schemes 13.33 and 13.34 employ the same starting material, an available[93] racemic decalin derivative. Both syntheses use similar methods to establish the required *trans* junction and the stereochemistry at C-3 and C-4. The stereochemistry of the ring fusion is established by a dissolving-metal reduction. The protonation generates the *trans* ring junction because the proton is delivered

92. P. Ma, V. S. Martin, S. Masamune, K. B. Sharpless, and S. M. Viti, *J. Org. Chem.* **47**, 1378 (1982); S. M. Viti, *Tetrahedron Lett.* **23**, 4541 (1982); J. M. Finan and Y. Kishi, *Tetrahedron Lett.* **23**, 2719 (1982).

93. Y. Kitahara, A. Yoshikoshi, and S. Oida, *Tetrahedron Lett.*, 1763 (1964).

Scheme 13.33. Aphidicolin Synthesis: B. M. Trost, Y. Nishimura, and K. Yamamoto[a]

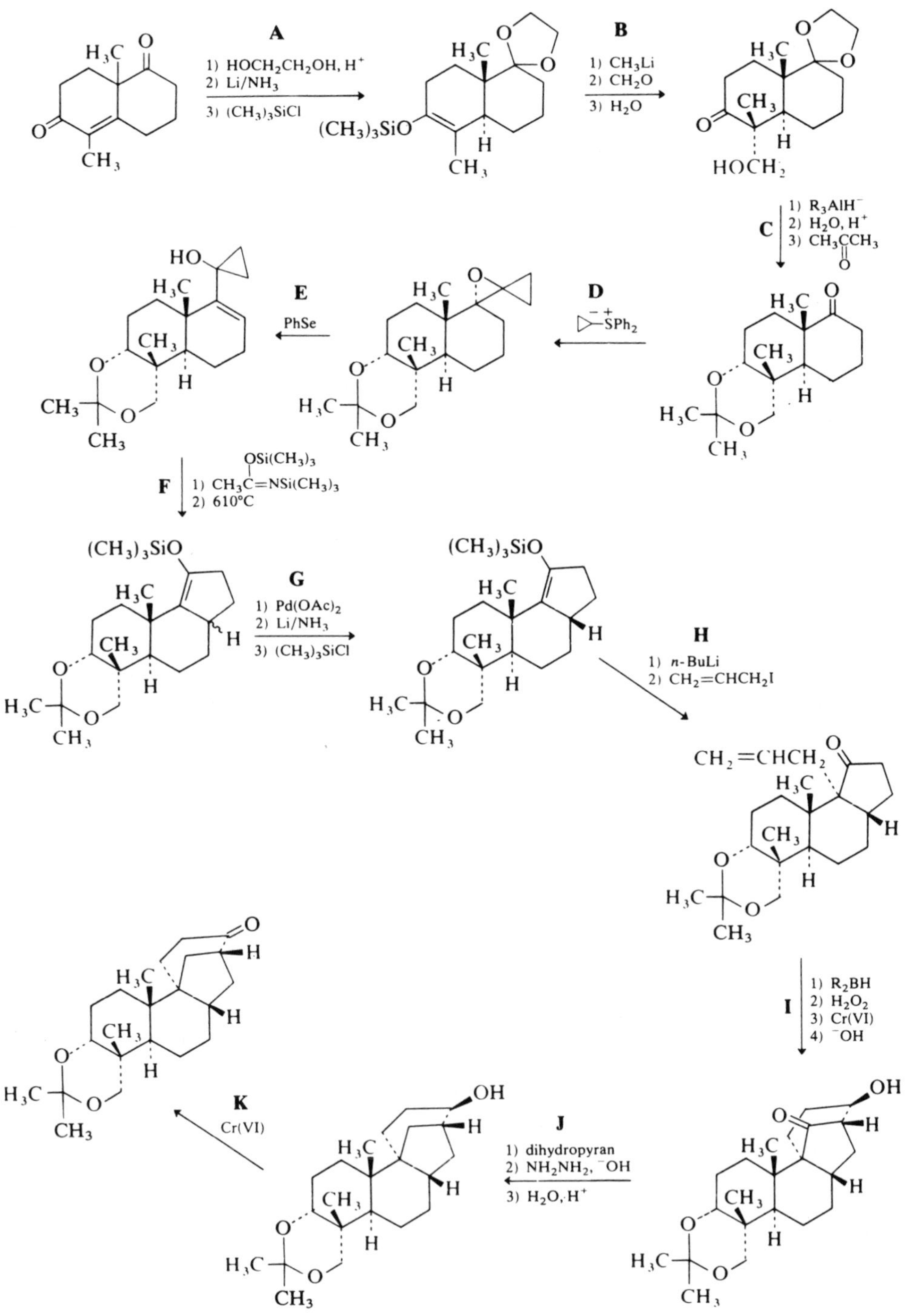

A: 1) $HOCH_2CH_2OH, H^+$ 2) Li/NH_3 3) $(CH_3)_3SiCl$

B: 1) CH_3Li 2) CH_2O 3) H_2O

C: 1) R_3AlH^- 2) H_2O, H^+ 3) CH_3CCH_3 (O)

D: cyclopropyl $\overset{-}{}\overset{+}{S}Ph_2$

E: PhSe

F: 1) $CH_3C(OSi(CH_3)_3){=}NSi(CH_3)_3$ 2) 610°C

G: 1) $Pd(OAc)_2$ 2) Li/NH_3 3) $(CH_3)_3SiCl$

H: 1) n-BuLi 2) $CH_2{=}CHCH_2I$

I: 1) R_2BH 2) H_2O_2 3) Cr(VI) 4) ^-OH

J: 1) dihydropyran 2) NH_2NH_2, ^-OH 3) H_2O, H^+

K: Cr(VI)

a. B. M. Trost, Y. Nishimura, and K. Yamamoto, *J. Am. Chem. Soc.* **101**, 1328 (1979).

Scheme 13.34. Aphidicolin Synthesis: J. E. McMurry, A. Andrus, G. M. Ksander, J. H. Musser, and M. A. Johnson[a]

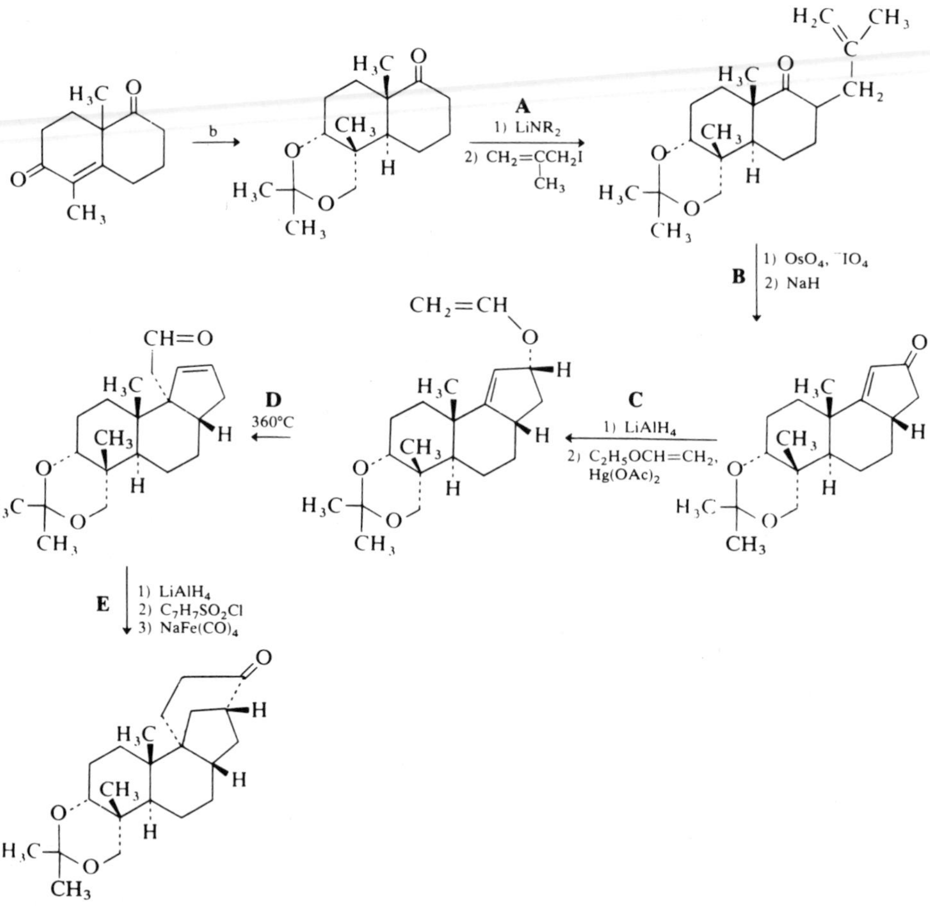

b

A
1) $LiNR_2$
2) $CH_2{=}C(CH_3)CH_2I$

B
1) OsO_4, $^-IO_4$
2) NaH

C
1) $LiAlH_4$
2) $C_2H_5OCH{=}CH_2$, $Hg(OAc)_2$

D
360°C

E
1) $LiAlH_4$
2) $C_7H_7SO_2Cl$
3) $NaFe(CO)_4$

a. J. E. McMurry, A. Andrus, G. M. Ksander, J. H. Musser, and M. A. Johnson, *J. Am. Chem. Soc.* **101**, 1330 (1979).
b. This transformation is carried out by a sequence of reactions similar to that in Scheme 13.33.

anti to the axial methyl group at C-10. The product is trapped as a silyl enol ether.

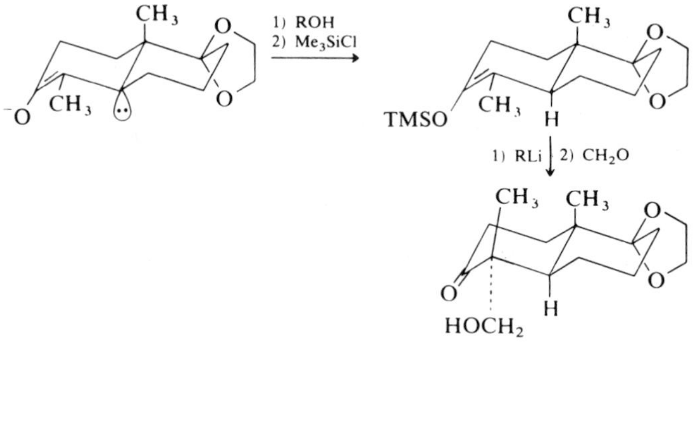

1) ROH
2) Me_3SiCl

1) RLi 2) CH_2O

On regeneration of the enolate, reaction with formaldehyde again occurs on the side of the ring opposite the axial methyl group. The stereochemistry of the alcohol group at C-3 is then established by use of a bulky reducing agent which approaches from the less hindered equatorial direction. The resulting diol is converted to an acetonide, which serves as a protective group throughout the synthesis. This sequence of reaction establishes the relative configuration at positions 3, 4, 5, and 10.

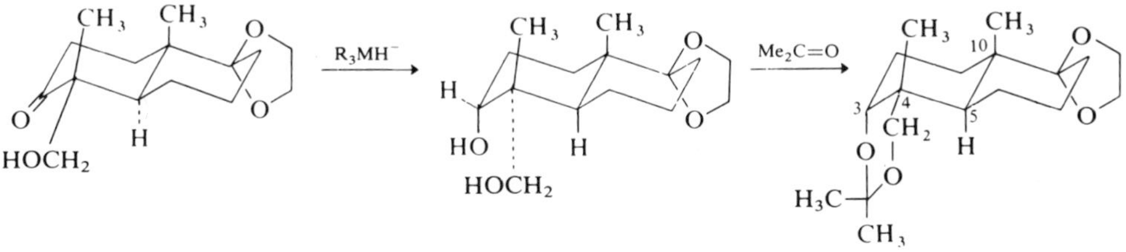

The methods for attachment of the third and fourth rings in the two syntheses differ and will be discussed separately. Scheme 13.33 uses an annulation sequence that had been shown to be a reasonably general way of constructing a cyclopentanone ring onto an existing ketone. The C-8—C-11 bond is established by a thermally induced vinylcyclopropane rearrangement.

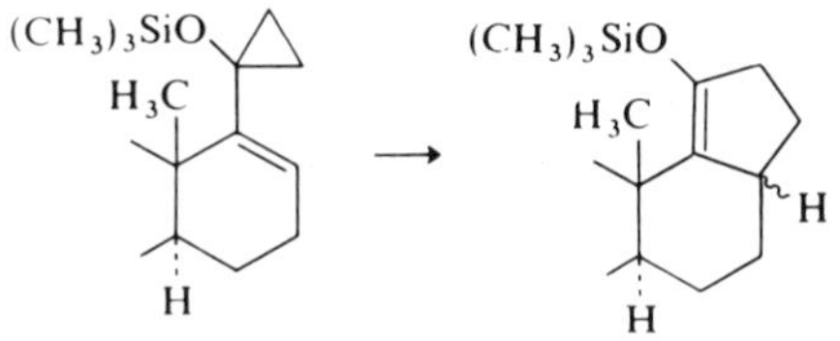

This reaction proved not to be very stereoselective, and an oxidation–reduction (step **G**) was necessary to establish the desired stereochemistry at C-8. The oxidation affords the α,β-unsaturated ketone, which is then reduced by Li/NH_3. The stereochemistry at C-8 is established by the β protonation of the reduction intermediate (see Section 5.5.1 for discussion of the stereochemistry of enone reductions). The dominant factor in controlling the stereochemistry is that protonation from the β-face, avoids forcing the C-8—C-11 bond into an axial conformation, where a diaxial interaction with the C-10 methyl group would result.

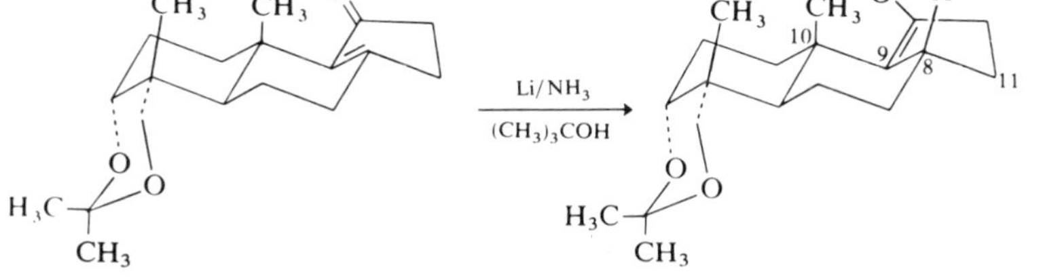

In step **H**, the enolate is regenerated, and attack by allyl iodide *anti* to the methyl group at C-10 generates the correct stereochemistry at C-9. Geometric constraints require that the final ring closure, which is carried out by an aldol condensation, proceed as shown in Scheme 13.33.

In the synthesis in Scheme 13.34, the first configuration that is established after construction of the decalin system is the one at C-8 in step **A**. An enolate alkylation is carried out with methallyl iodide. The observed, and desired, stereochemistry is governed by the C-10 methyl group, which blocks attack from the "top" side of the molecule. In step **B**, the five-membered ring is formed by intramolecular aldol condensation. The reduction of the enone (step **C**) establishes the configuration at C-13, and this chirality is subsequently transferred to C-9 by the intramolecular Claisen rearrangement in step **D**.

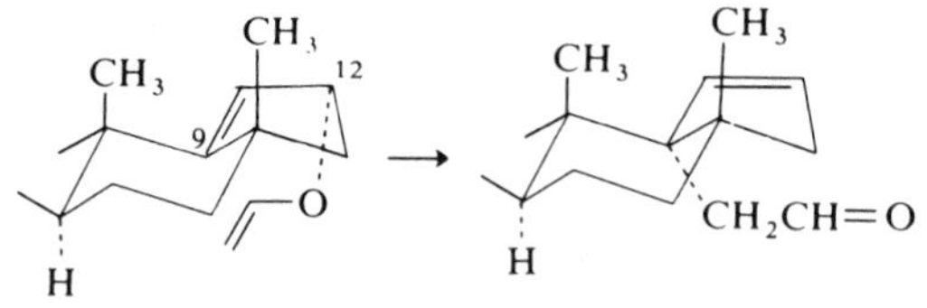

CH_3 CH_3 12 9 O H → CH_3 CH_3 $CH_2CH{=}O$ H

Again, geometric constraints require that the final C-12—C-16 bond formation result in the correct stereochemistry. The final step in Scheme 13.34 is an example of the use of the acylferrate reagent discussed in Section 8.4.

The final stages of the aphidicolin synthesis in Schemes 13.33 and 13.34 had already been developed during studies on the structure of the material.[94] The reaction sequence involves epoxide formation and hydrolysis of the epoxide to a diol.

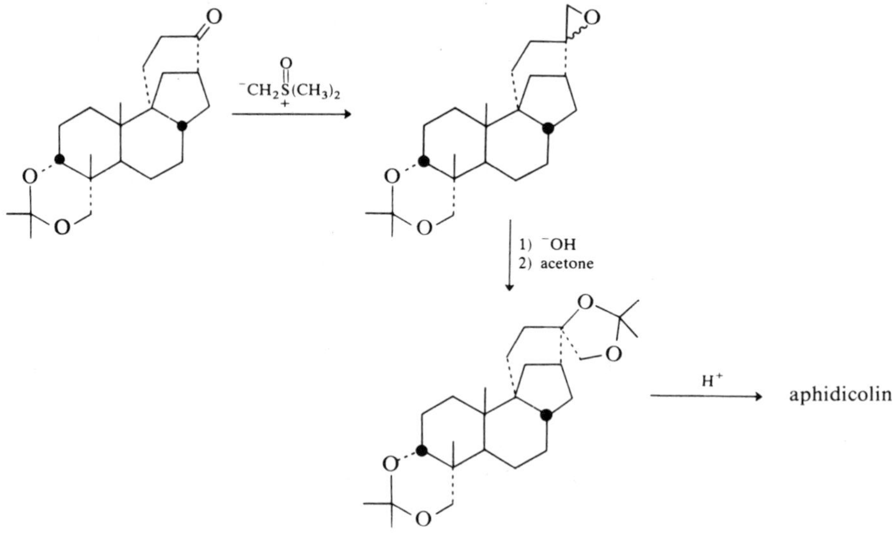

$^-CH_2\overset{+}{S}(O)(CH_3)_2$ → ; 1) ^-OH 2) acetone ; H^+ → aphidicolin

The epoxide formation step in this sequence is not very stereoselective and gives a mixture of isomers which must be separated. An alternative means for conversion

94. W. Dalziel, B. Hesp, K. M. Stevenson, and J. A. J. Harvis, *J. Chem. Soc., Perkin Trans. 1*, 2841 (1973).

of the ketone intermediate to aphidicolin has subsequently been developed. This method has the advantage of being stereoselective.[95] The protected ketone intermediate is first converted to the enol triflate and then carbonylated in methanol to give the ester (see Section 8.2 for discussion of this reaction). This unsaturated ester is epoxidized preferentially from the β face. Finally, the epoxide ring is opened by $LiAlH_4$ (LAH). Reduction of the ester group occurs in the same reaction to generate the diol structure.

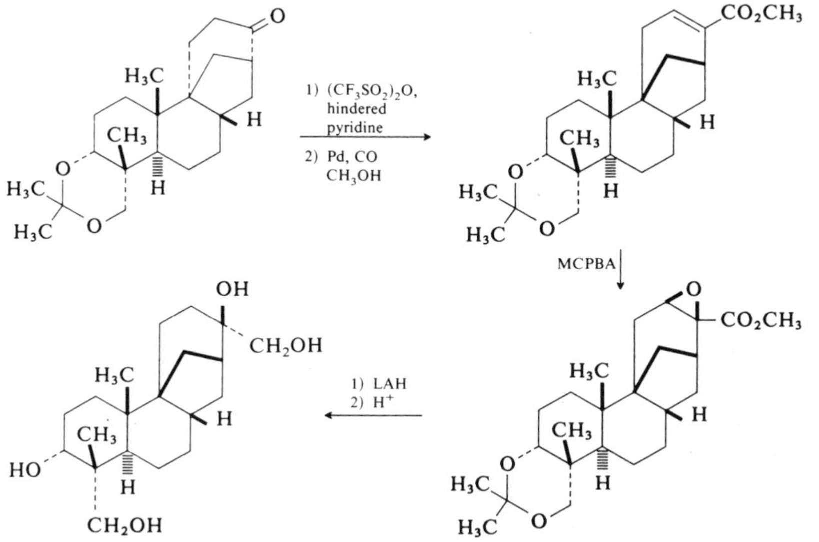
1) (CF3SO2)2O, hindered pyridine
2) Pd, CO CH3OH
MCPBA
1) LAH
2) H+
CO2CH3
CH2OH
OH
HO

In the aphidicolin synthesis in Scheme 13.35, a key bicyclic intermediate is established by a mercuric ion-induced polyene cyclization. Steps **A–D** serve to construct the necessary polyene. As was discussed in Section 10.1.1, the stereochemistry of such cyclizations is predictable on the basis of the polyene conformation.

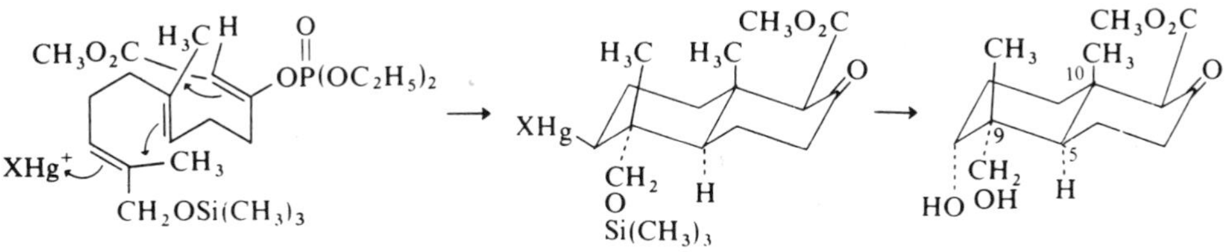
CH3O2C
OP(OC2H5)2
XHg+
CH2OSi(CH3)3
XHg
Si(CH3)3
HO OH

The relative configurations at C-4, C-5, and C-10 are established in the cyclization. The configuration at C-3 is established by converting it to a carbonyl group and then using a bulky reducing agent to introduce the axial hydroxyl. The six-membered ring encompassing C-13, C-12, C-16, C-15, and C-14 is constructed by a Robinson annulation in step **I**. The newly formed *spiro* ring becomes symmetrical in step **M**, so it does not present a stereochemical problem. The crucial element of

95. C. J. Rizzo and A. B. Smith III, *Tetrahedron Lett.* **29**, 2793 (1988).

Scheme 13.35. Aphidicolin Synthesis: E. J. Corey, M. A. Tius, and J. Das[a]

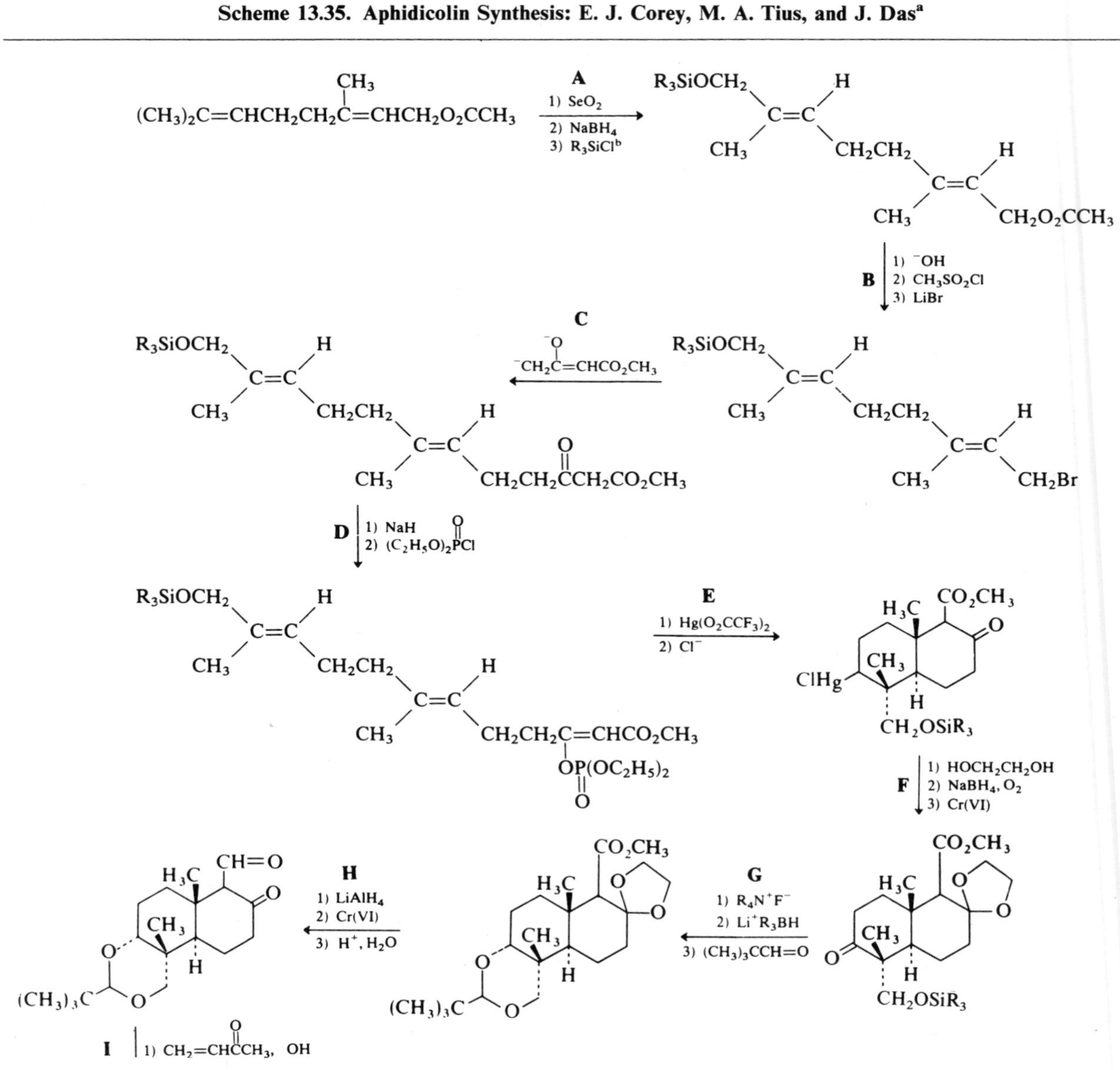

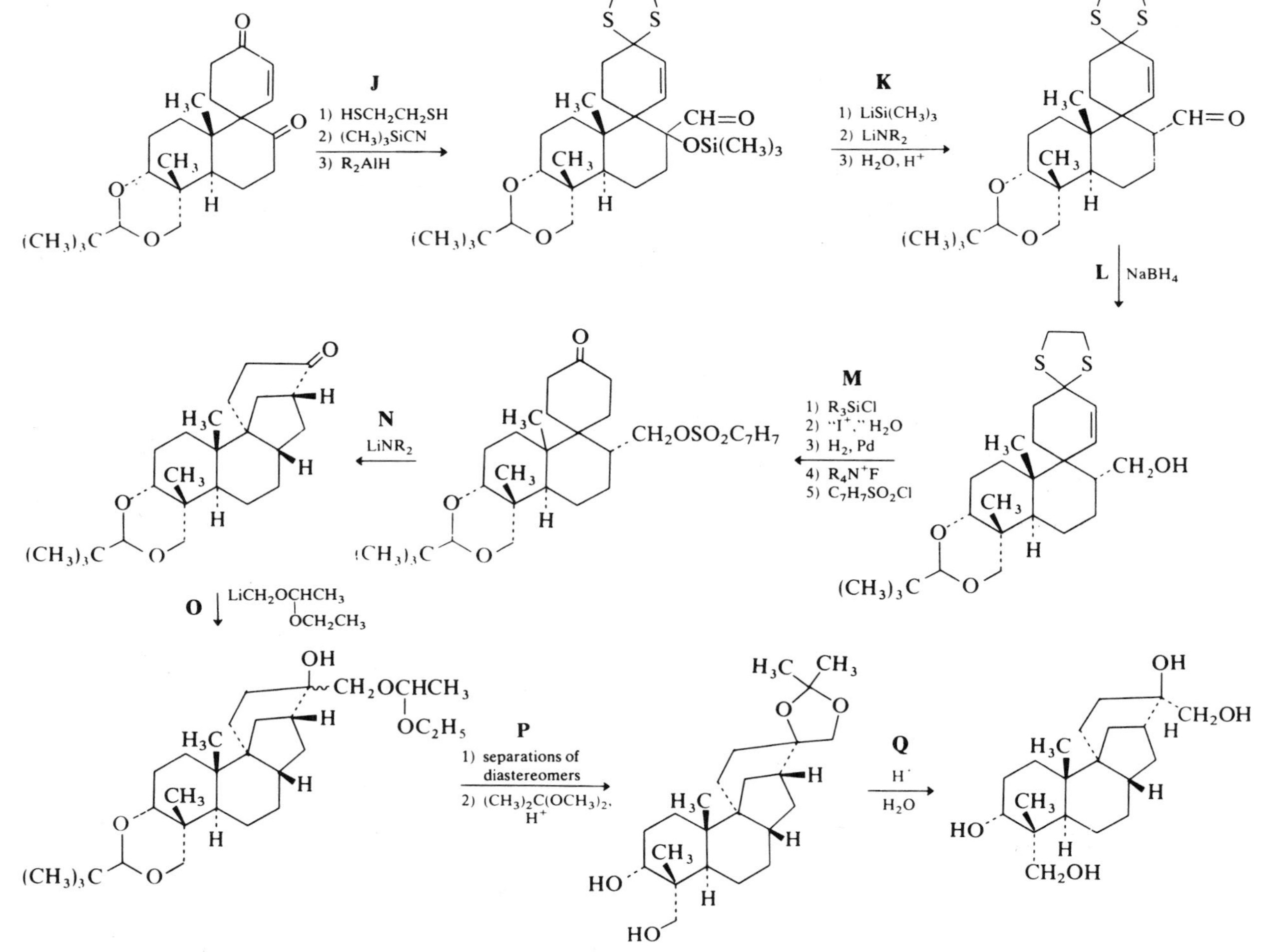

J 1) $HSCH_2CH_2SH$ 2) $(CH_3)_3SiCN$ 3) R_2AlH

K 1) $LiSi(CH_3)_3$ 2) $LiNR_2$ 3) H_2O, H^+

L $NaBH_4$

M 1) R_3SiCl 2) "I^+," H_2O 3) H_2, Pd 4) $R_4N^+F^-$ 5) $C_7H_7SO_2Cl$

N $LiNR_2$

O $LiCH_2OCHCH_3$ / OCH_2CH_3

P 1) separations of diastereomers 2) $(CH_3)_2C(OCH_3)_2$, H^+

Q H^+, H_2O

a. E. J. Corey, M. A. Tius, and J. Das, *J. Am. Chem. Soc.* **102**, 1742 (1980).
b. R_3Si = *t*-butyldimethylsilyl.

stereochemistry is that at C-8, which is established in steps **J** and **K** by a novel series of reactions invented to solve this particular problem. The ketone is first converted to a α-trimethylsilyloxynitrile. The initial cyanide attack probably occurs from the more open equatorial direction. The nitrile is then reduced to an aldehyde, and trimethylsilyllithium is added to the carbonyl group. This generates a β-hydroxysilane, which undergoes elimination.

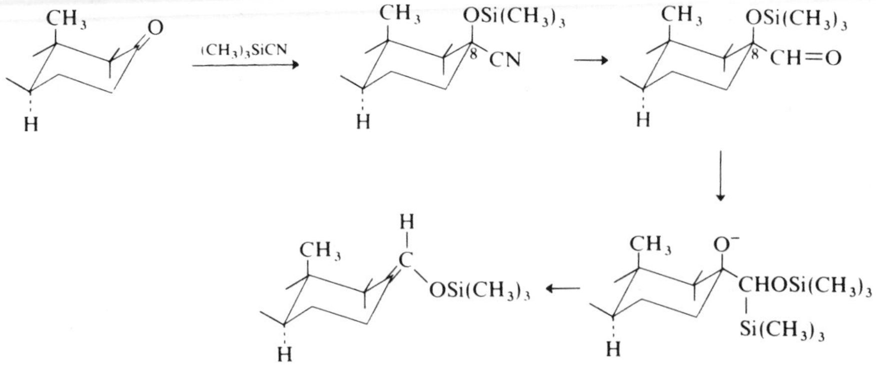

CH_3 O $(CH_3)_3SiCN$ CH_3 $OSi(CH_3)_3$ 8 CN CH_3 $OSi(CH_3)_3$ 8 CH=O H H H

H CH_3 C $OSi(CH_3)_3$ CH_3 O^- $CHOSi(CH_3)_3$ $Si(CH_3)_3$ H H

Because C-8 is reconverted to sp^2 hybridization, the stereochemistry up to this point is not crucial. The final stereochemistry at C-8 is established by protonation of the silyl enol ether. Evidently, the C-10 methyl group is the controlling factor.

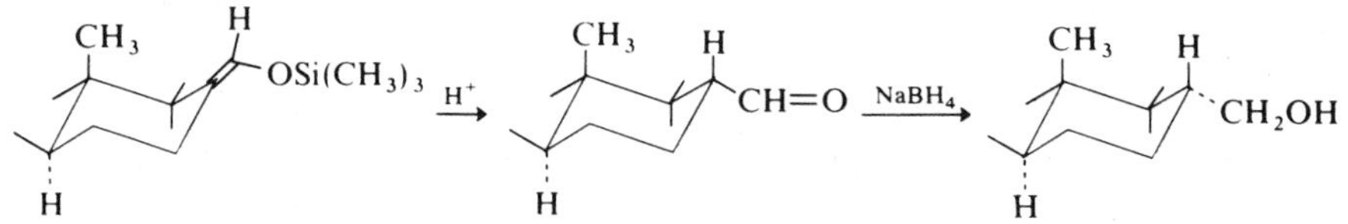

CH_3 H $OSi(CH_3)_3$ H^+ CH_3 H CH=O $NaBH_4$ CH_3 H CH_2OH H H H

After the aldehyde group is in place, it is reduced to the alcohol (step **L**). The next series of reactions (step **M**) first removes the ketal protecting group and then the carbon–carbon double bond. The primary hydroxyl group is protected as a silyl ether during this sequence. It is then deprotected and converted to the tosylate. In step **N**, the five-membered ring is constructed by an intramolecular alkylation. Careful control of this reaction so as to generate the *kinetic enolate* by use of a hindered base was crucial to obtaining the desired product. The synthesis in Scheme 13.35 was then completed by an independent method (steps **O–Q**), but this method, like the original one, is not highly stereoselective.

A fourth aphidicolin synthesis is outlined in Scheme 13.36. Steps **A** through **D** construct a key intermediate by a series of reactions, each of which has been discussed at earlier points in this text. Step **D** is a modified version of singlet oxygen oxidation which provides an enone as the product as the result of decomposition of the intermediate allylic hydroperoxide. Steps **E–J** are the crucial ones for elaborating the C and D rings. A Diels–Alder reaction in which an α,β-unsaturated ketone functions as the diene gives a pyran in step **E**. After using the ester substituent to build on a vinyl group, a thermal Claisen rearrangement (step **G**) gives a spiro ketone. This ketone is then converted to an α-diazoketone by a method discussed in Section 10.2.2. A photolytic Wolff rearrangement is then followed by a intramolecular [2 + 2] cycloaddition. The cyclobutanone formed by the cycloaddition is very acid-sensitive, and the silyl substituent controls the direction of the acid-catalyzed fragmentation. This sequence completes construction of the ring skeleton.

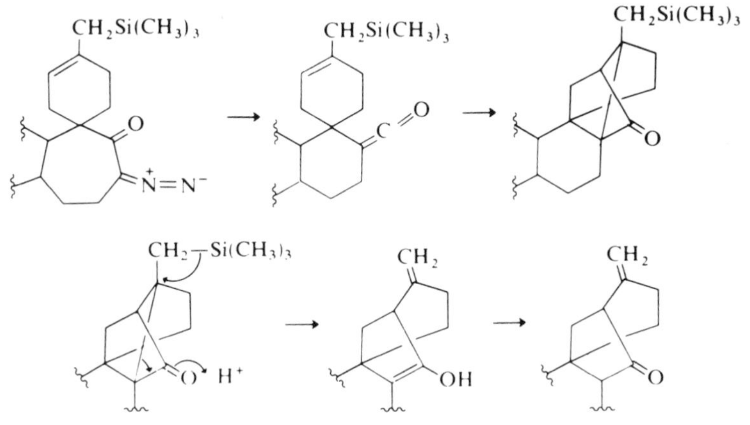

$CH_2Si(CH_3)_3$ $CH_2Si(CH_3)_3$ $CH_2Si(CH_3)_3$

CH_2—$Si(CH_3)_3$ CH_2 CH_2

The completion of the synthesis requires some functionalization and a change in stereochemistry at C-5. The C-16—C-17 hydroxylation is accomplished in step **L** and is stereoselective as the result of a steric effect by the large *t*-butyldimethylsilyl protective group. Step **M** accomplishes reductive removal of the extraneous hydroxyl group at C-11. Inversion at C-5 is accomplished by an oxidation to an enone, followed by lithium metal reduction. The stereochemistry is governed by the protonation of the enolate intermediate (as discussed in Section 5.5.1). The final functionalization at C-3 and C-4 is similar to the early steps in Schemes 13.33 and 13.34.

The synthesis in Scheme 13.37 begins with an intermediate decalin that also appears in Scheme 13.33 and 13.34. Steps **A** and **B** effect O-benzylation and then a Robinson annulation. Step **C** is the photocycloaddition of an allene to the enone system (see Section 6.4). The key steps for constructing the C/D ring structure proceed from this methylenecyclobutane. First, the methylene group is removed by ozonolysis. The resulting cyclobutane is first reduced and then treated with base. The base causes opening of the cyclobutane ring by a reverse aldol condensation. The ring-opened ketoaldehyde recyclizes to a less strained bicyclo[2.2.2]octanone.

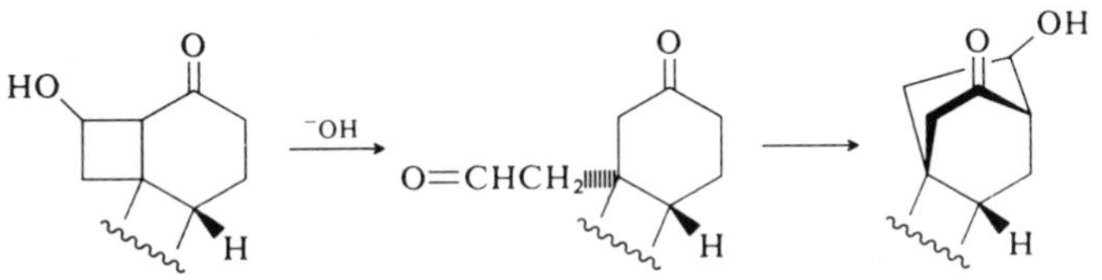

HO $\xrightarrow{^-OH}$ O=CHCH₂ → OH

The hydroxyl group is converted to a xanthate, and, after reduction of the carbonyl group, a Chugaev elimination is done (see Section 6.8.3). The final stage of the

Scheme 13.36. Aphidicolin Synthesis: R. E. Ireland, W. C. Dow, J. D. Godfrey, and S. Thaisrivongs[a]

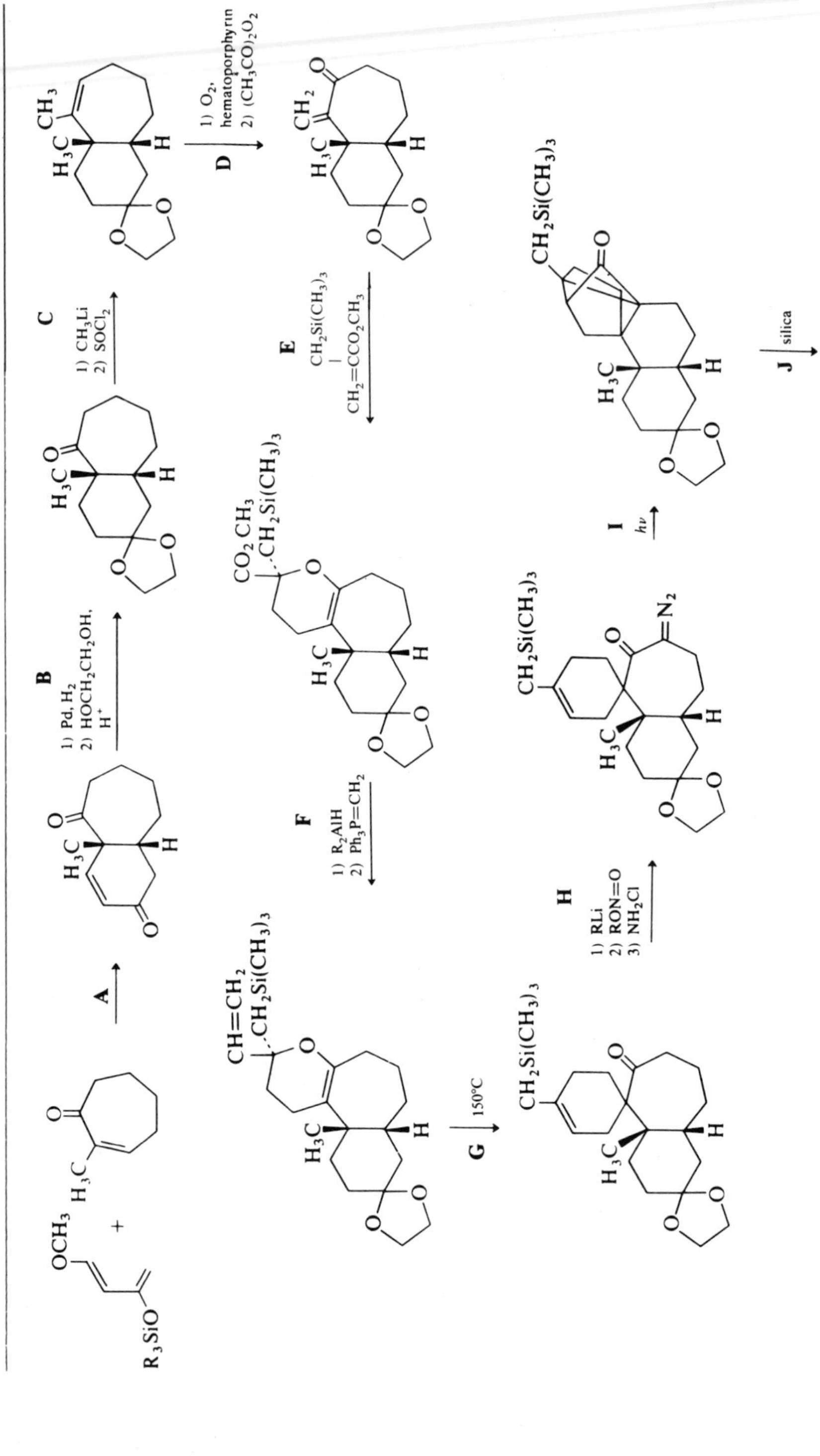
R3SiO
OCH3
H3C
A
B
1) Pd, H2
2) HOCH2CH2OH, H+
C
1) CH3Li
2) SOCl2
D
1) O2, hematoporphyrin
2) (CH3CO)2O2
E
CH2Si(CH3)3
CH2=CCO2CH3
F
1) R2AlH
2) Ph3P=CH2
G
150°C
H
1) RLi
2) RON=O
3) NH2Cl
I
hν
J
silica
CO2CH3
CH=CH2
CH2Si(CH3)3
N2

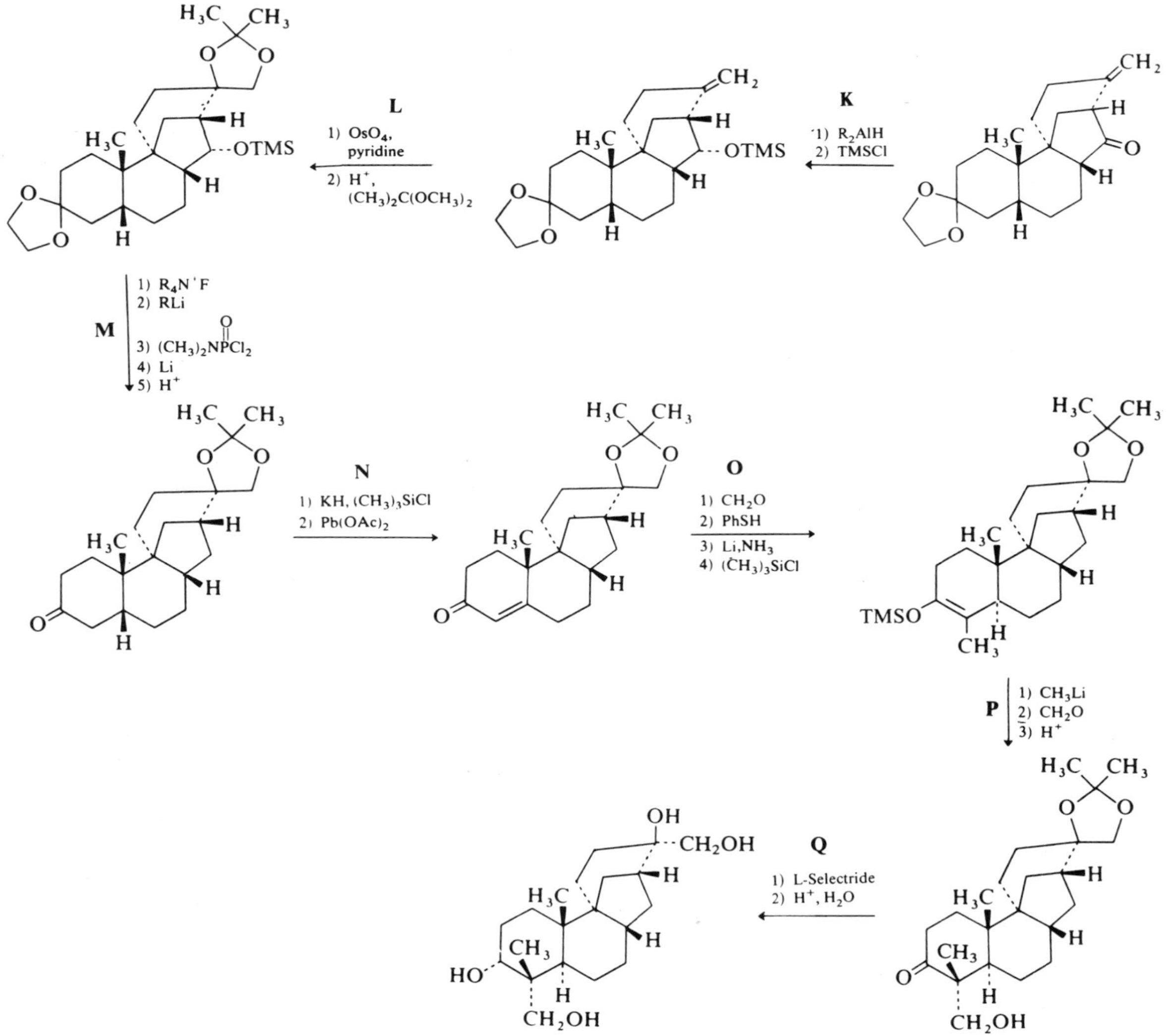

K
1) R_2AlH
2) TMSCl

L
1) OsO_4, pyridine
2) H^+, $(CH_3)_2C(OCH_3)_2$

M
1) $R_4N^+F^-$
2) RLi
3) $(CH_3)_2NP(O)Cl_2$
4) Li
5) H^+

N
1) KH, $(CH_3)_3SiCl$
2) $Pb(OAc)_2$

O
1) CH_2O
2) PhSH
3) Li, NH_3
4) $(CH_3)_3SiCl$

P
1) CH_3Li
2) CH_2O
3) H^+

Q
1) L-Selectride
2) H^+, H_2O

a. R. E. Ireland, J. D. Godfrey, and S. Thaisrivongs, *J. Am. Chem. Soc.* **103**, 2446 (1981); R. E. Ireland, W. C. Dow, J. D. Godfrey, and S. Thaisrivongs, *J. Org. Chem.* **49**, 1001 (1984).

Scheme 13.37. Aphidicolin Synthesis: R. M. Bettolo, P. Tagliatesta, A. Lupi, and D. Bravetti[a]

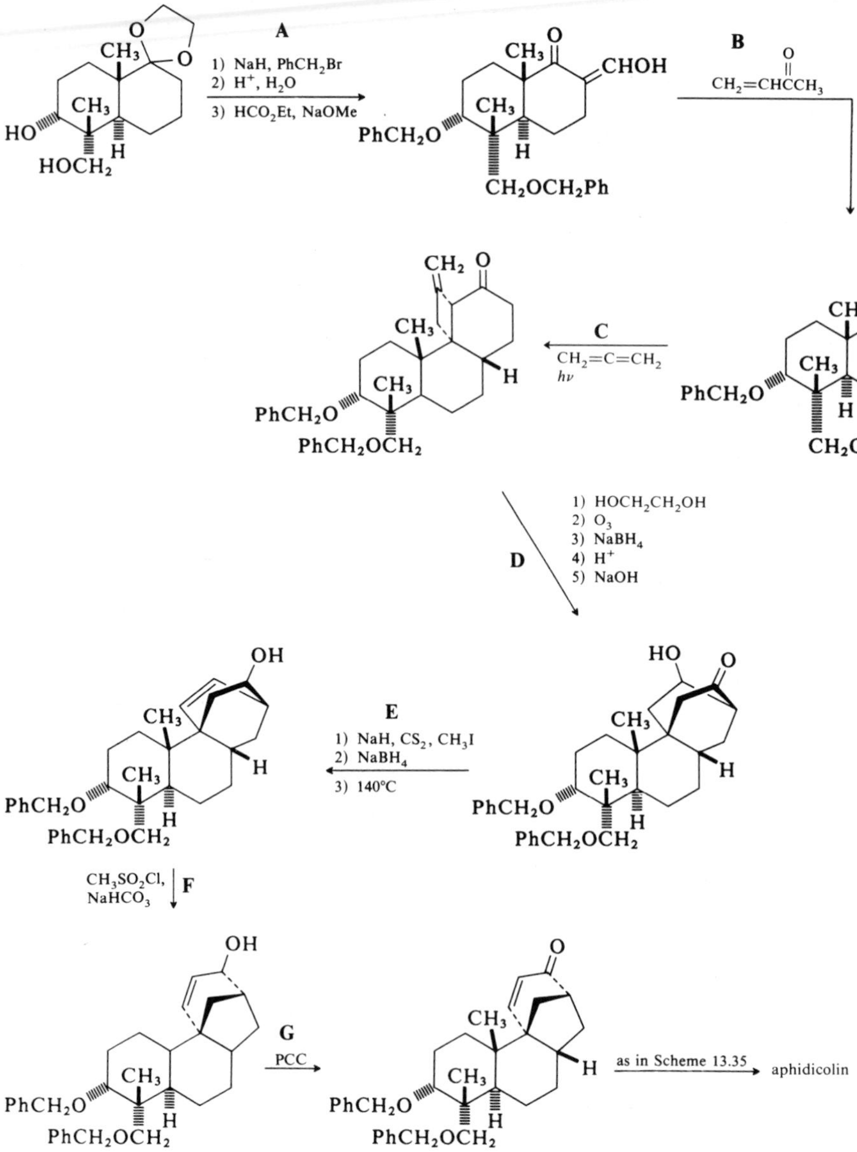

A: 1) NaH, $PhCH_2Br$; 2) H^+, H_2O; 3) HCO_2Et, NaOMe

B: $CH_2{=}CHCCH_3$ (C=O)

C: $CH_2{=}C{=}CH_2$, $h\nu$

D: 1) $HOCH_2CH_2OH$; 2) O_3; 3) $NaBH_4$; 4) H^+; 5) NaOH

E: 1) NaH, CS_2, CH_3I; 2) $NaBH_4$; 3) 140°C

F: CH_3SO_2Cl, $NaHCO_3$

G: PCC

as in Scheme 13.35 → aphidicolin

a. R. M. Bettolo, P. Tagliatesta, A. Lupi, and D. Bravetti, *Helv. Chim. Acta* **66**, 1922 (1983).

Scheme 13.38. Aphidicolin Synthesis: E. E. van Tamelen, S. R. Zawacky, R. K. Russell, and J. G. Carlson[a]

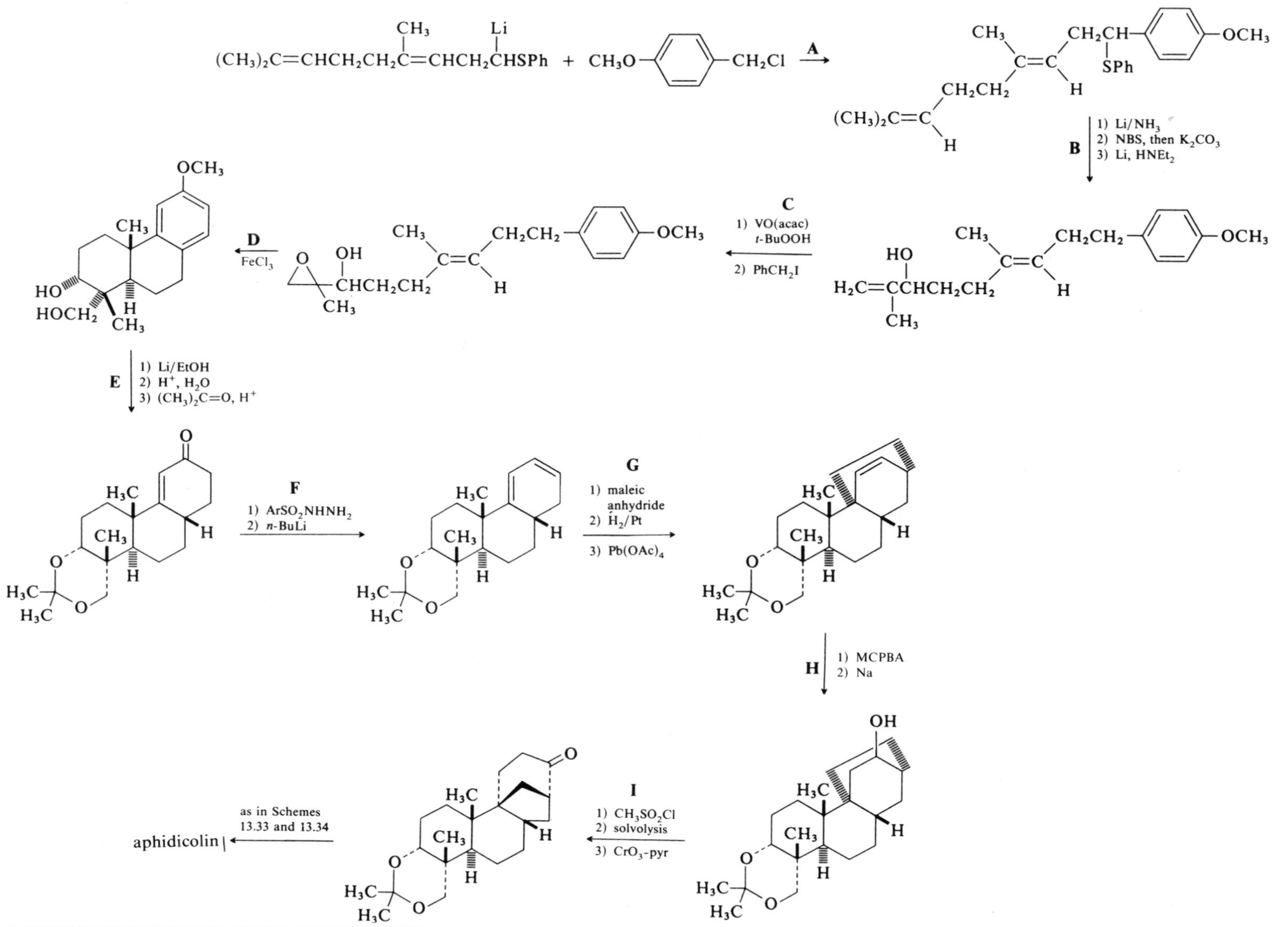

$(CH_3)_2C{=}CHCH_2CH_2C(CH_3){=}CHCH_2CH(Li)SPh + CH_3O{-}C_6H_4{-}CH_2Cl \xrightarrow{A}$

A: 1) Li/NH_3 2) NBS, then K_2CO_3 3) Li, $HNEt_2$ (step B)

C: 1) VO(acac), t-BuOOH 2) $PhCH_2I$

D: $FeCl_3$

E: 1) Li/EtOH 2) H^+, H_2O 3) $(CH_3)_2C{=}O$, H^+

F: 1) $ArSO_2NHNH_2$ 2) n-BuLi

G: 1) maleic anhydride 2) H_2/Pt 3) $Pb(OAc)_4$

H: 1) MCPBA 2) Na

I: 1) CH_3SO_2Cl 2) solvolysis 3) CrO_3-pyr

as in Schemes 13.33 and 13.34 → aphidicolin

a. E. E. van Tamelen, S. R. Zawacky, R. K. Russell, and J. G. Carlson, *J. Am. Chem. Soc.* **105**, 142 (1983).

Scheme 13.39. Aphidicolin Synthesis: R. A. Holton, R. M. Kennedy, H.-B. Kim, and M. E. Kraft[a]

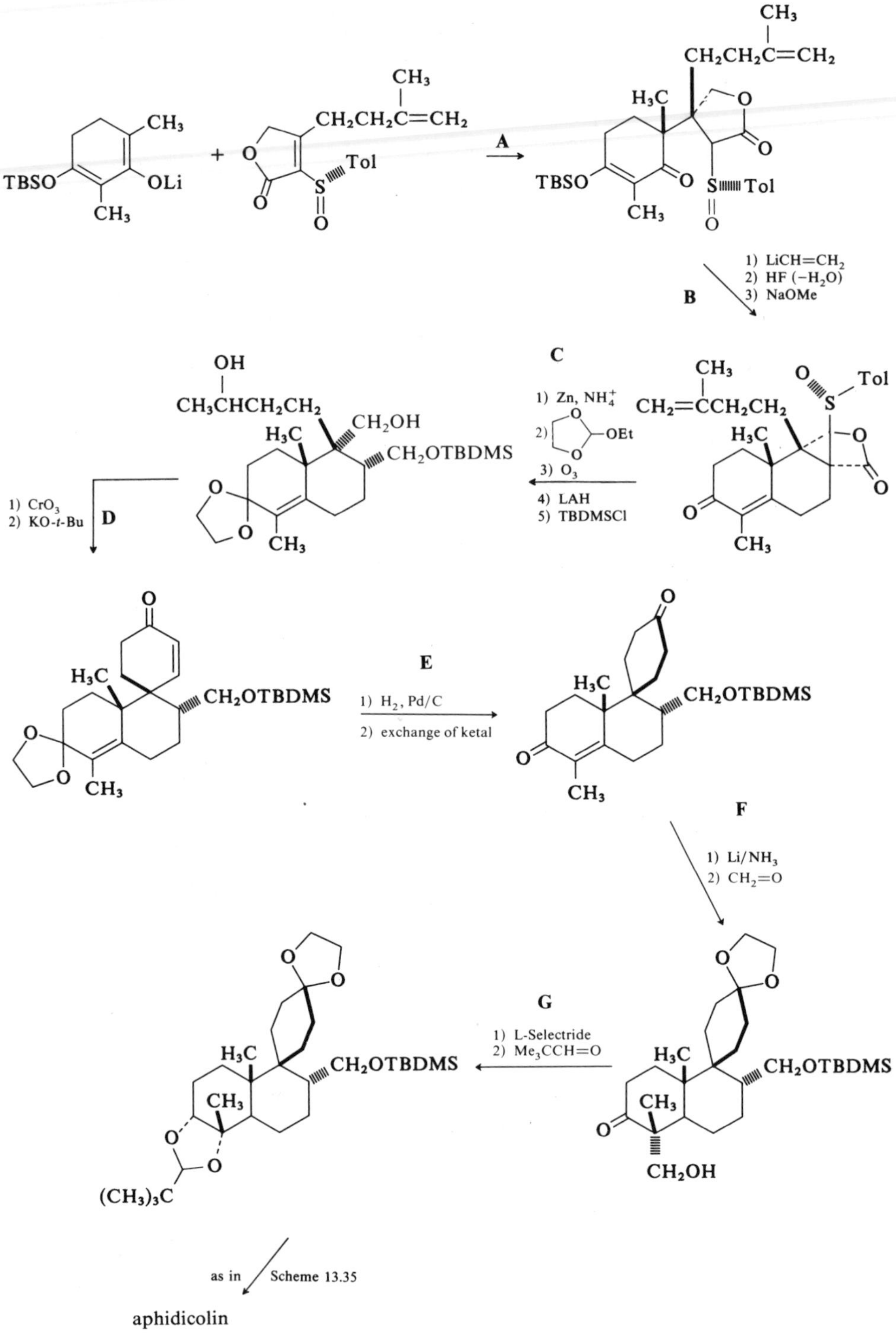

A

B
1) $LiCH{=}CH_2$
2) HF ($-H_2O$)
3) NaOMe

C
1) Zn, NH_4^+
2) (2-ethoxy-1,3-dioxolane)
3) O_3
4) LAH
5) TBDMSCl

D
1) CrO_3
2) KO-*t*-Bu

E
1) H_2, Pd/C
2) exchange of ketal

F
1) Li/NH_3
2) $CH_2{=}O$

G
1) L-Selectride
2) $Me_3CCH{=}O$

as in Scheme 13.35

aphidicolin

a. R. A. Holton, R. M. Kennedy, H.-B. Kim, and M. E. Kraft, *J. Am. Chem. Soc.* **109**, 1597 (1987).

synthesis in Scheme 13.37 is to rearrange the C/D rings from a [2.2.2] to the [3.2.1] structure found in aphidicolin. This is done by converting the remaining hydroxyl group to a mesylate. Under solvolytic conditions, the desired rearrangement occurs through an allylic cation.

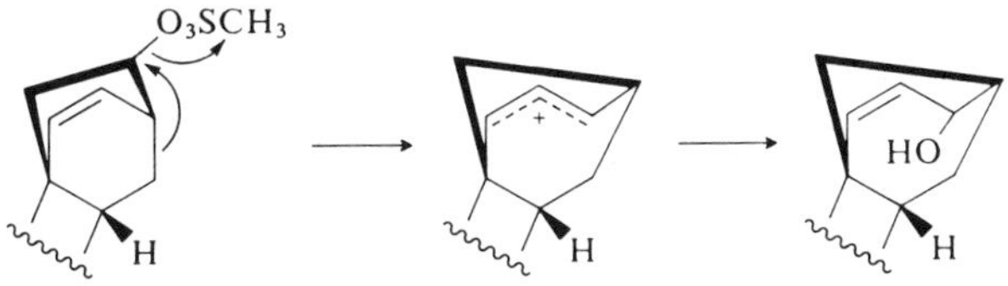

O_3SCH_3 → → HO

The final stages of this synthesis follows the route used in Scheme 13.35.

The synthesis in Scheme 13.38 combines elements of a biomimetic-type polyene cyclization with a rearrangement similar to that just described in Scheme 13.37. The early stages of the synthesis culminate in the construction of the allylic alcohol in step **B**. In step **C**, it is epoxidized using the stereoselective VO(*t*-BuOOH) method (Section 12.2.1). This epoxidation provides the key intermediate for the polyene cyclization. The mild Lewis acid $FeCl_3$ is used to promote the cyclization, which terminates in an electrophilic substitution of the methoxyphenyl substituent. Steps **E** and **F** convert the methoxyphenyl ring to a diene. This diene undergoes a Diels–Alder reaction in step **G**. After catalytic reduction of the double bond, the anhydride is subjected to an oxidative bis-decarboxylation (see Section 12.4.2 for discussion of this reaction). The resulting alkene is epoxidized, and the epoxide is reduced. A rearrangement is done at this point. The reaction is similar to that used in Scheme 13.37, except that it involves a saturated, rather than allylic, system. The final steps in the synthesis are those used in Schemes 13.33 and 13.34.

Each of the preceding syntheses of aphidicolin were conducted with racemic materials, so the final product was racemic aphidicolin. There has, to date, been one enantioselective synthesis of aphidicolin. The key step in establishing the absolute configuration of the product is the first one shown in Scheme 13.39. This is a Michael addition reaction carried out with an enantiomerically pure sulfoxide. The chiral sulfinyl substituent controls the stereochemistry at the two newly formed quaternary carbon atoms. The subsequent steps control the configuration of the other chiral centers relative to the original two stereocenters. The B-ring is closed in the last reaction of step **B** by a conjugate addition to a diene moiety. Step **D** involves oxidative generation of a ketoaldehyde, which undergoes aldol condensation to a *spiro*-cyclohexenone. After several functional group manipulations in steps **E–G**, this route intersects with that in Scheme 13.35.

General References

Protective Groups

T. W. Greene, *Protective Groups in Organic Synthesis*, Wiley, New York, 1981.
J. F. W. McOmie (ed.), *Protective Groups in Organic Chemistry*, Plenum, New York, 1973.

Synthetic Equivalent Groups

T. A. Hase (ed.), *Umpoled Synthons*, Wiley-Interscience, New York, 1987.
D. Lednicer, *Adv. Org. Chem. Methods Results* **8**, 179 (1972).
D. Seebach, *Angew. Chem. Int. Ed. Engl.* **18**, 239 (1979).

Synthetic Analysis and Planning

E. J. Corey and X.-M. Cheng, *The Logic of Chemical Synthesis*, Wiley, New York, 1989.
J.-H. Fuhrop and G. Penzlin, *Organic Synthesis: Concepts, Methods and Starting Materials*, Verlag Chemie, Weinheim, 1983.
T. Lindberg, *Strategies and Tactics in Organic Synthesis*, Academic Press, New York, 1984.
S. Warren, *Designing Organic Syntheses, A Programmed Introduction to the Synthon Approach*, Wiley, New York, 1978.
S. Warren, *Organic Synthesis: The Disconnection Approach*, Wiley, New York, 1982.

Stereoselective Synthesis

J. W. ApSimon, *Tetrahedron* **35**, 2797 (1979).
P. A. Bartlett, *Tetrahedron* **36**, 3 (1980).
G. M. Coppola and H. F. Schuster, *Asymmetric Synthesis*, Wiley-Interscience, New York, 1987.
B. Fraser-Reid and R. C. Anderson, *Fortschr. Chem. Org. Naturstoffe* **39**, 1 (1980).
S. Hanessian, *Total Synthesis of Natural Products, the Chiron Approach*, Pergamon, New York, 1983.
Y. Izumi and A. Tai, *Stereodifferentiating Reactions*, Academic Press, New York, 1977.
H. B. Kagan and J. C. Fiaud, *Top. Stereochem.* **10**, 175 (1977).
H. S. Mosher and J. D. Morrison, *Asymmetric Organic Reactions*, Second Edition, American Chemical Society, Washington, D.C., 1976.
S. Nogradi, *Stereoselective Syntheses*, Verlag Chemie, Weinheim, 1987.

Description of Total Syntheses

N. Anand, J. S. Bindra, and S. Ranganathan, *Art in Organic Synthesis*, Second Edition, Wiley-Interscience, New York, 1988.
J. ApSimon (ed.), *The Total Synthesis of Natural Products*, Vols. 1–7, Wiley-Interscience, New York, 1973–1988.
J. S. Bindra and R. Bindra, *Creativity in Organic Synthesis*, Academic Press, New York, 1975.
S. Danishefsky and S. E. Danishefsky, *Progress in Total Synthesis*, Meredith, New York, 1971.
I. Fleming, *Selected Organic Syntheses*, Wiley-Interscience, New York, 1973.

Problems

(*References for these problems will be found on page* 780.)

1. Indicate conditions which would be appropriate for the following transformations involving introduction or removal of protecting groups.

(a) CH_3, CH_3, OH, $(CH_3)_2CH$ → CH_3, CH_3, $OCH_2OCH_2CH_2OCH_3$, $(CH_3)_2CH$

(b)

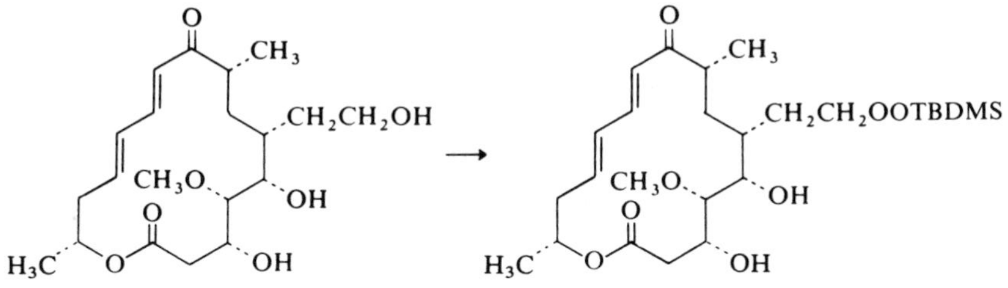

$\longrightarrow$

(c)

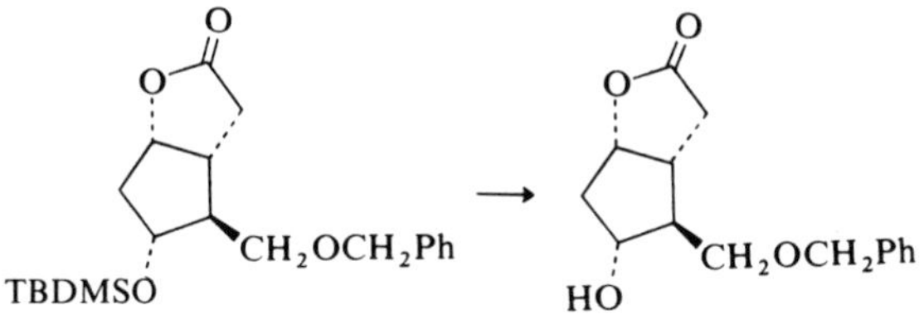

$\longrightarrow$

(d)

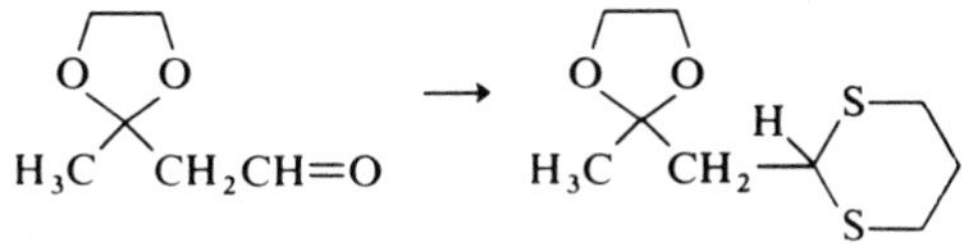

$\longrightarrow$

(e)

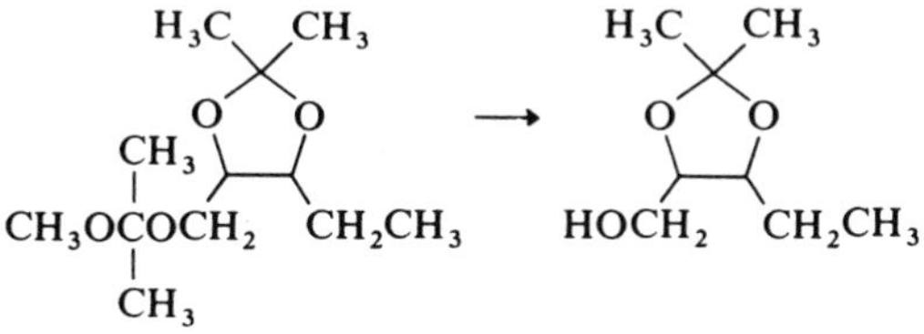

$\longrightarrow$

(f)

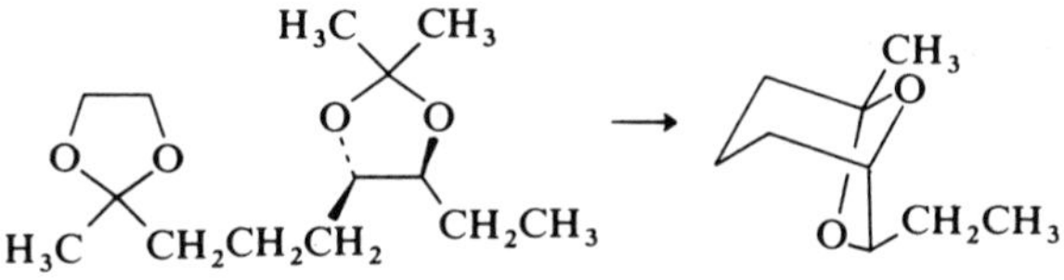

$\longrightarrow$

2. Indicate the product to be expected under the following reaction conditions.

(a)

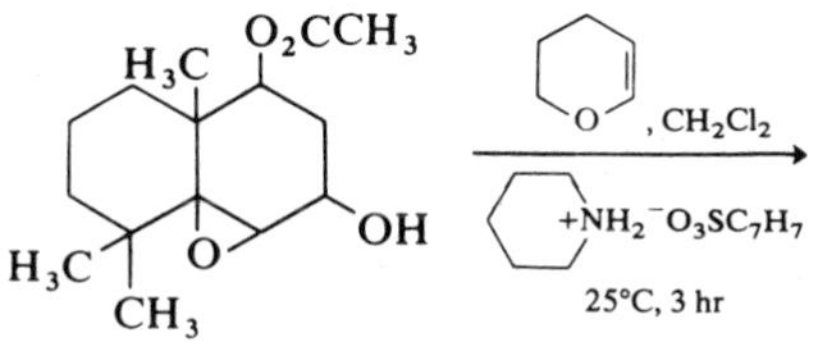

, CH_2Cl_2

$^+NH_2\ ^-O_3SC_7H_7$

25°C, 3 hr

(b)

$-CH_2C-C-CH_2OH + CH_2=CCH_3$ $\xrightarrow[CH_2Cl_2]{POCl_3}$

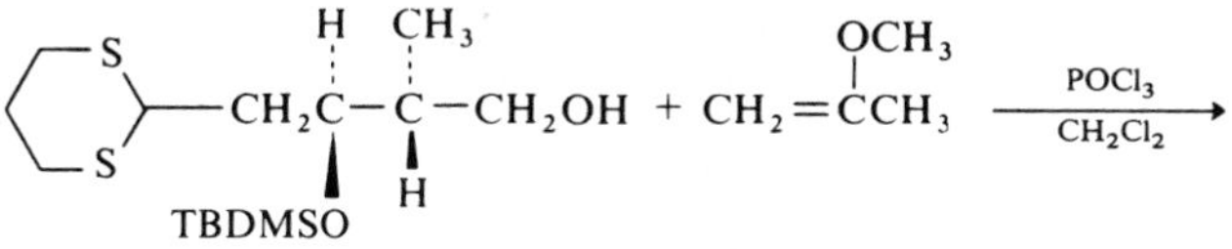

(c) $PhCH_2OCNHCH_2C-N$ (proline, CO_2H) $\xrightarrow[\text{ethanol}]{\text{Pd, cyclohexadiene}}$

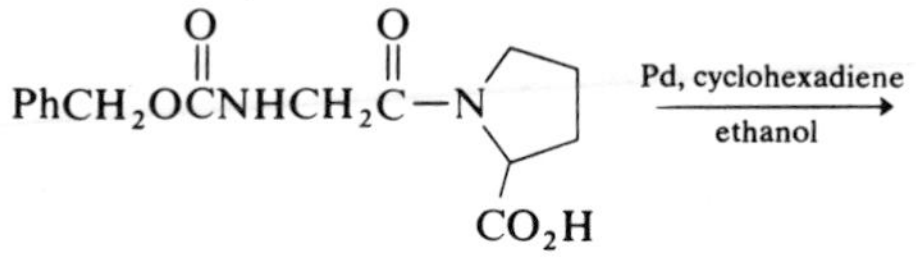

(d) NH_2, H, CO_2H, C, CH_2SH + $PhCH{=}O$ $\longrightarrow$

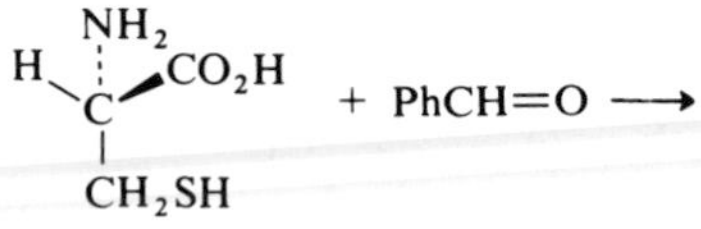

(e) $CH(SCH_2CH_3)_2$, H—C—OH, H—C—OH, HO—C—H, HO—C—H, CH_3 + CH_3CCH_3 (O) $\xrightarrow{CuSO_4}$ formula is $C_{13}H_{26}O_4S_2$

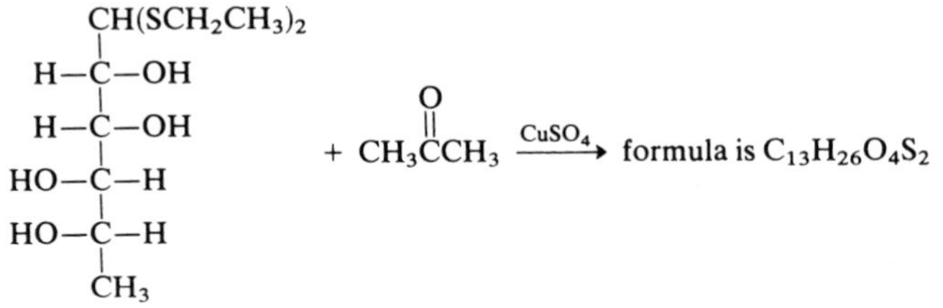

3. In each of the synthetic transformations shown, the reagents are appropriate, but the reactions will not be practical as they are written. What modification would be necessary to permit each transformation to be carried out to give the desired product?

(a) CH_3, O, H, H_3C, O, O, CH_3 $\xrightarrow{LiAlH_4}$ CH_3, O, H, H_3C, OH, CH_3, $CHCH_2OH$

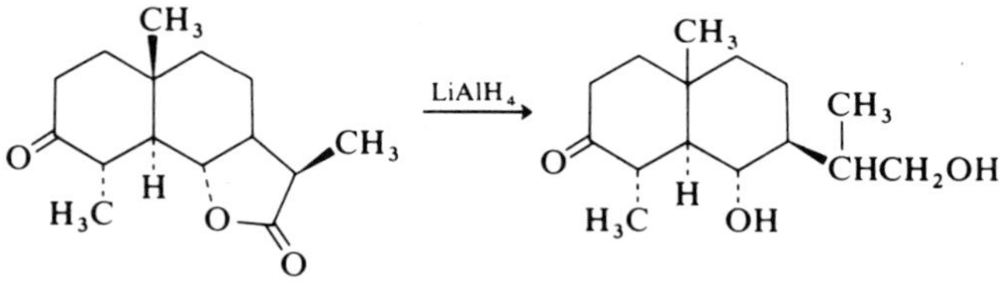

(b) H, OH, H_3C, O, H_3C, CH_3 $\xrightarrow[\text{2) } H_2NNH_2\text{, KOH}]{\text{1) } CrO_3\text{-acetone}}$ O, H_3C, H_3C, CH_3

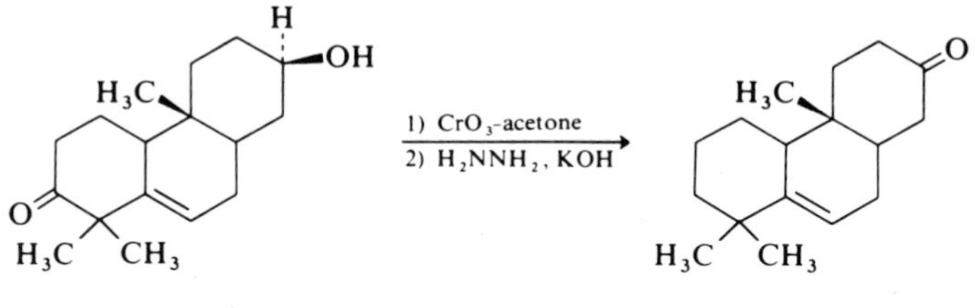

(c) H_3C, CH_3, H_3C, OH, $CH_2CH_2CH_2OH$ $\xrightarrow[\text{pyridine}]{POCl_3}$ H_3C, CH_3, H_3C, $CH_2CH_2CH_2OH$

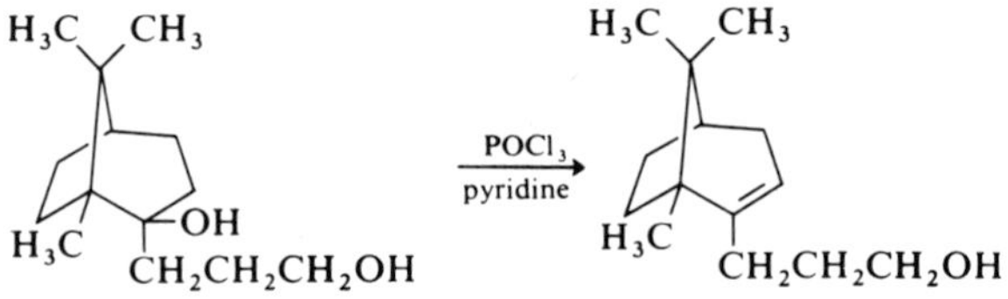

(d) CH_3, O, H_3C $\xrightarrow[NaNH_2]{CH_3I}$ H_3C, CH_3, O, H_3C

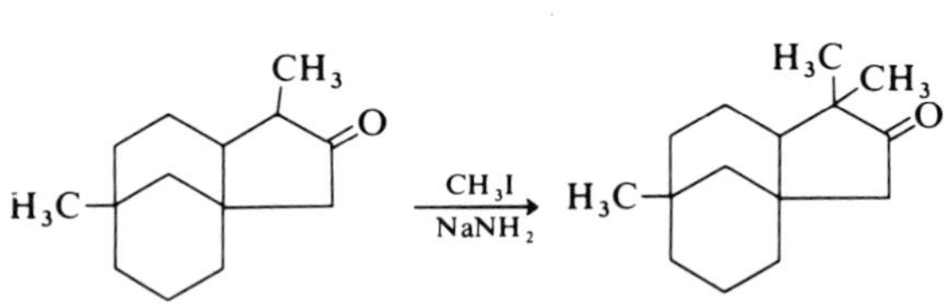

(e) [structure: $H_2NCH(HO_2C)$-thiazolidine with CH_3, CH_3, CO_2CH_2Ph] $\xrightarrow{(CH_3)_2CHN=C=NCH(CH_3)_2}$ [structure: H_2N-β-lactam, CH_3, CH_3, CO_2CH_2Ph]

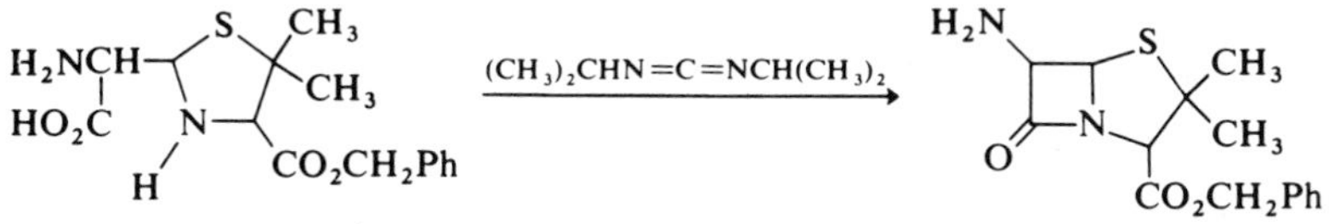

(f) [structure: H_3C bicyclic dione] $\xrightarrow{CH_3CH=PPh_3}$ [structure: H_3C, $CH_3CH=$ bicyclic ketone]

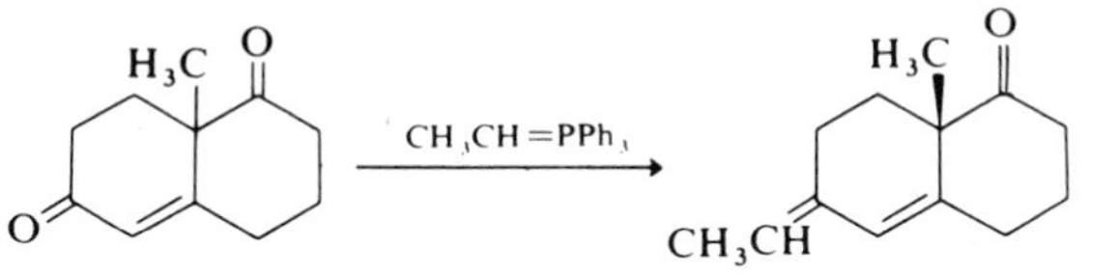

4. Under certain circumstances, each of the following groups can serve as a temporary protecting group for secondary amines by acting as a removable tertiary substituent. Suggest conditions which might be appropriate for subsequent removal of each group.

(a) $PhCH_2-$ (b) $CH_2=CHCH_2-$ (c) $CH_2=CH-$ (d) $PhCH_2OCH_2-$

5. Show how synthetic equivalent groups might be used to efficiently carry out the following transformations.

(a) $BrH_2C-C_6H_4-CH_2Br$ (meta) $\longrightarrow$ [cyclophane diketone structure]

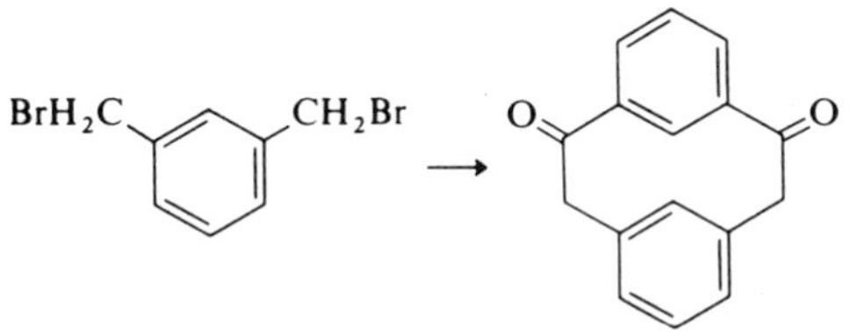

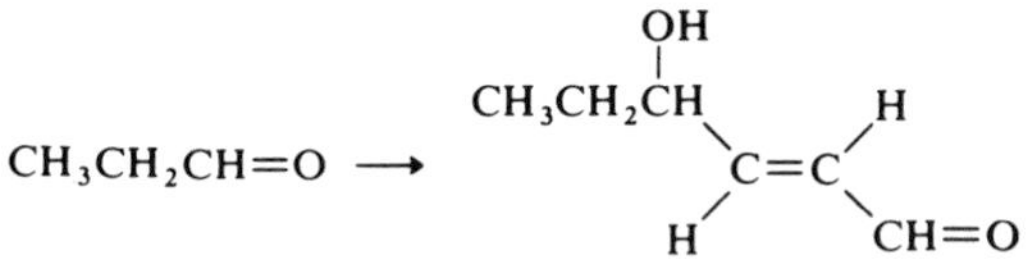

(b) $CH_3CH_2CH=O \longrightarrow CH_3CH_2CH(OH)CH=CHCH=O$ (trans)

(c) [structure: acetonide pyranoside ketone, OCH_3, H_3C, H_3C] $\longrightarrow$ [structure: acetonide pyranoside with OH and CH_2OH, OCH_3]

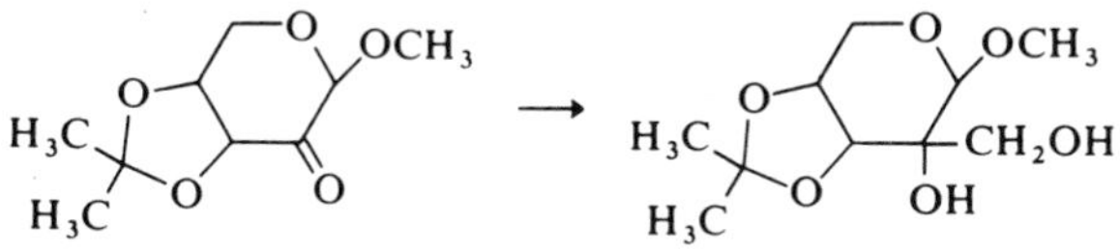

(d) $(CH_3)_2C=CHCH_2CH_2C(CH_3)(-O-)CH_2$ (epoxide) $\longrightarrow (CH_3)_2C=CHCH_2CH_2C(CH_3)=CHCH=O$

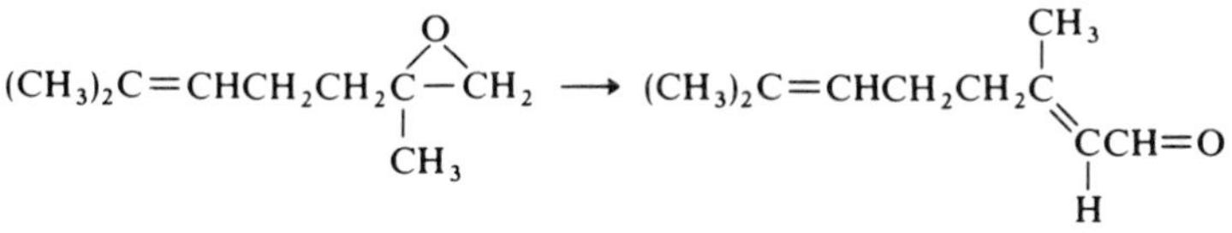

(e) [structure: bicyclic enone (hydroazulenone)] [structure: $CH_3C(=O)$–, OH substituted bicyclic]

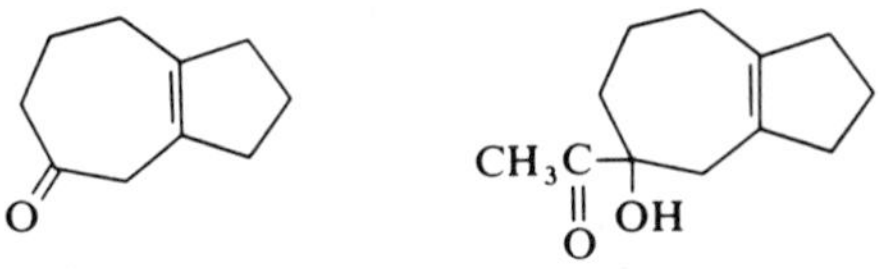

(f) 3-pyridinecarboxaldehyde (CH=O) → 3-pyridyl-$\overset{O}{\|}{C}CH_2CH_2CN$

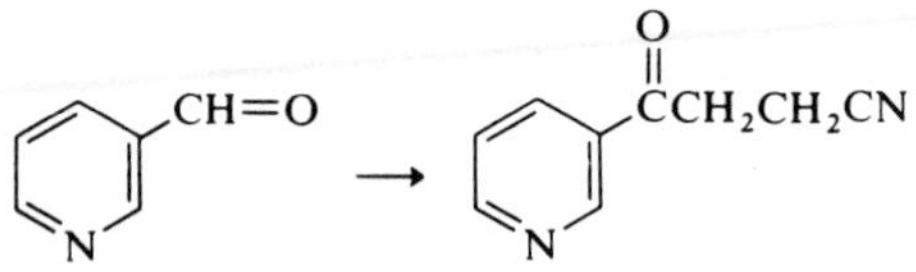

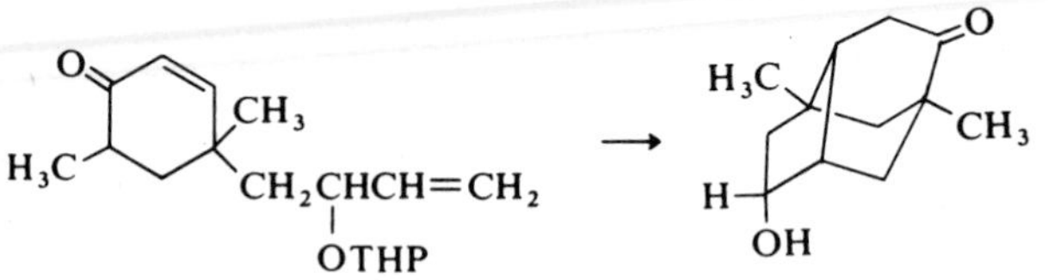

(g) O=, H_3C, CH_3, $CH_2CHCH=CH_2$, OTHP → H_3C, O, CH_3, H, OH

(h) C6H5–CH=O → C6H5–$\overset{O}{\|}{C}CH_2CH_2CH_2CH_3$

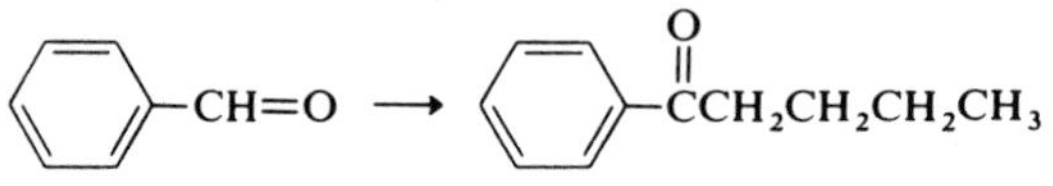

(i) CH_3, CH_2Cl, CH=O, CH_3, CH_3, CH_3 → CH_3, O, CH_3, CH_3, CH_3

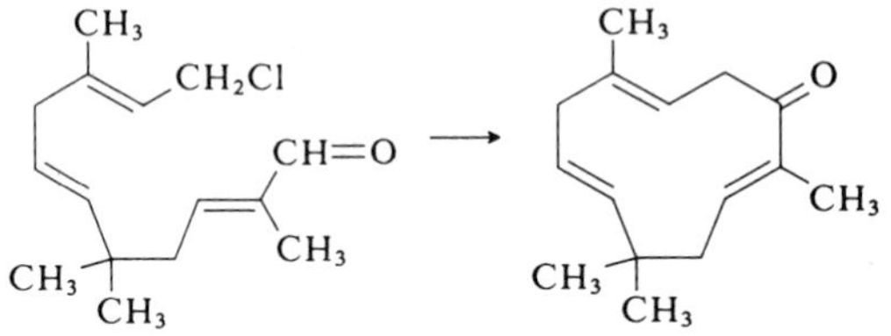

6. Indicate a reagent or short sequence that would accomplish each of the following synthetic transformations:

(a) $CH_3\overset{O}{\|}{C}CH=CH_2$ → $\overset{O}{\|}{C}CH_3$, O

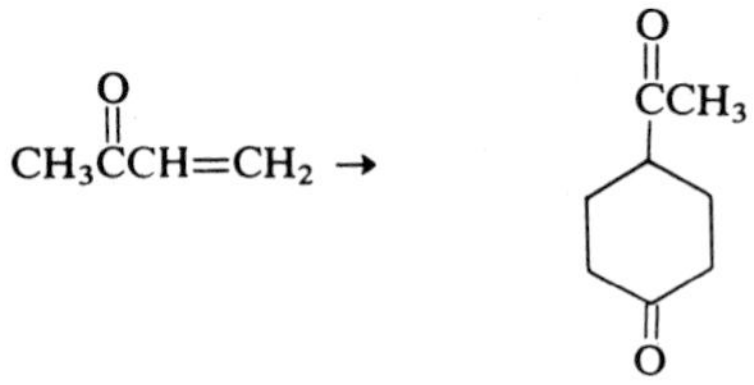

(b) $H_2C=C$, CH_3, $C(CH_3)_2$, OH → O, CH_3, $C(CH_3)_2$, OH

(c) cyclohexene oxide → OH, CH=CHCH=O

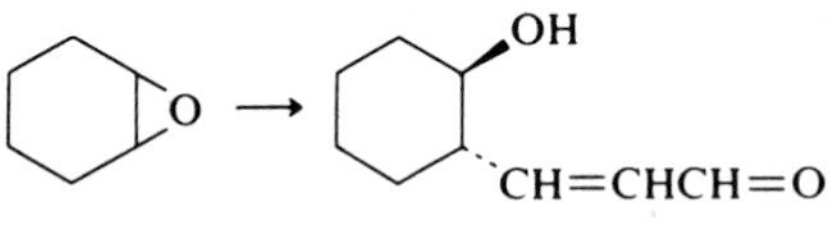

(d) O, Ph → H_3C, O, Ph

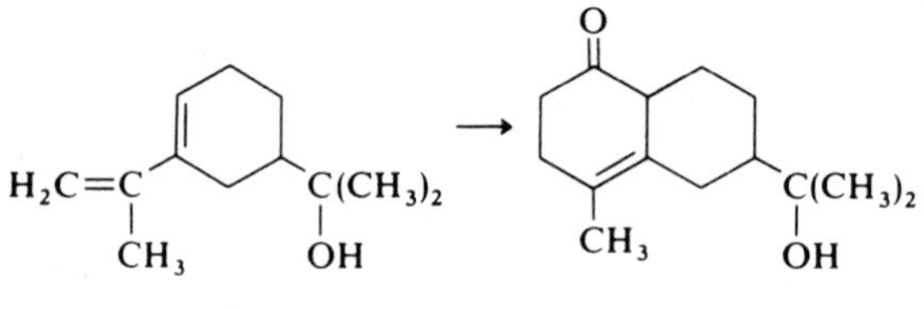

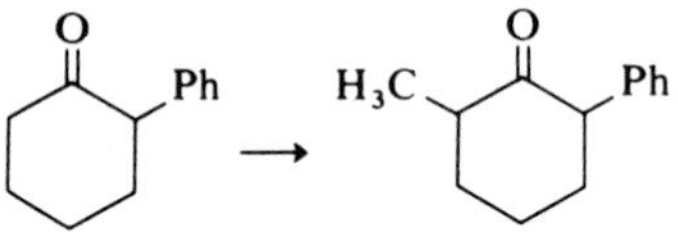

(e) $CH_3C(=CH_2)CH_2OH \rightarrow CH_3C(=CH_2)CH_2CH_2CH(OH)C(CH_3)=CH_2$

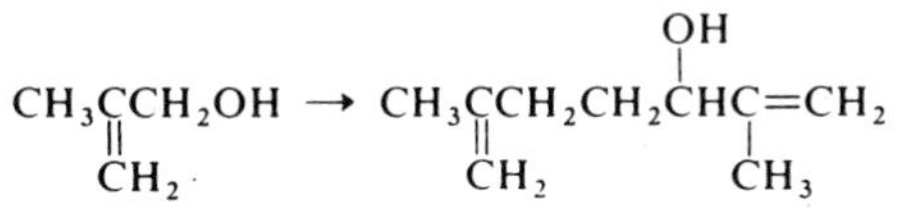

(f) [structure with CO_2CH_3] → [structure with O_2CCH_3]

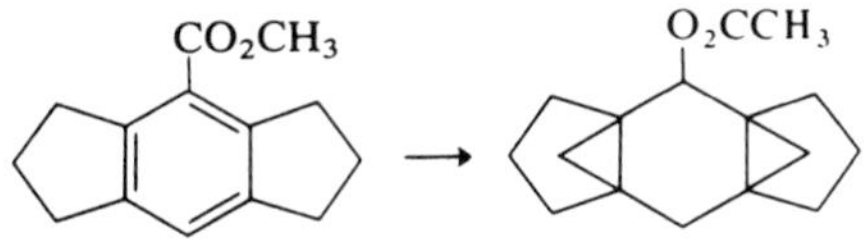

7. Indicate a reagent or short reaction sequence which could accomplish synthesis of the material shown on the left from the starting material on the right.

(a) [cyclopentane with CCH_3 (C=O) and $CH_2CH_2CH_2CH_3$] ⇒ [cyclopentenone]

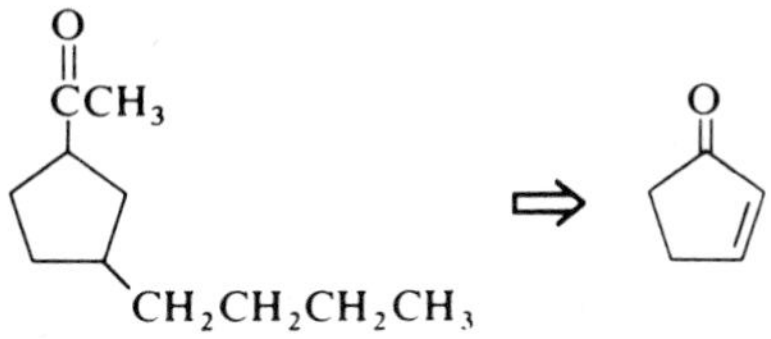

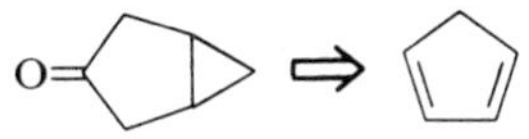

(b) [bicyclic ketone, O=] ⇒ [cyclopentadiene]

(c) [spiro bicyclobutene] ⇒ $C(CH_2Br)_4$

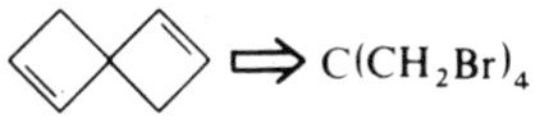

(d) [spiro enone with CH_3, CH_3] ⇒ [$Si(CH_3)_3$, H, C=C, H; HO, CH_3, CH_3]

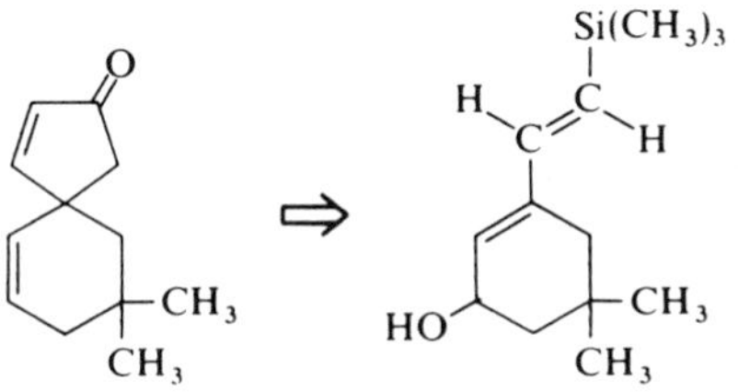

(e) $PhCH_2OCH_2CH_2C$ [chain with CH_3O, H, H, HO, H, HO, H, H, CO_2CH_3] ⇒ [$HOCH_2CH_2$, O, O, CH_3, CH_3O, O, CH_3]

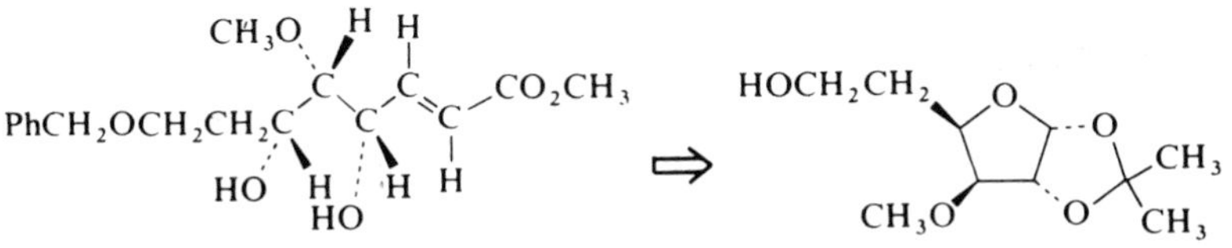

8. Because they are readily available from natural sources in optically pure form, carbohydrates are very useful starting materials for the synthesis of optically pure substances. However, the high number of similar functional groups present in carbohydrates requires versatile techniques for selective protection and selective reaction. Show how appropriate manipulation of protecting groups and/or selective reagents might be employed to effect the desired transformations.

(a) [$HOCH_2$, $HOCH$, O, $HOCH_2$, O, O, CH_3, CH_3] → [CH_2OH, O, $HOCH_2$, O, O, CH_3, CH_3]

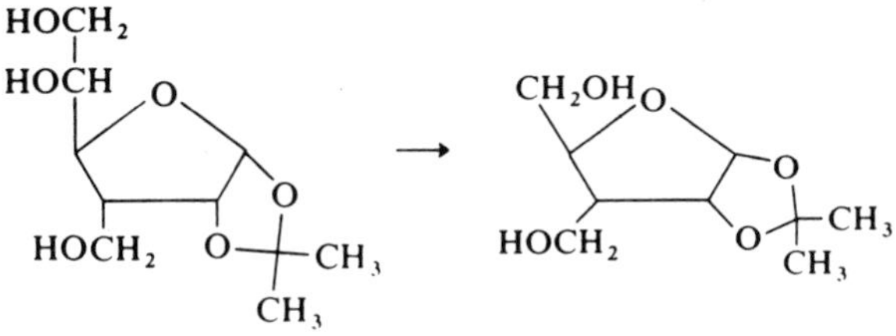

(b)

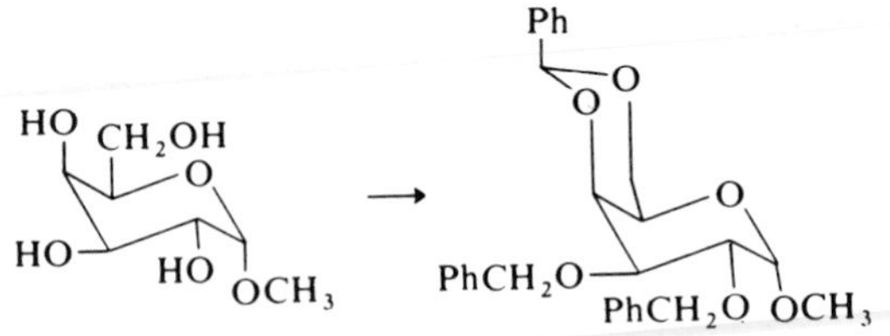

(c)

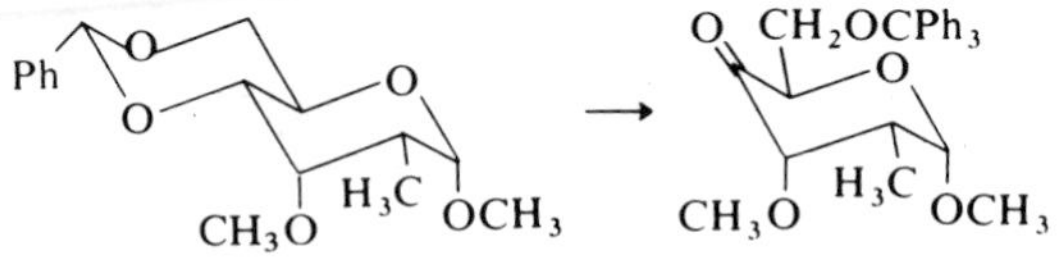

(d)

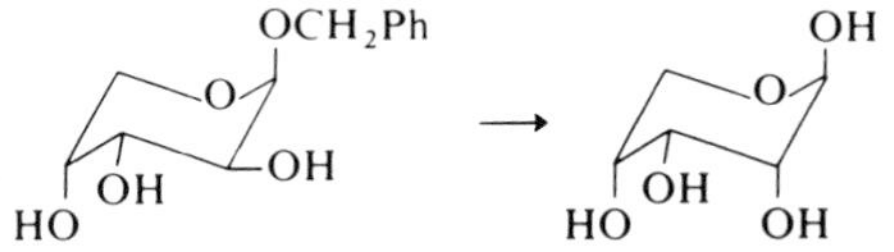

9. Synthetic transformations which are parts of total syntheses of natural products are outlined by a general retrosynthetic outline. For each retrosynthetic disconnection, suggest a reagent or short sequence of reactions which could accomplish the forward synthetic reaction. The proposed route should be diastereoselective but need not be enantioselective.

(a)

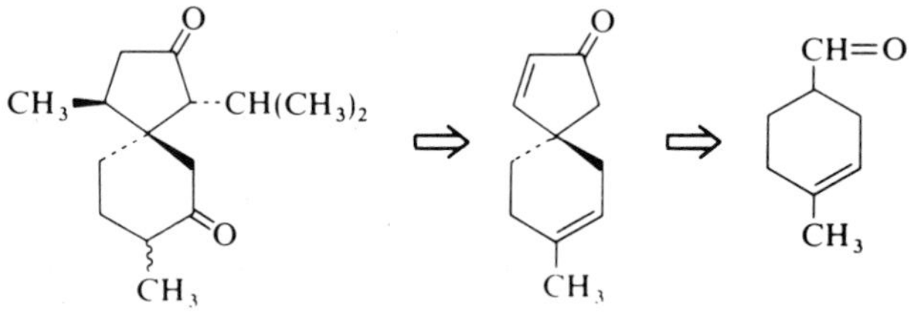

(b)

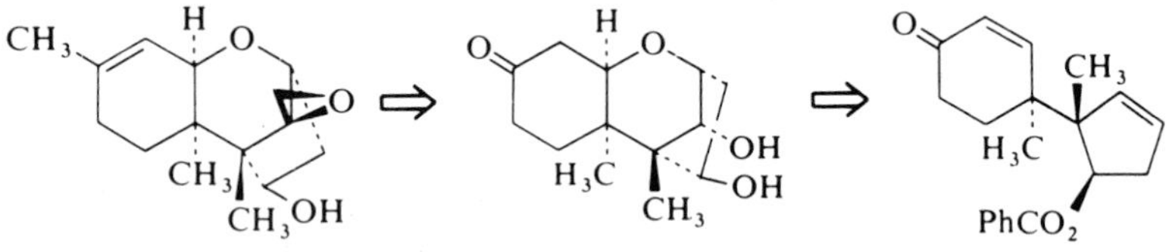

(c)

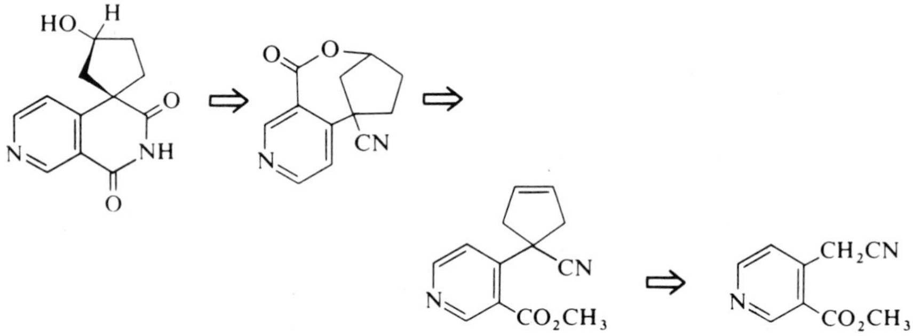

(d)

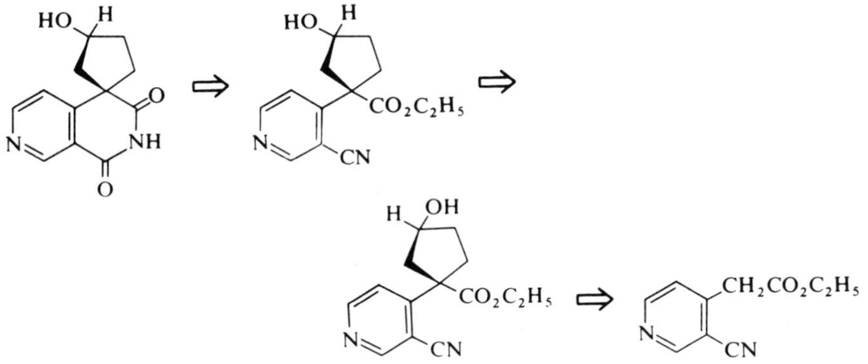

HO H ⇒ HO H $CO_2C_2H_5$ CN ⇒ H OH $CO_2C_2H_5$ CN ⇒ $CH_2CO_2C_2H_5$ CN

(e)

CH_3 H CH_3 ⇒ CH_3 H CH_3 OH ⇒ CH_3 $CH{=}CHCO_2C_2H_5$ CH_3 ⇓ CO_2H CH_3

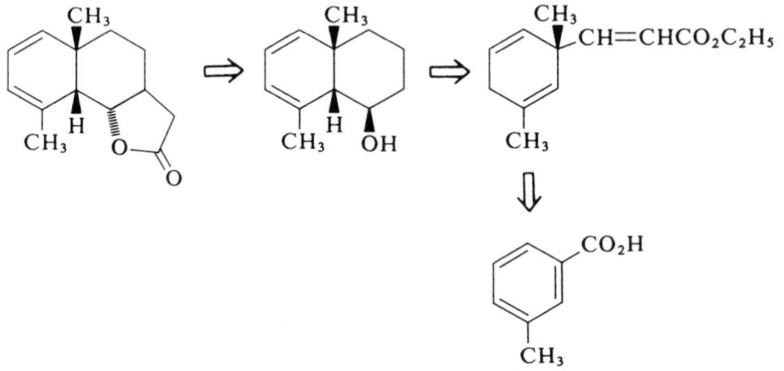

10. Diels–Alder reactions are attractive for many synthetic applications, particularly because of their predictable stereochemistry. There are, however, significant limitations on the type of compound which can serve as a dienophile or diene. As a result, the idea of synthetic equivalency has been exploited in this area. For each of the reactive dienophiles and dienes given below, suggest one or more transformations which might be carried out on a Diels–Alder adduct derived from it that would lead to a product not directly attainable by a Diels–Alder reaction. Give the structure of the diene or dienophile "synthetic equivalent" and indicate why the direct Diels–Alder reaction would not be possible.

Dienophiles

(a) $CH_2{=}CH\overset{+}{P}Ph_3$

(b) $CH_2{=}CH\overset{+}{S}(O^-)Ph$

(c) $CH_2{=}C(O_2CCH_3)CO_2C_2H_5$

Dienes

(d) $CH_2{=}C(OSi(CH_3)_3){-}CH{=}CHOCH_3$

(e) $CH_2{=}C(O_2CCH_3){-}C(SPh){=}CH_2$

11. One approach to the synthesis of enantiomerically pure materials is to start with an available homochiral material and effect the synthesis by a series of stereospecific reactions. Devise a sequence of reactions which would be appropriate for the following syntheses based on enantiomerically pure starting materials.

(a) 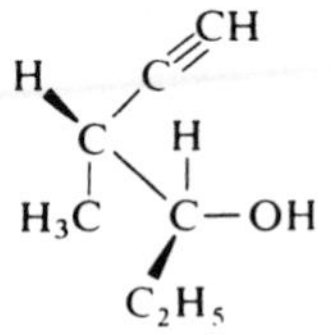 from

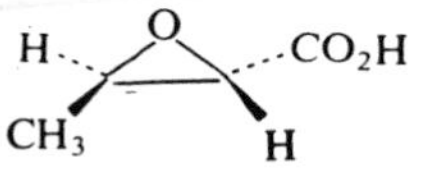

(b) 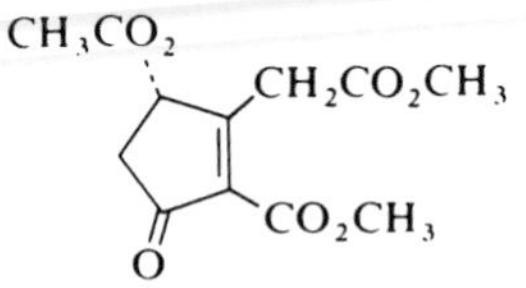 from

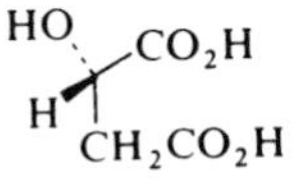

(c) 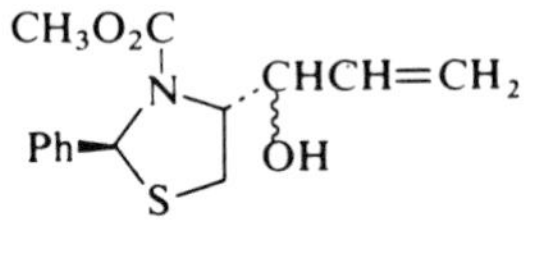 from

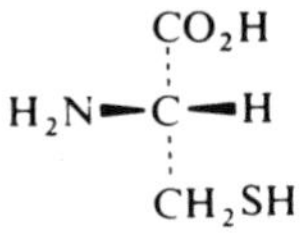

(d) 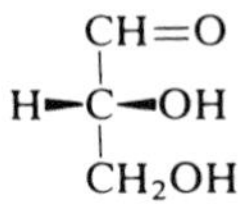 from

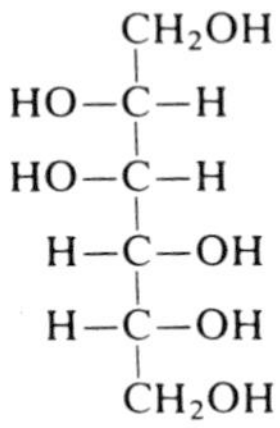

(e) 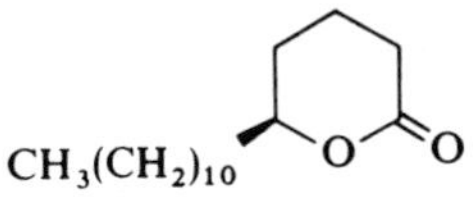 from

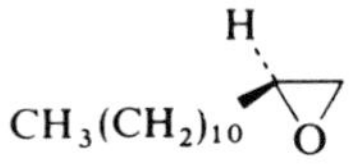

(f) 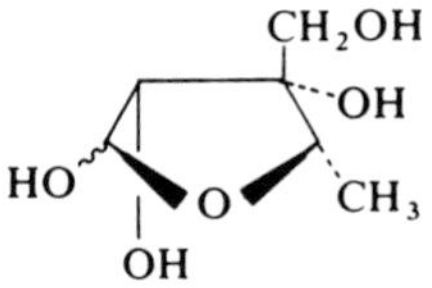 from

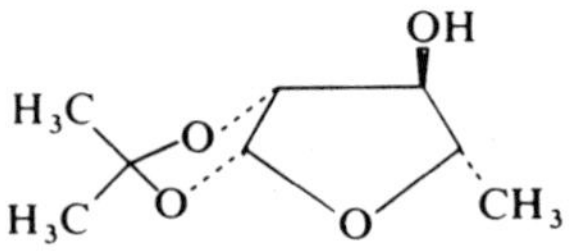

(g) 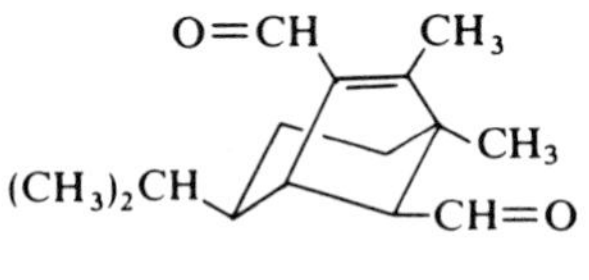 from

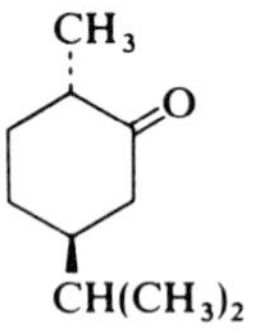

(h) 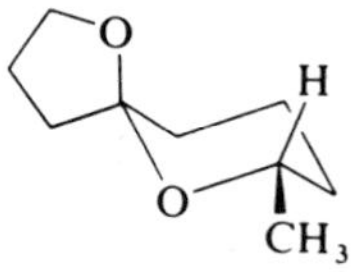 from

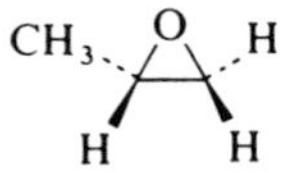

12. Several natural product syntheses are outlined in retrosynthetic form. Suggest a reaction or short reaction series which could accomplish each lettered transformation in the forward synthetic direction. The structures shown refer to racemic materials.

(a) Isotelekin

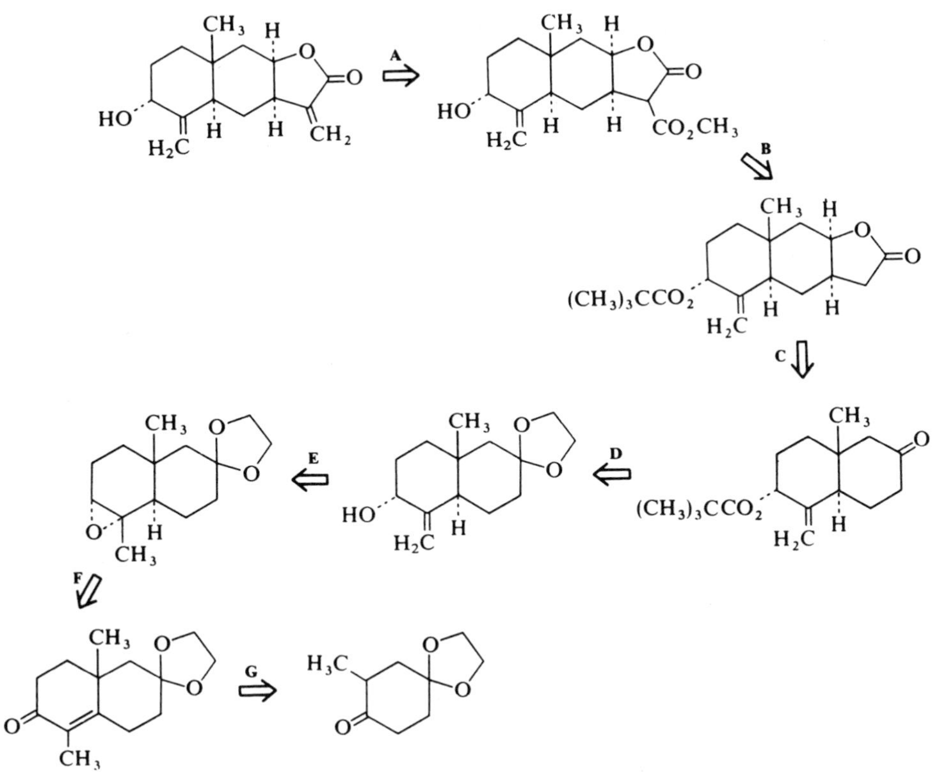

CH_3 H HO H_2C H H CH_2 $\stackrel{A}{\Rightarrow}$ CH_3 H HO H_2C H H CO_2CH_3 $\stackrel{B}{\Rightarrow}$ CH_3 H $(CH_3)_3CCO_2$ H_2C H H $\stackrel{C}{\Rightarrow}$ CH_3 $(CH_3)_3CCO_2$ H_2C H $\stackrel{D}{\Leftarrow}$ CH_3 HO H_2C H $\stackrel{E}{\Leftarrow}$ CH_3 O H CH_3 $\stackrel{F}{\Rightarrow}$ CH_3 O CH_3 $\stackrel{G}{\Rightarrow}$ H_3C O

(b) Disparlure

H $(CH_2)_9CH_3$ O H $(CH_2)_4CH(CH_3)_2$ $\stackrel{A}{\Rightarrow}$ THP—O H C—$(CH_2)_9CH_3$ HO C $(CH_2)_4CH(CH_3)_2$ H $\stackrel{B}{\Rightarrow}$ THP—O H···C—$(CH_2)_9CH_3$ HO C $CH_2CH_2CH{=}CHCH(CH_3)_2$ H $\stackrel{C}{\Rightarrow}$ THP—O H—C—$(CH_2)_9CH_3$ O O $\stackrel{D}{\Leftarrow}$ O C$(CH_2)_9CH_3$ O O $\stackrel{E}{\Leftarrow}$ CO_2H O O

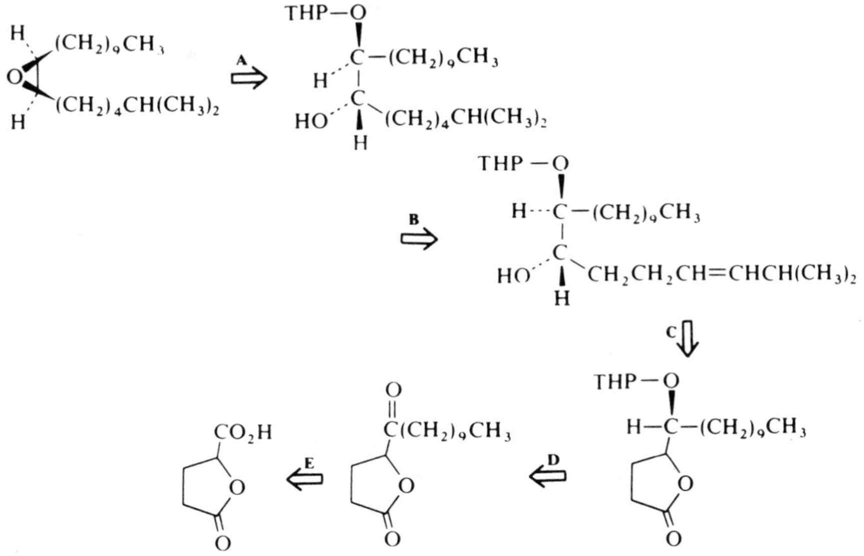

(c) Aromandrene

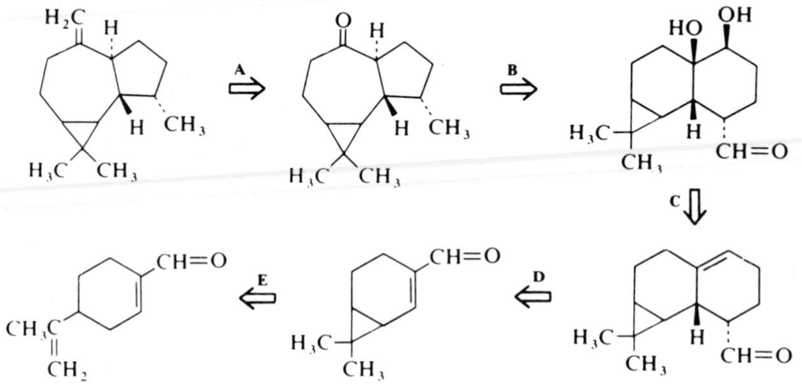

H_2C, H, H, CH_3, H_3C, CH_3 ⇨A O, H, H, CH_3, H_3C, CH_3 ⇨B HO, OH, H_3C, H, CH_3, CH=O

⇩C

CH_3C, CH_2, CH=O ⇦E H_3C, CH_3, CH=O ⇦D H_3C, H, CH_3, CH=O

(d) α-Bourbonene

CH_3, H, $(CH_3)_2CH$, H, H, CH_3 ⇨A CH_3, O, $(CH_3)_2CH$, H, CH_3 ⇨B H_3C, O, CCH_3, $(CH_3)_2CH$, H, CCH_3, O

⇩C

CH_3, CH_2CH_2, H, C=C, CHC, $H_5C_2O_2C$, CH_3, $CCO_2C_2H_5$, $(CH_3)_2CH$, H ⇦D CH_3, $CO_2C_2H_5$, $(CH_3)_2CH$, H, $CO_2C_2H_5$

E ⇩

$(CH_3)_2CHCH_2CH_2\overset{O}{\overset{\|}{C}}CH_3$

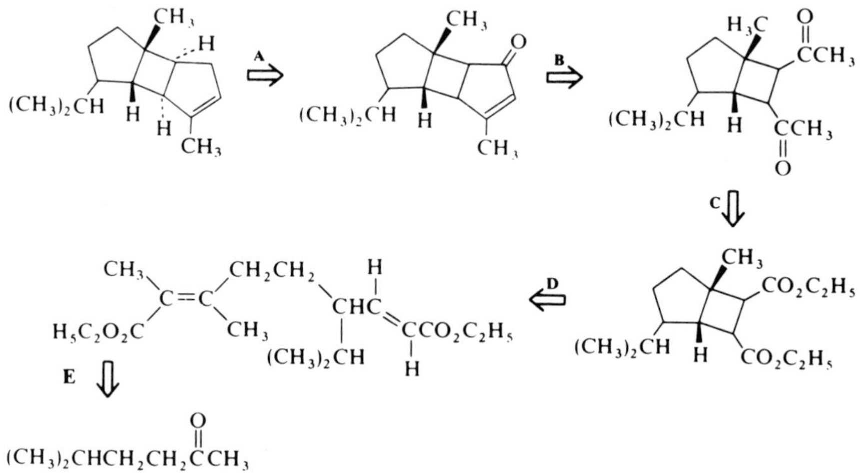

(e) Caryophyllene

H_3C, H, H_3C, H_3C, H, H_2C ⇨A H_3C, H, H_3C, CH_3, OH, H, HO ⇨B H_3C, H, H_3C, CH_3, H, CO_2CH_3, O, O

⇩C

O ⇦E H_3C, H, H_3C, H, O ⇦D H_3C, H, H_3C, CH_3, H, O, CO_2CH_3

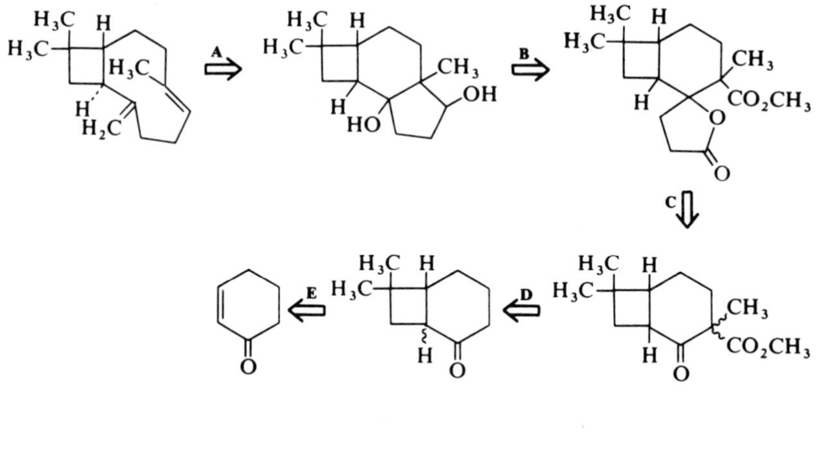

13. Perform a retrosynthetic analysis for each of the following molecules. Develop at least three outline schemes. Discuss the relative merits of the three schemes and develop a fully elaborated synthetic plan for the one you consider to be most promising.

(a) (b)

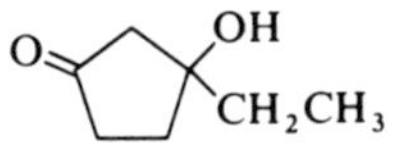

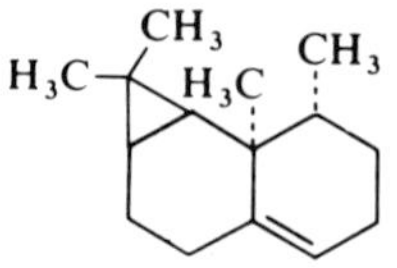

(c) 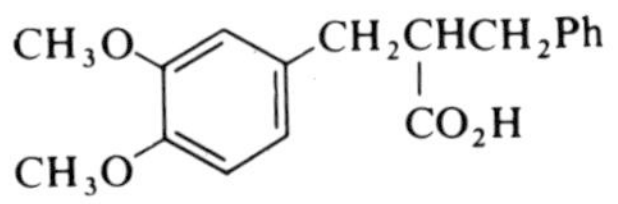 (d)

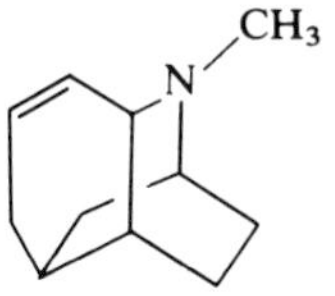

(e) 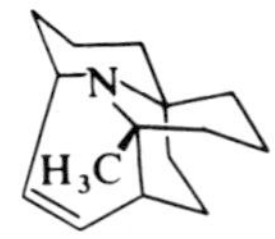 (f)

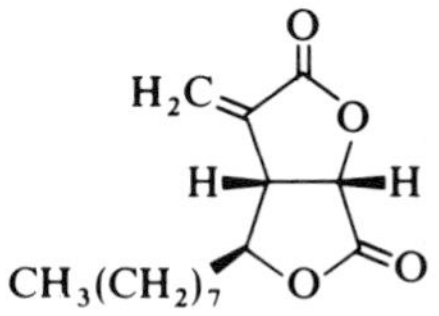

(g) 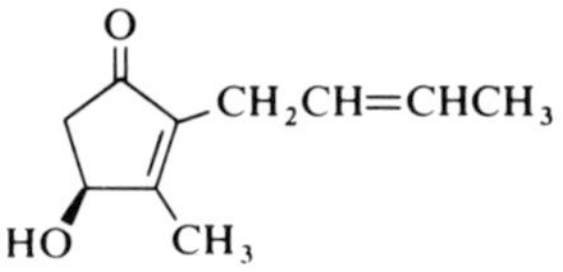 (h)

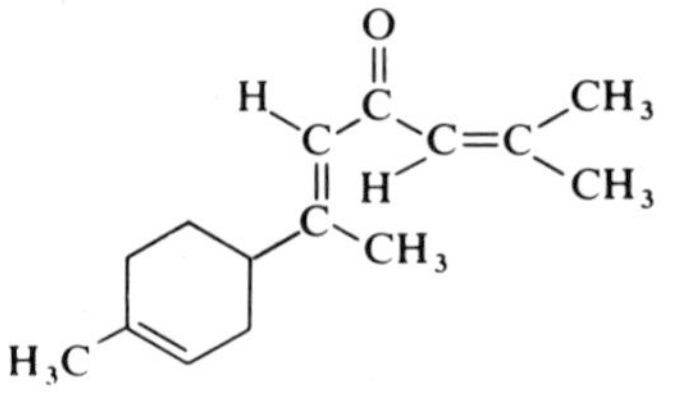

(i) 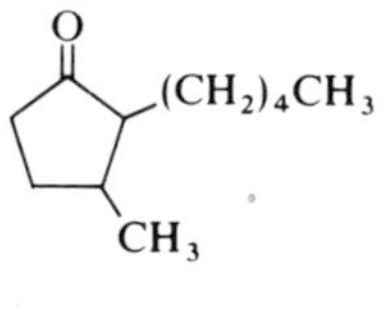 (j)

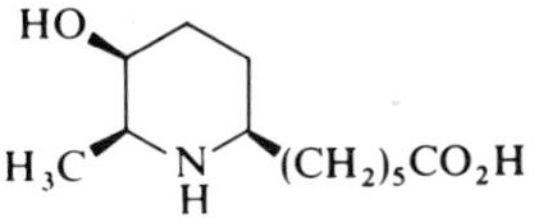

(k) 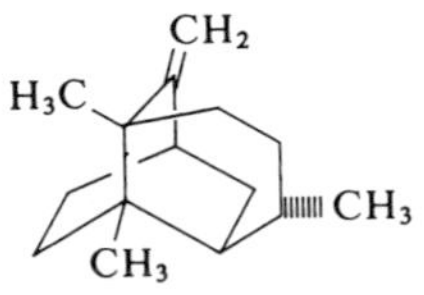 (l)

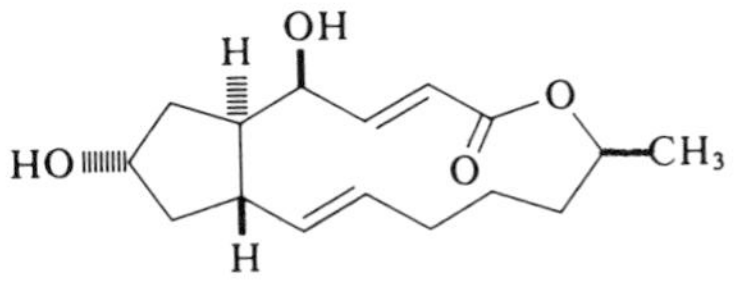

(m)

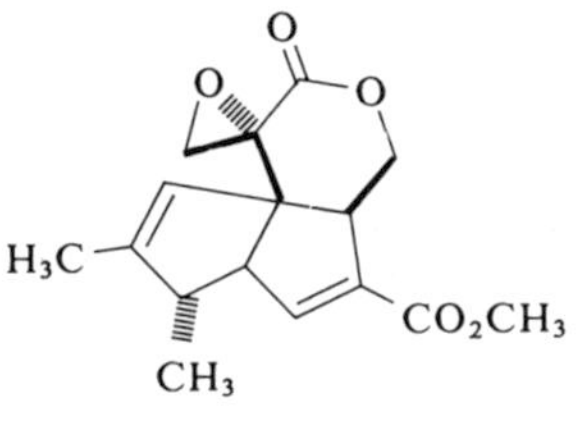

14. Suggest methods that would be expected to achieve a diastereoselective synthesis of the following compounds.

(a) 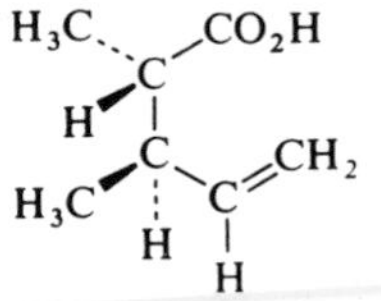 H_3C, CO_2H, H, H_3C, H, CH_2, H

(b) 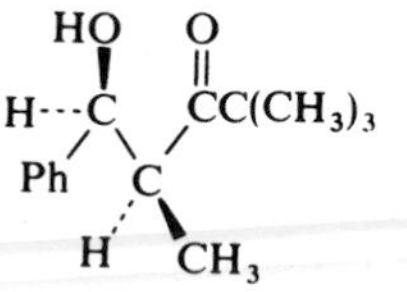 HO, O, H, $CC(CH_3)_3$, Ph, H, CH_3

(c) 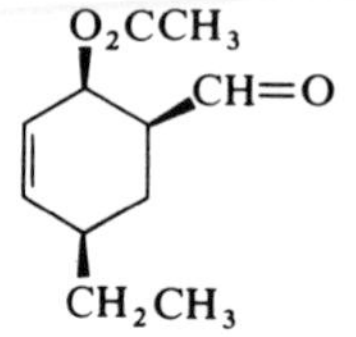 O_2CCH_3, $CH{=}O$, CH_2CH_3

(d) 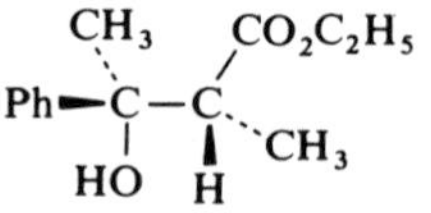 CH_3, $CO_2C_2H_5$, Ph—C—C, CH_3, HO, H

(e) 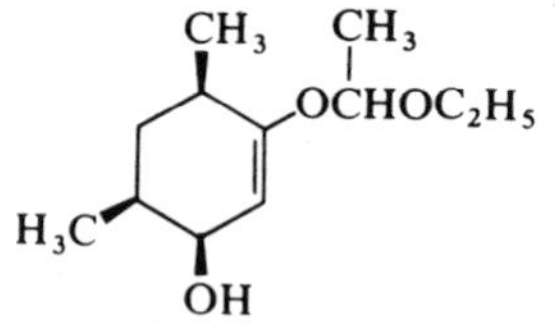 CH_3, CH_3, $OCHOC_2H_5$, H_3C, OH

15. Devise a route for synthesis of the desired compound in high enantiomeric purity from the suggested starting material.

(a) 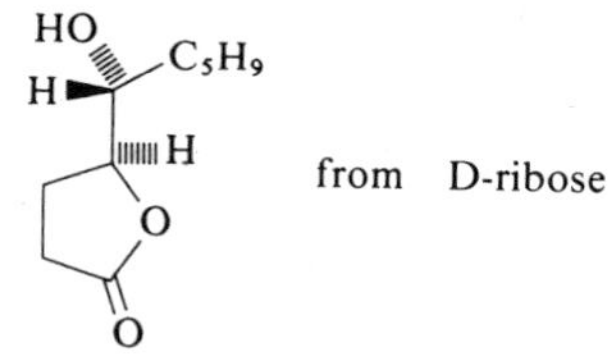 HO, C_5H_9, H, H, O, O from D-ribose

(b) OCH_2Ph, $ArSO_2N$ H, OCH_2Ph, O, $CO_2C_2H_5$ from $ArSO_2N$ H CH_2OH

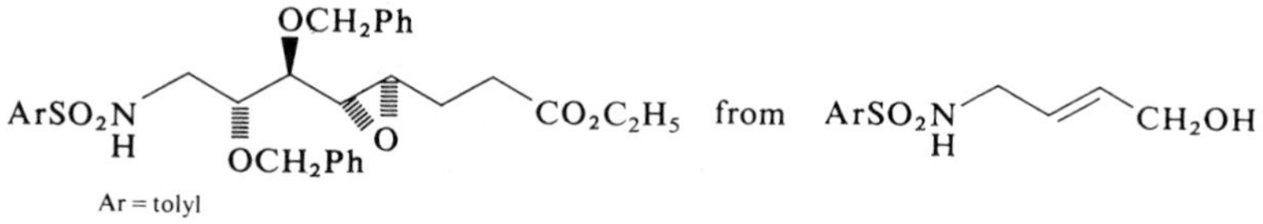

Ar = tolyl

(c) OCH_3, O, H_3C, O from H_3C—(cyclopentadiene)

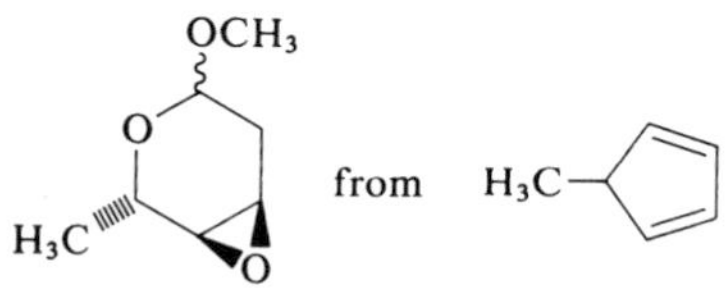

16. By careful consideration of transition state geometry using models, predict the *absolute configuration* of the major product for each reaction. Explain the basis of your prediction.

(a) 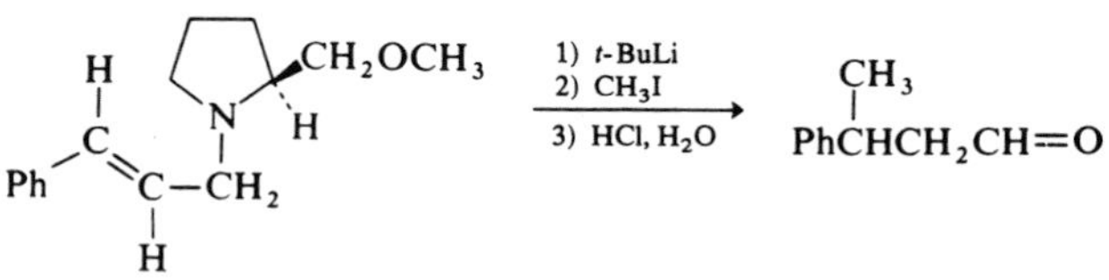 H, Ph, C, N, CH_2OCH_3, H, C—CH_2, H $\xrightarrow[\text{3) HCl, } H_2O]{\text{1) } t\text{-BuLi; 2) } CH_3I}$ $PhCH(CH_3)CH_2CH{=}O$

(b) $PhCH_2CH_2$–(oxazoline; O, N; Ph; CH_2OCH_3) $\xrightarrow[\text{3) } H^+, H_2O]{\text{1) } LiNR_2 \text{ 2) } CH_3CH_2CH_2CH_2I}$ $PhCH_2CH(CH_2CH_2CH_2CH_3)CO_2H$

(c) $(CH_3)_3C$–C(H)($CO_2C(CH_3)_3$)–N=(cyclohexylidene) $\xrightarrow[\text{2) } CH_3I]{\text{1) } LiNR_2}$ (2-methylcyclohexanone: O, CH_3)

(d) $PhC(=O)$–(oxazoline; O, N; Ph; CH_2OCH_3) $\xrightarrow[\text{2) } H^+, H_2O]{\text{1) } (CH_3)_2CHCH_2MgBr}$ $(CH_3)_2CHCH_2C(OH)(Ph)CO_2H$

(e) (H, Ph)C=C(CH_3)–C(=O)–N(pyrrolidine, CO_2H) $\xrightarrow[\text{2) } KOC(CH_3)_2]{\text{1) } Br_2}$ (H_3C, O, Ph, H epoxide)–C(=O)–N(pyrrolidine, CO_2H)

(f) $H_2C=CHCH=CHO_2C-C(Ph)(H)(OCH_3)$ + (5-hydroxynaphthoquinone: O, OH, O) $\xrightarrow[0°C]{B(O_2CCH_3)_3}$ (adduct: O, OH, O, OCHPh, OCH_3)

(g) $PhCH=CH$–(oxazoline; O, N; Ph; CH_2OCH_3) $\xrightarrow[\text{2) } H^+, H_2O]{\text{1) } C_2H_5Li}$ $PhCH(C_2H_5)CO_2H$

References for Problems

Chapter 1

1a. W. S. Matthews, J. E. Bares, J. E. Bartmess, F. G. Bordwell, F. J. Cornforth, G. E. Drucker, Z. Margolin, R. J. McCallum, G. J. McCollum, and N. E. Vanier, *J. Am. Chem. Soc.* **97**, 7006 (1975).
b. H. D. Zook, W. L. Kelly, and I. Y. Posey, *J. Org. Chem.* **33**, 3477 (1968).
2a. H. O. House and M. J. Umen, *J. Org. Chem.* **38**, 1000 (1973).
b. W. C. Still and M.-Y. Tsai, *J. Am. Chem. Soc.* **102**, 3654 (1980).
c. H. O. House and B. M. Trost, *J. Org. Chem.* **30**, 1341 (1965).
d. D. Caine and T. L. Smith, Jr., *J. Am. Chem. Soc.* **102**, 7568 (1980).
e. M. F. Semmelhack, S. Tomoda, and K. M. Hurst, *J. Am. Chem. Soc.* **102**, 7567 (1980).
f. R. A. Lee, C. McAndrews, K. M. Patel, and W. Reusch, *Tetrahedron Lett.*, 965 (1973).
g. R. H. Frazier, Jr., and R. L. Harlow, *J. Org. Chem.* **45**, 5408 (1980).
3a. M. Gall and H. O. House, *Org. Synth.* **52**, 39 (1972).
b. P. S. Wharton and C. E. Sundin, *J. Org. Chem.* **33**, 4255 (1968).
c. B. W. Rockett and C. R. Hauser, *J. Org. Chem.* **29**, 1394 (1964).
d. J. Meier, *Bull. Soc. Chim. Fr.*, 290 (1962).
e. M. E. Jung and C. A. McCombs, *Org. Synth.* **58**, 163 (1978).
f & g. H. O. House, T. S. B. Sayer, and C.-C. Yau, *J. Org. Chem.* **43**, 2153 (1978).
4a. J. M. Harless and S. A. Monti, *J. Am. Chem. Soc.* **96**, 4714 (1974).
b. A. Wissner and J. Meinwald, *J. Org. Chem.* **38**, 1967 (1973).
c. W. J. Gensler and P. H. Solomon, *J. Org. Chem.* **38**, 1726 (1973).
d. H. W. Whitlock, Jr., *J. Am. Chem. Soc.* **84**, 3412 (1962).
e. C. H. Heathcock, R. A. Badger, and J. W. Patterson, Jr., *J. Am. Chem. Soc.* **89**, 4133 (1967).
f. E. J. Corey and D. S. Watt, *J. Am. Chem. Soc.* **95**, 2302 (1973).
5. W. G. Kofron and L. G. Wideman, *J. Org. Chem.* **37**, 555 (1972).
6. C. R. Hauser, T. M. Harris, and T. G. Ledford, *J. Am. Chem. Soc.* **81**, 4099 (1959).
7a. N. Campbell and E. Ciganek, *J. Chem. Soc.*, 3834 (1956).
b. F. W. Sum and L. Weiler, *J. Am. Chem. Soc.* **101**, 4401 (1979).
c. K. W. Rosemund, H. Herzberg, and H. Schutt, *Chem. Ber.* **87**, 1258 (1954).
d. T. Hudlicky, F. J. Koszyk, T. M. Kutchan, and J. P. Sheth, *J. Org. Chem.* **45**, 5020 (1980).
e. C. R. Hauser and W. R. Dunnavant, *Org. Synth.* **40**, 38 (1960).
f. G. Opitz, H. Milderberger, and H. Suhr, *Justus Liebigs Ann. Chem.*, **649**, 47 (1961).
g. K. Wiesner, K. K. Chan, and C. Demerson, *Tetrahedron Lett.*, 2893 (1965).
h. K. Shimo, S. Wakamatsu, and T. Inoüe, *J. Org. Chem.* **26**, 4868 (1961).
i. T. A. Spencer, K. K. Schmiegel, and K. L. Williamson, *J. Am. Chem. Soc.* **85**, 3785 (1963).

j. G. R. Kieczkowski and R. H. Schlessinger, *J. Am. Chem. Soc.* **100**, 1938 (1978).
8a–d. E. D. Bergmann, D. Ginsburg, and R. Pappo, *Org. React.* **10**, 179 (1959).
e. L. Mandell, J. U. Piper, and K. P. Singh, *J. Org. Chem.* **28**, 3440 (1963).
f. H. O. House, W. A. Kleschick, and E. J. Zaiko, *J. Org. Chem.* **43**, 3653 (1978).
g. J. E. McMurry and J. Melton, *Org. Synth.* **56**, 36 (1977).
h. D. F. Taber and B. P. Gunn, *J. Am. Chem. Soc.* **101**, 3992 (1979).
i. H. Feuer, A. Hirschfeld, and E. D. Bergmann, *Tetrahedron* **24**, 1187 (1968).
j. A. Baradel, R. Longeray, J. Dreux, and J. Doris, *Bull. Soc. Chim. Fr.*, 255 (1970).
k. H. H. Baer and K. S. Ong, *Can. J. Chem.* **46**, 2511 (1968).
l. A. Wettstein, K. Heusler, H. Ueberwasser, and P. Wieland, *Helv. Chim. Acta* **40**, 323 (1957).
9a. E. Wenkert and D. P. Strike, *J. Org. Chem.* **27**, 1883 (1962).
b. S. J. Etheredge, *J. Org. Chem.* **31**, 1990 (1966).
c. R. Deghenghi and R. Gaudry, *Tetrahedron Lett.*, 489 (1962).
d. P. A. Grieco and C. C. Pogonowski, *J. Am. Chem. Soc.* **95**, 3071 (1973).
e. E. M. Kaiser, W. G. Kenyon, and C. R. Hauser, *Org. Synth.* **V**, 559 (1973).
f. J. Cason, *Org. Synth.* **IV**, 630 (1963).
g. S. A. Glickman and A. C. Cope, *J. Am. Chem. Soc.* **67**, 1012 (1945).
h. W. Steglich and L. Zechlin, *Chem. Ber.* **111**, 3939 (1978).
i. S. F. Brady, M. A. Ilton, and W. S. Johnson, *J. Am. Chem. Soc.* **90**, 2882 (1968).
j. R. P. Hatch, J. Shringarpure, and S. M. Weinreb, *J. Org. Chem.* **43**, 4172 (1978).
10. S. Masamune, *J. Am. Chem. Soc.* **86**, 288 (1964).
11. E. J. Corey, M. Ohno, R. B. Mitra, and P. A. Vatakencherry, *J. Am. Chem. Soc.* **86**, 478 (1964).
12. J. Fried, in *Heterocyclic Compounds*, R. C. Elderfield (ed.), Vol. 1, Wiley, New York, 1950, p. 358.
13. R. Chapurlat, J. Huet, and J. Druex, *Bull. Soc. Chim. Fr.*, 2446, 2450 (1967).
14a. F. Kuo and P. L. Fuchs, *J. Am. Chem. Soc.* **109**, 1122 (1987).
b. L. A. Paquette, H.-S. Lin, D. T. Belmont, and J. P. Springer, *J. Org. Chem.* **51**, 4807 (1986).
c. R. K. Boeckman, Jr., D. K. Heckenden, and R. L. Chinn, *Tetrahedron Lett.* **28**, 3551 (1987).
d. D. Seebach, J. D. Aebi, M. Gander-Coquot, and R. Naef, *Helv. Chim. Acta* **70**, 1194 (1987).
e. F. E. Ziegler, S. I. Klein, U. K. Pati, and T.-F. Wang, *J. Am. Chem. Soc.* **107**, 2730 (1985).
f. M. E. Kuehne, *J. Org. Chem.* **35**, 171 (1970).
g. D. A. Evans, S. L. Bender, and J. Morris, *J. Am. Chem. Soc.* **110**, 2506 (1988).
h. K. Tomioka, Y.-S. Cho, F. Sato, and K. Koga, *J. Org. Chem.* **53**, 4094 (1988).
i. K. Tomioka, H. Kawasaki, K. Yasuda, and K. Koga, *J. Am. Chem. Soc.* **110**, 3597 (1988).
15a. T. Kametani, Y. Suzuki, H. Furuyama, and T. Honda, *J. Org. Chem.* **48**, 31 (1983).
b. R. A. Kjonaas and D. D. Patel, *Tetrahedron Lett.* **25**, 5467 (1983).
c. D. F. Taber and R. E. Ruckle, Jr., *J. Am. Chem. Soc.* **108**, 7686 (1986).
d. M. Yamaguchi, M. Tsukamoto, and I. Hirao, *Tetrahedron Lett.* **26**, 1723 (1985).
e. D. L. Snitman, M.-Y. Tsai, D. S. Watt, C. L. Edwards, and P. L. Stotter, *J. Org. Chem.* **44**, 2838 (1979).
f. A. G. Schultz and J. P. Dittami, *J. Org. Chem.* **48**, 2318 (1983).
16. J. G. Henkel and L. A. Spurlock, *J. Am. Chem. Soc.* **95**, 8339 (1973).
17. M. S. Newman, V. DeVries, and R. Darlak, *J. Org. Chem.* **31**, 2171 (1966).
18. P. A. Manis and M. W. Rathke, *J. Org. Chem.* **45**, 4952 (1980).
19. F. D. Lewis, T.-I. Ho, and R. J. DeVoe, *J. Org. Chem.* **45**, 5283 (1980).

Chapter 2

1a. G. Ksander, J. E. McMurry, and N. Johnson, *J. Org. Chem.* **42**, 1180 (1977).
b. J. Zabicky, *J. Chem. Soc.*, 683 (1961).
c. G. Stork, G. A. Kraus, and G. A. Garcia, *J. Org. Chem.* **39**, 3459 (1974).
d. H. Midorikawa, *Bull. Chem. Soc. Jpn.* **27**, 210 (1954).
e. G. Stork and S. R. Dowd, *Org. Synth.* **55**, 46 (1976).
f. E. C. Du Feu, F. J. McQuillin, and R. Robinson, *J. Chem. Soc.*, 53 (1937).

g. E. Buchta, G. Wolfrum, and H. Ziener, *Chem. Ber.* **91**, 1552 (1958).
h. L. H. Briggs and E. F. Orgias, *J. Chem. Soc., C*, 1885 (1970).
i. J. A. Profitt and D. S. Watt, *Org. Synth.* **56**, 1984 (1977).
j. U. Hengartner and V. Chu, *Org. Synth.* **58**, 83 (1978).
k. E. Giacomini, M. A. Loreto, L. Pellacani, and P. A. Tardella, *J. Org. Chem.* **45**, 519 (1980).
l. N. Narasimhan and R. Ammanamanchi, *J. Org. Chem.* **48**, 3945 (1983).
m. M. P. Bosch, F. Camps, J. Coll, A. Guerro, T. Tatsouka, and J. Meinwald, *J. Org. Chem.* **51**, 773 (1986).
2a. M. W. Rathke and D. F. Sullivan, *J. Am. Chem. Soc.* **95**, 3050 (1973).
b. E. J. Corey, H. Yamamoto, D. K. Herron, and K. Achiwa, *J. Am. Chem. Soc.* **92**, 6635 (1970).
c. E. J. Corey and D. E. Cane, *J. Org. Chem.* **36**, 3070 (1971).
d. E. W. Yabkee and D. J. Cram, *J. Am. Chem. Soc.* **92**, 6328 (1970).
e. W. G. Dauben, C. D. Poulter, and C. Suter, *J. Am. Chem. Soc.* **92**, 7408 (1970).
f. P. A. Grieco and K. Hiroi, *J. Chem. Soc., Chem. Commun.*, 1317 (1972).
g. T. Mukaiyama, M. Higo, and H. Takei, *Bull. Chem. Soc. Jpn.* **43**, 2566 (1970).
h. I. Vlattas, I. T. Harrison, L. Tokes, J. H. Fried, and A. D. Cross, *J. Org. Chem.* **33**, 4176 (1968).
i. A. T. Nielsen and W. R. Carpenter, *Org. Synth.* **V**, 288 (1973).
j. M. L. Miles, T. M. Harris, and C. R. Hauser, *Org. Synth.* **V**, 718 (1973).
k. A. P. Beracierta and D. A. Whiting, *J. Chem. Soc., Perkin Trans. 1*, 1257 (1978).
l. T. Amatayakul, J. R. Cannon, P. Dampawan, T. Dechatiwongse, R. G. F. Giles, D. Huntrakul, K. Kusamran, M. Mokkhasamit, C. L. Raston, V. Reutrakul, and A. H. White, *Aust. J. Chem.* **32**, 71 (1979).
m. R. M. Coates, S. K. Shah, and R. W. Mason, *J. Am. Chem. Soc.* **101**, 6765 (1979).
n. K. A. Parker and T. H. Fedynyshyn, *Tetrahedron Lett.*, 1657 (1979).
o. M. Miyashita and A. Yoshikishi, *J. Am. Chem. Soc.* **96**, 1917 (1974).
p. W. R. Roush, *J. Am. Chem. Soc.* **102**, 1390 (1980).
q. L. Fitjer and U. Quabeck, *Synth. Commun.* **15**, 855 (1985).
r. A. Padwa, L. Brodsky, and S. Clough, *J. Am. Chem. Soc.* **94**, 6767 (1972).
s. W. R. Roush, *J. Am. Chem. Soc.* **102**, 1390 (1980).
t. C. R. Johnson, K. Mori, and A. Nakanishi, *J. Org. Chem.* **44**, 2065 (1979).
u. T. Yanami, M. Miyashita, and A. Yoshikoshi, *J. Org. Chem.* **45**, 607 (1980).
3a. K. D. Croft, E. L. Ghisalberti, P. R. Jefferies, and A. D. Stuart, *Aust. J. Chem.* **32**, 2079 (1971).
b. L. H. Briggs and G. W. White, *J. Chem. Soc., C*, 3077 (1971).
c. D. F. Taber and B. P. Gunn, *J. Am. Chem. Soc.* **101**, 3992 (1979).
d. G. V. Kryshtal, V. V. Kulganek, V. F. Kucherov, and L. A. Yanovskaya, *Synthesis*, 107 (1979).
e. S. F. Brady, M. A. Ilton, and W. S. Johnson, *J. Am. Chem. Soc.* **90**, 2882 (1968).
f. R. M. Coates and J. E. Shaw, *J. Am. Chem. Soc.* **92**, 5657 (1970).
g. K. Mitsuhashi and S. Shiotoni, *Chem. Pharm. Bull.* **18**, 75 (1970).
h. G. Wittig and H.-D. Frommeld, *Chem. Ber.* **97**, 3548 (1964).
i. R. J. Sundberg, P. A. Bukowick, and F. O. Holcombe, *J. Org. Chem.* **32**, 2938 (1967).
j. D. R. Howton, *J. Org. Chem.* **10**, 277 (1945).
k. Y. Chan and W. W. Epstein, *Org. Synth.* **53**, 48 (1973).
l. I. Fleming and M. Woolias, *J. Chem. Soc., Perkin Trans. 1*, 827 (1979).
m. F. Johnson, K. G. Paul, D. Favara, R. Ciabatti, and U. Guzzi, *J. Am. Chem. Soc.* **104**, 2190 (1982).
n. M. Ihara, M. Suzuki, K. Fukumoto, T. Kametani, and C. Kabuto, *J. Am. Chem. Soc.* **110**, 1963 (1988).
4a. W. A. Mosher and R. W. Soeder, *J. Org. Chem.* **36**, 1561 (1971).
b. M. R. Roberts and R. H. Schlessinger, *J. Am. Chem. Soc.* **101**, 7626 (1979).
c. J. E. McMurry and T. E. Glass, *Tetrahedron Lett.*, 2575 (1971).
d. D. J. Cram, A. Langemann, and F. Hauck, *J. Am. Chem. Soc.* **81**, 5750 (1959).
e. W. G. Dauben and J. Ipaktschi, *J. Am. Chem. Soc.* **95**, 5088 (1973).
f. T. J. Curphey and H. L. Kim, *Tetrahedron Lett.*, 1441 (1968).
g. K. P. Singh and L. Mandell, *Chem. Ber.* **96**, 2485 (1963).
h. S. D. Lee, T. H. Chan, and K. S. Kwon, *Tetrahedron Lett.* **25**, 3399 (1984).
i. J. F. Lavallee and P. Deslongchamps, *Tetrahedron Lett.* **29**, 6033 (1988).
5. T. T. Howarth, G. P. Murphy, and T. M. Harris, *J. Am. Chem. Soc.* **91**, 517 (1969).
6a. E. Vedejs, K. A. J. Snoble, and P. L. Fuchs, *J. Org. Chem.* **38**, 1178 (1973).
b. P. B. Dervan and M. A. Shippey, *J. Am. Chem. Soc.* **98**, 1265 (1976).

7a. E. E. Schweizer and G. J. O'Neil, *J. Org. Chem.* **30**, 2082 (1965); E. E. Schweizer, *J. Am. Chem. Soc.* **86**, 2744 (1984).
b. G. Büchi and H. Wüest, *Helv. Chim. Acta* **54**, 1767 (1971).
c. G. H. Posner, S.-B. Lu, and E. Asirvathan, *Tetrahedron Lett.* **27**, 659 (1986).
8. R. B. Woodward, F. Sondheimer, D. Taub, K. Heusler, and W. M. McLamore, *J. Am. Chem. Soc.* **74**, 4223 (1952).
9. G. Stork, S. D. Darling, I. T. Harrison, and P. S. Wharton, *J. Am. Chem. Soc.* **84**, 2018 (1962).
10. J. R. Pfister, *Tetrahedron Lett.*, 1281 (1980).
11. R. M. Jacobson, G. P. Lahm, and J. W. Clader, *J. Org. Chem.* **45**, 395 (1980).
12a. A. I. Meyers and N. Nazarenko, *J. Org. Chem.* **38**, 175 (1973).
b. W. C. Still and F. L. Van Middlesworth, *J. Org. Chem.* **42**, 1258 (1977).
13a. R. V. Stevens and A. W. M. Lee, *J. Am. Chem. Soc.* **101**, 7032 (1979).
b. C. H. Heathcock, E. Kleinman, and E. S. Binkley, *J. Am. Chem. Soc.* **100**, 8036 (1978).
14a. W. A. Kleschick and C. H. Heathcock, *J. Org. Chem.* **43**, 1256 (1978).
b. S. D. Darling, F. N. Muralidharan, and V. B. Muralidharan, *Tetrahedron Lett.*, 2761 (1979).
15a. M. Ertas and D. Seebach, *Helv. Chim. Acta* **68**, 961 (1985).
b. S. Masamune, W. Choy, F. A. J. Kerdesky, and B. Imperiali, *J. Am. Chem. Soc.* **103**, 1566 (1981).
c. C. H. Heathcock, C. T. Buse, W. A. Kleschick, M. C. Pirrung, J. E. Sohn, and J. Lampe, *J. Org. Chem.* **45**, 1066 (1980).
d. R. Noyori, K. Yokoyama, J. Sakata, I. Kuwajima, E. Nakamura, and M. Shimizu, *J. Am. Chem. Soc.* **99**, 1265 (1977).
e. D. A. Evans, E. Vogel, and J. V. Nelson, *J. Am. Chem. Soc.* **101**, 6120 (1979); D. A. Evans, J. V. Nelson, E. Vogel, and T. R. Taber, *J. Am. Chem. Soc.* **103**, 3099 (1981).
f. C. T. Buse and C. H. Heathcock, *J. Am. Chem. Soc.* **99**, 8109 (1977).
16. M. T. Reetz and A. Jung, *J. Am. Chem. Soc.* **105**, 4833 (1983).

Chapter 3

1a. M. E. Kuehne and J. C. Bohnert, *J. Org. Chem.* **46**, 3443 (1981).
b. B. C. Barot and H. W. Pinnick, *J. Org. Chem.* **46**, 2981 (1981).
c. T. Mukaiyama, S. Shoda, and Y. Watanabe, *Chem. Lett.*, 383 (1977).
d. H. Loibner and E. Zbiral, *Helv. Chim. Acta* **59**, 2100 (1976).
e. E. J. Prisbe, J. Smejkal, J. P. H. Verheyden, and J. G. Moffatt, *J. Org. Chem.* **41**, 1836 (1976).
f. B. D. MacKenzie, M. M. Angelo, and J. Wolinsky, *J. Org. Chem.* **44**, 4042 (1979).
g. A. I. Meyers, R. K. Smith, and C. E. Whitten, *J. Org. Chem.* **44**, 2250 (1979).
h. W. A. Bonner, *J. Org. Chem.* **32**, 2496 (1967).
i. B. E. Smith and A. Burger, *J. Am. Chem. Soc.* **75**, 5891 (1953).
j. W. D. Klobucar, L. A. Paquette, and J. F. Blount, *J. Org. Chem.* **46**, 4021 (1981).
k. G. Grethe, V. Toome, H. L. Lee, M. Uskokovic, and A. Brossi, *J. Org. Chem.* **33**, 504 (1968).
l. B. Neises and W. Steglich, *Org. Synth.* **63**, 183 (1984).
2. A. W. Friederang and D. S. Tarbell, *J. Org. Chem.* **33**, 3797 (1968).
3. H. R. Hudson and G. R. de Spinoza, *J. Chem. Soc., Perkin Trans. 1,* 104 (1976).
4a. L. A. Paquette and M. K. Scott, *J. Am. Chem. Soc.* **94**, 6760 (1972).
b. P. N. Confalone, G. Pizzolato, E. G. Baggiolini, D. Lollar, and M. R. Uskokovic, *J. Am. Chem. Soc.* **99**, 7020 (1977).
c. E. L. Eliel, J. K. Koskimies, and B. Lohri, *J. Am. Chem. Soc.* **100**, 1614 (1978).
d. H. Hagiwara, M. Numata, K. Konishi, and Y. Oka, *Chem. Pharm. Bull.* **13**, 253 (1965).
e. A. S. Kende and T. P. Demuth, *Tetrahedron Lett.*, 715 (1980).
f. P. A. Grieco, D. S. Clark, and G. P. Withers, *J. Org. Chem.* **44**, 2945 (1979).
5a, b. D. Seebach, H.-O. Kalinowski, B. Bastani, G. Crass, H. Daum, H. Dorr, N. P. DuPreez, V. Ehring, W. Langer, C. Nussler, H.-A. Oei, and M. Schmidt, *Helv. Chim. Acta* **60**, 301 (1977).
c. G. L. Baker, S. J. Fritschel, J. R. Stille, and J. K. Stille, *J. Org. Chem.* **46**, 2954 (1981).
d. D. Seebach, H.-O. Kalinowski, B. Bastani, G. Crass, H. Daum, H. Dorr, N. P. DuPreez, V. Ehring, W. Langer, C. Nussler, H.-A. Oei, and M. Schmidt, *Helv. Chim. Acta* **60**, 301 (1977).
e. M. D. Fryzuk and B. Bosnich, *J. Am. Chem. Soc.* **99**, 6262 (1977).
f. S. Hanessian and R. Frenette, *Tetrahedron Lett.*, 3391 (1979).
g. K. G. Paul, F. Johnson, and D. Favara, *J. Am. Chem. Soc.* **98**, 1285 (1976).

6a. P. Henley-Smith, D. A. Whiting, and A. F. Wood, *J. Chem. Soc., Perkin Trans. 1*, 614 (1980).
b. P. Beak and L. G. Carter, *J. Org. Chem.* **46**, 2363 (1981).
c. M. E. Jung and T. J. Shaw, *J. Am. Chem. Soc.* **102**, 6304 (1980).
d. P. N. Swepston, S.-T. Lin, A. Hawkins, S. Humphrey, S. Siegel, and A. W. Cordes, *J. Org. Chem.* **46**, 3754 (1981).
e. P. J. Maurer and M. J. Miller, *J. Org. Chem.* **46**, 2835 (1981).
f. N. A. Porter, J. D. Byers, A. E. Ali, and T. E. Eling, *J. Am. Chem. Soc.* **102**, 1183 (1980).
g. G. A. Olah, B. G. B. Gupta, R. Malhotra, and S. C. Narang, *J. Org. Chem.* **45**, 1638 (1980).
7a. A. K. Bose, B. Lal, W. Hoffman III, and M. S. Manhas, *Tetrahedron Lett.*, 1619 (1973).
b. J. B. Hendrickson and S. M. Schwartzman, *Tetrahedron Lett.*, 277 (1975).
c. J. F. King, S. M. Loosmore, J. D. Lock, and M. Aslam, *J. Am. Chem. Soc.* **100**, 1637 (1978); C. N. Sukenik and R. G. Bergman, *J. Am. Chem. Soc.* **98**, 6613 (1976).
d. R. S. Freedlander, T. A. Bryson, R. B. Dunlap, E. M. Schulman, and C. A. Lewis, Jr., *J. Org. Chem.* **46**, 3519 (1981).
8a. J. Jacobus, M. Raban, and K. Mislow, *J. Org. Chem.* **33**, 1142 (1968).
b. M. Schmid and R. Barner, *Helv. Chim. Acta* **62**, 464 (1979).
c. V. Eswarakrishnan and L. Field, *J. Org. Chem.* **46**, 4182 (1981).
d. R. F. Borch, A. J. Evans, and J. J. Wade, *J. Am. Chem. Soc.* **99**, 1612 (1977).
e. H. S. Aaron and C. P. Ferguson, *J. Org. Chem.* **33**, 684 (1968).
9. B. Koppenhoefer and V. Schuring, *Org. Synth.* **66**, 151, 160 (1987).
10. B. E. Watkins and H. Rapoport, *J. Org. Chem.* **47**, 4471 (1982).
11. A. Brändstrom, *Adv. Phys. Org. Chem.* **15**, 267 (1977); D. Landini, A. Maia, and A. Pampoli, *J. Org. Chem.* **51**, 5475 (1986).
12. R. M. Magid, O. S. Fruchey, W. L. Johnson, and T. G. Allen, *J. Org. Chem.* **44**, 359 (1979).
13. R. B. Woodward, R. A. Olofson, and H. Mayer, *Tetrahedron Suppl.* **8**, 321 (1966); R. B. Woodward and R. A. Olofson, *J. Am. Chem. Soc.* **83**, 1007 (1961); B. Belleau and G. Malek, *J. Am. Chem. Soc.* **90**, 1651 (1968).
14a. E. J. Corey, K. C. Nicolaou, and L. S. Melvin, Jr., *J. Am. Chem. Soc.* **97**, 654 (1975).
b. L. M. Beacham III, *J. Org. Chem.* **44**, 3100 (1979).
c. J. Huang and J. Meinwald, *J. Am. Chem. Soc.* **103**, 861 (1981).
d. P. Beak and L. G. Carter, *J. Org. Chem.* **46**, 2363 (1981).
15. T. Mukaiyama, S. Shoda, T. Nakatsuka, and K. Narasaka, *Chem. Lett.*, 605 (1978).
16. R. U. Lemieux, K. B. Hendriks, R. V. Stick, and K. James, *J. Am. Chem. Soc.* **97**, 4056 (1975).
17a. T. Mukaiyama, R. Matsueda, and M. Suzuki, *Tetrahedron Lett.*, 1901 (1970).
b. E. J. Corey and D. A. Clark, *Tetrahedron Lett.*, 2875 (1979).

Chapter 4

1a. N. Kharasch and C. M. Buess, *J. Am. Chem. Soc.* **71**, 2724 (1949).
b. I. Heilbron, E. R. H. Jones, M. Julia, and B. C. L. Weedon, *J. Chem. Soc.*, 1823 (1949).
c. A. J. Sisti, *J. Org. Chem.* **33**, 3953 (1968).
d. H. C. Brown and G. Zweifel, *J. Am. Chem. Soc.* **83**, 1241 (1961).
e. F. W. Fowler, A. Hassner, and L. A. Levy, *J. Am. Chem. Soc.* **89**, 2077 (1967).
f. A. Hassner and F. W. Fowler, *J. Org. Chem.* **33**, 2686 (1968).
g. A. Padwa, T. Blacklock, and A. Tremper, *Org. Synth.* **57**, 83 (1977).
h. I. Ryu, S. Murai, I. Niwa, and N. Sonoda, *Synthesis*, 874 (1977).
i. R. A. Amos and J. A. Katzenellenbogen, *J. Org. Chem.* **43**, 560 (1978).
j. P. Jacob III and H. C. Brown, *J. Am. Chem. Soc.* **98**, 7832 (1976).
k. H. C. Brown and G. J. Lynch, *J. Org. Chem.* **46**, 531 (1981).
l. R. C. Cambie, R. C. Hayward, P. S. Rutledge, T. Smith-Palmer, B. E. Swedlund, and P. D. Woodgate, *J. Chem. Soc., Perkins Trans. 1*, 180 (1979).
m. A. B. Holmes, K. Russell, E. S. Stern, M. E. Stubbs, and N. K. Welland, *Tetrahedron Lett.* **25**, 4183 (1984).
n. N. S. Zefirov, T. N. Velikokhat'ko, and N. K. Sadovaya, *Zh. Org. Khim.* (*Engl. Trans.*) **19**, 1407 (1983).
o. F. B. Gonzalez and P. A. Bartlett, *Org. Synth.* **64**, 175 (1985).

2. D. J. Pasto and J. A. Gontarz, *J. Am. Chem. Soc.* **91**, 2147 (1969).
3. D. J. Pasto and C. C. Cumbo, *J. Am. Chem. Soc.* **86**, 4343 (1964).
4. D. J. Pasto and J. A. Gontarz, *J. Am. Chem. Soc.* **93**, 6902 (1971).
5. R. Gleiter and G. Müller, *J. Org. Chem.* **53**, 3912 (1985).
6. G. Stork and R. Borch, *J. Am. Chem. Soc.* **86**, 935 (1964).
7. T. Hori and K. B. Sharpless, *J. Org. Chem.* **43**, 1689 (1978).
8a. E. Kloster-Jensen, E. Kovats, A. Eschenmoser, and E. Heilbronner, *Helv. Chim. Acta* **39**, 1051 (1956).
b. P. N. Rao, *J. Org. Chem.* **36**, 2426 (1971).
c. R. A. Moss and E. Y. Chen, *J. Org. Chem.* **46**, 1466 (1981).
d. J. M. Jerkunica and T. G. Traylor, *Org. Synth.* **53**, 94 (1973).
e. G. Zweifel and C. C. Whitney, *J. Am. Chem. Soc.* **89**, 2753 (1967).
f. R. E. Ireland and P. Bey, *Org. Synth.* **53**, 63 (1973).
g. W. I. Fanta and W. F. Erman, *J. Org. Chem.* **33**, 1656 (1968).
h. W. E. Billups, J. H. Cross, and C. V. Smith, *J. Am. Chem. Soc.* **95**, 3438 (1973).
i. G. W. Kabalka and E. E. Gooch III, *J. Org. Chem.* **45**, 3578 (1980).
j. E. J. Corey, G. Wess, Y. B. Xiang, and A. K. Singh, *J. Am. Chem. Soc.* **109**, 4717 (1987).
k. I. Nakatsuka, N. L. Ferreira, W. C. Eckelman, B. E. Francis, W. J. Rzeszotarski, R. E. Gibson, E. M. Jagoda, and R. C. Reba, *J. Med. Chem.* **27**, 1287 (1984).
l. W. Oppolzer, H. Hauth, P. Pfaffli, and R. Wenger, *Helv. Chim. Acta* **60**, 1801 (1977).
m. G. H. Posner and P. W. Tang, *J. Org. Chem.* **43**, 4131 (1978).
9a. E. J. Corey and H. Estreicher, *J. Am. Chem. Soc.* **100**, 6294 (1978).
b. E. J. Corey and H. Estreicher, *Tetrahedron Lett.*, 1113 (1980).
c. G. A. Olah and M. Nohima, *Synthesis*, 785 (1973).
10. D. J. Pasto and F. M. Klein, *Tetrahedron Lett.*, 963 (1967).
11. H. J. Reich, J. M. Renga, and I. L. Reich, *J. Am. Chem. Soc.* **97**, 5434 (1975).
12. H. C. Brown, G. J. Lynch, W. J. Hammar, and L. C. Liu, *J. Org. Chem.* **44**, 1910 (1979).
13. P. A. Bartlett and J. Myerson, *J. Am. Chem. Soc.* **100**, 3950 (1978).
14a. D. A. Evans, J. E. Ellman, and R. L. Dorow, *Tetrahedron Lett.* **28**, 1123 (1987).
b. K. C. Nicolaou, R. L. Magolda, W. J. Sipio, W. E. Barnette, Z. Lysenko, and M. M. Joullie, *J. Am. Chem. Soc.* **102**, 3784 (1980).
c. K. C. Nicolaou, W. E. Barnette, and R. L. Magolda, *J. Am. Chem. Soc.* **103**, 3472 (1981).
d. S. Knapp, A. T. Levorse, and J. A. Potenza, *J. Org. Chem.* **53**, 4773 (1988).
e. T. H. Jones and M. S. Blum, *Tetrahedron Lett.* **22**, 4373 (1981).
15a. W. T. Smith and G. L. McLeod, *Org. Synth.* **IV**, 345 (1963).
b. K. E. Harding, T. H. Marman, and D. Nam, *Tetrahedron Lett.* **29**, 1627 (1988).
c. S. Terahima, M. Hayashi, and K. Koga, *Tetrahedron Lett.*, 2733 (1980).
16. A. Toshimitsu, K. Terao, and S. Uemura, *J. Org. Chem.* **51**, 1724 (1986).
j. H. C. Brown and P. Heim, *J. Org. Chem.* **38**, 912 (1973).
k. R. O. Hutchins and N. R. Natale, *J. Org. Chem.* **43**, 2299 (1978).
l. M. R. Detty and L. A. Paquette, *J. Am. Chem. Soc.* **99**, 821 (1977).
m. C. A. Bunnell and P. L. Fuchs, *J. Am. Chem. Soc.* **99**, 5184 (1977).
n. Y.-J. Wu and D. J. Burnell, *Tetrahedron Lett.* **29**, 4369 (1988).
o. P. W. Collins, E. Z. Dajani, R. Pappo, A. F. Gasiecki, R. B. Bianchi, and E. M. Woods, *J. Med. Chem.* **26**, 786 (1983).
17. W. Oppolzer and P. Dudfield, *Tetrahedron Lett*, **26**, 5037 (1985); D. A. Evans, J. A. Ellman, and R. L. Dorow, *Tetrahedron Lett.*, **28**, 1123 (1987).
18. T. W. Bell, *J. Am. Chem. Soc.*, **103**, 1163 (1981).
19a. H. C. Brown and B. Singaran, *J. Am. Chem. Soc.*, **106**, 1794 (1984).
b. S. Masamune, B. M. Kim, J. S. Petersen, T. Sato, S. J. Veenstra and T. Imai, *J. Am. Chem. Soc.* **107**, 4549 (1985).

Chapter 5

1a. W. R. Roush, *J. Am. Chem. Soc.* **102**, 1390 (1980).
b. H. C. Brown, S. C. Kim, and S. Krishnamurthy, *J. Org. Chem.* **45**, 1 (1980).
c. G. W. Kabalka, D. T. C. Yang, and J. D. Baker, Jr., *J. Org. Chem.* **41**, 574 (1976).

d. J. K. Whitesell, R. S. Matthews, M. A. Minton, and A. M. Helbling, *J. Am. Chem. Soc.* **103**, 3468 (1981).
e. K. S. Kim, M. W. Spatz, and F. Johnson, *Tetrahedron Lett.*, 331 (1979).
f. M.-H. Rei, *J. Org. Chem.* **44**, 2760 (1979).
g. R. O. Hutchins, D. Kandasamy, F. Dux III, C. A. Maryanoff, D. Rolstein, B. Goldsmith, W. Burgoyne, F. Cistone, J. Dalessandro, and J. Puglis, *J. Org. Chem.* **43**, 2259 (1978).
h. H. Lindlar, *Helv. Chim. Acta* **35**, 446 (1952).
i. E. Vedejs, R. A. Buchanan, R. Conrad, G. P. Meier, M. J. Mullins, and Y. Watanabe, *J. Am. Chem. Soc.* **109**, 5878 (1987).
j. C. B. Jackson and G. Pattenden, *Tetrahedron Lett.* **26**, 3393 (1985).
2. D. C. Wigfield and D. J. Phelps, *J. Am. Chem. Soc.* **96**, 543 (1974).
3a. E. J. Corey, T. K. Schaaf, W. Huber, U. Koelliker, and N. M. Weinshenker, *J. Am. Chem. Soc.* **92**, 397 (1970).
b. E. J. Corey and R. Noyori, *Tetrahedron Lett.*, 311 (1970).
c. R. F. Borch, *Org. Synth.* **52**, 124 (1972).
d. D. Seyferth and V. A. Mai, *J. Am. Chem. Soc.* **92**, 7412 (1970).
e. R. V. Stevens and J. T. Lai, *J. Org. Chem.* **37**, 2138 (1972).
f. M. J. Robins and J. S. Wilson, *J. Am. Chem. Soc.* **103**, 932 (1981).
g. G. R. Pettit and J. R. Dias, *J. Org. Chem.* **36**, 3207 (1971).
h. P. A. Grieco, T. Oguri, and S. Gilman, *J. Am. Chem. Soc.* **102**, 5886 (1980).
i. M. F. Semmelhack, S. Tomoda, and K. M. Hurst, *J. Am. Chem. Soc.* **102**, 7567 (1980).
4a. F. A. Carey, D. H. Ball, and L. Long, Jr., *Carbohydr. Res.* **3**, 205 (1966).
b. D. J. Cram and R. A. Abd Elhafez, *J. Am. Chem. Soc.* **74**, 5828 (1952).
c. R. N. Rej, C. Taylor, and G. Eadon, *J. Org. Chem.* **45**, 126 (1980).
d. M. C. Dart and H. B. Henbest, *J. Chem. Soc.*, 3563 (1960).
e. E. Piers, W. deWaal, and R. W. Britton, *J. Am. Chem. Soc.* **93**, 5113 (1971).
f. A. L. J. Beckwith and C. Easton, *J. Am. Chem. Soc.* **100**, 2913 (1978).
g. D. Horton and W. Weckerle, *Carbohydr. Res.* **44**, 227 (1975).
h. R. A. Holton and R. M. Kennedy, *Tetrahedron Lett.* **28**, 303 (1987).
i. H. Iida, N. Yamazaki, and C. Kibayashi, *J. Org. Chem.* **51** 1069, 3769 (1986).
j. D. A. Evans and M. M. Morrissey, *J. Am. Chem. Soc.* **106**, 3866 (1984).
k. N. A. Porter, C. B. Ziegler, Jr., F. F. Khouri, and D. H. Roberts, *J. Org. Chem.* **50**, 2252 (1985).
l. G. Stork and D. E. Kahne, *J. Am. Chem. Soc.* **105**, 1072 (1983).
m. Y. Yamamoto, K. Matsuoka, and H. Nemoto, *J. Am. Chem. Soc.* **110**, 4475 (1988).
5a. D. Lenoir, *Synthesis*, 553 (1977).
b. J. A. Marshall and A. E. Greene, *J. Org. Chem.* **36**, 2035 (1971).
c. B. M. Trost, Y. Nishimura, and K. Yamamoto, *J. Am. Chem. Soc.* **101**, 1328 (1979); J. E. McMurry, A. Andrus, G. M. Ksander, J. H. Musser, and M. A. Johnson, *J. Am. Chem. Soc.* **101**, 1330 (1979).
d. R. E. Ireland and C. S. Wilcox, *J. Org. Chem.* **45**, 197 (1980).
e. P. G. Gassman and T. J. Atkins, *J. Am. Chem. Soc.* **94**, 7748 (1972).
f. A. Gopalan and P. Magnus, *J. Am. Chem. Soc.* **102**, 1756 (1980).
g. R. M. Coates, S. K. Shah, and R. W. Mason, *J. Am. Chem. Soc.* **101**, 6765 (1979); Y.-K. Han and L. A. Paquette, *J. Org. Chem.* **44**, 3731 (1979).
h. L. P. Kuhn, *J. Am. Chem. Soc.* **80**, 5950 (1958).
i. R. P. Hatch, J. Shringarpure, and S. M. Weinreb, *J. Org. Chem.* **43**, 4172 (1978).
j. T. Shono, Y. Matsumura, S. Kashimura, and H. Kyutoko, *Tetrahedron Lett.*, 1205 (1978).
6a, b. K. E. Wiegers and S. G. Smith, *J. Org. Chem.* **43**, 1126 (1978).
c. D. C. Wiegfield and F. W. Gowland, *J. Org. Chem.* **45**, 653 (1980).
7. D. Caine and T. L. Smith, Jr., *J. Org. Chem.* **43**, 755 (1978).
8. R. E. A. Dear and F. L. M. Pattison, *J. Am. Chem. Soc.* **85**, 622 (1963).
9a. S. Danishefsky, M. Hirama, K. Gombatz, T. Harayama, E. Berman, and P. F. Schuda, *J. Am. Chem. Soc.* **101**, 7020 (1979).
b. A. S. Kende, M. L. King, and D. P. Curran, *J. Org. Chem.* **46**, 2826 (1981).
c. A. P. Kozikowski and A. Ames, *J. Am. Chem. Soc.* **103**, 3923 (1981).
d. E. J. Corey, S. G. Pyre, and W. Su, *Tetrahedron Lett.* **24**, 4883 (1983).
e. T. Rosen and C. Heathcock, *J. Am. Chem. Soc.* **107**, 3731 (1985).
f. T. Fujisawa and T. Sato, *Org. Synth.* **66**, 121 (1987).

g. H. J. Liu and M. G. Kulkarni, *Tetrahedron Lett.* **26**, 4847 (1963).
10a. H. C. Brown and W. C. Dickason, *J. Am. Chem. Soc.* **92**, 709 (1970).
b. D. Seyferth, H. Yamazaki, and D. L. Alleston, *J. Org. Chem.* **28**, 703 (1963).
c. G. Stork and S. D. Darling, *J. Am. Chem. Soc.* **82**, 1512 (1960).
11a. R. F. Borch, *Org. Synth.* **52**, 124 (1972).
b. R. F. Borch, M. D. Bernstein, and H. D. Durst, *J. Am. Chem. Soc.* **93**, 2897 (1971).
c. A. D. Harmon and C. R. Hutchinson, *Tetrahedron Lett.*, 1293 (1973).
12. S.-K. Chung and F.-F. Chung, *Tetrahedron Lett.*, 2473 (1979).
13. D. F. Taber, *J. Org. Chem.* **41**, 2649 (1976).
14. D. R. Briggs and W. B. Whalley, *J. Chem. Soc., Perkin Trans. 1*, 1382 (1976).
15. C. M. Tice and C. H. Heathcock, *J. Org. Chem.* **46**, 9 (1981).
16a. N. J. Leonard and S. Gelfand, *J. Am. Chem. Soc.* **77**, 3272 (1955).
b. P. S. Wharton and D. H. Bohlen, *J. Org. Chem.* **26**, 3615 (1961); W. R. Benn and R. M. Dodson, *J. Org. Chem.* **29**, 1142 (1964).
c. G. Lardelli and O. Jeger, *Helv. Chim. Acta* **32**, 1817 (1949).
d. R. J. Petersen and P. S. Skell, *Org. Synth.* **V**, 929 (1973).
17a. N. M. Yoon, C. S. Pak, H. C. Brown, S. Krishnamurthy, and T. P. Stocky, *J. Org. Chem.* **38**, 2786 (1973).
b. D. J. Dawson and R. E. Ireland, *Tetrahedron Lett.*, 1899 (1968).
c. R. O. Hutchins, C. A. Milewski, and B. A. Maryanoff, *J. Am. Chem. Soc.* **95**, 3662 (1973).
d. M. J. Kornet, P. A. Thio, and S. I. Tan, *J. Org. Chem.* **33**, 3637 (1968).
e. C. T. West, S. J. Donnelly, D. A. Kooistra, and M. P. Doyle, *J. Org. Chem.* **38**, 2675 (1973).
f. M. R. Johnson and B. Rickborn, *J. Org. Chem.* **35**, 1041 (1970).
g. N. Akubult and M. Balci, *J. Org. Chem.* **53**, 3338 (1988).
18. H. Iida, N. Yamazaki, and C. Kibayashi, *J. Org. Chem.* **51**, 3769 (1986).
19a. H. C. Brown, G. G. Pai, and P. K. Jadhav, *J. Am. Chem. Soc.* **106**, 1531 (1984).
b. E. J. Corey, R. K. Bakshi, S. Shibata, C.-P. Chen, and V. K. Singh, *J. Am. Chem. Soc.* **109**, 7925 (1987).
c. M. Srebnik, P. V. Ramachandran, and H. C. Brown, *J. Org. Chem.* **53**, 2916 (1988).
20. L. A. Paquette, T. J. Nitz, R. J. Ross, and J. P. Springer, *J. Am. Chem. Soc.* **106**, 1446 (1984).
21. J. A. Marshall, *Acc. Chem. Res.* **13**, 213 (1980); J. A. Marshall and K. E. Flynn, *J. Am. Chem. Soc.* **106**, 723 (1984); J. A. Marshall, J. C. Peterson, and L. Lebioda, *J. Am. Chem. Soc.* **106**, 6006 (1984).

Chapter 6

1a. B. M. Trost, S. A. Godleski, and J. P. Genet, *J. Am. Chem.* **100**, 3930 (1978).
b. M. E. Jung and C. A. McCombs, *J. Am. Chem. Soc.* **100**, 5207 (1978).
c. L. E. Overman and P. J. Jessup, *J. Am. Chem. Soc.* **100**, 5179 (1978).
d. C. Cupas, W. E. Watts, and P. von R. Schleyer, *Tetrahedron Lett.*, 2503 (1964).
e. T. C. Jain, C. M. Banks, and J. E. McCloskey, *Tetrahedron Lett.*, 841 (1970).
f. G. Büchi and J. E. Powell, Jr., *J. Am. Chem. Soc.* **89**, 4559 (1967).
g. M. Raban, F. B. Jones, Jr., E. H. Carlson, E. Banuccci, and N. A. LeBel, *J. Org. Chem.* **35**, 1497 (1970).
h. H. Yamamoto and H. L. Sham, *J. Am. Chem. Soc.* **101**, 1609 (1979).
i. H. O. House, T. S. B. Sayer, and C.-C. Yau, *J. Org. Chem.* **43**, 2153 (1978).
j. M. C. Pirrung, *J. Am. Chem. Soc.* **103**, 82 (1981).
k. M. Sevrin and A. Krief, *Tetrahedron Lett.*, 187 (1978).
l. L. A. Paquette, G. D. Crouse, and A. K. Sharma, *J. Am. Chem. Soc.* **102**, 3972 (1980).
m. N.-K. Chan and G. Saucy, *J. Org. Chem.* **42**, 3838 (1977).
n, o. J. A. Marshall and J. Lebreton, *J. Org. Chem.* **53**, 4108 (1988).
p. R. L. Funk, W. J. Daily, and M. Parvez, *J. Org. Chem.* **53**, 4142 (1988).
q. J. D. Winkler, J. P. Hey, and P. G. Williard, *Tetrahedron Lett.* **29**, 4691 (1988).
r. F. A. J. Kerdesky, R. J. Ardecky, M. V. Lakshmikathan, and M. P. Cava, *J. Am. Chem. Soc.* **103**, 1992 (1981).

s. B. B. Snider and R. A. H. Hui, *J. Org. Chem.* **50**, 5167 (1985).
2a. W. Oppolzer and M. Petrzilka, *J. Am. Chem. Soc.* **98**, 6722 (1976).
b. A. Padwa and N. Kamigata, *J. Am. Chem. Soc.* **99**, 1871 (1977).
c. P. A. Jacobi, A. Brownstein, M. Martinelli, and K. Grozinger, *J. Am. Chem. Soc.* **103**, 239 (1981).
d. H. W. Gschwend, A. O. Lee, and H.-P. Meier, *J. Org. Chem.* **38**, 2169 (1973).
e. T. Kametani, M. Tsubuki, Y. Shiratori, H. Nemoto, M. Ihara, K. Fukumoto, F. Satoh, and H. Inoue, *J. Org. Chem.* **42**, 2672 (1977).
f. J. L. Gras and M. Bertrand, *Tetrahedron Lett.*, 4549 (1979).
g. H. Sato, M. Sakaguchi, Y. Fujimoto, T. Tatsuno, and H. Yoshioka, *Chem. Pharm. Bull.* **33**, 412 (1985).
3a. K. C. Brannock, A. Bell, R. D. Burpitt, and C. A. Kelly, *J. Org. Chem.* **29**, 801 (1964).
b. K. Ogura, S. Furukawa, and G. Tsuchihashi, *J. Am. Chem. Soc.* **102**, 2125 (1980).
c. E. Vedejs, M. J. Arco, D. W. Powell, J. M. Renga, and S. P. Singer, *J. Org. Chem.* **43**, 4831 (1978).
d. J. J. Tufariello and J. J. Tegeler, *Tetrahedron Lett.*, 4037 (1976).
e. W. A. Thaler and B. Franzus, *J. Org. Chem.* **29**, 2226 (1964).
f. B. B. Snider, *J. Org. Chem.* **41**, 3061 (1976).
g. L. A. Paquette, *J. Org. Chem.* **29**, 2851 (1964).
h. M. E. Monk and Y. K. Kim, *J. Am. Chem. Soc.* **86**, 2213 (1964).
i. B. Cazes and S. Julia, *Bull. Soc. Chim. Fr.*, 925 (1977).
j. D. L. Boger and D. D. Mullican, *Org. Synth.* **65**, 98 (1987).
4a. P. E. Eaton and U. R. Chakraborty, *J. Am. Chem. Soc.* **100**, 3634 (1978).
b. H. Hogeveen and B. J. Nusse, *J. Am. Chem. Soc.* **100**, 3110 (1978).
c. T. Oida, S. Tanimoto, T. Sugimoto, and M. Okano, *Synthesis*, 131 (1980).
d. J. N. Labovitz, C. A. Henrick, and V. L. Corbin, *Tetrahedron Lett.*, 4209 (1975).
e. W. Steglich and L. Zechlin, *Chem. Ber.* **111**, 3939 (1978).
f. F. D. Lewis and R. J. DeVoe, *J. Org. Chem.* **45**, 948 (1980).
g. S. P. Tanis and K. Nakanishi, *J. Am. Chem. Soc.* **101**, 4398 (1979).
h. S. Danishefsky, M. P. Prisbylla, and S. Hiner, *J. Am. Chem. Soc.* **100**, 2918 (1978).
i. T. Hudlicky, F. J. Koszyk, T. M. Kutchan, and J. P. Sheth, *J. Org. Chem.* **45**, 5020 (1980).
5. H. E. Zimmerman, G. L. Grunewald, R. M. Paufler, and M. A. Sherwin, *J. Am. Chem. Soc.* **91**, 2330 (1969).
6. C. J. Albisetti, N. G. Fisher, M. J. Hogsed, and R. M. Joyce, *J. Am. Chem. Soc.* **78**, 2637 (1956).
7. N. Shimizu, M. Ishikawa, K. Ishikura, and S. Nishida, *J. Am. Chem. Soc.* **96**, 6456 (1974).
8. R. Schug and R. Huisgen, *J. Chem. Soc., Chem. Commun.*, 60 (1975).
9. W. L. Howard and N. B. Lorette, *Org. Synth.* **V**, 25 (1973).
10. S. Danishefsky, M. Hirama, N. Fritsch, and J. Clardy, *J. Am. Chem. Soc.* **101**, 7013 (1979).
11. D. A. Evans, C. A. Bryan, and C. L. Sims, *J. Am. Chem. Soc.* **94**, 2891 (1972).
12. J. Wolinsky and R. B. Login, *J. Org. Chem.* **35**, 3205 (1970).
13. C. H. Heathcock and R. A. Badger, *J. Org. Chem.* **37**, 234 (1972).
14. B. J. Arnold, S. M. Mellows, P. G. Sammes, and T. W. Wallace, *J. Chem. Soc., Perkin Trans. 1,* 401 (1974); B. J. Arnold, P. G. Sammes, and T. W. Wallace, *J. Chem. Soc., Perkin Trans. 1,* 409 (1974).
15. B. Bichan and M. Winnik, *Tetrahedron Lett.*, 3857 (1974).
16a. R. A. Carboni and R. V. Lindsey, Jr., *J. Am. Chem. Soc.* **81**, 4342 (1959).
b. L. A. Carpino, *J. Am. Chem. Soc.* **84**, 2196 (1962); **85**, 2144 (1963).
17a. N. Ono, A. Kanimura, and A. Kaji, *Tetrahedron Lett.* **27**, 1595 (1986).
b. R. V. C. Carr, R. V. Williams, and L. A. Paquette, *J. Org. Chem.* **48**, 4976 (1983).
c. C. H. DePuy and P. R. Story, *J. Am. Chem. Soc.* **82**, 627 (1960).
d. S. Ranganathan, D. Ranganathan, and R. Iyengar, *Tetrahedron* **32**, 961 (1976).
18a. D. L. J. Clive, G. Chittattu, N. J. Curtis, and S. M. Menchen, *J. Chem. Soc., Chem. Commun.*, 770 (1978).
b. B. W. Metcalf, P. Bey, C. Danzin, M. J. Jung, P. Casara, and J. P. Veveri, *J. Am. Chem. Soc.* **100**, 2551 (1978).
c. T. Cohen, Z. Kosarych, K. Suzuki, and L.-C. Yu, *J. Org. Chem.* **50**, 2965 (1985).
d. T. Cohen, M. Bhupathy, and J. R. Matz, *J. Am. Chem. Soc.* **105**, 520 (1983).
e. R. G. Shea, J. N. Fitzner, J. E. Farkhauser, A. Spaltenstein, P. A. Carpino, R. M. Peevey, D. V. Pratt, B. J. Tenge, and P. B. Hopkins, *J. Org. Chem.* **51**, 5243 (1986).
f. R. L. Funk, P. M. Novak, and M. M. Abelman, *Tetrahedron Lett.* **29**, 1493 (1988).

g. R. M. Coates and C. H. Cummins, *J. Org. Chem.* **51**, 1383 (1986).
h. K. Ogura, S. Furukawa, and G. Tsuchihashi, *J. Am. Chem. Soc.* **102**, 2125 (1980).
i. H. F. Schmitthenner and S. M. Weinreb, *J. Org. Chem.* **45**, 3372 (1980).
j. R. A. Gibbs and W. H. Okamura, *J. Am. Chem. Soc.* **110**, 4062 (1988).
k. E. Vedejs, J. D. Rodgers, and S. J. Wittenberger, *J. Am. Chem. Soc.* **110**, 4822 (1988).
19a. D. J. Faulkner and M. R. Peterson, *J. Am. Chem. Soc.* **95**, 553 (1973).
b. N. A. LeBel, N. D. Ojha, J. R. Menke, and R. J. Newland, *J. Org. Chem.* **37**, 2896 (1972).
c. G. Büchi and H. Wüest, *J. Am. Chem. Soc.* **96**, 7573 (1974).
d. C. A. Henrick, F. Schaub, and J. B. Siddall, *J. Am. Chem. Soc.* **94**, 5374 (1972).
e. R. E. Ireland and R. H. Mueller, *J. Am. Chem. Soc.* **94**, 5897 (1972).
f. E. J. Corey, R. B. Mitra, and H. Uda, *J. Am. Chem. Soc.* **86**, 485 (1964).
g. J. E. McMurry and L. C. Blaszcak, *J. Org. Chem.* **39**, 2217 (1974).
h. W. Sucrow, *Angew. Chem. Int. Ed. Engl.* **7**, 629 (1968).
i. O. P. Vig, K. L. Matta, and I. Raj, *J. Indian Chem. Soc.* **41**, 752 (1964).
j. W. Nagata, S. Hirai, T. Okumura, and K. Kawata, *J. Am. Chem. Soc.* **90**, 1650 (1968).
k. H. O. House, J. Lubinkowski, and J. J. Good, *J. Org. Chem.* **40**, 86 (1975).
l. L. A. Paquette, G. D. Grouse, and A. K. Sharma, *J. Am. Chem. Soc.* **102**, 3972 (1980).
m. R. L. Funk and G. L. Bolton, *J. Org. Chem.* **49**, 5021 (1984).
n. A. P. Kozikowski and C.-S. Li, *J. Org. Chem.* **52**, 3541 (1987).
o. B. M. Trost and A. C. Lavoie, *J. Am. Chem. Soc.* **105**, 5075 (1983).
p. A. P. Marchand, S. C. Suri, A. D. Earlywine, D. R. Powell, and D. van der Helm, *J. Org. Chem.* **49**, 670 (1984).
q. M. Kodoma, Y. Shiobara, H. Sumitomo, K. Fukuzumi, H. Minami, and Y. Miyamoto, *J. Org. Chem.* **53**, 1437 (1988).
20a. J. J. Tufariello, A. S. Milowsky, M. Al-Nuri, and S. Goldstein, *Tetrahedron Lett.* **28**, 263 (1987).
b. G. H. Posner, A. Haas, W. Harrison, and C. M. Kinter, *J. Org. Chem.* **52**, 4836 (1987).
c. F. E. Ziegler, A. Nangia, and G. Schulte, *J. Am. Chem. Soc.* **109**, 3987 (1987).
d. M. P. Edwards, S. V. Ley, S. G. Lister, B. D. Palmer, and D. J. Williams, *J. Org. Chem.* **49**, 3503 (1984).
e. R. E. Ireland and M. D. Varney, *J. Org. Chem.* **48**, 1829 (1983).
f. D. J.-S. Tsai and M. M. Midland, *J. Am. Chem. Soc.* **107**, 3915 (1985).
g. E. Vedejs, J. M. Dolphin, and H. Mastalerz, *J. Am. Chem. Soc.* **105**, 127 (1983).
21. S. D. Burke, D. M. Armistead, and K. Shankaran, *Tetrahedron Lett.* **27**, 6295 (1986).
22a. K. Nomura, K. Okazaki, H. Hori, and E. Yoshii, *Chem. Pharm. Bull.* **34**, 3175 (1986).
b. Y. Tamura, M. Sasho, K. Nakagawa, T. Tsugoshi, and Y. Kita, *J. Org. Chem.* **49**, 473 (1984).
23a. C. Siegel and E. R. Thornton, *Tetrahedron Lett.* **29**, 5225 (1988).
b. D. P. Curran, B. H. Kim, J. Daugherty, and T. A. Heffner, *Tetrahedron Lett.* **29**, 3555 (1988).
c. H. Waldman, *J. Org. Chem.* **53**, 6133 (1988).

Chapter 7

1a. H. Neumann and D. Seebach, *Tetrahedron Lett.*, 4839 (1976).
b. P. Canonne, G. Foscolos, and G. Lemay, *Tetrahedron Lett.*, 155 (1980).
c. T. L. Shih, M. Wyvratt, and H. Mrozik, *J. Org. Chem.* **52**, 2029 (1987).
d. R. K. Boeckman, Jr., and E. W. Thomas, *J. Am. Chem. Soc.* **101**, 987 (1979).
e. G. M. Rubottom and C. Kim, *J. Org. Chem.* **48**, 1550 (1983).
f. S. L. Buchwald, B. T. Watson, R. T. Lum, and W. A. Nugent, *J. Am. Chem. Soc.* **109**, 7137 (1987).
g. T. Okazoe, K. Takai, K. Oshima, and K. Utimoto, *J. Org. Chem.* **52**, 4410 (1987).
h. J. W. Frankenfeld and J. J. Werner, *J. Org. Chem.* **34**, 3689 (1969).
i. E. R. Burkhardt and R. D. Rieke, *J. Org. Chem.* **50**, 416 (1985).
2. R. W. Herr and C. R. Johnson, *J. Am. Chem. Soc.* **92**, 4979 (1970).
3a. J. S. Sawyer, A. Kucerovy, T. L. Macdonald, and G. J. McGarvey, *J. Am. Chem. Soc.* **110**, 842 (1988).
b. T. Cohen and J. R. Matz, *J. Am. Chem. Soc.* **102**, 6900 (1980).
c. C. R. Johnson and J. R. Medich, *J. Org. Chem.* **53**, 4131 (1988).

d. B. M. Trost and T. N. Nanninga, *J. Am. Chem. Soc.* **107**, 1293 (1985).
e. T. Morwick, *Tetrahedron Lett.* **21**, 3227 (1980).
f. W. C. Still and C. Sreekumar, *J. Am. Chem. Soc.* **102**, 1201 (1980).
g. R. F. Cunio and F. J. Clayton, *J. Org. Chem.* **41**, 1480 (1976).
4a. J. J. Fitt and H. W. Gschwend, *J. Org. Chem.* **45**, 4258 (1980).
b. S. Aikyama and J. Hooz, *Tetrahedron Lett.*, 4115 (1973).
c. K. P. Klein and C. R. Hauser, *J. Org. Chem.* **32**, 1479 (1967).
d. B. M. Graybill and D. A. Shirley, *J. Org. Chem.* **31**, 1221 (1966).
e. M. M. Midland, A. Tramontano, and J. R. Cable, *J. Org. Chem.* **45**, 28 (1980).
f. W. Fuhrer and H. W. Gschwend, *J. Org. Chem.* **44**, 1133 (1979).
g. D. F. Taber and R. W. Korsmeyer, *J. Org. Chem.* **43**, 4925 (1978).
h. R. R. Schmidt, J. Talbiersky, and P. Russegger, *Tetrahedron Lett.*, 4273 (1979).
i. R. M. Carlson, *Tetrahedron Lett.*, 111 (1978).
j. R. J. Sundberg, R. Broome, C. P. Walters, and D. Schnur, *J. Heterocycl. Chem.* **18**, 807 (1981).
5a. M. P. Dreyfuss, *J. Org. Chem.* **28**, 3269 (1963).
b. P. J. Pearce, D. H. Richards, and N. F. Scilly, *Org. Synth.* **52**, 19 (1972).
c. U. Schöllkopf, H. Küppers, H.-J. Traencker, and W. Pitteroff, *Justus Liebigs Ann. Chem.* **704**, 120 (1967).
d. J. V. Hay and T. M. Harris, *Org. Synth.* **53**, 56 (1973).
e. F. Sato, M. Inoue, K. Oguro, and M. Sato, *Tetrahedron Lett.*, 4303 (1979).
f. J. C. H. Hwa and H. Sims, *Org. Synth.* **V**, 608 (1973).
6a. J. H. Rigby and C. Senanyake, *J. Am. Chem. Soc.* **109**, 3147 (1987).
b. K. Takai, Y. Kataoka, T. Okazoe, and K. Utimoto, *Tetrahedron Lett.* **29**, 1065 (1988).
c. E. Nakamura, S. Aoki, K. Sekiya, H. Oshino, and I. Kuwajima, *J. Am. Chem. Soc.* **109**, 8056 (1987).
d. H. A. Whaley, *J. Am. Chem. Soc.* **93**, 3767 (1971).
7. J. Barluenga, F. J. Fananas, and M. Yus, *J. Org. Chem.* **44**, 4798 (1979).
8a–b. W. C. Still and J. H. McDonald III, *Tetrahedron Lett.* **21**, 1031 (1980).
c. E. Casadevall and Y. Povet, *Tetrahedron Lett.*, 2841 (1976).
9. P. Beak, J. E. Hunter, Y. M. Jan, and A. P. Wallin, *J. Am. Chem. Soc.* **109**, 5403 (1987).
10. C. J. Kowalski and M. S. Haque, *J. Org. Chem.* **50**, 5140 (1985).
11. C. Fehr, J. Galindo, and R. Perret, *Helv. Chim. Acta* **70**, 1745 (1987).
12. M. P. Cooke, Jr., and I. N. Houpis, *Tetrahedron Lett.* **26**, 4987 (1985); E. Piers and P. C. Marais, *Tetrahedron Lett.* **29**, 4053 (1988).
13a. C. Phillips, R. Jacobson, B. Abrahams, H. J. Williams, and C. R. Smith, *J. Org. Chem.* **45**, 1920 (1980).
b. T. Cohen and J. R. Matz, *J. Am. Chem. Soc.* **102**, 6900 (1980).
c. T. R. Govindachari, P. C. Parthasarathy, H. K. Desai, and K. S. Ramachandran, *Indian J. Chem.* **13**, 537 (1975).
d. W. C. Still, *J. Am. Chem. Soc.* **100**, 1481 (1978).
e. E. J. Corey and D. R. Williams, *Tetrahedron Lett.*, 3847 (1977).
f. T. Okazoe, K. Takai, K. Oshiama, and K. Utimoto, *J. Org. Chem.* **52**, 4410 (1987).
g. M. A. Adams, A. J. Duggan, J. Smolanoff, and J. Meinwald, *J. Am. Chem. Soc.* **101**, 5364 (1979).
h. S. O. deSilva, M. Watanabe, and V. Snieckus, *J. Org. Chem.* **44**, 4802 (1979).
14. M. Kitamura, S. Suga, K. Kawai, and R. Noyori, *J. Am. Chem. Soc.* **108**, 6071 (1986); K. Soai, A. Ookawa, T. Kaba, and K. Ogawa, *J. Am. Chem. Soc.* **109**, 7111 (1987); M. Kitamura, S. Okada, and R. Noyori, *J. Am. Chem. Soc.* **111**, 4028 (1989).

Chapter 8

1a. C. Huynh, F. Derguini-Boumechal, and G. Linstrumelle, *Tetrahedron Lett.*, 1503 (1979).
b. N. J. LaLima, Jr., and A. B. Levy, Jr., *J. Org. Chem.* **43**, 1279 (1978).
c. A. Cowell and J. K. Stille, *J. Am. Chem. Soc.* **102**, 4193 (1980).
d. T. Sato, M. Kawasima, and T. Fujisawa, *Tetrahedron Lett.*, 2375 (1981).
e. H. P. Dang and G. Linstrumelle, *Tetrahedron Lett.*, 191 (1978).
f. B. M. Trost and D. P. Curran, *J. Am. Chem. Soc.* **102**, 5699 (1980).

g. D. J. Pasto, S.-K. Chou, E. Fritzen, R. H. Shults, A. Waterhouse, and G. F. Hennion, *J. Org. Chem.* **43**, 1389 (1978).
h. B. H. Lipshutz, J. Kozlowski, and R. S. Wilhelm, *J. Am. Chem. Soc.* **104**, 2305 (1982).
i. P. A. Grieco and C. V. Srinivasan, *J. Org. Chem.* **46**, 2591 (1981).
j. C. Iwata, K. Suzuki, S. Aoki, K. Okamura, M. Yamashita, I. Takahashi, and T. Tanaka, *Chem. Pharm. Bull.* **34**, 4939 (1988).
k. A. Alexakis, G. Cahiez, and J. F. Normant, *Org. Synth.* **62**, 1 (1984).
1. J. Tsuji, Y. Kobayashi, H. Kataoka, and T. Takahashi, *Tetrahedron Lett.* **21**, 1475 (1980).
m. W. A. Nugent and R. J. McKinney, *J. Org. Chem.* **50**, 5370 (1985).
n. R. M. Wilson, K. A. Schnapp, R. K. Merwin, R. Ranganathan, D. L. Moats, and T. T. Conrad, *J. Org. Chem.* **51**, 4028 (1986).
o, L. N. Pridgen, *J. Org. Chem.* **47**, 4319 (1982).
2a. B. H. Lipshutz, M. Koerner, and D. A. Parker, *Tetrahedron Lett.* **28**, 945 (1987).
b. B. H. Lipshutz, R. S. Wilheim, J. A. Kozlowski, and D. Parker, *J. Org. Chem.* **49**, 3928 (1984).
c. J. P. Marino, R. Fernandez de la Pradilla, and E. Laborde, *J. Org. Chem.* **52**, 4898 (1987).
d. C. R. Johnson and D. S. Dhanoa, *J. Org. Chem.* **52**, 1887 (1987).
3a. H. Urata, A. Fujita, and T. Fuchikami, *Tetrahedron Lett.* **29**, 4435 (1988).
b. Y. Itoh, H. Aoyama, T. Hirao, A. Mochizuki, and T. Saegusa, *J. Am. Chem. Soc.* **101**, 494 (1979).
c. P. G. M. Wuts, M. L. Obrzut, and P. A. Thompson, *Tetrahedron Lett.* **25**, 4051 (1984).
4a. R. J. Anderson, V. L. Corbin, G. Cotterrel, G. R. Cox, C. A. Henrick, F. Schaub, and J. B. Siddall, *J. Am. Chem. Soc.* **97**, 1197 (1975).
b. P. deMayo, L. K. Sydnes, and G. Wenska, *J. Org. Chem.* **45**, 1549 (1980).
c. Y. Yamamoto, H. Yatagai, and K. Maruyama, *J. Org. Chem.* **44**, 1744 (1979).
d. H. Shostarez and L. A. Paquette, *J. Am. Chem. Soc.* **103**, 722 (1981).
e. W. G. Dauben, G. H. Beasley, M. D. Broadhurst, B. Muller, D. J. Peppard, P. Pesnelle, and C. Suter, *J. Am. Chem. Soc.* **97**, 4973 (1975).
f. L. Watts, J. D. Fitzpatrick, and R. Pettit, *J. Am. Chem. Soc.* **88**, 623 (1966).
g. J. I. Kim, B. A. Patel, and R. F. Heck, *J. Org. Chem.* **46**, 1067 (1981).
h. J. A. Marshall, W. F. Huffman, and J. A. Ruth, *J. Am. Chem. Soc.* **94**, 4691 (1972).
i. H.-A. Hasseberg and H. Gerlach, *Helv. Chim. Acta* **71**, 957 (1988).
5a, b. B. O'Connor and G. Just, *J. Org. Chem.* **52**, 1801 (1987); G. Just and B. O'Connor, *Tetrahedron Lett.* **26**, 1799 (1985).
6. B. H. Lipshutz, R. S. Wilheim, J. A. Kozlowski, and D. Parker, *J. Org. Chem.* **49**, 3928 (1984); E. C. Ashby, R. N. DePriest, A. Tuncay, and S. Srivasta, *Tetrahedron Lett.* **23**, 5251 (1982).
7a. C. G. Chavdarian and C. H. Heathcock, *J. Am. Chem. Soc.* **97**, 3822 (1975).
b. E. J. Corey and D. R. Williams, *Tetrahedron Lett.*, 3847 (1977).
c. G. Mehta and K. S. Rao, *J. Am. Chem. Soc.* **108**, 8015 (1986).
8a. C. M. Lentz and G. H. Posner, *Tetrahedron Lett.*, 3769 (1978).
b. A. Marfat, P. R. McGuirk, R. Kramer, and P. Helquist, *J. Am. Chem. Soc.* **99**, 253 (1977).
c. L. A. Paquette and Y.-K. Han, *J. Am. Chem. Soc.* **103**, 1831 (1981).
d. A. Alexakis, J. Berlan, and Y. Besace, *Tetrahedron Lett.* **27**, 1047 (1986).
e. M. Sletzinger, T. R. Verhoeven, R. P. Volante, J. M. McNamara, E. G. Corley, and T. M. H. Liu, *Tetrahedron Lett.* **26**, 2951 (1985).
9a. E. J. Corey and E. Hamanaka, *J. Am. Chem. Soc.* **89**, 2758 (1967).
b. Y. Kitagawa, A. Itoh, S. Hashimoto, H. Yamamoto, and H. Zozaki, *J. Am. Chem. Soc.* **99**, 3864 (1977).
c. B. M. Trost and R. W. Warner, *J. Am. Chem. Soc.* **105**, 5940 (1983).
d. S. Brandt, A. Marfat, and P. Helquist, *Tetrahedron Lett.*, 2193 (1979).
10. R. H. Grubbs and R. A. Grey, *J. Am. Chem. Soc.* **95**, 5765 (1973).
11. H. L. Goering, E. P. Seitz, Jr., and C. C. Tseng, *J. Org. Chem.* **46**, 5304 (1981).
12. A. Marfat, P. R. McGuirk, and P. Helquist, *J. Org. Chem.* **44**, 1345 (1979).
13. N. Cohen, W. F. Eichel, R. J. Lopresti, C. Neukom, and G. Saucy, *J. Org. Chem.* **41**, 3505 (1976).
14a. R. J. Linderman, A. Godfrey, and K. Horne, *Tetrahedron Lett.* **28**, 3911 (1987).
b. H. Schostarez and L. A. Paquette, *J. Am. Chem. Soc.* **103**, 722 (1981).
c. Y. Yamamoto, S. Yamamoto, H. Yatagai, Y. Ishihara, and K. Maruyam, *J. Org. Chem.* **47**, 119 (1982).
d. T. Kawabata, P. A. Grieco, H.-L. Sham, H. Kim, J. Y. Jaw, and S. Tu, *J. Org. Chem.* **52**, 3346 (1987).

15a. A. Minato, K. Suzuki, K. Tamao, and M. Kumada, *Tetrahedron Lett.* **25**, 83 (1984).
b. E. R. Larson and R. A. Raphael, *Tetrahedron Lett.*, 5401 (1979).
c. M. C. Pirrung and S. A. Thomson, *J. Org. Chem.* **53**, 227 (1988).
d. J. Just and B. O'Connor, *Tetrahedron Lett.* **29**, 753 (1988).
e. M. F. Semmelhack and A. Yamashita, *J. Am. Chem. Soc.* **102**, 5924 (1980).
f. A. M. Echavarren and J. K. Stille, *J. Am. Chem. Soc.* **110**, 4051 (1988).
16a. J. E. Bäckvall, S. E. Byström, and R. E. Nordberg, *J. Org. Chem.* **49**, 4619 (1984).
b. M. F. Semmelhack and C. Bodurow, *J. Am. Chem. Soc.* **106**, 1496 (1984).
c. D. Valentine, Jr., J. W. Tilley, and R. A. Le Mahieu, *J. Org. Chem.* **46**, 4614 (1981).
d. A. S. Kende, B. Roth, P. J. SanFilippo, and T. J. Blacklock, *J. Am. Chem. Soc.* **104**, 5808 (1982).
17a. P. A. Bartlett, J. D. Meadows, and E. Ottow, *J. Am. Chem. Soc.* **106**, 5304 (1984).
b. M. Larcheveque and Y. Petit, *Tetrahedron Lett.* **28**, 1993 (1987).

Chapter 9

1a. N. Miyaura, K. Yamada, and A. Suzuki, *Tetrahedron Lett.*, 3437 (1979).
b. F. K. Sheffy, J. P. Godschalx, and J. K. Stille, *J. Am. Chem. Soc.* **106**, 4838 (1984).
c. P. Jacob III and H. C. Brown, *J. Am. Chem. Soc.* **98**, 7832 (1976).
d. D. Milstein and J. K. Stille, *J. Org. Chem.* **44**, 1613 (1979).
e. H. C. Brown and K. K. Wang, *J. Org. Chem.* **51**, 4514 (1986).
f. H. Yatagai, Y. Yamamoto, and K. Maruyama, *J. Am. Chem. Soc.* **102**, 4548 (1980).
g. R. Mohan and J. A. Katzenellenbogen, *J. Org. Chem.* **49**, 1234 (1984).
h. B. M. Trost and A. Brandi, *J. Org. Chem.* **49**, 4811 (1984).
i. H. C. Brown and T. Imai, *J. Am. Chem. Soc.* **105**, 6285 (1983).
j. H. C. Brown, N. G. Bhat, and J. B. Campbell, Jr., *J. Org. Chem.* **51**, 3398 (1986).
2a. H. C. Brown, M. M. Rogic, H. Nambu, and M. W. Rathke, *J. Am. Chem. Soc.* **91**, 2147 (1969); H. C. Brown, H. Nambu, and M. M. Rogic, *J. Am. Chem. Soc.* **91**, 6852 (1969).
b. H. C. Brown and R. A. Coleman, *J. Am. Chem. Soc.* **91**, 4606 (1969).
c. H. C. Brown and G. W. Kabalka, *J. Am. Chem. Soc.* **92**, 714 (1970).
d. G. Zweifel, R. P. Fisher, J. T. Snow, and C. C. Whitney, *J. Am. Chem. Soc.* **93**, 6309 (1971).
e. H. C. Brown and M. W. Rathke, *J. Am. Chem. Soc.* **89**, 2738 (1967).
f. H. C. Brown and M. M. Rogic, *J. Am. Chem. Soc.* **91**, 2146 (1969).
3. See references in Scheme 9.2.
4a. H. C. Brown, H. D. Lee, and S. U. Kulkarni, *J. Org. Chem.* **51**, 5282 (1986).
b. J. A. Sikorski, N. G. Bhat, T. E. Cole, K. K. Wang, and H. C. Brown, *J. Org. Chem.* **51**, 4521 (1986).
c. S. U. Kulkarni, H. D. Lee, and H. C. Brown, *J. Org. Chem.* **45**, 4542 (1980).
d. M. C. Welch and T. A. Bryson, *Tetrahedron Lett.* **29**, 521 (1988).
5a. D. R. McKean, G. Parrinello, A. F. Renaldo, and J. K. Stille, *J. Org. Chem.* **52**, 422 (1987).
b. L. Kuwajima and H. Urabe, *J. Am. Chem. Soc.* **104**, 6830 (1982).
c. A. Pelter, K. J. Gould, and C. R. Harrison, *Tetrahedron Lett.*, 3327 (1975).
d. A. Pelter and R. A. Drake, *Tetrahedron Lett.* **29**, 4181 (1988).
e. L. E. Overman and M. J. Sharp, *J. Am. Chem. Soc.* **110**, 612 (1988).
f. H. C. Brown and S. U. Kulkarni, *J. Org. Chem.* **44**, 2422 (1979).
6a. W. R. Roush, M. A. Adam, and D. J. Harris, *J. Org. Chem.* **50**, 2000 (1985).
b. S. J. Danishefsky, S. DeNinno, and P. Lartey, *J. Am. Chem. Soc.* **109**, 2082 (1987).
c. J. Hooz and D. M. Gunn, *J. Am. Chem. Soc.* **91**, 6195 (1969).
7a, b. H. C. Brown and N. G. Bhat, *J. Org. Chem.* **53**, 6009 (1988).
c. Y. Satoh, H. Serizawa, N. Miyzura, S. Hara, and A. Suzuki, *Tetrahedron Lett.* **29**, 1811 (1988).
d, e. H. C. Brown, D. Basavaiah, S. U. Kulkarni, N. Bhat, and J. V. N. Vara Prasad, *J. Org. Chem.* **53**, 239 (1988).
8a. J. A. Marshall, S. L. Crooks, and B. S. DeHoff, *J. Org. Chem.* **53**, 1616 (1988).
b. B. M. Trost and T. Sato, *J. Am. Chem. Soc.* **107**, 719 (1985).
9a. W. E. Fristad, D. S. Dime, T. R. Bailey, and L. A. Paquette, *Tetrahedron Lett.*, 1999 (1979).
b. E. Piers and H. E. Morton, *J. Org. Chem.* **45**, 4263 (1980).

c. J. C. Bottaro, R. N. Hanson, and D. E. Seitz, *J. Org. Chem.* **46**, 5221 (1981).
d. Y. Yamamoto and A. Yanagi, *Heterocycles* **16**, 1161 (1981).
e. M. B. Anderson and P. L. Fuchs, *Synth. Commun.* **17**, 621 (1987); B. A. Narayanan and W. H. Bunelle, *Tetrahedron Lett.* **28**, 6261 (1987).
f. A. Hosomi, M. Sato, and H. Sakurai, *Tetrahedron Lett.*, 429 (1979).
10. P. A. Grieco and W. F. Fobare, *Tetrahedron Lett.* **27**, 5067 (1986).
11. E. J. Corey and W. L. Seibel, *Tetrahedron Lett.* **27**, 905 (1986).
12a. J. A. Marshall, S. L. Crooks, and B. S. DeHoff, *J. Org. Chem.* **53**, 1616 (1988); J. A. Marshall and W. Y. Gung, *Tetrahedron Lett.* **29**, 3899 (1988).
b. E. Moret and M. Schlosser, *Tetrahedron Lett.* **25**, 4491 (1984).
c. L. K. Truesdale, D. Swanson, and R. C. Sun, *Tetrahedron Lett.* **26**, 5009 (1985).
d. B. M. Trost and T. Sato, *J. Am. Chem. Soc.* **107**, 719 (1985).
e. F. Bjorkling, T. Norin, C. R. Unelius, and R. B. Miller, *J. Org. Chem.* **52**, 292 (1987).
f. J. K. Stille and B. L. Groh, *J. Am. Chem. Soc.* **109**, 813 (1987).
g. Y. Yamamoto and A. Yanagi, *Heterocycles* **16**, 1161 (1981).
h. R. L. Funk and G. L. Bolton, *J. Org. Chem.* **49**, 5021 (1984).
i. L. E. Overman, T. C. Malone, and G. P. Meier, *J. Am. Chem. Soc.* **105**, 6993 (1983).
j. Y. Naruse, T. Esaki, and H. Yamamoto, *Tetrahedron Lett.* **29**, 1417 (1988).
k. J.-M. Fu, M. J. Sharp, and V. Sniekus, *Tetrahedron Lett.* **29**, 5459 (1988).
13a. H. C. Brown, T. Imai, M. C. Desai, and B. Singaran, *J. Am. Chem. Soc.* **107**, 4980 (1985).
b, c. H. C. Brown, R. K. Bakshi, and B. Singaran, *J. Am. Chem. Soc.* **110**, 1529 (1988).
d. H. C. Brown, M. Srebnik, R. K. Bakshi, and T. E. Cole, *J. Am. Chem. Soc.* **109**, 5420 (1987).

Chapter 10

1a. S. Julia and A. Ginebreda, *Synthesis*, 682 (1977).
b. R. Breslow and H. W. Chang, *J. Am. Chem. Soc.* **83**, 2367 (1961).
c. D. J. Burton and J. L. Hahnfeld, *J. Org. Chem.* **42**, 828 (1977).
d. D. Seyferth and S. P. Hopper, *J. Org. Chem.* **37**, 4070 (1972).
e. G. L. Closs, L. E. Closs, and W. A. Böll, *J. Am. Chem. Soc.* **85**, 3796 (1963).
f. L. G. Mueller and R. G. Lawton, *J. Org. Chem.* **44**, 4741 (1979).
g. F. G. Bordwell and M. W. Carlson, *J. Am. Chem. Soc.* **92**, 3377 (1970).
h. A. Burger and G. H. Harnest, *J. Am. Chem. Soc.* **65**, 2382 (1943).
i. E. Schmitz, D. Habish, and A. Stark, *Angew. Chem. Int. Ed. Engl.* **2**, 548 (1963).
j. R. Zurflüh, E. N. Wall, J. B. Sidall, and J. A. Edwards, *J. Am. Chem. Soc.* **90**, 6224 (1968).
k. M. Nishizawa, H. Takenaka, and Y. Hayashi, *J. Org. Chem.* **51**, 806 (1986).
l. H. Nishiyama, K. Sakuta, and K. Itoh, *Tetrahedron Lett.* **25**, 233 (1984).
m. H. Seto, M. Sakaguchi, and Y. Fujimoto, *Chem. Pharm. Bull.* **33**, 412 (1985).
n. D. F. Taber and E. H. Petty, *J. Org. Chem.* **47**, 4808 (1982).
o. B. Iddon, D. Price, H. Suschitzky, and D. J. C. Scopes, *Tetrahedron Lett.* **24**, 413 (1983).
p. A. Chu and L. N. Mander, *Tetrahedron Lett.* **29**, 2727 (1988).
q. G. E. Keck and D. F. Kachensky, *J. Org. Chem.* **51**, 2487 (1986).
r. D. H. R. Barton, J. Guilhem, Y. Hervé, P. Potier, and J. Thierry, *Tetrahedron Lett.* **28**, 1413 (1987).
2a. K. B. Wiberg, B. L. Furtek, and L. K. Olli, *J. Am. Chem. Soc.* **101**, 7675 (1979).
b. A. E. Greene and J.-P. Depres, *J. Am. Chem. Soc.* **101**, 4003 (1979).
c. R. A. Moss and E. Y. Chen, *J. Org.* **46**, 1466 (1981).
d. B. M. Trost, R. M. Cory, P. H. Scudder, and H. B. Neubold, *J. Am. Chem. Soc.* **95**, 7813 (1973).
e. T. J. Nitz, E. M. Holt, B. Rubin, and C. H. Stammer, *J. Org. Chem.* **46**, 2667 (1981).
f. L. N. Mander, J. V. Turner, and B. G. Colmbe, *Aust. J. Chem.* **27**, 1985 (1974).
g. P. J. Jessup, C. B. Petty, J. Roos, and L. E. Overman, *Org. Synth.* **59**, 1 (1979).
h. H. Dürr, H. Nickels, L. A. Pacala, and M. Jones, Jr., *J. Org. Chem.* **45**, 973 (1980).
i. G. A. Scheihser and J. D. White, *J. Org. Chem.* **45**, 1864 (1980).
j. M. B. Groen and F. J. Zeelen, *J. Org. Chem.* **43**, 1961 (1978).
k. R. C. Gadwood, R. M. Lett, and J. E. Wissinger, *J. Am. Chem. Soc.* **108**, 6343 (1986).
l. V. B. Rao, C. F. George, S. Wolff, and W. C. Agosta, *J. Am. Chem. Soc.* **107**, 5732 (1985).

m. Y. Araki, T. Endo, M. Tanji, J. Nagasawa, and Y. Ishido, *Tetrahedron Lett.* **28**, 5853 (1987).
n. M. Newcomb and J. Kaplan, *Tetrahedron Lett.* **28**, 1615 (1987).
o. G. Stork, P. M. Sher, and H.-L. Chen, *J. Am. Chem. Soc.* **108**, 6384 (1986).
p. E. J. Corey and M. Kang, *J. Am. Chem. Soc.* **106**, 5384 (1984).
q. G. E. Keck, D. F. Kachensky, and E. J. Enholm, *J. Org. Chem.* **50**, 4317 (1985).
r. A. DeMesmaeker, P. Hoffmann, and B. Ernst, *Tetrahedron Lett.*, **30**, 57 (1989).
3a. W. J. Hehre, J. A. Pople, W. A. Latham, L. Radom, E. Wasserman, and Z. R. Wasserman, *J. Am. Chem. Soc.* **98**, 4378 (1976); N. C. Baird and K. F. Taylor, *J. Am. Chem. Soc.* **100**, 1333 (1978); J. M. Bofill, J. Farrás, S. Olivella, A. Solé, and J. Vilarrasa, *J. Am. Chem. Soc.* **110**, 1694 (1988).
b. P. H. Mueller, N. G. Rondan, K. N. Houk, J. F. Harrison, D. Hooper, B. H. Willen, and J. F. Liebman, *J. Am. Chem. Soc.* **103**, 5049 (1981).
c, d. R. Gleiter and R. Hoffmann, *J. Am. Chem. Soc.* **90**, 5457 (1968).
4. C. D. Poulter, E. C. Friedrich, and S. Winstein, *J. Am. Chem. Soc.* **91**, 6892 (1969).
5. P. L. Barili, G. Berti, B. Macchia, and L. Monti, *J. Chem. Soc., C*, 1168 (1970).
6a. R. K. Hill and D. A. Cullison, *J. Am. Chem. Soc.* **95**, 2923 (1973).
b. A. B. Smith III, B. H. Toder, S. J. Branca, and R. K. Dieter, *J. Am. Chem. Soc.* **103**, 1996 (1981).
c. M. C. Pirrung and J. A. Werner, *J. Am. Chem. Soc.* **108**, 6060 (1986).
7. E. W. Warnhoff, C. M. Wong, and W. T. Tai, *J. Am. Chem. Soc.* **90**, 514 (1968).
8a. S. A. Godleski, P. von R. Schleyer, E. Osawa, Y. Inamoto, and Y. Fujikura, *J. Org. Chem.* **41**, 2596 (1976).
b. P. E. Eaton, Y. S. Or, and S. J. Branca, *J. Am. Chem. Soc.* **103**, 2134 (1981).
c. G. H. Posner, K. A. Babiak, G. L. Loomis, W. J. Frazee, R. D. Mittal, and I. L. Karle, *J. Am. Chem. Soc.* **102**, 7498 (1980).
d. T. Hudlicky, F. J. Koszyk, T. M. Kutchan, and J. P. Sheth, *J. Org. Chem.* **45**, 5020 (1980).
e. L. A. Paquette and Y.-K. Han, *J. Am. Chem. Soc.* **103**, 1835 (1981).
f. L. A. Paquette and R. W. Houser, *J. Am. Chem. Soc.* **91**, 3870 (1969).
9a. Y. Ito, S. Fujii, M. Nakatsuka, F. Kawamoto, and T. Saegusa, *Org. Synth.* **59**, 113 (1979).
b. P. Nedenskov, H. Heide, and N. Clauson-Kass, *Acta Chem. Scand.* **16**, 246 (1962).
c. L.-F. Tietze, *J. Am. Chem. Soc.* **96**, 946 (1974).
d. E. G. Breitholle and A. G. Fallis, *J. Org. Chem.* **43**, 1964 (1978).
e. E. Y. Chen, *J. Org. Chem.* **49**, 3245 (1984).
f. G. Mehta and K. S. Rao, *J. Org. Chem.* **50**, 5537 (1985).
g. T. V. Rajan Babu, *J. Org. Chem.* **53**, 4522 (1988).
h. W. D. Klobucar, L. A. Paquette, and J. P. Blount, *J. Org. Chem.* **46**, 4021 (1981).
i. F. E. Ziegler, S. I. Klein, U. K. Pati, and T.-F. Wang, *J. Am. Chem. Soc.* **107**, 2730 (1985).
j. T. Hudlicky, F. J. Koszyk, D. M. Dochwat, and G. L. Cantrell, *J. Org. Chem.* **46**, 2911 (1981).
k. R. E. Ireland, W. C. Dow, J. D. Godfrey, and S. Thaisrivongs, *J. Org. Chem.* **49**, 1001 (1984).
l. C. P. Chuang and D. J. Hart, *J. Org. Chem.* **48**, 1782 (1983).
10a. S. D. Larsen and S. A. Monti, *J. Am. Chem. Soc.* **99**, 8015 (1977).
b. S. A. Monti and J. M. Harless, *J. Am. Chem. Soc.* **99**, 2690 (1977).
c. F. T. Bond and C.-Y. Ho, *J. Org. Chem.* **41**, 1421 (1976).
d. E. Wenkert, R. S. Greenberg, and H.-S. Kim, *Helv. Chim. Acta* **70**, 2159 (1987).
e. B. B. Snider and M. A. Dombroski, *J. Org. Chem.* **52**, 5487 (1987).
f. G. A. Kraus and K. Landgrebe, *Tetrahedron Lett.* **25**, 3939 (1984).
11. L. Blanco, N. Slougi, G. Rousseau, and J. M. Conia, *Tetrahedron Lett.*, 645 (1981).
12. J. A. Marshall and J. A. Ruth, *J. Org. Chem.* **39**, 1971 (1974).
13. C. A. Grob, H. R. Kiefer, H. J. Lutz, and H. J. Wilkens, *Helv. Chim. Acta* **50**, 416 (1967).
14. M. P. Doyle, W. E. Buhro, and J. F. Dellaria, Jr., *Tetrahedron Lett.*, 4429 (1979).
15. R. Tsang, J. K. Dickson, Jr., H. Pak, R. Walton, and B. Fraser-Reid, *J. Am. Chem. Soc.* **109**, 3484 (1987).

Chapter 11

1a. L. Friedman and H. Shechter, *J. Org. Chem.* **26**, 2522 (1961).
b. E. C. Taylor, F. Kienzle, R. L. Robey, and A. McKillop, *J. Am. Chem. Soc.* **92**, 2175 (1970).

c. G. F. Hennion and S. F. de C. McLeese, *J. Am. Chem. Soc.* **64**, 2421 (1942).
d. J. Koo, *J. Am. Chem. Soc.* **75**, 1889 (1953).
e. E. C. Taylor, F. Kienzle, R. L. Robey, A. McKillop, and J. D. Hunt, *J. Am. Chem. Soc.* **93**, 4845 (1971).
f. G. A. Ropp and E. C. Coyner, *Org. Synth.* **IV**, 727 (1963).
g. M. Shiratsuchi, K. Kawamura, T. Akashi, M. Fujij, H. Ishihama, and Y. Uchida, *Chem. Pharm. Bull.* **35**, 632 (1987).
h. D. C. Furlano and K. D. Kirk, *J. Org. Chem.* **51**, 4073 (1986).
i. C. K. Bradsher, F. C. Brown, and H. K. Porter, *J. Am. Chem. Soc.* **76**, 2357 (1954).
2a. E. C. Taylor, E. C. Bigham, and D. K. Johnson, *J. Org. Chem.* **42**, 362 (1977).
b. P. Studt, *Justus Liebigs Ann. Chem.*, 2105 (1978).
c. T. Jojima, H. Takeshiba, and T. Kinoto, *Bull. Chem. Soc. Jpn.* **52**, 2441 (1979).
d. R. W. Bost and F. Nicholson, *J. Am. Chem. Soc.* **57**, 2368 (1935).
e. H. Durr, H. Nickels, L. A. Pacala, and M. Jones, Jr., *J. Org. Chem.* **45**, 973 (1980).
3a. C. L. Perrin and G. A. Skinner, *J. Am. Chem. Soc.* **93**, 3389 (1971).
b. R. A. Rossi and J. F. Bunnett, *J. Am. Chem. Soc.* **94**, 683 (1972).
c. M. Jones, Jr., and R. H. Levin, *J. Am. Chem. Soc.* **91**, 6411 (1969).
d. Y. Naruta, Y. Nishigaichi, and K. Maruyama, *J. Org. Chem.* **53**, 1192 (1988).
e. S. P. Khanapure, R. T. Reddy, and E. R. Biehl, *J. Org. Chem.* **52**, 5685 (1987).
f. G. Büchi and J. C. Leung, *J. Org. Chem.* **51**, 4813 (1986).
4a. M. P. Doyle, J. F. Dellaria, Jr., B. Siegfried, and S. W. Bishop, *J. Org. Chem.* **42**, 3494 (1977).
b. T. Cohen, A. G. Dietz, Jr., and J. R. Miser, *J. Org. Chem.* **42**, 2053 (1977).
c. M. P. Doyle, B. Siegfried, R. C. Elliot, and J. F. Dellaria, Jr., *J. Org. Chem.* **42**, 2431 (1977).
d. G. D. Figuly and J. C. Martin, *J. Org. Chem.* **45**, 3728 (1980).
e. E. McDonald and R. D. Wylie, *Tetrahedron* **35**, 1415 (1979).
f. M. P. Doyle, B. Siegfried, and J. F. Dellaria, Jr., *J. Org. Chem.* **42**, 2426 (1977).
g. A. P. Kozikowski, M. N. Greco, and J. P. Springer, *J. Am. Chem. Soc.* **104**, 7622 (1982).
h. P. H. Gore and I. M. Khan, *J. Chem. Soc., Perkin Trans. 1,* 2779 (1979).
i. A. A. Leon, G. Daub, and I. R. Silverman, *J. Org. Chem.* **49**, 4544 (1984).
j. S. R. Wilson and L. A. Jacob, *J. Org. Chem.* **51**, 4833 (1986).
k. S. A. Khan, M. A. Munawar, and M. Siddiq, *J. Org. Chem.* **53**, 1799 (1988).
l. J. R. Beadle, S. H. Korzeniowski, D. E. Rosenberg, B. J. Garcia-Slanga, and G. W. Gokel, *J. Org. Chem.* **49**, 1594 (1984).
5. B. L. Zenitz and W. H. Hartung, *J. Org. Chem.* **11**, 444 (1946).
6. T. F. Buckley III and H. Rapoport, *J. Am. Chem. Soc.* **102**, 3056 (1980).
7. G. A. Olah and J. A. Olah, *J. Am. Chem. Soc.* **98**, 1839 (1976).
8. E. J. Corey, S. Barcza, and G. Klotmann, *J. Am. Chem. Soc.* **91**, 4782 (1969).
9. E. C. Taylor, F. Kienzle, R. L. Robey, and A. McKillop, *J. Am. Chem. Soc.* **92**, 2175 (1970).
10. M. Essiz, G. Guillaumet, J.-J. Brunet, and P. Caubere, *J. Org. Chem.* **45**, 240 (1980).
11. S. P. Khanapure, L. Crenshaw, R. T. Reddy, and E. R. Biehl, *J. Org. Chem.* **53**, 4915 (1988).
12a. J. H. Boyer and R. S. Burkis, *Org. Synth.* **V**, 1067 (1973).
b. H. P. Schultz, *Org. Synth.* **IV**, 364 (1963); F. D. Gunstone and S. H. Tucker, *Org. Synth.* **IV**, 160 (1963).
c. D. H. Hey and M. J. Perkins, *Org. Synth.* **V**, 51 (1973).
d. K. Rorig, J. D. Johnston, R. W. Hamilton, and T. J. Telinski, *Org. Synth.* **IV**, 576 (1963).
e. K. G. Rutherford and W. Redmond, *Org. Synth.* **V**, 133 (1973).
f. M. M. Robinson and B. L. Robinson, *Org. Synth.* **IV**, 947 (1963).
g. R. Adams, W. Reifschneider, and A. Ferretti, *Org. Synth.* **V**, 107 (1973).
h. G. H. Cleland, *Org. Synth.* **51**, 1 (1971).
13a. R. E. Ireland, C. A. Lipinski, C. J. Kowalski, J. W. Tilley, and D. M. Walba, *J. Am. Chem. Soc.* **96**, 3333 (1974).
b. J. J. Korst, J. D. Johnston, K. Butler, E. J. Bianco, L. H. Conover, and R. B. Woodward, *J. Am. Chem. Soc.* **90**, 439 (1968).
c. K. A. Parker and J. Kallmerten, *J. Org. Chem.* **45**, 2614, 2620 (1980).
d. F. A. Carey and R. M. Guiliano, *J. Org. Chem.* **46**, 1366 (1981).
e. R. B. Woodward and T. R. Hoye, *J. Am. Chem. Soc.* **99**, 8007 (1977).
f. E. C. Horning, J. Koo, M. S. Fish, and G. N. Walker, *Org. Synth.* **IV**, 408 (1963); J. Koo, *Org. Synth.* **V**, 550 (1973).

14. T. F. Buckley III and H. Rapoport, *J. Am. Chem. Soc.* **102**, 3056 (1980).
15. H. C. Bell, J. R. Kalman, J. T. Pinhey, and S. Sternhell, *Tetrahedron Lett.*, 3391 (1974).
16. B. Chauncy and E. Gellert, *Aust. J. Chem.* **22**, 993 (1969); R. I. Duclos, Jr., J. S. Tung, and H. Rapoport, *J. Org. Chem.* **49**, 5243 (1984).
17. W. G. Miller and C. U. Pittman, Jr., *J. Org. Chem.* **39**, 1955 (1974).
18. W. Nagata, K. Okada, and T. Aoki, *Synthesis*, 365 (1979).

Chapter 12

1a. Y. Butsugan, S. Yoshida, M. Muto, and T. Bito, *Tetrahedron Lett.*, 1129 (1971).
b. E. J. Corey and H. E. Ensley, *J. Am. Chem. Soc.* **97**, 6908 (1975).
c. R. G. Gaughan and C. D. Poulter, *J. Org. Chem.* **44**, 2441 (1979).
d. E. Vedejs, D. A. Engler, and J. E. Telschow, *J. Org. Chem.* **43**, 188 (1978).
e. K. Akashi, R. E. Palermo, and K. B. Sharpless, *J. Org. Chem.* **43**, 2063 (1978).
f. A. Hassner, R. H. Reuss, and H. W. Pinnick, *J. Org. Chem.* **40**, 3427 (1975).
g. R. N. Mirrington and K. J. Schmalzl, *J. Org. Chem.* **37**, 2877 (1972).
h. K. B. Sharpless and R. F. Lauer, *J. Org. Chem.* **39**, 429 (1974).
i. J. A. Marshall and R. C. Andrews, *J. Org. Chem.* **50**, 1602 (1985).
j. R. H. Schlessinger, J. J. Wood, A. J. Poos, R. A. Nugent, and W. H. Parsons, *J. Org. Chem.* **48**, 1146 (1983).
k. R. K. Boeckman, Jr., J. E. Starett, Jr., D. G. Nickell, and P.-E. Sum, *J. Am. Chem. Soc.* **108**, 5549 (1986).
l. E. J. Corey and Y. B. Xiang, *Tetrahedron Lett.* **29**, 995 (1988).
m. D. J. Plata and J. Kallmerten, *J. Am. Chem. Soc.* **110**, 4041 (1988).
n. B. E. Rossiter, T. Katsuki, and K. B. Sharpless, *J. Am. Chem. Soc.* **103**, 464 (1981).
o. J. Mulzer, A. Angermann, B. Schubert, and C. Seilz, *J. Org. Chem.* **51**, 5294 (1986).
p. R. H. Schlessinger and R. A. Nugent, *J. Am. Chem. Soc.* **104**, 1116 (1982).
q. H. Niwa, T. Mori, T. Hasegawa, and K. Yamada, *J. Org. Chem.* **51**, 1015 (1986).
2a. J. P. McCormick, W. Tomasik, and M. W. Johnson, *Tetrahedron Lett.*, 607 (1981).
b. H. C. Brown, J. H. Kawakami, and S. Ikegami, *J. Am. Chem. Soc.* **92**, 6914 (1970).
c. R. M. Scarborough, Jr., B. H. Toder, and A. B. Smith III, *J. Am. Chem. Soc.* **102**, 3904 (1980).
d. B. Rickborn and R. M. Gerkin, *J. Am. Chem. Soc.* **90**, 4193 (1968).
e. J. A. Marshall and R. A. Ruden, *J. Org. Chem.* **36**, 594 (1971).
f. G. A. Kraus and B. Roth, *J. Org. Chem.* **45**, 4825 (1980).
g. T. Sakan and K. Abe, *Tetrahedron Lett.*, 2471 (1968).
h. K. J. Clark, G. I. Fray, R. H. Jaeger, and R. Robinson, *Tetrahedron* **6**, 217 (1959).
i. T. Kawabata, P. Grieco, H.-L. Sham, H. Kim, J. Y. Jaw, and S. Tu, *J. Org. Chem.* **52**, 3346 (1987).
j. P. T. Lansbury, J. P. Galbo, and J. P. Springer, *Tetrahedron Lett.* **29**, 147 (1988).
k. J. P. Marino, R. F. de la Pradilla, and E. Laborde, *J. Org. Chem.* **52**, 4898 (1987).
l. J. A. Marshall and R. A. Ruden, *J. Org. Chem.* **36**, 594 (1971).
m. J. E. Toth, P. R. Hamann, and P. L. Fuchs, *J. Org.* **53**, 4694 (1988).
3. E. L. Eliel, S. H. Schroeter, T. J. Brett, F. J. Biros, and J.-C. Richer, *J. Am. Chem. Soc.* **88**, 3327 (1966).
4. E. E. Royals and J. C. Leffingwell, *J. Org. Chem.* **31**, 1927 (1966).
5. W. W. Epstein and F. W. Sweat, *Chem. Rev.* **67**, 247 (1967).
6. D. P. Higley and R. W. Murray, *J. Am. Chem. Soc.* **96**, 3330 (1974).
7. R. Criegee and P. Günther, *Chem. Ber.* **96**, 1564 (1963).
8a. S. Isoe, S. Katsumura, S. B. Hyeon, and T. Sakan, *Tetrahedron Lett.*, 1089 (1971).
b. Y. Ogata, Y. Sawaki, and M. Shiroyama, *J. Org. Chem.* **42**, 4061 (1977).
c. F. G. Bordwell and A. C. Knipe, *J. Am. Chem. Soc.* **93**, 3416 (1971).
d. B. M. Trost, P. R. Bernstein, and P. C. Funfschilling, *J. Am. Chem. Soc.* **101**, 4378 (1979).
e. C. S. Foote, S. Mazur, P. A. Burns, and D. Lerdal, *J. Am. Chem. Soc.* **95**, 586 (1973).
f. J. P. Marino, K. E. Pfitzner, and R. A. Olofson, *Tetrahedron* **27**, 4181 (1971).
g. M. A. Avery, C. Jennings-White, and W. K. M. Chong, *Tetrahedron Lett.* **28**, 4629 (1987).

9a. P. N. Confalone, C. Pizzolato, D. L. Confalone, and M. R. Uskokovic, *J. Am. Chem. Soc.* **102**, 1954 (1980).
b. S. Danishefsky, R. Zamboni, M. Kahn, and S. J. Etheredge, *J. Am. Chem. Soc.* **103**, 3460 (1981).
c. J. K. Whitesell, R. S. Matthews, M. A. Minton, and A. M. Helbling, *J. Am. Chem. Soc.* **103**, 3468 (1981).
d. F. A. J. Kerdesky, R. J. Ardecky, M. V. Lakshmikanthan, and M. P. Cava, *J. Am. Chem. Soc.* **103**, 1992 (1981).
e. J. K. Whitesell and R. S. Matthews, *J. Org. Chem.* **43**, 1650 (1978).
f. R. Fujimoto, Y. Kishi, and J. F. Blount, *J. Am. Chem. Soc.* **102**, 7154 (1980).
g. S. P. Tanis and K. Nakanishi, *J. Am. Chem. Soc.* **101**, 4398 (1979).
h. R. B. Miller and R. D. Nash, *J. Org. Chem.* **38**, 4424 (1973).
i. R. Grewe and I. Hinrichs, *Chem. Ber.* **97**, 443 (1964).
j. W. Nagata, S. Hirai, K. Kawata, and T. Okumura, *J. Am. Chem. Soc.* **89**, 5046 (1967).
k. W. G. Dauben, M. Lorber, and D. S. Fullerton, *J. Org. Chem.* **34**, 3587 (1969).
l. E. E. van Tamelen, M. Shamma, A. W. Burgstahler, J. Wolinsky, R. Tamm, and P. E. Aldrich, *J. Am. Chem. Soc.* **80**, 5006 (1958).
m. S. D. Burke, C. W. Murtishaw, J. O. Saunders, J. A. Oplinger, and M. S. Dike, *J. Am. Chem. Soc.* **106**, 4558 (1984).
n. B. M. Trost, P. G. McDougal, and K. J. Haller, *J. Am. Chem. Soc.* **106**, 383 (1984).
10. W. P. Keaveney, M. G. Berger, and J. J. Pappas, *J. Org. Chem.* **32**, 1537 (1967).
11. B. M. Trost and K. Hiroi, *J. Am. Chem. Soc.* **97**, 6911 (1975).
12. E. C. Taylor, C.-S. Chiang, A. McKillop, and J. F. White, *J. Am. Chem. Soc.* **98**, 6750 (1976).
13a. F. Delay and G. Ohloff, *Helv. Chim. Acta* **62**, 2168 (1979).
b. R. Noyori, T. Sato, and Y. Hayakawa, *J. Am. Chem. Soc.* **100**, 2561 (1978).
14a. I. Saito, R. Nagata, K. Yubo, and Y. Matsuura, *Tetrahedron Lett.* **24**, 4439 (1983).
b. J. R. Wiseman and S. Y. Lee, *J. Org. Chem.* **51**, 2485 (1986).
c. H. Nishiyama, M. Matsumoto, H. Arai, H. Sakaguchi, and K. Itoh, *Tetrahedron Lett.* **27**, 1599 (1986).
15a. R. E. Ireland, P. G. M. Wuts, and B. Ernst, *J. Am. Chem. Soc.* **103**, 3205 (1981).
b. R. M. Scarborough, Jr., B. H. Toder, and A. B. Smith III, *J. Am. Chem. Soc.* **102**, 3904 (1980).
c. P. F. Hudrlik, A. M. Hudrlik, G. Nagendrappa, T. Yimenu, E. T. Zellers, and E. Chin, *J. Am. Chem. Soc.* **102**, 6894 (1980).
d. T. Wakamatsu, K. Akasaka, and Y. Ban, *J. Org. Chem.* **44**, 2008 (1979).
e. D. A. Evans, C. E. Sacks, R. A. Whitney, and N. G. Mandel, *Tetrahedron Lett.* 727 (1978).
f. F. Bourelle-Wargnier, M. Vincent, and J. Chuche, *J. Org. Chem.* **45**, 428 (1980).
g. J. A. Zalikowski, K. E. Gilbert, and W. T. Borden, *J. Org. Chem.* **45**, 346 (1980).
h. E. Vogel, W. Klug, and A. Breuer, *Org. Synth.* **55**, 86 (1976).
i. L. D. Spicer, M. W. Bullock, M. Garber, W. Groth, J. J. Hand, D. W. Long, J. L. Sawyer, and R. S. Wayne, *J. Org. Chem.* **33**, 1350 (1968).
j. B. E. Rossiter, T. Katsuki, and K. B. Sharpless, *J. Am. Chem. Soc.* **103**, 464 (1981).
k. L. A. Paquette and Y.-K. Han, *J. Am. Chem. Soc.* **103**, 1831 (1981).
l. T. Wakamatsu, K. Akasaka, and Y. Ban, *Tetrahedron Lett.*, 2755 (1977).
m. M. Muelbacher and C. D. Poulter, *J. Org. Chem.* **53**, 1026 (1988).
n. P. T. W. Cheng and S. McLean, *Tetrahedron Lett.* **29**, 3511 (1988).
o. A. B. Smith III and R. E. Richmond, *J. Am. Chem. Soc.* **105**, 575 (1983).
p. R. K. Boeckman, Jr., J. E. Starrett, Jr., D. G. Nickell, and P.-E. Sum, *J. Am. Chem. Soc.* **108**, 5549 (1986).
16. Y. Gao and K. B. Sharpless, *J. Org. Chem.* **53**, 4081 (1988).
17. C. W. Jefford, Y. Wang, and G. Bernardinelli, *Helv. Chim. Acta* **71**, 2042 (1988).

Chapter 13

1a. E. J. Corey, J.-L. Gras, and P. Ulrich, *Tetrahedron Lett.*, 809 (1976).
b. K. C. Nicolaou, S. P. Seitz, and M. R. Pavia, *J. Am. Chem. Soc.* **103**, 1222 (1981).
c. E. J. Corey and A. Venkateswarlu, *J. Am. Chem. Soc.* **94**, 6190 (1972).

d–f. H. H. Meyer, *Justus Liebigs Ann. Chem.*, 732 (1977).
2a. M. Miyashita, A. Yoshikoshi, and P. A. Grieco, *J. Org. Chem.* **42**, 3772 (1977).
b. E. J. Corey, L. O. Wiegel, D. Floyd, and M. G. Bock, *J. Am. Chem. Soc.* **100**, 2916 (1978).
c. A. M. Felix, E. P. Heimer, T. J. Lambros, C. Tzougraki, and J. Meienhofer, *J. Org. Chem.* **43**, 4194 (1978).
d. P. N. Confalone, G. Pizzolato, E. G. Baggiolini, D. Lollar, and M. R. Uskokovic, *J. Am. Chem. Soc.* **97**, 5936 (1975).
e. A. B. Foster, J. Lehmann, and M. Stacey, *J. Chem. Soc.*, 4649 (1961).
3a. D. M. Simonović, A. S. Rao, and S. C. Bhattacharyya, *Tetrahedron* **19**, 1061 (1963).
b. R. E. Ireland and L. N. Mander, *J. Org. Chem.* **32**, 689 (1967).
c. G. Büchi, W. D. MacLeod, Jr., and J. Padilla, *J. Am. Chem. Soc.* **86**, 4438 (1964).
d. P. Doyle, I. R. Maclean, W. Parker, and R. A. Raphael, *Proc. Chem. Soc.*, 239 (1963).
e. J. C. Sheehan and K. R. Henry-Logan, *J. Am. Chem. Soc.* **84**, 2983 (1962).
f. E. J. Corey, M. Ohno, R. B. Mitra, and P. A. Vatakencherry, *J. Am. Chem. Soc.* **86**, 478 (1964).
4a. B. ElAmin, G. M. Anantharamaiah, G. P. Royer, and G. E. Means, *J. Org. Chem.* **44**, 3442 (1979).
b. B. Moreay, S. Lavielle, and A. Marquet, *Tetrahedron Lett.*, 2591 (1977); B. C. Laguzza and B. Ganem, *Tetrahedron Lett.*, 1483 (1981).
c. J. I. Seeman, *Synthesis*, 498 (1977); D. Spitzner, *Synthesis*, 242 (1977).
d. H. J. Anderson and J. K. Groves, *Tetrahedron Lett.*, 3165 (1971).
5a. T. Hylton and V. Boekelheide, *J. Am. Chem. Soc.* **90**, 6987 (1968).
b. B. W. Erickson, *Org. Synth.* **53**, 189 (1973).
c. H. Paulsen, V. Sinnwell, and P. Stadler, *Angew. Dhem. Int. Ed. Engl.* **11**, 149 (1972).
d. S. Torii, K. Uneyama, and M. Isihara, *J. Org. Chem.* **39**, 3645 (1974).
e. J. A. Marshall and A. E. Greene, *J. Org. Chem.* **36**, 2035 (1971).
f. E. Leete, M. R. Chedekel, and G. B. Bodem, *J. Org. Chem.* **37**, 4465 (1972).
g. H. Yamamoto and H. L. Sham, *J. Am. Chem. Soc.* **101**, 1609 (1979).
h. K. Deuchert, U. Hertenstein, S. Hünig, and G. Wehner, *Chem. Ber.* **112**, 2045 (1979).
i. T. Takahashi, K. Kitamura, and J. Tsuji, *Tetrahedron Lett.* **24**, 4695 (1983).
6a. S. Danishefsky and T. Kitahara, *J. Am. Chem. Soc.* **96**, 7807 (1974).
b. P. S. Wharton, C. E. Sundin, D. W. Johnson, and H. C. Kluender, *J. Org. Chem.* **37**, 34 (1972).
c. E. J. Corey, B. W. Erickson, and R. Noyori, *J. Am. Chem. Soc.* **93**, 1724 (1971).
d. R. E. Ireland and J. A. Marshall, *J. Org. Chem.* **27**, 1615 (1962).
e. W. S. Johnson, T. J. Brocksom, P. Loew, D. H. Rich, L. Werthemann, R. A. Arnold, T. Li, and D. J. Faulkner, *J. Am. Chem. Soc.* **92**, 4463 (1970).
f. L. Birladeanu, T. Hanafusa, and S. Winstein, *J. Am. Chem. Soc.* **88**, 2315 (1966); T. Hanafusa, L. Birladeanu, and S. Winstein, *J. Am. Chem. Soc.* **87**, 3510 (1965).
7a. A. B. Smith III and W. C. Agosta, *J. Am. Chem. Soc.* **96**, 3289 (1974).
b. R. S. Cooke and U. H. Andrews, *J. Am. Chem. Soc.* **96**, 2974 (1974).
c. L. A. Hulshof and H. Wynberg, *J. Am. Chem. Soc.* **96**, 2191 (1974).
d. S. D. Burke, C. W. Murtiashaw, M. S. Dike, S. M. S. Strickland, and J. O. Saunders, *J. Org. Cyem.* **46**, 2400 (1981).
e. K. C. Nicolaou, M. R. Pavia, and S. P. Seitz, *J. Am. Chem. Soc.* **103**, 1224 (1981).
8a. E. M. Acton, R. N. Goerner, H. S. Uh, K. J. Ryan, D. W. Henry, C. E. Cass, and G. A. LePage, *J. Med. Chem.* **22**, 518 (1979).
b. E. G. Gros, *Carbohydr. Res.* **2**, 56 (1966).
c. S. Hanessian and G. Rancourt, *Can. J. Chem.* **55**, 1111 (1977).
d. R. R. Schmidt and A. Gohl, *Chem. Ber.* **112**, 1689 (1979).
9a. S. F. Martin and T. Chou, *J. Org. Chem.* **43**, 1027 (1978).
b. W. C. Still and M.-Y. Tsai, *J. Am. Chem. Soc.* **102**, 3654 (1980).
c. J. C. Bottaro and G. A. Berchtold, *J. Org. Chem.* **45**, 1176 (1980).
d. A. S. Kende and T. P. Demuth, *Tetrahedron Lett.*, 715 (1980).
e. J. A. Marshall and P. G. M. Wuts, *J. Org. Chem.* **43**, 1086 (1978).
10a. R. Bonjouklian and R. A. Ruden, *J. Org. Chem.* **42**, 4095 (1977).
b. L. A. Paquette, R. E. Moerck, B. Harirchian, and P. D. Magnus, *J. Am. Chem. Soc.* **100**, 1597 (1978).
c. P. S. Wharton, C. E. Sundin, D. W. Johnson, and H. C. Kluender, *J. Org. Chem.* **37**, 34 (1972).
d. S. Danishefsky, T. Kitahara, C. F. Yan, and J. Morris, *J. Am. Chem. Soc.* **101**, 6996 (1979).
e. B. M. Trost, J. Ippen, and W. C. Vladuchick, *J. Am. Chem. Soc.* **99**, 8116 (1977).

11a. E. J. Corey, E. J. Trybulski, L. S. Melvin, Jr., K. C. Nicolaou, J. A. Secrist, R. Lett, P. W. Sheldrake, J. R. Falck, D. J. Brunelle, M. F. Haslanger, S. Kim, and S. Yoo, *J. Am. Chem. Soc.* **100**, 4618 (1978).

b. K. G. Paul, F. Johnson, and D. Favara, *J. Am. Chem. Soc.* **98**, 1285 (1976).

c. P.N. Confalone, G. Pizzolato, E. G. Baggiolini, D. Lollar, and M. R. Uskokovic, *J. Am. Chem. Soc.* **97**, 5936 (1975).

d. E. Baer, J. M. Grosheintz, and H. O. L. Fischer, *J. Am. Chem. Soc.* **61**, 2607 (1939).

e. J. L. Coke and A. B. Richon, *J. Org. Chem.* **41**, 3516 (1976).

f. J. R. Dyer, W. E. McGonigal, and K. C. Rice, *J. Am. Chem. Soc.* **87**, 654 (1965).

g. E. J. Corey and S. Nozoe, *J. Am. Chem. Soc.* **85**, 3527 (1963).

h. R. Jacobson, R. J. Taylor, H. J. Williams, and L. R. Smith, *J. Org. Chem.* **47**, 3140 (1982).

12a. R. B. Miller and E. S. Behare, *J. Am. Chem. Soc.* **96**, 8102 (1974).

b. S. Iwaki, S. Marumo, T. Saito, M. Yamada, and K. Katagiri, *J. Am. Chem. Soc.* **96**, 7842 (1974).

c. G. Büchi, W. Hofheinz, and J. V. Paukstelis, *J. Am. Chem. Soc.* **91**, 6473 (1969).

d. M. Brown, *J. Org. Chem.* **33**, 162 (1968).

e. E. J. Corey, R. B. Mitra, and H. Uda, *J. Am. Chem. Soc.* **86**, 485 (1964).

13a. I. Fleming, *Selected Organic Syntheses*, Wiley, London, 1973, pp. 3–6; J. E. McMurry and J. Melton, *J. Am. Chem. Soc.* **93**, 5309 (1971).

b. R. M. Coates and J. E. Shaw, *J. Am. Chem. Soc.* **92**, 5657 (1970).

c. T. F. Buckley III and H. Rapoport, *J. Am. Chem.* **102**, 3056 (1980).

d. D. A. Evans, A. M. Golob, N. S. Mandel, and G. S. Mandel, *J. Am. Chem. Soc.* **100**, 8170 (1978).

e. E. J. Corey and R. D. Balanson, *J. Am. Chem. Soc.* **96**, 6516 (1974).

f. J. L. Herrmann, M. H. Berger, and R. H. Schlessinger, *J. Am. Chem. Soc.* **95**, 7923 (1973).

g. R. F. Romanet and R. H. Schlessinger, *J. Am. Chem. Soc.* **96**, 3701 (1974); R. A. LeMahieu, M. Carson, and R. W. Kierstead, *J. Org. Chem.* **33**, 3660 (1968); G. Büchi, D. Minster, and J. C. F. Young, *J. Am. Chem. Soc.* **93**, 4319 (1971).

h. J. H. Babler, D. O. Olsen, and W. H. Arnold, *J. Org. Chem.* **39**, 1656 (1974); R. J. Crawford, W. F. Erman, and C. D. Broaddus, *J. Am. Chem. Soc.* **94**, 4298 (1972).

i. C. S. Subramanian, P. J. Thomas, V. R. Mamdapur, and M. S. Chandra, *J. Chem. Soc., Perkin Trans. 1*, 2346 (1979).

j. S. Hanessian and R. Frenette, *Tetrahedron Lett.*, 3391 (1979).

k. E. Piers, R. W. Britton, and W. de Waal, *J. Am. Chem. Soc.* **93**, 5113 (1971); K. J. Schmalzl and R. N. Mirrington, *Tetrahedron Lett.*, 3219 (1970); N. Fukamiya, M. Kato, and A. Yoshikoshi, *J. Chem. Soc., Chem. Commun.*, 1120 (1971); G. Frater, *Helv. Chim. Acta* **57**, 172 (1974); K. Yamada, Y. Kyotani, S. Manabe, and M. Suzuki, *Tetrahedron* **35**, 293 (1979); M. E. Jung, C. A. McCombs, Y. Takeda, and Y. G. Pan, *J. Am. Chem. Soc.* **103**, 6677 (1981); S. C. Welch, J. M. Gruber, and P. A. Morrison, *J. Org. Chem.* **50**, 2676 (1985); S. C. Welch, C. Chou, J. M. Gruber, and J. M. Assercq, *J. Org. Chem.* **50**, 2668 (1985); H. Hagaiwara, A. Okano, and H. Uda, *J. Chem. Soc., Chem. Commun.*, 1047 (1985); G. Stork and N. H. Baird, *Tetrahedron Lett.* **26**, 5927 (1985).

l. E. J. Corey and R. H. Wollenberg, *Tetrahedron Lett.*, 4705 (1976); R. Baudouy, P. Crabbe, A. E. Greene, C. LeDrain, and A. F. Orr, *Tetrahedron Lett.*, 2973 (1977); A. E. Greene, C. LeDrian, and P. Crabbe, *J. Am. Chem. Soc.* **102**, 7583 (1980); P. A. Bartlett and F. R. Green, *J. Am. Chem. Soc.* **100**, 4858 (1978); T. Kitahara, K. Mori, and M. Matsui, *Tetrahedron Lett.*, 3021 (1979); Y. Köksal, P. Raddatz, and E. Winterfeldt, *Angew. Chem. Int. Ed. Engl.* **19**, 472 (1980); K. H. Marx, P. Raddatz, and E. Winterfeldt, *Justus Liebigs Ann. Chem.*, 474 (1984); C. LeDrain and A. E. Green, *J. Am. Chem. Soc.* **104**, 5473 (1982); T. Kitahara and K. Mori, *Tetrahedron* **40**, 2935 (1984); K. Nakatani and S. Isoe, *Tetrahedron Lett.* **26**, 2209 (1985); B. M. Trost and S. M. Mignani, *Tetrahedron Lett.* **27**, 4137 (1986); B. M. Trost, J. Lunch, P. Renault, and D. H. Steinman, *J. Am. Chem. Soc.* **108**, 284 (1986).

m. S. Danishefsky, M. Hirama, K. Gombatz, T. Harayam, E. Berman, and P. F. Schuda, *J. Am. Chem. Soc.* **101**, 7020 (1979); W. H. Parsons, R. H. Schlessinger, and M. L. Quesada, *J. Am. Chem. Soc.* **102**, 889 (1980); S. D. Burke, C. W. Murtiashaw, J. O. Saunders, and M. S. Dike, *J. Am. Chem. Soc.* **104**, 872 (1982); L. A. Paquette, G. D. Amis, and H. Schostarez, *J. Am. Chem. Soc.* **104**, 6646 (1982); M. C. Pirrung and S. A. Thompson, *J. Org. Chem.* **53**, 227 (1988); T. Ohtsuka, H. Shirahama, and T. Matsumoto, *Tetrahedron Lett.* **21**, 3851 (1983); D. E. Cane and P. J. Thomas, *J. Am. Chem. Soc.* **106**, 5295 (1984); D. F. Taber and J. L. Schuchardt, *J. Am. Chem. Soc.* **107**, 5289 (1985).

14a. R. E. Ireland, R. H. Mueller, and A. K. Willard, *J. Am. Chem. Soc.* **98**, 2568 (1976).
b. W. A. Kleschick, C. T. Buse, and C. H. Heathcock, *J. Am. Chem. Soc.* **99**, 247 (1977); P. Fellmann and J. E. Dubois, *Tetrahedron* **34**, 1349 (1978).
c. B. M. Trost, S. A. Godleski, and J. P. Genêt, *J. Am. Chem. Soc.* **100**, 3930 (1978).
d. M. Mousseron, M. Mousseron, J. Neyrolles, and Y. Beziat, *Bull. Chim. Soc. Fr.*, 1483 (1963); Y. Beziat and M. Mousseron-Canet, *Bull. Chim. Soc. Fr.*, 1187 (1968).
e. G. Stork and V. Nair, *J. Am. Chem. Soc.* **101**, 1315 (1979).
15a. R. D. Cooper, V. B. Jigajimmi, and R. H. Wightman, *Tetrahedron Lett.* **25**, 5215 (1984).
b. C. E. Adams, F. J. Walker, and K. B. Sharpless, *J. Org. Chem.* **50**, 420 (1985).
c. G. Grethe, J. Sereno, T. H. Williams, and M. R. Uskokovic, *J. Org. Chem.* **48**, 5315 (1983).
16a. H. Ahlbrect, G. Bonnet, D. Enders, and G. Zimmermann, *Tetrahedron Lett.*, 3175 (1980).
b. A. I. Meyers, G. Knaus, K. Kamata, and M. E. Ford, *J. Am. Chem. Soc.* **98**, 567 (1976).
c. S. Hashimoto and K. Koga, *Tetrahedron Lett.*, 573 (1978).
d. A. I. Meyers and J. Slade, *J. Org. Chem.* **45**, 2785 (1980).
e. S. Terashima, M. Hayashi, and K. Koga, *Tetrahedron Lett.*, 2733 (1980).
f. B. M. Trost, D. O'Krongly, and J. L. Balletire, *J. Am. Chem. Soc.* **102**, 7595 (1980).
g. A. I. Meyers, R. K. Smith, and C. E. Whitten, *J. Org. Chem.* **44**, 2250 (1979).

Index

acetals
  as carbonyl-protecting groups, 689
  reactions with
    allylic silanes, 468
    allylic stannanes, 478
acetoacetate carbanions
  acylation of, 90–91
  as enolate synthetic equivalents, 15
  O- versus C-alkylation of, 24
acetonides, as diol-protecting groups, 682–683
acid chlorides
  acylation of alcohols by, 145
  acylation of enolates by, 91
  decarbonylation of, 425–426
  halogenation of, 194–195
  preparation of, 145
  reaction with
    alkenyl silanes, 466
    allylic silanes, 466
    organocadmium compounds, 392
    organocopper compounds, 407
acyl anions, synthetic equivalents for, 692–695
acyl imidazolides, 147
  acylation of carbanion by, 90–91
  selective ester formation by, 683
acylation of
  alcohols, 145, 151–152
  alkenes, 495
  amines, 152–155
  aromatic rings, 580–584
  ester enolates, 84–91
  ketone enolates, 92–94
acylium ions
  in ene reactions, 495
  in Friedel–Crafts acylation reactions, 580
acyloin condensation, 263–265
acyloins: *see* ketones, α-hydroxy
acyloxazolinones
  aldol condensation reactions of, 73–74
  enantioselective alkylation of, 28–30
alane, 235
alcohols
  acylation of, 145, 151–152
  allylic
    from alkenes via selenides, 661
    from alkenes by selenium dioxide oxidation, 659–660
    from alkenes by singlet oxygen oxidation, 641–643
    from allylic selenides, 329
    from allylic sulfoxides by [2,3]-sigmatropic rearrangement, 329
    enantioselective epoxidation, 627–628, 731–732
    from epoxides by ring-opening, 638–639
    iodocyclization of monocarbonate esters of, 182
    oxidation by manganese dioxide, 619
    from β-oxido ylides, 98–99
    titanium-catalyzed epoxidation, 627–629
    vanadium-catalyzed epoxidation, 627
  benzylic, oxidation by manganese dioxide, 619
  conversion to alkyl halides, 122–128
  enantioselective synthesis of, 209–210
  in Friedel–Crafts alkylation, 579
  inversion of configuration of, 135
  oxidation, 615–624
    aluminum alkoxide-hydride acceptor, 623
    chromium(VI) reagents, 615–619
    chlorodimethylsulfonium salts, 621–622

alcohols (*cont.*)
oxidation (*cont.*)
dimethyl sulfoxide-based reagents, 620–623
manganese dioxide, 619
ruthenium tetroxide, 619
trichloroacetaldehyde, 624
preparation from
aldehydes and ketones by Grignard addition, 376, 378
aldehydes and ketones by reduction, 232–233
alkenes by hydroboration-oxidation, 204–205
alkenes by oxymercuration, 171–173
allylboranes and aldehydes, 457–460
epoxides and organocopper reagents, 405
epoxides by reduction, 247, 640
esters by Grignard addition, 376, 378
esters by reduction, 233
organoboranes by carbonylation, 446
protecting groups for, 678–686
in radical cyclization reactions, 544–545
reductive deoxygenation of, 253
aldehydes
aldol condensation of, 55–75
β-alkoxy, aldol condensation reactions of, 70–71
aromatic
by formylation, 583
by oxidation at methyl groups, 663
decarbonylation of, 425–426
enolates, alkylation of, 27–28
oxidation of, 652–653
preparation
from alcohols by oxidation, 615–624
from alkenes by ozonolysis, 646–647
from alkenylsilanes by epoxidation, 638
from alkyl halides by carbonylation, 426
from dihydro-1,3-oxazines, 38
from diols by oxidative cleavage, 647–648
from esters by partial reduction, 236–237
from organoboranes, 447–450
by palladium-catalyzed oxidation of alkenes, 416–417
by reaction of Grignard reagents with triethyl orthoformate, 377, 378
by reduction of nitriles, 237
protecting groups for, 689–691
reaction with
allylic boranes, 457–460
allylic silanes, 467–469
allylic stannanes, 476–478
organolithium compounds, 384
organomagnesium compounds, 376, 378
reduction by
hydride donors, 232–233
silanes, 249
α,β, unsaturated
α,β, unsaturated (*cont.*)
from alkenes by selenium dioxide oxidation, 659–660
from alkenyl silanes, 466
reactions with allylic silanes, 474
reactions with organoboranes, 452
unsaturated, cyclization by Lewis acids, 495
Alder rule, 285–286
aldol condensation, 55–75
of boron enolates, 66–68
cyclic transition state for, 61
directed aldol condensation, 60
intramolecular, 75–79
mechanism of, 55–58
mixed condensation with aromatic aldehydes, 58–60
of silyl enol ethers, 69
stereoselectivity of, 58, 61–65, 70–71
alkenes
acylation of, 494–495
addition of
alcohols, 171
carbenes, 523–528
halogens, 176–181
hydrogen halides, 167–171
nitrosyl chloride, 191
radicals, 545–551, 554
selenium reagents, 188–189, 661
silanes, 465
stannanes, 474
sulfenyl halides, 185–188
trifluoroacetic acid, 171
arylation by diazonium ions, 595–596
bromohydrins from, 179–180
cycloaddition reactions with
azides, 302
diazo compounds, 302–303
nitrile oxides, 306
nitrones, 305
epoxidation, 630–633
by nitriles and hydrogen peroxide, 631
by peroxycarboxylic acids, 630
stereoselectivity of, 630–631
in Friedel–Crafts alkylation, 579
hydration of, 170–171
hydroboration of, 200–203
enantioselective, 207–210
hydroformylation of, 425
hydrogenation of, 219–228
metal ion complexes of, 427
oxidation by
chromium(VI) reagents, allylic, 658–659
osmium tetroxide, 625–626
osmium tetroxide-periodate, 644
potassium permanganate, 624
potassium permanganate-periodate, 643
selenium dioxide, 659
singlet oxygen, 640–642

alkenes (*cont.*)
oxidation by (*cont.*)
Wacker reaction, 416–417
oxymercuration of, 171–176
ozonolysis of, 645–647
palladium-catalyzed carbonylation of, 421–422
palladium-catalyzed oxidation, 416–417
palladium-catalyzed reaction with aryl halides, 418–419
photocycloaddition reactions of, 310–315
preparation of
from alkenylboranes, 453–455
by alkylation of alkenyllithium reagents, 375
from alkynes by hydroboration, 211
from alkynes by reduction, 247–248
from carbonyl compounds by reductive dimerization, 262
from carboxylic acids by oxidative decarboxylation, 649–650
from dicarboxylic acids by *bis*-decarboxylation, 650
from β-hydroxysilanes by elimination, 102–103
from ketones by Lombardo's reagent, 391
from sulfones by Ramberg-Bäcklund rearrangement, 508
by thermal elimination reactions, 343–349
by Wittig reactions, 95–102
reactivity in cycloaddition reactions, 303-304
alkylation
of aldehydes, 27–28
of carboxylic acid dianions, 28
by conjugate addition reactions, 39–44
of dianions, 19–20
of dihydro-1,3-oxazine anions, 38
of enamines, 31–32
of enol silyl ethers, 494
of enolates
of aldehydes, 27–28
of cyclohexanone, 16–17
of decalone, 17–18
of esters, 28
intramolecular, 19
of N-acyl oxazolidinones, 28–30
solvent effects in, 20–23
Friedel-Crafts, 575–580
of hydrazones, 36–38
of imine anions, 34–36
of nitriles, 30
of oxazoline anions, 39
of phenols, 27
in tandem with organocopper conjugate addition, 409–410
alkynes
alkylation by organoboranes, 452–453
enantioselective, 457
halogenation of, 197–200
hydration of, 198–199
alkynes (*cont.*)
palladium-catalyzed reaction with halides, 421
partial hydrogenation of, 228
reactions with
organocopper reagents, 411
stannanes, 474–475
reduction by
dissolving metals, 256
lithium aluminum hydride, 247–248
allenes
electrophilic addition to, 195–197
preparation from
cyclopropenylidenes, 532
organocuprates and propargylic reagents, 405
π-allyl complexes
of nickel, 422–423
of palladium, 415–418
amides
alkylation of, 132–133
conversion to amines by Hofmann rearrangement, 537–539
N-iodo, radical reactions of, 544
preparation of, 146, 152–156
by Claisen rearrangement of O-allylic amide acetals, 327
from ketones by reaction with hydrazoic acid, 540
from nitriles, 155
from oximes by Beckmann rearrangement, 540
N-protecting groups for, 688–689
reduction by
alane, 238
diborane, 238
diisobutylaluminum hydride, 688
lithium aluminum hydride, 234
amine oxides
allylic, [2,3]-sigmatropic rearrangement of, 331
in oxidation of alkenes to diols, 626
oxidations of boranes by, 204–205
thermal elimination reactions of, 344
amines
alkylation of, 132
aromatic
ortho-alkylation of, 331
diazotization of, 588–589
enantioselective synthesis of, 210, 394
preparation from
amides, 234, 537–539
carboxylic acids by Curtius rearrangement, 537
carboxylic acids by Schmidt reaction, 539
imines by Grignard addition, 379
imines by reduction with sodium cyanoborohydride, 237
organoboranes, 205–206
phthalimide by alkylation, 132–133

amines (*cont.*)
protecting groups for, 686–689
reductive alkylation of, 250
amino acids, enantioselective synthesis of, 225–227
ammonium ylides, [2,3]-sigmatropic rearrangement of, 330
anilines, see amines, aromatic
anthracene
Diels–Alder reactions of, 297–298
antisynthetic transforms, 700
aphidicolin, synthesis of, 732–747
aromatic compounds
Birch reduction, 255–256
carbene addition reactions, 528
chromium tricarbonyl complexes of, 429–430
halogenation, 572–575
mercuration, 585–587
nitration, 571–572
nitrene addition reactions, 536
oxidation of substituents, 662–663
thallation, 587–588
aromatic substitution
by addition-elimination, 596–598
copper-catalyzed, 601–603
diazonium ions in, 588–596
electrophilic, 571–585
by elimination-addition, 599–601
by metalation, 585–588
palladium-catalyzed, 418–419
by the $S_{RN}1$ mechanism, 605–606
azides
acyl, in Curtius rearrangement, 537
alkyl
preparation by nucleophilic substitution, 131–132
reactions with organoboranes, 205–206
aryl
preparation from diazonium ions, 595
reaction with organoboranes, 205–206
aziridines, equilibria with azomethine ylides, 306–307
azo compounds, thermal elimination of nitrogen from, 339–340

Baeyer–Villiger oxidation, 654–655
Barbier reaction, 387
Barton deoxygenation, 252–253
Beckmann rearrangement, 540
benzene, chromium tricarbonyl complex of, 429–430
benzyne, 599–600
from 1-aminobenzotriazole, 600
from benzothiadiazole-1,1-dioxide, 600
by diazotization of *o*-aminobenzoic acid, 599–600
betaine, as intermediate in Wittig reaction, 95
bicyclo[2.2.1]heptadien-7-ones, decarbonylation of, 338
biogenetic-type synthesis, 82
Birch reduction, 255–256
bond disconnection, 700
borane, see diborane
boranes
alkyl
alcohols from, 204–205
amines from, 205–207
carbonylation of, 445–446
fragmentation reaction of, 511
halides from, 207
hydroboration by, 201–202
radical reactions of, 549
reactions of, 203–207, 445–464
oxidation of, 204–205
palladium-catalyzed coupling with halides, 462
preparation of, 443–444
thermal isomerization of, 203
alkenyl, palladium-catalyzed coupling with alkenyl halides, 462
alkynyl, reactions with aldehydes and ketones, 461
allyl
preparation of, 444
reactions with aldehydes and ketones, 457–460
aryl, preparation of, 444
halo
conversion to amines by azides, 207
hydroboration by, 202
borinate esters, 444
boron enolates
aldol condensation reactions of, 66–68
formation of
from α-diazo ketones, 67
from ketones, 67
boron tribromide, cleavage of ethers by, 141
boron trifluoride
cleavage of ethers by, 141
preparation of organoboranes from, 444
boronate esters, 444
β-allylic, 459–460
aryl, preparation from aryllilthium reagents and trimethyl borate, 461–462
bromides
alkyl, preparation from
alcohols, 123–124
carboxylic acids, 650
organoboranes, 207
aryl, preparation, 573–574, 591–594
see also halides
bromine, addition to alkenes, 177
bromohydrins, synthesis from alkenes, 179–180
N-bromosuccinimide

N-bromosuccinimide (*cont.*)
bromination of ketones by, 192–193
bromohydrins from alkenes by, 180–181

cadmium, organo- compounds
preparation, 392
reaction with acid chlorides, 392
carbanions
acylation of, 84–94
in aromatic $S_{RN}1$ substitution reactions, 606
conjugate addition reactions, 40–44
formation by deprotonation, 1–5
bases for, 3–5
of phosphine oxides, 102
phosphonate, 101
stabilization by substituents, 4
trimethylsilyl
olefin forming reactions of, 102–103
see also enolates
carbenes and carbenoid intermediates, 511–532
α-acyl, 533
addition reactions of, 522–528
generation of, 516–522
insertion reactions, 528–531
reactions with aromatic compounds, 528
rearrangement reactions of, 531–532
stereochemistry of addition reactions, 514
structures of, 512–516
substituent effects on reactivity and structure, 513–515
carbobenzyloxy groups, as protecting groups for amines, 230, 686
carbocations
fragmentation reactions of, 509–511
as intermediates
in addition of hydrogen halides to alkenes, 168
in chlorination of alkenes, 178
Friedel–Crafts alkylation, 575–580
polyene cyclization, 496–499
reactions with
alkenes, 493–494
unsaturated silanes, 494
unsaturated stannanes, 494
rearrangements of, 168, 178, 499–505
carbon acids, pK of, 3–4
carbonylation, palladium-catalyzed, 421–422
carboxylation, of ketones, 91–92
carboxylic acids
acylation reagents from, 145–151
alkylation of, 135
amides from, 152–154
cesium salts, alkylation of, 135
dianions of, alkylation of, 28
α-diazo, reaction with ketones, 505
enantioselective synthesis of, 28, 39
esterification by diazomethane, 134
carboxylic acids (*cont.*)
Hundsdiecker reaction, 650
α-hydroxy, oxidative decarboxylation to ketones, 651
β-keto, synthesis of, 91–92
oxidative decarboxylation of, 649–651
preparation from
aldehydes by oxidation, 652
alcohols by oxidation, 615–624
Grignard reagents and carbon dioxide, 379–380
methyl ketones by hypohalite oxidation, 658
protecting groups for, 691–692
pyridine-2-thiol esters as acylating agents, 148
reduction by
diborane, 238
trichlorosilane, 249
unsaturated, iodolactonization of, 181–182
cerium, organo- compounds
reactions with
hydrazones, 394
ketones, 394
chelation
in enolate of α-alkoxy esters, 326
in Grignard addition reactions of α-alkoxyketones, 387
chiral auxiliary, 702, 726, 728
chlorides
alkyl, preparation from
alcohols, 121–122
alkenes, 167–171
see also halides
chlorination, 574
of alkenes, 178
of alkynes, 197–200
aromatic, 574
2-chloro-3-ethylbenzoxazolium ion, conversion of alcohols to chlorides by, 126
2-chloroisoxazolium ion, activation of carboxylic acids by, 147–148
2-chloro-1-methylpyridinium ion, activation of carboxylic acids by, 147–148
chromium, π-complexes with aromatics, 429–430
chromium(VI) oxidants, 615–619
for oxidation of saturated hydrocarbons, 663
Claisen condensation, 84–88
Claisen rearrangement, 321–328
of O-allyl orthoesters, 322–323
of allyl vinyl ethers, 321–322
of ester silyl enol ethers, 323, 326–327
steric effects on rate, 326
Claisen–Schmidt condensation, 58–60
Clark–Eschweiler reductive alkylation, 250
Clemmensen reduction, 265
conjugate addition reactions, 39–46
kinetic conditions for, 43
of organocopper reagents, 403, 408–409

conjugate addition reactions *(cont.)*
in Robinson annulation reaction, 75–79
tandem alkylation and, 44
convergent synthesis, 700
Cope rearrangement, 316–321
catalysis by Pd(II) salts, 320
enantioselectivity of, 318
oxy-: *see* oxy-Cope rearrangement
stereoselectivity of, 317–318
copper(I) bromide
in preparation of organocopper reagents, 403–404
in Sandmeyer reaction, 591
copper(II) bromide, bromination of ketones by, 193
copper, organo- compounds, 401–414
addition to
alkynes, 411
acetylenic esters, 411
alkenyl, preparation from acetylenes, 411
cuprates
cyano, 402–403
mixed, 402–403
structure of, 402
2-thienyl, 402–403
as intermediates in, conjugate addition of Grignard reagents to enones, 401
organoboranes from, 444
preparation of, 402–404
reactions of with
acid chlorides, 407
allylic esters, 404–405
epoxides, 405
halides, 404
propargylic systems, 404
copper salts
as catalysts for nucleophilic aromatic substitution, 601–603
copper(I) trifluoromethanesulfonate
catalysis of alkene photocycloaddition by, 311
coupling of aryl halides by, 411
crown ethers, catalysis by, 129, 130
cuprates, see copper, organo-
Curtius rearrangement, 537
cyanohydrins
ethers of, as acyl anion equivalents, 693
reduction to aminomethyl carbinols, 504
cycloaddition reactions, 283–315
Diels–Alder, 284–299
1,3-dipolar, 300–307
of ketenes, 307–308
photochemical, 310–315
cyclobutadiene, iron tricarbonyl complex of, 428
cyclobutanes, preparation of, 308–310, 311–314, 375
cyclobutanones, synthesis of, 108, 307–308
cycloheptatrienylidene, 515–516
cyclohexanones
cyclohexanones *(cont.)*
N,N-dimethylhydrazone, alkylation of, 37
enamines of, 32
enolates of
intramolecular alkylation, 19
stereoselective alkylation, 16–17
cyclopentadienones, Diels–Alder addition reactions of, 338
cyclopropanes
divinyl, Cope rearrangement of, 320
preparation from
alkenes by carbene addition, 523–528
enones and sulfur ylides, 105
pyrazolines, 303–304, 340–341
cyclopropanones
conversion to oxyallyl ions, 307
as intermediates in Favorskii rearrangement, 507
cyclopropenes, from alkenyl carbenes, 532
cyclopropenylidene, 515–516

Danishefsky's diene, 296
Darzens reaction, 109
decalone, alkylation of enolates of, 17–18
decarbonylation
of acid chlorides, 425–426
of aldehydes, 425–426
of bicyclo[2.2.1]heptadien-7-ones, 338–339
decarboxylation
during amine-catalyzed condensations, 84
of β-keto acids, 15
of malonic acid derivatives, 15
oxidative, 649–651
diazenes, elimination of nitrogen from, 337
diazirines
carbenes from, 519
preparation of, 519
diazo compounds
additions to ketones, 504–505
from *N*-aziridinoimines, 719
carbenes from, 516–517
metal ion-catalyzed reactions of, 518–519, 526–527, 529–530
preparation of, 517–518
diazoketones: *see* ketones, diazo
diazomethane, in preparation of methyl esters, 134
diazonium ions
aromatic, 588–596
conversion to aryl azides, 595
conversion to aryl halides, 591–594
preparation of, 588–589
reaction with thiolates, 595
reductive dediazonation of, 590
phenols from, 591
as intermediates in Tiffeneau-Demjanov rearrangement, 504
diborane

diborane (*cont.*)
addition to alkenes, 200–203
as reducing agent, 235
reduction of
amides, 238
carboxylic acids, 238
epoxides, 640
dicyclohexylcarbodiimide, activation of carboxylic acids by, 147, 149, 153, 685
Dieckmann condensation, 88
Diels-Alder reaction, 284–298
catalysis by Lewis acids, 289, 300
of cyclopentadienones, 338
enantioselective, 294–295
intramolecular, 298–300
inverse electron demand, 286, 343
of pyridazines, 342
of pyrones, 342
of quinodimethanes, 296–297, 338
regioselectivity of, 286–287
stereoselectivity of, 284–286
transition state of, 285
dienes
Diels–Adler reactions of, 296–298
intramolecular photocycloaddition of, 311–312
preparation from
alkenyl halides and alkenyl boranes, 462
alkenyl halides and alkenyl Grignard reagents, 420
alkenyl halides by nickel-catalyzed coupling, 423
2,5-dihydrothiophene-1,1-dioxides, 337–338
reaction with singlet oxygen, 642
1,5-dienes
hydroboration of, 723
[3,3]-sigmatropic rearrangements of, 316–321
dienophiles, 289–295
masked functionality in, 289–290
as synthetic equivalent groups, 289–290
diethylaluminum cyanide, 45
2,5-dihydrothiophene-1,1-dioxides
preparation of dienes from, 337–338
preparation of quinodimethanes from, 338
diimide, generation and reduction by, 230–232
diisobutylaluminum hydride
reduction of
amides, 688
lactones, 237
nitriles, 235
unsaturated ketones, 239–240
dimethylaminopyridine, as acylation catalyst, 145, 147, 149
dimethylboron bromide, cleavage of ethers by, 141
dimethyl sulfoxide
in conversion of alkenes to bromohydrins, 179–180
in oxidation of alcohols, 620–621
dimethylsulfonium methylide, 103–107
dimethylsulfoxonium methylide, 103–107
diols
cleavage by
lead tetraacetate, 648–649
periodate, 643–644, 647–648
preparation by
epoxide ring-opening, 633–634
hydroxylation of alkenes, 624–626
reduction dimerization of carbonyl compounds, 261–262
monotosylates, rearrangement of, 501–503
protecting groups for, 682–683
rearrangements of, 501–503
α-diones
cleavage by periodate, 648
preparation by
oxidation of alkynes, 625
oxidation of enamino ketones, 642
oxidation of ketones by selenium dioxide, 657
dioxetanes, 642
dioxolanes
as carbonyl-protecting groups, 689–690
reactions with allylic silanes, 469
vinyl, as dienophiles, 293–294
diphenylphosphoryl azide, activation of carboxylic acids by, 153
dipolar cycloaddition reactions, 300–307
intromolecular, 305
regioselectivity in, 302–303
stereoselectivity in, 302
dipolarophiles, 301
1,3-dipoles, 300–301
dithianes
as acyl anion equivalent, 694, 712
lithiation of, 694
double stereodifferentiation, 70, 724

elimination reactions
carbenes from α, 512, 521–522
cheletropic, 336–339
of cyclic diazenes, 337
of β-hydroxysilanes, 102–103
thermal, 336–349
enamines
alkylation of, 31–32
conjuate addition to enones, 45–46
[2 + 2] cycloaddition reactions of, 309
of cyclohexanone, 32
halogenation of, 194
nucleophilicity, 31
preparation of, 31
enantioselective reactions
addition of allylic boranes to aldehydes, 459–460
addition of organocopper reagents to enones, 409

enantioselective reactions (*cont.*)
addition of organozinc reagents to aldehydes, 389
aldol condensations of N-acyl oxazolidinones, 73–74
aldol condensations involving double stereodifferentiation, 70
alkylation of N-acyl oxazolidinones, 28–30
alkylation of alkynes by organoboranes, 457
alkylation of hydrazones, 37–38
alkylation of oxazolines, 39
Cope rearrangement of 1,5-dienes, 318
epoxidation of allylic alcohols, 627–628
hydroboration, 207–210
hydrogenation, 223–227
ketones from organoboranes, 455–457
Robinson annulation reaction, 79
in synthesis of longifolene, 720
ene reaction, 332–336, 495
enol ethers
of β-diketones
reduction to enones, 240
lithiation of, 693–694
preparation from esters by Lombardo's reagent, 391
enol phosphate esters
from α-haloketones, 137
reduction of, 258
enol silyl ethers: *see* silyl enol ethers
enolates
of N-acyl oxazolidinones
enantioselective alkylation, 28–30
acylation of, 84–94
alkylation of, 11–19
by conjugate addition, 39–44
intramolecular, 25–26
O versus C, 23–27
in aromatic $S_{RN}1$ substitution reactions, 606
boron
aldol condensation reactions of, 66–68
formation from α-diazo ketones, 67
formation from ketones, 67
of cyclohexanones, stereoselective alkylation, 16–17
of decalones, stereoselective alkylation, 17–18
of esters
acylation of, 84–91
stereoselective formation, 323
formation of, 5–11
in competition with Grignad addition, 380
from enol acetates, 11
kinetic versus thermodynamic control of, 5–6
by reduction of α,β-enones, 11, 254–255
from trimethylsilyl enol ethers, 11
halogenation of, 194
of isopropyl phenyl ketone,
O- versus C-alkylation, 25
enolates (*cont.*)
magnesium
acylation of, 88–91
from α-bromoketones, 66
oxidation by
molecular oxygen, 655–657
MoO_5-pyridine-HMPA, 654
reactions with
benzene-chromium tricarbonyl complex, 430
trimethyl silyl chloride, 6–7
reactivity, effect of
crown ethers, 23
counter ion, 23
hexamethylphosphoric triamide, 23
solvents, 21–23
tetramethylethylenediamine, 23
in Robinson annulation reaction, 78–79
selenenylation of, 195
sulfenylation of, 195
of α,β-unsaturated ketones
alkylation, 26
protonation of, 26
zinc, preparation from α-haloketones, 389–390
enols
in aldol condensations, 55–58
in halogenation of ketones, 191–193
epoxides
carbenes from, 519–520
isomerization to ketones, 635
preparation from
alkenes by epoxidation, 630–631
allylic alcohols by epoxidation, 627–628
carbonyl compounds and sulfur ylides, 105–108
reaction with
organocopper reagents, 405
organolithium compounds, 384
reduction to alcohols, 247, 640
ring-opening reactions, 633–640
base-catalyzed, 638–639
hydrolysis, 633–635
with selenide ions, 639
trimethylsiyl, preparation of, 109–110
esters
acetylenic, addition of organocopper reagents to, 411
as alcohol-protecting groups, 683–686
α-alkoxy
aldol condensation of enolates, 71
Claisen rearrangement of silyl enol ethers, 326–327
enolates of, 326
condensation reactions of, 84–88
conversion to
amides, 155
enol ethers by Lombardo's reagent, 391
α-diazo
reaction with organoboranes, 450

esters (*cont.*)
α-diazo (*cont.*)
rhodium-catalyzed carbenoid reactions of, 529–530
enantioselective synthesis of, 28–30
enolates of
alkylation of, 28
stereoselective formation, 323
formate, formation of hydroxymethylene derivatives by, 93
β-keto
reaction with π-allylpalladium compounds, 417
synthesis by enolate acylation, 84–94
organozinc derivatives of, 391
preparation
by acylation of alcohols, 145–147
from aldehydes by oxidation, 652–653
by alkylation of carboxylate ions, 135
from alkyl halides by carbonylation, 426
from carboxylic acids using diazomethane, 134
from carboxylic acids by oxidative decarboxylation, 649–650
by Favorskii rearrangement, 506–508
by Fischer esterification, 151
from ketones by Baeyer-Villiger oxidation, 654–655
from organoboranes and α-haloacetate esters, 450
by palladium-catalyzed carbonylation, 421
pyrolysis of, 347–348
reaction with organomagnesium compounds, 376, 378
reduction
by calcium borohydride, 234
by lithium aluminum hydride, 233
by lithium borohydride, 234
β-sulfonyl, reaction with π-allylpalladium compounds, 417–418
α,β-unsaturated
addition of organocopper reagents to, 408
copper-catalyzed addition of Grignard reagents, 411
preparation by palladium-catalyzed carbonylation, 421
reaction with allylic silanes, 474
reduction of, 240
xanthate, pyrolysis of, 348–349
ethers
alkenyl: *see* enol ethers
allyl phenyl, Claisen rearrangement of, 328
allyl vinyl, Claisen rearrangement of, 321–322
cleavage of, 141–142

Favorskii rearrangement, 506–508
ferrocence, 429
Fischer esterification, 151–152
Fischer–Tropsch process, 425
fluoride ion
as catalyst for Michael reaction, 40
in reactions of allylic silanes, 469–470
fluorination, aromatic, 575
fluorine, addition to alkenes, 180–181
2-fluoro-1-methylpyridinium ion, in preparation of azides from alcohols, 131
fragmentation reactions, 509–511, 541, 716
free radicals: *see* radicals
Friedelcrafts acylation reactions, 580–584
intramolecular, 582
Friedel–Crafts alkylation reactions, 575–580
catalysts for, 579
chloromethylation, 582–583
intramolecular, 579–580
rearrangement during, 578
frontier orbitals
in Diels–Alder reactions, 284–287
in 1,3-dipolar cycloadditions, 302–303
in ketene cycloaddition reactions, 307–308
in radical reactions, 546–547

glycols: *see* diols
Grignard reagents: *see* magnesium, organo-compounds

halides
alkenyl
from alkenyl boranes, 210–211
nickel-catalyzed coupling of, 423
palladium-catalyzed coupling with alkenyl boranes, 462
palladium-catalyzed reaction with alkenyl stannanes, 480
palladium-catalyzed reaction with Grignard reagents, 420
palladium-catalyzed reaction with organolithium compounds, 420
palladium-catalyzed reaction with organozinc compounds, 420
alkyl
by addition of hydrogen halides to alkenes, 167–171
enantioselective synthesis of, 210
preparation from alcohols, 122–128
aryl
copper-catalyzed coupling, 411–412
nickel-catalyzed coupling of, 423–424
nickel-catalyzed coupling with Grignard reagents, 424
palladium-catalyzed alkenylation of, 418–419
palladium-catalyzed coupling with alkenyl stannanes, 480
palladium-catalyzed coupling with alkyl boranes, 462

halides (*cont.*)
aryl (*cont.*)
palladium-catalyzed coupling with arylboronic acids, 462
palladium-catalyzed reactions with organozinc compounds, 420
preparation from diazonium intermediates, 591-594
reductive dehalogenation of, 244-247, 250-251, 257-258
halogenation
of acid halides, 194-195
of alkenes, 176-181
stereoselectivity of, 177-178
of alkynes, 198-200
aromatic, 572-575
of ketones, 191-194
reagents for, 184-185
hard-soft-acid-base theory
application to enolate alkylation, 24
Heck reaction, 418-419
Hofmann-Loeffler reaction, 543-544
Horner-Wittig reaction, 102
hydrazones
diazo compounds from, 517
N,N-dimethyl
alkylation of anions of, 36-37
enantioselective alkylation of, 37-38
hydrolysis to ketones, 38
sulfonyl
diazo compounds from, 519
Shapiro reaction of, 266
Wolff-Kishner reduction of, 265
hydroboration, 200-203
of alkynes, 210-211
of 1,5-dienes, 723
enantioselective, 207-210
regioselectivity of, 200-201
stereoselectivity of, 201-202
hydroformylation, 425
hydrogenation, 219-230
catalysts for homogeneous, 222-223
enantioselective, 223-227
isomerization during, 222
stereoselectivity of, 220-222
hydrogenolysis, 230
hydrosilation, 465
hydroxymethylene derivatives
dianions of, alkylation, 93
synthesis of, 93
N-hydroxysuccinimide in activation of carboxylic acids, 153
hypohalites
acyl, as halogenating agents, 574, 575
ions, in oxidation of methyl ketones, 658

imidazolides: *see* acyl imidazolides
imines
imines (*cont.*)
anions of, 32-36
aldol condensation reactions of, 75
enantioselective alkylation of, 36
regioselective alkylation of, 35-36
condensation reactions of, 80-84
formation by rearrangement of alkyl nitrenes, 535
reactions with unsaturated silanes, 470, 472
reduction by sodium cyanoborohydride, 237
imino ethers
synthesis of, 133
iodides: *see* also halides
alkyl, preparation from alcohols, 125-126
aryl, preparation from diazonium ions, 594
iodination, aromatic, 575
iodine azide, as reagent, 192
iodine isocyanate, as reagent, 192
iodolactonization, 181-182
isobenzofurans as Diels-Alder dienes, 297
isoxazoles, from alkenes and nitrile oxides by cycloaddition, 306
isoxazolines, from alkenes and nitrones by cycloaddition, 305-306

Jones reagent, 617
juvabione, synthesis of, 703-712

ketals, vinyl, as dienophiles, 293-294
ketenes
cycloaddition reactions of, 307-308
dienophilic synthetic equivalent for, 292
as intermediates in diazoketone rearrangements, 533-534
β-ketoacids
decarboxylation of, 15
ketones
α-acetoxy
from enol acetates by epoxidation, 637
by oxidation with lead tetraacetate, 653
acylation of, 93-94
α-alkoxy
as by-products of Favorskii rearrangement, 507
reaction with Grignard reagents, 387
α-allyloxy, Claisen rearrangement of enolate, 327
Baeyer-Villiger oxidation of, 654-655
α-bromo
formation of, 191-193
reaction with phosphines and phosphites, 137
conversion to aminomethyl carbinols, 504
conversion to carboxylic acids by haloform reaction, 658
α-diazo

ketones (*cont.*)
α-diazo (*cont.*)
cyclization of unsaturated by boron trifluoride, 496
preparation of, 517–518
reaction with organoboranes, 450
rhodium-catalyzed carbenoid reactions of, 529–530
Wolff rearrangement, 532–534
enantioselective synthesis of, 37–38, 409, 455–457
enolates, stereoselective formation, 61–64
α-fluoro, 194
α-halo
Favorskii rearrangement of, 506–508
formation from alkenyl halides by epoxidation, 637
magnesium enolates from, 66
reaction with organoboranes, 450
zinc enolates from, 389
hindered, Wittig reaction of, 98
α-hydroxy, preparation of, 263–265, 637, 653–654, 655–656
oxidation of, 651–654
by selenium dioxide, 659
photocycloaddition reactions of, 315
preparation from
acid chlorides and organocadmium reagents, 392
acid chlorides and organocopper reagents, 407
acid chlorides and stannanes, 482–483
alcohols by oxidation, 615–624
alkenes by hydroboration-oxidation, 205
alkenes by palladium-catalyzed oxidation, 416–417
alkenyl silanes by epoxidation, 638
alkyl halides by carbonylation, 426, 484
alkynes by hydration, 198–199
aminomethyl carbinols by rearrangement, 504
aromatics by acylation, 583–585
carboxylic acids and organolithium reagents, 383, 385
epoxides by Lewis acid-catalyzed rearrangement, 635
α-hydroxy carboxylic acids by oxidative decarboxylation, 651
nitriles and Grignard reagents, 377, 379
organoboranes, 446–447
protecting groups for, 689–691
reactions with
allylic silanes, 466–467, 469–470
diazoalkanes, 504–505
organolithium compounds, 383–387
organomagnesium compounds, 376, 378
sulfur ylides, 103–108
trimethylsilyl cyanide, 504

ketones (*cont.*)
reduction of (by)
dissolving metals, 253, 261–262
Grignard reagents, 380
hydride-donor reagents, 232–233
hydrogen exchange, 249–250
silanes, 248–249
reductive deoxygenation of, 265–269
reductive dimerization of, 261–263
ring expansion of cyclic, 504–505
stereoselective reduction of, 241–244, 250
α,β-unsaturated
addition or organocopper reagents to, 408–409
from alkenyl mercury compounds and acid chlorides, 393
alkenyl stannanes and alkenyl trifluoromethanesulfonates by carbonylation, 484
deprotonation of, 7
enolates of, 26
photocycloaddition reactions of, 313–314
reactions with allylic silanes, 473–474
reactions with sulfur ylides, 105
reduction of, 239–240, 254–255
tandem conjugate addition-alkylation of, 409–410
Knoevenagel condensation, 83–84

lactams, by iodocyclization of O,N-trimethylsilyl imidates, 182
lactones
formation of, 148–150, 422, 544, 551
macrocyclic, 148–149
α-methylene, synthesis, 82
protection as dithioketals, 692
reduction of, 234
lead tetraacetate
cleavage of diols, 648–649
in oxidative cyclization of alcohols, 544–545
oxidative decarboxylation of carboxylic acids, 649–651
Lindlar's catalyst, 228
lithium organo- compounds
alkenyl
alkylation of, 375
preparation by Shapiro reaction, 374
alkylation of, 374–375, 384
alkynyl, 369
allylic, alkylation of, 375
benzylic, alkylation of, 375
configurational stability of, 372
organoboranes from, 444
preparation of, 368–373
from halides by halogen-metal exchange, 371–372
by lithiation, 369–371
from stannanes by metal-metal exchange, 373

lithium organo- compounds (*cont.*)
  preparation of (*cont.*)
    from sulfides by reduction, 373
  reaction with
    carbonyl compounds, 382–387
    carboxylic acids, 383
    halostannanes, 475
    trimethylsilyl chloride, 464
lithium tri-*t*-butoxyaluminum hydride, 235
lithium triethylborohydride, 247, 640
longifolene, synthesis of, 712–720

magnesium, organo- compounds
  alkylation of, 375–376
  alkynyl, 369
  allylic, isomerization of, 381
  benzyl, reaction with formaldehyde, 382
  copper-catalyzed conjugate addition of, 411, 412
  cyclopropylmethyl, ring-opening of, 381
  organoboranes from, 444–445
  preparation of, 366–367, 369
  reactions with
    aldehydes, 376, 378
    carbon dioxide, 379, 380
    esters, 376, 378
    halostannanes, 475
    imines, 379
    ketones, 376, 378
    triethyl orthoformate, 376, 378
    trimethylsilyl chloride, 464
  stereochemistry of, 367–368
  structure of, 366–368
malonic acids, decarboxylation of, 13
malonic ester anions
  acylation of, 91
  cyclization of »-haloalkyl, 13
  as enolate synthetic equivalents, 13
  reaction with π-allylpalladium compounds, 417
Mannich reaction, 80–82
Markownikoff's rule, 167
Meerwein arylation reaction, 595–596
Meerwein–Pondorff–Verley reduction, 249
mercurinium ion intermediate, 172
mercury organo- compounds
  α-acetoxy, 548
  aromatic, 585–587
  carbenes from, 522
  preparation of, 393, 547–548
  reaction with
    acid chlorides, 393
    halogens, 393
  reduction by sodium borohydride, 174, 548
mercury salts
  in aromatic mercuration, 586
  initiation of polyene cyclization by, 498
  in oxymercuration reactions, 171–176
  ring-opening of cyclopropanes by, 711
Michael reaction, 39–45
  intramolecular, in synthesis of longifolene, 713
Michaelis–Arbuzov reaction, 136
Mitsunobu reaction
  inversion of alcohol configuration by, 135
  in preparation of alkyl azides, 131–132
  in preparation of alkyl iodides, 126

nickel organo- compounds
  π-allyl complexes, 422, 428
  coupling of halides by, 423–424
  as intermediates in coupling of halides and Grignard reagents, 424
nitration, 571–572
  by acetyl nitrate, 572
  by nitronium salts, 572
  by trifluoroacetyl nitrate, 572
nitrenes, 512, 535–536
  alkyl, rearrangement of, 535
  aryl, rearrangement of, 535
  carboalkoxy, 536
  sulfonyl, 536
nitrenoid intermediate, 512
nitrile oxides, cycloaddition reactions, 306
nitriles
  alkylation, 30
  aromatic acylation by, 583
  conversion to primary amides, 155–156
  in epoxidation of alkenes, 631
  α-halo, reactions with organoboranes, 450
  preparation of
    from aryl halides, 601
    by conjugate addition of cyanide, 45
    by nucleophilic substitution, 130–131
  reaction with
    organomagnesium compounds, 377, 379
    Reformatsky reagent, 390
  reduction to aldehydes, 235
  α,β-unsaturated, addition of organocopper reagents to, 408
nitrate esters
  alkoxy radicals from, 545
  diazotization by, 589
nitroalkenes
  conjugate addition reactions of, 44–45
  as dienophiles, 292
nitrones, cycloaddition reactions, 305
nitroso compounds, from alkenes, 191
nitrosyl chloride, 191
Normant reagents, 411

Oppenauer oxidation, 623
orbital symmetry requirements
  Diels–Alder reaction, 284–287
  1,3-dipolar cycloaddition, 302–303
  [2 + 2] cycloaddition, 307–308
organocadmium compounds: *see* cadmium, organo-

organocerium compounds: *see* cerium, organo-
organocopper compounds: *see* copper, organo-
organolithium compounds: *see* lithium, organo-
organomagnesium compounds: *see* magnesium, organo-
organomercury compounds: *see* mercury, organo-
organonickel compounds: *see* nickel, organo
organopalladium compounds: *see* palladium, organo-
organothallium compounds: *see* thallium, organo-
organotin compounds: *see* stannanes
organozinc compounds: *see* zinc, organo-
ortho esters
  as carboxylic acid protecting groups, 691–692
  in Claisen rearrangement of allylic alcohols, 322
  reaction with Grignard reagents, 377–378
osmium tetroxide, 625–626, 637
oxalyl chloride
  in preparation of acid chlorides, 145
  in Swern oxidation, 621
oxaphosphetane, as intermediate in Wittig reaction, 95–96
oxazolidinones, see acyloxazolidinones
1,3-oxazine, dihydro, alkylation of anions, 38
oxazolines
  alkylation of anions, 39
  as carboxylic acid-protecting groups, 691
oxetanes, from alkene-carbonyl photocycloaddition, 315
oxirines, as intermediates in Wolff rearrangement, 532–533
oximes
  Beckmann rearrangement of, 540
  formation in nitrite ester photolysis, 545
oxy-Cope rearrangement, 321
  anionic, 321
  in synthesis of juvabione, 710
oxymercuration, 171–176
  stereoselectivity of, 175–176
oxygen
  reaction with
    enolates, 655–657
    radical intermediates, 174
  singlet
    generation of, 641
    lifetime of, 641
    reaction with alkenes, 641–643
ozonolysis, 645–647

palladium, organo- compounds
  π -allyl
    preparation of, 415
    reaction with enolates, 417–418
  formation by oxidative addition, 415–416, 418
  as reaction intermediates, 414–422
palladium, organo- compounds (*cont.*)
  as reaction intermediates (*cont.*)
    in conversion of alkenyl halides to esters by carbonylation, 421
    in coupling of halides and organometallic reagents, 420
    in reaction of aryl halides and alkenes, 418–419
Paterno–Buchi reaction, 315
periodate ion
  cleavage of diols, 644–645, 647–648
permanganate ion, oxidation of
  alkenes, 624, 643
  alkynes, 625
  aronatic side-chains, 662–663
peroxycarboxylic acids
  epoxidation of alkenes by, 630–631
phase transfer catalysis, 129, 130
phenolate anions, C- versus O-alkylation of, 27
phosphate esters
  aryl, reduction of, 258
  vinyl, reduction of, 258
phosphines, chiral, in hydrogenation catalysts, 223–227
phosphonate carbanions, Wittig reactions of, 101
phosphonate esters, preparation of, 136–137
phosphonium salts
  alkoxy, as intermediates in nucleophilic substitution, 124–125
  cyclopropyl, as synthetic equivalent groups, 697–698
  deprotonation of, 96
  preparation of, 96
  vinyl, as dienophiles, 293
phthalimide
  as amine-protecting group, 687
  in synthesis of amines, 132–133
pinacol rearrangement, 499–503
  in synthesis of longifolene, 714
polyolefin cyclization, 496–499, 737
Prelog–Djerassi lactone, stereoselective synthesis of, 720–732
protective groups, 677–692
  for amides
    2,4-dimethoxyphenyl, 688–689
    4-methoxyphenyl, 688–689
  amino-protecting groups, 686–689
    allyloxycarbonyl, 686–687
    amides as, 688
    t-butoxycarbonyl, 686
    carbobenzyloxy, 686
    phthalimides as, 687
    sulfonamides as, 687
    β,β,β-trichloroethyloxycarbonyl, 687
    trifluoroacetyl, 688
  carbonyl-protecting groups, 689–691
    acetals, 689
    dioxolanes, 689–690

protective groups (*cont.*)
carbonyl-protecting groups (*cont.*)
dithioketals, 690
oxathiolanes, 690
carboxylic acids
t-butyl esters, 691
ortho esters, 691–692
oxazolines, 691
β,β,β-trichloroethyl esters, 691
hydroxyl-protecting groups, 678–686
allyl, 682
benzyl, 681
t-butyl, 680
3,5-dimethoxybenzyl, 681
4-methoxybenzyl, 681
β-methoxyethoxymethyl, 670–680
methoxymethyl, 679–680
methylthiomethyl, 680
silyl ethers, 682
trichloroethyl carbonate esters, 685
tetrahydropyranyl, 678–679
triphenylmethyl, 680–681
pyrazolines
conversion to cyclopropanes, 303, 340–341
from dipolar cycloaddition reactions, 302–303
pyridazines, Diels–Alder reactions of, 342
pyridine-2-thiol, esters of
as acylating agents, 148–149
reaction with organomagnesium compounds, 377
pyridine-2-thione, N-hydroxy esters, radicals from, 542
pyridinium chlorochromate, 617
pyrones, Diels–Alder, addition reactions of, 342

quinodimethanes
from benzo[b]thiophene dioxides, 338
as Diels–Alder dienes, 296–297
quinones, as dienophiles, 289

radicals
alkoxy, 544–545
in aromatic substitution, 603–604
aryl
from N-nitrosoacetanilides, 604
reactions of, 603–605
cyclization of, 174, 246, 367, 551–555
regioselectivity and stereoselectivity in, 552
fragmentation reactionbs of, 555–558
generation of, 542–543
from halides, 542, 547, 553
from N-hydroxypyridine-2-thione esters, 542
by Mn(III) oxidation, 551
by reduction of organomercury compounds, 174, 543, 548
from selenides, 550, 555
from thiono esters, 253, 547, 550

radicals (*cont.*)
5-hexenyl, cyclization of, 174, 246, 367
as intermediates, 541–558
in preparation of organomanesium compounds, 367
intramolecular hydrogen abstraction by, 543–545
rearrangement reactions of, 555–558
substituent effects on reactivity, 546
trapping of
by alkenes, 554
by isonitriles, 553–554
by oxygen, 175
Ramberg–Bäcklund rearrangement, 508
Red–Al: *see* sodium bis(2-methoxyethoxy)aluminum hydride
resolution, in enantioselective synthesis, 701
Robinson annulation reaction, 75–79

Sandmeyer reaction, 591
Selectrides: see trialkylborohydrides
selenenyl halides, addition reactions with alkenes, 188, 661
selenides, preparation of, 347
selenium dioxide, 657, 659
selenolactonization, 188
selenoxides
allylic, [2,3]-sigmatropic rearrangements of, 329
in conversion of alkenes to allylic alcohols, 661
in conversion of epoxides to allylic alcohols, 639
preparation from selenides, 347
thermal elimination reactions of, 347
Shapiro reaction, 266, 374
Sharpless asymmetric epoxidation, 627–628, 731–732
[2,3]-sigmatropic rearrangements, 328–332
of allylic amine oxides, 331
of allylic ethers, 331–332
of allylic selenoxides, 329
of allylic sulfonium ylides, 329
of allylic sulfoxides, 329
of ammonium ylides, 330
of S-anilinosulfonium ylides, 331
[3,3]-sigmatropic rearrangements, 316–328
anionic oxy-Cope, 321
Claisen, 321–328
Cope, 316–321
of ester silyl enol ethers, 323
oxy-Cope, 320–321
of unsaturated iminium ions, 470
silanes
alkenyl, reactions with electrophiles, 465–466, 494
allylic
in polyene cyclizations, 498

silanes (*cont.*)
allylic (*cont.*)
reaction with electrophiles, 466–469, 494
reactions with α,β-unsaturated carbonyl compounds, 472–474
β-hydroxy, elimination reactions of, 102–103
synthesis of, 464
silyl enol ethers
aldol condensation reactions of, 69, 71
conversion to α-hydroxyketones by oxidation, 637, 653–654
halogenation of, 194
enolates from, 11
of ester enolates
Claisen rearrangement of, 323, 727–728
stereoselective formation, 323
as nucleophiles in conjugate addition reactions, 40, 44
oxidation of, 653–654
preparation from trimethylsilyl esters and Lombardo's reagent, 392
reactions with carbocations, 494
Simmons–Smith reagent, 526
sodium bis-(2-methoxyethoxy)aluminum hydride, 235
sodium cyanoborohydride, 235
sodium tetracarbonylferrate, 426
solvent effects
in enolate alkylation, 20–23
in nucleophilic substitution, 128–130
solvents, polar aprotic, 21
stannanes
alkenyl
palladium-catalyzed coupling with alkenyl trifluoromethanesulfonates, 480–481
palladium-catalyzed coupling with halides, 480
reactions with carbocations, 494
α-alkoxy
preparation of, 373, 475
reaction with organolithium compounds, 373
allylic
radical substitution reactions of, 550
reactions with acetals, 478
reactions with carbocations, 494
reactions with aldehydes, 476–478
reactions with thioacetals, 478
α-amino, preparation of, 475
halo
reactions with carbonyl compounds, 476
reactions with organometallic compounds, 475
metal-metal exchange reactions of, 373
palladium-catalyzed reactions with acid chlorideds, 482–483
synthesis of, 475
trialkyl, as hydrogen atom donors, 250–253
stereochemistry, control of in synthesis, 701–702
stereoselectivity of
addition of hydrogen halides to alkenes, 169–170
aldol condensation, 58–70
amine oxide pyrolysis, 344
Claisen rearrangement, 322–323
Cope rearrangement, 316–318
Diels–Alder reaction, 285–288
epoxidation of alkenes, 630–631
epoxidation of allylic alcohols, 627–628
hydroboration of alkenes, 201–202
hydrogenation of alkenes, 220–227
iodolactonization, 181–182
oxymercuration, 175–176
Wittig reaction, 98
sulfides, conversion to organolithium compounds, 373
sulfonamides
as amine protecting groups, 687
radical cyclization reactions of, 544
sulfonates
mono-, of diols, rearrangement of, 501–503
preparation from alcohols, 121–122, 135
reaction with Grignard reagents, 376
reduction of, 244–245
sulfones
α-halo, Ramberg–Bäcklund rearrangement of, 508
vinyl, as dienophiles, 292
sulfonium ylides, 103–107
[2,3-]sigmatropic rearrangement of, 329–330
sulfoxides
alkylation of, 136
α-alkylthio, as acyl anion equivalent, 694
allylic, [2,3]-sigmatropic rearrangement of, 329
β-keto, 94
alkylation of, 94
reductive desulfinylation of, 94
synthesis by acylation of, 94
vinyl, as dienophiles, 293
sulfoximines, reactions of, 108
Swern oxidation, 621
synthetic analysis, 699–701
synthetic equivalent groups, 692–699
in Diels–Alder reactions, 289

thallium, organo-compounds, preparation by electrophilic thallation, 587–588
tetrabromocyclohexadienone, as brominating reagent, 185, 192
thioamides, alkylation of, 136
thioketals, desulfurization of, 266
thiols, alkylation of, 136
Tiffeneau-Demjanov rearrangement, 504
tin, organo-compounds: *see* stannanes
titanium tetraisopropoxide, as catalyst for epoxidation of allylic alcohols, 627–628

trialkylborohydrides, as reducing agents, 235, 244, 640
triazenes, in conversion of carboxylic acids to esters, 134
tri-*n*-butyltin hydride
  in radical generating reactions, 542, 547, 553
  reductive dehalogenation by, 250–251
triethyl orthoformate
  reaction with Grignard reagents, 377, 378
trimethylsilyl iodide
  cleavage of esters by, 141
  cleavage of ethers by, 141
  generation *in situ,* 142
trifluoromethanesulfonates
  alkenyl
    palladium-catalyzed reaction with alkenyl stannanes, 480–481
    preparation from ketones, 481
  alkyl, preparation from alcohols, 122
trimethyloxonium tetralfuoroborate
  alkylation of amides by, 133
triphenylphosphine
  in Mitsunobu reaction, 126
  in Wittig reaction, 95

Ullman coupling reaction, 411–412
umpolung, 693

vanadyl acetylacetonate, as catalyst for epoxidation of allylic alcohols, 627
Vilsmeier–Haack reaction, 584–585

Wacker reaction, 416
Wilkinson's catalyst
  hydrogenation with, 222–223
  reduction of enones using triethylsilane, 240
Wittig reactions, 95–102
  intramolecular, 101–102
  stereoselectivity of, 98–99
Wittig rearrangement, 331–332
Wolff rearrangement, 532–534
Wolff–Kishner reduction, 265–266

xanthate ester pyrolysis, 348–349
X-ray structure
  ethylmagnesium bromide, 366
  lithium anion of *N*-phenylimine of methyl *t*-butyl ketone, 34
  lithium enolate of methyl *t*-butyl ketone, 22
  phenyllithium, 369
  potassium enolate of methyl *t*-butyl ketone, 23

ylide
  phosphorus, 95–98
    functionalized, 99
    β-oxido, 99
  sulfur, 103–108

zinc, organo- compounds
  in cyclopropanation by methylene iodide, 526
  enantioselective addition to aldehydes, 389
  preparation of, 388
  Reformatsky reaction of, 389–390